Families with Broad Distribution across the Northern Hemisphere

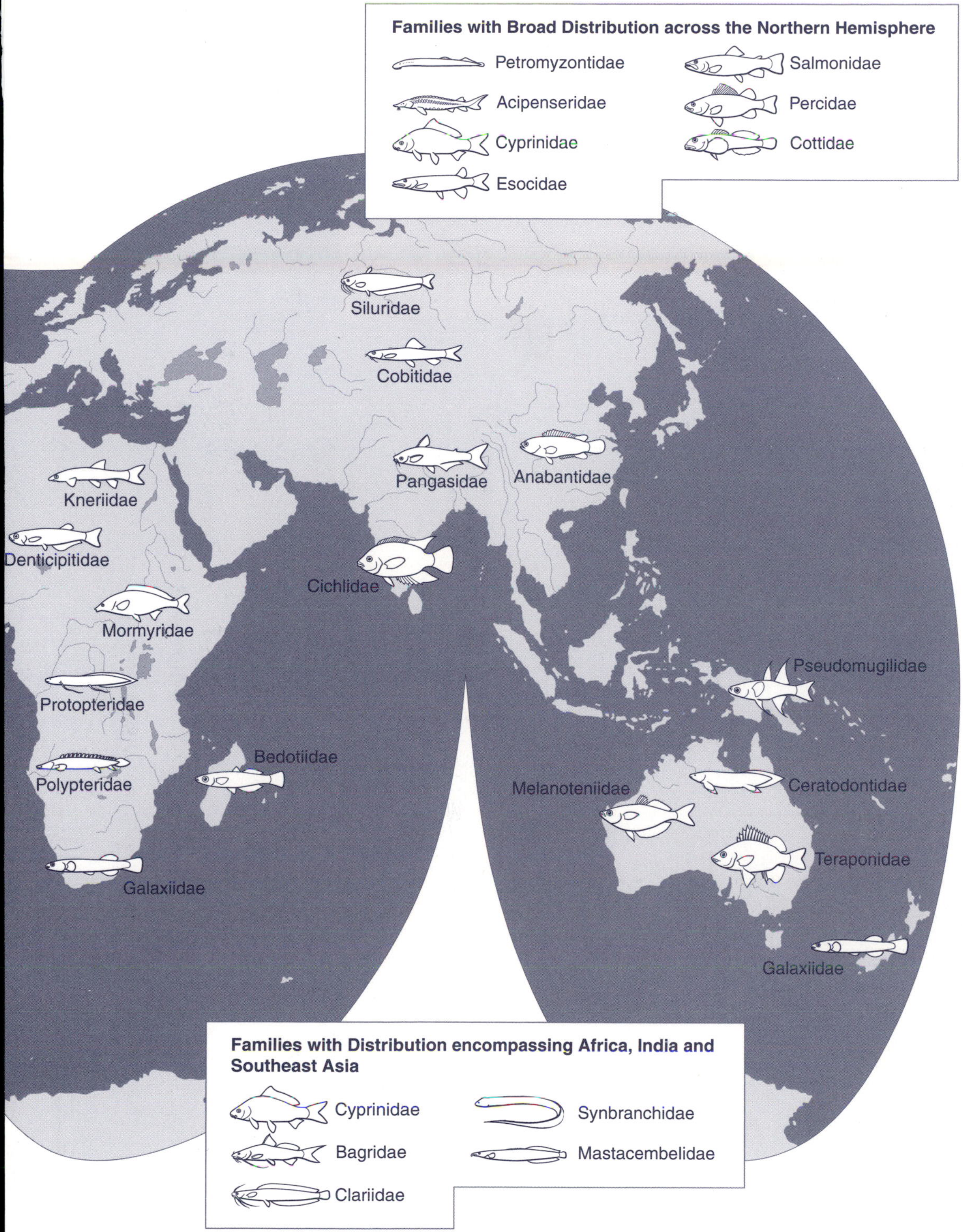

Petromyzontidae   Salmonidae

Acipenseridae   Percidae

Cyprinidae   Cottidae

Esocidae

Siluridae

Cobitidae

Kneriidae

Pangasidae   Anabantidae

Denticipitidae

Cichlidae

Mormyridae

Pseudomugilidae

Protopteridae

Bedotiidae

Polypteridae

Melanoteniidae   Ceratodontidae

Teraponidae

Galaxiidae

Galaxiidae

Families with Distribution encompassing Africa, India and Southeast Asia

Cyprinidae   Synbranchidae

Bagridae   Mastacembelidae

Clariidae

Course:**ZOO 456/556** Price:**131.70** Sem:**2068** MBI:**3777**

This book is property of Southeastern Textbook Rental

1. Check your e-mail for specific deadlines, penalties, hours, etc.
2. Last day to return textbooks with no penalties is the first business day following the last day of finals.
3. A fine will be assessed on books returned from the second business day through the fifth business day following final exams.
4. Books not returned by the fifth business day following the last day of finals will be assessed a purchase fee which is due and payable by the early fee payment deadline of the following semester. Failure to pay this debt will result in this account being turned over to an outside collection agency. Student is responsible for all related cost (collection/attorney fees in the amount of 33 1/3% of the principal, interest, late fees and related court cost).
5. Books are issued by barcode number. YOU MUST RETURN BOOK ISSUED TO YOU. Books with incorrect barcode will be confiscated and returned to the rightful renter.

BOND'S BIOLOGY OF FISHES

Third Edition

MICHAEL BARTON
Centre College

THOMSON

BROOKS/COLE

Australia • Canada • Mexico • Singapore • Spain
United Kingdom • United States

THOMSON
BROOKS/COLE

Bond's Biology of Fishes, third edition
Michael Barton

Publisher: Michelle Julet
Acquisitions Editor: Peter Adams
Development Editor: Shelley Parlante
Assistant Editor: Kari Hopperstead
Editorial Assistant: Kristin Lenore
Technology Project Manager: Travis Metz
Marketing Manager: Ann Caven
Project Manager, Editorial Production: Belinda Krohmer
Art Director: Lee Friedman
Print Buyer: Lisa Claudeanos
Permissions Editor: Kiely Sisk

Production Service: G & S Book Services
Text Designer: Andrew Ogus
Photo Researcher: Annette Coolidge
Copy Editor: Jan Six
Illustrator: John Norton
Cover Designer: Jean-Benoit Levy
Cover Image: © Ray Troll
Cover Printer: Phoenix Color Corp.
Compositor: G & S Book Services
Printer: RR Donnelley

For more information about our products, contact us at:
Thomson Learning Academic Resource Center
1-800-423-0563

For permission to use material from this text or product, submit a request online at **http://www.thomsonrights.com**. Any additional questions about permissions can be submitted by email to **thomsonrights@thomson.com**.

Library of Congress Control Number: 2005932707
ISBN: 0-120-79875-1

Thomson Higher Education
10 Davis Drive
Belmont, CA 94002-3098
USA

Asia (including India)
Thomson Learning
5 Shenton Way
#01-01 UIC Building
Singapore 068808

Australia/New Zealand
Thomson Learning Australia
102 Dodds Street
Southbank, Victoria 3006
Australia

Canada
Thomson Nelson
1120 Birchmount Road
Toronto, Ontario M1K 5G4
Canada

UK/Europe/Middle East/Africa
Thomson Learning
High Holborn House
50/51 Bedford Row
London WC1R 4LR
United Kingdom

Latin America
Thomson Learning
Seneca, 53
Colonia Polanco
11560 Mexico
D.F. Mexico

Spain (including Portugal)
Thomson Paraninfo
Calle Magallanes, 25
28015 Madrid, Spain

To my parents,
for making my education their highest priority.

To my wife Chris,
for making my well-being her highest priority.

To Carl and Lenora Bond,
a life partnership that has left its mark.

CONTENTS

Preface   xvi

PART ONE
AN INTRODUCTION
TO THE FISHES   1

1 · ICHTHYOLOGISTS AND WHAT THEY DO   3

A History of Ichthyology   4

Diversification of Ichthyological Research   12

Methods in Fish Biology   13

    Taxonomic Convention—Naming Fishes
    and Rules of Nomenclature   14

    Common Names   15

    Higher Classification of Fish Groups   15

    Centers for the Study of Fishes   15

2 · ORIGINS OF FISHES   19

Fishes—The First Vertebrates   20

Origins of Vertebrates   20

Major Groups of Living Fishes   25

    Jawless Fishes   27

    Cartilaginous Gnathostomes   27

    Bony Fishes   28

3 · THE ARCHITECTURE OF FISHES   31

Environmental Constraints
on the Design of Fishes   32

General Body Form and Fins   32

    The Head   33

    Body Form   35

    Topography of the Body   36

    Fins   37

Skin and Scales   44

    Epidermis   44

    Dermis   48

    Scales   48

Internal Support: The Skeletal System   48

    The Axial Skeleton   48

    Fins and Associated Appendicular
    Skeleton   60

Skeletal Muscle   64

General Relationships of Internal
Organs   64

    Alimentary Canal and Associated
    Structures   64

    Urogenital Organs   65

    The Heart   65

PART TWO
THE DIVERSITY OF FISHES 68

4 • EVOLUTION AND SYSTEMATICS 71

Foundations of Evolutionary Theory 72

Processes of Selection 72

    Speciation Phenomena in Fishes 74

    Ecological Character Displacement 79

    Macro- versus Microevolution 80

Systematic Studies: Tracing
Evolutionary Histories 80

    Application of Molecular and
    Computational Advances 81

An Integrated Approach to
Understanding Evolution 85

5 • JAWLESS FISHES

Early Vertebrate Experiments and
Descendant Forms 89

The First Fishes: Relationships
of Modern and Extinct Forms 90

The Hagfishes and Lampreys: A
Monophyletic Group? 90

    Hagfishes and Conodonts 92

    Order Myxiniformes 92

    Order Petromyzontiformes
    (Hyperoartia) 94

The "Ostracoderm" Fishes:
Proto-Gnathostomes? 96

    The †Pteraspidomorphi 97

    The †Cephalaspidomorphi 99

6 • THE GNATHOSTOMES

Fishes with Jaws 105

Origin of the Jaws 106

Gnathostomes at the Silurian/
Ordovician Transition 106

    Class †Placodermi 107

Class Chondrichthyes (Elasmo-
ranchiomorphi) 109

    Evolution of the Jaw Suspensorium
    in Gnathostome Fishes 109

    Subclass Euchondrocephali 111

    Subclass Elasmobranchii 114

7 • THE BONY FISHES (TELEOSTOMI)
Acanthodians and Sarcopterygians 129

Introduction to the Teleostomi 130

Class †Acanthodii 130

Class Sarcopterygii: Lobe-Finned
Fishes and the Tetrapods 131

    The †Onychodontiformes: Fishes
    of Uncertain Affinities 132

    Subclass Coelacanthimorpha 134

    Taxon Dipnomorpha 136

    Subclass †Rhizodontimorpha 138

    Subclass †Osteolepimorpha
    (†Rhipidistia) 138

The Origins of Tetrapods 138

8 • INTRODUCTION TO THE
ACTINOPTERYGIAN FISHES
Origins and Modern Diversity 147

Adaptations of the Actinopterygians 148

Origins of the Actinopterygii 148

Class Actinopterygii 149

    Subclass Chondrostei 149

    Subclass Neopterygii 156

9 • TELEOSTEAN FISHES
Early Members of the Most Diverse
Assemblage of Fishes 159

What Are the Teleosts? 160

Division Teleostei: Identifying Features 160

Extinct Primitive Teleosts 160

Order †Pholidophoriformes   160

Order †Leptolepidiformes   161

Order †Tselfatiformes   161

Order †Ichthyodectiformes   161

Diversity of Living Teleosts   161

Subdivision Osteoglossomorpha:
The "Bonytongues"   162

Subdivision Elopomorpha:
Tarpons, Ladyfishes, Eels, and Relatives   166

Subdivision Clupeomorpha:
Herrings and Relatives   172

10 · THE OSTARIOPHYSI

Masters of the Freshwater Domain   177

Ostariophysi as a Sister Group of
the Clupeomorpha?   178

Superorder Ostariophysi   178

Series Anotophysi   178

Series Otophysi   179

11 · THE EUTELEOSTEI

Argentiniform and Salmoniform Fishes   193

Classification of the Euteleostei   194

Order Argentiniformes   194

Order Salmoniformes   196

12 · FASTER AND DEEPER

Esociformes and Some Basal Neoteleosts
(Stomiiformes, Aulopiformes, Myctophiformes,
Lampridiformes, and Polymixiiformes)   205

Straddling the Neoteleostean Fence   206

Esociformes: Sister Group
of the Neoteleosts   206

Stop the Presses! Sister Group to the
Salmonoids After All?   206

Order Esociformes   206

Neoteleostei   207

Superorder Stenopterygii   207

Superorder Cyclosquamata   210

Superorder Scopelomorpha   212

Acanthomorph Fishes   212

Superorder Lampridomorpha   212

Superorder Polymixiomorpha   213

13 · THE PARACANTHOPTERYGII

Cods, Toadfishes, and Other Bottom-
Dwelling Teleosts   217

Paracanthopterygians at a Glance   218

Superorder Paracanthopterygii   218

Order Percopsiformes (Salmopercae)   218

Order Gadiformes (Anacanthini)   219

Order Amblyopsiformes (Anacanthini)   222

Order Ophidiiformes (Anacanthini)   222

Order Batrachoidiformes
(Anacanthini, Pediculati)   223

Order Lophiiformes (Anacanthini,
Pediculati)   224

14 · ACANTHOPTERYGIAN FISHES

Stephanoberyciformes, Zeiformes, and Beryciformes—
Possibly Percomorphs, Possibly Not   229

Introducing the Acanthopterygians   230

Percomorph Fishes   230

Order Stephanoberyciformes   230

Order Zeiformes (Zeomorphi)   232

Order Beryciformes (Berycomorphi)   233

15 · MORE ACANTHOPTERYGIANS

The Synbranchiformes, *Elassoma,* Gasterosteiformes,
Mugilomorpha, and Atherinomorpha   235

The Smegmamorpha: Housecleaning among
the Percomorphs, or Just a Spurious Notion?   236

Order Synbranchiformes   236

Order Gasterosteiformes   238

Series Mugilomorpha 239

  Order Mugiliformes 239

Series Atherinomorpha 240

  Order Atheriniformes 240

  Order Beloniformes (Synentognathi
or Exocoetoidei) 242

  Order Cyprinodontiformes (Microcyprini) 244

16 • PERCIFORM FISHES

Perches and a Whole Lot More 251

The Perciformes: An Embarrassment
of Riches 252

  Suborder Percoidei 252

  Suborder Labroidei 254

  Suborder Trachinoidei 255

  Suborder Pholidichthyoidei 257

  Suborder Blennioidei 257

  Suborder Zoarcoidei 257

  Suborder Notothenioidei 258

  Suborder Icosteoidei 258

  Suborder Gobiesocoidei 258

  Suborder Callionymoidei 259

  Suborder Gobioidei 259

  Suborder Kurtoidei 261

  Suborder Acanthuroidei 261

  Suborder Scombrolabracoidei 262

  Suborder Scombroidei 262

  Suborder Stromateoidei 263

  Suborder Anabantoidei 264

  Suborder Channoidei 264

17 • THE SPINY AND THE STRANGE

Scorpaenoid, Pleuronectiform, and Tetraodontiform Fishes 269

The Scorpaenoidei: Welcome to the
Perciformes—Again 270

Relationships of "Scorpaeniform" Fishes 270

  Suborder Scorpaenoidei 270

The Most Highly Derived Teleosts? 272

  Order Pleuronectiformes (Heterosomata) 272

  Order Tetraodontiformes (Plectognathi) 275

PART THREE
FORM AND FUNCTION 280

18 • THE INTEGUMENT OF FISHES

Skin, Scales, and Associated Structures 283

Integumentary Composition and
Embryonic Origins 284

  The Epidermis 284

  The Dermis 284

  Phylogeny of Scale Types 285

Hydrodynamics of Fishes 287

Color 291

  Occurrence of Color in Fishes 291

  Color Change 293

  Significance of Color 296

Bioluminescence 300

  Occurrence of Bioluminescence 300

  Production of Light 302

  Control of Luminescence 303

  Significance of Bioluminescence 303

Venomous Fishes 305

19 • MUSCLES, LOCOMOTION, AND BUOYANCY 311

Musculoskeletal System 312

  Musculature of the Trunk
and Appendages 312

  Neural Control of the
Musculoskeletal System 316

Modes of Locomotion   318

   Swimming   318

   Fin-Swimming Modes   326

   Swimming Speed   327

   Nonswimming Locomotion   329

The Swim Bladder and the
Regulation of Buoyancy   332

   Buoyancy: Gas Bladder and
   Retes, Fats, and Oils   332

PART FOUR
ADAPTATION
TO THE EXTERNAL
ENVIRONMENT   342

20 · THE SENSORY ARSENAL OF FISHES I

Vision   345

Integrating Sensory Biology with the
Ecology and Behavior of Fishes   346

General Morphology of Fish Eyes   346

Structure and Function
of Fish Visual Systems   348

   Adaptation of the Eye to the Optical
   Properties of Water   348

   Visual Cells and Pigments   348

   Focus, Accommodation, and
   Regulation of Light   352

Visual Adaptation in the Lives of Fishes   356

   Visual Adaptations of
   Deep-Ocean Fishes   357

   "Dual-Purpose" Eyes in Fishes   357

   Sensitivity to Light in the
   Ultraviolet Range   359

   Sensitivity to Polarized Light   359

   Some Alternative Photoreceptors   359

21 · THE SENSORY ARSENAL OF FISHES II

Systems for the Detection and
Production of Auditory, Mechanical,
and Electrical Stimuli   363

Structure of Auditory, Mechanosensory,
and Electrosensory Systems   364

   Auditory Systems: The Inner Ear   364

   The Lateral Line System   366

   Other Sensory Receptors   370

Function of Auditory, Mechanosensory,
and Electrosensory Systems   370

   Membranous Labyrinth   370

   Equilibrium   373

   Mechanosensory Function   374

   Electrosensory System Function   374

Production of Electricity   376

   Electrogenic Fishes   376

   Structure of Electric Organs   377

   Functions of Electric Organs   379

Sound Production in Fishes   380

   Nature of Sounds   380

   Significance of Sound Production   384

22 · THE SENSORY ARSENAL OF FISHES III

Olfaction, Taste, and Other Chemical Senses   389

Chemosensation in Fishes   390

Olfaction   390

   Structure of Receptor Organs   390

   Function and Significance of Olfaction   394

Taste   397

   Taste Receptors   397

   Function and Significance of Taste   397

Other Chemosensory Receptors   399

Coda: Relating Sensory Adaptations
to the Environment   400

PART FIVE
HOMEOSTATIC
MECHANISMS 404

23 • USE AND ACQUISITION OF FOOD 407

Metabolism and Growth 408

Energetic Concerns 408

Metabolism 408

Growth 409

Age and Growth Studies 411

Nutrition 411

Dietary Requirements of Fishes 412

Role of Hormones 413

The Alimentary Canal: Anatomical Features 414

The Oral Cavity and Pharynx 414

Esophagus, Stomach, and Intestine 418

Foraging Activity and Feeding in Fishes 421

Feeding in the Ecological Context 421

Detection and Selection of Food 421

Dietary Diversity and Foraging Activity 423

Functional Morphology of the
Feeding Apparatus 426

Agnathan Fishes 426

Chondrichthyan Fishes 426

Jaws and Branchial Apparatus
in Bony Fishes 427

Preparation of Food for Digestion 429

Ecomorphology and Feeding in Fishes 430

Digestion 430

The Digestive Process 430

Rates and Efficiency of Digestion 433

24 • CIRCULATION AND GAS EXCHANGE 441

Architecture of the Gills 442

Agnathan Fishes 442

Gnathostome Fishes 443

Gills as Gas Exchange Surfaces 446

The Branchial Sieve 446

Branchial Irrigation 449

Extrabranchial and Aerial Oxygen Uptake 451

Cutaneous Respiration 451

Alternatives to Gills: Air Breathing 451

Circulation 453

Vascular Components 454

Secondary Circulation 459

Muscle and Choroid Retes 459

The Blood of Fishes 461

Blood Compounds Dissolved in Plasma 461

Blood Cells 461

25 • OSMOTIC AND SOLUTE REGULATION 469

Osmoregulation 470

Water and Solutes—Maintaining
the Right Balance 470

Osmotic and Ionic Regulation in
Freshwater Fishes 471

Osmotic and Ion Regulation
in Marine Fishes 473

Diadromous and Other Euryhaline Fishes 478

Eggs and Larvae 479

Applications of Studies on
Osmoregulation to Fish Culture 480

Role of the Endocrine System in
Osmotic Regulation 480

The Urinary System 481

The Kidney 481

Role of the Kidneys in Excretion
and Osmoregulation 485

Evolutionary Considerations 488

Gills, Kidneys, and the Origins of Tetrapods 488

26 • NERVOUS AND ENDOCRINE SYSTEMS 493

The Central and Peripheral Nervous System 494

The Brain 494

Cranial Nerves 497

Spinal Cord and Nerves 498

The Endocrine System 500

The Hypothalamus 500

The Pineal Organ 501

The Pituitary Gland 501

Thyroid Gland 505

Interrenal Tissue 505

Chromaffin Tissue 506

Ultimobranchial Gland 506

Pancreas (Islets of Langerhans) 506

Gastroenteric Mucosa 507

Gonads 507

Corpuscles of Stannius 507

Other Organs with Endocrine Function 508

Neurosecretory Cells 508

Caudal Neurosecretory System 508

Pseudobranchial Neurosecretory Gland 508

Paraneurons 509

Natriuretic Peptides 509

Evolutionary and Environmental
Considerations 509

PART SIX
GENERATIONS OF FISHES 514

27 · REPRODUCTION AND DEVELOPMENT 517

Anatomy of the Reproductive System 518

Hagfishes 518

Lampreys 518

Sharks 518

Chimaeras (Ratfishes) 519

Bony Fishes 519

Function and Reproductive Patterns 520

Reproductive Strategy 521

Role of the Endocrine System 522

Semelparity and Iteroparity 522

Finding Mates 524

Egg Production in Oviparous Fishes 524

Egg Retention, Internal Incubation,
and Viviparity 526

Reproductive Guilds 530

Differences Between the Sexes 530

Hybridization 533

Selective Breeding 533

Embryonic and Early Development in Fishes 534

Embryology 534

Early Life History 536

28 · THE GENETICS OF FISHES 551

Introduction 552

Fundamental Concepts 552

Gene Expression 552

Mitosis and Meiosis 555

Mendelian Inheritance 555

Gene Mapping 565

Non-Mendelian Inheritance 565

Chromosomes 567

Chromosome and Arm Number in a Species 567

Induced Polyploidy 570

Hybrids 570

Sex Determination 572

Chromosomal Determination 572

Environmental Determination 573

Hermaphroditism 573

All-Female Species 574

Quantitative Genetics 574

Polygenic Traits 574

Heritability 575

Genetics and Phylogenetics 577

The "Big" Questions 579

Other Phylogenetic Applications 583

Detection and Study of Species 583

Evolution, Conservation, and Management 584

Population Genetics 584

Biogeography 585

Local Adaptation and Outbreeding
Depression 586

Genetics and Fisheries Management 587

Conservation Genetics 589

Recent Directions 589

Fish as Model Systems 589

Genetic Engineering 590

PART SEVEN
INTERACTIONS OF FISHES WITH THEIR ENVIRONMENT 598

29 • THE DISTRIBUTION OF FISHES 601

Introduction 602

Historical Geology and Fish Distribution 605

Stream Capture 605

Continental Drift 606

Glaciation 608

Distribution of Freshwater or Inland Fishes 609

Biogeographical Realms 609

Distribution of Marine Fishes 624

Pelagic Fishes 625

Shore and Shelf Fishes 628

Deep-Sea Fishes 633

30 • ECOLOGY OF FISHES I

An Introduction to Some Basic
Ecological Concepts 639

Understanding Ecology 640

Methods of Ecological Study 640

The Abiotic Environment 641

The Biotic Environment 641

The Abiotic Environment of Fishes 642

Temperature 643

Density and Viscosity of Water 644

Hydrostatic Pressure 644

Dissolved Substances 644

Transparency of Water 646

The Biotic Environment of Fishes 646

Energy: The Driving Force of
Ecological Interactions 646

Levels of Ecological Organization 648

Trophic Relationships in Fish Communities:
Eating and Being Eaten 651

Structure and Continuity in
Ecological Systems 655

Structuring the Fish Community 655

Spatial Integrity and Continuity 655

Temporal Continuity 656

Looking at the Big Picture:
Landscape Ecology, Metapopulations,
and Macroecology 657

Landscapes and Associated
Metapopulations 657

Macroecology: The Unification of
Ecology and Biogeography 657

31 • ECOLOGY OF FISHES II

Freshwater Fishes in Flowing Waters 663

Water—Abundant and Essential 664

Stream Environments in Temperate Zones 665

Upland Stream Environments 665

Temperate Lowland Rivers and Streams:
Habitat and Fish Assemblages 670

Tropical Rivers and Streams 672

Tropical Fish Assemblages 673

Energetics and Trophic Structure
of Riverine Food Webs 674

Trophic Adaptations in Stream Fishes 674

Riverine Trophic Structure and
the Role of Fishes 675

The Landscape Connection 677

32 · ECOLOGY OF FISHES III

Freshwater Fishes in Still Waters 681

Environmental Features
of Lentic Systems 682

Origins of Lakes 682

Lake Environments as
Habitat for Fishes 682

Temperate Lentic Environments 684

Tropical Lentic Environments 685

Fish Assemblages of Lentic Waters 686

Temperate Lakes 686

Tropical Lakes 688

Trophic Adaptations 688

Ecomorphology and Dietary Versatility 688

Species Flocks 689

Optimal Foraging Theory 689

Hypogean Systems: An Unusual
Aquatic Habitat 691

Inland Saline Waters and Associated
Fish Assemblages 695

33 · ECOLOGY OF FISHES IV

Coastal Marine Environments: The
Continental Shelf 701

Introduction to the Marine Realm 702

Coastal Marine Environments 702

The Continental Shelf 702

Pelagic Fish Fauna 703

Benthic Fish Fauna 704

Coral Reefs 705

34 · ECOLOGY OF FISHES V

Coastal Marine Environments: Beaches,
Estuaries, and Rocky Intertidal Shores 713

The Nearshore Environment 714

Protected Inshore Environments:
Bays and Estuaries 716

Temperate Estuarine Ecosystems and
Associated Fish Assemblages 716

Tropical Estuarine Ecosystems and
Associated Fish Assemblages 717

Exposed Inshore Environments: Surf Zones
of Sandy Beaches and Rocky Shores 722

Fish Assemblages of Surf Zones 722

The Fish Fauna of Rocky Shores 722

Seasonal Variation in Coastal
Fish Assemblages 726

Inshore Environments as Nurseries 726

35 · ECOLOGY OF FISHES VI

The Pelagic and Benthic Realm Beyond the
Continental Shelf; Polar Environments 731

Large Marine Ecosystems 732

The Open-Water Pelagic Realm 732

The Epipelagic Zone 733

The Mesopelagic Zone 736

The Bathypelagic Realm and Beyond 737

The Continental Slope and Abyssal Plain 740

Adaptations of Deep Benthic and
Benthopelagic Fishes 741

Deep Benthic Fish Assemblages 742

Marine Fishes of Polar Environments 743

The Nature of Polar Environments 743

The Polar Fish Fauna 744

36 · BEHAVIOR I

Getting Along in the Physical World 755

Behavior: The Ichthyological Perspective 756

xiv • Contents

Locomotor Responses to Stimuli 756

Phototaxis 757

Geotaxis 758

Electrotaxis and Magnetotaxis 758

Thigmotaxis 759

Rheotaxis 759

Chemotaxis 760

Homing and Migration 761

Homing Behavior 761

Migration 762

37 • BEHAVIOR II

Feeding, Fooling Around, and Finding Your
Friends: The Social World of Fishes 773

Feeding Behavior of Individuals 774

Detection of Food 774

Foraging Behavior 775

Application of Feeding Adaptations 775

Ecological Implications of
Foraging Behavior 776

Fish with Other Fish: Social Interactions 776

Games Fishes Play 776

Communication: Signals and
Social Behavior 777

Reproductive and Parental Behavior 778

Courtship and Breeding Behavior 779

Sexual Selection and
Sexual Dynamics 781

Parental Care 782

Shoals and Schools 783

Formation of Schools 783

Sensory Facilitation of Schooling 784

Adaptive Advantage of Schooling 784

Mixed-Species Shoals and Schools 786

Symbiosis 787

Symbiotic Relationships Among Fishes 787

Symbiotic Relationships
with Invertebrates 787

Learning in Fishes 791

Adaptive Behavioral Modification
and Its Applications 791

PART EIGHT
FISHES AND HUMANITY 798

38 • PARASITES AND DISEASES OF FISHES 801

Parasites as Extremely Symbiotic Creatures 802

Evolution of Parasitic Forms and Life Histories 802

Evolutionary Considerations and
Ecological Consequences 803

Parasite Communities 804

Survey of Parasitic Organisms and
the Diseases They Cause 804

Viruses 804

Eubacteria (Monera) 805

Protista 805

Fungi 807

Animalia 807

Impact of Parasites on Evolution and
Adaptation of Host Species 810

Dynamics of Host–Parasite Associations 810

Behavior and Susceptibility to Parasites 812

Parasites as Biological "Tags"
of Fish Populations 813

Nonparasitic Fish Diseases 814

Environmentally Induced Diseases 814

Dietary Diseases 814

Genetic Disorders 814

Fishes as Carriers of Human
Parasites and Diseases 815

Toxic Fishes 816

Ciguatera 816

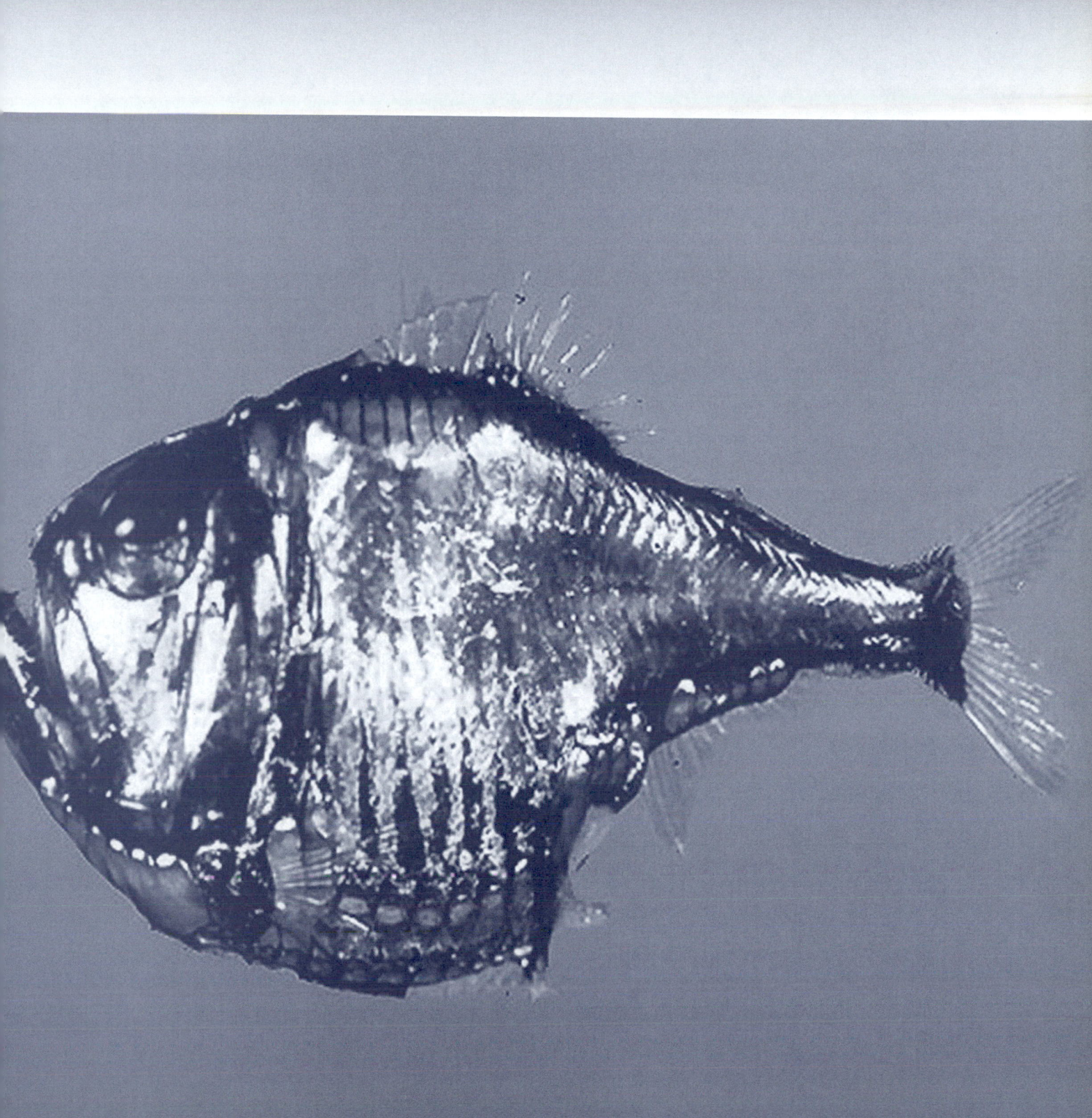

that it will appeal to anyone whose innate curiosity about this remarkable group of animals compels them to seek out a bit more information about fish biology. Academic training is not always a prerequisite for valuable contributions to biology. Charles Darwin gained tremendous insight into the workings of evolution by natural selection from sharing a pint at the pub with the members of the local pigeon fanciers' club. These enthusiasts were more than happy to introduce him to the techniques for the breeding of such unusual varieties as tumblers, trumpeters, and pouters. Fish biologists can learn much from the accountant who spends hours before her aquaria observing the life activities of her charges, or from the Nez Percé Indian out before dawn setting his gillnets for salmon in the Columbia River.

We all have an intuitive sense of what constitutes a fish. Fishes are, in many ways, similar to us, with a mouth at one end and an anus at the other, paired sense organs and appendages, and a vertebral column much like ours. Fishes are also distinctive in the ways in which they differ from us. They live in the water, so most of them depend on gills for gas exchange. Their paired appendages are fins, not bony limbs like ours. Whereas our skin is soft and pliable, fishes are covered with hard, bony scales. But when we investigate the nature of fishes in more detail, we must acknowledge the limits of our intuitive understanding. We learn of fishes that lack jaws, fishes in which some of the sense organs are not paired, fishes with virtually no vertebral column, fishes in which the appendages are bony like ours, and some fishes that lack appendages altogether. We discover fishes that breathe using lungs rather than gills. We learn that scales, though a general feature of the skin of fishes, are by no means universal.

It is the purpose of this first part to introduce the student of fishes to the basic features of fishes and how they arose in the evolution of animals. This part will also provide some insight into fishes by considering the origins and development of that branch of science devoted to the study of fishes—known as *ichthyology*. Our goal in presenting this book is to inspire future generations of biologists by introducing them to the wonderfully complex lives of fishes. Those with an interest purely in the fishes themselves may choose to pursue a career devoted to unraveling the mysteries of the evolution and interrelationships of fishes. Others may develop interests in specific disciplines of biology, such as behavior, ecology, or physiology. For these, it is hoped that understanding something of the biology of fishes will enable them to more fully appreciate that particular aspect of biology that will be the focus of their chosen profession. For those whose interests in fishes may not necessarily be career related, it is hoped that this book will help them to more fully understand why fishes captured their fancy in the first place. The authors enthusiastically endorse your continued attempts to pursue learning about fishes as you pursue them in a stream with a fly rod or through the glass of an aquarium.

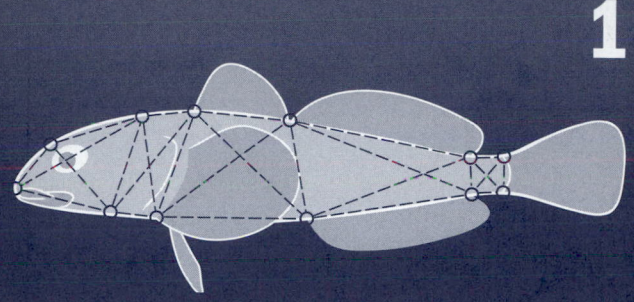

1

ICHTHYOLOGISTS AND WHAT THEY DO

A HISTORY OF ICHTHYOLOGY

DIVERSIFICATION OF ICHTHYOLOGICAL RESEARCH

METHODS IN FISH BIOLOGY

Taxonomic Convention — Naming Fishes and Rules of
  Nomenclature
  PhyloCode — The End of Taxonomy as We Know It?
Common Names
Higher Classification of Fish Groups
Centers for the Study of Fishes

In this chapter, we pay respects to our intellectual forebears—those biologists who laid the foundation of ichthyological study and who have contributed to its current course of development. The study of fishes is an ancient one, and fishes are especially significant as they have been integral in many ways to the formalization of vertebrate study. This formalization is manifested in the practices of taxonomy and its derivative discipline, systematics. *Taxonomy* provides us with a formal protocol for the identification of species. From this, we are able to pursue *systematic* study—that is, the reconstruction of evolutionary interrelationships. The sciences of taxonomy and systematics have developed through strict adherence to prescribed rules for measurement and characterization of specimens. The incorporation of molecular techniques has added another dimension to our understanding of the interrelationships of fishes. Learning how to recognize what constitutes a fish and, hence, how to distinguish among the different lineages of fishes is our entrée into the study of this fascinating group of vertebrates.

A HISTORY OF ICHTHYOLOGY

Our perception of what constitutes science is rooted among the ancient Greeks, who provided us with the disciplinary approach that served as the foundation for the formalization of scientific study. Before there was science, however, there was simple human need. The study of fishes undoubtedly had its origins in the simple desire of humans to feed, clothe, and equip themselves with useful tools. The earliest ichthyologists were hunters and gatherers who had learned how to obtain the most useful fishes, where to obtain them in abundance, and at what times of the year they might be most available. Knowing nothing of the systematic interrelationships of the salmonids, of the remarkable regulatory physiology that enables them to move from freshwater to the ocean and back, or of the phenomenally sensitive olfactory and visual systems that guide their migrations, the Native Americans of the northeast Pacific coast still learned enough of the life history of these fishes to develop a subsistence culture in which salmonids assumed an integral role.

The insights that early cultures had into the natural world, including its fishes, are manifested in the material objects that they left behind. Artistic expressions of this awareness range from the abstract renderings of many species still readily identifiable in the pottery of the Mimbres Indians of the American Southwest (Fig. 1.1; Jett and Moyle, 1986; Moyle and Moyle, 1991) to the highly realistic renderings in the tombs of ancient Egypt. One such painting, from the tomb of Ahanekht at Thebes, shows a fishing scene in which at least ten species, including damselfishes, elephantfishes, and Nile perch, are readily identifiable (Fig. 1.2). Priests of ancient Egypt apparently discouraged the consumption of aquatic resources—ironic in a culture so dependent on the riverine ecosystem of the Nile—yet the Nile perch (*Lates*) was also worshipped (Cuvier and Pietsch, 1995). Egyptian funerary practices included the embalming of animals as well as humans. Perhaps some of the earliest anatomical studies of fishes were conducted at this time.

Fish and fisheries are integral to the Judeo-Christian tradition as well. Moses, in the development of *kashruth* (the dietary laws followed by observant Jews), forbade the consumption of fishes lacking scales or appendages. The apostle Peter and his contemporaries probably harvested the same fishes that are the target species for today's fishery workers

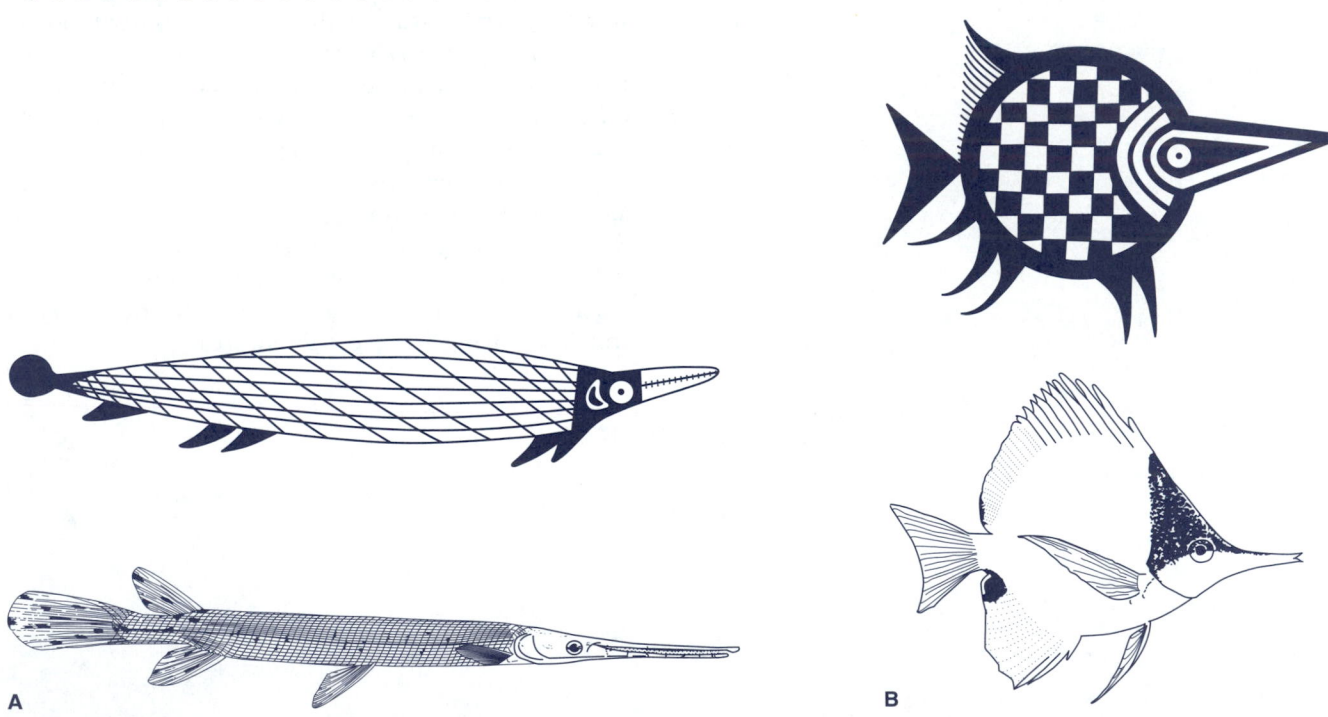

FIGURE 1.1

Illustrations of **A,** gar and **B,** butterflyfish from Mimbres pottery. Source: S. C. Jett and P. B. Moyle, 1986. The exotic origins of fishes depicted on prehistoric Mimbres pottery from New Mexico, *American Antiquity 51:* 688–720. Copyright © 1986 Society for American Archaeology.

FIGURE 1.2

Illustration of fishing scene from the tomb of Ahanekht, Egypt. Source: Cuvier and Pietsch, 1995.

on the Sea of Galilee, now known as Lake Kinneret. These include cyprinids of the genus *Barbus* and *Mirogrex*, cichlids of the genus *Sarotherodon*, and *Mugil cephalus* of the family Mugilidae.

Ichthyology became incorporated into formal scientific study by the greatest classical thinker of all, **Aristotle** (384–322 B.C.). Aristotle provided the earliest taxonomic classification of animals, in which 117 species of Mediterranean fishes could be described. He recognized the differences between fishes and marine mammals and provided anatomical descriptions of many species as well as original observations on their natural history. He recognized that male sharks could be distinguished by the presence of clasper organs and that certain sea basses change sex as they grow older. Some of Aristotle's students continued his ichthyological research. Theophrastus, for example, produced a treatise on amphibious fishes. The Romans, though less devoted to the pursuit of pure science, had a healthy appreciation of fishes and wrote extensively about them. The great naturalist Pliny the Elder, in compiling the works of Aristotle and other Greeks, included many observations of fishes, including such verifiable peculiarities as the sawfish (family Pristidae) and unverifiable ones such as mermaids. The impact of Aristotle on scholarship in general and on natural sciences in particular was such that no significant efforts were made to progress beyond Aristotelian dogma until the Renaissance.

With the dawn of the Renaissance, a renewed appreciation of scholarship brought about a more critical reading of the writings of Aristotle, and his stranglehold on the sciences was loosened. At this time, we can see perhaps the earliest development of two major approaches to the study of organisms. Carl Gans (1978) distinguished these as a "principles" approach and a "naturalist" approach. In the principles approach, the investigator is most concerned with understanding a basic biological principle. Certain organisms may be especially valuable in contributing to this understanding. This approach is inevitably reductionist, leading ultimately to the cellular and molecular fundamentals of life. The naturalist, on the other hand, is curious about the organisms themselves and how they interact with their environment.

Although the earliest naturalists may have been motivated simply by survival, the flowering of sciences in the Renaissance was fostered by individuals whose lives were governed by a passion for living things. The writings of three 16th-century scholars, **Hippolyte Salviani, Pierre Belon,** and **Guillaume Rondelet,** may be viewed as the beginnings of modern ichthyology. What makes their work so important was that it was based on actual observations, not just recitation of the works of the ancients, and included detailed, finely executed illustrations (Fig. 1.3). Rondelet's masterwork, *De Piscibus Marinum,* included descriptions of 244 species.

The incremental improvements in navigation and shipbuilding technology throughout the Middle Ages and the early Renaissance culminated in the era of exploration, colonization, and ultimately cultural hegemony by western Europeans in the 17th through 19th centuries. Expeditions of exploration and discovery—including such legendary voyages as James Cook's exploration of the Pacific and Charles Darwin's voyage of the *Beagle*—were generally well-conceived and -executed endeavors. A scientific approach to exploration necessitated that such voyages be staffed with

LE
VNZIEME LIVRE
DES POISSONS.

Des poiſſons plats, é premierement du
Turbot picquant.

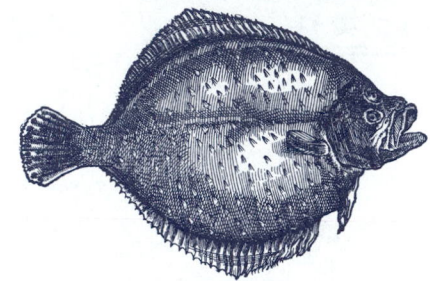

CHAPITRE I.

 Es poiſſons plats, qui ſont appellés en
Grec Πλατεῖς, en Latin *Plani*, ou *Lati*,
ſont ceux qui ſont tellement eſtendus en
long é en large que tous ſont tenures é
minces par toutes les extremités ; é ne
ſont eſtendus en dos é en ventre, mais
par les coſtés, é l'eſpine eſtédue par le milieu du dos, depuis la teſte iuſques à la
queúe, ne iette point ces branches com-
Hh   3

FIGURE 1.3

Illustration of flounder from Guillaume Rondelet's *Histoire entière des poissons* (1558). Source: G. Cuvier and T. W. Pietsch, 1995. Historical portrait of the progress of ichthyology, from its origins to our own time. Copyright 1995 Johns Hopkins University Press. Reprinted with permission.

naturalists who would collect and record the exotic flora and fauna from these far-flung destinations. Among the greatest of these early explorer-naturalists was **Georg Marcgrave** of Saxony (1610–1644) who did not live to witness the publication of his monumental *Natural History of the Fishes of Brazil* in 1648. Also deserving mention is naturalist **Mark Catesby's** (1679–1749) treatise, *Natural History of Carolina, Florida, and the Bahama Islands*, published from 1731 to 1743. This work included some of the earliest renderings of North American fishes.

Natural history collections became the fashion in Europe as ladies and gentlemen of substance and stature vied for the most exotic and unusual specimens for their collections. King Adolf Frederik of Sweden and his wife Lovisa Ulrika typified this enthusiasm for natural history, amassing a collection of more than 1,100 jars containing alcohol-preserved specimens (Åhlander et al., 1997; Fernholm and Wheeler, 1983). The more enlightened of these enthusiasts

made their collections available for scientific study. Such enthusiasm ultimately culminated in the patronage of natural history museums open to the public, such as the great Musée Nationale d'Histoire Naturelle in Paris. As beneficiaries of the massive effort of collection and documentation by the explorer-naturalists, these museums would emerge as centers of scholarship in the natural sciences.

The study of European fishes also advanced at this time. The British naturalists **John Ray** (1627–1705) and **Francis Willoughby** (1635–1672) published their *Historia Piscium* in 1686. Included in this work were descriptions of 420 species, 178 of them new, arranged in a provisional system of classification. It was the work of **Carolus Linnaeus** (1707–1778), the father of the modern science of taxonomy, that marked the beginning of a modern, systematic approach to the study of organisms, including fishes. Linnaeus, a professor at the University of Uppsala, was chiefly a botanist, but his close friend and colleague **Peter Artedi** (1705–1735) devoted his attention to the fishes. Linnaeus had a somewhat dubious reputation among his students. His house would be packed for his evening lectures, yet he was a hard taskmaster. Linnaeus would send students off to the most remote, poorly explored places in search of new specimens. Sometimes these unfortunates never returned (I recall, on a visit to Linnaeus's home in Uppsala, reading a letter from the widow of one of Linnaeus's lost protégés petitioning him for support, as she deemed him responsible for her dire financial straits).

Artedi's work, done in the context of Linnaeus's continuing refinement of the principles of taxonomy, merits his being considered the "father of ichthyology." Artedi recognized five orders: Malacopterygii, the soft-rayed fishes; Acanthopterygii, the spiny-rayed fishes; Branchiostegi, including the trunkfishes and anglerfishes; Chondropterygii, including the cartilaginous fishes, and the Plagiuri, or cetaceans. (The last order was an error that Artedi could have easily avoided by a closer reading of Aristotle.) Artedi developed standard methods for making counts and measurements of anatomical features that are still used today. One of Linnaeus's associates at that time was Albertus Seba, a wealthy pharmacist from Amsterdam. Seba had assembled an outstanding collection of natural history specimens—or "cabinet" as it was called at the time—and was to provide fish specimens for Artedi's work. Tragically, pleasure preceded work and, after an evening of socializing at Seba's residence, Artedi fell into one of Amsterdam's many canals and drowned at the age of 30.

Linnaeus acquired Artedi's notes and manuscripts, which became a core contribution to his masterwork, *Systema Naturae*. By the 10th edition of this work, published in 1758, Linnaeus had developed the taxonomic approach, including the binomial nomenclature still in use today. He revised the orders introduced by Artedi, not always for the better, by emphasizing the position of the pelvic fins. Fishes lacking pelvic fins were placed in the order Apodes; those with abdominal, thoracic, or jugular pelvic fins were named Abdominales, Thoracici, and Jugulares respectively. It should be noted that the contributions of Linnaeus and Artedi, important as they were in providing a foundation for modern taxonomic study, were not grounded in evolutionary theory. It would take almost a century before Charles Darwin would provide the intellectual foundation from which we would be able to perceive that the degree of similarity in taxonomic features was a consequence of phylogenetic relationship.

Two individuals near the dawn of the 19th century stand out for their heroic attempts to consolidate the ever-expanding body of ichthyological knowledge: **Marc Elisier Bloch** (1723–1799) of Berlin, and **Georges Cuvier** (1769–1832; Fig. 1.4) of Paris. Bloch's works were the standard references on fishes until Cuvier, with able assistance from his devoted pupil **Achille Valenciennes**, summarized just about all there was to know about fishes up to that time in his monumental *Histoire Naturelle des Poissons*. Published between 1828 and 1849, this work ran to 22 volumes. Descriptions of 4,055 species—2,311 new to science—were provided (Bauchot et al., 1997). It is one of the most ambitious treatises on natural history ever attempted, and Cuvier occupies a central niche in the pantheon of the great scientists of the western world. Cuvier recognized the contributions of one of his countrymen, **Bernard Germain Etienne de la Ville, comte de Lacepède** (1756–1826), as being especially important. Lacepède had to work under the most trying of circumstances. Deep in the throes of revolution and war, France was intellectually isolated from the rest of the world, making Lacepède's contributions, specifically his five-volume work, also titled *Histoire Naturelle des Poissons,* all the more remarkable.

The 17th and 18th centuries witnessed the dramatic expansion of European influence worldwide. Advances in shipbuilding and navigational technology allowed Europeans to visit the remotest corners of the globe. One of the pioneers in the study of tropical marine fishes was **Peter Forsskål**. Born in Finland in 1732 and trained by Linnaeus at the University of Uppsala, Forsskål participated in the first systematic exploration of the Red Sea region on a voyage commissioned by the Danish king. Tragically, most of the naturalists on that expedition, Forsskål included, succumbed to malaria while en route to Bombay (Mumbai). Much of his collection, mainly preserved as dried skins, survives and includes some of the earliest fish specimens from that region (Nielsen, 1993).

FIGURE 1.4
Baron Georges Cuvier (1769–1832). Source: Cuvier and Pietsch, 1995.

The scientific exploration of the Americas advanced our knowledge of the remarkable diversity of fishes to be found in the New World. One of the early students of American fishes was **Charles Alexandre LeSueur** (1778–1846), a student of Cuvier, who made early collections of the fishes of the Great Lakes and Saint Lawrence River regions. As the Atlantic seaboard was settled, regional studies of its flora and fauna began to appear. Among the earliest of these studies was *The Fishes of New York, Described and Arranged* by **Samuel Mitchell** (1764–1831). By the early 19th century, the vast North American interior was being explored and settled. Brilliant, adventurous, and restless individuals such as **John James Audubon** (1785–1851) and **Constantine Rafinesque** (1783–1840; Fig. 1.5) figure prominently in the faunal documentation of North America. Whereas Audubon is most remembered for his elegant portrayal of North American birds, his sometime traveling companion Rafinesque is perhaps best remembered for his detailed account of fishes, *Ichthyologia Ohiensis*, published in 1820. A true eccentric, who emigrated from Sicily to assume a teaching post at Transylvania University in Lexington, Kentucky, Rafinesque frequently failed to meet his teaching duties, as he would be out roving the wilderness and seining its streams for more fish specimens. An often-told tale of the adventures of Audubon and Rafinesque concerns Rafinesque using Audubon's prized Cremona violin to knock down bats in a cabin they shared. Audubon retaliated by providing Rafinesque with drawings of imaginary fishes, the descriptions of which Rafinesque promptly published.

FIGURE 1.5
Constantine Samuel Rafinesque-Schmaltz (1783–1840).
Source: Myers, 1964.

Another European who greatly enriched American science was **Louis Agassiz** (1807–1873; Plate 1.1). A Swiss scientist who had established his reputation through a series of studies on freshwater fishes and who pioneered the study of paleoichthyology, Agassiz was eventually lured to the United States to assume a teaching post at Harvard in 1846 (see Going Deeper).

At the close of the 18th century, the Pacific coast of North America was still largely unexplored. Spanish missionaries were colonizing the California coast, and they had other things on their mind than fishes. The Spanish had gained a foothold in the South, while the Russians were established in the North. The British were also interested in exploiting the abundant natural resources of this region. **Georg Steller** (1709–1746), a young German naturalist who accompanied the great Russian explorer Vitus Bering in his landmark expeditions to the Arctic, made some of the earliest observations of boreal fishes of the Pacific. **Johann Julius Walbaum** (1724–1800), a compiler of ichthyological literature, whose work was heavily criticized by his contemporaries (including Cuvier), nonetheless is noteworthy for providing the original descriptions of the great Pacific salmons of the genus *Oncorhynchus*. **Sir John Richardson**, who was also a student of the fishes of Australia and China, greatly expanded our understanding of the fish fauna of this region through his *Fauna Boreali-Americana*, published from 1831 to 1837.

The late 19th century witnessed a revolution in the structure of American college and university curricula, as the German model of instruction, which emphasized mastery of a fairly narrowly defined area of knowledge, began to take hold. Upstart universities like Cornell and Johns Hopkins paved the way in the development of a curricular model in which the student exercised a far greater degree of autonomy in choosing his course of study along the lines of chosen major fields. (The choice of gender here is deliberate and appropriate, as most colleges still had not opened their

GOING DEEPER • The Great Chain of Being: Louis Agassiz, David Starr Jordan, and Carl Hubbs

Even in these egalitarian times, we still take a measure of pride in our pedigree. Where we come from seems to count for something, both in the procreative and in the academic sense. I am especially proud that I can locate on my academic pedigree the three greatest zoologists of the last two centuries: Louis Agassiz, David Starr Jordan, and Carl Hubbs. The significance of these three individuals to education in general and to ichthyology in particular gives us pause for special consideration.

Louis Agassiz (1807–1873; Plate 1.1) was born in a small Swiss village on the shores of Lake Morat. As a youth, he was fascinated by the stream ichthyofauna of his native Switzer-

land, but his first opportunity at serious ichthyological work came in the course of his advanced studies at Munich. At the age of 21, Agassiz began a most ambitious project—the classification of the Brazilian fishes collected by Johan Baptiste von Spix (1781–1826). This work, *Selecta Genera et Species Piscium*, published in 1829, marked Agassiz's debut as a professional ichthyologist. In 1832, Agassiz moved to Paris, home of Baron Cuvier (Fig. 1.4) and the center of the zoological universe at the time. Cuvier and another important early ally, the eminent German naturalist Alexander von Humboldt, advanced the career of the rising young student of fishes. It was von Humboldt

who secured for him a professorship at Neuchâtel, where Agassiz commenced study on the rich stores of fossil fishes found in the mountains of his homeland. His treatise, *Recherches sur les poissons fossiles,* published from 1833 to 1843, increased the number of named fossil fishes to more than 1,700 and was a landmark in the study of paleoichthyology.

At about this time, Agassiz also pioneered the study of glaciers, and he was among the first to propose that the Northern Hemisphere was once covered with great ice sheets. Agassiz had attracted the attention of Harvard University, and in 1847 he was offered a professorship in zoology. At Harvard, Agassiz threw himself into

doors to women.) Within these majors, an array of elective courses would also be available. One of the greatest educators of this era happened to also be one of the greatest ichthyologists of the 20th century, **David Starr Jordan** (1851–1931; Plate 1.2). Jordan's contributions in the field of university education and fish biology are so significant that he merits separate consideration (see Going Deeper). Two of David Starr Jordan's best known collaborators were **Charles Henry Gilbert** (1859–1928), a meticulous worker who contributed tremendously to our knowledge of Pacific coast fishes, and **Barton Warren Evermann** (1853–1932), co-author of *Fishes of North and Middle America*. Others of the Stanford group included **John O. Snyder** (1867–1943), who studied the fishes of Mexico, Hawaii, and the freshwater fishes of the Pacific Coast and **Edwin C. Starks** (1867–1932), who published on fish osteology and on the fishes of Panama and Puget Sound.

By the dawn of the 20th century, universities were emerging as research powerhouses. Academicians trained in zoology joined their colleagues at the great museums in advancing ichthyological knowledge worldwide. Museum staffs indeed had their work cut out for them, as more and more collections continued to pour in from exotic locations. England, at the height of her colonial powers, was fortunate to have two enormously gifted ichthyologists on the staff of the British Museum at this time: **Albert Gunther** (1830–1914) and **Georges Boulenger** (1858–1937). Gunther's *Catalogue of the Fishes of the British Museum*, published from 1859 to 1870, was an eight-volume work with descriptions of more than 6,800 species and mention of nearly 1,700 others. Boulenger specialized in percoid fishes, producing several important monographs. He became the world's authority on the freshwater fishes of Africa. It seems that by the time of his retirement from the British Museum, Boulenger had had enough of fishes. He focused all his attention on growing roses and was engaged in producing a monograph on the European varieties at the time of his death.

Throughout the 19th and early 20th centuries, many other individuals made significant contributions to our understanding of fishes. Some of these deserve mention, if only with a few brief lines:

William O. Ayres (1817–1891). Ayres wrote on Pacific coast fishes of the United States. He was California's first ichthyologist.

Charles Girard (1822–1895). A native of France, Girard became an authority on the fishes of the western United States.

Spencer Fullerton Baird (1823–1887). Baird was the first Commissioner for the United States Fish Commission, the forerunner of the U.S. Fish and Wildlife Service. After retirement, he became Secretary of the Smithsonian Institution.

Edward Drinker Cope (1840–1897). An accomplished zoologist and paleontologist, Cope studied North American freshwater fishes. *Copeia*, the journal of the American Society of Ichthyologists and Herpetologists, is named in his honor.

---

his research, producing volumes on fishes and geology, and developing the zoological museum. A collecting trip to California resulted in a paper that introduced the ichthyological community to that most remarkable family of fishes, the viviparous surfperches (Embiotocidae). His most significant contributions, however, were in the field of pedagogy; he revolutionized the study of natural history. A forceful and eloquent speaker, Agassiz was perhaps the greatest popularizer of science in his time.

One of Agassiz's teaching innovations was the establishment of the Anderson School of Natural History at Penikese Island in Buzards Bay, Massachusetts. Some consider this institution the forerunner of the prestigious Marine Biological Laboratory at Woods Hole. At Penikese, students and teachers from around the country could gain firsthand contact with the ocean realm. Here, Agassiz convinced one of his summer school students, David Starr Jordan, that his future lay in fishes. One of the great ironies of the history of science is that Agassiz, for all his groundbreaking work in the biology of extinct and modern fishes and in geology, remained an adherent of the doctrine of special creation. His stubborn resistance to the tide of Darwinism sweeping the intellectual world in the latter half of the 19th century caused Agassiz to be viewed as somewhat of an anachronism by his younger colleagues. Unquestionably, though, Agassiz's contributions to science were enormous, and his inspirational touch landed on vast numbers of students, the most noteworthy of these undoubtedly being David Starr Jordan.

The second link in this great chain, David Starr Jordan (1851–1931; Plate 1.2), began his career as a student of Agassiz and a graduate of Cornell University, taking his master's degree there in 1872. After a brief stint as a professor of biology at Butler University in Indianapolis, Jordan was appointed professor of natural history at Indiana University in Bloomington. One of Jordan's talents was selecting the ablest

George Brown Goode (1851–1899). A pioneer in the study of deep-ocean fishes, Goode co-authored, with **Tarleton Bean** (1846–1915), *Oceanic Ichthyology—A Treatise on the Deep Sea and Pelagic Fishes of the World.*

Other selected students of fishes who completed their careers in the first part of the 20th century included:

Theodore Nicholas Gill (1837–1914). Gill described several orders of fishes. Jordan (1905) called him "the keenest interpreter of taxonomic facts yet known in the history of ichthyology."

Samuel Garman (1843–1927). Garman was a student of Agassiz at Harvard University; he became a specialist on elasmobranchs and deep-sea fishes.

Seth Meek (1859–1914). Meek published on freshwater fishes of Central America and the United States.

Carl H. Eigenmann (1863–1927). Born in Germany, and appointed professor of zoology at Indiana University by Jordan, Eigenmann later became Curator of Fishes at the Carnegie Museum. His research on such subjects as Pacific Coast fishes, embryology, blind cave fishes, and the freshwater fishes of South America led to his being referred to as one of the foremost ichthyologists of his time.

Bashford Dean (1867–1928). Dean was an authority on ancient fishes, and through his outstanding *Bibliography of Fishes* became best known as a compiler of ichthyological knowledge.

John Treadwell Nichols (1883–1958). Nichols studied the fishes of China and initiated the publication of *Copeia.*

By the middle of the 20th century, the study of fishes was flourishing in a number of institutions of higher learning throughout the world. A few ichthyologists stand out, not only for their research productivity, but also for their role as educators and instructors of future generations of fish biologists. Especially noteworthy are **Carl Hubbs** (1894–1979; Plate 1.3) and **Edward Raney** (1909–1984); the list of students who received their PhDs under the tutelage of these masters of fish biology includes many who are the acknowledged leaders in the profession today. Distinguished students of Carl Hubbs include **William Eschmeyer** of the California Academy of Sciences and editor of the recently published *Catalog of Fishes* (1998); **Reeve Bailey** and **Robert Rush Miller** (1916–2003), both Curators Emeriti at Michigan and authorities on North American freshwater fishes; and **George Losey**, Hawaii Institute of Marine Biology, an authority on the behavior and ecology of coral reef fishes. During his career at Cornell University, Edward Raney became one of the most respected ichthyologists of the past century, specializing in the systematics and ecology of freshwater fishes. A few of his students include **Royal Suttkus**, Emeritus Professor at Tulane University

colleagues with which to collaborate. One of these was Charles Henry Gilbert (1859–1928), who teamed with Jordan both at Butler and Indiana Universities to produce a series of outstanding works, culminating in their monumental *Synopsis of the Fishes of North America,* published in 1883. Jordan's talents in university administration led to his being appointed president of Indiana University in 1885. His arduous administrative duties did not diminish his ichthyological productivity, however.

In 1890, Leland Stanford, a Californian who had made millions in railroads, was visiting the most prestigious universities in the East, in hopes of luring one of their top administrators to the post of president for his newly established school in Palo Alto. After tragically losing their 16-year-old son to typhoid fever during a trip to Europe, the grieving Stanfords had decided to devote their philanthropic endeavors to the education of the children of the developing West. The presidency of a magnificent new university, endowed with Stanford's enormous resources and named in memory of his son, was an offer Jordan could not refuse. Jordan decamped for the sunnier climes of California and continued his stellar career in education and ichthyology. Another productive collaboration, this time with Barton Warren Evermann (1853–1932), resulted in the publication of perhaps Jordan's best known work, the four-volume treatise *Fishes of North and Middle America.* In all, Jordan's efforts resulted in the description of a total of 1,085 genera and more than 2,500 species of fishes. Late in his career as an educator, Jordan became a globe-trotting peace activist, publishing numerous articles and acting as the director of the World Peace Foundation.

In a letter written in 1924 to the celebrated paleontologist Roy Chapman Andrews, Jordan stated, "I came back from Japan last fall with the largest collection ever made there. . . . In order to work up the collection fully, I engaged the services of the ablest student I have had for the past 30 years, Carl Leavitt Hubbs, now curator of vertebrates in the University of Michigan." Indeed, Hubbs's (1894–1979; Plate 1.3)

and a specialist on the freshwater fishes of the Southeast; **C. Richard Robins**, Emeritus Professor at the University of Miami and author of numerous books on fishes; and **Bruce B. Collette** of the National Marine Fisheries Service, an authority on the systematics of pelagic marine fishes. These gentlemen are relatively easy to spot at ichthyology meetings, as they are inevitably surrounded by clusters of younger colleagues and graduate students, much like the bar jacks that accompany a large stingray as it makes its rounds of the coral reef.

Of course, not all ichthyology in the 19th and 20th centuries was being practiced in North America. A number of other countries have produced outstanding fish researchers of international stature in the past 200 years. A few of these include,

Francis Hamilton (1762–1829; Scotland). Hamilton described numerous fish species from India.

Felipe Poey y Aloy (1799–1891; Cuba). Poey y Aloy became an authority on the fishes of his homeland.

Johannes Müller (1808–1858; Germany). Müller was a comparative anatomist who revised some of Cuvier's work and defined several new groups of fishes.

Pieter Bleeker (1819–1878; the Netherlands). Bleeker provided much knowledge of the fishes of the East Indies.

Francis Day (1829–1899; England). Day published an important two-volume work on the fishes of India.

Franz Steindachner (1834–1919; Germany). Steindachner studied fishes from many countries, including Mexico, the United States, and Japan.

Lev Semënovich Berg (1876–1950; Russia). Author of *Freshwater Fishes of the U.S.S.R. and Adjacent Countries*, Berg was Russia's greatest ichthyologist and a leading scholar on the history of geographic exploration.

C. Tate Regan (1878–1943; England). A taxonomist at the British Museum of Natural History, Regan published a significant classification of fishes for the 14th edition of the *Encyclopaedia Britannica*.

John Richardson Norman (1899–1944; England). Norman studied flatfishes at the British Museum. He produced an important draft synopsis of orders, families, and genera of fishes that unfortunately never saw publication during his lifetime but remains a significant contribution to fish systematics.

J. L. B. Smith (1897–1968; South Africa). One of the African continent's greatest scientists, Smith is probably best known for his role in introducing the coelacanth (see Chapter 7) to the world.

Sunder Lal Hora (1896–1955; India). Hora was a scholar specializing in the systematics and life history of the fishes of India and a pioneer in the study of warm water aquaculture.

Peter Whitehead (1930–1992; England). Born in Africa and an authority on clupeoid fishes while at the

prolific career as an ichthyologist began at Stanford under the tutelage of Jordan and Charles Henry Gilbert. His first 13 papers on fishes, including further studies on Agassiz's viviparous surfperches, were published while he was still a student at Stanford. Soon after completing his studies at Stanford, Hubbs married Laura Clark, who would become an invaluable colleague and collaborator in much of his work. Singularly devoted to his work with fishes, Hubbs was generous with colleagues and collaborators who marveled at his inexhaustible ichthyological acumen. He could not stand ceremony, however, and shunned the graduation exercises where his master's and later doctoral degrees were awarded. Viewing his election to Phi Beta Kappa in his junior year at Stanford as an empty honor, he snubbed them as well.

With a PhD from the University of Michigan in hand, Hubbs became the first director of that state's Institute of Fisheries Research. Collaboration with Robert Rush Miller, later to become his son-in-law, produced a series of papers on the fishes of the American Southwest. Relocation to the Scripps Institution of Oceanography renewed Hubbs's interest in marine fishes, and a long series of studies soon followed, encompassing not only nearshore and deep-ocean fishes, but also birds, marine mammals, and aboriginal humans. This industrious, insightful, and irreverent student of fishes authored hundreds of papers, mentored many of the leading fish biologists of today, and is regarded as the greatest ichthyologist of the 20th century after Jordan.

My own mentor, Carl Bond, received his PhD under the direction of Karl Lagler (1912–1985) at the University of Michigan, where Reeve Bailey also served as his research advisor. Both Lagler and Bailey were doctoral students of Hubbs when he served on the faculty at Michigan. I was introduced to fishes as an undergraduate at the University of California, Los Angeles, by Boyd Walker (1917–2001), one of Hubbs's students at Scripps. Such an illustrious academic genealogy is no guarantee of success, of course, but it is a source of great pride to know that my educational legacy had its source in the minds of Agassiz, Jordan, and Hubbs.

British Museum of Natural History, Whitehead was one of the finest illustrators to put his hand to fishes. A curious and complex personality, he played a significant role in the rediscovery of original music manuscripts of Mozart and Beethoven that were stolen by the Nazis during World War II.

Norman B. Marshall (1915–1996; England). N. B. "Freddy" Marshall's enormous contributions made him the 20th century's leading authority on deep-sea ichthyofauna. Numerous books, many of them beautifully illustrated by his wife Olga, attest to the breadth and depth of his oceanographic knowledge.

Tokiharu Abe (1911–1996; Japan). For many years, Abe was one of Japan's most prolific ichthyologists. He specialized in the systematics of marine fishes.

Not all practitioners of the noble science of fish biology are able to devote themselves full-time to the task. When he was Crown Prince, **Emperor Akihito** (Plate 1.4) of Japan was a prolific researcher, specializing in the systematics of gobies. His ascent to the Chrysanthemum Throne has necessarily curtailed his ichthyological studies.

At the dawn of this millennium, it is hoped that at least half of the readers of this book will realize how tragically underrepresented their sex is in this recitation of the leading luminaries of ichthyology. Because women have only comparatively recently come into their own in the world of academe, there have been precious few who have gained celebrity in academic circles, including the study of fishes. This is not to say that there have not been noteworthy contributions by women. Balon et al. (1994) have produced an excellent overview of female contributions to fish biology.

Three women in particular merit special mention. **Ethylwynn Trewavas** (1900–1993) was a student of C. Tate Regan and went on to become a senior scientist at the British Museum, where she specialized in the systematics of African cichlids. **Rosemary Lowe-McConnell** (b. 1921) is widely recognized as the leading specialist on the tropical freshwater ichthyofauna, having published the definitive text on the subject (Lowe-McConnell, 1987). **Eugenie Clark** (b. 1922), the author of two popular autobiographies, *Lady with a Spear* and *The Lady and the Sharks*, and the subject of numerous television specials, is perhaps the most widely known ichthyologist today. A former student of Carl Hubbs, Clark became fascinated at an early age by the exploits of the pioneer deep-ocean researcher William Beebe. A specialist on the biology of sharks, she has probably logged more scuba hours than any other ichthyologist. Fortunately, the conditions of gender equality under which we presently work mean that it is no longer remarkable to be female and a fish biologist. Still, a survey of the 1998 directory of the American Society of Ichthyologists and Herpetologists shows that only about 17 percent of the membership is female.

If this admittedly cursory look at the history of ichthyology has whetted the reader's appetite, a few references will provide additional information. Jordan (1905) and Pietsch and Grobecker (1987) provide good historical overviews. Myers (1964), Dymond (1964), and Hubbs (1964a, 1964b) have detailed the history of ichthyology in North America. The extensive bibliography in Lindberg (1974) provides the names of ichthyologists active in the different world regions up through the late 1960s. Of particular value is the recently published historical portrait originally written by Cuvier and edited by Pietsch (1995).

DIVERSIFICATION OF ICHTHYOLOGICAL RESEARCH

The history of ichthyology, especially from the time of Linnaeus, is inextricably tied to the naturalist approach to the study of organisms. We cannot begin to appreciate the nature of an organism until we can accurately identify it and be able to reliably distinguish it from related forms. Taxonomy, therefore, is the mother of the formalized study of the natural history of organisms. The publication of Darwin's *Origin of Species,* fully a century after Linnaeus's 10th edition of *Systemae Naturae,* paved the way for an appreciation of the evolutionary relationships that subsume the taxonomic characterization of species. Hence, the branch of biology known as **systematics**, which discerns phylogenetic relationships based on the characteristics of a given group of species, arose from the earlier formalization of taxonomic technique.

Systematics and taxonomy, as well as life history studies, were the chief concerns of most of the aforementioned ichthyologists. As the science of ichthyology has matured, we see that it has diversified to include a variety of other disciplines. In the classical sense of the term, an ichthyologist is primarily focused on understanding the biology of the particular group of fishes that is his or her specialty. Researchers who follow the principles approach, using fishes to advance knowledge in a given area of biology, might not necessarily view themselves as ichthyologists. The proliferation of studies using zebrafish (*Danio rerio*) to investigate the fundamentals of developmental genetics (see Chapter 27, Going Deeper, for more details) is a good case in point. This may not strictly be ichthyological research, but our understanding of the biology of fishes is greatly enhanced when an organism that is discovered to be the ideal model for a given line of research just happens to be a fish. And so it is that fishes have made their contribution to the understanding of the biological realm from the level of biological molecules and cells to the level of the structure and function of ecosystems.

METHODS IN FISH BIOLOGY

A researcher who uses fishes to unlock some deep biological mystery should be well versed in the methodology of his own discipline, be it cell biology, physiology, or ecology. These individuals will also need to know sufficient fish biology to be able to adequately maintain their experimental subjects, but they may not necessarily be schooled in those fundamental issues of taxonomy and systematics on which the science of ichthyology is based. The serious student of fishes must first appreciate and understand the methodological foundations on which the systematic study of fishes rests. This methodology is based on the identification and recording of suitable characteristics used to define and distinguish species. Once a species is rendered distinguishable from other species, further investigations into its biology can be pursued.

Many features of the internal and external anatomy of a fish, such as vertebrae, fin spines, or muscle segments, are serially repetitive in a metameric sense (see Chapter 27 for a discussion of metamerism). Other features, such as scales or lateral line pores, are numerous but not necessarily metameric. Counts of these features, termed **meristic** characteristics, are extremely useful, because they are relatively stable and more or less independent of body size. **Morphometric** characteristics are measurable features of the topography of the fish and include such useful features as head length, eye length, distance between orbits, body depth, and length of fins. Such measurements are subject to the phenomenon of **allometric growth**, whereby features may change with increasing size of the individual.

The standard reference for procedures used in making meristic and morphometric measurements on fishes remains Hubbs and Lagler (1958). Strauss and Bond (1990) are also helpful in understanding the application of taxonomic technique. The construction of truss diagrams (Fig. 1.6) is helpful in quantitative comparisons of morphological features (Strauss and Bookstein, 1982; Strauss and Bond, 1990). Other, more subtle anatomical features may be useful in the identification of a given group of fishes. For example, the size and position of the swim bladder may not only be useful in characterizing a given species, but it may provide insight into the habitat as well. A large, well-developed swim bladder suggests that the fish spends most of its time up in the water column. The musculature and bones associated with the swim bladder might also suggest a role in acoustic communication (see Chapter 21).

Powerful statistical computation techniques are available for the analysis of discrete measurements, such as those obtained through meristic counts, and of data that may exist along a continuum, such as morphometric measurements of specimens of different sizes within a given fish species. Familiarity with such statistical procedures, such as analysis of variance (ANOVA), analysis of covariance (ANCOVA), and principal components analysis (PCA), is essential for the proper evaluation of taxonomic data.

Any taxonomic study will only be as good as the specimens used. Care must be taken in the acquisition, preservation, and handling of fish specimens in order to ensure that the specimens accurately represent the natural form of the creature. Of course, it goes without saying that fishes should not be collected unless the necessary permits are acquired. The gear for the collection of fishes may include traps or different kinds of nets. In freshwater, backpack or boat-mounted electroshockers are very effective. Photographs of specimens should be taken as soon after collection as possible. Color is

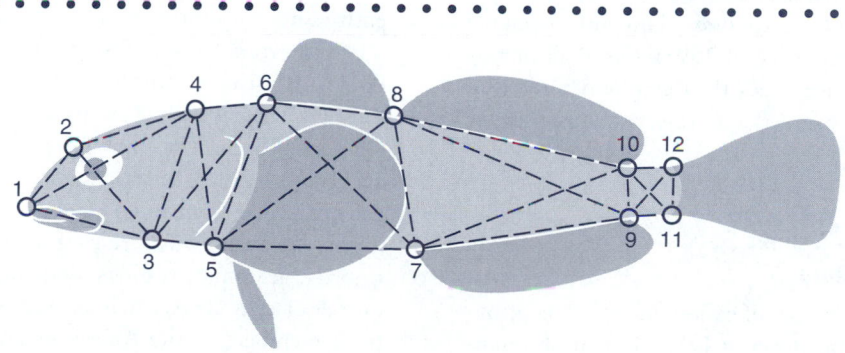

FIGURE 1.6
Example of truss diagram used in making morphometric measurements. Five truss cells (quadrilaterals) identified using 12 well-defined mid-sagittal landmarks. Adapted from Strauss and Bond, 1990.

often an important diagnostic tool, and it fades rapidly when specimens are preserved. Age and growth studies may necessitate the removal of scales, fin spines, or otoliths. Specimens to be kept for later study are preserved, usually in a buffered 5 to 10 percent Formalin[R] solution (Formalin[R] is the commercial name for a 40 percent formaldehyde solution). Preserved specimens are later rinsed thoroughly and stored in isopropyl or ethanol solutions.

Special collection and preservation techniques are necessary if the specimens are to be subjected to molecular analysis. Tissue samples should be preserved and stored in a medium that will prevent denaturing of component nucleic acids or proteins. Ethanol is preferable for this. (Enterprising fish biologists caught in the field without a source of ethanol have been known to resort to distilled beverages, such as rum, that have a sufficient alcohol content.)

Although counts and measurements remain the most frequently used diagnostic tools, other techniques may be brought to bear on the question of species identity and phylogenetic relationships. Because it lends itself so well to fossilization, skeletal tissue is the best morphological feature to use in reconstructing phylogenies—the evolutionary history of a given taxon can be deduced from comparisons of the skeletons of living forms with the fossils of extinct ones. The skeleton is a living scaffold to which the rest of the organism is attached. The size, shape, and presence or absence of certain bony elements provide insight into the biomechanical function of the whole animal. Getting at the skeleton can be a tricky procedure, requiring hours of patient dissection. Certain techniques are helpful. Soft tissues can be enzymatically dissolved or subjected to the scavenging actions of flesh-eating (dermestid) beetles. Small, delicate specimens are best prepared by rendering the soft tissues transparent through an enzymatic process and then staining the bone and cartilage (Taylor, 1967). The U.S. Geological Survey has published a valuable manual to promote quality control in the collection of taxonomic data and associated field methods (Walsh and Meador, 1998). Once collections are made and properly prepared, they can be deposited in a museum, where they can serve as reference material for further study.

· · · · · · · · · · · · · · · · · · · ·
Taxonomic Convention—Naming Fishes and Rules of Nomenclature

The Linnaean system of classification has been continuously refined since its inception in 1758. The most important building blocks of the system are the genus and species names—for example, *Lepomis gibbosus* (Linnaeus, 1758), the formal name of the pumpkinseed sunfish. The generic name is capitalized, but the species name is not; both names are printed in a type font (usually italics) different from the surrounding text; and the endings are governed by rules of Latin grammar regarding gender and case. The author and date of the name follow the name; parentheses indicate that the author originally named the species in a different genus. The date establishes the priority of the name. (This is important because many species were named more than once as fish were brought back to museums by different expeditions. The oldest available name is the one used, with few exceptions.) To be formally available, a name must have been proposed in a genus plus species context, using a generic name with a specific name not used before in that genus, and based on a type specimen (or, in older literature, specimens).

Linnaeus was the first to recognize the need for consistency in nomenclature, and the fact that his binomial system is still in use today reflects its inherent value. The designation of type specimens, though seemingly archaic, is still a standard museum convention. In Linnaeus's time, the potential for variation within species was not as widely recognized as it is today. A species description was thus based on **type specimens** that were taken to represent the entire species. One individual was designated the **holotype**—the single specimen on which the species description was based. Additional specimens used in the species description were designated **paratypes**. A small book of rules, the *International Code of Zoological Nomenclature* (International Trust for Zoological Nomenclature, British Museum [Natural History], London, 1985), is the final arbiter in questions of taxonomy and nomenclature. The International Committee on Zoological Nomenclature oversees the rules and keeps them up to date. This committee has the power to permanently establish a name for the sake of stability, when requested. Lately, the International Code has allowed for a greater variation in types to reflect the fact that species are not fixed and immutable entities but vary throughout their range.

The words chosen for species names may be Latin or Greek or Latinized words of any language representable in the Latin alphabet. For example, the scientific name of the Chinook salmon, *Oncorhynchus tshawytscha*, is composed of a Latinized Greek generic name meaning "hooknose" and a Latinized Tungusic or Chukchi Siberian salmon name. There are many other rules governing the application of priority, grammar, endings, and special problems. A committee of the American Fisheries Society, with assistance from members of the American Society of Ichthyologists and Herpetologists, has worked for years to ensure that the names of the more than 2,400 species of North American fishes are based on the best interpretation of the rules (Nelson et al., 2004). Occasionally, nomenclatural rules make it necessary to change a name in use.

PhyloCode—The End of Taxonomy as We Know It?

Some researchers have challenged the continued use of the Linnaean classification altogether, because it is not based on evolutionary relationships (see Withgott, 2000). They propose to institute a new nomenclature, termed *PhyloCode*, based on phylogenetic relationships. The advocates of this controversial new set of rules for nomenclature claim that its advantages include greater stability, as the naming of new taxa will not require previously discovered taxa to be renamed (http://www.ohiou.edu/phylocode).

Common Names

Common names for North American fishes have been gleaned from tradition or created by the Committee on Names of Fishes of the American Fisheries Society (Nelson et al., 2004) for the convenience of those who are uncomfortable with Latin names and also to provide continuity when occasional changes in formal classification are necessary. Some have argued that common names must reflect natural groups, the way formal names must. But that would destroy the traditional value, stability, and ecological uses of common names. (Ichthyologists certainly do not want to get in the same hot water that dinosaur biologists got in for promoting the change from the well-known name of *Brontosaurus* to *Apatosaurus*.) As an example of ecological application, the terms *trout* and *salmon* have been used to refer roughly to life history patterns—trouts usually do not migrate to the sea (except for anadromous rainbow trouts, cutthroat trouts, and some strains of brown trouts), and salmon usually do (except for kokanee and landlocked Atlantic salmon). These names are useful because of their widespread recognition; their inconsistent application can be mitigated by the use of scientific names.

Higher Classification of Fish Groups

The higher classification of fishes conforms to the Linnaean hierarchy of levels, of which two—*families* and *orders*—are traditionally held to be important. Here we must distinguish between fish groups and categories. The group names (e.g., Cottidae, Sphyrnidae) are intended to refer to real, evolved, and related entities in nature; the categories or levels (family, order) are not. The term **clade** (from the Greek *klados* = branch) has become widely used in systematic literature, as it denotes a group of organisms that share features by virtue of common descent. As more is learned about the

TABLE 1.1 FORMAT AND TERMINOLOGY FOR TAXONOMIC CLASSIFICATION (AFTER WEITZMAN, 1997)

Category	Standard Suffix
Class	
Division	
Subdivision	
Infradivision	
Group	
Subgroup	
Infragroup	
Section	
Subsection	
Infrasection	
Order	-iformes
Suborder	-oidei
Superfamily	-oidea
Family	-idae
Subfamily	-nae
Tribe	-ini

natural groups of fishes (e.g., Cichlidae, Otophysi, Elasmobranchii, Teleostei), the categories become more arbitrary and less relevant to scientific discussion. Nevertheless, they are useful for recording biodiversity in a general way. (In descending order, some categories are *kingdom, phylum, class, division, order, family, tribe, genus, species*, plus intermediate categories identified by the prefixes *super-*, as in *superfamily*, or *sub-*, as in *suborder*). Many workers prefer to end ordinal names with a standard ending, such as orders with *-iformes* and suborders with *-oidei*, but the practice is becoming less standardized (Table 1.1).

Centers for the Study of Fishes

Most colleges and universities have small teaching and research collections of zoological specimens. Some institutions have become centers for fish research, and their collections are among the largest and most comprehensive, reflecting their pioneering role in the global exploration of fishes. The Natural History Museum in London and the Musée Nationale d'Histoire Naturelle in Paris are two of the best-known research museums in the world.

In the United States, eight institutions figure most prominently in ichthyological research: the National Museum of Natural History (USNM, a branch of the Smithsonian Institution) in Washington, DC; the American Museum of Natural History (AMNH) in New York City; the Academy of Natural Sciences (ANSP) in Philadelphia; the Museum of Comparative Zoology (MCZ) at Harvard

University in Cambridge, Massachusetts; the University of Michigan Museum of Zoology (UMMZ) in Ann Arbor; the Field Museum of Natural History (FMNH) in Chicago; the California Academy of Sciences (CAS) in San Francisco; and the Natural History Museum of Los Angeles County (LACM). More than 24 million fish specimens are housed in just these eight institutions (Poss and Collette, 1995).

Some of these museums are affiliated with world–class aquaria, where the public can view fishes in all their glory. Excellent exhibits can be seen at the National Aquarium in Baltimore, the John G. Shedd Aquarium in Chicago, and the Steinhart Aquarium of the California Academy of Sciences in San Francisco. Improvements in the technology of fish maintenance, coupled with an increased interest in family-oriented recreational activities, have fueled a boom in aquarium building in the United States in the past decade. Metropolitan areas are turning to aquariums as a means of revitalizing downtown areas. Excellent exhibits focusing on regional fish faunas can be seen at the Monterey Aquarium, housed in an old cannery building in Monterey, California; the New England Aquarium in Boston; the Chattanooga Aquarium in Tennessee; and the Newport Aquarium across the Ohio River from Cincinnati.

KEY POINTS AND CONNECTIONS

- Insight into the ways of fishes dates back many centuries before the formalization of vertebrate study. This is evident from the artistic expression of early cultures as divergent as the ancient Egyptians and the aboriginal Americans.

- The ancient Greeks and Romans are credited with the development of a formal approach to scientific study, and much of this was directed at understanding the ways of fishes. Aristotle's early taxonomic work included numerous observations of the biology of fishes.

- As was the case with science in general, scholarship on fishes rapidly developed during the Renaissance. Advances in shipbuilding and navigation greatly expanded the horizons of European society, and the great explorer-naturalists of the 16th, 17th, and 18th centuries significantly enhanced our understanding of the biology of fishes.

- The emergence of institutions of higher learning and the great museums of natural history during the 19th and 20th centuries placed ichthyology in a new context—one in which we see biology, including the study of fishes, incorporated in the development of academic disciplines. David Starr Jordan exemplified the master academician whose scientific skill was directed toward the study of fishes.

- The development of special techniques for the collection and storage of fish specimens, and adherence to certain protocols for measuring,

identifying, and naming fishes has greatly enhanced our ability to make sense of this most diverse group of vertebrates.

FISH LINKS

http://www.fishbase.org Published by the International Center for Living Aquatic Resources Management, MCPO Box 2631, 0718 Makati City, Philippines. This is one of the most comprehensive online resources for fish biology. It has links with many other fish-related sites.

http://www2.biology.ualberta.ca/jackson.hp/IWR/index.php Ichthyology Web Resources, a site maintained by Keith L. Jackson at the University of Alberta. This site is one of the most extensive compilations of resources of interest to fish biologists.

http://www.calacademy.org/research/ichthyology/catalog/ Website for the Ichthyology Department at the California Academy of Sciences. Includes an online version of William Eschmeyer's *Catalog of Fishes*, published by the California Academy of Sciences in 1998.

http://www.flmnh.ufl.edu/fish/ A valuable site that communicates the breadth of ichthyological research being done at a major fish research facility, the Florida Museum of Natural History at the University of Florida.

www.ohiou.edu/phylocode/ PhyloCode is the new phylogenetic nomenclature proposed to supplement or possibly replace the conventional Linnaean system.

BUILDING AN ICHTHYOLOGY LIBRARY

References on the History of Ichthyology

Cuvier, G., and T. W. Pietsch (Trans., Ed.). 1995. *Historical portrait of the progress of ichthyology from its origins to our own time.* Johns Hopkins University Press, Baltimore.

Jordan, D. S. 1905. *A guide to the study of fishes.* Vols. 1 and 2. Henry Holt, New York.

———. 1922. *The days of a man: Being memories of a naturalist, teacher and minor prophet of democracy.* Vols. 1 and 2. World Book, Yonkers-on-Hudson, NY.

Pietsch, T. W., and W. D. Anderson, Jr. (Eds.). 1997. Collection building in ichthyology and herpetology. *Am. Soc. Ichthyol. Herpetol. Spec. Publ. 3.*

Valuable Texts Covering Techniques in Basic Taxonomy and Fish Biology

Blackwelder, R. E. 1967. *Taxonomy: a text and reference book.* John Wiley, New York.

Committee on the Use of Fishes in Research. 2004. *Guidelines for the use of fishes in research.* Available online at http://www.fisheries .org/html/Public_Affairs/Sound_Science/Guidelines2004.shtml (The need to guarantee the health and welfare of fishes that are experimental subjects in the lab and in the field makes this essential reading.)

Eschmeyer, W. N. 1998. Catalog of fishes, Vols. I–III. *Calif. Acad. Sci. Spec. Publ. 1,* Center for Biodiversity Research and Information, San Francisco. (Certainly one of the most ambitious ichthyological endeavors in the past century.)

Lagler, K. F. 1952. *Freshwater fishery biology*. W. C. Brown, Dubuque, IA. (Still one of the best practical ichthyology texts.)

Murphy, B. R., and D. W. Willis (Eds.). 1996. *Fisheries techniques* (2nd ed.). American Fisheries Society, Bethesda, MD.

Schreck, C. B., and P. B. Moyle (Eds.). 1990. *Methods for fish biology*. American Fisheries Society, Bethesda, MD.

• •

REFERENCES

• •

Åhlander, E., S.O. Kullander, and B. Fernholm. 1997. Ichthyological collection building at the Swedish Museum of Natural History, Stockholm, pp. 13–25. In Collection building in ichthyology and herpetology, T. W. Pietsch and W. D. Anderson (Eds.). *Am. Soc. Ichthyol. Herpetol. Spec. Publ. 3.*

Balon, E. K., M. N. Bruton, and D. L. G. Noakes. 1994. Women in ichthyology: An anthology in honour of ET, Ro, and Genie. *Env. Biol. Fishes 41:* 7–438.

Bauchot, M.-L., J. Daget, and R. Bauchot. 1997. Ichthyology in France at the beginning of the 19th century: The *Histoire Naturelle des Poissons* of Cuvier (1769–1832) and Valenciennes (1794–1865), pp. 27–80. In Collection building in ichthyology and herpetology, T. W. Pietsch and W. D. Anderson (Eds.). *Am. Soc. Ichthyol. Herpetol. Spec. Publ. 3.*

Berra, T. M. 1997. Some 20th-century fish discoveries. *Env. Biol. Fishes 50:* 1–12.

———. 2001. *Freshwater fish distribution.* Academic Press, San Diego.

Cuvier, G., and T. W. Pietsch (Trans., Ed.). 1995. *Historical portrait of the progress of ichthyology from its origins to our own time.* Johns Hopkins University Press, Baltimore.

Dymond, J. R. 1964. A history of ichthyology in Canada. *Copeia 1964:* 33–41.

Eschmeyer, W. N. 1998. Catalog of fishes, Vols. I–III. *Calif. Acad. Sci. Spec. Publ. 1.*

Fernholm, B., and A. Wheeler. 1983. Linnaean fish specimens in the Swedish Museum of Natural History, Stockholm. *Zool. J. Linn. Soc. 78:* 199–286.

Gans, C. 1978. All animals are interesting! *Amer. Zool. 18:* 3–9.

Horn, M. H. 1976. In honor of Carl L. Hubbs. *Bull. So. Calif. Acad. Sci. 75:* 57–59.

Hubbs, C. L. 1964a. History of ichthyology in the United States after 1850. *Copeia 1964:* 42–60.

———. 1964b. David Starr Jordan. *Syst. Zool. 13*(4): 195–200.

———, and K. F. Lagler. 1958. *Fishes of the Great Lakes region.* University of Michigan Press, Ann Arbor.

Jett, S. C., and P. B. Moyle. 1986. The exotic origins of fishes depicted on prehistoric Mimbres pottery from New Mexico. *Amer. Antiquity 51:* 688–720.

Jordan, D. S. 1905. *A guide to the study of fishes,* Vols. 1 and 2. Henry Holt, New York.

Kessel, E. L. (Ed.). 1970. *Festschrift* for George Sprague Myers: In honor of his sixty-fifth birthday. *Proc. Calif. Acad. Sci. 38.*

Lindberg, G. U. 1974. *Fishes of the world: A key to families and a checklist.* John Wiley, New York.

Lowe-McConnell, R. H. 1987. *Ecological studies in tropical fish communities.* Cambridge University Press, Cambridge, UK.

Moyle, P. B., and M. A. Moyle. 1991. Introduction to fish imagery in art. *Env. Biol. Fishes 31:* 5–23.

Myers, G. S. 1964. A brief sketch of the history of ichthyology in America to the year 1850. *Copeia 1964:* 33–41.

Nelson, J. S. 1994. *Fishes of the world* (3rd ed.). John Wiley, New York.

———, E. J. Crossman, H. Espinosa-Pérez, L. T. Findley, C. R. Gilbert, R. N. Lea, and J. D. Williams. 2004. *Common and scientific names of fishes from the United States, Canada, and Mexico* (6th ed.). American Fisheries Society, Bethesda, MD.

Nielsen, J. G. 1993. Peter Forsskål—a pioneer in Red Sea ichthyology. *Israel J. Zool. 39:* 283–286.

Norris, K. S. 1974. To Carl Leavitt Hubbs, a modern pioneer naturalist on the occasion of his eightieth year. *Copeia 1974:* 581–594.

Pietsch, T. W., and D. B. Grobecker. 1987. *Frogfishes of the world: Systematics, zoogeography, and behavioral ecology.* Stanford University Press, Stanford, CA.

Poss, S. G., and B. B. Collette. 1995. Second survey of fish collections in the United States and Canada. *Copeia 1995:* 48–70.

Strauss, R. E., and C. E. Bond. 1990. Taxonomic methods: Morphology, pp. 109–140. In *Methods for fish biology,* C. B. Schreck and P. B. Moyle (Eds.). American Fisheries Society, Bethesda, MD.

Strauss, R. E., and F. L. Bookstein. 1982. The truss: Body form reconstructions in morphometrics. *Syst. Zool. 31:* 113–135.

Taylor, W. R. 1967. An enzyme method of clearing and staining small vertebrates. *Proc. U.S. Nat. Mus. 122:* 1–17.

Walsh, S. J., and M. R. Meador. 1998. Guidelines for quality assurance and quality control of fish taxonomic data collected as part of the National Water Quality Assessment Program. *U.S. Geol. Surv. Water Resources Invest. 98-4239.*

Weitzman, S. H. 1997. Systematics of deep-sea fishes, pp. 43–77. In *Deep-sea fishes,* D. J. Randall and A. P. Farrell (Eds.). Academic Press, San Diego.

Withgott, J. 2000. Is it "So long, Linnaeus"? *Bioscience 50*(8): 646–651.

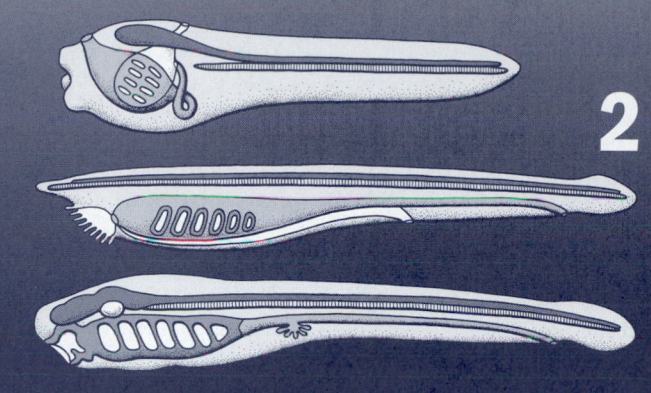

2

ORIGINS OF FISHES

FISHES—THE FIRST VERTEBRATES

ORIGINS OF VERTEBRATES

MAJOR GROUPS OF LIVING FISHES
Jawless Fishes
Cartilaginous Gnathostomes
The Elasmobranchs
The Holocephalans
Bony Fishes
Evolutionary Features

Fishes have an extraordinary evolutionary history, dating back more than 500 million years. As members of the deuterostome phylum Chordata, fishes share some affinities with echinoderms, hemichordates, and cephalochordates. Classical morphological studies, coupled with recent discoveries concerning the genetic mechanisms controlling development, have given us insight into the origins of the vertebrate body plan. Measured in terms of total species diversity, the fishes have been extravagantly successful in the adaptation of this body plan to the exigencies of aquatic life. Several lineages to which we have assigned the common name "fishes" have evolved over the last 500 million years. These have contributed in vastly different quantities to the modern fish fauna. Depending on one's perspective, the jawless fishes—the hagfishes and lampreys—constitute a single, monophyletic taxon or two taxa of disparate origins. More numerous are the chondrichthyans—cartilaginous fishes that have displayed remarkable morphological diversity throughout their long evolutionary history. By far the most diverse group is the osteichthyans—bony fishes that constitute the vast majority of modern fishes in the world's fresh and marine waters. This chapter will serve as a brief introduction to the origins and diversification of these lineages of aquatic vertebrates.

FISHES—THE FIRST VERTEBRATES

Before we delve into the question of just what constitutes a fish, we need to be sure that we understand what constitutes a vertebrate. We somewhat pejoratively refer to the fishes as "lower" vertebrates. In no way should this cause us to view fishes as somehow less adapted or less successful than the "higher" vertebrates—one of the main goals of this book is to convince you otherwise. Denoting fishes as "lower" vertebrates only refers to the fact that fishes appear as the original vertebrate group and, hence, the source of all other, "higher" vertebrates. Those animals to which we apply the term "fish" are the longest surviving group of vertebrates. They encompass groups that have had a continuous phylogenetic history of several hundred million years, as well as groups that can trace their ancestry back only to the Cretaceous period, when dinosaurs roamed the Earth (Table 2.1). Hence, when we refer to a fish as "modern," we only mean that it has survived into the present. Modern fishes thus represent all levels of phylogenetic advancement, from the ancient hagfishes to the much more recently evolved perches.

ORIGINS OF VERTEBRATES

Until recently, biologists classified all life into one of five *kingdoms*. Microbiologists studying prokaryotes that are believed to resemble the earliest life forms have provided convincing evidence that the so-called "Archaebacteria" should be split off from the rest of the Monera (Prokaryotae) and placed in their own separate kingdom. The kingdom **Animalia** is believed to have arisen from some protozoan lineage of the kingdom **Protista** that acquired an integrated, multicellular body form. From bilaterally symmetrical, **triploblastic** (i.e., possessing three embryonic tissue sources, the **ectoderm**, **mesoderm**, and **entoderm**) animals arose more complex forms that acquired an internal body cavity—the **coelom**—that was of tremendous adaptive significance. The coelom permitted an increase in internal space to house tissues and organs, provided an internal repository for wastes, and contributed to a more efficient hydrostatic skeleton, in which an arrangement of longitudinal and circular trunk muscles could work against this fluid-filled internal space. The coelomic cavity appeared more than once in the evolution of animals; in one lineage, the **eucoelomates**, the cav-

TABLE 2.1   REPRESENTATION OF SOME MAJOR GROUPS OF FISHES THROUGH TIME

Era	Period	Epoch	Dur[1]	BP[2]	Major Group
Cenozoic	Quaternary	Recent			
		Pleistocene	1.8	1.8	
	Tertiary	Pilocene	3.2	5	
		Miocene	19	24	
		Oligocene	14	38	
		Eocene	20	58	
		Palaeocene	7	65	
Mesozoic	Cretaceous		79	144	
	Jurassic		64	208	
	Triassic		37	245	
Palaeozoic	Permian		41	286	
	Carboniferous		74	360	
	Devonian		48	408	
	Silurian		30	438	
	Ordovician		67	505	
	Cambrian		65	570	

Major Group columns (left to right): Myxini, Pteraspidomorpha, Cephalaspidomorpha exc. petromyzonts, Petromyzontiformes, Placodermi, Elasmobranchii, Holocephali, Acanthodii, Actinopterygii, Coelacanthimorpha, Dipnoi

[1] Approximate duration in million of years.
[2] Approximate million of years before present at beginning of each period of epoch.

ity arose entirely within the mesoderm, such that mesodermally derived tissues formed a lining for the entire cavity.

Early worm-like eucoelomates diverged into two basic lineages, the *protostomes* and *deuterostomes.* Variation in the sets of genes controlling differentiation (see further on) suggests that the split between protostomes and deuterostomes could have been initiated during the Cambrian, almost 550 million years ago (Carroll, 1995; Erwin et al., 1997, Erwin, 1999). Some molecular studies, however, place this divergence from 630 million to almost one billion years ago—up to 400 million years earlier than indicated by the fossil record (Hausdorf, 2000; Lynch, 1999; Wang et al., 1999).

Protostomes are distinguished from deuterostomes primarily by differences in embryonic development, including the axis of development. **Protostomes** are so named because the **blastopore** (the early invagination in a ball of embryonic cells that ultimately contributes to the establishment of the triploblastic condition) occurs near what will become the mouth of the animal. In **deuterostomes**, on the other hand, the blastopore forms in the vicinity of the anus. The vast majority of animal species are protostomes, as this group includes such extraordinarily diverse phyla as the arthropods and the mollusks. The major deuterostome phyla are the **Echinodermata, Hemichordata,** and **Chordata,** the latter including the subphylum **Vertebrata**. Because of the fundamental similarities in the embryology of the deuterostomes, students enrolled in a course in vertebrate developmental biology will usually begin by studying the embryogenesis of sea urchins or starfish.

Investigations of the evolutionary relationships of the deuterostomes have documented that the echinoderms, hemichordates, and chordates constitute a monophyletic assemblage. One group of fossil deuterostomes, the **carpoids,** sometimes referred to as **calcichordates,** stylophorans, or mitrates (Fig. 2.1), is especially intriguing, as affinities with both echinoderms and chordates have been proposed for them (Jefferies, 1986, 1997; Peterson, 1994). Dominguez et al. (2002) have identified paired pharyngeal slits—a distinctly chordate feature—in a fossil carpoid. Pharyngeal gill slits are a defining feature of members of the phylum Hemichordata (a group that includes the worm-like acorn worms and the sessile, tentaculate pterobranchs) and of members of the phylum Chordata (Fig. 2.2).

In addition to the subphylum Vertebrata, the phylum Chordata includes the subphyla **Urochordata,** commonly known as *tunicates,* and the **Cephalochordata,** commonly known as *lancelets* or *amphioxus.* The urochordates, at first glance, suggest nothing that would demonstrate affinities with the vertebrates. Adult tunicates are mainly sessile filter feeders, although several forms exist that are modified for a planktonic existence. The term *tunicate* comes from the mantle or tunic that covers the epidermis of the animal. Filter feeding is accomplished using a perforated pharynx. When one looks at the diminutive larvae of tunicates, however,

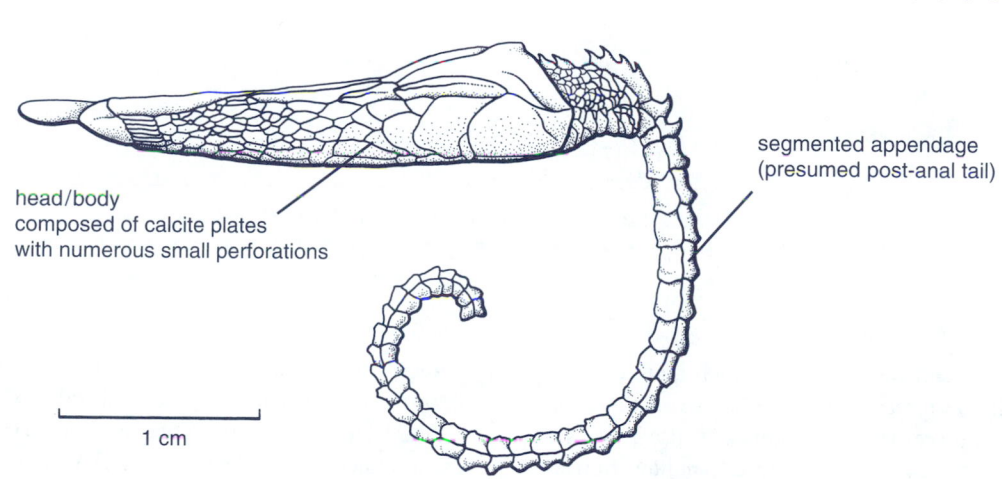

head/body
composed of calcite plates
with numerous small perforations

segmented appendage
(presumed post-anal tail)

1 cm

FIGURE 2.1

A carpoid (calcichordate), a possible protochordate group that dated from the Middle Cambrian to the Middle Devonian (520–370 MYA). Source: Jefferies, 1986.

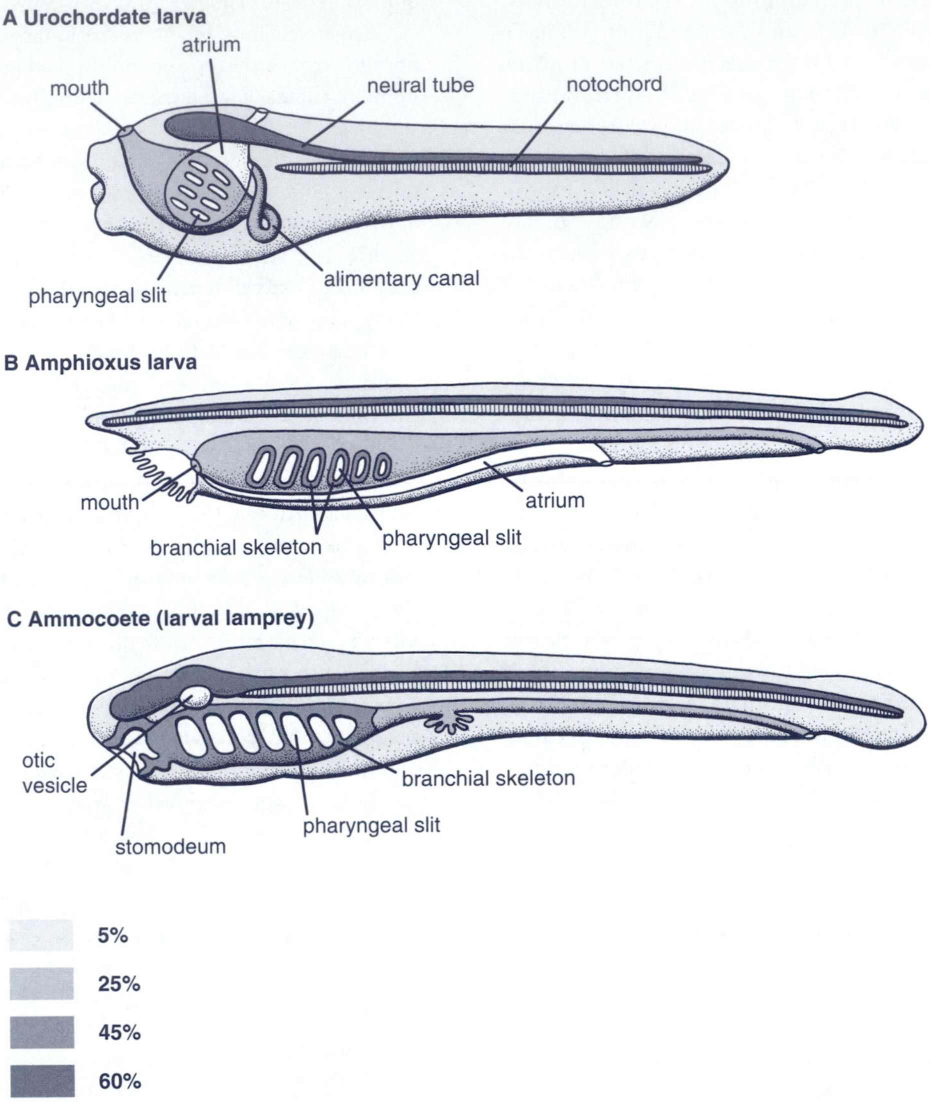

A Urochordate larva

atrium

mouth neural tube notochord

pharyngeal slit alimentary canal

B Amphioxus larva

mouth atrium

branchial skeleton pharyngeal slit

C Ammocoete (larval lamprey)

otic vesicle branchial skeleton

pharyngeal slit

stomodeum

5%

25%

45%

60%

FIGURE 2.2

Comparison of **A**, a larval tunicate with **B**, a cephalochordate, and **C**, the ammocoetes (larval) stage of a lamprey.

the chordate affinities of the group are much more evident. In addition to the perforations in the pharynx, a condition known as **pharyngotremy**, larval tunicates also possess a well-developed **notochord** with a dorsal, hollow **neural tube**. The latter two structures support the actions of a muscular post-anal tail region. These features provide the larval tunicate with the motility necessary for successful dispersal before eventually undergoing metamorphosis into the sessile

adult form. It is this combination of characteristics that serves to unify members of the phylum Chordata (Fig. 2.2).

Under what circumstances and in what manner did the earliest ancestors of the fishes arise? Walter Garstang (1928) proposed a theory that remains one of the most satisfying explanations of vertebrate origins. According to Garstang, vertebrates arose from tunicate-like ancestors that produced the larval body form seen in modern urochordates. Through a process he termed **paedomorphosis**, the adult stage was progressively delayed, and the creature persisted in the larval form. The maturation of the reproductive organs proceeded independently of the maturation of body form, and the creature eventually became sexually mature while still in the larval, juvenile body form. Hence, the sessile adult stage was dispensed with entirely, and the basic body form seen in fishes, with a perforated pharynx, notochord, dorsal spinal cord derived from the hollow neural tube, and a post-anal tail, or caudal, region, was established. In the subphylum Cephalochordata, the notochord persists as the main internal support for the animal. In this group, the head region is poorly defined, with the notochord extending from the tip of the tail to the tip of the snout (the name *amphioxus* literally means "sharp at both ends").

Cephalochordates share with vertebrates a number of other features, including a highly segmented body plan—a feature not evident among the tunicates but present in many protostome phyla. We use the term **metamerism** to refer to the serial repetition of both internal and external body features. In this way, the skeleton, trunk muscles, nervous system, and other anatomical features may be serially repeated in cephalochordates and vertebrates. The division of the nervous system into a somatic portion, controlling the segmental trunk structures, and a visceral system, controlling the internal soft organs—primarily the gut tube derivatives—is also a feature unique to the cephalochordates and vertebrates. Amphioxus, the best known being the lancelet (*Branchiostoma lanceolatum*), swims much like fishes, employing motor nerves from the spinal cord to stimulate the trunk muscles to contract in alternating fashion.

Most lancelets are sessile forms that burrow in the sediments of shallow marine waters. Here, they use their perforated pharynx to filter-feed much like the tunicates. Whereas tunicates and cephalochordates depend on ciliary action in the pharynx to generate a feeding current, vertebrates have developed a muscularized pumping mechanism. The cartilaginous gill arches or branchial basket seen in vertebrates possess a greater recoil elasticity that helps in the pumping action.

Two key features distinguish cephalochordates from vertebrates. Cephalochordates are small, with only modest oxygen requirements that can be largely met by diffusion across the undifferentiated body surface. The pharyngeal perforations are mainly a feeding mechanism. In vertebrates, these perforations become associated with the development of extensive gill surface areas to promote greater oxygen uptake. The other distinguishing feature is, of course, the acquisition of a cartilaginous or bony *vertebral column* to replace the notochord. We see varying degrees of

vertebral development among fishes, ranging from the lampreys and hagfishes with little if any traces of vertebrae to the perch-like fishes with well-developed vertebral structures. The embryonic vertebrate first establishes the notochord as an internal skeletal support. This will serve as a template, around which the vertebral elements will form. Vertebrates also develop a much more distinct head region. Whereas the notochord runs to the tip of the snout in cephalochordates, it terminates behind the skull in vertebrates (Fig. 2.2). It is for this reason that the cephalochordates are considered "acraniates," whereas vertebrates are classified as "craniates."

Although Garstang's hypothesis is not without its detractors (e.g., Ruben and Parrish, 1990), there is much evidence to support it. The tail of the larval tunicate consists of serially repeating bands of muscle—a good candidate for the metameric arrangement of muscles in the trunk of cephalochordates and vertebrates. A highly modified class of tunicates, the planktonic larvaceans, shows a paedomorphic body plan in which the adult retains the tail. Lacalli (1999) provided molecular evidence linking the development of the vertebrate trunk with tail structures seen in modern tunicates.

Recent studies on the developmental genetics of animals may prove instrumental in clarifying the role of paedomorphosis in chordate evolution. What we understand of the genetic mechanisms controlling the development and differentiation of animals leads us to believe that a dramatic modification of the body plan—as the paedomorphosis hypothesis suggests underlies the genesis of the vertebrate body plan—may be under the control of a few select gene families. These gene sets, termed **homeobox, homeotic**, or *hox* genes, could be common to several otherwise apparently unrelated lineages (see Chapter 28). In the late 19th century, an argument was advanced that the annelids shared fundamentally the same body plan as chordates, only inverted. This notion, much derided throughout most of the 20th century, actually may have some merit, as it is consistent with the discovery of genes specific for the dorsoventral patterning of the bilateral body plan. For example, vertebrates (deuterostomes) possess the gene *chordin*, which is responsible for patterning dorsal structures, whereas insects (protostomes) possess the gene *sog*, responsible for ventral determination. Geneticists view these two genes (as well as related ones responsible for patterning the opposite side in each animal) as fully homologous counterparts (Lacalli, 1996). Homologous gene sets are also implicated in the formation of structures such as the neural tube, long believed to have a common origin among vertebrates, cephalochordates, and urochordates (Shimeld, 1999). Indeed, the phenomenon of metamerism itself, long regarded as independently acquired by proto-stomes and deuterostomes, is now seen to be the product of similar *hox* gene clusters.

The modification of comparatively few regulatory genes may be the key to understanding major evolutionary events, such as the split between protostomes and deuterostomes, and the dramatic shift in body form that Garstang suggested took place in the ancestor of the modern chordate subphyla. Recently, researchers have discovered that the formation of the tail in tunicates is under the control of a single gene (Pennisi and Roush, 1997). The existence of such a gene, termed *manx* by its discoverers, would suggest that the tailed larval form is indeed an ancestral body type. The acquisition of the sessile adult phase would be viewed as a derived condition that coincided with the suppression of the *manx* gene.

The appearance of a more defined head region distinguishes the vertebrates from earlier forms such as the cepha-lochordates (Gans and Northcutt, 1983). Cephalochordates have been shown to possess only one *hox* gene cluster, whereas most vertebrates have been demonstrated to have four (Garcia-Fernandez and Holland, 1994). Duplication of the critical *hox* gene clusters is now believed to be the key to understanding the origins of the vertebrate head (Monastersky, 1996). The evolutionary history of the actinopterygians, especially the teleosts, appears to be one in which *hox* clusters have undergone additional proliferation and subsequent elimination of selected genes within the clusters (Málaga-Trillo and Meyer, 2001).

We have a much clearer understanding of the past 400 million years of vertebrate evolution, due largely to the existence of an extensive fossil record, than we do of the earliest origins of the vertebrates. The fortuitous discoveries of fossils of early soft-bodied chordates (see Going Deeper) give us a glimpse, at least, into our earliest origins.

GOING DEEPER · What Is the Oldest Fish?

Humans seem to have an intuitive need to be informed of life's superlatives, so it is inevitable that students of fishes would ask, "What is the world's oldest fish?" Answering this question is not easy, however, as two complications arise. One of them is the inadequacy of the fossil record in providing a comprehensive explanation of the early history of life on Earth. If we view life as a reel of film continuously unwinding for our pleasure and amazement, the fossil record constitutes only a few frames of the story that circumstances have placed at our disposal several million years after the events unfolded. The other complication is a problem of semantics—just what do we mean by the term *fish*? As it is a term of our own devising, we can apply it at our own discretion. For the sake of consistency, however, we should attempt to be as precise as possible in defining a fish, so that we can more readily distinguish them from their chordate ancestors. Lancelets, for example, are not normally termed "fishes," yet we do not seem to have a problem in designating hagfishes as "fishes," even though they might be considered a sister group to all other vertebrates. With these caveats in mind,

we can attempt to answer the question at hand.

Minute fossils known as **conodonts** (Plate 2.1) have long been useful to paleontologists as biostratigraphic indicators correlating with specific sedimentary sequences. What has not been understood until recently is their phylogenetic position, mainly because fossil conodonts were never observed in the context of the organisms that possessed them. All this changed with the fortuitous discovery of fossils from the Lower Carboniferous of the Edinburgh district, Scotland, in 1983. These fossils definitively showed conodonts to be tooth elements of an elongate, soft-bodied, vertebrate-like creature somewhat resembling modern hagfishes. Because conodonts appear in the fossil record as far back as the late Cambrian, they provide convincing evidence that vertebrates may have been well established by then (Briggs, 1992). Histochemical studies by Kemp and Nicoll (1996), however, suggest that conodonts are more closely related to cephalo-chordates than to vertebrates.

The so-called "Cambrian explosion" of approximately 550 million years ago, in which a vast array of new animal forms rapidly diversified,

was likely the consequence of the evolution of the capacity to mineralize tissue. Not only did this calcification confer significant adaptive advantages on corals, mollusks, arthropods, echinoderms, and chordates, but it resulted in tissue that would lend itself much more readily to fossilization. Because paleontologists are adept at reconstructing an entire animal from its skeletal framework, we have a very good idea what animals possessing calcified skeletal parts looked like, even though they may have long since disappeared. However, the constraints of fossilization have frustrated our attempts to understand much of the history of soft-bodied organisms, including vertebrate ancestors, in the era preceding the Cambrian—soft bodies just do not lend themselves to adequate fossilization. It is for this reason that the Lower to Middle Cambrian Burgess Shale formation of British Columbia is of such inestimable value. Here, conditions were just right for the fossil preservation of a spectacular array of soft-bodied creatures, some of which appear to be the earliest ancestors of the vertebrates. One such creature, the 40-millimeter *Pikaia* (Plate 2.2), may be the earliest known chordate (Gould, 1989).

MAJOR GROUPS OF LIVING FISHES

The term *fish* is an all-encompassing one, referring as it does to a diverse assemblage of vertebrate taxa (Fig. 2.3). The earliest classification systems relegated fishes to one class, *Pisces*. Throughout this book, we will refer mainly to Nelson's (1994) classification scheme (Table 2.2). The science of systematics marches on, however, and studies published since the appearance of Nelson's work have resulted in several revisions to the classification of fishes, and these revisions are included in the systematics chapters of this book. Many of these revisions can be attributed to the significant role that molecular biology currently plays in elucidating the evolutionary relationships of animals (see Chapter 4). The discovery and analysis of an increasing quantity of fossil material has also changed our perception of the evolutionary history of fishes.

In the evolution of fishes, two events are most noteworthy—the acquisition of mobile *paired appendages* and the development of functioning *jaws*. The development of jaws serves to distinguish two superclasses of aquatic vertebrates: the Agnatha and the Gnathostomata. The modern agnaths apparently arose from ancestors that predated the origin of paired appendages, whereas this feature is indicative of the gnathostomes. For our purposes, then, a *fish* is a vertebrate that possesses a cranium, is aquatic throughout its life, and hence relies mainly on gills for gas exchange. Even as broad a generalization as this does not account for such fishes as mudskippers (Gobiidae; see Chapter 16), which spend most of their time out of the water, or such amphibians as the mudpuppy (*Necturus*), which has gills and is thus confined to the water. If we also want to restrict our definition of a fish to a vertebrate in possession of jaws and paired appendages, this would exclude the lampreys and hagfish. Indeed,

Several startling fossil discoveries that have recently been made in the Chengjiang region of southern China may well push the origins of the fishes back into the early Cambrian. Shu et al. (1999) and Chen et al. (1999) have described genera of lamprey-like creatures, indicating fishes to be present at least 530 million years in the past. Indeed, the anatomical sophistication apparent in these fossils suggests that they might trace their origins as far back as 750 million years ago (Zimmer, 1999). One of these genera, *Haikouichthys* (Plate 2.3), appears to possess vertebral elements associated with its notochord and cephalic structures resembling the ammocoete larvae of lampreys (Shu et al., 2003a). Studies of fossils of the genus *Haik-ouella* (Plate 2.4), however, suggest that it is derived from even earlier deuterostome stock, as it possessed external gills but lacked pharyngeal gill pouches and showed no evidence of a notochord (Shu et al., 2003b).

The modern hagfishes and lampreys and the aforementioned fossil chordates such as *Pikaia* and *Haikouichthys* might be dismissed as merely "fish wannabes," because they lack the dermal accoutrements normally associated with true fishes. If this is the case, we must look elsewhere for the definitive earliest fish. Pierre-Yves Gagnier (1989) reported on a remarkably complete fossil of a creature that had the skin structure and general body shape of an early group of fishes known as the Arandaspida (see Chapter 5). This Bolivian fossil, *Sacabambaspis*, was dated at 470 million years old and was claimed at the time to be the oldest fish. That the fossil record of armored fishes may extend back even further has been suggested from some recent analyses of fossil fragments dating back 500 million years to the late Cambrian. Detailed scanning electron microscope studies of these fragments revealed an array of tubes and cavities unique to dentine, a mineralized tissue found in the dermal armor of ancient fishes, the scales of modern fishes, and in our own teeth. These studies concluded that the fragments were from the skin of a fossil fish, *Anatolepis*, first discovered in early Ordovician sediments of Spitsbergen, but now demonstrated to have been present during the late Cambrian (Smith et al., 1996). Another late Cambrian fish-like vertebrate with a somewhat differing dermal ultrastructure has also been discovered in Australia (Young et al., 1996). So, for now, it appears that *Anatolepis* has the nod as the world's oldest armored fish.

What should be obvious by now is that fishes, whatever we define them to be, have had a long and convoluted evolutionary history. One particularly interesting feature seen in the fossils of some of the very earliest fish-like chordates is the presence of well-developed paired fins running along the sides of their bodies—a seeming confirmation of Balfour's controversial "fin-fold" theory of the origin of paired fins (discussed in Chapter 5). Possessing the rudiments of paired fin structure, pharyngeal pouches with gills, trunk muscles arranged as segmented myomeres, and a notochord, these earliest fish-like chordates can confidently be placed as full participants in the Cambrian explosion that produced most of the major animal taxa 550 million years ago. A proliferation of armored forms soon followed, and these were apparently well established by the close of the Cambrian.

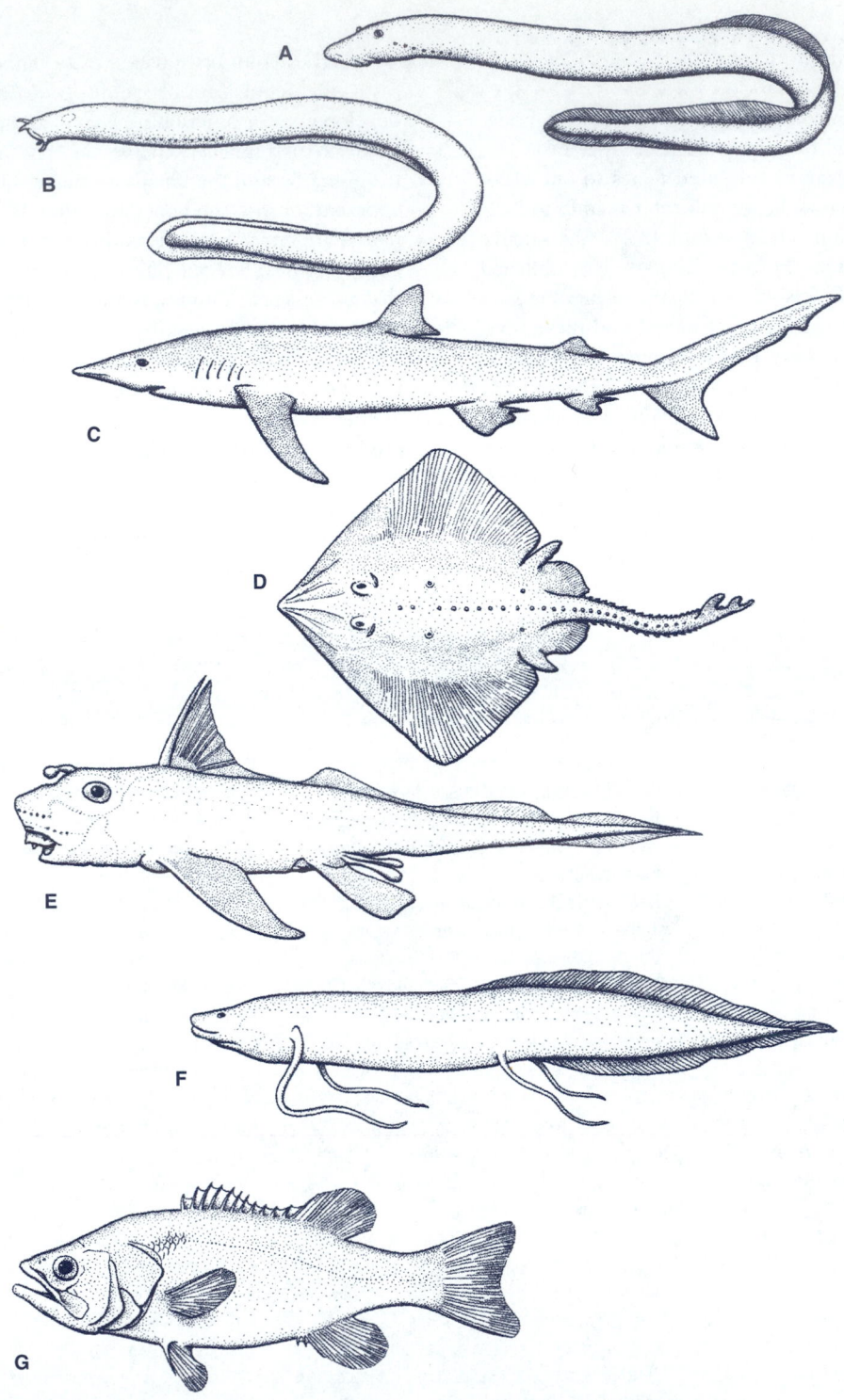

FIGURE 2.3
Examples of groups of living fishes. **A,** lamprey (class Cephalaspidomorphi, order Petromyzontiformes); **B,** hagfish (class Myxini, order Myxiniformes); **C,** shark (class Chondrichthyes, subclass Elasmobranchii); **D,** ray (skate; class Chondrichthyes, subclass Elasmobranchii); **E,** chimaera (class Chondrichthyes, subclass Holocephali); **F,** lungfish (class Sarcopterygii, superorder Ceratodontimorpha); **G,** teleost (class Actinopterygii, subclass Neopterygii).

TABLE 2.2   HIGHER CATEGORIES OF FISHES TO THE LEVEL OF DIVISION (AFTER NELSON, 1994)

Super class Agnatha
    Class Myxini (hagfishes)
    †Class Pteraspidomorphi (extinct armored fishes)
    Class Cephalaspidormorphi (lampreys, extinct relatives)
Super class Gnathostomata
    †Grade Placodermomorphi
    †Class Placodermi (placoderms)
    Grade Chondrichthiomorphi
    Class Chondrichthyes
        Subclass Holocephali (chimaeras)
        Subclass Elasmobranchii (sharks and rays)
    Grade Teleostomi (bony fishes)
    †Class Acanthodii (acanthodians)
    Class Sarcopterygii (lobe-finned fishes)
        Subclass Coelacanthimorpha (coelacanths)
        Subclass (unnamed)
            †Infraclass Porolepimorpha
            Infraclass Dipnoi (lungfishes)
    Class Actinopterygii (ray-finned fishes)
        Subclass Chondrostei (sturgeons and bichirs)
        Subclass Neopterygii
            Division Teleostei (most bony fishes)

the hagfish is regarded as a sister group of the vertebrates in some classification schemes, owing to the absence of any evidence of a vertebral column (Janvier, 1981). Although the phylogenetic relationship of the hagfish to the bony fishes is far more distant than the relationship of tetrapods to those same bony fishes, it is convenient to group this disparate assemblage of chordate animals together as fishes.

Jawless Fishes

Hagfishes and lampreys (Fig. 2.3 A, B) are both considered "agnathan" (i.e., without jaws) yet the extent to which they share a common phylogeny is debatable (see Chapter 5). Janvier (1981) considered the hagfishes to be protovertebrates or, more specifically, a sister group to all other vertebrate groups, yet this view has been somewhat modified (see Chapter 5). Living hagfishes (class **Myxini**, order **Myxiniformes**) have rudimentary eyes covered with thick skin and are elongate, marine predators and scavengers that lack paired fins and vertebrae. Their skeleton is entirely cartilaginous. The Myxiniformes consist of one family (**Myxinidae**) with 5 genera and about 60 species. These are found mainly in temperate seas, and some species are targeted by fisheries

because of the use of their skins for small leather items. They are characterized by barbels (slender sensory structures) around the mouth and the nasal opening. Species range in length from about 300 mm to one meter.

The lampreys (order **Petromyzontiformes**) comprise 41 extant species that range in length from 75 mm to more than a meter. All undergo a prolonged larval period that they spend buried in the sediments of stream bottoms, and they are seldom seen except during migrations. Larger species (some anadromous—see Chapter 36 for definition of anadromy) are "parasitic" and feed on fishes and, at times, on marine mammals. Application of the term *parasite* to lampreys is somewhat questionable, however. A true parasite must exist in a symbiotic state with its host; killing the host means the demise of the parasite. By attaching itself to a fish and sucking its blood and other body fluids, the lamprey is actually more of a specialized predator. The smaller species usually complete their entire life cycle in freshwater and are filter feeders during the larval stage. Rapid development of the gonads accompanies metamorphosis, and they do not feed as adults. Although all spawn in freshwater, many of the "parasitic" forms grow to maturity in the sea. Lampreys are eel-shaped and have a cartilaginous skeleton. They lack paired fins, and the adults have an oral suctorial disc set with keratinous (horn-like) teeth. They are found in the cold, temperate waters of higher latitudes, primarily in the Northern Hemisphere.

Cartilaginous Gnathostomes

Fishes of the class **Chondrichthyes** are characterized by the presence of jaws, paired appendages, and a skeleton composed of cartilage. Long a favorite of instructors of vertebrate anatomy, they display the vertebrate body plan at its simplest and most straightforward. The extant chondrichthyans can be readily divided into two groups—those possessing serial gill slits along the side of the head (the *Elasmobranchii*) and those with a single gill opening on each side of the head (the *Holocephali*). The class encompasses approximately 60 families, 185 genera, and upwards of 1,160 species; more than 95% of these are elasmobranchs (Compagno, 1999).

The Elasmobranchs

Sharks and rays are in the subclass **Elasmobranchii** ("plategill"). They are widely distributed, mainly in marine waters. Sharks evoke a more visceral response from people than any other group of fishes. Their unsavory reputation is largely unwarranted—only a few species engage in the predatory behavior that prompts such attention from the media. Many species, in fact, are subject to human predation. Elasmobranchs are the target of fisheries, not only for the harvest of their flesh, but also for other materials such as their hides for leather and their jaws and teeth for souvenirs. Whereas only about 30 people are killed each year by sharks, almost 800,000 tons of sharks and rays are taken in fisheries (Paxton and Eschmeyer, 1998). The spines of stingrays and other elasmobranchs are to be avoided, as they can cause painful injury, yet some aboriginal societies have made good use of these spines as sewing needles and spear tips. Sharks are generally large animals: The various species average about 2 m long. The absolute size champion among the fishes is the whale shark (*Rhincodon typus*), a plankton-feeding behemoth that reaches 12 meters and weighs as much as 12,000 kilograms, whereas the dwarf shark (*Squaliolus*) matures at about 200 mm. There are at least 376 and perhaps as many as 480 species of sharks (Compagno, 1991).

The elasmobranchs are an almost entirely marine group, although a few species, including the bull shark, which has been implicated in many shark attacks, enter freshwater. There are approximately 500 species of skates and rays, and like the sharks, they are widely distributed in tropical areas. One subfamily of stingrays, the Potamotrygoninae, is widely distributed in the Amazon River basin. Rays range in width from about 0.1 m to a "wingspan" of almost 7 m in the manta rays.

The Holocephalans

Chimaeras of the subclass **Holocephali** are related to the sharks and rays, and they share certain features of anatomy. Holocephali means "whole head," and it is as much a reference to the fusion of the palatoquadrate to the neurocranium seen in modern species (see Chapter 6) as it is to the presence of only a single gill aperture on either side of the head, which imparts a much more unified appearance to the head than in the elasmobranchs. The common name *chimaera* is a reference to a mythical beast composed of parts of several animals—the bizarre appearance of holocephalans seems to suggest such a creature. Their appearance has also caused them to be known as *rabbitfishes, ratfishes, elephant fishes,* and *ghost sharks.* Some species are used for food, and their oily livers produce a fine lubricant that once was commonly used by fishery workers. There are about 31 species in 6 genera. Except for the few found on the continental shelf, most live in the oceanic depths and are seldom seen or caught.

Bony Fishes

The remainder of the living fishes are distinguished by skeletons of bone or some combination of bone and cartilage. These are the bony fishes, grade **Teleostomi** or "Ostei-

chthyes"—by far the most diverse taxon of fishes. Nelson (1994) placed the count of known species of bony fishes at around 24,000, with FishBase weighing in at close to 28,000 of the 28,800 total fish species described to date. Eschmeyer (1990) estimated that the final tally may include as many as 55,000 named species and subspecies combined. Bony fishes represent about half of all living vertebrates, yet they are the least understood in terms of total species diversity and phylogenetic interrelationships.

One ancient lineage of bony fishes, including the lungfishes (**Dipnoi**), coelacanths (**Actinistia**), and their extinct relatives, is believed to have given rise to the tetrapods (see Chapter 7). Lungfishes and coelacanths (so named from fossils that were noted as having hollow rays) are classified as separate subclasses in the class **Sarcopterygii** ("fleshy-finned"), which also includes the tetrapods (Nelson, 1994). The living coelacanth (*Latimeria*) belongs to a subclass thought to have been extinct for millions of years until a specimen was captured near the coast of South Africa in 1938. A separate population has been identified almost 10,000 km to the east, off the coast of Sulawesi, Indonesia, after a researcher first discovered a specimen for sale at a local fish market (Erdmann et al., 1998; see Chapter 7). This remarkable animal retains many unique features that had been accurately deduced from the fossils of its extinct relatives, although the living species shows specialization for its particular deep-water mode of life, for instance, in its osmoregulatory adaptations (described in Chapter 25) and in the structure of its gas bladder.

The lungfishes are placed in the infraclass Dipnoi ("two-breathing"). They have well-developed lungs that allow them to live in water with low oxygen content. There is one living species of lungfish in South America, one in Australia, and four in Africa. All live in freshwater. Most reach sizes between 600 mm and 1 m in length, but an African species and the Australian species reach about 1.8 m.

Other bony fishes, of the class **Actinopterygii** ("ray-finned"), diversified along evolutionary lines distinct from those that produced terrestrial vertebrates, although some of the members of the subclass Chondrostei, the Polypteridae ("many fins") have some characteristics reminiscent of lungfishes. The Polypteridae were formerly placed in the infraclass Cladistia. This group includes a dozen or so species of African bichirs and reedfishes. Also among the Chondrostei are about 26 species of sturgeons and two species of paddlefishes.

The subclass **Neopterygii** ("new fins") contains the gars and the bowfin of North America and the division **Teleostei** ("complete" or "perfect bone"), which includes more than 20,000 species, of which diverse types are found in all oceans and in most freshwaters.

Bony fishes have been successful in just about all aquatic habitats. About 41 percent of the bony fishes are found in freshwater, with the rest either fully marine or at least tolerant of brackish water. They include brotulas found in the oceanic abyss and clingfishes living in the splash zone of the supratidal zone; killifishes and gobies living in the burrows of estuarine crabs; and catfishes and loaches withstanding the rushing torrents of the Andes and Himalayas. Also among the bony fishes are the pupfishes living in hot springs and the notothenioids living in Antarctic waters so cold that special glycoprotein compounds have evolved in the blood and tissues to prevent freezing (see Chapters 24 and 35). Even the deepest parts of the freshwater aquifer may be inhabited by fishes—the recovery of blind catfishes from deep artesian wells in Texas (see Chapter 10) attests to the habitat versatility of osteichthyan fishes.

Habitat versatility in bony fishes has been accommodated by an astounding diversity of body form and locomotor strategies. Bony fishes swim by many methods; walk and wriggle, both in and out of water; leap, glide, and even fly. They range in size from minuscule 1-cm gobioids to giant tunas, marlins, swordfishes, catfishes, and the arapaima (*Arapaima gigas*) of South America, with lengths exceeding 4 m. The colors of bony fishes rival those of the butterflies and birds; their shapes and postures are bewildering; and their modes of life and some of their anatomical and behavioral adaptations for feeding and breeding challenge the most creative of imaginations.

Evolutionary Features

The bony fishes include soft-rayed forms, which have retained anatomical characteristics that are considered primitive, as well as spiny-rayed forms, which show more derived features. Although some of the most primitive of the modern soft-rayed fishes, such as the gars, have heavy ganoid scales with a complex bony composition, most of them possess much simpler cycloid scales. Pelvic fins tend to be situated along the abdomen in more primitive bony fishes, and many show orbitosphenoid and mesocoracoid bones in the skull and pectoral girdle, respectively (see Chapter 3). Soft-rayed fishes include the bony-tongues or osteoglossomorphs, which are possibly the most primitive living teleosts (Lauder and Liem, 1983), tarpons, herrings, and eels. Trouts, minnows, catfishes, and their relatives also retain many of these primitive characteristics. Fishes with more derived characters include lanternfishes, flying fishes, cods, and the spiny-rayed groups such as perches, basses, scorpionfishes, and related forms. Many of these have spines in the fins, ctenoid scales, and pelvic fins in a thoracic position (see Chapter 3). The acquisition of derived characteristics is not necessarily a measure of success, however. Compara-

tively primitive groups, such as the herrings, are among the most abundant and diverse fishes in the ocean, and minnows and catfishes dominate freshwaters worldwide.

KEY POINTS AND CONNECTIONS

• As members of the phylum Chordata, fishes are characterized by a pharynx with perforations, a dorsal hollow neural tube, and the primordial internal skeletal structure known as the notochord. This body plan has been suggested to have arisen through a process of retention of larval and juvenile characteristics, known as paedomorphosis. The existence of families of regulatory gene sets known as homeotic genes suggests that a dramatic modification of animal form may be accomplished as a consequence of mutation at relatively few gene loci.

The nature of homeotic genes, and their role in vertebrate development and evolution, is discussed in more detail in Chapter 28.

• Although all fishes would be classified as "craniates," the hagfishes, lacking evidence of a vertebral column, have sometimes been classified as a sister taxon of the vertebrates. Lampreys are more definitely vertebrates, and they display greater habitat versatility—their life history typically includes a period of residence in the ocean, with spawning and larval stages spent in freshwaters.

• The class Chondrichthyes includes the sharks, skates, and rays of the subclass Elasmobranchii, and the chimaeras of the subclass Holocephali. The chondrichthyans are characterized by the presence of a cartilaginous skeleton. Although a few species of elasmobranchs penetrate freshwater drainages, the vast majority of the members of this class are confined to marine waters.

• The bony fishes ("Osteichthyes") are, by far, the most diverse group. This assemblage includes two distinct lineages: the lobe-finned Sarcopterygii, from which the tetrapods are believed to have arisen, and the ray-finned Actinopterygii. Only a handful of sarcopterygians have survived into modern times, whereas the vast majority of bony fishes are actinopterygians. The division Teleostei, encompassing most of the modern bony fishes, is widespread in both marine and freshwaters.

FISH LINKS

http://www.geocities.com/CapeCanaveral/Hall/1383/2TopCone.htm Web page of James Davison, collector of conodont fossils. This link includes basic information about conodonts and an extensive library of photographs.

REFERENCES

Briggs, D. E. G. 1992. Conodonts: A major extinct group added to the vertebrates. *Science 256:* 1285–1286.

Carroll, S. B. 1995. Homeotic genes and the evolution of arthropods and chordates. *Nature 376:* 479.

Chen, J.-Y., D.-Y. Huang, and C.-W. Li. 1999. An early Cambrian craniate-like chordate. *Nature 402:* 518–522.

Compagno, L. J. V. 1991. The evolution and diversity of sharks, pp. 15–22. In *Discovering sharks*, S. H. Gruber (Ed.). American Littoral Society, Highlands, NJ.

———. 1999. Systematics and body form, pp. 1–42. In *Sharks, skates, and rays: The biology of elasmobranch fishes*, W. C. Hamlett (Ed.). Johns Hopkins University Press, Baltimore.

Dominguez, P., A. G. Jacobson, and R. P. S. Jefferies. 2002. Paired gill slits in a fossil with a calcite skeleton. *Nature 417:* 841–844.

Erdmann, M. V., R. L. Caldwell, and M. K. Moosa. 1998. Indonesian "king of the sea" discovered. *Nature 395:* 335.

Erwin, D. H. 1999. The origin of body plans. *Amer. Zool. 39:* 617–629.

———, J. Valentine, and D. Jablonski. 1997. The origin of animal body plans. *Amer. Sci. 85:* 126–137.

Eschmeyer, W. N. 1990. Genera in a classification, pp. 435–495. In *Catalog of the genera of recent fishes*, W. N. Eschmeyer (Ed.). California Academy of Sciences, San Francisco.

Gagnier, P. 1989. The oldest vertebrate: A 470-million-year-old jawless fish, *Sacabambaspis janvieri*, from the Ordovician of Bolivia. *Nat. Geog. Res. 5*(2): 250–253.

Gans, C., and R. G. Northcutt. 1983. Neural crest and the origin of vertebrates: A new head. *Science 220:* 268–274.

Garcia-Fernandez, J., and P. W. H. Holland. 1994. Archetypical organization of the amphioxus *Hox* gene cluster. *Nature 370:* 563–566.

Garstang, W. 1928. The morphology of the Tunicata and its bearing on the phylogeny of the Chordata. *Quart. J. Microsc. Soc. 72:* 51–87.

Gould, S. J. 1989. *Wonderful life*. W. W. Norton, New York.

Hausdorf, B. 2000. Early evolution of the Bilateria. *Syst. Biol. 49*(1):130–142.

Janvier, P. 1981. The phylogeny of the Craniata, with particular reference to the significance of fossil "agnathans." *J. Vert. Paleontol. 1*(2): 121–159.

Jefferies, R. P. S. 1986. *The ancestry of vertebrates*. British Museum of Natural History, London.

———. 1997. A defence of the calcichordates. *Lethaia 30:* 1–10.

Kemp, A., and R. S. Nicoll. 1996. A histochemical analysis of biological residues in conodont elements. *Modern Geol. 20:* 287–302.

Lacalli, T. 1996. Dorsoventral axis inversion: A phylogenetic perspective. *BioEssays 18*(3): 251–254.

———. 1999. Tunicate tails, stolons, and the origin of the vertebrate trunk. *Biol. Rev. 74:* 177–198.

Lauder, G. V., and K. F. Liem. 1983. Patterns of diversity and evolution in ray-finned fishes, pp. 1–24. In *Fish neurobiology*, Vol. 1, R. G. Northcutt and R. E. Davis (Eds.). University of Michigan Press, Ann Arbor.

Lynch, M. 1999. The age and relationships of the major animal phyla. *Evolution 53*(2): 319–325.

Málaga-Trillo, E., and A. Meyer. 2001. Genome duplication and accelerated evolution of *Hox* genes and cluster architecture in teleost fishes. *Amer. Zool. 41:* 676–686.

Monastersky, R. 1996. Jump-start for the vertebrates: New clues to how our ancestors got a head. *Science News 149:* 74–75.

Nelson, J. S. 1994. *Fishes of the world* (3rd ed.). John Wiley, New York.

Paxton, J. R., and W. N. Eschmeyer (Eds.). 1998. *Encyclopedia of fishes*. Academic Press, San Diego.

Pennisi, E., and W. Roush. 1997. Developing a new view of evolution. *Science 277:* 34–37.

Peterson, K. J. 1994. The origin and early evolution of the Craniata, pp. 14–37. In Major features of vertebrate evolution, D. R. Prothero and R. M. Schoch (Eds.). *Short Courses in Paleontology no. 7.* University of Tennessee and the Paleontological Society, Knoxville.

Ruben, J. A., and J. K. Parrish. 1990. Antiquity of the chordate pattern of exercise metabolism. *Paleobiology 16:* 355–359.

Shimeld, S. M. 1999. The evolution of dorsoventral pattern formation in the chordate neural tube. *Amer. Zool. 39:* 641–649.

Shu, D.-G., H.-L. Luo, S. C. Morris, X.-L. Zhang, S.-X. Hu, L. Chen, J. Han, M. Zhu, Y. Li, and L.-Z. Chen. 1999. Lower Cambrian vertebrates from south China. *Nature 402:* 42–46.

———, S. C. Morris, J. Han, Z.-F. Zhang, K. Yasui, P. Janvier, L. Chen, X.-L. Zhang, J.-N. Liu, Y. Li, and H.-Q. Liu. 2003a. Head and backbone of the early Cambrian vertebrate *Haikouichthys*. *Nature 421:* 526–529.

———, S. C. Morris, Z.-F. Zhang, J.-N. Liu, J. Han, L. Chen, X.-L. Zhang, K. Yasui, and Y. Li. 2003b. A new species of yunnanozoan with implications for deuterostome evolution. *Science 299:* 1380–1384.

Smith, M. P., I. J. Sansom, and J. E. Repetski. 1996. Histology of the first fish. *Nature 380:* 702–704.

Wang, D. Y.-C., S. Kumar, and S. B. Hedges. 1999. Divergence time estimates for the early history of animal phyla and the origin of plants, animals and fungi. *Proc. Roy. Soc. Lond. B 266:* 163–171.

Young, G. C., V. N. Karatajute-Talimaa, and M. M. Smith. 1996. A possible late Cambrian vertebrate from Australia. *Nature 383:* 810–812.

Zimmer, C. 1999. Fossils give glimpse of old mother lamprey. *Science 286:* 1064–1065.

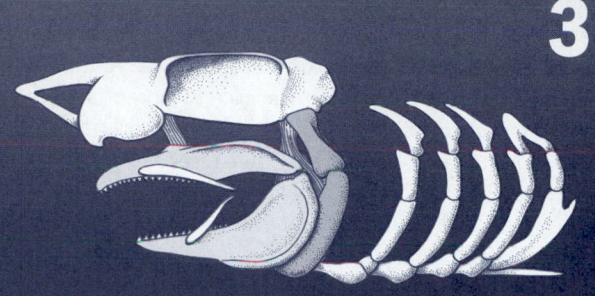

3

THE ARCHITECTURE
OF FISHES

ENVIRONMENTAL CONSTRAINTS ON THE DESIGN OF FISHES

GENERAL BODY FORM AND FINS

The Head
Body Form
Topography of the Body
Fins
  Median or Unpaired Fins
  Paired Fins

SKIN AND SCALES

Epidermis
Dermis
Scales

INTERNAL SUPPORT: THE SKELETAL SYSTEM

The Axial Skeleton
  The Skull
  The Vertebral Column
Fins and Associated Appendicular Skeleton
  Median Fins and Associated Skeleton
  Pectoral Fins and Associated Skeleton
  Fin Rays

SKELETAL MUSCLE

GENERAL RELATIONSHIPS OF INTERNAL ORGANS

Alimentary Canal and Associated Structures
Urogenital Organs
The Heart

In this chapter, we will discuss the essentials of what it means to be a fish. Our focus is on the basic structure of fishes; the functions of the various components that make up a fish will be treated in later chapters. It is a little difficult to appreciate just how fishes work, as they are adapted to existence in a medium that is foreign to us. The overall body form of fishes, diverse in its adaptation to a range of aquatic environments, has evolved in response to a medium that is much more viscous than the atmosphere. The skin of fishes is complex and features mineral depositions in the dermis, known as *scales*. Through these scales, fishes gain significant protection, yet their skin is not constructed with the threat of desiccation in mind, as it is in tetrapods. Compared to tetrapods, the skull of fishes is an extremely complex and kinetic device, supremely adapted to accomplish roles in feeding, respiration, and sensation. Whereas we must contend with the forces of gravity in our own locomotor adaptations, fishes need not bother with this constraint. Tetrapod locomotion has become focused on the appendages, whereas in fishes, it is primarily focused on the skeleton and associated muscles of the trunk. Although paired appendages first evolved in fishes, their role is primarily to stabilize and to direct forward momentum rather than to be the primary propulsive organs, as they are in terrestrial vertebrates. It is important to keep in mind, however, that in a group as diverse as the fishes, there will always be exceptions to the generalizations presented here. Just consider the diverse array of fishes that have become adapted, to varying degrees, to terrestrial conditions. From these, we learn that being a fish does not necessarily exclude one from a measure of participation in the terrestrial realm.

ENVIRONMENTAL CONSTRAINTS
ON THE DESIGN OF FISHES

The challenge of architecture is to create structures that fulfill specific design criteria but at the same time integrate successfully with the surrounding landscape. So it is with fishes—natural selection has shaped them to function in highly specific ways, and unless they become thoroughly integrated with their immediate surroundings, they will not survive. What is most striking about the biology of fishes is how the basic package, the architectural form, is so amenable to modification that fishes have successfully adapted to virtually every aquatic habitat—and to a few terrestrial environments as well.

Living on land as we do, it is a bit hard for us to conceive of the range of adaptations that permit existence in aquatic habitats. Compared to air, water is an extremely dense and viscous medium. As the density of their medium closely approximates that of their own tissues, fishes have little to fear from the forces of gravity. However, the density of fish tissue slightly exceeds that of water, so fishes must still exert a slight amount of energy to maintain themselves up in the water column. You may not be aware of the density of water until you try to lift a 10-gallon aquarium and discover that, when full, it weighs more than 83 pounds (or, for the non–metrically challenged, 37.85 liters weighing 37.85 kg). Such a dense medium exerts considerable force when moving—an important consideration for stream-dwelling fishes. Although the increase of water pressure with depth may constrain the vertical movements of fishes, it has not prevented many species from successfully inhabiting the deepest parts of the ocean.

Not only is water a physical force to be reckoned with, but its chemical properties have dictated the course of fish evolution. Water only marginally dissolves oxygen, and this solubility varies inversely with the temperature and concentration of dissolved solute. The design of the gills in fishes reflects the need to maximize oxygen extraction from a medium that has a characteristically low solubility to oxygen, and one needs to keep in mind that gas exchange is a *passive* process that relies on the maintenance of concentration gradients across diffusion surfaces (see Chapter 24). To cope with situations where the concentration of dissolved oxygen does not suffice, some fishes have evolved the capacity to breathe air.

Fishes must also strike a balance with their environment in the ions and other compounds that they exchange. Vertebrates are in possession of osmoregulatory capabilities that enable the maintenance of an internal ionic environment that is decidedly different from their external surroundings (see Chapter 25). The great 19th-century physiologist Claude Bernard developed the concept of the *milieu intérieur*—what we know today as **homeostasis,** the active regulation of a state of internal stability in the face of potential change in the external realm. Such active interplay with the environment also includes other phenomena, such as the perception of waterborne odors, the ingestion of nutrients available in the water, and, unfortunately, the effects of toxic dissolved substances.

The surface of a body of water is highly reflective, and most light that is absorbed will be rapidly extinguished in the first few meters of depth. In spite of this, fishes are, for the most part, highly visual creatures—that is, light plays an important role in their lives. Sound, on the other hand, is rapidly propagated through water; fishes have exploited this in the development of a broad range of sound-generating capabilities. It naturally follows that fishes have also developed astounding sensitivity to sonic disturbances propagated through the water. Although we may have some awareness of this through our own modest acoustic sensitivity, we are unable to fully appreciate the remarkable capacity that fishes have to both generate and detect electrical fields—a feature that vertebrates largely left behind when they moved onto the land.

The entire range of aquatic environments, both freshwater and marine, is potentially habitable by fishes. They abound in the open water and in close association with sandy and muddy bottoms, coral reefs, and rock outcrops. Fishes have evolved as inextricable features of the complex ecosystems in the hydrosphere. Their diversity is a testament to their success in exploiting the range of opportunities available to them. This chapter serves as an introduction to the morphology of fishes—the architecture that serves as the basis for their success. From this basic introduction, we will proceed to investigate in more detail the form, function, and diversity of fishes in later chapters. For the sake of expediency, we will forgo, for the time being, the intricacies of the phylogenetic relationships of living and fossil fishes. It will suffice for now to recognize modern fishes as comprising three basic lineages: *agnaths* (or jawless fishes), *cartilaginous fishes* (including sharks, skates, rays, and ratfishes), and *bony fishes.*

GENERAL BODY FORM AND FINS

The majority of fishes have a more or less streamlined body to minimize resistance as they move through the water. Usually, a head, trunk, and tail can be distinguished (Fig. 3.1), but fishes are different from most tetrapods in that they lack a neck region. Hence, head mobility independent of

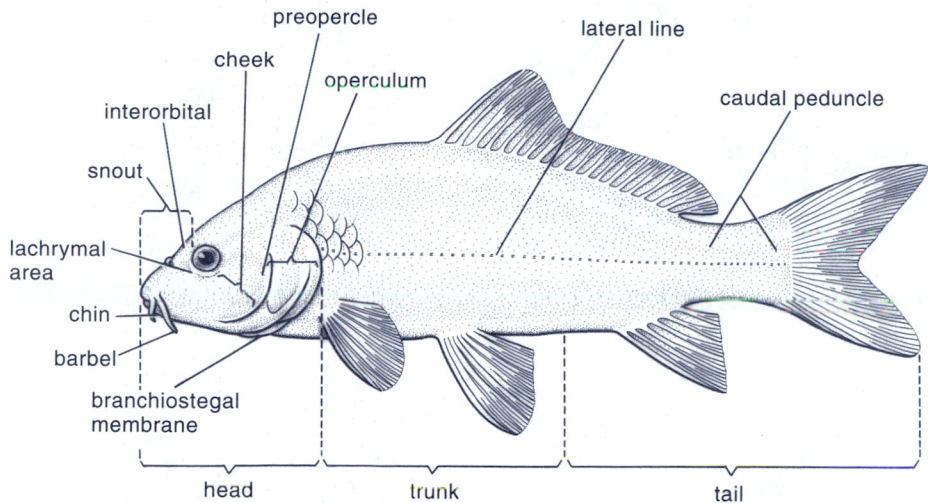

FIGURE 3.1
Diagram of bony fish, showing external features.

the rest of the body is not possible, save for a few unusual exceptions, such as the fossil placoderms (see Chapter 6) and the salamanderfish (*Lepidogalaxias salamandroides;* see Chapter 11). In most bony fishes, the *head* extends from the tip of the snout to the posterior edge of the **operculum**, a set of bones that covers the gills. In the lampreys, hagfishes, sharks, rays, and in those bony fishes—such as eels—in which the opercular cover is embedded in the skin, other points of reference for the three body regions must be used. The *trunk* or abdominal region houses the internal organs in a cavity termed the **coelom**. It begins at the posterior margin of the gill cavity and extends, in most fishes, to approximately the origin of the anal fin, but the true distinction between the trunk and tail is internal (see Vertebral Column). A subdivision of the coelom, termed the **pericardial cavity**, contains the heart.

The Head

Anatomical regions of the head of a bony fish (Fig. 3.1) include the **snout**, from the eye to the anterior tip of the upper jaw; the *operculum;* the *cheek*, between the eye and the angle of the preopercle; the **branchiostegal membrane**, below the operculum; the chin, or **mentum**; and the *interorbital*. For elasmobranch fishes, the head runs from the snout to the serial gill slits on the sides or ventral surfaces just anterior to the pectoral fins. The **lachrymal** region is below the anterior edge of the eye. The shape of the skull is very much determined by the mode of feeding and by the

prominence of the eyes. Whereas the eyes may be the most conspicuous feature of such nocturnal reef-dwellers as the aptly named bigeyes (Priacanthidae), they may be absent in species that inhabit the permanently darkened recesses of caves. Head-associated features, such as spines or sensory pores, are often useful in the systematic study of fish taxa.

The mouth is located near the anterior of the fish; however, its exact position may vary. It may be *inferior*, as in sturgeons and many elasmobranchs, *subterminal* as in dace, *terminal* as in trout, to *oblique* or even superior as in sandfishes (Fig. 3.2). Usually visible externally on bony fishes are the lower jaw bones and the paired upper jaw bones, the **premaxillae** and **maxillae**. Some species have **supramaxillary** bones attached to the maxillae. Protrusible mouths have independently evolved in cartilaginous fishes and in advanced bony fishes. In most of the spiny-rayed fishes and their relatives, ascending processes of the premaxillae keep those bones oriented as the processes slide down and forward from their resting place in the nasal region. Other types of jaw protrusion are seen in the carp-like fishes and sturgeons (Lauder, 1983). Many species have their lips or jaws bound to the snout or chin by a continuous bridge of skin, or **frenum**, so that the mouth is nonprotractile.

Barbels (Figs. 3.1, 3.2A), fleshy, elongate structures that carry tactile and chemosensory receptors, may be present around the nostrils and mouth and on chins. They may be minute and simple, or conspicuous and sometimes branched, as in some catfishes. They take the designation of the structure that bears them, such as *maxillary, mandibular,*

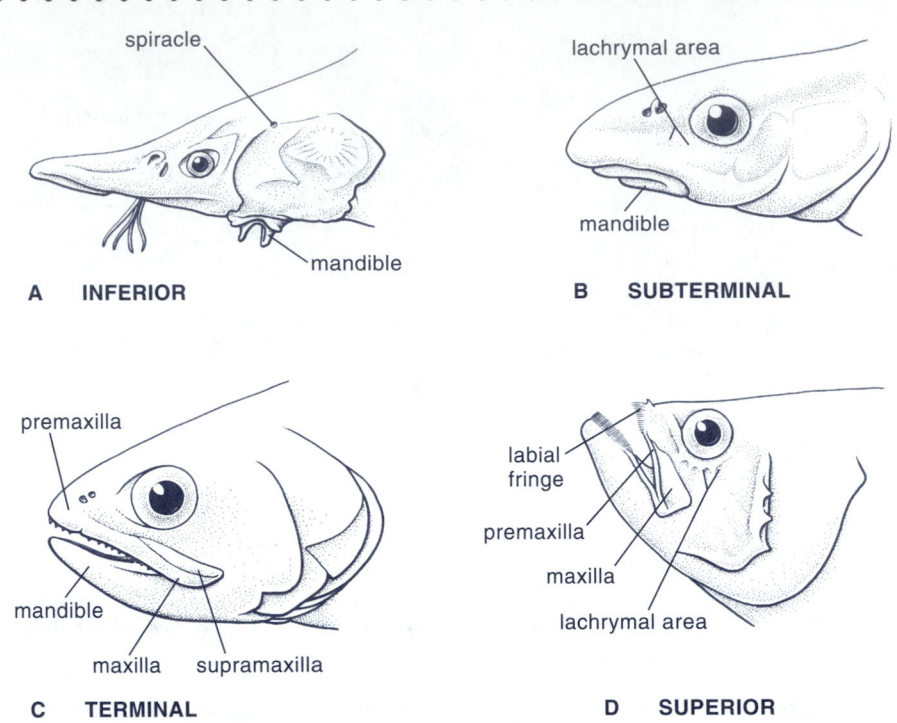

FIGURE 3.2

Examples of mouth positions in fishes: **A**, Inferior (sturgeons); **B**, subterminal (dace); **C**, terminal (trout); **D**, superior (sandfish). (*D* based on Jordan and Evermann, 1900.)

nasal, rostral, and *mental* (on chin) barbels. Similar to barbels but usually without special sensory functions are **cirri**, or various flaps of skin on the lips or other parts of the head. Many of these projections serve to obliterate the profile of the fish, making it less conspicuous (see Fig. 18.8B).

Lampreys have a series of fleshy *fimbria* surrounding the mouth, which is a jawless sucking disc (Fig. 3.3A). The jawless mouth and single nasal opening of hagfish have four barbels each. Sharks and rays may have oronasal grooves and labial folds in the mouth region (Fig. 3.3B). These structures aid in maintaining a flow of water through the nostrils for smell or taste as the animal moves or pumps water through the gills (Bell, 1993; see Chapter 24).

Spines may be among the most prominent features on the head of bony fishes (Fig. 3.4A). They are commonly found on the opercular bones, and they make some common fishes, such as yellow perch (*Perca flavescens*) and various species of sculpins (*Cottus*), difficult to handle. Head spines usually take their names from the bones that bear them, such as *opercular* or *parietal*, but they are sometimes named from their location: *preocular* or *nuchal*, for instance.

Sensory canals on the head (part of the lateral line sensory system discussed later) can be recognized by rows of pores or open grooves in the skin (Fig. 3.4B). Sensory organs in these canals respond to water movement and other mechanical stimuli (see Chapter 21).

The nostrils of living fishes, except for hagfishes, lungfishes, and some specialized bony fishes, have no internal openings to the oral cavity. There may be a blind sac on each side, with its single opening separated into incurrent and excurrent portions. Usually, the sac has anterior and posterior nares barely separated from each other. In some fishes, such as eels, the olfactory organ is a greatly expanded, tubelike cavity, with the nares widely separated (Fig. 3.5).

In rays, many sharks, and some primitive bony fishes such as the bichirs (*Polypterus*), sturgeons (*Acipenser*), and paddlefishes (*Polyodon*), an opening called the **spiracle** is found behind the eye (Fig. 3.2A). This aperture is the remnant of a full gill slit—originally between the mandibular and hyoid arches (Romer, 1970)—that has been reduced in size as a result of the modification of dorsal hyoid arch elements to serve as a suspensory apparatus for the mandibular

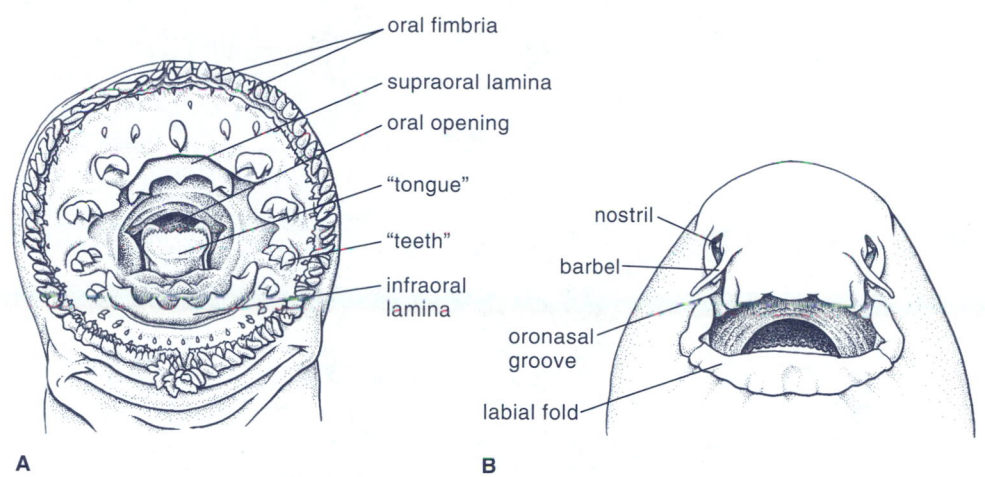

FIGURE 3.3

A, Oral disc of lamprey *Lampetra minima;* **B,** Ventral view of mouth and rostrum of shark *Chiloscyllium indicum* showing grooves and labinal folds.

arch. In bottom-dwelling rays and some benthic sharks, the mouth may be buried in the substrate, and the respiratory current is thus brought to the gills through the spiracles.

Body Form

Much about a fish's mode of existence can be inferred from its body form (Fig. 3.6). Tunas and mackerels exemplify the body form of fast-swimming, open-water fishes. This streamlined configuration, with an elliptical to round cross section and a narrow *caudal peduncle* just in front of the caudal fin, is called **fusiform.** This term is also often applied to the body shapes of fishes that are considerably more laterally compressed (**compressiform**) than the tunas, such as Pacific salmons (*Oncorhynchus*).

The compressiform shape is also seen in deep-bodied species that may not be in constant motion but may be capable of quick bursts of speed or sharp turns. Familiar fishes of this shape are sunfishes (genus *Lepomis* of the family Centrarchidae), snappers (Lutjanidae), porgies (Sparidae), and flounders (Pleuronectidae). Fishes that are flattened dorsoventrally are termed **depressiform.** Depressiform fishes include skates and rays, angel sharks (Squatinidae), toadfishes (Batrachoididae), and goosefishes (Lophiidae). Obviously, this shape suits the fish for life on the bottom, but the greatly flattened mantas (Mobulidae) and eagle rays (Myliobatididae), derived from bottom-living forms, have become adapted for graceful swimming through the water.

Eel-shaped fishes are called **anguilliform,** after the genus *Anguilla*, which includes the American and European eels. This term also refers to their distinctive mode of locomotion (see Chapter 19). Other descriptive terms used in connection with body form are **filiform,** for thread-shaped fishes such as snipe eels (Nemichthyidae); **taeniform,** for the ribbon-like shape of such fishes as gunnels (Pholidae), pricklebacks (Stichaeidae), and hairtails and cutlassfishes (Trichiuridae); **sagittiform,** for the somewhat arrow-like shape of pikes (Esocidae), gars (Lepisosteidae), and others; and **globiform,** exemplified by the rotund lumpsuckers (Cyclopteridae).

Of course, not all fishes have body forms that can be described by these convenient terms. Boxfishes and cowfishes (Ostraciidae), seahorses (Syngnathidae), and sea moths (Pegasidae; Fig. 3.7) are some examples of fishes with peculiar shapes. A familiar freshwater fish, the brown bullhead (*Ameiurus nebulosus*), is an example of a fish with a combination of shapes, having a depressed head, a body of round cross section, and a laterally compressed caudal region.

A body form often encountered in marine fishes—many from considerable depths—is that exemplified by the chimaeras (see Fig. 2.3E) and grenadiers (Macrouridae), which have a large head and forebody with a tapering afterbody and tail. This *chimaeriform* body, or one resembling it, can also be seen in some poachers (Agonidae), spiny eels (Halosauridae), and a few others. Some of these fishes hold the body still and swim by undulating their pectoral fins, but others swim by undulations of the body.

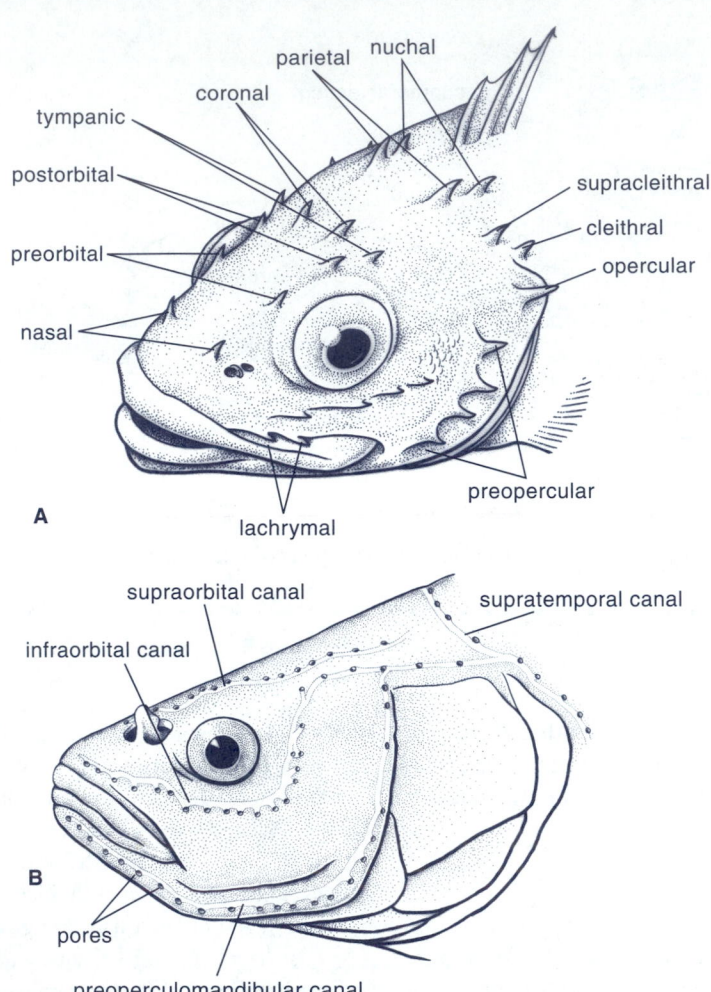

FIGURE 3.4
A, Illustration of head of rockfish (Sebastinae), showing spines (adapted from Hart, 1973); **B,** illustration of head of tui chub (Gila bicolor, family Cyprinidae), showing position of cephalic sensory canals and pores.

Topography of the Body

Some regions of the body are described by terms that aid in locating identifying features (Fig. 3.8). The dorsal surface, just behind the **occiput** (the posterior terminus of the skull), is called the **nuchal** region, sometimes characterized by a hump. The most anterioventral part of the body is usually the narrow *isthmus* that extends far forward below and between the gill openings. Posterior to this is the *breast*, and posterior to that is the *belly*. The narrow part of the body of the fish, just anterior to the caudal fin, is called the **caudal peduncle;** it functions to transmit the power of the lateral undulation of the trunk muscles to the tail.

Conspicuous along each side of many fishes is the trunk canal of the **lateral line system,** a continuation of the network of sensory canals on the head. The usually single lateral line may be an open groove in the skin, as in some chimaeras, or a row of pores in the skin or scales. Lines may be multiple, as in the greenlings (Hexagrammidae), or reduced in various ways (Coombs et al., 1988). Herrings (Clupeidae), for instance, lack an extended lateral line; it often appears only on a few anterior scales.

The gill openings of lampreys are in a lateral position, appearing as a row of seven nearly circular apertures. Hagfishes may have from 1 to 16 circular gill openings placed

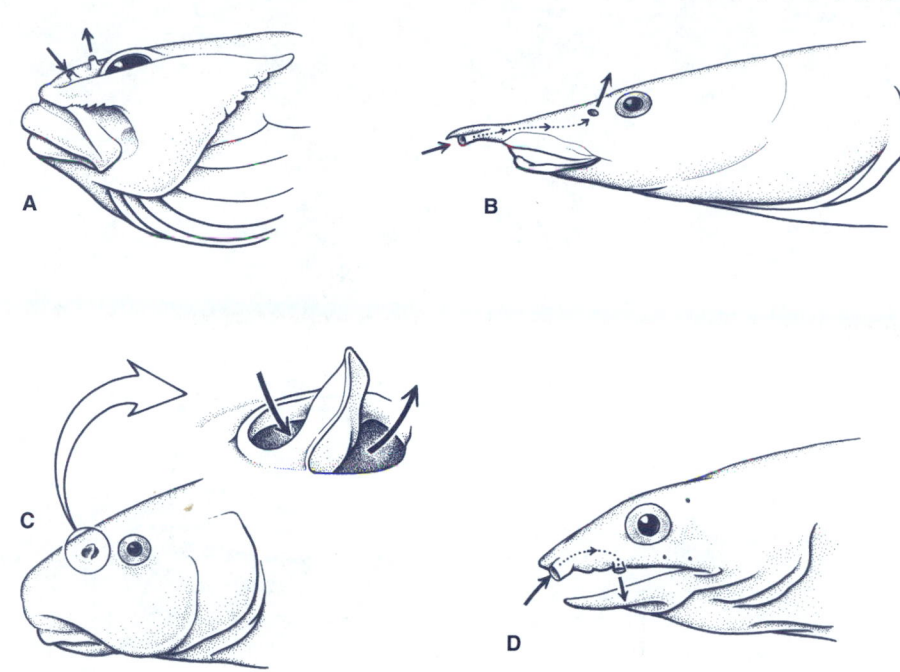

FIGURE 3.5
Representative nostrils; arrows show incurrent and excurrent apertures: **A,** Sculpin (Cottidae); **B,** spinyback (Mastacembelidae); **C,** typical bony fish nostrils divided by flap of skin (Catostomidae); **D,** worm eel (Myrophis).

well behind the head. In sharks and rays, five to seven individual gill openings occur in a series. Those of sharks are mostly lateral, whereas those of rays are mostly ventral. Gill openings of bony fishes and chimaeras are typically in a lateral position, just anterior to the pectoral girdle and covered by the operculum, but they may be placed well behind the operculum, as in eels, or in a ventral position, as in the swamp eels (Synbranchidae).

Fins

Fins, of course, are conspicuous features of the fish body. Fins are stiffened by structures called **rays,** which may be either soft and flexible or more rigid **spines** in the more derived fishes such as the percomorphs (see Chapter 14). The median or unpaired fins are the **dorsal, anal,** and **caudal** fins. These are in line with the axial skeleton and are supported by median elements associated with the vertebral column. The paired fins are the **pectoral** and **pelvic** (ventral) fins (see Fig. 3.8). They are supported, respectively, by the pectoral and pelvic girdles embedded in the musculature.

Median or Unpaired Fins

These are the dorsal fins, along the back; the caudal or tail fin; and the anal fins, located ventrally just behind the anus in most fishes (see Fig. 3.8).

The dorsal fin may extend the length of the back, be divided into two or three separate fins, or be single and small. The dorsal fin is lacking in some families, such as the gymnotid knifefishes of South America. In more advanced bony fishes, the anterior part of the dorsal fin (or the entire first dorsal fin if there are two) is supported by spines, which can be stiff and sharp or secondarily modified into flexible structures, as in the freshwater sculpins (*Cottus*). In some groups—for example, salmon and trout (Salmonidae), various catfishes, most characoids, and the lanternfishes (Myctophidae)—there is a small, fleshy, usually rayless, **adipose fin** on the dorsal part of the caudal peduncle.

The dorsal fins function in stabilization and in helping to achieve quick changes in direction, but they can be used in conjunction with the caudal and anal fins in braking. Many species that have dorsal fins extending the length of the back can move by undulating the fin, but some species

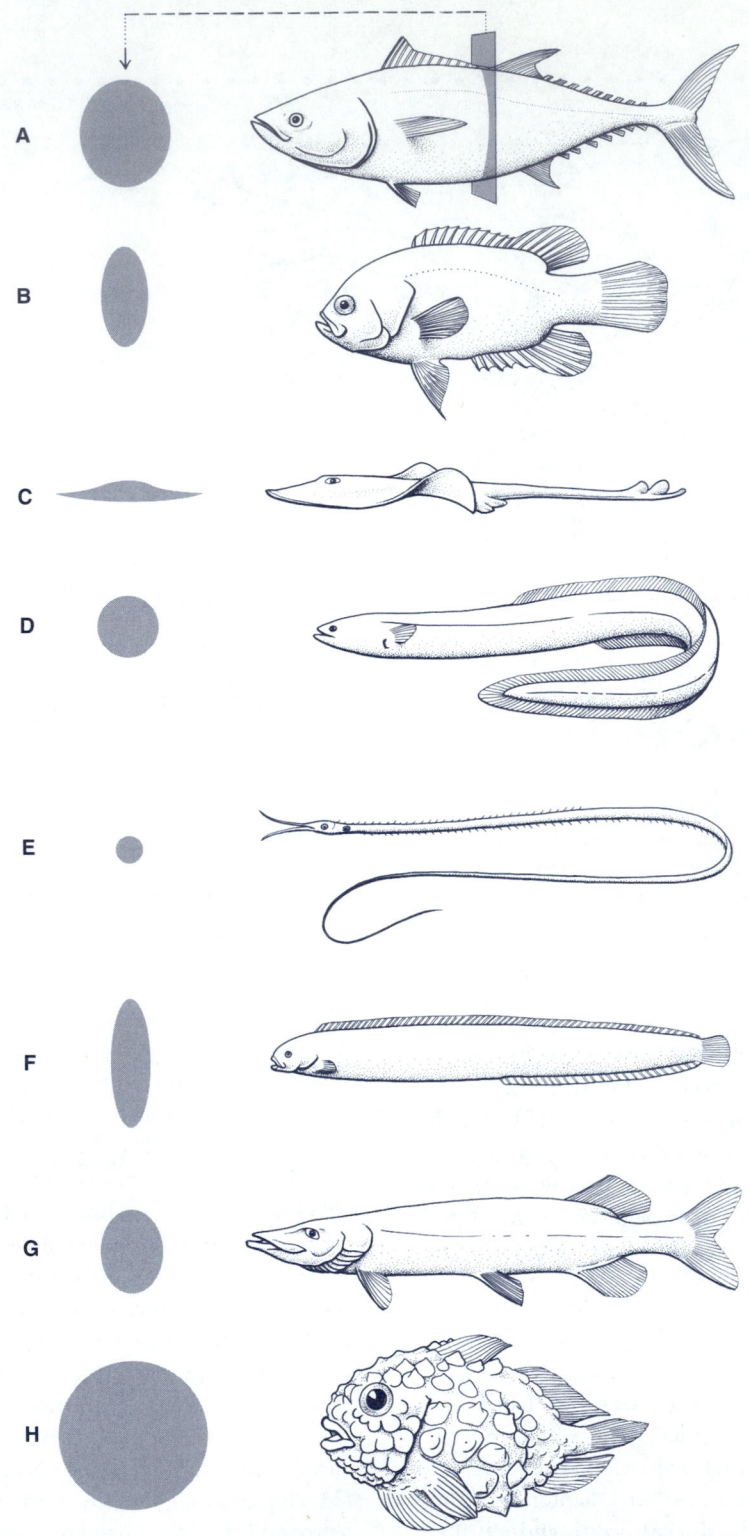

FIGURE 3.6

Representative body shapes in fishes, with typical cross section: **A,** Fusiform (tuna, Scombridae); **B,** compressiform (sunfish, Centrarchidae); **C,** depressiform (skate, Rajidae), dorsal view; **D,** anguilliform (eel, Anguillidae); **E,** filiform (snipe eel, Nemichthyidae); **F,** taeniform (gunnel, Pholidae); **G,** sagittiform (pike, Esocidae); **H,** globiform (lumpsucker, Cyclopteridae). (*H* based on Jordan and Evermann, 1900.)

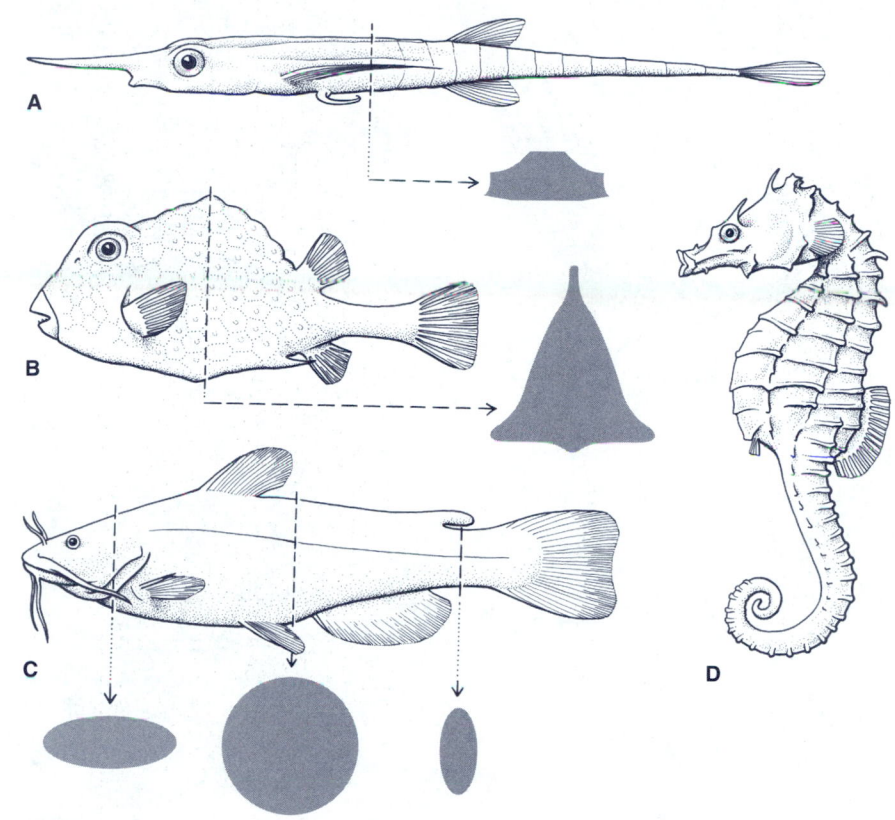

FIGURE 3.7
Example of unusual body shapes in fishes: **A,** Sea moth (Pegasidae); **B,** cowfish (Ostraciidae); **C,** bullhead
(Ictaluridae); **D,** seahorse (Syngnathidae). (*B* and *C* based on Jordan and Evermann, 1900.)

with short dorsals, such as pipefishes and seahorses, also use
the dorsal for locomotion (see Chapter 19).

Modified dorsal fins (Fig. 3.9) include the sucking disc
atop the head of the remoras (Echeneidae) that allows them
to cling to sharks, or to other large fishes, turtles, and cetaceans, and be carried along as hitchhikers. The angling apparatus of anglerfishes (Lophiiformes) is a modified dorsal
fin spine. Some species display a showy dorsal fin in intraspecific communication (see Chapter 37); the size and color
of dorsal fins are sexually dimorphic in many species.

In the bichirs (Polypteridae) of Africa (Fig. 3.10A),
the dorsal fin is divided into a unique series of finlets, each
consisting of a spine with a few soft rays attached along the
length of each spine. Most scombrids have a series of finlets
posterior to the dorsal fin (see Fig. 3.6A). These consist of
detached soft rays, usually branched and set in tough skin.

Several species of the cod family (Gadidae) have three dorsal fins that have soft rays only.

Most fishes have anal fins located posterior to the anus
(see Fig. 2.3). Fishes that lack an anal fin include lampreys,
chimaeras, skates, rays, some sharks (*Squalus, Somniosus,*
etc.), and a few bony fishes, including the king-of-the-
salmon (*Trachipterus*) and male pipefishes (*Syngnathus*).

The anal fin is generally short-based, but there are many
species whose anal fins exceed the dorsal fin in length. Some
have long bases, so that the anal fin stretches from the anus
to the caudal fin, even when the anus is located nearly under
the chin, as in the knifefishes (Gymnotidae and Rhamphichthyidae) of South America. Flounders (Pleuronectidae) and
gouramis (Osphronemidae) are compressiform fishes with
long-based anal fins. Only a comparatively few fishes, such
as cods (Gadidae), have more than one anal fin. Some, such

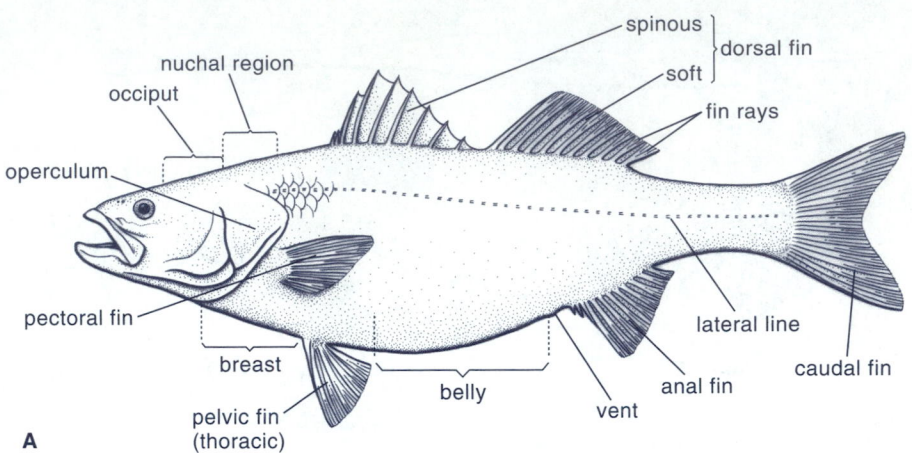

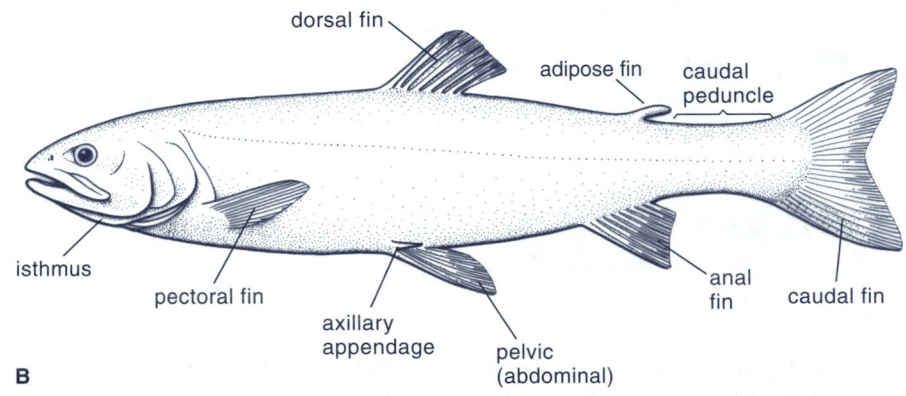

FIGURE 3.8
Body regions and fins: **A,** Spiny-rayed fish; **B,** soft-rayed fish.

as the jack mackerel (*Trachurus;* see Fig. 3.15C), have the anal fin spines separate from the soft rays, forming a small spinous anal fin. However, if anal fin spines are present, they are usually located at the anterior end of the single anal fin. **Finlets** posterior to the anal fin are present in the sauries (Scomberesocidae) and tunas, mackerels, and allied fishes (Scombridae).

Males of some species (notably of the Poeciliidae) have the anal fin modified into an intromittent organ, called a **gonopodium.** Some of the opisthoproctids (spookfishes) have the anal fin at the posterior terminus of the body, displacing the caudal fin upward.

Caudal fins appear in a variety of shapes, sizes, and kinds, and their internal structure often reflects phylogenetic relationships more than the other fins. Swimming habits may

be deduced to some extent from the caudal fin (Fig. 3.11). Fishes having a crescent-shaped (lunate) caudal fin and a narrow caudal peduncle are generally among the speediest of fishes and are capable of rapid, sustained motion. Many pelagic species have forked tails and are constantly on the move. Species with truncate, rounded, or emarginate caudal fins may be strong swimmers but are somewhat slower than those mentioned earlier. Fishes with small caudal fins or caudal fins that are continuous with the dorsal and anal fins tend to be weak swimmers, or may move by wriggling along the bottom.

Most familiar bony fishes have **homocercal** caudal fins, which externally appear to be symmetrical but are actually asymmetrical internally, because the tip of the vertebral axis turns upward, with most of the fin attached below it.

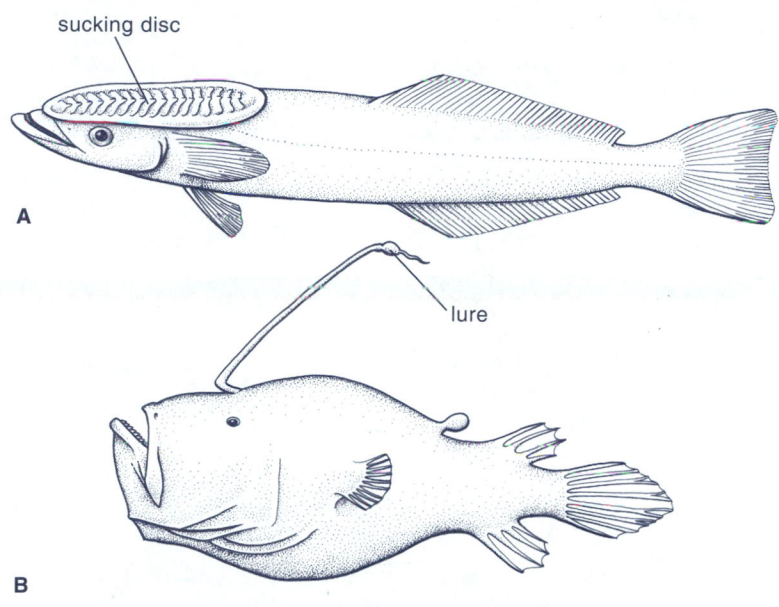

FIGURE 3.9
Modified dorsal fins: **A,** Sucking disc (remora, Echeneidae); **B,** fishing rod and lure (anglerfish, Ceratiidae).
(*B* based on Jordan and Evermann, 1900.)

Evidently, this is a character state derived from the markedly asymmetrical **heterocercal** (epicercal) fin structure of the sharks, sturgeons, and several extinct groups, in which the body axis turns upward posteriorly and most of the caudal fin is actually borne ventrally near the end of the tail. Intermediate stages can be seen in the gars (Lepisosteidae) and bowfin (*Amia*), in which the caudal fins are called *abbreviate heterocercal* because they are only slightly asymmetrical externally. On the upper edge of the caudal fin of gars, sturgeons, and many extinct fishes, there are a series of modified, elongate scales or *fulcra*. These are called "fringing fulcral scales" by Moy-Thomas and Miles (1971). They are present also on the leading edges of paired fins in some primitive bony fishes.

The homocercal tail of some of the basal teleosts differs from that seen in the derived forms in that, like the gar and bowfin, more than one vertebra is included in the upturned portion (Fig. 3.11G; Patterson, 1968).

The symmetrical tail of the cods and hakes (Gadidae) is called **isocercal**. The true caudal fin of cods is small and is borne on a symmetrical plate at the end of a tapering series of vertebrae. Most of what appears to be the caudal fin is actually composed of dorsal and anal fin elements (Fig. 3.11I).

Long, tapering, or whip-like tails are called **leptocercal**. Other symmetrical tails that come to a more abrupt point, as in lungfishes, are called **diphycercal**. In a few fishes, the caudal portion of the body is absorbed during development, so that the dorsal and anal fins bridge over the posterior terminus of the body in what is called a **gephyrocercal** tail. This is seen in the molas or ocean sunfishes (Molidae). Some of the earliest fossil fishes possessed a **hypocercal** tail, in which the notochord runs to the lower lobe of the caudal fin (see Chapter 5; Fig. 5.7).

Paired Fins
The pectoral fins of bony fishes, composed of soft rays only, are borne by the pectoral (shoulder) girdle, which in most fishes forms the posterior border of the gill cavity (see Fig. 3.25). The right and left halves of the pectoral girdle meet ventrally at the body midline and diverge dorsally, each half fastening more or less firmly to the posterior part of the skull. Pectoral fins are usually prominent; only a few groups lack them or have them reduced in size. Groups with less prominent pectoral fins are generally elongate, eel-shaped, or taeniform. Many such fishes are adapted for wriggling along the bottom or through vegetation.

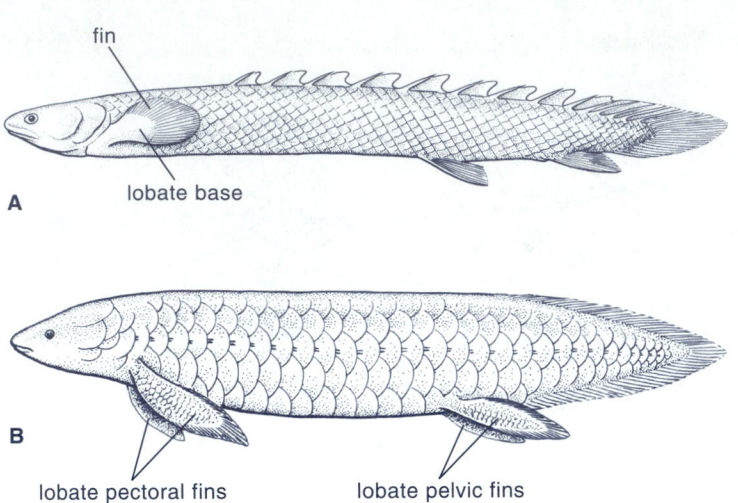

FIGURE 3.10

A, Bichir *Polypterus* showing unique dorsal fin configuration and example of lobate pectoral fins; **B,** lobate pectoral and pelvic fins of Australian lungfish *Neoceratodus*. (*B* based on Goodrich, in Lankester, 1909.)

The chondrichthyan fishes and the more primitive bony fishes usually have the pectoral fins placed low on the body, below the centers of buoyancy and mass (see Fig. 2.3C, E). In the bony fishes, these fins tend to have oblique bases (slanting downward and posteriorly), in part because of the presence of the mesocoracoid bone in the pectoral girdle. Although suitable for trimming the balance of the fish either at rest or in motion, these low-placed pectoral fins have less use in locomotion and braking than the pectoral fins of the more derived fishes, which lack the mesocoracoid bone, and in which the pectoral fins are generally placed higher on the body and have vertical bases (see Fig. 2.3G). These fins, which are closer to the centers of mass and buoyancy, are versatile and more efficient for use in locomotion, maneuvering, braking to sudden stops, aggressive displays, and other purposes (see Chapter 19). In some deep-bodied species, a shift in position of the gas bladder has accompanied the relocation of the pectoral fins. In some, the gas bladder is above the center of gravity, so that maintaining balance and locomotion by means of the pectoral fins are facilitated.

The low position of the pectoral fin in primitive fishes (and to a lesser extent in some advanced fishes) allows the fins to touch bottom and provide support in species living close to the substrate. These fins often bear numerous taste buds and touch receptors. The bichirs (*Polypterus*) bear the pectoral fins on a lobate base, with a skeleton peculiar to the group. The skeleton of the fin lobe consists of two rod-like bones with a cartilaginous plate between them, connected proximally to the scapula and coracoid. Distally, the rods and plate attach to ossified radials that bear the fin rays (Berg, 1947). This arm-like pectoral fin is called a **brachiopterygium** (see Fig. 3.10A). The coelacanth (*Latimeria*) and the Australian lungfish (*Neoceratodus;* see Fig. 3.10B) also have lobate pectoral fins, but of a different structure. Highly modified pectoral fins, supported by a central joined axis with no radials (fin ray supports) or fin rays, are seen in the African and South American lungfishes, *Protopterus* and *Lepidosiren*, respectively.

Some species have pectoral fins that aid them in remaining at the bottom or even on a vertical surface. Sisorid catfishes apparently use their pectoral fins as well as a pad between the fins in clinging to surfaces (Fig. 3.12A). The algae-eaters (Gyrinocheilidae) have pectoral fins (and mouths) modified for use in maintaining suction.

The African freshwater species *Pantodon buchholzi* is called the butterflyfish because of its pronounced upturned pectoral fins. These fishes, which can measure up to 10 cm, can make leaps of up to 2 m over the water surface. Unique among fishes is the pectoral apparatus of the South American characiform family Gasteropelecidae, commonly referred to as freshwater hatchetfishes or freshwater flying fishes (see Chapter 10). The bones of the pectoral girdle

PART ONE

AN INTRODUCTION TO THE FISHES

Fishes, living as they do in a medium entirely unlike the one in which we live, may seem at first glance less accessible than, say, birds or mammals. The irony of this does not escape the serious student of fishes, be he or she an academician, a fish hobbyist, or someone working in a fisheries-related industry—for fishes are by far the most successful vertebrate group. Of the approximately 50,000 extant vertebrate species, more than half are fishes. Living in aquatic environments with varying degrees of accessibility, fishes are far less understood than terrestrial vertebrates. It is not entirely accurate, however, to characterize fishes as solely aquatic; many families include species capable of terrestrial sojourns, and a few fishes have also taken wing, so to speak, to make limited excursions into the realm of the birds and the bats.

Whereas it is somewhat unusual for new species of birds or mammals to be discovered and described, research journals devoted to fishes publish many new species descriptions each year. Descriptions of 25 new species were published in *Copeia*, the journal of the American Society of Ichthyologists and Herpetologists, during 1999, for instance.

Certain families—usually of the more highly derived bony fishes known as the teleosts—are especially rich in species, including those new to science. Berra (1997, 2001) has observed that the greatest number of freshwater species described in the past quarter century are from just such rich families: Cichlidae (cichlids), Cyprinidae (minnows and relatives), Characidae (characins), Loricariidae (a catfish family), and Cyprinodontidae (killifishes). The families Gobiidae (gobies) and Serranidae (sea basses) likewise account for the greatest number of recently described marine species. It is not surprising that the majority of these newly described species are tropical freshwater forms from South America, Africa, and Asia.

Although this book is primarily intended for the student engaged in an academic study of fishes, the authors hope

Swift ("The Distribution of Fishes"). I thank them for providing revisions of these chapters for this edition.

I would also like to thank editorial staff members who transformed my sometimes inarticulate thoughts on fishes into the book that now rests in your hands. I thank Jeremy Hayhurst for nurturing this book in its early stages and Thomson senior biology editor Peter Adams for helping to bring it to completion. I also wish to acknowledge the advice and assistance of the Thomson editorial and production staff, including Kristin Lenore, Belinda Krohmer, Kiely Sisk, and Kari Hopperstead. Gretchen Otto of G&S Book Services and Annette Coolidge of Photopia were instrumental in getting electronic words and images into a more tactile format. And Centre College, through its Faculty Development Committee, provided financial support for the writing of this book.

Most of the visual quality of this book can be credited to John Norton, who illustrated both the second and current editions. John is a true artist who, with a few lines, can convey a wealth of information.

With this edition, color photographs are included for the first time. Many individuals have been most generous in providing photographs from their extensive collections. Although they are listed in the credit for each image, I would like to collectively thank them here. Thanks also to Ray Troll for providing one of his unique pieces of art for the cover of the book.

Several individuals gave freely of their time and expertise to provide critical commentary on the chapters: Kimberly A. Bjorgo, West Virginia University; Frank J. Bulow, Tennessee Technological University; Donald G. Buth, University of California, Los Angeles; Thomas G. Coon, Michigan State University; Lawrence R. Curtis, Oregon State University; Michael Fine, Virginia Commonwealth University; Malcolm Gordon, University of California, Los Angeles; Dennis C. Haney, Furman University; David C. Heins, Tulane University; Karen Martin, Pepperdine University; Hayden T. Mattingly, Tennessee Technological University; R. Glenn Northcutt, University of California, San Diego; Jean Porterfield, St. Olaf College; J. Michael Redding, Tennessee Technological University; Stacia Sower, University of New Hampshire; Melanie Stiassny, Columbia University; and Hong Y. Yan, Marine Research Station, Taiwan National Academy of Science.

I would like to extend special thanks to Hong Y. Yan. Through his intercession, I received the benefit of advice and comments from J. A. C. (Colin) Nicol (1915–2004) on one of the chapters. In like manner, I am grateful to Stacia Sower for contacting Aubrey Gorbman (1914–2003) and arranging for him to read one of the chapters. I am honored by the contributions that these two great researchers made to this book near the ends of their long and productive careers.

PREFACE

This book is, first and foremost, a tribute to its original author and my former mentor, Carl Bond. I have had the good fortune to have been associated with this book since its inception. As a graduate student of Carl's, I received acknowledgment for having read parts of the manuscript of the first edition, published in 1979. For the second edition, published in 1996, I contributed chapters on ecology and behavior. Now, Carl has entrusted the reputation of this classic work to me as I assume authorship for the third edition.

This edition reflects the growth and development of fish biology during the past decade. Molecular approaches to systematic study, in conjunction with traditional anatomical and paleontological investigations, have led to some significant changes in our perception of the evolutionary relationships of fishes. This edition consists of six chapters devoted to the topic of ecology and two on behavior, whereas the second edition had only one chapter for each topic. This extended coverage acknowledges the increasingly important contribution that fish studies have made to the maturation of these disciplines. A chapter devoted exclusively to parasites and diseases of fishes and a significant expansion of the chapter on fisheries and fishery science underscore our increased appreciation of fishes as essential natural resources. Integrated approaches to the study of fishes are recognized in the extensive cross-referencing of chapters throughout this edition.

My continuing education as a student of fishes has occurred over a remarkable interval of time in the history of ichthyology. Most "fishy types" of my generation, educated in the late 1960's and early 1970's, viewed ichthyology as a science on the verge of changing from a discipline focused mainly on anatomical, distributional, and natural history studies to one that was increasingly assuming an interdisciplinary perspective. We have witnessed the transformation of our field from one in which fishes were mainly studied for their own sake to one in which fishes have become recognized as important organisms for the elucidation of the entire spectrum of biological study, from molecules to ecosystems. The great ichthyologists of the last century are passing the baton to a host of molecular geneticists, developmental biologists, community ecologists, and behavioral biologists, who have all discovered that fishes can provide crucial insights into the way the world works. There is still the need, however, for a foundation on which to base these different disciplines. That is the intention of this book—to provide you with an introduction to the biology of fishes. A little understanding of the basic biology of this fascinating group of animals will provide a context for the discipline you eventually choose to pursue. Although the primary audience of this book is intended to be students who are investigating fishes in a junior- or senior-level college biology course, the authors recognize that learning is a lifelong pursuit. We hope that this book finds its way onto the bookshelf of anyone who professes an enthusiasm for fishes.

This book would not have been possible without the support and encouragement of many individuals. First, I would like to thank my wife, Chris. Being a fellow alumnus of the Fisheries and Wildlife Department at Oregon State University, as well as a colleague here at Centre College, she has been able to provide far more than the usual support and encouragement that a devoted spouse can offer. I would also like to thank my sons Erik and Geoffrey for providing sufficient distraction from work on this book; they continue to remind me that life is more than just fishes.

I owe special thanks to four great scholars of fishes who set me afloat on the academic stream. Boyd Walker (1917–2001) and Vladimir Walters (1927–1987) were my introduction to the world of fish biology at the University of California, Los Angeles. With Michael Horn as my mentor at California State University, Fullerton, I learned that the most interesting fishes were those that could survive between the tides of rocky shores. Finally, my deepest thanks and appreciation are extended to Carl Bond, who shepherded me through my years at O.S.U.

Chapters were contributed to the second edition by Anthony Gharrett ("The Genetics of Fishes") and Camm

Scombroid Poisoning 817

Tetrodotoxin Poisoning 817

Other Fish Poisons 818

39 · FISHES AND FISHERY RESOURCES

Their Use and Conservation 823

The Value of Fishes 824

Fishery Resources 824

A Chronology of Fishery Interests 824

Some Fishery Statistics 825

Freshwater and Anadromous Fisheries 825

The Great Lakes: Decline of a
Temperate Lake Fishery 829

Anadromous Fisheries 830

Tropical Freshwater Fisheries 831

Marine Fisheries 832

Pelagic Fisheries 832

Demersal Fisheries 833

The Science and Technology of Fisheries 834

Scientific Management of Fisheries 835

Recreational Fisheries 837

The Ornamental Fish Trade 837

Aquaculture 838

Conservation of Habitat and Biodiversity 840

Chemical Pollutants 840

Extinctions and the Impact of
Exotic Species 841

Hydrological Modifications 843

Global Environmental Perturbations 843

Are There Fish in Our Future? 844

Preserving Fish and Fisheries 844

Redefining Stewardship 845

Greek and Latin Word Roots and Terms 853

Glossary 855

Subject Index 861

Systematic Index 871

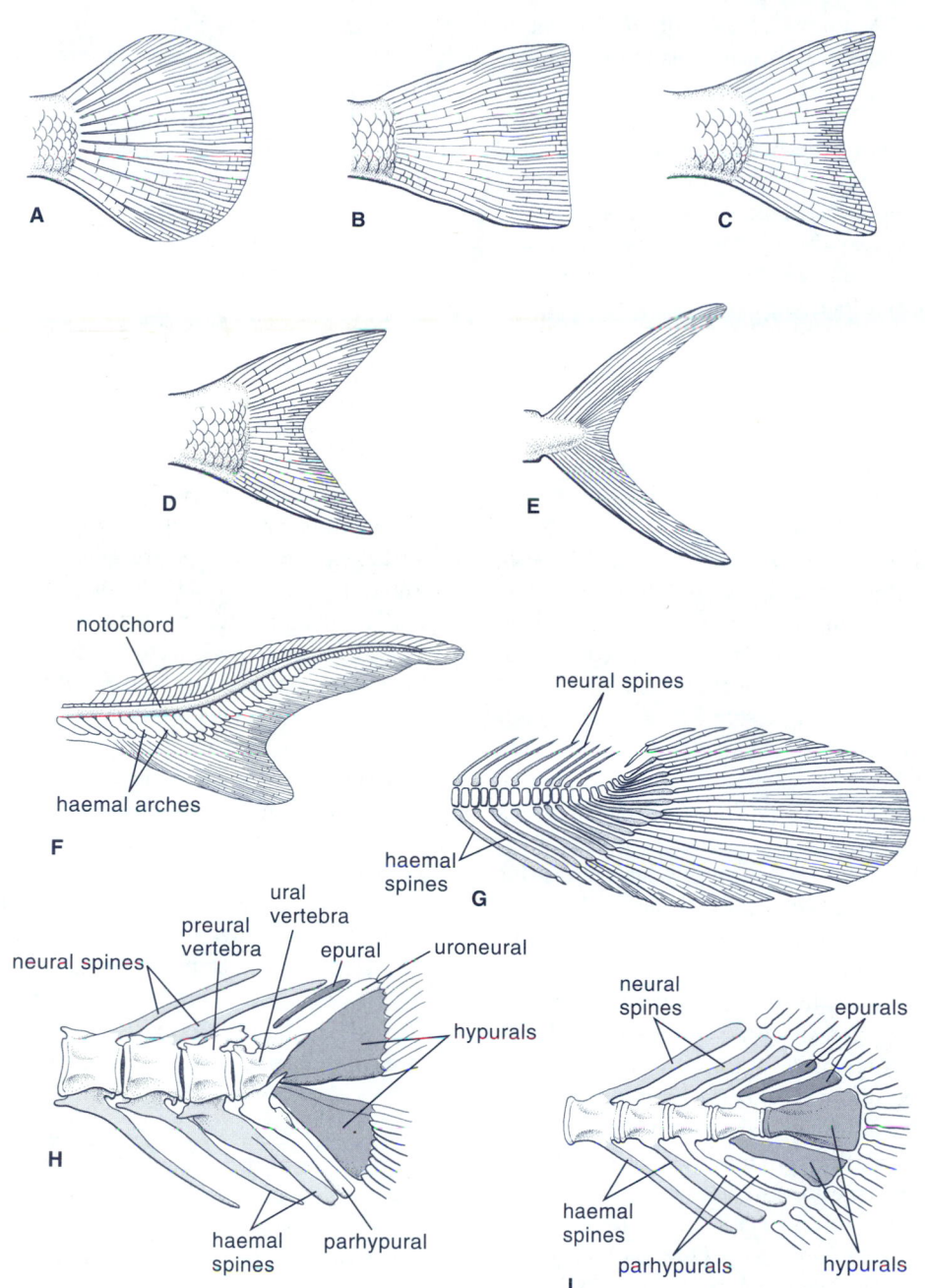

FIGURE 3.11

Representative types of caudal fins: **A,** rounded; **B,** truncate; **C,** emarginated; **D,** forked; **E,** lunate. Type of caudal fins, showing internal structure: **F,** heterocercal (sturgeon, Acipenseridae); **G,** abbreviate heterocercal (bowfin, Amiidae); **H,** homocercal (striped bass, Moronidae); **I,** isocercal (cod, Gadidae). (*A* based on Goodrich, 1930; *B* based on Jordan and Evermann, 1900.)

are expanded ventrally to give a somewhat hatchet-like appearance and to provide a broad area for the attachment of well-developed pectoral musculature. This powerful musculature enables these surface-dwelling fishes to leap clear of the water. A comparison with perhaps the best-known flying fishes, members of the marine family Exocoetidae (see Chapter 15), is instructive here. The enlarged pectoral musculature of gasteropelecids enables the generation of sufficient force to break through the water's surface and sustain flight—they are the only fishes known to engage in this kind of locomotion (Nelson, 1994). Exocoetids, on the other hand, are not deep-bodied; their broad pectoral fins are extended for gliding, and the glide may be sustained for prolonged periods through the rapid sculling action of the enlarged lower lobe of the caudal fin on the water's surface.

Catfishes have developed spine-like structures, consisting of consolidated soft rays, at the leading edge of the pectoral fins. Careless handling of bullhead or madtom catfishes (family Ictaluridae) will make one quickly realize the potency of this defensive weapon. Care must be taken especially with the madtoms, as these fishes produce mildly venomous secretions from glands associated with the spines. A few catfishes make use of these spines in terrestrial locomotion. The "walking catfish" (*Clarias batrachus*) is especially notorious. Accidentally introduced into Florida waters, where it has become somewhat of a pest, this native of Southeast Asia can successfully elude fish control efforts by climbing out of the water and walking away.

Many derived fishes are structurally specialized for certain habitats and ways of life, and their pectoral fins are often involved in the specialization (Fig. 3.12). Threadfins (Polynemidae; see Chapter 16) have pectoral fins that are each divided into two parts, the lower one consisting of several filaments that reach great lengths in some species. These filaments are thought to function as tactile organs and, when extended and fanned out, can enhance foraging efficiency. Other fishes with detached pectoral rays are the flying gurnards (Dactylopteridae), the searobins (Triglidae), and some of the stonefishes (Synanceidae). These families also have finger-like rays that are probably tactile in function.

Sexual dimorphism of the pectoral fins (and other fins) is fairly common among fishes, particularly in those that develop specialized behaviors, coloration, or other features that readily differentiate the sexes. Fishes with sexually dimorphic paired fins include the dragonets (Callionymidae), killifishes and pupfishes (Cyprinodontidae), gobies (Gobiidae), and suckers (Catostomidae).

Pelvic appendages of fishes are often smaller than the pectoral fins, more restricted in function, and subject to greater variation in placement (see Figs. 3.12 and 3.13). Elasmobranchs and primitive bony fishes are characterized by pelvic girdles and fins located in the abdominal region. The pelvic girdle is embedded in the ventral musculature and, unlike the pelvic girdle of tetrapods, is not connected to other skeletal elements. A few groups have the pelvic girdle moved forward, toward the pectoral girdle, but still lack contact between the bones. More derived bony fishes usually have thoracic pelvic fins, placed below or a little behind the pectoral fins, with a more or less firm connection between the pelvic and pectoral girdles. The pelvic fins of these groups usually have a spine and a few soft rays. The more primitive bony fishes tend to have many-rayed pelvic fins. In the cods and their relatives (Gadidae), some of the blennies (Blennioidei), toadfishes (Batrachoididae), and others, the pelvic fins are placed in a jugular position, anterior to the pectorals.

Pelvic fins are reduced or lost in many forms, especially in elongate fishes that wriggle along the bottom. Pelvic fins usually function in stabilizing and braking; they tend to be of less use than the pectoral fins in locomotion, except in some species of exocoetid flying fishes, in which the pelvic fins are used in gliding (see Chapter 15). Batfishes employ pelvic fins as well as pectorals in walking on the bottom. Several groups of fishes show modification and specialized functions of the pelvic fins. In the males of sharks, rays, and chimaeras, the pelvic fins are modified for use as intromittent organs in copulation. Many benthic species, such as sculpins, which live on hard surfaces, use pelvic fins to help hold them in place. In gobies (Gobiidae), clingfishes (Gobiesocidae), lumpfishes (Cyclopteridae), and algae-eaters (Gyrinocheilidae), pelvic fins have evolved into or have become incorporated into ventral sucking structures that aid the fish in holding on to the substrate (Fig. 3.13). The use of pelvic fins as tactile organs is exemplified by the gouramis, such as *Trichogaster*, an aptly named genus, as its pelvic fins are reduced and include an elongate, hair-like ray.

SKIN AND SCALES

As is the case in other vertebrates, the skin of fishes is made up of an outer *epidermis* and an inner *dermis* (Fig. 3.14; Van Oosten, 1957; Whitear 1986a, 1986b).

Epidermis

The epidermis is typically very thin, composed of only 10 to 30 layers of cells (an average thickness of about 250 μm) in most familiar fishes. Seahorses and their relatives may have only two or three layers of epidermal cells on the surface of their armor. The thickness of the epidermis on these fishes (about 20 μm) contrasts greatly with that found on the lips

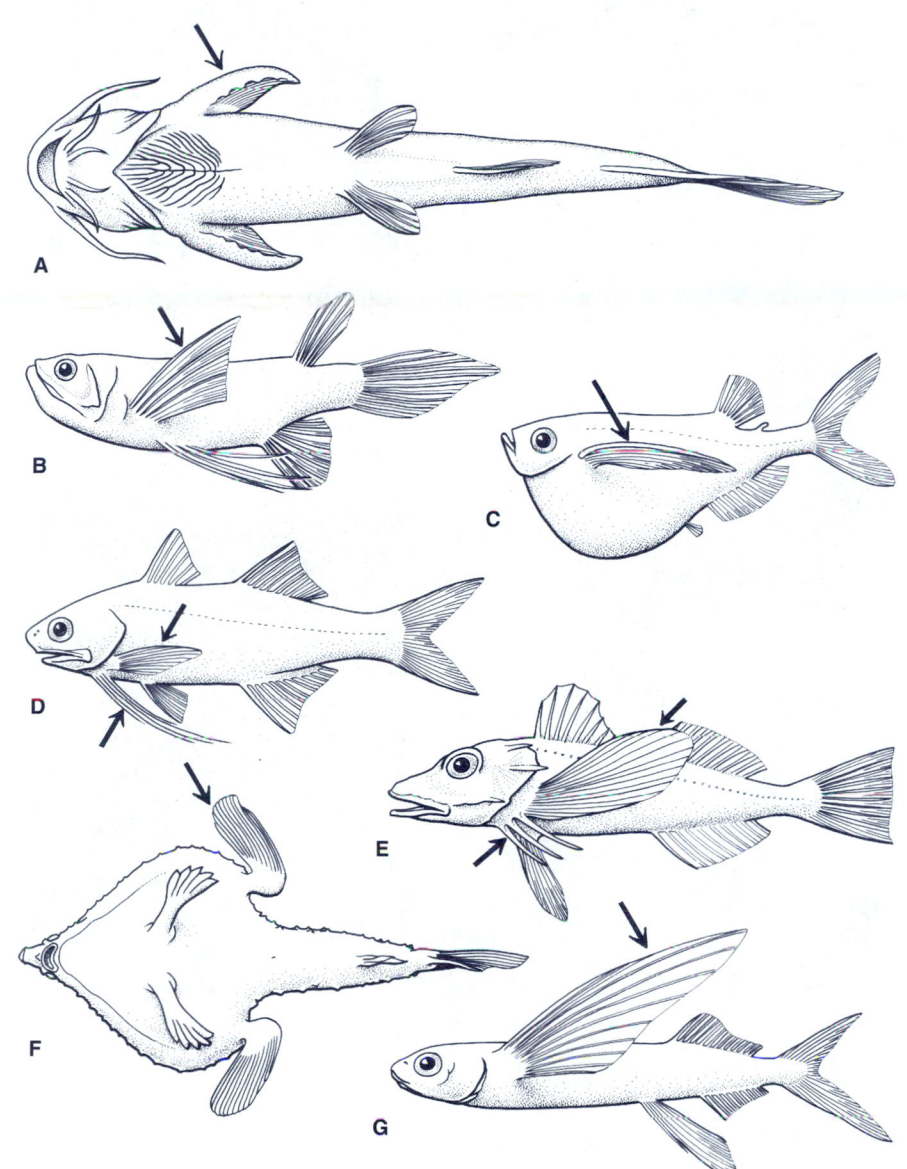

FIGURE 3.12

Examples of fish with modified pectoral fins; pectoral fins with arrows: **A,** Ventral view of sisorid catfish (*Glypto-thorax*); **B,** freshwater butterflyfish (*Pantodon*); **C,** hatchetfish *Gasteropelecus* **D,** threadfin (polynemidae); **E,** searobin (Triglidae); **F,** ventral view of batfish (Ogcocephalidae), with arm-like pectorals behind pelvics; **G,** flying fish (Exocoetidae). (B based on Herald, 1961; *D, E,* and *G* based on Jordan and Evermann, 1900.)

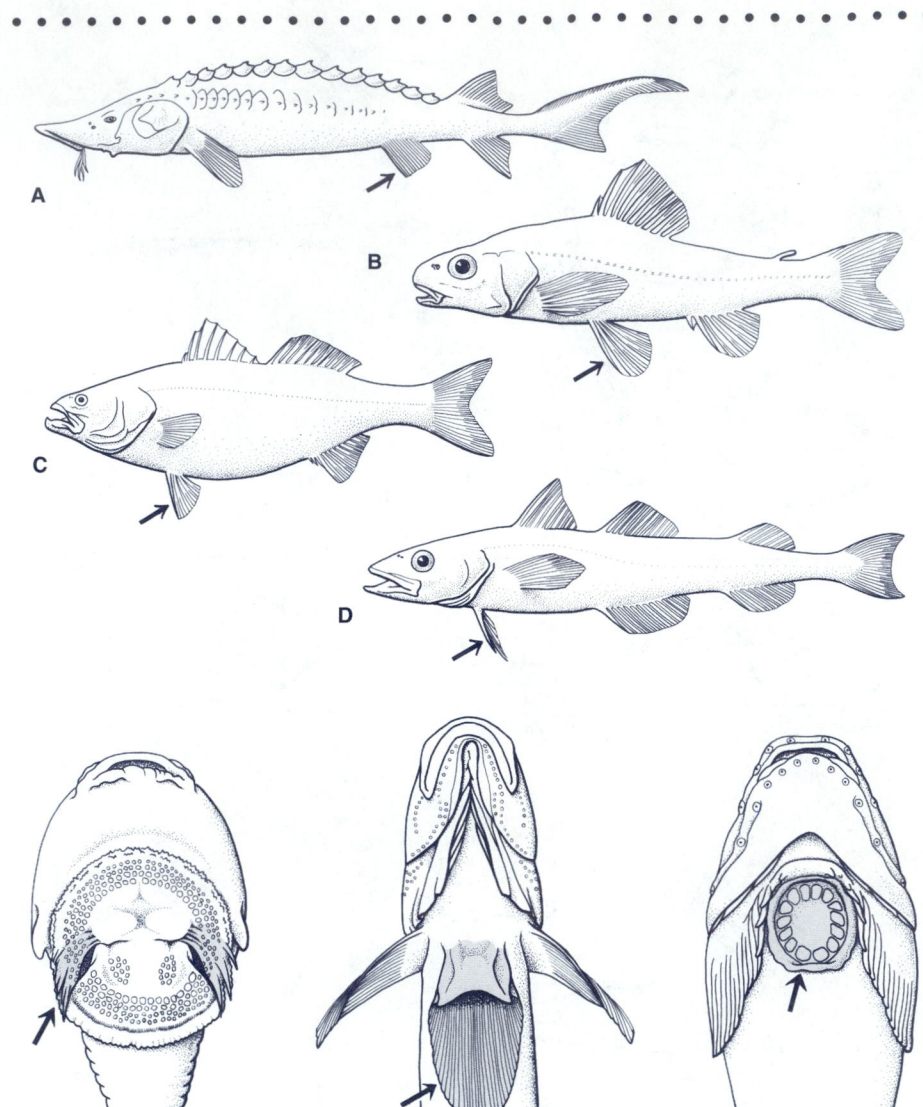

FIGURE 3.13
Examples of pelvic fin placement; pelvic fins with arrows: **A,** Abdominal (sturgeon, Acipenseridae); **B,** subabdominal (sand roller, Percopsidae); **C,** thoracic (bass, Moronidae); **D,** jugular (pollock, Gadidae). (Based on Jordan and Evermann, 1900.) Pelvic fins modified as sucking devices: **E,** Clingfish (Gobiesocidae); **F,** goby (Gobiidae); **G,** snailfish (Liparidae).

of sturgeons, which may be up to 3 mm thick. Exteriorly, the epidermis consists of squamous cells produced in a columnar germinative layer next to the dermis; these cells move outward, where they are eventually sloughed off. Except for the mucous covering, live cells of the epidermis are es-

sentially in contact with the medium, because no cornified layer is present, as it is in terrestrial vertebrates. However, in many species, a nonliving secretion of the epidermis, called a cuticle, covers the cells. The skin surface of lampreys consists of a thin cuticle, and a layer of cuticular secre-

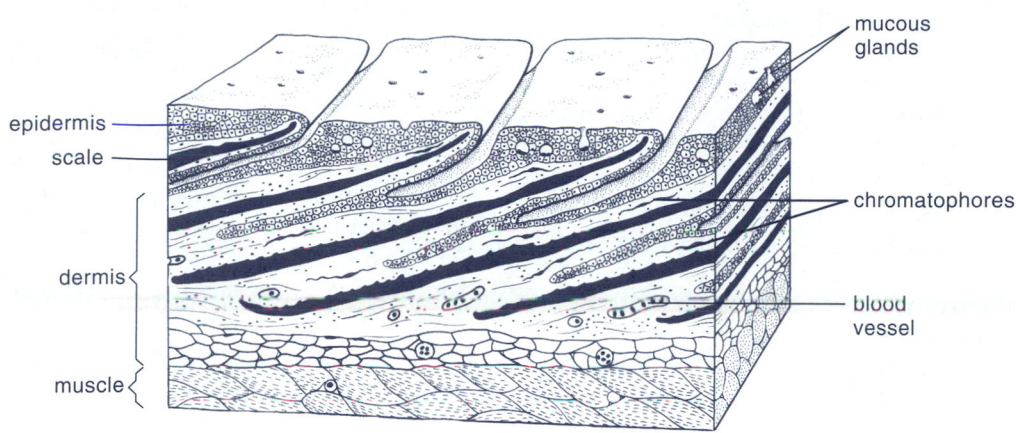

FIGURE 3.14
Section of fish skin. (Based on Wunder, 1936.)

tion may be associated with structures in other fishes that have contact with the bottom, such as the sucking disc of the clingfishes or the detached pectoral rays of gurnards. Some bony fishes, notably minnows (Cyprinidae) and their relatives, and certain salmoniform fishes secrete horny nuptial tubercles or pearl organs that cover part of the skin (see Fig. 27.7; Plate 10.2). These tubercles roughen the skin and provide friction during contact by breeding fishes.

Among the epidermal cells are unicellular mucus glands that discharge the mucus that forms the slimy outer coating of fishes. These glands appear to be of two types: those that discharge abruptly and refill, and those that produce slime gradually over a longer period. Mucus consists largely of glycoproteins that can absorb great amounts of water. The champions among mucus producers are the hagfishes, which have mucus glands with ducts that discharge slime and mucoprotein threads of considerable length (Whitear, 1986a). The oft-told story is of the recipe for hagfish jelly—add one hagfish to a bucket of water. The slime of fishes is largely protective in function. In addition to protecting the epidermis and making fishes difficult to grasp, slime can bind up and slough off particulate irritants and some heavy metal salts. Bacteria may be kept from the live epidermal cells by the mucus. Although mucus performs a lubricating function by helping fishes slip through the water, it appears to give only a slight advantage to most fishes and may function mainly in fast starts (Hoyt, 1975; Rosen and Cornford, 1971). Mucus can also precipitate certain suspended solids in muddy water. Special functions of mucus from the skin include its use as nest-building material in gouramis

(Osphronemidae) and as cocoon material in the African lungfishes (Protopteridae), which line their burrow with a desiccation-resistant covering during aestivation. When parrotfishes (Scaridae) rest at night, they surround themselves with a mucous envelope secreted from the oral cavity. The special, thick mucus of the clownfishes or anemonefishes (*Amphiprion, Dascyllus,* and *Premnas*) protects them from the stings of the anemones with which they cohabit (Lubbock, 1980). The mucus of snakeheads (Channidae) is used in western India to produce an extra strong building mortar (Antony, 1952).

When the epidermis is thin and consists of only a few layers, it is relatively simple in structure; a thicker epidermis may also contain nerve endings and pigment cells. Blood vessels are usually absent in the epidermis; nutritive materials diffuse through the intercellular matrix that holds the cells together. In some fishes, epidermal cells specialized for venom production are found associated with fin or head spines. Chimaeras (Holocephali), stingrays (Dasyatidae), stonefishes (Synanceinae), weevers (Trachinidae), and certain catfishes, including the aforementioned madtoms, are examples of fishes that possess various kinds of stinging spines. The venoms range from mildly irritating to painful, and some can be fatal to humans (Halstead, 1978).

The thickened epidermis of the sucking discs of clingfishes (Gobiesocidae) contains alveolar cells that form cushions that help mold the surface of the organ to the substrate. The light-producing organs (**photophores**) seen in many deepwater fish, such as members of the order Stomiiformes, are derived from epidermal cells (Whitear, 1986a).

Dermis

The dermis of fishes is much thicker than the epidermis—
it is here that the support network for the overlying epidermis resides. It is usually made up of two layers: a thin,
loose **stratum spongiosum**, just beneath the epidermis,
and a deeper **stratum compactum**. Most of the dermis
consists of fibrous connective tissue rather than actual cells.
Two sets of collagen fibers are disposed around the body in
opposing spirals, in such a manner that the fish can bend
without causing wrinkles in the skin (Whitear, 1986b).
A *subcutis* of connective tissue anchors the dermis to the musculature. The dermis contains pigment cells, blood vessels,
and nerves. The dominant feature of the dermis is the *scales*
or bony plates that assume a bewildering variety of shapes
and forms in fishes. In some fishes, especially those lacking
scales, the skin may be tough enough that it can be cured
to make leather products. The commercial fishery that has
developed for hagfishes supplies much of what passes for
"eelskin" in the leather trade. Shark skin, with its small,
rough scales, has a variety of specialized uses.

Scales

Most fishes have a covering of dermal scales. These may
be lacking, as in catfishes; embedded in the skin, as in
eels; or modified into bony plates or scutes (Fig. 3.15),
as in sturgeons (Acipenseridae), sticklebacks (Gasterosteidae), armored catfishes (including the Callichthyidae and
Loricariidae), and poachers (Agonidae). **Placoid** scales, or
dermal denticles, are typical of the sharks and their relatives. This type of scale consists of a basal plate, containing
some bone cells, that is buried in the skin, with a raised portion exposed (Fig. 3.15D). The overall structure is similar
to that of an elasmobranch tooth, with which these scales
are homologous, having a pulp cavity and tubules leading
into the dentine (Nelson, 1970). These denticles, with their
hard outer layer of vitrodentine, which is similar to tooth
enamel, make possible the use of dried shark skin as an abrasive similar to fine sandpaper.

Many extinct lobe-finned fishes have **cosmoid** scales,
with a layer of noncellular **cosmine** that lies beneath a very
thin outer layer of vitrodentine slightly different from that of
the placoid scale. Below the cosmine is a layer of vascularized
bone, called **isopedine,** and an inner layer of laminar bone.
Although the scales of the living coelacanth (*Latimeria*) are
said to be simplified cosmoid scales (Lagler et al., 1977),
Meinke (1982) concluded that coelacanths do not develop
cosmine. The scales of extant lungfishes are termed **elasmoid** (Whitear, 1986b).

Ganoid scales are encountered on bichirs and reedfish
(Polypteridae) and gars (Lepisosteidae) and, in modified

form, on the caudal fin of sturgeons (Acipenseridae) and
paddlefishes (Polyodontidae). Typically, the ganoid scale
has a rhomboid shape, with an anterior, peg-like extension overlapping the scale in front (Fig. 3.15E). The outer
layer of this scale is an acellular, enamel-like material called
ganoin, with a cosmine-like dentine layer beneath it. Gars,
however, lack this dentine layer (Kent and Carr, 2001).
Lamellar bone forms the basal plate (Whitear, 1986b). In the
extinct †Palaeonisciformes and the living bichirs and reedfish, the cosmine layer is perforated by tubules similar to
those present in cosmoid scales, and is underlain by a vascular area of transverse canals. This type of scale is sometimes
termed *palaeoniscoid*. The cosmine and tubules are reduced
in gars, sturgeons, and paddlefishes.

Scales of teleostean fishes are relatively simple, consisting of a mineralized surface layer and a thin, deeper layer of
collagenous tissue (Whitear, 1986b). They are quite thin when
compared with ganoid or cosmoid scales and lie in pockets of
the dermis, usually overlapping the scale behind in an im-
bricated manner. Scales of soft-rayed actinopterygians are
generally ovoid to nearly circular (subcircular) in shape and
lack spines or projections on the surface or posterior margin;
this type of smooth-rimmed scale is termed **cycloid**. More
derived teleosts tend to have **ctenoid** scales, with minute
spines on the exposed portions of the scales or in a comb-like
row on the posterior margin. Roberts (1993) has recognized
that the range of scale types that show spines transcends the
conventional meaning of the term *ctenoid*, so that care should
be used in evaluating the phylogeny of fishes based on the
presence of spines on scale surfaces (see Chapter 18).

The bony layer of the scale is usually characterized by
concentric ridges that represent growth increments during the life of the fish (Fig. 3.15F, G). Spacing and other
characteristics of the ridges (**circuli**) give biologists clues to
the life history of the individual fish. Year marks (**annuli**),
spawning marks, and signs of other developmental events
may be interpreted by a skilled scale reader. The innermost
part of the scale—the original locus of formation in the
larval fish—is called the **focus**. Lines called **radii** often lead
outward from the focus toward the edge of the scale. Scales
located over the lateral line are perforated by sensory pores.

INTERNAL SUPPORT: THE SKELETAL SYSTEM

The Axial Skeleton

The Skull

The skull (**syncranium**) of all vertebrates may be divided
into two discrete functional units: (1) the **neurocranium**

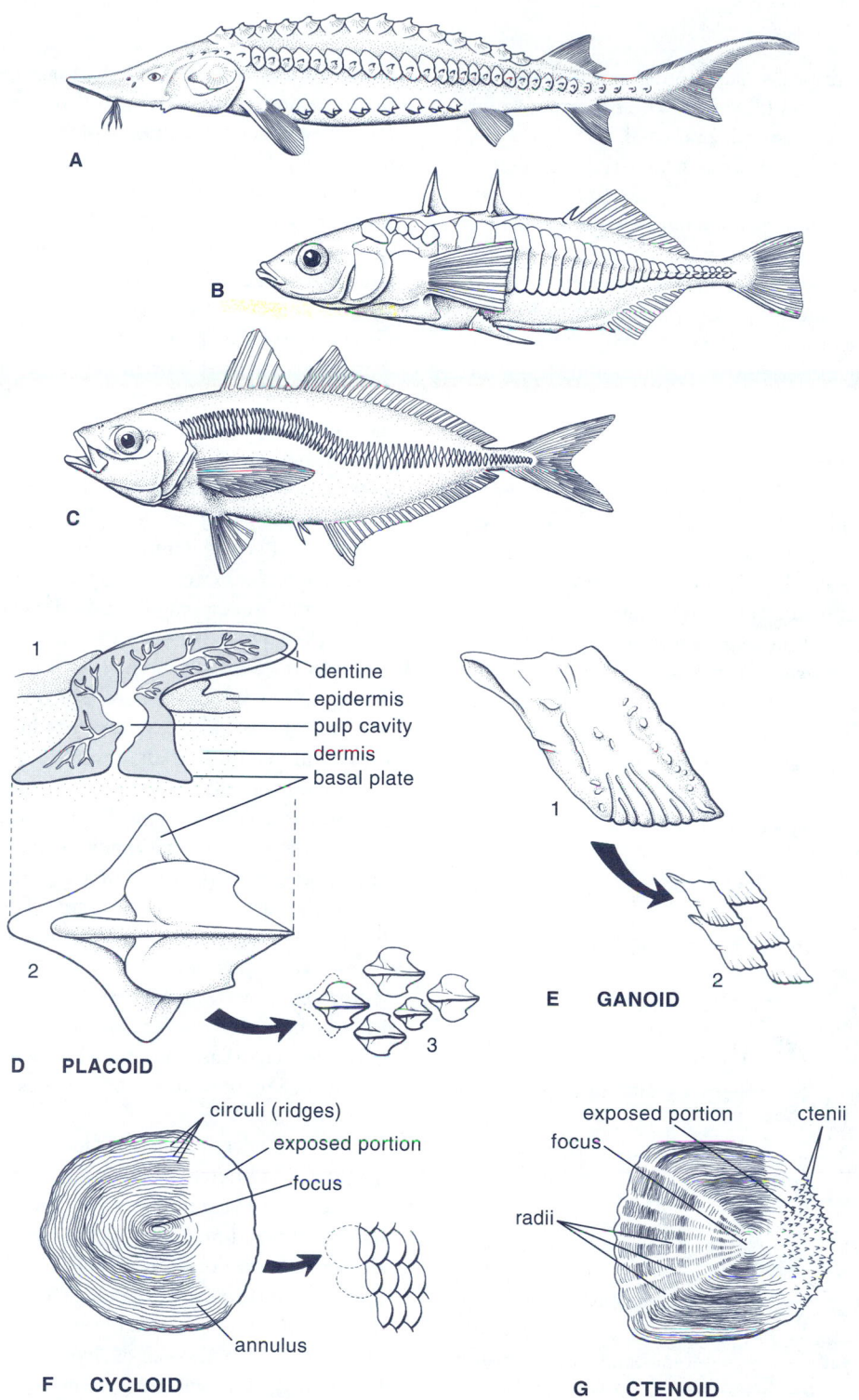

FIGURE 3.15

Examples of fishes with scutes: **A,** Sturgeon (Acipenseridae); **B,** stickleback (Gasterosteidae); **C,** jack (Carangidae). Example of types of scales (anterior to left): **D,** Placoid—1, sagittal section, 2, top view, 3, disposition of skin; **E,** ganoid—1, single scale, 2, disposition of fish; **F,** cycloid; **G,** ctenoid. (*A* and *B* based on Jordan and Evermann, 1900.)

(**braincase**), which supports, surrounds, and protects the brain and sense organs; and (2) the **splanchnocranium** or **branchiocranium** (the visceral, branchial, or gill arches), which supports the gills and from which the mandibular (jaws) and hyoid arches were derived. Note that many vertebrate morphology texts and most laboratory manuals refer to the elasmobranch neurocranium as the **chondrocranium**. It is more accurately termed the **chondroneurocranium** or **endocranium** (Compagno, 1988). The cranial units, whatever their nomenclature, differ not only in function, but also in embryological origin.

Development of the Skull

There is a general similarity among vertebrates in the embryonic development of the skull. Because there will be some discussion of the development of fish skulls later in this chapter, a generalized and abbreviated account of the process will be given here (Fig. 3.16). Hanken and Hall (1993) provide one of the most detailed treatments of the vertebrate skull.

It is the formation of the neurocranium, derived mainly from the embryonic **neural crest** (Gans, 1993; see Chapter 27) that distinguishes craniate chordates from the acraniates (see Chapter 2). The neurocranium forms at the anterior end of the notochord, beginning with the formation of two

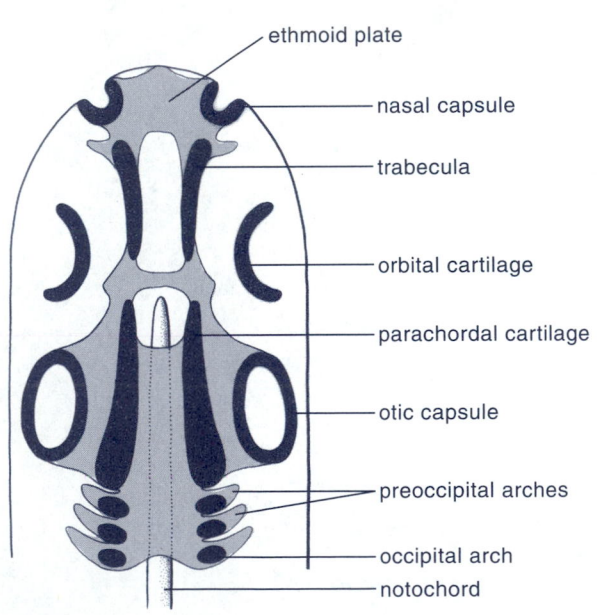

- ethmoid plate
- nasal capsule
- trabecula
- orbital cartilage
- parachordal cartilage
- otic capsule
- preoccipital arches
- occipital arch
- notochord

FIGURE 3.16
Generalized diagram of early stages of development of chondocranium. In *black,* separate cartilages; in *gray,* the formation of the ethmoid and basal plates.

parachordal cartilages. These cartilages, one on each side of the notochord and posterior to the forebrain, enlarge and form a structure, the **basal plate,** by fusing around the notochord. The basal plate enlarges and fuses with paired **occipital arch** cartilages that develop over the hindbrain, thus forming the rear wall of the neurocranium. Also uniting with the basal plate and occipital arch cartilages are paired **otic capsules** that form around the inner ears. The synotic tectum, a cartilage that forms over the posterior part of the brain, joins with the occipital arch cartilages. Anteriorly, two prechordal cartilages, called trabeculae, form, the anterior part from the neural crest, the posterior part from mesoderm (Balinsky, 1970). As the trabeculae grow, they fuse at their anterior ends to form the **ethmoid plate** and unite posteriorly with the developing basal plate.

The splanchnocranium, like the trabeculae just described, is derived from neural crest mesenchyme. This mesenchyme migrates ventrally between the future gill slits to form the precursors of what will become a series of skeletal bars that run between the gill slits to support and, in conjunction with the branchiomeric musculature, operate the gills. The ancestral number of these visceral arches is not known, but many systematists believe that there were eight, and that most or all of them functioned primitively in the support of gills.

In some fishes, a pair of **polar cartilages** forms between the trabeculae and joins them and the **parachordal cartilages,** so that three paired elements together contribute to the formation of the cranial floor. **Nasal capsules** form anteriorly from their respective cartilages, and **orbital cartilages** extend forward from the otic capsules. Antorbital processes develop from prechordal mesoderm. Enlargement and dorsal growth of these structures results in the cartilaginous cranium, which will be partially covered or replaced by bone in the bony fishes.

The Splanchnocranium and Jaws

The series of branchial arches that form around the pharyngeal region constitute the splanchnocranium. A primitive, undifferentiated branchial arch consists of four lateral paired elements (from dorsal to ventral, the **pharyngobranchial, epibranchial, ceratobranchial,** and **hypobranchial** elements) and a single, median, ventral element (the **basibranchial**) to which the paired hypobranchials articulate (see Fig. 3.18A). The splanchnocranium is associated with the support of gills, and in gnathostomes, the anterior two arches have been modified as elements of the jaws (see Fig. 6.3).

The first or anteriormost arch in living fishes is called the **mandibular arch** because of its contribution to the formation of the primary (i.e., primitive) upper and lower jaws (Schultze, 1993). As in the other arches, there are major

upper and lower elements on each side. In the course of evolution, the upper elements (**palatoquadrate cartilages**) of
the mandibular arch have slanted forward along the underside of the neurocranium and have joined anteriomedially to
complete the primitive upper jaw. Essentially, this remains
as the functional, tooth-bearing upper jaw of sharks and
rays, but it does not form the border of the mouth in bony
fishes, in which dermal skeletal elements surround and sup-
plant this original upper jaw.

The lower elements of the mandibular arch are called
the **mandibular** or **Meckel's cartilages**. These represent the lower element of the primary lower jaw, articulating with the palatoquadrate cartilage posteriorly and joining
together anteriomedially. In sharks and rays, the mandibular cartilage bears teeth and forms a complete mandible. In
bony fishes, it becomes ossified and still forms the jaw joint,
but again dermal bone elements cover Meckel's cartilage,
which is mostly reduced to a vestigial bit of cartilage hidden
beneath the dermal bones that now form the lower jaw. The
posterior ossification of the palatoquadrate cartilage is called
the **quadrate** bone; the ossification of Meckel's cartilage is
the **articular** bone.

The second arch in the series is the **hyoid arch**. The
upper element on each side, the **hyomandibular cartilage,**
has evolved in fishes into an important suspensory structure
that contributes to the suspension of the jaws and hyoid bar.
The **spiracle,** which is the remnant of the upper part of a
full primitive gill slit, is located between the mandibular and
hyoid arches. The lower element of the hyoid arch is the
ceratohyal cartilage, which becomes the supporting
structure of the median, **basihyal** cartilage. The remaining visceral arches typically support the gills and are called
branchial arches. The **ceratobranchial** element of the
fifth branchial arch of bony fishes is usually modified to bear
pharyngeal teeth.

The Dermatocranium

Fossils of the earliest fishes show a skull surrounded by a
dense envelope of dermal bone, with the neurocranium and
splanchnocranium remaining cartilaginous. The skulls of
living elasmobranch fishes provide a simplified introduction to skull anatomy, as the braincase (neurocranium) and
branchial arches (splanchnocranium) are the only components. In these fishes, the dermal component (the **dermatocranium**) consists only of the undifferentiated overlying
coat of denticles (the placoid scales) and the teeth, which are
considered homologous to the scales. As bony fishes evolved,
the dermal bone became more closely incorporated with the
neurocranium and splanchnocranium, and sites of ossification developed in association with all three components.

Skull of Living Agnaths

Hagfishes, considered by some to be a sister group of the
vertebrates (Janvier, 1981, 1996), have a cranium that appears to have been arrested at an early stage of development
(Fig. 3.17A). The skull "is made up of sinuous cartilaginous arches and plates to which muscles attach, and which
strengthen the walls of ducts" (Janvier, 1993). Trabeculae
are present, and there is a cartilaginous floor consisting of
the fused parachordals. The cartilaginous otic capsules are
fused to the parachordals, but the remainder of the sides
and top of the structure is membranous. Visceral arches are
not well developed in hagfishes. A rudimentary framework
of cartilage external to the pharynx and gill pouches serves
as a branchial skeleton, with a series of cartilages supporting the rasping organ (**lingual apparatus**) and its muscles.
The **velum,** which pumps water through the pharynx, has
a complex skeleton of cartilaginous rods.

The cranium of lampreys is more complete than that of
hagfishes (Fig. 3.17B). A partial roof for the brain is formed
anteriorly by extensions of the trabeculae. Posteriorly, there
are sidewalls, but only connective tissue covers most of the
brain. This rudimentary neurocranium, homologous with
that of other vertebrates, constitutes a minor part of the total
skull of lampreys. There are also dorsal cartilages supporting
the region anterior to the nasal opening, a series of cartilages
around the circular mouth, and a long, ventrally located lingual cartilage. Support of the gill region is provided by an
intricate **branchial basket** of cartilage that develops just
beneath the skin. Posterior extensions of the branchial basket
also cover and protect the heart. This basket is *not* homologous with the branchial skeleton of gnathostomes.

Skull of Elasmobranchs

In sharks, the chondroneurocranium usually forms a complete box, pierced in places by foramina and fenestrae for
the passage of nerves and blood vessels (Fig. 3.18). The olfactory and otic capsules are integral parts of this box. Anterior to the olfactory capsule is the **rostrum,** which may
be extremely elongated in some sharks (e.g., the sawshark,
Pliotrema; see Fig. 6.8A).

Rays may have large dorsal fontanelles, both in the rostral region and more posteriorly, so that the cranial roof appears to be incomplete. The rostrum in rays is often very
long. The extremes of length are seen in the eagle ray (*Myliobatis*), in which the rostrum is undeveloped, and in the
sawfish (*Pristis*), in which the rostrum may comprise nearly
one-third of the entire length of the fish.

In most elasmobranchs, there are prominent processes
anterior and posterior to the orbit (Fig. 3.18). Some species
show various crests and other sculpturing on the roof of the
cranium. Posteriorly, **occipital condyles** form a surface for

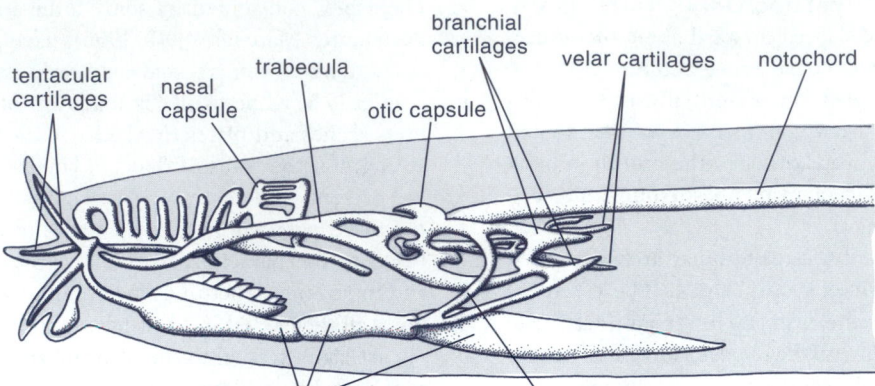

A   HAGFISH

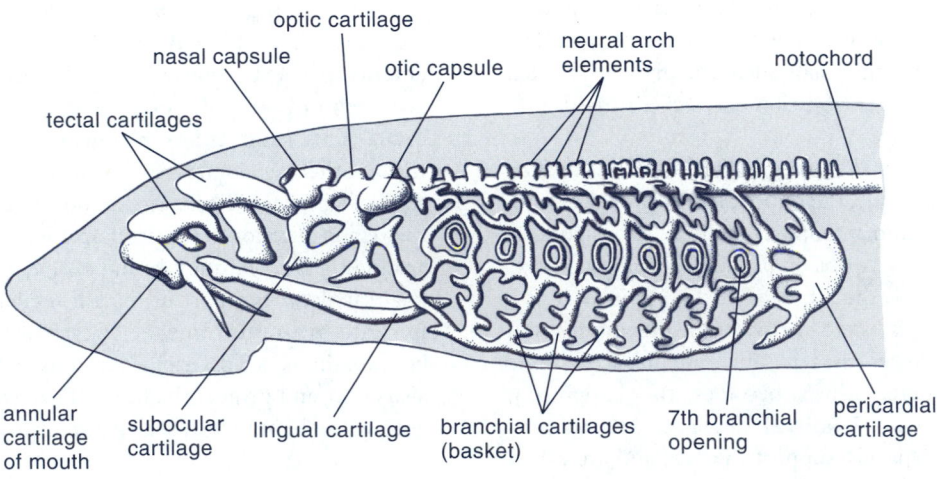

B   LAMPREY

FIGURE 3.17
Diagrams of skulls of living agnaths: **A,** hagfish; **B,** lamprey. (Based on Goodrich, in Lankester, 1909.)

articulation with the vertebral column, and the **foramen magnum** allows passage of the spinal cord.

Although modern elasmobranchs have virtually no bone in their skeletons, the cartilage may be calcified to some extent, even to the point of being as hard as bone. Prismatic calcifications of hydroxyapatite may strengthen major skeletal structures of adult elasmobranchs (Compagno, 1988), so that the surface of a dried chondrocranium has a mosaic appearance. Vertebral centra are often heavily calcified.

The jaws of elasmobranchs, as in other early gnathostome fishes, evolved as modifications of the anteriormost visceral arches (see Fig. 6.3); these have become greatly enlarged in some species. The palatoquadrate cartilage extends forward along the underside of the neurocranium, remaining free and movable in most species. It is usually attached to the cranium by means of the upper part of the hyoid arch—the *hyomandibular cartilage*—which articulates with the otic region. The lower jaw, or Meckel's cartilage, articulates with the posterior part of the palatoquadrate. The lower part of the hyoid arch is called the **anterohyal** (or *ceratohyal*) cartilage, and both it and the hyomandibular cartilage may bear *gill rays*—slender cartilaginous rods that strengthen

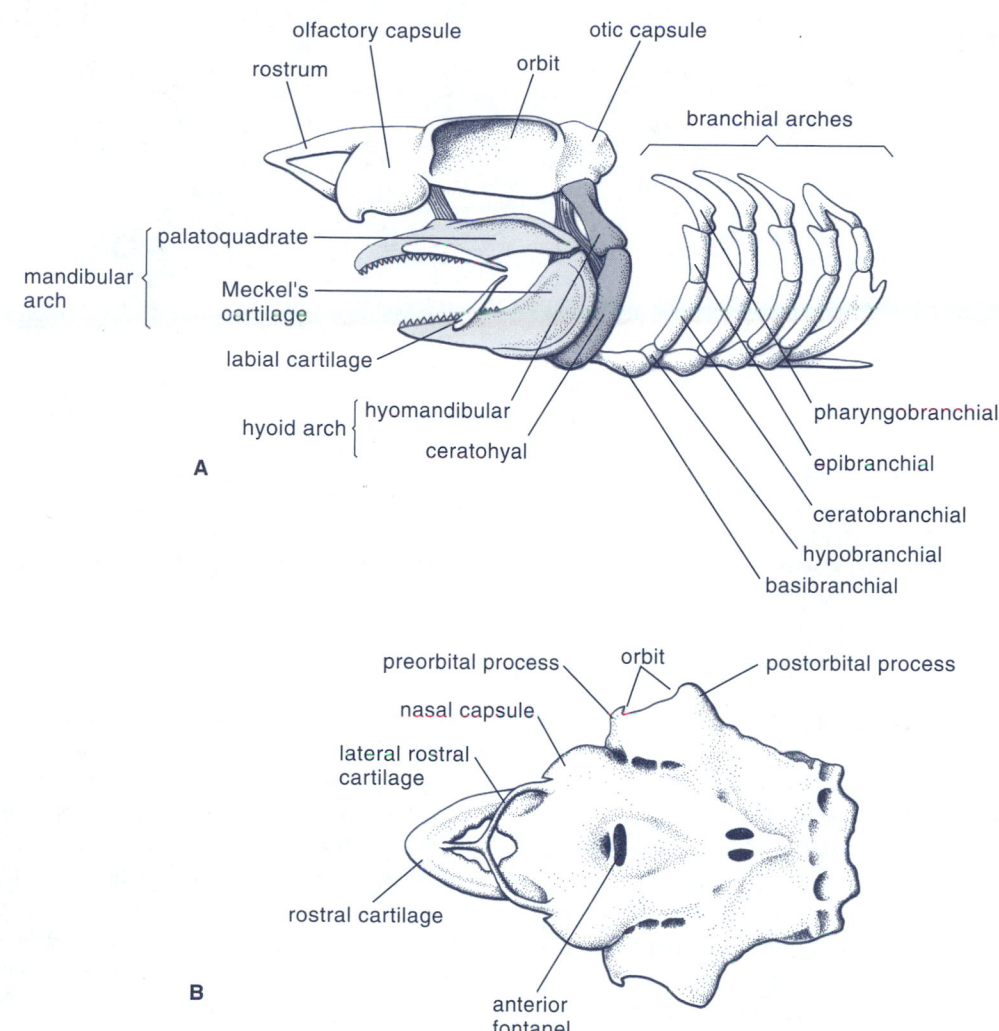

FIGURE 3.18
Diagram of elasmobranch skull. **A,** Lateral view, showing splanchnocranial elements (left side only). (Based on Bridge and Boulenger, 1904.) **B,** Dorsal view of shark skull.

the interbranchial septa from which gill tissue projects as lamellae. A basihyal cartilage connects the anterohyal cartilages ventrally. Chondrichthyan fishes show a range of involvement of the hyoid arch in jaw support. Most sharks and rays show a **hyostylic** condition, in which the palatoquadrate cartilage is free from the braincase and is buttressed posteriorly by the hyoid arch; this permits a considerable amount of jaw mobility (Fig. 3.19A). A few sharks show the **amphistylic** condition, in which the palatoquadrate is attached to both the neurocranium and the hyomandibular

(Fig. 3.19B). The evolution of jaw suspension mechanisms, resulting in a diversity of relationships between the cranium and visceral arches, is outlined in Chapter 6.

Vertebrates in which the upper jaw elements are fused to the cranium are termed **autostylic**. The chimaeras (Holocephali) have the palatoquadrate cartilage entirely fused to a cartilaginous neurocranium—a type of autostylic jaw suspension generally called **holostylic**. The hyomandibular cartilage serves no suspensory purpose for either upper or lower jaw. The hyoid arch is slightly

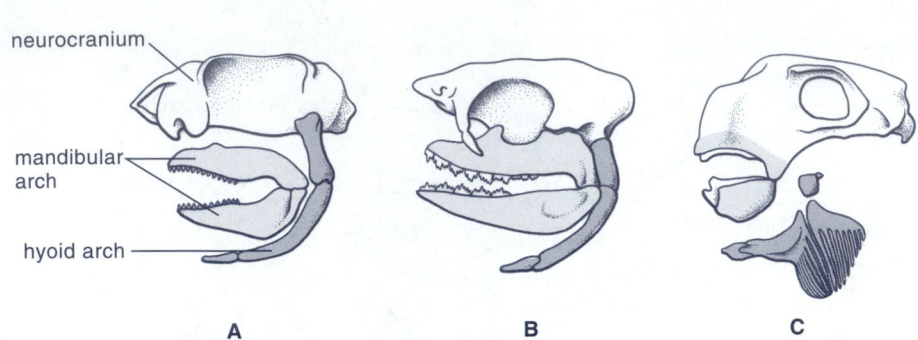

FIGURE 3.19
Illustrations showing relationship of mandibular arch to neurocranium in three types of jaw suspensions.
A, Hyostylic (upper jaw not firmly attached to neurocranium, with a ligamentous attachment to hyomandibular);
B, amphistylic (upper jaw attaches anteriorly to basal angle of neurocranium and posteriorly to postorbital process); **C,** holostylic (upper jaw fused to neurocranium). For clarity, the mandibular arch has been lowered in *A*; the hyoid arch has been flexed downward in *A* and *B and* has been lowered in *C*.

more modified than the branchial arches and bears gill rays (Fig. 3.19C).

There are five branchial arches in most sharks and rays, but a few species have six or seven. These arches typically consist of *pharyngobranchials, epibranchials, ceratobranchials,* and *hypobranchials.* Connective tissue might or might not hold pharyngobranchials to the roof of the pharynx, depending on the species. *Basibranchial cartilages* connect the hypobranchials on the ventral midline. Gill rays on the epibranchials and ceratobranchials support the gills on all but the last arch.

Skull of Bony Fishes

By far the most complex and kinetic vertebrate skulls are those of the bony fishes. Figures 3.20, 3.21, and 3.22 show the structures mentioned in this section, and frequent referral to these figures will be helpful. Good coverage of the ontogenetic development of the bony fish skull can be found in Weisel (1967) and Morris and Gaudin (1975). Gregory (1933, reprinted 2002) remains the classic treatise on the skull of bony fishes, and DeBeer (1937) and Schultze (1993) are also useful. Excellent treatment of the phases of skull development, structure, and diversity can be found in the three-volume work edited by Hanken and Hall (1993).

Some primitive bony fishes, such as paddlefishes and sturgeons, retain much of the cartilaginous neurocranium, with few ossifications. Others, such as the bowfin and salmonids, also retain much cartilage in the cranium, but in

the majority of bony fishes, the cartilage is mostly replaced by bone. Ossifications forming around and replacing cartilage are called **perichondral** and **endochondral cartilage bones,** respectively, whereas **membrane (dermal) bones** are formed in the dermis and are not preceded by a cartilage model. In teleosts, endochondral bones are especially prominent in the posterior region of the neurocranium. The ventral, unpaired **basioccipital bone** usually forms the occipital condyle, which articulates with the vertebral column. In perciform fishes and many other spiny-rayed groups, the lateral paired **exoccipital bones** contribute to the occipital condyle, forming, along with the basioccipital, a tripartite structure that articulates with the first vertebra. In most teleosts, the exoccipitals completely surround the spinal cord as it enters the skull through the *foramen magnum.* The dorsal, median, **supraoccipital bone,** in addition to forming part of the cranial roof, furnishes an anterior attachment surface for the epaxial trunk muscles. This bone may be extended into a crest to which the muscles attach. The crest reaches a large size in deep-bodied fishes with a large mass of muscle extending onto the skull, as in many advanced teleosts.

Much of the posterior part of the teleost skull consists of five endochondral bones that form in each **otic capsule,** protecting the membranous labyrinth. The largest of the complex are the paired **prootic bones,** which are anterior to the basioccipital and constitute a considerable portion of the

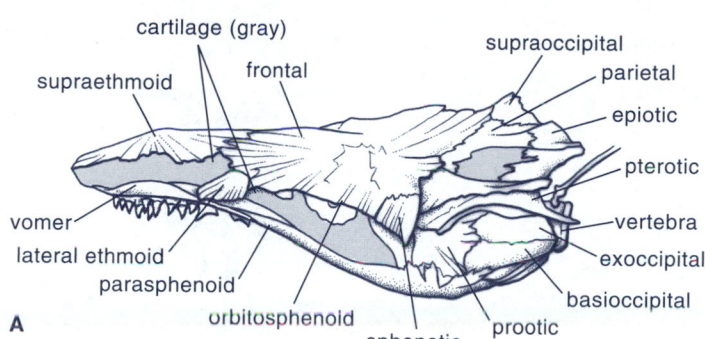

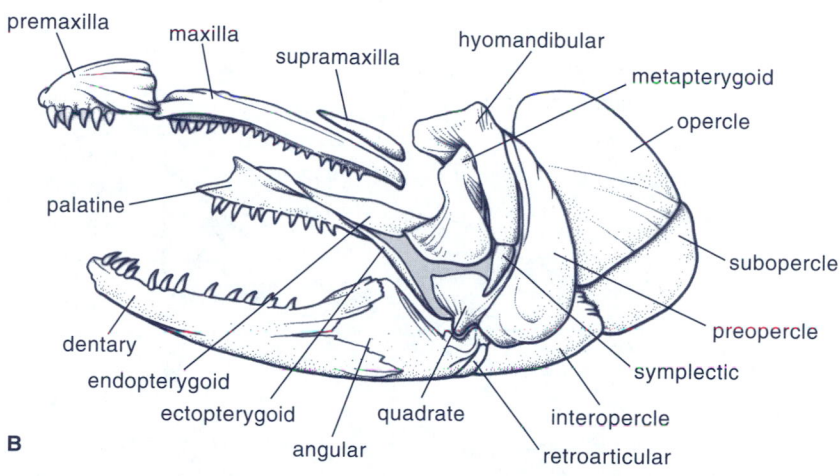

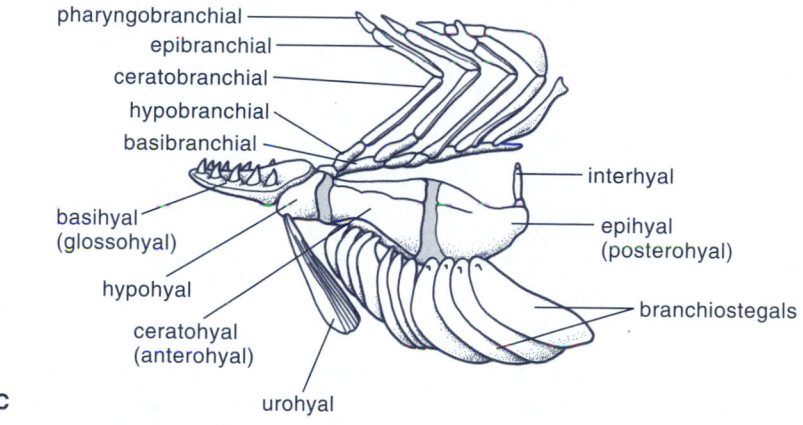

FIGURE 3.20
Skull bones of steelhead trout Oncorhynchus mykiss **A,** Neurocranium; **B,** jaws, suspensorium, and operculum (left side only); **C,** branchiohyoid apparatus (left side only). (Pharyngobranchials attach to ventral midline of neurocranium; interhyal attaches to hyomandibular posterior to symplectic.)

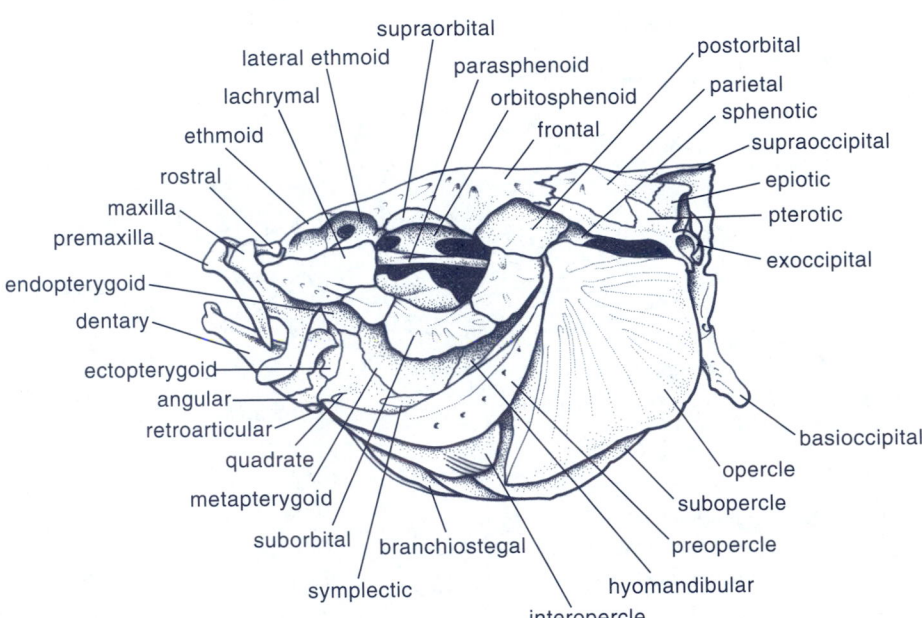

A   DORSAL VIEW

B   LATERAL VIEW

FIGURE 3.21
Skull of carp (*Cyprinus carpio*): **A,** Dorsal view of neurocranium; **B,** lateral view of skull.

lateral floor of the cranium in many species. Part of the posterior boundary of the orbit consists of the **sphenotic bone,** dorsal to the anterior part of the prootic. The sphenotic in part forms around the anterior semicircular canal of the inner ear. The **hyomandibular bone,** which supports the jaws, articulates with the sphenotic. The prominent ridges that usually mark the widest part of the cranium are formed by the **pterotic bones,** which ossify around the lateral semicircular canals and usually combine with a dermal element to produce a compound bone. The **epiotic bone,** which os-

sifies in part around the posterior semicircular canal, usually can be recognized as a process between the pterotic and the supraoccipital. The epiotic bone is the site of attachment of the pectoral girdle to the cranium. An **intercalar bone,** an ossification of a ligament, appears in the back wall of the cranium between the pterotic bone and the exoccipitals. This has replaced an endochondral bone (**opisthotic**) that has been lost by living bony fishes (Rojo, 1991).

Cartilage bones of the trabecular section of the cranium include the following: the paired **pterosphenoid**

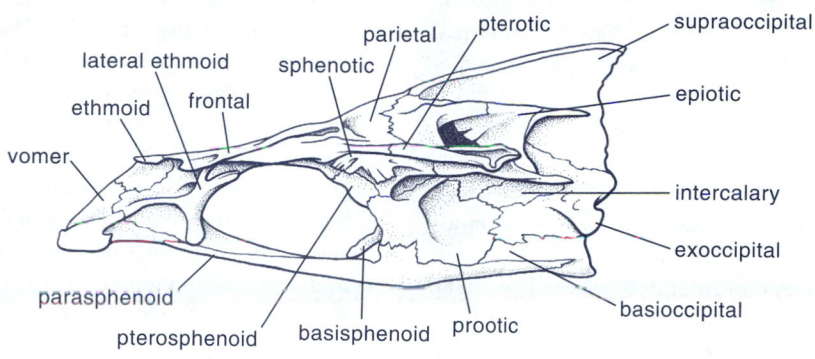

A   LATERAL VIEW

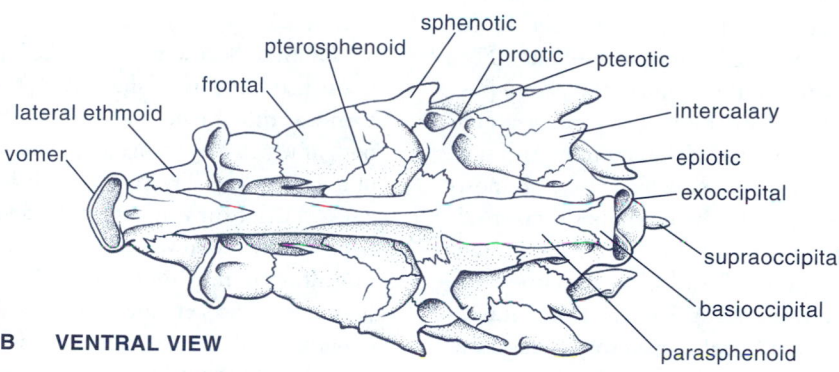

B   VENTRAL VIEW

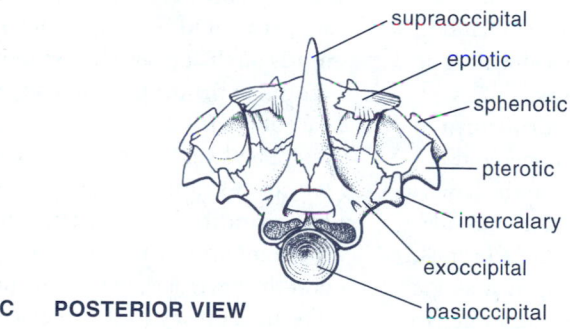

C   POSTERIOR VIEW

FIGURE 3.22

Neurocranium of percoid (based on *Morone saxatilis*): **A,** Lateral view; **B,** ventral view; **C,** posterior view.

bones, which form part of the posterior wall of the orbit, connecting with the prootic bones posteriorly; the median, **orbitosphenoid bone,** which forms a bony **interorbital septum,** through which the olfactory nerves pass in clupeiforms, salmoniforms, cypriniforms, and other basal bony fishes (this bone is absent in more derived teleosts); and the median **basisphenoid bone** in the posterior part of the orbit. The arms of the Y-shaped basisphenoid articulate with the prootics, and the shaft connects ventrally to the parasphenoid bone. Anterior to the orbit are paired, lateral **ethmoid bones,** which are of endochondral origin. The lateral ethmoids form a complex with dermal elements (**prefrontals**) in some fishes and are sites of attachment of the paired, dermal **lachrymal bones.**

The cartilage (endochondral) bones constitute the primary neurocranium, to which dermal bones are added. The dermal bones, also called *investing bones,* may include plates originating from scales, plates formed from coalescing tooth bases, and bones that form directly from membranes. The most posterior dermal bones of the skull are the paired **parietal bones,** which usually flank the median supraoccipital bone and constitute part of the roof of the cranium. Anterior to these are the large **frontal bones,** which make up most of the cranial roof. Parietals and frontals are both traversed by cephalic lateral line canals in many fishes. Typically, an unpaired ethmoid bone (**supraethmoid, dermal mesethmoid**) roofs the snout in front of the frontals. Paired **nasal bones,** which develop around cephalic sensory canals, are located on each side of the ethmoid. Ventrally, the **vomer** usually forms the anterior point of the neurocranium; it is often attached to the ethmoid in higher teleosts, but it is separate in the lower bony fishes. The vomer forms part of the roof of the mouth and often bears teeth. The long **parasphenoid** forms the ventral midline of the cranium, extending between the vomer and the basioccipital. The parasphenoid bears teeth in some of the teleosts.

A series of bones (**infraorbitals** or **circumorbitals**) partially surround the orbit, although the number and extent of these are reduced in many fishes. Infraorbitals enclose cephalic sensory canals and cover the head musculature. Usually the anterior bone of the group, the lachrymal, is larger than the others in the series. Others, such as the suborbital and postorbital bones, are named according to their positions with respect to the eye.

The suspension of the primary upper jaw and the primary and secondary lower jaws is by bones that form in the palatoquadrate cartilage. This cartilage is ossified in part as the **quadrate bone** at the posterior end. The quadrate is shaped somewhat like a quadrant of a circle, with the point downward. The lower jaw articulates with the point of the quadrate. Dorsally, the quadrate is attached to the

metapterygoid, another ossification of the palatoquadrate cartilage. This bone is instrumental in suspending the remainder of the primary upper jaw from the hyomandibular. Anterior to the quadrate and metapterygoid are two dermal bones that form along the lower edge of the palatoquadrate cartilage. One of these, the **endopterygoid,** stiffens the roof of the mouth. The other, the **ectopterygoid,** connects the quadrate and palatine, which is in the anterior part of the roof of the mouth, just behind and lateral to the head of the vomer. The palatine bone may have both endochondral and dermal components. If the dermal component is lacking, the bone is called the **autopalatine.**

In the upper jaw, the vomer and palatines, and sometimes the ectopterygoids, endopterygoids, and the parasphenoid, may bear teeth. However, the so-called secondary upper jaw, composed of the dermal **premaxillary** and **maxillary** bones, usually constitutes the main tooth-bearing (*dentigerous*) surface.

Within the lower jaw, Meckel's cartilage remains largely unossified in bony fishes, except for an anterior, **mentomeckelian** element and a posterior, **articular** element, both of which may form complexes with the dermal elements of the lower jaw. The major tooth-bearing bone of the lower jaw is the **dentary.** Between the dentary and the **quadrate,** from which the lower jaw is suspended, is the **angular** or **anguloarticular bone,** called the "*articular*" in most older literature. The **retroarticular bone,** consisting of endochondral and dermal elements, is on the posterior lower corner of the **angular;** it was often called the "angular" in older literature. A sensory canal runs through the angular and dentary bones.

In many nonteleost bony fishes, the lower jaw contains several more bones than the lower jaw of teleosts. Included are **prearticulars** on the inner surface; tooth-bearing **coronoids** on the upper edge; **splenials** and **postsplenials** just ventral to the dentary; and **supraangulars** in the posterior part of the jaw.

The hyoid arch becomes bone through multiple centers of ossification. The uppermost bone element is the **hyomandibular,** which articulates with the otic region of the cranium and acts as a suspension for the primary upper jaw, the lower jaw, the hyoid apparatus, and the operculum. The **metapterygoid** articulates with the anterior face of the hyomandibular. A peg-like bone, the **symplectic,** extends from the bottom of the hyomandibular to the quadrate. The **interhyal** attaches to the hyomandibular just behind the symplectic and suspends the remainder of the hyoid arch, which consists of the paired **posterohyal (epihyal), anterohyal (ceratohyal;** Nelson, 1969; Rojo, 1991), upper and lower **hypohyal** bones, and the unpaired **basihyal (glossohyal)** bone. The latter bears teeth in many fishes.

An unpaired bone, the **urohyal**, extends backward from the basihyals into the isthmus and constitutes the firm ventral connection between the head and trunk. The urohyal is a cartilage bone in sarcopterygians, but an ossification of a tendon in teleosts (Arratia and Schultze, 1990). Important dermal bones that connect with the posterohyal and anterohyal are the **branchiostegals**, which in some fishes protect the gills ventrally. In others, the branchiostegals stiffen a membrane that can be of greater importance than the operculum in pumping water over the gills. The **operculum**, which serves as a shield for the gills and as part of the branchial pump, is composed of four pairs of dermal bones. These are usually plate-like and are associated with the hyoid apparatus. The largest bone in the operculum is usually the **opercle**, which attaches by its anterodorsal corner to a condyle on the posterior edge of the hyomandibular. The **preopercle**, which carries a sensory canal, usually attaches along the hyomandibular for much of its length. The **interopercle** is below the preopercle, and the **subopercle** lies ventral to the opercle.

The branchial arches of teleosts are composed of a series of endochondral bones plus dermal tooth plates and gill rakers. The first three arches consist of a pharyngobranchial and an epibranchial in the upper section and a ceratobranchial and a hypobranchial in the lower part. The pharyngobranchial of the fourth arch is typically fused to that of the third arch or reduced, so that only the epibranchial and ceratobranchial are evident. The fifth arch is reduced further to one bone that may represent a ceratobranchial, and it is generally modified to bear pharyngeal teeth. A series of basibranchials is set between the left and right halves of the arches. These sometimes bear teeth, which appear just behind the teeth on the basihyal.

The Vertebral Column

In hagfishes and lampreys, the notochord persists without constriction (Goodrich, 1930). In hagfishes, the postcranial skeleton consists only of the notochord, which reaches anteriorly to the level of the midbrain, and the caudal fin skeleton (Janvier, 1996). The notochord of lampreys is flanked with primordial vertebral elements, the **arcualia** (Janvier, 1996). The notochord of elasmobranchs is constricted by cartilaginous **vertebral centra** so that, if extracted from the body intact, it would resemble a string of beads; the constricted portions would contrast markedly in diameter with the unconstricted portions that fit into the concavities of the **amphicoelous** (biconcave) centra. Some species have a single calcified cylinder formed within the centrum (the **cyclospondylous** condition), whereas others may possess two or more concentric cylinders (**tectospondylous**). In some, calcified radiating lamellae extend from the calcified cylinder, giving a somewhat star-shaped pattern in cross section (**asterospondylous**). Each centrum in the trunk of the elasmobranch has ventrolateral transverse processes (**basapophyses**), which bear the cartilaginous ribs. Dorsally, there is a **neural spine** surrounding the **neural canal**, through which the spinal cord runs. These neural arches consist of dorsal and ventral **intercalary plates** that alternate with basal dorsal plates (Daniel, 1934). In the tail region, the caudal vertebrae bear ventral **hemal arches** with spines.

Some basal bony fishes have a distinctive arrangement of the vertebrae. For instance, the bowfin (*Amia calva*) has two vertebrae in each body segment in the posterior section of the vertebral column, a condition called **diplospondyly**. Gars (Lepisosteidae) have **opisthocoelous** vertebrae, so called because they are concave posteriorly and convex anteriorly.

Although in some eels, the front and back surfaces of the centra are flat, and the blenniid *Andamia* has centra that are convex anteriorly, the typical teleost has ossified amphicoelous centra (concave anteriorly and posteriorly; Fig. 3.23) with the notochord filling the concavities. Basapophyses (**parapophyses**) are present, but might not be fused to the centra. Neural arches and spines are present, and the caudal vertebrae have hemal arches and spines. **Zygapophyses** can occur both anteriorly and posteriorly on the centra. The zygapophyses are generally small in fishes, with those of adjacent vertebrae not making contact, but they may be large and interlock in powerful swimmers such as the tunas and mackerels. The interlocking of the zygapophyses prevents excessive rotation of vertebrae, keeping them aligned in true dorsoventral position.

Ventral ribs (**pleural ribs**) usually attach to the vertebral basapophyses. **Intermuscular bones** that extend into the horizontal skeletogenous septum are often called **dorsal ribs**, regardless of whether they are borne on the centrum or on the pleural rib. Usually, the bones that lie in the myosepta (intermuscular bones) take their names from the structures that bear them; for example, those borne on the neural arch are called **epineurals**, those borne on the centra are **epicentrals**, and those on the ribs are called **epipleurals**. However, the terminology for these bones varies somewhat (see, e.g., Forey, 1973; Goodrich, 1930; Grassé, 1958). Johnson and Patterson (1993) pointed out that intermuscular bones are subject to much modification and that in spiny-rayed fishes, the epineurals may be displaced ventrally into the horizontal septum or onto the ribs, so that in such fishes, the bones that appear to be epipleurals are actually homologous with epineurals. Epineurals develop as bony growths on the neural arches, but in most fishes, they lose the bony connection and are attached by ligaments. Epipleurals and epicentrals are usually attached by ligaments, or may be rep-

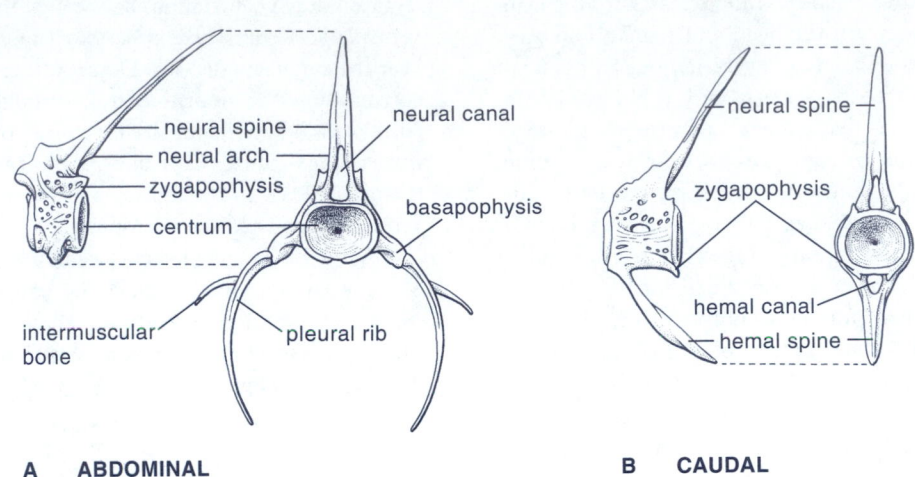

neural spine
neural arch
zygapophysis
centrum
neural canal
basapophysis

intermuscular bone
pleural rib

A   ABDOMINAL

neural spine
zygapophysis
hemal canal
hemal spine

B   CAUDAL

FIGURE 3.23
Vertebrae of teleost: **A,** Lateral and posterior views of abdominal vertebrae; **B,** lateral and posterior views of caudal vertebrae.

resented by ligaments (Johnson and Patterson, 1993). Intermuscular bones that are attached to the vertebrae by flexible or ligamentous joints may bend with the flexure of the musculature. A fossil fish (†*Chongichthys*) has sturdy, firmly attached epineural bones on the neural spines in the region of the dorsal fin (Arratia, 1982). This presumably derived condition must have limited flexion of the body somewhat. Intermuscular bones are more common in the more primitive bony fishes than in the higher groups.

The number of vertebrae tends to be fewer in more derived bony fishes. For instance, the salmons and trouts have around 60, and many of the perch-like fishes have from 25 to 35. Some elongate specialists, however, may have numerous vertebrae. Crestfishes (Lophotidae) have up to 200 vertebrae, and some species of snipe eels (Nemichthyidae) have up to 750 (Nelson, 1994). Lindsey (1975) noted that among related species, those that grow to a larger adult body size tend to have more vertebrae than those with a smaller adult size. He called this phenomenon **pleomerism,** meaning "many divisions." Latitudinal variation in vertebral number has also been observed in some families (Bailey and Gosline, 1955).

There is much modification of the vertebral column in the region of the caudal fin (see Fig. 3.11). In some of the primitive teleosts, the column may be upturned, with three or more progressively smaller centra involved (the upturned portion is called the **urostyle**). Below these, there is a supporting structure made up of about six **hypural** bones that appear to be modified hemal spines, while above the verte-

bral column there are one or more **uroneurals.** The hypural complex supports the rays of the caudal fin. In more advanced teleosts, the vertebral column ends in a *urostyle,* which is an upturned portion of the last vertebral centrum. The hypurals of higher teleosts are usually fused into larger plates, sometimes with one supporting the upper lobe of the caudal fin and one supporting the lower. Patterson (1968, p. 234) defined the teleosts as follows: "Actinopterygian fishes in which the vertebral centra are perichordally ossified, the lower lobe of the caudal fin is primitively supported by two hypurals articulating with a single centrum, and in which the neural arches are modified into elongate uroneurals, the anterior uroneurals extending forwards onto the preural centra."

Fins and Associated Appendicular Skeleton

Median Fins and Associated Skeleton
The median fins of clasmobranchs are supported by **basal cartilages** that are often segmented into **proximal, middle,** and **distal** elements (Fig. 3.24). In some sharks, the proximal elements may fuse into a single plate. In some flattened species, the basals may join with the neural spines.

Median fins of sarcopterygian (lobe-finned) fishes show a number of distinct features. The "caudal fin" of the coelacanth (*Latimeria*) is a case in point. What Nelson (1994) referred to as a three-lobed caudal fin is claimed by Uyeno (1991) to actually consist of a third dorsal fin and a second

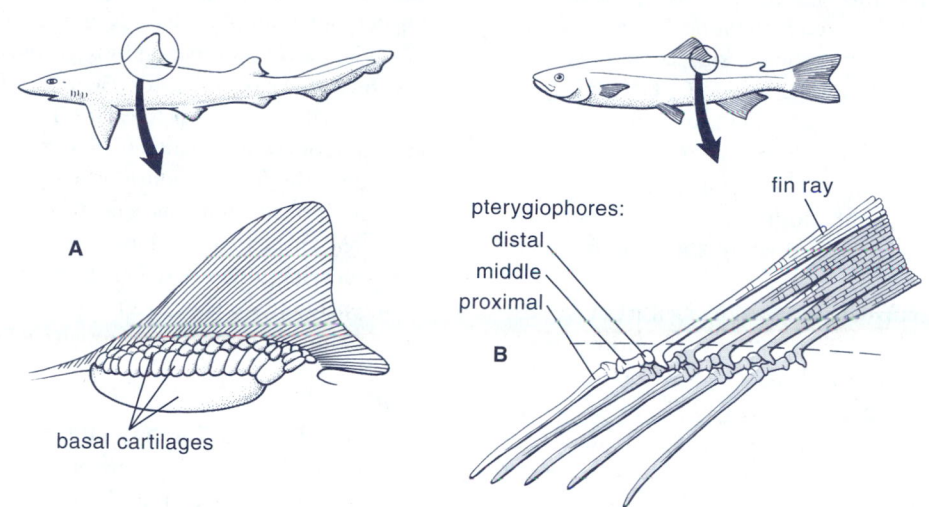

FIGURE 3.24

Skeletal supports of dorsal fin: **A,** Shark; **B,** bony fish. (Dashed lines show approximate body contour; *A* based on Goodrich, 1930.)

anal fin, in which the rays are borne on internal skeletal elements (the **pterygiophores**), that flank a small caudal fin (with rays not supported by pterygiophores) at the posterior terminus of the fish. The second dorsal fin and the first anal fin are lobate, extremely flexible, and important in the locomotion of the fish (see Chapter 7). In lungfishes, the median fins are supported by a series of basal elements that bear radials, which in turn bear fin rays, with the fin rays more numerous than the radials.

In teleosts, each ray of the median fins is typically supported by three pterygiophores—two ossified and one cartilaginous. Proximal pterygiophores are elongate, tapered bones set deeply into the medial skeletogenous septum, usually between the neural or hemal spines. Because of this, they are often called **interspinous bones;** those supporting the dorsal fin may be called **interneurals,** and those of the anal fin may be called **interhemals.** The middle pterygiophores are ossified and jointed flexibly to the proximal elements on one end and to the distal pterygiophores, if present, on the outer end. In soft-rayed fins, the cartilaginous distal pterygiophores usually fit between the bases of the two halves of the rays. In spinous fins, the distal pterygiophores may be lost or fused to the middle element. In extreme cases, fusion of the three elements is complete, and fin spines attach to single supporting structures.

Pterygiophores serve as attachment sites for epaxial muscles of the dorsal body musculature, and variation in mode of attachment is viewed as phylogenetically significant.

Mooi and Gill (1995) have determined that acanthomorph fishes can be classified into four main categories based on the mode of attachment of epaxial muscle to pterygiophores supporting the dorsal fin (see Chapter 15).

Pectoral Fins and Associated Skeleton

The pectoral fins of elasmobranchs are supported by a cartilaginous girdle consisting of an upper **scapular** section and a lower **coracoid** element (Fig. 3.25A). A small **suprascapular cartilage** may be present. In sharks, the two halves of the girdle are separate from each other dorsally and do not attach to the vertebral column. In rays, the two halves join each other or the vertebral column.

The elasmobranch pectoral fin consists of three basal cartilages; an anterior **propterygium**, a middle **mesopterygium**, and a posterior **metapterygium**. These articulate with the coracoid element. In rays, the articulation is horizontal, and the propterygium and metapterygium extend far forward and backward, respectively. Jointed **radials** attach to the basal cartilages and bear the fin rays at their distal ends.

The pelvic girdle of elasmobranchs is rather simple, consisting of a bar of cartilage crossing the ventral midline (Fig. 3.25B). Propterygia and metapterygia are borne on either side of the pelvic cartilage, and the enlarged metapterygia articulate with radial cartilages. The clasper cartilages that distinguish male chondrichthyans are formed from modified radials.

In the pectoral girdle of typical teleosts, the scapula and coracoid are ossified as endochondral bones, and part of their outer edges form the articular surface for the radials (**actinosts**) of the pectoral fin (Fig. 3.25C, D). This complex is applied to the inner surface of a secondary pectoral girdle consisting of a series of dermal bones. Actual attachment is to the **cleithrum,** usually the largest of the series. Cleithra meet at the ventral midline and extend upward toward the cranium. A series of **postcleithra** are present in most teleosts.

A **supracleithrum** attaches to the cleithrum and extends forward, where it attaches to the **posttemporal** bone, which is usually forked. If forked, the upper branch of the posttemporal attaches to the epiotic and the lower branch to the pterotic or the intercalar. One of the distinguishing features in the evolution of tetrapods is the development of a neck region, which permits greater mobility of the head relative to the pectoral girdle. This is made possible through the liberation of the pectoral girdle from these points of articulation at the back of the skull.

Most teleosts in which the pectorals are low and have oblique bases possess a **mesocoracoid** bone that forms a brace between the coracoid and the cleithrum. This is typical of soft-rayed fishes, such as herrings, salmons, carps, and catfishes, but is absent in most perciform teleosts—perches and basses, among others—that have vertical-based pectorals set higher on the sides.

The pelvic fin skeleton in teleosts is made up of plate-like **basipterygia**, one for each fin (Fig. 3.25E). These bones usually are joined to each other posteriorly and may meet anteriorly. Remnants of pterygiophores may be present where the fin rays join the basipterygium. The pelvic skeleton in fishes remains a fairly simple structure, unlike the pelvic girdle that develops in tetrapods. Here, the pelvic bones develop a direct articulation with the vertebral column, enabling a greater degree of support in animals that must counteract the forces of gravity while moving about.

Fin Rays

Fin rays of various groups of fishes differ in structure and origin. Fin rays of lampreys and hagfishes are simply rods of cartilage extending from the notochord, whereas elasmobranchs have fibrous, horny rays arising from the dermis. These structures, usually unbranched, are called **ceratotrichia** (*cerato* = "horn-like") because they are composed mostly of *keratin,* a scleroprotein that in tetrapods is incorporated in fingernails, horns, and other skin-associated structures. Other rays of a horny nature, composed of elastoidin fibers, are **actinotrichia** ("rod-hairs"), found in adipose fins and in the embryos of bony fishes. The

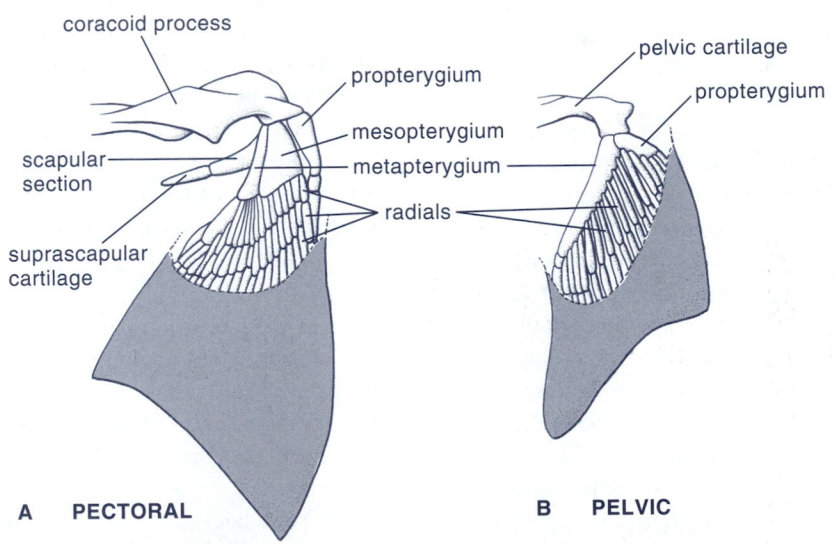

FIGURE 3.25
Skeletal supports of paired fins of shark: **A,** Ventral view of left pectoral; **B,** ventral view of left pelvic. (*A* and *B* based on drawings by John McKern.)

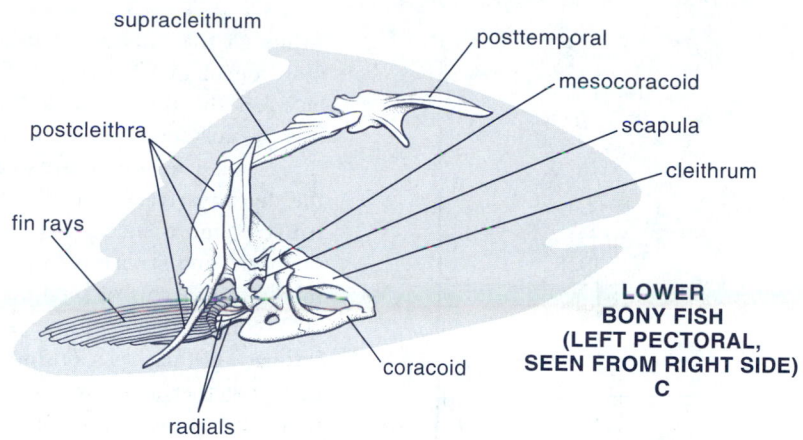

supracleithrum

posttemporal

mesocoracoid

scapula

postcleithra

cleithrum

fin rays

**LOWER
BONY FISH
(LEFT PECTORAL,
SEEN FROM RIGHT SIDE)
C**

coracoid

radials

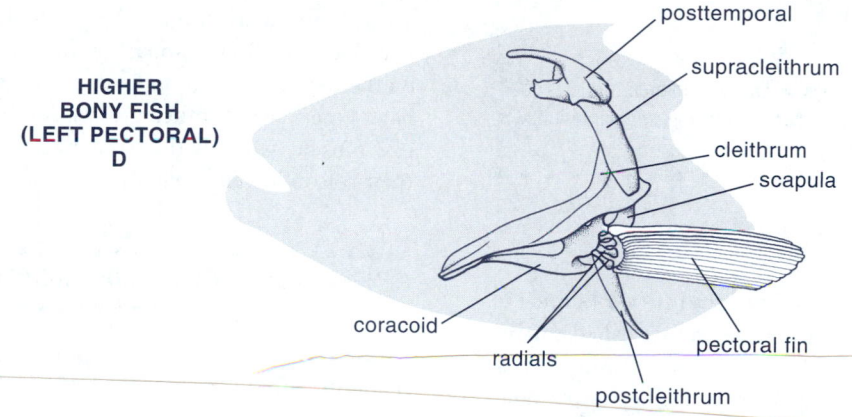

posttemporal

supracleithrum

**HIGHER
BONY FISH
(LEFT PECTORAL)
D**

cleithrum

scapula

coracoid

radials

pectoral fin

postcleithrum

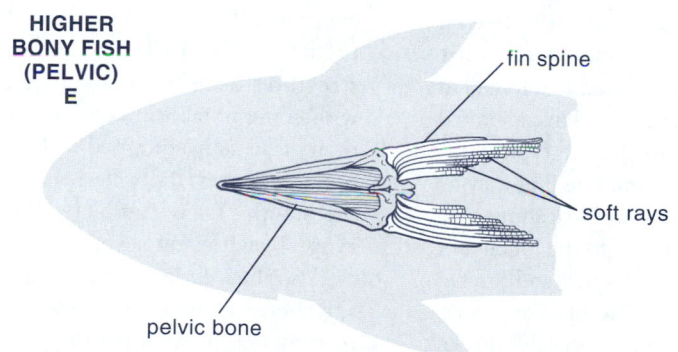

**HIGHER
BONY FISH
(PELVIC)
E**

fin spine

soft rays

pelvic bone

Skeleton of paired fins of teleost *Oncorhynchus mykiss:* **C,** Medial aspect of left pectoral bones. Skeletal supports of paired fins of teleost *Morone saxatilis:* **D,** Lateral view of left pectoral bones; **E,** ventral view of pelvic skeleton.

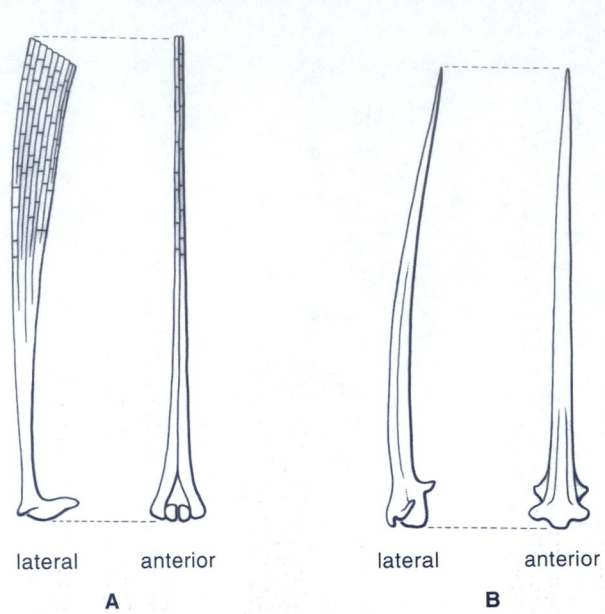

lateral   anterior   lateral   anterior

A   B

FIGURE 3.26
Comparison of soft and spinous rays: **A,** Lateral and anterior views of soft rays; **B,** same views of fin spine. Note branching, segmentation, and double construction of soft ray.

actinotrichia of the coelacanth (*Latimeria*) are very similar to those of bony fishes. Lungfishes have fin rays called **camptotrichia** that are different from the rays of other fishes. They are composed of an outer layer of flexible, fibrous bone over an uncalcified interior.

The dermal fin rays of bony fishes are called **lepidotrichia** (Whitear, 1986b) because they are believed to have arisen from rows of scales covering the horny, primitive fin rays. The lepidotrichia of the soft fins of most bony fishes are usually made up of several elements placed end to end in two closely apposed bilateral rows, so the rays have a jointed appearance and have two sides (Fig. 3.26). The typical unmodified soft ray is not only jointed and double, but may be branched. Some "soft" rays, such as those at the leading edge of the dorsal fin of a carp, are not really soft to the touch, but have been modified by the loss of the branched and jointed characteristics and have hardened into spine-like structures. True spines, such as those in the dorsal, anal, and pelvic fins of typical acanthopterygian fishes, are undivided, unbranched, typically hard, sharp structures, although these may be secondarily modified toward flexibility in some groups, such as sculpins. Lagler et al. (1962) claimed that the fin spines of acanthopterygians are formed from actinotrichia.

SKELETAL MUSCLE

Although the muscles of fishes will be covered in much more detail in Chapter 19, the fundamental relationship between the muscles and the skeletal system should at least be acknowledged here. Vertebrate muscle consists of three types: **smooth muscle**, primarily associated with the digestive tract and other visceral structures; **cardiac muscle**, more specifically the myocardial component of the heart; and **striated muscle**, the predominant tissue of the body in terms of total mass. Striated muscle works in close conjunction with the skeletal elements—indeed, the intimacy of their relationship is such that we often refer to them together as the **musculoskeletal system**. In comparing fishes with tetrapods, the basic pattern of locomotion is decidedly different, and hence the relative prominence of different muscles and their relationship to skeletal elements also differs. Fishes possess well-developed trunk muscles that they contract in a way that enables the caudal fin to act as a propulsive organ. Appendicular muscles are comparatively small. Tetrapods, on the other hand, rely mainly on the control of their appendages for locomotion, and they have experienced a much greater development of appendicular muscle, whereas the trunk muscles have diminished in their role in locomotion.

GENERAL RELATIONSHIPS OF INTERNAL ORGANS

Alimentary Canal and Associated Structures

Although the details of internal organ systems will be given in subsequent chapters, a general description of the viscera should be helpful here for the sake of orientation. This orientation will be aided by referral to Chapter 23, especially Figures 23.7–23.10. The alimentary canal and associated structures of fishes follow the generalized vertebrate plan with some notable exceptions. The anterior of the mouth cavity is usually equipped with oral valves that aid in pumping water over the gills by preventing backflow through the mouth. There can be teeth within the mouth and pharyngeal cavities on several bones other than the jawbones, as described earlier. Fishes, of course, retain the original vertebrate respiratory mode—the gills—that developed in conjunction with perforations in the pharyngeal region (see Chapter 2). As an adaptation for aquatic habitats in which insufficient dissolved oxygen is available (e.g., shallow coastal swamps), fishes have also developed a variety of accessory respiratory organs, including lungs. Arising from diverticula of the digestive tract and situated deep inside the body, where the threat of desiccation was minimized, lungs served as ideal respiratory structures in those vertebrates that made the transition from water to land.

Internally, the pharynx leads to a short gullet or *esophagus*, which is separated from the variously shaped stomach by a sphincter. There are certain groups of fishes that lack a true stomach, and in these, the esophagus communicates directly with the intestine. Beyond the distal end of the stomach, just past the pyloric sphincter, are blind sacs (**pyloric caeca**) opening off the intestine. These are found in most families of bony fishes, but not in the cartilaginous fishes.

The gut may be relatively straight and short, S-shaped, or variously elongated and coiled or folded, depending on the food habits of the species. Usually, the lumen of the posterior part of the gut is larger than that of the anterior part. In lampreys, the absorptive surface area of the intestinal lumen is enhanced through the presence of a longitudinal fold called the **typhlosole**. In elasmobranchs, the small intestine is readily distinguished from the large intestine because the latter contains an internal coil or helix, the **spiral valve**. Like the typhlosole, this increases the internal surface area and slows the passage of food for more efficient extraction of nutrients. Some lower bony fishes, such as bichirs, lungfishes, sturgeons, gars, and bowfin also have a spiral valve. In bony fishes, with few exceptions, the gut reaches the exterior at the anus or vent, but in elasmobranchs, the gut empties into a **cloaca**, which also receives the ducts from the urogenital system.

Associated with the alimentary tract are the liver and spleen, both of which are usually located near the stomach. Elasmobranchs and a few bony fishes have a discrete pancreas; otherwise it is associated with the liver to form a **hepatopancreas**.

Another organ associated with the alimentary canal is the **swim bladder** or **gas bladder**, which is derived from the digestive system and remains attached to it by a tube (**pneumatic duct**) in most soft-rayed fishes (see Chapter 19). Such fishes are called **physostomous**, from the Greek roots *physo* ("bladder") and *stom* ("mouth"). Most of the spiny-rayed fishes and their allies have lost the open connection except in the larval stages, when in many species air is gulped into the bladder. These fishes are called **physoclistous** (*clist* = "closed"). The gas bladder is typically a torpedo-shaped, thin-walled sac in the upper part of the body cavity, immediately below the kidney. Gas bladders vary considerably among the fishes in both shape and function.

Urogenital Organs

The dark red kidney tissue of bony fishes is typically located above the body cavity along the vertebrae, separated from the other viscera by membranes. Appearance and organization of the kidney material varies among fishes; thus, the prominent, large, pulpy kidneys of trout contrast greatly with the thinner organs of other actinopterygian fishes. The anterior part of the bony fish kidney, the **head kidney**, is specialized for blood cell formation and is usually more expanded than the posterior part. Elasmobranchs have more compact kidneys, usually located in the posterior part of the body cavity. Hagfish and lamprey kidneys are long and strap-like.

Typically, the urinary and reproductive ducts join to form the **urogenital sinus** in elasmobranchs. The sinus empties into the cloaca. Urinary bladders are absent in agnaths and cartilaginous fishes. In bony fishes, the ducts may join to form a single urogenital duct prior to reaching the exterior just behind the anus, or, depending on the species, may be arranged so that the left and right ducts of each system join to form a single tube (one urinary and one reproductive) that open to the exterior separately, the urinary duct posterior to the reproductive duct. The bladder of bony fishes usually forms as a posterior swelling of the urinary duct and is known as a **tubal bladder**, to distinguish it from the bladder that forms in a different manner in tetrapods.

Although there is much variation in their shape, size, and structure, the gonads of fishes are located along the dorsal part of the body cavity beneath the kidneys. Testes are usually much thinner than ovaries and lighter in color. Developing ovaries take on a yellowish coloration. Sperm ducts and oviducts may be more or less evident, depending on species, age, and stage of development.

· · · · ·
The Heart

The heart is placed in the pericardial cavity at the lower, anterior end of the body cavity, more or less below the gills and between the lower muscles and bones of the pectoral girdle. Elasmobranch hearts consist of a **sinus venosus, atrium, ventricle,** and a contractile **conus arteriosus.** Bony fish hearts have the conus reduced to a small, valve-bearing structure associated with an expansible basal section of the aorta called the **bulbus arteriosus.** The sinus venosus receives blood from the **ducts of Cuvier** and the hepatic veins (see Chapter 24).

· ·
KEY POINTS AND CONNECTIONS
· ·

• The head of fishes is defined as the region extending from the tip of the snout back to the gill slits in elasmobranchs or to the edge of the opercular region in bony fishes. The shape of the head is governed by modes of feeding and by the size and placement of the eyes. Protrusible jaw mechanisms, an important feeding adaptation, are conspicuous among chondrichthyan and osteichthyan fishes.

Chapter 23 elaborates on the ecomorphological adaptations of fishes, focusing on features of the head that accommodate feeding activities.

- A range of body forms are observed among fishes. The median and paired appendages are conspicuous features of the overall body form. The design of these appendages is closely correlated with mode of locomotion, especially in the design of the caudal fin, the chief locomotor appendage.

The role of appendages in locomotion is covered in Chapter 19.

- The thin epidermal layer of fish skin includes abundant mucus-secreting cells. Other epidermal structures include nuptial tubercles and photophores. Scales are the most conspicuous feature of the dermal layer, and a diversity of scale types are present. Primitive fishes have scales that are more robust, with a more complex mineral composition, whereas more derived bony fishes, such as the teleosts, have simplified scales, usually of the cycloid or ctenoid type.

The skin and associated structures are discussed in Chapter 18.

- The axial skeleton includes the skull and vertebral column. Skull design reflects phylogenetic history as well as feeding and respiratory adaptations. The skull is largely composed of two components, the neurocranium and the splanchnocranium. Bony fishes display extensive incorporation of dermal skeletal components into the skull to form a third component, the dermatocranium.

- Although the most primitive fishes display a well-developed notochord, the vertebral column is the chief axial skeletal support in most fishes. Amphicoelous (biconcave) vertebrae characterize most elasmobranchs, and bony fishes (teleosts). Structures associated with the vertebral centra include arches that develop to enclose the spinal cord or blood vessels, and lateral processes to serve as articulation points for ribs.

- Prominent within the body cavity are the digestive tract and urogenital organs, with the heart occupying an anterior subdivision of the body cavity.

Chapter 23 focuses on gut morphology and feeding adaptations, Chapter 25 provides information on the role of the kidneys in osmoregulation, and the structure and function of the reproductive organs is covered in Chapter 27.

FISH LINKS

http://speciesanalyst.net/fishnet/ FishNet is a distributed information system that links databases from a number of museums and other institutions that maintain fish collections. FishNet enables the efficient retrieval of information about specimens from partner institutions.

http://www.deepfin.org/ DeepFin is a research coordination network of fish biologists with interest in the issues of evolution and biodiversity of fishes. Their stated goal is "to establish the phylogenetic tree of all fishes, to decipher their evolutionary relationships."

BUILDING AN ICHTHYOLOGY LIBRARY

For this chapter, it is probably useful to consider the current crop of fish biology texts as well as some "classics," from which earlier generations of ichthyologists learned their trade. In addition to this book (which the authors wish to thank you for choosing), other helpful texts include,

Bone, Q., N. B. Marshall, and J. H. S. Blaxter. 1999. *Biology of fishes* (2nd ed.). Stanley Thornes, Cheltenham, UK.
Helfman, G. S., B. B. Collette, and D. E. Facey. 1997. *The diversity of fishes.* Blackwell Science, Malden, MA.
Moyle, P. B., and J. J. Cech, Jr. 2004. *Fishes: An introduction to ichthyology* (5th ed.). Prentice Hall, Upper Saddle River, NJ.

Earlier ichthyological standards include,

Jordan, D. S. 1905. *A guide to the study of fishes,* Vols. 1 and 2. Henry Holt, New York. (The first "modern" ichthyology text, and the first one to focus on North American fishes.)
Lagler, K. F., J. E. Bardach, and R. R. Miller. 1962. *Ichthyology.* John Wiley, New York. (The "gold standard" of ichthyology texts in the 1960s and 1970s.)
Marshall, N. B. 1966. *The life of fishes.* Universe Books, New York.
Nikolsky, G. V. 1961. *Special ichthyology.* Israel Program for Scientific Translations, Jerusalem. (Many of the classics in Russian ichthyology were translated in the 1960s and 1970s through this program.)
Norman, J. R., and P. H. Greenwood. 1975. *A history of fishes* (3rd ed.). Halstead Press, New York.

REFERENCES

Antony, A. C. 1952. Use of fish slime in structural engineering. *J. Bombay Nat. Hist. Soc. 50*(3): 682.
Arratia, G. 1982. *Chongichthys dentatus,* new genus and species, from the late Jurassic of Chile (Pisces: Teleostei: Chongichthyidae, new family). *J. Vert. Paleontol. 2*(2): 133–149.
———, and H. P. Schultze. 1990. The urohyal: Development and homology within osteichthyans. *J. Morphol. 203:* 247–282.
Bailey, R. M., and W. A. Gosline. 1955. Variation and systematic significance of vertebral counts in the American fishes of the family Percidae. *Misc. Publ. Mus. Zool. Univ. Mich. 93.*
Balinsky, B. I. 1970. *An introduction to embryology.* W. B. Saunders, Philadelphia.
Bell, M. 1993. Convergent evolution of nasal structure in sedentary elasmobranchs. *Copeia 1993:* 144–158.
Berg, L. S. 1940. Classification of fishes both recent and fossil. *Trav. Inst. Zool. Acad. Sci. URSS 5:* 87–517. Reprinted 1947, J. W. Edwards, Ann Arbor, MI.
Bridge, T. W., and G. A. Boulenger. 1904. Fishes. In *The Cambridge natural history,* Vol. 3, S. F. Harmer and A. E. Shipley (Eds.). MacMillan, London. Reprinted 1958, Wheldon and Wesley, Codicote, UK, and H. R. Engelmann (J. Cramer), Weinheim, Germany.

Compagno, L. J. 1988. *Sharks of the order Carchariniformes*. Princeton University Press, Princeton, NJ.

Coombs, S., J. Janssen, and J. F. Webb. 1988. Diversity of lateral line systems: Evolutionary and functional considerations, pp. 553–593. In *Sensory biology of aquatic animals*, J. Atema, R. R. Fay, A. N. Popper, and W. N. Tavolga (Eds.). Springer Verlag, New York.

Daniel, J. F. 1934. *The elasmobranch fishes*. Universtiy of California Press, Berkeley.

DeBeer, G. R. 1937. *The development of the vertebrate skull*. Clarendon Press, Oxford.

Forey, P. L. 1973. Relationships of elopomorphs, pp. 351–368. In Inter-relationships of fishes, P. H. Greenwood, R. S. Miles, and C. Patterson (Eds.). *Zool. J. Linn. Soc. 53* (Suppl. 1). Academic Press, New York.

Gans, C. 1993. Evolutionary origin of the vertebrate skull, pp. 1–35. In *The skull*, Vol. 2, J. Hanken and B. Hall (Eds.). University of Chicago Press, Chicago.

Goodrich, E. S. 1909. Cyclostomes and fishes. In *A treatise on zoology* (Part 9, Fasc. 1), R. Lankester (Ed.). Adam and Charles Black, London. Reprinted 1964, A. Asher, Amsterdam.

———. 1930. *Studies on structure and development of vertebrates*. Constable, London. Reprinted 1958, Dover, New York.

Grassé, P. P. (Ed.). 1958. Agnathes et poissons: Anatomie, ethologie, systematique. In *Traité de zoologie*, Vol. 133. Masson, Paris.

Gregory, W. K. 1933. Fish skulls. *Trans. Am. Phil. Soc. 23*: 75–481. Reprinted 2002, Krieger, Malabar, FL.

Halstead, B. W. 1978. *Poisonous and venomous marine animals of the world* (rev. ed.). Darwin Press, Princeton, NJ.

Hanken, J., and B. Hall (Eds.). 1993. *The skull*, Vols. 1, 2, and 3. University of Chicago Press, Chicago and London.

Hart, J. L. 1973. Pacific fishes of Canada. *Bull. Fish. Res. Bd. Can. 180*, Ottawa.

Herald, E. S. 1961. *Living fishes of the world*. Doubleday, Garden City, NY.

Hoyt, J. W. 1975. Hydrodynamic drag reduction due to fish slimes, pp. 653–672. In *Swimming and flying in nature*, Vol. 2, T. Y.-T. Wu, C. J. Brokaw, and C. Brennan (Eds.). Plenum Press, New York.

Janvier, P. 1981. The phylogeny of the Craniata, with particular reference to the significance of fossil "agnathans." *J. Vert. Paleontol. 1*(2): 121–159.

———. 1993. Patterns of diversity in the skull of jawless fishes, pp. 131–188. In *The skull*, Vol. 2, J. Hanken and B. Hall (Eds.). University of Chicago Press, Chicago and London.

———. 1996. *Early vertebrates*. Oxford University Press.

Johnson, G. D., and C. Patterson. 1993. Percomorph phylogeny: A survey of acanthomorphs and a new proposal. *Bull. Mar. Sci. 52*(1): 554–626.

Jordan, D. S. and B. W. Evermann. 1900. The fishes of North and Middle America. *Bull. U.S. Nat. Mus. 47*, Pt. 4.

Kent, G. C., and R. K Carr. 2001. *Comparative anatomy of the vertebrates* (9th ed.). McGraw-Hill, Boston.

Lagler, K. F., J. W. Bardach, and R. R. Miller, 1962. *Ichthyology*. John Wiley, New York.

Lagler, K. F., J. W. Bardach, R. R. Miller, and D. R. M. Passino. 1977. *Ichthyology* (2nd ed.). John Wiley, New York.

Lauder, G. V. 1983. Food capture, pp. 280–311. In *Fish biomechanics*, P. W Webb and D. Weihs (Eds.). Praeger, New York.

Lindsey, C. C. 1975. Pleomerism, the widespread tendency among related fish species for vertebral number to be correlated with maximum body length. *J. Fish. Res. Bd. Can. 32*: 2453–2469.

Lubbock, R. 1980. Why are clownfishes not stung by sea anemones? *Proc. Roy. Soc. Lond. B 207:* 35–61.

Meinke, D. K. 1982. A light and scanning electron microscope study of microstructure, growth, and development of the dermal skeleton of *Polypterus* (Pisces: Actinopterygii). *J. Zool. 197*(3): 355–382.

Mooi, R. D., and A. C. Gill. 1995. Association of epaxial musculature with dorsal-fin pterygiophores in acanthomorph fishes, and its phylogenetic significance. *Bull. Nat. Hist. Mus. Lond. (Zool.) 61*(2): 121–137.

Morris, S. L., and A. J. Gaudin. 1975. The cranial osteology of *Amphoistichus argenteus* (Pisces: Embiotocidae). *Bull. So. Cal. Acad. Sci. 75:* 29–38.

Moy-Thomas, J. A., and R. S. Miles. 1971. *Paleozoic fishes*. W. B. Saunders, Philadelphia.

Nelson, G. J. 1969. Gill arches and the phylogeny of fishes, with notes on classification of vertebrates. *Bull. Am. Mus. Nat. Hist. 141*(4): 475–552.

———. 1970. Pharyngeal denticles (placoid scales) of sharks, with notes on dermal skeleton of vertebrates. *Am. Mus. Nov. 2415:* 1–26.

Nelson, J. S. 1994. *Fishes of the world* (3rd ed.). John Wiley, New York.

Patterson, C. 1968. The caudal skeleton in lower pholidophoroid fishes. *Bull. Brit. Mus. Nat. Hist. (Geol.) 16*(5): 201–239.

Roberts, C. D. 1993. Comparative morphology of spined scales and their phylogenetic significance in the teleostei. *Bull. Mar. Sci. 52*(1): 60–113.

Rojo, A. L. 1991. *Dictionary of evolutionary fish osteology*. CRC Press, Boca Raton, FL.

Romer, A. S. 1970. *The vertebrate body* (4th ed.). W. B. Saunders, Philadelphia.

Rosen, M. W., and N. E. Cornford. 1971. Fluid friction of fish slimes. *Nature 234:* 49–51.

Schultze, H.-P. 1993. Patterns of diversity in the skills of jawed fishes, pp.189–254. In *The skull*, Vol. 2, J. Hanken and B. Hall (Eds.). University of Chicago Press, Chicago and London.

Uyeno, T. 1991. Observations on locomotion and feeding of released coelacanths, *Latimeria chalumnae*. *Env. Biol. Fishes 32:* 173–267.

Van Oosten, J. 1957. The skin and scales, pp. 207–244. In *The physiology of fishes*, M. E. Brown (Ed.). Academic Press, New York.

Weisel, G. F. 1967. Early ossification in the skeleton of the sucker (*Catostomus macrochelius*) and the guppy (*Poecilia reticulata*). *J. Morphol. 121:* 1–8.

Whitear, M. 1986a. Epidermis, pp. 8–38. In *Biology of the integument. 2. Vertebrates*, J. Berester-Hahn, A. G. Matoltsy, and K. S. Richards (Eds.). Springer Verlag, Berlin.

Whitear, M. 1986b. Dermis, pp. 39–64. In *Biology of the integument. 2. Vertebrates*, J. Berester-Hahn, A. G. Matoltsy, and K. S. Richards (Eds.). Springer Verlag, Berlin.

Wunder, W. 1936. *Physiologie der Süsswasserfische Mitteleuropas*. E. Schweizerbart, Stuttgart, Germany.

PART TWO

THE DIVERSITY
OF FISHES

Understanding a group of vertebrates as diverse as the fishes requires some fundamental appreciation of the basic principles of evolution and systematics. Evolution is the foundation on which biological thought is based. Nowhere are the fruits of evolution more evident than in the diversification that has taken place among the fishes. Charles Darwin contributed immensely to our understanding of evolution through his discovery that it proceeds through the process that we have come to define as *natural selection*. Selective processes result in the formation of new species and higher taxa and the extinction of others. The study of evolution continues to evolve, however; perhaps there are other, more subtle processes at work that have resulted in the range of diversity we observe among fishes. Why are some groups, such as the cichlid fishes, so incredibly diverse, whereas other families may only consist of a single species? Is this entirely the work of selective processes?

Systematics is the discipline that enables us to make sense of the evolutionary history of organisms. Until recently, systematic study entailed the detailed measurement of numerous anatomical features of the organism and the construction of phylogenies based on the degree of similarity of these features. *Cladistics* has emerged as the preferred methodology for systematic study. Systematics has also come to embrace other features of an organism besides its anatomy. Molecular, physiological, and behavioral characteristics may also be useful in the evaluation of evolutionary history.

This section of the book is largely devoted to investigation of the diversification and interrelationships of fishes. It is our hope that, after reading the chapters in this section, you will sit back and say to yourself, "Wow, I didn't realize there were so many fishes in the world!" Then you will dive back into the subsequent chapters in an attempt to understand what it is about fishes and the world they live in that could account for such a wonderful diversity of form and function.

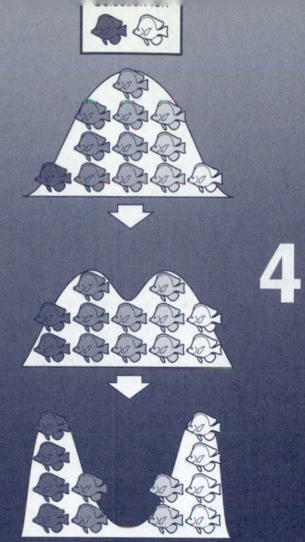

4

EVOLUTION AND SYSTEMATICS

FOUNDATIONS OF EVOLUTIONARY THEORY

PROCESSES OF SELECTION

Speciation Phenomena in Fishes
Ecological Character Displacement
Macro- versus Microevolution

SYSTEMATIC STUDIES: TRACING EVOLUTIONARY HISTORIES

Application of Molecular and Computational Advances
    Kinds of Molecular Data
    Applications of Molecular Data
    Computing Phylogenies

AN INTEGRATED APPROACH TO UNDERSTANDING EVOLUTION

Given the diversity of fishes and their dominance of the aquatic realm, it is not surprising that the practitioners of evolutionary and systematic biology have found both challenge and inspiration in their study. More than all other vertebrate taxa, the fishes have received the greatest attention from systematic biologists. Fish diversity, represented in the number both of modern species and of those in the fossil record, is largely the consequence of an evolutionary history that is closely associated with the vagaries of the drifting continents. Being for the most part confined to the water, fishes have experienced a distinctive evolutionary history, as aquatic habitats and their associated populations have become fragmented and isolated through time. This has provided systematic ichthyologists with a complex and convoluted evolutionary history to tease out of the extant as well as extinct forms. In this chapter, we will consider some of the basic principles of evolution and their application in systematic study with respect to the history of fishes.

FOUNDATIONS OF EVOLUTIONARY THEORY

The year 1859 was a momentous one in the history of human thought: Charles Darwin's (1809–1882) long-awaited masterpiece, the *Origin of Species,* was finally unleashed upon the world. Biology has become permanently transformed as a consequence. Proposing for the first time a plausible mechanism by which evolution proceeds, Darwin's work is the cornerstone on which the structure of modern biology rests. Central to Darwin's thesis is the phenomenon of **natural selection.** Darwin's great insight was in perceiving that organisms present a range of variation as a means of adaptation to the continual processes of change occurring in their environment.

Darwin was never able to provide an explanation for the genetic mechanism by which evolution by natural selection might proceed. At about the same time that the *Origin of Species* was shaking up the world of biological thought, the founding father of the modern science of genetics, Gregor Mendel (1822–1884), was laboring away in obscurity in a monastery in central Europe. One of the great ironies of biology is that Mendel never had the opportunity to share with Darwin his discovery of just the mechanism needed by Darwin. Mendel had discovered the principles of segregation of particulate traits and their independent assortment during mating, but he did not appreciate the evolutionary implications of his discovery. It was not until the dawn of the 20th century that Mendel's work was rediscovered and the evolutionary implications fully appreciated. The synthesis of Darwin's evolutionary principles with Mendelian genetics has been termed **neodarwinism.** The neodarwinian perspective holds that organisms constitute populations that have the potential for species formation. Genetic variation, which has its ultimate origins in random mutation, characterizes populations. Under conditions of isolation in space and time, these populations may form subspecies and eventually separate species. It is on the combination of genes in an organism that the processes of natural selection operate. Organisms are continually arranging and rearranging their gene combinations. Some of these combinations survive the winnowing processes of natural selection; others are consigned to oblivion. Although this modern synthesis has dominated evolutionary thinking since the 1940s, it is not without its critics. From her seminal studies on the origins of eukaryote organelles, Lynn Margulis has proposed that much of the evolutionary process can be attributed to the acquisition and merger of the genomes of quite unrelated organisms (see Margulis and Sagan, 2002).

PROCESSES OF SELECTION

Central to modern evolutionary theory is the action of selective forces operating on the range of variation among organisms. **Fitness** is the criterion by which we measure the success of an organism, the most practical measure of fitness being the number of viable offspring produced. Biologists recognize three kinds of selection operating on natural populations: *directional selection, stabilizing selection,* and *disruptive selection* (Fig. 4.1). With **directional selection,** the most fit organisms are at one extreme of the population distribution and, over time, the norm for the population shifts. With **stabilizing selection,** the most fit individuals represent the norm of the population, and the extremes are selected against. **Disruptive selection** is much the opposite, with the extremes being selected for at the expense of the norm of the population.

The agents of selection can be any number of environmental limiting factors, and these are usually distinguished as **physical factors,** such as temperature, salinity, or water flow velocity, and **biotic factors,** such as the presence of competitors or predators. An example of the combined influence of these two factors can be seen in the adaptations of a common species of stream-dwelling minnow in eastern North America, the central stoneroller (*Campostoma anomalum;* see Plate 10.2). Stonerollers get their name from their habit of moving gravel about during the preparation of a spawning nest. Physical factors, such as the texture and dimension of the gravel, are obviously important in determining the spawning success of this fish. Stream flow is also important, as flowing water is necessary to flush away sediment as the stoneroller prepares its nest. Stonerollers have a chisel-like lower lip that is used in the scraping of algae that grow on rocks and other submerged objects. Not only will the fish population be limited by the availability of algae, which itself will be influenced by the presence of suitable substrates on which to grow, but also by the presence of predators. Piscivorous bass (*Micropterus*) exert a significant influence on stoneroller habitat use—if bass are present, the standing crop of algae is larger owing to the decrease in grazing activity by stonerollers (Power, 1987; Power and Matthews, 1983; Power and Stewart, 1985).

Identifying selective forces in the environment is not as daunting a task as identifying the actual consequences of selection—actually witnessing the hand of evolution. David Reznick, John Endler, and their colleagues, through a series of elegant field studies on populations of guppies (*Poecilia reticulata*) living in streams in Trinidad, have documented

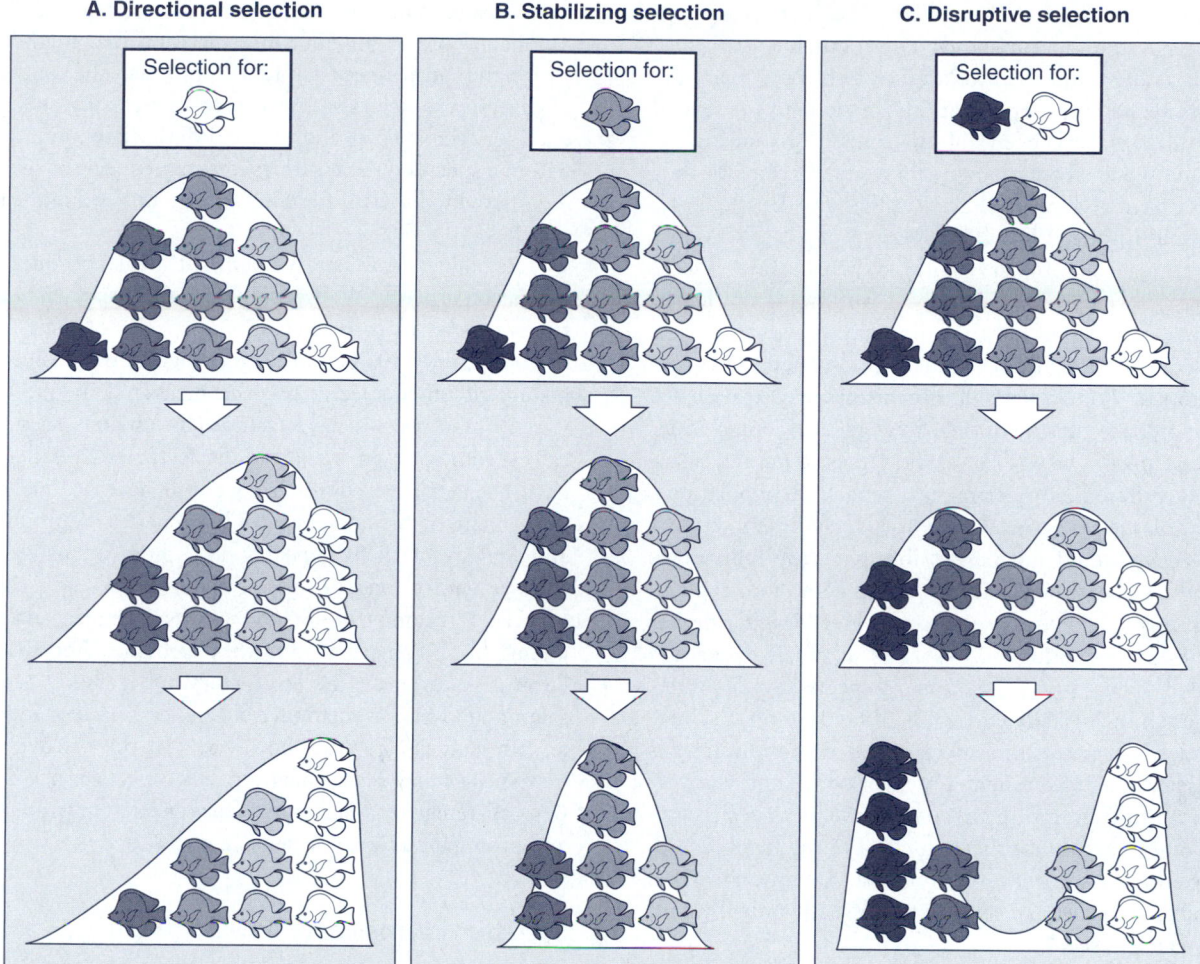

FIGURE 4.1
Illustration of three types of selection and their effects. Each curve represents the normal distribution of the trait in a population, and the arrows represent change in population genotypes and hence phenotypes through time. **A,** Directional selection—the adaptive norm changes as less fit genotypes are replaced with more fit ones. **B,** Stabilizing selection—intermediate genotypes are favored at the expense of extreme ones. **C,** Disruptive selection—extreme genotypes are selected at the expense of intermediate ones, occurring when a population exists in a heterogeneous environment. (Adapted from Volpe and Rosenbaum, 2000.)

such evolutionary changes. In the mountains of Trinidad, populations of guppies occupy different streams and encounter two potential predators. Killifish (*Rivulus hartii*) prey on smaller, immature guppies, but not on fully grown adults. Pike cichlids (*Crenicichla alta*), on the other hand, prefer larger guppies. As a consequence of this differential predation pressure, individual guppies that coexist with pike cichlids mature faster than those in killifish-dominated habitats and are thus sexually mature at a smaller body size. They reproduce more frequently and produce greater numbers of offspring. Through controlled breeding experiments, in which individuals were reared in the laboratory in the absence of predation, Reznick determined that the life history modifications indeed have a genetic basis and are not the result of unknown environmental variables. Reznick, Endler, and their colleagues have been monitoring manipulated guppy populations in the field for several years, and they have been able to actually document the evolutionary transformations in these populations as they have occurred. (Reznick, 1982; Reznick and Bryga, 1987; Reznick and Endler, 1982, 1990; Reznick et al., 1997).

The great evolutionary biologist Ernst Mayr (1904–2005) has provided us with the best working definition of a **species**—a group of interbreeding individuals forming a natural population that is reproductively isolated from other such groups. Populations of organisms may experience isolation and subdivision of their total genetic composition, their **gene pool**. Once isolated, the forces of natural selection may differentially act on the separated populations. The eventual result is the evolution of such differences that interbreeding would not be possible were the individuals from these isolated populations ever to come into contact. We call this outcome **speciation**. Ascertaining whether organisms are reproductively isolated or not is easier said than done. Consequently, defining a species by this criterion, though biologically correct, is difficult to do. Evaluating species in terms of their demonstrated **morphology** is much easier. This also validates the work of evolutionary biologists who are attempting to connect modern species with their evolutionary forebears manifested in the fossil record. In these situations, however, only those anatomical features that lend themselves to fossilization—usually only the hard parts—can be used.

If species are to be defined in terms of their reproductive isolation, we must ask the question, "How do populations become reproductively isolated?" Two kinds of mechanisms seem to account for most speciation events—*allopatric* ("different country") speciation and *sympatric* ("same country") speciation. In the Linnaean scheme, the species is the final category—the ultimate definition of the organism to be classified. The species is held to be more real and less arbitrary than other systematic categories. But we must acknowledge that the species exists both as an abstract concept, when discussed as the product of evolution, and as a real entity, in the form of the organism that is being classified.

For fishes, species have been described as reproductively independent lineages, diagnosed by different genetically based character states and ecological roles (Smith et al., 1995). A proposed new species is more readily accepted if it can be shown to possess several unique characters drawn from several lines of evidence beyond the morphological—including biochemical, behavioral, and ecological. A named

species that differs in only one feature from other species is suspect until someone demonstrates that it is not an ecophenotypic or polymorphic variation that might occur within the species (see Chapters 28 and 31). The congruence of diverse characters is the key to recognizing the difference between species and partly differentiated populations within a species. The entire spectrum of species—well differentiated to barely different—is encountered in the study of fishes.

As biologists, we study not only the final product of the speciation process, but the organisms that represent varying stages of species formation. Some species of fishes began to diverge only 10,000 years ago, whereas the evolutionary history of other species may be measured in millions of years. Some species have changed in obvious shapes and colors; some only in small, hidden features. Not all species have been discovered yet, so we still have an inadequate assessment of fish diversity. The genetics, reproduction, and ecology of all fish species need more intensive study. Species with a long history of intensive investigation are especially meaningful for the new systematic and other biological information that they provide. For example, current research on the life history and genetics of trout, salmon, and bass is changing our general understanding of all fishes. It is also true, however, that the discovery of a new species may reveal startling new aspects of fish biology (e.g., all-female and hermaphroditic species, or protandrous and protogynous species; see Chapter 28).

Speciation Phenomena in Fishes

When compared with other taxa, the rate with which speciation can occur in fishes appears to be much higher. In fishes, the time required for speciation is generally less than 0.3 million years, whereas that measured for island-dwelling birds and arthropods is in the range of 0.6 to 1.3 million years (McCune, 1997). Most fishes first set foot (or fin) on the road to speciation when their ancestral populations became isolated by natural barriers thousands or millions of years ago. This eventually culminated in the differentiation of genomes that had been subjected to differential selection by changing environments. This **allopatric** or **vicariant** ("replacement") model is consistent with the distribution patterns of North American fishes. Allopatric sister populations are found on opposite sides of geographic or ecological barriers. Examples of the allopatric model include the Olympic, central, eastern, and European mudminnows and the Alaska blackfish (*Umbra* and *Dallia;* Fig. 4.2I); four species of pikeminnows (*Ptychocheilus;* Fig. 4.2II); the lake suckers and their relatives (*Chasmistes,*

Deltistes, and *Xyrauchen;* Fig. 4.2III); and the darters of the *Etheostoma variatum* group (Fig. 4.2IV). The two subspecies of creek chubsuckers, *Erimyzon oblongus* (*E. o. oblongus* in eastern North America, and *E. o. claviformis* in the midwestern United States, Fig. 4.3) may represent speciation in progress. These two subspecies are distinct, except in the southeastern United States, where an overlap in morphological features exists.

Populations divided by natural barriers are often exposed to different environmental conditions. When this is the case, differential survival and reproduction of genetic variants changes the genetic makeup of the separated populations through time. A classic example of this allopatric speciation model is the species pairs on the Atlantic and Pacific sides of Panama. The isthmus emerged as land 3 to 5 million years ago, separating all the local marine fishes into pairs of populations (Jackson et al., 1996). Many of these have changed sufficiently to enable systematists to discriminate each member of one population from each member of the other, so they are recognized as distinct species (Collins, 1996).

The deglaciation of boreal habitats at the close of the Pleistocene created a host of opportunities for reinvasion of freshwater habitats by **anadromous** species—fishes that spawn in freshwater and migrate to the sea (see Chapters 31 and 36). Numerous examples demonstrating the evolution of endemic radiations of anadromous fishes are to be found among the petromyzontids, salmonids, osmerids, and gasterosteids. General features of these radiations include the repeated colonization of freshwater habitats and rapid phenotypic evolution (Bell and Andrews, 1997). This appears to be the consequence of an especially rapid rate of development of reproductive isolation in fishes inhabiting postglacial bodies of water (Schluter, 1996). When dispersal or hydrographic changes bring sister populations back into contact after a period of isolation, the different forms either merge or diverge genetically, depending on the amount of genetic change that occurred during separation and the environmental context at the time and place of secondary contact. A freshwater example of two species newly sympatric after long separation by glaciation is *Prosopium cylindraceum* (round whitefish) of northern and eastern North America and *P. williamsoni* (mountain whitefish) of western North America. These species are sympatric and distinct in northern British Columbia and the adjacent Northwest Territories (Scott and Crossman, 1972). In the Mississippi and Great Lakes drainages, a northern species, *Luxilus cornutus* (common shiner), and a southern species, *L. chrysocephalus* (striped shiner), established a zone of secondary contact near the southern margin of glaciation following the withdrawal of ice. These two species of *Luxilus* engage in frequent

hybridization, with different genetic results at different localities. Such a situation is sometimes termed **parapatric** speciation—where neighboring populations become distinct species adapted to local environmental conditions yet maintain a zone of hybridization along a common border.

In cases of parapatric speciation, when populations come into contact after a period of genetic separation, interbreeding may occur. If the intermediate individuals have lower fitness than the divergent parental types, selection may favor mate recognition systems that promote assortative mating (i.e., preference for mates belonging to the same genetic type). This process, theoretically important but largely undemonstrated in fishes, is called **reproductive character displacement**. It is recognized by the presence of greater distinctiveness between sympatric members of two populations than between allopatric members of the same two species. Divergent recognition cues in fishes may include body shapes, color patterns, courtship behaviors, courtship sounds, electrical signals, luminescent displays, olfactory cues, nest characteristics, and time and place of spawning.

Although allopatric speciation has been confirmed as the predominant mode of speciation (see Coyne and Orr, 2004), sympatric speciation has been invoked in a number of cases involving fishes. In the case of **sympatric** speciation, isolation and subdivision of the gene pool occurs even while the organisms are fully coexisting, in the absence of any geographic barriers. Sympatric speciation was long held in suspicion because few would acknowledge that gene pools could diverge in the absence of spatial isolation. An increasing body of evidence supports this contention, however (Tautz, 2003). In sympatric speciation, interruption of gene flow must occur by means other than that seen in allopatric speciation models. The sympatric speciation model might apply to colonists of different environments within a lake. Species may evolve by divergent resource use and reproduction, despite a limited exchange of genes in intermediate depths and habitats. Examples of this form of speciation can be inferred from the depth distributions of Great Lakes ciscoes (Koelz, 1929) and Lake Baikal sculpins (Kozhov, 1963). Undoubtedly the most dramatic example of this is the extensive formation of *species flocks* in the family Cichlidae that has taken place in the African Great Lakes. Greenwood (1984) defined a **species flock** as a group of closely related species that are derived from a common ancestor and are found in a common location, such as a single discrete body of water. Individual species living in the African Great Lakes are each adapted to different depths, substrates, temperatures, light, pressure, and food resources, and they have eventually become genetically separated from each other through the

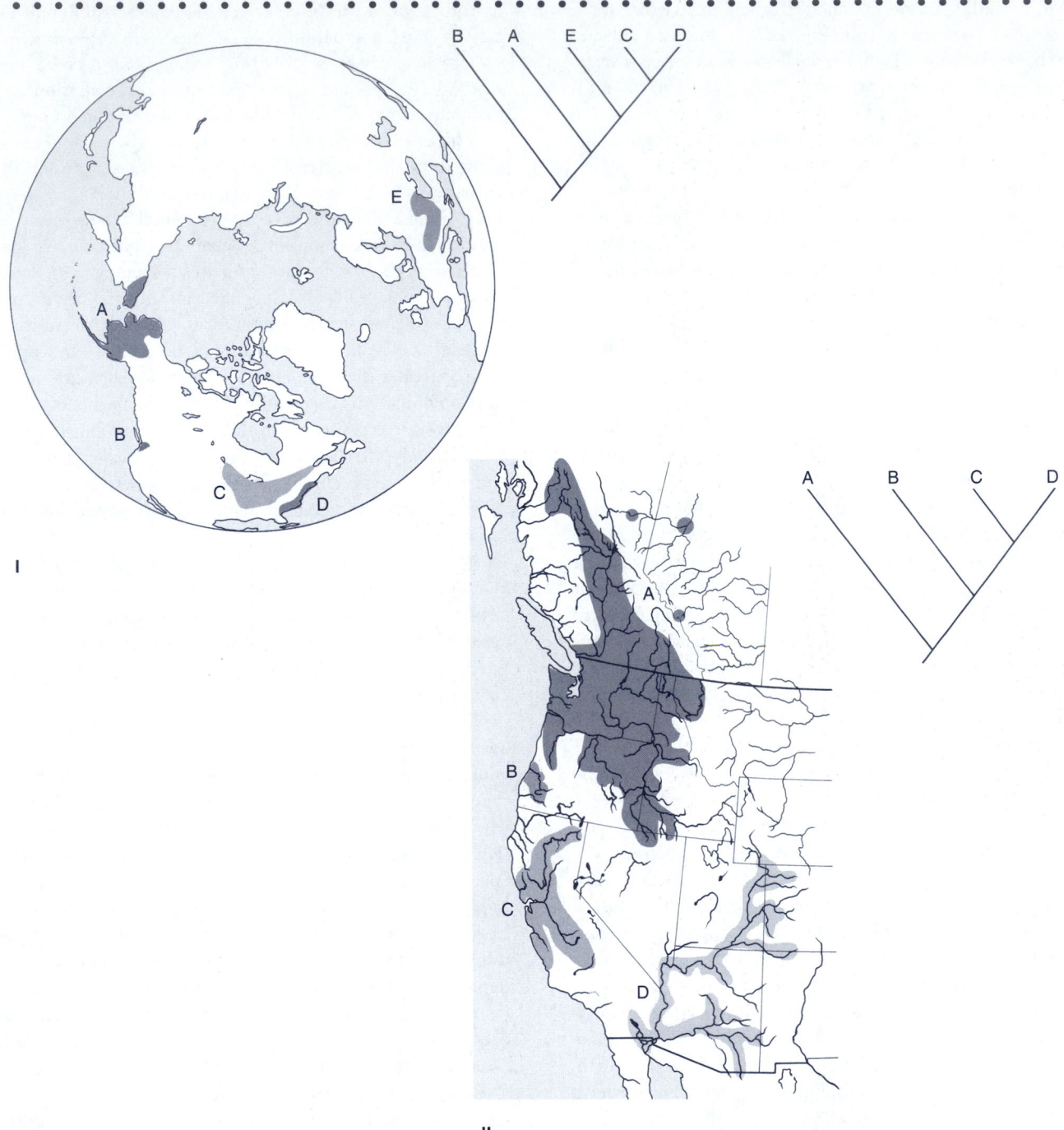

FIGURE 4.2

Examples of allopatric species distributions suggestive of vicariant speciation: **A,** Alaska blackfish (*Dallia pectoralis*); **B,** Olympic mud-minnow *(Novumbra hubbsi)*; **C,** central mudminnow (*Umbra limi*); **D,** eastern mudminnow (*U. pygmaea*); and *E*, European mudminnow (*U. krameri*). Four species of pikeminnow (*Ptychocheilus*): **A,** *P. oregonensis*; **B,** *P. umpquae*; **C,** *P. grandis*; **D,** *P. lucius*.

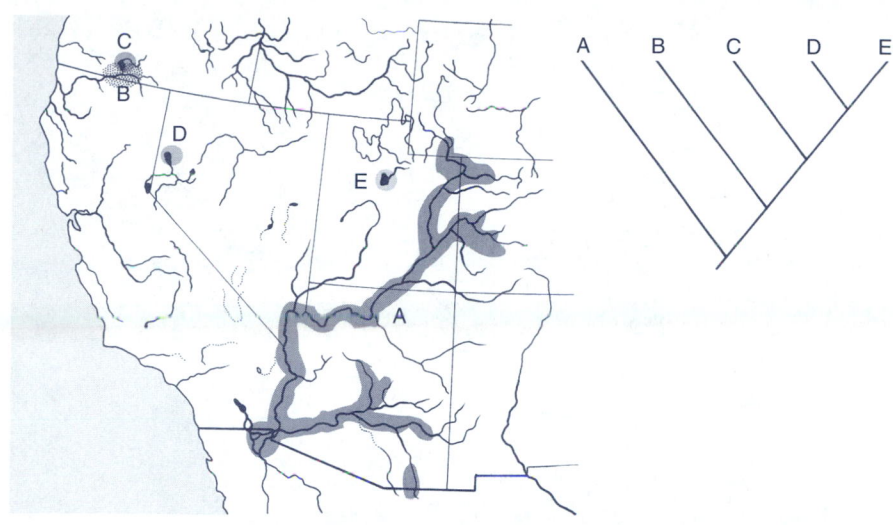

III

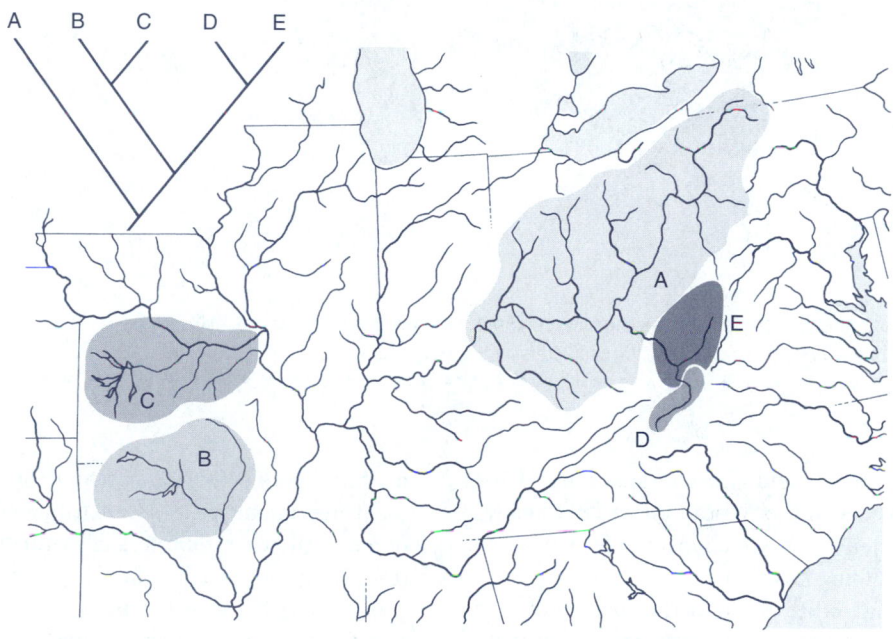

IV

River and lake suckers of the *Chasmistes* group: **A,** Razorback sucker (*Xyrauchen texanus*); **B,** Lost River sucker (*Deltistes luxatus*); **C,** shortnose sucker (*Chasmistes brevirostris*); **D,** cui-ui (*C. cujus*); **E,** June sucker (*C. liorus*). Five darters of the subgenus *Poecilichthys*: **A,** Variegate darter (*Etheostoma variatum*); **B,** Arkansas saddled darter (*E. euzonum*); **C,** Missouri saddled darter (*E. tetrazonum*); **D,** Kanawha darter (*E. kanawhae*); **E,** candy darter (*E. osburni*).

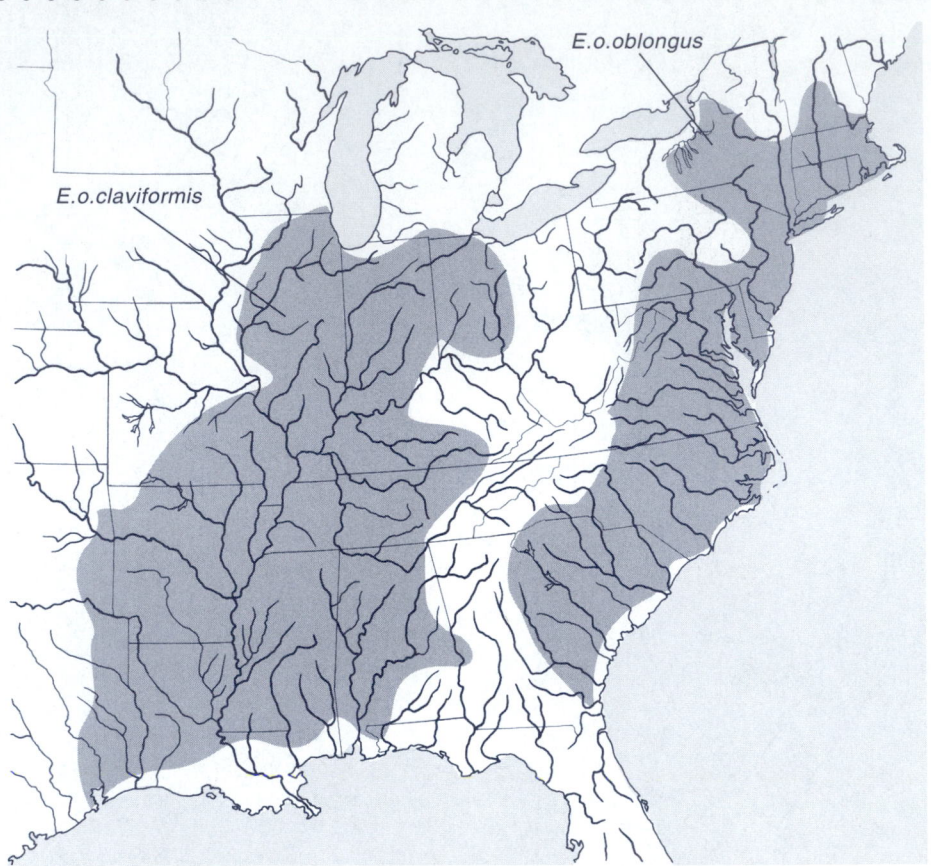

FIGURE 4.3
Distribution of eastern and central subspecies of creek chubsuckers (*Erimyzon*), which have not diverged to species status, judging from the lack of unique diagnostic characters for the two populations.

selection for different places and times of spawning. The peculiar geological history of the African Great Lakes may, however, have permitted extensive allopatric speciation as well (see Chapter 32, Going Deeper).

Mate selection and other reproductive behaviors are significant targets for the selective process. Speciation that begins with a change in the mate recognition system seems to characterize many North American freshwater fishes, such as small, colorful darters, minnows, pupfishes, and live-bearers. In many of these forms, the species differences are mainly in male breeding colors, implying that female choice of mates is an important part of the speciation process (Houde and Endler, 1990; Smith et al., 1995). A similar possibility has been suggested for the diverse flocks of closely related cichlid species in the African Great Lakes (Dominey, 1984; Galis and Metz, 1998). We term situations where reproductive success is influenced by success in

mating **sexual selection**. Examples include the competition for mates among males (see Chapter 37) and female choice of mates showing indicators of maturity and genetic success (e.g., large body size, bright colors, vigor, large territories, or good nests). Those features that make a male more desirable to a female may also make it more vulnerable to predation, however. For example, the sexually dimorphic traits that are functional in mate recognition systems in swordtails (Poeciliidae) are the same features preferred by their predators (Rosenthal et al., 2001).

One of the forces behind sexual selection (and consequent differentiation) is the struggle between the sexes to control reproduction. The "battle of the sexes" is more than just a colorful expression—it is a biological reality. Males and females do have different priorities that become part of their life histories. For the male, sex is all about maximizing the number of viable offspring that carry his genes. For the female, maximizing the output of offspring must be balanced by the need to conserve resources for the care of offspring just produced and to provide for future broods that might arise from liaisons with males that are even more fit than previous partners (Mestel, 1998). Sexual selection is especially useful in explaining processes of sympatric speciation—features that are the product of sexual selection can become rapidly amplified in a comparatively short period of time among co-occurring fish populations. Laboratory mating studies of Lake Malawi cichlids (see Chapter 32, Going Deeper) have demonstrated a strong preference of females for males from their own populations; such strong assortative mating is seen as a mode of incipient sympatric speciation (Knight and Turner, 2004).

Sticklebacks (Gasterosteidae) have long been considered definitive organisms for the study of behavior and speciation phenomena in fishes (see Chapters 15, 22, and 37). Boughman (2001) has demonstrated the role of sexual selection in the speciation process operating on several populations of recently diverged sticklebacks (*Gasterosteus*). The phenomenon of "sensory drive," in which adaptive changes in the perceptual system in females have an impact on mate choice, was investigated in several lakes in British Columbia characterized by different ambient light levels. Here, stickleback populations evolved differing sensitivities to red light, and males developed mating coloration that ranged from black to red. Females with greater light sensitivity in the red spectrum developed greater preference for redder males.

Speciation, although of crucial importance, is not a primary process in evolution. It is simply a byproduct of descent with modification in different populations. Speciation is the consequence of differential adaptive capacities, not a procedure driven by nature for some advantage. Differential adaptation and the resultant diversification of species have provided us with the astounding array of form and function we see in fishes. The most spectacular examples are those taxa that demonstrate evolutionary innovation in the matching of morphology and behavior to ecology, including the coral reef percoids, South American characins, South American catfishes, African cichlids, Holarctic minnows, North Pacific scorpaeniforms, inshore blennies and gobies, and Antarctic icefishes, to name just a few. An explosive diversification of species from ancestral stock is termed **adaptive radiation**. Among the fish taxa mentioned, we witness great flowerings of diversity that occur when new opportunities for trophic interactions are set in motion by unusually diverse spatial habitats. Coral reefs provide complex three-dimensional structure; African lakes provide depth gradients along fluctuating shorelines and islands. South American, North American, and Eurasian rivers provide millions of miles of branching fluvial gradients.

Inshore marine habitats are linear and heterogeneous, and the Antarctic shelf provides inshore gradients at a stable temperature that exclude most competitors. What each of these habitats provides is a spatially varying foundation on which the evolving fishes, among themselves, increase the trophic opportunities for specialization and proliferation.

The enhanced potential for speciation that is evident in fishes has led some to reconsider traditional notions of just what constitutes a species. Normally, we would view a species as something that comes about as a consequence of the selective processes previously discussed, survives for a period of time, and then goes extinct, never to reappear again. Turner (2002) proposed that our cladistic view of species as monophyletic entities may not always hold true—he proposed scenarios whereby species may be lost through hybridization, yet later are reestablished when conditions again favor the divergence of parental forms. This would be an application of **reticulate evolution**—a situation in which the genome of one species becomes incorporated into that of another. Admittedly, it is difficult to conceive of species disappearing and then reappearing at a later time or place. But if any group of animals can pull this off, fishes can.

* * * * * * * * * * * * * * * *

Ecological Character Displacement

The ecological differences and the associated morphological distinctions between fish species are often exaggerated by the enhanced growth and survival of individuals using a part of the resource base not shared with the competing species. That is, lower fitness of overlapping and competing individuals may select for divergence. This process is called **ecological character displacement** (Robinson and Wilson, 1994) and is ultimately responsible for species differences in size, jaws, teeth, gill rakers, habitat choice, and feeding behaviors. When this phenomenon is found to take place within a species, it is termed **trophic polymorphism**. It has been widely observed in northern temperate fishes, including salmonids, sticklebacks, and sunfishes. In most cases, the polymorphic feature is gill raker dimension, and it is associated with divergence into limnetic planktivores (longer gill rakers) and benthic macrocarnivores (shorter gill rakers). The two morphs of *Percichthys trucha* observed in lakes of temperate South America demonstrate an unusual departure. Here, a shallow water form having longer gill rakers is distinguished from a deeper water form having shorter gill rakers. In this case, both populations are benthic feeders, yet they specialize on different prey items. This is the first case of trophic polymorphism recorded in temperate South American waters (Ruzzante et al., 1998).

Variation in a morphological feature such as gill raker dimension is a demonstration of the fact that most or all significant evolutionary changes occur when one variant in a population has more success than the others (e.g., long and finely spaced gill rakers catch more food in the pelagic zone; short and broad gill rakers catch more food in the benthic zone). The genes responsible for the phenotypes that experience more successful survival and reproduction in their respective environments increase in proportion to their alternatives during subsequent generations.

Macro- versus Microevolution

Evolutionary biologists generally perceive the shifting allele frequencies in gene pools of populations—an evident manifestation of differential selective processes—as examples of **microevolution**. If these microevolutionary processes culminate in the reproductive isolation of the populations under consideration, speciation is said to have occurred. Studies have shown that such reproductive isolation—the necessary prerequisite for speciation—can occur very rapidly. Sockeye salmon (*Oncorhynchus nerka*) were first introduced into Lake Washington in the northwestern United States in 1937. In the ensuing time, amounting to fewer than 13 generations, divergent and genetically distinct populations using different reproductive environments (river spawning versus lake spawning) have arisen (Hendry et al., 2001).

But what of the formation of major groups of organisms—the phenomenon known as **macroevolution**? Over the vast expanse of geological time, can microevolutionary processes lead to the origin of major animal taxa? Most evolutionary biologists agree that this may be the case. Originally, it was presumed that such change was, as Darwin originally perceived it, a slow, gradual process inexorably leading to new species. In the 1950s, Ernst Mayr proposed an alternative mechanism in which speciation could occur at a greater rate. Mayr identified the phenomenon known as the **founder principle,** in which a subunit of an original population becomes genetically isolated and, containing only a fraction of the total genetic diversity of the ancestral population and experiencing an alien environment, undergoes rapid evolutionary divergence leading to new species. In the early 1970s, Niles Eldridge and Steven Jay Gould proposed a new theory of evolutionary rate that was more consistent with a fossil record that indicated long periods of evolutionary stasis followed by periods of rapid change (Eldridge and Gould, 1972). Their theory, known as **punctuated equilibrium,** suggests that the rate of speciation is not slow and constant but rather subject to dramatic shifts through time. This would help explain the paucity of transitional forms in the fossil record.

Molecular genetic studies have also revealed developmental processes that facilitate the rapid emergence of higher taxa. Modifications of comparatively few gene loci, which may be involved in critical developmental processes, may lead to dramatic modifications in body form. These modifications may confer reproductive isolation and promote speciation. A major phyletic line, such as lobe-finned, lung bearing fishes, would have arisen among the early bony fishes through an accumulation of traits that distinguish this particular outgroup, and they could have appeared in a comparatively short period of time.

Much remains to be understood about the patterns and processes of evolution. We can read the fossil record—comprehensive in some cases, sketchy in others—and, by observing the transformations experienced in modern populations, we can make reasonable inferences about how these mechanisms may have operated in the remote past. In the final analysis, though, we must remember the admonition of a 1973 essay by the great evolutionary geneticist Theodosius Dobzhansky: "Nothing in biology makes sense except in the light of evolution."

SYSTEMATIC STUDIES: TRACING EVOLUTIONARY HISTORIES

In formulating his principles of taxonomy, Carolus Linnaeus (see Chapter 1) had not foreseen the application that his work would have in elucidating the evolutionary relationships of organisms identified and classified using his approach. A brilliant naturalist but a man firmly grounded in the conventions of 18th-century biology, Linnaeus could or would not consider species as anything other than fixed, immutable entities. Darwin and his disciples have taught us otherwise. Taxonomy provides a foundation for the discipline known as **systematics**—the study of the evolutionary relationships of organisms. When species of a given taxon are identified and enumerated, according to the methodologies discussed in Chapter 1, it is possible to elucidate their phylogenetic interrelationships and reconstruct their evolutionary history. Through the identification of species, taxonomists also provide valuable information that researchers can use to assess biodiversity and, hence, the overall health of a given ecosystem. Until recently, the validity of a systematic analysis rested mainly on the persuasiveness of the argument of the researcher. In 1966, four ichthyologists (Greenwood, Rosen, Weitzman, and Myers, 1966) presented such an argument in their reconstruction of the evolutionary history of the teleostean fishes (see Chapter 9). Although many teleost taxa have experienced significant revision since the publication of this important work, it remains a foundation for the systematic study of fishes.

In the past few decades, systematic analysis has benefited from much more quantifiable approaches, including *phenetics* and *cladistics*. **Phenetic** analysis involves grouping organisms strictly on the number of characteristics they have in common. The emphasis is placed on the number of different characteristics used—the more the better. Evolutionary relationships, though not explicitly addressed, are implied, as those species sharing the greatest number of characteristics are assumed to be the most closely related. The problem lies in assessing the evolutionary significance of the characteristics used. A traditional, less quantifiable approach might consider a given feature, such as the presence or absence of a cartilaginous skeleton, to be of much greater evolutionary significance than the number of fin spines in a dorsal fin.

Willi Hennig, a German entomologist, revolutionized systematic biology when he introduced the approach known as **cladistics** (Hennig, 1966). Hennig claimed that the reconstruction of the phylogeny of a group of organisms depends on the recognition of shared derived characteristics—that is, features that differ from the ancestral condition. Hennig defined **apomorphies** as features derived or different from the ancestral form (i.e., "advanced" character states) and **plesiomorphies** as characters held in common with ancestral forms. **Synapomorphies** (literally, "forms derived away together") are shared derived characters, and **symplesiomorphies** are shared ancestral characters. **Autapomorphies** are specializations unique to only one taxon. Shared derived character states, pertaining to such features as number of spines, scale arrangements, or articulations among skeletal elements, are the features that may be useful in the diagnosis of a monophyletic group, or **clade,** and are thus the basis for constructing phylogenetic classifications. Symplesiomorphies provide no useful phylogenetic information, because a wide variety of unrelated forms may share a given ancestral character. For example, with respect to the phylogeny of fishes, the presence of a vertebral column is an example of a symplesiomorphy, as it is shared by a large number of descendant forms. However, the pattern of rib articulation with the vertebrae might be synapomorphic if a given clade can be uniquely distinguished from the ancestral form by a shared pattern of number and placement of ribs. Care must also be taken to recognize cases of **homoplasy**—similarity in features that is not due to common origin. The best known case is **convergent evolution,** where selection acts on unrelated forms, causing them to adapt in similar ways.

The goal of systematic study according to Hennig's principles is the development of a diagrammatic interpretation of phylogeny known as a **cladogram** or **phylogenetic tree**—a reconstruction through time of the evolutionary relationships of a given group of organisms. In such a depiction, a monophyletic group will be defined at a **node**—the point in the cladogram where one or more synapomorphies set the group apart from the rest of the organisms under consideration (Fig. 4.4). The procedure for developing a phylogeny for a given taxon starts with distinguishing primitive character states (plesiomorphies) from shared derived ones (synapomorphies). This is accomplished by comparison of the taxon in question with other taxa, termed **outgroups,** that fall outside the group in question. This procedure, termed **polarization of characters,** designates states as advanced, or **derived,** if they differ from the ancestral state prevalent among the diverse outgroups. For example, anal fin spines are a derived state; the absence of anal spines is an ancestral condition among fishes. The presence of pelvic, dorsal, and anal fin spines among the Acanthomorpha is evidence that acanthomorph families share a common ancestry derived from non-acanthomorphs that lacked fin spines (see Chapter 12). Phylogenetic relationships are affirmed by the **congruence** (consistency) of numerous independent characters in a cladistic analysis of all possible related groups (Fig. 4.5).

Care must be taken in the application of a particular character state and its claim to be apomorphic, lest it be taken out of a particular phylogenetic context. An apomorphic character can only be assessed relative to the clade in question, and some characters appear with great frequency among fishes. A case in point is the evolution of internal fertilization and viviparity. Viviparity appears to be a derived, apomorphic state, yet it appears in comparatively advanced forms, such as the poeciliids, as well as in primitive groups, such as elasmobranchs and coelacanths. Obviously, the placenta as a structure accommodating viviparity has been independently derived in a number of fish taxa, ranging from the most primitive to the most advanced. Studies have even demonstrated that an organ as complex as the placenta may have rapidly evolved multiple times within a single family—in this case the poeciliids—and even within a single genus (Reznick et al., 2002; see Chapter 15).

· ·
Application of Molecular and Computational Advances

A diversity of morphological attributes of fishes, including meristic and morphometric characters, coloration, and other features, have traditionally been used to delineate evolutionary relationships. Recently, the science of systematics has been revolutionized by two major developments—the application of molecular techniques and the development of powerful statistical computation tools. Two questions need to be addressed here: (1) What kinds of data can be derived from molecular studies? and (2) How can these be interpreted and applied to evolutionary study?

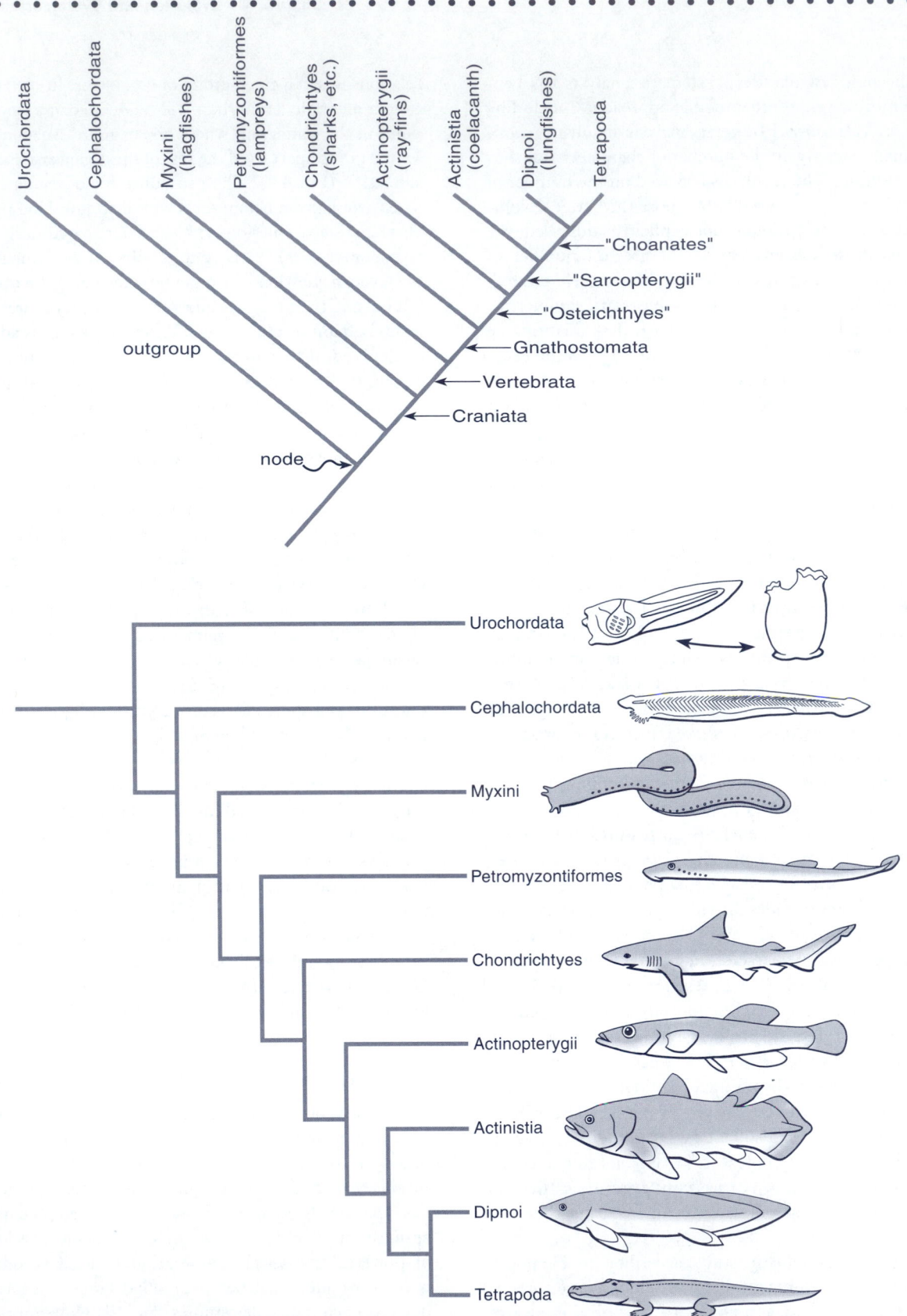

Urochordata

Cephalochordata

Myxini
(hagfishes)

Petromyzontiformes
(lampreys)

Chondrichtyes
(sharks, etc.)

Actinopterygii
(ray-fins)

Actinistia
(coelacanth)

Dipnoi
(lungfishes)

Tetrapoda

← "Choanates"

← "Sarcopterygii"

← "Osteichthyes"

← Gnathostomata

← Vertebrata

← Craniata

outgroup

node ⟶

Urochordata

Cephalochordata

Myxini

Petromyzontiformes

Chondrichtyes

Actinopterygii

Actinistia

Dipnoi

Tetrapoda

FIGURE 4.4
Different ways to represent cladograms, each with taxa indicated at terminal branch points.

Kinds of Molecular Data

Molecular systematists are practitioners of the "find 'em and grind 'em" approach to evolutionary investigations. Each individual organism is a repository of genetic information that informs us of its evolutionary history; all we need to do is get at the genome or investigate the genome products (see Chapter 28). Before the advent of the sophisticated techniques that enabled access to the genome itself, molecular biologists investigated the genome products, specifically the protein products coded for and assembled by the nucleic acids. Protein polymorphisms, termed **allozymes**, reflect genomic variation and can be useful in delineating populations. A technique known as **electrophoresis**, in which molecules are separated based on their charge and density characteristics, has had broad application in fish stock identification for several decades. The development of increasingly sophisticated methodologies and instrumentation has permitted direct access to the nuclear and mitochondrial genomes themselves—a boon to evolutionary investigations. Tiny quantities of gene sequences can be amplified using **polymerase chain reaction (PCR)**, and **restriction fragment length polymorphism (RFLP)** and **DNA–DNA hybridization** studies are used to measure genetic divergence within and among populations (Stepien and Kocher, 1997). One type of DNA polymorphism that has proven especially valuable in assessing genetic relatedness are the short, repeated sequences of DNA, known as **simple tandem repeat polymorphisms (STRPs)**, that occur at particular loci in the genome. Abbreviated sequences consisting of 2 to 9 base pairs, termed **microsatellites**, have proven to be especially useful molecular markers in the identification of fish populations, including commercial fisheries stocks (O'Connell and Wright, 1997).

The *mitochondrial* genome has a number of attributes that make it especially useful in phylogenetic studies. Mitochondrial genomes in fishes are clonally inherited—they are haploid and do not experience recombination. Mitochondrial DNA (**mtDNA**) is inherited only through the female line. It apparently evolves more rapidly than nuclear DNA, thus making for the ready identification of evolutionary processes even in closely related taxa (Stepien and Kocher, 1997). The **cytochrome** *b* gene is probably the most intensively studied component of the mitochondrial genome in fishes. Genetic markers that evolve more slowly, such as those in the nuclear DNA, are particularly useful in the derivation of higher phylogenetic relationships.

Mitochondrial genomic studies have proven especially valuable in the elucidation of the phylogenetic relationships of higher taxa among the teleosts (Inoue et al., 2001; Miya et al., 2001). Such studies have also revealed certain fundamental correlations between habitat and phylogeny.

Freshwater fishes, normally perceived as forming populations that are spatially disconnected, have comparatively deep phylogenetic histories (*depth* being an expression of mitochondrial differentiation), whereas marine fishes, inhabiting a more uniform environment, show comparatively shallow intraspecific phylogenetic structure (Avise, 2000). Exceptions are numerous, however. In the deep-sea bristlemouth species *Cylothone alba* (see Chapter 12), five distinct mtDNA types sampled from different oceanic regions have been identified where morphological differences are not apparent (Miya and Nishida, 1997).

Applications of Molecular Data

Molecular analysis can involve two basic kinds of information: *distance data* and *character data;* each can be used to answer different evolutionary questions (Moritz and Hillis, 1996). In the case of **distance data,** molecular difference is assessed as a single variable. This can then be expressed as a divergence in molecular sequences—for example, the base sequence of the DNA molecule—through time. The resultant **molecular clock** provides information pertinent to the time when two given groups may have diverged from their common ancestor. This notion is not without its detractors, however, as it rests on the assumption that the rate of mutation is constant across time and taxa.

Character data consist of a series of discrete variables (characters) with multiple states. These are instrumental in the assessment of the degree of similarity based on the number of characters held in common by the groups being studied. The analyses of distance and character data have made possible stock identification and the inference of phylogenies with an unprecedented degree of precision. "DNA fingerprinting" tests are so specific that the source stock for caviar imported into the United States can be precisely determined (see Chapter 8, Going Deeper). In a clever application of molecular biology, scientists were able to uncover fraud in a salmon fishing competition. Molecular techniques were used to determine that a fish submitted during the competition at a Finnish lake did not originate from lake stocks but was purchased at a fish market (Primmer et al., 2000).

Although classical systematists sometimes dispute the interpretation of data by their molecular colleagues, a greater insight into the phylogeny of a given group is achieved through the simultaneous assessment of molecular data with more traditional data sets derived from morphological studies—what has become known as a **"total evidence"** approach (Parker, 1997; Porterfield et al., 1999). Morphological study has the advantage of ready accessibility of a broad array of taxa available from museum collections. Museum specimens may not have been preserved in ways that enable biochemical analysis, but a large amount of taxonomic data can be obtained from them. Molecular approaches have the advantage of the availability of a far greater data set, often comprising hundreds or thousands of characters, from which phylogenies can be computed. The extent to which nucleotide substitution occurs in genomes permits the phylogenetic analysis of taxa so closely related as to be morphologically indistinguishable and of taxa so evolutionarily remote that few, if any, common morphological characters are available (Hillis and Wiens, 2000).

Combining morphological and molecular studies in a "total evidence" approach has proven most valuable, both in the assessment of the integrity of closely related fish stocks and in the determination of phylogenetic relationships of higher taxa. As a demonstration of the former, the examination of mitochondrial and nuclear DNA, combined with studies on scute morphology, permitted scientists to determine the evolutionary and dispersal history of two closely related species of North Atlantic sturgeons. Although the North American sturgeon *Acipenser oxyrhynchus* probably diverged from the European sturgeon *A. sturio* more than 15 million years ago, and contact between them has since been limited owing to geographic distance, molecular and morphological investigations indicate that *A. oxyrhynchus* crossed the Atlantic and successively replaced *A. sturio* over much of its European range sometime during the Middle Ages (Ludwig et al., 2002). In some cases, however, the morphological data conflict with molecular data, making the resolution of the phylogeny of a given group more difficult (see Hardman and Page, 2003).

The individual nature of an organism's molecular composition has enabled the broad application of **molecular pedigree analysis,** in which familial relationships in fish populations can be assessed (Wilson and Ferguson, 2002). Total evidence approaches, however, have facilitated the understanding of higher phylogenetic interrelationships, when applied to selected orders of fishes such as the Lampridiformes (Wiley et al., 1998; see Chapter 12), and have greatly enhanced our understanding of acanthomorph fishes as a whole (Wiley et al., 2000).

Computing Phylogenies

The principle of *parsimony* is the basis of phylogenetic hypothesis testing. Parsimony dictates that we accept the hypothesis of relationship that requires the fewest ad hoc hypotheses about character change. This is logically extended to the principle that the shortest cladistic tree that can be calculated from a complete data set is the best estimate of phylogenetic relationships (Fig. 4.5). Acceptance of the parsimony principle does not require a belief that evolution is parsimonious; rather, it requires that the data arrive at the

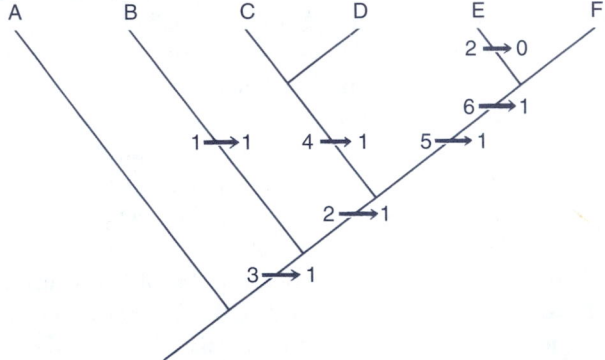

FIGURE 4.5

Matrix of six characters and six taxa, with resulting tree. Taxa *A* and *B* are outgroups; *C*, *D*, *E*, and *F* are the study group. The tree hypothesis is discovered by minimizing the number of ad hoc hypotheses about homoplasy—that is, finding the resolved tree that requires the minimum number of evolutionary steps. Characters *5* and *6* are correlated; all of the characters except *2* are congruent.

most accurate representation of that evolutionary history in the most efficient manner. Phylogenetic inference can proceed in one of two ways: (1) the definition of an **algorithm,** a sequence of steps that will culminate in the development of the phylogeny, best depicted in the form of phylogenetic trees or cladograms; and (2) the definition of a criterion for comparison of alternative phylogenies, with the intent of determining which one is better (Swofford et al., 1996).

Molecular studies, like classical morphometric and meristic studies, generate very large data sets that can only be handled with sophisticated computational tools. Each new generation of desktop or laptop computer is capable of handling ever larger data sets. Software development has coevolved with the increasingly sophisticated hardware, so that today's phylogenetic systematist has at hand an outstanding array of analytical methodologies. It is not surprising that ichthyologists, having to contend with numbingly diverse assemblages of fish species displaying a bewildering variety of characteristics, were among the first to develop software to search data sets for shared changes and to calculate phylogenetic trees. Two of the most widely used programs are HENNIG86 (developed by J. S. Farris) and PAUP (Phylogenetic Analysis Using Parsimony, developed by D. Swofford). Swofford et al. (1996) listed more than 40 software packages available to construct and evaluate phylogenies. Many of these are available on the Internet.

The Internet has greatly facilitated molecular phylogenetic studies by acting as a public repository for genetic data (Blake and Bult, 1996). For example, GenBank, the most frequently used gene repository, was initiated by the National Center for Biotechnology Information (NCBI) in 1988 with a few thousand gene sequences, and it had grown to more than 500,000 sequences by 1995. As of this writing, GenBank has more than 4 million gene sequences available, representing more than 58,000 organisms, some with their complete genome catalogued. With tools such as this at hand, molecular phylogenetics can be applied across the whole spectrum of life, from the assessment of variation in individual genes within a population to studies on the origin of phyla. As an example of the latter, an analysis of a diverse array of amino-acid sequences from mitochondrial genes available from the GenBank repository has suggested that the origin of protostome and deuterostome lineages should be pushed back to a period of time more than 600 million years ago, well before the so-called Cambrian explosion (Lynch, 1999). If this were the case, it would be much easier to understand how the earliest of the known fishes (discussed in Chapters 2 and 5) could have achieved the level of sophistication observed in fossils more than 550 million years old.

Patterson and Johnson (1997) sounded a cautionary note for those who wish to harness modern computational power in the study of organismal phylogeny: "One consequence of the cladistic revolution and the development of numerical cladistics is that a matrix and a computer-generated parsimony analysis are now the norm; if you do not provide them, referees will demand them. The larger the matrix, [the] . . . greater is the likelihood that error will creep in." (p. 404).

• •
AN INTEGRATED APPROACH TO UNDERSTANDING EVOLUTION
• •

The selective value of the characters assessed in phylogenetic study is evident from their survival through the ages. The marriage of Darwinian and Mendelian principles has given us the insight to appreciate the genetic mechanisms underlying the processes enabling some characters to prevail whereas others are consigned to oblivion. We must remain, however,

open to alternative interpretations in order to derive the most thorough explanation of the evolutionary process. The Darwinian revolution resulted in the rejection of an earlier mechanism of evolution, proposed by the 19th-century naturalist Jean Baptiste de Lamarck (1744–1829). To explain evolution, Lamarck proposed that the environment was capable of inducing certain characteristics in organisms that subsequently could be inherited by future generations.

It has become increasingly recognized that a creature can experience entirely different developmental and life histories depending on the environment it encounters (see Chapter 30). Perhaps because the most easily studied organismal models in genetics have been those in which the developmental process is rapid, and thus relatively impervious to environmental influence, we have tended to think of evolutionary processes as "hardwiring" developmental and life history features. Animals such as fruit flies and zebrafish, to mention two classic genetic models (see Chapter 27, Going Deeper), exemplify this bias (Bolker, 1995; Dusheck, 2002). Genetic mechanisms, however, are profoundly influenced by the environment. Consider, for example, the processes controlling sexual differentiation in labroids—the loss of a male causes the dominant female in his harem to change into a male (see Chapters 16 and 27).

Organisms produce certain proteins, termed "heat shock" or "chaperone" proteins, that bind to unstable proteins to help them maintain their conformation and thus their function (Feder, 1999). It has been suggested that heat shock proteins—so named because their role is especially appreciated in conditions of thermal stress (see Hightower et al., 1999)—can have a significant impact on the evolutionary process. Mutated proteins can be held in a stabilized and hence unexpressed state by these protein chaperones until certain environmental conditions cause them to release the mutation. In this manner, rapid evolutionary change can be induced by environmental conditions; "stored" genetic change is released all at once to cause a rapid and dramatic evolutionary modification (Holmes, 2002).

Biologists have long puzzled over the cause of latitudinal gradients in species diversity—why are there so many species near the equator, and progressively fewer as one moves toward the poles? The one constant in such latitudinal gradients is *temperature*, causing some researchers to suggest that high species diversity in the tropics is a consequence of higher temperatures inducing higher rates of mutation and hence a greater incidence of speciation (Rohde, 1992). In a comparison of mitochondrial genes in latitudinally separated species pairs of birds, Bromham and Cardillo (2003) found no significant effect of latitude on the rate of molecular evolution. But birds are endothermic animals; could the same be said of poikilothermic ("cold-blooded") organisms such as fishes? Application of molecular studies such as these to fishes could provide valuable insights into the relationships between speciation and biogeography (see Chapter 29).

As biology has become more focused on the study of genetic mechanisms, the idea of the genome as a set of instructions providing for a pattern of development that is a consequence of the evolutionary process yet immune from immediate environmental influence became ingrained in science education. Perhaps Lamarck perceived something that we are only recently becoming aware of—that the development and adaptation of an organism is the *shared* responsibility of the genome and of environmental influences. An interdisciplinary wedding of evolutionary, ecological, and developmental studies gives us insight into the ways in which organisms integrate genetic information with environmental cues, thus revealing to us the true dimensions of the evolutionary process.

KEY POINTS AND CONNECTIONS

- The agents of natural selection can be either physical factors of the environment or biotic factors. These two factors can combine to influence the distribution and abundance of fishes.
- Speciation in fishes can be allopatric, parapatric, or sympatric. Allopatric or vicariant speciation is consistent with the distribution patterns of most North American fishes, but sympatric speciation is likely to have occurred in situations where adaptive radiation has culminated in the formation of diverse species flocks, as in the African Great Lakes.

  The consequences of speciation, observed in the worldwide distribution and abundance of fishes, are treated in Chapter 29.

- The evolutionary relationships of fishes are greatly clarified through the application of the science of systematics. Willi Hennig introduced the concept of cladistics, which has had great influence on systematic studies, especially of fishes. According to Hennig's principles, evolutionary relationships are discerned through the presence of synapomorphies, or shared derived characters.

  In the subsequent chapters dealing with the systematics of fishes (5 through 17), the phylogenetic relationships of the taxa discussed are expressed in the form of cladograms, as is the convention in cladistic systematics.

- The emergence of molecular techniques for the assessment of evolutionary relationships has greatly enhanced the study of fishes. The genetic integrity of fish stocks and the phyletic relationships of taxa can be determined by measuring variation in genes or in their products. The enormous data sets generated from molecular studies can be analyzed using the most recent advances in computational technology, including online databases and software for the generation of phylogenies. *Molecular techniques are discussed in greater detail in Chapter 28.*

FISH LINKS

http://www.nmnh.si.edu/vert/fish.html Home page of the Division of Fishes at the Smithsonian Institution National Museum of Natural History.

http://research.amnh.org/ichthyology/index.html Home page of the Department of Ichthyology, American Museum of Natural History.

http://www.calacademy.org/research/ichthyology/ Home page of the Department of Ichthyology, California Academy of Sciences.

http://www.asih.org Home page of the American Society of Ichthyologists and Herpetologists.

http://phylogeny.arizona.edu/ Developed by D. R. Maddison and W. P. Maddison of the University of Arizona, The Tree of Life is a most ambitious project containing information about the phylogeny and biodiversity of all organisms.

http://www.ucmp.berkeley.edu/subway/specex.html This site was developed by the University of California Museum of Paleontology. A useful introduction to the principles and applications of cladistics is provided by the link "Journey Into Phylogenetic Systematics."

BUILDING AN ICHTHYOLOGY LIBRARY

A valuable introduction to the practice of taxonomy:

Winston, J. E. 2000. *Describing species: Practical taxonomic procedure for biologists.* Columbia University Press, New York.

Because appropriate nomenclature is essential:

Nelson, J. S., E. J. Crossman, H. Espinosa-Pérez, L. T. Findley, C. R. Gilbert, R. N. Lea, and J. D. Williams. 2004. Common and scientific names of fishes from the United States, Canada, and Mexico. *Amer. Fish. Soc. Spec. Publ. 29.* American Fisheries Society, Bethesda, MD.

Important texts on cladistic theory and molecular applications:

Ferraris, J. D., and S. R. Palumbi (Eds.). 1996. *Molecular zoology.* Wiley-Liss, New York.
Hennig, W. 1966. *Phylogenetic systematics.* University of Illinois Press, Urbana.
Hillis, D. M., C. Moritz, and B. Mable. 1996. *Molecular systematics* (2nd ed.), Sinauer, Sunderland, MA.

A few of the best introductions to evolutionary biology:

Avise, J. C. 2000. *Phylogeography: The history and formation of species.* Harvard University Press, Cambridge, MA.
Berra, T. 1990. *Evolution and the myth of creationism.* Stanford University Press, Stanford, CA. (As long as creationists insist on using the Bible as a biology text, keep this one close at hand. Also, it's written by one of our own, an ichthyologist.)
Coyne, J. A., and H. A. Orr. 2004. *Speciation.* Sinauer, Sunderland, MA.
Patterson, C. 1999. *Evolution* (2nd ed.). Cornell University Press, Ithaca, NY. (Colin Patterson was, at the time of his death in 1998, one of the world's leading fish paleontologists.)

Ridley, M. (Ed.). 1997. *Evolution*. Oxford University Press, Oxford. (A compendium of the classic works in evolution.)

Stearns, S. C., and R. F. Hoekstra. 2000. *Evolution: An introduction*. Oxford University Press, Oxford.

Volpe, E. P., and P. A. Rosenbaum. 2000. *Understanding evolution*. (6th ed.). McGraw-Hill, Boston.

• •

REFERENCES

• •

Avise, J. C. 2000. *Phylogeography: The history and formation of species*. Harvard University Press, Cambridge, MA.

Bell, M. A., and C. A. Andrews. 1997. Evolutionary consequences of postglacial colonization of fresh water by primitively anadromous fishes, pp. 323–363. In *Evolutionary ecology of freshwater animals*, B. Streit, T. Städler, and C. M. Lively (Eds.). Birkhäuser Verlag, Basel.

Blake, J. A., and C. J. Bult. 1996. Biological databases on the Internet, pp. 3–18. In *Molecular zoology: Advances, strategies, and protocols*, J. D. Ferraris and S. R. Palumbi (Eds.). Wiley-Liss, New York.

Bolker, J. A. 1995. Model systems in developmental biology. *BioEssays* 17(5): 451–455.

Boughman, J. W. 2001. Divergent sexual selection enhances reproductive isolation in sticklebacks. *Nature 411*: 944–948.

Bromham, L., and M. Cardillo. 2003. Testing the link between the latitudinal gradient in species richness and rates of molecular evolution. *J. Evol. Biol. 16*: 200–207.

Collins, T. 1996. Molecular comparisons of transisthmian species pairs: Rates and patterns of evolution. pp. 303–334. In *Evolution and environment in tropical America*, J. B. C. Jackson, A. F. Budd, and A. G. Coates (Eds.). University of Chicago Press, Chicago.

Coyne, J. A., and H. A. Orr. 2004. *Speciation*. Sinauer, Sunderland, MA.

Dominey, W. J. 1984. Effects of sexual selection and life history on speciation: Species flocks in African cichlids and Hawaiian *Drosophila*, pp. 231–249. In *Evolution of fish species flocks*, A. A. Echelle and I. Kornfield (Eds.). University of Maine Press, Orono.

Dobzhansky, T. 1973. Nothing in biology makes sense except in the light of evolution. *Amer. Biol. Teacher 35*: 125–129.

Dusheck, J. 2002. The interpretation of genes. *Nat. Hist. 111*(8): 52–59.

Eldridge, N., and S. J. Gould. 1972. Punctuated equilibria: An alternative to phyletic gradualism, pp. 82–115. In *Models in paleobiology*, T. J. M. Schopf (Ed.). Freeman, Cooper, San Francisco.

Feder, M. E. 1999. Organismal, ecological, and evolutionary aspects of heat-shock proteins and the stress response: Established conclusions and unresolved issues. *Amer. Zool. 39*: 857–864.

Galis, F., and J. A. J. Metz. 1998. Why are there so many cichlids? *Trends Ecol. Evol. 13*(1): 1–2.

Greenwood, P. H. 1984. What is a species flock? pp. 13–19. In *Evolution of fish species flocks*, A. E. Echelle and I. Kornfield (Eds.). University of Maine Press, Orono.

———, D. E. Rosen, S. H. Weitzman, and G. S. Myers. 1966. Phyletic studies of teleostean fishes with a provisional classification of living forms. *Bull. Am. Mus. Nat. Hist. 131*: 339–456.

Hardman, M., and L. M. Page. 2003. Phylogenetic relationships among bullhead catfishes of the genus *Ameirus* (Siluriformes: Ictaluridae). *Copeia 2003*: 20–33.

Hendry, A. P., J. K. Wenberg, P. Bentzen, E. C. Volk, and T. P. Quinn. 2001. Rapid evolution of reproductive isolation in the wild: Evidence from introduced salmon. *Science 290*: 516–518.

Hennig, W. 1966. *Phylogenetic systematics*. University of Illinois Press, Urbana.

Hightower, L. E., C. E. Norris, P. J. DiIorio, and E. Fielding. 1999. Heat shock responses of closely related species of tropical and desert fish. *Amer. Zool. 39*: 877–888.

Hillis, D. M., and J. J. Wiens. 2000. Molecules versus morphology in systematics: Conflicts, artifacts, and misconceptions, pp. 1–19. In *Phylogenetic analysis of morphological data*, J. J. Wiens (Ed.). Smithsonian Institution Press, Washington, DC.

Holmes, B. 2002. Ready, steady, evolve. *New Scientist 175*: 28–31.

Houde, A. E., and J. A. Endler. 1990. Correlated evolution of female mating preference and male color pattern in *Poecilia reticulata*. *Science 248*: 1405–1408.

Inoue, J. G., M. Miya, K. Tsukamoto, and M. Nishida. 2001. A mitogenetic perspective on the basal teleostean phylogeny: Resolving higher-level relationships with longer DNA sequences. *Mol. Phylogen. Evol. 20*: 275–285.

Jackson, J. B. C., A. G. Coates, and A. Budd. 1996. *Environmental and biological change in Neogene and Quaternary tropical America*. University of Chicago Press, Chicago.

Knight, M. E., and G. F. Turner. 2004. Laboratory mating trials indicate incipient speciation by sexual selection among populations of the cichlid fish *Pseudotropheus zebra* from Lake Malawi. *Proc. Roy. Soc. Lond. B 271*: 675–680.

Koelz, W. 1929. Coregonid fishes of the Great Lakes. *Bull. U.S. Bur. Fish. 43*: 297–643.

Kozhov, M. 1963. Lake Baikal and its life. *Monogr. Biol. 11*. W. Junk, The Hague.

Ludwig, A., L. Debus, D. Lieckfeldt, I. Wirgin, N. Benecke, I. Jenneckens, P. Williot, J. R. Waldman, and C. Pitra. 2002. When the American sturgeon swam east. *Nature 419*: 447–448.

Lynch, M. 1999. The age and relationships of the major animal phyla. *Evolution 53*: 319–325.

Margulis, L., and D. Sagan. 2002. *Acquiring genomes: A theory of the origins of species*. Basic Books, New York.

McCune, A. R. 1997. How fast is speciation? Molecular, geological, and phylogenetic evidence from adaptive radiations of fishes, pp. 585–610. In *Molecular evolution and adaptive radiation*, T. J. Givnsih, and K. J. Sytsma (Eds.). Cambridge University Press, Cambridge.

Mestel, R. 1998. The genetic battle of the sexes. *Nat. Hist. 107*(1): 44–49.

Miya, M., and M. Nishida. 1997. Speciation in the open ocean. *Nature 389*: 803–804.

———, A. Kawaguchi, and M. Nishida. 2001. Mitogenomic exploration of higher teleostean phylogenies: A case study for moderate-scale evolutionary genomics with 38 newly determined complete mitochondrial DNA sequences. *Mol. Biol. Evol. 18*: 1993–2009.

Moritz, C., and D. M. Hillis. 1996. Molecular systematics: Context and controversies, pp. 1–4. In *Molecular systematics* (2nd ed.), D. M. Hillis, C. Moritz, and B. K. Mable (Eds.). Sinauer, Sunderland, MA.

O'Connell, M., and J. M. Wright. 1997. Microsatellite DNA in fishes. *Rev. Fish Biol. Fisheries 7*: 331–363.

Parker, A. 1997. Combining molecular and morphological data in fish systematics: Examples from the Cyprinodontiformes, pp. 163–188. In *Molecular systematics of fishes*, T. D. Kocher and C. A. Stepien (Eds.). Academic Press, San Diego.

Patterson, C., and G. D. Johnson. 1997. Comments on Begle's "Monophyly and relationships of argentinoid fishes." *Copeia 1997*: 401–409.

Porterfield, J. C., L. M. Page, and T. J. Near. 1999. Phylogenetic relationships among fantail darters (Percidae: *Etheostoma*: *Catonotus*): Total evidence analysis of morphological and molecular data. *Copeia 1999:* 551–564.

Power, M. E. 1987. Predator avoidance by grazing fishes in temperate and tropical streams: Importance of stream depth and prey size, pp. 333–351. In *Predation*, W. C. Kerfoot and A. Sih (Eds.). University Press of New England, Hanover, NH.

———, and W. J. Matthews. 1983. Algae-grazing minnows (*Campostoma anomalum*), piscivorous bass (*Micropterus* sp.), and the distribution of attached algae in a small prairie-margin stream. *Oecologia 60:* 328–332.

———, and A. J. Stewart. 1985. Grazing minnows, piscivorous bass, and stream algae: Dynamics of a strong interaction. *Ecology 66:* 1448–1456.

Primmer, C. R., M. T. Koskinen, and J. Piironen. 2000. The one that did not get away: Individual assignment using microsatellite data detects a case of fishing competition fraud. *Proc. Roy. Soc. B (Biol. Sci.) 267:* 1699–1704.

Reznick, D. 1982. The impact of predation on life history evolution in Trinidadian guppies: Genetic basis of observed life history patterns. *Evolution 36:* 1236–1250.

———, and H. Bryga. 1987. Life history evolution in guppies (*Poecilia reticulata*) 1. Phenotypic and genetic changes in an introduction experiment. *Evolution 41:* 1370–1385.

———, and J. A. Endler. 1982. The impact of predation on life history evolution in Trinidadian guppies (*Poecilia reticulata*). *Evolution 36:* 160–177.

———, and———. 1990. Experimentally induced life history evolution in a natural population. *Nature 346:* 357–359.

———, M. Mateos, and M. S. Springer. 2002. Independent origins and rapid evolution of the placenta in the fish genus *Poeciliopsis. Science 298:* 1018–1020.

———, F. H. Shaw, F. H. Rodd, and R. G. Shaw. 1997. Evaluation of the rate of evolution in natural populations of guppies (*Poecilia reticulata*). *Science 275:* 1934–1937.

Robinson, B. W., and D. S. Wilson. 1994. Character release and displacement in fishes: A neglected literature. *Am. Nat. 144:* 596–627.

Rohde, K. 1992. Latitudinal gradients in species diversity—the search for the primary cause. *Oikos 65*(3): 514–527.

Rosenthal, G. G., T. Y. F. Martinez, F. J. García de León, and M. J. Ryan. 2001. Shared preferences by predators and females for male ornamentals in swordtails. *Am. Nat. 158:* 146–154.

Ruzzante, D. E., S. J. Walde, V. E. Cussac, P. J. Macchi, and M. F. Alonso. 1998. Trophic polymorphism, habitat and diet segregation in *Percichthys trucha* (Pisces: Percichthyidae) in the Andes. *Biol. J. Linn. Soc. 65:* 191–214.

Schluter, D. 1996. Ecological speciation in postglacial fishes. *Phil. Trans. Roy. Soc. Lond. B 351:* 807–814.

Scott, W. B., and E. L. Crossman. 1972. Freshwater fishes of Canada. *Fish. Res. Bd. Canada Bull. 184.*

Smith, G. R., J. Rosenfield, and J. Porterfield. 1995. Processes of origin and criteria for preservation of fish species, pp. 44–57. In Evolution and the aquatic ecosystem, J. Neilsen (Ed.). *Am. Fish. Soc. Spec. Publ.*, Bethesda, MD.

Stepien, C. A., and T. D. Kocher. 1997. Molecules and morphology in studies of fish evolution, pp. 1–11. In *Molecular systematics of fishes*, T. D. Kocher and C. A. Stepien (Eds.). Academic Press, San Diego.

Swofford, D. L., G. J. Olsen, P. J. Waddell, and D. M. Hillis. 1996. Phylogenetic inference, pp. 407–514. In *Molecular systematics* (2nd ed.), D. M. Hillis, C. Moritz, and B. K. Mable (Eds.). Sinauer, Sunderland, MA.

Tautz, D. 2003. Evolutionary biology: Splitting in space. *Nature 421:* 225–226.

Turner, G. F. 2002. Parallel speciation, despeciation, and respeciation: Implications for species definition. *Fish and Fisheries 3:* 225–229.

Volpe, E. P. and P. A. Rosenbaum. 2000. *Understanding evolution* (6th ed.). McGraw-Hill, Boston.

Wiley, E. O., G. D. Johnson, and W. W. Dimmick. 1998. The phylogenetic relationships of lampridiform fishes (Teleostei: Acanthomorpha), based on a total-evidence analysis of morphological and molecular data. *Mol. Phylogen. Evol. 10:* 417–425.

———,———, and———. 2000. The interrelationships of acanthomorph fishes: A total evidence approach using molecular and morphological data. *Biochem. Syst. Ecol. 28:* 319–350.

Wilson, A. J., and M. M. Ferguson. 2002. Molecular pedigree analysis in natural populations of fishes: Approaches, applications, and practical considerations. *Can. J. Fish. Aquat. Sci. 59:* 1696–1707.

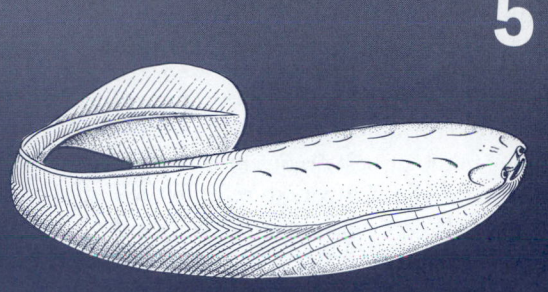

5   JAWLESS FISHES: EARLY VERTEBRATE EXPERIMENTS AND DESCENDANT FORMS

THE FIRST FISHES: RELATIONSHIPS OF MODERN AND EXTINCT FORMS

THE HAGFISHES AND LAMPREYS: A MONOPHYLETIC GROUP?

Hagfishes and Conodonts
Order Myxiniformes
Order Petromyzontiformes (Hyperoartia)

THE "OSTRACODERM" FISHES: PROTO-GNATHOSTOMES?

The †Pteraspidomorphi
    †Arandaspida
    †Astraspida
    †Heterostraci
    †Thelodonti
The †Cephalaspidomorphi
    Anaspida
    †Osteostraci
    †Galeaspida
    †Pituriaspida

I n this chapter, we will consider creatures that represent the earliest vertebrates and, hence, the first fishes. The fossil record has revealed a diversity of early "experiments" among the archaic fishes, yet there are only two surviving descendant forms of these first fishes—the hagfishes and the lampreys. Although they are similar in being jawless and lacking paired appendages, hagfishes and lampreys show a number of distinctive features. Some researchers have suggested that hagfishes may have descended from lineages that predate the evolution of vertebrates. Accumulating molecular evidence, however, points to a common origin for these two ancient groups of fishes. Although descended from the earliest fishes, hagfishes and lampreys also display a number of highly derived features of anatomy, physiology, and behavior. Hagfishes, for example, have a complex array of secretory cells in the skin that secrete copious quantities of slime. Lampreys have evolved an anadromous lifestyle that is possible only because of the evolution of remarkable physiological and behavioral adaptations. These features hint at what may have been a remarkable array of adaptations that existed hundreds of millions of years ago among the earliest fishes—features that we can only infer from their fortunately quite extensive fossil record.

THE FIRST FISHES: RELATIONSHIPS OF MODERN AND EXTINCT FORMS

The earliest vertebrates that appeared in the fossil record were creatures that lacked both jaws and well-developed paired appendages. One of the most vexing questions that students of vertebrate evolution must wrestle with is the relationship of these early fossil forms with the highly derived descendant forms of jawless fishes—the hagfishes and lampreys (Fig. 5.1). Various evolutionary scenarios have been mooted, yet a consensus opinion on the interrelationships of hagfishes, lampreys, and a host of long-extinct forms known only from the fossil record has yet to be achieved. In 1996, Philippe Janvier presented a most comprehensive classification of early fishes. By the time of the publication of his more recent phylogeny (Janvier, 2001), however, he had reconsidered the interrelationships of the modern agnathan fishes. The arrangement of the taxa presented here is therefore based largely on the phylogenies developed and amended by Janvier (1996, 2001) and by Nelson (1994). Extinct groups are identified by a dagger (†).

THE HAGFISHES AND LAMPREYS: A MONOPHYLETIC GROUP?

As discussed in Chapter 2, the hagfishes (Plate 5.1A), though an extant group of fishes, apparently date their ancestry back to the earliest origins of craniates. Janvier (1996) used the term **Hyperotreti** to refer to the hagfishes. Although Janvier (1996) considered hagfishes to be protovertebrates, they share with vertebrates a distinctive embryological feature, the **neural crest** (see Chapter 27). This epidermal thickening of the margins of the dorsal groove gives rise to gill arches; dermal skeletal elements, including scales and fin rays in fishes; eye structures, such as the lens and cornea; and pigment cells (Gans and Northcutt, 1983; Smith, 1993; Smith and Hall, 1990). Although comparative morphological analyses of extant agnaths and gnathostomes might suggest that modern cyclostomes are paraphyletic, as proposed by Janvier (1996, 2001; Fig. 5.2A), this is one case where the morphological evidence has not been confirmed by molecular studies. Janvier's (2001) phylogeny proposes three clades among living craniates: hagfishes, lampreys, and

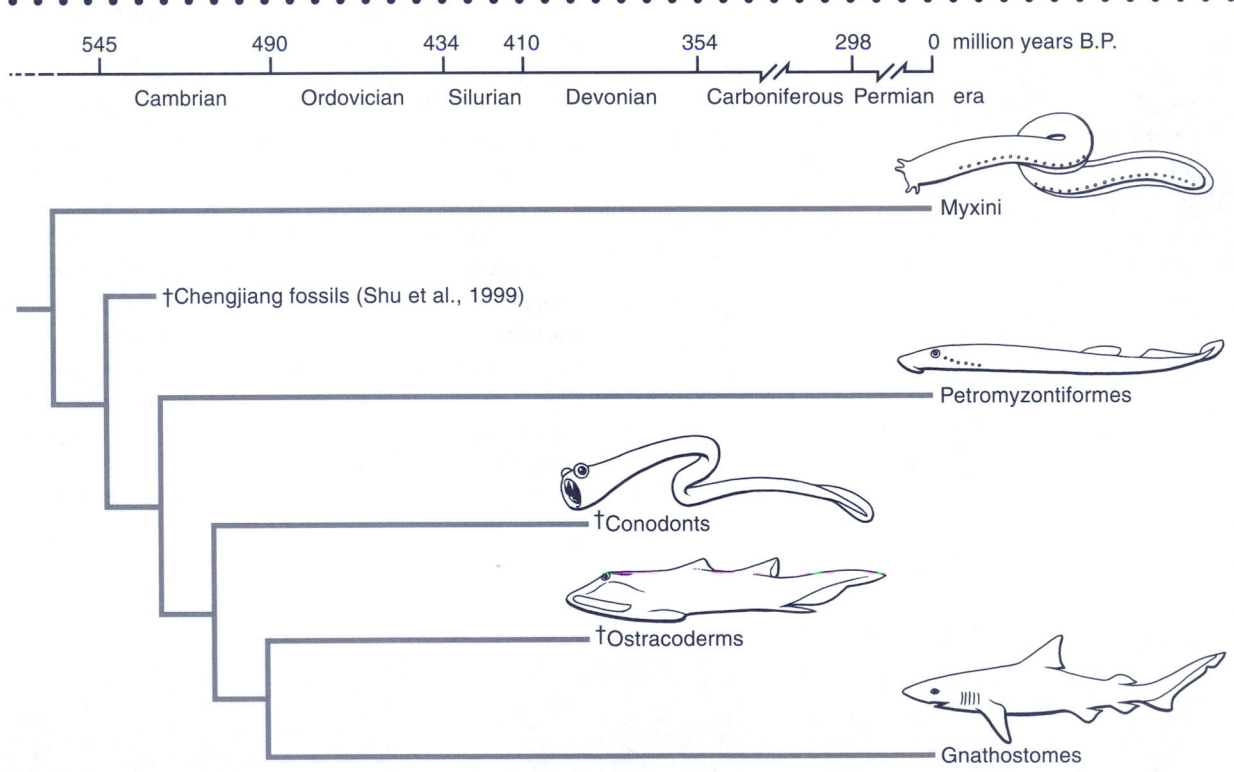

FIGURE 5.1
Tentative phylogeny of major fossil and living groups of vertebrates (From Janvier, 1999; Zimmer, 2000.)

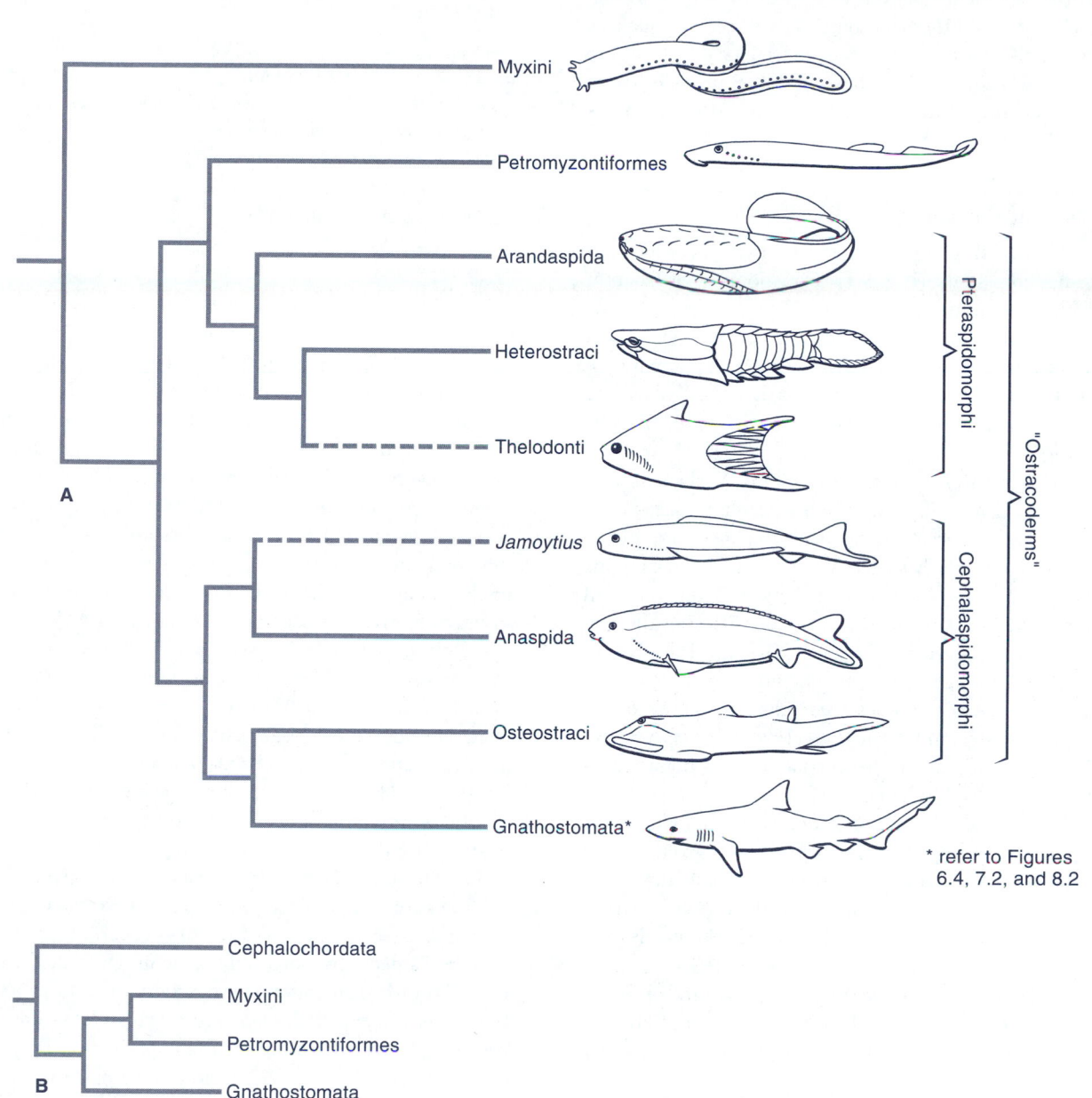

FIGURE 5.2

A, Cladogram showing phylogeny of living and fossil agnathan fishes (after Janvier, 2001); **B,** Phylogeny based on molecular studies that support monophyly of the cyclostomes (based on Mallatt et al., 2001, and Hedges, 2001).

gnathostomes, yet the analysis of nuclear protein-encoding genes (Hedges, 2001), ribosomal RNA gene sequences (Mallatt et al., 2001), and mitochondrial DNA (Delarbre et al., 2002; Furlong and Holland, 2002) all provide strong evidence for the monophyly of hagfishes and lampreys (Fig. 5.2B).

Hagfishes and Conodonts

Of special interest is the puzzling relationship between hagfishes and conodonts (Fig. 5.1). Conodonts (see Chapter 2) have long been known in the fossil record, but their affinities with the craniates were, until recently, uncertain at best. It was the discovery of a distinctly craniate fossil with conodont structures in the oral region that firmly fixed the conodonts as chordates (Aldridge et al., 1986, 1993; Briggs, 1992; Briggs et al., 1983). This conodont fossil shows definite craniate features, such as a notochord, V-shaped myomeres, a tail with fin rays, and, most peculiarly, enormous eyes. This last feature, together with the small size of the fossils (less than 1 cm) has suggested to some that conodonts were larval craniates. A relationship between conodonts and hagfishes based on similarities in the dentition of the two groups has been suggested (Krejsa and Slavkin, 1987; Krejsa et al., 1990). Although a fundamental phylogenetic relationship may exist between the fossil conodonts and the modern hagfishes, Aldridge and Donoghue (1998) did not perceive the tooth elements seen in fossil conodonts as homologous with the lingual teeth of hagfishes.

Aldridge et al. (1993) concluded that conodonts were vertebrates and probably a primitive sister group of the †Heterostraci. Histochemical analysis of conodont elements, however, points to an affinity with cephalochordates and not with hagfishes or vertebrates (Kemp and Nicoll, 1996). Forey and Janvier (1994) pointed out that the conodont fossils show no evidence of gill openings and observed that their hard tissues are unlike the enamel and bone of vertebrates, so they did not include the group into their vertebrate phylogeny.

One of the most important agnathan features seen in hagfishes that indicate a development different from that of the jaw-bearing vertebrates is the arrangement of the gills and the branchial skeleton. In both hagfishes and lampreys, the gill tissue (which is endodermal in origin) and the branchial arteries and nerves are internal to the branchial skeleton, which is a cartilaginous "basket" external to the gill pouches in the living forms. The branchial skeleton is fused to the neurocranium. Gill openings are pores, not slits. Vertebrae do not replace the notochord, and paired fins are absent.

Order Myxiniformes

DIAGNOSIS OF THE MYXINIFORMES
- Simple cartilaginous skeleton; jaws and paired fins absent.
- Single semicircular canal in inner ear.
- Rudimentary eyes lacking musculature.
- 1 to 16 pairs of external gill apertures.
- Lateral line system not developed.

Source: Nelson, 1994.

Striking video footage, taken during oceanographer Craig Smith's studies of the role of whale carcasses in the food chain of the ocean bottom (Dybas, 1999), depicts the role that hagfishes play in oceanic ecosystems. Dead or dying large vertebrates are attacked by hundreds of hagfishes as they sink to the ocean floor, until the carcass becomes a writhing mass of scavenging activity. These obviously primitive creatures occupy a significant niche in the world's oceans—one that they may have assumed at the dawn of their existence. Hagfishes live on soft bottoms of mud, silt, or clay. They have been taken in tidepools, however, as well as at depths greater than 5,000 m (Martini, 1998).

Only one definitive fossil hagfish, †*Myxinikela*, from the Carboniferous (Pennsylvanian) deposits of Illinois, has been discovered to date (Bardack, 1991, 1998). Another fossil from the same deposit, †*Gilpichthys*, was originally proposed to have hagfish affinities (Bardack and Richardson, 1977), but Bardack (1998) now questions this interpretation.

Because of their capacity to secrete copious amounts of slime from numerous glands arrayed on their ventrolateral surface (see Chapter 3; Bardack, 1991; Carroll, 1988), hagfishes are also known as *slime hags*. These glands contain both mucus-secreting cells and **thread cells** that produce protein filaments that probably serve to reinforce the slime layer. Hagfish filaments possess structural features akin to the protein fibers of the eukaryote cytoskeleton and have proven valuable in studies on their mechanical properties (see Fudge et al., 2003). The slime layer protects the skin surface of the hagfish during feeding bouts in carcasses. It may also serve to deter potential predators. Hagfishes are burrowing animals, and it has been proposed that the slime may serve to protect the skin while burrowing and to stabilize the burrow walls (Brodal and Fänge, 1963; Hardisty, 1979). Martini (1998) has provided details on the burrowing behavior of hagfishes and concluded that the slime secretion plays no role in burrowing.

Although hagfishes are primarily known as scavengers on large, moribund creatures that land on the ocean floor, their burrowing behavior puts them in contact with numerous potential prey items. Polychaete worms and shrimps have been found in their gut contents. Mallatt (1985) noted that the feeding apparatus of hagfishes is very efficient at pulling in worms by using rapid, lateral, pincerlike movements of the rows of keratinaceous dental cusps that border the sides of the mouth (Plate 5.1B). The similarity in the musculature and other structures of this lingual biting system has led Yalden (1985) to postulate that Myxinoidea could be included in the †Cephalaspidomorphi. Once a prey item is grabbed, the hagfish ties a knot in its tail and loops it forward on the body, using it as a brace to gain leverage when pulling at the prey. This is especially useful in tearing chunks of meat from a large carcass. This peculiar looping action is also used by the animal to clear slime from its skin surface. Once in the gut, ingested food items are wrapped in a mucus bag that is permeable to digestive enzymes and digested food. Once nutrients have been extracted from this bag, it is excreted as a wrapper around the feces.

The eyes of hagfishes are minute and completely covered by skin. Large tentacles surround the terminal nasal opening and the mouth. The structure of the inner ear of hagfishes is simpler than that of vertebrates, consisting of only a single semicircular canal and macula (Jorgensen, 1998; see Chapter 21 for details on sense organs).

Whereas lampreys may be found in freshwater, hagfishes are exclusively marine. They develop from a large, **meroblastic** egg (see Chapter 27 for details on embryology), and larvae have never been collected (Hardisty, 1979; Janvier, 1996). Their eggs (Fig. 5.3) may be more than 25 mm long and equipped with hooklike tendrils that can hold them together in a string or clump. Developing eggs are seldom seen, so that knowledge of hagfish embryology is limited (Gorbman and Tamarin, 1985). Walvig (1963) reported that one investigator found only 151 hagfish eggs over a 20-year period. The largest cluster contained 21 eggs.

The pharynx of hagfishes may consist of as few as 5 to as many as 16 pairs of gills, depending on the species. In some genera, the efferent branchial ducts are collected into a tube with a single external opening on each side. On the left side, a peculiar duct (the **pharyngocutaneous duct**), apparently homologous with the gills, originates at the posterior portion of the pharynx and communicates either with the left branchial duct or, in the South American *Neomyxine tridentiger*, directly to the exterior.

The cartilaginous skeleton of hagfishes is not well developed, with no rudiments of vertebrae and only a membranous roof in the skull. The arrangement of the central nervous system is distinct from that of other vertebrates, as the dorsal roots of the spinal nerves are united with the ventral roots.

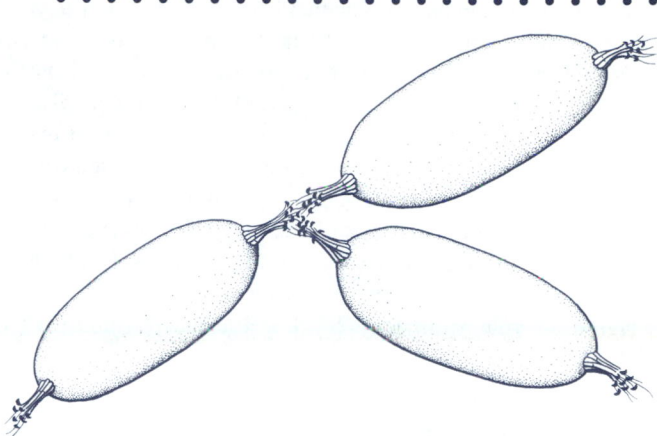

FIGURE 5.3
Eggs of hagfish, class Myxini (actual length ca. 30 mm).

Hagfishes are unique in possessing an array of contractile structures in their vascular system. The primary heart is located in the usual position for a heart, just behind the gills, and has three chambers. There are also paired **caudal hearts,** paired **cardinal vein hearts,** and a **portal vein heart** that pumps blood from an intestinal vein and the right jugular vein to the liver. These accessory hearts help reestablish blood flow after it leaves several sinuses. Hagfishes are unusual in having an extensive system of large venous sinuses and a well-developed dermal capillary network (Forster, 1998; Lomholt and Franko-Dossar, 1998; Welsch and Potter, 1998; see Chapter 24). To a lesser degree, lampreys also display sinuses between the arterial and venous systems. In both cases, this should not necessarily be interpreted as a primitive feature but rather as a highly specialized alternative to the typical vertebrate circulatory system, albeit one that pumps at the lowest pressure recorded in fishes (Forster, 1998).

Six genera of hagfishes are known, most of them from temperate waters. These are usually placed in the family **Myxinidae,** but those with multiple branchial openings are sometimes treated as a separate family, Eptatretidae, or as a subfamily, Eptatretinae (Nelson, 1994). There are more than 40 species; two of the best known are the Atlantic hagfish (*Myxine glutinosa*) and the Pacific hagfish (*Eptatretus stoutii*) of the eastern Pacific Ocean.

As one might expect, there is not much of a market for a fish that goes by the name "slime hag," although the Japanese (brave souls) consider them a food item. Their economic importance, centered in Asia, comes mainly from the use of their skin for leather items. The demand for "eelskin" products has resulted in the depletion of hagfish stocks in Asian waters, and there is concern for other stocks, especially off the coast of North America (Honma, 1998). Hagfishes are also considered a pest in the gill net or long-line fisheries as they mutilate the catch before it can be recovered. Although this activity might cause hagfishes to be viewed in a negative light by commercial fishery workers, their role as scavengers is vital to benthic marine ecosystems. Given their scavenging and burrowing habits, these "earthworms of the deep" play a significant role in nutrient recycling and substratum turnover (Martini, 1998).

Order Petromyzontiformes (Hyperoartia)

The Carboniferous rocks of Illinois and Montana have given us what little fossil evidence there is of the true lampreys. Only two genera of fossil lampreys, †*Mayomyzon* and †*Hardistiella*, are known. The presence of a piston cartilage in the lingual apparatus of †*Mayomyzon* suggests the early adoption of a "parasitic" mode of feeding. A third genus, †*Pipiscus*, was originally proposed to be a lamprey based on the presence of a sucking-type oral disc (Janvier, 1996). Bardack (1998), however, doubted that †*Pipiscus* is closely related to the lampreys. The Petromyzontiformes have been considered a sister group of the scaly anaspids, with the naked-skinned †*Jamoytius* interpreted as an intermediate form. The molecular evidence cited earlier indicates a closer affinity with hagfishes, however (Fig. 5.2B).

DIAGNOSIS OF THE PETROMYZONTIFORMES
- Cartilaginous skeleton; jaws and paired fins absent.
- Two semicircular canals.
- Naked, eellike body, with seven pairs of lateral gill openings.
- Eyes well developed in adult; not developed in larvae (ammocoetes).
  Source: Nelson, 1994.

Lampreys inhabit temperate seas and streams, occurring in both the Northern and Southern Hemispheres. The lampreys of the Northern Hemisphere are classified in the family Petromyzontidae, whereas two families, the Geotriidae and the Mordaciidae, are known from the Southern Hemisphere (Allen et al., 2002; Gill et al., 2003). The pouched lamprey (*Geotria australis*), with a broad distribution encompassing the Falkland Islands, South America, New Zealand, and Australia, is the sole geotriid lamprey. The short-headed lampreys of the genus *Mordacia* (family Mordaciidae) comprise two species from Australian freshwaters and one species confined to South America (Eschmeyer, 1998; Nelson, 1994). Using morphological features,

Gill et al. (2003) have confirmed the monophyly of the Petromyzontiformes and of the Northern Hemisphere species in the family Petromyzontidae. The Southern Hemisphere genera appear to represent highly specialized forms; their precise relationships have yet to be ascertained (Gill et al., 2003).

Modern lampreys have no paired fins (Plates 5.2 and 5.3; see Fig. 2.3); whether this is a basal or derived condition is difficult to assess. Like their once-presumed fossil relative †*Jamoytius*, their skin is lacking in mineralization, which, in all likelihood, has contributed to their paucity in the fossil record. Lampreys are eellike, with lateral eyes and a ventral mouth consisting of a circular disc set with horny teeth. The unpaired **nasohypophyseal** opening is between the eyes and leads to a blind pouch adjacent to the pituitary gland but lacking communication with the orobranchial chamber. The skeleton is cartilaginous and not well developed except for the skull and branchial region. No vertebral centra are developed, and the neural arches are rudimentary. Dorsal and caudal fins are present. Myotomes are not divided horizontally into epaxial and hypaxial muscles, as they are in jawed fishes.

Lampreys possess some peculiar internal features. Whereas hagfishes had only one semicircular canal in the inner ear, lampreys have developed two. The internal labyrinth has a ciliated epithelium, the left duct of Cuvier is absent, and the dorsal and ventral roots of the spinal nerves alternate and do not unite. In adults, the gills open into a respiratory tube that begins at the mouth, extends under the esophagus, and ends at the seventh pair of gills. In the eyeless, toothless larvae (called **ammocoetes;** Plate 5.2B; see Fig. 2.2), the gills open to the long pharynx, but this access is cut off posteriorly during metamorphosis, and a new esophagus forms. The gallbladder and bile ducts disappear in adults.

Some relationship to bony fishes may be indicated by the embryonic formation of a neural keel, instead of a neural tube as in sharks and rays. Furthermore, a **bulbus arteriosus** is present, instead of a conus arteriosus as in other cartilaginous fishes (see Chapter 24). These and other features suggest that lampreys, though agnathous like hagfishes, are possibly more closely related to gnathostomes (Forey, 1995; Forey and Janvier, 1994).

All lampreys have a long larval life. Very small eggs are deposited in nests made in the gravel bottoms of streams. The incubation period varies with temperature, but is usually two to four weeks. When the tiny larvae hatch, they drift to soft bottoms in pools and eddies and begin a life of filtering plankton and detritus at the mud–water interface. The ammocoete larvae spend about five years as blind, burrowing filter feeders, after which metamorphosis takes place, and they become predators on other fishes.

With regard to adult nutritional strategies, there are two types of lampreys, "parasites" and nonfeeding forms. After metamorphosis, the "parasitic" types feed by attacking fishes with their sucking mouths (Fig. 3.3A), rasping holes in the skin with their pistonlike tongues, and pumping out blood and body fluids—a process aided by the secretion of an anticoagulant from paired buccal or "salivary" glands. A few species ingest small fragments of flesh and viscera as well as scales and small bones (Beamish, 1980; Bond et al., 1983). This removal of tissue from the prey may be facilitated by the longitudinal laminae of "teeth" on the tongue closing together in a lateral biting motion (Yalden, 1985). The sight of a lamprey firmly attached to the side of a host fish leaves many with the impression that this is indeed a form of parasitism. However, a true parasite exists in a symbiotic state with its host and does not kill it. A more accurate perception of lampreys is as predators that engage in a prolonged and debilitating attack on their prey. If the host fish does not succumb to the attack outright, it is probably left in such a weakened state that infection through the wound is probable and death occurs later. The size of the host organism obviously is important in determining if lampreys are parasites or predators. Large fishes and even cetaceans are reported to experience lamprey attacks, but their body mass probably means they suffer little ill effect. Many "parasitic" lampreys are anadromous, spending their postmetamorphic growth period in salt water before returning to streams to spawn and die. These may reach a meter in length. Other "parasitic" lampreys remain in freshwater and may grow to half a meter or more, as do the landlocked populations of sea lamprey (*Petromyzon marinus*) in the Great Lakes. Some strictly freshwater species may reach adult size at 15 cm or even smaller (Kan and Bond, 1981).

The so-called "brook" lampreys confine their feeding to the larval stage, and after metamorphosis they spend a few months in hiding while their gonads mature. In typical semelparous fashion (see Chapter 27), they then spawn, usually at lengths of less than 20 cm, and die. In many instances, brook lampreys may share the same drainage with a "parasitic" sister species, forming so-called "paired" species. It is generally accepted that the nonfeeding species are derived from the predatory forms in the species pair. In some instances, more than one nonfeeding species has evolved from a predatory species (Hardisty, 1979; Potter, 1980), so that Vladykov and Kott (1979) have suggested using the term *satellite species* in place of "paired" species. Paired or satellite species have been reported in all lamprey genera except *Petromyzon*, *Caspiomyzon*, and *Geotria* (Hardisty and Potter, 1971). On the Pacific Coast of North America, there are the river lamprey (*Lampetra ayresi;* "parasite") and the closely related western brook lamprey (*L. richardsoni;* nonfeeding).

The western brook lamprey has a wider distribution than the river lamprey. It is found in many small creeks, both coastal and inland, whereas the river lamprey lives in larger streams, generally close to marine waters, where it feeds on herring and other fishes. Another Pacific Coast series consists of the widely distributed, "parasitic" Pacific lamprey (*Lampetra tridentata*) and its assumed derivatives, the nonfeeding Pit-Klamath brook lamprey (*L. lethophaga*) which is restricted to creeks in the upper reaches of the Klamath and Pit river systems, and the nonfeeding Kern brook lamprey (*L. hubbsi*) of the Kern river system.

The Arctic lamprey, *Lampetra japonica*, distributed from Siberia to Japan and Alaska, has at least five satellite species—two found in Asia and three in North America—and there are three pairs of species in the North American genus *Ichthyomyzon* (Potter, 1980). Other examples of satellite species include the following: *Lampetra fluviatilis* and the nonfeeding *L. planeri* of Europe; *Eudontomyzon danfordi* and the nonfeeding *E. vladykovi* and other brook lampreys of Eastern Europe; and *Mordacia mordax* and the nonfeeding *M. praecox* of Australia.

The economic value of lampreys is slight, even though they are used as food in some areas, and they have been used as a source for lightweight oil. Henry I, named "Beauclerc" ("fine scholar") in recognition of his scholarly acumen (he was reputedly the first Norman king to be fluent in English), is reported to have perished from eating too many lampreys. Petromyzontids can exert a profound negative impact on local environments and economies if they attack fishes of sport or commercial value. The chronicle of the invasion of the upper Great Lakes by the sea lamprey is a sad one. The valuable lake trout (*Salvelinus namaycush*) and lake whitefish (*Coregonus clupeaformis*) virtually disappeared as commercial species when the mortality due to the invading lamprey was superimposed on fishing mortality and deteriorating environmental conditions. In many other instances, the effect has not been as severe, but smaller lampreys, such as the chestnut lamprey (*Ichthyomyzon castaneus*) or the landlocked Pacific lampreys (*Lampetra tridentata*), are known to attack various freshwater game fishes. The smallest known "parasitic" lamprey, *L. minima*, is capable, apparently through strength of numbers, of killing fingerling trout and tui chub (*Gila bicolor*). Most adults are less than 80 mm long (Bond and Kan, 1973). This species, once thought to have become extinct through chemical treatment of Miller Lake in Oregon, was rediscovered in adjacent drainages (Lorion et al., 2000).

Larval lampreys are eaten by a variety of fishes, and are sometimes used as bait. Adult lampreys are excellent bait for sturgeon and have been found in the stomachs of other fishes, including sharks. Seasonally, they may form a great portion of the food of the California sea lions that live near the mouths of rivers that sustain runs of the Pacific lamprey.

· ·
THE "OSTRACODERM" FISHES: PROTO-GNATHOSTOMES?
· ·

About 470 million years ago, a small armored vertebrate (Fig. 5.4) was swimming lazily over a shallow, muddy ocean bottom covered with thousands of lamp shells (phylum Brachiopoda) and teeming with trilobites (phylum Arthropoda). This 20 cm long early fish thrived in the warm coastal shallows, possibly scraping and ingesting the abundant algae and microfauna or filtering detritus and microorganisms out of the water. An earthquake, a sudden buckling of the earth's crust, and a coastal river becomes diverted, finding a new opening to the ocean. Our little armored fish and thousands like it are rapidly exterminated, together with a host of invertebrate associates, by the sudden influx of freshwater. (It is the invertebrates that provide the most convincing evidence that these early fishes were marine and not freshwater.) Buried in the fine sediments, the fishes and invertebrates become fossilized and remain hidden until Pierre-Yves Gagnier, a researcher affiliated with the *Musée Nationale d'Histoire Naturelle* in Paris, discovers them near the village of Sacabamba in southern Bolivia in 1986. By naming his find †*Sacabambaspis janvieri,* Gagnier honors both the locale of their discovery and one of the world's authorities on early vertebrates, Philippe Janvier (Gagnier, 1989).

Gagnier discovered the earliest example of a type of fish commonly referred to as an **ostracoderm** (literally, "shell-skin"). It is one of a complex assemblage of early vertebrates that arose coincident with the evolution in many other animal phyla of the capacity to form mineralized tissues. This has left us with the opportunity to find the fossil remains of a number of these earliest armored fishes and to suggest possible phylogenies. These (see Figs. 5.1 and 5.2A) are especially helpful in drawing relationships not only between these early armored fishes and their soft-bodied ancestors mentioned in Chapter 2, but also between these ostracoderms and modern fishes, including those with and those without jaws.

It is easy to recognize the adaptive advantages of a hard, bony carapace, in view of the prevalence of large, predatory arthropods such as the eurypterid "water scorpions" that were abundant during the Silurian. Bone tissue in vertebrates contributes to basically two skeletal components—the *exoskeleton* and the *endoskeleton* (Kardong, 2002). The **endoskeleton** usually arises from **endochondral** or replacement bone—so named because the bone elements are first formed as cartilaginous precursors (see Chapter 3). The endochondral skeleton, originally formed in cartilage, was probably the earliest one to appear (Janvier, 1996). In the early ostracoderms, the **exoskeleton** (also known as the **dermal skeleton**) probably formed by direct ossification in skin tissues, while deeper skeletal elements remained as cartilage. The notochord remained as the chief axial support. Protection was not the only advantage to ossification. The dermal skeleton could have served as a reservoir for the calcium and phosphates necessary for metabolism. These earliest of fishes also appear to have had remarkable sensory capabilities, including mechanoreception and electroreception, as evidenced by the presence of canallike structures in the bone plates covering their bodies. Perhaps a hard carapace

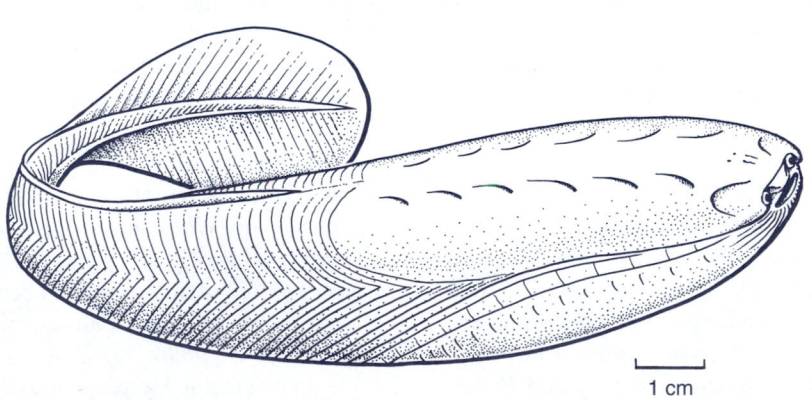

FIGURE 5.4
†*Sacabambaspis* from Ordovician rocks of southern Bolivia (actual size ca. 20 cm).

improved the insulation of electroreceptive devices (Carroll, 1988; Northcutt and Gans, 1983). It has been suggested that bony armor might have had a role in osmoregulation. The great renal physiologist Homer Smith advocated a freshwater origin for the vertebrates, contrary to the prevailing opinion and the fossil record that suggests otherwise. Smith (1932) reasoned that dermal armor could serve early freshwater fishes by restricting the osmotic uptake of water. Although this proposal has largely been discounted, it is interesting to note that the permeability of the exoskeleton of freshwater crustaceans is lower than that of marine crustaceans (Lockwood, 1962). As more of these early armored fishes are discovered in the fossil record, it is becoming apparent that the term "ostracoderm," once considered to denote a monophyletic assemblage, actually covers at least two distinct classes, the †**Pteraspidomorphi** and the †**Cephalaspidomorphi**. Janvier (2001) recognized that these armored fishes appear to have more in common with gnathostomes than with either the hagfishes or lampreys.

Among the ostracoderm fishes, Janvier (1996) recognized eight taxa: the †**Arandaspida**, †**Astraspida**, †**Heterostraci**, †**Anaspida**, †**Osteostraci**, †**Galeaspida**, †**Pituriaspida**, and †**Thelodonti**. These are somewhat tentatively distributed among the †Pteraspidomorphi and †Cephalaspidomorphi (Fig. 5.2A). The suffix *aspid* literally means "shield," and hence these taxa are distinguished largely on the basis of the characteristic morphology of their dermal armor, especially on the head. In addition to the aforementioned Ordovician sites in Bolivia, the rocks of the Amadeus Basin of Australia and the Harding Sandstone Formation in Colorado have yielded many valuable fossils of these earliest of vertebrates. Sansom et al. (2001) claimed that the diversification of vertebrates from the Late Cambrian through the Ordovician, including the ostracoderm and other agnathan lineages, was the result of environmental factors associated with sea level changes acting in concert with intrinsic genetic features of the animals—specifically the duplication of homeotic (*Hox*) gene clusters (see Chapters 2 and 28). Although Moy-Thomas and Miles (1971) placed the hagfishes among the cephalaspidomorphs, the affinities of hagfishes with the extinct ostracoderms seem remote at best. Janvier (1996) claimed affinities between lampreys and the extinct cephalaspids, but a revision of this phylogeny has removed the lampreys from the ostracoderm lineages (Janvier, 2001).

· · · · · · · · · · · ·
The †Pteraspidomorphi

The pteraspidomorphs are also known as the **Diplorhina** (literally "two nostrils"). Impressions in the dorsal head shields of fossils reveal a brain with two separate olfactory bulbs. It is believed that these must have been associated with separate nasal openings. No nasohypophyseal canal is present. Usually, there is only one pair of branchial openings, although there are several gill pouches, but the arandaspids have several individual gill openings. Bony armor lacking true bone cells is present.

†*Arandaspida*

This group includes the earliest known ostracoderm, †*Sacabambaspis*. Three other genera have been assigned to this taxon. The eyes of arandaspids are located at the extreme anterior of the head, where they and the presumed nostrils adjacent to them are separated by a T-shaped plate. Two large, roughly oval shields cover the dorsal and ventral sides of the head. The elaborate tubercles that ornament the head carapace are the primary means of differentiating the known genera. Although little is known of the internal anatomy of ostracoderms, impressions on the underside of the head shields suggest the presence of at least 10 branchial pouches per side. The structure of the exoskeleton shows the complexity typical of these early fishes. It has three layers: a **laminar** basal layer, a middle **cancellar** layer, and a superficial **spongy** layer, where the tubercles can be found. There is some evidence that portions of the endoskeleton may have been calcified as well.

†*Astraspida*

The astraspids are known from comparatively few fossils but, thanks to the excellent preservation properties of the sandstone of the Harding Formation, their anatomical features are known in some detail. Most distinctive to this group is the thick, glassy **enamel layer** that covers the tubercles. Sections through the tubercles have revealed the processes by which they are regenerated and replaced.

†*Heterostraci*

Heterostracan fossils have long been known to scientists but were not necessarily recognized as vertebrates. T. H. Huxley, the legendary 19th-century anatomist and chief defender of Darwinism, was the first to recognize them as fishes. Heterostracans reached their greatest development in the Upper Silurian and the Lower Devonian. A complex group consisting of a number of genera, the heterostracans are best exemplified by the genus †*Pteraspis* (Fig. 5.5A). Most were covered by bony plates that formed a shield over the anterior part of the body and scales on the posterior body, including the tail. Paired fins were lacking. Typically, the head was flat, the eyes lateral, and the mouth subterminal to slightly superior. In most specimens, the caudal fin shape is **hypocercal,** with the notochord entering the

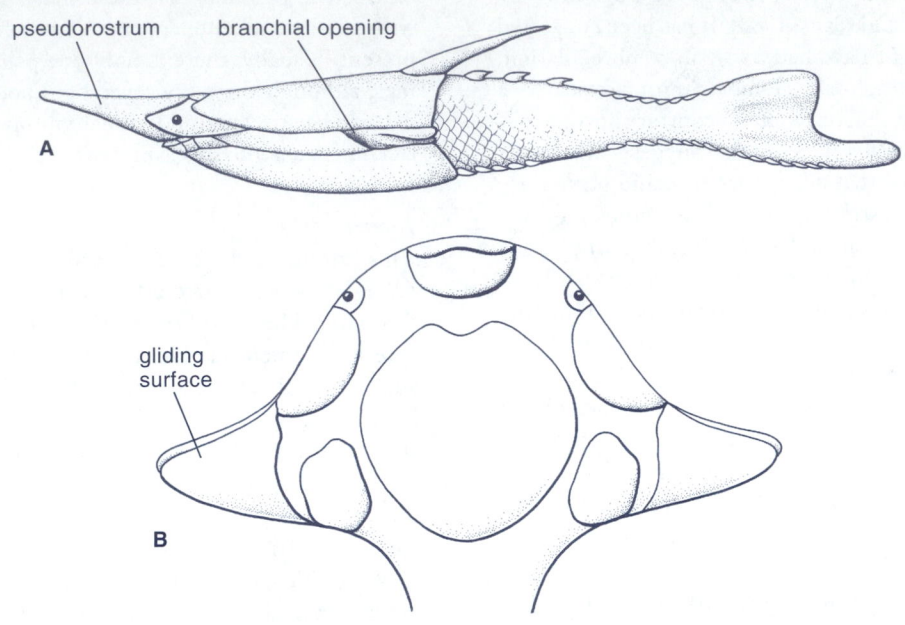

FIGURE 5.5

Representatives of the †Pteraspidomorphi: **A,** †*Pteraspis,* lateral view (actual size ca. 22 cm); **B,** Dorsal view of †*Pycnosteus,* with pronounced lateral "gliding surfaces" (actual size ca. 80 cm). (After Moy-Thomas and Miles, 1971.)

lower lobe, although some heterostracans show the heterocercal condition. Length was usually less than 300 mm, but some reached 1.5 m. One of the earliest of the heterostracans is †*Anglaspis,* which was almost completely covered by large bony plates anteriorly and large bony scales posteriorly—these scales being smaller on the almost symmetrical caudal fin, probably the most flexible section of the body. The single most distinctive feature of the group—one that they actually share with holocephalans and bony fishes—is a single common branchial opening.

Some genera have lateral projections from the head shield, forming what appear to be underwater "gliding" surfaces (Fig. 5.5B). One genus (†*Doryaspis*) had an anterior extension or *pseudorostrum* similar in appearance to the rostrum of the saw shark (*Pristiophorus*). Specializations such as tubular or dorsally placed mouths and stabilizing keels have led paleontologists to believe that the range of habits and habitats of heterostracans extended well beyond the bottom-feeding and -dwelling habits usually associated with ostracoderms.

†Thelodonti

Most thelodonts were small fishes, usually ranging from 100 to 200 mm, but fossils of †*Thelodus parvidens* are up to one meter long (Turner, 1986). They lived mainly during the Silurian and Devonian and differed from other ostracoderms in having a covering of small, denticlelike scales instead of plates or solid armor. Dorsal, anal, caudal (symmetrical to hypocercal), and lateral fins are present, and there appear to be eight or nine branchial sacs on each side, which open separately to the exterior. The body form and the scales of thelodonts suggest a monophyletic origin (Janvier, 1986; Wilson and Caldwell, 1993), but their actual phylogenetic status remains uncertain, and they are likely a paraphyletic group (Janvier, 1996). They have often been considered to be closely related to gnathostomes (Sansom et al., 2001).

The enigmatic "fork-tailed" thelodonts (Fig. 5.6) displayed prominent caudal fins, supported by 8 to 14 scaled lobes connected by a web that, in specimens from the Devonian, is also scaled. Wilson (1998) erected a new order, the

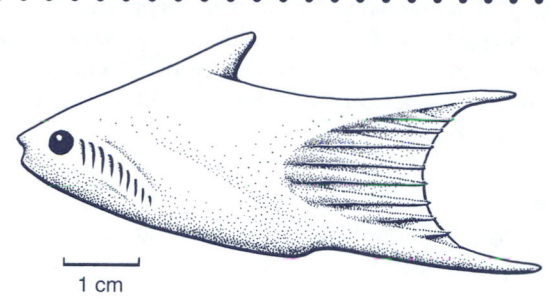

FIGURE 5.6
Representative of the †Thelodonti (Furcacaudiformes; actual size ca. 7 cm).

†Furcacaudiformes, to include the fork-tailed thelodonts; these agnaths are unusual in possessing a deeply laterally compressed body.

The †Cephalaspidomorphi

The cephalaspidomorphs are also known as the **Monorhina,** owing to the presence of a single nasal opening. Care must be taken in assigning phylogenetic significance

to monorhiny versus diplorhiny in the early evolution of vertebrates. Although monorhiny appears to be an ancestral condition in craniates, including hagfishes, it may have been secondarily derived in lampreys and several ostracoderms, as the ancestral condition for vertebrates was probably diplorhiny (Forey, 1995; Forey and Janvier, 1993).

Anaspida

The name *anaspid* is derived from the absence of a cephalic shield in these fishes. The anaspids were slender, somewhat laterally compressed fishes with multiple external branchial openings. Peculiar triradiate spines are often observed adjacent to these openings. A strongly hypocercal caudal fin is present, and an anal fin is seen in some fossils (Fig. 5.7A). Pectoral fins covered with minute scales and showing evidence of radial musculature also characterized the group. One of the most striking fossil anaspids, *†Jamoytius,* shows a well-developed pair of ventrolateral "fin folds" (Fig. 5.7B). In this and in other ostracoderm lineages, we find perhaps the earliest evidence of paired appendicular skeletal structure (see Going Deeper). Interestingly, *†Jamoytius* shows an absence of mineralized skeleton and, as such, has been considered by some to be a sister group of the lampreys (Forey, 1995; Janvier, 1996), notwithstanding the previously discussed molecular phylogenies

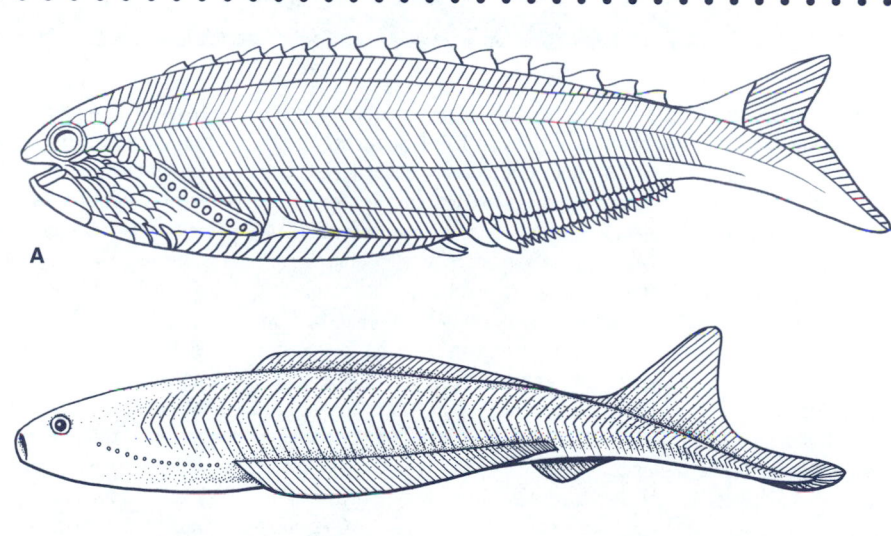

FIGURE 5.7
Representatives of the order †Anaspidiformes: **A,** *†Rhyncholepis* (actual size ca. 6 cm); **B,** *†Jamoytius* (actual size ca. 15 cm).

that support a sister group relationship of the lampreys and hagfishes.

†Osteostraci

Among the osteostracans is a large assemblage of fossils termed the Cornuata, so named for the large cornual processes that spread like wings from the cephalic shields (Fig. 5.8A,B). Several genera of noncornuate forms are also to be found in this taxon. A distinguishing feature of osteostracans is a shallow depression in the dorsal surface of the generally horseshoe-shaped cephalic shield. Termed a **cephalic field**, this structure is generally believed to have been associated with seismosensory or electrosensory capabilities. This is one group of ostracoderms in which an appendicular skeleton is evident. Just behind the cephalic shield can be found a pair of prominent pectoral fins, apparently associated with complex musculature. There is some evidence that these fins had an endoskeletal support as well. A small lobe that lies just ventral to the caudal fin, often interpreted as an anal fin, is also distinctive of the group. Osteostracan fossils reveal the degree to which soft tissue anatomy can be inferred from fossils. Cast in the underside of the cephalic shield, the brain cavity shows a division into

telencephalon, mesencephalon, metencephalon, and medulla oblongata (see Chapter 26). A number of canals, possibly representing nerve tracts, radiate away from the brain, some of them associating with the underside of the cephalic field (Fig. 5.8C).

†Galeaspida

The galeaspids were another ostracoderm group with a well-developed cephalic shield. Galeaspids were comparatively small (less than 20 cm) and had a large depression in the orobranchial cavity that communicated with a prominent medial dorsal opening. A pair of nasal openings also communicated with the orobranchial cavity. The pronounced medial cavity has been variously interpreted as similar to the blind nasohypophyseal cavity of lampreys and their extinct relatives or, more likely, homologous to the nasopharyngeal duct of hagfishes, as both communicate with the orobranchial region.

†Pituriaspida

A poorly understood group, known only from a few specimens collected in western Queensland, the pituriaspids also had a prominent cephalic shield. Being in possession of

GOING DEEPER · The Origin of Paired Appendages

Traditionally, we view the appearance of paired appendages as an evolutionary phenomenon that occurred concomitantly with the development of jaws. To a certain degree, this is true, as the earliest gnathostomes also appear to be the first vertebrates to show the paired appendicular form that we associate with the vertebrate body plan—anterior pectoral appendages and posterior pelvic ones. The earliest appendicular structures in vertebrates were thought to be unpaired, median fins (Coates, 1994). Paired appendages were believed to have first originated during the Ordovician and Silurian in the form of continuous fin folds (the fossil †Jamoytius is noteworthy for displaying a particularly well-developed continuous lateral fin fold; Fig. 5.7B) or discrete pectoral fins, as seen in anaspids and osteostracans (Fig. 5.8B). The recent discoveries of lower Cambrian chordate fossils that show definite lateral fin folds suggest, however,

that paired fins may actually predate median appendages (Janvier, 1999; Shu et al., 1999).

The earliest vertebrates, being aquatic organisms, were faced with the problems of maintaining stability as they swam (see Chapter 19). The development of serially homologous trunk musculature enabled the development of metachronal lateral undulations of the body as a means of forward propulsion. If the earliest vertebrates were burrowers or benthic creepers, the problems of maintaining stability in all three axes would be minimal. Once fishes became pelagic, however, they had to cope with problems of pitch, yaw, and roll before continuous swimming was possible (Nursall, 1962). Continuous fin folds or at least functioning pectoral appendages may have been the earliest remedies to stability, yet it is obvious that independently maneuverable paired structures set fore and aft were the best possible solution—one that

became a standard vertebrate feature, achieving its greatest degree of sophistication when vertebrates gained the land. The functional significance of paired appendages led early anatomists to speculate on the origins of these structures. According to Goodrich (1930), two theories vied for attention in the 19th century. Gegenbaur suggested that gill arches were the source of paired appendages, whereas Balfour argued that a continuous fin fold, originating at the dorsal midline and continuing around the body, splitting at the anus and ending just behind the gills, served as the source for both median and paired fin elements. Given the presence in the fossil record of examples of continuous lateral fin structures in the earliest vertebrates, Balfour's theory seemed to be the more credible.

The story of the origin of paired appendages appears to be another case in which regulatory

well-developed pectoral appendages and showing evidence of endochondral ossification, the pituriaspids are tentatively grouped with the galeaspids, osteostracans, and gnathostomes, which also display these features.

KEY POINTS AND CONNECTIONS

- The earliest of craniates, including the first vertebrates, are known to us from a diversity of fossils. Only two taxa, the hagfishes and the lampreys, have descended from these ancient lineages to be represented among the modern fish fauna. An absence of jaws and paired appendages in both of these groups is indicative of their affinities with the earliest fishes to appear in the fossil record.

- The hagfishes appear to predate the evolution of the vertebrates, although their proper phylogenetic position remains a subject of debate. A relationship of the hagfishes with the conodonts, a fossil group recently confirmed as being craniate, has also been proposed.

Included in Chapter 2 is a discussion of the earliest origins of the craniates.

- A diverse assemblage of agnathan fishes, commonly referred to as ostracoderms, are known from fossils as old as 470 million years. Their abundance in the fossil record is attributed to their capacity to form mineralized tissues in the dermis. At least two distinct classes of ostracoderms, the †Pteraspidomorphi and the †Cephalaspidomorphi, are known.

- Although lampreys appear to have features in common with some fossil cephalaspidomorphs, including an absence of scales or other bony inclusions in the skin, molecular studies have supported a sister group relationship between hagfishes and lampreys. The lampreys have evolved an unusual mode of feeding, considered by some to be a form of parasitism. The anadromous lifestyle seen among lampreys is indicative of the evolution of sophisticated osmoregulatory capabilities.

The excretory and osmoregulatory capabilities of hagfishes and lampreys are discussed in Chapter 25, and the evolution of endocrine function in these earliest of craniates is considered in Chapter 26.

BUILDING AN ICHTHYOLOGY LIBRARY

Essential to any library are the following books on the origins of fishes and texts dealing with the modern agnaths:

Ahlberg, P. E. (Ed.). 2001. *Major events in early vertebrate evolution: Paleontology, phylogeny, genetics, and development.* Taylor and Francis, London.
Hardisty, M. W. 1979. *The biology of cyclostomes.* Chapman and Hall, London.

gene sets, the *Hox* genes, have dominated the development of the vertebrate body plan (see Chapters 2 and 28). Primitive gene sets have been demonstrated to control the development of metameric (i.e., serially repeating) features of both invertebrate and vertebrate bodies. Understanding how the *Hox* genes function has enabled us to go beyond Balfour's theory to understand how the duplication of gene systems may have resulted in the formula for paired appendages that is ubiquitous among vertebrates. In this case, similar *Hox* gene sets were recruited to regulate the sequential development of fins in the pectoral and pelvic regions (Coates and Cohn, 1998; Shubin et al., 1997; Tanaka et al., 2002). This may explain why the pectoral and pelvic formula is so rigidly adhered to among vertebrates, yet Coates and Cohn (1998) claimed that the pelvic appendages, though appearing later than the pectorals, should not be interpreted as a direct serial homologue of them. For one, pelvic appendages are consistently anatomically simpler than pectorals. In their evaluation of the lateral fin fold hypothesis, Bemis and Grande (1999) concluded that it represents an "idealistic morphology" that is not consistent with modern phylogenetic approaches. As attractive as this theory may be, especially given the apparent evidence for it in the fossil record, it still remains a subject of controversy.

Interesting enough, Gegenbaur may be partially vindicated, as Tabin et al. (1999) have pointed out that the gene sets that control limb patterning are the same as those governing the development of the branchial arches. Studies on the control of gene expression in fin formation in the elasmobranch *Scyliorhinus canicula*, however, have suggested that the axis for fin development lies parallel to the body axis, supporting an origin of pectoral and pelvic appendages in paired fin folds (Tanaka et al., 2002). Whereas the number of appendages may vary widely among the arthropods, vertebrates are pretty much stuck with two pairs. One peculiar exception to this appears among the acanthodians, which show what appear to be numerous paired structures along their ventrolateral surface (see Fig. 7.1A). Manta rays (family Mobulidae) are sometimes considered another exception. Possessing an anterior pair of cephalic lobes that assist in feeding, one might claim them as the only living vertebrates with three pairs of functional appendages. These cephalic lobes are actually an anterior subdivision of the pectoral appendages (Nelson, 1994).

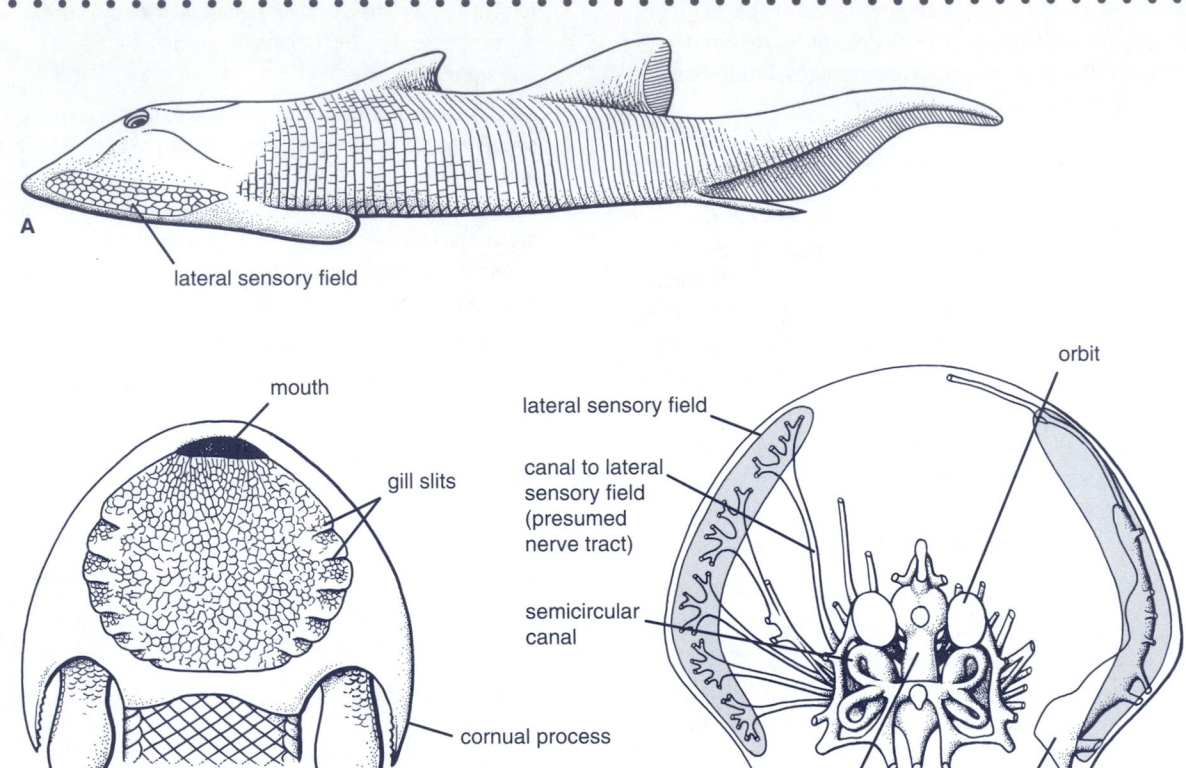

FIGURE 5.8

Representatives of †Osteostraci (order †Cephalaspidiformes): **A,** †*Ateleaspis* (actual size ca. 15 cm); **B,** Dorsal view of osteostracan head showing pectoral fins and lateral cornual processes; **C,** Cranial anatomy of osteostracan inferred from internal cast beneath head shield (ventral view).

————, and I. C. Potter. 1971. *The biology of lampreys,* 4 vols. Academic Press, New York.

Janvier, P. 1996. *Early vertebrates.* Oxford University Press, Oxford.

Jorgensen, J. M., J. P. Lomholt, R. E. Weber, and H. Malte (Eds.). 1998. *The biology of hagfishes.* Chapman and Hall, London.

Long, J. A. 1995. *The rise of fishes: 500 million years of evolution.* Johns Hopkins University Press, Baltimore. (Beautifully illustrated with numerous photographs and drawings.)

Moy-Thomas, J. A., and R. S. Miles. 1971. *Paleozoic fishes* (2nd ed.). Revised by R. S. Miles. Chapman and Hall, London.

REFERENCES

Aldridge, R. J., and P. C. J. Donoghue. 1998. Conodonts: A sister group to hagfishes? pp. 15–31. In *The biology of hagfishes,* J. M. Jorgensen, J. P. Lomholt, R. E. Weber, and H. Malte (Eds.). Chapman and Hall, London.

————, D. E. G. Briggs, E. N. K. Clarkson, and M. P. Smith. 1986. The affinities of conodonts—new evidence from the Carboniferous of Edinburgh, Scotland. *Lethaia 19:* 279–291.

————, D. E. G. Briggs, M. P. Smith, E. N. K. Clarkson, and N. D. L. Clark. 1993. The anatomy of conodonts. *Phil. Trans. Roy. Soc. Lond. B 340:* 405–421.

Allen, G. R., S. H. Midgeley, and M. Allen. 2002. *Field guide to the freshwater fishes of Australia.* Western Australian Museum, Perth.

Bardack, D. 1991. First fossil hagfish (Myxinoidea): A record from the Pennsylvanian of Illinois. *Science 254:* 701–703.

————. 1998. Relationships of living and fossil hagfishes, pp. 3–14. In *The biology of hagfishes,* J. M. Jorgensen, J. P. Lomholt, R. E. Weber, and H. Malte (Eds.). Chapman and Hall, London.

————, and E. S. Richardson, Jr. 1977. New agnathous fishes from the Pennsylvanian of Illinois. *Fieldiana (Geol.) 33:* 489–510.

Beamish, R. J. 1980. Adult biology of the river lamprey (*Lampetra ayresi*) and the Pacific lamprey (*Lampetra tridentata*) from the Pacific coast of Canada. *Can. J. Fish. Aquat. Sci. 37:* 1906–1923.

Bemis, W. E., and L. Grande. 1999. Development of the median fins of the North American paddlefish (*Polyodon spathula*), and a reevaluation of the lateral fin-fold hypothesis, pp. 41–68. In *Mesozoic fishes 2—Systematics and fossil record*, G. Arratia and H.-P. Schultze (Eds.). Verlag Dr. Friedrich Pfeil, München, Germany.

Bond, C. E., and T. T. Kan. 1973. *Lampetra (Entosphenus) minima* n. sp., a dwarfed parasitic lamprey from Oregon. *Copeia 1973:* 568–574.

———,———, and K. W. Myers. 1983. Notes on the marine life of the river lamprey, *Lampetra ayresi,* in Yaquina Bay, Oregon, and the Columbia River estuary. *Fish. Bull. 81*(1): 165–167.

Briggs, D. E. G. 1992. Conodonts: A major extinct group added to the vertebrates. *Science 256:* 1285–1286.

———, D. E. G. Clarkson, and R. J. Aldridge. 1983. The conodont animal. *Lethaia 20:* 1–14.

Brodal, A., and R. Fänge (Eds.). 1963. *The biology of* Myxine. Universitetsforlaget, Oslo.

Carroll, R. L. 1988. *Vertebrate paleontology and evolution.* Freeman, New York.

Coates, M. I. 1994. The origin of vertebrate limbs. *Development* (Suppl.): 169–180.

———, and M. J. Cohn. 1998. Fins, limbs, and tails: Outgrowths and axial patterning in vertebrate evolution. *BioEssays 20:* 371–381.

Delarbre, C., C. Gallut, V. Barriel, P. Janvier, and G. Gachelin. 2002. Complete mitochondrial DNA of the hagfish (*Eptatretus burgeri*): The comparative analysis of mitochondrial DNA sequences strongly supports the cyclostome monophyly. *Mol. Phylogen. Evol. 22:* 184–192.

Dybas, C. L. 1999. Undertakers of the deep. *Nat. Hist. 108*(9): 40–47.

Eschmeyer, W. L. (Ed.). 1998. Catalog of fishes, Vol. 3, Genera of fishes. *Calif. Acad. Sci. Spec. Publ. 1.*

Forey, P. 1995. Agnathans recent and fossil, and the origin of jawed vertebrates. *Rev. Fish Biol. Fisheries 5:* 267–303.

———, and P. Janvier. 1993. Agnathans and the origin of jawed vertebrates. *Nature 361:* 129–134.

———, and———. 1994. Evolution of the early vertebrates. *Am. Scientist 82:* 554–565.

Forster, M. E. 1998. Cardiovascular function in hagfishes, pp. 237–258. In *The biology of the hagfishes,* J. M. Jorgensen, J. P. Lomholt, R. E. Weber, and H. Malte (Eds.). Chapman and Hall, London.

Fudge, D. S., K. H. Gardner, V. T. Forsyth, C. Riekel, and J. M. Gosline. 2003. The mechanical properties of hydrated intermediate filaments: Insights from hagfish slime threads. *Biophys. J. 85:* 2015–2027.

Furlong, R. F., and P. W. Holland. 2002. Bayesian phylogenetic analysis supports monophyly of Ambulacraria and of cyclostomes. *Zool. Sci. 19:* 593–599.

Gagnier, P. Y. 1989. The oldest vertebrate: A 470 million-year-old jawless fish, *Sacabambaspis janvieri,* from the Ordovician of South America. *Nat. Geog. Res. 5:* 250–253.

Gans, C., and R. G. Northcutt. 1983. Neural crest and the origin of vertebrates: A new head. *Science 220:* 268–274.

Gill, H. S., C. B. Renaud, F. Chapleau, R. L. Mayden, and I. C. Potter. 2003. Phylogeny of living parasitic lampreys (Petromyzontiformes) based on morphological data. *Copeia 2003:* 687–703.

Goodrich, E. S. 1930. *Studies on the structure and development of vertebrates.* Constable, London. Reprinted 1958, Dover, New York.

Gorbman, A., and A. Tamarin. 1985. Early development of oral, olfactory, and adenohypophyseal structures of agnathans and its evolutionary implications, pp. 165–185. In *Evolutionary biology of primitive fishes,* R. E. Foreman, A. Gorbman, J. M. Dodd, and R. Olsson (Eds.). Plenum, New York.

Hardisty, M. W. 1979. *Biology of the cyclostomes.* Chapman and Hall, London.

———, and I. C. Potter. 1971. Paired species, pp. 249–277. In *The biology of lampreys,* Vol. 1. Academic Press, London.

Hedges, S. B. 2001. Molecular evidence for the early history of living vertebrates, pp. 119–134. In *Major events in early vertebrate evolution: Paleontology, phylogeny, genetics, and development,* P. E. Ahlberg, (Ed.). Taylor and Francis, London.

Honma, Y. 1998. Asian hagfishes and their fisheries biology, pp. 45–56. In *The biology of hagfishes,* J. M. Jorgensen, J. P. Lomholt, R. E. Weber, and H. Malte (Eds.). Chapman and Hall, London.

Janvier, P. 1986. Le nouvelles conceptions de la phylogenie de la classification des "Agnathes" et des Sarcoptérygiens, pp. 123–138. In Les poissons: Classification et phylogenèse, Y. Francois and M. L. Beuchot (Eds.). Oceanis 12(3).

———. 1996. *Early vertebrates.* Oxford University Press, Oxford.

———. 1999. Catching the first fish. *Nature 402:* 21–22.

———. 2001. Ostracoderms and the shaping of the gnathostome characters, pp. 172–186. In *Major events in early vertebrate evolution: Paleontology, phylogeny, genetics, and development,* P. E. Ahlberg, (Ed.). Taylor and Francis, London.

Jorgensen, J. M. 1998. Structure of the hagfish inner ear, pp. 557–563. In *The biology of hagfishes,* J. M. Jorgensen, J. P. Lomholt, R. E. Weber, and H. Malte (Eds.). Chapman and Hall, London.

Kan, T. T., and C. E. Bond. 1981. Notes on the biology of the Miller Lake lamprey *Lampetra (Entosphenus) minima. Northwest Science 55*(1): 70–74.

Kardong, K. 2002. *Vertebrates: Comparative anatomy, function, evolution* (3rd ed.). McGraw-Hill, New York.

Kemp, A., and R. S. Nicoll. 1996. A histochemical analysis of biological residues in conodont elements. *Modern Geol. 20:* 287–302.

Krejsa, R. J., and H. C. Slavkin. 1987. Earliest craniate teeth identified: The hagfish–conodont connection. *J. Dent. Res. 66* (Spec. Issue): 144.

———, D. Bringas, Jr., and H. C. Slavkin. 1990. A neontological interpretation of conodont elements based on agnathan cyclostome tooth structure, function, and development. *Lethaia 19:* 279–291.

Lockwood, A. P. M. 1962. The osmoregulation of Crustacea. *Biol. Rev. 37:* 257–305.

Lomholt, J. P., and F. Franko-Dossar. 1998. The sinus system of hagfishes—lymphatic or secondary circulatory system? pp. 259–272. In *The biology of hagfishes,* J. M. Jorgensen, J. P. Lomholt, R. E. Weber, and H. Malte (Eds.). Chapman and Hall, London.

Lorion, C. M., D. F. Markle, S. B. Reid, and M. F. Docker. 2000. Redescription of the presumed-extinct Miller Lake lamprey, *Lampetra minima. Copeia 2000:* 1019–1028.

Mallatt, J. 1985. Reconstructing the life cycle and the feeding of ancestral vertebrates, pp. 59–68. In *Evolutionary biology of primitive fishes,* R. E. Foreman, A. Gorbman, J. M. Dodd, and R. Olsson (Eds.). Plenum Press, New York.

———, J. Sullivan, and C. J. Winchell. 2001. The relationship of lampreys to hagfishes: A spectral analysis of ribosomal DNA sequences, pp. 106–118. In *Major events in early vertebrate evolution: Paleontology, phylogeny, genetics, and development,* P. E. Ahlberg (Ed.). Taylor and Francis, London.

Martini, F. H. 1998. The ecology of hagfishes, pp. 57–77. In *The biology of hagfishes,* J. M. Jorgensen, J. P. Lomholt, R. E. Weber, and H. Malte (Eds.). Chapman and Hall, London.

———, M. Lesser, and J. B. Heiser. 1997. Ecology of the hagfish, *Myxine glutinosa* L., in the Gulf of Maine: II. Potential impact on benthic communities and commercial fisheries. *J. Exp. Mar. Biol. Ecol. 214:* 97–106.

Moy-Thomas, J. A., and R. S. Miles. 1971. *Paleozoic fishes* (2nd ed.). Revised by R. S. Miles. Chapman and Hall, London.

Nelson, J. S. 1994. *Fishes of the world* (3rd ed.). Wiley, New York.

Northcutt, R. G., and C. Gans. 1983. The genesis of neural crest and epidermal placodes: A reinterpretation of vertebrate origins. *Quart. Rev. Biol. 58:* 1–28.

Nursall, J. R. 1962. Swimming and the origin of paired appendages. *Amer. Zool. 2:* 127–141.

Potter, I. C. 1980. The Petromyzontiformes, with particular reference to paired species. *Can. J. Fish. Aquat. Sci. 37:* 1595–1615.

Sansom, I. J., M. M. Smith, and M. P. Smith. 2001. The Ordovician radiation of vertebrates, pp. 156–171. In *Major events in early vertebrate evolution: Paleontology, phylogeny, genetics, and development,* P. E. Ahlberg (Ed.). Taylor and Francis, London.

Shu, D.-G., H.-L. Luo, S. C. Morris, X.-L. Zhang, S.-X. Hu, L. Chen, J. Han, M. Zhu, Y. Li, and L.-Z. Chen. 1999. Lower Cambrian vertebrates from south China. *Nature 402:* 42–46.

Shubin, N., C. Tabin, and S. Carroll. 1997. Fossils, genes, and the evolution of animal limbs. *Nature 388:* 639–648.

Smith, H. 1932. Water regulation and its evolution in fishes. *Quart. Rev. Biol. 7:* 1–26.

Smith, M. M. 1993. A developmental model for evolution of the vertebrate exoskeleton and teeth: The role of cranial and trunk neural crest. *J. Evol. Biol. 27:* 387–448.

———, and B. K. Hall. 1990. Development and evolutionary origins of vertebrate skeletogenic and odontogenic tissues. *Biol. Rev. 65:* 277–373.

Tabin, C. J., S. B. Carroll, and G. Panganiban. 1999. Out on a limb: Parallels in vertebrate and invertebrate limb patterning and the origin of appendages. *Amer. Zool. 39:* 650–663.

Tanaka, M., A. Münsterberg, W. G. Anderson, A. R. Prescott, N. Hazon, and C. Tickle. 2002. Fin development in a cartilaginous fish and the origin of vertebrate limbs. *Nature 416:* 527–531.

Turner, S. 1986. *Thelodus macintoshi* Stetson 1928, the largest known thelodont (Agnatha: Thelodonti). *Breviora 486.*

Vladykov, V. D., and E. Kott, 1979. Satellite species among the holarctic lampreys (Petromyzontidae). *Can. J. Zool. 57:* 860–867.

Walvig, F. 1963. The gonads and the formation of the sexual cells, pp. 530–580. In *The biology of* Myxine, A. Brodal and R. Fänge (Eds.). Universitetsforlaget, Oslo.

Welsch, U., and I. C. Potter. 1998. Dermal capillaries, pp. 273–283. In *The biology of hagfishes*, J. M. Jorgensen, J. P. Lomholt, R. E. Weber, and H. Malte (Eds.). Chapman and Hall, London.

Wilson, M. V. H. 1998. The Furcacaudiformes: A new order of jawless vertebrates with thelodont scales, based on articulated Silurian and Devonian fossils from northern Canada. *J. Vert. Paleontol. 18*(1): 10–29.

———, and M. W. Caldwell. 1993. New Silurian and Devonian fork-tailed thelodonts are jawless vertebrates with stomachs and deep bodies. *Nature 361:* 442–444.

Yalden, D. W. 1985. Feeding mechanisms as evidence for cyclostome monophyly. *Zool. J. Linn. Soc. 84:* 291–300.

Zimmer, C. 2000. In search of vertebrate origins: Beyond brain and bone. *Science 287:* 1576–1579.

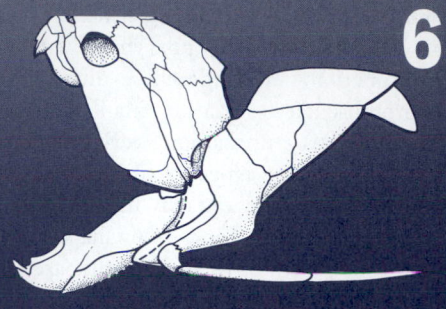

6

THE GNATHOSTOMES: FISHES WITH JAWS

ORIGIN OF THE JAWS

GNATHOSTOMES AT THE SILURIAN/ORDOVICIAN TRANSITION

Class †Placodermi
    Order †Rhenaniformes
    Order †Antiarchiformes (Pterichthyes)
    Order †Acanthothoraciformes (Palaeacanthaspidoidei)
    Order †Ptyctodontiformes
    Order †Petalichthyformes
    Order †Phyllolepiformes
    Order †Arthrodiriformes

CLASS CHONDRICHTHYES (ELASMOBRANCHIOMORPHI)

Evolution of the Jaw Suspensorium in Gnathostome Fishes
Subclass Euchondrocephali
    †Iniopterygii
    †Paraselachii
    Holocephali
    Superorder Holocephalimorpha
Subclass Elasmobranchii
    Cohort †Palaeoselachii
    Cohort Euselachii

The evolution of jaws among the early fishes undoubtedly ranks as an event that was crucial to the success of the vertebrates. Novel modes of feeding facilitated the successful invasion of new habitats, first in the aquatic realm and ultimately on land. The early Silurian witnessed a veritable explosion of gnathostome (jawed) fishes. Two of these taxa, the placoderms and the acanthodians, are no longer around, but they are well represented in the fossil record. Two other groups, the chondrichthyans and the bony fishes, constitute the vast majority of modern fish taxa. In this chapter, we will take a closer look at the origins of gnathostomes, with particular attention paid to the extinct placoderms and to both extinct and modern chondrichthyan fishes.

ORIGIN OF THE JAWS

In comparing the diversity of jawless vertebrates with those that possess jaws, it is obvious that the evolution of jaws, tied as it is to success in feeding, was of momentous importance. The early agnaths probably had somewhat restricted feeding strategies, primarily consisting of the filtration of small organisms out of the water or from the bottom muck. This is not to imply that macrocarnivory is the exclusive domain of the gnathostomes—witness the success of such specialized predator/scavengers as the hagfishes. The rudimentary jawlike apparatus of these agnaths is not considered homologous with the gnathostome jaw mechanism, nor does it possess the range of function seen in gnathostomes. The evolution of jaws, with all their diversity of form and function, permitted a tremendous expansion of feeding opportunities in vertebrates. Most of the early agnaths probably were bottom dwellers. With the new jaw mechanism, vertebrates greatly expanded the range of the aquatic realm available for exploitation. Fossil crinoids (sea lilies and feather stars of the phylum Echinodermata) that were contemporaries of early jawed fishes show greater evidence of arm regeneration than earlier fossils, possibly reflecting increased instances of nonlethal attack by predatory gnathostomes (Baumiller and Gahn, 2004).

As the early gnathostomes ranged further and faster in pursuit of prey, natural selection favored features that contributed to greater mobility and flexibility. Heavy, bony armor was deemphasized, and a greater premium was placed on the development of paired appendages for control and stability. Although the fossil record reveals that jaws and paired appendages did not appear at the same time—in some of the fossil agnaths discussed in Chapter 5 there is evidence of paired fin structure—they are generally viewed as synapomorphic characteristics within the gnathostomes.

The serial homology of jaws with the immediately posterior branchial arches has long been recognized (Gegenbaur, 1872; Goodrich, 1930; Gregory, 1929). Both jaws and branchial arches arise from neural crest elements in the embryo. It was reasoned that the evolution of jaws was a relatively straightforward process of enlargement of both the skeletal elements and their associated musculature: What was originally a gill arch became a fully functional biting and grasping device. The story is not as simple as that, however. For one, modern agnaths are in possession of **external branchial arches** that surround and enclose the gill pouches. Their homology with the **internal branchial arches** of gnathostomes is questionable (Mallatt, 1984), and hence the modern agnaths are peripheral to the story of jaw origins. Furthermore, gnathostome jaws are innervated by cranial nerve V (trigeminal), which in agnathans innervates an area anterior to the pharynx that is not associated with gill function. The inherent evolutionary conservatism of neuromuscular association facilitates the derivation of homologous structures among vertebrates.

There is one aspect of the story of the origin of jaws in which modern agnaths may figure significantly. Expression of homeotic (*Hox*) genes (see Chapters 2 and 28) occurs in the development of the mandibular arch, which contributes to the formation of the velum in lamprey embryos. Loss of *Hox* expression in the developing mandibular arch, as observed in gnathostomes, may have been a key event in the development of jaws (Cohn, 2002).

The question of the adaptive significance of transitional forms must also be addressed. Did "proto-jaws" have adaptive value before they were fully formed in the gnathostomes? Mallatt (1996) suggested that the jaws evolved first as a mechanism to promote gill ventilation rather than as a feeding device. According to his scenario, the proto-jaw developed as a mandibular arch with enlarged adductor muscles that could rapidly close and seal the oral cavity while accelerating the flow of water past the gills. Enlarged hypobranchial muscles attached to the mandibular arch would enable rapid mouth opening, and here lies the significance of this mechanism as a feeding device. The strong suction coming from the rapid opening of the mouth, now bordered by the mandibular arches, would bring in prey items that could be grasped by these same arches. Eventually, dentition became associated with the arches to improve prey grasping abilities, and the jaws subsequently commenced a remarkably complex evolutionary journey (Fig. 6.1). It is possible that the homeotic gene cluster replication associated with the evolution of the Craniata (see Chapter 2) may also be implicated in the modifications of the head that are associated with the evolution of jaws (Monastersky, 1996).

GNATHOSTOMES AT THE SILURIAN/ ORDOVICIAN TRANSITION

Studies have suggested that the ancestors of the gnathostomes had a poorly calcified endoskeleton and a body covering of tiny scales rather than heavy bony plates (Carroll, 1988; Smith and Hall, 1990). Among the agnathan fossil forms, the thelodonts appear as the most likely ancestors of the gnathostomes (van der Brugghen and Janvier, 1993; Wilson and Caldwell, 1993). At the close of the Ordovician, a moderate glaciation event occurred that resulted in a significant drop in sea level, as more water was taken up into the polar ice caps. The fossil record of this time records a decline

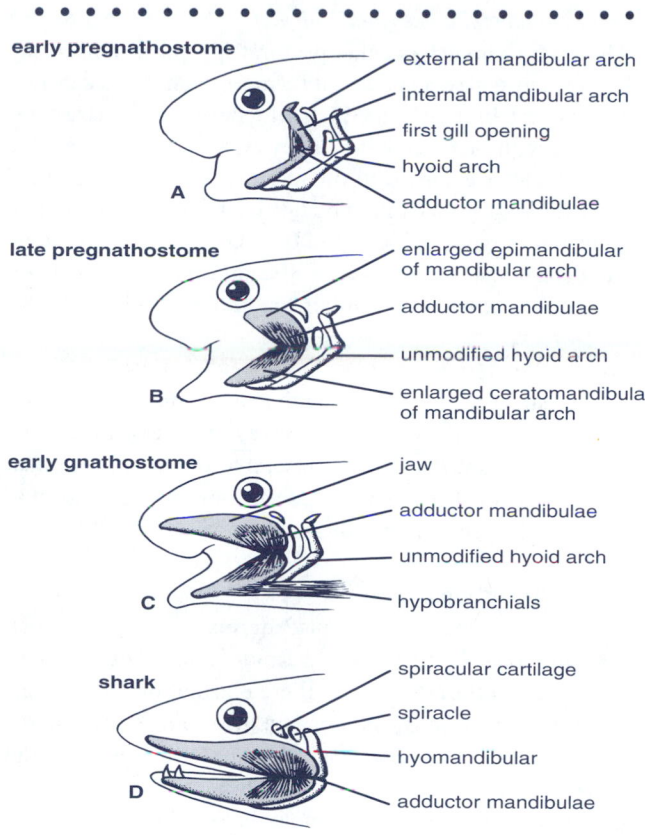

FIGURE 6.1

Possible scenario for the evolution of jaws in vertebrates. (From Mallatt, 1996.)

in species diversity that rapidly rebounded at the beginning of the Silurian (about 345 MYA). A proliferation of new species appeared, including both agnaths and gnathostomes (Janvier, 1996). Four groups of jawed vertebrates appeared at this time—placoderms, chondrichthyans (early shark relatives), acanthodians (sometimes referred to as "spiny sharks"), and the bony fishes (formerly known as the class Osteichthyes). The Early Silurian rocks of the low-lying islands offshore from Estonia have produced the earliest representatives of the acanthodians and actinopterygians. The acanthodians demonstrate uncertain affinities, but they are currently considered a sister group of the bony fishes. The actinopterygians, a group of bony fishes also known as the "ray-finned" fishes, become, as we shall see, the most diverse vertebrate taxon of all.

Nelson (1994) classified the vertebrates that bear jaws in the superclass **Gnathostomata**. Originating in the Silurian, or possibly as early as the late Ordovician (Sansom et al., 1996), gnathostomes flourished in the Devonian, and

several lineages survived into the Carboniferous and beyond, when most clades of jawless vertebrates dwindled. Gnathostomes include not only the modern fishes, but the tetrapods as well. Gnathostome fossils share derived characteristics that are not present in agnaths: Jaws are present; gill tissue, branchial arteries, and branchial nerves are external to the gill arches; pectoral and pelvic fins are present; the inner ear is characterized by three semicircular canals; and the branchial skeleton is not fused to the neurocranium. The notochord, persistent in early forms, is partially or completely replaced by vertebrae in modern gnathostomes.

Gnathostomes include two main groups of modern fishes, the cartilaginous fishes and the bony fishes. The cartilaginous fishes—sharks and rays (the elasmobranchs) and chimaeras—are placed in the class **Chondrichthyes**. The bony fishes are considered by Nelson (1994) to constitute the taxon **Euteleostomi**. That taxon is intermediate between a grade and a class and replaces the earlier term "Osteichthyes." There is disagreement on how some groups are associated, however. For instance, Jarvik (1980) and Lagios (1979, 1982) hold the somewhat unorthodox view that the lungfishes and the coelacanth (*Latimeria*) should be aligned with the elasmobranchs. Moreover, there are differing views on the phylogenetic affinities of the two major extinct lines of gnathostomes—the acanthodians and the placoderms. Earlier classifications presented acanthodians as members of the class †Placodermi. Jarvik (1980) considered them to have chondrichthyan affinities, whereas others consider them to be closely related to the bony fishes (Miles, 1973).

The justification of the association depends on the emphasis given to certain features of anatomy. Nelson (1994) has assigned the †Acanthodii to a monophyletic group, the Teleostomi, which includes the Actinopterygii and the Sarcopterygii. Most cladistic phylogenies place the acanthodians as a sister class to osteichthyans (i.e., actinopterygians and sarcopterygians; Pough et al., 1999). The phylogeny depicted in Figure 6.4 represents a consensus view on the diversification of the higher groups of gnathostomes.

Class †Placodermi

The placoderms are distinguished by the dermal armor that covers the head and anterior part of the trunk. They share features with bony fishes, sharks, rays, and chimaeras. Schaeffer and Williams (1977) perceived no close relationship with other major groups of fishes and expressed the opinion that the placoderms are a sister group of the remainder of the gnathostomes. Although some have claimed affinities between the placoderms and the chondrichthyan fishes, Maisey (1986) considered the placoderms a problematic group and did not include them with the Chondrichthyes. In the following treatment, the †Placodermi will be treated as a class distinct from the other major groups of gnathostomes.

Placoderms are rather diverse in structure and body form; some are sharklike, and others are flattened like rays. The bony armor is divided by a "neck joint" between the armor of head and body, so that, apparently, the head could be raised (Fig. 6.2A). Scales or small tessellated plates (**tesserae**) are present in many species. The endoskeleton is at least partially ossified. Pectoral and pelvic fins are present, and the caudal fin is heterocercal in most. The eyes are typically rather far forward. Placoderms are peculiar gnathostomes in that dentition, in the sense that we understand tooth structure, is usually not present. Primitive bony plates attached to the jaw cartilage formed shearing or biting surfaces. Recent studies on one of the more derived placoderm groups, the arthrodires (see further), indicated that they did indeed form teeth in the jaw, suggesting a more complex evolutionary history for dentition than was previously thought (see Chapter 23). Most placoderms are known from the Devonian, but some appeared in the Upper Silurian, and some persisted into the Carboniferous.

The following groups of placoderms (referred to here as orders) are given different taxonomic ranks by various authors; some appear as subclasses or classes. Authorities differ on the number of recognized taxa of placoderms. Nelson (1994) recognized the 7 orders discussed here. Denison's (1978) classification included 11 orders, whereas Janvier (1996) recognized more than 15 taxa.

Order †Rhenaniformes

This order is composed of depressiform fishes with terminal mouths that lack gnathal plates (Fig. 6.2B). Instead of toothlike structures on plates, these fossils show tubercles on the palatoquadrate, which is connected to the cranium by a hyomandibular, much like the jaw linkage in elasmobranchs. The mouth was apparently protrusible. The overall appearance of the †Rhenaniformes is much like that of rays, with broad, flat pectoral fins that border the rather robust body from the eyes to the pelvic fins. The tail is diphycercal. Eyes and nostrils are dorsal, and the nostrils are set almost between the eyes.

There are a few large plates on the head, but much of the head and all of the body and fins are covered by a mosaic of small, scalelike plates. Part of the cranium is ossified. Vertebral centra appear to be present, but they are actually fused neural and hemal arches (Carroll, 1988). Most of the fossils are small fishes, but some are a meter or more in length, and they were common in the Devonian (Long, 1995).

Order †Antiarchiformes (Pterichthyes)

This group has such peculiar pectoral appendages that some ichthyologists have considered them to constitute a separate class of vertebrates (Fig. 6.2C). The pectorals are large and covered with plates of bone. They articulate with the large body shield, are jointed in the middle, and have an ossified or calcified endoskeleton. The head is relatively small. The body of some species is covered by overlapping scales. Some had diverticula extending from the gill cavity into the body that may have been used in air breathing (Denison, 1978).

Order †Acanthothoraciformes (Palaeacanthaspidoidei)

These fishes are unusual in that the head armor of some species consists of large plates, whereas in others the dermal plates are so small that they resemble scales. These small depressiform fishes from the Lower Devonian are also distinguished by the absence of a neck joint.

Order †Ptyctodontiformes

This order contains small placoderms, usually less than 200 mm long. Head and body armor is not as extensive or as heavy as in the arthrodires; there is armor only on the anterior part of the body, and none on the snout. A plate on the cheek apparently covers the gill opening. In body form and in several other characteristics, including the tooth plates, the Ptyctodontiformes resemble holocephalans. Pectoral fins are large, as are the pelvic fins. As in holocephalans, prepelvic claspers are present, but here they are tipped with dermal scutes.

Order †Petalichthyformes

These are characterized by numerous head plates; scales on the body, on the pectoral fins, and in the snout region; and large lateral spines in front of the pectorals. The eyes are dorsal but anterior. The caudal fin is thought to be diphycercal.

Order †Phyllolepiformes

This order is represented by a single depressiform genus with a reduced number of armor plates. The nuchal plate on the head and the median dorsal body plate are enlarged. The neck joint does not appear to have been movable. The snout region seems to be unarmored.

Order †Arthrodiriformes

This order, also referred to as †Coccostei, contains most of the known placoderms (Fig. 6.2A). The arthrodires are characterized by a heavily armored head and forebody, and some have impressive, tusklike gnathal plates that form biting surfaces. The gills open between the head and the body armor, and the slit usually is covered by a bony plate.

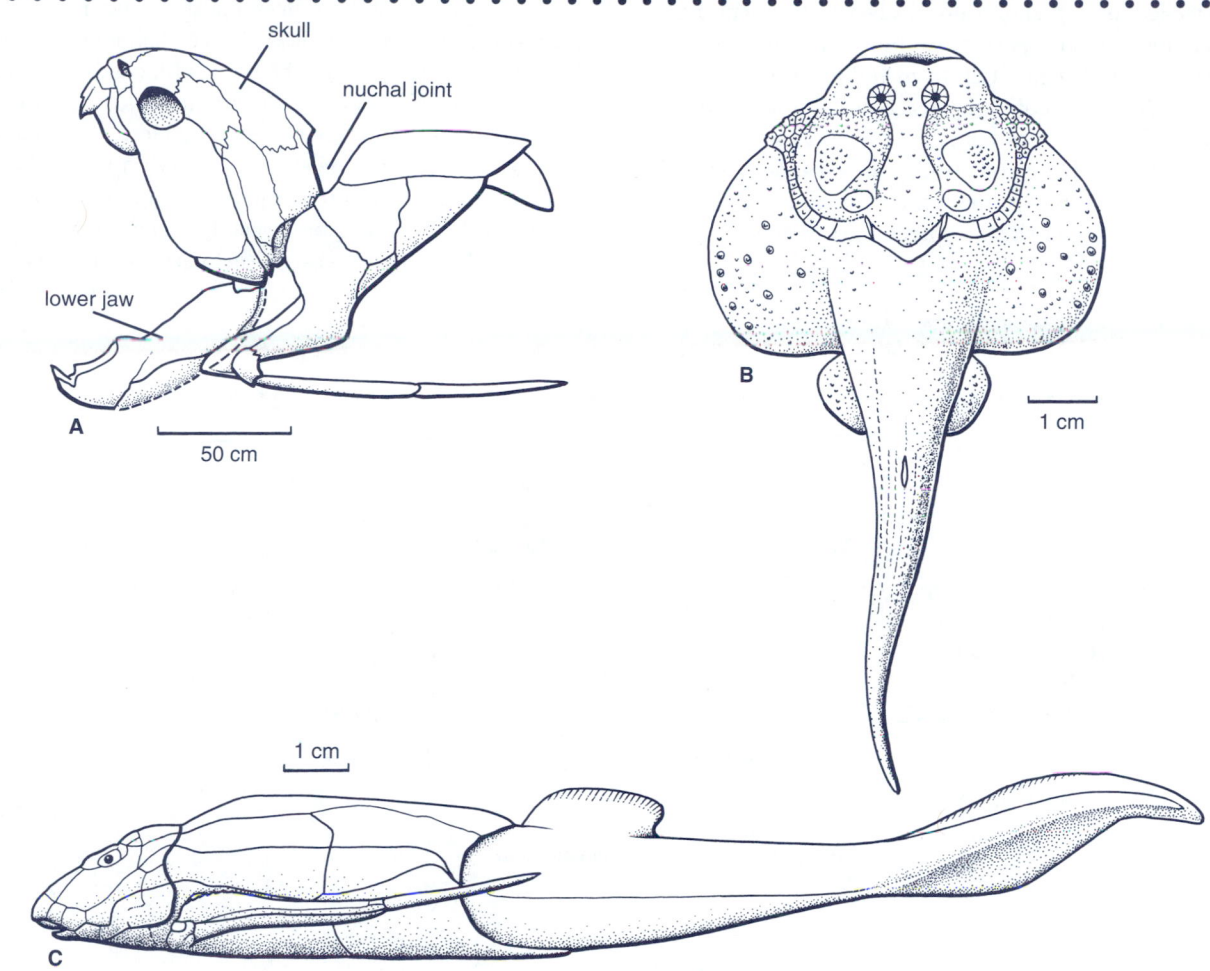

FIGURE 6.2

A, Representative of the arthrodires †*Coccosteus;* actual size ca. 35 cm); skull of †*Dunkleosteus,* showing how nuchal gap facilitates expansion of the gape (actual size ca. 100 cm); **B,** Dorsal view of representative of the order †Rhenaniformes †*Gemuendina* showing convergence with the batoids (actual size ca 12 cm); **C,** Representative of the †Antiarchiformes †*Bothriolepis;* actual size ca. 20 cm).

A few genera contain species that reached more than 6 m, but most specimens are considerably smaller.

CLASS CHONDRICHTHYES (ELASMOBRANCHIOMORPHI)

Evolution of the Jaw Suspensorium in Gnathostome Fishes

Gnathostome fishes have evolved a diversity of jaw configurations with respect to the relationship of the jaw (mandibular arch), jaw support elements (mainly the hyoid arch), and

the cranium (refer to Chapter 3, the Splanchnocranium and Jaws). In the **paleostylic** condition, seen in modern agnaths, arch elements are not directly associated with the skull in a manner that would enable them to function as jaws (Fig. 6.3). The earliest gnathostomes, such as the placoderms and acanthodians, display an **autostylic (euautostylic)** condition, in which the mandibular arch (the upper palatopterygoid and lower Meckel's cartilages) is suspended from the cranium, with no support derived from the hyoid arch. From this ancestral condition, a number of derivative forms arose. The **amphistylic** condition, in which the jaw is attached to the cranium anteriorly via a ligament from the palatoquadrate and posteriorly by the hyomandibular, is seen in some early chondrichthyan and teleostome fishes. Among the fossil chondrichthyans, a condition known as **autodiastylic** appears as ancestral to the elasmobranchs and holocephalans. Here, the hyoid arch is still not involved in jaw suspension, but it has developed elements (the cerato- and epihyal) to support a soft opercular covering (Grogan et al., 1999).

From the autodiastylic condition, the jaw support mechanisms diverged in the evolution of elasmobranchs and holocephalans (Fig. 6.3). The **hyostylic** condition evolved in the lineages of elasmobranchs culminating in the evolution of the modern sharks, skates, and rays. Here, the hyoid arch assumes a significant role in supporting the mandibular arch elements. Maisey (1980) pointed out that amphistyly is not qualitatively different from hyostyly, and perhaps should be considered a type of hyostyly, because the hyomandibular

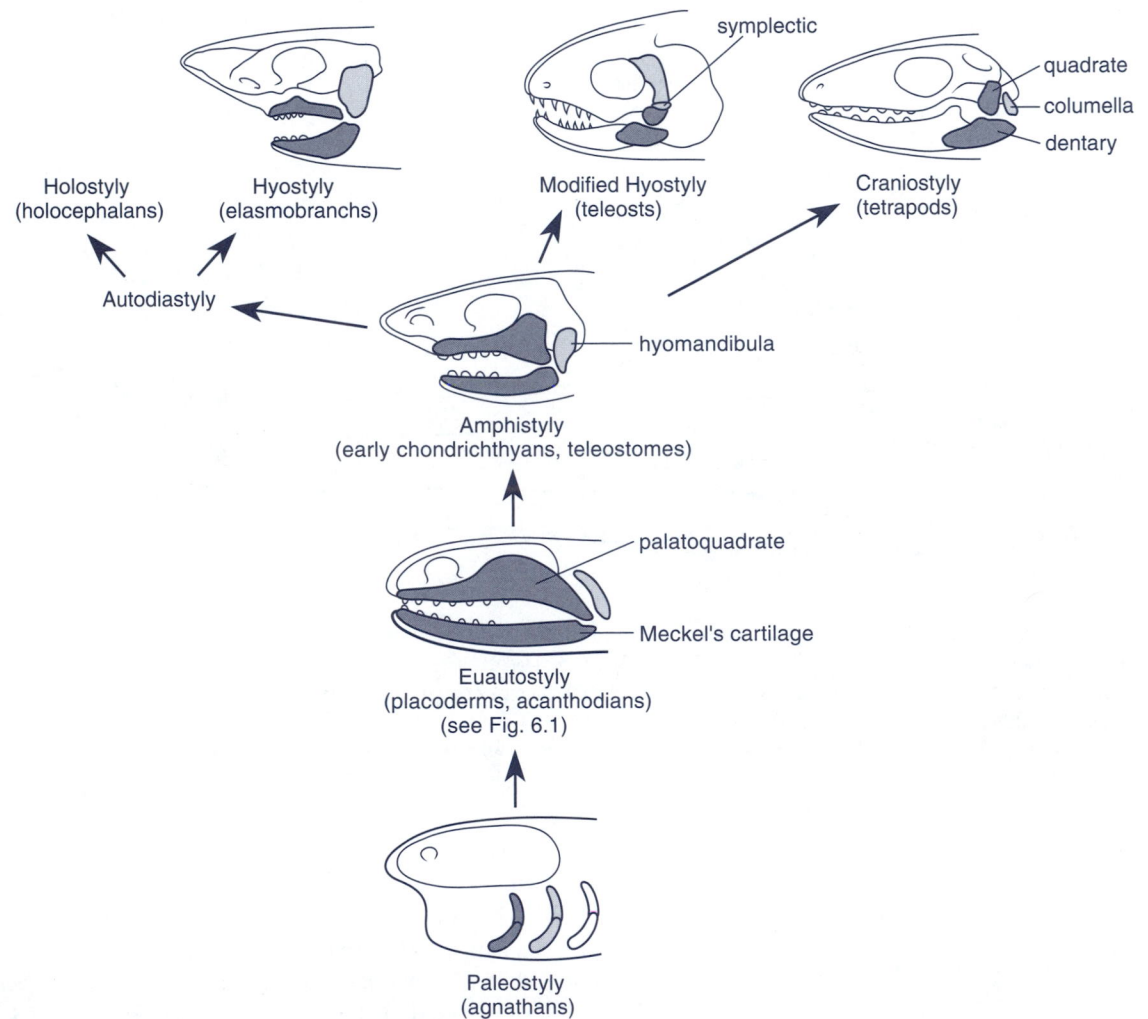

FIGURE 6.3

Evolution of jaw suspension in vertebrates. Mandibular arch elements are represented by dark shading; hyoid elements, including the symplectic, by light shading. (From Kardong, 1998; Grogan et al., 1999.)

aids in the suspension of both types. Most bony fishes show a modified version of the hyostylic condition. As discussed in Chapter 3, the skull of bony fishes incorporates significant dermal elements in the jaw, and the hyostylic condition is accordingly modified. From their analyses of the origins of chimaerid jaw suspensoria, Grogan et al. (1999) claimed that amphistyly is derived from hyostyly and that the autodiastylic condition is ancestral to all other gnathostome suspensorial designs.

There is a graded transition among species from small hyomandibular cartilages, with little suspensory function, to large hyomandibulars, with major suspensory function. Rays are only hyostylic, but most sharks at least have ligamentous connections between the palatoquadrate cartilage and the neurocranium. Connections, either ligamentous or firm, range in location in various species from rostral to postorbital positions. Some kinds of connections restrict movement of the palatoquadrate, whereas others allow ventral and forward movement.

The sharks that possess an orbital process on the palatoquadrate and show an attachment to the orbit include *Notorhynchus*, *Chlamydoselachus*, *Hexachus*, *Heptranchias*, the squaloids, pristiophoroids (sawsharks), and *Squatina*. Maisey (1980) called this kind of jaw suspension **orbitostylic** and suggested that sharks having such a suspension are more closely related to each other than to other sharks. All of these have ethmoidal or postorbital connections to the neurocranium as well as the hyomandibular and orbital connections.

The condition known as **holostyly** evolved from the autodiastylic euchondrocephalans and is demonstrated in the Holocephali. In this condition, the upper jaw is broadly fused to the cranium; yet the monophyly of the chondrichthyans, implying a close phylogenetic relationship between the sharks and chimaeroids, is still claimed (Fig. 6.4; Lund and Grogan, 1997; Grogan et al., 1999). Our classification here follows Grogan et al. (1999) in incorporating the Holocephali into their subclass Euchondrocephali. Nelson (1994) designated the Holocephali and Elasmobranchii as subclasses within the class Chondrichthyes—an arrangement that is probably more familiar to students of fishes.

· · · · · · · · · · · · · ·
Subclass Euchondrocephali

†Iniopterygii

The order †Iniopterygiformes contains Paleozoic (Carboniferous) fishes with the pectoral fin attached to the nuchal region (Fig. 6.5A) and characterized by prominent spines, often armed with hooks. Somewhat resembling the modern chimaeras (Holocephali), the iniopterygians have teeth more closely resembling those of elasmobranchs—the dentition consisting of denticles arranged in rows.

†Paraselachii

Included here is the order †Petalodontiformes (Petalodontida). These fishes are known from the Lower Carboniferous to the Upper Permian. They have been placed with the Holocephali by some authors. *†Janassa* is depressiform, with a prominent rostrum and large, horizontally oriented paired fins. *†Belantsea* (Fig.6.5B) is compressiform, with a deep body and a large head with high, rounded dorsal fins, the first of which is large and originates on the head above the orbits. The rounded pelvic fins are set near the caudal; there is no anal fin; and the caudal fin is structurally heterocercal but externally resembles a homocercal tail. Teeth are large and serrate, and the gills are placed posterior to the neurocranium.

Holocephali

Mention of the class Chondrichthyes inevitably brings to mind the more conspicuous members of the group—the sharks. A much more obscure but no less fascinating group are the chimaerids, most commonly referred to as *ratfishes* because of their long, slender tail. In these fishes, the skeleton is cartilaginous and the notochord persistent. As mentioned earlier, the skull displays a holostylic jaw suspension; branchial arches are all placed below the neurocranium; and the gill openings are covered by fleshy opercula. Except in a few fossil lines, the teeth are grinding plates with no enamel. In living forms, there is no spiracle and no cloaca, and the oviducts open to the exterior separately. There is usually a strong spine at the leading edge of the first dorsal fin, supported by a synarcual plate formed from neural arch elements.

Holocephali are known from the Upper Devonian to the present. A diverse assemblage of extinct forms, known as **bradyodonts** (so called because the teeth are thought to have slow growth and replacement), has been treated variously by modern authors. Depending on the evaluation of certain aspects of their structure, some are placed closer to sharks than to chimaeras. There are several extinct taxa that resemble holocephalans but are placed variously in and out of the group by paleoichthyologists. Many of the bradyodonts share a peculiar feature with modern chimaerids—a clasper-type appendage on top of the skull.

Holocephalans show many remarkable resemblances to the ptyctodontid placoderms and might be related to them. The ranking of groups within a classification scheme differs among students of the subclass. For instance, Patterson (1965) included many of the extinct groups as suborders

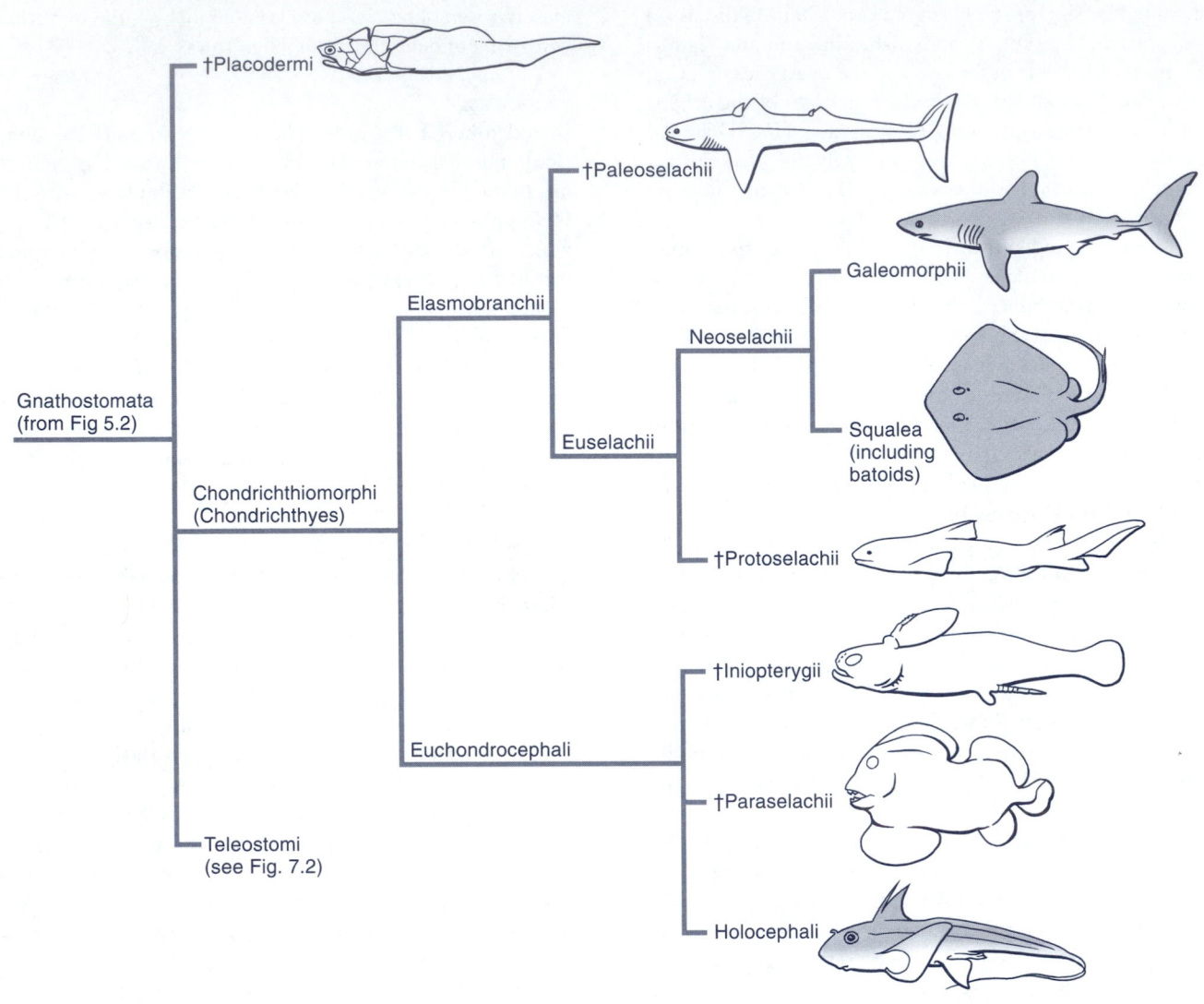

FIGURE 6.4

Phylogeny of gnathostome fishes, focusing on the Chondrichthyes. (From Nelson, 1994; Compagno, 1991; deCarvalho, 1996; Grogan et al., 1999.)

under Chimaeriformes, whereas Lund (1986) considered most of those groups as separate orders. The most recent revisions of the holocephalans (Didier, 1995; Lund and Grogan, 1997; Grogan et al., 1999) emphasized the evolution of the jaw suspension in deducing the phylogenetic relationships of the chondrichthyan fishes.

Superorder Holocephalimorpha
This group, distinguished by holostylic jaw suspension and dentition that consists of grinding plates, includes the modern Holocephali.

Order Chimaeriformes
This order contains all the modern chimaeras, which are known from the Lower Jurassic to the present. These living chimaerids are among the most bizarre of fishes, as indicated by their various common names. These names are based on the anatomical features or general appearance of the species and include the aforementioned *ratfish*, *spookfish*, *ghost shark*, *chimaera*, *rabbitfish*, and *elephantfish*. Modern species are usually placed in three families, all marine. Species are usually between 60 cm and 2 m long.

FIGURE 6.5
A, Representative of the †Iniopterygii (actual size ca. 30 cm; after Zangerl and Case, 1973); **B,** Representative of the order †Petalodontiformes †*Belantsea;* actual size ca. 29 cm; after Lund, 1989); **C,** representative of the order Chimaeriformes *Callorhynchus.*

DIAGNOSIS OF THE CHIMAERIFORMES

- Inferior placement of mouth, with respiratory water taken in mainly through nostrils.
- Two dorsal fins, with first preceded by a sharp, venomous spine.
- Caudal fin heterocercal (Callorhynchidae), diphycercal (Chimaeridae), or leptocercal (Rhinochimaeridae).
- Males with pelvic claspers.

Sources: Nelson, 1994; Lund and Grogan, 1997

Members of the family **Chimaeridae** are equipped with a frontal clasper on top of the head and a set of abdominal claspers (**tenacula**) in pockets just anterior to the pelvic fins. Fishes in this widespread family are found in moderate-depth or shallow water.

The **Rhinochimaeridae** are characterized by a long, depressed, and pointed snout, from which they take the name "longnose chimaeras." They have well-developed frontal claspers and reduced paired claspers. This family contains deepwater forms of wide distribution.

The plownose chimaeras (**Callorhynchidae**) have a peculiar snout that turns back on itself ventrally, forming a flattened appendage just in front of the mouth (Plate 6.1). Frontal claspers are also present. There is one genus, *Callorhynchus*, with about four species in shallow to moderately deep waters of the Southern Hemisphere, including the coastal waters of Argentina, Peru, Australia, and South Africa.

• • • • • • • • • • • • •
Subclass Elasmobranchii

Among the most intriguing fishes are the elasmobranchs—the modern sharks and rays and their fossil relatives. Many piscine superlatives—including the largest species, the most fearsome predators, and the most elegant swimmers—can be assigned to this subclass. Although our attention is captured by the large pelagic species, more than 60% of the named species inhabit the deep sea (Martin and Treberg, 2002). Elasmobranchs have cartilaginous endoskeletons, but the cartilage in many species is calcified. These calcifications may appear superficially on the endocranium in prismatic or granular form, or within the endoskeletal cartilage, as in the vertebrae of many species. Jaw suspension is hyostylic or amphistylic, and the branchial skeleton is posterior to the neurocranium. Most forms have **dermal denticles** in the form of **placoid scales** in the skin, but some have no scales. No operculum is present; there are five to seven separate gill openings on each side. Like the holocephalans, males have **pelvic claspers** that are inserted into the oviduct of the female during copulation. Fins are stiffened by horny rays called **ceratotrichia**. There is no gas bladder. A cloaca is present. Compagno (1999a) listed 394 described species of living elasmobranchs.

Recent discoveries of placoid scales in the Harding Sandstone of Colorado suggest that elasmobranchs could have originated as early as the late Ordovician (Sansom et al., 1996). By the Devonian, they constituted a diverse group, and this diversity has been the subject of numerous phylogenetic interpretations. Schaeffer (1967) pointed out that there are three general branches evident in the evolution of the elasmobranchs: a primitive **cladodont** branch (extinct); an intermediate and related **hybodont** group (extinct); and a more recent branch, with a few living forms occupying transitional positions between the hybodonts and the modern elasmobranchs. Modern elasmobranchs are all grouped by Compagno (1991, 1999b) into the subcohort Neoselachii (Table 6.1). Maisey (1984) termed the living elasmobranchs a "reasonably coherent group" and examined their possible monophyly. That the modern elasmobranchs are monophy-

letic has been confirmed in the most recent classifications of living forms (deCarvalho, 1996; Compagno, 1999a, 1999b).

Cladodonts are characterized by such features as teeth bearing an enlarged central cusp flanked by smaller cusps of the same conical shape (Fig. 6.6B), a persistent notochord, a short rostral section of the brain case, palatoquadrate articulation with the enlarged postorbital processes of the cranium, and long jaws reaching from the snout to behind the skull. Members of the cohort †Paleoselachii are examples of cladodonts. The paleoselachian *†Cladoselache* (Fig. 6.6A) is perhaps the best-studied Devonian shark (Janvier, 1996).

The teeth of hybodonts are variable, with some genera being similar to cladodonts in dentition and others having more flattened teeth, suitable for grinding (Fig. 6.6D). Among the most bizarre of vertebrate fossils are the tooth whorls attributed to the genus *Helicoprion* (Fig. 6.6E). Classified as eugeniodontid by Janvier (1996) and as a protoselachian hybodont by Kemp (1999), these fishes have been the subject of an ongoing debate over the proper placement of the tooth whorls on the body (Ellis, 2001). Hybodonts share most of the characteristics mentioned for the cladodonts, but they differ in the structure of the pectoral fin skeleton, possession of an anal fin, and reduction of caudal fin radials (Fig. 6.6C; Maisey, 1982). In some, the rostral portion of the skull is enlarged. Modern elasmobranchs (neoselachians) have vertebral centra that replace the notochord, and shorter jaws, which in some may be protrusible. The jaws have hyostylic or amphistylic suspension. The modern elasmobranchs have larger neural and hemal elements of the vertebrae than hybodonts. Schaeffer (1967) indicated that the living sharks—except for the transitional Chlamydoselachidae, Heterodontidae, and Hexanchidae—represent two main phyletic lines, the galeoids and the squaloids. The rays, or batoids, represent another line of modern elasmobranchs. DeCarvalho (1996) and Shirai (1996) recognized two phyletic lines, the divisions Galeomorphii and Squalea, with the batoids subsumed under the latter division (see Table 6.2).

Some investigators claim that paleozoic sharks represent a single premodern level of evolutionary organization, although there may be several types or designs (Zangerl, 1973). Nelson (1994) arranged the sharks into the following superorders: (1) †Cladoselachimorpha, including the cladodonts; (2) †Xenacanthimorpha, including the family †Xenacanthidae; and (3) Euselachii, comprising the fossil †Ctenacanthiformes and †Hybodontiformes and seven orders of living sharks and rays. Unlike the classification of sharks and rays by Compagno (1977, 1991, 1999b), Nelson (1994) did not separate those two groups into separate superorders. Other views of shark phylogeny and classification can be seen in Maisey (1984, 1986), Seret (1986), Cappetta (1987), and Carroll (1988).

TABLE 6.1   CLASSIFICATION OF SHARKS AND RAYS, INCLUDING DIAGNOSES OF MODERN ORDERS

Taxon	Diagnosis
Class Chondrichthyes	
Subclass Elasmobranchii	
Cohort †Paleoselachii	
Several orders of extinct Paleozoic sharks, including	
Order †Cladoselachiformes	
Order †Symmoriiformes	
Cohort Euselachii	
Subcohort †Protoselachii	
Order †Xenacanthiformes	
Order †Ctenacanthiformes	
Order †Hybodontiformes	
Subcohort Neoselachii (designated an infraclass by deCarvalho, 1996)	
Superorder †Palaeospinacomorphii	
Division Galeomorphii	
Superorder Heterodontoidea	
Order Heterodontiformes (horn sharks)	Two spiny dorsal fins; five gill slits; small spiracles; groove connects mouth with nostrils.
Superorder Galeoidea	
Order Orectolobiformes (carpet sharks, whale sharks)	Two dorsal fins, lacking spines; fifth gill slit overlaps fourth behind pectoral fin origin in many; mouth small, placed well in advance of eyes; nasoral groove present.
Order Lamniformes (mackerel sharks)	Two dorsal fins, lacking spines; five gill slits, many have last two above pectoral fin; small spiracles; large mouth extending well behind eye.
Order Carcharhiniformes (requiem sharks)	Two dorsal fins, lacking spines; five gill slits, with one to three over pectoral fin; spiracles present in some families; gill rakers absent.
Division Squalea	
Superorder Notidanoidea	
Order Hexanchiformes	Single dorsal fin, lacking spine; six or seven gill slits; small spiracles, situated well behind eye.
Suborders Chlamydoselachoidei (frill sharks) and Hexanchoidei (cow sharks)	
Superorder Echinorhinoidea	
Order Echinorhiniformes (bramble sharks)	Two small dorsal fins, lacking spines; fifth gill slit larger than others; minute spiracles situated well behind eyes; body covered with coarse denticles.
Superorder Squaloidea	
Order Squaliformes	Two dorsal fins, some with spines; five gill slits; spiracles present.
Suborders Squaloidei (dogfish sharks) and Dalatioidei (sleeper sharks)	
Superorder Hypnosqualea	
Order Squatiniformes (angel sharks)	Raylike body with dorsal eyes; two dorsal fins, lacking spines; no anal fin; five gill slits; terminal mouth; large spiracle.
Order Pristiophoriformes (saw sharks)	Elongate body with snout forming long flattened blade with unequal size teeth on each side; pair of long barbels; two dorsal fins with spines generally lacking; no anal fin; large spiracle.
Order Rajiformes (Superorder Rajomorphii in Compagno, 1999a, 1999b; see Table 6.2)	

Sources: Compagno, 1991, 1999b; Nelson, 1994; deCarvalho, 1996; McEachran et al., 1996.

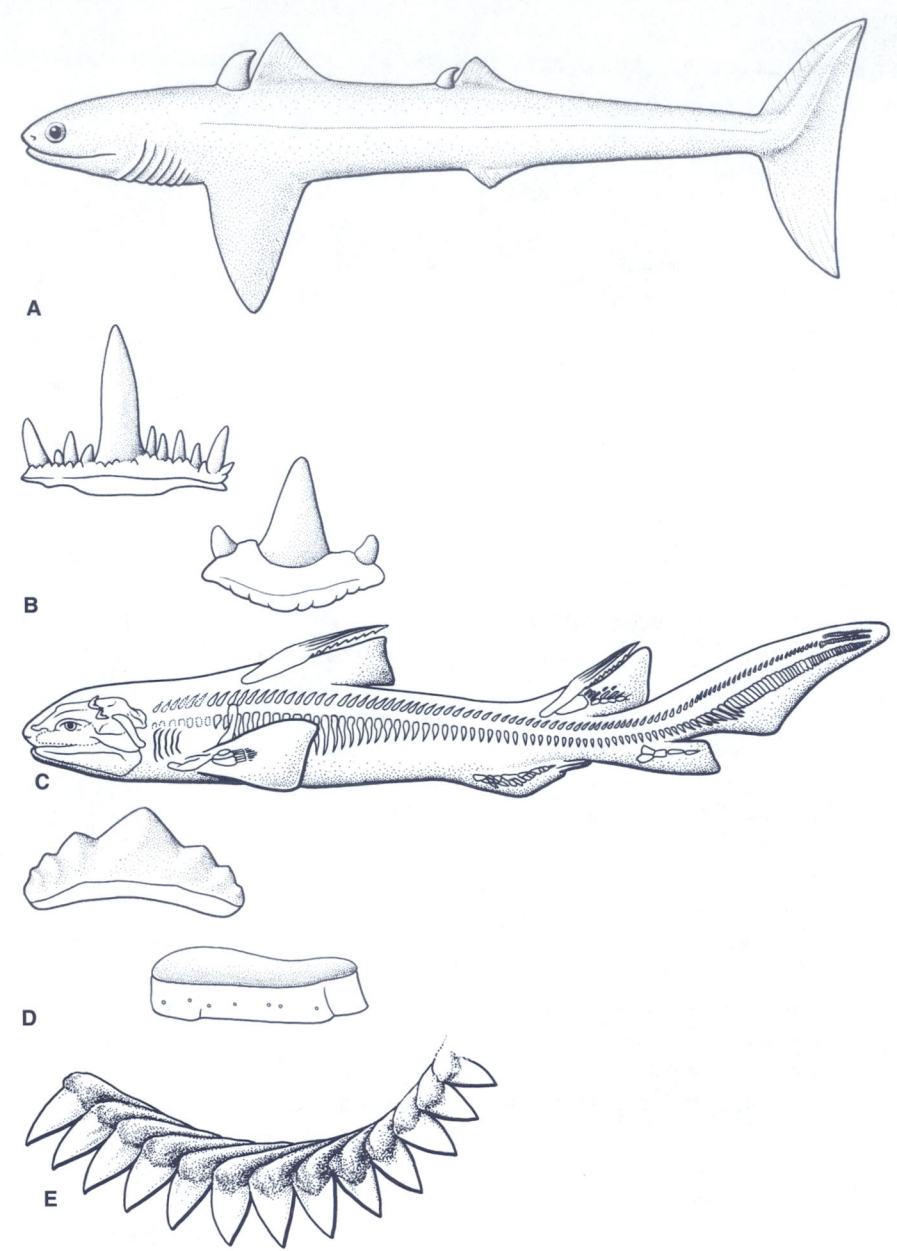

FIGURE 6.6
A, Representative of the order †Cladoselachiformes (†*Cladoselache*; actual size ca. 50 cm); **B,** Cladodont teeth;
C, Representative of the order †Hybodontiformes (†*Hamiltonichthys;* actual size ca. 28 cm); **D,** Hybodont teeth;
E,Tooth whorl of *Helicoprion*.

Compagno (1973, 1977, 1991) placed the sharks and rays into the subclass Elasmobranchii along with their paleozoic relatives. Paleozoic sharks are placed into the cohort †Palaeoselachii, which Compagno (1991) saw as comprising eight orders. Modern sharks and rays, and their extinct near relatives (xenacanths, ctenacanths, and hybodonts), are included in the cohort Euselachii. Compagno (1991, 1999b) did not follow the practice of splitting the living Euselachii into sharks and rays as lines of equal rank. Instead, he recognized four evolutionary lines, presented as three superorders of sharks and one superorder of rays. Our perspective on the early history of the sharks and allied forms will undoubtedly change in the coming years as additional fossil discoveries, such as the assemblage of sharks unearthed from Lower Carboniferous deposits in Montana (Lund, 1990, 1991), come under the close scrutiny of paleoichthyologists. In the following treatment, Compagno's (1991) arrangement is amended based on revisions of the higher classification of elasmobranchs by Shirai (1996) and deCarvalho (1996). Compagno (1999b) recognized the provisional nature of elasmobranch classification and summarized several problem areas requiring further clarification.

Cohort †Palaeoselachii

This group contains a number of forms that not only show primitive features when compared with the modern selachians, but show some remarkable adaptations as well. The order †Cladoselachiformes (Fig. 6.6A) contains Devonian species with broad-based paired fins, heavy dorsal fin spines, and branched cladodont teeth. The heterocercal caudal fin is lunate. The order †Symmoriiformes, from the Carboniferous, has cladodont teeth and lacks fin spines of the usual type. One family shows strange sexual dimorphism, in that the males of one species have an odd spinelike structure above the pectoral girdle. This appendage curves forward to form a flattened, denticle-covered blade over the top of the head, which is covered by similar spinous denticles. Zangerl (1984) claimed that the complex could have served to mimic a large mouth.

Cohort Euselachii

Under this group, Compagno (1991) listed two subcohorts. One, the †Protoselachii, includes extinct groups that are close to the modern level. Orders included are †Ctenacanthiformes, †Xenacanthiformes, and †Hybodontiformes (Fig. 6.6C).

The ctenacanths are known from the Devonian to the Triassic. They have developed mesopterygia and propterygia in their broad-based pectoral fins. The xenacanths existed from the Lower Devonian to the Upper Triassic, mostly in freshwater habitats. They have a long archipterygial axis supporting the pectoral fins, which are broad-based in some species. †Xenacanthus of the Permian was rather eel-shaped, with a diphycercal tail and a prominent occipital spine.

The hybodonts appear to be closely related to the ctenacanths, but have some modern features. The pectoral fins are supported by propterygia, mesopterygia, and metapterygia (see Fig. 3.25), and the fin radials, as in modern sharks, do not extend to the edges of the fins, but bear ceratotrichia. Hybodonts have a long history, extending from the Upper Devonian to the Tertiary.

Subcohort Neoselachii

This taxon (designated an infraclass by deCarvalho, 1996) comprises the modern sharks and rays, plus the closely related extinct sharks of the superorder †Palaeospinacomorphii, which existed from the Lower Jurassic to the early Tertiary. These sharks had long jaws and essentially cladodont teeth.

. . . .

Division Galeomorphii

Four orders of sharks are brought together under this taxon because of similarities in their cranial skeleton and in the structure of their pectoral fins. All have an anal fin and five gill openings. Most have dorsal fins, but only the Heterodontiformes have dorsal fin spines. Most of the familiar genera of sharks belong to this superorder, including the three known large sharks that are adapted to feed on plankton and small nekton. These are the whale shark (*Rhincodon typus*), the basking shark (*Cetorhinus maximus*), and the "megamouth" (*Megachasma pelagios*; Fig. 6.7A) from the Pacific. These three species have numerous small teeth in as many as 70 rows (Maisey, 1985) and are specialized for straining plankton, although the whale shark has been observed to take advantage of dense schools of small fishes. All other sharks have dental arrangements for grasping or cutting prey. The members of this superorder range from camouflaged, lurking benthic predators to the swiftest and most rapacious of pelagic sharks. Some of the specializations for pelagic life include buoyancy control, acute **acousticolateralis** sense, and control of internal temperature. Furthermore, these sharks have remarkable abilities in detecting electrical stimuli (see Chapter 21).

Superorder Heterodontoidea

Order Heterodontiformes

Only one genus, *Heterodontus* (Fig. 6.7B), of the family **Heterodontidae**, is included. Species are referred to as *horn sharks* because of the strong spines at the front of each dorsal fin. The generic name alludes to the variation of dentition in

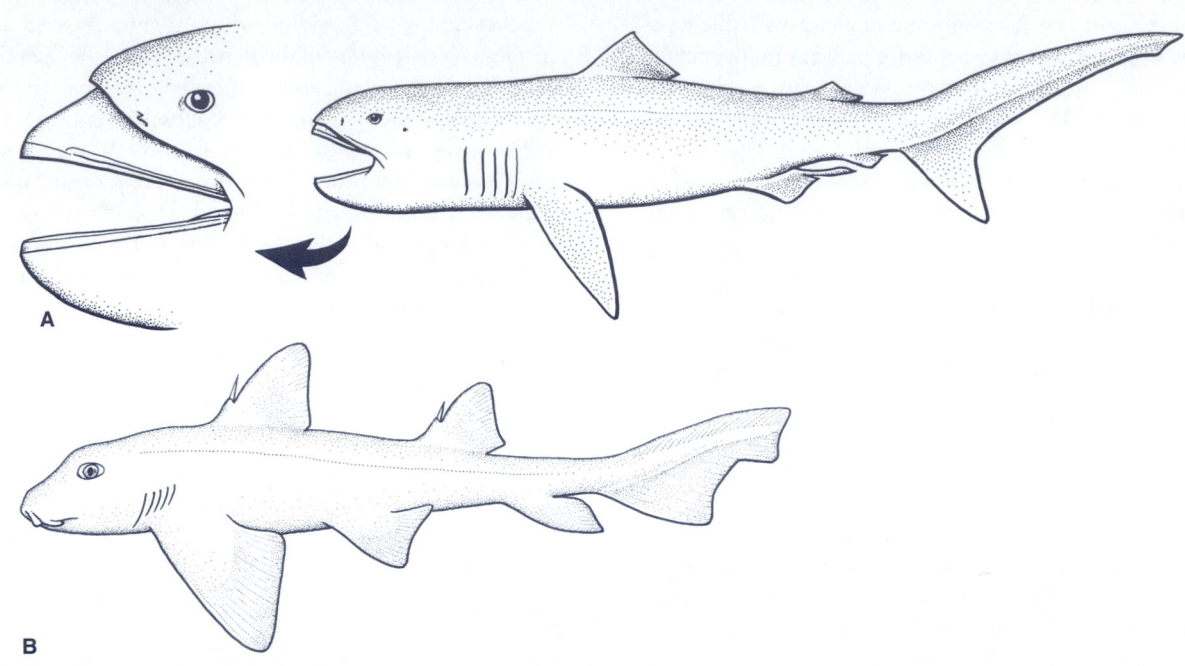

FIGURE 6.7
Representatives of the division Galeomorphii: **A,** Megamouth shark (*Megachasma pelagios*); **B,** Horn shark (*Heterodontus*).

the short and modified mouth. Anterior teeth are small and sharp, whereas those in the back of the jaws are molariform. The horn sharks are found in the Indo-Pacific in tropical to warm temperate waters. They may reach 1.5 m in length.

Heterodontus has often been considered to be closely related to such primitive sharks as hybodonts and ctenacanths. Some evidence for this involves the suspension of the upper jaw, which might be said to be structurally hyostylic but functionally amphistylic, according to Schaeffer (1967). Heterodontid teeth are known from the Jurassic.

Superorder Galeoidea

Order Orectolobiformes

The sharks of this group are usually placed into two families—the **Orectolobidae,** containing the carpet sharks, nurse sharks, zebra sharks, and wobbegons, and the **Rhincodontidae,** or whale sharks (Plate 6.3). In Compagno's (1991) arrangement, the Orectolobidae are restricted to a few genera, and the remaining fishes are placed in the following families: Parascyllidae, Brachaeluridae, Hemiscyllidae, Stegostomatidae, and Ginglymostomatidae. These sharks are found mainly in the tropical parts of the Indo-

Pacific, with most of the genera being present in Australian waters. One genus, *Ginglymostoma,* occurs in the Atlantic. For the most part, these are small sharks, reaching less than a meter long, but the Atlantic nurse shark has a maximum length of about 4.2 m, and others, such as the zebra shark (*Stegostoma*) and some species of *Orectolobus* may exceed 3 m long. The genus *Orectolobus* contains the strikingly marked wobbegons of Australia. Some of the carpet sharks in the western Pacific are known as "cat sharks," a name usually reserved for fishes of another order and family.

Rhincodon (= Rhiniodon) typus, the whale shark (Plate 6.3), is the largest living fish, attaining a length of 15 m. One specimen of 11.5 m was estimated to weigh about 12,000 kg. This sluggish giant is present in all tropical seas. Its mouth is terminal and broad. The whale shark feeds on great quantities of small, schooling fishes, such as herring, and on squid and planktonic crustacea. The mouth is equipped with numerous rows of small teeth, but most food items appear to be captured by straining through the fine gill rakers. The gill slits are very long and set rather high on the sides, partially above the pectoral fin. The body has a humpbacked appearance, and the caudal fin is very large. The color pattern is striking, consisting of yellowish or white spots on a gray to brown background. These huge animals are oviparous; the eggs are oval and surprisingly small (300 × 90 mm) considering the size of the shark. Whale sharks predictably congregate in select locations worldwide, such as the Sea of Cortez off Baja California, Ningaloo Reef in Western Australia, and the Galapagos Islands. Because of this, they are targeted by tourist operations seeking to provide an unusual animal encounter for recreational divers (Colman, 1997). Whale sharks also find themselves a casualty of the emergence of an affluent middle class in China, a country where shark fin soup is esteemed as an exemplification of conspicuous consumption. One fin from a whale shark is reported to sell for as much as $10,000 in China. Conservationists are campaigning for a worldwide ban on the heinous practice of "finning," in which sharks are caught, their fins cut off, and the fishes then thrown overboard to die.

Order Lamniformes

Usually, these are large and active sharks of shallow waters, with a few exceptions. Families included are **Odontaspididae** (= Carchariidae), the sand tigers, which are distributed in temperate to tropical waters and reach lengths of 3 m or more; **Pseudocarchariidae; Mitsukurinidae** (= Scapanorhynchidae), the goblin shark, a strange, primitive shark with a long, flat rostrum; **Alopiidae,** the thresher sharks; **Cetorhinidae,** the basking sharks; **Megachasmidae,** the megamouth; and **Lamnidae** (= Isuridae), the mackerel sharks. Naylor et al. (1997) have developed a phylogeny of lamniform sharks based on DNA sequence data. The lamnids are among the best known sharks, because of the large size and great appetite of some species. One lamnid, the white shark (*Carcharodon carcharias;* Plate 6.4), has been implicated in many fatal attacks on human beings. This is a giant among sharks—the largest mature male is recorded at 5 m, and the largest mature female at 7.1 m. This female had a body mass of 2,300 kg (Gottfried et al., 1996). Books that focus exclusively on the biology of this fearsome giant have been published (Ellis and McCosker, 1991; Klimley and Ainley, 1996) and well received owing to our fascination with this ichthyological superlative. An even larger extinct cousin, *Carcharodon megalodon,* prowled the Cenozoic seas. Reaching a total length of 15 m (Gottfried et al., 1996), this fish—the largest predatory shark that ever lived—would have made water sports a very dicey proposition. *Carcharodon carcharias* is found in tropical to temperate seas. In the eastern Pacific, it ranges as far north as 60° N to Petersburg, Alaska, and has attacked surfers in Oregon at temperatures around 13° C. It ranges at least to Newfoundland in the western North Atlantic. The family Lamnidae also includes the mako sharks of the genus *Isurus,* and the salmon and mackerel sharks (*Lamna*). The mako is

often sought as a big game fish and is taken commercially in some areas; *L. nasus,* the porbeagle, has some importance as a commercial fish in Norway and Iceland.

The goblin shark (*Mitsukurina owstoni*) is a deepwater fish that has been found in the Indian Ocean, the western Pacific off Japan, and the Atlantic near Portugal. It is characterized by a flat, elongate snout and protrusible jaws set with slender, sharp teeth. Some of these teeth were found embedded in a malfunctioning communications cable brought up from 1,300 m deep in the Indian Ocean. An interesting point to ponder is whether this act of selachian sabotage was provoked by the electronic activity within the cable or simply a consequence of feeding activity in the vicinity of the cable. The thresher sharks (*Alopias*) have the upper caudal lobe elongated, so that it comprises about one half of the total length of the fish. This great tail is used to herd the small, schooling fishes on which the sharks prey (Gubanov, 1972). When in shallow water, the tail slaps and splashes the surface. The several species of this genus are found in warm seas, and some are harvested commercially.

Unlike its swift and ferocious relatives, the gigantic basking shark is a slow-moving plankton feeder. It is close to the whale shark in reported maximum length, reaching 13.5 m. Its teeth, though numerous, are very small and would appear to have limited function. The gill rakers, however, are long and slender, constituting an excellent sieve for the small crustaceans on which the shark feeds. A long-held belief was that basking sharks, unable to sustain themselves during periods of low plankton density, shed their gill rakers, migrated offshore, and engaged in a form of hibernation (Parker and Boesman, 1954). Although the seasonal pattern of shedding gill rakers during the winter has been well documented, recent studies on the energetics of these large planktivores suggest that the threshold density of plankton needed to sustain them is significantly lower than previously believed (Sims, 1999). Not all individuals appear to shed their gill rakers; under certain conditions, the prey density may be sufficient to sustain activity throughout the year. Basking sharks are distributed in cold and temperate seas and have often been the object of harpoon fisheries for their liver oil.

A new species of giant planktivorous shark, *Megachasma pelagios* (the megamouth), was discovered off Hawaii in 1976, when the first specimen entangled itself in a parachutelike sea anchor being dragged by a research vessel. Its depth of capture was estimated to be about 160 m. The first specimen, about 4.2 m long, has a mouth about 1.2 m wide. Its teeth are small, and the gill rakers are adapted to strain plankton. The fish was described and named in 1983 (Taylor et al., 1983). Since the discovery of the first specimen, numerous others have been obtained from coastal waters off Brazil to Japan and Southeast Asia, and a book detailing the biology of this mysterious species has been published (Yano et al., 1997).

Order Carcharhiniformes

This is a large assemblage that includes eight families and about 40 genera, encompassing many of the more familiar species of sharks. Sharks of this order and of the order Squaliformes dominate the deep-sea shark fauna (Martin and Treberg, 2002). The families included are **Scyliorhinidae**, the cat sharks and spotted dogfishes; **Proscyllidae**, the finback cat sharks; **Pseudotriakidae**, false cat sharks; **Leptochariidae**, the barbeled hound sharks; **Triakidae**, the smooth hounds or smooth dogfishes; **Hemigaleidae**, the weasel sharks; **Carcharhinidae**, the requiem sharks; and **Sphyrnidae**, the hammerhead sharks.

Cat sharks are small sharks, often with striking color patterns, found in warm seas in many parts of the globe. These found at considerable depths are of drab coloration. The swell sharks of the genus *Cephaloscyllium* are capable of swallowing air and inflating their stomachs when brought out of the water. The inflated specimens then float until they can deflate themselves, a task that appears easy for some but harder for others, taking hours or days to accomplish. Smooth dogfishes are widespread shallow-water forms, some of which reach 2 m long. Familiar North American species are *Mustelus canis* of the Atlantic coast and *M. henlei* of the Pacific.

The requiem shark family (**Carcharhinidae**) contains several well-known medium to large species from tropical and temperate waters. Blacktip and whitetip sharks, so named because of their fin coloration, are of the genus *Carcharhinus* and are found in warm seas. One member of the group, the bull shark (*C. leucas*) of the western Atlantic, is found in the freshwaters of Lake Nicaragua, where it is known to make fatal attacks on bathers. The species is known to enter the Mississippi River. Similar species from Africa and India also enter freshwater. Another member of the family is the tiger shark (*Galeocerdo cuvieri*). This is a circumtropical species that reaches about 5.5 m and has some fame as a sport fish. The blue shark (*Prionace glauca;* Plate 6.5), is found in most warm and temperate waters. It has a slender body, a remarkable blue coloration, and is an active feeder, often attacking hooked salmon, to the dismay of the angler. The topes (*Galeorhinus*) are found in the Indo-Pacific and the eastern Atlantic. The soupfin shark (*Galeus zyopterus*) was once the target of a valuable fishery on the west coast of North America because of the vitamin A content of its liver oil. Production of synthetic vitamin A lowered the price, so that the fishery was abandoned.

The hammerhead sharks of the family **Sphyrnidae** were also sought for their vitamin-rich liver oil. These medium to large sharks are characterized by flat lateral expansions of the head, so that from above or below, the outline is that of the letter T. The eyes and nostrils are borne on the outer edge of the structure. Hammerhead sharks are confined to warm waters, and the family is circumtropical. *Sphyrna mokarran* has been measured at 5.4 m, and *S. zygaena* to over 4 m. They are reported to have hearty appetites and to feed on a variety of animals, including other hammerheads and the formidable stingrays.

Several hypotheses have been advanced to explain the function of the expanded head of these peculiar sharks. Some researchers have suggested that the flattened surface assists in making tighter turns. Studies have demonstrated a greater degree of head mobility in hammerheads when compared with other carcharhiniform sharks, thus lending some credence to the notion that their peculiar head contributes to hydrodynamic efficiency (Nakaya, 1995). Others have postulated that the positioning of nostrils and eyes at the margins of the cephalic lobes enhances olfactory and visual bilateral discrimination. Expanding and flattening a head already well endowed with electroreceptor organs may further enhance this sensory modality as well.

. . . .
Division Squalea

This taxon is based on similarities of cranial and pectoral anatomy and includes three orders of sharks with diverse appearance and dentition, one of which, the Hexanchiformes, has not usually been grouped with the others. Some orders include species with more than the usual five gill openings. Specializations include barbels and a tooth-studded rostrum in the sawshark (Fig. 6.8A), "cookie-cutting" dentition in *Isistius* (Fig 6.8B), luminosity in that genus and in *Etmopterus*, small body size in some genera, and large body size in the Greenland shark (*Somniosus*). DeCarvalho (1996) erected the division Squalea to include the superorder Hypnosqualea, which includes the orders Squatiniformes (angel sharks), Pristiophoriformes (saw sharks), and Rajiformes (skates and rays). Compagno's (1999a, 1999b) classification has the superorder Squalomorphii as including the orders Hexanchiformes (cow and frilled sharks), Squaliformes (dogfish sharks), and Pristiophoriformes, with the angel sharks in a separate superorder, the Squatinomorphii. Rays also constitute a separate superorder, the Rajomorphii. Although our framework for the general classification of elasmobranchs is largely based on deCarvalho (1996), the skates, rays, and allied forms (see Table 6.2) are treated here as a constituting their own superorder Rajomorphii (= Batoidea), in the sense of Compagno (1999a, 1999b).

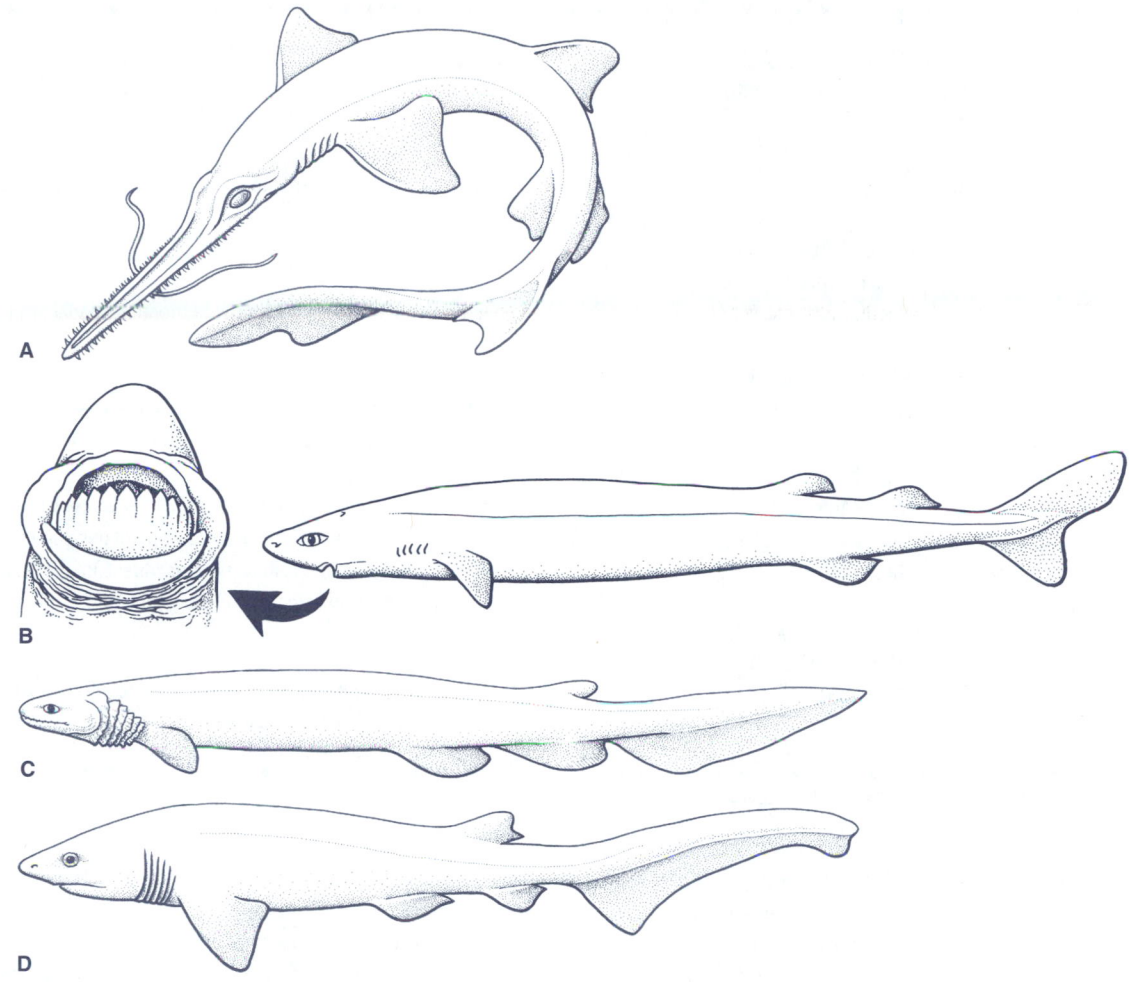

FIGURE 6.8

Representatives of the division Squalea: **A,** Sawshark (*Pliotrema*); **B,** cookie-cutter shark (*Isistius*); **C,** frill shark (*Chlamydoselachus*); **D,** Six-gill shark (*Hexanchus*).

Superorder Notidanoidea

Order Hexanchiformes

These sharks have six or seven gill arches and slits. There is an anal fin and a single dorsal fin without a spine. The suborder Chlamydoselachoidea has a single monotypic family, **Chlamydoselachidae,** consisting of the species *Chlamydoselachus anguineus*, the frill shark (Fig. 6.8C). This almost eel-shaped shark takes its name from the frilly extensions of the interbranchial septa, which overlap succeeding gill slits. Its notochord is unconstricted over most of the length of the trunk. The mouth is nearly terminal and is set with teeth having a broad base bearing three sharp cusps. These characters are similar to those of the extinct "cladodont" sharks, suggesting that the frill shark might represent a transitional form between primitive and modern sharks. This is a deep-water species known from several localities, reaching almost 2 m in length.

The suborder Hexanchoidei contains the family **Hexanchidae,** including the six-gill sharks (*Hexanchus;* Fig. 6.8D) and the "broad-headed" seven-gill shark (*Notorhynchus*); and the family **Heptranchidae** (the "pointy-headed" seven-gill sharks). These rather elongate, moderate- to large-sized sharks are also known as *cow sharks*. They have a subtermi-

Taxon	Diagnosis
chthyes	
ohort Neoselachii	
Superorder Rajomorphii (Batoidea)	
Order Torpediniformes	Branchial muscles modified into powerful electric organs; flaccid skin; well-developed caudal fin.
Suborders Torpedinoidei and Narcinoidei (electric rays)	
Order Pristiformes (sawfishes)	Snout drawn out into long flat blade, bordered by equal-sized teeth; sharklike body with two dorsal fins and well-developed caudal fin.
Order Rhiniformes (sharkfin guitarfishes or wedgefishes)	Two large falcate dorsal fins; origin of first dorsal over or anterior to pelvic fins; caudal fin with well-developed ventral lobe.
Order Rhinobatiformes (= Rhynchobatiformes; guitarfishes)	Two rounded (non-falcate) dorsal fins, origin of first well posterior to pelvic fin bases; ventral lobe of caudal fin not developed.
Order Rajiformes (skates)	Usually two small dorsal fins located near causal fin; tail very slender; most with prickles forming row on dorsal midline.
Order Myliobatiformes (stingrays— including butterfly rays, eagle rays, and manta rays)	Most with single small dorsal fin, lost in many species; most with prickles on dorsal midline; tail slender to whiplike, usually with barbed, venomous sting on dorsal surface.

Sources: Nelson, 1994; McEachran et al., 1996; Compagno, 1999a, 1999b.

nal mouth, and the teeth are mainly multicuspid, although more than one kind of tooth (heterodont condition) can be found in all species. The six-gill shark, *Hexanchus griseus*, reaches nearly 8 m and is widespread in temperate seas. Its food is usually herrings and other small fishes. It has been used in the manufacture of oil and meal.

Superorder Echinorhinoidea

Order Echinorhiniformes
The bramble shark, or alligator dogfish (*Echinorhinus*), is a robust shark that reaches 3 m. It is known mainly from warm seas, where it has been taken from depths as great as 900 m. Two species are described.

Superorder Squaloidea

Order Squaliformes
These sharks have five or six gill openings, two dorsal fins, often with a spine, and usually no anal fin. Two suborders, the Squaloidei and the Dalatioidei, are recognized. Perhaps the most widely recognized shark family, the **Squalidae** includes the spiny dogfish (*Squalus acanthias*), known to

legions of vertebrate anatomy students. It is abundant in temperate seas, reaching a length of about 2 m. Its flesh is edible, but its commercial value is not as high now as it used to be when it was sought for the vitamin-rich oils of the liver. Development of synthetic vitamins caused the decline of the dogfish fishery. This species might prove to have other health benefits, as it has been found to contain a broad-spectrum steroidal antibiotic called **squalamine** (Moore et al., 1993).

The dalatioid sharks include the Dalatiidae—the sleeper sharks—that reach more than 7 m. These are sluggish, cold-water animals that act both as predators and scavengers, but feed primarily on fishes. Also in this suborder is the smallest known shark, *Squaliolus laticaudus*, from the eastern Pacific near Japan and the Philippines; it reaches a length of only 15 cm. A closely related species from the Atlantic has been measured at 22 cm. Among the most peculiar of the dalatioids are the luminous sharks of the genus *Isistius*, including the Gulf dogfish (*I. plutodus*). These diminutive pelagic sharks show a remarkable specialization for cutting round plugs of flesh out of larger organisms. They have enlarged lower teeth and lips that aid in maintaining suction (Fig. 6.8B; see Plate 37.1). They attack moving prey head-on and allow the momentum of the larger animal to swivel them around, thus aiding in detaching the round piece of flesh. Because of this habit, they are called "cookie-cutter" sharks.

Superorder Hypnosqualea

Order †Protospinaciformes

This taxon is based on the genus †*Protospinax* from the Jurassic, a kind of "shark ray" that has some characteristics of both sharks and rays (Cappetta, 1987; deCarvalho and Maisey, 1996).

Order Pristiophoriformes

There is but one family in this order: **Pristiophoridae,** the sawsharks (Fig. 6.8A). The two genera, *Pristiophorus* (with five gill openings) and *Pliotrema* (with six gill openings) have the rostrum extended into a long, flat blade armed on each edge with teeth. There are two large barbels on the undersurface of the rostrum. These sharks are mainly found in the warm Indo-Pacific, but a rare species is known from the Bahamas. Fossils of the family are known from the Cretaceous.

Order Squatiniformes

This group contains one genus of depressiform fishes, with pectorals expanded forward but not fused with the head. The gill openings are mainly ventrolateral, and the spiracles are large, as in most batoids. They have two spineless dorsal fins set on the tail, no anal fin, and an essentially hypocercal caudal fin. These are the only living fishes to retain this ancient tailfin design. The order contains the single family **Squatinidae,** with one genus *Squatina,* the monkfishes or angel sharks. These are tropical to temperate in distribution, usually being found in shallow water. Despite their raylike appearance, their locomotion is sharklike, accomplished by movements of the tail. The largest species reaches about 2.4 m and a weight of 72 kg.

Superorder Rajomorphii (Batoidea)

DeCarvalho (1996) included the Rajiformes as members of the superorder Hypnosqualea, but Compagno (1999a, 1999b) regarded them as a distinct superorder, the Rajomorphii (Batoidea). The skates, rays, and related orders (Fig. 6.9) are recognized by a **depressiform** body, with the pectoral fins extending forward and fusing to the head, so that the five pairs of gill openings are ventral. There are several additional skeletal characteristics that distinguish them from sharks (Maisey, 1984): The suprascapulae are joined to each other over the vertebral column and either articulate with the column or fuse into a synarcual formed from the fusion of the anterior vertebrae. The posterior hypobranchial is in

contact with the shoulder girdle, either articulating with it or fusing to it. The palatoquadrate does not articulate with the neurocranium.

The rajomorphs are benthic predators, except for the mantas, which feed on plankton and small fishes in the open water. Feeding specializations in the group include the cephalic fins of the mantas, the crushing and grinding teeth of the eagle rays and others, and the great rostrum of the sawfish, with strong teeth set into sockets. Other specializations of note are the electric capabilities of the torpedoes, the stinging spines of stingrays, and the physiological modification of two families of stingrays that allow the invasion of freshwater. One of the most striking evolutionary adaptations was the transition of the order Myliobatiformes from benthic dwellers to pelagic giants that glide through the waters.

Order Torpediniformes

This order (Fig. 6.9A) consists of four families of electric rays—**Torpedinidae,** Hypnidae, Narcinidae, and Narkidae—in two suborders (McEachran et al., 1996). These have generally rounded discs, with the propterygia of the pectoral fins contacting the cranium anterior to the eyes. All have large electric organs in the disc on each side of the head (see Chapter 21). These organs are bundles of modified muscle tissue called **electroplaques,** with the component muscle cells termed **electrocytes.** They allow the fish to deliver powerful shocks that can stun prey or possibly discourage predators, although they are known to be eaten by sharks. Up to 200 volts have been recorded from large specimens. The shock may be powerful enough to render humans unconscious, as evidenced by the word roots of the family names, which all refer to sleep or numbness. Mediterranean species were used by the ancients as a form of electrotherapy for ailments such as arthritis and gout. Electric rays are found in tropical and temperate waters over a considerable depth range. Some of the deepwater forms, such as *Typhlonarke,* are blind. The largest species is thought to be *Torpedo nobiliana,* an Atlantic species that reaches 1.8 m in length.

Order Pristiformes

These are the sawfishes, family **Pristidae.** They are striking in their convergence with the sawsharks in general appearance, having the rostrum formed into a long, flat blade armed with teeth set in sockets. These are shallow-water fishes of warm seas, bays, and tropical rivers. There is a resident population in Lake Nicaragua. Some species are said to reach nearly 11 m long and a weight of about 2,400 kg.

The saw is used in feeding: While the sawfish moves through a school of fishes, lateral movements of this weapon

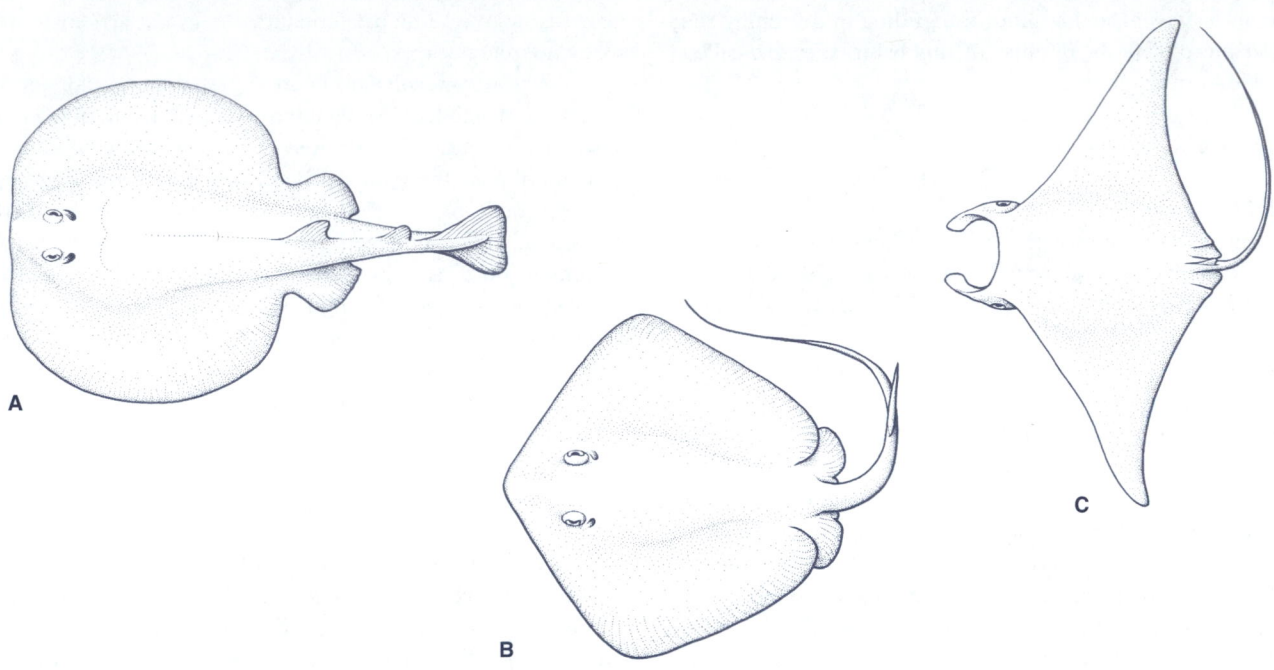

FIGURE 6.9
Representatives of the superorder Rajomorphii: **A,** Electric ray (*Torpedo*, order Torpediniformes); **B,** Stingray
(*Dasyatis*, order Myliobatiformes); **C,** Manta (*Manta birostris*, order Myliobatiformes).

GOING DEEPER   ·   The Barndoor Skate: An Incidental Casualty

Although extinction in freshwater and terrestrial ecosystems is, unfortunately, all too frequently documented, the conventional wisdom is that marine species, ranging over the vast and often inaccessible stretches of the oceanic realm, are essentially immune to extinction. Marine species, when compared with terrestrial species, do show a decreased extinction rate (McKinney, 1998). There has yet to be documented a case of extinction in historic times of a marine fish (Carlton et al., 1999). Even commercially significant species, subjected to intense fishery pressure that often culminates in a catastrophic population decline and the collapse of the fishery, have been observed to rebound in numbers when more stringent controls are placed on harvest. Regrettably, this may not be true for certain species, such as the barndoor

skate (*Raja laevis*). The barndoor skate is a veritable giant among the rajids, with a total body length up to 2 m and a width of more than 1 m. Barndoor skates slowly patrol the ocean bottom, consuming a diversity of invertebrates, including worms, shrimp, and lobsters, and several species of fishes, such as herring and menhaden (Raloff, 1999).

Elasmobranchs in general are quite susceptible to the effects of fishing, because they possess a suite of life history features—slow growth, late maturation, and low fecundity—that make them especially vulnerable (see Chapter 39). The pelagic longline fishery for such desirable species as swordfish and tuna has had a devastating impact on elasmobranchs. Baum et al. (2003) reported that Northwest Atlantic populations of several

species of large pelagic sharks have declined by as much as 75% in the past 15 years. Skates have been the target of limited commercial fisheries worldwide, but it is their incidental catch in trawl and longline fisheries for other bottom-dwelling species, such as cod, pollock, and flounder, that has had the most severe impact. A study by Casey and Myers (1998) suggested that the days of the barndoor skate are numbered—that this species may become the first documented case of extinction of a marine fish species.

Casey and Myers reviewed research vessel survey data from the Northwest Atlantic Ocean dating back to the early 1950s. They demonstrated a consistent decline in population numbers in the past decades—from an estimated total number of 600,000 individuals

kill and injure individuals that can be subsequently eaten. Fish impaled on the teeth of the saw are scraped off on the bottom and eaten. There are reports of serious injury and death resulting from bathers being in the way of a startled sawfish in the Ganges. Sawfishes are ovoviviparous but, fortunately for the mother, the rostrum and its teeth are soft until after birth.

Orders Rhiniformes and Rhinobatiformes (= Rhynchobatiformes)

Formerly these groups were considered a single order, the Rhinobatiformes, but McEachran et al. (1996) have tentatively proposed ordinal status for these two families of rather sharklike fishes, the **Rhinidae** and the **Rhinobatidae** (= Rhynchobatidae), generally referred to as guitarfishes. This name apparently is based on the appearance given by the flattened head, pectoral fins, and snout, followed by the tapering body and tail. These are fishes of shallow tropical and subtropical waters. They live on the bottom, feeding mainly on a variety of invertebrates. Most species reach about 1 m in length. Though considered edible, they are not highly esteemed as food fishes.

Order Rajiformes

This order contains batoids that lack some of the specialized features of the other orders and contains one family (**Rajidae**) with two subfamilies, the Rajinae and the Arhyncho-batinae (McEachran et al., 1996). Compagno (1999a, 1999b) elevated these two subfamilies to families within the Rajiformes. The rajids are remarkable for their high species diversity, with at least 165 species reported. A morphologically conservative group, they share a number of synapomorphies that corroborate a monophyletic origin (McEachran and Dunn, 1998). They have no stinging spines on the tail, no sawlike rostrum, and no large electric organs between the pectoral fin and the skull, although small electrical organs may be present. Rajids are fishes with the head, body, and pectoral fins combined into a flat disc, with a slender tail bearing very small dorsal and caudal fins. Most species are called *skates*. Most familiar are the skates of the subfamily Rajinae—a nearly cosmopolitan marine group, found from estuaries to the depths in warm and cold seas alike. Species range up to more than 2 m in length, but most do not grow more than 75 cm long. Unlike most of the other rays, the skates are oviparous. Their eggs are enclosed in horny cases, which are called "mermaid purses" in some localities. Recently, some intriguing studies have been published on the physiological and behavioral adaptations of embryonic skates within these enclosures (see Chapter 27). Some of the larger skates are commercially harvested in Europe. This fishery and the widespread incidental catch of skates may threaten some species with extinction on both sides of the Atlantic (see Going Deeper).

in the 1950s to less than 500 individuals in the 1970s. A decline in the mean size of individuals has also demonstrated their vulnerability to commercial fishery operations. Trawl fisheries are particularly devastating in their impact on bottom-dwelling fauna, especially large elasmobranchs (Carlton et al., 1999). Considering their estimated age at maturity (about 11 years) and fecundity, Casey and Myers (1998) concluded that the remaining population of barndoor skates experiences a fishery-related mortality well beyond what would enable the population to continue. Recent surveys conducted in conjunction with the development of a new bottom fishery for Greenland halibut (*Reinhardtius hippoglossoides*) have revealed previously unknown populations of barndoor skates living in waters deeper and further to the north of their reported range. The seasonal closure of trawl fisheries on other banks inhabited by barndoor skates apparently has contributed to their survival there as well.

Casey and Myers have argued that the only hope for the continued survival of the barndoor skate is for certain areas to be designated as fishery-free zones—a controversial position given the presence of several commercially desirable species in these areas. In March of 1999, two conservation organizations independently petitioned the National Marine Fisheries Service (NMFS) to designate *Raja laevis* an endangered species (Raloff, 1999). Such a designation would ensure federal protection for this species. A similarly worrisome trend has been documented for other species of large skates, such as the common skate (*Dipturus batis*) in the Atlantic. Fishes of the family Rajidae have demonstrated varying responses to exploitation, owing to their diversity in life history characteristics. Interesting enough, smaller species of skates, such as the little skate (*Raja erinacea*) and the thorny skate (*Raja radiata*), seem to have taken over the niche occupied by their larger brethren, and their numbers have increased. It would indeed be ironic if the first marine fish to be forced into extinction is one that was never the target of commercial exploitation but only an innocent bystander. Given what scientists are beginning to learn about the vulnerability of large elasmobranchs and other slow-growing species to human intrusion, the plight of the barndoor skate is likely to be visited upon other species.

Order Myliobatiformes

Rays of this order have large pectoral fins that combine with the head to form a broad disc, with a slender tail, usually having a stout stinging spine. The caudal and dorsal fins are reduced or absent. Included are six families: **Platyrhinidae, Zanobatidae, Hexatrygonidae, Urotrygonidae, Potamotrygonidae,** and **Dasyatidae.** The platyrhinids and zanobatids were originally classified as guitarfishes. The hexatrygonids are distinguished by their elongate, thin snouts.

These are all warm-water fishes, seldom entering cold temperate waters, and are usually found close to shore. The Potamotrygonidae are found in rivers of South America. Some dasyatids of the genera *Dasyatis* and *Himantura* appear to be permanent residents of freshwater in Africa, Asia, and New Guinea. The various stingrays and the butterfly rays live on the bottom, often concealing themselves in sand or other fine materials. Their food is shellfishes and bottom-living fishes. The tail spines are typically barbed and grooved along the edges. The venom produced in the groove can make a wound caused by the spine to be both painful and dangerous. The largest stingrays may reach a width of 2 m.

Many members of the Dasyatidae have become modified for a pelagic existence. The eagle rays and bat rays (*Aetobatus* and *Myliobatis*) and cownose rays (*Rhinoptera*) feed on the bottom, often dislodging bottom materials through the hydraulic action of the powerful movements of their large pectorals. Clams, oysters, and other invertebrates make up most of their food. The teeth of these rays are in the form of broad grinding plates. Locomotion is by "flying" movements of the winglike pectorals. The long, whiplike tail is usually held straight behind. Some species reach 1.2 m in width.

Mantas or devil rays (*Manta, Mobula;* Fig. 6.9C; Plate 6.6) have adapted to feeding on plankton and small, schooling fishes. They swim through the water by means of the wide, slender-tipped pectoral "wings," holding the mouth open. The peculiar **cephalic fins,** positioned on either side of the mouth, are used to guide food into the mouth. When curled into the spiral resting position, these fins give the impression of horns, hence the name "devil" rays. Although some species reach less than 1 m in width, others may reach several meters. A specimen of *Manta birostris* was measured at 6.6 m and is thought to have weighed more than 1,600 kg. Many of the dasyatids have a habit of leaping clear of the water and landing with a loud noise. Cartwheeling is another interesting behavior of these rays. Cownose rays are occasionally seen in more or less regularly oriented schools of up to 6,000 individuals.

KEY POINTS AND CONNECTIONS

- The evolution of the jaw in vertebrates facilitated the expansion of feeding opportunities and, hence, the exploitation of a greater variety of habitats. Jaws evolved from modifications of portions of the branchial arches—possibly as a consequence of improvements in the musculoskeletal control of gill ventilation. This resulted in the acquisition of the ability to apply suction in the act of feeding.

  Details on the role of jaws in prey capture are given in Chapter 23.

- Four gnathostome groups appeared during the early Silurian: placoderms, chondrichthyans, acanthodians, and bony fishes. Placoderms were characterized by heavy dermal armor covering the head and part of the trunk. Abundant during the Devonian, several orders of placoderms have been described from the fossil record. Chondrichthyans are distinguished by an endoskeleton composed entirely of cartilage.

- A diversity of jaw suspension configurations evolved among the gnathostomes. Chondrichthyans are especially noteworthy for the diversity of suspension designs present within the taxon. Two modern chondrichthyan taxa are known: the Holocephali and the Elasmobranchii. The modern holocephalans include the ratfishes and chimaeras.

- The elasmobranchs represent the greatest diversification of chondrichthyans, with a large number of fossil and modern forms represented. They share with the holocephalans the feature of pelvic claspers in the males. Dermal denticles in the form of placoid scales also characterize the elasmobranchs.

FISH LINKS

http://www.flmnh.ufl.edu/fish/Sharks/ISAF/ISAF.htm The International Shark Attack File is a compilation of all known shark attacks that is administered by the American Elasmobranch Society and the Florida Museum of Natural History. A useful site that communicates valuable information on an undesirable attribute of elasmobranchs.

http://www.elasmo-research.org The ReefQuest Centre for Shark Research, based in Vancouver, British Columbia, is "dedicated to shark and ray conservation through its scientific research and public education programs."

BUILDING AN ICHTHYOLOGY LIBRARY

Three excellent volumes that provide a helpful overview of research on early fishes:

Arratia, G., and G. Viohl (Eds.). 1996. *Mesozoic fishes: Systematics and paleoecology.* Verlag Dr. Friedrich Pfeil, München, Germany.
———, and H.-P. Schultze (Eds.). 1999. *Mesozoic fishes 2: Systematics and fossil record.* Verlag Dr. Friedrich Pfeil, München, Germany.

———, and A. Tintori (Eds.). 2004. *Mesozoic fishes 3: Systematics, paleoenvironments, and biodiversity*. Verlag Dr. Friedrich Pfeil, München, Germany.

Two important works on systematic ichthyology:

Greenwood, P. H., R. S. Miles, and C. Patterson (Eds.). 1973. *Interrelationships of fishes*. Academic Press, New York.
Stiassny, M. L. J., L. R. Parenti, and G. D. Johnson. 1996. *Interrelationships of fishes*. Academic Press, San Diego.

Three current works on the chondrichthyan fishes:

Carrier, J. C., J. A. Musick, and M. R. Heithaus (Eds.). 2004. *Biology of sharks and their relatives*. CRC Press, Boca Raton, FL.
Hamlett, W. C. (Ed.). 1999. *Sharks, skates, and rays: The biology of elasmobranch fishes*. Johns Hopkins University Press, Baltimore.
Klimley, A. P., and D. G. Ainley (Eds.). 1996. *Great white sharks: The biology of* Carcharodon carcharias. Academic Press, San Diego.

REFERENCES

Baum, J. K., R. A. Myers, D. G. Kehler, B. Worm, S. J. Harley, and P. A. Doherty. 2003. Collapse and conservation of shark populations in the Northwest Atlantic. *Science 299:* 389–392.
Baumiller, T. K., and F. J. Gahn. 2004. Testing predator-driven evolution with Paleozoic crinoid arm regeneration. *Science 305:* 1453–1455.
Cappetta, H. 1987. Chondrichthyes II. Mesozoic and Cenozoic Elasmobranchii, pp. 1–193. In *Handbook of paleoichthyology*, Vol. 3B. H.-P. Schultze (Ed.). Gustav Fischer Verlag, Stuttgart, Germany.
Carlton, J. T., J. B. Geller, M. L. Reaka-Kudla, and E. A. Norse. 1999. Historical extinctions in the sea. *Ann. Rev. Ecol. Syst. 30:* 515–538.
Carroll, R. L. 1988. *Vertebrate paleontology and evolution*. Freeman, New York.
Casey, J. M., and R. A. Myers. 1998. Near extinction of a large, widely distributed fish. *Science 281:* 690–692.
Cohn, M. J. 2002. Lamprey *Hox* genes and the origin of jaws. *Nature 416:* 386–387.
Colman, J. G. 1997. A review of the biology and ecology of the whale shark. *J. Fish Biol. 51:* 1219–1234.
Compagno, L. J. V. 1973. Interrelationships of living elasmobranchs, pp. 15–61. In *Interrelationships of fishes*, P. H. Greenwood, R. S. Miles, and C. Patterson (Eds.). Academic Press, New York.
———. 1977. Phyletic relationships of living sharks and rays. *Amer. Zool. 17:* 303–322.
———. 1991. The evolution and diversity of sharks, pp. 15–22. In *Discovering sharks*, S. H. Gruber (Ed.). American Littoral Society, Highlands, NJ.
———. 1999a. Checklist of living elasmobranchs, pp. 471–498. In *Sharks, skates, and rays: The biology of elasmobranch fishes*, W. C. Hamlett (Ed.). Johns Hopkins University Press, Baltimore.
———. 1999b. Systematics and body form, pp. 1–42. In *Sharks, skates, and rays: The biology of elasmobranch fishes*, W. C. Hamlett (Ed.). Johns Hopkins University Press, Baltimore.
deCarvalho, M. R. 1996. Higher-level elasmobranch phylogeny, basal squaleans, and paraphyly, pp. 9–34. In *Interrelationships of fishes*, M. L. J. Stiassny, L. R. Parenti, and G. D. Johnson (Eds.). Academic Press, San Diego.

deCarvalho, M. R., and J. G. Maisey. 1996. Phylogenetic relationships of the late Jurassic shark *Protospinax* Woodward 1919 (Chondrichthyes: Elasmobranchii), pp. 9–46. In *Mesozoic fishes: Systematics and paleoecology*, G. Arratia and G. Viohl (Eds.). Verlag Dr. Friedrich Pfeil, München, Germany.
Denison, R. 1978. Placodermi, pp. 1–128. In *Handbook of paleoichthyology*, Vol. 2. H.-P. Schultze (Ed.). Gustav Fischer Verlag, New York.
Didier, D. 1995. Phylogenetic systematics of extant chimaeroid fishes (Holocephali. Chimaeroidei). *Am. Mus. Novit. 3119:* 1–86.
Ellis, R. 2001. The *Helicoprion* mystery. *Nat. Hist. 110*(2): 76–77, 80.
———, and J. E. McCosker. 1991. *Great white shark*. Stanford University Press, Stanford, CA.
Gegenbaur, C. 1872. *Untersuchungen zur vergleichenden Anatomie der Wirbelthiere. III Das Kopfskelet der Selachier*. Engelmann, Leipzig, Germany.
Goodrich, E. S. 1930. *Studies on the structure and development of vertebrates*. Dover, New York (Reprinted 1958).
Gottfried, M. D., L. J. V. Compagno, and S. C. Bowman. 1996. Size and skeletal anatomy of the giant "megatooth" shark *Carcharodon megalodon*, pp. 55–66. In *Great white sharks: The biology of* Carcharodon carcharias, A. P. Klimley and D. G. Ainley (Eds.). Academic Press, San Diego.
Gregory, W. K. 1929. *Our face from fish to man*. Capricorn, New York (Reprinted 1965).
Grogan, E. D., R. Lund, and D. Didier. 1999. Description of the chimaerid jaw and its phylogenetic origins. *J. Morphol. 239:* 45–59.
Gubanov. Y. P. 1972. On the biology of the thresher shark (*Alopias vulpinus* Bonnaterre) in the northwest Indian Ocean. *J. Ichthyol. 12:* 591–600.
Janvier, P. 1996. *Early vertebrates*. Oxford University Press, Oxford.
Jarvik, E. 1980. *Basic structure and evolution of vertebrates*, Vols. 1 and 2. Academic Press, New York.
Kardong, K. V. 1998. *Vertebrates: Comparative anatomy, function, evolution*. WCB/McGraw-Hill, Boston.
Kemp, N. E. 1999. Integumentary system and teeth, pp. 43–68. In *Sharks, skates, and rays: The biology of elasmobranch fishes*, W. C. Hamlett (Ed.). Johns Hopkins University Press, Baltimore.
Klimley, A. P., and D. G. Ainley (eds.). 1996. *Great white sharks: The biology of* Carcharodon carcharias. Academic Press, San Diego.
Lagios, M. D. 1979. The coelacanth and the Chondrichthyes as sister groups: A review of shared apomorph characters and a cladistic analysis and reinterpretation, pp. 25–44. In The biology and physiology of the living coelacanth, J. E. McCosker and M. D. Lagios (Eds.). *Occ. Pap. Calif. Acad. Sci. 134.*
———. 1982. *Latimeria* and the Chondrichthyes as sister taxa: A rebuttal to recent attempts at refutation. *Copeia 1982:* 942–948.
Long, J. A. 1995. *The rise of fishes: 500 million years of evolution*. Johns Hopkins University Press, Baltimore.
Lund, R. 1986. The diversity and relationships of the Holocephali, pp. 97–106. In *Indo-Pacific fishes*, T. Uyeno, R. Arai, T. Tarniuchi, and K. Matsuura (Eds.). Ichthyological Society of Japan, Tokyo.
———. 1989. New petalodonts (Chondrichthyes) from the Upper Mississippian Bear Gulch limestone (Naimurian E_2b) of Montana. *J. Vert. Paleontol. 9:* 350–368.
———. 1990. Chondrichthyan life history styles as revealed by the 320 million years old Mississippian of Montana. *Env. Biol. Fishes 27:* 1–19.

————. 1991. Shadows in time—a capsule history of sharks, pp. 23–28. In *Discovering sharks*, S. H. Gruber (Ed.). American Littoral Society, Highlands, NJ.

————, and E. D. Grogan. 1997. Relationships of the Chimaeriformes and the basal radiation of the Chondrichthyes. *Rev. Fish Biol. Fisheries 7:* 65–123.

Maisey, J. G. 1980. An evaluation of jaw suspension in sharks. *Am. Mus. Novit. 706:* 1–19.

————. 1982. The anatomy and relationships of Mesozoic hybodont sharks. *Am. Mus. Novit. 2724:* 1–48.

————. 1984. Higher elasmobranch phylogeny and biostratigraphy. *Zool. J. Linn. Soc. 82:* 33–54.

————. 1985. Relationships of the megamouth shark, *Megachasma*. *Copeia 1985:* 228–231.

————. 1986. Heads and tails: A chordate phylogeny. *Cladistics 2:* 201–256.

Mallatt, J. 1984. Early vertebrate evolution: Pharyngeal structure and the origin of gnathostomes. *J. Zool. 204:* 169–183.

————. 1996. Ventilation and the origin of jawed vertebrates: A new mouth. *Zool. J. Linn. Soc. 117:* 329–404.

Martin, R., and J. Treberg. 2002. *Biology of deep-sea sharks: A review.* Proceedings of the International Congress on the Biology of Fish, Vancouver, Canada.

McEachran, J. D., and K. A. Dunn. 1998. Phylogenetic analysis of skates, a morphologically conservative clade of elasmobranchs (Chondrichthyes: Rajidae). *Copeia 1998:* 271–290.

————,————, and T. Miyake. 1996. Interrelationships of the batoid fishes (Chondrichthyes: Batoidea), pp. 63–84. In *Interrelationships of fishes*, M. L. J. Stiassny, L. R. Parenti, and G. D. Johnson (Eds.). Academic Press, San Diego.

McKinney, M. L. 1998. Is marine biodiversity at less risk? Evidence and implications. *Diversity and Distributions 4:* 3–8.

Miles, R. S. 1973. Relationships of acanthodians, pp. 63–104. In *Interrelationships of fishes*, P. H. Greenwood, R. S. Miles, and C. Patterson (Eds.). Academic Press, London.

Monastersky, R. 1996. Jump-start for the vertebrates: New clues to how our ancestors got a head. *Science News 149:* 74–75.

Moore, K. S., S. Wehrli, H. Roder, M. Rogers, J. N. Forest, Jr., D. McGrimmon, and M. Zazloff. 1993. Squalamine: An ammosterol antibiotic from the shark. *Proc. Natl. Acad. Sci. USA 90:* 1354–1358.

Nakaya, K. 1995. Hydrodynamic function of the head in the hammerhead sharks (Elasmobranchii: Sphyrnidae). *Copeia 1995:* 330–336.

Naylor, G. J. P., A. P. Martin, E. G. Mattison, and W. M. Brown. 1997. Interrelationships of lamniform sharks: Testing phylogenetic hypotheses with sequence data, pp. 199–218. In *Molecular systematics of fishes*, T. D. Kocher and C. A. Stepien (Eds.). Academic Press, San Diego.

Nelson, J. S. 1994. *Fishes of the world* (3rd ed.). John Wiley, New York.

Parker, H. W., and M. Boesman. 1954. The basking shark (*Cetorhinus maximus*) in winter. *Proc. Zool. Soc. Lond. 124:* 185–194.

Patterson, C. 1965. The phylogeny of the chimaeroids. *Phil. Trans. Roy. Soc. Lond. (Biol. Sci.) B 249:* 101–219.

Pough, F. H., C. M. Janis, and J. B. Heiser. 1999. *Vertebrate life* (5th ed.). Prentice Hall, Upper Saddle River, NJ.

Raloff, J. 1999. Skating to extinction? *Science News 155:* 280–282.

Sansom, I. J., M. M. Smith, and M. P. Smith. 1996. Scales of thelodont and shark-like fishes from the Ordovician of Colorado. *Nature 379:* 628–630.

Schaeffer, B. 1967. Comments on elasmobranch evolution, pp. 3–35. In *Sharks, skates and rays*, P. W. Gilbert, R. F. Matherson, and D. P. Rall (Eds.). Johns Hopkins University Press, Baltimore.

————, and M. Williams. 1977. Relationships of fossil and living elasmobranchs. *Amer. Zool. 17:* 293–302.

Seret, B. 1986. Classification et phylogénèse des chondrichthyes. *Oceanus 11:* 161–180.

Shirai, S. 1996. Phylogenetic interrelationships of neoselachians (Chondrichthyes: Euselachii), pp. 9–34. In *Interrelationships of fishes*, M. L. J. Stiassny, L. R. Parenti, and G. D. Johnson (Eds.). Academic Press, San Diego.

Sims, D. W. 1999. Threshold foraging behavior of basking sharks on zooplankton: Life on an energetic knife-edge? *Proc. Roy. Soc. Lond. B 266:* 1437–1443.

Smith, M. M., and B. Hall. 1990. Developmental and evolutionary origins of vertebrate skeletogenic and odontogenic tissues. *Biol. Rev. 65:* 277–374.

————, and Z. Johanson. 2003. Separate evolutionary origins of teeth from evidence in fossil jawed vertebrates. *Science 299:* 1235–1236.

Taylor, L. R., L. J. V. Compagno, and P. J. Strusaker. 1983. Megamouth—a new species, genus, and family of lamnoid shark (*Megachasma pelagios*, family Megachasmidae) from the Hawaiian Islands. *Proc. Calif. Acad. Sci. 43:* 87–110.

Van der Brugghen, W., and P. Janvier. 1993. Denticles in thelodonts. *Nature 364:* 107.

Wilson, M. V. H., and M. W. Caldwell. 1993. New Silurian and Devonian fork-tailed "thelodonts" are jawless vertebrates with stomachs and deep bodies. *Nature 361:* 442–444.

Yano, K., J. F. Morrissey, Y. Yabumoto, and K. Nakaya. 1997. *Biology of the megamouth shark*. Tokai University Press, Tokyo.

Zangerl, R. 1973. Interrelationships of early chondrichthyans, pp. 1–14. In *Interrelationships of fishes*, P. H. Greenwood, R. S. Miles, and C. Patterson (Eds.). Academic Press, London.

————. 1984. On the microscopic anatomy and possible function of the "spine-brush" complex of *Stethacanthus* (Elasmobranchii: Symmoriida). *J. Vert. Paleontol. 14:* 372–378.

————, and G. R. Case. 1973. Iniopterygia, a new order of chondrichthyan fishes from the Pennsylvanian of North America. *Fieldiana (Geol.) 6:* 1–67.

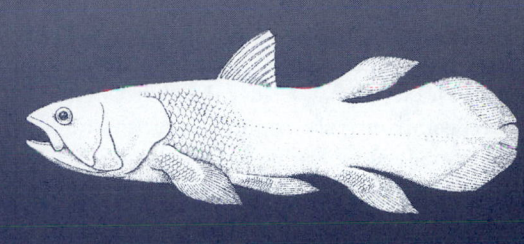

7

THE BONY FISHES (TELEOSTOMI): ACANTHODIANS AND SARCOPTERYGIANS

INTRODUCTION TO THE TELEOSTOMI

CLASS †ACANTHODII

CLASS SARCOPTERYGII: LOBE-FINNED FISHES AND THE TETRAPODS

The †Onychodontiformes: Fishes of Uncertain Affinities
Subclass Coelacanthimorpha
　Order Coelacanthiformes
Taxon Dipnomorpha
　Infraclass †Porolepimorpha
　Infraclass Dipnoi: Lungfishes
Subclass †Rhizodontimorpha
Subclass †Osteolepimorpha (†Rhipidistia)
　Order †Osteolepiformes

THE ORIGINS OF TETRAPODS

The evolutionary history of the gnathostome fishes suggests the early divergence of a number of lineages, such as the placoderms and chondrichthyans discussed in the previous chapter, and the early teleostomes, to be considered here. The teleostomes encompass a primitive gnathostome group, the acanthodians, and the group traditionally classified as "Osteichthyes"—a term that currently lacks taxonomic validity, as it probably refers to a polyphyletic group. The acanthodians have been referred to as "spiny sharks," in acknowledgement of their superficial resemblance to the chondrichthyans. The osteichthyans, or "bony fishes," experienced an early divergence into two major lineages—the sarcopterygians (the "lobe-finned" fishes) and the actinopterygians (the "ray-finned" fishes). Though abundant in the fossil record, very few sarcopterygians have survived into modern times. Those that are still around are a most intriguing group, exhibiting an assemblage of features that have caused them to be considered the closest relatives of the tetrapods.

INTRODUCTION TO THE TELEOSTOMI

If diversity and novelty equate to "success" in the evolution of vertebrates, then the teleostomes must move to the head of the class. Within this group are found (1) the extinct acanthodians, a strange archaic assemblage that some consider to be relatives of the sharks; (2) the sarcopterygians, from which the terrestrially adapted vertebrates (tetrapods) are derived; and (3) the actinopterygians, which will be the subject of several ensuing chapters. One feature that ties all these groups together is the existence of *bone* in the dermis and endoskeleton. In many classification schemes, the term "Osteichthyes" is used to denote the class of bony fishes, excluding the acanthodians. The latter have long been considered a separate lineage.

Nelson (1994) chose to abandon the use of "Osteichthyes" as a class and to adopt the use of **Teleostomi** as a grade embracing the remainder of the vertebrates, including the **†Acanthodii,** the **Sarcopterygii** (including the tetrapods), and the ray-finned fishes, or **Actinopterygii.** Primitive members of this grade share certain characters with the Elasmobranchii and †Placodermi: The caudal fin may be heterocercal; various groups may have spiracles; the heart may have valves in the conus arteriosus (see Chapter 3); the digestive tract may have an intestinal spiral valve and terminate in a cloaca (see Chapter 23); or the anus may be placed between the bases of the pelvic fin. For the most part, however, the caudal fin of bony fishes is homocercal; the spiracles, conus arteriosus, and cloaca are absent; and the anus is variously placed, usually just anterior to the anal fin. The ossification of bony fishes includes the significant incorporation of dermal bones in the head region (see Chapter 3, Figs. 3.20–3.22). There is usually a gas bladder or lung, but this might be lost in derived species. In the earliest acanthodians, the resemblance to sharks extends to the presence of serial gill slits with multiple gill covers (Janvier, 1996). In all other teleostomes, consolidation of gill tissue takes place, with a single gill slit beneath a series of opercular bones covering the gills. The septa that separated gills in fishes with serial gill slits are progressively reduced. Whereas gill arch levator muscles are absent among the Chondrichthyes, their presence among the sarcopterygians and actinopterygians is viewed as a synapomorphic (shared derived) character (see Chapter 4) that validates the taxon Teleostomi (Springer and Johnson, 2004).

CLASS †ACANTHODII

The acanthodians, commonly known as "spiny sharks" (the Latin name refers to the prominent median and paired spines that distinguish the group), probably represent an early invasion of pelagic habitats. The acanthodians trace their origins to the Lower Silurian and persisted until the Lower Permian. As such, they were contemporaries of a diversity of benthic ostracoderms. Under these circumstances, these early yet comparatively sophisticated gnathostome fishes probably evolved in response to greater feeding opportunities in the pelagic realm. This is evidenced by a somewhat elongate, fusiform body with a heterocercal tail, well-developed eyes set far forward in the skull, and a terminal to slightly subterminal mouth (Fig. 7.1A). Whereas most fossils show the short gill rakers characteristic of a predatory, macrocarnivorous lifestyle, some acanthodians possessed elongate gill rakers, suggestive of a life spent straining small invertebrates out of the water column. Non-imbricating small scales, with bony bases covered by dentine, are present. Scale growth was unique in that the crown of individual scales grew by the deposition of successive layers of dentine (Fig. 7.1B). Among the most distinguishing features seen in some acanthodians (e.g., †*Diplacanthus* and †*Climatius*) are the stout spines that formed the leading edges of all fin surfaces, save for the caudal fin. The potential significance of acanthodians in unraveling the mystery of the origin of paired appendages (as mentioned in Chapter 5) is seen in the best known genus, †*Climatius* (Fig. 7.1A). This genus shows two rows of ventrolateral spines between the pectoral and pelvic fins, suggestive to some of the remnants of primordial fin fold structures.

The gill clefts of acanthodians are covered by gill covers borne on the hyoid and branchial arches and by smaller plates subsidiary to those larger plates, but, in advanced forms, the hyoid gill cover is enlarged and appears to cover all the slits (Denison, 1979). The aforementioned †*Climatius* exemplifies the earliest of the acanthodians, as it shows the primitive condition of serial gill slits with multiple covers. The acanthodian endocranium is ossified, and dermal bone is present on the head. Some acanthodians show ossifications in the vertebral column as well. The teeth are unlike those of both the elasmobranchs and the other teleostomes, differing in microscopic structure and method of replacement. Tooth-bearing whorls apparently added new teeth toward the medial aspect of the whorl, whereas new teeth were added to the front of the jaw as it elongated with growth. Some species had nonreplaceable, isolated teeth on the mouth lining (Carroll, 1988), whereas others showed a tooth arrangement and pattern of replacement like that seen in sharks (Janvier, 1996).

There is some resemblance between the †Acanthodii and primitive actinopterygians in the structure of the ventral part of the neurocranium and in the development of an operculum. The relationships of the †Acanthodii are still very much uncertain. Widely divergent interpretations have

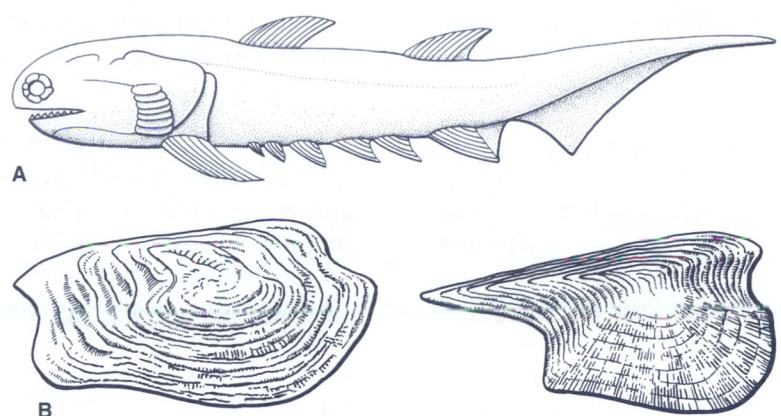

FIGURE 7.1
A, Representative of the †Acanthodii (actual size ca. 15 cm; after Watson, 1937); **B,** Scales from an acanthodian.

been made of their affinities by paleoichthyologists, who have variously considered them to be allied to placoderms, elasmobranchs, or actinopterygians. Denison (1979) stated that "it is necessary that these interpretations be made, but it is important that they be distinguished from facts." He "provisionally" accepted the proposal by Miles (1973) that acanthodians are aligned with the "Osteichthyes." The most current interpretation of their phylogeny places acanthodians as a sister group to the fishes originally grouped in the class "Osteichthyes" (Lauder and Liem, 1983; Maisey, 1986; Nelson, 1994). Acanthodians are sometimes divided into as many as seven orders, but most researchers (Moy-Thomas and Miles, 1971; Nelson, 1994; Janvier, 1996) have recognized three: †Climatiiformes, †Ischnacanthiformes, and †Acanthodiformes. Although most of these were small fishes, some exceptional fossils are as long as 2 m.

The earliest remains of acanthodians are mainly from marine deposits, and they seem to have been primarily a marine group. By the beginning of the Devonian, however, their habitat included freshwater (Denison, 1979). There is some evidence that species with long, heavy pectoral spines were bottom dwellers (Denison, 1979). Whether they possessed gas bladders is not known (Carroll, 1988). Fossilized gut contents have revealed that †Acanthodes fed on small invertebrates but took fishes occasionally. Fish remains have been found also in climatiids (Denison, 1979).

Zhu et al. (1999; Zhu and Ahlberg, 2001) have reported remarkable Early Devonian fossil finds from Asia that may have a dramatic impact on our interpretation of the origins of osteichthyan fishes. One fossil, named †Psarolepis, appears to be a basal sarcopterygian, yet it has features of the pectoral girdle and dentition like that of actinopterygians. Even more remarkable, it possesses stout appendicular spines much like those of the acanthodians. Both †Psarolepis and another fossil basal sarcopterygian, named †Achoania, possess an eyestalk—a feature previously considered unique to placoderms and chondrichthyans. Another Early Devonian fossil find, this time from southeastern Australia, appears to be a basal osteichthyan that more closely resembles the actinopterygians. The skull of this fossil also shows clear evidence of an eyestalk (Basden et al., 2000).

CLASS SARCOPTERYGII:
LOBE-FINNED FISHES AND THE TETRAPODS

Although they figure insignificantly in the total diversity of modern fishes, the sarcopterygian fishes are of enormous evolutionary importance, as they have the closest phylogenetic affinity with the tetrapods. Our focus here will be on the fishes, but the story of sarcopterygians would be incomplete unless some mention was made of the evolution of the tetrapods. The vast majority of sarcopterygian fishes are known only from the fossil record. Only seven (eight, if we accept the designation of a recently discovered coelacanth population as a distinct species—see Going Deeper) extant species have been described. The classification of sarcopterygians, especially in the context of their relationship to the tetrapods, has been a bone of contention for some time. The classification presented by Nelson (1994) is largely followed here. This includes the coelacanths (actinistians), lungfishes (dipnoans), four extinct lineages (†Onychodontiformes,

hizodontiformes, and †Osteolepifor-
ods, both extant and extinct. Cloutier
have contributed a phylogenetic analysis of ygians in which 13 clades are identified
and their interrelationships suggested (Fig. 7.2).

All sarcopterygians are distinguished by the presence
of fleshy, lobate fins, with a flexible internal skeletal support.
They also have an epichordal lobe in a caudal fin of varying
shape—heterocercal, heterodiphycercal, or diphycercal.
The latter two shapes often show an axial lobe, as in the
living coelacanth (*Latimeria*). Many ancestral lobefins had
two dorsal fins, as does *Latimeria*, but not the living lungfishes, whereas those closest to the tetrapods lacked a dorsal
fin altogether. Cosmine is typical of the dermal skeletons
of many of the extinct lobefins, and the cosmoid scales and

dermal bones of the fossil forms are characterized by pores
that lead to a canal system that researchers believe may have
been electrosensitive.

The †Onychodontiformes: Fishes of Uncertain Affinities

The †Onychodontiformes, also known as the †Struniiformes,
are a sarcopterygian lineage of uncertain affinities. They are
sometimes considered a sister group of the rest of the sarcopterygians (Janvier, 1996; Thomson, 1993). They possess distinct dentition, consisting of sigmoidal teeth that are housed
in deep internasal pits of the palate when the mouth is closed.
Only a few genera are known from the fossil record, including †*Strunius*, which is distinguished by a diphycercal tail
with an elongate axial notochordal lobe (Fig. 7.3A).

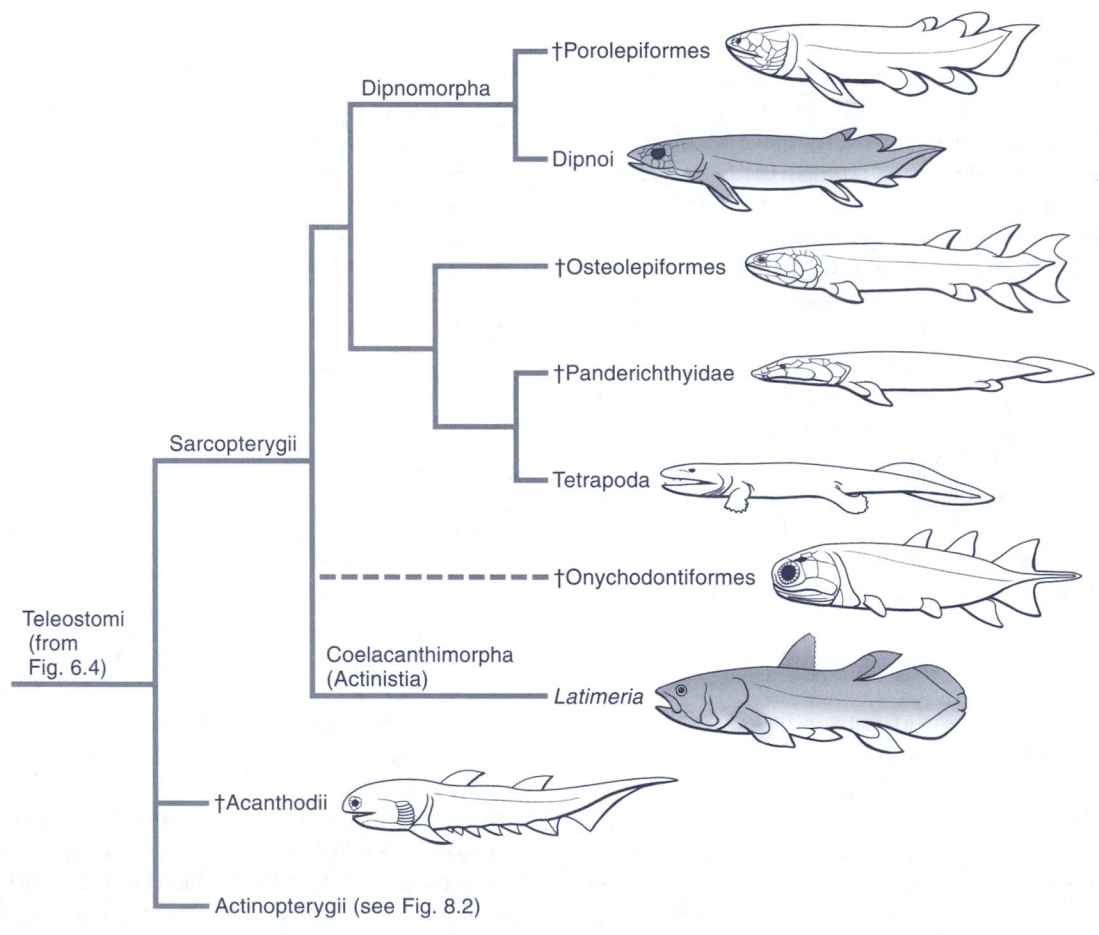

FIGURE 7.2
Phylogeny of the sarcopterygians, as suggested by Cloutier and Ahlberg, 1996.

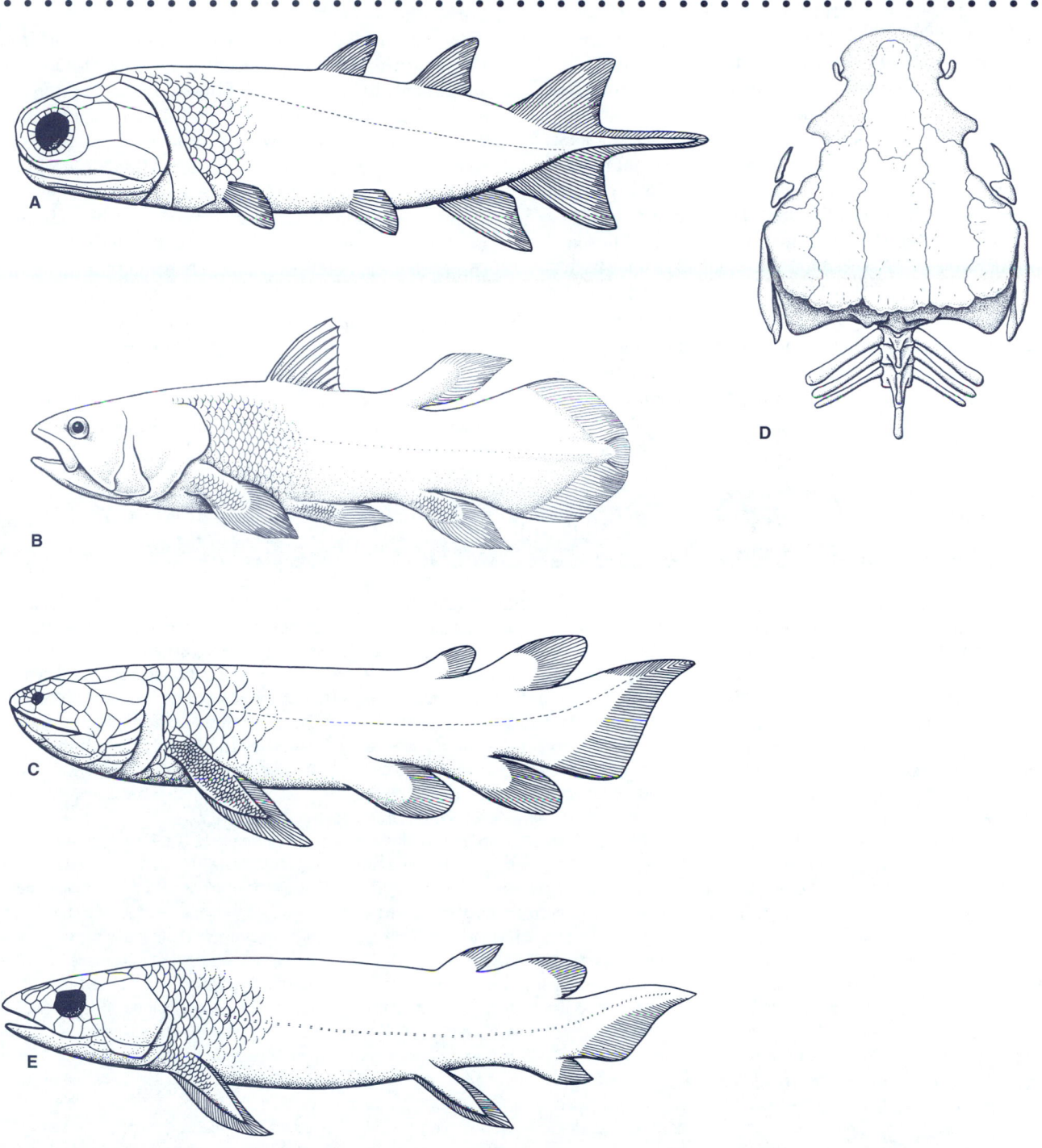

FIGURE 7.3
Representatives of the Sarcopterygii: **A,** Order †Onychodontiformes (*†Strunius;* actual size ca. 6 cm); **B,** Order Coelacanthiformes (*Latimeria chalumnae*); **C,** Order †Porolepiformes (*†Holoptychius;* actual size ca. 25 cm); **D,** Dorsal view of skull roof of *Neoceratodus;* **E,** Fossil dipnomorph (*†Dipterus;* actual size ca. 14 cm).

. . . .

Subclass Coelacanthimorpha

The living coelacanth and its extinct relatives differ from lungfishes in having the palatoquadrate separate from the neurocranium and having the skull divided into anterior (ethmoid) and posterior (otico–occipital) sections at a joint between the frontals and the parietals. A subcephalic muscle connecting the two parts can pull the anterior part ventrally in a biting motion (Carroll, 1988). Jaw suspension is hyostylic in most, but the hyomandibular is reduced in some representatives. The paired fins are paddlelike; some have a median axis with a proximal bone articulating by means of a ball-and-socket-like joint. A cloaca is absent. **Choanae** (see further), passageways that connect the nasal openings with the oral cavity, are present in some of the groups, and **cosmoid** scales are present in the extinct osteolepiforms and porolepiforms.

Order Coelacanthiformes

This order includes four extinct families plus the family **Coelacanthidae,** which includes several extinct genera and the only living members of the subclass, the genus *Latimeria* (Fig. 7.3B; see Going Deeper). Until the discovery of *Latimeria,* the coelacanths were known only as a comparatively homogeneous group from the Middle Devonian to the Upper Cretaceous. The eminent zoologist Louis Agassiz discovered and described the group in 1839 from a fossil tail fragment and was struck by the fact that the fin rays supporting the tail were hollow. He coined the name *Coelacanthus* ("hollow spine") to describe this unique group of fishes. The group had achieved a global distribution by the late Mesozoic, with fossils found in both marine and freshwater deposits. More than 120 fossil species in 47 genera have been described, and the fossils are found over a time span of 380 million years (Cloutier and Forey, 1991). They

---

GOING DEEPER · **The Story of the Coelacanth: A Twice-Told Tale of a Fish as Rare as a Full-Size Spare**

The story of the first discovery of the coelacanth (*Latimeria chalumnae*) has entered the annals of zoological folklore as arguably the greatest animal discovery of the past century. On December 22, 1938, the trawler *Nerine* docked at the South African port of East London with a very strange fish collected from a shark gill net set out by the crew. Fortunately, they had sufficient foresight to save any ichthyological oddities for Marjorie Courtenay Latimer, curator of the East London Museum. The fish had been taken in deep water off the mouth of the Chalumna River, just to the South of East London. Miss Latimer, sensing that this was a major find, dashed off a quick sketch to the brilliant but eccentric ichthyologist James L. B. Smith at Rhodes University in Grahamstown. The sketch was all that Smith needed to realize that this was indeed major news, as he recognized it as a specimen of a group of fishes believed to have been extinct for almost 80 million years. What made the find especially captivating was the recognition at the time that this "living fossil" could be a genuine ancestor of the tetrapods. In honor of the woman who brought the fish to his attention, and recognizing the location of its first discovery, Smith named the fish *Latimeria chalumnae* (Smith, 1939).

Smith instantly undertook an ambitious hunt for additional specimens. He recruited Eric Hunt, captain of a trading schooner, to spread the word among the locals. Fourteen years of searching finally paid off, and in 1952, the true home of the coelacanth was discovered nearly 2,000 km to the north, in the Comoros Islands. Local fishermen there occasionally landed one of these giants while fishing for *Ruvettus pretiosus,* the oil of which is valued as both a laxative and a mosquito repellant. The Comoros Islanders knew the coelacanth well—they even had a name for it: *ngombessa.* A second specimen was spirited out of the Comoros, much to the dismay of the French colonial government that administered the islands.

The French had their revenge. They imposed an embargo on coelacanth export, and French researchers Jacques Millot and Jean Anthony eclipsed Smith as the go-to guys on all matters coelacanth. In a bibliography of *Latimeria chalumnae,* published by Bruton (1991), more than 60 of their articles are cited. The title of Millot's first paper on the fish, "*Notre*" coelacanthe (Millot, 1953), pretty much expressed the sentiments of the French.

By the mid-1960s, more than 50 coelacanths had been recovered from the Comoros,

and the study of the fish progressed from anatomical descriptions to investigations of its physiology and biochemistry. The rarity of coelacanths and the depth at which they are found (usually ranging from 70 to 600 m; McCosker, 1979) made them difficult subjects for live study. At the time of publication of Bruton and Coutouvidis's (1991) inventory of all known specimens, a total of 172 specimens had been caught.

In the early 1970s, Hans Fricke of the Max Planck Institute began attempts at filming coelacanths in their natural habitat. Improvements in video technology have resulted in some striking images of coelacanths at work and at play. Remarkable discoveries were made, such as the tetrapodlike motions of the paired appendages, and the curious head-standing behavior mentioned earlier, and initial estimates of population size were attempted (Fricke et al., 1991; Fricke and Hissmann, 1994; Hissmann et al., 1998). In 1975, John McCosker of the California Academy of Sciences headed up a major coelacanth expedition to the Comoros—a live specimen exhibited at the Academy's Steinhart Aquarium would have been an unprecedented zoological event. The expedition did manage to collect two specimens, which were frozen, and

are thought by some to be derived from porolepiform stock (Andrews, 1973), but there is no consensus on that. There is evidence that the intracranial joint of the coelacanths is not homologous with that of the rhipidistians (Osteolepiformes; Bjerring, 1973; Thomson, 1967).

Coelacanthids usually have two external nostrils per side, but no choanae (see Fig. 7.7). Cosmine is lacking. There are two dorsal fins, the anterior one placed in the front half of the body. The caudal fin is diphycercal in all except one genus, and an axial lobe takes the form of a small **epicaudal** fin. The gas bladder or lung is calcified in extinct species.

Latimeria is large, typically about 2 m long, with reported lengths of more than 2.5 m. Weights of around 80 kg have been reported. The smallest free-living specimen captured was 42 cm long, barely longer than the estimated size at birth. Specimens of around 1.8 m have been estimated to be 11 years old. Life colors are brownish to bluish, with lighter spots because of unpigmented areas on scales. The scales are modified from the cosmoid type and consist of a dense isopedine plate and an outer mineralized layer to which denticles are fused. These denticles have a pulp cavity and dentine plus an outer layer that closely resembles enamel. There are reports of the Comoros Islands people using the scales like sandpaper to roughen innertubes being repaired.

The pectoral, pelvic, second dorsal, and anal fins are all lobate and flexible. The pectoral fin is capable of 180-degree rotation. The larger cartilaginous fin supports or "spines" are hollow, as in the fossil relatives. The coelacanth appears to be slow moving and spends much time hovering and drifting instead of rapidly pursuing prey. The second dorsal fin and the second anal fin are used in sculling motions, flexing to the same side at the same time and rotating as if to maintain

their tissues were made available to a number of researchers. The increased availability of fresh tissue specimens led to the publication of one of the first comprehensive reviews of coelacanth biology (McCosker and Lagios, 1979). Since then, Musick et al. (1991), Thomson (1991), and Forey (1998) have also published major works on the coelacanth. Popular accounts include Smith's (1956) memoir and a recent entry that brings the story up to date (Weinberg, 2000). Coelacanth fever has scientists expressing genuine concern for these amazing creatures—their rarity and popularity could result in the overexploitation of an already vulnerable species.

The coelacanth story took a strange and fascinating turn in 1997, when Arnaz Erdmann, while honeymooning with her husband Mark, a coral reef ecologist scouting out suitable field sites, spotted a strange but somehow familiar creature in a fish market in the seaport town of Manado in North Sulawesi, Indonesia. Fortunately, the Erdmanns snapped a quick photograph before the fish was sold. When word began to spread of this momentous find, the Erdmanns were compelled to return to Indonesia to continue their search for more specimens. In 1998, a barely alive second specimen caught in a gill net was delivered to them, and they were able to photograph the still-swimming fish before it expired. When they alerted colleagues of their discovery, the ichthyological community and the world at large was presented with the startling news that coelacanths appeared to be much more wide ranging than originally believed (Erdmann, 1999; Erdmann et al., 1998; Forey, 1998; Thomson, 1999). Here was a population that existed more than 6,000 km from the Comoros. French researchers, still smarting from the insult perpetrated by Smith, who had had the temerity to announce to the world the existence of a fish that had been making its home in a French territorial possession, quickly brought out a taxonomic description of the Indonesian population (Pouyaud et al., 1999). Aside from slight differences in coloration, the Indonesian coelacanths appeared very similar to those originally described from the Comoros Islands. Pouyaud et al. (1999) claimed sufficient differences in mitochondrial DNA sequences to assign a new species name, *Latimeria menadoensis*, to the Indonesian population, and their results have largely been substantiated by a more ambitious study (Holder et al., 1999). At the banquet following the 1999 meeting of the American Society of Ichthyologists and Herpetologists, a resolution was read honoring the Erdmanns for introducing to the world another coelacanth and excoriating "scientists from the land of Jean LaFitte and editors of otherwise respected journals" for denying the Erdmanns the courtesy of authoring the description of this new species. French researchers are conspicuously absent among the multiple authors who have recently published a paper describing the initial efforts at preserving the genome of *L. menadoensis* for future study via the generation of genomic bacterial artificial chromosome libraries (Danke et al., 2004).

The coelacanth has entered our consciousness as a prime example of both anachronism and elusiveness. *Time* magazine once described Richard Nixon as a "coelacanth of American anti-Communism." In recent television ads, Volkswagen has touted the availability of a full-size spare tire in their cars as being something as rare as a coelacanth. Surviving for millions of years in the ocean depths, the evolutionary conservatism displayed by this remarkable fish is the best example of the old maxim, "If it ain't broke, don't fix it."

trim and balance. The pectoral and pelvic fins are used in a synchronous, alternating sequence in slow swimming, resembling the manner in which tetrapods use their fore- and hindlimbs when walking (Fricke et al., 1987; Uyeno, 1991). The first dorsal fin is kept erect, as if used as a rudder. While foraging, coelacanths sometimes engage in a curious "head-standing" behavior, in which the body is held in a vertical position. Possibly, this facilitates the use of their acute electroreceptive capacity while searching for prey. Uyeno (1991) proposed a change in fin terminology, pointing out that the bulk of the tail complex is made up of a third dorsal fin and a second anal fin, and that the so-called "terminal lobe" is, in fact, the caudal fin. This epicaudal fin appears to function as a stabilizer when the fish is turning. The presence of a lateral line extending to the tip of the lobe indicates that it may also have a sensory function (Hissmann and Fricke, 1996).

Unlike that of extinct species, the swim bladder of *Latimeria* is not ossified. Instead, it is filled with fat—a modification for deepwater buoyancy control. It retains a closed connection with the ventral part of the pharynx. Because of the prominence of the swim bladder, the kidney has been displaced to a ventral and posterior position. The spiracular canal is closed, and internal nares are lacking. There is a large "rostral organ" with electroreceptive function beneath the skin of the snout (Bemis and Heatherington, 1982). Pleural ribs are lacking. The notochord is essentially a fluid-filled tube, with a thin layer of notochordal cells surrounded by heavier connective tissue. The intracranial joint allows the anterior part of the cranium to be flexed up and down during feeding or breathing. (Recall from Chapter 6 that a similar feature evolved in some placoderm fishes. The independent development of such a flexible joint in two vastly different lineages of fishes can be seen as an example of convergent evolution.)

Although the reproductive biology of *Latimeria* is not fully known, the species is ovoviviparous, even though the males do not have obvious intromittent organs. The 8- to 9-cm diameter eggs are said to be the largest shell-less eggs among all the fishes; large females have been found to carry up to 20 (Balon, 1991; Thomson, 1991). The gestation period was originally believed to be about 13 months, with the young born at a length of about 32 cm. The discovery of annuli in the scales of a newborn coelacanth suggests that the gestation period may last as long as 3 years—the longest known for any vertebrate (Froese and Palomares, 2000).

Latimeria is a lurking predator that lives on steep rock slopes. During the day, they seek refuge in caves. At night, they move out to prey on fishes and other nekton near the bottom (Uyeno and Tsutsumi, 1991). Stomach contents have included such deepwater fishes as lanternfish, cardinalfish, snappers (*Symphysanodon*), deep-sea witch eel,

a deepwater alfonsino (*Beryx decadactylus*), swell sharks, and cephalopods.

The osmoregulatory strategy pursued by coelacanths is quite similar to that seen in elasmobranchs (Griffith et al., 1974; Griffith and Pang, 1979; see Chapter 25), leading some to propose a closer affinity with the Chondrichthyes (Lagios, 1979, 1982), but this has not received wide acceptance (Compagno, 1979). There has been much speculation concerning the phylogenetic relationship of extant lobe-fin fishes, including *Latimeria*, to the tetrapods. Fritzsch (1987) has proposed that the basilar papillae of the inner ear of *Latimeria* are homologous with those seen in tetrapods. Gorr and Kleinschmidt (1993) demonstrated that the beta-hemoglobin chains in *Latimeria* appear more closely related to those of frogs than of other fishes, and Jarial et al. (1999) commented on the similarities between coelacanths and tetrapods in the structure of collagen fibrils. Homeotic (*Hox*) gene clusters (see Chapters 2 and 28) identified from the Indonesian coelacanth (*L. menadoensis*) confirm its tetrapod affinities (Koh et al., 2003). As we shall see later in this chapter, *Latimeria*, as a sarcopterygian, does have some phylogenetic affinities with tetrapods, but other groups are much closer to the ancestral stock of the tetrapods.

• • • • • • • • • • •
Taxon Dipnomorpha

Cloutier and Ahlberg (1996) and Janvier (1996) united the porolepimorph and dipnoan lungfishes in the taxon Dipnomorpha. Characters shared by the lungfishes and the extinct porolepiforms include the structure of the pectoral fins, which have a jointed axis, and the reduction of the ability to move the intracranial joint (Chang, 1991).

Infraclass †Porolepimorpha

Order †Porolepiformes

Jarvik (1980) suggested that members of this order, which lived from the Lower to the Upper Devonian, have given rise to the urodele amphibians—a proposal which has not gained much support. They were rather robust fishes (Fig. 7.3C), and the comparatively few genera that are known from the fossil record all seem to have been associated with near-shore marine environments (Janvier, 1996). Some had rhomboid scales containing cosmine; others had thin, round scales (Nelson, 1994).

Infraclass Dipnoi: Lungfishes

These fishes have received various placements in different systems of classification, having been considered a class, a subclass, an infraclass, and a series under the infraclass

Choanata. They differ from other bony fishes in having autostylic suspension of the upper jaw (the palatoquadrate is fused to the neurocranium; see Fig. 6.3). Furthermore, the teeth are fused into crushing plates. Maxillae, premaxillae, and coronoids are lacking. The connection to the esophagus is ventral. Internal nares (choanae) are present (see Fig. 7.7), as is a cloaca. The roofing bones of the skull share no known homologies with those of other bony fishes. They form a mosaic of numerous small bones in some extinct forms, usually with one or two relatively enlarged median plates at the posterior part of the braincase, and other, paired plates surrounding these. Students of the group have not applied names to the bones, but instead have designated them with letters of the alphabet. Figure 7.3D shows the skull roof of a modern lungfish, which has a reduced number of plates, with two elongated plates capping the median of the skull.

The paired fins of lungfishes consist of a long central axis with, in the Australian species, the fin rays disposed along it (see Fig. 3.10B; Plate 7.2). In the South American and African lungfishes, the fin rays are lacking, and the fins form long, slender appendages. A special pulmonary blood circulation is developed in living species, and the atrium of the heart is divided into left and right chambers by an incomplete septum. A spiral valve is present in the intestine. Like their coelacanth brethren, the lungfishes are capable of electroreception.

Fossil lungfishes (Fig. 7.3E) are found on all continents (Marshall, 1987), some from marine deposits. Nearly 60 nominal genera of extinct lungfishes are known, and as many as five extinct orders are recognized by some authorities. About half of the genera date back to the Devonian. Nelson (1994) grouped the lungfishes into two superorders, the extinct †Dipterimorpha and the Ceratodontimorpha, which includes the living species. A minute fossil fish, originally described as †Palaeospondylus gunni, has been recently identified as the larva of the dipterimorph †Dipterus valenciennesi. Extremely abundant in some deposits, these fossils are considered to represent the oldest definitive vertebrate larva (Thomson et al., 2003). Only orders with living representatives will be considered here.

Order Ceratodontiformes

This order contains the living Queensland lungfish (Neoceratodus forsteri) and an extinct genus †Ceratodus that had a broader distribution. Both are placed in the family **Ceratodontidae**, which is known from the Lower Triassic onward. N. forsteri is known from the Lower Cretaceous (Marshall, 1987). With a documented fossil record extending back almost 150 million years, the Queensland lungfish holds the distinction of being the world's oldest surviving vertebrate species (Allen et al., 2002).

Neoceratodus reaches nearly 2 m long and is a heavy-bodied fish, with large scales and paddlelike paired fins. The caudal fin is diphycercal. The species differs structurally from other living lungfishes by having a comparatively simple, unpaired lung, four pairs of gills, and a cartilaginous endocranium. Kemp (1999a, 1999b) has recently published a detailed study of the development of the skull of Neoceratodus. There are many differences in life history and habit between Neoceratodus and other lungfish species. The Australian species uses its gills to extract oxygen from the water. The lung can supplement oxygen uptake when activity increases or if the water becomes stagnant. No special nest is made; the eggs are laid among vegetation; and the young have no special external gills. Neoceratodus apparently is omnivorous, feeding on vegetation and on the many small forms of animal life that live among the plants, but it does not digest most plant material efficiently (Kemp, 1987). Unlike the lepidosireniform lungfishes that practice **aestivation**—a state of inactivity associated with seasonal drought conditions—the Australian lungfish frequents permanent bodies of water and is incapable of aestivation, but it can survive several weeks to a few months if kept moist in mud or vegetation (Kemp, 1987).

Geneticists have long been intrigued by the extremely large genomes of lungfishes, with Neoceratodus containing approximately 25 times the amount of genetic material of humans (Rock et al., 1996). A most instructive comparison is that between the genome size of lungfishes and the genome size determined for one of the most highly derived group of fishes, the Tetraodontiformes (see Chapter 17). Early vertebrate evolution has resulted in the duplication of a number of genes, including patterning gene sets, the homeotic or Hox genes (see Chapters 2 and 28). Studies of the genome of Neoceratodus, however, reveal that the duplication of Hox genes frequently observed in teleost fishes apparently occurred after the split between sarcopterygians and actinopterygians (Longhurst and Joss, 1999).

Order Lepidosireniformes

This order contains two families, Lepidosirenidae, from South America, and Protopteridae, from Africa (see Fig. 2.3F). Both have paired lungs, filamentous paired fins, and a membranous endocranium. Both families are known from the Upper Cretaceous onward, and consist of elongate fishes with fairly small scales. These species generally live in the shallows of larger bodies of water or in swamps that periodically dry up. Here, low concentrations of dissolved oxygen are frequently encountered, and the fishes in this order have evolved a much greater dependence on the lung for gas exchange. The gills are reduced and relatively ineffective compared to those of Neoceratodus. If the swamps dry up, both the African and the South American species can burrow into the muddy bottom and aestivate for several months, waiting for the next rainy season.

The family **Protopteridae** contains at least four species, all in the genus *Protopterus*. *Protopterus aethiopicus* reaches a length of 1.8 m, *P. dolloi* 1.3 m, *P. annectans* slightly less than 1 m, and *P. amphibius* is not known to reach more than 443 mm (Greenwood, 1987). *Protopterus annectans*, an African species, is known for the formation of an effective mucous cocoon, within which it can aestivate for seven or eight months (Greenwood, 1987). *Protopterus aethiopicus* has been maintained in its cocoon in the laboratory for as long as four years, far exceeding the normal aestivation period (Coates, 1937). Other species may aestivate in moist or water-filled burrows or, as in the case of females of *P. dolloi*, seek out open water during the dry season (Greenwood, 1987). There is fossil evidence of aestivation among other families of lungfishes as early as the Permian (Carroll, 1988).

When these fishes return to full activity after aestivation, nests are constructed, and breeding begins. The African species of *Protopterus* make simple holes near the edge of the swamp, whereas the South American *Lepidosiren* constructs a burrow. In both genera, the male guards the eggs and larvae. Larvae are held in place by a secretion from a cement organ on the breast region and have feathery external gills, similar to some species of salamanders.

Only one species, *Lepidosiren paradoxa*, is known from the family **Lepidosirenidae**. It is characterized by reduced paired fins, five gill arches, and the development, in breeding males, of feathery, gill-like structures on the pelvic fins. These seem to act as gills in reverse, releasing oxygen in the vicinity of the young. *Lepidosiren* feeds on animals, especially snails, but also consumes algae. Maximum length is about 1.25 m. The species appears to be very tenacious. One individual that arrived at Oregon State University with several cm of the tail missing not only healed the raw wound, but regenerated the section almost perfectly within two years.

Subclass †Rhizodontimorpha

The family †Rhizodontidae (Upper Devonian to Pennsylvanian) of the order †Rhizodontiformes is related to the osteolepiforms but lacks choanae (Long, 1989). This group is currently classified as a tetrapodomorph and a sister group to the panderichthyids (Elpistostegalia) and tetrapods (see Fig. 7.2; Cloutier and Ahlberg, 1996).

Subclass †Osteolepimorpha (†Rhipidistia)

These fishes existed from the Lower Devonian to the Lower Permian. They are choanate fishes with branched fin rays and an intracranial joint somewhat different from that of the coelacanths, in that the fifth cranial nerve passes through an opening posterior to the joint (Carroll, 1988).

Order †Osteolepiformes

Nelson (1994) claimed that this well-known group of Devonian and Carboniferous fishes includes at least seven families. These fishes differ from the porolepiforms in that they have a pineal foramen and that the pectoral fins are inserted low on the body. Two families, †Eusthenopteridae and †Panderichthyidae (= †Elpistostegalia) (Fig. 7.4A, B) have received much attention as close relatives to the ancestors of early tetrapods, especially because the skeleton of their paired fins (Fig. 7.5A–C) can be compared almost directly with that of the first tetrapods (Carroll, 1988; Coates and Clack, 1990; Colbert, 1980). The †Panderichthyidae have emerged as the most likely ancestors of the tetrapods (Ahlberg and Milner, 1994). Other families are †Osteolepidae, †Rhizodopsidae, †Tristichopteridae, †Canowindridae, and †Megalichthyidae. Some members of this order reached about 4 m in length. Recent investigations have questioned the validity of the order. Janvier (1996) retained the order †Osteolepiformes but removed the most likely tetrapod ancestors, the †Panderichthyidae. Ahlberg and Johanson (1998) claimed that the †Osteolepiformes are paraphyletic to the tetrapods and that the group has no taxonomic validity. In their revision of the classification of this group, they placed †*Panderichthys* as a sister taxon to the tetrapods.

THE ORIGINS OF TETRAPODS

One of the major contributions of the cladistic approach to phylogenetic systematics is that it has greatly clarified the relationship of terrestrial vertebrates to certain groups of fishes—specifically, the sarcopterygians. One of the reasons for discarding the term "Osteichthyes" as a class designating the bony fishes is that it reinforces the erroneous notion that actinopterygians and sarcopterygians, as constituent clades, share a greater phylogenetic affinity with each other than with tetrapods. We now recognize that sarcopterygian fishes such as the lungfishes are more closely related to us than they are to perches. In this context, then, consideration of the origins of terrestrial vertebrates takes on special meaning.

Until fairly recently, our view of the origin of tetrapods had certain groups of lobe-finned fishes hauling themselves out of the water in search of who knows what. This resulted in the eventual replacement of their "fishy" fins with limbs more appropriate for terrestrial locomotion. Studies on the comparative embryological development of paired appendages in chondrichthyans and bony fishes have confirmed

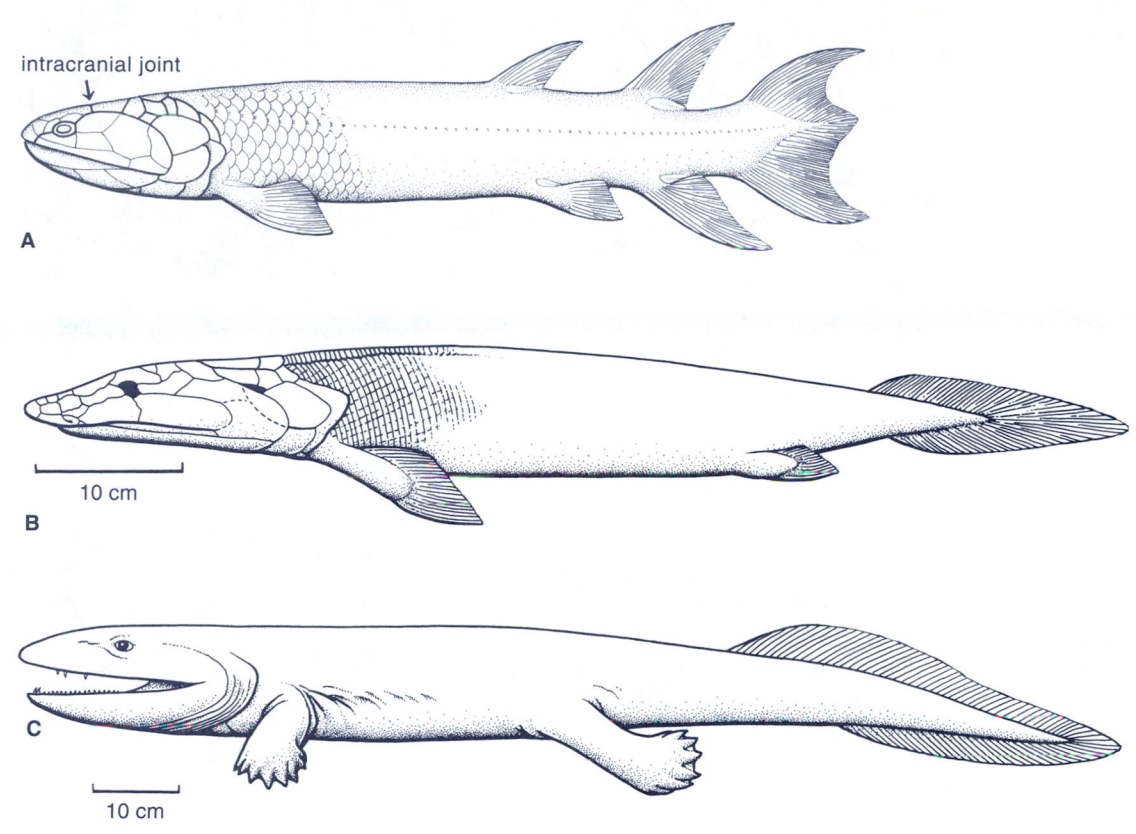

intracranial joint

10 cm

10 cm

A

B

C

FIGURE 7.4

Representatives of order †Osteolepiformes. **A,** †Eusthenopteridae (*†Eusthenopteron;* actual size ca. 40 cm);
B, †Panderichthyidae (*†Panderichthys;* actual size ca. 75 cm); **C,** Early tetrapod (*†Acanthostega;* actual size
ca. 125 cm).

that the pattern of appendicular muscle development in tetrapods is more akin to that seen in bony fishes. The genetic mechanism controlling the formation of paired fins in bony fishes, and hence of limbs in tetrapods, apparently predates the divergence of the sarcopterygians and the actinopterygians (Neyt et al., 2000). Recent studies, however, have indicated that our equation of the development of tetrapodal limbs among the bony fishes with the achievement of terrestriality is inaccurate. Clack (1997) observed that all Late Devonian tetrapod fossils are found in association with fishes. It is increasingly apparent that tetrapods had developed limbs long before they left the water and that their limbs were more suited to aquatic than terrestrial locomotion.

With limbs that may still have been useful in supporting the body in water, the tetrapodal gait may have arisen in these aquatic forms well before the transition to land (Shubin et al., 2004). The existence of tracks from the Early Devonian in Australia point to an even earlier origin for tetrapods (Long, 1990). One of the longest trackways of fossil footprints, located on a small island off the southwest coast of Ireland, shows no evidence of tail dragging, leading scientists to conclude that the animal was underwater (Westenberg, 1999). The dentition of the earliest tetrapods strongly suggests a piscivorous mode of existence. When these teeth are sectioned, they show a distinctive labyrinthine pattern of convolutions in the dentine—a pattern termed **labyrin-**

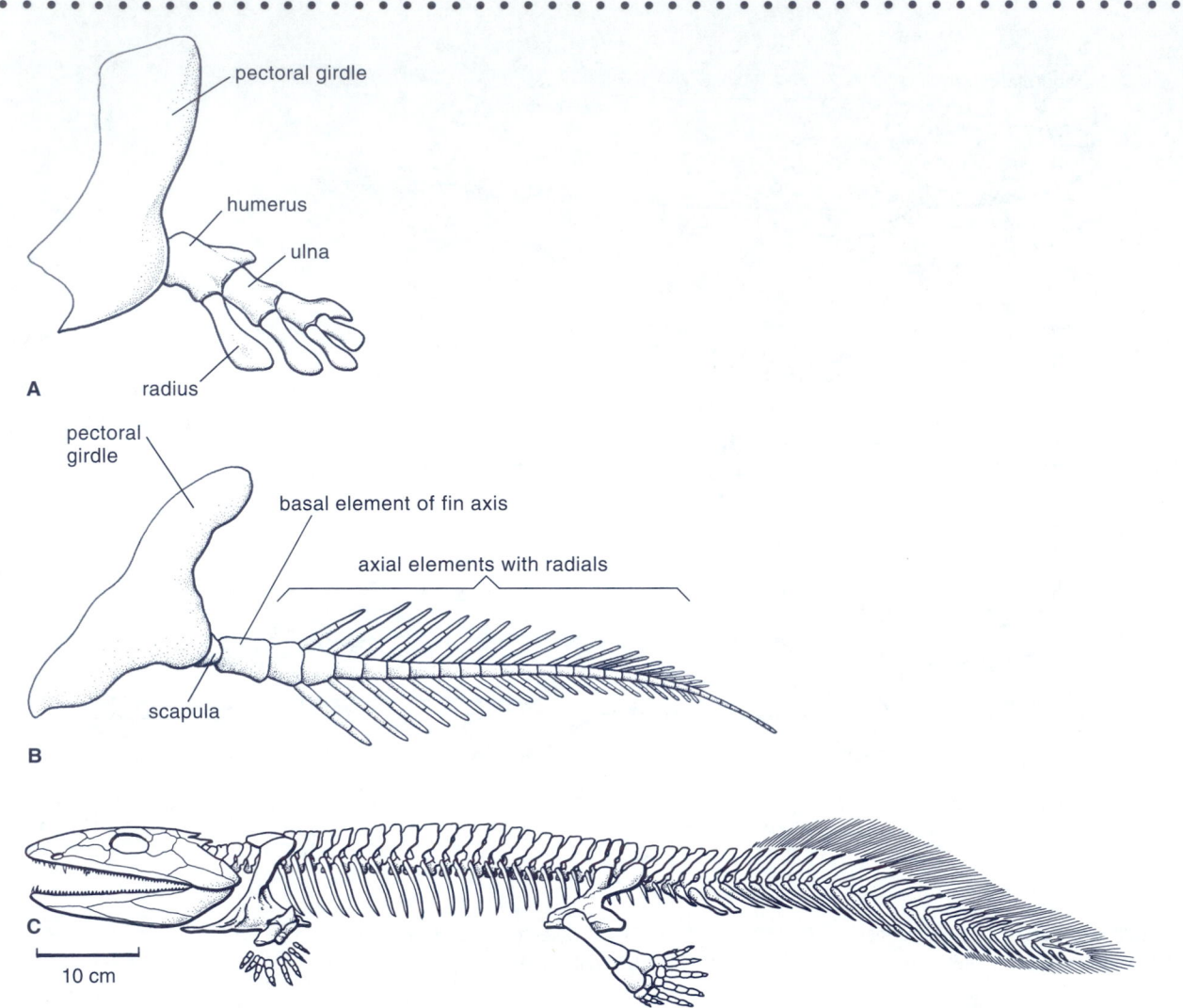

FIGURE 7.5
Pectoral skeleton of **A,** †*Eusthenopteron,* compared with **B,** Pectoral skeleton of modern lungfish, and **C,** Skeleton of †*Acanthostega,* showing appendages with multiple digits.

thodont or **polyplocodont.** This tooth type is also seen in osteolepiform fishes (Fig. 7.6).

The internal nostril, or *choana,* is integral to the story of terrestrial adaptation in vertebrates—with it, an alternative route for the intake of atmospheric gases becomes possible. With a choana, it is not necessary to open the mouth to breathe. In most bony fishes, the nasal openings are on the surface of the skull, and there is no choana. As mentioned earlier, the nasal capsule of the coelacanth is so configured, with both anterior and posterior openings on the side of the skull. In tetrapods, one of the nasal openings is in the roof of the mouth. The origin of the choana has long been a source of controversy, and this question is embedded in the larger problem of determining which lobe-finned lineage served as the ancestor of the tetrapods. Zhu and Ahlberg (2004) have shed new light on the subject in their discovery of what appears to be a transitional form in fossils of the lobefin †*Kenichthys.* This fossil fish demonstrates the emergence of the choana as a displaced posterior external nostril. The choana appears to have arisen from the pos-

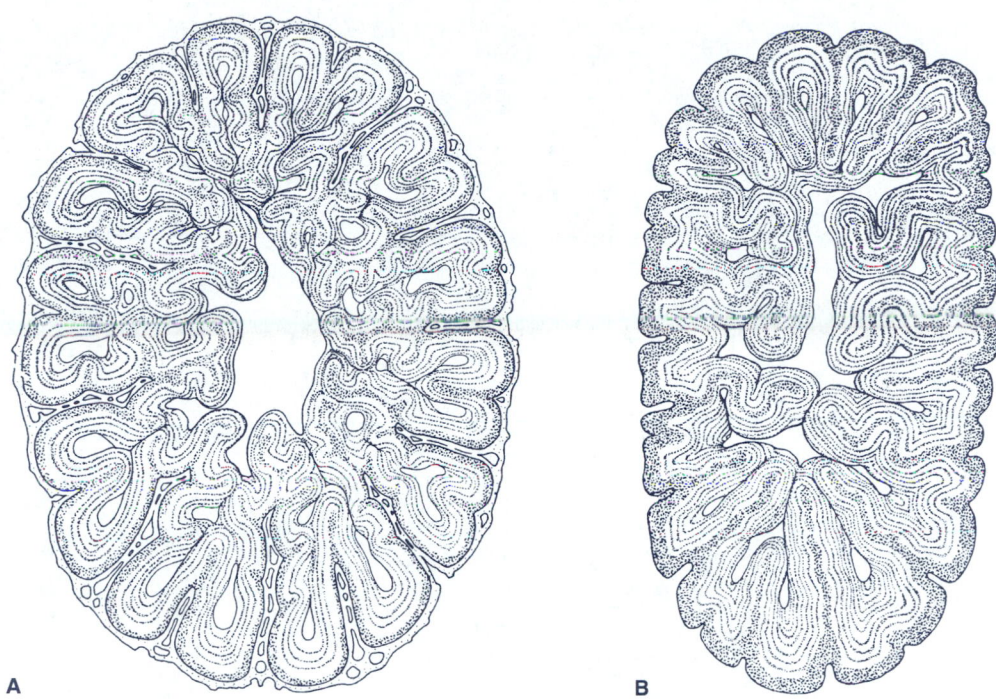

FIGURE 7.6

Labyrinthodont or polyplocodont dentition in **A,** the osteolepiform fish *Polyplocodus* and **B,** the temnospondyl-
ous tetrapod †*Benthosuchus.* (From B. J. Stahl, 1974. *Vertebrate history: Problems in evolution,* McGraw Hill, fig.
6.10, p. 218.)

terior nostril independently in both lungfishes and ancestral
tetrapods, and these choanae have come to occupy different
positions on the vertebrate palate (Fig. 7.7).

The best known early tetrapod, †*Ichthyostega,* was
originally believed to have been terrestrial. Other fossils,
including †*Acanthostega,* †*Hynerpeton,* and †*Elginerpeton,*
point to an aquatic existence in these first tetrapods (Ahlberg, 1995; Carroll, 1995). Grooves in the ceratobranchial
bones in †*Acanthostega,* for example, apparently accommodated branchial aortic arches carrying blood to gills (Coates
and Clack, 1991), and this feature was apparently shared
with †*Ichthyostega* (Clack et al., 2003). The grooved gill
bars, together with an unusual otic region that suggests an
ear highly specialized for hearing in water, strongly suggest
that †*Ichthyostega* was a much more aquatically adapted tetrapod than previously believed (Clack et al., 2003).

†*Elginerpeton* may be the oldest tetrapod known from
the fossil record. This Late Devonian creature shows a mosaic of panderichthyid and tetrapod features. Pentadactyly
was apparently not a feature of the first tetrapods. †*Acanthostega* showed eight toes on its front limbs (Fig. 7.5C;

Coates and Clack, 1990). †*Ichthyostega* had seven toes on its
hind limbs, whereas another Devonian tetrapod, †*Tulerpeton,* had six toes (Pough et al., 1999). The fossil fin of a Late
Devonian rhizodontid sarcopterygian, possibly of the genus
†*Sauripterus,* shows an array of radials that strongly suggest the digits of tetrapods. They are clearly not meant to
be load-bearing skeletal elements, however (Daeschler and
Shubin, 1998).

Rosen et al. (1981) reported that the lungfishes possessed a suite of characters that made them the most likely
sister group of early tetrapods, but Holmes (1985) refuted
many of their arguments. Coates and Clack's (1991) studies
of the branchial skeleton of †*Acanthostega* showed a close
resemblance to the fossil lungfish †*Chirodipterus,* however.
Jamieson (1991) mentioned that the spermatozoa of lungfishes suggest a relationship to both *Latimeria* and amphibians. Of the modern fishes, the dipnoans certainly appear
to be the most closely related to tetrapods (see Fig. 7.2).
Molecular phylogenetic data have supported the contention
that lungfishes are more closely related to tetrapods than are
coelacanths (Meyer, 1995). Zhu and Yu (2002) have reported

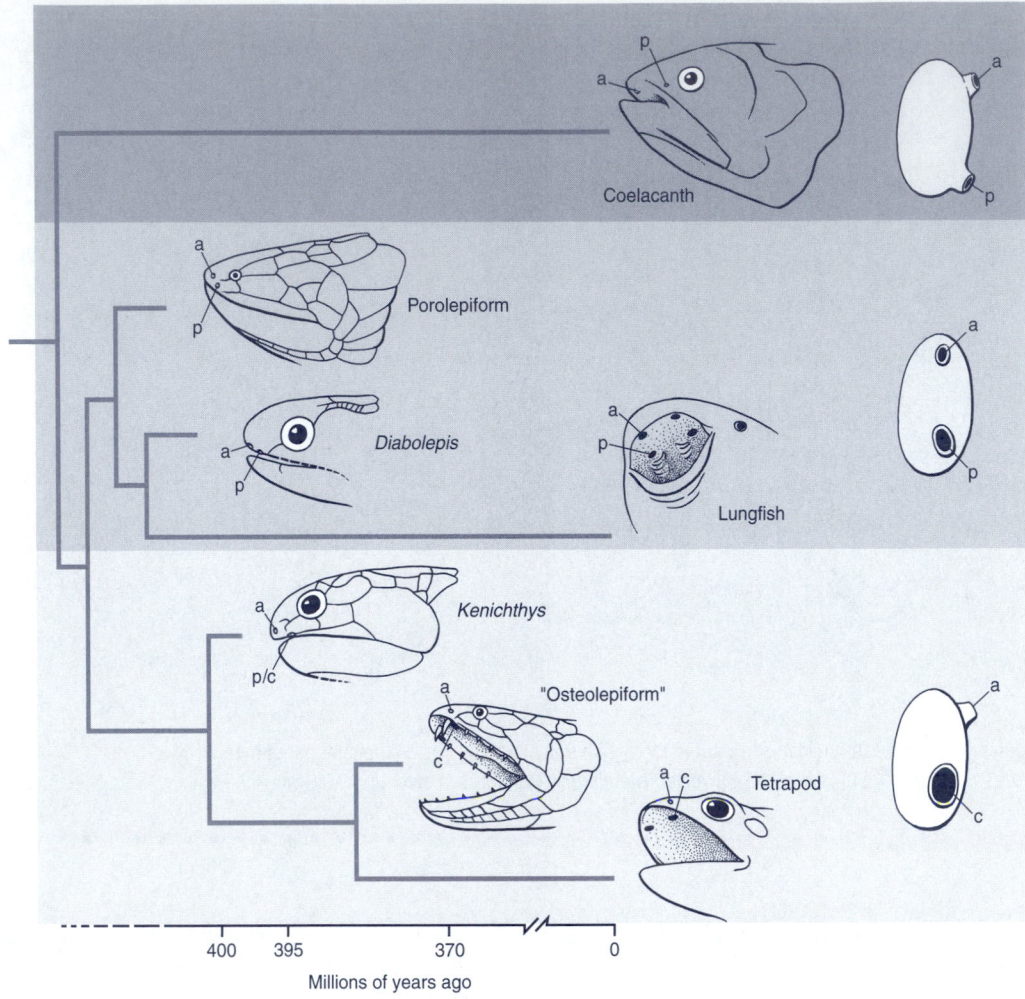

FIGURE 7.7

Interrelationships of fossil and extant sarcopterygians, showing the position of the choanae. †*Diabolepis* is an early lungfish that still shows external nostrils. In †*Kenichthys*, the posterior nostril is intermediate between that seen in an external nostril and a true choana. Ventral views of nasal capsules on the left side of the head are depicted for the coelacanth, lungfish, and tetrapod. *a*—anterior nostril; *p*—posterior nostril; *c*—choana. (From Janvier, 2004; Zhu and Ahlberg, 2004.)

on the discovery of a fossil sarcopterygian that possesses an eyestalk yet exhibits a character complex consistent with a fish that is close to the point at which lungfishes diverged from tetrapod ancestors.

The evidence for well-developed tetrapods roaming the shallows of the Late Devonian waterways has been well established in the past few years. We are much less sure, how-ever, of the events that surrounded the actual emergence of the tetrapods onto land. One Late Devonian genus, †*Hyner-peton*, appears to have been more terrestrially adapted than other early tetrapods (Daeschler et al., 1994). The evolution of the amniote condition, in which the embryo is surrounded by an array of extraembryonic membranes, is generally viewed as a definitive terrestrial adaptation. Modern terrestrial non-amniotes, such as frogs and salamanders, usually must return to water to breed. Although it has been proposed that the amniote condition evolved from certain embryo-retaining forms (recall that *Latimeria* gestates its eggs internally) as a means of enhancing maternal–fetal exchange, the widely held view that this actually represents a terrestrial innovation cannot be refuted (Laurin and Girondot, 1999).

A gap of more than 20 million years separates the earliest tetrapods from the first definitive terrestrial vertebrates. Paton et al. (1999) closed that gap slightly in reporting on the discovery of what appears to be one of the earliest amniotelike tetrapods. This fully terrestrial animal, named †*Casineria*, had a number of skeletal features consistent with locomotion on land. Included among these was the presence of pentadactyly. Another recently described tetrapod (†*Pederpes*) from this period (about 350 million years ago) possessed a fully pentadactylic foot but showed evidence of polydactyly in the presence of a supernumerary digit (Clack, 2002). From these crucial fossil finds, we might conclude that a reduction of the number of digits per limb to five provided for sufficient strength and stability to counteract the forces of gravity and get these creatures moving further and further away from the swamps from which they arose. The phylogenetic relationships of early tetrapods, reflecting the acquisition of increased terrestrial adaptations, have been explored by Ruta et al. (2003).

· ·
KEY POINTS AND CONNECTIONS
· ·

- The taxon Teleostomi encompasses the extinct acanthodians, the lobe-finned Sarcopterygii, and the ray-finned Actinopterygii. Acanthodians, sarcopterygians, and primitive actinopterygians share some of the characters seen in two other early gnathostome groups, the placoderms and elasmobranchs.

- The class †Acanthodii dates its origin to the early Silurian, at a time when the ostracoderms were dominant. Their elongate, somewhat fusiform bodies and heterocercal tails, coupled with a comparatively sophisticated jaw apparatus, probably facilitated greater penetration into pelagic habitats, especially in the marine realm.

- Although the sarcopterygians were a dominant group during the Paleozoic, with numerous orders represented, only two lineages, the coelacanths (Actinistia) and the lungfishes (Dipnoi) have survived into modern times. Until a few years ago, only one modern species of coelacanth, *Latimeria chalumnae*, inhabiting deep rocky regions off the southeast coast of Africa, was known. A second species, *L. menadoensis*, was described from specimens caught in the coastal waters of Indonesia.

- Only six species of modern lungfishes have survived, all of them inhabiting freshwater. Australia and South America are each home to a single species, whereas four species have been described from Africa. A remarkable physiological and behavioral adaptation is seen among the lungfishes of the order Lepidosireniformes (African and South American species). These fishes are able to enter a state of profound metabolic depression, known as aestivation, as a means to withstand periodic drought conditions.

The respiratory adaptations of aestivating fishes are discussed in Chapter 24.

- The evidence that is currently available supports an origin of the tetrapods from ancestral sarcopterygians. These ancestral forms appear to have shared many features with the lungfishes. Fossil evidence suggests that these earliest tetrapods were still fully aquatic creatures.

The circulatory and respiratory adaptations accommodating atmospheric gas exchange, including those found in sarcopterygians and hence most applicable to the evolution of tetrapods, are discussed in Chapter 24. Excretory and osmoregulatory adaptations of tetrapods are considered in Chapter 25.

· ·
BUILDING AN ICHTHYOLOGY LIBRARY
· ·

Your library ought to include essential scientific works on the coelacanth as well as recreational reading. Examples from each category are listed here, together with important texts on the lungfishes and the origins of the tetrapods.

Bemis, W. E., W. W. Burggren, and N. E. Kemp (Eds.). 1987. *The biology and evolution of lungfishes.* Alan R. Liss, New York.

Clack, J. A. 2002. *Gaining ground: The origin and evolution of the tetrapods.* Indiana University Press, Bloomington.

Forey, P. L. 1998. *History of coelacanth fishes.* Chapman and Hall, London.

McCosker, J. E., and M. D. Lagios (Eds.). 1979. The biology and physiology of the living coelacanth. *Occ. Pap. Calif. Acad. Sci. 134.*

Noakes, D. L. G. 1999. *Environmental Biology of Fishes,* Vol. 54. (This issue contains several articles pertaining to the discovery of the Indonesian species of coelacanth.)

Smith, J. L. B. 1956. *Old fourlegs. The story of the coelacanth.* Longmans, Green, London.

Weinberg, S. 2000. *A fish caught in time: The search for the coelacanth.* HarperCollins, New York.

Zimmer, C. 1998. *At the water's edge: Fish with fingers, whales with legs, and how life came ashore but then went back to sea.* Touchstone, Simon and Schuster, New York.

· ·
REFERENCES
· ·

Ahlberg, P. E. 1995. *Elginerpeton pancheni* and the earliest tetrapod clade. *Nature 373:* 420–425.

———, and A. R. Milner. 1994. The origin and early diversification of tetrapods. *Nature 368:* 507–514.

———, and Z. Johanson. 1998. Osteolepiformes and the ancestry of tetrapods. *Nature 395:* 792–794.

Allen, G. R., S. H. Midgeley, and M. Allen. 2002. *Field guide to the freshwater fishes of Australia.* Western Australian Museum, Perth.

Andrews, S. M. 1973. Interrelationships of crossopterygians, pp. 137–177. In *Interrelationships of fishes*, P. H. Greenwood, R. S. Miles, and C. Patterson (Eds.). Academic Press, London.

Balon, E. K. 1991. Prelude: The mystery of a persistant life form, pp. 9–13. In *The biology of* Latimeria chalumnae *and evolution of coelacanths*, J. A. Musick, M. N. Bruton, and E. K. Balon (Eds.). Kluwer, Dordrecht, the Netherlands.

Basden, A. M., G. C. Young, M. I. Coates, and A. Ritchie. 2000. The most primitive osteichthyan braincase? *Nature 403:* 185–188.

Bemis, W. E., and T. E. Heatherington. 1982. The rostral organ of *Latimeria chalumnae:* Morphological evidence of an electroreceptive function. *Copeia 1982:* 467–471.

Bjerring, H. C. 1973. Relationships of Coelacanthiformes, pp. 179–205. In *Interrelationships of fishes*, P. H. Greenwood, R. S. Miles, and C. Patterson (Eds.). Academic Press, London.

Bruton, M. N. 1991. Bibliography of the living coelacanth *Latimeria chalumnae*, with comments on publication trends. *Env. Biol. Fishes 32:* 403–433.

———, and S. E. Coutouvidis. 1991. An inventory of all known specimens of the coelacanth *Latimeria chalumnae*, with comments on trends in the catches. *Env. Biol. Fishes 32:* 371–390.

Carroll, R. L. 1988. *Vertebrate paleontology and evolution*. Freeman, New York.

———. 1995. Problems of the phylogenetic analysis of Paleozoic tetrapods. *Bull. Mus. Nat. Hist. Nat., Paris, 4e. Sér., 17:* 389–455.

Chang, M. M. 1991. Rhipidistians, dipnoans, and tetrapods, pp. 3–28. In *Origins of major groups of tetrapods: Controversies and consensus*, H.-P. Schultze, and L. Trueb (Eds.). Cornell University Press, Ithaca, NY.

Clack, J. A. 1997. Devonian tetrapod trackways and trackmakers: A review of the fossils and footprints. *Palaeogeography, Palaeoclimatology, Palaeoecology 130:* 227–250.

———. 2002. An early tetrapod from "Romer's Gap." *Nature 418:* 72–76.

———, P. E. Ahlberg, S. M. Finney, P. Dominguez Alonso, J. Robinson, and R. A. Ketcham. 2003. A uniquely specialized ear in a very early tetrapod. *Nature 425:* 65–69.

Cloutier, R., and P. E. Ahlberg. 1996. Morphology, characters, and the interrelationships of basal sarcopterygians, pp. 445–479. In *Interrelationships of fishes*, M. L. J. Stiassny, L. R. Parenti, and G. D. Johnson (Eds.). Academic Press, San Diego.

———, and P. L. Forey. 1991. Diversity of extinct and living actinistian fishes (Sarcopterygii), pp. 59–74. In The biology of *Latimeria chalumnae* and evolution of coelacanths, J. A. Musick, M. N. Bruton, and E. K. Balon (Eds.). *Env. Biol. Fishes 32* (1–4).

Coates, C. W. 1937. Slowly the lungfish gives up its secrets. *Bull. N.Y. Zool. Soc. 40:* 25–34.

Coates, M. I., and J. A. Clack. 1990. Polydactyl in the earliest known tetrapod limbs. *Nature 347:* 66–69.

———, and———. 1991. Fish-like gills and breathing in the earliest-known tetrapod. *Nature 352:* 234–235.

Colbert, E. H. 1980. *Evolution of the vertebrates* (3rd ed.). John Wiley, New York.

Compagno, L. J. V. 1979. Coelacanths: Shark relatives or bony fishes? pp. 45–52. In The biology and physiology of the living coelacanth, J. E. McCosker and M. D. Lagios (Eds.). *Occ. Pap. Calif. Acad. Sci. 134.*

Daeschler, E. B., and N. Shubin. 1998. Fish with fingers? *Nature 391:* 133.

———,———, K. S. Thomson, and W. W. Amaral. 1994. A Devonian tetrapod from North America. *Science 265:* 639–642.

Danke, J., T. Miyake, T. Powers, J. Schein, H. Shin, I. Bosdet, M. Erdmann, R. Caldwell, and C. T. Amemiya. 2004. Genome resource for the Indonesian coelacanth *Latimeria menadoensis*. *J. Exp. Zool. 301A:* 228–234.

Denison, R. 1979. Acanthodii, pp. 1–62. In *Handbook of paleoichthyology*, Vol. 5. H.-P. Schultze (Ed.). Gustav Fisher Verlag, Stuttgart, Germany.

Erdmann, M. V. 1999. An account of the first living coelacanth known to scientists from Indonesian waters. *Env. Biol. Fishes 54:* 439–443.

———, R. Caldwell, and M. K. Moosa. 1998. Indonesian "king of the sea" discovered. *Nature 395:* 335.

Forey, P. L. 1998. *History of the coelacanth fishes*. Chapman and Hall, London.

Fricke, H., O. Reinike, H. Hofer, and W. Nachtigall. 1987. Locomotion of the coelacanth *Latimeria chalumnae* in its natural environment. *Nature 329:* 331–333.

———, and K. Hissmann. 1994. Home range and migrations of the living coelacanth *Latimeria chalumnae*. *Mar. Biol. 120:* 171–180.

———,———, J. Schauer, O. Reinicke, L. Kasang, and R. Plante. 1991. Habitat and population size of the coelacanth *Latimeria chalumnae* at Grand Comoro. *Env. Biol. Fishes 32:* 287–300.

Fritzsch, B. 1987. Inner ear of the coelacanth fish *Latimeria* has tetrapod affinities. *Nature 327:* 153–154.

Froese, R., and M. L. D. Palomares. 2000. Growth, natural mortality, length–weight relationship, maximum length and length-at-first-maturity of the coelacanth *Latimeria chalumnae*. *Env. Biol. Fishes 58:* 45–52.

Gorr, T., and T. Kleinschmidt. 1993. Evolutionary relationships of the coelacanth. *Am. Sci. 81:* 72–82.

Greenwood, P. H. 1987. The natural history of lungfishes, pp. 163–179. In *The biology and evolution of lungfishes*, W. E. Bemis, W. W. Burggren, and N. E. Kemp (Eds.). Alan R. Liss, New York.

Griffith, R. W., and P. K. T. Pang. 1979. Mechanisms of osmoregulation in the coelacanth: Evolutionary implications, pp. 79–93. In The biology and physiology of the living coelacanth, J. E. McCosker and M. D. Lagios (Eds.). *Occ. Pap. Calif. Acad. Sci. 134.*

———, B. L. Umminger, B. F. Grant, P. K. T. Pang, and G. Pickford. 1974. Serum composition of the coelacanth, *Latimeria chalumnae* Smith. *J. Exp. Zool. 187:* 87–102.

Hissmann, K., and H. Fricke. 1996. Movements of the epicaudal fin in coelacanths. *Copeia 1996:* 606–615.

———,———, and J. Schauer. 1998. Population monitoring of the coelacanth (*Latimeria chalumnae*). *Conserv. Biol. 12:* 759–765.

Holder, M. T., M. V. Erdmann, T. P. Wilcox, R. L. Caldwell, and D. M. Hillis. 1999. Two living species of coelacanths? *Proc. Natl. Acad. Sci. USA 96:* 12616–12620.

Holmes, E. B. 1985. Are lungfishes the sister group of tetrapods? *Biol. J. Linn. Soc. 25*(4): 379–397.

Jamieson, B. G. M. 1991. *Fish evolution and systematics: Evidence from spermatozoa*. Cambridge University Press, Cambridge.

Janvier, P. 1996. *Early vertebrates*. Oxford University Press, Oxford.

———. 2004. Wandering nostrils. *Nature 432:* 23–24.

Jarial, M. S., L. R. Ganion, and B. A. Verhoestra. 1999. Ultrastructure of the collagen fibrils in the coelacanth. *J. Fish Biol. 55:* 1119–1122.

Jarvik, E. 1980. *Basic structure and evolution of vertebrates*, Vols. 1 and 2. Academic Press, New York.

Kemp, N. E. 1987. The biology of the Australian lungfish, *Neoceratodus forsteri*, pp. 181–198. In *The biology and evolution of lungfishes*, W. E. Bemis, W. W. Burggren, and N. E. Kemp (Eds.). Alan R. Liss, New York.

————. 1999a. Ontogeny of the skull of the Australian lungfish *Neocera-todus forsteri* (Osteichthyes: Dipnoi). *J. Zool. 248:* 97–137.

————. 1999b. Sensory lines and rostral skull bones in lungfish of the family Neoceratodontidae (Osteichthyes: Dipnoi). *Alcheringa 23:* 289–307.

Koh, E. G. L., K. Lam, A. Christoffels, M. V. Erdmann, S. Brenner, and B. Venkatesh. 2003. Hox gene clusters in the Indonesian coelacanth, *Latimeria menadoensis. Proc. Natl. Acad. Sci. USA 100:* 1084–1088.

Lagios, M. D. 1979. The coelacanth and the Chondrichthyes as sister groups: A review of shared apomorph characters and a cladistic analysis and reinterpretation, pp. 25–44. In The biology and physiology of the living coelacanth, J. E. McCosker and M. D. Lagios (Eds.). *Occ. Pap. Calif. Acad. Sci. 134.*

————. 1982. *Latimeria* and the Chondrichthyes as sister taxa: A rebuttal to recent attempts at refutation. *Copeia 1982:* 942–948.

Lauder, G. V., and K. F. Liem. 1983. The evolution and interrelationships of the actinopterygian fishes. *Bull. Mus. Comp. Zool. 150:* 95–197.

Laurin, M., and M. Girondot. 1999. Embryo retention in sarcopterygians, and the origin of the extra-embryonic membranes of the amniotic egg. *Ann. Sci. Nat. 3:* 99–104.

Long, J. A. 1989. A new rhizodontiform fish from the early Carboniferous of Victoria, Australia, with remarks on the phylogenetic position of the group. *J. Vert. Paleontol. 9:* 1–17.

————. 1990. Heterochrony and the origin of tetrapods. *Lethaia 23:* 157–166.

Longhurst, T. J., and J. M. P. Joss. 1999. Homeobox genes in the Australian lungfish *Neoceratodus forsteri. J. Exp. Zool. (Mol. Dev. Evol.) 285:* 140–145.

Maisey, J. G. 1986. Heads and tails: A chordate phylogeny. *Cladistics 2:* 201–256.

Marshall, C. R. 1987. A list of fossil and extant dipnoans, pp. 15–23. In *The biology and evolution of lungfishes,* W. E. Bemis, W. W. Burggren, and N. E. Kemp (Eds.). Alan R. Liss, New York.

McCosker, J. E. 1979. Inferred natural history of the coelacanth, pp. 17–24. In The biology and physiology of the living coelacanth, J. E. McCosker and M. D. Lagios (Eds.). *Occ. Pap. Calif. Acad. Sci. 134.*

————, and M. D. Lagios. 1979. The biology and physiology of the living coelacanth. *Occ. Pap. Calif. Acad. Sci. 134.*

Meyer, A. 1995. Molecular evidence on the origin of tetrapods and the relationships of the coelacanth. *TREE 10:* 111–116.

Miles, R. S. 1973. Relationships of acanthodians, pp. 63–104. In *Interrelationships of fishes,* P. H. Greenwood, R. S. Miles, and C. Patterson (Eds.). Academic Press, London.

Millot, J. 1953. "Notre" coelacanthe. *Rev. Madagascar, Tananarive 17:* 18–20.

Moy-Thomas, J. A., and R. S. Miles. 1971. *Paleozoic fishes* (2nd ed.). Revised by R. S. Miles. Chapman and Hall, London.

Musick, J. A., M. N. Bruton, and E. K. Balon (Eds.). 1991. *The biology of Latimeria chalumnae and evolution of coelacanths.* Kluwer, Dordrecht, the Netherlands.

Nelson, J. S. 1994. *Fishes of the world* (3rd ed.). John Wiley, New York.

Neyt, C., K. Jagla, C. Thisse, L. Haines, and P. D. Currie. 2000. Evolutionary origins of vertebrate appendicular muscle. *Nature 408:* 82–86.

Paton, R. L., T. R. Smithson, and J. A. Clack. 1999. An amniote-like skeleton from the Early Carboniferous of Scotland. *Nature 398:* 508–513.

Pough, F. H., C. M. Janis, and J. B. Heiser. 1999. *Vertebrate life* (5th ed.). Prentice Hall, Upper Saddle River, NJ.

Pouyaud, L., S. Wirjoatmodjo, I. Rachmatika, A. Tjakrawidjaja, R. Hadiaty, and W. Hadie. 1999. Une nouvelle espèce de coelacanthe. Preuves génétiques et morphologiques. *Comptes Rendus—Sér. III 322:* 261–267.

Rock, J., M. Eldridge, A. Champion, P. Johnston, and J. Joss. 1996. Karyotype and nuclear DNA content of the Australian lungfish, *Neoceratodus forsteri* (Ceratodidae: Dipnoi). *Cytogenet. Cell Genet. 73:* 187–189.

Rosen, D. E., P. L. Forey, B. G. Gardiner, and C. Patterson. 1981. Lungfishes, tetrapods, paleontology, and plesiomorphy. *Bull. Am. Mus. Nat. Hist. 167*(4): 159–276.

Ruta, M., M. I. Coates, and D. L. J. Quicke. 2003. Early tetrapod relationships revisited. *Biol. Rev. 78:* 251–345.

Shubin, N. H., E. B. Daeschler, and M. I. Coates. 2004. The early evolution of the tetrapod humerus. *Science 304:* 90–93.

Smith, J. L. B. 1939. A living fish of Mesozoic type. *Nature 143:* 455–456.

————. 1956. *Old fourlegs. The story of the coelacanth.* Longmans, Green, London.

Springer, V. G., and G. D. Johnson. 2004. Study of the dorsal gill-arch musculature of teleostome fishes, with special reference to the Actinopterygii. *Bull. Biol. Soc. Wash. 11.*

Stahl, B. J. 1974. *Vertebrate history: Problems in evolution.* McGraw-Hill, New York.

Thomson, K. S. 1967. Mechanisms of intracranial kinetics in fossil rhipidistian fishes (Crossopterygii) and their relatives. *Zool. J. Linn. Soc. 46:* 223–253.

————. 1991. *Living fossil: The story of the coelacanth.* W. W. Norton, New York.

————. 1993. The origin of the tetrapods, pp. 33–62. In Functional morphology and evolution, P. Dodson and P. Gingerich (Eds.). *Am. J. Sci. (Special Vol.) 293A.*

————. 1999. The coelacanth: Act three. *Am. Sci. 87:* 213–215.

————, M. Sutton, and B. Thomas. 2003. A larval Devonian lungfish. *Nature 426:* 833–834.

Uyeno, T. 1991. Observations on locomotion and feeding of released coelacanths, *Latimeria chalumnae. Env. Biol. Fishes 32:* 267–273.

————, and T. Tsutsumi. 1991. Stomach contents of *Latimeria chalumnae* and further notes on its feeding habits. *Env. Biol. Fishes 32:* 275–279.

Watson, D. M. S. The acanthodian fishes. *Phil. Trans. Roy. Soc. (Biol. Sci.) 228:* 49–146.

Weinberg, S. 2000. *A fish caught in time: The search for the coelacanth.* HarperCollins, New York.

Westenberg, K. 1999. From fins to feet. *Nat. Geog. 195*(5): 114–127.

Zhu, M., and P. E. Ahlberg. 2001. A primitive sarcopterygian fish with an eyestalk. *Nature 410:* 81–84.

————, and————. 2004. The origin of the internal nostril of tetrapods. *Nature 432:* 94–97.

————, and X. Yu. 2002. A primitive fish close to the common ancestor of tetrapods and lungfish. *Nature 418:* 767–770.

————,————, and P. Janvier. 1999. A primitive fossil fish sheds light on the origin of bony fishes. *Nature 397:* 607–610.

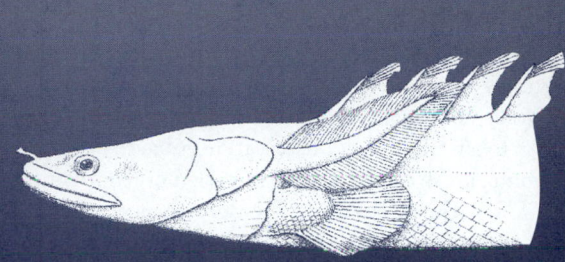

8 INTRODUCTION TO THE ACTINOPTERYGIAN FISHES: ORIGINS AND MODERN DIVERSITY

ADAPTATIONS OF THE ACTINOPTERYGIANS

ORIGINS OF THE ACTINOPTERYGII

CLASS ACTINOPTERYGII

Subclass Chondrostei
Order †Cheirolepiformes
Order Polypteriformes: Bichirs and Reedfish
Order †Palaeonisciformes
Order †Tarrasiiformes
Order †Phanerorhynchiformes
Order †Saurichthyiformes
Order Acipenseriformes
Subclass Neopterygii
Order †Pycnodontiformes
Order †Aspidorhynchiformes
Order †Pachycormiformes
Order Semionotiformes
Order Amiiformes

A most significant event in the early evolutionary history of the bony fishes was the divergence of the sarcopterygian (lobe-finned) fishes from the actinopterygian, or "ray-finned," fishes. What is puzzling is that the lobefins—so prevalent during the Devonian, and providing the ancestral stock for the evolution of tetrapods—are almost completely absent from the modern bony fishes. The actinopterygians, on the other hand, have experienced an unprecedented radiation and diversification. Most likely, this can be attributed to musculoskeletal developments that promoted versatility in feeding and locomotion. As we will see in this chapter, the modern fish fauna contains only a few representatives from the earliest lineages of actinopterygians. These survivors from an earlier time in the history of fishes nonetheless represent some fascinating adaptations to the aquatic realm. Although a few will readily enter marine waters, these evolutionary relics are most commonly associated with freshwater.

ADAPTATIONS OF THE ACTINOPTERYGIANS

In terms of the extent of their representation among the modern fish fauna, none of the groups we have considered so far can be considered to be especially successful. They are noteworthy for their peculiarity—a condition reflecting their affinity with forms long extinct. Only about 84 species of modern agnathan fishes are known. Chondrichthyans have experienced a modest diversification as marine carnivores, with more than 840 species counted among the modern fish fauna. The modern sarcopterygians—for all of the fascination they engender owing to their phylogenetic relationship to the tetrapods—are the barest remnant of a much greater assemblage known from the fossil record, with only 8 recognized species alive today. It is the actinopterygian fishes that fill the world's oceans, lakes, and rivers with such a diverse array of solutions to the problems of aquatic existence. The estimated 28,000 species of modern actinopterygians show a staggering diversity of forms, including both archaic species with lines of descent traced back to the earliest forms and more recent examples that have undergone explosive bouts of speciation. Two factors account for their success. The first is their capacity to thoroughly dominate freshwater habitats. Although the actinopterygians were not the first group to inhabit freshwater, they have dominated it to the extent that about 41 percent of the known species are exclusively freshwater forms (Horn, 1972). This is especially remarkable when one considers that only about 0.01 percent of the hydrosphere is composed of habitable lakes and rivers (see Chapters 31 and 32). The second factor is the amazing evolutionary plasticity of the actinopterygian body form. Fishes have become modified to fill just about every conceivable aquatic niche. Let us consider two of the most significant examples of such modification—feeding and locomotor adaptations.

Unlike the ancestral bony fishes, the actinopterygians possess a highly kinetic skull. The evolutionary trends we recognize among the actinopterygians are a reduction of bone mass in the skull and an increase in flexibility at points of articulation. This has permitted the development of a range of feeding adaptations, including protrusible jaws that can participate in suction-type feeding and allow fishes to get the jump on potential prey. An arsenal of dentition on a variety of bones in the gape and throat also contributes to feeding versatility.

Locomotor versatility was the consequence of a general reduction in the extent of dermal ossification and of the evolution of the homocercal caudal fin. Although some modern families, such as pipefishes (Syngnathidae) and trunkfishes (Ostraciidae), have evolved heavy dermal ossification, actinopterygians usually have thin scale coats, thus enabling a greater range of motion in trunk muscles and, hence, a greater array of locomotor strategies (see Chapter 19). With the heterocercal caudal fin, as seen in many archaic groups, including early actinopterygians, there is an asymmetry of forward thrust application, so that the paired appendages are called on to sustain lift. Some researchers have suggested that the homocercal caudal fin permits a more horizontal forward motion, thus decreasing the dependence on using the paired appendages as control surfaces. This—in conjunction with the neutral buoyancy conferred on the animal by the swim bladder—would enable the paired fins to function over a greater range of activities. It should be pointed out that not all investigators agree that symmetry of caudal fin surface necessarily correlates with uniformity in forward thrust (Lauder, 1989, 1994, 2000; see Chapter 19). Although fins may still be important in stabilizing forward motion, they may have greater importance in enabling quick turns. Other functions that fins may take on include walking along the sea floor as well as on land; sensing the environment; and producing sound. Freed from the job of constantly contributing lift, the pectoral fins may be folded flat against the body in situations where minimizing drag is essential.

ORIGINS OF THE ACTINOPTERYGII

The earliest evidence of actinopterygian fishes may be dated to a couple of recent fossil finds from the Early Devonian, dated at more than 400 million years ago. One of these, †*Psarolepis,* was discussed in Chapter 7 in the context of the origins of the sarcopterygians. The phylogenetic position of this fossil remains uncertain, as it possessed features of acanthodians, sarcopterygians, and actinopterygians. It may represent a basal group from which both sarcopterygians and actinopterygians evolved (Zhu et al., 1999). Another fossil from Early Devonian limestone of southeastern Australia also displays actinopterygian features, but its phylogenetic position remains uncertain (Basden et al., 2000). Isolated scales and dermal bone fragments are also known from the Early Devonian (Janvier, 1996). The fossil record of actinopterygians remains sketchy through the Silurian but reflects an increase in abundance during the Early Carboniferous (Janvier, 1996). Actinopterygians count among their number possibly the largest fish that ever existed—the genus †*Leedsichthys* (Fig. 8.1). Discovered in Jurassic British clay sediments, this planktivorous behemoth may have reached sizes in excess of 12 meters (Long, 1995).

A suite of characteristics define the actinopterygian fishes. The teeth are capped with a transparent mineralized tissue termed acrodine. The upper jaw is defined by an elongate, rearward projecting maxilla that becomes reduced

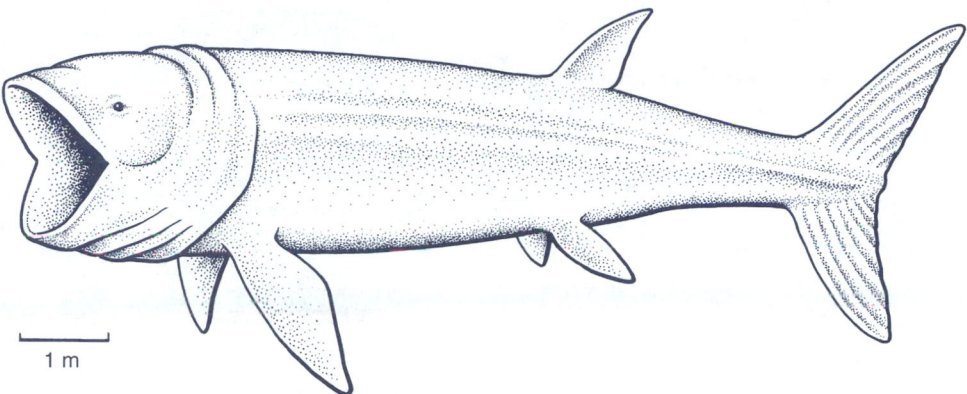

FIGURE 8.1
†*Leedsichthys,* a member of the pachycormids, an ancient group of actinopterygians of the Jurassic and Cretaceous.

in derived forms. A single dorsal fin is usually present and is frequently subdivided into two regions. The term *Actinopterygii* ("ray-finned") comes from the structure of the median and paired appendages, in which the fin surface is composed of slender rays that support intervening membranes. In the case of the median fins, the rays insert directly into the body, with no intervening basal points of articulation, as in the chondrichthyans. The paired appendages also insert directly onto internal skeletal supports and have no fleshy or bony structure to them. One peculiar order, the Polypteriformes, is an apparent exception to this appendicular characterization. The dermal skeleton covering the most primitive actinopterygians consists of heavy ganoid scales or isolated bony scutes. These give way to lighter, simpler cycloid and ctenoid scales, and some forms have divested themselves of scales entirely. As might be expected in so diverse a group, some of these characteristics may not be present. Acrodine, however, is present in all but a few groups (Janvier, 1996).

Actinopterygian fishes are also distinguished by the mode of formation of the cerebral hemispheres of the brain. Rather than forming through evagination, as occurs in other craniates, the cerebral hemispheres develop through a process of outward eversion of the lateral walls of the embryonic telencephalon (Northcutt, 1995; see Chapter 26).

CLASS ACTINOPTERYGII

Nelson (1994) divided the class Actinopterygii into two subclasses: the *Chondrostei,* including the living bichirs, sturgeons, paddlefishes, and several extinct groups; and the *Neopterygii.* The distinctive features of the bichirs strongly suggest that they ought to be considered a sister group of all other actinopterygians. The neopterygians include the gars and the bowfin and a diverse array of teleostean fishes, from bonytongues, tarpons, and herrings to trouts, percoids, and their derivatives, both living and extinct (Fig. 8.2).

Subclass Chondrostei

The chondrosteans share some ancestral characters such as spiracles; heterocercal or abbreviate heterocercal tails; and more fin rays than ray supports in the dorsal and anal fins, yet there is evidence that the group is not monophyletic (McCune and Schaeffer, 1986; Nelson, 1994).

Order †Cheirolepiformes
This order has a single family (†Cheirolepidae), known only from the Devonian. Large mouths, strongly heterocercal caudal fins, dorsal fins set far back over the anal fin, and small scales distinguish the order. Individuals of †*Cheirolepis* were up to 0.5 m long.

Order Polypteriformes: Bichirs and Reedfish
There is some difficulty in fitting the bichirs and reedfish of Africa into the chondrostean framework. Gardiner and Schaeffer (1989) proposed a phylogeny that suggests a relationship of †*Cheirolepis* and the Polypteriformes to the rest of the actinopterygians. Bartsch (1997) identified sarcopterygian features in the cranium of polypteriform fishes but still recognized them as basal members of the Actinopterygii. Molecular studies have confirmed the polypteriforms as the

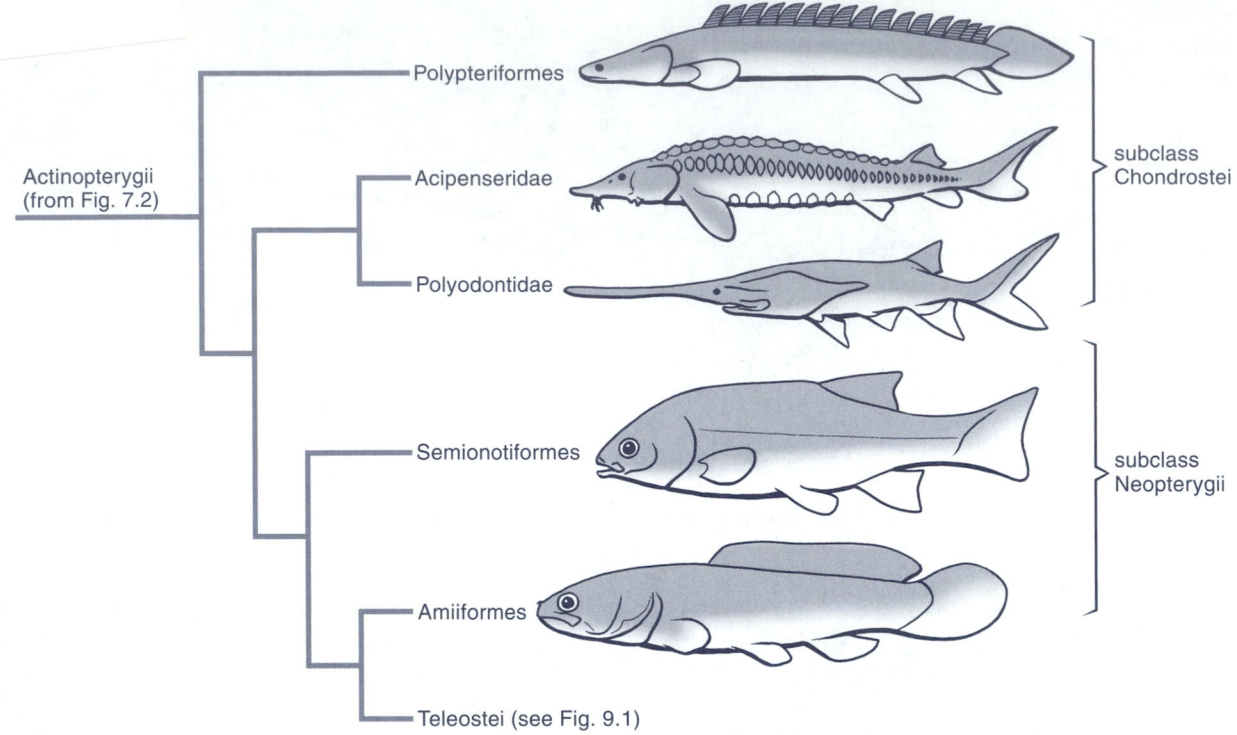

FIGURE 8.2

Phylogeny of the Actinopterygii. (From Janvier, 1996; Grande and Bemis, 1996.)

most basal group of actinopterygians (Meyer, 2003). It has long been known that polypteriform fishes are distinct from other actinopterygians in having four gill arches instead of the usual five that characterize gnathostome vertebrates. Britz and Johnson (2003) questioned the claims put forth by several researchers that the fourth gill arch is homologous with the fifth gill arch of other actinopterygians. Rather, they provided evidence that the fifth arch has been lost in polypteriform fishes, thus making their fourth arch homologous with the fourth arch of other actinopterygians.

Polypteriform fishes have a lunglike evagination originating from the ventral part of the esophagus that functions in gas exchange—an important adaptation given their propensity for living in oxygen-deficient waters. There are several other peculiarities of skeletal and soft anatomy that add to the belief that they should be placed in their own subgroup, outside or inside the Actinopterygii (Jarvik, 1980; Rosen

et al., 1981). They are often classified as Brachiopterygii or Cladistia. Depending on the authors of the classification system, they are placed variously as subclass, infraclass, or order (Benton, 1990; Carroll, 1988; Lauder and Liem, 1983a, 1983b; Nelson, 1994; Rosen et al., 1981).

The order Polypteriformes contains the single family **Polypteridae**, including the bichirs (*Polypterus;* see Fig. 3.10A) and the reedfish (*Erpetoichthys* [= *Calamoichthys*] *calabaricus*), known in the aquarium trade as the ropefish. The dorsal fin of these fishes is different from that of any other group; it consists of a series of separate small fins, each with one large, spinelike ray and one or more soft branches from the posterior edge of the spine (see Figs. 3.10A and 8.3A). The spines can make handling these fishes an unpleasant task. The lobate nature of the fins prompted Jordan (1923) to classify the order as "crossopterygians"—an archaic term used to identify members of the class Sarcopterygii.

DIAGNOSIS OF THE POLYPTERIFORMES
- Maxilla firmly bonded to skull.
- Spiracle present, but spiracular canal absent.
- Lobate pectoral fin supported by bony propterygium, metapterygium, and cartilaginous mesopterygium.
- Ganoid scales consisting of three layers—ganoin, cosmine, and isopedine.
- Spiral valve in intestine (like chondrichthyans).
- Heart with conus arteriosus (like chondrichthyans).
  Sources: Lauder and Liem, 1983a, 1983b; Nelson, 1994.

Polypterus includes about 10 species, all found in fresh-waters on the African continent. They are small- to medium-sized predators, usually maturing around 20 cm in length. *P. bichir* may reach a length of about 1 m, is sought as food, and is generally roasted in coals with the scales left on. The larvae have a rather amphibian appearance due to the large, feathery external gills (Fig. 8.3A). The single species of reedfish, *Erpetoichthys calabaricus* of West Africa, is long and slender, reaching about 90 cm in length (Plate 8.1). It lacks pelvic fins. It is somewhat amphibious in habit, being able to survive terrestrial conditions for up to an hour. This in itself

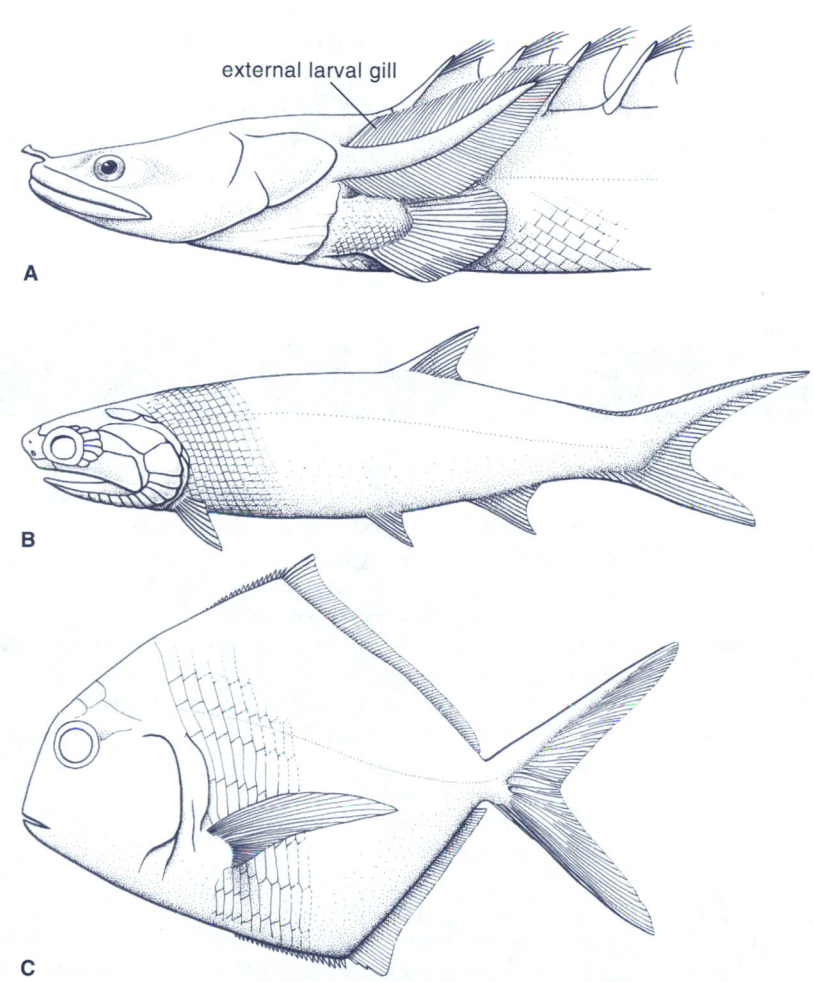

FIGURE 8.3
A, Anterior portion of *Polypterus,* showing external gills of juvenile. Representatives of the order †Palaeonisciformes; **B,** Palaeoniscoid (actual size ca. 15 cm); **C,** Platysomoid (actual size ca. 15 cm). (B after Romer, 1970; C after Moy-Thomas and Miles, 1971.)

poses an intriguing question: Assuming such terrestriality is facilitated by the availability of lungs, is it possible that the polypteriform mode of breathing can be correlated with that possibly used by the earliest tetrapods (see Chapter 7)? Brainerd et al. (1989) provided some interesting observations. Among these is the recording of negative pressures in the pleuroperitoneal cavities of two polypterid species, *Polypterus senegalus* and *Erpetoichthys calabaricus;* this indicates a mode of lung breathing that is distinct from other fishes and amphibians. Whereas lungfishes and amphibians usually fill the lung using positive pressure generated from muscle contractions in the buccal region, polypteroids inhale using **recoil aspiration**. This method uses the passive recoil gained from a rigid dermal jacket, composed of scales articulated by peg-and-socket joints into continuous rows. This does not appear to inhibit body flexibility, as especially demonstrated in the eellike *Erpetoichthys calabaricus*. Because the skin surfaces of the earliest tetrapods apparently had heavy scale coats, it has been suggested that these may have facilitated recoil aspiration as well (Brainerd et al., 1989). The evolution of lung filling by positive buccal pressure, as seen in modern amphibian lineages, may be a consequence of the loss of such heavy scale jackets.

Order †Palaeonisciformes

This order is represented by a number of extinct actinopterygian families. The diversity of fossil forms raises the possibility that this is not a monophyletic group. Nelson (1994) identified four suborders: †Palaeoniscoidei, †Redfieldioidei, †Platysomoidei, and †Dorypteroidei. These fishes had an evolutionary history from the Devonian into the Cretaceous and developed along several lines. Body forms ranged from the typical fusiform, as in palaeoniscoids (Fig. 8.3B) to the compressiform in platysomoids such as the †Chirodontidae (Fig. 8.3C), which had superficially symmetrical caudal fins and lacked pelvic fins. For the most part, these fishes were covered by ganoid scales and had large bony heads. The mouths were generally long, with conical teeth, and seem to have been designed mainly for catching and holding prey. Maxillaries were large and probably not movable. The platysomoids had more specialized mouths that were reduced in size and were provided, in some instances, with crushing teeth.

GOING DEEPER · Caviar

How ironic that the last word in gourmet food items is nothing more than the processed eggs of one of the most primitive fishes alive today. Caviar, long associated with royalty and *haute cuisine*, was not always such an exalted dish. In France, just before World War One, 40 centimes would buy you a couple loaves of bread or a kilogram of caviar. Ancient Persians are believed to have been the first consumers of caviar, which they called *chavjar* (loosely translated as "cake of power"). The Russian czars probably did the most to elevate caviar to the position it holds today, and the Caspian Sea region is home to the most legendary varieties of the delicacy.

The most cherished of all caviars came from the sterlet (*Acipenser ruthenus*), a rather diminutive sturgeon, whose size makes it attractive for public aquaria. The rarity of the sterlet has made its small golden eggs almost unobtainable today. Almost 90 percent of the world's caviar production comes from three species of sturgeon from the Caspian Sea region: the beluga (*Huso huso*), the oscietra or osetra (*Acipenser gueldenstaedti*), and the sevruga (*A. stellatus*). The United States accounts for almost 30 percent of the world's caviar consumption and is emerging as a leading producer of caviar. The world caviar market has undergone traumatic upheaval in the last decade, mainly because of the breakup of the former Soviet Union. Now there are five nations competing for the dwindling sturgeon resources of the Caspian: Kazakhstan, Turkmenistan, Azerbaijan, Russia, and Iran. In the absence of Soviet control, regulation of the fishery has all but disappeared, and poaching and smuggling are running rampant. Officials from the U.S. Fish and Wildlife Service believe that as much as 50 percent of the caviar imported into the United States comes from illegal fisheries. Pollution in the Caspian, mainly from the petroleum industry, is on the increase. In the early 1990s, Iranian caviar production peaked at around 300 tons per year. In 1999, it barely exceeded 100 tons. Recently, the United States and 142 other countries that are signatories to the Convention on International Trade in Endangered Species (CITES) have initiated steps to eliminate the illegal trade in caviar, in the hope that they can restore proper controls on a dangerously overexploited resource. All caviar that is imported into the United States is subjected to DNA fingerprinting at the National Fish and Wildlife Forensics Laboratory in Ashland, Oregon. If it does not conform to import declarations that specify species and source of caviar, the shipment is confiscated and the importer is subject to federal prosecution. The testing procedure for the caviar samples, which involves matching nucleotide sequences from the mitochondrial genome with a database of living sturgeon species, recently determined that almost one fourth of the caviar imported into the New York area was mislabeled (Birstein et al., 1998). In March of 2002, CITES concluded that beluga

Order †Tarrasiiformes

Fishes of this order had a diphycercal rather than heterocercal caudal fin that was continuous with elongate dorsal and anal fins. They are known from the Carboniferous.

Order †Phanerorhynchiformes

A single genus that resembles sturgeons is known from the Upper Carboniferous. Their fins were supported by a reduced number of fin rays.

Order †Saurichthyiformes

These fishes, from the Triassic and Jurassic, had an elongate body with the dorsal and anal fins set posteriorly, as in the gars. Their heads and mouths were elongate. Some species were up to one meter in length.

Order Acipenseriformes

This order includes the sturgeons and paddlefishes and related extinct forms. Its phylogenetic position remains unclear. Most morphological studies place the acipenseriform fishes as a sister group to the Neopterygii (Fig. 8.2), yet molecular studies favor a closer relationship of the sturgeons and paddlefishes with the gars and bowfin, to the exclusion of teleosts (Meyer, 2003).

Suborder †Chondrosteoidei

Chondrosteoid fossils are known from the Lower Triassic to the Lower Cretaceous. One family, the †Chondrosteidae, resembled sturgeons in general body shape, including a heterocercal tail and inferior mouth. The Errolichthyidae, probable relatives, had an almost terminal mouth.

Suborder Acipenseroidei

The sturgeons (Fig. 8.4A; Plate 8.2) of the family **Acipenseridae** are superficially distinguishable from paddlefishes by having bony scutes along the sides and back and four barbels on the underside of the rostrum, which is shorter than in the paddlefishes. A total of 25 extant species of sturgeon are recognized, including 17 within the genus *Acipenser* (Birstein and Bemis, 1997; Birstein and DeSalle, 1998). Osteological studies on fossil and recent forms and mitochondrial DNA sequencing have confirmed that *Scaphirhynchus*

stocks in the Caspian had recovered sufficiently to permit a limited fishery—a move that has both baffled and outraged conservationists, who have challenged the data that CITES used to make this decision (Stone, 2002).

Cultured sturgeon may be able to provide some relief from the fishery pressure exerted on natural populations. Osetra sturgeon are currently being farm-raised in San Francisco, California, and a caviar importer has recently initiated a sturgeon culture program with beluga, sevruga, and osetra sturgeons at a Florida aquaculture facility. Although only the roe from sturgeons can properly be referred to as caviar, other fish species produce palatable substitutes. North American paddlefish (*Polyodon spathula*) produces a quality product known in caviar circles as "Chattanooga beluga." One of the better "caviars" comes from the roe of the shad (*Alosa sapidissima*). The eggs of salmonids, including trout, salmon, and whitefish, are processed as caviar, as is the roe of lumpfish (Cyclopteridae). The Japanese produce a crunchy, delicate orange product from the eggs of flying fishes (Exocoetidae).

The processing of caviar is, as might be expected, a precise and painstaking procedure, which undoubtedly contributes to the mystique of the delicacy. The intact egg sac is removed from the freshly killed animal and placed on a sieve that is gently shaken to sort the eggs according to size. Once the eggs are sorted and cleaned, the caviar master—the keeper of the secret recipes for the best preservation procedures—will classify the eggs according to color and apply the appropriate quantities of highly purified salt. Salt cures and preserves the eggs, giving them their characteristic texture. Although Borax® is sometimes added to impart a softer, sweeter flavor, it is an illegal food additive in the United States and is therefore not present in caviar imported there. Caviar is then packed into its characteristic lacquered tins and shipped to distributors. Caviar has a limited shelf life, governed by the amount of salt used in processing, and should be kept very cold but not frozen. It should not be stored nor served in containers of metal—such as silver—that is prone to oxidation. A gold spoon is best to serve caviar, but if you have loaned out your collection of gold caviar spoons, glass, wood, or plastic may be used. Purists will consume caviar served unadorned on lightly buttered toast points, but any bland cracker will do. These individuals will regard as blasphemous the serving of caviar with hard-boiled eggs, sour crème, or chopped onion, as is commonly done. The only beverages permitted to accompany caviar are champagne or, better yet, a bracing shot of vodka from the bottle stored in your freezer. Just remember to raise your glasses in salute to the future health of the glorious animal that makes possible this most precious commodity.

and *Pseudoscaphirhynchus* are sister taxa to all other modern species, but that the validity of the genus *Huso* may be questionable (Birstein and DeSalle, 1998; Grande and Bemis, 1996; Mayden and Kuhajda, 1996). The genera *Scaphirhynchus* and *Pseudoscaphirhynchus* do not appear to form a monophyletic group, however (Birstein et al., 2002).

Sturgeons are distributed around the northern part of the Northern Hemisphere (**holarctic** distribution) and have marine, freshwater, and anadromous members. The waters of central and eastern Europe are especially noted for the numbers of species of sturgeon found there. Sturgeons are highly esteemed for their flesh and their roe, from which premium grades of caviar are made (see Going Deeper). Leather from sturgeons is used for bookbinding, and isinglass, a gelatinous extract from the swim bladder, is used to clarify beer. North American sturgeon fisheries are insignificant compared to the tonnage of sturgeon landed in eastern Europe and Iran.

The largest sturgeon is the beluga (*Huso*) of the Caspian and Black seas. It is reported to reach a length of about 9 m and a weight of 1,500 kg. Large specimens may be 100 years old and carry more than 7 million eggs. One of the most important commercial sturgeons is the Russian sturgeon (*Acipenser gueldenstaedti*) of the Caspian Sea and the Sea of Azov, reaching about 2.3 m long. The largest North American species is the white sturgeon (*Acipenser transmontanus*) of the Pacific coast. Lengths of 6 m and weights of about 850 kg have been reported. This giant is found in both fresh and salt water from southern California to Cook Inlet in Alaska and frequents large rivers such as the Snake, Columbia, Sacramento, and Fraser. The smallest members of the family are the strange-looking shovelnose sturgeons (Fig. 8.4A). *Scaphirhynchus platorhynchus* of the Mississippi reaches about 1 m long, whereas the small shovelnose (*Pseudoscaphirhynchus hermanni*) of the Amur Darya River in the former Soviet Union is reported to reach a maximum length of only 27 cm.

The food of sturgeons includes worms, crustaceans, and small fishes that can be sucked up by the greatly protrusible mouth. Beluga are reported to eat larger fishes and, occasionally, the young of the Caspian seal. A typical sturgeon's life history includes a migration from feeding grounds to breeding grounds in large rivers. Spawning takes place over gravel in fairly swift water. The demersal, adhesive eggs hatch after 3 to 5 days, and the larvae—about 1 cm long—drift downstream to suitable rearing areas in the river or the sea. Growth is slow, with many species reaching a length of only about 1.2 m after 10 years. Males reach sexual maturity earlier than the females. Medium-sized species may become mature at 8 to 12 years for the males and 10 to 15 years for the females.

Their requirement for clean water; their slow growth and late maturity; and the fact that the eggs are the most valuable product from these animals pose problems for fishery managers. The white sturgeon was once a prized commercial fish in California, Oregon, and Washington, but a virtually unregulated fishery—especially for the large females—has depleted the stock, and replacement is slow. The species constitutes a minor fishery at this time, and aquaculture programs have been initiated in California to supplement the dwindling populations. Caspian stocks of sturgeon are also under heavy fishing pressure, and large hatcheries have been erected in the former Soviet Union and Iran to supplement the yield of naturally spawned fish. In the case of the beluga, the destruction of spawning habitat is so extensive in the Danube and Volga rivers that hatchery managers cannot obtain enough brood stock for all of the hatcheries (Birstein et al., 1997).

Because the life history of sturgeons makes them especially vulnerable to overexploitation, several North American species currently receive federal protection or are the subject of intensive state management programs. The pallid sturgeon (*Scaphirhynchus albus*) and the Alabama sturgeon (*S. suttkusi*) are listed as "endangered." *Acipenser oxyrhynchus*, the Atlantic sturgeon, is listed as a species "of special concern," and a subspecies, the gulf sturgeon (*A. o. desotoi*) is considered "threatened." Also listed as "threatened" are the lake sturgeon (*A. fulvescens*) found in the Mississippi drainage and north through the Great Lakes, the Red River of the North, the St. Lawrence River, and Hudson Bay; and the shortnose sturgeon (*A. brevirostrum*) of fresh and marine waters of the Atlantic coast. Another vulnerable species, the green sturgeon (*A. medirostris*), is primarily marine and ranges from southern California to Alaska and westward to Asia, entering bays and rivers to spawn. The Kootenai River population of white sturgeon (*A. transmontanus*) is also classified as endangered. Most species of sturgeons—and the paddlefishes as well—should be considered threatened due to a combination of overexploitation and habitat degradation (Birstein et al., 1997; Waldman, 1995).

The paddlefishes (Fig. 8.4B; Plate 8.3) of the family **Polyodontidae** are characterized by an extremely long snout with two minute barbels. They lack the bony scutes of the sturgeons. There are two living species, *Polyodon spathula* of eastern North America and *Psephurus gladius* of the Yangtze River in China, but the origins of the family have been dated back to the Upper Cretaceous (Grande and Bemis, 1991). Both are found only in freshwater. *Psephurus* has a swordlike snout, relatively short gill rakers, and a protrusible mouth. Its food appears to be other fishes. Although some specimens have been reported as long as 7 m, authenticated records report a maximum size of about 4 m.

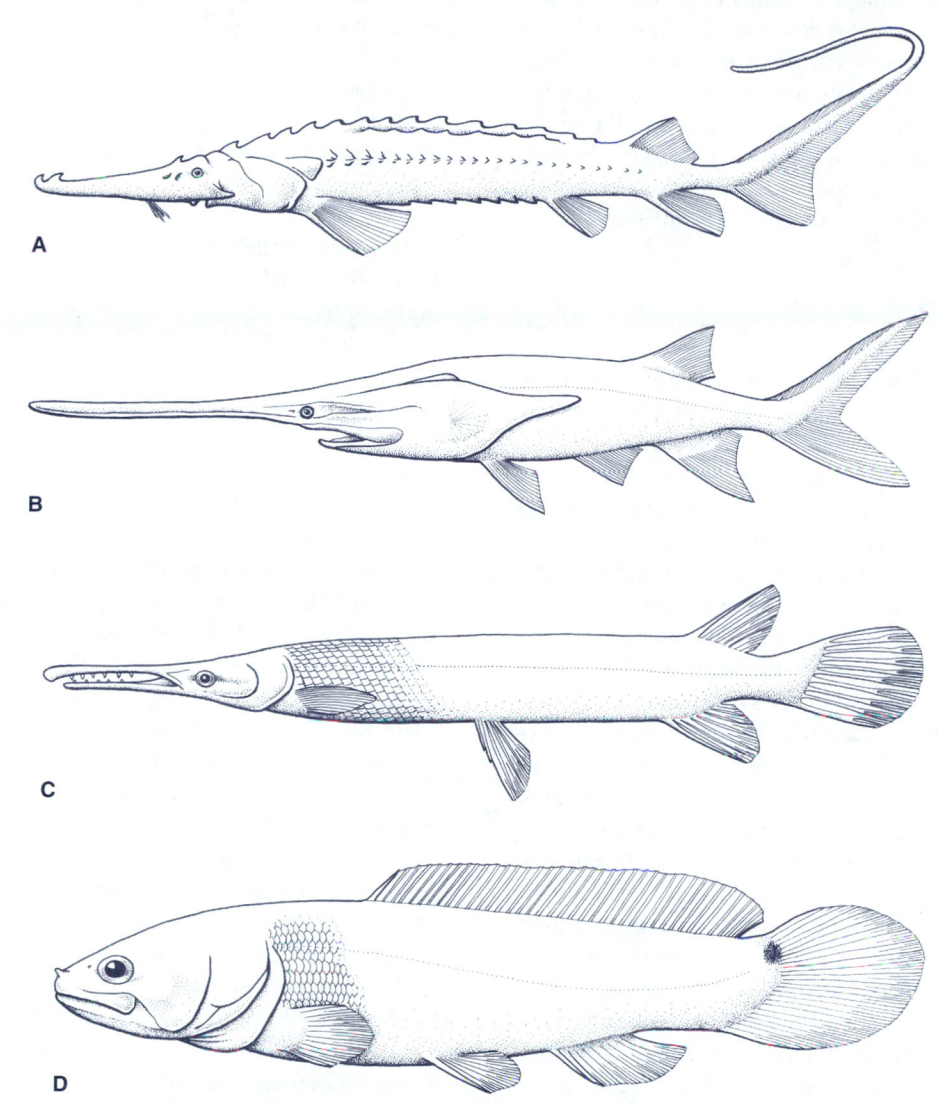

FIGURE 8.4
Representatives of the order Acipenseriformes: **A,** Shovelnose sturgeon (*Pseudoscaphirhynchus*); **B,** Paddlefish
(*Polyodon*). Representatives of **C,** Order Lepisosteiformes, gar (*Lepisosteus*); **D,** Order Amiiformes, bowfin (*Amia*).

Polyodon has a broad, paddle-shaped snout, long gill rakers, and a nonprotrusible mouth. Crustaceans, especially *Daphnia*, and other plankton are strained from the water by means of the long gill rakers. Both sturgeons and paddlefishes have an abundance of electroreceptors on the snout—these are similar to the ampullae of Lorenzini seen in elasmobranchs (see Chapter 21). In paddlefishes, the electroreceptors respond to the low-frequency emissions of

Daphnia (Wilkens et al., 1997, 2001; Wojtenek et al., 2001), and studies have shown that sensitivity to these emissions is enhanced by an optimal level of background noise (termed *stochastic resonance*), thus maximizing foraging success in areas of low visibility (Russell et al., 1999). *Polyodon* reaches a length of about 2 m, a weight of 76 kg, and spawns in swift rivers over gravel bars in the spring. Growth is more rapid than that of sturgeons. Although the range of the paddlefishes has been reduced by human activities, the species has become numerous in several reservoirs in the Mississippi drainage and forms the basis of a popular fishery based on snagging with treble hooks. Impoundment of the free-flowing stretches of these rivers has resulted in the ironic situation that this fishery can only be maintained through intensive artificial propagation and introductions, as most of the natural spawning habitat has been eliminated.

. . . .
Subclass Neopterygii

The gars, the bowfin, and the teleosts do not retain the full heterocercal tail and the spiracle that are typical of many chondrosteans, and only one living family has ganoid scales. The rays of the dorsal and anal fin correspond in number with the underlying fin ray supports. The gars and the bowfin share some structures not seen in the division Teleostei (see Chapter 9). There is a close correspondence between the multiple caudal fin rays and their supports. The ventral branches of the spinal nerve roots penetrate the lateral musculature and course ventrally outside the musculature. The bowfin, gars, and their fossil relatives were formerly placed in a group called "Holostei," which was proposed to represent a group intermediate between the Chondrostei and the Neopterygii. Another view of the relationship of these groups was the placement of gars into the division Ginglymodi of the Neopterygii, and the bowfin, their extinct relatives, and the teleosts into the division Halecostomi (the bowfin and closely related fossil genera in the subdivision Halecomorphi, and the teleosts in the subdivision Teleostei). A study of mitochondrial DNA sequences from gars, bowfin, and teleosts has actually lent support to the archaic concept of the "Holostei" by suggesting that the gars and the bowfin form a clade distinct from teleosts (Normark et al., 1991). In spite of the molecular data, Gardiner et al. (1996) concluded that the bowfin represents a sister group to the teleosts and the gars are a sister group of these two combined (Fig. 8.2). Different interpretations of the rather complex relationships of the extinct and extant neopterygians can be found in Patterson (1973), Patterson and Rosen (1977), Lauder and Liem, (1983a, 1983b), and Nelson (1994). Nelson (1994) chose to include in the subclass Neopterygii two orders with modern representatives and three extinct orders, followed by the monophyletic division Teleostei. The three extinct orders are briefly reviewed here, followed by the two orders with modern representatives.

Order †Pycnodontiformes
The †Pycnodontiformes were a group of shallow-water marine fishes that lived from the Upper Triassic to the Eocene. Their phylogenetic affinities are currently uncertain, although some researchers have placed them as a sister group to the teleosts.

Order †Aspidorhynchiformes
Existing from the Upper Jurassic to the Cretaceous, these elongate fishes somewhat resembled the needlefishes (Belonidae).

Order †Pachycormiformes
Living from the Jurassic to the Upper Cretaceous, the genera in this order included forms that closely resembled the teleosts, including the giant †Leedsichthys mentioned earlier.

Order Semionotiformes
This order includes (depending on the authority) two or three extinct families, comprising several genera, and the living gars, family **Lepisosteidae**. Gars are found in eastern North America from the Great Lakes region to Costa Rica, with one species reaching Cuba. Fossil gars date from the Upper Cretaceous of Europe and have been found in India, Africa, and North America.

Gars are elongate fishes with the body covered by heavy ganoid scales and the head with equally hard bone (Fig. 8.4C; Plate 8.4). Although many fishes resemble gars in having an elongate snout, gars are distinctive in that the nostrils reside at the tip rather than at the base of the snout—the position seen in paddlefishes and beaked teleosts such as the beloniform needlefishes (see Chapter 14). This condition arises as a consequence of elongation in the ethmoid bones of the skull, between the eyes and the nares. Both jaws are elongate and armed with several rows of strong, sharp teeth. Some of the upper teeth are borne on a series of infraorbital bones. The dorsal and anal fins are set far back, and the caudal fin is abbreviate heterocercal. Gars differ from other living fishes (except the blennioid genus *Andamia*) in having **opisthocoelous** vertebrae—these vertebrae are concave posteriorly and convex anteriorly. A spiral valve is present in the intestine, and the gas bladder is divided internally into interconnected chambers.

Gars live in quiet, often weedy waters and can usually be observed lying almost motionless near the surface. In addition to oxygen obtained through the gills, they can extract atmospheric oxygen using a well-vascularized swim bladder. Prey is captured by means of a rapid lateral strike with the jaws that follows a patient wait for the prey to come within range. According to Wiley (1976), there are two genera of gars, *Lepisosteus* with four species and *Atractosteus* with three. Most live in freshwater, but the alligator gar (*Atractosteus spatula*) and its close relatives may enter salt water. These are the largest of the gars, reaching 3 m in length. Gars have little economic value. They are used as food to some extent in the southern United States and Mexico, and one species is cultured in Cuba, and the ganoid scales are sometimes used as souvenirs and ornaments. The eggs are reported to be poisonous.

Order Amiiformes

Although this order includes a diverse group of fossil fishes, known from the Triassic to the Cretaceous, it has but one living representative, the North American bowfin (*Amia calva;* Fig. 8.4D; Plate 8.5) of the family **Amiidae** (Grande and Bemis, 1998). The bowfin shares key features of jaw operation (see Chapter 23) with teleosts (Liem, 1997), reinforcing the perception of them as sister groups (Fig. 8.2). The order has fossil representatives in Europe, Asia, and North America. *Amia* is found in eastern North America, from the Great Lakes south, and frequents warm, shallow water. Males construct and guard the nest—a depression made in aquatic vegetation—and court several females. After hatching, the young are cared for by the male, which sometimes broods the young in its mouth. The gas bladder is divided by internal septa and can function in aerial respiration, allowing the species to inhabit stagnant waters having a low oxygen concentration. The bowfin is found only in freshwater. The sexes differ in maximum adult size. Females may reach 1 m but 60 cm is a more common length, with males being somewhat smaller than the females. An additional expression of sexual dimorphism is the **ocellus**—a dark spot at the base of the male's caudal fin. The bowfin is more of a scientific curiosity—owing to its primitive nature—than an esteemed sport or food fish. Although it is known to put up a good fight on light tackle, the flesh is of poor quality, but the eggs are used in some places as a source of caviar. Fish managers take a dim view of the bowfin; its voracious habits have given it an undesirable reputation.

KEY POINTS AND CONNECTIONS

• Actinopterygian fishes are characterized by a distinctive dental composition and by fin surfaces composed of slender rays that support intervening membranes. Diversity of body form, especially as it pertains to feeding, and locomotor adaptations are undoubtedly keys to their success. An overall reduction and simplification of cranial bones results in a highly kinetic skull. The development of thinner scales and sophisticated musculoskeletal adaptations of the axial skeleton have resulted in a greater range of locomotor strategies.

The locomotor diversity present in fishes is considered in Chapter 19.

• The actinopterygian fishes have become especially successful in the invasion of freshwater habitats, and most surviving early actinopterygians are primarily freshwater inhabitants.

• Members of the subclass Chondrostei have archaic features, such as spiracles, a heterocercal caudal fin, and numerous fin rays in the fins. Members of the chondrostean order Polypteriformes possess a unique configuration of spines and soft rays in the dorsal fin and have lobate fins reminiscent of the sarcopterygians. The order Acipenseriformes includes some of the largest freshwater fishes—the sturgeons. Bony scutes are a conspicuous feature of sturgeons, and several species are highly esteemed as sources for caviar. Also in this order are two species known as paddlefishes—so named for their conspicuous snout.

• Among the subclass Neopterygii are two archaic families, the gars (Lepisosteidae) and the bowfin (Amiidae). These inhabitants of slow-moving, sometimes stagnant freshwaters of Central and North America are adept at breathing air, thanks to a well-vascularized swim bladder that functions as a lung.

A discussion of the modifications of the circulatory system and swim bladder that permit its use as a lung is included in Chapter 24.

FISH LINKS

http://www.wscs.info/index_home.html Homepage of the World Sturgeon Conservation Society—an organization dedicated to international cooperation on issues of management and conservation of sturgeons.

BUILDING AN ICHTHYOLOGY LIBRARY

Birstein, V. J., J. R. Waldman, and W. E. Bemis. 1997. *Sturgeon biodiversity and conservation.* Kluwer, Dordrecht, the Netherlands. (Reprinted from *Env. Biol. Fishes 48*(1–4).)

Bronzi, P., D. J. McKenzie, G. Arlati, and R. Rossi. 1999. Proceedings of the 3rd International Symposium on Sturgeons, Piacenza, Italy. *J. Appl. Ichthyol. 15.*

van Winkle, W. (Ed.). 2002. Biology, management, and protection of North American Sturgeon. *Amer. Fish. Soc. Symp. 28.* American Fisheries Society, Bethesda, MD.

Greenwood, P. H., R. S. Miles, and C. Patterson (Eds.). 1973. Interrelationships of fishes. *J. Linn. Soc. (Zool.) 53* (Suppl. 1). Academic Press, New York.

Saffron, I. 2002. *Caviar: The strange history and uncertain future of the world's most coveted delicacy.* Broadway Books, New York.

REFERENCES

Bartsch, P. 1997. Aspects of craniogenesis and evolutionary biology in polypteriform fishes. *Neth. J. Zool. 47:* 365–381.

Basden, A. M., G. C. Young, M. I. Coates, and A. Ritchie. 2000. The most primitive osteichthyan braincase? *Nature 403:* 185–188.

Benton, M. 1990. *Vertebrate paleontology.* Chapman and Hall, London.

Birstein, V. J., and W. E. Bemis. 1997. How many species are there within the genus *Acipenser? Env. Biol. Fishes 48:* 157–163.

————, and R. DeSalle. 1998. Molecular phylogeny of Acipenserinae. *Mol. Phylogen. Evol. 9:* 141–155.

————, W. E. Bemis, and J. R. Waldman. 1997. The threatened status of acipenseriform species: A summary. *Env. Biol. Fishes 48:* 427–435.

————, P. Doukakis, and R. DeSalle. 2002. Molecular phylogeny of Acipenseridae: Nonmonophyly of Scaphirhynchinae. *Copeia 2002:* 287–301.

————, ————, B. Sorkin, and R. DeSalle. 1998. Population aggregation analysis of three caviar-producing species of sturgeons and implications for the species identification of black caviar. *Conserv. Biol. 12:* 766–775.

Brainerd, E. L., K. F. Liem, and C. T. Sampler. 1989. Air ventilation by recoil aspiration in polypterid fishes. *Science 246:* 1593–1595.

Britz, R., and G. D. Johnson. 2003. On the homology of the posteriormost gill arch in polypterids (Cladistia, Actinopterygii). *Zool. J. Linn. Soc. 138:* 495–503.

Carroll, R. L. 1988. *Vertebrate paleontology and evolution.* Freeman, New York.

Gardiner, B. G., and B. Schaeffer. 1989. Interrelationships of lower actinopterygian fishes. *Zool. J. Linn. Soc. 97:* 135–187.

————, J. G. Maisey, and D. T. J. Littlewood. 1996. Interrelationships of basal neopterygians, pp. 117–146. In *Interrelationships of fishes*, M. L. J. Stiassny, L. R. Parenti, and G. D. Johnson (Eds.). Academic Press, San Diego.

Grande, L., and W. E. Bemis. 1991. Osteology and phylogenetic relationships of fossil and recent paddlefishes (Polyodontidae) with comments on the interrelationships of Acipenseriformes. *J. Vert. Paleontol. 11*(1 Suppl.): 1–121.

————, and————. 1996. Interrelationships of Acipenseriformes, with comments on "Chondrostei," pp. 85–115. In *Interrelationships of fishes*, M. L. J. Stiassny, L. R. Parenti, and G. D. Johnson (Eds.). Academic Press, San Diego.

————, and————. 1998. A comprehensive phylogenetic study of amiid fishes (Amiidae) based on comparative skeletal anatomy: An empirical search for interconnected patterns of natural history. *Soc. Vert. Paleontol. Mem. 4. Suppl. J. Vert. Paleontol. 18:* 1–690.

Horn, M. H. 1972. The amount of space available for marine and freshwater fishes. *Fish. Bull. 70:* 1295–1297.

Janvier, P. 1996. *Early vertebrates.* Oxford University Press, Oxford.

Jarvik, E. 1980. *Basic structure and evolution of vertebrates,* Vols. 1 and 2. Academic Press, New York.

Jordan, D. S. 1923. A classification of fishes, including families and genera as far as known. *Stanford Univ. Publ. Biol. Sci. 3.*

Lauder, G. V. 1989. Caudal fin locomotion in ray-finned fishes: Historical and functional analysis. *Amer. Zool. 29:* 85–102.

————. 1994. Caudal fin locomotion by teleost fishes: Function of the homocercal tail. *Amer. Zool. 34*(5): 13A (abstract).

————. 2000. Function of the caudal fin during locomotion in fishes: Kinematics, flow visualization, and evolutionary patterns. *Amer. Zool. 40:* 101–122.

————, and K. F. Liem. 1983a. The evolution and interrelationships of the actinopterygian fishes. *Bull. Mus. Comp. Zool. 150*(3): 95–197.

————, and————. 1983b. Patterns of diversity and evolution in ray-finned fishes, pp. 1–24. In *Fish neurobiology*, Vol. 1. R. G. Northcutt and R. E. Davis (Eds.). University of Michigan Press, Ann Arbor.

Liem, K. F. 1997. Developmental analysis of a functional complex that defines *Amia* and the Teleostei. *S. Afr. J. Sci. 93:* 523–528.

Long, J. A. 1995. *The rise of fishes: 500 million years of evolution.* Johns Hopkins University Press, Baltimore.

Mayden, R. L., and B. R. Kuhajda. 1996. Systematics, taxonomy, and conservation status of the endangered Alabama sturgeon, *Scaphirhynchus suttkusi* Williams and Clemmer (Actinopterygii, Acipenseridae). *Copeia 1996:* 241–273.

McCune, A. R., and B. Schaeffer. 1986. Triassic and Jurassic fishes: Patterns and diversity, pp. 171–181. In *The beginning of the age of dinosaurs.* Cambridge University Press, Cambridge.

Meyer, A. 2003. Recent advances in the (molecular) phylogeny of vertebrates. *Ann. Rev. Ecol. Syst. 34:* 311–338.

Moy-Thomas, J. A., and R. S. Miles. 1971. *Palaeozoic fishes.* W. B. Saunders, Philadelphia.

Nelson, J. S. 1994. *Fishes of the world* (3rd ed.). John Wiley, New York.

Normark, B. B., R. McCune, and R. G. Harrison. 1991. Phylogenetic relationships of neopterygian fishes, inferred from mitochondrial DNA sequences. *Mol. Biol. Evol. 8:* 819–834.

Northcutt, R. G. 1995. The forebrain of gnathostomes: In search of a morphotype. *Brain Behav. Evol. 46:* 275–318.

Patterson, C. 1973. Interrelationships of holosteans, pp. 233–305. In Interrelationships of fishes, P. H.. Greenwood, R. S. Miles, and C. Patterson (Eds.). *J. Linn. Soc. (Zool.) 53, Suppl. 1.* Academic Press, New York.

————, and D. E. Rosen. 1977. Review of ichthyodectiform and other Mesozoic teleost fishes and the theory and practice of classifying fossils. *Bull. Am. Mus. Nat. Hist. 158:* 81–172.

Romer, A. S. 1970. *The vertebrate body* (4th ed.). W. B. Saunders, Philadelphia.

Rosen, D. E., P. L. Forey, B. G. Gardiner, and C. Patterson. 1981. Lungfishes, tetrapods, paleontology, and plesiomorphy. *Bull. Am. Mus. Nat. Hist. 167:* 159–276.

Russell, D. F., L. A. Wilkens, and F. Moss. 1999. Use of behavioral stochastic resonance by paddle fish for feeding. *Nature 402:* 291–294.

Stone, R. 2002. Scientists deplore OK for sturgeon catch. *Science 295:* 2191.

Waldman, J. R. 1995. Sturgeons and paddlefishes: A convergence of biology, politics, and greed. *Fisheries 20:* 20–21, 49.

Wiley, E. O. 1976. The phylogeny and biogeography of fossil and recent gars (Actinopterygii: Lepisosteidae). *Mus. Nat. Hist. Univ. Kansas Misc. Publ. 64:* 1–111.

Wilkens, L. A., D. F. Russell, X. Pei, and C. Gurgens. 1997. The paddlefish rostrum functions as an electrosensory antenna in plankton feeding. *Proc. Roy. Soc. Lond. B 264:* 1723–1729.

————, B. Wettring, E. Wagner, W. Wojtenek, and D. Russell. 2001. Prey detection in selective plankton feeding by the paddlefish: Is the electric sense sufficient? *J. Exp. Biol. 204:* 1381–1389.

Wojtenek, W., X. Pei, and L. A. Wilkens. 2001. Paddlefish strike at artificial dipoles simulating the weak electric fields of planktonic prey. *J. Exp. Biol. 204:* 1391–1399.

Zhu, M., X. Yu, and P. Janvier. 1999. A primitive fossil fish sheds light on the origin of bony fishes. *Nature 397:* 607–610.

9 TELEOSTEAN FISHES: EARLY MEMBERS OF THE MOST DIVERSE ASSEMBLAGE OF FISHES

WHAT ARE THE TELEOSTS?

DIVISION TELEOSTEI: IDENTIFYING FEATURES

EXTINCT PRIMITIVE TELEOSTS

Order †Pholidophoriformes
Order †Leptolepidiformes
Order †Tselfatiformes
Order †Ichthyodectiformes

DIVERSITY OF LIVING TELEOSTS

Subdivision Osteoglossomorpha: The "Bonytongues"
    Order Hiodontiformes
    Order Osteoglossiformes
Subdivision Elopomorpha: Tarpons, Ladyfishes, Eels, and
Relatives
    Order Elopiformes
    Order Albuliformes
    Order Notacanthiformes
    Order Anguilliformes (Apodes)
    Order Saccopharyngiformes
Subdivision Clupeomorpha: Herrings and Relatives
    Order Clupeiformes

The most daunting task for anyone studying the systematics of vertebrates is to make sense of the phylogenetic history of that amazingly diverse assemblage of fishes known as the teleosts. The name *teleost* itself implies some ultimate resolution in the evolution of fishes (*teleost* literally means "perfect bone"), and indeed, the teleosts represent the fullest flowering of the ichthyological family tree. Here is a group of fishes that has successfully invaded every habitable place in the water, with many species even venturing out onto land. Teleosts have refined the sensory arsenal of fishes to the point that virtually no aquatic environment is uninhabitable to them. The electrogenic and electroreceptive abilities of some groups enable them to inhabit turbid rivers where vision alone does not suffice. Bioluminescence enables other groups to forage and communicate in the darkest depths of the ocean. Trophic versatility is another key to their success. Teleosts have developed a staggering array of feeding strategies, with herbivores, carnivores, detritivores, and even parasites well represented. In this and in subsequent chapters, we will review the best attempts by systematic ichthyologists to put some order to this fascinating assemblage of fishes.

WHAT ARE THE TELEOSTS?

Answering this question is indeed a daunting task, as it is necessary to provide an inclusive definition for more than 28,000 species. Obviously, fishes have achieved their amazing diversity in the world's waterways largely through the evolutionary accomplishments of this particular group. In Chapter 8 (see Fig. 8.2), we recognized the teleosts as a clade of neopterygians. The question that first needs to be answered is, "Does the term 'Teleostei' imply a monophyletic origin?" Teleostean fishes have long been recognized as possessing a suite of characters that implied monophyly. Yet Greenwood et al. (1966), in their masterful revision of the systematics of teleostean fishes, called into question the notion of monophyly. Would it not be more realistic to consider such a diverse group to have had multiple origins among earlier pholidophorid "holostean" ancestors? (Historians of fish systematics might view 1966 as a watershed year, as it marked both the publication of Greenwood et al. and Hennig's *Phylogenetic Systematics*.) Gosline (1965) was among the first to advocate the monophyly of teleosts, and DePinna (1996) provided convincing evidence for it by identifying 27 synapomorphic characters among both fossil and recent forms. Helfman et al. (1997) incorporated 22 characters to construct a cladogram showing the interrelationships of the major groups. Nelson (1994) proposed an extinct group, the †Pachycormiformes, as a sister group of the monophyletic teleosts. The teleosts trace their origins back to the Triassic, possibly as early as 220 million years ago, and have a rich fossil history though the Mesozoic and Cenozoic (Arratia, 1997). About 96 percent of all fishes are teleosts. Nelson (1994) grouped them into 38 orders, 426 families, and 4,064 genera.

DIVISION TELEOSTEI: IDENTIFYING FEATURES

Teleosts are defined by Patterson (1968) as actinopterygians with "the lower lobe of the caudal fin primitively supported by two hypurals articulating with a single centrum" and having elongate uroneurals, modified from neural arches (see Fig. 3.11), that provide stiffening for the upper lobe of the caudal fin. This gives an internal asymmetry to a caudal fin that appears to be symmetrical externally. Teleosts form a unique bone, the **urohyal,** from ossification of the sternohyoideus tendon (Arratia and Schultze, 1990). The premaxillary and maxillary bones develop a looser articulation with the braincase. As the premaxillary bone becomes more prominent in the gape of teleosts (see Figs. 3.20 and 3.21), they acquire the ability to greatly protrude their jaws, which is useful in generating suction and engulfing prey items during feeding. Recent teleosts have reduced the number of bones constituting each side of the lower jaw to three or four (Patterson and Rosen, 1977). Cycloid or ctenoid scales replace the ganoid scales seen in more primitive forms, and the spiral valve in the intestine is lost. The branchiostegal rays at the floor of the gill cavities in teleosts—a feature that has been demonstrated to be of phylogenetic significance (McAllister, 1968)—are modified to permit a greater sophistication in the control of the flow of water across the gills (Gosline, 1971). Patterson and Johnson (1995) claimed that the presence of **intermuscular bones**—segmental ossifications within the myosepta of the trunk muscles—is a distinctly teleostean feature, which they used to distinguish major lineages of teleosts.

EXTINCT PRIMITIVE TELEOSTS

DePinna (1996) proposed including the diverse assemblage of fossil forms that appear to be intermediate between basal lineages of teleosts and the gars and bowfin as members of the Teleostei. The few orders discussed here represent fossil forms that vary in their affinities with recent teleosts.

Order †Pholidophoriformes

This order, which is known from the Triassic to the Cretaceous (Berg, 1940), may be the ancestral group that gave rise to the osteoglossomorph and elopomorph lineages that diversified, like all other teleosts, during the Cretaceous.

The pholidophorids were small fishes (up to 40 cm), the fossils of which show vertebrae in various stages of evolutionary development. The cranial morphology of these fishes shares several specializations with that of modern teleosts (Patterson, 1975). They may have given rise to the leptolepidiforms and other primitive teleosts in the Triassic or Jurassic (Carroll, 1988; Nelson, 1994).

DIAGNOSIS OF THE †PHOLIDOPHORIFORMES
- Lower jaw resembles holosteans such as the gars and *Amia*.
- Most with ganoid scales and enamel on skull bones.
- Uroneurals are developed.
- Caudal fin resembles that of teleosts, with hypurals each bearing more than one fin ray.[1]
- Teleost characteristics include the presence of supramaxillae, a myodome[2] extending into the basioccipital, fused vomers, and the division of the premaxilla into a lateral, moveable portion and a more medial lateral dermethmoid.

[1] Feature shared with two other extinct groups, the †Leptolepidiformes and the †Ichthyodectiformes.

[2] A deep pit in the floor of the braincase where the extrinsic eyeball muscles originate.

Sources: Patterson, 1975; Patterson and Rosen, 1977.

Order †Leptolepidiformes

These fishes, known from the Triassic to the Cretaceous, are generally considered to be teleosts, although they show affinity with more primitive groups. Patterson (1975) showed that they are advanced over the pholidophorids and share many specializations of skull anatomy with other teleosts. Leptolepidiforms show greater vertebral ossification than the pholidophorids and have cycloid scales (Carroll, 1988). Fossils of adults generally do not exceed 5 cm in length.

Order †Tselfatiformes

This order, dating from the Cretaceous, shows affinities with clupeomorphs, elopomorphs, and especially osteoglossomorphs, with which they have been combined by some authorities (Patterson, 1967).

Order †Ichthyodectiformes

This order, known from the Jurassic and Cretaceous, includes early teleosts that show relationships to the basal living teleost groups and are placed by Nelson (1994) in the subdivision Osteoglossomorpha. They are unusual in having large uroneurals that extend laterally over the centra of the preural vertebrae. Furthermore, they have an endoskeletal bone (ethmopalatine) on the floor of the nasal cavity (Carroll, 1988; Nelson, 1994). Authorities recognize three to five families, all of which tend to be large, predaceous fishes. One genus, †Xiphactinus (†Portheus), included species as long as 4 m. One of the most famous fish fossils, unearthed from late Cretaceous rocks in Kansas and currently exhibited at the Sternberg Museum at Fort Hays State University, is a large specimen of †Xiphactinus that apparently died from ingesting a 2 m long ichthyodectid (†Gillicus) that remains perfectly preserved in the gut (Plate 9.1).

DIVERSITY OF LIVING TELEOSTS

Four major assemblages appear to have arisen among the teleosts. The more primitive representatives of all groups have some features in common, and similar features are seen in the body plans of many of the more advanced orders and suborders. There is a pronounced departure from a generalized "fishy" body plan that is expressed in an array of specialized forms well adapted for a variety of feeding and locomotion strategies (Gosline, 1971).

The more primitive teleosts typically have elongate, somewhat fusiform body shapes and have 50 to 60 vertebrae. Although some groups, such as eels and other specialized fishes, have many vertebrae and long flexible bodies,

more derived fishes tend to have shorter, deeper bodies with fewer (20 to 30) vertebrae. The primitive fishes tend to have single, short-based dorsal fins near the middle of the back, with all fins having only soft rays. More advanced fishes show a subdivision of the dorsal fin or elongate dorsals originating rather far forward. Spines usually add support for the anterior parts of dorsal, anal, and pelvic fins. The deepening of the body is usually accompanied by the encroachment of body musculature onto the top of the head, and the supraoccipital expands into a large crest as a point of origin for these muscles.

The pectoral fins of primitive teleosts resemble those seen in gars and the bowfin, typically being set low on the body, with the base slanting downward and backward or nearly horizontal in some. There are certain limitations of movement inherent in this positioning. These low pectoral fins, though useful in guiding and braking, are not as versatile as the pectoral fins of more derived forms that are set higher on the sides with a more vertical base. The latter position might be more suitable for locomotion and for maneuvering and braking. Many fishes are capable of hovering or even reversing direction using only their pectorals. Several modifications of the pectoral girdle are involved in the positioning of the fins, including changes in the relative sizes of bones and rotation of the axis of the girdle. The presence of the **mesocoracoid** bone in the pectoral girdle of primitive groups readily distinguishes them from more advanced ones. This bone forms an arch with the scapula and coracoid, bracing the fin base at an angle. The loss of the mesocoracoid bone in higher groups appears to allow the base of the pectoral to be aligned with the vertical axis of the girdle.

Pelvic fins in more primitive bony fishes are placed rather far back on the belly and are hence referred to as **abdominal** in position. The supporting bones—a reduced pelvic girdle—are not firmly connected to any other bony structure but are situated in the musculature of the body wall. More derived fishes have pelvic fins placed far forward on the breast region, below the pectorals. In this **thoracic** position, the pelvic girdle is attached to the lower portion of the pectoral girdle. A few groups of fishes have pelvic fins in an intermediate position, with no connection to the pectoral girdle, and some maintain the intermediate position with a ligamentous connection; these are variously called **subabdominal** or **subthoracic** pelvic fins. In a few instances, the pelvic fins are placed forward of the pectoral base in a **jugular** position—usually considered an advanced characteristic. In highly agile fishes, the vertical alignment of the origin of the spinous dorsal fin, the pectoral fins, and the thoracic pelvic fins add greatly to their ability to stop quickly or make tight turns. There is generally a change in the center of balance concomitant with the change in position of the paired fins, because the position of the gas bladder changes, moving farther forward and higher in the body in most advanced fishes, contributing to greater balance.

Ancestral fishes differ from more derived forms in the structure of the upper jaw. The outer edge of the jaw consists of both the premaxilla and the maxilla in basal actinopterygians. The premaxillae are seldom protractile, and both premaxillae and maxillae may bear teeth. On the other hand, the premaxillae form the upper border of the mouth in the derived forms. The maxillae are excluded from the actual border of the gape, are situated above the premaxillae, and do not bear teeth. The protrusible upper jaw is made possible by long, ascending processes of the premaxillae that slide along the anterodorsal part of the skull.

The orbitosphenoid bone forms a large part of the interorbital septum in many lower teleosts, but is lacking in most of the middle and higher teleosts. Scales of primitive teleosts are usually cycloid; those of derived ones are usually ctenoid. The hard, rough margins of most ctenoid scales may function to protect against injury and may possibly serve a hydrodynamic function, by holding a "boundary layer" of water next to the fish, or improve retention of mucus on the body surface (see Chapter 18).

The more primitive teleosts retain an open pneumatic duct from the alimentary canal to the gas bladder—a condition termed **physostomous**. This is in contrast to the **physoclistous** condition of more derived teleosts, in which the duct is absent or closed (see Chapter 24 for a discussion of gas bladder function). The pancreas is often a discrete gland in the more primitive forms (as it is in elasmobranchs) but is most usually diffuse in more derived fishes and may be associated with the liver to form a **hepatopancreas** in some of the most recently evolved forms.

The foregoing traits are some of the generally recognized differences between primitive and derived groups of teleosts. (Others will be mentioned later in the characterization of the different orders.) One of the difficulties in defining the interrelationships of the teleosts is the extent to which these characteristics occur in various combinations among the different orders. The four major monophyletic lines (subdivisions) of living teleosts are as follows:

1. The **Osteoglossomorpha**. This group includes bonytongues, elephantfishes (mormyrids), featherbacks, mooneyes, and their allies. Nelson (1994) recognized a single living order, **Osteoglossiformes**, and one extinct order, †Ichthyodectiformes. Some researchers (cf. Li and Wilson, 1996) consider the mooneyes and goldeyes to belong in a separate order, the **Hiodontiformes**.

2. The **Elopomorpha**. This group includes the tarpons, tenpounders, eels, and their relatives. Four extant orders are included: **Elopiformes, Notacanthiformes, Anguilliformes,** and **Saccopharyngiformes**. One extinct order, †Crossognathiformes, is considered to be closely related.

3. The **Clupeomorpha**. This group includes the herrings and their allied forms. There is a single living order, **Clupeiformes**, and an extinct order, †Ellimmichthyiformes.

4. The **Euteleostei**. The majority of the teleosts are included here. According to Nelson (1994), this subdivision comprises 32 orders, 391 families, 3,795 genera, and 22,262 species. In the decade since Nelson's important work was published, more than 5,000 new species have been described.

Some earlier classification systems featured a large, inclusive order, the "Clupeiformes" or "Isospondyli," that included various representatives of all four of the aforementioned groups. Ancestral characters held in common by many soft-rayed teleosts (symplesiomorphies) defined membership in this obsolete order.

• • • • • • • • • • • • • • • •
Subdivision Osteoglossomorpha: The "Bonytongues"

Greenwood et al. (1966) recognized the Osteoglossomorpha as a monophyletic group based on the shared presence of paired tendon bones on the hypobranchial and basibranchial and, most important, on the primary bite formed by the medial parasphenoid and the tongue, the literal translation of *osteoglossomorph* being "bony tongue." Osteoglossomorph fishes have been classified as a sister group of all other teleosts (Patterson and Rosen, 1977; Patterson, 1994), but Hilton (2001) asserted that care should be taken in assessing the phylogenetic significance of the tongue bite apparatus, as this character complex has lent itself to being misunderstood. Students of the group (Li and Wilson, 1996, 1999; Li et al., 1997a) have proposed the designation of the Hiodontiformes (the North American goldeye and mooneye) as a sister group to the rest of the osteoglossomorphs. A fossil family, the †Lycopteridae from the Jurassic and Cretaceous of Asia, is considered a sister group to both orders of extant osteoglossomorphs (Fig. 9.1).

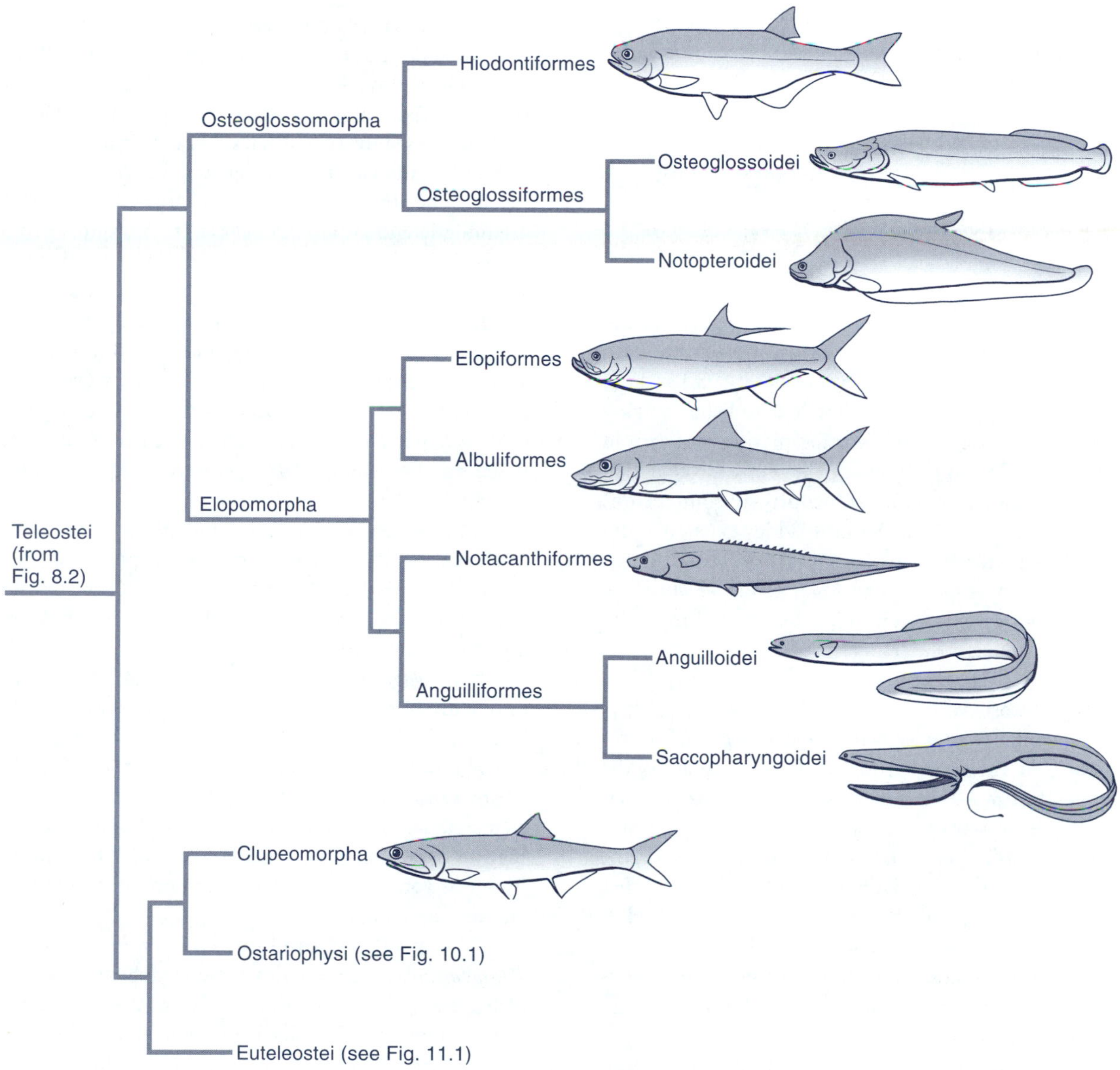

FIGURE 9.1
Phylogeny of the Teleostei showing interrelationships of the osteoglossomorph, elopomorph, and clupeomorph fishes. (From Forey et al., 1996; Li and Wilson, 1996.)

DIAGNOSIS OF THE HIODONTIFORMES AND OSTEOGLOSSIFORMES
- Absence of supramaxilla (in Hiodontiformes).[1]
- Primary bite between tongue and parasphenoid.
- Paired tendon bones on second hypobranchial or second hypobranchial and basibranchial.
- Intestine passes to the left of the stomach.[2]

[1] Synapomorphy that distinguishes the two orders.
[2] Intestine passes to the right in most fishes (except cyprinids and some atherinids).

Sources: Nelson, 1994; Li and Wilson, 1996.

Order Hiodontiformes

This order includes one family, the **Hiodontidae,** with two extant species. The mooneye (*Hiodon tergisius*) and the goldeye (*H. alosoides*) are silvery, herringlike fishes ranging from northeastern and central North America westward to Alberta and into British Columbia. They have a connection between the gas bladder and the ear (**otophysic connection**) that consists of diverticula of the gas bladder extending into the skull. They are rarely larger than 45 cm and have limited use as food and sport fishes. The fossil genus †*Eohiodon* is well known from several species reported from Eocene deposits in western North America (Li et al., 1997a).

Order Osteoglossiformes

This order contains species with morphological features characteristic of the most primitive teleosts, such as the bonytongues, yet it also includes fishes with extremely sophisticated electrosensory capabilities—the mormyrids (elephantfishes and their relatives). The anatomy of the Osteoglossiformes is essentially similar to that of other primitive bony fishes, and they were formerly included in a more inclusive interpretation of "Clupeiformes" or "Isospondyli." The osteoglossomorph bite discussed earlier is actually shared by other Mesozoic groups. Rather than using the dentary and upper jaw bones, the toothed tongue bone secures against toothed bones of the roof of the mouth. Usually, the parasphenoid is involved, but in some species the entopterygoids bear teeth as well. An extension from the parasphenoid forms a support for the entopterygoid. The structure is rare in teleosts, being found only in osteoglossomorphs and alepocephaloids (slickheads; Gosline, 1971). Usually, a tissue connection between the gas bladder and the ear is present. The Osteoglossiformes are thought to be a very ancient group. Some extinct families dating from the Jurassic have osteoglossomorph affinities, and some close relatives of extant families are known from the Cretaceous. The Eocene genus †*Phareodus* was the first fossil osteoglossomorph discovered in North America, and related forms have been identified in Australia (Li et al., 1997b).

Suborder Osteoglossoidei

This includes the families **Osteoglossidae** and Pantodontidae. Osteoglossids are found in Africa, South America, southeast Asia, and Australia. The subfamily Heterotidinae includes only two species—the arapaima or pirarucu of South America (*Arapaima gigas*) and *Heterotis niloticus* of Africa. The arapaima is one of the largest freshwater fishes, attaining lengths up to 2.5 m (Fig. 9.2A), and is prized as a food fish. Both *Heterotis* and *Arapaima* have air bladders with numerous small subdivisions that increase the area available for the absorption of oxygen. This apparently makes possible the use of atmospheric oxygen for respiration. *Heterotis* also has a special suprabranchial respiratory organ that aids in that function. *Heterotis* and *Arapaima* build large nests in shallow water and give protection to their young following hatching. The young of *Heterotis* are equipped with external gills.

The genera *Osteoglossum* and *Scleropages* are placed in the subfamily Osteoglossinae. Two species of *Osteoglossum*—*O. ferreirai*, the black arowana, and *O. bicirrhosum*, the silver arowana—are often imported from South America as aquarium fishes (Plate 9.2)

Scleropages jardini and *S. leicharti* are regarded as the only truly freshwater teleosts native to Australia. They are primary freshwater fishes (see Chapter 29), with no relatives known from salt water. The other fishes found in the freshwaters of Australia belong to families that are marine, anadromous, or catadromous, or have fossil relatives from marine deposits. *Scleropages formosus* of Thailand and the Malay region is apparently a mouth brooder, as are the species of *Osteoglossum*.

The family **Pantodontidae** contains one species, *Pantodon buchholzi* of Africa, which reaches a maximum size of less than 15 cm (see Fig. 3.12B). Males have a modified anal fin thought to function in internal fertilization. The eggs float at the surface, as do the fry after hatching. This peculiar little fish has subthoracic pelvic fins that are made up of long, separate, rather filamentous rays. The pectoral fins are relatively large and expanded and are characterized by expanded cleithra that support large pectoral muscles. These apparently give the fishes the ability to flap their pectoral fins like wings, which they do when leaping from the water. The common names *butterflyfish* and *freshwater flying fish* are derived from this habit, but this species should not be confused with the freshwater hatchetfishes—fishes of the characoid family Gasteropelecidae that also have enlarged pectoral girdles and are capable of launching themselves

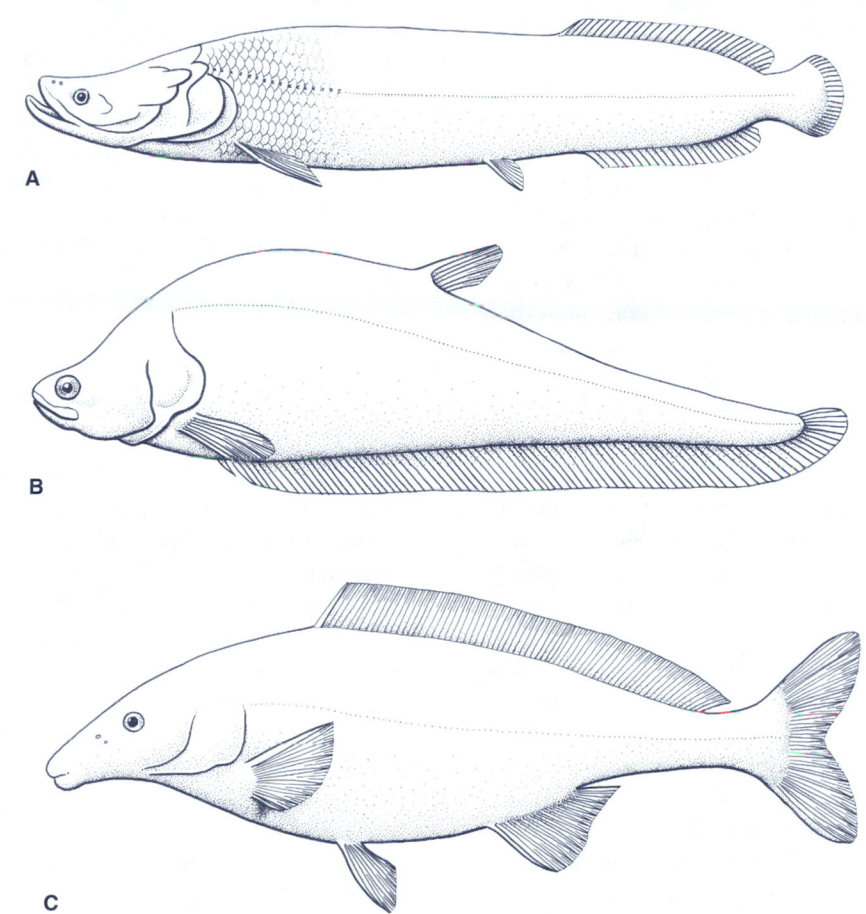

FIGURE 9.2
Representative of the suborder Osteoglossoidei: **A**, Arapaima (*Arapaima*). Representatives of the suborder Notopteroidei: **B**, Featherfin knifefish (*Notopterus*); **C**, Elephantfish (*Mormyrus*).

from the water. Both are popular aquarium fishes, but care should be taken never to leave the top of the tank open.

Suborder Notopteroidei

Included here are the Notopteridae (the featherbacks of southeast Asia, the Indo–Malayan archipelago, and Africa), Mormyridae, and Gymnarchidae ("electric" fishes of Africa).

The superfamily Notopteroidea includes the featherbacks (**Notopteridae**). These are rather strange appearing, elongate, compressed fishes with a long anal fin beginning in the anterior third of the body and extending to the caudal fin, with which it is confluent (Fig. 9.2B). A very small dorsal fin is placed about midway down the back, except in the

African genus *Xenomystis*, which lacks a dorsal fin. Some have a rather humped back and a concave dorsal head profile. Several species are used as food, and some enter the aquarium trade. The sophisticated use of the elongate anal fin for locomotion in this group makes them fascinating aquarium fishes to watch.

The superfamily Mormyroidea includes the Mormyridae, the elephantfishes, and the closely related Gymnarchidae. All members of these families inhabit African freshwaters. The mormyrids are sometimes placed in their own order, because of their remarkable electrogenic and electrolocation capabilities and their exceptionally large cerebellum. Whereas the brain in most fishes accounts for only

0.1 to 0.3 percent of body weight, in the mormyrids, it is an impressive 3 percent—greater than humans, who weigh in at about 2 percent (Nilsson, 1999).

Phylogenetic studies have deduced the interrelationships of "mormyriform" fishes by employing molecular techniques and focusing on the origins and diversification of the electric organs (Alves-Gomes, 1999; Alves-Gomes and Hopkins, 1997; Sullivan et al., 2000). Electrogenic fishes can be broadly classified into two groups—those that produce powerful emissions that can stun other organisms, and those that emit weak signals as a means of communication and orientation (see Chapters 21 and 36). Mormyrids fall into the latter category, with weak electric organs present in the caudal region. Mormyrids and notopterids have a peculiar inner ear structure, in that the sacculus and lagena are separated from the utriculus and the semicircular canals. The skull of mormyrids has small cavities occupied by vesicles that originate as part of the gas bladder in the early developmental stages. These features appear to be modifications enhancing the acoustic acuity of the inner ear system (see Chapter 21).

The family **Mormyridae** contains about 18 genera and almost 200 species. Molecular studies have confirmed two subfamilies, with the Petrocephalinae as a sister group of the rest of the genera that constitute the monophyletic subfamily Mormyrinae (Lavoué et al., 2000). Included among the mormyrids are some of the oddest looking fishes known from freshwater: the so-called elephantfishes of the genus *Gnathonemus*. These have the snout greatly elongated and curved downward, with a very small mouth at the tip, often equipped with a thick barbel. Many species have a more restrained shape, but all are characterized by a narrow caudal peduncle (Fig. 9.2C).

The single species of **Gymnarchidae**, *Gymnarchus niloticus*, has been studied extensively because it shares with mormyrids the capacity for orientation using generated weak electrical fields. They differ from mormyrids in appearance, being elongate and lacking the pelvic, anal, and caudal fins. The dorsal fin, which runs most of the length of the back, is used for locomotion forward or backward, by means of undulations that pass down the fin.

Subdivision Elopomorpha:
Tarpons, Ladyfishes, Eels, and Relatives

The elopomorphs are mostly marine and diverse in body form. The tarpons and bonefishes, game fish esteemed for their fighting abilities, and bizarre, rarely observed deep-sea eels can be found among the elopomorphs. These fishes share features in skeletal design, such as fused retroarticulars and angulars and prenasal and rostral ossicles, but it is the presence of the **leptocephalus** larvae that is the most unifying feature. Their adult morphology is as varied as the interpretations of the systematic interrelationships within the group (Forey, 1973a, 1973b; Forey et al., 1996; Lauder and Liem, 1983a, 1983b; Smith, 1984). With a fossil history that extends back to the Jurassic, the elopomorphs have been thought to be more primitive than the osteoglossomorphs, and the most primitive of the extant teleosts (Arratia, 1996). There are about 800 modern species in about 25 families.

Order Elopiformes

This order includes the tarpons and ladyfishes. Known from the Jurassic to the present, they were well represented among the Cretaceous ichthyofauna. Elopiform fishes possess many of the primitive features that distinguish lower from higher teleosts. These fishes show some relationships to the herringlike fishes and were sometimes placed with them in an order called "Isospondyli." However, they share the leptocephalus larval stage with the eellike fishes and are considered to be part of an evolutionary line that includes the eels, notacanths, and halosaurs. The leptocephalus larvae of various fishes may be ribbonlike or leaflike in general shape (see Fig. 27.11), and their metamorphosis to the juvenile body form usually involves significant shrinkage in length. The tissues of the larva are essentially translucent, and the teeth are prominent. Larvae of elopiforms differ from those of eels in having forked caudal fins (Richards, 1984; Fig. 9.3A).

DIAGNOSIS OF THE ELOPIFORMES
- Body usually slender and compressed.
- Gape bordered by premaxilla and toothed maxilla.
- Bony gular plate present between lower jaw bones.
- Ethmoidal commissure of the cephalic sensory canal system.
- Deeply forked caudal fin.
- Scales cycloid.
  Sources: Forey, 1973a, 1973b; Nelson, 1994.

The ladyfishes (tenpounders and machetes) of the family **Elopidae** and the tarpons of the family **Megalopidae** belong to this order. These are large, big-scaled fishes of warm seas. Aside from a good fight, they have nothing much to offer the sport fisher, as their flesh is not palatable. The ladyfish or "tenpounder" (*Elops saurus*), the best known fish of its genus (Plate 9.3), reaches somewhat less than 1 m in length, a weight of 14 kg, and is circumtropical. The tarpon (*Megalops atlanticus*; Fig. 9.3B), may reach 2.6 m in length and 150 kg, but the oxeye (*M. cyprinoides*) of the Indo-West Pacific usually does not exceed 1 m in length. The tarpons

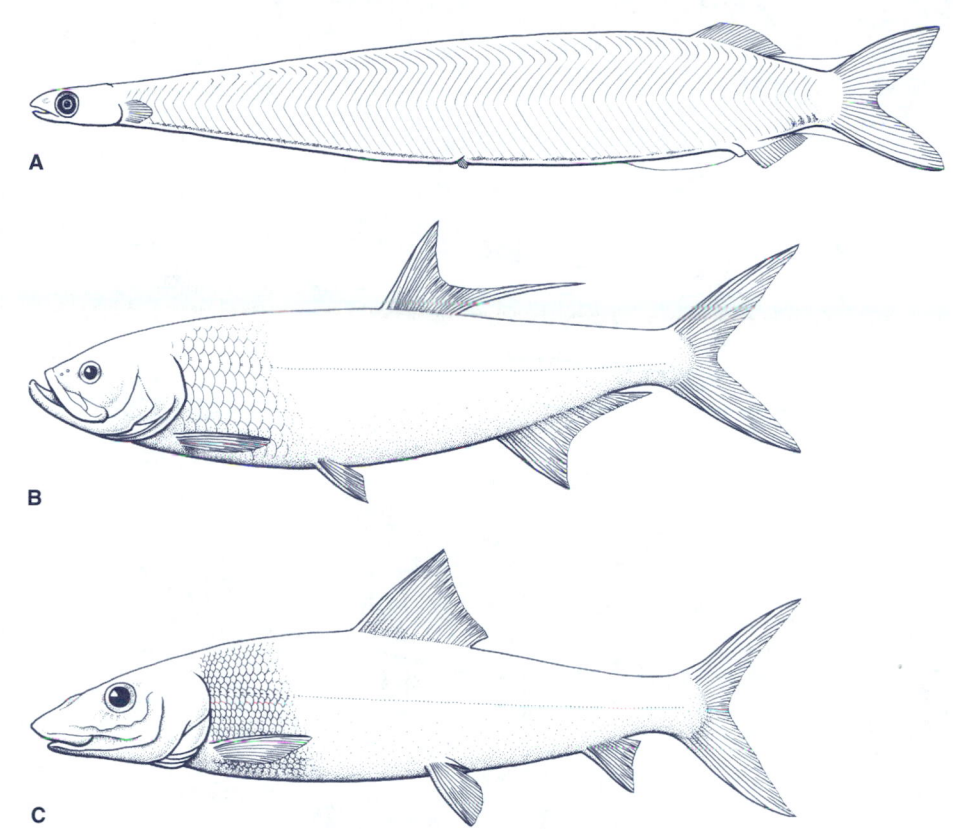

FIGURE 9.3
Representatives of the order Elopiformes: **A,** Leptocephalus larvae of *Elops;* **B,** Tarpon (*Megalops*). Representative of the order Albuliformes: **C,** Bonefish (*Albula*).

contrast somewhat with *Elops* because they possess a conus arteriosus, lack pseudobranchiae, and have a connection between the gas bladder and the otic region of the skull.

Order Albuliformes

The families **Albulidae** (bonefishes) and **Pterothrissidae** (the deep-sea bonefish) constitute the suborder Albuloidei. These are fishes with abdominal pelvic fins, low-based pectoral fins, and forked caudal fins. They are placed in the single family Albulidae by some. The bonefish (*Albula vulpes;* Fig. 9.3C) is a prized game fish of tropical and subtropical waters around the world. *Albula (= Dixonina) nemoptera* can be found in the Caribbean and along the Pacific coast of Central America. *Pterothrissus (= Isteus)* has at least two species; one lives in deep water off Japan, the other off West Africa. The albuloids are placed in the Anguilliformes by some authorities (Lauder and Liem, 1983a).

Order Notacanthiformes

Although Nelson (1994) considered these to constitute the suborder Notacanthoidei within the order Albuliformes, Forey et al. (1996) classified the notacanths as a separate order (Fig. 9.1). Notacanthiform fishes have high-based pectoral fins, abdominal pelvic fins, and thin, tapering tails with small or no caudal fins. The leptocephalus larvae of some of these fishes may exceed the adults in length. The order includes two families, the **Halosauridae** and the Notacanthidae. Halosaurids are widely distributed, meso- to bathypelagic fishes, occurring as deep as 2,000 m. They resemble true eels in being elongate and in having leptocephalus larvae. Although the abdominal pelvic fins and the cycloid scales indicate a comparatively primitive teleost, these are physoclistous fishes. The tail tapers to a point, without a caudal fin, but individuals may possess easily regenerated tail tips forming a pseudocaudal fin. They range to about 50 cm in

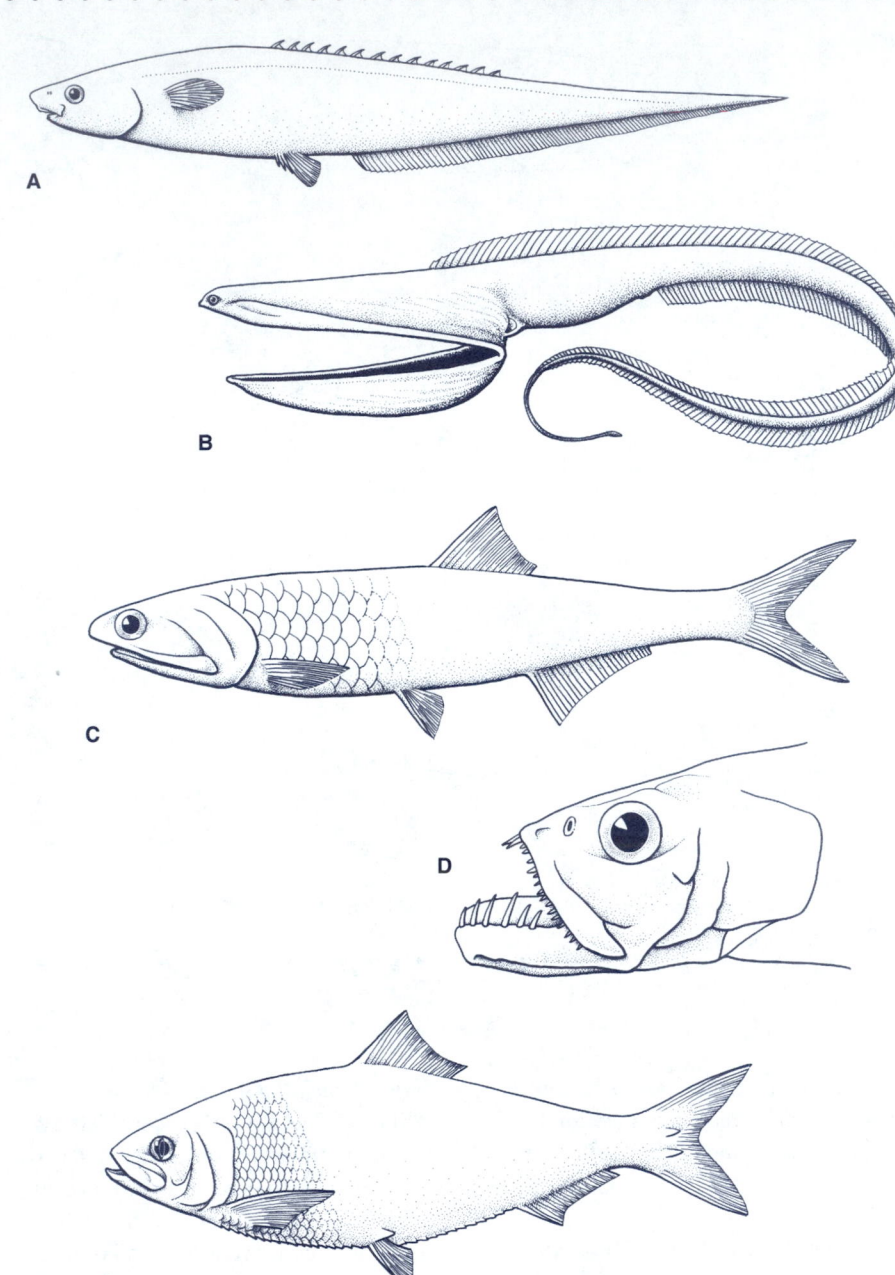

FIGURE 9.4
Representative of the order Notacanthiformes: **A,** Spiny eel (*Notacanthus*). Representatives of the order Anguilliformes: **B,** Gulper (*Eurypharynx*); Representatives of the order Clupeiformes: **C,** Anchovy (*Engraulis*); **D,** Head of wolf herring (*Chirocentrus*); **E,** Shad (*Alosa*).

length. Halosaurids are known from the fossil record as far back as the Upper Cretaceous.

The spiny eels of the family Notacanthidae (Fig. 9.4A) have a series of spines down the back, spines in the anal fin, and the upper jaw bordered by the premaxillae only. These eels live at depths down to 3,000 m and may have a world-wide distribution. Adults reach about 1.5 m.

Order Anguilliformes (Apodes)

This is a large order of marine and catadromous species, comprising about 20 families. They are frequently referred to as the "true" eels to distinguish them from the many families of elongate, "eellike" fishes. They are thought to have their origin in an albuloidlike ancestor, but they are so specialized that the relationships are obscure. The main link uniting the elopiforms and the eels is the leptocephalus larvae known from both groups.

DIAGNOSIS OF THE ANGUILLIFORMES

- Greatly elongated body, with vertebrae numbering up to 750.
- Mesocoracoid, opisthotic, symplectic, and posttemporals not present.
- Gill openings usually small and placed back from the edge of the concealed operculum.
- Pectoral girdle not connected to skull; pectoral fins absent in many.
- Pelvic girdle and fins absent.
- The dorsal and anal fins are confluent with the caudal fin.
- Swim bladder present and usually physostomous.
- Oviducts absent.
- Scales absent in many families.

  Sources: Robins, 1989; Nelson, 1994.

GOING DEEPER · Awesome Anguillids

When Aristotle opened up an eel and was unable to locate the gonads, he reasoned that the fish was generated spontaneously from the mud where it was frequently observed. The truth about these amazing fishes is no less bizarre. Completing their life cycle requires three metamorphic changes in body form and a journey of up to 6,000 km. Eels are the best known example of a **catadromous** lifestyle, spending most of their lives in freshwater and returning to the ocean to spawn once and die (**semelparity**). The family Anguillidae includes one genus, with about 15 species that range through the Atlantic and Pacific oceans. The two North Atlantic species, *Anguilla anguilla* and *A. rostrata*, have long been the focus of intensive study, as they are the subject of a valuable fishery—one that is in a perilous state of decline.

The European eel (*Anguilla anguilla*) is distributed in freshwater and estuaries from the Arctic Circle to the coast of Africa, through the Mediterranean Sea to the Black Sea and on islands from Iceland to the Azores and Madeira. The American eel (*A. rostrata*) ranges from northern South America to Labrador and, on rare occasions, Greenland. After spending four to eight years in freshwater, males have attained a length of about 60 cm and begin their long,

arduous journey back to the spawning grounds (see Chapter 36). Females of the European eel may not migrate until they have spent about 12 years in freshwater and have reached about 1.5 m long. American eel females migrate at about 1 m long. The life history of these eels was not understood until early in the 20th century. Leptocephalus larvae have long been known to scientists, but they were originally believed to be distinct species. In 1896, two Italian scientists collected leptocephali, held them in laboratory aquaria, and observed their remarkable transformation into young eels, known as **elvers**. Following this discovery, the Danish ichthyologist Johan Schmidt began the search to find where the leptocephali originated. After many years of patiently tracking the leptocephalus larvae at sea, Schmidt located the spawning grounds in a patch of the North Atlantic Ocean just east of the Bahamas called the Sargasso Sea. Here, he found that the European and American species spawn in adjacent but overlapping regions at a depth of about 400 m. The fragile, transparent leptocephalus larvae drift in the ocean currents for months or years. The Gulf Stream plays a pivotal role in the transport of leptocephalus larvae back to the vicinity of coastal streams. The journey of the American eel larvae takes about one

year, but the European eel larvae, carried north by the Gulf Stream across the north Atlantic, may take up to three years to reach the coastal drainages of Europe or North Africa.

Just what the leptocephali subsist on during their migration has long been a mystery, as all guts examined appeared to be empty (Moser, 1981). Mochioka and Iwamizu (1996) discovered gut contents in the leptocephali of many families of eels, including the families Congridae, Muraenidae, Muraenesocidae, Nettastomidae, and Ophichthidae. In all instances, the gut contents were fecal pellets and remains of the gelatinous "houses" of **larvaceans**—tunicates that secrete coverings in which they live and that they use as filtering devices during planktonic feeding. Apparently, the large teeth of the leptocephali are used to grasp and hold the larvacean while the liquid in the house is sucked out. The liquid contains fecal pellets and other organic material that can be used as food. Dissolved organic carbon may also serve as a source of nutrition (Otake and Maruyama, 1993). It is altogether possible that anguillid leptocephali in the North Atlantic have evolved a similar feeding strategy, focusing on planktonic tunicates. Leptocephali are able to sustain a prolonged larval stage and grow to consider-

Suborder Anguilloidei

Most of the eel families are in this suborder. Perhaps the best known are the catadromous eels of the family **Anguillidae** (see Fig. 3.6D; Plate 9.4). Anguillids have minute cycloid scales, a complete lateral line, and comparatively well-developed pectoral fins. Species are known from the Atlantic and Pacific (see Going Deeper), and they are a highly esteemed food fish in Europe and Asia. Molecular studies have revealed four clades of the genus *Anguilla* that correspond to their geographic ranges: Indo-Atlantic, Oceania, tropical Pacific, and Indo-Pacific (Aoyama et al., 2001). The morphological similarity and comparatively small measured genetic distance between clades shows the genus to be quite young (Bastrop et al., 2000). In the North Atlantic, two species are known, the European eel (*Anguilla anguilla*) and the American eel (*A. rostrata*). The two species are distinguished mainly by the number of vertebrae, with the American eel having about 107 vertebrae and the European eel about 114. Some biologists consider them to be a single species (Robins et al., 1991).

Suborder Muraenoidei

Eels of this suborder have no scales, and the lateral line is reduced. Members of the family **Chlopsidae** (Xenocongridae), the false morays, are found in tropical and subtropical waters. These are small, and some lack pectoral fins. Their cryptic behavior has resulted in some peculiar symbiotic associations. For example, *Kaupichthys nuchalis* of the Caribbean area is reported to hide inside sponges.

Some of the largest eels are the morays of the family **Muraenidae** (Plate 9.5). These are widespread in tropical and subtropical waters, especially where coral reefs abound. Although most species do not reach more than 1 m long, there are reports of some individuals as long as 3 m. Many morays are characterized by bold colors and patterns, and most have strong jaws and many sharp teeth. They have dorsal, anal, and caudal fins but lack pectoral fins and scales. Despite their fearsome reputation, most moray attacks occur when careless divers stick their hands in reef crevices without first looking to see if a moray is at home. Their sharp teeth can inflict painful wounds. The flesh of some

able size through a dramatic diminution of their metabolic rate as their size increases (Bishop and Torres, 1999).

As the leptocephalus larvae approach the coastal estuaries, they metamorphose into "**glass eels**" that begin to move shoreward under their own power or helped by longshore currents and tidal streams (Martin, 1995). The glass eels gradually acquire pigmentation and become elvers that traverse the estuaries and ascend the streams. After a journey that may take several hundred kilometers, the elvers are home and continue to grow and mature into the brownish green, slimy ("**yellow phase**") creatures that some find revolting but others find delectable.

After several years spent in freshwater, the maturing eels turn a silvery hue (termed **silver phase**), increase the size of their eyes, and begin the long trek back to the Sargasso Sea. This journey may include extensive terrestrial forays. Efficient cutaneous gas exchange and a serpentine aquatic locomotor strategy (see Chapter 19) that also serves them well on overland journeys make this possible. Undoubtedly, the migrating eels rely on a suite of sensory mechanisms and environmental cues to orient themselves during their journey. For example, studies have demonstrated that olfaction is critical in guiding the migration of both yellow-phase and silver-phase eels through estuaries and out to sea (Barbin, 1998; Barbin et al., 1998) and that they cue their movements to the ebb and flow of tides (Parker and McCleave, 1997). Tracking studies have shown that, once at sea, the migrating eels travel at depths ranging from 50 to 400 m. The eels appear to exhibit diurnal shifts in preferred depth, moving into shallower waters during the night (Tesch, 1978). *Anguilla anguilla* appears to move to even greater depths, as evidenced by its appearance in the diet of benthopelagic predators living as deep as 700 m (Reinsch, 1968). Ample stores of fat sustain the eels on this migration and provide energy for gonadal development as the journey ends (van Ginneken and van den Thillart, 2000). Fecundity estimates for *Anguilla* range as high as 19 million eggs per female (Barbin and McCleave, 1997). European eels may spend as long as a year at sea before reaching the Sargasso Sea. As might be expected for species with broad geographic ranges, there is much variation in the described life history parameters (Helfman et al., 1987; Jessop, 1998). For example, there is a clinal increase in elver length from south to north along the Atlantic coast of North America (Vladykov, 1966).

Although the catadromous lifestyle of eels strongly suggested a panmictic reproductive strategy, in which each species constituted a single, randomly mating population, this does not appear to be the case. Molecular studies have discounted the claim by some that Ameri-

morays has been known to carry ciguatera poisoning (see Chapter 38). There are about 15 genera, including *Muraena*, *Gymnothorax*, *Echidna*, and *Enchelycore*, and nearly 400 species in this family.

Suborder Congroidei

This comparatively diverse assemblage of eels includes nine families with almost 500 species (Nelson, 1994). The family **Congridae**, congers and garden eels (see Plate 36.1), includes some large members, such as *Conger conger*, which reaches 2.7 m and is an excellent food fish. Congers are marine and are found in tropical and temperate waters. The pike congers (**Muraenesocidae**) have narrow, long jaws with large teeth. Some species that reach more than 2 m are sought as food. They are found in the Indian and Pacific oceans, some in deep water.

Also included in this suborder is the tropical and temperate marine family **Synaphobranchidae**. These are commonly called the neck eels or cutthroat eels because of their gill openings that are confluent midventrally. The family contains the subfamily Simenchelinae, including only a single species, the pugnose or snubnose parasitic eel (*Simenchelys parasiticus*). These eels reportedly are capable of burrowing into and feeding on larger fishes—a behavior resembling that observed in hagfishes (Robins and Robins, 1976). In perhaps the only known instance of endoparasitism by a vertebrate, pugnose eels have been recovered from the heart of a mako shark (Caira et al., 1997).

Almost half of the congroid species are snake eels and worm eels of the family **Ophichthidae**. Most are tropical to subtropical marine fishes inhabiting shallow waters, but some enter coastal freshwaters.

Only about nine species of snipe eels (**Nemichthyidae**; see Fig. 3.6E) are known. These are characterized by elongate jaws that curve away from each other toward the tips. Snipe eels are filiform and greatly elongated, with up to 750 vertebrae. The long dorsal and anal fins are confluent with the very small caudal fin. These are deepwater eels that have been caught at depths from about 500 m to more than

can and European eels are actually the same species (see Chapter 36). Icelandic populations of *Anguilla*, however, appear to demonstrate some hybridization between European and American species (Avise et al., 1990). Recent data from samples of European eels from the Mediterranean, the Baltic, and the North Atlantic reveal a pattern of genetic structure strongly suggesting that mating is nonrandom and gene flow among populations is restricted (Wirth and Bernatchez, 2001).

The spawning area of the Japanese eel (*Anguilla japonica*) was not known until 1991. The species spawns in the summer, west of the Mariana Islands at about Lat. 15° N and Long. 140° E (Tsukamoto, 1992; Tsukamoto et al., 1989). This area is in the influence of the North Equatorial Current. From there, the leptocephalus larvae drift passively to the east at depths of 70 to 150 m until, at a length of about 20 mm, they begin diurnal vertical migrations that allow them to transfer, with the aid of trade winds, to the north-flowing Kuroshio Current (Kimura et al., 1993).

As seems to be the case with most commercial fisheries in the North Atlantic, the eel fishery is in serious trouble. Eel fishers working the Atlantic coastal rivers and streams, catching adult eels by setting out strings of baited traps, have reported a steady decline in the harvest of American eel over the past decade. Overfishing, blockage of migratory routes, loss of habitat, and pollution have all been suggested as causes for this decline (Haro et al., 2000). By the early 1980s, the European eel was extinct in more than 80 percent of the rivers on the Iberian Peninsula (Lobón-Cerviá, 1999). This near-catastrophic decline in the Iberian eel fishery is largely attributed to overharvesting of the elvers. A large elver fishery also exists in North America. During the spring elver season, thousands of kilograms of the young eels are caught to supply the insatiable eel culture industry that has mushroomed in Asia. Regrettably, the stringent regulation of elver harvesting has resulted in a black market supplied by poachers. The young elvers are shipped mainly to Japan or Taiwan, where they are fattened on specially prepared diets. When the pond-reared eels are of suitable size, the pond is drained and the eels harvested. Being the hardy and resilient creatures that they are, they can be shipped live to markets throughout Asia. Whereas Americans have yet to cultivate a taste for eel, Europeans enjoy them and Asians, especially the Japanese, revere them. Grilled *unagi* is a summer favorite among the diners in hundreds of restaurants in Tokyo that specialize in eel dishes. Ironic indeed, that the overharvesting of elvers to sustain the aquaculture industry may ultimately lead to the demise of the traditional eel fishers who have plied their trade for generations on both sides of the Atlantic.

7,000 m. They range in length from around 0.5 to 2 m, with most of the length in the tail.

Order Saccopharyngiformes

The gulpers and swallowers, eellike fishes of deep marine waters, possess some of the most bizarre anatomical modifications seen among vertebrates. They are scaleless and lack several bone elements, including the opercular bones and branchiostegals. The greatly elongated hyomandibular and the rest of the suspensorium project backward, so that in some species the mouth can be much longer than the head. The gills are small and plumelike. Some species are very elongate, reaching a length of 1.8 m. The leptocephali, recognizable by their deep bodies with V-shaped myomeres (blocks of trunk musculature; see Chapter 19) rather than the usually encountered W-shape, are found in shallow water (Böhlke, 1966; Robins, 1989).

Suborder Cyematoidei

Cyema and *Neocyema,* the bobtail snipe eels or arrow eels, are much shorter than the nemichthyid snipe eels, as if they have had the posterior part of the tail chopped off. These eels reach about 15 cm long. The dorsal and anal fins begin in the posterior half of the body and extend to the truncated end of the tail, where they extend backward like the feathers on a dart. As in the snipe eels, the long mouth cannot be fully closed. There are two species in the single family Cyematidae, found in deep water in all warm seas. *Cyema atrum* has been found as deep as 5,100 m.

Suborder Saccopharyngoidei

The three families that constitute this suborder are all deep sea in habitat and are greatly specialized, in part by the elimination of many anatomical features such as the swim bladder, scales, and ribs. The caudal and pelvic fins are reduced or absent. The hyoid arch is reduced to only the hyomandibular, and the gill arches are greatly reduced and situated well behind the head, with no bony or ligamentous connection to it. They lack the cleithrum, supracleithrum, lateral ethmoids, supraoccipital, orbitosphenoid, basisphenoid, and alisphenoids. The mouth is from three to seven times as long as the neurocranium and leads to a highly distensible pharynx that enables these fishes to engulf prey almost as large as they are (Fig. 9.4B).

The **Saccopharyngidae,** the swallowers or whiptail gulpers, contains nine species with large mouths and large, recurved teeth. They have extremely capacious stomachs and feed on fishes. The tail portion, which can reach a length of up to 2 m, has a complex luminescent organ that flashes like a red neon sign. Some forms with an attenuated tail still have as many as 300 vertebrae.

The gulper eels of the family **Eurypharyngidae** also have enormous mouths, but with small teeth (Fig. 9.4B; Plate 9.6). They reach a length of about 60 cm. Juveniles have a caudal fin, but this is lost in the adult. They have five complete gills and six gill clefts inside the "opercular" cavity. The stomach is nondistensible, and the food is mainly bathypelagic crustaceans. Members of the remaining family of the suborder, **Monognathidae,** lack the typical upper jaw bones—maxillaries and premaxillaries—but its members nonetheless are predators, probably feeding on shrimp. A rarely encountered family comprising about 14 species, they are known from only 70 specimens, most of them taken below 2,000 m (Nelson, 1994). They are among the most modified of fishes. A bifid spine is developed from the front part of the neurocranium and protrudes anteriorly from the roof of the mouth as a rostral fang (Bertelsen et al., 1989). This "ethmoid tooth" is hollow and equipped with a ducted gland. The males appear to change considerably at maturity, reabsorbing the mandible, losing the typical abdominal pouch, and greatly increasing the size of their olfactory organs, most likely to aid in the location of mates (Bertelsen et al., 1989). Length is up to 11 cm.

The fossil order †Crossognathiformes, known as early as the Lower Cretaceous, may be aligned with the eels. Three families are known from this marine order.

· ·
Subdivision Clupeomorpha: Herrings and Relatives

The herrings and anchovies make up the subdivision Clupeomorpha, consisting of the extinct order †Ellimmichthyiformes, with the living members in the order Clupeiformes. The swim bladder is extended forward in two branches that enter the skull and terminate in small vesicles within ossified bullae (small, somewhat globular structures), forming an otophysic connection with the utriculus of the ear. Clupeomorphs were sometimes considered relatives of the Elopiformes, but they lack some of the basal characteristics, such as the gular plate and conus arteriosus, encountered in those fishes. They show the usual complex of primitive teleostean features, such as fin surfaces composed only of soft rays and pelvic fins in an abdominal location. In the caudal fin, the urostyle is made up of the centrum of the terminal vertebrae and the first uroneural. The first hypural is separated from the urostyle (Lauder and Liem, 1983b). Many have **scutes** (see Chapter 3) that project posteriorly along the ventral midline. (Those that also have scutes along the dorsal midline are termed "double-armored.") Clupeomorphs introduced a new mode of jaw articulation, in which the angular is associated with the articular rather than with the retroarticular, and this appears in more derived teleosts.

Since the taxon was first described by Linnaeus, it has grown to include more than 80 genera and 350 species. The eccentric British ichthyologist Peter Whitehead, mentioned in Chapter 1, was perhaps the greatest student of the clupeomorphs, having published some of the best introductions to the group (Whitehead, 1985; Whitehead et al., 1988). Recent classifications of teleosts, emphasizing the dichotomous branching of different taxa, have grouped the clupeomorphs and the euteleosts ("higher" teleosts) together as the Clupeocephala (Patterson and Rosen, 1977). Johnson and Patterson (1996) proposed a new taxon, the **Otocephala,** to include the clupeomorphs and ostariophysans, and Lecointre and Nelson (1996) supported this in their summary of the morphological and molecular evidence. Molecular studies by Zaragüeta et al. (2002) have also supported a sister group relationship of the clupeomorphs with the Ostariophysi.

Order Clupeiformes

In terms of total biomass, the clupeiform fishes—especially the families Engraulidae and Clupeidae—are among the most significant fishery items harvested from the world's oceans (see Chapter 39). Why this is so has much to do with the predominance of planktivory among the Clupeiformes—feeding lower on the food chain contributes to their abundance. They trace their origins to the Upper Jurassic.

DIAGNOSIS OF THE CLUPEIFORMES

- Presence of a chamber in the pterotic bone with which some of the cephalic lateral line canals connect (recessus lateralis).[1]
- Part of the sensory canal system extends over the opercle and subopercle.
- Scales are cycloid and usually silvery and deciduous.
- Many with compressed, keellike bellies, with scutes on the midline.

[1]Except in the extinct order †Ellimmichthyiformes and the suborder Denticipitoidei.

Source: Nelson, 1994.

Suborder Denticipitoidei

This suborder contains one family, **Denticipitidae,** the tooth-head or denticle herring, in which a complete lateral line is present and the skull bones bear dermal denticles.

The only species, *Denticeps clupeoides,* is confined to West Africa, mainly Nigeria.

Suborder Clupeoidei

Fishes of this suborder lack the lateral line along the sides of the body and vary with respect to the degree of connection of the gas bladder with the gut. The caudal fin has 19 principal rays, contrasting with 16 in the Denticipitoidei.

The family **Engraulidae** contains the anchovies—small, silvery fishes rather rounded in cross-section and possessing a long snout and maxillary. The snout projects well beyond the lower jaw (Fig. 9.4C; Plate 9.7). Anchovies are usually less than 25 cm long, and most species live in tropical or subtropical waters. There are 15 to 20 genera and more than 100 species.

Like herrings, anchovies are the objects of great fisheries. One of the greatest is the anchoveta fishery off Peru, in the region of the Peru current, where tremendous biological production occurs. The anchoveta (*Engraulis ringens*) forms the basis for a fishery that has produced a harvest of up to 13 million metric tons in some years, in addition to the considerable numbers eaten by cormorants and other marine birds. Production of the anchoveta depends on the ocean currents. During the times of "*El Niño*" (also known as the El Niño Southern Oscillation, or ENSO), the coastal Peru Current that brings cold water north along the West coast of South America fades, and warmer equatorial waters intrude, causing an interruption of the enriching upwelling. The fishery and the birds suffer (see Chapter 39).

The ilishas, mostly of tropical and subtropical waters, have usually been considered a subfamily of the Clupeidae, but recent works have treated them as a separate family, the **Pristigasteridae** (Grande, 1982; Nelson, 1994; Whitehead, 1985; Whitehead et al., 1988). They have long-based anal fins and strong scutes on the ventral midline. Some species lack pelvic fins, and the genus *Raconda* has no dorsal fin. There are 9 genera and about 35 species, some of which, such as *Ilisha africanus* and *I. Elongatus,* are important in commercial fisheries.

The family **Chirocentridae** contains the wolf herrings or dorabs, *Chirocentrus dorab* and *C. nudus,* which differ in several respects from typical herrings. *C. dorab* may reach a length of 3.5 m and has numerous large, canine teeth (Fig. 9.4D). The gas bladder is divided internally by septa, and there is a bony appendage in the axilla of the pectoral fin. Wolf herrings are found in the Indo–Pacific and are used as food in some areas. Large specimens can be dangerous when handled.

The family **Clupeidae** includes the herrings, pilchards, shads, sardines, and similar fishes, and ranks as one of the most commercially important groups of fishes (see Chapter 39). Many species occur in dense schools, making them easier to capture, and their oil-rich flesh makes them the object of fisheries worldwide. In terms of the total tonnage of fishery products landed, the clupeids rank first—up to 20 million metric tons per year have been harvested in some years. The uses to which they are put are numerous—from food for humans and domestic animals to fertilizers and oils.

Some important members of the family are herring (*Clupea*), with species in both the north Atlantic and north Pacific (see Fig. 36.4); menhaden (*Brevoortia*) of the western Atlantic; and the American shad (*Alosa sapidissima;* Fig. 9.4E), which is sought for both food and sport. Others are the Pacific sardine (*Sardinops sagax*), historically one of the greatest fisheries of the California coast, depleted to the point of exterminating the fishery, and now staging a comeback; and the sprat (*Sprattus sprattus*) of the North and Baltic seas. In all, there are about 186 species of herrings in about 56 genera (Whitehead, 1985; Whitehead et al., 1988). Most are small fishes, seldom exceeding 0.5 m long , except *Palonia castelnaui*, a freshwater herring of South America, which reaches at least 1.5 m (personal communication, Dr. Barry Chernoff). The smallest known clupeomorph is also from South America: *Amazonsprattus scintilla*, described as a clupeid, reaches about 20 mm long, and is mature at about 15 mm (Roberts, 1984).

Clupeomorph life histories vary. Most are pelagic marine species, but many are found in freshwaters. Some—the shad and alewife (*Alosa*), for example—are anadromous. Pelagic forms spawn in the open water, whereas others attach eggs to seaweed or other substrates. From an ecological perspective, they are crucial links in aquatic food chains. Converting plankton into fish tissue, they are a critical food resource for larger piscivores.

KEY POINTS AND CONNECTIONS

- Although the question of teleostean phylogeny has yet to be completely resolved, the current consensus favors monophyly of the teleosts. Fully 96 percent of all named species of fishes are teleosts.

- Features that define the teleosts include the development of the urohyal bone and the modification of the premaxillary and maxillary bones to permit jaw protrusion. Jaw protrusion has greatly enhanced the feeding versatility of teleosts.

A discussion of the role of cranial skeletal features in the feeding strategies of teleosts can be found in Chapter 23.

- The teleosts encompass four major subdivisions: the Osteoglossomorpha, Elopomorpha, Clupeomorpha, and Euteleostei. Osteoglossomorphs, elopomorphs, and clupeomorphs are all comparatively primitive groups, whereas the vast majority of the modern teleosts are members of the subdivision Euteleostei.

- The Osteoglossomorpha include two orders in which the primary bite is formed from the medial parasphenoid and the tongue.

Included in Chapter 19 is a discussion of the unusual mode of locomotion used by members of the osteoglossomorph family Notopteridae. The electrical abilities of members of the families Mormyridae and Gymnarchidae are discussed in Chapters 21 and 36.

- Elopomorph fishes include tenpounders and tarpons of the order Elopiformes, bonefishes of the order Albuliformes, halosaurs and spiny eels of the order Notacanthiformes, and true eels of the order Anguilliformes. These primitive orders all share a distinctive larval form, the leptocephalus.

- The clupeomorph fishes include the herrings and related forms. In terms of total biomass, these mainly planktivorous fishes are the most significant living resource harvested from the world's oceans.

The significance of the world fishery for clupeomorph species is discussed in Chapter 39.

BUILDING AN ICHTHYOLOGY LIBRARY

Dixon, D. A. (Ed.). 2003. Biology, management, and protection of catadromous eels. *Amer. Fish. Soc. Symp. 33.* American Fisheries Society, Bethesda, MD.

Gosline, W. A. 1971. *Functional morphology and classification of teleostean fishes.* University Press of Hawaii, Honolulu. (Although the classification may be somewhat outdated, this book remains a classic synthesis of teleostean morphology and phylogeny.)

Grande, L. 1984. Paleontology of the Green River Formation, with a review of the fish fauna. *Bull. Geol. Surv. Wyoming 63.* (An excellent survey of the fossil fish fauna, including early teleosts, from one of the most important fossil-bearing regions of the world.)

Tesch, F. W. 1977. *The eel.* Chapman and Hall, London. (The 5th edition of this classic, edited by J. E. Thorpe, was published in 2003 by Iowa State University Press, Ames.)

REFERENCES

Alves-Gomes, J. A. 1999. Systematic biology of gymnotiform and mormyriform electric fishes: Phylogenetic relationships, molecular clocks, and rates of evolution in the mitochondrial rRNA genes. *J. Exp. Biol. 202:* 1167–1183.

———, and C. D. Hopkins. 1997. Molecular insights into the phylogeny of mormyriform fishes and the evolution of their electric organs. *Brain Behav. Evol. 49:* 324–351.

Aoyama, J., M. Nishida, and K. Tsukamoto. 2001. Molecular phylogeny and evolution of the freshwater eel, genus *Anguilla. Mol. Phylogen. Evol. 20:* 450–459.

Arratia, G. 1996. Reassessment of the phylogenetic relationships of certain Jurassic teleosts and their implications on teleostean phylogeny, pp. 219–242. In *Mesozoic fishes—systematics and paleoecology*, G. Arratia and G. Viohl (Eds.). Verlag Dr. Friedrich Pfeil, München, Germany.

———. 1997. Basal teleosts and teleostean phylogeny. *PalaeoIchthyologica 7:* 5–168.

———, and H.-P. Schultze. 1990. The urohyal: Development and homology within osteichthyans. *J. Morph. 203:* 247–282.

Avise, J. C., W. S. Nelson, J. Arnold, R. K. Koehn, G. C. Williams, and V. Thorsteinsson. 1990. The evolutionary genetic status of Icelandic eels. *Evolution 44:* 1254–1262.

Barbin, G. P. 1998. The role of olfaction in homing and estuarine migratory behavior of yellow-phase American eels. *Can. J. Fish. Aquat. Sci. 55:* 564–575.

———, and J. D. McCleave. 1997. Fecundity of the American eel *Anguilla rostrata* at 45° N in Maine, U.S.A. *J. Fish Biol. 51:* 840–847.

———, S. J. Parker, and J. D. McCleave. 1998. Olfactory cues play a critical role in the estuarine migration of silver-phase American eels. *Env. Biol. Fishes 53:* 283–291.

Bastrop, R., B. Strehlow, K. Jürss, and C. Sturmbauer. 2000. A new molecular phylogenetic hypothesis for the evolution of freshwater eels. *Mol. Phylogen. Evol. 14:* 250–258.

Berg, L. S. 1940. Classification of fishes both recent and fossil. *Trav. Inst. Zool. Acad. Sci. URSS 5:* 87–517. Reprinted 1947, J. W. Edwards, Ann Arbor.

Bertelsen, E., J. G. Nielsen, and D. G. Smith. 1989. Suborder Saccopharyngoidei: Families Saccopharyngidae, Eurypharyngidae, and Monognathidae, pp. 636–665. In Fishes of the western North Atlantic, Part 9, Vol. 1, Orders Anguilliformes and Saccopharyngiformes, E. B. Böhlke (Ed.). *Mem. Sears Found. Mar. Res.* Memoir, New Haven, CT.

Bishop, R. E., and J. J. Torres. 1999. Leptocephalus energetics: Metabolism and excretion. *J. Exp. Biol. 202:* 2485–2493.

Böhlke, J. E. 1966. Lyomeri, Eurypharyngidae, Saccopharyngidae, pp. 603–628. In Fishes of the western North Atlantic. *Mem. Sears Found. Mar. Res. 1(5).*

Caira, J. N., G. W. Benz, J. Borucinska, and N. Kohler. 1997. Pugnose eels, *Simenchelys parasiticus* (Synaphobranchidae) from the heart of a shortfin mako, *Isurus oxyrinchus* (Lamnidae). *Env. Biol. Fishes 49:* 139–144.

Carroll, R. L. 1988. *Vertebrate paleontology and evolution.* Freeman, New York.

DePinna, M. C. C. 1996. Teleostean monophyly, pp. 147–162. In *Interrelationships of fishes*, M. L. J. Stiassny, L. R. Parenti, and G. D. Johnson (Eds.). Academic Press, San Diego.

Forey, P. L. 1973a. Relationships of elopomorphs, pp. 351–368. In *Interrelationships of fishes*, P. H. Greenwood, R. S. Miles, and C. Patterson (Eds.). *Zool. J. Linn. Soc. 53* (Suppl. 1). Academic Press, New York.

———. 1973b. A revision of the elopiform fishes, fossil and recent. *Bull. Br. Mus. (Nat. Hist.) Geol. Suppl. 10:* 1–222.

———, D. T. J. Littlewood, P. Ritchie, and A. Meyer. 1996. Interrelationships of elopomorph fishes, pp. 175–191. In *Interrelationships of fishes*, M. L. J. Stiassny, L. R. Parenti, and G. D. Johnson (Eds.). Academic Press, San Diego.

Gosline, W. A. 1965. Teleostean phylogeny. *Copeia 1965:* 186–194.

———. 1971. *Functional morphology and classification of teleostean fishes.* University Press of Hawaii, Honolulu.

Grande, L. 1982. A revision of the fossil genus *Diplomystus* with comments on the interrelationships of clupeomorph fishes. *Am. Mus. Novitat. 2728:* 1–34.

Greenwood, P. H., D. E. Rosen, S. H. Weitzman, and G. S. Myers. 1966. Phyletic studies of teleostean fishes with a provisional classification of living forms. *Bull. Am. Mus. Nat. Hist. 131:* 339–456.

Haro, A., W. Richkus, K. Whalen, A. Hoar, W.-D. Busch, S. Lary, T. Brush, and D. Dixon. 2000. Population decline of the American eel. *Fisheries 25(9):* 7–16.

Helfman, G. S., B. B. Collette, and D. E. Facey. 1997. *The diversity of fishes.* Blackwell Science, Malden, MA.

———, D. E. Facey, L. S. Hales, Jr., and E. L. Bozeman, Jr. 1987. Reproductive ecology of the American eel. *Am. Fish. Soc. Symp. 1:* 42–56.

Hilton, E. J. 2001. Tongue bite apparatus of osteoglossomorph fishes: Variation of a character complex. *Copeia 2001:* 372–381.

Jessop, B. M. 1998. Geographic and seasonal variation in biological characteristics of American eel elvers in the Bay of Fundy area and on the Atlantic coast of Nova Scotia. *Can. J. Zool. 76:* 2172–2185.

Johnson, G. D., and C. Patterson. 1996. Relationships of lower euteleostean fishes, pp. 251–332. In *Interrelationships of fishes*, M. L. J. Stiassny, L. R. Parenti, and G. D. Johnson (Eds.). Academic Press, San Diego.

Kimura, K., K. Tsukamoto, and T. Sugimoto. 1993. *Migration of the Japanese eel larvae in the subtropical gyre-effect of the tradewind.* Paper presented at the Fourth Indo-Pacific Fish Conference (Bangkok). Abstr., p. 73 of program.

Lauder, G.V., and K.F. Liem 1983a. The evolution and interrelationships of the actinopterygian fishes. *Bull. Mus. Comp. Zool. 150:* 95-197.

———, and———. 1983b. Patterns of diversity and evolution in ray-finned fishes, pp. 1–24. In *Fish neurobiology*, Vol. 1. R. G. Northcutt and R. E. Davis (Eds.). University of Michigan Press, Ann Arbor.

Lavoué, S., R. Bigorne, G. Lecointre, and J.-F. Agnèse. 2000. Phylogenetic relationships of mormyrid electric fishes (Mormyridae; Teleostei) inferred from cytochrome *b* sequences. *Mol. Phylogen. Evol. 14:* 1–10.

Lecointre, G., and G. Nelson. 1996. Clupeomorpha, sister-group of Ostariophysi, pp. 193–207. In *Interrelationships of fishes*, M. L. J. Stiassny, L. R. Parenti, and G. D. Johnson (Eds.). Academic Press, San Diego.

Li, G.-Q., and M. V. H. Wilson. 1996. Phylogeny of Osteoglossomorpha, pp. 163–174. In *Interrelationships of fishes*, M. L. J. Stiassny, L. R. Parenti, and G. D. Johnson (Eds.). Academic Press, San Diego.

———, and———. 1999. Early divergence of Hiodontiformes sensu stricto in east Asia and phylogeny of some Late Mesozoic teleosts from China, pp. 369–384. In *Mesozoic fishes 2—systematics and fossil record*, G. Arratia and H.-P. Schultze (Eds.). Verlag Dr. Friedrich Pfeil, München, Germany.

———, and L. Grande. 1997a. Review of *Eohiodon* (Teleosteii: Osteoglossomorpha) from western North America, with a phylogenetic reassessment of Hiodontidae. *J. Paleontol. 71:* 1109–1124.

———, L. Grande, and M. V. H. Wilson. 1997b. The species of †*Phareodus* (Teleosteii: Osteoglossidae) from the Eocene of North America and their phylogenetic relationships. *J. Vert. Paleontol. 17:* 487–505.

Lobón-Cerviá, J. 1999. The decline of eel *Anguilla anguilla* (L.) in a river catchment of northern Spain 1986–1997. Further evidence for a critical status of eel in Iberian waters. *Arch. Hydrobiol. 144:* 245–253.

Martin, M. H. 1995. The effects of temperature, river flow, and tidal cycles on the onset of glass eel and elver migration into fresh water in the American eel. *J. Fish Biol. 46:* 891–902.

McAllister, D. E. 1968. Evolution of branchiostegals and classification of teleostome fishes. *Bull. Nat. Mus. Can. 221:* 1–239.

Mochioka, N., and M. Iwamizu. 1996. Diet of anguilloid larvae: Leptocephali feed selectively on larvacean houses and fecal pellets. *Mar. Biol. 125:* 447–452.

Moser, H. G. 1981. Morphological and functional aspects of marine fish larvae, pp. 90–131. In *Marine fish larvae*, R. Lasker (Ed.). University of Washington Press, Seattle.

Nelson, J. S. 1994. *Fishes of the world* (3rd ed.). John Wiley, New York.

Nilsson, G. 1999. The cost of a brain. *Nat. Hist. 108*(10): 66–73.

Otake, T. K., and K. Maruyama. 1993. Dissolved and particulate organic matter as possible food sources for eel leptocephali. *Mar. Ecol. Prog. Ser. 92:* 27–35.

Parker, S. J., and J. D. McCleave. 1997. Selective tidal stream transport by American eels during homing movements and estuarine migration. *J. Mar. Biol. Ass. U.K. 77:* 871–889.

Patterson, C. 1967. A second specimen of the Cretaceous teleost *Protobrama* and the relationships of the suborder Tselfatioidei. *Ark. Zool. 19:* 215–234.

———. 1968. The caudal skeleton in Lower Triassic pholidophorid fishes. *Bull. Brit. Mus. (Nat. Hist.) Geol. 16*(5): 201–239.

———. 1975. The braincase of pholidophorid and leptolepid fishes, with a review of the actinopterygian braincase. *Phil. Trans. Roy. Soc. Lond. B 269:* 275–579.

———. 1994. Bony fishes, pp. 57–84. In *Major features of vertebrate evolution*, R. S. Spencer (Ed.). University of Tennessee Press, Knoxville.

———, and G. D. Johnson. 1995. The intermuscular bones and ligaments of teleostean fishes. *Smithson. Cont. Zool. 559.*

———, and D. E. Rosen. 1977. Review of ichthyodectiform and other Mesozoic teleost fishes and the theory and practice of classifying fossils. *Bull. Am. Mus. Nat. Hist. 158:* 81–172.

Reinsch, H. H. 1968. Fund von Flußaalen, *Anguilla anguilla* L., im Nordatlantik. *Arch. Fisch Wiss. 19:* 62–63.

Richards, W. J. 1984. Elopiformes: Development, pp. 60–62. In *Ontogeny and systematics of fishes*, H. G. Moser, W. J. Richards, D. M. Cohen, M. P. Fahay, A. W. Kendall, Jr., and S. L. Richardson (Eds.). *Am. Soc. Ichthyol. Herpetol. Spec. Publ. 1.*

Roberts, T. R. 1984. *Amazonsprattus scintilla*, new genus and species from the Rio Negro, Brazil, the smallest known clupeomorph fish. *Proc. Calif. Acad. Sci. 43:* 317–321.

Robins, C. R. 1989. The phylogenetic relationships of the anguilliform fishes, pp. 9–33. In Fishes of the western North Atlantic, Part 9, Vol. 1, Orders Anguilliformes and Saccopharyngiformes, E. B. Böhlke (Ed.). *Mem. Sears Found. Mar. Res.*

———, and C. R. Robins. 1976. New genera and species of dysommine and synaphobranchine eels (Synaphobranchidae) with an analysis of the Dysomminae. *Proc. Acad. Nat. Sci. Phil. 127:* 249–280.

———, R. M. Bailey, C. E. Bond, J. R. Brooker, E. A. Lachner, R. N. Lea, and W. B. Scott. 1991. World fishes important to North Americans. *Am. Fish. Soc. Spec. Publ. 21.*

Smith, D. G. 1984. Elopiformes, Notacanthiformes, and Anguilliformes: Relationships, pp. 94–102. In *Ontogeny and Systematics of Fishes*, H. G. Moser, W. J. Richards, D. M. Cohen, M. P. Fahay, A. W. Kendall Jr., and S. L. Richardson (Eds.). Am. Soc. Ichthyol. Herpetol. Spec. Publ. 1.

Sullivan, J. P., S. Lavoué, and C. D. Hopkins. 2000. Molecular systematics of the African electric fishes (Mormyroidea: Teleostei) and a model for the evolution of their electric organs. *J. Exp. Biol. 203:* 665–683.

Tesch, F.-W. 1978. Horizontal and vertical swimming of eels during the spawning migration at the edge of the continental shelf, pp. 378–391. In *Animal migration, navigation, and homing*, K. Schmidt-Koenig and W. T. Keeton (Eds.). Springer Verlag, Berlin.

Tsukamoto, K., 1992. Discovery of the spawning area for Japanese eel. *Nature 356:* 789–791.

———, A. Umezawa, O. Taketa, N. Mochioka, and T. Kayihara. 1989. Age and birth date of *Anguilla japonica* leptocephali collected in western North Pacific in September, 1986. *Bull. Jpn. Soc. Sci. Fish. 55:* 1023–1028.

Van Ginneken, V. J. T., and G. E. E. J. M. van den Thillart. 2000. Eel fat stores are enough to reach the Sargasso. *Nature 403:* 156–157.

Vladykov, V. D. 1966. Remarks on the American eel (*Anguilla rostrata* LeSueur). Sizes of elvers entering streams; the relative abundance of adult males and females; and the present economic importance of eels in North America. *Verh. Int. Ver. Limnol. 16:* 1007–1017.

Whitehead, P. J. P. 1985. FAO species catalog. Clupeoid fishes of the world (suborder Clupeoidei), Part 1—Chirocentridae, Clupeidae, and Pristigasteridae. *FAO Fisheries Synopsis No. 125*, Vol. 7 (Pt. 1). FAO, Rome.

———, G. J. Nelson, and T. Wongratana. 1988. Clupeoid fishes of the world (suborder Clupeoidei). *FAO Fisheries Synopsis No. 125*, Vol. 7 (Pt. 2). FAO, Rome.

Wirth, T., and L. Bernatchez. 2001. Genetic evidence against panmixia in the European eel. *Nature 409:* 1037–1040.

Zaragüeta, B. R., S. Lavoué, A. Tillier, C. Bonillo, and G. Lecointre. 2002. Assessment of otocephalan and protacanthopterygian concepts in the light of multiple molecular phylogenies. *Comp. Rend. Acad. Sci. Sér. Biol. 325:* 1–17.

10

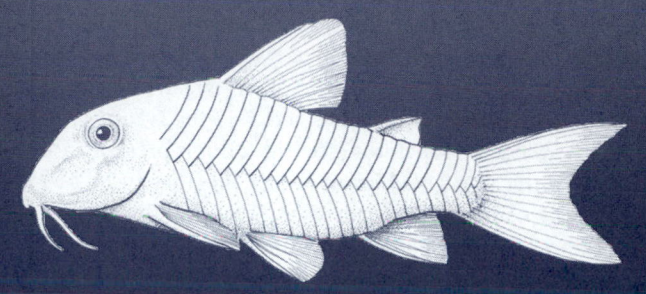

THE OSTARIOPHYSI: MASTERS OF THE FRESHWATER DOMAIN

OSTARIOPHYSI AS A SISTER GROUP OF THE CLUPEOMORPHA?

SUPERORDER OSTARIOPHYSI

Series Anotophysi
  Order Gonorynchiformes
Series Otophysi
  Order Cypriniformes
  Order Characiformes
  Order Siluriformes (Nematognathi)
  Order Gymnotiformes

The fishes of the superorder Ostariophysi thoroughly dominate freshwater habitats worldwide. How strange that in this vast and disparate group, only a few families are represented in the marine realm. Their mastery of the freshwaters of the world appears to have been facilitated by the development of an extraordinary suite of sensory capabilities. These include modifications of the anterior vertebrae and associated musculature to form a series of minute ossicles. When sound waves propagated through the water strike the swim bladder and set it to vibrating, this series of ossicles, termed the *Weberian apparatus,* transmits these vibrations to the inner ear—much like the series of ear bones found in the middle ear of mammals. Ostariophysan fishes are also characterized by their ability to secrete alarm substances when their skin surface is damaged. These substances signal conspecifics that danger is nearby. Extraordinary tactile and chemosensory capabilities are associated with the barbels that develop around the mouth of members of the order Siluriformes— the ostariophysan group that includes the catfishes. Perhaps the most extraordinary sensory adaptation seen among ostariophysans is the electrical communication witnessed among members of the order Gymnotiformes. Whereas most members of this order have evolved a sophisticated means of communication involving the discharge of weak electrical impulses, the electric eel has achieved notoriety as one of the few fish species that have developed the ability to emit electric discharges powerful enough to stun large prey.

OSTARIOPHYSI AS A SISTER GROUP OF THE CLUPEOMORPHA?

Most current classifications have grouped all of the remaining orders of teleostean fishes under the lineage (variably considered a cohort, subdivision, or infradivision) Euteleostei (Lauder and Liem, 1983; Nelson, 1994; Rosen, 1973). As mentioned in Chapter 9, the Clupeomorpha are now viewed as a sister group of the Ostariophysi. In Johnson and Patterson's (1996) revision of the euteleosts, the sister group relationship of the Ostariophysi and Clupeomorpha has been affirmed. Accordingly, Nelson's (1994) concept of what constitutes the subdivision Euteleostei differs from that of Johnson and Patterson (1996), as the sister group relationship with the Clupeomorpha necessarily excludes the Ostariophysi from inclusion in the Euteleostei (see Fig. 9.1).

SUPERORDER OSTARIOPHYSI

The fishes of this superorder are the dominant freshwater fishes of the world, accounting for about 64 percent of the freshwater ichthyofauna. A few species have successfully invaded marine waters, however. An extremely diverse assemblage, the more than 6,500 species of this superorder make up about 27 percent of the known species of fishes in the world (Nelson, 1994).

Ostariophysan fishes have modified anterior vertebrae and ribs, and most have a bony connection between the swim bladder and the ear, the **Weberian apparatus,** which aids in the reception of sound (see Chapter 21). The swim bladder is usually subdivided, with the smaller, anterior chamber covered by a silvery peritoneal tunic. Moreover, these fishes, except the gymnotoids (Fink and Fink, 1981), possess a distinctive alarm or fright substance, **"schreckstoff,"** a pheromone (Pfeiffer, 1977), which is released into the water by injury to the skin and produces a fright reaction in conspecifics. Many other groups of fishes, such as darters and gobies, have been demonstrated to exhibit alarm reactions to skin secretions (Smith, 1992).

Many of the ostariophysans have structures called **unculi,** or horny projections arising from single cells (Roberts, 1982). No other fishes are known to bear these structures, which are concentrated in areas of contact, such as lips, ventral surfaces, and sucking discs. Similar keratinaceous growths, termed **breeding tubercles,** are especially noticeable in many ostariophysans during the breeding season, but they are not as widespread among the Ostariophysi as the unculi are. These and other contact organs differ from the unculi in being multicellular in origin (Wiley and Collette, 1970).

The Ostariophysi are divided into two series (Fink and Fink 1981, 1996): the **Anotophysi,** in which the anterior vertebrae are only slightly modified; and the **Otophysi,** in which a Weberian apparatus (see Fig. 21.6) is present and in contact with the gas bladder and inner ear (Fig. 10.1). From a synthesis of anatomical, paleontological, and developmental studies, Chardon and Vandewalle (1997) have proposed an evolutionary history of the Weberian apparatus.

Series Anotophysi

The series Anotophysi includes only the order Gonorynchiformes (= Gonorhynchiformes).

Order Gonorynchiformes

DIAGNOSIS OF THE GONORYNCHIFORMES
- Mouth small, jaws without teeth.
- Anteriormost three vertebrae modified in what appears to be a forerunner of the Weberian apparatus.
- Epibranchial organ present.
- Cycloid (chanoids) or ctenoid (gonorynchoids) scales present.

  Sources: Nelson, 1994; Grande and Poyato-Ariza, 1999.

This order is considered the basal group of the Ostariophysi and has been confirmed as a monophyletic assemblage (Johnson and Patterson, 1997; Grande and Poyato-Ariza, 1999).

Suborder Chanoidei

This is a monotypic suborder, containing only the well-known milkfish (*Chanos chanos;* Fig. 10.2A; see Plate 10.1), family **Chanidae,** which is the object of pond culture in many areas of Southeast Asia. Modifications of the anterior vertebrae that parallel those observed in association with the development of the Weberian apparatus in members of the much more diverse series Otophysi have been described for *Chanos chanos* (Coburn and Chai, 2003). Five fossil genera are also included in this family (Grande and Poyato-Ariza, 1999). *Chanos chanos* can be found in coastal waters of the Indian and Tropical Pacific Oceans, where it reaches a length of about 1.8 m (Wheeler, 1975). A streamlined fish of marine or brackish water, it has a large, forked caudal fin and is capable of great leaps. Milkfish are efficient in converting their diet of microscopic phytoplankton into fish tissue. Young are collected for stocking in ponds by a variety of means, often by providing shade along the beach in shallow water, then using a small seine or dip net to capture the fry that gather there.

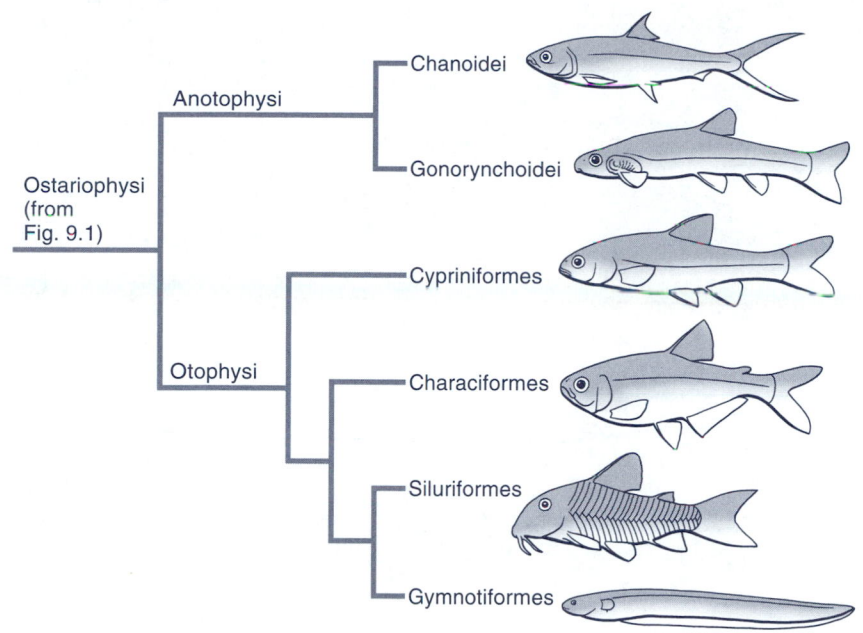

FIGURE 10.1
Phylogeny of the Ostariophysi (after Fink and Fink, 1996; Johnson and Patterson, 1996).

Culture methods differ from country to country, but they usually involve a complex of ponds, sometimes interconnected, to provide proper conditions for fry, fingerlings, and larger fishes being grown for the market. The Taiwanese are masters of milkfish culture, obtaining yields of up to 2,000 kg per hectare—especially remarkable considering the fact that special wintering ponds must be provided because of the generally cooler climate there.

Suborder Gonorynchoidei (= Gonorhynchoidei)

This suborder includes two families, the **Gonorynchidae** and the **Kneriidae** (Grande and Poyato-Ariza, 1999). Five species of *Gonorynchus*, commonly known as mousefishes, sand eels, or sandfishes, have been described on the basis of internal and external morphological features (Grande, 1999). *Gonorynchus* is a small marine fish of the Indo–Pacific (Fig. 10.2B). It is slender, has a pointed snout, a single barbel, and burrows into sandy bottoms. Whereas the milkfish possesses a swim bladder, it is absent in gonorynchoids. Species of *Gonorynchus* appear to be adapted to deep water; specimens from depths exceeding 700 m have been collected (Grande, 1999).

The kneriids are very small African freshwater fishes, some of which are paedomorphic. One species (*Kneria auriculata*) is called the "shell-ear" because of the peculiar

cup-shaped structure on the opercula of the males (Fig. 10.2C). Kneriids are airbreathers that can remain out of the water for some hours and also can wriggle up damp vertical surfaces in order to ascend streams. The shell-ear is believed to aestivate.

Series Otophysi

The otophysans are the dominant freshwater fishes in most parts of the world. This is a diverse group with more than 6,000 species. All have a well-developed Weberian apparatus between the swim bladder and the ear that aids in sound reception. The otophysans are varied in body form, size, habits, and habitat. Although recent authors have recognized two lineages based on the structure of the otophysic connection, there seems to be no consensus on the relationships or classification. Fink and Fink (1981) recognized the Cyprinophysi, which included the minnows, carps, and suckers of the order Cypriniformes, and the Characiphysi, which included the characins. In an update of their 1981 revision, Fink and Fink (1996) recognized the catfishes (order Siluriformes) and electric eels (order Gymnotiformes) as constituting a separate taxon, the Siluriphysi. Both Nelson (1994) and Fink and Fink (1996) recognized the four orders discussed next.

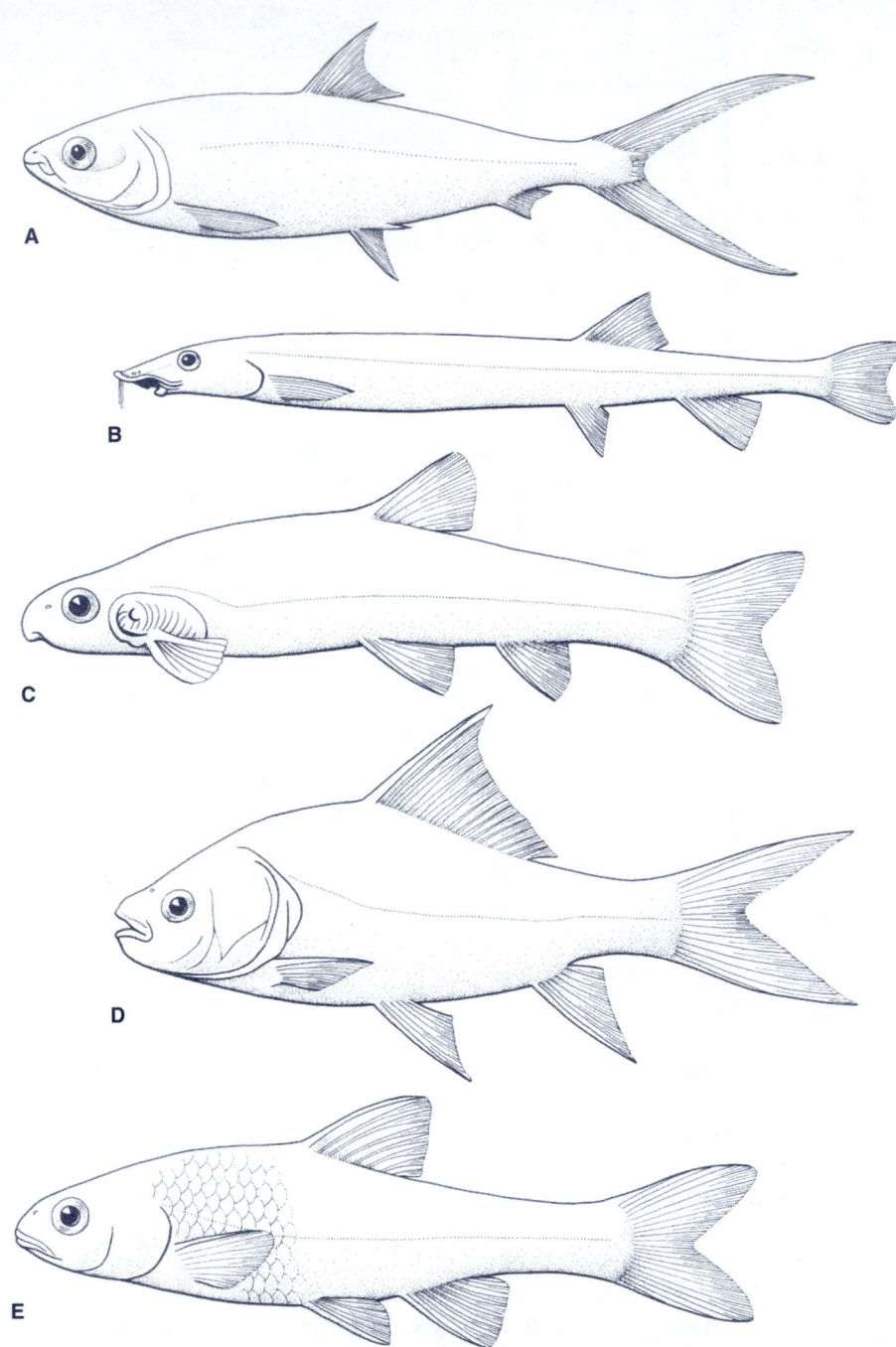

FIGURE 10.2
Representatives of the order Gonorynchiformes: **A,** Milkfish (*Chanos*); **B,** Sandfish (*Gonorynchus*); **C,** Shell-ear (*Kneria*). Representatives of the order Cypriniformes, family Cyprinidae: **D,** Catla (*Gibelion*); **E,** Shiner (*Notropis*).

Order Cypriniformes

DIAGNOSIS OF THE CYPRINIFORMES

- Jaws lacking teeth.
- Well-developed teeth on lower pharyngeal (fifth ceratobranchial) bones; teeth press against a cartilaginous pad borne on a posterior process of the basioccipital bone (see Fig. 23.6 C, D).
- Kinethmoid bone present between mesethmoid and ascending process of premaxillary bone.
- Orbitosphenoid, parietal, symplectic, and subopercular bones present.
- Many with barbels.
- Unfused third and fourth vertebrae of Weberian apparatus.
- Cycloid scales in most; head usually scaleless.
- Many with prominent intermuscular bones.
- True fin spines absent; some with stiffened, hard rays that resemble spines.
- Adipose fin absent (except in some loaches).
  Sources: Nelson, 1994; Fink and Fink, 1996.

The Cypriniformes includes the minnows, carps, loaches, suckers, and allied forms. They are native to Africa, Asia, Europe, and North America, and have been introduced in South America and Australia. They are fishes of great importance, both ecologically and economically.

The **Cyprinidae** (carps and minnows) appear to trace their origins to Southeast Asia, but now they are found in freshwaters worldwide. With more than 2,000 species grouped in approximately 340 genera, they are an extraordinarily successful family, with about half of the known species native to Asia (Banarescu and Coad, 1991). A quick perusal of the stock of any tropical fish store will reveal the significance of this family in the aquarium fish trade. Here can be found danios (see Chapter 27, Going Deeper), barbs, and rasboras. Cyprinids are abundant in lakes and streams of the Northern Hemisphere. Two species, the goldfish (*Carassius auratus*) and the carp (*Cyprinus carpio*) are particularly well known—in fact, notorious. Both have long been cultured, and numerous ornamental varieties have been produced. Because of its importance as a cultured food fish, the carp has been distributed almost worldwide from its original range in Eurasia. It has been the subject of selective breeding that has produced high-backed, deep-bodied individuals with few scales, rapid growth, and great efficiency in food utilization. The carp is well accepted as food in Europe and Asia and is used to some extent elsewhere. The stocking of carp in North American waters in the 19th century has led to its notoriety, as it is considered a pest that destroys the habitat of game fish while foraging, by stirring up the bottom and uprooting aquatic vegetation that native species use for nesting and cover. Although goldfish do not suffer as poor a reputation as carp, they have also been widely introduced. They are of some use for mosquito control where dissolved oxygen content is too low for other species. However, they have also been associated with the introduction of several harmful parasites that infect native populations of fishes (Lever, 1996).

Some of the cyprinids reach a large size, weighing up to 130 kg or more. The mahseer of India (*Tor tor*) is one of the largest. Other giants are *Catlocarpio siamensis* of Thailand and *Gibelion catla* of India, which is a favored food fish (Fig. 10.2D). Carps cultured in China include the bighead carp (*Hypophthalmichthys nobilis*); the silver carp (*H. molotrix*); and the grass carp (*Ctenopharyngodon idella*). The grass carp has been introduced in Europe and North America for use in the control of aquatic vegetation. It is now established in the Mississippi drainage and is commercially exploited in some places. Although effective as a weed control agent, the grass carp has also been implicated in the introduction of an Asian species of tapeworm to native cyprinids in North American drainages (Lever, 1996).

The phylogeny of the cyprinids has been the subject of a number of investigations (see. Cavender and Coburn, 1992; Chen et al., 1984; Howes, 1991). Cavender and Coburn recognized two phyletic lines within the Cyprinidae and designated these as the subfamilies Leuciscinae and Cyprininae. The leuciscines are broadly characterized as feeding with rapid swimming motions and employing mechanisms to lift the head while feeding, whereas cyprinines feed in a more sluggish manner, keeping the head rigid. Nelson (1994) recognized eight cyprinid subfamilies. Analysis of cytochrome *b* sequences has suggested that the modern genera constitute five separate lineages (Cunha et al., 2002).

Of the eight subfamilies recognized by Nelson (1994), the Leuciscinae include all approximately 270 species of minnows native to North America. Most of these are distributed east of the continental divide. Within the leuciscines, all save for the genus *Notemigonus* are grouped as a common assemblage, informally termed the phoxinines. The phoxinines themselves have been grouped into three clades: the western minnows, the chub group, and the shiners (including *Notropis;* Coburn and Cavender, 1992; Simons and Mayden, 1999). Most are small fishes that can serve as forage for larger predators. The most diverse genus is *Notropis*, with about 70 species (Fig. 10.2E; Mayden, 1989). The genus *Cyprinella* consists of about 25 species. Both of these are found mainly east of the Rocky Mountains. Another eastern North American genus that is often the most common fish of small streams is the stoneroller (*Campostoma*), so named from its habit of nudging and carrying stones about in the construction of nests (see Plate 10.2).

The largest of the North American minnows are in the genus *Ptychocheilus*. Originally referred to as *squawfish*, the preferred name is now *pikeminnow*. They are found in the Colorado, Sacramento, Columbia, and contiguous drainages, and they are reputed to be voracious predators. The Colorado pikeminnow (*P. lucius*) was reported by early ichthyologists to reach 1.5 m in length. The northern pikeminnow (*P. oregonensis*) now reaches about 75 cm, although its Pliocene-age ancestors in Oregon and Idaho reached 1.5 m.

The family **Gyrinocheilidae** includes fishes of the mountainous areas of Southeast Asia. They hold themselves in place in swift water by sucking onto the substrate with the mouth. Two opercular apertures on each side of the head facilitate gas exchange, the inhalant opening above and the exhalant opening below. They are distinctive among the Cypriniformes for the absence of pharyngeal teeth. Rather, they have numerous fine gill rakers useful for feeding on algae. These so-called "algae eaters" are a popular addition to the tropical fish hobbyist's tank, as these active fishes are continually scouring the tank sides and bottom, thus keeping algal blooms in check.

Suckers of the family **Catostomidae** are closely related to the minnows and are thought to have had their origin in Asia, even though only one primitive representative, *Myxocyprinus*, is found there now. Except for this species and for *Catostomus catostomus* (Fig. 10.3A), which has invaded Siberia from Alaska during the Pleistocene interglacial, the suckers are known only from North America. Harris and Mayden (2001) have contributed a reassessment of the phylogenetic relationships of catostomids based on molecular data.

Suckers have a single row of 16 or more pharyngeal teeth. Many of the members of the family have ventral mouths with thick, papillose lips, as exemplified by *Catostomus*, which is represented in most major drainages of North America. The largest of the genus is *C. luxatus*, the Lost River sucker of the Klamath drainage in Oregon and California, an endangered species that reaches about 1 m. Buffalofishes (*Ictiobus;* Fig. 10.3B) and the quillbacks and carpsuckers (*Carpiodes*) are large, carplike catostomids of the Mississippi and contiguous river systems. They are used as food and occasionally are cultured in the southern United States.

The family **Cobitidae** (= Cobitididae), the loaches, are small, slender fishes of Eurasia and Africa (Fig. 10.3C). They have three to six pairs of barbels around the mouth, and their gas bladder is encapsulated in bone. More than 100 species are known, in about 18 genera. The weatherfishes (*Misgurnus*) of European waters are well known because they react to changing barometric pressure with agitated swimming. Others (e.g., *Botia*) are kept in aquaria because of their striking color patterns and vigorous activity. Many species swallow air to supplement their oxygen supply. Those that pass the air through the anus are called "squeakers."

The family **Balitoridae**, formerly known as the Homalopteridae, contains the freshwater river loaches of Eurasia. These have at least three pairs of barbels close to the mouth, and a Weberian apparatus that differs in structure from that of the cobitids.

The balitorid subfamily Nemacheilinae includes several genera of small fishes from Eurasia. Some species, especially of the genus *Nemacheilus*, are exported from southeast Asia for use as aquarium fishes. A blind loach, *N. smithi*, is known from caves in Iran. One species lives in hot springs at 5,200 m elevation—the highest fish habitat known (Kottelat and Chu, 1988). Another species lives in caves at depths of up to 400 m below the surface of the ground.

The subfamily Balitorinae (flat loaches) contains fishes that are flattened on the ventral surface and generally have large pectoral and pelvic fins that are modified as adhesive organs. Most live in swift hill streams in southern Asia, including parts of India and China, Taiwan, and Borneo. Some have the gill openings reduced in size, and all have ventral mouths. There are usually three or four pairs of barbels around the mouth. Although they are often called "hill stream loaches," some live in other habitats. A blind, scaleless species of *Homalopterus* is known from caves in Thailand. There are nearly 30 genera and at least 120 species in this subfamily.

Order Characiformes

DIAGNOSIS OF THE CHARACIFORMES

- Well-developed teeth in jaw; diversity of tooth types reflects nutritional versatility.
- Pharyngeal teeth usually present, but not as well developed as in Cypriniformes.
- Orbitosphenoid, parietal, symplectic, and subopercular bones present.
- Barbels absent.
- Fins with soft rays only.
- First hypural of caudal fin separated from the compound ural centrum.
- Most with cycloid scales; some with ctenoid or ctenoid-like scales.
- Adipose fin present in most.
  Sources: Lauder and Liem, 1983; Nelson, 1994.

This tropical order contains the characins (Characidae) and their relatives. These are found in freshwater in both Africa

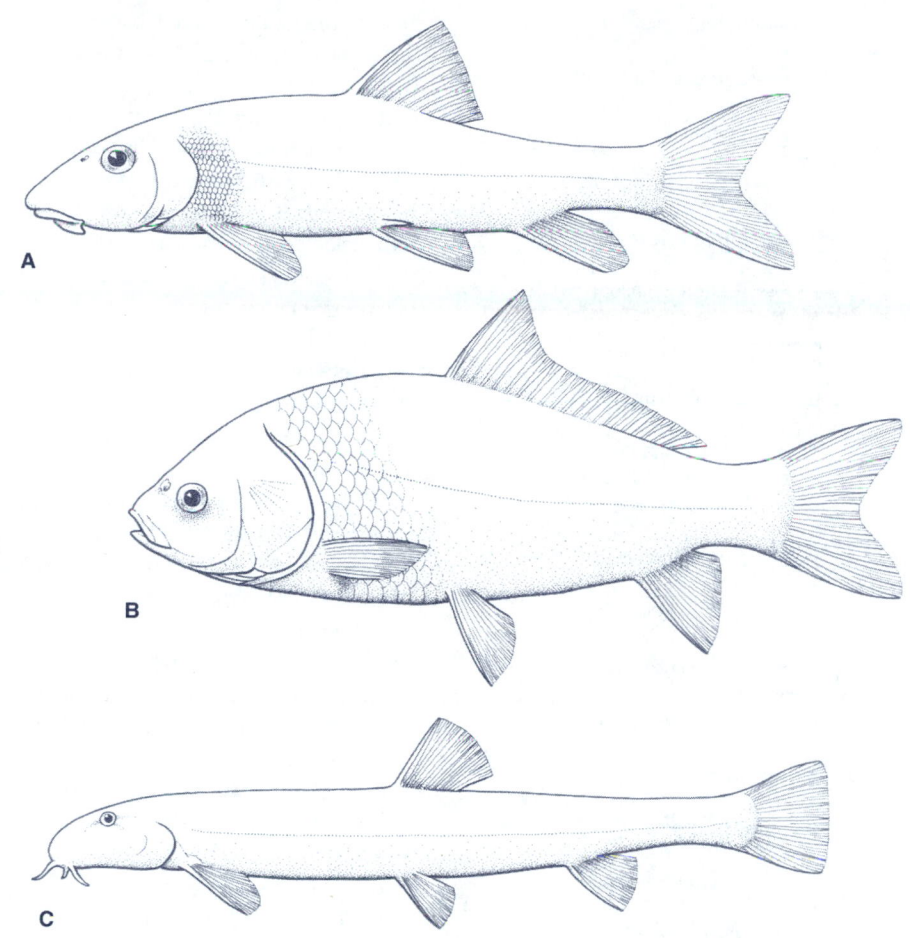

FIGURE 10.3

Representatives of the order Cypriniformes: **A,** Sucker (Catostomidae, *Catostomus*); **B,** Buffalofish (Catostomidae, *Ictiobus*); **C,** Loach (Cobitidae, *Cobitus*).

and South America, north through Central America and Mexico to the Rio Grande in Texas. Although the order is widely distributed in both Africa and South America, no genera are common to the two continents. About 25 genera and 200 species occur in Africa and more than 200 genera and more than 1,300 species in the Americas. Most authorities recognize several families—usually from 10 to 16 (Eschmeyer, 1990; Greenwood et al., 1966; Lauder and Liem, 1983)—but others prefer to regard most of these as subfamilies (Gosline, 1971; Robins et al., 1991). Comprehensive overviews of the phylogeny and classification of the Characiformes are presented by Vari (1998), Weitzman and Malabarba (1998), and Buckup (1998), while Ortí (1997) and Ortí and Meyer (1997) described phylogenetic approaches

derived from molecular studies. A tentative phylogeny of characiform fishes is presented in Fig. 10.4. The families listed here are those accepted by Nelson (1994).

The family **Citharinidae** of Africa is thought to be the most primitive of the characins (Fink and Fink, 1981). Most have ctenoid scales, and some have slightly protrusible upper jaws. Included in the citharinids is the subfamily Distichodontinae, which includes elongate predators and fin-eaters as well as compressiform herbivores and predators on insects and small crustacea. Members of the subfamily Citharininae, the moonfishes (Fig. 10.5A), are deep-bodied and have no teeth on the small maxillae. Reaching a size of about 85 cm in length and 2.5 kg in weight, moonfishes are esteemed food fishes.

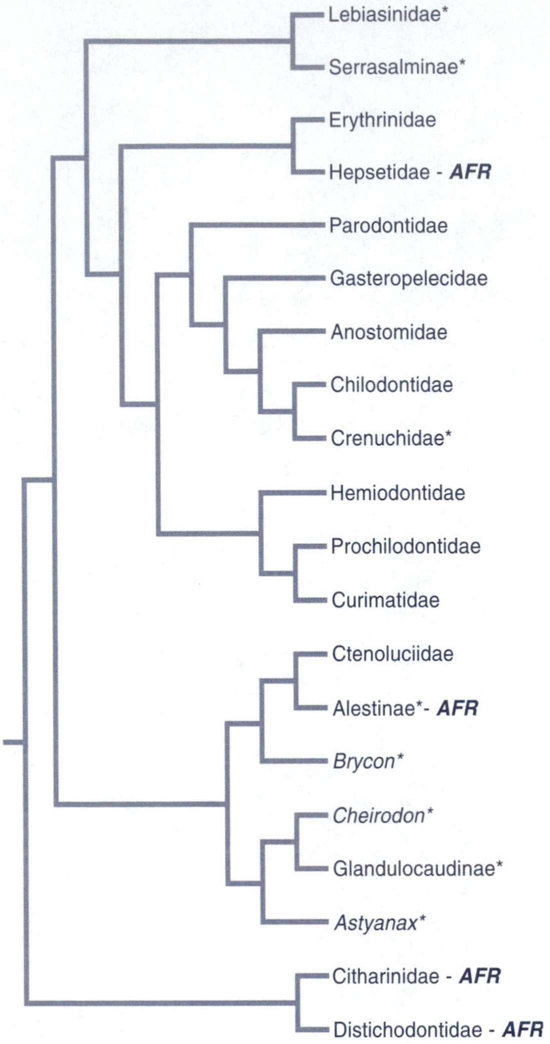

* Classified in family Characidae by Nelson (1994).

FIGURE 10.4
Phylogeny of the characiform fishes. *AFR* indicates African taxa (after Ortí, 1997).

Fishes of the South and Central American family **Hemiodontidae** usually lack teeth in the lower jaw. The hemiodontids include two subfamilies, the Parodontinae, which are algae-scraping, benthic fishes, and the Hemiodontinae, which are found in open water. Among the hemiodontines are the only characiform fishes with highly protrusible upper jaws. Members of the genus *Anodus* lack teeth in the jaws but have numerous gill rakers and other adaptations for feeding on plankton.

Fishes of the family **Curimatidae**, ranging from South America north to Costa Rica, are mainly detritivores. They have modifications of the gill chamber, such as loss of teeth on some pharyngeal tooth plates, and an epibranchial organ. The subfamily Curimatinae lacks jaw teeth. The subfamily Prochilodontinae has teeth in the jaws and a protractile mouth with large lips. The abundance and diversity of hemiodontids and curimatids in South American freshwaters reflects the significance of planktivory and detritivory in the trophic structure of these freshwater ecosystems (see Chapter 31).

Members of the South American families **Anostomidae** and **Lebiasinidae** are popular additions to home aquaria. Anostomids are known by tropical fish hobbyists as "headstanders," because of the peculiar head-down posture they assume while swimming.

The family **Erythrinidae** of South America includes some large-mouthed predators that can reach a length of 1 m. Some have lunglike gas bladders and are said to be capable of limited terrestrial forays.

The family **Ctenoluciidae**, the pike characins (see Plate 10.3), range from South America to Panama. These are large (up to 1 m), predaceous fishes with a body form that converges remarkably with that of the pikes (Esocidae; see Plate 12.1).

The family **Hepsetidae** of Africa is monotypic. *Hepsetus odoe* is a predaceous fish known as the Kafue pike. It reaches about 40 cm long. According to Géry (1977), this species spawns in a floating foam nest.

The family **Gasteropelecidae** of South America and Panama includes three genera, *Gasteropelecus*, *Carnegiella*, and *Thoracocharax*. Robins et al. (1991) listed nine species that are exported as aquarium fishes. These are known as freshwater hatchetfishes or freshwater flyingfishes. Their ability to leap clear of the water with great force, aided by their winglike pectoral fins and powerful pectoral muscles, is similar to the freshwater flyingfish described in Chapter 9.

The family **Characidae** contains numerous genera of diverse habits in both Africa and the Americas. There is little agreement among ichthyologists on the classification of the family. For instance, Géry (1977) and Nelson (1994) both listed 11 subfamilies, but only 4 of them coincided. Weitzman and Malabarba (1998) reflected on the challenge of systematic study of this family and concluded that they cannot be presently diagnosed as a monophyletic group.

Many small and colorful species of this family are kept in home aquaria and form the basis for a lucrative import-export trade. The tetras (Fig. 10.5B), such as *Hyphessobrycon* and *Hemigrammus*, are examples. The piranhas, native to drainages in northern South America, are characins of the subfamily Serrasalminae. They are notorious for their sharp

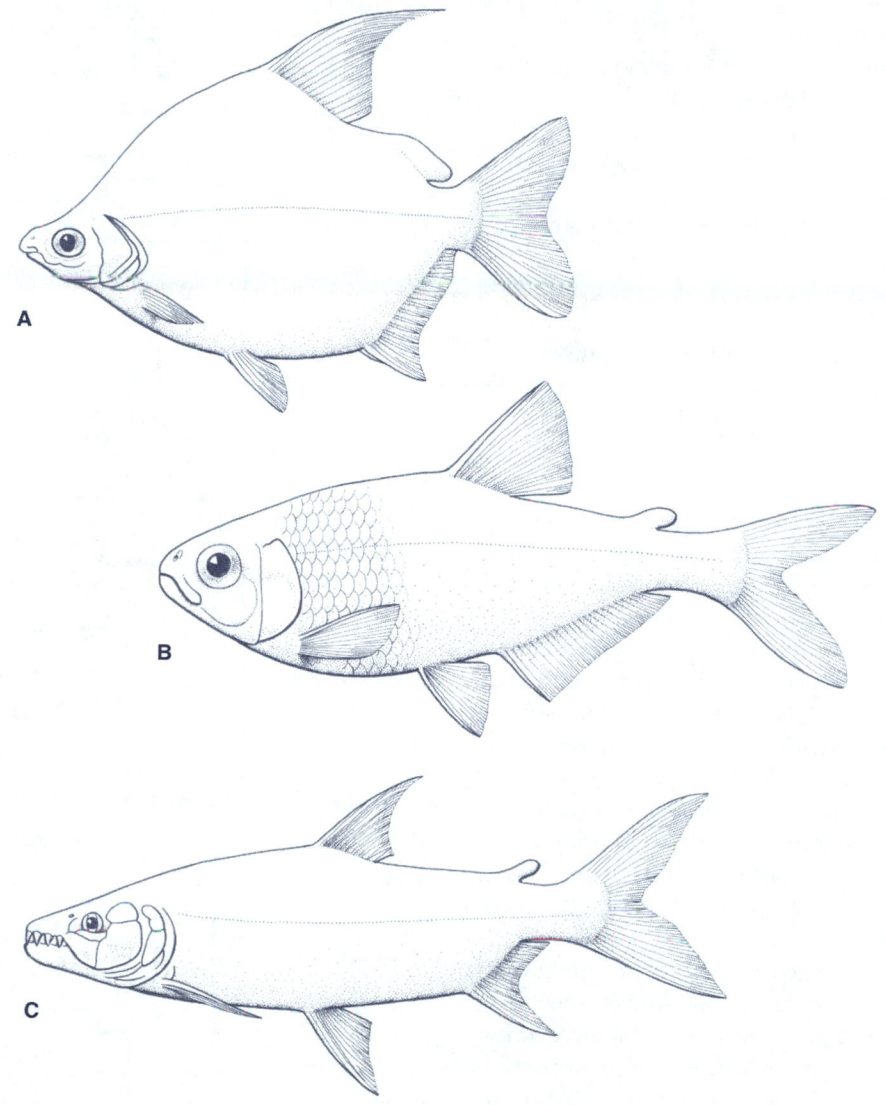

FIGURE 10.5
Representatives of the order Characiformes: **A,** Moonfish (Citharinidae); **B,** Tetra (Characidae); **C,** Tigerfish (Characidae, *Hydrocynus*).

teeth, strong jaws, and voracious feeding habits. Examples of genera of piranhas are *Serrasalmus* and *Pygocentrus* (see Plate 10.4). Although they are capable of inflicting painful bites, the reputation of piranhas is somewhat exaggerated. Although groups of red-bellied piranhas (*P. nattereri*) have been known to engage in a feeding frenzy when they encounter a bleeding animal, they are usually shy and retiring. Also to be found among the serrasalmines is the tambaqui (*Colossoma macropomum*), a gentle, herbivorous giant that weighs as much as 30 kg and is an esteemed food fish. Because of its fruit- and seed-eating habits, the tambaqui is a crucial link between terrestrial and freshwater ecosystems in the Amazonian rainforest. This species also shows great potential for fish farming (Araujo-Lima and Goulding, 1997). Another piranha relative, the silver-dollar fishes of the genus *Metynnis*, are among the most popular of aquarium fishes. The tigerfish (*Hydrocynus*) of Africa (Alestiinae) is another fierce predator (Fig. 10.5C), as is *Cynodon* of South America.

Order Siluriformes (Nematognathi)

DIAGNOSIS OF THE SILURIFORMES
- Teeth present on premaxillary, but absent on maxillary in most.
- Subopercle, symplectic, and parietal bones absent.
- Barbels around mouth well developed.
- Weberian apparatus incorporates five or more vertebrae; second through fourth or fifth vertebrae may be fused.
- Scales absent; skin bare or covered with bony plates, some of which may bear dermal denticles.
- Intermuscular bones absent.
- Dorsal and pectoral fins usually with soft rays modified into stiff spines; spines equipped with locking mechanism; some with venom glands associated with spines.
- Adipose fin usually present.

  Sources: Nelson, 1994; Teugels, 1996.

The order Siluriformes, the catfishes, is a widespread and economically important group of fishes. Among the most distinctive features of catfishes are the barbels around their mouths. Taste receptors, which occur on the head and trunk of catfishes, are especially abundant on the barbels. Three types of barbels—mandibular, maxillary, and nasal—may be found among the Siluriformes. The maxillary barbel is associated with a mobile palatine-maxillary system that consists of the maxillary and palatine bones together with specialized muscles and ligaments that permit considerable mobility (Gosline, 1975). Diogo and Chardon (2001) discussed the evolution and adaptation of this important sensory tool in catfishes. Fossil catfishes are known from the Paleocene and were diverse by the Eocene.

Teugels (1996) recognized 33 families of catfishes, including 416 genera and more than 2,500 species. More than 60 percent of these are native to Central and South America. Catfishes are known from all continents except Antarctica, where they occur as fossils. Those found in Australian freshwaters belong to the marine families Ariidae and Plotosidae—the former widespread in warm seas, the latter in the Indo-Pacific. Most of the other catfish families are strictly freshwater fishes. The phylogenetic relationships of the Siluriformes are summarized in Fig. 10.6. Phylogenetic studies using molecular techniques have focused particularly on the families Clariidae and Ictaluridae; it is no coincidence that these are among the most economically important families (Volckaert and Agnèse, 1996).

The family **Diplomystidae** (velvet catfishes) of South America is considered to contain the most primitive living catfishes (Campos et al., 1997; Teugels, 1996). They are unique among catfishes in having well-developed maxillae bearing teeth. The fifth vertebra is not fused or firmly connected to the fourth, as in the remaining families. There are

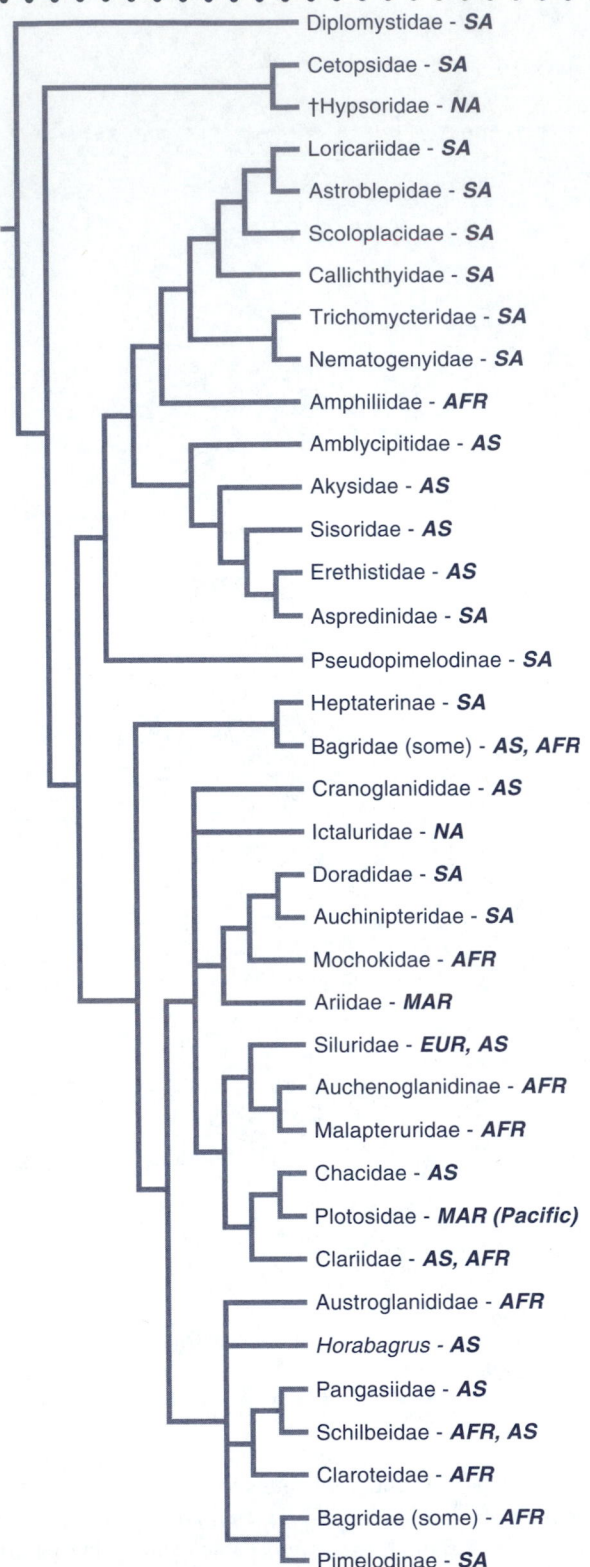

FIGURE 10.6
Phylogeny of the major taxa of catfishes, excluding the recently discovered *Lacantunia enigmatica*. *NA* = North America, *SA* = South America, *AFR* = Africa, *AS* = Asia, *EUR* = Europe, *MAR* = marine (after dePinna, 1998; Teugels, 1996).

four, perhaps five, species, of which *Diplomystes* (= *Diplomyste*) *chilensis* and *Olivaichthys viedmensis* may be the best known.

Recently, a new family, genus, and species of catfish, *Lacantunia enigmatica,* was discovered in the Río Usumacinta of southern Mexico. The species name is most appropriate, as this fish has a most puzzling assemblage of features, suggesting it may be among the most ancient of catfishes. Systematic relationships with other New World families are not apparent (Rodiles-Hernández et al., 2005). The authors marveled that a fish of such size (approaching 0.5 m in length), which is commonly fished by the local people, could have escaped scientific detection for so long. Given its recent discovery, the phylogenetic position of this family (**Lacantuniidae**) is not considered in Fig. 10.6.

Nelson (1994) identified 15 families of catfishes from South America. Among the largest of these in terms of body size are the **Pimelodidae**—long-whiskered catfishes that can grow to a length of 3 m. These large predators are important food fishes in the Amazon Basin (Barthem and Goulding, 1997). One assemblage of South American catfish families is considered monophyletic because they share gas bladders encapsulated in bone; denticles (termed **odontodes**) on the skin; armor or fin rays that resemble placoid scales; and common features in the structure of the first pectoral ray (dePinna, 1998). This assemblage, sometimes classified in the suborder Loricarioidea, includes a few families that lack dermal armor. Included among these is the family **Trichomycteridae**, the pencil catfishes. The trichomycterids include a number of genera that are sometimes viewed as parasites, because they attack the gill cavities of larger fishes. Members of the genus *Vandellia* (the candirus) are apparently attracted to nitrogenous wastes produced by fish gills and have consequently been known to enter the urinary opening of mammals, including humans. A word of caution—when swimming in the Amazon Basin, always leave the water to urinate (see Chapter 37).

Armored loricarioids include the **Callichthyidae** (plated catfishes; Fig. 10.7A); **Scoloplacidae,** the spiny dwarf catfishes; and **Loricariidae,** the suckermouth catfishes. Species from many of the aforementioned families, including the Aspredinidae, Doradidae, Pimelodidae, Callichthyidae, and Loricariidae, are popular additions to the home aquarium.

One of the best-known families of Old World catfishes is the **Siluridae,** the sheatfishes. These elongate predators are some of the largest fishes to be found in freshwater. The European wels (*Silurus glanis*) may reach a weight of 130 kg or more and a length of 3 m (Fig. 10.7B), whereas the Asian silurid *Wallagonia attu* may reach lengths of 2 m. Another catfish family known for the size of some of its members is

the **Pangasiidae** of Asia. Two genera are described in the family (Roberts and Vidthayanon, 1991). The genus *Pangasius* includes one of the largest freshwater fishes, *P. gigas,* weighing as much as 300 kg and measuring up to 3 m. The other genus, *Helicophagus,* contains only two species and is aptly named from its diet of mollusks. The closely related Schilbeidae occur in both Asia and Africa, as does the family Bagridae.

Some families of catfishes are able to extract oxygen from the atmosphere and hence engage in terrestrial activity. Air-breathing fishes can be found in the families **Clariidae** of Asia and Africa and **Heteropneustidae** (= Saccobranchidae) of Asia. Fishes in the clariid family are equipped with an arborescent accessory breathing organ in the gill chamber, and the gill cavity of the Heteropneustidae communicates with large, paired air sacs in the musculature of the body. These families are placed together in the Clariidae by some ichthyologists because the two groups have similar structures except for the air-breathing apparatus. Included among the clariids is the notorious "walking catfish" (*Clarias batrachus*). Albino varieties were once imported into the United States from their native Thailand for the tropical fish trade. Because of their terrestrial capabilities (see Chapter 24), they rapidly spread throughout South Florida waterways, becoming one of the most conspicuous of the many exotic fish species that have become established in North America (see Chapter 39). This species is an ichthyological version of the classic example of selection for melanic forms that was first described for moths in industrialized areas of Europe. In the case of *Clarias,* the visibility and vulnerability of the albinic forms to predators ensured a rapid reversion to naturally dark forms in waterways where they became naturalized.

The African family **Malapteruridae** contains two electrogenic species, one of which, *Malapterurus electricus,* can deliver a severe shock. Another African family, the **Mochokidae,** contains some species that habitually swim upside down (see Plate 10.5). Some of these have reverse countershading, with dark bellies and lighter dorsal surfaces. The upside-down swimming may aid in feeding on plankton in the surface layer and may also help in obtaining oxygen from well oxygenated water near the air–water interface (Chapman et al., 1994).

The North American freshwater catfishes, Ictaluridae, which were native to western North America from Eocene to Pliocene, are now native only east of the Rocky Mountains, but they have been introduced and are now established in most parts of the continent with a suitably warm climate. The blue catfish (*Ictalurus furcatus*) is the giant of the group, weighing up to 70 kg. The channel catfish (*I. punctatus*) may reach more than 20 kg, and the flathead catfish (*Pylodictis*

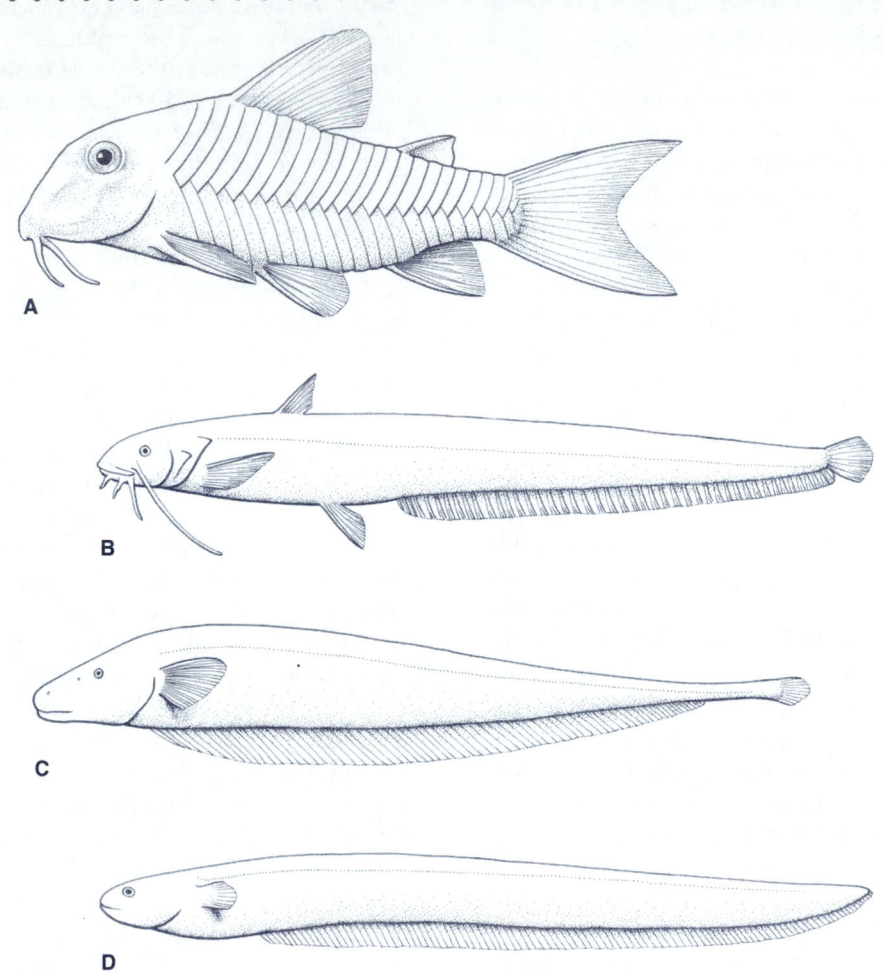

FIGURE 10.7
Representatives of the order Siluriformes: **A,** Armored catfish (Callichthyidae); **B,** Wels (*Silurus*). Representatives of the order Gymnotiformes: **C,** Knifefish (*Apteronotus*); **D,** Electric eel (*Electrophorus*).

olivaris) is nearly as large. These large ictalurids are the subject of one of the most peculiar forms of fishing, known as "noodling." Noodling is about as basic as fishing can be, as the only tackle involved is one's hands. The noodler gropes around beneath banks or undercuts where a catfish is suspected to hide. When the fish attempts to bite or swallow the noodler's hand, the fish is simply hauled in. Many states outlaw this practice. Not only is it a potentially dangerous activity, but conservationists argue that hand-fishing favors the removal of the largest individuals—the ones that could make the greatest reproductive contribution to the population. The channel catfish has long been a favored food fish and the object of a commercial fishery. It is cultured extensively throughout the warmer parts of the United States.

Modern methods of fish culture; the availability of dry, pelleted fish food; and consumer demand have made the rearing of channel catfish one of the most profitable of aquaculture enterprises.

The bullhead catfishes (*Ameiurus*, formerly in *Ictalurus*) are well known to anglers and fish biologists; their native distribution encompasses streams and lakes of North America east of the continental divide. Hardman and Page (2003) incorporated molecular techniques in a reassessment of the phylogenetic relationships within this genus.

The madtom catfishes (*Noturus*) have venom glands associated with their pectoral spines and can cause painful wounds. Taylor (1969) and Burr and Stoeckel (2000) have published definitive works on these small, secretive fishes.

There are three ictalurid genera with blind species: *Trogloglanis* and *Satan* from Texas, and *Preitella* from northern Mexico. *Trogloglanis* and *Satan* were discovered in artesian wells drilled to depths of more than 500 m (Sneegas and Hendrickson, 2003). Living as they do in conditions of extreme hydrostatic pressure, they lack swim bladders (Hubbs and Bailey, 1947).

Order Gymnotiformes

DIAGNOSIS OF THE GYMNOTIFORMES

- Palatines not ossified; ectopterygoids absent.
- Pelvic girdle and fins absent.
- Dorsal fin absent, but some may have an elongate, fleshy filament on the back that resembles an adipose fin.
- Anal fin usually begins just behind the anus and stretches to the end of the tail.
- The caudal fin is usually lacking.
- Anus and urogenital apertures placed far forward, usually in advance of the pectorals.

  Sources: Nelson, 1994; Albert and Campos-da-Paz,1998.

The Gymnotiformes (New World knifefishes) are elongate fishes of South and Central American freshwaters (Fig. 10.7C). Because of features they share with the catfishes, Fink and Fink (1981) placed them in the order Siluriformes. One feature they hold in common with catfishes is an abundance of electroreceptors on the body surfaces.

The electrogenic and electroreceptive capabilities of gymnotiform fishes are especially well known. Two classes of electroreceptors have been identified in gymnotiform fishes—one is sensitive to low-frequency emissions and is useful in prey detection, whereas the other detects high-frequency emissions and is used for orientation and social communication (see Chapters 21 and 36).

Two suborders of gymnotiform fishes are generally recognized (Albert and Campos-da-Paz, 1998; Campos-da-Paz and Albert, 1998; Nelson, 1994). Pronounced sexual dimorphism, in which mature males display greatly hypertrophied facial features, is known in some species of gymnotiform fishes (Cox Fernandes et al., 2002).

Suborder Sternopygoidei

This suborder contains weakly electric fishes with highly developed modes of electric communication and electrolocation. Families usually recognized include **Sternopygidae, Rhamphichthyidae, Hypopomidae,** and **Apteronotidae.** Many of these have a bizarre appearance; some

bear a superficial resemblance—except for the dorsal and anal fins, of course—to *Gymnarchus* or the mormyrids, the African electric fishes discussed in Chapter 9.

Suborder Gymnotoidei

This suborder includes the families **Gymnotidae** and **Electrophoridae.** The latter family is monotypic, including only the electric eel, *Electrophorus electricus* (Fig. 10.7D), which is a large, elongate, but heavy-bodied fish with about 240 vertebrae. The electric eel has about 80 percent of its body in the caudal region, which is largely made up of electric organs (see Chapter 21 for a discussion of the structure of electrogenic tissue). This fish can be considered an aquatic "stun gun," as its electric organs are capable of producing powerful discharges. Emissions of up to 650 volts have been recorded, although the average is about 350 volts. The buccopharyngeal cavity of the electric eel has become modified for air breathing—a useful adaptation for the swampy lowlands normally inhabited by this species.

- -

KEY POINTS AND CONNECTIONS

- -

- Members of the superorder Ostariophysi are currently considered a sister group to the Clupeomorpha. The group is divided into the Anotophysi, in which only rudimentary modification of the anterior vertebrae is apparent, and the Otophysi, in which these vertebrae are modified to form the Weberian apparatus.

An illustration of the Weberian apparatus, as it appears in suckers of the family Catostomidae, is shown in Chapter 21.

- Although the vast majority of ostariophysan fishes are freshwater inhabitants, members of the anotophysan families Chanidae and Gonorynchidae inhabit marine waters.

- Included among the most diverse orders of ostariophysan fishes are the Cypriniformes, a wide-ranging order that is most diverse in southeast Asia but is well represented in the north temperate freshwater ichthyofauna; the Characiformes, dominant members of the South American freshwater ichthyofauna, but prevalent in Africa as well; and the Siluriformes, which are abundant worldwide. Two families of siluriform catfishes, the Ariidae and Plotosidae, represent the only significant invasion of marine waters by otophysan fishes.

The distinctive pharyngeal dentition of cypriniform fishes is discussed in Chapter 23.

- The order Gymnotiformes includes some of the most unusual fishes found in freshwater. Their electroreceptive and electrogenic capabilities are legendary, culminating in the evolution of one of the most sophisticated communication modes known among animals, involving the emission and detection of weak electrical signals.

Chapters 21 and 36 provide discussions of the structure and function of electrical organs and their role in fish behavior.

FISH LINKS

http://bio.slu.edu/mayden/cypriniformes/home.html Cypriniformes Tree of Life: Dr. Richard Mayden of Saint Louis University is
directing a major research effort focusing on the evolutionary relationships, classification, and diversity of this extraordinary order of fishes.

http://silurus.acnatsci.org/All Catfish Species Inventory: The mission
of this research program dedicated to all things catfish is succinctly
stated: "To facilitate the discovery, description, and dissemination of
knowledge of all catfish species by a global consortium of taxonomists
and systematists."

BUILDING AN ICHTHYOLOGY LIBRARY

A definitive work on the cyprinids:

Winfield, I. J., and J. S. Nelson. 1991. *Cyprinid fishes: Systematics, biology,
and exploitation.* Chapman and Hall, London.

The following four books are indicative of the evolutionary,
ecological, and economic significance that ostariophysans
play in the Neotropics:

Araujo-Lima, C., and M. Goulding. 1997. *So fruitful a fish: Ecology,
conservation, and aquaculture of the Amazon's tambaqui.* Columbia
University Press, New York.
Barthem, R., and M. Goulding. 1997. *The catfish connection: Ecology, mi-
gration, and conservation of Amazon predators.* Columbia University
Press, New York.
Goulding, M. 1980. *The fishes and the forest: Explorations in Amazonian
natural history.* University of California Press, Berkeley.
Malabarba, L. R., R. E. Reis, R. P. Vari, Z. M. S. Lucena, and C. A. S.
Lucena (Eds.). 1998. *Phylogeny and classification of neotropical fishes.*
EDIPUCRS, Porto Alegre, Brazil.

REFERENCES

Albert, J. S., and R. Campos-da-Paz. 1998. Phylogenetic systematics of
Gymnotiformes with diagnosis of 58 clades: A review of available
data, pp. 419–446. In *Phylogeny and classification of neotropical
fishes,* L. R. Malabarba, R. E. Reis, R. P. Vari, Z. M. S. Lucena, and
C. A. S. Lucena (Eds.). EDIPUCRS, Porto Alegre, Brazil.
Banarescu, P., and B. W. Coad. 1991. Cyprinids of Eurasia, pp. 127–155.
In *Cyprinid fishes. Systematics, biology, and exploitation,* I. J.
Winfield and J. S. Nelson (Eds.). Chapman and Hall, London.
Barthem, R., and M. Goulding. 1997. *The catfish connection: Ecology, mi-
gration, and conservation of Amazon predators.* Columbia University
Press, New York.
Buckup, P. A. 1998. Relationships of the Characidiinae and phylogeny
of characiform fishes (Teleostei: Ostariophysi), pp. 123–144.In
Phylogeny and classification of neotropical fishes, L. R. Malabarba,
R. E. Reis, R. P. Vari, Z. M. S. Lucena, and C. A. S. Lucena (Eds.).
EDIPUCRS, Porto Alegre, Brazil.
Burr, B. M., and J. N. Stoeckel. 2000. The natural history of madtoms
(genus *Noturus*), North America's diminutive catfishes. *Am. Fish.
Soc. Symp. 24:* 51–101.
Campos, H., G. Arratia, and C. Cuevas. 1997. Karyotypes of the most
primitive catfishes (Teleostei: Siluriformes: Diplomystidae). *J. Zool.
Syst. Evol. Res. 35:* 113–119.

Campos-da-Paz, R., and J. S. Albert. 1998. The gymnotiform "eels" of
tropical America: A history of classification and phylogeny of the
South American electric knifefishes (Teleostei: Ostariophysi:
Siluriphysi), pp. 401–417. In *Phylogeny and classification of neotropi-
cal fishes,* L. R. Malabarba, R. E. Reis, R. P. Vari, Z. M. S. Lucena,
and C. A. S. Lucena (Eds.). EDIPUCRS, Porto Alegre, Brazil.
Cavender, T., and M. M. Coburn. 1992. Phylogenetic relationships of
North American Cyprinidae, pp. 293–327. In *Systematics, historical
ecology, and North American freshwater fishes,* R. L. Mayden (Ed.).
Stanford University Press, Palo Alto, CA.
Chapman, L. J., L. Kaufman, and C. A. Chapman. 1994. Why swim upside down? A comparative study of two mochokid catfishes. *Copeia
1994:* 130–155.
Chardon, M., and P. Vandewalle. 1997. Evolutionary trends and possible
origin of the Weberian apparatus. *Neth. J. Zool. 47*(4): 383–403.
Chen, X. L., P. Q. Yue, and R. D. Lin. 1984. Major groups within the
family Cyprinidae and their phylogenetic relationships. *Acta Zoo-
taxon. Sin. 9:* 424–440.
Coburn, M. M., and T. M. Cavender. 1992. Interrelationships of North
American cyprinid fishes, pp. 328–373. In *Systematics, historical
ecology, and North American freshwater fishes,* R. L. Mayden (Ed.).
Stanford University Press, Palo Alto, CA.
———, and P. Chai. 2003. Development of the anterior vertebrae of
Chanos chanos (Ostariophysi: Gonorhynchiformes). *Copeia 2003:*
175–180.
Cox Fernandes, C., J. G. Lundberg, and C. Riginos. 2002. Largest of all
electric-fish snouts: Hypermorphic facial growth in male *Apteronotus hasemani* and the identity of *Apteronotus anas* (Gymnotiformes:
Apteronotidae). *Copeia 2002:* 52–61.
Cunha, C., N. Mesquita, T. E. Dowling, A. Gilles, and M. M. Coelho.
2002. Phylogenetic relationships of Eurasian and American
cyprinids using cytochrome b sequences. *J. Fish Biol. 61:*
929–944.
dePinna, M. C. C. 1998. Phylogenetic relationships of neotropical
Siluriformes (Teleostei: Ostariophysi): Historical overview and
synthesis of hypotheses, pp. 279–330. In *Phylogeny and classification
of neotropical fishes,* L. R. Malabarba, R. E. Reis, R. P. Vari, Z. M. S.
Lucena, and C. A. S. Lucena (Eds.). EDIPUCRS, Porto Alegre,
Brazil.
Diogo, R., and M. Chardon. 2001. Adaptive transformation of the
palatine-maxillary system in catfish: Increased mobility of the maxillary barbel, pp. 367–383. In *Sensory biology of jawed fishes: New
insights,* B. G. Kapoor and T. J. Hara (Eds.). Science Publishers,
Enfield, NH.
Eschmeyer, W. N. 1990. Genera in a classification, pp. 435–495. In *Cata-
log of the genera of recent fishes,* W. N. Eschmeyer (Ed.). California
Academy of Sciences, San Francisco.
Fink, S. V., and W. L. Fink. 1981. Interrelationships of the ostariophysan
fishes (Teleostei). *Zool. J. Linn. Soc. 72:* 297–353.
———. 1996. Interrelationships of ostariophysan fishes (Teleostei), pp.
209–249. In *Interrelationships of fishes,* M. L. J. Stiassny, L. R.
Parenti, and G. D. Johnson (Eds.). Academic Press, San Diego.
Géry, J. 1977. *Characoids of the world.* TFH, Neptune City, NJ.
Gosline, W. A. 1971. *Functional morphology and classification of teleostean
fishes.* University Press of Hawaii, Honolulu.
———. 1975. The palatine-maxillary mechanism in catfishes with com-
ments on the evolution and zoogeography of modern siluroids. *Occ.
Pap. Calif. Acad. Sci. 120:* 1–31.
Grande, T. 1999. Revision of the genus *Gonorynchus* Scopoli, 1777
(Teleostei: Ostariophysi. *Copeia 1999:* 453–469.

———, and F. J. Poyato-Ariza. 1999. Phylogenetic relationships of fossil and recent Gonorynchiform fishes (Teleostei: Ostariophysi). *Zool. J. Linn. Soc. 125:* 197–238.

Greenwood, P. H., D. E. Rosen, S. H. Weitzman, and G. S. Myers. 1966. Phyletic studies of teleostean fishes with a provisional classification of living forms. *Bull. Am. Mus. Nat. Hist. 131:* 339–456.

Hardman, M., and L. M. Page. 2003. Phylogenetic relationships among bullhead catfishes of the genus *Ameirus* (Siluriformes: Ictaluridae). *Copeia 2003:* 20–33.

Harris, P. M., and R. L. Mayden. 2001. Phylogenetic relationships of major clades of Catostomidae (Teleostei: Cypriniformes) as inferred from mitochondrial SSU and LSU rDNA sequences. *Mol. Phylogen. Evol. 20:* 225–237.

Howes, G. J. 1991. Systematics and biogeography: An overview, pp. 1–33. In *Cyprinid fishes: Systematics, biology, and exploitation,* I. J. Winfield and J. S. Nelson (Eds.). Chapman and Hall, London.

Hubbs, C. L., and R. M. Bailey. 1947. Blind catfishes from artesian waters of Texas. *Occ. Pap. Mus. Zool. Univ. Mich. 499.*

Johnson, G. D., and C. Patterson. 1996. Relationships of lower euteleostean fishes, pp. 251–332. In *Interrelationships of fishes,* M. L. J. Stiassny, L. R. Parenti, and G. D. Johnson (Eds.). Academic Press, San Diego.

———. 1997. The gill-arches of gonorynchiform fishes. *S. Afr. J. Sci. 93:* 594–600.

Kottelat, M., and X. L. Chu. 1988. Review of *Yunnanitus* with descriptions of a miniature species flock and six new species from China (Cypriniformes: Homalopteridae). *Env. Biol. Fishes 23:* 65–93.

Lauder, G. V., and K. F. Liem. 1983. The evolution and interrelationships of the actinopterygian fishes. *Bull. Mus. Comp. Zool. 150:* 95–197.

Lever, C. 1996. *Naturalized fishes of the world.* Academic Press, San Diego.

Mayden, R. L. 1989. Phylogenetic studies of North American minnows, with emphasis on the genus *Cyprinella* (Teleostei: Cypriniformes). *Univ. Kansas Mus. Nat. Hist. Misc. Publ. No. 80.*

Nelson, J. S. 1994. *Fishes of the world* (3rd ed.). Wiley, New York.

Ortí, G. 1997. Radiation of characiform fishes: Evidence from mitochondrial and nuclear DNA sequences, pp. 219–243. In *Molecular systematics of fishes,* T. D. Kocher and C. A. Stepien (Eds.). Academic Press, San Diego.

———, and A. Meyer. 1997. The radiation of characiform fishes and the limits of resolution of mitochondrial ribosomal DNA sequences. *Syst. Biol. 46*(1): 75–100.

Pfeiffer, W. 1977. The distribution of fright reaction and alarm substances in fishes. *Copeia 1977:* 517–539.

Roberts, T. R. 1982. Unculi (horny projections arising from single cells), an adaptive feature of the epidermis of ostariophysan fishes. *Zool. Scripts 11:* 55–76.

———, and C. Vidthayanon. 1991. Systematic revision of the Asian catfish family Pangasidae, with biological observations and descriptions of three new species. *Proc. Acad. Nat. Sci. Phil. 143:* 97–144.

Robins, C. R., R. M. Bailey, C. E. Bond, J. R. Brooker, E. A. Lachner, R. N. Lea, and W. B. Scott. 1991. World fishes important to North Americans. *Am. Fish. Soc. Spec. Publ. 21.*

Rodiles-Hernández, R., D. A. Hendrickson, J. G. Lundberg, and J. M. Humphries. 2005. *Lacantunia enigmatica* (Teleostei: Siluriformes), a new and phylogenetically puzzling freshwater fish from Mesoamerica. *Zootaxa 1000:* 1–24.

Rosen, D. E. 1973. Interrelationships of higher euteleostean fishes, pp. 397–513. In Interrelationships of fishes, P. H. Greenwood, R. S. Miles, and C. Patterson (Eds.). *J. Linn. Soc. (Zool.) 53, Suppl. 1.* Academic Press, New York.

Simons, A. M., and R. L. Mayden. 1999. Phylogenetic relationships of North American cyprinids and assessment of homology of the open posterior myodome. *Copeia 1999:* 13–21.

Smith, G. R. 1992. Phylogeny and biogeography of the Catostomidae, freshwater fishes of North America and Asia, pp. 778–826. In *Systematics, historical ecology, and North American freshwater fishes,* R. L. Mayden (Ed.). Stanford University Press, Palo Alto, CA.

Sneegas, G. W., and D. A. Hendrickson. 2003. Extreme catfishes. *Amer. Currents 29*(1): 1–2.

Taylor, W. R. 1969. A revision of the catfish genus *Noturus* Rafinesque with an analysis of higher groups of Ictaluridae. *U.S. Nat. Mus. Bull. 282.*

Teugels, G. G. 1996. Taxonomy, phylogeny, and biogeography of catfishes (Ostariophysi, Siluroidei): An overview. *Aquat. Living Resour. 9:* 9–34.

Vari, R. P. 1998. Higher level phylogenetic concepts within Characiformes (Ostariophysi), a historical review, pp. 111–122. In *Phylogeny and classification of neotropical fishes,* L. R. Malabarba, R. E. Reis, R. P. Vari, Z. M. S. Lucena, and C. A. S. Lucena (Eds.). EDIPUCRS, Porto Alegre, Brazil.

Volckaert, F. A., and J.-F. Agnèse. 1996. Evolutionary and population genetics of Siluroidei. *Aquat. Living Resour. 9:* 81–92.

Weitzman, S. H., and L. R. Malabarba. 1998. Perspectives about the phylogeny and classification of the Characidae (Teleostei: Characiformes), pp. 161–170. In *Phylogeny and classification of neotropical fishes,* L. R. Malabarba, R. E. Reis, R. P. Vari, Z. M. S. Lucena, and C. A. S. Lucena (Eds.). EDIPUCRS, Porto Alegre, Brazil.

Wheeler, A. 1975. *Fishes of the world: An illustrated dictionary.* MacMillan, New York.

Wiley, M. L., and B. B. Collette. 1970. Breeding tubercles and contact organs in fishes: Their occurrence, structure, and significance. *Bull. Am. Mus. Nat. Hist. 143:* 145–216.

11

THE EUTELEOSTEI: ARGENTINIFORM AND SALMONIFORM FISHES

CLASSIFICATION OF THE EUTELEOSTEI

Order Argentiniformes
   Suborder Argentinoidei
   Suborder Alepocephaloidei
Order Salmoniformes
   Suborder Osmeroidei
   Suborder Salmonoidei

Modern cladistic approaches to classification have resulted in a game of musical chairs for many taxa of teleostean fishes. This is especially evident in revisions that have tackled the two orders of fishes discussed in this chapter. Recent classifications have retained the taxon Protacanthopterygii to include the orders Argentiniformes and Salmoniformes. Argentiniform fishes are predominantly found in the deep ocean. They are distinguished by an unusual branchial structure, the *crumenal organ,* that apparently functions in feeding. Much better known are the Salmoniformes—an order that contains several species that figure significantly in sport and commercial fisheries. Dominated by anadromous forms, the Salmoniformes present an intriguing evolutionary and biogeographical history, and their significance to many North American aboriginal societies cannot be overstated.

CLASSIFICATION OF THE EUTELEOSTEI

In his 1994 treatise, Joseph Nelson considered an assemblage of comparatively basal teleosts to compose the superorder **Protacanthopterygii**. This taxon has been the subject of numerous interpretations and, as Nelson (1994) stated, "continues to be unstable, largely because the many characters exhibit a mosaic distribution, show reduction, are otherwise highly modified, or are primitive for the euteleosts." The removal of the ostariophysans from the euteleosts and their resolution as a sister group of the clupeomorphs, as discussed in Chapter 10, resolves some of the problems associated with the higher classification of euteleosts. However, many problems still remain in the elucidation of the interrelationships of this complex and confusing assemblage of bony fishes. In Nelson's (1994) classification, the superorder Protacanthopterygii comprises the orders Esociformes (including the pikes), Osmeriformes (including the smelts and deepwater argentinoids), and Salmoniformes (including the trouts and salmons). Johnson and Patterson (1996) have tackled many of the thorny problems associated with the higher classification of these groups and have presented a classification somewhat different from that of Nelson (1994), yet they retained the taxon Protacanthopterygii to encompass the two orders discussed here. Ishiguro et al. (2003) critically evaluated the validity of the Protacanthopterygii from the perspective of the mitochondrial genome. They concluded that it is a polyphyletic basal assemblage, located somewhere between more basal teleosts, such as the elopomorphs, and the more derived neoteleosts (see Chapter 12). We will follow the phylogeny proposed by Johnson and Patterson (1996), as summarized in Fig. 11.1. Most noteworthy in this classification is the designation of the Esociformes as a sister group of the neoteleosts rather than allied with the Osmeriformes and Salmoniformes (see Chapter 12).

Order Argentiniformes

The argentiniform fishes include about 160 species of widespread, deep-sea fishes, variously arranged in up to nine—but usually five to seven—families (Begle, 1992; Eschmeyer, 1990; Nelson, 1994). The **crumenal organ** (also known as the **epibranchial organ**) is a distinctive feature of argentiniform fishes and apparently functions for the consolidation of small prey.

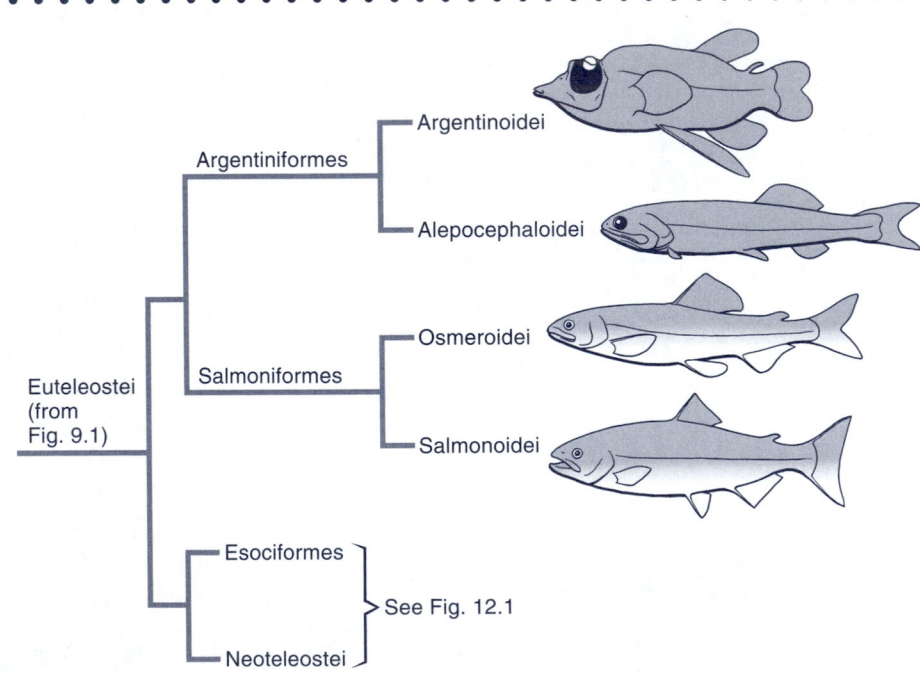

FIGURE 11.1

Phylogeny of lower euteleostean fishes showing relationship of the Esociformes to the Argentiniformes and Salmoniformes (after Johnson and Patterson, 1996).

DIAGNOSIS OF THE ARGENTINIFORMES
- Crumenal organ present.*
- Posterior gill arches modified.
- Ventral displacement of first two to four epineural bones.*
- Caudal fin forked; lowermost fin ray of upper caudal lobe supported by both caudal median cartilages.*
- Photophores present in some species.
- Swim bladder physoclistous if present.
  * Supports monophyly of order.
  Source: Johnson and Patterson, 1996.

Suborder Argentinoidei

This suborder contains four families, the Argentinidae (argentines), Bathylagidae (deep-sea smelts), Opisthoproctidae (barreleyes; Fig. 11.2A), and Microstomatidae (pencil-smelts), which are included as members of the Argentinidae by Robins et al. (1991). Argentinoids are characterized by silvery color and an absence of teeth on the maxilla. An adipose fin is usually present, and the dorsal fin is close to the middle of the body. Some species of argentines are harvested commercially.

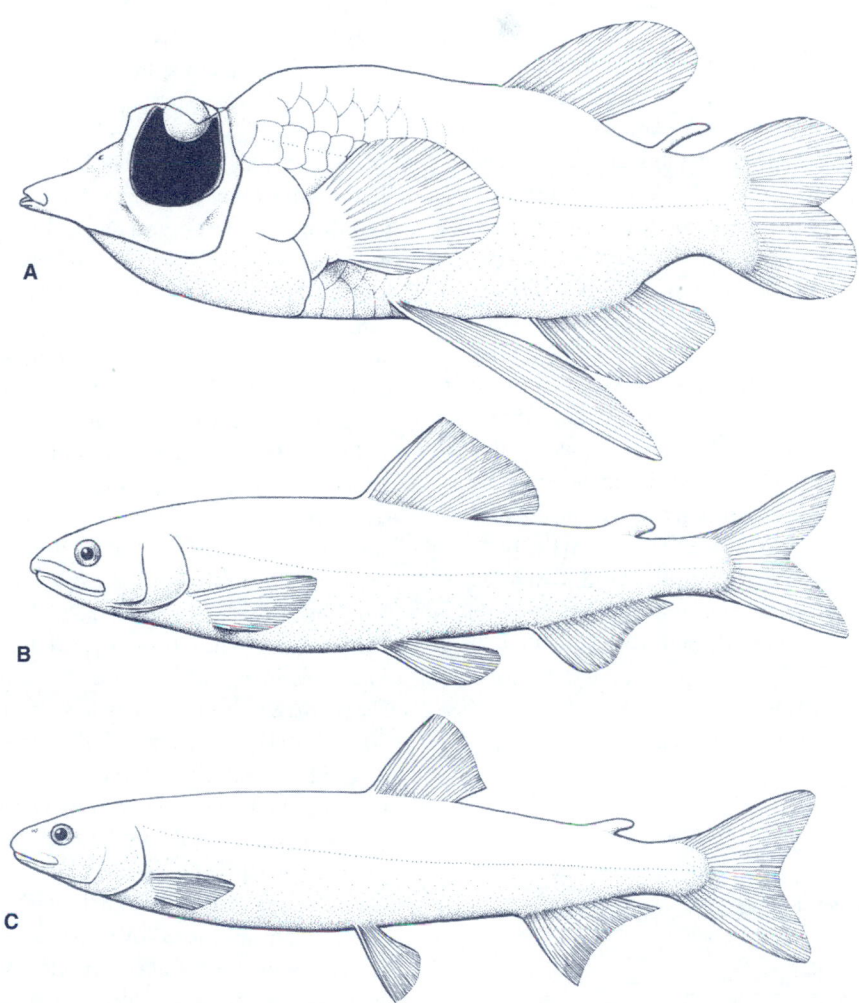

FIGURE 11.2

A, Representative of suborder Argentinoidei: barreleye (Opisthoproctidae, *Macropinna*); B, Representative of suborder Osmeroidei: ayu (Osmeridae, *Plecoglossus*); C, Representative of superfamily Galaxioidea: "southern grayling" (*Prototroctes maraena*).

Suborder Alepocephaloidei

These are dark-colored fishes of the deep sea. They have the dorsal fin set over the anal fin, and they lack an adipose fin and gas bladder. Families usually included in this suborder are the slickheads (Alepocephalidae), the apparently primitive Leptochilichthyidae, and the tubeshoulders (Platytroctidae = Searsidae). The common name *tubeshoulder* is derived from the opening of a gland that lies beneath the cleithrum and produces a luminous substance. Begle (1992) considered all the Alepocephaloidei to belong in the single monophyletic family Alepocephalidae, without erecting any subfamilies.

Order Salmoniformes

From a cultural and commercial perspective, the salmoniform fishes are a most significant group. Ranging as they do across both marine and freshwater habitats, they have a complex and fascinating evolutionary history. The order Salmoniformes (*sensu* Greenwood et al., 1966) was formerly considered the basal group of a more inclusive superorder Protacanthopterygii, but more recent phylogenetic studies have raised serious doubts about the composition of this superorder, and many former members have been placed elsewhere (see Fink and Weitzman, 1982; Rosen, 1973). Fink (1984a) expressed the opinion that the superordinal name Protacanthopterygii was no longer useful. He considered Salmoniformes to be coextensive with the family Salmonidae (Fink, 1984b). Nelson (1994) recognized the orders Salmoniformes and Osmeriformes, restricting the former to the single family Salmonidae. Begle (1991) avoided the term Salmoniformes. Here, we follow Johnson and Patterson (1996) in recognizing the order Salmoniformes as consisting of two suborders, the Osmeroidei and the Salmonoidei.

DIAGNOSIS OF THE SALMONIFORMES*
- Separate dermethmoid and supraethmoid bones.
- Epipleural bones absent.
- Tripartite occipital condyle.
- Nuptial tubercles.
- Scales lacking radii.
- Anadromous migratory behavior.
  - * All these features support monophyly of the order.
  - Source: Johnson and Patterson, 1996.

It should be noted here that Begle's (1991, 1992) contributions, though widely cited in this systematic consideration of salmoniform fishes, have been heavily criticized by Johnson and Patterson (1996, 1997) and Patterson and Johnson (1997), who viewed his conclusions with great suspicion.

Suborder Osmeroidei

Begle (1992) proposed a series of characters defining the osmeroids, including a vomer with a short shaft, ventral condyle on the pelvic girdle, absence of the orbitosphenoid, and reduction of the articular. Of these, Johnson and Patterson (1996) noted that the last two characters also distinguish esocoids. An absence of teeth on the gill rakers also distinguishes most osmeroids. Many of these fishes have an adipose fin, which usually contains a small cartilage (Matsuoka and Iwai, 1983). A few emit a "cucumber odor," caused by *trans*-2-*cis*-6-nonadienal. This has been noted in osmerids, retropinnids, argentinids, and a few other fishes (McDowall et al., 1993) and apparently has originated independently more than once (Begle, 1991). This group contains diadromous or freshwater fishes of both the Northern and Southern Hemispheres. They are small—usually less than 30 cm long.

Superfamily Osmeroidea

The family **Osmeridae** contains the smelts, small fishes found in both fresh and salt water of temperate and cold parts of the Northern Hemisphere. There are seven genera and 11 species of smelts. They are slender, silver-sided fishes of delicate flavor and are popular as food fishes. Most species prefer to spawn on sand or small gravel. Some, such as the eulachon (*Thaleichthys pacificus;* see Plate 11.1), are anadromous, whereas others spawn on ocean beaches at high tide, usually in areas of some freshwater seepage. Large congregations of spawners make it easy to capture them with dip nets. The Holarctic capelin (*Mallotus villosus*) supports a commercial fishery in the North Atlantic, where the annual catch formerly exceeded a million metric tons.

The ayu (*Plecoglossus altivelis*) of Japan, China, and Korea usually has been placed in the monotypic family Plecoglossidae, but it is closely related to the smelts, especially *Osmerus,* and is now placed with them in the Osmeridae (Begle, 1991; Johnson and Patterson, 1996). This species (Fig. 11.2B) has a row of large, rather square-cut, chisel-like teeth on the maxillaries and dentaries (Howes and Sanford, 1987). The ayu feeds on diatoms and associated organisms growing on the rocks in river bottoms.

The ayu is an annual fish, without overlapping generations. They spawn in lower parts of rivers in fall and early winter. Some reproductive plasticity has been documented, with large females spawning once and smaller ones spawning twice within a two-week period (Iguchi and Tsukamoto, 2001). The young are carried to sea and return to the streams in late winter and early spring. Their growth is rapid, so that they are large enough to sustain a fishery by early summer. Also known as "sweetfish," the ayu is sought as a delicacy in Japan, where it was the target of the celebrated cormorant fishery. Tethered cormorants, with rings around their throats to prevent them from swallowing their prey, were allowed to catch small fish. The birds were then pulled back into the boat, and the captured fish removed. Another fishing method, called *tomozuri,* involves the use of a previously captured live ayu, which is introduced, along with snagging hooks, into the territory of another fish. When the resident fish comes to protect its territory, it is snagged.

The family **Salangidae** (icefishes, noodlefishes, or glassfishes), considered by many ichthyologists to be closely related to the osmerids, were aligned with the galaxioids by Begle (1991). Johnson and Patterson (1996) considered them to be members of the tribe Salangini, subfamily Osmerinae, family Osmeridae. They are small (10 cm), slender, transparent fishes with a flattened head. Partly because of their resemblance to larval smelt, they are thought to be neotenic (Nelson, 1994). They are distributed in marine and freshwaters along the Asian coast from the Greater Sunda Islands to the Amur River. During their spawning runs into freshwater, they are taken in commercial quantities. Like the ayu, they are short-lived. The genus *Sundasalanx* is sometimes placed in its own family, Sundasalangidae. It lacks an adipose fin, has some osteological peculiarities, and the females of one species can mature at less than 15 cm (Roberts, 1981, 1984).

Superfamily Galaxioidea

These are fishes of temperate waters of the Southern Hemisphere. Two families are currently recognized: **Retropinnidae** and **Galaxiidae**. Whereas Nelson (1994) recognized the Lepidogalaxiidae as a separate family, Johnson and Patterson (1996) considered the lepidogalaxiines and galaxiines as subfamilies within the Galaxiidae. In the family Retropinnidae, only the right gonad is present, and a short, horny keel is developed in front of the anus (McDowall, 1990). The family includes the Southern Hemisphere "grayling" (*Prototroctes maraena;* Fig. 11.2C) of Australia and Tasmania (another species of the genus, *P. oxyrhynchus,* native to New Zealand, is now considered extinct), and the southern smelts of the genera *Retropinna* (see Plate 11.2) of Australia, Tasmania, and New Zealand and *Stokellia* of New Zealand.

The southern grayling (*Prototroctes maraena*) deposits numerous (30,000 to 60,000 per female) small, demersal eggs in freshwater during the southern autumn. The larvae apparently drift to the sea, where they remain until spring, when they return to freshwater. Individuals are usually less than 30 cm long, but lengths of more than 45 cm have been noted. The southern smelts superficially resemble the osmerids but have the dorsal fin far back, near the adipose fin. They are translucent fishes, usually less than 13 cm long, and are often called cucumberfish because of their distinctive

odor—a feature they share with the osmerids. Although the chemical responsible for this odor has been identified as *trans-2-cis-*6-nonadienal, its functional significance remains a mystery (McDowall et al., 1993). Some species are apparently confined to freshwater, but others may be anadromous, ascending streams from the sea during the southern spring and summer (McDowall, 1990). Spawning takes place mainly in summer. Newly hatched fish drift to sea and return upriver when 5 to 6 cm long. On the upstream migration, these colorless young are often captured, along with other species, and used as food. These larval fishes are called "whitebait" and are cooked in patties. The spawning males have numerous pearl organs or nuptial tubercles. Some species reach maturity in one year.

The southern grayling is not closely related to the salmonoid graylings (*Thymallus*) of the Northern Hemisphere, but was given the name by early settlers because of its resemblance. This fish has declined in numbers in Australia and Tasmania. With the extinction of *P. oxyrhynchus* in New Zealand, the Prototroctinae are now a monotypic subfamily. Changes in habitat due to land use and the introduction of exotic species have been implicated in the decline, but McDowall (1990) claimed that the southern grayling has declined in areas that have not been so affected.

The subfamily Lepidogalaxiinae was considered monotypic by Nelson (1994), consisting only of the salamanderfish (*Lepidogalaxias salamandroides*). Although it superficially resembles a smelt, the galaxiid *Lovettia* has many features in common with *Lepidogalaxias,* and Johnson and Patterson (1996) have included it in the subfamily Lepidogalaxiinae. Members of *Lovettia* are anadromous; the young drift to sea soon after hatching and return in a year. They are very slender and transparent as they enter freshwater but darken as they mature. In mature males, the anus and urogenital opening are located in the anterior part of the abdomen.

The salamanderfish (see Plate 11.3) is restricted in distribution to southwestern Australia. Although resembling the galaxiids, it is readily distinguished from them by the presence of scales, fused frontal bones, and greatly modified cephalic lateral line pores. Once considered a relative of the esocoids (Rosen, 1974), it is currently classified with the galaxiids as an osmeroid (Begle, 1991; Nelson, 1994). Molecular phylogenetic studies (Waters et al., 2000) provide evidence that *Lepidogalaxias* is not a galaxiid and also have suggested that the proposed esocoid affinity may have merit. A tiny fish, *Lepidogalaxias* reaches sexual maturity at 40 mm. The dorsal fin is placed posteriorly, as in pikes, mudminnows, and galaxiids. Caudal fin rays are reduced to nine in number and are unbranched. The anal fin of the male has highly modified skeletal supports and fin rays, and is covered by large scales set in thick skin (Rosen, 1974).

This species is unusual in that it has no oculomotor muscles. The eye itself is immobile, being attached to the socket. The head, unlike the heads of most fishes, can be moved downward and laterally (McDowall and Pusey, 1983).

Lepidogalaxias can aestivate in dry soil during droughts, although it does not appear to have accessory air-breathing organs (Berra et al., 1989). Pusey (1989) noted that it loses very little water during the first several weeks of aestivation and does not accumulate much urea. Significant mortality among aestivating individuals has been recorded as the habitat dries up. Reproduction follows aestivation, and apparently females in which lipid reserves drop below a critical value do not spawn. Males die after spawning at the age of about one year. Some females die following their first spawning, but others survive to spawn again the next year (Pusey, 1990).

The subfamily Galaxiinae was once considered a separate order (Berg, 1940) but has been variously placed into a suborder Galaxioidei or retained in the Salmonoidei. Here it is included in the Osmeroidei, as treated by Begle (1991) and Johnson and Patterson (1996). McDowall (1999) used caudal skeletal morphology as a means of distinguishing different genera in the subfamily. These are mostly small fishes, less than 30 cm long. The subfamily is viewed as comprising two tribes: Aplochitonini (genus *Aplochiton* of southern South America and the Falkland Islands) and Galaxiini (genus *Galaxias* and five other genera that range throughout the Southern Hemisphere but are concentrated in southeastern Australia and Tasmania; Johnson and Patterson, 1996).

Some of the galaxiines are diadromous. *Aplochiton marinus* of South America is believed to be an ocean spawner. The Galaxiinae are rather small freshwater or catadromous fishes of Australia, New Zealand, Tasmania, and the southern tips of Africa and South America. The most widespread and speciose genus is *Galaxias* (see Plate 11.4). *Galaxias maculatus* has one of the broadest distributions of any freshwater fish, being found in Tasmania, New Zealand, Chile, and the Falkland Islands. As might be expected, it shows extensive variation in mitochondrial DNA sequences among these disparate populations (Waters and Burridge, 1999).

Galaxias maculatus spawns in vegetation along the shore at high tide and leaves the eggs to incubate above the level of the sea until the next extreme high tide two weeks later. Eggs hatch when they are again covered by water, and the larvae swim into the ocean, later to ascend the rivers. Landlocked populations are reported to spawn in tributaries to lakes on freshets that subside and leave the eggs on shore. The eggs hatch during a subsequent freshet. Landlocked *Galaxias maculatus*, locally called *puye*, occur in enormous concentrations in Chilean lakes.

Suborder Salmonoidei

Salmonoids are soft-rayed and mostly physostomous, although the swim bladder may be absent in some. There are no connections of the swim bladder with the ear (i.e., no otophysic connection). An adipose fin is often present. The suborder is often considered a basal one, from which several of the higher groups could have evolved. One of the most striking evolutionary features of the salmonoids is their tetraploid karyotype (see Chapter 28). Salmonoids probably date as far back as the Cretaceous, but the oldest known fossil is of the genus †*Eosalmo* from the Eocene (Wilson and Li, 1999). Nelson (1994) recognized only one family, the Salmonidae.

The family **Salmonidae** includes the trouts and salmons of the subfamily Salmoninae, the whitefishes of the family Coregoninae (sometimes considered a separate family, the Coregonidae), and the graylings of the family Thymallinae, all native to the Northern Hemisphere (Stearley and Smith, 1993). All have an adipose fin. Salmonids have an unusual feature of their reproductive anatomy, in that ducts do not develop in association with the testes or ovaries. In this feature, they resemble the much more primitive cyclostomes. All retain a large proportion of cartilage in the cranium. Coregonines are viewed as a sister group to the thymallines and salmonines (Fig. 11.3; Johnson and Patterson, 1996; Wilson and Li, 1999). Salmonids are among the most valuable food and game fishes, with a long history of significant cultural association with humans (see Going Deeper).

The genus *Salmo* contains the Atlantic salmon (*S. salar;* see Plate 11.5) and the brown trout (*S. trutta*). The Pacific salmons and trouts make up the genus *Oncorhynchus* (Smith and Stearley, 1989). The name *Oncorhynchus* literally means "hooked nose," as it refers to the pronounced curvature seen in the upper and lower jaws (termed **kype**) of males on reaching sexual maturity. The six Pacific salmons are found in the North Pacific—five of these along the coast of North America, and all six in Asia. Some species—for example, the coho salmon (*O. kisutch*) and Chinook salmon (*O. tshawytscha*)—have been successfully introduced into the Great Lakes, where they currently are the basis of a sport fishery. The rainbow or steelhead trout (*O. mykiss*) of the Pacific coasts of North America and Siberia, and the brown trout of Europe, are currently sought primarily for sport, although they both are used in aquaculture. Both have been transplanted to temperate parts of the Southern Hemisphere. The Atlantic salmon is a prized sport fish. As its dwindling populations continue to be fished commercially in some parts of the North Atlantic, it has become the basis of a remarkably successful aquaculture industry, although one that is not without some controversy (see Chapter 39).

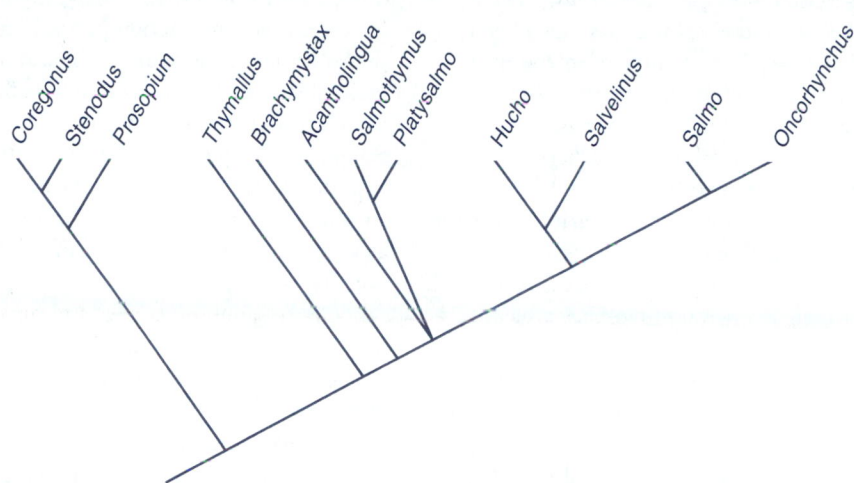

FIGURE 11.3

Phylogeny of the salmonid fishes (family Salmonidae; after Stearley and Smith, 1993).

Various species of trouts and salmons, or races within these species, may be anadromous, feeding and growing in the ocean but spawning and spending a portion of their early life in freshwater. Among the species of Pacific salmons, individuals die soon after spawning—a life history strategy referred to as **semelparity** (see Chapter 27). However, in the rainbow trout and the cutthroat trout (*O. clarki*), the hormonal changes that cause mortality in the Pacific salmons are usually not as severe and are typically reversed. Members of the genus *Salmo* generally do not die following spawning. This reproductive strategy, typical of most fishes, is termed **iteroparity**. The phenomenon of semelparity is a fascinating topic, as the selective factors that contribute to its evolution remain largely unknown. Energetically speaking, semelparity represents a maximum reproductive investment in a single spawning event as opposed to spreading the investment out over several spawning seasons. Studies by Crespi and Teo (2002) have suggested that certain features of the salmonid reproductive life history, such as long-distance migrations that contribute to increased mortality between spawning intervals, and the development of increased egg size, which would enhance juvenile survival, were crucial in the evolution of the semelparous condition.

The Chinook (chinook) salmon (*O. tshawytscha*) is the largest of the Pacific salmons, reaching a maximum weight of about 57 kg. This magnificent fish is of great value as a commercial fish, and it supports a major sport fishery. Once, it ascended the Columbia River and its tributaries,

migrating upstream for distances of more than 1,500 km, but now hydroelectric and irrigation dams have severely impeded access to spawning areas. It has been proposed (Nelson et al., 2003) that "chinook" be changed to a proper noun (i.e., capitalized) in recognition that this species was named after the Native American tribe of that name, although the Chinook Tribe has, to date, not received recognition from the federal government. As a cultural entity that is historically established does not depend on the federal government for validation, we will henceforth use the proper noun for this species.

The coho salmon (*O. kisutch*) is more numerous than the Chinook salmon in most areas, so it supports a larger sport fishery. The coho is well adapted to short coastal streams and lower tributaries of large river systems for spawning purposes and, thus, has not been affected as severely as the Chinook by dam construction. However, it has been affected by changes in habitat partly due to deforestation. This has changed the stream flow patterns, altered water temperatures, and allowed siltation of spawning areas. Moreover, there have been deleterious effects on the fitness of certain races by hybridization with hatchery stocks. The ease with which the species can be reared in modern hatcheries has made possible the sustaining of coho runs in depleted streams, even though the genetic integrity of the native stocks may be at risk. The introduction of the coho into the Great Lakes has provided a remarkable sport fishery, especially in Michigan. The salmons find ample food in the large stocks of alewife (*Alosa pseudoharengus*), which are abundant in the lakes. Chinook salmon have also been stocked in the Great Lakes, and the ability of introduced species of *Oncorhynchus* to reproduce in streams draining into the Great Lakes has contributed to the survival of the fishery. Overall, the combination of intense fishery pressure from both sport and commercial interests and habitat degradation has dealt a severe blow to the integrity of the Pacific salmon populations. Pacific salmon have disappeared from 40 percent of their historical breeding ranges, and in more than two thirds of their historical range, they are considered threatened or endangered (Committee on Protection and Management of Pacific Northwest Anadromous Salmonids, 1996).

The Pacific salmon with the most prized flesh is the sockeye salmon (*Oncorhynchus nerka;* see Plate 11.6 A, B). The sockeye, unlike the piscivorous coho and Chinook salmons, depends heavily on pelagic crustaceans for food. It forms the basis for a short yearly season of intensive commercial fishing. In contrast to the precarious situation of the other salmon species along the coasts of Washington, Oregon, and California, sockeye salmon stocks are plentiful in Canadian and Alaskan waters, especially Bristol Bay. Because this salmon normally spends its early life in lakes, the best runs are into rivers with large lake systems. The great runs that entered the Fraser River of Canada early in the 20th century were severely compromised by railroad construction that resulted in massive amounts of rock being dumped into the Hell's Gate Canyon in British Columbia. In 1914, this construction precipitated a massive rock slide that devastated the salmon runs. The torrential current at Hell's Gate was a tremendous challenge to migrating fish even without the increase in gradient caused by the slide. Sound management policies, including the construction of fish passage facilities, have done much to restore the sockeye runs in Canada. The landlocked form of the sockeye, known as **kokanee,** is valuable as a game fish in lakes of the Pacific Northwest.

Chum salmon (*O. keta*) and pink salmon (*O. gorbuscha*) are sought as commercial fishes and have only minor importance as game fishes. In earlier times, however, they were an important subsistence item for coastal tribes of Native Americans (see Going Deeper). In most parts of their ranges, these two species ascend rivers only a short distance to spawn, and the young drift to the ocean immediately after emerging from their gravel nests. In the other species, the young spend from a few months to two years in freshwater. The pink salmon is remarkable in that the life cycle is almost invariably two years, and some streams have heavy runs in even-numbered years. The odd-year and even-year populations have evolved some morphological distinctions during the thousands of years of chronological isolation. Hatchery production of pink salmon is carried out successfully in Siberia and southeast Alaska. In some years, the Alaska operations have produced so many fish that prices have been depressed.

GOING DEEPER · Salmonoids and the Native American Fishery

Technology-based societies, often mistakenly referred to as "civilized," tend to romanticize the nature of the relationship of aboriginal cultures with the resources on which they depend. The image of the American Indian in the minds of Whites is often that of the bronzed warrior in some mystical communion with the natural world as he slowly paddles his canoe along the lakeshore. Such peoples are held to develop a reverence for nature that, by some mysterious power, prevents them from ever abusing the gifts provided to them by the Great Spirit above. In fact, Native Americans in the past centuries were no different from the rest of humanity: They needed food and shelter and exploited the resources available to them in ways that often had a profound impact on their environment. Nowhere is this more evident than in the ways in which aboriginal Americans obtained and used fishes in North American waters before the arrival of Europeans. Fishes, especially salmonoids, constituted the main source of protein in the diet of aboriginal Americans of the Pacific coast and the Great Lakes regions. The Indians of present-day coastal Alaska, British Columbia, Washington, Oregon, and northern California harvested enormous quantities of Pacific salmon. The oil-rich eulachon was also valued by the Northwest coastal Indians. Great Lakes Indians harvested large quantities of lake trout and whitefishes. Although other species, such as sturgeons and herrings, also figured significantly in the catch of Native Americans, our focus here is on the impact that aboriginal fishing had on salmonines, especially species of *Oncorhynchus*. The chronology of fishery impacts on a resource can be conveniently divided into an aboriginal phase — our focus here — and an industrial phase, which will be considered in some detail in Chapter 39.

Native Americans had developed a variety of ingenious techniques for enhancing their harvest of fishes. In the Great Lakes region, Chippewa women fashioned basswood, nettle, and other natural fibers into scoop and gill nets. When the whitefish were migrating, two-man crews in canoes would work the rapids where they congregated. While one steadied the

The chars of the genus *Salvelinus* differ in coloration from the trout and salmon in having light spots on a darker background instead of dark spots on a light background. There are differences in cranial osteology as well. A biological difference is the season of spawning. Trout tend to spawn in late winter, spring, or early summer, whereas chars spawn mainly in autumn. Chars are freshwater and anadromous fishes found in cold waters of the Northern Hemisphere. In common usage, they are usually called trout. Included in the genus are the Dolly Varden (*S. malma;* see Plate 11.9) and the brook trout (*S. fontinalis*). The lake trout (*S. namaycush*) supported a large commercial fishery in the Great Lakes prior to the spread of the sea lamprey, which feeds on them. The bull trout (*S. confluentus*) formerly occupied many of the cold waters of the Columbia River system and ranged from northern California north through western Canada to Alaska, but now is disappearing from parts of its range because of land use practices and introduction of the brook trout. Dolly Varden and bull trout largely exist in a parapatric distribution (see Chapter 4), yet they are sympatric over some parts of their range, and in these areas they are known to form hybrids (Redenbach and Taylor, 2002).

Other genera in the Salmoninae are the huchens (*Hucho*), which are large, voracious fishes of cold waters of Europe and Asia, and the lenok (*Brachymystax*) of Asia (Stearley and Smith, 1993). Several other genera (*Salvethymus, Salmothymus, Acantholingua, Parahucho*) native to Eurasia have been proposed but disputed. In their phylogenetic study of salmonines, Phillips and Oakley (1997) used molecular data to validate the status of at least one of these disputed genera, *Parahucho.*

Within the Coregoninae, genera include *Coregonus*, the whitefishes, which are holarctic food fishes once abundant in the Great Lakes; *Stenodus*, the inconnu, a large, predatory whitefish of arctic Asia and North America; and *Prosopium*, a holarctic genus that includes the Rocky Mountain whitefish (*P. williamsoni;* see Plate 11.10). The graylings (*Thymallus*) constitute the only genus in the Thymallinae. Graylings are attractive, holarctic fishes with long, colorful dorsal fins and are sought over much of their range by anglers (see Plate 11.11). The southernmost limit of the natural range of the Arctic grayling (*T. arcticus*) barely extends into the continental United States; native populations in Michigan and Montana have been greatly diminished, but they have been reintroduced in mountainous areas in some northern states.

KEY POINTS AND CONNECTIONS

- The order Argentiniformes includes five to seven families of primarily deep-ocean fishes that are all distinguished by the presence of an epibranchial or crumenal organ—a structure that apparently facilitates the accumulation of small prey items.

- The order Salmoniformes includes two diverse suborders of fishes that are broadly distributed in higher latitudes. The suborder Osmeroidei includes the smelts of the superfamily Osmeroidea. These generally

canoe, the other would use a large scoop net to collect the fish. With cedar floats and stone weights, gill nets could be set in the lake, even under ice during the winter. Ice fishing with spears and lures was also practiced (Doherty, 1990). Aboriginal peoples of the Northwest Pacific coast also used a variety of natural materials in the pursuit of their fisheries. Bull kelp (*Nereocystus luetkeana*), when dried and braided, made a tough, monofilament-like line. Whale sinew, local nettle, and cedar bark also made excellent materials for weaving nets or making fishing line. The proficiency of women in producing fishing gear extended to the contribution of their own long hair in the making of braided leaders for fishing lines.

Pacific coastal salmon fishers would cut an access channel through a kelp bed to channel incoming schools of salmon. They would station their canoes along the channel and harpoon the fish as they made their way to the river mouth. The most efficient harvesting of salmon took place as the fishes concentrated during their upstream migrations. Strategically placed traps (see Plate 11.7) and weirs (fences set to guide the migrating fish into traps; see Plate 11.8) were used with great efficiency, and gill nets were often deployed from adjacent shorelines in quiet backwaters where salmon rested. Dams and weirs exhibited the greatest environmental impact, as they could modify the stream course and prevent runs from reaching the spawning

beds. They were the most efficient manner of harvesting, however. Once the harvest was complete, the dams and weirs would be dismantled, lest the rising waters carry them away. Undoubtedly the most dramatic fishery technique was employed along the steep canyons of the middle Columbia, where the water tumbled through the narrow, basaltic cliffs. Here, Indian fishers erected wooden platforms above the torrents and speared or netted salmon as they struggled through the rapids.

Fishing was an arduous and time-consuming task. The subdivision of tasks according to sex was a natural consequence of the fact that men, with their greater size and strength, were most adept at catching, whereas women focused on

small fishes occur in dense concentrations in coastal waters; many species are food fishes highly esteemed for their delicate flavor and texture.

- Whereas the osmeroids are primarily associated with temperate waters of the Northern Hemisphere, the fishes of the superfamily Galaxioidea appear as Southern Hemisphere analogs of the osmeroids.
- The suborder Salmonoidei includes the salmons, trouts, chars, and whitefishes. Salmons and trouts possess a body form well adapted for inhabiting flowing waters; anadromy figures significantly in their life histories. Their popularity as food and game fishes has resulted in widespread introductions worldwide. The whitefishes are lake inhabitants with a circumpolar distribution, whereas the graylings also inhabit large river systems and lakes of the far North. Both have been historically important as food fishes, and the beauty and fighting abilities of graylings have given them legendary status among fly fishers.

The ecology of salmonids as dominant members of flowing freshwaters in high latitudes is discussed in Chapter 31. Their complex evolutionary history is considered in Chapters 4 and 28, whereas Chapter 27 offers a discussion of reproductive adaptations in anadromous fishes.

- -
BUILDING AN ICHTHYOLOGY LIBRARY
- -

Because of the significance of the salmonids, many books have been written about them. Two of the best ones follow:

Groot, C., and L. Margolis (Eds.). 1991. *Pacific salmon life histories.* UBC Press, Vancouver.

Mills, D. H. 1989. *Ecology and management of Atlantic salmon.* Chapman and Hall, London.

Life history information on salmoniform fishes is given in several regional fish guides. Three excellent ones follow:

Hart, J. L. 1973. Pacific fishes of Canada. *Fisheries Research Board of Canada Bull. 180.*

McDowall, R. M. 1990. *New Zealand freshwater fishes: A natural history and guide* (2nd ed.). Heinemann Reed, Auckland.

Scott, W. B., and E. J. Crossman. 1973. Freshwater fishes of Canada. *Fisheries Research Board of Canada Bull. 184.* (New edition published in 1998 by Galt House, Oakville, Ontario.)

A well-illustrated work on Native American fishery techniques:

Stewart, H. 1977. *Indian fishing: Early methods on the Northwest coast.* University of Washington Press, Seattle.

- -
REFERENCES
- -

Begle, D. P. 1991. Relationships of the osmeroid fishes and the use of reductive characters in phylogenetic analysis. *Syst. Zool.* 40(1): 33–53.

———. 1992. Monophyly and relationships of the argentinoid fishes. *Copeia 1992:* 350–366.

Berg. L. S. 1940. Classification of fishes, both recent and fossil. *Trav. Inst. Zool. Acad. Sci. URSS 5:* 87–517 (Reprinted 1947, J. W. Edwards, Ann Arbor, MI).

Berra, T. M., D. M. Sever, and G. R. Allen. 1989. Gross and histological morphology of the swimbladder and lack of respiratory structures in *Lepidogalaxias salamandroides,* an aestivating fish from Western Australia. *Copeia 1989:* 850–856.

the processing of the catch. Large quantities of fish were cooked by boiling or roasting, and much of the catch was preserved by drying and smoking. Properly cured, salmon could last for several months before spoiling. As preservation techniques improved, salmon became available throughout the year, and fishery pressure consequently increased. Indian fishers at the Dalles rapids on the Columbia River developed a valuable food product called *salmon pemmican.* By the time the salmon reached the Dalles, they had lost much of their fat content. The much leaner meat dried quicker and was less prone to spoilage. Women would pulverize the freshly caught fish and set them out to dry in the warm, desiccated air of the Columbian Plateau.

Because it stayed edible for a much longer period of time, pemmican became a valuable item for trade with Indian tribes from the interior.

The number of aboriginal peoples populating the Pacific Northwest before contact with Europeans has been estimated at about 200,000 (Boyd, 1990). Anthropologist Robert Shalk (in Taylor, 1999) estimated their annual harvest of salmon in the Columbia River Basin at more than 41 million pounds, or about 4.5 to 6.3 million fish out of a total run of 11 to 16 million. This is comparable to the total quantity harvested when the industrial fishery was well underway at the close of the 19th century. The diverse tribal groups that peopled California's Central Valley drainage had access to a greater variety of resources and, hence, were not as dependent on fishes. Still, a population estimated at about 110,000 individuals managed to harvest and consume an estimated 8.5 million pounds of Chinook salmon, exceeding the peak commercial harvests in the Sacramento-San Joaquin River system in the late 19th century (Yoshiyama, 1999). So Indians could harvest enormous quantities of fish yet exert little if any lasting impact, whereas industrial fisheries have all but wiped out salmonid populations in the Columbia River drainage. An understanding of the Native American approach to the fishery may provide some insight. Indians fished anadromous populations throughout their range, harvesting in coastal waters and along

Boyd, R. T. 1990. Demographic history, 1774–1874, pp. 135–148. In *Handbook of North American Indians, Vol. 7, Northwest coast,* W. Suttles (Ed.). Smithsonian Institution, Washington, DC.

Clausen, J. 2000. One fish, two fish. *The Nation 270*(3): 22–24.

Committee on Protection and Management of Pacific Northwest Anadromous Salmonids. 1996. *Upstream: Salmon and society in the Pacific Northwest.* National Academy Press, Washington, DC.

Crespi, B. J., and R. Teo. 2002. Comparative phylogenetic analysis of the evolution of semelparity and life history in salmonid fishes. *Evol. 56*(5):1008–1020.

Doherty, R. 1990. *Disputed waters: Native Americans and the Great Lakes fishery.* University Press of Kentucky, Lexington.

Eschmeyer, W. N. 1990. Genera in a classification, pp. 435–495. In *Catalog of the genera of recent fishes,* W. N. Eschmeyer (Ed.). California Academy of Sciences, San Francisco.

Fink, W. L. 1984a. Basal euteleosts: Relationships, pp. 202–206. In Ontogeny and systematics of fishes, H. G. Moser, W. J. Richards, D. M. Cohen, M. P. Fahay, A. W. Kendall, Jr., and S. L. Richardson (Eds.). *Am. Soc. Ichthyol. Herpetol. Spec. Publ. 1.*

———. 1984b. Stomiiformes: Relationships, pp. 181–184. In Ontogeny and systematics of fishes, H. G. Moser, W. J. Richards, D. M. Cohen, M. P. Fahay, A. W. Kendall, Jr., and S. L. Richardson (Eds.). *Am. Soc. Ichthyol. Herpetol. Spec. Publ. 1.*

———, and S. H. Weitzman. 1982. Relationships of the stomiiform fishes (Teleostei) with a description of *Diplophos. Bull. Mus. Comp. Zool. 150*(2): 31–93.

Greenwood, P. H., D. E. Rosen, S. H. Weitzman, and G. S. Myers. 1966. Phyletic studies of teleostean fishes with a provisional classification of living forms. *Bull. Am. Mus. Nat. Hist. 131:* 339–456.

Howes, G. J., and C. P. J. Sanford. 1987. Oral ontology of the ayu, *Plecoglossus altivelis,* and comparisons with the jaws of other salmoniform fishes. *Zool. J. Linn. Soc. 89*(2): 133–169.

Iguchi, K., and Y. Tsukamoto, 2001. Semelparous or iteroparous: Resource allocation tactics in the ayu, an osmeroid fish. *J. Fish Biol. 58:* 520–528.

Ishiguro, N. B., M. Miya, and M. Nishida. 2003. Basal euteleostean relationships: A mitogenomic perspective on the phylogenetic reality of the "Protacanthopterygii." *Mol. Phylogen. Evol. 27:* 476–488.

Johnson, G. D., and C. Patterson. 1996. Relationships of lower euteleostean fishes, pp. 251–332. In *Interrelationships of fishes,* M. L. J. Stiassny, L. R. Parenti, and G. D. Johnson (Eds.). Academic Press, San Diego.

—————————. 1997. Comments on Begle's "Monophyly and relationships of argentinoid fishes." *Copeia 1997:* 401–409.

Matsuoka, M., and T. Iwai. 1983. Adipose fin cartilage found in some teleostean fishes. *Jpn. J. Ichthyol. 30*(1): 37–46.

McDowall, R. M. 1990. *New Zealand freshwater fishes: A natural history and guide* (2nd ed.). Heinemann Reed, Auckland.

———. 1999. Caudal skeleton in *Galaxias* and allied genera. *Copeia 1999:* 932–939.

———, and B. J. Pusey. 1983. *Lepidogalaxias salamandroides* Mees—a redescription, with natural history notes. *Rec. West. Aus. Mus. 11*(1): 11–23.

———, B. M. Clark, G. J. Wright, and T. G. Northcote. 1993. *Trans*-2-*cis*-6-nonadienal: The cause of cucumber odor in osmerid and retropinnid smelts. *Trans. Amer. Fish. Soc. 122*(1): 144–147.

Nelson, J. S. 1994. *Fishes of the world* (3rd ed.). Wiley, New York.

———, E. J. Crossman, H. Espinosa-Perez, L. T. Findley, C. R. Gilbert, R. N. Lea, and J. D. Williams. 2003. The "Names of Fishes" list, including recommended changes in fish names: Chinook salmon for chinook salmon, and *Sander* to replace *Stizostedion* for the sauger and walleye. *Fisheries 28*(7): 38–39.

Patterson, C., and G. D. Johnson. 1997. The data, the matrix and the message: Comments on Begle's "Relationships of the osmeroid fishes." *Syst. Biol. 46:* 358–365.

the freshwater routes. The industrial fishery concentrated its activity solely along the lower stretch of the river. Indian fishing gear was adapted to local conditions, whereas the industrial fishing techniques focus on maximizing the take. Aboriginal fisheries were strictly regulated by ceremonial strictures and kinship ties, especially in the Northwest, where dependence on the resource was greater. An inadvertent form of conservation, this ensured moderation in the harvest. Although aboriginal Americans did modify rivers and streams to enhance the harvest, their impact was nowhere as drastic as the modern White impact on the landscape—an impact that all too often has had nothing to do with the harvest of fish.

In the late 1960s, the highly publicized "fish-in" became the focus of the emerging civil rights movement among Native Americans. These confrontations with state and federal fishery managers and the law enforcement community culminated in a lawsuit filed by Yakima fisherman David Sohappy. This and related litigation challenged the federal government to honor treaties made with the various tribes in the 1850s that guaranteed them access to the fishery. The landmark Boldt Decision of 1974 was the most famous of a series of rulings that had the effect of guaranteeing Native Americans up to 50 percent of the annual harvest. Currently, the Columbia River Inter-Tribal Fish Commission—a confederation of four Columbia River treaty tribes—is spearheading the movement in the Pacific Northwest to dismantle the massive dams that have converted the once magnificent and free-flowing Columbia River into a series of lakes. Tribal fishery experts believe that such drastic measures are needed to resurrect the declining runs of salmon (Clausen, 2000). Native Americans realize that the injustices of the past were not fully compensated by the favorable court decisions in the 1970s. Not only have they reclaimed the right to the fishery, they have become key players in the attempt to resurrect the devastated salmon fisheries of the Pacific Northwest.

Phillips, R. B., and T. H. Oakley. 1997. Phylogenetic relationships among the Salmoninae based on nuclear and mitochondrial DNA sequences, pp. 145–162. In *Molecular systematics of fishes*, T. D. Kocher and C. A. Stepien (Eds.). Academic Press, San Diego.

Pusey, B. J. 1989. Aestivation in the teleost fish *Lepidogalaxias salamandroides* Mees. *Comp. Biochem. Physiol. 92A*(1): 137–138.

———. 1990. Seasonality, aestivation, and the life history of the salamanderfish *Lepidogalaxias salamandroides*. *Env. Biol. Fishes 29:* 15–26.

Redenbach, Z., and E. B. Taylor. 2002. Evidence for historical introgression along a contact zone between two species of char (Pisces: Salmonidae) in northwestern North America. *Evolution 56:* 1021–1035.

Roberts, T. R. 1981. Sundasalangidae, a new family of minute freshwater salmoniform fishes from southeast Asia. *Proc. Calif. Acad. Sci. 42*(9): 295–302.

———. 1984. Skeletal anatomy and classification of the neotenic Asian salmoniform superfamily Salangoidea (icefishes or noodlefishes). *Proc. Calif. Acad. Sci. 43*(13): 179–220.

Robins, C. R., R. M. Bailey, C. E. Bond, J. R. Brooker, E. A. Lachner, R. N. Lea, and W. B. Scott. 1991. World fishes important to North Americans. *Am. Fish. Soc. Spec. Publ. 21.*

Rosen, D. E. 1973. Interrelationships of higher euteleostean fishes, pp. 397–513. In Interrelationships of fishes, P. H. Greenwood, R. S. Miles, and C. Patterson (Eds.). *J. Linn. Soc. (Zool.) 53 Suppl. 1.* Academic Press, New York.

———. 1974. Phylogeny and zoogeography of salmoniform fishes and relationships of *Lepidogalaxias salamandroides. Bull. Am. Mus. Nat. Hist. 153*(2): 265–326.

Smith, G. R., and R. F. Stearley. 1989. The classification and scientific names of rainbow and cutthroat trouts. *Fisheries 14*(1): 4–10.

Stearley, R. F., and G. R. Smith. 1993. Phylogeny of the Pacific trouts and salmons (*Oncorhynchus*) and genera of the family Salmonidae. *Trans. Am. Fish. Soc. 122:* 1–33.

Taylor, J. E., III. 1999. *Making salmon: An environmental history of the Northwest fisheries crisis.* University of Washington Press, Seattle.

Waters, J. M., and C. P. Burridge. 1999. Extreme intraspecific mitochondrial DNA sequence divergence in *Galaxias maculatus* (Osteichthys: Galaxiidae), one of the world's most widespread freshwater fish. *Mol. Phylogen. Evol. 11*(1): 1–12.

———, J. A. López, and G. P. Wallis. 2000. Molecular phylogenetics and biogeography of galaxiid fishes (Osteichthys: Galaxiidae): Dispersal, vicariance, and the position of *Lepidogalaxias salamandroides. Syst. Biol. 49*(4): 777–795.

Wilson, M. V. H., and G.-Q. Li. 1999. Osteology and systematic position of the Eocene salmonid †*Eosalmo driftwoodensis* Wilson from western North America. *Zool. J. Linn. Soc. 125:* 279–311.

Yoshiyama, R. M. 1999. A history of salmon and people in the Central Valley region of California. *Rev. Fish. Sci. 7*(3–4): 197–239.

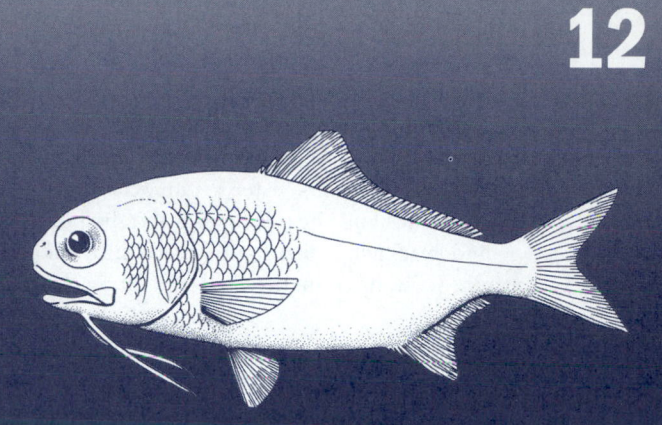

12

FASTER AND DEEPER: ESOCIFORMES AND SOME BASAL NEOTELEOSTS (STOMIIFORMES, AULOPIFORMES, MYCTOPHIFORMES, LAMPRIDIFORMES, AND POLYMIXIIFORMES)

STRADDLING THE NEOTELEOSTEAN FENCE

ESOCIFORMES: SISTER GROUP OF THE NEOTELEOSTS

Stop the Presses! Sister Group to the Salmonoids After All?
Order Esociformes

NEOTELEOSTEI

Superorder Stenopterygii
Order Stomiiformes
Order Ateleopodiformes
Superorder Cyclosquamata
Order Aulopiformes
Superorder Scopelomorpha
Order Myctophiformes

ACANTHOMORPH FISHES

Superorder Lampridomorpha
Order Lampridiformes (Allotriognathi)
Superorder Polymixiomorpha
Order Polymixiiformes

In this chapter, we will consider fishes from two significant lineages: the Esociformes, an order that includes some of the most highly esteemed temperate freshwater game fishes; and an assemblage of orders that we would consider basal members of the taxon known as Neoteleostei. Three orders of basal neoteleosts—the Stomiiformes, Aulopiformes, and Myctophiformes—represent the most extensive radiations of teleosts into the mesopelagic and bathypelagic realms of the ocean. When one considers the success of other basal teleostean groups in the deep sea—such as the gulpers and swallowers of the order Anguilliformes, and the argentines and related forms of the order Argentiniformes—an interesting pattern emerges: The invasion of the deep pelagic realms was largely accomplished by comparatively primitive teleosts that developed an array of remarkable adaptations enabling them to survive and flourish in what we might consider one of the harshest and most forbidding regions of the ocean. Although the deep benthic realm is inhabited by an assemblage of fishes that is taxonomically quite distinct from the aforementioned orders (members of the orders Gadiformes and Ophidiiformes, for example, are especially prevalent on the deep ocean floor), these also are representatives of groups that appeared comparatively early in the evolution of teleosts. Also to be considered here are the basal acanthomorph fishes of the orders Lampridiformes and Polymixiiformes. These broadly distributed marine pelagic fishes are the first groups that we have considered to possess true fin spines.

STRADDLING THE NEOTELEOSTEAN FENCE

Included in this chapter is a diverse grouping of bony fishes that appears to straddle a major divide in the phylogeny of the teleosts. On one side are the Esociformes—included by some with the salmoniform fishes in the superorder Protacanthopterygii. The protacanthopterygian designation represented an earlier view that esociform fishes were comparatively "primitive" teleosts. As discussed hereafter, that perception has changed of late. On the other side are basal members of the assemblage known as **neoteleosts**. This divergence is also evident in the habitats of the Esociformes and of this assemblage of basal neoteleosts. The Esociformes are freshwater fishes, with a north temperate to polar distribution. The basal neoteleosts discussed in this chapter are, for the most part, pelagic marine fishes of moderate to extreme depths.

ESOCIFORMES: SISTER GROUP OF THE NEOTELEOSTS

As mentioned in Chapter 11, the esociform fishes have led a somewhat peripatetic existence, first finding a home with the salmon and smelts in the Protacanthopterygii, but more recently being consigned to a group (the **Neognathi**) that includes the Esociformes and the rest of the teleosts (Neoteleostei; see Fig 12.1; Johnson and Patterson, 1996). The widely held view that the esocids are the most "primitive" euteleosts (Lauder and Liem, 1983; Nelson, 1994) has been challenged by Johnson and Patterson (1996), who proposed a sister group relationship between the Esociformes and the neoteleosts. In a survey of the modes of tooth attachment in the jaws and pharyngeal bones of a diversity of teleosts (see Chapter 23), Fink (1981) identified a type in which the teeth are hinged and depressible; this feature is shared by esociform and neoteleost fishes, uniting the two groups in a sister group relationship, together with an absence of cells in skeletal tissue (Johnson and Patterson, 1996).

Stop the Presses! Sister Group to the Salmonoids After All?

In considering the phylogenetic relationships of a group of vertebrates as diverse as the fishes, it is to be expected that our understanding of these relationships may never be complete. Novel interpretations, based on molecular evidence, often serve to reinforce traditional morphological studies, but they may sometimes challenge them. Such is the case with the sister group relationships of the Esociformes. Although our classification of teleosts, including esociform relationships, follows that put forth by Johnson and Patterson (1996), recent molecular evidence has challenged this

interpretation. Based on the analysis of DNA sequences from both mitochondrial and nuclear genomes, López et al. (2004) asserted an unambiguous sister group relationship between the Esociformes and the salmonoids. So the pikes, mudminnows, and blackfish, remaining uncertain of their heritage, continue to wander in the desert of unresolved phylogenies.

Order Esociformes

These fishes (see Fig. 3.6G) constitute a well-recognized group, usually accorded ordinal rank but sometimes considered a suborder (Berg, 1940) or superorder (Jamieson, 1991). Though not necessarily the most primitive of teleosts, they have an ancient origin in North American freshwaters dating back to the Cretaceous (Wilson et al., 1992).

DIAGNOSIS OF THE ESOCIFORMES
- Loss of orbitosphenoid and mesocoracoid bones.*
- Absence of teeth in the maxillary bones.*
- Presence of pro- and preethmoid bones.
- Scales on cheek and operculum.
- Fins of soft rays only.
- Dorsal and anal fins set well back on body.
- Physostomous.
  * Characters shared with more advanced neoteleosts.
  Sources: Lauder and Liem, 1983; Johnson and Patterson, 1996.

The family **Esocidae** contains the pikes (*Esox;* see Plate 12.1), medium- to large-sized carnivorous fishes of lakes and slow rivers. Pikes are voracious carnivores, with their snout and jaws extended into a long, flattened mouth set with sharp teeth. They are the quintessential ambush predator, with an elongate body and medial fins set well back for optimal thrust and stability in a straight-line attack (see Chapter 19, Going Deeper). The northern pike (*E. lucius*) is one of the most widely distributed freshwater fishes, with a circumpolar distribution encompassing North America, Europe, and Asia (see Fig. 29.5). It is a large fish; individuals weighing more than 25 kg have been reported. Even larger, but not as widely distributed, is the muskellunge (*E. masquinongy*), found in the upper Mississippi drainage, the Great Lakes, and in some contiguous drainages. The "muskie" is one of North America's greatest trophies for sportfishers. The Amur pike (*E. reicherti*) is found in Siberia. Also included in this genus are the pickerels—smaller versions of the esocid body plan.

Mudminnows of the family **Umbridae** (see Plate 12.2) are found in both North America and Europe. Unlike the pikes, these are small fishes, usually less than 15 cm long.

Mudminnows prefer very slow water—usually bogs, stagnant ditches, and streams of low gradient. They will hide in the mud when disturbed, and one species (*Umbra limi*) is reported to survive dry periods by burrowing into the bottom sediments.

The family **Dalliidae,** represented by the Alaska blackfish (*Dallia pectoralis*) is usually included in the Umbridae, but it has many characteristics that appear to distinguish it from that family. *Dallia pectoralis* is found on the Chukot Peninsula of Siberia and in Alaska, where it inhabits slow streams, lakes, and bogs. The winters are long and cold in these areas, so that the blackfish must be inactive for a great part of the year, usually passing the coldest part of the winter buried in the bottom. Some sphagnum bogs freeze to the bottom, and the blackfish are sometimes immobilized in ice. They can be frozen externally, but if their internal body temperature does not drop low enough to crystallize the body fluids, they can survive. They appear to withstand oxygen-deficient periods in summer by using atmospheric oxygen. Blackett (1962) reported rather slow growth for the species—165 mm at age three. Although blackfish usually do not exceed 20 cm, aboriginal peoples have used them as food for sled dogs.

The phylogeny presented in Figure 12.1 reflects Nelson's (1994) arrangement for the higher taxa, but it also includes an alternative phylogeny suggesting different relationships among the esocids. Based on molecular studies, López et al. (2000) proposed reassignment of *Dallia* and *Novumbra* to the Esocidae, with the three species of *Umbra* in the family Umbridae. *Esox* and *Novumbra* are placed in the subfamily Esocinae, and *Dallia* becomes the sole member of the Daliinae (Fig. 12.1).

NEOTELEOSTEI

The remaining teleosts make up the monophyletic **Neoteleostei,** a group that encompasses seven superorders but itself is not a formal taxonomic category (Nelson, 1994). Neoteleosts include the superorders Stenopterygii, Cyclosquamata, Scopelomorpha, Lampridomorpha, Polymixiomorpha, Paracanthopterygii, and Acanthopterygii (Fig. 12.1). All these superorders, save for the Stenopterygii, constitute the clade Eurypterygii (Johnson, 1992; Rosen, 1973).

Certain skeletal and muscular developments among the neoteleosts have contributed to a greater flexibility and versatility, especially when it comes to the manipulation of food items. The first vertebra of neoteleosts articulates with three skeletal elements of the back of the head—the basioccipital, left, and right exoccipitals—rather than just with the basioccipital, as seen in more primitive fishes. This

tripartite condyle, typical for the neoteleosts, has also been identified in some salmonids, however (Johnson and Patterson, 1996). Neoteleosts have developed a **retractor dorsalis** muscle (= **retractor arcus branchialium**) that originates on the vertebral column and inserts on the pharyngobranchials. This muscle allows those fishes to exercise great control over the manipulation of food by the pharyngeal jaws (Lauder and Liem, 1983). Neoteleosts also have a **rostral cartilage** that lies between the neurocranium and the premaxillaries, and many have hinged teeth (Fink, 1981).

Superorder Stenopterygii

Order Stomiiformes

The Stomiiformes are a widely distributed, deep-sea group with a long history of phylogenetic study (Fink, 1984, 1985; Fink and Fink, 1986; Fink and Weitzman, 1982; Harold and Weitzman, 1996). Stomiiformes were formerly placed as a suborder within the Salmoniformes, but they have been reclassified largely because of their neoteleostean characteristics. They were presented by Rosen (1973) as a primitive sister group of the rest of the neoteleosts. Harold and Weitzman (1996) argued for the subdivision of the order into two monophyletic infraorders—the Gonostomata and the Photichthya. These, for the most part, coincide with the subordinal designations of Nelson (1994) described further on. Harold and Weitzman (1996) and Harold (1998) proposed that the gonostomatid genera *Diplophos* and *Manducus* together constituted a sister group of all other stomiiform fishes.

DIAGNOSIS OF THE STOMIIFORMES
- Gape bordered by premaxilla and maxilla.
- Adductor mandibulae muscle medially subdivided, resulting in two insertion points.*
- Greatly enlarged posterior branchiostegal rays.*
- Scales, when present, are cycloid.
- Bioluminescent organs common.
- Physostomous swim bladder, with distinctive posterior location of rete mirable.*
- Fin surfaces composed of soft rays.
- Some with small adipose fin.
  * Indicates support of monophyly for the group.
  Sources: Rosen, 1973; Harold and Weitzman, 1996.

Within the order Stomiiformes, the body shapes range from the deep-bodied hatchetfishes to the anguilliform dragonfishes. Some genera include species with prominent chin barbels. In bioluminescent species, photophores are arranged in various patterns, and most have photophores that shine into the eye (Marshall, 1979). The photophore

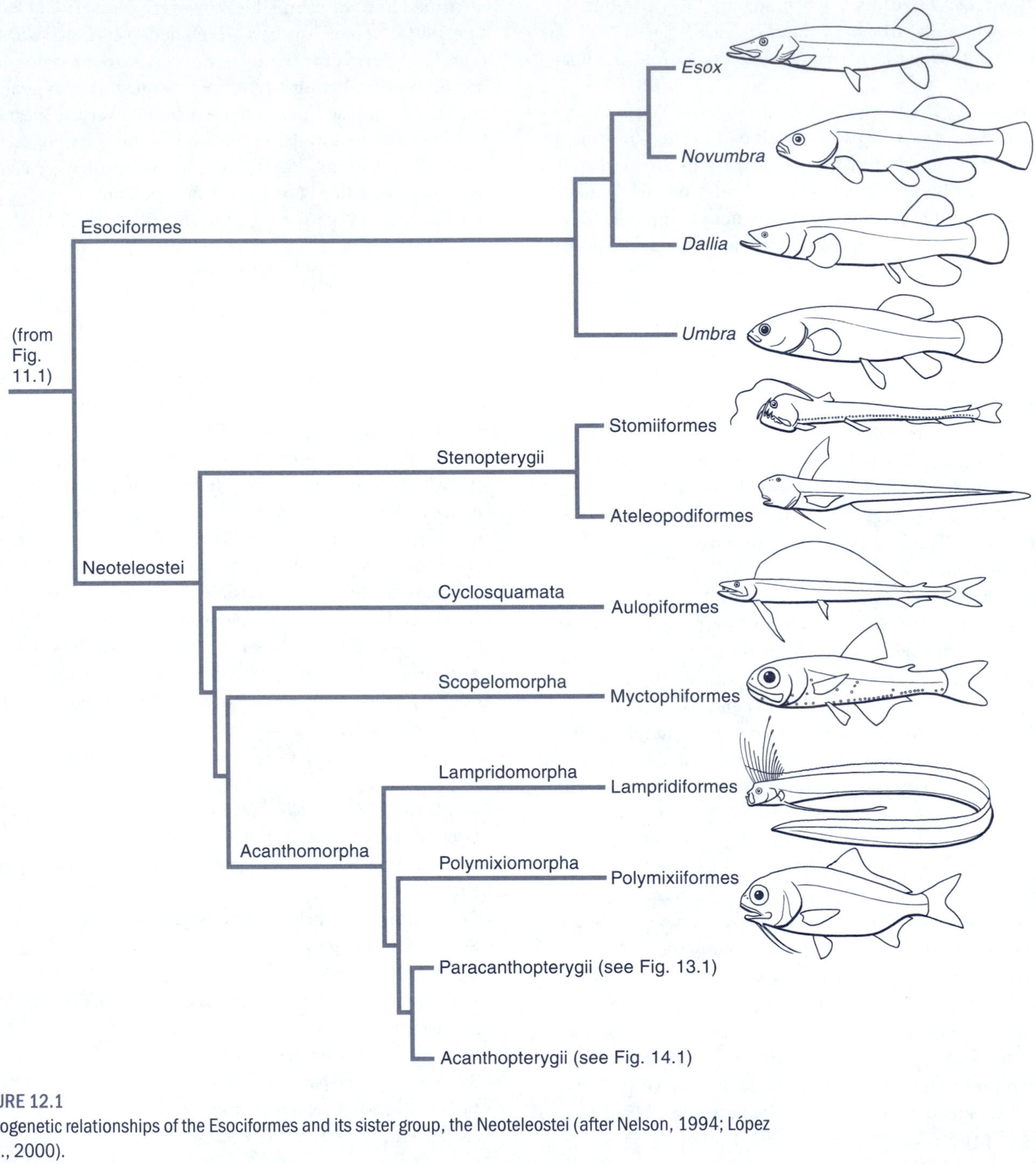

FIGURE 12.1
Phylogenetic relationships of the Esociformes and its sister group, the Neoteleostei (after Nelson, 1994; López et al., 2000).

arrangement is a valuable systematic tool, as it distinguishes the various subgroups (Weitzman, 1997).

Suborder Gonostomatoidei

This suborder contains the lightfishes and bristlemouths (**Gonostomatidae**) and the marine hatchetfishes (**Sternoptychidae**). Gill rakers are present (Nelson, 1994), and all have photophores, usually with ducts.

The gonostomatids are elongate fishes with ventral series of photophores that extend onto the isthmus and branchiostegals. Their mouths are large, with the maxillary extending far back beyond the eye. There are between 25 and 30 species in six genera. They live from the mesopelagic to the bathypelagic zone, and they have adaptations in coloration, feeding apparatus, and buoyancy mechanisms to suit that specific habitat. Members of the genus *Cyclothone* (Fig. 12.2A) are the most widespread of the deep pelagic fishes (Marshall, 1979), and are among the most abundant vertebrates (Ahlstrom et al., 1984). Harold (1998) and Miya and Nishida (2000) have contributed revisions focusing on the two most common genera, *Gonostoma* and *Cyclothone*.

The 10 genera of **Sternoptychidae** are divided into two subfamilies, the Maurolicinae (pearlsides and their relatives) and the Sternoptychinae (marine hatchetfishes; Nelson, 1994). The maurolicines have a robust body shape but are not highly compressed. They have small ventral photophores and a small mouth. Marine hatchetfishes are bizarre in appearance, with extremely compressed bodies and large, usually oblique mouths (see Plate 12.3). Scales consist of elongate, narrow plates arranged vertically on the sides. Photophores are comparatively large, distributed in series on the lower lateral surfaces and directed ventrally. These are open-water fishes of the lower epipelagic or mesopelagic realms and, like the pearlsides, seldom exceed 75 mm in length.

Suborder Photichthyoidei

This group includes the **Photichthyidae** (= Phosichthyidae; Eschmeyer and Bailey, 1990) and the **Stomiidae**, in which Fink (1985) and Nelson (1994) included the following six subfamilies: Chauliodontinae (viperfishes; Fig. 12.2B; see Plate 12.4); Stomiinae (scaly dragonfishes or barbeled

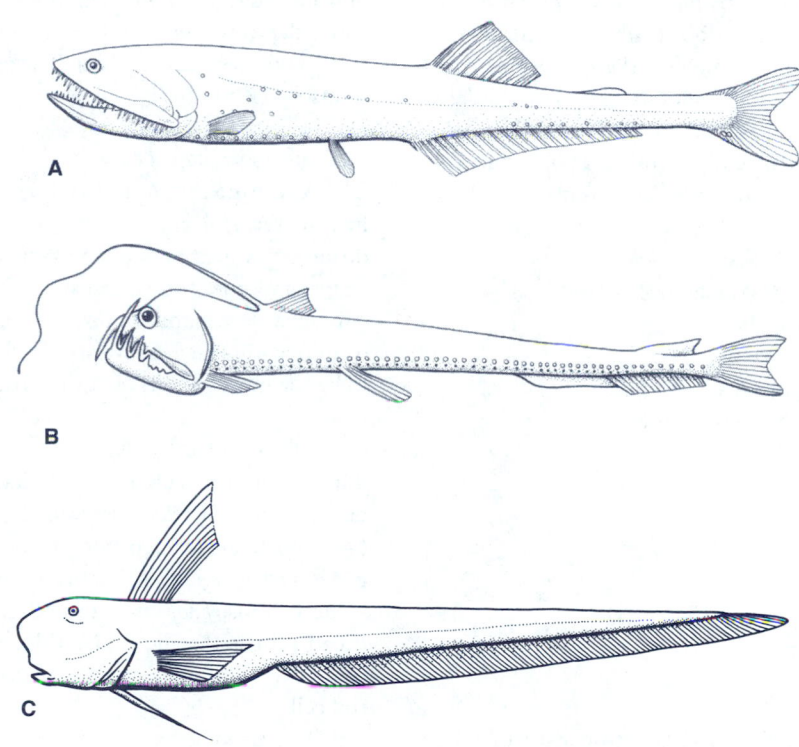

FIGURE 12.2
Representatives of the order Stomiiformes: **A,** Bristlemouth (Gonostomatidae, *Cyclothone*); **B,** Viperfish (Stomiidae, *Chauliodus*); **C,** Representative of the order Ateleopodiformes (*Ateleopus*).

dragonfishes); Astronesthinae (snaggletooths); Melanosto-miinae (scaleless dragonfishes); Idiacanthinae (stalkeyes); and Malacosteinae (loosejaws).

There are seven genera and about 20 species of photich-thyids known from the Pacific, Indian, and Atlantic oceans. These "lighthousefishes," as they are sometimes called, are mesopelagic and migrate at night from water of about 200 to 300 m to shallower depths. They are rather small fishes without chin barbels, and they have two rows of photophores along their lower sides to the origin of the anal fin, then a single row from there to the caudal fin. Fishes of the pho-tichthyid genus *Vinciguerria*, along with the gonostomatid *Cyclothone*, are among the most abundant vertebrates on Earth (Ahlstrom et al., 1984).

The fishes of the family Stomiidae are mostly small, predatory fishes, reaching about 150 mm in length. Some have extremely large teeth for their size and are bizarre in appearance. All show special adaptations for life in the deep sea (Marshall, 1979; Randall and Farrell, 1997).

Order Ateleopodiformes

The sole family in this order, the **Ateleopodidae** (Fig. 12.2C), contains four genera and about 12 species of fishes from moderate depths of tropical and warm temperate seas. Their relationships have been subject to different interpretations (Berg, 1940; Eschmeyer, 1990; Olney et al., 1993; Nelson, 1994). Nelson (1994) elevated them to ordinal status. These are elongate fishes with a short head and trunk and a long tail, so that they resemble chimaeras (Chimaeridae) or grenadiers (Macrouridae) in general body shape.

DIAGNOSIS OF THE ATELEOPODIFORMES
- Several cranial bones, including the orbitosphenoid and basisphenoid, lacking.
- Dentition poorly developed.
- Short, sometimes high dorsal fin.
- Elongate anal fin confluent with small caudal fin.
- Jugular, sometimes filamentous pelvic fins.
- Swim bladder absent.

  Source: Nelson, 1994.

. . . .
Superorder Cyclosquamata

Order Aulopiformes

This order was erected by Rosen (1973) to consist of the "Iniomi" of Gosline et al. (1966), but excluding the Mycto-phiformes. The placement of these fishes in a separate order has been accepted by many (Eschmeyer, 1990; Fink,1984; Nelson, 1994; Weitzman, 1997), but some (Johnson, 1982;

Okiyama, 1984) remain to be convinced. Nelson (1994) and Baldwin and Johnson (1996) recognized four modern sub-orders, comprising 43 genera. The aulopiform fishes are known from the Cretaceous onward.

DIAGNOSIS OF THE AULOPIFORMES
- Maxillaries excluded from border of nonprotrusible mouth.
- Greatly elongated second pharyngobranchial; uncinate process of second epibranchial contacts third pharyngobranchial.
- Fin rays soft; abdominal pelvic fins.
- Adipose fin usually present.
- Many with photophores.
- Swim bladder, if present, physoclistous.

  Sources: Lauder and Liem, 1983; Nelson, 1994.

Suborder Giganturoidei

The giganturoids include two deepwater families, the Bathysauridae, classified as a subfamily of the Synodontidae by Nelson (1994) but included in this suborder by Baldwin and Johnson (1996), and the Giganturidae. The latter are commonly known as telescopefishes because of their tubular, forward-directed eyes. Tubular eyes are seen in a number of deepwater, piscivorous families. They are believed to enhance depth perception (Gartner et al., 1997). A new genus *Bathysauroides* is also included in this suborder (Baldwin and Johnson, 1996).

Suborder Aulopoidei

The **Aulopidae** (= Aulopodidae) are the threadsails or flag-fins of warm marine waters (see Plate 12.5). Some of the dozen or so members of the genus *Aulopus* live in shallow water, and one, the "Sergeant Baker" (*A. purpurissatus*) of Australia, is caught by inshore anglers and is a food fish of minor importance. Nelson (1994) considered the family to be the most primitive of the order.

Suborder Chlorophthalmoidei

This suborder contains the families Chlorophthalmi-dae (greeneyes), Scopelarchidae (pearleyes), Notosudidae (= Scopelosauridae; paperbones), and Ipnopidae, includ-ing *Ipnops* ("grideyes"), and *Bathypterois* (tripodfishes or spiderfishes). *Ipnops* has strange, lensless eyes consisting of broad retinas that lie under thin, transparent bones of the roof of the skull. The retinas are yellow and contain mainly rod cells.

The tripodfishes (*Bathypterois*) have elongated, modi-fied rays in the paired fins and in the anal and caudal fins. They have been observed to support themselves on the muddy bottom at great depths using their pelvic and caudal fins as a tripod (see Plate 12.6).

Suborder Alepisauroidei

The families Synodontidae, Paralepididae (barracudinas), Anotopteridae (daggertooths), Evermannellidae (sabertooth fishes), Omosudidae (omosudids), and Alepisauridae (lancetfishes) make up this suborder. The Pseudotrichonotidae are also classified in this suborder by Nelson (1994), based on a perceived relationship with the Synodontidae (Johnson, 1992).

Members of the families **Anotopteridae** (daggertooths) and **Alepisauridae** (lancetfishes) are elongate, scaleless, predatory fishes of the deep epipelagic to mesopelagic regions. The Anotopteridae include only one species, *Anotopterus pharao*. It is a fearsome predator that can sever the spine of prey species with a single slashing blow or remove strips of flesh from prey too large to consume whole (Welch and Pankhurst, 2001). The two species of alepisaurid lancetfishes also have large mouths and fearsome teeth, but they are distinguished by a prominent dorsal fin (Fig. 12.3A). Individuals up to 2 m long are occasionally found stranded on the beaches of the Pacific Northwest during the spring months, startling anglers who have never seen such a fearsome-looking fish.

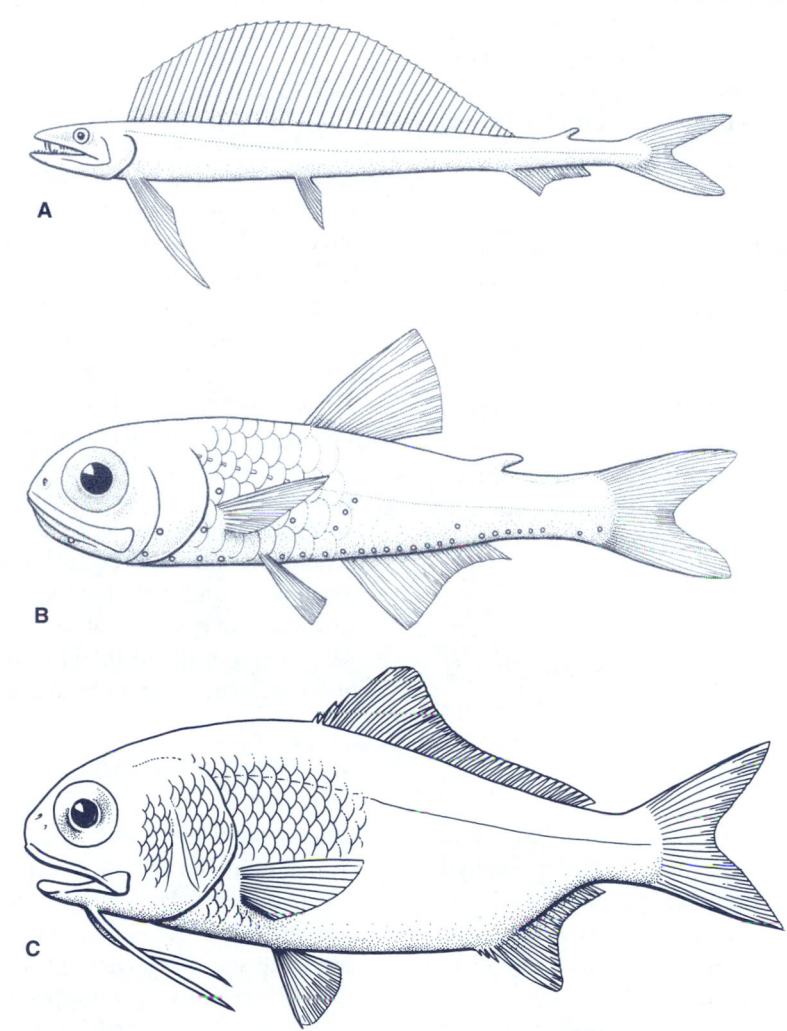

FIGURE 12.3

A, Representative of the order Aulopiformes: lancetfish (Alepisauridae, *Alepisaurus*); **B,** Representative of the order Myctophiformes: lanternfish (Myctophidae); **C,** Representative of the order Polymixiiformes: beardfish (*Polymixia*).

The family **Synodontidae** (lizardfishes; see Plate 12.7) are mostly shore fishes of warm waters, but some are pelagic. Many of them have a strong, somewhat musky odor. The related *Harpadon nehereus*, or Bombay duck, is a colorless, translucent schooling fish. It is common in the northern Indian Ocean, where it is harvested, dried, and, because of its strong flavor, used in the making of sauces.

Superorder Scopelomorpha

Rosen (1973) classified all remaining teleosts in the clade Ctenosquamata. This clade is composed of the superorders Scopelomorpha (the myctophiform fishes) and Acanthomorpha, which includes the lampridiomorphs and polymixiomorphs, discussed in this chapter, and the paracanthopterygians and acanthopterygians considered in Chapters 13 through 17. Johnson (1992) believed this clade to be monophyletic because of similarities in the structure of the dorsal gill arches.

Order Myctophiformes

The order Myctophiformes has been considered a member of more inclusive groups, such as the Iniomi (Gosline et al., 1966), but the current definition of this order is that it contains only the Myctophidae (lanternfishes) and the Neoscopelidae (blackchins).

DIAGNOSIS OF THE MYCTOPHIFORMES

- Mouth bordered by premaxillaries; maxillaries excluded from gape.
- Retractor dorsalis muscle subdivided, with medial head attaching to third pharyngobranchial and lateral head inserting on fourth pharyngobranchial.
- Mesocoracoid absent.
- Enlarged, toothed third upper pharyngobranchial.
- Swim bladder, if present, physoclistous; many species with fat-filled bladders.
- Adipose fin present.

   Sources: Rosen, 1973; Lauder and Liem, 1983.

Diverse fossil myctophiforms are known as early as the Cretaceous. Stiassny (1996) affirmed both the monophyly of the Myctophiformes, as originally proposed by Rosen (1973), and his own claim that they constituted a sister group of the species-rich Acanthomorpha (see Fig. 12.1).

Lanternfishes of the family **Myctophidae** (Fig. 12.3B; see Plate 12.8) are small fishes found in all oceans from the surface to the depths, although most of the nearly 250 species live in waters shallower than 1,000 m. They have relatively large eyes and numerous photophores, each species apparently with its own patterns. Within a species, photophore patterns are sexually dimorphic.

Important genera are *Diaphus*, *Lampanyctus*, *Myctophum*, and *Tarletonbeania*. These fishes are noted for their diel vertical migrations, moving toward or actually to the surface at night and back into the aphotic zone by day. Although they have no direct commercial importance, they are key players in oceanic food webs. Collectively, they constitute a tremendous biomass that contributes indirectly to world fisheries by converting plankton into food for commercially important fishes higher up the food chain. The **Neoscopelidae** are somewhat larger than the lanternfishes and generally have smaller eyes. They are black in color, and most have fewer photophores than lanternfishes. They live at depths ranging from 700 to 2,000 m in most seas. There are five or six species, some of which are widespread in deep water.

ACANTHOMORPH FISHES

At this juncture, the enormous contributions that Donn Eric Rosen has made to systematic ichthyology should be obvious; what a tragedy that ichthyology lost one of its most brilliant practitioners at the peak of his career! Rosen (1973) conceived of the Acanthomorpha as a taxon uniting the polymixiomorphs, paracanthopterygians, and acanthopterygians, mainly on the basis of true fin spines in the dorsal, anal, and pelvic fins of many members. As such, the acanthomorphs, consisting of about 300 families and more than 14,000 species, represent the culmination of teleostean phylogeny. Stiassny (1986) contributed a number of features that affirmed the group as monophyletic. Johnson and Patterson's (1993) major assessment of the acanthomorphs also confirmed their monophyly, and a total evidence analysis, combining molecular and morphological data, lent further credence to this view (Wiley et al., 2000). The origin of acanthomorphs can be traced to the Upper Cretaceous (Patterson, 1993).

Superorder Lampridomorpha

Order Lampridiformes (Allotriognathi)

At one time, the order Lampridiformes was believed to be part of the series Percomorpha, but they are now considered a primitive sister group of all other acanthomorphs (Johnson and Patterson, 1993; Olney et al., 1993; see Fig. 12.1). A total evidence analysis of morphological and molecular characteristics has confirmed the monophyly of the order (Wiley et al., 1998). Lampridiforms are marine, and the order includes a number of deepwater forms. Fossils are known from the Eocene. This order includes the opah and the greatly elongate and ribbonlike oarfishes and ribbonfishes, plus some smaller fishes with uncertain affinities.

DIAGNOSIS OF THE LAMPRIDIFORMES
- Mesocoracoid and opisthotic bones absent, orbitosphenoid present in some.
- Highly protrusible jaw with unique association between premaxillaries and maxillaries; premaxillary excludes maxillary from gape.
- Scales cycloid or absent.
- Mostly soft-rayed, but Veliferidae may have one or two modified fin spines in dorsal and anal fins.
- Thoracic pelvic fins.
- Swim bladder physoclistous.

Sources: Johnson and Patterson, 1993; Olney et al., 1993.

Suborder Lamproidei

This suborder includes the widely distributed opahs (**Lampridae**, genus *Lampris*). Two species are known: *L. guttatus*, which is found worldwide in the pelagic realm, and *L. immaculata*, known only from higher latitudes of the Southern Hemisphere. Because of their greatly compressed, ovate body, the opahs are also known as *moonfish*. Opahs are noted for their color pattern of blue or blue-gray on the back, silver on the sides, reddish silver on the belly, and red jaws and fins, all with an overlay of silver or whitish spots. They may reach about 2 m in length and may weigh up to 270 kg.

Suborder Veliferoidei

This group includes only the family **Veliferidae**, called *sailbearers* because of the very large dorsal and anal fins. The body is compressed as much as in the opah.

Suborder Trachipteroidei

This group includes fishes that are greatly compressed but are also elongate (**taeniform**). Many have bizarre coloration or fin shapes, so that they are thought to be the basis of sea serpent stories. The Lophotidae, or crestfishes, have a crest that extends forward on the head and bears the anterior part of the long dorsal fin. The Radiicephalidae contains one species that is similar to the crestfishes. The ribbonfishes (**Trachipteridae**) also have an elongate dorsal fin, as do the oarfishes (**Regalecidae**). The anal fin is vestigial in radiicephalids and is entirely lacking in trachipterids and regalecids (Olney, 1984). In the regalecids, the first several dorsal rays may be very high, and the elongate pectoral fins have earned one species (*Regalecus glesne*; see Plate 12.9) the name *oarfish*. *Trachipterus altivelus*, the king-of-the-salmon, occurs on the Pacific coast of North America. *T. arcticus*, the dealfish, is found in the Atlantic. *Regalecus glesne* is a widespread pelagic species, occurring in all oceans. This fish reaches 8 m long, has a thin compressed body, and a dorsal fin that might look like the mane of a horse.

Suborder Stylephoroidei

This group contains the single family **Stylephoridae**, the tube-eyes, which live at greater depths than do the other members of the order Lampridiformes. They derive their common name from the telescopic eyes that may point upward or forward (Nelson, 1994). The lower lobe of their caudal fin is elongate, as are the first two rays of the dorsal fin. The mouth opening is small and at the end of a tubular snout, but the oral cavity can be rapidly increased in volume (up to nearly 40-fold) as the fish suction-feeds on plankton (Pietsch, 1978). Presently, there is only one known species (*Stylephorus chordatus*), which reaches about 30 cm.

Superorder Polymixiomorpha

Order Polymixiiformes

These are tropical marine fishes, called barbudos or beardfishes because of their long barbels. They are placed in one family, the **Polymixiidae**, whose taxonomic position is not certain. It has been placed among the Paracanthopterygii, the Perciformes, or in an order of its own. Robins et al. (1991) and Eschmeyer (1990) retained it in the Beryciformes. Stiassny (1986) and Stiassny and Moore (1992) placed it in an uncertain position as a sister group of the remainder of the Acanthomorpha.

DIAGNOSIS OF THE POLYMIXIIFORMES
- Truncated posterior supramaxillary bones present.
- Anterior branchiostegals modified to support pair of hyoid barbels.
- Palatovomerine ligament that passes between lateral maxillary processes.
- Two sets of intermuscular bones — epineurals and epipleurals.*
- Dorsal fin with four to six spines; anal fin with four spines.

  * Unique among the acanthomorphs.

Sources: Stiassny, 1986; Patterson and Johnson, 1995; Baldwin and Johnson, 1996.

The five species of *Polymixia* live in mid-depths in the western Pacific and the Atlantic (Fig. 12.3C). The well-developed chin barbels in this enigmatic group undoubtedly function as sensory devices; similar structures have independently developed in other groups, such as the codfishes, goatfishes, and a number of catfish families.

KEY POINTS AND CONNECTIONS
- Several groups of comparatively primitive teleosts are discussed in this chapter. Fishes of the order Esociformes are considered a sister group of the rest of the taxa considered here; both groups share

a common dental feature of having hinged, depressible teeth. The absence of cells in the skeletal tissues also unites these two groups.

- The best known esociform fishes are the pikes and pickerels—voracious predators with elongate bodies and large, toothy mouths. Smaller members of this order include the mudminnows (Umbridae) and the blackfish (Dalliidae)—fishes adapted to cold, stagnant, swampy backwaters in the far northern latitudes.

A discussion of the esocid body form as representing the optimum for rapid predatory attack is provided in the Going Deeper box in Chapter 19.

- Two superorders of teleosts—the Stenopterygii and the Scopelomorpha—include orders that are especially successful in deep-ocean environments. Among the stenopterygians, members of the order Stomiiformes—particularly the lightfishes and bristlemouths of the family Gonostomatidae and the hatchetfishes of the family Sternoptychidae—are especially prevalent in the depths of the mesopelagic. The scopelomorph order Myctophiformes includes the lanternfishes (Myctophidae), another family that dominates the deep epipelagic and mesopelagic regions. Lanternfishes are well known for their extensive diurnal vertical migrations, sometimes in excess of several hundred meters. As might be expected, bioluminescence is characteristic of these deep-ocean fishes.

The structure and function of bioluminescent organs, or photophores, is discussed in Chapter 18, whereas the significance of gonostomatids and myctophids in deep-ocean environments is considered in Chapter 35.

BUILDING AN ICHTHYOLOGY LIBRARY

A few books should be mentioned here, as they deal with the biology of deep ocean fishes, several groups of which have been discussed in this chapter:

Randall, D. J., and A. P. Farrell (Eds.). 1997. *Deep-sea fishes*, Vol. 16, Fish Physiology Series. Academic Press, San Diego.

During his career at the British Museum, Norman Marshall (1915–1996) became one of the best-known ichthyologists. His books are classic works on deep-sea biology.

Marshall, N. B. 1954. *Aspects of deep sea biology.* Hutchinson, London.
———. 1971. *Explorations in the life of fishes.* Harvard University Press, Cambridge, MA.
———. 1980. *Deep-sea biology: Developments and perspectives.* Garland STPM Press, New York.

REFERENCES

Ahlstrom, E. H., J. Richards, and S. H. Weitzman. 1984. Families Gonostomatidae, Sternoptychidae, and associated stomiiform groups: Development and relationships, pp. 184–198. In *Ontogeny and systematics of fishes*, H. G. Moser, W. J. Richards, D. M. Cohen,

M. P. Fahay, A. W. Kendall, Jr., and S. L. Richardson (Eds.). *Am. Soc. Ichthyol. Herpetol. Spec. Publ. 1.*

Baldwin, C. C., and G. D. Johnson. 1996. Interrelationships of Aulopiformes, pp. 355–404. In *Interrelationships of fishes*, M. L. J. Stiassny, L. R. Parenti, and G. D. Johnson (Eds.). Academic Press, San Diego.

Berg, L. S. 1940. Classification of fishes, both recent and fossil. *Trav. Inst. Zool. Acad. Sci. URSS 5:* 87–517 (Reprinted 1947, J. W. Edwards, Ann Arbor, MI).

Blackett, R. F. 1962. Some phases in the life history of the Alaskan blackfish *Dallia pectoralis. Copeia 1962:* 124–130.

Eschmeyer, W. N. 1990. Genera in a classification, pp. 435–495. In *Catalog of the genera of recent fishes*, W. N. Eschmeyer (Ed.). California Academy of Sciences, San Francisco.

———, and R. M. Bailey. 1990. Genera of recent fishes, pp. 7–433. In *Catalog of the genera of recent fishes*, W. N. Eschmeyer (Ed.). California Academy of Sciences, San Francisco.

Fink, W. L. 1981. Ontogeny and phylogeny of tooth attachment modes in teleost fishes. *J. Morphol. 167:* 167–184.

———. 1984. Stomiiformes: Relationships, pp. 181–184. In Ontogeny and systematics of fishes, H. G. Moser, W. J. Richards, D. M. Cohen, M. P. Fahay, A. W. Kendall, Jr., and S. L. Richardson (Eds.). *Am. Soc. Ichthyol. Herpetol. Spec. Publ. 1.*

———. 1985. Phylogenetic relationships of the stomiid fishes (Teleostei: Stomiiformes). *Misc. Pub. Mus. Zool. Univ. Michigan No. 171.*

———, and S. V. Fink. 1986. A phylogenetic analysis of the genus *Stomias*, including the synonymization of *Macrostomias. Copeia 1986:* 494–503.

———, and S. H. Weitzman. 1982. Relationships of the stomiiform fishes (Teleostei) with a description of *Diplophos. Bull. Mus. Comp. Zool. 150*(2): 31–93.

Gartner, J. V., Jr., R. E. Crabtree, and K. J. Sulak. 1997. Feeding at depth, pp. 115–193. In *Deep-sea fishes*, Vol. 16, Fish Physiology Series, D. J. Randall and A. P. Farrell (Eds.). Academic Press, San Diego.

Gosline, W. A., N. B. Marshall, and G. W. Mead. 1966. Order Iniomi. Characters and synopsis of families, pp. 1–18. In Fishes of the western North Atlantic. *Mem. Sears Found. Mar. Res. 1*(5).

Harold, A. S. 1998. Phylogenetic relationships of the Gonostomatidae (Teleostei: Stomiiformes). *Bull. Mar. Sci. 62:* 715–741.

———, and S. H. Weitzman. 1996. Interrelationships of stomiiform fishes, pp. 333–353. In *Interrelationships of fishes*, M. L. J. Stiassny, L. R. Parenti, and G. D. Johnson (Eds.). Academic Press, San Diego.

Jamieson, B. G. M. 1991. *Fish evolution and systematics: Evidence from spermatozoa.* Cambridge University Press, Cambridge, UK.

Johnson, G. D. 1992. Monophyly of the euteleostean clades—Neoteleostei, Eurypterygii, and Ctenosquamata. *Copeia 1992:* 8–25.

———, and C. Patterson. 1993. Percomorph phylogeny: A survey of acanthomorphs and a new proposal. *Bull. Mar. Sci. 52*(1): 554–626.

———, and———. 1996. Relationships of lower euteleostean fishes, pp. 251–332. In *Interrelationships of fishes*, M. L. J. Stiassny, L. R. Parenti, and G. D. Johnson (Eds.). Academic Press, San Diego.

Johnson, R. K. 1982. Fishes of the families Evermannellidae and Scopelarchidae: Systematics, morphology, interrelationships, and zoogeography. *Fieldiana (Zool.) N.S. 12:* 1–252.

Lauder, G. V., and K. F. Liem. 1983. The evolution and interrelationships of the actinopterygian fishes. *Bull. Mus. Comp. Zool. 150*(3): 95–197.

López, J. A., P. Bentzen, and T. W. Pietsch. 2000. Phylogenetic relationships of esocoid fishes (Teleostei) based on partial cytochrome *b* and 16S mitochondrial DNA sequences. *Copeia 2000:* 420–431.

————, W.-J. Chen, and G. Orti. 2004. Esociform phylogeny. *Copeia 2004:* 449–464.

Marshall, N. B. 1979. *Developments in deep-sea biology.* Blandford Press, Poole, Dorset, UK.

Miya, M., and M. Nishida. 2000. Molecular systematics of the deep-sea fish genus *Gonostoma* (Stomiiformes: Gonostomatidae): Two paraphyletic clades and resurrection of *Sigmops. Copeia 2000:* 378–389.

Nelson, J. S. 1994. *Fishes of the world* (3rd ed.). Wiley, New York.

Okiyama, M. 1984. Myctophiformes: Relationships, pp. 254–259. In Ontogeny and systematics of fishes, H. G. Moser, W. J. Richards, D. M. Cohen, M. P. Fahay, A. W. Kendall, Jr., and S. L. Richardson (Eds.). *Am. Soc. Ichthyol. Herpetol. Spec. Publ. 1.*

Olney, J. E. 1984. Lampridiformes: Development and relationships, pp. 368–379. In Ontogeny and systematics of fishes, H. G. Moser, W. J. Richards, D. M. Cohen, M. P. Fahay, A. W. Kendall, Jr., and S. L. Richardson (Eds.). *Am. Soc. Ichthyol. Herpetol. Spec. Publ. 1.*

————, G. D. Johnson, and C. G. Baldwin. 1993. Phylogeny of lampridiform fishes. *Bull. Mar. Sci. 52*(1): 137–169.

Patterson, C. 1993. An overview of the early fossil record of acanthomorphs. *Bull. Mar. Sci. 52*(1): 29–59.

————, and G. D. Johnson. 1995. The intermuscular bones and ligaments of teleostean fishes. *Smith. Cont. Zool. 559.*

Pietsch, T. W. 1978. The feeding mechanism of *Stylephorus chordatus* (Teleostei: Lampridiformes): Functional and ecological implications. *Copeia 1978:* 255–262.

Randall, D. J., and A. P. Farrell (Eds.). 1997. *Deep-sea fishes,* Vol. 16, Fish Physiology Series. Academic Press, San Diego.

Robins, C. R., R. M. Bailey, C. E. Bond, J. R. Brooker, E. A. Lachner, R. N. Lea, and W. B. Scott. 1991. Common and scientific names of fishes from the United States and Canada (5th ed.). *Am. Fish. Soc. Spec. Publ. 20.*

Rosen, D. E. 1973. Interrelationships of higher euteleostean fishes, pp. 397–513. In Interrelationships of fishes, P. H. Greenwood, R. S. Miles, and C. Patterson (Eds.). *J. Linn. Soc. (Zool.) 53 Suppl. 1.* Academic Press, New York.

Stiassny, M. L. J. 1986. The limits and relationships of the acanthomorph teleosts. *J. Zool. (Lond.) B1:* 411–460.

————. 1996. Basal ctenosquamate relationships and the interrelationships of the myctophiform (scopelomorph) fishes, pp. 405–426. In *Interrelationships of fishes,* M. L. J. Stiassny, L. R. Parenti, and G. D. Johnson (Eds.). Academic Press, San Diego.

————, and J. A. Moore. 1992. A review of the pelvic girdle of acanthomorph fishes, with comments on hypotheses of acanthomorph relationships. *Zool. J. Linn. Soc. 104:* 209–242.

Weitzman, S. H. 1997. Systematics of deep-sea fishes, pp. 43–77. In *Deep-sea fishes,* Vol. 16, Fish Physiology Series, D. J. Randall and A. P. Farrell (Eds.). Academic Press, San Diego.

Welch, D. W., and P. M. Pankhurst. 2001. Visual morphology and feeding behavior of the daggertooth. *J. Fish. Biol. 58*(5): 1427–1437.

Wiley, E. O., G. D. Johnson, and W. W. Dimmick. 1998. The phylogenetic relationships of lampridiform fishes (Teleostei: Acanthomorpha), based on a total-evidence analysis of morphological and molecular data. *Mol. Phylogen. Evol. 10:* 417–425.

————. 2000. The interrelationships of acanthomorph fishes: A total evidence approach using molecular and morphological data. *Biochem. Syst. Ecol. 28:* 319–350.

Wilson, M. V. H., D. B. Brinkman, and A. G. Newman. 1992. Cretaceous Esocoidei (Teleostei): Early radiation of the pikes in North American fresh waters. *J. Paleontol. 66*(5): 839–846.

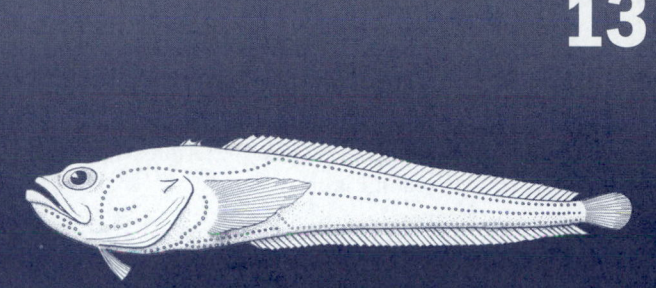

13 THE PARACANTHOPTERYGII: CODS, TOADFISHES, AND OTHER BOTTOM-DWELLING TELEOSTS

PARACANTHOPTERYGIANS AT A GLANCE

SUPERORDER PARACANTHOPTERYGII

Order Percopsiformes (Salmopercae)
Order Gadiformes (Anacanthini)
Order Amblyopsiformes (Anacanthini)
Order Ophidiiformes (Anacanthini)
    Suborder Ophidioidei
    Suborder Bythitoidei
Order Batrachoidiformes (Anacanthini, Pediculati)
Order Lophiiformes (Anacanthini, Pediculati)
    Suborder Lophioidei
    Suborder Antennarioidei
    Suborder Chaunacoidei
    Suborder Ogcocephalioidei
    Suborder Ceratioidei

The paracanthopterygian fishes include a diverse assemblage of teleosts with a complex and confusing evolutionary history. The name *Paracanthopterygii* (*para* = "like") suggests that this group resembles but is not quite the same as the acanthopterygian, or "spiny-finned," teleosts. Earlier taxonomic revisions have moved major groups of teleosts in and out of the Paracanthopterygii, and the general consensus of students of the group is that it is probably not a natural phylogenetic assemblage. As the taxon stands now, it includes small freshwater forms as well as large marine forms found at all depths of the ocean. Although the best known ones, such as the cods, conform to the "typical" fish profile, others are bizarre looking creatures with greatly enlarged heads or are elongate and more closely resemble eels. Paracanthopterygians, as a whole, are a group of teleosts that have evolved a close association with benthic substrates, either resting directly on them or hovering just above them while foraging for animals hiding in the sediments.

PARACANTHOPTERYGIANS AT A GLANCE

Although a few paracanthopterygian fishes inhabit North American freshwaters, most are benthic and benthopelagic marine fishes. Among the paracanthopterygians inhabiting the continental shelf areas of the world's oceans are the cods and their allies of the order Gadiformes—among the most important of the commercially harvested fishes (see Chapter 39). Elongate, eellike forms, especially adapted to living in the deepest reaches of the world's oceans or in caves, are to be found among the Ophidiiformes; and a bizarre array of bottom-dwelling, large-mouthed predators constitute the orders Batrachoidiformes and Lophiiformes. The paracanthopterygian fishes present an especially challenging problem to modern fish systematists; arguments still persist over what groups constitute definitive members of the taxon.

Among Donn Eric Rosen's strengths as an ichthyologist was his choice of collaborators. He and his colleagues (Greenwood et al., 1966) recognized the mostly soft-ray finned superorder Paracanthopterygii as distinct from the Acanthopterygii, which were characterized by much more durable spines. In 1969, he copublished a major work with the renowned British paleoichthyologist Colin Patterson, in which they sought to clarify the relationships of a number of teleost orders. Rosen and Patterson (1969) suspected that the members of the orders Percopsiformes, Gadiformes, Batrachoidiformes, Lophiiformes, and Gobiesociformes constituted a monophyletic assemblage, and they demonstrated that a number of features—particularly the jaw musculature and the caudal skeleton—did indeed indicate monophyly.

Since the superorder was first defined by Greenwood et al. (1966), it has been subjected to modification and reinterpretation by several students of the group (Fraser, 1972; Patterson and Rosen, 1989; Rosen, 1985; Rosen and Patterson, 1969; Stiassny, 1986). Parenti (1993) has suggested that the paracanthopterygians may have a sister group relationship with another large assemblage of teleosts, the Atherinomorpha, which are characterized by the predominance of flexible spines and soft rays.

Rosen (1985) eventually expressed reservations about whether the Paracanthopterygii, as he understood them to be, could be accepted as a monophyletic group. Patterson and Rosen (1989) presented a cladogram that characterized the superorder and its included lineages, eliminating the Gobiesociformes and the Zoarcoidei, which were originally included by Greenwood et al. (1966); the Polymixiiformes, which Rosen and Patterson (1969) had added, and other groups (e.g., Indostomiformes, Gobioidei) that other authors had included. The remaining orders that are presented here

as belonging to the superorder Paracanthopterygii are those that appear to represent some consensus among ichthyologists, even though there are unresolved questions regarding the relationships of the fishes included in the Percopsiformes, Ophidiiformes, and Batrachoidiformes. Further study will no doubt result in phylogenetic interpretations that differ from those summarized in Figure 13.1.

SUPERORDER PARACANTHOPTERYGII

Order Percopsiformes (Salmopercae)

The Percopsiformes are endemic to North American freshwaters. An extinct suborder, †Sphenocephaloidei, is known from marine deposits from the Cretaceous. Murray and Wilson (1999) considered the Percopsiformes to be monophyletic if the Amblyopsidae were removed. Their phylogeny, in which the amblyopsids are given separate, ordinal status, is followed here.

DIAGNOSIS OF THE PERCOPSIFORMES

- Subocular shelf, orbitosphenoid, and basisphenoid are absent.
- Maxillary excluded from the border of the mouth (gape).
- Six branchiostegal rays present.
- Ctenoid scales present.
- Flexible spines in the dorsal and anal fins; single supraneural behind the first or second neural spine; subabdominal or subthoracic pelvic fins; some members with an adipose fin.
- Caudal fin with 16 branched rays; caudal fin is supported by two plates that are the result of the fusion of hypurals.
- Physoclistous gas bladder.

Sources: Lauder and Liem, 1983; Murray and Wilson, 1999; Nelson, 1994; Rosen and Patterson, 1969.

Members of the family **Percopsidae** have an adipose fin. The family consists of only two species: *Percopsis omiscomaycus*, the troutperch of eastern North American drainages (Fig. 13.2A), and *P. transmontana*, the sand roller of the Columbia River system. These are small fish of still or slow streams. *Percopsis* seems to be nocturnal in habit, remaining in deep water or hiding during the day and moving into shallow water at night. The family is known from the Late Palaeocene (Murray and Wilson, 1996).

The monotypic family **Aphredoderidae** consists of the pirate perch (*Aphredoderus sayanus*), which lacks an adipose fin, has subthoracic pelvic fins, and has the urogenital opening in a jugular position. (The family name Aphredoderidae literally means "excrement throat".) It is a freshwater species that lives in sluggish lowland streams along

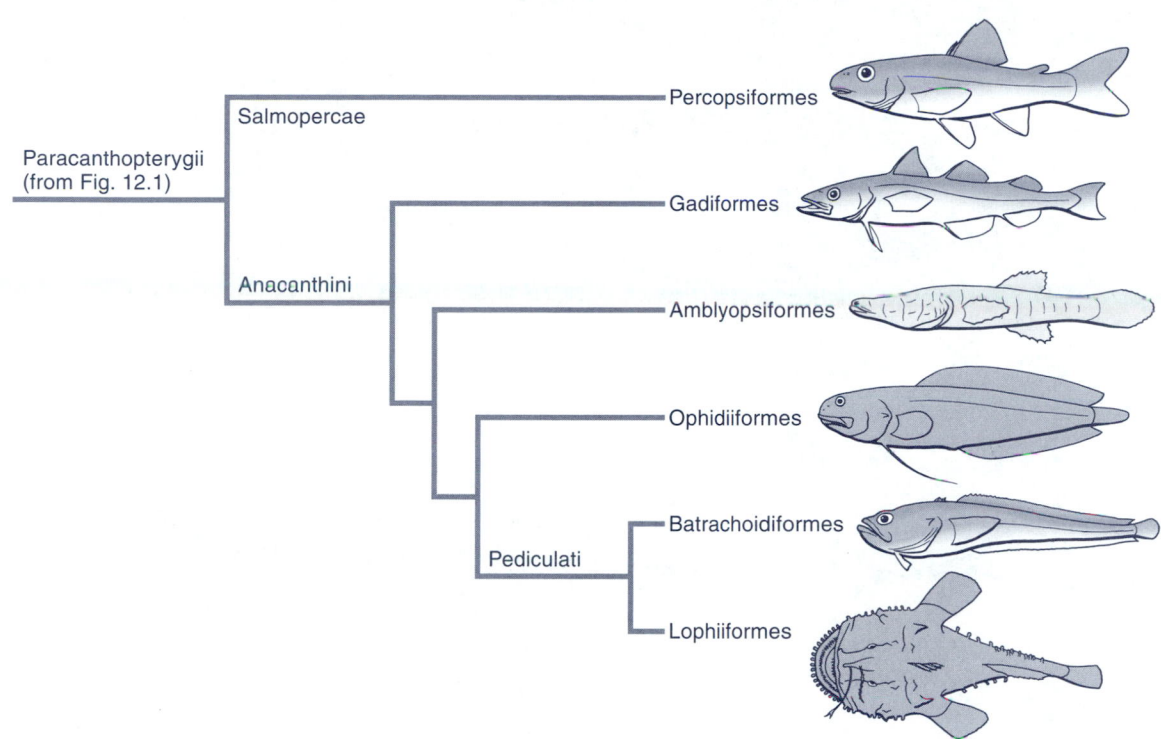

FIGURE 13.1
Phylogeny of paracanthopterygian fishes (after Patterson and Rosen, 1989; Murray and Wilson, 1999).

the Gulf and Atlantic coasts and in the Mississippi drainage of the United States. Fletcher et al. (2004) discovered that the peculiar placement of the urogenital opening facilitates an unusual mode of reproduction. Female individuals thrust their heads into the narrow spaces of underwater root masses and release their eggs; the males congregate at these spawning sites and likewise enter the spawning crevices headfirst to release their sperm. Following the act of spawning, the males were observed defending the oviposition sites by obstructing their entrances. The forward placement of the urogenital opening facilitates the headfirst egg release and fertilization.

Order Gadiformes (Anacanthini)

Patterson and Rosen (1989) referred to the fishes of this order as "core paracanthopterygians." They exhibit a complex of primitive and derived characteristics. The relationships within the order have been subject to much study and interpretation in recent years, and various arrangements of

suborders and families have been proposed. Important references include Cohen (1984, 1989), Cohen et al. (1990), Fahay and Markle (1984), and Siebert (1990). This order contains the cods and their allies, including the grenadiers, hakes, and burbots (Table 13.1). With few exceptions, these are marine fishes and are known in the fossil record from the beginning of the Paleocene.

DIAGNOSIS OF THE GADIFORMES
- The premaxillaries exclude the maxillaries from the gape.
- Absence of the orbitosphenoid and mesocoracoid.
- Cycloid scales usually present.
- Fins with soft rays; thoracic or jugular pelvic fins.
- Isocercal or leptocercal caudal fin; accessory bones ("X" and "Y" bones) present between the neural spines and haemal spines of preural vertebrae in many genera; first neural spine in close contact or joined to the crest of the supraoccipital, except in *Muraenolepis*.
- Physoclistous gas bladder.

Sources: Cohen, 1984; Fahay and Markle, 1984; Patterson and Rosen, 1989.

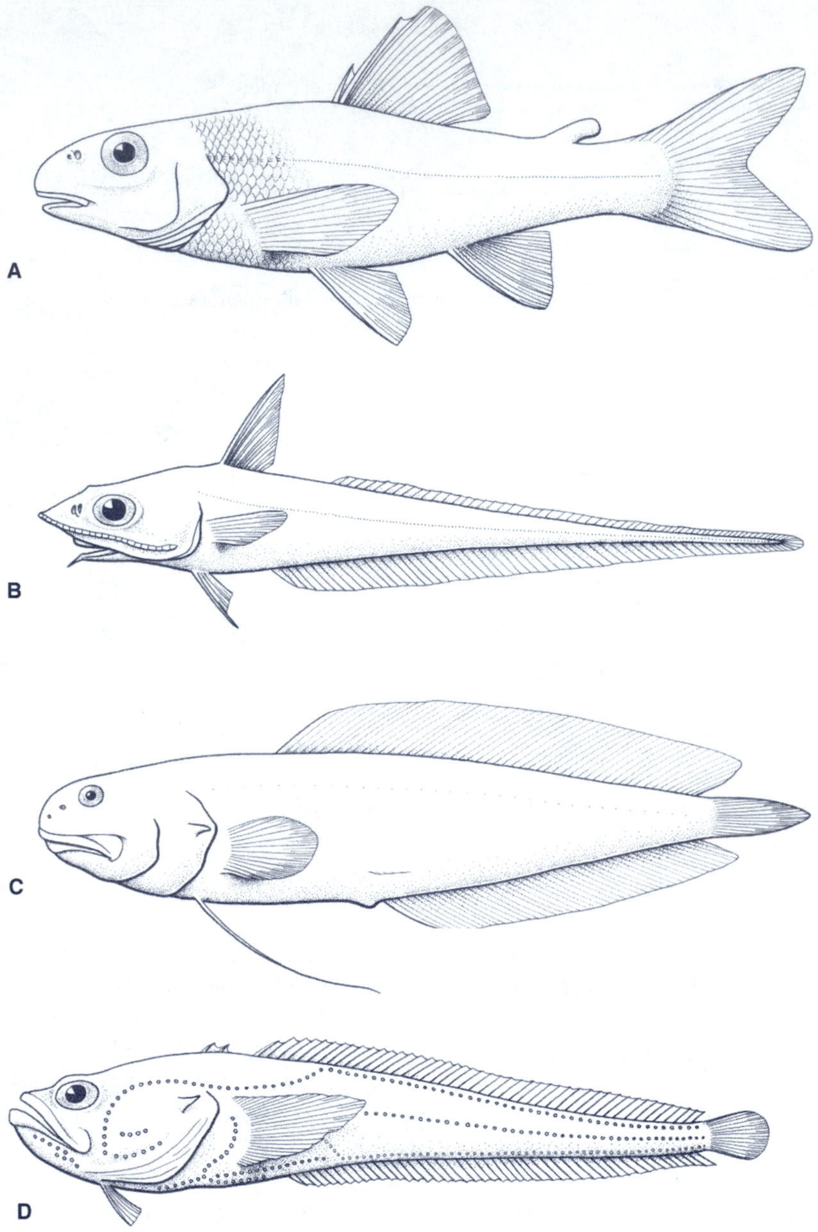

FIGURE 13.2

A, Representative of Order Percopsiformes: troutperch (*Percopsis omiscomaycus*); **B,** Representative of order Gadiformes: grenadier (*Macrourus*); **C,** Representative of order Ophidiiformes: livebearing brotula (*Brotulina*); **D,** Representative of order Batrachoidiformes: midshipman (*Porichthys*).

TABLE 13.1   FAMILIES OF GADIFORM FISHES

Family	Common Name	Type	Distribution
Ranicipitidae	tad-pole cod	Monotypic (*Raniceps raninus*)	Northeast Atlantic Ocean
Euclichthyidae	eucla cod	Monotypic (*Euclichthys polynemus*)	Marine waters off Australia and New Zealand
Macrouridae	grenadiers or rattails		Worldwide deep marine
Steindachneriidae	luminous hake	Monotypic (*Steindachneria argentea*)	Gulf of Mexico and tropical Atlantic
Moridae	codlings		Worldwide deep marine
Melanonidae	pelagic cods		Southern Atlantic and Pacific Oceans
Macruronidae	straptails, southern hakes		Southern Hemisphere marine
Bregmacerotidae	codlets	Monogeneric (*Bregmaceros*)	Widespread in tropical and subtropical seas
Muraenolepididae	eel cods		Cold marine waters of the Southern Hemisphere
Phycidae	phycid hakes, rocklings		Atlantic Ocean and elsewhere in Southern Hemisphere
Merluciidae	merluciid or silver hakes		Worldwide marine
Gadidae	cods		Arctic, Atlantic, and Pacific marine, with one Holarctic freshwater species

Although there is active, ongoing work on the phylogeny of gadiforms, there seems to be no general agreement on the relationships of the currently recognized families. Markle (1989) used a classification that had four suborders: Ranicipitoidei, Melanonoidei, Macrouroidei, and Gadoidei. Nelson (1984) recognized the suborders Muraenolepidoidei, Gadoidei, and Macrouroidei, but in a later edition (Nelson, 1994), he declined to identify suborders of gadiform fishes owing to the differences of opinion regarding their classification. Eschmeyer (1990, 1998) also did not delimit suborders, and we shall follow suit here.

Among the gadiform fishes, the grenadiers of the family **Macrouridae** are especially prominent members of the deep-ocean ichthyofauna. These are elongate fishes with a large head, often with large eyes. The trunk of the body is short, with 10 to 16 abdominal vertebrae; the long, tapering body is fringed by long dorsal and anal fins and ends in a leptocercal tail that lacks a distinct caudal fin (Fig. 13.2B). Some may have a spinous second dorsal ray and ctenoid scales. Photophores containing luminescent symbiotic bacteria are present in many species, and drumming muscles are associated with the gas bladder of the males of certain species. Macrourids are benthopelagic fishes, usually found at depths from 200 to 2,000 m (Marshall and Iwamoto, 1973). Howes (1989) suggested that certain genera—*Bathygadus, Gadomus,* and *Trachyrhincus*—usually placed in this family should be considered to be more closely related to gadids.

The abundance of some species of grenadiers has made them a likely candidate for commercial exploitation (see Going Deeper, Chapter 35).

The **Moridae,** or codlings, are deep-sea fishes with an unusual otophysic connection, consisting of very large diverticula from the gas bladder that pass through wide foramina in the exoccipitals.

Closely related to the cods are the hakes of the family **Merlucciidae.** Robins et al. (1991) included these fishes in the Gadidae. These are commercial fishes of the Atlantic, eastern Pacific, and the seas around southern New Zealand and South Africa. The Pacific hake (*Merluccius productus*) has become a major commercial species in the northeastern Pacific Ocean. Though not nearly as desirable a food fish as their relatives the cods and the pollock, it supported a modest fishery that the Russians eventually developed into a much more significant one. The species is known in the fishery industry as "Pacific whiting."

Members of the family **Gadidae** have two and sometimes three dorsal fins. The caudal fin is of the "pseudocaudal" type, in that it consists mostly of dorsal and anal elements, with a very small "true" caudal making up the central part. Many cods have diverticula of the gas bladder in connection with the inner ear, which apparently increase the acuity of hearing.

Members of the subfamily Gadinae are believed to have originated in the North Atlantic. Several species are found in the waters of the North Atlantic and Arctic Oceans, but only three are seen in the eastern North Pacific Ocean. Carr et al. (1999) have determined that the three genera native to the Pacific coastal waters represent distinct phylogenetic lineages that entered the Pacific independently. Molecular studies have revealed the Gadidae to be the most derived of the gadiform families (Bakke and Johansen, 2002).

The rocklings (some species referred to as hakes) and the burbot have usually been placed in the Gadidae as the subfamily Lotinae, but are now considered by some authors to constitute a separate family, the Lotidae (Cohen, 1984; Eschmeyer, 1990, 1998). Some of the lotine genera include *Brosme*, the cusks; *Molva*, the European ling, and *Ciliata*, the rocklings. The burbot (*Lota lota;* see Plate 13.3) is a Holarctic fish of freshwater.

From a cultural and economic standpoint, no fish family has had more significance for the development of northern European and North American society than the Gadidae (see Fig. 3.13D). Commercial fishers know members of this family as cods, haddocks, pollock, and whitings. All have commercial value as food or as a source of high-quality fish meal. Cod fishing in the North Atlantic has been pursued by many nations since the middle ages. Cod fishery encampments were among the earliest European settlements in North America. The object of most of this effort has been the Atlantic cod (*Gadus morhua*), which may reach a length of 1.2 m and a weight of 45 kg. Longlining has been the primary fishery method employed over the years, but trawling, especially since the advent of steam-powered vessels, forms the basis of the fisheries today. Also, gill nets are employed in the cod fisheries of New England. Landings of the species in recent years have been nearly two million metric tons per year, but the fishery is presently in a serious state of decline (see Chapter 39).

Another important commercial gadid in the North Atlantic is the haddock (*Melanogrammus aeglefinus*), which has also suffered from overfishing by trawlers. The species reaches about 1.1 m long and 16 kg. The pollock (*Pollachius virens*) and the blue whiting (*Micromesistius poutassou*) are additional important commercial species. The Pacific Ocean counterpart of the Atlantic cod, the Pacific cod (*G. macrocephalus*), was once fished in Canadian and Alaskan waters by sailing vessels operating from San Francisco ports. Landings of the walleye pollock (*Theragra chalcogramma*) from the Pacific Ocean and Bering Sea usually exceed six million metric tons annually—currently the highest for the cod family. In the summer of 2000, a United States federal judge issued an injunction effectively closing a vast portion of the North Pacific ground fishery in one of the toughest-ever enforcements of the Endangered Species Act. Criticized for inadequate protection of endangered populations of Stellar's sea lion (*Eumetopias jubata*), the National Marine Fisheries Service has been mandated to develop a more effective protection plan. Until that time, only the sea lions may prey on the cod and pollock in the restricted zone, which includes up to 80 percent of the Gulf of Alaska.

• • • • • • • • • • • • • • • • • •
Order Amblyopsiformes (Anacanthini)

The Amblyopsiformes comprise a single family (**Amblyopsidae**) that includes the cave fishes and the swampfish (*Chologaster cornuta*) of the southeastern part of the United States. All save for the swampfish are inhabitants of subterranean waters. Like the pirate perch, these fishes have spines in their dorsal and anal fins and have the anus positioned far forward. Unlike their former relatives among the Percopsiformes, the amblyopsids have cycloid scales. In the swampfish, the eyes are reduced but not covered by skin, and a peculiar appendage, which may have a role in courtship behavior, develops on the dorsal surface of the snout in males (Ross and Rohde, 2003). The spring cavefish (*C. agassizi*) has extremely reduced eyes, and members of the genera *Amblyopsis*, *Speoplatyrhinus*, and *Typhlichthys* (see Plate 13.4) are blind. The lateralis system of the cavefishes is well developed and includes lines of superficial neuromasts on the head and body. They have cycloid scales, and, in species with pelvic fins, the fins are abdominal. The remarkable suite of adaptations that permit amblyopsids and other fishes to inhabit the darkest recesses of caves is further discussed in Chapter 32.

• • • • • • • • • • • • • • • • • •
Order Ophidiiformes (Anacanthini)

These fishes were once included with the Perciformes, then considered to be a suborder of the Gadiformes. Most recent treatments have accorded them ordinal status within the Paracanthopterygii (Cohen and Nielsen, 1978; Gordon et al., 1984; Nelson, 1994). The order contains a variety of fishes, some elongate, with tapering bodies and long dorsal and anal fins that are confluent with the caudal fin in many genera. Dorsal and anal pterygiophores are more numerous than the vertebrae adjacent to the fins. Pelvic fins have small spines in some species and are usually jugular to mental, so that they resemble elongate barbels.

Ophidiiforms are mostly benthic or benthopelagic, with few freshwater representatives. There may be as many as 400 species, some of which live at great depths.

Suborder Ophidioidei

Members of this suborder are oviparous, with the anterior nostril positioned some distance from the lip. The cusk eels and brotulas of the family **Ophidiidae** (Fig. 13.2C) range from the shallows to great depths in most seas. There are about 50 genera and more than 165 species, usually arranged in four subfamilies. Most are small, but members of the genus *Genypterus* found in Australia, South America, and South Africa may grow to more than a meter long. *Genypterus reedi* and *G. chilensis*, both found in Chilean waters, are prized as food fishes.

The subfamily Neobythitinae includes some rather brightly colored littoral fishes, but others are dark, deep-water forms. *Abyssobrotula galatheae* is known as the deepest living fish, having been captured at 8,470 m in the Puerto Rico trench. *Acanthonus armatus* is a benthopelagic species with a huge head containing a large, fluid-filled cavity. The low specific gravity of the fluid provides buoyancy.

The pearlfishes of the family Carapidae lack caudal fins; their dorsal and anal fins are elongate, and the body tapers posteriorly to a point (see Plate 13.5). The anus is jugular, probably an adaptation for living with the posterior part of the body inside invertebrate hosts. Larvae are greatly elongate and bear a bannerlike, elongated dorsal ray (**vexillum**) just behind the head. There are two subfamilies and seven genera, with 31 species (Markle and Olney, 1990). The subfamily Carapinae, which are small, thin-bodied fishes, have no pelvic fins and have members that live commensally (some parasitically)—primarily in sea cucumbers, but they may inhabit secondary hosts such as bivalve molluscs (Paredes-Ríos and Balart, 2000). Trott (1970) proposed a phylogeny of carapids in which parasitism evolved in conjunction with the loss of fins. In this scheme, the subfamily Pyramodontinae, which includes species that are deeper bodied and have pelvic fins, were perceived as a basal group. They are nearly circumtropical, predatory fishes with large anterior teeth and a protractile upper jaw.

Suborder Bythitoidei

In this group, the anterior nostril is quite close to the upper lip, and the males have an intromittent organ. In some species, the caudal fin is separated from the dorsal and anal fins. The family **Bythitidae** contains two subfamilies of viviparous (live-bearing) brotulas that are widely distributed in marine waters. There are about 80 species in 28 genera (Gordon et al., 1984). Bythids frequent caves and crevices of coral reefs. The genus *Lucifuga* contains blind species that live in freshwater caves in Cuba. Other members of the genus (with minute but functional eyes) are found in lime-stone sinks in the Bahamas (see Plate 13.6). Another genus, *Ogilbia*, has species in freshwater caves of Yucatan and in brackish waters of the Galapagos Islands (Nelson, 1994).

Order Batrachoidiformes (Anacanthini, Pediculati)

This order contains the toadfishes and midshipmen—squat, broad-mouthed, bottom fishes of tropical and temperate seas. They are related to the Lophiiformes, and Patterson and Rosen (1989) considered these two orders to constitute the clade Pediculati.

DIAGNOSIS OF THE BATRACHOIDIFORMES
- Flattened cranium with parasphenoid joining the frontal bones; gape formed from both premaxillaries and maxillaries, but only premaxillaries bear teeth.
- Ribs absent.
- Radials of pectoral fins elongate.
- Jugular pelvic fins with one spine and two or three soft rays.
- Short, spinous dorsal fin, bearing two to four spines, is rostral to a long, soft dorsal fin.

Source: Patterson and Rosen, 1989.

Batrachoidiform fishes have only three pairs of gills—their quiescent lifestyle and consequent low metabolic rates mean that they can get by with one fewer pair than other bony fishes. Sharp spines may be present in the dorsal fin and on the edge of the operculum. Venom is produced in glands at the bases of these spines in certain genera, some of which (*Thalassophryne* and *Daector*) have hollow spines through which a potent venom can flow (see Chapter 39).

Many species can produce a variety of sounds with special muscles that attach to and can vibrate the gas bladder. These sounds have been described as hoots, grunts, and boat whistle blasts (see Chapter 21). There is only one family, **Batrachoididae**, divided into three subfamilies (Collette, 1966; Smith, 1952). There are about 19 genera and 64 species.

The subfamily Batrachoidinae contains the toadfishes, which are shallow-water marine fishes of most tropical and warm temperate seas. They have no venom glands and no photophores. There are three spines in the first dorsal fin. Most have barbels and skin flaps around the mouth, and members of some genera, such as *Batrachomoeus*, have remarkable camouflage that suits them for life on coral. Some species display bold and brilliant coloration, as befits their coral reef habitat; the aptly named *Sanopus splendidus* is a good example (see Plate 13.7). The genus *Opsanus* is represented on the Gulf and Atlantic coast of North America.

Members of the subfamily Porichthyinae are found on the continental shelf of North and South America. Their first dorsal fin consists of two spines, there are no scales, and some species, such as the Atlantic midshipman (*Porichthys plectrodon*), have venom glands (Hoese and Moore, 1977). They have multiple lateral lines (Hart, 1973; Nelson, 1994). The genus *Porichthys* takes the name *midshipman* from the several rows of photophores that course along the body like buttons on a midshipman's uniform (Fig. 13.2D). These are among the few shallow-water fishes that produce light. The genus *Aphos* of the Pacific coast of South America lacks photophores.

The subfamily Thalassophryninae contains venomous toadfishes. In contrast to the solid dorsal and opercular spines of the other batrachiform fishes, these have two sharp, hollow spines in the first dorsal fin and one on the operculum. Venom glands in association with the spines allow the injection of venom. They do not have photophores, and the lateral line is single or lacking. These are fishes mainly of warm waters of Central and South America, with a few freshwater species, especially in the Amazon.

• •
Order Lophiiformes (Anacanthini, Pediculati)

DIAGNOSIS OF THE LOPHIIFORMES
- Head with large mouth and reduced gill openings.
- Ribs absent.
- Pectoral fins with long, narrow radials.
- Pelvic fins, if present, anterior to pectoral fins; composed of a single spine and four or five soft rays.
- Caudal fin borne on single hypural plate.
- Swim bladder, when present, is physoclistous.
  Sources: Pietsch, 1984; Rosen and Patterson, 1969.

This order includes the highly derived anglerfishes and frogfishes. The spinous dorsal fin of these bizarre fishes may be modified into a fishing lure that is placed above the head. The anteriormost elements form a sort of fishing rod, the **illicium,** that often bears a flap or a bulbous lure, the **esca,** at the tip. If the esca is bitten off by the potential prey, it can be regenerated. In many species, the paired fins are modified into limblike structures that aid in locomotion along the bottom. The order is divided into five suborders (Pietsch and Grobecker, 1987).

Suborder Lophioidei
This suborder consists of one family, **Lophiidae,** the anglers or goosefishes (see Plate 13.8). These fishes have a large, flat head and a wide mouth set with sharp, depressible teeth. They have jugular pelvic fins with one spine and five soft rays.

Goosefishes are widely distributed from cold waters of the Northern Hemisphere (Barents Sea) to the tropics. The genus *Lophius*, with eight species, is perhaps the best known, as it is commercially harvested. Fishery marketers, in their infinite wisdom, decided that the fish-buying public would turn away from anything with the name "goosefish," so the British name "monkfish" is used. Monkfish livers are also sold under the name "ankimo." *Lophius americanus* is the common Atlantic coast species in the Americas; *L. piscatorius* and *L. budegassa* are European species. Members of the genus *Lophius* may reach 1.3 m. Lophiids lie concealed on the bottom and attract prey by moving their angling apparatus. *Lophius americanus* deposits up to 2.6 million eggs in a veil of gelatinous material secreted by the ovaries. The veils can be up to 10 m long and nearly a meter wide (Breder and Rosen, 1966). Other genera are *Lophiomus* (one species), *Lophioides* (13 species), and *Sladenia* (three species).

Suborder Antennarioidei
Included here are the small frogfishes, batfishes, and their allies. The pelvic fin is set well in advance of the pectoral fin. Scales are absent, but the skin bears spinules or denticles. The body is laterally compressed.

The family **Antennariidae,** the frogfishes, is a circum-tropical group of 12 genera and 41 species with a concentration of species in the Indo-Australian archipelago (Pietsch and Grobecker, 1987). They are primarily benthic, shallow-water fishes, but a few species are found deeper than 100 m, and one (*Antennarius nummifer*) has been taken as deep as 293 m (Pietsch and Grobecker, 1987). Their slow, lumbering gait belies an amazing efficiency at feeding, in which they can "inhale" comparatively large prey species more rapidly and with a larger increase in oral cavity volume than any other fish (Grobecker and Pietsch, 1979; Pietsch and Grobecker, 1990).

One of the best known frogfishes is the sargassumfish (*Histrio histrio;* see Plate 13.9). Individuals of this species may live far from land in floating mats of *Sargassum* (see Chapter 35), depositing their eggs among the weeds in gelatinous veils or rafts that are much like smaller versions of those produced by the goosefishes. *Histrio* and some species of *Antennarius* are known to inflate themselves by swallowing air, but Pietsch and Grobecker (1987) remarked that inflation is usually in response to much "poking and manipulation."

The monotypic **Tetrabrachiidae** (Pietsch, 1981) are found from the Moluccas islands to northern Australia. *Tetrabrachium ocellatum* has highly modified pectoral fins that are divided into upper and lower sections, the lower part bound to the body by a membrane. Likewise, the **Lophichthyidae** contain but one species, *Lophichthys boschmai*, of New Guinea. The **Brachionichthyidae** of southern Australia have a spinous dorsal fin set forward on the head, with the three spines connected by membranes. As many as seven species, some undescribed, are known (Nelson, 1994).

Suborder Chaunacoidei

This suborder contains the sea toads or coffinfishes of the family **Chaunacidae,** which are found in warm and temperate seas at depths from 90 to more than 2,000 m (Caruso, 1989). They have a very short illicium, with an esca covered with cirri. The head is large and globose, and the gill openings are placed well behind the pectorals. There are two genera: *Chaunax*, which has about 12 species, and *Bathychaunax*, with two species (Caruso, 1989). Nelson (1994) grouped the chaunacoids with the following two suborders in a single suborder, the Ogcocephalioidei.

Suborder Ogcocephalioidei

This suborder contains the single family **Ogcocephalidae,** the batfishes (see Plate 13.10), which Pietsch and Grobecker (1987) indicated are closely related to the ceratioids. The batfishes are bizarre in the extreme, displaying a dorsoventrally compressed body with scales that are modified into tubercles or bucklers (Bradbury, 1967). The illicium is retractable, and they can walk over the bottom on their pectoral and pelvic fins (Nelson, 1994). According to Bradbury (1967, 1988), there are nine genera and more than 60 species of batfishes, distributed in tropical oceans.

Suborder Ceratioidei

This large group is made up of 11 families of anglerfishes. Although the larvae are planktonic in the epipelagic realm, the adults are most frequently found in the depths of the meso- and bathypelagic zones. They are small fishes with large mouths (see Plate 13.11). The females are larger than the males. Pelvic fins are absent. The illicium usually carries a lure containing luminescent bacteria. Some species bear barbels that have autogenic luminescent systems (Marshall, 1979). Videotape observations of individuals of the deep-dwelling genus *Gigantactis* (whipnose anglerfishes, of the family Gigantactinidae) have shown them engaging in a peculiar foraging behavior that involves drifting upside down with the current. The illicium is extended in a slight arc out in front of the fish, and the esca is positioned a few centimeters above the bottom (Moore, 2002). Bertelsen (1951) remains the classic reference on the suborder. There are about 149 species in 35 genera (Bertelsen, 1984; Nelson, 1994).

In some ceratioid families, the males are parasitic on the females, attaching firmly with their jaws and eventually becoming, in some cases, not much more than testicular tissue dependent on the female blood supply for all nutrients (see Chapter 27). Obligate parasitic males are known in the families Ceratiidae, Neoceratiidae, and Linophrynidae (including Aceratiidae and Photocarynidae). Males of the Caulophrynidae may be facultative parasites. In the Melanocetidae, Himantolophidae, and Oneirodidae, the males are known to be free swimming, but they have a remarkably large olfactory apparatus and jaws suited to clamping onto the female.

KEY POINTS AND CONNECTIONS

• The Paracanthopterygii are a taxon that encompasses a large and varied assemblage of primarily benthic marine fishes. Recent systematic studies have indicated that this taxon is probably not monophyletic.

• Two orders, the Percopsiformes and the Amblyopsiformes, once thought to be more closely related, are quiet-water inhabitants of streams and subterranean waters in North America.

• The order Gadiformes includes among its members perhaps the most commercially significant food fishes in the world—the cods and their relatives. Molecular phylogenetic studies provide evidence that the family Gadidae is perhaps the most derived gadiform family.

Refer to Chapter 39 for a discussion of the gadiform fisheries.

• The order Ophidiiformes encompasses an assemblage of elongate, eellike fishes with an extraordinary range of depths and habitats. Included in the order are cusk eels and brotulas, with representatives that inhabit crevices in freshwater caves and coral reefs as well as species that are among the deepest known fishes.

Refer to Chapter 35 for a discussion of fishes characteristic of deep-ocean environments.

• Fishes of the order Batrachoidiformes are squat, seemingly ungainly in appearance, but effective in ambush predation. Being fishes of limited mobility, they display a commensurate diminution of gill structure. The batrachoidiform fishes are benthic fishes of shallow waters. Many produce powerful venom from glands associated with hollow spines as a means of warding off predators. Extraordinary structural and physiological modifications of the swim bladder—involving an intimate interplay of muscular, neuronal, and endocrine systems—have evolved for the purposes of communication, especially during mating.

The Going Deeper box in Chapter 21 provides a summary of the role of vocalization in the complex reproductive life histories of batrachoidids.

• Fishes of the order Lophiiformes resemble the Batrachoidiformes in possessing limited mobility and a squat, ungainly body morphology.

Members of this highly derived order are distinguished by the modification of dorsal fin spines as fishing lures. Shallow-water members of the order have a highly cryptic body profile and coloration that renders them nearly invisible among the rocks and vegetation where they are normally found. Limblike modifications of the paired fins enable them to move about along the bottom in lieu of well-developed swimming capabilities. Some members have perfected their predatory lifestyle in the deep pelagic realm of the ocean. In this inhospitable environment, some have evolved a most unusual reproductive strategy, in which males are profoundly reduced in size and assume a parasitic existence on their female partners.

• •

REFERENCES

• •

Bakke, I., and S. Johansen. 2002. Characterization of mitochondrial ribosomal RNA genes in gadiformes: Sequence variations, secondary structural features, and phylogenetic implications. *Mol. Phylogen. Evol.* (electronic publ., Aug. 14, 2002).

Bertelsen, E. 1951. The ceratioid fishes. *Dana Rep. 39:* 1–276.

———. 1984. Ceratioidei: Development and relationships, pp. 325–335. In Ontogeny and systematics of fishes, H. G. Moser, W. J. Richards, D. M. Cohen, M. P. Fahay, A. W. Kendall, Jr., and S. L. Richardson (Eds.). *Am. Soc. Ichthyol. Herpetol. Spec. Publ. 1.*

Bradbury, M. G. 1967. The genera of batfishes. *Copeia 1967:* 399–422.

———. 1988. Rare fishes of the deep-sea genus *Halieutopsis:* A review with descriptions of four new species (Lophiiformes: Ogcocephalidae). *Fieldiana Zool. (N.S.) 44:* 1–22.

Breder, C. M., Jr., and D. E. Rosen. 1966. *Modes of reproduction in fishes.* Natural History Press, Garden City, NY.

Carr, S. M., D. S. Kivlichan, P. Pepin, and D. C. Crutcher. 1999. Molecular systematics of gadid fishes: Implications for the biogeographic origins of Pacific species. *Can. J. Zool. 77:* 19–26.

Caruso, J. H. 1989. Systematics and distribution of the Atlantic chaunacid anglerfishes (Pisces: Lophiiformes). *Copeia 1989:* 153–165.

Cohen, D. M. 1984. Gadiformes: Overview, pp. 259–265. In Ontogeny and systematics of fishes, H. G. Moser, W. J. Richards, D. M. Cohen, M. P. Fahay, A. W. Kendall, Jr., and S. L. Richardson (Eds.). *Am. Soc. Ichthyol. Herpetol. Spec. Publ. 1.*

———. 1989. Papers on the systematics of gadiform fishes. *Nat. Hist. Mus. Los Angeles Co. Sci. Ser. 32.*

———, and J. G. Nielson. 1978. Guide to the identification of genera of the fish order Ophidiiformes with a tentative classification of the order. *NOAA Tech. Rep. NMFS Circ. 147.*

———, T. Inada, T. Iwamoto, and N. Scialabba. 1990. Gadiform fishes of the world (Order Gadiformes). An annotated and illustrated catalogue of cods, hakes, grenadiers and other gadiform fishes known to date. FAO Species Catalog, Vol. 10. *FAO Fish. Synop. 125.*

Colette, B. B. 1966. A review of the venomous toadfishes, subfamily Thalassophryninae. *Copeia 1966:* 846–864.

Eschmeyer, W. N. 1990. Genera in a classification, pp. 435–495. In *Catalog of the genera of recent fishes,* W.N. Eschmeyer (Ed.). California Academy of Sciences, San Francisco.

———. 1998. Catalog of fishes. Vol. 3, Genera of fishes. *Calif Acad. Sci. Spec. Publ. 1.*

Fahay, M. P., and D. F. Markle. 1984. Gadiformes: Development and relationships, pp. 265–283. In Ontogeny and systematics of fishes,

H. G. Moser, W. J. Richards, D. M. Cohen, M. P. Fahay, A. W. Kendall, Jr., and S. L. Richardson (Eds.). *Am. Soc. Ichthyol. Herpetol. Spec. Publ. 1.*

Fletcher, D. E., E. E. Dakin, B. A. Porter, and J. C. Avise. 2004. Spawning behavior and genetic parentage in the pirate perch (*Aphredoderus sayanus*), a fish with an enigmatic reproductive morphology. *Copeia 2004:* 1–10.

Fraser, T. H. 1972. Some thoughts about the teleostean fish concept—the Paracanthopterygii. *Jpn. J. Ichthyol. 19:* 232–242.

Gordon, D. J., D. F. Markle, and J. E. Olney. 1984. Ophidiiformes: Development and relationships, pp. 308–319. In Ontogeny and systematics of fishes, H. G. Moser, W. J. Richards, D. M. Cohen, M. P. Fahay, A. W. Kendall, Jr., and S. L. Richardson (Eds.). *Am. Soc. Ichthyol. Herpetol. Spec. Publ. 1.*

Greenwood, P. H., D. E. Rosen, S. H. Weitzman, and G. S. Myers. 1966. Phyletic studies of teleostean fishes with a provisional classification of living forms. *Bull. Am. Mus. Nat. Hist. 131:* 339–456.

Grobecker, D. B., and T. W. Pietsch. 1979. High-speed cinematographic evidence for ultrafast feeding in antennariid anglerfishes. *Science 205:* 1161–1162.

Hart, J. L. 1973. Pacific fishes of Canada. *Fish. Res. Bd. Can. Bull. 180,* Ottawa.

Hoese, H. D., and R. H. Moore. 1977. *Fishes of the Gulf of Mexico.* Texas A&M University Press, College Station, TX.

Howes, G. J. 1989. Phylogenetic relationships of macrouroid and gadoid fishes based on cranial myology and arthrology, pp. 113–128. In Papers on the systematics of gadiform fishes, D. M. Cohen (Ed.). *Nat. Hist. Mus. Los Angeles Co. Sci. Ser. 32.*

Lauder, G. V., and K. F. Liem. 1983. The evolution and interrelationships of the actinopterygian fishes. *Bull. Mus. Comp. Zool. 150*(3): 95–197.

Markle, D. F. 1989. Aspects of character homology and phylogeny of the Gadiformes, pp. 59–88. In Papers on the systematics of gadiform fishes, D. M. Cohen (Ed.). *Nat. Hist. Mus. Los Angeles Co. Sci. Ser. 32.*

Markle, D. F., and J. E. Olney. 1990. Systematics of the pearlfishes (Pisces: Carapidae). *Bull. Mar. Sci. 47*(2): 269–410.

Marshall, N. B. 1979. *Developments in deep-sea biology.* Blandford Press, Poole, Dorset, UK.

———, and T. Iwamoto. 1973. Fishes of the western North Atlantic. Family Macrouridae. *Sears Found. Mar. Res. Memoir (Yale Univ.) 1*(6): 496–665.

Moore, J. A. 2002. Upside-down swimming behavior in a whipnose anglerfish (Teleostei: Ceratioidei: Gigantactinidae). *Copeia 2002:* 1144–1146.

Murray, A. M., and M. V. H. Wilson. 1996. A new Palaeocene genus and species of percopsiform (Teleostei: Paracanthopterygii) from the Paskapoo Formation, Smoky Tower, Alberta. *Can. J. Earth Sci. 33:* 429–438.

———, and———. 1999. Contributions of fossils to the phylogenetic relationships of the percopsiform fishes (Teleostei: Paracanthopterygii): Order restored, pp. 397–411. In *Mesozoic Fishes 2: Systematics and fossil record,* G. Arratia and H.-P. Schultze (Eds.). Verlag Dr. Friedrich Pfeil, München, Germany.

Nelson, J. S. 1994. *Fishes of the world* (3rd ed.). Wiley, New York.

Paredes-Ríos, G. A., and E. F. Balart. 2000. Corroboration of the bivalve, *Pinna rugosa,* as a host of the Pacific pearlfish *Encheliophis dubius* (Ophidiiformes: Carapidae), in the Gulf of California, México. *Copeia 2000:* 521–522.

Parenti, L. R. 1993. Relationships of atherinomorph fishes (Teleostei). *Bull. Mar. Sci. 52:* 170–196.

Patterson, C., and D. E. Rosen. 1989. The Paracanthopterygii revisited: Order and disorder, pp. 5–36. In Papers on the systematics of gadiform fishes, D. M. Cohen (Ed.). *Nat. Hist. Mus. Los Angeles Co. Sci. Ser. 32.*

Pietsch, T. W. 1981. The osteology and relationships of the anglerfish genus *Tetrabrachium* with comments on lophiiform classification. *Fish. Bull. 79:* 387–419.

Pietsch, T. W. 1984. Lophiiformes: Development and relationships, pp. 320–325. In Ontogeny and systematics of fishes, H. G. Moser, W. J Richards, D. M. Cohen, M. P. Fahay, A. W. Kendall, Jr., and S. L. Richardson (Eds.). *Am. Soc. Ichthyol. Herpetol. Spec. Publ. 1.*

———, and D. B. Grobecker. 1987. *Frogfishes of the world.* Stanford University Press, Stanford, CA.

———, and ———. 1990. Frogfishes: Masters of aggressive mimicry, these voracious carnivores can gulp prey faster than any other vertebrate predator. *Sci. Am. 262*(6): 96–103.

Robins. C. R., R. M. Bailey, C. E. Bond, J. R. Brooker, E. A. Lachner, R. N. Lea, and W. B. Scott. 1991. Common and scientific names of fishes from the United States and Canada (5th ed.). *Am. Fish. Soc. Spec. Publ. 20.*

Rosen, D. E. 1985. An essay on euteleostean classification. *Am. Mus. Novit. 2827.*

———, and C. Patterson. 1969. The structure and relationships of the paracanthopterygian fishes. *Bull. Am. Mus. Nat. Hist. 141*(3): 357–474.

Ross, S. W., and F. C. Rohde. 2003. Life history of the swampfish from a North Carolina stream. *Southeast. Nat. 2*(1): 105–120.

Siebert, D. J. 1990. Book review: Papers on the systematics of gadiform fishes by D. M. Cohen (Ed.), 1989. *Copeia 1990:* 889–893.

Smith, J. L. B. 1952. The fishes of the family Batrachoididae from South and East Africa. *Ann. Mag. Nat. Hist. Ser. 12*(5): 313–339.

Stiassny, M. L. J. 1986. The limits and relationships of the acanthomorph teleosts. *J. Zool. (Lond.) B1:* 411–460.

Trott, L. B. 1970. Contributions to the biology of carapid fishes (Paracanthopterygii: Gadiformes). *Univ. Calif. Publ. Zool. 89.*

14 ACANTHOPTERYGIAN FISHES: STEPHANOBERYCIFORMES, ZEIFORMES, AND BERYCIFORMES — POSSIBLY PERCOMORPHS, POSSIBLY NOT

INTRODUCING THE ACANTHOPTERYGIANS

PERCOMORPH FISHES

Order Stephanoberyciformes
Order Zeiformes (Zeomorphi)
Order Beryciformes (Berycomorphi)
    Suborder Trachichthyoidei
    Suborder Berycoidei
    Suborder Holocentroidei

In this brief chapter, we will first consider key features of the most diverse assemblage of teleosts—the spiny-finned Acanthopterygii. The Percomorpha constitute the most diverse assemblage of acanthopterygian fishes, yet there is some disagreement as to just what constitutes a percomorph. Three orders of basal acanthopterygian fishes originally classified as percomorphs but now considered "pre-percomorph" are considered here. These taxa are small in species composition, but large in phylogenetic significance, having led a somewhat peripatetic existence among the myriad of classification systems proposed for teleosts in the last few decades. They are exclusively marine and are especially well represented among the deep-ocean ichthyofauna. A couple of families, such as the flashlight fishes (Anomalopidae) and the squirrelfishes (Holocentridae) are common in shallow waters, however.

INTRODUCING THE ACANTHOPTERYGIANS

Acanthopterygian fishes constitute more than half of the named species of bony fishes. Although fish systematists generally recognize them as a coherent group, their bewildering diversity poses a major challenge for students of taxonomy and evolution. They are, for the most part, characterized by hard, sharp spines in the dorsal, anal, and pelvic fins. Acanthopterygians typically have two distinct dorsal fins, the first being composed of spines and the second of soft rays. Spines are also found in the pelvic and anal fins. Baudelot's ligament, which supports the pectoral skeleton via its attachment to the supracleithrum, originates on the basioccipital in acanthopterygians, rather than on the first vertebrae, as in most lower teleosts (Nelson, 1994). Johnson and Patterson (1993) highlighted three derived characteristics as diagnostic of acanthopterygians: the presence of pelvic spines, the reduction in size and number of free pelvic radials, and the presence of an anteromedial process of the pelvic bone. A remarkable diversification of fin structure accompanies a range of locomotor strategies. A physoclistous gas bladder is well developed in those taxa that hold position in the water column, but this may be reduced or absent in bottom-dwelling forms (see Chapter 19).

The acanthopterygian upper jaw is protrusible in most families; increased jaw mobility is facilitated by a well-developed ascending process of the premaxilla. The maxilla tends to be excluded from the gape. The retractor dorsalis muscle (= retractor arcuum branchialium) is generally inserted only on the third pharyngobranchial (Rosen, 1973). This is consistent with an advanced state of development of pharyngeal dentition—a feature that has contributed to a diversification of feeding strategies. With some exceptions, acanthopterygian orders have ctenoid scales.

PERCOMORPH FISHES

In their overview of percomorph phylogeny, Johnson and Patterson (1993) succinctly stated the problem inherent in addressing the evolution of this bewildering assemblage of teleostean fishes: The characteristics that unambiguously define the group are yet to be identified. The orders discussed in this chapter and the next differ in the degree to which they conform to the perception of what constitutes a percomorph. One thing is for certain—there sure are a lot of them! Nelson (1994) included about 230 families, more than 2,140 genera, and 12,000 species in the series Percomorpha—roughly half of all named fish species. Although they are a predominantly marine group, one of the key features that have contributed

to percomorph diversity is the success of the group in inhabiting a variety of freshwater environments. Some families—the Cichlidae, for example—demonstrate remarkable genetic plasticity that has culminated in the evolution of tremendous species diversity (see Chapter 32, Going Deeper).

In some classification systems (e.g., Nelson, 1994), percomorphs are distinguished from the series Atherinomorpha (see Chapter 15) in that they have a fourth pharyngobranchial, which the atherinomorphs lack, and they usually retain more than two infraorbital bones. The mechanism for upper jaw protrusion differs from that of the atherinomorphs (Lauder and Liem, 1983). Stiassny (1990) presented the percomorphs as monophyletic because the two halves of the pelvic girdle are fused medially.

There are many problems yet to be solved in the matter of the relationships among the percomorphs. Accordingly, there are different theories and opinions concerning the classification of these fishes. The mullets (see Chapter 15) have been removed from the Percomorpha into the series Mugilomorpha by Stiassny (1990), but she pointed out that the placement of this group is not certain, because the mugiloids have an essentially percomorph pelvic girdle. Johnson and Patterson (1993) presented a view of percomorph relationships that is somewhat different from that proposed by Nelson (1994). They excluded the Stephanoberyciformes, Beryciformes, and Zeiformes from the Percomorpha and suggested new designations for higher taxa. This is the classification as represented in Figure 14.1.

Order Stephanoberyciformes

The Stephanoberyciformes are marine fishes, with some living in the deep sea. Included in this order are the bigscale fishes or ridgeheads (**Melamphaidae**) that are widespread in the bathypelagic realm.

DIAGNOSIS OF THE STEPHANOBERYCIFORMES
- Bones of skull very thin; teeth absent in palate.
- Orbitosphenoid, subocular shelf, and supramaxillaries absent.
- Cycloid scales in some families; others with spines on scales, or lacking scales.
- Single dorsal fin, set well back on body in most species; dorsal spines, if present, are weak.
- Pelvic fins abdominal to jugular.
  Sources: Ebeling and Weed, 1973; Moore, 1993; Nelson, 1994.

The **Gibberichthyidae** or gibberfishes include a single genus (*Gibberichthys*) with only two species. The larvae and prejuveniles of *Gibberichthys* pass through what is termed a **kasidoron** stage, during which they have long, strangely branched pelvic fins (Fig. 14.2; Keene and Tighe, 1984).

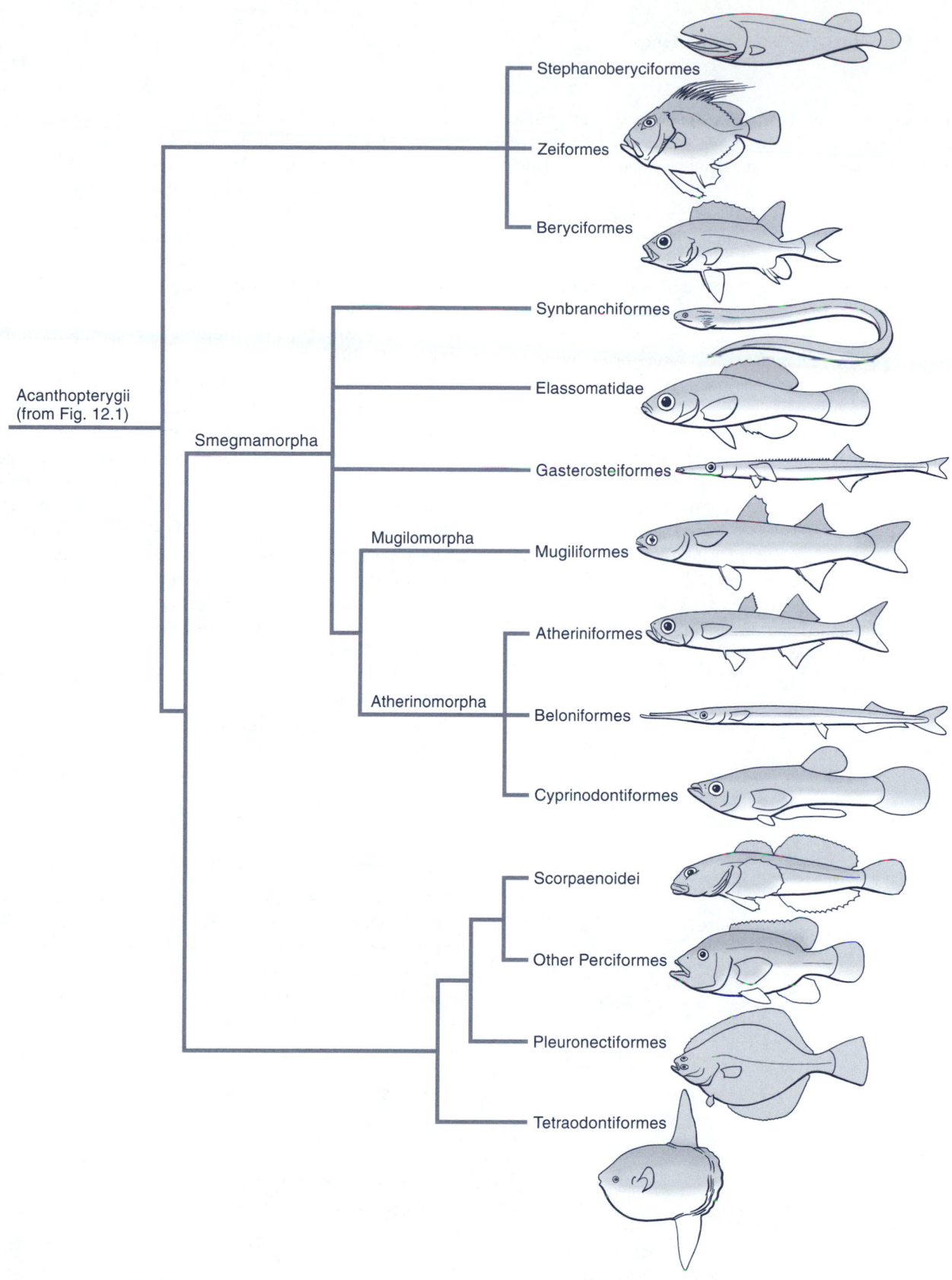

Stephanoberyciformes

Zeiformes

Beryciformes

Synbranchiformes

Elassomatidae

Gasterosteiformes

Mugiliformes

Atheriniformes

Beloniformes

Cyprinodontiformes

Scorpaenoidei

Other Perciformes

Pleuronectiformes

Tetraodontiformes

Mugilomorpha

Atherinomorpha

Smegmamorpha

Acanthopterygii
(from Fig. 12.1)

FIGURE 14.1
Phylogenetic relationships of acanthopterygian fishes (after Johnson and Patterson, 1993).

It has been suggested that these appendages may mimic pelagic siphonophores or *Sargassum* weed; whether they are luminescent or not has yet to be demonstrated (deSylva and Eschmeyer, 1977).

The **Stephanoberycidae** or pricklefishes resemble the melamphaids in having several short spines preceding the caudal fin above and below. Their common name comes from the scale coat that may be covered with small spines.

Three families of stephanoberyciform fishes are known as whalefishes, from their generally cetacean body form and cavernous mouths. They are open-water forms found in most oceans. None of them have fin spines. The two known species of redmouth whalefishes (family **Rondeletiidae**, genus *Rondeletia*) have the pores of the lateralis system of the body disposed in several vertical rows over a smooth skin. The single species of red whalefish, *Barbourisia rufa* (family **Barbourisiidae**), in contrast to the redmouth whalefishes, has spiny skin (see Plate 14.1). There are nine genera and about 35 species of flabby whalefishes (family **Cetomimidae**). These have no scales and no pelvic fins, and some species have rudimentary eyes. Colgan et al. (2000) have confirmed the three families of whalefishes as forming a monophyletic group.

• •

FIGURE 14.2
Kasidoron prejuvenile of gibberfish (Gibberichthyidae). (After D. P. DeSylva and W. N. Eschmeyer, 1977. Systematics and biology of the deep-sea fish family Gibberichthyidae: A senior synonym of the family Kasidoroidae. *Proc. Calif. Acad. Sci. Ser. 4, 41*: 215–231.)

• •

The families Mirapinnidae and Megalomycteridae include strange oceanic fishes that some ichthyologists have thought to be closely related to the Beryciformes. They were once placed in an order of their own (Bertelsen and Marshall, 1956), and some recent researchers consider them to belong among the Lampridiformes (Eschmeyer, 1990; Nelson, 1984; Rosen and Patterson, 1969). Nelson (1994), basing the placement on the findings of Moore (1993) and others, considered these to be stephanoberyciforms.

The family **Mirapinnidae** is now considered to have two subfamilies, the hairyfish (Mirapinninae) and the tapetails (Eutaeniophorinae; Nelson 1994). Only one species of hairyfish (*Mirapinna esau*) is known. It is scaleless and has skin covered with small, villous projections that give the appearance of fur. Its pectoral fins are placed high on the sides, well behind the head, and the large pelvic fins are set about where pectoral fins are usually seen. No mature specimens have been captured; all have been less than 50 mm long. The subfamily Eutaeniophorinae contains three species of elongate fishes that are called tapetails because of a long, ribbon-like extension of the caudal fin in juveniles. Both subfamilies are found in the deep waters of the Atlantic, Indian, and Pacific Oceans.

The family **Megalomycteridae,** which was formerly placed in its own suborder, has four genera and five species. These fishes are known as largenose fishes, because of their pronounced nasal openings leading to highly developed olfactory lamellae. They are represented in both the Atlantic and Pacific.

• • • • • • • • • • • • • •

Order Zeiformes (Zeomorphi)

This order includes the dories and related families (suborder Zeoidei); Nelson (1994) provisionally placed boarfishes (suborder Caproidei) also within this order. Zeiform fishes are widespread in tropical and temperate seas. They are found in deeper coastal waters near continental shelves and slopes, or around seamounts. Fossils are known from the Eocene.

DIAGNOSIS OF THE ZEIFORMES

• Body form generally deeply compressed.

• Orbitosphenoid, subocular shelf, and supramaxilla absent; mouth is protrusible, greatly so in some genera.

• Ctenoid scales.

• Thoracic pelvic fins, sometimes with a spine and 5 to 10 soft-rays.

• Spinous anal fin sometimes forms a nearly distinct fin anterior to the soft-rays.

Source: Heemstra, 1980.

The family **Zeidae** (the dories) are fishes of moderate depths and are often targeted by commercial fisheries. The John Dory (*Zeus faber*) has a round, dark spot on each side, and some fishermen of various European countries refer to it as "St. Peter's fish," perpetuating a legend that the marks are prints of the saint's thumb and forefinger (see Plate 14.2). The species may reach a maximum length of nearly 1 m. Other families in the order are the Grammicolepididae (diamond dories), Parazenidae, Macrurocyttidae, and Oreosomatidae.

The family **Caproidae** (the boarfishes), which are noted for their greatly protractile mouths, do not appear to be closely related to the other fishes placed in the order Zeiformes (Rosen, 1973). Some authors (Heemstra, 1980; Johnson and Patterson, 1993) have suggested that these fishes might better be placed in another group, possibly the percomorphs.

Rosen (1984) suggested that the Zeiformes were allied with the Tetraodontiformes, but that view has not been widely accepted. More recently, it has been proposed that the Zeiformes form a monophyletic group with the Gadiformes—a proposal derived largely from molecular studies and not supported by the widely disparate morphologies of the two orders (Miya et al., 2001; Wiley et al., 2000). In their revision of the Zeiformes, Tyler et al. (2003) viewed the gadiform–zeiform relationship as suspect. Although Nelson (1994) provisionally included the caproids as a suborder of the Zeiformes, Tyler et al. (2003) did not. Although Tyler et al. affirmed that the Zeiformes (excluding the Caproidei) form a monophyletic group, the interrelationship of the zeiform fishes, the caproids, and the tetraodontiform fishes remains a thorny issue yet to be resolved.

Order Beryciformes (Berycomorphi)

This order embraces the roughies, alfonsinos, soldierfishes, and their allies. These marine fishes are generally classified as basal acanthopterygians and as a primitive sister group of the Perciformes (Lauder and Liem, 1983) or of the Perciformes and Smegmamorpha (Johnson and Patterson, 1993, see Fig. 14.1). The beryciform fishes have displayed considerable diversity as far back as the Cretaceous. Nelson (1994) recognized three suborders.

DIAGNOSIS OF THE BERYCIFORMES

- Orbitosphenoid, subocular shelf, and supramaxillaries present.
- Ctenoid or cycloid scales.
- Spines in the dorsal, anal, and pelvic fins.
- Thoracic or subthoracic pelvic fins with one spine and 6 to 13 soft rays.
- Caudal fin with a few procurrent spines, both dorsally and ventrally.
- Most with physoclistous swim bladder, but physostomous or absent in some.

Sources: Lauder and Liem, 1983; Johnson and Patterson, 1993; Nelson, 1994.

Suborder Trachichthyoidei

Families in this suborder include the deep-dwelling **Diretmidae** (spinyfins) and **Anoplogastridae** (fangtooths). Also included in the suborder are the **Anomalopidae** (flashlightfishes or lanterneyes) that occur in tropical and subtropical waters mainly in the Indo–Pacific region. This is one of the few bioluminescent shallow-water groups (see Chapter 18). The photophores in these fishes are subocular and are filled with luminous bacteria. The light can be revealed or concealed by everting or retracting the photophore or by advancing or withdrawing a shutterlike membrane (see Plate 14.3).

The **Monocentridae** or pineconefishes of the Indo-Pacific region have large scales that are firmly united to one another and prominent dorsal fin spines that are alternately oriented to right and left. There are two photophores on the ventral surface of the mandible. There are two genera, *Monocentrus* and *Cleidopus*. The **Trachichthyidae** (roughies and slimeheads) now include the family Korsogasteridae. Some of the roughies contribute to an important midwater commercial fishery in the South Pacific (see Going Deeper box, Chapter 35). The orange roughy (*Hoplostethus atlanticus*) is the target of one of the most intensive deepwater fisheries, particularly off the coast of Australia and New Zealand. Firm of flesh and relatively energetic, *Hoplostethus* challenges our perception of the physiology and adaptations of species normally associated with the nutrient-poor ocean depths.

Suborder Berycoidei

Zehren (1979) and Nelson (1994) considered this suborder to include only the **Berycidae** (alfonsinos and redfishes), which appear to be the most primitive members of the order. These are usually bright red fishes of moderate depths that have a widespread distribution. The family is composed of two genera, *Beryx* and *Centroberyx*.

Suborder Holocentroidei

A well-known family is the **Holocentridae,** which contains the big-eyed, brilliantly colored soldierfishes and squirrelfishes of tropical coral reefs (see Plate 14.4). These are active at night and can be found hiding in various shelters during the day. Studies of ribosomal DNA sequences by Colgan et al. (2000) have provided insight into the phylogenetic relationships of the stephanoberyciform and beryciform fishes. The phylogeny that Colgan et al. proposed excludes the Holocentridae from the Beryciformes but places them in a clade of non-beryciforms that includes the Gempylidae, Zeidae, and Atheriniformes.

KEY POINTS AND CONNECTIONS

- More than half of the named species of fishes are members of the spiny-finned taxon Acanthopterygii. These fishes are usually distinguished by two distinct dorsal fins, with the first composed of stiff, sharp spines and the second composed of soft, flexible rays.

- Of the acanthopterygians, the Percomorpha are the most diverse taxon, especially in the marine realm. Many percomorphs, however, have been successful in invading a range of freshwater habitats.

For a discussion of the most successful example of percomorph radiation into freshwater habitats, the cichlid fishes of the African Great Lakes, refer to the Going Deeper box in Chapter 32.

- Three orders of basal acanthopterygians—the Stephanoberyciformes, the Zeiformes, and the Beryciformes—are widespread in deep-ocean environments. Two beryciform families, the holocentrid squirrelfishes and the anomalopid flashlightfishes are shallow-water inhabitants but avoid the daylight hours. The flashlightfishes display remarkable bioluminescent capabilities—with light from a well-developed organ at the base of the eyes—that are useful for scanning the immediate environment during nocturnal forays.

For more details concerning the bioluminescence exhibited by flashlightfishes, refer to Chapter 18.

BUILDING AN ICHTHYOLOGY LIBRARY

An essential volume, still available from the University of Miami for $35.00. Don't delay! Buy today!

Johnson, G. D., and W. D. Anderson (Eds.). 1993. Proceedings of the symposium on the phylogeny of Percomorpha, June 15–17, 1990, in Charleston, SC, at the 70th annual meeting of the American Society of Ichthyologists and Herpetologists. *Bull. Mar. Sci. 52*(1).

REFERENCES

Bertelsen, E., and N. B. Marshall. 1956. The Mirapinnati, a new order of teleost fishes. *Dana Rep. 42:* 1–34.

Colgan, D. J., C.-G. Zhang, and J. R. Paxton. 2000. Phylogenetic investigations of the Stephanoberyciformes and Beryciformes, particularly whalefishes (Euteleostei: Cetomimidae), based on partial 12S rDNA and 16S rDNA sequences. *Mol. Phylogen. Evol. 17*(1):15–25.

DeSylva, D. P., and W. N. Eschmeyer. 1977. Systematics and biology of the deep-sea fish family Gibberichthyidae, a senior synonym of the family Kasidoroidae. *Proc. Calif. Acad. Sci. 41*(6): 215–231.

Ebeling, A. W., and W. H. Weed, III. 1973. Fishes of the western North Atlantic. Order Xenoberyces (Stephanoberyciformes). *Sears Found. Mar. Res. Memoir (Yale Univ.) 1*(6): 397–478.

Eschmeyer, W. N. 1990. Genera in a classification, pp. 435–495. In *Catalog of the genera of recent fishes*, W. N. Eschmeyer (Ed.). California Academy of Sciences, San Francisco.

Heemstra, P. C. 1980. A revision of the zeid fishes (Zeiformes: Zeidae) of South Africa. *J. L. B. Smith Inst. Ichthyol. Bull. 41.*

Johnson, G. D., and C. Patterson. 1993. Percomorph phylogeny: A survey of acanthomorphs and a new proposal. *Bull. Mar. Sci. 52*(1): 554–626.

Keene, M. J., and K. A. Tighe. 1984. Beryciformes: Development and relationships, pp. 383–292. In Ontogeny and systematics of fishes, H. G. Moser, W. J. Richards, D. M. Cohen, M. P. Fahay, A. W. Kendall, Jr., and S. L. Richardson (Eds.). *Am. Soc. Ichthyol. Herpetol. Spec. Publ. 1.*

Lauder, G. V., and K. F. Liem. 1983. The evolution and interrelationships of the actinopterygian fishes. *Bull. Mus. Comp. Zool. 150*(3): 95–197.

Miya, M., A. Kawaguchi, and M. Nishida. 2001. Mitogenomic exploration of higher teleostean phylogenies: A case study for moderate-scale evolutionary genomics with 38 newly determined complete mitochondrial DNA sequences. *Mol. Biol. Evol. 18:* 1993–2009.

Moore, J. A. 1993. Phylogeny of the Trachichthyiformes (Teleostei: Percomorpha). *Bull. Mar. Sci. 52*(1): 14–136.

Nelson, J. S. 1984. *Fishes of the world* (2nd ed.). Wiley, New York.

———. 1994. *Fishes of the world* (3rd ed.). Wiley, New York.

Rosen, D. E. 1973. Interrelationships of higher euteleostean fishes, pp. 397–513. In Interrelationships of fishes, P. H. Greenwood, R. S. Miles, and C. Patterson (Eds.). *J. Linn. Soc. (Zool.) 53 Suppl. 1.* Academic Press, New York.

———. 1984. Zeiformes as primitive plectognath fishes. *Am. Mus. Novit. 2782.*

Rosen, D. E., and C. Patterson. 1969. The structure and relationships of the paracanthopterygian fishes. *Bull. Am. Mus. Nat. Hist. 141*(3): 357–474.

Stiassny, M. L. J. 1990. Notes on the anatomy and relationships of the bedotiid fishes of Madagascar, with a taxonomic revision of the genus *Rheocles* (Atherinomorpha: Bedotiidae). *Am. Mus. Novit. 2979.*

Tyler, J. C., B. O'Toole, and R. Winterbottom. 2003. Phylogeny of the genera and families of zeiform fishes, with comments on their relationships with tetraodontiforms and caproids. *Smithson. Cont. Zool. 618.*

Wiley, E. O., G. D. Johnson, and W. W. Dimmick. 2000. The interrelationships of acanthomorph fishes: A total evidence approach using molecular and morphological data. *Biochem. Syst. Ecol. 28:* 319–350.

Zehren, S. J. 1979. The comparative osteology and phylogeny of the Beryciformes (Pisces: Teleostei). *Evol. Monogr. (Univ. Chicago) 1.*

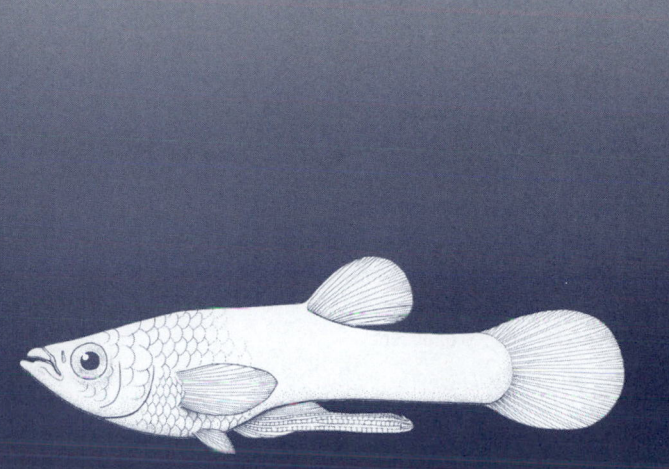

THE SMEGMAMORPHA: HOUSECLEANING AMONG THE PERCOMORPHS, OR JUST A SPURIOUS NOTION?

Order Synbranchiformes
   Suborder Synbranchoidei
   Suborder Mastacembeloidei
   Suborder Elassomatoidei
Order Gasterosteiformes
   Suborder Gasterosteoidei
   Suborder Syngnathoidei

SERIES MUGILOMORPHA

Order Mugiliformes

SERIES ATHERINOMORPHA

Order Atheriniformes
   Superfamily Atherinopsoidea
   Superfamily Atherinoidea
Order Beloniformes (Synentognathi or Exocoetoidei)
   Suborder Adrianichthyoidei
   Suborder Exocoetoidei (Belonoidei)
Order Cyprinodontiformes (Microcyprini)
   Suborder Aplocheiloidei
   Suborder Cyprinodontoidei

Several orders of acanthopterygians have been proposed to constitute a distinctive clade, the Smegmamorpha. This taxon incorporates several orders of teleosts that were previously considered to have little in common. Smegmamorphs are fishes of freshwater, estuarine, and coastal marine habitats. The Synbranchiformes are eellike, tropical freshwater fishes that can tolerate conditions of low dissolved oxygen because of their air-breathing abilities. The pygmy sunfishes of the family Elassomidae had long been classified as members of the North American sunfish family Centrarchidae. Rather than being diminutive centrarchids, however, they have features that suggest that they should be grouped with other smegmamorphs. The Gasterosteiformes are peculiar fishes characterized by a body encased in bony armor or with rows of bony scutes. Sticklebacks (family Gasterosteidae) have been the subject of classic behavioral studies and are also well known among evolutionary biologists for the extraordinary degree of localized morphological variation exhibited among populations. The Mugilomorpha and Atherinomorpha include families that are primarily adapted to inshore coastal marine and estuarine waters, with several species having become adapted to freshwater. In these environments, they can usually be found at or near the water's surface.

THE SMEGMAMORPHA: HOUSECLEANING AMONG THE PERCOMORPHS, OR JUST A SPURIOUS NOTION?

As mentioned in Chapter 13, Parenti (1993) proposed the possibility of a sister group relationship between the paracanthopterygians and the atherinomorphs. Johnson and Patterson (1993) have proposed an alternative phylogeny in constructing, as part of the Percomorpha, the clade **Smegmamorpha** to include mugilomorphs, atherinomorphs, the pygmy sunfishes (*Elassoma*), sticklebacks, and swamp eels. Nelson (1984) originally classified the mugilioids as a suborder of the Perciformes, but in his most recent edition (Nelson, 1994), he placed the mugilomorphs as a series with greater affinity to the atherinomorphs. The Smegmamorpha are diagnosed primarily by the first epineural originating on the parahypophysis or lateral process of the centrum of the first vertebrae. Johnson and Patterson (1993) discussed a number of other shared derived characters that distinguish this clade.

It is obvious that much systematic work remains to be done before the proper phylogenetic interrelationships of the Smegmamorpha can be ascertained. Although Johnson and Patterson (1993) presented compelling arguments for the validity of the clade, especially as it pertains to the relationship between atherinomorphs and mugilomorphs, the Smegmamorpha may not represent a monophyletic group (Wiley et al., 2000). From the phylogeny constructed from an exhaustive study of teleostome gill arch musculature (Springer and Johnson, 2004), Springer and Orrell (2004) argued for the rejection of the Smegmamorpha, claiming it does not constitute a monophyletic clade. Perhaps it is the creative etymology that most appeals, as the name *Smegmamorpha* is derived from the first letters of the six taxa recognized (including both suborders of the Synbranchiformes—the **S**ynbranchioidei and the **M**astacembeloidei—**E**lassomatidae; **G**asterosteiformes; **M**ugilomorpha; and **A**therinomorpha); the resulting acronym means "cleansing agent" in Greek and Latin. Springer and Orrell's (2004) reservations notwithstanding, the phylogeny presented in Figure 14.1 reflects this classification.

Order Synbranchiformes

This order contains the swamp eels (Synbranchoidei) and the spiny-backed eels (Mastacembeloidei). These are elongate fishes of subtropical and tropical freshwater, although some species are known to enter waters of very low salinity. Travers's (1984) examination of the phylogenetic relationships of the group resulted in the placement of the mastacembeloid spiny-backed eels (formerly included in the Perciformes) into this order.

DIAGNOSIS OF THE SYNBRANCHIFORMES

- Scales, if present, cycloid.
- Pectoral girdle well behind head and not connected to cranium; posttemporal bone reduced or absent.
- Pelvic girdle and fins absent.
- Caudal fin reduced or absent.

Suborder Synbranchoidei

- Scales are lacking except in the genus *Monopterus*.
- Gills greatly reduced; gill openings are confluent ventrally.
- Pectoral fins absent.
- Dorsal and anal fins are reduced to a ridge; rays absent.
- Swim bladder and ribs absent.

Suborder Mastacembeloidei

- Scales reduced or absent.
- Dorsal and anal fin spines lacking in Chaudhuriidae or forming isolated series in Mastacembelidae.
- Physoclistous swim bladder present.

Source: Travers, 1984.

Suborder Synbranchoidei

Species of synbranchoid eels show a range of gill structure, with some species (possibly the most primitive) possessing four pairs of fully developed gill arches, whereas others show varying degrees of reduction in gill tissue. The latter species primarily extract oxygen from the atmosphere, using their highly distensible pharynx or intestine. Many are capable of making terrestrial sojourns. The order contains one family (**Synbranchidae**) with 15 species known from the tropics. Some live in caves in Africa and Yucatan; most are found in freshwater or brackish-water swamps.

Synbranchus is the most widespread genus, found in Asia, Africa, South America, and Mexico. The rice eel (*Monopterus albus;* Fig. 15.1A) of southeast Asia and Indonesia can spend the dry season in holes in the bottom when the swamps dry up. The larvae of *M. albus* can live in waters of low oxygen concentration because they have a thin skin with many capillaries in which the main flow of blood is opposite the current of the water that the pectoral fins bring from the relatively oxygen-rich surface layer (Liem, 1981). Lauder and Liem (1983) suggested a close relationship between the Synbranchiformes and the snakeheads of the suborder Channoidei (see Chapter 16).

Suborder Mastacembeloidei

This suborder contains 75 species, commonly known as the spiny eels. These are freshwater or brackish-water fishes of Africa and Asia. The best known representatives are in the family **Mastacembelidae** (Fig. 15.1B). In this family, the anterior nostrils form tubes at the tip of the elongate snout

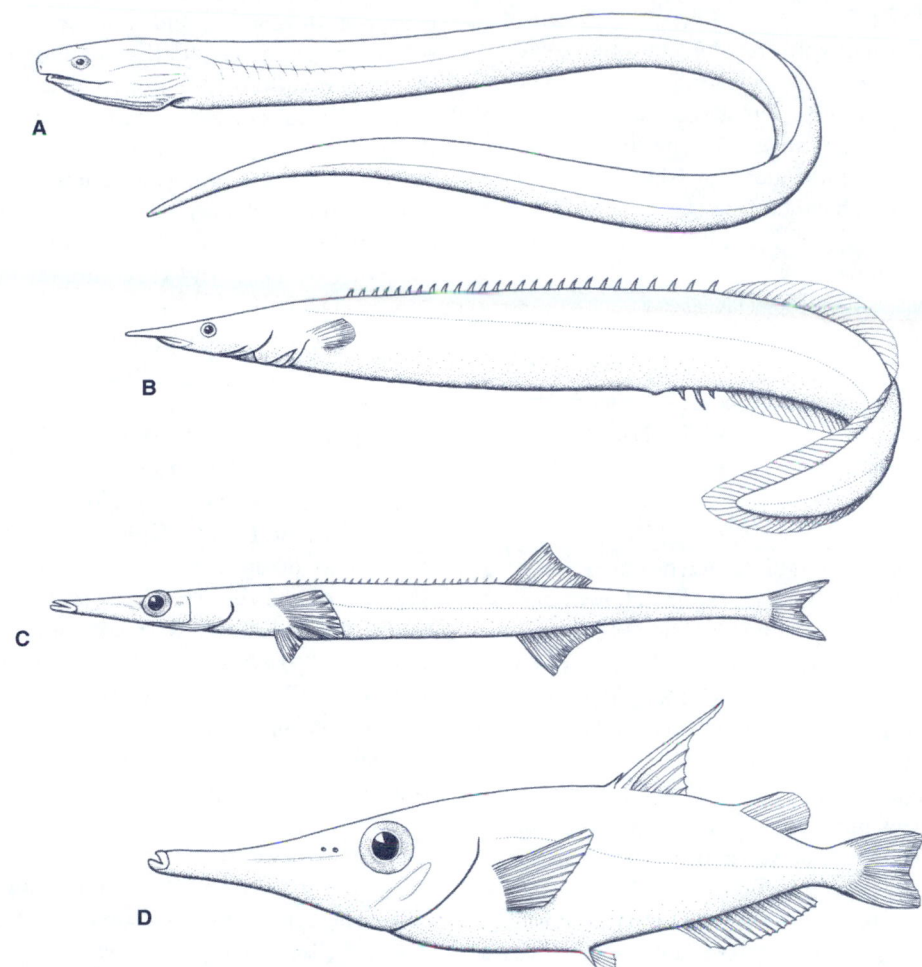

FIGURE 15.1
Representatives of the order Synbranchiformes: **A,** Synbranchidae; **B,** Mastacembelidae. Representatives
of the order Gasterosteiformes: **C,** Tubesnout (*Aulorhynchus*); **D,** snipefish (Macrorhamphosidae).

and open on each side of a tentacle. Many mastacembelids
live in swamps and depend, at least in part, on atmospheric
oxygen.

Members of the family **Chaudhuriidae** lack the isolated spines of Mastacembelidae and have the rostral appendage reduced. They are sometimes placed into a separate
order or suborder. They are small freshwater fishes found in
China and Southeast Asia.

Suborder Elassomatoidei
The pygmy sunfishes (*Elassoma;* see Plate 15.1) form the
family **Elassomatidae**. These fishes have formerly been
placed with the Centrarchidae, although this group differs
from the centrarchids in many characters (Branson and

Moore, 1962; Johnson, 1984; Wiley et al., 2000). Nelson
(1994) retained the group in the Perciformes, but Johnson
(1984) believed that *Elassoma* should not be placed in the
suborder Percoidei. Johnson and Patterson (1993) placed
the family in the Smegmamorpha (see Fig. 14.1).

These diminutive fishes are less than 5 cm long. They
have no lateral line of sensory pores and five or fewer dorsal spines. The scales are cycloid. All species occur only in
slow-flowing, heavily vegetated freshwaters of southeastern United States. Of the six known species, three have
broad geographic distributions, whereas the three others
are much more restricted, to the point that they have been
recommended for threatened or endangered status (Quattro
et al., 2001).

Order Gasterosteiformes

This order encompasses such a diversity of forms that some systems of classification have apportioned them into two or three separate orders. One feature that they all appear to have in common is an elongate snout. Among the fishes in this order are sticklebacks, tubesnouts, pipefishes, seahorses, seamoths, trumpetfishes, and cornetfishes.

DIAGNOSIS OF THE GASTEROSTEIFORMES
- Orbitosphenoid, basisphenoid, and supramaxillaries are absent; many with elongated snouts.
- Most with dermal armor.
- Pelvic fins abdominal to subthoracic; pelvic bones not connected to the pectoral girdle; some with well-developed pelvic spines.
- Many with stout, well-developed dorsal spines.
  Sources: Pietsch, 1978; Fritzsche, 1984.

Most species are marine and prefer warm waters, but a few species, especially the sticklebacks, reside in freshwater. Fossils are known from the Eocene. Fritzsche (1984) summarized the systematic challenges of the group and reviewed the classification schemes of Greenwood et al. (1966) and Banister (1970). The classification within the order as presented here follows that of Pietsch (1978) and Nelson (1994), but we recognize Johnson and Patterson's (1993) classification of the order as a member of the Smegmamorpha (see Fig. 14.1). It should be noted, however, that mitochondrial genomic analysis of a broad array of teleostean lineages has revealed a surprisingly close affinity of the gasterosteiform fishes with members of the order Scorpaeniformes (Miya et al., 2001).

Suborder Gasterosteoidei

This group includes the sticklebacks and their close relatives. The family **Aulorhynchidae** contains the tubesnouts of the North Pacific. *Aulorhynchus flavidus* has about 18 to 20 separate spines along the back (Fig. 15.1C). The **Hypopty-chidae** (sand eels) of the western Pacific resemble the sticklebacks but lack free spines (Ida, 1976).

The family **Gasterosteidae** contains the sticklebacks of northern temperate waters—among the most intensely studied of all fishes for their remarkable morphological variation, ecological adaptations, behavior, and physiology. The genus *Gasterosteus*, the threespine sticklebacks (see Fig. 3.15B), includes both freshwater and anadromous forms, the latter having a much greater covering of bony plates along the sides (Gibson, 2005). Sticklebacks exhibit a tremendous degree of localized variation in body and fin morphology throughout the Northern Hemisphere (Bell, 1984, 1994; Walker, 1997). Peichel et al. (2001) have

demonstrated that variation in such features as spine length, bony plate formation, and number of gill rakers are controlled by genetic factors that map to independent regions of the chromosomes. Through detailed analysis of gene sequences, Colosimo et al. (2005) have demonstrated that the parallel evolution of numerous reduced-plate morphs is attributable to repeated selection for alleles of a specific signaling protein known to influence skin-associated development. At least for freshwater populations of *G. aculeatus* along the Northeastern Pacific rim, the numerous instances of parallel evolution of the reduced-plate morphs appear to be derived from a single genetic basis.

The fascinating reproductive behavior of members of the genus *Gasterosteus*, involving intensive bouts of nest building and courtship by the males, has been the subject of much study (see Chapter 37). The gasterosteoids secrete a gluelike substance from their kidneys that is used by males to construct nests out of plant materials. The ninespine sticklebacks of the genus *Pungitius* range from temperate freshwaters into the Arctic, where they are found in both marine and freshwater. Members of this genus and the related genus *Spinachia*, the fifteenspine sticklebacks, build nests well off the bottom in vegetation. Other genera are *Apeltes*, the fourspine sticklebacks, and *Culaea*, the brook sticklebacks. Gasterosteids are seldom longer than 10 cm, although *Spinachia spinachia* is reported to reach 18 cm.

Suborder Syngnathoidei

According to Pietsch (1978), the seamoths of the family Pegasidae belong in this suborder. Some investigators, however, place them in their own order Pegasiformes. The **Pegasidae** (see Fig. 3.7A) are among the most peculiar of fishes. Their body is covered by sculptured bone, with the tail enclosed in bony rings. The small mouth opens below an odd rostrum composed of modified nasal bones. Pectoral fins are broad, fanlike, and oriented horizontally. These are small shore fishes of the Indo-Pacific.

Also in this suborder are the pipefishes and seahorses of the family **Syngnathidae**, peculiar little fishes with no pelvic fins and a body enclosed in bony rings. The pipefishes (*Syngnathus*) are elongate fishes, but the seahorses (*Hippocampus*) have the head flexed ventrally, so that it is at a right angle to the body (see Fig. 3.7D). The tail is flexible and prehensile. Both pipefish and seahorse males are equipped with brood pouches, in which the eggs are incubated. Although most seahorse species are believed to practice monogamy during the breeding season, molecular genetic studies of the pipefish (*Syngnathus typhle*) have shown that several females may contribute to the brood gestated by a given male (Jones et al., 1999). These fishes are often cryptically colored, with a shape obscured by numerous leaflike lobes. Perhaps the best examples are the weedy seadragon (*Phyllopteryx foliatus*) and the leafy seadragon (*Phycodurus eques*) of Australian coastal waters (see Plate 15.3).

The family **Solenostomidae,** the ghost pipefishes, are less well known than the syngnathids. They have a long, tubelike snout, large pectoral and caudal fins, and a high, flexible, spinous dorsal fin. The female incubates the eggs between the pelvic fins. The single genus (*Solenostomus*) is found in the Indo-Pacific and consists of three species (Orr and Fritzsche, 1993). Seahorses and their relatives have long been esteemed in Asian cultures, where they are used in traditional medicines. Dried seahorses are popular souvenirs, and thousands of live specimens make their way into the home aquarium trade. The trade in these species is considered a threat to several populations worldwide (Lourie et al., 1999).

The family **Indostomidae** of the freshwaters of Southeast Asia is sometimes placed into its own order. Its exact placement may be open to question, but it has features in common with the Gasterosteiformes and probably belongs with this order (Greenwood, 1966; Pietsch, 1978). These curious little fishes have free spines before the dorsal fin, and the slender body is enclosed in bony rings, so it combines some of the characteristics of both sticklebacks and pipefishes. The Indostomidae were long believed to be a monotypic family, but two new species have recently been described (Britz and Kottelat, 1999).

The trumpetfishes of the family **Aulostomidae** are circumtropical. They have a series of isolated spines along the back, and a very long snout (see Plate 37.2). They reach a length of about 60 cm. The family **Fistulariidae,** or cornetfishes, are also circumtropical and may approach 2 m in length. They are extremely slender and have a long filament formed from the middle rays of the caudal fin (see Plate 15.4). The sides of the long snout are equipped with sharp ridges. Large specimens have been observed to spend up to half an hour stalking a school of small goatfishes. During this time, the fish moved so slowly as to seem almost motionless.

The family **Centriscidae,** found in the Indo-Pacific region, includes the shrimpfishes (called razorfishes in some areas). They have a thin body with a sharp ventral edge and an armor of plates. Their locomotion is usually by means of undulating fins, and they generally maintain a vertical position, often with the head down (see Plate 15.5). They are found in the Indo-Pacific region, often associated with coral reefs.

The snipefishes (**Macrorhamphosidae**) are also Indo-Pacific coral reef associates and, like the shrimpfishes, may swim mainly with their head down (Fig. 15.1D). The body is quite deep, the snout long, and they possess a large dorsal fin spine.

SERIES MUGILOMORPHA

The mugilomorphs are another example of a fish taxon that has had some difficulty finding a home among the percomorphs. DeSylva (1984) claimed that the fishes that we consider here to be members of the series Mugilomorpha were related to the sphyraenoids (barracudas) and the polynemoids (threadfins). Stiassny (1990) and Stiassny and Moore (1992), however, proposed that the Mugilomorpha should be a series (a taxon between superorder and order) with a sister group relationship to the Atherinomorpha. These authors, however, noted that mugilomorphs share with holocentrids and higher percomorphs a union of the pelvic girdle elements that is lacking in the atherinomorphs and other basal percomorph clades. Stiassny (1993) mentioned several other features of the pelvic girdle that give evidence of the mugilomorphs as a group distinct from the atherinomorphs. In their challenge to the validity of the clade Smegmamorpha, Springer and Orrell (2004) cited accumulating evidence for a monophyletic Mugilomorpha–Atherinomorpha sister group assemblage. They resurrected the name **Percesoces,** first used by Edward Drinker Cope (1871; see Chapter 1) for this sister group.

Order Mugiliformes

This order consists of a single family, the mullets (**Mugilidae**). These shallow-water marine and brackish-water fishes are especially adapted for feeding on and processing small particles and benthic ooze. Modifications of the gut, including a muscular, gizzardlike structure in the anterior part of the stomach, esophageal gastric glands, and an elongate intestine, facilitate their detritivorous habits (see Chapter 23).

DIAGNOSIS OF THE MUGILIFORMES
- Poorly developed teeth in jaws; well-developed gill rakers and pharyngeal apparatus for filter feeding.
- Scales cycloid or intermediate between cycloid and ctenoid.
- Lateral line on the trunk may be poorly developed or absent.
- Pectoral fins placed high on body; pelvic fins subabdominal.
- Ligamentous or bony connection of pelvic bones to cleithrum of pectoral girdle absent.
- Well separated spinous and soft dorsal fins.
Sources: DeSylva, 1984; Nelson, 1994.

The mullets are widespread in tropical to warm temperate climates. There are nearly 100 species in about a dozen genera, including *Agonostomus, Liza, Rhinomugil,* and *Mugil.* One species, the striped mullet (*Mugil cephalus;*

see Plate 15.6), is nearly circumtropical. Mullets are an esteemed food fish; not only are they harvested from the wild, but they are cultured in ponds as well. In most places where mullet culture is practiced (e.g., Hong Kong and Taiwan), the young fish or "seed" are gathered from natural spawning areas or are simply allowed to flow with the tide into ponds, which are then closed off. Taiwanese mullet culturists have learned how to produce mullet juveniles through artificial propagation.

A final note on the popular culture front: In the film *Cool Hand Luke*, a term of derision directed at an individual with long, shaggy hair was "mullethead." Henceforth, a mullet has come to mean not only the highly esteemed fish, but also a style of haircut that makes the statement, "business up front, party in the back."

SERIES ATHERINOMORPHA

Atherinomorphs share many features—including the placement of pectoral, dorsal, and pelvic fins; terminal to dorsal position of the mouth; and similarity in the disposition of olfactory lamellae—that suggest convergent adaptations to life near the water's surface. The series includes such well-known surface dwellers as the silversides and rainbowfishes (Atheriniformes or "atherinoids"), the flyingfishes and their relatives (Beloniformes), and the killifishes and topminnows (Cyprinodontiformes). Atherinomorphs have an upper jaw structure that is distinct from that of other acanthopterygians—the absence of the ball-and-socket articulation between the palatine and maxilla prevents the premaxillaries from being locked in the protruded position (Nelson, 1994). Cycloid scales predominate. Relatively large demersal eggs with many oil droplets and long, adhesive chorionic filaments that can entangle in vegetation or other structures characterize the group (Rosen and Parenti, 1981). Approximately 20 families constitute the series Atherinomorpha, with about 170 genera and 1,100 species included.

Order Atheriniformes

The silversides, priapiumfishes, and their relatives have yet to find a secure home among the numerous cladograms drawn up by systematic ichthyologists. Silversides have some features in common with the mullets, and formerly they were placed with them in a suborder of the Perciformes (Regan, 1929) or in the separate order Mugiliformes (Berg, 1940; Lagler et al., 1977). Priapiumfishes have been placed variously with the Cyprinodontiformes, Mugiliformes, or in their own order Phallostethiformes.

DIAGNOSIS OF THE ATHERINIFORMES
- Two dorsal fins, the first weak but composed of true spines.
- Pelvic fins are small, subthoracic to anterior in position.
- Both ctenoid and cycloid scales.
- Lateral line poorly developed or absent.

Sources: Nelson, 1994; Saeed et al., 1994.

Saeed et al. (1994) have proposed the division of the Atheriniformes into two superfamilies, Atherinopsoidea and Atherinoidea. Dyer (1997) and Dyer and Chernoff (1996), however, proposed an alternative classification in which the family Atherinopsidae is separated from the rest of the Atheriniformes. The classification by Saeed et al. (1994) is followed here.

Superfamily Atherinopsoidea
The group commonly referred to as the silversides, once thought to be the monophyletic family Atherinidae, appears to consist of distinctly different New World and Old World forms (Saeed et al., 1994). New World silversides, the **Atherinopsidae** are small fishes, mostly marine, with some freshwater representatives. Most have a prominent silver band along each side and lack a lateral line. They are found along most tropical and warm temperate coasts of the New World and often invade freshwater in the absence of an established freshwater fauna. The genus *Chirostoma* contains the charal and pejerreys of the plateaus of Mexico and parts of South America, where they are important food fishes. *Labidesthes sicculus* is a freshwater species of North America. One of the best known of the atherinopsids is the California grunion (*Leuresthes tenuis;* see Going Deeper), fossils of which date from the Middle Eocene. *Atherinopsis californiensis* is one of the largest silversides, reaching a length of more than 30 cm. Typical of many atherinomorph fishes, the eggs have filaments that attach to vegetation or to other objects.

Two other families, the **Notocheiridae** and the **Isonidae,** are also included among the Atherinopsoidea (Saeed and Ivantsoff, 1991; Saeed et al., 1994). The notocheirids are commonly referred to as *surf silversides* because of their abundance in the turbulent zone where waves break; they are nearly transparent, comparatively deep bodied, with a pronounced keel along the belly.

Superfamily Atherinoidea
The family **Atherinidae** includes several genera of silversides, such as *Atherina* (Fig. 15.2A) and *Atherinomorus,* that are primarily known from coastal and inland waters of Europe, Australia, New Guinea, and elsewhere in the Indo-Pacific. The family **Bedotiidae** consists of the genera

Bedotia and *Rheocles* (Stiassny, 1990). These small (about 8 cm), colorful fishes are native to the inland waters of Madagascar. They were considered plesiomorphic atherinomorphs by Stiassny (1990) and Parenti (1993), but not by Saeed et al. (1994) or Dyer and Chernoff (1996). Whatever their final placement, bedotiids are well characterized by a remarkable thickening of the centra, dorsal, and haemal spines in the posteriormost vertebrae.

The **Melanotaeniidae** (rainbowfishes) are a family of freshwater fishes native to Australia, with a few species in New Guinea. A colorful group, they are among the most popular of aquarium fishes. The closely related family **Pseudomugilidae** are commonly called blue-eyes (see Plate 15.8). Some of these can enter salt water and have a fairly broad coastal distribution in Australia. The **Telmatherinidae** contains the Celebes rainbowfish (*Telmatherina ladigesi*)

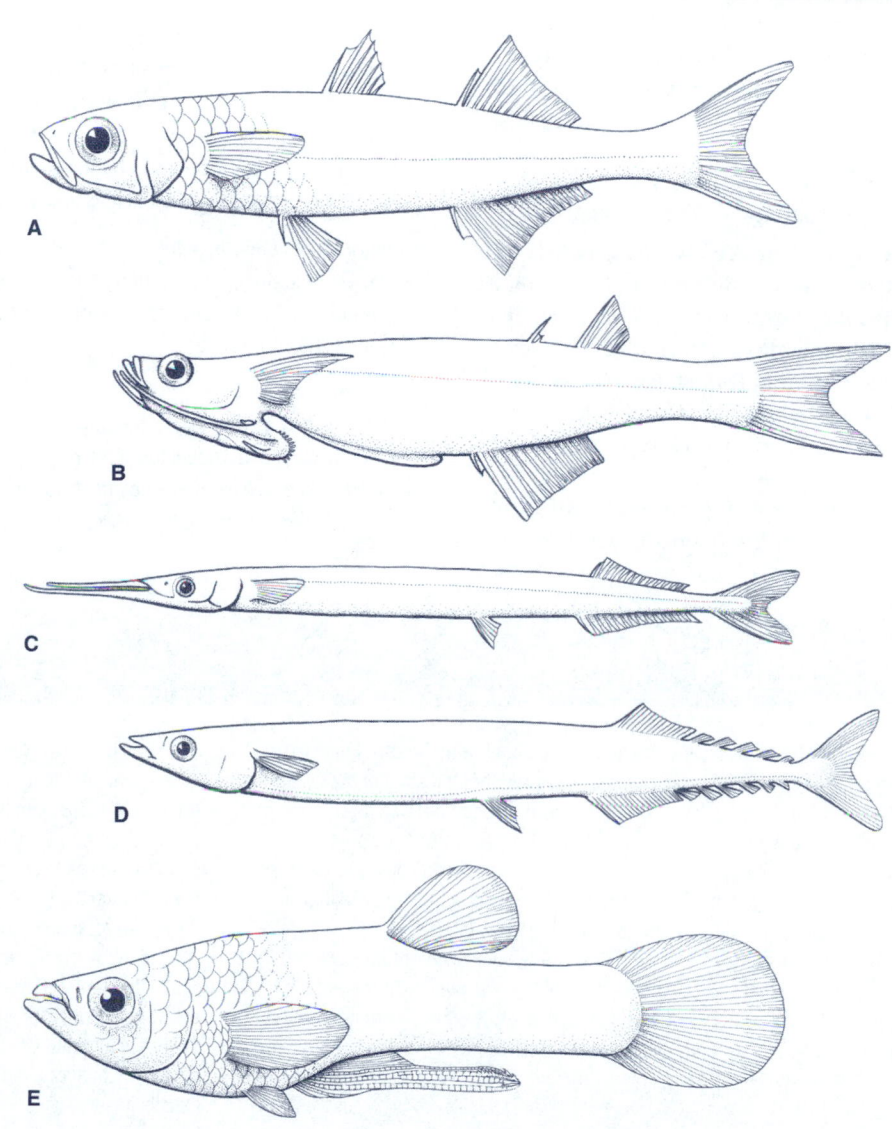

FIGURE 15.2
Representatives of the order Atheriniformes: **A**, Silverside (Atherinidae, *Atherina*); **B**, Priapiumfish (Phallostethidae, *Neostethus*). Representatives of the order Beloniformes: **C**, Needlefish (*Belone*); **D**, Saury (*Scomberesox*). Representative of the order Cyprinodontiformes: **E**, Guppy (*Poecilia*).

of Sulawesi, a beautifully colored fish with elongate dorsal and anal fins, which is prized as an aquarium fish.

The **Phallostethidae** (including the Neostethidae) of Thailand, the Philippines, and the Indo–Malayan archipelago are called *priapiumfishes* because of the strange intromittent organ of the male. (In Greco–Roman mythology, Priapus, son of Aphrodite and Dionysus, was the god of the male genitalia.) This device is situated in the region of the isthmus and is supported by a skeletal structure made up of elements from the pelvic and pectoral girdles, plus the first pair of ribs. The anus and genital pore are jugular in position (Fig. 15.2B). Parenti (1987) and Dyer and Chernoff (1996) considered the Dentatherinidae from the western Pacific to be closely related to the Phallostethidae, but Ivantsoff et al. (1987) argued that they are more closely related to Old World atherinids.

• •

Order Beloniformes (Synentognathi or Exocoetoidei)

Beloniform fishes are active surface dwellers, mostly of warm seas. Some species enter freshwater, and some may also be found in temperate marine waters. This order includes the ricefishes (Adrianichthyidae), flyingfishes (Exocoetidae), halfbeaks (Hemiramphidae), sauries (Scomberesocidae), and needlefishes (Belonidae). Fossil beloniforms are known from the Eocene. There are nearly 40 genera and 180 living species in the order.

Using a combination of morphological and biochemical (mitochondrial and nuclear DNA sequence) data, Lovejoy (2000) has provided a novel phylogeny of the beloniform fishes, in which the Exocoetidae are nested within the Hemiramphidae and the Scomberesocidae are nested within the Belonidae.

DIAGNOSIS OF THE BELONIFORMES

- Orbitosphenoid and mesocoracoid absent.
- Nasal organs variously reduced.
- Lower pharyngeal bones fused.
- Lateral line, if present, situated very low on the body.
- Fins with soft rays.
- Pectoral fins set high on the sides.
- Pelvic fins located abdominally, with six rays.
- Single dorsal fin set far back over anal fin (sauries have series of dorsal and anal finlets).
- Caudal fin usually has only 13 branched rays, with the lower lobe longer and with more principal rays than the upper.

Suborder Adrianichthyoidei

- Vomer, metapterygoid, ectopterygoid, or supracleithrum absent.
- Expanded articular surface of the fourth epibranchial.
- Lateral line canal absent.

Suborder Exocoetoidei

- Platelike process on the posterior, ventral part of the basioccipital.
- Median lower pharyngeal tooth plate.
- Elongation of jaw elements occurs at some point in early life history.

Source: Rosen and Parenti, 1981; Nelson, 1994.

GOING DEEPER • Those Weird and Wacky Grunions

At some time or another, we all fall victim to malicious duplicity from our peers, especially in our younger years, when we are much less critical in our judgments. How many of you have been on a snipe hunt? You know the drill: Provided with a gunny sack, you are sent out in the woods in the middle of the night in pursuit of this mythical beast. It only takes one trip to realize that you have been had. When I first heard about the grunion during my childhood in Southern California, it sounded like a locally adapted version of the snipe hunt. A most implausible scenario: You show up at the beach, bucket in hand, in the dead of night, preferably after midnight during a full moon, and you scoop up hundreds of silvery fish that have rushed up the beach with the surf. Much to my surprise, I learned that the story of the California grunion (*Leuresthes tenuis*) was no snipe hunt. As an undergraduate at UCLA, I learned much more about this fascinating fish from the master of "grunionology"—my ichthyology teacher, Boyd Walker (1952, 1959).

Silversides (families Atherinopsidae and Atherinidae), living as they do at the water's surface, have evolved spawning rhythms cued by the lunar cycles that govern the height of the tides. Such lunar periodicity results in a closely synchronized spawning event that ensures maximum fertilization and hence survival of larvae. From March through August, depending on the latitude, the spawning intensity of the Atlantic silverside (*Menidia menidia*) peaks in the several days following a new or full moon (Middaugh, 1981). In the shallow estuaries that are home to *Menidia*, this lunar periodicity enables the fish to congregate and spawn in the high intertidal zone, where they secure the eggs to vegetation by long threads. Timing the spawning run to coincide with the predicted time when high tide is at its peak probably enables the fish to deposit its eggs when tidal currents are at a minimum. Calmer waters undoubtedly contribute to successful fertilization. This lunar sensitivity has also been demonstrated in the laboratory with populations of another species of *Menidia* (*M. beryllina*) that may occur in freshwaters far from marine tidal influence (Sherrill

Suborder Adrianichthyoidei

This suborder was defined by Rosen and Parenti (1981) and contains only the family **Adrianichthyidae.** These are native to Asia and the East Indies. The ricefishes and medakas of the genus *Oryzias* are found in fresh and brackish waters of Southeast Asia, the East Indies to Timor and Celebes, Luzon and north to Japan. Medakas have, for many years, been a valuable species for genetic study.

Fertilized eggs are held for some time in a cluster near the genital pore of the female before they are eventually brushed off on a suitable substrate. The harvesting of eggs from the genital pore is easily accomplished in experimental situations. *Adrianichthys* and *Xenopoecilus* of the freshwaters of Celebes are large-jawed fishes with internal fertilization and eggs that hatch on extrusion. In *Horaichthys* of India, the gonopodium of the males is deflected to the right and the genital orifice of the females to the left.

Suborder Exocoetoidei (Belonoidei)

The eggs of many species bear long tendrils or filaments. Collette et al. (1984) mentioned that the eggs of the fishes in this group have no or only minute oil droplets and that the presence of a beak supports Rosen and Parenti's (1981) division of the order into the Adrianichthyoidei (no beak) and Exocoetoidei (with beak).

The family **Belonidae** contains the needlefishes, which are slender, elongate fishes with the jaws protruded into a long, beaklike mouth set with sharp, fine teeth (Fig. 15.2C).

The largest members of the family may reach more than 1.5 m in length. Needlefishes are noted for their inclination to jump from the water at the slightest provocation, sometimes with gruesome results. A young snorkeler in Florida was nearly killed when a houndfish (*Tylosurus crocodilus*) leapt from the water and impaled her in the neck. In Hawaii, a surfer was killed when a *Platybelone* struck him in the eye and penetrated his brain. Schools may scatter with great leaps at the swift approach of a boat. Needlefishes are attracted to light and have been known to make leaps at lanterns set out during nightly fishing expeditions. Multiple independent adaptations to freshwater apparently have occurred among the needlefishes (Lovejoy and Collette, 2001). *Strongylura kreffti*, a native of New Guinea and the only freshwater member of the genus, makes an interesting aquarium fish, providing it is the sole inhabitant of the tank and a supply of feeder fish is readily available.

The family **Scomberesocidae** contains the sauries (Fig. 15.2D), which usually have a short beak and a series of mackerel-like finlets behind the dorsal and anal fins. They are also known to leap from the water. *Cololabis saira* of the North Pacific, a plankton feeder, occurs in large schools and is the object of a commercial fishery.

The **Hemiramphidae,** or halfbeaks, are surface-dwelling planktivores also noted for their jumping ability. The lower lobe of the caudal fin is larger than the upper lobe. In this family, the lower jaw is greatly extended, but the upper jaw is fairly short (see Plate 15.9). Lovejoy's (2000)

and Middaugh, 1993). A Florida species of *Menidia* (*M. peninsulae*) spawns only below the mean low-water tide level. Its eggs are therefore subject to greater predation (Middaugh and Hemmer, 1987).

The California grunion (*Leuresthes tenuis*) is the best known of the intertidally spawning atherinomorphs. Like *Menidia,* its spawning activities are closely cued to lunar and tidal cycles. Grunions inhabit the surf zone of sandy beaches, however, so they must contend with wave surge in addition to the tidal regime. What results is one of the more fascinating examples of sex on the beach. Grunion runs reach their peak usually two to six days after a new or full moon and occur late at night during spring and summer months.

Right at the peak of high tide, males and females rush up the beach with an incoming wave surge and are stranded by the thousands. The female wriggles into the wet sand and deposits her eggs, while the male wraps himself around her and fertilizes the eggs (see Plate 15.7). The next wave surge will carry the parents back out to sea. The eggs remain buried in the moist sand until the next extreme high tide comes along about two weeks later. If, for some reason, the larvae miss that appointment, their emergence can be delayed an additional two weeks. At this time, contact with the agitated water stimulates hatching. Spawning and hatching are examples of *semilunar* periodicity, occurring on an approximately 14-day cycle. Gonadal development

in *Leuresthes* is on an 18-day cycle, however. Similar spawning behavior has been observed in the related Gulf of California grunion (*L. sardina*). Unlike the California grunion, this species makes spawning runs both at night and during the day. The false grunion, *Atherinops* (*Colpichthys*) *regis,* also spawns at high tide during the daytime along the Pacific coastline of Mexico. Of the many examples that have evolved among fishes for placing eggs in environments where the threat of predation is minimized, the grunion is one of the most spectacular. And quite tasty as well! Many a night has been spent around a campfire on a Southern California beach passing around a skillet of fried grunion and a jug of wine while waiting for the sunrise.

study refuted the view that the hemiramphids, by only developing the lower jaw, are somehow paedomorphic derivatives of the needlefishes, which show elongation of both the upper and lower jaws during development. Lovejoy also proposed that the freshwater halfbeaks of the Indo–West Pacific are actually more closely related to the needlefishes and sauries, whereas the rest of the family retains its affinity with the flyingfishes.

The flyingfishes (**Exocoetidae**) are commonly observed gliding over the water's surface in many warm seas. The "two-wing" flyingfishes are genera in which the gliding surfaces are composed of the pectoral fins. Genera with short pectoral fins, such as *Fodiator* and *Parexocoetus,* have limited glide capabilities, whereas long finned genera, such as *Exocoetus,* can sustain glides up to 25 m. The "four-wing" flyingfishes, such as the genus *Cypselurus,* employ both pectoral and pelvic fins for lift (see Plate 15.10). The four-wing flyingfishes can sustain their flight by rapid sculling of the elongated lower lobe of the caudal fin on the water surface. In this manner, flights of up to 200 m are possible (Paxton and Eschmeyer, 1998). Speeds of up to 90 km/hr are not unusual during flights (Rayner, 1981). The spotted flyingfish (*Cheilopogon pinnatibarbatus*) is the largest known species, reaching lengths of 50 cm and weights of more than 1 kg. The California flyingfish (*Cypselurus californicus*) reaches a length of about 45 cm, whereas the smallest species, *Parexocoetus mento,* reaches only 14 cm.

Order Cyprinodontiformes (Microcyprini)

This order includes the killifishes, topminnows, and their relatives. Cyprinodontiforms are widely distributed, occurring naturally on all continents except Australia and Antarctica. They are especially abundant in Central America, where about one third of the known species occur. They are small, usually found in estuarine or freshwater environments, and include about 850 species in about 110 genera (Costa, 1998). Two suborders are recognized: the Aplocheiloidei and the Cyprinodontoidei.

DIAGNOSIS OF THE CYPRINODONTIFORMES
- Orbitosphenoid, mesocoracoids, and basisphenoids absent.
- Maxillaries are excluded from the border of the mouth by the premaxillaries.
- Lateral line canal system is incomplete or not well formed.
- Abdominal pelvic fins with soft rays.
- Single dorsal fin set behind the middle of the body.
- Caudal fin with varying number of rays, never forked.

Suborder Aplocheiloidei
- Metapterygoid present.
- Three basibranchials.

- Pelvic fin bases close together.
- First two dorsal radials each supporting a fin ray.

Suborder Cyprinodontoidei
- Metapterygoid absent.
- Two basibranchials.
- Pelvic fin bases not close together.
- One dorsal ray articulating with the first two radials.
- Upper and lower hypurals usually fused to the last vertebral centrum.

Sources: Nelson, 1994; Parenti, 1981; Costa, 1998.

Suborder Aplocheiloidei

These are small, freshwater fishes ranging from Florida to Argentina in the Americas, but inhabiting tropical Africa and the southern part of Asia as well. Parenti (1981) arranged them in two families, the Aplocheilidae and the Rivulidae, but Nelson (1984, 1994), Robins et al. (1991), and Eschmeyer (1990, 1998) listed the latter as a subfamily of the Aplocheilidae. Costa (1998) recognized two families and provided evidence for monophyly of the suborder.

Aplocheiloid fishes are remarkable for the ability of the embryos to enter a stage of developmental arrest known as **diapause** (Hrbek and Larson, 1999; see Chapter 27). This is an especially useful adaptation among the so-called "annual" fishes that deposit eggs in the mud of temporary ponds and die when the ponds dry up. During the ensuing rainy season, the eggs hatch and repopulate the ponds. A complex history of the origins of annualism, in which the feature arose prior to the isolation of Neotropical from Old World populations and was subsequently lost in species that have become adapted to permanent waters, has been proposed (Murphy and Collier, 1997; Murphy et al., 1999).

The family **Aplocheilidae** contains numerous species that are popular in the aquarium trade. The aplocheilids are fishes of Africa and southern Asia and include panchaxes and lyretails of the genus *Aphiosemion;* panchaxes of the genera *Epiplatys* and *Pachypanchax;* and the nothos of *Nothobranchius.*

The **Rivulidae** are distributed from Florida to Argentina and include such aquarium fishes as the genera *Rivulus, Pterolebias,* and *Cynolebias.* The mangrove rivulus (*Rivulus marmoratus;* see Plate 15.11) was previously thought to be rare throughout its range, until Taylor (1988, 1990) discovered that its primary microhabitat was the abandoned burrows of the land crab *Cardisoma guanhumi.* This species is remarkable for being hermaphroditic and capable of self-fertilization (Harrington, 1961). Functional males are sometimes encountered, however.

Several new species of rivulids have been discovered in the past decade, prompting a systematic revision of the group

(Costa, 2004; Hrbek et al., 2004). Costa (2004) proposed that *R. marmoratus* should be reclassified as a member of a new subfamily, the Cryptolebiatinae, and hence should be named *Cryptolebias marmoratus*.

Suborder Cyprinodontoidei

These fishes are distributed in freshwaters and estuarine habitats throughout much of the temperate and tropical parts of the world, but they are not native to Australia and New Guinea. Many members of this suborder are brightly colored and enter the aquarium fish trade. Phylogenies of the cyprinodontiformes advanced by Parenti (1981) and Costa (1998) both endorsed the distinction of cyprinodontoids from aplocheiloids, but they differed in the definition of the phylogenetic interrelationships of the cyprinodontoid families. For example, Parenti held the Profundulidae to be a sister group of the rest of the cyprinodontoid families, whereas Costa proposed the Profundulidae and Goodeidae as taxa jointly forming a sister group to the Fundulidae.

Members of the families **Profundulidae** and **Fundulidae** are commonly referred to as *killifishes*. The profundulids contain a single genus (*Profundulus*) and are known from Central America. Killifishes of the genus *Fundulus*—a significant group, as they are frequently employed in bioassays and other experimental procedures—are placed in the family Fundulidae by Parenti (1981), but are retained in the Cyprinodontidae by Robins et al. (1991). Other genera include *Adinia*, *Leptolucania*, and *Lucania*. The rainwater killifish (*Lucania parva*) of the Atlantic coast and the Gulf of Mexico has, for some unknown reason, been introduced to saline lakes in Utah and to parts of the Pacific coast. Fundulids have a disjunct distribution in North America; *Fundulus* species inhabiting the coastal drainages of western North America appear to be a sister clade to all other species of fundulids (Wiley, 1986; Bernardi, 1997).

The four-eyed fishes of the family **Anablepidae** actually have only two eyes, but each pupil is divided into an upper section and a lower one. The eyes protrude from the top of the head, so that the upper half can be above the surface of the water as the fish swims in the usual position just below the surface, allowing light from above the water to reach the retina. Simultaneously, the lower section of the pupil allows light from below the surface to enter the eye (see Plate 15.11; see Fig. 20.6A; see Chapter 20). Another peculiarity of the family is the asymmetrical orientation of the intromittent organ of the male and the genital opening of the female. About 60% of the males have the intromittent organ oriented to the right; the remainder are sinistral. Fortunately, the reverse is true for the orientation of the genital orifice of the females. Four-eyed fishes are found from southern Mexico to northern South America.

Jenynsia, a genus of southern Brazil and northern Argentina, is placed with the Anablepidae by Parenti (1981) and Eschmeyer (1998) but retained as a separate family *Jenynsiidae* by Robins et al. (1991). As in the anablepids, the gonopodium and the female genital aperture are asymmetrically oriented either to right or left. Ghedotti (1998, 2000) supported the monophyly of the families Anablepidae and Poeciliidae, uniting them in the superfamily Poecilioidea.

The family **Poeciliidae** contains the live-bearing tooth carps—guppies, mollies, and swordtails. Males are characterized by a gonopodium used for insemination of the females (see Chapter 27). Certain egg-laying species are included in the family by Parenti (1981) and Parenti and Rauchenberger (1989), who considered the live-bearing Poeciliidae as constituting a distinct subfamily. Livebearers are found primarily in the warmer parts of North and South America and are among the most popular of aquarium fishes. The guppy (*Poecilia reticulata*; Fig. 15.2E) is one of the easiest livebearers to maintain and breed. Some members of the family are known as *mosquitofish* because of their habit of feeding close to the water's surface on insect larvae. The North American *Gambusia affinis* and *G. holbrooki* have been introduced to many areas of the world to aid in insect control, but the insect control value of these introductions has been more than offset by their detrimental environmental impact. The *Gambusia* species compete with and are predators on various native species.

The genera *Poecilia* and *Poeciliopsis* are noted for the existence of unisexual (female) "species" produced by hybridization or by the activation of eggs by sperm that does not contribute genetic material to the developing eggs (Schultz, 1989; Wetherington et al., 1989). Reznick et al. (2002) provided evidence that the placenta—integral to the evolution of the viviparous state—may have evolved multiple times in *Poeciliopsis*.

Parenti (1981) and Nelson (1994) included the subfamilies Goodeinae and Empetrichthyinae in the family **Goodeidae**. Robins et al. (1991), however, retained the oviparous empetrichthyines—springfishes (*Crenichthys*) and poolfishes (*Empetrichthys*)—in the family Cyprinodontidae. The Goodeinae, commonly referred to as *splitfins* because of a separation of the anterior rays of the anal fin in males, are viviparous topminnows endemic to the plateau of central Mexico. Developing embryos absorb nutrients through branched filaments called **trophotaeniae**, which are attached at the anal region. The slightly separated anterior lobe of the anal fin seen in males may serve as a gonopodium. In the aquarium trade, the various members of the subfamily are called *allotoca*, *goodea*, *characodon*, and *skiffia*. The springfishes and poolfishes are endemic to isolated bodies of water in southern Nevada, where their status is threatened.

They are entirely lacking in pelvic bones and practice external fertilization.

The family **Cyprinodontidae** contains oviparous killifishes of several genera. They range from the coastal waters of northern South America north throughout the United States, but are also found in the West Indies, the Mediterranean region including the Nile, and other parts of Africa.

In the late 1960s, the Devil's Hole pupfish (*Cyprinodon diabolis*), which is known from just one small spring in Nevada, became emblematic of the emerging environmental movement. When the spring that constituted the entire habitat for the species was threatened by pumping of water from the aquifer, it became the center of a legal contest that ultimately served to bring the plight of several threatened species to the attention of the public. Ultimate victory was achieved with the passage of the **Endangered Species Act** in 1973. Pupfishes of the genus *Cyprinodon* have become especially diverse in a number of localities, including the deserts of the American Southwest (see Chapter 31), the Caribbean and the Bahamas, and Mexico (Barton, 1999; Bunt et al., 1999; Echelle and Echelle, 1992, 1998; Holtmeier, 2001; Humphries and Miller, 1981; Turner and Liu, 1977). From their analysis of the variation in mitochondrial DNA sequences, Echelle et al. (2005) have traced the origins of the genus *Cyprinodon* and its subsequent dispersal. In some of the localities where pupfishes are found, salinities may exceed that of full-strength seawater, making the members of this genus among the most euryhaline of all vertebrates (see Chapter 25).

KEY POINTS AND CONNECTIONS

- In introducing the clade Smegmamorpha, Johnson and Patterson (1993) proposed that a number of seemingly unrelated orders of acanthopterygian teleosts might share a common phylogenetic history. The smegmamorphs encompass a diverse array of small to medium-sized fishes that are primarily distributed in coastal marine and estuarine waters, with many families being well represented among the freshwater ichthyofauna of the world.

- Two rather distinct suborders of eellike freshwater fishes, the Synbranchoidei (swamp eels) and the Mastacembeloidei (spiny eels), constitute the order Synbranchiformes. Of the two suborders, the swamp eels are especially well adapted for existence in stagnant waters; their gills may be rudimentary, but their skin is well vascularized for cutaneous gas exchange, and some possess specialized organs for atmospheric gas exchange.

  For a discussion of the respiratory adaptations of the Synbranchiformes and other air-breathing fishes, see Chapter 24.

- Fishes of the order Gasterosteiformes are among the most bizarre in body morphology. An elongate snout is characteristic of members of the order; many have bodies encased in bony armor or have rows of bony scutes along the lateral surfaces. The trumpetfishes (Aulostomidae) and cornetfishes (Fistulariidae) are elongate predators on small fishes in coral reef environments. The seamoths (Pegasidae), seahorses (Syngnathidae), and shrimpfishes (Centriscidae) are examples of gasterosteiform fishes with a body encased in bony armor. The fascinating reproductive behavior of the sticklebacks of the family Gasterosteidae has made them perhaps the most watched freshwater fishes in existence.

  For discussions of the speciation phenomenon as it pertains to the evolution of feeding adaptations and body form in sticklebacks, refer to Chapters 4 and 23. The reproductive behavior of sticklebacks is considered in more detail in Chapter 37.

- The Mugilomorpha include a single family, the mullets (Mugilidae). Broadly distributed in warm marine and estuarine waters, these fishes show a number of dietary adaptations for the consumption of detritus and other particulate matter.

  Dietary adaptations of mullets and other species that consume particulate matter are discussed in Chapter 23.

- Fishes of the series Atherinomorpha show adaptations for a life spent at or near the water's surface, such as the position of the mouth and fins. Atherinomorph fishes have successfully invaded a number of freshwater habitats worldwide; some of the most popular fishes in the tropical fish trade are members of this taxon. A diversity of reproductive adaptations is seen among the atherinomorphs, including species that are livebearing. Their ability to successfully invade inland waters ranging from freshwater streams, lakes, and ponds to hypersaline lagoons is possible because of their strong osmoregulatory capabilities. Some species have become adapted to ephemeral waters that periodically dry up by producing eggs that can temporarily interrupt their development—a condition known as diapause. One of the most unusual visual adaptations seen in all fishes is exhibited by the "four-eyed" fishes of the cyprinodontiform family Anablepidae. With each eye subdivided into upper and lower portions, this fish is capable of viewing simultaneously the waters beneath it and the surroundings above the water's surface.

  A discussion of the unusual reproductive adaptations of atherinomorphs, including live-bearing forms, can be found in Chapters 27 and 28. The remarkable osmoregulatory adaptations of the cyprinodontiform fishes are considered in Chapter 25, and the structure and function of the eyes of surface-dwelling atherinomorphs, including the four-eyed fishes, is mentioned in Chapter 20.

FISH LINKS

http://www.aka.org/ Website of the American Killifish Association, dedicated enthusiasts of this most attractive and intriguing group of fishes.

BUILDING AN ICHTHYOLOGY LIBRARY

Bell, M. A. 1994. *The evolutionary biology of the threespine stickleback.* Oxford University Press, Oxford.

Dawson, C. E. 1985. *Indo-Pacific pipefishes (Red Sea to the Americas).* Gulf Coast Reseach Laboratory, Ocean Springs, MS.

Huber, J. H. 1992. *Review of* Rivulus, *ecobiogeography, relationships.* Société Française d'Ichthyologie, Paris.

Meffe, G. K., and F. F. Snelson, Jr. (Eds.), 1989. *Ecology and evolution of livebearing fishes (Poeciliidae).* Prentice Hall, Englewood Cliffs, NJ.

Wildekamp, R. H., and B. R. Watters. 1993–1996. *A world of killies: Atlas of the oviparous cyprinodontiform fishes of the world,* 3 vols. American Killifishes Association, Mishawaka, IN.

The texts mentioned below are valuable for their discussions of the adaptations of fishes, especially cyprinodontiform species, to the arid conditions of the American Southwest.

Blackwelder, E., C. L. Hubbs, R. R. Miller, and E. Antevs. 1948. The Great Basin: With emphasis on glacial and postglacial times. *Bull. Univ. Utah Biol Ser. 10*(7). University of Utah, Salt Lake City.

Hubbs, C. L., R. R. Miller, and L. C. Hubbs. 1974. Hydrographic history and relict fishes of the north-central Great Basin. *Mem. Calif. Acad. Sci. 7.* California Academy of Sciences, San Francisco.

Naiman, R. J., and D. L. Soltz. 1981. *Fishes in North American deserts.* Wiley, New York.

REFERENCES

Banister, K. E. 1970. The anatomy and taxonomy of *Indostomus paradoxus* Prashad and Mukerji. *Bull. Br. Mus. Nat. Hist. (Zool.) 19*(5): 179–209.

Barton, M. G. 1999. Studies on the biology of inland fishes of the Bahamas: Adaptations of cyprinodont fishes. *Proc. 8th Symp. Nat. Hist. Bahamas* (abstr.).

Bell, M. A. 1984. Evolutionary phenetics and genetics. The threespine stickleback, *Gasterosteus aculeatus* and related species, pp. 431–528. In *Evolutionary genetics of fishes,* B. J. Turner (Ed.). Plenum, New York.

———. 1994. *The evolutionary biology of the threespine stickleback.* Oxford University Press, Oxford.

Berg, L. S. 1940. Classification of fishes both recent and fossil. *Trav. Inst. Zool. Acad. Sci. USSR 5*: 87–517 (reprinted 1947, J. W. Edwards, Ann Arbor, MI).

Bernardi, G. 1997. Molecular phylogeny of the Fundulidae (Teleostei, Cyprinodontiformes) based on the cytochrome b gene, pp. 189–197. In *Molecular systematics of fishes,* T.D. Kocher and C.A. Stepien (Eds.). Academic Press, San Diego.

Branson, B. A., and G. A. Moore. 1962. The lateralis components of the acousticolateralis system in the sunfish family Centrarchidae. *Copeia 1962*: 1–108.

Britz, R., and M. Kottelat. 1999. Two new species of gasterosteiform fishes of the genus *Indostomus* (Teleostei: Indostomidae). *Ichthyol. Explor. Freshwat. 10*(4): 327–336.

Bunt, T. M., B. J. Turner, D. Duvernell, C. Holtmeier, and M. Barton. 1999. Molecular evidence for reproductive isolation between two sympatric trophic morphs of San Salvador pupfish (*Cyprinodon*). *Proc. 8th Symp. Nat. Hist. Bahamas* (abstr.).

Collette, B. B., G. E. McGowen, N. V. Parin, and S. Mito. 1984. Beloniformes: Development and relationships, pp. 335–354. In Ontogeny and systematics of fishes, H. G. Moser, W. J. Richards, D. M. Cohen, M. P. Fahay, A. W. Kendall, Jr., and S. L. Richardson (Eds.). *Am. Soc. Ichthyol. Herpetol. Spec. Publ. 1.*

Colosimo, P. F., K. E. Hosemann, S. Balabhadra, G. Villarreal, Jr., M. Dickson, J. Grimwood, J. Schmutz, R. M. Myers, D. Schluter, and D. M. Kingsley. 2005. Widespread parallel evolution in sticklebacks by repeated fixation of ectodysplasin alleles. *Science 307*: 1928–1933.

Cope, E. D. 1871. Contribution to the ichthyology of the Lesser Antilles. *Trans. Am. Phil. Soc. New Ser. 14*(3): 445–483.

Costa, W. J. E. M. 1998. Phylogeny and classification of the Cyprinodontiformes (Euteleostei: Atherinomorpha): A reappraisal, pp. 537–560. In *Phylogeny and classification of neotropical fishes,* L. R. Malabarba, R. E. Reis, R. P. Vari, Z. M. S. Lucena, and C. A. S. Lucena (Eds.). EDIPUCRS, Porto Alegre, Brazil.

———. 2004. Relationships and redescription of *Fundulus brasiliensis* (Cyprinodontiformes: Rivulidae) with description of a new species and notes on the classification of the Aplocheiloidei. *Ichthyol. Explor. Freshwat. 15*: 105–120.

DeSylva, D. P. 1984. Mugiloidei: Development and relationships, pp. 530–533. In Ontogeny and systematics of fishes, H. G. Moser, W. J. Richards, D. M. Cohen, M. P. Fahay, A. W. Kendall, Jr., and S. L. Richardson (Eds.). *Am. Soc. Ichthyol. Herpetol. Spec. Publ. 1.*

Dyer, B. S. 1997. Phylogenetic revision of Atherinopsinae (Teleostei, Atherinopsidae), with comments on the systematics of the South American freshwater fish genus *Basilichthys* Girard. *Misc. Publ. Mus. Zool. Univ. Mich. 185:* 1–64.

———, and B. Chernoff. 1996. Phylogenetic relationships among atheriniform fishes (Teleostei, Atherinomorpha). *Zool. J. Linn. Soc. 117:* 1–69.

Echelle, A. A., and A. F. Echelle. 1992. Mode and pattern of speciation in the evolution of inland pupfishes of the *Cyprinodon variegatus* complex (Teleostei: Cyprinodontidae): An ancestor-dependant hypothesis, pp. 691–709. In *Systematics, historical ecology, and North American freshwater fishes,* R. L. Mayden (Ed.). Stanford University Press, Stanford, CA.

———, and ———. 1998. Evolutionary relationships of pupfishes in the *Cyprinodon eximius* complex (Atherinomorpha: Cyprinodontiformes). *Copeia 1998:* 852–865.

———, E. W. Carson, A. F. Echelle, R. A. van den Bussche, T. E. Dowling, and A. Meyer. 2005. Historical biogeography of the New-World pupfish genus *Cyprinodon* (Teleostei: Cyprinodontidae). *Copeia 2005:* 320–339.

Eschmeyer, W. N. 1990. Genera in a classification, pp. 435–495. In *Catalog of the genera of recent fishes,* W. N. Eschmeyer (Ed.). California Academy of Sciences, San Francisco.

———. (Ed.). 1998. Catalog of fishes, Vol. 3, Genera of fishes. *Calif. Acad. Sci. Spec. Publ. 1.*

Fritzsche, R. A. 1984. Gasterosteiformes: Development and relationships, pp. 398–405. In Ontogeny and systematics of fishes, H. G. Moser, W. J. Richards, D. M. Cohen, M. P. Fahay, A. W. Kendall, Jr., and S. L. Richardson (Eds.). *Am. Soc. Ichthyol. Herpetol. Spec. Publ. 1.*

Ghedotti, M. J. 1998. Phylogeny and classification of the Anablepidae (Teleostei: Cyprinodontiformes), pp. 561–581. In *Phylogeny and classification of neotropical fishes,* L. R. Malabarba, R. E. Reis, R. P. Vari, Z. M. S. Lucena, and C. A. S. Lucena (Eds.). EDIPUCRS, Porto Alegre, Brazil.

————. 2000. Phylogenetic analysis and taxonomy of the poecilioid fishes (Teleostei: Cyprinodontiformes). *Zool. J. Linn. Soc. 130:* 1–53.

Gibson, G. 2005. The synthesis and evolution of a supermodel. *Science 307:* 1890–1891.

Greenwood, P. H., D. E. Rosen, S. H. Weitzman, and G. S. Myers. 1966. Phyletic studies of teleostean fishes with a provisional classification of living forms. *Bull. Am. Mus. Nat. Hist. 131:* 339–456.

Harrington, R. W., Jr. 1961. Oviparous hermaphroditic fish with internal self fertilization. *Science 134:* 1749–1750.

Holtmeier, C. L. 2001. Heterochrony, maternal effects, and phenotypic variation among sympatric pupfishes. *Evol. 55:* 330–338.

Hrbek, T., and A. Larson. 1999. The evolution of diapause in the killifish family Rivulidae (Atherinomorpha, Cyprinodontiformes): A molecular phylogenetic and biogeographic perspective. *Evol. 55:* 1200–1216.

————, C. Pereira de Deus, and I. P. Farias. 2004. *Rivulus duckensis* (Teleostei; Cyprinodontiformes): New species from the Tarumã Basin of Manaus, Amazonas, Brazil, and its relationship to other neotropical Rivulidae. *Copeia 2004:* 569–576.

Humphries, J. M., and R. R. Miller. 1981. A remarkable species flock of pupfishes, genus *Cyprinodon,* from Yucatán, México. *Copeia 1981:* 52–64.

Ida, H. 1976. Removal of the family Hypoptychidae from the suborder Ammodytoidei, order Perciformes, to the suborder Gasterosteoidei, order Sygnathiforms. *Jpn. J. Ichthyol. 23:* 33–42.

Ivantsoff, W., B. Said, and A. Williams. 1987. Systematic position of the family Dentatherinidae in relationship to Phallostethidae and Atherinidae. *Copeia 1987:* 649–658.

Johnson, G. D. 1984. Percoidei: Development and relationships, pp. 464–498. In Ontogeny and systematics of fishes, H. G. Moser, W. J. Richards, D. M. Cohen, M. P. Fahay, A. W. Kendall, Jr., and S. L. Richardson (Eds.). *Am. Soc. Ichthyol. Herpetol. Spec. Publ. 1.*

————, and C. Patterson. 1993. Percomorph phylogeny: A survey of acanthomorphs and a new proposal. *Bull. Mar. Sci. 52*(1): 554–626.

Jones, A. G., G. Rosenqvist, A. Bergland, and J. C. Avise. 1999. The genetic mating system of a sex-role-reversed pipefish (*Syngnathus typhle*): A molecular inquiry. *Behav. Ecol. Sociobiol. 46:* 357–365.

Lagler, K. F., J. W. Bardach, R. R. Miller, and D. R. M. Passino. 1977. *Ichthyology* (2nd ed.). Wiley, New York.

Lauder, G. V., and K. F. Liem. 1983. The evolution and interrelationships of the actinopterygian fishes. *Bull. Mus. Comp. Zool. 150*(3): 95–197.

Liem, K. F. 1981. Larvae of air-breathing fishes as countercurrent flow devices in hypoxic environments. *Science 211:* 1177–1179.

Lourie, S. A., J. C. Pritchard, S. P. Casey, S. K. Truong, H. J. Hall, and A. C. J. Vincent. 1999, The taxonomy of Vietnam's exploited seahorses (family Syngnathidae). *Biol. J. Linn. Soc. 66:* 231–256.

Lovejoy, N. R. 2000. Reinterpreting recapitulation: Systematics of needlefishes and their allies (Teleostei: Beloniformes). *Evolution 54:* 1349–1362.

————, and B. B. Collette. 2001. Phylogenetic relationships of New World needlefishes (Teleostei: Belonidae) and the biogeography of transitions between marine and freshwater habitats. *Copeia 2001:* 324–338.

Middaugh, D. P. 1981. Reproductive ecology and spawning periodicity of the Atlantic silverside, *Menidia menidia* (Pisces: Atherinidae). *Copeia 1981:* 766–776.

————, and M. J. Hemmer. 1987. Reproductive ecology of the tidewater silverside, *Menidia peninsulae* (Pisces: Atherinidae) from Santa Rosa Island, Florida. *Copeia 1987:* 727–732.

Miya, M., A. Kawaguchi, and M. Nishida. 2001. Mitogenomic exploration of higher teleostean phylogenies: A case study for moderate-scale evolutionary genomics with 38 newly determined complete mitochondrial DNA sequences. *Mol. Biol. Evol. 18:* 1993–2009.

Murphy, W. J., and G. E. Collier. 1997. A molecular phylogeny for aplocheiloid fishes (Atherinomorpha, Cypronodontiformes): The role of vicariance and the origins of annualism. *Mol. Biol. Evol. 14:* 790–799.

————, J. E. Thomerson, and G. E. Collier. 1999. Phylogeny of the neotropical killifish family Rivulidae (Cyprinodontiformes, Aplocheiloidei) inferred from mitochondrial DNA sequences. *Mol. Phylogen. Evol. 13:* 289–301.

Nelson, J. S. 1984. *Fishes of the world* (2nd ed.). Wiley, New York.

————. 1994. *Fishes of the world* (3rd ed.). Wiley, New York.

Orr, J. W., and R. A. Fritzsche. 1993. Revision of the ghost pipefishes, family Solenostomidae (Teleostei: Syngnathoidei). *Copeia 1993:* 168–172.

Parenti, L. R. 1981. A phylogenetic and biogeographic analysis of cyprinodontiform fishes (Teleostei, Atherinomorpha). *Bull. Am. Mus. Nat. Hist. 168:* 335–557.

————. 1987. Phylogenetic aspects of tooth and jaw structure of the medaka, *Oryzias latipes,* and other beloniform fishes. *J. Zool. (Lond.) 211:* 561–572.

————. 1993. Relationships of atherinomorph fishes (Teleostei). *Bull. Mar. Sci. 52:* 170–196.

————, and M. Rauchenberger. 1989. Systematic overview of the Poeciliinae, pp. 3–12. In *Ecology and evolution of livebearing fishes (Poeciliidae),* G. K. Meffe and F. F. Snelson, Jr. (Eds.). Prentice Hall, Englewood Cliffs, NJ.

Paxton, J. R., and W. N. Eschmeyer. 1998. *Encyclopedia of fishes* (2nd ed.). Academic Press, San Diego.

Peichel, C. L., K. S. Nereng, K. A. Ohgi, B. L. E. Cole, P. F. Colosimo, C. A. Buerkle, D. Schluter, and D. M. Kingsley. 2001. The genetic architecture of divergence between threespine stickleback species. *Nature 414:* 901–905.

Pietsch, T. W. 1978. Evolutionary relationships of the sea moths (Teleostei: Pegasidae) with a classification of gasterosteiform families. *Copeia 1978:* 517–529.

Quattro, J. M., W. J. Jones, J. M. Grady, and F. C. Rohde. 2001. Gene–gene concordance and the phylogenetic relationships among rare and widespread pygmy sunfishes (genus *Elassoma*). *Mol. Phylogen. Evol. 18*(2): 217–226.

Rayner, J. M. V. 1981. Flight adaptations in vertebrates, pp. 137–172. In Vertebrate locomotion, M. H. Day (Ed.). *Symp. Zool. Soc. Lond. 48.* Academic Press, New York.

Regan, C. T. 1929. Fishes, pp. 305–329. In *Encyclopaedia Britannica* (14th ed.), Vol. 9. Encyclopedia Britannica, London.

Reznick, D. N., M. Mateos, and M. S. Springer. 2002. Independent origins and rapid evolution of the placenta in the fish genus *Poeciliopsis. Science 298:* 1018–1020.

Robins. C. R., R. M. Bailey, C. E. Bond, J. R. Brooker, E. A. Lachner, R. N. Lea, and W. B. Scott. 1991. Common and scientific names of fishes from the United States and Canada (5th ed.). *Am. Fish. Soc. Spec. Publ. 20.*

————, and L. R. Parenti. 1981. Relationships of *Oryzias,* and the groups of atherinomorph fishes. *Am. Mus. Novit. 2719:* 1–25.

Saeed, B., and W. Ivantsoff. 1991. *Kalyptatherina,* the first telmatherinid genus known outside Sulawesi. *Ichthyol. Explor. Freshwat. 2*(3): 227–238.

———, ———, and L. E. L. M. Crowley. 1994. Systematic relationships of atheriniform families within Division I of the series Atherinomorpha (Acanthopterygii) with relevant historical perspectives. *Voprosi Ikhtiol. 34*(4): 1–32.

Schultz, R. J. 1989. Origins and relationships of unisexual poeciliids, pp. 69–87. In *Ecology and evolution of livebearing fishes*, G. K. Meefe and F. F. Snelson, Jr. (Eds.). Prentice Hall, Englewood Cliffs, NJ.

Sherrill, M. T., and D. P. Middaugh. 1993. Spawning periodicity of the inland silverside, *Menidia beryllina* (Pisces: Atherinidae) in the laboratory: Relationship to lunar cycles. *Copeia 1993:* 522–528.

Springer, V. G., and G. D. Johnson. 2004. Study of the dorsal gill-arch musculature of teleostome fishes with special reference to the Actinopterygii. *Bull. Biol. Soc. Wash. 11.*

———, and T. M. Orrell. 2004. Appendix: Phylogenetic analysis of 147 families of acanthomorph fishes based primarily on dorsal gill-arch muscles and skeleton. *Bull. Biol. Soc. Wash. 11.*

Stiassny, M. L. J. 1990. Notes on the anatomy and relationships of the bedotiid fishes of Madagascar, with a taxonomic revision of the genus *Rheocles* (Atherinomorpha: Bedotiidae). *Am. Mus. Novit. 2979.*

———. 1993. What are grey mullets? *Bull. Mar. Sci. 52:* 197–219.

———, and J. A. Moore. 1992. A review of the pelvic girdle of acanthomorph fishes, with comments on hypotheses of acanthomorph intrarelationships. *Zool. J. Linn. Soc. 104:* 209–242.

Taylor, D. S. 1988. Observations on the ecology of the killifish *Rivulus marmoratus* (Cyprinodontidae) in an infrequently flooded mangrove swamp. *Northeast Gulf Sci. 10*(1): 63–68.

———. 1990. Adaptive specializations of the cyprinodont fish *Rivulus marmoratus. Fla. Sci. 53*(3): 239–248.

Travers, R. A. 1984. A review of the Mastacembeloidei, a suborder of synbranchiform teleost fishes. Part II: Phylogenetic analysis. *Bull. Br. Mus. (Nat. Hist.) Zool. 47:* 83–150.

Turner, B. J., and R. K. Liu. 1977. Extensive interspecific genetic compatibility in the New World killifish genus *Cyprinodon. Copeia 1977:* 259–269.

Walker, B. W. 1952. A guide to the grunion. *Calif. Fish Game 38*(3): 409–420.

———. 1959. The timely grunion. *Nat. Hist. 68*(6): 302–307.

Walker, J. A. 1997. Ecological morphology of lacustrine threespine stickleback *Gasterosteus aculeatus* L. (Gasterosteidae) body shape. *Biol. J. Linn. Soc. 61:* 3–50.

Wetherington, J. D., R. A. Schenck, and R. C. Vrijenhoek. 1989. The origins and ecological success of unisexual *Poeciliopsis*: The frozen niche-variation model, pp. 259–275. In *Ecology and evolution of livebearing fishes (Poeciliidae)*, G. K. Meffe and F. F. Snelson, Jr. (Eds.). Prentice Hall, Englewood Cliffs, NJ.

Wiley, E. O. 1986. A study of evolutionary relationships of *Fundulus* topminnows (Teleostei: Fundulidae). *Am. Zool. 26:* 121–130.

———, G. D. Johnson, and W. W. Dimmick. 2000. The interrelationships of acanthomorph fishes: A total evidence approach using molecular and morphological data. *Biochem. Syst. Ecol. 28:* 319–350.

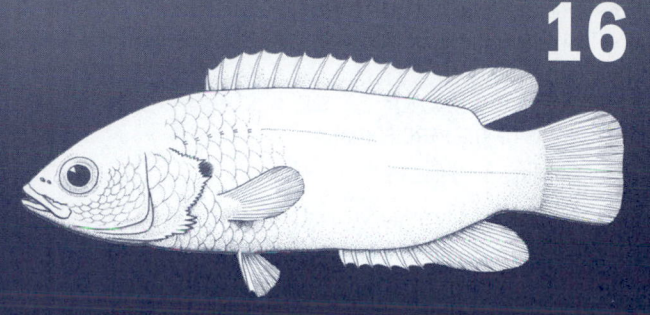

16

PERCIFORM FISHES: PERCHES AND A WHOLE LOT MORE

THE PERCIFORMES: AN EMBARRASSMENT OF RICHES

Suborder Percoidei
Suborder Labroidei
Suborder Trachinoidei
Suborder Pholidichthyoidei
Suborder Blennioidei
Suborder Zoarcoidei
Suborder Notothenioidei
Suborder Icosteoidei
Suborder Gobiesocoidei
Suborder Callionymoidei
Suborder Gobioidei
Suborder Kurtoidei
Suborder Acanthuroidei
Suborder Scombrolabracoidei
Suborder Scombroidei
Suborder Stromateoidei
Suborder Anabantoidei
Suborder Channoidei
    Anabantoid and Channoid Fishes Reconsidered

Although the order Perciformes gets its name from the perches, the undistinguished appearance of its namesake in no way represents the diversity of form and function seen in this, the largest order of vertebrates. Perciform fishes dominate shallow-water marine habitats, although they can be found in all types of environments. The diversity of the order poses a real challenge to students of fish systematics, and most agree that the order is probably polyphyletic. Nelson (1994) listed 18 suborders of perciform fishes, with three suborders—the Percoidei, Labroidei, and Gobioidei—accounting for more than three quarters of the described species. Many of the best-known food and game fishes are percoids. The labroids have evolved a distinctive and highly effective jaw apparatus that has conferred on them an extraordinary array of feeding adaptations. The diminutive size and cryptic behaviors of some perciform fishes, such as the gobioids and blennioids, have made them especially successful in exploiting the highly irregular topography of coral reef environments. At the opposite end of the percomorph size spectrum are the large and powerful tunas and billfishes of the suborder Scombroidei—fishes of the open waters of the epipelagic zone. If there is a niche to be filled in the aquatic realm, a perciform fish is there to fill it. Some have even spilled over into the terrestrial realm, skipping across coastal mudflats or crawling overland from one pond to another.

THE PERCIFORMES: AN EMBARRASSMENT OF RICHES

The fishes somewhat tentatively grouped under the taxon Perciformes should be approached with caution—even a little trepidation—by students of systematic ichthyology. They are a daunting group indeed. The Perciformes are by far the largest vertebrate order, with nearly 9,300 species in 148 families reported by Nelson (1994), with many more new species discovered and described in the past decade. Perciform fishes exhibit such diversity of form and function that they undoubtedly will be the last frontier of systematic ichthyology. Their diversity encompasses minute gobies and monstrous tunas and swordfishes; slender snake mackerels and quillfishes and deep-bodied butterflyfishes and spadefishes; fishes of the deep sea and fishes of the surface layers. Coral reef environments are home to some of the most diverse perciform families, including the gobies (Gobiidae), wrasses (Labridae), blennies (Blenniidae), and damselfishes (Pomacentridae). Although perciforms are the predominant group in marine waters, thousands of species inhabit fresh and brackish water.

In recent years, there has been a shift in systematic studies of the teleosts from the investigation of phylogenetic histories within families to more ambitious studies in which the intended goal is to attempt to make sense of the higher levels of fish classification (cf. Johnson and Patterson, 1993; Miya et al., 2001; Mooi and Gill, 1995; Wiley et al., 2000). Our understanding of the history of the Perciformes—an order of fishes that is far from being confirmed as monophyletic—should benefit greatly from this trend. Molecular studies, as discussed elsewhere in this book (see Chapter 4), have provided a new perspective in our understanding of the phylogenetic relationships of major taxa. Morphological investigations also continue to provide insight. One approach has been to focus on the evolutionary modification of a selected feature in fishes. Examples of this approach include McAllister's (1968) treatise on the variation and volution of the branchiostegals in teleosts, Mooi and Gill's (1995) study of epaxial musculature and pterygiophores in acanthomorph fishes, and investigations of the dorsal gill-arch musculature and skeleton of actinopterygians by Springer and Johnson (2004) and Springer and Orrell (2004).

The Perciformes include most of the spiny-rayed fishes. Fossil evidence of perciform fishes is sparse in late Cretaceous strata, but definitive fossils are known from the early Tertiary period (Patterson, 1993). The fossil record indicates that most modern suborders were present by the Eocene, suggesting a comparatively rapid rate of evolution (Carroll, 1988). Opinions differ as to the total number of suborders of perciform fishes; usually, about 20 are included. The subordinal designations used here largely reflect those outlined by Nelson (1994), with additional perspectives provided by Johnson (1993).

DIAGNOSIS OF THE PERCIFORMES

- Adults with acellular bone.
- Orbitosphenoid and mesocoracoid absent.
- Maxillary is excluded from the gape by the premaxillaries.
- Branchiostegal rays usually number six or seven; four of these are placed on the outer surface of the upper portion of the ceratohyal.
- Spines associated with fins.
- Ctenoid scales usually present.
- Vertebrae number 24 or fewer.
- Pectoral fins usually placed high on the side, with almost vertical fin bases.
- Pelvic fins thoracic or jugular, with the pelvic girdle usually connected to the cleithra; pelvic fins typically have one spine and five (or fewer) soft rays.
- Dorsal fin commonly subdivided into an anterior spiny fin and a posterior fin composed of soft rays; adipose fin never seen.
- The caudal fin has 17 or fewer principal rays.
- Swim bladder, if present, physoclistous. (Fishes of the family Bathyclupeidae are a notable exception.)
  Sources: Lauder and Liem, 1983a; Johnson and Patterson, 1993; Nelson, 1994.

In spite of the numerous features shared by all members of the Perciformes, none appear unique to the order, causing Lauder and Liem (1983a) to claim the order to be "clearly polyphyletic" and Johnson and Patterson (1993) to state that they "have found nothing to indicate that Perciformes . . . are monophyletic."

Suborder Percoidei

This immense group contains more than 70 families and about 2,900 species. Although monophyly has been demonstrated for several families, the suborder, as currently understood, "is undoubtedly polyphyletic" (Johnson, 1993). The percoids are mostly marine, but several families have successfully invaded freshwater. Members of some of the families are important as food fishes, others as sport fishes, and many are aquarium favorites. While retaining a fairly conservative basic body plan (in many ways the archetype of what a fish ought to look like), percoids have achieved an astounding diversity of ecological, behavioral, and reproductive adaptations. Space permits only a cursory glance at the percoids, focusing on some of the better known families.

Among the freshwater representatives of the Percoidei are the black basses and sunfishes of the family **Centrarchidae**. These are small to medium-sized fishes of North America and are especially abundant in the southern half of the United States. The largemouth bass (*Micropterus salmoides;* see Plate 16.1) is probably the best known game fish with worldwide popularity—the result of its widespread introduction outside of its native range. There are probably more trophy largemouths mounted on home and office walls across the United States than there are in the lakes. The world record for a largemouth bass is for a beast caught in Georgia in 1932 that weighed 22 lbs. 4 oz. (more than 10 kg). Also included in the more than 30 species of centrarchids are the crappies (*Pomoxis*) and several members of the genus *Lepomis,* of which the bluegill (*L. macrochirus*) is probably the best known; it is a good panfish and is often cultured with the largemouth bass in ponds. The only sunfish native to the Pacific drainages of North America is the Sacramento perch (*Archoplites interruptus*)—a representative of a genus that has been found in Pliocene deposits of Oregon, Idaho, and Washington. The pygmy sunfishes of the family Elassomatidae were once considered centrarchids, but they are currently not even considered members of the Perciformes (see Chapter 15).

Although the family **Percidae** is present in Europe and Asia, it reaches its greatest development in eastern North America, where there are about 115 species, mostly darters of the genus *Etheostoma* (see Plate 16.2). The familiar yellow perch (*Perca flavescens;* see Plate 16.3) of North America has a close relative in the Eurasian *P. fluviatilis.* The North American sauger (*Stizostedion canadense*) and walleye (*S. vitreum*) have an Old World congener in *S. lucioperca.* The latter two genera are highly esteemed food and sport fishes. Nelson et al. (2003) claimed that the generic name *Sander* has priority over *Stizostedion;* the aforementioned species are thus to be considered *Sander canadensis* and *S. vitreus* (modification of species names is in conformation with the masculine gender of *Sander*).

Other percoid families with important food fishes include the **Lutjanidae** (snappers), **Sciaenidae** (drums and croakers), **Carangidae** (jacks and pompanos), and **Bramidae** (pomfrets). The numerous sea basses, hamlets, and groupers of most warm seas are in the family **Serranidae.** The related temperate perches (**Percichthyidae**) are found in both fresh and marine waters of the Southern Hemisphere. The temperate basses of the family **Moronidae** are placed with the percichthyids by many ichthyologists, but Johnson (1984) did not believe there is a close relationship between these two taxa. Some moronids, such as the popular striped bass (*Morone saxatilis;* see Plate 16.4) of North America, are anadromous. The moronid *Dicentrarchus labrax* of Europe

and northern Africa is a commercial fish caught in estuaries and along rocky shores.

The **Centropomidae** (snooks) are found in tropical to subtropical marine, brackish, and—occasionally—freshwaters in the Pacific, Atlantic, and Indian oceans. Snooks of the genus *Centropomus* are popular food and game fishes. Although Nelson (1994) considered fishes of the genera *Lates* and *Psammoperca* to be centropomids, Mooi and Gill (1995) separated them into their own family, the **Latidae.** The genus *Lates* includes the giant Nile perch (*L. nilotica*) and the barramundi perch (*L. calcarifer;* also known as Asian seabass) from Australia. Nile perch were introduced into Lake Victoria, with disastrous consequences for the native ichthyofauna (see Chapter 32). The **Grammatidae** (basslets) of the tropical West Atlantic include small, colorful fishes that are popular in marine home aquaria. **Priacanthidae** (bigeyes) are nocturnal, small to medium-sized fishes (up to 50 cm), mostly of warm waters. Some species are circumtropical. Many species are red or purplish in color. There are four genera, including *Priacanthus* and *Cookeolus.* The **Apogonidae** (cardinalfishes) also are mostly nocturnal, warm-water fishes. They are small, usually not much more than 20 cm long, and are noted for their oral incubation of eggs. There are about 300 species, more than a third of these in the genus *Apogon* (see Plate 16.5). Some species are found in freshwaters of Pacific islands.

The helmet gurnards (family **Dactylopteridae**) are characterized by a large, bony head, with large spines extending backward from the lower part of the operculum, and greatly enlarged pectoral fins, with the first few rays short and separated from the rest of the fin (see Plate 16.6). These features give the helmet gurnards a superficial resemblance to the searobins of the scorpaenoid family Triglidae (which are sometimes also called gurnards; see Chapter 17). Those separate pectoral rays and the pelvic fins are used in "walking" over the substrate.

These tropical marine fishes are often referred to as "flying gurnards," because they have been said to leap from the water and glide. This is highly unlikely, as the fins seem too weak and the body too robust for such activity. They often swim off the bottom, spread their pectoral fins, and "glide" through the water to a gentle landing some distance from where they started. Wheeler (1975) mentioned the display of the pectoral fins by individuals that have been startled. Some members have a bony connection, by means of parapophyses, between the gas bladder and the cranium—an adaptation that might aid in hearing. They have some features in common with various gasterosteiform fishes (Pietsch, 1978), but Johnson and Patterson (1993) rejected the notion that the dactylopterids might have an origin among the Gasterosteiformes.

The flying gurnards have long been classified among the scorpaenoid fishes, based chiefly on the presence of a posterior extension of the infraorbital bone (see Chapter 17). This so-called "suborbital stay" is now considered of doubtful homology with that of the scorpaenoids. Imamura (2000) has proposed that the flying gurnards be united with the tilefishes (formerly the family **Malacanthidae**) to constitute the enlarged family Dactylopteridae. The tilefishes are marine fishes of the Pacific, Indian, and Atlantic oceans. *Lopholatilus chamaeleonticeps,* the ocean whitefish of the North Atlantic, once supported a large commercial fishery, until a natural incursion of cold water greatly diminished the population. The fishery was nonexistent from 1882 until 1915, and only incidental catches of this once prevalent species are now realized.

Once considered a separate order intermediate between the clupeiforms and the galaxiids (Berg, 1940), the fishes of the deep-ocean family **Bathyclupeidae,** which retain a pneumatic duct and have their maxillaries included in the gape, are now placed among the percoids.

The **Polynemidae** contains the threadfins, which are tropical shore fishes. The snout is prominent, reaching well beyond the large mouth, and the pectoral fins are composed of an upper part, rather typical of percoid pectoral fins, and a lower section composed of four to eight long filaments that apparently serve as tactile organs. The upper rays attach to the first two actinosts; the third bears no fin rays; and the fourth supports the filamentous rays. The dorsal fins are widely separated. Threadfins have been observed to swim a spiral course, up one piling and down the next, with the pectoral filaments fanned out, presumably to detect prospective food items over about a 40 cm wide area. Examples of genera are *Polynemus, Polydactylus* (see Plate 16.7), and *Eleutheronema;* species of the last genus may reach 2 m in length. The history of the classification of this group is reviewed by deSylva (1984a). Berg (1940) and Lindberg (1974) considered the polynemids to be a separate order, and Johnson (1993) suggested an affinity between the polynemids and the drums of the family **Sciaenidae.** The drums are another highly esteemed food fish commonly found in inshore marine waters and estuaries, with some species invading freshwaters. Their name derives from their sound-producing ability, in which the swim bladder is used as a resonating chamber (see Chapter 21).

· · · · · · · · · · ·
Suborder Labroidei

Fully 15 percent of all living fish species are to be found in the suborder Labroidei. This suborder has traditionally been united by the presence of a remarkable pharyngeal jaw mechanism that has conferred on the group an extraordinary versatility in feeding capacities. According to Kaufman

and Liem (1982), the suborder Labroidei is defined by three characters:

- The fifth ceratobranchials are fused into a single bone.
- The bones of the upper pharyngeal jaws are in direct contact with the base of the cranium.
- The esophageal sphincter muscle is present as a single continuous sheet with no subdivisions.

Stiassny and Jensen (1987) agreed with this diagnosis and added additional characters:

- The lower pharyngeal jaw has a ventral keel for muscle attachment.
- The lower pharyngeal jaw is suspended directly from the neurocranium by a "muscle sling" made up of several component muscles.
- The lower pharyngeal jaw is structurally united into a single unit.

Prior to these studies, the suborder was considered to contain only three families: the Labridae (wrasses), Scaridae (parrotfishes), and Odacidae (greenbones). Kaufman and Liem (1982) and Stiassny and Jensen (1987) included the Scaridae and Odacidae in the Labridae and added the following pharyngognathous families to the suborder Labroidei: Cichlidae, Embiotocidae, and Pomacentridae. These families were formerly placed in the suborder Percoidei. The perception that the Labroidei constitute a natural group has been challenged by Streelman and Karl (1997), whose studies of nuclear DNA have suggested that the distinctive pharyngeal jaw configuration of the labroids independently evolved in more than one lineage.

The **Labridae** (see Plate 16.8) are one of the largest marine families, with nearly 60 genera and about 500 species. They are distributed in most tropical to warm temperate seas and are usually brightly colored, small species, but there are species that are as long as 3 m (Wheeler, 1975). They have sharp, heavy teeth in the front of the mouth in addition to the crushing pharyngeal jaws. Although labrids in a given area may show a diversity of jaw and dental adaptations, these do not necessarily correlate to differences in feeding strategy (Clifton and Motta, 1998). Many of the smaller species are "cleaners" that pick parasites from other species (see Plate 16.13A). Many species experience sex changes at some point in their life histories (Warner and Robertson, 1978; see Chapter 27).

Nelson (1994), Richards and Leis (1984), and Eschmeyer (1998) considered the families Scaridae and Odacidae to be labroids. The **Scaridae** are medium-sized to large (ca. 1.2 m and 70 kg) tropical and subtropical fishes usually associated with coral reefs. Their anterior teeth are fused into a parrotlike beak, with which they scrape attached vegetation or bite off chunks of coral. More than 80 species have been identified. The **Odacidae** are fishes of Australia and New Zealand. They have nonprotractile jaws, with fused teeth forming a cutting edge. There are about 12 species.

The **Cichlidae** are an important freshwater family, which ranges through warm freshwaters from India, Africa, South America, and Central America north to the Rio Grande River. Farias et al. (2000) have published a comprehensive investigation of cichlid phylogeny, incorporating both molecular and morphological data. The presence of cichlids on the southern continents and on islands such as Madagascar and Sri Lanka suggests either that their ancestors could withstand brackish or even marine waters or that their distribution patterns date from the Cretaceous, when those landmasses were closer together.

Only one genus (*Etroplus*) occurs in southern Asia, where the pearl spot (*E. suratensis*) of India is taken as a food fish. In the Americas, the Rio Grande perch (*Herichthys cyanoguttatum*) is the most northerly representative of the family. Some members of the New World genera *Apistogramma*, *Astronotus* and *Cichlasoma* enter the aquarium trade, as do many others in the family. The red oscar (*Astronotus ocellatus*) has long been a popular aquarium fish; given its predatory habits, however, it is not a very good member of a community tank (see Plate 16.9). The freshwater angelfish (*Pterophyllum scalare*) and the discus (*Symphysodon discus*) are also longtime favorites of aquarists.

Owing to their extraordinary diversification in lacustrine situations (see Chapter 32, Going Deeper), the African cichlids remain a classic study in speciation events. Two major lineages comprise the African cichlids: the Tilapiinae and the Haplochrominae. These lineages appear to have separated approximately eight million years ago (Nagl et al., 2001). The tilapiines have been especially important in fish culture. A well-known food fish is *Oreochromis mossambicus*, which was one of the first cichlids to be introduced outside of its native range. Its ease of culture and food conversion efficiency point to its potential in aquaculture. A hardy and adaptable genus, it has been widely introduced, with undesirable consequences in some places (Lever, 1996). Because these fishes are highly prolific, they will quickly overcrowd a pond, resulting in stunting of the population. Other cultured species are *O. nilotica* and *Tilapia zilli*.

In some of the East African Great Lakes, there are "species flocks" of haplochromine cichlids, usually of the genus *Haplochromis* and its close relatives. There is some evidence that numerous species and some new genera have evolved in certain lakes from a few founding species. Some of these flocks include well over 100 species, showing a remarkable range of trophic adaptations. The diversity of

nutritional strategies includes sand plowers, detritivores, herbivores, insectivores, carnivores, eye biters, fin biters, scale eaters, thieves of eggs and larvae, and other specialties. One predatory species from Lake Malawi, *Nimbochromis livingstonii*, feigns death by lying motionless on its side. Its mottled coloration effectively mimics the decaying flesh of a dead fish (see Plate 16.10). This entices smaller fishes that are effectively dispatched when they venture too close (Barlow, 2000). Even among species with similar feeding habits, subtle differences in reproductive behavior have apparently allowed the development of closely related forms that may occupy only slightly different niches in the same lake—an example of sympatric speciation. Courtship and nesting activities in cichlids follow set patterns of behavior that have been of great interest to ethologists. Many cichlids are oral incubators; others build nests in the substrate. Kornfield and Smith (2000) have authored an excellent review of the evolutionary biology of cichlids.

The **Embiotocidae** (surfperches) are coastal fishes of the North Pacific, with most of the species ranging from California northward. Two species occur in Japan. One Californian species, *Hysterocarpus traski*, lives in freshwater. These are viviparous fishes that give birth to precocial young. In at least one species, newborn males are sexually mature. It is indeed curious that some labroids, such as the cichlids, achieve an extraordinary degree of speciation, whereas others, such as the embiotocids, are comparatively species-poor, with only 23 known species. Studies on the molecular phylogeny of embiotocids have revealed that, in spite of these profound differences in species numbers, the degree of genetic differentiation within the embiotocids and the cichlids is very similar (Bernardi and Bucciarelli, 1999).

The **Pomacentridae** are small, tropical, marine fishes, usually of shallow waters around reefs. The family includes the damselfishes and anemonefishes. Using mitochondrial DNA sequence data, Tang (2001) has confirmed the monophyly of the family. Pomacentrids may be the dominant members of reef communities in some areas. They are often conspicuous because of their numbers and their coloration. Their compact size belies a most belligerent disposition, as many species guard small patches of the reef environment, chasing away intruders that are sometimes much larger. Herbivorous species of damselfishes may tend a garden of algae within their territory.

Suborder Trachinoidei

This group includes the weeverfishes, stargazers, and their allies. Trachinoids are generally not as elongate as the blennies, but they bear structural resemblances to those fishes. An important feature is the jugular pelvic fins. The composition of this suborder has been subject to various interpretations. Some ichthyologists (Gosline, 1968; Rosenblatt, 1984) placed them with the blennies, whereas others considered them a separate suborder but left some families in the Blennioidei and included some usually placed within the Percoidei. The stargazers of the family **Uranoscopidae** (see Plate 16.11) are marine fishes with venomous spines at the edge of the opercle. Venom glands at the base of the spines can deliver poison through grooves in the spine. *Astroscopus* has electric organs, derived from

extrinsic eye muscles, behind the eyes, capable of discharging 50 volts.

The weevers of the family **Trachinidae** are also venomous, having opercular and dorsal spines equipped with venom-producing tissues (Fig. 16.1A). Although the effect of the weever's poison on humans is severe, it may be less so than that of the stargazers, which is known to cause death.

Most families of trachinoids have the habit of concealing themselves in sand or in other soft bottom materials. Their eyes are generally placed on top of their heads, their mouths are in a superior position, and there are usually

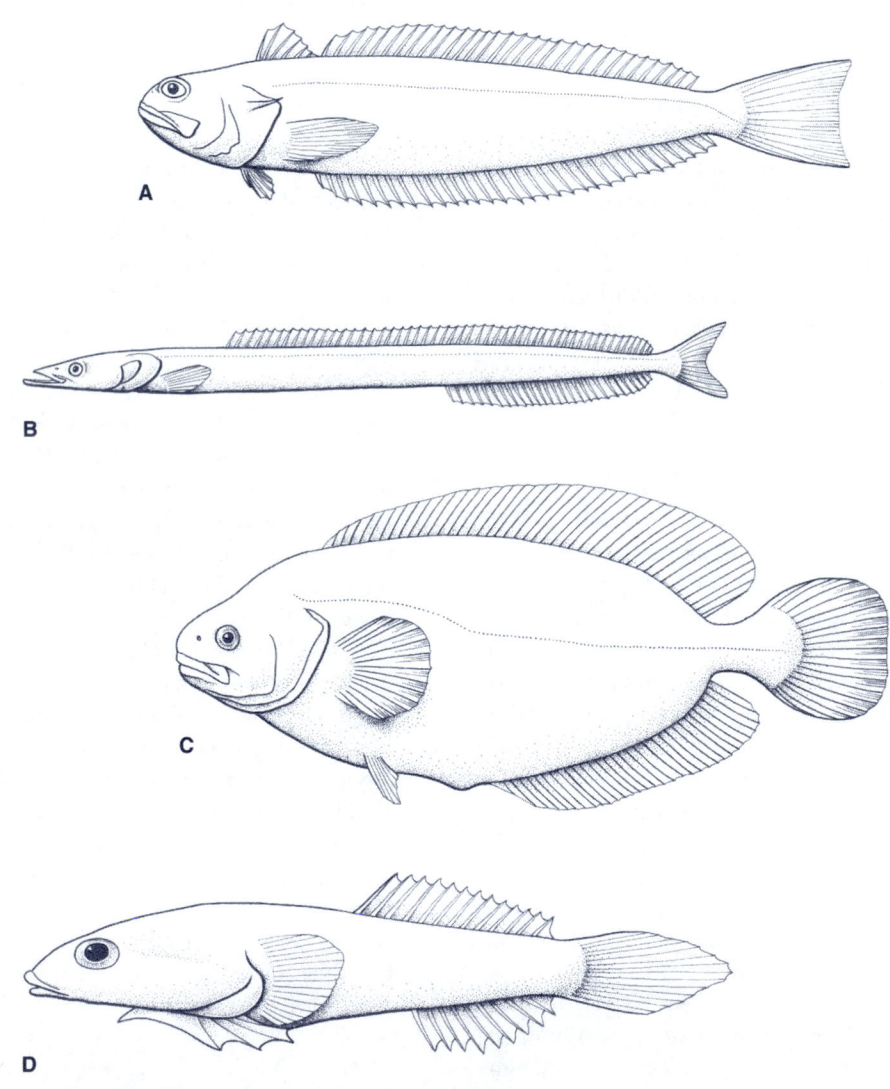

FIGURE 16.1

Representatives of the order Perciformes: **A,** Weever (*Trachinus*); **B,** Sand lance (*Ammodytes*); **C,** Juvenile ragfish (*Icosteus*); **D,** Clingfish (*Gobiesox*).

fringes or flaps that prevent the intake of sand with the respiratory water. Some of these families are **Trichodontidae,** the sandfishes of the North Pacific; **Trichonotidae,** the sand divers; and **Leptoscopidae,** of Australia and New Zealand. The **Dactyloscopidae,** the sand stargazers of tropical America, have similar adaptations, but they are now classified as blennioids (Doyle, 1998; Springer, 1993). Other trachinoid families include the **Pinguipedidae** (sandperches; formerly Mugiloididae; Rosa and Rosa, 1987), **Cheimarrhichthyidae** (torrent fishes; placed in the preceding family by some investigators), **Percophidae** (duckbills), and **Creediidae** (sandburrowers).

The bent-tooths (**Champsodontidae**) and the swallowers (**Chiasmodontidae**—deepwater fishes that can swallow fishes larger than themselves) are considered by Nelson (1994) to be trachinoid but of dubious affinities. Johnson (1993) and Mooi and Johnson (1997) suggested that the champsodontids may be related to scorpaeniform fishes.

The sand lances of the family **Ammodytidae** are small, slender, marine fishes with a protruding lower jaw and forked caudal fin (Fig. 16.1B). They lack a gas bladder, their scales are cycloid, and the pelvic fins are absent or jugular in location. Sand lances are burrowing forms, found mainly in temperate to cool seas, and may be important as food for predatory species. (In my morphology class, this seems to be one of the most common species recovered from the stomachs of dogfish sharks during student dissections.) They are harvested in northern Japan for use as food in fish culture.

Suborder Pholidichthyoidei

This is an Indo-Pacific group, containing only the **Pholidichthyidae,** or convict blennies. There are probably only two species of these small, dark, elongate fishes. They are reported to live in burrows excavated in soft substrates (Trnski et al., 1989). Springer and Freihofer (1976) have proposed a possible relationship of the convict blennies to the suborder Blennioidei.

Suborder Blennioidei

This group contains a large variety of marine fishes with jugular pelvic fins, in which each radial of the dorsal and anal fins corresponds with a neural or haemal spine. Most of the blennioids are rather elongate, with many being either taeniform or eel-shaped. Springer (1993) proposed monophyly for the blennioids based on several character complexes, including features of the pectoral and pelvic fins and associated skeletal supports, gill arches, and caudal fin elements. Analysis of mitochondrial DNA sequences has largely confirmed Springer's (1993) classification and the

monophyly of the "blenniiform" suborders (Blennioidei, Notothenoidei, and Zoarcoidei; Stepien et al., 1997). The six families included here are represented by nearly 130 genera and close to 700 species.

The **Blenniidae** (combtooth blennies, scaleless blennies) include numerous small shore fishes of tropical and subtropical seas. They are common in the intertidal zone, and one genus, *Salarias,* contains species that will leap from one tide pool to another when disturbed (see Plate 16.12). The **Clinidae** contain the familiar kelpfishes of the Pacific coast of North America as well as numerous blennies from other shores. These are quite often perchlike in appearance. The **Tripterygiidae,** or threefins, including the cockabullies of Australia, are similar in appearance to the foregoing two families and, like them, are noted for the male's care of the eggs during incubation (see Chapter 37).

Other families included in the Blennioidei by Nelson (1994) and Eschmeyer (1998) are the **Labrisomidae,** **Chaenopsidae** (pikeblennies), and **Dactyloscopidae** (sand stargazers). One chaenopsid species, *Hemiemblemaria simulus,* is commonly called the *wrasse blenny* because of its remarkable resemblance to young individuals of the bluehead (*Thalassoma bifasciatum*), a wrasse common to the Caribbean. Blueheads are a cleaner fish commonly seen on Caribbean reefs (see Chapter 37). *Hemiemblemaria,* unlike most blennioids, swims up off the bottom and, in this manner, mimics the cleaning behavior of its model. This may enable it to approach and prey on fishes that mistake it for a bluehead. Another species of wrasse, *Labroides dimidiatus* establishes cleaning stations on Indo-Pacific reefs and also has a blennioid mimic, *Aspidontus taeniatus* (see Plates 16.13A, B).

Suborder Zoarcoidei

These are blenny-like marine fishes that were formerly classified with the blennies, then placed with the gadiforms and ophidiiforms (Greenwood et al., 1966), but are now considered to constitute a separate suborder (Eschmeyer, 1998; Nelson, 1994). In a brief overview of the history of the classification of the suborder, Anderson (1984) concluded that the zoarcoids are allied with the Blennioidei. Most of the fishes in this suborder are elongate, with long dorsal and anal fins and more or less compressed bodies, but their body form ranges from the filiform quillfishes to the rather deep-bodied prowfishes. The members of the suborder have a single nostril on each side.

The **Zoarcidae** (eelpouts) resemble cusk eels in appearance and somewhat in habit. Many are viviparous. There are about 45 genera and, according to Anderson (1984), probably close to 200 species. Most eelpouts are small, but *Macrozoarces* reaches about 1 m. They are typically deepwater species, with some found deeper than 5,000 m, although some species live in the intertidal zone. The family is widespread, especially in cold northern waters. The blood of some Antarctic zoarcids has a glycoprotein that inhibits ice formation in the tissues—a feature common to many species that live in extremely cold waters (see Chapter 24).

The **Bathymasteridae** (ronquils) are fishes found only in the North Pacific, and according to Matarese (1989) are closely related to the **Stichaeidae,** or pricklebacks, which are found mainly in the North Pacific but also occur in the North Atlantic and Arctic oceans. They typically live in shallow coastal waters. The **Cryptacanthodidae,** or wrymouths, contain a few species in the North Atlantic and North Pacific. The giant wrymouth (*Cryptacanthodes giganteus*) reaches a length of nearly 1.2 m. Members of the family **Pholidae,** the gunnels (see Plate 16.14), are shallow-water fishes of the North Atlantic and the North Pacific. Pricklebacks and gunnels are common intertidal inhabitants of rocky shores, especially in the North Pacific (see Chapter 34).

The **Anarhichadidae,** or wolffishes, contain at least one species that is taken commercially, *Anarhichas lupus.* The wolf eel (*Anarrhichthys ocellatus*) of the North Pacific, reaches 2.4 m in length and is noted for its strong dentition of canine teeth in the front of the jaws and wide molariform teeth in back. Its diet is primarily shellfish, and the species is often found in the traps of Pacific coast crab fishers.

The **Ptilichthyidae** (quillfish), **Scytalinidae** (graveldivers) and **Zaproridae** (prowfish) are all monotypic families from the North Pacific.

Suborder Notothenioidei

This suborder contains five families that were sometimes aligned with the Blennioidei (Gosline, 1968; Stevens et al., 1984) but were given subordinal status by Nelson (1984, 1994) and Eschmeyer (1990, 1998). Ritchie et al. (1997), Bargelloni et al. (2000), and Derome et al. (2002) have produced phylogenies of the suborder based on mitochondrial gene sequences. About 100 species live in Antarctic or subantarctic seas, with a few entering freshwaters of Australia. The **Bovichtidae** (= Bovichthyidae; Eschmeyer and Bailey, 1990; Robins et al., 1991b), or thornfishes, are found in Australia, New Zealand, and southern South America. **Nototheniidae** (see Plate 16.15), the Antarctic cods or cod icefishes, occur in Australia, Tasmania, New Zealand, and the Antarctic. The plunderfishes (**Harpagiferidae**), spiny fishes that are also referred to as *Antarctic sculpins,* range north to the Falkland islands and southern South America. Antarctic dragonfishes (**Bathydraconidae**) are found in both deep and shallow water of the Antarctic. Icefishes (**Channichthyidae**) of the seas surrounding Antarctica have extremely

reduced amounts of hemoglobin in the blood. Like the eelpouts, some of these Antarctic families have glycoprotein antifreeze in their blood as an adaptation against freezing in supercooled waters (see Chapter 24).

Suborder Icosteoidei

The monotypic ragfish family (**Icosteidae**) is found in deep water in the North Pacific. *Icosteus aenigmaticus* (Fig. 16.1C) is a flabby, limp fish with reduced scales. The young have pelvic fins, but those fins disappear in adults. The species reaches a length of more than 2 m. Berg (1940) and Gosline (1971) considered the Icosteidae to constitute a separate order.

Suborder Gobiesocoidei

The Gobiesocoidei were formerly considered to be a part of the Paracanthopterygii but were eliminated from that superorder by Patterson and Rosen (1989). The taxon contains the single family **Gobiesocidae.** It is now treated as a perciform suborder (Nelson, 1994), but it has been considered a perciform derivative by some (Allen, 1984) and is placed as a pre-perciform group by others (Eschmeyer and Bailey, 1990; Eschmeyer, 1998; Robins et al., 1991). In his critique of paracanthopterygian phylogenetic studies, Gill (1996) suggested that the gobiesocids may again find themselves allied with the paracanthopterygians. The limits and placement of the group are not agreed on by recent authors. The monograph of Briggs (1955) remains the definitive reference for the group.

The suborder and family Gobiesocidae, including the formerly recognized family Alabetidae (= Cheilobranchidae; Springer and Fraser, 1976), is characterized by a joint between the interoperculum and the epihyal, a joint between the supracleithrum and the cleithrum, and a greatly modified heart, like a strongly bent tube in which the atrium and ventricle lie side by side with the atrium on the left side. The common cardinal veins are expanded as large chambers, and the sinus venosus is reduced (Lauder and Liem, 1983b).

The gobiesocids are small, soft-rayed, scaleless fishes of tropical and temperate shores; some enter freshwater. Some species of the South African genus *Chorochismus* reach about 30 cm. At least one South American species of *Tomicodon* is used as food.

Most gobiesocids have a powerful suctorial disc on the ventral surface below the pectorals (see Fig. 3.13E; Fig. 16.1D), hence the common name of "clingfishes." This disc consists of the modified pelvic fins and skin folds and can exert enough suction that the specimens are sometimes very difficult to remove from a smooth surface. The Northern clingfish (*Gobiesox maeandricus*) uses its adhesive disc to gain leverage while prying limpets from rocks, thus exploiting a food resource unavailable to many other intertidally occurring predators (Drucker, 1991). *Sicyaces sanguineus* of Chile is somewhat amphibious and seeks food intertidally. In its preferred habitat—the splash zone of wave-washed shores—the sucking disc is especially useful.

The northern clingfish ranges from southern California to Alaska. Other North American genera are *Acyrtops* and *Rimicola*. The genus *Alabes* contains small, eellike fishes from the seas of Australia. Their sucking disc is greatly reduced, if evident at all, and the gill openings are confluent beneath the head, as in some other members of the family.

Suborder Callionymoidei

The dragonets consist of the **Callionymidae** (Fig. 16.2A) and **Draconettidae,** both of warm seas. They are highly colored fishes that are noted for the large, showy fins of the males.

Suborder Gobioidei

This extremely diverse suborder consists of eight families with about 270 genera and more than 2,100 species, mostly of shallow, tropical marine waters, but with about 200 species in freshwater (Nelson, 1994). Wang et al. (2001) have published a phylogeny of gobioids based on mitochondrial ribosomal sequences.

Gobioids exhibit a diversity of lifestyles. Some can spend considerable time out of the water and obtain food on mudflats. There are several so-called *amphidromous* species (see Chapter 36) that spawn in freshwater, with the larvae drifting out to sea. Here, the larvae experience a planktonic phase of one to six months before reascending the rivers. This lifestyle has enabled amphidromous gobies to become the dominant ichthyofauna of insular river systems, as seen in the Indo–Pacific and Caribbean (Keith, 2003). A few gobioid species are symbiotic with burrowing creatures; some of these burrow dwellers, such as *Typhlogobius*, are blind. Some act as "cleaner" fishes that remove ectoparasites from other fishes. The pelvic fins of some families are united into a sucking disc that can be used to adhere to the substrate. Nowhere is this of more adaptive value than in the Hawaiian species that use this adhesive organ to support them as they climb up the waterfalls of torrential mountain streams (Schoenfuss and Blob, 2003).

Springer (1983) listed the following synapomorphies for the Gobioidei:

- Parietals are lacking.
- A cartilage at the anterior of the pelvic girdle fits between the lower ends of the cleithra.

- The dorsal end of the symplectic does not meet the dorsal end of the interhyal.
- The first basibranchial is cartilaginous.

Johnson and Brothers (1993) mentioned other synapomorphies, including the following:

- Hypurals 1 and 2 are fused, as are hypurals 3 and 4.
- The uppermost pectoral ray articulates with the dorsalmost actinost rather than with the scapula.
- Supraneurals and the basisphenoid are lacking.

The loach goby family (**Rhyacichthyidae**) is found in freshwater from China to New Guinea. They have flat heads, an inferior mouth, and the paired fins and lower part of the body are modified into a sucking disc. As such, they are aptly named, with more than a superficial resemblance to the balitorid river loaches (see Chapter 10). These, the only gobies with a lateral line, are generally considered to be the most primitive of the gobioid fishes (Wang et al., 2001). The only genus, *Rhyacichthys*, has two species.

The family **Odontobutidae** of eastern Asia contains three genera of freshwater fishes that share some characteristics of the pectoral girdle with the Rhyacichthyidae. They apparently are close to the Eleotridae (Nelson, 1994).

The **Eleotridae** (= Eleotrididae) are sometimes combined with the Gobiidae, although they are distinguished by the presence of separate pelvic fins. Commonly referred to as sleepers, these are usually bottom fishes, but some live in midwater and some are pelagic. *Gobiomorus dormitor* of Central America reaches a length of nearly 60 cm. Members of the genera *Oxyeleotris* and *Bunaka* of southeast Asia and the Indo–Australian areas, respectively, may reach lengths of 50 cm or more. *O. marmoratus* is a prized food fish.

The **Gobiidae** are the largest family in the suborder and include species from all warm seas, some tropical freshwaters, and a few temperate marine and estuarine localities. According to Nelson (1994), there are about 212 genera and nearly 1,900 species in this family, making it the largest family of marine fishes. These are mainly small fishes, mostly less than 10 cm long. One of the smallest vertebrates is a goby (*Trimmatom nanus*) of the Chagos archipelago (Indian Ocean), in which mature females reach no more than about 10 mm (see Plate 16.16). A marine goby of the Marshall islands (*Eviota zonura*) is nearly as small. Some small gobies of Southeast Asia and the Philippines occur in great enough numbers that they can be harvested, mixed with salt, and fermented to make *bagoong*, a sauce that is eaten with vegetables or rice. The bagoong fishery targets larval and postlarval individuals up to about 25 mm on their migration from marine waters into rivers.

FIGURE 16.2
Representatives of the order Perciformes: **A,** Dragonet (*Callionymus*); **B,** Forehead brooder (*Kurtus*);
C, butterfish (*Peprilus*); **D,** Climbing perch (*Anabas*).

The skipping gobies or mudskippers were formerly considered to be in a separate family, the Periophthalmidae (Berg, 1940). Nelson (1994) placed them in the family Gobiidae, subfamily **Oxudercinae**. These fishes inhabit tropical shore areas with soft bottoms and are usually seen at the water's edge or on the mud, rocks, or mangrove roots along the shore. They can pull themselves along with their arm-like pectoral fins or can flip around with great rapidity by flexing their body (see Plate 16.17; see Chapter 19).

The eyes of *Periophthalmus* are set high on the head, the gill cavity is expanded, and the skin is very vascular, so that it functions in respiratory exchange. Mudskippers can live out of water for more than 30 hours if not subjected to excessive heat or desiccation. In high humidity, their ability to stay out of water may be limited by the inability to excrete nitrogenous wastes rather than the inability to obtain oxygen. One mudskipper, *Boleophthalmus chinensis*, is cultured as a food fish in China and Taiwan. Other genera are *Oxuderces*, *Apocryptes*, and *Periophthalmodon*.

The subfamily **Sicydiinae** includes many freshwater species. They are distributed around the world in warm waters and include genera such as *Awaous*, *Sicydium*, and *Sicyopterus*, noted earlier for their ability to ascend waterfalls.

The eel gobies or eellike gobies, formerly placed in the family Gobioididae, are nearly circumtropical in fresh, brackish, and marine waters. The burrowing gobies (*Trypauchen*) occur in fresh, brackish, and marine waters from Africa to Japan and the Philippines.

On the Pacific coast of the continental United States, there are 12 native gobies, including *Typhlogobius californiensis*, the blind goby, and *Gillichthys mirabilis*, the longjaw mudsucker, which is a popular bait fish capable of withstanding prolonged exposure to atmospheric conditions. There are at least 58 gobiids on the Atlantic Coast of the United States, ten of the genus *Coryphopterus* and nine of *Gobionellus*.

Exotic goby species have been introduced into California from Asia and into the Great Lakes from the Black and Caspian Seas. In both cases, the fishes probably arrived in the ballast water of ships.

Other families of Gobioidei include the **Kraemeriidae**, sand gobies or sandfishes, which range from Hawaii to the Indian Ocean; and the **Microdesmidae**, or wormfishes, which are circumtropical (Thacker, 2000). The **Xenisthmidae**, an Indo-Pacific group, are considered a subfamily of the Gobiidae by Springer (1983) but are listed as a family by Eschmeyer and Bailey (1990), Eschmeyer (1998), and Nelson (1994).

The family **Schindleriidae** contains three species. These are tiny, larvoid (paedomorphic), marine fishes of the tropical Pacific that resemble the Ammodytoidei, but also bear a resemblance to the larvae of various other fishes, including microdesmids. They are usually less than 3 cm long. *Schindleria praematura* is among the most common surface fishes in parts of the tropical Pacific. These fishes were formerly placed in a separate suborder Schindleroidei, but Johnson and Brothers (1993) demonstrated their relationship to the Gobioidei. Watson and Walker (2004) recently described the third species of *Schindleria*, *S. brevipinguis*, claiming it to be the smallest known vertebrate. With the males maturing at 7 mm and the largest specimen measuring only 8.4 mm, these tiny translucent fishes apparently live no more than two months.

Suborder Kurtoidei

The family **Kurtidae** contains the single genus *Kurtus*, with two species of nurseryfishes or forehead brooders (Fig. 16.2B). These are tropical fishes in which the egg clusters are hung on hooks developed from the supraoccipital of the males. The ribs of these fishes form a tubular, ossified structure, which encloses the gas bladder. *Kurtus* is found in both marine and freshwater habitats in the Indo-Pacific. This family is placed in the suborder Percoidei by Eschmeyer and Bailey (1990), but it is placed in the Kurtoidei here following Lauder and Liem (1983b), Nelson (1994), and Eschmeyer (1998). Johnson (1993) mentioned several features, including distinctive, gridlike sensory papillae on the head and body, and the configuration of the dorsal gill arch elements, that suggest an affinity with the cardinalfishes of the family Apogonidae.

Suborder Acanthuroidei

Except for the pelagic *Luvarus*, this suborder contains fishes typical of reef habitats. They are characterized by 11 synapomorphies, most involving the skeleton (Tyler et al., 1989). Included among these are the following:

- Four or five branchiostegals are present.
- The premaxillaries and maxillaries are bound together, so that movement of the upper jaw is limited.
- The first neural spine is fused to its vertebral centrum.
- The lachrymal and second infraorbital are joined loosely if at all.

The family **Siganidae**, spinefoots or rabbitfishes, often placed in its own suborder, is distinguished by two spines and three soft rays in each pelvic fin. The dorsal and anal spines of rabbitfishes are venomous. Species of *Siganus* are largely herbivorous and show promise in aquaculture.

The **Luvaridae** were accepted as a member of the suborder Scombroidei prior to the study by Tyler et al. (1989), in which cladistic analysis refuted the relationship to the scombroids and confirmed the relationship to the Acanthuroidei. The family has only one species, the louvar (*Luvarus imperialis*), an oceanic fish of warm latitudes noted for its red fins and pink body color. It has been reported to reach a length of about 1.8 m.

The surgeonfishes and tangs of the family **Acanthuridae** are common and colorful inhabitants of coral reefs. *Acanthurus* has an erectile, sharp, forward-pointing blade on each side of the caudal peduncle. Other genera with folding spines are *Zebrasoma*, *Paracanthurus*, and *Ctenochaetus*. Fishes of the Indo–Pacific genus *Naso* (the unicornfishes) have a long, spikelike projection extending forward from the forehead (see Plate 16.18). They have one or two fixed spines on each side of the caudal peduncle. The related *Prionurus* has three or more such spines.

The family **Zanclidae** contains the graceful and colorful Moorish idol (*Zanclus cornutus*), which is found around coral reefs in the Pacific and Indian oceans. Its striking color pattern of black bars on a yellow-white background makes it a favorite with aquarists and divers. Fishes of this genus have no spines or retractible blades on the caudal peduncle. Johnson (1993) supported the inclusion of the deep-bodied, coastal marine spadefishes (**Ephippidae**) and scats (**Scatophagidae**) among the acanthuroids. Some species of scats enter freshwater and are popular fishes in the home aquarium.

· · · · · · · · · · · · · · ·
Suborder Scombrolabracoidei

The monotypic family **Scombrolabracidae** contains the widespread, deep-sea black mackerel (*Scombrolabrax heterolepis*), which has been classified among the Scombroidei, Percoidei, or Trichiuroidei by various authors, although Roule (1922) first indicated that it should be considered a distinct suborder. This fish bears a resemblance to members of the family Gempylidae, but has such percoid characters as protractile premaxillaries and serrations on the opercle and preopercle. The skeleton has fewer (30) vertebrae than most scombroids, and it shows a procurrent spur on the lowermost principal caudal ray, which allies it with the percoids (Johnson, 1975). Bond and Uyeno (1981) pointed out that *Scombrolabrax* is distinguished from all other fishes by virtue of its unique vertebral configuration that justifies placing the family in its own suborder: The fifth through the twelfth vertebral centra are hollowed out bilaterally into ventrally opening bulbous bullae, into which evaginations of the gas bladder fit.

· · · · · · · · · · · · · · ·
Suborder Scombroidei

This suborder includes the barracudas, snake mackerels, mackerels, tunas, billfishes, and their relatives, many species of which are valuable in commercial and sport fisheries. They are open-water fishes especially adapted for powerful swimming (see Chapter 19, Going Deeper). The scombroids have been the subject of a number of systematic interpretations and a diversity of opinions as to what families constitute the suborder. All of the members, however, share the common feature of nonprotractile premaxillaries with maxillaries firmly attached to them (Lauder and Liem, 1983b; Nelson, 1994).

Collette et al. (1984) recognized six families in the suborder: Scombrolabracidae, Gempylidae, Trichiuridae, Xiphiidae, Istiophoridae, and Scombridae. Johnson (1986) recognized the Sphyraenidae, Gempylidae, and Scombridae, which he divided into eight tribes. Robins et al. (1991) considered the Gempylidae to be part of the Trichiuridae. Nelson (1994) listed the following families in the suborder: Sphyraenidae, Gempylidae, Trichiuridae, Scombridae, and Xiphiidae.

The barracudas of the family **Sphyraenidae** were once considered to be closely related to the mullets (deSylva 1984b, 1984c), but Johnson (1986) considered the barracudas to be a primitive sister group to the rest of the scombroids. The barracudas have large mouths and strong, sharp teeth set in sockets. The pectoral fins are set low, and the lateral line is well developed. As in the mullets, the dorsal fins are well separated. These are fierce, medium-sized to large predators of warm seas. One species, the great barracuda (*Sphyraena barracuda;* see Plate 16.19) reaches a length in excess of 2 m. Barracudas are palatable fishes but, like many tropical species, they have been suspected of causing a potentially severe toxic reaction known as "fish poisoning" or **ciguatera** (see Chapter 38).

The **Gempylidae**, escolars or snake mackerels, are usually found in deep water, but some are found near the surface. They are usually equipped with long, strong teeth and have a slender, streamlined form. There are usually finlets behind the dorsal and anal fins. The caudal fin is forked. Pelvic fins are reduced to a single spine in some species, but have one spine and five rays in others. One species, *Ruvettus pretiosus*, the oilfish, reaches about 2 m in length (see Chapter 7, Going Deeper).

The **Trichiuridae**, hairtails and cutlassfishes, have exceptionally strong teeth and a compressed body with a long, tapering tail, with that of *Trichiurus* ending in a fine point. The caudal fin, if present, is small and forked. Pelvic fins are rudimentary or absent. Hairtails are harvested as food in many tropical countries.

The systematics of the **Scombridae**, as might be expected from a family containing so many commercial and game species, have been well studied. This family includes the tunas and mackerels (see Fig. 3.6A); these are mostly swift-moving species of the surface waters, usually in warm seas, and are noted for their wide-ranging migrations. Morphological studies have subdivided the scombrids into two subfamilies, the Scombrinae and the monotypic Gasterochismatinae, although molecular investigations have cast some doubt on this classification (Collette et al., 2001). Many tunas maintain a body temperature several degrees higher than the surrounding water as an adaptation for enhanced metabolic efficiency (see Chapter 24).

Mackerels of the genus *Scomber* may reach only a few kilograms in weight, but they are valuable commercial fishes. Examples are *S. scombrus* of the Atlantic and *S. japonicus* of the Pacific. Even smaller species are in the genus *Rastrelliger,* which is of great commercial value in some tropical areas such as the Gulf of Thailand.

The largest tuna is one of the three species commonly referred to as bluefin—the Atlantic bluefin tuna (*Thunnus thynnus*), which reaches 4 m in length and may weigh up to 800 kg. Electronic tagging studies have confirmed the presence of two populations; although they intermingle in foraging grounds in the North Atlantic Ocean, one population spawns in the Gulf of Mexico and the other in the Mediterranean Sea (Block et al., 2005). Generally perceived as the world's most valuable fish, a single bluefin may fetch a price of more than one million yen (about ten thousand dollars at the current exchange rate) in the Japanese market, as this species is highly esteemed for the making of sashimi. The desirability of bluefin tuna has resulted in a number of captive rearing operations being developed worldwide (Farwell, 2001). A smaller yet highly desirable species on the Pacific Coast of North America is *T. alalunga,* the albacore. The yellowfin tuna (*T. albacares*) has been historically the most important species in the Pacific tuna fishery, but it is captured mainly south of the United States. The skipjack tuna (*Katsuwonus pelamis*), though small, is an important commercial species, especially in the Pacific. Other genera in the family include *Sarda,* the bonitos; *Auxis,* the frigate mackerels; *Scomberomorus,* the Spanish mackerels; and *Acanthocybium,* the wahoo.

Two related families of scombroid fishes are termed *billfishes* after the greatly elongated upper jaw that is a defining feature. For the present, we will consider the billfishes to be scombroids, although molecular studies have questioned this placement, suggesting that they may belong in a separate suborder (Collette et al., 2001). The family **Xiphiidae** contains the swordfish (*Xiphias gladius*), a wide-ranging pelagic predator. The swordfish has a smooth, bladelike rostrum that apparently is used in lateral slashing at prey; it makes up about one third the length of the fish. The bill apparently can also be effectively used as a spear for impaling prey. Adults lack scales and teeth. There is no pelvic fin girdle and no pelvic fin. This is a prized food fish and is sought by harpooners, longliners, and sportfishers. (Those of you who saw the film *The Perfect Storm* may take comfort in knowing that no swordfish were sacrificed in making the movie. All of the animals were "animatronic" robots—even the entrails scattered on the boat deck were carefully modeled after the internal anatomy of swordfishes!) Swordfishes have been known to attack and pierce small boats. Maximum size is about 540 kg at a length of nearly 5 m.

Billfishes of the family **Istiophoridae** are among the most popular of the large marine game fishes. These fishes have a rough, rounded bill, pelvic fins, and retain scales and teeth as adults. The first dorsal fin is elongate, and is very high in the sailfish (*Istiophorus;* see Plate 16.20). Spearfishes and the white and striped marlins are in the genus *Tetrapturus;* the blue marlin is in the genus *Maikara.* Blue marlin may reach a weight of about 640 kg, and the black marlin (*Istiompax*) has been recorded at nearly 710 kg.

Billfishes and swordfishes have a remarkable brain heater that elevates the temperature of their brains and eyes. The heater is developed from the superior rectus eye muscle, part of which is modified for thermogenesis rather than contraction (Block, 1987; see Chapter 24).

Suborder Stromateoidei

This suborder is composed of marine fishes that have papillose lateral sacs extending from the pharynx or esophagus behind the gill arches. Many of the species habitually associate with and perhaps feed on large jellyfishes. Pelvic fins are subthoracic to jugular and are absent in the adults of some species. The lachrymals are expanded, so that the maxillaries are mostly hidden by them, and the scales are cycloid to weakly ctenoid (Haedrich, 1967). There are six families in this suborder (Horn, 1984). Robins et al. (1991a) presented all the members of the suborder as belonging to the single family Stromateidae, but Eschmeyer (1998) recognized six families.

Amarsipidae is a monotypic family of the tropical Indo-Pacific. The pelagic *Amarsipus carlsbergi* is apparently known only from larvae and juveniles (Nelson, 1994). The medusafishes of the family **Centrolophidae** are distributed widely in tropical and temperate seas. Their common name is derived from the habit of swimming beneath some of the large jellyfishes. *Icichthys lockingtoni* is found in the North Pacific; *Centrolophus niger* occurs in the northeast Atlantic. Other genera include *Psenopsis*, *Tubbia*, and *Seriolella*. The man-of-war fishes or driftfishes of the family **Nomeidae** have teeth in the esophageal sacs and retain their pelvic fins as adults. Some members of the genus *Nomeus* associate with large jellyfishes and swim among their tentacles. The family **Ariommatidae** has one genus, *Ariommus*, with six species in deep waters of warm seas.

Squaretails of the family **Tetragonuridae** are widely distributed in tropical and subtropical areas, with some species ranging into temperate waters. The smalleye squaretail (*Tetragonurus cuvieri*) ranges from British Columbia to Australia and New Zealand.

The **Stromateidae**, the butterfishes or "white pomfrets," are known as good food fishes. They have teeth in the expanded esophagus, and some lack pelvic fins as adults. Species are well distributed in warm temperate and tropical inshore waters. Examples of genera are *Pampus*, *Stromateus*, and *Peprilus* (Fig. 16.2C).

Suborder Anabantoidei

This suborder includes the climbing perches and gouramies. Various ichthyologists have differing views of the composition of the suborder; some (Robins et al., 1991b) chose to simplify the arrangement by recognizing only one family (Anabantidae), whereas Lauder and Liem (1983b) and Eschmeyer (1998) recognized five. In this treatment, we follow Nelson (1994) by including five families in this group.

These fishes are distinguished by a labyrinth organ that is developed from the upper part of the first gill arch and occupies much of the gill chamber. The bones of the arch are expanded and folded so as to present a great surface area in a small space. Oxygen can be extracted from air trapped in this structure, so these fishes are at home in warm waters that may be very low in oxygen (see Chapter 24). They are freshwater fishes of Asia and Africa. Many species are known for their remarkable territorial courtship and nesting behavior.

The family **Anabantidae** contains the climbing perches of the genera *Anabas* of Asia and *Ctenopoma* of Africa. *Anabas* is equipped with stout spines on the operculum that aid in pulling the fish along over the ground when it migrates to a suitable habitat during the dry season (Fig. 16.2D).

The kissing gourami is in the family **Helostomatidae**. The kissing action for which this species is so well known is probably a form of sparring. The giant gourami (*Osphronemus goramy*) is placed in the **Osphronemidae**. It is the largest of the gouramies, reaching 60 cm, and is a favorite food fish in Southeast Asia, where it is the object of fish culture. Other favorite aquarium gouramis, such as *Trichopsis*

and *Trichogaster*, are in the **Belontiidae**, a family that also includes the Siamese fighting fish (*Betta splendens*) and the paradise fish (*Macropodus*). The eggs of most anabantoids are incubated at the surface, some floating of their own accord, with others placed in bubble nests that are attended by the male. In investigations of the correlation between phylogeny and the evolution of egg care in *Betta*—specifically, the transition from bubble nest building to mouthbrooding—Rüber et al. (2004) deduced a complex pattern of evolution of brooding and egg care, in which mouthbrooding apparently evolved in more than one clade (see Chapter 37).

The family **Luciocephalidae** includes one small species, the pikehead (*Luciocephalus pulcher*), of the Indo-Malayan region. It is distinguished by the absence of dorsal and anal fin spines and a gas bladder, the presence of a simple suprabranchial organ, and a highly protractile mouth. It captures food by lunging forward and extending the premaxillary forward up to about one third the length of the head. Suction is not necessarily involved (Lauder and Liem, 1981).

Suborder Channoidei

The snakeheads of the family **Channidae** (= Ophiocephalidae) are sometimes placed into their own suborder because of the unique structure of the air-breathing organ, subabdominal pelvic fins, and lack of spines. Snakeheads are much more elongate than anabantoids; some species approach 1 m in length. They are voracious predators, and though prized as food, they cannot be cultured with food fishes vulnerable to predation. A great advantage of snakeheads is that they can be held alive in the market for days if kept properly moist. There are two genera, *Channa* of Asia (about 18 species) and *Parachanna* of Africa (3 species). Five species of snakeheads have reportedly been introduced into North American waters, with three of these apparently established as reproducing populations (Courtenay and Williams, 2004). The northern snakehead (*Channa argus;* see Plate 16.21) has appeared at a number of locations throughout the southern United States. Because of their voracious predatory habits and their ability to survive prolonged emersion, snakeheads should be considered exotic species of great concern.

Anabantoid and Channoid Fishes Reconsidered

Based on structural features of their gill arches and associated musculature, a modified classification of anabantoid and channoid fishes has recently been proposed (Springer and Johnson, 2004; Springer and Orrell, 2004). These authors have introduced a new order, the **Anabantomorpha**, to include the anabantoids and channoids as members that possess the aforementioned suprabranchial labyrinth organs, plus the members of the family Nandidae (sometimes separated into three separate families, the Nandidae, Badidae, and Pristolepidae). All of these share the feature of teeth on the parasphenoid bone.

KEY POINTS AND CONNECTIONS

- The spiny-rayed fishes of the order Perciformes, the largest order of vertebrates, represent such a diversity of form and function that the order is probably not a monophyletic assemblage.

  Refer to Chapter 4 for a discussion of the "total evidence" approach to fish systematics as it applies to attempts to place the perciform fishes in the context of teleostean evolution.

- Almost 3,000 species currently constitute the suborder Percoidei. Among the most conspicuous fishes in shallow marine and freshwaters, they are targeted by sportfishers worldwide who value their flavor and fighting ability.

- The labroid fishes have evolved a distinctive pharyngeal jaw mechanism that enables a diversity of feeding strategies. Fishes of the family Cichlidae exemplify the successful radiation of labroids into freshwater habitats, whereas the wrasses of the family Labridae and the parrotfishes of the family Scaridae are especially conspicuous members of coral reef communities. Wrasses are particularly well known as cleaners of other fishes, especially in coral reef communities.

  Refer to the Going Deeper box in Chapter 32 for a discussion of the explosive speciation of cichlid fishes in the African Great Lakes and the Going Deeper box in Chapter 37 to revisit the phenomenon of cleaning symbiosis.

- Many members of the suborder Trachinoidei possess venomous spines on the operculum and dorsal fin. Most are cryptic fishes that bury themselves in sandy or muddy ocean bottoms.

  A discussion of venomous fishes is found in Chapter 18.

- The suborders Blennioidei, Zoarcoidei, and Notothenioidei appear to be a monophyletic assemblage; they are primarily benthic fishes inhabiting coastal marine waters. Blennioids are somewhat elongate and well adapted for hiding in crevices and burrows, especially in coral reef habitats. The zoarcoids are even more elongate, almost eel-like, and some species are found in deep water. The notothenioids are primarily known for the remarkable diversity they have achieved in Antarctic coastal waters.

  The diversification of the Antarctic notothenioids is discussed in the context of the ecology of polar waters in Chapter 35.

- The clingfishes of the suborder Gobiesocoidei have had an interesting classification history, as the group was recently considered to be allied with the paracanthopterygians. Their extreme morphological modification, including pelvic fins modified into a thoracic adhesive disc, has made it difficult to properly assess their phylogenetic affinities. Nonetheless, this distinctive adaptation enables them to adhere to rocky intertidal substrates that commonly experience severe turbulence.

  The ichthyofauna of rocky shores, including the gobiesocoid clingfishes, is discussed in Chapter 34.

- The gobies of the suborder Gobioidei form one of the largest groups of perciform fishes, yet their small size and generally cryptic behavior often render them inconspicuous. In most gobies, the pelvic fins are fused into a cup-shaped sucking disc.

- The tunas and billfishes of the suborder Scombroidei are among the largest perciform fishes. They are generally inhabitants of open waters, where their size and speed make them top-level carnivores of the pelagic realm. The barracudas, recent additions to the suborder after long having been considered relatives of the mullets, are fearsome predators in warm coastal waters.

  Adaptations of fishes of the epipelagic realm, including the scombroids, are covered in Chapter 35.

FISH LINKS

http://cichlidresearch.com A comprehensive site maintained by Ronald Coleman, a cichlid researcher on the faculty at California State University, Sacramento.

BUILDING AN ICHTHYOLOGY LIBRARY

Barlow, G. W., and G. C. Williams. 2000. *The cichlid fishes: Nature's grand experiment in evolution.* Perseus, Cambridge, UK.
Craig, J. F. 2000. *Percid fishes: Systematics, ecology, and exploitation.* Blackwell, Oxford.
Keenleyside, M. H. A. (Ed.). 1991. *Cichlid fishes—behavior, ecology, and evolution.* Chapman and Hall, London.

REFERENCES

Allen, L. G. 1984. Gobiesociformes: Development and relationships, pp. 629–636. In Ontogeny and systematics of fishes, H. G. Moser, W. J. Richards, D. M. Cohen, M. P. Fahay, A. W. Kendall, Jr., and S. L. Richardson (Eds.). *Am. Soc. Ichthyol. Herpetol. Spec. Publ. 1.*
Anderson, M. E. 1984. Zoarcidae: Development and relationships, pp. 578–582. In Ontogeny and systematics of fishes, H. G. Moser, W. J. Richards, D. M. Cohen, M. P. Fahay, A. W. Kendall, Jr., and S. L. Richardson (Eds.). *Am. Soc. Ichthyol. Herpetol. Spec. Publ. 1.*
Bargelloni, L., S. Marcato, L. Zane, and T. Patarnello. 2000. Mitochondrial phylogeny of notothenioids: A molecular approach to Antarctic fish evolution and biogeography. *Syst. Biol. 49*(1): 114–129.
Barlow, G. W. 2000. *The cichlid fishes: Nature's grand experiment in evolution.* Perseus, Cambridge, UK.
Berg, L. S. 1940. Classification of fishes both recent and fossil. *Trav. Inst. Zool. Acad. Sci. USSR 5:* 87–517 (Reprinted 1947, J. W. Edwards, Ann Arbor, MI).
Bernardi, G., and G. Bucciarelli. 1999. Molecular phylogeny and speciation of the surfperches (Embiotocidae, Perciformes). *Mol. Phylogen. Evol. 13*(1): 77–81.
Block, B. A. 1987. Billfish brain and eye heater: A new look at non-shivering heat production. *News Physiol. Sci. 2:* 208–214.

————, S. L. H. Teo, A. Walli, A. Boustany, M. J. W. Stokesbury, C. J. Farwell, K. C. Weng, H. Dewar, and T. B. Williams. 2005. Electronic tagging and population structure of Atlantic bluefin tuna. *Nature 434:* 1121–1127.

Bond, C. E., and T. Uyeno. 1981. Remarkable changes in the vertebrae of perciform fish *Scombrolabrax* with notes on its anatomy and systematics. *Jpn. J. Ichthyol. 28*(3): 259–262.

Briggs, J. C. 1955. A monograph of the clingfishes (order Xenopterygii). *Stanford Ichthyol. Bull. 6:* 1–224.

Carroll, R. L. 1988. *Vertebrate paleontology and evolution.* W. H. Freeman, New York.

Clifton, K. B., and P. J. Motta. 1998. Feeding morphology, diet, and ecomorphological relationships among five Caribbean labrids (Teleostei, Labridae). *Copeia 1998:* 953–966.

Collette, B. B., T. Potthoff, W. J. Richards, S. Ueyanagi, J. L. Russo, and Y. Nishikawa. 1984. Scombroidei: Development and relationships, pp. 591–619. In Ontogeny and systematics of fishes, H. G. Moser, W. J. Richards, D. M. Cohen, M. P. Fahay, A. W. Kendall, Jr., and S. L. Richardson (Eds.). *Am. Soc. Ichthyol. Herpetol. Spec. Publ. 1.*

————, C. Reeb, and B. A. Block. 2001. Systematics of the tunas and mackerels (Scombridae), pp. 5–33. In *Tuna: Physiology, ecology, and evolution,* B. A. Block and E. D. Stevens (Eds.). Academic Press, San Diego.

Courtenay, W. R., and J. D. Williams. 2004. Snakeheads (Pisces, Channidae)—a biological synopsis and risk assessment. *U.S. Geol. Surv. Circ. 1251.*

Derome, N., W.-J. Chen, A. Dettai, C. Bonillo, and G. Lecointre. 2002. Phylogeny of Antarctic dragonfishes (Bathydraconidae, Notothenioidei, Teleostei) and related families based on their anatomy and two mitochondrial genes. *Mol. Phylogen. Evol. 24*(1): 139–152.

DeSylva, D. P. 1984a. Polynemoidei: Development and relationships, pp. 540–541. In Ontogeny and systematics of fishes, H. G. Moser, W. J. Richards, D. M. Cohen, M. P. Fahay, A. W. Kendall, Jr., and S. L. Richardson (Eds.). *Am. Soc. Ichthyol. Herpetol. Spec. Publ. 1.*

————. 1984b. Mugiloidei: Development and relationships, pp. 530–533. In Ontogeny and systematics of fishes, H. G. Moser, W. J. Richards, D. M. Cohen, M. P. Fahay, A. W. Kendall, Jr., and S. L. Richardson (Eds.). *Am. Soc. Ichthyol. Herpetol. Spec. Publ. 1.*

————. 1984c. Sphyraenoidei: Development and relationships, pp. 534–540. In Ontogeny and systematics of fishes, H. G. Moser, W. J. Richards, D. M. Cohen, M. P. Fahay, A. W. Kendall, Jr., and S. L. Richardson (Eds.). *Am. Soc. Ichthyol. Herpetol. Spec. Publ. 1.*

Doyle, K. D. 1998. Phylogeny of the sand stargazers (Dactyloscopidae: Blennioidei). *Copeia 1998:* 76–96.

Drucker, E. G. 1991. Mechanics and function of adhesion by the northern clingfish during predation on limpets. *Amer. Zool. 35*(5): Abstr. 18A.

Eschmeyer, W. N. 1990. Genera in a classification, pp. 435–495. In *Catalog of the genera of recent fishes,* W. N. Eschmeyer (Ed.). California Academy of Sciences, San Francisco.

————. 1998. Catalog of fishes, Vol. 3, Genera of fishes. *Calif Acad. Sci. Spec. Publ. 1.*

————, and R. M. Bailey. 1990. Genera of recent fishes, pp. 7–433. In *Catalog of the genera of recent fishes,* W. N. Eschmeyer (Ed.). California Academy of Sciences, San Francisco.

Farias, I., G. Orti, and A. Meyer. 2000. Total evidence: Molecules, morphology, and the phylogenetics of cichlid fishes. *J. Exp. Zool. (Mol. Dev. Evol.) 288:* 76–92.

Farwell, C. J. 2001. Tunas in captivity, pp. 391–412. In *Tuna: Physiology, ecology, and evolution,* B. A. Block and E. D. Stevens (Eds.). Academic Press, San Diego.

Gill, A. C. 1996. Comments on an intercalar path for the glossophayryngeal (Cranial IX) nerve as a synapomorphy of the Paracanthopterygii and on the phylogenetic position of the Gobiesocidae (Teleostei: Acanthomorpha). *Copeia 1996:* 1022–1029.

Gosline, W. A. 1968. The suborders of perciform fishes. *Proc. U.S. Natl. Mus. 124:* 1–78.

————. 1971. *Functional morphology and classification of teleostean fishes.* University Press of Hawaii, Honolulu.

Greenwood, P. H., D. E. Rosen, S. H. Weitzman, and G. S. Myers. 1966. Phyletic studies of teleostean fishes with a provisional classification of living forms. *Bull. Am. Mus. Nat. Hist. 131:* 339–456.

Haedrich, R. L. 1967. The stromateoid fishes: Systematics and a classification. *Bull. Mus. Comp. Zool. Harv. Univ. 135:* 31–139.

Horn, M. H. 1984. Stromateoidei: Development and relationships, pp. 620–628. In Ontogeny and systematics of fishes, H. G. Moser, W. J. Richards, D. M. Cohen, M. P. Fahay, A. W. Kendall, Jr., and S. L. Richardson (Eds.). *Am. Soc. Ichthyol. Herpetol. Spec. Publ. 1.*

Imamura, H. 2000. An alternative hypothesis on the phylogenetic position of the family Dactylopteridae (Pisces: Teleostei), with a proposed new classification. *Ichthyol. Res. 47:* 203–222.

Johnson, G. D. 1975. The procurrent spur: An undescribed perciform caudal character and its phylogenetic implications. *Occ. Pap. Calif. Acad. Sci. 121:* 1–23.

————. 1984. Percoidei: Development and relationships, pp. 464–498. In Ontogeny and systematics of fishes, H. G. Moser, W. J. Richards, D. M. Cohen, M. P. Fahay, A. W. Kendall, Jr., and S. L. Richardson (Eds.). *Am. Soc. Ichthyol. Herpetol. Spec. Publ. 1.*

————. 1986. Scombroid phylogeny: An alternative hypothesis. *Bull. Mar. Sci. 39:* 1–41.

————. 1993. Percomorph phylogeny: Progress and problems. *Bull. Mar. Sci. 52:* 3–28.

————, and E. B. Brothers. 1993. *Schindleria:* A paedomorphic goby (Teleostei: Gobioidei). *Bull. Mar. Sci. 52:* 441–471.

————, and C. Patterson. 1993. Percomorph phylogeny: A survey of acanthomorphs and a new proposal. *Bull. Mar. Sci. 52:* 554–626.

Kaufman, L. S., and K. F. Liem. 1982. Fishes of the suborder Labroidei (Pisces: Perciformes): Phylogeny, ecology, and evolutionary significance. *Breviora 472.*

Keith, P. 2003. Biology and ecology of amphidromous Gobiidae of the Indo-Pacific and Caribbean regions. *J. Fish Biol. 63:* 831–847.

Kornfield, I., and P. F. Smith. 2000. African cichlid fishes: Model systems for evolutionary biology. *Ann. Rev. Ecol. Syst. 31:* 163–196.

Lauder, G. V., and K. F. Liem. 1981. Prey capture by *Luciocephalus pulcher:* Implication for models of jaw protrusion in teleost fishes. *Environ. Biol. Fish. 6:* 257–268.

————, and ————. 1983a. The evolution and interrelationships of the actinopterygian fishes. *Bull. Mus. Comp. Zool. 150*(3): 95–197.

————, and ————. 1983b. Patterns of diversity and evolution in ray-finned fishes, pp. 1–24. In *Fish neurobiology,* Vol. 1, R. G. Northcutt and R. E. Davis (Eds.). University of Michigan Press, Ann Arbor.

Lever, C. 1996. *Naturalized fishes of the world.* Academic Press, San Diego.

Lindberg, G. U. 1974. *Fishes of the world: A key to families and a checklist.* Wiley, New York.

Matarese, A. C. 1989. Phylogenetic relationships of the ronquils (Perciformes: Bathymasteridae). *Prog. Abstr. 65th Ann. Meeting ASIH,* p. 115 (abstr.).

McAllister, D. E. 1968. Evolution of branchiostegals and classification of teleostome fishes. *Nat. Mus. Can. Bull. 221.*

Miya, M., A. Kawaguchi, and M. Nishida. 2001. Mitogenomic exploration of higher teleostean phylogenies: A case study for moderate-scale evolutionary genomics with 38 newly determined complete mitochondrial DNA sequences. *Mol. Biol. Evol. 18:* 1993–2009.

Mooi, R. D., and A. C. Gill. 1995. Association of epaxial musculature with dorsal-fin pterygiophores in acanthomorph fishes, and its phylogenetic significance. *Bull. Nat. Hist. Mus. Lond. (Zool.) 61*(2): 121–137.

———, and G. D. Johnson. 1997. Dismantling the Trachinoidei: Evidence of a scorpaenoid relationship for the Champsodontidae. *Ichthyol. Res. 44*(2): 143–176.

Nagl, S., H. Tichy, W. E. Mayer, I. E. Samonte, B. J. McAndrew, and J. Klein. 2001. Classification and phylogenetic relationships of African tilapiine fishes inferred from mitochondrial DNA sequences. *Mol. Phylogen. Evol. 20*(3): 361–374.

Nelson, J. S. 1984. *Fishes of the world* (2nd ed.). Wiley, New York.

———. 1994. *Fishes of the world* (3rd ed.). Wiley, New York.

———, E. J. Crossman, H. Espinosa-Perez, L. T. Findley, C. R. Gilbert, R. N. Lea, and J. D. Williams. 2003. The "Names of Fishes" list, including recommended changes in fish names: Chinook salmon for chinook salmon, and *Sander* to replace *Stizostedion* for the sauger and walleye. *Fisheries 28*(7): 38–39.

Patterson, C. 1993. An overview of the early fossil record of acanthomorphs. *Bull. Mar. Sci. 52:* 29–59.

———, and D. E. Rosen. 1989. The Paracanthopterygii revisited: Order and disorder, pp. 5–36. In Papers on the systematics of gadiform fishes, D. M. Cohen (Ed.). *Nat. Hist. Mus. Los Angeles Co. Sci. Ser. 32.*

Pietsch, T. W. 1978. Evolutinary relationships of the sea moths (Teleostei: Pegasidae) with a classification of gasterosteiform families. *Copeia 1978:* 517–529.

Richards, W. J., and J. M. Leis. 1984. Labroidei: Development and relationships, pp. 542–547. In Ontogeny and systematics of fishes, H. G. Moser, W. J. Richards, D. M. Cohen, M. P. Fahay, A. W. Kendall, Jr., and S. L. Richardson (Eds.). *Am. Soc. Ichthyol. Herpetol. Spec. Publ. 1.*

Ritchie, P. A., S. Lavoué, and G. Lecointre. 1997. Molecular phylogenetics and the evolution of Antarctic notothenioid fishes. *Comp. Biochem. Physiol. 118A*(4): 1009–1025.

Robins, C. R., R. M. Bailey, C. E. Bond, J. R. Brooker, E. A. Lachner, R. N. Lea, and W. B. Scott. 1991a. Common and scientific names of fishes from the United States and Canada (5th ed.). *Am. Fish. Soc. Spec. Publ. 20.*

Robins, C. R., R. M. Bailey, C. E. Bond, J. R. Brooker, E. A. Lachner, R. N. Lea, and W. B. Scott. 1991b. World fishes important to North Americans. *Am. Fish. Soc. Spec. Publ. 21.*

Rosa, I. L., and R. S. Rosa. 1987. *Pinguipes* Cuvier and Valenciennes and Pinguipedidae Gunther, the valid names for the fish taxa usually known as *Mugiloides* and Mugiloididae. *Copeia 1987:* 1048–1051.

Rosenblatt, R. H. 1984. Blennioidei: Introduction, p. 551–552. In Ontogeny and systematics of fishes, H. G. Moser, W. J. Richards, D. M. Cohen, M. P. Fahay, A. W. Kendall, Jr., and S. L. Richardson (Eds.). *Am. Soc. Ichthyol. Herpetol. Spec. Publ. 1.*

Roule, L. 1922. Description de *Scombrolabrax heterolepis* nov. gen. nov. sp., poisson abyssal nouveau de l'Île Madère. *Bull. Inst. Oceanogr. (Monaco) 408:* 1–8.

Rüber, L., R. Britz, H. H. Tan, P. K. L. Ng, and R. Zardoya. 2004. Evolution of mouthbrooding and life-history correlates in the fighting fish genus *Betta*. *Evol. 58:* 799–813.

Schoenfuss, H. L., and R. W. Blob. 2003. Kinematics of waterfall climbing in Hawaiian freshwater fishes (Gobiidae): Vertical propulsion at the aquatic–terrestrial interface. *J. Zool. (Lond.) 261:* 191–205.

Springer, V. G. 1983. *Tyson belos,* new genus and species of western Pacific fish (Gobiidae, Xenesthminae), with discussion of gobioid osteology and classification. *Smithson. Cont. Zool. 390.*

———. 1993. Definition of the suborder Blennioidei and its included families (Pisces: Perciformes). *Bull. Mar. Sci. 52*(1): 472–495.

———, and T. H. Fraser. 1976. Synonymy of the fish families Cheilobranchidae (= Alabetidae) and Gobiesocidae, with descriptions of two new species of *Alabes. Smithson. Cont. Zool. 234.*

———, and W. C. Freihofer. 1976. Study of the monotypic fish family Pholidichthyidae (Perciformes). *Smithson. Cont. Zool. 216.*

———, and G. D. Johnson. 2004. Study of the dorsal gill-arch musculature of teleostome fishes with special reference to the Actinopterygii. *Bull. Biol. Soc. Wash. 11.*

———, and T. M. Orrell. 2004. Appendix: Phylogenetic analysis of 147 families of acanthomorph fishes based primarily on dorsal gill-arch muscles and skeleton. *Bull. Biol. Soc. Wash. 11.*

Stepien, C. A., A. K. Dillon, M. J. Brooks, K. L. Chase, and A. N. Hubers. 1997. The evolution of blennioid fishes based on an analysis of mitochondrial 12S rDNA, pp. 245–270. In *Molecular systematics of fishes,* T. D. Kocher and C. A. Stepien (Eds.). Academic Press, San Diego.

Stevens, E. G., W. Watson, and A. C. Matarese. 1984. Notothenoidea: Development and relationships, pp. 561–564. In Ontogeny and systematics of fishes, H. G. Moser, W. J. Richards, D. M. Cohen, M. P. Fahay, A. W. Kendall, Jr., and S. L. Richardson (Eds.). *Am. Soc. Ichthyol. Herpetol. Spec. Publ. 1.*

Stiassny, M. L. J., and J. S. Jensen. 1987. Labroid interrelationships revisited: Morphological complexity, key innovations, and the study of comparative diversity. *Bull. Mus. Comp. Zool. 151*(5): 269–319.

Streelman, J. T., and S. A. Karl. 1997. Reconstructing labroid evolution with single-copy nuclear DNA. *Proc. R. Soc. Lond. B 264:* 1011–1020.

Tang, K. L. 2001. Phylogenetic relationships among damselfishes (Teleostei: Pomacentridae) as determined by mitochondrial DNA data. *Copeia 2001:* 591–601.

Thacker, C. 2000. Phylogeny of the wormfishes (Teleostei: Gobioidei; Microdesmidae). *Copeia 2000:* 940–957.

Trnski, T., J. M. Leis, and P. Wirtz. 1989. Pholidichthyidae—convict blennies, engineerfishes, pp. 259–261. In *The larvae of Indo-Pacific shorefishes,* J. M. Leis, and T. Trnski (Eds.). University of Hawaii Press, Honolulu.

Tyler, J. C., G. D. Johnson, I. Nakamura, and B. B. Collette. 1989. Morphology of *Luvarus imperialis* (Luvaridae), with a phylogenetic analysis of the Acanthuroidei (Pisces). *Smithson. Cont. Zool. 485.*

Wang, H.-Y., M.-P. Tsai, J. Dean, and S.-C. Lee. 2001. Molecular phylogeny of gobioid fishes (Perciformes: Gobioidei) based on mitochondrial 12S rRNA sequences. *Mol. Phylogen. Evol. 20*(3): 390–408.

Warner, R. R., and D. R. Robertson. 1978. Sexual patterns in the labroid fishes of the western Caribbean. I: The wrasses (Labridae). *Smithson. Cont. Zool. 254:* 1–27.

Watson, W., and H. J. Walker. 2004. The world's smallest vertebrate, *Schindleria brevipinguis,* a new paedomorphic species in the family Schindleriidae (Perciformes: Gobioidei). *Rec. Aust. Mus. 56:* 139–142.

Wheeler, A. 1975. *Fishes of the world: An illustrated dictionary.* MacMillan, New York.

Wiley, E. O., G. D. Johnson, and W. W. Dimmick. 2000. The interrelationships of acanthomorph fishes: A total evidence approach using molecular and morphological data. *Biochem. Syst. Ecol. 28:* 319–350.

17 THE SPINY AND THE STRANGE: SCORPAENOID, PLEURONECTIFORM, AND TETRAODONTIFORM FISHES

THE SCORPAENOIDEI: WELCOME TO THE PERCIFORMES—AGAIN

Relationships of "Scorpaeniform" Fishes
Suborder Scorpaenoidei

THE MOST HIGHLY DERIVED TELEOSTS?

Order Pleuronectiformes (Heterosomata)
    Suborder Psettodoidei
    Suborder Pleuronectoidei
Order Tetraodontiformes (Plectognathi)
    Suborder Triacanthoidei
    Suborder Tetraodontoidei

We close out our discussion of the diversity of fishes with three primarily marine taxa that demonstrate, to varying degrees, extensive modifications of the teleostean body plan. Perciform affinities are seen in two of these groups, the Scorpaenoidei and the Pleuronectiformes. Long considered a separate order of fishes, the scorpaenoids are now considered by some to be a perciform suborder. This diverse assemblage comprises a number of primarily benthic families and includes such familiar fishes as the sculpins and the rockfishes. The pleuronectiform flatfishes, though exhibiting some perciform affinities, display an extraordinary modification of their body plan, with the adoption of profound asymmetry. This is manifested in a benthic-adapted fish that lies on one side, with both eyes shifted over to the other side. Fishes of the order Tetraodontiformes show an extreme modification of the teleostean body plan, suggesting that they might be considered the most highly derived of the bony fishes. Rigid, with comparatively few vertebrae, they primarily rely on movements of their pectoral, dorsal, anal, or caudal fins for locomotion.

THE SCORPAENOIDEI: WELCOME TO THE PERCIFORMES—AGAIN

This fairly diverse assemblage of acanthopterygian fishes, once considered a separate order, the Scorpaeniformes, may now be considered a suborder of perciform fishes. Thus, the classification of scorpaeniform fishes has come full circle since Berg (1940) placed them among the Perciformes as the suborder Cottoidei. Gosline (1971) and Lauder and Liem (1983) regarded them as perciform derivatives, but Washington et al. (1984), Nelson (1994), Robins et al. (1991), and Eschmeyer (1998) considered them "pre-perciform." Mooi and Gill (1995) investigated the variation in the mode of attachment of epaxial muscles to dorsal fin pterygiophores in a broad range of acanthomorph fishes. Among their conclusions was the recommendation that the scorpaeniform fishes be considered a suborder within the Perciformes. Mooi and Gill (1995) pointed out several features that suggested a percoid affinity for the scorpaenoid fishes, observing that the general resemblance of many scorpaenoids to the percoid basses of the family Serranidae may be more than superficial.

Relationships of "Scorpaeniform" Fishes

Although many authors have acknowledged the so-called "scorpaeniform" fishes to have perciform affinities, relationships within the group have been subject to considerable revision. Imamura (1996) proposed that the flatheads of the suborder Platycephaloidei and the fishes of the suborder Scorpaenoidei form a monophyletic group; with the inclusion of the scorpaeniform fishes among the Perciformes, the flatheads are now considered a taxon within the suborder Scorpaenoidei (Mooi and Gill, 1995). Three families, including the **Platycephalidae,** are commonly referred to as flatheads. Some of these Indo-Pacific benthic marine fishes are highly esteemed as food fishes.

Suborder Scorpaenoidei

The suborder Scorpaenoidei is a rather large and important group that includes almost 1,300 species of predominantly marine benthic fishes.

DIAGNOSIS OF THE SCORPAENOIDEI
- Large head, with a posterior extension of the third infraorbital bone forming a suborbital stay that runs from the orbit to the preoperculum.
- Head and body often covered with bony plates or spines.
- Venom glands often associated with spines.
- Large, broad-based pectoral fins.

- Caudal fin usually rounded or truncate, rarely forked.
- Type I epaxial musculature morphotype (with partially separate muscle mass or muscle fiber slips inserting on processes on the radials of the pterygiophores*).
* A feature held in common with many perciform families.
Source: Lauder and Liem, 1983; Mooi and Gill, 1995.

The **suborbital stay** is a prominent feature that unites members of the suborder, as is a caudal fin skeleton with two platelike hypurals sutured to the terminal centrum (Lauder and Liem, 1983). Scorpaenoids typically have a large head, with numerous prominent spines, and large, broad-based pectoral fins. Many scorpaenoids have venomous spines; some are among the most dangerous animals in the ocean.

The largest family in the suborder Scorpaenoidei is the **Scorpaenidae** (see Fig. 3.4A), the scorpionfishes and their relatives, which Nelson (1994) divided into 12 subfamilies, with almost 400 species in about 60 genera. These are marine fishes, found mostly in temperate and tropical waters in the Pacific and Indian Oceans. The rockfishes and redfishes (Sebastinae) are well-known commercial fishes of both the North Pacific and the North Atlantic. A few more species are known from the South Atlantic. The most species-rich genus (*Sebastes*) includes at least 110 species distributed in the North Pacific Ocean and the Gulf of California (Eschmeyer et al., 1983; Love et al., 2002; Nelson, 1994). Eschmeyer (1998) has elevated the subfamily to comprise its own family, the Sebastidae, but Love et al. (2002) favored retaining the rockfishes within the family Scorpaenidae. They are locally known to sport fishers on the Pacific coast as "rock cod," "sea bass," and "snapper." The Acadian redfish (*Sebastes fasciatus*) and the deepwater redfish (*S. mentella*), both of the Atlantic, are marketed under the name "ocean perch." The Pacific Ocean perch (*Sebastes alutus*) is another important species. In recent years, commercially exploited species of rockfishes have suffered serious declines in abundance due to overfishing (see Chapter 39).

The scorpionfishes found in warmer waters are mostly colorful and well provided with spines, cirri, and fleshy flaps, giving many of them a bizarre appearance, which, however, renders them virtually indistinguishable from the substrate on which they rest (see Plate 17.1). Most scorpaenids have venom-producing tissues along the dorsal fin spines. Wounds caused by these spines are very painful.

The subfamily Pteroinae, typified by the tropical genus *Pterois,* the lionfishes and turkeyfishes, are quite venomous and can cause severe illness in persons punctured by the dorsal spines. In spite of this, they are a hardy and popular addition to marine aquaria (see Chapter 39).

The Indo-Pacific stonefishes of the family **Synanceidae** take their common name from their resemblance to the rocks among which they live. They are reputed to produce the strongest venom of all fishes in glands at the bases of the dorsal spines. This venom is discharged through grooves in the spines and can be fatal to humans (see Chapter 18).

The **Triglidae**, or searobins (see Fig. 3.12E), are characterized by bony-plated heads and large pectoral fins, with a few of the lower rays detached as separate, finger-like tactile and chemoreceptive organs. The pelvic fins are relatively large and strong and aid in "walking" along the bottom—the similarities of members of this family to the gurnards have been noted (see Chapter 16). Some species of this widely distributed family are commercially harvested.

The family **Anoplopomatidae** contains the sablefish (*Anoplopoma fimbria*) and the skilfish (*Erilepis zonifer*), both of the North Pacific. The sablefish, known also as "black cod," has very oily flesh and is the subject of a commercial fishery.

The family **Hexagrammidae** includes the greenlings (*Hexagrammos*), distinguished by an unusual lateral line system consisting of several canals running along the body (Fig. 17.1A). Also in the family are the pelagic Atka mackerel (*Pleurogrammus monopterygius*) and the bottom-dwelling ling cod (*Ophiodon elongatus*). Those that are discouraged from eating ling cod because of the green tinge of the un-cooked flesh are missing a treat; this fish is highly prized by sportfishers.

The **Cottidae** (sculpins) are usually large-headed fishes with large, fanlike pectoral fins (Fig. 17.1B), and many are characterized by strong spines on the gill cover. They often have bony plates on the skin, and they seldom have more than a few rows of scales. They are typically marine fishes of the temperate and cold waters of the Northern Hemisphere. There are about 70 genera and 300 species (Nelson, 1994). Freshwater members are mainly in the genus *Cottus*, whose species are common bottom-dwelling inhabitants of cold streams. Whereas the percid darters have become the most diverse benthic fishes in streams east of the Rocky Mountains in North America, sculpins seem to have occupied that niche in drainages of the Pacific slope. Sculpins are widespread in Pacific coast inshore marine waters and coastal streams, but only a few species are seen in freshwaters of central and east-ern North America.

Three families, the Cottocomephoridae, Comephoridae, and Abyssocottidae have become particularly associated with the remarkable ecology and biogeographical history of Lake Baikal in Siberia (see Chapters 29 and 32). The **Cotto-comephoridae** are a relatively large Lake Baikal group, with at least eight genera and 24 species. Nelson (1994) grouped these as a subfamily within the Cottidae. Most species tend to be heavy-bodied, benthic forms, but some exploit open-water habitats in what is the oldest and deepest inland body of water on Earth.

FIGURE 17.1
Representatives of the suborder Scorpaenioidei: **A**, Greenling (*Hexagrammos*); **B**, Sculpin (*Cottus*).

The **Comephoridae**, or Baikal oilfishes, are viviparous pelagic fishes. The two species of *Comephorus* inhabit deep water but swim near to the surface at night. They are nearly colorless and have large pectoral fins. Unlike most fishes in the suborder, they have a caudal fin that is truncate to deeply emarginate. The sex ratio is said to be heavily skewed toward females (Wheeler, 1975).

The **Abyssocottidae** are a generally deepwater family, consisting of six genera and 20 species (Nelson, 1994). In his classic work on the Lake Baikal sculpins, Taliev (1955) classified the abyssocottids within the family Cottidae.

The **Hemitripteridae** are a marine group with eight species primarily known from the North Pacific, although one species, the sea raven (*Hemitripterus americanus*) is known from the Atlantic. The hemitripterids have been proposed to be a sister group of the poachers (family Agonidae) described further on (Washington et al., 1984; Yabe, 1985).

The "blob" sculpins or "fatheads" of the genera *Ebinania, Neophrynichthys,* and *Psychrolutes* were retained in the Cottidae by Robins et al. (1991), but Nelson (1982, 1994) recognized the group as a separate family, the fathead sculpins (**Psychrolutidae**). These are more or less tadpole-shaped marine sculpins of the Pacific, Indian, and Atlantic oceans (see Plate 17.2). Some species are quite small, but others, such as the flabby, globose *Psychrolutes phrictus* of the eastern North Pacific may reach 70 cm and nearly 10 kg.

The family **Agonidae** contains the poachers and alligatorfishes, curious armored fishes of cold marine waters. They are present in the Arctic, Antarctic, and North Atlantic oceans, along the coasts of southern South America, and are especially numerous in the North Pacific.

The **Cyclopteridae** contains the lumpfishes, whereas the snailfishes are now considered to constitute a separate family, the **Liparidae** (= Liparididae; Kido, 1988). Most of the fishes in these two families have their pelvic fins modified into a round suctorial disc, with which they cling to rocks and vegetation (see Fig. 3.13G). They have a worldwide distribution in cold marine waters, where they range from the intertidal zone to depths of more than 7,000 m. Snailfishes have a thin, usually scaleless skin, with a loose, gelatinous layer just beneath. This gives the impression of a fish wearing a skin suit that is a few sizes too large for it. Lumpfishes produce a palatable caviar that is dyed black to more closely resemble that of sturgeons (see Chapter 8).

THE MOST HIGHLY DERIVED TELEOSTS?

In most systems of teleost classification, the flatfishes and the tetraodontiform fishes are the last two orders considered—the perception being that these constitute the most highly derived groups of bony fishes. Herein lies the problem: Both of these primarily marine orders display extraordinary anatomical modifications that result in unique lifestyles, especially where feeding and locomotion are considered. But does the fact that they are usually the last taxa considered in systematic treatments mean that they should both be the last to "climb aboard" the cladogram? Do their extreme anatomical modifications warrant their being considered the most derived taxa from a phylogenetic perspective? Or is their true phylogeny nested somewhere among the other groups we have considered? The extent to which these two orders depart from the "traditional" perciform body plan makes the determination of their phylogenies especially problematic. They do not appear to share a common ancestry—so why should they share the final chapter of this systematic treatment of fishes? The anatomical modifications evident in each order do present a convincing argument for their being considered the "ultimate" teleosts, however.

Of the two orders, flatfishes are much more likely to form a monophyletic group with other perciform fishes (Joseph Nelson, personal communication). What sets them apart is the acquisition of a most unusual metamorphic event during their ontogeny. Flatfishes display one of the most extraordinary modifications of the teleost body plan when they shift from a bilaterally symmmetrical larva to an extremely asymmetrical juvenile and adult. During this metamorphosis, the eye from one side migrates to the other side, and the fish assumes a benthic existence lying on its "blind" side. Much modification of the skeleton and other tissues accompanies this dramatic transformation. The end result is an entirely novel application of lateral undulations of the trunk muscles for the purpose of locomotion; the fish swims with its laterally compressed body oriented horizontally instead of vertically.

The tetraodontiform fishes are generally viewed as the most advanced teleosts; their generally rigid body is propelled using sophisticated appendicular motions (see Chapter 19). This shift from a locomotor mode that relies heavily on the lateral undulation of trunk muscles working against a flexible vertebral column to one in which the vertebral column is comparatively rigid somewhat parallels the evolutionary history of tetrapod locomotion, where locomotor activity likewise becomes more focused on the appendages and less on motions of the trunk.

In addition to the usual prey available to fishes, the Tetraodontiformes, by virtue of the evolution of powerful jaws with well-developed dentition, are able to consume animals such as corals, urchins, and arthropods that are otherwise protected from predation by virtue of their hard skeletal enclosures.

Order Pleuronectiformes (Heterosomata)

The flatfishes are mostly marine, benthic carnivores, common to most coasts. A few enter freshwater, and some are found at great depths in the oceans. They are important food fishes.

DIAGNOSIS OF THE PLEURONECTIFORMES
- Adults not bilaterally symmetrical; body with extreme lateral compression.
- Dorsal and anal fins with long bases; the dorsal fin usually extends to the cranium.
- Branchiostegal rays usually number six or seven.
- Scales cycloid, ctenoid, or with tubercles.
- Swim bladder rarely present.

Source: Nelson, 1994.

Early in their development, after a period of bilateral symmetry, flatfishes begin side-swimming, and an eye migrates from what becomes the bottom side to what becomes the upper side. In the starry flounder (*Platichthys stellatus*, family **Pleuronectidae**), this metamorphosis begins 27 to 104 days after egg fertilization, and eye migration requires about five days. Starry flounders are one of the few pleuronectiform species that are bimodal in terms of which side the fish lies upon (Fig. 17.2A). A distinct geographic trend is observable, with about half of the individuals from the coast of California to southeast Alaska being right-eyed (**dextral**), with left-eyed (**sinistral**) fish becoming increasingly common in populations from the Alaskan Peninsula. All individuals from the Japanese coast are left-eyed (Hart, 1973). Policansky (1982) found that the direction of eye migration in the starry flounder is genetically based and involves one allele. Starry flounders therefore provide a good demonstration of the fact that, although some pleuronectiform families may be predominantly dextral whereas others are sinistral, there is *no* real phylogenetic significance to the direction of eye migration. The condition has apparently arisen several times in the evolution of flounders and soles (Berendzen and Dimmick, 2002; Chapleau, 1993).

Generally, in flatfishes, the side turned toward the bottom is blind and mostly lacks pigmentation. The upper side is often cryptically colored and capable of rapid color change, allowing some species to become almost invisible to predators or prey (see Plate 17.3A). They are usually aided in this camouflage by the sediments in which they are partially buried or which they may distribute over themselves. There is much variation among the nearly 570 extant species, so that the classification of the flatfishes presents problems that seem to prevent ichthyologists from reaching a consensus on the relationships within the group.

A history of the classification of the Pleuronectiformes shows evidence of much disagreement concerning the number and arrangement of suborders and families in the group (Ahlstrom et al., 1984; Berg, 1940; Chapleau, 1993; Cooper and Chapleau, 1998; Eschmeyer and Bailey, 1990; Lindberg, 1974; Nelson, 1994). Nelson's (1994) classification is consistent with current views in its recognition of two suborders, the Psettodoidei and the Pleuronectoidei. Molecular studies by Berendzen and Dimmick (2002) confirmed this subdivision and affirmed the monophyly of the order.

Suborder Psettodoidei

This suborder is composed of a single monogeneric family, **Psettodidae**, in which spinous rays are seen in the dorsal, pelvic, and anal fins, and in which the dorsal fin does not extend onto the head. The pelvic fins are almost symmetrical, and the eyes may be on either the right or the left side. Because the psettodids show the least amount of movement of the eye during metamorphosis, they are generally considered the most basal of the flatfishes. *Psettodes* has several percoid characteristics, causing some ichthyologists to suggest a relationship with the Serranidae, but Chapleau (1993) stated that there is little evidence for such a relationship. The genus *Psettodes* is found in marine waters of tropical Africa, the Red Sea, and the Indo–West Pacific. Phylogenetic studies based on muscle and skeletal features (Hoshino, 2001) and the aforementioned molecular studies by Berendzen and Dimmick (2002) have both confirmed the status of the psettodids as a sister group to all other flatfishes.

Suborder Pleuronectoidei

Fishes in this suborder have no spines in the dorsal or anal fins. Berendzen and Dimmick (2002) provided evidence for the existence of three clades among the pleuronectoids. Fishes of the families Citharidae, Cynoglossidae, Samaridae, Soleidae, *Trinectes* (Achiridae), and *Poecilopsetta* (Poecilopsettidae) appear to constitute one distinct clade. The families Bothidae and part of the family Paralichthyidae constitute another clade, and the third clade consists of the Pleuronectidae and the other members of the Paralichthyidae. What is most apparent in this classification and in that proposed by Hoshino (2001) is the amount of work still to be done before a thorough understanding of the phylogenetic interrelationships of this complex and confusing group can be achieved.

Although the fishes of the family **Citharidae** are unique among the pleuronectoids in having pelvic spines, the family is probably not monophyletic (Hensley and Ahlstrom, 1984). Citharids are found in the Indo-Pacific, the Mediterranean, and West Africa.

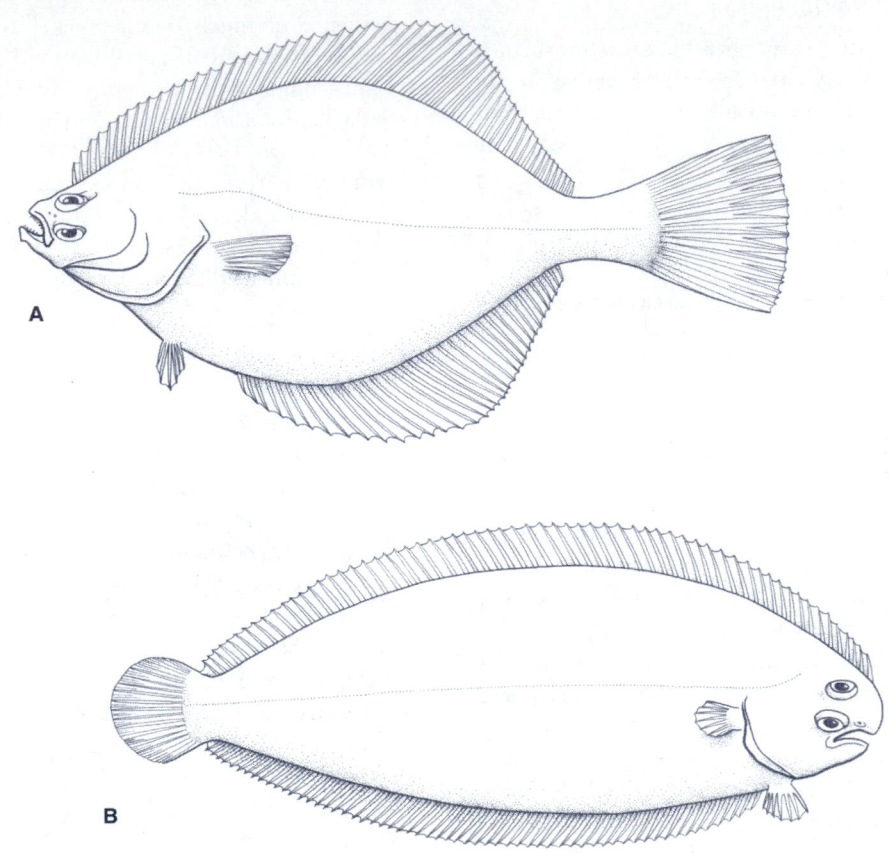

FIGURE 17.2
Representatives of the order Pleuronectiformes: **A,** Starry flounder (*Platichthys*); **B,** Sole (*Solea*).

The family **Samaridae** includes about 20 species, found mainly in deeper waters of the tropical and subtropical Indo-Pacific. This family was formerly considered to be part of the Pleuronectidae (Chapleau, 1993).

The dextral **Achiridae** (American soles) are fishes of the Americas, and are found both in rivers and in marine habitats, especially in South America. There are nine genera and about 28 species.

The family **Soleidae** includes the true soles. In these dextral fishes, the mouth is not terminal but shifted toward the side that rests on the bottom (Fig. 17.2B). Members of the Soleidae tend to prefer warm waters, but a few are found in temperate seas. Some species are known to enter rivers, especially in Asia, Australia, and Africa. The skin of the Moses sole (*Pardachirus marmoratus*) secretes a substance that acts as a powerful predator deterrent (see Chapter 18). A European species (*Solea solea*) possesses a black-tipped pectoral fin that is raised when the fish is theatened. This

action may mimic the erection of venomous dorsal fin spines by the sympatric trachinid weeverfish (see Chapter 16).

Tonguefishes of the family **Cynoglossidae,** which are considered by some ichthyologists to be soleids, are found primarily in tropical and subtropical seas. Chapleau (1988) confirmed them to be a monphyletic group. These are left-eyed, slender fishes that have the dorsal and anal fins confluent with the caudal fin. They are used as food wherever large enough species exist. Some species are no more than 50 cm long. Most North American species are in the genus *Symphurus* and are mostly too small to be taken commercially.

The family **Bothidae** includes the lefteye flounders— species with both eyes on the left side. This widespread group consists of 21 genera (Hensley, 1986) and nearly 120 species, only a few of which are found in North America. In many species, the eyes are widely separated, with the males of some species showing a wider separation than the females. Examples include *Bothus lunatus*, the peacock flounder, and

B. ocellatus, the eyed flounder, both of which range through-out the Caribbean. These species exemplify the remarkable cryptic coloration capabilities of the flatfishes (see Plates 17.3A, B). Some slender-bodied tropical species are noted for their large mouths. Examples are *Chascanopsetta lugubris* and *Kamoharaia megastoma,* in which the mandibles are longer than the head.

The **Scophthalmidae** were recognized as a family by Hensley and Ahlstrom (1984). This sinistral (left-eyed) family contains deep-bodied species from the Black and Mediterranean seas and the North Atlantic. The turbot (*Psetta maxima*), which reaches a meter in length, is a prized food fish in Europe. The proper phylogenetic position of this family remains unresolved (Berendzen and Dimmick, 2002; Chapleau, 1993).

The family **Pleuronectidae,** or righteye flounders, has representatives in most seas, from the arctic regions to southern Australia. It includes more than 90 species that are nearly all dextral. The family is diverse, and four or five subfamilies are generally recognized. Chapleau and Keast (1988) recommended that the subfamilies be elevated to families.

The subfamily Pleuronectinae comprises about 60 species, many of which are well-known commercial fishes, such as the halibuts of the genus *Hippoglossus*. These are the largest of the flatfishes, the Atlantic halibut (*H. hippoglossus*) weighing more than 300 kg and the Pacific halibut (*H. stenolepis*) attaining about 210 kg. Native Americans gill-netting near Sitka, Alaska, caught a Pacific halibut that was estimated to weigh more than 500 kg (http://www.halibut.net).

Some pleuronectines are called "soles" even though they do not show the characteristics of Soleidae. Examples are the petrale sole (*Eopsetta jordani*), the rex sole (*Glyptocephalus zachirus*), and the rock sole (*Lepidopsetta bilineata*). These so-called "soles" are generally smaller than the flounders and halibuts, but the species that are large enough to sustain a fishery are highly prized as food. The yellowfin sole (*Limanda aspera*) has been targeted by commercial fisheries in the Bering Sea. The plaice (*Pleuronectes platessa*) is an important commercial fish in Europe. The starry flounder (*Platichthys stellatus*) is a favored sport and food fish of the Pacific coast of North America.

Another problematic family, the **Paralichthyidae,** is likely to be polyphyletic but shares some affinities with the pleuronectids (Berendzen and Dimmick, 2002). These sinistral fishes include several food fishes of the coasts of the Atlantic, Indian, and Pacific oceans. Within this family, one group includes the genus *Paralichthys*. This genus comprises well-known species such as *P. californicus,* the California halibut; *P. olivaceus,* the olive flounder of the western Pacific; and *P. dentatus,* the summer flounder of the western Atlantic. The other paralichthyid assemblage includes the widespread

genus *Citharichthys* (sanddabs and whiffs), which includes a species that ascends rivers.

Order Tetraodontiformes (Plectognathi)

The molas, boxfishes, puffers, and their relatives all have a small, terminal mouth with strong jaws and incisors or a sharp beak composed of modified jaw elements. If the simplification of genetic composition is deemed a significant indicator of phylogenetic advancement, then the tetraodontiformes are possibly the most highly derived of all the fishes. The sequencing of the genome of the pufferfish (*Takifugu rubripes*) has revealed an extremely compact genome that is similar to that of humans in the total number of genes but has eliminated most of the repetitive duplications of DNA that fill out the human genome (Aparicio et al., 2002). Comparison of the genome of the freshwater pufferfish (*Tetraodon nigroviridis*) with that of humans has proven instrumental in confirming that the duplication of the whole genome has occurred in the evolution of actinopterygians (see Chapter 28) and in reconstructing the ancestral vertebrate genome (Jaillon et al., 2004).

DIAGNOSIS OF THE TETRAODONTIFORMES
- Maxillae and premaxillae fused or at least strongly bound together.
- Nasals, parietals, and suborbitals absent.
- Palatines and hyomandibulars are ankylosed to the skull.
- Gill openings restricted; operculum and subopercular are reduced in size; preoperculum is elongate in many.
- Scales are usually modified into plates or spines.
- Swim bladder present in all except the molids.

Source: Tyler, 1980.

The tetraodontiform fishes are also characterized by a pronounced reduction in vertebral number. This affects the degree to which the trunk can flex—an important feature in the generation of escape responses (see Chapter 19). Limited vertebral flexion may therefore be correlated with the evolution of an array of mechanical and chemical defensive mechanisms among tetraodontiform fishes, including sharp spines with frictional locking mechanisms; hard, bony carapaces; the capacity to inflate; and toxic secretions. Inflation capacity appears to have independently evolved twice in tetraodontiform fishes, whereas the rigid carapace has evolved four times (Brainerd and Patek, 1998).

Winterbottom (1974) and Friel and Wainwright (1997) based their classifications of the order on myological features, whereas Tyler's (1980) classification is based on osteological characteristics. Early life history characteristics have also been used to discern phylogenetic relationships (Leis, 1984). Most of the differences in the phylogenies erected by various authors concern the taxonomic ranking of groups as either families or subfamilies and the placement of the Balistidae and Ostraciidae, which Winterbottom (1974) placed in the suborder Tetraodontoidei. Both Eschmeyer (1998) and Nelson (1994) recognized the nine families discussed here.

The origins of the Tetraodontiformes have been the subject of considerable investigation. The order is well represented in the fossil record, possibly dating back as early as the Late Cretaceous (Sorbini, 1979). The exceptional preservation of a broad array of fossils from Eocene sites in northern Italy has provided important information on the early history of this highly derived group (Tyler and Santini, 2002). There are about 100 genera and 340 living species, with only about 20 recorded from freshwater (Nelson, 1994). The Acanthuroidei and the Zeiformes (see Chapter 14) have both been advanced as sister taxa to the Tetraodontiformes, which Nelson (1994) considered a "post-perciform" group.

Suborder Triacanthoidei

This suborder contains the spikefishes and triplespines. They have a large, locking pelvic fin spine and one or two pelvic soft rays. The dorsal fin usually has six strong spines. Two families, the **Triacanthodidae** and the **Triacanthidae**, constitute this suborder. Whereas the triacanthodid spikefishes range from the western Atlantic through the Indo-Pacific, the triacanthid triplespines are known only from the Indo-Western Pacific, although several fossil triacanthid taxa are known from Europe (Santini and Tyler, 2002a).

Suborder Tetraodontoidei

Nelson (1994) has divided this diverse group of fishes into three superfamilies. The pelvic spines in fishes of this suborder are lacking or very small.

Superfamily Balistoidea

This group includes fishes often referred to as leatherjackets, because of the appearance of the skin in which the scales are often embedded. Prominent dorsal spines, with a mechanism that permits the first spine to be locked in an erect position, also characterize the balistoideans. Families in the group are all marine, from the Atlantic, Pacific, and Indian oceans. The families are the triggerfishes (**Balistidae**; see Plate 17.4) and the filefishes (**Monacanthidae**). The Monacanthidae were considered a subfamily of the Balistidae by Winterbottom (1974) and Robins et al. (1991). They are most diverse in the coastal tropical waters of Australia. The largest member of the superfamily Balistoidea is the meter-long scrawled filefish (*Aluterus scriptus*) of tropical and subtropical waters.

Superfamily Ostracoidea

The boxfishes and trunkfishes (**Ostraciidae**) are the only members of the superfamily Ostracoidea. The ostraciids, like the aforementioned balistoideans, are mainly shore fishes of the tropics, some with brilliant coloration. Members of both these superfamilies have separate, well-developed teeth. The trunkfishes and boxfishes are enclosed in a bony carapace so that only the eyes, jaws, and fins are mobile. Extensive vertebral fusion, resulting in a reduction of overall vertebral numbers, is observed in the boxfishes (Klassen, 1996).

Superfamily Tetraodontoidea

This superfamily includes fishes that are usually covered by spines and can fill their stomach (or a sac that evaginates from the anterior part of the stomach) with air or water, so that an individual can inflate itself like a spiny balloon. The stomach of the balloonfish (*Diodon holocanthus*) is a simple sac that has apparently lost its digestive function (Brainerd, 1994). Inflation is initiated by coordinated cycles of buccal expansion and compression, brought about by rotation of the hyoid apparatus and pectoral girdle; coordinated activity of the pyloric and cardiac sphincters help sustain the inflated state. The collagen fibers of the dermis are configured to accommodate the dramatic increase in body volume and, in the case of the spiny puffers, support the erectile spines (Brainerd, 1994; Wainwright et al., 1995). These fishes have no true teeth; the sharp edges of the jawbones form a beaklike structure useful for crushing heavily armored invertebrates. The parrotfishes of the family Scaridae are convergent in displaying a similar beak, although in their case, it is composed of fused teeth and not of the jawbones.

The family **Triodontidae** contains but one species, the threetooth puffer (*Triodon macropterus*) of seas from Japan to Indonesia.

The **Tetraodontidae**, or puffers, produce a potent neurotoxin called *tetrodotoxin* (= *tarichatoxin*), which can be fatal (see Chapter 18). Tetrodotoxin is apparently produced by symbiotic bacteria and is most concentrated in the liver, skin, and ovaries. The Japanese delicacy *fugu* is prepared from the flesh of the puffer, but great care must be taken to avoid tissues containing concentrations of the toxin. Tetrodotoxin poisoning can bring on profound states of inertia; the catatonic states of "zombies" from the Caribbean *vodou* culture are reported to be induced by ingestion of preparations from the skin of puffers. Puffers are mostly marine and are found in tropical and subtropical waters of the Indian, Pacific, and Atlantic oceans. Some species live in the freshwaters of southern Asia and Africa.

Members of the family **Diodontidae** (porcupinefishes; considered part of the Tetraodontidae by Robins et al., 1991), have longer spines on the skin than puffers (see Plate 17.5). Their inflated and dried skins are familiar curios in many tropical areas.

The family **Molidae** contains the molas. In these fishes of coastal waters and the open ocean, the early larvae resemble those of other tetraodontiform fishes, but as the fish develops, the caudal portion becomes restricted. This results in one of the most peculiar of all body forms seen among the teleosts (Fig. 17.3A, B; see Plate 17.6). These fishes have evolved a unique "pseudocaudal" fin at their posterior margin; it is composed of soft dorsal- and anal-fin rays that have migrated to the end of the animal together with supporting pterygiophores from the immediately adjacent dorsal and anal fins (Santini and Tyler, 2002b; Fig. 17.4). There are three monotypic genera: *Masturus* and *Mola* are closely related and together form a sister clade to *Ranzania* (Santini and Tyler, 2002b). Although the family is mostly tropical and subtropical, *Mola mola* ranges into temperate waters. This species, commonly called the ocean sunfish, may weigh up to 1,000 kg. It is often seen at the surface, sometimes on its side as if basking in the sun. Molas range to a depth of more than 350 m and feed on a variety of invertebrates, including jellyfishes, and some small fishes. Molas may be the most fecund fishes in existence, as the mature females can produce as many as 300 million eggs. Out in the open ocean, substrates on which to attach and associate are rare. Consequently, molas are usually heavily parasitized, both externally and internally.

The molas are generally considered to be the most derived members of what we have already observed to be the most highly derived order of teleosts. It is therefore particularly interesting to observe the extent to which their adaptation to an open-ocean mode of existence has resulted in many skeletal elements being replaced by cartilage. A peculiar form of phylogenetic symmetry is therefore apparent, as molas bring fishes full circle back to the earliest cartilaginous forms, such as the hagfishes, lampreys, and sharks.

KEY POINTS AND CONNECTIONS

- The scorpaenoids are a diverse assemblage of perciform fishes that had, until recently, been considered a separate order, the Scorpaeniformes. They are generally large-headed fishes with prominent spines in the fins and on opercular surfaces. Venom glands are often associated with the stout, sharp spines of scorpaenoids, and this suborder includes some of the most dangerous fishes. A prominent suborbital stay, a skeletal element that runs beneath the orbit, also characterizes the group.

  A discussion of the properties of fish venoms, including those of scorpaenoid fishes, is included in Chapter 18.

- The flatfishes of the order Pleuronectiformes are among the most morphologically modified of all fishes. Their bilaterally symmetrical larvae undergo a dramatic transformation, resulting in an extremely compressed body with the eyes situated on one side of the body. The most primitive of the flatfishes are in the suborder Psettodoidei,

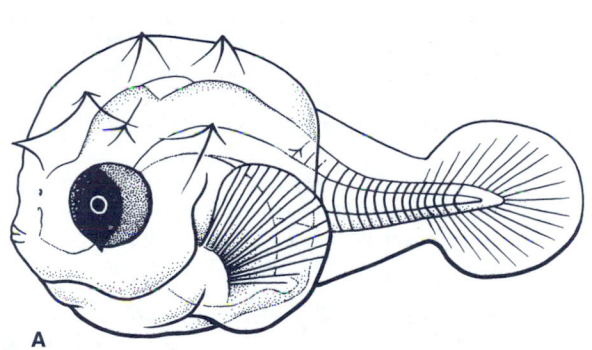
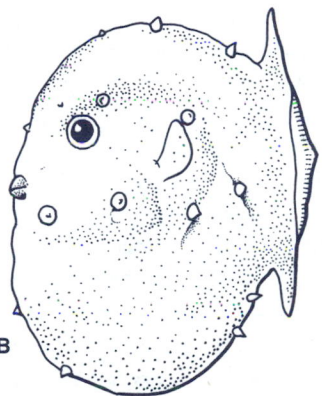

FIGURE 17.3
Transformation of larval *Mola mola*: **A,** Larva at about 1.84 mm (from Sanzo, 1939); **B,** Larva at 37 mm (from Reuvens, 1894). Note prominent body spines in early larval stage.

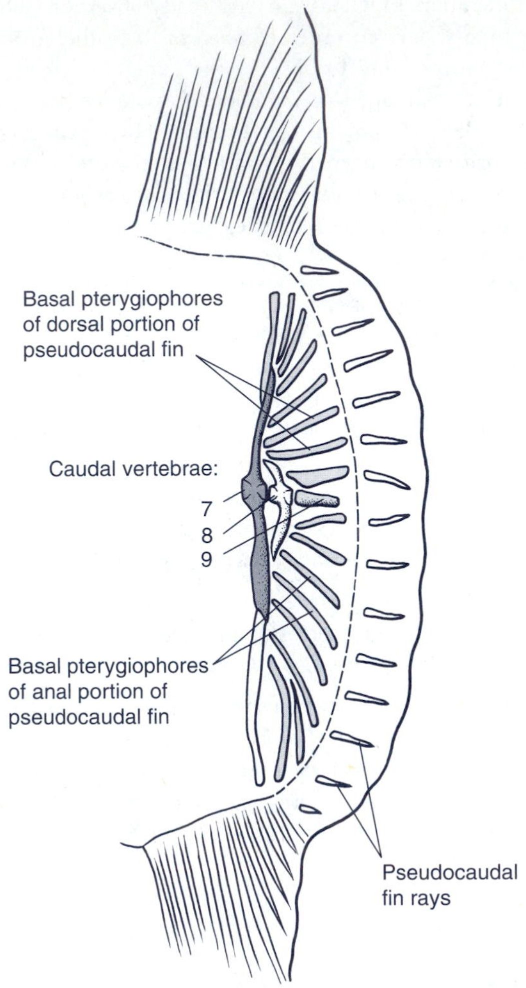

Basal pterygiophores
of dorsal portion of
pseudocaudal fin

Caudal vertebrae:

7
8
9

Basal pterygiophores
of anal portion of
pseudocaudal fin

Pseudocaudal
fin rays

FIGURE 17.4
Skeletal structure of the "pseudocaudal" fin of *Mola mola*.

characterized by retention of spines in the fins and a greater degree of remaining symmetry. Pleuronectiformes show features suggesting that they may form a monophyletic group with the Perciformes. Many species of these benthic inhabitants are among the most highly esteemed food fishes.

• Tetraodontiform fishes may be the most derived of all the bony fishes. These fishes generally have a small mouth with powerful jaws equipped with sharp teeth or tooth plates derived from the jawbones. The number of vertebrae is reduced, and these fishes generally keep their body fairly rigid during swimming, relying on movements of the fins for locomotion.

The ostraciiform, balistiform, and tetraodontiform modes of locomotion, as discussed in Chapter 19, are named after families of tetraodontiform fishes.

BUILDING AN ICHTHYOLOGY LIBRARY

An outstanding and highly readable introduction to one of the most commercially significant families of ocean fishes:

Love, M. S., M. Yoklavich, and L. Thorsteinson. 2002. *The rockfishes of the Northeast Pacific.* University of California Press, Berkeley.

REFERENCES

Ahlstrom, E. H., K. Amaoka, D. A. Hensley, H. G. Moser, and B. Y. Sumida. 1984. Pleuronectiformes: Development, pp. 640–670. In Ontogeny and systematics of fishes, H. G. Moser, W. J. Richards, D. M. Cohen, M. P. Fahay, A. W. Kendall, Jr., and S. L. Richardson (Eds.). *Am. Soc. Ichthyol. Herpetol. Spec. Publ. 1.*

Aparicio, S., J. Chapman, E. Stupka, and 38(!) co-authors. 2002. Whole-genome shotgun assembly and analysis of the genome of *Fugu rubripes. Science 297:* 1301–1310.

Berendzen, P. B., and W. W. Dimmick. 2002. Phylogenetic relationships of pleuronectiformes based on molecular evidence. *Copeia 2002:* 642–652.

Berg, L. S. 1940. Classification of fishes both recent and fossil. *Trav. Inst. Zool. Acad. Sci. USSR 5:* 87–517 (Reprinted 1947, J. W. Edwards, Ann Arbor, MI).

Brainerd, E. L. 1994. Pufferfish inflation: Functional morphology of post-cranial structures in *Diodon holocanthus* (Tetraodontiformes). *J. Morphol. 220:* 243–261.

———, and S. N. Patek. 1998. Vertebral column morphology, C-start curvature, and the evolution of mechanical defenses in tetraodontiform fishes. *Copeia 1998:* 971–984.

Chapleau, F. 1988. Comparative osteology and intergeneric relationships of the tongue soles (Pisces: Pleuronectiformes: Cynoglossidae). *Can. J. Zool. 66:* 1214–1232.

———. 1993. Pleuronectiform relationships: A cladistic reassessment. *Bull. Mar. Sci. 52:* 516–540.

———, and A. Keast. 1988. A phylogenetic reassessment of the monophyletic status of the family Soleidae, with comments on the suborder Soleoidei (Pisces: Pleuronectiformes). *Can. J. Zool. 66:* 1214–1232.

Cooper, J. A., and F. Chapleau. 1998. Monophyly and intrarelationships of the family Pleuronectidae (Pleuronectiformes) with a revised classification. *Fish. Bull. 96:* 686–726.

———. 1998. Catalog of fishes, Vol. 3, Genera of fishes. *Calif. Acad. Sci. Spec. Publ. 1.*

———, and R. M. Bailey. 1990. Genera of recent fishes, pp. 7–433. In *Catalog of the genera of recent fishes,* W. N. Eschmeyer (Ed.). California Academy of Sciences, San Francisco.

———, E. S. Herald, and H. Hammann. 1983. *A field guide to Pacific coast fishes of North America.* Houghton Mifflin, Boston.

Friel, J. P., and P. C. Wainwright. 1997. A model system of structural duplication: Homologies of adductor mandibulae muscles in tetraodontiform fishes. *Syst. Biol. 46(3):* 441–463.

Gosline, W. A. 1971. *Functional morphology and classification of teleostean fishes.* University Press of Hawaii, Honolulu.

Hensley, D. A. 1986. Current research on Indo-Pacific bothids. In *Indo-Pacific fish biology,* T. Uyeno, R. Arai, T. Taniuichi, and K. Matsuura (Eds.). Ichthyological Society of Japan, Tokyo.

———, and E. H. Ahlstrom. 1984. Pleuronectiformes: Relationships, pp. 670–687. In Ontogeny and systematics of fishes, H. G. Moser, W. J. Richards, D. M. Cohen, M. P. Fahay, A. W. Kendall, Jr., and S. L. Richardson (Eds.). *Am. Soc. Ichthyol. Herpetol. Spec. Publ. 1.*

Hoshino, K. 2001. A new hypothesis of intrarelationships of the Pleuronectiformes (Teleostei) based on mycological and osteological characters. *Abstr. 81st Ann. Meeting Am. Soc. Ichthyol. Herpetol.,* Pennsylvania State University.

Imamura, H. 1996. Phylogeny of the family Platycephalidae and related taxa (Pisces: Scorpaeniformes). *Species Diversity 1:* 123–233.

Jaillon, O., and 60(!) co-authors. 2004. Genome duplication in the teleost fish *Tetraodon nigroviridis* reveals the early vertebrate proto-karyotype. *Nature 431:* 946–957.

Kido, K. 1988. Phylogeny of the family Liparididae, with the taxonomy of the species found around Japan. *Mem. Fac. Fish. Hokkaido Univ. 35*(1): 125–256.

Klassen, G. J. 1996. Fusion complex of abdominal vertebrae 1–5 in boxfishes (Tetraodontiformes: Ostraciidae): Reinterpreting character evolution. *Copeia 1996:* 859–865.

Lauder, G. V., and K. F. Liem. 1983. Patterns of diversity and evolution in ray-finned fishes, pp. 1–24. In *Fish neurobiology,* Vol. 1, R. G. Northcutt and R. E. Davis (Eds.). University of Michigan Press, Ann Arbor.

Leis, J. M. 1984. Tetraodontiformes: Relationships, pp. 459–463, In Ontogeny and systematics of fishes, H. G. Moser, W. J. Richards, D. M. Cohen, M. P. Fahay, A. W. Kendall, Jr., and S. L. Richardson (Eds.). *Am. Soc. Ichthyol. Herpetol. Spec. Publ. 1.*

Lindberg, G. U. 1974. *Fishes of the world: A key to families and a checklist.* Wiley, New York.

Love, M. S., M. Yoklavich, and L. Thorsteinson. 2002. *The rockfishes of the Northeast Pacific.* University of California Press, Berkeley.

Mooi, R. D., and A. C. Gill. 1995. Association of epaxial musculature with dorsal-fin pterygiophores in acanthomorph fishes, and its phylogenetic significance. *Bull. Nat. Hist. Mus. Lond. (Zool.) 61*(2): 121–137.

Nelson, J. S. 1982. Two new South Pacific fishes of the genus *Ebinania* and contributions to the systematics of Psychrolutidae (Scorpaeniformes). *Can. J. Zool. 60*(6): 1470–1504.

———. 1994. *Fishes of the world* (3rd ed.). Wiley, New York.

Policansky, D. 1982. The asymmetry of flounders. *Sci. Am. 246*(5): 116–122.

Reuvens, C. L. 1894. Remarks on the genus *Orthragoriscus. Notes Leyden Mus. 16:* 126–130.

Robins. C. R., R. M. Bailey, C. E. Bond, J. R. Brooker, E. A. Lachner, R. N. Lea, and W. B. Scott. 1991. Common and scientific names of fishes from the United States and Canada (5th ed.). *Am. Fish. Soc. Spec. Publ. 20.*

Santini, F., and J. C. Tyler. 2002a. Phylogeny and biogeography of the extant species of triplespine fishes (Triacanthidae, Tetraodontiformes). *Zool. Scripta 31:* 321–330.

———, and———. 2002b. Phylogeny of the ocean sunfishes (Molidae, Tetraodontiformes), a highly derived group of teleost fishes. *Ital. J. Zool. 69:* 37–43.

Sanzo, L. 1939. Rarissimi stadi larvali di Teleostei. *Arch. Zool. Ital. 26:* 121–151.

Sorbini, L. 1979. Segnalazione di un plettognato Cretacico. *Plectocretacicus* nov. gen. *Boll. Mus. Civ. St. Nat. Verona 6:* 1–4.

Taliev, D. N. 1955. *Sculpin fishes of Baikal (Cottidae).* Akad. Nauk, USSR, Moscow and Leningrad.

Tyler, J. C. 1980. Osteology, phylogeny, and higher classification of the fishes of the order Plectognathi (Tetraodontiformes). *NOAA Tech. Rep. NMFS Circ. 434.*

———, and F. Santini. 2002. Review and reconstructions of the tetraodontiform fishes from the Eocene of Monte Bolca, Italy, with comments on related Tertiary taxa. *Mus. Civ. Stor. Nat. Verona, Stud. Ric. Giac. Terz. Bolca 9:* 47–119.

Wainwright, P. C., R. G. Turingan, and E. L. Brainerd. 1995. Functional morphology of pufferfish inflation: Mechanisms of the buccal pump. *Copeia 1995:* 614–625.

Washington, B. B., W. N. Eschmeyer, and K. M. Howe. 1984. Scorpaeniformes: Relationships, pp. 438–447. In Ontogeny and systematics of fishes, H. G. Moser, W. J. Richards, D. M. Cohen, M. P. Fahay, A. W. Kendall, Jr., and S. L. Richardson (Eds.). *Am. Soc. Ichthyol. Herpetol. Spec. Publ. 1.*

Wheeler, A. 1975. *Fishes of the world: An illustrated dictionary.* MacMillan, New York.

Winterbottom, R. 1974. The familial phylogeny of the Tetraodontiformes (Acanthopterygii: Pisces) as evidenced by their comparative myology. *Smithson. Contr. Zool. 155.*

Yabe, M. 1985. Comparative osteology and myology of the superfamily Cottoidea (Pisces: Scorpaeniformes), and its phylogenetic classification. *Mem. Fac. Fish. Hokkaido Univ. 32*(1): 1–130.

PART THREE

FORM
AND FUNCTION

From our initial focus on fish diversity, it is hoped that the reader has gained an insight into the dramatic capacity of fishes to adapt to the aquatic realm and will appreciate the extent to which fishes demonstrate a diversity of solutions to the problems of aquatic existence. Embedded within this diversity, however, are unifying features that are common to most if not all fishes. Our investigation of form and function in fishes considers the contributions of two key structural components: the integument and the musculoskeletal system. For fishes in general, the skin acts as an interface between their internal environment and the external, aquatic realm. Scales may be considered ubiquitous among fishes, yet there many instances where fishes have divested themselves of this feature. In some habitats, scales are not necessary, or their presence may actually be a hindrance. Among the functions we attribute to the skin, locomotion may not immediately come to mind. However, by contributing to the genesis of a smooth profile, offering minimal frictional resistance, and by providing a tough encasement firmly attached to the muscles, the skin plays a key role in locomotion.

Fishes have evolved a range of locomotor adaptations, and all of these modes reflect the need to move about in a medium that is much more viscous than our own. A stout-bodied, bottom-dwelling fish, however, obviously resorts to a distinctly different locomotor strategy than an elongate surface dweller. The biomechanical aspects of locomotion in fishes demonstrate the complex interplay of integument, muscles, and skeleton. Although we will explore locomotion and biomechanics in the specific contexts of the contribution of each of these structural features, we must be mindful of the degree to which these systems operate as an integrated unit.

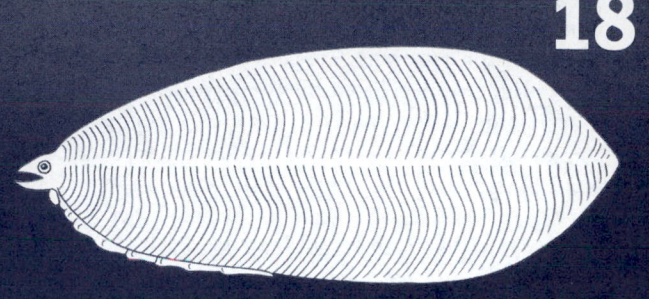

18 THE INTEGUMENT OF FISHES: SKIN, SCALES, AND ASSOCIATED STRUCTURES

INTEGUMENTARY COMPOSITION AND EMBRYONIC ORIGINS

The Epidermis
The Dermis
Phylogeny of Scale Types

HYDRODYNAMICS OF FISHES

COLOR

Occurrence of Color in Fishes
Color Change
  Control of Color Change
  Optical Filters
Significance of Color
  Concealment
  Advertisement
  Disguise
  Mimicry

BIOLUMINESCENCE

Occurrence of Bioluminescence
Production of Light
  Light from Symbiotic Bacteria
  Nonbacterial Luminescence
Control of Luminescence
Significance of Bioluminescence
  Concealment (Crypsis)
  Advertisement
  Disguise
  Special Considerations

VENOMOUS FISHES

The integument is one of the largest organs of the vertebrate body and includes an astounding diversity of structures associated with the multiple tasks that skin accomplishes for the animal housed within. Much of the specialized function of the skin arises from its role as the primary interface between the organism and its environment. Skin must be both protective and resilient. It serves as the first line of defense against pathogen invasion, while at the same time governing the passage of substances essential to life functions. The skin also serves as the location for the array of sensory structures that animals have evolved in order to perceive the environment in which they live. For fishes, the usual vertebrate sensory and communicative arsenal of vision, olfaction, hearing, taste, and touch includes integumentary components specifically modified for an aquatic existence. Additional modalities, such as electroreception and bioluminescence, have greatly enhanced the adaptation of some fishes to the aquatic realm. These, too, incorporate components of integumentary origin.

Cells and glands found in the skin are capable of producing powerful secretions that make some fishes unpalatable or dangerous to attack. Several species of fishes are considered among the most dangerous animals in the world, owing to the potency of the venom produced by their skin glands. These glands are often found in association with sharp, stiff spines that function as effective venom delivery systems.

The presence of such a diversity of integumentary structures devoted to sense reception, communication, or protection speaks to the role that skin plays in behavior. Visual detection of skinborne colors, olfactory and gustatory detection of chemical secretions originating from the skin, and the extremely sophisticated electrocommunicative abilities observed in several fish taxa all demonstrate the role of the integument in fish behavior.

INTEGUMENTARY COMPOSITION AND EMBRYONIC ORIGINS

As mentioned in Chapter 3, the skin of vertebrates is typically a complex bilayer, composed of an epidermis and a dermis (see Fig. 3.14). The constraints of locomotion in a medium that is much more viscous than our own confer special properties on the skin of fishes—properties that contribute to minimizing resistance to forward momentum (see Going Deeper). The most conspicuous feature of the skin of fishes is the array of scales of diverse shapes and sizes that arise from the dermis. (The basic scale types have been covered in some detail in Chapter 3.) Although somewhat dated, the reviews by Hawkes (1974a, 1974b), Bullock and Roberts (1974), and Roberts and Bullock (1976) are still valuable introductions to the subject of the integumentary composition of fishes and the associated pathological conditions.

The Epidermis

Most vertebrate skin precursors have been established by the end of the neurulation stage of the embryo (Kardong, 2002). The embryonic ectoderm gives rise to a thin layer of epidermal cells that form the outermost skin covering (see Fig. 3.14). Vertebrates are distinctive in demonstrating the development of clusters of ectodermal cells that arise adjacent to the neural groove. These are termed neural crest cells, and they give rise to a diverse assemblage of tissues, including skeletal components, muscles, and glands. Skin-associated structures, such as pigment cells and the cornea of the eye, also arise from the neural crest (Holland et al., 2001; Kent and Carr, 2001). In fishes, two types of cells are seen in the epidermis: epidermal cells and unicellular glands (Kardong, 2002). Epidermal cells form the different layers of the stratified epidermis and may be cuboidal or columnar in shape. These cells may contain numerous secretory vesicles that contribute to the formation of the mucous layer typical of fish skin. Interspersed among the epidermal cells are several types of unicellular glands. Included in these are granular cells and goblet cells, which contribute to the mucous layer, and club cells, which produce alarm secretions (Fig. 18.1). Granular cells are especially abundant in the skin of lampreys, yet this group appears to lack the goblet cells that are normal skin constituents in elasmobranchs and bony fishes. Obviously, then, the epidermis is usually covered with a slimy mucous layer contributed by different cellular components, but in some fishes, such as the lampreys, a thin, nonliving cuticle may be secreted over the epidermis (see Chapter 3). Some fishes, such as the wrasses (Labridae) and the parrotfishes (Scaridae) secrete an elaborate mucous cocoon from large goblet cells. This cocoon, in which the fish resides each night, is believed to act as a predator deterrent. Videler et al. (1999) have investigated the biochemical properties of parrotfish cocoons and have determined them to have antibiotic properties as well.

The epidermis also gives rise to the hard, acellular layer of enamel that coats the teeth and sometimes the scales of fishes. Ganoin, seen in the scales of primitive actinopterygians, is enamel-like in its origins and composition. Perhaps the greatest distinction that can be made between the skin of fishes and that of terrestrial vertebrates is the degree of development of keratinized epidermal structures. The scales of terrestrial amniotes, such as the snakes and lizards, are good examples of the role that keratin plays in conferring desiccation resistance on the skin of tetrapods, yet Frolich (1997) suggested that the low, crawling body posture of the first tetrapods may have prompted increased keratinization as a means of protection against abrasion. This is not to say that keratin formation is absent altogether in fishes. There are numerous examples of keratinized features, such as the distinctive "teeth" that line the oral cavity of lampreys, the scraping edge of the jaws in some herbivorous cyprinids, and the nuptial (breeding) tubercles seen in many groups of fishes (see Plate 10.2; see Fig. 27.7).

One of the most unusual forms of symbiosis known in fishes has been discovered to exist between the mahi mahi (*Coryphaena hippurus*) and strange, spinous cells that are closely associated with the epidermis. Because of their extensive cytoplasmic interdigitations among epidermal cells, these symbionts were once believed to be highly modified epidermal cells themselves. However, analysis of DNA sequences from these cells has revealed them to be distinct organisms existing in a bizarre, poorly understood commensal relationship with their host individuals (Langdon et al., 1995).

The Dermis

The histological composition of the dermis has been described in Chapter 3. The dermis of the dorsal region arises from embryonic dermatomes that differentiate from the mesodermal somites. The dermis of the flanks and belly arise from lateral plate mesoderm. The collagen of one of the layers of the dermis—the stratum compactum—is organized into plies that wrap in spirals around the body. This permits the skin to flex without wrinkling. This dermal layer is highly elasticized and behaves much like a rubber band. When a fish flexes its trunk while swimming, the skin that is stretched on one side of its body stores energy that contributes to moving its tail in the opposite direction.

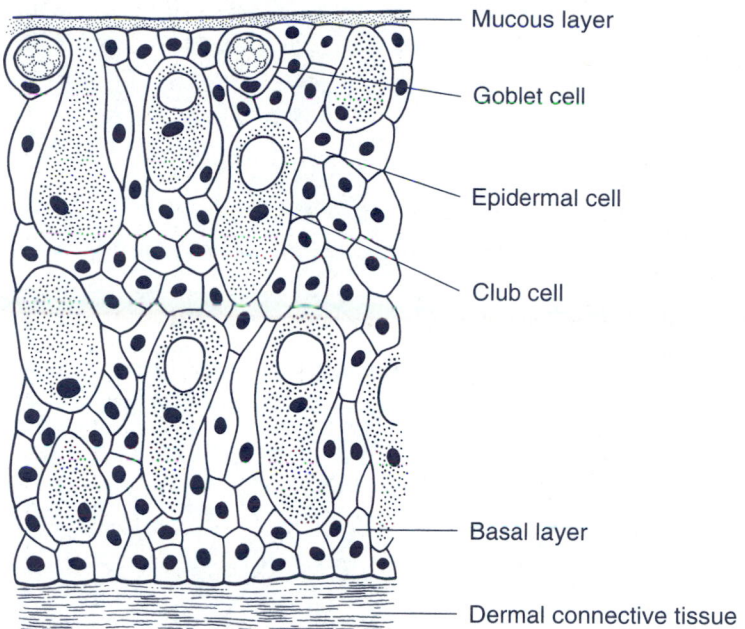

Mucous layer

Goblet cell

Epidermal cell

Club cell

Basal layer

Dermal connective tissue

FIGURE 18.1
Cellular composition of the epidermis of a bony fish.

In fishes such as the elasmobranchs, in which the axial skeleton is not well developed, the skin is an essential locomotor organ, against which the trunk muscles are acting to propel the animal forward. The most conspicuous feature of the dermis in fishes are the scales that make up much of its bulk.

Phylogeny of Scale Types

Unlike the scales of tetrapods, which are keratinaceous and largely arise from the epidermis, the scales of fishes are of dermal origin. Figure 3.15 illustrates the range of scale types seen in modern fishes. Early ostracoderms and placoderms had a complex scale coat, with both epidermal (enameloid) and dermal (bony) contributions (Fig. 18.2). Typically, the surface of this exoskeleton was ornamented with tiny tubercles that consisted of a surface layer of enamel covering a layer of dentin (or dentine). Both spongy and compact (lamellar) bones formed the base for the exoskeleton (Fig. 18.2A). Modern agnaths represent a dramatic departure from this ancestral skin design, as there is an absence of mineralization.

In chondrichthyans, the dermal bone is absent, but dentin and enamel layers persist in the form of **placoid** scales (dermal denticles). Placoid scales (Fig. 18.2B) form in the dermis but project through the epidermis to form a tough, prickly skin surface that has unusual hydrodynamic properties, discussed in the next section and in the Going Deeper box. The scale coat is scarce to absent in the holocephalans, but their well-developed fin spines are considered to be homologous structures, differing from placoid scales in their size and in the fact that they continue to grow throughout the life of the animal (Kemp, 1999). Among the batoids, the denticles are more restricted in their distribution, but several species are characterized by the development of tracts of large thorns or tubercles (Deynat, 1998; Deynat and Séret, 1996; Miyake et al., 1999). Fin spines, including the venomous caudal spines of some batoids, and teeth, including the bizarre lateral teeth on the rostrum of the batoid sawfishes (Pristidae) and selachian sawsharks (Pristiophoridae), are all considered homologous to placoid scales. The rostral teeth of the pristids remain constant in number and continue to grow throughout the life of the animal, reaching sizes of up to four inches in large specimens. The rostral teeth of the pristiophorids are nongrowing, but they are replaced by successively larger teeth as the animal grows (Kemp, 1999).

Among bony fishes, the scales have been distinguished as either *rhomboid* or *elasmoid* (Kent and Carr, 2001). **Rhom-**

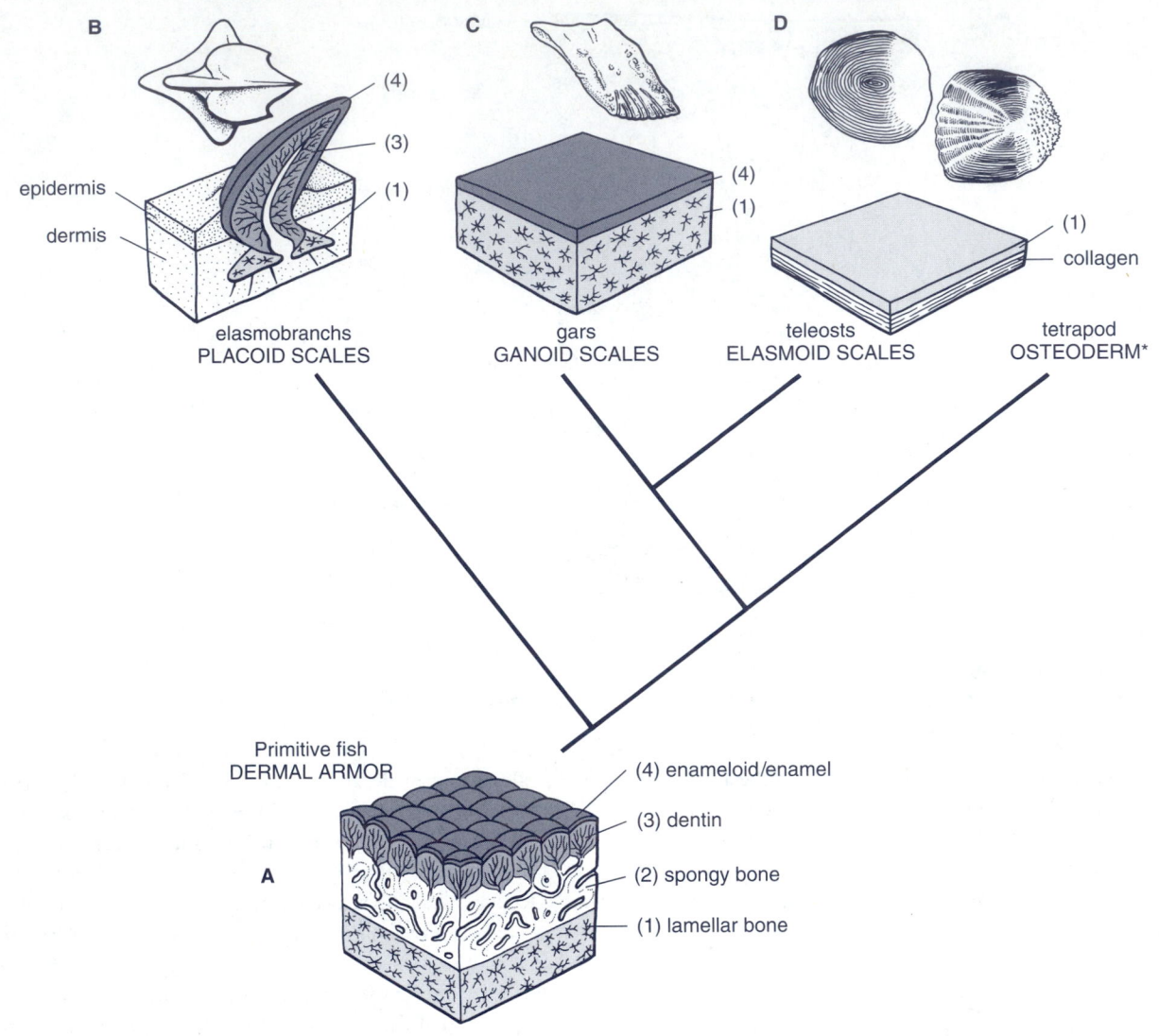

FIGURE 18.2
Phylogeny of scale types, emphasizing dermal contributions among fishes. **A,** Section of dermal armor of primitive fish, showing (1) lamellar bone; (2) spongy bone; (3) dentin; (4) enameloid/enamel. **B,** Section of elasmobranch placoid scale; (3) plus (4) constitute a denticle. **C,** Section of ganoid scale. **D,** Teleost elasmoid scales. *Osteoderms are remnants of dermal seals found in some tetrapods, such as caecilians and crocodilians. (Modified from Kent and Carr, 2001.)

boid scales typically have a diamond shape and form a distinctive, pavementlike skin covering, as exemplified by the gars (Fig. 18.2C). Mineral composition defines these rhomboid scales as **ganoid,** named for **ganoin**—the distinctive, enamel-like surface coating on the scale. The bulk of the scale is composed of dermal bone, however. An unmineralized layer, the **ganoin membrane,** separates the ganoin from the overlying epidermis (Sire, 1994). In addition to the gars, the extinct paleoniscoid fishes and the modern polypterids are characterized by ganoid scales. Gars and polypterids share the same processes of ganoin formation, which are similar to the processes of enamel deposition in teeth (Sire, 1994). Another rhomboid type, the **cosmoid** scale, is seen in fossil sarcopterygians. It is characterized by a distinctive canal system that penetrates the dentine and enamel layers.

The cycloid and spinous scales that characterize most bony fishes are examples of **elasmoid** scales. Ever since their first application in the determination of age and growth in carp (*Cyprinus carpio*) more than one hundred years ago, elasmoid scales have been enormously important in life history studies of fishes (Carlander, 1987). Modern techniques, including the use of transmission and scanning electron microscopy (Dingerkus and Koestler, 1986; Roberts, 1993; Sire, 1994; Yamada and Watabe, 1979), have provided great insight into the formation and differentiation of elasmoid and other scale types in fish skin. Compared to the scales described for more primitive bony fishes, elasmoid scales are much simpler in composition, consisting of a thin layer of dense bone overlying a fibrous plate composed of collagen (Fig. 18.2D). Living coelacanths have scales that appear elasmoid, but with their exposed surfaces covered with placoid-like denticles (Elliott, 2000).

Cycloid scales are present in all major lineages of teleosts, and spined scales are almost as widely distributed. Only the relatively small clades Osteoglossomorpha and Salmonoidei are lacking in some form of spined scale. Roberts (1993) recognized three types of spined scales among the teleosts: (1) **crenate**, with simple indentations or projections along the scale margin; (2) **spinoid**, with spines continuous with the main body of the scale; and (3) true **ctenoid**, where the spines are separate from the main body of the scale (Fig. 18.3). Among the ctenoid scales, Roberts (1993) identifies the category *transforming ctenoid* (Fig. 18.3D) as a scale type that is highly variable in size and shape but appears to have a consistent general configuration. It is unique to, and thus supports the monophyly of, the percomorphs. Individual species may possess both cycloid and spined scales. Some flounders, for example, have cycloid scales on the side that lies on the substrate, but the eyed upper side possesses ctenoid scales. Scombroids achieve their hydrodynamic efficiency through a very smooth skin with minute scales (see Going Deeper). Larval swordfishes (Xiphiidae), however, are heavily armored, with well-developed spinoid scales, apparently to confer better protection against predators. As these fishes mature, the scales are not reabsorbed, but rather they become deeply embedded in the thickening dermis (Govoni et al., 2004).

According to the dietary laws of the Jewish faith, the presence of scales (and fins) is what determines whether a fish may be consumed or not (kosher or *kashrut*). The range of variation in scale types, however, makes the determination of what constitutes a kosher fish a difficult proposition. Jewish scholars of the Torah affirm that only fishes with cycloid and ctenoid scales conform to the dietary laws. Dr. James Atz of the American Museum of Natural History has served as a consultant to the Jewish community on such matters;

with his assistance, a list of fishes deemed kosher has been made available (http://www.kashrut.com/articles/fish/).

HYDRODYNAMICS OF FISHES

The skin surface of fishes is but one component in an integrated body form that has developed to permit many fishes to pursue the path of least resistance through the water. Investigators studying the hydrodynamics of fish swimming have speculated that a vortex sheet forms behind the trailing edge of the vertical fins and is absorbed by the succeeding fins, the last dorsal or the anal fin passing the vortex sheet on to the caudal fin, from which it is shed from the fish (Blake, 1983; Webb, 1978; Weihs, 1989). This can establish a flow pattern that resembles that caused by a continuous fin but with less friction drag (Blake, 1983). One promising technique for investigating drag-reducing mechanisms involves analyzing the visualized wake of a swimming fish using **digital particle image velocimetry** (DPIV): An experimental tank in which a fish is swimming is seeded with minute, neutrally buoyant, fluorescent spheres. These particles are illuminated by a laser as they flow past the swimming fish, and the flow is imaged using a high-resolution video camera (Fig. 18.4; Anderson et al., 2001; Drucker and Lauder, 1999, 2001; Lauder, 2000; Wolfgang et al., 1999).

Most fishes that swim at moderate to fast speeds are well streamlined. They have a somewhat conical head that, with the pectoral region, forms a comparatively short "entering" section forward of the point of greatest body diameter. This is followed by an "afterbody" that tapers to the caudal peduncle. As a body of such shape thrusts against the water, there is typically positive pressure on the entering section and slightly negative pressure along the afterbody. Part of the energy expended by the swimming fish goes into overcoming the **pressure drag** of any turbulent wake that is formed behind the caudal fin. There is also **frictional drag,** caused by the passage of the water over the skin of the fish. A thin *boundary layer* of water exists next to the skin. Water in this layer can flow smoothly over the skin in a laminar fashion in very small fish, or in larger fish as they move slowly. As size and speed increase, the laminar flow in the boundary layer changes to turbulent flow and increases drag. If the turbulent layer separates from the skin, a turbulent wake can result, and drag increases further.

Scientists studying the hydrodynamics of swimming fishes predict the nature of the boundary layer by calculating the **Reynolds number** (*Re*), which is an expression of the ratio of inertial forces to viscous forces (Birkhoff, 1950; Blake, 1983; Walters, 1963). This involves the length of the body (*L*), velocity (*V*), and the kinematic viscosity

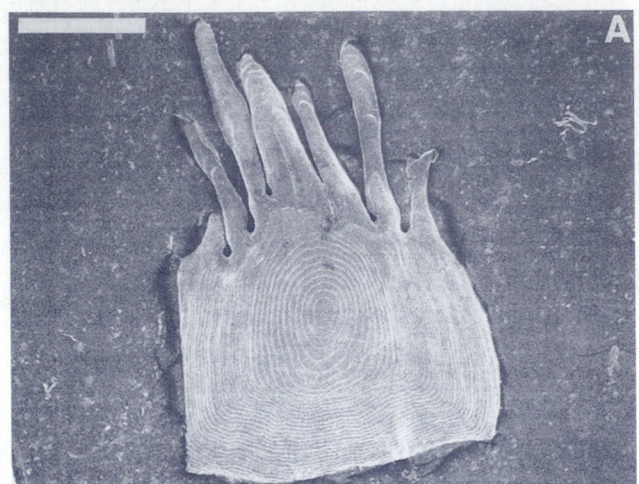
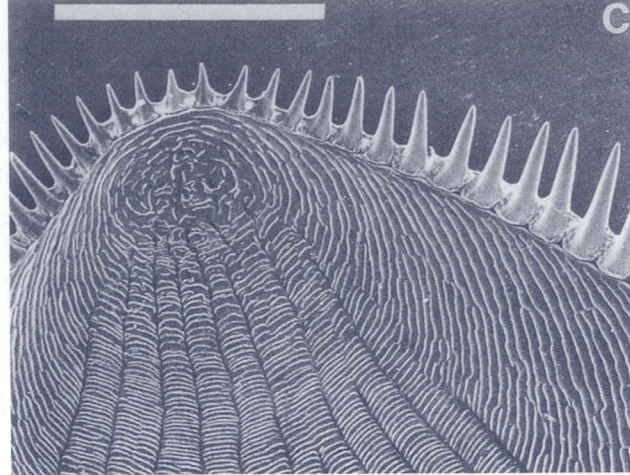
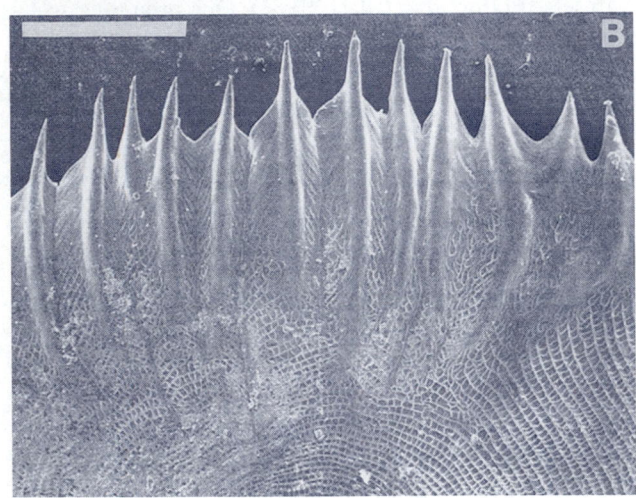
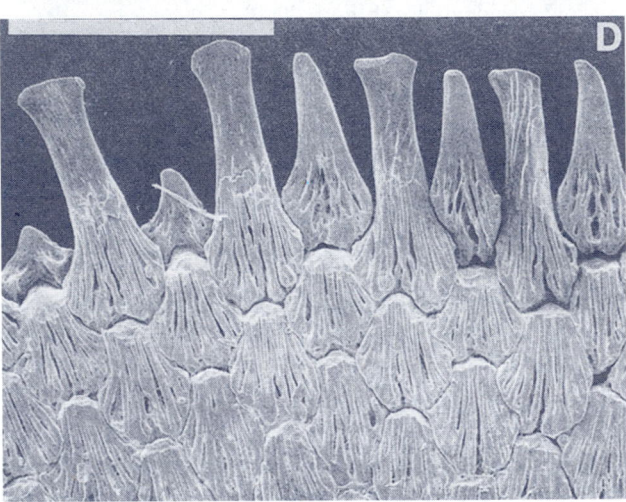

FIGURE 18.3
Scanning electron micrographs showing variation in spiny scale types among teleost fishes: **A,** Crenate scale, as seen in the aulopiform *Bathypterois quadrifilis.* **B,** Spinoid scale, as seen in the myctophid *Notoscopelus japonicus.* **C,** Ctenoid scale, with ctenii on scale periphery, as seen in the gobiid *Yongeichthys nebulosus.* **D,** Transforming ctenoid scale of *Epinephelus nigritus;* the transforming ctenoid scale is distinct in having marginal rows of complete spines and submarginal rows of truncated spines. (From Roberts, 1993.)

(ν; Videler, 1993, defined this as the ratio of dynamic viscosity over density) of water: $Re = LV/\nu$ (Videler, 1993; Yates, 1983). *Re* is a dimensionless number, which can be less than one, but most often is in the thousands or millions for fishes of appreciable size. Blake (1983) stated that for streamlined bodies under steady flow, the boundary layer changes from laminar to turbulent as the Reynolds number increases from 5×10^5 to 5×10^6, passing through a transitional or unstable flow during which turbulence increases.

 ⊛ Drag on the swimming fish is proportional to the square of the velocity, so a fish that must swim rapidly must exert an appreciable amount of energy to maintain its speed. Fishes known for their fast swimming have evolved means of reducing drag and means to sustain energy output for their comparatively high sustained speeds. The drag on a flexible, somewhat compliant fish body can be two to five times the drag on a rigid body of the same size and shape. This drag is increased by the undulations of the swimming fish's body, as the flow crosses over from one side to the other and causes separation of the boundary layer. Some fishes minimize that effect by using what is called "kick-and-glide" or "burst-and-coast" swimming, in which they accelerate to a given speed or swim up to a given depth, then hold their body rigid and glide for a distance. In fishes that are heavier than

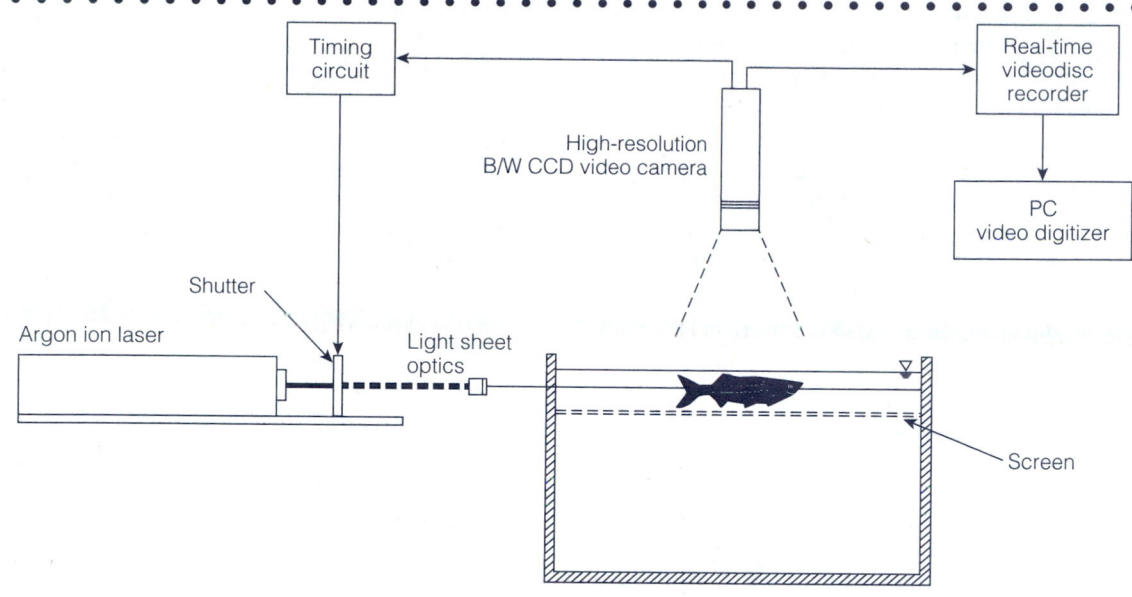

FIGURE 18.4
Illustration of experimental apparatus used for digital particle image velocimetry (DPIV). (Reproduced courtesy of M. S. Triantafyllou, Massachusetts Institute of Technology.)

water, the pectoral fins are used as gliding planes to produce lift, but the decreasing speed allows them to sink, so their progression is based on swimming up and gliding down. The saithe (*Gadus virens*) is reported to accelerate up at about 10 body lengths per second (Ls^{-1}) then coast down, achieving an average speed of 5 Ls^{-1} (Videler and Weihs, 1982). Those fishes with nearly neutral buoyancy can maintain a nearly constant depth in kick-and-glide swimming. A gliding fish generally has about one third the drag of a steadily swimming fish at the same speed (Videler, 1981; Weihs, 1974).

The tunas and tunalike scombroids have excellent morphological characteristics for reducing drag (see Chapters 18 and 19, Going Deeper). The maximum body depth is usually at least 60 percent of the body length behind the anterior point (Weihs and Webb, 1983). Their streamlined bodies are shaped so that the generation of adverse pressure gradients is delayed and laminar flow is maintained. They present a clean surface that aids in maintaining a laminar or at least an attached turbulent flow along the skin. The body is held nearly straight during swimming, as most lateral movement is confined to the caudal peduncle and the high-aspect-ratio caudal fin. The caudal peduncle is equipped with flexible finlets that are thought to direct the flow across the peduncle in a manner that minimizes separation of the boundary layer (Aleev, 1969; Lindsey, 1978). The peduncular keels (see Fig. 19.8B) might aid in maneuvering. Ships

with such keels are reported to have a much tighter turning radius than those not so equipped (Watts, 1960). Some investigators believe that the corselet of scales that roughens the forebody of some tunas in a characteristic pattern helps maintain an attached, though turbulent boundary layer, for any attached boundary layer produces less drag than if that layer were separated. There seem to be conflicting views as to whether the water flow that results from ram irrigation (see Chapter 24; *ram* is a distinct type of gill operation) of the gills helps keep an attached turbulent boundary layer or encourages separation. Some non-scombroids have slots behind the opercular opening that would appear to improve the efficiency of the water flow past the gills.

The roughened corselet of scales observed in tunas is but one example of the hydrodynamic roughening of fish skin. The coarse skin of elasmobranchs may also confer greater hydrodynamic efficiency (see Going Deeper). Ctenoid scales with spines that protrude through the mucous coating are encountered in a variety of teleosts. The roughness of any protrusions (Fig. 18.5) that project through the boundary layer can cause small vortices to form and encourage a turbulent boundary layer to remain attached, so that drag will be reduced. Some examples of fishes that have special modifications of their scales to provide a roughened or patterned exterior are the roughscale pomfret (*Taractes asper*), the catalufas (Priacanthidae), the squaretails

(*Tetragonurus*), and the oilfish (*Ruvettus pretiosus*). The oilfish has spaces beneath the skin filled with seawater that is injected into the boundary layer as the fish undulates its body (Bone, 1972). This could aid in keeping the boundary layer attached. A different type of injection system has been reported in some ribbonfishes (Trachypteridae). Canals in the skin possibly accept boundary layer water in areas of high pressure and release it farther along the body in areas of low pressure (Walters, 1963). This system could have the effect of helping maintain a laminar boundary layer.

The role of mucus in the reduction of frictional drag in some fishes has been investigated (Hoyt, 1975; Rosen and Cornford, 1971). Mucus may reduce drag by reducing the viscosity of water as it flows past the body. Most of the experiments involved study of the flow of water mixed with mucus in a tube. There is, however, much variation in the effectiveness of slime from different species. The demonstration of the effectiveness of the mucus in lowering the viscosity of water in a pipe does not necessarily prove that the mucus can lower the frictional drag on a swimming fish (Videler, 1993). Experiments with newly killed fish and wax models, however, have suggested that mucus does play an important role in overall drag reduction in at least some species of fishes (Daniel, 1981).

Some studies have suggested that an overall reduction in drag may also be achieved through schooling behavior. Weihs (1973) claimed that fishes in schools, swimming in a suitable geometric pattern—usually a diamond shape—can increase swimming efficiency for all but the leaders. Zuyev and Belyayev (1970) have observed that the tail-beat frequency of leaders in a school exceeds that of the followers. In spite of such observations, however, the hydrodynamic advantages of schooling remain a subject of some controversy (see Chapter 37). Studies on swimming trout

GOING DEEPER • Dare to Build a Better Fish I: Lessons for the "Thorpedo"

The British zoologist James Gray was the first to recognize that large, rapidly swimming creatures, such as fishes and dolphins, have no business achieving the speeds they do given their power output (Gray, 1957). Thus was born Gray's paradox. Swimming fishes exemplify this paradox because they can achieve a velocity well in excess of what they should be able to given their available muscle mass. The secret appears to be the control of the drag-inducing vortices generated by the body moving through the water. The shape and texture of the body surface will determine whether the passage of the water is smooth (laminar flow) or becomes disturbed (turbulent flow). When the flow becomes turbulent, drag-inducing vortices are generated. The creative manipulation of these vortices appears to be the way to resolve Gray's paradox. Much of that manipulation involves the skin surface of the animal.

There appears to be an inverse correlation between locomotor efficiency and scale size in fishes. Large scales would hinder the lateral mobility of the trunk and tail. Hence, large, swift-moving species, such as the tunas, have very small scales or are scaleless altogether on the most flexible parts of the body. As mentioned earlier, however, scombroids possess a corselet of well-developed scales on the less mobile pectoral region of the body. This may confer a hydrodynamic advantage. The swiftest pelagic sharks, such as the lamnids, may also gain some advantage from their body covering of minute dermal denticles. The skin surface of sharks may be one of the most sophisticated mechanisms of drag reduction known among animals—something that has attracted the attention of Olympic swimmers and aeronautical engineers.

Two different kinds of drag exert a negative impact on forward momentum as a body moves through the water. Form or pressure drag results when orderly flow separates along a moving body. Streamlined body shapes, as discussed in the Going Deeper box in Chapter 19, serve to minimize this type of drag. Surface drag ("skin friction") is attributed to forces that retard the flow of water across the body. Elite swimmers often shave their body hair in order to minimize surface drag. But here is where the story gets interesting.

From the earlier discussion of drag reduction, recall that a roughened texture, such as the skin of a shark, appears to *enhance* the velocity at which a body can move through the water. The placoid scales of some species of pelagic sharks produce a series of minute grooves (Fig. 18.13) aligned with the fluid flow along the body of the swimming animal. Apparently, this grooved texture creates tiny vortices that act to hold the flow of water closer to the skin surface and hence reduce turbulence. Although some are skeptical of this phenomenon (Vogel, 1996), the principle appears to work in some applications: Swimwear manufacturers have seized on the concept as the Next Big Thing in performance enhancement. Elite swimmers like America's Jenny Thompson and Lenny Krayzelburg and Australia's Ian Thorpe (the "Thorpedo") were swimming in the 2000 Olympics in full body suits. Body suits are used by some swimmers to streamline the body profile and reduce the rippling of the skin and muscles that might contribute to drag. But the Fastskin[tm] body suit, designed by Speedo[tm], is something new altogether. Speedo[tm] claims that a significant reduction in the coefficient of drag is achieved by mimicry of the skin of sharks. The tiny resin riblets of the body suit create microturbulence in the boundary layer, which inhibits flow separation. Water stays closer to the body, and surface drag is reduced. Some detractors claim that

have shown that fishes are capable of reducing muscle activity through the exploitation of vortices generated by schoolmates or from flowing water (Liao et al., 2003).

COLOR

Occurrence of Color in Fishes

Only a few fishes lack or nearly lack skin pigment. Cave-dwelling species, which live in the dark, are notable examples, generally appearing pale to white or slightly colored by blood or other body fluids. Fish larvae are often unpigmented or may have pigments deposited in restricted places, such as the head or yolk sac. The larvae of some smelts and gunnels, for example, may remain virtually trans-

parent up to a length of 2 cm or more, and eel leptocephali, depending on the species, may reach 15 cm or more while still transparent (Fig. 18.6A, B). Studies on the pattern of pigment deposition in vertebrates have provided valuable insights into understanding the overall regulation of morphogenesis. In these as in many other aspects of vertebrate development, zebrafishes have proven to be most valuable objects of study (see Chapter 27, Going Deeper; Milos and Dingle, 1978a, 1978b; McClure, 1999).

Pigment is mostly contained in special cells called **chromatophores** that arise from the neural crest and thus share a number of features with nerve cells (Meyer-Rochow, 2001). Coloration in the internal organs, skin, flesh, or bones in some species can also be independent of specialized cells. For instance, adults of the lamprey (*Geotria australis*) maintain two dorsolateral blue-green stripes caused by the deposition of biliverdin (a green bile pigment) during their

the performance advantage of these sophisticated body suits comes from increased buoyancy—something that is prohibited in competitive swimming—resulting from the application of water-resistant materials. While Speedo™ continues to tout the success of their revolutionary suit, kinesiologists remain skeptical (Sanders et al., 2001; Zempel, 2001). The suits appear to be catching on among the top athletes in competitive swimming, however, with more of them opting to wear them for the events of the 2004 Olympics in Athens (Krieger, 2004).

Whether or not the Fastskin™ actually represents a technological breakthrough should not detract from the hydrodynamic efficiency of the shark skin after which it is modeled. Some placoid scales have another feature that enables sharks to easily slip through the water. The grooved surfaces of the denticles of several species of pelagic sharks sit on raised pedestals that apparently act to absorb turbulence, thus further reducing pressure drag (see Fig. 3.15D). Racing yachts have used this design in the construction of their distinctive winged keels (Helfman et al., 1997). In 1987, an America's Cup sailboat with a hull textured to mimic the skin of sharks was so successful that hull texturing became outlawed. The aeronautical industry has focused on this distinctive feature of the scales of sharks in hopes that it will lead to improvements in aircraft aerodynamics. The placoid denticles are anchored to the skin by fibers attached to tiny muscles, so that sharks might be actively reducing drag by adjusting their denticles in response to water pressure and turbulence. Engineers at the California Institute of Technology have produced an active drag reduction system modeled after the skin of sharks (Gupta et al., 1996). Perhaps the airplane of the future will be covered with minute airfoils, modeled after placoid scales, that will be

capable of fine-tuning aerodynamics—making it less likely that your in-flight beverage will move from your fold-out tray into your lap.

200µm

FIGURE 18.13
Scanning electron micrograph of the skin of a pelagic shark, a model for the Speedo™ Fastskin™ body suit. (Photo courtesy of John Mansfield / Microbeam Analysis Society.)

© John Mansfield / Microbeam Analysis Society

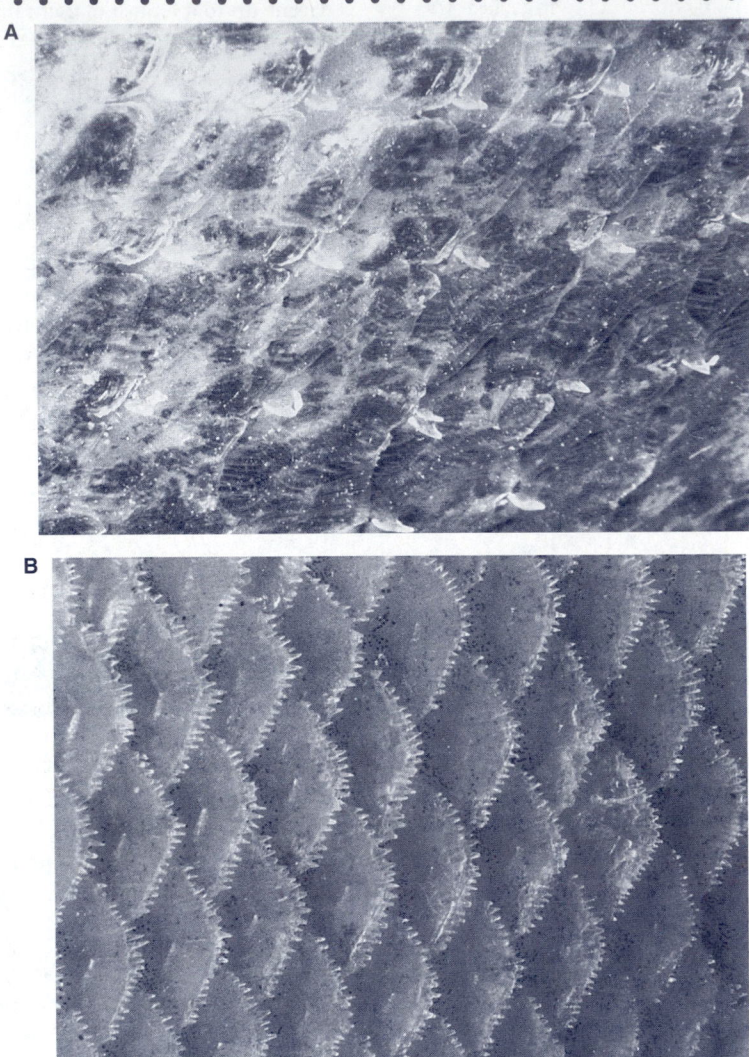

FIGURE 18.5
Rough surfaces on fishes: **A,** Spiny scales of rough pomfret (*Taractes asper*). **B,** Strongly ctenoid scales
of popeye catalufa (*Pristigenys serrula*).

marine life. Although blue pigment is not common, a blue carotenoprotein is found in the muscles, fin bases, or walls of the alimentary canal of some tropical fish larvae. Chromatophores are named for the color they impart or for the color of the pigment they carry (Fingerman, 1965; Fox, 1957; Fox and Vevers, 1960; Fujii, 1969). Colors imparted by cells are of two kinds: those due to pigments (**biochromes**), and those due to the reflection of light from a colorless, mirror-like surface and refraction by the tissues (structural colors or **schematochromes**; Simon, 1971). These two types of colors are often encountered in combination. Chromatophores classified by color include **melanophores,** contain-

ing a black or brown pigment (*melanin*); **erythrophores,** containing reddish pigments (*carotenoids* and *pteridines*); **xanthophores,** containing yellow pigments (*carotenoids*); **leucophores,** containing white *purines,* usually guanine and hypoxanthine, in the form of small crystals that can be moved within the cytoplasm; and **iridophores,** containing *purines,* mostly guanine, in large, nonmotile crystals. Iridophores in some fishes have been shown to be motile (Kasakawa et al., 1989). Cells carrying more than one pigment are called **compound chromatophores.** A type of melanophore (**phaeomelanophore**) that carries red melanin is found in some cichlids (Avtalion and Reich, 1989).

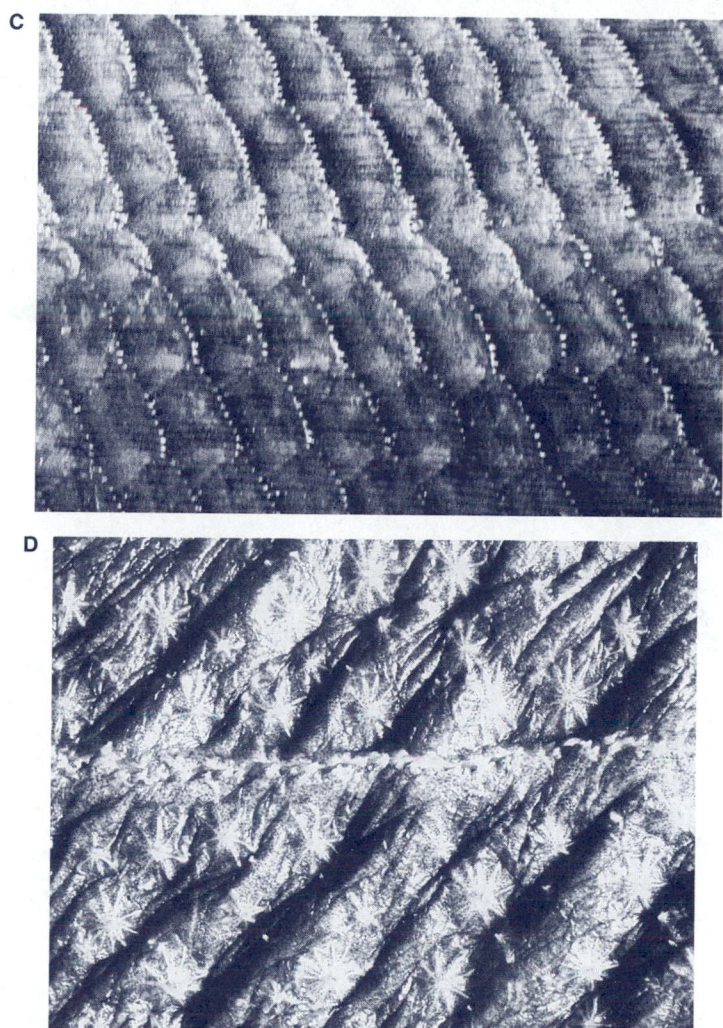

FIGURE 18.5
Rough surfaces on fishes: **C**, Ridged and strongly ctenoid scales of smalleye squaretail (*Tetragonurus cuvieri*).
D, Stellate, spinous scales of prickly shark (*Echinorhinus cookei*).

Chromatophores are located mainly in the dermis, but sometimes they may be in the epidermis or may even be hypodermal (Hawkes, 1974b). Chromatophores are also found in the peritoneum (lining of the body cavity) and sometimes around parts of the central nervous system.

Color Change

Color changes involving biochromes, and some changes involving structural colors, depend on the movement of pigments within the chromatophore; these short-term and often rapid changes are called *physiological* or *neural* color changes. Long-term color changes may be due to increases in the number of chromatophores and the consequent general alteration of pigmentation distribution within cells, referred to as *morphological* color change.

Pigment occurs in the organelles of cells—sometimes, as in the case of the purines, in crystalline form. Melanin is carried in organelles called *melanosomes;* erythrophores contain *pterinosomes* bearing red pteridines. Chromatophores are usually characterized by irregular cell boundaries and dentritically branched processes, although finely radiate and other cell shapes are known (Fig. 18.6C–E). When the pigment-bearing organelles are aggregated in the center of a cell, the

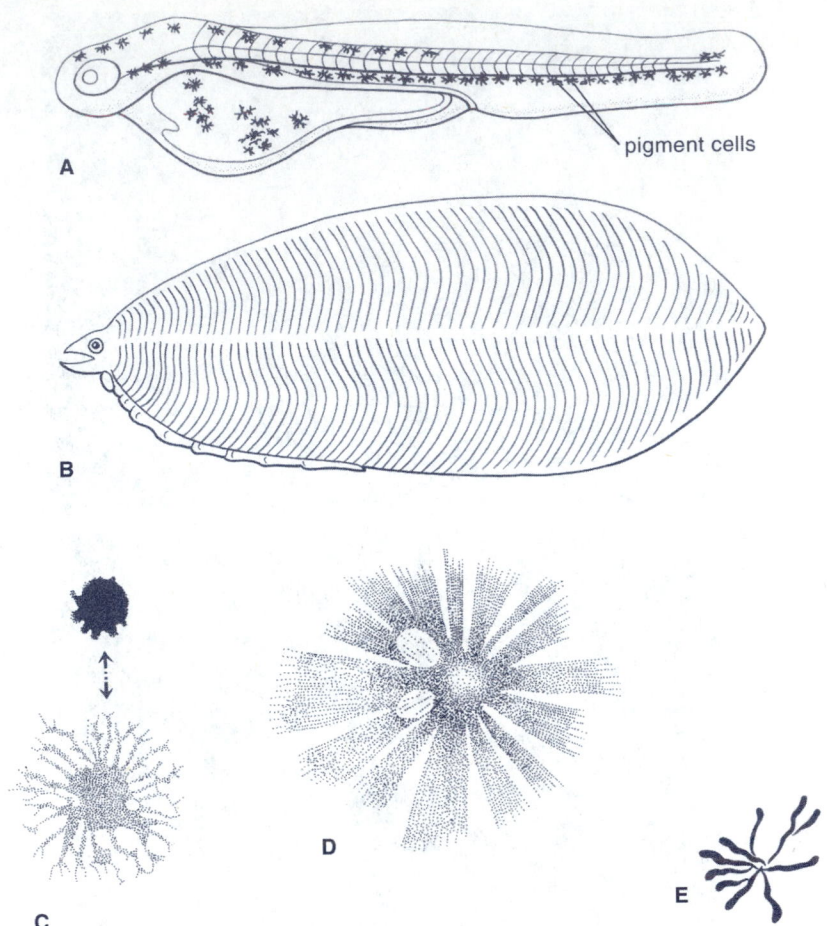

pigment cells

FIGURE 18.6
Larvae of fishes: **A,** Showing scattered chromatophores; **B,** Unpigmented leptocephalus larva of eel
(*Thallassenchelys coheni*); **C,** Dendritic chromatophore, showing pigment aggregated (*top*) and dispersed
(*bottom*); **D,** Chromatophore with pigment in finely radiate pattern; **E,** Pigment in coarsely radiate pattern
(after Wunder, 1936).

skin appears pale; when the organelles are dispersed through-out the cell, the skin takes on the color of the chromatophore. Various hues are made possible by combinations of different chromatophores overlying each other or by the action of com-pound chromatophores, which can carry separate organelles with pigments of different colors (Fujii, 1969, 1993).

Of special interest are the structural colors caused by the iridophores. Many open-water species have a continu-ous subdermal sheet of iridophores on the lateral and ventral sides of the body. This **stratum argenteum,** as it is called, is made up of several layers of cells, so much of the light—up to 80 percent—striking the side of the fish is reflected. Fur-thermore, there are usually some iridophores in the dermis.

If all these diminutive mirrors are parallel to the surface of the body, the fish generally appears uniformly silvery, but if the reflecting crystals are oriented differentially, the appar-ent color of an individual may be dark when viewed from above and light when seen from below (Denton and Nicol, 1966). This is the mechanism by which *countershading* is achieved (see further on).

In the neon tetra (*Paracheirodon innesi*), changes in color reflectivity can be caused by cytoplasmic movement between reflective plates within the iridophores. The plates are like playing cards in a deck (Lythgoe and Shand, 1983). The neon tetra has both physiologically active and physi-ologically inactive iridophores (Lythgoe and Shand, 1982).

Physiologically active iridophores are present in the irides-cent lateral stripe along the body, and this stripe shifts in color in response to ambient light intensity. Here, layers with a high refractive index alternate with layers of low refractive index. The high refractive index layers are thin intracellular crystals composed of guanine and hypoxanthine. Interspersed among these are the low refractive index layers, which are cytoplasmic in nature. Each iridophore contains a multilayered stack of about 20 crystal and cytoplasm pairs (Lythgoe and Shand, 1989). Color modulation arises from changes in light intensity falling on the iridophore itself. A rhodopsin-like molecule identified within the iridophores may figure significantly in this color modulation (Lythgoe et al., 1984).

Leucophores are dendritic, light-reflecting chromatophores that are present in the Cyprinodontiformes and probably in related groups (Fujii, 1993). The chemical composition of the reflecting material, which reflects visible light in all directions, is not known. The reflecting substance is found in organelles termed *leucosomes* (Fujii, 1993).

Control of Color Change

The aggregation and dispersal of pigments in biochrome chromatophores appear to be under both hormonal and neural control. Although in some fishes, one or the other of these may dominate (Naitoh et al., 1985), most fishes probably combine both control processes. The neuronal nature of chromatophores is also evident in the degree to which they exhibit direct sensitivity to a number of environmental stimuli, including light, electromagnetic fields, osmolarity, and pH (Meyer-Rochow, 2001).

The pituitary gland secretes **melanocyte-stimulating hormone (MSH)** and apparently other substances that cause the movement of the melanosomes resulting in a change in color. Not all species are affected in the same manner by MSH, which in most species causes the pigment to disperse and the skin to darken. In some instances, however, the same hormone causes aggregation in one species and dispersal in another. Some researchers have suggested that other hormones in addition to MSH that affect chromatophores are secreted by various parts of the pituitary (Abbot, 1973). A **melanophore-concentrating hormone (MCH)** of hypothalamic origin appears to be active in a number of teleost species (Nagai et al., 1986).

Epinephrine (adrenalin), a derivative of the amino acid tyrosine that has widespread and profound effects on a number of cell and tissue sites in the vertebrate body, generally concentrates pigment in fish melanophores. The same effect can be caused in many species, but not in all, by **melatonin**, a secretion of the pineal body. Furthermore, there is some evidence that the pseudobranch (see Chapter 24)

may be active in secreting or activating a hormone-like substance involved in aggregating pigment in chromatophores.

The nervous control of chromatophores appears to involve the release of neurotransmitter substances by the neurons innervating the cells. Some evidence points to the existence of double innervation of chromatophores, so that one set of fibers releases a transmitter that causes the dispersal of pigment, and another set produces a neurotransmitter that causes pigment aggregation. Leucophores, melanophores, and to some extent erythrophores and xanthophores are under nervous control. Iridophores are under the control of the sympathetic nervous system in at least some species (Kasukawa et al., 1987).

The intracellular transport of pigment granules appears to be facilitated by the mechanochemical actions of microtubular "motors." The operation of these motors and hence the direction of pigment movements appears to be regulated by intracellular concentrations of cyclic adenosine monophosphate (cAMP) and calcium ions (Kotz and McNiven, 1994).

Color changes may result from a fish's response to a changed background color in its surroundings, or they may be due to any one of many responses to social, behavioral, or chemical stimuli (Fujii, 1969, 1993). Numerous fishes are known to react to the color of their environment by altering their body color to match it. Shallow-water fishes apparently take cues from the light penetrating the surface of the water and the light reflected from the bottom or other background (Walls, 1963). The neuronal mediation of coloration response appears to involve a comparison of the light from the two sources. Light from the surface impinges on the lower part of the retina, whereas that reflected from the bottom strikes the upper part, and the eyes of fishes appear to be specialized to respond to this differential illumination. Responses may be rapid, taking only a few seconds, or they may require hours (Odiorne, 1957). Darkening or paling can be effected by light striking the pineal body in many fishes. There is evidence that other parts of the central nervous system can respond to light, because paling responses have been induced by exposure to light in individuals whose eyes and pineal have been rendered nonfunctional.

Morphological color changes can often be identified in the various life history stages of fish. For instance, young trout and salmon are usually colored to resemble their stream habitat and may have vertical bars (**parr marks**) on the sides, but as they physiologically prepare to enter the ocean, they develop pelagic coloration—becoming silvery on the sides and blue or green on the back. Later, as they return to freshwater for breeding, the pelagic coloration is obliterated and a nuptial coloration, often involving very bright colors, appears.

As eels of the genus *Anguilla* mature and prepare to enter the ocean migratory phase, they become silvery. Pankhurst and Lythgoe (1982) found that the maturing European eel (*A. anguilla*) lost xanthophores and experienced a decrease in the amount of the reflective purines guanine and hypoxanthine, but that reflecting layers were reorganized to produce the silvery camouflage. Changes in color can be brought on by changes in diet, especially if carotenoids are involved (Watiporn, 1988). The diet during the larval stage of flounders has been implicated in the absence of pigment. Juveniles who were fed rotifers or brine shrimp (*Artemia*) showed a high incidence of albinism (Seikei, 1989). In some instances, parasitism has caused a change in pigmentation (Ward, 1988). Loss of iridophores can be induced by retaining fishes under constant illumination in a light container (Fries, 1958). Keeping fishes on a dark background results in an increase of melanophores (Odiorne, 1957). Melanogenesis seems to be controlled by the pituitary and is brought about by the action of the adrenocorticotropic hormone (ACTH; Fujii, 1969).

Optical Filters

All fishes do not experience the same spectrum of colors in their environment, because the quality and intensity of light striking the retina of the eye can be influenced by color in the lens or cornea (see Chapter 20). Some cichlids have a yellow pigment, which apparently acts as an optical filter, in the lens and cornea. The deep-sea hatchetfish (*Argyropelecus*) has yellow lenses (McFall-Ngai et al., 1986). Muntz (1983) suggested that yellow lenses in marine fishes might render the camouflage of photophores in other fishes less efficient, because such lenses absorb less of the bioluminescence than of the surface light, so the light of the photophores would be brighter than the background.

Greenlings (Hexagrammidae) have corneal filters that respond to a change in light with a change in density of pigmentation (Gnyubkin, 1989). Other fishes with corneal filters include sculpins and wrasses (Gnyubkin and Gamburtseva, 1981).

· · · · · · · · · · ·
Significance of Color

The significance of coloration can only be determined in the context of the environment in which that color is being expressed. The development of inexpensive, field-compatible spectroradiometric instrumentation that enables an objective, quantitative assessment of animal coloration has permitted more sophisticated interpretation of the role of color in the lives of fishes (Endler, 1990). Pigment is important to fishes for many reasons. For fishes that live in shallow water, pigment may be effective in protecting them from damage by ultraviolet radiation. McArdle and Bullock (1987) reported that severe losses of Atlantic salmon (*Salmo salar*) at Irish hatcheries were due in part to solar ultraviolet radiation. Ex-

posure of Chinook salmon (*Oncorhynchus tshawytscha*) fertilized eggs and fry to various artificial lights indicated that there was damage from ultraviolet light (Dey and Damkaer, 1990). Pelagic marine species may also be adversely affected by increased ultraviolet radiation (see Chapter 39).

Color may function also in predator avoidance, by concealment, disguise, or mimicry; in predator deterrence, by the advertisement of toxins or other noxious attributes; and in mating behavior.

Concealment

A common method of concealment among fishes is **obliterative countershading,** in which the dorsal surface is dark whereas the sides and belly are pale. Many open-water species show the type of countershading that is sometimes referred to as *pelagic coloration,* with dark backs of the general hue of the water as seen from above (usually blue or green), and silver sides and belly. With this arrangement, the animal blends in with the darker background of the water's depths when viewed from above, and the light color of the sides lightens the shadows on the flanks, so that the entire fish can appear to have a uniform color, which renders it more obscure when viewed from the sides.

Some smelts and silversides are translucent but have a reflective layer around the body cavity and another around the red lateral muscles. In both instances, the silver does not extend over the top, so the tissues are essentially countershaded, like the fish itself. Some brightly colored reef fishes are countershaded in that the upper sides and back are black or some other dark tone, whereas lighter, brighter patterns cover the flanks and belly. Several species of butterflyfishes, including *Chaetodon lunula,* have been observed turning their dark backs toward approaching predators—an act that should render a conspicuous fish less so. Some of the black on the sides of this species serves to conceal an apparent social signal of yellow, which replaces the black during aggressive encounters (Hamilton and Peterman, 1971).

Dark pigment in the peritoneum or gut wall of deep-sea fishes might serve to conceal the bioluminescent emissions of their ingested prey. The blue pigment in the gut wall of certain tropical fish larvae could mask the presence of the orange or red crustacea on which they feed (Herring, 1967).

Concealment can be aided by an overall resemblance to the substrate or background. The transparency of larvae may make them less visible to predators (McFall-Ngai, 1990). Some species have hues and patterns similar to those of the bottom (demersal coloration), whereas some other patterns bear a general resemblance to vegetation. Many species can change color or pattern to match their surroundings. Cutthroat trouts (*Oncorhynchus clarkii*) that live in streams that course beneath a dense forest canopy tend to be very dark and heavily spotted; those that frequent open riffles are lighter; and those living in ponds opalescent with colloidal clay can be ghostly pale. Much attention has been given to the flounders and other flatfishes (see Plate 17.3), but various blennies, sculpins, scorpionfishes, and other fishes are capable of rapid color change (see Chapter 37).

Many demersal fishes have flaps, cirri, and irregular outlines that aid in concealment. Fishes that generally resemble plants include snailfishes (*Liparis*), pipefishes and seahorses (Syngnathidae), pricklebacks (Stichaeidae), and gunnels (Pholidae). Many bottom-dwelling fishes, such as the marine sculpins (Cottidae), stonefishes (Synanceinae), and stargazers (Uranoscopidae), show a color resemblance to a substrate that consists of a combination of stones, shells, and algae.

Disruptive coloration (Cott, 1940; Muntz, 1990)—consisting of stripes, bars, ocelli, and other markings—may cause a fish to appear conspicuous out of its accustomed habitat. In its usual habitat, however, those markings tend to break up the outline of the individual or to make the eyes or other readily recognizable features less prominent (Fig. 18.7). In both schooling and nonschooling species, disruptive markings can present a confusing pattern. The sight of a large aggregation of horizontally striped fishes such as threadfins (Polynemidae) in the brightness of a coral lagoon might challenge the focal abilities of a predator's eyes. Eye stripes and "hoods" of dark color that cover the top or anterior of the head and the eye are common in fishes. Elongate fishes usually have horizontal eye stripes, and those with deep bodies or blunt heads tend to show vertical stripes or stripes that follow the contour of the head.

Advertisement

Advertisement by bright colors or conspicuous patterns may serve many purposes. Schooling species have markings that probably aid individuals in staying with their own kind. Even countershaded species, such as sardines and tunas, may have species-specific rows of spots or series of stripes that are arranged so as not to detract from the obliterative shading to any great degree, but still allow species recognition. Cleanerfishes, which remove ectoparasites from other fishes, usually have distinctive coloration and markings, so that they are readily recognized by larger species and are allowed to approach with little danger of being eaten. Differential coloration of the sexes aids in the recognition and attraction of potential mates. Color may figure significantly in reproductive success, as in the Pecos pupfish (*Cyprinodon pecosensis*), in which Kodric-Brown (1983) found that darker males mated with more than three times as many females per hour than lighter males. With respect to the relationship between color and behavior, it is important to recognize the functional differences between colors that are part of the permanent palette of fishes and those that appear only during reproduction (Kodric-Brown, 1998).

Nicoletto (1991) has proposed that the density of carotenoid pigment is an indicator of male vigor in the guppy (*Poecilia reticulata*). As carotenoids must be obtained from the diet, they constitute a resource limiting the expression of sexual coloration, and hence they influence reproductive success (Grether et al., 1999; Olson and Owens, 1998; see Chapter 37). Such a correlation between carotenoid availability, immunocompetence (indicating overall health and vigor of the

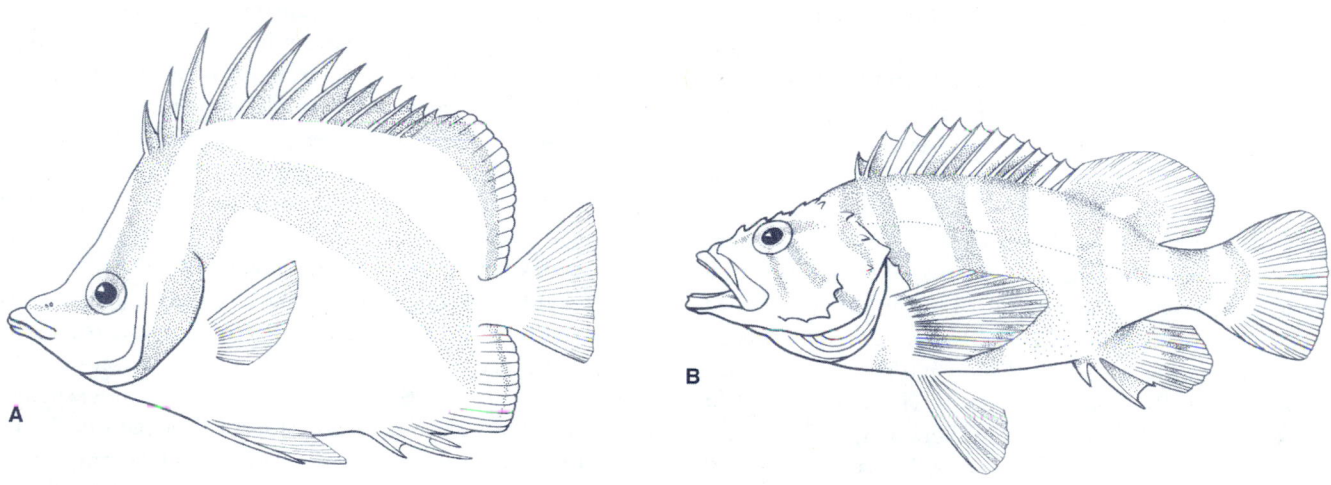

FIGURE 18.7

Examples of fishes with disruptive pigment patterns: **A,** Scythe butterflyfish (*Chaetodon falcifer*); **B,** Treefish (*Sebastes serriceps*). (*B* after Miller and Lea, 1972.)

individual), and reproductive success has been demonstrated in bird species (Blount et al., 2003; Faivre et al., 2003).

Changes in color and behavior may coincide. One example in a schooling species is the rudderfish (*Kyphosus elegans*), in which certain individuals change their color pattern from stripes to spots, round up stragglers from the school, and discourage members of other schools from feeding in their school's area. During their brief stints of such activity, the spotted individuals are in marked contrast to the schooling individuals. There are numerous instances where color functions to advertise mood or to predict the fighting ability of an individual in intraspecific encounters (Maynard, 1988).

Often, bright colors and bold patterns advertise unpalatability or venomous capacity (**aposematic coloration**). Weeverfishes (*Trachinus*), which are generally cryptically colored, erect their dorsal fin, displaying a conspicuous black mark, to warn potential predators. Conspicuous examples of fishes with warning coloration are the lionfishes (*Pterois*), whose bold coloration asserts their highly venomous nature, and the sabertoothed blenny (*Meiacanthus nigrolineatus*), which has venom glands associated with enlarged teeth in the lower jaw. Some surgeonfishes show a distinctive bright spot at the location of their formidable peduncular spines.

Disguise

Many fish species are effective at disguising themselves, not only with cryptic or distinctive color patterns, but also with a body shaped to resemble other objects; this is often coupled with an appropriate behavior. An adaptive advantage has apparently accrued to species that bear a close likeness to objects in the environment that elicit neutral reactions in predators or prey. Mimicry is a special case that merits separate consideration (see next subsection). Most disguises among fishes cause them to resemble plants (Fig. 18.8; see Plate 13.9). The leaffish of South America (*Monocirrhus polyacanthus*) has a barbel at the chin, simulating the stem of a leaf, and has coloration, body shape, and postural behavior that complete the illusion. The young of a labrid (*Hemipteronotus pavo*) has a "stem" consisting of the first dorsal fin, and its posture and coloration resemble a floating or drifting frond of seaweed. If color is involved in mimicry, it is termed **chromatic mimesis** (Fujii, 1993). Other fishes that mimic vegetation are the batfishes (*Platax*) and the young of a carangid (*Trachinotus falcatus*), both of which combine color and posture in the deceit. Other genera with leaflike young include *Lobotes* and *Oligoplites*. The naked sole (*Gymnachirus melas*) looks like ragged vegetation by means of color and serrate dorsal and anal fins. The filefish (*Aluterus*), holding position with its mouth to the substrate and its slender tail upward, resembles eelgrass. Juveniles of many fishes are

suitably colored and shaped to resemble floating or drifting plant debris, but one of the most remarkable instances is seen in the ephippid *Chaetodipterus*, native to mangrove estuaries, whose young resemble the blackened, sunken seed pods of mangroves.

An interesting adaptive coloration is the possession of large, realistic eyespots (**ocelli**), especially in the posterior part of the fish. These spots are believed to misdirect predators or to advertise unpalatability (Neudecker, 1989). When coupled with disruptive bars or stripes that make the fish's real eyes inconspicuous, these eyespots effectively disguise the direction in which the fish is heading and may be confusing to a predator (see Chapter 37). About 91 percent of the species in the butterflyfish genus *Chaetodon* have eye camouflage (often a dark bar across the eye), and nearly half combine this with false eyespots, usually not on the head. Neudecker (1989) reported that the false eyespots of butterflyfishes are usually situated in a part of the body that, if injured, would allow a high probability of recovery and survival after an attack. Most eyespots of nonschooling species are in the posterior third of the fish, mostly on the caudal peduncle or the dorsal fin. Spots in the anterior third tend to be high on the body, just below the formidable dorsal spines. Ocelli may be signals involved in intraspecific behavior or may be camouflage or disruptive coloration. They could be considered instances of mimicry, in that they might resemble the eyes of other organisms.

Mimicry

Mimicry may serve many functions among various animals, and several types of mimicry have been recognized, especially among insects. Mimicry is not widespread among fishes, and many species engage in mimicry only as juveniles. Fishes exemplify the classical mimic–model relationship, with mimics tending to be rarer than their models. Eagle and Jones (2004) identified examples of mimicry among several families of coral reef fishes; they distinguished species that engaged in mimicry as a means of facilitating feeding, discouraging predation, or both. Fishes that mimic other species (*models*) that are distasteful or otherwise avoided (or at least not preferred) by predators are engaging in **Batesian mimicry**. Thus, a sole living among venomous weeverfish may lift a black pectoral fin to mimic the warning signal given by the weeverfish model's dorsal fin, or a blenny may assume the coloration of a cleaner wrasse model and be able to live near predators with a reduced probability of being eaten. Young eeltail catfish (*Plotosus anguillarus*), themselves venomous, may aggregate to resemble sea anemones.

A variation on Batesian mimicry, **aggressive mimicry**, occurs when one species mimics another to gain resources

FIGURE 18.8
Examples of fishes with structure and coloration resembling vegetation: **A,** Young of tripletail (*Lobotes*), with general resemblance to mangrove leaf; **B,** Mosshead warbonnet (*Chirolophus nugator*), with cirri and flaps resembling marine algae; **C,** Sargassumfish (*Histrio histrio*).

from the model. An example involves the wrasse (*Labroides dimidiatus;* see Plate 16.13A), which has access to larger fishes as an ectoparasite cleaner, and one of the saber-toothed blennies (*Aspidontus taeniatus;* see Plate 16.13B), which mimics the wrasse and can thereby feed on skin torn from unsuspecting larger fishes. With experience, however, victimized fishes learn to distinguish the cleaner from the outlaw. Another example involves two characoid fishes of South America: *Probolobus heterostomus,* a scale eater, closely resembles in shape and coloration some species of *Astyanax*. This mimicry allows the scale eater to school freely with *Astyanax* and prey on their scales (Sazima, 1977). Likewise, some fry-eating (paedophagous) cichlids of Lake Malawi (*Cyrtocara*) are capable of mimicking the color pattern of species they are preying on and can make the change from, for instance,

a longitudinal stripe to a lateral row of spots within seconds (McKaye and Kocher, 1983).

The display of **lures** to entice prey toward the mouth is a special kind of mimicry, usually involving modified body parts, that enhances predation. The angler catfish (*Chaca chaca*) uses its maxillary barbels for such purposes. The lures of lophiiform anglerfishes may be luminescent; resemble worms or other edible invertebrates; or, as in a species of *Antennarius,* closely resemble a fish, complete with eye spots and structures resembling fins (Pietsch and Grobecker, 1978).

In **Mullerian mimicry,** two or more dangerous or unpalatable species assume similar warning coloration. This benefits both species, as predators learn to associate the warning coloration with unpalatability. This is a rare phenomenon among fishes; the closest qualifying instance

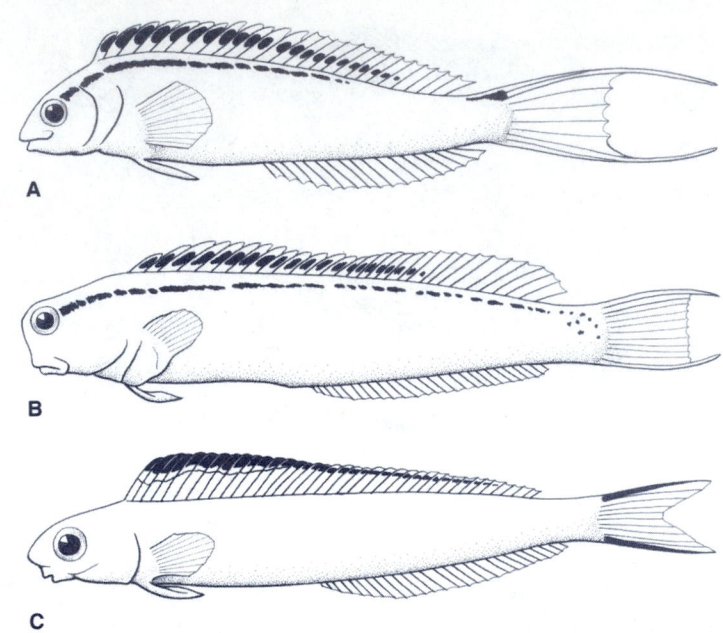

FIGURE 18.9
Mimicry among blennioid fishes: **A,** *Meiacanthus nigrolineatus,* a venomous species; **B,** *Ecsenius gravieri,* a nonaggressive species; **C,** *Plagiotremus townsendi,* an aggressive species (after Springer and Smith-Vaniz, 1972).

involves some of the saber-toothed blennies (Fig. 18.9). The venomous *Meiacanthus nigrolineatus* is a model not only for the nonaggressive *Ecsenius gravieri* (a Batesian mimic) but also for *Plagiotremus townsendi* (a Mullerian mimic), a fierce biter that, like *M. nigrolineatus,* is usually rejected by predators (Springer and Smith-Vaniz, 1972).

Some mouth-brooding cichlids display an interesting mimicry that ensures fertilization of the eggs, which are taken into the female's mouth after deposition and before fertilization. The male has a series of egglike spots ("dummy eggs") on his anal fin. He brings these to the attention of the female in the area where she has been depositing eggs, and as she attempts to pick them up, he releases sperm, fertilizing the eggs carried in her mouth.

BIOLUMINESCENCE

Occurrence of Bioluminescence

The ability to produce light has been noted in at least 45 families of fish (Table 18.1). Most of these are teleosts; only two families of elasmobranchs and no lampreys, hagfishes,

or nonteleost bony fishes are known to be luminous. Among the teleosts, a wide range of families, from the soft-rayed osmeriforms to the perciforms and lophiiforms, display bioluminescence. The lanternfishes (Myctophidae) have more luminescent genera and species than any other, but spookfishes (Opisthoproctidae), grenadiers (Macrouridae), batfishes (Ogcocephalidae), anglerfishes (Ceratioidei), and stomatioids (Stomiiformes) have numerous light-bearing representatives, and in some regions the Stomiiformes may outnumber lanternfishes.

There are a few bioluminescent fishes that are permanent residents of shallow water, including midshipmen (Batrachoididae), flashlightfishes (Anomalopidae), and slipmouths (Leiognathidae), but most bioluminescent species live at moderate to great depths, and many move into surface waters as part of a nightly feeding migration (Morin, 1981). Most bioluminescent fishes inhabit the mesopelagic realm of the ocean, usually at depths ranging from 300 to 1,000 m. At some localities, up to 66 percent of the fish species and more than 50 percent of the individuals collected are luminescent. Estimates of the total number of luminescent mesopelagic fishes range from 600 to 700 species (Marshall, 1979). The light organs and modes of light

TABLE 18.1   FAMILIES REPORTED TO CONTAIN BIOLUMINESCENT FISHES

Order	Family	Type of Photophore Bacterial	Type of Photophore Self-luminous
Squaliformes	Squalidae		√
	Dalatiidae		√
Torpediniformes	Torpedinidae		√
Anguilliformes	Congridae	√	
	Saccopharyngidae		√?
Clupeiformes	Engraulidae		√
Osmeriformes	Opisthoproctidae	√	
	Bathylagidae		√
	Alepocephalidae		√
	Platytroctidae		√
Stomiiformes	Gonostomatidae		√
	Sternoptychidae		√
	Stomiidae*		√
	Phosichthyidae		√
Aulopiformes	Chlorophthalmidae	√	
	Scopelarchidae		√
	Paralepididae		√
	Evermannellidae		√
Myctophiformes	Neoscopelidae		√
	Myctophidae		√
Gadiformes	Macrouridae	√	
	Moridae	√	
	Steindachneriidae		√
Batrachoidiformes	Batrachoididae		√
Lophiiformes	Melanocetidae	√	
	Himantolophidae	√	
	Diceratiidae	√	
	Oneirodidae	√	
	Ceratiidae	√	
	Gigantactidae	√	
	Centrophyrnidae	√	
	Linophyrnidae	√	
	Thaumatichthyidae	√	
Beryciformes	Anomalopidae	√	
	Monocentridae	√	
	Trachichthyidae	√	
Perciformes	Acropomatidae	√	
	Apogonidae	√	√
	Leiognathidae	√	
	Sciaenidae		√
	Pempheridae		√
	Chiasmodontidae		?

Source: Based on Herring and Morin, 1978; Morin, 1981; Herring, 1982; and Pietsch, personal communication to Bond.
* Includes Astronesthinae, Chauliodontinae, Idiacanthinae, Melanostominae, and Malacosteinae.

production are more diverse in fishes than in any other marine animals (Herring, 1982).

. . . .
Production of Light

Light production usually takes place in special organs called **photophores** (Fig. 18.10). Light is chemically produced through a chemical reaction of the enzyme **luciferase** with a heterocyclic phenol, **luciferin**, in the presence of an oxygen source and adenosine triphosphate (ATP).

Although most luminous fishes produce their own light (are self-luminous), many depend on symbiotic luminous bacteria nurtured in special, glandlike structures (see Table 18.1). At least one species of fish depends on luciferin obtained through its diet for luminescence. The plainfin midshipman (*Porichthys notatus*) of the Pacific Coast of North America is one of the few shallow-water species with self-luminous photophores, of which it has about 700. In the southern part of its range, it is luminescent, but north of San Francisco, where its occurrence is discontinuous, it is not. However, luminescence can be induced in individuals from Puget Sound after the dietary intake of luciferin from the ostracod *Vargula hilgendorfi* (Thompson et al., 1988; Thompson and Tsuji, 1989). Luminescence of this dietary nature occurs also in *Parapriacanthus* (Pempheridae) and in some genera of Apogonidae (Haneda, 1986).

Light from Symbiotic Bacteria

Most luminous fishes of shallow water have only bacterial photophores, usually in the region of the eye or gut (Morin, 1981), but fish from deep water may have bacterial light organs in other parts of the body. Herring (1977) reported that bacterial bioluminescence is found in relatively few fishes, as compared with the great numbers of species that are self-luminous, and that there are usually no more than four bacterial luminous organs per individual. There are about 65 genera with bacterial photophores and 130 genera that are self-luminous (Hastings and Morin, 1991).

The bacterial luminous organs usually open to the exterior. Certain bacteria appear to be restricted to given groups of fishes. Some are said to be species specific, although *Photobacterium leiognathi* is reported to live with members of

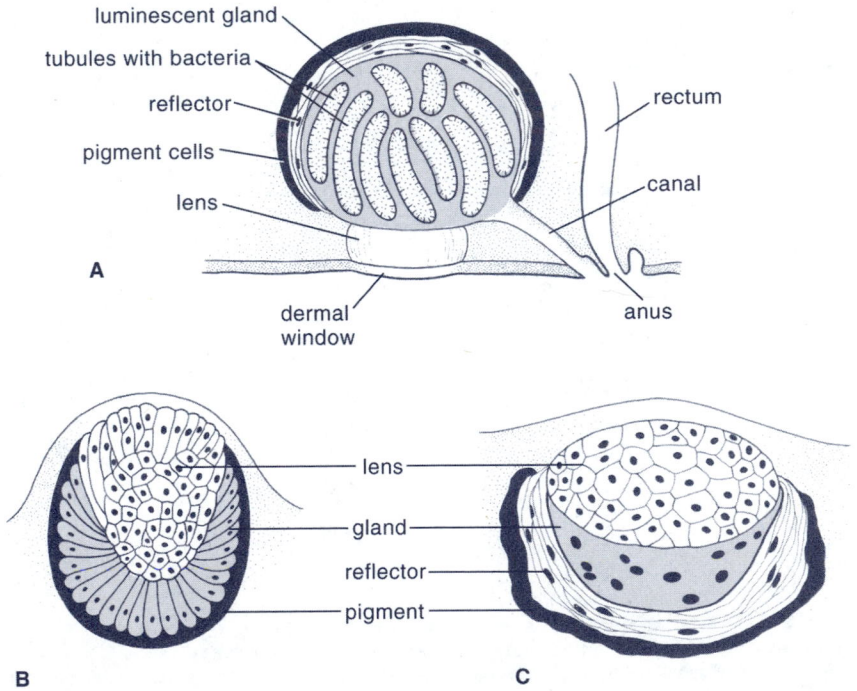

FIGURE 18.10
Illustrations of examples of photophores: **A**, Bacterial photophore, in which luminous bacteria are nurtured in tubelike structures. **B**, Self-luminous organ with lens and pigment sheath. **C**, Self-luminous organ with lens, reflector, and pigment sheath.

the Leiognathidae, Percichthyidae, Apogonidae, and squids (Fukasawa et al., 1988). The pineconefish (*Monocentrus japonicus*) harbors the luminous bacterium *Vibrio fisheri*.

Luminous structures can be associated with a variety of body parts. Grenadiers (Macrouridae), some cods (Moridae), steindachneriid hakes (Steindachneriidae), and roughies (Trachichthyidae) are examples of groups with luminescent structures along the ventral surface, close to the anus. In some of these, gelatinous tissue under the skin allows the light to diffuse over a relatively wide area. Some of the light glands release luminous materials that can spread along the exterior of the fish. Slipmouths (Leiognathidae) have a bacterial light gland around the esophagus, a lens that concentrates the light and directs it into the gas bladder, and a pigmented shutter that can conceal the light. A species of the leiognathid genus *Gazza* shows luminescence from the mouth as well as from the ventral surface and can project luminescence as a discrete beam (McFall–Ngai and Dunlap, 1982). Other fishes with bacterial luminescence associated with the digestive system, anus, or ventral musculature are the spookfishes (Opisthoproctidae), lanternbellies (Acropomatidae), and the genus *Siphamia* of the cardinalfishes (Apogonidae). Most of these can light up their thoracic or ventral areas because of internal reflectors and translucent tissue under the skin.

The flashlightfishes (Anomalopidae) bear bacterial light organs under the eyes (see Plate 14.3). The two best known species are *Photoblepharon palpebratum* and *Anomalops kaptotron*, both of which are shallow-water fishes of the Indo-West Pacific. *Anomalops* has been observed to flash its light on and off in a rather regular manner by rotating its light organ inward and down into a black-pigmented pocket under the eye. *Photoblepharon* shows a more steady light but can conceal it by drawing a fold of black tissue over it. Most members of the suborder of deep-water anglerfishes (Ceratioidei) are luminous, with bacterial photophores on the barbels, the esca, and elsewhere on the body (see Chapter 13). Foran (1991) claimed that two taxa of deep-water fishes in which bioluminescence is especially well represented, the myctophids and the Stomiiformes, used bacterial symbionts. DNA hybridization studies, however, have verified that the luminescence in myctophid and stomiiform fishes is not bacterial in origin (Haygood et al., 1994).

Nonbacterial Luminescence

Self-luminous fishes commonly possess a series of light organs along the ventral aspect of the body that direct the light downward. However, a few have dorsal photophores, and many of the stomiiforms have an abundance of small, simple organs along the dorsal surface that enable the entire fish to be silhouetted in light. Some have light organs within the mouth as well as on the jaws. There may be numerous (1,000 or more) photophores per individual, and more than one type of light organ on an individual. Photophores are not always open to the exterior and are not usually associated with the gut in these fishes (Herring, 1977).

Light organs vary from very small, unpigmented structures, such as those on the back and fins of the stomiiforms, to very complex ones, with the glandular portion surrounded by a reflector that directs the light through a lens, which can concentrate the beam as it is emitted. Some photophores have irislike structures that can control the amount of light emitted (Herring and Morin, 1978).

Members of the family Platytroctidae (= Searsidae) have a light gland above the pectoral fin with an opening to the surface. Apparently, luminous material can be released voluntarily from the gland. Similar glands are known in the ceratioids. This type of luminescence is called *extracellular* and contrasts with the intracellular type, in which the luminous material is confined to the cells (**photocytes**) that produce the light.

Although certainly not a natural phenomenon, the development of transgenic fluorescent fishes merits at least a brief mention. Zebrafish (see Chapter 27, Going Deeper) have been genetically modified with the insertion of genes from a jellyfish (*Aequorea*) and a sea anemone (*Discosoma*) that produce green fluorescent and red fluorescent protein, respectively. Under visible light, the zebrafish gives off a fluorescent glow that becomes transformed into brilliant coloration when viewed under ultraviolet light. Originally developed as pollution monitors, these fish are now available to home aquarists (http://www.glofish.com).

Control of Luminescence

In the anomalopids, some stomiiforms, and the pineconefishes (Monocentridae), control of the display of light is indirect, by concealing or screening the luminous tissue. In most self-luminous bony fishes, however, the photophores are innervated and appear to be under direct nervous control. Injections of epinephrine usually cause activity of luminous tissue in most species tested, but the exact relationships of the nervous control and the mechanism of hormonal control have yet to be clarified (Hastings and Morin, 1991).

Significance of Bioluminescence

According to Morin (1981), luminescence functions in obtaining prey (by luminescent predators), in evading predators (by luminescent prey), and in communication between conspecifics. Concealment, advertisement, and disguise are all applications of luminescence.

The spectra of bioluminescent light emissions fall into two main categories: (1) blue to blue-green and (2) red, with most species emitting in the blue range of the spectrum (Montgomery and Pankhurst, 1997). Blue emissions are perceived as most useful to deep-ocean fishes for purposes of communication. Red emissions are useful for both shallow- and deep-water fishes as sources of illumination for investigating the surrounding environment. The aforementioned anomalopid flashlightfishes best exemplify this. The value of bioluminescence in identification and communication is obvious when one recognizes that about two thirds of deep oceanic fishes produce light. The identification of specific selective advantages is difficult for most species because of their habitats. The functions of light are thought to parallel in part the functions of color, and of course color is involved in bioluminescence. Many species have color filters built into their light organs, so that the emitted light may be of different wavelengths. Montgomery and Pankhurst (1997) provided several examples of visual systems of deep-dwelling fishes that represent specific adaptations for the detection of bioluminescence.

Concealment (Crypsis)

McFall-Ngai (1990) reported that pelagic fishes can gain concealment by three major types of crypsis: transparency, reflection of light, counterillumination, or combinations of these. According to the principle of **counterillumination,** the placement of photophores and other luminous tissue in a ventral position in most luminous species of mid-water fishes enables the fish to match the ambient light coming from above, so that predators hunting from below are less likely to discern the fish's silhouette. There is evidence that various lanternfishes and other species with ventral photophores regulate the intensity of their light emission according to the light they perceive from above (Young and Ropper, 1977). Many luminescent species, including representatives of the Myctophidae, Opisthoproctidae, and Stomiiformes, apparently adjust their luminescence by comparing the light intensity from above with the light emanating from photophores that shine into their eyes (Lawry, 1974). Some species are known to switch on ventral photophores when illuminated from above. The color filters in the ventral light organs of hatchetfishes have a transmission band close to 480 nm that matches blue-green light, which has great penetration into sea water (Denton, 1970). McFall-Ngai and Morin (1991) found that leiognathids, which have bacterial photophores around the esophagus, increased the intensity of their luminescence as illumination from above was increased, although not in direct proportion to the increase in ambient light.

McAllister (1967) has suggested that the dense pigmentation of the peritoneum or stomach of a deep-sea predator serves mainly to guard against the lights of recently swallowed prey shining through the body wall and betraying the location of the predator.

Advertisement

Communication involving luminescence is active in the reproduction of some species. Midshipmen (*Porichthys*) are known to display the light from photophores during courtship (Wheeler, 1975), and other fishes might do the same. Some lanternfishes have sexually dimorphic patterns of luminous organs; such different patterns might serve in the recognition of mates (Lagler et al., 1977).

There are several other advantages that might accrue to fishes that advertise themselves with light. Species recognition could aid in keeping schools together or could aid individuals of nonschooling species in maintaining territories. Bioluminescent communication can involve more than one species or even other taxa of fishes (Herring, 1990).

Sudden displays of light by a single fish or by an aggregation of small fishes might serve to startle or confuse predators. Defense by confusing visually oriented predators is considered by Hastings and Morin (1991) as probably the most important function of bioluminescence in fishes. The Atlantic midshipman (*Porichthys plectrodon*), possessor of a venomous spine, is known to flash its lights upon approach of predators, thus warning them away.

Several predators, including stomiiforms and anglerfishes, have luminous organs or tissue close to the mouth, in the mouth, or on barbels or illicia (see Chapter 13). In some or all of these cases, prey could be attracted to the mouth by a show of photophores.

Disguise

Some photophores tend to disguise fishes by creating a resemblance to other objects in the environment. If the aforementioned baits closely resemble some organism that is habitually eaten by the fish's prospective prey, this might be considered disguise. Some investigators have suggested that the luminous tissue of certain fishes mimics various luminous invertebrates. Also, perhaps a school of small luminous fish might take on the appearance of a single organism large enough to be intimidating to potential predators.

Special Considerations

The placement of light organs on the head, so that objects in the visual field of the fish are lit, permits the illumination of prey species; the feeding observations of stomiiform fishes appear to confirm this. Some melanostomiines combine a red-sensitive retina (see Chapter 20) with a large, red, post-orbital photophore that allows them to see nearby prey, especially those with red pigment (Denton, 1970). Because few deep-sea animals have red-sensitive retinas, the red spotlight of this stomiiform is not conspicuous (Marshall, 1979). One species of loosejaws (Malacosteinae) is known to have red and green photophores and visual pigments sensitive to both those colors (O'Day and Fernandez, 1974). If an individual of a prey species illuminates a nearby predator, other members of the species could avoid it. According to the "**burglar alarm theory**," foraging on bioluminescent prey stimulates light production as a means of exposing the predators to those that they might feed on. Studies of the feeding ecology of the midshipman (*Porichthys notatus*) have revealed a trophic cascade of bioluminescence in which the burglar alarm theory is applied. The midshipman is dependent on bioluminescent prey as a source of luciferin. Midshipmen rise up in the water column at night to feed on luminescent prey items, such as the aforementioned ostracod *Vargula*, but they are also sensitive to light emitted by dinoflagellates. Because of this, they can home in on nonluminescent predators of the dinoflagellates as well, thus broadening their dietary range during feeding intervals (Mensinger, 1995).

The peculiar "cookie cutter" shark (*Isistius brasiliensis*) may actually employ its counterillumination abilities as a means to attract the large fishes and cetaceans on which it feeds (see Chapter 37, Plate. 37.1). A patch of skin on the ventral surface near the head lacks bioluminescent photophores, which causes this region to appear conspicuous to large predators viewing the shark from below. It has been proposed that this heavily pigmented region lacking photophores acts as a lure. When large predators approach, *Isistius* can make a quick grab at a chunk of flesh, turning the potential predator into prey (Widder, 1998). In some species, there are probably multiple uses of the photophores in predation, intraspecific communication, and avoidance of predators.

VENOMOUS FISHES

Many species of fishes have evolved integumentary features that provide protection through noxious secretions. Poisonous and venomous fishes apparently enjoy an ecological advantage, in that predators may be injured or killed by their stings or poisons. Some species, such as the Moses sole (*Pardachirus marmoratus*), release toxins from glands scattered in the skin; these serve as an effective predator deterrent. In other species, toxic glands are highly localized and associated with spines that facilitate the introduction of venom into the victim. A number of fish families contain species capable of inflicting painful stings that combine mechanical injury with the release of venom (Halstead, 1978; Maretic, 1988). The stingrays of the families Dasyatidae and Potamotrygonidae are among the most notorious of stinging fishes. The generally larger pelagic dasyatids—the eagle rays, cownose rays and mantas (Myliobatinae)—have stings and venom-producing tissue, but they are not often implicated in injury to humans. The stingers of the rays are stiff, with serrated edges (Fig. 18.11A). The spine is located in an integumentary sheath, and the venom is mainly located in the epidermis of this sheath that overlies the spine (Fig. 18.11B, C). When the spine is thrust into the flesh of a victim, the skin sheath is broken, and the venom is released into the wound.

The spines of stingrays are usually situated close to the thick, muscular base of the tail. When the ray is forced to defend itself—for example, against a person stepping on the disc—the tail is curled quickly over the back and the spines are thrust at the offender. The wounds resulting from these stings are painful and sometimes dangerous. The victim is usually in danger from secondary infections, tetanus, and gangrene, as well as from the effects of the venom.

Other cartilaginous fishes with venomous spines are the dogfish shark (*Squalus*) and the chimaeras. In these fishes, the venom glands are along the dorsal spines, and the danger of being stung is not as great as with the stingrays. Most punctures come from careless handling. The venom of sharks and chimaeras is not as dangerous as that of stingrays.

Among the bony fishes, certain catfishes, weeverfishes, surgeonfishes, scorpionfishes, stargazers, rabbitfishes, toadfishes, and the sabertooth blennies are known to be venomous, and several other groups, including the Carangidae and Scatophagidae, are reported to have venom associated with spines.

Catfishes have venom glands in the skin sheathing the dorsal and pectoral spines, and some groups have axillary venom glands that supply their secretions to the exterior of the pectoral spine. Most catfish stings are painful but not dangerous, but some species cause edematous swelling and gangrene at the wound. The family Plotosidae, an Indo-Pacific group containing both fresh- and saltwater members, contains the most dangerous venomous catfishes—some capable of causing death in humans (Halstead, 1978). Some of the catfishes that are commonly kept in home aquaria can inflict severe stings that occasionally result in numbness and shock; included are the shovelhead catfishes of the family Pimelodidae, the thorny catfishes of the family Doradidae, and the air-breathing clariid catfishes. Air-breathing catfishes of the families Heteropneustidae and Clariidae can

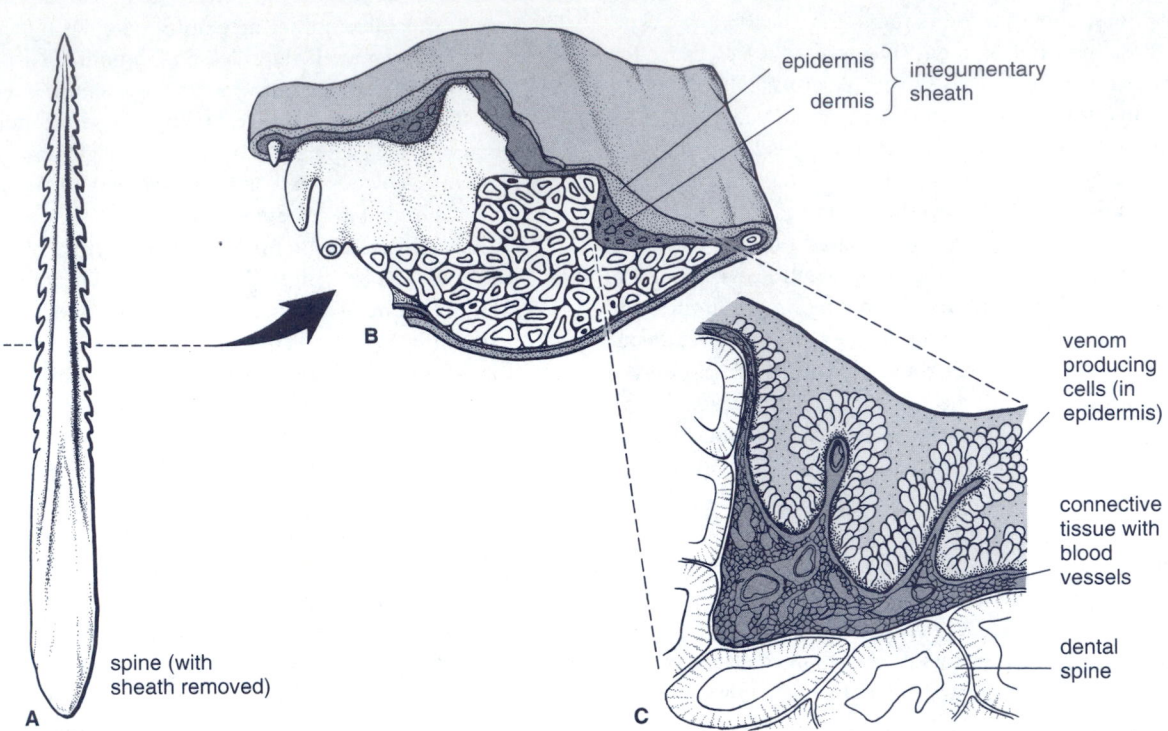

FIGURE 18.11
Stinger and associated venom-producing cells in a stingray (*Urolophus halleri*): **A,** Isolated stinger. **B,** Cross section through stinger. **C,** Section showing stinger and associated integumentary sheath (after Russell, 1969).

be especially aggressive stingers. Venomous catfish species can be found among the Bagridae, Siluridae, and the marine Ariidae. What the diminutive madtom catfishes (*Noturus*)—members of the only freshwater catfish family native to North America, the Ictaluridae—lack in size, they amply compensate for with potent venom glands at the base of their pectoral spines.

Catfishes react to being grabbed or restrained by lashing violently from side to side. Usually, the pectoral and dorsal spines are locked in an erect position during this activity, so they can pierce the attacker. Care should be exercised when handling live catfishes of any kind. Usually, they can be grasped directly behind the pectoral fins with reduced danger of being stung.

Toadfishes (Batrachoididae; see Plate 13.7) are found mainly in warm coastal waters of most seas. One subfamily, the Thalassophryninae, has members with hollow opercular and dorsal fin spines surrounded by venom glands (Halstead, 1970). Venomous toadfishes occur in the tropical eastern Pacific and western Atlantic. Two freshwater species are

known from South America. Their venom is not regarded as being as dangerous as that of other teleosts such as the scorpaenoids or trachinids.

Fishes of the family Scorpaenidae are widespread along tropical and temperate shores. Members of the family generally have venom glands in grooves along the dorsal, anal, and pelvic spines (Roche and Halstead, 1972). In some genera, such as *Sebastes*, the venom is not virulent, and the spines mainly inflict a painful wound. In other genera, such as *Scorpaena* (see Plate 17.1) and the tropical *Pterois*, the venom is more powerful (that of *Pterois* sometimes killing humans). Species of *Pterois*—Indo-Pacific coral reef inhabitants commonly known as turkeyfish or lionfish—are extremely colorful, making them popular additions to marine aquaria. The aquarium fish trade is responsible for a number of introductions of exotic species, possibly including *Pterois volitans* (see Chapter 39). This species has recently been recorded at several localities along the southeastern U.S. coastline and on Bermuda. Included among these sightings are juveniles—evidence that breeding populations may

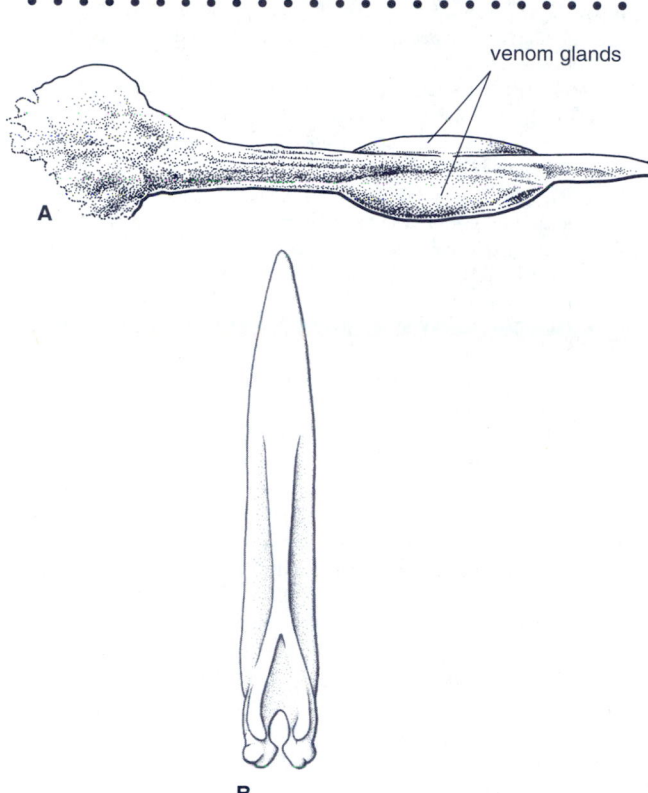

venom glands

A

B

FIGURE 18.12
Stinging spines in teleosts: **A,** Dorsal spine of the stonefish (*Synanceia horrida*), showing fusiform venom glands lying alongside the spines. **B,** Grooved dorsal spine of venomous weeverfish (family Trachinidae).

have become established (Paula Whitfield, National Centers for Coastal Ocean Science, personal communication).

The scorpaenoid family Synanceidae (stonefishes) has very large venom glands associated with the dorsal, anal, and pelvic spines. The ducts of the glands run in a groove to a point near the tip of the spine (Fig. 18.12A). Stonefish venom is extremely dangerous, and as the fishes are so well camouflaged, there are many recorded instances of people stepping on them and being severely affected by the venom. This tropical Indo–Pacific family is responsible for numerous deaths. The aptly-named species *Synanceia horrida* is regarded by many to be the world's most venomous fish.

Weeverfishes (Trachinidae; see Fig. 16.1A) are among the fishes that have venomous opercular spines as well as venomous fin spines. Weeverfishes are found from the North Sea south into the Mediterranean and have considerable contact with fishery workers and divers throughout their range. They are said to attack when disturbed, or even without provocation, striking with the bladelike opercular

spine (Maretic, 1988). Both fin and opercular spines are deeply grooved, as in most venomous fishes (Fig. 18.12B). The venom-producing tissue is in the grooves. Few fatalities result from weever stings, but permanent damage to joints can result (Maretic, 1988).

Unlike terrestrial vertebrates—snakes in particular—very few fishes appear to have a venomous bite. Morays have been reported to have poisonous fangs, but this appears not to be the case. Some authorities have claimed that venom-producing tissue is present in the skin of their palate. Regardless of any purported venomous capacity, the bite of a moray eel is very painful and highly prone to infection. The sabertooth blennies (*Meiacanthus*) have large canine teeth in the lower jaw. Venom glands are associated with these grooved teeth, so that their bite is a most unpleasant experience (Fishelson, 1974).

KEY POINTS AND CONNECTIONS

• The integument of fishes consists of a complex arrangement of tissue layers from two distinct embryonic sources; the epidermis arises from the ectoderm, and the dermis arises from the mesoderm. Neural crest cells contribute a number of cell types to the skin of fishes. Scales, which are among the most characteristic features of the skin of fishes, arise within the dermis.

The origins of dermal armor and scale types in the context of the early evolution and diversification of fishes are discussed in Chapters 5 through 7.

• Integumentary composition is important in the consideration of general body form and swimming in fishes. Integumentary features such as mucus secretion and scale microanatomy may be integral to the maintenance of hydrodynamic efficiency in swift-swimming fishes.

The discussion of the topics of body form and hydrodynamic efficiency as they pertain to fish locomotion continues in Chapter 19.

• The visual context of fish skin is manifested in the actions of cellular components that impart and control color. Also to be considered is the special adaptive feature of bioluminescence that is widely observed among deepwater fishes.

The adaptive value of bioluminescence must be considered in the context of the visual systems of deep-ocean fishes, which is discussed in Chapter 20. The applications of bioluminescence are further explored in Chapter 35, where the adaptations of deep-ocean fishes are also discussed.

• Many species of fishes possess integumentary cells that produce potent toxins. Most venomous fishes couple this venom-secreting capacity with sharp spines to create an effective venom delivery apparatus, but comparatively few species appear to possess a venomous bite.

Toxicity in fish tissues extends beyond integumentary venom glands. Other forms of fish toxicity are discussed in Chapter 38.

REFERENCES

Abbot, F. S. 1973. Endocrine regulation of pigmentation in fish. *Am. Zool. 13:* 885–894.

Aleev, Y. G. 1969. *Function and gross morphology in fish.* Israel Program for Scientific Translations, Jerusalem.

Anderson, E. J., W. R. McGillis, and M. A. Grosenbaugh. 2001. The boundary layer of swimming fish. *J. Exp. Biol. 204:* 81–102.

Avtalion, R. R., and L. Reich. 1989. Chromatophore inheritance in red tilapias. *Bamidgeh (Isr. J. Aquacult.) 41*(3): 98–104.

Birkhoff, G. 1950. *Hydrodynamics.* Princeton University Press, Princeton, NJ.

Blake, R. W. 1983. *Fish locomotion.* Cambridge University Press, Cambridge, UK.

Blount, J. D., N. B. Metcalfe, T. R. Birkhead, and P. F. Surai. 2003. Carotenoid modulation of immune function and sexual attractiveness in zebra finches. *Science 300:* 125–127.

Bone, Q. 1972. Buoyancy and hydrodynamic functions in the castor oil fish, *Ruvettus pretiosus* (Pisces: Gempylidae). *Copeia 1972:* 78–87.

Bullock, A. M., and R. J. Roberts. 1974. The dermatology of marine teleost fish. I. The normal integument. *Oceanogr. Mar. Biol. Ann. Rev. 13:* 383–411.

Carlander, K. D. 1987. A history of scale age and growth studies of North American freshwater fish, pp. 3–14. In *Age and growth in fish,* R. C. Summerfelt and G. E. Hall (Eds.). Iowa State University Press, Ames.

Cott, H. B. 1940. *Adaptive coloration in animals.* Methuen, London.

Daniel, T. L. 1981. Fish mucus: In situ measurements of polymer drag reduction. *Biol. Bull. 160:* 376–382.

Denton, B. J. 1970. On the organization of reflecting surfaces in some marine animals. *Phil. Trans. Roy. Soc. Lond. B 258:* 285–313.

———, and J. A. C. Nicol. 1966. A survey of reflectivity in silvery teleosts. *J. Mar. Biol. Assoc. UK 46:* 655–722.

Dey, D. B., and D. M. Damkaer. 1990. Effects of spectral irradiance on the early development of chinook salmon. *Prog. Fish.Cult. 53*(3): 141–154.

Deynat, P. P. 1998. Le revêtement cutané des raies (Chondrichthyes, Elasmobranchii, Batoidea). II. Morphologie et arrangement des tubercules cutanés. *Ann. Sci. Nat. 3–4:* 155–172.

———, and B. B. Séret. 1996. Le revêtement cutané des raies (Chondrichthyes, Elasmobranchii, Batoidea). I. Morphologie et arrangement des denticules cutanés. *Ann. Sci. Nat. Zool. 17*(2): 65–83.

Dingerkus, G., and R. J. Koestler. 1986. Application of scanning electron microscopy to the study of shark dermal denticles. *Scan. Electr. Microsc. 2:* 513–519.

Drucker, E. G., and G. V. Lauder. 1999. Locomotor forces on a swimming fish: Three dimensional vortex wake dynamics quantified using digital particle image velocimetry. *J. Exp. Biol. 202:* 2393–2412.

———, and———. 2001. Wake dynamics and fluid forces of turning maneuvers in sunfish. *J. Exp. Biol. 204:* 431–442.

Eagle, J. V., and G. P. Jones. 2004. Mimicry in coral reef fishes: Ecological and behavioral responses of a mimic to its model. *J. Zool. (Lond.) 264:* 33–43.

Elliott, D. G. 2000. Integumentary system, pp. 95–108. In *The laboratory fish,* G. A. Ostrander (Ed.). Academic Press, San Diego.

Endler, J. A. 1990. On the measurement and classification of colour in studies of animal colour patterns. *Biol. J. Linn. Soc. 41:* 315–352.

Faivre, B., A. Grégoire, M. Préault, F. Cézilly, and G. Sorei. 2003. Immune activation rapidly mirrored in a secondary sexual trait. *Science 300:* 103.

Fingerman, M. 1965. Chromatophores. *Physiol. Rev. 45:* 296–339.

Fishelson, L. 1974. Histology and ultrastructure of the recently found toxic gland in the fish *Meiacanthus nigrolineatus* (Blenniidae). *Copeia 1974:* 386–392.

Foran, D. 1991. Evidence of luminous bacterial symbionts in the light organs of myctophid and stomiiform fishes. *J. Exp. Zool. 259:* 1–8.

Fox, D. L. 1957. The pigments of fishes, pp. 367–385. In *The physiology of fishes,* M. B. Brown (Ed.). Academic Press, New York.

Fox, H. M., and H. G. Vevers. 1960. *The nature of animal colors.* Macmillan, New York.

Fries, E. F. B. 1958. Iridescent white reflecting chromatophores (Antaugophores) iridoleucophores in certain teleost fishes, particularly in *Bathygobius. J. Morphol. 103:* 203–254.

Frolich, L. M. 1997. The role of the skin in the origin of amniotes: Permeability barrier, protective covering, and mechanical support, pp. 327–352. In *Amniote origins. Completing the transition to land,* S. S. Sumida and K. L. M. Martin (Eds.). Academic Press, San Diego.

Fujii, R. 1969. Chromatophores and pigments, pp. 307–353. In *Fish physiology,* Vol. III, W. S. Hoar and D. J. Randall (Eds.). Academic Press, New York.

———. 1993. Coloration and chromatophores, pp. 535–562. In *The physiology of fishes,* D. H. Evans (Ed.). CRC Press, Boca Raton, FL.

Fukasawa, S., T. Suda, and S. Kubota. 1988. Identification of luminous bacteria isolated from the light organ of the fish *Acropoma japonicum. Agric. Biol. Chem. 52*(1): 285–286.

Gnyubkin, V. F. 1989. Response of pigmented cornea of whitespotted greenling to changes in illumination. *Biol. Morya Vladivostok 1:* 25–32.

———, and A. G. Gamburtseva. 1981. Morphological variation in the coloration of the cornea of the eye of a fish. *J. Ichthyol. 21*(1): 175–181.

Govoni, J. J., M. A. West, D. Zivotofsky, P. R. Bowser, and B. B. Collette. 2004. Ontogeny of squamation in swordfish, *Xiphias gladius. Copeia 2004:* 391–396.

Gray, J. 1957. How fishes swim. *Sci. Am. 197*(2): 48–54.

Grether, G. F., J. Hudon, and D. F. Millie. 1999. Carotenoid limitation of sexual coloration along an environmental gradient in guppies. *Proc. Roy. Soc. Lond. B 266:* 1317–1322.

Gupta, B., R. Goodman, F. Jiang, and Y.-C. Tai. 1996. Analog VLSI system for active drag reduction. *IEEE Micro. 16*(5): 53–59.

Halstead, B. W. 1970. *Poisonous and venomous marine animals of the world,* Vol. 3. U.S. Government Printing Office, Washington, DC.

———.1978. *Poisonous and venomous marine animals of the world* (rev. ed.). Darwin Press, Princeton, NJ.

Hamilton, W. J., III, and R. M. Peterman. 1971. Countershading in the colourful reef fish, *Chaetodon lunula:* Concealment, communication or both. *Anim. Behav. 19*(2): 357–364.

Haneda, Y. 1986. On a new type of luminous fishes and squids, ingested luminescence, pp. 838–839. In Indo-Pacific fish biology, T. Uyeno, R. Arai, T. Taniuchi, and K. Matsuura (Eds.). *Proc. 2nd Int. Congr. Indo-Pac. Fishes.* Ichthyological Society of Japan, Tokyo.

Hastings, J. W., and J. G. Morin. 1991. Bioluminescence, pp. 131–170. In *Neural and integrative animal physiology: Comparative animal physiology* (4th ed.), C. L. Prosser (Ed.). Wiley-Liss, New York.

Hawkes, J. W. 1974a. The structure of fish skin I: General organization. *Cell Tissue Res. 149:* 147–158.

———. 1974b. The structure of fish skin II: The chromatophore unit. *Cell Tissue Res. 149:* 159–172.

Haygood, M. G., D. B. Edwards, G. Mowlds, and R. H. Rosenblatt. 1994. Bioluminescence of myctophid and stomiiform fishes is not due to bacterial luciferase. *J. Exp. Zool. 270:* 225–231.

Helfman, G. S., B. B. Collette, and D. F. Facey. 1997. *The diversity of fishes.* Blackwell, Malden, MA.

Herring, P. J. 1967. The pigments of plankton at the sea surface. Aspects of marine zoology. *Symp. Zool. Soc. Lond. 19:* 215–235.

———.1977. Bioluminescence of marine organisms. *Nature 276:* 788–793.

———. 1982. Aspects of the bioluminescence of fishes. *Oceanogr. Mar. Biol. Ann. Rev. 20:* 415–470.

———. 1990. Bioluminescent communication in the sea, pp. 245–264. In *Light and life in the sea,* P. J. Herring, A. K. Campbell, M. Whitfield, and L. Maddox (Eds.). Cambridge University Press, Cambridge, UK.

———, and J. G. Morin. 1978. Bioluminescence in fishes, pp. 273–329. In *Bioluminescence in action,* P. J. Herring (Ed.). Academic Press, London.

Holland, P. W. H., H. Wada, M. Manzanares, R. Krumlauf, and S. M. Shimeld. 2001. The origin of the neural crest, pp. 33–39. In *Major events in early vertebrate evolution: Palaeontology, phylogeny, genetics, and development,* P. E. Ahlberg (Ed.). Taylor and Francis, London.

Hoyt, J. W. l975. Hydrodynamic drag reduction due to fish slimes, pp. 653–672. In *Swimming and flying in nature,* Vol. 2, T. Y.-T. Wu, J. Brokaw, and C. Brennan (Eds.). Plenum Press, New York.

Kardong, K. V. 2002. *Vertebrates: Comparative anatomy, function, evolution* (3rd ed.). McGraw-Hill, Boston.

Kasukawa, H., N. Oshima, and R Fujii. 1987. Mechanism of light reflection in blue damselfish motile iridophore. *Zool. Sci. 4:* 243–257.

Kemp, N. E. 1999. Integumentary system and teeth, pp. 43–68. In *Sharks, skates, and rays: The biology of elasmobranch fishes,* W. C. Hamlett (Ed.). Johns Hopkins University Press, Baltimore.

Kent, G. C., and R. K. Carr. 2001. *Comparative anatomy of the vertebrates* (9th ed.). McGraw-Hill, Boston.

Kodric-Brown, A. 1983. Determinants of male reproductive success in pupfish (*Cyprinodon pecoensis*). *Anim. Behav. 31*(1): 128–137.

———. 1998. Sexual dichromatism and temporary color change in the reproduction of fishes. *Amer. Zool. 38:* 70–81.

Kotz, K. J., and M. A. McNiven. 1994. Intracellular calcium and cAMP regulate directional pigment movements in teleost erythrophores. *J. Cell Biol. 124:* 463–474.

Krieger, K. 2004. Do pool sharks swim faster? *Science 305:* 636–637.

Lagler, K. F., J. W. Bardach, R. R. Miller, and D. R. M. Passino. 1977. *Ichthyology* (2nd ed.). Wiley, New York.

Langdon, J. S., A. Masters, T. Thorne, and S. Wilton. 1995. Bizarre organism from the skin of mahi mahi, *Coryphaena hippurus* L. (Teleostei: Coryphaenidae). *J. Fish Diseases 18:* 481–494.

Lauder, G. V. 2000. Function of the caudal fin during locomotion in fishes: Kinematics, flow visualization, and evolutionary patterns. *Amer Zool. 40:* 101–122.

Lawry, J. V. 1974. Lantern fish compare downwelling light and bioluminescence. *Nature 247:* 155–157.

Liao, J. C., D. N. Beal, G. V. Lauder, and M. S. Triantafyllou. 2003. Fish exploiting vortices decrease muscle activity. *Science 302:* 1566–1569.

Lindsey, C. C. l978. Form, function, and locomotory habits in fish, pp. l–l00. In *Fish physiology,* Vol. VII, W. S. Hoar and D. J. Randall (Eds.). Academic Press, New York.

Lythgoe, J. N., and J. Shand. 1982. Change in spectral reflexions from the iridophores of the neon tetra. *J. Physiol. 325:* 23–34.

———, and———. 1983. Diel color changes in the neon tetra *Paracheirodon innesi.* *Env. Biol. Fishes 8*(4): 249–254.

———, and———. 1989. The structural basis for iridescent colour changes in dermal and corneal iridophores in fish. *J. Exp. Biol. 141:* 313–325.

———, ———, and R. G. Foster. 1984. Visual pigment in fish iridocytes. *Nature 308:* 83–84.

Maretic, Z. 1988. Fish venoms, pp. 445–476. In *Handbook of natural toxins,* Vol. 3, Marine toxins and venoms, A. T. Tu (Ed.). Marcel Dekker, New York.

Marshall, N. B. 1979. *Developments in deep-sea biology.* Blandford Press, Poole, Dorset, UK.

Maynard, D. J. 1988. Status signaling and the social structure of juvenile coho salmon. *Diss. Abst. Int. Pt. Biol. Sci. Eng. 48*(12).

McAllister, D. E. 1967. The significance of ventral bioluminescence in fishes. *J. Fish. Res. Bd. Can. 24*(3): 537–554.

McArdle, J., and A. M. Bullock. 1987. Solar ultraviolet radiation as a causal factor of "summer syndrome" in cage-reared Atlantic salmon, *Salmo salar* L.: A clinical and histopathological study. *J. Fish Diseases 10:* 255–264.

McClure, M. 1999. Development and evolution of melanophore patterns in fishes of the genus *Danio* (Teleostei: Cyprinidae). *J. Morphol. 241:* 83–105.

McFall-Ngai, M. J. 1990. Crypsis in the pelagic environment. *Am. Zool. 30*(l): 175–188.

———, and P. V. Dunlap. 1982. Three new modes of luminescence in the leiognathid fish *Gazza minuta:* Discrete luminescence. *Mar. Biol. 73*(3): 227–237.

———, and J. G. Morin. 1991. Camouflage by disruptive illumination in leiognathids, a family of shallow-water bioluminescent fishes. *J. Exp. Biol. 156:* 119–137.

———, J. F. Crescitelli, J. Childress, and J. Horowitz. 1986. Patterns of pigmentation in the eye lens of the deep-sea hatchet fish *Argyropelecus affinis* Garman. *J. Comp. Physiol. A159:* 791–800.

McKaye, K. R., and T. Kocher. 1983. Head ramming behavior by three paedophagous cichlids in Lake Malawi, Africa. *Anim. Behav. 31*(l): 206–210.

Mensinger, A. F. 1995. Ecomorphological adaptations to bioluminescence in *Porichthys notatus.* *Env. Biol. Fishes 44:* 133–142.

Meyer-Rochow, V. B. 2001. Fish chromatophores as sensors of environmental stimuli, pp. 317–334. In *Sensory biology of jawed fishes: New insights,* B. G. Kapoor and T. J. Hara (Eds.). Science Publishers, Enfield, NH.

Miller, D. J., and R. N. Lea. 1972. Guide to the coastal marine fishes of California. *Calif. Fish Bull. 157.*

Milos, N., and A. D. Dingle. 1978a. Dynamics of pigment pattern formation in the zebrafish, *Brachydanio rerio* I. Establishment and regulation of the lateral line melanophore stripe during the first eight days of development. *J. Exp. Zool. 205:* 205–216.

———, and———. 1978b. Dynamics of pigment pattern formation in the zebrafish *Brachydanio rerio* II. Lability of lateral line stripe formation and regulation of pattern defects. *J. Exp. Zool. 205:* 217–224.

Miyake, T., J. L. Vaglia, L. H. Taylor, and B. K. Hall. 1999. Development of dermal denticles in skates (Chondrichthys, Batoidea): Patterning and cellular differentiation. *J. Morphol. 241:* 61–81.

Montgomery, J., and N. Pankhurst. 1997. Sensory physiology, pp. 325–349. In *Deep-sea fishes,* D. J. Randall and A. P. Farrell (Eds.). Academic Press, San Diego.

Morin, J. G. 1981. Bioluminescent patterns in shallow tropical marine fishes, pp. 569–574. In The reef and man, E. D. Gomez, C. E. Birkeland, R. W. Buddenmeier, R. E. Johannes, J. A. Marsh, Jr., and R. T. Tsuda (Eds.). *Proceedings of the 4th International Symposium,* Vol. 2. Marine Science Center, Quezon City, Philippines.

Muntz, W. R. A. 1983. Bioluminescence and vision, pp. 217–238. In *Experimental biology at sea*, A. G. MacDonald and I. G. Priede (Eds.). Academic Press, London.

———. 1990. Stimulus, environment and vision in fishes, pp. 491–511. In *The visual system of fish*, R. Douglas and M. Djamgos (Eds.). Chapman and Hall, London.

Nagai, M., N. Oshima, and R. Fujii. 1986. A comparative study of melanin-concentrating hormone (MCH) action on teleost melanophores. *Biol. Bull. 171:* 360–370.

Naitoh, T., A. Morikawa, and Y. Omura. 1985. Adaptation of a common freshwater goby, yoshinobori, *Rhinogobius branneus* Temminck and Schlegel to various backgrounds including those containing different sizes of black and white checkerboard pattern. *Zool. Sci. 2*(l): 59–63.

Neudecker, S. 1989. Eye camouflage and false eyespots: Chaetodontid responses to predators. *Env. Biol. Fishes 25:* 143–157.

Nicoletto, P. F. 1991. The relationship between male ornamentation and swimming performance in the guppy, *Poecilia reticulata. Behav. Ecol. Sociobiol. 28*(5): 365–370.

O'Day, W. T., and H. R. Fernandez. 1974. *Aristostomias scintillans* (Malacosteidae): A deep-sea fish with visual pigment apparently adapted to its own bioluminescence. *Vision Res. 14:* 545–550.

Odiorne, J. M. 1957. Color changes, pp. 387–401. In *The physiology of fishes*, Vol. 2, M. F. Brown (Ed.). Academic Press, New York.

Olson, V. A., and I. P. F. Owens. 1998. Costly sexual signals: Are carotenoids rare, risky, or required? *TREE 13:* 510–514.

Pankhurst, N., and J. M. Lythgoe. 1982. Structure and colour of the integument of the European eel *Anguilla anguilla* (L.). *J. Fish. Biol. 21*(3): 279–296.

Pietsch, T., and D. Grobecker. 1978. The compleat angler: Aggressive mimicry in an antennariid anglerfish. *Science 201:* 369–370.

Roberts, C. D. 1993. Comparative morphology of spined scales and their phylogenetic significance in the teleostei. *Bull. Mar. Sci. 52:* 60–113.

Roberts, R. J., and A. M. Bullock. 1976. The dermatology of marine teleost fish. II. Dermatopathology of the integument. *Oceanogr. Mar. Biol. Ann. Rev. 14:* 227–246.

Roche, E. T., and B. W. Halstead. 1972. The venom apparatus of California rockfishes (family Scorpaenidae). *Calif. Dept. Fish Game Bull. 156.*

Rosen, M. W., and N. E. Cornford. 1971. Fluid friction of fish slimes. *Nature 234:* 49–51.

Russell, F. E. 1969. Poisons and venoms, pp. 401–449. In *Fish physiology* W. S. Hoar and D. J. Randall (Eds.). Academic Press, New York.

Sanders, R., B. Rushall, H. Toussaint, J. Stager, and H. Takagi. 2001. *Bodysuit yourself: But first think about it.* http://www.rohan.sdsu.edu/dept/coachsci/swimming/bodysuit/fiveauth.htm

Sazima, I. 1977. Possible case of aggressive mimicry in a neotropical scale-eating fish. *Nature 170:* 510–512.

Seikei, T. 1989. Albinism of hatchery reared flounder (*Poralichthys olivaceus*) as a result of deformation of asymmetrical development of skin structure, p. 489 (abstr.). In The early life history of fish, 3rd ICES Symposium, I. H. S. Blaxter, J. C. Gamble, and H.V. Westernhagen (Eds.). *Cons. Int. Explor. Mer 191.*

Simon, H. 1971. *The splendor of iridescence: Structural colors in the animal world.* Dodd and Mead, New York.

Sire, J.-Y. 1994. Light and TEM study of nonregenerated and experimentally regenerated scales of *Lepisosteus oculatus* (Holostei) with particular attention to ganoine formation. *Anat. Rec. 240:* 189–207.

Springer, V., and W. Smith-Vaniz. 1972. Mimetic relationships involving fishes of the family Blenniidae. *Smithson. Contr. Zool. 112.*

Thompson. E. M., and F. I. Tsuji. 1989. Two populations of the marine fish *Porichthys notatus*, one lacking in luciferin essential for bioluminescence. *Mar. Biol. 102:* 161–165.

———, B. G. Nafpaktitis, and F. I. Tsuji. 1988. Dietary uptake and blood transport of *Vargula* (crustacean) luciferin in the bioluminescent fish, *Porichthys notatus. Comp. Biochem. Physiol. 89A*(2): 203–209.

Videler, J. J. 1981. Swimming movements, body structure, and propulsion in cod, *Gadus morhua*, pp. 1–27. In Vertebrate locomotion, M.H. Day (Ed.). *Symp. Zool. Soc. Lond. 48.* Academic Press, New York.

———. 1993. *Fish swimming.* Chapman and Hall, London.

———, and D. Weihs. 1982. Energetic advantages of burst-and-coast swimming of fish at high speeds. *J. Exp. Biol. 97:* 169–178.

———, G. J. Geertjes, and J. J. Videler. 1999. Biochemical characteristics and antibiotic properties of the mucous envelope of the queen parrotfish. *J. Fish Biol. 54:* 1124–1127.

Walls, G. L. 1963. *The vertebrate eye and its adaptive radiation.* Hafner, New York.

Walters, V. 1963. The trachypterid integument and an hypothesis on its hydrodynamic function. *Copeia 1963:* 260–270.

Ward, P. I. 1988. Sexual dichromatism and parasitism in British and Irish freshwater fish. *Anim. Behav. 36*(4): 1210–1215.

Watiporn, P. 1988. Effects of carotenoid pigments from different sources on color changes of fancy carp, *Cyprinus carpio.* Abstr., Master of Science thesis (Fish. Sci.). *Notes Fac. Fish. Kasetsart Univ. 10:* 1.

Watts, E. H. l960. The relationship of fish locomotion to the design of ships, pp. 27–40. In Vertebrate locomotion. *Symp. Zool. Soc. Lond. 5.*

Webb, P. W. 1978. Hydrodynamics: Nonscombroid fish, pp. 189–237. In *Fish physiology*, Vol. VII, W. S. Hoar and D. J. Randall (Eds.). Academic Press, New York.

Weihs, D. 1973. Hydromechanics of fish schooling. *Nature 245:* 48–50.

———. 1974. Energetic advantages of burst swimming in fish. *J. Theoret. Biol. 48:* 215–229.

———. 1989. Design features and mechanics of axial locomotion in fish. *Amer. Zool. 29:* 151–160.

———, and Webb, P. W. 1983. Optimization of locomotion, pp. 339–371. In *Fish biomechanics*, P. W. Webb and D. Weihs (Eds.). Praeger, New York.

Wheeler, A. 1975. *Fishes of the world: An illustrated dictionary.* Macmillan, New York.

Widder, E. A. 1998. A predatory use of counterillumination by the squaloid shark, *Isistius brasiliensis. Env. Biol. Fishes 53:* 267–273.

Wolfgang, M. J., J. M. Anderson, M. A. Grosenbaugh, D, K. P. Yue, and M. S. Triantafyllou. 1999. Near-body flow dynamics in swimming fish. *J. Exp. Biol. 202:* 2303–2327.

Wunder, W. 1936. *Physiologie der Süsswasserfische Mitteleuropas.* E. Schweizerbart, Stuttgart.

Yamada, J., and N. Watabe. 1979. Studies on fish scale formation and resorption I. Fine structure and calcification of the scales in *Fundulus heteroclitus* (Atheriniformes: Cyprinodontidae). *J. Morphol. 159:* 49–66.

Yates, G. T. 1983. Hydromechanics of body and caudal fin propulsion, pp. 177–213. In *Fish biomechanics*, P. W. Webb and D. Weihs (Eds.). Praeger, New York.

Young, R. E., and C. E. Ropper. 1977. Intensity regulation of bioluminescence during countershading in living midwater animals. *Fish. Bull. 75:* 239–252.

Zempel, C. 2001. The emperor's new swimsuit. *Swimmer Jan./Feb. 2001:* 54–57.

Zuyev, G. V., and V. V. Belyayev. 1970. An experimental study of the swimming of fish in groups as exemplified by the horsemackerel (*Trachurus mediterraneus ponticus* Aleev.). *J. Ichthyol. 10:* 545–548.

19

MUSCLES, LOCOMOTION, AND BUOYANCY

MUSCULOSKELETAL SYSTEM

Musculature of the Trunk and Appendages
Neural Control of the Musculoskeletal System

MODES OF LOCOMOTION

Swimming
Anguilliform Swimming
Subcarangiform Swimming
Carangiform Swimming
Thunniform Swimming
Ostraciiform Swimming
Fin-Swimming Modes
Locomotion Using Dorsal and Anal Fin Undulation
Locomotion Using Oscillation of Dorsal and Anal Fins
Locomotion Using the Pectoral Appendages
Swimming Speed
Nonswimming Locomotion

THE SWIM BLADDER AND THE REGULATION OF BUOYANCY

Buoyancy: Gas Bladder and Retes, Fats, and Oils
Vertical Movement
Gas Resorption
Gas Secretion

We tend to think of fishes as neutrally buoyant and hence excused from the laws of gravity, unlike tetrapods, in which locomotion involves counteracting gravitational forces. Although fishes may not be required to expend significant amounts of energy counteracting the forces of gravity, they must exert a significantly greater force against the water as they move through it. Consider the energy you expend in swimming the length of a pool compared to what you would expend by simply walking alongside it. This is what makes swimming such a beneficial form of exercise: Any movement in relation to the water demands the expenditure of large amounts of energy, because water is about 800 times as dense as air, and thus significant energy must be expended to push through it, especially if appreciable speed is required, as for escape or pursuit.

The maintenance of a vertical position in a fluid environment also requires the expenditure of energy, either in the constant adjustments made to fin and body surfaces as the fish swims or through buoyancy adjustments made with the swim bladder. Some of the most remarkable adaptations among the fishes are those seen in large pelagic species that are capable of making rapid adjustments in vertical position while moving through the water with extraordinary velocity. Fishes must work to change their vertical position or to maintain their position in currents. Not all fishes are neutrally buoyant, however; many spend virtually their entire lives on the bottoms of lakes, streams, or the ocean. Among these are many species that remain quiescent until their prey is in the vicinity, then strike with a swiftness that can only be observed with the aid of high-speed cinematography.

In this chapter, we will explore the morphological and physiological features of fishes that permit them to master the horizontal and vertical dimensions of their aquatic realm. We will consider the musculo-skeletal features that facilitate movement and the diversity of locomotor strategies that result. We will also investigate the role that the swim bladder plays in the maintenance of neutral buoyancy.

MUSCULOSKELETAL SYSTEM

The best place to begin a discussion of the locomotor capacities of fishes is with the integrated system that makes movement possible—the musculoskeletal system. Muscles operate in partnership with a skeletal support system, which may include the vertebral column—or its evolutionary and developmental predecessor, the notochord—and the tough skin covering to which the trunk muscles are firmly anchored. It is especially interesting to consider the role of the notochord, as it represents the original internal support structure of early swimming vertebrates. Long et al. (2002) have made in vitro assessments of the contribution of the hagfish notochord to the maintenance of body stability during swimming.

Musculature of the Trunk and Appendages

The skeletal musculature of fishes consists mainly of the large muscles of the trunk and tail. Other, smaller muscles are associated with the bones of the jaws, the branchial arches, and the fins. The trunk musculature consists of a series of blocks of striated muscle fibers, called **myomeres** or **myotomes** (the term *myotome* is more commonly used to denote that portion of the embryonic somites that will differentiate into axial skeletal muscle), separated by sheets of connective tissue, called **myosepta** or **myocommata**. Gnathostome fishes display a remarkably consistent design in which each myoseptum consists of six specifically arranged tendons; these tendons are apparently not present among the agnathans (Gemballa et al., 2003). Myomeres and myosepta are examples of the **metamerism**, or serial repetition of body parts, that is a defining attribute of most higher animals.

Most fishes have both red and white muscle cells (**myofibers**). The red fibers are oriented more or less parallel with the body axis, whereas the white fibers may deviate as much as 45° from the body axis (Videler, 1993). The myomeres are folded so that, just under the skin, their outer edges resemble the letter W tipped on its side (Fig. 19.1). In lampreys, the angles of flexing of the myomeres are slight, especially anteriorly—but in sharks and bony fishes, the bends are sharper and more evident. The modification and folding of the myomeres are so elaborate that a short and simple description is difficult. Suffice it to say that they

greatly enhance the contractile force that can be applied to the genesis of forward momentum in the fish. A careful dissection of the trunk musculature of a fish will reveal the extent of myomere complexity—performing this dissection at your favorite seafood restaurant is a sure way to impress your dining companions with your ichthyological acumen.

In lampreys, the myomeres extend forward from their edges on the sides of the body to their origins on the axial skeleton. In sharks and bony fishes, however, the myomeres extend posteriorly toward the axial skeleton from the regions of the backward-pointing upper and lower flexures visible on the surface, each fitting inside another, so that a cross section of the trunk or tail cuts through several myomeres on each side, showing the myosepta as concentric lines (Fig. 19.2A, B). There are two myomeres per vertebral centrum in some fishes, but because of the folded pattern, a given myomere, depending on the species, might have an overall anterior–posterior span of 3 to 12 intervertebral joints (Wainwright, 1983). Each myomere is typically divided into four or more portions by myosepta. A vertical septum (mid-dorsal in the abdominal region) separates the muscles into bilateral left and right halves. On each side of the body, a main horizontal septum at the level of the vertebral column divides the myotomes into **epaxial** (upper) and **hypaxial** (lower) muscles. Less evident horizontal septa further subdivide the lateral muscles, so that posterior to the body cavity, there are usually four recognizable muscle bundles on each side (Wainwright, 1983). The myomeres connect externally to the skin and internally to the median or vertical septum and to the myosepta and horizontal septa (Videler, 1977). Lampreys and hagfishes do not have horizontal septa.

Lamprey myomeres are characterized by flattened white (fast) muscle fibers surrounded by slow fibers and intermediate fibers. Along the sides of most fishes, just under the skin, lie the lateral superficial muscles, which are usually dark in color (Fig. 19.3A, D). This **red muscle** is used in normal, sustained swimming activity and is fatigue resistant at slow or cruising speed. In highly mobile fishes, such as the tunas, red muscle is more extensive than in sedentary fishes (Fig. 19.2C). Red muscle is extremely well vascularized for efficient oxygen supply and, in some cases, for thermoregulation (see Chapter 24). The remainder of the lateral musculature, predominantly **white muscle**, is used for sudden bursts of rapid swimming, such as during escape or the capture of prey (Fig. 19.3A, B, D, E). White muscle fibers operate primarily in an anaerobic mode when engaging in burst activity. Pelagic cephalopods (squids; *Loligo*), which demonstrate many remarkable morphological and ecological convergences with fishes (Packard,

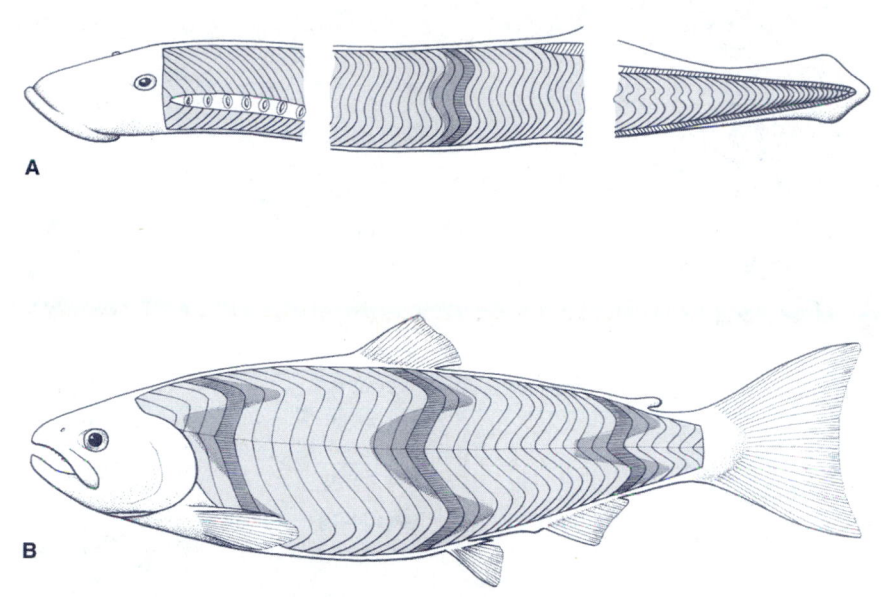

FIGURE 19.1
Lateral body musculature. **A,** Lamprey, showing myomere patterns in anterior, middle, and posterior sections; **B,** Illustration of myomere patterns in salmon (myomere number reduced). Full extent of selected myomeres shown in dark blue. (*B* based on Greene and Greene, 1914.)

1972; Rodhouse and White, 1995), show a similar differentiation of slow aerobic and fast anaerobic muscles in the mantle—a structure important in cephalopod locomotion (Mommsen et al., 1981).

Anteriorly, the body musculature connects to the pectoral girdle and head; posteriorly, the connection is to the caudal fin or to caudal fin tendons (Lindsey, 1978). The main horizontal septum is composed of two thin layers of tendons. The outer ends of the tendons connect to the superficial red muscle. The inner ends of one sheet attach to the posterior half of the vertebral centra, and the tendons of the other sheet course forward and attach to the front of the centra. Tendons form at the ends of the anterior and posterior cones of the myomeres, and are well developed in strong swimmers, but are weak or short in most bony fishes. Elasmobranchs have well-developed tendons, and the scombroid fishes have especially strong tendons, particularly in the caudal region (Fierstine and Walters, 1968).

The track of fibers in thin, superficial red muscles is usually parallel to the body axis, but the orientation of fibers in deeper muscle is usually curved, so that they form angles of as much as 40° with the body axis. Alexander (1969) has demonstrated that series of fibers form spiral tracks along the body. This was done by examining the origins and inser-

tions of individual fibers on successive myocommata. A set of helices is formed in each of the arms of the recumbent W that generally describes the shape of the myomeres. Apparently, the intricate relationships between the shape of the myomeres, with their cone-in-cone arrangement, and the orientation of the fibers allows the fibers to contract at the same rate regardless of their distance from the vertebral column. The muscles shorten about 5 percent on contraction regardless of their position in the myomere. The force of contraction is thereby transferred to the skeleton efficiently, so that maximum power output is gained.

About half of a fish's weight is locomotor muscle (Bone, 1978). Red fibers usually constitute a minor part of the lateral musculature—0.5 to 10 percent—although in tunas and other active pelagic fishes, the red fibers may make up nearly 30 percent of the muscle (Greer-Walker and Pull, 1975; Johnston, 1981). In the tunas and their close relatives, and in some of the swift sharks, the red musculature extends from the lateral position to the vertebral column (Fig. 19.2C). In higher vertebrates, different fiber types intermingle in a given single muscle, but in fishes there is usually a much clearer distinction between red and white muscle fibers (Bone et al., 1999). Some fishes, however, may show a mixture of red and white fibers. Some

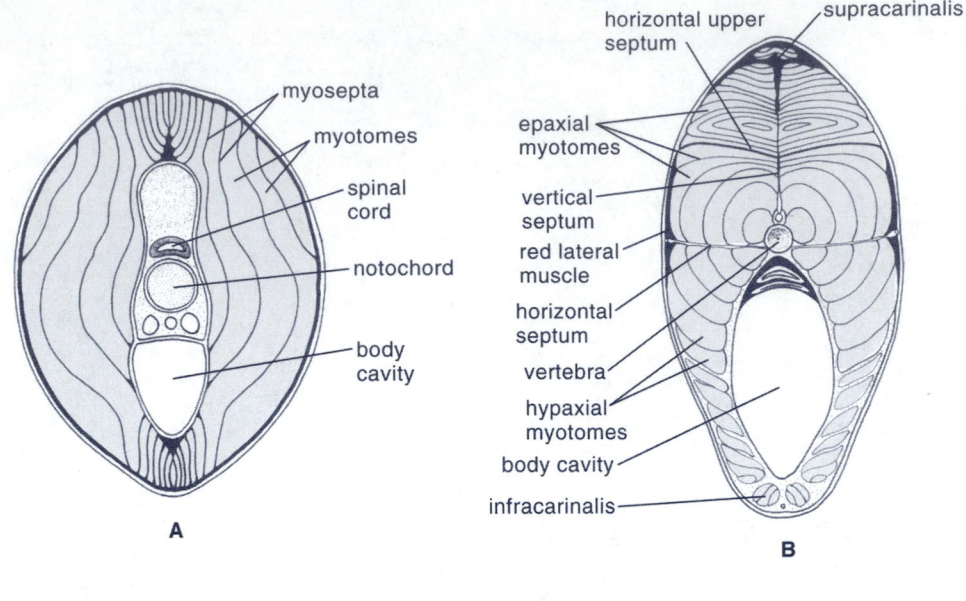

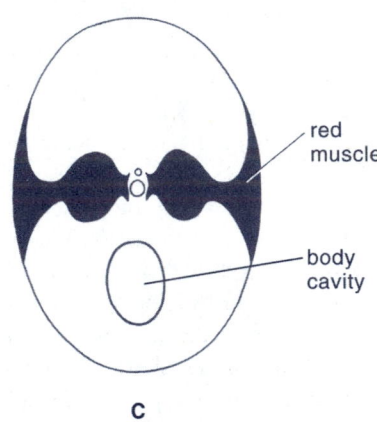

FIGURE 19.2
Illustration of body musculature in cross section. **A,** Lamprey (*Lampetra tridentata*); **B,** Chinook salmon
(*Oncorhynchus tshawytscha*); **C,** Illustration showing approximate extent of red muscle (black) in skipjack
tuna (*Katsuwonus pelamis*).

species, including various salmonids and cyprinids, have intermediate muscle fibers, called "pink fibers," somewhat concentrated just inside the red musculature (Fig. 19.3D), but also scattered among the white fibers (Coughlin and Rome, 1996; Gill et al., 1982). Johnston (1983) reported that five muscle fiber types have been recognized in the spotted dogfish (*Scyliorhinus canicula*) and four in the perch (*Perca*).

The red (**slow** or **tonic**) fibers and the white (**fast** or **twitch**) fibers differ in several ways related to their respective functions (Bone, 1978, 1989; Johnston, 1981). Red muscle fibers, which are capable of prolonged activity in sustained cruising, are of small diameter and are well supplied with lipids and some glycogen. They are also well endowed with peripheral capillaries. Red muscle operates primarily aerobically and contains large stores of myoglobin, which impart the reddish color and serve as oxygen storage sites. This ensures aerobic metabolic activity even when oxygen consumption exceeds the ability of the

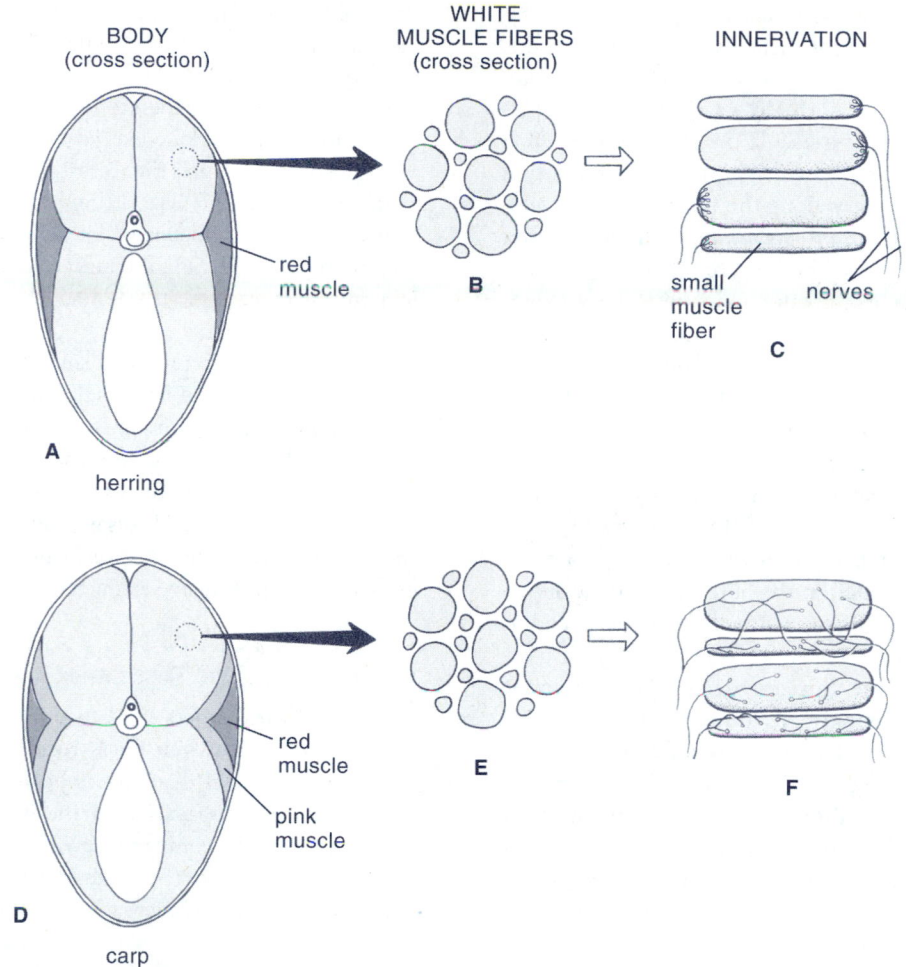

BODY
(cross section)

**WHITE
MUSCLE FIBERS**
(cross section)

INNERVATION

red
muscle

B

small
muscle
fiber

nerves

C

A

herring

red
muscle

E

F

pink
muscle

D

carp

FIGURE 19.3

A, Cross section of herring (Clupeidae) with red muscle shaded; **B,** Disposition of large and small white muscle fibers; **C,** Terminal innervation of muscle fibers. **D,** Cross section of carp (Cyprinidae), showing position of red muscle (heavy shading) and "pink" muscle (light shading); **E,** Disposition of muscle fibers; **F,** Multiple innervation of muscle fibers. (Redrawn with modification from Bone et al., 1978.)

circulatory system to supply oxygen via hemoglobin. Abundant mitochondria (about 16 to 35 percent by volume) provide oxidative biochemical pathways for the ATP production that consumes the reserves of lipids in the cells. The white fibers, on the other hand, have little lipid, low mitochondrial volume (up to 8 percent), and few peripheral capillaries. White fibers are mostly of larger diameter, have little or no myoglobin, and use anaerobic glycolysis to produce ATP, although there is a range of aerobic capacity of white muscle among fishes.

As might be expected, the amount of red muscle in a given species depends on its mode of swimming. Greer-

Walker and Pull (1975), who examined cross sections of several fishes cut about two-thirds back toward the tail, found that 5 to 15 percent of the cross sectioned area in most species consisted of red muscle. A few families, including Sparidae, Clupeidae, Carangidae, and Scombridae, exceeded 15 percent. Some, including those that mainly use their fins for swimming, had 0 to 5 percent red muscle in the cross section. The ratfishes (Chimaeridae) and wrasses (Labridae) are good examples of fishes with lower quantities of red muscle. Gill et al. (1989) showed that the percentage of cross-sectional area made up of red muscle in five freshwater fishes sectioned behind the anal fin varied with habitat and life history. The yellow perch (*Perca*), which cruises in schools, had a red muscle composition of about 15 percent, whereas two species of esocids, which are lurking predators, had only 4 to 5.5 percent of red fibers. A surfperch (*Cymatogaster aggregata*), which normally swims by means of its pectoral fins, had pectoral musculature composed of 90 to 95 percent red fibers (Webb, 1973, 1975). The power output of red muscle may also vary along the body axis. For example, in skipjack tuna (*Katsuwonus pelamis*), the total power output of red muscle located in the anterior of the body was approximately 40 percent greater than posteriorly (Syme and Shadwick, 2002).

The winglike pectoral fins of skates and most other rays have muscle bundles arranged in two layers—deep and superficial—both above and below the elongate fin rays. The deep bundles, next to the fin rays, originate on the rays or on the pectoral girdle and attach via tendons to more distal parts of the fin rays. The superficial fibers originate on the pectoral girdle or on connective tissue of the deep fiber tendons. Red muscle is located in the outer layer of the superficial bundles, with red fibers of small diameter extending into the deep bundles.

The muscles of the fins are derived from selected embryonic myotomes but usually do not correspond with body segments in adults. **Carinal muscles** on the dorsal and ventral midlines serve mainly as protractors and retractors for the dorsal and anal fins (Fig. 19.4). The caudal fin musculature is dorsoventrally asymmetrical, matching the skeletal asymmetry of that structure (Lauder, 1982, 1989). Important elements of the caudal musculature are the lateral superficial muscles, which can move the fin laterally; various flexors that attach to the fin rays; and the hypochordal longitudinal muscle that connects the hypural skeleton to the fin rays of the dorsal portion of the caudal fin. This complex system can curve the fin rays and change the span of the fin.

As might be expected, caudal fin design is especially critical in initiating and sustaining lift and forward momentum. The conventional perception has been that heterocercal caudal fins, because of the asymmetry of the dorsal lobe relative to the ventral one, generate lift and torque forces that have to be counteracted by the body and pectoral fins, whereas the homocercal fin, possessing symmetry of the dorsal and ventral lobes, correspondingly generates symmetrical forces. Although the heterocercal caudal fin of sharks appears to conform to this model (Lauder, 2000; Wilga and Lauder, 2002), that of sturgeons does not (Lauder, 2000). Studies on the use of pectoral fins by sharks confirm their role in stabilizing the body, yet researchers disagree as to whether they contribute to lift or not (Fish and Shan-

nahan, 2000; Wilga and Lauder, 2002). The pectoral fins of swimming sturgeons generate no lift unless they are actively moved dorsally or ventrally (Wilga and Lauder, 1999). The homocercal caudal fins of fishes as divergent as bluegills and tunas also appear to generate lift, as the lateral excursions of the dorsal caudal lobe exceed those of the ventral lobe, and the speed at which the dorsal lobe travels exceeds that of the ventral lobe. These actions are possible through the evolution of the hypochordal longitudinal muscle, which is unique in being the only intrinsic caudal muscle whose line of action is at an angle to the body axis (Gibb et al., 1999; Lauder, 1989, 2000).

Each median fin ray has a set of erector and depressor muscles. Soft fins also have inclinators capable of bending the rays. The musculature of the paired fins (Fig. 19.4B, C) includes abductors, adductors, and arrectors, with some fibers attaching on individual fin rays or their basals to give the fin great flexibility. This is especially true in some fishes, such as surfperches and wrasses, where the pectoral fins are used as major propulsive elements.

Neural Control of the Musculoskeletal System

Tonic and twitch fibers tend to differ in their patterns of innervation, with twitch fibers receiving axons of a larger diameter. As axonal diameter is positively correlated with the velocity of propagation of the neural impulse, this is consistent with the comparatively rapid velocity of contraction in twitch fibers (Schmidt-Nielsen, 1997). Typically, white fibers receive innervation at their ends (**focal innervation**; Fig. 19.3C), whereas red muscles demonstrate **multiple innervation**, with multiple motor endplates along the length of the fiber. **Terminal innervation** is at one or both ends, and **dual innervation** is by two axons at adjacent sites at one or both ends of a muscle fiber (Bone, 1989). Dual innervation, which may facilitate the coordination of rapid fiber contractions by the synchronous firing of nerve impulses, appears to be the primitive mode of innervation of vertebrate twitch fibers (Bone, 1999). Tonic muscle fibers that have multiple innervation and do not conduct action potentials in their outer membranes have been described in various teleosts and may be present in the shark genus *Scyliorhinus* (Bone, 1989).

The innervation of muscle fibers varies among the different phyletic lineages of fishes (Agarkov et al., 1976; Bone, 1988, 1989; Johnston, 1981). Lampreys have only a portion of their fast muscle fibers innervated, as these fibers are coupled electrically to each other and to the intermediate fibers. The fast fibers have focal innervation and the slow fibers have multiple innervation in both lampreys and hag-

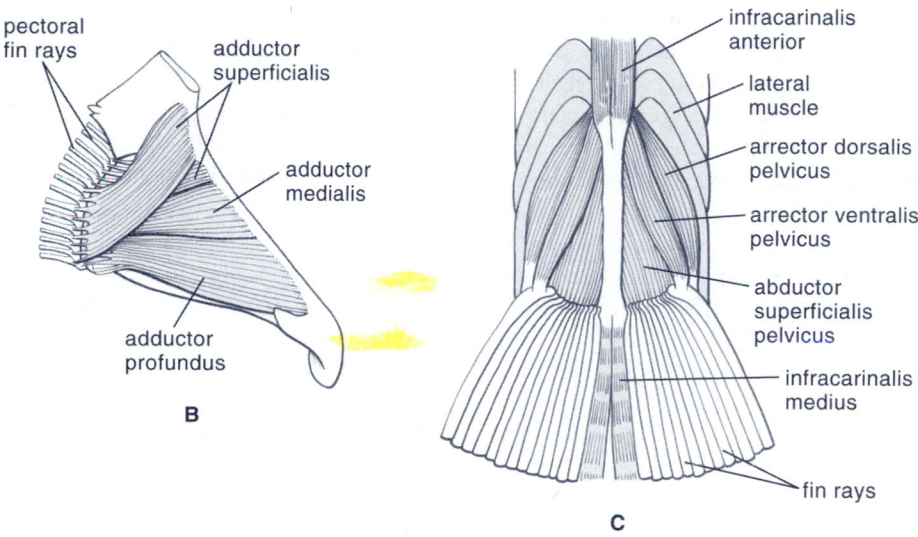

FIGURE 19.4
A, Illustration of inclinator muscles of dorsal fin; **B,** Medial aspect of pectoral fin muscles; **C,** Ventral view of pelvic fin muscles. (*A* based on Greene and Greene, 1914; *B* redrawn from Winterbottom, 1974; *C* based on Greene and Greene, 1914.)

fishes, but in the former, two axons supply each of the innervated fast fibers (Bone, 1989). In elasmobranchs, white muscle fibers are usually innervated terminally or focally. Non-teleostean bony fishes have terminal innervation of white muscle. Included among the lower teleostean groups with terminal innervation are the Clupeiformes, Anguilliformes, Gonorynchiformes, Hiodontidae, and Alepocephalidae. Some catfish families also show terminal innervation. Acanthopterygians have multiple innervation (Hudson, 1969; Johnston, 1981). Nerve branches from two or more

spinal nerves run in the myosepta and extend fibers out onto the surface of the myomeres, where sometimes complex nerve endings innervate the muscle fibers. The multiple innervation of white fibers in higher teleosts (Fig. 19.3F) may permit their use in slow, sustained cruising as well as burst swimming (Bone et al., 1999). Alternate neuromotor pathways—the *C-start* and the *S-start* (described later)— are apparently used in the two primary locomotor modes used in escape response or prey attack, as observed in the muskellunge (*Esox masquinongy;* Hale, 2002).

Measurement of electrical activity in muscle tissue (**electromyogram** or **EMG**) has revealed that white muscle fibers propagate action potentials, but that red fibers show local, nonpropagated activity (Bone et al., 1978). In slow, sustained swimming, only red muscle fibers are used by most fishes, although in the species having pink or intermediate fibers, some of these fibers may be recruited in sustained locomotion. At moderate speeds, the recruitment of pink fibers increases. If burst speed is necessary, the white or fast fibers are used (Bone et al., 1978; Johnston, 1983). Because of their oxidative metabolism and ample capillary network, the red muscle fibers are efficient in slow contraction and can operate for long periods without fatiguing. Pink muscle has high levels of both aerobic oxidative and anaerobic glycolytic enzymes, is fairly fast in contraction speed, and resists tiring more than white muscle, maintaining efficiency at a high rate of work.

Burst speed using the white muscle can last only a few minutes in most fishes before fatigue sets in. Trout are reported to use 50 percent of their stored muscle glycogen in about 15 seconds. The conversion of glycogen to lactate is rapid, but in the white muscle, the lactate level decreases very slowly, taking up to 18 hours. Lactate can be oxidized in the gills and apparently can supply part of the energy used in ion exchange. Recovery in the comparatively well-vascularized red muscle is relatively rapid, taking an hour or less (Batty and Wardle, 1979; Bone, 1975; Johnston, 1981).

MODES OF LOCOMOTION

Swimming

Modes of swimming, whether mainly by body movement or by fin action, have been studied at least from the time of Aristotle. Sophisticated techniques such as EMG recording and high-speed cinematography enable scientists to obtain a clearer understanding of the swimming of fishes, although not all pieces of the puzzle are yet in place. Digital particle image velocimetry (DPIV; described in Chapter 18) is a valuable technique for the study of fish swimming when combined with the analysis of muscle kinematics. Typically, fish swim by generating serial waves of contractions of the myomeres from head to tail, resulting in undulations of the body. The series of muscle contractions alternates from one side to the other, forming the curvature that pushes against the water to generate forward thrust (Weihs, 1989). For most fishes, a relatively constant cross section of red muscle along the trunk results in a uniform distribution of power generation along the length of the fish (Shadwick et al., 1998). Fins are involved in swimming as stabilizers, rud-

ders, and brakes and, in many species, as a means of propulsive locomotion. Fin-powered locomotion is supplemental in some species, but forms the major mode of swimming in others. Deep-bodied fishes displaying the more derived acanthopterygian pattern of fin placement (pectorals high up on the body and pelvics shifted anteriorly) are generally held to have greater maneuverability than more fusiform fishes with more basal fin patterns. This is not always the case, however: The goldfish (*Carassius auratus*), having a more fusiform body and less derived fin placement, was demonstrated to swim faster and turn through sharper angles than deeper-bodied silver dollar fish (*Metynnis hypsauchen*) and angelfish (*Pterophyllum scalare;* Schrank et al., 1999).

When considering locomotion in any animal, it is convenient to refer to its **gait**, which has been defined by Webb (1998) as "a unique combination of muscle-propulsor behavior which is recruited more or less discretely while moving over a portion of the total performance range of an animal." Although there are many fishes that employ more than one type of gait—along a continuum between those that use full body undulations and those that use only movement of fin surfaces—a series of categories describing swimming modes has been useful. With propulsor systems incorporated as components of gaits, two simple categories can be defined: (1) body and caudal fin swimming (**BCF gaits**), providing greater power, useful for high-speed swimming and fast starts; and (2) median and paired fin swimming (**MPF gaits**), used at slower speeds (Blake, 2004; Webb, 1998). The degree of specialization for a given category of locomotion tends to result in performance limitations in another. Studies on swimming performance of certain species (Korsmeyer et al., 2002) may focus on the transition from one gait to another. EMG measurements reveal characteristic muscle contraction patterns for these swimming modes, but confirm a continuum of contractile modes, with anguilliform and thunniform modes at opposite ends of the kinematic spectrum (Knower et al., 1999). Categories such as *anguilliform* and *thunniform* are named after fish taxa that exemplify that particular mode of swimming, but they have no phylogenetic significance. The terms used here follow Lindsey (1978) and Breder (1926). Dominant swimming modes are discussed hereafter, and a detailed summary of the range of gaits is shown in Table 19.1.

Anguilliform Swimming

The mode of swimming seen in lampreys, hagfishes, some sharks, eels, larvae of many species, some elongate flatfishes, some zoarcoids, and other thin-bodied fishes is called **anguilliform,** after the eels, the best known practitioners of the mode (Fig. 19.5A). An interesting parallel to the evolution of swimming modes in fishes is seen in Mesozoic

TABLE 19.1   FISH PROPULSION SYSTEMS

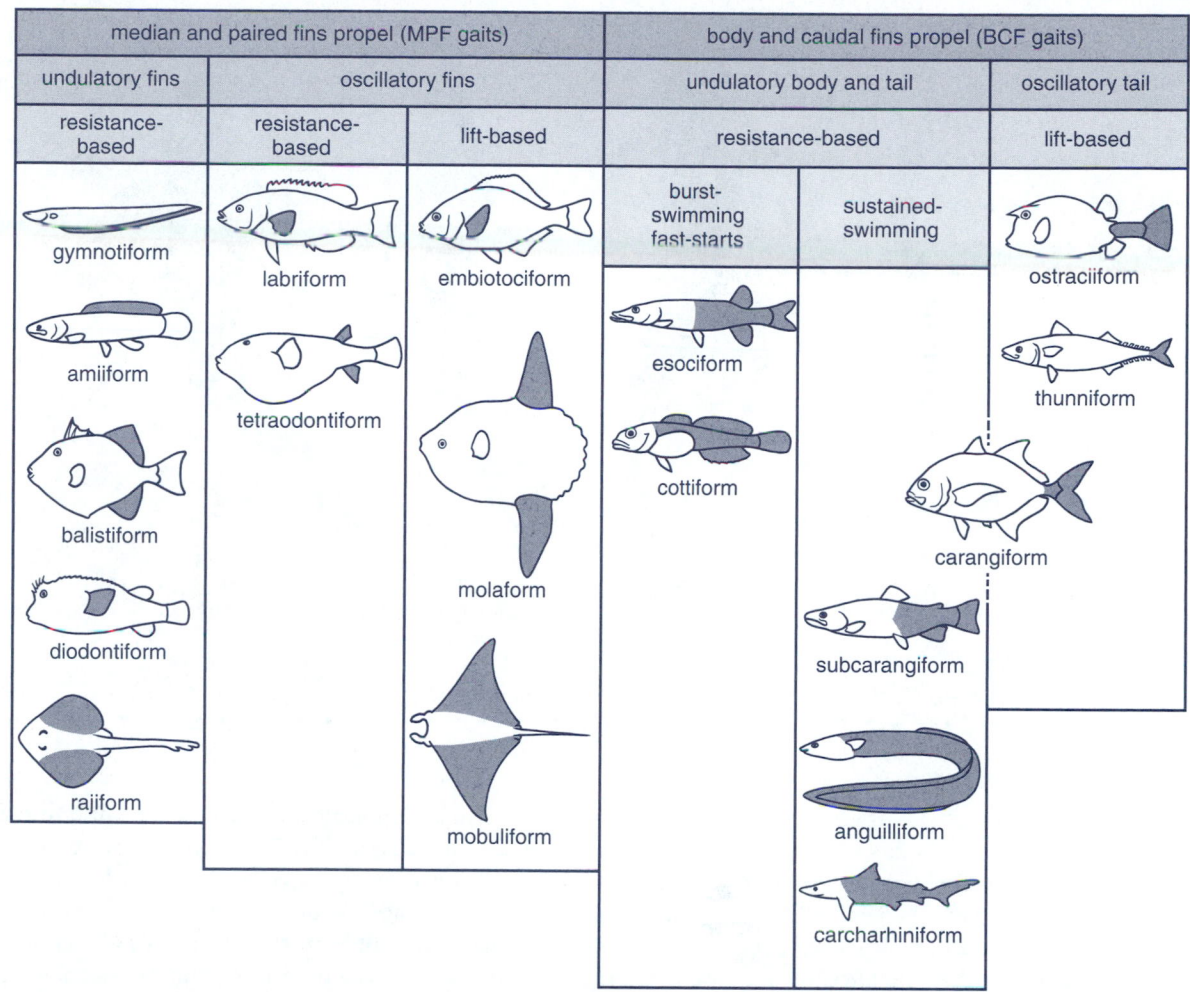

median and paired fins propel (MPF gaits)			body and caudal fins propel (BCF gaits)		
undulatory fins	oscillatory fins		undulatory body and tail		oscillatory tail
resistance-based	resistance-based	lift-based	resistance-based		lift-based
gymnotiform	labriform	embiotociform	burst-swimming fast-starts / esociform, cottiform	sustained-swimming / carangiform, subcarangiform, anguilliform, carcharhiniform	ostraciiform, thunniform
amiiform	tetraodontiform	molaform			
balistiform					
diodontiform					
rajiform		mobuliform			

Gaits are distinguished as either MPF (median and paired fins) or BCF (body and caudal fin), and fin/body motion is distinguished as either undulatory or oscillatory. When oscillating propulsors are used like oars, they are termed *resistance-based,* and when used as wings, they are termed *lift-based*. Shaded areas indicate parts of body or fins involved in gaits (after Webb, 1998).

ichthyosaurs. By the Cretaceous, a thunniform mode had evolved among the best known of these aquatic reptiles, but the earliest forms that first appeared in the early Triassic apparently swam in an anguilliform fashion (Motani et al., 1996). In anguilliform swimming, the entire body undulates, and more than one wave is present at once. The specific wavelength, which is the undulation's wavelength divided by the body length, is less than one (Blake, 1983). Fishes that engage in anguilliform locomotion demonstrate a range of variation in kinematic parameters, such as the amplitude and velocity of contractile waves propagated along the body (Gillis, 1996). Many of the eels, pricklebacks, loaches, and other fishes that swim in the anguilliform fashion are benthic in habit but swim off the bottom for short periods.

As in other fishes, red muscle is exclusively used during slow swimming, and white muscle fibers are recruited with increasing speeds. At slow speeds, the anterior muscles are virtually inactive, but as speed increases, the enhanced recruitment of anterior muscles has been demonstrated (Gillis, 1998). Many practitioners of anguilliform locomotion are able to reverse the direction of muscle contractions in order to swim backwards. In eels of the genus *Anguilla*, certain features of muscle kinematics, such as the amplitude profile of the waveform as it moves along the body,

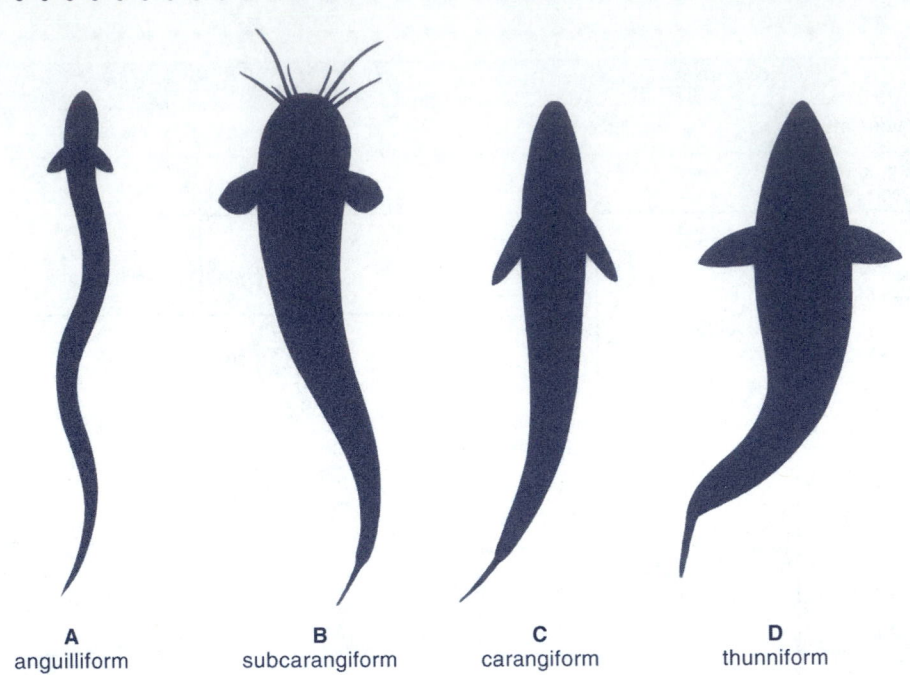

A	B	C	D
anguilliform	subcarangiform	carangiform	thunniform

FIGURE 19.5
Four modes of swimming: **A**, Anguilliform; **B**, Subcarangiform; **C**, Carangiform; **D**, Thunniform.

have been demonstrated to differ between forward and backward locomotion (D'Aout and Aerts, 1999). Using anguilliform locomotion, eels of the genus *Anguilla* migrate thousands of kilometers when moving to their spawning grounds. The power output of silver-phase (migrating) eels has been demonstrated to significantly increase over that of yellow-phase (nonmigrating) individuals (Ellerby et al., 2001a).

Some pelagic fishes have the slender, elongate shape typical of eellike swimmers. Sauries (Scomberesocidae) and sand lances (Ammodytidae) are examples of surface and shallow-water species that swim in the anguilliform fashion. Needlefishes (Belonidae) are unique among anguilliform swimmers in possessing prominent fins that are useful in making rapid darting motions to capture prey and escape predators (Liao, 2002). Mesopelagic eelpouts are also reported to use the anguilliform mode, and there is little reason to doubt that other mesopelagic and bathypelagic fishes of slender shape do the same. There are several pelagic fishes of elongate shape among the Stomiiformes, especially in the genera *Chauliodus, Stomias,* and *Idiacanthus,* and in the subfamily Melanostomiinae. The deep-pelagic anguilliform snipe eels, gulpers, and swallowers are also of greatly elongate shape. Other examples of pelagic fishes that

seem morphologically suited to anguilliform swimming can be found among the Myctophiformes. Anguilliform swimmers with compressed bodies or with dorsal and anal fins that significantly increase the height or span of the fish are more efficient than those with cylindrical bodies or those that are tapered toward the tail. Fishes using this mode are generally slower than those using the subcarangiform mode, which is next in the sequence of swimming modes, and into which the anguilliform mode grades.

Subcarangiform Swimming

Fishes with a thick forebody have reduced anterior flexibility, and their undulations are mainly confined to the posterior part of the body (Fig. 19.5B), so that usually less than one wavelength is present. The designation of this mode of swimming as **subcarangiform** indicates that it is advanced beyond the anguilliform in terms of achievable velocity but is not up to the level of speed and efficiency seen in fast swimmers exemplified by the jacks of the family Carangidae. Subcarangiform swimmers are typically of compressed, fusiform shape, but may vary from fusiform with a nearly rounded caudal peduncle to rather compressiform species. Many have a rounded forebody with little flexibility and a more compressed midbody and caudal peduncle.

The caudal fins of subcarangiform swimmers tend to be truncate, rounded, or emarginate, but some stream-living catfishes and cyprinids, for example, have deeply emarginate to forked caudal fins. The caudal fin is of great importance to most fishes that swim in non-anguilliform modes. The caudal fin has considerable versatility and flexibility and is used for significant and subtle changes in direction and power. Loss of the caudal fin in most active subcarangiform swimmers handicaps the individual in maneuvering and in making fast starts and turns, but the performance in straight-line swimming in some species is not greatly hindered (Lindsey, 1978). Many species have median fins that effectively increase the body depth and aid in the production of forward thrust. Sometimes these fins can be depressed for rapid straight swimming and erected to accomplish tight turns. This is true especially for spinous dorsal fins that are set rather far forward. In some species, the dorsal fins are divided into two or three sections and are not completely depressible. The undulatory nature of the swimming is quite evident, and usually the arrangement of the vertical fins as stabilizers is not efficient enough to prevent the head from yawing back and forth as the fish progresses. Yawing is less pronounced in subcarangiform swimmers than in those that employ the anguilliform mode. Subcarangiform swimmers include trouts, cods, goldfishes, basses, and many other familiar fishes.

Carangiform Swimming

In the **carangiform** mode, the anterior one half to two thirds of the body is not very flexible and is bent only slightly during swimming. The flexure that provides forward thrust develops mainly in the back third of the body (Fig. 19.5C), so that usually up to one half wavelength will be present (Blake, 1983). The posterior part of the body usually tapers sharply to a narrow caudal peduncle and then flares to a strongly forked or even lunate caudal fin. This shape avoids the possibility of strong thrust in advance of the caudal fin, which transmits the locomotor power to the medium. The length of the waveform developed on the body is usually less than one half of the body length in most carangiform swimmers. The caudal fin, being forked or lunate, has a high vertical span in relation to its area and has what is termed a high **aspect ratio,** which is the square of the span of the fin divided by the surface area of the fin (Fierstine and Walters, 1968). Aspect ratios among the carangids typically have values of about 3.5. This is somewhat greater than that seen in subcarangiform swimmers, such as largemouth bass (1.5) or brown trout (about 2). Some minnows with forked caudal fins have aspect ratios higher than 2.0, and adult Pacific salmons have ratios of about 2.7, approaching the carangiform range. Usually, the caudal fins with high aspect

ratios are less flexible than those with lower aspect ratios and cannot be controlled as effectively by intrinsic muscles. The functions of guiding and adjusting small direction changes, as seen in the more flexible tail fins of the subcarangiform swimmers, have been taken over by other fins. Although some carangiform swimmers, such as the herrings, show few specializations, others may have their pelvic fins and a spinous dorsal fin moved forward, close to a vertical line run through the origin of the pectorals; may have one or more finlets following the dorsal and anal fins; or may have lateral keels developed on the caudal peduncle. The carangiform type of swimming is found in a variety of fish families, from herrings and sardines to more advanced fishes including the eponymous carangids (jacks) and some of the scombrids.

Thunniform Swimming

The **thunniform** mode of locomotion (Fig. 19.5D) has received a great amount of attention from scientists because of the speeds achieved and the interesting adaptations of the thunniform swimmers (see Going Deeper). Relatively few species swim in this fashion, but they include some of the largest teleosts, as well as sharks of appreciable size. There are remarkable physiological and circulatory adaptations involved with sustained swimming in these fishes, such as the heat-conserving rete that increases body heat up to as much as 20°C above ambient (see Chapter 24; Carey and Gibson, 1983) and modifications for streamlining. A remarkable degree of convergence in the development of the thunniform body design is seen in such remotely related fishes as the lamnid sharks and scombrid tunas. This convergence extends even to the architecture of the myomeres and their associated tendons (Donley et al., 2004).

This mode of swimming may be an improvement in efficiency over the carangiform mode, in that the great bulk of the body musculature is used mainly to generate rapid lateral movements of the caudal fin rather than undulations of the body (Shadwick et al., 1999). The extremely narrow caudal peduncle and the high, thin caudal fin impart tremendous thrust without causing much yawing movement of the body or head. The aspect ratio of the caudal fin ranges from about 4.0 to more than 8.5 in tunas, and up to 10 in marlins and sailfishes. Analysis of scombroid phylogenies indicates that the high-aspect-ratio caudal fin has evolved at least three times—in billfishes, in the wahoo (*Acanthocybium*), and in the tunas (Westneat and Wainwright, 2001). Some control of the span of the caudal fin can be exercised in the tunas, but the stiff caudals of the marlins and sailfishes can be changed little by intrinsic muscles. Likewise, the lamnid sharks that swim in this mode have caudal fins with a virtually fixed span. The caudal fins of the thunniform swimmers constitute a powerful propeller that can drive the body forward at great speed or can efficiently move the fish over long distances at moderate speeds. About 90 percent of the locomotor thrust is contributed by the caudal fin. In tunas, the vertebral column is relatively stiff, as is the caudal peduncle. There is, however, a joint at the anterior end of the peduncle that allows flexure. Scombrids have oblique tendons that extend posteriorly from the myoseptal fibers of epaxial and hypaxial myotomes (over three to seven vertebrae, depending on the species) before inserting on a vertebra. The first peduncular vertebra is the site of the posteriormost attachment of posterior oblique tendons (Fierstine and Walters, 1968; Westneat et al., 1993). By means of this connection, the powerful body musculature

GOING DEEPER · Dare to Build a Better Fish II: Pedal to the Metal and Fin to the Flow

Those of us with biological training who do not mind a bit of grease under our fingernails cannot help but seek out the biological influences in the construction of those marvelous pieces of machinery that enable us to get from one place to another. Fishes, especially those specifically adapted for extremely rapid locomotion, provide valuable models for understanding the constraints imposed on the design of automobiles, boats, and airplanes. I am not ashamed to admit that for me, fast cars are one of life's essentials. In the world of high-performance cars, it's all about acceleration and top speed.

Two forms of automobile racing have evolved with these goals in mind. Drag racing pits one car against another in a contest to see who can cover one quarter of a mile in the shortest time. A "top-fuel" dragster (Fig. 19.6A) can cover that distance in about four and a half seconds and reach a terminal velocity of more than 325 miles per hour. Here, the quickest driver takes home the trophy. The other "holy grail" of auto racing is absolute top speed. Currently, Royal Air Force Flight Lieutenant Andy Green holds this record at 763.035 miles per hour—the choice of a jet aircraft pilot to drive was a good one, as this was the first time an automobile achieved supersonic speeds. Clearly, the design criteria for a car built to cover one quarter of a mile in the shortest possible time are vastly different from those for a car intended to attain maximum velocity over a far greater distance. In the ichthyological realm, this contrast in design priorities is evident in a comparison of fishes adapted for a quick launch, such as the esocid pikes, with those adapted for sustained high-speed mobility, such as the scombroids.

Top-fuel dragsters, like pikes, are elongate, with a rearward concentration of propulsive elements. Their overall silhouettes are somewhat similar in emphasizing a posterior concentration of mass (see Fig. 3.6G). In a dragster, the engine and driver are located as far to the rear as possible for maximum traction and stability. Although the rear wheels and tires are enormous for purposes of traction, the front wheels are barely wide enough to keep the car on track as the driver hurtles toward the finish line. The massive engine, rear wheels, and driver account for most of the weight of the dragster, as the weight of all other components, including the critical steering mechanism, is kept at a minimum. The pike (see Fig. 3.6G, Plate 12.1) follows a similar design. The posterior placement of the dorsal and anal fins permits them to combine with the large caudal fin surface to generate maximum thrust. Weihs (1973) pioneered the application of hydrodynamic theory

A

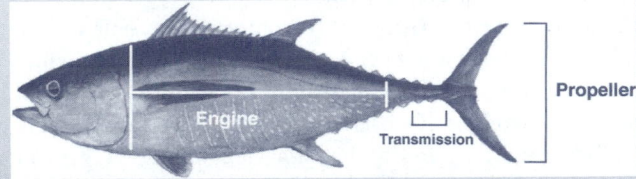

B

C

Propeller

Engine

Transmission

© Petersen Automotive Museum 2001.

FIGURE 19.6
A, A top-fuel dragster piloted by legendary driver Shirley Muldowney; **B,** A streamlined tuna (Thunnus albacares) showing mechanical arrangement of components of propulsion system (Photo courtesy of Company of Biologist Ltd.); **C,** Photograph of a land speed racer constructed from the belly tank of a World War II fighter plane.

can flex the stiff caudal peduncle from side to side (Lindsey, 1978). There is another joint at the end of the peduncle, between that structure and the caudal fin. The great lateral tendon extends along both sides of the peduncle and inserts on the caudal fin ray bases. Various tendons extend past this postpeduncular joint (Lindsey, 1978). The caudal fin is greatly stiffened by the overlap of the hypural plate (ural fan) with the strong bases of the fin rays (Weihs, 1989). In species with a bony peduncular keel, the large tendon that has its origin in the extensions of the myosepta of the posterior part of the body musculature runs over the keel much as a line runs over a pulley (Lindsey, 1978). This great lateral tendon and other tendons from various groups of posterior myomeres insert on the bases of the

to fish design. He predicted that esocids, with their large, posteriorly positioned fin surfaces combined with high-amplitude tail movements, would generate superior straight-line attack performance. Kinematic studies of both C-start and S-start performances of pikes have confirmed this. Acceleration rates during C-starts exceed those achieved during S-starts, demonstrating pikes to be extremely effective in predator avoidance as well (Frith and Blake, 1991, 1995). Fishes such as pikes, with dorsal and anal fins set close to the caudal fin, enjoy a "double tail" effect in increasing the speed of rapid starts (Weihs and Webb,1983; Weihs, 1989). Frith and Blake (1991) confirmed this in demonstrating that the caudal fin generates about three quarters of the thrust in a fast start, with the remaining one quarter contributed by the dorsal and anal fins. Muscle accounts for the majority of the pike's body mass (55 to 60 percent), and it is anaerobically fueled to maximize power output for short bursts of activity (Webb, 1978). The skin is thin, to minimize the mass of nonessential components (Webb, 1984). The modern racing car is skinned with aluminum or with the lightest of composite materials—sit on one and you will leave a lasting impression, both on the car and on the pit crew, who will politely but firmly escort you to the parking lot. Like the top-fuel dragster, the pike is designed for rapid acceleration and short bursts of straight-line performance. A dragster is not the best choice for the cut-and-thrust of everyday freeway driving, nor is the pike very effective in maneuvering about in structurally heterogeneous aquatic habitats.

The illustration on the cover of the August 1999 issue of the *Journal of Experimental Biology* (Vol. 202) explicitly acknowledged the merit of studying the biology of fishes from the perspective of a mechanical engineer (Fig. 19.6B). Here, we see the muscle mass of the trunk of a yellowfin tuna (*Thunnus albacares*) as the engine and its stiff, narrowed caudal peduncle as the transmission delivering power to the rigid, lunate caudal fin that is characteristic of scombroids. The fusiform body shape, with the maximum depth about midway between the head and the tail, and the head slightly blunted relative to the tail, is, from a hydrodynamic perspective, the most efficient shape for cutting through the water with a minimum of resistance. This reduces the formation of drag-inducing vortices along the body. The small, highly flexible finlets along the dorsal and ventral sides of the caudal peduncle in tunas and their relatives appear to form a latticework that works to maintain laminar flow in the region of the fish just aft of its greatest depth (Aleev, 1969, 1977; Altringham and Shadwick, 2001; Lindsey, 1978). By maintaining water flow along the keel, the finlets appear to enhance the development of a caudal fin vortex, thus increasing thrust (Altringham and Shadwick, 2001). In billfishes, the greatly elongate preorbital "spear" gives the head a much more tapered profile. An apparent hydrodynamic consequence of carrying this lethal weapon about—it is indeed used for prey capture—is a pronounced forward shift of the maximum body depth (see Plate 16.20).

It was not long after the Wright brothers first took to the air that aeronautical engineers recognized the aerodynamic advantages of the fusiform shape. The belly tanks carrying extra fuel that were slung beneath the wings of World War II fighters very closely approximate the profile of a tuna. Just after the war, these tanks could be had for a pittance, and hot rodders soon recognized that they had an ideal body for a car designed with terminal velocity in mind (Fig. 19.6C). A wind-cheating shape, combined with a powerful motor and a steering mechanism that is only needed to keep the car going in a straight line, was a good start on a land speed record attempt.

As mentioned in Part I of this story (see Chapter 18, Going Deeper), the resolution of Gray's paradox is fundamental to understanding the biomechanics of aquatic locomotion. It is not surprising that the U.S. Navy is very interested in the resolution of this paradox, as it might provide insight into more efficient surface and underwater propulsion systems. The Office of Naval Research (ONR) is currently investigating the applications of Autonomous Undersea Vehicles (AUVs) in oceanographic research and military surveillance. These devices currently use rotating propellers in their propulsion systems, just like conventional watercraft. If an alternative propulsion system were available, what would it look like? Because fishes, especially swift ones, like tunas and other scombroids, can apparently overcome the limits imposed by drag, as identified in Gray's paradox, why not model propulsion systems on the motions of a swimming fish? Researchers at the Massachusetts Institute of Technology have done just that: RoboTuna! This remarkable device (Fig. 19.7) is a sophisticated recreation of the complex system of muscles, bones, and tendons that make up the genuine article. The first-generation RoboTuna has a set of eight rigid "vertebrae" connected with low-friction ball bearing joints. Lateral flexion is generated by a set of servomotors that operate an elaborate system of pulleys and cables. That

caudal fin rays (Fig. 19.8). The placement of the tendons in relation to the joints is such that the **angle of attack** of the fin—the angle between the fin surface and the direction of movement of the fish—will be close to optimum for the generation of thrust.

In many thunniform swimmers, the fins other than the caudal can be depressed into grooves or recesses that make the surface of the fish "clean" from the standpoint of streamlining. Jaw bones fit neatly into fairings, and the bulge of the eye is streamlined by "adipose eyelids." The sclera of the eye (see Chapter 20) of scombroids is well ossified, possibly to prevent distortion of the eyeball during swift cruising. The rather stiff and, in some instances, large pectoral fins can be extended and employed to provide lift and steering. The dorsal fin, which in tunas has a high anterior portion well in advance of the middle of the fish, can be erected and used in quick maneuvering.

In addition to the mechanical and hydrodynamic adaptations just described, scombroids display a number of remarkable physiological accommodations for an extremely high-energy mode of living. Muscle contractions measured in tunas are unlike those observed in other fishes. Most fishes generate forward momentum by coordinated serial waves of muscle contraction that generate body undulations. As described earlier, tunas are distinctive in having masses of red, aerobic muscle fibers deep in the body, where other fishes would only have white fibers (see Fig. 19.2C). The deep red fibers contract somewhat out of phase in relation to the more superficial fibers. This produces a large degree of shear between superficial and deep fibers, resulting in the generation of appreciably more work by the muscles—work that is focused at the caudal peduncle and fin rather than in the trunk where the muscle is concentrated (Katz et al., 2001).

To maximize the delivery of oxygen and fuels, the muscle capillaries of tunas form dense, branching networks resembling those seen in the flight muscles of birds. Capillary dimensions per unit volume of mitochondria in scombroids, however, do not approach those seen in the intensely aerobic muscles of high-performance endotherms such as birds and bats (Mathieu-Costello et al., 1996). Weber and

RoboTuna can indeed recreate the type of flow field generated by its biological counterpart is verified by the fact that it also demonstrates Gray's paradox—RoboTuna experiences greater drag when stationary than when it is swimming. A spinoff project has been the development of a robotic pike that can replicate the rapid acceleration characteristic of esocids. Can a submarine that undulates its way through the water be too far off? Mitsubishi Heavy Industries of Japan is apparently more interested in animatronic applications of robotic fish, as they have developed a remote-controlled robotic sea bream and are planning to build a robotic coelacanth (Hadfield, 1999).

In his insightful analysis of the relationship between form and swimming in fishes, Webb (1984) constructed a triangular plane of functional morphology. At the center of the triangle is a surfperch, representing a locomotor generalist. At one corner is a pike, representing optimal development for acceleration; at another corner is a tuna, representing sustained high-speed cruising. If we wanted to complete our analogy of fishes as the living embodiment of automotive design, we would seek a correlate for the butterflyfish that occupies the third corner of the triangle. The short but deep-bodied members of the family Chaetodontidae are ideally designed for maximum maneuverability in and around coral reefs, at the expense of rapid starts or high speeds. In the world of motor sports, ranging from go-karts to Formula One, such agility makes for some of the most exciting forms of racing. But for the sake of those for whom the automobile remains just an underappreciated appliance,

it is probably best to just leave it at pikes and tunas and save the chaetodontids for another time.

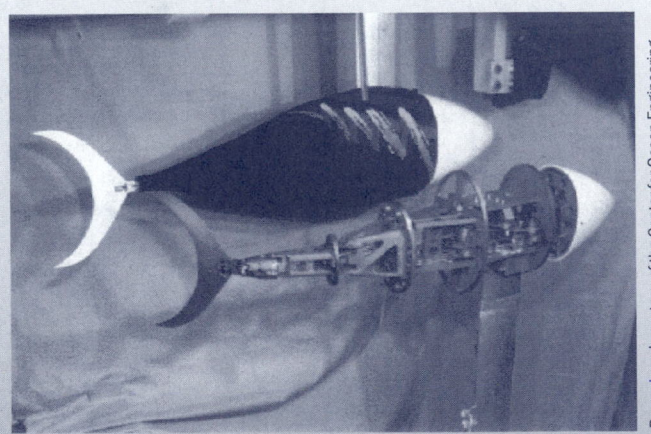

FIGURE 19.7
RoboTuna—a robotic fish engineered to replicate the swimming motions of a fish (photograph courtesy of Michael Triantafyllou, Dept. of Ocean Engineering, Massachusetts Institute of Technology).

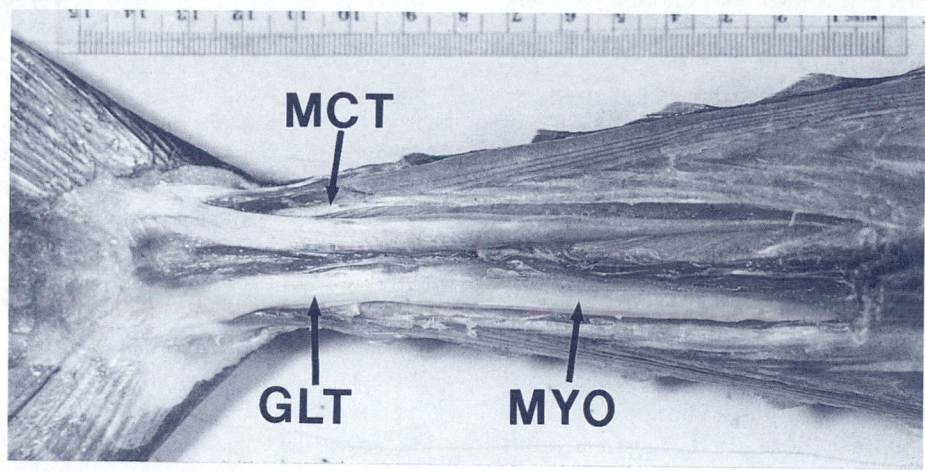

A

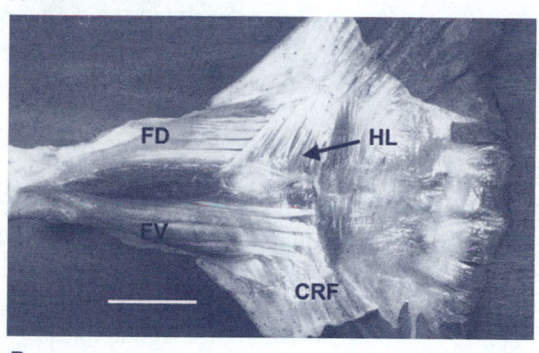

B

FIGURE 19.8

A, *Scomberomorus cavalla* caudal peduncle dissected to show great lateral tendon (GLT), median caudal tendon (MCT), and modified posterior myomeres (MYO); **B,** Caudal fin musculature of blue marlin, *Makaira nigricans*; deep dissection after removal of great lateral tendon. CRF: modified candal ray flaps; FD: flexor dorsalis muscle; FV: flexon ventralis muscle; HL: hypochordal longitudinalis muscle. Scale bar-5dm. (Photographs courtesy of Dr. Mark W. Westneat.)

Haman (1996) predicted that high-performance fishes such as the tunas should show similar pathways for metabolic fuel consumption as those seen in endurance-adapted mammals. These would reflect the consumption of a greater proportion of proteins and lipids, and fewer carbohydrates, and an increased reliance on intramuscular fuel stores in order to circumvent the problems associated with the transmembrane transport of circulatory fuel stores.

Tunas experience extremely rapid rates of growth. This reduces the vulnerability of the juveniles to predation, thus increasing their rate of survival to adulthood (Dickson, 1996). It has been suggested that the highly efficient physiological mechanisms used by scombroids to deliver oxygen and metabolic fuels to muscles are not so much an adaptation to enable sustained high-speed cruising as one enabling this rapid growth rate. Enhanced rates of gonadal growth, digestion,

and clearance of accumulated lactic acid in exhausted muscle also are possible because of the efficiency of oxygen and fuel delivery systems (Brill, 1996).

Ostraciiform Swimming

The **ostraciiform** mode of swimming is named for the boxfishes of the family Ostraciidae, which cannot bend their bone-covered body at all. They swim by oscillating a slightly flexible caudal fin on the end of a short caudal peduncle, so that practically all the forward thrust is contributed by the caudal fin, although in some species each stroke of the caudal fin is countered by simultaneous strokes of the dorsal and anal fins. The caudal fin is swung from side to side by alternate contractions of the myomeres on either side, and the undulation of the tail in a subcarangiform manner accompanies burst swimming (Gordon et al., 1996). The drag on species that resort to this mode of swimming is many times greater than on streamlined fishes, and none of them are fast swimmers. Awkward as the rigid carapace may appear to be while swimming, it is instrumental in maintaining stability through the generation of vortices that act as self-correcting trimming forces (Bartol et al., 2003, 2005). Boxfish swimming is actually quite complex, as up to three gaits, involving different combinations of all fin surfaces, have been recorded (Hove et al., 2001). In slow swimming, they commonly use their dorsal and anal fins, but they also employ their pectoral fins for thrust generation.

Electric rays, because of their laterally inflexible bodies and short caudal portion, use a sculling motion in swimming, and those trichiurids with caudal fins, judging from observations on *Aphanopus* (Bone, 1971), might use this method in very slow swimming, such as in stalking prey. Hatchetfishes (Sternoptychidae) have been observed swimming in a similar manner. Compared to other modes of swimming, ostraciiform locomotion is comparatively rare.

Fin-Swimming Modes

Although most fishes can swim by undulating at least part of the body, the caudal peduncle, and the caudal fin, many use their pectoral fins or the dorsal and anal fins as their typical means of locomotion at slow or moderate speeds. These fishes usually hold their body very straight while moving. There are some possible reasons for this, such as minimizing the activity that prey can see when being approached head-on, or preventing the distortion of electric fields in weakly electric fishes that monitor distortions in such fields (Nanjappa et al., 2000; see Chapters 10, 21, and 36).

Fin-swimming modes can be distinguished as incorporating either **undulation,** in which waves are propagated along fin surfaces that are broadly attached to the fish's body, or **oscillation,** in which fins with narrower bases are sculled back and forth. Although fishes do not always conform exactly to the modes that are described, some habitually use a certain fin or combinations of fins, so that descriptive terms can be applied. As is the case for the previously described locomotor strategies, these terms are usually derived from the names of fishes that typify a given style of fin swimming.

Locomotion Using Dorsal and Anal Fin Undulation

A great variety of fishes generate waves of undulation in both their dorsal and anal fins for locomotion. Because this mode of swimming is developed to a high degree in the triggerfishes and filefishes, it is called **balistiform** (Lighthill and Blake, 1990). The body is held straight, and undulatory waves traveling down the vertical fins move the fish either forward or backward, depending on the direction of the waves. Some variation of this mode is seen in such groups as eels, trumpetfishes (Aulostomidae), percoids, and flatfishes.

Many fishes have dorsal fins that extend the length of the back and can swim by sending undulatory waves along that fin to propel the body either forward or backward, depending on the direction of the waves. This mode of swimming is called **amiiform,** after the bowfin (*Amia calva*). Other species with long dorsal fins and small or absent anal fins that swim in this manner are found among the ribbonfishes, the hairtails and scabbard fishes (Trichiuridae), and in some sculpins, especially *Nautichthys*. The latter has a high spiny dorsal fin, which is slanted forward at about a 45° angle when the soft-rayed dorsal fin is used for amiiform swimming. Other examples include the African *Gymnarchus* and some of its relatives in the family Mormyridae. These are electric fishes that commonly maintain a straight posture except when swimming rapidly. The pipefishes and seahorses have dorsal fins with short bases in relation to their body length, but these flexible fins can undulate with great speed, causing the fishes to move as if powered by an invisible propeller. Although the pectoral fins usually take part in the locomotion, these fishes are thus said to swim in the amiiform manner.

Other electric fishes that use a single, broad-based fin for locomotion are the gymnotoids of South America, but in these, the anal fin is used. The freshwater knifefishes of Africa and Southeast Asia (Notopteridae) also swim in this manner. This mode is called **gymnotiform** (Lighthill and Blake, 1990).

Locomotion Using Oscillation of Dorsal and Anal Fins

Puffers, along with some of the triggerfishes, can swim by sculling the dorsal and anal fins, moving both fins simultaneously toward the same side, in what is called the **tetraodontiform** mode. *Latimeria* was reported by Locket (1980) to use only the second dorsal and anal fins in slow swimming. The two fins are moved simultaneously in the same direction, the complete sculling stroke requiring twisting of the lobate fins. The pectoral fins are used for minor adjustments. Observation of *Latimeria* by Fricke et al. (1987) indicated that the animal commonly drifts with currents and uses paired fins for guidance and stabilization. The caudal fin is used for fast starts, and both paired and unpaired fins are used in slow swimming, with the paired fins alternating in a tetrapodal form of locomotion (see Chapter 7). As mentioned earlier, ostraciid boxfishes may engage in oscillation of the dorsal and anal fins in conjunction with movements of the caudal fin.

Locomotion Using the Pectoral Appendages

Although the rajomorph (batoid) fishes are a monophyletic group (see Chapter 6), they exhibit a diversity of locomotor patterns, ranging from axial flexion involving the body and tail (as in the sawfishes and guitarfishes) to undulation or oscillation of the pectoral fins (Rosenberger, 2001). Pectoral fin undulation, as practiced by most skates and rays, is usually termed **rajiform** locomotion. Among the rajomorphs, undulatory motion of the pectoral fins grades into oscillatory motion, as seen in manta rays and eagle rays, that seems more like the wing movements of birds. Rajiform undulation of the pectoral fins is characterized by a higher average number of waves per fin length (averaging 1.3 for the Atlantic stingray, *Dasyatus sabina*), whereas fin oscillators swim with a lower average number of waves per fin length (0.4 for the cownose ray, *Rhinoptera bonasus;* Rosenberger, 2001). Bone (1999) has provided information on the muscular adaptations associated with the range of locomotion observed in elasmobranchs.

Among the perchlike fishes, many species employ narrow-based pectorals as oars or paddles, moving them simultaneously or alternately to provide a propulsive force. In most of these fishes, the base of the pectoral is high on the side and angled forward, so that the base is not quite vertical. This placement aids in feathering the fin on its forward stroke. Twisting the fin on the power stroke causes it to behave like a propeller, in that the angles of attack vary along the length of the fin (Westneat, 1996). This mode is called **labriform,** after the wrasses. Many non-percoids also swim in this manner, including, for example, some species of characins and the mesopelagic beryciform *Anoplogaster*. For slow swimming, the eelpout (*Melanostigma pammelas*) is reported to swim by means of alternating strokes of the pectoral fins.

Obviously, pectoral fin shape varies considerably among the labriform swimmers. Some of the percoids have long, slightly falcate pectoral fins that provide greater swimming performance (Walker and Westneat, 2002). Others, such as the sculpins and snailfishes, use their shorter, broad-based fins to engage in labriform swimming.

. . . .
Swimming Speed

Because of the density of the medium through which they must move, fishes cannot match the sustained speeds that some terrestrial and aerial animals can attain. Fishes can achieve maximum velocity for only a few seconds or minutes before their white muscles are fatigued and rest is required. High burst speeds are typically employed in escape responses, pursuit of prey, or overcoming strong currents.

Many laboratory studies have been conducted to determine the **critical swimming speed** (U_{crit})—that is, the maximum speed at which a fish can swim for a selected interval of time (Jones et al., 1974). These tests have most often been carried out by placing the experimental animals in tubes through which water can be pumped at desired speeds, and then increasing the speed at regular intervals until the fishes fail to swim (Beamish, 1978). Fishes tested have primarily been subcarangiform swimmers of the family Salmonidae, and most of them have been less than 20 cm long. Typical swimming intervals range from 5 to 30 minutes, and velocities range from about 2.5 to 10 body lengths per second (Ls^{-1}). This type of study can be useful in assessing the ability of fishes to swim under stress from pollutants, lowered dissolved oxygen, or disease.

In burst acceleration, fishes often make what investigators have called *C-starts* and *S-starts*. The **C-starts** are "startle" responses involving the **Mauthner neurons**—two large neurons that originate in the brain and send axons down the spinal cord, where they communicate with motor neurons running to the trunk muscles. The electrical synapses employed in impulse transmission by Mauthner neurons enable a much shorter response time than that possible in neurons using neurotransmitter chemicals (see Chapter 26). In C-starts, the fish abruptly bends its caudal region to one side to form a C or L shape, then swings it back in the first of a short series of power strokes (Fig. 19.9A). This type of start may give the fish a direction different from its original course, so that correction is required. C-starts are primarily used as a means of escape from predators, whereas S-starts are useful in prey capture. In **S-starts,** the body forms an S shape (Fig. 19.9B), and the direction of the forward motion is maintained (Frith and Blake, 1995). More elongate fishes are more inclined to launch themselves in an S-start. The increase in swimming speed can be accomplished by an increase in frequency or amplitude of body undulation or tail stroke, usually both, although at least one species, the jack mackerel (*Trachurus*), has been reported to maintain constant amplitude while increasing the frequency (Hunter and Zweifel, 1971).

Burst speeds and critical swimming speeds (Table 19.2) have been measured in some fishes by a variety of methods, ranging from simply attaching a speedometer to a fishing line to sophisticated methods involving high-speed cinematography or sonar. Swimming performance has been measured also by forcing fish to swim in flumes or tunnels through which water is pumped at known speeds, and in circular "doughnut" tanks that are rotated at selected speeds. Burst speed in anguilliform swimmers seems to be generally much slower than in subcarangiform, carangiform,

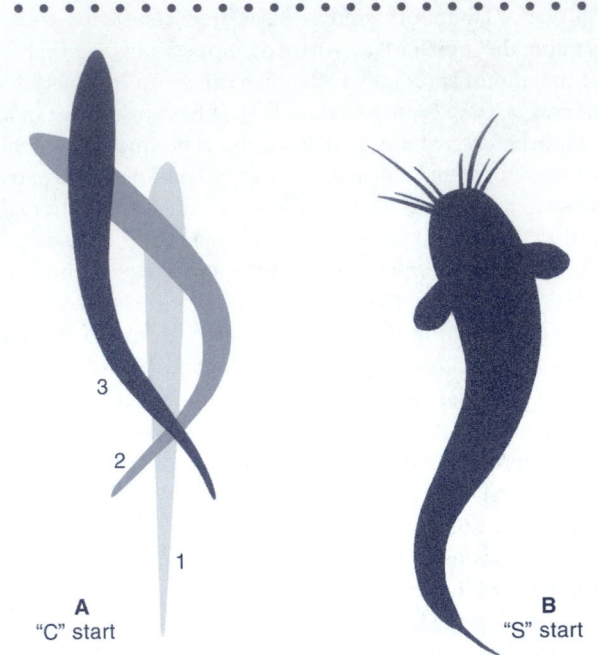

FIGURE 19.9
Fast starts: **A,** Fish in C-start—light shading: fish at rest; medium shading: fish in C flexion; dark shading: fish in first power stroke; **B,** Fish (*Clarias*) in S-start.

or thunniform swimmers, although comparatively few data are available for eellike swimmers. Whether the speed is expressed in relative terms, such as lengths per second (Ls^{-1}), or in absolute terms (cm s^{-1}), the anguilliform swimmers are apparently slower than most of the other swimming types. Burst speed in *Anguilla vulgaris* has been reported at about 2 Ls^{-1}. The eelpout (*Zoarces viviparus*) can swim at about 3 Ls^{-1}, and the flounder (*Platichthys flesus*) has been reported to swim at nearly 4 Ls^{-1} (Beamish, 1978). A small midwater eelpout (*Melanostigma*) has been clocked at 4 to 5 Ls^{-1} for a few minutes (Belman and Anderson, 1979). Small fishes can cover more lengths per second than larger ones, so of the aforementioned fishes, the eel swimming at 2 Ls^{-1} was traveling at about 115 cm s^{-1}, whereas the eelpout and gunnel, at 3 Ls^{-1}, were moving at 30 cm s^{-1} or less.

One might expect that, given their size and the frictional resistance of the medium through which they travel, larval fishes would not be very quick. This is not the case, however. Larvae of coral reef fishes measuring 1–2 cm have been recorded swimming at a mean speed of 13.7 Ls^{-1}, with some achieving velocities as high as 34 Ls^{-1} (Leis and McCormick, 2002). Putting this into perspective, the world record for a 100-m freestyle swimming event is just over 48 seconds (less than 2 Ls^{-1}); an Olympic swimmer moving at about 41 Ls^{-1} could cover the 100 m in less than 4 seconds!

Much attention has been given to the subcarangiform swimmers in the family Salmonidae. Several members of the genera *Salmo* and *Oncorhynchus* have burst speeds from 6 to 10 Ls^{-1} and actual speeds of 300 to more than 500 cm s^{-1}. Rainbow trout have been clocked at more than 1,000 cm s^{-1} (about 36 km h^{-1}) in an instantaneous burst of speed, but when followed over 10-second bursts, the speed was about 200 cm s^{-1}. Other subcarangiform swimmers, such as carp, suckers, and cods, show burst speeds of 100 to 300 cm s^{-1}, swimming at 5 to 8 Ls^{-1} (Beamish, 1978).

Among carangiform swimmers, bluefish (*Pomatomus saltatrix*) can swim at about 13 Ls^{-1} or 200 cm s^{-1}, and the jack mackerel (*Trachurus symmetricus*) has shown speeds up to 7.5 Ls^{-1} or about 210 cm s^{-1} in a 28-cm fish (Hunter, 1971; Hunter and Zweifel, 1971). This species closely approaches the thunniform mode and can maintain locomotion at near-burst velocities for a few hours.

Tunas and their close relatives that swim in the thunniform manner are the fastest fishes. The wahoo (*Acanthocybium solandri*) has been clocked to more than 20 Ls^{-1} and 2,100 cm s^{-1}, or about 75 km h^{-1}. A sailfish (*Istiophorus*) was once estimated at 84 km h^{-1} from the rate at which a hooked fish stripped line from a reel. Wardle et al. (1989) observed swimming speeds of 1,200 to 3,200 cm s^{-1} in caged bluefin tuna (*Thunnus thynnus*). This equates to 0.3 to 1 Ls^{-1} and 3.6 to 10.8 km h^{-1}.

Sustained speeds of fishes are difficult to observe and measure in the field. Methods used to measure or estimate sustained speeds include following schools, following individuals tagged with sonic tags, or tagging and recapturing migrating individuals. Because of the long distances that some species must travel while on migrations, the efficient use of stored energy is of prime importance, so their speed is geared as closely to their resting metabolism rate as possible. In a study of several species of fish (using individuals of 700 g or less), Belokopytkin and Shul'man (1989) determined that energy cost was highest at 0.5 Ls^{-1} and at 6 to 7 Ls^{-1} and faster, with the lowest expenditure of energy in most of the species studied at 2 to 3 Ls^{-1}. The migration speeds of a variety of species, from sharks to percoids and flatfishes, have been measured at less than 1 Ls^{-1}. Some of the Pacific salmons travel at close to 2 Ls^{-1}. Using ultrasonic tracking devices, the average swimming speed of sockeye salmon has been estimated at 1 Ls^{-1} (2.4 km h^{-1}), which is close to the most efficient speeds of about 0.82 Ls^{-1} (1.8 km h^{-1}) determined in the laboratory (Quinn, 1988). Bluefin tuna tagged in the Gulf of Mexico migrated to Norway in an interval of time that was calculated to equal 760 cm s^{-1}, or about 2.25 km h^{-1} (Wardle et al., 1989).

TABLE 19.2   BURST AND CRITICAL SWIMMING SPEEDS OF SELECTED FISHES

Species	Ls^{-1}	$cm\ s^{-1}$
Anguilliform mode		
Pholis gunnelus	3	30
Anguilla vulgaris	2	115
Zoarces viviparus	ca. 3	18–21
Subcarangiform mode		
Oncorhynchus kisutch	ca. 6–10	287–533
O. mykiss	ca. 3–4	186–226
Carassius auratus	9.4–11	74–200
Danio rerio (U_{crit})[1]	15.5	56
Gadus merlangus	5–9	70–180
Carangiform mode		
Clupea harengus	ca. 6–10	67–131
Scomber scombrus	5–9	190–300
Pomalobus psuedoharengus	14–16	400–500
Trachurus mediterraneus	16.4	258
Labriform mode		
Scarus schlegeli (U_{crit})[2]	2.8–3.7	66–82
Thunniform mode		
Acanthocybium solandri	18.4	2,100
Thunnus albacares	6–21	523–2,072
T. thynnus[3]	0.3–1	1,200–3,200
Balistiform mode		
Rhinecanthus aculeatus[3]	3.2–5.0	55–93

Source: Adapted from Beamish, 1978.
[1] Plaut, 2000; [2] Koremeyer et al., 2002; [3] Wardle et al., 1989. Plaut (2000) provided a summary of critical swimming speeds for a number of small fishes.

In discussing maximum swimming speeds, Wardle and He (1988) and Wardle et al. (1989) reported on the **stride length** in fishes—that is, the distance covered with one oscillation of the tail, which is governed by the twitch contraction time of the white lateral muscle. In *Scomber scombrus*, which has a stride of nearly one body length at a tail beat of 18 Hz, and a resulting speed of 550 cm s⁻¹, twitch contraction time was measured at 26 ms, and a top speed of 590 cm s⁻¹ was predicted. In a bluefin tuna of 2.26 m, twitch contraction time was 50 ms, and tail beat frequency was 10 Hz. At an average stride of 0.65 body length (= 1.46 m), a speed of 54 km h⁻¹ could be achieved. Theoretically, if a stride of 1 body length could be used at high speed, the resulting velocity could be 81 km h⁻¹.

Nonmigrating fishes traveling in schools tend to move faster than migrating fishes. Apparently, migrating fish must maintain a speed that is energetically efficient over long distances. Herring swim from less than 1 to nearly 8 Ls^{-1} while schooling. Schooling tuna are known to swim from 2 to 15 Ls^{-1}, and there is at least one report of 21 Ls^{-1} (Beamish, 1978).

Nonswimming Locomotion

Although swimming is the primary means of locomotion among fishes, there are many other ways in which fishes move from one place to another. Some of these are more than interesting oddities—they enable a diverse array of fishes to exploit niches from which obligatory swimmers are excluded. Some of the modes of travel have been described as follows (Lindsey, 1978): wriggling (snakelike progression), both in the water and on land; using the pectoral fins as crutches (*crutching*); flipping; skipping; sucking-and-hitching; gliding; flying; and drifting passively.

Probably, all the fishes that wriggle through mud or in and around coral or through the interstices of other hard bottom materials are capable of swimming, but their ability to move in a snakelike manner by forcing bends of their bodies against the bottom allows them to live a much different life from the swimmers in the water above them. The slender and colorful snake eels of the family Ophichthyidae, which number about 250 species, are excellent examples, as are other eel families, including the morays (Muraenidae), congers (Congridae), and the extremely slender spaghetti eels (Moringuidae). Spaghetti eels have assumed a fossorial existence, burrowing headfirst through sandy or muddy sediments. Other marine examples of fishes that engage in snakelike wriggling include the hagfishes (Myxinidae), cusk eels (Ophidiidae), gunnels (Pholidae), pricklebacks (Stichaeidae), eelpouts (Zoarcidae), wolf eels (Anarhichadidae), and another fossorial group, the graveldivers (Scytalinidae).

Freshwater fishes that move by wriggling include the larvae of lampreys, the polypterid ropefish (*Erpetoichthys calabaricus*), the eel catfishes (*Channalabes*) of the family Clariidae, some loaches (Cobitidae), the electric "eel" or knifefish (*Electophorus electricus*), and the members of the Synbranchidae.

The catadromous eels (Anguillidae) are good examples of fishes that can move effectively in a snakelike manner on land as well as in the water; their extensive migrations may, indeed, entail a bit of overland travel. Electromyographic recordings on the white muscle of eels (*Anguilla rostrata*) swimming in water versus those moving on land have revealed dramatically different patterns of muscle recruitment during locomotion (Gillis, 2000), and muscle power output decreased significantly during terrestrial locomotion (Ellerby et al., 2001b). Sojourns out of the water are known in other groups as well. Some synbranchids are known to travel overland, and some pholids and stichaeids remain hidden among rocks and vegetation in the intertidal area at low tide and do a fair job of wriggling away when disturbed. Many other fishes are capable of terrestrial locomotion, using the trunk, fins, and other structures that are specially modified. The climbing perch (*Anabas;* see Fig. 16.2D) has a series of backward-directed spines on the operculum. These spines assist in gripping the substrate as the fish travels overland.

Mudskippers (Gobiidae, subfamily Oxudercinae; see Plate 16.17, Fig. 20.6C) have highly modified pectoral fins that are armlike, with "elbows" separating a rather stiff upper part from a more flexible and expanded lower section. In terrestrial locomotion, the pelvic fins give support as the pectoral fins are both extended forward. The pectoral fins are then pulled back, and the fish proceeds after the manner of a person using crutches. *Clarias*, the walking catfish, moves on land by extending the pectoral fins and bending the body to thrust first one pectoral spine and then the other forward. Many fishes can move by flipping, using their tail to propel themselves forward. This mode might seem haphazard, but such fishes as the sleepers (Eleotridae), gobies (Gobiidae), and clinids (Clinidae) can be very precise in their vigorous jumps, leaving a shallow pool and flipping quickly over mudflats or rocks to dive into some other haven. Fishes with ventral suckers (see Fig. 3.13E–G) or

pads that can be used to adhere to the substrate can usually move over wet surfaces by attaching, hitching the body forward a short distance, and reattaching. Homalopterids and sisorid catfishes (see Fig. 3.12A) are examples of fishes that can overcome artificial barriers such as low concrete weirs in this manner. Lampreys, with their powerful oral sucking discs (see Fig. 3.3A), can move up wet vertical surfaces at waterfalls or dams by this suck-and-hitch method.

There are a few fishes that "walk" on the bottom or climb aquatic vegetation. In some of these, the pectoral fins are modified by having the lower rays separate and stiffened, so that they can be used to pull the fish along in a crutching or crawling fashion. Examples are the sea robins (Triglidae; see Fig. 3.12E) and certain genera of stonefishes, such as *Inimicus* and *Choridactylus*. The batfishes (Ogcocephalidae; see Fig. 3.12F, Plate 13.10) have strong pelvic fins placed in advance of the laterally situated, armlike pectorals, which are quite strong and are bent as if they have elbows. With these paired fins, plus the short-based, mobile anal fin, these fishes can walk over the bottom as a veritable quintiped (Lindsey, 1978). The pelvic fins of skates (Rajidae) have projections of the anterior lobe that permit a similar form of locomotion termed "punting" (Koester and Spirito, 2003). Some relatives of the batfishes, in the family Antennariidae, use their armlike pectoral fins to climb around in vegetation. The sargassumfish (*Histrio histrio;* see Fig. 18.8C; Plate 13.9) is one of the best known examples. Various flatfishes can use undulations of their dorsal and anal fins to crawl along the bottom. Tongue soles (Cynoglossinae) have been described as traveling in a millipede-like manner, leaving obvious tracks in the substrate (Clarke and Pearcy, 1968).

Many strong swimmers can leap significant distances. Milkfish (*Chanos*), mullets (*Mugil*), needlefishes (Belonidae), tunas (Scombridae), and various silversides (Atherinidae and Atherinopsidae) are known to be great jumpers. Eagle rays (Myliobatidae) and mantas (Mobulidae) often leap and return to the water with their flat body cupped so as to make a loud noise. Leaping activity can serve to escape predators, to descend on floating prey, to remove attached parasites, or for reasons unknown, including "play" activity. Fishes ascending streams may jump over low barriers or falls. The salmonids are noted for such leaping ability.

Several species of fishes have been reported to skip over the water's surface by taking short jumps or by applying the force of their fins or body to the surface and skipping without full submergence. Individuals that are injured or in danger from predators sometimes skitter across the water. Lindsey (1978) has reported an interesting set of morphological modifications related to such skittering in the genus *Chela*. These Asian cyprinids can bend the head upward to nearly a 90° angle relative to the dorsal surface. This "neck bending" is accompanied by a downward thrust of the pectoral fins, so that if the fish is at the surface, a rapid succession of neck bendings aids in skipping across the water.

The needlefishes (Belonidae) and the halfbeaks (Hemirhamphidae) include many species that move along the surface with short, successive jumps or actually keep their body above the surface with rapid sweeps of the caudal fin. Some of these species have elongated lower lobes of the caudal fin. Some have enlarged pectoral fins and are capable of gliding short distances. *Euleptorhamphus longirostris* has been given as an example of a halfbeak that can glide (Myers, 1950).

Gliding flight is best developed in the Exocoetidae, or flyingfishes (see Fig. 3.12G; Plate 15.9), some of which can make flights that cover 50 m at speeds of up to 2,500 cm s^{-1} (about 90 km h^{-1}; Rayner, 1981). These flights are generally close to the water's surface, but heights of more than 5 m can be reached. Flights can be prolonged by propulsion with the enlarged lower lobe of the caudal fin. Flyingfishes accelerate from swimming speed to gliding speed by rapid vibration of the caudal fin once the body is free of the water. Single flights of up to 400 m have been recorded. Most exocoetids have only the pectoral fins greatly enlarged, with moderate enlargement of the pelvic fins, which are used for maintaining trim or balance. In "four-winged" species, the pelvic fins are very large and add greatly to the area of gliding membrane (see Chapter 14). This decreases the wing loading (the weight supported by the unit area of gliding surface) and can result in slower flight, as the speed of gliding is proportional to the square root of the wing loading (Rayner, 1981).

The freshwater hatchetfishes (Gasteropelecidae) of South America, also called "freshwater flyingfishes" (see Fig. 3.12C) not only have enlarged pectoral fins, but they have pectoral muscles that make up as much as 25 percent of the weight of the fish. These fishes are known to leap as far as 5 m (Rayner, 1981). The butterflyfish (*Pantodon buchholzi*) of African freshwaters (see Fig. 3.12B) has enlarged pectoral fins and muscles arranged to move the pectoral fins forcefully up and down (Blake, 1983; Greenwood and Thompson, 1960). Aerial sojourns of up to 2 m have been recorded, and Rayner (1981) reported that *Pantodon* can catch flying insects. Although the pectoral dexterity of this remarkable little fish suggests that it could engage in flapping its pectoral fins, true directional flight has yet to be demonstrated.

Other "flying" fishes that should be mentioned are the flying gurnards (Dactylopteridae). These bony-headed, rather clumsy-looking fishes have enlarged pectoral fins and are known to be primarily benthic. The pectoral fins have been described as being too weak to support the fish in gliding flight, and reports of gliding appear to be mistaken (Blake, 1983). Nonetheless, there are records of dactylopte-

rids being found on the decks of boats, and reports of their leaping from one tank to another when held in captivity.

"Hitchhikers" among the fishes are mainly lampreys (Petromyzontidae) and remoras (Echeneidae). Trichomycterids and other catfishes that parasitize larger fishes should also qualify as hitchhikers. In some western North American rivers, lampreys attach to salmon on their upstream spawning migration. Remoras attach to large fishes, turtles, and whales and are carried along wherever their hosts travel.

Passive drifting is the mode of transportation of the larvae and floating eggs of many species. Ocean sunfishes (Molidae) of great size have been observed drifting passively at the surface, and the young of tripletails (Lobotidae) are known to emulate drifting mangrove leaves. Adult tripletails are often observed floating on their sides at the surface (Manooch, 1984). Fishes with weak powers of locomotion and fishes that can gather food with a minimum of swimming may stay in a moving water mass and, although their activity is directed only toward food gathering, may ride the current across oceans. The sargassum fish (*Histrio histrio*) is one of the best known examples of a fish species exhibiting a close association with drifting seaweed.

The cephalopod squids mentioned earlier have evolved a sophisticated form of jet propulsion that involves forcing water through a nozzle created by the mantle cavity. In fishes, however, the forceful ejection of water out of the gill chamber through the opercular opening does not appear to be a significant locomotor mode (Lindsey, 1978). In many instances, however, this action does generate forward momentum. In observing fish in aquaria, one can see them compensate for the force created by opercular jets by movement of the fins.

Scientists who have made observations from submersibles at mid-ocean depths have noted various species in apparent lethargy oriented with the head up (Barham, 1971; Clarke and Haedrich, 1968; Clarke and Pearcy, 1968; Clarke and Rosenblatt, 1968), including members of the families Bathylagidae, Gonostomatidae, Myctophidae, Paralepididae, Nemichthyidae, Regalecidae, and Trichiuridae. Although the last four families were noted as using movement of the fins or body to maintain position, the myctophids apparently overcame their slight negative buoyancy by strong pumping contractions of the operculum every two to four seconds (Barham, 1971). There is speculation—and some observational evidence—that myctophids may migrate to the surface and back by means of opercular movements (Barham, 1971); they have been seen oriented head downward during the time that migration back to depth occurs. Moreover, fishes with capacious gill cavities—various sculpins, for example—eject water forcibly when making a fast start. The ejection of water from the opercular openings may have some importance in reduction of drag.

THE SWIM BLADDER AND THE REGULATION OF BUOYANCY

One feature that distinguishes most vertebrates is the development of a medial evagination in the anterior part of the embryonic gut tube. In sarcopterygians and tetrapods, this becomes the lung, whereas in actinopterygians, it becomes a **swim bladder** (or *gas bladder*). Vertebrate morphologists disagree about whether the lung and the swim bladder constitute homologous structures, but their similarities are evident: Both form as outpocketings of the embryonic gut; both are gas-filled bags, and both share similar musculature and patterns of innervation (Kardong, 2002). Whereas lungs are paired, ventral structures, swim bladders are single structures occupying the dorsal part of the body cavity of fishes. Of course, their primary function is quite different, with lungs obtaining oxygen for respiration, whereas swim bladders are primarily organs for the maintenance of buoyancy. Berenbrink et al. (2005) have reconstructed the evolutionary history of oxygen-secreting ability in vertebrates, including its application to both swim bladder and eye function (see Chapter 24). This evolution of gas secretion ability involved a complex interplay of vascular modifications, including countercurrent exchangers, and various blood parameters, such as the Bohr and Root effects, buffering capacity, and ion exchange functions in red blood cell membranes.

In its simplest configuration, the swim bladder is a torpedo-shaped organ, yet many variations exist among fishes. Minnows and carps (Cyprinidae) have anterior and posterior chambers connected by a sphincter. The swim bladder of the featherbacks (Notopteridae) is divided into lateral halves, with the two chambers communicating anteriorly. In the herrings (Clupeidae), the swim bladder has a posterior opening to the exterior near the anus, through which gas may be voided. Some fishes have posterior extensions of the gas bladder reaching beyond the body cavity. In viviparous perches (Embiotocidae), for example, the swim bladder extends along the ventral surface of the vertebrae, whereas in the hairtails (*Trichiurus*), the posterior extension of the swim bladder runs along the concave anterior face of the first interhaemal bone, used for anal fin support.

The functions of the gas bladder include buoyancy maintenance, sound production, and sound reception. Some primitive actinopterygians use the swim bladder as a lung (see Chapter 24). When used for the exchange of respiratory gases, the swim bladder is usually compartmentalized and heavily vascularized, as in the bowfins and gars. For larval fishes, it is believed that a critical time period exists, even for physoclistous species, in which they must access the air–water interface or submerged gas bubbles in order to obtain and ingest a small air bubble for initial gas

bladder inflation. Larval zebrafishes reared in experimental chambers where they were denied access to the air–water interface were much less successful at gas bladder inflation (Goolish and Okutake, 1999).

The most sophisticated swim bladder designs are seen in those fish that use them for vocalization (see Chapter 21). Drums and croakers (Sciaenidae)—so named for the sounds they generate—have unusual, sexually dimorphic gas bladders, with variously shaped sacs (Fig. 19.10) or branching caeca arranged along each side of the organ. Sound production is usually accomplished by special vibratory muscles attached at or near the gas bladder. Gas bladders also serve in sound reception by acting as a resonator.

Buoyancy: Gas Bladder and Retes, Fats, and Oils

Although fishes may engage their body and fin shapes in the generation of lift as well as forward momentum, the gas bladder is the major means by which they maintain vertical position in the water column. The tissues of a fish generally have densities greater than water. Scales and bones have a specific gravity of about 2.0, and other tissues are in the 1.05 to 1.10 range. Fats and oils have a specific gravity of about 0.90 to 0.93 and thus tend to decrease the overall density of fishes. The combination of all tissues in a fish without a gas bladder or some other device for maintaining buoyancy results in an overall specific gravity in the range of 1.06 to 1.09. Because freshwater has a specific gravity of 1.0 and the salt water of the oceans about 1.026, such fishes must exert continuous effort to provide enough hydrodynamic lift to prevent sinking.

In bottom-dwelling fishes, such as darters (*Etheostoma*), most flatfishes (Pleuronectiformes), and sculpins (*Cottus*), which do not require neutral buoyancy, the gas bladder is frequently absent. Various bathypelagic fishes have lost the gas bladder (achieving buoyancy by other means), and the gas bladder is also absent in the agnaths and the cartilaginous fishes. Flying gurnards (Dactylopteridae), which are robust fishes that live mostly on the bottom, gain lift from their large pectorals and a small gas bladder when they swim off the substrate.

Because they lack gas bladders, sharks and rays have an overall negative buoyancy. This presents no problem for the bottom-dwelling species, but pelagic ones must swim constantly to prevent sinking. Large pectoral fins provide lift in many species of sharks, and hammerhead and bonnethead sharks use their distinctive head for hydrodynamic purposes (Bone, 1988; Nakaya, 1995). Rays gain both lift and resistance to sinking from their broad pectoral fins that can act like parachutes. Some pelagic fishes in which the gas bladder is small or lacking depend heavily on hydrodynamic

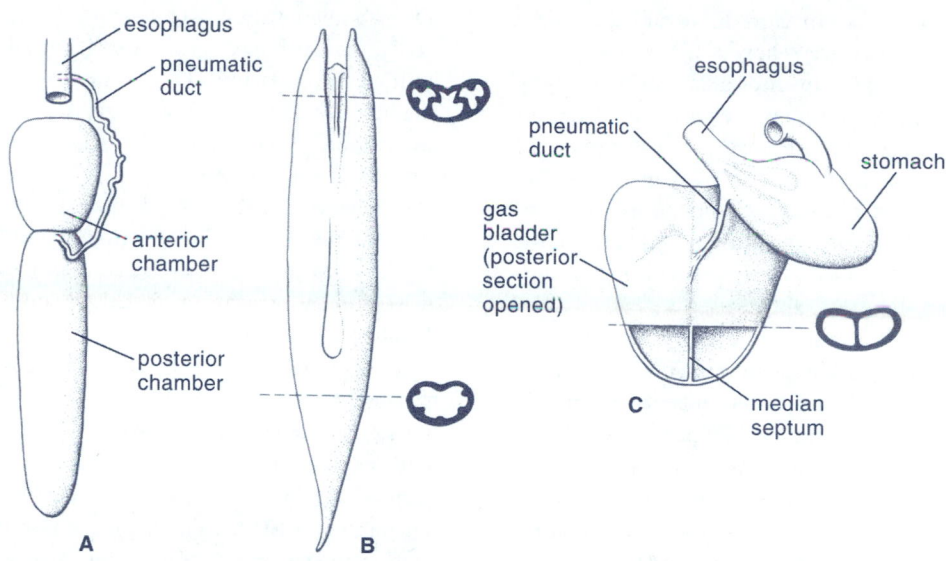

FIGURE 19.10
Examples of gas bladders, ventral views. **A,** Sucker (Catostomidae), showing long and crooked pneumatic duct; **B,** Seatrout (*Cynoscion*) showing anterior and posterior chambers in cross section; **C,** Channel catfish (Ictaluridae) opened to show median septum (stomach displaced to side).

lift from pectoral fins, caudal keels, and the angle of attack of their bodies.

Although the buoyancy gain from the fats and oils present in most bony fishes is negligible, these lipids can be of importance in exceptionally oily species or in fishes with special inclusions of lipids in the body cavity, liver, or bones. Phleger (1998) distinguished two categories of lipids with respect to their role in buoyancy maintenance. Some lipids, such as wax esters and **squalene,** directly contribute to buoyancy, whereas others, such as the triacylglycerols, function primarily in energy storage but make a small contribution to the overall reduction of body density. Many elasmobranchs are aided in maintaining buoyancy by the inclusion of lipids, including squalene, in their large livers (Bass and Ballard, 1972; Beldridge, 1972). Squalene has a specific gravity of about 0.86 and is metabolically inert, so that its only function appears to be buoyancy maintenance (Phleger, 1998). Some sharks have livers that hold up to 30 percent of the entire weight of the individual, and up to 90 percent of the weight of their liver may be oil (Bone, 1988). When these fishes are deprived of food, certain liver compounds are not broken down, apparently because of their critical role in buoyancy maintenance (Satchell, 1999).

Squalene is present in small amounts in several bony fishes, including *Latimeria,* but most bony fishes that de-

pend on lipids for buoyancy have wax esters with a specific gravity of approximately 0.86 in or around the gas bladder, in the skin, in the musculature, and even inside the cranium, as in the orange roughy (*Hoplichthys atlanticus;* Phleger and Grigor, 1990). In *Latimeria,* many myctophids, gempylids, and probably other mesopelagic fishes, up to 15 percent of their wet weight is composed of wax esters. The oilfish (*Ruvettus*), whose tissues are perfused with oil, has a density of 0.87 and is nearly neutrally buoyant. The eulachon, whose generic name (*Thaleichthys*) means "fat fish," contains about 20 percent lipid, including some squalene. This fish was formerly called the "candlefish" because a string wick could be placed in a dried fish to make a crude candle. Lipids may be of relatively greater importance in those species having reduced bones and watery flesh.

The pelagic *Mola mola,* which sometimes is seen floating at the surface, has no gas bladder, is remarkably watery, and has thin, light bones—features seen in numerous deep-sea fishes. *Mola* has a rather tough cartilaginous layer beneath the skin. The body fluids of marine fishes contribute a small amount of lift due to the fact that their osmotic concentration is about half that of seawater. In a fish as large as the mola, this may contribute significantly to density reduction. With as much as 86 percent of their tissue mass composed of water, marine fishes gain a buoyancy advantage over their freshwater brethren.

Fishes that live near the surface, in midwater, or freely swimming close to the bottom gain an advantage from a swim bladder because they are relieved of the necessity of maintaining a chosen depth by muscular effort. Gaining and maintaining neutral buoyancy over a range of depths is physiologically challenging because the water pressure changes about 1 atmosphere for every 10 m increase in depth. As a fish descends, the increasing pressure will compress the gas bladder, and without compensatory action, the contribution of the compressed bladder to neutral buoyancy will decrease.

An ascending fish faces the opposite problem: The decrease of pressure on the gas bladder allows it to expand, and if no "relief valve" is provided, the other internal organs will be crowded out by the expanding gas bladder. Although this is not a natural occurrence, bottom-dwelling fishes that are hooked and rapidly brought to the surface by sportfishers often have their stomachs protruding from their mouths. Sometimes, in a forced ascent, the buoyancy of the expanding gas bladder can become great enough that the fish cannot overcome it to swim back to its accustomed depth. Furthermore, dragging a fish up from a great enough depth can result in a ruptured gas bladder. In their natural vertical movements, fishes generally do not move rapidly through a sufficient depth range to bring about more than a 25 percent change in gas bladder volume. At great depths, the change in pressure in terms of percentages is small for each 10-meter change in depth. For instance, ascending from 100 to 90 m increases the volume of the gas bladder only by 10 percent. Those species moving through greater depth ranges are specially equipped with means of emptying and filling the gas bladder (Marshall, 1979; Steen, 1970).

The size, structural modifications, and placement of gas bladders reflect the ecology and habits of fishes. Benthic fishes with small or no gas bladders depend on friction with the bottom materials, wedging into tight places, digging in, or holding on with special suction structures to maintain position. Those fishes that swim up to capture food above the bottom can derive some benefit from a small gas bladder that ranges from 0.3 to 5 percent of their volume. Most bathypelagic species found below 1,000 m lack gas bladders, but some benthic fishes found as deep as 7,000 m may have small gas bladders.

Marine fishes that live above the bottom but confine their activities to narrow depth ranges usually have gas bladders of about 5 to 5.6 percent of their volume, compared to about 7 to 10.6 percent in freshwater species. These fishes can hold themselves motionless with comparative ease, using slight fin or body movements to counteract weak currents. Many of these species may be slightly heavier than water and sink slowly between adjustments of position. Deep-bodied fishes usually have the gas bladder placed above their center of gravity, so that little effort is required to hold the body upright. In some slender fishes, the swim bladder is below the center of gravity, and the fish must use fin motion to remain upright. These species will go "belly up" when anesthetized.

Species of fish adapted to flowing (*lotic*) water tend to have smaller gas bladders than still-water (*lentic*) fishes, as the less buoyant condition is favorable to maintaining a given station in the stream. In related stream fishes, those species habitually living in the swiftest water have less capacious gas bladders than those living in slower currents. Furthermore, in laboratory experiments, stream species reared in still water proved to have larger gas bladders than those reared in a current (Gee, 1968, 1972). Within a given species, there is a general capability to adjust buoyancy to suit the current flow encountered. Such adjustment may be made to maintain a station in swift water (reduction of buoyancy) or to relieve the muscular effort required for maintaining position as the current velocity decreases (increase of buoyancy). Over the course of the life history of some species, the relative volume of the gas bladder changes considerably. In the longnose dace (*Rhinichthys cataractae*), the young live in slower flowing water at stream margins and have a comparatively large gas bladder. As the fish become older and move into swift currents, the gas bladder grows at a much slower rate than the body of the individual (Gee, 1974). In certain anadromous trout and salmon, the young are adapted to a stream life for periods of up to three years or more. During this time, the young are less buoyant than they will be when they begin their downstream migration to the sea. Becoming more buoyant in a current thus facilitates dispersal.

Vertical Movement

Conspicuous in the lives of many fishes is vertical migration (see Chapter 35). This can take the form of rapid, spontaneous changes in depth in pursuit of escaping prey or periodic, nocturnal migrations that will bring fishes into closer proximity with preferred food items. Naturally, the resolution of the physiological problems associated with depth change involves modifications to the gas bladder. The speed with which many scombroids make changes in depth may exceed the capacity of a swim bladder to make buoyancy compensations. Consequently, many powerful scombroids lack a gas bladder and govern their depth mainly by rapid swimming, using the pectoral fins for lift when necessary. The larger the pectoral fins, the less speed is needed to maintain the lift. Examples of scombroids without gas bladders are the Spanish mackerels and their close relatives (*Scomberomorus*), skipjacks (*Euthynnus*), and mackerels (*Scomber*). Larger tunas and their xiphioid relatives (billfishes and swordfishes) have gas bladders and consequently can rely on a slower swimming speed than smaller species, although their vertical movements near the surface are restricted unless their gas bladders have extremely tough walls or special adaptations for rapid compensation.

The billfishes, some of which are among the largest teleosts, can make rapid vertical excursions and compensate rather quickly for changing pressure because of the nature of their gas bladders. These are divided into numerous small cells, each of which has its own gas secretion and absorption glands (Robins, 1974), so that the pressure can be adjusted rapidly. The bluefish (*Pomatomus*) also has the capability of rapid secretion and absorption of gas (Alexander, 1972).

The yellowfin tuna (*Thunnus albacares*) is unusual in that its gas bladder is not inflated in fishes of about 2 kg, and these small fishes have a specific gravity of about 1.09. Quick growth and inflation of the bladder reduces the specific gravity of a 10-kg fish to about 1.05.

Many mesopelagic fishes perform regular vertical migrations, so that the greatest concentrations of the species will approach the surface during the night, but will be at depths between 400 m and more than 500 m in the daytime. Others may move from greater depths to a nocturnal depth of about 150 m. The lanternfish family (Myctophidae) includes perhaps the largest number of vertically migrating species. The dragonfishes, viperfishes, scaly dragonfishes, scaleless dragonfishes, black or stalkeye dragonfishes, and snaggletooths—all different subfamilies in the family Stomiidae—are also known for their extensive vertical migrations. Some families have members that migrate vertically but do not reach the surface. Included are the bigheads (Melamphaidae) and the hatchetfishes (Sternoptychidae). Certain epipelagic fishes, such as *Clupea harengus* and other herrings, are known to go as deep as 150 m and return to the surface in the course of a day. Herrings are typically not neutrally buoyant (Blaxter and Batty, 1984), but they can expel gas through a pore behind the anus.

Many of these vertical migrants have gas-filled bladders and face the difficulties inherent in moving the captured gas through pressure changes of 15 to 50 atmospheres or more. The problems are greatest for those fishes that move all the way to the surface. Some species have avoided these problems in part by incorporating materials of low specific gravity in the body. The gas bladder may be partially or completely filled with fat, as in the black scabbardfish and certain macrourids (rattails), or reduced in size and invested with fat, as in some lanternfishes and anglemouths. The gas bladders of most deep benthopelagic fishes, such

as the macrourids, contain a fatty foam rich in cholesterol and oxygen bubbles (Phleger, 1998). Some lipids associated with the gas bladder of lanternfishes have a specific gravity close to that of squalene. For the most part, however, gas is maintained in the bladder by special structures that can secrete or absorb gas (Satchell, 1991; Scholander, 1957). Of course, some physostomous fishes retain the ability to release gas through the pneumatic duct or, in the herrings, through a posterior duct opening behind the anus, but even these must secrete gas into the lumen of the gas bladder if a reasonable hydrostatic balance is to be maintained. There is little evidence indicating that a perfect hydrostatic balance is kept throughout vertical migration. Some species of lanternfishes (e.g., *Notoscelopus kroyeri* and *Tarletonbeania crenularis*) are negatively buoyant (Marshall, 1979).

Gas Resorption

Gas bladders are nearly impervious to gas leakage because of the complex structure of their walls. They are usually composed of four layers, the outer one consisting of densely woven but elastic fibers. The next layer is of more loosely organized fibers, and the inner two layers are smooth muscle and epithelium. In many species, the wall is made more impervious to gases by a layer of guanine crystals just below the outer elastic-fiber layer. Tiny overlapping platelets may aid in reducing leakage in the gas bladders of certain poeciliids. The wall thickness is usually in the range of 50 to 300 μm, but can be greater, as in the black scabbardfish (*Aphanopus carbo*), which has walls up to 1.5 mm thick (Bone, 1971). The gas bladder in this species is also tightly enclosed by the ribs (Howe et al., 1980). The tough wall composition of the swim bladders of many species makes their removal fairly easy. Gas bladders can be removed by careful dissection and, if gently inflated while they dry, will make useful specimens for demonstrating gas bladder morphology.

The problem of removing gas from a gas-tight sac is overcome by special areas in the wall, where a rich bed of capillaries can be exposed to the lumen of the gas bladder. Gas resorption involves the diffusion from a high gas tension in the bladder to a lower tension in the blood, at a rate governed by the tensions, temperature, area of the capillary bed, and the rate at which the blood flows through the bed. The capillaries are usually disposed in a subcircular or oval area in the dorsal wall of the gas bladder and can be isolated from the lumen during the nonabsorbent phase of operation by a sphincter. Because of its shape, this structure is usually called the **oval**. Relaxation of the sphincter and contraction of radial muscles in the oval bring about maximum exposure of the capillaries to the gases in the bladder. There are exceptions to the arrangement described, however: In some fishes, the capillaries cannot be isolated, but resorption can be prevented by constriction of the capillaries or by thickening of the epithelial lining by muscle contractions.

In the eels of the genus *Anguilla*, the pneumatic duct is modified for the resorption of gas (Fig. 19.11A). In several percoid fishes, pipefishes, sticklebacks, and others, a diaphragm separates the posterior, gas-resorbing part of the gas bladder from the anterior, gas-secreting section. Such fishes are called **euphysoclistic,** to distinguish them from **paraphysoclistic** fishes, in which the gas-secreting and gas-diffusing areas are not well separated (Fig. 19.11B). Gas resorption in many euphysoclists is accompanied by the contraction of the gas-secreting chamber and may involve anterior displacement of the diaphragm as well as expansion of the resorbent chamber.

The veins leading from the oval or other resorbent structure lead into the cardinal vein in most physoclists. In the eel, blood leaves the resorbent structure via the pneumatic duct vein, which proceeds directly to the heart. Because the blood with its load of gas thus must pass through the gills before being distributed to the systemic circulation, there is an opportunity for excess gas to be diffused into the surrounding water. Considering the rapid ascents made by lanternfishes moving from 500 m to the surface, the resorption process might seem too slow for the proper adjustment of gas bladder volume to be made. Certainly, the rates of resorption calculated for freshwater fishes would not allow for a rapid ascent. However, anatomical studies of the gas resorption apparatus in lanternfishes have revealed large areas of capillaries in relation to the volume of the gas bladder and capacious arteries and veins serving the oval. The capillaries of the oval are separated from the lumen of the bladder by tissue no thicker than 1 μm. Furthermore, most vertical migrants are small—usually less than 100 mm—and have a small gas bladder in relation to their size, as their buoyancy is often aided by inclusion of lipids.

Gas Secretion

Filling the gas bladder on descent cannot be as easily accomplished as the removal of gas during the ascent. Gas must somehow be secreted from the lower pressure in the blood into the higher pressure in the gas bladder. Oxygen, for instance, found essentially at a pressure of 0.2 atmospheres or less in the blood, must be secreted into the gas bladder against a pressure gradient up to hundreds of atmospheres. The apparatus by which the gas secretion is accomplished consists of closely associated bundles of arterial and venous capillaries running counter to each other. Blood goes to the gas bladder through the arterial capillar-

ies, circulates through the specialized bed of capillaries from which the gases are secreted into the gas bladder, and then flows away through the venous capillaries. These bundles of capillaries form the **rete mirabile** or "wonderful net." This remarkable organ apparently evolved four separate times in the teleosts, and has become secondarily eliminated several times (Berenbrink et al., 2005). The parallel alignment of the small vessels forms an efficient countercurrent structure that serves to concentrate gases at the site of the **gas gland.** Depending on the species, there may be from a few hundred to more than 200,000 capillaries in the rete.

The process of gas concentration is aided by the secretion of lactic acid into the blood by the gas gland. This acid increases the partial pressure of carbon dioxide in the blood by releasing CO_2 from bicarbonate, but its greatest effects are on the partial pressure of oxygen. Acidification of the blood promotes the **Bohr effect**—higher partial pressures are required to keep oxygen bound to hemoglobin. The **Root effect** is also operative as the pH is lowered—the quantity of oxygen that can combine with hemoglobin, regardless how high the partial pressure, is decreased (see Chapter 24 for further discussion of the Bohr and Root effects). The Root effect is of great importance in the release of oxygen into the gas bladder. There is, moreover, a "salting out" effect—the lactate ions decrease the amount of gas that can be held in solution (Scholander, 1954; Scholander and Van Dam, 1954). In vitro culture of gas gland cells has greatly enhanced our understanding of their function (Prem and Pelster, 2001; Sötz et al., 2002).

Because of the aforementioned effects, the blood leaving the gas gland through the venous capillaries of the rete has greater partial pressures of gases, especially oxygen, than the blood in the arterial capillaries. The two sets of capillaries are intimately associated, so that a cross section of the rete would show them in a checkerboard or mosaic pattern. The capillaries of the rete are very long in relation to other capillaries. The great length and number of capillaries allow ample opportunity for the oxygen at high partial pressure in the venous capillaries to diffuse into the blood in the arterial capillaries. Eventually, the pressure in the small vessels is higher than that in the lumen of the gas bladder, so that bubbles of gas are released at the gas gland, serving to inflate the bladder.

The structure and dimensions of retes and gas glands differ among phyletic lines and also with respect to habitat affiliation. Some of the upper mesopelagic lanternfishes have retial capillaries that are 1 to 2 mm long, whereas those living as deep as 1,000 m may have retes up to 7 mm long. *Anguilla* has a rete of about 4 mm. The golden redfish (*Se-*

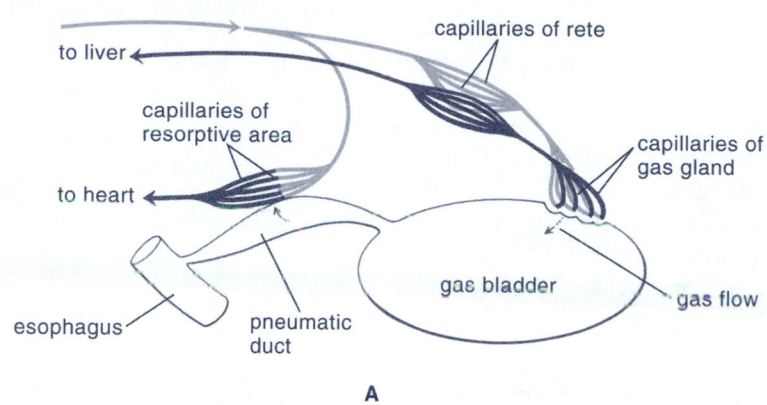

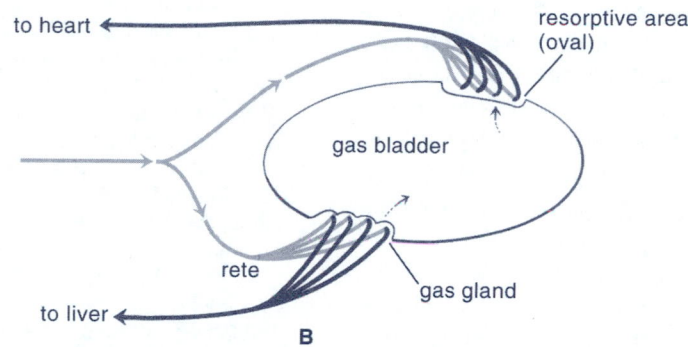

FIGURE 19.11

A, Illustration of the gas bladder of *Anguilla,* showing relationships of rete and gas gland and resorptive area on pneumatic duct; **B,** Diagram of gas bladder, rete, and resorptive oval of physoclistous fish.

bastes norvegicus), which lives down to about 600 mm has retes of 7 to 10 mm in fishes 30 to 45 cm long. In benthopelagic grenadiers (Macrouridae), those species that are commonly found shallower than about 600 m have retes of 6 mm or less, and those found deeper than 2,000 m have retes longer than 25 mm. Relatives living at intermediate depths have retes of intermediate lengths (Marshall, 1979). Many of the Macrouridae have from two to six retes. Multiple retes and gas glands are common among deep-sea fishes.

The internal structure of gas glands differs among species. The glandular epithelium is a specialized part of the inner gas bladder lining and may be single-celled, multicellular, or folded. Capillaries penetrate between folds or, in some species, enter giant, specialized cells.

The gases contained in the gas bladders of shallow-water fishes usually consist mainly of nitrogen and oxygen,

in the proportions found in the atmosphere—about 80 percent nitrogen and 20 percent oxygen. Members of the Salmonidae tend to have greater proportions of nitrogen, although in some species, the gas secreted into the gas bladder may be high in oxygen, which is later exchanged for nitrogen by diffusion. Salmonids do not have well-developed retes or gas glands, so that gas enters the lumen of the bladder over a wide area of the bladder wall. Gases contained in the gas bladders of deep-sea fishes contain higher proportions of oxygen than nitrogen. A deep-sea eel, *Synaphobranchus,* has been noted as having 75.1 percent oxygen, 20.5 percent nitrogen, 3.1 percent carbon dioxide, and 0.4 percent argon in the gas bladder. Various deepwater species, including lanternfishes, have from 76 percent to 88 percent oxygen in the gas bladder, but lanternfishes captured at the surface show about 43 percent oxygen.

KEY POINTS AND CONNECTIONS

- As a key component in the locomotor system of fishes, the skeletal musculature of the trunk consists of a serial arrangement of blocks termed myomeres. These are separated by connective tissue myosepta and are arranged in complex, broadly overlapping units. Skeletal muscle fibers fall into two general categories: white or "twitch" fibers, used for rapid contraction, as in burst swimming; and red or "tonic" fibers, used for sustained locomotor activity. The vascular supply, biochemical composition, and patterns of innervation of each muscle type are profoundly different and reflect specific musculoskeletal applications.

The skeletal muscles of the head that function in feeding and respiration are discussed in Chapter 23.

- The manner in which animals use their musculoskeletal system for purposes of locomotion has been termed their "gait." For fishes, gaits can be broadly classified as body and caudal fin swimming (BCF) gaits or median and paired fin swimming (MPF) gaits. Specific locomotor modes are described, using terminology largely derived from the taxa of fishes that best characterize the relevant modes. Many fishes also engage in modes of locomotion that do not involve swimming.

- An important component of the overall locomotor strategy of fishes is the means to achieve neutral or close to neutral buoyancy. For many fishes, especially chondrichthyans, this is achieved using lipid depositions that reduce their overall specific gravity. For bony fishes, the swim bladder contributes to neutral buoyancy through the selective resorption or secretion of gas. The swim bladder uses sophisticated physiological processes, involving specialized structures such as the rete mirabile, oval, and gas gland, for transferring gas to and receiving gas from the circulatory system.

The role of the swim bladder in sound production is described in Chapter 21; its role in respiratory gas exchange is discussed in Chapter 24. Swim bladder function in deep-ocean fishes that engage in extensive vertical migrations is mentioned in Chapter 35.

FISH LINKS

http://www.mcz.harvard.edu/fish/Home page for the Ichthyology Department of the Museum of Comparative Zoology at Harvard University. Includes a link to the laboratories of George V. Lauder, a leader in studies on the functional morphology of fishes.

http://biomech.ucsd.edu/research.htm Robert Shadwick, of the Shadwick Biomechanics Laboratory at the University of California, San Diego, uses an engineering-based approach to the study of the energetics and mechanics of fish swimming.

http://web.mit.edu/towtank/www/Tuna/tuna.html Home page of David Barrett's RoboTuna project at the Massachusetts Institute of Technology.

http://web.mit.edu/towtank/www/Pike/pike.html John Muir Kumph's robotic pike project, also at MIT.

BUILDING AN ICHTHYOLOGY LIBRARY

Aleev, Y. G. 1969. *Function and gross morphology in fish.* Israel Program for Scientific Translations, Jerusalem.

Alexander, R. M. 1970. *Functional design in fishes.* Hutchinson, London.

Block, B. A., and E. D. Stevens. 2001. *Tuna: Physiology, ecology, and evolution.* Academic Press, San Diego.

Hoar, W. S., and Randall, D. J. (Eds.). 1978. *Fish physiology,* Vol. VII, Locomotion. Academic Press, New York.

Lauder, G. V., and J. H. Long, Jr. (Eds.). 1996. Aquatic locomotion: New approaches to invertebrate and vertebrate biomechanics. *Amer. Zool.* 36(6).

REFERENCES

Agarkov, G. B., Y. N. Varich, and K. A. Snezhina. 1976. Innervation of the locomotor organs in some fish. *J. Hydrobiol.* 12(2): 57–59.

Aleev, Y. G. 1969. *Function and gross morphology in fish.* Israel Program for Scientific Translations, Jerusalem.

Aleev, Y. G. 1977. *Nekton.* Junk, The Hague.

Alexander, R. McN. 1969. The orientation of muscle fibers in the myomeres of fishes. *J. Mar. Biol. Assoc. UK 49:* 263–190.

———. 1972. The energetics of vertical migration by fishes. *Symp. Soc. Exp. Biol. 26:* 273.

Altringham, J. D., and R. E. Shadwick. 2001. Swimming and muscle function, pp. 313–344. In *Tuna: Physiology, ecology, and evolution,* B. A. Block and E. D. Stevens (Eds.). Academic Press, San Diego.

Barham, E. G. 1971. Deep-sea fishes—lethargy and vertical orientation, pp. 100–118. In *Proceedings of the International Symposium on Biology of Sound Scattering in the Ocean,* G. B. Farquhar (Ed.). Maury Center for Ocean Science, Department of the Navy, Washington, DC.

Bartol, I. K., M. Gharib, D. Weihs, P. W. Webb, J. R. Hove, and M. S. Gordon. 2003. Hydrodynamic stability of swimming in ostraciid fishes: Role of the carapace in the smooth trunkfish *Lactophrys triqueter* (Teleostei: Ostraciidae). *J. Exp. Biol. 206:* 725–744.

———, ———, P. W. Webb, D. Weihs, and M. S. Gordon. 2005. Body-induced vortical flows: A common mechanism for self-corrective trimming control in boxfishes. *J. Exp. Biol. 208:* 327–344.

Bass, A. H., and J. A. Ballard. 1972. Buoyancy control in the shark *Odontaspis taurus* (Rafinesque). *Copeia 1972:* 594–595.

Batty, R. S. and Wardle, C. S. 1979. Restoration of glycogen from lactic acid in the anaerobic swimming muscle of plaice, *Pleuronectes platessa* L. *J. Fish Biol. 15:* 509–519.

Beamish, F. W. H. 1978. Swimming capacity, pp. 101–187. In *Fish physiology,* Vol. VII, W.S. Hoar and D. J. Randall (Eds.). Academic Press, New York.

Beldridge, H. D., Jr. 1972. Accumulation and function of liver oil in Florida sharks. *Copeia 1972:* 306–325.

Belman, B. W., and Anderson, M. E. 1979. Aquarium observations on feeding by *Melanostigma pammelas* (Pisces: Zoarcidae). *Copeia 1979:* 366–369.

Berenbrink, M., P. Koldkjær, O. Kepp, and A. R. Cossins. 2005. Evolution of oxygen secretion in fishes and the emergence of a complex physiological system. *Science 307:* 1752–1757.

Blake, R.W. 1983. *Fish locomotion.* Cambridge University Press, Cambridge, UK.

———. 2004. Fish functional design and swimming performance. *J. Fish Biol.* 65: 1193–1222.

Blaxter, J. H. S., and R. S. Batty. 1984. The herring swimbladder: Loss and gain of gas. *J. Mar. Biol. Assoc. UK* 64: 441–459.

Bone, Q. 1971. On the scabbard fish *Aphanopus carbo. J. Mar. Biol. Assoc. UK* 51: 219–255.

———. 1975. Muscular and energetic aspects of fish swimming, pp. 493–528. In *Swimming and flying in nature*, T. Y.-T. Wu, C. J. Brokaw, and C. Brennan, (Eds.). Plenum Press, New York.

———. 1978. Locomotor muscle, pp. 361–424. In *Fish physiology*, Vol. VII, W. S. Hoar and D. J. Randall (Eds.). Academic Press, New York.

———. 1988. Muscles and locomotion, pp. 99–141. In *Physiology of elasmobranch fishes*, T. J. Shuttleworth (Ed.). Springer Verlag, Berlin.

———.1989. Evolutionary patterns of axial muscle systems in some invertebrates and fish. *Amer. Zool.* 29: 5–18.

———. 1999. Muscular system: Microscopical anatomy, physiology, and biochemistry of elasmobranch muscle fibers, pp. 115–143. In *Sharks, skates, and rays: The biology of elasmobranch fishes*, W. C. Hamlett (Ed.). Johns Hopkins University Press, Baltimore.

———, Kiceniuk, J., and Jones, D. R. 1978. On the role of the different fibre types in fish myotomes at intermediate swimming speeds. *Fish. Bull.* 76: 691–699.

———, N. B. Marshall, and J. H. S. Blaxter. 1999. *Biology of fishes* (2nd ed.). Stanley Thornes, Cheltenham, UK. (Reprinted 2nd ed., Chapman and Hall, 1995).

Breder, C. M. 1926. The locomotion of fishes. *Zoologica (N.Y.)* 4: 159–297.

Brill, R. W. 1996. Selective advantages conferred by the high performance physiology of tunas, billfishes, and dolphin fish. *Comp. Biochem. Physiol.* 113A(1): 3–15.

Carey, F. G., and Q. H. Gibson. 1983. Heat and oxygen exchange in the rete mirabile of the bluefin tuna, *Thunnus thynnus. Comp. Biochem. Physiol.* 74A(2): 333–342.

Clarke, W. D., and Haedrich, R. L. 1968. Dive 218, pp. 14–15. In *Gulfview—Diving Log*, May 27–June 12, 1967, R. D. Gaul and W. D. Clarke (Eds.). Gulf Universities Research Corp., Ocean Springs, MS.

———, and W. G. Pearcy. 1968. Dive 217, pp. 10–14. In *Gulfview—Diving Log*, May 27–June 12, 1967, R. D. Gaul and W. D. Clarke (Eds.). Gulf Universities Research Corp., Ocean Springs, MS.

———, and Rosenblatt, R. H. 1968. Dive 219, pp. 16–17. In *Gulfview—Diving Log*, May 27–June 12, 1967, R. D. Gaul and W. D. Clarke (Eds.). Gulf Universities Research Corp., Ocean Springs, MS.

Coughlin, D. J., and L. C. Rome. 1996. The roles of pink and red muscle in powering steady swimming in scup, *Stenotomus chrysops. Amer. Zool.* 36: 666–677.

D'Aout, K., and P. Aerts. 1999. A kinematic comparison of forward and backward swimming in the eel *Anguilla anguilla. J. Exp. Biol.* 202: 1511–1521.

Dickson, K. 1996. Rapid rates of growth, metabolism, and swimming in juvenile tunas, pp. 115–119. In *High performance fish II: Symposium proceedings*, International Congress on Biology of Fishes, San Francisco State University, July 14–18, 1996. American Fisheries Society, Bethesda, MD.

Donley, J. M., C. A. Sepulveda, P. Konstantinidis, S. Gemballa, and R. E. Shadwick. 2004. Convergent evolution in mechanical design of lamnid sharks and tunas. *Nature* 429: 61–65.

Ellerby, D. J., I. L. Spierts, and J. D. Altringham. 2001a. Slow muscle power output of yellow- and silver-phase European eels (*Anguilla anguilla* L.): Changes in muscle performance prior to migration. *J. Exp. Biol.* 204: 1369–1379.

———, ———, and———. 2001b. Fast muscle function in the European eel (*Anguilla anguilla* L.) during aquatic and terrestrial locomotion. *J. Exp. Biol.* 204: 2231–2238.

Fierstine, H. L., and Walters, V. 1968. Studies in locomotion and anatomy of scombroid fishes. *Mem. So. Calif. Acad. Sci.* 6: 1–31.

Fish, F. E., and L. D. Shannahan. 2000. The role of the pectoral fins in body trim of sharks. *J. Fish Biol.* 56: 1062–1073.

Fricke, H., O. Reinicke, H. Hofer, and W. Nachtigall. 1987. Locomotion of the coelacanth *Latimeria chalumnae* in its natural habitat. *Nature* 329: 331–333.

Frith, H. R., and R. W. Blake. 1991. Mechanics of the startle response in the northern pike, *Esox lucius. Can. J. Zool.* 69: 2831–2839.

———, and———. 1995. The mechanical power output and hydromechanical efficiency of northern pike (*Esox lucius*) fast-starts. *J. Exp. Biol.* 198: 1863–1873.

Gee, J. H. 1968. Adjustment of buoyancy by longnose dace (*Rhinichthys cataractae*) in relation to velocity of water. *J. Fish. Res. Bd. Can.* 25: 1485–1496.

———. 1972. Adaptive variation in swimbladder length and volume in dace, genus *Rhinichthys. J. Fish. Res. Bd. Can.* 29: 119–127.

———. 1974. Behavioral and developmental plasticity of buoyancy in the longnose, *Rhinichthys cataractae*, and blacknose, *R. atratulus* (Cyprinidae) dace. *J. Fish. Res. Bd. Can.* 31: 35–41.

Gemballa, S., L. Ebmeyer, K. Hagen, T. Hannich, K. Hoja, M. Rolf, K. Treiber, F. Vogel, and G. Weitbrecht. 2003. Evolutionary transformations of myoseptal tendons in gnathostomes. *Proc. Roy. Soc. Lond. B* 270: 1229–1235.

Gibb, A. C., K. A. Dickson, and G. V. Lauder. 1999. Tail kinematics of the chub mackerel *Scomber japonicus:* Testing the homocercal tail model of fish propulsion. *J. Exp. Biol.* 202: 2433–2447.

Gill, H. S., A. H. Weatherly, and T. Bhesania. 1982. Histochemical characterization of myotomal muscle in the bluntnose minnow, *Pimephales notatus* Rafinesque. *J. Fish. Biol.* 21: 205–214.

———, ———, R. Lee, and D. Legere. 1989. Histochemical characterization of myotomal muscle of five teleost species. *J. Fish. Biol.* 34: 375–386.

Gillis, G. B. 1996. Undulatory locomotion in elongate aquatic vertebrates: Anguilliform swimming since Sir James Gray. *Amer. Zool.* 36: 656–665.

———. 1998. Neuromuscular control of anguilliform locomotion: Patterns of red and white muscle activity during swimming in the American eel *Anguilla rostrata. J. Exp. Biol.* 201: 3245–3256.

———. 2000. Patterns of white muscle activity during terrestrial locomotion in the American eel (*Anguilla rostrata*). *J. Exp. Biol.* 203: 471–480.

Goolish, E. M., and K. Okutake. 1999. Lack of gas bladder inflation by the larvae of zebrafish in the absence of an air–water interface. *J. Fish Biol.* 55: 1054–1063.

Gordon, M. S., I. Plaut, and D. Kim. 1996. How puffers (Teleostei: Tetraodontidae) swim. *J. Fish Biol.* 49: 319–328.

Greene, C. W., and C. H. Greene. 1914. The skeletal musculature of the king salmon. *Bur. Comm. Fish. Doc. 796* (Republished 1915 in *Bur. Comm. Fish Vol. 33* for 1913).

Greenwood, P. H., and Thompson, K. S. 1960. The pectoral anatomy of *Pantodon buchholzi* Peters (a freshwater flying fish) and the related Osteoglossidae. *Proc. Zool. Soc. Lond.* 135: 283–301.

Greer-Walker, M. and Pull, G. A. 1975. A survey of red and white muscle in marine fish. *J. Fish Biol.* 7: 295–300.

Hadfield, P. 1999. The fish you never have to feed. *New Scientist 162:* 17.

Hale, M. E. 2002. S-start and C-start escape responses of the muskellunge (*Esox masquinongy*) require alternative neuromotor mechanisms. *J. Exp. Biol. 205:* 2005–2016.

Hove, J. R., L. M. O'Bryan, M. S. Gordon, P. W. Webb, and D. Weihs. 2001. Boxfishes (Teleostei: Ostraciidae) as a model system for fishes swimming with many fins: Kinematics. *J. Exp. Biol. 204:* 1459–1471.

Howe, K. M., D. L. Stein, and C. E. Bond. 1980. First records off Oregon of the pelagic fishes *Paralepis atlantica, Gonostoma atlanticum,* and *Aphanopus carbo,* with notes on the anatomy of *Aphanopus carbo. Fish. Bull. 77:* 700–703.

Hudson, R. C. L. 1969. Polyneural innervation of the fast muscles of the marine teleost *Cottus scorpius* L. *J. Exp. Biol. 50:* 47–67.

Hunter, J. R. 1971. Sustained speed of jack mackerel, *Trachurus symmetricus. Fish. Bull.* 69(2): 267–271.

———, and J. R. Zweifel. 1971. Swimming speed, tail beat frequency, tail beat amplitude, and size in jack mackerel, *Trachurus symmetricus,* and other fishes. *Fish. Bull. 69:* 253–267.

Johnston, I. A. 1981. Structure and function of fish muscles, pp. 71–113. In Vertebrate locomotion, M. H. Day (Ed.). *Symp. Zool. Soc. Lond. 48.* Academic Press, New York.

———. 1983. Dynamic properties of fish muscle, pp. 36–67. In *Fish biomechanics,* P. W. Webb and D. Weihs (Eds). Praeger, New York.

Jones, D. R., J. W. Kiceniuk, and O. S. Bamford. 1974. Evaluation of the swimming performance of several fish species from the MacKenzie River. *J. Fish. Res. Bd. Can. 31:* 1641–1647.

Kardong, K. V. 2002. *Vertebrates: Comparative anatomy, function, evolution* (3rd ed.). McGraw-Hill, Boston.

Katz, S. L., D. A. Syme, and R. E. Shadwick. 2001. Enhanced power in yellowfin tuna. *Nature 410:* 770–771.

Knower, T., R. E. Shadwick, S. L. Katz, J. B. Graham, and C. S. Wardle. 1999. Red muscle activation patterns in yellowfin (*Thunnus albacares*) and skipjack (*Katsuwonus pelamis*) tunas during steady swimming. *J. Exp. Biol. 202:* 2127–2138.

Koester, D. M., and C. P. Spirito. 2003. Punting: An unusual mode of locomotion in the little skate *Leucoraja erinacea* (Chondrichthyes: Rajidae). *Copeia 2003:* 553–561.

Korsmeyer, K. E., J. F. Steffensen, and J. Herskin. 2002. Energetics of median and paired fin swimming, body and caudal fin swimming, and gait transition in parrotfish (*Scarus schlegeli*) and triggerfish (*Rhinecanthus aculeatus*). *J. Exp. Biol. 205:* 1253–1263.

Lauder, G. V. 1982. Structure and function in the tail of the pumpkinseed sunfish (*Lepomis gibbosus*). *J. Zool. (Lond.) 197:* 483–495.

———. 1989. Caudal fin locomotion in ray-finned fishes: Historical and functional analysis. *Amer. Zool. 29:* 85–102.

———. 2000. Function of the caudal fin during locomotion in fishes: Kinematics, flow visualization, and evolutionary patterns. *Amer. Zool. 40:* 101–122.

Leis, J. M., and M. I. McCormick. 2002. The biology, behavior, and ecology of the pelagic, larval stage of coral reef fishes, pp. 171–199. In *Coral reef fishes: Dynamics and diversity in a complex ecosystem,* P. F. Sale (Ed.). Academic Press, San Diego.

Liao, J. C. 2002. Swimming in needlefish (Belonidae): Anguilliform locomotion with fins. *J. Exp. Biol. 205:* 2875–2884.

Lighthill, M. J., and R. Blake. 1990. Biofluid dynamics of balistiform and gymnotiform locomotion. Part I. Biological background and analysis of elongated body theory. *J. Fluid Mech. 212:* 183–207.

Lindsey, C. C. 1978. Form, function, and locomotory habits in fish, pp. 1–100. In *Fish physiology,* Vol. VII, W. S. Hoar and D. J. Randall (Eds.). Academic Press, New York.

Long, J. H., Jr., M. Koob-Emunds, B. Sinwell, and T. J. Koob. 2002. The notochord of hagfish *Myxine glutinosa:* Visco-elastic properties and mechanical functions during steady swimming. *J. Exp. Biol. 205:* 3819–3831.

Manooch, C. S., III. 1984. *Fisherman's guide: Fishes of the southeastern United States.* North Carolina State Museum of Natural History, Raleigh.

Marshall, N. B. 1979. *Developments in deep-sea biology.* Blandford Press, Poole, Dorset, UK.

Mathieu-Costello, O., R. W. Brill, and P. W. Hochachka. 1996. Structural basis for oxygen delivery: Muscle capillaries and manifolds in tuna red muscle. *Comp. Biochem. Physiol. 113A*(1): 25–31.

Mommsen, T. P., J. Ballantyne, D. MacDonald, J. Gosline, and P. W. Hochachka. 1981. Analogues of red and white muscle in squid mantle. *Proc. Nat. Acad. Sci. USA 78:* 3274–3278.

Motani, R., H. You, and C. McGowan. 1996. Eel-like swimming in the earliest ichthyosaurs. *Nature 382:* 347–348.

Myers, G. S. 1950. Flying of the half beak, *Euleptorhamphus. Copeia 1950:* 320.

Nakaya, K. 1995. Hydrodynamic function of the head in the hammerhead sharks (Elasmobranchii: Sphyrnidae). *Copeia 1995:* 330–336.

Nanjappa, P., L. Brand, and M. J. Lannoo. 2000. Swimming patterns associated with foraging in phylogenetically and ecologically diverse American weakly electric teleosts (Gymnotiformes). *Env. Biol. Fishes 58:* 97–104.

Packard, A. 1972. Cephalopods and fish: The limits of convergence. *Biol. Rev. 47:* 241–307.

Phleger, C. F. 1998. Buoyancy in marine fishes: Direct and indirect role of lipids. *Amer. Zool. 38:* 321–330.

———, and M. R. Grigor. 1990. Role of wax esters in determining buoyancy in *Hoplostethus atlanticus* (Beryciformes: Trachichthyidae). *Mar. Biol. 105*(2): 229–233.

Plaut, I. 2000. Effects of fin size on swimming performance, swimming behavior and routine activity of zebrafish *Danio rerio. J. Exp. Biol. 203:* 813–820.

Prem, C., and B. Pelster. 2001. Swimbladder gas gland cells cultured on permeable supports regain their characteristic polarity. *J. Exp. Biol. 204:* 4023–4029.

Quinn, T. P. 1988. Estimated swimming speeds of migrating adult sockeye salmon. *Can. J. Zool. 66:* 2160–2163.

Rayner, J. M. V. 1981. Flight adaptations in vertebrates. pp. 137–172. In Vertebrate Locomotion, M. H. Day (Ed.). *Symp. Zool. Soc. Lond. 48.* Academic Press, New York.

Robins, C. R. 1974. Billfish biology: Facts for the fisherman. Addendum to The International Marine Angler 36(5): 1–4.

Rodhouse, P. G., and M. G. White. 1995. Cephalopods occupy the ecological niche of epipelagic fish in the Antarctic Polar Frontal Zone. *Biol. Bull. 189:* 77–80.

Rosenberger, L. J. 2001. Pectoral fin locomotion in batoid fishes: Undulation versus oscillation. *J. Exp. Biol. 204:* 379–394.

Satchell, G. H. 1991. *Physiology and form of fish circulation.* Cambridge University Press, Cambridge, UK.

———. 1999. Circulatory system: Anatomy of the periphera circulatory system. pp. 218–237. In *Sharks, skates, and rays: The biology of elasmobranch fishes,* W. T. Hamlett (Ed.). Johns Hopkins University Press, Baltimore.

Schmidt-Nielsen, K. 1997. *Animal physiology: Adaptation and environment* (5th ed.). Cambridge University Press, Cambridge, UK.

Scholander, P. F. 1954. Secretion of gases against high pressures in the swim bladder of deep sea fishes. II. The rete mirabile. *Biol. Bull. 107:* 260–277.

———.1957. The wonderful net. *Sci. Am. 196*(4): 96–107.

———, and L. Van Dam. 1954. Secretion of gases against high pressures in the swim bladder of deep sea fishes I. Oxygen dissociation in blood. *Biol. Bull. 107:* 247–259.

Schrank, A. J., P. W. Webb, and S. Mayberry. 1999. How do body and paired-fin positions affect the ability of three teleost fishes to maneuver around bends? *Can. J. Zool. 77:* 203–210.

Shadwick, R. E., J. F. Steffensen, S. L. Katz, and T. Knower. 1998. Muscle dynamics in fish during steady swimming. *Amer. Zool. 38:* 755–770.

———, S. L. Katz, K. E. Korsmeyer, T. Knower, and J. W. Covell. 1999. Muscle dynamics in skipjack tuna: Timing of red muscle shortening in relation to activation and body curvature during steady swimming. *J. Exp. Biol. 202:* 2139–2150.

Sötz, E., H. Niederstätter, and B. Pelster. 2002. Determinants of intracellular pH in gas gland cells of the swimbladder of the European eel *Anguilla anguilla. J. Exp. Biol. 205:* 1069–1075.

Steen, J. B. 1970. The swim bladder as a hydrostatic organ, pp. 414–443. In *Fish physiology,* Vol. IV, W. S. Hoar and D. J. Randall (Eds.). Academic Press, New York.

Syme, D. A., and R. E. Shadwick. 2002. Effects of longitudinal body position and swimming speed on mechanical power of deep red muscle from skipjack tuna (*Katsuwonus pelamis*). *J. Exp. Biol. 205:* 189–200.

Videler, J. J. 1977. Mechanical properties of fish tail joints, pp. 183–194. In *Physiology of movement—biomechanics,* W. Nachtigall (Ed.). Gustav Fischer Verlag, Stuttgart.

———. 1993. *Fish swimming.* Chapman and Hall, London.

Wainwright, S. A. 1983. To bend a fish, pp. 68–90. In *Fish biomechanics,* P. W. Webb and D. Weihs (Eds.). Praeger, New York.

Walker, J. A., and M. W. Westneat. 2002. Performance limits of labriform propulsion and correlates with fin shape and motion. *J. Exp. Biol. 205:* 177–187.

Wardle, C. S., and P. He. 1988. Burst swimming speeds of mackerel, *Scomber scombrus* L. *J. Fish Biol. 32:* 471–478.

———, J. J. Videler, T. Arimoto, J. M. Franco, and P. He. 1989. The muscle twitch and the maximum swimming speed of giant bluefin tuna, *Thunnus thynnus* L. *J. Fish Biol. 35:* 129–137.

Webb, P. W. 1973. Kinematics of pectoral fin propulsion in *Cymatogaster aggregata. J. Exp. Biol. 59:* 697–710.

———. 1975. Efficiency of pectoral-fin propulsion of *Cymatogaster aggregata,* pp. 573–584. In *Swimming and flying in nature,* Vol. 2, T. Y.-T. Wu, C. J. Brokaw, and C. Brennan (Eds.), Plenum Press, New York.

———. 1978. Hydrodynamics: Nonscombroid fish, pp. 189–237. In *Fish physiology,* Vol. VII, W. S. Hoar and D. J. Randall (Eds.). Academic Press, New York.

———.1984. Form and function in fish swimming. *Sci. Am. 251*(1): 72–82.

———. 1998. Swimming, pp. 3–24. In *The physiology of fishes* (2nd ed.) D. H. Evans (Ed.). CRC Press, Boca Raton, FL.

Weber, J.-M., and F. Haman. 1996. Pathways for metabolic fuels and oxygen in high performance fish. *Comp. Biochem. Physiol 113A*(1): 33–38.

Weihs, D. 1973. Hydromechanics of fish schooling. *Nature 245:* 48–50.

———. 1989. Design features and mechanics of axial locomotion in fish. *Amer. Zool. 29:* 151–160.

———, and P. W. Webb. 1983. Optimization of locomotion, pp. 339–371. In *Fish Biomechanics,* P. W. Webb and D. Weihs (Eds.). Praeger, New York.

Westneat, M. W. 1996. Functional morphology of aquatic flight in fishes: Kinematics, electromyography, and mechanical modeling of labriform locomotion. *Amer. Zool. 36:* 582–598.

———, W. Hoese, C. A. Pell, and S. A. Wainwright. 1993. The horizontal septum: Mechanisms of force transfer in locomotion of scombrid fishes (Scombridae, Perciformes). *J. Morphol. 217*(2): 183–204.

———, and S. A. Wainwright. 2001. Mechanical design for swimming: Muscle, tendon, and bone, pp. 271–311. In *Tuna: Physiology, ecology, and evolution,* B. A. Block and E. D Stevens (Eds.). Academic Press, San Diego.

Wilga, C. D., and G. V. Lauder. 1999. Locomotion in sturgeon: Function of the pectoral fins. *J. Exp. Biol. 202:* 2413–2432.

———, and———. 2002. Function of the heterocercal tail in sharks: Quantitative wake dynamics during steady horizontal swimming and vertical maneuvering. *J. Exp. Biol. 205:* 2365–2374.

Winterbottom, R. 1974. The familial phylogeny of the Tetraodontiformes (Acantopterygii: Pisces) as evidenced by their comparative myology. *Smithson. Cont. Zool. 155.*

PART FOUR

ADAPTATION TO THE EXTERNAL ENVIRONMENT

In Chapters 3 and 30, we discuss some of the properties of water as a medium in which fishes live. It is these properties that make it somewhat difficult for us to fully appreciate the adaptive capacity of fishes: They live in an environment that is alien to us. Yet it is just that fact that makes fishes such fascinating subjects for study. Water can be viewed as the context in which fishes have evolved, adapted, and presently carry out their lives. The adaptation of fishes to their watery environment can be viewed from the perspective of the function of the array of sensory systems that detect and relay information. Water is the medium that delivers the message to fishes, so to speak. Understanding the structure and function of sensory systems that have evolved specifically to work in an aquatic environment is essential to understanding what fishes are all about. Our approach, therefore, is to consider how fishes relate to their environment by investigating the operation of the diverse array of sensory systems seen among the fishes.

Water is also the medium that delivers nutrition to fishes. Much of the sensory capacity of fishes is directed at the location and ingestion of food. We will consider feeding adaptations in this section, as it seems natural to consider this aspect of the biology of fishes as a demonstration of the adaptive use of their sensory systems. Like so many other aspects of fish biology, however, feeding cannot be understood from just a single perspective. We can fully appreciate the role that feeding plays only if we consider its morphological, behavioral, and ecological contexts.

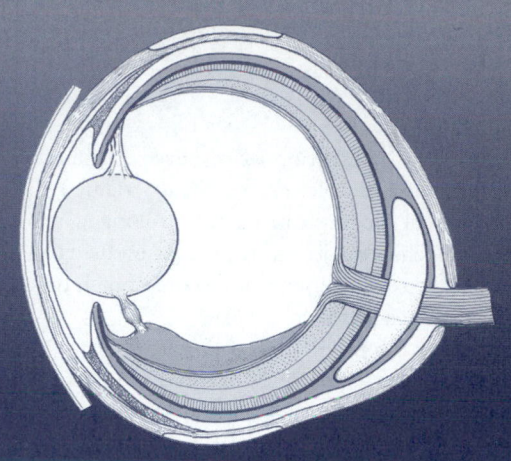

20

THE SENSORY ARSENAL OF FISHES I: VISION

INTEGRATING SENSORY BIOLOGY WITH THE ECOLOGY AND BEHAVIOR OF FISHES

GENERAL MORPHOLOGY OF FISH EYES

STRUCTURE AND FUNCTION OF FISH VISUAL SYSTEMS

Adaptation of the Eye to the Optical Properties of Water
Visual Cells and Pigments
Focus, Accommodation, and Regulation of Light
   Focus and Accommodation
   Regulation of Light

VISUAL ADAPTATION IN THE LIVES OF FISHES

Visual Adaptations of Deep-Ocean Fishes
"Dual-Purpose" Eyes in Fishes
Sensitivity to Light in the Ultraviolet Range
Sensitivity to Polarized Light
Some Alternative Photoreceptors

One only has to open one's eyes underwater to gain an appreciation for the extraordinary visual capacities displayed by fishes. The aquatic realm, which to us appears as an unfocused blur of poorly defined colors and shadows, is to fishes clear, sharp, and well resolved. The vertebrate eye is a marvel of optical engineering, and fishes have adapted it for effective vision in virtually all aquatic habitats. Light is available in the aquatic realm in a broad spectrum of wavelengths and intensities. Fishes have modified their eyes to maximize visual sensitivity in the well-lit, highly transparent waters of tropical coral reefs; in fast-flowing streams turbid with sediment; and even in the vast oceanic depths where sunlight never penetrates, but bioluminescent organisms make vision a useful adaptation. As with all adaptations, the structure and function of the fish eye is best understood in an evolutionary context. The photoreceptive structures of the earliest vertebrates were undoubtedly simple structures designed for the translation of light stimuli into neural signals. From such humble beginnings arose a diverse array of eye designs. In some environments, vision has become the primary sensory modality, whereas in others, its use is effectively supplemented by other sensory systems. For this reason, vision also needs to be investigated in the context of the entire sensory arsenal available to fishes. This chapter begins with a brief consideration of the importance of sensory biology in the study of the ecology and behavior of fishes. Our focus then shifts to the structure and function of the visual systems of fishes and their applications.

INTEGRATING SENSORY BIOLOGY
WITH THE ECOLOGY AND BEHAVIOR OF FISHES

Elsewhere in this book are chapters (Chapters 30–37) devoted to the ecology and behavior of fishes—aspects of the lives of fishes that are challenging to cover in a comprehensive fashion in the few chapters devoted to these subjects. Nowhere does fish biology benefit more from an integrative approach than in considering the sensory and neural processes that permit the environmental adaptations that are considered in the discussions of fish ecology and behavior. Although the sensory systems will be discussed as individual entities, we should always acknowledge their unified application in the context of the physically and biologically complex environments that fishes inhabit.

The role of coloration in fishes (see Chapter 18) can only be appreciated in the context of the evolution of complementary visual systems (Yokoyama and Yokoyama, 1996). Aspects of the behavior of fishes can best be understood by assessing the comparative sensory inputs resulting in a given suite of behaviors. The now classical studies on breeding in guppies and sticklebacks have affirmed the vital role that vision plays in reproductive behavior, yet we must also acknowledge the roles that minute quantities of pheromones, or body and fin vibrations during courtship, play in achieving successful reproduction. Visual detection of color or other bodily features provides information on a given individual's condition and hence affect the inclination to mate. The role of nonvisual means of detecting reproductive condition, though not as well understood, are likely to prove essential to the processes of sexual selection that govern the evolution of reproductive behaviors (Sargent et al., 1998).

Prey location, attack, and consumption rely on a thorough integration of the entire sensory arsenal of fishes, and the dominant sensory modality may shift at different points in the sequence of events associated with predation. In the muskellunge (*Esox masquinongy*), for example, vision is primarily used to locate potential prey, and information from the lateral line becomes increasingly important as the prey is approached and attacked. Individuals in which the lateral line system has been experimentally suppressed will significantly alter their mode of prey approach (New et al., 2001).

Schooling is another behavioral adaptation in which we assume vision is a prerequisite for the highly cohesive and coordinated actions associated with this essential aspect in the lives of many species of fishes. Yet studies on the schooling performance of saithe (*Pollachius virens*) have revealed that the lateral line system plays a much greater role than previously recognized (Partridge and Pitcher, 1980).

An integrated approach to the study of sensory biology also recognizes the potential for temporal variation in sensory use—especially in the acquisition of resources. Antarctic nototheniid fishes, for example, are seasonally denied the opportunity for visual feeding during the prolonged winter nights. Hydromechanical stimulus perception using the lateral line and tactile sense perception become much more important at these times (Janssen, 1996).

The reference to a sensory "arsenal" may appear a bit militaristic, yet the term is an appropriate one, given the challenges that must be met daily by fishes in order to ensure their survival. As we will see, this arsenal contains some of the most sophisticated devices available among vertebrates for the detecting and processing of vital environmental information that enables successful feeding, breeding, and other life processes. One feature that we are becoming increasingly aware of is the degree to which the individual components of this arsenal work in concert to achieve optimal adaptation to the environment.

GENERAL MORPHOLOGY OF FISH EYES

The shape and structure of the eyes of fishes vary tremendously in response to habitat requirements, yet all eyes conform to the basic vertebrate plan. The major features of the eye are an anterior chamber, an iris, a lens, and a vitreous chamber containing the vitreous humor and lined by the retina (Fig. 20.1). The entire structure is covered by the sclerotic coat, a tough covering of connective tissue that is transparent in the region where the eye is in contact with the water. The transparent section of the sclera is called the cornea. The sclera of elasmobranchs and teleosts may be stiffened by cartilaginous structures or, in the case of the latter, by scleral ossicles (see Chapter 19).

The eye is generally nearly spherical, but usually the corneal surface is flattened, so that the spherical lens is nearly in contact with the cornea. Some fishes have nonspherical lenses. Stingray lenses are flattened to the extent that their equatorial diameter exceeds their axial diameter by 18 percent. In many sharks, the equatorial diameter of the lenses exceeds the axial diameter by 12 to 16 percent. Some deep-water teleosts have slightly flattened lenses, and *Trachipterus* (Trachipteridae) has an equatorial lens diameter greater than the axial diameter (Sivak, 1990). Some highly compressed species, such as butterflyfishes, have extremely flattened eyes—to the point that they are described as disc–like (Bauchot et al., 1989). A pigmented, vascular choroid layer separates the retina and the sclera. The choroid layer is continuous with the iris and prevents the blurring of images caused by internal reflection of the retinal images back to the retinal photoreceptors (Ali and Klyne, 1985). A choroid

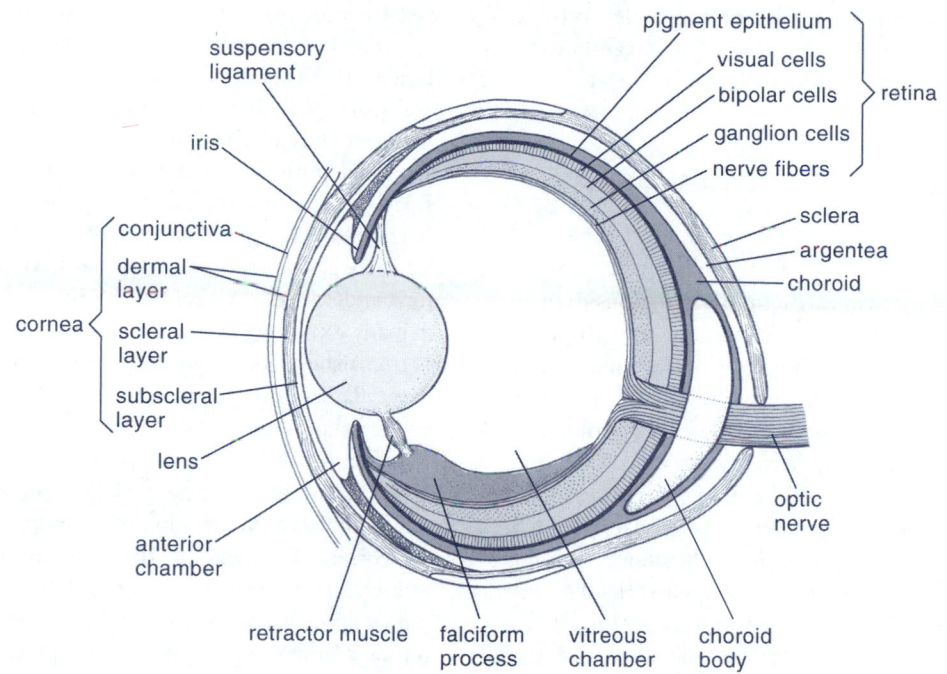

FIGURE 20.1

Vertical section through a fish eye (not drawn to scale), showing the relationships of its parts.

body or "gland"—actually a rete mirabile—is prominent in the choroid of many teleosts (see Chapter 24).

Lampreys differ from other fishes in that their eye has no **circumorbital sulcus**—the invagination of skin around the outward portion of the eyeball. Instead, the skin is continuous over the eye, forming a dermal cornea or "spectacle" separate from the scleral cornea. The lamprey eye is rather solidly attached to the rim of the eye socket; the eyeball lacks any cartilaginous or bony reinforcement, although the eye socket has some cartilage medially and ventrally (Nicol, 1989).

The slightly flattened lens of the elasmobranch eye is suspended by a band of gelatinous tissue that attaches around the equator of the lens. This suspensory material has a relatively robust dorsal portion. Some species have a ventral papilla, called the *pseudocampanule,* that aids in the suspension of the lens and may contain smooth muscle fibers that can protract the lens (Nicol, 1989).

The lens in bony fishes is suspended mainly by a dorsal ligament, but is attached by soft material all around its circumference. The retractor muscle of the teleost lens, which is ventral to the lens and is attached to the anterior end of the falciform process, is an outgrowth of the choroid that extends through a fissure in the retina (Fig. 20.1). The retina

consists of a pigmented epithelium; the visual receptor cells (*rods* and *cones*); *bipolar cells* that connect the visual cells to the *ganglion cells,* which are closest to the vitreous humor; and the nerve fibers leading to the optic nerve. *Horizontal cells* with large cell bodies make connections between visual cells; *amacrine cells,* which lack axons, form horizontal connections between ganglion cells (Hawryshyn, 1992; Wagner, 1990).

Cone photoreceptors have short outer segments that are conical, whereas rods have long, cylindrical ones. The inner segments of the cones approach the outer segments in size, whereas the inner segments of the rods are smaller than the outer segments (Fernald, 1993). Cones in teleosts differ in length, with the longest cones sensitive to long light waves and the shortest to short light waves (Fernald, 1993).

Eyelids, except for those of some elasmobranchs, are not well developed in fishes. The eyelids of most sharks move very little, but in some genera (e.g., *Ginglymostoma, Galeorhinus,* and *Cephaloscyllium*), the lids can partially or completely cover the eye (Nicol, 1989). Some sharks have a third eyelid, the **nictitating membrane**, which is attached at the anteroventral aspect of the eye and fits beneath the lower lid, from which it is developed. It can be moved upward and obliquely backward to cover and protect the surface of the eye. Coverage is complete in some carcharhinids. The so-called adipose eyelids of some sharks and some teleosts (e.g., herrings and tunas) are immobile and serve to streamline the slight bulge of the eye beyond the surface of the head.

STRUCTURE AND FUNCTION OF FISH VISUAL SYSTEMS

Adaptation of the Eye to the Optical Properties of Water

The eyes of fishes show many structural adaptations to their visual environment. Fishes live in a medium whose optical properties are much different from those of the atmosphere. Depending on the angle of incidence of light, a calm water surface can reflect 80 percent or more of the light striking it. If the water surface is agitated, there is great variation in the transmission of light, regardless of the angle of incidence. The bending of light rays entering water (approximately 48.6°) is such that a fish in water with a perfectly smooth surface views objects above the water through a circle ("Snell's window") subtended by a 97.2° cone above each eye (Fig. 20.2). Nearly all objects from horizon to horizon appear in the circle, which is surrounded by the reflective undersurface seen beyond the limits of the cone. In turbulent waters, the circular window in the surface is broken up, and light is transmitted through ever-changing patterns (Walls, 1942).

Light that enters water is absorbed rapidly but differentially, with red, for instance, attenuating rapidly and blue penetrating to greater depths. Absorption is defined by Loew and McFarland (1990) as "the change of electromagnetic energy into some other form," and light that is redirected by reflection, refraction, or diffraction is said to be *scattered*. Researchers have suggested that the spectral sensitivity of the eyes of the earliest fishes, such as the ostracoderms, may have broadened as competition forced the invasion of midwater habitats by previously benthic forms (Hawryshyn, 1998). The various species of fishes must cope with life in a variety of habitats, including the bright surface, brilliant coral reefs, dimly lit caves, sheltered forest streams, dark bogs, and ocean depths where sunlight cannot penetrate and the only light is that of bioluminescent creatures. Some even invade the land or commonly peer above the water's surface, thus having to resolve the problems of vision in the very different medium of air. It is obvious that fishes have evolved visual systems that can be of use in a diversity of habitats. Many fishes, however, live where light is minimal to nonexistent and vision is not the primary sense. Other senses, such as olfaction, mechanosensory, or electrosensory lateral line systems, are of greater importance to these fishes. Yet even in the ocean depths, where surface light does not penetrate, vision may still be significant in the lives of fishes because of bioluminescence (see Chapter 18).

The eyes of most fishes are placed on the sides of the head so that they have wide lateral fields of vision. This lateral placement does not preclude binocular vision in certain segments of the visual field. Some species (Fig. 20.3) specialize in the inspection of more restricted parts of their surroundings and, for example, have their eyes set forward (*Gigantura* and others) or upward (*Argyropelecus* and others) for binocular vision. Some species have the eyes positioned for a wider field of vision below (*Hypopthalmus*) or above, as in many topminnows that feed from the surface and in many bottom fishes that have no need of vision below. Some bottom-living flatfishes and sand stargazers (Dactyloscopidae) have eyes on short stalks (Fig. 20.3E).

In most fishes, the reception of a large field of view is achieved by the placement of the spherical lens so that it bulges through the opening of the pupil and nearly touches the cornea. The lens can thus gather light from nearly all of each lateral field. The light-gathering ability of fish eyes can be appreciated by considering the relationships of their physical dimensions to optical properties. One of these is **Matthiessen's ratio**, which is the focal length divided by the lens radius. This ratio commonly ranges in various species from about 2.17 to 2.56, with the ratio for a given species being fairly constant. Drawing an analogy with a single lens reflex (SLR) camera, the lens aperture is expressed as the f number (Matthiessens's ratio divided by two; the lower the number, the greater the aperture diameter). In the eyes of fishes, this is about 1.1 to 1.3—comparable to a fast camera lens; a short focal length allows for sufficient depth of field.

The cornea of teleosts is usually made up of four layers (Lythgoe, 1975a). Externally, there is a multicellular epithelial layer, and inside that there is a collagenous stroma that is separated from the inner layer (*endothelium*) by the thin *Descemet's membrane*. The cornea has a refractive index approximating that of water, whereas the lens is constructed to have an effective refractive index of about 1.67 (Munz, 1971). Fish lenses studied by Fernald and Wright (1983) proved to have a gradient of refractive index, with the less compact outer cells showing a refractive index of 1.38 and the cells near the center having an index of 1.56. These authors pointed out that because light curves in a spherical lens, the gradient of refractive index makes a sharp focus possible.

Visual Cells and Pigments

Visual cells of fishes include rods, single cones, and double cones (including "identical twin" cones), with subtypes in various species (some have both long and short single

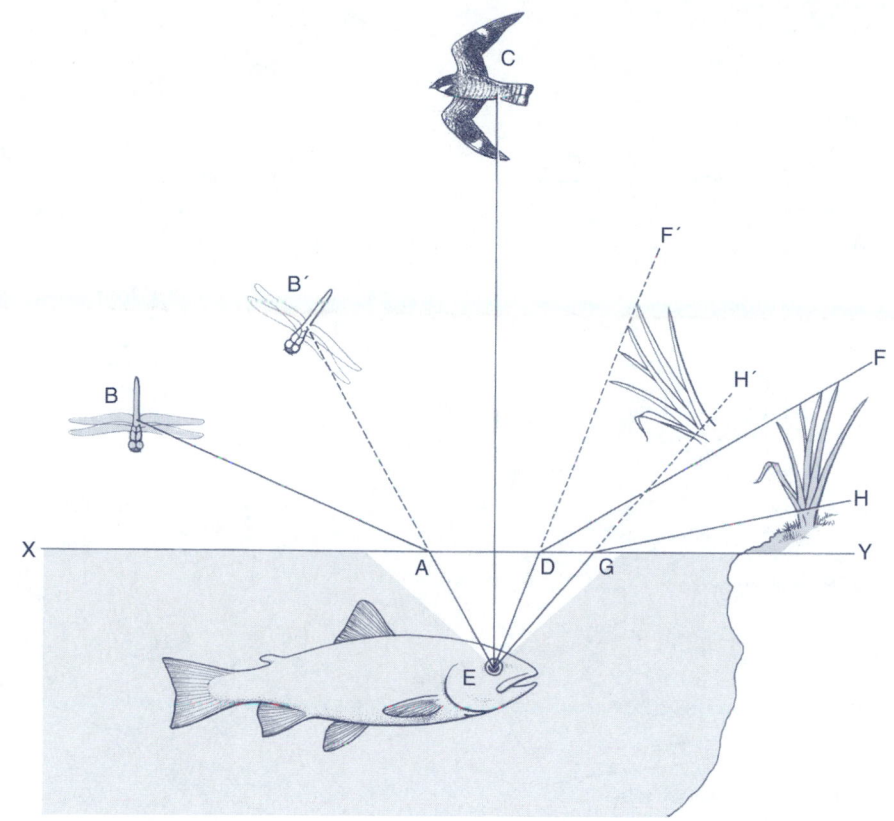

FIGURE 20.2
Visual field of a fish in relationship to the refraction of light entering with a perfectly flat surface (XY). Because of the bending of the light rays, the fish's eye (E) does not receive light striking the surface above the shaded area. The bird at C, directly above E, is seen in its actual position. The insect at B is perceived as if at B', and the angles EDF and EGH cause the plant to be seen as if the top were at F' and the bottom at H'.

cones). Molecular phylogenetic analyses of visual pigments (*opsin*—see later) enabling color vision in vertebrates indicate that five distinct genetic lineages arose well before the teleosts diverged from other vertebrate lineages; these lineages are evident in fishes as primitive as the lampreys (Bowmaker, 1998; Collin and Trezise, 2004). Color vision, which requires at least two classes of cones, each sensitive to different portions of the electromagnetic spectrum, is well documented among fishes (Hawryshyn, 1998). The cones in some species can respond to blue, green, or red, yet comparatively few vertebrates possess fully trichromatic color vision. This variety indicates that certain cells are specialized to respond to specific wavelengths and light intensities. The discrimination of wavelengths has been demonstrated in the goldfish by behavioral training (Neumeyer, 1986). Hawryshyn and McFarland (1987) used conditioned response to demonstrate that cones sensitive to ultraviolet (see

later), green, and red wavelengths showed evidence of sensitivity to polarized light (see later), whereas blue-sensitive cones showed no such sensitivity. The presence of pigment that absorbs ultraviolet light has been shown in the Japanese dace (*Tribolodon*) by microspectrophotometric analysis (Harosi and Hashimoto, 1983).

Although rods dominate in the retinas of deep-dwelling fishes, cones are known in deepwater members of the following families: Chlorophthalmidae, Dirctmidae, Agonidae, Omosudidae, Notosudidae, Scorpaenidae, and Zoarcidae. However, only double cones are present in these families. Ontogenetic changes in the number and type of cones have been demonstrated in the splitnose rockfish (*Sebastes diploproa*), a species that has shallow-living larvae and small juveniles that stay near the surface for about a year prior to moving to deep water (Boehlert, 1978, 1979). As this species grows and the individuals migrate to deep water, and their vision adapts to

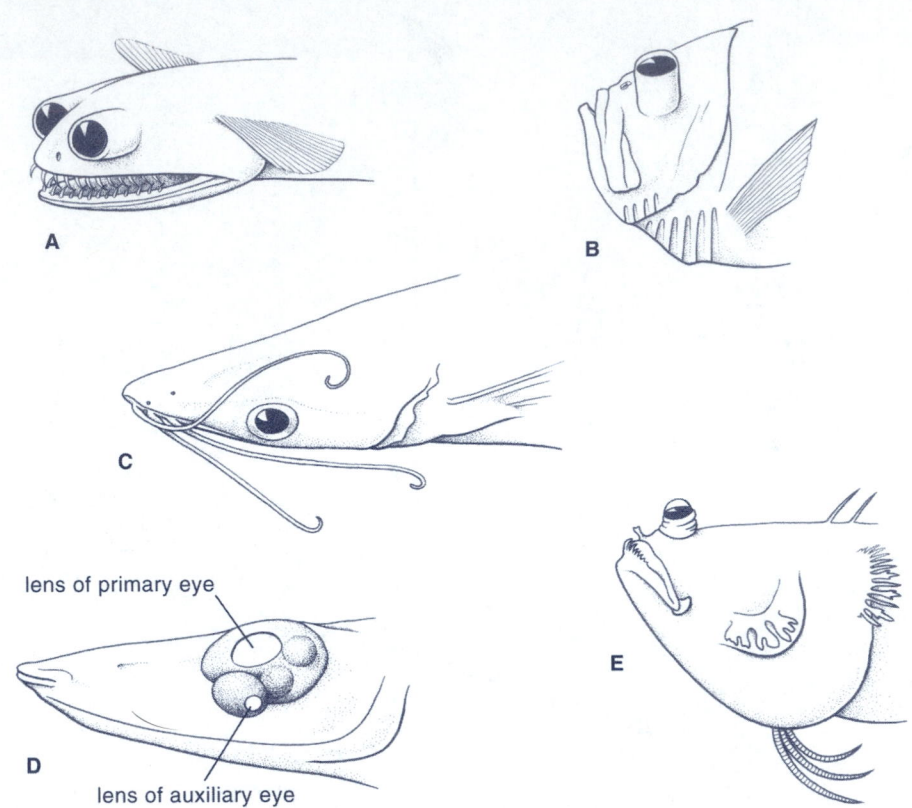

lens of primary eye

lens of auxiliary eye

FIGURE 20.3
Examples of fishes with unusual eyes: **A**, *Gigantura*; **B**, *Argyropelecus*; **C**, *Hypophthalmus* (eyes directed obliquely downward); **D**, *Bathylychnops*; **E**, *Dactyloscopus*.

a dimmer environment, cones decrease in density and rods increase. Single cones disappear, apparently by fusion into double cones.

Cones are set in regular patterns in many species and are often arranged in a mosaic of squares (Wagner, 1990). Other patterns may approximate rows, circles, crosses, or triangles. This regular arrangement ensures that the most important parts of the retina are supplied with each type of cell specialized for the various wavelengths of importance to the vision of the species (Lythgoe, 1979). In the surface-dwelling guppies (*Poecilia reticulata*), the lower part of the retina is dominated by green-sensitive cones, best suited for detecting dark objects against light coming from above, whereas the upper part of the retina is rich in red-, green-, and blue-sensitive cones to maximize color discrimination of objects in the water. When male guppies are courting, they always present in front of or below the female (Levine and MacNichol, 1982).

Rods function in dim light and outnumber the cones in fishes with duplex retinas (i.e., retinas with both rods and cones). Many elasmobranchs and deep-sea teleosts have either pure-rod retinas or retinas with only a few cones. Several elasmobranch species (representing about 10 families) that have duplex retinas show rod–cone ratios of 5:1 to 100:1, with most between 7:1 and 13:1. In species that commonly feed in dim light, numerous rods may connect with only a few bipolar cells, which in turn connect to a single ganglion cell. This convergent adaptation aids in summing the visual input and increases the retina's sensitivity in subdued light. Several groups of fishes, including lampreys, cusk eels, congers, and various deep-sea families, have two to several layers of visual cells (multiple-bank retinas) that could increase the efficiency of light interception. In some species, there are more layers of cells at the fovea than in other parts of the retina. In *Chauliodus*, the number of retinal layers increases with the size of the fish, and this is correlated with shifts to a deeper habitat (Locket, 1980). The number of rods in nocturnal or deep-sea species can reach up to 20 million per mm^2. Some of these fishes have retinas that absorb 90 percent or more of the light striking them, so the threshold at which they can detect light is probably lower than that of humans, in which the retina absorbs about 30 percent of the blue-green light striking it.

The photosensitive pigments in visual cells consist of an apoprotein (*opsin*) combined with a light–sensitive *chromophore* (Hawryshyn, 1998). Two basic types are seen in fishes—**rhodopsin,** based on retinal from vitamin A$_1$, and **porphyropsin,** based on an aldehyde of vitamin A$_2$ (3-dehydroretinal; Bowmaker, 1990; Munz, 1971). In the case of rhodopsin, for example, the absorption of photons initiates a complex biochemical sequence that converts rhodopsin into retinal + opsin. This conversion has been termed "bleaching," as the free retinal and opsin do not absorb light in the visible spectrum (Withers, 1992). This conversion changes the membrane properties of the photoreceptor cells, and this promotes the depolarization of the associated neurons, relaying a signal to the brain that results in the sensation of vision. Given the diversity of color environments in which vertebrates are found, it is not surprising that a diverse array of opsin gene sequences have evolved to best accommodate the visual capacities of animals to their respective environments (Yokoyama and Yokoyama, 1996).

Freshwater fishes generally have mainly porphyropsin in their retinas, and marine fishes have mainly rhodopsin, but many freshwater species and a few marine species have both. Elasmobranchs have rhodopsin almost exclusively, as do many cyprinids and a few other teleosts. Changes in pigments and pigment ratios have been documented in several species in response to the quality of seasonal light, temperature, hormone level, life history stage, and diet (Beatty, 1984; Bowmaker and Kunz, 1987; Muntz and Mouat, 1984). For instance, some species of anadromous fishes—for example, *Petromyzon marinus, Morone americana* (Ali and Klyne, 1985), and *Oncorhynchus* (Muntz, 1971)—may have a preponderance of one pigment over the other at certain life history stages. Lampreys change from porphyropsin in their freshwater phase to rhodopsin in the ocean, and Pacific salmon are known to change from rhodopsin to porphyropsin on their spawning migration. Among freshwater fishes, there is much variation in the proportions of the two pigments, and these proportions vary in given species with age, season, and temperature (Beatty, 1984). An example of ontogenetic change in spectral sensitivity is seen in young pollock. As they move from shallow to deep water at a length of about 50 mm, a progressive change in the maximum absorbance of light (λ_{max}) from the violet to the blue range takes place (Shand et al., 1988).

Most fishes are thought to have color vision. Behavioral experiments have demonstrated color sensitivity (Douglas and Hawryshyn, 1990), and Ali and Klyne (1985) claimed that colorblindness in fishes has yet to be demonstrated. Bowmaker et al. (1988) claimed that fishes of the genus *Malacosteus* have the potential for color vision even though their retinas possess only rods. This may extend to other deep-sea genera with paired visual pigments (Bowmaker, 1990). Pigment pairs arise because the same opsin can form two different pigments, depending on whether it binds with a retinal from vitamin A$_1$ or A$_2$.

Visual pigments of fishes respond to light of a considerable range of wavelengths. Maximum absorbance (λ_{max}) ranges from about 360 nm (UV; Harosi and Hashimoto, 1983) to 625 nm (red; Levine and MacNichol, 1979). The actual spectral sensitivity, however, may be quite broad. The Atlantic cod (*Gadus morhua*), for example, has double cones with λ_{max} of 517 nm and single cones with λ_{max} of 446 nm, but has a range of spectral sensitivity from about 400 nm to more than 650 nm (Bowmaker, 1990). From the standpoint of the absorbance of wavelengths, Bowmaker (1990) considered marine fishes to fall into three broad habitat categories—deep ocean, coastal, and tropical coral reef (Table 20.1).

The usual maximum absorption shown by the visual pigments of marine species ranges from 477 to 522 nm (Munz, 1971). Deep-sea species (Category A) tend to have pigments that maximize vision in a dimly lighted, blue environment. Some of these apparently have porphyropsin along with rhodopsin. Most of the deep-dwelling species studied by Partridge et al. (1988) had pigments ranging in maximum absorption from 475 to 488 nm. Two species had two visual pigments each, one with λ_{max} of 466 and 500 nm, the other with 478 and 485 nm. Levine and MacNichol (1979) pointed out that fishes living in the same general environment will have different spectral absorbance values depending on microhabitat, niche, and other factors. Species from shallow waters (Categories B and C), however, tend toward maximum absorbance at shorter wavelengths than those of benthic or crepuscular species.

Most freshwater species have pigments with maximum wavelength absorptions (λ_{max}) from 498 to 535 nm, but the portion of that range differs by species, habitat, food habits, and age. Levine and MacNichol (1979) proposed four overlapping visual categories of freshwater fishes, in which visual sensitivity reflects habitat, feeding behavior, and time of activity (Table 20.2).

Among the Group I fishes (see Table 20.2), the four-eyed fish (*Anableps anableps*) has a most unusual visual system that simultaneously accommodates aquatic and terrestrial vision (see later). This fish has single cones with λ_{max} at about 409 nm, but has two types of double cones. The

TABLE 20.1 VISUAL CATEGORIES OF MARINE FISHES GROUPED ACCORDING TO HABITAT (FROM BOWMAKER, 1990)

Category	Habitat	Spectral sensitivity
A	Deep ocean	Visual pigments adapted to dim blue light; some red sensitivity to bioluminescent emissions; some with yellow lenses to enhance sensitivity to photophore emissions
B	Coastal	Sensitivity to abundant light in the blue-green range; λ_{max} = 440–460 nm (violet to blue) and 520–540 nm (green)
C	Tropical coral reef	Strong blue light; λ_{max} of 495 nm (blue-green) and 513–530 nm (green); many species with yellow corneas—facilitates attenuation of strong blue light

components of one type of double cone possess dissimilar λ_{max} values of about 463 and 576 nm. In the other type of double cone, both components have the same λ_{max} value of 576 nm. The absence of blue-sensitive pigments in Group III fishes may enhance visual acuity and hence prey detection in conditions of low light (Bowmaker, 1990).

Hawryshyn (1992) pointed out that many fishes, including carp, have four types of cones—short cones, with a peak sensitivity at about 460 nm; middle cones, with λ_{max} at about 530 nm; long cones, with λ_{max} at about 600 nm; and ultraviolet cones, with λ_{max} at around 380 nm. There is considerable overlap in sensitivity, with long cones absorbing relatively well in the UV range. Cones sensitive to ultraviolet showed a greater response to vertically polarized light than the middle and long cones, which were better at sensing horizontal polarization (Hawryshyn, 1992).

Focus, Accommodation, and Regulation of Light

Focus and Accommodation

Several species of fishes cannot focus images sharply because of *spherical aberration* in their lenses (Kreuzer and Sivak, 1984). Light transmitted through the periphery of the lens tends to be focused at a slightly different focal point from the light transmitted through the center of the lens, thereby creating a blurred image. One might expect that the lenses of visually oriented predators, such as trout and pike, might show less spherical aberration than that observed in fishes that are more dependent on chemical senses, such as bullhead catfishes and carps.

Chromatic aberration, resulting from light of different wavelengths focusing at different distances from the lens, is small in fishes because of the gradient of the refractive index in the lens, ranging from about 2 percent of the focal length in cichlids to 5.3 percent in the goldfish, with most fishes studied at 4 to 5 percent. This aberration is probably not large enough to cause serious problems in vision (Fernald, 1990; Sivak, 1990).

Lampreys differ from other fishes in the mechanism for accommodation to near and distant vision. The lens is not suspended from the interior of the eyeball, as in other fishes, but rather is held in place by the pressure of the fluid in the vitreous cavity. In accommodating for distant vision, the lens is forced back by the contraction of a muscle that flattens the cornea.

Although elasmobranchs are said to accommodate by moving the lens forward, Somiya and Tamura (1973) noted no lens movement or deformation of the eyeball in the four elasmobranchs they studied. Sivak (1990) cited work showing accommodative ability in two sharks and a stingray. In some fishes, including skates, some rays, mudskippers, and some flatfishes, there is a type of inactive accommodation (Ali and Klyne, 1985). The shape of the eyeball diverges considerably from the spherical, placing the lower part of the retina closer to the lens, so that distant objects above the level of the eye are in focus and close objects at the level of the eye or below it are also in focus. This "ramp retina" allows the fish to accommodate, if necessary, by changing the position of the head. Many elasmobranchs can reduce the pupil to a very small aperture (or two or more separated small apertures) when in bright light. This probably creates an effect similar to a

TABLE 20.2   VISUAL CATEGORIES OF FRESHWATER FISHES GROUPED ACCORDING TO HABITAT, FEEDING BEHAVIOR, AND TIME OF
ACTIVITY (AFTER LEVINE AND MACNICHOL, 1979)

Group number	Habitat	Spectral sensitivity and cone characteristics
I	Shallow water, diurnal	Single cones with maximal absorption down to λ of 410 nm; double cones with maximal absorbance up to 580 nm (yellow)
II	Diurnal midwater generalists	Eyes sensitive to blue and green; single cones with λ_{max} around 460 nm (blue); double cones with λ_{max} at 540 nm (green); other red-sensitive double cones ($\lambda_{max} = 580-629$ nm); some with small single cones that absorb in the UV range (355–390 nm)
III	Crepuscular midwater predators	Green-sensitive (500–540 nm) single cones; red-sensitive double cones
IV	Crepuscular and nocturnal benthic	Rods present, absorbing maximally at 540 nm; two types of single cones— one with λ_{max} around 530–540 nm, the other at more than 600 nm

pinhole camera, giving reasonably good focus to both close and far objects. In the barrel-shaped eyes of certain deep-sea species, the immobile lens does not allow for accommodation, but the retina is specialized by division into two parts, so that objects at two different distances can be seen in focus.

Although the lens of the typical teleost eye is spherical, or nearly so, the vitreous chamber is not, giving the retina an ellipsoid shape (see Fig. 20.1). As a result, relatively distant objects that are lateral to the fish are in focus, and objects close to the fish are not. Objects that are close and right in front of the fish in the binocular field are in better focus than more distant objects. Accommodation by teleosts to distant vision in the anterior field is accomplished by moving the lens posteriorly by means of the retractor lentis muscle. The accommodative movements of teleosts are generally nasal–temporal (Fernald and Wright, 1985), but some species move the lens slightly toward the back of the eye.

Fishes differ in their ability to accommodate. Most teleosts have been found to have well-developed, triangular lens muscles that enable effective accommodation. Although the members of a few freshwater fish families, such as the centrarchids and channids, have well-developed musculature to permit accommodation, eye muscles are less developed in most freshwater fishes, resulting in only moderate powers

of accommodation. The lens muscle appears to be of small diameter in many freshwater species, and in some species with poor accommodation (*Anguilla,* some catfishes) the muscle is not highly developed. Among marine species, *Mugil cephalus* was noted as having a lens muscle with negligible function, and a few other species showed little or no accommodation (Somiya and Tamura, 1973).

Generally, elasmobranchs have been thought to be *hypermetropic* (farsighted) and teleosts *myopic* (nearsighted) because of their respective mechanisms for accommodation. Retinoscopy (study of the refraction of the eye) by several investigators has demonstrated that many teleosts are apparently slightly *emmetropic* (farsighted), having relatively small refractive errors. However, because some researchers have measured to the nearest surface of the retina and some have measured through to the back, there is still doubt about the degree of significance of this error. Judging from the distribution and architecture of visual cells in the retina, the most acute vision is aimed at anterior objects. In fishes with a duplex retina, cones are most numerous in a retinal area in the posterior (temporal) part of the eye, where images from the anterior field of view are focused. In deep-sea fishes with rod-only retinas, the rods of the posterior area of the retina have elongated, light-sensitive sections.

Several species of teleosts, including shallow- and deep-water marine species and some freshwater fishes, have been shown to possess a **fovea**—a small depression in the posterior part of the retina at the site of the greatest concentration of visual cells. The fovea, depending on the species, may be shallow and saucer shaped, or deeper with the shape of a trumpet bell. The shallow shape is said to aid in maintaining a visual fix, and the deeper type can refract the light through the convex sides to a great number of visual cells and, because of its optical properties, may aid the fish in judging distance in monocular fields (Harkness and Bennet-Clark, 1978). Unusual, steep-sided foveas with up to 28 superimposed banks of rods are seen in the alepocephalid *Bajacalifornia drakei*. These foveas appear to be specifically adapted for the precise location of bioluminescent prey (Montgomery and Pankhurst, 1997).

In many teleosts, the iris is shaped to allow the greatest possible oblique view forward (Fig. 20.4), even to the point of providing an opening large enough to let light coming from the lateral field strike the anterior part of the retina without passing through the lens (Munk and Frederickson, 1974; Munz, 1971; Walls, 1942). Again referring to our SLR camera analogy, the eyes of fishes function at such low *f* stops that the iris cannot keep aberrant light from reaching the retina, and some species have irises with diameters larger than the diameter of the lens (Fernald, 1993). The effect of these **aphakic apertures** on vision is still under investigation. However, in deep-sea fishes, in which aphakic apertures commonly occur around the lens, these **apertures** are thought to increase the illumination of the central part of the retina, especially if light arises from obliquely placed sources under conditions near the threshold of vision. If the lensless aperture is forward of the lens, the posterior part of the retina receives the extra illumination. This appears to be an optimal design for fishes with visual systems dedicated almost entirely to the detection of bioluminescence (Montgomery and Pankhurst, 1997).

Regulation of Light

Eyes in vertebrates are distinguished as being either *photopic* or *scotopic*. **Photopic** eyes are designed to operate at high light intensity, with a maximum of visual acuity and color vision. **Scotopic** eyes operate at lower light levels, as in deep-dwelling or nocturnal fishes, and minimize the threshold of detectable light at the expense of acuity and color perception (Withers, 1992). Photopic eyes are rich in cones, whereas scotopic eyes have a greater concentration of rods for vision at low light levels. Tiny oil droplets are sometimes incorporated into the cones to improve the channeling of light to the visual pigments.

The regulation of light as it enters or after it enters the eye and the adaptation to light or dark is accomplished by several means: (1) Retinomotor mechanisms can move pigment or visual cells to regulate photoreception; (2) contractile irises can reduce the amount of light admitted; (3) pigment in the cornea, iris, or lens can filter light; (4) a reflective tapetum lucidum in the choroid or retina can reflect light back through the visual cells; and (5) from a behavioral standpoint, the fish can swim to or away from the source of light.

Retinomotor Mechanisms

Light and dark adaptation in many teleosts is accomplished by retinomotor movements of pigment and visual cells (Fig. 20.5). Pigment cells in the outer layer of the retina con-

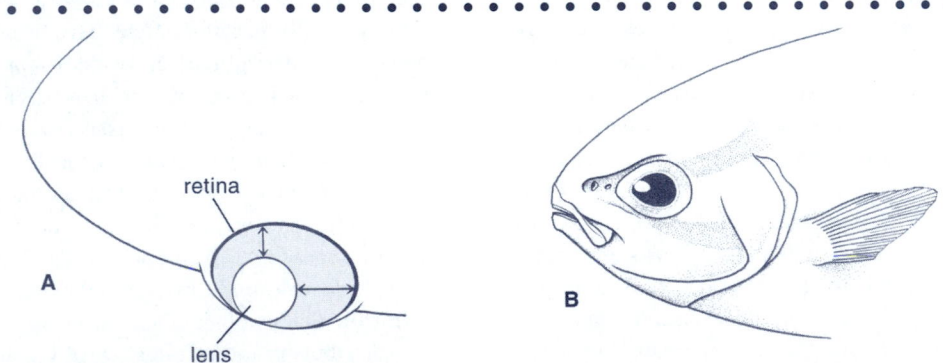

FIGURE 20.4
A, Eyeball shape and placement of lens in relation to retina in many teleosts. Close objects in the anterior field can be in sharp focus, whereas the lateral field is adapted to more distant vision. **B,** Elliptical eye shape (Girellidae) that allows for a large anterior field of vision by a laterally placed eye. Directly in front of the eye is a groove that facilitates forward vision.

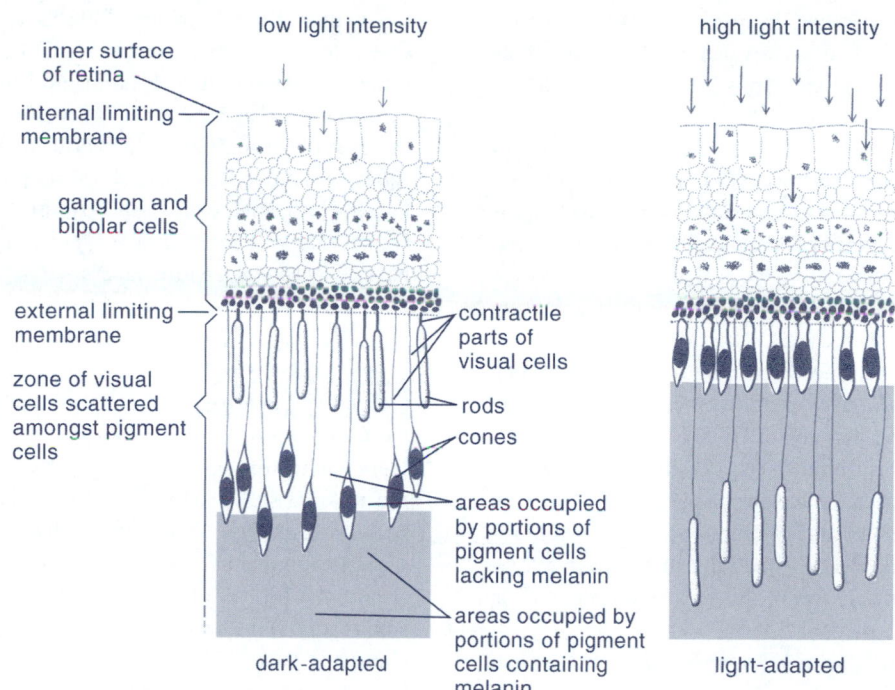

low light intensity

high light intensity

inner surface of retina

internal limiting membrane

ganglion and bipolar cells

external limiting membrane

zone of visual cells scattered amongst pigment cells

contractile parts of visual cells

rods

cones

areas occupied by portions of pigment cells lacking melanin

areas occupied by portions of pigment cells containing melanin

dark-adapted

light-adapted

FIGURE 20.5

Motion of rods, cones, and pigment in the retina of teleosts. In light adaptation (*right*), the rods are moved away from the light and are protected by the forward movement of pigment.

tain processes through which melanin can move to or from the outer parts of the visual cells. Under bright illumination, the eye adapts by the movement of melanin toward the visual cells and the movement of the outer segments of the rods into the pigmented area, where they are shielded from the light. In dim light, the pigment is drawn back, and the contractile or myoid part of the rods pulls the rods away, allowing the receptors to be exposed to light. The movement of the cones is opposite that of the rods, but the cones are not usually hidden by the pigment. In wrasses (Labridae), a red pigment is prominent in the pigment cells, which also contain melanin. Cylinders of pigment are extended to screen the rods and long cones. In light adaptation (photopic vision), the extension of the red pigment, which mainly absorbs rays under 560 nm, does not allow any other light but red from reaching the long single cones and rods (Fineran and Nicol, 1974). In dark adaptation (scotopic vision), the rods shorten and are not screened by the red cylinders. The time required for pigment movement in the tapetum (see later) of sharks or for the shifting of pigment and visual cells in teleosts is considerable—about 2 hours for advancing or receding in the shark tapetum, and about 30 minutes for light adaptation and one hour or more for dark adaptation

in teleosts such as *Oncorhynchus*. Retinomotor activity is not only influenced by ambient light; it is also influenced by circadian rhythms (Yokoyama and Yokoyama, 1996).

Contractile Irises

Most elasmobranchs but few teleosts have contractile irises that can control the amount of light entering the eye (Munz, 1971). Contraction of the iris is usually more rapid than dilation. Some teleost species have mobile pupils under nervous control, but the eels of the genus *Anguilla*, like the elasmobranchs, appear to have irises that contract in direct response to light (Seliger, 1962; Young, 1933). The pupillary operculum of rays may have a fringe with up to a dozen projections that form several small apertures in a crescent- or U-shaped configuration around the lateral and ventral aspects of the pupil, effectively diminishing the entering light but maintaining a greater visual field (Murphy and Howland, 1991).

Among the teleosts, stargazers, armored catfishes (Loricariidae), and many flatfishes (particularly bothids) have a specialized iris that forms a pupillary operculum that can expand to leave a thin, U-shaped aperture and cut off most of the light reaching the pupil. The ability of loricariid catfishes to adjust the diameter of their irregularly shaped pupils in response to illumination apparently is not associated with visual accommodation; rather, it appears to be an adaptation that renders these benthic creatures less conspicuous by minimizing the size of their pupils (Douglas et al., 2002).

Pigment in Cornea or Lens

Some fishes have pigment in the cornea or lens that acts as a filter to eliminate certain light, especially short wavelengths of light. Some benthic species in several families have specially constructed, iridescent corneas that protect the eyes from strong down-welling light without greatly affecting light entering the eye from the lateral field of view (Lythgoe, 1975a). Iridescence is produced when light passes "from one medium through a thin layer of another which has a different refractive index" (Lythgoe, 1975b, p. 263). There are about six types of iridescent corneas, all with lamellae whose thickness is less than the wavelength of light. Because of their regular arrangement and thinness, they refract and reflect light to cause iridescence when illuminated from above; this reduces the light that enters the eyes (Lythgoe, 1975a, 1975b).

The semiterrestrial blenny *Dialommus fuscus* has a photosensitive pigment in its nonconstricting iris and in its aqueous humor, which turns dark when the fish is out of water in bright light (Stevens and Parsons, 1980). This change in optical density apparently protects the retina from excessive illumination in air. Gnyubkin (1989) showed that the yellow cornea of *Hexagrammos stelleri* reacts to the level of illumination by dispersing or aggregating pigment, but the reaction is slow, requiring about 100 minutes.

Reflective Tapetum Lucidum

The "eyeshine" of fishes, including sharks and *Latimeria*, can be caused by various layers of reflective material associated with the choroid or retina or, in some fishes, with the cornea (Best and Nicol, 1980; Cohen, 1990; Locket, 1974). A choroidal tapetum lucidum is present in most elasmobranchs, and a retinal tapetum lucidum is present in many freshwater teleosts (Best and Nicol, 1980). These tapeta are reflective layers that usually involve guanine crystals as reflectors, which are as efficient as a good mirror. A fibrous type of choroidal tapetum, consisting of shiny, tendonlike structures is seen in some marine fishes (Muntz, 1971). The tapetum lucidum acts to reflect light that has passed through the visual cells back into the system and is an effective adaptation for sight in dim light. In sharks, the cells in which the guanine crystals provide a reflective surface are parallel to the plane of the retina throughout the fundus of the eye but are oblique peripherally, so they are always perpendicular to the light entering the eye, thus increasing their efficiency as reflectors. Melanin is associated

with the guanine reflectors and can migrate along the reflectors to mask the reflectivity. This has two possible functions: One is to adapt the eye for vision in bright light, whereas the other may be to reduce eyeshine and thus make the fish less conspicuous. Extraneous light is prevented from entering the eyeball through its walls by the **stratum argenteum,** a reflective layer outside the choroid. This is of importance in fishes with translucent tissue around the orbits, such as the catfish genera *Ompok* and *Kryptopterus*. Irises usually also have reflective or opaque layers.

VISUAL ADAPTATION IN THE LIVES OF FISHES

Dependence on sight varies greatly among fishes, as does the relative size of the eyes. Sight-feeding, diurnal fishes such as trout, bass, and sunfishes have prominent eyes, with a diameter equal to about one fifth or one sixth the length of the head. The eyes of the pike, a diurnal predator, only seem small (about one tenth the head length) in proportion to its elongated head. Crepuscular or nocturnal fishes that hunt by sight and deep-sea fishes that are at least partly sight dependent tend to have eyes that are one third to one half their head length. Wide pupils and tremendous numbers of rods (up to 20 million per mm^2) in the retina aid in the increase of sensitivity in such eyes (Marshall, 1979). Those nocturnal or deepwater fishes that depend largely on olfaction, taste, the mechanosensory lateral line, or electroreception tend to have reduced eyes. These small-eyed fishes include such freshwater fishes as mormyrids, gymnarchids (see Chapters 9 and 21 for details on the remarkable electrical capabilities of these fishes), and various catfishes that are nocturnal or found in turbid waters. In the marine environment, bathypelagic fishes, such as gulper eels, snipe eels, and ceratioid anglerfishes, and many demersal species have relatively small eyes (Marshall, 1966).

The quality of eyes of the same size can differ markedly from species to species. The number, disposition, and types of visual cells; connections of the cells to the optic neurons; mechanisms for accommodation; and effectiveness of the tapetum lucidum all determine the efficiency of the eye. In a study of 31 shallow-water species off New Zealand, Pankhurst (1989) found that based on photoreceptors and eye morphology, fishes of different ecological functions or distributions had differing visual structures and abilities. Carnivores had larger eyes than herbivores relative to the size of the body. Diurnal species appeared to have better visual acuity than nocturnal species but, based on their retinal features, the nocturnal species had better sensitivity to light. Relatively large eyes were present in small nocturnal species and in planktivores. Studies of the visual systems of Hawaiian coral reef fishes have demonstrated the value of an integrated approach, in which spectral sensitivity is correlated with the spectral characteristics of fish coloration and the visual features of the coral reef environment, including the microhabitats of the ichthyofaunal inhabitants (Losey et al., 2003; Marshall et al., 2003a, 2003b).

Visual Adaptations of Deep-Ocean Fishes

Some of the most remarkable adaptations of eyes are seen in marine fishes living at moderate to great depths (see Chapter 35). A fundamental difference in visual needs and their associated adaptations can be seen between mesopelagic and bathypelagic species. Mesopelagic fishes show the greatest variation in eye design; operating in a region of downwelling daylight, they must be able to discriminate shapes in dim ambient light, but also be able to detect point-source illumination from bioluminescence. Although bathypelagic species are operating in regions of absolute darkness save for bioluminescence, their generally smaller and less sensitive eyes are still well adapted for the detection of point-source bioluminescent emissions (Warrant and Locket, 2004). In some deep-dwelling fishes, the parts of the eye retain approximately the same proportions as those of shallow-water species but are enlarged and occupy a greater portion of the head. In others, the eye is tubular, with a greatly enlarged lens and a large pupillary opening admitting light to the large lens, which focuses on a relatively small retina. In these greatly modified eyes, as in virtually all fish eyes, the distance from the center of the lens to the retina (focal length) is about 2.55 (in some as low as 2.10) times the lens radius, conforming to Matthiessen's ratio (Walls, 1942), so their optical properties may be similar to more normal fish eyes. These deepwater eyes are fixed, lack mechanisms for accommodation, and mostly receive clear images from one direction only. Some have two effective parts of the retina, or an accessory retina, and can form images coming from two different directions. Usually, a kind of accessory lens (lens pad) aids in directing light to the accessory retina. The genus *Ipnops* lacks eyes but has visual cells located on top of the flattened head (Marshall, 1979).

Some deepwater fishes, including hatchetfishes (*Argyropelecus*), scaleless dragonfishes (*Echiostoma*), spinyfins (*Diretmus*), greeneyes (*Chlorophthalmus*), and pearleyes (*Scopelarchus*) have yellow lenses. Although a yellow filter reduces the amount of light reaching the retina, it may function to make photophores that emit the light that matches the dim daylight seem brighter, so that functions involving detection of bioluminescence, such as intraspecific communication or predation, can be better accomplished (Muntz, 1976). Somiya

(1979) suggested that the yellow filter functions mainly in reducing short-wavelength light in species that have some special strategy for use of the dim, longwave light. For instance, *Echiostoma barbatum* has yellow lenses, red photophores, and red-sensitive visual pigment. Objects such as potential prey or predators illuminated by emissions from its photophores would thus be made more visible. Other genera that have red photophores and red-sensitive visual pigments are *Pachystomias* and *Malacosteus*. Because most deep-sea fishes are not sensitive to red light, and thus cannot see the red photophores, these three genera have an advantage in illuminating potential prey without being seen clearly (Munk, 1982).

Fishes that apparently cannot focus a sharp image on the retina because of a flattened lens were reported by Munk (1984). These include five species of anglerfishes and the gulper eel. Some mormyrids are known to have flattened lenses. These small, optically nonadjusted eyes may be best used in detecting movement (Munk, 1984). An eelpout from the Bering Sea (*Opaeophacus acrogeneius*) that lives near the limit of light penetration has a strange modification of the eye lens. There is a vertical trench in the lens, filled with a jellylike material that is suspected to have a refractive index less than that of the lens, so that light striking the trench would be distributed over a wide area of the retina. Although this might diminish visual acuity, it might enhance sensitivity to movement of objects silhouetted against the dim light (Bond and Stein, 1984). The sand lance (*Limnichthytes fasciatus*) has a flattened lens, yet it achieves a remarkable degree of refractive power from its highly convex cornea. This fish lies concealed in the sand, with only its eyes protruding. From this position, it lunges out at small crustaceans. Much like those of a chameleon, the eyes of the sand lance move independently of each other while scanning the seabed for prey. Rapid adjustments to the convex cornea, using striated corneal muscles, enable each eye to change its focus and to view objects at varying distances. In this way, depth perception is made possible in the absence of stereoscopic vision (Pettigrew and Collin, 1995; Pettigrew et al., 1999).

"Dual-Purpose" Eyes in Fishes

In at least three cases among vertebrates, visual sensitivity has been enhanced by what appear to be attempts to duplicate eyes beyond the two normally found—in a sense, the evolution of "four-eyed" fishes (Schwab et al., 2001). Probably the best known example are the three species of the genus *Anableps* (see Chapter 15). The blennioid fishes of the closely related genera *Mnierpes* and *Dialommus* are another example. The third case is that of the mesopelagic species

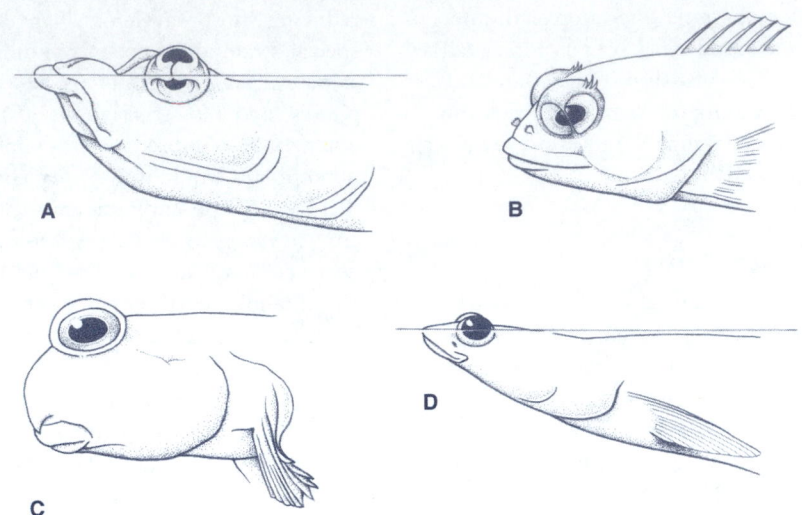

FIGURE 20.6

Fishes with eyes adapted for aerial vision: **A**, *Anableps*, in which the eye is modified to have an aerial (above) and an aquatic (below) aperture; **B**, *Dialommus*, in which the angled cornea with two flat surfaces causes double images on the retina in aerial vision; **C**, *Periophthalmus*, which spends much of its time completely out of water; **D**, *Rhinomugil corsula*, which often swims with its eyes above the water surface.

Bathylychnops exilis (Fig. 20.3D) of the family Opisthoproctidae. Fishes in this family have large, upward-directed eyes, but *B. exilis* has in addition small auxiliary eyes that have developed near the ventral base of the primary eyes; these are oriented downward.

In the first two aforementioned cases, eye subdivision is an accommodation to permit aerial vision—these fishes have eyes greatly modified to enable both aerial and aquatic vision. The four-eyed fishes (*Anableps*) of Central America and northern South America swim at the surface, with the upper half of the eye exposed to the air (Fig. 20.6A; see Plate 15.11). The iris is modified so that two flaps divide the pupil of the eye at the level of the water surface. The lens is egg shaped, so that light entering from above the water's surface passes through the short axis, which compensates for the added refraction of light by the aerial cornea, and light entering from the water passes through the long axis (Schwassman and Kruger, 1965). The upper part of the cornea is thicker and is curved more than the lower part. Whereas Charman and Tucker (1973) reported that the "normal" cornea of the goldfish has a refractive index of about 1.33, Sivak (1976) reported the refractive index of the *Anableps* cornea to be 1.51—greater than that reported from the corneas of other fishes. Although the ultrastructure of the photoreceptors in the upper part of the retina is similar to that in the lower part, the ventral retina is thicker and contains

more cells (Borwein and Hollenberg, 1973). *Anableps* must submerge its eyes frequently to prevent drying.

The labrisomid blennies (*Dialommus fuscus;* Fig. 20.6B) of the Galapagos Islands and *Mnierpes macrocephalus*, from the Pacific coast of Central America, have eyes that are divided by vertical septa. The two corneal surfaces are flat and are angled at 110° relative to each other. When these fishes are out of the water, they can focus clearly on objects because of the 1.0 index of refraction of their eye fluids, but the two angled corneal surfaces would be expected to result in the formation of dual images. Under water, the fishes would perceive only a single image (Graham, 1997; Graham and Rosenblatt, 1970; Stevens and Parsons, 1980).

Some flyingfishes have three flattened surfaces on the cornea, but Ali and Klyne (1985) believed that there is little difference between the image perceived in the air and the image formed under water. Mudskippers (*Periophthalmus*), which spend much of their time on mudflats or among mangrove roots, have prominent eyes set high on the head (Fig 20.6C; see Plate 16.17). The lens of the eye is flattened more than in most other fishes, so that aerial vision is good. A pocket below the eye carries moisture, and the eye can be retracted into this pocket to prevent drying. An Indian mullet (*Rhinomugil corsula*) is convergent with *Anableps* in its body and head shape and the placement of its eyes (Hora, 1938; Walls, 1963). *R. corsula* can hold its eyes above water as it swims at the water surface (Fig. 20.6D). Its aerial vision appears to be acute, for the slightest movement of an observer is sufficient to send this nimble fish below the surface, to pop up again at a safe distance.

The archerfish (*Toxotes jaculator*), although it keeps its eyes below the surface, has excellent aerial vision and can squirt water at insects and other small prey with great accuracy. The eyes of the archerfish are little modified. With a narrow (14°) field of binocular vision and no fovea, it is believed to compensate with some form of "range finder" in the brain (Walls, 1963). Archerfish will gather at the site of an insect prey on a twig just above the water surface and commence "shooting" until the insect is dislodged. The first one to arrive at the site where the insect hits the water claims the prize. Recent studies have verified the "range finder" hypothesis as it pertains to the fish's ability to precisely determine where the insect will land. Much like the outfielder who gauges his or her position to be right under a fly ball, an archerfish will shift its body axis within 100 ms after dislodging its prey to align with the precise location where the insect will land. Although the outfielder is constantly observing the trajectory of the fly ball, the archerfish moves to the point of capture without any additional visual feedback (Rossel et al., 2002).

Sensitivity to Light in the Ultraviolet Range

Visual sensitivity in the ultraviolet (UV) range is known among all classes of vertebrates (Zhang, 2003), especially tropical reef fishes (Losey et al., 2003). Using molecular techniques, Shi and Yokoyama (2003) were able to reconstruct the phylogeny of UV-sensitive visual pigments in vertebrates. The discovery that many teleost species possess visual pigments with a peak absorption around 360 nm indicates widespread visual capabilities in the UV range (Losey et al., 1999). Losey et al. (1999) suggested roles for these pigments in communication, including mate choice. UV-sensitive retinal cone cells are known from *Dascyllus albisella*, a planktivorous pomacentrid that inhabits coral reef areas. Ontogenetic changes in UV sensitivity have been recorded in *Dascyllus;* increased accumulation of UV-blocking compounds in the eyes accompany their maturation from juveniles closely associated with coral heads to adults that forage further afield (Losey et al., 2000). Of 22 species of pelagic marine fish larvae sampled in inshore waters of the Pacific Northwest, 18 possessed ultraviolet and violet visual pigments—adaptive features that appear to be conducive to planktivory (Britt et al., 2001). UV sensitivity appears to diminish with smoltification in anadromous salmonids but becomes reestablished with sexual maturity (Deutschlander et al., 2001).

Sensitivity to Polarized Light

The ability to sense polarized light occurs in many fishes, although its functional advantage to the fish is not definitely known (Cameron and Pugh, 1991; Dill, 1971; Hawryshyn, 1992). Sensitivity to polarized light in animals may have applications entirely within the underwater world, at the water's surface, and in viewing the overlying atmosphere. Multiple adaptive applications suggest a diversity of neuronal accommodations in the development of this sensory capability (Wehner, 2001). It has been suggested that the paired cones of fishes operate in the detection of polarized light to improve their visual acuity in turbid waters by filtering out the light scattered from suspended particles (Bains, 1996). There is speculation that the navigation of migratory species is in part facilitated by their sensitivity to polarized light. Some features of living organisms, such as fish scales, polarize light under water; this confers on them a biological signature that might make them more readily visible against a background of inanimate objects.

Some Alternative Photoreceptors

Many fishes find themselves in certain environments where vision is of limited value. The widespread adaptation of fishes to cave environments (see Chapter 32) usually has resulted in the diminution of their visual capacities. Blindness does not necessarily mean a complete lack of sensitivity to light, however, for photoreceptors may exist in the skin of some blind species and are associated with the central nervous system of many others. The Somalian cavefish (*Phreatichthys andruzzii*) lacks optic nerves and has markedly reduced optic lobes, yet still demonstrates photosensitivity (Ercolini and Berti, 1975). Photoreceptors in the caudal region of larval lampreys (ammocoetes) appear to facilitate the initiation of a burrowing response when the animal is exposed to light. The rudimentary eyes of hagfishes are not especially light sensitive, but in one species, *Eptatretus burgeri*, light-sensitive areas are known from the skin of the tail and in the pale stripe that extends down the back (Patzner, 1978).

The most important site of extraocular photoreception is the pineal organ and its associated structures (see Chapter 26). Histological studies have demonstrated secretory cells in the pineal bodies of various species and in the parapineal of *Latimeria* (Locket, 1980; McNulty, 1981). Electrophysiological studies have shown receptor potentials produced in the pineal nerve following stimulation by light. The role of the pineal complex in regulating chromatophores has been demonstrated in many experiments. Cave-dwelling species initiate swimming movements away from a light source stimulating the pineal region. Most fishes fall into one of three categories regarding illumination of the pineal region (Breder and Rasquin, 1950): Fishes of Category 1 have a transparent or translucent tissue "window" covering the pineal complex and usually react positively to light. Category 2 fishes have an opaque covering over the pineal and usually react negatively to light. Category 3 fishes can control the entry of light to some extent by the action of chromatophores above the pineal complex; these fishes vary in their reaction to light.

The pineal window in some sharks can allow the transmission of up to seven times as much light as adjacent parts of the head. In some species, the threshold for the detection of light at this site is below that perceived as moonlight. In addition to a pineal window, tunas have a tubelike, translucent structure that directs light to the dorsal part of the brain. About 25 percent of the incident light can be transmitted to the vicinity of the apparently photosensitive pineal. An Indian catfish (*Heteropneustes fossilis*) has a lens-like structure in the pineal window. *H. fossilis* is a nocturnal species, but it is thought that the combination of the "lens," the pineal window, and a pineal fossa allows a sufficient concentration of light onto the pineal to stimulate photoendocrine function (Srivastava and Srivastava, 1991).

Experiments with the aforementioned eyeless Somalian cavefish (*Phreatichthys*) have shown that its central nervous system is sensitive to light. Sensitivity of the pineal to light is involved in thermoregulatory behavior, according to experiments by Kavaliers (1980), in which white suckers (*Catostomus commersoni*) with the pineal shielded spent significantly more time in lighter and warmer parts of the experimental environment than did individuals whose pineals were unshielded. The suckers appeared to show both a fast (neural) and a slower (hormonal) response. These and a host of other studies have clearly demonstrated the integrative capacity of the visual senses—fishes are made more complete because of the way in which their visual senses integrate with other sensory modalities to better connect them with their watery world.

KEY POINTS AND CONNECTIONS

- Although conforming to the basic structure of the vertebrate eye, the eyes of fishes display a dramatic diversity in design in response to the range of aquatic habitats in which they are found. Some distinctive features are also observed. Scleral ossicles reinforce the eyes of some swift-moving fishes, and several species possess a choroid body. Eyelids are seen only among elasmobranchs, which may also have a nictitating membrane.

  The function of the choroid body is discussed in Chapter 24.

- The retinal photoreceptors of fishes include rods, single cones, and double cones; their spectral sensitivities reflect the particular habi-

tats of the fishes. The diversity of cone structure is evidence of the extensive development of color vision among fishes.

The photic properties of different fish habitats are discussed in greater detail in Part VII, focusing on habitats and ecology. Chapter 35, in which the range of illumination of pelagic waters and the consequent visual adaptations of fishes are further discussed, is especially helpful.

- An analogy can be drawn between the light-gathering properties of fish eyes, such as the dimensions of the lens, and the lens of a camera. A diversity of means to regulate light entering the eye, including retinomotor mechanisms; pigments in the lens, cornea, or iris; and a reflective tapetum lucidum, are observed among fishes. Few teleosts have contractile irises, however.

- Specific correlates of visual capacity with the habits and habitats of fishes can be made. Visual adaptations to diet and feeding behavior and time of maximal activity (diurnal vs. nocturnal or crepuscular) are readily discernable, as are the adaptive advantages of sensitivity to UV and polarized light. Some of the most unusual eyes are seen among deep-ocean fishes and in the few surface-dwelling species in which aerial vision is made possible through the "subdivision" of their eyes.

FISH LINKS

http://www.hawaii.edu/loseylab/ The research laboratory of Dr. George Losey of the Hawai'i Institute of Marine Biology. Dr. Losey currently researches the influence of ultraviolet light on marine fishes.

BUILDING AN ICHTHYOLOGY LIBRARY

Ali, M. A., and M. A. Klyne. 1985. *Vision in vertebrates.* Plenum Press, New York.

Douglas, R., and M. B. A. Djamgoz (Eds.). 1990. The visual system of fish. Chapman and Hall, London.

Kageyama, C. J. 1999. What Fish See: Understanding Optics and Color Shifts for Designing Lures and Flies. Frank Amato, Portland, OR.

An interesting book that applies studies on fish vision to the design of fishing lures; Kageyama is an optometrist, lure designer, and avid steelhead sportfisher.

von der Embde, G., J. Mogdans, and B. G. Kapoor (Eds.). 2004. *The senses of fish: Adaptations for the reception of natural stimuli.* Springer Verlag, Berlin.

REFERENCES

Ali, M. A., and M. A. Klyne. 1985. *Vision in vertebrates.* Plenum Press, New York.

Bains, S. 1996. Sunfish shows the way through the fog. *Science 272:* 653.

Bauchot, R., A. Thomot, and M. L. Bauchot. 1989. The eye muscles and their innervation in *Chaetodon trifasciatus* (Pisces, Teleostei, Chaetodontidae). *Env. Biol. Fishes 25*(1–3): 221–233.

Beatty, D. D. 1984. Visual pigments and the labile scotopic visual system of fish. *Visual Pigment Biochem. 24:* 1563–1573.

Best, A. C. G., and J. A. C. Nicol. 1980. Eyeshine in fishes. A review of ocular reflectors. *Can. J. Zool. 58*(6): 945–956.

Boehlert, G. W. 1978. Intraspecific evidence for the function of single and double cones in the teleost retina. *Science 202:* 309–311.

———. 1979. Retinal development in postlarval through juvenile *Sebastes diploproa:* Adaptations to a changing photic environment. *Rev. Can. Biol. 38*(4): 265–280.

Bond, C. E., and D. L. Stein. 1984. *Opaeophacus acrogeneius,* a new genus and species of Zoarcidae (Pisces: Osteichthyes) from the Bering Sea. *Proc. Biol. Soc. Wash. 97:* 522–525.

Borwein, B., and M. J. Hollenberg. 1973. The photoreceptors of the "four-eyed" fish, *Anableps anableps* L. *J. Morphol. 140*(4): 405–441.

Bowmaker, J. K. 1990. Visual pigments of fishes, pp. 81–107. In *The visual system of fish,* R. H. Douglas and M. B. A. Djamgoz (Eds.). Chapman and Hall, London.

———. 1998. Evolution of colour vision in vertebrates. *Eye 12:* 541–547.

———, and Y. W. Kunz. 1987. Ultraviolet receptors, tetrachromatic colour vision and retinal mosaics in brown trout (*Salmo trutta*): Age-dependent changes. *Vision Res. 27:* 2101–2108.

———, H. J. A. Dartnall, and P. J. Herring. 1988. Long-wave sensitive visual pigments in some deep-sea fishes: Segregation of "paired" rhodopsins and porphyropsins. *J. Comp. Phys. A 163*(5): 685–698.

Breder, C. M., Jr., and P. Rasquin. 1950. A preliminary report on the role of the pineal organ in the control of pigment cells and light reactions in recent teleost fishes. *Science 111:* 10–12.

Britt, L. L., E. R. Loew, and W. N. McFarland. 2001. Visual pigments in the early life stages of Pacific northwest marine fishes. *J. Exp. Biol. 204:* 2581–2587.

Cameron, P. A., and B. N. Pugh. 1991. Double cones as a basis for a new type of polarization vision in vertebrates. *Nature 353:* 161–164.

Charman, W. N., and J. Tucker. 1973. The optical system of the goldfish eye. *Vision Res. 13:* 1–8.

Cohen, J. L. 1990. Vision in elasmobranchs, pp. 463–490. In *The visual system of fish,* R. H. Douglas and M. B. A. Djamgoz (Eds.). Chapman and Hall, London.

Collin, S. P., and A. E. O. Trezise. 2004. The origins of colour vision in vertebrates. *Clin. Exp. Optometry 87:* 217–223.

Deutschlander, M. E., D. K. Greaves, T. J. Haimberger, and C. W. Hawryshyn. 2001. Functional mapping of ultraviolet photo-sensitivity during metamorphic transitions in a salmonid fish, *Oncorhynchus mykiss. J. Exp. Biol. 204:* 2401–2413.

———, , and ———. 1982. Color vision in fishes. *Sci. Am. 246:* 108–117.

Dill, P. A. 1971. Perception of polarized light by yearling sockeye salmon. *J. Fish. Res. Bd. Can. 28:* 1319–1322.

Douglas, R. H., and C. W. Hawryshyn. 1990. Behavioral studies of fish vision: Analysis of visual capabilities, pp. 373–418. In *The visual system of fish,* R. H. Douglas and M. B. A. Djamgoz (Eds.). Chapman and Hall, London.

———, S. P. Collin, and J. Corrigan. 2002. The eyes of suckermouth armoured catfish (Loricariidae, subfamily Hypostomus): Pupil response, lenticular longitudinal spherical aberration and retinal topography. *J. Exp. Biol. 205:* 3425–3433.

Ercolini, A., and R. Berti. 1975. Light sensitivity experiments and morphology studies of the blind phreatic fish *Phreatichthys andruzzii* Vinciguerra from Somalia. *Monit. Zool. Ital. 6:* 29–43.

Fernald, R. D. 1990. The optical system of fishes, pp. 45–62. In *The visual system of fish,* R. H. Douglas and M. B. A. Djamgoz (Eds.). Chapman and Hall, London.

———. 1993. Vision, pp. 161–189. In *The physiology of fishes,* D. H. Evans (Ed.). CRC Press, Boca Raton, FL.

———, and S. E. Wright. 1983. Maintenance of optical quality during crystalline lens growth. *Nature 301:* 618–620.

———, and ———. 1985. Growth of the visual system in the African cichlid fish, *Haplochromis burtoni:* Accommodation. *Vision Res. 25:* 163–170.

Fineran, B. A., and J. A. C. Nicol. 1974. Studies on the eyes of New Zealand parrot fishes (Labridae). *Proc. Roy. Soc. Lond. B 186:* 217–247.

Gnyubkin, V. F 1989. Response of pigmented cornea of whitespotted greenling to changes in illumination. *Biol. Morya Vladivostok 1:* 25–32.

Graham, J. B. 1997. *Air-breathing fishes: Evolution, diversity, and adaptation.* Academic Press, San Diego.

———, and R. H. Rosenblatt. 1970. Aerial vision: Unique adaptation in an intertidal fish. *Science 168:* 586–588.

Harkness, L., and H. C. Bennet-Clark. 1978. The deep fovea as a focus indicator. *Nature 272:* 814–816.

Harosi, F. I., and Y. Hashimoto. 1983. Ultraviolet visual pigment in a vertebrate: A tetrachromatic cone system in the dace. *Science 222:* 1021–1023.

Hawryshyn, C. W. 1992. Polarization vision in fish. *Amer. Sci. 80:* 164–175.

———. 1998. Vision, pp. 345–374. In *The physiology of fishes* (2nd ed.) D. H. Evans (Ed.). CRC Press, Boca Raton, FL.

———, and W. N. McFarland. 1987. Cone photoreceptor mechanisms and the detection of polarized light in fish. *J. Comp. Physiol. A 160:* 459–465.

Hora, S. L. 1938. Notes on the biology of the fresh-water grey mullet *Mugil corsula* H., with observations on the probable mode of origin of aerial vision in fishes. *J. Bombay Nat. Hist. Soc. 40:* 62–68.

Janssen, J. 1996. Use of the lateral line and tactile senses in feeding in four Antarctic nototheniid fishes. *Env. Biol. Fishes 47:* 51–64.

Kavaliers, M. 1980. The pineal organ and circadian rhythms of fishes, pp. 631–643. In *Environmental physiology of fishes,* M. A. Ali (Ed.). Plenum Press, New York.

Kreuzer, R. O., and J. G. Sivak. 1984. Spherical aberration of the fish lens: Interspecies variation and age. *J. Comp. Physiol. A 154:* 415–422.

Levine, J. S., and E. F. MacNichol, Jr. 1979. Visual pigments in teleost fishes: Effects of habitat, microhabitat, and behavior on visual system evolution. *Sens. Process. 3*(2): 95–131.

Locket, N. A. 1974. The choroidal tapetum lucidum of *Latimeria. Proc. Roy. Soc. Lond. B186:* 281–290.

———. 1980. Variation of architecture with size in the multiple-bank retina of a deep-sea teleost, *Chauliodus sloani. Proc. Roy. Soc. Lond. B208:* 223–242.

Loew, E. R., and W. N. McFarland. 1990. The underwater visual environment, pp. 1–43. In *The visual system of fish,* R. H. Douglas and M. B. A. Djamgoz (Eds.). Chapman and Hall, London.

Losey, G. S., T. W. Cronin, T. H. Goldsmith, D. Hyde, N. J. Marshall, and W. N. McFarland. 1999. The UV visual world of fishes: A review. *J. Fish Biol. 54:* 921–943.

———, P. A. Nelson, and J. P. Zamzow. 2000. Ontogeny of spectral transmission in the eye of the tropical damselfish, *Dascyllus albisella* (Pomacentridae), and possible effects on UV vision. *Env. Biol. Fishes 59:* 21–28.

———, W. N. McFarland, E. R. Loew, J. P. Zamzow, P. A. Nelson, and N. J. Marshall. 2003. Visual biology of Hawaiian coral reef fishes. I. Ocular transmission and visual pigments. *Copeia 2003:* 433–454.

Lythgoe, J. N. 1975a. The structure and phylogeny of iridescent corneas in fishes, pp. 253–262. In *Vision in fishes: New approaches in research*, M. A. Ali (Ed.). Plenum Press, New York.

———. 1975b. The iridescent cornea of the sand goby *Pomatoschistus minutus* (Pallas), pp. 263–277. In *Vision in fishes: New approaches in research*, M.A. Ali (Ed.). Plenum Press, New York.

———. 1979. *The ecology of vision*. Clarendon Press, Oxford.

Marshall, N. B. 1966. *The life of fishes*. Universe Books, New York.

———. 1979. *Developments in deep-sea biology*. Blandford Press, Poole, Dorset, UK.

Marshall, N. J., K. Jennings, W. N. McFarland, E. R. Loew, and G. S. Losey. 2003a. Visual biology of Hawaiian coral reef fishes. II. Colors of Hawaiian coral reef fishes. *Copeia 2003:* 455–466.

———, K. Jennings, W. N. McFarland, E. R. Loew, and G. S. Losey. 2003b. Visual biology of Hawaiian coral reef fishes. III. Environmental light and an integrated approach to the ecology of reef fish vision. *Copeia 2003:* 467–480.

McNulty, J. A. 1981. A quantitative morphological study of the pineal organ in the goldfish, *Carassius auratus. Can. J. Zool. 59:* 1312–1325.

Montgomery, J., and N. Pankhurst. 1997. Sensory physiology, pp. 325–349. In *Deep-sea fishes*, D. J. Randall and A. P. Farrell (Eds.). Academic Press, San Diego.

Munk, O. 1982. Cones in the eye of the deep-sea teleost *Diretmus argenteus. Vision Res. 22:* 179–181.

———.1984. Non-spherical lenses in the eyes of some deep-sea teleosts. *Arch. Fischereiwiss. 34*(2–3): 145–153.

———, and R. D. Frederickson. 1974. On the function of aphakic apertures in teleosts. *Vidensk. Medd. Dan. Naturhist. Foren. Kbh. 137:* 65–94.

Muntz, W. R. A. 1976. On yellow lenses in mesopelagic animals. *J. Mar. Biol. Assoc. UK 56:* 963–976.

———, and G. S. V. Mouat. 1984. Annual variations in the visual pigments of brown trout inhabiting lochs providing different light environments. *Visual Pigment Biochem. 24:* 1575–1580.

Munz, F. W. 1971. Vision: Visual pigments, pp. 1–32. In *Fish physiology*, Vol. V., W. S. Hoar and D. J. Randall (Eds.). Academic Press, New York.

Murphy, C. J., and H. C. Howland. 1991. The functional significance of crescent-shaped pupils and multiple pupillary apertures. *J. Exp. Zool. Suppl. 5:* 22–28.

Neumeyer, C. 1986. Wavelength discrimination in goldfish. *J. Comp. Physiol. A 158:* 203–213.

New, J. G., L. Alborg Fewkes, and A. N. Khan. 2001. Strike feeding behavior in the muskellunge, *Esox masquinongy:* Contributions of the lateral line and visual sensory systems. *J. Exp. Biol. 204:* 1207–1221.

Nicol, J. A. C. 1989. *The eyes of fishes*. Clarendon Press, Oxford.

Pankhurst, N. W. 1989. The relationship of ocular morphology to feeding modes and activity periods in shallow marine teleosts from New Zealand. *Env. Biol. Fishes 26:* 201–211.

Partridge, B. L., and T. J. Pitcher. 1980. The sensory basis of fish schools: Relative roles of lateral line and vision. *J. Comp. Physiol. 135:* 315–325.

Partridge, J. C., S. N. Archer, and J. N. Lythgoe. 1988. Visual pigments in the individual rods of deep-sea fishes. *J. Comp. Physiol. A 162:* 543–550.

Patzner, R. A. 1978. Experimental studies on the light sense in the hagfish *Eptatretus burgeri* and *Paramyxine atami. Helgolander Wiss. Meeresunters. 31*(1–2): 180–190.

Pettigrew, J. D., and S. P. Collin. 1995. Terrestrial optics in an aquatic eye: The sandlance, *Limnichthytes fasciatus* (Creediidae, Teleostei). *J. Comp. Physiol. A 177:* 397–408.

———, S. P. Collin, and M. Ott. 1999. Convergence of specialized behaviour, eye movements and visual optics in the sandlance (Teleostei) and the chameleon (Reptilia). *Curr. Biol. 9:* 421–424.

Rossel, S., J. Corlija, and S. Schuster. 2002. Predicting three-dimensional target motion: How archer fish determine where to catch their dislodged prey. *J. Exp. Biol. 205:* 3321–3326.

Sargent, R. C., V. N. Rush, B. D. Wisenden, and Y. Y. Yan. 1998. Courtship and mate choice in fishes: Integrating behavioral and sensory ecology. *Amer. Zool. 38:* 82–96.

Schwab, I. R., V. Ho, A. Roth, T. N. Blankenship, and P. G. Fitzgerald. 2001. Evolutionary attempts at 4 eyes in vertebrates. *Trans. Am. Ophthalm. Soc. 99:* 145–157.

Schwassman, H. O., and L. Kruger. 1965. Experimental analysis of the visual system of the four-eyed fish (*Anableps microlepis*). *Vision Res. 5:* 269–281.

Seliger, H. H. 1962. Direct action of light in naturally pigmented muscle fibers. I. Action spectrum for contraction in eel iris sphincter. *J. Gen. Physiol. 46:* 333–342.

Shand, J., J. C. Partridge, S. N. Archer, G. W. Potts, and J. N. Lythgoe. 1988. Spectral absorbance changes in the violet/blue sensitive cones of the juvenile pollack, *Pollachius pollachius. J. Comp. Physiol. 163:* 699–703.

Shi, Y., and S. Yokoyama. 2003. Molecular analysis of the evolutionary significance of ultraviolet vision in vertebrates. *Proc. Nat. Acad. Sci. USA 100*: 8308–8313.

Sivak, J. G. 1976. Optics of the eye of the "four-eyed fish" (*Anableps anableps*). *Vision Res. 16:* 531–534.

———. 1990. Optical variability of the fish lens, pp. 63–80. In *The visual system of fish*, R. H. Douglas and M. B. A. Djamgoz (Eds.). Chapman and Hall. London.

Somiya, H. 1979. 'Yellow lens' eyes and luminous organs of *Echinostoma barbatum* (Stomiatoidei, Melanostomiatidae). *Jpn. J. Ichthyol. 25*(4): 269–272.

———, and T. Tamura. 1973. Studies on the visual accommodation in fishes. *Jpn. J. Ichthyol. 20*(4): 193–206.

Srivastava, G., and C. B. L. Srivastava. 1991. A lens-like specialization for photic input in the pineal window of an Indian catfish, *Heteropneustes fossilis. Experientia 47:* 698–700.

Stevens, J. K., and K. E. Parsons. 1980. A fish with double vision. *Nat. Hist. 89*(1): 62–67.

Wagner, H. J. 1990. Retinal structure of fishes, pp. 109–158. In *The visual system of fish*, R. H. Douglas and M. B. A. Djamgoz (Eds.). Chapman and Hall, London.

Walls, G. L. 1942. The vertebrate eye and its adaptive radiation. *Cranbrook Inst. Sci. Bull. 19.*

———. 1963. *The vertebrate eye and its adaptive radiation*. Hafner, New York.

Warrant, E. J., and N. A. Locket. 2004. Vision in the deep sea. *Biol. Rev. 79:* 671–712.

Wehner, R. 2001. Polarization vision—a uniform sensory capacity? *J. Exp. Biol. 204:* 2589–2596.

Withers, P. C. 1992. *Comparative animal physiology*. Saunders, Fort Worth, TX.

Yokoyama, S., and R. Yokoyama. 1996. Adaptive evolution of photoreceptors and visual pigments in vertebrates. *Ann. Rev. Ecol. Syst. 27:* 543–567.

Young, J. Z. 1933. Comparative studies on the physiology of the iris. I. Selachians. *Proc. Roy. Soc. Lond. B 112:* 228–241.

Zhang, J. 2003. Paleomolecular biology unravels the evolutionary mystery of vertebrate UV vision. *Proc. Nat. Acad. Sci. USA 100:* 8045–8047.

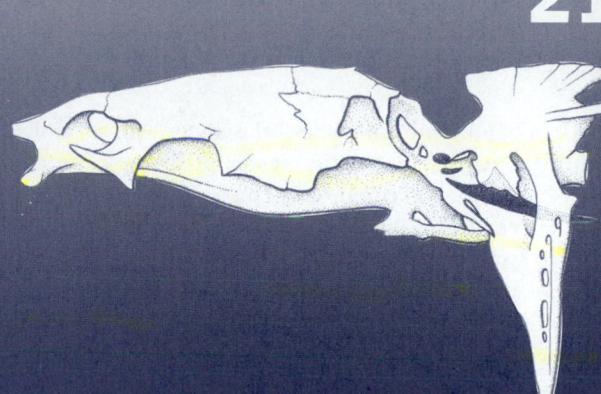

21

THE SENSORY ARSENAL OF FISHES II: SYSTEMS FOR THE DETECTION AND PRODUCTION OF AUDITORY, MECHANICAL, AND ELECTRICAL STIMULI

STRUCTURE OF AUDITORY, MECHANOSENSORY, AND ELECTROSENSORY SYSTEMS

Auditory Systems: The Inner Ear
The Lateral Line System
  Mechanosensory Components
  Electrosensory Components
Other Sensory Receptors
  Nociception — Can Fishes Experience Pain?

FUNCTION OF AUDITORY, MECHANOSENSORY, AND ELECTROSENSORY SYSTEMS

Membranous Labyrinth
  Sound Reception
  Sound Location
Equilibrium
  Angular Acceleration
  Gravistatic Function
Mechanosensory Function
Electrosensory System Function

PRODUCTION OF ELECTRICITY

Electrogenic Fishes
Structure of Electric Organs
Functions of Electric Organs

SOUND PRODUCTION IN FISHES

Nature of Sounds
  Stridulation
  The Gas Bladder and Sound Production
  Miscellaneous Sources of Sound
Significance of Sound Production

ishes are attuned to a world of vibration, both electromagnetic and mechanical. As aquatic creatures, living in a medium through which these kinds of stimuli are propagated with particular speed and efficiency, they have evolved special sensitivities that we are unable to appreciate. In many ways, water is superior to air as a medium for the transmission of both acoustic and electrical stimuli. Sound travels through water at a speed ranging from 1,500 to 1,540 m sec^{-1} (the variation reflects ambient temperature and salinity)—nearly five times the speed of sound through air. Sound can travel for long distances in water with little attenuation; sound waves can be reflected off the bottom, off the surface, and off density boundaries caused by differences in temperature or salinity. While penetrating the tissues of fishes, sonic vibrations stimulate a variety of receptors, and these signals, when transduced into nerve impulses, enable the recipient fish to make appropriate locomotor and behavioral responses. The transmission of electromagnetic emissions is a function of the amount of ions dissolved in the water. Electroreception is a valuable tool that fishes use to gain information about their immediate environment. As the likeliest sources of electrical emissions are other animals, electroreception is valuable as a means of detecting and capturing prey, avoiding predators, or communicating with conspecifics. Although terrestrial vertebrates have evolved vocalization capabilities derived from the movement of air through their respiratory passageways, fishes have developed other means of sonic communication, using a variety of their anatomical features, including the swim bladder, skeletal elements, and associated musculature specifically modified for the generation of sounds.

STRUCTURE OF AUDITORY, MECHANOSENSORY, AND ELECTROSENSORY SYSTEMS

To the physics student, acoustic vibrations are pressure variations accompanied by a longitudinal oscillation of the particles through which the vibration travels. For most fishes, the frequency range of interest is between 50 and 1,000 Hz (Schellart and Wubbels, 1998). Being dependent on a conductive medium, electrical activity is invariably associated with aquatic organisms. Fishes are the perfect organisms in which to study the adaptation to and exploitation of these two physical phenomena. Mechanosensory function in fishes is the domain of the octavolateralis system. This system consists of three quite different components—the auditory, equilibrium, and lateral line systems—that all use a common type of receptor cell, the hair cell (Schellart and Wubbels, 1998). Electrosensory capacity is also generally included as a component of the octavolateralis system. The term *octavolateralis* is an acknowledgment of the common use of the octaval (eighth or auditory) cranial nerve and the lateral line nerves for processing sensory information, but does not necessarily imply that the three systems share common evolutionary origins (Popper and Fay, 1999). In fishes, the octavolateralis system includes the inner ear (fishes have no middle or outer ear), the neuromasts and canals that make up the lateral line system, and the ampullary and tuberous organs of the electrosensory lateral line. The functions of the ear appear to be primarily balance and sound reception. The organs of the lateral line respond mainly to the displacement of water and to pressure. The ampullae and tuberous organs sense electrical fields and biologically generated electrical signals.

Auditory Systems: The Inner Ear

Vertebrate auditory systems arose very early in the evolution of vertebrates. In some cases, the basic tasks of the inner ear, such as the discrimination of acoustic features, the localization of sound sources, and the analysis of the frequency of acoustic emissions, are accomplished by fundamentally similar structures that have become modified in the adaptation of vertebrates to specific habitats. In other instances, the acoustic "soundscape" may have dictated novel means of detection and processing of sensory input (Fay and Popper, 2000). In the late 19th century, the concept of the "acousticolateralis" system emerged from the belief that the ear arose as an invagination of the anterior part of the lateral line. Although a common embryonic origin and innervation pattern for the ear and lateral line system has since been disproved (Popper and Fay, 1999), the two systems share many structural and functional features.

In chondrichthyan fishes, the inner ear is embedded in the cartilaginous otic capsule of the neurocranium (see Chapter 3). It is composed of a triangular membranous sac, the vestibule, and a membranous labyrinth composed of three semicircular canals. The ear of bony fishes is composed of the osseous labyrinth, a cavity including ducts within the bones of the otic capsule, and the membranous labyrinth within the osseous structure (Lewis et al., 1985). The membranous labyrinth typically consists of two or three more or less distinct chambers—the utriculus, sacculus, and lagena—and three semicircular canals (Fig. 21.1). The vestibule of chondrichthyans is also subdivided into utriculus, sacculus, and lagena. The canals and the utriculus constitute the pars superior of the organ and are mainly involved in balance and the detection of angular acceleration, and the sacculus and the lagena are the pars inferior and are mainly involved in hearing (Fay and Popper, 1980; Popper, 1983; Popper and Fay, 1999). A few fishes use the utriculus also in sound detection. In the herrings, for example, hearing is enhanced by projections of the gas bladder that run forward adjacent to the utriculus. These two parts of the inner ear are nearly separated in some minnows (Cyprinidae) and are completely separate in featherfin knifefishes (Notopteridae) and some gobies (Gobiidae). The size of the various parts of the ear varies among groups or species; in the bowfin (*Amia*) and Ostariophysi, the lagena is larger than the sacculus, but the converse is true for most other groups. The semicircular canal system is reported to be larger than the pars inferior in the flyingfishes (Exocoetidae) and goosefishes (*Lophius;* Platt and Popper, 1981).

The hard, calcareous deposits in the inner ear, termed otoliths, are essential for both hearing and maintenance of balance in fishes. These structures consist of crystalline calcium carbonate embedded in a protein matrix. A recently discovered gene, named *starmaker*, has been implicated in their morphogenesis (Söllner et al., 2003). Otoliths differ greatly in size and shape among species and are often so distinctive that they can be used for species identification, even in larval fishes, and for phylogenetic analysis (Assis, 2003; Chen and Yan, 2002). The accretion of material to otoliths as they grow is usually in a regular pattern of layers, so that age and other life history features of individuals of many species can be assessed by the study of the otoliths using appropriate methods, including such sophisticated techniques as scanning electron microscopy and X-ray tomography (Hamrin et al., 1999; Radtke, 1984; Waldron and Gerneke, 1997). Analysis of isotope signatures in the otoliths of the weakfish (*Cynoscion regalis*) has demonstrated that this species possesses a strong homing instinct, so that several reproductively discrete populations exist (Thorrold et al., 2001). Because of their calcareous nature, otoliths may

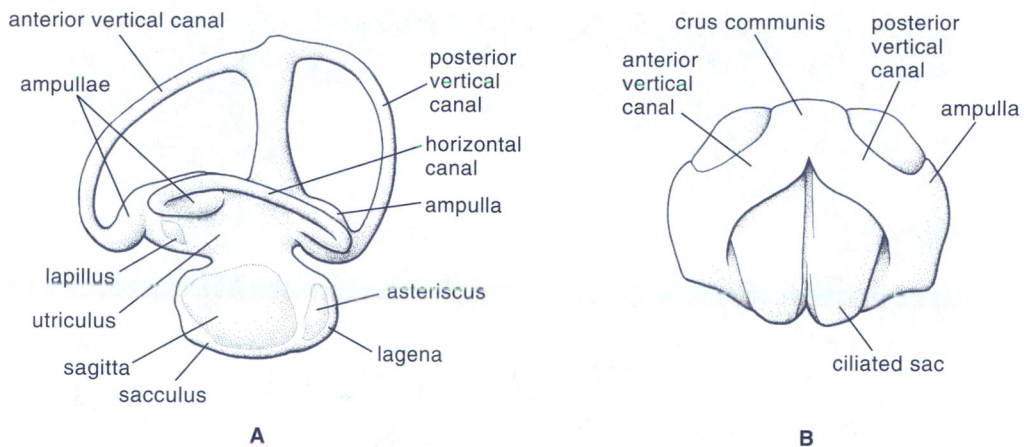

FIGURE 21.1
Membranous labyrinth of **A,** Cutthroat trout (*Oncorhynchus clarkii*) and **B,** Pacific lamprey (*Lampetra tridentata*), both showing the left labyrinth from left.

become preserved as part of the sedimentary history of the Earth. In this capacity, their chemical composition has been used to reconstruct past climate history (Andrus et al., 2002; Ivany et al., 2000; Patterson, 1999; see Chapter 30).

In sharks, rays, and chimaeras, an **endolymphatic duct** runs from the vestibule to an opening on top of the head. In bony fishes, this duct is abbreviated or lacking. The utriculus is the place of attachment of the three semicircular canals, each of which is oriented in a different plane in relation to the others: One is horizontal, situated on the lateral aspect of the utriculus; the other two are vertical, one posterior and one anterior, at right angles to each other, placed so that each is at an approximately 45° angle to the axis of the body as viewed from above. The vertical component of the labyrinth, to which the semicircular canals are attached dorsally—termed the **crus communis**—is continuous with the utriculus. Lewis et al. (1985) reported that in some elasmobranchs, the semicircular canals do not connect to the utriculus.

Each semicircular canal has an ampulla at its junction with the utriculus (the ampulla of the horizontal canal is at the anterior connection). Rising from the floor of each ampulla is an eminence called **a crista,** on which is a gelatinous ridge or **cupula,** covering hairlike, ciliary extensions of the hair cells, which are the receptor cells of the inner ear. The sensory organ (**macula**) consists of a group of sensory hair cells and supporting cells (Montgomery, 1988; Schellart and Wubbels, 1998).

The crista and cupula form what is regarded as a flexible diaphragm in the ampulla (Platt, 1983). Movement of the head in any direction will cause the **endolymph** of the

canals to deform one or more of the cupulae, bending and thus stimulating the hair cells of the maculae.

The sacculus is attached ventrally to the utriculus; the lagena, which attaches to the posterior part of the sacculus, may be well delineated but is not distinct in many species. In all three of these organs, the hair cells are located on epithelial maculae, on which the otoliths lie. The maculae consist of a basement membrane that supports numerous hair cells, each with its bundle of filaments. Each bundle has a long, true cilium (**kinocilium**), with several shorter filaments (**stereocilia**) that are actually microvilli, graded in length, alongside it (Platt, 1983; Popper and Coombs, 1980; Schellart and Wubbels, 1998).

The otoliths of elasmobranchs consist of calcareous granules (**otoconia**) in a soft matrix and have been reported to include mineral particles, such as sand grains, that enter through the endolymphatic duct (Montgomery, 1988). These exogenous particles are called **otarena**. In most bony fishes, the otoliths (Fig. 21.2) are hard structures, and those of most species have a characteristic shape and size. These are generally held in place by a gelatinous membrane. The otoliths of the utriculus, sacculus, and lagena are called, respectively, the **lapillus, sagitta,** and **asteriscus.** In the labyrinth of lampreys, elasmobranchs, and many bony fishes, there is a macula (*macula neglecta*) without an otolith (Lewis et al., 1985).

Hair cells on the maculae are disposed in groups with similar orientation, so that the kinocilium is on the same side of each of the cells. Because of the contours, shapes, and placement of the various maculae and the pattern of polarity,

FIGURE 21.2
Photograph of medial aspect of otolith of rainbow trout (*Oncorhynchus mykiss*), showing growth rings.
(Photograph courtesy of John McKern.)

there are hair cells oriented in different planes in relation to the axis of the body (Fay and Popper, 1980; Lewis et al., 1985; Platt and Popper, 1981; Popper and Fay, 1999; Schellart and Wubbels, 1998). The saccular macula of teleosts can have four or more regions of different hair cell orientation; the lagenar macula generally has two regions of opposite orientation (Platt et al., 1989; Popper, 1983; Popper and Fay, 1999). The cilia bundles of the hair cells differ in length and arrangement in various sensory areas. Those in the semicircular canals have kinocilia that are much longer than those in the maculae, and the stereocilia range up to about half the length of the kinocilium. The bundles of cilia in the otolithic organs of several species of bony fishes studied by Platt and Popper (1981) generally varied in length and in relative lengths of the kinocilia and stereocilia.

The membranous labyrinth of hagfishes consists of a lower chamber that has a single macula, associated with a membrane that carries mineral inclusions (mostly calcium phosphate, or apatite, with some calcium carbonate), which serve the function of an otolith (Lewis et al., 1985). Above the lower section, a continuous canal forms an arch. This has been called a single semicircular canal, but it has also been considered to represent both the anterior and posterior vertical canals, because ampullar swellings with annular cristae appear near the junctions with the pars inferior. Ross (1963) noted a tiny endolymphatic duct.

In lampreys, the labyrinth is more complicated. The common macula is functionally divided into anterior horizontal, vertical, and posterior horizontal parts that appear to correspond to the sacculus, utriculus, and lagena (see later;

Lowenstein, 1971), but it is still covered by an "otolith" made up of calcareous crystals, mostly apatite. There are two vertical semicircular canals, with divided cavities. Two ciliated sacs, which constitute a large portion of the labyrinth, are not seen in any other vertebrate. The function of these structures is not well known.

The otolithic maculae and ampullar maculae of the inner ear are innervated by cranial nerve VIII (the auditory, acoustic, or octaval nerve; see Chapter 26). Each hair cell is in contact at its base with an afferent and an efferent nerve fiber ending. The ratio of hair cells to neurons entering the maculae in the burbot (*Lota lota*) and in the bowfin (*Amia calva*) is about 10 to 1 (Popper, 1983) and can be far more in other species (Popper and Fay, 1999). Although their mode of action is poorly understood, the efferent neurons are known to modulate the responses of the hair cells (Popper and Fay, 1999).

The Lateral Line System

Whereas mechanoreception in terrestrial vertebrates is restricted to the sense of hearing—the inner ear transduction of air pressure waves from propagated sounds—mechanoreception in aquatic vertebrates is much more of a "whole-body" process, involving receptor cells at numerous locations along the body surface. What we term the **lateral line system** is a complex and sophisticated array of hair cells and associated structures that serve to transduce and relay the stimulus to the nerve cells. These hair cells constitute the mechanosensory component, whereas the electrosensory component consists of specialized **electroreceptor cells**. Both mechanoreceptors and electroreceptors are innervated by lateral line nerves (Wullimann, 1998). Zebrafishes (*Danio rerio*) have become the established model in investigations of vertebrate genetics and morphogenesis (see Chapter 27, Going Deeper), including investigations of the development and innervation of the lateral line (Dambly-Chaudière et al., 2003).

Mechanosensory Components

Lateral line organs may be free **neuromasts** on the skin or in pits, or they may be located in canals or grooves on the head and body (Fig. 21.3). The lateral line canals typically open to the surface through pores penetrating the bones or scales (Figs. 21.3 and 21.4). The cephalic lateral line canal courses through a distinct subset of skull bones. Usually, these canals through the bones are of small diameter, but in some groups, such as the drum family (Sciaenidae), they are spacious. Different sizes of canals in some fishes may allow for tuning to different frequencies (Denton and Gray, 1988, 1989; Platt et al., 1989). The lateral line canal system of the coelacanth (*Latimeria*) appears to consist of a combination of features reminiscent of the canal systems of ancient vertebrates and those of more modern fishes (Hensel and Balon, 2001).

Herrings and their close relatives (Clupeidae) are peculiar in that they possess a tubelike intracranial space called the **lateral recess (recessus lateralis)**, which connects the head canals of the lateral line system with the inner ear and anterior bullae of the gas bladder. It is situated between the pterotic and frontal bones above and the prootic bone below. A membrane separates the lateral recess from the labyrinth of the auditory organ. Extremely small changes in position of the membrane, due to changing pressure, can cause a flow within the perilymph around the labyrinth of the inner ear (Blaxter et al., 1989).

Typical neuromast receptor organs of the lateral line system are found not only in canals (**canal neuromasts**; Fig. 21.3) but also on the skin (**superficial neuromasts** or **pit organs**). These organs may differ in size, in the number of hair cells, and in the number and morphology of stereocilia in each hair cell. Superficial neuromasts are of several types, some sitting flush with the epidermis with hair cells extending into the water, some in pits either in the dermis or epidermis, and some on papillae (see Coombs et al., 1988, for terminology). Superficial neuromasts are usually arranged in series, either as accessory lines oriented with the lateral line canals or as replacement lines (or groups) that take the place of evolutionarily lost canals (Coombs et al., 1988). In larval sculpins, all neuromasts are superficial, but many become enclosed in canals as the fish matures (Jones and Janssen, 1992).

The hair cells of a neuromast are covered by a single cupula. Adjacent hair cells within a neuromast usually show opposing orientations (e.g., with the kinocilia oriented headward in some and tailward in the others). Water movement striking the side of the fish sets up an impulse in the fluid of the lateral canal. This causes slight deformation of the cupulae, thus bending the hair cells, which elicits a response transmitted into the nerve innervating the neuromast. Lateral line canals are hydrodynamic detectors that respond to acceleration (not sound) close to the source of vibration (Kalmijn, 1989).

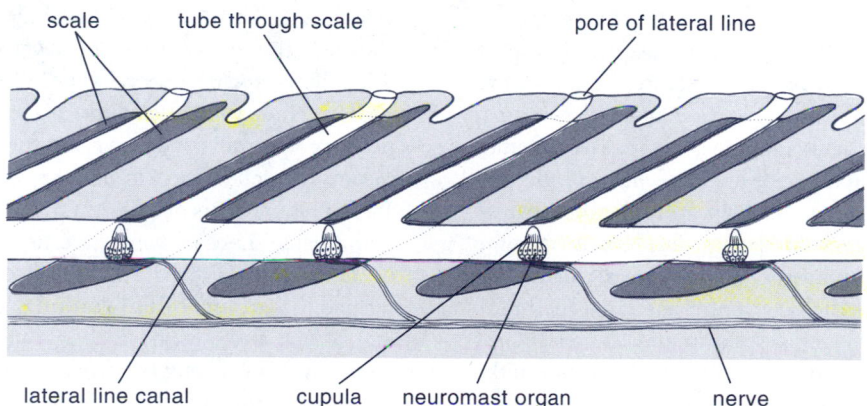

FIGURE 21.3
Relationship of lateral line canal to scales in a typical teleost. The drawing represents a horizontal section, with the thickness of the scales and the size of the lateral line exaggerated.

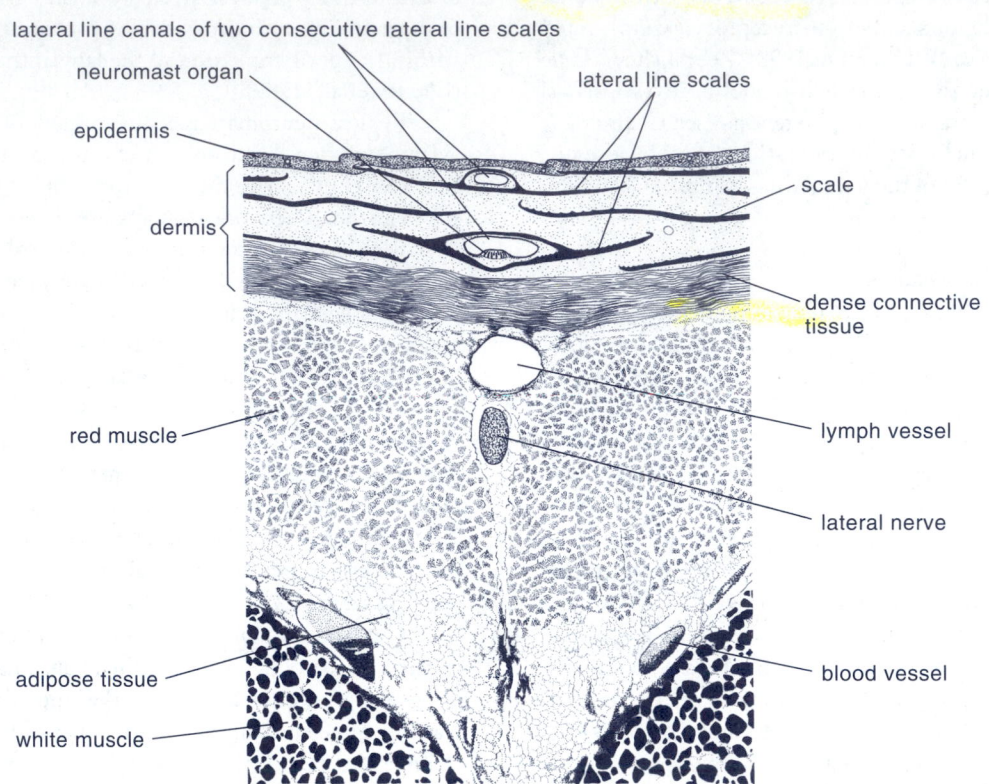

FIGURE 21.4
Relationship of lateral line to scales, skin, and muscles.

Electrosensory Components

The ability to respond to weak electrical fields is a plesiomorphic (see Chapter 4) character of vertebrates that was present in the earliest fishes, lost in the basal actinopterygians, but independently developed in at least two teleostean taxa (Wullimann, 1998). Presumed electroreceptors have been noted in fossils of agnaths, acanthodians, osteostracans, lungfishes, and rhipidistians (Hopkins, 1983). Alves-Gomes (2001) has reviewed the development of electrogenic and electroreceptive capacities in teleosts. Among extant fishes, electroreception is known in lampreys, elasmobranchs, some primitive nonteleosts, and some teleosts—particularly the osteoglossomorphs and the ostariophysans (von der Emde and Schwarz, 2001). It is interesting to note that electroreception does not appear to be present in hagfishes, nor in the closest extant relatives of the teleosts, the gars (*Lepisosteus*) and the bowfin (*Amia calva*). Mormyrids and gymnotoids are most remarkable among the teleosts because their electroreceptive capacities are used in conjunction with their ability to generate weak electrical fields for purposes of orientation and communication. Among these, the electric eel (*Electrophorus electricus*) is undoubtedly the most notorious of the electrogenic fishes, owing to the potency of its electrical discharge. Von der Emde (1998) provided a phylogeny of fishes showing the distribution of electroreceptive capability.

There are two general types of electroreceptors: **ampullary** and **tuberous.** Early Palaeozoic fishes apparently possessed ampullary electroreceptors, and they are known in lampreys, chondrichthyans, sarcopterygians, chondrosteans, and four orders of teleosts (Alves-Gomes, 2001). In elasmobranchs, these receptors are usually termed **ampullae of Lorenzini** (Fig. 21.5). Specialized sensory structures, termed the **organs of Fahrenholz,** are found in lungfishes and bichirs and have been shown to be homologous to the ampullae of Lorenzini (Northcutt, 1986). Ampullary organs are found in other nonteleost fishes, such as the sturgeons and paddlefishes. Among teleosts, ampullary organs are known in catfishes, gymnotiform knifefishes, xenomystine knifefishes (Notopteridae), and mormyriforms (Jorgensen, 1989; Zakon, 1986). Lampreys have electroreceptors, but they are

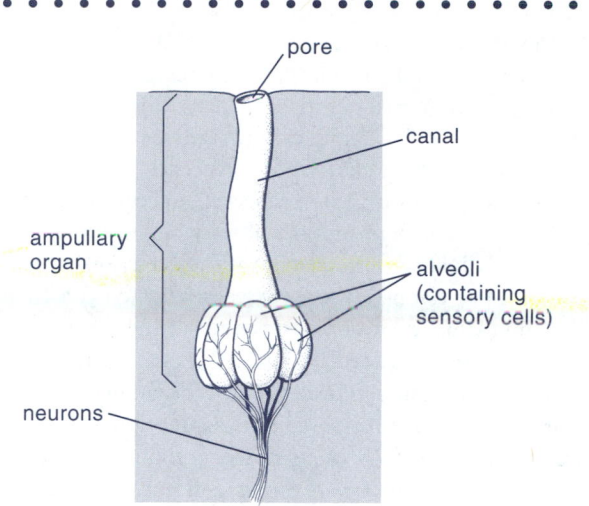

FIGURE 21.5
Ampulla of Lorenzini from an elasmobranch fish.

not of the ampullary type. Instead, they are end buds that formerly were thought to be taste or lateral line receptors. These receptor cells are set in the epidermis with the apical ends (of which there are up to 25) extending to the surface. Electroreceptors are somewhat concentrated on the head of lampreys, but they are distributed sparsely on the body as well (Ronan, 1986).

In elasmobranchs, the ampullae are usually distributed in clusters on the head and on the pectoral fins of skates. These clusters take their names from their respective placements and innervation—supraorbital, buccal, hyoid, and mandibular. At the surface of the skin are pores that open into the canal or tubule of each ampulla. In the Rajidae and many other rays, the canals are both large in diameter (about 1 mm) and long in relation to those of sharks, and the canals spread widely from the clusters. In freshwater stingrays, such as the Potamotrygonidae of South America or *Dasyatis garouaensis* of Africa, the ampullae are smaller and less complex than in their marine relatives. They are not arranged in clusters but are distributed over the body. The canals are short, so that the microampullae are nearly at the surface (Raschi and Makanos, 1989; Zakon, 1986). Canal length in ampullae is apparently an evolutionary response to the differing resistivity of the fish body in relation to the freshwater or salt water medium. In salt water, the body fluids are more resistant than the medium, and a long duct aids in increasing the potential along the receptor organ. In freshwater, the canals are shorter.

A conductive, gelatinous material fills the canal of ampullary organs, but the walls are electrically resistive (Montgomery, 1988) or at least passive (Zakon, 1986). The composition of this mucopolysaccharide gel may vary along the length of the canal (Whitehead et al., 1999). The sensory cells are set into the epithelium of the walls of the ampullae, with a kinocilium (in elasmobranchs) or apical microvilli (in teleosts) exposed to the lumen of the ampulla (Bullock, 1981). The receptor cells are separated by supporting cells, which are arranged so that current will not flow between adjacent cells but through them. There are hundreds of sensory cells per ampulla (Zakon, 1986). Whitehead et al. (1999) identified two types of receptor cells in the ampullae of the catfish *Arius graeffei*, an inhabitant of fresh, brackish, and fully marine waters of northern Australia. Electroreceptors are well distributed over the head and body in electroreceptive bony fishes, except in sturgeons, paddlefishes, bichirs, and the South American and African lungfishes, in which they are found on the head only. Ampullary organs have not been demonstrated in *Latimeria*, but there is a large rostral organ that is suspected to be electrosensory (Northcutt, 1986; Northcutt and Bemis, 1993).

In the electroreceptors found in mormyroids, gymnotoids, and various other freshwater fishes that are ampullary in nature, each ampulla has a canal about 100 to 200 μm long leading to the surface. The other main category of electroreceptor is the tuberous organ, which typically has no opening to the exterior. The "canals," if present, are filled with loose epithelial cells (Zakon, 1986). Tuberous electroreceptors appear to be more sensitive to higher electrical frequencies than ampullary ones, and they are most closely associated with the weakly electrogenic mormyrid and gymnotiform fishes. Consequently, these types of receptors are used in the detection of **electric organ discharge (EOD;** see later). The differences in the histology of tuberous organs found in various gymnotiforms and mormyrids have resulted in the identification of different kinds of tuberous receptors, including knollenorgans, gymnotomasts, and mormyromasts (Kramer, 1990; Szabo, 1974; Zakon, 1986). The electromotor corollary discharge pathways that link electric organ discharge with electroreception specifically suppress the sensitivity of the knollenorgans in mormyrids (von der Emde, 1998). Mormyrids, therefore, show an extremely sophisticated filtering device that prevents the brain from self-stimulating through the highly sensitive knollenorgans when electrical signals are generated. In some respects, mormyrids have the most sophisticated of the electroreceptive and electrogenic systems: Two separate pathways from receptor organs into the brain are maintained—one for electrocommunication and the other for electrolocation (von der Emde, 1998).

The means by which electroreceptive fishes process sensory information appear to be remarkably complex and subtle. Osteoglossomorph fishes appear to be "prewired" for the transmission of electrosensory information. Using a fluorescent neuronal tracer, Wullimann and Roth (1994) determined that the non-electroreceptive osteoglossomorph *Pantodon buchholzi* possessed the same telencephalocerebellar neuronal pathway as that used for electroreception by mormyrids. Refining sensory input is possible through the aforementioned generation of electromotor corollary discharge. Sensory processing can best be understood as the generation of expectations or predictions about sensory inputs and the removal of these expected inputs from the incoming flow of sensory information (Bullock, 1988). This permits novel sensory input to be more readily distinguished. Octavolateral sensory processing embodies this within the cerebellum (or cerebellumlike structures) of electroreceptive fishes (Bell et al., 1997a, 1997b, 1999). Here, an adaptive form of sensory processing occurs as learned predictions about sensory input are generated but then deleted from the actual sensory input, thus permitting unpredicted inputs to be more readily discerned. Removing the predictable or redundant maximizes the information content of the sensory input.

. . . .
Other Sensory Receptors

Additional cutaneous sensory receptors found in some elasmobranchs are the **vesicles of Savi,** whose function is still under investigation. These are found in the snout region and along the anterior edge of the pectoral fins of electric rays, mostly on the ventral surface. Similar vesicles are found in some other families of rays (Barry and Bennett, 1989). Usually, there are about 200 vesicles, with only 30 to 40 on the dorsal surface of the snout. According to Barry and Bennett (1989), these are mechanoreceptors sensitive to vibrations up to 350 Hz (cycles per second).

Additional organs of possible proprioceptive function are the spiracular organs of sharks, rays, and such primitive bony fishes as the sturgeons and bichirs (Barry and Bennett, 1989) and the subcutaneous corpuscular pressure receptors seen in sharks (Bone and Blaxter, 1999). **Proprioceptors r**espond to changes in position, balance, and the use of muscles, yet muscle proprioceptors are poorly represented in fishes (see Chapter 22, Coda). These organs, where present, are similar in structure to the typical lateral line sensory organs.

The physical stimuli of touch and temperature are primarily received by fishes through free nerve endings in the skin. Although the ampullae of Lorenzini have been shown to react markedly to changes in temperature, this property has nothing to do with their primary function. Furthermore, some sharks, searobins, and eels have corpusclelike structures that may be specialized touch receptors.

Nociception—Can Fishes Experience Pain?

One question concerning the sensory biology of fishes that has received considerable attention, largely due to activism on the part of animal rights groups, is whether fishes can perceive pain or not. Animal rights groups are opposed to angling, as they claim that fishes experience pain from being hooked. Research on the presence of nociceptors in fishes is taken as evidence in support of this contention. Sneddon et al. (2003) defined **nociception** as the detection of noxious stimuli. They claimed that their studies, which consisted of the administration of noxious stimuli to the mouth region of rainbow trout, showed evidence of awareness of pain and discomfort by fishes. University of Wyoming researcher James Rose strongly opposed this contention. He claimed that behavioral responses to noxious stimuli are separate from the perception of pain—that nociception is a response to a broader array of threatening stimuli. Rose (2002) asserted that fishes, lacking the higher centers of neural processing in the cerebral cortex that are responsible for fear or suffering, are thus unable to feel pain. One thing is certain, however—until we can hear it from the fishes themselves, we will be unable to provide a definitive answer to this question.

. . . .
FUNCTION OF AUDITORY, MECHANOSENSORY, AND ELECTROSENSORY SYSTEMS
. . . .

The inner ear and lateral line systems of fishes work in concert to provide the individual with a broad array of mechanosensory information, but they differ in the type of stimuli to which they respond. The inner ear is active in the detection of changes in position due to acceleration in any plane and in the detection of position changes with respect to gravity (equilibration); it is also active in the detection of sound. The mechanosensory component of the lateral line responds mainly to the movement of water over the receptors and, as such, is more sensitive to stimuli that arise closer to the individual. Water movements may arise from many sources, such as currents, the activity of other animals, or movement of the fish itself. Whereas the inner ear is innervated by cranial nerve VIII (octaval), the lateral line receives innervation from three to five different cranial nerves (Coombs and Montgomery, 1999).

. . . .
Membranous Labyrinth

Sound Reception

Acoustic energy in water is expressed in the form of compression waves that are accompanied by particle displacement—the latter component being sensed by the mechanoreceptive lateral line. Because water is so much denser than air (about

800 times as dense), a greater amount of energy is required to generate sound in water; but when the sound is propagated, it travels at about 4.5 times the speed of sound in air (approximately 1,500 m sec^{-1} versus 330 m sec^{-1}) and is not rapidly attenuated. Low-frequency sound propagation is relatively easier in water than the propagation of high frequencies. Fishes lacking specialized mechanisms for the enhancement of acoustic signals depend on sensory cells in the inner ear that detect the displacement of fluid particles in the endolymph. Because of the limited amplitude resulting from particle motion, acoustic "generalists" tend to have a comparatively narrow range of frequencies that they can sense and also to have higher hearing thresholds (Hong Yan, Marine Research Station, Academia Sinica, personal communication). Fishes classified as acoustic "specialists" possess structures such as the otophysic connections discussed hereafter; these are used to sense compression waves transmitted to the inner ear. These forms can thus detect sounds by sensing both compression waves and particle displacement.

Many natural physical processes cause sound in water. Earthquakes, volcanism, winds and the water movements caused by them, rain, movement of bottom materials by currents or waves, and action of ice all provide background noise. Anthropogenic sound may be evident in certain places and may consist of noise from ships and submarines and from industrial operations both on shore and in the sea. Biological sound, either incidental or deliberate, may arise from animals. In addition to fishes, mollusks, crustaceans, and marine mammals are all sound producers. Sounds from all biological sources may have significance for fishes (e.g., in finding mates or prey or in escaping from predators). Dolphins may even be able to disable prey fishes with sound (Norris and Mohr, 1983).

At one time, it was widely believed that fishes were deaf. However, careful study has revealed that fishes are sensitive to sounds and that, in most species, the sonic energy is received by the sacculus and lagena (Hawkins, 1973; Jenkins, 1989). Compression waves cause movement of the maculae in relation to the otoliths that rest on them. Because fishes have a density very close to that of water, pressure waves cause them to vibrate along with the water. The heavier otoliths do not vibrate at the same time or rate as the rest of the fish, so there is movement in relation to the beds of hair cells on which the otoliths lie, and the sensory cilia are bent and stimulated.

Responses to a wide range of frequencies have been measured in fishes, but the response is related to the auditory equipment of the various species. Generally, species with a functional connection between the gas bladder and the ear (otophysic connection) have greater sensitivity than those without such a connection and respond to a greater range of frequencies. With some exceptions, such as the hardhead catfish (*Arius felis*), otophysan members of the Ostariophysi (see Chapter 10) appear to have exceptional powers of hearing in terms of both sensitivity and range of frequency (Popper and Coombs, 1980). In these fishes, the Weberian apparatus, which consists of bones modified from the first few vertebrae and their processes, forms a connection between the anterior chamber of the air bladder and the labyrinth (Fig. 21.6). The ossicles through which these vibrations are conducted are the tripus (a crescent-shaped structure in contact with the anterior part of the gas bladder) and the smaller intercalarium, scaphium, and claustrum. The claustrum is in contact with the walls of the membranous labyrinth, which in these fishes is modified so that the left and right organs coalesce posteriorly to form the sinus impar ("impaired sinus"). In otophysans, the lagena and the lagenar otolith (asteriscus) are larger than the sacculus and the saccular otolith (sagitta), but the sensory epithelium of the sacculus may be as extensive as that of the lagena (Platt and Popper, 1981).

Examples of otophysan fishes are the goldfish (*Carassius auratus*) and the loach (*Nemacheilus barbatula*), which are known to respond to sounds up to about 3,500 Hz, and the minnow (*Phoxinus*), which in various experiments has responded to a range of frequencies from 20 to 7,000 Hz. Another otophysan, the brown bullhead (*Ameiurus nebulosus*), has been reported to have absolute frequency limits of 60 to 10,000 Hz (Poggendorf, 1976). Sensitivity is usually greatest at much lower frequencies than at the upper limits. For instance, the goldfish and brown bullheads have hearing thresholds in the 200 to 1,000 Hz range (Hawkins, 1981; Tavolga, 1971).

In a variety of fishes, there are direct connections between the gas bladder and the ear. The codlings or deep-sea cods (Moridae) have large branches of the gas bladder in contact with the skull. In the mormyrid fishes, small portions of the gas bladder become separated during early development and are enclosed within the skull in contact with the ear. Although the mormyrids are undoubtedly best known for the role of EOD and electroreception in communication and orientation, they can also generate sound and display remarkable acoustic abilities. The acoustic sensitivity of mormyrids is significantly enhanced by the presence of these small, gas-filled *tympanic bladders* associated with the sacculus in the inner ear (Fletcher and Crawford, 2001; Yan and Curtsinger, 2000).

Herrings and anchovies have gas-filled bullae, attached to diverticula of the gas bladder, in close connection with the utricular area of the ear. A membrane within each bulla separates the gas from a section of the bulla filled with perilymph, which is in intimate contact with the utriculus. An elastic

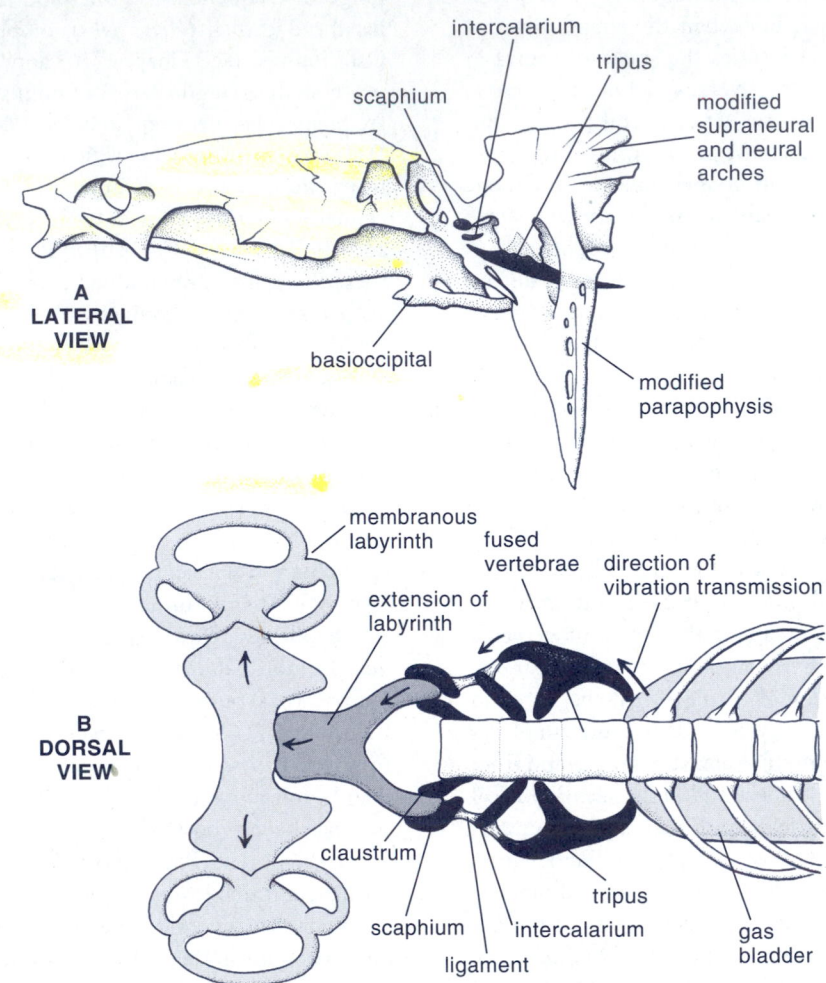

FIGURE 21.6
Weberian apparatus. **A,** Skull of *Catostomus,* showing the modified first three vertebrae and their relationship to the left Weberian ossicles (in black, claustrum not shown); **B,** Illustration of the relationship of the Weberian ossicles (in black) to gas bladder, vertebrae (shown here unmodified and unfused), and the extension of the sinus impar of the membranous labyrinth.

thread extends from the membrane to the ear. The recessus lateralis is also in close contact with the utriculus. Herrings respond to frequencies from 0.01 to 1,000 Hz—a range that extends from depth-related pressure changes to high-frequency sound (Blaxter, 1981; Denton and Blaxter, 1976). Shad (*Alosa sapidissima*) can detect sounds from 100 Hz to 180 kHz. The evolution of sensitivity to ultrasonic emissions apparently has enabled shad to detect echolocating odontocete predators, such as porpoises (Mann et al., 1998). Ultrasound sensitivity appears to be largely restricted to clupeiform fishes—possibly just to members of the subfamily Alosinae (Higgs et al., 2004; Mann et al., 2001). Some soldierfishes (Holocentridae) of the genus *Myripristis* have a close connection of the gas bladder with the ear and exceed many Otophysi in their response to high frequencies. They are able to detect frequencies up to about 3,000 Hz (Coombs, 1981). Featherfin knifefishes (Notopteridae) have an otophysic connection but do not have especially acute hearing. Other families with otophysic connections are the tarpons (Megalopidae), porgies (Sparidae), and bigeyes (Priacanthidae). Drums (Sciaenidae) and triggerfishes (Balistidae) have their gas bladder in close relationship to the skull. A strange adaptation for contact between the gas bladder and the skeleton is seen in the escolars (Scombrolabracidae), in which several vertebrae develop hollows into which diverticula of the gas bladder extend. Whether this has a relationship to hearing is unknown (Bond and Uyeno, 1981).

Species with air-breathing chambers in the head, such as climbing perch and snakeheads, maintain a bubble of air in the suprabranchial cavity close to the auditory region. This bubble could act as a resonator. Liem (1967) noted a tympanum on the exoccipital foramen of the saccular bulla of *Luciocephalus*. This is near the suprabranchial cavity. Moreover, there are cranial lobes of the gas bladder that are in close proximity to the suprabranchial cavity. Gas bladders can increase the sensitivity of fishes to sounds and increase the range of frequencies detected (Platt et al., 1989). A gas-filled suprabranchial cavity might have a similar effect.

Fishes without a close connection between the gas bladder and the ear usually respond to a narrower range of frequencies than those with an otophysic linkage. The highest frequencies causing responses in these fishes are mostly below 1,200 Hz, often in the 300- to 600-Hz range (Tavolga, 1971). Responses to sound are elicited either by conditioning the fishes to show a specific behavior on receiving a sound or by inserting an electrode around the auditory nerve and measuring microphonic potentials resulting from stimulation of the ear by sound. Another approach is the measurement of acoustically evoked potentials using the auditory brainstem response (ABR) protocol (Kenyon et al., 1998). Here, auditory responses can be measured in a less invasive manner, as the recording electrodes are placed on the skin covering the skull in the region of the medulla.

Species without an otophysic connection appear to be less responsive to acoustic stimuli at higher frequencies. In one study that involved implanting electrodes in the ears of a tilapia and a channel catfish, both species had similar sensitivities to underwater vibrations up to 600 Hz. The response of the tilapia gradually diminished above that frequency, but was still evident at 900 Hz. The response of the catfish went as high as 4,000 Hz before diminishing. Deflation of the gas bladder made little difference in the tilapia, but the catfish lost some sensitivity at higher frequencies. Removal of gas from the gas bladder is known to reduce the sensitivity to sound at a given frequency, so sounds must be of higher amplitude to be heard. Placing a small, gas-filled balloon in contact with the head of a fish that naturally lacks a gas bladder will increase both its sensitivity to sound and its highest frequency of response (Blaxter, 1981). Mauthner neuron–mediated escape responses (see Chapter 26) have also been identified as initiated via transduction of acoustic pressure

stimuli from the gas bladder (Canfield and Eaton, 1990). Although the gas bladder is perceived as being valuable in the detection of hydrostatic cues important to the maintenance of proper depth, fishes lacking a gas bladder, such as the dogfish (*Scyliorhinus canicula*), have been demonstrated to be capable of hydrostatic pressure detection via hair cells (Fraser and Shelmerdine, 2002).

Fishes of various types have been shown to distinguish tones, with ostariophysans responding to frequency differences from 3 to 5 percent over the range of 200 to 500 Hz. Fishes other than ostariophysans are less able to discriminate frequencies, and some, such as sharks and eels, respond only to frequency differences of 50 percent or more. Fishes can also distinguish the intensity of sound.

Sound Location

Although early experimentation with directional hearing seemed to show that orientation to near-field sounds received by the lateral line was possible, that ability has been doubted (Schuijf and Buwalda, 1980). More recent research on the ability to localize sound in the far field has shown that directional hearing involves the detection of particle displacement or velocity as well as pressure. Terrestrial vertebrates might be seen as having an advantage in the detection of point sources of acoustic stimuli, as the lower velocity of sound propagated through the air would enhance lateral discrimination. Because the distance between the ears is comparatively small in fishes, especially larvae and juveniles, directional hearing through the differential perception of signals from the right versus the left ear is not possible (Schellart and Wubbels, 1998). Because of the differential orientation of the maculae and the orientation of hair cells in directional groups, researchers have theorized that stimulation of both ears can result in output that, when compared by the central nervous system, allows the fish to determine the direction from which the sound originated (Popper, 1983; Popper et al., 1988; Schellart and Wubbels, 1998). Sharks and some bony fishes have been observed to swim toward a sound source on many occasions (Corwin, 1981; Myrberg and Nelson, 1991).

• • • • • •
Equilibrium

Angular Acceleration

This is sometimes referred to as *dynamic equilibrium* and is detected by the ampullar maculae of the semicircular canals. The hair cells in that sensory epithelium, like the hair cells in other parts of the inner ear and the mechanical lateral line system, are pear shaped to cylindrical, and each has a single kinocilium that is longer, larger, and more complex than the numerous shorter stereocilia that form a sloping series, with the longest next to the kinocilium (Fig. 21.7). Bending the ciliary bundle toward the kinocilium inhibits the firing frequency of an associated neuron, whereas bending away from the kinocilium is excitatory. The vertical canals respond to rotation in any direction, but there appears to be the possibility of greater stimulation by movement on horizontal axes (Lowenstein, 1971).

Gravistatic Function

Detection of gravity, or *static equilibrium*, involves the utriculus in most fishes, although there is evidence that the utriculus is also active in sound perception in some groups (such as the herrings) and that in others, the lagena or part of the sacculus are involved in maintenance of equilibrium. Impulses from the utriculus control a series of reflexes that govern posture. Some that are easiest to observe are the curling of fins and the rolling of eyes as a live specimen is held at an angle. Change of position in relation to gravity is apparently detected by means of the deformation of sensory hairs as the utricular otolith (lapillus) shifts. If a fish is rotated, there will be initial stimulation of semicircular canal cristae as well as of the utricular maculae, but when the endolymph of the canals has stabilized, the otoliths will continue to deform the sensory hairs. Platt and Popper (1981) have reported that such a great diversity exists among fishes that generalizations about the division of function among parts of the inner ear and about a "typical" teleost ear may be untenable.

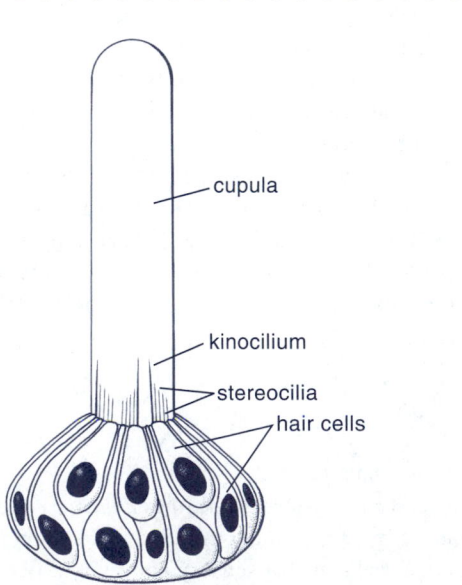

FIGURE 21.7
Lateral line organ (neuromast).

Mechanosensory Function

Neuromast receptor organs respond to deformation by mechanical stimuli, particularly water movement in relation to the fish. Thus, currents from whatever source (e.g., the fish swimming), water displacements caused by other organisms, and the small displacements caused by sources of sound at close range can be detected. The responses of canal and superficial neuromasts have been studied in cichlids and the rainbow trout (Münz, 1989). The canal neuromasts extend their sensitivity to higher frequencies than the superficial neuromasts. In the trout, canal neuromasts respond to water acceleration, and superficial neuromasts respond to water velocity. In the cichlids, superficial neuromasts respond to water velocity, but both types respond to water acceleration (Münz, 1989). The frequency range of 50 to 100 Hz seems to be optimal for reception by the lateral line, although it can respond to frequencies from below 10 Hz to about 500 Hz (Coombs and Janssen, 1989; Kalmijn, 1989).

Coombs and Montgomery (1999) provided a summary of lateral-line-mediated behaviors that included schooling, predator avoidance, hydrodynamic imaging, courtship communication, and feeding that involves the detection of prey on the surface or beneath the surface. Several studies (Coombs, 1999; Janssen and Corcoran, 1998; Janssen et al., 1999) have demonstrated the role of the lateral line in the detection of prey. The lateral line has been demonstrated to be useful for size discrimination in particulate feeders, such as the zooplanktivorous alewife (*Alosa pseudoharengus*), when prey densities preclude filter feeding (Janssen et al., 1995). The lateral line would seem to be especially important among species of nocturnal habit or species living in caves or the deep ocean. The latter two groups show extensive modification of the lateral line canals and organs.

Electrosensory System Function

The weak electric organs of mormyroids and gymnotiforms were once referred to as *pseudoelectric*, probably due to a lack of instruments sensitive enough to measure the discharges, although as early as 1841 they were assumed to have an electrical function. Once the discharges were confirmed, about 1880, the search began for the function of the organs. Matching the production of electrical fields with the presence of receptors responsive to the fields required clever experimentation and solid scientific reasoning. Because some electroreceptors were found to be sensitive to changes in temperature, they were first considered to be thermoreceptors.

Some apparently non-electrogenic fishes (sharks, rays, catfishes, and others) possess the electrosensory ampullae of Lorenzini or similar electrosensitive organs. These fishes are said to have **passive electrosensory systems,** because they react to externally generated electric stimuli only. In **active electrosensory systems,** the electric fishes generate stimuli to which their own tuberous electroreceptors are sensitive. The receptors in active systems can also react to external stimuli. Weakly electric fishes are capable of perceiving both the resistive and the capacitive components of objects in their environment. This imparts a sort of "color" to an object that is thus detectable in dark environments that preclude the use of vision (von der Emde and Schwarz, 2001).

Electroreceptors are located in the skin, in patterns that vary with the species. In general, ampullary organs, equipped with a canal opening to the exterior, are responsive to low-frequency stimuli of 0.1 to 50 Hz over long periods. These are called *tonic receptors,* in contrast to the tuberous organs, which do not open to the exterior and are sensitive to higher frequencies up to 2,000 Hz. The tuberous organs are *phasic receptors,* are not sensitive to direct current, and become insensitive to prolonged stimuli. Electroreceptors are innervated by the lateral line nerves.

The tuberous organs, found only in Gymnotiformes and Mormyriformes, show structural differences that apparently reflect differences in function. Some fishes have the ability to sense differences in waveform and frequency, and some have their receptors programmed for low or high thresholds.

Sources of electrical stimuli are both biological and environmental—caused not only by the activity of electrical organs or by other processes (such as secretory or muscular activity) of aquatic animals, but also by movements of water masses, atmospheric processes, and various geological and electrochemical processes.

Functions of electroreception include but may not be restricted to the location of objects, communication, and navigation (Hopkins, 1983; Kramer, 1990). Fishes with active electrical systems hold their bodies straight and swim by undulating the dorsal, anal, or pectoral fins. Some species move backward equally well as they move forward. The tail-first approach, seen often in gymnarchids, is probably advantageous in that the posterior sections of the electric organs are somewhat isolated from the rest of the body, so that maximum current density is set up around the tip of the tail. Distortions of the field by objects are thus maximized. Moreover, only the tail is exposed to the possible dangers of a new situation. The straight posture allows fishes to establish a symmetrical electric field around their bodies, and the extent of the field is governed by the resistance of the water and the nature of the electric discharge. The field approximates a dipole field in some species and in the young of other species. In apteronotids and rhamphichthyids, the anterior three quarters of the fish act as a distributed source of current. The tail, where the ends of the electric organs are

close to the surface and are not surrounded by body fluids, acts more like a point source (Fig. 21.8). Objects encountered in the field—whether they are good or poor conductors—distort the field. The high-frequency receptors sense the change in impedance, and the fish reacts appropriately to the information received. Although the range of this system is not great and probably operates best within only a few centimeters of the fish (Kramer, 1990), it is apparently of significance to nocturnal fishes and to those fishes that live in turbid waters.

Communication between electric fishes may have significance in such aspects of life as reproduction and the spacing of individuals. The range of communication may be from less than 50 cm to nearly 7 m, depending on water conditions and the species involved. Some species can quickly change the frequency of their pulses to avoid "jamming" by interfering frequencies. This change is termed the **jamming avoidance response (JAR;** Heiligenberg and Rose, 1985; Viete and Heiligenberg, 1991).

Passive electrolocation allows fishes to react to the fields that emanate from living organisms in the water and from inanimate sources. The activity of muscles establishes very small alternating current (AC) fields, and electric organs set up larger fields. Direct current (DC) fields from organisms originate from potentials involving the body fluids and the surrounding medium. Wounds are reported to strengthen the DC fields, so that a wounded organism can be detected electrically at a greater distance than can an uninjured organism.

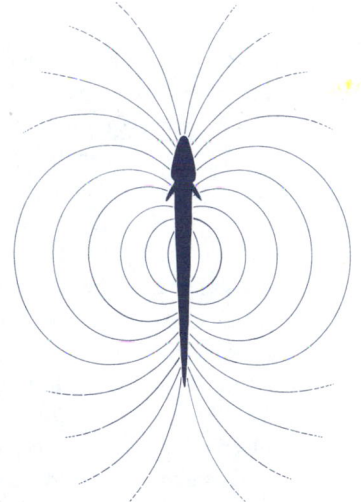

FIGURE 21.8
Representation of the electric field around a gymnotoid fish.

Certain sharks have been shown to locate flatfish concealed by covers that allowed the passage of an electrical field but prevented the passage of odors (Kalmijn, 1971). Many electrosensory species react to magnets or to other inanimate sources of electrical fields, and sharks will readily seek out a dipole field from electrodes set out to imitate the field around a flatfish.

Some skates have been shown to be sensitive to the Earth's magnetic field, and Kalmijn (1978) was successful in training stingrays to respond to magnetic fields. Elasmobranchs apparently can detect electrical gradients as low as 5×10^{-4} millivolt/cm (Feng, 1991), which is low enough that they can use the gradient caused by their swimming in the magnetic field of the Earth for orientation. The use of their sensitivity to weak electrical and magnetic fields for orientation is suspected in salmon and eels and is supported by some experimental evidence.

Walker et al. (1982) trained yellowfin tuna to discriminate between two Earth-strength magnetic fields, ambient and altered. The gradients were equal but opposite. Although this species has no ampullary organs, it has a concentration of magnetite in the ethmoid region that could respond to magnetic fields (see Chapter 36). Magnetite is known also from the otoliths of certain elasmobranchs (Vilches-Troya et al., 1984) and from the olfactory lamellae of rainbow trout (Diebel et al., 2000).

PRODUCTION OF ELECTRICITY

Electrogenic Fishes

Electric discharges in fishes are of two distinct types—strong and weak. Three taxa of fishes have long been known to generate electrical shocks powerful enough to stun other organisms. Physicists had to unravel the mysteries of electricity before its generation in the tissues and organs of fishes could be understood, however. The electric rays of the family Torpedinidae are believed to have been used by ancient Roman physicians in an early form of electrotherapy. The electric catfish (*Malapterurus electricus*) was featured in Egyptian hieroglyphics as early as 4,750 years ago, and has an Arabic name translating as "father of thunder." In South America, the electric eel (*Electrophorus electricus*), the most powerful producer of electricity among fishes, was undoubtedly well known by the indigenous cultures long before its discovery by European explorers. The electrogenic capabilities of weakly electric fishes, however, were confirmed only after instrumentation sensitive enough to study them became available, and the strong electrical capability of the electric stargazers (*Astroscopus*) was not recognized until the 20th century.

Electric fishes are a diverse assemblage, yet they share certain convergent features: They are generally slow-moving or sedentary, are active at night or in murky waters of low visibility, and have thickened skin that serves as a good insulator. Most have reduced eyes, and some electric rays are blind. Generally, the cerebellum is enlarged, greatly so in the mormyrids (Nelson, 1994; Nilsson, 1999). Electric organ discharge (EOD) is useful for purposes of orientation and navigation in waters where other senses, such as vision, might be impaired. It is also valuable as a means of communication among conspecifics. Electric organs are present in six orders of fishes, suggesting that the ability to produce electricity evolved independently at least six times—possibly more, depending on the phylogeny accepted (Feng, 1991). Strongly electric fishes are the electric rays (Torpedinidae, 10 genera), electric catfish (Malapteruridae, *Malapterurus*), electric eel (Electrophoridae, *Electrophorus electricus*), and the electric stargazers (Uranoscopidae, *Astroscopus*). The electric rays, or torpedoes, are widespread in marine waters, some living at considerable depths. They are benthic and slow, and some of the larger species, such as *Torpedo nobiliana*, are capable of delivering a shock of 220 volts. Electric catfish live in the murky waters of African rivers. They are known to reach a length of about 1 m and to produce shocks of 350 volts. The electric eel, an Amazonian gymnotiform species, is a sluggish fish like the previous two and lives in water of low visibility. This relatively large fish (at least one specimen was measured at nearly 3 m long) can generate pulses of up to 650 volts, but typical discharges are on the order of 350 volts. Electric stargazers, marine fishes of the western Atlantic, are sand burrowers that can deliver up to 50 volts.

Weakly electric fishes are in families that are mostly found in tropical freshwaters, but one group is marine. All are either benthic or semibenthic and rather sluggish. The skates (Rajidae), well known and nearly cosmopolitan in the oceans, include numerous species of small to moderate size. Mormyrids, the elephantfishes and their relatives, are freshwater fishes of Africa, as are their close relatives, the gymnarchids. The mochokid (synodontid) catfishes are also from Africa. Many of these fishes are nocturnal in habit, and most live in waters of relatively low visibility. In the freshwaters of South America, the knifefishes of the order Gymnotiformes (which also includes the electric eel) form a group of weakly electric fishes that are strikingly convergent with the African mormyriforms (Alves-Gomes, 1999). This order includes the families Gymnotidae, Apteronotidae, Sternopygidae, Hypopomidae, and Rhamphichthyidae. Some taxonomists consider the order to comprise but one broad family, the Gymnotidae (Robins et al., 1991).

Structure of Electric Organs

Electric organs consist of blocks of modified striated muscle fibers, termed **electroplaques.** The cells of electroplaques, termed **electrocytes,** are usually flattened, with innervation on one side. They are arranged to allow a summation of potentials from the depolarization of membranes that permits the generation of a current. Electrocytes are typically thin (about 10 to 30 μm in torpedoes) and waferlike and are arranged in bundles or stacks (Fig. 21.9). One surface of a typical electrocyte is heavily innervated, and the opposite face is irregular, with numerous papillalike projections. Gelatinous material surrounds the bundles or columns of electric cells, and the electric organs are rich in blood vessels, nerves, and connective tissue. The electrocytes of skates are of two types, flat and cup-shaped. They are not packed tightly, and in some species they retain the striations of muscle cells. The South American ghost knifefishes (Apteronotidae) are unusual in that their electrogenic capacity has evolved from modified spinal neurons, and their electric cells of muscle origin have apparently been lost (Bass, 1986; Bennett, 1970, 1971). These enlarged spinal neurons pass forward after entering the electric organ and then loop back; they can reach more than 100 μm in diameter in both the forward- and backward-running sections (Bennett, 1971).

Among the strongly electric fishes (Fig. 21.10A–D), the electric eel has three separate electric organs, forming a large part of its bulk. The hypaxial section of the long caudal region is made up mostly of the main electric organ, with a smaller one, the **organ of Hunter**, running along its ventral surface. The **organ of Sachs** is posterior to the main organ. These organs have formed from axial musculature, and the electrocytes are ribbonlike, flattened anteroposteriorly, and extend from the medial septum out toward the skin. A large adult may have more than 100,000 electrocytes in the main organ on each side, as there can be up to 6,000 vertical arrays of up to 25 ribbonlike cells. The posterior surfaces of the electrocytes are innervated from spinal nerves. Current flow in the organ is from back to front (Bone and Marshall, 1982), with reverse flow in the surrounding water. The organ of Sachs produces weak pulses of about 10 volts, and the frequency depends on the activity of the fish, from a few pulses per minute at rest to 30 per second while active. The organ of Hunter apparently is capable of generating both strong and weak pulses (Kramer, 1990).

In the electric catfish, the electric organ is derived from pectoral musculature. It is only a few millimeters thick and lies in the skin between the epidermis and a fatty layer that covers most of the body musculature (Grassé, 1958). The several million electrocytes are disclike, about 1 mm in diameter, with a short stalk on the posterior, innervated face. Each side of the organ is innervated by branches from a large neuron in the anterior part of the spinal cord. Current flow is anterior to posterior.

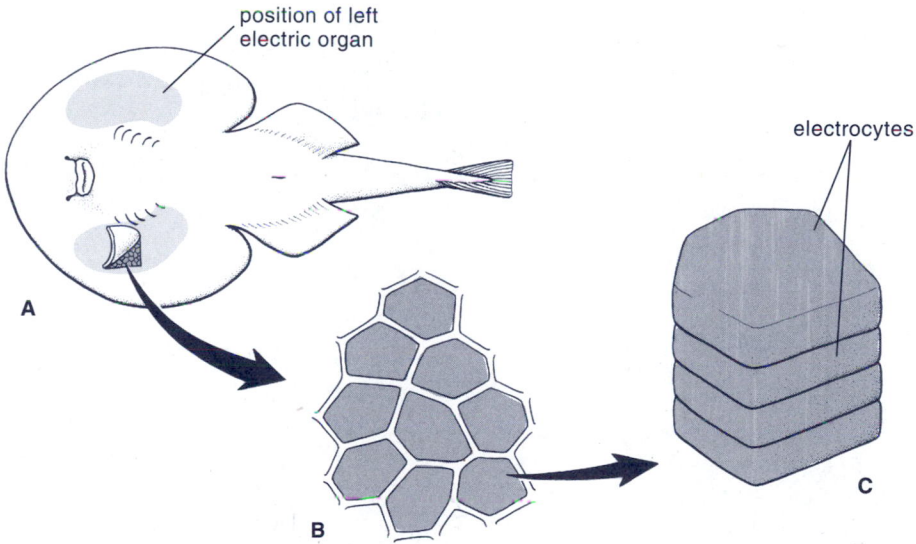

FIGURE 21.9

A, Ventral view of electric ray (*Torpedo*), showing position of electric organs; **B,** Shape of electrocytes as viewed ventrally; **C,** Diagram of arrangement of one "stack" of electrocytes.

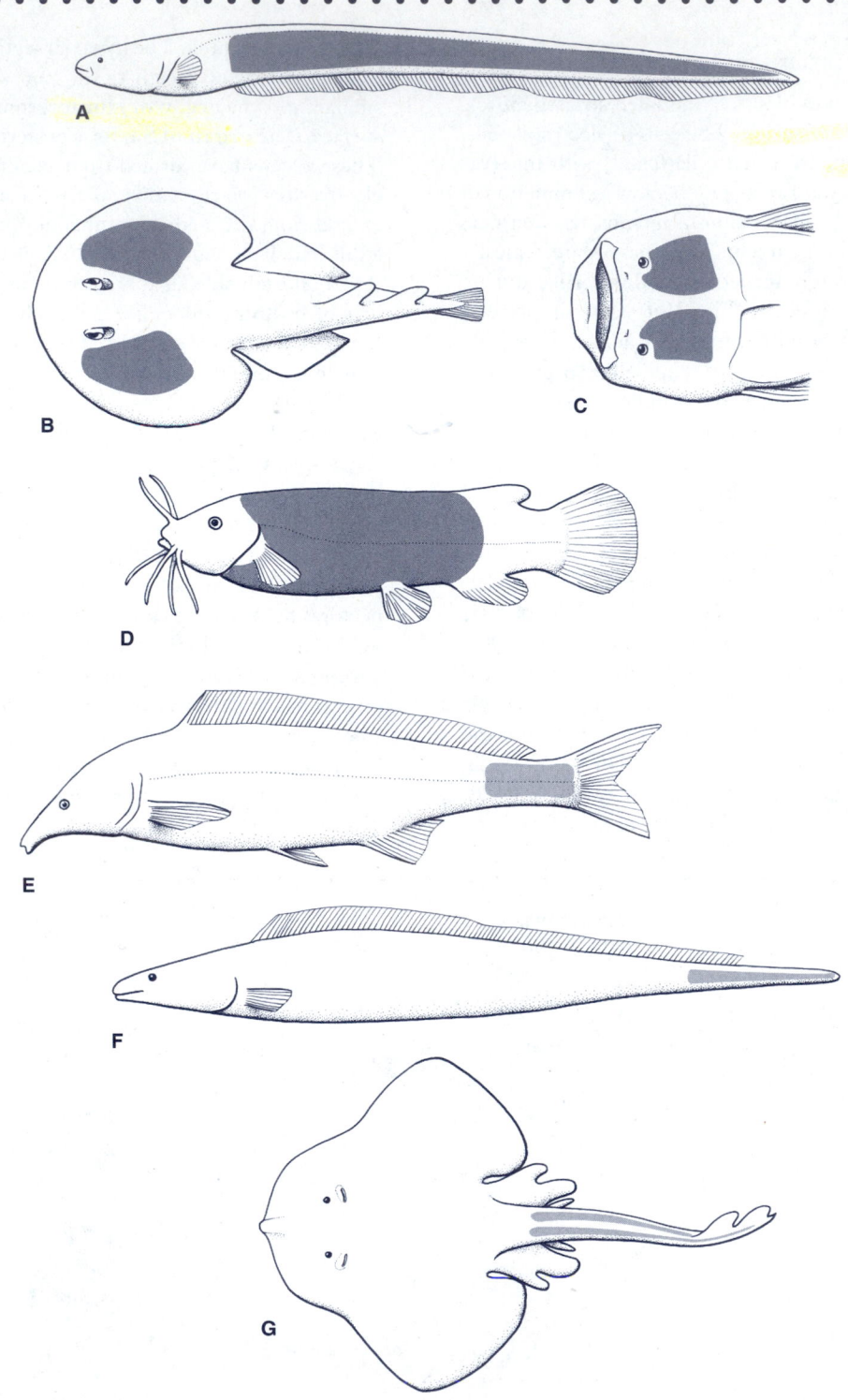

FIGURE 21.10

Positions of electrical organs (shaded regions) in electric fishes. Strongly electric fishes: **A,** Electric eel (*Electrophorus*); **B,** Electric ray (*Torpedo*); **C,** Stargazer (*Astroscopus*); **D,** Electric catfish (*Malapterurus*). Weakly electric fishes: **E,** Mormyridae; **F,** Gymnarchidae; **G,** Rajidae.

Electric rays have a large, kidney-shaped electric organ on each side of their disc, adjacent (lateral) to the head and branchial region. The columns of hexagonal to roughly circular electrocytes that make up the organ are oriented vertically and reach from the dorsal to the ventral surface. The electrocytes, which reach 7 mm in diameter in some species, are derived from branchial musculature. Current flow is ventral to dorsal. The genus *Narcine* has an accessory organ at the posterior part of each main organ.

Like the electric rays, the stargazers (*Astroscopus*) have dorsoventrally flattened electric organs in the head region. These relatively small organs are situated posterior to the eye and are derived from extrinsic eye muscles. The flattened electrocytes may be as much as 5 mm in diameter. Current direction is dorsal to ventral.

Weakly electric fishes tend to have one or more elongate electric organs along each side in the caudal peduncle (Fig. 21.10E–G). Mormyrids have two columns of cells on each side of the caudal peduncle. *Gymnarchus* has four thin columns per side in the posterior half of the caudal region, and skates are equipped with one pair of organs running most of the length to the tail. In *Gymnotus*, the electric organ extends from below the head to the tip of the tail, coursing along the ventral aspect of the body. In addition to elongate, ventral main organs, some gymnotoids (*Steatogenys, Gymnorhamphichthys*) have small accessory organs under the skin of the chin region. *Steatogenys* has another small organ in the pectoral region. Apteronotids have nerve-derived lateral electric organs reaching from above the pectoral fin to the base of the caudal fin. Current flow is forward in *Gymnarchus* and backward in *Raja*, and both directions are possible in the gymnotoids and in some mormyrids (Bennett, 1971). Mochokid (synodontid) catfishes have an electric organ apparently derived from a sonic muscle above the gas bladder (Hagedorn et al., 1990).

Some weakly electric fishes (gymnotiforms and mormyrids) have larval electric organs as well (Bass, 1986). These are similar in structure to the adult organs, but they do not develop into the organs found in the adults. The larval organs reach from near the head to the caudal peduncle, whereas the adult organs are generally more posterior in location.

Functions of Electric Organs

The function of strong electric organs appears to be the stunning of prey and the discouragement of intruders or predators. The use of electricity to obtain prey has been observed in the electric rays, and its use for this purpose in other species seems probable, considering the circumstances under which these electric species live. All are secretive, living near the bottom, in situations that probably allow prey to approach closely without alarm (sit-and-wait predators). Although the stargazer can be especially well concealed, allowing small crustaceans or fish to move onto the sand under which the predator is buried, there is doubt that the electric discharge is of sufficient duration to stun prey (Pickens and McFarland, 1964). Electric catfishes (*Malapterurus*) are known to feed on schooling clupeids and schilbeids, which would seem to be agile enough to escape from the sluggish predators. The use of the electric organs to stun several prey fish at a time was inferred in a study of the food habits of the catfish (Sagua, 1979). Rankin and Moller (1986) reported that *M. electricus* used electrical discharges against intruders in defense of shelter sites, especially against members of other species.

Electric discharges in the weakly electric fishes function both in the electrolocation of nearby objects and in communication with other fishes. As might be expected, electric organ discharge is closely linked neuronally with electroreception. Not only can mormyrid fishes detect their discharges via their own electroreceptors, but corollary discharge pathways link the electrosensory lobes of the brain with the motor nuclei initiating EOD (Bell and von der Emde, 1995; Bell et al., 1995, 1999).

Intraspecific communication seems to be the main function of EOD in the skates (Bratton and Ayers, 1987). The African mormyroids (Mormyridae and Gymnarchidae) and New World knifefishes (Gymnotiformes) live mostly in turbid waters, and some are nocturnal in habit, so vision might be of limited use. Other senses—hearing, olfaction, and mechanoreception—must aid their orientation to their surroundings, as in other nocturnal species, but the possession of a system to generate and detect electrical fields sets these animals apart. For the most part, these electric fishes hold their bodies rigid and straight, depending on undulations of their fins for propulsion. The straight posture ensures symmetry in the electrical field that they generate (see Fig. 21.8). Interference with or distortion of the field by nearby objects can be detected by electroreceptors. Moreover, electric fields can be detected by conspecifics and by fishes of other species and may indicate species identity, sex, age, and reproductive state. Electrolocation forms the basis of a sophisticated communication system (see Chapter 36).

The electric organ discharge pattern is characteristic of a given species, with frequencies and other features of the pulses or waves differing from one species to another. Although many electric fishes have the ability to vary their EOD, oscillograph tracings show characteristic, species-specific shapes and amplitudes. The EODs of mormyrids and most gymnotoids are pulsed. *Gymnarchus niloticus* and some gymnotoids, such as *Apteronotus, Eigenmannia,* and *Sternopygus*, have an EOD in the form of a wave. Mochokids can produce discharges continuously or in bursts. Various mormyrids discharge from one to six pulses per second as a rule, but can accelerate their discharge up to about 130 pulses per second. Output in mormyrids generally ranges from about 9 to 16 volts. *Gymnarchus* is reported to operate at about 4 to 7 volts, discharging at about 300 pulses per second. Investigators have demonstrated a wide range of pulse frequencies in gymnotoids—from 2 up to approximately 1,000 per second.

Control of electrical discharges in weakly electric fishes appears to be both neural and hormonal and involves a "pacemaker nucleus" located in the brainstem (Dye and Meyer, 1986; Keller et al., 1991). The frequency of discharge can be altered by the administration of steroids. Androgens decrease the frequency of discharge in *Sternopygus* and *Apteronotus* when given in large doses. On administration of low doses by daily injections or by implantation, testosterone and dihydrotestosterone (DHT) can bring about large decreases of frequency in *Sternopygus*, whereas a slight increase is caused by estradiol. Long-term effects in *Apteronotus* are reversed from those in *Sternopygus*, in that a decrease in frequency is caused by estrogen and an increase by DHT. This is apparently the basis for the specificity of EOD depending on age and reproductive state (Dye and Meyer, 1986).

SOUND PRODUCTION IN FISHES

Nature of Sounds

Sound-producing structures among fishes include teeth, skeletal elements, muscles, and the gas bladder, and the sounds made by fishes have been described by a great variety of terms. Sounds of schooling fish swimming have been called rustles or roars. Stridulation produces sounds reminiscent of clicks, rasps, or scratches when not aided by the gas bladder, and croaks, grunts, and knocks when the gas bladder acts as a resonator. The frequency of stridulatory sounds in the freshwater drum (*Aplodinotus grunniens*) can range from 150 to 2,000 Hz (Schneider and Hasler, 1960), although the gas bladder–aided sounds are generally below 1,000 Hz, whereas unaided stridulation usually produces frequencies in the 1,000- to 4,000-Hz range.

Sounds made by vibrating the gas bladder have been described as hoots, boops, grunts, yelps, knocks, and croaks, with the toadfishes (*Opsanus*) being known for their "boat whistle" sounds (see Going Deeper). Gas bladder sounds are harmonic and usually of low frequency, from 40 to 250 Hz, with the great majority in the 75- to 100-Hz range, but some

---

GOING DEEPER · Sex for a Song: Vocal Adaptations and Phenotypic Plasticity Among the Batrachoidids

The batrachoidid toadfishes (see Plate 13.7) include several species that are primarily coastal and benthic in their habitat. With their broad, flattened heads, wide mouths, and slender tails, their body form is that of an ambush predator. These hardy bottom dwellers, capable of surviving prolonged exposure to atmospheric conditions, have long been a favorite of researchers investigating the physiological adaptations of marine fishes (Pärt et al., 1999). During the mating season, usually from late spring to summer, male toadfishes will occupy a hole or crevice to which they will attempt to attract gravid females. The males create a diverse range of grunts, croaks, or whistles. Whereas agonistic encounters may result in short, sharp grunts, hums and whistles are used to attract females. The courtship hums of *Porichthys* can be so loud that they keep people living on houseboats awake at night—not unlike that noisy couple living in the apartment above you! Females attracted to the nesting site lay several large, adhesive eggs that are zealously guarded by the males. The males continue to guard the hatchlings until their yolk sac is absorbed and they assume a more mobile existence.

Studies on the reproductive biology of toadfishes represent an ideal example of the benefits of an integrated approach to the study of fishes. Ecologists, behavioral biologists, and neurobiologists all have collaborated to provide insight into one of the most fascinating and complex reproductive strategies known among fishes. A key feature of this reproductive strategy is the sounds that toadfishes make to attract mates. Toadfishes, especially the midshipman (*Porichthys*), also exemplify the benefits of the evolution of alternative reproductive strategies—an adaptation well represented among the teleosts (see Chapters 27 and 37).

As described elsewhere in this chapter, the toadfishes have evolved a most sophisticated set of muscles that can vibrate the gas bladder at extraordinarily high frequencies—approximately 50 times faster than locomotor muscles (Fine et al., 2001; Rome et al., 1996). Similar muscle adaptations are also responsible for the generation of sound from the rattle of the rattlesnake. The piercing "boat whistle" call of the male oyster toadfish (*Opsanus tau*) must be intoxicating to a female, yet, strangely, only lower frequency components of the call overlap the fish's best auditory capability (Yan et al., 2000). Because the sonic muscles are used predominantly in mating behavior, seasonal hypertrophy might be expected. This has been demonstrated in the haddock (*Melanogrammus*

sciaenids can produce frequencies up to 2,000 Hz. Some sounds, such as the "chatter" detected in hydrophone surveys of estuarine waters of the southeastern United States, have been attributed both to stridulatory actions of pharyngeal teeth in weakfish (*Cynoscion regalis*) and to gas bladder sonic muscles in cusk eels (*Ophiodon marginatum;* Sprague and Luczkovich, 2001).

Stridulation

The grinding, snapping, or rubbing together of teeth is the most common type of stridulation among fishes (Tavolga, 1971). Pharyngeal teeth appear to be important sound producers, especially because they are close enough to the gas bladder that the sounds can be amplified (Takemura, 1984). In the cichlid *Oreochromis* (*Tilapia*) *mossambica*, special muscles run from the occipital region and the first vertebra to the upper pharyngeals; in this case, sounds made in the absence of the gas bladder were essentially of the same amplitude as those produced by the intact animal (Lanzing, 1974). Jaw teeth are often used in sound production; some filefishes (Monacanthidae) have ridges, apparently effective in stridulation, on the backs of the front teeth, and various perciform fishes can snap their teeth together, making a sharp sound.

Other mechanisms of stridulation include the movement of fin spines against their sockets in catfishes (Fig. 21.11A), triggerfishes, filefishes, sticklebacks, surgeonfishes, and others; contact between the first dorsal interspinous bone and modified neural spines in some sisorid catfishes; and friction between other skeletal parts in triggerfishes, seahorses, and clownfishes.

The Gas Bladder and Sound Production

Incidental sounds may be made by the release of air from the lungs or other cavities used in air breathing. Some fishes, such as the Atlantic eel and some catfishes, cause sounds by the release of gas from the gas bladder. Such sounds are common, but their importance is unknown. Both Pacific and Atlantic herring (*Clupea pallasii* and *C. harengus,* respectively) produce distinctive bursts of sonic emissions from swim bladder gases released via the anus (Wilson et al., 2003). Yes, indeed—fish farts appear to have communicative value! The emissions were primarily at night, when visual communication is likely impaired, and the rates of emission increased with increases in density of individuals. The authors of this study displayed an appreciation of the humor of the situation by naming these emissions Fast Repetitive Tick (FRT) sounds.

aeglefinus; Templemen and Holder, 1958) and the weakfish (*Cynoscion regalis;* Connaughton et al., 1997). No seasonal cycle of sonic neuromuscular development has been detected in the toadfish, however (Johnson et al., 2000).

The structure and function of the sonic emission system is integral to the evolution of alternative mating tactics and associated morphology in the plainfin midshipman (*Porichthys notatus;* see Fig. 13.2D). For several years, researchers in the laboratory of Andrew Bass have investigated the alternative mating tactics of this species. Bass and his colleagues discovered a pronounced sexual dimorphism in the size of the gas bladder and the development of the sonic musculature (Bass and Marchaterre, 1989a, 1989b). Yet more surprising, these differences extend to alternative reproductive forms that have evolved

among the males (Bass, 1996; Brantley et al., 1993). Michael Fine had earlier discovered a very similar dimorphism in the oyster toadfish (*Opsanus tau;* Fine et al., 1984).

Two morphs, Type I and Type II, have been identified in the male midshipman. Type I males build nests and attract females with their incessant singing. When threatened, they also respond with short, sharp grunts. Type II males are a form of "sneaker" or "satellite" male (see Chapter 37). They do not build nests or attract females. Instead, they will wait for an opportune moment to invade a Type I male's nest while the female is there and get in a few quick bouts of sperm release. Type I and Type II males are dramatically different in a number of features. Type I males are significantly larger and heavier than Type II males—on the average about twice as large and eight times heavier—but Type II males reach

sexual maturity earlier and have a ratio of gonad to body mass up to eight times that of Type I males (Bass, 1996). The coloration of Type II males more closely resembles that of females.

As one might expect, there are significant disparities in the sonic muscle/gas bladder complex responsible for vocalizations and the neuronal systems that control them. The ratio of vocal muscle to body mass is up to six times greater in Type I fish, and the muscle fibers themselves show significantly greater development (Bass, 1996). Bass and his colleagues used a proteinaceous compound, *biocytin,* to trace the neuronal pathways controlling vocalization. They discovered that the circuitry was largely similar between males and females but that the neurons of Type I males were up to three times larger when compared to Type II males and females. Similar patterns of sexual dimorphism

Most gas bladder sounds are due to the generation of vibrations. Several species are equipped with muscles that vibrate the gas bladder directly or indirectly. Some trigger-fishes vibrate the gas bladder by rubbing or drumming the pectoral fin against an area where the gas bladder is close to the body wall. With more than 150 species, the aptly named drums and croakers of the family Sciaenidae are among the most diverse of the sound-producing fishes. In some sciaenid genera, both males and females have sonic muscles associated with the gas bladder, whereas other genera show sonic muscles only in males (Hill et al., 1987). The freshwater drum (*Aplodinotus*) produces sound by vibrating special muscles of the body walls that attach to broad tendons stretching over the gas bladder, whereas another sciaenid genus (*Micropogon*) has tendons attached directly to the gas bladder (Schneider and Hasler, 1960).

Several species have muscles that originate on the skull or vertebral column and insert either on the gas bladder itself or on ribs or other structures associated with it. Certain fishes of the families Brotulidae, Macrouridae, Serranidae, Priacanthidae, Holocentridae, Scorpaenidae, and Teraponidae have been noted as having such arrangements. One of the best-developed structures is the "elastic spring mechanism" seen in certain catfishes (*Galeichthys, Bagre*). This consists of plates formed from the first few vertebrae that are placed in contact with the dorsal part of the gas bladder wall. Muscles stretching between the skull and this springy apparatus can vibrate it rapidly, setting up audible vibrations of the gas bladder (Tavolga, 1971).

Intrinsic sonic muscles are incorporated into the walls of the gas bladder in several species of toadfishes (Batrachoididae), in searobins (Triglidae), in flying gurnards (Dactylopteridae) and in the zeid *Zeus faber*. In most of these fishes, the bladder tends to be divided into left and right chambers; it is heart shaped in the toadfishes (Fig. 21.11B) and nearly completely divided in gurnards. Usually, an internal diaphragm divides the bladder into anterior and posterior sections. The diaphragm is typically perforated by an opening surrounded by a sphincter muscle, and its configuration undoubtedly is associated with a distinctive manner of sound production.

Typically, the sonic muscles of fishes are red and tonic in nature (see Chapter 19); the grunters of the family Teraponidae appear to be an exception. Well vascularized, well supplied with myoglobin, and resistant to fatigue, sonic musculature is capable of generating the highest frequencies of contraction of any vertebrate muscle. Tavolga (1971) mentioned frequencies as high as 100 Hz (i.e., contractions per second), and Rome et al. (1996) have determined that the "boat whistle" mating call of the toadfish (*Opsanus tau*) is generated using muscle contractions in the range of 200 Hz. Extremely rapid calcium recycling and crossbridge attachment/detachment rates in the muscle fibers have been recorded in association with this enhanced rate of muscle contraction.

in neuronal pathways and sonic musculature have been observed in oyster toadfish (Fine et al., 1984, 1990a), suggesting that the mating system of *Opsanus* is largely similar to that of *Porichthys*. In an attempt to find hormonal correlates of the two male forms (Type I vs. Type II), Bass and his colleagues have focused on the hormonal cascade that culminates in the sexual differentiation and maturation of the gonads. Testosterone was detected in both males and females, with Type II males having the highest levels, followed by females, with Type I males having the lowest amounts. Another form of testosterone that is unique to teleosts, *11-ketotestosterone,* exists at much higher levels in Type I males, however, and seems to be the key hormone supporting courtship behavior, including vocalizations. The enzyme *aromatase,* which catalyzes the conversion of testosterone to estrogen, exists in much greater concentrations in Type II males. This enzyme, which is active in the vocal centers of the central nervous system, appears to be a key player in the expression of brain circuitry that determines whether a fish becomes a singer or not (Schlinger et al., 1999). Researchers in Michael Fine's laboratory have identified neurons in the brain of *Opsanus* that bind steroids, including testosterone and estrogen. Forebrain sites that, when electrically stimulated, evoke the "boat whistle" sound contain steroid-binding neurons (Fine and Perini, 1994; Fine et al., 1990b, 1996).

The story of the reproductive biology of the oyster toadfish and the plainfin midshipman demonstrates that the generation of a given phenotype is the result of a number of diverse influences—physical and biotic aspects of the environment, phylogenetic history, and ontogenetic processes. Bass (1998) recognized this in his conceptualization of the phenotype of the organism as being manifested by the combined influences of behavioral characters, structural characters, and the environment. When the behavioral ecologist, neurobiologist, and anatomist all sit before a tank watching the courtship of the midshipman, each one probably sees something a little bit different, but their combined vision gives the truest picture of the means by which these sexy little beasts ensure their survival.

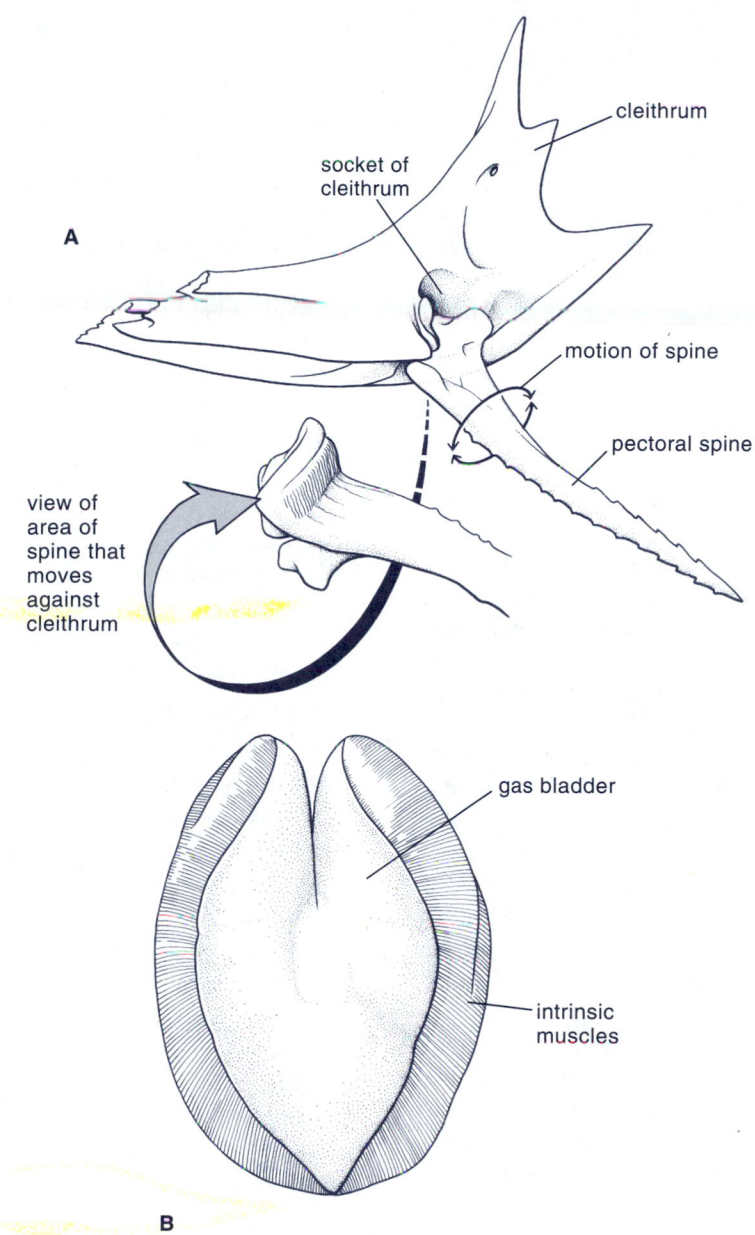

FIGURE 21.11

A, Left cleithrum and pectoral spine of sea catfish (Ariidae) showing roughened flange with which stridulatory sounds are made; **B,** Gas bladder of toadfish (*Opsanus*) with musculature along outer edges.

Miscellaneous Sources of Sound

The act of swimming can cause the production of sound because of the movement of parts of the skeleton against each other and the turbulence caused by movement through the water. Large schools of fish have been noted to generate sounds as they veer and turn (Moulton, 1960). The release of bubbles because of reduced pressure during rapid ascents probably causes sounds. Fishes also cause sounds in breaking the surface of the water, either for feeding or for other purposes. Sportfishers are attracted to the exact spot where the quarry is feeding by hearing the smacking sound of the bluegill or the splash of the trout. Many know the double splash of the leaping carp. Eagle rays are known to make prodigious leaps that result in loud slaps as they fall back to the water.

Feeding activities are often noisy. Parrotfishes (Scaridae), wrasses (Labridae), surfperches (Embiotocidae), and many other fishes that crush molluscs or crustaceans with their strong teeth in the jaws or throat make much incidental sound. The opening and closing of oral and opercular valves in large fishes can be audible, as can the release of air from the mouth or anus of airbreathers. The squeaker (*Misgurnus fossilis*) uses the alimentary canal as a respiratory surface—it gets its name from the sounds made as respiratory gas is expelled.

Significance of Sound Production

Although incidental sounds are of significance to fish, in that such noises may be detected by potential prey or predators, most sounds produced volitionally by fishes appear to play a role in communication and are generated with the purpose of gaining some adaptive advantage for the sender of the sound. Myrberg (1981) classified fishes into six groups based on the manner in which they use perceived sound: predators, prey, mates, companions, competitors for mates, and competitors for other resources.

Reproductive behavior is accompanied by sounds in several marine and freshwater species. In most instances, the males seem to generate most of the sounds and are usually better equipped to do so than the females. In freshwater, sounds are used in courtship—or at least in the presence of prospective spawning partners—in such families as the sunfishes (Centrarchidae), minnows (Cyprinidae), cichlids, characids, and anabantids. Several marine fishes, including cods, seahorses, gobies, blennies, damselfishes, hamlets, and parrotfishes, are also known to use sounds as part of courtship (Lobel, 1991, 1992). Some of the most remarkable courtship sounds are made by toadfishes (*Opsanus*), in which the male produces growls, grunts, and the aforementioned "boat whistles" (see Going Deeper). The naked goby (*Gobiosoma bosci*) produces sounds under various stimuli, but only during the reproductive season (Mok, 1981). The male of the freshwater drum (*Aplodinotus grunniens*) also produces sounds only during the spawning season, but it does not develop sound-producing capabilities until the third year of life. Females of *Pomacentrus partitus* can locate nest sites by the courtship sounds of the males and seem to be able to distinguish among and evaluate the males by their sounds (Myrberg et al., 1986).

A variety of species emit sounds when defending or maintaining territories. Several marine percomorph families, including soldierfishes, gobies, wrasses, damselfishes, and grunters, engage in this behavior. Other defensive sound producers include sea catfishes, toadfishes, triggerfishes,

and cods. Among the freshwater fishes, the cichlids are notable for making defensive and territorial sounds.

Alarm or startle sounds made when disturbed have been attributed to nearly 40 families of fishes (Myrberg, 1981). The sounds are mostly produced internally, but some are made as a consequence of swimming. The general perception is that such sounds serve to startle and momentarily confuse potential predators, enabling prey to make good their escape. These sounds might also act as an alarm signal to conspecifics or to other fishes receiving them. Sound signals are thought to be of significance to schooling fishes for helping to maintain contact and, hence, school cohesion. Gudgeons (*Gobio gobio*), which commonly live in small groups, become quiet when held alone in aquaria, but produce characteristic creaking sounds when in groups (Ladich, 1988).

KEY POINTS AND CONNECTIONS

- The octavolateralis system of fishes consists of three components—the auditory, equilibrium, and lateral line systems. All use a common receptor cell—the hair cell—and electrosensory capacity is also generally included as an octavolateral component.

- The lateral line system and the inner ear of fishes share many structural and functional features, but they have been demonstrated to have separate embryonic origins.

- The inner ear of fishes is located in the otic region of the skull and consists of a series of membranous sacs and associated semicircular canals. The inner ear functions both in hearing and the maintenance of balance.

- Mechanoreception in fishes is a "whole-body" experience, involving an array of receptor cells located on the body surface. Mechanoreception in the lateral line is accomplished through the use of hair cells existing as free neuromasts on the surface of the skin, in pits, or in canals or grooves on the head and body. The electrosensory component consists of specialized electroreceptor cells of two types—ampullary and tuberous. Both mechanoreceptor and electroreceptor cells are innervated by lateral line nerves.

- Fishes with an otophysic connection—a functional connection between the gas bladder and the ear—tend to have greater sensitivity to acoustic stimuli than those lacking such a connection. A special type of otophysic connection, the Weberian apparatus, is found in most of the members of the predominantly freshwater Ostariophysi. *The phylogenetics of ostariophysan fishes is covered in Chapter 10.*

- The capacity to generate electrical discharges from specialized tissues is known from six different orders of fishes. Electric organs consist of blocks of modified striated muscle fibers, termed electroplaques. Electrical discharges are of two types—strong and weak. The few groups of fishes capable of producing strong electrical discharges use them to capture prey and to ward off intruders.

- Weakly electric fishes use their electric organ discharge (EOD) to orient themselves in turbid or nocturnal waters where other senses, such as vision, might be impaired. Weak EOD is also valuable as a means of communication.

- The mormyroids and gymnotiforms are among the best known of the weakly electric fishes. They tend to swim primarily with fins and hold their bodies rigid in order to minimize the distortion of the generated electrical fields.

Electrotactic responses of fishes are discussed in Chapter 36. Additional details of the phylogenetic significance of electroreceptive and electrogenic capabilities in mormyroids are given in Chapter 9, whereas the phylogeny of the gymnotiform fishes is covered in Chapter 10.

- Many fishes are capable of producing sound, either through the rubbing of hard skeletal parts or by specialized muscles that act on the swim bladder. Sonic musculature, working in conjunction with the swim bladder, is an important component of the reproductive behavioral repertoire of many fish species.

FISH LINKS

http://www.life.umd.edu/biology/popperlab/ Home page of Arthur Popper's aquatic bioacoustics research program at the University of Maryland.

http://www.nbb.cornell.edu/neurobio/sisneros/teambass/ teambass.htm Home page of Andrew Bass, one of the foremost experts on toadfish mating behavior.

http://nelson.beckman.uiuc.edu/ Home page of Mark Nelson's Electrosensory Signal Processing Lab at the University of Illinois, Urbana-Champaign. Dr. Nelson's lab studies the sensory biology of weakly electric fishes.

BUILDING AN ICHTHYOLOGY LIBRARY

Bullock, T. H., and W. Heiligenberg (Eds.). 1986. *Electroreception.* Wiley, New York.

Fay, R. R., and A. N. Popper (Eds.). 1999. *Comparative hearing. Fish and amphibians.* Springer Verlag, New York.

Kapoor, B. G., and T. J. Hara. 2001. *Sensory biology of jawed fishes: New insights.* Science Publishers, Enfield, NH.

Journal of Experimental Biology, Vol. 202(10). (1999). This issue is devoted to current research in electric organ discharge and electroreception in fishes.

Moller, P. 1995. *Electric fishes: History and behavior.* Chapman and Hall, London.

REFERENCES

Alves-Gomes, J. A. 1999. Systematic biology of gymnotiform and mormyriform electric fishes: Phylogenetic relationships, molecular clocks, and rates of evolution in the mitochondrial rRNA genes. *J. Exp. Biol. 202:* 1167–1183.

———. 2001. The evolution of electroreception and bioelectrogenesis in teleost fish: A phylogenetic perspective. *J. Fish Biol. 58:* 1489–1511.

Andrus, C. F. T., D. E. Crowe, D. H. Sandweiss, E. J. Reitz, and C. S. Romanek. 2002. Otolith $\Delta^{18}O$ record of mid-Holocene sea surface temperatures in Peru. *Science 295:* 1508–1511.

Assis, C. A. 2003. The lagenar otoliths of teleosts: Their morphology and its application in species identification, phylogeny, and systematics. *J. Fish Biol. 62:* 1268–1295.

Barry, M. A., and M. V. L. Bennett. 1989. Specialized lateral line receptor systems in elasmobranchs: The spiracular organs and vesicles of Savi, pp. 591–606. In *The mechanosensory lateral line*, S. Coombs, P. Görner, and H. Münz (Eds.). Springer Verlag, New York.

Bass, A. H. 1986. Electric organs revisited: Evolution of a vertebrate communication and orientation organ, pp. 13–70. In *Electroreception*, T. H. Bullock and W. Heiligenberg (Eds.). Wiley, New York.

———. 1996. Shaping brain sexuality. *Amer. Sci. 84:* 352–363.

———. 1998. Behavioral and evolutionary neurobiology: A pluralistic approach. *Amer. Zool. 38:* 97–107.

———, and M. A. Marchaterre. 1989a. Sound generating (sonic) motor system in a teleost fish (*Porichthys notatus*): Sexual polymorphism in the ultrastructure of myofibrils. *J. Comp. Neurol. 286:* 141–153.

———, and———. 1989b. Sound generating (sonic) motor system in a teleost fish (*Porichthys notatus*): Sexual polymorphisms and general synaptology of a sonic motor nucleus. *J. Comp. Neurol. 286:* 154–169.

Bell, C. G., and G. von der Emde. 1995. Electric organ corollary discharge pathways in mormyrid fish II. The medial juxtalabar nucleus. *J. Comp. Physiol. A 177:* 463–479.

———, K. Dunn, C. Hall, A. Caputi. 1995. Electric organ discharge pathways in mormyrid fish I. The mesencephalic command associated nucleus. *J. Comp. Physiol. A 177:* 449–462.

———, V. Z. Han, Y. Sugawara, and K. Grant. 1997a. Synaptic plasticity in a cerebellum-like structure depends on temporal order. *Nature 387:* 287–281.

———, D. Bodznick, J. Montgomery, and J. Bastian. 1997b. The generation and subtraction of sensory expectations within cerebellum-like structures. *Brain Behav. Evol. 50* (Suppl. 1): 17–31.

———, V. Z. Han, Y. Sugawara, and K. Grant. 1999. Synaptic plasticity in the mormyrid electrosensory lobe. *J. Exp. Biol. 202:* 1339–1347.

Bennett, M. V. L. 1970. Comparative physiology: Electric organs. *Ann. Rev. Physiol. 32:* 471–528.

———. 1971. Electric organs, pp. 347–491. In *Fish physiology*, Vol. V, W. S. Hoar and D. J. Randall (Eds.). Academic Press, New York.

Blaxter, J. H. S. 1981. The swimbladder and hearing, pp. 61–71. In *Hearing and sound communication in fishes*, W. N. Tavolga, A. N. Popper, and R. R. Fay (Eds.). Springer Verlag, New York.

———, J. C. Gamble, and H. V. Westernhagen. 1989. The early life history of fish. *Rapp. Proc. Verb. Réunions Cons. Int. Explor. Mer 191.*

Bond, C. B., and T. Uyeno. 1981. Remarkable changes in the vertebrae of the perciform fish *Scombrolabrax* with notes on its anatomy and systematics. *Jpn. J. Ichthyol. 28:* 259–262.

Bone, Q., and N. B. Marshall. 1982. *Biology of fishes.* Blackie, Glasgow, UK.

———, and J. H. S. Blaxter. 1999. *Biology of fishes* (2nd ed.). Stanley Thornes, Cheltenham, UK.

Brantley, R. K., M. A. Marchaterre, and A. H. Bass. 1993. Androgen effects on vocal muscle structure in a teleost fish with inter- and intrasexual dimorphisms. *J. Morphol. 216:* 305–318.

Bratton, B. O., and J. L. Ayers. 1987. Observations on the electric organ discharge of two skate species (Chondrichthyes: Rajidae) and its relationship to behaviour. *Env. Biol. Fishes 20:* 241–254.

Bullock, T. H. 1981. Comparisons of the electric and acoustic senses and their central processing, pp. 525–570. In *Hearing and sound communication in fishes*, W. N. Tavolga, A. N. Popper, and R. R. Fay (Eds.). Springer Verlag, New York.

———. 1988. The comparative neurology of expectation: Stimulus acquisition and neurobiology of anticipated and unanticipated input, pp. 269–284. In *Sensory biology of aquatic animals*, J. Atema, R. R. Fay, A. N. Popper, and W. N. Tavolga (Eds.). Springer Verlag, New York.

Canfield, J. G., and R. C. Eaton. 1990. Swimbladder acoustic pressure transduction initiates Mauthner-mediated escape. *Nature* 347: 760–762.

Chen, L. S., and H. Y. Yan. 2002. The relative distribution of otoliths as a means of larval fish identification. *Zool. Stud.* 41(2): 144–152.

Connaughton, M. A., M. L. Fine, and M. H. Taylor. 1997. The effects of seasonal hypertrophy and atrophy on fiber morphology, metabolic substrate concentration and sound characteristics of the weakfish sonic muscle. *J. Exp. Biol.* 200: 2449–2457.

Coombs, S. 1981. Interspecific differences in hearing capabilities for select teleost species, pp. 173–178. In *Hearing and sound communication in fishes*, W. N. Tavolga, A. N. Popper, and R. R. Fay (Eds.). Springer Verlag, New York.

———. 1999. Signal detection theory, lateral-line excitation patterns and prey capture behavior of mottled sculpin. *Anim. Behav.* 58: 421–430.

———, and J. Janssen. 1989. Peripheral processing by the lateral line system of the mottled sculpin (*Cottus bairdi*), pp. 299–319. In *The mechanosensory lateral line*, S. Coombs, P. Görner, and H. Münz (Eds.). Springer Verlag, New York.

———, and J. C. Montgomery. 1999. The enigmatic lateral line system, pp. 319–362. In *Comparative hearing: Fish and amphibians*, R. R. Fay and A. N. Popper (Eds.). Springer Verlag, New York.

———, ———, and J. F. Webb. 1988. Diversity of lateral line systems: Evolutionary and functional considerations, pp. 553–593. In *Sensory biology of aquatic animals*, J. Atema, R. R. Fay, A. N. Popper, and W. N. Tavolga (Eds.). Springer Verlag, New York.

Corwin, J. T. 1981. Audition in elasmobranchs, pp. 81–105. In *Hearing and sound communication in fishes*, W. N. Tavolga, A. N. Popper, and R. R. Fay (Eds.). Springer Verlag, New York.

Dambly-Chaudière, C., D. Sapède, F. Soubiran, K. Decorde, N. Gompel, and A. Ghysen. 2003. The lateral line of zebrafish: A model system for the analysis of morphogenesis and neural development in vertebrates. *Biol. Cell 95*: 579–587.

Denton, E. J., and J. H. S. Blaxter. 1976. The mechanical relationships between the clupeid swimbladder, inner ear and lateral line. *J. Mar. Biol. Ass. UK 56*: 787–807.

———, and J. A. B. Gray. 1988. Mechanical factors in the excitation of the lateral lines of fishes, pp. 595–617. In *Sensory biology of aquatic animals*, J. Atema, R. R. Fay, A. N. Popper, and W. N. Tavolga (Eds.). Springer Verlag, New York.

———, and ———. 1989. Some observations on the forces acting on neuromasts in fish lateral line canals, pp. 229–263. In *The mechanosensory lateral line*, S. Coombs, P. Görner, and H. Münz (Eds.). Springer Verlag, New York.

Diebel, C. E., R. Proksch, C. R. Green, P. Nielson, and M. M. Walker. 2000. Magnetite defines a vertebrate magnetoreceptor. *Nature* 406: 299–302.

Dye, J. C., and J. H. Meyer. 1986. Central control of the electric organ discharge in weakly electric fish, pp. 71–102. In *Electroreception*, T. H. Bullock and W. Heiligenberg (Eds.). Wiley, New York.

Fay, R. R., and A. N. Popper. 1980. Structure and function in teleost auditory systems, pp. 3–42. In *Comparative studies of hearing in vertebrates*, A. N. Popper and R. R. Fay (Eds.). Springer Verlag, New York.

———, and ———. 2000. Evolution of hearing in vertebrates: The inner ears and processing. *Hear. Res.* 149(1–2): 1–10.

Feng, A. S. 1991. Electric organs and electroreceptors, pp. 319–334. In *Comparative animal physiology: Neural and integrative animal physiology* (4th ed.), C. L. Prosser (Ed.). Wiley, New York.

Fine, M. L., and M. A. Perini. 1994. Sound production evoked by electrical stimulation of the forebrain in the oyster toadfish. *J. Comp. Physiol. A 174*: 173–185.

———, D. Economos, R. Radtke, and J. R. McClung. 1984. Ontogeny and sexual dimorphism of the sonic motor nucleus in the oyster toadfish. *J. Comp. Neurol.* 225: 105–110.

———, N. M. Burns, and T. M. Harris. 1990a. Ontogeny and sexual dimorphism of the sonic muscle in the oyster toadfish. *Can. J. Zool.* 68: 1374–1381.

———, D. A. Keefer, and H. Russel-Mergentahl. 1990b. Autoradiographic localization of estrogen-concentrating cells in the brain and pituitary of the oyster toadfish. *Brain Res.* 536: 207–219.

———, F. A. Chen, and D. A. Keefer. 1996. Autoradiographic localization of dihydrotestosterone and testosterone concentrating neurons in the brain of the oyster toadfish. *Brain Res. 709*: 65–80.

———, K. L. Malloy, C. B. King, S. L. Mitchell, and T. M. Cameron. 2001. Movement and sound generation by the toadfish swim bladder. *J. Comp. Physiol. A187*: 371–379.

Fletcher, L. B., and J. D. Crawford. 2001. Acoustic detection by sound-producing fishes (Mormyridae): The role of gas-filled tympanic bladders. *J. Exp. Biol.* 204: 175–183.

Fraser, P. J., and R. L. Shelmerdine. 2002. Fish physiology: Dogfish hair cells sense hydrostatic pressure. *Nature 415*: 495–496.

Grassé, P. P. (Ed.). 1958. Agnathes et poissons: Anatomie, éthologie, systématique. In *Traité de zoologie*, Vol. 13, 3 parts. Masson, Paris.

Hagedorn, M., M. Womble, and T. B. Finger. 1990. Synodontid catfish: A new group of weakly electric fish. *Brain Behav. Evol. 35*: 268–277.

Hamrin, S. F., E. Arneri, P. Doering-Arjes, H. Mosegaard, A. Patwardhans, A. Sasov, M. Schatz, D. van Dyck, H. Wickström, and M. van Heel. 1999. A new method for three-dimensional otolith analysis. *J. Fish Biol. 54*: 223–225.

Hawkins, A. D. 1973. The sensitivity of fish to sounds. *Oceanogr. Mar. Biol. Ann. Rev. 11*: 291–340.

———. 1981. The hearing abilities of fish, pp. 109–133. In *Hearing and sound communication in fishes*, W. N. Tavolga, A. N. Popper, and R. R. Fay (Eds.). Springer Verlag, New York.

Heiligenberg, W., and G. Rose. 1985. Neural correlates of the jamming avoidance response (JAR) in the weakly electric fish *Eigenmannia*. *Trends Neurosci.* 8: 442–449.

Hensel, K., and E. K. Balon. 2001. The sensory canal systems of the living coelacanth, *Latimeria chalumnae*: A new instalment. *Env. Biol. Fishes 61*: 117–124.

Higgs, D. M., D. T. T. Plachta, A. K. Rollo, M. Singheiser, M. C. Hastings, and A. N. Popper. 2004. Development of ultrasound detection in American shad (*Alosa sapidissima*). *J. Exp. Biol. 207*: 155–163.

Hill, G. L., M. L. Fine, and J. A. Musick. 1987. Ontogeny of the sexually dimorphic sonic muscle in three sciaenid species. *Copeia 1987*: 708–713.

Hopkins, C. D. 1983. Functions and mechanisms in electroreception, pp. 215–259. In *Fish neurology*, Vol. I, R. G. Northcutt and R. E. Davis (Eds.). University of Michigan Press, Ann Arbor.

Ivany, L. C., W. P. Patterson, and K. C. Lohmann. 2000. Cooler winters as a possible cause of mass extinctions at the Eocene/Oligocene boundary. *Nature 407:* 887–890.

Jenkins, D. B. 1989. The utricle in *Ictalurus punctatus*, pp. 73–78. In *Hearing and sound communication in fishes*, W. N. Tavolga, A. N. Hopper, and R. R. Fay (Eds.). Springer Verlag, New York.

Janssen, J., and J. Corcoran. 1998. Distance determination via the lateral line in the mottled sculpin. *Copeia 1998:* 657–662.

———, W. R. Jones, A. Whang, and P. E. Oshel. 1995. Use of the lateral line in particulate feeding in the dark by juvenile alewife (*Alosa pseudoharengus*). *Can. J. Fish. Aquat. Sci. 52:* 358–363.

———, V. Sideleva, and H. Biga. 1999. Use of the lateral line for feeding in two Lake Baikal sculpins. *J. Fish Biol. 54:* 404–416.

Jones, W. R., and J. Janssen. 1992. Lateral line development and feeding behavior in the mottled sculpin, *Cottus bairdi* (Scorpaeniformes: Cottidae). *Copeia 1992:* 485–492.

Jorgensen, J. M. 1989. Evolution of octavolateralis sensory cells, pp. 115–149. In *The mechanosensory lateral line*, S. Coombs, P. Görner, and H. Münz (Eds.). Springer Verlag, New York.

Kalmijn, A. J. 1971. The electric sense of sharks and rays. *J. Exp. Biol. 55:* 371–383.

———. 1978. Electric and magnetic sensory world of sharks, skates, and rays, pp. 507–528. In *Sensory biology of sharks, skates, and rays*, B. S. Hodgson and R. F. Mathewson (Eds.). Office of Naval Research, Arlington, VA.

———. 1989. Functional evolution of lateral line and inner ear sensory systems, pp. 187–215. In *The mechanosensory lateral line*, S. Coombs, P. Görner, and H. Münz (Eds.). Springer Verlag, New York.

Keller, C. H., M. Kawasaki, and W. Heiligenberg. 1991. The control of pacemaker modulations for social communication in the weakly electric fish *Sternopygus. J. Comp. Physiol. A 169:* 441–450.

Kenyon, T. N., F. Ladich, and H. Y. Yan. 1998. A comparative study of hearing ability in fishes: The auditory brainstem response approach. *J. Comp. Physiol. A 182:* 307–318.

Kramer, B. 1990. *Electrocommunication in teleost fishes.* Springer Verlag, Berlin.

Ladich, F. 1988. Sound production by the gudgeon, *Gobio gobio* L., a common European freshwater fish (Cyprinidae, Teleostei). *J. Fish Biol. 32:* 707–715.

Lanzing, W. J. R. 1974. Sound production in the cichlid *Tilapia Mossambica* Peters. *J. Fish. Biol. 6:* 341–347.

Lewis, E. R., E. L. Leverenz, and W. S. Bialek. 1985. *The vertebrate inner ear.* CRC Press, Boca Raton, FL.

Liem, K. F. 1967. A morphological study of *Luciocephalus pulcher*, with notes on gular elements in other recent teleosts. *J. Morphol. 121:* 103–134.

Lobel, P. S. 1991. Mating strategies of coastal marine fishes. *Oceanus 34:* 19–26.

———. 1992. Sounds produced by spawning fishes. *Env. Biol. Fishes 33:* 351–358.

Lowenstein, D. 1971. The labyrinth, pp. 207–240. In *Fish physiology*, Vol. V, W. S. Hoar and D. J. Randall (Eds.). Academic Press, New York.

Mann, D. A., Z. Lu, M. C. Hastings, and A. N. Popper. 1998. Detection of ultrasonic tones and simulated dolphin echolocation clicks by a teleost fish, the American shad (*Alosa sapidissima*). *J. Acoust. Soc. Am. 104:* 562–568.

———, D. M. Higgs, W. N. Tavolga, M. J. Souza, and A. N. Popper. 2001. Ultrasound detection by clupeiform fishes. *J. Acoust. Soc. Am. 109:* 3048–3054.

Mok, H. K. 1981. Sound production in the naked goby, *Gobiosoma bosci* (Pisces, Gobiidae)—a preliminary study, pp. 447–455. In *Hearing and sound communication in fishes*, W. N. Tavolga, A. N. Popper, and R. R. Fay (Eds.). Springer Verlag, New York.

Montgomery, J. C. 1988. Sensory physiology, pp. 79–98. In *Physiology of elasmobranch fishes*, T. J. Shuttleworth (Ed.). Springer Verlag, Berlin.

Moulton, J. M. 1960. Swimming sounds and the schooling of fishes. *Biol. Bull. Woods Hole 119:* 210.

Münz, H. 1989. Functional organization of the lateral line periphery, pp. 285–297. In *The mechanosensory lateral line*, S. Coombs, P. Görner, and H. Münz (Eds.). Springer Verlag, New York.

Myrberg, A. A., Jr. 1981. Sound communication and interception in fishes, pp. 395–425. In *Hearing and sound communication in fishes*, W. N. Tavolga, A. N. Popper, and R. R. Fay (Eds.). Springer Verlag, New York.

———, and D. R. Nelson. 1991. The behavior of sharks: What have we learned? pp. 92–100. In Discovering sharks, S. H. Gruber (Ed.), *Am. Littoral Soc. Spec. Publ. 14.* Highlands, NJ.

———, M. Mohler, and J. D. Catala. 1986. Sound production by males of a coral reef fish *(Pomacentrus partitus):* Its significance to females. *Anim. Behav. 34:* 913–923.

Nelson, J. S. 1994. *Fishes of the world* (3rd ed.). Wiley, New York.

Nilsson, G. 1999. The cost of a brain. *Nat. Hist. 108*(10): 66–73.

Norris, K. S., and B. Mohr. 1983. Can odontocoetes debilitate prey with sound? *Am. Nat. 122*(1): 85–104.

Northcutt, R. G. 1986. Electroreception in nonteleost bony fishes, pp. 257–285. In *Electroreception*, T. H. Bullock and W. Heiligenberg (Eds.). Wiley, New York.

———, and W. E. Bemis. 1993. Cranial nerves of the coelacanth, *Latimeria chalumnae* (Osteichthyes: Sarcopterygii: Actinistia), and comparisons with other Craniata. *Brain Behav. Evol. 42*(Suppl. 1): v–x, 1–76.

Pärt, P., C. M. Wood, K. M. Gilmour, S. F. Perry, J. Laurent, J. Zadunaisky, and P. J. Walsh. 1999. Urea and water permeability in the ureotelic gulf toadfish (*Opsanus beta*). *J. Exp. Zool. 283:* 1–12.

Patterson, W. P. 1999. Oldest isotopically characterized fish otoliths provide insight to Jurassic continental climate of Europe. *Geology 27:* 199–202.

Pickens, P. E., and W. N. McFarland. 1964. Electric discharge and associated behaviour in the stargazer. *Anim. Behav. 12:* 362–367.

Platt, C. 1983. The peripheral vestibular system of fishes, pp. 89–123. In *Fish neurobiology*, R. G. Northcutt and R. E. Davis (Eds.). University of Michigan Press, Ann Arbor.

———, and A. N. Popper. 1981. Fine structure and function of the ear, pp. 3–38. In *Hearing and sound communication in fishes*, W. N. Tavolga, A. N. Popper, and R. R. Fay (Eds.). Springer Verlag. New York.

———, ———, and R. R. Fay. 1989. The ear as part of the octavolateralis system, pp. 633–651. In *The mechanosensory lateral line*, S. Coombs, P. Görner, and H. Münz (Eds.). Springer Verlag, New York.

Poggendorf, D. 1976. The absolute threshold of learning in the bullhead (*Ameiurus nebulosus*) and contributions to the physics of the Weberian apparatus of the Ostariophysi, pp. 147–181. In *Sound reception in fishes*, W. N. Tavolga (Ed.). Dowdin, Hutchinson, and Ross, Stroudsburg, PA.

Popper, A. N. 1983. Organization of the inner ear and auditory processing, pp. 125–178. In *Fish neurobiology*, R. G. Northcutt and R. Davis (Eds.). University of Michigan Press, Ann Arbor.

———, and S. Coombs. 1980. Auditory mechanisms in teleost fishes. *Amer. Sci. 68*(4): 429–440.

————, and R. R. Fay. 1999. The auditory periphery in fishes, pp. 43–100. In *Comparative hearing: Fish and amphibians*, R. R. Fay and A. N. Popper (Eds.). Springer Verlag, New York.

————, P. H. Rogers, W. M. Saidel, and M. Cox. 1988. Role of the fish ear in sound processing, pp. 687–710. In *Sensory biology of aquatic animals*, J. Atema, R. R. Fay, A. N. Popper, and W. N. Tavolga (Eds.). Springer Verlag, New York.

Radtke, R. 1984. Scanning electron microscope evidence for yearly growth zones in giant bluefin tuna, *Thunnus thynnus*, otoliths from daily increments. *Fish. Bull. 82:* 434–440.

Rankin, H. C., and P. Moller. 1986. Social behavior of the African electric catfish, *Malapterurus electricus*, during intra- and interspecific encounters. *Ethology 73*(3): 177–190.

Raschi, W., and L. A. Makanos. 1989. The structure of the ampullae of Lorenzini in *Dasyatis garouaensis* and its implications on the evolution of freshwater electroreceptive systems, pp. 101–111. In *Eighth International Symposium on Morphological Sciences*, W. C. Hamlett and B. Tota (Eds.). Rome, Italy.

Robins, C. R., R. M. Bailey, C. E. Bond, J. R. Brooker, E. A. Lachner, R. N. Lea, and W. B. Scott. 1991. World fishes important to North Americans. *Am. Fish. Soc. Spec. Publ. 21.*

Rome, L. C., S. A. Syme, S. Hollingworth, S. L. Lindstedt, and S. M. Baylor. 1996. The whistle and the rattle: The design of sound producing muscles. *Proc. Nat. Acad. Sci. USA 93:* 8095–8100.

Ronan, M. 1986. Electroreception in cyclostomes, pp. 209–224. In *Electroreception*, T. H. Bullock and W. Heiligenberg (Eds.). Wiley, New York.

Rose, J. D. 2002. The neurobehavioral nature of fishes and the question of awareness and pain. *Rev. Fisheries Sci. 10*(1): 1–38.

Ross, D. M. 1963. The sense organs of *Myxine glutinosa* L., pp. 150–160. In *The biology of* Myxine, A. Brodal and R. Fänge (Eds.). Universitetsforlaget, Oslo.

Sagua, V. O. 1979. Observations on the food and feeding habits of the African electric catfish *Malapterurus electricus* (Gmelin). *J. Fish Biol. 15:* 61–69.

Schellart, N. A. M., and R. J. Wubbels. 1998. The auditory and mechanosensory lateral line system, pp. 283–312. In *The physiology of fishes* (2nd ed.), D. H. Evans (Ed.). CRC Press, Boca Raton, FL.

Schlinger, B. A., C. Greco, and A. H. Bass. 1999. Aromatase activity in the hindbrain vocal control region of a teleost fish: Divergence among males with alternative reproductive tactics. *Proc. Roy. Soc. Lond. B 266:* 131–136.

Schneider, H., and A. D. Hasler. 1960. Laute und lauterzeugung beim Süsswassertrommler *Aplodinotus grunniens* Rafinesque (Sciaenidae: Pisces). *Z. Vergl. Physiol. 4:* 499–517.

Schuijf, A., and R. J. A. Buwalda. 1980. Underwater localization— A major problem in fish acoustics, pp. 43–78. In *Comparative studies in hearing in vertebrates*, A. N. Popper and R. R. Fay (Eds.). Springer Verlag. New York.

Sneddon, L. U., V. A. Braithwaite, and M. J. Gentle. 2003. Do fishes have nociceptors? Evidence for the evolution of a vertebrate sensory system. *Proc. Roy. Soc. Lond. B 270:* 1115–1121.

Söllner, C., M. Burghammer, E. Busch-Nentwich, J. Berger, H. Schwarz, C. Riekel, and T. Nicolson. 2003. Control of crystal size and lattice formation by *starmaker* in otolith biomineralization. *Science 302:* 282–286.

Sprague, M. W., and J. J. Luczkovich. 2001. Do striped cusk-eels *Ophiodon marginatum* (Ophidiidae) produce the "chatter" sound attributed to weakfish *Cynoscion regalis* (Sciaenidae)? *Copeia 2001:* 854–859.

Szabo, T. 1974. Anatomy of the specialized lateral line organs of electroreception, pp. 13–58. In *Handbook of sensory physiology*, Vol. III, Pt. 3, A. Fessard (Ed.). Springer Verlag, Berlin.

Takemura, A. 1984. Acoustical behavior of the freshwater goby *Odontobutis obscura. Bull. Jpn. Soc. Sci. Fish. 50:* 561–564.

Tavolga, W. N. 1971. Sound production and detection, pp. 135–205. In *Fish physiology*, Vol. V, W. S. Hoar and D. J. Randall (Eds.). Academic Press, New York.

Templemen, W., and V. M. Holder. 1958. Variation with fish length, sex, stage of sexual maturity and season in the appearance and volume of the drumming muscles of the swimbladder in the haddock *Melanogrammus aeglefinus* (L.). *J. Fish. Res. Bd. Can. 15:* 355–390.

Thorrold, S. R., C. Latkoczy, P. K. Swart, and C. M. Jones. 2001. Natal homing in a marine fish metapopulation. *Science 291:* 297–299.

Viete, S., and W. Heiligenberg. 1991. The development of the jamming avoidance response (JAR) in *Eigenmannia:* An innate behavior indeed. *J. Comp. Physiol. A 169:* 15–33.

Vilches-Troya, J., R. F. Dunn, and D. P. O'Leary. 1984. Relationship of the vestibular hair cells to magnetic particles in the otolith of the guitarfish *(Rhinobatos productus)* sacculus. *J. Comp. Neurol. 236:* 489–494.

Von der Emde, G. 1998. Electroreception, pp. 313–343. In *The physiology of fishes* (2nd ed.), D. H. Evans (Ed.). CRC Press, Boca Raton, FL.

————, and S. Schwarz. 2001. Detection of electric signals in jawed fishes, pp. 161–180. In *Sensory biology of jawed fishes: New insights*, B. G. Kapoor and T. J. Hara (Eds.). Science Publishers, Enfield, NH.

Waldron, M. E., and D. A. Gerneke. 1997. Comparison of two scanning electron microscope techniques for examining daily growth increments on fish otoliths. *J. Fish Biol. 50:* 450–454.

Walker, M. M., A. E. Dizon, and J. L. Kirschvink. 1982. Geomagnetic field detection by yellowfin tuna, pp. 755–758. In *Oceans 82 Conference Record: Industry, government, education.* Partners in Progress, Washington, DC.

Whitehead, D. L., I. R. Tibbetts, and L. Y. M. Daddow. 1999. Distribution and morphology of the ampullary organs of the salmontail catfish, *Arius graeffei. J. Morphol. 239:* 97–105.

Wilson, B., R. S. Batty, and L. M. Dill. 2003. Pacific and Atlantic herring produce burst pulse sounds. *Proc. Roy. Soc. Lond. B* (Suppl. 3): *Biology Letters S95.*

Wullimann, M. F. 1998. The central nervous system, pp. 245–282. In *The physiology of fishes* (2nd ed.), D. H. Evans (Ed.). CRC Press, Boca Raton, FL.

————, and G. Roth. 1994. Descending telencephalic information reaches longitudinal torus and cerebellum via the dorsal preglomerular nucleus in the teleost fish, *Pantodon buchholzi:* A case of neural preadaptation? *Brain Behav. Evol. 44:* 338–352.

Yan, H. Y., and W. S. Curtsinger. 2000. The otic gasbladder as an ancillary auditory structure in a mormyrid fish. *J. Comp. Physiol. A 186:* 595–602.

————, M. L. Fine, N. S. Horn, and W. E. Colón. 2000. Variability in the role of the gasbladder in fish audition. *J. Comp. Physiol. A 186:* 435–445.

Zakon, H. H. 1986. The electroreceptive periphery, pp. 103–156. In *Electroreception*, T. H. Bullock and W. Heiligenberg (Eds.). Wiley, New York.

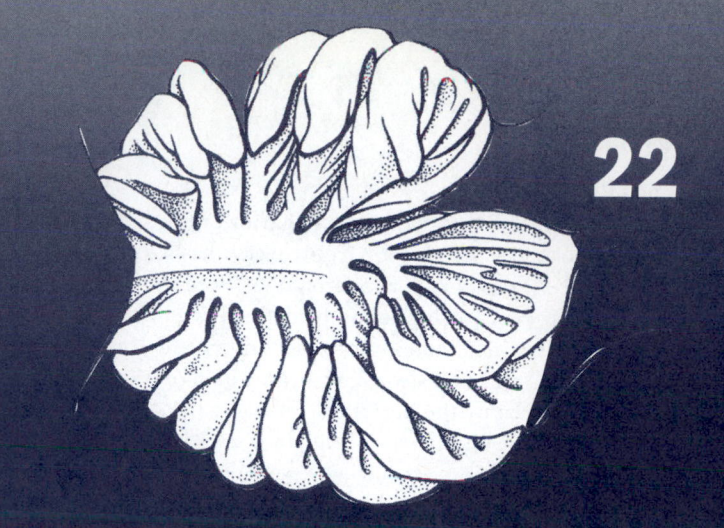

22

THE SENSORY ARSENAL OF FISHES III: OLFACTION, TASTE, AND OTHER CHEMICAL SENSES

CHEMOSENSATION IN FISHES

OLFACTION

Structure of Receptor Organs
Function and Significance of Olfaction
    Olfactory Thresholds
    Detection of Food
    Orientation
    Other Behavior

TASTE

Taste Receptors
Function and Significance of Taste
    Sapid Substances
    Taste Thresholds
    Relationship to Feeding and Other Behavior

OTHER CHEMOSENSORY RECEPTORS

CODA: RELATING SENSORY ADAPTATIONS TO THE ENVIRONMENT

Chemosensation in fishes is primarily associated with olfaction and taste (gustation). Olfaction in fishes is quite different from the sense of smell observed in terrestrial organisms, as fishes are detecting features of waterborne substances of differing solubility, often with remarkable sensitivity. Olfactory cues greatly enhance fishes' ability to orient to their environment. Olfactory information figures significantly in communication—a good case in point being the role of pheromones—and is often associated with learned behavior in fishes. Taste, on the other hand, appears to function more for the purposes of immediate gratification—nowhere more apparent than in the location and ingestion of food. The senses of olfaction and taste are somewhat restricted in their location—not surprisingly, this means the head and mouth. The skin of fishes also possesses additional chemosensory capabilities. The so-called "common chemical sense," may relay information about the concentration of selected ions. Solitary chemosensory cells, which appear to have a function similar to the cells that compose taste buds, are also broadly distributed in the epidermis of fishes.

CHEMOSENSATION IN FISHES

Chemosensation is possibly the earliest sensory modality, having evolved more than 500 million years ago (Hara, 1994). For vertebrates, the primary chemosensory pathways are olfaction and taste (gustation). Chemosensory processes in the aquatic environment differ from those taking place in the terrestrial environment in two fundamental ways: First, rather than detecting airborne molecules, fishes sense molecules in solution. Second, these molecules, being transported in water rather than air, are dispersed much more slowly. Consequently, solubility is a far more important feature than volatility for molecules that are detected in an aquatic environment, and it is dissolved molecules that figure significantly in the behavioral responses of fishes (Hara, 1992a). Dissolved substances may originate far from the fishes and be carried by currents or, more slowly, by diffusion to the receptor organs. The excellent solvent properties of water (see Chapter 30) allow a myriad of organic and inorganic substances to be carried in solution, and even though many molecules are only sparingly soluble, they may still be detected by the acute olfactory and gustatory organs of fishes. Among the ecological applications of chemical communication, Zimmer (2000) included predation, courtship and mating, kinship recognition, and habitat colonization.

Fishes are well equipped for sensing chemicals in solution. Many have taste buds distributed over the body, but particularly in the epidermis of the head, and some have especially sensitive olfactory organs. Taste is important at close range, whereas olfaction may be of greater importance in locating the sources of more distant stimuli. In the absence of confounding currents, a gradient of dissolved material will exist that may allow orientation to the source of the odor or taste. Currents or turbulence may alter the state of the gradient and influence the capacity for orientation (Kleerekoper, 1969).

Olfaction interfaces with behavior in fishes in four main areas: feeding, homing migration, reproduction, and fright reaction (Hara and Zielinski, 1989). Taste is of primary importance in food selection and consumption, but it may be involved in other aspects of behavior. Fishes are known to detect and recognize the smell of prey (Little, 1983), predators (Rehnberg and Schreck, 1986), individuals of their own species (Todd et al., 1967), and even of specific small streams (Hasler and Wisby, 1951). The reaction of fishes to odors depends not only on the species involved, but can also involve the previous learning experiences of the individual.

Semiochemicals have been defined as chemicals that mediate interactions between organisms. Two categories of semiochemicals have been identified: **allelochemicals,** which have an interspecific effect, and **pheromones,** which act intraspecifically, affecting the behavior or physiology of members of the same species. Controversy has surrounded the precise definition of a pheromone (Sorenson, 1992). For example, could a compound be defined as a pheromone if it is merely a metabolic by-product and not produced as a specialized communication signal? Reproductive behavior is often triggered by the detection—usually by olfaction—of pheromones. The detection of pheromones may be of importance in homing behavior. In tetrapod vertebrates, pheromonic detection is of sufficient importance that a secondary olfactory system, the vomeronasal system, apparently evolved specifically to aid in the location of a receptive mate (Firestein, 2001).

Despite the variation observed in chemosensory systems in aquatic and terrestrial animals, the molecular (protein) configuration of cellular receptor sites is remarkably consistent. The gene families coding for the assembly of these proteins have been identified for many vertebrate and invertebrate species. At least nine olfactory receptor gene systems, each arising from a single ancestral gene, have been identified as most likely present in the common ancestor of the fishes and tetrapods; representatives of eight of these nine gene complexes are still observable in modern fish species (Niimura and Nei, 2005). Olfactory receptor genes may number up to 100 in zebrafish and catfish, but they may include as many as 1,000 in the mouse or rat. The significance of olfaction is evident from the observation that as much as 4 percent of the eukaryote genome may be given over to the coding of olfactory systems (Firestein, 2001; Mombaerts, 1999). The biochemistry of the transduction of chemosensory stimuli remains poorly understood, but recent research has provided insights into the nature of receptor sites and their interaction with waterborne chemicals that elicits the sensation of smell or taste.

OLFACTION

Structure of Receptor Organs

The organs of olfaction in fishes are associated with the nostrils, located in sacs or pits on the anterior part of the head, usually directly in front of the eye in bony fishes. In elasmobranchs, the nostrils are usually in front of the mouth. The olfactory receptor cells are in the epithelium lining the sacs. This epithelium is columnar and is underlain by a basement membrane. In addition to sensory epithelial cells, the olfactory epithelium includes ciliated nonsensory cells, supporting (sustentacular) cells, and mucous cells. Olfactory receptor cells differ from other sensory cells in that the receptor cells themselves are neurons that extend an axon directly to the brain. The receptor cells are located on series of lamellae (folds) that are disposed in the epithelial lining of the nasal sac. The epithelium of the lamellae contains two types of receptor cells: **ciliated receptor cells** and **microvillous receptor cells** (Caprio, 1988). The segregation of microvillous and ciliated sensory cells in fish olfactory systems may be a precursor to the origin of the vomeronasal system, which consists of microvillous receptor cells capable of pheromone detection, in tetrapods (Hara, 1992c). Water must flow through the sac so that dissolved material can come into contact with the sensory cells. Cells with a distinctive rod-shaped protrusion are also seen and are considered a distinct receptor cell type (Zeiske et al., 1992).

The ciliated receptor cells are somewhat elongate, and a small swelling at their distal end reaches the lumen of the olfactory sac into which the cilia extend (Hara, 1971, 1975). The sensory cells taper at their bases to form very thin (ca. 0.2 μm) axons. These are grouped into fascicles that extend to the olfactory bulb, where they form synapses with dendrites of mitral cells. Fish species differ in the size and number of cilia per olfactory receptor cell. There are about 20 cilia per cell in the burbot (*Lota lota*), and these appear to be 20 to 30 μm long (Gemne and Döving, 1969). Ciliary movements observed in vitro have shown no regular pattern; bending and straightening occur in seemingly random sequences. The relationship of the cilia and their movements to the olfactory sense has not yet been fully described, but they seem to be the site where chemical substances come into contact with the appropriate nervous tissue, eliciting the sensation of smell (Cagan, 1984). The binding of odorants to receptor sites on the cilia appears to be a reversible reaction. This causes a conformational change in the receptor sites that elicits the generation of action potentials in the sensory cell (Brand and Bruch, 1992; Hara, 1992c). The number of receptor cells per mm² of olfactory epithelium is usually in the range of 50,000 to 95,000, but densities up to 500,000 cells/mm² have been reported (Caprio, 1984; Yamamoto, 1982).

In cyclostomes, elasmobranchs, and a variety of bony fishes (including some catfishes, cyprinoids, cods, and mormyroids), the olfactory bulb—a part of the olfactory lobe of the brain—is adjacent to the olfactory organ. In many of these fishes, the olfactory bulb is close to the rest of the brain; but in elasmobranchs, *Latimeria*, and some teleosts, the bulb is distant from the forebrain. In *Squalus*, the olfactory bulb is so closely associated with the olfactory epithelium that there is no discrete olfactory nerve (Kent, 1992). The olfactory bulb of teleosts is a concentrically laminated structure similar to that seen in higher vertebrates. Four layers

are present: a superficial olfactory nerve layer, a glomerular layer, a mitral cell layer, and a deep internal cell layer (Satou, 1992). The axons of the large cells of the mitral cell layer of the olfactory bulb contribute to the olfactory tract, which extends to the olfactory lobe. Each tract has two bundles, lateral and medial. The length of the olfactory nerve is a function of the positions of the olfactory organ and bulb relative to the olfactory lobe of the brain. Three configurations have been recognized:

1. Bulbs close to nares; the olfactory nerves are hence very short, but a long olfactory tract connecting the bulb to the olfactory lobe is present (elasmobranchs, most ostariophysans; Fig. 22.1A).

2. Bulbs close to hemispheres of the forebrain, so that long olfactory nerves are employed (most teleosts, including *Anguilla*, *Esox*, and *Oncorhynchus*; Fig. 22.1B).

3. Bulb is intermediate in position (as in *Gymnothorax*, *Coryphaena*, and some gadiform and characoid species; Hara, 1975; Hara and Zielinski, 1989; Kleerekoper, 1969; Meisami, 1991).

The thin olfactory nerve fibers conduct impulses more slowly (0.2 m sec⁻¹) than most other vertebrate axons (Meisami, 1991). The identity of the olfactory cells as ganglionic cells, and their direct connection with the brain, has led some morphologists to conclude that the olfactory receptor is essentially a primitive structure that has remained relatively unchanged (Kleerekoper, 1969).

The olfactory epithelium, which is ectodermal in origin and developed from the nasal placodes, is typically arranged as a series of olfactory lamellae on a longitudinal ridge, called the **raphe**, forming what is usually termed the **olfactory rosette** in most bony fishes (Zeiske et al., 1992). In elasmobranchs, the raphes are transverse and support olfactory lamellae on each side of each raphe. There is great variation in the arrangement of olfactory lamellae (Fig. 22.2), and the number of olfactory lamellae varies greatly among fishes (Hara, 1993; Yamamoto, 1982). Some, such as clingfishes and seahorses, have none. Others have but one or two, and familiar fishes (including salmonids, minnows, and pikes) have fewer than 20. Various eels have numerous olfactory lamellae; *Anguilla*, for instance, has around 90, but a greater number has been reported for the porgy (*Haplopagrus guentheri*), which has about 230 (Kleerekoper, 1969). In some species, the number of lamellae increases as the individual grows. **Microsmatic** fishes, with fewer lamellae (and presumably fewer receptor cells), generally have weaker powers of olfaction. However, although the relationship seems somewhat obscure (Hara, 1971), olfactory powers are probably related more to the area of the olfactory epithelium

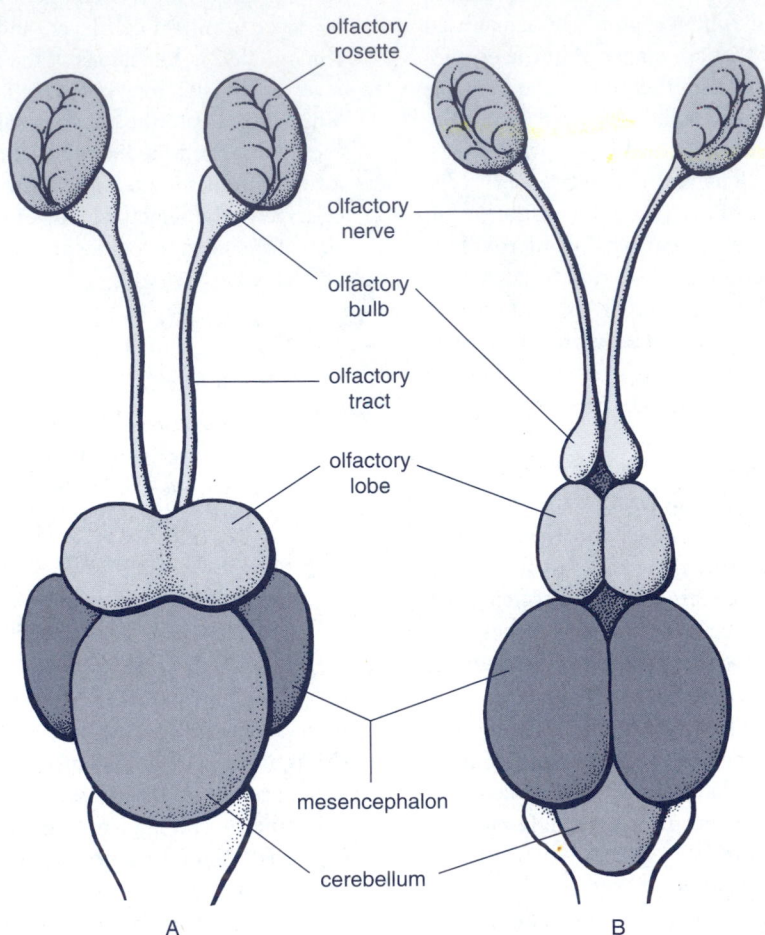

FIGURE 22.1
Dorsal view of brain: **A,** Catfish (*Parasilurus asotus*); **B,** Trout (*Oncorhynchus mykiss;* after Hara, 1975).

in relation to the surface area of the body and to the disposition of receptor cells in the olfactory epithelium. The paucity of olfactory lamellae in sticklebacks (*Gasterosteus*) does not preclude a role for olfaction in mating behavior (see Chapter 37).

The general observation has been made that fishes with oval rosettes that have an intermediate number of olfactory lamellae have intermediate powers of olfaction, and that those with long rosettes (macrosmatic fishes) have great olfactory acuity. Macrosmatic fishes tend to have continuously distributed olfactory epithelium on the lamellae, with a dense packing of sensory and ciliated nonsensory cells (Yamamoto, 1982). However, Zeiske et al. (1992) found no correlation between the number of receptor cells and olfactory acuity.

The majority of fishes have only external nares, which do not communicate with the pharynx. Typically, there is an anterior and posterior naris on each side (see Fig. 3.5). Fishes that have a communication between the external sacs and the oral cavity or pharynx include lungfishes, hagfishes, some eels, and stargazers (Uranoscopidae; Atz, 1952a, 1952b). The internal nares are not homologous in these diverse forms, because they develop differently. The lungfishes are considered to have true internal nares (choanae). Unusual anatomy is seen in *Tetraodon* and *Diodon*, in which the olfactory epithelium is situated in perforated "nasal lobes" or on bifid "nasal tentacula" protruding from the surface of the head at the usual position of olfactory organs (Kleerekoper, 1969). Some anglerfishes have tentacular nostrils (Marshall, 1967).

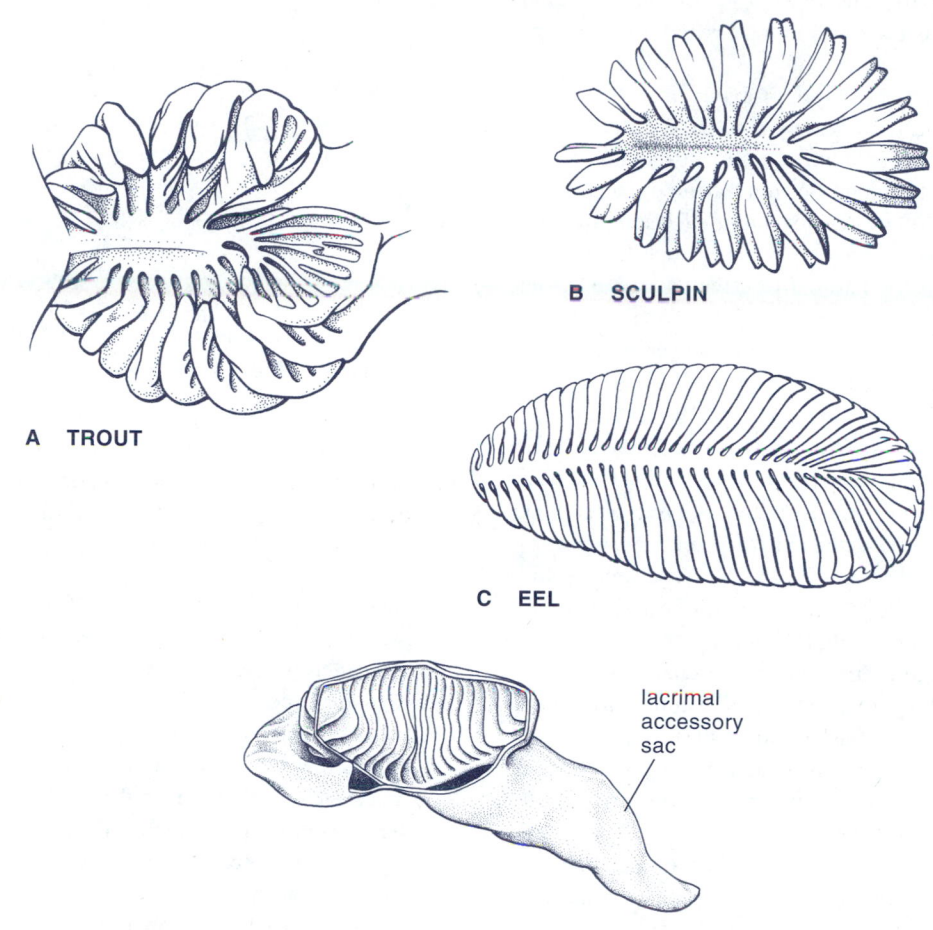

A TROUT

B SCULPIN

C EEL

lacrimal
accessory
sac

D EUROPEAN FLOUNDER

FIGURE 22.2
Dorsal view of olfactory rosette: **A,** Trout (*Oncorhynchus mykiss*), showing concave rosette with fingerlike processes; **B,** Staghorn sculpin (*Leptocottus armatus*), convex rosette; **C,** American eel (*Anguilla rostrata*), flat rosette; **D,** European flounder (*Pleuronectes flesus*), with accessory sacs. In eels, the incurrent water moves from left to right of the figure; in the others, the water impinges near the middle of the rosette (*D* after Kleerekoper, 1969).

For chemical substances to come into contact with the olfactory epithelium in the olfactory sacs, the sacs must be irrigated with water. Respiratory movements can facilitate the irrigation of the olfactory sacs in those fishes with internal nares, but maintaining a flow of water in most fishes requires some special hydraulic engineering. Circulation of water within the nasal sac is accomplished in one of three ways: (1) by forward movement of the fish in relation to the water (ram irrigation), (2) by the action of cilia in the sac or extensions of it, and (3) by pumping effected by direct or indirect constriction of the sensory nasal sac and accessory sacs.

Fishes that depend on their forward motion or on water currents for the movement of water through the olfactory organ usually have flaps or ridges behind the anterior nares to guide the water over the olfactory epithelium. These cutaneous structures are easily seen in common freshwater fishes such as trout, minnows, and suckers. Sharks and rays are notable for their occasionally elaborate complement of flaps and grooves associated with the nares (and mouth;

see Fig. 3.3). These flaps are arranged so that they accept a current of water over the olfactory epithelium as the animal swims or, in some species, as it pumps a respiratory current (Bell, 1993).

In eels, some catfishes, and probably in other fishes having long nasal sacs and widely separated anterior and posterior nostrils, cilia are important in moving water through the system, although the contraction of facial muscles that would facilitate the movement of water into the olfactory sacs may also be involved. Many such fishes have tubular extensions of the nares. In nettostomatid eels, the anterior nostrils are tubular, but the posterior nostrils are pores or slits, usually opening in front of the eye. In the genus *Nettenchelys,* however, the posterior nostril opens behind the head at a distance of up to two times the length of the head, depending on the species (Smith and Castle, 1982).

Latimeria has tubular anterior nostrils and slitlike posterior nostrils in front of the eyes. There is a five-sectioned, complicated rosette bearing the olfactory lamellae. Cilia apparently move water through the nasal cavity. Bichirs and reedfishes have remarkably complex nasal organs, with an anterior tube and a valvular posterior nostril. Both nasal cavities are divided into large and small chambers, which house multiple rows of lamellae. Kinocilia apparently move water through the organ (Zeiske et al., 1992).

Modifications for pumping water in and out of the olfactory chamber in a cyclic sequence are common among bony fishes, but can be seen also in the lampreys. Lampreys have a single olfactory opening, leading to an olfactory rosette that could either represent a fusion of two organs of smell or be a consequence of the lack of the organ splitting into two parts embryologically. The olfactory nerve is doubled in cyclostomes (Kleerekoper, 1969). However, the nasal portion of the canal is continuous with a blind tube called the nasohypophyseal canal or pouch that runs posteriorly under the brain and ends between the brain and the pharynx. Contractions of the branchial muscles apply pressure to the pouch and cause rhythmic emptying and refilling. As water flows into the pouch, some is shunted into the olfactory organ by a valve at its entrance (Kleerekoper, 1969).

In bony fishes, accessory sacs that are continuous with the nasal sac are commonly located under the skin in the region lateral to the nostrils, and some species have such sacs under the skin between the nasal organs. Respiratory and other movements involving the jaws and dermal bones of the face affect the volume of the nasal and accessory sacs and cause water to flow in and out over the olfactory epithelium as the sacs are compressed and relaxed by these movements. The anterior nostril is usually incurrent and the posterior one excurrent (Fig. 22.3). Some bony fish groups have a single nostril on each side. This can result from the loss of

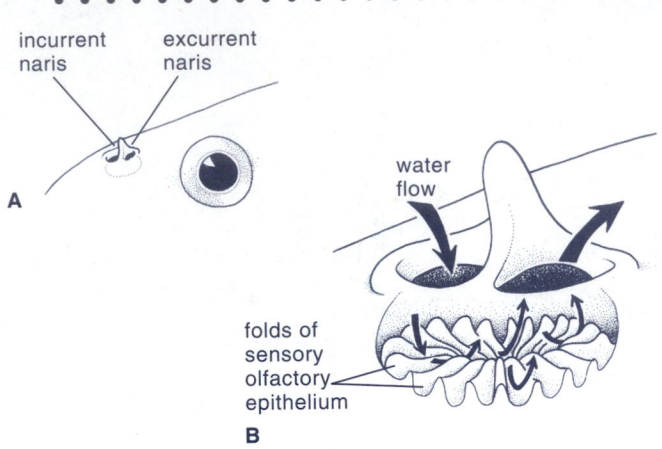

FIGURE 22.3
A, Position of left nares of Cyprinidae; **B,** Water flow over olfactory rosette of Cyprinidae.

the posterior naris, as in eelpouts (Zoarcidae) and snailfishes (Liparidae), or from the coalescence of anterior and posterior nares, as in sticklebacks (Gasterosteidae). Other groups having single nostrils are the cichlids (Cichlidae) and damselfishes (Pomacentridae).

Olfactory stimuli are specifically defined as those chemical stimuli that are transmitted to the central nervous system via cranial nerve I (the olfactory nerve), whereas gustatory stimuli are transmitted via cranial nerves VII (facial), IX (glossopharyngeal), or X (vagus; Hara, 1994). Associated with the olfactory apparatus in vertebrates is the terminal nerve system (see Chapter 26). Although there is no direct evidence that this system has a chemosensory function (Meredith and White, 1987), experiments with various animals, including goldfish, have suggested a relationship to reproductive function, especially in the detection of pheromones (Meisami, 1991, Yamamoto, 1982). However, Fujita et al. (1991) referred to the terminal nerve as having no known function and demonstrated that the olfactory system is responsible for chemosensory responses to pheromones in the male goldfish.

Function and Significance of Olfaction

Olfactory Thresholds

Early research on olfactory thresholds involved behavioral experiments, in which fishes trained by reward or punishment were conditioned to select or avoid water that held the chosen chemical, then were tested on more and more dilute concentrations. More modern methods may involve conditioned heart rate, electroencephalograms, recordings from nerve tracts, and other electrophysiological methods. Thresholds from various sources for a number of fish species and chemical substances have been tabulated by Kleerekoper (1969), Little (1983), Caprio (1984), Meisami (1991), and Hara (1992b). Examples of these thresholds are presented in Table 22.1. Although lower thresholds have been reported, Hara and Zielinski (1989) reported that the lowest olfactory threshold in fish yet identified by modern electrophysiological means is 10^{-13} M for a preovulatory pheromone (17-α, 20-β dihydroxy-4-pregnen-3-one) in goldfish. Fishes show generally low thresholds for amino acids, many of which contribute to food odors. Ictalurid catfishes show low thresholds (10^{-9} to 10^{-6} M) for amino acids (Hara, 1992b; Little, 1983).

Eels of the genus *Anguilla* have especially acute olfaction and have been used successfully in experiments involving conditioned responses to extremely dilute solutions of food extracts or other substances. For example, only one molecule of β–phenylethyl alcohol (10^{-18} M) in the olfactory sac has been reported to cause a conditioned response in a trained eel (Kleerekoper, 1969). The olfactory organ of the eel is relatively large and is equipped with many folds in the epithelium, which, unlike that of most fishes, is pigmented.

Döving et al. (1980) found that Arctic char (*Salvelinus alpinus*) and Arctic grayling (*Thymallus arcticus*) responded to the odors of bile acids at thresholds much lower than those for amino acids. They suggested that two different olfactory receptors might be involved in sensing the two different kinds of acids, because the response could be measured only from the medial part of the olfactory bulb when bile acids were presented and only from the lateral part when amino acids were presented. Fish generally have low olfactory thresholds for steroid hormones (Hara, 1992b).

Olfaction may also play a key role in osmoregulation. Both migratory fishes (such as salmonids) and nonmigratory freshwater fishes (such as cyprinids) have been demonstrated to possess acute olfactory sensitivity to environmental calcium and, to a lesser degree, sodium. It has been suggested that this enables such fishes to recognize osmotically optimal environments and contributes to the maintenance of internal calcium homeostasis (Hubbard et al., 2002).

Detection of Food

The detection of food is a major function of olfaction in most fishes and may be of special importance in species that feed in dim light or search through bottom materials or vegetation for edible objects. Among bony fishes, most active, open-water predators are primarily sight-oriented hunters (e.g., sharks); even among these species, however, olfaction, along with other chemical senses, is of great importance. It is well known that sharks can detect baits or wounded fish by olfactory cues. Some sharks lost the ability to find food placed into a tank with them when both their nostrils were plugged with cotton, but sought and found it when only one nostril was occluded, although the pattern of search differed from that followed when both nostrils were clear. In many of these experiments, the possibility of sighting the food was removed by concealing it in cheesecloth or in some other substance through which odor could penetrate. In other experiments, sharks were blinded but still would perform the typical figure-eight search pattern in locating food (Tester, 1963).

Some of the most interesting studies have involved the introduction of extracts, dilutions, or washes of various substances into experimental tanks with sharks. Extracts of fish flesh—especially with a considerable oil content—caused search and feeding activity even when presented in very dilute concentrations. Rotting flesh appeared to repel sharks to some extent. In studying the response of sharks to living fish, Tester (1963) found that water flowing over uninjured but distressed or excited fish caused a greater response in

TABLE 22.1   OLFACTORY THRESHOLDS OF FISHES FOR VARIOUS SUBSTANCES

Substance	Fish	Threshold	Source
Amino acids	hagfish	10^{-6} to 10^{-5} M	Döving and Holmberg, 1974
L-methionine	lemon shark	10^{-8} to 10^{-7} M	Zeiske et al., 1986
Amino acids	catfishes	10^{-9} to 10^{-6} M	Caprio, 1984
Methionine	grayling	1.3×10^{-6} M	Döving et al., 1980
Bile acids	grayling	6.3×10^{-9} M	Döving et al., 1980
Bile acids	goldfish	10^{-9} M	Meisami, 1991
Steroids	goldfish	10^{-13} to 10^{-12} M	Meisami, 1991
Phenylethyl alcohol	eel (*A. anguilla*)	2.9×10^{-20} M	Little, 1983
Phenylethyl alcohol	rainbow trout	10^{-9} M	Little, 1983
Sucrose	minnow (*Phoxinus*)	1.2×10^{-3} M	Little, 1983

the test animals than did water flowing over quiescent fish. Starved sharks were more responsive than well-fed specimens, some detecting the odor of food at a concentration of 1×10^{-4} ppm—a dilution that may be stronger than the undetermined threshold level. Amino acid thresholds as low as 1×10^{-14} M have been reported for elasmobranchs (Montgomery, 1988), and Hueter and Gilbert (1991) mentioned studies indicating that blacktip and reef sharks could sense extracts of grouper flesh at 1 part per 10 billion.

For obvious reasons, there has been interest in the attractiveness of human odors to sharks. Human urine appears to be detected by some sharks but causes no particular activity, whereas human blood attracts some of the species tested. In one series of experiments, human sweat caused what Tester (1963) termed "aversion" in sharks at about 1 ppm. Repulsion of sharks by odors has been studied for many years, in the hope that a suitable repellent can be found. Certain dyestuffs and copper acetate have shown limited promise, but research on the shark-repelling poison (*pardaxin*) found in a tropical flatfish (*Pardachirus marmoratus*) has led to the discovery that certain surfactants appear to be effective in repelling sharks (see Chapter 18).

Fishes other than sharks and rays that are known to orient to food by olfaction include lampreys, hagfishes, African lungfishes, and many teleosts. Among the latter are numerous minnows, catfishes, eels, perches, wrasses, and cods. Field observations and examination of the relative size and degree of development of olfactory organs and olfactory centers have led investigators to believe that certain plankton feeders of both shallow and deep waters and some swift predators orient toward food at least partially by olfaction.

Orientation

Orientation in the environment is achieved partly through the sense of smell in many fishes—undoubtedly in more than are presently recognized. Some minnows can be taught to distinguish between the odors of streams that differ in geology or in organic components such as prevalent species of aquatic plants, indicating that they might be able to recognize localities by very dilute olfactory cues (Hasler and Wisby, 1951). A variety of fishes, including several salmonid species and some centrarchids, including sunfishes (*Lepomis*), apparently can locate their home areas by means of the olfactory sense. The ability to do so has been demonstrated for Pacific salmon (*Oncorhynchus*), which can return to the stream locality from which they migrated months or years earlier (Hasler and Scholz, 1983).

Olfaction plays a prominent part in the salmon's homing behavior during the spawning migration. Field experiments have shown that blinded salmon taken from a spawning stream and displaced downstream can make the correct choices of tributaries and return to their home stream if allowed full use of their olfactory organs, but distribute in a random manner if their nostrils are plugged (Wisby and Hasler, 1954).

Electroencephalographic studies have shown that strong impulses can be recorded from the olfactory bulb of salmon stimulated by home stream waters, although some fish reacted to non-home natural waters as well (Oshima et al., 1969). Imprinting salmon smolts with the organic compounds morpholine or phenylethyl alcohol, which do not occur naturally, and then, at the time of spawning migration, placing the appropriate attractant (morpholine or phenylethyl alcohol) in a stream other than the one that the smolts descended has resulted in attracting the returning adults to the non-home stream (Cooper and Hirsch, 1982). Salmon apparently depend on olfactory cues in nonreproductive homing as well as reproductive homing (Stabell, 1992). Studies by Allan Scholz at Eastern Washington University revealed that thyroid hormone levels are integral to the ability of salmon to imprint on odors. Researchers investigating the molecular mechanisms of olfactory imprinting have discovered that imprinting salmon with phenylethyl alcohol triggers enhanced cyclic guanosine monophosphate (cGMP) production in olfactory neurons. This compound acts as an intracellular messenger that facilitates signal transmission (Barinaga, 1999).

Studies by Selset and Döving (1980) showed that mature Arctic char showed a preference for water that contained the odor of smolts from their home stream population. The ability of several species of fishes, including salmonids, to recognize the odor of their own kin has been reviewed by Olsén (1992). There are two hypotheses involved in the relationship of olfaction to homing. One is that the fishes respond to an imprint of the odors of soils, vegetation, and other such materials present in the stream from which they migrated. The second is that there is an innate response to pheromones released by fish of their own particular strain in the stream from which they came. Pheromones important to homing are believed to originate in the liver and are released with the feces (Stabell, 1992).

Other Behavior

Olfaction has been implicated or suspected to be of importance in many other aspects of fish behavior (Little, 1983). In reproductive behavior, in addition to homing to the spawning stream, the sense of smell is important in the location of mates, the triggering of certain phases of the spawning act, the recognition of young, and territorial defense. Pheromones related to reproductive behavior (see Chapter 27) can be detected at very low thresholds. Olfaction is involved in the social behavior of fishes in many ways, including the recognition of members of the same species, collectively (in schools) or individually (see Chapter 37).

The detection and avoidance of predators is of great importance. In some fishes, especially the ostariophysans, the presence of predators is indirectly noted by the sensing of alarm substances released from injured individuals of conspecific and closely related fishes. Minute amounts of a fright substance ("**schreckstoff**" in German) issuing from damaged mucous cells of a wounded fish cause an almost immediate fright reaction and retreat in members of the same species. A dilution of skin extract to about 0.02 part per trillion can be sensed (Little, 1983). A fright reaction is caused in Pacific salmon by rinses or extracts of mammalian skin containing L-serine (Brett and MacKinnon, 1954). This may represent a recognition of the odors of potential terrestrial predators such as foraging bears. There is evidence that minnows that have survived attacks by predators show fright reactions when exposed to the odor of the predator species (Little, 1983). Sensitivity to alarm substances can be significantly degraded by environmental contaminants, such as cadmium (Scott et al., 2003).

For decades, researchers have pursued the development of a chemical compound that could serve as an effective shark repellent. A most promising development in this area was presented at the 2004 Annual Meeting of the American Society of Ichthyologists and Herpetologists. Eric Stroud and his colleagues (Stroud et al., 2004) have isolated semiochemicals from extracts of decayed shark carcasses. A 200-ml dose of a mixture of these compounds was effective in clearing an aggregation of Caribbean reef sharks (*Carcharhinus perezei*) from an experimental feeding area.

TASTE

Taste Receptors

Whereas the olfactory system is integrated with the capacity for learning, as in recognizing food odors, geographic locations, or the pheromones of conspecifics, the sense of taste (gustation) is devoted to the instinctual recognition of food sources (Sorenson and Caprio, 1998). The receptors of the gustatory sense are called *taste buds,* which are groups of specialized epithelial cells. Taste buds are somewhat oblate or pear-shaped structures, made up of about 100 to 150 elongate cells, including basal cells and supporting cells in addition to the receptor cells (Fig. 22.4). The role of the basal cells is not clear. There may be chemical synapses between the sensory cells and the basal cells, which in turn may synapse with afferent nerve fibers. There is a possibility that the basal cells are mechanoreceptors. Taste buds, which are about 30 to 80 μm long and 20 to 50 μm wide, are set in the epithelium (above the basement membrane) with the apical end at the surface of the epithelium (Caprio, 1984; Kapoor et al., 1975). The apical ends of the receptor cells have microvilli that protrude from their surface. Histologically, there are two types of cells involved, pale sensory cells and dark supporting cells (Meisami, 1991; Reutter, 1982, 1992).

Taste buds are generally concentrated on the mouth and in the pharyngeal region and associated gill arches of fishes. Some species, notably carps and salmonids, have a great concentration of taste buds on a specialized palatal organ, which may be of principal importance in gustation (Hara, 1971). Many species, including catfishes, have external taste buds on specialized structures such as barbels, elongate fin rays, and certain areas of the body surface. Some catfishes have taste buds over much of the body. The distribution of external taste buds differs between species. In cyprinids, they are more numerous toward the head than toward the tail and increase from dorsal to ventral (Gomahr et al., 1992). Atema (1971) reported concentrations of taste buds of up to 50 per mm^2 on the surfaces of gill arches in *Ameiurus natalis*, and a total of around 175,000 on the entire body surface. Taste buds commonly occur on the lips and head of bottom-feeding fishes, but this is by no means universal (Livingston, 1987). Harvey and Batty (1998) studied the distribution of taste buds in early life history stages of the cod. Taste buds first appeared on the snout and lips, and developed their greatest concentrations on the barbel and pelvic fin surfaces—consistent with the feeding behavior of these benthopelagic fishes.

Innervation of the taste buds of the oropharyngeal cavity is usually by the glossopharyngeal (IX) and vagus (X) nerves. The facial (VII) nerve innervates the external taste buds and, in some fishes, some buds at the anterior part of the mouth (Caprio, 1984; Hara and Zielinski, 1989). In the catfish, the gustatory system innervated by the facial nerve detects distant chemical stimuli, whereas that innervated by the vagal nerve is used for discriminatory behaviors accompanying selective ingestion (Hara, 1994).

Function and Significance of Taste

Sapid Substances

The variety of stimuli that can be detected by olfaction is much broader than those that can be recognized through gustation, although both systems respond to amino acids and nucleotides (Sorenson and Caprio, 1998). Some sharks that have a keen sense of taste have been observed to prefer certain food fishes in taste tests (Tester, 1963) and are known to react to substances that are described by humans as bitter and sour. Some sharks react to salt, seemingly through the gustatory sense. Although there has been progress in

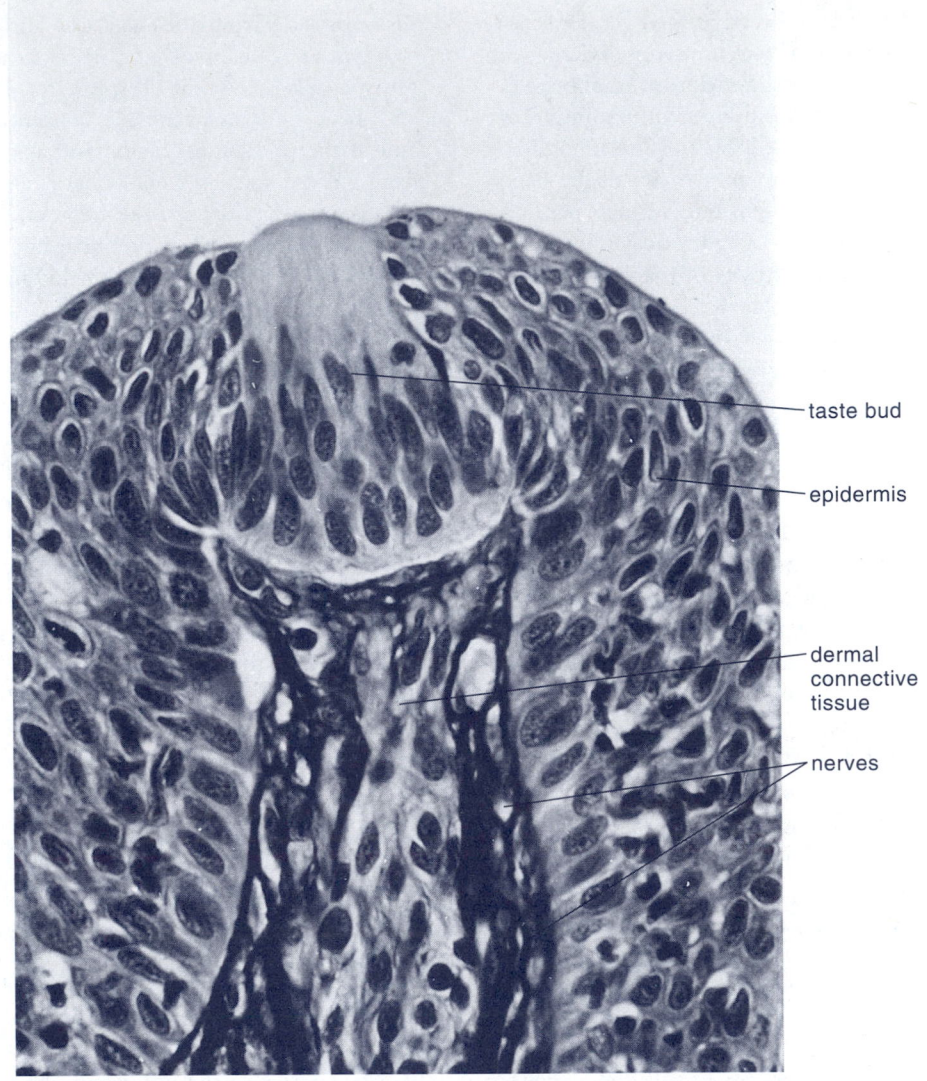

FIGURE 22.4
Taste bud of juvenile rainbow trout (*Oncorhynchus mykiss*). (Photomicrograph courtesy of Joseph Wales.)

research on the sensory physiology of sharks in the last three decades (Hueter and Gilbert, 1991), comparatively little research has been carried out on their sense of taste.

Much more is known of the gustatory capabilities of bony fishes. Various species are known to react to the four categories of taste known to humans (bitter, sweet, salt, and sour) as well as to many other tastes of greater biological importance. Through electrophysiological studies and training for conditioned responses, various teleosts have been shown to respond to numerous substances placed into contact with the taste receptors (Marui and Caprio, 1992).

A gustatory response is elicited by numerous substances, including amino acids, aliphatic acids, nucleotides, many salts, quinine and related bitter materials, as well as saliva, milk, and extracts of earthworms, silkworm pupae, and other food items. Gustatory responses to amino acids are by far the best studied (Hara, 1994). The **palatal organ** of the carp responds strongly to carbon dioxide (Hara, 1971). This thick, muscular pad in the anterior part of the oral cavity in cyprinids has dense aggregations of taste buds that permit highly localized assessment of the palatability of ingested organic particles (see Chapter 23).

According to the most current hypothesis concerning gustation, stimulus molecules elicit the sensation of taste through their interaction with certain receptors on the cell surface. This leads to the activation of second-messenger cascades, culminating in an increase in the level of intracellular calcium ions or other factors that bring about a depolarization of the receptor cell, initiating the release of neurotransmitters and the generation of action potentials in the postsynaptic taste fibers. Nitric oxide (NO) has been demonstrated to be a significant neurotransmitter substance. The stimulation of taste receptors with amino acids brings about a significant increase in the activity of nitric oxide synthase, necessary for the synthesis of NO, thus demonstrating its potential role in taste perception (Huque and Brand, 1994).

In testing the responses of single nerve fibers, investigators have noted some specialization among the receptors; some receptors react to many tastes, but others only to specific tastes or combinations (Kiyohara et al., 1975; Tucker, 1983). Carp appear to have seven, and puffers at least three different kinds of receptors.

Taste Thresholds

Experimental evidence has shown that taste receptors on various parts of the body have different sensitivities and thresholds, and that there are notable specific differences among fishes. Even within the same species, taste responses can differ between strains. For example, Japanese carp reacted more strongly than did a Swedish strain to quinine and extracts of worms and silkworm pupae, whereas the Swedish strain responded only weakly to quinine but much more strongly to sucrose than did the Japanese strain (Konishi and Zotterman, 1963).

The thresholds at which fish can respond to tastes vary from species to species but, generally speaking, fishes react to lower concentrations of sapid substances than humans. In the minnow (*Phoxinus*), for instance, the threshold reported for sucrose is 1.2×10^{-5} M, and that for fructose is 1.6×10^{-5} M (Little, 1983). The Mexican blind cavefish (*Astyanax mexicanus*) is reported to have a much greater taste sensitivity than minnows—up to thousands of times greater for the four basic taste qualities (Hara, 1971). *Phoxinus* can detect sodium chloride at 4×10^{-5} M, and *Ameiurus* can taste quinine at about 1×10^{-4} M. Nurse and lemon sharks are known to react to the chemical betaine at concentrations as low as 1×10^{-9} M (Carr, 1982).

Amino acids are detected by fishes at low concentrations and often elicit feeding responses. Johnson et al. (1990) obtained electrophysiological responses to several amino acids by *Tilapia zillii*, a herbivore, at a test concentration of 1×10^{-6} M, but no thresholds were sought. Caprio (1984) and Marui and Caprio (1992) tabulated thresholds for amino acids in several fishes. For instance, channel catfish can detect L-alanine and L-arginine at 1×10^{-9} to 1×10^{-11} M. *Pseudorasbora* can detect L-alanine and proline at about the same levels.

Relationship to Feeding and Other Behavior

Experiments with the chemical stimulation of feeding behavior in a variety of fishes, including the percoids *Lagodon* and *Orthopristis* (Carr, 1982), have disclosed that amino acids and the compound betaine are responsible for much of the feeding stimulus. Species vary in their responses to extracts of marine organisms and to artificial mixtures.

Olfaction and gustation are both important in sensing more or less distant sources of stimuli, which may be of great importance in food location (see Chapter 23) and various reproductive activities (see Chapter 27). Sharks are well known for sensing food odors from distances of several kilometers (Hueter and Gilbert, 1991), but taste may be involved in this as well as in the eventual selection and ingestion of food. Bullheads (*Ameiurus*) are known to rely heavily on the external gustatory sense in finding food at a distance (Bardach et al., 1967a), and the same appears to be at least partially true of other fishes with numerous taste buds over the body and fins. In most fishes, the taste receptors on the lips, in the mouth, and on the branchial arches are instrumental in the final detection of food items, in initiating the reflexes involved in seizing and swallowing, and in the rejection of unwanted items (Atema, 1971; Hara, 1994). Some function of taste is suspected in the courtship of certain fishes—cichlids and gouramis, for instance—because of the mouth and fin contact observed during mating behavior. The recognition of young may be in part dependent on taste, although olfaction may be of greater importance.

OTHER CHEMOSENSORY RECEPTORS

Although olfaction and taste are undoubtedly the chemical senses of greatest importance to fishes, there are other senses that might serve important functions. A *common chemical sense*, attributed to free nerve endings in the skin, has been suggested as a sensor of solutions of salts, acids, and alkaline materials (Whitear, 1992). This sense has been studied very little, and its true significance is not well known.

Also poorly understood is the function of solitary chemosensory cells (SCCs), which are structurally similar to the receptor cells in taste buds and have been described in the epidermis of virtually all aquatic vertebrates, including some larval amphibians (Kotrschal, 1999). These cells form a synaptic contact at their base with an afferent neuron and possess a single, stout microvillus at their apical end (Sorenson and Caprio, 1998). These cells may share innervation with cranial nerves that carry information from the taste buds, but they also may be innervated by spinal nerves. The cytology of these bipolar cells, often found at concentrations of more than 1,000 per mm^2, has been described by Whitear (1992). Kotrschal (1991) found densities of 2,000 to 4,000 SCCs per mm^2 in some European cyprinids. Two genera of benthic, intertidally occurring gadoids, termed rocklings (*Ciliata* and *Gaidropsarus*), have become models for SCC research because of the extreme density of these unusual chemosensory cells located on the anterior dorsal fins of these genera. Up to six million SCCs may be present on the dorsal fin rays of *Gaidropsarus mediterraneus* (Whitear, 1992).

The searobins (Triglidae), which show a well-developed chemical sense located in the modified lower pectoral rays, have no taste buds in the skin of those rays, but apparently have solitary chemosensory cells (Silver and Finger, 1984). These cells are innervated by spinal nerves (Bardach et al., 1967b; Finger, 1982). Fishes with modified fin rays that bear taste buds innervated by cranial nerves include the hakes of the genus *Urophycis,* the rocklings (in which fin ray taste buds are restricted to the anteriormost ray of the anterior dorsal fin), and the gouramis (Belontiidae). The fins of rocklings receive input from sensory spinal nerves as well as from the facial nerve (Whitear and Kotrschal, 1988). It has been proposed that SCCs function as generalized detectors of current-borne stimuli, but do not function in the determination of the source of stimuli, and that highly specialized versions of the SCC, as seen in the rocklings, may function in predator detection. This is evident from the observation that this SCC system responds to a much narrower range of substances—specifically, secretions produced by other fishes (Kotrschal, 1995, 1999).

The apparent response of pit organs and free neuromasts of the lateral line system to certain ions has been reported by Katsuki and Yanagisawa (1982), but Whitear (1992) doubted the validity of these studies.

CODA: RELATING SENSORY ADAPTATIONS TO THE ENVIRONMENT

Fishes are indeed extraordinary for the range of aquatic habitats that they inhabit. Environments as diverse as the wave-swept rocky intertidal shores, the uniformly cold and stable depths of the bathypelagic realm, and the moving and still waters of small headwater streams and large rivers present a vast array of sensory stimuli for resident fishes. Given the diversity of environmental stimuli, the evolutionary

divergence of sensory systems is not surprising. Sensory adaptation to a given environment must also be considered in the context of evolutionary history. Skates and flounders may co-inhabit a given patch of muddy ocean bottom, but the sensory arsenals on which they depend must be considered in light of their unique phylogenies.

Kotrschal (1999) has provided valuable insights into the range of sensory adaptations in fishes by comparing the brains and sensory mechanisms of fishes from a range of habitats. Whereas mesopelagic fishes possess moderate-sized brains, with well-developed optic tecta, the brains of bathypelagic fishes are among the smallest relative to body size. The olfactory and acousticolateralis systems are well developed in fishes living in the uniformly cold and dark ocean depths. Although the gustatory sense appears minimal in bathypelagic forms, those feeding on or near the bottom (e.g., macrourids) appear to show greater dependence on the sense of taste. In the well-lit and sometimes turbulent, epipelagic inshore and offshore environments, vision is of much greater importance. Whereas the retinas of deep-dwelling fishes are rod-dominated, those of fishes in shallower water are cone-dominated, providing for sophisticated color vision. Retinal specializations in shallow-water forms permit the perception of polarized and ultraviolet light as well. Chemosensation would appear to be of limited use in highly turbulent environments, especially for the perception of objects at a distance. Pheromonal communication is still an integral component in the lives of many shallow-water species, however. Likewise, acousticolateralis systems would seem to be of limited use in environments that experience extremes in wave or current action. Some intertidally occurring fishes (e.g., stichaeids) show sophisticated and highly branched lateral line systems, however.

The aforementioned generalizations are developed primarily from comparisons of shallow-water marine species with deep-dwelling forms. Kotrschal (1999) noted similar patterns in freshwater taxa, such as the cyprinids and cichlids —benthic inhabitants tend to have well-developed chemosensory capacities, especially taste, but less developed visual and acousticolateralis lobes in the brain.

Chemosensation has been demonstrated to be instrumental in the induction of phenotypic changes in the crucian carp (*Carassius carassius*)—possibly the first example of induced morphological defense known in vertebrates. In the presence of chemical cues from piscivores, such as the northern pike (*Esox lucius*) or the perch (*Perca fluviatilis*), the body of the carp developed greater depth (Brönmark and Miner, 1992; Brönmark and Pettersson, 1994). Experiments have demonstrated that the handling time of deeper bodied carp by pike is greater, and, in preference studies, pike preferred more fusiform carp (Nilsson et al., 1995). The deeper body morph occurs with significantly greater frequency in lakes inhabited by piscivores (Poléo et al., 1995).

Although fishes display remarkable versatility in sensory adaptation to environmental stimuli from a range of aquatic habitats, they are also noteworthy for apparently lacking a particular sensory modality: *Proprioceptors* that monitor the contractile status of muscles are poorly developed in fishes (Bone et al., 1999). The perception of body position and posture is apparently of much more significance in tetrapodal locomotion.

KEY POINTS AND CONNECTIONS

- Chemosensation is an ancient sensory modality that is chiefly manifested in fishes in the form of taste and olfaction. The operation of these senses is dependent on the solubility and dispersal ability of odorant and sapid molecules in the water. In general, fishes display extreme sensitivity to waterborne substances, and this is especially useful in feeding and interspecies communication.

- Olfactory receptors are usually located in the anterior part of the head, in recesses known as the nares. Olfactory receptor cells are modified neurons located on series of lamellae that constitute the epithelial lining of the nasal sac. Two types of receptor cells — ciliated receptors and microvillous receptors — have been identified.

- Olfaction has been demonstrated to be essential in food location, identification of conspecifics for mating purposes, territorial defense, and recognition of young. Several classical studies with salmonids have demonstrated their extremely acute olfactory sense to play an essential role in recognizing their natal streams from great distances. Communication using pheromones is also widespread among fishes.

  Orientation in response to chemical cues (chemotaxis) is integral to the study of fish behavior. Aspects of chemotaxis, including a most remarkable application in the long-distance migration of salmonids, are discussed in Chapter 36.

- Whereas olfaction has been integrated with learning capacity in fishes, as in associating distinct smells with predators or prey, the gustatory sense operates at a more instinctual level, primarily enabling the rapid identification of food sources. Taste receptor cells are generally present in the form of taste buds that are localized on the mouth and pharyngeal epithelium. Taste buds may also form dense concentrations on special sensory structures, the barbels, which surround the mouth in many species.

  The ways in which the different sensory modalities are used during feeding are discussed in Chapter 23.

- A common chemical sense, attributed to free nerve endings in the epidermis, enables the detection of certain ions. Fishes also possess solitary chemosensory cells that are structurally similar to the receptor cells of taste buds. These receptor cells may form dense aggregations in certain areas of the epidermis, such as the fin rays.

- The adaptation of fishes to a diversity of environments is reflected in the differential emphasis placed on their sensory systems. For example, fishes inhabiting deep oceans have especially well-developed olfactory and acousticolateralis systems instead of a developed sense of vision. In habitats where vision is important, fishes may differ in the degree to which their retinas are dominated by rods or cones — cones being especially prevalent in the retinas of fishes inhabiting shallow waters.

BUILDING AN ICHTHYOLOGY LIBRARY

Hara, T. J. (Ed.). 1992. *Fish chemoreception.* Chapman and Hall, London.

Kleerekoper, H. 1969. *Olfaction in fishes.* Indiana University Press, Bloomington.

REFERENCES

Atema, J. 1971. Structures and functions of the sense of taste in the catfish (*Ictalurus natalis*). *Brain Behav. Evol. 4:* 273–294.

Atz, J. W. 1952a. Internal nares in the teleost *Astroscopus. Anat. Rec. 113:* 105–116.

———. 1952b. Narial breathing in fishes and the evolution of internal nares. *Q. Rev. Biol. 27:* 366–376.

Bardach, J. E., J. H. Todd, and R. Crickmer. 1967a. Orientation by taste in fish of the genus *Ictalurus. Science 155:* 1276–1278.

———, M. Fujiya, and A. Holl. 1967b. Investigation of external chemoreceptors of fishes, pp. 641–665. In *Olfaction and taste II,* T. Hayashi (Ed.). Pergamon Press, Oxford.

Barinaga, M. 1999. Salmon follow watery odors home. *Science 286:* 705–706.

Bell, M. 1993. Convergent evolution of nasal structure in sedentary elasmobranchs. *Copeia 1993:* 144–158.

Bone, Q., N. B. Marshall, and J. H. S. Blaxter. 1999. *Biology of fishes* (2nd ed.). Stanley Thornes, Cheltenham, UK.

Brand, J. G., and R. C. Bruch. 1992. Molecular mechanisms of chemosensory transduction: Gustation and olfaction, pp. 126–149. In *Fish chemoreception,* T. J. Hara (Ed.). Chapman and Hall, London.

Brett, J. R., and D. MacKinnon. 1954. Some aspects of olfactory perception in migrating adult coho and spring salmon. *J. Fish. Res. Bd. Can. 11:* 310–318.

Brönmark, C., and J. G. Miner. 1992. Predator-induced phenotypical change in body morphology in crucian carp. *Science 258:* 1348–1350.

———, and L. B. Pettersson. 1994. Chemical cues from piscivores induce a change in morphology of crucian carp. *Oikos 70:* 396–402.

Cagan, R. H. 1984. Olfactory recognition in rainbow trout, pp. 285–299. In *Comparative physiology of sensory systems,* L. Bolis, R. D. Keynes, and S. H. P. Maddrell (Eds.). Cambridge University Press, Cambridge, UK.

Caprio, J. 1984. Olfaction and taste in fish, pp. 257–284. In *Comparative physiology of sensory systems,* L. Bolis, R. D. Keynes, and S. H. P. Maddrell (Eds.). Cambridge University Press, Cambridge, UK.

———. 1988. Peripheral filters and chemoreceptor cells in fishes, pp. 313–338. In *Sensory biology of aquatic animals,* J. Atema, R. R. Fay, A. N. Popper, and W. N. Tavolga (Eds.). Springer Verlag, New York.

Carr, W. E. S. 1982. Chemical stimulation of feeding behavior, pp. 259–274. In *Chemoreception in fishes*, T. J. Hara (Ed.). Elsevier, Amsterdam.

Cooper, J. C., and P. J. Hirsch. 1982. The role of chemoreception in salmonid homing, pp. 343–362. In *Chemoreception in fishes*, T. J. Hara (Ed.). Elsevier, Amsterdam.

Döving, K. B., and K. Holmberg. 1974. A note on the function of the olfactory organ of the hagfish *Myxine glutinosa. Acta Physiol. Scand. 91:* 430–432.

———, R. Selset, and G. Thommesen. 1980. Olfactory sensitivity to bile acids in salmonid fishes. *Acta Physiol. Scand. 108:* 123–131.

Finger, T. E. 1982. Somatotopy in the representation of the pectoral fin and free fin rays in the spinal cord of the sea robin, *Prionotus carolinus. Biol. Bull. 163:* 154–161.

Firestein, S. 2001. How the olfactory system makes sense of scents. *Nature 413:* 211–218.

Fujita, I., P. W. Sorenson, N. B. Stacey, and T. Hara. 1991. The olfactory system, not the terminal nerve, functions as the primary chemosensory pathway mediating responses to sex pheromones in male goldfish. *Brain Behav. Evol. 38:* 313–321.

Gemne, G., and K. B. Döving. 1969. Ultrastructural properties of primary olfactory neurons in fish (*Lota lota* L.). *Am. J. Anat. 126:* 457–476.

Gomahr, A., M. Palzenberger, and K. Kotrschal. 1992. Density and distribution of external taste buds in cyprinids. *Env. Biol. Fishes 33:* 125–134.

Hara, T. J. 1971. Chemoreception, pp. 79–120. In *Fish physiology*, Vol. V, W. S. Hoar and D. J. Randall (Eds.). Academic Press, New York.

———. 1975. Olfaction in fish. *Prog. Neurobiol. 5:* 271–335.

———. 1992a. *Fish chemoreception*. Chapman and Hall, London.

———. 1992b. Overview and introduction, pp. 1–12. In *Fish chemoreception*, T. J. Hara (Ed.). Chapman and Hall, London.

———. 1992c. Mechanisms of olfaction, pp. 150–170. In *Fish chemoreception*, T. J. Hara (Ed.). Chapman and Hall, London.

———. 1993. Chemoreception, pp. 191–218. In *The physiology of fishes*, D. H. Evans (Ed.). CRC Press, Boca Raton, FL.

———. 1994. Olfaction and gustation in fish: An overview. *Acta Physiol. Scand. 152:* 207–217.

———, and B. Zielinski. 1989. Structural and functional development of the olfactory organ in teleosts. *Trans. Am. Fish. Soc. 118*(2): 183–194.

Harvey, R., and R. S. Batty. 1998. Cutaneous taste buds in cod. *J. Fish Biol. 53:* 138–149.

Hasler, A. D., and W. J. Wisby. 1951. Discrimination of stream odors by fishes and relation to parent stream behavior. *Am. Nat. 85:* 223–238.

———, and A. T. Scholz. 1983. *Olfactory imprinting and homing in salmon: Investigations into the mechanism of the imprinting process.* Springer Verlag, Berlin.

Hubbard, P. C., P. M. Ingleton, L. A. Bendell, E. N. Barata, and A. V. M. Canário. 2002. Olfactory sensitivity to changes in environmental [Ca^{2+}] in the freshwater teleost *Carassius auratus:* An olfactory role for the Ca^{2+}-sensing receptor? *J. Exp. Biol. 205:* 2755–2764.

Hueter, R. E., and P. W. Gilbert. 1991. The sensory world of sharks. *Underwater Nat. 19–20:* 48–55.

Huque, T., and J. G. Brand. 1994. Nitric oxide synthase activity of the taste organ of the channel catfish, *Ictalurus punctatus. Comp. Biochem. Physiol. 108B:* 481–486.

Johnson, P. B., H. Zhou, and M. A. Adams. 1990. Gustatory sensitivity of the herbivore *Tilapia zillii* to amino acids. *J. Fish Biol. 35:* 387–393.

Kapoor, B. G., H. E. Evans, and R. A. Pevzner. 1975. The gustatory system in fish. *Adv. Mar. Biol. 13:* 53–108.

Katsuki, Y., and K. Yanagisawa. 1982. Chemoreception in the lateral line, pp. 227–242. In *Chemoreception in fishes*, T. J. Hara (Ed.). Elsevier, Amsterdam.

Kent, G. C. 1992. *Comparative anatomy of the vertebrates.* Mosby, St. Louis.

Kiyohara, S., I. Hidaka, and T. Tamura. 1975. Gustatory response in the puffer. 2. Single fiber analyses. *Bull. Jpn. Soc. Sci. Fish 41*(4): 383–391.

Kleerekoper, H. 1969. *Olfaction in fishes.* Indiana University Press, Bloomington.

Konishi, J., and Y. Zotterman. 1963. Taste functions in the carp. *Acta Physiol. Scand. 52:* 150–161.

Kotrschal, K. 1991. Solitary chemosensory cells: Taste, common chemical sense or what? *Rev. Fish Biol. Fish. 1:* 3–22.

———. 1995. Ecomorphology of solitary chemosensory cell systems in fish: A review. *Env. Biol. Fishes 44:* 143–155.

———. 1999. Sensory systems, pp. 126–142. In *Intertidal fishes: Life in two worlds*, M. H. Horn, K. L. M. Martin, and M. A. Chotkowski (Eds.). Academic Press, San Diego.

Little, E. E. 1983. Behavioral function of olfaction and taste in fish, pp. 351–376. In *Fish neurobiology*, Vol. 1, R. G. Northcutt and R. E. Davis (Eds.). University of Michigan Press, Ann Arbor.

Livingston, M. E. 1987. Morphological and sensory specializations of five New Zealand flatfish species, in relation to feeding behavior. *J. Fish Biol. 31:* 775–795.

Marshall, N. B. 1967. The olfactory organs of bathypelagic fishes, pp. 57–70. In *Aspects of marine zoology*, N. B. Marshall (Ed.). Academic Press, New York.

Marui, T., and J. Caprio. 1992. Teleost gustation, pp. 171–198. In *Fish chemoreception*, T. J. Hara (Ed.). Chapman and Hall, London.

Meisami, E. 1991. Chemoreception, pp. 335–434. In *Neural and integrative animal physiology*, C. L. Prosser (Ed.). Wiley, New York.

Mombaerts, P. 1999. Seven transmembrane proteins as odorant and chemosensory receptors. *Science 286:* 707–711.

Montgomery, J. C. 1988. Sensory physiology, pp. 79–98. In *Physiology of elasmobranch fishes*, T. J. Shuttleworth (Ed.). Springer Verlag, Berlin.

Niimura, Y., and M. Nei. 2005. Evolutionary dynamics of olfactory receptor genes in fishes and tetrapods. *Proc. Nat. Acad. Sci. USA 102:* 6039–6044.

Nilsson, P. A., C. Brönmark, and L. B. Pettersson. 1995. Benefits of a predator-induced morphology in crucian carp. *Oecologia 104:* 291–296.

Olsén, K. H. 1992. Kin recognition in fish mediated by chemical senses, pp. 229–248. In *Fish chemoreception*, T. J. Hara (Ed.). Chapman and Hall, London.

Oshima, K., W. E. Hahn, and A. Gorbman. 1969. Electroencephalographic olfactory responses in adult salmon to waters traversed in the homing migration. *J. Fish. Res. Bd. Can. 26:* 2123–2133.

Poléo, A. B. S., S. A. Øxnevad, K. Østbye, E. Heibo, R. A. Anderson, and L. Asbjørn-Vøllestad. 1995. Body morphology of crucian carp *Carassius carassius* in lakes with or without piscivorous fish. *Ecography 18:* 225–229.

Rehnberg, B. G., and C. B. Schreck. 1986. The olfactory L-serine receptor in coho salmon: Biochemical specificity and behavioral response. *J. Comp. Physiol. 159:* 61–67.

Reutter, K. 1982. Taste organ in the barbel of the bullhead, pp. 77–91. In *Chemoreception in fishes*, T. J. Hara (Ed.). Elsevier, Amsterdam.

———. 1992. Structure of the peripheral gustatory organ, represented by the siluroid fish *Plotosus lineatus* (Thunberg), pp. 60–78. In *Fish chemoreception*, T. J. Hara (Ed.). Chapman and Hall, London.

Satou, M. 1992. Synaptic organization of the olfactory bulb and its central projection, pp. 40–59. In *Fish chemoreception*, T. J. Hara (Ed.). Chapman and Hall, London.

Scott, G. R., K. A. Sloman, C. Rouleau, and C. M. Wood. 2003. Cadmium disrupts behavioural and physiological responses to alarm substance in juvenile rainbow trout (*Oncorhynchus mykiss*). *J. Exp. Biol. 206:* 1779–1790.

Selset, R., and K. B. Döving. 1980. Behavior of mature anadromous char (*Salmo alpinus* L.) towards odorants produced by smolts of their population. *Acta Physiol. Scand. 108:* 113–122.

Silver, W. L., and T. E. Finger. 1984. Electrophysiological examination of a non-olfactory, non-gustatory chemosense in the searobin, *Prionotus carolinus*. *J. Comp. Physiol. A 154:* 167–174.

Smith, D. G., and P. H. J. Castle. 1982. Larvae of the nettastomid eels: Systematics and distribution. *Dana Rep. 90:* 1–44.

Sorenson, P. W. 1992. Hormones, pheromones, and chemoreception, pp. 199–228. In *Fish chemoreception*, T. J. Hara (Ed.). Chapman and Hall, London.

———, and J. Caprio. 1998. Chemoreception, pp. 375–405. In *The physiology of fishes* (2nd ed.), D. H. Evans (Ed.). CRC Press, Boca Raton, FL.

Stabell, O. B. 1992. Olfactory control of homing behaviour in salmonids, pp. 249–270. In *Fish chemoreception*, T. J. Hara (Ed.). Chapman and Hall, London.

Stroud, E. M., M. M. Herrmann, and S. H. Gruber. 2004. Semiochemicals as shark repellants—identification and behavioral responses. *Abstr. Ann. Meeting Amer. Soc. Ichthyol. Herpetol.*, Norman, OK, 26–31 May, 2004.

Tester, A. L. 1963. Olfaction, gustation, and the common chemical sense in sharks, pp. 255–282. In *Sharks and survival*, P. W. Gilbert (Ed.). D. C. Heath, Boston.

Todd, J. H., J. Atema, and J. E. Bardach. 1967. Chemical communication in social behavior of a fish, the yellow bullhead (*Ictalurus natalis*). *Science 158:* 672–673.

Tucker, D. 1983. Fish chemoreception: Peripheral anatomy and physiology, pp. 311–349. In *Fish neurobiology*, Vol. 1, R. G. Northcutt and R. E. Davis (Eds.). University of Michigan Press, Ann Arbor.

Whitear, M. 1992. Solitary chemosensory cells, pp. 103–125. In *Fish chemoreception*, T. J. Hara (Ed.). Chapman and Hall, London.

———, and K. Kotrschal. 1988. The chemosensory anterior dorsal fin in rocklings (*Gaidropsarus* and *Ciliata*, Teleostei, Gadidae): Activity, fine structure and innervation. *J. Zool. (Lond.) 216:* 339–366.

Wisby, W. J., and A. D. Hasler. 1954. Effect of olfactory occlusion in migrating silver salmon (*Oncorhynchus kisutch*). *J. Fish. Res. Bd. Can. 11:* 472–478.

Yamamoto, M. 1982. Comparative morphology of the peripheral olfactory organ in teleosts, pp. 39–59. In *Chemoreception in fishes*, T. J. Hara (Ed.). Elsevier, Amsterdam.

Zeiske, E., J. Caprio, and S. H. Gruber. 1986. Morphological and electrophysiological studies on the olfactory organ of the lemon shark, *Negaprion brevirostris* (Poey), pp. 381–391. In Indo-Pacific fish biology. *Proceedings of the 2nd International Conference on Indo-Pacific Fishes*, T. Uyeno, R. Arai, T. Taniuchi, and K. Matsuura (Eds.). Ichthyological Society of Japan, Tokyo.

———, B. Theisen, and H. Breuker. 1992. Structure, development, and evolutionary aspects of the peripheral olfactory system, pp. 13–39. In *Fish chemoreception*, T. J. Hara (Ed.). Chapman and Hall, London.

Zimmer, R. K. 2000. Importance of chemical communication in ecology. *Biol. Bull. 198:* 167.

PART FIVE

HOMEOSTATIC
MECHANISMS

Modern physiological study can be said to have its origins in the work of the great 19th-century scientist Claude Bernard. Bernard was the first to recognize that animal function is dependent on the ability to maintain stability in the *milieu intérieur,* or internal environment. Bernard recognized that animals are capable of maintaining this milieu intérieur constant, and that this constancy is essential to normal physiological function. By the early 20th century, the hierarchical organization of organisms—from cells to tissues to organs to organ systems to the whole animal—had become recognized as the key to understanding physiological function. Walter Cannon coined the term **homeostasis** to refer to the tendency of animals to maintain the constancy of the internal environment in the face of fluctuation in the external environment. We have come to recognize that **homeostasis** is an active process: Animals are obtaining and expending energy in the attempt to maintain a stable internal environment, which is thus optimized in terms of the broad array of biochemical and physiological functions that define living systems.

Survival in stressful environments often depends on an animal's ability to minimize fluctuations in ion concentrations, osmotic pressure; pH, and dissolved oxygen. For endothermic animals, this capacity also extends to the ability to maintain a narrow range of internal temperatures. Fishes, for the most part, do not include thermoregulation as part of their homeostatic processes. In this section, we will explore the physiological functions—specifically, gas exchange, circulation, excretion, and osmoregulation—that sustain an individual fish. None of these systems can be fully understood in isolation. All of them are integrated in into a functional whole. For example, although we will consider the circulatory system mainly in its role in oxygen uptake and delivery, it serves, of course, myriad other duties, including the transport of nutrients and endocrine secretions. It is critical to the maintenance of the internal constancy of the chemical environment of the cells and tissues. Communication among tissues and organs is essential to this integration. For this reason, we will also explore the functions of the nervous and endocrine systems—two systems that, in quite different ways, enable the total integration of the organism into a functional unit.

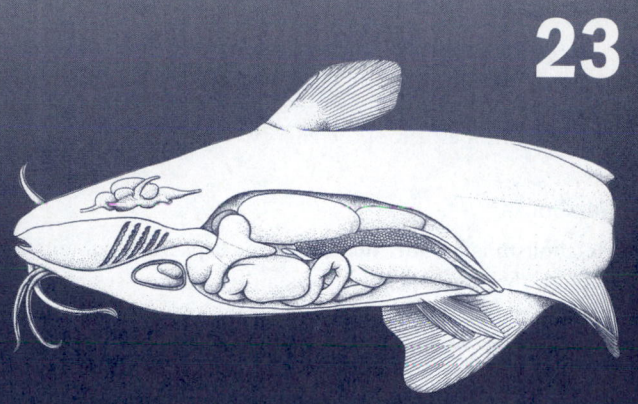

23

USE AND ACQUISITION OF FOOD

METABOLISM AND GROWTH

Energetic Concerns
Metabolism
Growth
Age and Growth Studies

NUTRITION

Dietary Requirements of Fishes
  Protein
  Carbohydrates
  Lipids
  Vitamins
  Minerals
Role of Hormones

THE ALIMENTARY CANAL: ANATOMICAL FEATURES

The Oral Cavity and Pharynx
  Teeth
  Pharynx
Esophagus, Stomach, and Intestine
  Esophagus
  Stomach
  Intestine
  Pyloric Caeca
  Liver
  Pancreas
  Spleen

FORAGING ACTIVITY AND FEEDING IN FISHES

Feeding in the Ecological Context
Detection and Selection of Food
Dietary Diversity and Foraging Activity
  Macrocarnivores
  Microcarnivores
  Herbivores
  Omnivores

FUNCTIONAL MORPHOLOGY OF THE FEEDING APPARATUS

Agnathan Fishes
Chondrichthyan Fishes
Jaws and Branchial Apparatus in Bony Fishes
  Suction Feeding
  Gill Rakers and Pharyngeal Teeth
Preparation of Food for Digestion
Ecomorphology and Feeding in Fishes

DIGESTION

The Digestive Process
  Gut Morphology and Digestion
  Enzymatic Activity
Rates and Efficiency of Digestion

Fishes are no different than any other organism in that much of their daily activity focuses on the acquisition of food. Food supplies energy and essential materials that make bodily maintenance, growth, and the elaboration of reproductive tissue possible. Like all other organisms, fishes budget their energy intake, allocating portions for specific needs. A portion of the ingested food is expended simply in the maintenance of the individual. Energy over and above this amount can be used in the processes of growth— both of the whole body and of the reproductive tissues essential for the perpetuation of the species. The conversion of energy for the purposes of maintenance and growth is what we refer to as *metabolism*. Fishes are especially fascinating for the diversity of feeding strategies they exhibit. Although most fishes are carnivores, just about every other feeding strategy is also found, especially among the bony fishes. Some carnivores are adapted for the consumption of very small prey, such as plankton; others ingest large prey. Omnivores, herbivores, and detritivores round out the picture of nutritional adaptations of fishes to their aquatic realm. In this chapter, we will investigate the fundamentals of metabolism and growth, consider the nutritional requirements of fishes, and gain a greater appreciation for the range of dietary strategies displayed by fishes.

METABOLISM AND GROWTH

Energetic Concerns

It has long been recognized that there is a definite correlation between global patterns of species richness and areas of high productivity. The availability of abundant and diverse food resources has resulted in the evolution of a greater diversity of species able to use these resources. **Energetics** is that branch of biology that is concerned with the ways in which energy (the unit of measurement of which is the **calorie**, or in larger applications, the *kilocalorie*) is apportioned in the growth and maintenance of individuals, populations, and communities. It seeks to explain the nature of ecosystems from the perspective of the energy that is made available for their development and maintenance. A simple energetic model applicable to fishes, which depicts the apportioning of available energy, is illustrated in Table 23.1.

The needs of the organism that must be met before energy in appreciable amounts can be used for growth include the standard metabolism (see later) of the resting animal, any swimming or other muscle activity over the resting condition, and the energy of what is called **specific dynamic action** (**SDA**; Roberts and Bullock, 1989). SDA includes energy used in the deamination of amino acids that are not used in growth and energy used in the digestion and assimilation of food. This energy is also called the "heat of nutrient metabolism" (Smith, 1989). SDA has been studied in a few fishes. The amount of energy attributed to SDA in the bioenergetics of the cutthroat trout has been estimated to range from about 14 to about 47 percent of the energy consumed by individual fish. Usually, SDA in fishes ranges from 10 to 18 percent of the energy consumed.

Metabolism

Metabolism is broadly defined as the capacity to obtain and convert chemical energy to sustain the processes of maintenance and growth of cells and tissues. Metabolism can be measured by calorimetric methods—that is, by measuring the total heat production of the organism. A simpler yet more efficient method is to measure the total amount of oxygen consumed in oxidative processes (see later). For endothermic organisms, the **basal metabolic rate (BMR)** is measured when the animal is in an inactive state and is under no stress. The animal must also be in a postabsorptive state, as digestive activity will increase the metabolic rate as a consequence of SDA. For poikilothermic organisms such as fishes, in which the resting metabolic rate will be a function of the ambient temperature, the **standard metabolic rate (SMR)** is measured. Because fishes are inevitably going to be engaged in a modest level of activity while their metabolic rate is being measured, an alternative measure, termed the **routine metabolic rate (RMR)**, is used (Hill and Wyse, 1989). Routine metabolic rates for some familiar freshwater fishes are shown in Table 23.2.

Oxygen consumption increases dramatically with activity, expressed as the **active metabolic rate (AMR)**, up to the point beyond which the fish cannot extract any more oxygen from the water. The history of the individual in regard to acclimatization to the environment is important to oxygen consumption, as are circadian and seasonal cycles.

Some fishes can resort to anaerobic metabolism when using stored carbohydrates in burst or prolonged swimming beyond their aerobic capacities (Beamish, 1980), and some appear to be in the anaerobic mode even in moderate swimming (Duthie,1982). Some fishes, such as the cyprinid goldfish, carps, and *Rasbora*, can withstand prolonged

TABLE 23.1   ENERGETIC MODEL FOR THE APPORTIONING OF ENERGY IN AN INDIVIDUAL FISH

Variable	Definition	Equation
DE	energy that is apparently digestible	$DE = IE - FE$
IE	intake of energy in food	
ME	energy that can be metabolized	$ME = IE - (FE + UE + ZE)$
FE	waste energy in feces	
UE	waste energy in urine	
ZE	energy excreted through gills	
RE	energy retained as growth of useful products, such as gametes	$RE = ME - HE$
HE	total heat production, including basal metabolism, activity, and specific dynamic action	

Sources: Smith, 1989; National Research Council, 1981, 1983.

TABLE 23.2   ROUTINE METABOLIC RATES MEASURED IN SELECTED NORTH AMERICAN FRESHWATER FISHES

Species	Temperature (°C)	O_2 Consumption (mg kg^{-1} hr^{-1})	Source
Carassius auratus	10	15.7	Beamish and Mookherji, 1964
	20–22	30–160	Beamish and Mookherji, 1964
	32–35	127–262	Beamish and Mookherji, 1964
Cyprinus carpio	10	17	Beamish, 1964
	20	48	Beamish, 1964
	30	104	Beamish, 1964
Salmo trutta	10	81	Beamish, 1964
	20	128	Beamish, 1964
	30	282	Beamish, 1964
Salvelinus alpinus	9.6	170	Christiansen et al., 1991
Cottus sp.	15	92–157	Bond (personal data)
	25	150–264	Bond (personal data)
Morone saxatilis*	8–24	44–218	Kruger and Brocksen, 1978
Sander vitreus (Stizostedion vitreum)	20–25	231–277	Cai and Summerfelt, 1992

* Also found in marine waters.

deprivation of oxygen—a testament to their anaerobic capabilities. Goldfish are most unusual in their accumulation of ethanol rather than the usual lactate as an end product of anaerobic metabolism (van den Thillart et al., 1983). Anaerobic metabolism incurs an **oxygen debt,** more appropriately termed **excess post-exercise oxygen consumption (EPOC).** The use of the latter term recognizes that the increased oxygen consumption immediately following a bout of exercise is not necessarily a direct consequence of the need to eliminate accumulated lactic acid, as was once believed.

Maintaining an optimum water quality for ichthyofauna is important because of the possible combined effects of poor conditions. When the dissolved oxygen content of the water is lowered, for instance, the fish responds by increasing the rate of gill ventilation. This added activity requires a greater consumption of oxygen, so the individual can be placed in the position of attempting to extract a greater amount of oxygen from a smaller supply. A rise in temperature or an increase in carbon dioxide content could combine with the lowered oxygen and increased activity to make the situation intolerable.

Metabolism can be measured by direct methods, such as the measurement of heat production, or by indirect methods, such as the measurement of the amount of oxygen consumed or carbon dioxide produced. The indirect methods, especially measuring oxygen consumption, are the most efficient and reliable means of measuring the rates of metabolic activity, especially for aquatic organisms. A **respirometer** is a device commonly used to measure metabolic rate

expressed as the amount of oxygen consumed. This is usually done by measuring the dissolved oxygen concentration as it enters a fish's mouth and as it exits the opercular chamber. The difference represents the amount of oxygen extracted by the test organism, which is a function of a variety of factors, including species, size, temperature, nutritional state, and level of activity. Fishes are **ectothermic,** meaning that their source of body temperature is outside of the body, and **poikilothermic,** meaning that their body temperature reflects ambient conditions (some remarkable exceptions to this are discussed in Chapter 24). As ectothermic poikilotherms, fishes have a measurable metabolic rate that is generally less than that of tetrapods. Recent studies, however, have suggested that, when temperature and body size are taken into account, virtually all living organisms conform to a universal resting metabolic rate (Gillooly et al., 2001). In an elegant demonstration of the integration of physiological and ecological study, Brown et al. (2004) have developed a theory of metabolism demonstrating that the rates at which resources are taken up and allocated ultimately control ecological processes at all levels from the individual to the biosphere.

Growth

Growth, here defined as the elaboration of tissue resulting in an increase in body mass, is inextricably tied to the metabolic activities of the organism. The most commonly used descriptor of growth is the **condition factor** (k = weight/length^3). Although not directly measuring growth, a number of indices (e.g., glycine uptake by scales, RNA/DNA ratios) can illustrate the metabolic correlates of growth (Busacker and Adelman, 1987; Busacker et al., 1990). Growth is one aspect of the overall metabolic function of an organism. Consumed energy in excess of that needed for basic maintenance can be applied to growth or to the elaboration of gametes.

In vertebrates, the majority of acquired mass during growth takes the form of white muscle. Almost two thirds of the protein synthetic activity experienced during growth in fishes is directed toward myofibril construction (Mommsen, 2000). Most fishes continue to grow throughout their lives (**indeterminate growth**), but individuals of a given species have a genetically determined potential for reaching a characteristic maximum size under the most favorable circumstances. Some gobies never reach more than a few centimeters in length, but sharks and tunas, for instance, may reach several meters in length and weigh hundreds of kilograms. The characteristic upper size for a species is reached in a relatively short time in short-lived species, but it may be attained only after decades in long-lived species. Under less favorable environmental conditions, when a greater proportion of the energy budget must be allocated to maintenance costs, fishes may reach a size smaller than that genetically attainable for the species (an extreme example of this, described for brook char populations in extremely oligotrophic lakes, is discussed in Chapter 32), or they may fail to develop gonadal tissue. Growth in fishes generally follows a sigmoid pattern of increase in size with age, as illustrated in Figure 23.1. The curve represents growth from hatching to the maximum possible size in a given environment. Actually, fish growth is usually more irregular than the idealized curve shows. Growth is usually faster in warm weather than in cold weather and may decrease during migrations or spawning, sometimes even becoming negative, when metabolic demands exceed the food energy intake (Fig. 23.2). In addition to annual fluctuations, fish growth may occur in "stanzas" during the normal life history, with each stanza being defined by a sigmoid curve showing a decrease in growth rate before the resumption of more rapid growth with the entry into the next stanza. Growth stanzas generally result from physical, physiological, or ecological changes and can be represented by incremental marks (e.g., annuli) in skeletal tissue. A migratory fish might end its first growth stanza and begin its second by moving from a stream into a lake or into the ocean, and begin a third when it is large enough to feed on other fishes rather than on small invertebrates.

Factors involved in irregular growth or in the limitation of growth may be environmental or may be concomitants of the fish's physiology. Environmental factors include water

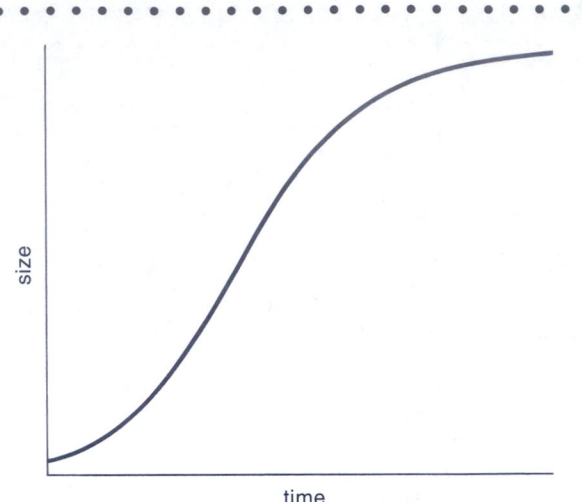

FIGURE 23.1
Idealized sigmoid growth curve.

quality, which embraces such entities as dissolved oxygen, carbon dioxide, ammonia, salinity, pH, and chemical pollutants; and other physical factors, such as temperature and light, both of which change with the season and can exert effects either directly on the fish or on the abundance, composition, and availability of food. Physiological factors that can influence growth include age, sexual maturity, and state of health. Behavioral factors involved might include spawning activity, migrations, and the defense of territory. The production of gametes and reproductive activity can require significant proportions of ingested energy. Waiwood and Majkowski (1984) estimated that cod (*Gadus morhua*) between ages 6 and 10 used about 19 percent of total food biomass in reproduction, and older cods between ages 11 and 15 used about 38 percent.

Brett (1979) summarized the effects of environmental factors on the growth of fish, discussing these in terms of the factors presented by Fry (1971): (1) *controlling factors*, such as pH and temperature; (2) *limiting factors*, such as oxygen; (3) *masking factors*, such as salinity; and (4) *directing factors*, which may be environmental factors such as light or temperature that induce genetically established responses in fishes.

Generally, food is most available and the temperature is most conducive for growth during spring, summer, and fall in temperate areas, but most species do not grow at a constant rate during these periods. Summer-spawning species, such as the bluegill, may put on most of the year's growth in the spring and then grow slowly through the spawning period and the warmest part of the summer. Spring spawners grow best following spawning, but their growth may slow during

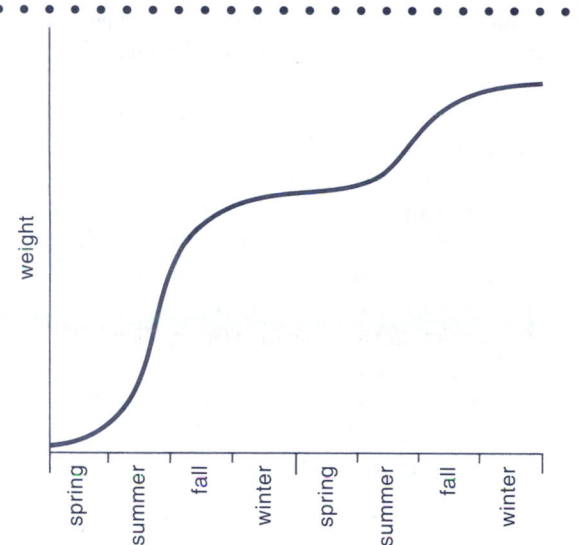

FIGURE 23.2
Theoretical growth curve of fish in temperate climate, showing cessation of growth in winter.

the hot summer. Some, such as the largemouth bass (*Micropterus salmoides*), feed heavily in late summer and early fall.

Age and Growth Studies

The study of growth is not only of great scientific interest but is also of considerable practical importance in the management of fisheries. Growth may be studied in terms of nutritional and bioenergetic considerations, with the expectation that the knowledge gained will aid in growing cultured species faster while minimizing the expenses associated with their maintenance. Or growth may be considered in relation to age for various species or stocks within species. Information on the influence of population size on growth is useful in establishing catch regulations. Comparing size at first maturity among genetically similar stocks can provide insight into possible limitations on growth in various bodies of water. Various relationships of length to weight are used by fishery managers to compare the general condition of fishes from separate stocks, of fishes from the same stocks at different times of the year, or of fishes from different bodies of water.

Age and growth studies of fishes use several methods. Length frequencies plotted for large samples of individuals generally show a multimodal distribution, with the modes representing prominent lengths of each year class. This method can be useful for short-lived species or for the first few years of life of long-lived species, but it has limited application for slow-growing fishes that live a decade or longer.

Many marine species, such as the scorpaenid rockfishes, live to advanced ages. The shortraker rockfish (*Sebastes borealis*) is known to live up to 120 years, and the rougheye rockfish (*S. aleutianus*) reaches 140 years (Beamish and McFarlane, 1987).

In many species, the interpretation of recognizable marks left periodically in scales or other hard structures, such as otoliths (see Chapter 21), can result in the determination of age. In those species that lay down compressed circuli in the scales during winter or other periods of seasonal slow growth and form annuli (see Fig. 3.15), age can be determined. Then, assuming that there is a relationship between the growth of the scale and the growth of the fish, the measurement of a selected scale dimension, such as the radius of each annulus and the radius to the edge of the scale, can allow a back calculation resulting in an estimated length of the fish at each annulus. For long-lived species, age determination and back calculation of length are better accomplished from cross- ections of otoliths, fin rays and spines, opercular bones, or vertebrae (Beamish and McFarlane, 1987; Carlander, 1987; Lagler, 1956; Royce, 1972).

Validation of the results of such studies can be accomplished by research on individuals that are captured, measured, marked, released, and later recaptured. Comparison of scales or fin ray sections taken at initial capture and at recapture of the same individual can be useful to validate or invalidate the use of these methods for the estimation of age and growth for the species involved. Marks can consist of clipped fins or numbered metal or plastic tags to identify individual fishes. Other marks can involve the injection or feeding of fluorescent chemicals, such as oxytetracycline, which leave deposits in bone or other calcified structures. These marks can be identified by the use of ultraviolet light after the marked structure is removed from the fish and sectioned.

Age and growth studies in fishes also include such techniques as the measurement of RNA/DNA ratios (Bulow, 1987), uptake of radioactive amino acids (Adelman, 1987), and aminoacetic acid (Busacker and Adelman, 1987). Caillet and Radtke (1987) and Caillet et al. (1986) reported the use of electron microprobe analysis of calcium and phosphorus in the vertebrae of sharks to verify the seasonal deposition of growth bands. This might prove a useful method for studying long-lived, slow-growing species—some sharks are known to live up to 100 years (Pike, 1990).

NUTRITION

Fishes, like other animals, require the common components of food—proteins, carbohydrates, fats, minerals, and vitamins. Specific requirements and optimum levels of these nutrients in diets have been studied in only a few species, usually of economic importance. Most information available on fish nutrition pertains to species that are reared in captivity, such as the Japanese eel (*Anguilla japonica*), several members of the families Salmonidae and Cyprinidae, the channel catfish (*Ictalurus punctatus*), and the red seabream (*Pagrus major*, family Sparidae). Because fish culturists are continually seeking means of producing more and better fishes at lower costs, there is a practical value in studying fish nutrition. Even though species may be specific in their requirements, the knowledge gained from studying one species can contribute to understanding the nutritional needs of other species.

Because fishes are ectothermic, their metabolism is directly influenced by temperature; some species are adapted to cold waters and others to warm waters, each showing characteristic changes in metabolic rate over their tolerance range. Furthermore, fishes of various species, especially carnivores, can use proteins and fats as energy sources more efficiently than mammals.

. . . .
Dietary Requirements of Fishes

Protein

Compared with other vertebrates, fishes generally require more dietary protein (Bowen, 1987; Horn, 1998). This is especially so in species that are subject to intensive aquaculture. The protein requirements of salmonid fishes have been extensively studied. These requirements differ with temperature; less protein is required at low temperatures than at higher ones. Wilson (1989) determined protein requirements in prepared diets for Chinook and coho salmon to be 40 percent of the diet; for sockeye salmon, 45 percent; and for rainbow trout, 40 to 45 percent. The natural foods of trout have a dietary protein content as low as 12 percent, yet this appears to be sufficient (Brown, 1957). Perhaps these differences reflect an enhanced dietary requirement in response to the stress of rearing in confined aquaculture situations.

Channel catfish production diets contain 32 to 36 percent protein. Common carp diets contain similar amounts. Diets for tilapias have from 28 to 32 percent protein, although young tilapia may grow best when fed 50 to 56 percent protein (Lovell, 1989; Wilson, 1989). The juveniles of the red seabream (*Pagrus major*) and the buri or yellowtail (*Seriola quinqueradiata*) are estimated to need 55 percent protein in the diet.

Carbohydrates

The inclusion of carbohydrates in the diet can spare some protein for use in growth rather than for energy expended in activity. Carbohydrates are usually much more inexpensive

than proteins, so there is an economic advantage if they can be fed to cultured fishes. When one considers the quantity of fish meal that may be incorporated into prepared diets, we might also conclude that carbohydrates have less impact on ocean fish stocks (see Chapter 39).

Omnivorous species, such as carp, can digest carbohydrates better than trout or other carnivores and can have a higher percentage of carbohydrates in the diet. Carnivorous species may show nutritional disorders if they are fed an excess of digestible carbohydrates. Salmonids usually respond to high levels of dietary carbohydrate by depositing an excess of glycogen in the liver and excess fat in the liver and kidneys. Trout are naturally diabetic and will retain high levels of glucose in the blood when fed excess carbohydrates (Committee on Animal Nutrition, 1993).

Lipids

Fats provide an energy source for fishes, but they can be used only in limited amounts. If fed in excess, fat will infiltrate the liver and possibly cause death. Fats differ greatly in digestibility—those with high melting points (*saturated fats*) are difficult for fishes to digest. When digestible fats are used in balanced diets, some of the dietary protein is spared for growth or other purposes (Sargent et al., 1989). Oxidized (rancid) fats are harmful in fish diets; fishes that are fed oxidized fats have been noted to develop fatty degeneration of the liver (Halver, 1989). Some naturally occurring oils are known to contain substances that are toxic to some fishes. For instance, cottonseed oil contains cyclopropene fatty acids, which are harmful to trout if fed in excess. For fishes, as for other vertebrate classes, certain fatty acids are deemed essential, meaning that they must be supplied in their diets (Kanazawa, 1985).

Vitamins

As in other vertebrates, vitamins are an essential dietary item. As might be expected, the most thorough studies have been made of salmonids, especially trout (*Oncorhynchus mykiss, Salmo trutta, Salvelinus fontinalis*) and the Chinook and coho salmons (*O. tshawytscha, O. kisutch*). Essential vitamins for salmon and trout include ascorbic acid (C), thiamine (B_1), riboflavin (B_2), pyridoxine (B_6), vitamin B_{12}, biotin (H), choline, folic acid (folacin), inositol, niacin, pantothenic acid, tocopherol (E), and vitamins K and A. Generally, studies with the aforementioned species and with other cultured species, such as channel catfish, carp, Japanese eel, red seabream, and buri (*Seriola quinqueradiata*), have shown similar qualitative needs for vitamins, but deficiency symptoms differ from species to species, and their quantitative requirements may vary (Halver, 1989).

A deficiency of ascorbic acid in trout can cause abnormal spinal curvature (scoliosis, lordosis; Fig. 23.3) and internal bleeding. Poor or no growth is a consequence of withholding several vitamins, and a high mortality rate accompanies a deficiency of tocopherol, biotin, thiamine, and especially pyridoxine. Salmonine populations of the Great Lakes experience early mortality syndrome (EMS) owing to a deficiency in dietary thiamine; in humans, this deficiency results in the disorder known as *beriberi*. EMS has been shown to result from a diet consisting mainly of alewives (*Alosa pseudoharengus*), a species shown to contain high concentrations of the thiamine-degrading enzyme thiaminase (Brown et al., 2005; Honeyfield et al., 2005). Excesses of some vitamins are also detrimental. For instance, an excess of vitamin A (hypervitaminosis A) causes pathological changes in trout, including an enlarged liver and abnormal bone growth.

Minerals

There has been some difficulty in assessing the dietary requirements of minerals and other trace elements in fishes, because of their ability to absorb elements directly from the water. Calcium, chloride, cobalt, phosphorus, strontium, and sulfate can be extracted from the water by trout and probably by other species as well. Furthermore, many elements are required in such small amounts that quantitative assessment is difficult (Lall, 1989; Piper et al., 1982).

Minerals and related substances required in appreciable quantities by most animals are calcium, phosphorus, magnesium, potassium, sulfur, and chloride. Many other elements are known to affect the health and growth of animals, even though they are present only in trace amounts. Some that are considered essential in fish are iodine, iron, copper, manganese, molybdenum, cobalt, zinc, selenium, and fluorine (Committee on Animal Nutrition, 1993). Others that

are less known but may be essential in the diet are nickel, vanadium, silicon, arsenic, lead, cadmium, bromine, and tin (Lall, 1989).

Calcium and phosphorus are important to bone growth, and a deficiency of either can result in abnormal skeletal development. Both elements have other important roles in metabolism. Most of the required calcium can be absorbed directly from the water (Lall, 1989) at the gills, mouth lining, and fins, but a low percentage (0.34 percent per kilogram of supplied diet) is required in the food. Phosphorus is absorbed by fishes, but the low concentration in water means that most must be supplied in the diet. The phosphorus requirement in the diet is about 0.4 percent per kilogram (Lall, 1989). Magnesium is essential to the proper development of bone and to normal growth and appetite. Freshwater fishes need about 0.05 percent magnesium in their diet.

Iron is essential for cellular respiration and is usually sufficiently supplied to fish in both natural and artificial diets. Iodine is necessary for the production of hormones from the thyroid gland, and a deficiency of this element in the diet can cause goiter (Lall, 1989).

Role of Hormones

Hormones are integral to the regulation of rates of metabolism and growth (see Chapter 26). As fishes are largely carnivorous creatures, in which protein ingestion is of paramount importance, the endocrine regulatory mechanisms might be fundamentally different from those of mammals, in which monitoring glucose levels is essential to the regulation of growth and metabolism (Silverstein et al., 2000). Neuropeptide Y (NPY), a neuropeptide commonly found

FIGURE 23.3
Rainbow trout (*Oncorhynchus mykiss*), showing severe scoliosis resulting from diet lacking ascorbic acid.

in the brain and pancreas, is a powerful stimulator of feeding behavior in fishes, as is b-endorphin (Mommsen, 1998; Silverstein et al., 2000). Thyroid hormones are thought to bring about an increase in fish growth by also increasing appetite and by increasing the efficiency of food conversion. There is evidence that a combination of 17-methyltestosterone and thyroid hormone might enhance the growth rates of hatchery salmonids (Donaldson et al., 1979). The integration of growth and metabolism is perceived in the establishment of an insulin/insulinlike growth factor (IGF) axis. This route of regulation of growth and metabolism is but one of many that may act to influence the synthesis and release of growth hormone (GH) by the pituitary gland (Mommsen, 1998). The "diabetic goby" has become a valuable research tool in the elucidation of the endocrine regulation of growth and metabolism: Longjaw mudsuckers (*Gillichthys mirabilis*) subjected to isletectomy (removal of the endocrine pancreatic tissue) provide diabetic models for endocrine study (Kelley et al., 2000; Mommsen, 1998).

Physiological changes influenced by hormones affect the growth rate during migrations, spawning, and wintering—all natural components of the life cycle. Hormonal influence can be seen in the differential growth of the sexes. In some species, males are of much smaller maximum size than females. This occurs in many species that have internal fertilization, but the most extreme examples of sexually dimorphic size are the ceratioid anglerfishes, in some of which the males parasitize the females (see Fig. 35.1C; Plate 35.3). Fertilization is external in the ceratioids. In those species in which the males are larger than the females, size appears to be related to behavior, as the males are involved in building or guarding nests or in other activities that require size and stamina.

Hormones are also involved in the response of fishes to stressful situations. Increased cortisol secretion is a commonly observed response to stress (Hazon and Balment, 1998). Most fishes grow at a slower rate under stresses such as overcrowding. Even when sufficient food is made available, fishes held in overcrowded ponds grow poorly—an immediate manifestation of physiological stress that is revealed through changes in hormone levels, especially elevated levels of cortisol, the chief stress response hormone (Jobling, 1994). Chronic exposure to cortisol can have deleterious effects. Serotonin, a neurotransmitter released from the brain of stressed fishes, initiates a cascade that culminates in the release of cortisol into the blood. Interesting enough, feeding fishes a diet enhanced with L-tryptophan—a biosynthetic precursor of serotonin—appears to counteract the stress effects of cortisol and to render the fish less aggressive. The suggestion has been made that serotonin may act

not only to initiate the release of cortisol, but to suppress it as well (Lepage et al., 2002; Winberg et al., 2001). Fishes subjected to chronic stress, such as overcrowding, show elevated plasma growth hormone levels; stunted fishes, however, show tissue insensitivity to growth hormone (Mommsen, 1998).

THE ALIMENTARY CANAL: ANATOMICAL FEATURES

The Oral Cavity and Pharynx

Adaptations for a bewildering array of feeding strategies are seen in fishes, and these adaptations naturally involve the size and placement of the mouth (see Chapter 3) and the size and kind of teeth in the mouth or pharynx. Most fishes have the mouth at or very near the tip of the snout; but numerous bottom feeders, such as suckers or sturgeons, have subterminal or inferior mouths. Superior mouths, possessed by relatively few fishes, are a specific adaptation for capturing food from the surface or for ambushing overhead prey from a benthic vantage point. The shape and position of the mouth provide a clue to feeding habits, especially when considered along with the size and placement of teeth. Until recently, it was believed that teeth came about only after jaws evolved. The placoderms, among the earliest gnathostomes (see Chapter 6), were not believed to have had true teeth. However, Smith and Johanson (2003) claimed that true teeth composed of dentine are observed in more derived lineages of placoderms, suggesting that dentition may have arisen more than once in the evolution of gnathostomes. Their claim has not gone unchallenged, however (Burrow, 2003).

Teeth

The structure of fish dentition suggests a complex evolutionary history; what we would consider true teeth have obviously evolved several times among the fishes. Agnathans have teeth that are unique in being composed of a keratinaceous material. The highly variable dentition of elasmobranchs is considered homologous with their placoid scales, as they share the same composition (see Chapter 3). As these teeth are constantly being lost during use, they are continually replaced from nested rows lining the inside of the jaw. At least one elasmobranch species, the bamboo shark (*Chiloscyllium plagiosum*), has enhanced its feeding opportunities through a novel modification to its dentition. The sharp, spiky teeth seen in this species are useful for the ingestion of fishes and squid. These teeth are attached via broad, flexible ligaments that permit the teeth to flatten down on the jaw, also permitting the ingestion and crushing of hard-shell prey items such as crabs (Ramsay and Wilga, 2004).

The tips of actinopterygian teeth have a transparent cap of hard, enamel-like material, the **acrodine**. It is perhaps the most structurally consistent feature of the teeth of bony fishes, only being absent in those groups in which the teeth are modified into crushing plates or are lost altogether (Janvier, 1996). Attachment of teeth to the supporting bone falls into four general types (Fink, 1981). Type 1 is the most primitive and features a strong, mineralized connection between tooth and jaw or pharyngeal bone. This pattern is seen in bichirs, paddlefishes, gars, bowfins, basal teleosts, and a few higher fishes. In Type 2 attachment, common to many teleosts, mineralization is incomplete, and the tooth is connected to the jaw by collagen. Type 3 teeth, found mainly in the Stomiiformes, are hinged and depressible, so that captured prey can be moved toward the esophagus but prevented from escaping when the teeth are erected. The base is not fully mineralized anteriorly in the area of the hinge. In Type 4 attachment, collagen attaches the tooth to the posterior part of the base and acts as a hinge. When the tooth depresses, the anterior edge lifts off the base, exposing the pulp cavity. This type is found in pikes, some stomi-iforms, and several groups of more derived teleosts.

Continuous replacement of the teeth characterizes most bony fishes. The primitive manner of tooth replacement, termed *extraosseous*, is one in which the teeth are regenerated from soft tissue outside of the bone to which they will become attached. An *intraosseous* form of replacement, in which the teeth form within sockets in the bone, has apparently independently evolved in at least three teleostean clades (Trapani, 2001).

Pikes (Esocidae), handsawfishes (Alepisauridae), and many sharks are equipped with large mouths and prominent, sharp teeth that identify them as predators on rather large prey that can be swallowed whole. Some sharks have dental arrangements that enable them to bite large chunks out of animals that are too big to swallow. Some characins, such as the piranhas (*Serrasalmus*), and other teleosts, such as barracudas (*Sphyraena*), may do the same. A variety of deep-sea predators have daggerlike teeth that help them grasp relatively large prey and hold it until it can be swallowed (see Plate 35.1).

Many large-mouthed fishes that have small teeth or no teeth at all in the mouth may be equipped with other structures that can hold prey or strain plankton out of the water. Pads of small, conical or cardiform teeth on the jaws or on several bones (such as the vomer and the palatines) are seen in many species that are opportunistic in capturing a variety of prey. The largemouth bass (*Micropterus salmoides*) and

many species of catfishes are examples of successful predators with small teeth.

Fishes with specialized feeding habits may depart from the usual dental configurations in remarkable fashion (Fig. 23.4). Wolf eels (*Anarrhichthys*), which habitually feed on shelled invertebrates, have strong, canine teeth in the front of the jaws for grasping prey and blunt molars for crushing the shells. Lungfishes have tooth plates that are used for holding or crushing prey. Parrotfishes (Scaridae) can bite off chunks of coral with a beaklike structure formed by the fusion of the front teeth.

Many butterflyfishes (Chaetodontidae) have small mouths at the end of a slender snout—an arrangement that is useful in removing food items from crevices. Motta (1982, 1984a, 1984b, 1985) discovered that the protrusion of the mouth in the genus *Chaetodon* involves three different, complex couplings of bones. High-speed video analysis of the feeding of another chaetodontid, the forcepsfish (*Forcipiger longirostris*), has demonstrated a remarkable ability to rapidly protrude its already elongated snout. This enables these fishes to approach and strike their prey from a greater distance (Ferry-Graham and Lauder, 2001; Ferry-Graham et al., 2001). The teeth of butterflyfishes contain iron. Those species that feed on the hardest prey have more iron in their teeth than those that feed on somewhat softer materials (Motta, 1987). Slipmouths (Leiognathidae), dories (Zeidae), mojarras (Gerridae), and several other families are capable of protruding their mouths an exceptional distance to siphon in prey.

Teeth are borne on several of the head and face bones. In the upper jaw, this includes the premaxillary and the maxillary in most of the soft-rayed teleosts, but in the more derived taxa, the maxillary usually does not bear teeth. Additional teeth are commonly seen on the vomer and the palatines. Many species bear teeth on the pterygoids and the parasphenoid. In the lower jaw, the dentaries are usually the main toothed bones, but teeth may be present on the tongue (glossohyal) and the basibranchials (Fig. 23.5).

Pharynx

The gills lie just behind the oral cavity in the pharynx. There are typically four pairs of gills in bony fishes, but sharks and rays may have gills on five to seven arches (see Chapter 24). The gill arches may be equipped with inwardly directed projections called **gill rakers** that aid in filtering food from the water (Fig. 23.6). Fishes that consume large prey may have gill rakers that are few in number and small, but many carry rough prominences or denticles that aid in holding and swallowing prey. Plankton feeders usually have an extensive straining sieve formed of long, slender gill rakers. Among the salmonids, the variation in gill raker morphology is

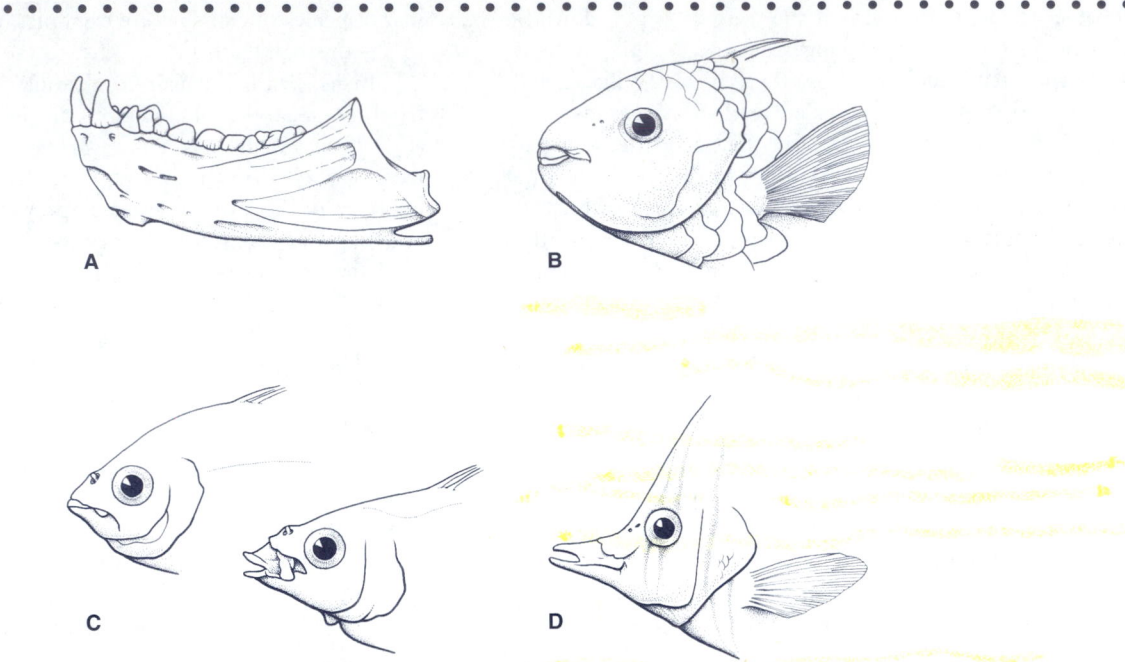

FIGURE 23.4

Examples of teeth and mouths: **A,** Dentary bone of wolf eel (*Anarrhichthys*); **B,** Beaklike teeth of parrotfish (Scaridae); **C,** Protrusible mouth of slipmouth (Leiognathidae); **D,** Small, specialized mouth of butterflyfish (Chaetodontidae).

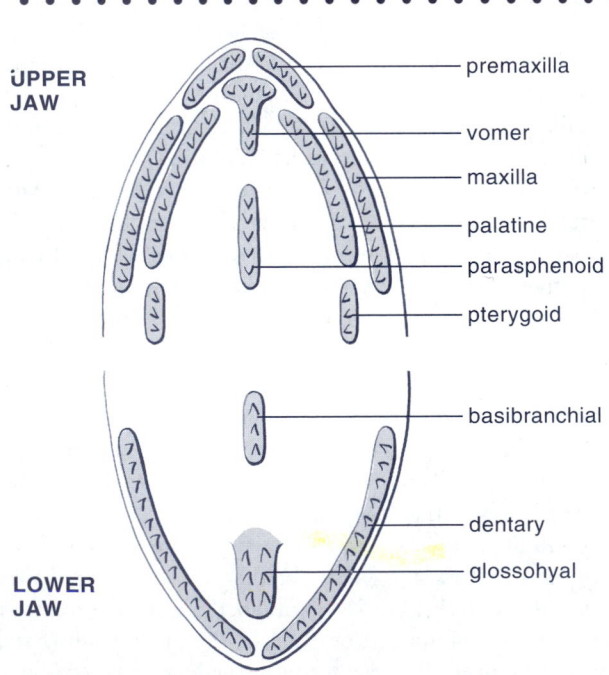

FIGURE 23.5

Location of tooth-bearing bones in the mouth of bony fishes.

apparent when comparing Chinook salmon (*Oncorhynchus tshawytscha*), which feed on herring and other small fishes, with sockeye salmon (*O. nerka*), which feed mainly by filtering plankton. Variation in gill raker morphology may also reflect a trophic polymorphism that exists within a given species. Again, salmonids are among the best examples of this (see Chapters 4 and 32).

An **epibranchial** or **crumenal organ** that may function in the collection and concentration of small food particles is present in certain groups of fishes that feed on plankton and similar materials. This organ is epibranchial or postbranchial in position. It is found, for instance, in the plankton-eating herrings (*Clupea;* Gibson and Ezzi, 1985), the herbivorous milkfish (*Chanos*), the suborder Stromateidoidei, and a number of deep-sea fishes, including the four-eyed fish *Bathylychnops* (see Fig. 20.3D; Stein and Bond, 1985). Undoubtedly contributing to the success of the family Cyprinidae in freshwater is the degree of modification of oral and branchial structures for the ingestion and sorting of suspended materials. The **palatal organ** present on the roof of the pharynx in cyprinids is thought to be involved primarily in the selective retention of food particles. Studies have identified especially dense aggregations

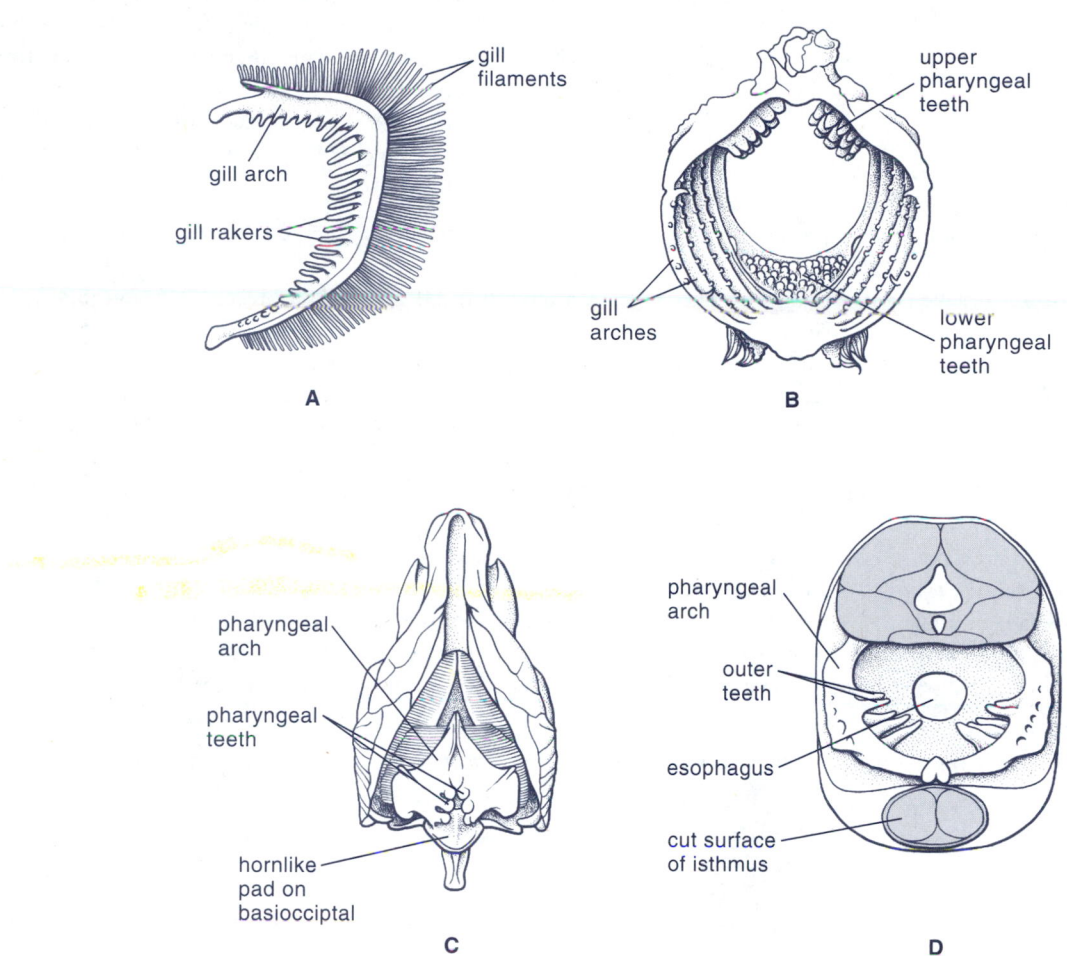

FIGURE 23.6
Examples of gill rakers and pharyngeal teeth: **A,** Illustration of gill arch with rakers and gills; **B,** Anterior view of gill arches and pharyngeal teeth of surfperch (Embiotocidae); **C,** Ventral view of the pharyngeal region of a carp (*Cyprinus*), with pharyngeal arch displaced anteriorly to expose the basioccipital pad; **D,** Anterior aspect of pharyngeal teeth of the pikeminnow (*Ptychocheilus*), cross section behind the last gill arch with musculature and other soft tissue removed.

of taste buds in association with the palatal organ, suggesting that its role may primarily be chemosensory during the sorting of suspended organic and inorganic materials during feeding (see Chapter 22; Callan and Sanderson, 2003).

The fifth gill arches of bony fishes are usually reduced to a single lower element (the fifth ceratobranchial) on each side. This bone bears teeth that bite against opposing teeth borne on the upper elements (pharyngobranchials) of one or all of the four branchial arches. In minnows (Cyprinidae) and suckers (Catostomidae), the lower pharyngeal bones bite against a pad borne on an extension of the basioccipital bone. The monophyletic group Labroidei includes the damselfishes (Pomacentridae), cichlids (Cichlidae), surfperches (Embiotocidae), wrasses (Labridae), parrotfishes (Scaridae), and the cales and weed-whitings (Odacidae); these are examples of fishes with pharyngeal jaws (**pharyngognathous** teleosts). The fifth ceratobranchials are fused and bear teeth in these families (Lauder and Liem, 1983). In the cichlids, the lower pharyngeals bite against the second to fourth pharyngobranchials (Liem, 1991). Pharyngeal teeth are varied in size and shape, ranging from small, conical points to grinding plates (see Fig. 23.6).

Esophagus, Stomach, and Intestine

Esophagus

In general, the esophagus in fishes is a short tube that is highly distensible, so that relatively large objects can be swallowed, but microphagous fishes have less distensible tubes than those of predatory fishes. Esophageal walls are generally equipped with both circular and longitudinal muscles, which provide peristaltic activity to enable swallowing. The lining of the esophagus consists of stratified epithelium and columnar epithelium, with numerous mucous cells or glands. Taste buds are probably present in all species, and may be located in other areas than the digestive tract (see Chapter 22). Gastric glands appear in the posterior part of the esophagus in some mullets (Mugilidae) and sculpins (Cottidae). In more primitive bony fishes, the esophagus is the site of the connection of the swim bladder with the alimentary canal via the pneumatic duct (see Chapters 3 and 19).

Several modifications of the esophagus are known. The butterfishes (Stromateidae) and their close relatives have muscular sacs connected to the esophagus. In some stromateid genera (*Pampus, Nomeus*), these esophageal sacs are lined with teeth, which are attached to thin bones in the walls of the sacs. The sacs of stromateids serve various functions in different species, such as mucus production, food storage, or preparation of food by trituration (grinding or crushing). In some fishes, such as the Alaskan blackfish (*Dallia pectoralis*) and the rice eel (*Monopterus alba*), the esophagus is modified for respiration.

Stomach

In most fishes, a stomach is present, varying in shape and structure according to the diet of the various species. Usually, the stomach is a bent, more or less muscular tube shaped like a U or V (Fig. 23.7A). Another common form is a bag-shaped stomach, with anterior openings from the esophagus and to the gut. Heavy-walled, gizzardlike stomachs are found in the gizzard shad (*Dorosoma*), mullets (*Mugil*), and a few others (Fig 23.7B).

The stomach is lacking in lampreys, hagfishes, chimaeras, and some bony fishes, including minnows (Cyprinidae), pipefishes (Syngnathidae), sauries (Scomberesocidae), and parrotfishes (Scaridae). In these fishes, gastric glands are absent, and the esophagus empties directly into the intestine.

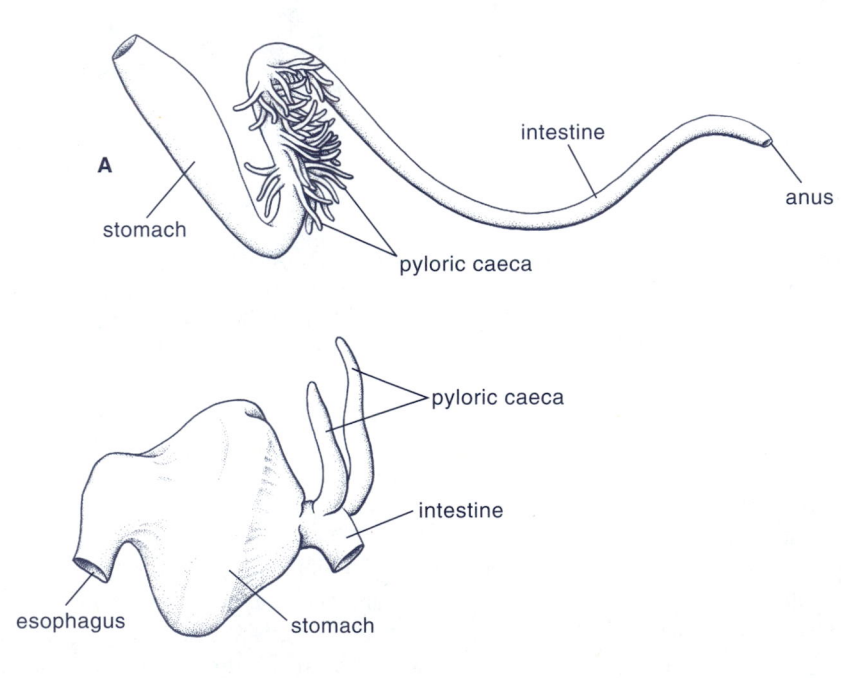

FIGURE 23.7
Examples of stomachs and pyloric caeca (anterior to left): **A,** Stomach, caeca, and intestine of trout (Salmonidae); **B,** Stomach and pyloric caeca of mullet (Mugilidae).

Hagfishes have a sphincter between the branchial region and the digestive gut.

There is evidence that some fishes, especially elasmobranchs, are capable of fully everting their stomach out their mouths as a means of ridding themselves of ingested noxious materials. Drugs that induce vomiting in other animals are capable of inducing such stomach-rinsing behavior (Sims et al., 2000).

Intestine

The intestinal epithelium includes cells termed **enterocytes** that serve as the site for absorption of digested materials. Movement of nutrients across the cell membrane of enterocytes may be an active or a passive process and may involve the participation of specialized carrier molecules (Horn, 1998). The length of the intestine in fishes is generally correlated with the overall digestibility of the ingested foodstuffs (Smith, 1989). Carnivores generally have short guts (see Fig. 23.10; see Table 23.3). Herbivorous fishes and fishes that eat detritus and mud may have guts several times their body length. These long intestines are usually folded into distinct patterns in the body cavity (Fig. 23.8).

Omnivorous fishes have guts of intermediate length (Fig. 23.9). A simple fold, sometimes referred to as a **typhlosole**, is present in the intestine of lampreys. An interesting contrast is seen in the guts of modern agnaths. The intestine of predatory lampreys, which feed on the blood and juices or finely divided flesh of their prey, is extremely thin-walled and can be greatly distended, but when empty it appears as a thin cord. The hagfishes ingest larger pieces of their prey, and the hagfish intestine has a thick wall and an extensively folded lining.

Sturgeons, lungfishes, polypterids, *Latimeria*, the bowfin (*Amia*), and gars (*Lepisosteus*), and the sharks and related cartilaginous fishes possess a **spiral intestine** (or **spiral valve**; Fig. 23.10). Typhlosoles and spiral valves serve to increase the absorptive surface along the length of the intestine. In most fishes, the vent or anus represents the posterior opening of the gut. A few families, such as the herrings (Clupeidae), have an opening to the exterior from the swim bladder, posterior to the anus. Other than a few rare exceptions (such as the female pipefishes of the genus *Nerophis*), only the sharks, rays, lungfishes, and male coelacanths have a **cloaca,** which receives the end opening of the gut as well as the ducts of the urinary and genital systems (although the cloaca is present in most tetrapods, it is absent in most mammals).

Organs that develop from the embryonic gut tube that eventually becomes the digestive tract include the pyloric caeca, liver, pancreas, and swim bladder; with the exception of the swim bladder, these remain attached to and associated with the gut.

Pyloric Caeca

On the intestine of most bony fishes, just beyond the pyloric end of the stomach, there may be one to many blind sacs, called **pyloric caeca.** A few groups, such as topminnows (Cyprinodontidae), pikes (Esocidae), and some catfishes (Ictaluridae), lack these structures. The bichir (*Polypterus*) has only one pyloric caecum, the yellow perch (*Perca flavescens*) has three, and in the flatfishes (Pleuronectiformes), the pyloric caeca usually number no more than five. In other groups, such as mackerels (Scombridae), salmons (Salmonidae), and snailfishes (Liparidae), the number of these caeca may be 200 or more. Caeca of different species vary considerably in size, state of branching, and connection with the gut. In the sturgeons (Acipenseridae), the many caeca form a large mass, but only a single duct leads to the intestine. In salmon, on the other hand, each caecum communicates directly with the gut (see Fig. 23.7A). The functions of the pyloric caeca probably involve both digestion and absorption by increasing the intestinal absorptive surface area without

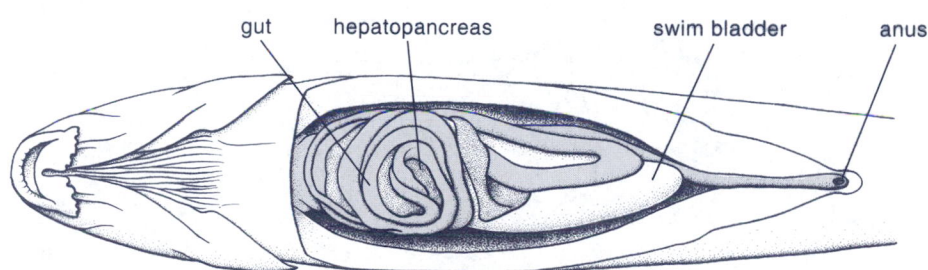

gut   hepatopancreas   swim bladder   anus

FIGURE 23.8
Elongate gut of sucker (*Catostomus macrocheilus*), a microphagous fish.

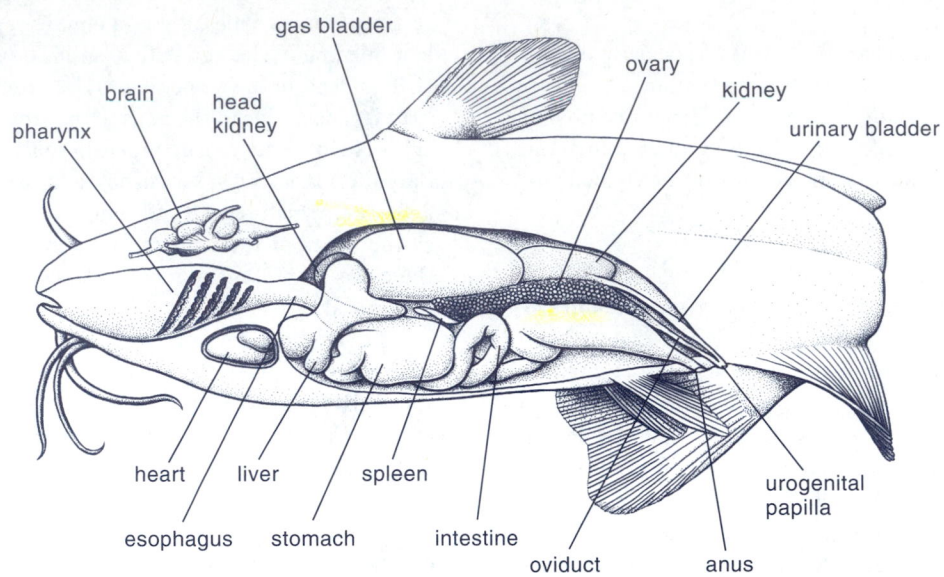

FIGURE 23.9
Bullhead (*Ameiurus*), head sectioned slightly to left of midline, body cavities opened to show relative positions of internal organs. (Note that the head kidney is separated from the renal kidney by the gas bladder.)

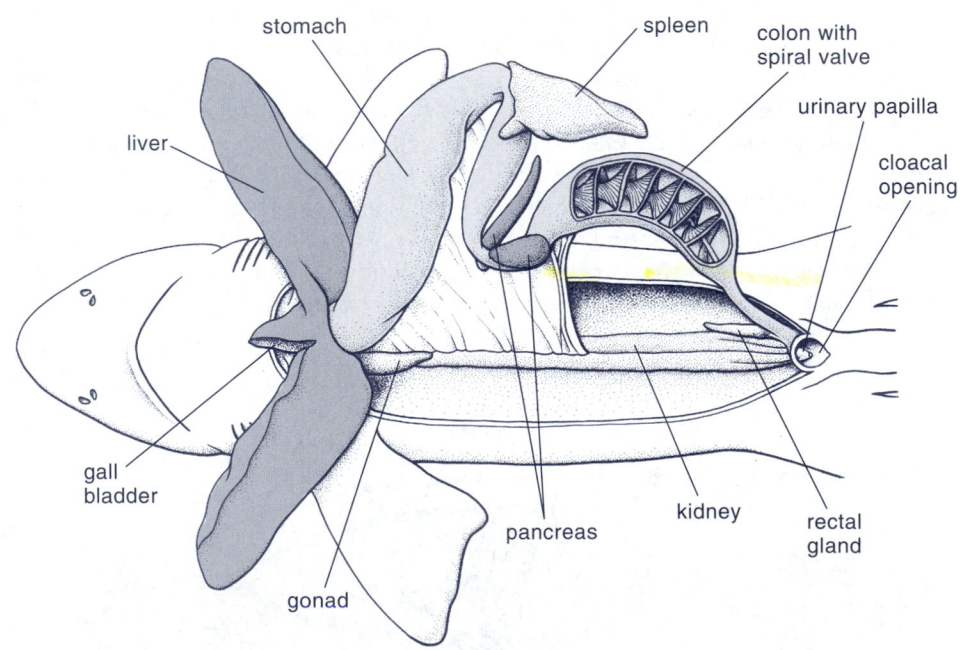

FIGURE 23.10
Illustration of viscera of shark (spiral valve opened to show internal structure). (Based on Daniel, 1934.)

increasing the length of the intestine. Retrograde peristaltic contractions have been observed in the digestive tract of the Atlantic halibut (*Hippoglossus hippoglossus*). These appear to be the means of filling the pyloric caeca with chyme (digesting food mass) and mixing it with digestive secretions (Rønnestad et al., 2000). Digestive enzymes, including trypsin, have been isolated from the pyloric caeca of many species.

Liver

The liver is a large gland in all fishes, but sharks and rays have especially large livers that may constitute up to 30 percent of the total body mass (see Fig. 23.10). The liver usually lies over or partially surrounds the stomach. It is typically bilobed, but it may have only one lobe, as in salmon, or three, as in mackerels and *Squalus*. In hagfishes, the liver is in two distinct parts, with separate ducts leading to the gall bladder. Adult lampreys have no bile ducts or gall bladder, but in most other fishes, the gall bladder is present and functions to store liver secretions. Ordinarily, one hepatic duct originates from each lobe of the liver and joins the cystic duct from the gallbladder to form the bile duct. Vertebrate liver function includes bile secretion and glycogen storage, in addition to several other biochemical processes. Pelagic sharks have especially large livers, and their fatty composition is believed to contribute to their buoyancy (see Chapter 19).

Pancreas

The pancreas secretes several enzymes that are active in digestion. Furthermore, the pancreatic islets have the endocrine function of producing insulin (see Chapter 26). Hagfishes have a small pancreas, with several ducts that empty into the bile duct. Lampreys have pancreatic tissue scattered throughout the liver and intestinal wall—a configuration that may reflect an exclusively endocrine function. Among the bony fishes, the pancreatic tissue is usually diffuse and located in or around the liver. This is especially true of the pancreas of more derived taxa of fishes, in which the pancreas and liver are combined into a *hepatopancreas*. The lungfishes (*Protopterus*) and many of the soft-rayed bony fishes have a discrete pancreas. In sharks and rays, the pancreas is a compact organ, usually consisting of two distinct lobes: a dorsal lobe, located posterodorsally to the stomach and duodenum; and a ventral lobe, lying in the curve of the duodenum. The pancreatic duct may reach the small intestine separately from the bile duct, as in the sharks, or it may discharge into the bile duct, as in the gar (*Lepisosteus*) and lungfish (*Protopterus*).

Spleen

The spleen is usually recognized as a dark red, often pyramidal structure lying on or behind the stomach, to which it attaches by a bandlike ligament. In some fishes, such as the lungfishes, it is absent. Although the spleen is associated with the digestive organs, it has no digestive function, but rather is instrumental in blood cell formation. The function of red blood cell destruction has also been ascribed to the spleen of more derived lineages of bony fishes. A diffuse aggregation of tissue along the intestine that functions like a spleen is observed in lampreys and hagfishes.

FORAGING ACTIVITY AND FEEDING IN FISHES

Feeding in the Ecological Context

Feeding adaptations must be considered in the context of the trophic interrelationships of members of a given community. Figure 23.11 shows a simplified diagram of the generalized trophic relationships of fishes. A more specific depiction of trophic participation, showing the feeding interactions of fishes as members of a warm temperate, estuarine community, is given in Figure 34.2. The impact of feeding interrelationships on the structuring of ecological communities and the behavioral implications of feeding are among the topics covered in Chapters 30–37. A key feature that has contributed to the success of fishes is the diversity of feeding strategies that has permitted them to dominate aquatic ecosystems.

Detection and Selection of Food

Feeding is carried out daily by most fishes and may be the most frequent of voluntary activities. Flight from enemies, reproduction, migration, and many other activities might be occasional or periodic, but feeding is usually part of the daily routine, and foraging in some species may require extended periods of time. Most fishes show some sort of morphological or behavioral specialization for a particular type of food, but they may also demonstrate considerable versatility in their response to the availability of different food items. It has generally been perceived that the body composition of prey species closely reflects the dietary requirements of their carnivorous predators, so that foraging behavior is dictated by prey availability and not necessarily by their nutritional composition. Herbivores and omnivores, operating in a heterogeneous nutritional environment, will exhibit a greater ability to select food based on their specific nutritional requirements. When the diet of predators

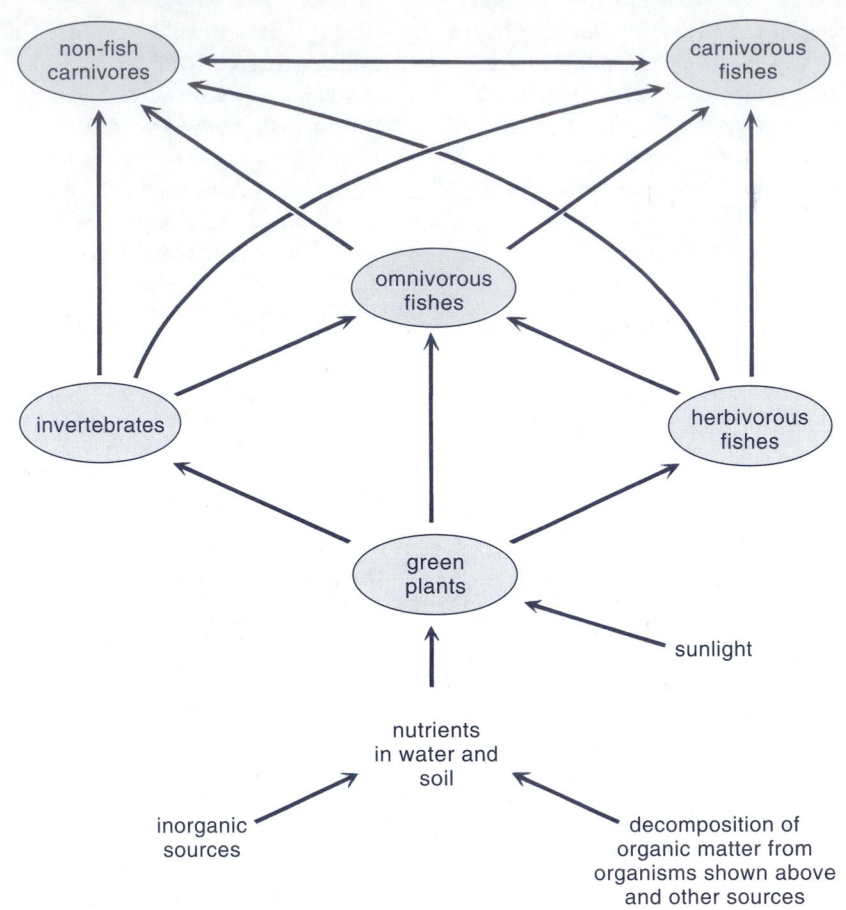

FIGURE 23.11

Simplified diagram of hypothetical trophic relationships that include fishes.

in some arthropod communities is manipulated to render them protein- or lipid-deficient, they exhibit compensatory nutrient selection in response to their specific nutritional requirements (Mayntz et al., 2005). It should not come as too great a surprise if a similar strategy is discovered among fishes, as carnivory is their predominant nutritional mode.

Competition can be reduced or avoided through adjustments in feeding practices. As Liem (1980a) remarked, even specialists can be jacks-of-all-trades if necessary to find food. Consider how the different species in your home aquarium, each one showing distinctive feeding morphologies or behaviors, all readily adapt to a diet of flake food. Usually, when individuals of a given species are presented with two choices of food, they will feed on the more abundant food source until a threshold of intake is realized, at which time the fishes switch to the alternative food source. Diet switching is integral to our understanding of optimal foraging theory

(see Chapter 37), but dietary shifts may occur that appear not to represent optimal foraging (Gerking, 1994).

Fishes will bring to bear their powers of locomotion and all manner of specialized structures in the attempt to maximize their food intake. The detection of possible food items at a distance can be through chemical senses, the eyes, the auditory organs, the lateral line system, or the electrosensory system (see Chapters 20–22). Most species are visual feeders; examples range from most fish larvae to large-eyed pelagic predators, such as scombroids. In many species of fishes, the position of the eyes in the head reflects particular feeding habits.

Olfaction and taste both play major roles in feeding. Many species that appear to feed mainly by sight are known also to have acute olfaction. Fishes such as lungfishes, eels, spiny eels (Mastecembelidae), and others have well-developed olfactory systems, as do sharks, rays, and many deep-sea fishes. Through olfaction, many species get their first signal that food is in the vicinity and then are able to follow a chemical gradient to the source. The bilateral nature of the olfactory organs and other sensory systems enables the lateral discrimination of environmental cues—a necessary component of gradient following.

Taste, along with the tactile sense, appears to be significant in the final selection of food and its retention for preparation and swallowing. Although taste buds are typically in the mouth, the distribution of external taste buds over the skin of a wide variety of species is well known (see Chapter 22). Many fishes with barbels, including catfishes and cyprinids, are examples. Fishes appear to separate food from unwanted detritus at the level of the pharynx. Taste buds are abundant on the gill arches, gill rakers, epibranchial organ, and the tissue surrounding the pharyngeal teeth. In many instances, unwanted material is ejected through the gill openings after being subjected to a final test by sense receptors in the pharynx. Large particles are usually ejected through the mouth with a "coughing" action. Gill rakers, pharyngeal teeth and bristles, epibranchial organs, and special pharyngeal musculature all serve mechanical functions in retaining or rejecting ingested material. The black surfperch (*Embiotoca jacksoni*) has pharyngeal muscles that enable it to sort small food organisms from other indigestible particles (Schmitt and Holbrook, 1984).

The amount of food ingested per day and the amount of time spent feeding depend on many factors. Active predators, with high metabolic rates, require a greater energy intake than do sessile fishes. If a predator feeds habitually on small organisms, a considerable amount of time must be spent gathering prey in order to meet its energy requirements. However, if the predator can catch and swallow large organisms, it might satisfy itself with one or two captures per day. Filter feeders foraging in waters with a high concentration of biomass might also be able to gain sufficient nutrition in a relatively short period of time. Because metabolic rate varies with temperature (Jobling, 1994), cold-water predators that endure winter conditions should require less food than warm-water predators, such as tunas, that avoid cold water. Deep-sea predators (such as anglerfishes and gulpers) in uniformly frigid waters usually subsist on infrequently available, large meals (Marshall, 1966).

Daily and seasonal temperature fluctuations affect food intake in most species. The predator, feeding on the bodies of other animals, such as insects, crustaceans, squids, and other fishes, is ingesting high-protein, high-calorie food that provides nutritional requirements without much bulk. On the other hand, the herbivores or detritivores (see later) must ingest large quantities of less concentrated foodstuffs, sometimes including a large proportion of indigestible

material. Consequently, feeding activity in these fishes requires longer periods of time.

Some species feed mainly by sight and are active by day, although peaks of feeding activity may occur in the morning and evening. Other fishes that depend more on chemical senses can feed effectively in twilight (*crepuscular feeding*) or at night, so they may be most active in the early morning and late evening. Feeding activity may change in response to seasons, cycles of migratory or reproductive activity, age, and size.

In relative terms, small individuals consume more per day in relation to their body weight than large individuals. The daily food intake of juvenile brown trout weighing 5 g at 15°C is about 11 percent of their body weight per day. The same species at 50 g at the same temperature eats 6.6 percent of body weight per day; whereas fishes at 500 g under the same conditions have a food intake of 4.3 percent body weight per day (Jobling, 1994). Many fishes show a fairly consistent average daily food intake despite fluctuations when feeding is measured over an extended period. Furthermore, food intake appears to be adjusted to the nutritive value of available food, so that greater amounts of less nutritious foods are eaten.

Dietary Diversity and Foraging Activity

Most species of fishes are predatory, feeding on live animals or parts thereof. But in some habitats, especially in the tropics, anywhere from 10 to 20 percent of the species present—or nearly half of the individuals—may depend primarily on plant material for food. Although some species are highly specialized and thus restricted by their feeding apparatus to a narrow range of foods, most fishes tend toward opportunism and will take advantage of a diversity of foods. Although some species may be considered as mainly carnivores or herbivores for the purpose of some discussions, they may often manifest more omnivorous tendencies.

Ecological classifications often categorize fishes by their foods and feeding strategies, so that the role of a given species in an aquatic community can be elucidated (Gerking, 1994; Hyatt, 1979; Keenleyside, 1979). A broad classification of feeding types would group fishes into four categories: macrocarnivores, microcarnivores, herbivores, and omnivores.

Macrocarnivores

Benthic invertebrate fauna provides a significant proportion of the diet of macrocarnivorous fishes. For the most part, smaller organisms such as worms, mollusks, aquatic insects (including larvae and pupae), and crustaceans constitute the main food source for most carnivores, whereas larger organisms are safe from all but large or specialized fishes.

Piscivorous fishes vary in the extent to which they practice cannibalism. In most, it represents only a small proportion of their diet. But in some groups—the gadoid hakes (*Merluccius*) and pollocks (*Theragra*), for instance—cannibalism can account for 40 to 70 percent of the diet (Juanes, 2003).

Some benthic predators live on the bottom and hunt by searching and capturing individual organisms as they are encountered. Others simply remain still and capture invertebrate prey that comes close. Many stream-dwelling fishes subsist mainly on terrestrial insects that land on the water. Many bottom feeders swim above the substrate and search by sight for prey, and a few others supplement sight feeding by other senses. Nocturnal foragers may use vision only minimally and rely instead on taste, olfaction, the lateral line system, tactile sense, and electroreceptors (if present) to locate prey. Barbels or elongate fins carrying organs of touch and taste may aid in the search.

Predaceous fishes may seek prey individually, in schools, or in the company of other species. Pelagic piscivores, such as sharks, most salmon species, tunas, jacks, and dolphinfishes, rely mainly on vision to pursue and capture prey. In some instances, groups of predators appear to engage in cooperative hunting by herding prey against the shore or into tightly packed schools or "balls," thus preventing effective escape. The blackfin reef shark (*Carcharhinus melanopterus*) has even been observed leaving the water to consume fishes that have been herded onto the shore (Wetherbee, 1990).

Many benthic predators, such as sculpins, stonefishes, groupers, morays, and flatfishes, use cryptic coloration and lie in wait in a hiding place or on the substrate; unsuspecting prey that venture too close learn too late that sessile, bottom-dwelling fishes can often move with lightning speed and unerring accuracy. Benthic lophiiforms, such as the goosefishes, batfishes, and frogfishes, employ lures (**illicia**) to draw prey close to their mouths. The ceratioid anglerfishes and the viperfishes (*Chauliodus*) have elaborate lures (some of which are luminescent) developed from the dorsal fin, for the attraction of prey. The squarehead or angler catfishes (*Chaca*) use wormlike barbels at the corners of their broad mouths as lures.

Some predators, such as electric rays (Torpedinidae) and some species of stargazers (Uranoscopidae), are aided in food capture by powerful electric organs that stun or immobilize prey. The electric catfish (*Malapterurus electricus*) also appears to use electric organs to stun schools of the clupeids and schilbeid catfishes on which it feeds (Sagua, 1979).

Benthic predators, such as flatfishes, may stalk prey along the bottom before attacking. Pelagic predators are generally neutrally buoyant species and can hang nearly motionless in the water as they gradually make an approach on their chosen prey. Gars (*Lepisosteus*), pikes (*Esox*), trumpetfishes (*Aulostomus*), and some piscivorous cichlids are slender-bodied (compressiform) predators that disclose little view of their actual size as they slowly approach their prey head-on. Trumpetfishes are known to shadow larger fish and then dash out to capture prey (see Plate 37.2).

Cleaning symbiosis is a special kind of seemingly mutualistic relationship, in which a larger fish provides nourishment for smaller species (see Chapter 37). In some tropical areas, there are fishes that mimic the cleaners and thus gain nourishment by biting off parts of fins or skin. The ichthyoborids of Africa are known as *fin biters* because they pursue a parasitic existence by feeding on the fins of other species. Scale-eating fishes are known from both South America and Africa. Certain characoids of South America and several cichlids of the African lakes have dentition adapted for removing scales from other fishes. The slender piranha (*Serrasalmus elongatus*) is reported to eat both scales and parts of fins, as well as some flesh.

Certain cichlids of African lakes are termed **paedophagous**, because they eat the young of other species. They dislodge broods of fry from the mouths of females, either by ramming the hyoid region, as in the genus *Cyrtocara* (McKaye and Kocher, 1983) or by engulfing the snout of the female and sucking out the young (Wilhelm, 1980).

Lampreys are generally termed parasitic feeders, but they really behave more like predators. A lamprey that is small in comparison to its host behaves as a parasite—it can suck out fluids without killing the host; however, those lampreys that are close to the host in size, or those (e.g., *Lampetra ayresi*) that rasp out and ingest viscera, inevitably kill their victims rather quickly (Beamish, 1980). Several years ago, the sea lamprey (*Petromyzon marinus*) in the Great Lakes was calculated to cause up to 56 percent of the mortality in lake trout (*Salvelinus namaycush*) older than nine years (Swanson and Swedberg, 1980) and up to 75 percent of the mortality in whitefish (*Coregonus;* Spangler et al., 1980).

Although lampreys usually discard a lifeless victim, Beamish (1980) presented evidence of continued feeding on dead prey. One small species (*Lampetra minima*), previously thought to be extinct (see Chapter 5), commonly remained with prey until all soft tissue was consumed (Kan and Bond, 1981). Hagfishes are mostly **saprophagous**, feeding as scavengers, although benthic invertebrates have been found in the guts of some species. Certain catfishes of the families Cetopsidae and Trichomycteridae, from the Amazon, apparently parasitize the gills of pimelodid catfishes and other large species, causing a flow of blood on which they feed.

Microcarnivores

Planktivorous fishes vary in their pursuit practices—some are relatively passive and nonselective in the ingestion of particulate food items that are filtered using the gill rakers, whereas others exhibit greater selectivity and spend greater amounts of time in the active pursuit of more elusive individuals (Lazzaro, 1987). Janssen (1978) noted that alewifes (*Alosa pseudoharengus*) switch their feeding modes based on the abundance and activity of prey species. Such shifts may be related to optimal foraging, so that a reasonable return of energy gained for energy expended can be realized (Crowder, 1985). Shifts in feeding modes may entail a modification in locomotor patterns. Young Sacramento perch (*Archoplites interruptus*) feeding on individual plankton used a leisurely approach in capturing passive prey but used a fast start (from a slightly S-shaped body flexion) and pursuit to capture evasive prey. The energy cost of capturing evasive prey was eight times that of the capture of passive prey (Vinyard, 1982).

Herbivores

There are few or no fishes that are lifelong herbivores, because their larvae and juveniles usually begin life feeding on zooplankton or extremely small benthic and pelagic animals. Furthermore, species that feed on filamentous algae or vascular plants also ingest whatever animals—snails, crustaceans, or insect larvae—are attached to the vegetation. Nonetheless, there are many species that, as adults, depend on the intake of plant material and show appropriate structural modifications for gathering, processing, and digesting it.

It should be apparent that planktivorous fishes may ingest significant amounts of phytoplankton, either deliberately or incidental to their capture of zooplankton. Those species that feed on phytoplankton engage in filter feeding using long, fine, and closely spaced gill rakers. Some herrings (*Sardinella*, *Dorosoma*), menhaden (*Brevoortia*), cyprinids (*Hypophthalmichthys*), and cichlids are feeders on phytoplankton.

Many species browse on large filamentous or thallose algae or on vascular plants. The grass carp (*Ctenopharyngodon*), noted for its use in aquatic weed control, can ingest huge quantities of vegetation, which it breaks loose with its rather hard, toothless mouth and shreds with its pharyngeal teeth. The barb (*Puntius javanicus*) is another browser on aquatic vegetation. Other herbivores include species of the South American genus *Myleus*, commonly known as redhooks; the African characoids of the genus *Distichodus;* and certain cichlids of the genera *Oreochromis* and *Tilapia*. In marine waters, the milkfish (*Chanos chanos*) is a commercially significant herbivorous species, apparently feeding on phytoplankton and filamentous algae. In northern temperate, rocky coastal waters, the adults of the mosshead sculpin (*Clinocottus globiceps*) and two of the larger species of stichaeids (*Xiphister mucosus* and *Cebidichthys violaceus*) feed mainly on algae (Barton, 1982; Grossman, 1986; Yoshiyama et al., 1996). Rabbitfishes (Siganidae) are browsers on filamentous algae, and various reef fishes, such as triggerfishes, sea chubs, damselfishes, and surgeonfishes, also use algae as food. Some surgeonfishes—for example, the orangespot surgeonfish (*Acanthurus olivaceus*)—have thick-walled, gizzardlike stomachs for processing coarse vegetation (Wheeler, 1975).

Numerous species in both marine and freshwater habitats are adapted to scrape films of diatoms and other vegetation from substrates. This grazing is accomplished by means of modified jaw edges in some species and by teeth in others. Some fishes have only the lower jaw modified and scrape by a unidirectional, chiseling action, whereas others employ both jaws in scraping. In some of these fishes, the teeth are bristlelike, whereas in others, they are chisel-like. Examples of grazers and scrapers are the chiselmouth (*Acrocheilus*), the ayu (*Plecoglossus*), various suckers (Catostomidae), the stoneroller (*Campostoma*), various cichlids (*Pseudotropheus*, *Gephyrochromis*), parrotfishes (Scaridae), and suckermouth catfishes of the family Loricariidae.

Some highly specialized herbivorous strategies have developed in the Amazon basin. Many characiform fishes there commonly feed on fruit, nuts, and flowers of plants growing over the waterways or in the seasonally flooded forests.

Omnivores

Omnivorous fishes commonly consume a variety of foods, both plant and animal, in their daily fare. The common carp (*Cyprinus carpio*) is perhaps the best example of a fish that will eat various kinds of animals, including fishes, but eats plant material as well. The channel catfish is one of many other species with a wide taste in foods. Although the rainbow trout is primarily known as a carnivore, most studies of its food habits have disclosed filamentous algae in its stomach.

Although they may be specialized for a narrow (*stenophagous*) diet, some species turn to a wider choice of food when the opportunity arises. A variety of herbivores can be caught with animal bait. The chiselmouth, marvelously adapted for scraping the substrate, can be captured on a hook baited with an earthworm, on a small spinner, or even on a floating artificial fly. Ayu, with maxillary and mandibular teeth set on the outer surface of the bones and aligned to form long scraping edges, adapt quickly to feeding on float-ing pellets in hatcheries. Tui chub (*Gila bicolor*) are ordinar-ily pickers of small prey and algae among vascular plants or along the bottom, but they will move to the surface by the thousands to feed on unusually large hatches of insects. The prickly sculpin (*Cottus asper*) is well adapted to a benthic life, but small individuals of 5 or 6 cm will swim well off the bottom to capture individual *Daphnia*.

Many of the cichlids of the African Great Lakes have teeth, jaws, and pharyngeal teeth that are morphologically specialized for feeding on a narrow range of foods, but they nonetheless are opportunistic and will take advantage of abundant and easily obtained foods (Greenwood, 1974; Liem, 1980a). McKaye and Marsh (1983) studied the feed-ing behavior of two algae-scraping cichlids, *Petrotilapia tri-dentiger* and *Pseudotropheus zebra*, in Lake Malawi. Although the territorial males of *P. tridentiger* engaged almost exclu-sively in algae scraping, the females, juveniles, and nonter-ritorial males switched frequently to eating zooplankton, as did both sexes of *P. zebra*. Because the males do not leave their territories, their ability to specialize on the algae avail-able within their restricted range is of great importance (McKaye and Marsh, 1983).

Consumption of freshly dead or partially decomposing organic matter, termed **detritivory**, might be classified as a distinctive type of omnivorous strategy. In ocean areas and lakes where detritus may accumulate over soft bottoms, de-tritivores may be abundant, especially in the tropics. Detri-tivorous fishes, such as the characin *Prochilodus scrofa*, may have elaborate, coiled intestines characterized by regional differentiation of cells and tissues (Nachi et al., 1998). A detritivorous feeding strategy, coupled with an extremely broad salinity tolerance and pugnacious nature, may be the key to the broad distribution and abundance of the sheeps-head minnow (*Cyprinodon variegatus*) in the Americas.

FUNCTIONAL MORPHOLOGY OF THE FEEDING APPARATUS

Agnathan Fishes

In a group of vertebrates as phylogenetically and ecologi-cally diverse as the fishes, one can expect tremendous varia-tion in the design of the feeding apparatus. The diversity of feeding structures is even observable between the two sur-viving agnathans, the hagfishes and the lampreys. Hagfishes lack the oral disc of the lampreys and have their lingual teeth arranged bilaterally so that, when everted, the lingual teeth and the teeth in the roof of the mouth can "bite" the prey and tear a hole in the skin or pull off pieces of flesh. Hagfishes

can also use their "knot tying" behavior (see Chapter 5) to aid in feeding (Hardisty, 1979).

In larval lampreys, a dorsal hood surrounds the oral opening. The oral cavity is filled with a fine, many-branched filter that retains fine particles. Adult lampreys have an oral disc set with horny teeth and a "tongue" that can act to cut and rasp because of its longitudinal and transverse tooth laminae. At the same time that it rasps, it acts as a siphon pump that, as it retracts in the rasping motion, reduces pres-sure and serves to suck out fluids and tissues from the prey.

Chondrichthyan Fishes

Jawed fishes show great variety in the structure and func-tion of the feeding apparatus. Remarkable evolutionary ad-vances are seen from the early gnathostomes to the teleosts, and even within the Actinopterygii, greater complexity and efficiency in the feeding mechanism is evident in the evolu-tion of more derived lineages (Lauder, 1982). In the earliest gnathostomes, the jaws appear to have had few moving parts. Probably, some could elevate the cranium by the function of the intracranial joint at the same time that the mandible was lowered, with little or no lateral movement possible in the lower jaw and no forward movement possible in the upper jaw. Some of the living sharks with amphistylic or orbito-stylic jaw suspension (*Chlamydoselachus, Heptranchias*) may approach this arrangement. Most modern sharks have a type of hyostylic jaw suspension that allows the upper jaw to swing forward and down (see Chapter 6). Wilga et al. (2000) suggested that the ancestral mechanism of jaw depression in gnathostomes, as witnessed in the modern Chondrichthyes, involved the coracomandibularis muscle, with the coraco-hyoideus enabling the depression of the hyoid arch. The nurse shark (*Ginglymostoma cirratum*) has especially well-developed coracohyoid and coracobranchial muscles for de-pressing the hyoid and branchial arches, which, when work-ing with a comparatively small mouth filled with small teeth, facilitates suction feeding by this bottom feeder (Motta and Wilga, 1999). Although the cranial muscles involved in the depression and elevation of the lower jaw are similar among the different taxa of elasmobranchs, those involved in up-per jaw protrusion and retraction show a more varied evo-lutionary history (Wilga et al., 2001). The absence of pha-ryngeal dentition in chondrichthyan fishes might suggest less sophisticated approaches in the handling of prey items, specifically the ability to separate out edible from inedible items. Dean et al. (2005) argue that the elaborate cranial musculature seen in chondrichthyans enables sophisticated prey handling as well as prey capture. Batoids, for example, have highly developed lower jaw musculature that enables sorting and processing of the hard shells of benthic prey.

Jaws and Branchial Apparatus in Bony Fishes

The numerous moving parts of the jaws, the branchiohyoid apparatus, and the opercular series require complex sets of muscles that may vary significantly in structure among fishes of different phylogenetic lineages and ecological habits. Accompanying the development of jaws in gnathostomes were specialized forms of myosin in the muscles associated with the jaws. Myosin is the motor protein associated with the thick filaments in striated muscles. The muscles in the jaws of gnathostomes, from sharks to mammals, have a specialized form of masticatory myosin that confers rapid contractile force and power (Hoh, 2002). Examples of these muscles are shown in Figure 23.12. The **adductor mandibulae,** which closes the mouth, is the largest of the head muscles, attaining a proportionately large size in those fishes that have crushing teeth or bite chunks out of prey organisms. For instance, the formidable appearance of the wolf eel (*Anarrhichthys*) is partially attributed to its large and well-developed jaw muscles; the posterior portion of the cranium is greatly compressed and smooth, providing a large area of attachment for the adductor mandibulae. The tetraodontiform fishes, which are among the most derived of fish taxa (see Chapter 17), display a range of subdivisions of the adductor mandibulae that involve the duplication of pre-existing muscles. The sophistication in food handling afforded by such a subdivision of the jaw muscles has also been observed in loricarioid catfishes and parrotfishes of the family Scaridae (Friel and Wainwright, 1999). Damselfishes (Pomacentridae) have an extremely unusual biting mechanism—one that receives assistance from the muscles and bones of the pectoral girdle. When the pectoral girdle is protracted, it makes contact with the lower pharyngeal jaw, causing its elevation and contributing to the total biting force that can be exerted on prey items (Galis and Snelderwaard, 1997).

The opercular musculature is well developed in most fishes, especially those of sedentary habits, because of the role of the operculum in irrigation of the gills. Bottom fishes, such as sculpins and catfishes, generally have better developed branchiostegal muscles than do active species; swift fishes, such as tunas, that depend on continuous swimming movement for gill oxygenation (ram irrigation or ventilation) tend to have smaller opercular and branchiostegal muscles.

Among the Osteichthyes, jaw depression is primarily enabled by the sternohyoideus muscle—considered homologous to the coracohyoideus of the Chondrichthyes (Wilga et al., 2000). Also witnessed among the bony fishes is a progression from comparatively simple, elongate jaws, which mainly catch and hold prey, to suction feeding aided by protrusible jaws. Lungfishes are unusual in having developed

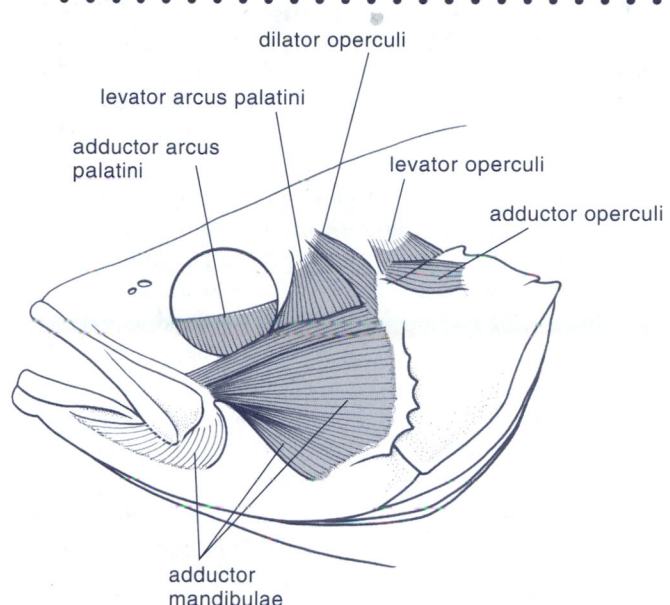

FIGURE 23.12
Examples of jaw and opercular muscles in *Sebastes melanops*.

effective suction feeding without protrusible jaws (Bemis, 1987). The myctophiform genus *Aulopus* possibly represents jaw protrusibility at its incipient stages (Gosline, 1981). The evolutionary sequence leading to protrusible jaws involved the development of movable maxillaries and premaxillaries, enlargement of the premaxillaries, exclusion of maxillaries from the gape, and development of a long ascending process of the premaxillaries. Lauder (1983) remarked that the feeding apparatus of fishes can consist of more than 30 movable bony parts and more than 50 muscles, whereas Choi (2002) claimed that feeding can involve as many as 60 bones and 80 muscles.

Generally, in primitive bony fishes, there is little movement in the premaxillary, and the head of the maxillary is attached to the neurocranium just posterior to the premaxillary. In opening the mouth, the contraction of the epaxial body muscles elevates the neurocranium, and the contraction of the muscles along the ventral aspect of the head (mainly the sternohyoideus, but including, in some fishes, the geniohyoideus and hypaxial body muscles) lowers the mandible. In *Latimeria*, neurocranial elevation is facilitated by the intracranial joint (Lauder, 1980a). In *Amia* and the teleosts, there is an additional coupling that serves to open the mouth. This is powered by the levator operculi muscles that raise the opercular apparatus, which is coupled to the mandible through the interoperculomandibular ligament.

Another coupling is seen in derived teleosts, which have an interoperculohyoid ligament (Lauder, 1982). The ligamentous coupling between the maxillary and the mandible causes the maxillary to pivot on its connection to the neurocranium and to swing downward and forward, thus forming the lateral edges of the open mouth.

Usually, the mouth-opening sequence is accompanied by the lowering of the floor of the mouth by the depression of the hyoid apparatus. During this time, the opercular opening is closed. This can lower the pressure in the orobranchial cavity and aid in the capture of food. Opercular expansion contributes to the suction but usually occurs only as the mouth begins to close.

The closure of the mouth is accomplished mainly by the adductor mandibulae. In *Amia*, hyoid depression occurs after mouth closure (Lauder, 1979). Additional information on mouth function in actinopterygians can be found in Lauder (1980b), Lauder and Liem (1980), and Rand and Lauder (1981).

Suction Feeding

In cypriniforms and in other derived lineages of teleosts, the mobile premaxilla makes jaw protrusion possible, and suction feeding is well developed. Depending on the species, the protrusion of the premaxilla can (1) add to the velocity of approach to the prey, (2) allow the fish to "reach" for food, (3) aid in forming a more efficient tunnel for suction, and (4) if left protruded while the mandible is adducted, allow the mouth to close faster. Different mechanisms have evolved for the protrusion of the premaxilla in different groups of fishes (Motta, 1984b). A common type involves a ligamentous coupling of the maxilla to the mandible so that, as the mandible is depressed, the maxilla is partly rotated and a process on its head engages the premaxilla and forces it forward. In some advanced forms, the premaxilla is attached to the mandible, so that there is a direct pull on it as the mandible is depressed.

The sequence of events during the suction feeding of a percoid (*Gymnocephalus cernua*), as reported by Elshoud-Oldenhave and Osse (1976), consists of a preparatory phase, in which the volume of both oral and branchial cavities is decreased, followed by three subsequent phases:

Phase I—Abduction pressure is exerted with the mouth closed.

Phase II—The mouth is opened, and the oral and branchial chambers are enlarged.

Phase III—The mouth is closed and compressed to force water and ingested material posteriorly; water is forced out by the compression of the posterior chamber, and food is retained on the gill rakers.

During the suction phase, the mouth cavity is said to be like a truncated cone with the base at the posterior, with pressure becoming more negative as the wider base is approached (Liem, 1990). Studies by Grubich (2001) have suggested that the neuromuscular mechanisms that permit the *lateral* expansion of the oral cavity are an essential—but heretofore overlooked—aspect of the evolution of suction feeding in bony fishes.

The speed at which small prey can be engulfed by various fishes using suction feeding is usually very rapid. Most species can complete oral expansion in less than 100 msec, and many require less than 50 msec. Small prey can be engulfed in as little as 4 msec by the frogfish (*Antennarius;* Grobecker and Pietsch, 1979) or about 5 msec by the swamp eel (*Monopterus;* Liem, 1980b). *Antennarius* can complete the entire feeding sequence in as little as 25 msec. Grobecker and Pietsch (1979) suggested that such ultrafast feeding is common for sit-and-wait predators with large mouths. Sanford and Wainwright (2003) have developed a sophisticated technique for investigating the kinematics of suction feeding: They implanted tiny piezoelectric ceramic crystals in different regions of the fish's mouth; these crystals are capable of receiving ultrasonic signals and emit ultrasonic pulses when distorted. This method has proven to be much more precise than conventional video recording techniques—kinematic functions that are otherwise not visible are detectable using this procedure.

Gill Rakers and Pharyngeal Teeth

Gill rakers exhibit considerable diversity in shape and size, usually reflecting the diet of the species involved (Fig. 23.13). The branchial sieve created by the gill rakers is probably not a passive filter in fishes that eat plankton or other small organisms. The space between adjacent gill arches can be controlled, and gill rakers and the grooves between them are controlled by musculature that can govern the ability to retain particles of differing sizes. In certain cyprinids, the palatal organ works in concert with the gill rakers to govern the dimensions of the "branchial slit" that allows water to pass from the mouth through the branchial sieve (Hoogenboezem et al., 1991). Studies by Sanderson et al. (2001) have indicated that filter-feeding fishes are not using elongate gill rakers for trapping suspended particles. The elaborate gill raker systems that have evolved in several groups apparently are used in a crossflow filtration process that moves food particles away from the gill raker surfaces and toward the esophagus. Apparently, not all fishes require gill rakers for the retention of very small particles. The cichlid *Tilapia galilaea* was able to ingest and retain particles as small as 0.07 mm after its gill rakers and microbranchiospines were surgically removed (Drenner et al., 1987).

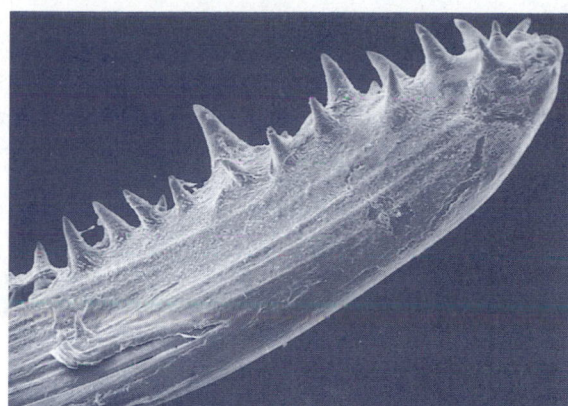

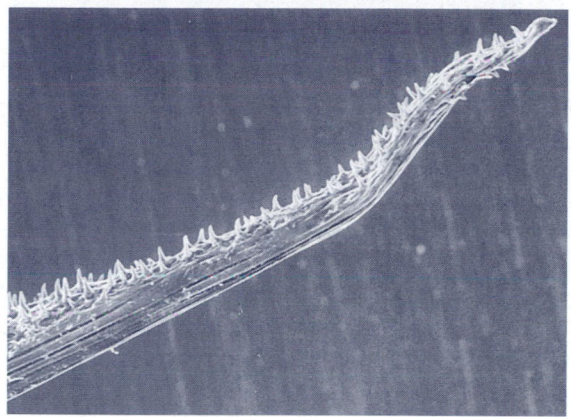

FIGURE 23.13
A, Distal half of gill raker of bocaccio (*Sebastes paucispinis*), a piscivore, showing coarse teeth; **B,** Tip of gill raker of pygmy rockfish (*S. wilsonsi*), a crustacean eater, showing numerous fine teeth. (Scanning electron micrographs courtesy of A. H. Soeldner and G. Pequeno.)

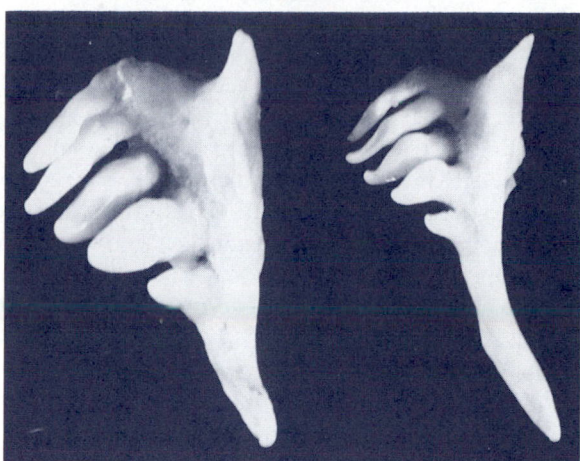

FIGURE 23.14
Pharyngeal teeth of tui chub (*Gila bicolor*), *left*, spring-living race that feeds on gastropods and large insects; *right*, lacustrine race that feeds on zooplankton and small insects. The specimens were of equal length.

In many species, the pharyngeal teeth function simply to aid the oral teeth and the gill rakers in holding prey and perhaps in forcing prey into the esophagus. In some species, the pharyngeal teeth demonstrate structural modifications for crushing, grinding, shearing, or other actions relevant to a particular type of diet. Cypriniform fishes exemplify the development of pharyngeal teeth at the expense of dentition in the gape (see Chapter 10). Modifications to the pharyngeal teeth in response to local differences in food resources may even occur within a single species (Fig. 23.14).

Preparation of Food for Digestion

Many predatory fishes, such as the largemouth bass (*Micropterus salmoides*), have small, cardiform teeth on relatively broad pads both in the mouth and on the pharyngeal bones. These teeth act to catch and hold prey and to aid in swallowing it, but they do not prepare the prey for the digestive process. Other predators, such as salmon or trout, have rows of sharp teeth that commonly tear or break the skin of the prey at the time of capture or during the swallowing process. The pikeminnows (*Ptychocheilus*) have no teeth in the mouth but are equipped with strong pharyngeal teeth, which tear the prey as it is forced into the esophagus. A few predators, such as sharks and piranhas, bite pieces out of the prey, with the grisly practice of exsanguination (i.e., death by blood loss) by the white shark being among the most notorious of feeding practices (see Chapter 37).

Although a number of fishes (chimaeras, lungfishes, wolffishes, croakers, and others) have tooth plates or molariform teeth in the mouth, with which they can grind or mash food, the pharyngeal teeth appear to be the major apparatus used for mastication among the teleosts. Some of the strongest development of pharyngeal teeth can be seen in species that habitually feed on shellfish or coral. Many of these, such as the surfperches (Embiotocidae), have pharyngeal mills that can triturate the food organisms past recognition. Some herbivorous fishes that feed on macroalgae and aquatic plants must thoroughly tear or grind the vegetation, so that digestive enzymes can act on the cell contents. The grass carp (*Ctenopharyngodon idella*) has relatively long, rough-edged pharyngeal teeth that intermesh while tearing the soft plants on which the species feeds.

Mucous cells or glands are present in the mouth and pharynx of most fishes, but the greatest mucus production is usually in the esophagus. Mucus aids in the retention of small particles at the gill rakers, facilitates the process of swallowing by lubricating large particles, and may aid in holding finely divided particles together for swallowing. Butterfishes (Stromateidae) and their close relatives have expanded esophageal sacs for food storage. They have teeth in the lining of the sac to triturate the food, and the sacs are well equipped with mucous glands. The digestive process appears to begin in the esophagus of certain mullets (Mugilidae) and sculpins (Cottidae) that have gastric glands in the posterior section.

Ecomorphology and Feeding in Fishes

The morphological studies described earlier have provided important insights into the relationship between anatomical structure and feeding actions in fishes. The past several years have witnessed the emergence of a new field of study: **ecomorphology,** which seeks to relate all aspects of the functional morphology of animals with their ecological performance. This approach has been especially fruitful when applied to the diversity of feeding types represented in a particular community—specifically, in the correlation of fish skull kinetics to diet and feeding behavior (Liem, 1993). Specific examples of ecomorphological adaptation have already been considered in the evolution of morphological variation that accompanies trophic polymorphism within numerous species (see Chapter 4). The sticklebacks (Gasterosteidae) are among the best known examples demonstrating the evolution of morphological variation among populations (Bell and Foster, 1994; Grant, 1994; Walker, 1997). In the threespine stickleback (*Gasterosteus aculeatus;* see Plate 15.2), body morphologies favoring quick, evasive action (maximum acceleration and maneuverability) characterize populations coexisting with predators. Differences in locomotor behavior associated with feeding may account for the observation that limnetic, zooplanktivorous populations of *G. aculeatus* have more elongate snouts than populations that feed on benthos (Walker, 1997). Both intra- and interspecific variations in morphology that correlate with features of aquatic habitats have been documented by a number of authors (see Wood and Bain, 1995).

Broad morphological correlates with feeding behavior have been observed in numerous fish communities. Mouth shape and total area correlated with trophic level for 18 species of Mediterranean fishes (Karpouzi and Stergiou, 2003). Assemblages of distantly related West African stream fishes can be morphologically grouped according to feeding type: (1) detritivores have elongated guts and dorsally positioned eyes; (2) piscivores have large bodies and mouths; (3) surface feeders have dorsally oriented mouths; and (4) benthic feeders have laterally placed eyes and small relative gut lengths (Hugueny and Pouilly, 1999). As might be expected, the cichlids (see Going Deeper, Chapter 32) have also proven worthy candidates for the ecomorphological approach to feeding studies (see Winemiller et al., 1995).

Ecomorphological approaches have also been applied to investigations of ontogenetic shifts in feeding behavior and prey preferences. Differential shifts in feeding mode, reflecting changes in oral anatomy, have been observed to accompany ontogenetic shifts in diet in two coexisting species of predatory percoids, pinfish (*Lagodon rhomboides*) and snook (*Centropomus undecimalis;* Luczkovich et al., 1995).

DIGESTION

The Digestive Process

A novel approach to studying the digestive processes of animals is to employ **reactor theory,** as modeled by chemical engineers to optimize the efficiency of a chemical process for the production of large quantities of materials. When applied to the digestive process, reactor theory can be used to determine the effect of such variables as nutrient quality and rate of intake on the rates of hydrolysis and absorption. Such applications can provide insight into the evolution of digestive processes (Jumars, 2000a, 2000b). For fishes, as for all other animals, the "bottom line" is the maximization of nutrient absorption from ingested foodstuffs. In most fishes, the chemical digestive process begins in the stomach, which differs from the esophagus in the composition of its walls and in the type of glands in its mucosal lining. In addition to secreting a protective mucus coating and pepsin (a protease), the glands of the vertebrate stomach secrete hydrochloric acid, which maintains the pH of the stomach contents in a range suitable for the action of pepsin. Elasmobranchs appear to be the earliest vertebrates capable of secreting gastric acids (Ballantyne, 1997). Pepsin shows a peak of activity at a pH of about 2, and pH values between about 1.5 and 4 are common in the stomachs of predatory or insectivorous species. The stomachs of some herbivorous fishes produce acidic secretions with a pH range suitable for the digestion of algae (Horn, 1989). Pepsin is not found in cyclostomes, chimaeras, lungfishes, and many teleosts that lack a stomach (Stevens, 1988).

Fishes secrete other gastric enzymes in addition to pepsin (Fänge and Grove, 1979). In some fishes, there are apparently proteases that are optimally active at a pH from 3 to 5. There are also enzymes in the stomach that act on foods other than proteins. Chitinase has been obtained from insectivores and from species that commonly feed on crustaceans and, hence, must be able to break down their chitinous exoskeletons. Amylase has been identified in some clupeids, and lipase in clupeids and the Mozambique tilapia (*Oreochromis mossambicus*). The flow of gastric juices is initiated by the act of feeding and especially by the distension of the stomach wall. Secretion is, to some extent, under the control of the vagus nerve.

The stomach generally acts to store food and to initiate digestion by mixing the food mass with the gastric juices. Depending on the food habits of the species, the stomach can be large and distensible or small and capable of passing small food items along during an extended feeding period each day. The stomachs of some mullets (Mugilidae), the milkfish (*Chanos chanos*), some herrings (Clupeidae), and some characins (Characidae)—all microphagous fishes—are modified into gizzards. In many of these, the gizzard involves only the pyloric part of the stomach, and the secretory function of the stomach is diminished. In the gizzard shad (*Dorosoma cepedianum*), the gizzard is divided into cardiac and pyloric sections. Many microphagous fishes have unspecialized stomachs, or none at all. There are great differences in the shape and development of the stomach, even among closely related species or species with similar food habits. The motility of the stomach is, in many instances, related to the degree of fullness, so that food is removed more rapidly from a full stomach than from one that is partially full. As in mammals, the release of **cholecystokinin** from endocrine cells of the intestine has been demonstrated to induce gallbladder motility, resulting in the release of bile into the duodenum, while simultaneously suppressing stomach motility and gastric emptying (Olsson et al., 1999).

As noted earlier, a variety of fishes lack a true stomach. Stomachs are not recognized in lampreys, hagfishes, chimaeras (Holocephali), lungfishes (Dipnoi), some gobies (Gobiidae), minnows (Cyprinidae), suckers (Catostomidae), sauries (Scomberesocidae), pipefishes (Syngnathidae), wrasses (Labridae), and in at least some of the members of the following families: Cobitidae, Cyprinodontidae, Poeciliidae, Atherinidae, Belonidae, Mugilidae, Cichlidae, Scaridae, Blenniidae, Callionymidae, and Gobiesocidae. Some of these may maintain some gastric function, because gastric glands have been noted in pipefishes and in the belonid needlefish *Xenentodon* (Fänge and Grove, 1979). There are several hypotheses explaining the loss of the stomach in various kinds of fishes, one being that the condition is neotenic, inasmuch as larval fishes generally lack a stomach. Given the low concentrations of chlorine available in freshwater, the ability to digest food in a completely alkaline gastric environment would seem to be advantageous, because the fishes would be freed from the burden of producing hydrochloric acid. In the absence of acid in the digestive tract of the cunner (*Tautogolabrus*), whole molluscs ingested by the fish are known to have survived the half-day journey from ingestion to defecation. Predatory fishes lacking a stomach often have an expanded portion of the intestine, in which large morsels can be stored while undergoing digestion. These expansions are often mistaken for stomachs, as they serve as a region for digestion.

The pyloric caeca, if present, branch from the intestine near the pyloric end of the stomach (Fig. 23.7). There are conflicting reports on whether digestive enzymes are produced by the caeca. Although some histological studies of fishes have indicated that enzyme-secreting cells are absent from the pyloric caeca (Fänge and Grove, 1979), the caeca are considered by Prosser and DeVillez (1991) to be sources of both carbohydrases and proteinases. Glass et al. (1987; as cited by Prosser and DeVillez, 1991) reported the presence of digestive enzymes (trypsin, amylase) in the caeca, but attributed the secretion of these enzymes to the diffuse pancreas. Smith (1989) ascribed no secretory function to the pyloric caeca. The caeca may have a function in absorption as well as in increasing the area of digestive surface. Pyloric caeca have been shown to be active in converting fatty alcohols into fatty acids (Cowey and Sargent, 1979).

Carnivores may expend great amounts of energy in the capture of food, yet digestion and assimilation are accomplished with comparative ease. Herbivores, on the other hand, expend little energy in gathering food, yet the processing of highly digestion-resistant plant materials presents special challenges to the digestive system. The structural carbohydrates composing plant tissue are highly impervious to the digestive enzymes that are endogenous to the vertebrate digestive system, and a variety of alternative measures must be employed. Marine fishes from temperate shores have been shown to be able to assimilate significant amounts of organic carbon from macroalgae, however (Edwards and Horn, 1982; Horn et al., 1985; Horn and Gibson, 1990). Herbivorous fishes may optimize their digestive efficiency by preferentially feeding on selected algal species or by feeding at times when photosynthetic activity is at a maximum (Choat and Clements, 1998). The king angelfish (*Holacanthus passer*) possesses an expanded portion of the intestinal tract containing microorganisms capable of algal digestion (Martínez-Díaz and Pérez-España, 1999). In this fashion, some herbivorous fishes resemble horses, as both are "hindgut fermenters," possessing expanded hindgut chambers where plant materials can be degraded by microbially assisted fermentation processes. One of the largest and most unusual prokaryotes discovered to date is *Epulopiscium fishelsoni*, an apparent intestinal symbiont in the guts of acanthurid surgeonfishes from both the Indo-Pacific and Caribbean regions (Clements and Bullivant, 1991; Fishelson et al., 1985). Microbial fermentation processes, involving symbiotic microorganisms, greatly facilitate the processes of digestion and absorption in herbivorous vertebrates, yet the role of intestinal symbionts in the digestive processes of herbivorous fishes has yet to be clarified (Choat and Clements, 1998; Clements and Choat, 1995). Perhaps the most unlikely source of plant-derived nutrients for vertebrates is wood. Although a few mammals, such as beavers, are known to consume woody materials, the only fishes that are known to consume wood are the loricariid catfishes. This capability appears to be facilitated by a host of symbiotic microorganisms (Nelson et al., 1999).

Gut Morphology and Digestion

As mentioned earlier, the length of the teleost intestine is correlated with dietary preferences. Detritivorous and herbivorous fishes, which habitually ingest a large proportion of indigestible material with their food, appear to have the longest relative gut length, and in these fishes, the gut is usually flexed or coiled in an elaborate manner. The added length increases the retention time of the food and allows more efficient processing of digestion-resistant materials. Table 23.3 shows a comparison of the relative gut lengths of several species.

Intestinal length increases allometrically in relation to body size, and this must be taken into account when making correlations between diet and intestinal length (Kramer and Bryant, 1995a, 1995b). In many species, the length (and

thus the volume and absorptive area) of the gut increases by a factor greater than the increase in body length—large adult individuals have a markedly greater gut length to body length ratio than do small individuals.

Enzymatic Activity

Digestion and absorption proceed in the intestine in a neutral to alkaline environment. The enzymes involved are secreted by the pancreas, intestinal mucosa, and possibly by the pyloric caeca. The types and amounts of enzymes present in the digestive system of a given species are related to the general food habits of that species (Fänge and Grove, 1979). Hagfishes are unique in their digestive processes. Food in the hagfish gut is enclosed in a mucoid bag secreted by the intestinal epithelium. Digestive enzymes can penetrate the bag, and the residual mass that is not absorbed is excreted in the mucoid wrapper (Pough et al., 1999). The pancreatic tissue is probably the source of many of the digestive enzymes, but the diffuse nature of the pancreas in most fishes makes it a challenge to locate the exact source of the enzymes. One of the most important proteases, trypsin, is secreted by the pancreas. Trypsin is an endopeptidase, which breaks up polypeptides at the bonds adjacent to specific amino acids (phenylalanine and tyrosine, in this case). Other endopeptidases, chymotrypsin and elastase, are produced by fish pancreatic tissue. Exopeptidases, which break up large peptides and remove some terminal amino acids from the chains, are also secreted by the pancreas of fishes (Stevens, 1988), as are

TABLE 23.3 RATIO OF INTESTINE LENGTH (L) TO BODY LENGTH (B) IN SELECTED FISHES

Species	L/B	Diet	Source
Elopichthys bambusa (kanyu)	0.63	Piscivorous	Kapoor et al. (1975)
Ptychocheilus oregonensis (northern pikeminnow)	0.78	Piscivorous	Weisel (1962)
Pyrrhulina filamentosa (pyrrhulina)	1.0	Insectivorous	Kapoor et al. (1975)
Xiphister mucosus (rock prickleback)	0.80	Herbivorous (macroalgae)	Barton (1982)
X. atropurpureus (black prickleback)	0.62	Carnivorous	Barton (1982)
Cebidichthys violaceus (monkeyface prickleback; >80 mm)	1.01	Herbivorous (macroalgae)	Barton (1982)
C. violaceus (<80 mm)	0.56	Omnivorous	Barton (1982)
Xenocharax spilurus (xenocharax)	2.0	Plants and invertebrates	Kapoor et al. (1975)
Carassius auratus (goldfish)	2.3	Phytoplankton, plants, detritus	McVay and Kaan (1940)
Citharinus congicus (citharinus)	4.0	Microphagous	Kapoor et al. (1975)
C. citharus (citharinus)	6.0–7.5	Microphagous	Kapoor et al. (1975)
Chondrostoma nasus (nase)	2.0	Herbivorous	Junger et al. (1989)
Ctenopharyngodon idella (grass carp)	2.5	Herbivorous	Kapoor et al. (1975)
Hypophthalmichthys molotrix (silver carp)	13.0	Phytoplankton	Kapoor et al. (1975)
Labeo horie (labeo)	15–20	Algae, detritus	Kapoor et al. (1975)
Catostomus catostomus (longnose sucker)	2.29	Benthivorous	Weisel (1962)
Centropristis striata (black sea bass)	0.71	Piscivorous	Blake (1930)
Prionotus carolinus (northern sea robin)	0.89	Benthivorous	Blake (1936)
Mulloides auriflamma (goatfish)	1.03	Benthivorous	Al-Hussaini (1946)

amylase and lipases. Chimaeras, and possibly other fishes, produce chitinase in the pancreas, although this enzyme is more commonly secreted by the gastric mucosa.

Surface epithelial cells of the intestine in vertebrates secrete a variety of enzymes that act on carbohydrates, but knowledge of the intestinal enzymes is by no means complete. Maltase is known from the intestines of the ayu (*Plecoglossus altivelis*), various salmonids, the common carp, and the sparid *Pagrus*. In addition, the carp appears to secrete sucrase and cellobiase (Stevens, 1988).

The liver secretes bile, which is stored in the gall-bladder. Bile aids in the digestion of fatty materials by emulsifying them—lipases and esterases cannot hydrolyze fats that are not in solution or nearly so.

· · · · · · · · · · · · ·
Rates and Efficiency of Digestion

Digestive rates are variable, depending on the type of food-stuff, the species of fish, the temperature, and the amount of food ingested. There is some evidence that small fishes of a given species digest food more rapidly than larger individuals. Because various investigators have used different test species, different methods, and different test foods, there is some difficulty in comparing the results of digestion rate studies. The temperature at which digestion is taking place is reported in these experiments, but this has not always been discussed in relation to the ecology of the species studied. Temperature influences processes such as the rate of secretion and the activity of digestive enzymes, the absorption rate of the digested food, and the muscular activity of the digestive tract. The amount of food fed to fishes has an effect on their rate of digestion; usually, a large meal is digested at a more rapid rate than a small one. Rates of gastric evacuation apparently decrease with increasing activity. This may possibly be a consequence of shunting blood away from the gut when the oxygen demands of the skeletal muscles increase (Jobling, 1981). Some results of digestion rate studies are shown in Table 23.4.

TABLE 23.4   TIME NECESSARY TO EMPTY DIGESTIVE TRACT FOR SELECTED FISHES

Species	Temperature (°C)	Time to 100% empty (hr)	Reference
Salmonidae			
Oncorhynchus mykiss	15	40	Grove et al. (1978)
Esocidae			
Esox lucius	12	72	Lane and Jackson (1969)
Cyprinidae			
Pimephales promelas	20	12–24	Lane and Jackson (1969)
Carassius auratus	20	60–72	Lane and Jackson (1969)
Cyprinus carpio	23	48	Lane and Jackson (1969)
Gibelio catla	28–30	18–54	Renade and Kewalramani (1967)
Cirrhina mrigala	28–30	18–60	Renade and Kewalramani (1967)
Ameiuridae			
Ameiurus nebulosus	20	60	Lane and Jackson (1969)
Embiotocidae			
Phanerodon furcatus	23–26	10–12	Bray and Ebeling (1975)
Centrarchidae			
Micropterus salmoides	20	60	Lane and Jackson (1969)
Cottidae			
Cottus gobio	10	100	Western (1971)
Gobiidae			
Gobius cobitis	14, 20	45.4, 27.8*	Horn and Gibson (1990)
Bleniidae			
Lipophrys pholis	11, 264	7.4, 18.4*	Horn and Gibson (1990)
Parablennius sanguinolentus	14, 20	66.3, 24.5*	Horn and Gibson (1990)
Scombridae			
Katsuwonus pelamis	23–36	14	Magnuson (1969)
Cichlidae			
Tilapia nilotica	25	7–15	Moriarty (1973)

* Gastric evacuation times at respective temperatures indicated.
Source: Table adapted in part from Fänge and Grove, 1979.

Smith (1989) indicated that carnivores have a relatively slow food passage in comparison to herbivorous fishes. Examples include the virtually complete digestion of salmon fry in 24 hours (at 15°C) by Dolly Varden (*Salvelinus malma*) and a similar time for the total digestion of food by rainbow trout (at 9–10°C). Tunas empty their gut in about 14 hours. Sharks generally are slower in digesting. Wetherbee (1990) stated that the lemon shark (*Negaprion brevirostris*) will feed for about 11 hours and then not eat for about 32 hours.

The determination of gut passage rate and assimilation frequency in herbivorous fishes is especially problematic (Choat and Clements, 1998). The herbivorous grass carp passes food in about 8 hours. Rabbitfishes (*Siganus*), following a meal of algae, empty the gut in 3 hours or less; juveniles begin defecation as early as 1.5 hours after ingestion (Bryan, 1975). In the sea chubs (*Kyphosus*), which practice microbial fermentation, the gut takes 20 hours or more to empty (Rimmer and Wiebe, 1987).

Digestion efficiency in carnivores ranges between 70 and 90 percent; but in herbivorous species, it usually ranges between 40 and 50 percent. Smith (1989) cited studies that indicated a higher digestibility of fats and protein in the diets of herbivores, however. These digestive efficiencies ranged from about 65 to 70 percent for lipids and from about 60 to 71 percent for proteins. Cai and Curtis (1989) reported a digestion efficiency of about 50 percent for grass carp fed aquatic plants (*Elodea*), and a digestive efficiency of 67 to 68 percent for the same species fed commercial catfish food.

Only a portion of the energy ingested is available for metabolism and growth, for several reasons. First, not all of the energy may be in materials that are digestible by the fish, or some materials may otherwise escape assimilation. This portion of the food energy is passed out of the alimentary canal as feces. The materials absorbed through the intestinal wall contain some energy in nitrogenous compounds that cannot be metabolized and are excreted by the gills or kidneys. Estimates of the energy in fecal and other wastes are usually in the range of 15 to 20 percent of the energy ingested. The remaining materials are available for necessary metabolism and growth.

Diet formulation for aquaculture situations, as discussed earlier, is an especially important area of fish research. Trout of about 15 cm held in 14°C water are fed dry diets at about 2 percent of body weight per day. Smaller fish are fed more, and larger fish less; more food is given in warmer water, and less in colder water. Channel catfish held in 25°C water are fed at the rate of 2.5 to 3 percent of body weight per day. Conversion efficiency of food into fish is usually from 1.6 to 2.5 kg of dry food to 1 kg of fish produced. Because not all of the energy ingested is used by the fish, large-scale aquaculture operations can have potentially serious environmen-tal impacts. The fecal waste produced by large numbers of fishes living in a confined space that is subsequently released into the environment represents a significant input of materials and energy to be consumed elsewhere in the ecosystem, resulting in aquatic pollution (see Chapter 39).

KEY POINTS AND CONNECTIONS

- Ingested food represents energy and materials available for maintenance and growth in fishes. Maintenance of life activities is expressed by the measurement of metabolism. For fishes, a frequently used measure of metabolic activity is the routine metabolic rate (RMR), in which a metabolic indicator, such as oxygen consumption, is measured while the test specimen is engaged in a modest amount of activity.

- Growth is defined as the elaboration of tissue resulting in an increase in body mass. Most fishes demonstrate indeterminate growth, as they continue to add tissue mass throughout their lives. Ingested energy that is not consumed in the metabolic maintenance of the individual can be applied to growth and to the elaboration of gametes.

  The production of gametes and hence offspring is the subject of Chapter 27.

- A diverse array of techniques for studying age and growth in fishes are available. Among the most valuable of these methods has been the measurement of bone deposition in various skeletal elements, such as scales, fin spines, opercula, and otoliths.

- Fishes are fairly typical of vertebrates in terms of their dietary requirement of proteins, carbohydrates, lipids, vitamins, and minerals. Given the preponderance of carnivory among fishes, it is not surprising that fishes, when compared with other vertebrates, generally require more dietary protein. Omnivorous species tend to have a higher percentage of carbohydrates in their diet. Fats are a useful energy source for fishes, but they can only be used in limited amounts.

- The role of hormones in growth and metabolism must be understood in the context of the overall metabolic framework of fishes. As fishes are largely carnivores, endocrine regulatory mechanisms governing the ingestion and processing of protein are probably more significant than the monitoring of glucose levels, as seen in mammals.

  Endocrine function, especially as it relates to metabolism and growth, is discussed in Chapter 26.

- Given their dietary diversity, it is not surprising that fishes possess an extraordinary array of adaptations of the mouth and alimentary canal. A variety of dentition types have evolved to facilitate the grasping and handling of different types of prey. One of the most significant developments seen among the bony fishes is the evolution of protrusible jaws that enable suction feeding. Bony fishes also possess distinctive pyloric caeca, situated near the pyloric region of the stomach. Intestinal length varies with diet—herbivorous and omnivorous fishes tend to have longer, more elaborate intestinal tracts.

- An ecomorphological approach to assessing feeding and digestion in fishes is useful, as it enables the association of the structure and function of the feeding apparatus with ecological aspects, such as the role of a given species in the overall trophic structure of the fish community.

An introduction to the trophic structure of fish communities is provided in Chapter 30.

FISH LINKS

http://ecomorphology.mlml.calstate.edu/ Web page of Lara Ferry-Graham, formerly a research associate in Peter Wainwright's laboratory at the University of California, Davis (http://www.fishlab.ucdavis.edu/). Ferry-Graham and Wainwright study the functional morphology of feeding in fishes.

BUILDING AN ICHTHYOLOGY LIBRARY

Gerking, S. D. 1994. *Feeding ecology of fish.* Academic Press, San Diego.
Gregory, W. K. 1933. Fish skulls: A study of the evolution of natural mechanisms. *Trans. Am. Phil. Soc. 23*(2).

Given the intimate association between skull morphology and feeding behavior, this is an especially valuable text. It was reprinted in 2002 and is available from Kreiger, Malabar, FL.

Shulman, G. E., and R. M. Love. 1999. *Advances in marine biology,* Vol. 36, *The biochemical ecology of marine fishes.* Academic Press, San Diego.

This volume includes several worthwhile contributions on bioenergetics and growth of fishes.

REFERENCES

Adelman, I. R. 1987. Uptake of radioactive amino acids as indices of current growth rate of fish: A review, pp. 65–79. In *Age and growth of fish,* R. C. Summerfelt and G. E. Hall (Eds.). Iowa State University Press, Ames.

Al-Hussaini, A. H. 1946. The anatomy and histology of the alimentary tract of the bottom feeder, *Mulloides auriflamma* (Forsk.). *J. Morphol. 28:* 121–154.

Ballantyne, J. S. 1997. Jaws: The inside story. The metabolism of elasmobranch fishes. *Comp. Biochem. Physiol. 118B*(4): 703–742.

Barton, M. G. 1982. Intertidal vertical distribution and diets of five species of central California stichaeoid fishes. *Calif. Fish Game 68*(3): 174–182.

Beamish, F. W. H. 1964. Respiration of fishes with special emphasis on standard oxygen consumption. II. Influence of weight and temperature on respiration of several species. *Can. J. Zool. 42:* 177–188.

———, and P. S. Mookherji. 1964. Respiration of fishes with special emphasis on standard oxygen consumption. I. Influence of weight and temperature on respiration of goldfish *Carassius auratus* L. *Can. J. Zool. 42:* 161–175.

Beamish, R. J. 1980. Adult biology of the river lamprey (*Lampetra ayresi*) and the Pacific lamprey (*Lampetra tridentata*) from the Pacific coast of Canada. *Can. J. Fish. Aquat. Sci. 37:* 1906–1923.

———, and G. A. McFarlane. 1987. Current trends in age determination methodology, pp. 15–42. In *Age and growth of fish,* R. C. Summerfelt and G. E. Hall (Eds.). Iowa State University Press, Ames.

Bell, M. A., and S. A. Foster. 1994. *The evolutionary biology of the three-spine stickleback.* Oxford University Press, Oxford.

Bemis, W. E. 1987. Feeding systems of living Dipnoi: Anatomy and function, pp. 249–275. In *The biology and evolution of lungfishes,* W. B. Bemis, W. W. Burggren, and N. B. Kemp (Eds.). Alan R. Liss, New York.

Blake, I. H. 1930. Studies on the comparative histology of the digestive tube of certain teleost fishes. I. A predaceous fish, the sea bass (*Centropristis striatus*). *J. Morphol. 50:* 39–70.

———. 1936. Studies on the comparative histology of the digestive tube of certain teleost fishes. III. A bottom-feeding fish, the sea robin (*Prionotus carolinus*). *J. Morphol. 60:* 77–120.

Bowen, S. H. 1987. Dietary protein requirements of fishes—a reassessment. *Can. J. Fish. Aquat. Sci. 44:* 1995–2001.

Bray, R. N., and A. W. Ebeling. 1975. Food, activity, and habitat of three "picker" type microcarnivorous fishes in the kelp forests of Santa Barbara, California. *Fish. Bull. 73*(4): 815–829.

Brett, J. R. 1979. Environmental factors and growth, pp. 599–675. In *Fish physiology,* Vol. VIII, *Bioenergetics and growth,* W. S. Hoar, D. J. Randall, and J. R. Brett (Eds.). Academic Press, New York.

Brown, J. H., J. F. Gillooly, A. P. Allen, V. M. Savage, and G. B. West. 2004. Toward a metabolic theory of ecology. *Ecology 85:* 1771–1789.

Brown, M. E. 1957. Experimental studies on growth, pp. 351–400. In *The physiology of fishes,* M. E. Brown (Ed.). Academic Press, New York.

Brown, S. B., J. D. Fitzsimons, D. C. Honeyfield, and D. E. Tillitt. 2005. Implications of thiamine deficiency in Great Lakes salmonines. *J. Aquat. Anim. Health 17:* 113–124.

Bryan, P. G. 1975. Food habits, functional digestive morphology, and assimilation efficiency of the rabbitfish *Siganus spinus* (Pisces, Siganidae) on Guam. *Pac. Sci. 29:* 269–277.

Bulow, F. J. 1987. RNA–DNA ratios as indicators of growth in fish: A review, pp. 45–64. In *Age and growth of fish,* R. C. Summerfelt and G. E. Hall (Eds.). Iowa State University Press, Ames.

Burrow, C. J. 2003. Comment on "Separate evolutionary origins of teeth from evidence in fossil jawed vertebrates." *Science 300:* 1661.

Busacker, G. P., and I. R. Adelman. 1987. Uptake of '4C-glycine by fish scales (in vitro) as an index to current growth rate, pp. 355–357. In *Age and growth in fish,* R. C. Summerfelt and G. E. Hall (Eds.). Iowa State University Press, Ames.

———, ———, and E. M. Goolish. 1990. Growth, pp. 363–387. In *Methods for fish biology,* C. B. Schreck and P. B. Moyle (Eds.). American Fisheries Society, Bethesda, MD.

Cai, Y., and R. C. Summerfelt. 1992. Effects of temperature and size on oxygen consumption and ammonia excretion by walleye. *Aquaculture 104:* 127–138.

Cai, Z., and L. Curtis. 1989. Effects of diet on consumption, growth, and fatty acid composition in young grass carp. *Aquaculture 81:* 47–60.

Caillet, G. M., and R. L. Radtke. 1987. A progress report on the electron microprobe analysis technique for age determination and verification in elasmobranchs, pp. 359–369. In *Age and growth of fish,* R. C. Summerfelt and G. E. Hall (Eds.). Iowa State University Press, Ames.

Caillet, G. M., R. L. Radtke, and B. A. Weldon. 1986. Elasmobranch age determination and verification: A review, pp. 345–360. In *Proceedings of the 2nd International Conference on Indo-Pacific Fishes*, T. Uyeno, R. Arai, T. Taniuchi, and K. Matsuura (Eds.). Ichthyological Society of Japan, Tokyo.

Callan, W. T., and S. L. Sanderson. 2003. Feeding mechanisms in carp: Crossflow filtration, palatal protrusions, and flow reversals. *J. Exp. Biol. 206:* 883–892.

Carlander, K. D. 1987. A history of scale age and growth studies of North American freshwater fish, pp. 3–16. In *Age and growth of fish*, R. C. Summerfelt and G. E. Hall (Eds.). Iowa State University Press, Ames.

Choat, J. H., and K. D. Clements. 1998. Vertebrate herbivores in marine and terrestrial environments: A nutritional ecology perspective. *Ann. Rev. Ecol. Syst. 29:* 375–403.

Choi, C. 2002. How fish hook fish. *J. Exp. Biol. 205:* 2202.

Christiansen, J. S., E. H. Jørgensen, and M. Jobling. 1991. Oxygen consumption in relation to sustained exercise and social stress in Arctic charr (*Salvelinus alpinus* L.). *J. Exp. Zool. 260:* 149–156.

Clements, K. D., and S. Bullivant. 1991. An unusual symbiont from the gut of surgeonfishes may be the largest known prokaryote. *J. Bacteriol. 173:* 5359–5362.

———, and J. H. Choat. 1995. Fermentation in tropical marine herbivorous fishes. *Physiol. Zool. 68:* 355–378.

Committee on Animal Nutrition. 1993. *Nutrient requirements of fish.* Board of Agriculture, National Resarch Council, National Academy Press, Washington, DC.

Cowey, C. B., and J. R. Sargent. 1979. Nutrition, pp. 1–69. In *Fish physiology*, Vol. VIII, *Bioenergetics and growth*, W. S. Hoar, D. J. Randall, and J. R. Brett (Eds.). Academic Press, New York.

Crowder, L. B. 1985. Optimal foraging and feeding mode shifts in fishes. *Env. Biol. Fishes 12:* 57–62.

Daniel, J. F. 1934. *The elasmobranch fishes.* University of California Press, Berkeley.

Dean, M. N., C. D. Wilga, and A. P. Summers. 2005. Eating without hands or tongue: Specialization, elaboration and the evolution of prey processing mechanisms in cartilaginous fishes. *Biol. Letters 1:* 357–361.

Donaldson, E. M., U. H. M. Fagerlund, D. A. Higgs, and J. R. McBride. 1979. Hormonal enhancement of growth, pp. 455–597. In *Fish physiology*, Vol. VIII, *Bioenergetics and growth*, W. S. Hoar, D. J. Randall, and J. R. Brett (Eds.). Academic Press, New York.

Drenner, R. W., G. L. Vinyard, K. D. Hambright, and M. Gophen. 1987. Particle ingestion by *Tilapia galileo* is not affected by the removal of gill rakers and microbranchiospines. *Trans. Am. Fish. Soc. 116:* 272–276.

Duthie, G. G. 1982. The respiratory metabolism of temperature-adapted flatfish at rest and during swimming activity and the use of anaerobic metabolism at moderate swimming speeds. *J. Exp. Biol. 97:* 359–373.

Edwards, T. W., and M. H. Horn. 1982. Assimilation efficiency of a temperate-zone intertidal fish (*Cebidichthys violaceus*) fed diets of macroalgae. *Mar. Biol. 67:* 247–253.

Elshoud-Oldenhave, M. J. W., and J. W. M. Osse. 1976. Functional morphology of the feeding system in the ruff—*Gymnocephalus cernua* (L. 1758)—(Teleostei, Percidae). *J. Morphol. 150:* 399–422.

Fänge, R., and D. Grove. 1979. Digestion, pp. 161–260. In *Fish physiology*, Vol. VIII, *Bioenergetics and growth*, W. S. Hoar, D. J. Randall, and J. R. Brett (Eds.). Academic Press, New York.

Ferry-Graham, L. A., and G. V. Lauder. 2001. Aquatic prey capture in ray-finned fishes: A century of progress and new directions. *J. Morphol. 248:* 99–119.

———, P. C. Wainwright, C. D. Hulsey, and D. R. Bellwood. 2001. Evolution and mechanics of long jaws in butterflyfishes (family Chaetodontidae). *J. Morphol. 248:* 120–143.

Fink, W. L. 1981. Ontogeny and phylogeny of tooth attachment modes in teleost fishes. *J. Morphol. 167:* 167–184.

Fishelson, L., W. L. Montgomery, and A. A. Myrberg. 1985. A unique symbiosis in the gut of tropical herbivorous surgeonfishes (Acanthuridae: Teleostei) from the Red Sea. *Science 229:* 4951.

Friel, J. P., and P. C. Wainwright. 1999. Evolution of complexity in motor patterns and jaw musculature of tetraodontiform fishes. *J. Exp. Biol. 202:* 867–880.

Fry, F. E. J. 1971. The effect of environmental factors on the physiology of fish, pp. 1–98. In *Fish physiology*, Vol. VI, W. S. Hoar and D. J. Randall (Eds.). Academic Press, New York.

Galis, F., and P. Snelderwaard. 1997. A novel biting mechanism in damselfishes (Pomacentridae): The pushing up of the lower pharyngeal jaw by the pectoral girdle. *Neth. J. Zool. 47(4):* 405–410.

Gerking, S. D. 1994. *Feeding ecology of fishes.* Academic Press, San Diego.

Gibson, R. N., and I. A. Ezzi. 1985. Effect of particle concentration on filter- and particulate-feeding in the herring *Clupea harengus. Mar. Biol. 88(2):* 109–116.

Gillooly, J. F., J. H. Brown, G. B. West, V. M. Savage, and E. L. Charnov. 2001. Effects of size and temperature on metabolic rate. *Science 293:* 2248–2251.

Gosline, W. A. 1981. The evolution of the premaxillary protrusion system in some teleostean fish groups. *J. Zool. (Lond.) 193:* 11–23.

Grant, P. R. 1994. Ecological character displacement. *Science 266:* 746–747.

Greenwood, P. H. 1974. The cichlid fishes of Lake Victoria, East Africa: The biology and evolution of a species flock. *Bull. Br. Mus. (Nat. Hist.) Zool.* (Suppl. 6): 1–134.

Grobecker, D. B., and T. W. Pietsch. 1979. High-speed cinematographic evidence for ultrafast feeding in antennariid angler fishes. *Science 205:* 1161–1162.

Grossman, G. D. 1986. Food resource partitioning in a rocky intertidal fish assemblage. *J. Zool. (Lond.) B1:* 317–355.

Grove, D. J., L. Lozoides, and J. Nott. 1978. Satiation amount, frequency of feeding, and gastric emptying rate in *Salmo gairdneri. J. Fish Biol. 12:* 507–516.

Grubich, J. R. 2001. Prey capture in actinopterygian fishes: A review of suction feeding motor patterns with new evidence from an elopomorph fish, *Megalops atlanticus. Amer. Zool. 41:* 1258–1265.

Halver, J. E. 1989. The vitamins, pp. 31–109. In *Fish nutrition* (2nd ed.), J. E. Halver (Ed.). Academic Press, San Diego.

Hardisty, M. W. 1979. *Biology of the cyclostomes.* Chapman and Hall, London.

Hazon, N., and R. J. Balment. 1998. Endocrinology, pp. 441–463. In *The physiology of fishes* (2nd ed.), D. H. Evans (Ed.). CRC Press, Boca Raton, FL.

Hill, R. W., and G. A. Wyse. 1989. *Animal physiology* (2nd ed.). Harper and Row, New York.

Hoh, J. F. Y. 2002. Review: "Superfast" or masticatory myosin and the evolution of jaw-closing muscles of vertebrates. *J. Exp. Biol. 205:* 2203–2210.

Honeyfield, D. C., J. P. Hinterkopf, J. D. Fitzsimons, D. E. Tillitt, J. L. Zajicek, and S. B. Brown. 2005. Development of thiamine deficiencies and early mortality syndrome in lake trout by feeding experimental and feral fish diets containing thiaminase. *J. Aquat. Anim. Health 17:* 4–12.

Hoogenboezem, W., J. G. M. van den Boogart, F. A. Sibbing, E. H. R. R. Lammens, A. Terlouw, and J. W. M. Osse. 1991. A new model of particle retention and branchial sieve adjustment in filter-feeding bream (*Abramis brama*, Cyprinidae). *Can. J. Fish. Aquat. Sci. 48*(1): 7–18.

Horn, M. H. 1989. Biology of marine herbivorous fishes. *Oceanogr. Mar. Biol. Ann. Rev. 27:* 167–272.

———. 1998. Feeding and digestion, pp. 43–63. In *The physiology of fishes* (2nd ed.), D. H. Evans (Ed.). CRC Press, Boca Raton, FL.

———, and R. N. Gibson. 1990. Effects of temperature on the food processing of three species of seaweed-eating fishes from European coastal waters. *J. Fish Biol. 37:* 237–247.

———, M. A. Neighbors, M. J. Rosenberg, and S. N. Murray. 1985. Assimilation of carbon from dietary and nondietary macroalgae by a temperate-zone intertidal fish, *Cebidichthys violaceus* (Girard) (Teleostei: Stichaeidae). *J. Exp. Mar. Biol. Ecol. 86:* 241–253.

Hugueny, B., and M. Pouilly. 1999. Morphological correlates of diet in an assemblage of West African freshwater fishes. *J. Fish Biol. 54:* 1310–1325.

Hyatt, K. D. 1979. Feeding strategy, pp. 71–119. In *Fish physiology*, Vol. VIII, *Bioenergetics and growth*, W. S. Hoar, D. J. Randall, and J. R. Brett (Eds.). Academic Press, New York.

Janssen, J. 1978. Feeding-behaviour repertoire of the alewife, *Alosa pseudoharengus*, and the ciscoes, *Coregonus hoyi* and *C. artedii. J. Fish. Res. Bd. Can. 35:* 249–253.

Janvier, P. 1996. *Early vertebrates*. Oxford University Press, Oxford.

Jobling, M. 1981. The influences of feeding on the metabolic rate of fishes: A short review. *J. Fish Biol. 18:* 385–400.

Jobling, M. 1994. *Fish energetics*. Chapman and Hall, London.

Juanes, F. 2003. The allometry of cannibalism in piscivorous fishes. *Can. J. Fish. Aquat. Sci. 60:* 594–602.

Jumars, P. A. 2000a. Animal guts as ideal chemical reactors: Maximizing absorption rates. *Am. Nat. 155:* 527–543.

———. 2000b. Animal guts as nonideal chemical reactors: Partial mixing and axial variation in absorption kinetics. *Am. Nat. 155:* 544–555.

Junger, H., K. Kotrschal, and A. Goldschmid. 1989. Comparative morphology and ecomorphology of the gut in European cyprinids (Teleostei). *J. Fish Biol. 34:* 315–336.

Kan, T. T., and C. E. Bond. 1981. Notes on the biology of the Miller Lake lamprey *Lampetra* (*Entosphenus*) *minima. Northw. Sci. 55:* 70–74.

Kanazawa, A. 1985. Essential fatty acid and lipid requirement of fish, pp. 281–298. In *Nutrition and feeding in fish*, C. B. Cowey, A. M. Mackie, and J. G. Bell (Eds.). International Symposium on Feeding and Nutrition in Fish, Aberdeen, UK.

Kapoor, B. G., H. Smit, and I. A. Verighina. 1975. The alimentary canal and digestion in teleosts. *Adv. Mar. Biol. 13:* 109–239.

Karpouzi, V. S., and K. I. Stergiou. 2003. The relationship between mouth size and shape and body length for 18 species of marine fishes and their trophic implications. *J. Fish Biol. 62:* 1353–1365.

Keenleyside, M. H. A. 1979. *Diversity and adaptation in fish behavior*. Springer Verlag, Berlin.

Kelley, K. M., J. T. Haigwood, R. Flores, and G. S. Nicholson. 2000. Integrating metabolism and growth: A view from the diabetic goby. *Abstr. Proc. 4th Int. Symp. Fish Endocrinol.*, Seattle, WA.

Kramer, D. L., and M. J. Bryant. 1995a. Intestine length in the fishes of a tropical stream: 1. Ontogenetic allometry. *Env. Biol. Fishes 42:* 115–127.

———, and———. 1995b. Intestine length in the fishes of a tropical stream: 2. Relationships to diet—the long and short of a convoluted issue. *Env. Biol. Fishes 42:* 129–141.

Kruger, R. L., and R. W. Brocksen. 1978. Respiratory metabolism of striped bass, *Morone saxatilis* (Walbaum), in relation to temperature. *J. Exp. Mar. Biol. Ecol. 31:* 55–66.

Lagler, K. F. 1956. *Freshwater fishery biology* (2nd ed.). Wm. C. Brown, Dubuque, IA.

Lall, S. P. 1989. The minerals, pp. 219–257. In *Fish nutrition* (2nd ed.), J. E. Halver (Ed.). Academic Press, San Diego.

Lane, T. H., and H. M. Jackson. 1969. Voidance time for 23 species of fish. *Invest. Fish Control No. 33.*

Lauder, G. V. 1979. Feeding mechanics in primitive teleosts and in the halecomorph fish *Amia calva. J. Zool. (Lond.) 187:* 543–578.

———. 1980a. The role of the hyoid apparatus in the feeding mechanism of the coelacanth *Latimeria chalumnae. Copeia 1980:* 1–9.

———. 1980b. Evolution of the feeding mechanism in primitive actinopterygian fishes: A functional anatomical analysis of *Polypterus, Lepisosteus*, and *Amia. J. Morphol. 163:* 283–317.

———. 1982. Patterns of evolution in the feeding mechanism of actinopterygian fishes. *Amer. Zool. 22:* 275–285.

———. 1983. Food capture, pp. 280–311. In *Fish biomechanics*, P. W. Webb and D. Weihs (Eds.). Praeger, New York.

———, and K. F. Liem. 1980. The feeding mechanism and cephalic myology of *Salvelinus fontinalis:* Form, function, and evolutionary significance, pp. 365–390. In *Charrs, salmonid fishes of the genus Salvelinus*, E. K. Balon (Ed.). Junk, The Hague.

———, and———. 1983. The evolution and interrelationships of the actinopterygian fishes. *Bull. Mus. Comp. Zool. 150*(3): 95–197.

Lazzaro, X. 1987. A review of planktivorous fishes: Their evolution, feeding behaviors, selectivities, and impacts. *Hydrobiology 146:* 97–167.

Lepage, O., O. Tottmar, and S. Winberg. 2002. Elevated dietary intake of L-tryptophan counteracts the stress-induced elevation of plasma cortisol in rainbow trout (*Oncorhynchus mykiss*). *J. Exp. Biol. 205:* 3679–3687.

Liem, K. F. 1980a. Adaptive significance of intra- and interspecific differences in the feeding repertoires of cichlid fishes. *Amer. Zool. 20:* 295–314.

———. 1980b. Air ventilation in advanced teleosts: Biochemical and evolutionary aspects, pp. 57–91. In *Environmental physiology of fishes*, M. A. Ali (Ed.). Plenum Press, New York.

———. 1990. Aquatic versus terrestrial feeding modes: Possible impacts on the trophic ecology of vertebrates. *Amer. Zool. 30:* 209–221.

———. 1991. Functional morphology, pp. 129–150. In *Cichlid fishes—behaviour, ecology and evolution*, M. H. A. Keenleyside (Ed.). Chapman and Hall, London.

———. 1993. Ecomorphology of the teleostean skull, pp. 422–452. In *The skull*, Vol. 3, J. Hanken and B. K. Hall (Eds.). University of Chicago Press, Chicago.

Lovell, R. T. 1989. Diet and fish husbandry, pp. 549–604. In *Fish nutrition* (2nd ed.), J. E. Halver (Ed.). Academic Press, San Diego.

Luczkovich, J. J., S. F. Norton, and R. G. Gilmore, Jr. 1995. The influence of oral anatomy on prey selection during the ontogeny of two percoid fishes, *Lagodon rhomboides* and *Centropomus undecimalis. Env. Biol. Fishes 44:* 79–95.

Magnuson, J. J. 1969. Digestion and food consumption by skipjack tuna *Katsuwonus pelamis. Trans. Am. Fish. Soc. 98:* 379–392.

Marshall, N. B. 1966. *The life of fishes*. Universe Books, New York.

Martínez-Díaz, S. F., and H. Pérez-España. 1999. Feasible mechanisms for algal digestion in the king angelfish. *J. Fish Biol. 55:* 692–703.

Mayntz, D., D. Raubenheimer, M. Salomon, S. Toft, and S. J. Simpson. 2005. Nutrient-specific foraging in invertebrate predators. *Science 307:* 111–113.

McKaye, K. R., and T. Kocher. 1983. Head ramming behavior by three paedophagous cichlids in Lake Malawi, Africa. *Anim. Behav. 31(1):* 206–210.

———, and A. Marsh. 1983. Food switching by two specialized algae-scraping cichlid fishes in Lake Malawi, Africa. *Oecologia 56(2–3):* 245–248.

McVay, J. A., and H. W. Kaan. 1940. The digestive tract of *Carassius auratus. Biol. Bull. 78(1):* 53–67.

Mommsen, T. P. 1998. Growth and metabolism, pp. 65–97. In *The physiology of fishes* (2nd ed.), D. H. Evans (Ed.). CRC Press, Boca Raton, FL.

———. 2000. Growth and metabolism in fishes: Who's the bear and where's the bull? *Abstr. Proc. 4th Int. Symp. Fish Endocrinol.,* Seattle, WA.

Moriarty, D. J. W. 1973. The physiology of digestion of blue-green algae in the cichlid fish *Tilapia nilotica. J. Zool. (Lond.) 171:* 25–39.

Motta, P. J. 1982. Functional morphology of the head of the inertial suction feeding butterflyfish, *Chaetodon miliaris* (Perciformes, Chaetodontidae). *J. Morphol. 174:* 283–312.

———. 1984a. Tooth attachment, replacement, and growth in the butterflyfish, *Chaetodon miliaris* (Chaetodontidae, Perciformes). *Can. J. Zool. 62(2):* 183–189.

———. 1984b. Mechanics and functions of jaw protrusion in teleost fishes: A review. *Copeia 1984:* 1–18.

———. 1985. Functional morphology of the head of Hawaiian and mid-Pacific butterflyfishes (Perciformes, Chaetodontidae). *Env. Biol. Fishes 13:* 253–276.

———, and C. D. Wilga. 1999. Anatomy of the feeding apparatus of the nurse shark, *Ginglymostoma cirratum. J. Morphol. 241:* 33–60.

Nachi, A. M., F. J. Hernandez-Blazquez, R. L. Barbieri, R. G. Leite, S. Ferri, and M. T. Phan. 1998. Intestinal histology of a detritivorous (iliophagous) fish *Prochilodus scrofa* (Characiformes, Prochilodontidae). *Ann. Sci. Nat. 2:* 81–88.

National Research Council. 1981. *Nutrient requirements of coldwater fishes.* National Academy Press, Washington, DC.

———. 1983. *Nutrient requirements of warmwater fishes.* National Academy Press, Washington, DC.

Nelson, J. A., D. A. Wubah, M. E. Whitmer, E. A. Johnson, and D. J. Stewart. 1999. Wood-eating catfishes of the genus *Panaque:* Gut microflora and cellulolytic enzyme activities. *J. Fish Biol. 54:* 1069–1082.

Olsson, C., G. Aldman, A. Larsson, and S. Holmgren. 1999. Cholecystokinin affects gastric emptying and stomach motility in the rainbow trout *Oncorhynchus mykiss. J. Exp. Biol. 202:* 161–170.

Pike, C. S., III. 1990. Uncovering the ages of sharks and its importance in fisheries management, pp. 109–111. In Discovering sharks, S. H. Gruber (Ed.), *Am. Littoral Soc. Spec. Publ. 14,* Highlands, NJ.

Piper, R. G., I. B. McElwain, L. E. Orme, J. P. McCraren. L. G. Fowler, and J. R. Leonard. 1982. *Fish hatchery management.* U.S. Fish and Wildlife Service, Washington, DC.

Pough, F. H., C. M. Janis, and J. B. Heiser. 1999. *Vertebrate life* (5th ed.). Prentice Hall, Upper Saddle River, NJ.

Prosser, C. L., and E. J. DeVillez. 1991. Feeding and digestion, pp. 205–229. In *Environmental and metabolic animal physiology,* C. L. Prosser (Ed.). Wiley-Liss, New York.

Ramsay, J. D., and C. D. Wilga. 2004. Dual function shark teeth. *Abstr. Ann. Meeting Soc. Integ. Comp. Biol.,* New Orleans.

Rand, D. M., and G. V. Lauder. 1981. Prey capture in the chain pickerel, *Esox niger:* Correlations between feeding and locomotor behavior. *Can. J. Zool. 59:* 1072–1078.

Renade, S. S., and H. G. Kewalramani. 1967. Studies on the rate of food passage in fish intestines. *FAO Fish. Rep. 44:* 349–358.

Rimmer, D. W., and W. J. Wiebe. 1987. Fermentative microbial digestion in herbivorous fishes. *J. Fish Biol. 31:* 229–236.

Roberts, R. J., and A. M. Bullock. 1989. Nutritional pathology, pp. 423–473. In *Fish nutrition* (2nd ed.), J. E. Halver (Ed.). Academic Press, San Diego.

Rønnestad, I., C. R. Rojas-Garcia, and J. Skadal. 2000. Retrograde peristalsis: A possible mechanism for filling the pyloric caeca? *J. Fish Biol. 56:* 216–218.

Royce, W. F. 1972. *Introduction to the fishery sciences.* Academic Press, New York.

Sagua, V. O. 1979. Observations on the food and feeding habits of the African electric catfish *Malapterurus electricus* (Gmelin). *J. Fish Biol. 15:* 61–69.

Sanderson, S. L., A. Y. Cheer, J. S. Goodrich, J. D. Graziano, and W. T. Callan. 2001. Crossflow filtration in suspension-feeding fishes. *Nature 412:* 439–441.

Sanford, C. P. J., and P. C. Wainwright. 2003. Use of sonomicrometry demonstrates the link between prey capture kinematics and suction pressure in largemouth bass. *J. Exp. Biol. 205:* 3445–3457.

Sargent, J., R. J. Henderson, and D. H. Tocher. 1989. The lipids, pp. 153–218. In *Fish nutrition* (2nd ed.), J. E. Halver (Ed.). Academic Press, San Diego.

Schmitt, R. J., and S. J. Holbrook. 1984. Ontogeny of prey selection by black surfperch *Embiotoca jacksoni* (Pisces: Embiotocidae): The roles of fish morphology, foraging behavior, and patch selection. *Mar. Ecol. Prog. Ser. 18(3):* 225–239.

Silverstein, J. T., V. Bondareva, J. B. K. Leonard, and E. M. Plisetskaya. 2000. Neuropeptide regulation of feeding in fish. *Abstr. Proc. 4th Int. Symp. Fish Endocrinol.,* Seattle, WA.

Sims, D. W., P. L. R. Andrews, and J. Z. Young. 2000. Stomach rinsing in rays. *Nature 404:* 566.

Smith, L. S. 1989. Digestive functions in teleost fishes, pp. 332–423. In *Fish nutrition* (2nd ed.), J. E. Halver (Ed.). Academic Press, San Diego.

Smith, M. M., and Z. Johanson. 2003. Separate evolutionary origins of teeth from evidence in fossil jawed vertebrates. *Science 299:* 1235–1236.

Spangler, G. R., D. S. Robson, and H. A. Regier. 1980. Estimates of lamprey-induced mortality in whitefish, *Coregonus clupeiformis. Can. J. Fish. Aquat. Sci. 39:* 2146–2158.

Stein, D. L., and C. E. Bond. 1985. Observations on the morphology, ecology, and behaviour of *Bathylychnops exilis* Cohen. *J. Fish Biol. 27:* 215–228.

Stevens, C. E. 1988. *Comparative physiology of the vertebrate digestive system.* Cambridge University Press, London.

Swanson, B. L., and D. V. Swedberg. 1980. Decline and recovery of the Lake Superior Gull Island Reef lake trout (*Salvelinus namaycush*) and the role of sea lamprey (*Petromyzon marinus*) predation. *Can. J. Fish. Aquat. Sci. 39:* 2074–2080.

Trapani, J. 2001. Position of developing replacement teeth in teleosts. *Copeia 2001:* 35–51.

Van den Thillart, G., V. B. Henegouwen, and F. Kesbeke. 1983. Anaerobic metabolism of goldfish *Carassius auratus* (L.): Ethanol and CO_2 excretion rates and anoxia tolerance at 20, 10, and 5°C. *Comp. Biochem. Physiol. 76A:* 295–300.

Vinyard, G. L. 1982. Variable kinematics of Sacramento perch (*Archoplites interruptus*) capturing evasive and non-evasive prey. *Can. J. Fish. Aquat. Sci. 39:* 208–211.

Waiwood, K., and J. Majkowski. 1984. Food consumption and diet composition of cod, *Gadus morhua*, inhabiting the southwestern Gulf of St. Lawrence. *Env. Biol. Fishes 11:* 63–78.

Walker, J. A. 1997. Ecological morphology of lacustrine threespine stickleback *Gasterosteus aculeatus* L. (Gasterosteidae) body shape. *Biol. J. Linn. Soc. 61:* 3–50.

Weisel, C. F. 1962. Comparative study of the digestive tract of a sucker, *Catostomus catostomus*, and a predaceous minnow, *Ptychocheilus oregonense*. *Am. Mid. Nat. 68:* 334–346.

Western, J. R. H. 1971. Feeding and digestion in two cottid fishes, the freshwater *Cottus gobio* and the marine *Enophrys bubalis*. *J. Fish Biol. 3:* 225–246.

Wetherbee, B. 1990. Feeding biology of sharks, pp. 74–76. In Discovering sharks, S. H. Gruber (Ed.), *Am. Littoral Soc. Spec. Publ. 14,* Highlands, NJ.

Wheeler, A. 1975. *Fishes of the world: An illustrated dictionary.* Macmillan, New York.

Wilga, C. D., P. C. Wainwright, and P. J. Motta. 2000. Evolution of jaw depression mechanisms in aquatic vertebrates: Insights from Chondrichthyes. *Biol. J. Linn. Soc. 71:* 165–185.

Wilhelm, W. 1980. The disputed feeding behavior of a paedophagous haplochromine cichlid (Pisces) observed and discussed. *Behavior 74:* 310–323.

Wilson, R. P. 1989. Amino acids and proteins, pp. 111–151. In *Fish nutrition* (2nd ed.), J. E. Halver (Ed.). Academic Press, San Diego.

Winberg, S., Ø., Øverli, and O. Lepage. 2001. Suppression of aggression in rainbow trout (*Oncorhynchus mykiss*) by dietary L-tryptophan. *J. Exp. Biol. 204:* 3867–3876.

Winemiller, K. O., L. C. Kelso-Winemiller, and A. L. Brenkert. 1995. Ecomorphological diversification and convergence in fluvial cichlid fishes. *Env. Biol. Fishes 44:* 235–261.

Wood, B. M., and M. B. Bain. 1995. Morphology and microhabitat use in stream fish. *Can. J. Fish. Aquat. Sci. 52:* 1487–1498.

Yoshiyama, R. M., W. D. Wallace, J. L. Burns, A. L. Knowlton, and J. R. Welter. 1996. Laboratory food choice by the mosshead sculpin, *Clinocottus globiceps* (Girard) (Teleostei: Cottidae), a predator of sea anemones. *J. Exp. Mar. Biol. Ecol. 204:* 23–42.

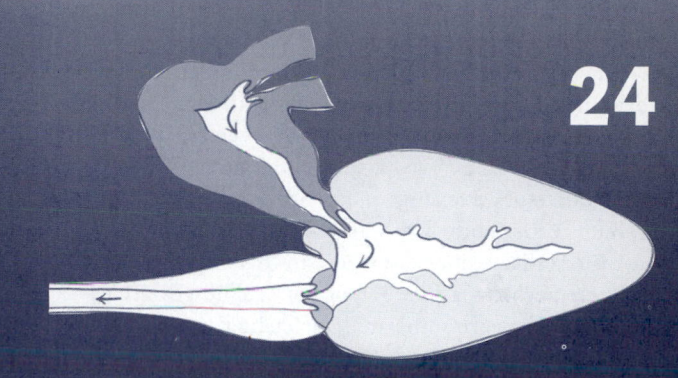

24

CIRCULATION
AND GAS EXCHANGE

ARCHITECTURE OF THE GILLS

Agnathan Fishes
Gnathostome Fishes
  Cartilaginous Fishes
  Bony Fishes

GILLS AS GAS EXCHANGE SURFACES

The Branchial Sieve
  Gill Filaments and Lamellae
Branchial Irrigation

EXTRABRANCHIAL AND AERIAL OXYGEN UPTAKE

Cutaneous Respiration
Alternatives to Gills: Air Breathing

CIRCULATION

Vascular Components
  The Heart
  Branchial Arteries
  Blood Flow in Gills
  Postbranchial Circulation
Secondary Circulation
Muscle and Choroid Retes

THE BLOOD OF FISHES

Blood Compounds Dissolved in Plasma
  Antifreeze Proteins
Blood Cells
  Bohr and Root Effects
  Blood Cells and Immune Responses

The term *respiration* is derived from the Latin word meaning "to breathe" or "to exhale." In this sense, the word is most appropriate in discussing the topic at hand—the exchange of gases between the organism and its environment. The term is also used by cell biologists to define the biochemical processes that result in the generation of energy necessary to drive cellular processes. These two forms of respiration are, of course, inextricably tied together. In complex organisms such as fishes, the first definition of respiration refers to the acquisition of oxygen and the release of carbon dioxide by specialized tissues. This gas exchange is a necessary prerequisite for the energy-yielding biochemical reactions that proceed at the cellular level.

Gas exchange is a function of the available surface area, measured as cm^2; metabolic activity is a function of the total mass or volume of an organism, measured as cm^3. Very small creatures (e.g., bacteria and protists) do not need specialized gas exchange surfaces, as their requirements for gas exchange are minimal and can be met across their undifferentiated surfaces. As organisms get larger, multicellular, and generally more complex, their metabolic demands increase by a power of 3 (reflecting their mass or volume), but their available surface area for gas exchange only increases by a power of 2. In effect, the surface to volume ratio decreases as organisms get larger, and we soon approach a limit where undifferentiated surfaces will not suffice for gas exchange. The solution to this dilemma takes the form of specialized gas exchange surfaces. For aquatic creatures, these surfaces take the form of gills, whereas terrestrial creatures, exploiting the much greater abundance of oxygen in the atmosphere, have acquired lungs—from certain ancestral fishes, of course. Fishes are an especially interesting case study among the vertebrates. Here, gills that develop from pharyngeal perforations are the predominant gas exchange surface, but in the sarcopterygians, lungs have evolved to enable access to atmospheric gases.

As animals got larger and more complex, the sites for uptake of oxygen and release of carbon dioxide became further removed from the tissue sites where the oxygen was delivered and the carbon dioxide taken away. Animals evolved circulatory systems as a means to optimize this uptake and delivery process, although the functions of the circulatory system include far more than gas exchange. In the pharyngeal gill system, we see an intimate association of epithelium specifically modified to accommodate gas exchange and circulatory tissue specifically modified for the rapid movement of oxygen into and carbon dioxide out of the body. We have combined the discussions of gas exchange and circulation here in this chapter, because it is not possible to consider the structure and function of one without the other.

ARCHITECTURE OF THE GILLS

Fishes live in a medium in which the availability of oxygen is limited because of its comparatively low solubility (Table 24.1). Unlike terrestrial organisms, which can avail themselves of an atmosphere that is rich and uniform in oxygen content, fishes must contend with an environment in which the solubility of oxygen varies with temperature and content of dissolved substances (Table 24.2). The hearts of fishes, consisting of multiple chambers, pump only deoxygenated blood to the gills. We have a wealth of information on the structure and function of the circulatory and respiratory systems of fishes that have evolved to meet the challenges of gas exchange and the distribution of dissolved substances to the tissues and organs. Special adaptations include the composition of the blood, the morphology of the circulatory apparatus, behavioral responses to oxygen levels,

TABLE 24.1   SOLUBILITIES OF THE MAIN ATMOSPHERIC GASES AT 15°C AND 1 ATM PRESSURE

Gas	Solubility (ml gas l^{-1} freshwater)
Nitrogen	16.9
Oxygen	34.1
Carbon dioxide	1,019.0

Source: Schmidt-Nielsen, 1997.

and the structure and function of the gills and other respiratory surfaces. Some of the more interesting adaptations involve the direct use of atmospheric oxygen. The lungfishes (Dipnoi), having evolved a special pulmonary circuit to accommodate gas exchange using the lungs, differ markedly from the other bony fishes in this respect.

From discussions elsewhere in this book, we know that the gills perform a number of roles. The gill arches bear structures that are important in feeding (see Chapter 23). Specialized cells in the gills are important in osmoregulation, and the gills serve as the main route for the excretion of nitrogenous wastes (see Chapter 25). In this chapter, we will focus on the role of the gills as gas exchange devices.

Agnathan Fishes

Agnaths are faced with unique constraints, in that feeding and respiration must be able to be carried out simultaneously without the benefit of jaws (Rovainen, 1996). Current theories on the evolution of jaws (see Chapter 6) propose that jaws originally evolved to facilitate gill irrigation and were later adapted for purposes of feeding.

The gill tissue of lampreys and hagfishes is arranged as a series of radiating ridges inside expanded pouches that are internal to the branchial skeleton. This skeleton consists of an elaborate, basketlike arrangement that has considerable elasticity. The pouches are flattened anteroposteriorly and are somewhat separated from each other. Depending on the species, hagfishes may have 5 to 15 pairs of gill pouches that open internally into an elongate pharynx. External openings are separate in some genera (Fig. 24.1A), but in others, some or all of the excurrent tubes from the gills may converge at a collecting tube, which conveys the excurrent respiratory water to an external pore on each side. Hagfishes usually have the respiratory apparatus set well back behind the head. They are unique in possessing a **pharyngocutaneous duct** that connects the pharynx with the exterior, opening on the left side behind the last branchial opening or, in *Myxine*, into the single gill opening.

Lampreys have seven gill pouches per side, each opening separately to the exterior via a short tube (Fig. 24.1B). Internally, they open into a special respiratory tube or "pharynx" beneath the esophagus. In larval lampreys (ammocoetes), the pharynx is continuous anteriorly with the oral cavity and posteriorly with the esophagus. It is similar to the pharynx of hagfishes, but during the metamorphosis of the ammocoetes, the pharynx disconnects from the esophagus posteriorly, with an ensuing separation of the two tubes. Anteriorly, the entrance to the respiratory tube is guarded by a **valvular velum**. This special respiratory tube is unique among vertebrates.

TABLE 24.2   EFFECT OF TEMPERATURE AND SALINITY ON SOLUBILITY OF OXYGEN IN WATER

Temperature ($°C$)	Solubility in freshwater (ml $O_2\, l^{-1}$)	Solubility in (35‰) seawater (ml $O_2\, l^{-1}$)
0	10.29	7.97
10	8.02	6.35
15	7.22	5.79
20	6.57	5.31
30	5.57	4.46

Source: Schmidt-Nielsen, 1997.

Gill irrigation in the cyclostomes (lampreys and hagfishes) is accomplished by contracting muscles around the branchial area forcing water out of the several gill pouches. The elastic recoil of the cartilaginous branchial basket aids in filling the pouches. Water can enter and leave the individual pouches through the separate external openings of lampreys and of those hagfishes that possess separate openings. It can also enter through the mouth or through the pharyngocutaneous duct, which opens just behind the last gill pouch of hagfishes. Gill irrigation in hagfishes is accomplished in part by a velar pump in the velar chamber, just posterior to the mouth. This scroll-like structure creates a current by alternately rolling tightly and unrolling (Hardisty, 1979; Johansen and Strahan, 1963; Malte and Lomholt, 1998). Despite the fact that the skin is well vascularized in hagfishes, the cutaneous uptake of oxygen is not evident. Rather, the dense capillary beds found in the skin appear to facilitate oxygen delivery to the skin while the animal is burrowing in frequently anoxic sediments (Malte and Lomholt, 1998).

Gnathostome Fishes

In the jawed fishes, the gill tissue occurs in the form of filaments or ridges on interbranchial septa that are borne on the gill arches (Fig. 24.1C and D). Both the septa and the gills are external to the branchial skeleton, in contrast with the agnaths. In gnathostomes, the branchial apparatus is concentrated into a smaller proportion of the body than in the lampreys and hagfishes.

Cartilaginous Fishes

The interbranchial septa of sharks and rays extend to the body wall, so that each branchial chamber is entirely separated from the others and each has its own opening (*gill slit*) to the exterior (Fig. 24.1C). Sharks typically have five openings, but six or seven openings are seen in some groups. Cartilaginous branchial rays extend from the gill arch outward within the septum. Each arch and septum bears a series of gill lamellae on both the anterior and posterior surfaces. The gill lamellae on one side of a septum constitute a hemibranch or "half-gill." The two hemibranchs on a gill arch are called a holobranch. In most sharks, the posterior hemibranch of the hyoid arch is present in the first gill pocket, so that an odd number of hemibranchs occurs on each side—9, 11, or 13, depending on whether the fish has five, six, or seven gill pouches. A remnant of the mandibular gill, called the mandibular or spiracular pseudobranch, is associated with the spiracle, which is anterior to the functional gills in most species and represents a remnant of the pharyngeal opening between the mandibular and hyoid arches (see Chapter 3; Goodrich, 1930; Laurent and Dunel-Erb, 1984).

Chimaeras (Chimaeriformes) have four branchial arches and four gill pouches covered by a fleshy operculum. Thus, the gill septa do not extend to the body wall and are only slightly longer than the gill filaments. Adult chimaeras have no spiracle, and the pseudobranch is absent. There are even numbers of hemibranchs on each side: A hyoid hemibranch is followed posteriorly by holobranchs on the first, second, and third branchial arches, and an anterior hemibranch on the fourth branchial arch. In general, the branchial apparatus of chimaeras is not as long as that of the sharks.

Bony Fishes

In bony fishes, the gill septa are progressively reduced (Fig. 24.1E, F, and G). Some of the more basal groups, such as sturgeons (Acipenseridae) and gars (Lepisosteidae), have slightly reduced septa, with the tips of the gill lamellae extending beyond as free filaments (Fig. 24.1F). In teleosts, the septa become greatly reduced to no more than small ridges along each gill arch, from which the gill tissue extends as long filaments (Fig. 24.1G). The gill apparatus is thus more compact than in the sharks, rays, and chimaeras. The gill arches are closely apposed, and the entire chamber is covered on each side by the bony operculum. The elimination of the septa results in greater respiratory efficiency because the flow of water through the secondary lamellae is facilitated.

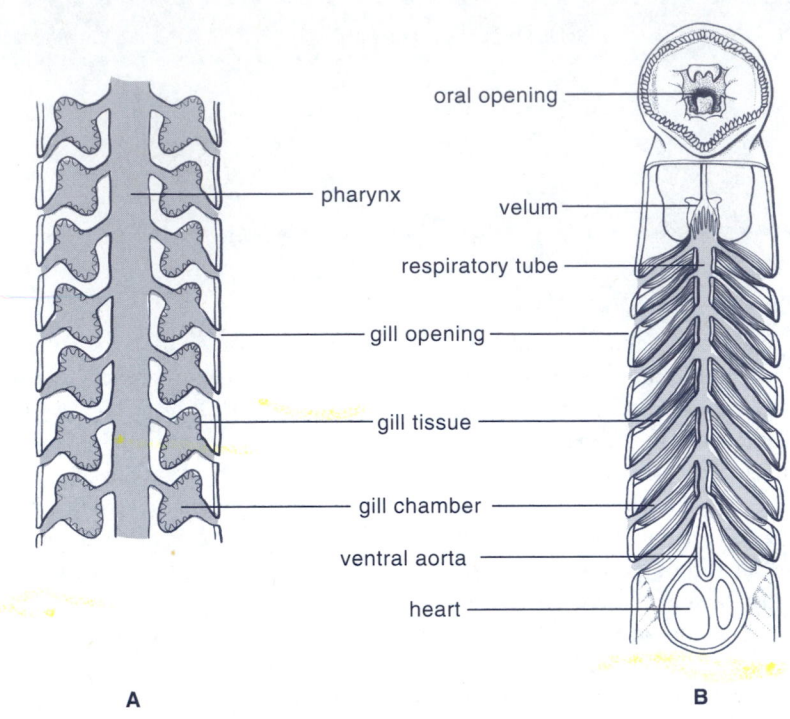

A

B

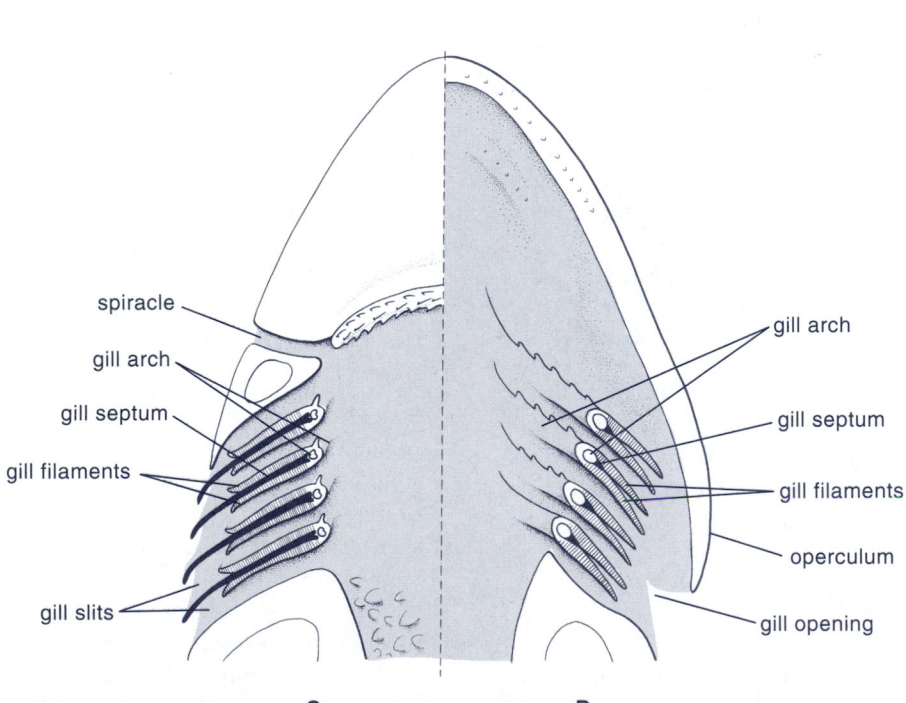

C   D

FIGURE 24.1

Arrangement of gills in frontal section: **A,** Hagfish (*Eptatretus*); **B,** Lamprey (*Lampetra*); **C,** Shark; **D,** Bony fish.

(*Continued*)

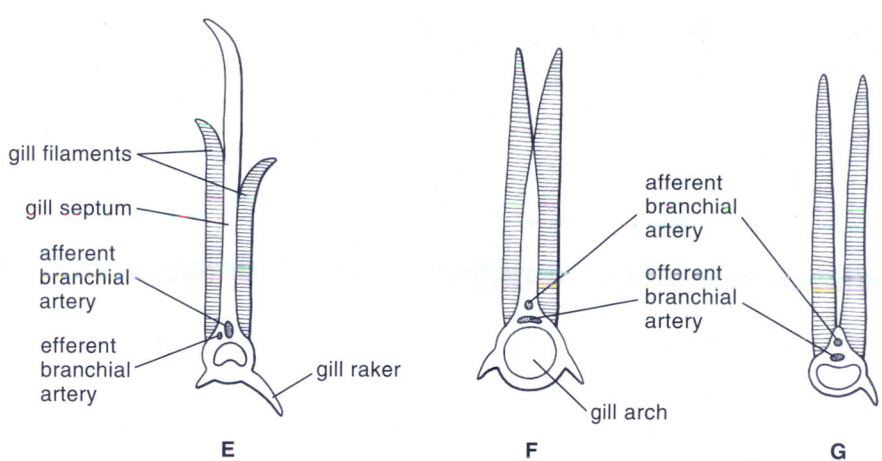

FIGURE 24.1 (*Continued*)
Relationship of branchial septum and gill tissue: **E**, Shark; **F**, Sturgeon; **G**, Teleost (*C* based on Weichert, 1965).

Bony fishes typically retain a pseudobranch at the site of the hyoid arch and holobranchs on each of the four branchial arches. In basal forms, such as sturgeons, gars, and coelacanths (*Latimeria*), a hyoidean hemibranch as well as a pseudobranch is retained. The deep-sea eel *Eurypharynx* is the only known teleost with five complete gill arches. Reduction of the gills has occurred in some air-breathing teleosts and others. Some members of the family Synbranchidae (swamp eels) have only one well-developed holobranch.

The pseudobranch is evidently the remnant of the primitive gill of the mandibular (or first visceral) arch (Kent, 1992; Romer, 1970). The mandibular pouch and slit may be retained as the spiracle in many sharks, all rays, and some primitive bony fishes, such as sturgeons, paddlefishes, and bichirs. For bottom-dwelling fishes, in which the mouth may be buried in the sediments, the spiracle may function as a point of entry for the respiratory current (see Chapter 3). Although sturgeons have a spiracle, most of the water actually enters the buccal cavity, to be passed across the gills from an aperture on the dorsal surface of the operculum (Kardong, 1998). In tetrapods, the spiracle evolved into the middle ear cavity and incorporates the hyomandibular bone as one of the bones (the *columella* or *stapes*) of the middle ear. The pseudobranch is associated with the spiracle in the sharks, rays, and sturgeons, but in the gars, the hyoidean hemibranch and the pseudobranch are adjacent on the inner side of the opercular base at the anterior portion of the branchial chamber. In the teleosts, that is the usual location of the pseudobranch (Fig. 24.2).

Although the pseudobranch closely resembles a functional gill in some species, it receives only oxygenated blood and therefore cannot function in respiratory gas exchange, but it may serve that function in the leptocephalus larvae of eels (Laurent and Dunel-Erb, 1984). Moreover, in most bony fishes, the pseudobranch is reduced to the appearance of a glandular organ that may be situated beneath the skin. The pseudobranch may be involved in the function of the choroid rete, which secretes oxygen into the retina of the eye (Beatty, 1975; Pelster and Randall, 1998). Blood flows directly from the pseudobranch, which is rich in carbonic anhydrase, to the ophthalmic artery. Although the contribution of the pseudobranch to the concentration of oxygen by the choroid is not known, the removal of the pseudobranch lowers the concentration of oxygen in the eye (Satchell, 1991). In some species, it has been shown that the secretion of gas into the gas bladder may be facilitated by the secretions of the pseudobranch (Wittenberg and Haedrich, 1974). According to Laurent and Dunel-Erb (1984), the functions of the pseudobranch remain to be ascertained, but may include a role in pressure detection.

Among the lungfishes, the greatest modification of the gill apparatus is seen in the African lungfishes (Protopteridae). These fishes retain a hyoidean hemibranch, have no gills on the first or second branchial arches, and have holobranchs on the third and fourth branchial arches and an anterior hemibranch on the fifth. Juveniles of the African lungfish develop external gills in association with the last three branchial arches—a feature that originally led them to

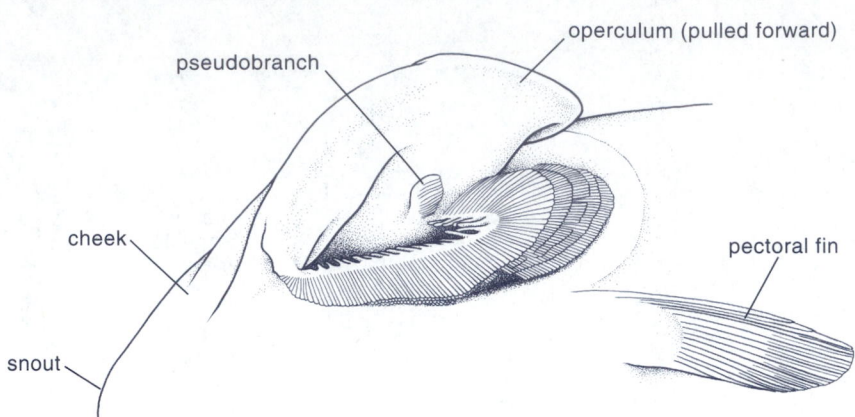

FIGURE 24.2
Position of pseudobranch in a trout (Salmonidae).

be mistaken for amphibians. The Australian lungfish (*Neoceratodus*) has a hyoid hemibranch and four holobranchs. None of the lungfishes retains a pseudobranch or spiracle. Males of the South American lungfish (*Lepidosiren paradoxa*) develop heavily vascularized filaments on the pelvic fins during the breeding season that seem to function like gills in reverse, as they apparently release oxygen to the developing embryos in the nest.

GILLS AS GAS EXCHANGE SURFACES

The Branchial Sieve

The effective exchange of gases in a gill-breathing fish depends on bringing the blood and respiratory water into close apposition on either side of a membrane through which the gases can diffuse. The uptake of oxygen by the blood is maximized if the blood and the water flow in opposition to each other—a system known as countercurrent exchange. In fishes, countercurrent exchange is realized by pumping water across a finely subdivided set of surfaces on the gills, the *branchial sieve*, while blood is pumped through internal spaces in the gills in the opposite direction. Keeping in mind that the movement of respiratory gases is always passive, in response to concentration gradients, moving the flow of respiratory water in the direction opposite to the flow of blood always ensures that a higher concentration of dissolved oxygen exists in the respiratory water than in the blood at the point of diffusion.

Gill Filaments and Lamellae

Each gill arch bears a number of filaments, so that the entire respiratory surface of an individual may consist of several hundred filaments—the actual number varies with factors such as size and general habits of the species. Active fishes generally have more filaments than sessile species (Hughes, 1984). About 300 to 500 filaments may be found on the gills of small, bottom-dwelling darters (Percidae; Branson and Ulrikson, 1967), whereas a perch of the same family weighing 30 g has about 1,500. A mackerel (Scombridae) weighing 800 g has about 2,400 filaments—typical of highly active families of teleosts. Each gill filament bears numerous secondary lamellae at right angles to the long axis of the filament. Scanning electron microscopy has been instrumental in investigations of the ultrastructure of gill lamellae (Olson, 1996b). The lamellae, distributed on both the upper and lower surfaces, are delicate ridges with thin walls. These walls constitute the barrier between the blood and the surrounding water and have three layers of cells: a relatively thick epithelial layer, a basement membrane, and pillar cells (Satchell, 1991). The thickness of these walls (the respiratory membrane through which gases must diffuse, or the water–blood barrier) varies with the mode of life of the different species. Tunas and their close relatives have very thin lamellar walls of about 0.53 to 1.0 μm. Most bony fishes have walls of 2 to 4 μm thickness, and some demersal species have respiratory membranes 5 to 6 μm thick. Most elasmobranchs for which data are available have lamellar thicknesses of 5 to 11 μm (Hughes, 1984).

The total number of lamellae in marine fish species studied by Hughes (1966) ranged from 52,000 to 689,000.

The total number reported depended on the number of filaments and the count of lamellae per unit length of filament and on the size of the fish. Slow-moving fishes usually have from 10 to 20 lamellae per mm, whereas active fishes have 30 to 40 per mm. Most bony fishes are in the 15 to 30 per mm range. Some air-breathing species have fewer lamellae, with a water–blood barrier 10 µm thick in *Anabas testudineus*, whereas the air–blood barrier of the aerial respiratory apparatus is 0.21 µm. In the mudskippers (*Periophthalmus*), the water–blood barrier of the gill lamellae is thin because the gills, moistened by water held in the branchial chamber, are the site of gas exchange (Hughes, 1984). Among the mudskippers, more effective mechanisms for obtaining oxygen while out of the water have precluded the need for highly developed gill surfaces (Graham, 1997).

A large surface area for the exchange of respiratory gases is generated by numerous filaments bearing small but numerous lamellae. As with other respiratory features, active and slower-moving fishes differ with respect to the surface area of the gill. Some examples of gill area estimates are given in Table 24.3. When the published gill surface area measurements were converted to values expected for a fish weighing 200 g (de Jager and Dekkers,1975), it was revealed that scombrids generally have gill areas greater than 1,000 mm²/g, with various tunas having 1,500 to 3,500 mm²/g. Other active pelagic fishes have gill surface areas of 500 to 1,000 mm²/g, and most bony fishes for which measurements were available were found to be in the range of 150 to 350 mm²/g.

The effectiveness of the gill as a surface for the exchange of respiratory gases depends on the contact made with the water being pumped through the system. In each hemibranch, the lamellae of adjacent filaments are juxtaposed to form tiny channels through which water is forced (Fig. 24.3). The filaments are equipped with muscles that hold the tips of the filaments of the posterior hemibranchs of each arch against the tips of the filaments of the anterior hemibranch on the following arch, so that all water must pass through the lamellar channels (Fig. 24.4). Dramatic differences can be observed in a comparison of lamellar architecture and flow dynamics in the moderately active largemouth bass with that in the highly active skipjack tuna (Table 24.4; Langille et al., 1983).

As the volume in the oral and branchial cavities changes with respiratory movements, there is a compensatory change in the space occupied by the gill mass due to the action of the gill arch and gill filament musculature. The tiny abductor and adductor muscles that span the gill filaments control their position—pulling the filaments of a gill arch closer together allows more water to bypass the gas exchange surfaces. The tips of the filaments from adjacent arches remain in contact most of the time and are separated briefly, at least in some species, during part of each opercular cycle. The tips are also separated during coughing, and during bypassing of excessive water flow.

In agnaths and elasmobranchs, the filaments are bound to the gill septa, but the secondary lamellae stop short of the septa (Fig. 24.5), so passages or canals are formed next to them. Water passes through the interlamellar channels to these canals along the septa and then toward the gill slit. Although this arrangement allows a countercurrent flow of blood and water, it may not be as effective as the countercurrent flow regime seen in teleosts (Butler and Metcalfe, 1988).

Internally, the secondary lamellae of the gill filaments are divided into numerous, capillary-sized channels by pillar or pilaster cells that are disposed in more or less regular rows. A marginal channel rims each lamella (Fig. 24.6). These small spaces receive blood from capillaries branching from the afferent arterioles of the filaments and pass the blood, counter to the flow of water outside the lamellae, to the efferent arterioles. There is a central sinus in the filament, through

TABLE 24.3   GILL SURFACE AREA EXPRESSED IN ARBITRARY UNITS PER GRAM OF BODY WEIGHT IN SELECTED SPECIES OF MARINE TELEOSTS

Common name	Family	Activity level	Relative gill surface area
Goosefish	Lophiidae	low	51
Toadfish	Batrachoididae	low	137
Summer flounder	Paralichthyidae	low	268
Puffer	Tetraodontidae	moderate	505
Eel	Anguillidae	moderate	902
Sea trout	Sciaenidae	moderate	1,253
Menhaden	Clupeidae	high	1,685
Butterfish	Stromateidae	high	1,725
Mackerel	Scombridae	high	2,551

Source: Figure 1.6 in Schmidt-Nielsen, 1997, based on Gray, 1954.

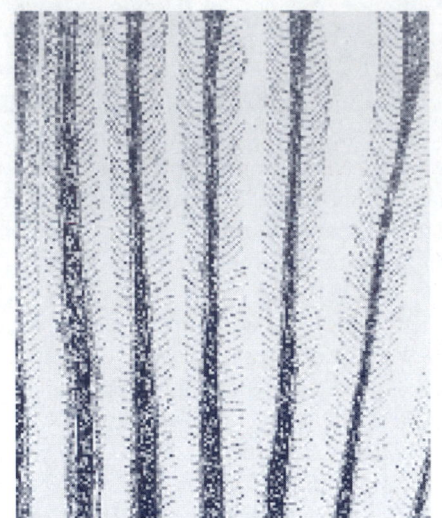

A

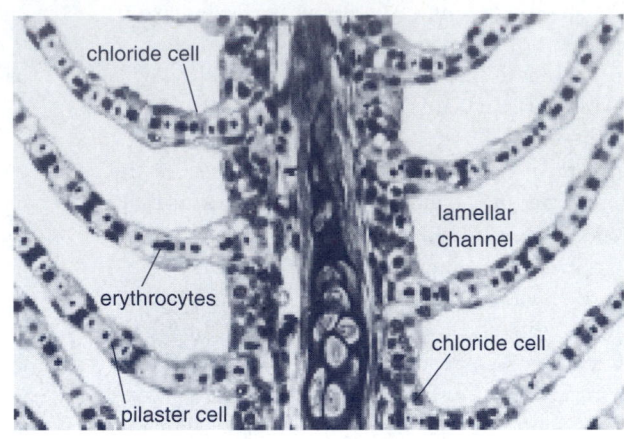

chloride cell

lamellar channel

erythrocytes

chloride cell

pilaster cell

B

FIGURE 24.3

Photomicrographs of sections along gill filaments of rainbow trout, showing **A,** Arrangement of secondary lamella to form numerous channels with great surface area; **B,** Pilaster cells, chloride cells, and erythrocytes, (Photographs courtesy of Professor Joseph H. Wales.)

mouth cavity

gill area

opercular cavity

operculum

gill opening

esophagus

ventral view of head
(partially sectioned)

blood vessels

gill arches

water flow

gill filaments

FIGURE 24.4

Frontal section through gill area of teleost, showing tips of filaments from adjacent arches held together so water flow must cross filaments through lamellar channels. Lamellae, which are at right angles to the axes of the filaments, are not shown.

TABLE 24.4   NUMBER OF LAMELLAR CHANNELS (PORES) AND FLOW VELOCITIES MEASURED IN LARGEMOUTH BASS
AND SKIPJACK TUNA

Species	Body mass (g)	Number of pores	Velocity of flow at rest (cm³ sec⁻¹)	Velocity of flow when active (cm³ sec⁻¹)
Largemouth bass (*Micropterus salmoides*)	837	302,700	2.15	17.6
Skipjack tuna (*Katsuwonus pelamis*)	1,667	7,186,000	24.6	143

Source: Langille et al., 1983.

which, some investigators believe, the blood can be bypassed without passing through the lamellae. Some investigators have claimed that blood is shunted through the central cavity or around the tips of the filaments under conditions of abundant dissolved oxygen, but others believe that no shunting occurs and that the only control of blood flow through the filament is by the muscles of the arterioles. The shunting of blood around the gill tissue is necessary in air-breathing fishes while they are in oxygen-poor water, or else they would lose oxygen from the blood to the water. In these cases, only enough blood to take care of ammonia excretion and ion exchange must go through the relatively thick-walled lamellae (Bone and Marshall, 1982).

Branchial Irrigation

The respiratory pump that forces the water across the gills consists of the buccopharyngeal cavity, plus all the mechanisms for opening, enlarging, and constricting it, and the parabranchial cavity (between the gills and the operculum), which can be enlarged or constricted by the action of the operculum and the branchiostegals (Saunders, 1961). The coordinated action of these two cavities produces a continuous flow of water across the gills. As the mouth is opened and water is sucked in by the enlarging of the buccal cavity, the parabranchial cavity is rapidly enlarged, but with its opercular opening sealed. This causes a negative pressure that draws water across the gills. When the mouth is closed, the

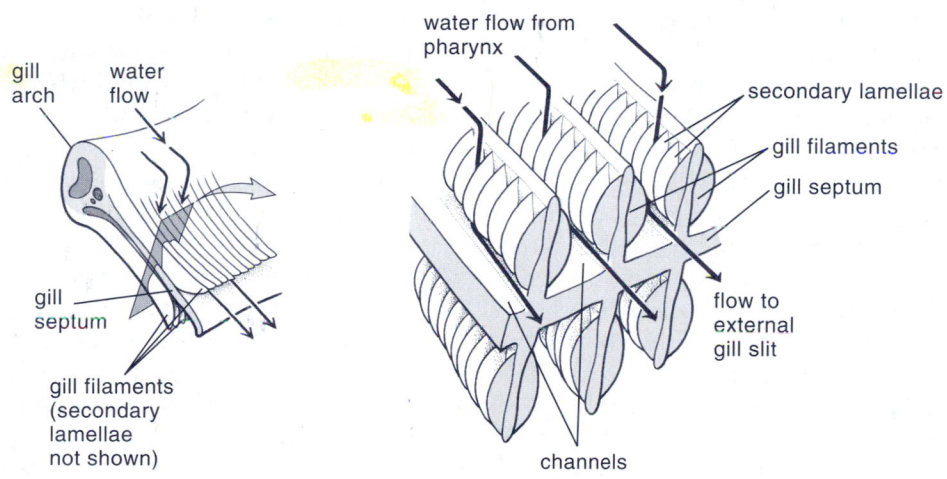

FIGURE 24.5

Illustration of gill filaments and secondary lamellae of dogfish (*Squalus acanthias*), showing path of respiratory water between lamellae, then through channels next to gill septum.

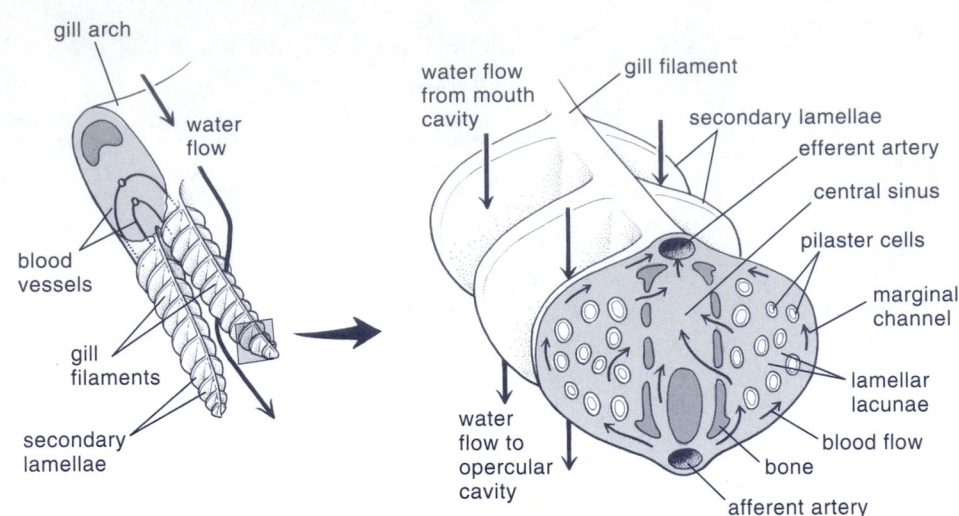

FIGURE 24.6
Illustration of section through gill filament of a bony fish at level of secondary lamella. Arrows depict flow of blood through spaces (lacunae) among pilaster cells.

oral valves prevent the escape of the water past the lips, and the water is forced across the gills, through the parabranchial cavity (which is now constricting), and exits via the opercular opening.

In hagfishes, respiratory water is pumped to the gills by a velar pump, situated in a chamber posterior to the buccal cavity. Water enters the chamber through the nostril and is forced toward the gill pouches by the action of paired, scroll-like structures that unroll into the velar chamber from the top and then force the water out of the chamber as they rapidly roll up again, with the edge of the roll in contact with the lateral walls of the chamber. This action also sucks water into the chamber from the nostril (Hardisty, 1979).

Adult lampreys apparently can draw a respiratory current through the mouth when free swimming, but they usually depend on a tidal action to move water in and out through the external branchial apertures. This mode is necessary when attached to prey or other objects. The branchial region can be constricted to force water out of the gills, and the elasticity of the branchial skeleton then greatly aids in expanding the gill chambers to draw water in. A valvular arrangement carries the incoming water to the medial side of the gill pouch, so that it can pass through the gill lamellae counter to the flow of blood. Larval lampreys use a velar pump in moving respiratory water, in addition to contractions of the branchial basket.

Baglioni (1908) was the first to propose a classification of the branchial pumps of teleosts that recognized the relationships between the habits of fishes and their respiratory apparatus. Most teleosts conform to this classification, in which a more sessile lifestyle is correlated with the increased use of the branchiostegal apparatus. Pelagic species pump respiratory water mainly through opercular movements. Some swift pelagic species irrigate the gills by holding the mouth open as they swim—a technique sometimes referred to as **ram ventilation,** but more appropriately termed **ram irrigation,** as it involves the movement of water across a respiratory surface. Fusion of the gill filaments and, in some cases, the lamellae is associated with the acquisition of ram irrigation in swift-moving fishes. Although sharks and scombroids are similar in the evolution of many features promoting an active pelagic lifestyle (see Chapter 19, Going Deeper) this does not extend to the aforementioned modification of the gill filaments. Comparisons of swift-moving lamnid sharks with less active species revealed the inherently conservative nature of the gill structure of chondrichthyans compared with teleosts (Benz, 1984).

Less active species that spend considerable amounts of time resting on the bottom tend to show greater development of the branchiostegal component of the branchial pump and combine opercular and branchiostegal movements in gill irrigation (Gosline, 1971; McAllister, 1968; see Chapter 9). Benthic species depend greatly on the branchiostegal apparatus and may have the opercular elements reduced.

Some swift species, such as tunas and billfishes, that depend on ram irrigation have nearly lost the ability to irrigate the gills by pumping. The branchial apparatus is highly modified in some species—the tips of the filaments of adjacent arches may be fused, or tissue connections may be present between branchial arches. Ram irrigation usually takes place at speeds of more than 1 km h^{-1} and has been calculated to take less than 1 percent of the total energy expended by a swimming skipjack tuna. Several marine fishes, including some benthic forms, can ram irrigate if their speeds reach 1.5 to 2 km h^{-1} (Roberts, 1975). Some stream fishes hold the mouth and opercula open to irrigate the gills passively while maintaining position in swift water. Some stream fishes, such as algae eaters (*Gyrinocheilus*), cling to the substrate while allowing the swift current to force water over the gills.

The volume of water pumped over the gills (*respiratory volume*) varies with factors such as morphology, size, temperature, carbon dioxide content of the water, oxygen content, and activity. The number and amplitude of respiratory movements increase as the level of activity increases or the dissolved oxygen content of the water decreases, so that the respiratory volume can be optimized. Relationships of breathing rate, amplitude of movement, and volume can be seen for the rainbow trout, in which the normal breathing rate is about 80 min^{-1}. With exercise, the rate was found to increase to about 100 min^{-1}, but the respiratory volume increased from 594 to 3,042 ml min^{-1} kg^{-1}. The efficiency of oxygen removal from the water is usually lower at high irrigation rates, because not all the water pumped through the gills comes in close enough contact with the lamellae for a sufficient time. However, the blood passing through the gills is usually 85 to 95 percent saturated with oxygen (Saunders, 1961, 1962; Shelton, 1970). Under favorable conditions, fishes can remove about 85 to 90 percent of the dissolved oxygen from water passing over the gills counter to the flow of blood. Such efficient removal rates usually occur when the amount of dissolved oxygen is high and the respiratory volume is low. More typically, the range is from 50 to 60 percent; and under conditions of low dissolved oxygen, high temperatures, and increased respiratory volume, the extraction rate may fall to 10 to 20 percent or even lower. Beyond the routine metabolic rate that may be determined for a given species, the actual metabolic rate depends on many factors, including size, temperature, pH, concentration of dissolved oxygen or carbon dioxide, and salinity.

Cutaneous Respiration

Most fishes are capable of absorbing some oxygen from the water through the skin, although in sharks and many bony fishes, this uptake might not be enough even to satisfy the local cutaneous oxygen requirement. Many species can obtain from 5 to 30 percent of their required oxygen through the skin (Feder and Burggren, 1985a). In most fishes, cutaneous respiration is of importance mainly during periods of low activity or of relatively low temperatures. Many amphibious fishes depend on the cutaneous absorption of oxygen for significant proportions of their respiratory requirements (Graham, 1997). Eels of the genus *Anguilla* obtain about 12 percent of their O_2 requirements through the skin; gobies of the genera *Boleophthalmus* and *Periophthalmus* cutaneously obtain 36 and 48 percent of their oxygen, respectively; and *Neochanna* (the New Zealand mudfish) obtains 43 percent (Feder and Burggren, 1985b). The reedfish (*Erpetoichthys*), though covered by heavy ganoid scales, obtains about 32 percent of its oxygen via the skin (Sacca and Burggren, 1982). Larval fish obtain oxygen via diffusion through the skin, and some, such as the larvae of *Monopterus albus*, have a countercurrent arrangement, in which the blood in the skin flows forward counter to the water propelled posteriorly by the pectoral fins (Bone et al., 1995). Some larvae have gill filaments that extend out of the gill opening, and special external gills are found in the larvae of polypterids and lepidosireniform lungfishes.

Alternatives to Gills: Air Breathing

The ability to extract oxygen from the atmosphere is an adaptation that evolved among the earliest bony fishes and occurs as a convergent specialization in many teleost groups. This adaptation is encountered in fishes that inhabit tropical swamps and beaches and in species such as the Alaskan blackfish (*Dallia pectoralis*) that can be found in freezing Arctic bogs. Overall, air breathing is more common among warm-water fishes than among temperate or cold-water types, and is most common among the tropical swamp dwellers.

Air breathing by means of lungs is an ancient characteristic of fishes, having probably originated in oxygen-poor environments during the Late Silurian and Early Devonian periods. Although lungs may have arisen in hypersaline Silurian seas (Packard, 1974), they were common among freshwater fishes of the Devonian swamps, and they persist today in lungfishes and other nonteleosts such as *Lepisosteus*, *Amia*, and *Polypterus*. Fishes may use a lung or gas bladder for the uptake of atmospheric oxygen, but a number of other structures also have been modified for that purpose.

Obviously, waters that are chronically deficient in dissolved oxygen can be inhabited by fishes only if those fishes can derive their oxygen from an alternative source. Throughout the tropics, swamps of high organic content and heavy vegetative cover support year-round populations of fishes, some of which are obligate air breathers. Swamps and streams that provide adequate dissolved oxygen supplies during part of the year, but that stagnate and even dry up at other times, may maintain a complement of specialized fishes that are facultative or obligate air breathers. These fishes can cope with the drying of the water either by burrowing and aestivating or by moving overland to more permanent bodies of water. Some mountain streams of the tropics support species that can withstand the torrents of the rainy season and can later adapt to the oxygen-poor pools of the dry season.

There are some fishes that expose themselves to air even though the surrounding water contains sufficient dissolved oxygen. Certain intertidally occurring blennioids and zoarcoids remain secreted beneath rocks or damp vegetation as the tide recedes and returns, living in a dewatered but damp habitat for 2 or 3 hours every tidal cycle (see Chapter 34). Here, they survive by maintaining metabolic rates in air comparable to or less than the rates measured while submerged (Bridges, 1988). Eels (*Anguilla*) expand the living space available to them by moving overland through wet vegetation to isolated ponds and lakes. Species of walking catfishes owe their broad geographical distribution in part to overland forays that place them in new bodies of water. The spread of the exotic walking catfish (*Clarias batrachus*) in Florida is an example. Several species of air breathers, including the blennoid *Dialommus*, swamp eels, and mudskippers, actively seek food while out of the water. The mudskippers are at home on sunny mudflats, where they move about freely, engaging in aggressive displays and other social behavior (Graham, 1997).

Adaptation to the aerial mode of respiration demands that air be brought into contact with highly vascularized tissue of considerable surface area. Usually, some cavity in the head or body is modified for the purpose of oxygen uptake. In some instances, an existing cavity has become modified; in others, a new cavity is formed; and in yet others, existing structures are modified to provide the requisite surface area. Furthermore, air-breathing species must be equipped to carry on the functions of osmoregulation, release of carbon dioxide, and the excretion of ammonia. The pumping of blood through very thin gill tissue to facilitate one or more of these functions could lead to a loss of oxygen from comparatively oxygen-rich blood to the oxygen-poor water. The

development of thicker gill lamellae, reduced branchial irrigation, and other modifications have aided in overcoming these problems. Fishes have apparently developed shunt pathways in the gills that allow the oxygenated blood to be routed through without coming into contact with hypoxic water (Burggren and Roberts, 1991; Randall, 1985; Steen and Kruysse, 1964).

Many bony fishes that commonly spend time out of the water absorb atmospheric oxygen through the skin. The eels (*Anguilla*), while in air, absorb 30 to 66 percent of their required oxygen via the skin (Berg and Steen, 1966; Feder and Burggren, 1985b). Examples of amphibious species that absorb much of their oxygen from the air through the skin include *Boleophthalmus* (43 percent); *Erpetoichthys* (41 percent); and *Periophthalmus* (76 percent; Feder and Burggren, 1985b). Other species listed by Johansen (1970) as obtaining significant amounts of oxygen from the air through the skin are the longjaw mudsucker (*Gillichthys mirabilis*) and the bluntnose knifefish (*Hypopomus brevirostris*), although the latter also has gills modified for air breathing. An eleotrid (*Dormitator latifrons*) has vascularized tissue on the top of the head that can absorb oxygen from the air when swimming just beneath the water surface.

Most air-breathing species rhythmically or periodically empty and fill a specialized cavity with air, so that the atmospheric oxygen can come into contact with vascularized tissue in that cavity (Liem, 1980). In the case of the electric eel (*Electrophorus electricus*), the cavity is the mouth and pharynx, where the lining is folded and otherwise modified to provide a large surface rich in blood vessels (Johansen, 1970). The mouth lining of some of the swamp eels, including *Synbranchus*, plays a significant part in respiration, although it is not much modified for an increase of absorptive surface area. The common carp brings a bubble of air into contact with a specialized part of the palate when in oxygen-poor water. A similar behavior has been observed in several species of gobies (Gee and Gee, 1991, 1995). The pharyngeal walls of mudskippers are reported to be a respiratory surface in air in addition to the gills and skin (Gordon et al., 1969; Teal and Carey, 1967).

In several species, including the synbranchoid *Amphipnous cuchia* and the snakeheads (Channidae), the walls of the pharynx are enlarged by diverticula (Hughes et al., 1974). The branchial chamber is the site of respiratory epithelium in many air breathers. Some, such as the walking catfishes (*Clarias*) and the labyrinth fishes of the suborder Anabantoidei, including the climbing perch (*Anabas*), the pikehead (*Luciocephalus pulcher*), and the three families commonly known as the gouramies, have parts of their gill arches modified into firm structures of large surface area bearing a respiratory epithelium. These respiratory organs are in an enlarged cavity above the gills and consist of a number of folded and crenellated plates. The corresponding organ of air-breathing catfishes develops from the second and fourth gill arches and is arborescent in nature, taking the name *gill tree*. An Asian catfish (*Saccobranchus fossilis*) has a pair of saclike diverticula leading from the branchial chamber into the lateral musculature. The vascular organization of the accessory respiratory organs of *Clarias* is very similar to that seen in the gill filaments (Olson et al., 1995).

One of the most remarkable adaptations for the use of atmospheric oxygen is the modification of parts of the alimentary canal for respiratory purposes. The swamp eel (*Monopterus albus*) has a modified esophagus that is used in aerial respiration (Liem, 1967). Some South American armored catfishes (e.g., *Plecostomus* and *Ancistrus*) use the stomach as a respiratory organ. Many loaches (Cobitidae) and several of the armored catfishes are intestinal breathers, using a large section of the gut exclusively as a "lung."

Although bony fishes apparently developed the lung and gas bladder early in their evolutionary history, no such feature is found among the cyclostomes or elasmobranchs. The lungs of lungfishes, like the lungs of tetrapods, connect with ducts from the ventral wall of the alimentary tract, whereas the gas bladders of physostomous actinopterygians have a dorsal pneumatic duct. The lung tissue of lungfishes somewhat resembles that of amphibians. It represents the most extensive development of a gastric diverticulum (lung or gas bladder) as a respiratory device. In the African and South American species, the lung is divided into right and left sections, whereas the Australian lungfish has an undivided lung. All have developed special pulmonary circulation, derived from the posterior (sixth) branchial arch (Burggren and Johansen, 1987).

Many physostomes use the gas bladder for aerial respiration (Liem, 1989). Some, such as the reedfish, obtain up to 40 percent of their oxygen through their lunglike gas bladder (Sacca and Burggren, 1982). The bowfin (*Amia calva*) has been shown to possess two distinct breathing modes—one in which inhalation is for purposes of respiration, and one for purposes of buoyancy regulation (Hedrick and Jones, 1993, 1999; Hedrick et al., 1994). Other fishes that breathe air include several osteoglossids, including the arapaima (*Arapaima gigas*), mudminnows (*Umbra*), the tarpon (*Megalops*), and the aimara (*Hoplerythrinus unitaeniatus*). The latter is a characoid of South American swamps; both it and the arapaima have gas bladders that have been secondarily modified to resemble a lung. The Clupeomorpha, a taxon noted for a gas bladder that has evolved a distinctive otophysic connection (see Chapter 9), is also noteworthy for the absence of species that have developed respiratory function involving the gas bladder (Liem, 1989).

Several air breathers have lost the capability to keep themselves supplied with oxygen from the water, even when in well-aerated situations, so that, if restrained from reaching the surface, they soon die. Such *obligate air breathers* include the South American and African lungfishes, the arapaima, the electric eel, the snakeheads, and the South American armored catfishes of the genus *Hoplosternum*. Magid and Babiker (1975) reported that large walking catfishes of the species *Clarias lazera* cannot survive even in oxygen-saturated water if prevented from breathing air. Many other species must rely on air breathing during periods of low dissolved oxygen or during excursions out of the water, but their gills can maintain their respiratory needs only if the dissolved oxygen content of the water is high enough (Johansen, 1970).

Those species that aestivate, spending the dry season buried in the mud, in burrows, or even in mucous cocoons, as in the case of the lungfishes, must be able to maintain themselves by breathing air over periods lasting for months. In addition to the African and South American lungfishes, fishes known to aestivate during summers or dry seasons include the bowfin (*Amia calva;* see Plate 8.5), the central mudminnow (*Umbra limi*), the walking catfishes (*Clarias*), the salamanderfish (*Lepidogalaxias salamandroides;* see Plate 11.3), the mudfishes of New Zealand (*Neochanna*), the swamp eels (*Synbranchus, Amphipnous*), the snakeheads (Channidae; see Plate 16.21), and the climbing perches (*Anabas*).

Most of the comparatively few marine fishes capable of aerial respiration are species that inhabit intertidal shores (Bridges, 1988; Martin and Bridges, 1999). The amphibious Chilean clingfish (*Sicyases sanguinus*), described in Chapter 34, is one of the best known examples. Tarpon, whose young are commonly found in brackish or freshwater, have a lunglike gas bladder. Although lacking air bladders or lungs, some sharks of the genus *Chiloscyllium* have been observed to gulp air.

CIRCULATION

The circulatory system can be conceived of as having two major components—the blood that serves as the means to transport dissolved gases, nutrients, and a host of other compounds essential for normal function, and the system of vessels that convey the blood to and from tissues and organs. These vessels consist of **arteries** that carry blood away from the heart, **veins** that carry blood back to the heart, and beds of **capillaries** that serve as exchange interfaces with the tissues. The walls of the capillaries have the thickness of a single cell, minimizing the diffusion distance between the blood and the tissues being supplied. In most fishes, the heart pumps deoxygenated blood to the gills. Here, the blood receives oxygen and is then conveyed to tissues, where the oxygen is released and carbon dioxide is picked up. Lungfishes show the development of an additional shunt, derived from the sixth aortic arch, in which a significant portion of the blood is diverted to a lung for oxygenation and then returned to the heart. Here, the heart is required to simultaneously pump oxygenated and deoxygenated blood. This system of paired circulatory shunts was an essential adaptation in the evolution of terrestrial vertebrates, although it apparently first evolved among fishes as an accommodation for existence in swampy waters deficient in dissolved oxygen.

Vertebrate circulation is also characterized by *portal systems*—venous systems that have capillary beds at either end. All vertebrates, save for hagfishes and mammals, have a **renal portal system,** in which blood collected from capillary beds in the tail flows through the renal portal vein to capillary beds in the kidneys. In fishes, blood from the renal portal system may enter the kidneys via segmental veins in the body wall, pelvic fins, or the posterior of the abdominal cavity (Kardong, 1998). Thus, for most vertebrates, the renal portal system is one of two vascular networks supplying the kidneys. The **hepatic portal system** drains the digestive tract and conveys blood to the liver, thus forming a direct route for the transport of absorbed nutrients to processing and storage sites in the liver. A third portal system is the tiny **hypophyseal portal system,** which connects the hypothalamus of the brain with the pituitary. This vascular network, which serves to convey neurosecretions from the brain to sites in the pituitary, is apparently absent in most fishes (see Chapter 26).

Vertebrates have also developed a **lymphatic system** as a derivative of the circulatory system. This system of blind-ended vessels primarily serves to recover blood fluids that leak out into tissues and return them to circulatory flow.

• • • • • • • • • • •
Vascular Components

The Heart
The vertebrate vascular system is characterized by a single main pump, the *heart,* although some fishes have evolved additional peripheral pumps to facilitate blood flow. Deoxygenated blood returning to a typical fish heart will first encounter a thin-walled **sinus venosus,** then an **atrium,** followed by a heavily muscularized **ventricle,** and, depending on the species, either a contractile **conus arteriosus** or an elastic **bulbus arteriosus** (see Fig. 24.7). The heart is contained within a subdivision of the body cavity, the **pericardial cavity,** which is located just ventral to the gills. Although the various groups of fishes have many features of the heart in common, there are distinguishing features that warrant considering the groups separately.

Although cardiac output is comparable to other fishes, the recorded blood pressure of hagfishes (Myxini) is lower than that of any vertebrate. A large proportion of the blood (approximately 30 percent of the total volume) is retained in a venous sinus system and may circulate at a lower rate than the central blood volume (Forster, 1998; Randall, 1970). Hagfishes actually have two structures composed of cardiac muscle—a **systemic heart,** like all other vertebrates, and a **portal heart** that is unique to hagfishes. The portal heart pumps blood from the gut through the two lobes of the liver and on to the systemic heart (Forster, 1998; Satchell, 1991). The hagfish systemic heart is the most primitive among the fishlike craniates. It has a well-developed sinus venosus that is partially divided into anterior and posterior portions by a fold of tissue. Blood flows from the sinus venosus to the atrium and then empties into the ventricle through a narrow passage. The ventricle pumps blood into a noncontractile but elastic portion of the ventral aorta, the bulbus arteriosus, a structure also seen in bony fishes (see Fig. 24.7). A conus arteriosus appears to be lacking in hagfishes; the **semilunar valves** that prevent backflow from the bulbus arteriosus are set into the walls of the ventricle itself. The elastic bulbus arteriosus helps to maintain a constant flow of blood to the gills. In addition to systemic and portal hearts, hagfishes have additional hearts of different design that assist in the propulsion of blood at characteristically low pressures. A **cardinal heart** in the head pumps venous blood toward the systemic heart. Because this pump is powered by extrinsic musculature, it may not fit the definition of a true heart. The **caudal heart,** located near the end of the tail, is most active when the animal stops swimming. Additional features of the hagfish heart that are often considered primitive are the absence of neural control and its relative insensitivity to acetylcholine and adrenalin. Forster (1998) suggested that these features could be interpreted as "specialized" rather than "primitive."

The heart in the lampreys (Petromyzontidae) is large in comparison to that of most fishes. The sinus venosus is a small, vertical, tubular structure. The atrium overlies the ventricle, and a bulbus arteriosus is present at the base of the ventral aorta. As in the hagfish, a conus arteriosus is lacking. Only the right common cardinal vein (duct of Cuvier) is present and empties into the atrium.

In elasmobranchs, the sinus venosus, atrium, ventricle, and conus arteriosus are well developed (Fig. 24.7, top), although the sinus venosus has very little cardiac muscle and may not be as important in filling the atrium as in bony fishes. The muscle of the ventricle has a compact outer layer (*compacta*) and an inner, spongy layer (*spongiosa*). The compact layer brings about higher efficiency in the function of the ventricle (Tota, 1989). In those sharks capable of maintaining their body temperatures significantly above the surrounding water (discussed in a later section), the compacta may comprise more than 40 percent of the mass of the ventricle, whereas in nonthermogenic sharks, it may make up as little as 15 percent. The larger compact layer in these "warm-blooded" sharks may be functionally related to the need for greater vascular output (Tota, 1989). Unlike lampreys and hagfishes, both right and left common cardinal veins are developed. The bulbus arteriosus is absent.

The hearts of bony fishes are more variable in structure than those of the other groups because of the great evolutionary diversity in living forms. The typical bony fish heart (Fig. 24.7, bottom) has a thin-walled sinus venosus that receives blood from the ducts of Cuvier and the hepatic veins. The blood empties into the atrium from the sinus venosus

and is pumped to the muscular ventricle. Very few bony fishes have a compact layer in the ventricle. From the ventricle, the blood is pumped into the bulbus arteriosus (or, in some primitive species, the conus arteriosus), through which it passes into the ventral aorta. Sturgeons and paddlefishes (Acipenseriformes), bichirs and reedfishes (Polypteriformes), gars (Lepisosteidae), bowfins (Amiidae), and some basal teleosts (such as the tarpons) retain a contractile conus arteriosus, with two or more rows of internal valves. Lungfishes have a conus arteriosus that retains some proximal cardiac muscle and valves, but in all three genera of lungfishes, the conus bends sharply and twists 270° (Burggren and Johansen, 1987).

In most teleosts, the conus arteriosus is reduced, nonmuscular, and bears only one set of valves between the ventricle and the nonmuscular, elastic bulbus arteriosus. The bulbus arteriosus may appear as a small, white dilation in a dissected fish, but when the heart is pumping, the bulbus expands to the size of the ventricle. The bulbus arteriosus is reported to be more than 30 times more expansible than

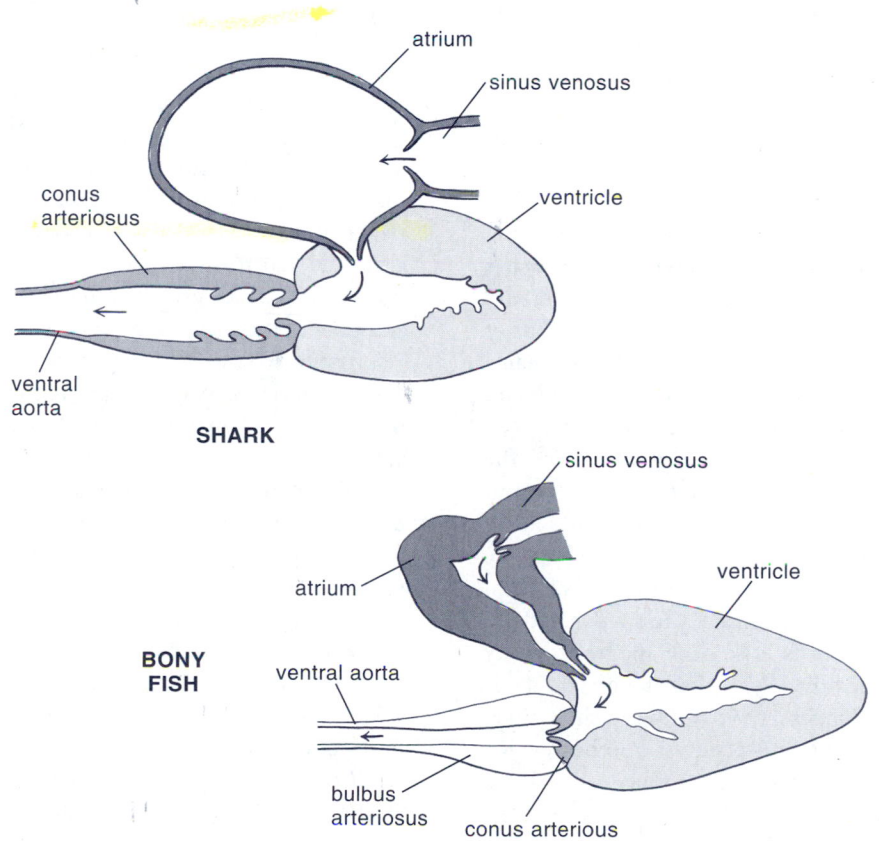

FIGURE 24.7
Comparison of the heart of a shark with that of a bony fish.

the aorta of humans (Licht and Harris, 1973) and can expand 700 percent in carp (Satchell, 1991). Blood is prevented from flowing back into the ventricle by valves located within the reduced conus arteriosus. Thus, the elastic walls of the bulbus maintain pressure on the blood flowing to the gills, maintaining an almost continuous flow, in contrast to the pulsatile flow from a conus arteriosus (Johansen and Gesser, 1986; Satchell, 1971). Teleosts living in temperate waters have been observed to have higher blood pressures than species living in cold waters at high latitudes, such as notothenioids. This has been correlated with differences in the tissue composition of the bulbus arteriosus, reflecting differences in circulatory demands (Icardo et al., 2000).

The atrium of the lungfish heart is partially divided into two parts by an incomplete septum (the **pulmonalis fold**); thus, the lungfishes are described as having a functional three-chambered heart similar to that of amphibians. The right chamber is generally larger than the left. In the South American lungfish (*Lepidosiren paradoxa*), the two sides are nearly completely separated by an atrioventricular cushion, which also serves as a valve between the atrium and the ventricle (Burggren and Johansen, 1987). The right division of the atrium receives deoxygenated blood via the sinus venosus, and the left side receives oxygenated blood from the lung via the pulmonary vein. Virtually complete separation of these two blood supplies is thought to be maintained through the atrium, and mixing in the ventricle is minimized by another incomplete partition. *Lepidosiren* has the best developed separation of the right and left halves of the heart, and the Australian lungfish (*Neoceratodus forsteri*) has the least developed—not surprising given its greater dependence on gills for gas exchange. The conus arteriosus of lungfishes is provided with a peculiar spiral fold that maintains the separation of oxygenated and deoxygenated blood supplies coming from the ventricle (much like that seen in the conus of amphibians). In both *Lepidosiren* and the African lungfishes (*Protopterus*), the fold starts in the ventral proximal wall of the conus arteriosus and continues along its length. It meets a second fold arising from the opposite wall farther along in the organ, which continues to maintain the separation of oxygenated and deoxygenated blood. In *Neoceratodus,* the spiral fold is little developed proximally but sufficiently developed distally so that the blood flow is divided (Burggren and Johansen, 1987). Oxygenated blood is guided mainly to the first and second gill arches (which lack gill tissue) and thence to the dorsal aorta.

Branchial Arteries

The **ventral aorta** of fishes extends forward beneath the pharynx. In lampreys, the aorta remains single to the fourth gill pouch, dividing into right and left branches at the septum between this pouch and the third pouch. Eight pairs of **afferent branchial arteries** branch from the aorta and enter the walls of the gill pouches (Fig. 24.8A).

Afferent branchial arteries in elasmobranchs arise from the single ventral aorta and enter each of the branchial arches to supply blood to the holobranchs borne by these arches and to the hyoid hemibranch (Figs. 24.8B, 24.1C). There are five pairs of branchial arteries in most rays and sharks—more in those having six or seven gill slits. Teleosts are similar to the elasmobranchs in this respect, except that the afferent branchial artery leading to the hyoid hemibranch is absent. Sturgeons more closely resemble the sharks, and some nonteleost actinopterygians are intermediate between the sharks and the bony fishes in their patterns of branchial vessels. In lungfishes, the ventral aorta is short, so the afferent branchial arteries branch from the conus arteriosus close to the heart.

In the gill arch, the afferent branchial arteries give rise to arterioles that supply capillaries or open spaces (*lacunae*) in the gill lamellae. After passing through the gills, the oxygenated blood flows through **efferent branchial arteries** to the **dorsal aorta**.

Blood Flow in Gills

The heart provides the major force to move the blood through the ventral aorta, up the afferent branchial arteries, and into the gill filaments. There are two pathways that blood can follow in the gill lamellae: the arterioarterial and the arteriovenous pathways (Nilsson, 1986). In the arterioarterial pathway, blood passes through the lamellae, is oxygenated, and proceeds to the efferent branchial artery and then to the suprabranchial artery, which in teleosts develops into the carotid artery anteriorly and into the dorsal aorta posteriorly. In the arteriovenous pathway, blood can pass from the filamental arteries to arteriovenous anastomoses and thence to the central filamental venous system and to possibly nutritive vascular beds in the gill tissue. Three distinct blood pathways have been described in the gill filaments of the skipjack tuna (*Katsuwonus pelamis*), a high-energy scombrid. The respiratory pathway responsible for gas exchange is arterioarterial. The interlamellar and nutrient pathways are arteriovenous. The interlamellar pathway has no known function, whereas the nutrient pathway appears to be integral to nonrespiratory homeostatic processes in the gill filaments (Olson et al., 2003).

Contractions of the ventricle generate a considerable pressure that is attenuated as the blood is forced through the intralamellar channels and into the dorsal aorta via the efferent branchial arteries. Farrell (1991) tabulated blood pressures of various fishes from several sources. Examples of systolic pressures in the ventral aorta of various species

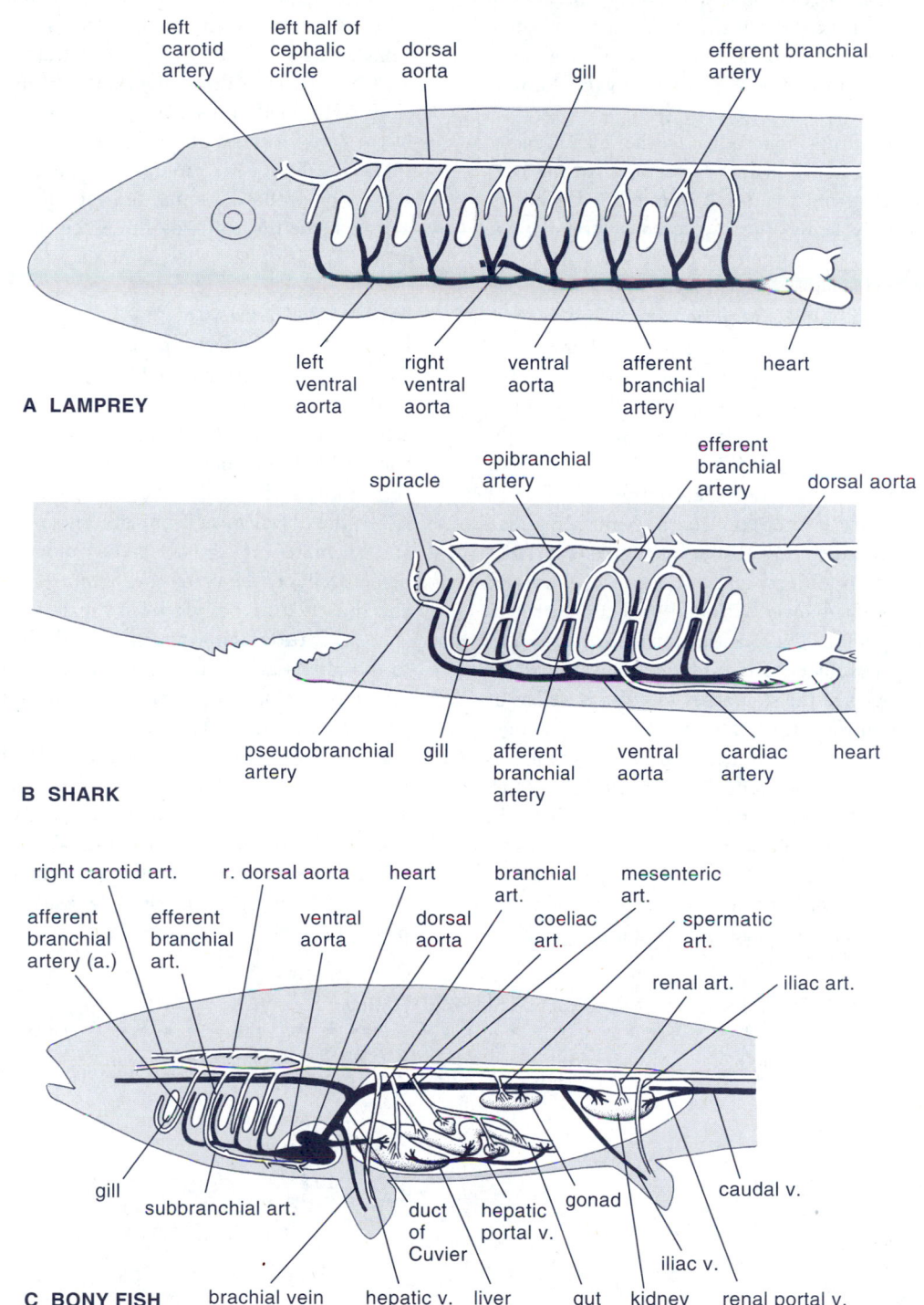

FIGURE 24.8
Circulatory pathways in fishes. **A,** Branchial arteries of a lamprey; **B,** Branchial arteries of a shark; **C,** Bony fish, showing branchial circulation and other major blood vessels.

are given in Table 24.5. The bulbus arteriosus, because of its elasticity (or the conus arteriosus in elasmobranchs and primitive bony fishes, because it is muscularized), maintains a positive pressure on the blood even though the diastolic ventricular pressure may drop to zero.

The pressure drops markedly as the blood passes through the gills. Pressure in the dorsal aorta of the hagfish (*Myxine*) ranges from about 2 to 14 mm Hg, the higher pressures apparently being due to the contraction of the gill pouches. Systolic pressure in the dorsal aorta of various elasmobranchs has been measured at 25 to 40 percent of that of the ventral aorta. The gill contractions superimpose pressure on the pulse caused by the heart. The pressure in the dorsal aorta is usually 40 to 50 percent of that in the ventral aorta in bony fishes, but some, such as the carp, maintain from 50 to 70 percent of the ventral aortic pressure in the dorsal aorta (Ngan et al., 1974). The recorded dorsal aortic pressures of the carp, 22 to 32 mm Hg, contrast with the low figures of 8 to 12 for the Antarctic icefish (Notothenioidei).

The volume of blood flow through the gills is variable, depending in part on gill resistance. Control of the blood flow through the gills involves both neuronal and hormonal factors (Nilsson, 1986). The position of filaments is controlled by striated abductor and adductor muscles, which can change the angles of the structures rapidly, as well as by smooth adductor musculature. There are vascular sphincters at the bases of the filamental efferent arteries that may govern blood flow to some extent (Nilsson, 1986). Another factor in gill blood flow, of course, is the rate at which the heart pumps blood. Typical cardiac outputs for bony fishes are in the range of 15 to 20 ml kg^{-1}min^{-1}, but outputs from 5 to 100 ml kg^{-1} min^{-1} have been reported. Elasmobranchs have cardiac outputs in the range of 20 to 25 ml kg^{-1} min^{-1}.

Although the heart supplies the major force that maintains blood flow in the dorsal aorta, systemic arteries, and veins, other structures may assist in the flow, such as the gill contractions in *Myxine* mentioned earlier. Serial contractions of the body musculature aid the blood flow, and one of the consequences of heart contraction inside a nonelastic pericardium is a drop in sinoatrial pressure that hastens venous blood flow. Among the special structures that help maintain the flow of body fluids are the lymph hearts and caudal hearts of hagfishes, some sharks, and eels.

Postbranchial Circulation

Lampreys and hagfishes have a single, median dorsal aorta and a peculiar section called the *cephalic circle* in the region of the first gill pouch or anterior to it, from which major arteries supply the head region with blood. In most jawed fishes, the dorsal aorta is paired anteriorly, with extensions continuing to the head as the internal carotid arteries. Posteriorly, the dorsal aorta is unpaired through the trunk region and continues into the tail as the caudal artery, through the hemal arches of the vertebrae. Arteries supplying the viscera and musculature branch from the aorta and caudal artery.

The **caudal vein** runs through the hemal arches ventral to the caudal artery. In lampreys and hagfishes, the caudal vein splits to form the paired **posterior cardinal veins**. In jawed fishes, the caudal vein enters the kidneys through the renal portal system. The postcardinal veins receive blood from the kidneys and gonads and from the musculature as they run forward to join the **common cardinal veins (ducts of Cuvier)**. The precardinal veins, the subclavian veins, and the jugular veins enter the ducts of Cuvier. Blood from the ducts enters the sinus venosus, which also receives blood from the hepatic portal system, draining the digestive

TABLE 24.5 SYSTOLIC PRESSURES IN VENTRAL AORTA FOR SELECTED FISHES

Species	Pressure (mm Hg)	Source
Hagfishes		
Myxine sp.	6.8	Satchell, 1991
Myxine sp.	17	Farrell, 1991
Eptatretus cirrhatus	10.8	Satchell, 1991
Elasmobranchs		
Squalus sp.	30	Farrell, 1991
Raja sp.	16	Farrell, 1991
Teleosts		
Oncorhynchus mykiss (rested)	45	Farrell, 1991
O. mykiss (exercised)	72	Farrell, 1991
Ictalurus sp.	40	Farrell, 1991
Gadus morhua	38	Farrell, 1991
Ophiodon elongatus	39	Farrell, 1991

tract and routing blood through capillaries in the liver (Fig. 24.8C).

Venous blood return in elasmobranchs and some bony fishes (except the Acanthopterygii) is facilitated in part by caudal pumps or caudal hearts (Satchell, 1991). Caudal pumps, as in some sharks, consist of left and right caudal sinuses and a series of vessels equipped with valves. As the fish swims, the bending of the caudal fin to the left and right will alternately compress the sinuses and force blood from the tail toward the caudal vein. The caudal heart of sharks is arranged in such a way that the caudal sinus can be compressed by the serial contractions of muscle in the tail, so that blood is pumped forward even when the fish is not swimming. The teleost caudal heart is made up of two chambers, one on each side of the caudal fin skeleton, connected by a foramen (Satchell, 1991). The skeletal muscles that operate these hearts originate on the vertebrae and insert on the hypural plate. In the eels (*Anguilla*), blood from the caudal fin and cutaneous veins is received into the right side of the caudal heart and is pumped first to the left side and then into the caudal vein. In some species, blood flows into both chambers from the tail and is pumped from both into the caudal vein, although there is a connection between the two chambers.

In elasmobranchs, venous blood return is aided by the *hemal arch pump,* which consists of a series of valves at the entry of each segmental vein into the caudal vein. Blood flow through the valves is aided by the contractions of the myotomes during locomotion (Satchell, 1991).

Secondary Circulation

The lymphatic system is usually described as a part of the vertebrate circulatory system that consists of a set of vessels that originate blindly in tissues. Their primary function is to return fluids that have leaked into the tissue spaces back to the cardiovascular system. The lymphatic system has remained an elusive entity in fishes. Several researchers have questioned the existence of a true lymphatic system in teleosts and instead have proposed a **secondary circulatory system** in its place (Olson, 1996a, 1998). Satchell (1991) presented a description of the secondary circulation in fishes. This system parallels the primary circulatory system; it receives blood through anastomosing short vessels arising from the dorsal aorta, the efferent branchial arteries, and segmental arteries. In some species, the mouths of these interarterial vessels have microvilli that extend out and screen most of the erythrocytes. This system was regarded as lymphatic until it was determined that its source of blood is the primary arteries and that blood cells are sometimes passed through its vessels (Satchell, 1991).

Most of the vessels of the secondary system are involved with the skin, although they are also found beneath the epithelium of the gut and in the mouth cavity. In the skin, the system is found in a network of capillaries under the epidermis of the exposed portions of the scales. These capillaries form from arteries flowing posteriorly beneath each scale, and they carry the almost-clear blood forward over the scale and join veins under the preceding and overlapping scale. The blood in this system is pumped toward the heart by the caudal heart in species that have this organ (Vogel, 1985). The extensive sinus system of hagfishes is thought to be comparable to the teleost secondary vascular system (Lomholt and Franko-Dossar, 1998). Although the secondary system may contain more than one half of the blood plasma in some species, its function and importance are not well understood. It is thought to function in nutrient transport and osmoregulation (Olson, 1996a, 1998; Satchell, 1991).

Muscle and Choroid Retes

Retes are networks of intertwining blood vessels that, because of their close juxtaposition, facilitate exchange between blood flowing into the rete and blood leaving the rete. Probably the best known example of a rete is the one associated with the gas bladder (see Chapter 19), but there are other retial systems in fishes elsewhere in the circulatory system. Among fishes that are considered to be elite swimmers, including representatives of the sharks, tunas, and billfishes, a form of regional endothermy, sometimes referred to as **heterothermy**, has evolved, in which certain regions of the body operate at temperatures significantly higher than the surrounding waters. Two categories of heterothermic conditions have been identified—those involving trunk muscles that maintain elevated temperatures and those involving muscle-derived tissue modified for warming the blood going to the brain (Katz, 2002).

Katz (2002) maintained that selective forces did not favor the development of trunk muscles specifically for the generation of heterothermic conditions; rather, heterothermy was a "happy accident." Nonetheless, the rete system associated with the trunk muscles is integral to the maintenance of elevated temperatures. Retial systems involved with trunk muscles are found in the scombroids and in the lamniform (families Lamnidae and Alopidae) sharks—both groups in which most members must swim constantly and vigorously to stay at a given depth. This rete is associated with the lateral red muscle—a site of nearly constant physiological activity that results in a core temperature several degrees higher than the medium in which the fish swims. Fishes are somewhat unusual among vertebrates because they show a clear distinction of fiber types in the different muscle masses. The red muscle, which is surrounded by white muscle (see Chapter 19), is supplied with blood by large, cutaneous arteries, and drained by lateral cutaneous veins. A rete is imposed between those blood vessels and the red muscle, forming a countercurrent system that functions in the exchange of heat (Fig. 24.9A, B). The heat generated by the activity of the red muscle is conserved deep within the fish (Fig. 24.9B; Carey et al., 1971; Graham and Dickson, 2001). Tunas as small as 207 mm have been shown to be capable of elevating their muscle temperatures as much as 3°C above ambient temperature (Dickson, 1994). In some primitive tunas, such as the skipjacks and the frigate mackerels, the rete is located in enlarged hemal arches (Schaefer, 1985).

Within the viscera of some lamnoid sharks and certain tunas, there is a retial heat exchanger that elevates the temperature in the body cavity. In the bluefin and bigeye tunas, this rete is formed from branches of the coeliomesenteric arteries. In sharks, the visceral rete forms from pericardial arteries (Carey et al., 1971).

In the extraocular muscles (specifically, the superior rectus) of the eyes of swordfishes and billfishes, there exists a mass of brown tissue that is rich in cytochrome *c*, much like the thermogenic "brown fat" seen in several small mammals. An associated countercurrent heat exchanger maintains a constant temperature of about 28°C in the eyes and brain. Many vertebrates produce heat through rapid and comparatively uncoordinated contractions of the muscles (shivering), yet the scombroid brain heater is a modified muscle that produces heat without contractions, much like the aforementioned brown fat bodies. The blood supply for the rete comes from the carotid artery (Block, 1987). Scombroids, being highly mobile visual predators, have eye muscles that are very active, as the eyes are in motion much of the time. The presence of such thermogenic muscle in close proximity to the brain may have served as an important precursor in the evolution of mechanisms to permit a constant brain temperature while the fishes experience the dramatic temperature changes that accompany their extensive vertical migrations (Carey, 1982). This extraocular heating system has been demonstrated to significantly improve visual resolution (see Chapter 20), enhancing the detection of rapidly moving prey by large oceanic predators such as sharks, tunas, and billfishes (Fritsches et al., 2005).

Another retial system is found in the choroid of the eyes of teleosts and the bowfin. The **choroid rete** forms a horseshoe-shaped body around the optic nerve and serves to maintain a hyperoxic intraocular partial pressure of oxygen (Pelster and Randall, 1998). A reconstruction of the evolutionary history of gas-secreting mechanisms in vertebrates suggests that the evolution of the choroid rete occurred only once in fishes and predates the evolution of a similar system

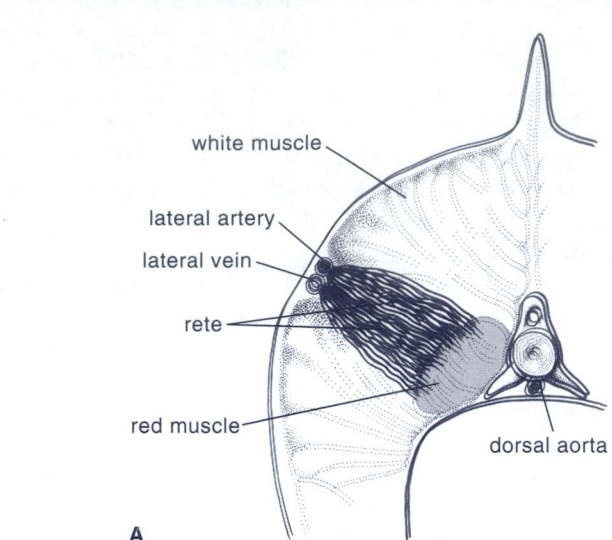

A

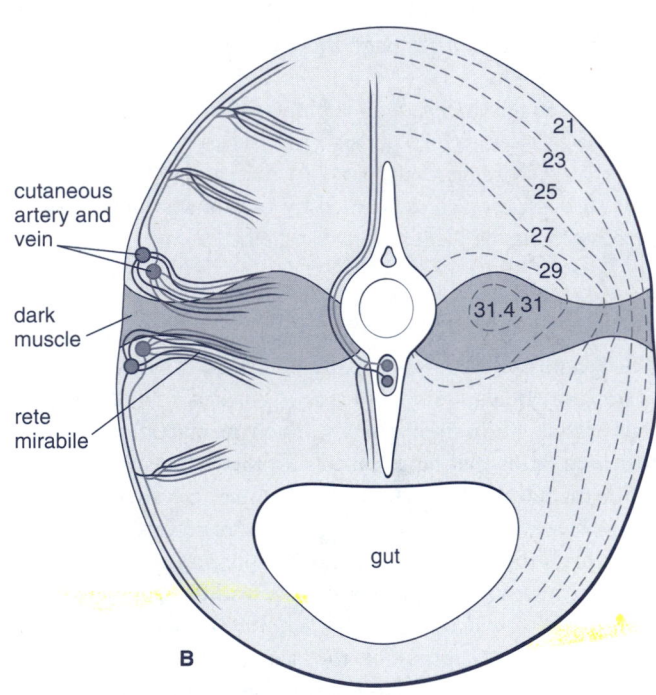

B

FIGURE 24.9

A, Representation of a partial cross section of porbeagle shark, showing relationships of lateral blood vessels and rete to deep-seated red musculature. **B,** Arrangement of blood vessels and rete in bluefin tuna (*left*) and the resultant thermal profile, expressed in °C (*right*), in the muscle tissues (after Carey, 1973).

associated with gas bladder function (see Chapter 19) by some 100 million years (Berenbrink et al., 2005). Predators that depend heavily on vision have large choroid retes and intraocular partial pressures that may exceed 1,300 mm Hg. Sedentary fishes may have partial pressures amounting to only one quarter of that figure. Elasmobranchs do not have choroid retes and may have an intraocular partial pressure of oxygen as low as 30 mm Hg.

THE BLOOD OF FISHES

Blood Compounds Dissolved in Plasma

In addition to its role as a transporter of dissolved gases, the blood transports a variety of other materials, including inorganic ions and a number of organic constituents, such as glucose, hormones, vitamins, and several plasma proteins. These proteins, some of which are involved in immune responses, may include two forms of alpha globulin, two forms of beta globulin, gamma globulin, albumin, transferrin, and others. These proteins buffer against pH changes and aid in the maintenance of osmotic pressure in body fluids. The osmotic concentration of blood in fishes varies according to habitat and osmoregulatory capabilities. The osmolality of bony fish blood ranges from somewhat below 200 milliosmoles (mOsm) in freshwater fish to more than 400 mOsm in marine species (see Chapter 25). Sodium and chloride ions are the main contributors to the total ion concentration, with lesser contributions from potassium, calcium, magnesium, urea, and free amino acids.

Antifreeze Proteins

The freezing point depression (Δ_{fp}) of bony fish blood ranges from about −0.6°C in freshwater fishes to about −0.75°C in marine species. Some Arctic fishes live in waters of −1.7°C; Antarctic waters can be as cold as −1.86°C. Fishes that live in such habitats are protected from freezing by blood glycoproteins (glycopeptides), which may account for one half or more of the osmolality of the blood. These antifreeze compounds function by binding to ice crystals, thus inhibiting the expansion of the ice front in tissues. The three-dimensional structure of many of these antifreeze proteins has been determined, so that the actual sites where they bind to ice are known (Davies et al., 2002). In Arctic fishes, the Δ_{fp} ranges down to −1.0°C, so those fishes living in the coldest water have supercooled body fluids and, with few exceptions, will freeze if brought into contact with ice. A similar situation occurs in the Antarctic, where most species have freezing points of −0.9°C to −1.54°C. Certain notothe-niids freeze only at temperatures lower than −2.2°C because

of elevated levels of glycopeptides in the blood (Eastman, 1993). The antifreeze of one species of nototheniid consists of eight separate glycopeptides that have a broad range of molecular weights (2,600 to 33,700 daltons). The antifreeze activity correlates positively with molecular weight (Eastman, 1993). Many temperate and polar fishes are known to seasonally adjust the osmolality of their blood (Duman and DeVries, 1974a, 1974b). One such fish is the winter flounder (*Pseudopleuronectes americanus*), which can withstand temperatures as low as −1.6°C (Fletcher, 1977). Some fishes increase the blood concentration of NaCl when adapting to cold; others increase the concentration of organic compounds (DeVries, 1971; Feeney and Hofman, 1973).

Blood Cells

Cellular constituents of the blood include the red blood cells, or **erythrocytes**, and the white cells, or **leukocytes**. Unlike those of mammals, the erythrocytes of fishes and other ectotherms contain the full complement of cellular organelles, including nucleus, mitochondria, and endoplasmic reticulum (Weber and Jensen, 1988). Erythrocytes obtain their characteristic color from **hemoglobin**, which is made up of the colorless protein *globin* and the iron-containing, red-yellow pigment *heme*. The heme component is the binding site for the oxygen. The affinity of hemoglobin for oxygen is controlled in part by nucleoside triphosphates, such as adenosine triphosphate and guanosine triphosphate (Fänge, 1992). The binding of oxygen to the hemoglobin is reversible, depending on the ambient partial pressure of oxygen.

The hemoglobin of lampreys and hagfishes is much like the myoglobin found in muscle tissue, in that it has only one chain (monomeric). Three monomeric forms exist in hagfishes; upon deoxygenation, these form heterodimers or heterotetramers (Müller et al., 2003). Hagfish hemoglobin has a molecular weight of 16,500 to 17,000 (Satchell, 1991); it is sometimes considered to represent a transitional form between invertebrate and vertebrate hemoglobins (Müller et al., 2003). As in most other vertebrates, the hemoglobin molecules of elasmobranchs and bony fishes are tetrameric—with four peptide chains—and have molecular weights of about 66,000 to 68,000 (Fänge, 1992; Satchell, 1991). Compared to other vertebrates, however, fishes exhibit a far greater diversity of hemoglobin structure—not surprising considering the diversity of habitats in which they are found (Pérez et al., 1995). Although the heme units of fish hemoglobins appear to be the same in different species, the proteins may differ among species and within species, so that more than one type of hemoglobin can be found in some species (Satchell, 1991). Differences in hemoglobin structures apparently aid in physiological adaptation to different environments (Fänge, 1992). Hemoglobins may differ in many features, such as the composition of amino acids, affinity for oxygen, electrophoretic mobility, and the extent of the Bohr effect (see later). Some salmonids may have up to 18 different hemoglobins during their life cycle (Satchell, 1991).

Only a few fishes lack hemoglobin; for instance, some channichthyids of the Antarctic and the leptocephalus larvae of eels have colorless blood, in which the oxygen is transported dissolved in the plasma. Channichthyids compensate for the lack of hemoglobin by living at low temperatures and maintaining heavily vascularized skin and fins for cutaneous gas exchange. They have increased their circulatory volume and show a rapid movement of blood through the respiratory system (Hemmingson, 1991). Lacking hemoglobin has not compromised their ability to lead an active life, as most are semipelagic predators with metabolic rates comparable to fishes in similar habitats that possess hemoglobin (Kock and Everson, 1997).

The nucleated erythrocytes of most fishes are usually oval; relatively few species have spherical cells. Lampreys have spherical red cells about 9 μm in diameter. Elasmobranchs have large, ovoid erythrocytes, their length ranging from 20 to 27 μm and their width from around 14 to 20 μm. The erythrocytes of bony fishes generally range from 12 to 14 μm in length and 8.5 to 9.5 μm in width, but lungfishes have unusually large erythrocytes, about 36 μm long. Some deep-sea teleosts with blood vessels of exceptionally small diameter have nonnucleated red cells that are about 5.5 μm long and 2.5 μm wide (Fänge, 1992). Examples are the sternoptychid genera *Maurolicus* and *Valencienellus* and the phosichthyid genus *Vinciguerria*. There is much speculation concerning the adaptive advantages of specific sizes of erythrocytes in vertebrates. In teleosts, an inverse relationship exists between erythrocyte size and aerobic swimming ability—increased oxygen transport and diffusion being facilitated by the increase in surface to volume ratio afforded by reduced cell size (Lay and Baldwin, 1999). Erythrocyte size may also be a function of capillary diameter. Snyder and Sheafor (1999) claimed that increased capillary diameter evolved in response to the need for reduced circulatory resistance accompanying the evolution of a pulmonary circuit. This would explain the unusually large erythrocytes observed in lungfishes and amphibians.

With notable exceptions, fishes have a smaller blood volume than other vertebrates; the blood volume usually ranges between 2 and 4 ml/100 g in bony fishes, compared with volumes of 6 ml/100 g or more in mammals (Lagler et al., 1977). Lampreys appear to have greater blood volumes than these (about 8.5 ml/100 g), and hagfishes, at 17.5 to 20 ml/100 g, have one of the highest blood volumes reported

among the craniates (Olson, 1998). Elasmobranchs are reported to have blood volumes of 6 to 8 ml/100 g (Satchell, 1971). Salmonids approach the blood volumes of elasmobranchs, with from 5 to more than 7 ml/100 g (Smith, 1966). Tunas (Scombridae) have a high blood volume, ranging from about 8 ml/100 g in a 9-kg fish to 13 ml/100 g in a 4.5-kg fish; smaller individuals are reported to have even greater blood volumes. Some investigators have suggested a phylogenetic regression in blood volume, from hagfishes with the highest blood volume to teleosts with the lowest (Olson, 1998; Satchell, 1971). The more derived bony fishes possess a more efficient circulatory system and thus need less blood for the transport of oxygen and other materials. Generally, there is an inverse relationship between the size of red blood cells and their number per unit volume of blood, with sharks and rays having fewer than half a million cells per cubic millimeter. However, some gobies may have similar counts—for instance, *Gobius exanthonemus* is reported to have an erythrocyte count of 4.25×10^5/mm³. An Antarctic fish, *Trematomus*, has from 6.6 to 8.0×10^5/mm³. Most bony fishes have red cell counts of 1 to 3×10^6/mm³, with a majority under 2×10^6/mm³, but some active marine fishes have higher numbers, ranging from 4 to 6×10^6/mm³.

The percentage of the blood volume that consists of red cells is called the **hematocrit** and is correlated with the red cell count. Humans have hematocrits of about 47 percent. Hagfishes have about 13 percent, elasmobranchs usually have hematocrits under 25 percent, and the spiny dogfish (*Squalus acanthias*) has about 13 percent. Most teleosts are in the 20 to 30 percent range; some marine species that apparently require large oxygen-carrying capacity—for instance, Atlantic mackerel (*Scomber scombrus*), bluefin tuna (*Thunnus thynnus*), and Atlantic herring (*Clupea harengus*)—have hematocrits of 51 to 52.5 percent (Satchell, 1991).

The hemoglobin concentration in fish blood, expressed in g/100 ml, is usually 7 to 10. The numbers of red blood cells, and consequently the hematocrit and hemoglobin concentration, can vary with season, temperature, and the nutritional state and health of the fish. Circadian changes have been noted in some species (Riggs, 1970). Blood hemoglobin and measured hematocrit have been observed to increase as a consequence of migration in American shad (*Alosa sapidissima;* Leonard and McCormick, 1999).

Blood cell formation (**hematopoiesis**) occurs at several sites in fishes. The spleen is usually the most important site of erythrocyte formation. The kidney produces both red blood cells and various leukocytes (Satchell, 1991). In elasmobranchs, the **organ of Leydig,** most often associated with the wall of the alimentary canal (commonly along the esophagus), is a site of leukocyte formation. In some elasmobranchs, there is similar tissue (*epigonal organ*) associated

PLATE 1.1
Louis Agassiz
(1807–1873).
Source: Hubbs, 1964.

PLATE 1.2
David Starr Jordan
(1851–1931).
Source: Hubbs, 1964a.

PLATE 1.3
Carl Leavitt Hubbs (1894 –1979).
Reproduced courtesy of © Scripps
Institution of Oceanography Archives
(SIO).

PLATE 1.4
Crown Prince (now Emperor)
Akihito examining some fish
specimens with George Sprague
Myers (1967). *Smithsonian
Institution Archives (SIA) record
Unit 7317, George Sprague Myers
Papers. Used with permission.*

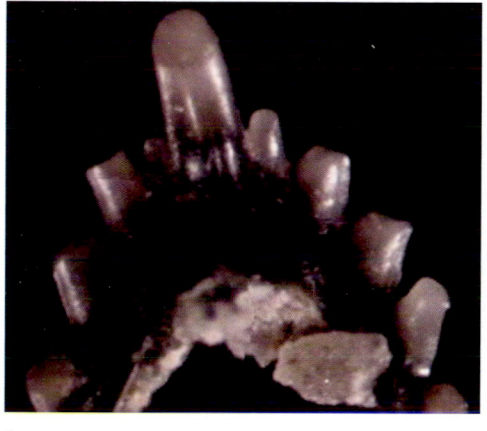

A

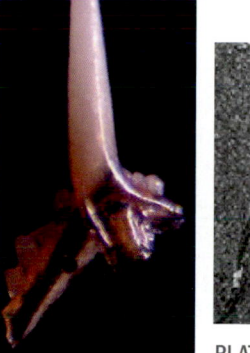

B

PLATE 2.1
Fossil conodonts, approximately 450 MYA. A, *Plectodina furcata* from the
Liepers Shale, Nashville, Tennessee. B, *Phragmodus undatus* from Dayton, Ohio.
Scale bar represents .5mm. *Photographs courtesy of James Davison.*

PLATE 2.2
Early chordates, *Pikaia*, from the Burgess Shale, British Columbia.
*Photograph from Department of Paleobiology, National Museum of
Natural History.*

PLATE 2.3
An early fish-like chordate, *Haikouichthys*, from Cambrian deposits of the Chengjiang region,
South China. Fossil is approximately 25 mm in length. *Reproduced with permission of Degan
SHU (Northwest University, Xi'an, China).*

PLATE 2.4
An early deuterostome (protochordate?),
Haikouella, from the Cambrian deposits of the
Chengjiang region, South China. *Reproduced with
permission of Degan SHU (Northwest University,
Xi'an, China).*

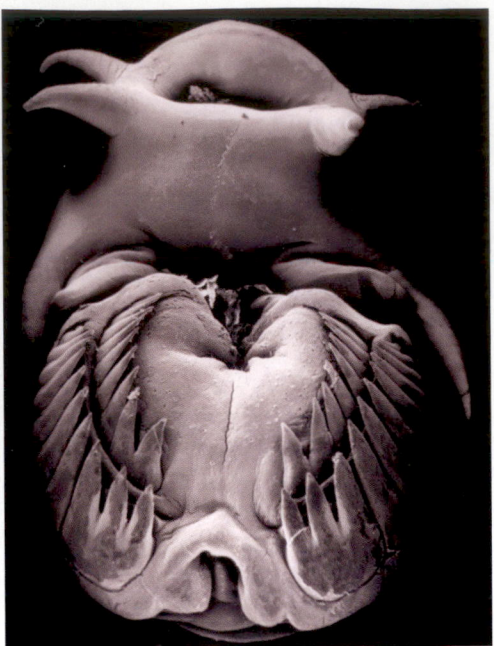

A

B

PLATE 5.1
A, Pacific hagfish (*Eptatretus stoutii*). © *Johnny Jensen/www.jjphoto.dk.* **B**, Rows of dental cusps in mouth of Pacific hagfish.
Photograph courtesy of Aubrey Gorbman. Reproduced with permission of Claudia Gorbman.

• •

A

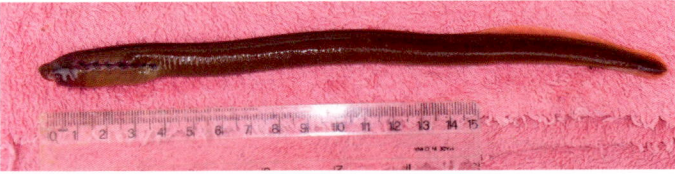

B

PLATE 5.2
A. Adult sea lamprey (*Petromyzon marinus*). © *Beck P. Kent /Animals Animals - Earth Sciences.* B. ammocoete stage of western brook lamprey (*Lampetra richardsoni*). © *Damon Goodman.*

• •

PLATE 5.3
Australian short-head lamprey (*Mordacia mordax*). © *Ross Felix.*

• •

PLATE 6.1
Plownose chimaera (*Callorhynchus milii*). *Courtesy of Erik Tveskov.*

PLATE 6.2
Whale shark (*Rhincodon typus*). © *Phillip Colla/Oceanlight.com.*

PLATE 6.3
Great white shark (*Carcharodon carcharias*). © *Phillip Colla/Oceanlight.com.*

PLATE 6.4
Blue shark (*Prionace glauca*). © *Phillip Colla/Oceanlight.com.*

PLATE 6.5
Manta ray (*Manta birostris*) with remora (family Echeneidae) attached below the gill slits. © *Phillip Colla/Oceanlight.com.*

PLATE 7.1
Australian lungfish (*Neoceratodus forsteri*).
© *Johnny Jensen/www.jjphoto.dk.*

PLATE 8.3
Paddlefish (*Polyodon spathula*). *Reproduced with permission of Patrick O'Neil and the Alabama Fisheries Association.*

PLATE 8.1
Ropefish (*Erpetoichthys calabaricus*).
© *Johnny Jensen/www.jjphoto.dk.*

PLATE 8.4
Longnose gar (*Lepisosteus osseus*). *Reproduced with permission of Patrick O'Neil and the Alabama Fisheries Association.*

PLATE 8.2
Alabama sturgeon (*Scaphirhynchus suttkusi*). *Reproduced with permission of Patrick O'Neil and the Alabama Fisheries Association.*

PLATE 8.5
Bowfin (*Amia calva*). *Reproduced with permission of Patrick O'Neil and the Alabama Fisheries Association.*

PLATE 9.1

Fossil of †*Xiphactinus audax* that has swallowed a †*Gillicus*. From Green River Formation, Wyoming. *Photograph courtesy of Jerry Choate, Sternberg Museum of Natural History, Fort Hays State University.*

PLATE 9.2

Arowana (*Osteoglossum bicirrhosum*). © *Johnny Jensen/www.jjphoto.dk.*

PLATE 9.3

Ladyfish (*Elops saurus*). © *Marcel Karssies.*

PLATE 9.4

European eel (*Anguilla anguilla*). *Courtesy of Steffen Zienert.*

PLATE 9.5
Green moray (*Gymnothorax funebris*). © *Phillip Colla/Oceanlight.com.*

PLATE 9.6
Umbrella mouth gulper eel *(Eurypharynx pelecanoides).*
Image provided courtesy of the NORFANZ partners—Australia's National Oceans Office and CSIRO and New Zealand's Ministry of Fisheries and NIWA. For more information on the voyage, visit www. oceans.gov.au /norfanz.

PLATE 9.7
Northern anchovy *(Engraulis mordax). Courtesy of Milton Love.*

PLATE 10.1
Milkfish *(Chanos chanos). Courtesy of Kwang-Tsao Shao.*

PLATE 10.2
Largescale stoneroller (*Campostoma oligolepis*); male showing breeding colors and nuptial tubercles. *Reproduced with permission of Patrick O'Neil and the Alabama Fisheries Association.*

PLATE 10.3
Pike characin (*Boulengerella maculata*).
Photograph courtesy of Aleksei Saunders.

PLATE 10.4
Red-bellied piranha (*Pygocentrus nattereri*).
©*Johnny Jensen/www.jjphoto.dk.*

PLATE 10.5
Upside-down catfish (*Synodontis nigriventris*). The name refers to the darkened belly that provides reverse countershading consistent with its upside-down mode of swimming.
© *Johnny Jensen/www.jjphoto.dk.*

PLATE 11.1
Eulachon (*Thaleichthys pacificus*). *National Oceanic & Atmospheric Administration (NOAA).*

PLATE 11.2
Australian smelt (*Retropinna semoni*).© *Ross Felix.*

PLATE 11.3
Salamanderfish (*Lepidogalaxias salamandroides*). © Tim Berra.

PLATE 11.4
Western galaxias (*Galaxias occidentalis*). © Klaus Busse.

PLATE 11.5
Atlantic salmon (*Salmo salar*).
© R. M. McDowall.

A

PLATE 11.6
A, Sockeye salmon, male in reproductive colors
(*Oncorhynchus nerka*). © Ernest Keeley.
B, Sockeye salmon showing ocean-adapted coloration.
© Johnny Jensen/www.jjphoto.dk.

B

PLATE 11.7
Native American salmon trap on an Alaskan river, circ 1867. *Courtesy of the U.S. Bureau of Fisheries, National Oceanic & Atmospheric Administration(NOAA).*

PLATE 11.8
Yakima Indians fishing at Celilo Falls, Columbia River, circa 1940. © *Corbis.*

PLATE 11.9
Dolly Varden (*Salvelinus malma*). *USGS, Western Fisheries Research Center.*

PLATE 11.10
Mountain whitefish (*Prosopium williamsoni*). Photograph courtesy of Ernest Keeley, Idaho State University.

PLATE 11.11
Arctic grayling *(Thymallus arcticus). Photograph courtesy of Ernest Keeley, Idaho State University.*

PLATE 12.1
Northern pike (*Esox lucius*). © *Johnny Jensen/www.jjphoto.dk.*

PLATE 12.2
Eastern mudminnow (*Umbra pygmaea*).
© *Johnny Jensen/www.jjphoto.dk.*

PLATE 12.3
Lovely hatchetfish (*Argyropelecus aculeatus*). *Courtesy of Don Flescher.*

PLATE 12.4
Viperfish (*Chauliodus sloani*), showing distinctive
photophore array. It has captured a hatchetfish
(*Argyropelecus*). *Photograph courtesy of
Professor Francesco Costa.*

PLATE 12.5
Royal flagfin (*Aulopus filamentosus*).
*Photograph provided courtesy
of Pedro M. N. Cambraia Duarte.*

PLATE 12.6
Tripodfish (*Bathypterois longifilis*). © NORFANZ. This image has been provided courtesy of the NORFANZ partners--Australia's National Oceans Office and CSIRO and New Zealand's Ministry of Fisheries and NIWA. For more information on the voyage, visit www.oceans.gov.au /norfanz.

PLATE 12.7
Inshore lizardfish (*Synodus foetens*).
Courtesy of Don Flescher.

PLATE 12.8
Lanternfish (*Gonichthys barnesi*), with characteristic photophores along body. © NORFANZ. This image has been provided courtesy of the NORFANZ partners--Australia's National Oceans Office and CSIRO and New Zealand's Ministry of Fisheries and NIWA. For more information on the voyage, visit www. oceans.gov.au /norfanz.

PLATE 12.9
A large oarfish (*Regalecus glesne*) captured in Sydney Harbor, Australia, 1954. *AP/ Wide World Photos.*

PLATE 13.1
Burbot (*Lota lota*). *Photograph courtesy of Ernest Keeley, Idaho State University.*

PLATE 13.2
Cavefish (*Typhlichthys subterraneus*). © *William Pflieger.*

PLATE 13.3
Pearlfish (*Carapus bermudensis*). *Courtesy of Don Flescher.*

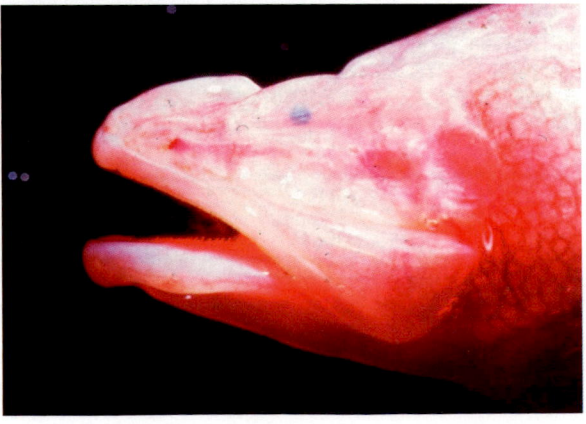

PLATE 13.4
Representative of suborder Bythitoidei, family Bythitidae (*Lucifuga spelaotes*). *Courtesy of Dennis W. Williams and Jill Yager.*

PLATE 13.5
Coral toadfish (*Sanopus splendidus*).
Courtesy of Anita C. Floyd.

PLATE 13.6
Goosefish (*Lophius americanus*). *Courtesy of Don Flescher.*

PLATE 13.7
Sargassumfish (*Histrio histrio*). © *Dave Cook.*

PLATE 13.8
Shortnose batfish (*Ogcocephalus nasutus*). *Courtesy of Don Flescher.*

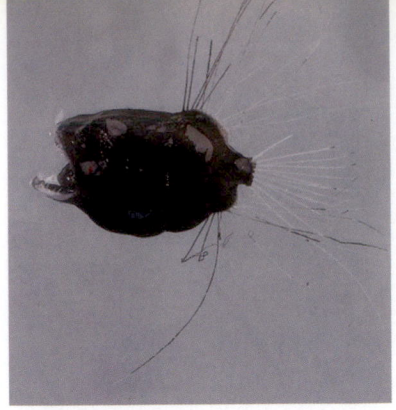

PLATE 13.9
Mossfish (*Caulophryne jordani*). Image provided courtesy of
the NORFANZ partners - Australia's National Oceans Office
and CSIRO and New Zealand's Ministry of Fisheries and NIWA.
For more inforamtion on the voyage., visit www.oceans.gov.
au /norfanz

PLATE 14.1
Red whalefish (*Barbourisia rufa*). *Courtesy of Dr. Keiichi Matsuura.*

PLATE 14.2
John Dory (*Zeus faber*). *Courtesy of Robert Patzner.*

PLATE 14.3
Atlantic flashlightfish (*Kryptophanaron alfredi*).
Photograph provided by J. E. Randall.

PLATE 14.4
Hawaiian squirrelfish (*Sargocentron xantherythrum*). *Photograph provided by J. E. Randall.*

PLATE 15.1

Banded pygmy sunfish (*Elassoma zonatum*). Reproduced with permission of Patrick O'Neil and the Alabama Fisheries Association.

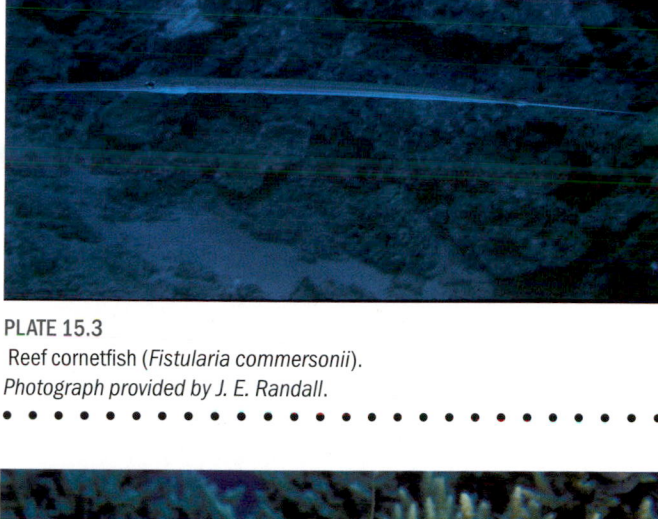

PLATE 15.3

Reef cornetfish (*Fistularia commersonii*). *Photograph provided by J. E. Randall.*

PLATE 15.2

Leafy seadragon (*Phycodurus eques*). *Photograph courtesy of Andrew Bowie.*

PLATE 15.4

Shrimpfish (*Aeoliscus*) in characteristic head-down posture. *Photograph provided by J. E. Randall.*

PLATE 15.5

Striped mullet (*Mugil cephalus*). *Photograph provided by J. E. Randall.*

PLATE 15.6
Male California grunion (*Leuresthes tenuis*) wrapped around female laying eggs in sand. *Photograph provided by Mike Brock.*

PLATE 15.7
Pacific blue-eye (*Pseudomugil signifer*).
© *Johnny Jensen/www.jjphoto.dk.*

PLATE 15.8
Halfbeak, sea garfish (*Hyporhamphus melanochir*). *Courtesy of Kwang-Tsao Shao.*

PLATE 15.9
Juvenile four-winged flyingfish (*Cypselurus*).
Courtesy of Kwang-Tsao Shao.

PLATE 15.10
Mangrove rivulus (*Rivulus marmoratus*), hermaphroditic form. © Mike Brock.

PLATE 15.11
Striped four-eyed fish (*Anableps anableps*) © RN Dr. Roman Slaboch.

PLATE 16.2
Greenside darter (*Etheostoma blennioides*). *Reproduced with permission of Patrick O'Neil and the Alabama Fisheries Association.*

PLATE 16.3
Yellow perch (*Perca flavescens*). *Reproduced with permission of Patrick O'Neil and the Alabama Fisheries Association.*

PLATE 16.1
Largemouth bass (*Micropterus salmoides*). *Reproduced with permission of Patrick O'Neil and the Alabama Fisheries Association.*

PLATE 16.4
Striped bass (*Morone saxatilis*). *Reproduced with permission of Patrick O'Neil and the Alabama Fisheries Association.*

PLATE 16.5
Flamefish (*Apogon maculatus*). *Photograph provided by J. E. Randall.*

PLATE 16.6
Flying gurnard (*Dactylopterus volitans*).
Courtesy of Robert Patzner.

PLATE 16.7
Threadfin (*Polydactylus plebeius*).
Photograph provided by J. E. Randall.

PLATE 16.8
Head of Spanish hogfish (*Bodianus rufus*)
showing protruding teeth characteristic of
labrids. *Courtesy of Dr. Robert Patzner.*

PLATE 16.9
Red oscar (*Astronotus ocellatus*).
© *Johnny Jensen/www.jjphoto.dk.*

PLATE 16.10
Predatory cichlid (*Nimbochromis livingstonii*) with coloration that mimics decaying flesh.© *George J. Reclos.*

PLATE 16.11
Lancer stargazer (*Kathetostoma albigutta*).
Photograph courtesy of Don Flescher.

PLATE 16.12
Jeweled blenny (*Salarias fasciatus*). *Photograph provided by J. E. Randall.*

A

B

PLATE 16.13
A, Striped cleaner wrasse (*Labroides dimidiatus*), the model. B, Mimic blenny (*Aspidontus taeniatus*), the mimic.
Photographs provided by J. E. Randall.

PLATE 16.14
Rockweed gunnels (*Apodichthys fucorum*). The variation in color permits them to blend in with a range of vegetation types.
Photograph courtesy of Milton Love.

PLATE 16.15
Antarctic notothenioid (*Cygnodraco mawsoni*, family Bathydraconidae).
Photograph courtesy of Joseph Eastman.

PLATE 16.16
This minute goby (*Trimmatom nanus*) is possibly the smallest vertebrate in the world.
Photograph courtesy of the Royal Ontario Museum and Richard Winterbottom.

PLATE 16.17
Silverlined mudskipper (*Periopthalmus argentilineatus*).
Photograph provided by J. E. Randall.

PLATE 16.18
Whitemargin unicornfish (*Naso annulatus*) showing prominent forehead spike. © *Johnny Jensen/www.jjphoto.dk.*

PLATE 16.19
Great barracuda (*Sphyraena barracuda*). © *Johnny Jensen/www.jjphoto.dk.*

PLATE 16.20
Sailfish (*Istiophorus platypterus*). *Photograph provided by J. E. Randall.*

PLATE 16.21
Northern snakehead (*Channa argus*). This specimen was prepared using a freeze-drying process developed by Hofinger Tier-Präparationen (http://www.praeparator.com/). *Wolfgang Hofinger, Hofinger-Tierpräparationen.*

PLATE 17.1
California scorpionfish (*Scorpaena guttata*). *Photograph courtesy of Milton Love.*

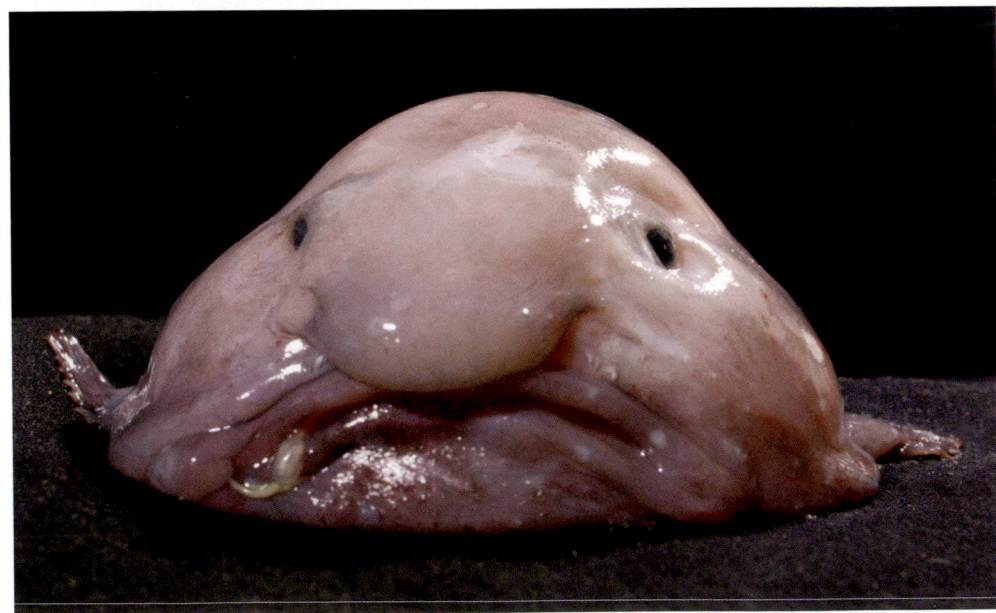

PLATE 17.2
Deepwater fathead sculpin (*Psychrolutes*). Note the parasitic copepod attached to its jaw. *Image provided courtesy of the NORFANZ partners-—Australia's National Oceans Office and CSIRO and New Zealand's Ministry of Fisheries and NIWA. For more information on the voyage, visit www.oceans.gov.au/norfanz. ©NORFANZ, National Oceans Office.*

A

B

PLATE 17.3

Cryptic coloration in members of the flounder genus *Bothus*: A, Cryptic coloration in the peacock flounder (*Bothus lunatus*). While following an individual of this species for about half an hour, the author observed numerous color changes that perfectly camouflaged the fish as it moved from sandy bottoms to rocky ones with varying degrees of vegetative cover.
B, Eyed flounder (*Bothus ocellatus*). *Photographs courtesy of Robert Patzner.*

PLATE 17.4
Queen triggerfish (*Balistes vetula*). *Courtesy of Jeppe Kolding.*

PLATE 17.5
Porcupinefish (*Diodon hystrix*).
© *Johnny Jensen/www.jjphoto.dk.*
© *Andy Pearson & Johnny Jensen.*

PLATE 17.6
Ocean sunfish (*Mola mola*).
Photograph courtesy of the National Oceanic and Atmospheric Administration (NOAA).

A

B

C

PLATE 33.1
Fish species known to engage in cryptic burrowing behavior: A, Sand diver (*Synodus intermedius*). ©2005 Paul Humann/Marinelifeimages.com. B, Southern stargazer (*Astroscopus y-graecum*). *Photograph courtesy of Michele Warren.* C, California halibut (*Paralichthys californicus*).
© *Mark Conlin. www.markconlin.com*

A

B

C

D

PLATE 33.2

Examples of fish species commonly associated with kelp forests of the northeast Pacific Ocean: A, Kelp greenling (*Hexagrammos decagrammus*). *Photograph courtesy of Marc Chamberlain.* B, Señorita (*Oxyjulis californica*). *Photograph courtesy of Milton Love.* C, Kelp rockfish (*Sebastes atrovirens*). *Photograph courtesy of Milton Love.* D, Giant kelpfish (*Heterostichus rostratus*). *Photograph courtesy of Jay Carroll.*

• •

A

B

C

PLATE 33.3

Diversity of jaw and tooth structure in coral reef fishes: A, Beaked coralfish (*Chelmon rostratus*) use their elongate snouts with small terminal mouths to forage on small invertebrates secreted among the crevices in coral reefs. ©*Johnny Jensen/www.jjphoto.dk.* B, Stoplight parrotfish (*Sparisoma viride*) using its sharp incisors to graze on algae. *Photograph © 2005 Paul Humann/ Marinelifeimages.com.* C, Great barracuda (*Sphyraena barracuda*) showing its fearsome dentition. *Photograph © 2005 Paul Humann/Marinelifeimages. com.*

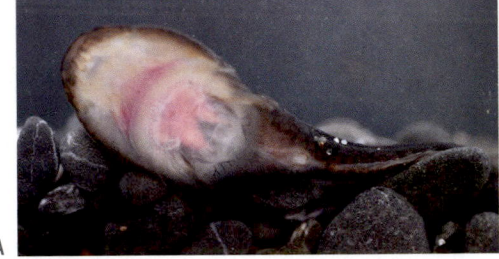

A

B

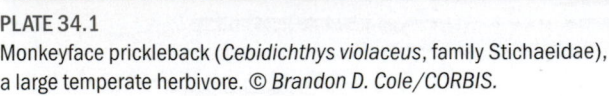

PLATE 34.1

Monkeyface prickleback (*Cebidichthys violaceus*, family Stichaeidae), a large temperate herbivore. © *Brandon D. Cole/CORBIS.*

PLATE 34.2

A, Northern clingfish (*Gobiesox maeandricus*); B, Ventral adhesive disc formed from modified thoracic pelvic fins. *Photographs courtesy of Michael Horn.*

PLATE 34.3

Variation in the cryptic coloration patterns of a rocky intertidal fish, the crevice kelpfish *(Gibbonsia montereyensis)*. *Photograph courtesy of Milton Love.*

● ●

A

B

C

D

PLATE 35.1

Toothy mesopelagic fishes: A, humpback anglerfish *(Melanocetus johnsonii)*; B, fangtooth *(Anoplogaster cornuta)*; C, viperfish *(Chauliodus sloani)*; D, scaleless black dragonfish *(Melanostomias)*. Note the prominent luminescent chin barbel used to lure prey as well as the luminescent organs just below the eye and at the bases of the teeth. *These images © NORFANZ. Images have been provided courtesy of the NORFANZ partners—Australia's National Oceans Office and CSIRO and New Zealand's Ministry of Fisheries and NIWA. For more information, visit http://www.oceans.gov.au/norfanz.*

● ●

PLATE 35.2
Female deep-sea anglerfish parasitized by two males. © *Peter David/ Getty Images, Inc.*

PLATE 35. 3
An epibenthic Antarctic notothenioid (*Trematomus eulepidotus*, family Nototheniidae). *Photograph courtesy of Joseph Eastman.*

PLATE 35.4
Gills of *Cryodraco antarcticus* (Channichthyidae) revealing paucity of respiratory pigments in blood. *Photograph courtesy of Joseph Eastman.*

PLATE 36.1
A colony of spotted garden eels (*Heteroconger hassi*). Large numbers of these congrid eels live together in areas with a fine sand bottom that is most conducive to their burrowing activity. They face into the current and pick out small organisms from the water flowing by. *Photograph courtesy of J. E. Randall.*

PLATE 37.1
Teeth of the cookie-cutter shark *Isistius plutodus*. *Photograph courtesy of Carl Bento,* © *Australian Museum.*

PLATE 37.2
Trumpetfish (*Aulostomus maculatus*) shadowing a Spanish hogfish (*Bodianus rufus*) while hunting for small fishes.
Photograph © *2005 Ned DeLoach/Marinelifeimages.com.*

PLATE 37.3
Male yellowface pikeblenny (*Chaenopsis limbaughi*) using erect dorsal fin and branchiostegal membranes in display for nearby female.
Photograph © 2005 Ned DeLoach/Marinelifeimages.com.

PLATE 37.4
Juvenile banded butterflyfish (*Chaetodon striatus*) with pronounced ocellus (eye spot) and eye obscured with eye bar.
Photograph © 2005 Ned DeLoach/ Marinelifeimages.com.

PLATE 37.5
Bar jack (*Caranx ruber*) shadow-feeding with southern stingray (*Dasyatis americana*). © Doug Perrine/SeaPics.com.

A

B

PLATE 37.6

A, Female *Lampsilis cardium* displaying fish-mimic mantle flaps. B, Young red-eye bass (*Micropterus coosae*) attacking mantle mimics. *Photograph courtesy of W. R. Haag.*

PLATE 37.7

Medusafish (*Nomeus gronovii*)— an intimate associate of cnidarians, especially the Portuguese man-of-war (*Physalia physalia*). *Photograph courtesy of Don Flescher.*

PLATE 37.8

Clownfish (*Amphiprion melanopus*) lurking among the tentacles of a sea anemone. © *Erik Schlogl.*

PLATE 38.1

Furuncle, a localized inflammation observed in cases of furunculosis. *Courtesy of Tore Håstein and the National Veterinary Institute, Oslo.*

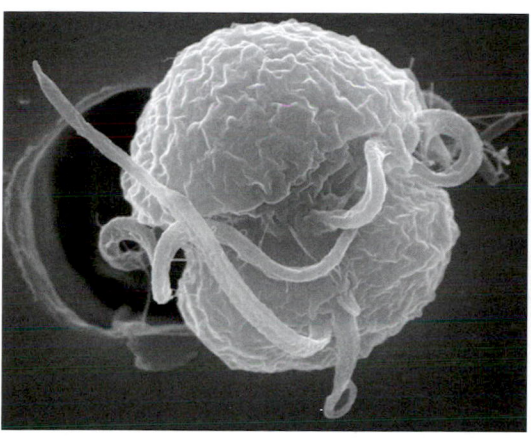

A

B

PLATE 38.2

A, Scanning electron micrograph of *Pfiesteria*. B, Fishes that have suffered an attack by *Pfiesteria*.
Source: Center for Applied Aquatic Ecology, North Carolina State University.

PLATE 38.3

View of aboral pole of *Trichodina*, showing denticular ring. *Dr. Josef Brief and the Marine Biology laboratory (MBL)/Woods Hole Oceanographic Institution.*

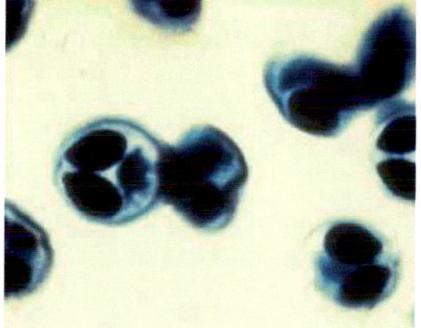

A

B

PLATE 38.4

Spores of Myxobolus cerebralis released into water by infected fish. B, Infected trout (on left) showing effects of whirling disease. Information about whirling disease can be obtained from the Whirling Disease Initiative website (http://whirlingdisease.montana.edu /education / lifecycle.htm). *AP Wide Word Photos/Denver Post, Gaylon Wampler.*

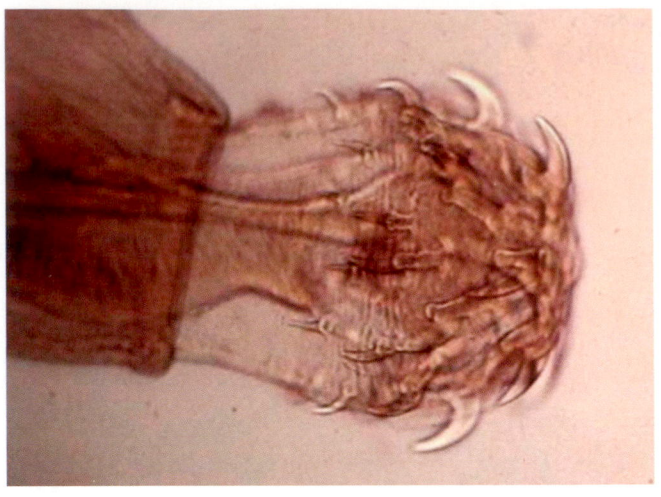

PLATE 38.5
Acanthocephalus showing everted spiny proboscis used to anchor into host tissues. *Joseph Carl Fox, PhD and the Center for Veterinary Sciences, Oklahoma State University.*

PLATE 38.6
The crustacean ectoparasite *Argulus. Courtesy of Ian Russon.*

PLATE 38.7
Cymothoid isopod attached to blackbar soldierfish (*Myripristis jacobus; Photograph © 2005 Paul Humann/ Marinelifeimages.com.*).

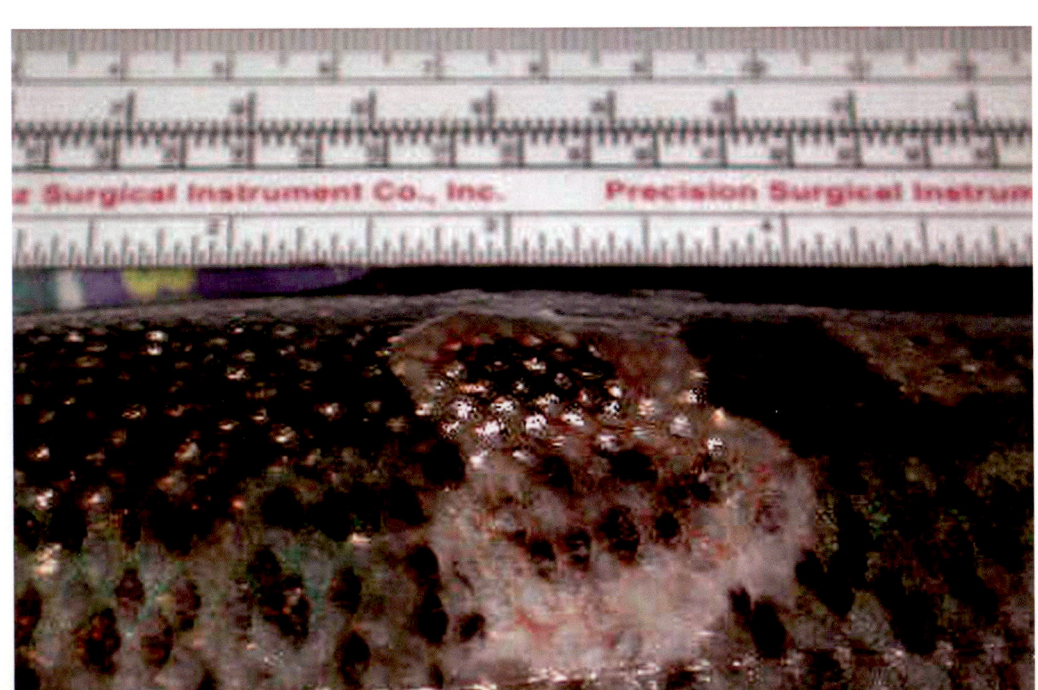

PLATE 38.8
Gross appearance of Atlantic salmon skin lesions from Atlantic Salmon Papillomatosis virus infection. *Maine DIF&W Department of Inland Fisheries and Wildlife.*

with the gonads. Similar tissue may occur in various places in fishes—in the gut wall, orbit, meninges, base of the cranium, and in the cranium above the hindbrain in teleosts. The thymus is a site of lymphocyte production as well. In lampreys, larval hematopoiesis occurs mainly in the typhlosole, with some occurring in the nephric fold. In adult lampreys, the "fat column" that extends along the dorsal surface of the spinal cord is the site of hematopoiesis (Potter et al., 1982).

The oxygen capacity of fish blood includes both oxygen carried in solution and oxygen carried in combination with hemoglobin in the erythrocytes. The Antarctic blackfin icefish (*Chaenocephalus aceratus*), which has no hemoglobin, has a reported blood oxygen capacity of 0.45 to 1.08 ml O_2/100 ml (Hemmingson and Douglas, 1972), whereas most teleosts have capacities in the 8 to 12 ml O_2/100 ml range. Very active fishes, such as the pelagic scombroids, and species adjusted to oxygen-poor waters have blood oxygen capacities as high as 20 ml O_2/100 ml. The oxygen capacity of the blood of sharks and rays is typically less than that of teleosts, usually ranging from 3.5 to 6 ml O_2/100 ml. When erythrocytes of fishes are exposed to hypoxia, they swell rapidly and remain swollen until the return of normoxia—a response that apparently enhances blood oxygen affinity (Weber and Jensen, 1988).

The actual oxygen content of the blood depends on many factors, including the partial pressure of oxygen in the water, the partial pressure of carbon dioxide, pH, temperature, and the activity of the fish (Fry, 1947). Normally, blood from the dorsal aorta is 85 to 95 percent saturated, whereas venous blood usually carries oxygen at 30 to 60 percent saturation. Trout undergoing strenuous exercise have been reported to experience a complete depletion of oxygen in the venous blood as it returns to the heart.

Bohr and Root Effects

The relationships among CO_2, pH, and the oxygen affinity of the blood are of special interest. One of these relationships, the Bohr effect or *Bohr shift* (Burggren et al., 1991), involves a decreased affinity of hemoglobin for oxygen at low pH due to the altered configuration of the hemoglobin molecule by binding hydrogen ions. This is manifested in the shift of the oxygen dissociation curve to the right (Fig. 24.10A). This augments the offloading of O_2 at respiring tissues, where CO_2 concentration is high. Because CO_2 is rapidly lost at the gill, H^+ dissociates from hemoglobin, allowing the effective loading of O_2. This effect is prominent in fishes that are adapted to habitats with high oxygen content and low CO_2. A distinct advantage exists for those species. At the gills, in the presence of a low partial pressure of CO_2, the blood can easily load oxygen even at low partial pressure; then, in the tissues, at a higher partial pressure of CO_2,

oxygen can be released independently of its partial pressure. This is one feature in which the Antarctic channichthyids may be at a disadvantage. Lacking hemoglobin, they are unable to experience the Bohr effect, which may limit their capacity for sustained swimming (Kock and Everson, 1997). A disadvantage of the Bohr effect in fishes adapted to low CO_2 and high oxygen conditions is that if the CO_2 content of the medium rises, increasingly higher dissolved oxygen content becomes necessary to facilitate loading of the hemoglobin. Fishes such as the bullhead catfishes (*Ameiurus*), which exhibit low activity and are adapted to slow-water habitats, where low pH and decreased oxygen content are normal, exhibit a diminished Bohr effect (Moyle and Cech, 1982).

A variation on the Bohr effect is the Root effect, which is unique to fish hemoglobins and mostly present in teleost fishes with gas bladders (Pelster and Randall, 1998). Root-effect hemoglobins were a necessary prerequisite for the evolution of the vascular modifications, such as the rete mirabile, that enable the secretion of gases at the swim bladder and the choroid of the eye (Berenbrink et al., 2005). The Root effect is manifested as a decrease in the oxygen saturation capacity of the blood with rising partial pressure of CO_2. A decrease in pH will render fish blood incapable of becoming 100 percent saturated with oxygen regardless of the partial pressure of oxygen (Fig. 24.10B). This has been demonstrated in experiments in which oxygen partial pressures up to 140 atmospheres were used. This effect promotes oxygen release to the swim bladder and the choroid rete of the eye. Although the Root effect may limit oxygen uptake at the gills, it is important in maintaining elevated oxygen partial pressures during muscle capillary transit, thus enhancing oxygen delivery to tissues (Brauner and Randall, 1996). The molecular basis for the Root effect has been summarized by Jensen et al. (1998).

Blood Cells and Immune Responses

Adaptive immunity is a feature unique to the vertebrates. The array of molecules involved in immune responses appears to have rapidly diversified early in the evolution of vertebrates (Hughes and Yeager, 1997). In fishes, as in other vertebrates, white blood cells (*leukocytes*) are integral to immune function (Iwama and Nakanishi, 1996). Leukocytes are not as numerous as red cells and usually number fewer than 1.5×10^5/mm³ in most fishes (Mulcahy, 1970). Overall, however, the leukocyte count of fishes far exceeds that found in humans (Satchell, 1991). The range within a single species may be great; counts for the common carp (*Cyprinus carpio*), for example, have been reported as ranging from about 3.2×10^4/mm³ to 1.46×10^5/mm³. There are four kinds of white blood cells: **granulocytes, thrombocytes, lymphocytes,** and **monocytes.** In hagfishes, only

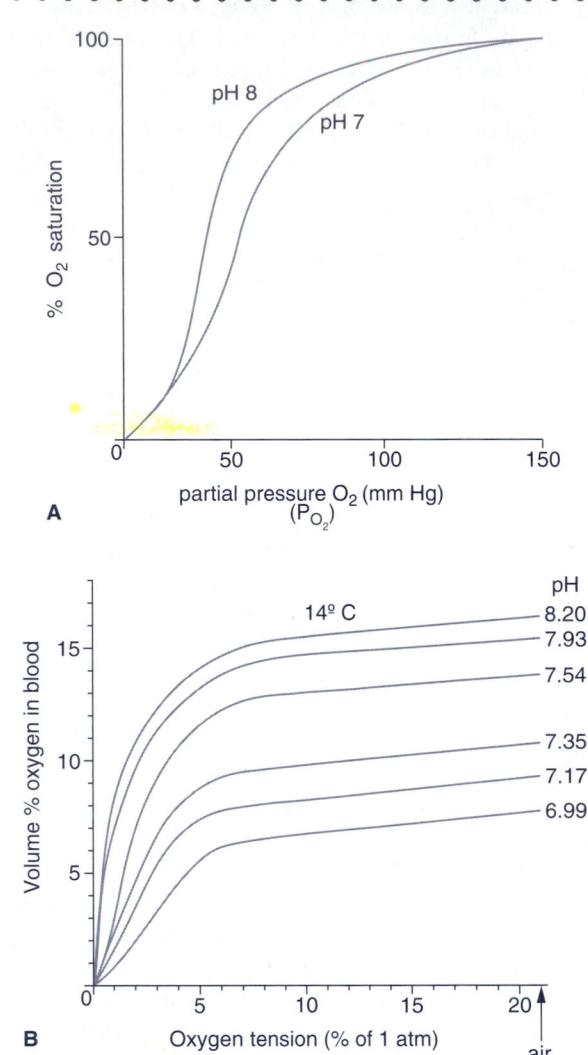

FIGURE 24.10

A, The Bohr effect, showing the impact of blood pH on oxygen dissociation curve. At a selected partial pressure of oxygen, the percentage saturation will be lower at pH 7 than at pH 8. **B,** The Root effect, showing the impact of blood pH on the saturation capacity of hemoglobin. As pH decreases, the hemoglobin responds with a diminished capacity to bind oxygen (after Steen, 1963).

a single kind of granulocyte is known, whereas lampreys and gnathostomes show greater variation in granulocyte type (Zapata et al., 1996). Granulocytes include four types of cells, named for their staining properties: neutrophils and three different types of eosinophils (Satchell, 1991). Neutrophils are common in most species, but eosinophils are not always present. Granulocytes are phagocytic, involved

in combating disease, and they may increase in number when a fish suffers a bacterial infection.

Thrombocytes are involved in blood clotting; they carry a compound that promotes the conversion of prothrombin to thrombin and they are more numerous than the other leukocytes in many marine fishes, constituting about half the total. Clotting in fishes is extremely rapid.

Lymphocytes include the phagocytic macrophages, plasma cells, and small lymphocytes, which may be active in protein production. More than 90 percent of the white blood cells in carp and trout can be lymphocytes. Monocytes are mononucleated macrophages.

KEY POINTS AND CONNECTIONS

- Fishes are faced with the challenge of extracting oxygen, which has a comparatively low solubility in water, from their environment and delivering it to tissues in adequate amounts to sustain life processes. For most fishes, the heart pumps only deoxygenated blood, which is routed to the gills for oxygenation before being delivered to the tissues and organs.

- Gills represent an extraordinary enhancement of epithelial surface area for the purposes of optimizing the quantity of dissolved oxygen that can be extracted from the water. Whereas agnathan and elasmobranch fishes display multiple gill pouches, chimaeriform and bony fishes have the gill tissue consolidated beneath a single covering, usually referred to as the operculum. Fishes use the musculature of the head region to pump water across the gill tissue; dissolved oxygen extraction is maximized through the use of a countercurrent flow regime, in which the flow of blood runs opposite to the flow of water across the gills.

It is important to recognize that gill tissue and its associated skeletal support functions not only in gas exchange, but also in feeding (see Chapter 23) and osmoregulation (see Chapter 25). Countercurrent exchange is also employed in the circulatory network (rete mirabile) supplying the gas bladder, which is discussed in Chapter 19.

- Fishes living in environments with diminished availability of dissolved oxygen have evolved a number of strategies for obtaining oxygen from the atmosphere. These include lungs and highly modified swim bladders, cutaneous gas exchange, and the use of highly modified regions in the branchial chambers.

Habitats in which fishes might encounter hypoxic conditions are discussed in Chapters 32 and 34.

- For most fishes, the heart is the sole means of propelling the blood through the veins and arteries. It consists of several chambers—deoxygenated blood first enters the sinus venosus, is passed to the atrium, then to the ventricle, and exits to the gills via the conus (or bulbus) arteriosus. Blood oxygenated at the gills is then passed via a dorsal aorta to arteries supplying the various tissues and organs.

• As in all other vertebrates, the blood of fishes serves as the means to deliver dissolved oxygen to the tissues. Respiratory pigments (hemoglobin) enhance the quantity of oxygen that can be transported. The blood also transports nutrients, endocrine substances, and agents of immune response. Vertebrate hemoglobin exhibits the Bohr effect, which facilitates the loading and unloading of oxygen. The hemoglobin of fishes is unique in demonstrating a related phenomenon, the Root effect, in which oxygen saturation capacity is diminished with increasing concentration of CO_2 or decreasing pH.

The freezing resistance properties of the blood of fishes living in polar latitudes and the remarkable hemoglobin-free blood of some Antarctic species are discussed in Chapter 35.

• •
BUILDING AN ICHTHYOLOGY LIBRARY
• •

Perry, S. F., and B. Tufts (Eds.). 1998. *Fish respiration* (Fish physiology series, Vol. 17). Academic Press, San Diego.

Satchell, G. H. 1991. *Physiology and form of fish circulation.* Cambridge University Press, Cambridge, UK.

• •
REFERENCES
• •

Baglioni, S. 1908. Den Atmungmechanismus den Fische. *Z. Allgem. Physiol. 7:* 177–282.

Beatty, D. D. 1975. The role of the pseudobranch and choroid rete mirabile in fish vision, pp. 673–678. In *Vision in fishes: New approaches in research,* M. A. Ali (Ed.). Plenum Press, New York.

Benz, G. W. 1984. On the conservative nature of the gill filaments of sharks. *Env. Biol. Fishes 10:* 111–116.

Berenbrink, M., P. Koldkjær, O. Kepp, and A. R. Cossins. 2005. Evolution of oxygen secretion in fishes and the emergence of a complex physiological system. *Science 307:* 1752–1757.

Berg, T., and J. B. Steen. 1966. Regulation of ventilation in eels exposed to air. *Comp. Biochem. Physiol. 18:* 511–516.

Block, B. 1987. Billfish brain and eye heater: A new look at non-shivering heat production. *News Physiol. Sci. 2:* 208–214.

Bone, Q., and N. B. Marshall. 1982. *Biology of fishes.* Blackie, Glasgow, UK.

———, ———, and J. H. S. Blaxter. 1995. *Biology of fishes* (2nd ed.). Blackie, Glasgow, UK.

Branson, B. A., and G. U. Ulrikson. 1967. Morphology and histology of the branchial apparatus in percid fishes of the genera *Percina, Etheostoma,* and *Ammocrypta* (Percidae: Percinae: Etheostomatini). *Trans. Am. Microsc. Soc. 86:* 371–389.

Brauner, C. J., and D. J. Randall. 1996. The interaction between oxygen and carbon dioxide movements in fishes. *Comp. Biochem. Physiol. 113A:* 83–90.

Bridges, C. R. 1988. Respiratory adaptations in intertidal fish. *Amer. Zool. 28:* 79–96.

Burggren, W. W., and K. Johansen. 1987. Circulation and respiration in lung fishes (Dipnoi), pp. 217–236. In *The biology and evolution of lungfishes,* W. E. Bemis, W. W. Burggren, and N. E. Kemp (Eds.). Alan R. Liss, New York.

———, and J. L. Roberts. 1991. Respiration and metabolism, pp. 353–435. In *Comparative animal physiology: Environmental*

and metabolic animal physiology (4th ed.), C. L. Prosser (Ed.). Wiley-Liss, New York.

———, B. McMahon, and D. Powers. 1991. Respiratory functions of blood, pp. 437–508. In *Comparative animal physiology: Environmental and metabolic animal physiology* (4th ed.), C. L. Prosser (Ed.). Wiley-Liss, New York.

Butler, P. J., and J. D. Metcalfe. 1988. Cardiovascular and respiratory systems, pp. 1–47. In *Physiology of elasmobranch fishes,* T. J. Shuttleworth (Ed.). Springer Verlag, Berlin.

Carey, F. G. 1973. Fishes with warm bodies. *Sci. Amer. 228*(2): 36–44.

———. 1982. A brain heater in swordfish. *Science 216:* 1327–1329.

———, J. M. Teal, J. W. Kanwisher, K. D. Lawson, and J. Beckett. 1971. Warm-bodied fish. *Amer. Zool. 11*(1): 137–145.

Davies, P. L., J. Baardsnes, M. J. Kuiper, and V. K. Walker. 2002. Structure and function of antifreeze proteins. *Phil. Trans. Roy. Soc. Lond. B 357:* 927–935.

de Jager, S., and W. J. Dekkers. 1975. Relations between gill structure and activity in fish. *Neth. J. Zool. 25:* 276–308.

DeVries, A. L. 1971. Freezing resistance in fishes, pp. 157–190. In *Fish physiology,* Vol. VI, W. S. Hoar and D. J. Randall (Eds.). Academic Press, New York.

Dickson, K. A. 1994. Tunas as small as 207 mm fork length can elevate muscle temperatures significantly above ambient water temperature. *J. Exp. Biol. 190:* 79–93.

Duman, J. G., and A. L. DeVries. 1974a. Freezing resistance in winter flounder *Pseudopleuronectes americanus. Nature 247:* 237–238.

———, and———. 1974b. The effects of temperature and photo-period on antifreeze production in cold water fishes. *J. Exp. Zool. 190*(1): 89–98.

Eastman, J. T. 1993. *Antarctic fish biology.* Academic Press, San Diego.

Fänge, R. 1992. Fish blood cells, pp. 1–54. In *Fish physiology,* Vol. XII, Part B, W. S. Hoar, D. J. Randall, and A. P. Farrell (Eds.). Academic Press, San Diego.

Farrell, A. P. 1991. Circulation of body fluids, pp. 509–558. In *Comparative animal physiology: Environmental and metabolic animal physiology* (4th ed.), C. L. Prosser (Ed.). Wiley-Liss, New York.

Feder, M. E., and W. W. Burggren. 1985a. Skin breathing in vertebrates. *Sci. Amer. 253*(5): 126–142.

———, and———. 1985b. Cutaneous gas exchange in vertebrates: Design, patterns, control and implications. *Biol. Rev. Cambridge Phil. Soc. 601:* 1–45.

Feeney, R. E., and R. Hofman. 1973. Depression of freezing point by glycoproteins from an Antarctic fish. *Nature 243:* 357–359.

Fletcher, G. I. 1977. Circannual cycles of blood plasma freezing point and Na^+ and Cl^- concentrations in Newfoundland winter flounder (*Pseudopleuronectes americanus*): Correlation with water temperature and photoperiod. *Can. J. Zool. 55:* 789–795.

Forster, M. E. 1998. Cardiovascular function in hagfishes, pp. 237–258. In *The biology of hagfishes,* J. M. Jørgensen, J. P. Lomholt, R. E. Weber, and H. Malte (Eds.). Chapman and Hall, London.

Fritsches, K. A., R. W. Brill, and E. J. Warrant. 2005. Warm eyes provide superior vision in swordfishes. *Current Biol. 15:* 55–58.

Fry, F. E. J. 1947. Effect of the environment on animal activity. University of Toronto, Biology Series, *SS/Publ. Ontario Fish. Res. Lab. 68:* 1–62.

Gee, J. H., and P. A. Gee. 1991. Reactions of gobioid fishes to hypoxia: Buoyancy and aquatic surface respiration. *Copeia 1991:* 17–28.

———, and———. 1995. Aquatic surface respiration, buoyancy control, and the evolution of air-breathing in gobies (Gobiidae: Pisces). *J. Exp. Biol. 198:* 79–89.

Goodrich, B. S. 1930. *Studies on the structure and development of vertebrates.* Constable, London (Reprinted 1958, Dover, New York).

Gordon, M. S., I. Boetius, D. H. Evans, R. McCarthy, and L. C. Oglesby. 1969. Aspects of the physiology of terrestrial life in amphibious fishes. 1. The mudskipper, *Periophthalmus sobrinus. J. Exp. Biol. 50:* 141–149.

Gosline, W. A. 1971. *Functional morphology and classification of teleostean fishes.* University Press of Hawaii, Honolulu.

Graham, J. B. 1997. *Air-breathing fishes: Evolution, diversity, and adaptation.* Academic Press, San Diego.

———, and K. A. Dickson. 2001. Anatomical and physiological specializations for endothermy, pp. 121–165. In *Tuna: Physiology, ecology, and evolution,* B. A. Block and E. D. Stevens (Eds.). Academic Press, San Diego.

Gray, I. E. 1954. Comparative study of the gill area of marine fishes. *Biol. Bull. 107:* 219–225.

Hardisty, M. W. 1979. *Biology of the cyclostomes.* Chapman and Hall, London.

Hedrick, M. S., and D. R. Jones. 1993. The effects of altered aquatic and aerial respiratory gas concentrations on air-breathing patterns in a primitive fish (*Amia calva*). *J. Exp. Biol. 181:* 81–94.

———, and———. 1999. Control of gill ventilation and air-breathing in the bowfin *Amia calva. J. Exp. Biol. 202:* 87–94.

———, S. L. Katz, and D. R. Jones. 1994. Periodic air-breathing behavior in a primitive fish revealed by spectral analysis. *J. Exp. Biol. 197:* 429–436.

Hemmingson, E. A. 1991. Respiratory and cardiovascular adaptations in hemoglobin-free fish: Resolved and unresolved problems, pp. 191–203. In *Biology of Antarctic fish,* G. di Prisco, B. Maresca, and B. Tota (Eds.). Springer Verlag, Berlin.

———, and E. L. Douglas. 1972. Respiratory and circulatory responses in a hemoglobin-free fish, *Chaenocephalus aceratus,* to changes in temperature and oxygen tension. *Comp. Biochem. Physiol. 43A:* 1031–1043.

Hughes, A. L., and M. Yeager. 1997. Molecular evolution of the vertebrate immune system. *BioEssays 19:* 777–786.

Hughes, G. M. 1966. The dimensions of fish gills in relation to their functions. *J. Exp. Biol. 45:* 177–195.

———. 1984. General anatomy of the gills, pp. 1–72. In *Fish physiology,* Vol. X, Part A, W. S. Hoar and D. J. Randall (Eds.). Academic Press, Orlando, FL.

———, B. R. Singh, R. N. Thakur, and J. S. D. Munshi. 1974. Areas of the air breathing surfaces of *Amphipnous cuchia* (Ham.). *Proc. Ind. Nat. Sci. Acad. B 40:* 379–392.

Icardo, J. M., E. Colvee, M. C. Cerra, and B. Tota. 2000. The bulbus arteriosus of stenothermal and temperate teleosts: A morphological approach. *J. Fish Biol. 57*(Suppl. A): 121–135.

Iwama, G., and T. Nakanishi. 1996. *The fish immune system: Organism, pathogen, and environment.* Academic Press, San Diego.

Jensen, F. B., A. Fago, and R. E. Weber. 1998. Hemoglobin structure and function, pp. 1–40. In *Fish respiration,* S. F. Perry and B. Tufts (Eds.). Academic Press, San Diego.

Johansen, K. 1970. Air breathing in fishes, pp. 361–411. In *Fish physiology,* Vol. IV, W. S. Hoar and D. J. Randall (Eds.). Academic Press, New York.

———, and H. Gesser. 1986. Fish cardiology: Structural, haemodynamic, electromechanical and metabolic aspects, pp. 71–85. In *Fish physiology: Recent advances,* S. Nilsson and S. Holmgren (Eds.). Croom Helm, London.

———, and R. Strahan. 1963. The respiratory system of *Myxine glutinosa* L., pp. 352–371. In *The biology of* Myxine, A. Brodal and R. Fänge (Eds.). Universitetsforlaget, Oslo.

Kardong, K. V. 1998. *Vertebrates: Comparative anatomy, function, evolution.* WCB/McGraw-Hill, Boston.

Katz, S. L. 2002. Review: Design of heterothermic muscle in fish. *J. Exp. Biol. 205:* 2251–2266.

Kent, G. C. 1992. *Comparative anatomy of the vertebrates.* Mosby-YearBook, St. Louis, MO.

Kock, K.-H., and I. Everson. 1997. Biology and ecology of mackerel icefish, *Champsocephalus gunnari:* An Antarctic fish lacking hemoglobin. *Comp. Biochem. Physiol. 118A:* 1067–1077.

Lagler, K. F., J. W. Bardach, R. R. Miller, and D. R. M. Passino. 1977. *Ichthyology* (2nd ed.). Wiley, New York.

Langille, L., E. D. Stevens, and A. Anantaraman. 1983. Cardiovascular and respiratory flow dynamics, pp. 92–137. In *Fish biomechanics,* P. W. Webb and D. Weihs (Eds.). Praeger, New York.

Laurent, P., and S. Dunel-Erb. 1984. The pseudobranch morphology and function, pp. 285–323. In *Fish physiology,* Vol. X, Part B, W. S. Hoar and D. J. Randall (Eds.). Academic Press, Orlando, FL.

Lay, P. A., and J. Baldwin. 1999. What determines the size of teleost erythrocytes? Correlations with oxygen transport and nuclear volume. *Fish Physiol. Biochem. 20:* 31–35.

Leonard, J. B. K., and S. D. McCormick. 1999. Changes in haematology during upstream migration in American shad. *J. Fish Biol. 54:* 1218–1230.

Licht, J. H., and W. S. Harris. 1973. The structure, composition and elastic properties of the teleost bulbus arteriosus in the carp *Cyprinus carpio. Comp. Biochem. Physiol. 46A:* 699–708.

Liem, K. F. 1967. Functional morphology of the integumentary, respiratory and digestive systems of the synbranchoid fish *Monopterus albus. Copeia 1967:* 375–388.

———. 1980. Air ventilation in advanced teleosts: Biochemical and evolutionary aspects, pp. 57–91. In *Environmental physiology of fishes,* M. A. Ali (Ed.). Plenum Press, New York.

———. 1989. Respiratory gas bladders in teleosts: Functional conservatism and morphological diversity. *Amer. Zool. 29:* 333–352.

Lomholt, J. P., and F. Franko-Dossar. 1998. The sinus system of hagfishes—lymphatic or secondary circulatory system? pp. 259–272. In *The biology of hagfishes,* J. M. Jørgensen, J. P. Lomholt, R. E. Weber, and H. Malte (Eds.). Chapman and Hall, London.

Magid, A. M. A., and M. M. Babiker. 1975. Oxygen consumption and respiratory behavior of three Nile fishes. *Hydrobiologia 46:* 359–367.

Malte, H., and J. P. Lomholt. 1998. Ventilation and gas exchange, pp. 223–234. In *The biology of hagfishes,* J. M. Jørgensen, J. P. Lomholt, R. E. Weber, and H. Malte (Eds.). Chapman and Hall, London.

Martin, K. L. M., and C. R. Bridges. 1999. Respiration in water and air, pp. 54–78. In *Intertidal fishes: Life in two worlds,* M. H. Horn, K. L. M. Martin, and M. A. Chotkowski (Eds.). Academic Press, San Diego.

McAllister, D. E. 1968. Evolution of branchiostegals and classification of teleostome fishes. *Bull. Nat. Mus. Can. 221:* 1–239.

Moyle, P. B., and J. J. Cech, Jr. 1982. *Fishes: An introduction to ichthyology.* Prentice Hall, Englewood Cliffs, NJ.

Mulcahy, M. F. 1970. Blood values in the pike *Esox lucius* L. *J. Fish Biol. 2:* 203–209.

Müller, G., A. Fago, and R. E. Weber. 2003. Water regulates oxygen binding in hagfish (*Myxine glutinosa*) hemoglobin. *J. Exp. Biol.* 206: 1389–1395.

Ngan, P. V., K. Hamamori, I. Hanyu, and T. Hibiya. 1974. Measurement of blood pressure of carp. *Jpn. J. Ichthyol.* 21(1):1–8.

Nilsson, S. 1986. Control of gill blood flow, pp. 86–101. In *Fish physiology: Recent advances*, S. Nilsson and S. Holmgren (Eds.). Croom Helm, London.

Olson, K. R. 1996a. The secondary circulation in fish: Anatomical organization and physiological significance. *J. Exp. Zool.* 275: 172–185.

———. 1996b. Scanning electron microscopy of the fish gill, pp. 31–45. In *Fish morphology: Horizon of new research*, J. S. Datta Munshi and H. M. Dutta (Eds.). Science Publishers, Lebanon, NH.

———. 1998. The cardiovascular system, pp. 129–154. In *The physiology of fishes* (2nd ed.), D. H. Evans (Ed.). CRC Press, Boca Raton, FL.

———, T. K. Ghosh, P. K. Roy, and J. S. D. Munshi. 1995. Microcirculation of gills and accessory respiratory organs of the walking catfish *Clarias batrachus. Anat. Rec.* 242: 383–399.

———, H. Dewar, J. B. Graham, and R. W. Brill. 2003. Vascular anatomy of the gills in a high energy demand teleost, the skipjack tuna (*Katsuwonus pelamis*). *J. Exp. Zool. 297A:* 17–31.

Packard, G. C. 1974. The evolution of air-breathing in Paleozoic gnathostome fishes. *Evolution 28:* 320–325.

Pelster, B., and D. Randall. 1998. The physiology of the Root effect, pp. 113–139. In *Fish respiration*, S. F. Perry and B. Tufts (Eds.). Academic Press, San Diego.

Pérez, J., K. Rylander, and M. Nirichio. 1995. The evolution of multiple hemoglobins in fishes. *Rev. Fish Biol. Fisheries 5:* 304–319.

Potter, I. C., Lord R. Percy, D. L. Barber, and D. J. Macey. 1982. The morphology, development and physiology of blood cells, pp. 233–292. In *The biology of lampreys*, M. W. Hardisty and I. C. Potter (Eds.). Academic Press, London.

Randall, J. E. 1970. The circulatory system, pp. 133–172. In *Fish physiology*, Vol. IV, W. S. Hoar and D. J. Randall (Eds.). Academic Press, New York.

Randall, J. E. 1985. Shunts in fish gills, pp. 71–82. In *Cardiovascular shunts: Phylogenetic, ontogenetic and clinical aspects*, K. Johansen and W. Burggren (Eds.). Munksgaard, Copenhagen.

Riggs, A. 1970. Properties of fish hemoglobins, pp. 208–252. In *Fish physiology*, Vol. IV, W. S. Hoar and D. J. Randall (Eds.). Academic Press, New York.

Roberts, J. L. 1975. Active branchial and ram gill ventilation in fishes. *Biol. Bull. 148*(1): 85–105.

Romer, A. S. 1970. *The vertebrate body* (4th ed.). W. B. Saunders, Philadelphia.

Rovainen, C. M. 1996. Feeding and breathing in lampreys. *Brain Behav. Evol. 48:* 297–305.

Sacca, R., and W. W. Burggren. 1982. Oxygen uptake in air and water in the air-breathing reedfish *Calamoichthys calabaricus:* Role of skin, gills and lungs. *J. Exp. Biol. 97:* 179–186.

Satchell, G. H. 1971. *Circulation in fishes*. Cambridge University Press, Cambridge, UK.

———. 1991. *Physiology and form of fish circulation*. Cambridge University Press, Cambridge, UK.

Saunders, R. L. 1961. The irrigation of the gills in fishes. I. Studies of the mechanism of branchial irrigation. *Can. J. Zool. 39:* 677–683.

———. 1962. The irrigation of the gills in fishes. II. Efficiency of oxygen uptake in relation to respiratory flow activity and concentrations of oxygen and carbon dioxide. *Can. J. Zool. 40:* 817–862.

Schaefer, K. M. 1985. Body temperatures in troll-caught frigate tuna, *Auxis thazard. Copeia 1985:* 231–233.

Schmidt-Nielsen, K. 1997. *Animal physiology: Adaptation and environment* (5th ed.). Cambridge University Press, Cambridge, UK.

Shelton, G. 1970. The regulation of breathing, pp. 293–359. In *Fish physiology*, Vol. IV, W. S. Hoar and D. J. Randall (Eds.). Academic Press, New York.

Smith, L. S. 1966. Blood volumes of three salmonids. *J. Fish. Res. Bd. Can. 23*(9): 1439–1446.

Snyder, G. K., and B. A. Sheafor. 1999. Red blood cells: Centerpiece in the evolution of the vertebrate circulatory system. *Amer. Zool. 39:* 189–198.

Steen, J. B. 1963. The physiology of the swimbladder of the eel *Anguilla vulgaris*. I. The solubility of gases and the buffer capacity of the blood. *Acta. Physiol. Scand. 58:* 124–137.

———, and A. Kruysse. 1964. The respiratory function of teleostean gills. *Comp. Biochem. Physiol. 12:* 127–142.

Teal, J. M., and F. G. Carey. 1967. Skin respiration and oxygen debt in the mudskipper *Periophthalmus sobrinus. Copeia 1967:* 677–679.

Tota, B. 1989. Myoarchitecture and vascularization of the elasmobranch heart ventricle, pp. 122–135. In Evolutionary and contemporary biology of elasmobranchs, W. C. Hamlett and B. Tota (Eds.), *J. Exp. Zool. Suppl. 2.*

Vogel, W. O. P. 1985. The caudal heart of fishes: Not a lymph heart. *Act. Anat. 121:* 41–45.

Weber, R. E., and F. B. Jensen. 1988. Functional adaptations in hemoglobins from ectothermic vertebrates. *Ann. Rev. Physiol. 50:* 161–179.

Wittenberg, J. B., and R. L. Haedrich. 1974. The choroid rete mirabile of the fish eye. II. Distribution and relation to the pseudobranch and to the swimbladder rete mirabile. *Biol. Bull. 146:* 137–156.

Zapata, A. G., A. Chibá, and A. Varas. 1996. Cells and tissues of the immune system of fish, pp. 1–62. In *The fish immune system: Organism, pathogen, and environment*, G. Iwama and T. Nakanishi (Eds.). Academic Press, San Diego.

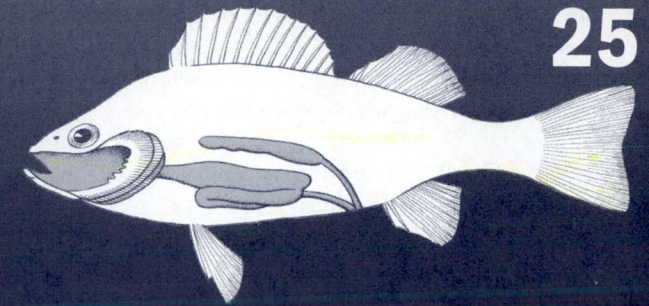

25

OSMOTIC AND SOLUTE REGULATION

OSMOREGULATION

Water and Solutes—Maintaining the Right Balance
Osmotic and Ionic Regulation in Freshwater Fishes
  Lampreys in Freshwater
  Elasmobranchs in Freshwater
  Freshwater Bony Fishes
Osmotic and Ion Regulation in Marine Fishes
  Hagfishes
  Lampreys in Salt Water
  Elasmobranchs
  Holocephali
  Latimeria
  Marine Teleosts
Diadromous and Other Euryhaline Fishes
Eggs and Larvae
Applications of Studies on Osmoregulation to Fish Culture
Role of the Endocrine System in Osmotic Regulation

THE URINARY SYSTEM

The Kidney
  General Features of the Kidney
  Hagfishes
  Lampreys
  Chondrichthyes
  Bony Fishes
Role of the Kidneys in Excretion and Osmoregulation

EVOLUTIONARY CONSIDERATIONS

Gills, Kidneys, and the Origins of Tetrapods

One of the more intriguing mysteries concerning the origin and diversification of fishes is whether they evolved in fully marine conditions, freshwater environments, coastal brackish waters, or possibly all of these. As discussed in Chapter 5, the fossil evidence points to a marine origin for the earliest vertebrates. Hagfishes, as the oldest extant craniates, are fully marine, yet lampreys, also with ancient origins, are best known from those species that freely move between freshwater and seawater.

Fishes show a range of adaptability to the ionic concentrations that they encounter in their environments. Some, such as the hagfishes, have only a limited capacity to selectively regulate their tissue ionic concentrations. For them, this is not of great concern, living as they do in the ocean depths, where fluctuation in ionic composition is at a minimum. But what about the teleosts that live in bays and estuaries? Tidal flushing and seasonal fluctuation in freshwater input may impose significant osmoregulatory challenges. Many migratory fish species will, in the course of their life histories, move from stream to ocean or from ocean to stream. In fishes, several tissue sites may function as the interface between the internal environment of the animal and its external environment. Included in these are the skin, the gut, and especially the gills and the kidneys. The regulation of body ionic concentration is inextricably tied to the mechanisms for the formation and release of urine—the task of the kidneys. But, as we shall see, the function of the kidneys in fishes is decidedly different from that in tetrapods. In this chapter, we will explore the tissues and organs that have evolved in fishes for the purpose of regulating body water content and ionic concentration and also for facilitating the removal of excess ions and the by-products of metabolism.

OSMOREGULATION

Water and Solutes—Maintaining the Right Balance

Some of the most crucial adjustments that fishes of all kinds must make in their respective environments concern the maintenance of proper water and salt balances in their tissues. With few exceptions, fishes are confined to the aquatic realm. Some lamprey species are known to make limited overland forays during their migrations. Although the Dipnoi and a few dozen species of bony fishes are capable of a limited amphibious existence, no chondrichthyans show this capability (Davenport and Sayer, 1993). Within the aquatic realm, fishes have achieved astounding success in both freshwater and the marine environment, yet each of these environments presents osmotic challenges that must be overcome. For freshwater fishes, failure to maintain the proper ion balance would result in the lethal dilution of body fluids, whereas active ion regulation is also necessary for marine fishes to prevent dehydration. Few fishes have internal salt concentrations that closely match the water in which they swim, so they must employ cellular and physiological processes to prevent the excessive gain or loss of water. The body fluids of freshwater fishes have a higher osmotic concentration than their medium, whereas marine species have more dilute body fluids than the seawater surrounding them. The fact that fishes have succeeded in inhabiting virtually the entire liquid hydrosphere means that they have experienced remarkable success in overcoming the inevitable challenges of existence in a medium with an osmotic concentration that is significantly different from that of their own body fluids.

Before continuing, a brief review of the concepts and terminology of salt and water balance is in order. One gram mole of a substance in 1 kg of solution is called a *molar* solution, whereas 1 gram mole per kg of solvent is referred to as a *molal* solution. In expressing osmotic activity, which depends on the number of nondissociated molecules and ions per unit volume or weight of solvent, the term **osmol (Osm)** is used. One gram mole per kg of water of a substance that does not dissociate can be said to have an **osmolality** of 1 osmol kg^{-1} and exerts an osmotic pressure of 22.4 atmospheres. Compounds that do dissociate have a higher osmolality, corresponding to the degree of dissociation. One mole of sodium chloride per kilogram has an osmolality close to 2 Osm kg^{-1}, because the compound dissociates nearly completely in solution. In dealing with the body fluids of animals, it is convenient to use smaller units than osmols, so the **milliosmol (mOsm)** is commonly used.

Osmotic concentration can be expressed also in terms of the **freezing point depression** (Δ_{fp}) of aqueous solutions.

A molal solution (1,000 mOsm) freezes at −1.86°C, which approximates the freezing point of seawater of average salinity. The salinity of seawater is expressed as **parts per thousand (ppt or ‰)** which equates to grams of solute per kilogram water. The mean salinity of the world's oceans is about 32 ‰. Table 25.1 shows comparisons between freezing point depression and salinity.

A solution with a lower amount of salt per unit volume than a solution to which it is being compared is said to be **hypo-osmotic** to the more concentrated solution, which in turn, of course, is **hyperosmotic** to the less concentrated solution. If the solutions have the same osmotic pressure, they are said to be **isosmotic**.

Although diffusion will be regarded in simple terms for the sake of this discussion, it is a complex process, involving such things as the nature of the membranes being penetrated, the concentration of solutes, and the electrical charges of particles. The discussion here will be mainly concerned with the net results of the process.

For terrestrial vertebrates, the kidney is the chief organ that regulates ion and water content, and to this function has been added the task of excretion of the nitrogenous byproducts of metabolism. For fishes, living for the most part in an aqueous environment, several additional epithelial tissue sites are involved in osmotic and ion regulation. Chief among these are the gills and the integument. The gut, urinary bladder, and specialized tissues, such as the elasmobranch rectal gland, also participate in osmoregulation. It is important to remember, however, that osmotic and ionic regulation is a totally integrated process in animals— the maintenance of homeostasis involves tissue sites for the detection of ion and water levels, neuronal and endocrine means of communication of osmotic state, and the actual tissues involved in the movements of water and solutes.

Given their diversity, it should not come as a surprise that fishes exhibit a range of osmoregulatory capabilities.

TABLE 25.1   FREEZING POINT DEPRESSION OF WATER AT SELECTED SALINITIES

Salinity (‰)	Δ_{fp} (°C)	Osmolality (mOsm kg^{-1})
5	−0.29	155
10	−0.58	312
15	−0.87	444
20	−1.13	608
25	−1.45	780
30	−1.72	925
32 (seawater)	−1.86	1,000
35	−2.03	1,091
40	−2.35	1,263

We can evaluate this capacity for osmoregulation graphically by assessing the effect of change in the external ionic environment on the internal ionic environment (Fig. 25.1). Animals that exhibit minimal change in internal ion and water levels when challenged by fluctuations in environmental ion concentrations are said to be strong **osmoregulators**. If an animal, such as a freshwater teleost, regulates its internal ionic composition above that of its environment, it is said to be a **hyper-osmoregulator;** if it regulates below its environmental ionic concentration, as would a marine teleost, it is termed a **hypo-osmoregulator**. Many fish species, such as anadromous lampreys and salmon or catadromous eels, regularly travel between freshwater and seawater in the course of their lives (see Chapter 36). These fishes engage in both hyper- and hypo-osmoregulation, depending on whether they are in freshwater or seawater. As the external ionic environment changes, however, some species do not compensate, and their internal osmoconcentration changes—these species are known as **osmoconformers**. Most animals are capable of regulating specific ions, however. In general, the distribution of the major fish taxa is an accurate reflection of their ability to maintain consistency in their internal ion and water composition. Hagfishes are osmoconformers, but they live in the ocean depths where they experience only minimal fluctuations of salinity. Lampreys, on the other hand, freely migrate between freshwater and the ocean. Chondrichthyan fishes have made only limited forays into freshwater, whereas the bony fishes have become well established in waters ranging from full-strength seawater (and sometimes hypersaline conditions) to freshwaters of widely varying ionic content.

Osmotic and Ionic Regulation in Freshwater Fishes

Because the osmotic concentration of typical fish blood, as expressed in mOsm kg^{-1}, is in the range of about 265 to 325 ($\Delta_{fp} = -0.5$ to $-0.61°C$), freshwater fishes are hyperosmotic to their medium and tend to gain water by diffusion through any semipermeable surface. If unchecked or uncompensated, this inward diffusion would result in a serious dilution of the body fluids. Waterproofing the body by means of a thick scale covering, bony armor, substantial quantities of connective tissue in the skin, or copious secretion of mucus might afford some protection (see discussion of ostracoderms in Chapter 5), but any site that maintains circulation of blood near the surface of the skin will also provide a conduit for the passage of water. The gills, obviously, cannot be waterproofed, and they provide ample surface for the inward diffusion of water as well as gases.

The osmotic uptake of water in excess of the needs of the organism is counteracted by driving the water out by some other means. This is accomplished by the kidney and its vascular supply. Blood is provided to the kidney by the renal artery and the renal portal system (see Chapter 24). The elevated pressure of blood from the renal artery permits ultrafiltration to take place as the blood passes through the capillaries of the glomeruli (see ensuing discussion of the kidneys). This results in the formation of a dilute urine, which is passed out of the body. The extensive ciliation of the kidney tubules enables them to propel copious amounts of filtrate. Blood leaving the glomerulus passes through capillaries surrounding the kidney tubules and returns to the general circulation via the renal vein. Blood from the renal portal vein also joins this network of capillaries around the tubules. Mainly divalent ions are released into the filtrate as the blood courses along these capillaries (Satchell, 1991).

Several researchers (Agarwal and John, 1975; Curtis and Wood, 1991; Evans, 1980; Marshall, 1995) have reported on the osmoregulatory capacity of the urinary bladder in freshwater and euryhaline (i.e., having a broad tolerance for fluctuating salinities; see Chapter 34) teleosts. Osmotic permeability of the bladder walls is low, but Na$^+$ and Cl$^-$ are actively reabsorbed through the walls.

The urine formed by freshwater fishes is quite dilute, with osmotic concentrations in various species and conditions

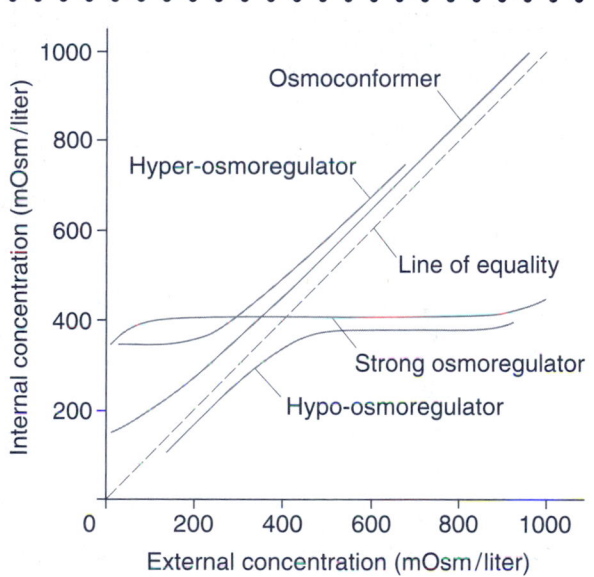

FIGURE 25.1
General patterns of plasma osmotic regulation demonstrated in the major taxa of fishes. Points where a change in external osmotic conditions results in an equal change of internal osmotic conditions fall along the line of equality. Animals that maintain internal osmoconcentration above this line are said to be hyper-osmoregulators, whereas those that osmoregulate below this line are termed hypo-osmoregulators. (From Gordon, 1977.)

ranging from about 16 to 55 mOsm kg^{-1} (Δ_{fp}=−0.03 to −0.09°C). The urine contains only small amounts of nitrogenous compounds, such as uric acid, creatine, and ammonia, as the bulk of nitrogenous waste is excreted via the gills (see discussion of kidney evolution later in this chapter).

Although the concentration of salt in urine is low, its copious flow causes a significant amount of salt to be lost. Salts are also lost by diffusion from the body. These losses are compensated by salt intake from food and by active absorption through the gills. The uptake of Cl$^-$ at the gills involves an exchange with HCO$_3^-$. Na$^+$ apparently moves across the cell membranes through a conductive channel created by protons secreted by a H$^+$-ATPase pump located in the gill epithelial cells (Karnaky, 1998). Hence, ion uptake is coordinated with acid−base balance in fishes and involves specialized cells (**chloride cells;** see Fig. 24.3) located in the gill epithelium (Claiborne, 1998; Karnaky, 1998; Perry, 1997). Chloride cells are characterized by an ovoid nucleus at the base of the cell, numerous mitochondria, and an extensive system of tubules (Fig. 25.2). Chloride cells have been identified in the gills of both freshwater and marine fishes, yet their function is more closely associated with salt secretion in seawater in euryhaline and marine fishes. It has been suggested that the proliferation of chloride cells on the gill

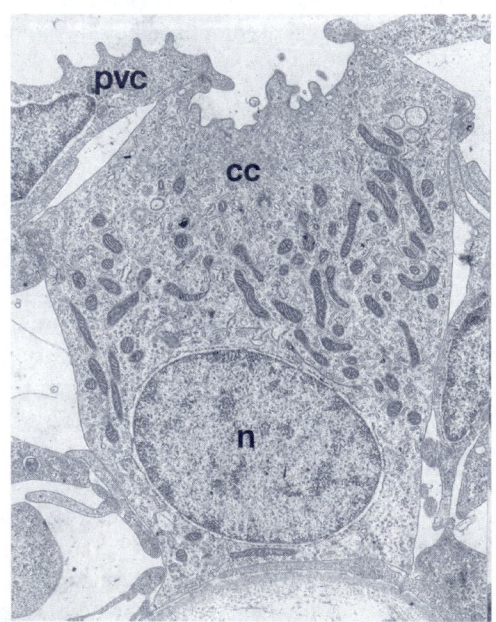

FIGURE 25.2
Transmisssion electron micrograph of a chloride cell (*cc*) in tilapia (*Oreochromis mossambica*). *n*—nucleus; *pvc*—adjacent pavement cell (Source: Perry, 1997).

lamellae of freshwater fishes that are experiencing osmotic challenge may impede the free diffusion of dissolved gases. Fishes may compensate for this by an increased ventilation rate or increased binding affinity of the hemoglobin (Perry, 1998). Some freshwater teleosts have been shown to drink water even though this requires the kidney to remove even greater amounts of excess water (Evans, 1980).

Lampreys in Freshwater

Although lampreys have not been the subject of as much physiological study as elasmobranchs and bony fishes, their capabilities as osmoregulators have been explored (see Table 25.3). About 70 percent of the body fluid is intracellular, about 7 percent is in the plasma, and the remainder is interstitial. The skin and gills of lampreys are permeable to water, but research findings on the extent of their permeability have been variable. The river lamprey (*Lampetra fluviatilis*) is reported to absorb water at a rate equivalent to about one third its body weight per day. The brook lamprey (*L. planeri*) apparently absorbs water at about twice the rate of the river lamprey (Morris, 1972).

The normal urine flow in freshwater lampreys is apparently not well known because of difficulties with laboratory methodology. Many values shown in the literature, ranging up to 360 ml kg^{-1} day^{-1} for river lamprey, are considered high by some authorities, but all researchers agree that the flow of the very dilute urine is copious. Osmolarity of *L. fluviatilis* urine in freshwater has been reported as ranging from about 20 to 38 mOsm l^{-1} (Evans, 1993; Youson, 1981). The amount of sodium excreted in the urine is about 117 mole h^{-1}; chloride is excreted at about 5 mole h^{-1}. Only small amounts of nitrogenous substances are excreted by the kidney. Most nitrogenous excretion is at the gills.

Salts lost by lampreys in the urine and through the body surface are balanced by salts in food and by direct absorption of salts from the water. The gills actively remove chlorides from the water and release them into the blood. Sodium, potassium, and calcium enter through the gills when made available as chlorides.

Elasmobranchs in Freshwater

In spite of their somewhat limited osmoregulatory capabilities, a number of elasmobranchs enter water of low salinity or even freshwater, and the freshwater stingrays (*Potamotrygon*) of the Amazon and some African and Asian dasyatids are confined to freshwater. Obviously, excess water would enter the body if they maintained the same osmotic concentration in the body fluids as that maintained by their marine counterparts. Some marine species can be acclimated to about half-strength seawater. They adjust by reducing the amounts of chloride, urea, and trimethylamine oxide

(TMAO) in the blood until the osmotic concentration is reduced to about 600 mOsm kg^{-1}. In the face of this reduced osmotic concentration, their hypo–osmotic state results in increased urine flow (Evans, 1993).

Elasmobranchs that often move from the sea into large tropical rivers include the smalltooth sawfish (*Pristis pectinata*) and the bull shark (*Carcharhinus leucas*), sometimes called the freshwater or Nicaragua shark. These species can reduce the osmotic concentration of their blood to a range of 485 to 550 mOsm l^{-1}, retaining only about 100 to 180 mmole l^{-1} urea in the blood. Urine flow in the smalltooth sawfish has been measured at an average of 10.4 ml kg^{-1} hr^{-1} (Thorson, 1967). The rectal gland of the bullshark regresses while in freshwater by decreasing the number of glandular tubules. The freshwater stingrays (*Potamotrygon*) have nearly broken the urea habit after a long history as a freshwater genus; their blood contains only slightly more than 1 mmole l^{-1} of urea. Their osmoregulation is essentially like that of freshwater teleosts, and the rectal gland is nonfunctional.

Freshwater Bony Fishes

The permeability of bony fishes to water is generally less than that of lampreys, although some bony fishes have been noted as absorbing up to one third of their body weight per day (Fig. 25.3A). Scales and other armor in the bony fishes aid in retarding water uptake, as armor might have done in the extinct ostracoderm relatives of the lampreys. The eel (*Anguilla*) is often singled out as having a nearly impervious skin. Eel skin is reported to be so dense that it represents about 10 percent of the body weight. For bony fishes, there is evidence that most of the water absorbed comes through the gills (Black, 1957; Conte, 1969). The body water of freshwater teleosts makes up about 70 to 75 percent of their body weight. Sturgeons, paddlefishes, gars, and the bowfin generally have a similar proportion of water. In freshwater teleosts, intracellular water accounts for about 60 percent (55 to 63 percent) of the total body water; about 12 to 16 percent is extracellular water, and plasma accounts for about 2 percent of body water (Thorson, 1961). Figures given by Parry (1966) are somewhat higher: 74 to 80 percent as intracellular water, and 2.5 to 3 percent of the body fluids in plasma.

Ionic plasma concentrations of selected freshwater, marine, and euryhaline teleosts are given in Table 25.2. The osmotic concentration of freshwater fish blood is usually around 300 mOsm kg^{-1}, but at least one species, the reedfish (*Erpetoichthys calabaricus*) has a dilute plasma of 199 mOsm kg^{-1} (Lutz, 1975a, 1975b). Many anadromous species in freshwater may maintain concentrations around 325 mOsm l^{-1}. Bony fishes in freshwaters move relatively large amounts of water through the kidneys. Urine flow varies among species and with the environment, but several studies have indicated

flows between 50 and 150 ml kg^{-1} day^{-1}. Urine production rates as low as 16 ml kg^{-1}day^{-1} have been measured for the pike, and as high as 330 ml kg^{-1}day^{-1} for the goldfish (Hickman and Trump, 1969). The osmotic concentration of the urine of freshwater bony fishes is usually between 30 and 40 mOsm l^{-1}. A range of 20 to 80 mOsm l^{-1} is given by Bone and Marshall (1982).

As might be expected, environmental pH can have a significant effect on gill morphology and function, specifically affecting ion transport capacity (Goss et al., 1995). Teleosts that inhabit the naturally acidic, ion-deficient Amazonian "blackwaters" (see Chapter 31) are an interesting exception. Here, the ion flux via the gills is remarkably insensitive to ambient pH (Gonzalez and Preest, 1999; Wilson et al., 1999).

The ionic composition of the urine of freshwater fishes varies greatly among species and conditions; only small amounts of nitrogenous compounds are excreted in the urine of freshwater fishes. In addition to urea and ammonia, which are excreted mainly through the gills, the urine may contain creatine, uric acid, amino acids, and creatinine (Hickman and Trump, 1969). Mention should be made of the African lungfishes (*Protopterus*), which produce urea during aestivation and store it in the blood, producing no urine for several months (Forster and Goldstein, 1969; Smith, 1930).

As in lampreys, the loss of salts in the urine and by diffusion in bony fishes is compensated by salt uptake from ingested food and by absorption through the gills. The uptake of sodium is in part related to the excretion of ammonia at the gills. Ion exchange mechanisms operating in branchial cells facilitate the exchange of ammonia for sodium and of bicarbonate for chloride. Under suitable conditions, Na$^+$ is also exchanged for H$^+$. The exchange rate for sodium in freshwater species is very small compared to the exchange that occurs in salt water species. Only about one percent of the exchangeable sodium in the body of the euryhaline threespine stickleback (*Gasterosteus aculeatus*) is exchanged per hour in freshwater. This rate is 20 times that for the same species in seawater (Mactz, 1974).

Osmotic and Ion Regulation in Marine Fishes

The osmotic concentration of salts in the blood of marine fishes (except hagfishes) is significantly less than that of seawater (see Tables 25.2–25.4). Bony fishes adapted to marine environments maintain blood at an osmolality of about 380 to 470 mOsm kg^{-1} ($\Delta_{fp} = -0.7$ to $-0.87°C$; see Table 25.2). The salt content of elasmobranch blood is slightly higher than this but forms only a part of the blood's osmolality, the remainder being due to the retention of nitrogenous substances, specifically urea and trimethylamine oxide (TMAO). For marine fishes, the maintenance of proper

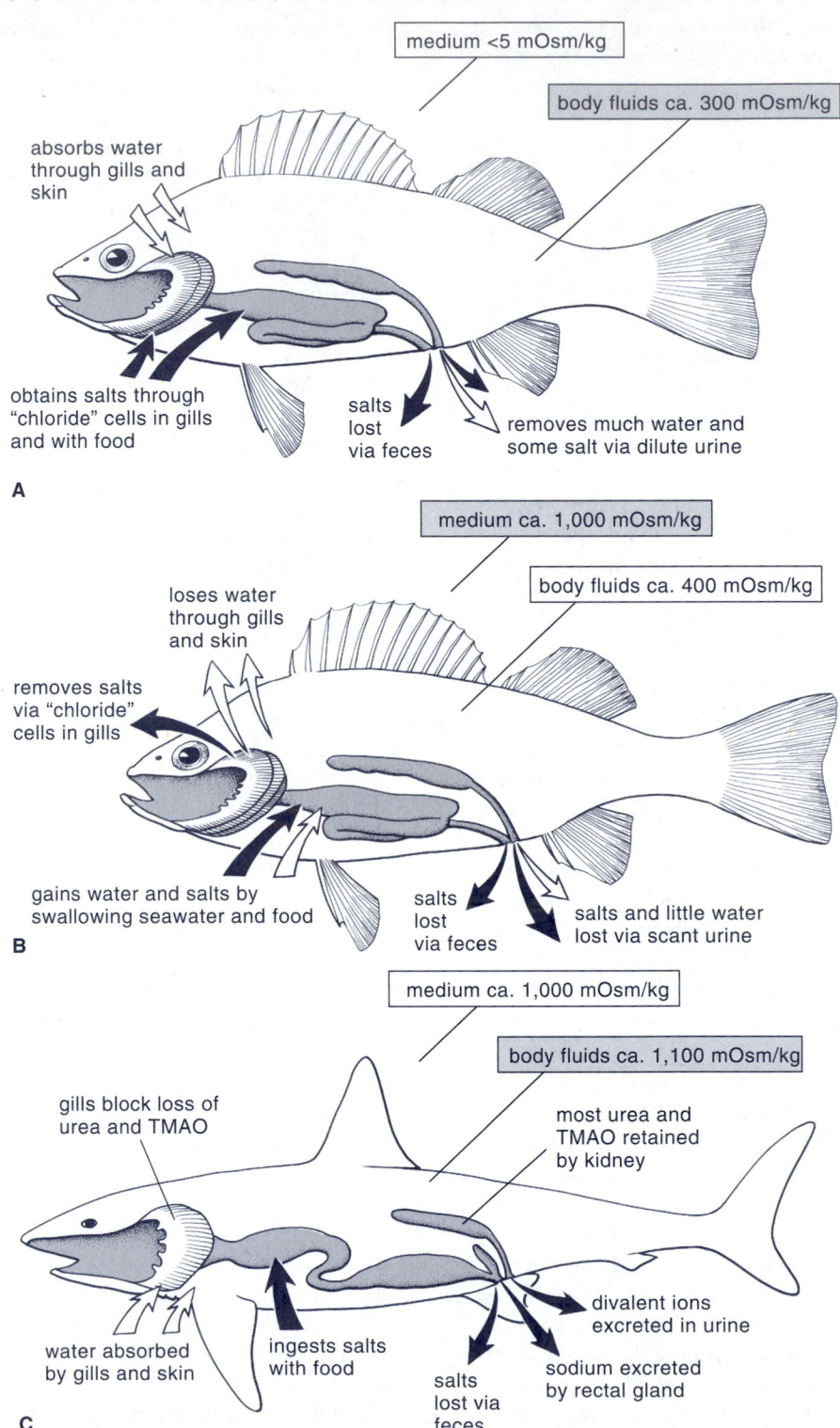

medium <5 mOsm/kg

body fluids ca. 300 mOsm/kg

absorbs water
through gills and
skin

obtains salts through
"chloride" cells in gills
and with food

salts
lost
via feces

removes much water and
some salt via dilute urine

A

medium ca. 1,000 mOsm/kg

body fluids ca. 400 mOsm/kg

loses water
through gills
and skin

removes salts
via "chloride"
cells in gills

gains water and salts by
swallowing seawater and food

salts
lost
via feces

salts and little water
lost via scant urine

B

medium ca. 1,000 mOsm/kg

body fluids ca. 1,100 mOsm/kg

gills block loss of
urea and TMAO

most urea and
TMAO retained
by kidney

water absorbed
by gills and skin

ingests salts
with food

salts
lost via
feces

divalent ions
excreted in urine

sodium excreted
by rectal gland

C

FIGURE 25.3

Osmoregulation and ion and water flux in fishes: **A,** Freshwater teleost; **B,** Marine teleost; **C,** Marine elasmobranch.

TABLE 25.2   PLASMA IONIC CONCENTRATIONS (mOsm kg^{-1}) OF SELECTED FRESHWATER, MARINE, AND EURYHALINE TELEOSTS

Species	Na	Cl	K	Mg	Ca	SO$_4$	Total
Cyprinus carpio (freshwater)	130	125	2.9	1.2	2.1	–	274
Lophius piscatorius (marine)	180	196	5.1	2.5	2.8	2.7	452
Pleuronectes flesus (euryhaline; in freshwater)	124	132	2.9	–	2.7	–	240
P. flesus (euryhaline; in seawater)	142	168	3.4	–	3.3	–	297

Source: Karnaky, in Evans, 1998.

water and salt balance requires different mechanisms from those used by freshwater fishes. There are at least three methods evident among marine fishes: (1) hagfishes are essentially osmoconformers; (2) elasmobranchs and *Latimeria* osmoregulate by retaining urea; and (3) the remaining bony fishes and lampreys osmoregulate using specialized epithelial cells, particularly in the gills.

Hagfishes

Hagfishes are characterized by a very permeable skin and virtually lack the ability to regulate sodium chloride, so that exposure to water of higher or lower salinity than seawater results in rather rapid changes in the osmotic concentration of their body fluids. Selected ions, especially divalent ions, are regulated by hagfishes, however (Table 25.3). The body fluids of hagfishes are practically isosmotic to their medium, although some investigators have shown a slight hypertonicity (Robertson, 1963). The ionic content of hagfish plasma differs from that of seawater in that Ca^{2+}, Mg^{2+}, and SO_4^{2-} are in lower concentrations and Na^+ is in higher concentration. The concentration of Cl^- in hagfish plasma has been reported as lower than that of seawater by some investigators and as higher than seawater by others. The concentration of K^+ is reported to be nearly the same as in seawater. The freezing point depression of hagfishes has been determined to range from -1.74 to $-1.98°C$ in full-strength seawater (Robertson, 1963). Urea appears to be present in very small amounts. Divalent ions and K^+ are excreted in the scant urine.

Lampreys in Salt Water

While they are in salt water equivalent to one half to full-strength seawater, river lampreys (*Lampetra fluviatilis*) and sea lampreys (*Petromyzon marinus*) can maintain their body fluids at about 285 to 330 mOsm kg^{-1}. As the body fluids are hypo-osmotic to the medium, diffusion tends to remove water from the body through the gills and skin. To recoup this loss, lampreys swallow seawater (50 to 220 ml kg^{-1} day^{-1}), which is absorbed from the gut.

Also absorbed from the gut are monovalent ions, mainly Na^+ and Cl^-, which are in excess of the lamprey's needs. Divalent ions from the seawater are not absorbed by the gut and are eliminated with the feces. Divalent ions are excreted also in the meager but concentrated urine. The excess chloride is excreted through the gills, apparently through the chloride cells described earlier.

Elasmobranchs

Osmoregulation in elasmobranchs (Fig. 25.3C) involves bolstering the osmotic concentration due to salts in the blood by the retention of urea and smaller amounts of other nitrogenous compounds. Urea, an end product of nitrogen metabolism, is produced through glutamine-dependent synthesis in the liver (Acher et al., 1999) and is excreted only in relatively small amounts via the urine of sharks and rays. As the glomerular filtrate passes along the kidney tubule, special segments resorb much (70 to 90 percent) of the urea, so that the blood contains about 350 mmole l^{-1} urea in a typical marine elasmobranch. The urea content of the blood

TABLE 25.3   PLASMA IONIC CONCENTRATIONS (mOsm kg^{-1}) OF AGNATHAN FISHES

	Na	Cl	K	Mg	Ca	SO$_4$	Total
Seawater	439	513	9.3	50	9.6	26	1050
Myxine glutinosa (Atlantic hagfish)	486	508	8.2	12	5.1	3	1035
Petromyzon marinus (sea lamprey)	156	159	32	7	3.5	–	333
Lampetra fluviatilis (river lamprey—freshwater)	120	104	3.9	2	2.5	–	272

Source: Karnaky, in Evans, 1998.

is usually given as 2 to 2.5 percent. The urine of marine elasmobranchs usually contains about 100 mmole l⁻¹ of urea. Trimethylamine oxide (TMAO), another nitrogenous waste product, appears in the blood at about 70–100 mmole l⁻¹ and is therefore of secondary importance to the osmolality of the blood. It is reabsorbed in the kidney tubule, and its concentration in the urine is about 10 mmole l⁻¹. The gills appear to be the major site of urea efflux (Karnaky, 1998). The importance of TMAO and other methylamines in the body and intracellular fluids may be mainly to counteract the destabilizing action of urea on proteins (Shuttleworth, 1988). Kelly and Yancey (1999) have reported a correlation between TMAO concentration in body tissues and depth of occurrence: Deeper-dwelling elasmobranchs, teleosts, and crustaceans all show higher concentrations of TMAO when compared with shallower-dwelling relatives. Kelly and Yancey suggested that increased TMAO concentrations may increase buoyancy, possibly counteract the protein destabilization brought on by increased pressure, or may simply be a consequence of dietary differences.

The concentration of the various solutes, mostly sodium, chloride, urea, and TMAO, in the blood of elasmobranchs produce a combined osmolality of about 1,000 to 1,100 mOsm kg⁻¹ in animals living in seawater of about 930 to 1,030 mOsm kg⁻¹. Thus, sharks and rays typically have an osmotic concentration in the blood about 50 to 100 mOsm kg⁻¹ higher than the concentration of the medium. Being hyperosmotic to seawater, elasmobranchs tend to gain water by diffusion. This influx is mainly through the gills, as the skin is nearly impervious. This excess water is excreted as urine, the flow of which is typically from 1 to 1.5 ml kg⁻¹ h⁻¹. Salts enter the marine elasmobranchs by diffusion and via ingested food because their body fluids have a more dilute concentration of salts than the medium. Furthermore, the drinking of

seawater has been confirmed in some elasmobranchs, but at a rate of only one tenth to one fifth that of teleosts.

Salts are excreted via two main pathways. The urine is the most important medium of excretion of divalent ions, the concentration of magnesium and phosphate in urine being more than 30 times their concentration in plasma, and the concentration of sulfate being nearly 140 times that of plasma. Sodium and chloride account for most of the osmotic concentration of the urine, but appear in nearly the same concentrations as those found in plasma (Table 25.4).

Sodium is excreted by the rectal gland, an organ that was of unknown function until about 1960 (Burger, 1962; Burger and Hess, 1960). The normal concentration of NaCl in rectal gland fluid is nearly twice that of plasma, although the osmolality of the two fluids is the same. Studies have supported the contention that the rectal gland functions in the regulation of intravascular fluid volume rather than osmoregulation (Karnaky, 1998). Small amounts of potassium and even smaller amounts of calcium and magnesium are excreted by this gland. The urea concentration of rectal gland fluid is about 4 percent of the concentration in the plasma. The volume of flow from the rectal gland is variable, usually 0.5 to 0.8 ml kg⁻¹ h⁻¹. When the rectal gland is experimentally rendered nonfunctional, the kidneys appear to compensate by releasing more copious urine (Shuttleworth, 1988). The model for salt secretion in the rectal gland is considered the same as that seen in seawater-adapted teleost chloride cells, and the genes for several transport-related proteins involved in rectal gland NaCl secretion have been sequenced (Karnaky, 1998).

Elasmobranchs possess chloride cells in the gills, which would suggest an active ion exchange at this site (Bone and Marshall, 1982; Kirschner, 1991). Although ions apparently can be lost and gained by the gills, Shuttleworth (1988)

TABLE 25.4 CONCENTRATIONS OF SELECTED SOLUTES IN PLASMA AND URINE OF THE SPINY DOGFISH (*SQUALUS ACANTHIAS*)

Solute	Concentration (mmole kg⁻¹)		
	Seawater	Plasma	Urine
Sodium	440	250	240
Potassium	9	4	2
Calcium	10	3.5	3
Magnesium	50	1.2	40
Chloride	490	240	240
Sulfate	25	0.5	70
Phosphate	0	0.97	33
Urea	0	350	100
TMAO	0	70	10
Osmolality	930 mOsm l⁻¹	1,000 mOsm l⁻¹	800 mOsm l⁻¹

TMAO = trimethylamine oxide.
Source: Hickman and Trump, 1969.

believed that any net elimination is insignificant compared to the amounts released by the kidneys and the rectal gland.

Holocephali

Chimaeras retain a high concentration of urea in the blood and maintain their internal osmotic pressure at or slightly above that of the surrounding seawater. They differ from sharks and rays in the relative proportions of solutes in their blood, as more salt and less urea and TMAO are retained (Table 25.5).

Although a discrete rectal gland like that seen in elasmobranchs has not been identified in holocephalans, it is considered by various investigators to exist in primitive form (Fänge and Fugelli, 1962). Secretory cells similar to those of rectal glands are present along with ducts entering the rectum.

Latimeria

This ancient marine fish osmoregulates in much the same manner as chondrichthyans. It retains urea and TMAO in the blood, and these nitrogenous solutes are more important in keeping a high osmotic concentration than they are in sharks, rays, and chimaeras (Table 25.6). Sodium chloride concentration in the blood of *Latimeria* is less than that in the cartilaginous fishes. The figures for Na⁺ and Cl⁻ approximate those seen in teleosts, but are much lower than the values reported for chondrichthyans. Urea and TMAO are higher in *Latimeria* than in the cartilaginous fishes. Although the reported range of osmolality varies, the usual level is slightly hypo-osmotic, and *Latimeria* must drink seawater to maintain its fluid level. The comparatively small gill surface area probably serves to minimize water loss. The coelacanth possesses a rectal gland, which is a likely site for salt secretion (Griffith and Pang, 1979). The kidney does not provide a high concentration of salt to be voided into the urine, and chloride cells in the gills are not numerous.

The similarity in osmoregulation between elasmobranchs and *Latimeria* is not indicative of a close phylogenetic relationship (Griffith and Pang, 1979). *Latimeria* differs

TABLE 25.5 CONCENTRATION OF SELECTED SOLUTES IN SERUM AND URINE OF *HYDROLAGUS*

Constituent	Concentration (mmole l⁻¹)	
	Serum	Urine
Na⁺	300	162
Cl⁻	306	268
Urea	245	51.6
TMAO	5.5	–
Osmolality	897 mOsm l⁻¹	844 mOsm l⁻¹

TMAO = trimethylamine oxide.
Source: Read, 1971.

TABLE 25.6 CONCENTRATION OF SELECTED SOLUTES IN SERUM OF *LATIMERIA*

Constituent	Serum concentration (mmole l⁻¹)
Na⁺	197
Cl⁻	87
Urea	377
TMAO	122
Osmolality	923 – 1,181 mOsm l⁻¹

TMAO = trimethylamine oxide.
Source: Griffith et al., 1974.

from chondrichthyans in that it is unable to reabsorb urea in the kidney tubules. This is a feature that *Latimeria* shares with the crab-eating frog (*Rana cancrivora*). Amphibians are almost exclusively restricted to freshwater, and the crab-eating frog, an inhabitant of mangrove estuaries of Southeast Asia, is an exception to the rule. It survives in an osmotically challenging environment by retaining far greater quantities of urea than any other amphibian.

Marine Teleosts

Marine teleosts (Fig. 25.3B) have slightly less total body water than freshwater teleosts and have a blood osmolality that normally ranges from 380 to 450 mOsm kg⁻¹ in an external environment of 800 to 1,200 mOsm kg⁻¹. Thus, compensation for the threat of desiccation is the rule here. Some species have been shown to lose from 30 to 60 percent of their water intake by osmosis. This loss is compensated for largely by ingesting seawater. The rate of drinking varies with the species, and within species it varies with salinity: The higher the salinity, the greater the rate of seawater intake. Marine species commonly ingest from 7 to more than 35 percent of their body weight in water per day (Conte, 1969; Johnson, 1973; Kirschner, 1991). From 60 to 80 percent of the ingested water is absorbed through the alimentary canal, beginning in the esophagus, where *Anguilla*, for example, absorbs a significant proportion of the swallowed water (Conte, 1969; Kirschner, 1991). Absorbed with the water are the monovalent ions Na⁺, K⁻, and Cl⁻. Divalent ions remain mostly in the gut; usually, less than 20 percent of those swallowed are absorbed.

The excess monovalent ions imbibed with the water are excreted mainly through the gills via chloride cells that resemble the salt-secreting cells of other animals (Evans et al., 1999; Kirschner, 1991; Vickers, 1961). These cells are concentrated in the gills, but in many species they appear also in the mouth lining and the skin of the inner surface of the operculum as well as in the skin of the head and anterior body (Marshall et al., 1997). The tubules observed in chloride cells (see Fig. 25.2) are continuous with the basal surface of the cell and are rich in Na^+, K^+, and adenosine triphosphatase (ATPase). More than one type of chloride cell, with distinctive cell morphology and location relative to the gill vasculature, has been identified in the gills of teleosts (Karnaky, 1998). Each ion-secreting cell has a cavity at its apex, opening to the exterior, and chloride is secreted from these cavities (Bone et al., 1995). Sodium chloride likely enters via connections with adjacent cells, but Na^+ is kept at a low level in the chloride cell by an Na^+/K^+-activated ATPase pump located on the plasma membrane at the base of the cell (Maetz, 1974). As Na^+ is removed from the cell at its base, it moves paracellularly around the chloride cell and into the fluid environment. The Cl^- is extruded from the apex of the cell, perhaps as a vesicle, into the water.

Guppies (*Poecilia reticulata*) adapted to salt water showed an increase of chloride cells not only on gill tissue, but in the mouth, the inner lining of the operculum and even in pockets under the scales (Vickers, 1961). In studies involving the immunohistochemical localization of Na^+/K^+ ATPase activity in the gills of freshwater- and seawater-adapted strains of guppies, enhanced immunoreactivity for Na^+/K^+ ATPase and increases in chloride cell size and number were observed in newborn guppies from seawater-adapted mothers, suggesting a maternal influence on osmoregulatory capacity in developing offspring (Shikano and Fujio, 1999).

Knowledge of the molecular biology of the transport mechanisms has advanced to the point that genes coding for the pump proteins involved have been identified and sequenced (Karnaky, 1998). Sodium excretion balances the intake through the gut, and the sodium–potassium exchange at the chloride cell appears to be important for the loss of sodium. Studies on the flounder (*Platichthys flesus*) showed that no sodium was excreted through the gills in potassium-free seawater (Maetz, 1969). Gill chloride secretion has been discovered not to be the exclusive domain of the chloride cells, however. Cultures of fish gill respiratory cells have also been demonstrated to possess chloride secretion capabilities, thus challenging the conventional model that claimed chloride cells to be unique in their osmoregulatory function (Avella and Ehrenfeld, 1997).

A peculiar variation on the theme of teleost osmoregulation is seen in the cobbler (*Cnidoglanis macrocephalus*), a plotosid catfish that apparently depends on a special salt gland (dendritic organ) for the maintenance of salt balance. The organ is external, is attached just posterior to the urogenital opening, and has an ultrastructure similar to that of the rectal gland of sharks. Ligature of the organ results in severe ion imbalance (Kowarsky, 1973).

Much remains to be learned about the mechanisms of ion exchange in fishes. Electron microscopy has revealed much about the ultrastructure of the secretory cells (Sakurai et al., 1997). Current research focuses on investigations of the biochemical pathways involved in the energetics of ion regulation and on the elucidation of the genomic components responsible for the synthesis of the necessary enzymes and other proteinaceous components involved in the processes.

Diadromous and Other Euryhaline Fishes

Many species are capable of living in both freshwater and salt water. **Diadromous** fishes include both marine and freshwater environments in their life histories. **Anadromous** fishes are hatched in freshwater, subsequently move to the sea to feed and grow, and then return to freshwater to spawn. For **catadromous** species, it is just the opposite. Many other species have a broad tolerance for salinity and can move freely between fresh and salt water. All of these must be able to adjust their osmoregulatory mechanisms, sometimes quite rapidly, depending on the speed with which they change habitat or the habitat changes around them. Diadromous species generally undergo progressive changes that may alter their appearance as well as their physiology (Fontaine, 1975). At a given time, depending on the stage of life history, these fishes are adapted specifically to one medium or the other, and do not usually change back rapidly. Some euryhaline species, however, can make rapid osmotic adjustments as they travel through a range of salinities.

The young of several anadromous species of salmonids, upon attaining a characteristic minimum size (often 10 to 15 cm in length), undergo a transformation from the freshwater **parr** stage to the salt-tolerant **smolt** stage. The critical size range, which in nature might not be reached for one, two, or more years, depending on the species, can be reached in only a few months of rearing in a hatchery. The fishes change to a silvery color, become slimmer in form, and show tendencies to migrate downstream into the ocean. If they are restrained from migrating, the salt tolerance of some species— such as the rainbow trout (*Oncorhynchus mykiss*)—regresses after several weeks. Chum salmon (*O. keta*) and pink salmon (*O. gorbuscha*) migrate seaward while they are very young and are salt-tolerant soon after the yolk sac is absorbed. This salt tolerance does not normally regress.

When diadromous species are adapted to salt water, the osmotic concentration of their blood is higher than in the same species adapted to freshwater. Examples of a few species are given in Table 25.7 (see also Table 25.2). For diadromous and otherwise euryhaline teleosts, it is not clear if the sites in the gill epithelium used for the uptake of salts during freshwater existence are the same as those involved in the excretion of salts during inhabitation of seawater. The Na^+/K^+ ATPase activity associated with ion transport across the gill epithelium is subject to both slow modulation via endocrine control and rapid modulation, possibly via cyclic AMP–mediated phosphorylation involving protein kinases (Tipsmark and Madsen, 2001).

Adaptation of euryhaline fishes to salt water generally requires drinking of the medium. Fishes such as rainbow trout and eels, which swallow little or no water while in freshwater, may drink about 4 to 15 percent of their body weight per day in salt water. The Mozambique tilapia (*Oreochromis mossambicus*) may swallow nearly 30 percent of its body weight per day (Johnson, 1973). Drinking rate increases with temperature (Maetz, 1974). Many species respond to the saline medium by changes in kidney function. The glomerular filtration rate may diminish dramatically, and the tubular reabsorption of water usually increases. Urine flow decreases to 10 percent or less of the urine flow in freshwater. The blood of a fish that commences hypo-osmoregulation usually stabilizes within 1.5 to 5 days.

The osmoregulatory capacity of some species is truly remarkable. The inanga (*Galaxias maculatus*) lives in salinities from less than 1 to 49 ppt and can be acclimated to 62 ppt. The Mozambique tilapia has been acclimated to 69 ppt (Chessman and Williams, 1975). Many members of the Poeciliidae and Cyprinodontidae can tolerate high salinities. The sheepshead minnow (*Cyprinodon variegatus*), a widely distributed cyprinodont in the New World that inhabits brackish coastal ponds and lagoons, has been found in salinities of more than 142 ‰—about four and one half times the salinity of full-strength seawater (Martin, 1972; Simpson and Gunter, 1956).

Many marine fishes, especially from tropical families, frequently move into and out of freshwater, some at certain life history stages, but some in a more spontaneous fashion. These fishes must be able to switch abruptly from conserving water to filtering out large volumes through the kidney, and must turn from the excretion of excess salts to their conservation. Several marine families have given rise to vicarious freshwater forms (see Table 29.1). The glomeruli of kidneys (see later discussion) are important for the filtration of excess water, and a few of these forms seem to be unlikely candidates for freshwater inhabitation, as they have few or no glomeruli. For example, there are freshwater species of pipefishes and toadfishes. These species apparently are quite impermeable and can increase their urine flow through a mechanism that is not well known. A marine toadfish (*Opsanus tau*) that can adapt to low salinity has been noted as having a plasma osmolality of 392 mOsm kg^{-1} in seawater and 250 mOsm kg^{-1} in freshwater (Lahlou et al., 1969). This species apparently engages in a pulsatile release of urea from the gills, while at the same time keeping its permeability to water unchanged (Pärt et al., 1999).

TABLE 25.7   COMPARISON OF PLASMA OSMOLALITY OF SELECTED DIADROMOUS FISHES IN FRESHWATER AND SEAWATER

Species	Plasma concentration (mOsm kg^{-1})	
	Freshwater	Seawater
Petromyzon marinus	280	333
Oncorhynchus tshawytscha	304	350
Salmo salar	328	344
Anguilla sp.	350	430

Eggs and Larvae

Eggs of fishes, at the time of deposition, are essentially isosmotic to the body fluids of the female. Hagfish eggs are nearly isosmotic with the environment as well. In elasmobranchs, there are sufficient concentrations of urea and TMAO in the eggs of oviparous species that they do not face an osmotic challenge when they are deposited. These embryos, as well as those of ovoviviparous species, maintain their levels of urea and TMAO through the sometimes lengthy incubation and are apparently not at an osmotic disadvantage (Shuttleworth, 1988). On the other hand, the eggs and sperm of teleosts are generally placed in environments with either higher or lower osmotic concentrations. Under conditions that are normal for the species involved, exposure of the gametes to a medium differing in osmolality exerts no ill effects. Furthermore, the results of experimentation with gametes of salmon, herring, and flounders have shown that fertilization in some species can occur over a broad range of salinities. The percentage of successful fertilization decreases at low (<15 ppt) salinity in the marine spawners and at high (>24 ppt) salinity in the salmon. Most freshwater species have low rates of fertilization and hatching in saline water (Holliday, 1969). Male pipefishes (Syngnathidae) that gestate their eggs in brood pouches and frequently live in brackish or freshwaters are capable of osmoregulation of the brood pouch fluid. This ensures that the eggs will develop in an ionically stable environment (Quast and Howe, 1980).

Fertilized eggs, although they imbibe water through the chorion, are somehow able to regulate their salt concentration to a great extent. There appears to be some effect of the salinity at which fertilization takes place on the subsequent ability of the eggs to develop optimally. This seems to be due to some initial influence by the salinity of the medium on the physical properties of the perivitelline fluid.

There are some examples of incubation and hatching at unusual salinities. The sheepshead minnow (*Cyprinodon variegatus*) has been known to hatch at 110 ppt, the desert pupfish (*C. macularius*) and the fourbeard rockling (*Enchelyopus cimbrius*) at 70 ppt, and the herring (*Clupea harengus*) at 60 ppt (Holliday, 1969). Chum salmon (*Oncorhynchus keta*) eggs, when transferred at the eyed stage to full-strength seawater, showed 50 percent survival through hatching, although the alevins did not survive (Kashiwagi and Sato, 1969). The eggs of a few primary freshwater teleosts are known to incubate and hatch normally at salinities up to 5 ppt.

Applications of Studies on Osmoregulation to Fish Culture

Knowledge of the impact of environmental salinity on egg and larval stages and of the relationships between osmoregulation and growth in salmonids and various other fishes used in aquaculture is of great practical importance.

By incubating eggs and rearing fry of rainbow trout at several salinities, it was determined that a hypertonic medium caused decreased hatching and increased mortality of alevins. Fry grew slower and showed elevated metabolic rates at hypertonic salinities. Fry reared in isotonic water showed no increase in growth rate (Morgan and Iwama, 1990).

Newly hatched larvae of some commercially significant marine species (herring, flounder) can tolerate broad ranges of salinity, but this tolerance diminishes with age. Alevins of chum salmon can survive in quarter-strength seawater one day after hatching, but they do not adapt well to full-strength seawater until the yolk sac is absorbed at 60 or more days after hatching. Cells that are rich in microtubular structures, resembling the chloride cells of mature fishes, appear in the epidermis of larval herring. Possibly osmoregulation is begun in such cells prior to the full development of the gills, gut, and kidneys (Holliday, 1969). The appearance of chloride cells in the gills of larval fishes several days before the development of secondary lamellae suggests that the gills may first function as osmoregulatory devices in fishes and secondarily become gas exchange structures (see Chapter 24). This has profound implications for investigations of the original function of gills in vertebrates (Rombough, 1999).

The culture of salmon and trout in net cages set in bays and fjords has certain advantages over rearing in freshwater, so the time or growth stage at which the young fishes can be transferred to a saline environment without undue mortality or altered metabolism must be known. Research on Atlantic salmon smolts has shown that a loss of appetite and a slowing of growth ensues after transfer to seawater, and those effects last beyond the shift of smolts to salt water osmoregulation. Their metabolism changed from a ten-dency to deposit lipids while in freshwater to a tendency to deposit proteins after transfer and adaptation to salt water (Usher et al., 1991). Numerous other studies have addressed parr–smolt transformation and freshwater–seawater transfer (Hansen et al., 1989).

Role of the Endocrine System in Osmotic Regulation

Although hormones are likely to be of great importance for osmoregulation in lampreys and hagfishes, these primitive fishes have not been studied as extensively as the gnathostomes. As discussed in Chapter 26, arginine vasotocin and small amounts of corticosteroids may figure significantly in the osmoregulation in agnaths. Adrenaline and arginine vasotocin have been determined to function as diuretics in lampreys (Rankin et al., 1983). Recent studies have identified natriuretic peptides as also important for osmoregulation in agnaths (see Chapter 26).

Many questions regarding the hormonal control of osmoregulation in elasmobranchs are still to be answered. In 1997, the American Elasmobranch Society held its first symposium on elasmobranch endocrinology; among the contributed papers was that of Acher et al. (1999), who reported on the unique, urea-based osmoregulatory mechanism of elasmobranchs and the diversity of hormones that have evolved in support of this system. A peptide called **rectin** appears to be a factor in the control of secretion of the rectal glands of *Squalus*, *Scyliorhinus*, and *Raja*. Adenosine stimulates secretion in rectal glands, and somatostatin may be inhibitory of rectal gland secretion. There is evidence that steroid hormones are also important in the regulation of the rectal gland and of urea content and permeability to water. Thyroid hormones and prolactin are involved as well (Shuttleworth, 1988).

Much knowledge has accumulated concerning the relationship of the endocrine secretions to osmoregulation in bony fishes, especially teleosts. The complex phylogeny of teleostean species is apparent in the diversity of responses demonstrated in experimental studies of endocrine function, but the evidence that hormones play a key role in osmoregulatory processes is incontrovertible. The conventional wisdom concerning osmoregulation in bony fishes has been that prolactin facilitates adaptation to freshwater, whereas cortisol facilitates seawater adaptation. Recent studies have suggested that cortisol may be important for both freshwater and seawater adaptation and that the growth hormone/insulinlike growth factor I axis may also figure significantly in teleost osmoregulation (McCormick, 2001).

Secretions of the pituitary gland are probably both directly and indirectly involved in the control of salt and water balance. Prolactin has been shown to decrease the permeability of membranes and to effect sodium retention in freshwater fishes (Avella et al., 1990; Lahlou, 1980; Young et al., 1989). Coho salmon smolts adapted to salt water and then transferred to freshwater increase their plasma prolactin concentrations up to 10 times those maintained in salt water. At the same time, plasma osmotic pressure changes from about 360 mOsm kg^{-1} to about 325 mOsm kg^{-1}. Coho smolts adapted to freshwater and then transferred to salt water reduce their plasma prolactin levels rapidly and change their osmolality to about 360 mOsm kg^{-1} (Avella et al., 1990).

Atlantic salmon (*Salmo salar*) undergo a decrease in plasma prolactin during the parr–smolt transformation (Prunet and Boeuf, 1989). Prunet et al. (1985) showed that the plasma prolactin in the rainbow trout decreased from 10–15 ng ml^{-1} to 3–5 ng ml^{-1} after one day following their transfer to seawater. Reciprocal transfer resulted in the restoration of the higher levels. In the euryhaline Mozambique tilapia, blood and pituitary prolactin levels were shown to be lower after transfer to salt water (Nicoll et al., 1981).

Arginine vasotocin influences kidney function and sodium permeability in both marine and freshwater species (Hickman and Trump, 1969). Chondrichthyans apparently show a much greater diversity of neurohypophyseal oxytocinlike secretions than what has been recorded in other vertebrate groups (Acher et al., 1999). Growth hormone is thought to influence the onset of salt tolerance in the young of anadromous salmonids. It shows effects on salinity preference as well as tolerance, but its exact role is not known (Butler, 1966; Holmes and Donaldson, 1969; Johnson, 1973). Bolton et al. (1987) noted that growth hormone influenced osmoregulation in the rainbow trout by reducing plasma levels of calcium, sodium, and magnesium.

Adrenocorticotropin and thyrotropin, secreted by the pituitary gland, may influence osmoregulation by stimulating secretions of the respective target glands. The thyroid gland is known to be active in migrating diadromous species. Thyroid hormones—specifically thyroxine (T4)—have been implicated in the development of osmoregulatory abilities in metamorphosing flounders (Schreiber and Specker, 1999). The interrenal tissue produces adrenocortical steroids that act on renal and extrarenal systems to aid in regulating the body fluids in both marine and freshwater species, but again details of this activity are largely lacking. Cortisol may be the most important steroid involved. Redding et al. (1991) implicated cortisol in osmoregulatory processes of juvenile coho salmon during adaptation to salt water. Plasma concentrations of sodium were lower in both fresh and salt water after injection with cortisol, and concentrations of potassium were higher. Cortisol had direct effects on the number and biochemical characteristics of chloride cells in Mozambique tilapia opercular membranes when these membranes were

removed and held in vitro (McCormick, 1990). The effect on restoring numbers of chloride cells was dose-dependent. Cortisol has a part in the control of absorption of water by the alimentary canal in the seawater phase of eels. It also has been shown to act on the permeability of the urinary bladder in salt water adapted fish. In his review of corticosteroids in osmoregulation, Dharmamba (1979) referred to cortisol as "the dominant hormonal factor" involved in the osmoregulation of marine-adapted fishes. Wendelaar Bonga (1993) also noted the importance of cortisol in the osmoregulation of fishes transferred from seawater to freshwater. The differentiation of chloride cells and their associated Na$^+$/K$^+$ ATPase activity has been demonstrated to be influenced by cortisol, growth hormone, insulinlike growth factor I, and thyroid hormones (McCormick, 1995).

Studies on osmoregulation in eels (*Anguilla*) have revealed the corpuscles of Stannius to play a role in ion and water balance. When sexually maturing "silver" eels (see Chapter 9, Going Deeper) were transferred to seawater, they increased the osmotic concentration of their blood serum, mainly by increases of Na$^+$ and Cl$^-$. On removal of the corpuscles of Stannius (*stanniectomy*), there was a decrease in the concentration of Na$^+$ and increases of Ca^{2+} and K$^+$. When eels with their corpuscles removed were injected with a preparation of corpuscles, their serum concentrations of Ca^{2+} and K$^+$ were reduced and Na$^+$ was increased (Holmes and Donaldson, 1969). When freshwater-adapted glass eels and elvers are challenged with seawater, they rapidly increase their drinking rate (Birrell et al., 2000). Stanniectomy influences the water balance in eels by also enhancing the drinking rate (van der Heijden et al., 1999).

The gill tissue of fishes also plays a role in osmoregulatory processes, in ways more subtle than those already mentioned. The extensive vascularization of gill tissue enables it to function as an important site for endocrine metabolism—the activation or deactivation of hormones. For example, the pillar cells of the gills contain the enzyme responsible for the activation of angiotensin (ANG) by converting it from ANG I to ANG II. Natriuretic peptides are also cleared at sites in the gills (Olson, 1998).

THE URINARY SYSTEM

The Kidney

When one considers the broad evolutionary spectrum of vertebrates and the diverse habitats in which they may be found, the structure and function of kidneys, with their myriad adaptations of form and physiology, are challenging topics to adequately communicate in a few paragraphs, even if we restrict our considerations to fishes. Various modes of life in fresh and salt water have resulted in the structural adjustment of the kidneys to accommodate changing function. In vertebrates, the kidneys share embryonic origins with the organs of reproduction—hence the term **urogenital system.** In some fishes, this association is evident, but in bony fishes, the systems have become quite distinct, with little common structure or function.

General Features of the Kidney

Although the coelacanth *Latimeria* is a notable exception in having a ventral, unpaired kidney, the kidneys of most fishes are slender, elongate, dark red organs extending along the dorsal aspect of the body wall, just ventral to the vertebrae (see Figs. 23.9, 23.10). When the viscera are removed from the body cavity, the kidneys can be seen through the peritoneum. Kidneys are paired, but usually placed close together in most bony fishes; their fusion along the midline is not uncommon. Excretory function is concentrated in the posterior section. The anterior part of the kidney is subject to modification both in structure and in function. In male elasmobranchs, chimaeras, and nonteleost bony fishes, such as sturgeons, gars, and bowfins, the anterior part of the kidney is involved in reproductive function. In most teleosts, the anterior part of the kidney has a concentration of lymphoid and hematopoietic tissue, with chromaffin (suprarenal) and interrenal tissue distributed along the postcardinal veins (Daniel, 1934; Goodrich, 1930; Romer and Parsons, 1978; Wake, 1979).

The structural unit of the kidney is the **nephron** or **kidney tubule,** which consists of a **renal corpuscle** and a **convoluted tubule,** the latter leading to ducts that convey urine to the exterior (see Figs. 25.4, 25.6). The renal corpuscle is made up of the double-walled **Bowman's capsule** and a **glomerulus**—a mass of capillaries coiled within the capsule. The lumen between the walls of Bowman's capsule is continuous with the remainder of the tubule, which consists of several segments (Romer and Parsons, 1978). Examples of tubule structure will be given later.

In the vertebrate embryo, the first and anteriormost kidney portion to develop is the **pronephros;** a most distinctive feature of this early kidney type is its connection to the coelomic cavity via the **nephrostome** (Fig. 25.4A). This part of the kidney usually degenerates with the successive development of more posterior kidney tubules, but in teleosts, it may participate in the formation of blood cells (**hematopoiesis;** Ganassin et al., 2000). The pronephros, with segmentally arranged tubules, opening by means of ciliated nephrostomes to the abdominal and pericardial cavities, is present during the embryonic development of fishes and persists in modified functional form in hagfishes and in larval lampreys. In both of these groups, the nephrostomes empty into the pericardial coelom. A few bony fishes, such as certain lanternfishes (Myctophidae), eelpouts (Zoarcidae), and clingfishes (Gobiescocidae), retain a pronephros (Weichert, 1965), which is always anterior to the opisthonephros, separated from it in hagfishes but intergrading in most others. A kidney retaining both pronephric and opisthonephric elements is some-

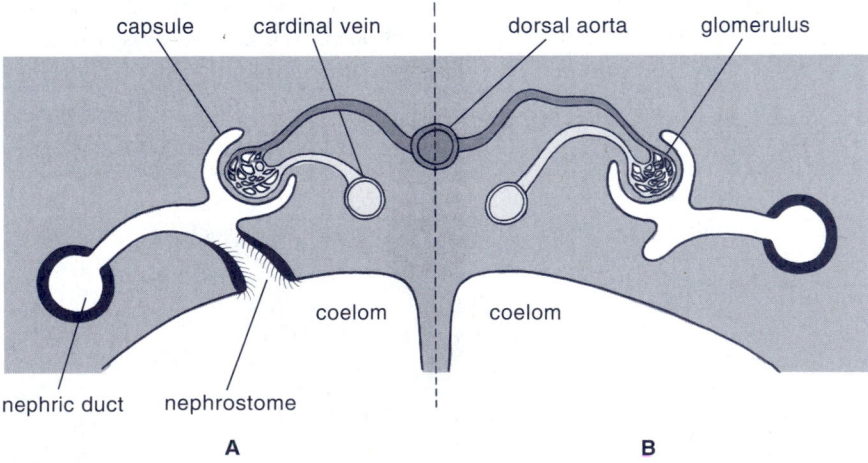

FIGURE 25.4

Comparison of pronephric versus opisthonephric kidney types: **A,** Pronephros; **B,** Opisthonephros. (Based on Goodrich, in Lankester, 1909.)

times called a **holonephros,** but others reserve this term for a kidney in which each trunk segment has a tubule.

The excretory kidney of most fishes is an **opisthonephros,** which is derived from the embryonic **mesonephros** (Hildenberg, 1988; Kardong, 1998; Fig. 25.4B). In amniotes, the mesonephros gives way to the **metanephros,** which becomes the functional kidney. In an opisthonephros, the segmental formation of kidney tubules becomes obscured, and the tubules become concentrated posteriorly. Hagfishes, in which the essential features of the mesonephros are retained (Fänge, 1963), are an exception. Here, the renal corpuscle is present, and the nephrostomes are absent. The amphibian kidney is also an opisthonephros.

Each kidney is provided with a duct draining posteriorly, either to a juncture with its fellow to form a median duct or to a bladder or sinus. The major duct draining each kidney is called the **archinephric** (or **nephric) duct** in most fishes, although a different urinary duct draining the kidney has been developed by the elasmobranchs and chimaeras.

Hagfishes

The hagfishes retain a pronephric kidney (Fänge, 1963). These are small, paired, and situated some distance anteriorly to the opisthonephros and dorsally to the heart on the pericardial wall. Usually, one large renal corpuscle (*glomus*) consisting of three glomeruli occurs posteriorly in each pronephros. There appears to be no connection of the corpuscles and tubules to a functional archinephric duct or to any other excretory duct. The nephrostomes enter the pericardial coelom, which is continuous with the body cavity, and the ciliated tubules connect with the pronephric vein. The function of the pronephros in the adult hagfish is problematic, but the organ is thought to be periodically hematopoietic and may have a lymphoid function. The functional, opisthonephric kidney of hagfishes is long and slender, extending nearly the length of the body cavity. The 30 or more renal corpuscles are arranged irregularly along the length of the kidney. The tubules lack cilia and are quite simple in structure, consisting of a short neck segment and a short proximal segment that empties into the archinephric duct or "ureter" (Fänge, 1963). A urinary sinus receives the archinephric ducts and communicates to the exterior through a urogenital papilla.

Lampreys

Lamprey larvae begin life with paired, pronephric kidneys that function through the prolarval stage. When that stage ends and the burrowing ammocoete stage begins, a pair of larval, opisthonephric kidneys develop posteriorly; these eventually take over as the functional kidneys as the pronephric kidneys lose their function. The larval kidneys apparently do not contribute to the formation of the adult kidney. They degenerate and are replaced with an entirely new structure during metamorphosis (Youson, 1981).

The long, straplike, opisthonephric kidneys of adult lampreys are suspended from the dorsal body wall. They are somewhat comma-shaped in cross section, with a dorsal portion tapering to a thin lower edge (Youson, 1981), along which the archinephric duct courses. In most species, the bulk of the organ consists of a single, elongate renal corpuscle with numerous tubules. This compound glomerulus or **glomus** has no typical Bowman's capsule; however, the capsule is present in the larvae, which have several renal corpuscles, with the posterior ones combining into glomeruli (Youson, 1981; Youson and McMillan, 1970). *Lampetra fluviatilis* is estimated to have 2,300 to 2,800 functional glomeruli (Rankin et al., 1983).

The tubule of the lamprey kidney consists of a ciliated neck segment; a proximal segment, divided into convoluted and straight portions; an intermediate segment; a distal segment, with straight and convoluted parts; and a collecting segment. There are two types of intermediate segments: a short type and a longer type that may be involved in urine formation. The collecting tubules communicate with the archinephric duct, which has muscle tissue in the walls, possibly for the expulsion of urine. The archinephric ducts from each side merge and enter a urogenital sinus, which opens to the exterior via a urogenital papilla.

Chondrichthyes

The kidneys of chimaeras are somewhat similar to those of sharks and rays, but the males have multiple archinephric ducts that reach the urogenital sinus separately (Hickman and Trump, 1969). The point of release of urinary wastes also differs slightly, in that the opening in chimaeras is behind the anus (Jollie, 1973).

The opisthonephric kidneys of sharks and rays are usually flattened, band-shaped or strap-shaped structures that are wider posteriorly (Fig. 25.5). Shapes may vary; in some species, the kidneys are lobate along the lateral edges. This lobate form is especially prominent in rays, in which the kidney may be confined to the posterior part of the body cavity in females. The anterior part of the kidney is usually reduced in female sharks as well. In male elasmobranchs, the anterior part of the kidney is relatively enlarged and functions as part of the genital system (Callard, 1988; Goodrich, 1909).

The nephrons of the anterior part of the kidney in male chondrichthyans may become modified to secrete seminal fluid, which is received by the archinephric duct. The archinephric duct is usually converted into a sperm duct for the transport of sperm and seminal fluid, but in some species, it receives urine from separate urinary ducts. In other species, collecting tubules convey the urine to urinary ducts

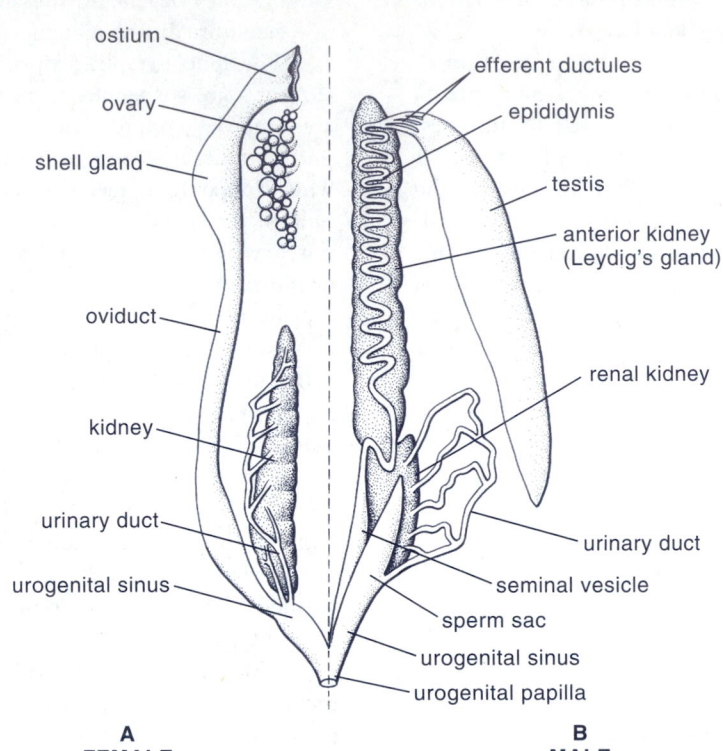

ostium

ovary

shell gland

oviduct

kidney

urinary duct

urogenital sinus

efferent ductules

epididymis

testis

anterior kidney
(Leydig's gland)

renal kidney

urinary duct

seminal vesicle

sperm sac

urogenital sinus

urogenital papilla

**A
FEMALE**

**B
MALE**

FIGURE 25.5
The urogenital organs of the shark: **A,** Female; **B,** Male. (Based on Goodrich, in Lankester, 1909.)

that have no connection with the archinephric duct. In male elasmobranchs, the archinephric duct (sperm duct) usually enlarges into a seminal vesicle that empties into the urogenital sinus along with the urinary duct. In females, the archinephric ducts typically drain into a urinary sinus.

The kidney tubules of sharks and rays consist of a long neck segment; an initial proximal segment, composed of two cytologically distinct regions; a second, more homogeneous proximal segment; and a distal segment, followed by a collecting tubule. In chimaeras, the neck tubule is shorter and the glomeruli are not as heavily vascularized, but overall the structure bears a resemblance to the kidney of *Squalus* (Hickman and Trump, 1969).

Bony Fishes
Sturgeons and paddlefishes have elongate kidneys, extending the length of the body cavity (Goodrich, 1930; Romer, 1970). The pronephric portion, or head kidney, is included. The kidneys are fused posteriorly, but they taper forward separately. Anteriorly, the kidney of the male is associated with the genital system, receiving many efferent ducts from the testes. Many of the tubules are modified for sperm trans-

port. The nephric ducts from each side meet posteriorly in an expanded, bladderlike section. In the bichir (*Polypterus*), it is the posterior section of the kidney that is most intimately associated with the testis, but there is no connection of the sperm duct with the archinephric duct short of the urogenital sinus.

In the bowfin (*Amia calva*), gars (*Lepisosteus*), and lungfishes (*Neoceratodus, Protopterus,* and *Lepidosiren*), the relationship between the testes and the kidneys resembles that of sturgeons and paddlefishes. In the lungfishes, the nephric ducts from the left and right kidneys run separately to the cloaca, except in the males of *Protopterus*, in which the nephric ducts unite before entering the cloaca. A *urinary caecum* or bladder is present in lungfishes (Romer and Parsons, 1978; Wake, 1987).

Teleosts, being more diverse than other fish groups and having broader geographic and ecological distribution, show a wide range of kidney morphologies. The nephrons of teleosts (Fig. 25.6A) are generally composed of the following elements: the glomerulus, a neck segment, a two-part proximal segment, an intermediate segment, a distal segment, and the collecting tubule. Freshwater fishes typically have

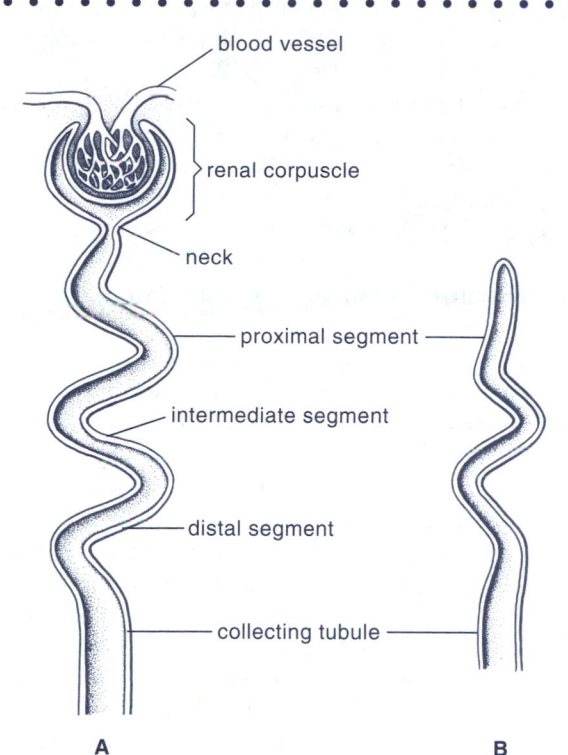

blood vessel

renal corpuscle

neck

proximal segment

intermediate segment

distal segment

collecting tubule

A   B

FIGURE 25.6
Components of two types of nephrons found in bony fishes: **A,** Glomerular; **B,** Aglomerular (as found in some marine fishes).

all of these components, permitting maximal reabsorption of NaCl (Karnaky, 1998). Marine fishes usually lack some of these components, especially the distal segment. Actually, the only nephron structures that all teleosts have in common are the second proximal segment and the collecting tubule. Marine fishes, which lose water across many body surfaces besides the kidneys, usually have smaller and fewer glomeruli than freshwater species (Hickman and Trump, 1969; Karnaky, 1998). Some have lost the glomeruli altogether (Fig. 25.6B).

At present, kidney structure and function has been thoroughly investigated only in relatively few species, so that a comprehensive system for the classification of kidney design is not possible. Ogawa (1961, 1962), however, made a noble attempt in proposing a classification based on the configuration or shape of marine teleost kidneys and suggested general relationships to phylogeny and habitat, as some of the proposed types are selectively encountered in freshwater species.

Ogawa's *Type I* kidney is exemplified in salmons and trouts (Salmonidae), herrings (Clupeidae), and the ayu

(*Plecoglossus*). These fishes have kidneys fused throughout their lengths, and the opisthonephros is continuous with the head kidney, which is the nonrenal remnant of the embryonic pronephros (Romer, 1970). The head kidney may be somewhat expanded laterally, especially in the salmonids. It consists of lymphoid tissue, with some suprarenal tissue included, and has lost the typical renal function. This type of nephron is mostly typical of marine fishes, although many clupeoids and salmonoids live in freshwater. The glomerular capsule is large (ca. 85–105 μm), as in freshwater species, but the neck segment and distal segment are lacking.

Many minnows (Cyprinidae), at least one loach (*Misgurnus*), some catfishes, and most eels have kidneys that are fused to each other anteriorly and posteriorly, but are separate through the midsection (Ogawa's *Type II*). The head kidney is usually expanded laterally. These fishes have large glomerular capsules (ca. 60–95 μm), and all segments are present in the freshwater species and in the marine catfish genus *Plotosus*. Morays (*Gymnothorax*) lack the distal segment.

Ogawa recognized his *Type III* in perchlike fishes (Percoidei), gobies (Gobioidei), barracudas (*Sphyraena*), blennies (Blennioidei), the medaka (*Oryzias*), the snakehead (*Channa*), lanternfishes (Myctophidae), mackerels (Scombridae), sculpins (Cottidae), flounders (Pleuronectidae), and others. In these fishes, the functional kidneys are found only posteriorly, and the head kidneys are well differentiated from the opisthonephric kidney. The glomerular capsule is relatively small (ca. 40–70 μm), and the distal segment is absent in the marine forms. *Channa* and *Oryzias*, both of which are freshwater forms, retain the distal segment.

Pipefishes and seahorses have narrow kidneys connected only at the most posterior end, and the head kidney is not developed. This structure represents Ogawa's *Type IV.* The nephron is greatly reduced, with only the second portion of the proximal segment and the collecting segment present.

Type V kidneys are completely separated from one another. The head kidney is developed, and may even retain pronephric glomeruli in some anglerfishes (Lophiiformes). This type also occurs in puffers (*Fugu, Canthigaster*) and boxfishes (*Ostracion*) and other closely related fishes. Some species are reported to lack glomeruli; the goosefish (*Lophius piscatorius*), the sargassumfish (*Histrio histrio*), and the porcupine fish (*Diodon*) are examples. These fishes retain only the second proximal segment and the initial collecting tubule. Species that have glomeruli lack the distal segment.

Role of the Kidneys in Excretion and Osmoregulation

The kidneys play a major role in the excretion of surplus ions and metabolic by-products. The role of the kidney in osmoregulation centers on the processes of urine formation

486 · Part Five Homeostatic Mechanisms

and excretion. The urine of freshwater fishes is much less concentrated than the blood and body fluids, whereas the urine of marine species is significantly more concentrated. Because the loop of Henle is not present in the fish nephron, fishes, unlike amniotes, are not able to form urine that is more concentrated than the blood plasma. Even so, the nephric structure in fishes includes a diversity of sites for the secretion and reabsorption of a number of dissolved substances (Fig. 25.7).

The glomerulus is a filter that allows blood plasma containing dissolved materials to pass into the space between the walls of Bowman's capsule, and thence into the kidney tubule. Blood cells and large molecules, such as proteins, cannot pass the filter. There would be no osmoregulatory advantage gained by freshwater fishes if the filtrate removed from the blood at the glomerulus were to be excreted with its normal complement of salts. The advantage is gained by excreting urine that is more dilute than the plasma. As the fluid passes down the tubule, substances are reabsorbed at specific locations. Glucose is reabsorbed in the proximal tubule, and salts are reabsorbed in the distal tubule, the walls of which are impermeable to water in many fishes. Further physiological advantage is gained in some fishes by the excretion of small amounts of waste nitrogenous compounds in the urine.

The kidneys of marine teleosts, as in freshwater fishes, are the major site for the excretion of divalent ions, but their role as a water pump is diminished because of the extrarenal losses of water. Consequently, the glomeruli are generally smaller and fewer than in freshwater fishes and elasmobranchs (Table 25.8). About 30 species of marine teleosts have lost their glomeruli completely to minimize renal loss of water (Figs. 25.6B; 25.7F). These species produce urine entirely by tubular secretion (Karnaky, 1998). Species lacking glomeruli include gulpers (Saccopharyngidae), seahorses and pipefishes (Syngnathidae), dragonets (Callionymidae), scorpionfishes (Scorpaenidae), poachers (Agonidae), sculpins (Cottidae), puffers (Tetraodontidae), toadfishes (Batrachoididae), clingfishes (Gobiesocidae), anglerfishes (Lophiidae), frogfishes (Antennariidae), and batfishes (Ogcocephalidae). In addition to the diminution, degeneration, or loss of glomeruli, marine fishes also tend to lose the distal segment of the kidney tubule (Hickman and Trump, 1969; Karnaky, 1998).

Urine flow in marine teleosts is scant, usually amounting to 1 to 2 percent of the body weight per day. In some species, most of the water in the urine enters through the kidney tubule. In others, water is filtered out by the glomerulus, and most of it is subsequently reabsorbed by the tubule. The osmotic concentration of urine is slightly less than that of the blood. The urinary bladder is involved in adjusting the concentration of the urine. It has greater osmotic permeability than that of freshwater teleosts and does not absorb salts.

Principal divalent ions in the urine of marine fishes are, in decreasing order of concentration: (1) sulfate, at concentrations sometimes exceeding 300 times that in the plasma; (2) magnesium, at 50 to more than 100 times the plasma concentration; and (3) calcium, at 4 to 10 times the plasma concentration. Phosphate occurs in concentrations less than that of calcium. There apparently is active secretion of magnesium, sulfate and, sometimes, phosphate into the tubule. In some marine species, calcium and magnesium are precipitated as salts in the tubular lumen and thus no longer influence the osmotic gradient. Chloride may be virtually absent from the urine in some species, but can appear in concentrations of up to 85 percent of the plasma concentration in others, especially when an individual is in a diuretic state (Black, 1957; Hickman and Trump, 1969).

Freshwater fishes produce dilute urine at rates up to 14 ml h^{-1} kg^{-1} (Hickman and Trump, 1969). Production of urine by marine fishes is about 10 percent of that or less. In freshwater fishes, the osmolality of urine is typically in the range of about 20 to 37 mOsm l^{-1}, and in marine fishes, the urine has an osmolality in the range of 300 to 400 mOsm l^{-1}, approximating the osmotic concentrations of the blood and body fluids (Hickman and Trump, 1969).

The kidney of hagfishes has been termed *atubular*, as it consists of large, segmentally arranged glomeruli that drain to the medial side of a primitive archinephric duct (Fig. 25.7A). The epithelium of the archinephric duct is functionally analogous to the first proximal tubule section in gnathostomes. Hagfishes are considered osmoconformers, with little capacity to regulate their plasma osmotic concentrations at levels significantly different from their seawater environment. Even so, urea, K$^+$, PO$_4^{3-}$, SO$_4^{2-}$, and Mg^{2+} are secreted during filtrate formation in the kidneys (Karnaky, 1998).

Compared with the simple nephric structure of hagfishes, the lamprey nephron has a complexity that resembles that seen in higher vertebrates (Karnaky, 1998). Lampreys in seawater apparently excrete NaCl at sites other than the kidney, whereas in freshwater-adapted species, almost all of the filtered Na$^+$ and Cl$^-$ is reabsorbed by the tubules (Fig. 25.7B).

The comparatively sophisticated kidneys of chondrichthyans are specialized to retain urea and trimethylamine oxide (TMAO) to the extent that, like the osmoconforming hagfishes, they have blood plasma that is nearly isosmotic to seawater. The nephrons of marine elasmobranchs are capable of recovering more than 85 percent of the urea that is filtered at the glomerulus (Lacy and Reale, 1995).

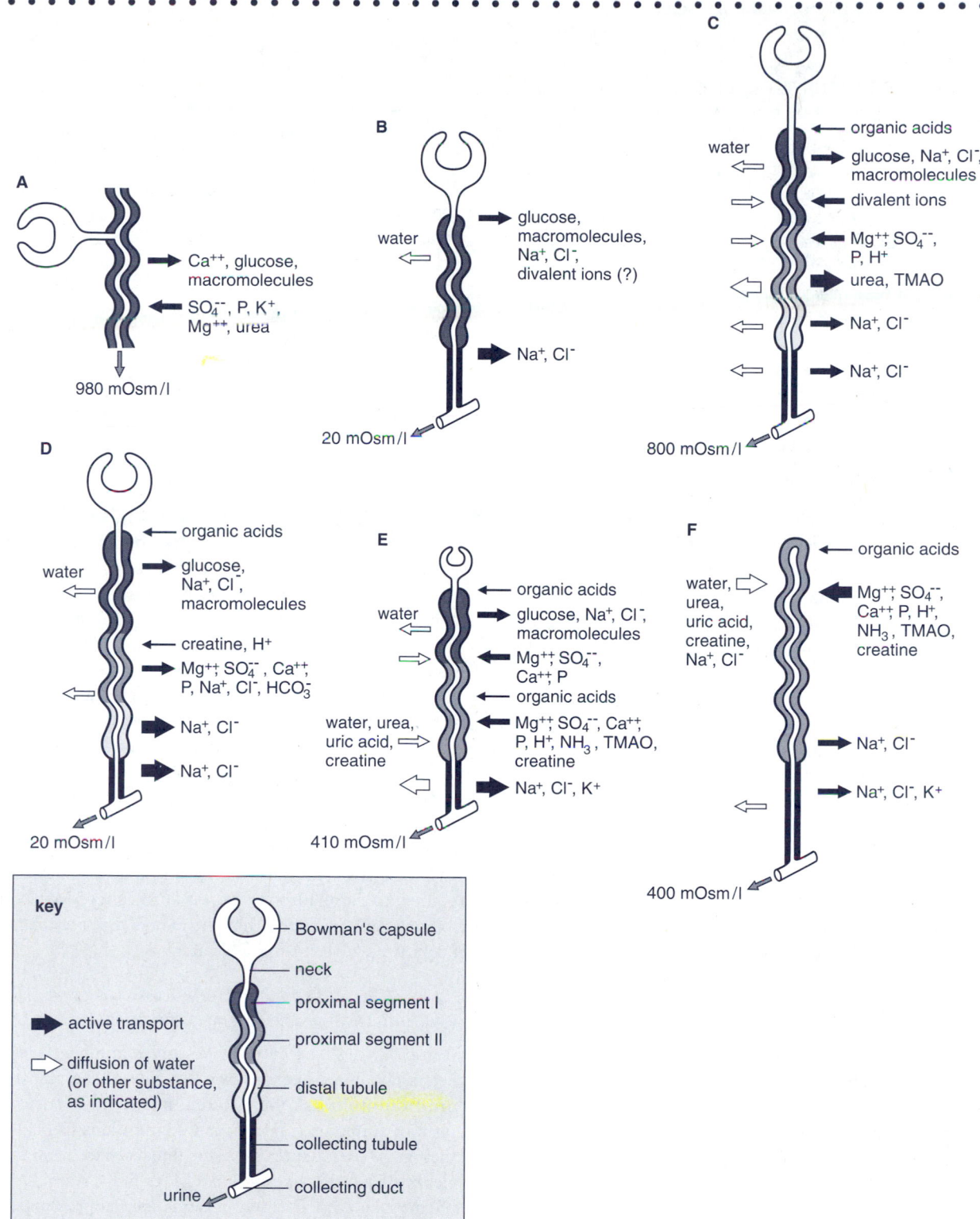

FIGURE 25.7

Schematic representation of the structure and function of the nephron in **A,** Hagfish; **B,** Lamprey in freshwater; **C,** Marine elasmobranch; **D,** Freshwater teleost; **E,** Marine glomerular teleost; **F,** Marine aglomerular teleost. Morphological differentiation of the nephron is shown, and tubular secretion and reabsorption of the compounds indicated. Solid arrows indicate active transport and open arrows indicate passive diffusion. Arrows pointing away from the nephron indicate movement of solutes from the nephron filtrate to the blood; arrows pointing toward the nephron indicate movement of solutes from the blood to the nephron filtrate. (based on Hickman and Trump, 1969.)

TABLE 25.8 NUMBER AND SIZE OF GLOMERULI IN SELECTED FISHES

Species	Weight (g)	Number of glomeruli in one kidney	Average diameter of glomeruli (µm)	Relative volume (mm³) of glomeruli per m² of body surface
Elasmobranchs				
Mustelus canis	485	4,400	185	60.93
Raja erinacea	1,060	1,200	190	10.38
Marine Teleosts				
Gadus morhua callarias	670	16,250	37	1.49
Lutjanus griseus	544	31,860	55	10.99
Pseudopleuronectes americanus	160	5,300	50	3.14
Freshwater Teleosts				
Ameiurus nebulosus	89	18,160	100	126.5
Cyprinus carpio	221	24,310	82	50.8
Perca flavescens	116	4,870	102	30

Source: Nash, 1931.

This unusual osmoregulatory feature of the chondrichthyans renders them capable of gaining salts from their environment (Fig. 25.7C). The elasmobranch kidney does not appear to be a major source of NaCl secretion and, with urine flows at approximately 25 percent of the glomerular filtration rate, a net tubular reabsorption of water is apparent (Evans, 1993).

Freshwater bony fishes have well-developed glomeruli that can filter much water from the blood (Fig. 25.7D). Some of this water is reabsorbed from the proximal tubule, but most is passed as urine. Monovalent ions are reabsorbed in both the proximal and distal tubule segments. In marine bony fishes, the glomeruli tend to be smaller, with some fishes (see earlier) classified as *aglomerular* because they lack these structures altogether (Fig. 25.7 F). The proximal tubule segment is long, with the first section adjacent to the glomerulus functioning much like the proximal segment of freshwater fishes. The more distal part of the proximal segment secretes divalent ions (mainly Mg^{2+} and SO_4^{2-}) into the tubule.

Among the bony fishes, the sarcopterygians present some unusual departures from the actinopterygian norm in terms of kidney structure and function. In many respects, they are more like chondrichthyans. The coelacanth (*Latimeria chalumnae*) retains urea and TMAO. Studies on the urine composition of coelacanths reveal that the kidney secretes these nitrogenous compounds as well as divalent ions and some carbohydrates, but plays no role in NaCl excretion (Griffith et al., 1976).

EVOLUTIONARY CONSIDERATIONS

Researchers have long held the view that fishes originated in the ocean and later spread into freshwater. The fossil record, which shows the earliest known fishes placed in a marine context (see Chapters 2 and 5), substantiates this view. The fact that both the cephalochordates and the hagfishes are marine, with hagfishes possessing little capacity to osmoregulate, also lends support to this assertion. The preeminent vertebrate paleontologist A. S. Romer, however, claimed that the earliest fossil fishes were from freshwater deposits (Romer and Grove, 1935), thus supporting the assertion of the distinguished physiologist Homer Smith that the glomerular design of the vertebrate kidney, specialized as it is for clearing water from the blood plasma, pointed to a freshwater origin for vertebrates. Griffith (1994) has attempted to reconcile these differing viewpoints by proposing that the first vertebrates were possibly anadromous, claiming that this is consistent with much of the fossil record and with physiological function in modern fishes. Griffith pointed to the anadromous lifestyle of lampreys as evidence for this and suggested that the conodonts (see Chapter 2) may have evolved a similar strategy.

Gills, Kidneys, and the Origins of Tetrapods

In most fishes, the kidney has only a minimal role in the excretion of nitrogenous wastes. Most fishes are termed ammoniotelic, as they excrete nitrogenous waste in the form of ammonia (NH_3) or ammonium ions (NH_4^+) across their gills (see Chapter 24). One unusual exception is a species of tilapia that lives in Lake Magadi, Kenya. The waters of this lake are extremely alkaline (pH approximately 10) and highly buffered, preventing the normal excretion of ammonia. Consequently, all nitrogenous wastes are excreted in the form of urea (Walsh, 1998; Wright and Land, 1998). Apparently, freshwater fishes differ from marine species in the proportion of ammonia versus ammonium ions excreted, with a greater proportion of nitrogenous waste transported across the gills as ammonium ions in marine species (Wilkie, 1997). Although teleosts are primarily ammoniotelic, marine species also tend to excrete a proportion of nitrogen in the form of urea (Jobling, 1994).

The term **ureotelic** is used to denote fishes that excrete primarily urea. Chondrichthyans and sarcopterygians are good examples of ureotelic fishes. Most of this efflux in chondrichthyans is across the gills, although a small portion of urea leaves after being filtered by the kidney—the enhanced tubular reabsorptive capacity of the kidneys promotes high plasma concentrations of urea and TMAO (Lacy and Reale, 1995; Karnaky, 1998).

With the loss of gills in tetrapods, the kidney becomes the sole route for excretion. Some tetrapods, such as amphibians and mammals, remain ureotelic, whereas others, including the reptiles and birds, excrete primarily uric acid, and are hence termed **uricotelic**.

As discussed in Chapter 7, there is compelling evidence that the first tetrapods were fully aquatic forms. Janis and Farmer (1999) sounded a cautionary note in response to the enthusiasm for a fully aquatic origin for tetrapods, however. They observed that, whereas †*Acanthostega* may have been gill-bearing, other genera, such as †*Ichthyostega* and †*Tulerpeton* must have had terrestrial ancestors owing to their loss of gills. Janis and Farmer asserted that gills would not have been lost in fully aquatic creatures owing to their value in the excretion of nitrogenous waste and ion regulation. They claimed that the fossil tetrapods showing no evidence of gills must have arisen from terrestrially adapted forms, in which the kidneys had taken over the role of nitrogen excretion.

• •
KEY POINTS AND CONNECTIONS
• •

- The success of fishes in both marine and freshwater habitats is attributable to their ability to maintain their internal ionic composition at levels that are quite different from the environment in which they live. Hagfishes are an exception, in that their ionic composition closely resembles the marine environment that they inhabit. Osmoregulation refers to the ability to maintain a relatively constant internal ionic composition—many fishes can do this even when the ionic composition of their environments is subject to extreme fluctuation. Being unable to osmoregulate, hagfishes are termed osmoconformers.

The physical and chemical properties of aquatic habitats that might affect the osmoregulatory capabilities of fishes are discussed in Chapter 30.

- The osmotic concentration of the blood and body fluids of a typical freshwater fish ranges between 265 and 325 mOsm kg^{-1}. Freshwater fishes are therefore hyperosmotic relative to their medium and tend to gain water by diffusion. This they counteract by excreting copious quantities of dilute urine. The uptake of salts is accomplished by chloride cells located mainly in the gill epithelium. Whereas lampreys and bony fishes have become well established in freshwater, only a few species of chondrichthyan fishes are known to invade freshwater habitats.

- Marine fishes are generally hypo-osmotic relative to their environment. Faced with an osmotic loss of water, they compensate by ingesting seawater and excreting surplus ions. Chloride cells in the gill tissue are useful here. Members of the Chondrichthyes and the coelacanth are distinctive in having an osmoregulatory strategy that involves the retention of significant quantities of urea and TMAO to maintain an elevated blood and body fluid osmolality. In spite of this shared mechanism and the fact that both groups possess a rectal gland for the regulation of ion levels or body fluid volume, these two groups do not appear to be closely related. Marine teleosts are also effective hypo-osmoregulators, maintaining their body ionic content at levels less than half that of full-strength seawater. Among the most remarkable fishes are those species termed euryhaline because they can effectively osmoregulate over a broad range of salinities. Such varying conditions are experienced by diadromous species during the course of their migrations and by inhabitants of bays and estuaries that may experience fluctuation in local environmental salinities.

- The endocrine system plays a crucial role in osmoregulation in fishes. Osmoregulatory influence is in evidence along the entire endocrine axis, from pituitary secretions to hormones released by organs such as the thyroid. Arginine vasotocin and corticosteroids appear to have an osmoregulatory role in agnaths. Steroid hormones, among other endocrine secretions, appear to influence water permeability and figure significantly in the function of the rectal gland in chondrichthyan fishes. Among the bony fishes, prolactin is known to facilitate freshwater adaptation, whereas cortisol has been implicated in adaptation to both marine and freshwater environments. Prolactin also has been shown to influence sodium retention in freshwater fishes, and arginine vasotocin appears to influence kidney function and sodium permeability in both marine and freshwater species.

The osmoregulatory role of endocrine secretions and their source tissues and organs is discussed in further detail in Chapter 26.

- In most fishes, the kidneys are closely paired, elongate, dark red organs that run just ventral to the vertebrae. The kidney is composed of varying numbers of functional subunits termed nephrons. Nephrons have an important role in the regulation of ionic composition; their structure and function vary among the different fish taxa. Differences in nephric structure and function may also reflect habitat—not surprising, given the decidedly different roles that kidneys play in fishes inhabiting freshwater versus marine environments.

Kidneys and the associated ducts for draining urinary wastes may be modified for the transmission of gametes in some fishes (see Chapter 27).

FISH LINKS

http://www.science.mcmaster.ca/~woodcm/ Home page for the
laboratory of Christopher Wood. Dr. Wood and his associates study
the transport of ions, nutrients, and respiratory gases in fishes and
crustaceans.

BUILDING AN ICHTHYOLOGY LIBRARY

Evans, D. H. (Ed.). 1998. *The physiology of fishes* (2nd ed.). CRC Press,
Boca Raton, FL.
Wood, C. M., and T. J. Shuttleworth (Eds.). 1995. *Cellular and molecular
approaches to fish ionic regulation.* Academic Press, San Diego.

REFERENCES

Acher, R., J. Chauvet, M.-T. Chauvet, and Y. Rouille. 1999. Unique
evolution of neurohypophyseal hormones in cartilaginous fishes:
Possible implications for urea-based osmoregulation. *J. Exp. Zool.
284:* 475–484.
Agarwal, S., and P. A. John. 1975. Functional morphology of the urinary
bladder in some teleostean fishes. *Forma Functio 8*(2): 19–26.
Avella, M., and J. Ehrenfeld. 1997. Fish gill respiratory cells in culture:
A new model for Cl⁻-secreting epithelia. *J. Membrane Biol. 156:*
87–97.
———, G. Young, P. Prunet, and C. B. Schreck. 1990. Plasma prolactin
and cortisol concentrations during salinity challenges of coho
salmon (*Oncorhynchus kisutch*) at smolt and post-smolt stages.
Aquaculture 91: 359–372.
Birrell, L., G. Cramb, and N. Hazon. 2000. Osmoregulation during the
development of glass eels and elvers. *J. Fish Fiol. 56:* 1450–1459.
Black, V. S. 1957. Excretion and osmoregulation, pp. 163–205. In
The physiology of fishes, Vol. 1, M. E. Brown (Ed.). Academic Press,
New York.
Bolton, J. P., N. L. Collie, H. Kawauchi, and T. Hirano. 1987. Osmo-
regulatory actions of growth hormone in rainbow trout (*Salmo
gairdneri*). *J. Endocrinol. 112:* 63–68.
Bone, Q., and N. B. Marshall. 1982. *Biology of fishes.* Blackie, Glasgow,
UK.
———, N. B. Marshall, and J. H. S. Blaxter. 1995. *Biology of fishes*
(2nd ed.). Blackie, Glasgow, UK.
Burger, J. W. 1962. Further studies on the function of the rectal gland in
the spiny dogfish. *Physiol. Zool. 35:* 205–217.
———, and W. N. Hess. 1960. Function of the rectal gland in the spiny
dogfish. *Science 131:* 670–671.
Butler, D. G. 1966. Effect of hypophysectomy on osmoregulation in the
European eel (*Anguilla anguilla* L.). *Comp. Biochem. Physiol.
18:* 773–781.
Callard, G. V. 1988. Reproductive biology, Part B, pp. 292–317. In
Physiology of elasmobranch fishes, T. J. Shuttleworth (Ed.). Springer
Verlag, Berlin.
Chessman, B. C., and W. D. Williams. 1975. Salinity tolerance and osmo-
regulatory ability of *Galaxias maculatus* (Jenyns) (Pisces, Salmoni-
formes, Galaxiidae). *Freshw. Biol. 5:* 135–140.
Claiborne, J. B. 1998. Acid–base regulation, pp. 177–198. In *The physi-
ology of fishes* (2nd ed.), D. H. Evans (Ed.). CRC Press, Boca
Raton, FL.

Conte, F. P. 1969. Salt secretion, pp. 241–292. In *Fish physiology,* Vol. I,
W. S. Hoar and D. J. Randall (Eds.). Academic Press, New York.
Curtis, B. J., and C. M. Wood. 1991. The function of the urinary bladder
in vivo in the freshwater rainbow trout. *J. Exp. Biol. 155:* 567–583.
Daniel, J. F. 1934. *The elasmobranch fishes.* University of California Press,
Berkeley.
Davenport, J., and M. D. J. Sayer. 1993. Physiological determinants of
distribution in fish. *J. Fish Biol. 43*(Suppl. A): 121–145.
Dharmamba, M. 1979. Corticosteroids and osmoregulation in fishes.
Proc. Ind. Nat. Sci. Acad. B45: 515–525.
Evans, D. H. 1980. Osmotic and ionic regulation by freshwater and
marine fishes, pp. 93–122. In *Environmental physiology of fishes,*
M. A. Ali (Ed.). Plenum Press, New York.
———. 1993. *The physiology of fishes.* CRC Press, Boca Raton, FL.
———. 1998. *The physiology of fishes* (2nd ed.). CRC Press, Boca
Raton, FL.
———, P. M. Piermarini, and W. T. W. Potts. 1999. Ionic transport in
the fish gill epithelium. *J. Exp. Zool. 283:* 641–652.
Fänge, R. 1963. Structure and function of the excretory organs of
myxinoids, pp. 516–529. In *The biology of* Myxine, A. Brodal
and R. Fänge (Eds.). Universitetsforlaget, Oslo.
———, and K. Fugelli. 1962. Osmoregulation in chimaeroid fishes.
Nature 196: 689.
Fontaine, M. 1975. Physiological mechanisms in the migration of marine
and amphihaline fish. *Adv. Mar. Biol. 13:* 241–355.
Forster, R. P., and L. Goldstein. 1969. Formation of excretory products,
pp. 313–350. In *Fish physiology,* Vol. I, W. S. Hoar and D. J.
Randall (Eds.). Academic Press, New York.
Ganassin, R. C., K. Schirmer, and N. C. Bols. 2000. Cell and tissue
culture, Chapter 38. In *The laboratory fish,* G. K. Ostrander (Ed.).
Academic Press, San Diego.
Gonzalez, R. J., and M. R. Preest. 1999. Ionoregulatory specializations
for exceptional tolerance of ion-poor, acidic waters in the neon tetra
(*Paracheirodon innesi*). *Physiol. Biochem. Zool. 72:* 156–163.
Goodrich, E. S. 1909. Cyclostomes and fishes. In *A treatise on zoology,*
Part 9, Fasc. 1, R. Lankester (Ed.). Adam and Charles Black, Lon-
don (Reprinted 1964, A. Asher, Amsterdam).
———. 1930. *Studies on the structure and development of vertebrates.*
Constable, London (Reprinted 1958, Dover, New York).
Goss, G., S. Perry, and P. Laurent. 1995. Ultrastructural and mor-
phometric studies on ion and acid–base transport processes in
freshwater fish, pp. 257–284. In *Cellular and molecular approaches
to fish ionic regulation,* C. M. Wood and T. J. Shuttleworth (Eds.).
Academic Press, San Diego.
Griffith, R. W. 1994. The life of the first vertebrates. *BioScience 44:*
408–417.
———, B. L. Umminger, B. F. Grant, P. K. T. Pang, and G. E. Pickford.
1974. Serum composition of the coelacanth, *Latimeria chalumnae*
Smith. *J. Exp. Zool. 187:* 87–102.
———, ———, ———, ———, L. Goldstein, and G. E. Pickford.
1976. Composition of the bladder urine of the coelacanth *Latimeria
chalumnae. J. Exp. Zool. 196:* 371–380.
———, and P. K. T. Pang. 1979. Mechanisms of osmoregulation in the
coelacanth: Evolutionary implications, pp. 79–93. In The biology
and physiology of the living coelacanth, J. E. McCosker and M. D.
Lagios (Eds.), *Occ. Pap. Calif. Acad. Sci. 134.*
Hansen, L. P., W. C. Clarke, R. L. Saunders, and J. E. Thorpe. 1989.
Salmonids smoltification III: Special issue. *Aquaculture 82:* 1–390.
Hickman, C. P., and B. F. Trump. 1969. The kidney, pp. 91–239. In *Fish
physiology,* Vol. I, W. S. Hoar and D. J. Randall (Eds.). Academic
Press, New York.

Hildenberg, M. 1988. *Analysis of vertebrate structure* (3rd ed.). Wiley, New York.

Holliday, F. G. T. 1969. The effects of salinity on the eggs and larvae of teleosts, pp. 293–311. In *Fish physiology*, Vol. I, W. S. Hoar and D. J. Randall (Eds.). Academic Press, New York.

Holmes, W. N., and E. M. Donaldson. 1969. The body compartments and the distribution of electrolytes, Part 1, pp. 1–89. In *Fish physiology*, Vol. I, W. S. Hoar and D. J. Randall (Eds.). Academic Press, New York.

Janis, C. M., and C. Farmer. 1999. Proposed habitats of early tetrapods: Gills, kidneys, and the water–land transition. *Zool. J. Linn. Soc. 126*: 117–126.

Jobling, M. 1994. *Fish bioenergetics.* Chapman and Hall, London.

Johnson, D. W. 1973. Endocrine control of hydromineral balance in teleosts. *Amer. Zool. 13*: 799–818.

Jollie, M. 1973. *Chordate morphology.* Robert E. Krieger, Huntington, NY.

Kardong, K. V. 1998. *Vertebrates: Comparative anatomy, function, evolution.* WCB/McGraw-Hill, Boston.

Karnaky, K. J., Jr. 1998. Osmotic and ionic regulation, pp. 157–176. In *The physiology of fishes* (2nd ed.), D. H. Evans (Ed.). CRC Press, Boca Raton, FL.

Kashiwagi, M., and R. Sato. 1969. Studies on the osmoregulation of the chum salmon, *Oncorhynchus keta* (Walbaum). I. The tolerance of eyed period eggs, alevins and fry of the chum salmon to seawater. *Tohoku J. Agr. Res. 20*(1): 41–47.

Kelly, R. H., and P. H. Yancey. 1999. High contents of trimethylamine oxide correlating with depth in deep-sea teleost fishes, skates, and decapod crustaceans. *Biol. Bull. 196*: 18–25.

Kirschner, L. B. 1991. Water and ions, pp. 13–107. In *Comparative animal physiology: Environmental and metabolic animal physiology* (4th ed.), C. L. Prosser (Ed.). Wiley-Liss, New York.

Kowarsky, J. 1973. Extra-branchial pathways of salt exchange in a teleost fish. *Comp. Biochem. Physiol. 46A*: 477–486.

Lacy, E. R., and E. Reale. 1995. Functional morphology of the elasmobranch nephron and retention of urea, pp. 107–146. In *Cellular and molecular approaches to fish ionic regulation*, C. M. Wood and T. J. Shuttleworth (Eds.). Academic Press, San Diego.

Lahlou, B. 1980. Les hormones dans l'osmorégulation des poissons, pp. 201–240. In *Environmental physiology of fishes*, M. A. Ali (Ed.). Plenum Press, New York.

———, I. W. Henderson, and W. H. Sawyer. 1969. Renal adaptations by *Opsanus tau*, a euryhaline aglomerular teleost, to dilute media. *Am. J. Physiol. 216*: 1266–1272.

Lutz, P. L. 1975a. Osmotic and ionic composition of the polypteroid *Erpetoichthys calabaricus. Copeia 1975*: 119–123.

Lutz, P. L. 1975b. Adaptive and evolutionary aspects of the ionic content of fishes. *Copeia 1975*: 369–373.

Maetz, J. 1969. Seawater teleosts: Evidence for a sodium–potassium exchange in the branchial sodium-excreting pump. *Science 166*: 613–615.

———. 1974. Aspects of adaptation to hypo-osmotic and hyper-osmotic environments, pp. 1–167. In *Biochemical and biophysical perspectives in marine biology*, Vol. 1, D. C. Malins and J. R. Sargent (Eds.). Academic Press, New York.

Marshall, W. S. 1995. Transport processes in isolated teleost epithelia: Opercular epithelium and urinary bladder, pp. 1–23. In *Cellular and molecular approaches to fish ionic regulation*, C. M. Wood and T. J. Shuttleworth (Eds.). Academic Press, San Diego.

———, S. E. Bryson, P. Darling, C. Whitten, M. Patrick, M. Wilkie, C. M. Wood, and J. Buckland-Nicks. 1997. NaCl transport and ultrastructure of opercular epithelium from a freshwater-adapted euryhaline teleost, *Fundulus heteroclitus. J. Exp. Zool. 277*: 23–37.

Martin, F. D. 1972. Factors influencing the local distribution of *Cyprinodon variegatus* (Pisces: Cyprinodontidae). *Trans. Am. Fish. Soc. 101*: 89–93.

McCormick, S. D. 1990. Cortisol directly stimulates differentiation of chloride cells in tilapia opercular membrane. *Am. J. Physiol. 259*: R857–R863.

———. 1995. Hormonal control of gill Na+, K+, −ATPase and chloride cell function, pp. 285–315. In Fish physiology, Vol. XIV, *Ionoregulation: Cellular and molecular approaches*, C. M. Wood and T. J. Shuttleworth (Eds.). Academic Press, New York.

———. 2001. Endocrine control of osmoregulation in teleost fish. *Amer. Zool. 41*: 781–794.

Morgan, J. D., and G. K. Iwama. 1990. The energetics of ion regulation in freshwater resident and anadromous juvenile rainbow trout (*Oncorhynchus mykiss*). *Bull. Aquacult. Assoc. Can. 90*(4): 57–60.

Morris, R. 1972. Osmoregulation, pp. 193–239. In *The biology of lampreys*, Vol. 2, M. W. Hardisty and I. C. Potter (Eds.). Academic Press, London.

Nicoll, C. S., S. W. Wilson, R. Nishioka, and H. A. Bern. 1981. Blood and pituitary prolactin levels in tilapia (*Sarotherodon mossambicus*; Teleostei) from different salinities as measured by a homologous radioimmunoassay. *Gen. Comp. Endocrinol. 44*: 365–373.

Ogawa, M. 1961. Comparative study of the external shape of the teleostean kidney with relation to phylogeny. *Sci. Rep. Tokyo Kyoiku Daigaku B 10*: 61–68.

Ogawa, M. 1962. Comparative study on the internal structure of the teleostean kidney. *Sci. Rep. Saitama Univ. B4*(2): 107–131.

Olson, K. R. 1998. Hormone metabolism by the fish gill. *Comp. Biochem. Physiol. 119A*: 55–65.

Parry, G. 1966. Osmotic adaptation in fishes. *Biol. Rev. 41*: 392–444.

Pärt, P., C. M. Wood, K. M. Gilmour, S. F. Perry, P. Laurent, J. Zadunaisky, and P. J. Walsh. 1999. Urea and water permeability in the ureotelic gulf toadfish (*Opsanus beta*). *J. Exp. Zool. 283*: 1–12.

Perry, S. F. 1997. The chloride cell: Structure and function in the gills of freshwater fishes. *Ann. Rev. Physiol. 59*: 325–347.

———. 1998. Relationships between branchial chloride cells and gas transfer in freshwater fish. *Comp. Biochem. Physiol. 119A*(1): 9–16.

Prunet, P., and G. Boeuf. 1989. Plasma prolactin levels during smoltification in Atlantic salmon, *Salmo salar. Aquaculture 82*: 297–305.

———, G. Boeuf, and L. M. Houde. 1985. Plasma and pituitary prolactin levels in rainbow trout during adaptation to different salinities. *J. Exp. Zool. 235*: 187–196.

Quast, W. D., and N. R. Howe. 1980. The osmotic role of the brood pouch in the pipefish *Syngnathus scovelli. Comp. Biochem. Physiol. 67A*: 675–678.

Rankin, J. C., I. W. Henderson, and J. A. Brown. 1983. Osmoregulation and the control of kidney function, pp. 66–88. In *Control processes in fish physiology*, J. C. Rankin, T. J. Pitcher, and R. T. Duggan (Eds.). Wiley, New York.

Read, L. J. 1971. Body fluids and urine of the holocephalan *Hydrolagus colliei. Comp. Biochem. Physiol. 39A*: 185–192.

Redding, J. M., R. Patiño, and C. B. Schreck. 1991. Cortisol effects on plasma electrolytes and thyroid hormones during smoltification in coho salmon, *Oncorhynchus kisutch. Gen. Comp. Endocrinol. 81*: 373–382.

Robertson, J. D. 1963. Osmoregulation and ionic composition of cells and tissues, pp. 504–515. In *The biology of* Myxine, A. Brodal and R. Fänge (Eds.). Universitetsforlaget, Oslo.

Rombough, P. J. 1999. The gill of fish larvae. Is it primarily a respiratory or an ionoregulatory structure? *J. Fish Biol. 55(Suppl. A):* 186–204.

Romer, A. S. 1970. *The vertebrate body* (4th ed.). W. B. Saunders, Philadelphia.

———, and B. H. Grove. 1935. Environment of the early vertebrates. *Am. Midl. Nat. 16:* 805–856.

———, and T. S. Parsons. 1978. *The vertebrate body: Shorter version.* W. B. Saunders, Philadelphia.

Sakurai. T., T. Hatae, T. Ichimura, and T. Ishida. 1997. Scanning electron microscopic study of the cytoplasmic tubules in lamprey chloride cells. *J. Elect. Micros. 1:* 93–95.

Satchell, G. H. 1991. *Physiology and form of fish circulation.* Cambridge University Press, Cambridge, UK.

Schreiber, A. M., and Specker, J. L. 1999. Metamorphosis in the summer flounder, *Paralichthys dentatus:* Thyroid status influences salinity tolerance. *J. Exp. Zool. 284:* 414–424.

Shikano, T., and Y. Fujio. 1999. Changes in salinity tolerance and branchial chloride cells of newborn guppy during freshwater and seawater adaptation. *J. Exp. Zool. 284:* 137–146.

Shuttleworth, T. J. 1988. Salt and water balance—extrarenal mechanisms, pp. 171–199. In *Physiology of elasmobranch fishes,* T. J. Shuttleworth (Ed.). Springer Verlag, Berlin.

Simpson, D. G., and G. Gunter. 1956. Notes on habits, systematic characters, and life histories of Texas salt water Cyprinodontes. *Tulane Stud. Zool. 4:* 115–134.

Smith, H. W. 1930. Metabolism of the lungfish, *Protopterus aethiopicus. J. Biol. Chem. 88:* 97–130.

Thorson, T. B. 1961. Partitioning of body water in Osteichthyes: Phylogenetic implications in aquatic vertebrates. *Biol. Bull. 120:* 238–254.

———. 1967. Osmoregulation in freshwater elasmobranchs, pp. 265–270. In *Sharks, skates, and rays,* P. W. Gilbert, R. F. Mathewson, and O. P. Rall (Eds.). Johns Hopkins University Press, Baltimore.

Tipsmark, C. K., and S. S. Madsen. 2001. Rapid modulation of Na+/K+ −ATPase activity in osmoregulatory tissues of a salmonid fish. *J. Exp. Biol. 204:* 701–709.

Usher, M. L., C. Talbot, and F. B. Eddy. 1991. Effects of transfer to seawater on growth and feeding of Atlantic salmon smolts. *Aquaculture 94:* 309–326.

van der Heijden, A. J. H., P. M. Verbost, M. J. C. Bijvelds, W. Atsma, S. E. Wendelaar Bonga, and G. Flik. 1999. Effects of sea water and stanniectomy on branchial Ca2+ handling and drinking rate in eel (*Anguilla anguilla* L.). *J. Exp. Biol. 202:* 2505–2511.

Vickers, T. 1961. A study of the so-called "chloride-secretory" cells of the gills of teleosts. *Q. J. Microsc. Sci. 60:* 507–518.

Wake, M. H. (Ed.). 1979. *Hyman's comparative anatomy* (3rd ed.). University of Chicago Press, Chicago.

———, 1987. Urogenital morphology of dipnoans, with comparisons to other fishes and to amphibians, pp. 199–216. In *The biology and evolution of lungfishes,* W. E. Bemis, W. W. Burggren, and M. E. Kemp (Eds.). Alan R. Liss, New York.

Walsh, P. J. 1998. Nitrogen excretion and metabolism, pp. 199–214. In *The physiology of fishes* (2nd ed.), D. H. Evans (Ed.). CRC Press, Boca Raton, FL.

Weichert, C. K. 1965. *Anatomy of the chordates* (3rd ed.). McGraw-Hill, New York.

Wendelaar Bonga, S. E. 1993. Endocrinology, pp. 469–502. In *The physiology of fishes,* D. H. Evans (Ed.). CRC Press, Boca Raton, FL.

Wilkie, M. P. 1997. Mechanisms of ammonia excretion across fish gills. *Comp. Biochem. Physiol. 118A:* 39–50.

Wilson, R. W., C. M. Wood, R. J. Gonzalez, M. L. Patrick, H. L. Bergman, A. Narahara, and A. L. Val. 1999. Ion and acid–base balance in three species of Amazonian fish during gradual acidification of extremely soft water. *Physiol. Biochem. Zool. 72:* 277–285.

Wright, P. A., and M. D. Land. 1998. Urea production and transport in teleost fishes. *Comp. Biochem. Physiol. 119A:* 47–54.

Young, G., B. T. Bjornsson, P. Prunet, R. Lin, and H. A. Bern. 1989. Smoltification and seawater adaptation in coho salmon (*Oncorhynchus kisutch*): Plasma prolactin, growth hormone, thyroid hormones, and cortisol. *Gen. Comp. Endocrinol. 74:* 335–345.

Youson, J. H. 1981. The kidneys, pp. 191–261. In *The biology of lampreys,* M. W. Hardisty and I. C. Potter (Eds.). Academic Press, London.

———, and D. B. McMillan. 1970. The opisthonephric kidney of the sea lamprey of the Great Lakes, *Petromyzon marinus* L. I. The renal corpuscle. *Am. J. Anat. 127:* 207–232.

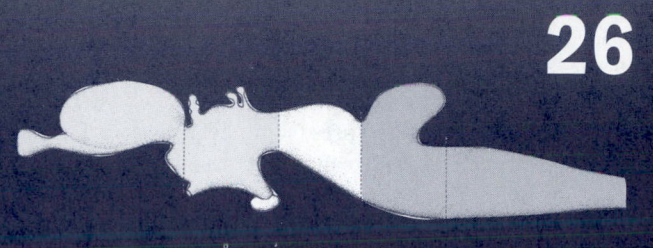

26

NERVOUS AND ENDOCRINE SYSTEMS

THE CENTRAL AND PERIPHERAL NERVOUS SYSTEM

The Brain
Cranial Nerves
   Lateral Line Nerves
Spinal Cord and Nerves

THE ENDOCRINE SYSTEM

The Hypothalamus
The Pineal Organ
The Pituitary Gland
   Agnaths
   Chondrichthyes
   Bony Fishes
   Hormones Released by the Adenohypophysis in Fishes
   Neurohypophyseal Hormones
Thyroid Gland
Interrenal Tissue
Chromaffin Tissue
Ultimobranchial Gland
Pancreas (Islets of Langerhans)
Gastroenteric Mucosa
Gonads
Corpuscles of Stannius
Other Organs with Endocrine Function
   Kidney
   Thymus

NEUROSECRETORY CELLS

Caudal Neurosecretory System
Pseudobranchial Neurosecretory Gland
Paraneurons
Natriuretic Peptides

EVOLUTIONARY AND ENVIRONMENTAL CONSIDERATIONS

Nowhere in the study of the vertebrate body plan is an integrative approach more necessary than in the consideration of the two systems that unite all the other systems and enable the organism to function as a whole—the nervous and endocrine systems. Although we will focus here on the structure and function of the components of these two systems, we must acknowledge that their function cannot be fully understood except in the context of the other tissues and organs in vertebrates. Structurally, the vertebrate nervous system is a discrete entity, consisting of bundles of sensory neurons that relay impulses from sense receptors to centralized processing areas in the brain, spinal cord, and peripheral ganglia, from which motor neurons in turn relay motor impulses to muscles and glands that communicate via neuromuscular junctions. The endocrine system is something altogether different, as it consists of tissue components from a variety of sources. Although diverse in embryonic origins, these tissues are functionally united in their mode of action— the release of chemical signals that travel to target tissues via the circulatory system—and their degree of communication that enables the various tissues and organs of the body to function as a cohesive whole. Although they were long held to be distinct systems, with the nervous system functioning mainly for rapid response to stimuli, and the endocrine system more involved in sustained maintenance, researchers are coming to recognize that the nature of the two systems is less distinct than previously thought. Orientation to the external environment and maintenance of internal homeostasis involve complex integrations of neuronal and hormonal functions and highly specialized and localized tissue responses that do not clearly fit into either neuronal or hormonal categories.

THE CENTRAL AND PERIPHERAL NERVOUS SYSTEM

Structurally, the vertebrate nervous system is usually divided into the **central nervous system** (**CNS**), which includes the brain and spinal cord, and the **peripheral nervous system** (**PNS**), which includes sensory (afferent) nerve tracts transmitting signals from sense receptors to the CNS and motor (efferent) tracts leaving the CNS. The CNS is responsible for integrating and processing afferent inputs and initiating efferent outputs. The PNS also includes ganglia peripheral to the CNS that function as satellite processing centers.

The Brain

A critical development in the evolution of vertebrates was the transition from a filter-feeding organism to one that was capable of moving water across pharyngeal perforations with a muscular pump—a feature that distinguishes the craniates from cephalochordates (see Chapter 2). With the development of a new and complex head, specifically modified for simultaneous feeding and gas exchange, and possessing a full complement of sensory organs, the brain became a significantly more complex entity. This development appears to have coincided with the duplication of genomic elements that is believed to have been instrumental in the evolution of vertebrates (Northcutt, 1996).

As with other vertebrates, the brains of fishes lie in the lumen of the **neurocranium**, protected by cartilage or bone, the membranous surrounding **meninges** and cerebrospinal fluid, and a fatty matrix that surrounds the brain and fills much of the cranial cavity. Contrary to the elaborate meningeal arrangement seen in tetrapods (having two or three layers of tissue—dura mater, pia mater, and arachnoid), most fishes have only a single layer (**primitive meninx**). Cranial nerves leave the brain and pass through foramina of the skull to their respective target tissues or organs. The spinal cord, with which the brain is continuous, leaves the cranium posteriorly through the **foramen magnum**. There is a great deal of variation in brain morphology among fishes, even though fish brains all follow a basic vertebrate plan (Fig. 26.1; Igarashi and Kamiya, 1972; Romer and Parsons, 1978). Although some differences in brain morphology are of phylogenetic significance, other differences are apparently due to the degree of development of different sensory and motor functions (Northcutt and Davis, 1983; Tuge et al., 1968). Differences in the relative sizes of fish brains are also evident. The brains of the two most primitive craniates, hagfishes and lampreys, differ strikingly in overall morphology. The brains of lampreys are slender and elongated and possess a well-developed system of ventricles; lampreys possess the lowest ratio of brain to body weight of all craniates. The hagfish brain is more compressed, with a much reduced system of ventricles; hagfishes possess a brain to body weight ratio that is comparable to that of teleosts and urodele amphibians (Wicht, 1996). Differences in the relative sizes of parts of the brain among fishes are often due to the comparative differences in the development of the special senses (sight, hearing, touch, smell, taste, and mechanosensory and electrosensory lateral line).

The brains of some fishes, including the modern coelacanth, weigh considerably less than 0.1 to 0.3 percent of the total body weight, whereas the relative brain weight of many cyprinid minnows may be twice that. Some sharks, with their large olfactory lobes and bulbs, have relatively large brain to body weight ratios (Northcutt, 1989a; Smeets et al., 1983). Elasmobranchs in general have been noted for having comparatively large brains for ectothermic animals. Ion pumping, measured by Na^+/K^+ ATPase activity, is the single most energy-consuming process in the brain. Although the average brain mass in elasmobranchs is up to three times that seen in teleosts, the Na^+/K^+ ATPase activity is only about one third of that measured in teleosts (Nilsson et al., 2000). The largest relative size among fish brains is that of the mormyrids (Mormyridae) of Africa, in which the brain may be more than 3 percent of body weight (see Chapter 9).

In embryonic fishes, the brain first develops in three sections (Fig. 26.2): the **forebrain** (**prosencephalon**), **midbrain** (**mesencephalon**), and **hindbrain** (**rhombencephalon**), as in other vertebrates, and then undergoes further differentiation (Kent, 1992; Nieuwenhuys, 1962; Romer and Parsons, 1978). The mesencephalon and rhombencephalon together constitute the **brain stem** or **truncus cerebri** (Nieuwenhuys and Pouwels, 1983). Using outgroup analysis, Northcutt (1995) has generated a morphotype (the ancestral configuration) of the gnathostome forebrain. This is the site of the olfactory sense. Its anterior part (**telencephalon**) is characterized by a pair of primary olfactory centers, the **olfactory bulbs,** from which olfactory nerves extend to the olfactory organ. Caudal to the olfactory bulbs, the telencephalon swells into what are often called "olfactory lobes" (Kent, 1992). These are usually larger than the bulbs and are mainly concerned with non-olfactory functions. In most bony fishes, the bulbs are situated just anterior to the lobes, but in elasmobranchs and in certain catfishes, carp, and cods, the olfactory bulbs are adjacent to the olfactory organ, and a long olfactory tract separates them from the lobes (Bernstein, 1970). In such species, and especially in sharks and rays, the bulbs may be comparatively large. Part of the telencephalon is developed as the **cerebrum** in elasmobranchs and bony fishes, although in the latter, the cerebral hemispheres are prominent only among the more primitive members. The cerebrum of lungfishes resembles

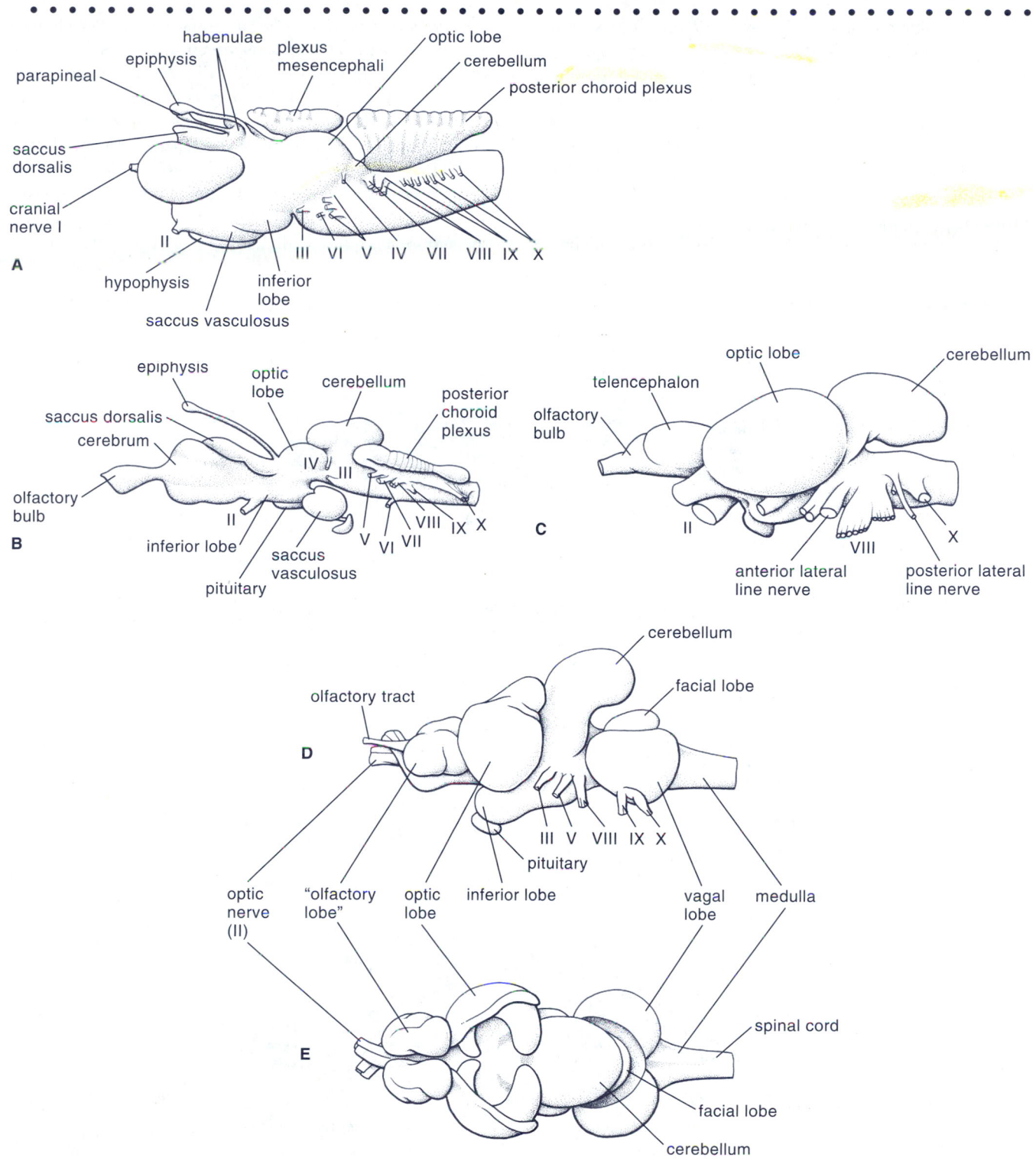

FIGURE 26.1

View of brain: **A,** Lamprey (*Petromyzon*); **B,** Shark (*Somniosus*); **C,** Pacific salmon (*Oncorhynchus*);
D, Carp (*Cyprinus*), lateral view; **E,** Carp, dorsal view (*A* and *B* after Romer, 1970).

that of amphibians (Romer and Parsons, 1978). The posterior part of the forebrain is the **diencephalon**, usually set off by a constriction from the telencephalon. The **pineal organ**, which in many fishes is sensitive to light, arises from the roof of the diencephalon in elasmobranchs and bony fishes (Bernstein, 1970; Ekström and Meissl, 1997). Lampreys and hagfishes have both a parapineal and pineal organ. The **hypothalamus**, in the floor of the forebrain, is closely associated with the adjacent pituitary gland. The hypothalamus, in its role as controller of the pituitary gland, is the main link between nervous and endocrine functions in vertebrates. Also in this area is the **saccus vasculosus**, which is a vascular evagination with thin walls (Fig. 26.4). This structure is found only in fishes. Its function is not definitely known, but it is lined with hair cells, and its sensory fibers go to other brain centers, including the hypothalamus (Kent, 1992).

The **optic lobes** are the most prominent feature of the midbrain (mesencephalon) and are especially large in visually oriented fishes. In hagfishes, which have vestigial eyes, the midbrain is quite small. The optic lobes of lungfishes are

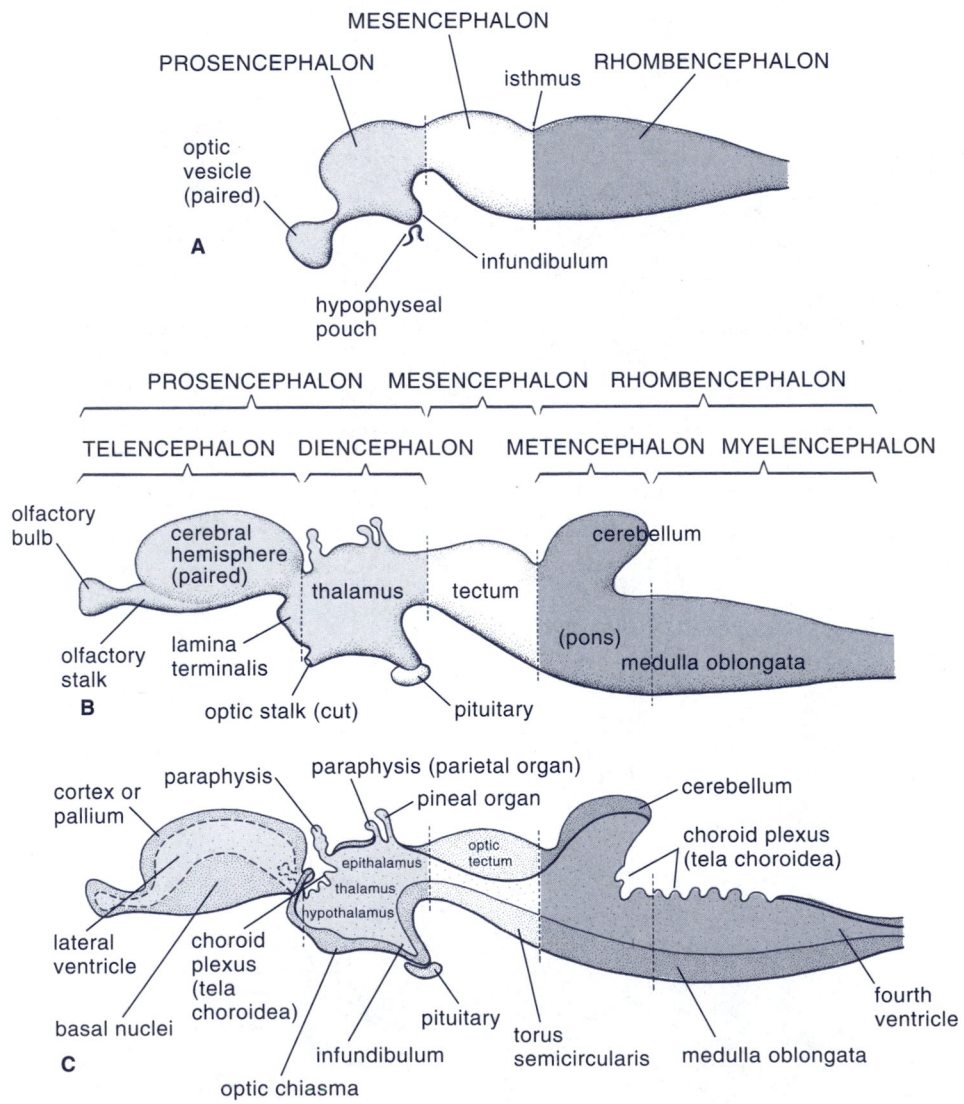

FIGURE 26.2
Stages in the development of the brain: **A,** Three main divisions developed; **B,** More advanced stage; **C,** Sagittal section of *B* (after Romer, 1970).

fused into a single median structure. Vision is a major function of the mesencephalon, but the midbrain also has other functions, including the facilitation of learning and the relay of sensory messages to motor responses. Afferent nerve fibers from olfactory and gustatory centers run to the roof of the mesencephalon or optic tectum. The **torus semicircularis** is developed in the floor of the midbrain and is especially large in fishes that have well-developed auditory and lateral line senses. Immediately ventral to the midline of the tectum lies the **torus longitudinalis.** This group of neurons, which responds to photostimulation and appears to receive input from oculomotor neurons controlling eye movements, is present only in actinopterygians except for the polypteriform fishes (Wullimann, 1994).

The major feature of the **metencephalon** (the anterior part of the rhombencephalon) of elasmobranchs and bony fishes is the **cerebellum.** In bony fishes, the cerebellum is divided into two major sections: the **valvula cerebelli,** extending rostrally below the optic tectum; and the **corpus cerebelli,** extending anteriodorsally. The valvula are large in catfishes and minnows and reach a very large size in mormyrids (Nieuwenhuys and Pouwels, 1983). The size of the cerebellum is variable, being especially well developed in large sharks. In the family Mormyridae, which have exceptional auditory and electrosensory systems, the cerebellum is large and complex and may overlie the forebrain. Catfishes, mackerels, and tunas usually have enlarged cerebella. In lampreys, the cerebellum is very small, and the structure is not recognized at all in hagfishes. It may be that the virtual absence of a cerebellum in agnaths is correlated with the fact that they do not engage in sustained swimming as much as teleosts (Stacia Sower, personal communication). Although the cerebellum is derived from the rhombencephalon, it is not considered a part of the brain stem by Nieuwenhuys and Pouwels (1983), even though it is "strongly interconnected." The brain stem consists of the rhombencephalon and mesencephalon and holds the origins and terminations of most of the cranial sensory nerves (Nieuwenhuys and Pouwels, 1983).

The cerebellum supports the coordination of movement, muscle tone, and posture or balance. Many fishes with enlarged cerebella are fast swimmers, but catfishes and mormyrids are slower moving yet have this structure enlarged, so the integration of certain sensory information by the cerebellum also appears to be an important function in some types of fishes. Mormyrids are noted for their "electrical" sense, and catfishes are equipped with organs for the reception of electrical impulses.

At the point where the corpus cerebelli meets the medulla, there are lateral swellings called **eminentiae granulares,** which may be enlarged in some species (such as in certain catfishes and trichiurid hairtails). The eminentia granularis appears to be an electroereceptive part of the mormyrid brain (Hopkins, 1983).

The most posterior part of the brain, the **myelencephalon,** is composed chiefly of the **medulla oblongata,** which in gnathostomes includes somatic and visceral sensory and motor areas. Several cranial nerves arise from the medulla oblongata, and impulses to and from spinal nerves are also relayed here. Various parts of the medulla are enlarged according to the sensory function and habits of fishes. For instance, certain suckers (Catostomidae) and minnows (Cyprinidae) have characteristic enlargements ("lobes") at the roots of the seventh, ninth, and tenth cranial nerves.

A pair of large neurons, called **Mauthner cells,** is found in the floor of the medulla of most fishes and aquatic amphibians (Diamond, 1971; Hardisty, 1979; Kuhlenbeck, 1975; Zottoli, 1978, 1981). These function in startle and escape reactions (see Chapter 19; Eaton, 1991; Eaton et al., 1995; Zottoli et al., 1995). The dendrites are associated with acoustic centers, and the axons travel the length of the spinal cord to connect with motor neurons to trunk and caudal musculature (Zottoli, 1978). These cells are joined by an electrical synapse. Because the synaptic delay is so much shorter in electrical synapses than in the more common chemical synapses, the coordinated motor response to stimuli is much more rapid. Mauthner cells are lacking in several groups of fishes, including anglerfishes (Lophiiformes), eels (Anguilliformes), pipefishes, seahorses (Syngnathidae) and others (Bone and Marshall, 1982; Zottoli, 1978, 1981). The aforementioned groups are not especially well known for speed and agility—locomotor features that would be enhanced through Mauthner cells. Cryptic coloration and behavior work better for these groups than a rapid escape response. For some anglerfishes, however, these limited locomotor abilities do not compromise their feeding efficiency—they are capable of extremely rapid generation of suction for prey inhalation (see Chapter 13).

• • • • • • • •
Cranial Nerves

The arrangement of the cranial nerves reflects the conservatism inherent in the evolution of the central nervous system, because the anatomy and the aggregate functions of the cranial nerves are remarkably uniform among the vertebrate classes. These nerve tracts consist of **sensory (afferent)** neurons, relaying information from peripheral sense organs to the brain, and **motor (efferent)** neurons, which arise in the brain and innervate muscles and glands. The nerves can also be classified as **somatic,** if they innervate somatic tissues such as skeletal muscle or skin, or **visceral,** if they innervate visceral tissues such as involuntary (smooth) muscle or glands. They can also be designated as **special**, if they are associated with highly localized senses such as sight, hearing, olfaction, or taste, or **general**, if they are concerned with sensory or motor functions in a broader distribution. Cranial nerves are conventionally named using Roman numerals, based on their relative position; although they were originally described according to the mammalian configuration, these designations work reasonably well for other vertebrates, including fishes. In the mammalian configuration, 12 cranial nerves are identified, but as many as 25 have been identified in some craniates (Kent and Carr, 2001). Each nerve may be associated with a multitude of functions, as it can carry both sensory and motor neurons innervating a variety of sites. Cranial nerves can also be grouped by function.

As luck would have it, once the designation of cranial nerves with Roman numerals was formalized, a new tract was discovered lying anterior to cranial nerve I. Rather than overturn convention, this *terminal nerve* was designated cranial nerve 0; it is apparently sensory in function. The *olfactory* (I), *optic* (II), *lateral line* (including anterior, middle, and posterior nerve tracts) and *acoustic* (VIII) nerves are sensory. The *oculomotor* (III), *trochlear* (IV), and *abducens* (VI) nerves are motor nerves; and the *trigeminal* (V), *facial* (VII), *glossopharyngeal* (IX) and *vagus* (X) nerves have both sensory and motor fibers present. Including the terminal nerve (0), 11 cranial nerves, plus the unnumbered lateral line nerves, are present in fishes. Actually, the ganglion of the *profundus* nerve (numbered V_1 in some texts) is fused to the trigeminal ganglion in lampreys, hagfishes, and lungfishes, but the two are not fused in other fishes (Northcutt and Bemis, 1993), so that they appear as separate nerves.

Fishes lack distinct *accessory* and *hypoglossal* nerves (cranial nerves XI and XII in amniotes). Neurons with the homologous function of the accessory nerve of other vertebrates are found in association with the vagus (X). Hypoglossal neurons are associated with nerves at the anterior end of the spinal cord. The cranial nerves found in fishes and the functions of their component neurons are summarized in Table 26.1.

Lateral Line Nerves

These are unnumbered cranial nerves that are placed so closely to other nerves that they have been considered components of those nerves, but Northcutt (1989b) has pointed out that a more plausible hypothesis is that the lateral line nerves are a separate series of cranial nerves.

The anterior lateral line nerve (ALLN) travels with cranial nerves V and VII and has two or three ganglia, depending on the species. Branches from these ganglia innervate neuromasts on the anterior part of the head. The middle lateral line nerve (MLLN), if present, innervates a pit line in bony fishes. The posterior lateral line nerve (PLLN) may travel with cranial nerve X and serves mainly the corporal neuromasts, including the lateral line canal (McCormick, 1983; Northcutt, 1989b; Song and Northcutt, 1991).

Spinal Cord and Nerves

The spinal cord, with only a few exceptions, is continuous with the medulla oblongata and extends to the end of the vertebral column. It is essentially a hollow tube, but the central canal is of small diameter compared to the thick walls. Around the central canal, making a pattern that in cross-section looks similar to a pair of butterfly wings, is **gray matter**, composed of unmyelinated (i.e., lacking the fatty, insulating wrapping provided by glial cells) nerve fibers running longitudinally. In lampreys and hagfishes, all the nerve fibers are unmyelinated, and the spinal cord is flattened dorsoventrally. The spinal cord of the lamprey is characterized by 8 to 12 giant *Müller's fibers* on each side (Hardisty, 1979; Kuhlenbeck, 1975; Rovainen, 1978). These are somatic motor axons that run the length of the spinal cord from Müller cell bodies in the brain stem.

The length of the spinal cord can vary significantly. In the ocean sunfishes (Molidae), it does not extend much farther than the hindbrain, and in the goosefish (*Lophius*), the cord is shortened, but a long terminal filament extends posteriorly from it.

The paired spinal nerves are arranged segmentally and arise from the gray matter as dorsal and ventral roots that merge and then typically branch into three parts. The dorsal root has a ganglion outside the spinal cord (dorsal root ganglion). Dorsal and ventral branches (or rami) serve the axial muscles and skin, whereas a visceral branch (ramus) supplies the internal organs. Lampreys differ from the other fishes in lacking the connection between the dorsal and ventral roots. In these forms, the dorsal roots originate opposite the myosepta, and the ventral roots opposite the myotomes.

The dorsal roots of spinal nerves in fishes carry somatic and visceral afferent fibers (and some visceral efferent fibers). Somatic and visceral efferent fibers enter the spinal cord through the ventral roots. Visceral efferent components of the cranial and spinal nerves contribute to the **autonomic nervous system** (sympathetic and parasympathetic components), which is involved in the control of smooth muscle and certain glands. Bony fishes have a chain of interconnected, segmentally arranged ganglia that are peripheral to the central nervous system. Sympathetic ganglia are found in an irregular series in the trunk region of elasmobranchs. Parasympathetic fibers are largely associated with cranial nerves, almost entirely within the vagus (X) nerve.

TABLE 26.1   CRANIAL NERVES OF VERTEBRATES AND ASSOCIATED FUNCTIONS IN FISHES

Cranial Nerve	*Location*	*Function*
0 (Terminal)	Projects to telencephalon; associated with olfactory nerve; distributed within olfactory bulb. Not apparent in lampreys (Eisthen and Northcutt, 1996).	Integration of sensory and autonomic functions related to reproductive physiology and behavior (Demski and Northcutt, 1983; Demski and Schwanzel-Fukuda, 1987; Fujita et al., 1991). Cells containing luteinizing hormone releasing hormone (LHRH), essential for reproductive function, have been identified in ganglion cells of the terminal nerve (Schwanzel-Fukuda, 1999). Functions in chemosensation.
I (Olfactory)	Runs from the olfactory epithelium in the olfactory organ to the olfactory lobe of the telencephalon.	
II (Optic)	Projects from the ganglion cell layer of the retina to the optic tectum. (Although classified as a cranial nerve, it develops as an extension of the brain during the development of the eyes.) Relays visual information, including color vision and sensitivity to light in the ultraviolet and visible spectra.	
III (Oculomotor)	Projects from mesencephalon to the inferior oblique and the superior, inferior, and medial rectus muscles of the eye.	Functions as a somatic motor nerve, controlling extrinsic eye muscles and visceral muscles of iris and ciliary body.
IV (Trochlear)	Projects from mesencephalon to superior oblique muscle of the eye.	Functions as a somatic motor nerve, controlling extrinsic eye muscle.

V (Trigeminal) Divided into three branches:
    Opthalmic
    Maxillary
    Mandibular
	Projects to metencephalon from various receptor sites.	Carries information from taste buds, tactile, and thermal receptors.
		    Somatic sensory nerve.
		    Somatic sensory nerve.
		    Somatic sensory fibers from jaw; motor fibers from muscle derivatives of first(mandibular) visceral arch.

VI (Abducens)	Projects from anterior medulla oblongata (myelencephalon) to lateral rectus of eye.	Functions as a somatic motor nerve, controlling extrinsic eye muscle.

VII (Facial) Divided into three branches:
    Superficial ophthalmic
    Buccal
    Hyomandibular
	Projects to myelencephalon from various receptors in head and body.	Different branches supply taste receptors on head, body tactile receptors, and control of head muscles derived from second (hyoid) visceral arch.

VIII (Acoustic)	Projects to myelencephalon from inner ear.	Acoustic reception.
IX (Glossopharyngeal)	Projects to myelencephalon from sense receptors, and motor tracts to muscles of first gill slit (3rd visceral arch).	Dorsal group of branches serves proprioceptors* and small portion of lateral line (Kent and Carr, 2001); branchial branches associated with taste organs of pharynx and branchial muscles.
X (Vagus)	Multiple branches from myelencephalon to posterior four gill slits, numerous internal organs.	Diverse sensory and motor functions.

* Proprioceptors are receptors located in muscles, tendons, and joints; they function to provide information on body position.

THE ENDOCRINE SYSTEM

The endocrine system used to be conceived of as an aggregate of glandular structures that enabled cells and tissues to communicate through substances (**hormones**) released into the circulatory system. Recently, the perception of endocrine function has broadened to include "any substance that operates at the cellular level, generated either externally or internally, which conveys to that cell a message to stop, start, or modulate a cellular process" (Norman and Litwack, 1997). The classical perception of endocrine function was that it involved the release of compounds that had an effect on tissues at some distance. Much more localized tissue responses result from **autocrine or paracrine** secretions. Autocrine secretions produce responses in the cells producing the secretions themselves, whereas paracrine signals elicit local action in their immediate neighborhood through their release by cells into the extracellular medium.

The endocrine system of fishes is comparable to that of higher vertebrates, but some endocrine tissues do not form discrete glands in fishes, and the sites of these tissues may be different from the sites of secretion in tetrapods. However, for the most part, similar hormones are produced. Moreover, fishes possess some additional endocrine tissues, such as the caudal neurosecretory system and the Stannius corpuscles, that do not have homologues in higher vertebrates. Following is a description of the endocrine glands of fishes,

with the general location of the tissue and a brief mention of the function of the secretions. The general locations of the endocrine glands are shown in Figure 26.3.

The Hypothalamus

As described earlier, the hypothalamus is located in the lower part of the diencephalon. It contains vital neurosecretory pathways for the initiation of endocrine responses to environmental stimuli. The functional relationship between the hypothalamus and the pituitary gland is currently defined in the context of the **hypothalamo–pituitary axis,** which consists of neurosecretory neurons in the hypothalamus and the pituitary gland, plus all other glands and target tissues under their direct control (Norris, 1997). In most fishes, including the teleosts, the hypothalamus consists of pronounced, evaginated lobes (Butler, 2000). Secretions arising from various regions of the hypothalamus influence a variety of pituitary functions. In some animal groups, these neurosecretions from the hypothalamus reach the pituitary via the hypothalamo–hypophyseal portal system (see Chapter 24). However, this vascular component is absent in agnaths and teleosts (Gorbman, 1995; Sower, 1998).

Aggregates of cell bodies that participate in the same neural circuits are traditionally termed **nuclei,** and the hypothalamus in fishes includes a diversity of these nuclei. The **in-**

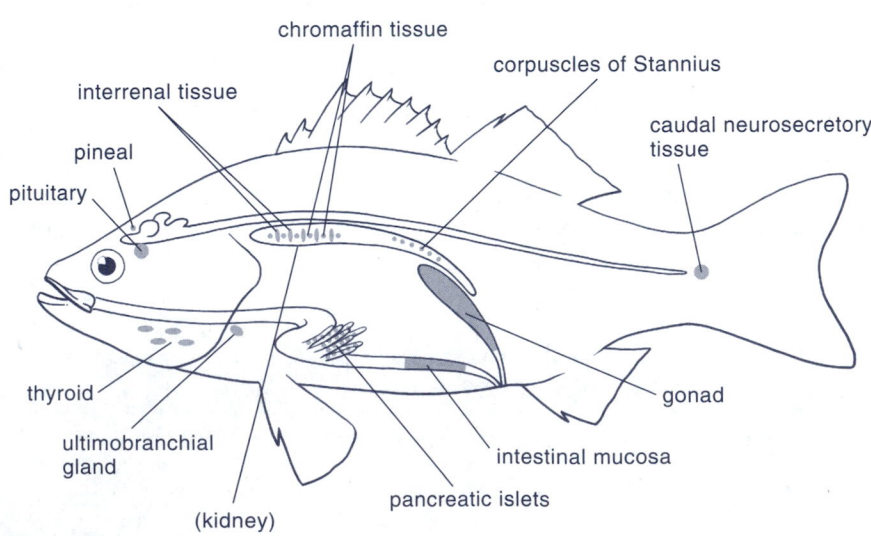

FIGURE 26.3
Approximate locations of tissues and organs known to have endocrine function in a bony fish.

fundibular cells of the cerebral vesicles of cephalochordates are considered homologous to the hypothalamus that appears in craniates (Northcutt, 1996). Four hypothalamic nuclei have been identified in chondrichthyans, and actinopterygians exhibit the greatest variation among gnathostomes (Northcutt, 1995). Auditory, mechanosensory, electroreceptor, and gustatory centers have all been identified as providing inputs to the hypothalamus (Northcutt, 1995). Electrical stimulation of the hypothalamus elicits aggressive and feeding behaviors; this suggests that the hypothalamus possesses highly integrative functions and has extensive neuronal communication with other areas of the brain (Demski, 1983).

The Pineal Organ

In vertebrates, a number of evaginations may be derived from the roof of the diencephalon, including the *parietal eye,* the so-called "third eye" (Ekström and Meissl, 1997; Fenwick, 1970), and the *pineal organ* or *epiphysis*. With light-sensitive cells, similar to those found in the retina, and a rudimentary lens positioned beneath a cornea derived from the overlying epidermis, the parietal eye of some reptiles may be quite a sophisticated photoreceptor (Kardong, 1997). Lampreys have both pineal and parietal organs, both with photoreceptive capabilities. The pineal organ is prominent in elasmobranchs and bony fishes, but the parietal organ is absent or rudimentary at best. In most bony fishes, a thickening of the pineal epithelium and an enhanced vascularization of the structure indicates a glandular function (Belsare, 1974). However, there is great variation among fishes, even within particular families, in the manifestation of the pineal organ's photoreceptive capacity or secretory capacity, or both.

The chief secretion of the pineal organ is the peptide **melatonin,** which aggregates melanin granules in the skin melanophores (see Chapter 18) of amphibians and has been shown to have a similar effect in some fishes. The diurnal pigmentation cycle of lamprey ammocoetes (dark by day, light by night) involves the action of melatonin aand other, nonendocrine entities. The phototransduction cascade linking the reception of light to the release of melatonin is believed to have features in common with retinal phototransduction. Chief among these are the accumulation of cyclic adenosine monophosphate (cAMP) in darkness and its diminution during light exposure, and the fact that sodium ions are necessary for photoreception. Excitatory amino acids, like glutamate or aspartate, are believed to be the neurotransmitters operating in the pineal organ (Ekström and Meissl, 1997).

The pineal organ is known to contain and release other materials, such as various peptides and arginine vasotocin (Kavaliers, 1980). The secretion of melatonin and serotonin, whose concentrations change during annual reproductive cycles, suggests a reproductive role for the pineal organ, chiefly as a regulator of seasonal rhythmicity. Serotonin is known to stimulate the release of gonad releasing hormones in fishes (Redding and Patiño, 1993). In addition to its role in gonadal maturation and pigmentation, the pineal gland may also be involved in the control of circadian locomotor rhythms, behavioral thermoregulation, growth, and metabolism (Ekström and Meissl, 1997; Tamura and Hanyu, 1980). Removal of the pineal organ from fishes can bring about changes in growth and can result in the stimulation of the pituitary and thyroid glands (Vodicnik et al., 1978, 1979). Pinealectomy can also disrupt circadian activity cycles (Kavaliers, 1980).

The Pituitary Gland

This gland, also referred to as the **hypophysis,** is located beneath the diencephalon (Fig. 26.3) and may be associated with the saccus vasculosus. It has a dual embryonic origin: A part of it arises from an evagination of the floor of the embryonic brain, which meets an ectodermal component growing upward from the dorsal part of the embryonic mouth cavity (stomodeum). The stomodeal component forms an early pouchlike structure (**Rathke's pouch**) in lower fishes, but begins as a solid structure in higher fishes. This component eventually forms the **adenohypophysis.** The neural (diencephalic) component of the adult pituitary gland, called the **neurohypophysis,** is closely associated with the hypothalamus.

Hypothalamic control of the neurohypophysis and adenohypophysis is mediated by specialized neurosecretory cells. Adenohypophyseal secretions are controlled by peptide neurosecretory substances from the hypothalamus classified as hormones but also known as *releasing factors*. In teleosts, these factors are released from hypothalamic neurons that enter the adenohypophysis directly and innervate it. An example of this is the neuroendocrine release, by gonadotropin-releasing hormone (GnRH), of gonadotropic hormones from the pituitary gland. Receptor sites for GnRH have been identified in the adenohypophysis. In addition to GnRH, other neurosecretions stimulate or inhibit the release of adenohypophyseal hormones (Sower, 1997, 1998; Sower and Kawauchi, 2001). Hagfishes and lampreys depend on simple diffusion between the brain and adenohypophysis to bring neurosecretory hypothalamic factors to the adenohypophysis (Nozaki et al., 1994; Sower, 1998).

There are commonly two parts of the neurohypophysis of fishes (and other vertebrates): the **infundibular stalk** and the **posterior lobe (pars nervosa)**. The remainder of the pituitary complex (i.e., the adenohypophysis or *pars*

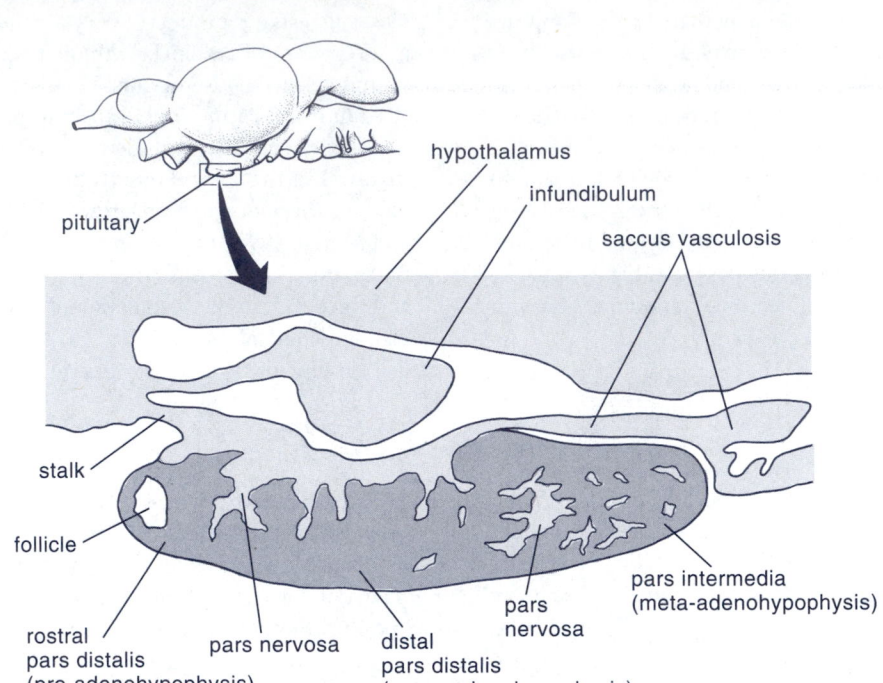

hypothalamus

infundibulum

saccus vasculosis

pituitary

stalk

follicle

pars intermedia
(meta-adenohypophysis)

rostral
pars distalis
(pro-adenohypophysis)

pars nervosa

distal
pars distalis
(meso-adenohypophysis)

pars
nervosa

FIGURE 26.4
Sagittal section of the pituitary gland of rainbow trout (*Oncorhynchus mykiss*).

buccalis), formed from the stomodeal component, consists of histologically distinct areas that have more or less functionally equivalent parts in the mammalian pituitary gland.

The adenohypophysis is divided into an anterior section, called the **pars distalis,** and a posterior section, or **pars intermedia** (*meta-adenohypophysis*), which is most intimately related to the pars nervosa (Fig. 26.4). In many fishes, the pars intermedia is completely fused with the neurohypophysis and forms what is referred to as the *neuro-intermediate lobe* (Gorbman et al., 1983; Holmes and Ball, 1974; Matty, 1985; Pickford and Atz, 1957; Schreck and Scanlon, 1977; Sower, 1998).

Agnaths

The role of the pituitary gland in agnathan endocrine systems is an especially attractive area of research, as it is believed that much can be learned about the origins of the vertebrate pituitary systems by studying these early forms. The distinguished comparative endocrinologist Aubrey Gorbman (1914–2003) stood out as a pioneer in the field of evolutionary endocrinology. Sower, Kawauchi, and their colleagues have isolated

and characterized several hormones from the brain and pituitary gland of lampreys, including those with essential roles in reproduction (Sower, 1998; Sower and Kawauchi, 2001). Their work has demonstrated that the function of the hypothalamo–hypophyseal axis in these primitive fishes is broadly similar to that seen in gnathostomes.

The pituitary glands of lampreys and hagfishes are considered to be more primitive than those of elasmobranchs and bony fishes (Fig. 26.5A–E). The gland is flattened and less complex than that of gnathostomes, yet resembles it in being composed of a neurohypophysis and an adenohypophysis (Sower, 1998). In hagfishes, the neurohypophysis appears as a tubular projection from the brain, whereas in lampreys it is poorly developed, being little more than a thin plate of cells in association with nerve fibers. The adenohypophysis of lampreys is divided into rostral, distal, and posterior sections, although these may not be homologous with those of gnathostome fishes. In hagfishes, the adenohypophysis is undivided, and there is no pars intermedia. Although several types of secretory cells are present, they are not clearly segregated by functional type. The agnathan pituitary gland probably does not secrete as full a range of hormones as in higher forms, and the divisions may not be comparable to those of other fishes. Pituitary hormones that have been isolated so far include adrenocorticotropin (ACTH), two forms of melanophore stimulating hormone (MSH), and two hormones of unknown function (Sower and Kawauchi, 2001).

Chondrichthyes

Although considered to be among the most primitive gnathostome fishes, elasmobranch fishes possess a suite of remarkable physiological adaptations that has stimulated much research into their endocrine control mechanisms (Gelsleichter and Manire, 1999). In the sharks and rays, the adenohypophysis has a thin forward extension reaching or nearly reaching the optic chiasma (Matty, 1985), and the pars nervosa is mixed with the pars intermedia to form a neuro-intermediate lobe (Gorbman et al., 1983). The pituitaries of sharks and rays are peculiar in having a small ventral lobe, attached to the pars distalis by a short stalk, and receiving a direct arterial supply (Fig. 26.5B). A structure identified in holocephalans as the *Rachendachhypophyse* (German for "throat roof hypophysis") may be comparable to the ventral lobe of elasmobranchs (Sathyanesan, 2005). In the Holocephali, it is detached and lies outside the chondrocranium in the roof of the mouth. Sharks and rays have a median eminence–like structure, with a hypophyseal portal system. There is also some evidence of this portal system in chimaeras.

Bony Fishes

There are considerable differences among the bony fishes in the shape and functional organization of the pituitary gland (Figs. 26.4, 26.5C–E).

Coelacanthiformes

The pituitary of *Latimeria* has an elongate ventral or rostral lobe of the pars distalis that is reminiscent of the ventral lobe of sharks and rays. This is one of the characters that has caused speculation about a close relationship between the two groups (Lagios, 1975, 1982; Fig. 26.5C, D). A median eminence and a hypophyseal portal system are present.

Dipnoi

In the lungfishes, the pituitary complex is more compact than in the groups previously covered and has a similarity to the pituitary of amphibians (Lagios, 1982; Matty, 1985; Romer and Parsons, 1978). A pars nervosa is formed at the posterior part of the neurohypophysis, and the pars inter-

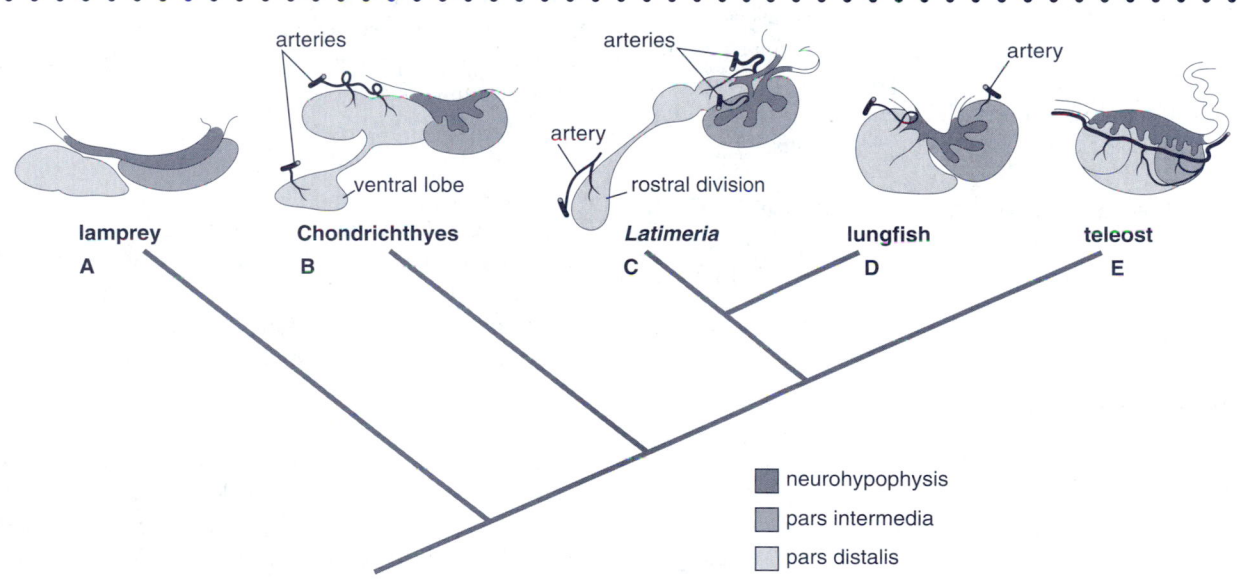

FIGURE 26.5
Phylogeny of pituitary structure in fishes: **A,** Lamprey; **B,** Chondrichthyan; **C,** *Latimeria;* **D,** Lungfish; **E,** Teleost (after Lagios, 1982; Kardong, 1997).

media is distinct. Furthermore, the median eminence and the portal system are developed. There is no saccus vasculosus in the brain of lungfishes.

Nonteleost Actinopterygians

The bichirs, reedfishes, sturgeons, paddlefishes, gars, and the bowfin all are characterized by a median eminence, constituting a portal system between the pars distalis and the floor of the infundibulum (Matty, 1985). As in teleosts, a saccus vasculosus is present. The pituitaries of these fishes are relatively compact, not divided into separate lobes, and the secretory cells are not divided into functional types. In the bichirs, a vestige of Rathke's pouch persists as an orohypophyseal duct—a condition also noted in at least one teleost, *Tenualosa (Hilsa) ilisha*, a herring of the Indian Ocean.

Teleosts

In the teleosts, the median eminence as such is lacking, and the adenohypophysis contains afferent branches from the neurohypophysis. There are many differences among the teleosts in the shape and functional organization of the pituitary complex (Gorbman et al., 1983). The glands are usually somewhat elongate, but they may be roughly globular, as in salmonids and cyprinids (Figs. 26.4, 26.5E). In the swamp eel (*Monopterus albus*), the adenohypophysis completely surrounds the neurohypophysis, except at the connection with the infundibulum (O and Chan, 1974). In most teleosts, the pituitary lies close to the hypothalamus, but in some it is on a short stalk. A notable case is the goosefish (*Lophius*), in which the gland is on a long stalk. In the goby *Lepidogobius*, the pituitary is surrounded by the hypothalamus.

Hormones Released by the Adenohypophysis in Fishes

All of the adenohypophyseal hormones are proteins. The following list summarizes the adenohypophyseal hormones known from fishes:

Adrenocorticotropin (ACTH)

This hormone stimulates the production of steroids in the interrenal tissue, which is homologous to the adrenal cortex of higher vertebrates.

Gonadotropic Hormones (GTH)

In fishes, there are two distinct gonadotropic hormones, which are considered equivalents to the follicle stimulating hormone (FSH) and luteinizing hormone (LH) seen in higher vertebrates. Fishes do not have well-differentiated beta cells (producing FSH) and gamma cells (producing LH), as in higher vertebrates, but some fishes have cells that

may approximate them. Fish gonadotropins consist of an alpha subunit and a beta subunit. The alpha subunits of the FSH and LH equivalents are the same; thyroid stimulating hormone (TSH) shares the same alpha subunit as well. The differences between FSH, LH, and TSH lie in the structure of the beta subunit.

The gonadotropic hormones stimulate the production of eggs and sperm and promote the production of steroids from other endocrine sites. GTH are released in response to gonadotropin releasing hormone (GnRH) secreted by the hypothalamus (Sower, 1990, 1998). Endocrine regulation of reproductive processes is further discussed in Chapter 27.

Growth Hormone (GH)

Growth hormone, also known as *somatotropin*, promotes growth, as both appetite and food conversion appear to increase as a result of treatment with GH. Growth hormone stimulates the liver to produce insulinlike growth factors (see later). The similarity in the peptide sequences of growth hormone, prolactin, and somatolactin suggests that they derive from a common ancestor (Mommsen, 1998).

Melanophore Stimulating Hormone (MSH)

This hormone acts on melanophores, causing aggregation or dispersal of pigment granules (see Chapter 18). In mammals, ACTH and MSH are derived from a common precursor molecule, *pro-opiomelanocortin* (POMC). This molecule is processed in different ways in the adenohypophysis, resulting in production of the final forms of ACTH and MSH; a similar system has been identified in fishes (Hazon and Balment, 1998).

Prolactin

This hormone acts on the kidney, gills, gut, and urinary bladder for the regulation of osmotic balance. It has been shown to affect the permeability of various tissues to water and to modify levels of sodium. Other effects involve xanthophore pigment dispersion and fat metabolism.

Somatolactin

This hormone was recently discovered in teleost fishes. Although its function remains unclear, it apparently plays a role in the smoltification in salmonids and in steroidogenesis in gonadal tissue (Norris, 1997).

Thyrotropin (TSH)

This hormone stimulates growth and secretions by the thyroid gland. As described earlier, it is structurally similar to gonadotropins.

Neurohypophyseal Hormones

In the various groups of fishes, several peptides are released by the neurohypophysis. Of these, **arginine vasotocin (AVT)** appears to be present and active in all species. Gorbman et al. (1983) listed the following neurohypophyseal peptides for the fishes: (1) cyclostomes: AVT only; (2) sharks: AVT, valitocin, and aspartocin; (3) rays: AVT and glumitocin; (4) chimaeras: AVT and oxytocin; (5) lungfishes: AVT and mesotocin; and (6) teleosts: AVT and isotocin. AVT appears to be involved in osmoregulation, aiding in maintaining the proper water balance. It is known to increase blood pressure. The neurohypophyseal peptides act on smooth muscle, and some of them (oxytocin, isotocin, and mesotocin) cause the constriction of branchial muscle.

Although the pituitary is obviously instrumental in governing many vital activities of the other endocrine glands, a strange cobalt-blue mutant of the rainbow trout (*Oncorhynchus mykiss*) lives to an age of at least five years with almost no pituitary tissue. This form appears occasionally from broods in Japanese hatcheries, but it is sterile because of abnormal oogenesis and spermatogenesis. It has some metabolic disorders, but the thyroid gland and the interrenal tissue appear normal (Yamazaki, 1974).

Thyroid Gland

The typical thyroid of fishes consists of separate epithelial follicles, each walled by a single layer of secretory cells and usually located in the region of the pharynx or heart. The space within the follicles contains a colloidal liquid material (Matty, 1985). The location and distribution of the thyroid follicles differ among fishes; those of lampreys and hagfishes are found along the ventral aorta in the branchial region, whereas in elasmobranchs, lungfishes, and *Latimeria*, the follicles form more or less compact glands ventral to the pharynx (Wendelaar Bonga, 1993). The thyroid follicles of many bony fishes are associated with the surface of the heart, ventral aorta, and the lower parts of the branchial arteries. A more compact type of thyroid gland has been described in a variety of bony fishes, such as the mudskipper (*Periophthalmus*), swordfish (*Xiphias*), parrotfishes (Scaridae), and a few others. In some fishes, such as the goldfish (*Carassius auratus*) and the platyfishes (*Xiphophorus*), thyroid follicles may also occur in the head kidney (i.e., the anterior hematopoietic section of the kidney) in addition to the usual subpharyngeal location.

The iodine-containing thyroid hormones (**thyroxine** and **tri-iodothyronine**) appear to have a variety of physiological effects in fishes. Many of these actions are not yet well understood, but it is certain that they are important in regulating the pigmentation of the skin and eye. There is also evidence that they affect the rate of oxygen consumption, promote the deposition of guanine in the skin, and influence carbohydrate and nitrogen metabolism. Thyroid hormones have long been known to be instrumental in the metamorphosis of amphibians, and thyroxine (T4) has an equally dramatic effect in promoting the transformation of symmetrical larval flatfishes into the extraordinarily asymmetrical juveniles that they become. In addition to this role in flatfish metamorphosis, thyroid hormones also appear to influence their osmoregulatory capacities (Huang et al., 1998; Schreiber and Specker, 1998, 1999). Involvement in osmoregulation has also been reported by Wendelaar Bonga (1993) and Hazon and Balment (1998), who reported that thyroid hormones appear to be involved in the initiation of migratory behavior and changes in osmoregulatory function in salmon as they move into seawater. Furthermore, effects on motor activity, skeletal growth, the maturation of gonads (Bern and Nishioka, 1985), and the function of the central nervous system have been noted. Leatherland (1982) emphasized that much evidence of the effects of thyroid hormones among fishes is contradictory, possibly because of species differences in the complex actions of the hormones or, in part, due to extraneous, unaccountable factors.

Interrenal Tissue

In elasmobranchs, the interrenal tissue, which is homologous with the adrenal cortex of higher vertebrates, is organized into glands situated between the posterior regions of the kidneys. The interrenal tissue of bony fishes is usually associated with the head kidney (pronephros), appearing as cells or groups of cells scattered there, especially along the cardinal veins. Cells similar to adrenocortical cells are found in the walls of the cardinal veins of lampreys, yet only small amounts of corticosteroids have been identified in agnaths (Hazon and Balment, 1998). Little is known of interrenal function in these animals (Matty, 1985). Although the regulation of the interrenal tissue is usually mediated by the action of ACTH, it appears that gonadotropins may be involved as well (Schreck et al., 1989).

The secretions of the interrenal tissue are steroids, most notably **cortisol, corticosterone,** and **cortisone.** Elasmobranchs produce a distinctive corticosteroid (1α-hydroxycorticosterone), whereas both freshwater and marine teleosts produce mainly cortisol (Hazon and Balment, 1998). The adrenocorticosteroids appear to exert some control over osmoregulatory processes, acting on the kidney, gills, and gastrointestinal tract. Cortisol treatment of the larvae of tilapia (*Oreochromis mossambica*) enhances their adaptation to conditions of elevated salinity (Lin et al., 1999). The metabolism of proteins and carbohydrates is affected by the corticosteroids, especially in such fishes as the Pacific salmons (*Oncorhynchus*), which make lengthy migrations while fasting and must use muscle protein in order to gain sufficient energy to complete their travels and the ensuing spawning process.

Secretions of the interrenal cells are important in the stress response of fishes. Stressful events, such as injury, crowding, or abrupt change from one container to another, can cause a rapid elevation of corticosteroids, which may peak in 1 to 24 hours, sometimes sooner (Schreck, 1981).

Chromaffin Tissue

This tissue is homologous with the adrenal medullary tissue of higher vertebrates, but the organization of the interrenal and chromaffin tissues into a compact gland is known in only one family of fishes (Cottidae). Usually, the chromaffin cells of bony fishes are distributed along the postcardinal veins, and they may intermingle to some extent with interrenal cells. In elasmobranchs, chromaffin tissue is associated with the sympathetic ganglia and the dorsal aorta anterior to the interrenal tissue. A separation of the two tissues is seen also in the cyclostomes, in which the chromaffin cells appear as strands along the dorsal aorta. Chromaffin cells are derived embryologically from postganglionic cells of the sympathetic nervous system, some cells of which also produce noradrenaline.

Chromaffin cells secrete the **catecholamines**: adrenaline and noradrenaline (epinephrine and norepinephrine), which are important in "fight or flight" stress responses in fishes (Schreck, 1981). The control of heart rate, blood pressure, blood flow through the gills (Wahlqvist, 1980; Wahlqvist and Nilsson, 1980), concentration of melanin in melanophores, and dilation of the pupils all have been ascribed to catecholamines.

Ultimobranchial Gland

In bony fishes, this gland develops from an embryonic gill pouch and is located below the esophagus near the sinus venosus, often on or closely associated with the pericardium. In elasmobranchs, the gland is on the left side of the midline, beneath the pharynx. It secretes the hormone **calcitonin,** which is involved in the inhibition of bone resorption in mammals and is thought to be involved in calcium metabolism in fishes. Experiments with rainbow trout (Fouchereau-Peron et al., 1986) have indicated that calcitonin is involved in the regulation of adaptation to seawater. The gland does not occur in cyclostomes.

Pancreas (Islets of Langerhans)

Modern agnaths are unique among craniates in having the exocrine and endocrine functions of the pancreas distinct and separated. Whereas the exocrine elements are present in the gut epithelium, the endocrine pancreatic function is located in an aggregate of submucosal islets, termed the **islet organ** (Youson, 2000). The pancreatic islets of bony fishes are usually dispersed in the connective tissue around the pyloric caeca, small intestine, spleen, and gall bladder. Some teleosts have bodies of pancreatic tissue gathered into "Brockmann bodies," some of which can produce both endocrine and exocrine secretions (Gorbman et al., 1983). A few species have a compact mass of pancreatic tissue on or near the gall bladder. The islet tissue is found in the walls of the intestine in lampreys. In sea lamprey ammocoetes, the islets are located in the gut epithelium at the junction of the bile duct. As the bile duct degenerates during metamorphosis, part of it becomes a caudal endocrine pancreas. An anterior (cranial) pancreas develops from the larval pancreas (Youson and Elliott, 1989). The cranial pancreas is on the dorsal wall of the esophagus where it joins the intestine (Barrington, 1972). The endocrine pancreas of hagfishes consists of follicles at the junction of the bile duct and the intestine (Matty, 1985). Elasmobranchs have a discrete pancreas that includes the islets. Four types of endocrine secreting cells have been identified in the fish pancreas. These produce **insulin, glucagon, somatostatin,** and **pancreas-peptide.** A fifth cell type of possible endocrine function has been detected in the European bass *Morone* (= *Dicentrarchus*) *labrax* (Carrillo et al., 1986).

The secretions of the pancreas have important roles in governing metabolism. Insulin is involved in the synthesis of protein, the conversion of glucose into glycogen, and in the production of fat from carbohydrate sources. Insulin is a member of a larger family of polypeptides that share similarities in peptide sequences. A related molecule, **insulinlike growth factor (IGF)**, is secreted by the liver and is known to play an important role in growth and tissue differentiation in fishes (Duan, 1997; Mommsen, 1998). Glucagon is an insulin antagonist and causes glucose in its stored form (*glycogen*) to enter the blood. Aside from releasing glucose from glycogen, glucagon is lipolytic (Plisetskaya, 1990). Somatostatin inhibits gastrointestinal motility in the rainbow trout and may slow the emptying of the stomach (Chen and Hale, 1992). Somatostatin may also be involved in the release of glucose from the liver in salmonids (Eilertson et al., 1991) and has been shown to inhibit the release of growth hormone in the eel *Anguilla* (Suzuki et al., 1990). Two types of somatostatin cells have been noted in mullets by Lozano and Agulleiro (1986).

Gastroenteric Mucosa

Several peptides with apparent endocrine regulatory function are found in the intestinal mucosal cells and in the nerves associated with the gut mucosa. Although studies have revealed these compounds to be present in fishes, their exact roles are not well understood (Holmgren et al., 1986).

At least 16 types of endocrine intestinal cells have been recorded in fishes (Rawdon and Andrews, 1990). The secretions of these cells appear to be involved mainly in regulating the secretions or motility of the digestive tract. **Gastrin** and related substances, which occur in lampreys, sharks, and bony fishes, cause the secretion of pepsin and hydrochloric acid in the stomach. Glucagon, somatostatin, and insulin are produced in the gut as well as in the pancreatic cells. **Substance P,** found in sharks and bony fishes, has been shown to have an excitatory effect on stomach and gut muscle (Holmgren, 1985; Holmgren et al., 1985). Other peptides that might be involved in the stimulation of gut motility are bombesin and serotonin. Some that may be involved in excitatory effects on gastric tissue are enkephalin and neurotensin. Bombesin might stimulate the secretion of stomach acid, and vasointestinal peptide (VIP) may inhibit gut motility (Wendelaar Bonga, 1993).

Gonads

The sex organs of both sexes (see Chapter 27) secrete steroids that are important in the manifestation of courtship, nest building, and other aspects of reproductive behavior, as well as in the differentiation of gonads, development and maintenance of secondary sexual characteristics, and gametogenesis. Steroids are of importance in the reproductive cycle of sharks (Tsang and Callard, 1987a, 1987b), lampreys (Linville et al., 1987), and bony fishes (Redding and Patiño, 1993). The ovary produces **estrogens,** the actions of which have not been well studied in fishes. Investigations have shown positive relationships between ovarian secretions and receptivity to males and the development of secondary sexual characteristics. Secretions (**pheromones**) that are released into the environment (see Chapters 36 and 37) and attract male fishes to females are also under the influence of estrogens (Stacey, 1981, 1987; Stacey and Sorensen, 1991).

The testes produce **androgens,** especially testosterone and 11-ketotestosterone. Other hormones that have been isolated from the testes include dehydroepiandrosterone and androstenedione. Many studies have shown androgens to be of great importance in the sexual behavior and spawning activity of male fishes. Postfertilization exposure of eggs to increased levels of maternally derived testosterone has been demonstrated to affect the subsequent development and sex differentiation of embryos and larvae. Interesting enough, females apparently also can influence the development of their offspring through transfer of the stress hormone cortisol directly to the yolk of eggs (McCormick, 1999). The presence of eggs in mouthbrooding cichlids apparently triggers a decrease in the concentrations of androgens and of the estrogen estradiol (Kishida and Specker, 2000; Specker and Kishida, 2000).

In those species exhibiting alternative reproductive strategies, alternative endocrine states are in evidence. As in the toadfishes described earlier (see Chapter 21, Going Deeper), two types of reproductive morphs exist among bluegills (*Lepomis macrochirus*). Nest-building, territory-defending males are larger than an alternative morph that exhibits female mimicry and sneaker spawning behavior (see Chapter 37). During the time of nest building and spawning, the larger males have significantly higher levels of circulating 11-ketotestosterone, yet the testosterone levels of the two morphs are similar (Kindler et al., 1989). Similar findings have been reported for the stoplight parrotfish (*Sparisoma viride*), another species exhibiting male dimorphism (Cardwell and Liley, 1991). In general, elevated levels of 11-ketotestosterone appear to correlate with the aggressive behavior necessary to establish and defend territories (Hourigan et al., 1991; Pankhurst and Kime, 1991; Sikkel, 1993).

Corpuscles of Stannius

The corpuscles of Stannius are found in the opisthonephric kidney of holosteans and teleosts. They vary in position among species, being found dorsally, dorsolaterally, or ventrolaterally; they are seldom arranged symmetrically. The featherbacks (*Notopterus*) have only a single corpuscle near the head kidney. The corpuscles of Stannius may be highly vascularized and lobulated (Belsare, 1973). In eight teleost species studied by Krishnamurthy and Bern (1971), the corpuscles were found to have prominent autonomic innervation.

Because removal of the corpuscles of Stannius brings about changes in plasma composition, their secretions are thought to be involved in osmoregulation (Henderson and Jones, 1973). A decline in sodium and a rise in potassium salts produce histological changes in the corpuscles. Pang and Pang (1986) showed that the removal of the Stannius corpuscles from the killifish *Fundulus heteroclitus* resulted in hypercalcemia. There appears to be a relationship between the adrenal cortex and the activity of the corpuscles of Stannius, because injections of corticosteroids bring about nuclear hypertrophy and other evidence of stimulation in the cells of the corpuscles. The secretion

stanniocalcin probably acts as a calcium-channel blocker to limit the entry of calcium into the chloride cells (Wendelaar Bonga, 1993).

Other Organs with Endocrine Function

Kidney

In fishes, the hormone **renin** appears to be secreted by certain granular cells (*juxtaglomerular cells*) located in the glomerular arterial supply and apparently derived from arterial cells. Renin in circulation forms **angiotensin** by acting on a plasma polypeptide precursor (*angiotensinogen*). Angiotensin is active in osmoregulation through sodium retention by the kidneys (Brown et al., 1990; Matty, 1986; see Chapter 25). The **renin–angiotensin system** (**RAS**) operates to counteract hypotension and fluid depletion. Although the RAS has been demonstrated in chondrichthyans and bony fishes (Olson, 1998), until recently it was not known to exist among the agnaths. The demonstration of its function in river lampreys (*Lampetra fluviatilis*) indicates that the RAS has been conserved as a feature of vertebrate osmoregulation for more than 500 million years (Brown et al., 2005).

Thymus

The thymus gland, much like other endocrine tissues in vertebrates, takes its origin from the branchial pouches of fishes and is generally found above the branchial chamber or pockets in lampreys, sharks, and bony fishes. The thymus appears to have a role primarily in the production of T-cells and lymphocytes, which are associated with the immune system of fishes (see Chapter 24).

NEUROSECRETORY CELLS

Implicit in our discussions of endocrine structure and function is the recognition of a special category of cells that act as an integrator of neural and endocrine functions. As such, these neurosecretory cells can act as a mediator that synchronizes endocrine activity in response to environmental cues. These cells are essentially neurons found in association with the central nervous system that also possess the secretory features found in endocrine cells. Depolarization of the cell membrane, brought about through synaptic association with other nerve cells, prompts the release of peptides or amines that behave as endocrine substances. The neurohypophysis and the chromaffin cells previously discussed demonstrate such neurosecretory cell function as a mediator of other endocrine activity. For example, neurons

in the brain stimulate the neurosecretory cells of the hypothalamus that project into the neurohypophysis to release the polypeptide gonadotropin releasing hormone (GnRH), which in turn prompts the endocrine cells of the adenohypophysis to release hormones that target endocrine activity in the gonads (Sower, 1997).

Caudal Neurosecretory System

The caudal neurosecretory system was first described in the eel *Anguilla japonica* (Enami, 1955). Since then, it has been identified in a number of elasmobranchs and bony fishes. Near the termination of the spinal cord in sharks, rays, teleosts, and some other bony fishes, such as *Lepisosteus* and *Polypterus,* are found enlarged secretory neurons known as **Dahlgren cells.** These appear to be of two types (Bhatt and Negi, 1987). Spontaneous bursts of activity have been recorded in the Dahlgren cells (Brierley et al., 2001). The axons of these neurosecretory cells terminate in a capillary bed that appears to function in the storage and release of secretions (Jaiswal and Belsare, 1973). In teleosts, the capillary network is contained in a well-defined neurohemal structure, called the **urophysis,** which in some species can be paired or on a stalk (Matty, 1985). The Dahlgren cells of this complex, which also includes the terminal filament of the spinal column, are the site of the production and release of the peptides **urotensin I** and **II** and possibly others, which resemble arginine vasotocin. Although the exact biological activity of the hormones is not established, experimentation has produced evidence that they appear to influence water balance and sodium regulation in some species (Matty, 1985).

In a study of the cyclic activity of this gland in an Indian catfish, Sharma and Sharma (1975) noted that stored materials disappeared during the breeding season, leading them to surmise that the caudal secretory system is involved in the reproductive cycle. Urotensin II increases during the spawning season in the white sucker (*Catostomus commersoni*) and has been shown to promote contractions in smooth muscle in the urogenital systems of some fishes (Hazon and Balment, 1998; Matty, 1985). The role of the caudal neurosecretory system in the production of compounds that behave as pheromones has also been suggested (Richards, 1974).

Pseudobranchial Neurosecretory Gland

A pseudobranchial neurosecretory organ has been identified in the anterior aortic arches of several Asian air-breathing fishes (Srivastava et al., 1981). Although its function is unclear, it may resemble that of the paraneurons described next.

Paraneurons

Paraneurons are a diverse and diffuse collection of cells with both receptor and endocrine secretory sites on their plasma membranes. Like other neurosecretory cells, they release peptides and amines, yet their function is poorly understood. **Merkel cells** are one kind of paraneuron that has been described from the epithelia of teleosts. Their distribution on oral surfaces, barbels, and fins suggests that they have a sensory function and may mediate other tissue responses through their neurosecretions. Mechanoreceptors and electroreceptors have also been implicated as sites for the release of neuropeptides. Neuroendocrine cells have been identified in the gills of some fishes, the lungs of dipnoans (*Protopterus*), and the swim bladders of air-breathing fishes (*Polypterus*), suggesting a role in the monitoring of dissolved gas concentrations (Zaccone et al., 1999).

Natriuretic Peptides

Natriuretic peptides are a category of compounds that were first isolated from myocardial tissue (deBold et al., 1981). Since then, they have been detected in a variety of tissues, including the brain, gonads, gut, and kidney, where they may have either a paracrine or an endocrine function (Hazon and Balment, 1998; Wendelaar Bonga, 1993). They are broadly represented in fishes, occurring in agnaths, chondrichthyans, and bony fishes, and their function includes the inhibition of the reabsorption of sodium and other cations from urine (Takei, 2000). Their presence in the brain points to their role as a neurotransmitter as well. Immunohistochemical studies of hagfish brains have indicated dense concentrations of natriuretic binding sites in the telencephalon and diencephalon (Donald et al., 1999). The function of these peptides appears to be in the neuroendocrine regulation of osmoregulatory and cardiovascular homeostasis.

EVOLUTIONARY AND ENVIRONMENTAL CONSIDERATIONS

Current research on endocrine systems has led to significant discoveries that are pertinent to evolutionary and environmental biology. Mention was made earlier of the studies on brain development in agnaths that have lent support to the role of gene duplication in vertebrate evolution. Studies by Sower (1998) have suggested that such gene duplication is also evident in the evolution of genes that regulate the synthesis of brain and pituitary hormones.

Optimization of the adaptation of an organism to its environment is fundamental to our understanding of evolutionary processes; it appears that the vertebrate endocrine system incorporates optimization strategies—specifically as they pertain to the problem of ensuring that a signal emitted by a sender is optimally tuned to the receptive capacities of the receiver. A number of studies have demonstrated that adaptive shifts in sense reception are correlated with changes in hormone levels (Zakon, 2004), and batrachoidid fishes have provided some compelling evidence for this (see Chapter 21, Going Deeper). Sisneros et al. (2004) have demonstrated that the application of steroid reproductive hormones to female midshipman fish (*Porichthys notatus*) enhances their sensitivity to the courtship calls of the male—a remarkable demonstration of a vertebrate's capacity to fine-tune communication at a time when it is most essential, during the reproductive season. The authors presented molecular evidence for the existence of estrogen receptors on the cell membranes of the auditory epithelial cells themselves.

Since the publication of Colborn and Clement's (1992) book on adverse environmental impacts on normal endocrine function, much toxicological research has been devoted to the topic of endocrine disruption. Certain chemicals can disrupt normal endocrine function by acting in one of two ways. Some industrial pollutants can mimic the effect of natural endocrine substances, greatly diminishing the adaptive capacity of the affected individual. Much attention has been devoted to compounds that mimic the effect of estrogens (Colborn and Thayer, 2000).

Industrial pollutants may also affect endocrine pathways by disrupting the flow of information that would ultimately lead to the production of endocrine substances. Waterborne pesticides have been demonstrated to have such an effect. Male Atlantic salmon (*Salmo salar*) exposed to triazine pesticides showed a diminished response to the pheromones released by females that are essential in inducing physiological and behavioral development in preparation for breeding (Moore and Lower, 2000; Moore and Waring, 1998). If the endocrine system is in any way the target of industrial pollutants, their adverse impact may be apparent at levels far below mandated water quality standards.

KEY POINTS AND CONNECTIONS

- The vertebrate nervous system is typically divided into central and peripheral components. Fishes demonstrate a number of neural specializations reflecting their adaptation to an aquatic existence. The evolution of the craniate condition has resulted in a significantly enlarged and more complex brain. In some fishes, the relative brain weight is especially great owing to the development of specialized capabilities. Enhanced olfactory capability in elasmobranchs is reflected in large olfactory lobes and bulbs. The cerebellum becomes enlarged in actively swimming fishes. Although mormyrid fishes are not especially strong swimmers, their cerebellum is particularly well developed to accommodate their extraordinary electroreceptive and acoustic sensory capabilities. Mauthner cells are found in the spinal cord of fishes. Using electrical synapses, these cells permit rapid startle and escape responses.

The neural control of locomotion, including the function of Mauthner neurons, is covered in Chapter 19. The sense receptors associated with the nervous system are discussed in Chapters 20-22.

• The vertebrate endocrine system has been traditionally viewed as a collection of glands that release signaling substances into the circulatory system in order to affect target tissues and organs at some distance. The perception of endocrine function has been broadened to include an array of sophisticated and subtle cell signaling and modulating actions, often having an impact at close range. Fishes display the axis of endocrine activity typical of other vertebrates, with some specializations. Examples of endocrine function unique to fishes include the corpuscles of Stannius, found in the kidney, which influence osmoregulation; and the caudal neurosecretory system, which appears to function in water balance and reproduction.

The role of endocrine secretions in osmoregulation is discussed in more detail in Chapter 25.

FISH LINKS

http://www.unh.edu/biochemistry/sower/index.html Stacia Sower's lab at the University of New Hampshire studies the origins of vertebrate endocrine function by focusing on the earliest of craniates—lampreys and hagfishes.

REFERENCES

Barrington, E. J. W. 1972. The pancreas and intestine, pp. 135–169. In *The biology of lampreys*, Vol. 2, M. W. Hardisty and I. C. Potter (Eds.). Academic Press, London.

Belsare, D. K. 1973. Comparative anatomy and histology of the corpuscles of Stannius in teleosts. *Z. Mikrosk. Anat. Forsch.* 87(4): 445–456.

Belsare, D. K. 1974. Morphology of the pineal organ in some carps. *Zool. Beitr.* 20(1): 47–54.

Bern, H. A., and R. S. Nishioka. 1985. Endocrine control of salmonid development and seawater adaptation, p. 7. In *Advances in aquaculture and fisheries research: A California sea grant symposium*, May 1983, K. B. Anderson (Ed.). California Sea Grant College Program, Davis.

Bernstein, J. J. 1970. Anatomy and physiology of the central nervous system, pp. 1–90. In *Fish physiology*, Vol. IV, W. S. Hoar and D. J. Randall (Eds.). Academic Press, New York.

Bhatt, S. D., and U. Negi. 1987. Caudal neurosecretory system in the crucian carp. *Matsya 12–13*: 52–61.

Bone, Q., and N. B. Marshall. 1982. *Biology of fishes*. Blackie, Glasgow, UK.

Brierley, M. J., A. J. Ashworth, J. R. Banks, R. J. Balment, and C. R. McCrohan. 2001. Bursting properties of caudal neurosecretory cells in the flounder *Platichthys flesus*, in vitro. *J. Exp. Biol.* 204: 2733–2739.

Brown, J. A., C. J. Gray, and S. M. Taylor. 1990. Direct effects of angiotensin II on glomerular ultrastructure in the rainbow trout, *Salmo gairdneri*. *Cell Tissue Res.* 260: 315.

———, C. S. Cobb, S. C. Frankling, and J. C. Rankin. 2005. Activation of the newly discovered cyclostome renin–angiotensin system in the river lamprey *Lampetra fluviatilis*. *J. Exp. Biol.* 208: 223–232.

Butler, A. B. 2000. Nervous system, pp. 331–355. In *The laboratory fish*, G. K. Ostrander (Ed.). Academic Press, San Diego.

Cardwell, J. R., and N. R. Liley. 1991. Androgen control of social status in males of a wild population of stoplight parrotfish, *Sparisoma viride* (Scaridae). *Horm. Behav. 25*: 1–18.

Carrillo, M., S. Zanuy, H. Duve, and A. Thorpe. 1986. Identification of hormone-producing cells of the endocrine pancreas of the sea bass, *Dicentrarchus labrax*, by ultrastructural immunocytochemistry. *Gen. Comp. Endocrinol. 61*: 287–301.

Chen, C. B., and E. Hale. 1992. Effect of somatostatin on intragastric pressure and smooth muscle contractility of the rainbow trout, *Oncorhynchus mykiss* Walbaum. *J. Fish Biol. 40*: 545–556.

Colborn, T., and C. Clement (Eds.). 1992. *Chemically-induced alterations in sexual and functional development: The wildlife/human connection*. Princeton Scientific, Princeton, NJ.

———, and K. Thayer. 2000. Aquatic ecosystems: Harbingers of endocrine disruption. *Ecol. Appl. 10*: 949–957.

deBold, A. J., H. B. Borenstein, A. T. Veres, and H. Sonnenberg. 1981. A rapid and potent natriuretic response to intravenous injection of atrial myocardial extract in rats. *Life Sci. 28*: 89–94.

Demski, L. S. 1983. Behavioral effects of electrical stimulation of the brain, pp. 317–359. In *Fish neurobiology, Vol. 2*, Higher brain areas and functions, R. E. Davis and R. G. Northcutt (Eds.). University of Michigan Press, Ann Arbor.

———, and R. G. Northcutt. 1983. The terminal nerve: A new chemosensory system in vertebrates? *Science 220*: 435–437.

———, and M. Schwanzel-Fukuda. 1987. The terminal nerve (nervus terminalis) structure, function, and evolution. *Ann. New York Acad. Sci. 519*.

Diamond, J. 1971. The Mauthner cell, pp. 265–346. In *Fish physiology*, Vol. V, W. S. Hoar and D. J. Randall (Eds.). Academic Press, New York.

Donald, J. A., T. Toop, and D. H. Evans. 1999. Natriuretic peptide binding sites in the brain of the Atlantic hagfish, *Myxine glutinosa*. *J. Exp. Zool. 284*: 407–413.

Duan, C. 1997. The insulin-like growth factor system and its biological actions in fish. *Amer. Zool. 37*: 491–503.

Eaton, R. C. (Ed.) 1991. Neuroethology of the Mauthner system. *Brain Behav. Evol. 37*: 250–332.

———, J. G. Canfield, and A. L. Guzik. 1995. Left–right discrimination of sound onset by the Mauthner system. *Brain Behav. Evol. 46*: 165–179.

Eilertson, C. D., P. K. O'Connor, and M. A. Sheridan. 1991. Somatostatin-14 and somatostatin-25 stimulate glycogenolysis in rainbow trout, *Oncorhynchus mykiss*, liver incubated in vitro: A systemic role for somatostatins. *Gen. Comp. Endocrinol. 82*: 192–205.

Eisthen, H. L., and R. G. Northcutt. 1996. Silver lampreys (*Ichthyomyzon unicuspis*) lack a gonadotropin-releasing hormone and FMRF amide-immunoreactive terminal nerve. *J. Comp. Neurol. 370*: 159–172.

Ekström, P., and H. Meissl. 1997. The pineal organ of teleost fishes. *Rev. Fish Biol. Fisheries 7:* 199–284.

Enami, M. 1955. Caudal neurosecretory system in the eel (*Anguilla japonica*). *Gunma J. Med. Sci. 4:* 23–36.

Fenwick, J. C. 1970. The pineal organ, pp. 91–108. In *Fish physiology,* Vol. III, W. S. Hoar and D. J. Randall (Eds.). Academic Press, New York.

Fouchereau-Peron, M., Y. Arlot-Bonnemains, M. S. Mouktar, and G. Milhaud. 1986. Adaptation of rainbow trout (*Salmo gairdneri*) to sea water: Changes in calcitonin levels. *Comp. Biochem. Physiol. 82A:* 83–87.

Fujita, I., P. W. Sorenson, N. B. Stacey, and T. Hara. 1991. The olfactory system, not the terminal nerve, functions as the primary chemosensory pathway mediating responses to sex pheromones in male goldfish. *Brain Behav. Evol. 38:* 313–321.

Gelsleichter, J., and C. A. Manire. 1999. Introduction to the proceedings of the first symposium on elasmobranch endocrinology, held at the thirteenth annual meeting of the American Elasmobranch Society (AES). *J. Exp. Zool. 284:* 473–474.

Gorbman, A. 1995. Olfactory origins and evolution of the brain–pituitary endocrine system: Facts and speculation. *Gen. Comp. Endocrinol. 97:* 171–178.

———, W. W. Dickhoff, S. R. Vigna, N. B. Clark, and C. L. Ralph. 1983. *Comparative endocrinology.* Wiley-Interscience, New York.

Hardisty, M. W. 1979. *Biology of the cyclostomes.* Chapman and Hall, London.

Hazon, N., and R. J. Balment. 1998. Endocrinology, pp. 441–463. In *The physiology of fishes* (2nd ed.), D. H. Evans (Ed.). CRC Press, Boca Raton, FL.

Henderson, I. W., and I. C. Jones. 1973. Hormones and osmoregulation in fishes. *Ann. Inst. Michel Pacha 5(2):* 69–235.

Holmes, R. L., and J. N. Ball. 1974. *The pituitary gland: A comparative account.* Cambridge University Press, Cambridge UK.

Holmgren, S. 1985. Substance P in the gastrointestinal tract of *Squalus acanthias. Mol. Physiol. 8:* 119.

———, D. J. Grove, and S. Nilsson. 1985. Substance P acts by releasing 5-hydroxytryptamine from enteric neurons in the stomach of the rainbow trout, *Salmo gairdneri. Neuroscience 14:* 683.

Hopkins, C. D. 1983. Functions and mechanisms in electroreception, pp. 215–259. In *Fish neurology,* Vol. I, R. G. Northcutt and R. E. Davis (Eds.). University of Michigan Press, Ann Arbor.

Hourigan, T. F., M. Nakamura, Y. Nagahama, K. Yamauchi, and E. G. Grau. 1991. Histology, ultrastructure, and in vitro steroidogenesis of the testes of two male phenotypes of the protogynous fish, *Thalassoma duperrey* (Labridae). *Gen. Comp. Endocrinol. 83:* 193–217.

Huang, L., A. M. Schreiber, B. Soffientino, D. A. Bengtson, and J. L. Specker. 1998. Metamorphosis of summer flounder (*Paralichthys dentatus*): Thyroid status and the timing of gastric gland formation. *J. Exp. Zool. 280:* 413–420.

Igarashi, S., and T. Kamiya. 1972. *Atlas of the vertebrate brain.* University Park Press, Baltimore.

Jaiswal, A. O., and D. K. Belsare. 1973. Comparative anatomy and histology of the caudal neurosecretory system in teleosts. *Z. Mikrosk. Anat. Forsch. 87:* 589–609.

Kardong, K. V. 1997. *Vertebrates: Comparative anatomy, function, evolution* (2nd ed.). WCB/McGraw-Hill, Boston.

Kavaliers, M. 1980. The pineal organ and circadian rhythms of fishes, pp. 631–643. In *Environmental physiology of fishes,* M. A. Ali (Ed.). Plenum Press, New York.

Kent, G. C. 1992. *Comparative anatomy of the vertebrates.* Mosby-YearBook, St. Louis, MO.

———, and R. K. Carr. 2001. *Comparative anatomy of the vertebrates* (9th ed.). McGraw-Hill, New York.

Kindler, P. M., D. P. Philipp, M. R. Gross, and J. M. Bahr. 1989. Serum 11-ketotestosterone and testosterone concentrations associated with reproduction in male bluegill (*Lepomis macrochirus:* Centrarchidae). *Gen. Comp. Endocrinol. 75:* 446–453.

Kishida, M., and J. L. Specker. 2000. Paternal mouthbrooding in the black-chinned tilapia, *Sarotherodon melanotheron* (Pisces: Cichlidae): Changes in gonadal steroids and potential for vitellogenin transfer to larvae. *Horm. Behav. 37:* 40–48.

Krishnamurthy, V. O., and H. A. Bern. 1971. Innervation of the corpuscles of Stannius. *Gen. Comp. Endocrinol. 16:* 162–165.

Kuhlenbeck, H. 1975. *The central nervous system of vertebrates, Vol. 4,* Spinal cord and deuterencephalon. Karger, Basel.

Lagios, M. D. 1975. The pituitary gland of the coelacanth *Latimeria chalumnae* Smith. *Gen. Comp. Endocrinol. 25:* 126–146.

———. 1982. *Latimeria* and the Chondrichthyes as sister taxa: A rebuttal to recent attempts at refutation. *Copeia 1982:* 942–948.

Leatherland, J. F. 1982. Environmental physiology of the teleostean thyroid gland: A review. *Env. Biol. Fishes 7:* 83–110.

Lin, G. R., C. F. Weng, J. I. Wang, and P. P. Hwang. 1999. Effects of cortisol on ion regulation in developing tilapia (*Oreochromis mossambicus*) larvae on seawater adaptation. *Physiol. Biochem. Zool. 72:* 397–404.

Linville, J. E., L. H. Hanson, and S. A. Sower. 1987. Endocrine events associated with spawning behavior in the sea lamprey (*Petromyzon marinus*). *Horm. Behav. 21:* 105–117.

Lozano, M. T., and B. Agulleiro. 1986. Immunocytochemical and ultrastructural study of the endocrine pancreas of *Mugil auratus* and *Mugil saliens* L. (Teleostei). *J. Submicrosc. Cytol. 18(1):* 85–98.

Matty, A. J. 1985. *Fish endocrinology.* Croom Helm, London.

———. 1986. Nutrition, hormones and growth. *Fish Physiol. Biochem. 2:* 141–150.

McCormick, C. A. 1983. Organization and evolution of the octavolateralis area of fishes, pp. 179–213. In *Fish neurobiology,* Vol. 1, R. G. Northcutt and R. E. Davis (Eds.). University of Michigan Press, Ann Arbor.

McCormick, M. I. 1999. Experimental test of the effect of maternal hormones on larval quality of a coral reef fish. *Oecologia 118:* 412–422.

Mommsen, T. P. 1998. Growth and metabolism, pp. 65–97. In *The physiology of fishes* (2nd ed.), D. H. Evans (Ed.). CRC Press, Boca Raton, FL.

Moore, A., and N. Lower. 2000. The impact of two pesticides on olfactory mediated endocrine function in mature male Atlantic salmon parr. *Abstr. 4th Int. Symp. Fish Endocrinol.,* University of Washington, Seattle.

———, and C. P. Waring. 1998. Mechanistic effects of a triazine pesticide on reproductive endocrine function in mature male Atlantic salmon (*Salmo salar* L.) parr. *Pest. Biochem. Physiol. 62(1):* 41–50.

Nieuwenhuys, R. 1962. Trends in the evolution of the actinopterygian forebrain. *J. Morphol. 111:* 69–88.

———, and E. Pouwels. 1983. The brain stem of actinopterygian fishes, pp. 25–87. In *Fish neurobiology,* Vol. I, R. G. Northcutt and R. E. Davis (Eds.). University of Michigan Press, Ann Arbor.

Nilsson, G., M. H. Routley, and G. M. C. Renshaw. 2000. Low mass-specific brain Na$^+$/K$^+$ ATPase activity in elasmobranch compared to teleost fishes: Implications for the large brain size of elasmobranchs. *Proc. Roy. Soc. Lond. B (Biol. Sci.) 267:* 1335–1339.

Norman, A. W., and G. Litwack. 1997. *Hormones*. Academic Press, San Diego.

Norris, D. O. 1997. *Vertebrate endocrinology* (3rd ed.). Academic Press, San Diego.

Northcutt, R. G. 1989a. Brain variation and phylogenetic trends in elasmobranch fishes, pp. 83–100. In *Evolutionary and contemporary biology of elasmobranchs*, W. C. Hamlett and B. Tota (Eds.). Alan R. Liss, New York.

———. 1989b. Phylogeny and innervation of lateral lines, pp. 17–78. In *The mechanosensory lateral line*, S. Coombs, P. Görner, and H. Münz (Eds.). Springer Verlag, New York.

———. 1995. The forebrain of gnathostomes: In search of a morphotype. *Brain Behav. Evol. 46:* 275–318.

———. 1996. The agnathan ark: The origin of craniate brains. *Brain Behav. Evol. 48:* 237–247.

———, and W. E. Bemis. 1993. Cranial nerves of the coelacanth, *Latimeria chalumnae* (Osteichthyes: Sarcopterygii: Actinistia), and comparisons with other Craniata. *Brain Behav. Evol. 42*(Suppl. 1): v–x, 1–76.

———, and R. E. Davis. 1983. Telencephalic organization in ray-finned fishes, pp. 203–236. In *Fish neurobiology*, R. G. Northcutt and R. E. Davis (Eds.). University of Michigan Press, Ann Arbor.

Nozaki, M., A. Gorbman, and S. A. Sower. 1994. Diffusion between the neurohypophysis and the adenohypophysis of lampreys, *Petromyzon marinus*. *Gen. Comp. Endocrinol. 96:* 385–391.

O, W., and T. H. Chan. 1974. A cytological study on the structure of the pituitary gland of *Monopterus albus* (Zuiew). *Gen. Comp. Endocrinol. 24:* 208–222.

Olson, K. R. 1998. The cardiovascular system, pp. 129–154. In *The physiology of fishes* (2nd ed.), D. H. Evans (Ed.). CRC Press, Boca Raton, FL.

Pang, P. K. T., and R. K. Pang. 1986. Hormone and calcium regulation in *Fundulus heteroclitus*. *Amer. Zool. 26:* 225–234.

Pankhurst, N. W., and D. E. Kime. 1991. Plasma sex steroid concentration in male blue cod, *Parapercis colias* (Bloch and Schneider) (Pinguipedidae), sampled underwater during the spawning season. *Aust. J. Mar. Freshw. Res. 42:* 129–137.

Pickford, G., and J. W. Atz. 1957. *The physiology of the pituitary gland of fishes*. New York Zoological Society, New York.

Plisetskaya, E. M. 1990. Endocrine pancreas of teleosts. *J. Exp. Zool. Suppl. 4:* 53–57.

Rawdon, B. B., and A. Andrews. 1990. Vertebrate gut endocrine cells: Comparative and developmental aspects, pp. 504–509. In *Progress in comparative endocrinology*, A. Epple, C. G. Scares, and M. A. Stetson (Eds.). Wiley-Liss, New York.

Redding, J. M., and R. Patiño. 1993. Reproductive physiology, pp. 503–534. In *The physiology of fishes*, D. H. Evans (Ed.). CRC Press, Boca Raton, FL.

Richards, I. S. 1974. Caudal neurosecretory system: Possible role in pheromone production. *J. Exp. Zool. 187:* 405–408.

Romer, A. S. 1970. *The vertebrate body* (4th ed.). W. B. Saunders, Philadelphia. ·

———, and T. S. Parsons. 1978. *The vertebrate body: Shorter version*. W. B. Saunders, Philadelphia.

Rovainen, C. M. 1978. Müller cells, "Mauthner" cells, and other identified reticulospinal neurons in the lamprey, pp. 245–269. In *Neurobiology of the Mauthner cell*, D. Faber and H. Korn (Eds.). Raven Press, New York.

Sathyanesan, A. G. 2005. The hypophysis and hypothalamo–hypophyseal system in the chimaeroid fish *Hydrolagus colliei* (Lay and Bennett) with a note on their vascularization. *J. Morphol. 116:* 413–449.

Schreck, C. B. 1981. Stress and compensation in teleostean fishes: Response to social and physical factors, pp. 295–321. In *Stress and fish*, A. D. Pickering (Ed.). Academic Press, New York.

———, and P. F. Scanlon. 1977. Endocrinology in fisheries and wildlife. *Fisheries 2*(3): 20–27.

———, C. Bradford, M. S. Fitzpatrick, and R. Patiño. 1989. Regulation of the interrenal of fishes: Nonclassical control mechanisms. *Fish Physiol. Biochem. 7:* 259–265.

Schreiber, A. M., and Specker, J. L. 1998. Metamorphosis in the summer flounder (*Paralichthys dentatus*): Stage-specific developmental response to altered thyroidal status. *Gen. Comp. Endocrinol. 111:* 156–166.

———, and———. 1999. Metamorphosis in the summer flounder, *Paralichthys dentatus:* Thyroid status influences salinity tolerance. *J. Exp. Zool. 284:* 414–424.

Schwanzel-Fukuda, M. 1999. Nervus terminalis, pp. 350–357. In *Encyclopedia of reproduction*, Vol. 3, E. Knobil and J. D. Neill (Eds.). Academic Press, San Diego.

Sharma, S., and A. Sharma. 1975. A note on the caudal neurosecretory system and seasonal changes observed in the urophysis of *Rita rita* (Bleeker). *Can. J. Zool. 53:* 357–360.

Sikkel, P. 1993. Changes in plasma androgen levels associated with changes in male reproductive behavior in a brood cycling marine fish. *Gen. Comp. Endocrinol. 89:* 229–237.

Sisneros, J. A., P. M. Forlando, D. L. Deitcher, and A. H. Bass. 2004. Steroid-dependent auditory plasticity leads to adaptive coupling of sender and receiver. *Science 305:* 404–407.

Smeets, W. J., A. J. R. Nieuwenhuys, and B. L. Roberts. 1983. *The central nervous system of cartilaginous fishes*. Springer Verlag, Berlin.

Song, J., and R. G. Northcutt. 1991. Morphology, distribution and innervation of the lateral line receptors of the Florida gar, *Lepisosteus platyrhincus*. *Brain Behav. Evol. 37:* 10–37.

Sower, S. A. 1990. Neuroendocrine control of reproduction in lampreys. *Fish Physiol. Biochem. 8:* 365–374.

———. 1997. Evolution of GnRH in fish of ancient origins, pp. 27–49. In *GnRH Neurons: Gene to behavior*, I. S. Parhar and Y. Sakuma (Eds.). Brain Shuppan, Tokyo.

———. 1998. Brain and pituitary hormones of lampreys, recent findings and their evolutionary significance. *Amer. Zool. 38:* 15–38.

———, and H. Kawauchi. 2001. Update: Brain and pituitary hormones of lampreys. *Comp. Biochem. Physiol. B 129:* 291–302.

Specker, J. L., and Kishida, M. 2000. Mouthbrooding in the black-chinned tilapia, *Sarotherodon melanotheron* (Pisces: Cichlidae): The presence of eggs reduces androgen and estradiol levels during paternal and maternal parental behavior. *Horm. Behav. 38:* 44–51.

Srivastava, C. B. L., A. Gopesh, and M. Singh. 1981. A new neurosecretory system in fish, located in the gill region. *Experientia 37:* 850–851.

Stacey, N. E. 1981. Hormonal regulation of female reproductive behavior in fish. *Amer. Zool. 21:* 305–316.

———. 1987. Roles of hormones and pheromones in fish reproductive behavior, pp. 28–69. In *Psychobiology of reproductive behavior*, D. Crews (Ed.). Prentice Hall, Englewood Cliffs, NJ.

———, and P. W. Sorensen. 1991. Function and evolution of fish hormonal pheromones, pp. 109–135. In *Biochemistry and molecular biology of fishes*, Vol. 1, P. W. Hochachka and T. P. Mommsen (Eds.). Elsevier, New York.

Suzuki, R., M. Kishida, and T. Hirano. 1990. Growth hormone secretion during long term incubation of the pituitary in the Japanese eel, *Anguilla japonica*. *Fish Physiol. Biochem. 8:* 159–165.

Takei, Y. 2000. Diverse osmoregulatory actions of natriuretic peptides in eels. *Abstr. 4th Int. Symp. Fish Endocrinol.*, University of Washington, Seattle.

Tamura, T., and I. Hanyu. 1980. Pineal sensitivity in fishes, pp. 477–496. In *Environmental physiology of fishes*, M. A. Au (Ed.). Plenum Press, New York.

Tsang, P. C. W., and I. P. Callard. 1987a. Morphological and endocrine correlates of the reproductive cycle of the aplacental viviparous dogfish, *Squalus acanthias. Gen. Comp. Endocrinol. 66:* 182–189.

———, and ———. 1987b. Luteal progesterone production and regulation in the viviparous dogfish, *Squalus acanthias. J. Exp. Zool. 241:* 377–383.

Tuge, H., K. Uchinashi, and H. Shimamura. 1968. *An atlas of the brains of fishes of Japan.* Tzukiji Shokan, Tokyo.

Vodicnik, M. J., R. E. Kral, V. L. de Vlaming, and L. W. Crim. 1978. The effects of pinealectomy on pituitary and plasma gonadotropin levels in *Carassius auratus* exposed to various photoperiod-temperature regimes. *J. Fish Biol. 12:* 187–196.

———, J. Olcese, G. Delahunty, and V. de Vlaming. 1979. The effects of blinding, pinealectomy and exposure to constant dark conditions on gonadal activity in the female goldfish, *Carassius auratus. Env. Biol. Fishes 4:* 173–177.

Wahlqvist, I. 1980. Effects of catecholamines on isolated systemic and branchial vascular beds of the cod, *Gadus morhua. J. Comp. Physiol. 137:* 139–143.

———, and S. Nilsson. 1980. Adrenergic control of the cardiovascular system of the Atlantic cod, *Gadus morhua*, during "stress." *J. Comp. Physiol. 137:* 145–150.

Wendelaar Bonga, S. E. 1993. Endocrinology, pp. 469–502. In *The physiology of fishes*, D. H. Evans (Ed.). CRC Press, Boca Raton, FL.

Wicht, H. 1996. The brains of lampreys and hagfishes: Characteristics, characters, and comparisons. *Brain Behav. Evol. 48:* 248–261.

Wullimann, M. F. 1994. The teleostean torus longitudinalis: A short review on its structure, histochemistry, connectivity, possible function and phylogeny. *Eur. J. Morphol. 32:* 235–242.

Yamazaki, F. 1974. On the so-called "cobalt" variant of rainbow trout. *Bull. Jpn. Soc. Sci. Fish. 40*(1): 17–25.

Youson, J. H. 2000. The agnathan enteropancreatic endocrine system: Phylogenetic and ontogenetic histories, structure, and function. *Amer. Zool. 40:* 179–199.

———, and W. M. Elliott. 1989. Morphogenesis and distribution of the endocrine pancreas in adult lampreys. *Fish Physiol. Biochem. 7:* 125–131.

Zaccone, G., A. Mauceri, L. Ainia, S. Fasulo, and A. Licata. 1999. Paraneurons in the skin and gills of fishes, pp. 417–441. In *Ichthyology: Recent research advances*, D. N. Saksena (Ed.). Science Publishers, Enfield, NH.

Zakon, H. 2004. Heeding the hormonal call. *Science 305:* 349–350.

Zottoli, S. J. 1978. Comparative morphology of the Mauthner cell in fish and amphibians, pp. 13–45. In *Neurobiology of the Mauthner cell*, D. S. Faber and H. Korn (Eds.). Raven Press, New York.

———. 1981. Electrophysiological and morphological characterization of the winter flounder Mauthner cell. *J. Comp. Physiol. 144:* 541–553.

———, A. P. Bentley, B. J. Prendergast, and H. I. Rieff. 1995. Comparative studies on the Mauthner cell of teleost fish in relation to sensory input. *Brain Behav. Evol. 46:* 151–164.

PART SIX

GENERATIONS
OF FISHES

The next two chapters are devoted to topics that have captured their fair share of attention from both scientists and the general public. Studies of developmental processes and mechanisms of heredity are at the forefront of contemporary biological research—understandable given their biomedical applications. Although our reason for considering these topics here is primarily to enhance our overall understanding of the biology of fishes, we would be remiss if we did not acknowledge the value of fishes as models in basic biomedical research. Several species of fishes have become the organisms of choice for elucidating the mysteries of genetic and reproductive processes as they apply to vertebrates. Perhaps no other field of biology is progressing as rapidly as the study of the mechanisms that control reproduction, development, and inheritance. This rapid progress has come about largely through the discovery and refinement of techniques that enable us to access these phenomena at the molecular level. However, a thorough understanding of these processes still depends on our ability to apply our understanding of molecular behavior to its consequences for cell structure and function, the organization of these cells into intact organisms, and ultimately the interaction of these organisms with each other and their environment.

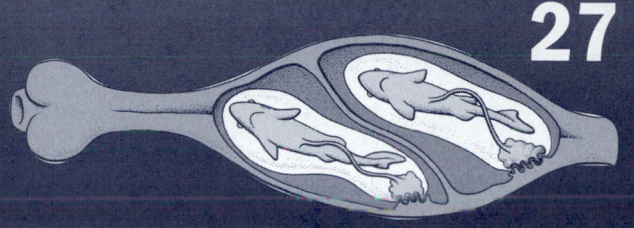

27

REPRODUCTION
AND DEVELOPMENT

ANATOMY OF THE REPRODUCTIVE SYSTEM

Hagfishes
Lampreys
Sharks
Chimaeras (Ratfishes)
Bony Fishes

FUNCTION AND REPRODUCTIVE PATTERNS

Reproductive Strategy
Role of the Endocrine System
Semelparity and Iteroparity
Finding Mates
Egg Production in Oviparous Fishes
Egg Retention, Internal Incubation, and Viviparity
Reproductive Guilds
Differences Between the Sexes
Hybridization
Selective Breeding

EMBRYONIC AND EARLY DEVELOPMENT IN FISHES

Embryology
  Parthenogenesis
Early Life History
  Adaptations of Larval Fishes
  Ecology of Early Life History Stages

Fishes are no different from other life forms in that a significant portion of their time and energy is devoted to the perpetuation of the species. The embryogenesis of the reproductive organs of fishes is typical of vertebrates, yet the different taxa vary in structural features of the testes and ovaries. These reproductive structures may originate as paired organs that are supported in the body cavity by mesenteries, but in many species, the gonads are single structures. Fishes vary dramatically in their reproductive adaptations; some fishes place most of their reproductive effort into the production of large numbers of eggs that, once released into the environment, never again receive the attention of the parents. Others, producing fewer eggs, invest great amounts of time in the construction of nests and the care of fertilized eggs and larvae. Many groups of fishes demonstrate the ultimate form of parental care of early life history stages, in that the female retains the fertilized eggs and gestates them internally; in a few species, this may even become the responsibility of the male.

Great variation is also seen in the embryogenesis of fishes. Some embryos have few yolk reserves stored in the egg and hence develop and hatch rapidly, whereas others, with greater deposits of nutritional materials, undergo a more prolonged embryological development. In reproduction, as in so many other aspects of the lives of fishes, a diverse array of strategies has evolved to ensure the success of the species.

ANATOMY OF THE REPRODUCTIVE SYSTEM

In vertebrate organisms, the urinary system invariably shares some components with the tissues that constitute the reproductive organs. Both share similar embryological origins; hence, the term **urogenital system** is often used to describe these shared components. The genital portion is comprised of gonads—the **testes** and **ovaries**—and the ducts that convey their products (eggs and sperm) to the outside. Gonads arise in the embryo from a portion of the lining of the coelomic epithelium adjacent to the developing kidneys, known as the **genital ridge**.

The gonads of fishes are usually elongate structures suspended by mesenteries from the dorsal aspect of the abdominal cavity. Their relationship to the kidneys and their associated ducts varies widely among groups. **Primordial germ cells**, destined to develop via meiosis into egg and sperm cells, arise outside of the embryonic gonads and migrate—usually through amoeboid motion—to the developing gonads. At this stage, the primordial germ cells have yet to differentiate into male or female gametes. Aquaculture researchers have succeeded in implanting donor primordial germ cells labeled with a fluorescent protein from newly hatched individuals of one species, rainbow trout (*Oncorhynchus mykiss*), into the peritoneal cavity of hatchlings of a different recipient species, the masu salmon (*O. masou*). These donor germ cells were later able to differentiate into functional eggs and sperm in the recipient gonad. The potential advantage of this procedure for aquaculturists rests in the fact that rainbow trout take up to two years to reach sexual maturity, whereas masu salmon only require one year (Takeuchi et al., 2004).

Hagfishes

In these early craniates, a single, elongate gonad is anchored from the ventral side of the gut (Brodal and Fänge, 1963; Sower and Gorbman, 1999). The gonad is contained in a membrane, with the comparatively small testes at the posterior end, and the much larger ovary taking up most of the length of the membrane; gonoducts are lacking in both sexes. In rare instances of bisexuality, both gonads occupy the same membrane (Sower and Gorbman, 1999). In *Myxine*, few males have been collected. The paucity of fully male forms may be attributed to their loss of feeding behavior; hence, they do not show up in baited traps. In *Eptatretus*, the sex ratio is close to normal, although a tendency toward hermaphroditism is seen in some females (Hardisty, 1979). The hagfish testis has been described as nodular and spiral in structure (Morisawa, 1995; Patzner, 1982). For lack of a sperm duct, the sperm are released directly into the body cavity. The spermatozoa of hagfishes are characterized by a pronounced middle section, containing numerous elongate mitochondria. Thus, they actually more closely resemble sperm cells seen in those vertebrates, including teleosts, that practice internal fertilization (Morisawa, 1995).

Unlike the lampreys, which deposit numerous small eggs, hagfishes produce comparatively few eggs. These eggs are large, with tough shells and hooks for attachment to each other or to the substrate (see Fig. 5.2). Oviducts are lacking. Both eggs and sperm reach the exterior by passing through one of two abdominal pores that open into a sinus, which communicates with a cloaca (see Chapter 3). No copulatory organs are present, and the mode of fertilization remains to be described (Patzner, 1998).

Lampreys

In lampreys, the gonads are single, suspended by a peritoneal fold (**mesorchium** in the male, **mesovarium** in the female), and extend along most of the length of the body cavity. They tend to be to the right of the intestine, and they have no connection with the kidneys. In immature specimens, they appear as thin, lobulate structures, but at maturity, they may fill virtually the entire body cavity, crowding the other viscera. Although testes and ovaries can be identified in larval lampreys, the environment appears to play a significant role in the actual determination of sex in the adult (Lowartz and Beamish, 2000). The gut, which may be greatly distended during the feeding stage, becomes small and nonfunctional in mature fishes and may be almost completely hidden by the ripe ovary or testis. The single ovary of lampreys is the result of the fusion of embryonic primordia (Hoar, 1969). No sperm ducts or oviducts are present; at spawning, both the eggs and sperm are shed into the body cavity, from which they exit through paired abdominal pores (which open before spawning occurs) to the urogenital sinus (Hardisty, 1979). A prominent urogenital papilla is developed in mature specimens of males. Eggs are slightly larger than a millimeter in diameter at extrusion. **Fecundity**—the number of eggs produced by the female—depends largely on body size in lampreys. Small, nonparasitic species may produce from 400 to 9,000 eggs, but large anadromous species deposit hundreds of thousands—124,000 to 260,000 in *Petromyzon marinus* (Hardisty, 1971).

Sharks

The testes of elasmobranchs are paired and are usually placed anteriorly in the body cavity, suspended dorsally by means of a mesorchium (Daniel, 1934; Goodrich, 1909, 1930; Romer, 1970). Often, the right testis is larger than the left. Sperm discharges into a central canal network that communicates with the anterior part of the kidney through efferent ducts traversing the mesorchium (see Fig. 25.5). The front part of the kidney is modified into a glandular **epididymis,** where the archinephric duct receives the **ductuli efferentes** and runs posteriorly via a coiled path. The passage through the epididymis may occur during the maturation of sperm. Just posterior to the efferent ductules that drain the testis, the kidney is modified into **Leydig's gland** (see Fig. 25.2); here, the tubules secrete a seminal fluid into the archinephric duct. As the archinephric duct runs posteriorly, it courses along the functional kidney as the **vas deferens** and then enlarges into a seminal vesicle from which a **sperm sac** opens dorsally. The vesicles and sperm sacs open into the **urogenital sinus,** which in turn empties into the **cloaca.** From the cloaca, sperm enter the grooves of the claspers, through which they can be transferred to the female. Associated with the claspers, under the skin of the pelvic fin and abdomen, are glandular sacs called **siphons** that secrete a lubricating fluid.

The ovaries of elasmobranchs are paired, but the left one may be greatly reduced in size in some species. Like the testes, the ovaries are placed well anteriorly in the body cavity, and each is suspended by a mesovarium. The oviducts receive the eggs anterior to the ovaries, usually through a common mouth or funnel (see Fig. 25.5). Eggs released into the coelom proceed into this funnel and then travel down the oviduct to the region of the **shell gland (nidamental gland),** where fertilization occurs and a horny shell or membrane is secreted (Romer, 1970). In oviparous (egg-laying) species, the shell is tough and protects the developing embryo. In ovoviviparous or viviparous (live-bearing) species, the shell is slight or vestigial, and the young develop in the posterior, uterine portion of the oviduct.

Chimaeras (Ratfishes)

In male ratfishes, the testes are compact structures placed forward in the body cavity (Dean, 1906). The testes of *Hydrolagus colliei* in early maturity are about 4.5 cm long and weigh about 13 g (Stanley, 1961). The reproductive system differs from that of sharks in the complexity of the network of efferent ducts traversing the mesorchium. These tubules join so that a smaller number (four to seven) then enter the testis canal, which is in the mesorchium at the base of the testis (Stanley, 1961). Furthermore, the posterior part of the archinephric duct expands into a structure called an **ampulla (or vesicula seminalis).** This is glandular in nature, is partially compartmentalized in some species, and has been described as a receptacle for the maturation of the sperm in the spermatophores coming from the epididymis.

The ampulla receives a number of ducts from the posterior part of the kidney.

In males, the urogenital pore opens medially behind the anus; adjacent to the opening are pelvic claspers that convey the seminal fluid to the genital openings of the female. Anterior to the anus is a pair of abdominal claspers that apparently function during copulation. Female chimaeras are unique in that the two oviducts open separately to the exterior, not connecting to each other or to the urinary system. A shell gland and a uterine portion are present in the oviducts.

Bony Fishes

In most bony fishes, the testes are whitish, lobate organs lying along the gas bladder, although in some groups (such as salmonids) the organs appear smooth. In most groups, there is no connection between the reproductive and urinary systems, and there are separate openings to the exterior for the two systems, with the urinary pore posterior to the genital pore. In some species, the sperm ducts connect with the urinary system in a **urogenital sinus,** located at the posterior end of the body cavity. Primitive bony fishes may differ from this plan (Fig. 27.1; Goodrich, 1909, 1930; Hoar, 1969).

In lungfishes, the testes are elongate and may stretch the length of the body cavity. A longitudinal collecting duct or network of ducts lies along the medial edge of the testis, and in the posterior section efferent ducts connect with the kidney. In *Protopterus,* the central ducts from each testis merge to form a median structure posterior to the testes, and the ducts from this tube enter the kidneys.

In the archaic *Polypterus,* the testes are closely bound to the kidneys, and although the sperm ducts enter that organ, they do not join the urinary system until the urogenital sinus is reached. The testes of the holostean *Amia* reside in the anterior part of the body cavity and are closely associated with the kidneys. A longitudinal duct is situated along the medial edge of the testes, and numerous efferent ducts extend from this to the anterior part of the kidney. Other archaic forms, such as *Acipenser* and *Lepisosteus,* are similar, but their testes are more elongate and communicate with a greater length of the kidney.

The testes of teleosts are of two types. Most have a structure in which spermatogenesis occurs along the length of a lobule, with a central lumen into which the sperm are shed. In atheriniforms, the spermatogenesis is confined to the distal end of the tubule (Billard et al., 1982; Grier, 1981; Jamieson, 1991).

Fish spermatozoa are varied in shape and structure (Jamieson, 1991). Most sperm of bony fishes have roundish heads, usually about 2 to 5 μm, that are made up mainly of the nucleus and numerous mitochondria. Usually, there is a

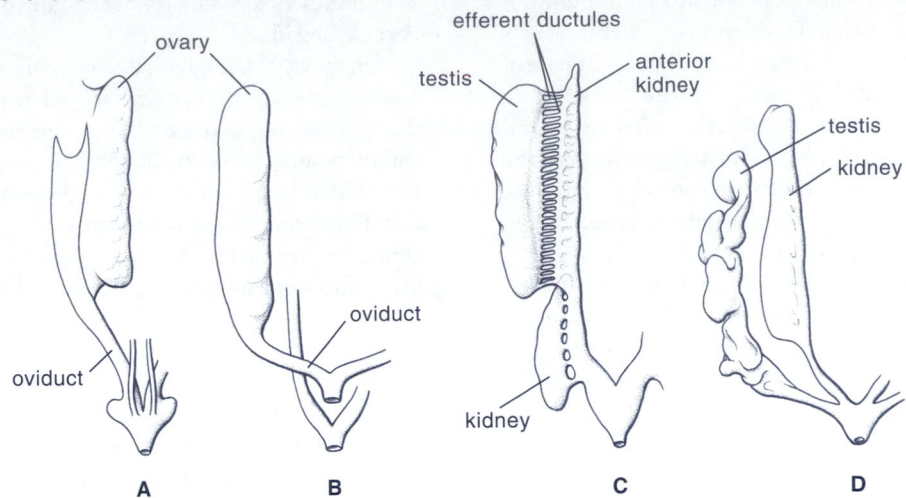

FIGURE 27.1

Gonads and reproductive ducts in bony fishes: **A,** Bowfin female (Amiidae); **B,** Representative teleost female; **C,** Bowfin male (Amiidae); **D,** Representative teleost male (Source: Goodrich, 1909).

single flagellum for propulsion, but a few species have two flagella. The spermatozoa of agnaths, chondrichthyans, lungfishes, *Latimeria*, bichirs, and sturgeons tend to have long, slender heads that are capped with an **acrosome**, which contains enzymes that enable the sperm cell to digest its way through the egg membranes.

In contrast to the chondrichthyans, teleosts usually have oviducts that are continuous with the covering of the ovaries, so that the ova are not shed into the body cavity. This is called the **cystovarian** condition (Hoar, 1969). This saccular ovary–oviduct system arises in one of two ways during development. Some species show a condition in which two folds along the edge of the genital ridge meet and merge to enclose a hollow, which becomes the central cavity of the ovary and extends behind it as the oviduct. In others, there is lateral growth of the edge of the genital fold, so that it curls upward to fuse with the body wall. This process also captures a bit of the coelom, which then becomes the cavity of the ovary and oviduct. Eggs shed from the oviducts reach the exterior via a pore between the anus and the urinary pore. Most nonteleosts show the **gymnovarian** condition, in which the ovaries open into the body cavity and the ova are conveyed through an open funnel into the oviduct. A few teleosts have gymnovarian ovaries; these include the Osteoglossiformes, Anguillidae, loaches of the genus *Misgurnus*, Salmonidae, and Galaxiidae. The nonteleost gars (Lepisosteidae) do not release their eggs into the body cavity (Hoar, 1969). Tetrapods also demonstrate the gymnovarian condition, shedding their eggs into the body cavity to be received by the funnel of the oviduct.

In synchronous spawners—those species that come into the reproductive state and spawn simultaneously—the weight of the ovary may be as much as 40 percent of the total body mass, but in nonsynchronous spawners, it is considerably less (Tyler and Sumpter, 1996). The ova of fishes are usually a few millimeters in diameter, but some may be exceptionally large, especially in chondrichthyans and in *Latimeria*. The number of eggs deposited usually varies with the size of the female, but ovoviviparous or viviparous species (see further) tend to produce fewer eggs than oviparous species, with exceptions. The ovaries of bony fishes are usually well separated, but partial or complete fusion of the right and left organs can be seen in some percoids. In the largemouth bass and some darters, the ovaries join posteriorly to produce a V-shaped structure. The ovaries of the yellow perch (*Perca flavescens*) are so completely fused that they give the appearance of a single organ. This ovary is fused to the body wall just posterior to the anus, and eggs are extruded when this area ruptures, so that the oviducts are not functional. The rupture of the body wall heals soon after oviposition (Parker, 1942). A few teleosts, such as needlefishes (*Strongylura*), have single ovaries.

FUNCTION AND REPRODUCTIVE PATTERNS

The investment of energy in reproduction is referred to as the **reproductive effort** (Kamler, 1992). There are several ways in which reproductive effort can be measured, including the number of eggs per female (*fecundity*) or the biomass of eggs per female, but one of the most useful is the **gonado-somatic index,** or the weight of the gonads expressed as a percentage of the body weight. There have been numerous studies on reproductive effort based on the energy budget, which is expressed as follows (Ricker, 1968; Wootton, 1990):

$$C = P_g + P_r + R + U + F$$

where C is the total energy consumed, P_g is the energy used in the growth of the body, P_r is that used in reproduction, R is the energy of metabolism, U is energy lost as urine and through other nonfecal excretory processes, and F is the energy contained in the feces. Through this energy budget calculation, it becomes possible to understand how some fishes adapt to living in stressful environments. In such cases, where C is constant, the elevation of metabolism (R) necessary for survival may preclude energy being available for reproduction (P_r). The allocation of resources to reproduction goes well beyond the energy necessary for gonadal development. Resources must be available for the development of secondary sexual characteristics and reproductive behavior as well (Wootton, 1990).

Reproductive Strategy

Reproductive strategy in fishes includes all those reproductive traits that optimize the number of offspring (Wootton, 1984). Reproductive traits are variations in the strategy made to respond to environmental fluctuations. Important traits are fecundity according to size and age, reproductive age, size of gametes, reproductive behavior, seasonal timing of reproduction, sex change, and the number of times spawning occurs in the life of the female (*parity*).

Fishes are well known for their high potential fecundity, with most species releasing thousands to millions of eggs annually (Hoar, 1969). The environment takes its toll on eggs and hatched young. Charles Darwin recognized that his theory of evolution by natural selection rested on the premise that species must often produce large numbers of offspring in order to ensure that some will survive to the age at which they in turn can reproduce and perpetuate the species. Thus, the minimum requirement of reproduction, if a species is to maintain itself in stable numbers, is the eventual replacement of each spawning pair by an equally successful pair. Stability of population numbers is seldom actually achieved, and the numbers fluctuate depending on the pressures of the environment. Fluctuations may be episodic or cyclic, depending on these environmental factors, and not all species in an area will be equally affected by the same environmental changes.

Fish species have evolved reproductive methods and attendant physiology that allow them to be successful under a great variety of conditions. The entire repertoire of morphological, physiological, ecological, and behavioral adaptations—the overall approach to reproduction—is included in the reproductive strategy. Strategies may include the production of great numbers of eggs, as seen in the gadids (see Chapter 39), or fewer eggs with greater opportunity for survival. Among the least fecund of freshwater teleosts are the madtom catfishes of the family Ictaluridae. These diminutive and mostly nocturnal cavity nesters may produce clutches of only 25 to 35 eggs, but they are among the largest individual eggs produced by freshwater fishes (Burr and Stoeckel, 1999). Reproductive strategies must ensure the survival of a portion of the eggs—through force of numbers, concealment, protection of nests, or retention in the body. Strategies must place the earliest feeding stage of the young in the proximity of ample and suitable food sources and must ensure that the juvenile fishes have eventual access to the living space of the adults. Time and location of reproduction are generally of great importance.

Within a species there may be annual, latitudinal, or altitudinal differences in the timing of spawning, and these differences may be tied to cycles of temperature, light, or seasonal rainfall. There are species that live in environments that offer favorable conditions for eggs and young over much of the year. Such species may have an extended spawning season, in which females release eggs in small numbers over an extended period. Extended spawning activity can occur near the equator, where seasonal change is slight; among deep-sea species, in which young do not depend on the seasonal abundance of food in upper layers; or in other thermally constant environments. Stein and Pearcy (1982) have reported evidence that some North Pacific macrourids are reproductively active throughout the year. The Borax Lake chub (*Gila boraxobius*) of southeastern Oregon lives in a thermal lake that undergoes only moderate seasonal changes in temperature. The species shows a peak spawning period in the spring but is reproductively active year round (Williams and Bond, 1983). The placement of eggs or young in the optimal place at the right time is due to the response of the endocrine system to environmental cues such as temperature and light, so that gametes are matured, and spawning migrations are undertaken (Lam, 1983). In some species, reproductive readiness may be brought on by an increasing photoperiod and warm temperatures, and spawning may be brought on by a shortened photoperiod and decreasing temperatures. In other species, the converse is true. Billard (1981) has recognized that reproduction for most species of teleosts falls into one of three categories:

Group I, in which gametogenesis is completed in summer and fall during declines in temperature and photoperiod, and in which spawning is performed in the cold season;

Group II, in which gametogenesis begins in the fall but is arrested during winter, so that maturation and spawning are completed in spring or summer;

Group III, in which gametogenesis takes place and is completed during times of increasing temperature and photoperiod, and spawning takes place in spring or summer.

Role of the Endocrine System

Fishes are particularly receptive to exogenous cues that initiate gametogenesis and the onset of reproductive behavior. The endocrine system (see Chapter 26) acts as the mediator in translating environmental cues into morphological, physiological, or behavioral changes in fishes as they prepare for reproductive activity. In some species, light may be more important than temperature, whereas others may be more responsive to temperature. Water flow and flooding, salinity fluctuation, lunar and tidal cycles, and the availability of food may all have reproduction-related impacts on the endocrine system.

The hypothalamus appears to be the site of release of **gonadotropin releasing hormone (GnRH;** Peter, 1981; Sundararaj, 1981). **Gonadotropic hormones (GTH),** which are secreted by the proximal pars distalis of the pituitary, promote the development of eggs and sperm and stimulate the production of androgenic and estrogenic steroids, which control sexual behavior and the development of secondary sexual characteristics. Secondary sexual characteristics may also be induced by proteinaceous hormones. For example, **luteinizing hormone (LH;** see Chapter 26) and **human chorionic gonadotropin (hCG)** promote elongation of the ovipositor in the bitterling (*Rhodeus,* see Fig. 38.1; Weber and Grau, 1999). In temperate and other fishes that spawn once per year, there is an annual cycle of endocrine activity in what has come to be known as the *hypothalamic–pituitary–gonad axis* (Sower and Kawauchi, 2001) that regulates preparation for spawning, spawning, and a quiescent postspawning period. A daily cycle of endocrine activity is also present in some fishes. The pineal body (and eye) plays a role in the regulation of the reproductive cycle, registering changes in diurnal light cycles as the season progresses (Vodicnik et al., 1978, 1979). Possible pathways and control mechanisms are shown in Figure 27.2.

Fish culturists have long known how to influence or control the maturation of brood stock in hatcheries or fish farms. By controlling photoperiods, temperature, and, for some species, water flow, fishes can be brought into spawning condition earlier or later than the normal reproductive season. More direct control of reproductive readiness can be obtained by the injection of gonadotropic hormones or hormone preparations that promote maturation and spawning (Donaldson and Hunter, 1983). A variety of substances has been used, including carp pituitary extract, partially purified salmon gonadotropin, and various mammalian gonadotropins (Lam, 1982). An effective substance for many species is hCG, as in the aforementioned studies on the bitterling. GnRH has been shown to be essential in the induction of spawning of the milkfish (*Chanos chanos*) and mullets (Mugilidae). The development of synthetic analogs of GnRH has been a great aid to aquaculturists working with a variety of species in many countries. Analogs of GnRH by themselves are not effective in bringing about ovulation in salmonids (Lin and Peter, 1986), so salmon gonadotropin is used in conjunction with GnRH to induce spawning in salmon (Sower et al., 1982). Analogs of GnRH are effective in inducing ovulation in several cyprinids (Lin and Peter, 1986) and some percoids (Matsuyama et al., 1993).

One often overlooked consequence of the release of pollutants into waterways is the impact that some of these introduced compounds can have on reproductive processes owing to the fact that they can effectively mimic the actions of endocrine substances. Bleached Kraft Mill Effluent (BKME), an effluent of pulp mills, has been demonstrated to mimic the effect of androgens in promoting the development of male secondary sexual characteristics in mosquitofish (*Gambusia;* Bortone et al., 1989; Sumpter, 1997).

Semelparity and Iteroparity

Individuals of **semelparous (monocyclic)** species make a one-time investment in the future of the species, in that they spawn only once before they die (Cole, 1954). The best known example—that of the Pacific salmons (*Oncorhynchus*)—underscores the fact that semelparity is most frequently observed in fishes that undertake extensive migrations that take them to the ocean to exploit greater feeding and growth opportunities before returning to freshwater to spawn and die (Crespi and Teo, 2002; Young, 1999). Steroids from **interrenal tissue** (the cells of which are specifically located in an adrenal cortex in mammals; see Chapter 26) have been implicated in the "programmed death" of semelparous fishes such as the Pacific salmons and the lampreys (Young, 1999). Obviously, the development of semelparity is most likely when conditions are sufficiently stable that reasonable reproductive success is assured, or under conditions that ensure that some compensatory mechanism can make up for the failures brought on by a fluctuating environment. The risk of failure would seem to be greatest in annual fishes, in which there is no overlap of generations (e.g., the ayu, *Plecoglossus altivelis*). In that diadromous species,

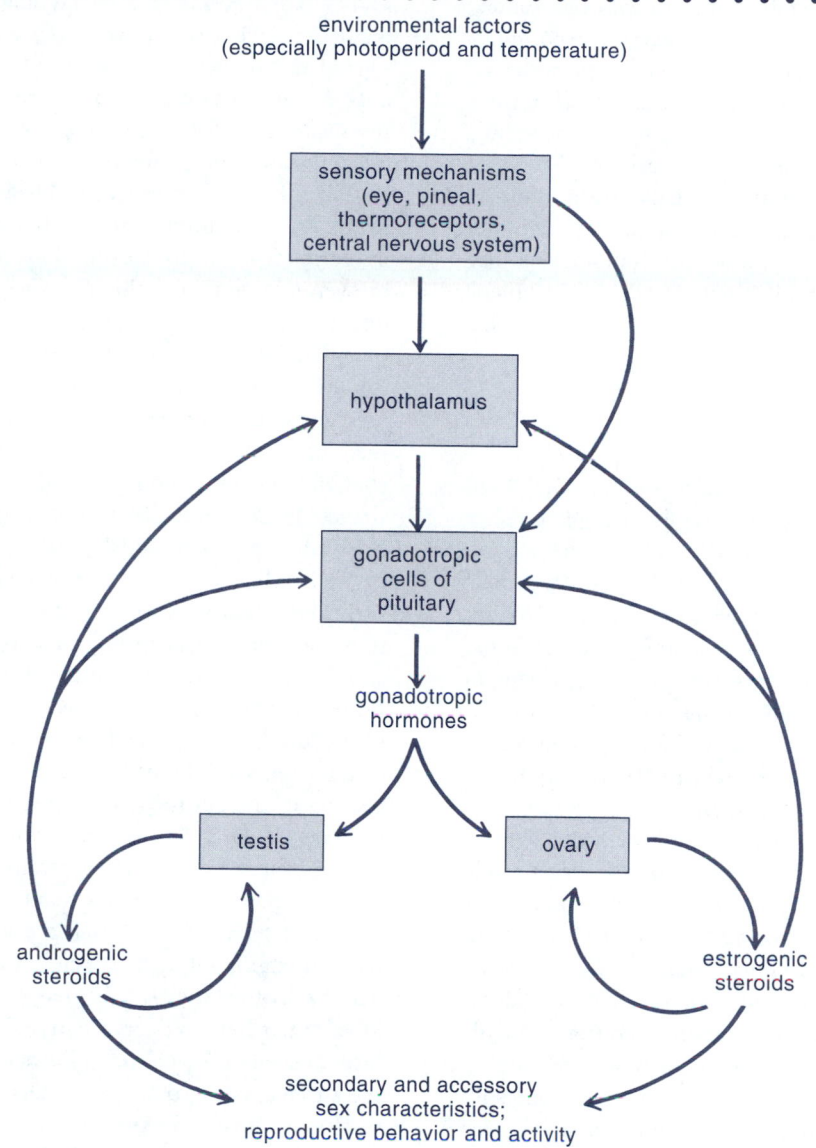

environmental factors
(especially photoperiod and temperature)

sensory mechanisms
(eye, pineal,
thermoreceptors,
central nervous system)

hypothalamus

gonadotropic
cells of
pituitary

gonadotropic
hormones

testis   ovary

androgenic   estrogenic
steroids   steroids

secondary and accessory
sex characteristics;
reproductive behavior and activity

FIGURE 27.2
Possible relationships among environmental factors, receptors, endocrine organs, and reproductive activity.

straying of the young as they return from the ocean could compensate for the failure of reproduction in any given stream. Pacific salmon have overlapping generations, caused by variation in the number of years spent in freshwater and in the marine environment. Moreover, there is a certain amount of straying among returning adults.

Iteroparous (polycyclic) species, in which individuals spawn several times during a lifetime, would appear to have a better chance of succeeding in an unpredictable environment, in that failure in one spawning season has a chance of being compensated for in some succeeding spawning season.

Long-lived species would be expected to have a greater chance at eventual success than short-lived species. Up to the onset of senescence, the older, larger females tend to increase the number of eggs they release, so that a decline in the number of individuals is at least partially offset by an increase in eggs. An example is the Borax Lake chub mentioned previously. Few females of this dwarf species live into the fourth year of life, but those that do grow to about three times the length of the rest of the female population and carry up to 80 times more eggs than are developed by most of the spawning population (Williams and Bond, 1983).

An interesting example of the adjustment of reproduction strategies to the prevailing environment was reported in the British sculpin (*Cottus gobio*). Northern populations of females living in unproductive, acid moorland waters had a long life span, matured at age two (some reached nine years), and spawned once a year. Fewer than 5 percent of the females from the more productive, southern chalk stream populations reached two years of age. They matured at age one and were capable of spawning at least four times during that year (Fox, 1978). A similar adjustment to the environment has been noted for the American shad (*Alosa sapidissima;* Carscadden and Leggett; 1975).

Finding Mates

Central to reproduction is that males and females come together on spawning or mating grounds. Schooling fishes and species in which males and females migrate to the same area may have no problems in getting the sexes together. Mate selection in fishes can depend on visual cues, involving color, pattern, size, or placement of photophores; on sonic cues; or on chemical signals (**pheromones**; see Chapters 26, 36, and 37). Many demonstrations leave no doubt that pheromones produced by the ovary or kidney of females attract males (Liley, 1982; Yambe et al., 1999). Sex hormones (steroids and prostaglandins and their metabolites) are most frequently implicated in reproductive pheromonal activity (Stacey and Sorensen, 1999). Male lampreys release a bile acid, possibly through the gills, that acts as a potent sex hormone by inducing searching and preference behavior in female lampreys (Li et al., 2002). The characid *Corynopoma riisei* (of the aptly named subfamily Glandulocaudinae) demonstrates an unusual mode of pheromone secretion—glands on the caudal peduncle of males apparently secrete a mucopolysaccharide that acts as a chemical signal during courtship (Atkins and Fink, 1979). There is circumstantial evidence that pheromones are important in the lives of some deepsea fishes—not surprising, given the challenges of locating potential mates among widely dispersed individuals in the dark depths of the ocean. Males of the benthopelagic halosaurs, the gonostomatid genus *Cyclothone*, and ceratioid anglerfishes have enlarged olfactory organs. Investigators have suggested that such males make use of their oversize scent organs in locating females (Marshall, 1979).

Egg Production in Oviparous Fishes

Eggs of **oviparous** (egg-laying) bony fishes are usually small, typically between 1.5 and 3 mm in diameter, but there is a tremendous range in size. Kamler (1992), in reviewing the data on egg size in fishes, noted that the eggs of the frill

shark (*Chlamydoselachus*), which range from 90 to 97 mm in diameter, have a wet weight 34 million times that of the eggs of the viviparous surfperch (*Cymatogaster*), which are only 0.3 mm diameter. Some fairly large eggs are seen in the madtom catfishes, in which the egg diameter is often greater than 3 mm—huge for such small fishes (Burr and Stoeckel, 1999). Some species of trout and salmon have egg diameters exceeding 5 mm, and ariid catfishes commonly have eggs from 15 to 25 mm. The eggs of a given species, or even of a given female, may vary in size with such factors as race, population, age, or nutrition of an individual (Kamler, 1992). The elongate, yolk-laden eggs of hagfishes (see Fig. 5.2) are up to 30 mm long, and oviparous sharks, rays, and chimaeras deposit eggs of 60 to 70 mm in egg cases that are up to 300 mm long. Some of the larger egg cases contain two eggs. Chondrichthyan egg cases are of various shapes, from spindlelike to purselike (they are sometimes referred to as "mermaids' purses"), as shown in Figures 27.3A and B. The distinctive shape of the egg cases apparently optimizes water flow across the surface of the case, thus facilitating gas exchange between the environment and the embryonic fish contained within (Zimmer, 1999). Embryonic little skates (*Raja erinacea*) further enhance water circulation by placing an embryonic extension at the tip of their tail into one of the horns of their egg case. Rapid beating of the tail helps to pull oxygenated water into the egg case (Koob and Summers, 1996; Leonard et al., 1999). The eggs of bony fishes are typically spherical, but elongate eggs are known in anchovies, some gobies, and clownfishes.

Most fishes lay eggs that are heavier than water (**demersal eggs**), but many species produce buoyant eggs that may be hydrostatically adjusted by oil inclusions, by imbibed water held in a large perivitelline space, or by a high ratio of surface to volume. This permits the eggs to float at the surface, or at some intermediate depth, depending on the species involved. **Aquaporins**—proteins that compose water channels in cell membranes—have been shown to be instrumental in the hydration of pelagic fish eggs, which permits the eggs to maintain positive buoyancy (Fabra et al., 2005). The pelagic eggs of some species drift freely; others attach to each other or to vegetation by means of tendrils (Fig. 27.4). Tendrils, hooks, or other attachment devices also occur on demersal eggs and are present on elasmobranch and hagfish eggs. Demersal eggs may be adhesive and deposited in clumps, such as the eggs of sculpins and darters, which stick together through the incubation period, or they may be attached singly to some substrate. Some are temporarily adhesive, like the eggs of trout and salmon. Such eggs adhere to the substrate and to each other for a short period and then separate. The practical advantage of temporary adhesion for a species that constructs gravel nests in

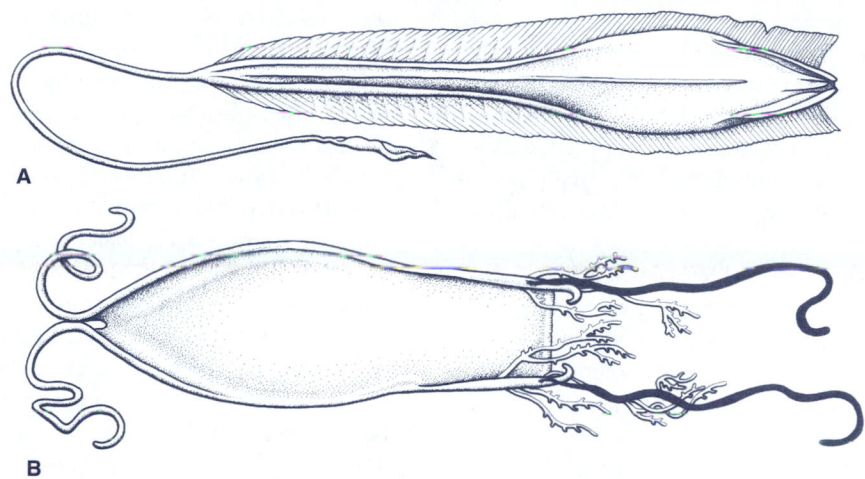

FIGURE 27.3
A, Egg case of chimaera (*Hydrolagus*); **B,** Egg case of shark (Scyliorhinidae).

running water and takes several minutes to cover the eggs is obvious.

Some fishes engage in mass spawning, with no pairing of individuals. Numerous males and females release gametes together in a suitable environment. The numbers of eggs are high, and they may be left to drift in the open water or to be carried by the turbulence of a stream, allowed to settle on the substrate, or released so that they may adhere to vegetation. Spawning often ensues after migration to a suitable site, often against a current that will carry eggs and larvae back to a nursery area. Some herrings are exemplary of open-sea mass spawners. Others migrate to shore areas, and some, like the shad, are anadromous. Beach spawning, in which fishes move into the extreme shallows under tidal influence, in order to place their eggs beyond the reach of most predators, is known in at least six teleost families (Martin and Swiderski, 2001). The most remarkable of these are several species of the well-known grunion (*Leuresthes*), which ride waves far up the beach during high tides in order to lay their eggs in essentially terrestrial environments (see Chapter 15, Going Deeper).

Polyandrous spawning is exhibited by many species in both freshwater and marine situations. Males position themselves on the spawning ground and surround females as the latter swim into their midst. Eggs and sperm are released simultaneously, sometimes with violent activity on the part of the spawners. In some suckers (Catostomidae), the activity consists of vigorous vibration, strong enough to dislodge the

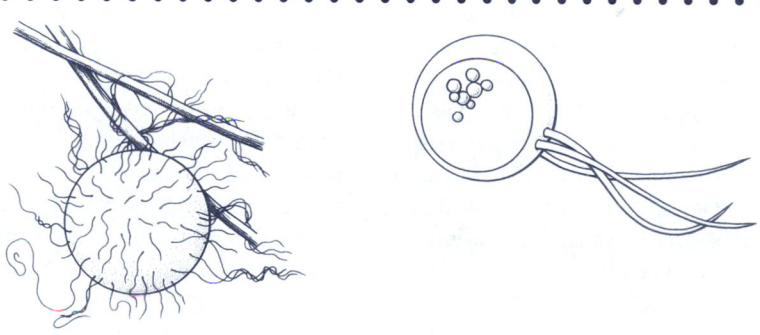

FIGURE 27.4
Teleost eggs with filaments and tendrils.

stones of the substrate and allow the eggs to sift into crevices. The eggs of yellow perch (*Perca flavescens*) are enclosed
in a gelatinous rope that often festoons vegetation or debris
in the spawning area. The sequestering of eggs in a nest,
either adhering to the substrate or enclosed in a gelatinous
matrix, has consequences for the oxygenation of the developing embryos (Strathmann and Hess, 1999). Many polyandrous spawners of the open ocean leave their eggs to drift
in the currents.

Pairing is common in many species, such as tunas and
carangids, which spawn in open water or over unprepared
sites on the bottom. Other species pair for spawning after
one or both members prepare a site for the reception of the
eggs. Preparation may range from merely fanning silt from
stones to the construction of elaborate nests. Some fishes,
such as salmon and trout, bury their eggs in a gravel nest,
termed a **redd,** that is quickly abandoned.

Blumel (1979) reviewed parental care among fishes,
noting more than 80 families of bony fishes in which some
kind of parental care was observed. The guarding of eggs
was the most common form of parental care in most families,
followed by nest building or cleaning of the spawning substrate and fanning eggs. Other behaviors listed among the 15
types of parental care include internal gestation, oral brooding, burying of eggs, and splashing water on eggs deposited
above the water line—a practice seen in the lebiasinid characin *Copella arnoldi*, commonly known as the "splash tetra."

Nest building is known in lungfishes, the bowfin (*Amia*),
osteoglossiforms, minnows, catfishes, sunfishes, snakeheads,
and many others. In numerous species, from primitive to
derived groups, the nest is guarded by one or both parents.
Parental care consists largely of warding off potential predators, ensuring adequate circulation of water across the developing eggs, and removing diseased eggs.

A few species leave their eggs in the care of other animals.
Some snailfishes (*Careproctus*) deposit their eggs in the gill
chambers of crabs of the genera *Paralithodes* and *Lopholith-
odes*. The tubesnout (*Aulichthys*) places its eggs in the peri-
branchial cavity of ascidians, and the bitterlings (*Rhodeus* and
close relatives) introduce their eggs into the mantle cavity of
freshwater mussels (see Fig. 37.1B), so that the eggs are incubated in the bivalve's chamber. *Percilia gillissi* of Chilean lakes
places its eggs in the outlet canals of a freshwater sponge.

In a number of pairing species, the eggs are carried and
protected by one of the parents after external fertilization.
Eggs are carried in the mouth or branchial cavity in about 10
families, but the practice is most prevalent in the Cichlidae,
in which several species are mouthbrooders. The breeding
habits and behaviors of cichlids are of great interest to
scientists and aquarists alike and have been the subject
of considerable study (Barlow, 2000; Keenleyside, 1991).

At least 17 species of cardinalfishes (Apogonidae) and 13 of
sea catfishes (Ariidae) are mouthbrooders (Breder and Rosen,
1966). Other mouthbrooding families are the Osteoglossidae, Liparidae, Opisthognathidae, Anabantidae, Belontiidae, Luciocephalidae, and Malapteruridae. The female amblyopsids carry their eggs in the branchial cavity. The ariid
catfish *Tachysurus* is reported to incubate its eggs intesti-
nally (Blaxter, 1969). Some banjo catfishes (Aspredinidae)
and some loricariid catfishes carry their eggs embedded
in the skin of the lower surfaces of the body and fins. The
males of some pipefish and seahorse species (Syngnathidae)
bear eggs on the skin, but in most species of this family, the
males carry eggs in brood pouches. In their close relatives,
the Solenostomidae, the female is equipped with a brood
pouch formed by the pelvic fins (Orr and Fritzsche, 1993).
The males of *Kurtus* have a hooklike structure projecting
forward from the forehead, from which the clustered eggs
hang during incubation.

Some oviparous fishes fertilize their eggs internally.
This practice is encountered, for example, among skates,
chimaeras, many sharks, some characins, catfishes, and rockfishes. Egg retention for limited durations after fertilization
had taken place in the follicle or ovarian lumen ultimately
culminated in the conditions of *ovoviviparity* and true *viviparity*, the latter perhaps best exemplified in the poeciliids
(see Chapter 15).

Egg Retention, Internal Incubation, and Viviparity

For fish species that practice some form of retention of their
fertilized eggs, a continuum of reproductive adaptations exist, ranging from the deposition of internally fertilized eggs
in the cleavage stage to the release of large, well-nourished
juveniles or young adults. Several possible advantages accrue
to fishes that practice internal gestation. The first of these
is *protection*. The eggs and embryos are safe from predators except those large enough to overwhelm the female.
They are protected from adverse water conditions; desiccation, anoxia, and injurious temperatures are not dangerous
unless the female is unable to escape these conditions. The
young are protected against mortality caused by drifting
away from suitable rearing areas as eggs or larvae. Another
possible advantage might be the *conservation of energy*, although careful study would be necessary to confirm energy
saving in specific instances. Certainly, nest building is un-
necessary, and the size of the male can be reduced. Large
numbers of eggs are generally not required, and usually
there is no need for long migrations to a specific breeding
site. Usually, fertilization is ensured, so that few eggs are
wasted. Live birth, in its different manifestations, may also be
of some advantage in species dispersal, as a single pregnant

female might be able to extend the range of a species. Similarly, the survival of one pregnant female after a catastrophe might allow the restoration of a population.

The terms *ovoviviparity* and *viviparity* have been applied to a range of reproductive adaptations involving internal gestation. In the broadest sense, "viviparity" includes all conditions in which hatched young are released from the female. **Ovoviviparous** species incubate their eggs and release their young without providing any maternal source of nourishment other than the egg yolk. Hamlett (1999) referred to this condition as *aplacental yolk sac viviparity*. Wourms (1981) favored the term **lecithotrophy** to describe the condition in which eggs are retained until hatching and all nourishment is derived from the reserves in the yolk, and **matrotrophy** to indicate nutritional dependence on the mother during gestation.

Lecithotrophy and matrotrophy are widespread among elasmobranchs (Dulvey and Reynolds, 1997; Wourms, 1977, 1981). Wourms (1981) recognized either condition in 40 of the 98 families accepted in Compagno's (1990) classification of elasmobranchs. Hamlett (1999) classified all elasmobranchs practicing internal gestation as "viviparous" but distinguished variations on the theme. Oviparous elasmobranchs produce relatively large eggs, with tough, structurally complex shells. Egg laying is viewed as the ancestral condition among elasmobranchs, with live bearing having evolved independently several times (Dulvey and Reynolds, 1997). About 10 percent of living shark species have developed **placentae**—uterine compartments in the oviducts that provide nutrient and gas exchange, transport of other molecules such as immunoglobulins, and osmoregulation. In these cases, the embryonic yolk sac contributes to the placenta

(Fig. 27.5). Elasmobranchs lacking such features are termed **aplacental.** Aplacental species may still engage in nutrient transfer to embryos through other specialized structures, such as the **trophonemata**—long uterine villi that increase the surface area for nutrient secretion and dissolved gas exchange (Hamlett, 1999). In bony fishes, there are about 15 live-bearing families—including the Latimeriidae, Zoarcidae, Bythitidae, Ophidiidae, Aphyonidae, Hemirhamphidae, Goodeidae, Jenynsiidae, Anablepidae, Poeciliidae, Comephoridae, Embiotocidae, Labriosomidae, and Clinidae.

As mentioned earlier, some oviparous species have set the stage for viviparity by fertilizing their eggs internally and releasing them at an early stage of development. Internal fertilization requires a modification of behavior and the development in males of a means of introducing spermatozoa into the genital orifice of females. Spermatophores have developed in the Poeciliidae and Horaichthyinae; those of the latter are barbed for sure attachment. Some of the most remarkable intromittent organs among fishes are seen in oviparous species (Fig. 27.6). Chimaeras, skates, and oviparous sharks have pelvic **claspers,** whereas the anal fin is modified into a **gonopodium** in the egg-laying, live-bearing poeciliids; the Anablepidae; and in the subfamily Horaichthyinae of the family Adrianichthyidae. Members of the Phallostethidae have elaborate, highly asymmetrical structures developed from the pelvic girdle for use in clasping and inseminating the female. Some sculpins have fleshy genital papillae that reach a large size. The freshwater butterflyfish (*Pantodon buchholzi*) has a hollow tube on the leading edge of the anal fin.

From the short-term retention of eggs to their internal incubation and hatching is only a short step. According to Wourms (1981), there are 14 families of bony fishes that

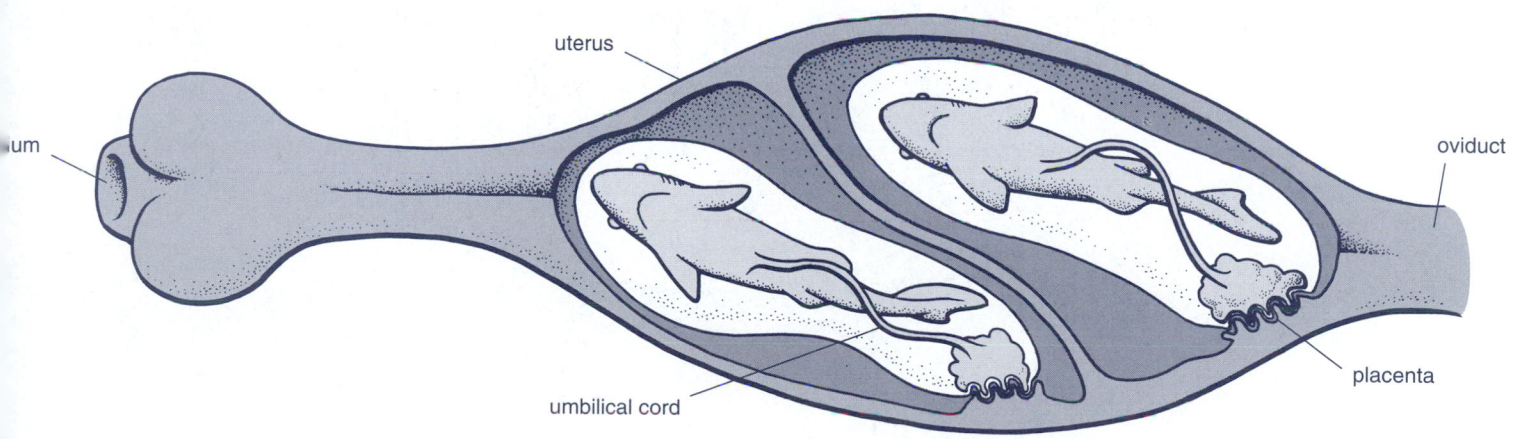

FIGURE 27.5
Development of uterine compartment in placental shark (redrawn from Hamlett et al., 1993).

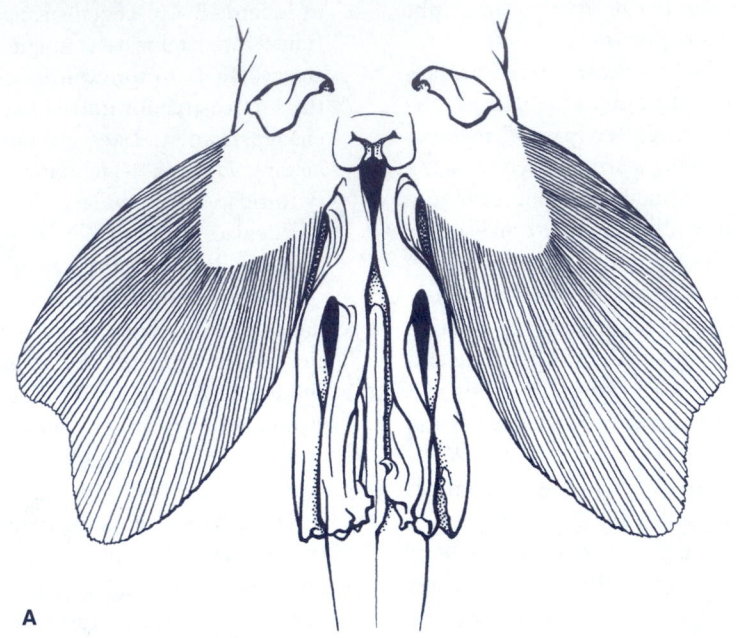

A

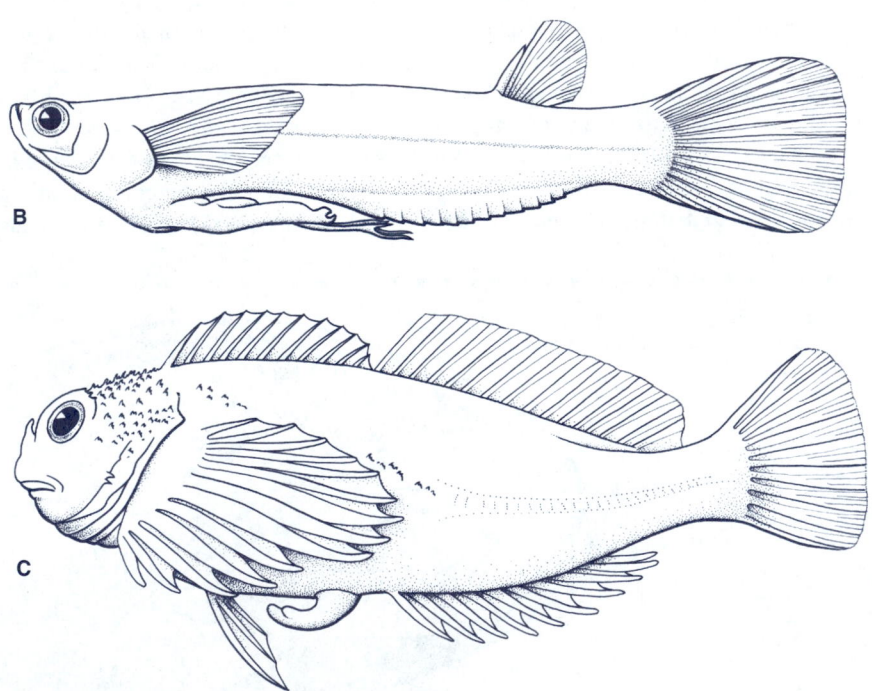

B

C

FIGURE 27.6
Examples of male oviparous fishes with large intromittent organs: **A,** Pelvic claspers of a chimaera (*Hydrolagus*); **B,** *Horaichthys;* **C,** *Clinocottus.*

gestate their eggs internally to some degree (Latimeriidae, Ophidiidae, Bythitidae, Aphyonidae, Hemirhamphidae, Goodeidae, Anablepidae, Poeciliidae, Embiotocidae, Zoarcidae, Clinidae, Labrisomidae, Scorpaenidae, and Comephoridae).

The term **larviparous** has sometimes been applied to the condition of giving birth to larvae. Many species of rockfishes (Scorpaenidae = Sebastidae) are ovoviviparous, releasing newly hatched larvae that weigh less than the eggs from which they developed. The rockfishes show little sexual dimorphism and have no particular specialization of the ovary, where the eggs remain during incubation. In an ophidiid genus (*Dinematichthys*), the eggs hatch from the ovarian follicles and develop into advanced larvae in the lumen of the ovary.

Another type of mainly lecithotrophic viviparity is seen in most Poeciliidae, in which the young are retained until the juvenile stage is reached (Thibault and Schultz, 1979). The young remain in the follicles of the ovary, with sufficient yolk for development. In some species, including those of *Gambusia* and *Xiphophorus*, the pericardial sac expands into a hoodlike or straplike structure that folds over the embryo's head; this structure is thought to have an absorptive function. Mother–embryo transfer of nutrients in poeciliids has been experimentally demonstrated, suggesting an increased role of matrotrophy in species previously considered primarily lecithotrophic. When mothers were injected with radiolabeled leucine, assays of their embryos revealed the uptake of the labeled compound (Marsh-Matthews et al., 2001).

Many poeciliid species are capable of retaining more than one brood at a time. This is called **superfetation** and is a result of the continued maturation of eggs during the development of previously fertilized eggs. Some species can store live sperm for several months, so that successive batches of oocytes can be fertilized as they mature. There may be as many as five broods at once in *Poeciliopsis prolifica* (Thibault and Schultz, 1979). In what may be the ultimate demonstration of superfetation, a South African member of the Clinidae (*Clinus superciliosus*) may harbor up to 12 broods at once (Veith, 1980).

The coelacanth (*Latimeria chalumnae*) has the largest eggs of all bony fishes—8.5 to 9.0 cm in diameter. Because of the large egg size and because of evidence of live-bearing reproduction in fossil coelacanths, it was long believed that the species would be viviparous. This was proven in 1975 by the discovery of five advanced young, up to 33 cm long, in the oviduct of a specimen at the American Museum of Natural History (Thomson, 1991). Although yolk is believed to be the primary source of nutrient in developing *Latimeria*, this species is considered matrotrophic, as it appears capable of producing uterine secretions for the purpose of nourishing young (Balon, 1991a, 1991b; Wourms, 1981; Wourms et al., 1988). It has been suggested that the extraembryonic

membranes that support terrestrial reproduction in amniotes (the amnion, chorion, and allantois) may have had their origins as structures facilitating mother–embryo exchange in sarcopterygian fishes. Given the development of embryonic retention adaptations in modern sarcopterygians such as *Latimeria*, it is an appealing proposal, yet one that is currently not substantiated (Laurin and Girondot, 1999).

Matrotrophic viviparous fishes nourish their developing embryos through a variety of adaptations—most involving the secretion of nutritive materials by the female, but some including the formation of pseudoplacentae or placentae. In some aplacental species, the developing young subsist exclusively on the yolk for a time, and then the yolk is supplemented or supplanted by secretions ("uterine milk") of the female and, in a few species, by material from dead eggs and embryos. Any secretions, cells, and cell debris other than blood that come from the female and are available to the embryo are called **histotroph**. In several species of sharks, the nutritive secretions are taken in through the mouth or the spiracles and swallowed. In aplacental species, the trophonemata that secrete the nutrient fluid are elongate and extend through the spiracles of the embryo into the gut. In placental sharks, there are three phases of nutrition according to Hamlett (1989, 1999), who applied the following terms: **vitellogenesis,** the phase in which yolk is the energy source; **histotroph secretion,** the phase involving nutrition secreted by the female; and **hematotrophic placentation.** Placental sharks are found in the families Carcharhinidae and Sphyrnidae. The placental condition is not known among the batoids. Yolk sac placentae, formed by the close apposition and eventual interdigitation of thin tissues of the yolk sac and the uterine oviduct (Fig. 27.5), are present in such carcharhinid sharks as *Mustelis canis, Carcharhinus falciformis*, some species of *Scoliodon, Prionace, Hemigaleus,* and in some hammerheads of the genus *Sphyrna* (Wourms, 1977). Placental species vary in embryonic dependence on the placentae (Dodd, 1983). In some, the early development of the embryo is supported mainly by yolk, and a placenta is not developed for several months. In some sharks, an umbilical cord is provided, with delicate vascular structures called **appendiculae**, which may function in the absorption of nutrient materials and perform other functions (Hamlett, 1989). In other species, placentation occurs early in gestation and is responsible for most of the embryonic nutrient. An example is *Scoliodon laticaudatus*, which is believed to have the smallest eggs (1 mm diameter) of any elasmobranch (Hamlett, 1989).

A bizarre type of matrotrophy is known in some families of sharks, mainly lamnoids. Advanced embryos devour newly ovulated eggs (**oophagy**) or their own weaker siblings (**adelphophagy;** Wourms, 1981). Considering the lifestyle of most lamnoids, this intrauterine cannibalism is good training for the future. Adelphophagy has also been reported in more benign species, such as the ophidioid genus *Cataetyx*. The young of *Latimeria* are believed to reach their large size at birth not only because of the large yolk but also, as mentioned earlier, because they may ingest egg debris and histotroph secretions (Balon, 1991a, 1991b) or possibly smaller embryos (Balon, 1984).

There is a great variety of adaptations for the nourishment of embryos among the viviparous teleosts (Hoar, 1969; Wourms, 1981; Wourms et al., 1988). Many of these adaptations involve supplemental structures of the female or the embryo. In the eelpout (*Zoarces viviparus*), the only special absorptive structure other than the skin is the hypertrophied posterior section of the intestine. Some other examples of specialized structures are the expanded pericardial sac in some Poeciliidae and the follicular pseudoplacentae in others; the expansion and specialization of the fin membranes in Embiotocidae; the structures called **trophotaeniae** that radiate from the anal region of the embryos of most Goodeidae and some ophidioids; and the great expansions of the gut in Anablepidae. Some embiotocid species have enlarged hindguts that protrude from the anus of developing young. Some of these specializations allow the retention of young until an advanced stage is reached. In an extreme case, the males of the embiotocid *Cymatogaster aggregata* are born sexually mature (Weibe, 1968), even though this species has eggs that are among the smallest known in fishes—about 0.3 mm (Eigenmann, 1894). Placenta-like connections are formed in some teleosts and batoids. A "branchial placenta," in which trophonemata or similar structures invade the branchial chamber of the embryo and form close associations with the gill tissue, has been described for *Jenynsia* (Jenynsiidae) and *Gymnura* (Dasyatidae; Wourms, 1981).

Reproductive Guilds

The tremendous diversity of their reproductive methods has inspired several attempts to classify fishes from ecological or behavioral perspectives. Balon (1975a, 1975b, 1981, 1984) proposed a classification that eventually encompassed 34 reproductive guilds. The guilds are arranged in three sections. The first section includes the *nonguarders*, and has two subsections: *open substratum spawners*, which leave their eggs exposed, and *brood hiders*, which conceal their eggs. The first subsection ranges from pelagic spawners to those that spawn on rock and gravel, plants, sand, and on land. Fishes of the second subsection hide their eggs in sand, rock and gravel, cavities, and in invertebrates. The second section, the *guarders*, includes *substratum choosers* and *nest spawners*. Substrata chosen include open water, rocks, plants, and

above-water sites. Nests range from froth to rock and gravel, plants, and sand holes, and may even include animals such as sea anemones. The third section are the *egg bearers*, including those that bear eggs (and sometimes larvae) *externally* on structures such as brood pouches, the mouth, or the branchial cavity, or internally, as in lecithotrophic or placental situations. Although the full reproductive habits of only a small fraction of the fishes are known, and although Balon's classification may be incomplete, or may not provide for some exceptional or versatile species, the guilds form a framework that is useful in summarizing the bewildering array of breeding habits in fishes.

Differences Between the Sexes

With many important exceptions (to be discussed later), the individuals of most species of fish are **dioecious** or **gonochoristic**—that is, they function as either male or female throughout their adult life. In the **hermaphroditic** condition, on the other hand, an individual produces both eggs and sperm at some stage of its development. Although the establishment of sex initially depends on the sex chromosomes—designated X and Y for most fishes—the actual phenotypic expression of sex is a more complex and subtle phenomenon, in which the environment, especially the temperature, may play a significant role (Conover and Heins, 1987a, 1987b; Conover and Kynard, 1981; see Chapter 28).

Although the sexes have very similar appearances in many species, sexual dimorphism or dichromatism is common in fishes and may be especially well marked in species with internal fertilization or elaborate reproductive behavior. The differences between the sexes may involve secondary sexual characteristics that are necessary for the accomplishment of copulation, oviposition, or incubation (*requisite* characteristics), or they may be so-called *accessory* characteristics, which may not be directly involved in the mechanics of reproduction but are important for recognition, courtship, or other reproductive behavior. Placing the onset of the development of secondary sexual characteristics in the context of genetic and physiological control mechanisms, especially as it pertains to the specific impact of steroid hormones, is especially challenging (Emerson, 2000). Requisite secondary sexual characteristics include the claspers of elasmobranchs and the various gonopodia of male phallostethiforms and cyprinodontiforms, such as the hemirhamphids and embiotocids. Even oviparous families, such as the sculpins, may have species with large genital papillae (Fig. 27.6C). Brood pouches and specialized ovipositors are also requisite characteristics.

Accessory secondary sexual characteristics are many and varied and are usually sexually dimorphic. Many structures and colors change with the reproductive state of the individual, whereas others are more or less permanent throughout the year. It is important to distinguish the environmental factors that influence temporary color changes associated with the onset of reproductive behaviors from those that have resulted in the evolution of more permanent coloration features that distinguish the sexes (Kodric-Brown, 1998). The males of many species are more brightly colored than the females and may have larger fins and bolder markings. **Sexual dichromatism** is seen in salmon and trout, in which the males are more colorful; in the bowfin, in which the male has a caudal ocellus; and in many minnows, characins, cichlids, and others.

Sexual dimorphism is often seen in fishes. Longer fins are characteristic of the males of many fishes; suckers, gobies, dragonets, and climbing perches are examples. The more prominent colors and larger fins of males can be significant in courtship or in aggressive displays toward rivals, or the fins can, in certain species, aid in holding the spawners together. This is especially true in species that have nuptial tubercles or contact organs on the fins or body. These structures are prominent on the fins and scales of suckers and reach large sizes on the heads of certain minnows (see Plate 10.2; Fig. 27.7). They are most common in species that spawn in flowing water. *Nuptial tubercles* are formed from both keratinized and nonkeratinized epidermal cells. The horny caps of the keratinized type are often pointed. Whitefish, grayling, smelts, ayu, retropinnids, kneriids, phractolaemids, most cyprinoid families, a few characoids, the mochokid catfishes, and percids all have nuptial tubercles. *Contact organs* are small, bony, spinelike structures that are usually associated with scales or fin rays; they are present in needlefishes, certain cyprinodontoids, characins, and sculpins (Wiley and Collette, 1970).

Size differences between the sexes are evident in many species. The reproductive pattern of the species determines which sex is the larger, but commonly the female, carrying the bulky eggs, is larger than the male. A notable exception is the lungfish *Protopterus aethiopicus*, in which the male is reported to be twice the length of the female (Greenwood, 1987). In cyprinid species in which the males engage in combat or guard nests, or in which a high level of sperm competition exists, the males will be equal in size to or larger than the females (Parker, 1992; Pyron, 1996; Snelson, 1972). The greatest disparity in size occurs in the ceratioid anglerfishes, in which the female can be many times larger than the male (see Plate 35.3). In certain species, the male ceratioid grasps the female's skin with his teeth, literally grows onto her, and becomes a testis-filled parasite, available for service at spawning time. Pietsch (1976) reviewed the reproductive strategies of ceratioids and showed that the males

of Ceratiidae, Linophymidae, and probably Neoceratiidae are obligate sexual parasites. The males of Caulophrynidae and the oneirodid genus *Leptacanthichthys* are thought to be facultative parasites. Sexual parasitism is not known in the remainder of the suborder.

Hermaphroditic fishes have attracted attention as a rich resource of physiological and genetic information as well as for their general interest (Atz, 1964). Occasional hermaphrodites are found in many gonochoristic species as an abnormality, but there are numerous species that are normally hermaphroditic; a couple are even capable of self-fertilization. **Synchronous** (or **simultaneous**) **hermaphrodites** have ripe ovaries and testes at the same time, but usually spawn with one or more other individuals, alternately taking the role of male and female. Although the belted sandfish (*Serranus subligarius*), a marine species of Florida, has been observed fertilizing its own eggs in captivity (Yamamoto, 1969), the mangrove rivulus (*Rivulus marmoratus*) is the only species known to be capable of self-fertilizing in the wild. Synchronous hermaphrodites are known from the following families: Chlorophthalmidae, Alepisauridae, Paralepididae, Evermannellidae, Cyprinodontidae, Seffanidae, Maenidae, and Labridae. Individuals of other families may occasionally be hermaphrodites. A familiar example is the striped bass (*Morone saxatilis*), in which occasional hermaphrodites are seen.

Sequential hermaphrodites are either first male (**protandrous**) or first female (**protogynous**; Yamamoto, 1969). Many of the species begin life with undifferentiated gonads that contain both male and female elements. The protandrous condition is known in members of the Gonostomatidae, Serranidae, Sparidae, Maenidae, Labridae, Platycephalidae, Pomacentridae, and Centropomidae. Protogynous hermaphrodites are known in the families Scaridae, Sparidae, Emmelichthyidae, Synbranchidae, Serranidae, Maenidae, and Labridae. In the Labridae, there are species in which some males do not pass through the female stage (*primary males*) as well as *secondary males* that change to male from female. The secondary males are brightly colored, but the primary male may be dull, or at least colored like the females. In one species, the two types of males exhibit different spawning behaviors—the primary males spawning in groups with a single female, and the secondary males pairspawning with females in turn. Bidirectional sex change, in which an individual has the ability to change sex more than once in its lifetime, has been described among coral-dwelling gobies of the genus *Gobiodon* (Munday et al., 1998).

The evolution of alternative reproductive strategies within a given species may be viewed as an adaptive advantage, as it confers greater reproductive versatility on the species (see Chapters 26 and 37; Alonzo and Warner, 2000;

FIGURE 27.7
Nuptial tubercles on: **A,** scales, anal fin, and caudal fin of sucker (*Catostomus*); **B,** Scales, head, and pectoral fin of peamouth (*Mylocheilus*).

Henson and Warner, 1997). In the bluegill sunfish (*Lepomis macrochirus*) for example, three mating tactics have been identified: parental, sneaker, and satellite. *Parentals* delay maturation, thus growing to a larger size. They establish nests and become territorial. The other two forms are considered "cuckolders" that engage in sperm competition with the parentals. *Sneakers* mature early and engage in ambush tactics to introduce their sperm to females; when larger and older, they become *satellites* and sneak their sperm in by mimicking females (Gross, 1984, 1991; Neff et al., 2003). It has been suggested that the filial cannibalism (consumption of one's own offspring) observed in bluegill sunfish may be a means of selectively removing cuckolder offspring from the nest (Neff, 2003).

Sex reversal and hermaphroditism are controlled by the endocrine system, which is genetically programmed in normally hermaphroditic species to act on the gonads in response to the proper stimuli—internal, external, or both. The experimental administration of androgens to genetically female fish has changed them into functional males, and the administration of estrogens has changed genetic males into functional females (Guerrero, 1979). The breeding of sex-reversed, functional males (that have no Y chromosome) to normal females can result in all-female progeny. This can be significant to fishery management, especially when no reproduction of early maturing pond fishes is desired or when a useful but potentially troublesome species, such as the grass carp (*Ctenopharyngodon idella*) is to be stocked.

Naturally occurring all-female species are known among the Poeciliidae; the best known example is probably the amazon molly (*Poecilia formosa*)—named after the mythical all-female tribe of warriors. This fish originated from the hybridization of other species of *Poecilia* (**hybridogenesis**) and is a permanently diploid species that produces diploid eggs (Schultz, 1973). The species is perpetuated through matings with males of other species, which contribute sperm that triggers the development of the eggs but does not contribute any genetic material (a condition known as **gynogenesis**). Gynogenesis results in the production of all-female offspring. Other gynogenetic species are known, and some have been artificially produced in the laboratory. All known species of unisexual fishes are hybrids, and it is likely that the same holds true for all unisexual amphibians and squamate reptiles (Vrijenhoek, 1994).

Experimental production of all-female broods has been accomplished with several species for various purposes. Some laboratories "clone" females of a species by activating eggs with sperm made impotent by irradiation or some other means and then restoring the diploid chromosome set contributed by the female by treating the activated eggs with pressure or temperature shocks to inhibit the first or, if desired, the second meiotic division (Streisinger et al., 1981). This induces the retention of the second polar body (Chevassus, 1983).

Hybridization

Fish hybridization in nature has been recognized for many years, and hybrids have long been produced in laboratories and hatcheries (see Chapter 28). Schwartz (1972, 1981) has documented the extensive research about the topic. Natural hybrids can occur by two or more species spawning in proximity at the same time. Fish spermatozoa remain active for many seconds to a few minutes, and sperm from a spawning pair or group closely upstream can invade the nest or spawning area of another group downstream. If the sperm and eggs are of different species, hybrids may result.

Hybridization involving the pairing of different species is known, but usually there are reproductive barriers—behavioral, physical, or temporal—that will preserve the integrity of species, especially those living sympatrically. Many species of cichlids in Lake Victoria can interbreed and produce viable offspring, yet their sexual selection for distinct color morphs reinforces assortative mating and, consequently, promotes species diversification. Recent eutrophication in the lake has resulted in increased turbidity. This has had the effect of relaxing these selective pressures, and species diversity has concomitantly declined (Seehausen et al., 1997).

Selective Breeding

Aquaculturists, whether they are involved in pond culture, hatchery operations for stock enhancement, salmon ranching, breeding of ornamental species, or other activities, are constantly seeking to maintain or improve the quality of their product. The maintenance of genetic variation in captive "wild" animals requires attention to known genetic principles (see Chapter 28) and requires the use of an adequate number of breeders and the minimization of certain types of inbreeding (Frankel and Soulé, 1981), so that some selection of breeders must be practiced. The maintenance of high quality in semi-wild stocks (i.e., ensuring sufficient genetic variation), such as those used in salmon ranching—in which brood stock is selected from fishes returning from the ocean—is somewhat difficult. Although large numbers of individuals are involved, they have been subjected to the artificial selection of hatchery life, and the identification of family lines requires sophisticated methods. Most selective breeding in fishes is accomplished with captive brood stock, from which selected matings can be made. Selective breeding in domesticated stocks may target specific features of a given species, but the overall goal is increased productivity (Gjedrem, 1983; Gjerde, 1993; Kirpichnikov, 1981). Fish are selected for resistance to the adverse environmental factors encountered in artificial surroundings, including diseases; extremes of temperature, dissolved oxygen, pH, and other physical and chemical factors.

Selection can be accomplished by inspecting prospective breeders for desirable phenotypic attributes—with little or no knowledge of the genotype—and choosing those that appear to have the best balance of the characteristics wanted. Thus, selective breeding requires that the stocks from which the prospective breeders are to be selected are subject to the same environmental factors, so that any differences among individuals will have less chance of being environmentally induced. Naturally, inbreeding is a consequence of selective breeding programs; because inbreeding generally leads to reduced performance, it is to be avoided

(Gall, 1983; Kincaid, 1983). Crossing unrelated strains of a species is often used to increase heterozygosity and produce **heterosis** (also known as **hybrid vigor**)—a particularly significant feature in breeding programs with carp (*Cyprinus carpio;* Kirpichnikov, 1981).

Artificial selection programs favor individuals that can adapt to the consequences of life at high densities, including stress brought on by the crowding and buildup of metabolic wastes in rearing facilities. Fish can be selected for rapid growth on either manufactured or natural food; for efficiency in weight gain per unit of food consumed; for increased fecundity; for delayed or accelerated sexual maturity; and for color, squamation, body conformation, and many other attributes that could make culture easier, increase production, and enhance the acceptability of the product in the market. Kirpichnikov (1981) suggested that selection can be used to improve anadromous or marine fishes. He mentioned faster growth of young, increased fertility, disease resistance during freshwater life, better growth and survival in the marine habitat, and shorter duration of marine life in anadromous species. Mention is made in Chapter 39 of programs for the production of improved breeds of Atlantic salmon (*Salmo salar*) and gilthead seabream (*Sparus aurata*) that exemplify the sophistication of applied genetic studies in fish culture. The dramatic increase in genetic engineering of aquaculture species has led to some concerns about possible adverse impacts on the gene pools of naturally occurring stocks, however.

EMBRYONIC AND EARLY DEVELOPMENT IN FISHES

Embryology

Studies on the development of fishes have greatly benefited from an integrative approach, in which the roles of genetics, evolution, and environment are all perceived as instrumental in the understanding of embryogenesis. A modular approach, in which the developing embryo is partitioned into functional or organizational subunits, has proven valuable in the elucidation of evolutionary mechanisms at work in vertebrate developmental processes (Bolker, 2000; Gilbert and Bolker, 2001; Wilson and Maden, 2005). Studies on the development of fishes have confirmed a common genetic basis for much of the patterning of the vertebrate embryo. Investigations of

GOING DEEPER • Zebrafish: The New *Drosophila*

In 2000, the scientific community was abuzz with the news that two competing teams of researchers, the publicly funded International Human Genome Sequencing Consortium, and a private corporation, Celera Genomics, had completed a first draft of the human genome. This achievement has been touted as the most significant development in biology since Watson and Crick's elucidation of the structure of the DNA molecule almost 50 years earlier. One of the participants in the Human Genome Project, The Sanger Centre in Cambridge, England, has turned its attention to sequencing the genome of a 4-cm long cyprinid fish that has long been popular in the aquarium trade, *Danio* (= *Brachydanio*) *rerio* (Hamilton-Buchanan, 1822), the zebrafish (Roush, 1997; Vogel, 2000a). Just as T. H. Morgan revolutionized genetics and developmental biology with his studies of the fruit fly (*Drosophila melanogaster*) early in the last century, the zebrafish has become the definitive vertebrate model for the study of genetic processes controlling development. Invertebrates, such as the easily maintained *Drosophila* and the nematode worm *Caenorhabditis elegans,* have made extraordinary contributions to our understanding of genetic processes owing to the identification of a vast number of mutant forms that highlight key processes in development. Until fairly recently, however, a suitable vertebrate model was not available. The laboratory mouse is less than satisfactory, as its development is confined to the mother's uterus. The zebrafish has emerged as the ideal candidate.

The zebrafish is native to sluggish coastal streams of eastern India, with the Ganges River as its type locality. Hardy and prolific, zebrafishes are a favorite of tropical fish hobbyists, who have long enjoyed their active yet peaceful nature (Axelrod and Schultz, 1955; Sterba, 1966). Males tend to be slightly slimmer than females, and they are known to form long-lasting pair bonds (Paxton and Eschmeyer, 1998). Molecular phylogenetic studies have revealed that the zebrafish is more closely related to the pearl danio (*Danio albolineatus*) than to the giant danio (*D. aequipinnatus*) and that species of the genus *Rasbora* are more closely related to the danios than are species of the genera *Puntius* and *Cyprinus* (Meyer et al., 1993). Analysis of morphological characters has indicated that the genus *Danio* is, in fact, paraphyletic, consisting of two natural groups of species (Fang, 2003).

The transformation of the zebrafish into the "zebrafish system" began about 20 years ago in the laboratory of George Streisinger at the University of Oregon. Streisinger recognized the inherent advantages to working with the zebrafish. Here was a fish with a long history in

the behavior of specific genes, however, have revealed that fishes have pursued a number of distinct evolutionary pathways. For example, paired appendages were a key innovation in the evolution of the vertebrate body form (see Chapter 5, Going Deeper). Investigations of the molecular genetics of limb formation and secondary elimination among vertebrates have revealed that the mechanism resulting in the loss of pelvic fins in some populations of sticklebacks is different from that resulting in the same condition in pufferfishes (Tanaka et al., 2005).

In normal fertilization, only one sperm enters the egg through a tiny passage in the chorion, known as the **micropyle.** In the pink salmon, the external funnel is about 15 μm but narrows to about 1 μm as the tube passes through the chorion (Depêche and Billard, 1994). Even if more than one sperm enters an egg—a condition known as **polyspermy,** common in elasmobranchs and a few bony fishes—only one sperm is involved in the actual event of fertilization (Ginzburg, 1972). The great majority of fishes release sperm into the water in the vicinity of the eggs. Because the sperm of freshwater species lives only a short time following release into the water, it is important that the fishes bathe the eggs in a heavy concentration of sperm as the eggs are being extruded. Fish sperm can be held alive for a short time in physiological (0.9 percent) saline solution. Sperm of marine fishes is generally viable for a longer time than sperm of freshwater species. Fish culturists are interested in methods of long-term sperm storage, and techniques are constantly being improved. Sperm of several species has been successfully stored for several days at temperatures near freezing in suitable conditions (Stoss, 1983). Sperm frozen in appropriate supporting media at the temperature of liquid nitrogen has yielded some success in the fertilization of fish eggs (Stoss, 1983). Cryopreservation of sperm with liquid nitrogen has been accomplished with a number of species, including carp (Lubzens et al., 1993), the wels catfish (*Silurus glanis;* Linhart et al., 1993), yellow perch (Ciereszko et al., 1993), rainbow trout (Wheeler and Thorgaard, 1991), and others (Harvey, 1993).

The development of fertilized fish eggs follows much the same pattern as in other vertebrates. Most fish eggs are **macrolecithal** (or **telolecithal**), meaning that they are endowed with a dense concentration of yolk, but there is much variation among fishes regarding the actual amount of yolk. A few species have a small egg, with only a small quantity of yolk (**microlecithal**). The paucity of yolk permits cleavage

the aquarium fish trade. It is easily kept, has a short generation time (2 to 3 months) and produces large clutches of eggs (100 to 200 per mating). Because the larvae are transparent, genetic anomalies in early developmental stages are readily discernible (Detrich et al., 1999). Streisinger and his associates cloned the zebrafish, developed mutagenic procedures, and developed the protocols for genetic mapping and assessing the genetic mechanisms governing development. In 1994, John Postlethwait and his team of researchers at the University of Oregon Neuroscience Institute published the first gene map of the zebrafish based on meiotic recombination (Postlethwait et al., 1994, 1999). Nobel Laureate Christiane Nüsslein-Volhard at the Max Planck Institute for Developmental Biology, and Wolfgang Driever at Massachusetts General Hospital, both veteran *Drosophila* geneticists, completed a laborious screening of almost two million embryos to isolate and characterize 2,000 developmental mutants that are essential for genetic mapping studies (Kahn, 1994). The December 1996 issue of *Development,* issued as a separate volume (123), was devoted entirely to research involving these mutants. The zebrafish genome has proven itself invaluable in countless genetic studies. For example, the zebrafish has provided tremendous insight into cell signaling pathways and the origins of cell fate, the patterning of the embryo in early development, and the role of *hox* gene clusters—sets of regulatory genes that are widespread among animal phyla (see Chapters 2 and 28; Amores et al., 1998; Fishman, 2001; Heide et al., 1994; Wilson et al., 1993). The genetic basis of human disorders such as obesity and bone disease has been investigated using the zebrafish system (Vogel, 2000b).

The zebrafish has fostered the emergence of interdisciplinary approaches to biological research, as it provides model systems for a broad array of studies from developmental physiology to behavioral ecology. For example, boron is known to be essential for growth in vascular plants, yet its role in animals remains elusive. Recent studies using zebrafish embryos have demonstrated boron to be essential for normal embryogenesis (Rowe and Eckhert, 1999). Transgenic zebrafish have also proven their value in toxicological studies (Carvan et al., 2000). Zebrafishes have been used to assess the role of habitat complexity in aggressive interactions and monopolization of food resources (Basquill and Grant, 1998). The potential for the application of this wonderful new tool in biological research is currently limited only by the imagination of the researcher. Who knows what this little fish from India will tell us next?

of the entire egg during early embryogenesis (**holoblastic cleavage**). This is the case in the primitive cephalochordates (amphioxus), lampreys, some sturgeons, and lungfishes. For these fishes, the cleavage is unequal, with larger cells at the vegetal pole, where the yolk is concentrated.

In some fishes with macrolecithal eggs, the cytoplasm is thinly distributed around the relatively large yolk, whereas others have the cytoplasm concentrated at the animal pole (opposite the vegetal pole). The cytoplasm forms a polar cap at the site of the nucleus following fertilization, and this begins to divide, forming the embryo on the surface of the yolk (**meroblastic cleavage**). The evolution of meroblastic cleavage in teleosts is correlated with an overall decrease in egg size, with increases in egg size appearing to have independently developed in several lineages (Collazo, 1996; Collazo et al., 1994).

As in other vertebrates, **gastrulation** in fishes signals the early differentiation of the embryo into the germ layers (**ectoderm, mesoderm, and endoderm**). Vertebrate embryos are classically conceived of as having three germ layers (**triploblastic**), yet Hall (2000) argued that the **neural crest** is an independently developed germ layer, making vertebrates uniquely "quadroblastic." During gastrulation in sharks and teleosts, the mass of dividing cells known as the **blastoderm** grows over and eventually engulfs the yolk to become the yolk sac. A thickening of cells at the margin of the blastoderm becomes the **embryonic shield**, which is destined to become the embryo proper. The metameric arrangement of the embryo becomes visible with the appearance of the **mesodermal somites** (Fig. 27.8A–D). Gastrulation in teleosts is perceived as radically different from more primitive bony fishes, such as sturgeons, and from amphibians (Bolker, 1994; Collazo et al., 1994). Teleosts and lampreys share a peculiarity of embryology, in that the central nervous system forms by the hollowing of a **medullary keel** instead of by the formation of medullary (neural) folds, as in other vertebrates. Another interesting peculiarity in the development of some bony fishes is the suspension of embryogenesis (**diapause**) experienced by the eggs of certain "annual" cyprinodontids, which deposit fertilized eggs in the bottoms of drying ponds. These eggs do not complete their development until the ponds hold water again. *Nothobranchius* and *Aphyosemion* are genera with annual species in which the eggs might not hatch for several months.

Usually, the length of the incubation period is governed by temperature. Within the optimum range for normal development, the incubation period shortens as the temperature increases. For instance, trout and salmon eggs will hatch in about 50 days at 10°C, but at 2°C, incubation requires about six months. The eggs of the common carp incubate normally at temperatures of from 15 to 30°C, and hatching occurs in about a week at the lower temperature, but is shortened by several hours up to one day at the higher temperature. Within the genetic capability of the species, the incubation temperature influences the meristic features of the individual. These are features that are primitively tied to **metamerism**—the serial repetition of internal and external body features (refer to Fig. 27.8D for the formation of mesodermal somites)—and include vertebrae and myomeres, but also scales and fin rays. As a general rule, individuals of a selected batch of eggs that are incubated at low temperature will have more vertebrae, scale rows, and fin rays than their siblings incubated at higher temperatures. There are a few instances, however, in which the opposite has been found.

Hatching in many bony fishes is aided by secretions of special glands on the head or inside the mouth. These secretions are generally enzymatic in nature and weaken or even liquefy the chorion.

Parthenogenesis

Although the development of a fish egg normally begins on fertilization by a spermatozoon, it has been experimentally induced without sperm, especially by various chemical and physical techniques. These artificially activated eggs usually do not develop normally (Blaxter, 1969). **Parthenogenesis** (the development of unfertilized eggs) does occur naturally in some fishes (Echelle and Mosier, 1981; Solar et al., 1991). There are two types of parthenogenesis: **androgenesis,** in which the egg is activated by sperm and develops without any contribution from the egg nucleus; and **gynogenesis,** in which sperm activates the egg but contributes no genetic material to the developing embryo (see Chapter 28).

Early Life History

Largely because the subject is of such great importance to fishery scientists, the published literature on early life histories of fishes is growing tremendously. The following references can provide an entrée to the literature: Balon (1985a, 1985b); Copp et al. (1999); Hoar and Randall (1988a, 1988b); Kamler (1992); Lasker (1981a); Moser et al. (1984); Russell (1976); and Webb (1999). Hatching in fishes has been identified as proceeding in one of two ways: (1) mechanical disruption of the shell by chewing or locomotor activities of the emerging larva, or (2) dissolution of the shell by embryonic secretions. The importance of the second method in fishes has been confirmed through numerous studies that have identified cells on the surface of fish embryos that secrete hatching enzymes (Yamagami, 1988). Newly hatched fishes of oviparous species may be tiny, undeveloped creatures destined to undergo considerable additional growth and differentiation, as in the lampreys, or they may be essentially

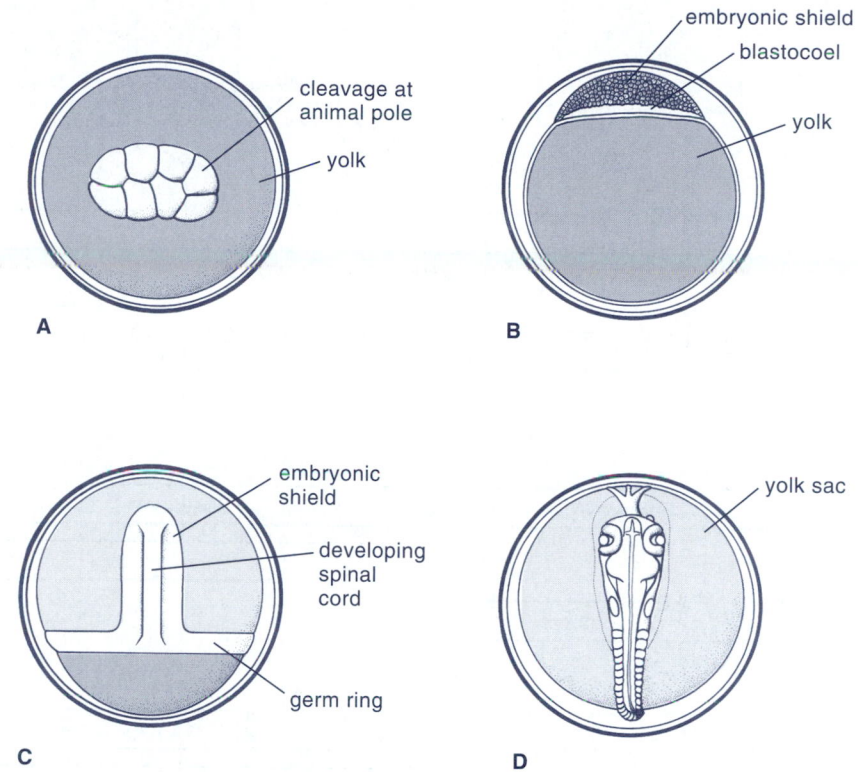

FIGURE 27.8

Examples of embryonic stages in teleosts: **A,** Cleavage; **B,** Blastula (sectioned); **C,** Embryonic shield; **D,** Organogeny with development of somites.

small replicas of the adults, as in oviparous sharks, rays, hagfishes, and some bony fishes. The larvae of freshwater teleosts generally are less distinctive and show a gradual transition to the juvenile stage, whereas those of marine species, usually with planktonic eggs and larvae, show a number of unusual adaptations and a pronounced metamorphosis into juveniles (Matarese et al., 1989; Webb, 1999).

Discussions of early life history stages of fishes have suffered somewhat from a lack of standardized terminology, and the diverse nature of very young fishes seems to prevent the ready acceptance of any standardization of terms. Early development is usually divided among the egg, larval, and juvenile developmental stages (Fig. 27.9; Kendall et al., 1984). According to Kendall et al. (1984), the egg stage is divided into early (ending with blastopore closure), middle (ending with freeing of the tailbud from the yolk), and late (ending with hatching) subdivisions. Following hatching, there is a transitional stage, the *yolk sac larva* (Fig. 27.10A), which ends with the absorption of the yolk sac. The next developmental subdivision is the *preflexion larva*, which ends

when the notochord begins to flex upward (Fig. 27.10B). This is followed by the *flexion* and *postflexion* larval subdivisions, following which the larva begins metamorphosis in the transitional stage, which ends when the fish has its full complement of fin rays and loses its larval characteristics. It then acquires juvenile characteristics.

Some workers consider the embryonic period to last until hatching or parturition, and others consider it to last until the young begin to feed for themselves. Hubbs (1943) proposed the simplest and most straightforward terminology, in which larval stages are defined as those beyond the embryo but still distinct from the juvenile, which is essentially a small version of the adult. Hubbs recognized two larval stages: the **prolarva,** which retains a yolk sac, and the **postlarva,** which has absorbed the yolk sac but is still unlike the juvenile stage. The upright-swimming larvae of flatfishes or the leptocephali of eels are examples of postlarvae. If yolk-bearing larvae transform directly into a juvenile, as is the case in many salmonids and certain sculpins, these larvae are called **alevins.** They are the "sac fry" of salmonid culturists.

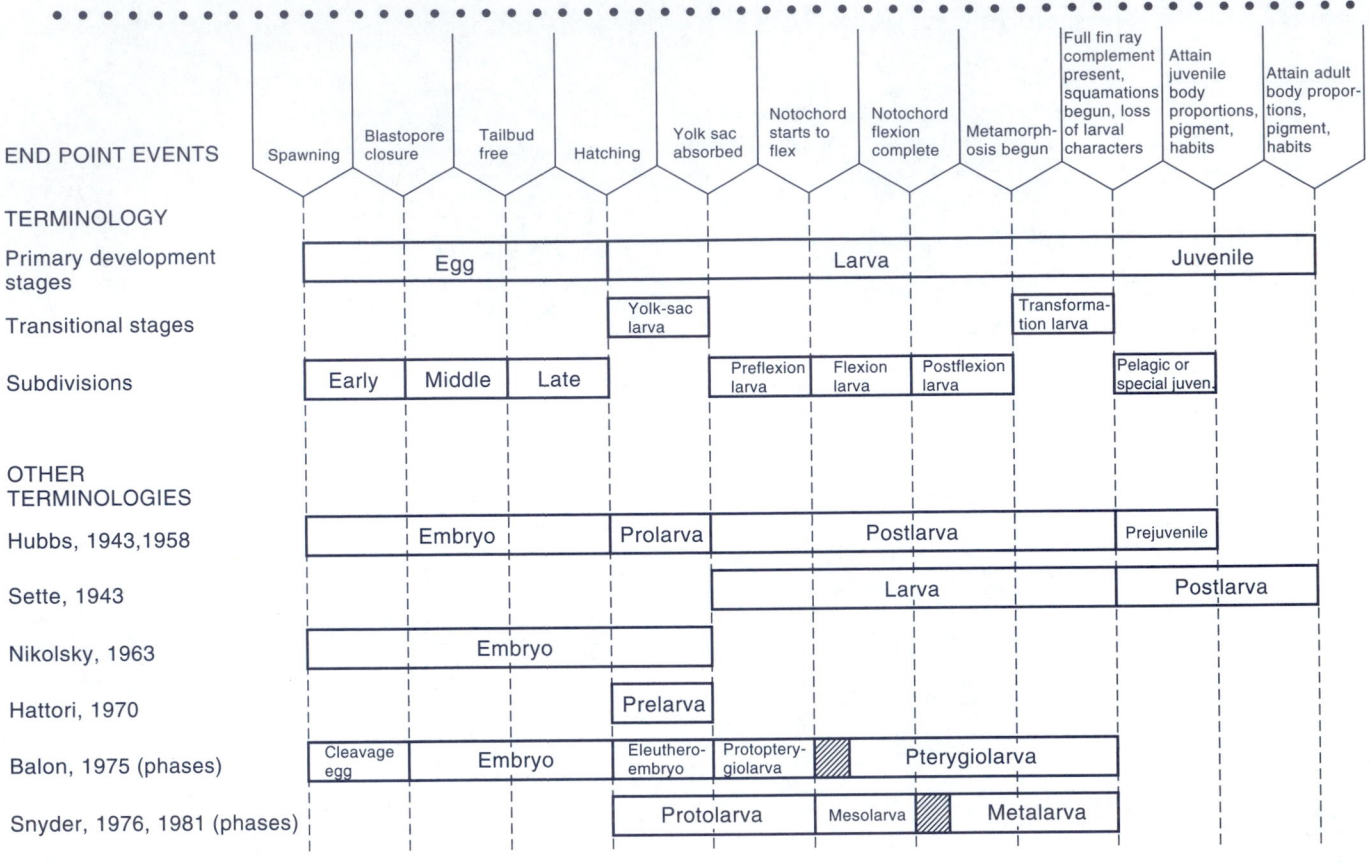

FIGURE 27.9
Terminology of life history stages of fishes (Source: Kendall et al., 1984).

A terminology suggested by Balon (1975b) divides the embryonic period into three phases—cleavage egg, embryo, and *eleutheroembryo* ("free embryo"), which is free from the egg. The term "free embryo" is not universally accepted. A newly hatched fish may be termed a *larva* (Kamler, 1992), or *larva* may simply refer to the stage between hatching and metamorphosis (Webb, 1999). The early larval phase, featuring undifferentiated fin folds, is called *apterolarva* by Balon. The later larval phase, with fin rays forming, is then called *pterolarva* (Balon, 1985b).

Balon (1979, 1981, 1985b, 1986) discussed the theory of **saltatorial ontogeny** in fishes, which questions whether a typical fish life history comprises a succession of barely noticeable small changes or a stepwise series of longer, steady states interspersed with rapid changes in function and form. He recognized three life history models—indirect, transitory, and direct—and divided the saltatorial ontogeny into up to five periods, and these periods into phases and steps. The embryo period has three phases in all three models:

cleavage, with three steps; embryo phase, with four steps; and free embryo, with three phases. The indirect model has a larval period that consists of the fin-fold phase and the fin-formed phase. The transitory model has what Balon termed the alevin period and phase, whereas the direct model has no larval or alevin period. The juvenile period is common to all three models, but in the transitory model, it is divided into parr, smolt, and juvenile phases. Other, subsequent periods in the three models are adult and senescent.

Adaptations of Larval Fishes

A marvelous array of adaptations to the environment is seen among fish larvae. Some of these adaptations involve structures and shapes that are entirely unlike those in the juvenile stage and require an extensive metamorphosis. Lampreys, for example, undergo great internal changes as well as some obvious external modifications during metamorphosis. They lose the functional gallbladder and bile ducts, grow a new esophagus as the respiratory tube disconnects from

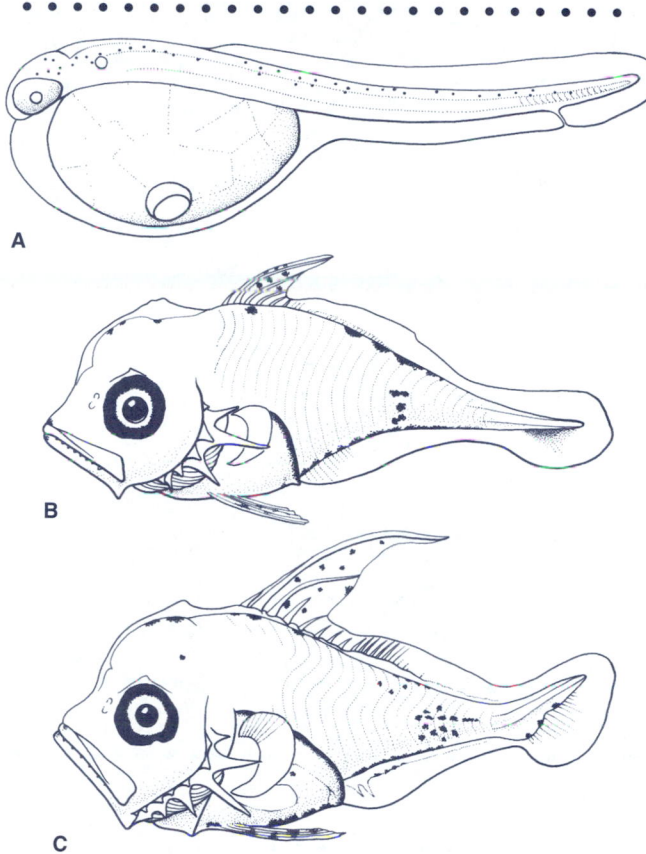

FIGURE 27.10

Examples of fish larvae: **A,** Yolk sac larva of *Brevoortia patronus* (2.6 mm); **B,** Preflexion larva of *Selene vomer* (3.2 mm); **C,** Postflexion larva of *Selene vomer* (3.9 mm) (Sources: *A* from Hettler, 1984; *B* and *C* from Aprieto, 1974).

the alimentary canal, gain functional eyes, lose the oral hood and filtering sieve (replacing them with an oral disc set with horny teeth), and acquire larger medial fins. The larval pronephric kidney is replaced by the adult mesonephros (opisthonephros; Hardisty, 1979). Eels change from the toothy,

leaflike, transparent leptocephalus (Fig. 27.11), to transparent "glass eels," to the juvenile or elver (Castle, 1984; Tesch, 1977). Both lampreys and eels shrink considerably in length from larva to juvenile. Examples of remarkable metamorphosis are numerous, including the following: the flatfishes, which change the entire architecture of the head as the eye on one side migrates to the other side (Ahlstrom et al., 1984); the molas, in which most of the caudal region of the body is lost during larval development, so that a normal caudal fin does not form (see Figs. 17.3, 17.4); and the swordfishes, which as larvae have pronounced, toothy jaws (Fig. 27.12). As larvae, many fishes—including the molas, swordfishes, and many other perciform and scorpaenid genera—have very large spines in relation to their size (Fig. 27.12). These spines are thought to represent a protection against predation (Moser, 1981).

Many larvae show special adaptations for respiration, ranging from highly vascular fin folds, pectoral fins, or yolk sacs to the featherlike true external gills of *Polypterus, Protopterus,* and *Lepidosiren* (Goodrich, 1930). Gill filaments project from the gill openings of several species in the embryonic or larval stages of fishes, including mormyroids, loaches, and some elasmobranchs. The larvae of *Protopterus annectans* begin to breathe air at 23 to 27 mm (Greenwood, 1987). Vision-enhancing adaptations are seen in some larvae that have elliptical eyes, believed to have better rotation around the long vertical axis than round eyes, so that a larger field of vision can be attained, whereas others have eyes mounted on stalks (Fig. 27.13). Some larvae, such as those of *Protopterus, Lepidosiren, Amia,* and *Lepisosteus,* have attachment or adhesive organs.

Many pelagic larvae are specially modified for flotation, so that they maintain a specific depth or range of depth. Persistent fin folds, gelatinous enclosures for the body, or elaborate fins providing enhanced surface area may enhance buoyancy in planktonic forms (Moser, 1981). In the larvae of many fishes, sinking is retarded by a high ratio of surface to volume. In these larvae, the fins may be of exceptionally large size, the fin rays or guts may extend into long, trailing filaments (see Figs. 14.2, 27.14), or the body may be

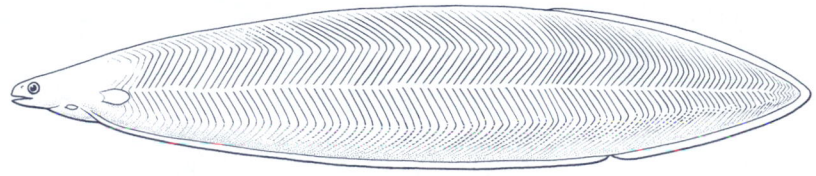

FIGURE 27.11

Leptocephalus larva of *Anguilla*.

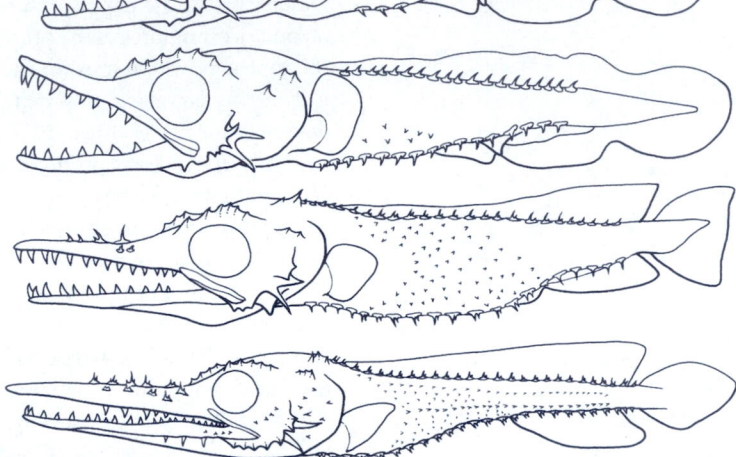

FIGURE 27.12
Spinous larvae of swordfish, *Xiphius gladius*. (Source: Potthoff and Kelley, 1982.)

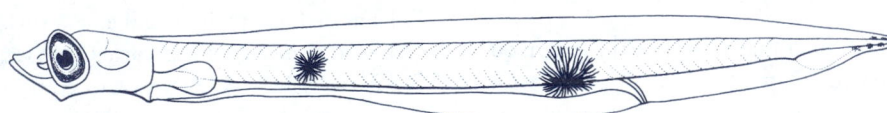

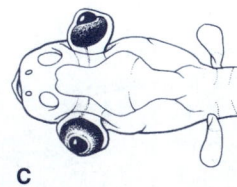

FIGURE 27.13
Examples of larvae with modified eyes: **A,** *Symbolophorus californiense,* with elliptical eyes on short stalks (9.6 mm); **B,** *Leuroglossus schmidti,* with elliptical eyes (11.7 mm); **C,** Dorsal view of head of *L. schmidti* (SL = 14.6 mm). (Sources: *A* from Moser and Ahlstrom, 1974; *B* and *C* from Dunn, 1983.)

covered with spines—although those spines may have more importance in deterring predators. Oil globules are effective in conferring hydrostatic balance and are common in drifting larvae. The inclusion of a large proportion of water in the flesh, as found among leptocephali, is another common flotation device.

The formation of pigmentation in fishes has attracted the attention of developmental biologists interested in understanding the underlying genetic mechanisms controlling pattern determination in vertebrates (see McClure, 1999). Distinctive patterns of melanophore deposition (see Chapter 18) occur in the larvae of several groups of fishes. Mostly, fishes with heavy pigmentation, either black or yellow, live close to the surface and may require protection from ultraviolet light (Moser, 1981). Some patterns of pigmentation may prevent the refraction of light from gut contents or from gas bladders. Others may aid in breaking up body outlines or otherwise making the larvae less noticeable to predators (Fig. 27.15).

In exceptional instances, fishes with larval characteristics mature sexually and reproduce. The retention of larval or juvenile characteristics in the mature organism is known as **neoteny**. Sexual maturation while still in the larval or juvenile form is known as **progenesis** or **paedomorphosis**. This phenomenon is common in the icefishes (Salangidae) of the coasts of China and Korea. Certain sauries and needlefishes are neotenic, and a tendency toward this condition is seen in some brook lampreys, in which significant gonadal development occurs while the individual is still in the ammocoete stage. Some of the best examples of neotenic fishes are the members of the gobioid genus *Schindleria* of the central Pacific Ocean (see Chapter 16). These tiny, transparent fishes reach about 20 mm in length and retain several larval characteristics, including functional pronephric kidney tubules, opercular gills, and a heart in which the atrium remains behind the ventricle instead of being folded over it, as normally occurs later in development (Brunn, 1940; Gosline, 1959; Watson et al., 1984).

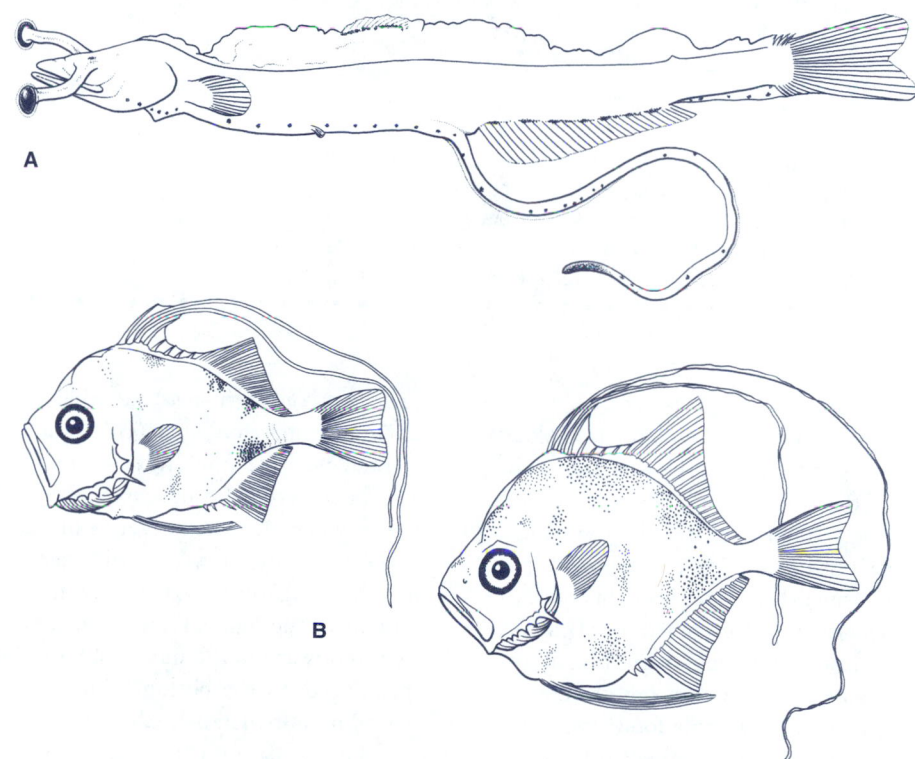

FIGURE 27.14
Examples of larvae with filamentous structures: **A,** *Myctophum aurolateratum* with stalked eyes and filamentous gut (26 mm); **B,** *Selene vomer* with filamentous fin rays (7.7 mm and 9.0 mm). (Sources: *A* from Moser and Ahlstrom, 1974; *B* from Aprieto, 1974.)

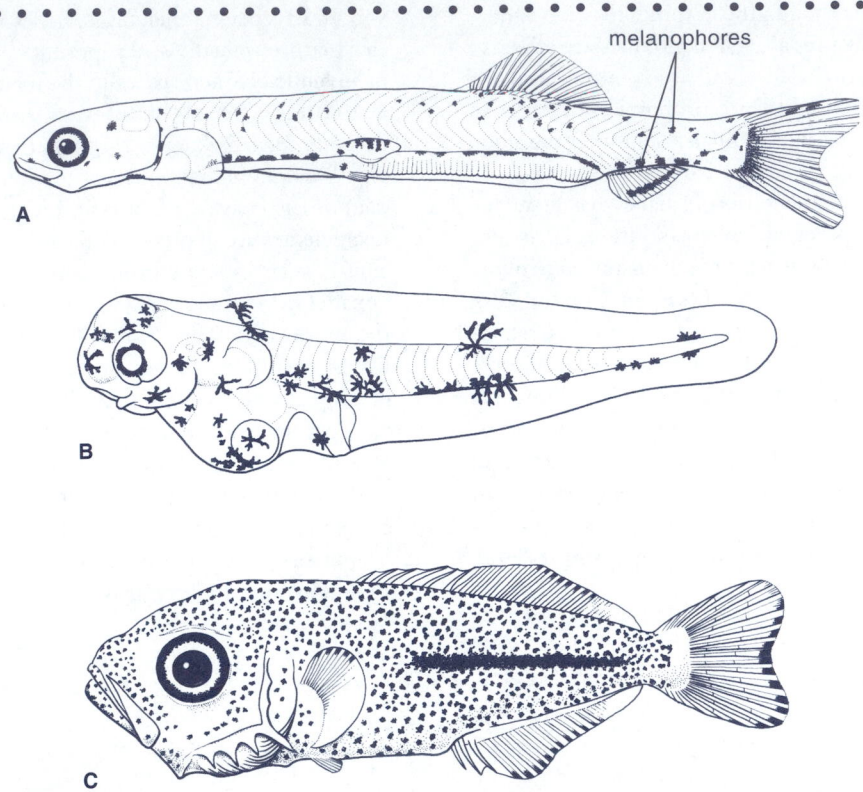

FIGURE 27.15
Examples of melanophore patterns in larvae: **A,** Larva of gulf menhaden (*Brevoortia patronus;* 16.5 mm),
showing melanophores over intestine and gas bladder as well as body and fins. Note post-anal series above
anal fin. **B,** Larva of white croaker (*Genyonemus lineatus;* 2.38 mm), showing a melanophore pattern that
may tend to break up the body outline; **C,** Heavily pigmented larva of leatherjacket (*Oligoplites saurus;*
SL = 5.7 mm). (Sources: *A* from Hettler, 1984; *B* from Watson, 1982; *C* from Aprieto, 1974.)

Ecology of Early Life History Stages

Space permits only a cursory look at the complex ecology
of the early life history stages of fishes. The proceedings of
the Fisheries Society of the British Isles 1997 symposium
on ichthyoplankton ecology (Geffen et al., 1997) provide a
good introduction to current research on the ecology of pelagic larval fishes. For marine fishes, a stable ocean environment appears to be necessary for optimum survival (Lasker,
1981b). Unusual or unseasonable storms, disruption of upwelling patterns, and changes in currents can change the
water quality or the abundance of suitable food organisms.
The young of freshwater species can be affected by many
weather-related circumstances, such as scouring floods and
fluctuations in water temperature or water level.

Although yolk sac larvae and early feeding larvae
may seem capable of only minor directed movements, experiments have shown that the larvae of walleye pollock
(*Theragra chalcogramma*) are able to move in relation to
environmental gradients (Olla and Davis, 1990). Larvae
showed positive phototaxis by moving horizontally from
a darkened area of a tank to an area of low light intensity,
but they demonstrated negative phototaxis under high light
intensity. They showed a daily periodicity in response to
light, moving toward the surface in the dark and toward the
bottom in the light. They avoided surface turbulence and
moved upward in the water column when chilled water was
introduced on the bottom. Phototactic responses are discussed in more detail in Chapter 37.

The transition from dependence on yolk to the ingestion of foods obtained from the environment has been called
one of the critical periods in the life of a fish (Kamler, 1992),
although Miller et al. (1988) suggested that the increased
mortality at this time may simply be due to the increased
vulnerability of smaller larvae to starvation. Assuming that

the larva has successfully navigated the hazardous waters of early life history, avoiding predation and transport to environments unfavorable to further development, it now initiates the ingestion of small, digestible particles that will supplement and eventually replace the yolk. Larval condition at this critical time rests largely on the ability to rapidly convert ingested food into biomass. A variety of assessment techniques, including measurement of the RNA/DNA ratio, C/N ratio, relative lipid content, or otolith growth indices, have been employed in the determination of larval condition (Suthers, 2000; Westernhagen et al., 1998).

How and where larvae feed is limited by the developmental status of the sensory systems and the feeding apparatus, including gape size, and by the hydrodynamic constraints of food capture (Webb, 1999). Food for fish larvae at first feeding must be small, because many larvae do not exceed 3 mm long at that stage. The initial food of larvae may include phytoplankton, copepods, ciliates, and mollusc larvae. Phytoplankton usually serves as a food source for only a brief time, but the northern anchovy (*Engraulis mordax*) can feed on dinoflagellates for up to 20 days after hatching, although then its growth rate is depressed (Hunter, 1981). Zooplankton, especially copepods, make up the most important segment of the food of planktonic larvae. The width of prey taken at the beginning of feeding for most clupeoid larvae is 50 to 100 µm, but the upper limit increases to about 200 µm as the larvae grow from about 3–4 mm to 7–10 mm. Larvae of piscivorous fishes, such as mackerels, tend to feed on larger prey than that eaten by the clupeoids and may become cannibalistic at 5 to 10 mm (Hunter, 1981; Nishimura and Hoshino, 1999).

Density of prey is of great importance to the survival of fish larvae at the onset of feeding. The locomotor powers of larvae are not great, and prey is detected at one body length or less during early feeding. Laboratory studies cited by Hunter (1981) indicated that densities of 220 to 4,000 copepod nauplii (a larval stage of crustaceans) per liter are necessary for 50 percent survival of fish larvae stocked at densities from 0.2 to 50 individuals per liter. Densities of nauplii in the open ocean are usually less than 40 individuals per liter. In enclosed waters, such as estuaries and lagoons, average copepod concentrations can be 200 individuals per liter or more. Zooplankton density in freshwater has also been found to correlate with the growth and survival of fish larvae (Welker et al., 1994). The distribution of plankton is by no means uniform, so that fish larvae that are found in dense patches of plankton are generally well fed (Hunter, 1981). Predation on eggs, larvae, and juveniles is of great importance to the survival of fish to maturity (Hunter, 1981; Nellen, 1986). Many pelagic animal taxa are capable of ingesting drifting eggs and yolk sac larvae, and some can pursue and catch swimming larvae. Predators include cnidarians (including siphonophores and chondrophores); comb jellies (Ctenophora); squids; pelagic hyperiidian and calanoid copepods; euphausid shrimp; arrowworms (Chaetognatha); tunicates, such as salps, doliolids, and pyrosomes; and fishes, which are probably the most important predators (Nellen, 1986).

Vulnerability to predation depends on many factors. Where and when the eggs are deposited is of great importance. Many fishes are nocturnal spawners and spawn in areas of drift or current, which gives pelagic eggs a chance to disperse before being exposed to diurnal predators. Diurnal spawners tend to release their eggs into currents or in parts of the ocean with few large planktonic predators (Hunter, 1981). Fishes that deposit demersal eggs that hatch into pelagic larvae usually spawn where currents will aid in the dispersal of the larvae.

As mentioned earlier, many species, both marine and freshwater, guard demersal eggs at least until hatching; and some guard demersal young, so that predation is minimized. In some of these fishes, such as the freshwater sculpins (in which the young are not guarded for any length of time), the yolk sac larvae find refuge under stones downstream from the nest stone. The alevins of trout and salmon remain buried in the gravel nests until the yolk is nearly used up and they are able to swim and seek food.

The size of eggs and larvae is important, in that small invertebrate predators may not be able to ingest larger eggs and larvae. Rapid growth of larvae will place them beyond the abilities of smaller predators to ingest them. The maturation of sensory organ systems aids in the detection of predators. Escape from predators is facilitated by the development of the locomotor apparatus so that the larvae can swim faster and longer. Nellen (1986) emphasized the importance of predation by fishes on eggs and larvae and advanced the hypothesis that high fecundity and cannibalism are of considerable importance in the ecology and recruitment of fishes.

Critical to fisheries managers is an understanding of the duration of planktonic larval life. For pelagic species with broad ranges across the open ocean, the widespread dispersal of larvae is not inconsistent with the distribution of the juveniles and adults. But what of species with more restricted distributions, such as coral reef fishes? How does a pelagic, planktonic larval life history permit juveniles and adults to ultimately settle in discrete coastal habitats, such as bays and estuaries or coral reefs? Some studies have suggested that the conventional wisdom that pelagic larval life necessarily means dispersal far from the location of birth is not necessarily correct. Larval coral reef fishes can be recruited to the very same reefs from which they were spawned (Jones et al., 1999; Swearer et al., 1999).

KEY POINTS AND CONNECTIONS

• As in other vertebrates, gamete production in fishes is accomplished by the testes and ovaries, which are suspended in the body cavity. The shared developmental history of the gonads and the urinary system is evident from their common designation as the urogenital system. The means to channel gametes out of the body vary significantly among the major taxa of fishes. Whereas agnaths lack ducts dedicated to the transmission of gametes, chondrichthyan and osteichthyan fishes generally have such ducts. Chondrichthyan and some bony fishes share one feature of reproductive design with terrestrial vertebrates — the eggs are released into the coelomic cavity before being received and channeled down the oviduct.

The "uro" component of the urogenital system is discussed in Chapter 25.

• Reproductive effort refers to the investment of energy in reproduction, whereas reproductive strategy refers to those features that optimize the number of offspring produced. Reproductive strategy encompasses all aspects of the lives of fishes, including morphological, physiological, ecological, and behavioral adaptations. Compared with other vertebrate classes, fishes are known for their high reproductive output, measured as fecundity. Mate selection is facilitated by visual or sonic cues or by chemical signals.

As an example of the integration of morphological, physiological, and behavioral contributions to reproductive strategy, the role of sound production and reception in reproduction is discussed in Chapter 21. Mating behavior in fishes is discussed in Chapter 37.

• Although most fishes are oviparous, many groups practice egg retention as a means of enhancing survival of the young. This may take the form of ovoviviparity (lecithotrophy), in which the gestating egg still receives nutrient from the yolk, or true viviparity (matrotrophy), in which nourishment comes from maternal contributions. In the latter condition, fishes may develop a form of placenta, across which nutrients may pass from mother to embryo.

Brooding and parental care in fishes are described in Chapter 37.

• Although most fishes are classified as gonochoristic, several families include members that are known to be hermaphroditic. Hermaphroditism may be sequential, in which the animal may sexually mature as one sex, only to change sex later in life. A few fishes are simultaneous hermaphrodites, in which both testes and ovaries develop at the same time. Self-fertilization in such cases is extremely rare, however.

The genetic aspects of sex determination, including hermaphroditism, are discussed in Chapter 28.

• Studies on the development of fishes have revealed features that are typical of vertebrate embryogenesis, but have also revealed that fishes have pursued a number of distinct evolutionary pathways. Gastrulation signals the onset of differentiation of tissues and organs and continues with the development and differentiation of mesodermally derived somites. Early life history stages are critical, as it is during this time that fishes may be most vulnerable to environmental perturbations.

The relationship between spawning success and the recruitment of commercially significant fish stocks is discussed in Chapter 39.

FISH LINKS

http://zfin.org/zf_info/zfbook/zfbk.html Developed by Monte Westerfield at the Institute of Neuroscience, University of Oregon, this—the definitive site on zebrafish—is an online version of Westerfield (1995).

BUILDING AN ICHTHYOLOGY LIBRARY

Chambers, R.C., and E.A. Trippel (Eds.). 1997. *Early life history and recruitment in fish populations.* Chapman and Hall, London.

Detrich, H. W., III, M. Westerfield, and L. I. Zon (Eds.). 1999. *The zebrafish: Biology,* Vol. 59, Methods in cell biology. Academic Press, San Diego.

———, ———, and———. 1999. *The zebrafish: Genetics and genomics,* Vol. 60, Methods in cell biology. Academic Press, San Diego.

Hall, B. K., and M. H. Wake (Eds.). 1999. *The origin and evolution of larval forms.* Academic Press, San Diego.

Hoar, W. S., and D. J. Randall (Eds.). 1988. *Fish physiology,* Vol. XI, The physiology of developing fish, Part A (eggs and larvae). Academic Press, San Diego.

Hoar, W. S., and D. J. Randall (Eds.). 1988. *Fish physiology,* Vol. XI, The physiology of developing fish, Part B (viviparity and posthatching juveniles). Academic Press, San Diego.

Kunz, Y. W. 2004. *Developmental biology of teleost fishes,* Vol. 28, Fish and Fisheries Series. Springer Verlag, Berlin.

REFERENCES

Ahlstrom, E. H., K. Amaoka, D. A. Hensley, H. G. Moser, and B. Y. Sumida. 1984. Pleuronectiformes: Development, pp. 640–670. In Ontogeny and systematics of fishes, H. G. Moser, W. J. Richards, D. M. Cohen, M. P. Fahay, A. W. Kendall, Jr., and S. L. Richardson (Eds.), *Am. Soc. Ichthyol. Herpetol. Spec. Publ. 1.*

Alonzo, S. H., and R. R. Warner. 2000. Female choice, conflict between the sexes and the evolution of male alternative reproductive behaviours. *Evol. Ecol. Res. 2:* 149–170.

Amores, A., A. Force, Y.-L. Yan, L. Joly, C. Amemiya, A. Fritz, R. K. Ho, J. Langeland, V. Prince, Y.-L. Wang, M. Westerfield, M. Ekker, and J. H. Postlethwait. 1998. Zebrafish *hox* clusters and vertebrate genome evolution. *Science 282:* 1711–1714.

Aprieto, V. L. 1974. Early development of five carangid fishes of the Gulf of Mexico and the south Atlantic coast of the United States. *Fish. Bull. 72:* 415–443.

Atkins, D. L., and W. L. Fink. 1979. Morphology and histochemistry of the caudal gland of *Corynopoma riisei* Gill. *J. Fish Biol. 14:* 465–469.

Atz, J. W. 1964. Intersexuality in fishes, pp. 145–232. In *Intersexuality in vertebrates including man,* C. N. Armstrong and A. J. Marshall (Eds.). Academic Press, New York.

Axelrod, H. R., and L. P. Schultz. 1955. *Handbook of tropical aquarium fishes.* McGraw-Hill, New York.

Balon, E. K. 1975a. Reproductive guilds of fishes: A proposal and definition. *J. Fish. Res. Bd. Can. 32:* 821–864.

———. 1975b. Terminology of intervals in fish development. *J. Fish Res. Bd. Can. 32:* 1663–1670.

———. 1979. The theory of saltation and its application to the ontogeny of fishes: Steps and thresholds. *Env. Biol. Fishes 4:* 97–101.

———. 1981. Saltatory processes and altricial to precocial forms in the ontogeny of fishes. *Amer. Zool. 21:* 573–576.

———. 1984. Patterns in the evolution of reproduction styles in fishes, pp. 35–53. In *Fish reproduction: Strategies and tactics*, G. W. Potts and R. J. Wooten (Eds.). Academic Press, London.

———. 1985a. *Early life histories of fishes: New developmental, ecological, and evolutionary perspectives.* Junk, Dordrecht, the Netherlands.

———. 1985b. The theory of saltatory ontogeny and life history models revisited, pp. 18–28. In *Early life histories of fishes*, E. K. Balon (Ed.). Junk, Dordrecht, the Netherlands.

———. 1986. Saltatory ontogeny and evolution. *Riv. Biol.—Biol. Forum 79:* 151–190.

———. 1991a. Probable evolution of the coelacanth's reproductive style: Lecithotrophy and orally feeding embryos in cichlid fishes and in *Latimeria chalumnae. Env. Biol. Fishes 32:* 249–265.

———. 1991b. Prelude: The mystery of a persistent life form, pp. 9–13. In *The biology of* Latimeria chalumnae *and evolution of coelacanths*, J. A. Musick, M. N. Bruton, and E. K. Balon (Eds.). Kluwer, Dordrecht, the Netherlands.

Barlow, G. W. 2000. *The cichlid fishes: Nature's grand experiment in evolution.* Perseus, Cambridge, MA.

Basquill, S. P., and J. W. A. Grant. 1998. An increase in habitat complexity reduces aggression and monopolization of food by zebrafish, *Danio rerio. Can. J. Zool. 76:* 770–772.

Billard, R. 1981. The reproductive cycle in teleost fish (Le cycle reproducteur chez les poissons Teleostéens). *Cah. Lab. Hydrobiol. Montereau 12:* 43–56.

———, A. Fustier, C. Weil, and B. Breton. 1982. Endocrine control of spermatogenesis in teleost fish. *Can. J. Fish. Aquat. Sci. 39:* 65–79.

Blaxter, J. H. S. 1969. Development: Eggs and larvae, pp. 177–252. In *Fish physiology*, Vol. III, W. S. Hoar and D. J. Randall (Eds.). Academic Press, New York.

Blumel, L. 1979. Paternal care in the bony fishes. *Quart. Rev. Biol. 54:* 149–161.

Bolker, J. A. 1994. Comparison of gastrulation in frogs and fish. *Amer. Zool. 34:* 313–322.

———. 2000. Modularity in development and why it matters to evo-devo. *Amer. Zool. 40:* 770–776.

Bortone, S. A., W. P. Davis, and C. M. Bundrick. 1989. Morphological and behavioral characters in mosquitofish as potential bioindication of exposure to kraft mill effluent. *Bull. Env. Contam. Toxicol. 43:* 370–377.

Breder, C. M., Jr., and D. E. Rosen. 1966. *Modes of reproduction in fishes.* Natural History Press, Garden City, NY.

Brodal, A., and R. Fänge (Eds.). 1963. *The biology of* Myxine. Universitetsforlaget, Oslo.

Brunn, A. F. 1940. A study of the fish *Schindleria* from South Pacific waters. *Dana Rep. 21:* 1–12.

Burr, B. M., and J. N. Stoeckel. 1999. The natural history of madtoms (genus *Noturus*), North America's diminutive catfishes. Catfish 2000. *Proc. Am. Fish. Soc. Symp. 24:* 51–101.

Carscadden, J. E., and W. C. Leggett. 1975. Life history variations in populations of American shad, *Alosa sapidissima* (Wilson), spawning in tributaries of the St. John River, New Brunswick. *J. Fish Biol. 7:* 595–609.

Carvan, M. J., III, T. P. Dalton, G. W. Stuart, and D. W. Nebert. 2000. Transgenic zebrafish as sentinels for aquatic pollution. *Ann. N. Y. Acad. Sci. 919:* 133–147.

Castle, P. H. J. 1984. Notacanthiformes and Anguilliformes: Development, pp. 62–102. In Ontogeny and systematics of fishes, H. G. Moser, W. J. Richards, D. M. Cohen, M. P. Fahay, A. W. Kendall, Jr., and S. L. Richardson (Eds.), *Am. Soc. Ichthyol. Herpetol. Spec. Publ. 1.*

Chevassus, B. 1983. Hybridization in fish. *Aquaculture 33:* 245–262.

Ciereszko, A., L. Ramseyer, and K. Dabrowski. 1993. Cryopreservation of yellow perch semen. *Prog. Fish. Cult. 55:* 261–264.

Cole, L. C. 1954. The population consequences of life history phenomena. *Quart. Rev. Biol. 29:* 103–137.

Collazo, A. 1996. Evolutionary correlations between early development and life history in plethodontid salamanders and teleost fishes. *Amer. Zool. 36:* 116–131.

———, J. A. Bolker, and R. Keller. 1994. A phylogenetic perspective on teleost gastrulation. *Am. Nat. 144:* 133–152.

Compagno, L. J. V. 1990. Relationships of the megamouth shark, *Megachasma pelagios* (Lamniformes: Megachasmidae), with comments on its feeding habits, pp. 357–380. In Elasmobranchs as living resources, H. L. Pratt, Jr. (Ed.), S. H. Fisheries Service, *NOAA Tech Rep. 90.*

Conover, D. O., and S. W. Heins. 1987a. Adaptive variation in environmental and genetic sex determination in a fish. *Nature 326:* 496–498.

———and———. 1987b. The environmental and genetic components of sex ratio in *Menidia menidia* (Pisces: Atherinidae). *Copeia 1987:* 732–743.

———, and B. E. Kynard. 1981. Environmental sex determination: Interaction of temperature and genotype in a fish. *Science 213:* 577–579.

Copp, G. H., V. Kováč, and K. Hensel (Eds.). 1999. When do fishes become juveniles—looking beyond metamorphosis to juvenile development. *Env. Biol. Fishes 56(1–2).*

Crespi, B. J., and R. Teo. 2002. Comparative phylogenetic analysis of the evolution of semelparity and life history in salmonid fishes. *Evolution 56:* 1008–1020.

Daniel, J. F. 1934. *The elasmobranch fishes.* University of California Press, Berkeley.

Dean, B. D. 1906. Chimaeroid fishes and their development. *Pub. Carnegie Inst. Washington 32:* 1–195.

Depêche, J., and R. Billard. 1994. *Embryology in fish, a review.* Société Française d'Ichtyologie, Paris.

Detrich, H. W., III, M. Westerfield, and L. I. Zon. 1999. Overview of the zebrafish system, pp. 3–10. In *The zebrafish: Biology*, H. W. Detrich, III, M. Westerfield, and L. I. Zon (Eds.). Academic Press, San Diego.

Dodd, J. M. 1983. Reproduction in cartilagenous fishes (Chondrichthyes), pp. 31–95. In *Fish physiology*, Vol. IX, Part A, W. S. Hoar, D. J. Randall, and E. M. Donaldson (Eds.). Academic Press, New York.

Donaldson, E. M., and G. A. Hunter. 1983. Induced final maturation, ovulation, and spermiation in cultured fish, pp. 351–403. In *Fish physiology*, Vol. IX, Part B, W. S. Hoar, D. J. Randall, and E. M. Donaldson (Eds.). Academic Press, New York.

Dulvey, N. K., and J. D. Reynolds. 1997. Evolutionary transitions among egg-laying, live-bearing and maternal inputs in sharks and rays. *Proc. Roy. Soc. Lond. B 264:* 1309–1315.

Dunn, J. R. 1983. Development and distribution of the young of northern smoothtongue, *Leuroglossus schmidti* (Bathylagidae), in the northeast

Pacific, with comments on the systematics of the genus *Leuroglossus* Gilbert. *Fish. Bull. 81:* 23–40.

Echelle, A. A., and D. T. Mosier. 1981. All-female fish: A cryptic species of *Menidia* (Atherinidae). *Science 212:* 1411–1413.

Eigenmann, C. H. 1894. On the viviparous fishes of the Pacific coast of North America. *Bull. U.S. Fish. Comm. 12:* 381–478.

Emerson, S. B. 2000. Vertebrate secondary sexual characteristics— physiological mechanisms and evolutionary patterns. *Am. Nat. 156*(1): 84–91.

Fabra, M., D. Raldúa, D. M. Power, P. M. T. Deen, and J. Cerdà. 2005. Marine fish egg hydration is aquaporin-mediated. *Science 307:* 545.

Fang, F. 2003. Phylogenetic analysis of the Asian cyprinid genus *Danio* (Teleostei, Cyprinidae). *Copeia 2003:* 714–728.

Fishman, M. C. 2001. Zebrafish—the canonical vertebrate. *Science 294:* 1290–1291.

Fox, P. J. 1978. Preliminary observations on different reproduction strategies in the bullhead (*Cottus gobio* L.) in northern and southern England. *J. Fish Biol. 12:* 5–11.

Frankel, O. H., and M. E. Soulé. 1981. *Conservation and evolution.* Cambridge University Press, Cambridge, UK.

Gall, G. 1983. Genetics of fish: A summary of conclusions. *Aquaculture 33:* 383–394.

Geffen, A. J., J. M. Fives, and J. E. Thorpe (Eds.). 1997. Ichthyoplankton ecology: The Fisheries Society of the British Isles annual symposium. *J. Fish Biol. 51*(Suppl. A).

Gilbert, S. F., and J. A. Bolker. 2001. Homologies of process and modular elements of embryonic construction. *J. Exp. Zool. 291:* 1–12.

Ginzburg, A. S. 1972. *Fertilization in fishes and the problem of polyspermy.* Academy of Sciences of the USSR. Translation by Israel Program for Scientific Translations, Jerusalem.

Gjedrem, T. 1983. Genetic variation in quantitative traits and selective breeding in fish and shellfish. *Aquaculture 33:* 51–72.

Gjerde, B. 1993. Breeding and selection, pp. 187–208. In *Salmon aquaculture,* K. Heen, R. L. Monohan, and F. Utter (Eds.). Blackwell, Oxford.

Goodrich, E. S. 1909. Cyclostomes and fishes. In *A treatise on zoology,* Part 9, Fasc. 1, R. Lankester (Ed.). Adam and Charles Black, London (Reprinted 1964, A. Asher, Amsterdam).

———. 1930. *Studies on the structure and development of vertebrates.* Constable, London (Reprinted 1958, Dover, New York).

Gosline, W. A. 1959. Four new species, a new genus, and a new suborder of Hawaiian fishes. *Pac. Sci. 13:* 67–77.

Greenwood, P. H. 1987. The natural history of lungfishes, pp. 163–179. In *The biology and evolution of lungfishes,* W. E. Bemis, W. W. Burggren, and N. E. Kemp (Eds.). Alan R. Liss, New York.

Grier, H. J. 1981. Cellular organization of the testis and spermatogenesis in fishes. *Amer. Zool. 21:* 345–357.

Gross, M. R. 1984. Sunfish, salmon, and the evolution of alternative reproductive strategies and tactics in fishes, pp. 55–75. In *Fish reproduction: Strategies and tactics,* G. W. Potts and R. J. Wootton (Eds.). Academic Press, London.

———. 1991. Evolution of alternative reproductive strategies: Frequency-dependent selection in male bluegill sunfish. *Phil. Trans. Roy. Soc. Lond. B 332:* 59–66.

Guerrero, R. D. 1979. Use of hormonal steroids for artificial sex reversal of *Tilapia,* pp. 512–514. In Symposium on hormonal steroids in fish, B. I. Sundararaj and S. V. Goswami (Eds.), *Proc. Ind. Nat. Acad. Sci. B45*(5).

Hall, B. K. 2000. The neural crest as a fourth germ layer and vertebrates as quadroblastic and triploblastic. *Evol. Dev. 2:* 3–5.

Hamlett, W. C. 1989. Evolution and morphogenesis of the placenta in sharks. *J. Exp. Zool.* Suppl. 2: 35–52.

———. 1999. *Sharks, skates, and rays: The biology of elasmobranch fishes.* Johns Hopkins University Press, Baltimore.

———, A. M. Eulitt, R. L. Jarrell, and M. A. Kelly. 1993. Uterogestation and placentation in elasmobranchs. *J. Exp. Zool. 266:* 347–367.

Hardisty, M. W. 1971. Gonadogenesis, sex differentiation, and gametogenesis, pp. 295–359. In *The biology of lampreys,* Vol. I, M. W. Hardisty amnd I. C. Potter (Eds.). Academic Press, London.

———. 1979. *Biology of the cyclostomes.* Chapman and Hall, London.

Harvey, B. 1993. Cryopreservation of fish spermatozoa, pp. 175–179. In *Genetic conservation of salmonoid fishes,* J. G. Cloud and G. H. Thorgaard (Eds.). Plenum Press, New York.

Heide, K. A., E. T. Wilson, C. J. Cretekos, and D. J. Grunwald. 1994. Contribution of early cells to the fate map of the zebrafish gastrula. *Science 265:* 517–520.

Henson, S. A., and R. R. Warner. 1997. Male and female alternative reproductive behaviors in fishes: A new approach using intersexual dynamics. *Ann. Rev. Ecol. Syst. 28:* 571–592.

Hettler, W. F. 1984. Description of eggs, larvae, and early juveniles of gulf menhaden, *Brevoortia patronus,* and comparisons with Atlantic menhaden, *B. tyrannus,* and yellowfin menhaden, *B. smithi. Fish. Bull. 82:* 85–95.

Hoar, W. S. 1969. Reproduction, pp. 1–72. In *Fish physiology,* Vol. III, W. S. Hoar and D. J. Randall (Eds.). Academic Press, New York.

———, and D. J. Randall (Eds.) 1988a. *Fish physiology,* Vol. XI. Academic Press, San Diego.

———, and———. 1988b. *Fish physiology,* Vol. XII. Academic Press, San Diego.

Hubbs, C. L. 1943. Terminology of early stages of fishes. *Copeia 1943:* 260.

Hunter, J. R. 1981. Feeding ecology and predation of fish larvae, pp. 34–77. In *Marine fish larvae,* R. Lasker (Ed.). University of Washington Press, Seattle.

Jamieson, B. G. M. 1991. *Fish evolution and systematics: Evidence from spermatozoa.* Cambridge University Press, Cambridge, UK.

Jones, G. P., M. J. Milicich, J. Emslie, and C. Lunow. 1999. Self-recruitment in a coral reef fish population. *Nature 402:* 802–804.

Kahn, P. 1994. Zebrafish hit the big time. *Science 264:* 904–905.

Kamler, E. 1992. *Early life history of fish: An energetic approach.* Chapman and Hall, London.

Keenleyside, M. H. A. 1991. Parental care, pp. 191–308. In *Cichlid fishes: behavior, ecology and evolution,* M. H. A. Keenleyside (Ed.). Chapman and Hall, London.

Kendall, A. W., Jr., E. H. Ahlstrom, and H. G. Moser. 1984. Early life history stages of fishes and their characters, pp. 11–22. In Ontogeny and systematics of fishes, H. G. Moser, W. J. Richards, D. M. Cohen, M. P. Fahay, A. W. Kendall, Jr., and S. L. Richardson (Eds.), *Am. Soc. Ichthyol. Herpetol. Spec. Publ. 1.*

Kincaid, H. L. 1983. Inbreeding in fish populations used for aquaculture. *Aquaculture 33:* 215–227.

Kirpichnikov, V. S. 1981. *Genetic basis of fish selection.* Springer Verlag, New York.

Kodric-Brown, A. 1998. Sexual dichromatism and temporary color changes in the reproduction of fishes. *Amer. Zool. 38:* 70–81.

Koob, T. J., and A. Summers. 1996. On the hydrodynamic shape of little skate (*Raja erinacea*) egg capsules. *Bull. Mt. Desert Is. Biol. Lab. 35:* 108–111.

Lam, T. J. 1982. Applications of endocrinology to fish culture. *Can. J. Fish. Aquat. Sci. 39:* 111–137.

———. 1983. Environmental influences on gonadal activity in fish, pp. 65–116. In *Fish physiology,* Vol. IX, Part B, W. S. Hoar,

D. J. Randall, and E. M. Donaldson (Eds.). Academic Press, New York.

Lasker, R. (Ed.). 1981a. *Marine fish larvae.* University of Washington Press, Seattle.

———. 1981b. The role of a stable ocean in larval fish survival and subsequent recruitment, pp. 80–87. In *Marine fish larvae,* R. Lasker (Ed.). University of Washington Press, Seattle.

Laurin, M., and M. Girondot. 1999. Embryo retention in sarcopterygians, and the origin of the extra-embryonic membranes of the amniotic egg. *Ann. Sci. Nat. 3:* 99–104.

Leonard, J. B. K., A. P. Summers, and T. J. Koob. 1999. Metabolic rate of embryonic little skate, *Raja erinacea* (Chondrichthyes: Batoidea): The cost of active pumping. *J. Exp. Zool. 283:* 13–18.

Li, W., A. P. Scott, M. J. Siefkes, H. Yan, Q. Liu, S.-S. Yun, and D. A. Gage. 2002. Bile acid secreted by male sea lamprey that acts as a sex pheromone. *Science 296:* 138–141.

Liley, N. R. 1982. Hormones and reproductive behavior in fishes, pp. 73–116. In *Fish physiology,* Vol. III, W. S. Hoar and D. J. Randall (Eds.). Academic Press, New York.

Lin, H. R., and R. E. Peter. 1986. Induction of gonadotropin secretion and ovulation in teleosts using LHRH analogs and catecholaminergic drugs: A review, pp. 667–670. In *The first Asian fisheries forum,* J. L. MacLean, L. B. Dizon, and L. V. Hosillos (Eds.). Asian Fisheries Society, Manila, Philippines.

Linhart, O., R. Billard, and J. P. Proteau. 1993. Cryopreservation of European catfish (*Silurus glanis* L.) spermatozoa. *Aquaculture 115:* 347–359.

Lowartz, S. M., and F. W. H. Beamish. 2000. Novel perspectives in sexual lability through gonadal biopsy in larval sea lampreys. *J. Fish Biol. 56:* 743–757.

Lubzens, E., S. Rothbard, and A. Hadani. 1993. Cryopreservation and viability of spermatozoa from the ornamental Japanese carp (nishikigoi). *Bamidgeh (Isr. J. Aquacult.) 45*(4): 169–174.

Marshall, N. B. 1979. *Developments in deep-sea biology.* Blandford Press, Poole, Dorset, UK.

Marsh-Matthews, E., P. Skierkowski, and A. DeMarais. 2001. Direct evidence for mother-to-embryo transfer of nutrients in the livebearing fish *Gambusia geiseri. Copeia 2001:* 1–6.

Matarese, A. C., A. W. Kendall, Jr., D. M. Blood, and B. M. Vinter. 1989. Laboratory guide to early life history stages of Northeast Pacific fishes. *NOAA Tech. Rep. NMFS 80.*

Matsuyama, M., M. Hamada, M. Ashitani, M. Kashiwagi, T. Iwai, K. Okuzawa, H. Tanaka, and H. Kagawa. 1993. Development of LHRH a copolymer pellet polymerized by ultraviolet and its application for maturation in red sea bream *Pagrus major* during the non-spawning season. *Bull. Jpn. Soc. Sci. Fish. 59:* 1361–1369.

McClure, M. 1999. Development and evolution of melanophore patterns in fishes of the genus *Danio* (Teleostei: Cyprinidae). *J. Morphol. 241:* 83–105.

Meyer, A., C. H. Biermann, and G. Ortí. 1993. The phylogenetic position of the zebrafish (*Danio rerio*), a model system in developmental biology: An invitation to the comparative method. *Proc. Roy. Soc. Lond. B 252:* 231–236.

Miller, T. J., L. B. Crowder, J. A. Rice, and E. A. Marshall. 1988. Larval size and recruitment mechanisms in fishes: Toward a conceptual framework. *Can. J. Fish. Aquat. Sci. 45:* 1657–1670.

Morisawa, S. 1995. Fine structure of spermatozoa of the hagfish *Eptatretus burgeri* (Agnatha). *Biol. Bull. 189:* 6–12.

Moser, H. G. 1981. Morphological and functional aspects of marine fish larvae, pp. 90–131. In *Marine fish larvae,* R. Lasker (Ed.). University of Washington Press, Seattle.

———, and E. H. Ahlstrom. 1974. Role of larval stages in systematic investigations of marine teleosts: The Myctophidae, a case study. *Fish. Bull. 72:* 391–413.

———, W. J. Richards, D. M. Cohen, M. P. Fahay, A. W. Kendall, Jr., and S. L. Richardson (Eds.). 1984. Ontogeny and systematics of fishes. *Am. Soc. Ichthyol. Herpetol. Spec. Publ. 1.*

Munday, P. L., M. J. Caley, and G. P. Jones. 1998. Bi-directional sex change in a coral-dwelling goby. *Behav. Ecol. Sociobiol. 43:* 371–377.

Neff, B. D. 2003. Paternity and condition affect cannibalistic behavior in nest-tending bluegill sunfish. *Behav. Ecol. Sociobiol. 54:* 377–384.

———, P. Fu, and M. R. Gross. 2003. Sperm investment and alternative mating tactics in bluegill sunfish (*Lepomis macrochirus*). *Behav. Ecol. 14:* 634–641.

Nellen, W. 1986. A hypothesis on the fecundity of bony fish. *Meeresforsch. 31:* 75–89.

Nishimura, K., and N. Hoshino. 1999. Evolution of cannibalism in the larval stage of pelagic fish. *Evol. Ecol. 13:* 191–209.

Olla, B. L., and M. W. Davis. 1990. Effects of physical factors on the vertical distribution of larval walleye pollock *Theragra chalcogramma* under controlled laboratory conditions. *Mar. Ecol. Prog. Ser. 63:* 105–112.

Orr, J. W., and R. A. Fritzsche. 1993. Revision of the pipefishes family Solenostomidae. *Copeia 1993:* 168–182.

Parker, G. A. 1992. The evolution of sexual size dimorphism in fish. *J. Fish Biol. 41*(Suppl. B): 1–20.

Parker, J. B. 1942. Some observations on the reproductive system of the yellow perch (*Perca flavescens*). *Copeia 1942:* 223–226.

Patzner, R. A. 1982. Die Reproduktion der myxinoiden. Ein vergleich von *Myxine glutinosa* und *Eptatretus burgeri. Zool. Anz. (Jena) 208:* 132–144.

———. 1998. Gonads and reproduction in hagfishes, pp. 378–395. In *The biology of hagfishes,* J. M. Jørgensen, J. P. Lomholt, R. E. Weber, and H. Malte (Eds.). Chapman and Hall, London.

Paxton, J. R., and W. N. Eschmeyer. 1998. *Encyclopedia of fishes* (2nd ed.). Academic Press, San Diego.

Peter, R. E. 1981. Gonadotropin secretion during reproductive cycles in teleosts: Influence of environmental factors. *Gen. Comp. Endocrinol. 45:* 294–305.

Pietsch, T. W. 1976. Dimorphism, parasitism, and sex: Reproductive strategies among deepsea ceratioid anglerfishes. *Copeia 1976:* 781–793.

Postlethwait, J., S. Johnson, C. N. Midson, W. S. Talbot, M. Gates, E. W. Ballenger, D. Africa, R. Andrews, T. Carl, J. S. Eisen, S. Horne, C. B. Kimmel, M. Hutchinson, M. Johnson, and A. Rodriguez. 1994. A genetic linkage map for the zebrafish. *Science 264:* 699–703.

———, A. Amores, A. Force, and Y.-L. Yan. 1999. The zebrafish genome, pp. 149–163. In *The zebrafish: Genetics and genomics,* H. W. Detrich, III, M. Westerfield, and L. I. Zon (Eds.). Academic Press, San Diego.

Potthoff, T., and S. Kelley. 1982. Development of the vertebral column, fins and fin supports, branchiostegal rays, and squamation of the swordfish, *Xiphias gladius. Fish. Bull. 80:* 161–186.

Pyron, M. 1996. Sexual size dimorphism and phylogeny in North American minnows. *Biol. J. Linn. Soc. 57:* 327–341.

Ricker, W. E. 1968. *Methods for assessment of fish production in fresh waters. IBP handbook No. 3.* Blackwell, Oxford.

Romer, A. S. 1970. *The vertebrate body* (4th ed.). W. B. Saunders, Philadelphia.

Roush, W. 1997. A zebrafish genome project? *Science 275:* 923.

Rowe, R. I., and C. D. Eckhert. 1999. Boron is required for zebrafish embryogenesis. *J. Exp. Biol. 202:* 1649–1654.

Russell, F. S. 1976. *The eggs and planktonic stages of British marine fishes.* Academic Press, London.

Schultz, R. J. 1973. Origin and synthesis of a unisexual fish, pp. 207–211. In *Genetics and mutagenesis of fish,* J. H. Schroder (Ed.). Springer Verlag, New York.

Schwartz, F. J. 1972. World literature to fish hybrids with an analysis by family, species, and hybrid. *Gulf Coast Res. Lab. Publ. 3:* 1–328.

———. 1981. World literature to fish hybrids with an analysis by family, species, and hybrid: Supplement 1. *NOAA Tech. Rep. NMFS SSRF-750:* 1–507.

Seehausen, O., J. J. M. van Alphen, and F. Witte. 1997. Cichlid fish diversity threatened by eutrophication that curbs sexual selection. *Science 277:* 1808–1811.

Snelson, F. F., Jr. 1972. Systematics of the subgenus *Lythrurus,* genus *Notropis* (Pisces: Cyprinidae). *Bull. Fla. State Mus. Biol. Sci. 17:* 1–92.

Solar, I. I., E. M. Donaldson, and D. Douville. 1991. A bibliography of gynogenesis and androgenesis in fish. *Can. Tech. Rep. Fish. Aquat. Sci. 1788.*

Sower, S. A., and A. Gorbman. 1999. Agnatha, pp. 83–90. In *Encyclopedia of reproduction,* Vol. 1, E. Knobil and J. D. Neill (Eds.). Academic Press, San Diego.

———, and H. Kawauchi. 2001. Update: brain and pituitary hormones of lampreys. *Comp. Biochem. Physiol. B 129:* 291–302.

———, C. B. Schreck, and E. M. Donaldson. 1982. Hormone induced ovulation of coho salmon (*Oncorhynchus kisutch*) held in seawater and freshwater. *Can. J. Fish. Aquat. Sci. 39:* 627–632.

Stacey, N., and P. Sorensen. 1999. Pheromones, fish, pp. 748–755. In *Encyclopedia of reproduction,* Vol. 3, E. Knobil and J. D. Neill (Eds.). Academic Press, San Diego.

Stanley, H. P. 1961. *Studies on the genital systems and reproduction in the chimaeroid fish* Hydrolagus colliei *(Lay and Bennett).* PhD thesis, Oregon State University.

Stein, D. L., and W. G. Pearcy. 1982. Aspects of reproduction, early life history, and biology of macrourid fishes off Oregon. *USA Deep Sea Res. 29:* 1313–1329.

Sterba, G. 1966. *Freshwater fishes of the world.* Pet Library, London.

Strathmann, R. R., and H. C. Hess. 1999. Two designs of marine egg masses and their divergent consequences for oxygen supply and desiccation in air. *Amer. Zool. 39:* 253–260.

Streisinger, G., C. Walker, N. Dower, D. Knauber, and F. Singer. 1981. Production of clones of homozygous zebrafish *Brachydanio rerio.* *Nature 291:* 293–296.

Stoss, J. 1983. Fish gamete preservation and spermatozoan physiology, pp. 305–350. In *Fish physiology,* Vol. IX, Part B, W. S. Hoar, D. J. Randall, and E. M. Donaldson (Eds.). Academic Press, New York.

Sumpter, J. P. 1997. Environmental control of fish reproduction: A different perspective. *Fish Physiol. Biochem. 17:* 25–31.

Sundararaj, B. I. 1981. Reproductive physiology of teleost fishes: A review of present knowledge and needs for future research. *FAO ADCP/Rep. 381/16,* Rome.

Suthers, I. 2000. Significance of larval condition: Comment on laboratory experiments. *Can. J. Fish. Aquat. Sci. 57:* 1534–1536.

Swearer, S. E., J. E. Caselle, D. W. Lea, and R. R. Warner. 1999. Larval retention and recruitment in an island population of a coral-reef fish. *Nature 402:* 799–802.

Takeuchi, Y., G. Yoshizaki, and T. Takeuchi. 2004. Surrogate broodstock produces salmonids. *Nature 430:* 629–630.

Tanaka, M., L. A. Hale, A. Amores, Y.-L. Yan, W. A. Cresko, T. Suzuki, and J. H. Postlethwait. 2005. Developmental genetic basis for the evolution of pelvic fin loss in the pufferfish *Takifugu rubripes.* *Dev. Biol. 281:* 227–239.

Tesch, F. W. 1977. *The eel.* Chapman and Hall, London.

Thibault, R. E., and R. J. Schultz. 1979. Reproductive adaptations among viviparous fishes (Cyprinodontiformes: Poeciliidae). *Evolution 32:* 320–333.

Thomson, K. S. 1991. *Living fossil: The story of the coelacanth.* W. W. Norton, New York.

Tyler, C. R., and J. P. Sumpter. 1996. Oocyte growth and development in teleosts. *Rev. Fish Biol. Fisheries 6:* 287–318.

Veith, W. J. 1980. Viviparity and embryonic adaptations in the teleost *Clinus superciliosus.* *Can. J. Zool. 58:* 1–12.

Vodicnik, M. J., R. E. Kral, V. L. de Vlaming, and L. W. Crim. 1978. The effects of pinealectomy on pituitary and plasma gonadotropin levels in *Carassius auratus* exposed to various photoperiod-temperature regimes. *J. Fish Biol. 12:* 187–196.

———, J. Olcese, G. Delahunty, and V. de Vlaming. 1979. The effects of blinding, pinealectomy and exposure to constant dark conditions on gonadal activity in the female goldfish, *Carassius auratus.* *Env. Biol. Fishes 4:* 173–177.

Vogel, G. 2000a. Sanger will sequence zebrafish genome. *Science 290:* 1671.

———. 2000b. Zebrafish earns its stripes in genetic screens. *Science 288:* 1160–1161.

Vrijenhoek, R. C. 1994. Unisexual fish: Model systems for studying ecology and evolution. *Ann. Rev. Ecol. Syst. 25:* 71–96.

Watson, W. 1982. Development of eggs and larvae of the white croaker, *Genyonemus lineatus* Ayres (Pisces: Sciaenidae), off the southern California coast. *Fish. Bull. 80:* 403–417.

———, A. C. Matarese, and E. G. Stevens. 1984. Trachinoidea: Development and relationships, pp. 554–561. In Ontogeny and systematics of fishes, H. G. Moser, W. J. Richards, D. M. Cohen, M. P. Fahay, A. W. Kendall, Jr., and S. L. Richardson (Eds.), *Am. Soc. Ichthyol. Herpetol. Spec. Publ. 1.*

Webb, J. F. 1999. Larvae in fish development and evolution, pp. 109–158. In *The origin and evolution of larval forms,* B. K. Hall and M. H. Wake (Eds.). Academic Press, San Diego.

Weber, G. M., and E. G. Grau. 1999. Prolactin in nonmammals, pp. 51–60. In *Encyclopedia of reproduction,* Vol. 4, E. Knobil and J. D. Neill (Eds.). Academic Press, San Diego.

Weibe, J. P. 1968. The reproductive cycle of the viviparous seaperch *Cymatogaster aggregatus* Gibbons. *Can. J. Zool. 46:* 1221–1234.

Welker, M. T., C. L. Pierce, and D. H. Wahl. 1994. Growth and survival of larval fishes: Roles of competition and zooplankton abundance. *Trans. Am. Fish. Soc. 123:* 703–717.

Westernhagen, H. V., C. Freitas, G. Fürstenberg, and J. Willführ-Nast. 1998. C/N data as an indicator of condition in marine fish larvae. *Arch. Fish. Mar. Res. 46*(2): 165–179.

Wheeler, P. A., and G. H. Thorgaard. 1991. Cryopreservation of rainbow trout semen in large straws. *Aquaculture 93:* 95–100.

Wiley, M. L., and B. B. Collette. 1970. Breeding tubercles and contact organs in fishes: Their occurrence, structure, and significance. *Bull. Am. Mus. Nat. Hist. 143:* 145–216.

Williams, J. E., and C. E. Bond. 1983. Status and life history notes on the native fishes of the Alvord Basin, Oregon and Nevada. *Great Basin Naturalist 43:* 409–420.

Wilson, E. T., K. A. Helde, and D. J. Grunwald. 1993. Something's fishy here—rethinking cell movements and cell fate in the zebrafish embryo. *Trends Genet. 9:* 348–352.

Wilson, L., and M. Maden. 2005. The mechanisms of dorsoventral patterning in the vertebrate neural tube. *Dev. Biol. 282:* 1–13.

Wootton, R. J. 1984. Introduction: Strategies and tactics in fish reproduction, pp. 1–12. In *Fish reproduction: Strategies and tactics,* G. W. Potts, and R. J. Wootton (Eds.). Academic Press, London.

———. 1990. *Ecology of teleost fishes.* Chapman and Hall, New York.

Wourms, J. P. 1977. Reproduction and development in chondrichthyan fishes. *Amer. Zool. 17:* 379–410.

———. 1981. Viviparity: The maternal–fetal relationship in fishes. *Amer. Zool. 21:* 473–515.

———, B. D. Grove, and J. Lombardi. 1988. The maternal–embryonic relationship in viviparous fishes, pp. 1–134. In *Fish physiology,* Vol. XI, Part B. W. S. Hoar and D. J. Randall (Eds.). Academic Press, San Diego.

Yamagami, K. 1988. Mechanisms of hatching in fish, pp. 447–499. In *Fish physiology,* Vol. XI, The physiology of developing fish, Part A, Eggs and larvae, W. S. Hoar and D. J. Randall (Eds.). Academic Press, San Diego.

Yamamoto, T. 1969. Sex differentiation, pp. 117–175. In *Fish physiology,* Vol. III, W. S. Hoar and D. J. Randall (Eds.). Academic Press, New York.

Yambe, H., M. Shindo, and F. Yamazaki. 1999. A releaser pheromone that attracts males in the urine of mature female masu salmon. *J. Fish Biol. 55:* 158–171.

Young, G. 1999. Migration, fish, pp. 234–244. In *Encyclopedia of reproduction,* Vol. 3, E. Knobil and J. D. Neill (Eds.). Academic Press, San Diego.

Zimmer, C. 1999. The mystery of the mermaid's purse. *Nat. Hist. 108*(6): 24–25.

28

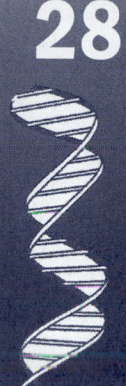

THE GENETICS OF FISHES

INTRODUCTION

FUNDAMENTAL CONCEPTS

Gene Expression
Mitosis and Meiosis
Mendelian Inheritance
Gene Mapping
Non-Mendelian Inheritance

CHROMOSOMES

Chromosome and Arm Number in a Species
Induced Polyploidy
Hybrids

SEX DETERMINATION

Chromosomal Determination
Environmental Determination
Hermaphroditism
All-Female Species

QUANTITATIVE GENETICS

Polygenic Traits
Heritability

GENETICS AND PHYLOGENETICS

The "Big" Questions
Other Phylogenetic Applications
Detection and Study of Species

EVOLUTION, CONSERVATION, AND MANAGEMENT

Population Genetics
Biogeography
Local Adaptation and Outbreeding Depression
Genetics and Fisheries Management
Conservation Genetics

RECENT DIRECTIONS

Fish as Model Systems
Genetic Engineering

A. J. GHARRETT

Genetics is intrinsic to all facets of the biology of fishes. It is the basis of their morphology and physiology, behavior and ecology, and systematics and evolution. From a narrow perspective, the discipline of genetics studies mechanisms for the inheritance of traits and for sex determination. From a broader perspective, however, genetics examines the variation within and among species and the processes by which genes encoded in the DNA of an organism are expressed. Some genes are responsible for an organism's short-term physiological responses, and others for a population's adaptation to interannual and global changes in the biotic and physical environment. Intraspecific genetic differences, which provide a buffer for environmental changes and allow both populations and species to evolve, are a focus for conservation and management biologists. Genetic divergence can lead to speciation. The delineation of species uses differences in sizes, shapes, colorations, or particular structures; and systematics is based on characteristics shared by evolutionary lineages. All of those features are determined by expression of the fish's genes, and most are related to the success of a species in its ecological niche.

INTRODUCTION

The mechanisms and concepts of genetics are the bases for every aspect of biology; moreover, they unite all biological fields into a cohesive whole. The particular aspects of biology for which genetic principles provide explanations are inheritance, gene expression, and evolutionary processes. *Inheritance* is the process by which genetic information is transmitted from parents to offspring; it explains why offspring generally resemble their parents and why individuals within a species share many characteristics. *Gene expression* is the means by which genetic information carried by an individual is revealed; it provides the mechanisms for ontogenetic progression and physiological response. *Evolutionary* processes provide insight into the causes of genetic divergence, which may result in speciation. Systematic descriptions of species ordinarily attempt to follow the evolutionary relationships among species (**phylogeny**), which are based on the descent from a common ancestor—in other words, individuals in a taxon share a genetic history. Applications of genetic methods to understanding the biology of and unraveling the relationships among fishes are developing rapidly. During the past decade, genetic methods have found increasing application to many facets of the biology of fishes. The use of multiple genetic approaches to address particular questions is a hallmark of the increasing role of genetics in studies of fishes. This chapter focuses on the principles of genetics as they pertain to fishes and on the genetic methods that are currently used or are being developed to study fish biology.

Inheritance is accomplished by the transmission of *deoxyribonucleic acid* (*DNA*) sequences from parent to offspring. DNA is carried in each cell of an individual and encodes the potential structures, functions, and even behaviors of each individual. The resulting individual is the outcome of genetic expression in a particular environmental context. Genetic variation arises from changes in the DNA sequence (**mutations**). Mutations provide genetic alternatives, which may be acted on by natural selection. Most gene changes are eliminated because they are not as effective in their contribution to the species' overall ability to meet and solve environmental challenges as the original gene. An occasional mutation, however, provides an even better solution, which is manifested by an increased ability to contribute offspring to subsequent generations—that is, the mutation increases **fitness**. Because the environment experienced by a population continually changes, both in year-to-year weather fluctuations and in longer-term climate cycles, there is usually no single best genetic solution for a population. As a result of this environmental uncertainty, populations usually display genetic diversity, which reflects the previous environmental history of the population and may ensure that at least a portion of the population will be able to withstand future environmental extremes. Genetic diversity is essential for the persistence of species because it provides raw materials for natural selection and may allow the species to evolve (Stearns, 1992).

Over time and space, species are not fixed entities; rather, they are dynamic and reflect the results of responses to evolutionary pressures that have acted earlier on populations of the species. To comprehend why there are so many species of fish (at least 24,618; Nelson, 1994) and to understand the origins—and perhaps destinies—of a particular fish species, one must have an appreciation for the genetic basis of and the related evolutionary processes that underlie fish diversity. Every aspect of the genetics of fish—from the molecular structure of genes and proteins to the fish's interaction with the biosphere—plays a role in understanding ichthyology.

FUNDAMENTAL CONCEPTS

Gene Expression

Genes are expressed when the information encoded in the DNA sequences (the **genotype**) is manifested in the organism. The result, which is often influenced by the environment, is the **phenotype**. Gene expression may be observed at many levels, ranging from the synthesis of a particular biomolecule to the exhibition of complex behaviors. The expression of the information encoded in the genetic material of an organism results in the phenotype that we recognize as characteristic of a particular species. The most fundamental level of organization of a **genome** (an organism's entire set of genetic information) is the sequence of nitrogenous bases or **nucleotides** in the DNA molecule. The four nucleotides common to DNA are *adenine* (**A**), *thymine* (**T**), *guanine* (**G**), and *cytosine* (**C**). DNA is a very long, double-stranded helical structure of nucleotides, held together by hydrogen bonds between the **A-T** and **G-C** pairs of the complementary strands. The amount of DNA carried in each molecule is enormous—more than one billion nucleotide pairs. Base pairing stabilizes the molecule and provides a mechanism for accurately **replicating** the information carried on the molecule: Each strand can direct the construction of new double-stranded molecules identical in sequence to the original. The result is two precise duplicates of the genome of an individual, which are needed for cell division and reproduction (Fig. 28.1).

The information encoded in the DNA base sequence includes both structural genes and regulatory genes. The DNA sequence in structural genes specifies a sequence of

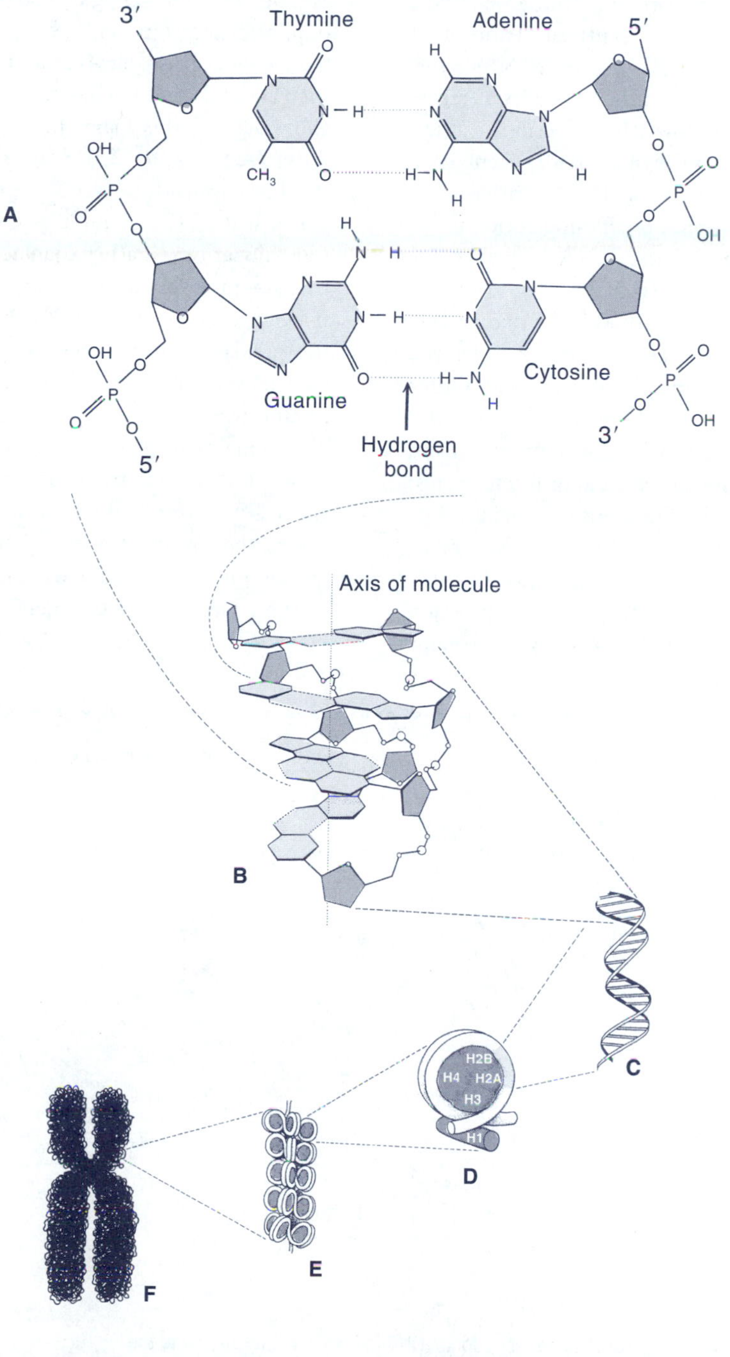

FIGURE 28.1

Deoxyribonucleic acid (DNA) is double-stranded and constructed from the deoxyribonucleotide triphosphate subunits possessing the nitrogenous bases adenine (A), guanine (G), cytosine (C), and thymine (T).
A, Hydrogen bonding between A-T and G-C base pairs holds the two strands together. **B,** The stack of base pairs is located at the center of the molecule; and **C,** Alternating deoxyribose sugars and phosphates connect the base pairs and wind helically around the outside. **D,** The DNA is packed into the nucleus of a eukaryote by histone proteins (H1, H2A, H2B, H3, and H4), around which it winds. **E,** The histones and other chromosomal proteins organize the DNA into **F,** Chromosomes.

ribonucleic acid (RNA) nucleotides, which in turn determines the order of amino acids in a protein. Numerous RNA molecules are important in protein synthesis (ribosomal RNAs, or rRNAs; and transfer RNAs, or tRNAs) and other housekeeping tasks. Protein synthesis occurs at **ribosomes** in the cytoplasm and follows the directions carried by messenger RNA (mRNA) copies of structural gene base sequences coded in the nuclear DNA. RNA synthesis, or **transcription,** is the construction of ribonucleotide polymers that are complementary to the DNA sequence, or **gene,** which specifies the structure (Fig. 28.2).

A nucleotide sequence signal, aptly called a **promoter,** precedes each structural gene, marking the site at which the RNA polymerase is to begin transcribing the gene. Other regulatory genes determine when and how long a particular structural gene will be expressed. During development, entire batteries of genes must be turned on or off at different stages; differentiation of cells during development results in different tissues. Although erythrocytes (unlike our own erythrocytes, they are nucleated in fishes!) and muscle cells carry the same genetic information, the differential expression of their genetic information results in a variety of proteins that produce cells possessing very different functions and capabilities. Physiological adaptation of individuals may require changes in gene expression later in life to optimize their chances for survival and reproduction. Examples of such adaptations are changes in osmotic competence, which allow diadromous fishes to move between salt- and freshwater (see Chapter 25), and sex changes of nonsimultaneous hermaphrodites (see Chapter 27). The organization of genes within the genome is complicated. In fact, genome reorganization—rather than accumulated changes in single genes—probably accounts for many of the differences between species (King and Wilson, 1975). The organization of the regulatory and structural genes and many other interspersed sequences in the DNA is a second level of genomic organization.

Gene expression can be detected at a number of levels. These levels trace the process of expression, starting with the mRNA transcribed from a particular gene; continuing to the protein translated from that message; then to a simple phenotype, such as coloration, that may be a result of the expression of that gene; or to more complex phenotypes, like size or fecundity, which result from the combined

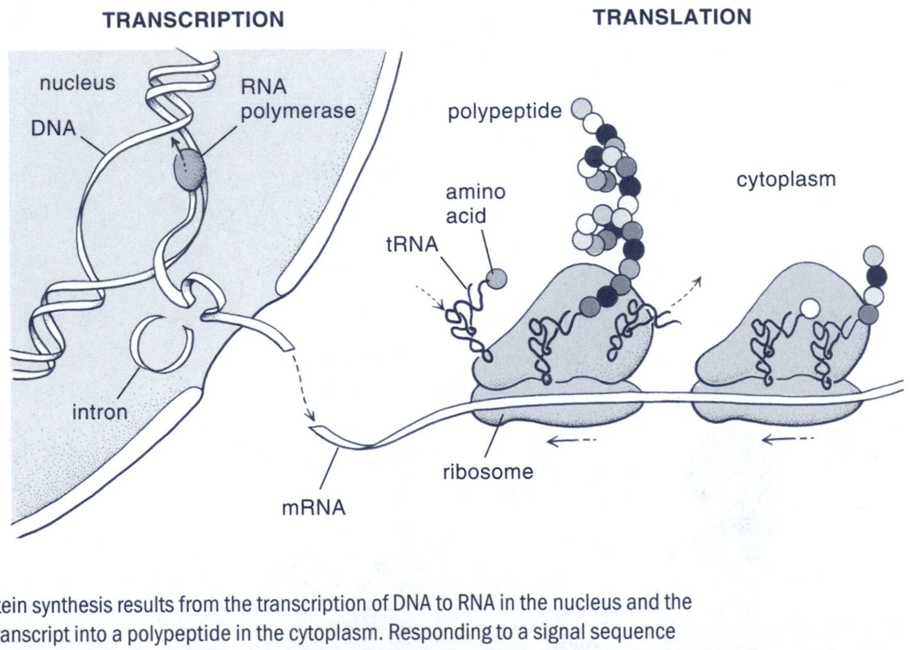

TRANSCRIPTION   **TRANSLATION**

FIGURE 28.2
Gene expression as protein synthesis results from the transcription of DNA to RNA in the nucleus and the translation of the RNA transcript into a polypeptide in the cytoplasm. Responding to a signal sequence (*promoter*) preceding the structural gene, the RNA polymerase catalyzes the synthesis of RNA, using one strand of the DNA as a template. Following transcription, the RNA is processed, which includes the excision of introns. The resulting messenger RNA (mRNA) is transported across the nuclear membrane into the cytoplasm, where the instructions it carries in its nucleotide sequence are translated into an amino acid sequence by ribosomes, using transfer RNAs (tRNAs) to mediate the positioning of the amino acids specified by the instructions in the sequence.

expression of a number of genes—these complex traits are referred to as **polygenic traits.**

Mitosis and Meiosis

Packing the enormous amount of DNA found in each very small cell creates some problems in cell division. For example, the armored catfish (*Corydoras elegans*) has about 6 picograms (10^{-12} grams) of DNA in each cell (Hinegardner and Rosen, 1972), which is about 5.5 billion nucleotide pairs, or 1.8 meters (6 feet!) of DNA. A typical cell is about 10 μm in diameter; and the DNA is localized within the cell in a much smaller organelle, the nucleus. Just think of the difficulty you would have unraveling 43 kilometers (27 miles) of very thin spaghetti packed into a 25-cm (10-inch) diameter bowl. Worse, while unraveling it, you must split it into two pieces lengthwise! This happens to the DNA at each cell division, during which it is accurately replicated and apportioned to daughter cells. This is possible because DNA is organized into separate **chromosomes**—another level of genome organization beyond the nucleotide sequence and gene order (see Figs. 28.1 and 28.3). The DNA of the armored catfish is distributed among 50 chromosomes (Hinegardner and Rosen, 1972), each of which has about 3.6 cm (1.4 inches) of DNA. Chromosomal proteins organize the DNA into highly compact structures during cell division (Fig. 28.3). Morphologically, chromosomes have **arms** that are joined at a **centromere**—the location to which spindle fibers attach during cell division (Fig. 28.4). The set of chromosomes in an organism is called the **karyotype** (e.g., Fig. 28.3); the chromosomes are described by their size and by the position of the centromere (Fig. 28.4). Genes are arranged linearly along chromosomes, always in the same order in a given species. As a result of this colinearity, a DNA sequence or gene for a particular trait is often conceptualized as its location on a chromosome, or its **locus** (plural **loci**). Maps of the relative positions of genes on their chromosomes have been developed for well-studied species like mouse, fruit fly, and tomato. More recently, maps of molecular markers have been developed for several fish species, such as the zebrafish (*Danio rerio*), medaka (*Oryzias latipes*), rainbow trout (*Oncorhynchus mykiss*), and tilapia (*Oreochromis niloticus*).

Most fishes, like most other vertebrates, are **diploid** (2N), which means that they possess two sets of matching chromosomes (one from each parent in sexual species) and, therefore, two copies of each gene (locus). During **mitosis**—normal proliferative or equational cell division—DNA replicates so that each chromosome, now a pair of DNA molecules, separates, and one of each newly formed chromosome is distributed to each new daughter cell. Each daughter cell

thereby receives a full, diploid complement of chromosomes (see Figs. 28.3B and 28.5A).

In vertebrates, gamete production (sex cell formation) includes the nuclear process known as **meiosis,** which is a modified mitotic process. One of the modifications is that one meiosis involves two cell divisions. As in mitosis, DNA replication occurs prior to the first cell division. However, the products of chromosomal (and DNA) replication, called **sister chromatids,** remain joined at a single centromere during the first division. Unlike mitosis, the **homologous** chromosomes (i.e., those carrying genes for the same traits) join together at metaphase to form **tetrads** or **bivalents,** which are the four copies of a particular chromosome. During this first division (Meiosis I), the homologous chromosomes (pairs of sister chromatids) are drawn to opposite poles of the cell, so that homologous chromosomes are segregated. This division is referred to as the *reduction division,* because only one copy of each diploid pair is transferred to each meiotic product. The second division (Meiosis II) is referred to as the *equational division,* because the two sister chromatids that separate at this stage derive from a single parental chromosome. During Meiosis II, the centromere of each pair of sister chromatids replicates, and each member of the resulting chromosome pair moves toward the opposite pole of the cell. The products are **haploid** (N) gametes (Fig. 28.5B), each of which bears a single complement of genes. When two gametes unite at fertilization to form a **zygote,** the diploid (2N) complement is restored.

Mendelian Inheritance

The rules for inheritance, or *Mendel's laws,* describe the expression of a simple trait, specified by a single locus, among the progeny of diploid parents. These rules reflect the segregation of chromosomes (and their genes) as a result of meiosis. An example of a simple Mendelian trait is albinism in the medaka (*Oryzias latipes;* Yamamoto, 1969). Mendel's first law describes the behavior of two alleles at a single locus. **Alleles** at a locus are alternative sets of information for a particular trait. Each individual carries two alleles at a locus—one on each of a homologous chromosome pair. If both alleles at a locus in one individual are indistinguishable (e.g., **AA, aa,** or **A'A'**), the organism is referred to as **homozygous.** If the two alleles it carries are distinguishable (e.g., **Aa, AA',** or **A'a**), an individual is labeled **heterozygous.** Of course, the meaning of "indistinguishable" depends on the level of resolution used—that is, how the trait itself is detected. For example, at a color-determining locus, two alleles may be indistinguishable based on the resulting color, but they may differ in actual amino acid or DNA sequence. Therefore, an individual may be homozygous for the color

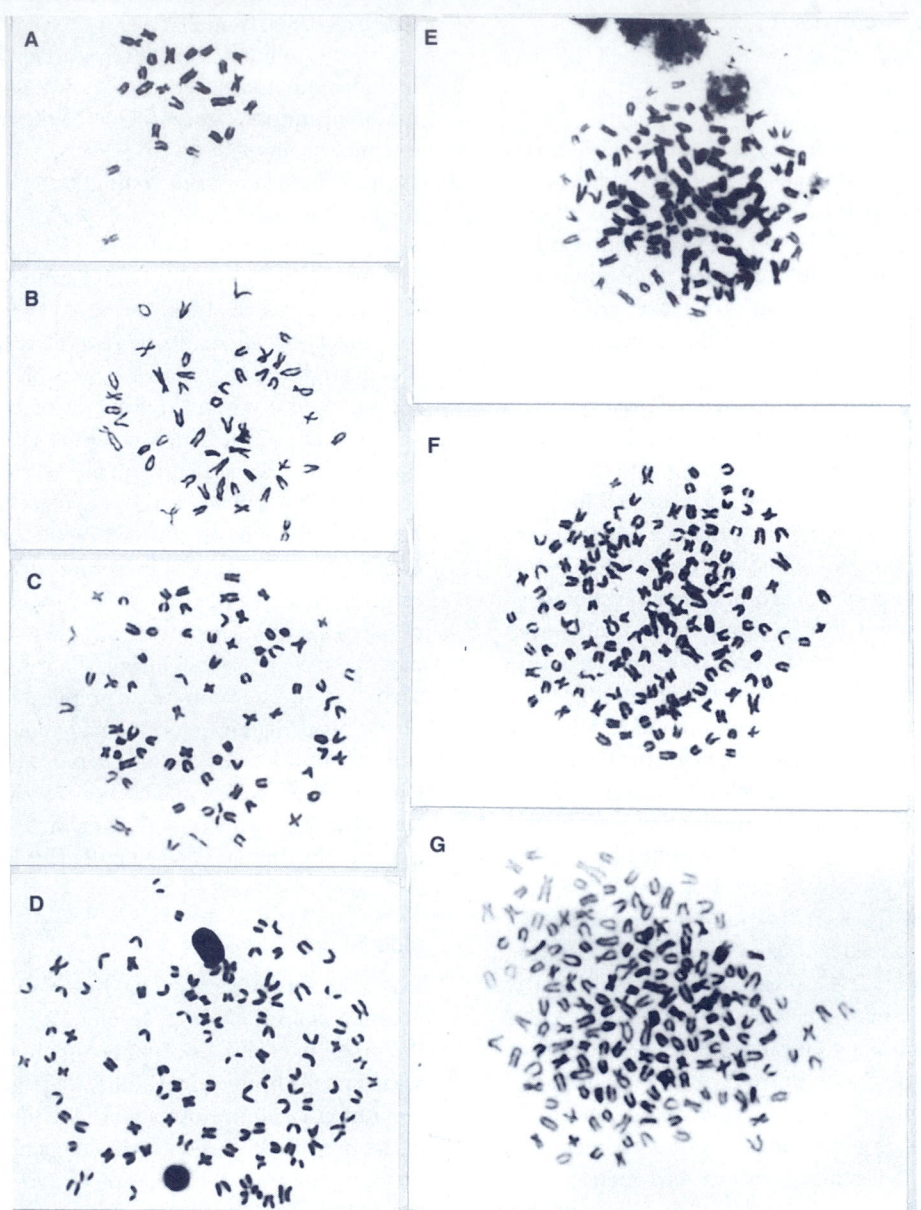

FIGURE 28.3

Karyotypes of metaphase **A**, Haploid; **B**, Diploid; **C**, Triploid; **D**, Tetraploid; **E**, Pentaploid; **F**, Hexaploid; and **G**, Heptaploid loach (*Misgurnus anguillicaudatus*). The haploid spread is from an embryo, the heptaploid spread from fry cells, and the others from 5- to 7-month-old fishes. (From Matsubara et al., 1995.)

allele, but heterozygous for the amino acid or DNA sequence alleles. The actual allelic composition of an individual is its *genotype*. The result of the allele combination that is actually observed in the individual is the *phenotype*. **Mendel's first law** predicts that the two copies of a trait in each diploid (2N) parent are distributed equally among its haploid (N)

gametes—that is, an offspring has a 50:50 chance of receiving one particular copy.

Often, the effect of one allele overrides the effect of the second allele, or is **dominant**; for instance, pigmentation in medaka is dominant over albinism. If the **A** allele is dominant, the phenotype of heterozygous individuals (e.g., **Aa**)

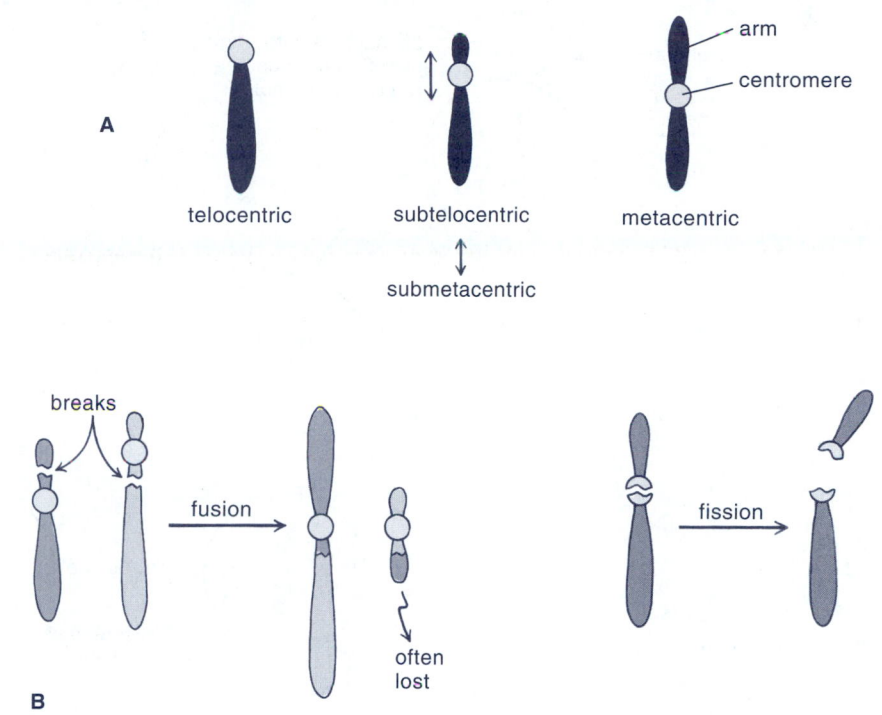

FIGURE 28.4
A, Chromosomes are categorized according to their configuration. They are *metacentric* if the ratio of the longer arm to the shorter arm is between 1.0 and 1.7, *submetacentric* if the ratio is between 1.7 and 3.0, *subtelocentric* for ratios between 3.0 and 7.0, and *telocentric* if the ratio is greater than 7.0. **B,** Changes in chromosome number can result from fusion through reciprocal translocations of chromosomes (*center*) or from fission of a chromosome (*right*). Telocentric chromosomes are counted as a single arm; fission and fusion change the chromosome number without changing the chromosome arm number.

will resemble that of the homozygous dominant (**AA**) ones. Alleles like **a,** whose phenotype appears only in individuals homozygous for that allele (**aa**), are referred to as **recessive.** The genotype of normally pigmented medaka possess at least one allele for normal pigmentation (**AA** or **Aa**), whereas the albino medaka genotype has two recessive alleles that specify no pigmentation. Additional examples of traits exhibiting dominance or recessive expression include coloration in the channel catfish (*Ictalurus punctatus;* albinism is recessive), common carp (*Cyprinus carpio;* recessive phenotypes are blue, gold, and grey), guppies (*Poecilia reticulata;* recessive phenotypes are blond, gold, and albino), and rainbow trout (*Oncorhynchus mykiss;* albinism is recessive); spotting in jewel tetras (*Hyphessobrycon eques*), green swordtails (*Xiphophorus helleri*), and platyfish (*Xiphophorus maculatus* has at least 9 alleles); and spinal abnormalities in guppies and Japanese medaka (*Oryzias latipes;* summarized in Tave, 1986).

The progeny of matings between an albino (**aa**) and a homozygous normal (**AA;** orange-red) medaka (*O. latipes*) carry one allele from each parent, so they are genotypically heterozygous (**Aa**), but phenotypically they resemble the pigmented parent. Among the progeny of matings between heterozygous fish, all three possible genotypes will occur (**AA** in 1/4, **Aa** or **aA** in 1/2, and **aa** in 1/4), and both normal and albino phenotypes will be apparent, but in a 3:1 ratio of dominant (normal; **AA, Aa,** and **aA**) to recessive (albino; **aa**) types (Fig. 28.6).

In other modes of inheritance that involve a single locus, heterozygotes are distinguishable either because they are intermediate between the two homozygous types (**partially dominant**) or because they exhibit the characteristics of each homozygous type (**codominant**). One example of partial dominance is coloration in the Siamese fighting fish (*Betta splendens*); one homozygote is steel blue, the heterozygote is blue, and the other homozygote is green

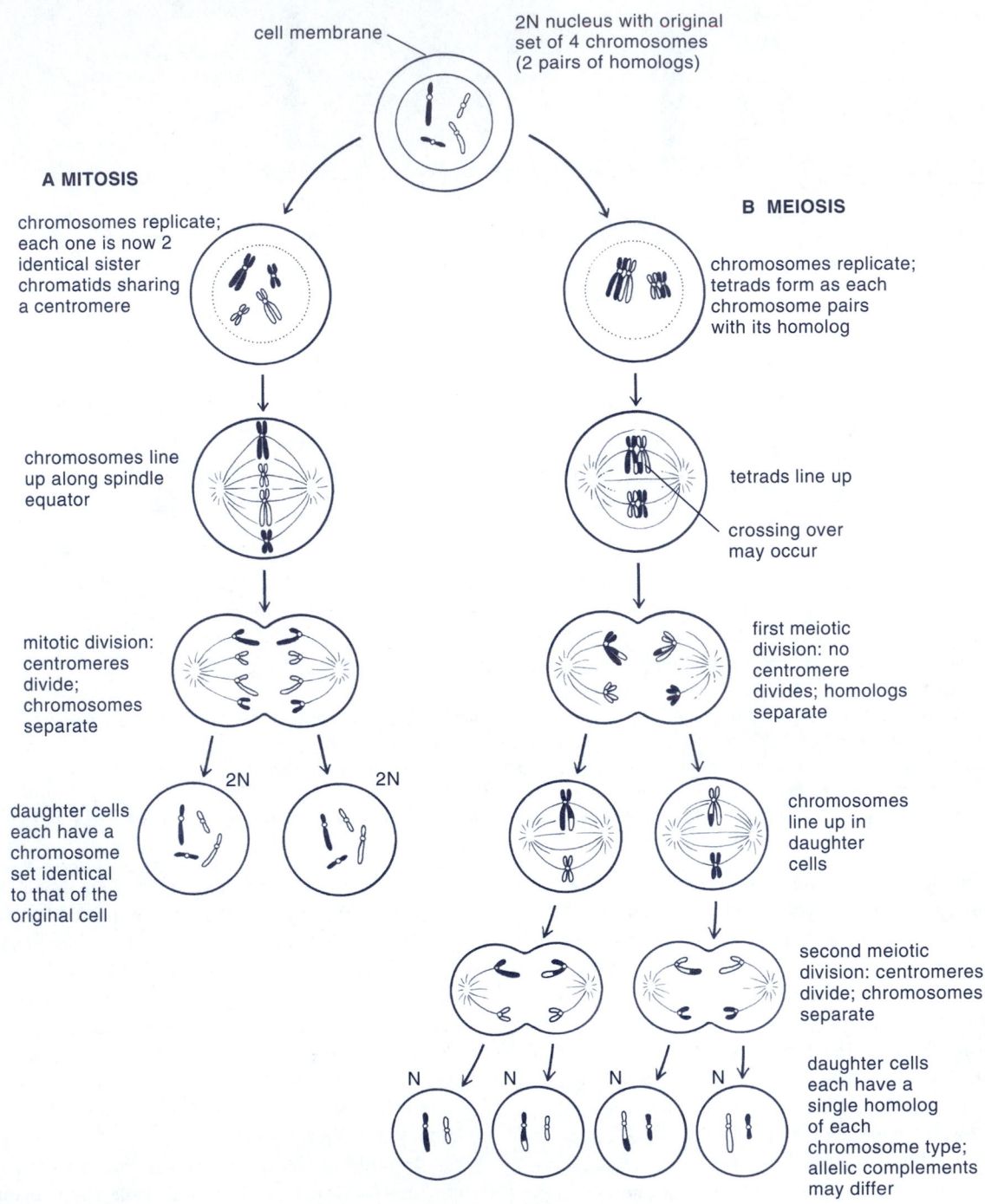

cell membrane

2N nucleus with original
set of 4 chromosomes
(2 pairs of homologs)

A MITOSIS

chromosomes replicate;
each one is now 2
identical sister
chromatids sharing
a centromere

B MEIOSIS

chromosomes replicate;
tetrads form as each
chromosome pairs
with its homolog

chromosomes line
up along spindle
equator

tetrads line up

crossing over
may occur

mitotic division:
centromeres
divide;
chromosomes
separate

first meiotic
division: no
centromere
divides; homologs
separate

2N   2N

daughter cells
each have a
chromosome
set identical
to that of the
original cell

chromosomes
line up in
daughter
cells

second meiotic
division: centromeres
divide; chromosomes
separate

N   N   N   N

daughter cells
each have a
single homolog
of each
chromosome type;
allelic complements
may differ

FIGURE 28.5

A, Mitosis in eukaryotes is the mechanism ensuring that each product of cell division has a complete
complement of information (chromosomes) identical to that of the parent cell. **B,** Meiosis in eukaryotes is the
process that reduces the diploid complement of chromosomes and ensures that each gamete has a complete
haploid complement.

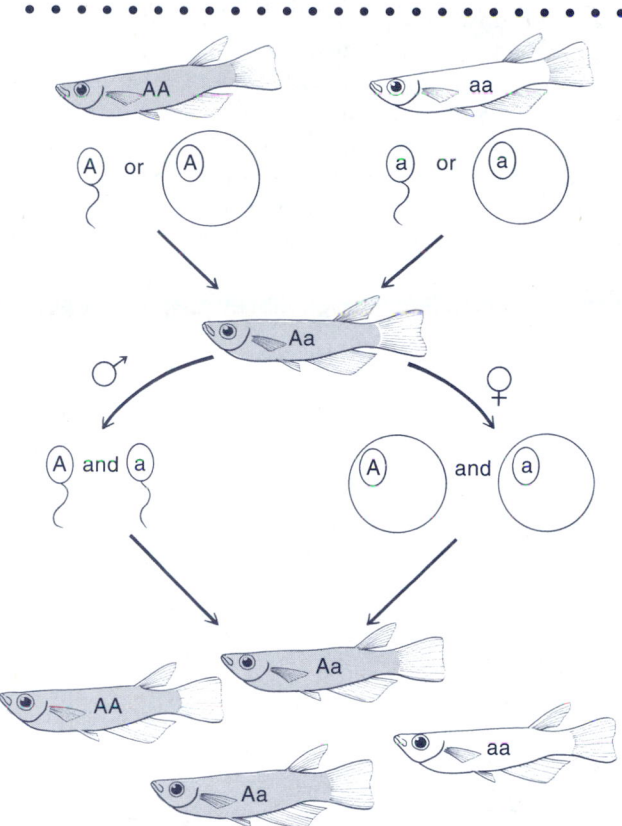

FIGURE 28.6
Inheritance of albinism in medaka follows Mendel's law of segregation. The phenotypic frequency in the second generation of a cross of a homozygous pigmented fish and an albino fish is 3 : 1.

(Wallbrunn, 1958). A second example of partial dominance involves spotting in brown trout (*Salmo trutta;* Skaala and Jorstad, 1988). Normal (**SS**) brown trout mated with a fine-spotted trout (**S'S'**) produce offspring (**SS'**) with an intermediate phenotype. Crosses between heterozygous individuals result in 1/4 normal (**SS**), 1/2 intermediate (**SS'** and **S'S**), and 1/4 fine-spotted (**S'S'**; Fig. 28.7).

Codominance can be observed in some variants of proteins that can be detected electrophoretically. These allelic variants result from amino acid substitutions (mutations) that create small electrical charge differences between their protein products. The differently charged proteins can be separated in an electric field by a technique termed **electrophoresis.** Electrophoretically detectable gene products of alleles at a locus are referred to as **allozymes.** Allozyme alleles differing by a single amino acid (frequently the result of a single DNA nucleotide change) can often be resolved. Protein electrophoresis ordinarily uses a starch or polyacrylamide gel as a support medium. Histochemical stains

specific for the activity of a particular enzyme are used to detect the location of that enzyme on the gel (Murphy et al., 1990). Individuals that are homozygous for an allele at a locus produce a single electrophoretic band, but heterozygous individuals produce multibanded patterns, reflecting the different allozyme products (Fig. 28.8). Whereas no more than a handful of simple Mendelian traits may be available for morphological characteristics in a species, it is relatively easy to identify electrophoretically dozens of protein coding loci. More than 30 loci have been used in electrophoretic studies of a wide variety of fish species, including poeciliids, salmonids, cyprinids, mugilids, clinids, and gasterosteids (Buth, 1984a; Buth and Haglund, 1994; Campton and Mahmoudi, 1991; Haglund et al., 1993; May and Johnson, 1990; Mayden and Matson, 1992; Morizot, 1990; Morizot et al., 1991b; Stepien and Rosenblatt, 1991; Wood and Mayden, 1992). Although potentially more powerful DNA-based methods (based on both mitochondrial and nuclear genes) have been developed and applied to many questions involving the genetics and the biology of

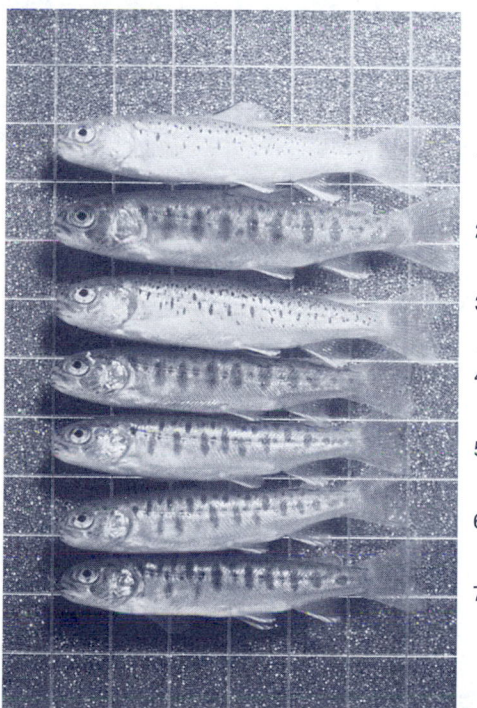

FIGURE 28.7
Inheritance of the spotting pattern in *Salmo trutta* exhibits partial dominance; the phenotypes of both homozygous types are partially expressed in the heterozygote. From the top, 1 and 3 are homozygous for fine spotted, 2 and 4 are normal, and 5, 6, and 7 are heterozygous. (From Jorstad et al., 1991.)

fishes, allozyme analysis remains important, both because it is an alternative tool and because of the sheer abundance of data available for many species. Good science demands that all potential tools be evaluated for suitability in addressing a question.

Mendel's second law extends the behavior of alleles at a locus to the inheritance of more than one trait. Mendel observed that what takes place at one locus does not depend on what happens at another locus—that is, the loci behave independently. Mendel's first law predicts that in crosses between 2 heterozygotes (**AaBb**), 3/4 of the progeny will have the dominant phenotype (**A-** is either **AA** or **Aa**) for the first trait; 3/4 of those (3/4 × 3/4, or 9/16 of the total) will also have the phenotype that is dominant for the second trait and be **A-B-**. The other 1/4 of the **A-** offspring (3/4 × 1/4, or 3/16 of the total) will be recessive (**A-bb**)

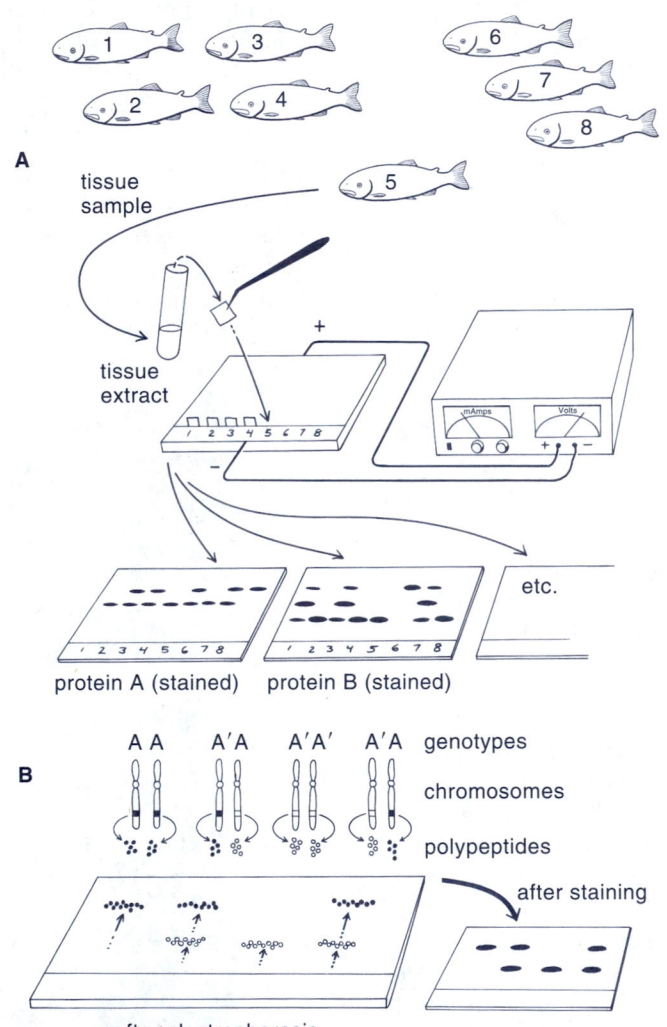

FIGURE 28.8

A, In protein electrophoresis, genetically determined differences in the expression of proteins are examined. Tissues are sampled from individual fishes, cells of the tissues are broken, and samples of the cytoplasm are introduced by filter-paper wicks into a supporting matrix of starch or a similar substance. **B,** An electrical field causes the proteins (charged particles) to move through the matrix. Differently charged (genetically different) polypeptides migrate at different rates (solid and unfilled circles in the figure represent differently charged gene products). The positions of the proteins in the matrix are determined by histochemical stains. (From Gharrett and Utter, 1982.)

for the second trait. The remaining 1/4 of the progeny are recessive for the first trait (**aa**). Of the **aa** progeny, 3/4 are dominant for the second trait (1/4 × 3/4, or 3/16, are **aaB-**), and the remainder (1/4 × 1/4, or 1/16) are double homozygous recessives (**aabb**). The phenotypic ratio expected as a result of the independent assortment of alleles at two loci (9:3:3:1) applies to traits expressed in a dominant–recessive relationship and would be reflected by the cross-hatched portions of Figure 28.9 if **A'** and **B'** were recessive alleles for loci located on different chromosomes. Much of the data available from fishes, however, involves codominant alleles (e.g., allozymes) that permit us to infer genotypes for individual fishes. The genotypic ratios expected for a cross between double heterozygotes is 1:2:1:2:4:2:1:2:1, which is far more informative than the 9:3:3:1 ratio (see Fig. 28.8)!

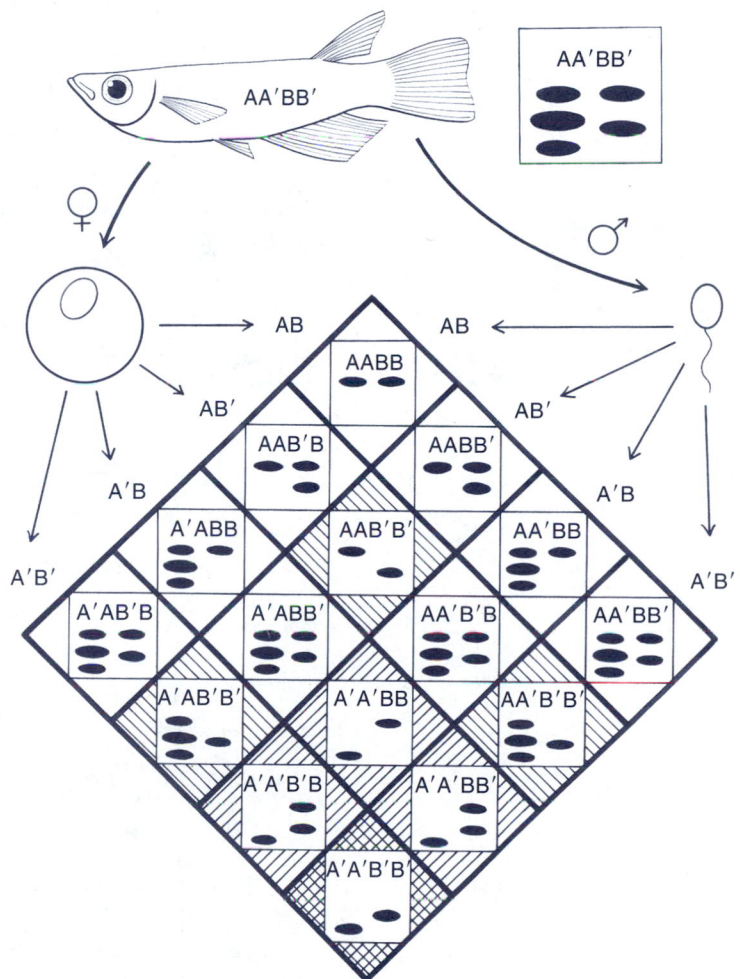

FIGURE 28.9

The expected phenotypic and genotypic frequencies that result from crossing two individuals, each of which is heterozygous for two electrophoretically detectable traits. The genotype and the electrophoretic pattern of the parents and of each progeny type are shown. One of the electrophoretic patterns results from a monomer (a single gene product—polypeptide—that forms a functional enzyme) and shows one (homozygote BB or B'B') or two (heterozygote BB') bands. The other results from a dimer (two gene products—polypeptides—that must aggregate to form a functional enzyme) and shows one (homozygote AA or A'A') or three (heterozygote AA') bands. Electrophoretically detectable traits are often codominant. Diagonal cross-hatching indicates which progeny would be undetectable if the traits were not codominant and shows the 9:3:3:1 ratio predicted by Mendel's law of independent assortment.

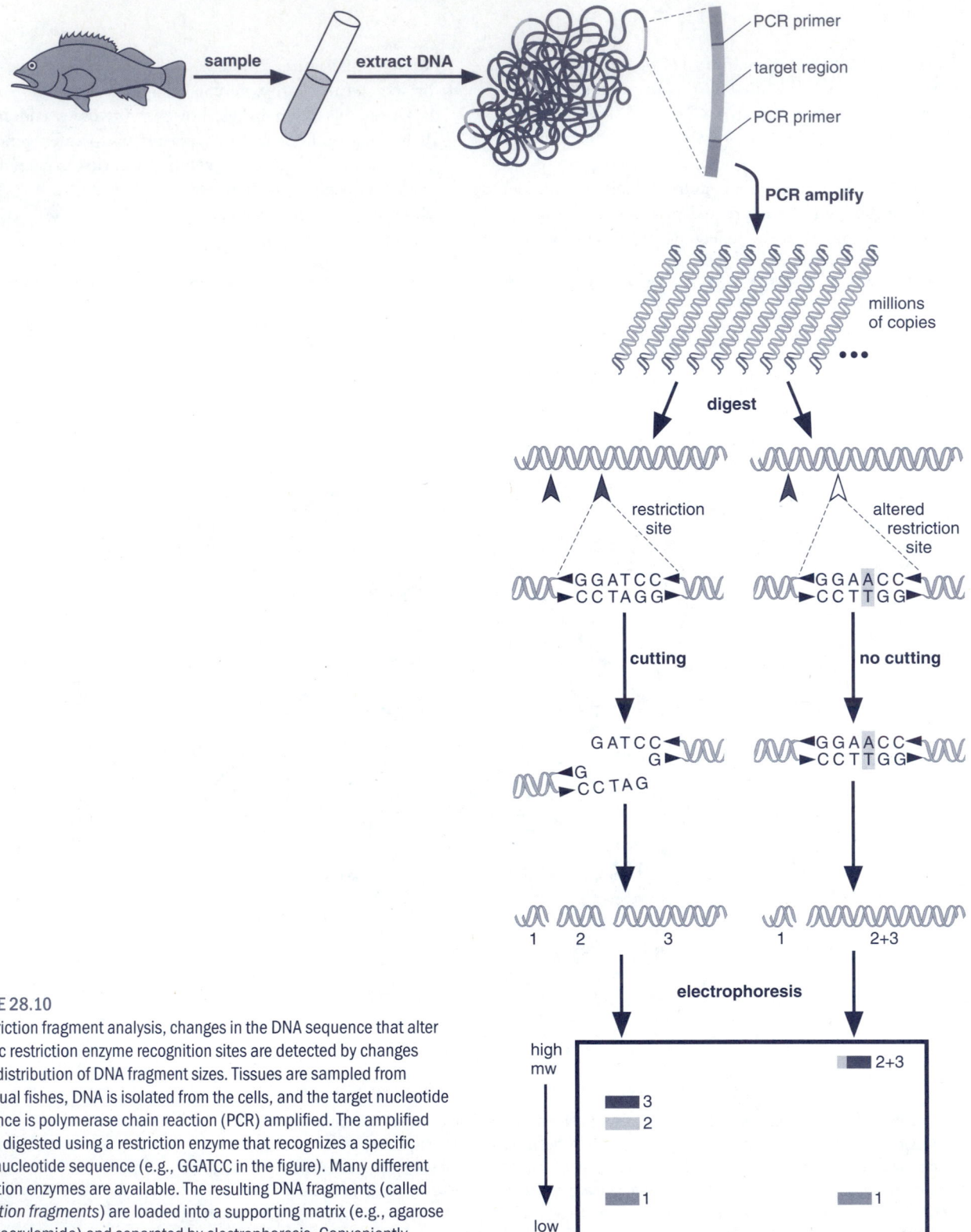

FIGURE 28.10

In restriction fragment analysis, changes in the DNA sequence that alter specific restriction enzyme recognition sites are detected by changes in the distribution of DNA fragment sizes. Tissues are sampled from individual fishes, DNA is isolated from the cells, and the target nucleotide sequence is polymerase chain reaction (PCR) amplified. The amplified DNA is digested using a restriction enzyme that recognizes a specific short nucleotide sequence (e.g., GGATCC in the figure). Many different restriction enzymes are available. The resulting DNA fragments (called *restriction fragments*) are loaded into a supporting matrix (e.g., agarose or polyacrylamide) and separated by electrophoresis. Conveniently, electrophoresis separates DNA fragments according to (the logarithm of) their molecular weight.

During the past decade, methods developed to detect allelic differences in DNA nucleotide sequences have been applied extensively to questions involving nearly every facet of fish genetics. This genetic variation can be detected directly by sequencing target DNA fragments isolated by cloning or amplified by polymerase chain reaction (PCR) techniques (reviewed in Wright, 1993) or detected indirectly by **restriction fragment length polymorphism (RFLP)** analysis (Figs. 28.10 and 28.11). RFLP analysis is done by digesting isolated DNA sequences with *restriction endonucleases*—enzymes that recognize short but specific sequences in the DNA and cleave the DNA. Restriction endonuclease digestion repeatably produces a set of fragments whose length is determined by the nucleotide sequence. An alteration (mutation) in one nucleotide of an endonuclease recognition site will prevent its cleavage and alter the restriction fragment set. After endonuclease digestion, the restriction fragments can be electrophoretically separated by size in agarose or polyacrylamide gels. The positions of the DNA fragments in the gel can be detected using DNA-specific dyes, radioisotopes, immunochemicals, or the ability of the fragments to base-pair with labeled DNA that has complementary nucleotide sequences. The restriction fragment data are not directly applied to most questions, but they can be used to map the location in the DNA sequence of the restriction sites that produce them. Analyses are usually conducted using the presence or absence of the restriction sites as binary characters.

Alleles of **microsatellite** loci are other frequently used DNA markers (Wright, 1993). Microsatellites are tandem repeats of short (generally two to four) nucleotides (see Chapter 4). The number of repeats can range from a few to more than a hundred. Allelic differences at microsatellite loci result from processes such as slippage (slipped-strand mispairing) during DNA transcription and crossovers that occur within the microsatellite locus, but do not align perfectly, and yield products that have a different number of repeats between the nonrepeated flanking sequences (unequal crossovers). Microsatellites are detected using PCR to amplify the repeated sequences lying between the flanking areas. The sizes of the microsatellite alleles are estimated from their electrophoretic mobility, as described for RFLP analyses (Fig. 28.12). Many PCR microsatellite alleles are relatively small (100–400 base pairs) and are resolved by an automatic DNA sequencer.

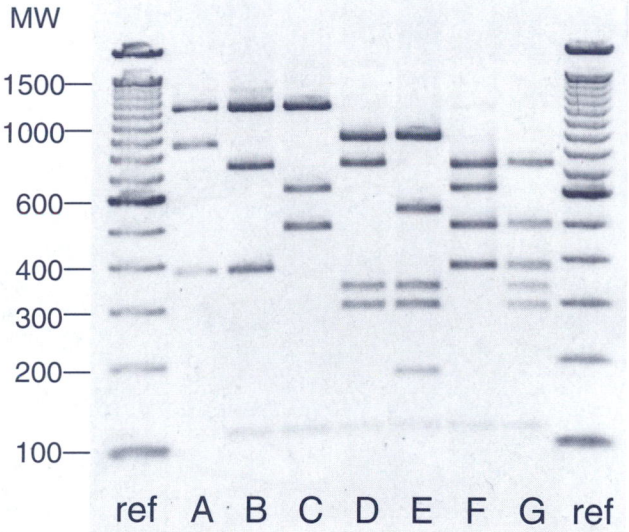

FIGURE 28.11

Fragment patterns from restriction endonuclease *Bst*N I digestion of the region of rockfish (*Sebastes*) mtDNA that codes the NADH dehydrogenase-3 and -4 protein subunits. Numbers on the left side of the figure are the molecular weights (in base pairs) of reference fragments. Lanes A–G are different species: **A**, *S. ciliatus;* **B**, *S. alutus;* **C**, *S. ruberrimus;* **D**, *S. melanops;* **E**, *S. flavidus;* **F**, *S. maliger, S. babcocki,* and others; and **G**, *S. brevispinis.* The group of species that share type F can be distinguished using other restriction enzymes. (From Gharrett et al., 2001.)

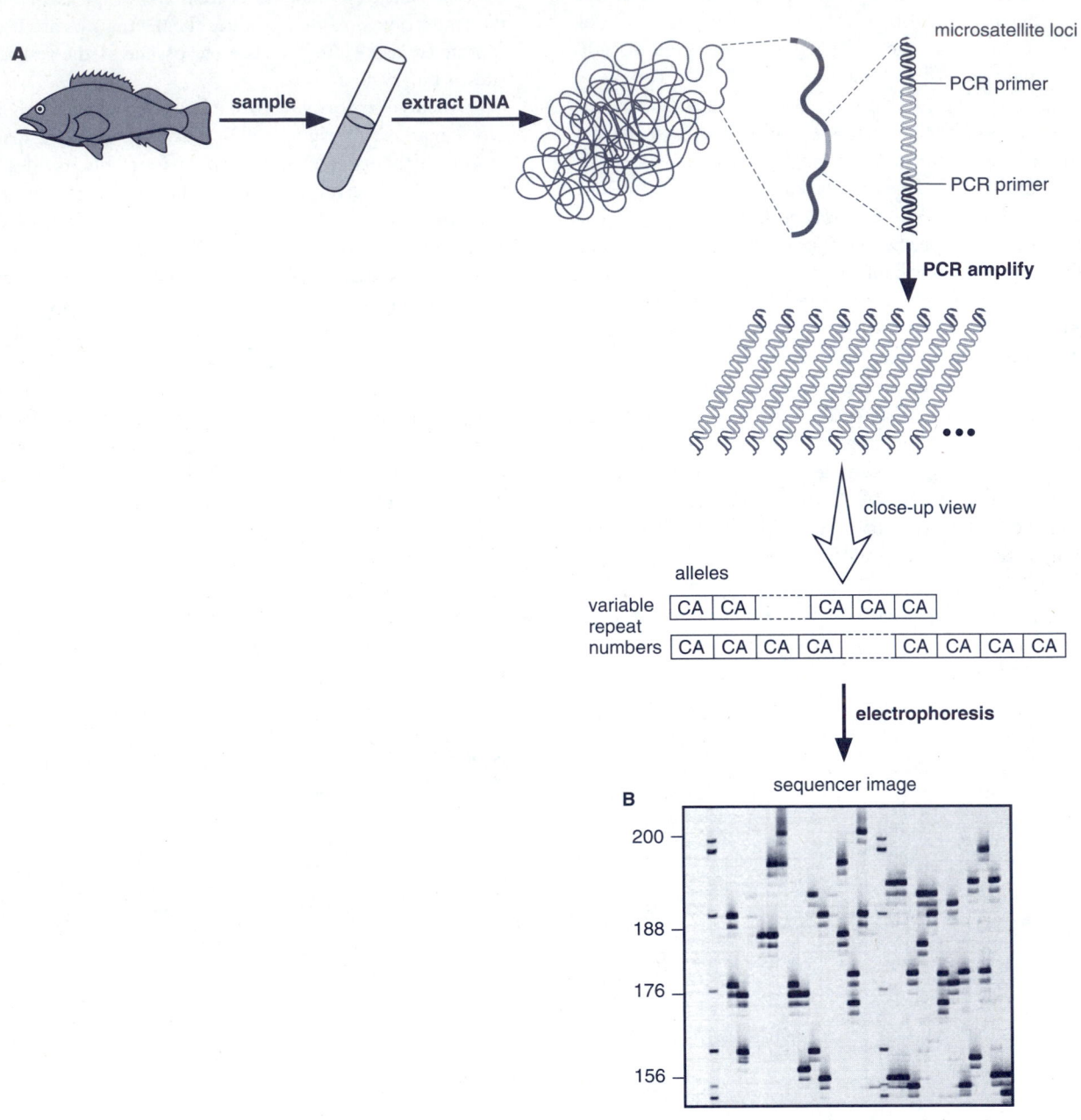

FIGURE 28.12

Allelic differences at microsatellite loci result from different numbers of tandem repeats of very short (ca. 2 to 4) nucleotides. **A,** Tissues are sampled from individual fishes, DNA is isolated from the cells, and the target microsatellite locus is PCR amplified. The sizes of the alleles at the locus are estimated by electrophoresis, often using an automated sequencer. **B,** In the sequencer image, the numbers on the left indicate the sizes of the alleles. Vertical lanes 13 and 30 are size references. Lanes 14, 18, 26, and 29 are faint. Can you see the 19 different alleles in this sample of boccacio (*Sebastes paucispinus*)?

Many additional loci can be detected using other DNA-based methods. Some methods use known DNA sequences to design primers that target specific regions of DNA. Other PCR methods amplify arbitrary anonymous sequences (i.e., the location and function are unknown). For example, **randomly amplified polymorphic DNAs (RAPDs;** Williams et al., 1990) are resolved using random oligonucleotide primers; and **anonymous fragment length polymorphisms (AFLPs;** Vos et al., 1995) are sequences between specific restriction endonuclease recognition sites (these and other methods are reviewed in Palumbi, 1996). Many of the methods that amplify anonymous sequences produce alleles that have a dominant–recessive (i.e., dominant alleles amplify and recessive alleles do not amplify) relationship. The disadvantage in the amount of information that can be obtained from this dominant–recessive relationship may be offset by the large number of loci that can often be resolved.

If Mendel's laws were all there were to genetics, we would find them cited as facts (though important ones) buried in biology texts. In fact, Mendel's laws describe the normal situation, and the rest of the genetics of inheritance elaborates on these rules, describes the underlying mechanisms, or considers the exceptions.

Gene Mapping

One important exception to Mendel's laws occurs because during meiosis, the distribution of genes to daughter cells is mediated by chromosomes. As a result, traits encoded by genes that are located close together on the same chromosome are **linked**. Alleles for linked traits generally are inherited as a unit, violating Mendel's second law (independent assortment). Linkage can be detected from deviations in the phenotypic or genotypic ratios predicted by Mendel's second law.

Recall that although the order of genes on homologous chromosomes is identical, the chromosomes often carry different alleles for those genes. Consequently, for a species to adapt to a dynamic environment, a mechanism that can produce different combinations of alleles on a chromosome would be advantageous. In fact, such a mechanism, called **recombination,** does exist, and it takes place during Meiosis I, when homologs are assembled as bivalents. The likelihood of recombination between two loci depends largely on their physical separation on a chromosome; the further they are apart, the more often recombination will occur between them. The frequency of recombination observed in offspring can be used to infer the distances separating the loci. Allozyme locus linkage maps have been made for a variety of fishes, including some salmoniform, atheriniform, and perciform species (May and Johnson, 1990; Morizot, 1990). An example of an allozyme map for *Xiphophorus* that assigns 56 loci, mostly protein coding loci, to 17 linkage groups is shown in Figure 28.13. More recently, detailed DNA marker–based linkage maps have been constructed for rainbow trout (*Oncorhynchus mykiss;* 476 markers; Young et al., 1998; and 208 markers; Sakamoto et al., 2000), zebrafish (*Danio rerio;* 414 markers; Postlethwait et al., 1994; 705 markers; Knapik et al., 1998; and 1053 markers; Hukriede et al., 1999), and medaka (*Oryzias latipes;* 633 markers; Naruse et al., 2000).

Linkage maps based on or including structural gene loci have advantages over noncoding markers such as microsatellites, RAPDs, and AFLPs, because they provide opportunities to compare the genome structures of different taxa. Although the structural gene loci mapped so far reflect only a small but growing number of the genes in fishes (estimated at tens of thousands), there are numerous linkage groups that are conserved not just among piscine species, but among vertebrates in general, even after 400 million years of divergence! These conserved linkage groups are referred to as **syntenies**. A recent map for medaka assigned 630 molecular and 4 phenotypic markers to 24 linkage groups (Naruse et al., 2000). The markers included DNA sequences that expressed genes involved in the embryological development of vertebrate morphology (*HOX* genes) and antigen (major histocompatibility; MHC) expression. Mammals have a single cluster of MHC genes, which includes both Class I and Class II genes. In zebrafish and medaka, two MHC Class I loci map to one linkage group (chromosome or chromosome region), but the MHC Class II locus maps to a second linkage group. Because the medaka (a beloniform) and the zebrafish (a cypriniform) are members of different taxa (Acanthopterygii and Ostariophysi, respectively), it seems likely that the ancestral teleost had a similarly dispersed MHC organization. However, these genes appear to have arisen in jawed fishes, because the agnathan sea lamprey (*Petromyzon marinus*) has a different system for adaptive immune responses that is based on highly diverse, leucine-rich repeats at a single locus, which form the variable portion of the receptors involved in antigen recognition (Pancer et al., 2004). Examination of additional nuclear genes and inclusion of more taxa will eventually enable us to chart the changes in genomic structure that accompanied (or directed) vertebrate evolution.

Non-Mendelian Inheritance

Most traits result from expression of diploid nuclear genes; however, another set of genes—mitochondrial genes—is inherited independently of nuclear genes. The *mitochondrion* is a subcellular organelle that carries the electron transport apparatus responsible for aerobic metabolism in eukaryotes.

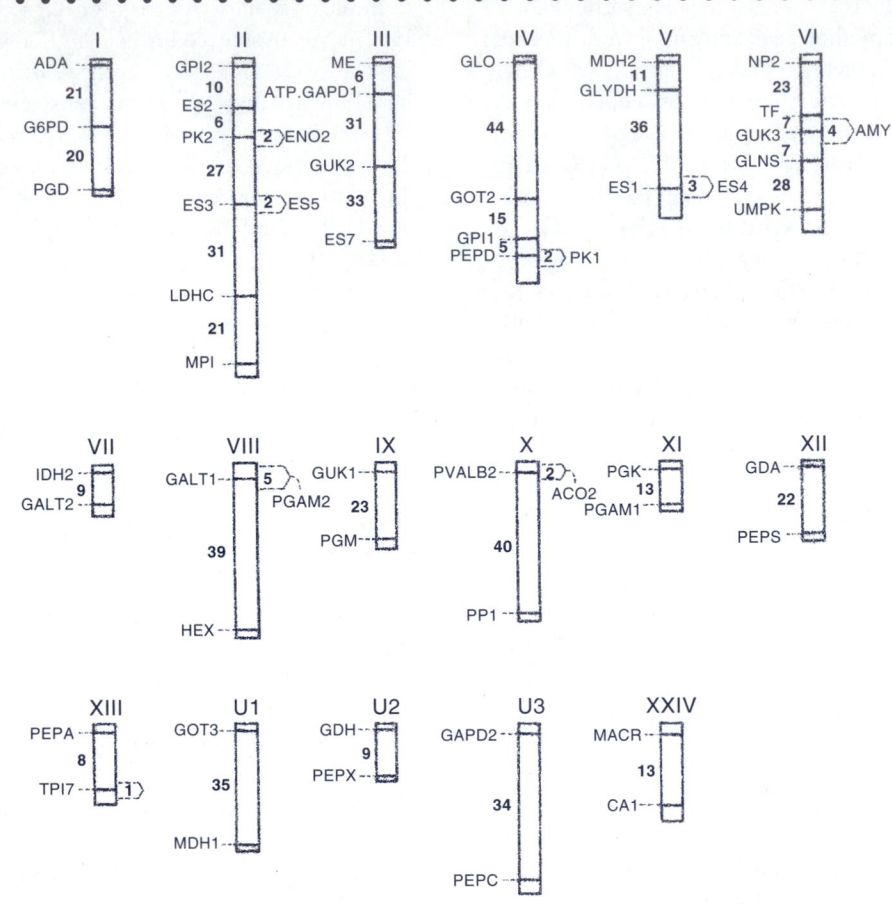

FIGURE 28.13

The linkage map of *Xiphophorus*. Abbreviations (e.g., G6PD) represent protein loci. Roman numerals designate linkage groups, and arabic numerals indicate recombination percentages (Morizot et al., 1991b).

Mitochondria in fish possess a DNA molecule—a circle of about 16,500 base pairs (reviewed in Meyer, 1993) that specifies several mitochondrial proteins (the rest of the mitochondrial proteins are specified by nuclear genes and are transported in from the cytoplasm) as well as some of its own protein synthesis machinery, including rRNAs and tRNAs (Fig. 28.14). Unlike nuclear DNA, mitochondrial DNA (mtDNA) is haploid (N). Although the number varies considerably, each cell carries about 1,000 mitochondria. During reproduction in vertebrates, most or all of the mitochondria received by an offspring come from the yolk of the egg—that is, mtDNA is derived almost entirely as a clone from the female parent.

Mitochondrial DNA comparisons have become an important tool both for phylogenetic and population genetic studies in fishes (see Chapter 4). One advantage of mtDNA is that the rate of change in nucleotide sequences of the mi-

tochondrion appears to be higher than the rate of change in many nuclear DNA sequences (Brown et al., 1979). Other advantages are that mtDNA is a relatively simple molecule (small and haploid) and can be easily isolated. Because there is no known mechanism for recombination between mtDNA molecules from different cells, each unique mtDNA sequence (clone) is treated as a single, complex allele rather than an assemblage of alleles.

The methods used to study mtDNA are similar to those used for studying nuclear DNA sequences, except that a particular nuclear sequence must be resolved from an enormous background of other sequences. The relatively simple techniques used in mtDNA analyses have been extended to studies of a broad range of fish species. Species studied include agnathan (Docker et al., 1999) and chondrichthyan (Rasmussen and Arnason, 1999) species as well as osteichthyans, including most of the major groups, ranging from

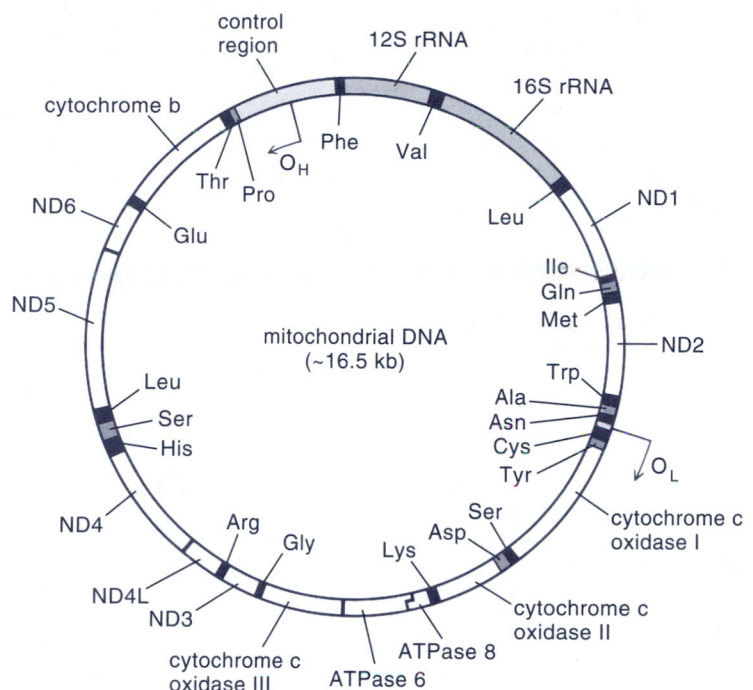

FIGURE 28.14
The mitochondrial DNA genome includes genes for structural proteins, transfer RNAs, and ribosomal RNAs.

the South American lungfish (*Lepidosiren paradoxa;* Meyer and Wilson, 1990) and coelacanths (*Latimeria chalumnae* and *L. menadoensis;* Holder et al., 1999) to butterflyfishes (Chaetodontidae; McMillan et al., 1999), rockfishes (*Sebastes;* Rocha-Olivares et al., 1999), and American eels (*Anguilla rostrata;* Avise et al., 1986). The entire mtDNA genome has been sequenced in numerous species. Rasmussen and Arnason (1999) examined the relationships between gnathostome species and amniote tetrapods. Their study included the common dogfish (*Scyliorhinus canicula*), star-spotted dogfish (*Mustelus manazo*), spiny dogfish (*Squalus acanthias*), starry skate (*Raja radiata*), loach (*Crossostoma lacustre*), carp (*Cyprinus carpio*), rainbow trout (*Oncorhynchus mykiss*), cod (*Gadus morhua*), coelacanth (*Latimeria chalumnae*), bichir (*Polypterus ornatipinnis*), and lungfish (*Protopterus dolloi*).

CHROMOSOMES

Chromosome and Arm Number in a Species

A constant number of chromosomes is usually characteristic of a species. The range of chromosome numbers reported for different fish species ranges from fewer than 20 (16 for the chocolate gourami, *Sphaerichthys osphromenoides*) to well over 400 (446 estimated for the schizothorazine cyprinid *Diptychus dipogon*, which is polyploid; see Table 28.1). Very small **microchromosomes** are found in the more ancient lineages Agnatha, Elasmobranchii, and Chondrostei. Because of the difficulty in counting them, it is likely that even larger chromosome complements exist.

A characteristic chromosome number may vary when chromosome arms rearrange without changing the amount of genetic material. For example, a metacentric chromosome may result from the **fusion** (translocation) of two telocentric chromosomes, or two telocentric chromosomes may result from the **fission** of a metacentric chromosome (see Fig. 28.4). As a result of such rearrangement, the chromosome numbers observed in some species vary over a small range, although the *chromosome arm number* remains constant. For example, the diploid chromosome number in rainbow trout varied from 58 to 64, but the chromosome arm number in each case was 104 (Thorgaard, 1976, 1983).

Among closely related species, the chromosome number is usually similar, but an even greater similarity is often observed for the chromosome arm number. In Salmonidae, the chromosome number varies between 52 and 84, but the arm number is much more restricted, ranging between 100 and 108. Other similarities in arm number can be seen, for example, among centrarchids and percids (Table 28.1). Scrutiny of the number of chromosome arms characteristic of a species suggests that many osteichthyan species have about 50 chromosome arms, whereas others have about 100. Although the chromosome number or chromosome arm number may be similar among various species of cyprinids, closer scrutiny reveals centromere rearrangements, which indicate chromosomal differentiation.

Species that have a larger number of chromosome arms also tend to have about twice as much DNA (Hinegardner and Rosen, 1972), which led Ohno (1970) to propose that some lineages arose through the complete duplication of a chromosome set (e.g., 2N to 4N). An event in which the entire chromosome complement is duplicated is called **polyploidy**. Such an event might result from failure in one cycle of cell division in an early embryo. If the embryo came from parents of the same species, this is called

TABLE 28.1   CHROMOSOME AND ARM NUMBERS (WHEN AVAILABLE) FOR A VARIETY OF PISCINE SPECIES (CALTON AND DENTON, 1974; CHIARELLI AND CAPANNA, 1973; GOLD ET AL., 1980; KIRPICHNIKOV, 1981; YU AND YU, 1990)

Taxon	2N	Arms
Agnatha		
Myxiniformes		
Eptatretus stouti	48*	48
Myxine glutinosa	42*	42
Petromyzontiformes		
Petromyzon marinus	164–168*	–
Gnathostomata		
Chondrichthyes		
Holocephali		
Chimaeriformes		
Hydrolagus colliei	58*	–
Elasmobranchii		
Squaliformes		
Squalus acanthias	62*	–
Rajiformes		
Narcine brasiliensis	28	50
Raja radiata	98*	104
Osteichthyes		
Acipenseriformes		
Scaphirhynchus platorynchus	112*	–
Polyodon spathula	120*	–
Acipenser naccarii	239±7*	–
Lepisosteiformes		
Lepisosteus platostomus	68*	–
Osteoglossiformes		
Hiodon alosoides	50	90
Anguilliformes		
Anguilla rostrata	38	–
Clupeiformes		
Alosa pseudoharengus	48	48
Clupea pallasi	52	60
Cypriniformes		
Astyanax mexicanus	50	90
Cyprinus carpio	104	168
Diptychus dipogon	est. 446	–

(Continued)

autopolyploidy; if the parents were from different species, **allopolyploidy.**

Cell DNA content, chromosomal complements, and duplicated loci that are electrophoretically and immunologically similar but map to different linkage groups (called **isoloci;** Allendorf and Thorgaard, 1984) indicate that the species in the family Salmonidae probably descend from an autotetraploid ancestor (Allendorf and Thorgaard, 1984; Buth, 1983). In contrast, members of the family Catostomi-

dae appear to be descended from an allotetraploid forebear (Ferris, 1984; Uyeno and Smith, 1972). Other taxa that include polyploid species are some old world cyprinids, probably the silurids, acipenseriforms, and possibly a single perciform (reviewed in Buth, 1983). In the oriental loach (*Misgurnus anguillicaudatus*), naturally occurring diploid (50 chromosomes), triploid (75 chromosomes), and tetraploid (100 chromosomes) individuals have been observed (Arai et al., 1991; see Fig. 28.3).

TABLE 28.1   CONTINUED

North American Cyprinidae	48–52	80–100
North American Catostomidae	96–100	–
Siluriformes		
North American Ictaluridae	40–72	62–94
Esociformes		
Esox	50	50
Umbra limi	22	44
Dallia pectoralis	78	118
Salmoniformes		
Coregonus	80	92–108
Oncorhynchus	52–74	104–106
Prosopium	64–82	100
Salvelinus	80–84	100–102
Myctophiformes		
Synodus lucioceps	46	76
Myctophidae	48	48
Gadiformes		
Pollachius virens	40	50
Cyprinodontiformes		
North American Cyprinodontidae	32–48	48–52
North American Poecilidae	46–69	46–69
Nothobranchius rachowi	16 or 18	–
Gasterosteiformes		
North American Gasterosteidae	42–46	54–78
Scorpaeniformes		
Hexagrammus octogrammus	48	48
Cottus pygmaeus	48	48
Perciformes		
North American Centrarchidae	46–48	46–48
North American Percidae	48	48–94
Mugil cephalus	48	48
Mugil curema	28	48
Sphaerichthys osphromenoides	16	30
Pleuronectiformes		
Citharichthys spilopterus	28	48
many others	48	48

* Indicates the presence of microchromosomes, which may be incompletely enumerated.

Induced polyploidy

Intentional manipulation of the number of sets of chromosomes characteristic of an organism (**ploidy manipulation**) began early in the 20th century (Ihssen et al., 1990). Ploidy manipulation is accomplished by interfering with normal fertilization or cell division. If cell division is interrupted at an early stage of embryogenesis, when cell division is synchronous, it is possible to double the ploidy of the embryo, say from 2N to 4N. Processes that disrupt cell division and increase ploidy include heat shock or cold shock, pressure shock, and some chemical or antibiotic treatments (Thorgaard and Allen, 1987). All these treatments interfere with spindle formation, preventing the distribution of chromosomes to the poles of the spindle apparatus, subsequent cytokinesis, or both. A ploidy increase can also be induced by polar body retention. Treatments that interfere with cell division can also cause a zygote to retain its second polar body. Knowing this, it is possible to manipulate the ploidy in fishes much as has been done with plants in agricultural applications that, for example, increase growth rates, sizes, and yields.

Common products of ploidy manipulation are all-female stocks, triploid (3N) stocks, and tetraploid (4N) stocks. All-female stocks are produced by stimulating fertilization with sperm whose DNA has been rendered useless (e.g., by ultraviolet light treatment) without inactivating its motility or its ability to recognize a micropyle. After fertilization, the zygote is heat shocked or pressure shocked to induce polar body retention and produce a diploid complement (the sperm's complement having been inactivated; Fig. 28.15A). In many fish species, the female has two X chromosomes (sex-determining chromosomes); all the resultant embryos of those species are female **gynogens,** which means that all the chromosomes are maternally derived (Fig. 28.15A). An all-female stock can have several advantages in an aquacultural context. For developing a brood stock, the number of eggs is usually limiting. For some species, like pink salmon (*Oncorhynchus gorbuscha*), the roe may at times be more valuable than the flesh, so an all-female stock is considerably more valuable. Finally, females often do not exhibit the pronounced secondary sex characteristics observed in males, and are of greater economic value. An all-female stock can be maintained by using steroids to reverse the sex of some of the females. In species in which males are heterogametic (see Sex Determination section), these sex-reversed phenotypic males have only X chromosome–carrying sperm and produce only female offspring.

Triploids are produced in a similar manner similar to gynogens, but using normal (not inactivated) sperm (Fig. 28.15B). Triploid embryos have one paternal chromosome complement and two maternal complements (Fig. 28.15B). Tetraploids are produced using heat shock or pressure shock after normal fertilization to block one of the early cleavage divisions. Triploids can also be produced by crossing diploid (2N) and tetraploid (4N) individuals: Haploid (N) and diploid (2N) gametes combine to produce a triploid (3N) zygote.

Polyploid fish have application in aquaculture and stocking, because fish that possess an odd complement of chromosomes (3N, 5N, etc.) often are sterile or nearly so. An odd ploidy ordinarily disrupts meiotic pairing and may arrest gametogenesis, particularly oogenesis. If oogenesis is partially or completely arrested at an early stage, the individual may not sexually mature on schedule (or at all) and may convert energy that ordinarily would be destined for gamete production into growth (Thorgaard and Allen, 1987). Delayed maturation can lead to a prolonged life span in **semelparous** species (which spawn once and then die) like the Pacific salmon (see Chapter 27). Early maturation results in revenue losses from precocious males and deteriorating flesh quality in maturing fishes, especially males. To avoid this problem, all-female triploid fish have been used in some Atlantic salmon pen-rearing operations. In addition to their sterility, triploids may outperform diploids after the age of maturation, when energy that is ordinarily devoted to gamete production in diploid fish may be directed toward additional growth in triploid individuals. Stocking sterile triploids into barren systems may produce large fish without the concern of their establishing a self-perpetuating population. Herbivorous sterile triploid grass carp (*Ctenopharyngodon idella*) are expected to reduce the abundance of the aquatic plant hydrilla (*Hydrilla verticillata*) in waters infested by this weed in the southern United States without the risk of establishing an exotic species. The hope of producing nonmaturing, trophy-sized fish has stimulated experimental releases of triploid Chinook salmon (*Oncorhynchus tshawytscha*) in the Great Lakes. Another important reason for using sterile fish in some applications is that the genes of sterile fish cannot introgress into and disrupt wild gene pools.

Although triploids produced by ploidy manipulation are mostly sterile, ploidy mosaicism occurs in the gonads of some triploids, which suggests that some unusual—possibly tripolar—cell divisions occur that can produce diploid cells (Teplitz et al., 1994). If absolute sterility is required, the manipulated species must be carefully evaluated before it is used.

Hybrids

The nature of the chromosomal complement in a species can be responsible for its reproductive isolation. If the alignment of chromosomes in a hybrid fails during meiosis,

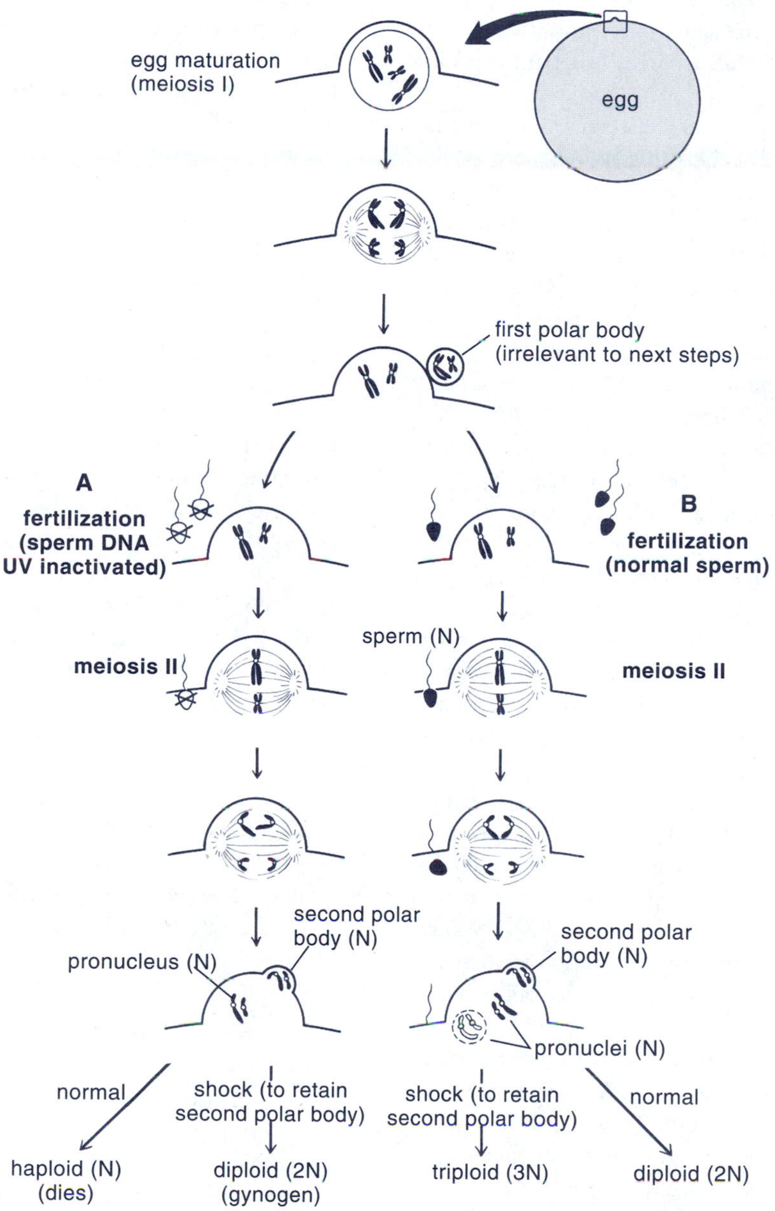

FIGURE 28.15

Chromosome manipulation is accomplished by disrupting cell division at fertilization or very early during development. **A,** Production of all-female individuals (*gynogenesis*) is done by using sperm whose DNA has been inactivated with ultraviolet light or ionizing radiation and then shocking the fertilized egg to induce the retention of the second polar body produced during oogenesis. The resulting individual has only maternal chromosomes. **B,** Triploidy can be induced by using sperm whose DNA has not been inactivated.

because the chromosomal organizations of the two parental species differ, the hybrid may be unable to produce functional gametes and will be sterile. The results of hybridization vary from cross to cross (Chevassus, 1983); hybrids between some species of *Xiphophorus* are often fertile (because many species share the same chromosome and arm number; Morizot et al., 1991a), whereas other interspecies hybrids are nonviable.

Natural hybridization occurs between some sympatric species (see Chapter 27). Dowling and Moore (1984, 1985) observed extensive hybridization between the cyprinid sibling species *Luxilus* (formerly *Notropis*) *cornutus* and *L. chrysocephalus* in the Midwest, but because the hybrids are less fit, the two species are able to maintain separate gene pools. Hybridization can also occur between endemic and introduced species. In Texas, smallmouth bass (*Micropterus dolomieu*), northern largemouth bass (*M. salmoides salmoides*), and Florida largemouth bass (*M. salmoides floridanus*) have been stocked in regions endemic to the Guadalupe bass (*M. treculi*). Genetic surveys have indicated substantial introgression from the introduced bass species into the endemic Guadalupe bass populations (Morizot et al., 1991b).

Rainbow trout (*Oncorhynchus mykiss*) have been stocked extensively in western North America. Where native cutthroat trout (*O. clarki*) are endemic, hybridization often takes place. Hybridization that occurs as a result of such introductions can jeopardize the genetic integrity of endemic species. If habitat loss or degradation (e.g., logging, dams, urbanization) also affects the species, the combined effect may threaten the existence of the endemic species in that area (Campton, 1987). Recently, hybrids have been observed between pink (*O. gorbuscha*) and Chinook (*O. tshawytscha*) salmon in the Laurentian Great Lakes. Neither species is native to the Great Lakes (Rosenfield et al., 2000). Natural hybrids between these species are not common in their native ranges; this illustrates that introductions can produce unexpected results.

Hybridization resulting from ill-advised transfers of fishes may jeopardize some native stocks, but some purposeful hybridizations may be useful. As with sterile triploid fish, sterile hybrids have advantages for recreational "put and take" fisheries and aquaculture. Because the increase in abundance of a sterile population of fish is under the control of the manager or aquaculturist, competition for limited food resources can be avoided, and fishes can realize their full growth potential. The use of sterile hybrid sunfishes in farm ponds (Childers, 1967) and hybrid tilapias in intensive culture are examples of such an application. As with induced triploids, diploid hybrids may not be completely sterile. For example, Simon and Noble (1968) observed survivals of be-

tween 0.09 and 2.1 percent for offspring of F_2 hybrids and back crosses with F_1 hybrids between pink (*Oncorhynchus gorbuscha*) and chum (*O. keta*) salmon.

The concepts of hybridization and polyploidization are brought together in triploid hybrids, which are produced by interspecific fertilization followed by polar body retention. Triploid hybrids, like many interspecies hybrids and triploids, would be expected to be sterile, especially because both hybridization and ploidy manipulation often produce sterility. Moreover, there may be an opportunity for combining the advantageous traits of two species. For example, triploid hybrids between pink (*Oncorhynchus gorbuscha*) and Chinook (*O. tshawytscha*) salmon result in a fish that has the early seawater tolerance of the pink salmon, but the size and flesh quality nearer that of the Chinook salmon. An added bonus is that this fish has a faster growth rate than either species (Joyce et al., 1994).

Although ploidy manipulation provides a useful aquacultural tool, caution must be exercised before embarking on a program that uses hybrid or polyploid sterile fish. First, it must be determined that the surviving fish is indeed hybrid or polyploid (Chevassus, 1983). Reports have been made of triploid, gynogenic, and androgenic (i.e., the genome is wholly paternally derived, possibly by multiple insemination) fishes among presumed hybrids. Second, the fish must be unequivocally sterile—incapable of interbreeding with other hybrids or of backcrossing with indigenous species. Third, although the hybrids and polyploids may not be able to produce viable offspring, they may still participate in spawning activities and thus reduce the productivity of the wild population, as has been purposely done to control the Mediterranean fruit fly in California. Finally, the potential ecological effects of releasing a new organism must be considered carefully.

SEX DETERMINATION

Most fish species have separate sexes, although there are a number of unisexual and hermaphroditic species. **Unisexual** fishes are all female and reproduce clonally (asexually), whereas **hermaphroditic** individuals produce both eggs and sperm, either simultaneously or sequentially. A variety of sex-determining mechanisms have been discovered in fishes.

Chromosomal Determination

The most common form of sex determination results from sex-determining genes carried on particular chromosomes. Female humans, for example, carry two X chromosomes, and male humans carry an X chromosome and a Y chromosome. These chromosomes are homologous pairs during meiosis. The Y chromosome carries male-determining genes, but the X chromosome carries many other essential genes that have no alleles on the Y chromosome. Many species of fishes (e.g., salmonids) share this means of sex determination, although the degree of physical divergence that is observed between human X and Y chromosomes is not always observed.

Some fish species share with many avian species an alternative chromosomal sex-determining mechanism. In this mechanism, the female has two different chromosomes—referred to as W and Z—whereas the male has two Z chromosomes. This mode of sex determination is observed in some poeciliids, some *Tilapia* species, and the fourspine stickleback (*Apeltes quadracus*).

The sex that produces the gametes that carry both sex chromosomes (i.e., X and Y chromosomes or W and Z chromosomes) is referred to as the **heterogametic** sex. The heterogametic sex can be determined cytologically if the sex chromosomes have differentiated sufficiently (e.g., sockeye salmon, *Oncorhynchus nerka*, and rainbow trout, *O. mykiss*; Thorgaard, 1977, 1978). Another method for determining the heterogametic sex is to make gynogenic females, reverse the sex of some of them with steroids to produce phenotypic males, and then cross the two sexes. If normal males of the species are heterogametic (XY), then all the offspring will be females ($XX_\female$ by $XX_{phenotypic\ \male} \rightarrow XX_\female$ offspring). However, if the females are heterogametic (WZ), then the progeny will be a mixture of males and females ($WZ_\female$ by $WZ_{phenotypic\ \male} \rightarrow ZZ_? + WZ_\female + WW_\male$ offspring). The heterogametic sex can also be determined from crosses between normal and sex-reversed fish. Note that ZZ zygotes may not survive.

Some species have more complicated chromosome-based sex determination mechanisms. The freshwater goby (*Gobionellus shufeldti*), for example, has multiple X chromosomes (Pezold, 1984), the characiform *Parodon affinis* has multiple W chromosomes (Filho et al., 1980), and southern platyfish (*Xiphophorus maculatus*) have a combination of X, Y, and Z chromosomes (Gordon, 1946). Tave (1986) described 10 sex-determining systems in fishes.

Environmental Determination

Our ability to reverse the sex of fish by using steroids suggests that, in some species of fishes, similar changes may occur naturally and may be regulated physiologically (i.e., hormonally). In most species of fishes, sex is chromosomally determined, and the sex ratio is 1:1, because the heterogametic sex produces nearly equal numbers of X– and Y–

(or W– and Z–) carrying gametes. However, in addition to sex-determining genes on X, Y, or W chromosomes, some species have autosomal chromosomes that influence sex determination (Kosswig, 1964). The sex of the atherinid Atlantic silverside (*Menidia menidia*) is governed by both the genotype of the fish and the environment (Conover and Kynard, 1981). During the summer, the sex ratio varies. In spring and early summer, 70 to 80 percent of the fishes are female; but later in the summer, the proportion is nearer to 50 percent or less. Differential mortality has been ruled out, and experiments in which developing embryos were subjected to different temperature treatments have indicated that in the silverside, the temperature experienced during development influences sex determination. There also appears to be a genetic component, because the progeny of different females react differently to the temperature treatments. Another species in which sex determination is influenced by temperature is *Poeciliopsis lucida* (Schultz, 1993; Sullivan and Schultz, 1986).

Hermaphroditism

A number of coral reef fishes are sequential hermaphrodites. These fishes spend part of their lives as one sex and part as the other (see Chapter 27). The change is usually triggered behaviorally. The most common form of hermaphroditism is **protogyny**, in which females turn into males (Warner, 1984). At least 14 families have protogynous species, and protogyny is especially common in the families Labridae, Scaridae, and Serranidae. Some harem-living species, like the cleaner wrasse (*Labroides dimidiatus*), have a rigid social hierarchy. The cleaner wrasse feeds by cleaning the skin, mouth, and gills of larger fishes at "cleaning stations." Each male has about 5 or 6 females in his harem. When the male is lost, the dominant female rapidly changes sex and takes over the harem. Within 10 days, the new male is actively producing sperm. In species that have a less rigid social structure, like the saddleback wrasse (*Thalassoma duperreyi*), the change from female to male is determined by the relative number and size of conspecifics in the vicinity (Ross et al., 1983; Ross et al., 1990). Protogyny is often found when being a large male is advantageous. Large males can better defend breeding sites or harems and, as a result, have the opportunity to inseminate many more eggs than they could produce as females.

A less common form of sequential hermaphroditism is **protandry**—a change from male to female. The Sparidae, Pomacentridae, and Muraenidae have protandrous species. The clown fishes or anemonefishes (*Amphiprion*) live near or within stinging anemones, which serve as their protector and provider of food. These fishes are generally found in pairs, a larger female and a smaller male. The loss of the female may result in the male becoming a female and joining with a smaller male (Warner, 1984). In social structures where mates are paired, larger females can produce more eggs, so protandry would be advantageous.

Some simultaneous hermaphrodites have been reported. The hamlets (*Hypoplectrus*) found in the Caribbean alternate male and female roles during their mating. Many deep-sea fishes, which rarely have high densities, are simultaneous hermaphrodites. Because encountering another member of the same species may be a rare event for such dispersed fishes, hermaphroditism provides the advantage of enabling these fishes to spawn with any mature conspecific they meet. The mangrove rivulus (*Rivulus marmoratus*) is the only species known to be capable of internal self-fertilization (Kweon et al., 1998).

All-Female Species

Several species of fish are exclusively female. Most of these species, such as the Amazon molly (*Poecilia formosa*) belong to the family Poeciliidae, which are livebearers. Unisexual species often arise as interspecies hybrids, which have developed mechanisms to avoid genetic recombination (expected in Meiosis I) and so keep their ancestral genome intact. Development of ova is triggered by sperm from males of another species (often one of the progenitor species), which are enticed into inseminating the ova. Of course, the male genome does not become part of the zygote. Analysis by protein electrophoresis has suggested that the unisexual Texas silverside (*Menidia clarkhubbsi*), an oviparous species, probably originated from hybridization between the inland silverside (*M. beryllina*) and the tidewater silverside (*M. peninsulae;* Echelle et al., 1983). It is possible that the Texas silverside arose numerous times from separate hybridization events.

QUANTITATIVE GENETICS

Many traits, including life history characteristics such as fecundity, size, and developmental and maturation timing, are subject to natural selection but are determined by the combined action of many loci and are influenced by the environment. These traits would be expected to reflect adaptive differences among populations that inhabit different environments. Natural selection on polygenic traits is probably the most important evolutionary force that natu-

ral populations experience. Understanding how polygenic traits respond to selection is essential to understanding evolutionary processes. Moreover, these are the kinds of traits that ordinarily are of interest to aquaculturists. Since early times, long before Darwin and Mendel, humans have been empirical animal breeders, who first domesticated animals and then selected these animals for desirable attributes like size, shape, and temperament. Sometimes, these efforts were successful, and sometimes they were not, but today the legacy of these efforts is reflected in the large variety of canine and bovine breeds. Comparable results can also be achieved with fishes—with the advantage that we now understand the process and have some predictive tools that can help determine the feasibility of selective breeding and the amount of effort that would be required to achieve a particular goal.

Polygenic Traits

Many of the agricultural advances that have been made in the 20th century have resulted from directed, artificial selection programs. The potential for similar improvements exists for aquatic species, but relatively little has yet been done to tap that potential. Many of the economically important traits in fishes result from the combined expression of alleles at many different loci, and these traits are referred to as **polygenic** or **metric** traits.

Although each individual locus contributing to a polygenic trait obeys Mendel's laws, it is not possible to follow its individual effects from parent to offspring, because those effects are obscured by the actions of the other loci and contributing environmental effects. Polygenic traits can be measured and have statistical properties: The mean and variance of the trait can be estimated for a sample of individuals. As a result, the demonstration of the genetic basis for the inheritance of a polygenic trait requires statistical analysis—hence the term **quantitative genetics**. The environment also influences the expression of polygenic traits: Even monozygotic ("identical") twins are not perfect replicates. The ultimate expression (i.e., the phenotype), therefore, results from the interpretation of the available genetic information (the genotype) in the context of the particular environment experienced by the individual, or conceptually:

Heritability

The objective of quantitative genetics is to determine the relative importance of the contributions of genotypic and environmental influences on the phenotype. For a group of individuals (e.g., a population), the variation observed in the magnitude of a particular trait (V_p) will in part be due to genetic determinants (V_G) and in part to environmental influences (V_E), so we can refine our conceptual relationship into an equation that partitions the total phenotypic variance:

$$V_P = V_G + V_A$$

The **heritability (in the broad sense;** H^2) of a particular trait is the proportion of the total phenotypic variability (V_p) that is genetic in nature (Fig. 28.16A):

$$H^2 = \frac{V_G}{V_P} = \frac{V_G}{(V_G + V_E)}$$

Alleles at a locus can have dominant, partially dominant, or recessive effects on a phenotype. Moreover, alleles at different loci can interact to produce **epistatic** effects. The overall genetic variation (V_G) accounts for all of these effects, but it can be partitioned into the genetic variation that results from considering each allele as independently contributing an increment toward the phenotype (remember, it is metric) and the variation that is attributable to dominance effects at a locus or to an interaction between loci. These are called the **additive** variation (V_A) and the **nonadditive** variation (V_N), respectively. The additive variation represents the portion of the genetic variation that can be predictably used for selective breeding, and the portion of the total phenotypic variation (V_p) that is additive genetic variation (V_A) is referred to as **heritability in the narrow sense** (h^2):

$$h^2 = \frac{V_A}{V_P} = \frac{V_A}{(V_G + V_E)} = \frac{V_A}{(V_A + V_N + V_E)}$$

Two basic approaches are used to estimate heritabilities—correlation between relatives and realized heritability.

The idea underlying the **correlation between relatives** is that related individuals possess some alleles that are identical (i.e., perfect duplicates of the DNA sequence), either because one relative provided the genetic information to the other (e.g., parent to offspring) or because the information was inherited from a common ancestor (e.g., siblings). Similarities between the phenotypes of related individuals (statistical correlations) that are not observed between unrelated individuals reflect genetic influence. For example, an individual receives one half of its genes from

each parent. Therefore, parent and offspring share one half of the same genetic information (barring mutations), and phenotypic similarities between parent and offspring that are not observed between the offspring and the population in general reflect heritability. Care must be exercised in selecting a particular experimental design, because factors such as egg quality or rearing conditions—which produce environments that are common to members of a family but differ between families—can bias the results. Heritabilities estimated from breeding experiments can be used to predict the extent to which breeding individuals at a phenotypic extreme (**artificial selection**) can alter the phenotype of a population.

Realized heritability is an empirical way of estimating heritability. In this method, the mean phenotype of the progeny of parents selected from one tail of a phenotypic distribution is compared to the mean of the parents relative to the mean of the population as a whole. The extent to which the progeny mean differs from the mean of the population as a whole but resembles the mean of the parents determines the realized heritability:

$$h^2_{\ realized} = \frac{(\bar{x}_{progeny} - \bar{x}_{population})}{(\bar{x}_{parents} - \bar{x}_{population})}$$

If the mean of the progeny is the same as the mean of the population, it is clear that the variation observed is entirely environmental. If, on the other hand, the mean of the progeny equals that of the parents, all the variation observed is genetic (Fig. 28.16B). A prediction of the change in the mean of a population in response to selective breeding can be made by rearranging the equation for realized heritability using heritability estimates from previous experiments.

Heritability estimates have been made in cultured species, including channel catfish (*Ictalurus punctatus*), several tilapia species, mosquitofish (*Gambusia affinis*), common carp (*Cyprinus carpio*), several salmonid species, and the velvet belly shark (*Etmopterus spinax*), for traits as diverse as size at a particular age, number of vertebrae, DDT tolerance, number of pyloric caeca, and disease resistance (summarized in Tave, 1986).

Although little directed selective breeding has been done with aquacultural species (compared to agricultural species), most cultured strains have experienced genetic selection resulting from culture. The very successful Norwegian Atlantic salmon (*Salmo salar*) broodstock was domesticated from an initial mixture of several Norwegian stocks. Many domestic rainbow trout (*Oncorhynchus mykiss*) strains derive from the McCloud River in California, and many farmed channel catfish (*Ictalurus punctatus*) strains originated from

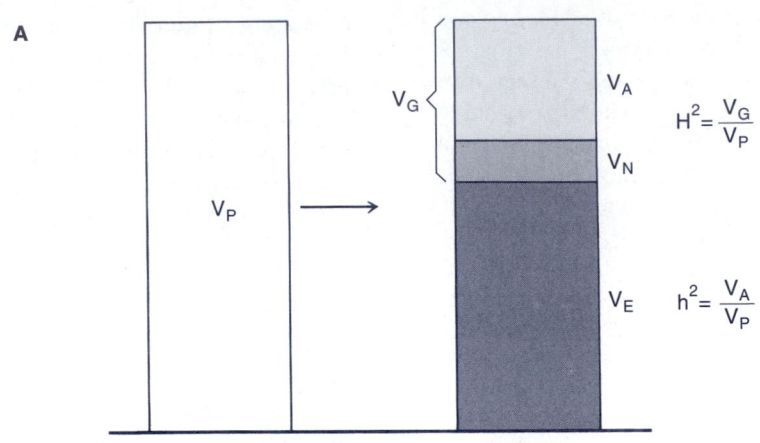

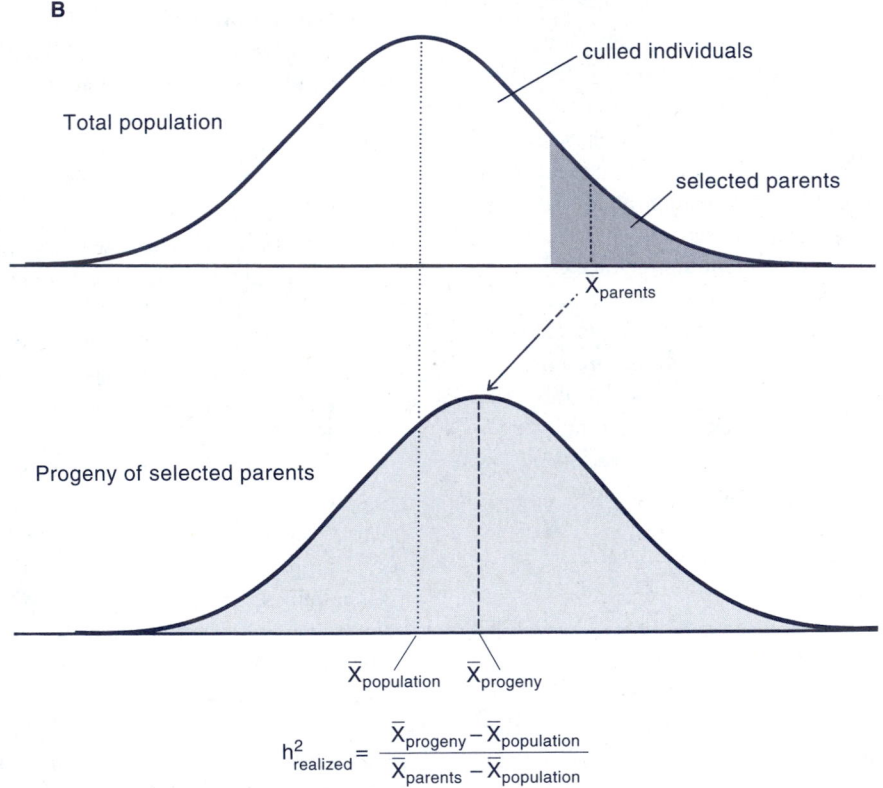

FIGURE 28.16

A, The total phenotypic variability observed in a population can be partitioned into genetic ($V_G = V_A + V_N$) and environmental (V_E) components. From these components, it is possible to estimate the extent to which the traits are determined genetically—that is, are heritable. **B,** Response to selection can be estimated from the change in the population mean of progeny produced by selected parents.

the Red River in Oklahoma. Often, the domestication process is accompanied by considerable divergence among domestic strains that results from unintentional selection by culturists. One indication of the involvement of genetic selection in the domestication process is that both domesticated rainbow trout and catfish have faster growth rates and higher survival rates in the hatchery environment than wild strains do. Domesticated stocks also exhibit different behaviors than wild fish, such as increased susceptibility to anglers (Dunham, 1986).

Directed broodstock improvement is quite possible in intensive aquaculture, where the environment can be closely controlled. A wide variety of traits hold potential for genetic improvement in an aquacultural setting. Siitonen and Gall (1989) advanced the spawning time of a captive rainbow trout stock an average of nearly seven days per generation for six generations. From breeding experiments, Iwamoto and his colleagues estimated that the heritability for weight during freshwater rearing of intensively cultured coho salmon (*Oncorhynchus kisutch*) was between 0.2 and 0.6. This means that between 20 and 60 percent of the size variation observed is attributable to additive genetic variation, which can be selected for. If we assume an intermediate value (e.g., 0.4) and use the largest 20 percent of the fish as breeders, we can rearrange the equation for realized heritability and predict an increase in the average size (response to selection) of about 30 percent per generation (Iwamoto et al., 1982).

Selective breeding has assumed an increasingly important role in fish farming (see Chapter 39). As competition increases, farmers look for production advantages, such as faster growth, resistance to specific diseases, and increased food conversion efficiency, and for improved product quality traits, such as flesh color—all of which respond to selective breeding (e.g., Gjedrem, 2000). Because carnivorous fish are two to three times as efficient as swine and poultry in producing edible food for humans, we can expect fish farming to become an increasingly important source of protein for the world's expanding human population. Several farmed cold-water species, including rainbow trout, coho salmon, Atlantic salmon, and Arctic char, have responded to directed selection in commercial applications (Gjedrem, 2000; Hershberger et al., 1990; Parsons, 1998), and programs to genetically improve warm-water species such as tilapia (Hulata et al., 1986), channel catfish (Wolters, 1993), and carp (Moav and Wohlfarth, 1976) for aquaculture have been underway for more than a decade. Recently, a new catfish variety (USDA 103) has been developed. Whereas the strain now used by farmers requires 18 to 24 months to reach market size, USDA 103 grows to market size 10 to 20 percent faster, which will be advantageous to producers (Weaver-Missick, 2001).

Molecular and quantitative genetics can be combined in some aquacultural applications. Although (by definition) quantitative traits are influenced by numerous loci, the contributing loci do not usually contribute equally to the expression of the trait. In some cases, a small number of loci may contribute disproportionately to phenotypic variation. A Mendelian locus (e.g., allozyme or microsatellite) that is closely linked to a disproportionately contributing quantitative locus can be used as a marker for the trait if linkage disequilibrium exists between the alleles at the marker and the quantitative loci and if the molecular markers and quantitative alleles are correlated. Such a locus is referred to as a **quantitative trait locus (QTL)**. QTLs can be used for **marker assisted selection**. For example, microsatellite loci QTLs were identified for high temperature tolerance in families of rainbow trout (*Oncorhynchus mykiss*) derived from backcrosses of hybrids between two lines—one selected for tolerance to high temperature and the other selected for tolerance to low temperature (Danzmann et al., 1997). Different QTLs were associated with high temperature tolerance in each hybrid family, and epistasis was observed between markers within families.

When one considers the scope of application of selective breeding of agricultural species, it is clear that the potential for genetic improvement of aquacultural species has only begun to be tapped. In contrast to many of the organisms used in agriculture, many wild populations of potentially culturable aquatic species still exist. Each population possesses inherent genetic variability, which is essential to the long-term success of the species; and in the near future, that genetic variation may also be invaluable in improving and maintaining aquacultural strains.

GENETICS AND PHYLOGENETICS

A focal point of phylogeny (and evolution) is that existing species are descended from a common ancestor. The nucleotide sequences in the DNA carry the "operating instructions" for guiding the development of an organism. Those instructions have been passed down from the common ancestor, modified by occasional DNA nucleotide changes, insertions or deletions of sequences, and chromosomal rearrangements. Sequence changes accumulate over time, but the "operating instructions" remain homologous to the ancestral sequence and preserve much of the original nucleotide sequence. Because genes thus carry the historical information of a taxon, the nucleotide sequences and

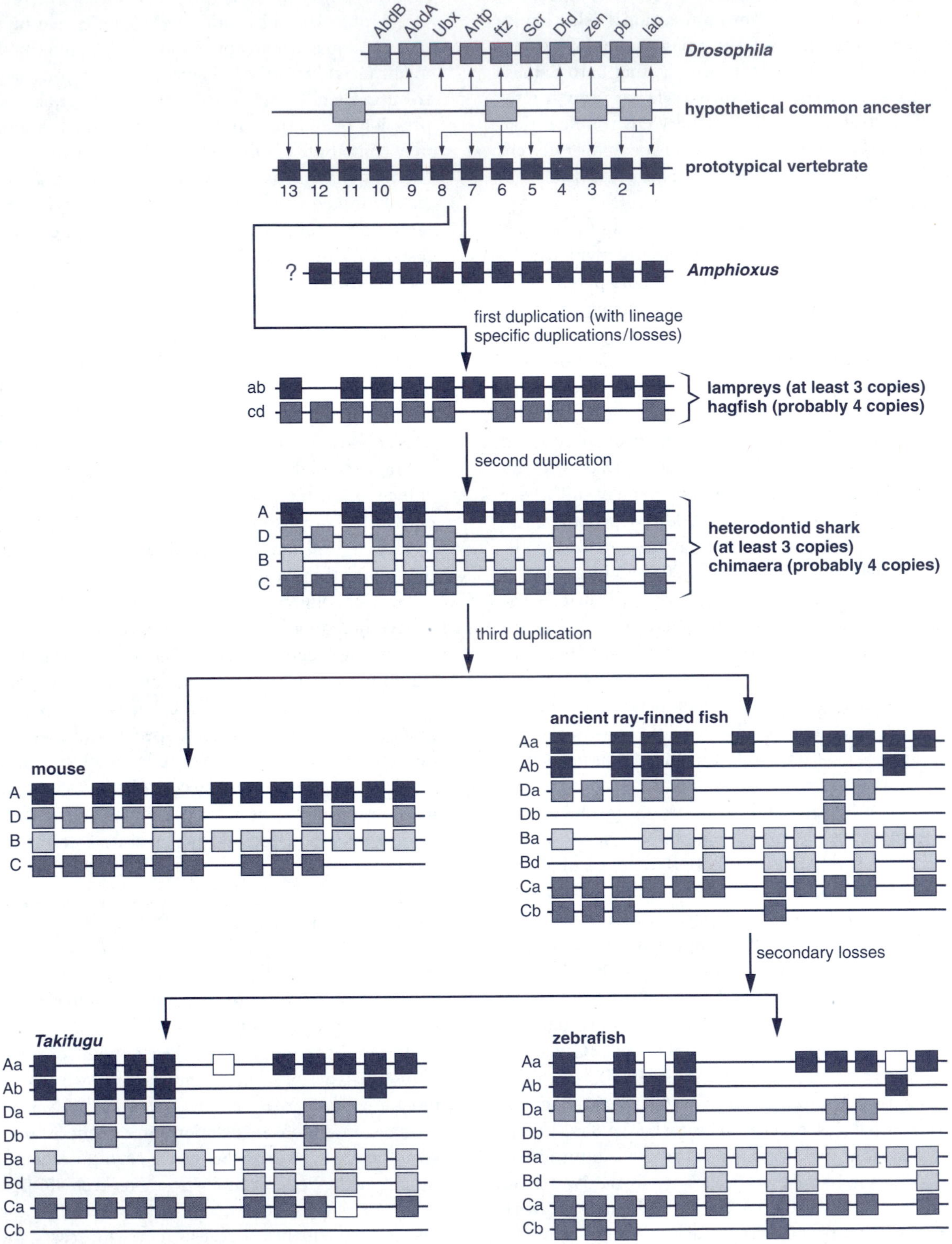

AbdB
AbdA
Ubx
Antp
ftz
Scr
Dfd
zen
pb
lab
Drosophila
hypothetical common ancester
prototypical vertebrate
13 12 11 10 9 8 7 6 5 4 3 2 1
?
Amphioxus
first duplication (with lineage specific duplications/losses)
ab
cd
lampreys (at least 3 copies)
hagfish (probably 4 copies)
second duplication
A
D
B
C
heterodontid shark (at least 3 copies)
chimaera (probably 4 copies)
third duplication
mouse
A
D
B
C
ancient ray-finned fish
Aa
Ab
Da
Db
Ba
Bd
Ca
Cb
secondary losses
Takifugu
Aa
Ab
Da
Db
Ba
Bd
Ca
Cb
zebrafish
Aa
Ab
Da
Db
Ba
Bd
Ca
Cb

gene products, such as allozymes, can provide phylogenetic information. Most historical systematics work has used structural features (e.g., the presence or absence of jaws, the number of gill slits, the location of paired fins) as phenotypic characteristics to assess common ancestry. Ancestral forms preserved in the fossil record provide much of the historical record.

Because the fossil record is often sketchy, incomplete, or nonexistent, many questions about higher levels of systematic relationships remain unanswered. The careful comparison of DNA sequences, which ultimately provide instructions for the homologous structures, provides a molecular paleontological tool to address some of the questions that cannot be answered by existing fossil and comparative structural data; and this tool permits us to approach and to examine some previously intractable questions. Although it is not yet feasible to compare the complete DNA sequences of several billion base pairs of two organisms, it is possible to focus on a limited number of sequences. There is an enormous variety of DNA sequences in a genome. Some genes code for structures that tolerate little modification, and their sequences are strongly conserved. For example, recognizable homology exists between the rRNA sequences of eukaryotes and prokaryotes. In contrast, microsatellite loci are highly variable, often displaying as many as 100 alleles within a single population.

• • • • • • • • • • •
The "Big" Questions

The two big questions in vertebrate evolution are (1) from which ancestors did vertebrates originate, and (2) from which vertebrate line did tetrapods (see Chapter 7) emerge? During the past decade, molecular data have substantially advanced our understanding of the emergence and divergence of vertebrates, but the origin of tetrapods remains unclear. Some of the most exciting work involves studies of genes that regulate the embryonic development of the anterior–posterior axis and appendicular structures in both invertebrates and vertebrates. These genes, called *HOX* genes (short for *homeobox genes;* sometimes designated *hox* or *Hox,* conventions differing between species) specify transcription factors (i.e., DNA-binding proteins), which serve as switches that turn on and off the expression of genes involved in cellular differentiation (see Chapter 2 and Going Deeper, Chapter 27). The expression of *HOX* genes themselves is also controlled by interactions with other regulatory molecules (Ruddle et al., 1999).

HOX genes occur in all animals that have bilateral symmetry; and the closely related *ParaHox* genes (Brooke et al., 1998), which probably originated near the time of the evolutionary emergence of metazoa, occur in all animals, including sponges and Cnidaria (reviewed in Finnerty, 2001). Many invertebrates (e.g., polychaetes and arthropods) have a cluster of 9 or 10 *HOX* genes, although some invertebrates, such as nematodes, have fewer genes, probably as a result of losses (Aboobaker and Blaxter, 2003). The comparison of *HOX* gene nucleotide sequences from invertebrates and vertebrates shows that these genes have been highly conserved and suggests that the complement of 13 *HOX* genes that are observed in vertebrate clusters arose by tandem duplications from a smaller number of ancestral genes (Fig. 28.17). The homology between invertebrate and vertebrate *HOX* genes and the colinearity of their gene order predates the protostome–deuterostome split (Kappen et al., 1989; Ruddle et al., 1994)!

In marked contrast to invertebrates, which have a single cluster of *HOX* genes, vertebrates have at least two and sometimes as many as seven or more *HOX* clusters. The organization of *HOX* genes within the clusters is strongly conserved, and the genes are arranged in the order in which they are deployed during development. The cephalochordate amphioxus (*Branchiostoma*) appears to be the nearest modern invertebrate relative to vertebrates (Stokes and Holland, 1998). The amphioxus and vertebrate species

FIGURE 28.17

HOX cluster organization in chordates, and a model for the evolution of *HOX* clusters in vertebrates (from Amores et al., 2004; Brooke et al., 1998; Málaga-Trillo and Meyer, 2001; Ruddle et al., 1999). Boxes represent the presence of certain *HOX* genes in a cluster. Shading reflects the presumed evolutionary relationships between clusters: Darkly shaded clusters presumably diverged from lightly shaded ancestral boxes. The first duplication and the second duplication produced the two different darkly and lightly shaded boxes (Málaga-Trillo and Meyer, 2001; Ruddle et al., 1999). Empty boxes are unexpressed pseudogenes. Ray-finned fishes have undergone a third duplication. Genes that have been lost from the clusters are represented by spaces; some species appear to have lost entire clusters.

share a number of morphological features, such as a hollow dorsal nerve cord, segmented muscle blocks (somites), and a notochord (see Chapter 2); however, the amphioxus, like other invertebrates, has only a single cluster of *HOX* genes. Genome mapping shows that the multiple *HOX* clusters that are observed in vertebrates are associated with different linkage groups, and homologies between the different clusters strongly suggest that two polyploidy (Postlethwait et al., 1998) events have occurred since the divergence of the vertebrate and cephalochordate (amphioxus) lineages (see Fig. 28.17), although another interpretation of those data is that many segmental duplications have occurred over time (Hughes et al., 2001). It has been presumed that the first polyploidy event provided the genetic raw material necessary for the emergence of vertebrates (Ohno, 1970). A second round of polyploidy probably preceded the divergence of gnathostomes (jawed vertebrates) from agnaths. Although lampreys appear to have three or four *HOX* clusters, and gnathostomes also have at least four *HOX* clusters, comparisons of lamprey *HOX* sequences to those of gnathostomes has indicated that the additional agnathan clusters arose independently by duplication after the split between jawed and jawless lineages (Force et al., 2002, Fried et al., 2003).

A third polyploidy event occurred in the ancestor of actinopterygian (ray-finned) fishes, increasing the number of clusters to at least seven, although some species (e.g., the pufferfish, *Takifugu rubripes*) have since lost *HOX* clusters, and loci within *HOX* clusters have been lost or have become nonfunctional pseudogenes in other species (Málaga-Trillo and Meyer, 2001). No data are yet available for the number of *HOX* clusters in the salmonids and catostomids, which underwent an additional genome duplication prior to species divergence. The sarcopterygians, which include the lobe-finned fishes and tetrapods have the four *HOX* clusters common to all gnathostomes (Málaga-Trillo and Meyer, 2001). Amusingly, DNA sequences from humans are often used to represent this "fish" lineage.

By moving or altering the sequences that regulate *HOX* gene expression and the sequences on which they act, the function of a specific *HOX* gene in an organism can be completely altered, although the *HOX* gene itself remains unchanged. For example, in the mouse, the *Hoxd3* and *Hoxa3* genes occupy the same position in different *HOX* clusters. Mutants of either gene result in defective development. Transgenic replacement of defective *Hoxa3* genes with a normal *Hoxd3* corrects the problem (Greer et al., 2000). The *Hoxa3* and *Hoxd3* structural genes thus appear to be interchangeable; the sequences that regulate their expression determine the timing and extent of their expression during development. We know that at least some of these regula-

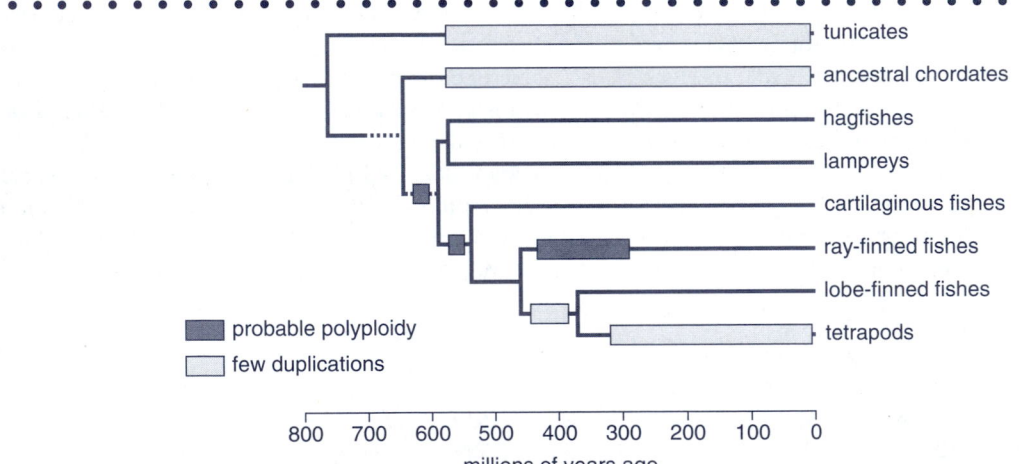

FIGURE 28.18
Phylogeny of chordates, with approximate times of emergence (Kumar and Hedges, 1998) and polyploidy events (Málaga-Trillo and Meyer, 2001; Prince, 2002; Robinson-Rechavi et al., 2004), based on molecular data. Data from Robinson-Rechavi et al. (2004) have indicated that few duplications occurred in some lineages. No firm data are available for the emergence of Cephalochordata.

tory sequences are also conserved, because *HOX* genes in genomic fragments of amphioxus that are introduced into mouse embryos are expressed during the ontogeny of the mouse's head in tissues (e.g., neural crest cells) that do not exist in amphioxus embryos (Manzanares et al., 2000).

The sequential polyploidy events that have been suggested by the increasing numbers of *HOX* clusters probably enabled the emergence of vertebrates and the divergence of gnathostomes from agnaths. These genome duplication events are supported by data collected for many other nuclear protein-coding genes (Robinson-Rechavi et al., 2004; Sharman and Holland, 1996). The third duplication, which took place in ray-finned fishes but not in lobe-finned fishes, has also been supported by data from many other genes (Vandepoele et al., 2004; Wittbrodt et al., 1998; Fig. 28.18).

DNA sequence analyses have also contributed to our present understanding of the origins of vertebrates and the relationships among vertebrates and within and among groups of fishes. Included among the data sets that have been used to evaluate these relationships are mitochondrial DNA (mtDNA), 28S and 18S ribosomal DNA (rDNA) sequences, and many protein-coding nuclear genes. Ribosomal (18S rDNA) and mtDNA sequences have also contributed to our present understanding of the origins of and relationships among vertebrates. Analyses of 18S rDNA have indicated that cephalochordates and vertebrates are a sister group to hemichordates and echinoderms (Cameron et al., 2000). Furthermore, both mtDNA and 18S rDNA analyses have placed the amphioxus as a sister taxon to other chordates (Cameron et al., 2000; Wada and Satoh, 1994). Analyses of 18S and 28S rRNA genes and numerous nuclear protein-coding genes (Mallatt and Sullivan, 1998; Stock and Whitt, 1992; Takezaki et al., 2003) have indicated that agnaths (lampreys and hagfish) are a monophyletic sister clade to gnathostomes and that chondrichthyans are a sister clade to the other jawed fishes (Robinson-Rechavi et al., 2004; Takezaki et al., 2003).

Analysis of sequence data and inference of relationships often is not a simple process. First, to estimate relationships among a number of species, a third species (an *outgroup*) is needed to provide a contrast (see Chapter 3). To be useful, the outgroup should be a sister taxon to all the other taxa being compared, which means that in a tree depicting their relationships, all the species of interest will cluster *before* the branch that includes the outgroup joins the tree. Selection of an appropriate outgroup that is not too distantly related is critical. Second, caution must be exercised in interpreting results based on only one or a few sources of sequence data, which may not truly represent the divergence of the

genome as a whole. Third, the analysis of sequence data is not always straightforward, because the effect of mutations varies enormously from position to position in the nucleotide sequence—that is, not all nucleotide sites are equally informative, and uninformative sites can obscure interpretations. Finally, different tree-drawing methods can infer different relationships.

Some or all of these factors may explain occasional unexpected results. For example, using a broad set of nucleotide sequences, it is clear that chondrichthyans are a sister taxon to bony fishes (Takezaki et al., 2003). Consequently, a chondrichthyan outgroup has been used in studies of the relationships among groups of bony fishes (e.g., Inoue et al., 2001). Unexpectedly, however, comparisons of mtDNA sequences of a variety of bony fishes and several chondrichthyan species that used either echinoderms or lampreys as an outgroup placed the chondrichthyan species cluster as a sister taxon with actinopterygian species, which together were clustered *under* the coelacanths, bichirs, lungfishes, and tetrapods. Neither the lungfishes nor the coelacanths were a sister group to tetrapods (Rasmussen and Arnason, 1999). If these results reflect the true historical relationships among fishes, interpretations such as the ancestral state of elasmobranch gill slits and cartilaginous skeleton would have to be revised! Clearly, caution must be exercised in interpreting molecular data, and conclusions that differ radically from results based on other kinds of data (e.g., morphology) require reevaluation. However, careful applications of molecular methods provide the only means to resolve many questions about the relationships among fishes that have not been solved by other methods.

The second "big" question—from which vertebrate line did tetrapods emerge?—remains unresolved. A study based on 44 protein-coding nuclear genes included sequence data for 10,404 amino acid positions from DNA sequences of mammals, birds, amphibians, coelacanths, lungfishes, ray-finned fishes, and cartilaginous fishes (which were used as the outgroup; Takezaki et al., 2004). The three different methods used to draw phylogenetic trees failed to agree on the relationships among tetrapods, coelacanths, and lungfishes (Fig. 28.19). Where do we go from here? Many genome projects are progressing; and more powerful analytical methods are being developed in support of those projects. Furthermore, by analyzing sequence data from taxa for which relationships are well defined by other criteria (e.g., morphology), the choices of appropriate sequence data can be improved. Finally, data from additional loci will improve resolution. Takezaki et al. (2004) estimated that the amino acid sequence data from at least 200 loci will be

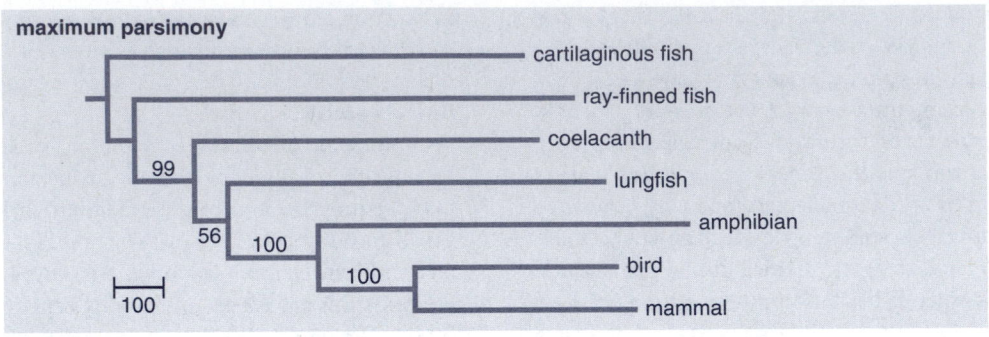

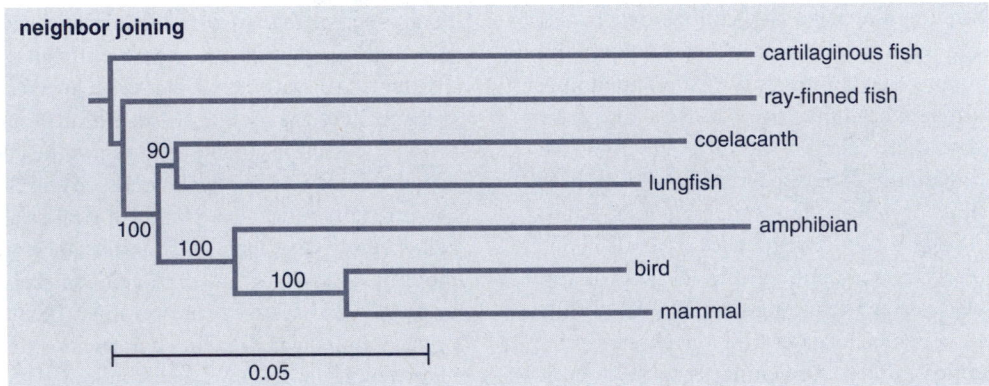

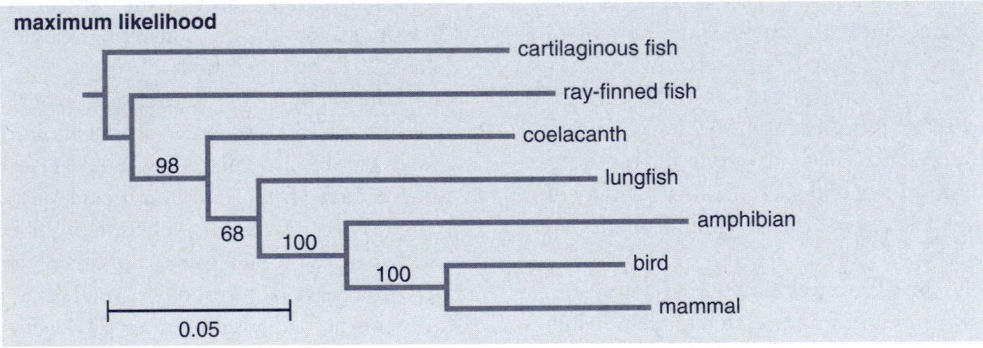

FIGURE 28.19

Three phylogenetic trees that infer the origins of tetrapods (Takezaki et al., 2004). All three trees are derived from the amino acid sequences of the same 44 protein-coding nuclear genes. The trees differ in the algorithms used to draw them: maximum parsimony, neighbor joining, and maximum likelihood; the different algorithms produce different scales. The numbers at the nodes indicate the confidence of the node. Note that the split of coelacanths, lungfishes, and tetrapods not only differs among trees, but that there is lower statistical confidence in two of the trees.

needed. At the rate nucleotide sequence data are accumulating, it should not be long before those data are available and we can learn more about our family tree.

Other Phylogenetic Applications

Genetic analysis provides a powerful tool that can be applied to a myriad of questions about relationships among taxa. Some genes are less conservative than the rRNA and *HOX* genes; at the extreme, there are sequences carried by most eukaryotes that appear to have little or no function and that can diverge freely through mutation and recombination. The variety of genes and the degree to which their sequences are conserved make it necessary to choose the genes carefully for examining divergence between two taxa. Several examples of these diverse applications follow.

Many of the cichlid species that emerged recently in the East African Great Lakes can be discriminated by mtDNA variation (Meyer et al., 1990; Schliewen et al., 1994), but the mtDNA of some species that have clearly diverged morphologically has not yet diverged genetically (Moran and Kornfield, 1993).

Sequences of the mitochondrial genes for cytochrome b and NADH dehydrogenase subunit 3 have been used to resolve phylogenetic questions about the lamprey genus *Lampetra* (Docker et al., 1999). Lamprey taxonomy is challenging, because there are few distinctive morphological characters, and species identification and classification primarily use adult dentition. Comparison of 11 lamprey species, representing three subgenera (*Lampetra*, *Lethenteron*, and *Entosphenus*), resulted in confirmation of the monophyletic relationship of the genus, but reassigned *L. hubbsi* from the subgenus *Entosphenus* to the subgenus *Lampetra*. The study also showed that the subgenera *Lampetra* and *Lethenteron* were not clearly delineated and supported the hypothesis that the three nonparasitic species were derived from parasitic lineages.

The revision of the classification of the family Salmonidae (Stearley and Smith, 1993; see Chapter 11) is a good example of the application of available morphological, paleontological, and molecular data in determining the coancestry of a group of species. In addition to the 119 morphological characters used to examine 33 extant and 4 fossil taxa, the study included molecular data available for mitochondrial DNA and chromosome complements. One important result of this review was the reclassification of Pacific drainage trouts (formerly *Salmo*) as *Oncorhynchus*.

Allozymes also provide important phylogenetic information. The large number of cyprinid species, the simi-larity and sympatry of many species, and their occasional hybridization often confuse their taxonomic status. Allozyme data corroborated the separation of the Rio Grande silvery minnow (*Hybognathus amarus*) from the Mississippi River (*H. argyritus*) and Atlantic slope (*H. regius*) forms of the silvery minnow (Cook et al., 1992). The level of reproductive isolation between the common shiner (*Luxilus cornutus*) and the striped shiner (*L. chrysocephalus*) has been estimated from characteristic electrophoretic differences between the species (Dowling and Moore, 1984, 1985).

Detection and Study of Species

Genetic information can also be used to confirm or resolve new species or to detect naturally occurring hybrids. In 1997 and 1998, two coelacanth fishes were collected off Indonesia, nearly 10,000 km from the Comoros Islands where *Latimeria chalumnae* was described. Preliminary reports described nine morphological and meristic differences between *L. chalumnae* and the newly described *L. menadoensis*. Reexamination of the data from *L. menadoensis* compared to published data for four of the characters of *L. chalumnae* showed no differences, but the comparison of sequences for about 4,800 base pairs of the mtDNA genome verified *L. menadoensis* as a new species (Holder et al., 1999; see Chapter 7).

Speciation takes place when populations diverge to the extent that a reproductive barrier develops. In some cases, such barriers are incomplete, and interspecific hybrids occur naturally. Even when barriers to reproduction are complete, it may be difficult to distinguish morphologically between closely related species. In some instances, allozyme or DNA sequence differences can be used to identify the species of a specimen. Such markers may provide phylogenetic information (Buth, 1984b), resolve new species, detect naturally occurring hybrids, or identify larval forms that have not been adequately described with traditional taxonomic techniques.

The all-female atherinid *Menidia clarkhubbsi*, studied by Echelle et al. (1983), is a cryptic species that is found in brackish waters together with *M. beryllina* and *M. peninsulae*. The species was recognized and subsequently described because its allozyme profile differed from both of the other species with which it was observed. This example of speciation is special in that it conflicts with our concept of a single common ancestor, because this all-female species appears to have emerged in more than one location from hybridization between two similar species.

Another example of the detection of cryptic species resulted from the observation of fixed differences at allozyme loci in Hawaiian bonefishes (*Albula*). Those observations led to a thorough morphological examination, in which two species possessing different morphotypes that clearly corresponded to the two different genetic profiles were identified (Shaklee and Tamaru, 1981).

DNA sequence comparison and other genetic methods provide a powerful phylogenetic tool, and some of the gaps in the fossil record may be closed using the living record encoded in the DNA sequence of modern species. It is important to understand, however, that genetic methods are not a panacea for fish systematics. Nucleotide sequence data are just one of several tools that can be brought to bear on a systematic question, and the results of some sequence comparisons may be inconclusive or may even conflict with results from other sequence comparisons. Genetic information is not a replacement for traditional, morphology-based methods. Only by using *all* the tools available will we be able to formulate the most accurate phylogeny of fishes; even with all the tools possible, it is unlikely that we will be able to unequivocally answer all the questions we now have. Moreover, the resolution of some phylogenetic mysteries will undoubtedly reveal new ones.

EVOLUTION, CONSERVATION, AND MANAGEMENT

Whereas the diversity of species records the results of evolutionary processes, the genetic variation that exists within and between fish populations provides a snapshot of the evolutionary process itself. Evolution acts at the population level. Individuals either survive or perish, but the population evolves in response to changes in natural selection pressures that result from interannual environmental fluctuations and long-term changes. Population divergence results from the adaptation of populations to local environmental differences and from random changes in the genetic complement of a population. The amount and nature of the genetic variation within a population—and, by extension, within a species—contributes to the resilience of a population or species to environmental changes. Most adaptively important characteristics are polygenic traits. Responses to selection that can be achieved with artificial selection on a genetically determined trait in a cultured population also occur in wild populations in response to environmental (selective) pressures. Individuals possessing phenotypes that increase their ability to contribute progeny under prevailing environmental conditions contribute more genes. If the phenotype has a genetic component (i.e., is heritable),

natural selection will increase the incidence of the favored phenotype in the population.

Population Genetics

Although most of the traits that are acted on by natural selection are polygenic traits—which express complex phenotypes, such as life history characteristics—polygenic traits are very difficult to study in wild populations. Consequently, the majority of studies of fish population genetics have focused on single gene characters, such as those that can be resolved by using protein electrophoresis or by surveying mitochondrial or nuclear DNA sequence variations.

Although alleles for some of these traits produce phenotypes that are favored by natural selection or are closely linked to loci that flourish, many alleles (although there is controversy about the extent) probably have little or no influence on fitness; these are referred to as **neutral alleles**. Neutral or nearly neutral traits can reveal information about the movements of fish between populations (**migration** or **gene flow**) and about the extent of genetic diversity within and among populations. Reproductively isolated populations tend to diverge genetically over time. This divergence results from both natural selection and from random processes (**random genetic drift**) that reflect errors in sampling the alleles transmitted between generations. The degree of missampling is inversely proportional to the size of the population—that is, divergence due to random genetic drift occurs more rapidly in small populations. Flipping an unbiased coin provides a good analogy to random genetic drift. Although the chances of observing tails on any one toss is 50:50, in a single toss, you only see the one *or* the other, because a coin landing on its edge is not allowed. If the coin is tossed 10 times, only 24.6 percent of the time will you see exactly five tails. At larger sample sizes (i.e., more tosses) the chance of observing tails exactly one half of the time decreases because there are many more possible outcomes, but, on the average, the proportion of tails observed gets closer and closer to 50 percent as the number of tosses increases.

Population genetics theory predicts that the divergence among populations will reflect the balance between gene flow (leading to homogenization of populations) and random genetic drift (leading to divergence of populations) for selectively neutral loci. One of the most frequent uses of allozyme data by fish geneticists has been to examine the genetic structure of populations and to use the extent of genetic divergence to infer and quantify dispersal patterns. In a study of ten marine shore species of southern California and Baja California, Waples (1987) observed that genetic divergence was inversely correlated with presumed dispersal ability. For example, the estimates of gene flow among black perch (*Embiotoca jacksoni*) populations are low, consistent with this fish being a livebearer and not having marine larvae. In contrast, the estimates of gene flow were quite high for the halfmoon (*Medialuna californiensis*), which has larvae that live offshore for several months and that may be widely dispersed by oceanographic influences. Estimates of gene flow for species whose larvae are found mostly inshore, such as the woolly sculpin (*Clinocottus analis*) and the island kelpfish (*Alloclinus holderi*), were intermediate.

· · · · · · ·
Biogeography

Genetic divergence among populations is also used to infer biogeographic relationships (see Chapter 29). A wealth of studies have included the northern studfish (*Fundulus catenatus;* Grady et al., 1990), the threespine stickleback (*Gasterosteus aculeatus;* Buth and Haglund, 1994), Pacific herring (*Clupea pallasi;* Grant and Utter, 1984), Pacific salmons (*Oncorhynchus;* Beacham et al., 1988; Gharrett et al., 1987; Utter et al., 1989), bluegill sunfish subspecies (*Lepomis macrochirus macrochirus* and *L. macrochirus purpurescens;* Avise and Smith, 1974), and the Tennessee shiner (*Notropis leuciodus;* Mayden and Matson, 1992). Genetically, Pacific herring comprise two obvious races: Bering Sea and northern Pacific Ocean. Grant and Utter (1984) speculated that the differences between these populations reflected repeated Pleistocene glaciations along the Alaskan coast, particularly when the Aleutian chain separated the two bodies of water. Genetic divergence is also observed among the widely separated populations of the threespine stickleback (*Gasterosteus aculeatus*) found in the Atlantic and Pacific basins (Buth and Haglund, 1994). Another, more divergent threespine stickleback group includes most of the samples collected from the Japanese Archipelago.

Biogeographic studies have been advanced dramatically by applications of DNA sequence-based data, because such data introduce a temporal element. Sequence (and RFLP) data provide a time frame, because changes in the sequence (mutations) accumulate one at a time, and the molecular data intrinsically carry a basis for constructing a family tree of genes, called a **gene genealogy** or **gene tree**. Although more divergent taxa may exhibit **homoplasy** (i.e., separate independent origins of the same trait) for a particular gene or sequence, the choice of a more conserved sequence can often resolve that problem. Even though it may be difficult to recognize the root for a gene tree (i.e., determine the ancestral lineage), genes that are contiguous in the tree (e.g.,

AA and AF in Fig. 28.20) are probably more closely related (have diverged more recently) than genes that are distal on the tree (e.g., AA and E or M in Fig. 28.20). In contrast, the set of alleles at an allozyme locus cannot be ordered unequivocally on a temporal scale based on their relative mobilities. That is, there is no dependable way to distinguish between older (ancestral) alleles and recently derived alleles without making inferences based on presumed ancestral relationships. Avise (2000) coined the term **phylogeography,** which couples the concept of gene trees with the geographic distribution of the genes. Phylogeographic studies help us to understand the modern distributions of a species and to interpret the signatures that historic geological processes—such as advances and retreats of glaciers, changes in sea level, and tectonic activity—have inscribed in the population structure.

Mitochondrial DNA provides an especially useful source of data for phylogeographic studies. Because there is virtually no recombination in mitochondria, each molecule (or lineage) carries its own mutational history. Moreover, because mtDNA is haploid (a maternally inherited clone), it is more sensitive to random drift than nuclear genes, and thus population divergence is more rapid.

Avise (1992) summarized studies of 19 marine, freshwater, and coastal species of fishes, birds, and invertebrates in the southeastern United States. In many freshwater and marine species, he observed deep "clefts" in the gene trees that corresponded to the geographic separation of Gulf of Mexico versus Atlantic Ocean populations. This geographic separation is reinforced by present-day oceanographic conditions: The prevailing current flows north and splits at the southern tip of Florida, continuing northward along both the eastern and western Florida coasts. Repeated glacial advances and retreats during the Pleistocene also influenced modern distributions. Although Pleistocene glaciers did not reach Florida, the glacial drop in sea level greatly increased the size of the Florida peninsula and decreased the size of the Gulf of Mexico. Moreover, the climate was more arid, reducing the extent of estuarine and saltmarsh habitat, and the climate was cooler during glacial maxima. Reduced and isolated estuaries may have contributed to the separation of Atlantic and Gulf of Mexico populations.

Another example of the use of phylogeography to unravel the history of a species involves the pink salmon (*Oncorhynchus gorbuscha;* Churikov and Gharrett, 2002). Pink salmon have fixed, two-year life cycles. Recently, there has been virtually no gene flow between broodlines; even- and odd-year broodlines in the same drainage have quite divergent genetic compositions. The geographic distribution

FIGURE 28.20

A minimum spanning network, showing mutational relationships among pink salmon mtDNA haplotypes (from Churikov and Gharrett, 2002). The haplotypes were determined from restriction endonuclease analyses of PCR-amplified fragments that included 97 percent of the mtDNA genome. In all, nearly 10 percent of the mtDNA genome was surveyed by the restriction analyses. The haplotypes represent both odd (darkly shaded) and even (lightly shaded) broodyear samples, each sampled from four populations (10 fishes from each) distributed around the Pacific rim from Southeast Alaska to the Kamchatka Peninsula. The size of the circle reflects the number of each haplotype observed.

of mtDNA haplotypes in their northern range is consistent with the population expansions and declines that would be expected from Pleistocene glacial advances and retreats. The gene tree (see Fig. 28.20) also indicates that, although even- and odd-year populations are generally distinct, their divergence is relatively recent—more recent than the total divergence reflected by the span of the gene tree. One explanation of the pink salmon mtDNA tree is that a breakdown in the two-year pink salmon life cycle occurs periodically, as has been observed following their colonization of the Laurentian Great Lakes (Wagner and Stauffer, 1982).

One of the important lessons that phylogeography has taught us is that genetic similarities between populations do not necessarily mean that the populations are connected by gene flow. An alternative explanation is that the populations were colonized from the same source but subsequently be-

came isolated. Their similarity is historical, not a result of recent processes; but their separation has probably been too recent for measurable divergence to occur.

Local Adaptation and Outbreeding Depression

The life histories of fishes are molded by natural selection (see Chapter 30). The success of a population depends on its ability to coordinate important life history events—such as behavior, timing of spawning, rate of embryonic development, and many other traits—with environmental events. The most successful fishes are those that pass on the most genes to the next generation. The genetic composition of a population may be culled and groomed to produce phenotypes that thrive in the local environment—a process termed **local adaptation**. This process is complex, because the population encounters both interannual and global variation in the environment and may receive immigrants from other populations. Moreover, the most successful genotypes often vary from generation to generation (Geiger et al., 1997).

Local adaptation, which has been well documented in fish populations (see reviews by Taylor, 1991; and Carvalho, 1993), involves an enormous number of loci and alleles. There are probably many different successful combinations among a huge number of poorer combinations. A possible consequence is that the hybridizing of well-adapted populations may lead to a decrease of fitness, rather than "hybrid vigor." The loss of fitness in hybrids between two populations of a species is termed **outbreeding depression**. If the alleles act additively, a decline would be expected in F_1 hybrids, producing what is termed **ecological outbreeding depression**. An intermediate spawning time in hybrids between populations possessing earlier and later spawning times might be less successful than either parental type in its own parental environment, but thrive in an intermediate environment. In some instances, outbreeding depression does not occur until the F_2 generation. This condition, termed **genetic outbreeding depression**, will happen if epistatic interactions between alleles at different loci are important determinants of fitness. In genetic outbreeding depression, complexes of interacting alleles remain intact in the F_1 generation but are disrupted by assortment (Mendel's second law) in the second generation (Fig. 28.21). Outbreeding depression has been demonstrated in largemouth bass (Philipp and Claussen, 1995) and pink salmon (Gharrett et al., 1999; Gilk et al., 2004). Epistatic QTLs were associated with high temperature tolerance in rainbow trout (*Oncorhynchus mykiss;* Danzmann et al., 1999). Not only was the epistatic mechanism shown in this study; but different families had different QTLs for the same trait, demonstrating that different gene combinations can evolve to meet similar environmental challenges.

Genetics and Fisheries Management

The differences in allozyme frequencies among populations provide markers that can be used for stock identification. Species of Pacific salmon (*Oncorhynchus*) are intercepted by fisheries in marine waters as they migrate to their natal streams to spawn, but before they have segregated into discrete stocks. The management of other species, like the witch flounder (*Glyptocephalus cynoglossus*) and the Atlantic cod (*Gadus morhua*), is also complicated by mixed-stock harvests (Fairbairn, 1981; Ruzzante et al., 1996). One of the goals of fisheries managers is the continued productivity of the har-

vested species, which in turn depends on the productivity of each contributing population. Mixed-stock harvesting can result in different exploitation rates for different component populations. In the extreme, some populations run the risk of severe overharvesting or extirpation (see Chapter 39).

For some species, naturally occurring genetic markers exist that can be used to estimate the stock composition of a catch and provide managers with information on stock abundances (Grant et al., 1999; Pella and Milner, 1987; Shaklee and Bentzen, 1998; Shaklee et al., 1999; Ward, 2000). When naturally occurring differences do not exist, it may be possible to genetically mark a population by altering the allozyme frequencies—for example, by selective breeding (Gharrett and Seeb, 1990; Skaala et al., 1990). Allozyme markers have also been used to assess the success of stocking of walleye (*Stizostedion vitreum vitreum-Sander vitreus v.;* Schweigert et al., 1977).

Although allozymes can be used to obtain information about the genetic structure of populations in a species, local adaptation primarily involves response to selection of quantitative traits such as life history characteristics. An example of local adaptation is the orientation in the water current of newly emerged sockeye salmon (*Oncorhynchus nerka*) fry. Sockeye usually rear in freshwater lakes for one or more years before their seaward migration. Fry emerging from redds in tributary or outlet streams must navigate to the rearing lake. Experiments have shown that newly emerged fry from populations that spawn in outlet streams move against the current, whereas fry from populations that spawn in tributaries move with the current (Raleigh, 1967; see Chapter 36).

There also appears to be a relationship between stream temperature and the timing of spawning in sockeye salmon. The timing of fry emergence is critical to the survival of salmon, because the window of food availability is limited; however, the rate of embryonic development, which determines when the fry can emerge, depends largely on the incubation temperature; and a longer incubation period is required in colder water. The obvious solution to the problem is to coordinate the timing of spawning with the incubation temperature and the timing of emergence. Natural selection favors phenotypes that adopt such a strategy, and sockeye salmon in the Fraser River system in British Columbia exhibit a strong relationship between spawning time and average incubation temperature (Brannon, 1987). The duration of embryological development from fertilization to seaward migration in pink salmon (*Oncorhynchus gorbuscha*), which leave freshwater soon after leaving the gravel, has a quantitative genetic component (Hebert et al., 1998). The survival of pink salmon depends on timing their emigration to meet favorable estuarine conditions.

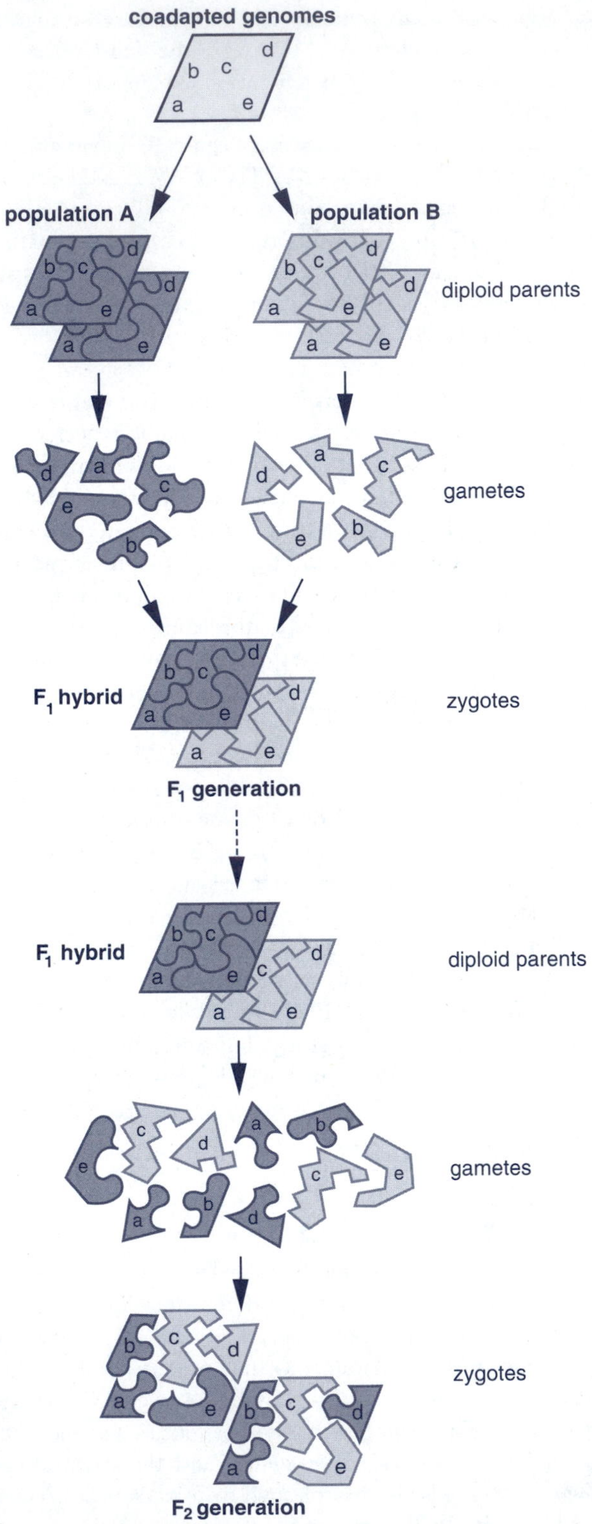

FIGURE 28.21

A conceptual example of outbreeding depression resulting from disrupted coadapted genomes. Multiple interacting (epistatic) alleles at several loci (a – e) are selected jointly to produce a well-adapted phenotype (coadapted genome) for a particular local environment or environment series encountered by a fish population. Alternative co-adapted genomes may be possible (Populations A and B). If two populations possessing different coadapted allele complexes are hybridized, the F_1 hybrid individuals will receive two complete sets of coadapted alleles, which may or may not affect the fitness of the F_1 population. However, the independent assortment of alleles in the second generation disrupts the coadapted genomes in individuals of the F_2 and later generations. Disruption of coadapted genomes by outbreeding may reduce population fitness and hence result in outbreeding depression (from Gharrett and Smoker, 1993).

Conservation Genetics

Conservation geneticists are concerned with maintaining both intra- and interpopulation variation, the loss of which poses a threat of extinction to a species (Nelson and Soulé, 1987). Genetic changes occur naturally over time as a result of environmental changes; however, humans have the ability to cause large changes in the genetic structure of fish populations in a relatively short time, as a result of harvest practices, urbanization and population growth, or hatchery practices (Goodman, 1990; Hindar et al., 1991; Riddell, 1993; Skaala et al., 1990; Thorpe, 1993). The erosion of genetic diversity in a species compromises that species' ability to adapt to normal environmental changes and, perhaps, threatens its existence. Worldwide, the estimate of the present-day extinction rate of species is near an all-time high (in evolutionary time!); the difference is that this rate now results largely from human activities (see Chapter 39).

Recognizing the crisis, the U.S. Congress enacted the Endangered Species Act in 1973 to provide a means for conserving endangered species and protecting the ecosystems on which they depend. In the act, the term *species* is broadly applied to include subspecies and other distinct groupings. To clarify the use of *species* in this context, the term **evolutionarily significant unit (ESU)** was developed to describe a population that was substantially isolated reproductively from other conspecifics and that represented an important component in the evolutionary legacy of the species (Waples, 1991). Controversy over the identification of specific ESUs persists, but the concept has been instrumental in listing a number of endangered ESUs, including Redfish Lake sockeye salmon (*Oncorhynchus nerka*) in Idaho, coastal cutthroat trout (*O. clarki clarki*) in Washington state, and Chinook salmon (*O. tshawytscha*) in the Snake River (Waples, 1995). Important among the criteria for defining an ESU is the genetic structure of the species. Genetic differences can establish that little gene flow occurs between the population of interest and other populations of the species.

Strong genetic divergence—deep clefts in gene trees—provides indicators of genetically distinct lineages. Variation at microsatellite loci shows that the coastal cutthroat trout in Washington State have a strong structure based on their natal creeks and that there is little gene flow between populations (Wenburg et al., 1998). We have also seen that a lack of divergence in molecular genetic traits does not necessarily indicate a lack of divergence in important adaptive traits. Consequently, other information—such as geographical features that reinforce isolation, unique phenotypic or life history characteristics (e.g., run timing or morphology), and indications of unusual or distinctive adaptation to the habitat—

is taken into consideration. For example, the Redfish Lake sockeye salmon are unique because they are the most southern anadromous sockeye population and because they spawn 1,500 km from the ocean at an altitude of 2,000 m—features unequalled by any other sockeye salmon population in the world. Their isolation was unquestioned, because the nearest sockeye population was 900 km away (Waples, 1995). A combination of differences in the spawning habitat of Chinook salmon in the Snake River, consistent genetic differences between Snake River populations and Columbia River populations downstream, and differences in the spawning times of the fall versus the spring/summer Chinook salmon led to the listing of two Snake River ESUs.

DNA analysis methods provide an approach for studying the history of fish species, because DNA can be extracted from archived scale samples and even from the surface of some otolith samples (Hutchinson et al., 1999). A comparison of microsatellite variation in present-day endangered populations of Danish Atlantic salmon with that in scales taken 60 years ago showed small changes in allele frequencies but a sharp reduction in the number of distinct alleles observed (Nielsen et al., 1997). These results are consistent with a decline in effective population size and the resulting erosion of genetic variability. Retrospective studies of archived specimens (if they have not been fixed in formalin, which degrades DNA) provide a powerful tool for conservation geneticists and stress the importance of genetic archives.

RECENT DIRECTIONS

Fishes as Model Systems

Fishes are the oldest, most numerous, and most diverse vertebrates, but they share many aspects of structure, function, and gene expression with higher vertebrates. Both the similarities and differences between fishes and other vertebrates are sources of insight into vertebrate biology. Comparative studies of vertebrate genes and gene complexes have improved our knowledge of their expression and function (reviewed by Powers, 1991). For example, information from fishes has been important for studies of the immune system, of hormones such as vasotocin and insulin, and of the precursor of egg yolk protein, vitellogenin (Chan et al., 1993; Heierhorst et al., 1993; Lazier and MacKay, 1993; Litman et al., 1990; Marchalonis and Schluter, 1990; Pohajdak et al., 1993). From comparative studies, researchers can extract information about the evolution of a gene or gene system, including the variation among the genes of different species that may be necessary for their appropriate function in different biological or ecological contexts. Moreover, as the regulation of gene expression depends on the organization of genes in complex expression systems, differences in gene organization can provide clues to the modes of expression.

The comparison of the immune systems of vertebrates provides an example in which data from fishes reveal alternative gene organizations. Vertebrates share many features of their immune system that do not occur in invertebrates. Of particular interest are the circulating antibodies that are characteristic of all vertebrates, including the agnaths. The organization of the gene complexes that produce circulating antibodies, however, varies. Studies of the horn shark (*Heterodontus francisci*) have revealed a different and, presumably, ancestral arrangement of the battery of genes that produce immunoglobulins (Litman et al., 1990). The gene organization of more advanced fishes, like the ladyfish (*Elops saurus*), rainbow trout (*Oncorhynchus mykiss*), and channel catfish (*Ictalurus punctatus*), resembles that of higher vertebrates (Pohajdak et al., 1993).

Because of the similar gene complement and the relative ease of culture in some species, fishes serve as important models for some kinds of research. Both genetic and physiological experiments can be performed on fish that cannot be done with humans. For example, species of the genus *Xiphophorus* have been particularly important in cancer research. In some crosses, melanomas develop from genes that control pigment patterns (Vielkind et al., 1989). These crosses provide a model for understanding the genetics of tumor production. Other cancer-inducing genes are being studied in other fish species, including rainbow trout (*Oncorhynchus mykiss*), Atlantic tomcod (*Microgadus tomcod*), winter flounder (*Pseudopleuronectes americanus*), and goldfish (*Carassius auratus;* reviewed in Van Beneden, 1993).

The genome structures of vertebrates vary enormously, ranging from the polyploid species, which have multiple copies of genes, to species that possess few redundant genes or highly repeated sequences. The pufferfish (*Takifugu rubripes*) has the smallest known genome of any vertebrate — less than 15 percent the size of the human genome (Brenner et al., 1993). Keeping in mind that most of the genes required for their existence and function are shared by all vertebrates, the "streamlined" genome of the pufferfish thus offers an excellent model for learning about the organization and expression of vertebrate genomes. Sequencing and analyzing this compact genome is one approach to learning about the organization of more complicated (e.g., human) genomes, which are cluttered with sequences that appear to be superfluous. The synteny (see Gene Mapping earlier) of some batteries of genes is highly conserved between the pufferfish and humans (Brunner et al., 1999; Gellner and

Brenner, 1999; Thrower et al., 1996): Eight genes that were found in 68,000 base pairs in the pufferfish occupied more than 600,000 base pairs in humans (Brunner et al., 1999). The genetics of the development of the zebrafish (*Danio rerio*) have received much attention, and many developmental mutations have been identified. Recently, several zebrafish mutants with "human" diseases have been found (Zon, 1999). Because the genome of the zebrafish has been mapped, these developmental mutants make the zebrafish a model for studying human diseases (see Chapter 27, Going Deeper).

Genetic Engineering

To many people, the term *genetic engineering* conjures up an image of mad scientists transplanting genes from one organism to another (*transgenics*) in exotic combinations to produce "monsters" (see Chapter 39). Although there is potential for abuse, there is also potential for societal benefits, such as enhanced protein production and superior food quality. Because of the potential risks, transgenic research and product development in many countries are stringently regulated to prevent applications that might become ecological or genetic threats.

What most people do not realize is that transgenic animals are not easy to produce. Moreover, the domestication of plants and animals and most of the agricultural advances made in the last century are nontransgenic products of genetic engineering; natural and artificial selection both engineer the genetic compositions of affected populations. The development of transgenic engineering provides another powerful tool that, as with all tools, should be used with care. The development of molecular techniques has made genetic engineering possible and now provides exciting possibilities for developing aquacultural products.

An entire industry has emerged that is based on genetically modified bacteria that produce biochemicals. For example, the human insulin gene has been cloned into bacteria, which now produce insulin that is identical to that produced by humans and is used to treat diabetes. The process involves introducing a functional gene from an exogenous source and inducing the recipient organism to express that gene. A number of experiments have been conducted to construct transgenic fishes. The results of those experiments have indicated that transgenic modification of fishes is possible, but not necessarily easy. Moreover, it is not always possible to predict the outcome of such an experiment.

In a successful and permanent gene transfer, a gene (i.e., DNA) must be introduced into the nucleus of a recipient cell and be integrated into its genome, transmitted to its progeny, and expressed in the progeny. Although the process is conceptually simple, there are many practical challenges to overcome. The introduction of the genetic material is ordinarily accomplished by microinjection or by electroporation—analogous to electrophoresis through the cell membrane (Cloud, 1990; Inoue et al., 1990). Ova or embryos are ordinarily used, because there are only one or a few nuclei, the nuclei are accessible, and the tissue has not irreversibly differentiated into cell lines other than the germ line. Unfortunately, there is little control over *where* the gene integrates into the genome, and the expression of the introduced gene depends somewhat on the location at which it inserts into the recipient's genome. Furthermore, the mere integration of a gene into the recipient genome is no guarantee that it will be expressed, because gene expression is a result of regulation that ordinarily involves sequences that signal the position of the gene and the circumstances under which it is to be expressed. Therefore, most genes that are introduced include a **promoter** sequence, which—as its name indicates—promotes the expression of the gene (see Fig. 28.2). The expression of many eukaryote genes is also enhanced or hindered by complex interactions with DNA-binding proteins called **enhancement factors,** such as those involved in *HOX* gene expression. Even if the gene is successfully integrated, it may disrupt an essential portion of the genome, or it may not end up in the germ line and be inherited.

The promise of genetic manipulation is the ability to design organisms that have characteristics not found in nature. This promise of genetic engineering also raises serious concerns. Engineered organisms—whether planned or unplanned—may have a competitive advantage over native species and displace them or even disrupt the ecosystem. Because of the unpredictability and the potential adverse effects, caution must be exercised in developing and testing transgenic fishes (Kapuscinski and Hallerman, 1990).

Genes coding for growth hormones have been successfully introduced into fish, and the expression of those genes has been documented. In many cases, faster growth has been observed. For example, a Chinook salmon (*Oncorhynchus tshawytscha*) growth hormone gene, driven by an ocean pout (*Macrozoarces americanus*) antifreeze promoter, has been successfully introduced into Chinook and coho (*O. kisutch*) salmon and rainbow trout (*O. mykiss;* Devlin et al., 1995). Tilapia growth hormone, driven by human cytomegalovirus regulatory sequences, has been introduced into tilapia (*Oreochromis hornorum;* Martínez et al., 1999). The transgenic coho salmon experienced a 10-fold increase in average size at 15 months of age compared with controls. In all these species, some of the fish were mosaics, expressing the gene in some tissues but not in others. However, some

of the fishes were transgenic, and in tilapia, the introduced genes were followed into the fourth generation. Much more work is required to ensure that there are no additional unexpected deleterious results, such as sterility, lethargy, and weakness, which have accompanied growth hormone gene transfer in swine (Marx, 1988).

In another transgenic application, a gene for an antifreeze protein, which is synthesized by some species of Arctic fishes, like the winter flounder (*Pseudopleuronectes americanus*), has been successfully introduced into Atlantic salmon (*Salmo salar*), with the hope that it might increase the survival of pen-reared fish in very cold waters.

There is an obligation for the geneticist and pisciculturist to thoroughly examine the potential genetic and ecological effects of culturing transgenic fishes where they may escape or of releasing them into natural systems. It is difficult to predict the possible negative results that may occur, but a study by Muir and Howard (1999) has suggested that a trait (like size) that increases mating success might give the transgenic fish an advantage in a wild system. In that system, even if the progeny of the transgenic line are less viable, the transgene (which they refer to as a "Trojan" gene) may flourish in the population because of sexual selection and ultimately erode the fitness of the population.

Although there are caveats to genetic engineering, the potential for increasing protein production to meet the world's growing demand in the face of dwindling resources will keep this technology in the foreground. The economic potential of transgenic organisms is boundless; only our imagination limits the possibilities. If we apply our knowledge wisely, during the next few decades, we should be able to benefit from natural production and also enjoy exciting new aquacultural products.

· ·
KEY POINTS AND CONNECTIONS
· ·

• The information carried in the DNA of a fish directs the processes by which a single fertilized cell differentiates into specialized cells and tissues that develop into a complex organism. In particular, homeobox (*HOX*) gene clusters regulate the embryonic development of the anterior-posterior axis and appendicular structures in fishes. *HOX* and other genes are important to our understanding the genetic basis of structure in fishes.

The morphology of fishes is presented in Chapter 2. Reproduction and embryological development in fishes are described in Chapter 27.

• Until recently, the systematics of fishes has been based almost exclusively on morphology. Molecular methods provide another set of tools to examine fish phylogeny. In particular, DNA sequence analyses have developed rapidly and have provided insights into relationships and evolutionary events that could not be addressed by morphology alone. For example, the *HOX* gene clusters provide evidence of episodes of complete genome duplication that were probably instrumental in the emergence of vertebrates and gnathostomes. These observations are currently being evaluated using data from other genes.

An introduction to fish classification is presented in Chapter 3. The origin of fishes is considered in Chapter 4, and the emergence of gnathostomes is examined in Chapter 5.

• A variety of factors has produced the existing population genetic structures of species. Over longer times, these factors have also led to speciation. Notable factors are climatological, geological, and oceanographic features that change over time and lead to the fragmentation and isolation of populations and to subsequent population expansions. Furthermore, fishes exhibit a variety of behaviors that can lead to reproductive isolation, including homing to natal grounds for spawning (e.g., salmonids) and mating behaviors.

The distribution of fishes is considered in Chapter 29, migration is discussed in Chapter 36, and fish behavior is examined in Chapters 36 and 37.

• As the human population continues to grow, the demand for high-quality protein is also increasing. That demand has led to severe depletions in many of the world's harvested fish stocks as a result of overfishing. The demand for high-quality protein is being met to some extent by aquaculture, but in the face of an inexorably increasing human population, it is unlikely that many stocks will be able to recover and likely that many others will be jeopardized. Human population growth has also led to habitat loss as a result of urbanization, habitat degradation, extraction of water resources, and pollution.

The use of fishes as resources and their conservation are discussed in Chapter 39.

• •
REFERENCES
• •

Aboobaker, A. A., and M. L. Blaxter. 2003. *Hox* gene loss during dynamic evolution of the nematode cluster. *Current Biol. 13:* 1–4.

Allendorf, F. W., and G. A. Thorgaard. 1984. Tetraploidy and the evolution of salmonid fishes, pp. 1–53. In *Evolutionary genetics of fishes,* B. J. Turner (Ed.). Plenum Press, New York.

Amores, A., T. Suzuki, Y.-L. Yan, J. Pomeroy, A. Singer, C. Amemiya, and J. H. Postlethwait. 2004. Developmental roles of puffer *Hox* clusters and genome evolution in ray-fin fish. *Genome Res. 14:* 1–10.

Arai, K., K. Matsubara, and R. Suzuki. 1991. Karyotype and erythrocyte size of spontaneous tetraploidy and triploidy in the loach *Misgurnus anguillicaudatus. Nippon Suisan Gakkaishi 57:* 2167–2172.

Avise, J. C. 1992. Molecular population structure and the biogeographic history of a regional fauna: A case history with lessons for conservation biology. *Oikos 63:* 62–76.

———. 2000. *Phylogeography: The history and formation of species.* Harvard University Press, Cambridge, MA.

———, and M. H. Smith. 1974. Biochemical genetics of sunfish. I. Geographic variation and subspecific intergradation in the bluegill, *Lepomis macrochirus. Evolution 28:* 42–56.

———, G. S. Helfman, N. C. Saunders, and L. S. Hales. 1986. Mitochondrial DNA differentiation in North Atlantic eels: Population genetic consequences of an unusual life history pattern. *Proc. Natl. Acad. Sci. USA 83:* 4350–4354.

Beacham, T. D., R. E. Withler, C. B. Murray, and L. W. Barner. 1988. Variation in body size, morphology, egg size, and biochemical genetics of pink salmon in British Columbia. *Trans. Am. Fish. Soc. 117:* 109–126.

Brannon, E. L. 1987. Mechanisms stabilizing salmonid fry emergence timing, pp. 120–124. In Sockeye salmon (*Oncorhynchus nerka*) population biology and future management, H. D. Smith, L. Margolis, and C. C. Wood (Eds.), *Can. Spec. Publ. Fish. Aquat. Sci. 96.*

Brenner, S., G. Elgar, R. Sanford, A. Macrae, B. Venkatesh, and S. Aparicio. 1993. Characterization of the pufferfish (*Fugu*) genome as a compact model vertebrate genome. *Nature 366:* 265–268.

Brooke, N. M., J. Garcia-Fernàndez, and P. W. H. Holland. 1998. The *ParaHox* gene cluster is an evolutionary sister of the *Hox* gene cluster. *Nature 392:* 920–922.

Brown, W. M., M. George, Jr., and A. C. Wilson. 1979. Rapid evolution of animal mitochondrial DNA. *Proc. Natl. Acad. Sci. USA 76:* 1967–1971.

Brunner, B., T. Todt, S. Lenzzner, L. Stout, U. Schilz, H.-H. Ropers, and V. M. Kalscheuer. 1999. Genomic structure and comparative analysis of nine *Fugu* genes: Conservation of synteny with human chromosome Xp22.2–p22.1. *Genome Res. 9:* 437–448.

Buth, D. G. 1983. Duplicate isozyme loci in fishes: Origins, distribution, phyletic consequences, and locus nomenclature, pp. 381–400. In *Isozymes: Current topics in biological and medical research,* Vol. 10, Genetics and evolution, M. C. Rattazzi, J. G. Scandalios, and G. W. Whitt (Eds.). Alan Liss, New York.

———. 1984a. Allozymes of the cyprinid fishes: Variation and application, pp. 561–590. In *Evolutionary genetics of fishes,* B. J. Turner (Ed.). Plenum Press, New York.

———. 1984b. The application of electrophoretic data in systematic studies. *Ann. Rev. Ecol. Syst. 15:* 501–522.

———, and T. R. Haglund. 1994. Allozyme variation in the *Gasterosteus aculeatus* complex, pp. 61–84. In *The evolutionary biology of the threespine stickleback,* M. A. Bell and S. A. Foster (Eds.). Oxford University Press, Oxford.

Calton, M. S., and T. E. Denton. 1974. Chromosomes of the chocolate gourami: A cytogenetic anomaly. *Science 185:* 618–619.

Cameron, C. B., J. R. Garey, and B. J. Swalla. 2000. Evolution of the chordate body plan: New insights from phylogenetic analyses of deuterostome phyla. *Proc. Natl. Acad. Sci. USA 97:* 4469–4474.

Campton, D. E. 1987. Natural hybridization and introgression in fishes: Methods of detection and genetic interpretations, pp. 161–192. In *Population genetics in fishery management,* N. Ryman and F. M. Utter (Ed.). University of Washington Press, Seattle.

———, and B. Mahmoudi. 1991. Allozyme variation and population structure of striped mullet (*Mugil cephalus*) in Florida. *Copeia 1991:* 485–492.

Carvalho, G. R. 1993. Evolutionary aspects of fish distribution: Genetic variability and adaptation. *J. Fish Biol. 43*(Suppl. A): 53–73.

Chan, S. J., Q.-P. Cao, S. Nagamatsu, and D. F. Steiner. 1993. Insulin and insulin-like growth factor genes in fishes and other primitive chordates, pp. 407–418. In *Biochemistry and molecular biology of fish,* Vol. 2, Molecular biology frontiers, P. W. Hochachka and T. P. Mommsen (Eds.). Elsevier, Amsterdam.

Chevassus, B. 1983. Hybridization in fish. *Aquaculture 33:* 245–262.

Chiarelli, A. B., and E. Capanna. 1973. Checklist of fish chromosomes, pp. 206–232. In *Cytotaxonomy and vertebrate evolution,* A. B. Chiarelli and E. Capanna (Eds.). Academic Press, New York.

Childers, W. F. 1967. Hybridization of four species of sunfishes (Centrarchidae). *Ill. Nat. Hist. Surv. Bull. 27:* 159–214.

Churikov, D., and A. J. Gharrett. 2002. Comparative phylogeography of the two pink salmon broodlines: An analysis based on mitochondrial DNA genealogy. *Mol. Ecol. 11:* 1077–1101.

Cloud, J. G. 1990. Strategies for introducing foreign DNA into the germ line of fish. *J. Reproduct. Fert. Suppl. 41:* 107–116.

Conover, D. O., and B. E. Kynard. 1981. Environmental sex determination: Interaction of temperature and genotype in a fish. *Science 213:* 577–579.

Cook, J. A., K. R. Bestgen, D. L. Propst, and T. L. Yates. 1992. Allozymic divergence and systematics of the Rio Grande silvery minnow, *Hybognathus amarus* (Teleostei: Cyprinidae). *Copeia 1992:* 36–44.

Danzmann, R. G., T. R. Jackson, and M. M. Ferguson. 1999. Epistasis in allelic expression at upper temperature tolerance QTL in rainbow trout. *Aquaculture 173:* 45–58.

Devlin, R. H., T. Y. Yesaki, E. M. Donaldson, S. J. Du, and C.-L. Hew. 1995. Production of germline transgenic Pacific salmonids with dramatically increased growth performance. *Can. J. Fish. Aquat. Sci. 52:* 1376–1384.

Docker, M. F., J. H. Youson, R. J. Beamish, and R. H. Devlin. 1999. Phylogeny of the lamprey genus *Lampetra* inferred from mitochondrial cytochrome *b* and ND3 gene sequences. *Can. J. Fish. Aquat. Sci. 56:* 2340–2349.

Dowling, T. E., and W. S. Moore. 1984. Level of reproductive isolation between two cyprinid fishes, *Notropis cornutus* and *N. chrysocephalus. Copeia 1984:* 617–628.

————, and————. 1985. Evidence for selection against hybrids in the family Cyprinidae (genus *Notropis*). *Evolution 39:* 152–158.

Dunham, R. A. 1986. Selection and crossbreeding responses for cultured fish, pp. 391–400. In *X Breeding programs for swine, poultry, and fish,* G. E. Dickerson and R. K. Johnson (Eds.). University of Nebraska Board of Regents, Lincoln.

Echelle, A. A., A. F. Echelle, and C. D. Crozier. 1983. Evolution of an all-female fish, *Menidia clarkhubbsi* (Atherinidae). *Evolution 37:* 772–784.

Fairbairn, D. J. 1981. Which witch is which? A study of the stock structure of witch flounder (*Glyptocephalus cynoglossus*) in the Newfoundland region. *Can. J. Fish. Aquat. Sci. 38:* 782–794.

Ferris, S. D. 1984. Tetraploidy and the evolution of catostomid fishes, pp. 55–93. In *Evolutionary genetics of fishes,* B. J. Turner (Ed.). Plenum Press, New York.

Filho, O. M., L. A. C. Bertollo, and P. M. G. Junior. 1980. Evidences for a multiple sex chromosome system with female heterogamety in *Apareiodon affinis* (Pisces, Parodontidae). *Caryologia 33:* 83–89.

Finnerty, J. R. 2001. Cnidarians reveal intermediate stages in the evolution of *hox* clusters and axial complexity. *Amer. Zool. 41:* 608–620.

Force, A., A. Amores, and J. H. Postlethwait. 2002. *Hox* cluster organization in the jawless vertebrate *Petromyzon marinus. J. Exp. Zool. 294:* 30–46.

Fried, C., S. J. Prohaska, and P. F. Stadler. 2003. Independent *hox*-cluster duplications in lamprey. *J. Exp. Zool. 299B:* 18–25.

Geiger, H. J., W. W. Smoker, L. A. Zhivotovsky, and A. J. Gharrett. 1997. Variability of family size and marine survival in pink salmon (*Oncorhynchus gorbuscha*) has implications for conservation biology and human use. *Can. J. Fish. Aquat. Sci. 54:* 2684–2690.

Gellner, K., and S. Brenner. 1999. Analysis of 148 kb of genomic DNA around the *wnnnt1* locus of *Fugu rubripes. Genome Res. 9:* 251–258.

Gharrett, A. J., A. K. Gray, and J. Heifetz. 2001. Identification of rockfish (*Sebastes* spp.) by restriction site analysis of the mitochondrial ND-3/ND-4 and 12S/16S rRNA gene regions. *Fish. Bull. 99:* 49–62.

————, and J. E. Seeb. 1990. Practical and theoretical guidelines for genetically marking fish populations. *Am. Fish. Soc. Symp. 7:* 407–417.

————, S. M. Shirley, and G. R. Tromble. 1987. Genetic relationships among populations of Alaskan chinook salmon (*Oncorhynchus tshawytscha*). *Can. J. Fish. Aquat. Sci. 44:* 765–774.

————, and W. W. Smoker. 1993. A perspective on the adaptive importance of genetic infrastructure in salmon populations to ocean ranching in Alaska. *Fish. Res. 18:* 45–58.

————, W. W. Smoker, R. R. Reisenbichler, and S. G. Taylor. 1999. Outbreeding depression in hybrids between odd- and even-broodyear pink salmon. *Aquaculture 173:* 117–129.

————, and F. M. Utter. 1982. Scientists detect genetic differences. *Sea Grant Today 12*(2): 3–4.

Gilk, S. E., I. A. Wang, C. L. Hoover, W. W. Smoker, S. G. Taylor, A. K. Gray, and A. J. Gharrett. 2004. Outbreeding depression in hybrids between spatially separated pink salmon, *Oncorhynchus gorbuscha,* populations: Marine survival, homing ability, and variability in family size. *Env. Biol. Fishes 69*(1–4): 289–298.

Gjedrem, T. 2000. Genetic improvement of cold-water fish species. *Aquacult. Res. 31:* 25–33.

Gold, J., W. J. Karel, and M. R. Strand. 1980. Chromosome formulae of North American fishes. *Prog. Fish Cult. 42:* 10–23.

Goodman, M. L. 1990. Preserving the genetic diversity of salmonid stocks: A call for federal regulation of hatchery programs. *Env. Law 20:* 111–116.

Gordon, M. 1946. Interchanging genetic mechanisms for sex determination in fishes under domestication. *J. Heredity 37:* 307–320.

Grady, J. M., R. C. Cashner, and J. S. Rogers. 1990. Evolutionary and biogeographic relationships of *Fundulus catenatus* (Fundulidae). *Copeia 1990:* 315–323.

Grant, W. S., J. L. García-Marín, and F. M. Utter. 1999. Defining population boundaries for fishery management, pp. 27–72. In *Genetics of sustainable fisheries management,* S. Mustafa (Ed.). Blackwell, Oxford.

————, and F. M. Utter. 1984. Biochemical population genetics of Pacific herring (*Clupea pallasi*). *Can. J. Fish. Aquat. Sci. 41:* 856–864.

Greer, J. M., J. Peutz, K. R. Tomas, and M. R. Capecchi. 2000. Maintenance of functional equivalence during paralogous *Hox* gene evolution. *Nature 403:* 661–665.

Haglund, T. R., D. G. Buth, and R. Lawson. 1993. Allozyme variation and phylogenetic relationships of Asian, North American, and European populations of the ninespine stickleback, *Pungitius pungitius,* pp. 438–452. In *Systematics, historical ecology, and North American freshwater fishes,* R. L. Mayden (Ed.). Stanford University Press, Stanford, CA.

Hebert, K. P., P. L. Goddard, W. W. Smoker, and A. J. Gharrett. 1998. Quantitative genetic variation and genotype by environment interaction of embryo development rate in pink salmon (*Oncorhynchus gorbuscha*). *Can. J. Fish. Aquat. Sci. 55:* 2048–2057.

Heierhorst, J., K. Leideris, and D. Richter. 1993. Vasotocin neuropeptide precursors and genes of teleost and jawless fish, pp. 339–356. In *Biochemistry and molecular biology of fish*, Vol. 2, Molecular biology frontiers, P. W. Hochachka and T. P. Mommsen (Eds.). Elsevier, Amsterdam.

Hershberger, W. K., J. M. Meyers, W. C. McAuley, and A. M. Saxton. 1990. Genetic changes in growth of coho salmon (*Oncorhynchus kisutch*) in marine netpens produced by ten years of selection. *Aquaculture 85:* 187–197.

Hindar, K., N. Ryman, and F. Utter. 1991. Genetic effects of cultured fish on natural fish populations. *Can. J. Fish. Aquat. Sci. 48:* 945–957.

Hinegardner, R., and D. E. Rosen. 1972. Cellular DNA content and the evolution of teleostean fishes. *Am. Nat. 106:* 621–644.

Holder, M. T., M. V. Erdman, T. P. Wilcox, R. L. Caldwell, and D. M. Hillis. 1999. Two living species of coelacanths? *Proc. Natl. Acad. Sci. USA 96:* 12616–12620.

Hughes, A. L., J. da Silva, and R. Friedman. 2001. Ancient genome duplications did not structure the human *Hox*-bearing chromosomes. *Genome Res. 11:* 771–780.

Hukriede, N. A., and 15 coauthors. 1999. Radiation hybrid mapping of the zebrafish genome. *Proc. Natl. Acad. Sci. USA 96:* 9745–9750.

Hulata, G., G. W. Wohlfarth, and A. Haley. 1986. Mass selection for growth rate in the Nile tilapia (*Oreochromis niloticus*). *Aquaculture 57:* 177–184.

Hutchinson, W. F., G. R. Carvalho, and S. I. Rogers. 1999. A nondestructive technique for recovery of DNA from dried fish otoliths for subsequent molecular genetic analysis. *Mol. Ecol. 8:* 891–894.

Ihssen, P. E., L. R. McKay, I. McMillan, and R. B. Phillips. 1990. Ploidy manipulation and gynogenesis in fishes: Cytogenetic and fisheries applications. *Trans. Am. Fish. Soc. 119:* 698–717.

Inoue, J. G., M. Miya, K. Tsukamoto, and M. Nishida. 2001. A mitogenomic perspective on the basal teleostean phylogeny: Resolving higher-level relationships with longer sequences. *Mol. Phylogenet. Evol. 20:* 275–285.

Inoue, K., S. Yamashita, J. Hata, S. Kabeno, S. Asada, E. Nagahisa, and T. Fujita. 1990. Electroporation as a new technique for producing transgenic fish. *Cell Different. Dev. 29:* 123–128.

Iwamoto, R. N., A. M. Saxton, and W. K. Hershberger. 1982. Genetic estimates for length and weight of coho salmon during freshwater rearing. *J. Heredity 73:* 187–191.

Johnson, J. B., T. E. Dowling, and M. C. Belk. 2004. Neglected taxonomy of rare desert fishes: Congruent evidence for two species of leatherside chub. *Syst. Biol. 53*(6): 841–855.

Jorstad, K. E., O. Skaala, and G. Dahle. 1991. The development of biochemical and visible genetic markers and their potential use in evaluating interaction between cultured and wild fish populations. *ICES Mar. Sci. Symp. 192:* 200–205.

Joyce, J. E., W. W. Smoker, R. Heintz, and A. J. Gharrett. 1994. Survival to fry and seawater tolerance of diploid and triploid hybrids between chinook (*Oncorhynchus tshawytscha*), chum (*O. keta*), and pink salmon (*O. gorbuscha*). *Can. J. Fish. Aquat. Sci. 51*(Suppl. 1): 25–35.

Kappen, C., K. Schugart, and F. H. Ruddle. 1989. Two steps in the evolution of the Antennapedia-class vertebrate homeobox genes. *Proc. Natl. Acad. Sci. USA 95:* 5459–5463.

Kapuscinski, A. R., and E. M. Hallerman. 1990. Transgenic fish and public policy: Anticipating environmental impacts of transgenic fish. *Fisheries 15:* 2–11.

King, M.-C., and A. C. Wilson. 1975. Evolution at two levels in humans and chimpanzees. *Science 188:* 107–116.

Kirpichnikov, V. S. 1981. *Genetic bases of fish selection*. Springer Verlag, New York.

Knapik, E. W., A. Goodman, M. Ekker, M. Chevrette, J. Delgado, S. Neuhass, N. Shimoda, W. Driever, M. C. Fishman, and H. J. Jacob. 1998. A microsatellite genetic linkage map for zebrafish (*Danio rerio*). *Nature Genet. 18:* 338–343.

Kosswig, C. 1964. Polygenic sex determination. *Experientia 20:* 190–199.

Kumar, S., and B. Hedges. 1998. A molecular timescale for vertebrate evolution. *Nature 392:* 917–920.

Kweon, H.-S., E.-H. Park, and N. Peters. 1998. Spermatozoon ultrastructure in the internally self-fertilizing hermaphroditic teleost, *Rivulus marmoratus* (Cyprinodontiformes, Rivulidae). *Copeia 1998:* 1101–1106.

Lazier, C. B., and M. E. MacKay. 1993. Vitellogenin gene expression in teleost fish, pp. 391–405. In *Biochemistry and molecular biology of fish*, Vol. 2, Molecular biology frontiers, P. W. Hochachka and T. P. Mommsen (Eds.). Elsevier, Amsterdam.

Litman, G. W., C. T. Amemiya, R. N. Haire, and M. J. Shamblott. 1990. Antibody and immunoglobulin diversity. *BioScience 40:* 751–757.

Málaga-Trillo, E., and A. Meyer. 2001. Genome duplications and accelerated evolution of *Hox* genes and cluster architecture in teleost fishes. *Amer. Zool. 41:* 676–686.

Mallatt, J., and J. Sullivan. 1998. 28S and 18S rDNA sequences support monophyly of lampreys and hagfishes. *Mol. Biol. Evol. 15:* 1706–1718.

Manzanares, M., H. Wada, N. Itasaki, P. A. Trainor, R. Krumlauf, and P. W. H. Holland. 2000. Conservation and elaboration of *Hox* gene regulation during evolution of the vertebrate head. *Nature 408:* 854–857.

Marchalonis, J., and S. F. Schluter. 1990. Origins of immunoglobulins and immune recognition molecules. *BioScience 40:* 758–768.

Martínez, R., A. Arenal, M. P. Estrada, F. Herrera, V. Huarta, J. Vázquez, T. Sánchez, and J. de la Fuente. 1999. Mendelian transmission, transgene dosage and growth phenotype in transgenic tilapia (*Oreochromis honorum*) showing ectopic expression of homologous growth hormone. *Aquaculture 173:* 271–283.

Marx, J. L. 1988. Gene-watcher's feast served up in Toronto. *Science 242:* 32–33.

Matsubara, K., K. Arai, and R. Suzuki. 1995. Survival potential and chromosomes of progeny of triploid and pentaploid females in the loach (*Misgurnus anguillicaudatus*). *Aquaculture 131:* 37–46.

May, B., and K. R. Johnson. 1990. Composite linkage map of salmonid fishes (*Salvelinus*, *Salmo*, and *Oncorhynchus*), pp. 4.151–4.159. In *Genetic maps, locus maps of complex genomes* (5th ed.), S. J. O'Brien (Ed.). Cold Spring Laboratory Press, New York.

Mayden, R. L., and R. H. Matson. 1992. Systematics and biogeography of the Tennessee shiner, *Notropis leuciodus* (Cope) (Teleostei: Cyprinidae). *Copeia 1992:* 954–968.

McMillan, W. O., L. A. Weigt, and S. R. Palumbi. 1999. Color pattern evolution, assortative mating, and genetic differentiation in brightly colored butterflyfishes (Chaetodontidae). *Evolution 53:* 247–260.

Meyer, A. 1993. Evolution of mitochondrial DNA in fishes, pp. 1–38. In *Biochemistry and molecular biology of fish*, Vol. 2, Molecular biology frontiers, P. W. Hochachka and T. P. Mommsen (Eds.). Elsevier, New York.

———, and A. C. Wilson. 1990. Origin of tetrapods inferred from their mitochondrial DNA affiliation to lungfish. *J. Mol. Evol. 31:* 359–364.

———, T. D. Kocher, P. Basasibwaki, and A. C. Wilson. 1990. Monophyletic origin of Lake Victoria cichlid fishes suggested by mitochondrial DNA sequences. *Nature 347:* 550–553.

Moav, R., and G. W. Wohlfarth. 1976. Two way selection for growth rate in the common carp (*Cyprinus carpio* L.). *Genetics 82:* 83–101.

Moran, P., and I. Kornfield. 1993. Retention of an ancestral polymorphism in the mbuna species flock (Teleostei: Cichlidae) of Lake Malawi. *Mol. Biol. Evol. 10:* 1015–1029.

Morizot, D. C. 1990. Use of fish gene maps to predict ancestral vertebrate genome organization, pp. 207–234. In *Isozymes: Structure, function, and use in biology and medicine,* Z. I. Ogita and C. L. Markert (Eds.). Wiley-Liss, New York.

———, S. W. Calhoun, L. L. Clepper, M. E. Schmidt, J. H. Williamson, and G. J. Carmichael. 1991a. Multispecies hybridization among native and introduced centrarchid basses in central Texas. *Trans. Am. Fish. Soc. 120:* 283–289.

———, S. A. Slaugenhaupt, K. D. Kallerman, and A. Chakravarti. 1991b. Genetic map of fishes of the genus *Xiphophorus* (Teleostei: Poeciliidae). *Genetics 127:* 399–410.

Muir, W. M., and R. D. Howard. 1999. Possible ecological risks of transgene organism release when transgenes affect mating success: Sexual selection and the Trojan gene hypothesis. *Proc. Natl. Acad. Sci. USA 96:* 13853–13856.

Murphy, R. W., J. W. Sites, Jr., D. G. Buth, and C. H. Haufler. 1990. Proteins I: Isozyme electrophoresis, pp. 45–126. In *Molecular systematics,* D. M. Hillis and C. Moritz (Eds.). Sinauer, Sunderland, MA.

Naruse, K., and 18 coauthors. 2000. A detailed linkage map of medaka, *Oryzias latipes:* Comparative genomics and genome evolution. *Genetics 154:* 1773–1784.

Nelson, J. S. 1994. *Fishes of the world* (3rd ed.). Wiley, New York.

Nelson, K., and M. Soulé. 1987. Genetical conservation of exploited fishes, pp. 345–368. In *Population genetics in fishery management,* N. Ryman and F. M. Utter (Eds.). University of Washington Press, Seattle.

Nielsen, E. E., M. M. Hansen, and V. Loeschcke. 1997. Analysis of microsatellite DNA from old scale samples of Atlantic salmon *Salmo salar:* A comparison of genetic composition over 60 years. *Mol. Ecol. 6:* 487–492.

Ohno, S. 1970. *Evolution by gene duplication.* Springer Verlag, New York.

Palumbi, S. R. 1996. Nucleic acids II: The polymerase chain reaction, pp. 205–247. In *Molecular systematics* (2nd ed.), D. M. Hillis, C. Moritz, and B. K. Mable (Eds.). Sinauer, Sunderland, MA.

Pancer, Z., C. T. Amemiya, G. R. A. Ehrhardt, J. Ceitlin, G. L. Gartland, and M. Cooper. 2004. Somatic diversification of variable lymphocyte receptors in the agnathan sea lamprey. *Nature 430:* 174–180.

Parsons, J. 1998. Status of genetic improvement in commercially reared stocks of rainbow trout. *World Aquacult. 29*(1): 44–47.

Pella, J. J., and G. B. Milner. 1987. Use of genetic marks in stock composition analysis, pp. 247–276. In *Population genetics in fishery management,* N. Ryman and F. M. Utter (Eds.). University of Washington Press, Seattle.

Pezold, F. 1984. Evidence for multiple sex chromosomes in the freshwater goby, *Gobionellus shufeldti* (Pisces: Gobiidae). *Copeia 1984:* 235–238.

Philipp, D. P., and J. E. Claussen. 1995. Fitness and performance differences between two stocks of largemouth bass from different river drainages within Illinois. *Am. Fish. Soc. Symp. 15:* 236–243.

Pohajdak, B., B. Dixon, and G. R. Stuart. 1993. Immune system, pp. 191–205. In *Biochemistry and molecular biology of fish,* Vol. 2, Molecular biology frontiers, P. W. Hochachka and T. P. Mommsen (Eds.). Elsevier, Amsterdam.

Postlethwait, J.H., and 14 coauthors. 1994. A genetic linkage map for zebrafish. *Science 264:* 699–703.

Postlethwait, J. H., and 28 coauthors. 1998. Vertebrate genome evolution and the zebrafish gene map. *Nature Genet. 18:* 345–349.

Powers, D. A. 1991. Evolutionary genetics of fish. *Adv. Genet. 29:* 119–228.

Prince, V. 2002. The *Hox* paradox: More complex(es) than imagined. *Dev. Biol. 249:* 1–15.

Raleigh, R. F. 1967. Genetic control in the lakeward migrations of sockeye salmon (*Oncorhynchus nerka*) fry. *J. Fish. Res. Bd. Can. 24:* 2613–2622.

Rasmussen, A.-S., and U. Arnason. 1999. Molecular studies suggest that cartilagenous fishes have a terminal position in the piscine tree. *Proc. Natl. Acad. Sci. USA. 96:* 2177–2182.

Riddell, B. E. 1993. Spatial organization of Pacific salmon: What to conserve?, pp. 23–41. In *Genetic conservation of salmonid fishes,* J. G. Cloud and G. H. Thorgaard (Eds.). Plenum Press, New York.

Robinson-Rechavi, M., B. Boussau, and V. Laudet. 2004. Phylogenetic dating and the characterization of gene duplications in vertebrates: The cartilaginous fish reference. *Mol. Biol. Evol. 21:* 580–586.

Rocha-Olivares, A., R. H. Rosenblatt, and R. D. Vetter. 1999. Molecular evolution, systematics, and zoogeography of the rockfish subgenus *Sebastosomus* (*Sebastes,* Scorpaenidae) based on mitochondrial cytochrome *b* gene and control region sequences. *Mol. Phylogenet. Evol. 11:* 441–458.

Rosenfield, J. A., T. Todd, and R. Greil. 2000. Asymmetric hybridization and introgression between pink salmon and chinook salmon in the Laurentian Great Lakes. *Trans. Am. Fish. Soc. 1229:* 670–679.

Ross, R. M., T. F. Hourigan, M. M. F. Lutnesky, and I. Singh. 1990. Multiple simultaneous sex changes in social groups of a coral-reef fish. *Copeia 1990:* 427–433.

———, G. S. Losey, and M. Diamond. 1983. Sex change in a coral-reef fish: Dependence of stimulation and inhibition on relative size. *Science 221:* 574–575.

Ruddle, F. H., C. T. Amemiya, J. L. Carr, C.-B. Kim, C. Ledje, C. S. Shashikant, and G. P. Wagner. 1999. Evolution of chordate *HOX* gene clusters. *Ann. N.Y. Acad. Sci. 870:* 238–248.

———, J. L. Bartels, K. L. Bentley, C. Kappen, M. T. Murtha, and J. W. Pendleton.1994. Evolution of *HOX* genes. *Ann. Rev. Genet. 28:* 423–442.

Ruzzante, D. E., C. T. Taggart, D. Cook, and S. Goddard. 1996. Genetic differentiation between inshore and offshore Atlantic cod (*Gadus morhua*) off Newfoundland: Microsatellite DNA variation and antifreeze level. *Can. J. Fish. Aquat. Sci. 53:* 634–645.

Sakamoto, T., R. G. Danzmann, K. Gharbi, P. Howard, A. Ozaki, S. K. Khoo, R. A. Woram, N. Okamoto, M. M. Fergison, L.-E. Holm, R. Guyomard, and B. Hoyheim. 2000. A microsatellite linkage map of rainbow trout (*Oncorhynchus mykiss*) characterized by large sex-specific differences in recombination rates. *Genetics 155:* 1331–1345.

Schliewen, U. K., D. Tautz, and S. Paabo. 1994. Sympatric speciation suggested by monophyly of crater lake cichlids. *Nature 368:* 629–632.

Schultz, R. J. 1993. Genetic regulation of temperature-mediated sex ratios in the livebearing fish *Poeciliopsis lucida. Copeia 1993:* 1148–1151.

Schweigert, J. F., F. J. Ward, and J. W. Clayton. 1977. Effects of fry and fingerling introductions on walleye (*Stizostedion vitreum vitreum*) production in West Blue Lake, Manitoba. *J. Fish. Res. Bd. Can. 34:* 2142–2150.

Shaklee, J. B., T. D. Beacham, L. Seeb, and B. A. White. 1999. Managing fisheries using genetic data: Case studies from four species of Pacific salmon. *Fish. Res. 43:* 45–78.

———, and P. Bentzen. 1998. Genetic identification of stocks of marine fish and shellfish. *Bull. Mar. Sci. 62:* 589–621.

———, and C. S. Tamaru. 1981. Biochemical and morphological evolution of Hawaiian bone fishes (*Albula*). *Syst. Zool. 30:* 125–146.

Sharman, A. C., and P. W. H. Holland. 1996. Conservation, duplication, and divergence of developmental genes during chordate evolution. *Neth. J. Zool. 46:* 47–67.

Siitonen, L., and G. A. E. Gall. 1989. Response to selection for early spawn date in rainbow trout, *Salmo gairdneri. Aquaculture 78:* 153–161.

Simon, R. C., and R. E. Noble. 1968. Hybridization in *Oncorhynchus* (Salmonidae). I. Viability and inheritance in artificial crosses of chum and pink salmon. *Trans. Am. Fish. Soc. 97:* 109–118.

Skaala, O., G. Dahle, K. E. Jorstad, and G. Naevdal. 1990. Interactions between natural and farmed fish populations: Information from genetic markers. *J. Fish Biol. 36:* 449–460.

———, and K. E. Jorstad. 1988. Inheritance of the fine-spotted pigmentation pattern of brown trout, *Salmo trutta* L. In Trouts in streams and lakes. *Polish Arch. Hydrobiol. 35:* 3–4.

Stearley, R. F., and G. R. Smith. 1993. Phylogeny of the Pacific trouts and salmon (*Oncorhynchus*) and genera of the family Salmonidae. *Trans. Am. Fish. Soc. 122:* 1–33.

Stearns, S. C. 1992. *The evolution of life histories.* Oxford University Press, New York.

Stepien, C. A., and R. H. Rosenblatt. 1991. Patterns of gene flow and genetic divergence in the northeastern Pacific Clinidae (Teleostei: Blennioidei), based on allozyme and morphological data. *Copeia 1991:* 875–896.

Stock, D. W., and G. S. Whitt. 1992. Evidence from 18S ribosomal RNA sequences that lampreys and hagfishes form a natural group. *Science 257:* 787–789.

Stokes, M. D., and N. D. Holland. 1998. The lancelet. *Am. Scient. 86:* 552–560.

Sullivan, J. A., and R. J. Schultz. 1986. Genetic and environmental basis of variable sex ratios in laboratory strains of *Poeciliopsis lucida. Evolution 40:* 152–158.

Takezaki, N., F. Figueroa, Z. Zaleska-Rutczynska, and J. Klein. 2003. Molecular phylogeny of early vertebrates: Monophyly of the agnathans as revealed by sequences of 35 genes. *Mol. Biol. Evol. 20:* 287–292.

———, F. Figueroa, Z. Zaleska-Rutczynska, N. Takahata, and J. Klein. 2004. The phylogenetic relationship of tetrapod, coelacanth, and lungfish revealed by the sequences of forty-four nuclear genes. *Mol. Biol. Evol. 21:* 1512–1524.

Tave, D. 1986. *Genetics for fish hatchery managers.* AVI, Westport, CN.

Taylor, E. B. 1991. A review of local adaptation in Salmonidae, with particular reference to Pacific and Atlantic salmon. *Aquaculture 98:* 185–207.

Teplitz, R., J. Joyce, S. I. Doroshov, and B. H. Min. 1994. A preliminary ploidy analysis of diploid and triploid salmonids. *Can. J. Fish. Aquat. Sci. 51*(Suppl. 1): 38–41.

Thorgaard, G. H. 1976. Robertsonian polymorphism and constitutive heterochromatin distribution in chromosomes of the rainbow trout (*Salmo gairdneri*). *Cytogenet. Cell Genet. 17:* 174–184.

———. 1977. Heteromorphic sex chromosomes in male rainbow trout. *Science 196:* 900–902.

———. 1978. Sex chromosomes in the sockeye salmon: A Y–autosome fusion. *Can. J. Genet. Cytol. 20:* 349–354.

———. 1983. Chromosomal differences among rainbow trout populations. *Copeia 1983:* 650–662.

———, and S. K. Allen. 1987. Chromosome manipulation and markers in fishery management, pp. 319–331. In *Population genetics in fishery management,* N. Ryman and F. M. Utter (Eds.). University of Washington Press, Seattle.

Thorpe, J. E. 1993. Impacts of fishing on genetic structure of salmonid populations, pp. 67–80. In *Genetic conservation of salmonid fishes,* J. G. Cloud and G. H. Thorgaard (Eds.). Plenum Press, New York.

Thrower, M. K., and 13 coauthors. 1996. Conservation of synteny between the genome of the pufferfish (*Fugu rubripes*) and the region on chromosome 14 (14q24.2) associated with familial Alzheimer disease (*AD3* locus). *Proc. Natl. Acad. Sci. USA 93:* 1366–1369.

Utter, F., G. Milner, G. Staahl, and D. Teel. 1989. Genetic structure of chinook salmon, *Oncorhynchus tshawytscha,* in the Pacific Northwest. *Fish. Bull. U.S. 87:* 239–264.

Uyeno, T., and G. R. Smith. 1972. Tetraploid origin of the karyotype of catostomid fishes. *Science 175:* 644–646.

Van Beneden, R. J. 1993. Oncogenes, pp. 113–136. In *Biochemistry and molecular biology of fish,* Vol. 2, Molecular biology frontiers, P. W. Hochachka and T. P. Mommsen (Eds.). Elsevier, Amsterdam.

Vandepoele, K., W. De Vos, J. S. Taylor, A. Meyer, and Y. Van de Peer. 2004. Major events in the genome evolution of vertebrates: Paranome age and size differ considerably between ray-finned fishes and land vertebrates. *Proc. Natl. Acad. Sci. USA 101:* 1638–1643.

Vielkind, J. R., K. D. Kallman, and D. C. Morizot. 1989. Genetics of melanomas in *Xiphophorus* fishes. *J. Aquat. Anim. Health 1:* 69–77.

Vos, P., R. Hogers, M. Bleeker, M. Riejans, T. Van de Lee, M. Hornes, A. Frijters, J. Pot, J. Peleman, M. Kuiper, and M. Zabeau. 1995. A new technique for DNA fingerprinting. *Nucl. Acids Res. 23:* 4407–4414.

Wada, H., and N. Satoh. 1994. Details of the evolutionary history from invertebrates to vertebrates, as deduced from the sequences of 18S rRNA. *Proc. Natl. Acad. Sci. USA 91:* 1801–1804.

Wagner, W. C., and T. M. Stauffer. 1982. Distribution and abundance of pink salmon in Michigan tributaries of the Great Lakes, 1967–1980. *Trans. Am. Fish. Soc. 111:* 523–526.

Wallbrunn, H. M. 1958. Genetics of the Siamese fighting fish, *Betta splendens. Genetics 43:* 289–298.

Ward, R. D. 2000. Genetics in fisheries management. *Hydrobiologia 420:* 191–201.

Waples, R. S. 1987. A multispecies approach to the analysis of gene flow in marine shore fishes. *Evolution 41:* 385–400.

———. 1991. Pacific salmon, *Oncorhynchus* spp., and the definition of "species" under the Endangered Species Act. *Mar. Fish. Rev. 55:* 11–22.

———. 1995. Evolutionarily significant units and the conservation of biological diversity under the Endangered Species Act. *Am. Fish. Soc. Symp. 17:* 8–27.

Warner, R. W. 1984. Mating behavior and hermaphroditism in coral reef fishes. *Am. Scient. 72:* 128–136.

Weaver-Missick, T. 2001. Purrfecting the catfish. *Agricult. Res. 49:* 16–18.

Wenburg, J. K., P. Bentzen, and C. J. Foote. 1998. Microsatellite analysis of genetic population structure in an endangered salmonid: The coastal cutthroat (*Oncorhynchus clarki clarki*). *Mol. Ecol. 7:* 733–749.

Williams, J., G. K. Kubelik, K. J. Livak, J. A. Rafalski, and S. V. Tingey. 1990. DNA polymorphisms amplified by arbitrary primers are useful as genetic markers. *Nucl. Acids Res. 18:* 6531–6535.

Wittbrodt, J., A. Meyer, and M. Schartl. 1998. More genes in fish? *BioEssays 20:* 511–515.

Wolters, W. 1993. Channel catfish breeding and selection programs: Constraints and future prospects, pp. 82–95. In *Selective breeding of fishes in Asia and the United States*, K. L. Main and B. Reynolds (Eds.). Oceanic Institute, Makapuu Pt., Honolulu, HI.

Wood, R. M., R. L. Mayden. 1992. Systematics, evolution, and biogeography of *Notropis chlorocephalus* and *N. lutipinnis. Copeia 1992:* 68–81.

Wright, J. M. 1993. DNA fingerprinting of fishes, pp. 57–91. In *Biochemistry and molecular biology of fishes*, Vol. 2, Molecular biology frontiers, P. W. Hochachka and T. P. Mommsen (Eds.) Elsevier, New York.

Yamamoto, T. 1969. Inheritance of albinism in the medaka, *Oryzias latipes*, with special reference to gene interaction. *Genetics 62:* 797–809.

Young, W. A., P. A. Wheeler, V. H. Coryell, P. Keim, and G. A. Thorgaard. 1998, A detailed linkage map of rainbow trout produced using doubled haploids. *Genetics 148:* 839–850.

Yu, X.-Y., and X.-Y. Yu. 1990. A schizothoracine fish species, *Diptychus dipogon*, with a very high number of chromosomes. *Chromosome Inf. Serv. 48:* 17–18.

Zon, L. I. 1999. Zebrafish: A new model for human disease. *Genome Res. 9:* 99–100.

PART SEVEN

INTERACTIONS OF FISHES WITH THEIR ENVIRONMENT

In previous sections, we have investigated the architecture, evolutionary history, and diversification of fishes and the ways in which they function in the aquatic realm. We have investigated the topic of function in terms of the means by which fishes sense the environment and exploit its resources and in terms of the processes by which they maintain themselves. In doing so, we have focused our attention on the structure and function of the whole organism and its component parts. We are now ready to take our studies of fishes to another level by investigating the nature of their interaction with the environment and with one another.

The topic of biogeography is integral to these considerations, as it is concerned with the distribution of organisms across the Earth's surface and with the circumstances that led to their being where they are. Biogeography is a marriage of the disciplines of biology and geology, because the distribution and abundance of life very much reflects the changes that have taken place in the Earth's crust over

the past hundreds of millions of years. Just what the consequences of such changes have been for the distribution of marine and freshwater fishes is the subject of the following chapter.

Ecology (from the Greek word *oikos,* meaning "home" or "household") has traditionally been defined as "the study of the relations of organisms to one another and to their surroundings." (Ricklefs and Miller, 1999). As such, ecology is more a study of processes than of objects, although we need to have a thorough understanding of the nature of the objects in order to properly interpret their ecological interactions. Ecology provides us with a means to grasp the diversity and complexity of the natural world and helps us to understand the fundamental interconnectedness of all life.

Ecology as a discipline in its own right has its origins early in the 20th century, but it is rooted in the detailed observations of countless naturalists dating back to the origins of our own species. From these observations came the recognition that certain species of organisms form natural associations,

each with a distinctive assemblage of component creatures. As ecology has evolved as a science, different aspects of the discipline developed particular schools of ecological thought. Some researchers, such as the brilliant Odum family (including Howard T., Eugene P., and William H. Odum), emphasized the continuity of ecological processes in studies of energy flow, whereas others, such as C. J. Krebs, emphasized factors influencing distribution and abundance (Krebs, 1994).

Understanding ecology necessitates some means of organizing the world of fishes into manageable components, as has been done in the ensuing chapters of this book. Such categorization of environments has to be somewhat arbitrary, however, as fishes more often than not refuse to recognize boundaries that are artificially imposed on what is in actuality a continuum of habitats. Gobies, for example, are one of the most diverse suborders of fishes found in coral reef environments, with about half of the family's known species occurring there (Paxton and Eschmeyer, 1998). Many goby species, however, can be found in freshwater streams, ponds, springs, and lakes. Hagfishes, on the other hand, are pretty much restricted to the deep-ocean environment. Some fishes, such as the salmons, inhabit one kind of environment at one phase in their life history, then shift to another one at a later phase.

Although our focus will be on the aquatic realm, we must recognize that a full understanding of aquatic ecology necessitates consideration of how and where the aquatic realm interfaces with the terrestrial realm. In Chapter 30, we will become acquainted with the sometimes surprising ways in which fishes are integral to terrestrial ecological processes. By investigating the relationship of fishes to their environment, we should be able to gain some insight into the ways in which humans affect the environments of fishes, sometimes for the better but, unfortunately, far too often for the worse.

Integral to our understanding of how fishes interact with their environment is an appreciation of their behavior. Evolution has equipped fishes with adaptive modes of behavior that enable optimization of resource use and hence maximization of their reproductive potential. Much of the behavioral repertoire of fishes is directed at other fishes—what we would generalize as social behavior. We will consider several aspects of the behavior of fishes that have enabled them to exist as fully integrated members of ecological communities.

29

THE DISTRIBUTION OF FISHES

INTRODUCTION

HISTORICAL GEOLOGY AND FISH DISTRIBUTION

Stream Capture
Continental Drift
Glaciation

DISTRIBUTION OF FRESHWATER OR INLAND FISHES

Biogeographical Realms
The Holarctic Region
The Palearctic Region
The Nearctic Region
Middle America
The Neotropical Region
The Oriental Region
The Ethiopian Region
The Australian Region

DISTRIBUTION OF MARINE FISHES

Pelagic Fishes
Polar Waters
North Temperate Waters
Tropical Waters
Shore and Shelf Fishes
Tropical and Temperate Regions
Polar Benthic and Shelf Fishes
Deep-Sea Fishes

CAMM C. SWIFT

Fishes have a long history of significant contribution to the study of both the recent and the ancient biogeography of our planet. Evolutionarily and geologically older and more numerous than all the other vertebrates combined, fishes are distributed literally everywhere that water exists on the planet. The majority of fishes can tolerate only a narrow range of salinity, however, and are physiologically restricted to either marine or freshwater. This means that many freshwater fish species are isolated in one or a few freshwater drainages and cannot disperse through the ocean to colonize adjacent drainages. The relatively large size of fishes and their lack of protection against desiccation make them unlikely to be transported long distances by the wind or carried alive by birds or other animals to colonize distant waters. Thus, freshwater fishes rely on adequate connections between lakes, rivers, and swamps taking place over geological time scales of thousands or millions of years as mountains rise, glaciers come and go, and sea levels rise and fall. Continental areas, such as the southeastern United States, have been relatively stable for millions of years. Drainage alterations over time have produced large numbers of species that are still being intensively studied. The more geologically active western United States has far fewer freshwater fishes, and more extinction has occurred there in the past. In a similar vein, the tropical and temperate regions have more fish species than the polar regions, the latter having been subjected to extreme cooling in the past. Since the last great extinction event on the planet more than 65 million years ago, about 35,000 species of fishes have developed. These fishes provide an exhaustive and detailed data set for analyzing a variety of biological and geological changes during the Earth's history.

INTRODUCTION

The distribution of fishes can be analyzed at many scales. Short-distance colonization of new habitats over ecological time can occur over a few days to a few years, whereas movement over tens to hundreds of miles with changing land use patterns occurs over decades to centuries (Matthews, 1998; Trautman, 1981; Welcomme, 1979). Similar dispersal has taken place over greater distances with the retreat of the glaciers over thousands to tens of thousands of years (Matthews, 1998; see Chapter 4). Finally, the evolution of distinct species, both living and fossil, has taken place over millions of years on one or more continents (Grande and Bemis, 1999; Humphries and Parenti, 1999; Mayden, 1992; Springer, 1982).

The distribution patterns related to longer geological time scales have figured prominently in the classical analyses of the global distributions of fishes (Boulenger, 1905; Briggs, 1974, 1995; Darlington, 1957; de Beaufort, 1951; Gunther, 1880; Jordan, 1901, 1928; Myers, 1938, 1941, 1949, 1951). Past events have included both the speciation and diversification of the organisms and the simultaneous geological changes (vicariant events) that may or may not have caused their splitting into multiple taxa, whether species, genera, or even higher categories.

Before the mid-1960s, dispersal by the active movement of fishes was widely assumed to be almost the only mechanism explaining the distribution of fishes between continents and landmasses separated by the ocean. Geologists reinforced the idea that the continents had been in or near their present positions for most if not all the time required for the evolution of the world's fish fauna (Fig. 29.1). This required fish ancestors either to have dispersed across areas of the present continents that are no longer inhabited or no longer present, or to have had greater salinity tolerance, enabling them to migrate across marine waters. Such tolerance would be required (indeed, indispensable) for crossing areas of ocean between continents.

Fishes have been classified by their salinity tolerance, as tabulated in Table 29.1. This table also includes several terms for the various patterns of fish migration, and these are discussed fully in Chapter 36. For distributional and biogeographical purposes, freshwater or inland fish families have been classified as either primary, secondary, or peripheral. **Primary** freshwater fishes are essentially intolerant of all but the lowest salinities—usually less than about 10 parts per thousand (‰)—and then only briefly. The ocean is usually about 35 ‰. Primary freshwater fishes are thought not to cross marine barriers. **Secondary** freshwater fishes are a little more tolerant of salinity and are expected to toler-

ate some salinity near the coast and thus to have the ability to cross narrow marine barriers. **Peripheral** is a collective term including a variety of fish that regularly live in waters of different salinities at various stages in their life history. This includes many kinds of coastal and migratory fishes. Finally, **euryhaline** fishes can freely tolerate marine to freshwater salinities at all life stages rather than at just certain stages, as many of the other categories (see Chapter 30). Primary and secondary freshwater fishes have been considered better candidates (or the only ones to consider) for biogeographic studies, as they did not cross salt water barriers today or, presumably, in the past (McDowall, 1988; Myers, 1949; Patterson, 1975; Sparks and Smith, 2005). These distinctions have become less important for several reasons.

First, many species in the "salt-tolerant" families (e.g., species of the sculpin genus *Cottus*, family Cottidae, or the killifish genus *Fundulus*, family Fundulidae) are as restricted to freshwater as many "true" or primary families of freshwater fishes. In the western United States, Australia, Madagascar, and elsewhere in the world, many freshwater habitats are dominated by freshwater fish species from secondary, diadromous, and peripheral families, like the petromyzontids, salmonids, cottids, atherinids, and cichlids. These have begun to be inclusively labeled as "inland" fishes (Hartel et al., 2002; Moyle, 1976, 2002; Ross, 2001; Stiassny and Raminosoa, 1994; Wydowski and Whitney, 1979, 2003). The inland fish fauna of Madagascar has only "salt-tolerant" families and lacks native primary freshwater fishes (Stiassny and Raminosoa, 1994). Allen et al. (2002a) divided the Australian "freshwater fishes" into two groups, primary freshwater families—including the primary, secondary, and peripheral fishes of Table 29.1—and estuarine freshwater families of marine species wandering into freshwater occasionally. Remarkably, some inland fishes that are isolated in the southwestern desert of the United States have higher salinity tolerances than some marine fishes (Soltz and Naiman, 1978).

A second reason why the distinction of freshwater families is no longer held to be crucial was provided by Cohen (1970), who demonstrated that many marine fishes were as continental as the so-called freshwater fishes, and thus their distribution could be equally informative. Cohen (1970) showed that about 42 percent of the world's fish species are restricted to freshwater. Of the remainder, 39 percent are continental—that is, they live in shallow cold or warm waters on the continental shelves. This adds up to 80 percent of fish species that can be called continental. Other exclusively offshore benthic and pelagic oceanic species, from both warm and cold water, constitute only about 19 percent of the fish species, even though they may contribute tremendous

RIVER KEY

NEARCTIC
1. Yukon
2. Mackenzie
3. Columbia
4. Sacramento/ San Joaquin
5. Colorado
6. Rio Grande
7. Missouri
8. Mississippi
9. Ohio
10. Tennessee
11. Mobile Bay Basin

NEOTROPICAL
1. Magdalena
2. Orinoco
3. Amazon
4. São Francisco
5. Paraguay
6. Paraná
7. Uruguay
8. Colorado

ETHIOPIAN
1. Senegal
2. Niger
3. Nile
4. Congo
5. Zambezi
6. Orange

PALAEARTIC
1. Rhine
2. Danube
3. Dnieper
4. Volga
5. Tigris
6. Ob
7. Yenisey
8. Lena
9. Amur
10. Yellow

ORIENTAL
1. Indus
2. Krishna
3. Godavari
4. Ganges
5. Brahmaputra
6. Salween
7. Yangzi
8. Mekong
9. Irrawaddy

AUSTRALIAN
1. Fly
2. Darling
3. Murray

LAKES AND INLAND SEAS

NEARCTIC
a. Great Lakes

ETHIOPIAN
a. Rift Lakes

PALAEARCTIC
a. Black Sea
b. Caspian Sea
c. Lake Baikal

NEOTROPICAL
a. Lake Titicaca

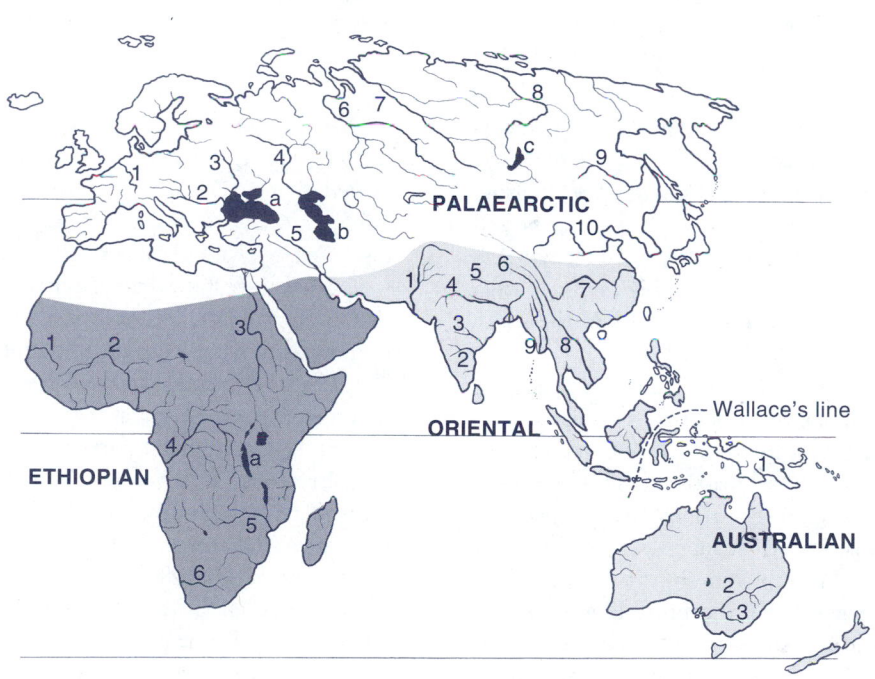

FIGURE 29.1

The biogeographical regions, with the continents and major rivers and lakes of today.

TABLE 29.1 DIVISIONS OF FRESHWATER FISH GROUPS

Division	Remarks
I. Primary	Groups with little or no tolerance for seawater; such as lungfishes, paddlefishes, bichirs, pikes, minnows, catfishes (except plotosids and ariids), characins, centrarchids, living osteoglossids.
	Archaeoliminic: originating in freshwaters and always so confined (Patterson, 1975).
	Telolimnic: confined to freshwaters at present, but less so in the past.
II. Secondary	Groups usually restricted to freshwater but with enough salt tolerance so that members can enter the ocean and sometimes cross narrow salt water barriers; garpikes, killifishes, livebearers, cichlids.
III. Diadromous	Migratory between freshwater and sea for the purpose of breeding (McDowall, 1988).
	1. Anadromous: diadromous fishes which spend most of their life in the sea and mature. When fully grown they return to freshwater to breed.
	2. Catadromous: diadromous fishes which spend most of their life in freshwater and go to the sea as adults to breed.
	3. Amphidromous: regularly migrating between freshwater and the sea for purposes other than breeding.
	A. Marine: spawning in marine water, with larvae and juvenile stages briefly in freshwater.
	B. Freshwater: spawning in freshwater with larvae and juvenile stages; temporarily marine before returning to freshwater.
IV. Vicarious	Nonmigratory species of otherwise marine groups living in freshwater, such as burbot, brook silverside, tule perch.
V. Complementary	Groups with close marine relatives that dominate freshwaters in the absence of primary and/or secondary fishes, such as the melanotaeniids of Australia-New Guinea.
VI. Sporadic	Fish that go back and forth between each medium and can breed in either medium; some anchovies, mullets, snappers, and gobies.
VII. Peripheral	A term coined to include categories III–VI above, namely all species that move between fresh and salt water at some or many life stages.
VIII. Euryhaline	Estuarine fishes that often and freely go between marine and freshwater, differing from most of the above categories, which usually are capable of changing mediums only at particular life stages.

* After Myers, 1938, 1949, 1951; McHugh, 1967; Patterson, 1975; McDowall, 1988.

biomass in the expanses of the ocean. Only a small percentage—less than 0.5 percent of all fish species—can tolerate a variety of salinities (Cohen, 1970; McDowall, 1988; Nelson, 1994). Even the small numbers of salt-tolerant estuarine and diadromous species are often restricted to the "offshore estuarine zones" that form near coasts with high rainfall, such as the North Sea, Panama Bay, Bering Sea, and the south China Sea (Haedrich, 1983; McHugh, 1967).

Third, geologists have verified and confirmed the reality of plate tectonics and continental drift, forcing the realization that fishes have been carried on continents over geological time. In other words, dispersal through the ocean was not necessary to explain their distributions. In retrospect, the fishes and other organisms have become part of the evidence that the continents had been moving, but these facts alone were not sufficient to overcome the prevailing geological paradigm. Early consideration of these factors led Rosen (1975) to declare that fish species need only be considered continental (rather than oceanic) to be useful for biogeographical studies. Indeed, even oceanic fishes can be informative as indicators of the history of the ocean basins (Haedrich, 1997; White, 1994; Zahuranec, 2000).

In this chapter, the fishes that live primarily in lakes, streams, springs, and marshes on land will be referred to as *inland* or *freshwater fishes*. Many of the terms noted in Table 29.1, and their predecessors, which were extensively discussed by McDowall (1988), have not acquired widespread usage despite their utility in the past. Those terms that apply to highly migratory fishes have long maintained currency and will be discussed further in Chapter 36.

The size and diversity of freshwater fish faunas in a descriptive sense is still an active area of study on all continents. Associations of particular fish species often define faunal regions. The borders of such regions help identify areas of past speciation events resulting from dispersal or vicariance (Banarescu, 1990, 1992, 1994; Greenwood, 1983; Hocutt and Wiley, 1986; Mayden, 1992). **Dispersal** involves active movement by the fish taxon from one place to another, whereas **vicariance** means that some alteration of the landscape created a barrier between two populations, leading to differentiation into two or more taxa. Documentation of fish faunas is particularly important where faunas are being heavily afflicted by humans and may be partially or completely lost. Rapid changes are taking place both in less studied areas, such as South America or Indonesia, and in better known regions, like the eastern United States, Japan, and Europe (Lundberg et al., 2000; Moyle and Leidy, 1992; Moyle and Williams, 1990; Stiassny and de Pinna, 1994; Warren and Burr, 1994). Many new species are still awaiting discovery in the poorly known areas, and even in the "backyards" of the best known areas, such as the United States (Burr and Mayden, 1992; Buth and Mayden, 2001; Near and Benard, 2004; Smith et al., 2002). Urgency has been injected into this search by the potential for the "homogenization" of fish faunas by the increasing artificial movement and establishment of nonnative species (Rahel, 2002).

The methods of analyzing fish distribution and biogeography fall into three general categories. First, measures of *diversity*—usually the number of resident species per some definable habitat unit, such as stream, lake, reef, country, continent, or hemisphere—can be compared with one another for a variety of similarities and differences. These methods can merely compare diversity without regard to the species involved, or they can compare various levels of relationships based on shared species versus nonshared species. Usually, these methods cluster units of habitat, such as river drainages, offshore islands, or continents, according to their sharing of species. The more similar the fish faunas, the closer they cluster or group together. Many such *faunal regions* or areas of the world were intuitively recognized and have been increasingly quantified statistically. Comparisons of the biological characteristics of the same species within different communities of other species (with partially or completely different competitors, predators, habitat, etc.) provide a rich source of data for comparative biology.

The second methodology, analyzing the origin of biological features, requires the incorporation of *phylogenetic* classifications of fishes. The relative position of a taxon on a phylogenetic tree provides the framework for interpreting changes and separating earlier from later developments (see Chapter 4). Fossils can enrich the data set by providing past distributional information and estimates of the minimum time involved for the appearance of novelties. A particular taxon, biological feature, or evolutionary event has to be at least as old as the fossil with the attribute (though the attribute could easily be older as well).

Finally, if phylogenetic hypotheses are available for more than one group of fishes, the *congruence* (i.e., correspondence) of the phylogenetic trees over a particular geographical area can strengthen an overall hypothesis of change in a regional fish fauna. These methods and their applications have been presented with particular reference to fishes by Lundberg and McDade (1990), Mayden and Wiley (1992), Stiassny and de Pinna (1994), and Mayden and Wood (1995). General and comprehensive works on bio-

geography are Humphries and Parenti (1999), Avise (2000), and Lomolino et al. (2005).

Smaller, less mobile kinds of fishes usually have narrower distributions, simply because they cannot disperse as rapidly. Thus, they often are more subject to endemism. **Endemism** is the restriction and uniqueness of certain organisms to a particular place, indicating that they have been isolated for a long time. The greater numbers and the smaller geographic ranges of small species permit a more detailed interpretation at finer scales than is possible with larger, more active fishes with fewer species and larger ranges. Thus, in eastern North America, darters (Percidae), stream minnows (Cyprinidae), madtoms (Ictaluridae, genus *Noturus*), and pygmy sunfishes (Elassomatidae), inform biogeographical study at finer scales than the larger bodied suckers (Catostomidae) or sunfishes (Centrarchidae; Mayden, 1992). The larger but fewer high-level or top predators, such as gars (Lepisosteidae) and black basses (*Micropterus*), are expected to more effectively inform biogeographical study at larger or coarser geographic scales (Swift et al., 1986).

Endemism is also greater in small-sized marine fishes, such as gobies (Gobiidae), damselfishes (Pomacentridae), pipefishes (Syngnathidae), cardinalfishes (Apogonidae), and blennies (Blenniidae). Also, because their eggs are attached to the substrate or because they use other types of parental care (Randall, 1998), the reduced number of offspring and the low adult dispersal ability keeps these fishes more restricted in distribution. Smaller species also usually have shorter lives, and, in an equivalent time period, will produce more generations than larger, long-lived species. Thus, evolutionary changes can proceed faster in smaller species. Of course, more small species can fit into a given habitat than larger ones, so diverse faunas are usually dominated by many small species— typically less than about 15 cm as adults.

HISTORICAL GEOLOGY AND FISH DISTRIBUTION

Stream Capture

Geological changes in river and stream drainages over time have been a primary factor in the speciation of fishes. Streams constantly erode surrounding canyons and can interact with the course of neighboring drainages. Proceeding laterally or upstream, the action of water lowers the divide between stream courses, until one becomes diverted or captured by another. The uplifting of mountain ranges accelerates this process. Volcanic eruptions or landslides can block a lake's outlet, create a new lake, or change the direction of stream flow. Such *stream captures* transfer some portion of an aquatic fauna into a new drainage system.

In the middle and upper parts of large drainage systems, these transfers usually occur in one direction— that is, a community of fishes in one drainage enters and mixes with a community of another system. An early axiom of fish biogeographers arose from the observation that each individual species in groups of related upland fishes often occurred in adjacent drainages across mountain divides. Thus, several groups of small darters (Percidae) and minnows (Cyprinidae) occur in adjacent drainages in the Appalachian mountains. These streams may drain to the Atlantic Coast, the Gulf of Mexico, or the interior Tennessee system. Such examples also occur in the Rocky Mountains with cutthroat trout (Salmonidae), in the Andes mountains with parasitic catfishes (Trichomycteridae), and in the Himalayas with hill stream fishes (Balitoridae).

In lowland expanses of associated marshes and wetlands, transfers can go both ways because fewer barriers to movement exist. Lowered sea levels have allowed currently separate river systems to join farther out on the continental shelf. For example, separate rivers now draining into Chesapeake and San Francisco bays were each a tributary of a larger river down the middle of each bay during glacial low levels of the sea about 10,000 years ago. These past connections are reflected in shared freshwater fish species that are now isolated from each other by marine or estuarine conditions.

Continental Drift

Today, the concept of continental drift is widely accepted. The Earth's crust, or *lithosphere*, is about 100 km thick and consists of a few major plates and several smaller ones. These tectonic plates float and move over an underlying *mantle* of viscous or thickly plastic material. Convection currents in the mantle send new rock to the surface along ridges in the oceans. The sea floor spreads away from these ridges, usually in two directions more or less perpendicular to the ridge.

One ridge runs down the middle of the Atlantic Ocean from north to south. At one time, the South America–Africa and North America–Europe pairs of continents were close together if not connected. They have been moving apart for 112 to 150 million years (Lundberg et al., 1998). The diverging eastern coastlines of North and South America have been stable in relation to sea level, with only slow changes in drainage patterns, over this time. Such coasts are called *passive margins* because little or no tectonic activity has taken place in or near them. The oldest passive continental margins, with less extreme climatic and geological disruption, usually have the largest inland or freshwater fish faunas today.

In contrast, along the western coasts of North and South America, the continental plates are converging or colliding. The Pacific Plate, largely underlying the Pacific Ocean, is slowly moving into the North American Plate, pushing up the western mountain ranges, like the Sierra Nevada and the coastal ranges of California, Oregon, and Washington. In a like manner, the Nazca Plate is colliding with South America, uplifting the Andes Mountains. Geologically disruptive mountain building has caused more radical changes in the landscape over shorter periods of time than on the east coasts of these continents. Moreover, the mountains have increasingly prevented rainfall from reaching the interior, leading to the aridity of the North American southwest. This explains the smaller native freshwater fish faunas in the western or Pacific drainages of North America (Minckley et al., 1986; Moyle, 2002; Smith et al., 2002) and South America (Malabarba et al., 1998). These western continental edges are called *active margins*, because of the mountain-building activity present.

The Himalayas formed as the Indian subcontinent collided with Asia, and they are continuing to uplift. As the Indian continent approached in the Paleocene and Oligocene, a broad, shallow sea gave way to the precursors of the present-day Ganges (eastward flowing) and Indus (westward flowing) river systems. As the mountains developed, they increasingly blocked monsoon storms from the Indian Ocean. Progressively, rainfall was restricted to the southern sides of these mountains, and the northern sides, much like the interior of the western United States, became progressively more arid.

About 225 million years ago, from the Later Paleozoic to the Early Mesozoic, the continents were combined into a "supercontinent" called Pangaea (Fig. 29.2). This large landmass began to split apart in the Jurassic, about 180 million years ago. The first separation between the future North America and Africa, about 150 million years ago, resulted in a northern and southern tier of future continents, separated by the westward-flowing Tethys Seaway. The northern component (Laurasia) would become North America, Europe, and Asia, and the southern one (Gondwana) would develop into South America, Africa, Antarctica, Madagascar, India, and Australia. By the late Cretaceous, about 90 million years ago, South America, Africa, Antarctica–Australia–New Guinea, and India–Madagascar had separated as distinct landmasses. In the Eocene (50–60 million years ago), India had separated from Madagascar, and both were large islands. Australia had separated from Antarctica, and New Zealand from Australia. By the Mid-Miocene (about 15 million years ago) the continents were near their present positions, with India in contact with Asia, and Africa and South America closely approaching Arabia and central America, respectively. Shallow seaways still separated the latter two pairs of continents (see Fig. 29.2). The present configuration of the continents is shown in Figure 29.1.

Islands have been natural laboratories for studying fish distributions. Many large islands with a history of contact with continental landmasses often have similar freshwater or inland fishes. Examples are Sri Lanka (Ceylon) close to India; New Guinea and Tasmania close to Australia; the British Isles near Europe; Trinidad near Venezuela; and Long Island near New York state. These are called *continental islands.* Another category of islands—*oceanic islands*—lack a history of connection to continental areas. In such cases, the resident fishes or their ancestors must have dispersed over narrow to wide expanses of ocean, have been carried on continents or their fragments, or have been derived from marine fishes. Thus the oceanic islands making up New Zealand have about 30

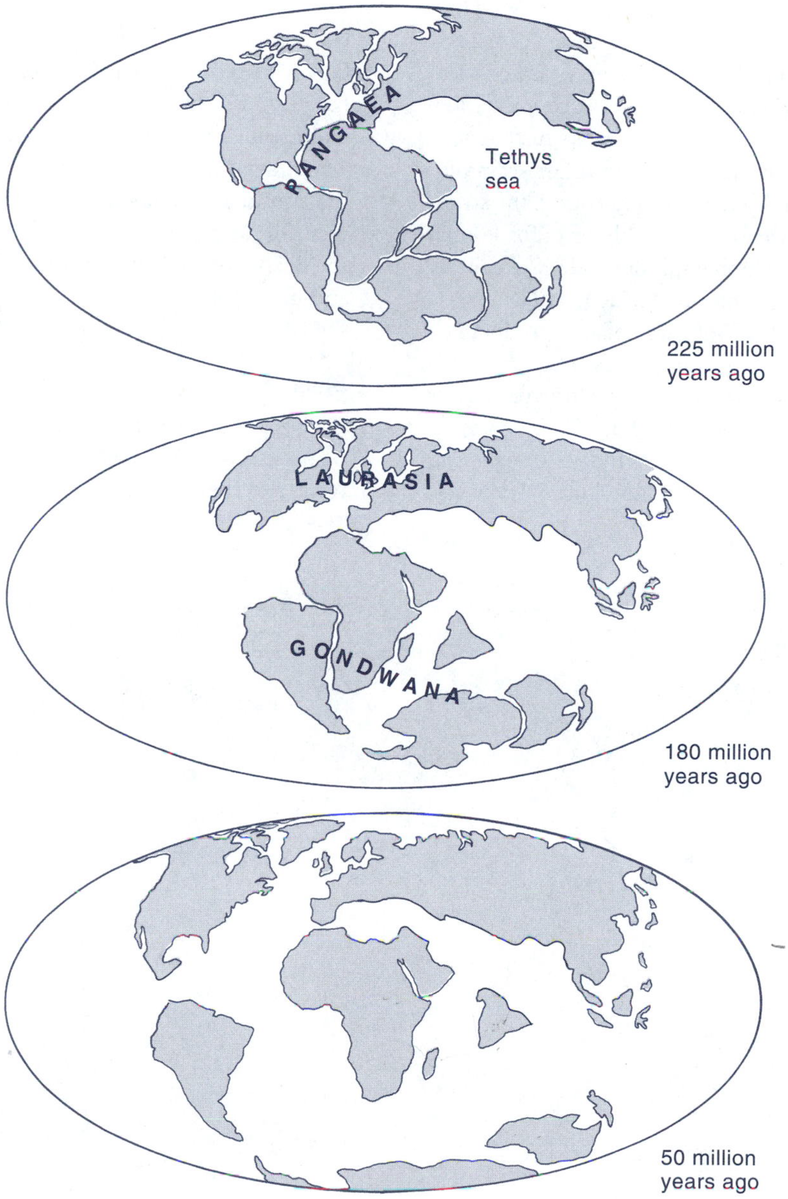

FIGURE 29.2
General changes in the configuration of the continents for the last 225 million years.

species of freshwater fishes. This is about the same number as the United Kingdom, a continental island group. The low number in the United Kingdom is due to its geologically recent emergence from glacial ice, whereas the low number in New Zealand is due to its distance over the ocean from other landmasses (McDowall, 1990).

The Hawaiian Islands are a classical example of oceanic islands. They formed over "hot spots" beneath the Earth's crust. Volcanic islands and seamounts (submerged mountains that may have been islands during lower sea stands) were formed from rising material from the Earth's mantle as they passed over the hot spot. A linear series of islands resulted, forming the greater Hawaiian chain of islands and seamounts. The currently emergent islands are only a few tens of thousands of years old, but the most distant seamounts to the northwest are close to 80 million years old. Although the current islands are relatively young, the string of preceding islands has provided the mechanism for a much longer period of faunal development independent of contact with continental areas. The volcanic Galapagos Islands are a product of a "triple junction" of three plates, producing a series of oceanic islands. Other examples of oceanic islands are Bermuda in the north Atlantic; the California islands off southern California and northern Baja California; the Cocos–Keeling islands in the southern Indian Ocean; and many of the islands of the open Pacific, from Micronesia to the eastern Pacific islands, such as Cocos (Costa Rica), Clipperton (France), and the Juan Fernandez (Chile) islands. Most of these islands are hundreds of kilometers from the nearest continent. Others, like the southern California islands, are only tens of kilometers offshore but separated by deep ocean gaps.

The mantle of the plates sinks back under continental areas, creating the deep trenches of the ocean. Older sea floor material has been destroyed by this process, and the oldest sea floor is less than about 70 millions years old, from the Late Cretaceous. This contrasts with the terrestrial continents, where rocks exist that are dated at more than one billion years old. The fossil record indicates that a deep-ocean habitat for fishes did not become available until the Late Cretaceous. Earlier fishes apparently were entirely shallow water, shelf or littoral inhabitants. During the Cretaceous, the continents reached positions that allowed the development of deep-ocean circulation; before this time, much of the deeper ocean was intermittently or continuously anoxic and uninhabitable by fishes.

· · · · ·
Glaciation

Glaciation has shaped the distribution of fishes since the Mid- to Late Miocene. Antarctic glaciation began at this time, and the Northern Hemisphere glaciation started later

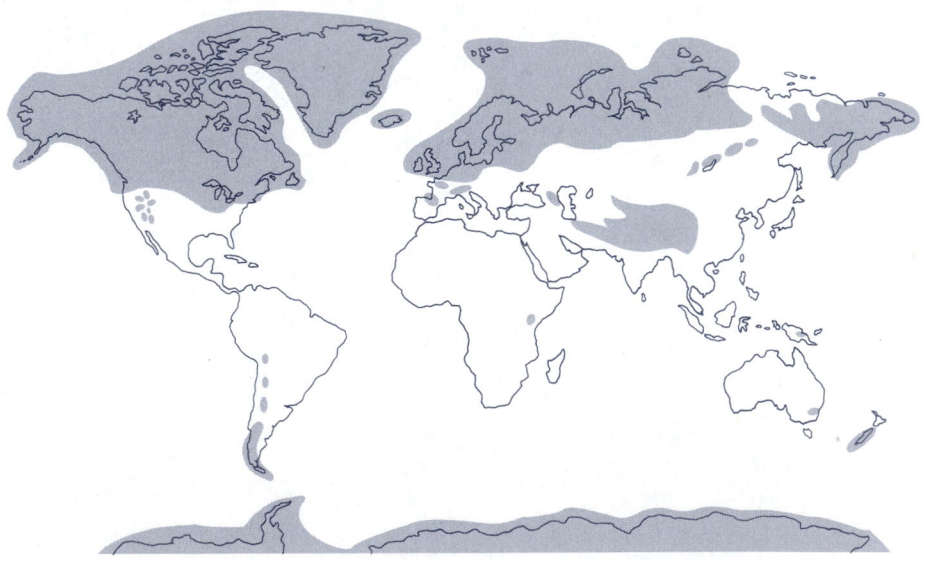

FIGURE 29.3
Extent of Pleistocene glaciation, which expanded over the northern continents several times in the last 1.5 million years.

in the Pliocene. Besides covering much of the land area with ice, the water taken up in glaciers lowered sea levels by tens of meters or more. Significant lowering of sea levels occurred in the Miocene, again in the Pliocene, and several times in the Pleistocene. Figure 29.3 shows the approximate extent of ice cover during the Pleistocene on the northern continents, southern Chile and Argentina, and southern New Zealand. At the same time, the continental shelves were exposed as much as 100 meters below present-day sea levels. Large river valleys in the San Francisco and Chesapeake bays and in the English Channel cut canyons into the emergent continental shelves. These are the submarine canyons of today on the continental margins. Peninsular Florida was twice as wide as today, and lowland habitat was even more extensive. A broad, emergent landmass of low relief extended across the Bering Sea, connecting present-day Alaska with Russia. Similar land bridges emerged between Australia and both Tasmania to the south and New Guinea to the north. Many of the islands of northwestern Indonesia also became connected to each other and to the southeast Asian landmass during these low sea stands. Conversely, during the Early Miocene (before the onset of glacial cycles) and during interglacials (when the glaciers were reduced or absent), the ocean was as much as 20 to 30 meters higher than at present. This submerged more than half of peninsular Florida, all of Long Island (New York), and flooded much of the Los Angeles Plain and Central Valley of California. This created an inland sea in the Central Valley of California. Many other lowland areas also would have disappeared under shallow ocean waters, such as the lower portions of many of the major lowland rivers of the world.

DISTRIBUTION OF FRESHWATER OR INLAND FISHES

Continental freshwater fishes demonstrate several patterns common to many other organisms on the globe. Diversity is highest in the tropical regions, and lowest in the polar regions, apparently because the increased (warmer) metabolism allows more species to occupy a given habitat or divide up the available resources. The latitudinal data for riverine fishes worldwide conform to this explanation (Allen et al., 2002b). Most freshwater fish habitat is in streams (lotic waters; see Chapter 31), and the greatest diversity in individual drainages usually occurs somewhere in the middle reaches. Diversity declines both in the uppermost tributaries and at the downstream interface with the marine environment. In the headwaters, fishes depend on allochthonous elements (material coming from outside the stream, such as insects, fruit, etc.; see Chapter 31). Downstream, greater water volumes (larger habitat size), generally higher temperatures,

and increased nutrients result in more habitat complexity in the water from autochthonous sources of energy (i.e., from the water itself; Lowe-McConnell, 1975).

Though less diverse, estuarine species are often very numerous in terms of individuals, and estuaries are among the most productive habitats on Earth. In fully marine conditions, the number of species goes up again. The reduced diversity at the freshwater–salt water interface is due to the greater fluctuation in abiotic factors (salinity, temperature, turbidity, etc.) controlling community structure more than biotic factors (predation, competition, etc.; Dunson and Travis, 1991, 1994). In a few coastal regions of the world with karst topography (i.e., bedrock of calcium carbonate limestone) and high rainfall, the resulting high calcium ion content physiologically permits marine fish to invade more inland habitats (see Chapter 32). The usual boundary between freshwater and salt water species is less distinct, such as in the "spring rivers" of Florida (Nordlie, 1990; Swift et al., 1986) and the northwest coast of Madagascar (Kiener, 1965; Stiassny and Raminosoa, 1994).

Biogeographical Realms

Today, each continent (except Antarctica and the continent-sized Greenland) has a distinctive freshwater fish fauna, with a unique combination of families, genera, and species. Some taxa are exclusive (or endemic) to one or more continents, and only a few are found on all continents. The classic zonation of the world is based on three realms with six regions. The realms are Arctogea (also called Megagea by Darlington, 1957), Neogea, and Notogea (see Fig. 29.1). **Arctogea** is divided into four regions as follows: (1) the **Ethiopian** region, consisting of Africa except for the Atlas mountain region in the northernmost strip along the Mediterranean Sea; (2) the **Oriental** region, including tropical Asia south of the Himalayas and extending out on to the Indonesian Islands to Wallace's Line; (3) the **Palearctic** region, including temperate and cold Eurasia north of the Himalayas, the northern third of Arabia, and about the northern tenth of Africa; and (4) the **Nearctic** region, including most of North America south through about the northern half of Mexico. The second realm is **Neogea,** consisting of the **Neotropical** region, from southern Mexico through Central and South America. The third realm is **Notogea,** consisting of the **Australian** region northward to include New Guinea and the Indonesian islands southwest of Wallace's Line. The term **Holarctic** is often used for the combined Nearctic and Palearctic regions spanning the complete polar, boreal, and temperate northern hemisphere.

These faunal regions roughly correspond to the boundaries of the continents and major landmasses. The boundaries between them typically have some intermixing of faunas. Between the Oriental and Australian regions, **Wallace's Line** reflects a relatively narrow oceanic gap. However, the deep ocean here means that a gap still existed during low sea stands. Wallace's Line separates Bali and Lompok—a sharp break in the distribution of several freshwater fish families that do not penetrate farther southwestward (Berra, 2001; Stiassny and Raminosoa, 1994). However, one group formerly known only west of Wallace's Line, the atherinomorph family Phallostethidae, was discovered recently in Sulawesi just east of the line (Parenti and Louie, 1998). According to geologists, the part of Sulawesi with the new species formerly was part of a single landmass west of Wallace's Line. The landmass split, and the eastern portion joined with and formed the western section of Sulawesi. The ancestral population of the new species likely was divided by the splitting of this landmass rather than by dispersal across an ocean barrier. Parenti and Louie (1998) convincingly argued that for fishes, Wallace's Line could be redrawn *through* Sulawesi, rather than between it and another landmass to the west.

Matthews (1998) statistically analyzed world fish distributions and found strong support for the classical subdivisions into realms and regions. His clustering of similarity among 52 areas worldwide was based on the world's major drainage basins and the distributional data of Berra (1981). The dominant clusters aligned closely with the classical realms and regions. The northern edge of Africa was not treated separately, so it did not cluster with the Palearctic region, as would be expected with analysis at a finer scale.

An area of southern South America (southern Chile and Argentina, together called *Patagonia*) also remained a distinct cluster, because it lacked many of the South American tropical fish groups and had some cold-water endemic species. One of Matthews's clusters united Madagascar and the islands of the West Indies, because they share just two families, Cichlidae and Cyprinodontidae, and have few other freshwater fishes. New Zealand also has a very small freshwater fish fauna, causing it to cluster as distinct from the rest of the Australian region. One fish family, the Cyprinidae, stood out because it spanned a much larger area than other freshwater fish families. Matthews attributed this to two divergent strategies among the "primary" freshwater fishes. The Cyprinidae (and other cypriniform families) had fewer species that were more widespread (discussed later; see Fig. 29.6), and the catfishes (Siluriformes) and characins (Characiformes) each had diversified into more families in smaller continental areas (see Going Deeper).

The Holarctic Region

The fishes of the Holarctic region—that is, the Nearctic and Palearctic regions combined—are very similar, and a few families are distributed across most of both, including the Percidae, Cyprinidae, and Esocidae. In both North America and Eurasia, the faunas are much richer in the southeastern portions of the regions and less diverse farther to the north, northwest, and west. This north–south difference is relatively abrupt in eastern North America and Eurasia, where the Pleistocene glaciation extirpated many freshwater fishes

GOING DEEPER • The Equivalence of Fish Taxa

At this point, the prudent student should be aware of the lack of quantitative measures of what constitutes a species, a genus, or a family of fishes. Is a family (genus, subfamily, etc.) of the Characiformes equal, in some scientific sense, to a family of the Cypriniformes? Currently, the answer to this question is either "no," "maybe," or "We have an intuitive idea that they are comparable, but we lack strong quantitative measures or definitions of these groups." Some systematists may think that the Cyprinidae could be divided into more families. For example, until the fish classification of Greenwood et al.

(1966), far fewer families were recognized in the Characiformes. Even at the species level of fish classification — often considered the more basic, elemental, and most objective taxonomic level — arguments can often develop. Many measures and assessments of diversity are based on the species level, because the naming of higher levels of classification is perceived as more subjective. Considerable recent scientific scrutiny of the concepts and definitions of species has developed from an increased interest in conservation. As more species and populations of fishes (and other organisms) have become

imperiled, it is imperative to scientifically and legally determine what group exactly must be protected. This has strengthened the scientific definition and integrity of what constitutes a species and has provided mechanisms to prioritize the importance of groups of fishes for both scientific and conservation purposes (Mayden and Wood, 1995; Moritz, 2002; Waples, 1995). This rigor has not been applied to higher levels of classification, but it would be desirable when making comparisons for biogeographical purposes.

Elsewhere in this chapter, we have noted the distribution of lampreys at the colder northern

or pushed them southward. Many taxa have not reinvaded northward in the thousands of years since the last glacial cycle (see Banarescu, 1986, 1990, 1992, for Europe; see Hocutt and Wiley, 1986; Matthews, 1998, for North America). For example, all members of the freshwater fauna in England must have arrived in the last ten thousand years or so. In North America, an intermingling of northern and southern forms occurs along the southern front of glacial advance, across New York state, Pennsylvania, Ohio, Indiana, and Illinois, in both the Great Lakes and Mississippi drainages. In eastern Europe and Asia, the glaciated areas largely drain north and have a smaller fauna of salmonids, whitefishes, esocids, and cottids than the unglaciated areas draining south, with mostly cypriniforms, percids, and catfishes. Areas of intermingling of northern and southern fish faunas occur at the western and eastern ends of the Palearctic region, namely in the Baltic, Eastern Mediterranean, and Black Sea drainages in eastern Europe, and in the Amur River basin in eastern Asia. In both of these areas, the cold-water and warmer (temperate) fish groups mix. Some unique groups, such as the species of the genera *Rhodeus, Huso,* and *Misgurnus,* are disjunct—that is, some representatives occur in each mixing area, but they are absent from much of the intervening expanse of Asia (Banarescu, 1990, 1992, 1994; Berg, 1949).

Six freshwater fish families are found only in the Holarctic region: Polyodontidae, Esocidae, Umbridae, Dalliidae, Catostomidae, and Percidae, along with a few other, diadromous families, such as the Petromyzontidae, Salmonidae,

Osmeridae, and Acipenseridae. The two living members of the Polyodontidae, or paddlefishes, are found in the Yangtze River of China and in the Mississippi River and associated drainages of North America (Fig. 29.4A). Fossils also occur in the northwestern United States and northeastern Asia (Grande and Bemis, 1991). The ancient sturgeon subfamily Scaphirhynchinae is found in eastern North America and in tributaries of the internally draining Sea of Aral, east of the Mediterranean and Black seas (Mayden and Kuhajda, 1996).

The Esocidae have a wide distribution, covering almost the complete northern, glaciated parts of the Holarctic region (Fig. 29.5). A great deal of this area is covered by one species, the northern pike (*Esox lucius*), in a circumpolar distribution. Three other migratory or salt-tolerant species are also Holarctic, widely and almost continuously distributed, namely the rainbow smelt (*Osmerus mordax*), the fourhorn sculpin (*Myoxocephalus quadricornis*), and threespine stickleback (*Gasterosteus aculeatus*). The freshwater cod or burbot (*Lota lota*) is also distributed circumglobally, primarily in northern glaciated regions, somewhat like the northern pike.

The mudminnows (Umbridae) have a disjunct distribution (Fig. 29.4B; see Fig. 4.2I), *Umbra krameri* occurs in the Danube watershed of southern Europe, and the two American species live in the Laurentian Great Lakes region (*Umbra limi*) and along Atlantic and northeast Gulf of Mexico drainages (*Umbra pygmaea*). These species probably date from a transatlantic connection in the Early Tertiary (Patterson, 1981).

and southern ends of the globe (*antitropical distribution*). Some researchers have recognized one family of lampreys (Petromyzontidae) with one subfamily in the northern hemisphere (Petromyzontinae) and two in the southern (Mordaciinae and Geotriinae); others elevate these groups to the family level and recognize three families of lampreys. Both sets of researchers agree that the two southern groups are more closely related to each other than they are to the northern group. Thus, both would also agree that only one ancestral lamprey had to travel between the poles to explain the distribution

(based on current knowledge). If, however, evidence supported a hypothesis that one of the southern families was closer to the northern family, other possible combinations of the timing of speciation events would be possible. At least two ancestral divisions across the equator could reasonably be postulated.

In either of the aforementioned cases, the analyses would be similar regardless of which classification was used. If Matthews (1998) had used subfamilies of the Cyprinidae rather than families, we would expect them to reinforce the patterns determined from the whole suite

of fishes analyzed. However, the distinction of a single, widespread family Cyprinidae might be lost. On the other hand, more endemism would be exposed if the subfamilies were found to be more narrowly distributed to particular landmasses. Until we have a more rigorously scientific idea of what constitutes a "family" of fishes, the comparisons will remain relative. In any case, most researchers today are more interested in the actual interrelationships—and the opportunities for scientific hypothesis testing that they provide—than in arguments over the naming of the hierarchical levels of classification.

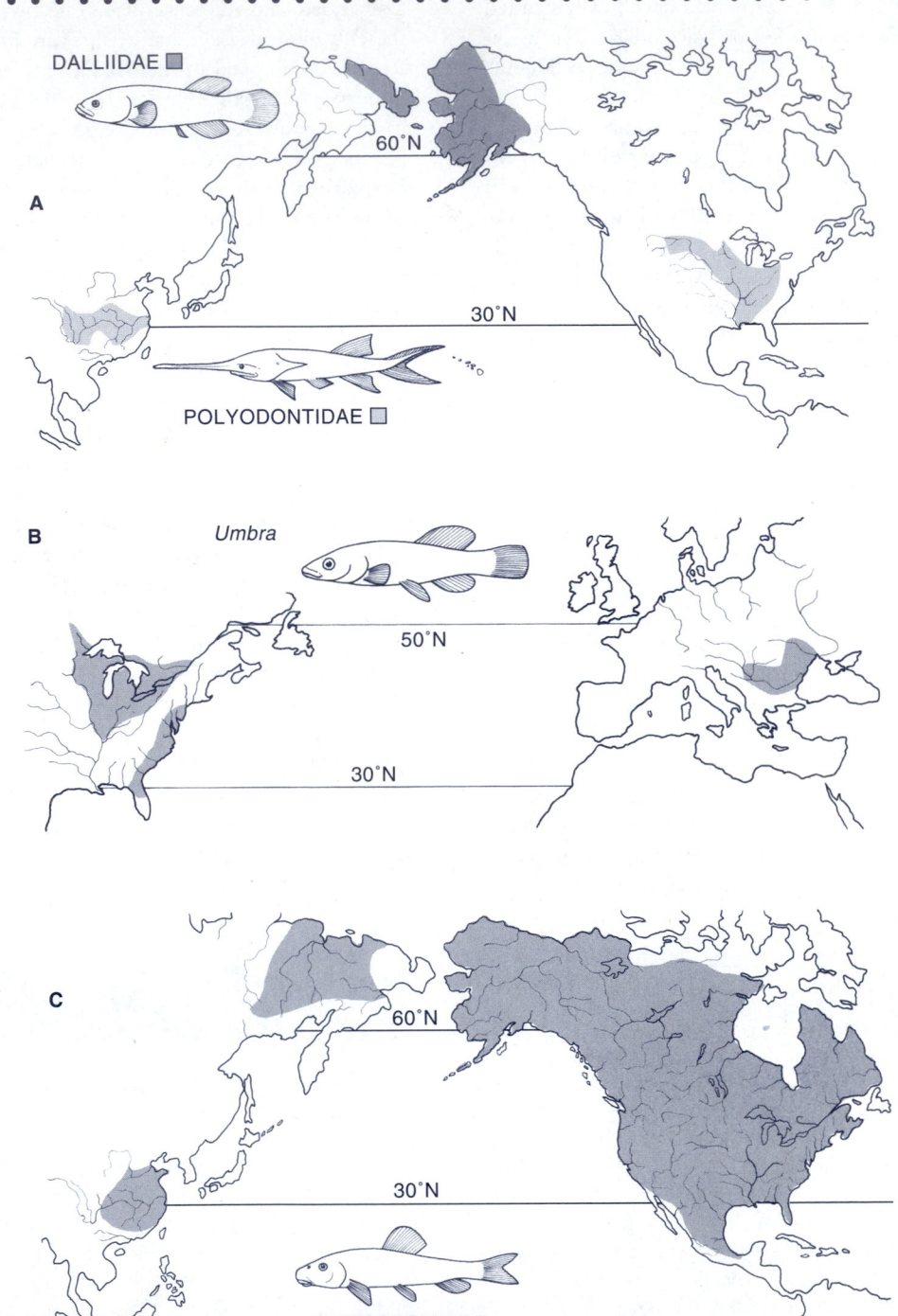

FIGURE 29.4
Distribution maps of the families **A,** Dalliidae; **B,** Polyodontidae; and **C,** Catostomidae, illustrating relationships between Asia and North America across the Pacific.

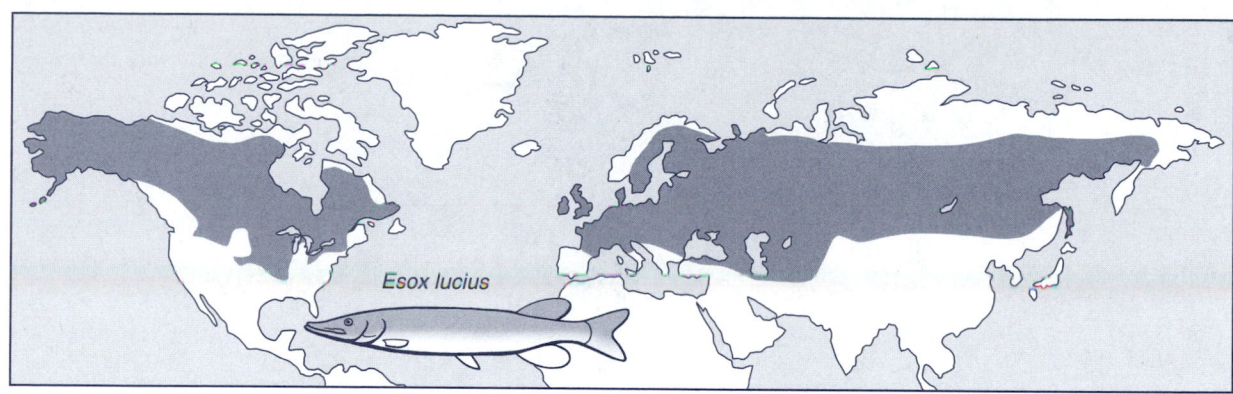

FIGURE 29.5

Distribution of the northern pike (*Esox lucius*) throughout the Holarctic region, mostly on recently glaciated terrain (after Grande, 1999).

The Dalliidae consist of three or four species, *Dallia pectoralis* in Alaska and two or three species across the Bering Strait in extreme northeastern Russia, which were only recently separated by the submergence of the Bering Strait in the last ten thousand years or so (see Fig. 4.2I). Given the fluctuations of glacial activity during the Pleistocene, this area has been exposed at least four or five times in the last million or so years. Thus, the fishes could have repeatedly come back into direct contact across the present marine gap.

The sucker family (Catostomidae) also is split by the Bering Strait, with a recent crossing in the longnose sucker (*Catostomus catostomus*), which is found on both sides of the Bering Strait but is considered the same species (Fig. 29.4C). A much older connection is required to account for the occurrence of one primitive sucker (*Myxocyprinus asiaticus*) in southeast China and the remaining members of the family almost throughout North America (Fig. 29.4C). This old connection is probably similar in age to that of the paddlefishes—namely, early Cenozoic. Known Eocene and Paleocene fossils of suckers occur in northwestern North America and northeastern China (Cavender, 1986).

The Percidae have two large-bodied genera, *Sander* (formerly *Stizostedion*) and *Perca*, with different species in the old and new worlds. Three other genera with a small number of species occur in the old world, and several genera of small-bodied "darters" include at least 150 species (Burr and Mayden, 1992; Wiley, 1992) in the eastern half of North America, primarily in the United States. The more northern, large-bodied genera represent recent connections

with Asia, and the rich eastern North American fauna has an older ancestry, dating from the early closeness of Europe and eastern North America (Collette and Banarescu, 1977; Wiley, 1992). A few taxa of other anadromous groups also show a transatlantic pattern, including the shads (*Alosa*) and the temperate basses (*Morone*) in eastern North America, and *Dicentrarchus* in western Europe.

The Cyprinidae or minnows (closely related to the suckers, Catostomidae) are the only other "primary" freshwater fish family with a Holarctic distribution, but they also extend into the Ethiopian and Oriental regions. They are absent from the Neotropical and Australian regions, stopping at Wallace's Line in Asia and just north of the isthmus of Tehuantepec in Mexico (Fig. 29.6). Like the percids, the cyprinids are much more diverse in eastern North America. The North American drainages west of the Rocky Mountains (and the continental drainage divide) have far fewer minnows and completely lack several families, such as the percids, amiids, and lepisosteids. Much of the speciation of the percids and of many cyprinids occurred entirely in the Nearctic region.

The killifish family (Fundulidae) is exclusively Nearctic and extends from the northern Laurentian Great Lakes region southward into the tropical parts of central Mexico (Yucatan) on the Atlantic coast, the northern edge of Cuba in the Caribbean, and the southern tip of Baja California on the Pacific side, but not into the Gulf of California. The ecologically similar pupfishes (Cyprinodontidae) extend into South America and north Africa around to the eastern

• •

CYPRINIDAE

FIGURE 29.6
Distribution of the family Cyprinidae, exceptional in its occurrence in the Nearctic, Palearctic, Oriental, and Ethiopian regions.

• •

Mediterranean and the margins of the northern Arabian Sea (Turkey).

The anadromous lampreys (Petromyzontidae) and sturgeons (Acipenseridae) are widespread in the Holarctic, and both families have some freshwater species on each continent. The catadromous anguillid eels are found throughout the Holarctic in tributaries of the north Atlantic, Indian, and western Pacific, but are absent from tributaries of the eastern Pacific Ocean. The anguillids also extend into the southern Hemisphere in the Indian and western Pacific oceans. Other diadromous or estuarine groups in the Nearctic include the Salmonidae, Osmeridae, Gasterosteidae, and Cottidae.

The Palearctic Region

The Palearctic region shares its freshwater fishes with other continental landmasses, but two endemic, salt-tolerant fish groups occur there as well: the amphidromous ayu (*Plecoglossus altivelis*) of Japan and Taiwan, and the noodlefishes (Salangidae) along the east coast of Asia. Two endemic families of cottoid fishes, the Comephoridae and the Abyssocottidae, occur in the ancient Lake Baikal in central Asia (see Fig. 29.1). This lake has a species flock of about 24 mostly interrelated species in these two families, both included in the family Cottidae by some workers (Kinziger et al., 2005; Kozhov, 1963; Sideleva, 2003). The lake occupies a mountainous area that has been undergoing tectonism for millions of years, and several older lake basins occupy the current

drainage. The endemic species of the remaining lake may represent multiple earlier source lakes rather than in situ speciation in only one lake.

The Palearctic region shares a number of freshwater families with tropical Africa and the Asian tropics of the Oriental region. It is richest in Cyprinidae (see Fig. 29.6) and related families (Cypriniformes). These and several other families show a hiatus in distribution across the Arabian peninsula, where the drying climate in the last few million years has eliminated populations. In other cases, separate populations may date from the much earlier separation of India from Africa. The Asian populations would be derived from fish carried on the Indian landmass from Africa to Asia. The loaches (Cobitidae) occur in the Palearctic (including Europe), Oriental, and Ethiopian regions. Climbing perches (Anabantidae), snakeheads (Channidae), spring swamp eels (Mastacembelidae), and the labyrinth catfishes (Clariidae) range from tropical (i.e., Oriental) Asia throughout the transitional area into temperate (Palearctic) Asia (Fig. 29.7). The hill stream catfishes (Sisoridae) from tropical and temperate Asia are absent from Africa. The old world catfishes, or sheatfishes (Siluridae), are shared by the Oriental and Palearctic regions. The flat and hill stream loaches (Balitoridae, formerly Homalopteridae) are found in several drainages circumscribing the Himalayan mountains. Thus, these fishes inhabit both the Palearctic region to the north and the Oriental region to the south by virtue of their adaptation to the changing drainage patterns of these steep mountain streams (Banarescu, 1992).

The Nearctic Region

The Nearctic region is characterized by several relict groups of archaic fishes. The primitive teleost family Amiidae has one living freshwater species, the bowfin (*Amia calva*), in eastern North America. Fossils of this and related families date from the Jurassic and the Cretaceous of Europe, Asia, and South America (Grande and Bemis, 1998). Cenozoic fossils of the genus *Amia* and its relatives (subfamily Amiinae) have been found throughout the Holarctic region. Thus, either a transpacific or a transatlantic relationship may exist. As no fossils of *Amia* have been found after the Miocene outside of North America, it is truly a "living fossil" (Grande and Bemis, 1998, 1999). A similar fossil record exists for the more primitive gars (Lepisosteidae), known from several living species in eastern North America, Mexico, and the Caribbean (Wiley, 1976). However, the absence of fossils in the eastern Palearctic region favors a transatlantic relationship. The Hiodontidae, with two living North American species, are related to the Notopteridae of Asia and Africa. However, all fossil hiodontids occur in western North America and northeastern Asia, indicating a transpacific relationship for this family (Cavender, 1986; Wilson and Williams, 1992). Three families of primitive acanthopterygians (Percopsiformes)— the Percopsidae (troutperches; 2 species), Aphredoderidae (pirate perch; 1 species), and Amblyopsidae (North American cave and spring fishes; 12 species)—are exclusively North American as living species. A distant relative of the living forms, the fossil genus *Sphenocephalus*, has been found in the Cretaceous of Europe; otherwise, all other Cretaceous and Cenozoic records of Percopsiformes are from North America (Wilson and Murray, 1996).

Two speciose families of freshwater fishes that are endemic to North America or the Nearctic region are the sunfishes and freshwater basses (Centrarchidae; about 30 species) and the North American catfishes and madtoms (Ictaluridae; about 45 species). Only one species from these two families is found west of the continental divide of North America: the Sacramento perch (*Archoplites interruptus*), a centrarchid that occurred in the Sacramento–San Joaquin and associated drainages of central California (Moyle, 2002). It was completely extirpated in its native range, but thrives in several alkaline natural and artificial lakes and reservoirs in eastern California and western Nevada (Moyle, 2002). Both Centrarchidae and Ictaluridae have an extensive fossil record in the western United States (Smith et al., 2002). In the eastern United States, the centrarchids and catfishes (except the madtoms) are most diverse in the lowlands of

the Mississippi Valley and the southeastern lowlands of the United States. The madtoms, small ictalurid catfishes of the genus *Noturus,* occur more in the uplands in small streams. Some madtoms show a central highlands distribution. Such species are divided across the present Mississippi River valley, both in the uplands of the Ozark Plateau in Arkansas and Missouri to the west and in the southern part of the Appalachian mountains and the Cumberland Plateau to the east. These two uplands have remained south of Pleistocene glaciation and have had a relatively stable existence for most of the Cenozoic.

The Percidae, Cyprinidae, and Catostomidae are particularly numerous in eastern North America, and often one species in a genus or subgenus will be found in each of several major tributaries of rivers in one or both of these upland areas (Boschung and Mayden, 2004; Mayden and Wood, 1995; Near and Benard, 2004; Strange and Burr, 1997). The approximately 22 upland species of *Nothonotus* darters (Percidae) epitomize this trend, with species in most major drainages of the lower Great Lakes, in several drainages on each side of the Appalachian Mountains, and in both the Ozark and Ouachita uplands west of the Mississippi River; there is also a relict southern species in eastern tributaries of the lower Mississippi River valley, as noted below (Near and Keck, 2005). Mayden et al. (1992) estimated that there were approximately 1,050 species of North American native freshwater fishes, some yet to be described (Fig. 29.8). The Mississippi River drainage as a whole has close to 400 species; the Mobile Bay system of drainages to the Gulf of Mexico has about 240 species. The independent drainages along the Atlantic coast and west of the Mississippi River along the Gulf of Mexico have fewer species. A strong division exists between the faunas east and west of the Appalachians (e.g., eastern and western chubsuckers; see Chapter 4, Fig. 4.2III). Avise (2004) illustrated the close congruence of the genetic dichotomy in the eastern and western populations of spotted sunfishes (from Bermingham and Avise, 1986) with the faunal dichotomy between the eastern and western drainages, based on a cluster analysis of their fish faunas (from Swift et al., 1986). The lowlands around northern Florida and southern Georgia have permitted some interchange of lowland fishes between the eastern and western sides of the Appalachians, but this mountain range has remained a strong barrier for many freshwater fishes for much of the Cenozoic. Low-elevation streams that do not have tributaries above Miocene sea levels have considerably fewer species, and the Florida peninsula has the fewest freshwater fishes (Swift et al., 1986; Fig. 29.8).

In the Nearctic region, many present fish distributions were influenced by the advance and retreat of the glacial ice.

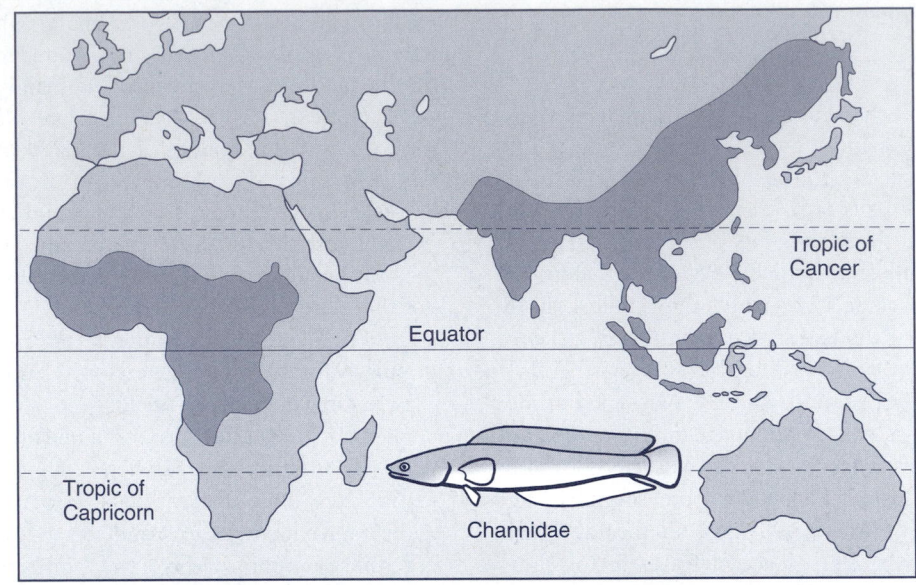

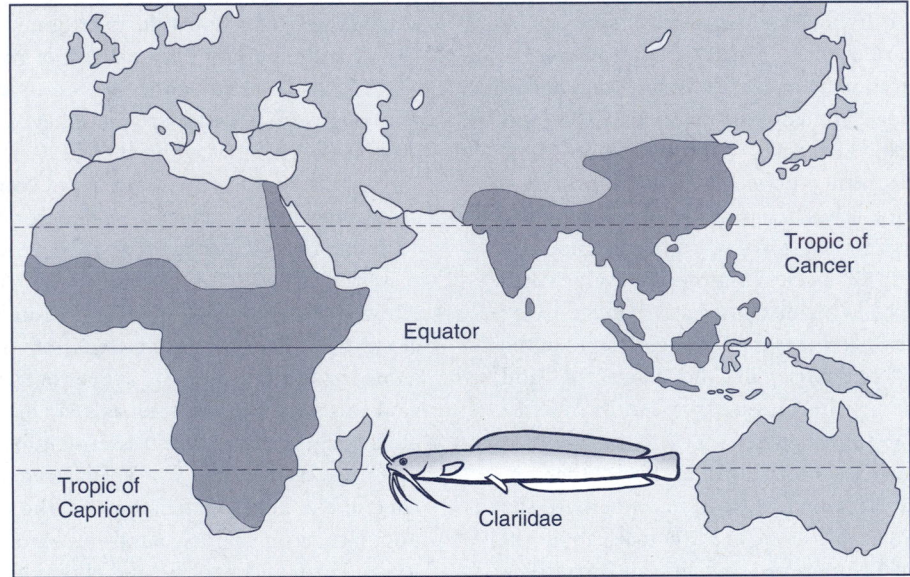

FIGURE 29.7
Freshwater fish families spanning the Palearctic and Ethiopian regions, but not crossing Wallace's Line into the Australian region.

Glacial advances are represented today by relict populations of otherwise northern species in the south. The bayou darter (*Etheostoma rubrum*), the southern redbellied dace (*Phoxinus erythrogaster*), and the bluntface shiner (*Cyprinella camura*) all occur in small, cool tributaries of the Mississippi River in western Mississippi, whereas their closest relatives are much farther north and at higher elevations. Glacial retreat allowed these species to invade drainages northward, via routes that changed with the hydrology of the Great Lakes, draining at times into the Atlantic, into Hudson Bay, and into the greater Mississippi drainage via three or four different points, including the western tip of Lake Superior, the southern end of Lake Michigan (the Illinois River outlet), and the southwestern end of Lake Erie. The Mississippi

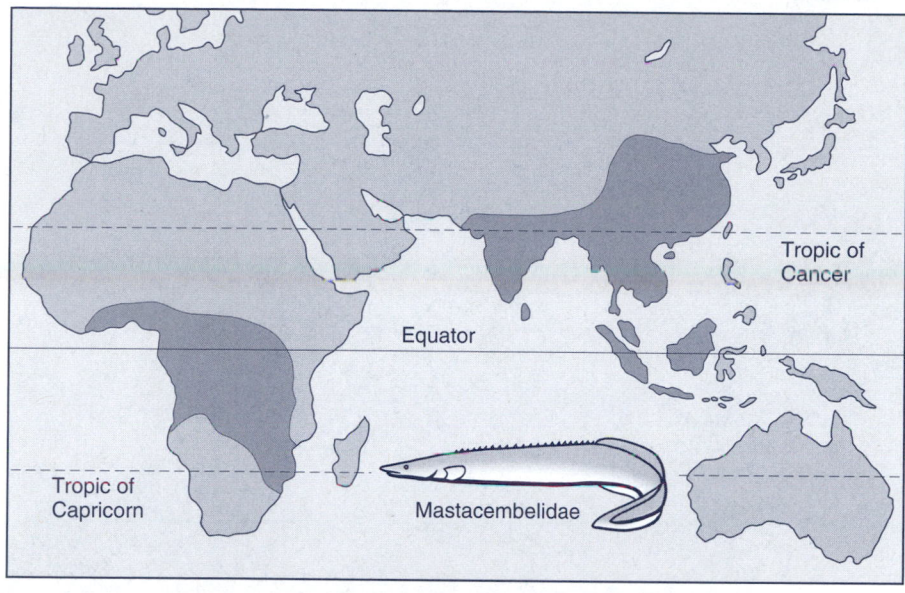

Mastacembelidae

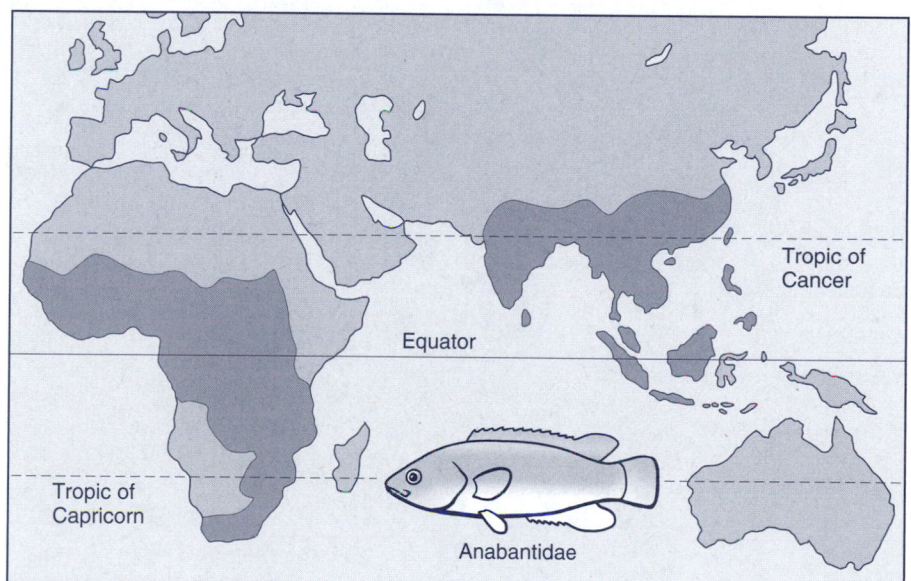

Anabantidae

drainage source area (or refugium) provided more than 80 percent of the invading species (Crossman and McAllister, 1986; Matthews, 1998; Trautman, 1981; Underhill, 1986).

During glacial advances, the western United States and northern Mexico were much wetter, large lakes dominated the basin and range provinces, and they had larger fish faunas. However, these faunas were still much smaller than those of the eastern United States (Smith et al., 2002). Only a few species closely related to the eastern North American fauna are present in the western fauna, which is dominated by distinctive groups of cyprinids, catostomids, cottids, sticklebacks, and salmonids. These species are relicts of the previous, wetter climatic period. During this glacial period, the interior watersheds of Idaho, Oregon, Utah, and Nevada

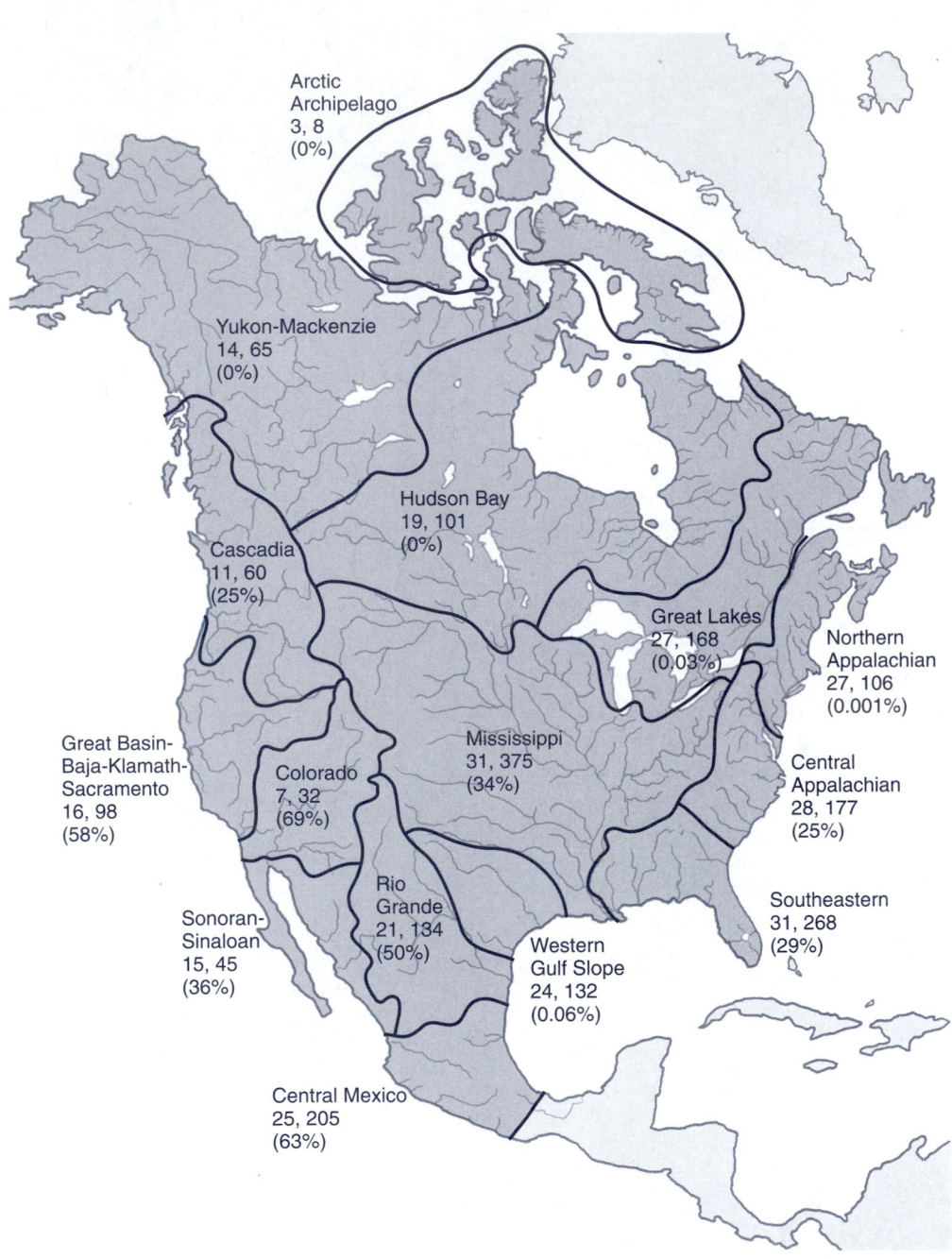

FIGURE 29.8

The distribution of freshwater fishes by major drainages in North America (after Burr and Mayden, 1992). The numbers represent the number of families, species, and percentage endemism at the species level, respectively.

drained toward the southwest across California, variously into the Sacramento–San Joaquin and Klamath drainages. Later interior flows went southward, from the Snake River across the present day Bonneville Basin into the Colorado River. During this same period, the Colorado plateau was uplifting as the Bonneville and Lahontan Basins expanded, isolating the major tributaries of the Colorado River from each other. Genera like *Rhinichthys*, *Lepidomeda*, and *Pantosteus* have a distinct species or recognizable taxon in each of the several major tributaries of the Colorado River drainage. Cognate species were isolated in major western drainages. For example, one species each of the minnow genus *Ptychocheilus* became isolated in the Colorado, Columbia, Umpqua, and Sacramento–San Joaquin drainages (see Fig. 4.2II).

During the last million years or so (the Pleistocene), some northern drainages have been diverted toward the Snake River and Columbia River drainages. A large area including eastern California, southeastern Oregon, Utah, and Nevada became an internal basin with no outlet to the ocean (the Great Basin). Several distinct forms of cutthroat trouts are distributed among the mountainous tributaries of what are now dry or hypersaline lakes. Some groups, like the stickleback, were once common in the interior but became completely extinct east of the Sierra Nevada mountains (Bell, 1994). Fossil evidence shows that they were present until a few thousand years ago in the area of the present Mojave Desert of southern California. Even the surviving fishes, like the western lake suckers, now show relict distributions (see Fig. 4.2III). Adaptation to the severe conditions of the interior deserts resulted in a morphological convergence toward a few surviving types. This relative homogeneity confused earlier attempts at classification, and molecular genetics is providing a valuable data set for unraveling the genealogical relationships (Smith et al., 2002). Even speciose genera that range across the whole continent show a strong dichotomy across the continental divide, such as the dichotomy that occurs in eastern and western clades of pupfishes (*Cyprinodon*, family Cyprinodontidae; Echelle et al., 2005) and in many of the freshwater sculpins (*Cottus*, family Cottidae; (Kinziger et al., 2005).

Middle America

Middle America (i.e., Central America and the southern part of Mexico) combines an interchange of North and South American faunas with local endemism (Bussing, 1985, 1998; Lundberg, 1998; Miller et al., 2005; Fig. 29.9). Between North and South America, during most of the Cenozoic, a changing combination of islands (associated with an equally complex interaction of tectonic plates) occupied the area of the present Middle American isthmus and Caribbean islands. From about 11 to 4 million years ago, a permanent

land connection was established. The earlier islands isolated fishes and resulted in widespread endemism (Fig. 29.9C) among species of cichlids, characids, and pimelodids, which were otherwise mostly South American. North American species of centrarchids, cyprinids, and ictalurids extend only into Mexico and extreme northern Middle America. As in western North America, more species of centrarchids and ictalurids occurred in central Mexico during past well-watered periods (Cavender, 1986).

Several families in the freshwaters of Middle America originated there (and thus are endemic), such as the Goodeidae, Profundulidae (Fig. 29.9D), Poeciliidae, and Anablepidae. The freshwater and marine silversides (Atherinopsidae) also have several endemic species in the freshwaters of Middle America, as well as coastal marine species farther north and south in North and South America (Dyer, 1998; White, 1986). As shown in Figure 29.9, Middle America has faunal mixing (Fig. 29.9A), considerable endemism associated with past tectonic activity (Fig 29.9C), several faunal areas (Fig. 29.9B), and phylogenetic relationships with extralimital areas of North America (Fig. 29.9D). The Antillean Islands have a small fauna, totaling about 71 freshwater fishes—predominantly on the larger islands of Cuba, Hispaniola, and Jamaica, with fewer on smaller and more distant islands (Burgess and Franz, 1989). Poeciliids constitute 46 of the species known.

The Neotropical Region

Until the aforementioned connection with Middle America was established, South America was isolated by ocean for a long time. Lundberg et al. (1998) estimated the separation of Africa (Ethiopian region) from South America across the south Atlantic Ocean at 112 to 150 million years ago. The basic configuration of South America has been in existence for most of the Cenozoic, with a spine of elevated mountains along the western edge. The eastern two thirds of the continent consists largely of lowlands. Two Late Mesozoic or older uplifted but more stable areas are found north and south of the main lower Amazon River drainage of today. In the early Cenozoic, the lowlands adjacent to the eastern base of the Andes were at times occupied by inland seas of varying extent. These seas opened to the north near the mouth of the present-day Magdelena River and to the south near the mouth of today's Paraná River (see Fig. 29.1). These seas were intermittently replaced or bordered by freshwater lakes in their upper reaches. During some periods, a long river existed flowing from northern Argentina north along the east side of the Andes to the Caribbean Sea. Thus, most of the Amazon drainage flowed north, and today's lower Amazon River was a relatively small drainage. In the Late Miocene, a lower barrier was breached, and most of the

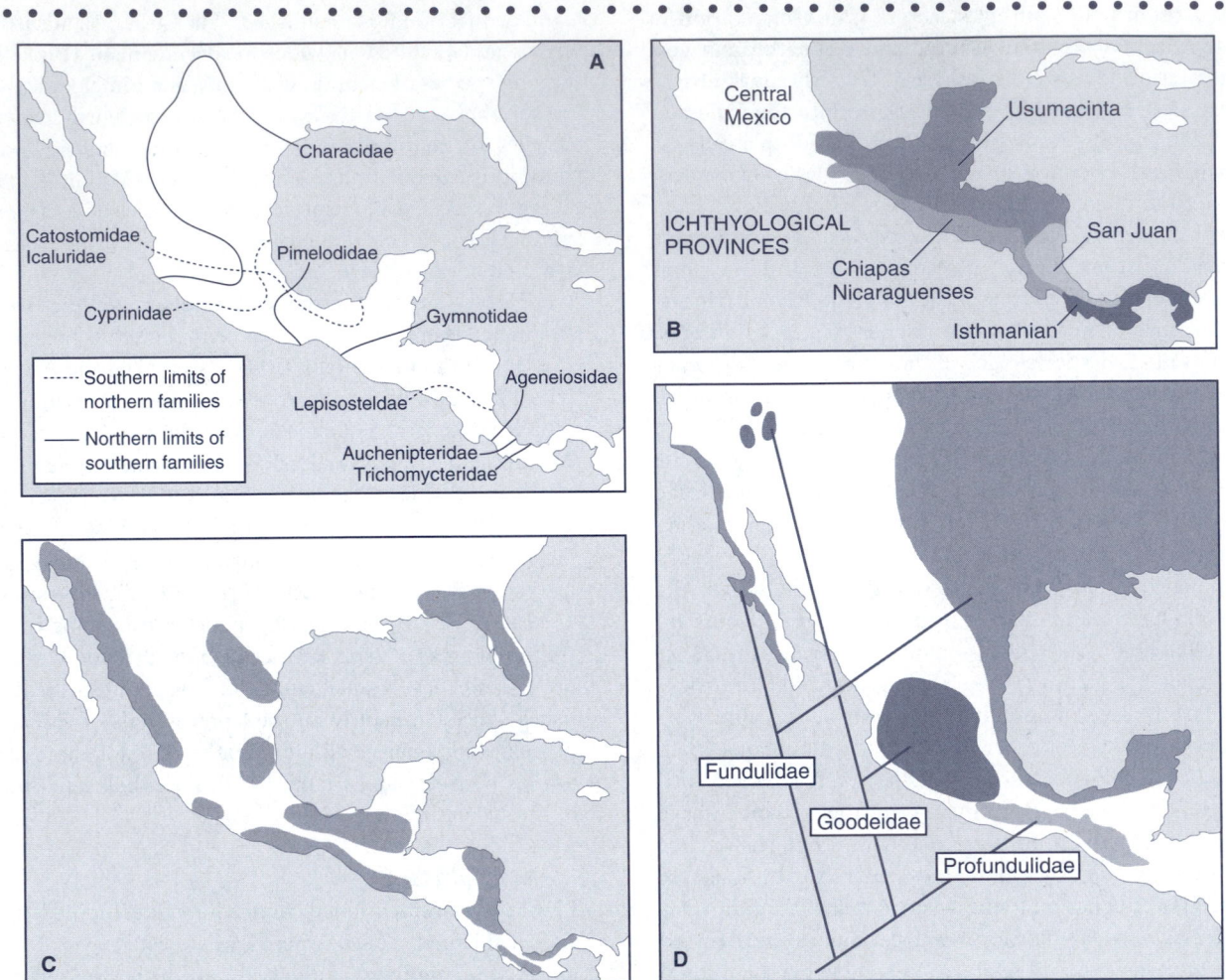

FIGURE 29.9

Four ways to analyze the freshwater fish distributions of Middle America: **A,** The mixing of North American families southward and South American families northward in this area of faunal interchange. **B,** Division of drainages into provinces based on the sharing of species within provinces. **C,** Areas of endemism, with distinct species and genera not found elsewhere, indicating isolation from other areas, sometimes on islands. **D,** The cladistic interrelationships of three families, Fundulidae (medium shading), Goodeidae (dark shading), and Profundulidae (light shading), providing a hypothesis for the evolutionary divisions (vicariances) that occurred leading to these current distributions. (*A* after Regan, 1906–1908; Miller, 1966; Lomolino et al., 2005; *B* after Bussing, 1998; *C* after Bussing, 1985; *D* after Costa, 1998).

interior drainage began to flow into the much-enlarged main Amazon River of today about eight million years ago. The Orinoco and Paraná drainages also became larger, acquiring some parts of the earlier north- and south-flowing system through the middle of the Cenozoic. No evidence exists for a prehistoric large river outlet to the Pacific from the Late Cretaceous onward (Lundberg et al., 1998). This means that the few Amazonian fishes in the small, steep Pacific drainages must have been transferred by stream captures. Lundberg et al. (1998) concluded that the great diversity of South American fishes has developed over much of the Cenozoic and, in fact, was largely in place by Miocene times. The Pliocene and Pleistocene saw much rearrangement of the fauna but no large-scale evolution of new species, as has been postulated for other organisms of the Amazonian fauna.

As recently as the previous edition of this book (1996), estimates of the number of South American freshwater or inland fish species ranged from 2,000 to 5,000, with the latter figure considered possibly too high. A few years later, Vari and Malabarba (1998) agreed with Schaefer's (1998) estimate of about 8,000 species. If the total number of fish species in the world reaches an estimated 35,000, then the fishes of the Neotropical realm will be 23 percent of this total! A majority of these species are found in the Amazon River drainage, but with large faunas in the Orinoco, Magdalena, and Paraná systems as well (see Fig. 29.1). Many sets of cognate species of large river fishes occur in three or four of these drainages. One each of three or four related species can occur in each drainage. A particularly diverse example is the characiform genus *Creagrutus* (Vari and Harold, 2001), with more than 50 species; each species is restricted to one or a few major tributaries of the Amazon, Orinoco, and Paraná systems and several other minor drainages. The Magdalena River has fewer species, as some fish groups documented in the fossil record have been lost (Lundberg et al., 1986).

Other South American drainages lying between the largest ones on the Atlantic and Caribbean coast have fewer species, and the fewest species are found in the steep drainages that flow to the Pacific. The small drainages of the southeastern coast of Brazil have been isolated for a long time and often have the basal (or primitive) sister taxon of groups that are otherwise widespread in the Amazon system. This suggests an early dichotomy or vicariance between this southeastern area and most of the rest of the continent.

The Neotropical fish fauna is richest in Characiformes and Siluriformes, with the Cichlidae a distant third in number of species. The weakly electric fishes (Gymnotiformes), with five or six families, are also rich, with about 150 species. Fourteen families of catfishes are also endemic to the region, along with several families of characiforms. The Characiformes and the Cichlidae are shared with Africa, but the faunas in these groups appear to be monophyletic on each continent. This means that a common ancestor gave rise to the whole fauna on each continent after an initial, early sharing of one or a few ancestral taxa.

Several other freshwater families are closely related to marine groups and probably evolved in the early Cenozoic in the inland shallow seas and large lakes of the region. Geologists have interpreted these early habitats as similar to today's Caspian Sea, with lowered salinity and a mixture of marine, estuarine, and freshwater fishes. However, the Neotropical inland sea was more tropical than the Caspian area of today (Lundberg et al., 1998). Freshwater stingrays (Potamotrygonidae), herrings (Clupeidae), anchovies (Engraulidae), needlefishes (Belonidae), croakers (Sciaenidae), and flatfishes (Soleidae) have from a few to many species in the major South American rivers. A phylogenetic study of the needlefishes by Lovejoy and Collette (2001) postulated three independent invasions of freshwater in South America and a fourth in Middle America.

The spine of the higher Andes mountains is inhabited by a small fauna of cold-tolerant fishes—primarily the trichomycterid and loricariid catfishes and, in the south, the cyprinodontid genus *Orestias*. Ancient high-altitude lakes, uplifted in the Early Miocene in Bolivia and Peru, including Lake Titicaca, have a large species flock of orestiine cyprinodontids that evolved in the absence of predatory fish species (Parenti, 1984). These species are distributed more in relation to tectonic events than by drainage basin. Southward, as noted earlier, Patagonia is a distinctive temperate region, with the most primitive living catfish, *Diplomystes*, and several species of percicthyids.

The Oriental Region

The Oriental region is less precisely bounded by the ocean, with broad land connections to the Palearctic and the Ethiopian regions. Several large river systems flow from the Palearctic southward into the Oriental region east of the Himalayas. They empty into tropical oceans (see Figs. 29.1 and 33.3). The Himalayas are a distinct boundary to the north, and the dry areas of the Middle East form a prominent hiatus to the west. The Oriental region extends from about eastern Iran through tropical Asia, with a northern transitional zone just south of the Yangtze River in China. Taiwan, the Philippines, and Indonesia are included as far as Wallace's Line, where a sharp boundary exists between the Oriental and Australian regions, as noted previously. Pleistocene low sea stands allowed many river systems that are presently separated (see Fig. 29.1) to connect with each other. Today, each major island west of Wallace's Line has a subset of fishes and shares many species with the other islands (Roberts, 1989) and with the mainland of Indochina (Rainboth, 1996a, 1996b).

The Oriental region is rich in ostariophysan fishes, both cypriniforms and siluriforms. Some cyprinid genera from subfamilies such as Bariliinae, Barbinae, and Garrinae are shared with Africa. Rainboth (1991) was able to subdivide southeast Asia (Oriental and Palearctic regions) into several fauna areas based on their cypriniform fishes, with separations both east–west between major rivers and north–south between the temperate and tropical areas. Catfish families endemic to the Oriental region include the Amblycipitidae, Chacidae, Cranoglanididae, and Pangasiidae. The catfish families Bagridae, Siluridae, and Sisoridae are shared with the Palearctic region, and the families shared with Africa include the Clariidae and Schilbeidae (see Fig. 29.7). The bagrid catfishes (like the cypriniform loaches, Cobitidae) are found in all three regions (Palearctic, Oriental, and Ethiopian). As already noted, the hill stream fishes are centered in the mountainous areas separating the Oriental from the Palearctic region. The Luciocephalidae and five or six other anabantoid families are exclusively Oriental. Other inland fish groups in the Oriental region but with wider distributions are the Osteoglossidae and Synbranchidae (shared with the Ethiopian, Australian, and Neotropical regions), Notopteridae (shared with Africa), Nandidae (shared with Africa and South America), Mastacembelidae, Chaudhuriidae, and Channidae (shared with the Ethiopian region and temperate Asia), and Toxotidae (shared with Australia).

The Ethiopian Region

The Ethiopian region is essentially Africa, Madagascar, and the southern two thirds of Arabia. However, the northwestern edge of Africa, including the Atlas mountains, is included in the Palearctic region (see Fig. 29.1). The Ethiopian region is almost entirely surrounded by ocean, except for the border with the Palearctic region, which is dominated by the arid Sahara Desert and Arabia. Central Africa is tropical; the northern and southern extremities are temperate. Tropical rainforests lie near the equator, dominated by the Zaire and parts of the upper Nile river drainages. The Nile also drains some of the central Rift Lakes with a dryer climate. In the Miocene and Pliocene, large, isolated interior drainage basins held lakes that facilitated the dispersal of fishes across the continent. Drainages that reached the ocean were smaller than today. From the Miocene onward, "rifting," or the internal separation of the continental landmass, created large lakes in east central Africa (see Going Deeper, Chapter 32). The drainage flows were altered, creating larger rivers that now drain all the way to the ocean (Greenwood, 1983). The longevity of the Rift Lakes and their varied geological history has resulted in the presence today of 16 major lakes from 28 to 69 square kilometers in size and from tens of thousands to perhaps as much as 10 million years old (Fryer and Iles, 1972), and many other minor lakes. The number of cichlid fish species per lake ranges from 30 up to more than 300, with species still being discovered. Several species of mastacembelid eels and clariid catfishes have evolved in these systems as well. These are the only lakes in the world with a fish faunal diversity approaching or perhaps exceeding that of similar-sized marine shore environments (Greenwood, 1983; Martens, 1997; Thieme et al., 2005).

Several fish families are endemic to Africa. Two archaic families of primitive bony fishes include the lungfish family Protopteridae and the chondrostean family Polypteridae, or bichirs. Archaic teleost families that are also endemic to Africa include the toothed herrings (Denticipidae), the osteoglossoid butterflyfishes (Pantodontidae), the notopteroid families Gymnarchidae and Mormyridae (sometimes included in one family), and the closely related gonorynchiform families Phractolaemidae and Kneriidae. The remaining endemic families are characiforms (Distichodontidae, Citharinidae, Ichthyoboridae, and Hepsetidae) and siluriforms (Amphiliidae, Malapteruridae, and Mochokidae). As already noted, several other families are shared with the Oriental and Palearctic regions, often even the same genera, indicating recent faunal connections—Mid-Cenozoic or Miocene or later. At least two other groups shared with South America—Cichlidae and Characiformes—date from much older connections (Lundberg, 1993; Lundberg et al., 1998). The Synbranchidae, Nandidae, and Aplocheilidae occur in the Ethiopian, Oriental, and sometimes the Australian regions. Lungfishes (Dipnoi) also have living representatives in South America, Africa, and Australia, but not in the Oriental region. Lungfish fossils from South America, Africa, and Madagascar complicate the relationships, and the divisions between phylogenetic lines may predate the split of these continents in the Late Cretaceous. Both the lungfishes and the synbranchids represent worldwide tropical archaic freshwater fishes that must date back to the Cretaceous divisions of the continents.

Disregarding desert areas, the fish fauna of Africa is poorest in the highlands of Ethiopia and the southern tip of Africa. Cyprinids, cichlids, and some catfish families are the most typical representatives. Ethiopia has some loaches (Cobitidae) related to fishes to the north and east in the Oriental and Palearctic regions, and the South African Cape area has anabantids and aplocheilids that are also related to the Oriental region. Tropical West Africa has the richest riverine fauna, centered on the Zaire (Congo) River basin, and some of these species range west into independent Atlantic drainages on the southern side of southern West Africa to Senegal. The Nile is also rich in fishes, but less so than the Zaire. These comparisons depend on the extent to which the Great Rift Lake faunas are considered part of these drainages or are separated out as distinct from the development of the more typical riverine fish faunas. The lakes variously drain internally, with no outlet to the sea (at least today), or empty into the upper parts of the Nile, Congo, and Zambezi rivers. Usually they empty over falls or rapids that do not allow riverine fishes to invade upstream. Many of the fishes of the lakes are hypothesized to be derived originally from a few riverine ancestors, and far fewer lake species subsequently invaded the rivers. Nonetheless, much of the faunal relationship between these large drainages in Africa relates to the interconnections between large, prehistoric, internally closed lake drainages covering most of central and northern Africa. The current Rift Lakes represent a small fraction of the lakes that were originally present (Greenwood, 1983; Roberts, 1975). Roberts and Greenwood divided the fish faunas of Africa into about 12 zoogeographic regions. Skelton (1996) divided the southern third of Africa, from the Zambezi River southward, into five regions, dominated by a distinctly endemic temperate fauna of small cyprinids called redfins. Greenwood (1983) estimated the fauna of Africa to comprise about 2,500 species, and Daget et al. (1986) and Teugels et al. (1994) estimated approximately 3,000 species for the continent—considerably more than the Nearctic (North America) or the Palearctic (Eurasia) regions, about the same as the Oriental region, but less than half the number projected for South America. The evolution of the exceptionally rich fauna of cichlids in the African lakes is discussed in Chapters 4 and 32.

Madagascar, which is part of the Ethiopian region, has about 40 species of inland fishes in the "secondary" or "peripheral" freshwater fish families (see Table 29.1); 84 percent of these are endemic, attesting to the long isolation of this large land area (Stiassny and Raminosoa, 1994). Madagascar was separated from Africa in the Jurassic, earlier than the separation of Africa from South America, but it did not separate from India until the Early Cenozoic. Evidence for this later separation is documented by the closely related cichlid genus *Etroplus* in southern India and the endemic *Paretroplus* in Madagascar. These two genera are relatively primitive, basal members of the very speciose family Cichlidae. As primitive sister groups, they provide valuable information for determining the changes elsewhere in the phylogenetic tree of the family (Stiassny and de Pinna, 1994). Although the fauna of Madagascar is small, Stiassny and de Pinna pointed out that it also contains several other relatively primitive taxa, such as the atherinomorph family Bedotiidae (a Madagascar endemic), the mugilomorph genus *Agonostomus* (shared with Middle America), and the catfish family Anchariidae (a Madagascar endemic), basal to the speciose and mostly marine catfish family Ariidae. The recent phylogenetic analysis by Sparks and Smith (2004) found that the rainbowfishes of Madagascar and the Australia–New Guinea region are more closely related than either is to African taxa. This supports the scenario of these two areas remaining in contact after their earlier separation from Africa, rather than some kind of oceanic dispersal.

The Australian Region

Most of the freshwater fishes of Australia and New Guinea are in families that are marine or estuarine in the rest of the world. Continental connections of this region have been lacking since before the Late Cretaceous. As in Madagascar, major groups of continental freshwater fishes, such as minnows, loaches, catfishes, anabantoids, and channids are

all lacking. Allen et al. (2002a) documented about 206 inland freshwater fish species—about a 10 percent increase in the 10 years since Allen's (1991a) earlier account. At least a few more species will probably be found in the future. Even with the addition of about 61 marine invaders, the inland fish fauna is small for such a large land area. More than 100 of the inland species occur in rivers of the wetter, northern and eastern coastal margins (see Fig. 29.1). The western drainages only have 12 or so species. Much of the western half of the interior desert has no known fishes. The eastern half of this desert—the internal drainage of Lake Eyre—has no outlet to the ocean today. This interior desert has 20 or so fish species, including gobies, atherinids, and eel-tailed catfishes, which are adapted to extremes of high temperature and salinity. Three archaic freshwater fish species occur in the well-watered northeastern part of the continent—namely, the Australian lungfish (*Neoceratodus forsteri;* in Australia only) and two osteoglossids, *Scleropages leichardti* and *S. jardinii.* The latter is shared with the Fly River region of New Guinea, and osteoglossids are found in the Ethiopian and Neotropical regions as well (Fig. 29.10). The well-watered areas are also rich in hardyheads and silversides (Atherinidae), rainbowfishes (Melanotaeniidae), blue-eyes (Pseudomugilidae), pygmy perches (Nannopercidae), gobies (Gobiidae), and gudgeons (Eleotridae). Several genera and a few species of these northern fishes are shared with New Guinea to the north, and many species in Australia, New Guinea, and Tasmania are endemic to the Australian region (Allen, 1991b; Allen et al., 2002a).

Tasmania to the south has a much smaller fauna of southern temperate fishes, some of which represent antitropical elements. Thus, the southern taxa of lampreys, family Geotriidae (one species) and Mordacidae (two species), occur in Tasmania (see Going Deeper). Small, freshwater, spiny-rayed, perchlike fishes of the families Percichthyidae, Nannopercidae, and Gadopsidae together contribute 19 species to the Australian inland fish fauna. The temperate galaxiids have their greatest diversity in Tasmania, with 17 species (11 of them endemic). Fewer galaxiids occur in southern South America (4 species), and one species is found in Southern Africa. The most primitive galaxiids are found in Tasmania, suggesting that some of the earliest members of the group occurred there. The timing of the distribution of the Galaxiidae among the southern continents (a Gondwana distribution) is confounded by the salinity tolerance of some species. Possibly, the marine life history stage was lost more than once to produce freshwater taxa (Waters et al., 2000). The closer phylogenetic relationship between the species of Australia and South America compared to Africa corresponds with the known earlier separation of Africa from these two continents. The galaxiids epitomize the concept

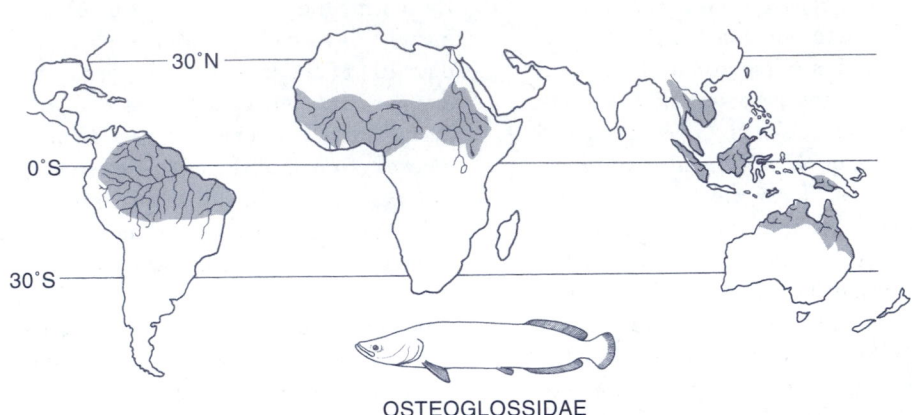

FIGURE 29.10
Distribution of the Osteoglossidae, reflecting a Gondwana distribution among the southern continents.

of a Gondwana distribution among these three southern continents (and, before the Miocene, probably a fourth or fifth: Antarctica) and New Zealand. As noted earlier, New Zealand has a small freshwater fish fauna of about 30 species (McDowall, 1990) of mostly marine-derived galaxiids, gobiids, and eleotrids.

The freshwater bovichthyids also occur in temperate and boreal Australia, and they are considered a primitive sister group to the notothenioids that are endemic largely to colder Antarctic and southern ocean waters. As with Madagascar, the temperate Australian–Tasmanian freshwater fish fauna is small but important because it includes the primitive sister groups to larger numbers of related fish species (see Chapter 4).

DISTRIBUTION OF MARINE FISHES

The breakup of the continents also created the separate ocean basins. The subdivision of the oceans and the appearance of colder, polar regions initiated the deep circulation of ocean waters. Just as mountains cause prevailing winds to drop their moisture, continents divert the flow of ocean currents, causing the upwelling of deeper, colder, and often nutrient-laden ocean water. This expanded the potential fish habitat in the ocean to greater depths. The westward-flowing, tropical Tethys Sea was closed off in the Mid-Cenozoic (Late Oligocene to early Miocene) by the Arabian peninsula.

From the Miocene onward, the Mediterranean Sea level fluctuated drastically. It mostly dried out and became hypersaline for a period during the Miocene, with the closure of the Straits of Gibraltar. The straits opened again in the Late Miocene, but the Mediterranean remained separated from the Red Sea to the east. The Red Sea was forming by "rifting"—the separation of two continental plates.

For most of the Cenozoic, ocean currents flowed westward across what was to become southern Middle America, until 3 to 10 million years ago, when the isthmus of Panama was closed off. This closure deflected warm water to the north. This led to the development of the Gulf Stream, which pushed warmer oceanic conditions farther north in the eastern Atlantic Ocean. Today's oceans and their prevailing currents are shown in Figure 33.3.

The barriers between the Mediterranean and the Red Sea in the Middle East and between the Caribbean Sea and the Pacific Ocean in Panama were both artificially breached with canals, to shorten transit distances for ships. Little exchange of fishes has occurred in Panama, because the canal is elevated by locks over mountains and is primarily freshwater (Rubinoff and Rubinoff, 1971). In the Middle East, however, the sea-level Suez Canal has allowed about 50 species of marine and estuarine fishes to invade west from the Red Sea into the Mediterranean. Few if any species have been documented invading in the opposite direction (Golani, 1993). The geologically recent extinction events in the Mediterranean left it with a reduced (depauperate) marine fish community. With more empty ecological niches to fill, Red Sea fishes were easily accommodated in the eastern Mediterranean (see Chapter 39). No Mediterranean species has been able to find a niche among the more diverse Red Sea fauna so far.

Many cold- and cooler-water fishes occur on both sides of the tropics, in a distribution pattern termed **antitropical** (Hubbs, 1952). If species are not as widely separated within the tropics (i.e., found between the tropics of Cancer and Capricorn) but are still absent in the mid-equatorial region, the distribution is called **antiequatorial** (Randall, 1982). We have already noted some freshwater fishes, such as the lampreys and galaxiids, that show an antitropical distribution. Other prominent antitropically distributed marine families include the Scorpaenidae, Cottidae, Zoarcidae, Agonidae, Oplegnathidae, Girellidae, Clinidae, Heterodontidae, Atherinopsidae, and some Pomacentridae. These distributions of marine fishes can be attributed to at least four causes (Randall, 1998): (1) species or their ancestors were able to cross the tropical regions during cooler periods in the past, when conditions of cool upwelling existed in tropical areas; (2) some fishes, such as the horn sharks (Heterodontidae), occur in deeper, cooler waters in the tropics and may be antitropical only in shallow water, actually occurring continuously in deep water; (3) landmasses, islands, or seamounts that provided past stepping stones for crossing the tropics may no longer exist; and (4) intervening populations have been extirpated, and recolonization is no longer possible. The first two possible causes seem more likely for the distantly separated antitropical species, and the second two probably apply more often to antiequatorial ones.

Salinity also affects the distribution of marine fishes—particularly those that are intolerant of lower salinities. The mouths of the major rivers of the world, such as the Indus, Zaire, Amazon, Mississippi, Orinoco, and Mekong, inhibit the development of coral reefs and other fully marine faunas. Many tropical reef fishes of the Caribbean have the same or cognate species much farther south in coral reefs on the southeastern coast of Brazil (Rocha, 2003), but a broad hiatus exists north and south of the mouth of the Amazon River. Large areas of low salinity form in the offshore estuarine zones of the northern polar region and adjacent to land areas with high rainfall and freshwater runoff (McHugh, 1967). Lower salinity areas in the temperate and boreal zones are dominated by salmonids, herrings (Clupeidae), and anchovies (Engraulidae). In the tropical areas, such as on the west coast of Panama, and at the mouths of major rivers like the Indus, Ganges, Mekong, Congo, and Amazon, anchovies, herrings, croakers and drums (Sciaenidae), ariid catfishes, jacks (Carangidae), and several other, less numerous

families occur. Many of these are virtually absent on marine offshore islands that lack estuarine influence (Randall, 1998; Smith-Vaniz et al., 1999).

As already noted, very few fish species are adapted specifically to low salinities, but a fauna of such specialists—particularly gobies (Gobiidae)—derived from marine ancestors occurs in the Black, Caspian, and eastern Mediterranean seas, where low salinities are the norm. In the Black and Caspian seas, salinities are low enough that many freshwater species of cyprinids also occur (Berg, 1949; Zenkevitch, 1963). During glacial advances into the eastern Mediterranean, some low-salinity specialists dispersed westward from the Black Sea into the eastern Mediterranean. When higher salinities returned, several species of brackish-adapted gobies were left widely scattered in small, coastal lagoons and rivermouths around the margins of the Mediterranean Sea (Miller, 1973).

The cyprinodontid genus *Aphanius* was split into a freshwater species in the Arabian peninsula and a euryhaline species in the eastern Indian Ocean by the closing of the Tethys Sea in the Miocene (Hrbek and Meyer, 2003). Later, several species evolved due to tectonic events in the eastern Mediterranean region, and one species also became widespread around the Mediterranean. In coastal California, the brackish-water tidewater goby (*Eucyclogobius newberryi*) also has widely isolated populations, with strong genetic differentiation between them, indicating isolation of up to a few million years (Dawson et al., 2002; Swenson, 1999).

Pelagic Fishes

The upper layers of the oceans—the epipelagic and mesopelagic zones—have also been divided up into faunal provinces based on the distribution of fishes (and other organisms). These zones correspond to the physical attributes of water masses, such as temperature, depth, current flow, and others (cf. Figs. 29.11, 33.3). Northern zones coincide with the oceanic provinces, but the broad tropical zone can be further subdivided based on the presence of oceanic currents that influence species distributions. Despite the naïve expectation that species in the homogeneous ocean basins would be little differentiated, studies like those of Zahuranec (2000) have disclosed 17 separate, largely nonoverlapping distributions for 17 species of lanternfishes (genus *Nannobrachium*). A similar situation was found for the pelagic liparid genus *Psednos*, with about 25 species (Chernova and Stein, 2002), and for the mesopelagic genus *Eustomias* (Clarke, 2001). Such distributions are prevalent in smaller fishes, but they would not apply as much to larger, mobile species like tunas.

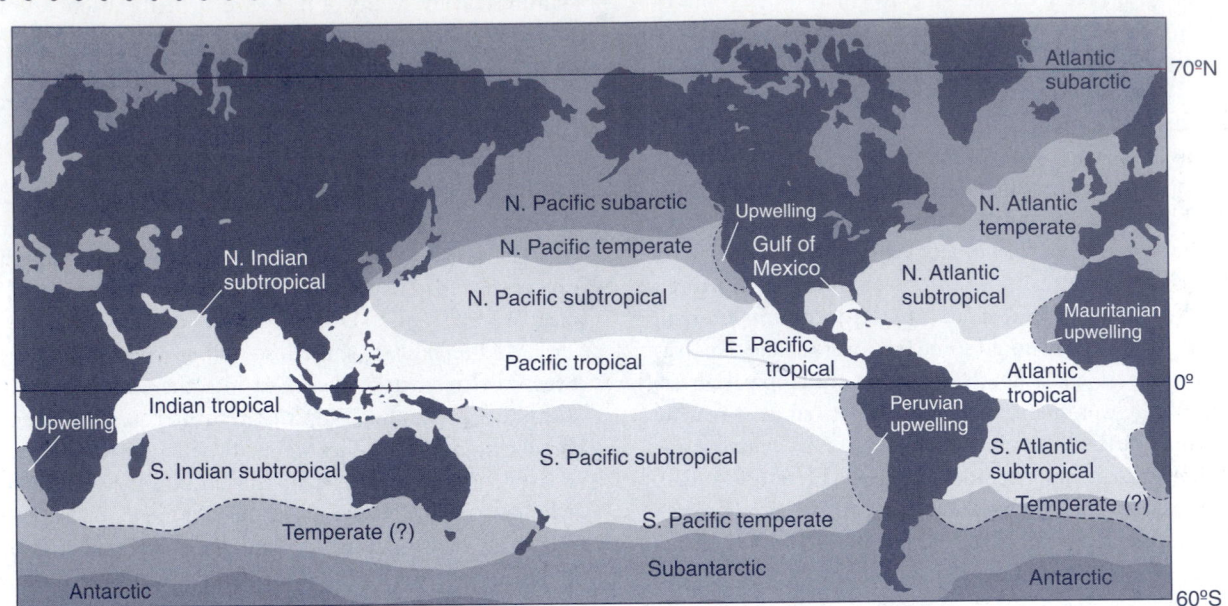

FIGURE 29.11
Faunal regions of the pelagic faunas of the upper layers of the ocean (after Zahuranec, 2000).

Polar Waters

In the Arctic and Antarctic regions (see Chapter 35), marine waters range from nearly −2°C in winter to 5 or 6°C in summer. Much of these areas can be covered with ice for most of the year. In the Arctic, the cold water extends to about the 60th parallel into the northern Bering Sea. On the Atlantic side, the boundary of the Arctic water begins at the southeastern corner of Labrador, at about the 50th parallel, extends up to Iceland, and then continues northeast over Norway to a bit east of the North Cape, at about the 70th parallel. This displacement of the cold-water boundary is due to the influence of the Labrador Current in the west and the northeastward drift of the warmer Gulf Stream toward northern Europe (see Fig. 33.3). The warmer but saltier Atlantic water is denser than the colder but fresher surface water of the Arctic Ocean. The more saline Atlantic water sinks and moves northeast into the Arctic. In the boundary area, the bottom is 500 to 1,000 meters at the shallowest, and considerable exchange (of both water and fishes) occurs between the Atlantic and Arctic Oceans. The Bering Sea gap with the north Pacific is much shallower, from 50 to 200 meters, and much less exchange is possible. The Arctic Ocean is about 35 to 40 percent continental shelf, compared to 8 percent for the world's oceans as a whole. One central basin reaches about 6,000 m in depth, but except for the cods (Gadiformes), typical deep-water fishes are lacking.

Only a few pelagic species occur in the shallow connection between the north Atlantic and the north Pacific. These include cold-water species, such as the anadromous Pacific salmon, and less migratory salmonids like the sheefish (*Stenodus leucichthys*) and chars (*Salvelinus*). In the Pacific, one hexagrammid, the Atka mackerel (*Pleurogrammus monopterygius*), is also pelagic, as are the juveniles of other greenlings. The skilfish (*Erilepis zonifer*), the young of the sablefish (*Anoplopoma fimbria*), and the walleye pollock (*Theragra chalcogramma*), a member of the cod family (Gadidae), are also open-water fishes. Some of these species extend southward along the eastern Pacific in cool, upwelling waters.

South of the Antarctic convergence, at about the 60th parallel in the South Atlantic (Fig. 29.12), water remains less than 5°C even in the summer. This area supports a small pelagic fauna—primarily representatives of the mesopelagic family Myctophidae, with a few Aulopiformes, Stomiiformes, and Beryciformes. Both McGinnis (1982) and Zahuranec (2000) recognized five subdivisions of the Antarctic Seas, more or less concentrically arranged from the mainland outward, based on the distribution of lanternfishes (Myctophidae).

North Temperate Waters

In the North Atlantic, a polar pelagic fauna is lacking, and a slightly more southern boreal fauna exists that spans the boreal and temperate zones of the North Atlantic. Along

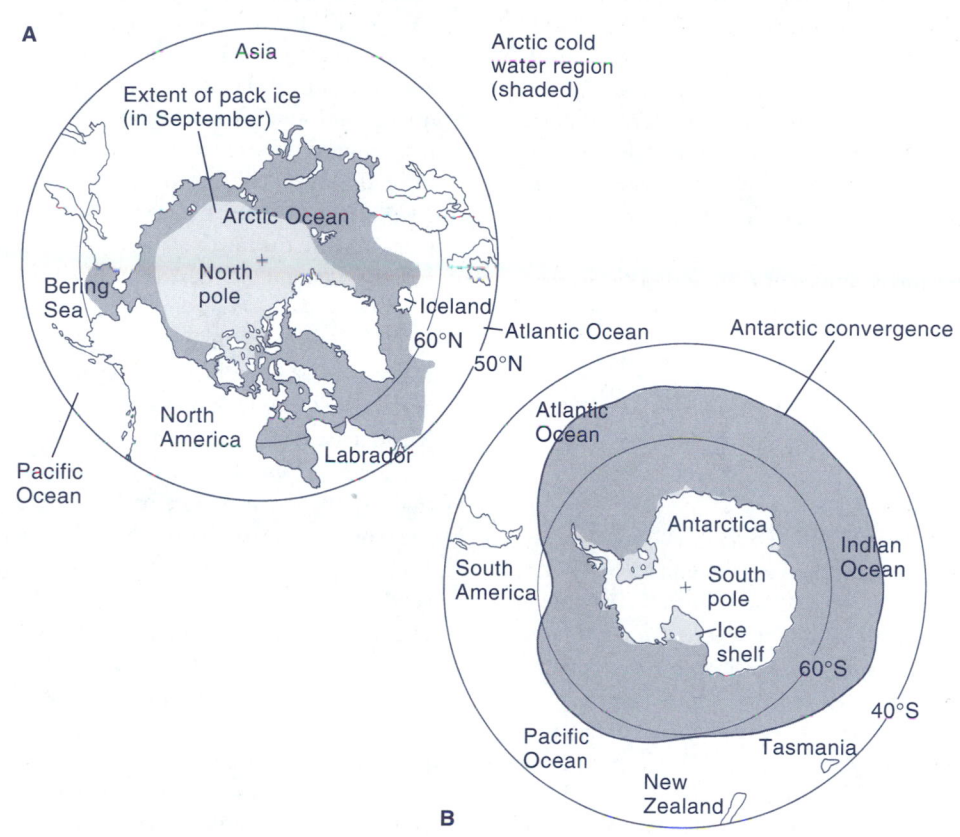

FIGURE 29.12
Oceanic subdivisions of the polar regions: **A,** Arctic; **B,** Antarctic.

with the boreal region in the Pacific (from about the 34th to the 42nd parallel N), wide-ranging species from both north and south occur, including sardine (*Sardinops sagax*), northern anchovy (*Engraulis mordax*), medusafish (*Icichthys lockingtoni*), blue shark (*Prionace glauca*), albacore (*Thunnus alalunga*), and bluefin tuna (*T. orientalis*) in the Pacific; and mackerel shark (*Lamna nasus*), opah (*Lampris guttatus*), Atlantic mackerel (*Scomber scombrus*), bluefin tuna (*Thunnus thynnus*), swordfish (*Xiphias gladius*), and mola (*Mola mola*) in the Atlantic. The presence of tuna is seasonal and timed according to their migrations, which span much of each ocean (see Chapter 36). The north temperate waters are inhabited by tunas (*Thunnus*), flyingfishes (Exocoetidae), sauries (Scomberesocidae), opah (Lamprididae), and some marlins (Istiophoridae), as well as many species of meso-pelagic fishes, such as lanternfishes (Myctophidae), bristle-mouths (Gonostomatidae), hatchetfishes (Sternoptychidae), and others. The Gulf Stream makes the warm temperate region much wider in Europe (from the English Channel to

Senegal) than on the North American side (from Cape Hatteras, North Carolina, southward to the middle of Florida).

Tropical Waters
The pelagic tropical regions are larger and more distinct than the cooler regions, and their boundaries are outlined in Figure 29.11. The pattern of oceanic currents and their deflection by the continents renders the tropical regions somewhat the inverse of the temperate regions—namely, much wider on the western side of the Atlantic and Pacific oceans than on the eastern side. At least one pelagic species pair is split between these two wide tropical areas on the western margins of major continental areas. The flyingfish *Fodiator rostratus* occurs along the coasts of western Europe and Africa, and the second and only other species in the genus (*F. acutus*) occurs along the western coasts of North, Central, and South America. Both occur in the tropical and subtropical waters of their respective areas. They are among the most primitive of the flyingfishes, so they may be

relicts of an old distribution (Parin and Belyanina, 2002). The other flyingfishes have a distribution similar to that of many mesopelagic fishes.

In the tropical Atlantic, a distinctive "Sargasso" fauna exists in the large, closed gyre of water in the central north Atlantic just east of the Caribbean Sea, the Bahamas, and Florida (see Fig. 36.3). This community consists of a floating epifauna associated with drifting clumps of the alga *Sargassum* and includes pipefishes, the sargassum fish (*Histrio histrio*) and the juveniles of many other fishes, particularly jacks (Carangidae) and stromateids (Dooley, 1972; see Chapter 35).

Shore and Shelf Fishes

Tropical and Temperate Regions

Shore and shelf fishes are usually benthic or, if pelagic, are restricted to the continental shelves (see Chapter 33). The continental shelf is usually defined offshore to the 100-meter depth contour. Often, this is about where the continental shelf abruptly drops off into the much steeper continental slopes. Randall (1998) defined reef and shore fishes as those down to 200 meters deep, perhaps because many of his study sites were oceanic islands that lacked a distinct continental shelf and dropped off steeply. Often, species that are mostly pelagic but wander into or over shore and shelf habitats are not included. Most of the true shelf and reef species have narrower distributions than pelagic species.

Many larger nektonic shelf and shore fishes show strong temperature preferences and move seasonally with water masses. Thus, warm-water fishes move up the Atlantic and Pacific coasts of North America in the spring and summer, following the Gulf Stream on the Atlantic side and the Davidson Current on the Pacific side. The Gulf Stream veers to the northeast, whereas the Davidson Current is abruptly deflected in southern California upon meeting the cold, southward-flowing California Current. This encounter sets up the Southern California Countercurrent, which circulates in a large area of southern California known as the Southern California Bight. In Australia, warm currents flow southward along the eastern coasts in the summer, also bringing down warm-water fishes that disappear in the winter. Particularly on the western coasts of North and South America, the well-known El Niño phenomenon, which takes place every few years, can carry tropical warm-water fishes far to the north or south, respectively. California populations of fishes like the blue shark or the California moray eel are largely or completely recruited from larvae or juveniles that disperse north from reproducing populations living farther south. Strong El Niños are often followed by strong La

Niñas—a reverse phenomenon, almost like a rebound, of strong cold flows in the opposite direction. Then, cooler-water species are carried south to California or north to Peru and Ecuador. In the northern Gulf of Mexico and along the middle and south Atlantic coasts, many tropical fishes come inshore and colonize reefs and other habitats in summer, only to be decimated during the cold winter months (Hastings, 1979). In some cases, permanent populations live offshore at greater depth during the winter. These movements make the coastwise boundaries between faunal regions indistinct.

Areas of cool-water upwelling lie along the west coast of North America in California and Baja California, and along the southwest coast of South Africa. Such areas allow cold- and warm-water faunas to exist in close proximity. Thus, in Magdalena Bay, on the outer Pacific coast of southern Baja California, tropical marine fishes and mangroves are found in the shallow, warm inner bay. Cool upwelling areas occur at the mouth of the bay, with kelp beds and distinctly temperate fishes. On a longer time scale—probably for at least thousands of years—cold-water fishes have become distributed in pockets of upwelling on the mainland and offshore islands from southern California southward into Baja California. Conversely, in central and northern California, the ocean is always cold. However, the upper ends of the larger bays get warmer in summer, and these areas hold populations of warm-water fishes that are relicts of earlier warm periods. Both north and south, this alternation has provided a complex intertwining of northern and southern faunal elements (Hastings, 2000; Hubbs, 1961).

Horn and Allen (1978) analyzed the distribution of California coastal marine fishes and found that the greatest species diversity was located between 32 to 42 degrees north latitude—that is, approximately the southern two thirds of California. They quantified what had been observed for a long time—namely, that many groups of fishes had more species overlapping in distribution in central and southern California than either farther north or farther south. Love et al. (2002) mapped the number of rockfish (*Sebastes*) species found at various points along the Pacific coast of North America. Of about 70 total species, only in the southern third of California were as many as 55 to 60 species recorded. Farther north into northern California and south into the upper third of Baja California, 45 to 55 species occurred. Only 1 to 5 species occurred at the northern extreme of the Aleutian Islands and near the southern end of Baja California. The species diversity in southern California has several causes. First, the confluence of northern and southern currents in an area of complex island and basin landforms provided an opportunity for the isolation and evolution of local, endemic species. Second, the confluence of currents brings together species from both the north and the south. Finally, these conditions have been present for millions of years. Far fewer species of rockfishes occur in the western north Pacific or southern South America (antitropically). The northeastern Pacific is where many of the species originated or evolved. The north Pacific is also rich in species of cottids, embiotocids, and agonids.

Three subdivisions, each with endemic elements, can be recognized in the tropical eastern Pacific: (1) the Gulf of California; (2) the Mexican zone, from the mouth of the Gulf of California to Nicaragua; and (3) the Panamic zone, from Nicaragua to northern Peru. The isolated nature of the tropical eastern Pacific and the number of isolated offshore islands (Clipperton, Cocos, Galapagos, Revillagigedo) make it one of the greatest endemic areas for marine shore fishes for its size in the world (Hastings and Robertson, 2001). About 85 percent of the species in this region are endemic. Very few species are shared with the Caribbean, only a short distance away across the isthmus of Panama. Moreover, only about 125 species of shore fishes are shared with the western Pacific. These transpacific shore fishes have been able to cross the eastern Pacific Barrier—perhaps the longest, most significant barrier to marine fish dispersal today (Lessios et al., 1997). Most of these species came eastward from the western Pacific, but some have gone westward from the offshore islands of the eastern Pacific (Robertson et al., 2005).

The distributions of marine shore fishes on the Atlantic coast of North America are similar to those of the Pacific, as the temperature divisions are similar (see Figs. 29.11 and 33.3) The tropical families and genera of the Caribbean, south Florida, and the Bahamas are similar to those in the Gulf of California, but the species are different (Thomson et al., 2000). The tropical western Atlantic has more species than the eastern Pacific, mostly because the complex, well-developed coral reefs of the Caribbean are very rare or nonexistent in the eastern Pacific. The size of the Caribbean coral reef and rocky shore fish fauna is second only to that of the Indo-West Pacific in size. As noted earlier, tropical species often do not survive the winter farther north. Thus, of the Atlantic coast of northern Florida is a transitional area (Smith-Vaniz et al., 1999). The warm temperate (or subtropical) region extends farther north, through about Cape Hatteras, North Carolina, and has been called the Carolinian Province, Carolinian Bight, or the South Atlantic Bight. Like the Southern California Bight, it is a zone of mixed warm- and cool-water faunas. North of the Carolinas and Virginia, the fish fauna is smaller and made up of species adapted to cooler waters. Even farther north, Cape Cod is often designated the southern boundary for strictly cold-water fishes—somewhat analogous to Point Conception in southern California, but considerably farther north.

Peninsular Florida, like the Baja California peninsula, has been a barrier to the dispersal of marine shore fishes. Atlantic coastal populations of some temperate marine shore fishes extend southward to northern Florida, and a disjunct population occurs across the peninsula in the northern Gulf of Mexico. In other cases, the two species or subspecies completely circumscribe the Florida peninsula to the south. They occur along the northeast coast of Florida as two overlapping species, or they interbreed as subspecies. Pleistocene sea levels were only slightly higher than present levels. Thus, the Florida peninsula would not have been breached by marine fish. During glacial low sea stands, cooler water may have allowed these species to extend southward, around the southern tip of Florida. It is also known that a clear oceanic breach in northern Florida existed earlier during the Miocene and Pliocene; thus, these speciation events could have occurred much earlier as well. This split has affected marine species of *Cynoscion*, *Brevoortia*, *Hypsoblennius*, *Fundulus*, and *Menidia*, as well as anadromous species such as *Alosa*, *Morone*, and *Acipenser*.

The coasts of Europe and Africa show similar subdivisions by temperature (see Figs. 29.11 and 33.3). The Gulf Stream makes the temperate zone very wide, from north of the British Isles to northern Africa, including the Mediterranean. Further south, opposite the "hump" of Africa, the Mauritanian upwelling presents cool water, as does a similar upwelling even farther south in Southwest Africa. Between these two upwelling areas, in the tropical zone, the Zaire (Congo) River empties into the Atlantic Ocean, inhibiting coral reef development. Tropical reef fishes are poorly developed and are found mostly in the Gulf of Guinea on some offshore islands (Robins, 1964). Several genera of Caribbean marine fishes have one or a few species in the Gulf of Guinea. In South Africa, the cold upwelling of Southwest Africa abruptly meets with the warm tropical waters from the eastern side of South Africa. This combination produces a large marine fish fauna in South Africa (Beckley et al., 2002), estimated at about 2,200 species, representing about 15 percent of the estimated number of marine shore fishes worldwide.

The tropical shelf habitat is much better developed on the eastern side of Africa, in the Indian Ocean, and extends from South Africa up to the Red Sea, the Persian Gulf, and across the western part of the northern Arabian Sea to the freshwater influence of the Indus River delta. The northern half of the Indian Ocean is warm, tropical, and continuous with the string of large tropical islands extending eastward from Indonesia through New Guinea and northern Australia. Farther west, the large West Pacific region consists of numerous islands that extend westward to the Hawaiian and Easter islands—a vast expanse of the globe. Randall (1998) estimated 2,800 species of shore fishes in the East Indian region, including Indonesia, New Guinea, and the Philippines. Randall and Lim (2000) estimated 3,365 species of shore fishes from the South China Sea, and Allen and Adrim (2003) estimated 2,057 species of coral reef fishes for Indonesia. These estimates cover roughly equal areas, with extensive overlap, and clearly coral reef fish predominate. Randall's (1998) explanation for this diversity included the long continuity of warm, equitable conditions near the equator; the geological complexity that subdivides the oceanic areas, where the isolation of populations may lead to speciation; the dispersal of extralimital species into the area; and the mixture of diverse habitats, including mangrove swamps, estuaries, soft substrates, and coral reefs. Furthermore, during low sea level stands, the area would have been even more strongly subdivided. Randall noted that the number of species diminished from the central area in both eastern and western directions. The western limits are represented by the Hawaiian Islands (about 566 species, of which 23 percent are endemic) and Easter Island (about 126 species, of which 22 percent are endemic), with the highest percent-

ages of endemic species for offshore islands. The southern limits include the Great Barrier Reef (about 1,300 species; Smith-Vaniz et al., 1999) to the southern limit of tropical habitats at Lord Howe Island (about 433 species). To the west of the East Indies, in the tropical Indian Ocean, about 750 species of shore fishes occur on the Chagos archipelago and about 900 in the Maldives. About 1,100 species occur in the Red Sea, and about 1,170 occur in Natal, on the southeastern coast of Africa (Randall, 1998)—both about the same number of species as the Great Barrier Reef, or about the same as the whole Caribbean! These numbers are likely to increase, but probably not by more than 20 percent or so. Myers (1999) noted distance, lack of dispersal ability, and lack of habitat diversity on offshore islands as factors accounting for the sharp drop in numbers of species across Micronesia. The Micronesian islands extend from the Asian continental plate to the islands on the Pacific plate. Springer (1982) demonstrated many distinctions among fish species distributions relative to the Pacific plate (Figs. 29.13 and 29.14). The Pacific plate has fewer species. Some species

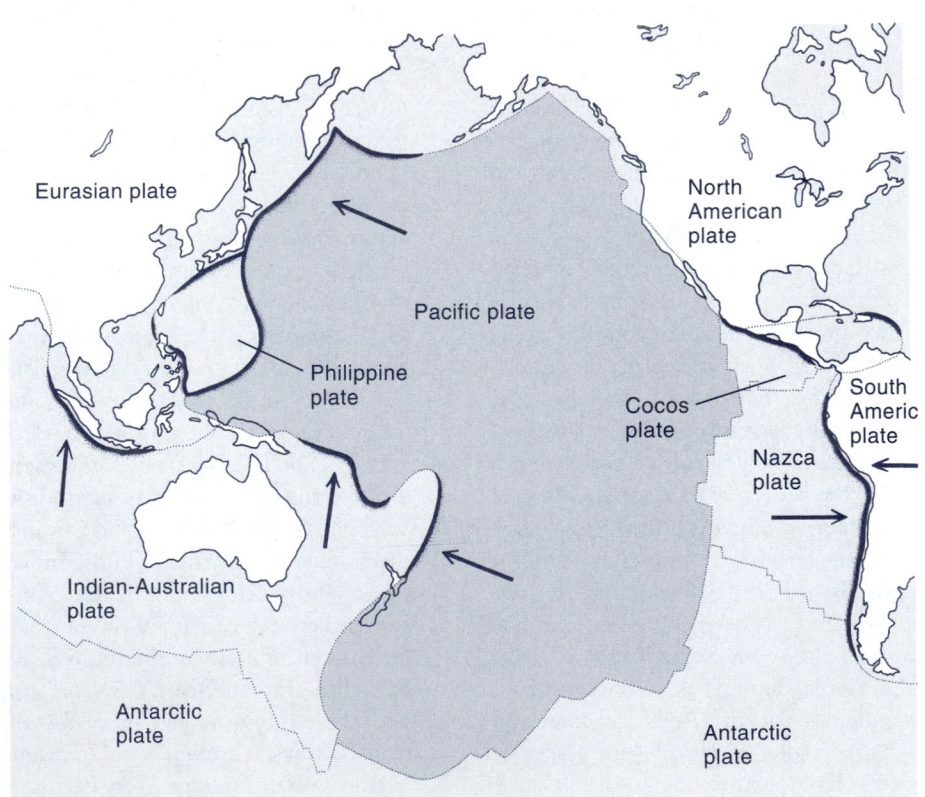

FIGURE 29.13
Boundaries and direction of movement of the major tectonic plates of the Pacific Ocean (after Springer, 1982).

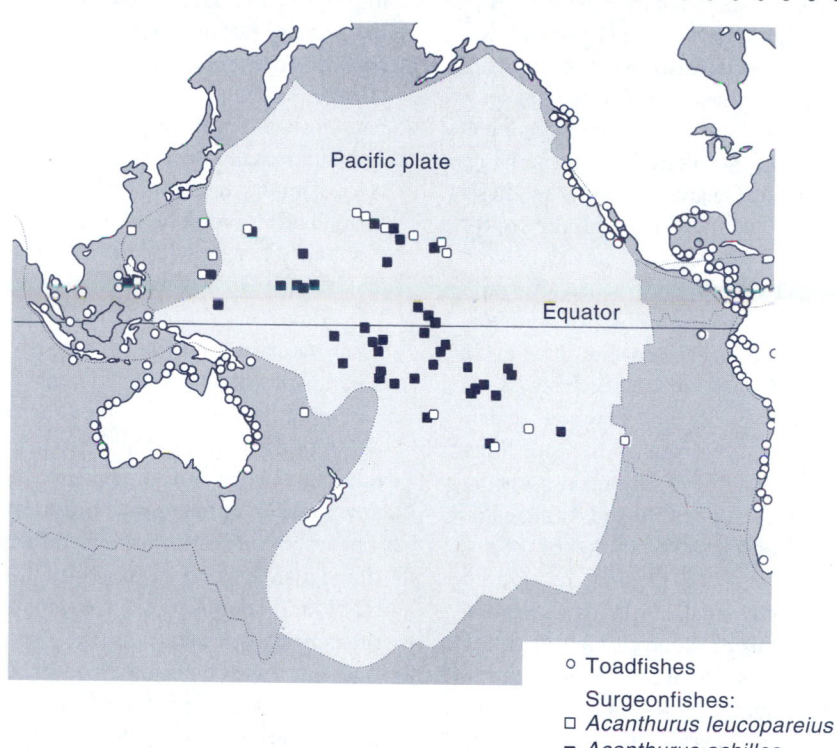

Pacific plate

Equator

○ Toadfishes
Surgeonfishes:
□ *Acanthurus leucopareius*
■ *Acanthurus achilles*

FIGURE 29.14

Representative distributions of marine shore fishes in relation to Pacific plates: non-Pacific plate toadfishes (*open circles*); Pacific plate endemic surgeonfishes (*solid squares*), and the antiequatorial distribution of one surgeonfish (*Acanthurus leucopareius; open squares; after Springer, 1982*).

only occur on the surrounding continental plates, and others are restricted to the Pacific plate, implying the influence of some larger past landmass that led to this dichotomy.

To the north, Taiwan has an estimated 2,200 species of shore fishes according to Randall (1998), and the Ryukyu Islands of Japan have an estimated minimum of 2,000 species. The Ryukyu islands constitute only about the southern one sixth of the country of Japan, but they have extensive coral reefs. Like California, Japan has a northward-flowing, warm Kuroshio Current and a corresponding, southward-flowing cold Oyashio Current. Weaker warm-water currents affect the western coast of Japan and coastal Korea and China. Two circular currents (gyres) are maintained farther north in the basins of the Sea of Japan. Two more basins are present in the Yellow Sea and the South China Sea. The presence of several distinct, large landmasses and intervening deep-ocean areas make the fish distributions of the western Pacific more complex and richer than anywhere else in the world. Numerical analysis of the distributions of 2,983 shore fish species of the Indo-Malay-Philippine archipelago by

Carpenter and Springer (2005) identified the central Philippines as the marine area with the highest diversity in the world. A lesser peak of diversity occurred between Malaysia and Sumatra. Such areas are important scientifically for the study of biological complexity at its maximum and justify strong conservation efforts to avoid losing such richness.

Nakabo (2002) reported about 3,863 fish species from Japan, including about 175 freshwater or inland species and a few hundred deep-water or offshore marine species. Because Japan extends from the rich North Pacific boreal and temperate fauna southward into the tropical East Indian region, 2,500 to 3,000 species of shore and shelf fishes occur there. This contrasts to the similar distance from Alaska to the southern tip of Baja California, where approximately 1,500 to 2,000 species of marine shore fishes occur (Mecklenburg et al., 2002; Miller and Lea, 1976; Thomson et al., 2000).

Four isolated oceanic islands or archipelagos at the northern or southern extremities of the tropics have acquired marine shore fish faunas by long-distance dispersal (Grove and Lavenberg, 1997; Randall, 1998; Smith-Vaniz et al.,

1999): (1) the Hawaiian Islands (North Pacific, 566 species); (2) Bermuda (North Atlantic, 433 species); (3) Lord Howe Island (southwestern Pacific, east of Australia; 434 species); and (4) the Galapagos Islands (southeastern Pacific, 444 species). They range in size from the small, essentially single platform of Bermuda (about 1,200 square km) to the large archipelagos of Hawaii and the Galapagos. None of these islands have a known history of connections with the continents. All are at least several million years old, and they have fish faunas that are closely related to the continental areas upstream from them in the oceanic currents. The two singleton islands (Lord Howe and Bermuda) have relatively low levels of endemism—less than 6 and 4 percent, respectively (Randall, 1998; Smith-Vaniz et al., 1999). The Hawaiian chain and the Galapagos have 20 to 30 percent endemism. This is due to the greater opportunity for isolation on multiple islands that have been changing in position and number for much of the last half of the Cenozoic. The number of species is similar at all four of these sites, and they may be close to saturated with species. If so, the smaller islands (Bermuda and Lord Howe, lacking endemism) acquired their diversity largely by immigration, whereas the archipelagos (Hawaii and Galapagos, with high endemism) also generated new endemic species in roughly the same period of time.

The dispersal distance to reach these islands may have been less than previously thought. Intervening seamounts possibly provided more "stepping stones" when they were exposed or at shallower depths during low sea stands (Hubbs, 1959). Moore et al. (2004) and Wilson and Kaufman (1987) have documented such dispersal with records of Eastern Atlantic shelf fishes from mid-Atlantic seamounts. Randall (1998) noted that repeated visits to Easter Island have disclosed the possible extinction of a few species and colonization by others over a few decades. Continued observations should be able to determine if the fauna is increasing or decreasing. Alternatively, the fish fauna may be saturated, and the number of new species arriving must be balanced by others being extirpated.

Polar Benthic and Shelf Fishes

Benthic, northern, cold-water fishes of the shore and shelves of the Arctic polar region are dominated by sculpins (Cottidae), smelts (Osmeridae), snailfishes (Liparidae), eelpouts (Zoarcidae), and cods (Gadidae). Several distinctive fish species occur, such as the Arctic sculpin (*Myoxocephalus scorpioides*) and the Arctic flounder (*Liopsetta glacialis*), and a few species, such as the Arctic char (*Salvelinus alpinus*) and the fourhorn sculpin (*Myoxocephalus quadricornis*) occur widely in both marine and freshwater habitats. The lack of deep water has limited the number of species, and Andriachev and Chernova (1994) tabulated only 415 species, includ-

ing freshwater and diadromous forms—only about half the number of both marine and freshwater species that are known from California. Glaciation has caused dramatic changes during the Pleistocene and possibly eliminated many fishes. Since the glacial maximum about 10,000 years ago and actually observed since the late 1800s, warming of Arctic waters has allowed northward movement of many boreal cods (Gadidae), flounders (Pleuronectidae), herrings (Clupeidae), skates (Rajidae), and other fishes (Perry et al., 2005; Zenkevitch, 1963).

Antarctica differs from the Arctic in having a deep-ocean fauna in addition to a shelf fauna (see Chapter 35). The history of the Antarctic fish fauna began with its glaciation in the Early Miocene (Eastman, 1993). The central, polar location of the Antarctic landmass, completely separated from the other continents, allows a simple, eastward-flowing current to encircle the continent almost completely. The weight of the Antarctic landmass has depressed its continental shelves down to about 500 meters—several times that found on most other continents, and twice the depth of the continental shelves in the Arctic. No intertidal, estuarine, or freshwater habitats occur, but fishes use the underside of the ice as a habitat (*cryopelagic* fishes). Several species that are normally shelf fishes have adapted to a pelagic existence. Depressions from 500 to 1,200 meters deep on the Antarctic shelf support deep-water-adapted notothenioids (*pseudobathyal* fishes) rather than the more typical bathyal fishes found in the rest of the world's oceans. As the margins of the continent present minimal barriers to fish movement, few zoogeographical regions are defined. The East Antarctic Province encircles about 80 percent of the continent, and the West Antarctic Province encompasses the Antarctic peninsula extending toward South America. The remaining four Antarctic zoogeographical regions coincide with islands or island groups (South Georgia, Marion-Crozet, Kerguelen-Heard, and Macquarie) around the periphery of the south polar ocean (Fig. 29.12B).

The shelf fish fauna is overwhelmingly dominated by the distinctive, cold-adapted notothenioids—a spiny-rayed group only known from the Antarctic and nearby islands and extremities of the southern three continents (see Chapter 16). The remaining shelf fishes are related to cold-water fishes that occur widely elsewhere in the world's oceans, such as snailfishes (Liparidae; about 60 Antarctic species), eelpouts (Zoarcidae; 22 species), and skates (Rajidae; 8 species). Eastman (1993) listed about 180 shelf or shore fishes out of a total marine fauna of 274 species from the Antarctic region, allowing for the deeper shelf. Some species are shared with the southern ends of South America, Australia, New Zealand, and the southern islands mentioned earlier. Notothenioids represent about 50 percent of these shelf species, and 70 to 90 percent of the shelf families are endemic, indicating a long isolation of the Antarctic region and the southern ocean fauna (Gon and Heemstra, 1990). This number of species in a relatively homogeneous environment indicates that the Antarctic had greater geographic complexity in the past, leading to speciation of isolated populations (Bargelloni et al., 2000).

Deep-Sea Fishes

As already noted, the deep-sea fauna is relatively small but widespread. The species can be pelagic or benthic. The zonation of bathypelagic fishes is similar to that of the mesopelagic species discussed earlier. The mesopelagic species are the most diverse compared to the epipelagic or bathypelagic communities. Like the pelagic species, the greatest diversity of benthic deep-sea fishes lies at middle depths (Haedrich, 1997). Thus, more species of deep-sea benthic fishes occur on the lower slope (around 1,500 meters depth) than on the shallower shelf or in the deeper rise and bathyal regions (see Fig. 33.1 and Chapter 35 for definitions of the deep-sea zones). The difficulty of extensively sampling the deep ocean bottom means that the distribution of benthic fishes is poorly known. However, the former assumption of widespread uniformity is giving way to a recognition of regional differences among benthic species, whose differentiation may rival that of the pelagic species (Haedrich, 1997).

The most diverse pelagic zone (the *mesopelagic*) is dominated by relatively primitive fishes such as lanternfishes (Myctophidae), hatchetfishes (Sternoptychidae), viperfishes (Chauliodontidae), and dragonfishes (Stomiidae), whereas the benthic fishes hark from more advanced groups such as cods (Gadidae), eelpouts (Zoarcidae), brotulas (Ophidiidae), slickheads (Alepocephalidae), and various species of sharks, rays, and chimaeras. The chimaeras (Holocephali) are almost exclusively deep-sea fishes, with a few species entering shelf waters in the colder oceanic areas of the world. The more primitive mesopelagic species composition suggests that such habitat was available and was invaded earlier than the benthic deep sea. This means that the reliable oxygenation of the water progressed from shallow to deep water, the deepest waters becoming oxygenated later. As with many beginnings of the modern fish fauna, this appears to date from Middle to Late Cretaceous.

KEY POINTS AND CONNECTIONS

- Fish distributions contribute greatly to the study of biogeography on many levels and time scales. Comparing distribution patterns with phylogenetic trees helps unravel the histories of both terrestrial and aquatic habitats and communities. The very numerous and widespread modern fish fauna provides an abundance of opportunities to answer questions about the history of the Earth.

- Almost half (42 percent) of the world's fishes are freshwater fishes, and 39 percent live in shallow, warm or cold water on the continental shelves. This leaves slightly less than 19 percent of the fish species in the 70 percent of the world covered by the deep ocean. An additional 1 percent or fewer fish species have the ability to go back and forth between fresh and salt water at some life stage.

- The largest and most diverse freshwater fish faunas have developed where larger landmasses have remained stable and relatively warm (but not too hot), such as the southeastern United States, Amazonian South America, central Africa, and southeastern Asia.

- Within these large, stable areas, occasional (geologically speaking) stream captures and changing sea levels have generated many of the species present today throughout the Cenozoic era (the last 65 million years).

- The Amazon and its associated river basins have the richest freshwater fish fauna in the world, followed by Asia, Africa, and North America.

- The largest and most diverse marine fish communities are also in shallow, warm (tropical) areas, but with unstable geological histories, providing information on the isolation of islands and the separation of ocean basins.

- The most diverse marine shore and shelf faunas are in the warm Indo-Western Pacific region, followed by the Caribbean Sea.

- The North Pacific has a rich cold-water marine fauna, but the North Atlantic and the Mediterranean have far fewer species, due to Pleistocene extinctions.

- The Antarctic also has a small fauna, but it is almost entirely endemic to the southern oceans. The Arctic marine fauna is composed of only a few widespread species from the North Atlantic and North Pacific.

- Land areas intervening between major continents, such as Middle America and the Arabian peninsula, have small freshwater faunas of a few endemic species and a mixture of fishes from the adjacent, larger landmasses.

- The onset of Miocene glaciation eliminated freshwater fishes from Antarctica, but both freshwater and marine fish groups remain that show Gondwana distributions on the southern extremities of Australia, South America, and Africa.

- The northern continents have been much more continuous over time, and several faunal exchanges have occurred from the Cretaceous onward. Pleistocene glaciations in the northern hemisphere have also caused extinctions, and the limited degree of reinvasion has left glaciated areas with significantly fewer freshwater fishes.

- North America has been a refugium for several primitive freshwater fish groups, including the gars (Lepisosteidae), bowfin (Amiidae), pirate perches (Aphredoderidae), and troutperches (Percopsidae).

- Primitive tropical freshwater fishes are found today in the tropical portions of one or more continents, including the lungfishes (Dipnoi; three families), swamp eels (Synbranchidae), bichirs (Polypteridae), and bonytongues (Osteoglossidae).

- The apparently homogeneous open oceans consist of about 20 faunal regions defined by pelagic fishes. Benthic deep-sea fishes will probably be found to have greater differentiation in their distribution than presently known.

- Shore fish zones can be divided into cold or boreal, temperate, subtropical, and tropical elements, but the prevailing north–south trending continental boundaries and the vagaries of prevailing currents blur the distinctions among these regions. Barriers between major marine environments, such as the Arabian, Floridian, and Baja California peninsulas, the isthmus of Panama, the mouths of the largest rivers, and cold upwelling areas, have led to the emergence of many distinct species of fishes.

- On a worldwide scale, the active rearrangement of the continents over the last 65 million years has subdivided the oceans, separated the continents, enabled fishes to invade the deep ocean, and helped create the largest number of fish species the world has ever known.

- The emphasis on defining biogeographical zones has given away to a dynamic understanding of dispersal, phylogenetic relationships, and differentiation among taxa in relation to time, geography, and evolutionary change.

BUILDING AN ICHTHYOLOGY LIBRARY

Berra, T. M. 2001. *Freshwater fish distribution.* Academic Press, San Diego.

Hocutt, C. H., and E. O. Wiley (Eds.). 1986. *The zoogeography of North America freshwater fishes.* Wiley, New York.

Humphries, C. J., and L. R. Parenti. 1999. *Cladistic biogeography. Interpreting patterns of plant and animal distributions* (2nd ed.). Clarendon Press, Oxford, UK.

Lomolino, M. V., B. R. Riddle, and J. H. Brown. 2005. *Biogeography* (3rd ed.). Sinauer, Sunderland, MA.

Mayden, R. L. (Ed.) 1992. *Systematics, historical ecology, and North American freshwater fishes.* Stanford University Press, Stanford, CA.

REFERENCES

Andriachev, A. P., and N. V. Chernova. 1994. Annotated list of fish-like vertebrates and fishes of the Arctic Seas and adjacent waters. *Vopr. Ikhtiol. 34:* 435–456 (Russian).

Allen, G. R. 1991a. *Freshwater fishes of Australia.* TFH, Neptune, NJ.

———. 1991b. *Field guide to the freshwater fishes of New Guinea* (Publ. No. 9). Christensen Research Institute, Madang, Papua New Guinea.

———, and M. Adrim. 2003. Coral reef fishes of Indonesia. *Zool. Stud. 42(1):* 1–72.

———, S. G. Midgley, and M. Allen. 2002a. *Field guide to the freshwater fishes of Australia.* Western Australian Museum, Perth.

Allen, P., J. H. Brown, and J. F. Gillooly. 2002b. Global biodiversity, biochemical kinetics, and the energetic-equivalence rule. *Science 297:* 1545–1548.

Avise, J. C. 2000. *Phylogeography: The history and formation of species.* Harvard University Press, Cambridge, MA.

———. 2004. *Molecular markers, natural history, and evolution* (2nd ed.). Sinauer, Sunderland, MA.

Banarescu, P. 1986. Chapter 5. Zoogeography and history of the freshwater fish fauna of Europe, pp. 88–107. In *The freshwater fishes of Europe,* Vol. I, Pt. II, J. Holcik (Ed.). Aula Verlag, Wiesbaden, Germany.

———. 1990. *Zoogeography of freshwaters,* Vol. I. Aula Verlag, Wiesbaden, Germany.

———. 1992. *Zoogeography of freshwaters,* Vol. II. Aula Verlag, Wiesbaden, Germany.

———. 1994. *Zoogeography of freshwaters,* Vol. III, Distribution and dispersal of freshwater animals in Africa, Pacific areas, and South America. Aula Verlag, Wiesbaden, Germany.

Bargelloni, L., S. Marcato, L. Zane, and T. Patarnello. 2000. Mitochondrial phylogeny of notothenioids: A molecular approach to Antarctic fish evolution and biogeography. *Syst. Biol. 49(1):* 114–129.

Beckley, L. E., P. A. Hulley, and P. H. Skelton. 2002. Synoptic overview of marine ichthyology in South Africa. *Mar. Freshw. Res. 53:* 99–105.

Bell, M. A. 1994. Palaeobiology and evolution of threespine stickleback, Chapter 15, pp. 438–471. In *The evolutionary biology of the stickleback,* M. A. Bell and S. A. Foster (Eds.). Oxford University Press, Oxford.

Berg, L. S. 1949. *Freshwater fishes of the USSR and adjacent countries* (4th ed.), Vol. 3. Israel Program for Scientific Translations, Jerusalem.

Bermingham, E., and J. C. Avise. 1986. Molecular zoogeography of freshwater fishes in the southeastern United States. *Genetics 113:* 939–965.

Berra, T. M. 1981. *An atlas of distribution of the freshwater fish families of the world.* University of Nebraska Press, Lincoln.

———. 2001. *Freshwater fish distribution.* Academic Press, San Diego.

Boschung, H. T., and R. L. Mayden. 2004. *Fishes of Alabama.* Smithsonian Institution, Washington, DC.

Boulenger, G. A. 1905. The distribution of African freshwater fishes. *Nature 72:* 413–421.

Briggs, J. C. 1974. *Marine zoogeography.* McGraw-Hill, New York.

———. 1995. *Global biogeography.* Elsevier, Amsterdam.

Burgess, G. H., and R. Franz. 1989. Zoogeography of the Antillean freshwater fish fauna, pp. 263–304. In *Biogeography of the West Indies. Past, present, and future,* C. A. Woods (Ed.). Sandhill Crane Press, Gainesville, FL.

Burr, B. M., and R. L. Mayden. 1992. Phylogenetics and North American freshwater fishes, pp. 18–75. In *Systematics, historical ecology, and North American freshwater fishes,* R. L. Mayden (Ed.). Stanford University Press, Stanford, CA.

Bussing, W. A. 1985. Patterns of distribution of the central American ichthyofauna, pp. 453–473. In *The great American interchange,* F. G. Stehli and S. D. Webb (Eds.). Plenum Press, New York.

———. 1998. Peces de las agues continentals de Costa Rica—Freshwater fishes of Costa Rica. *Revista Biol. Trop. 46*(Suppl. 2).

Buth, D. G., and R. L. Mayden. 2001. Allozymic and isozymic evidence for polytypy in the North American catostomid genus *Cycleptus. Copeia 2001:* 899–906.

Carpenter, K. E., and V. G. Springer. 2005. The center of the center of marine shore fish biodiversity: The Philippine Islands. *Env. Biol. Fishes 73:* 467–480.

Cavender, T. M. 1986. Review of the fossil history of North American freshwater fishes, pp. 699–724. In *Zoogeography of North American freshwater fishes,* C. H. Hocutt and E. O. Wiley (Eds.). Wiley, New York.

Chernova, N. V., and D. L. Stein. 2002. Ten new species of *Psednos* (Pisces, Scorpaeniformes: Liparidae) from the Pacific and North Atlantic Oceans. *Copeia 2002:* 755–778.

Clarke, T. A. 2001. Pelagic fishes of the genus *Eustomias,* subgenus *Dinematochirus* (Stomiidae), in the Indo-Pacific with the description of twelve new species. *Copeia 2001:* 683–699.

Cohen, D. M. 1970. How many recent fishes are there? *Proc. Calif. Acad. Sci. (Ser. 4) 38:* 341–346.

Collette, B. B., and P. Banarescu. 1977. Systematics and zoogeography of the fishes of the family Percidae. *J. Fish. Res. Bd. Can. 34:* 1450–1463.

Costa, W. J. E. M. 1998. Phylogeny and classification of the cyprinodontiformes (Euteleostei: Atherinomorpha): A reappraisal, pp. 537–560. In *Phylogeny and classification of neotropical fishes,* L. R. Malabarba, R. E. Reis, R. P. Vari, Z. M. S. Lucena, and C. A. S. Lucena (Eds.). Editoria Universitaria, Pontificia Universidade Catolica do Rio Grande do Sul, Porto Alegre, Brazil.

Crossman, E. J., and D. A. McAllister. 1986. Zoogeography of freshwater fishes of the Hudson Bay drainage, Ungava Bay, and the Arctic archipelago, pp. 53–104. In *Zoogeography of North American freshwater fishes,* C. C. Hocutt and E. O. Wiley (Eds.). Wiley, New York.

Daget, J., J. P. Gosse, and D. F. E. Thys van den Audenaerde (Eds.). 1986. *Checklist of the freshwater fishes of Africa.* ORSTOM, MRAC, Paris.

Darlington, P. J., Jr. 1957. *Zoogeography.* Wiley, London.

Dawson, M. N., K. D. Louie, M. Barlow, D. K. Jacobs, and C. C. Swift. 2002. Comparative phylogeography of sympatric sister species, *Clevelandia ios* and *Eucyclogobius newberryi* (Teleostei, Gobiidae), across the California Transition zone. *Mol. Ecol. 11:* 1065–1075.

de Beaufort, L. F. 1951. *Zoogeography of the land and inland waters.* Sidgwick and Jackson, London.

Dooley, J. K. 1972. Fishes associated with the pelagic sargassum complex, with a discussion of the sargassum community. *Contrib. Mar. Sci. Univ. Texas 16:* 1–30.

Dunson, W. A., and J. Travis. 1991. The role of abiotic factors in community organization. *Am. Nat. 138:* 1067–1091.

———, and———. 1994. Patterns in the evolution of physiological specialization in salt-marsh animals. *Estuaries 17*(1A): 102–110.

Dyer, B. S. 1998. Phylogenetic systematics and historical biogeography of the neotropical silverside family Atherinopsidae (Teleostei: Atheriniformes), pp. 519–536. In *Phylogeny and classification of neotropical fishes,* L. R. Malabarba, R. E. Reis, R. P. Vari, Z. M. S. Lucena, and C. A. S. Lucena (Eds.). Pontificia Universidade Catolica do Rio Grande do Sul, Porto Alegre, Brazil.

Eastman, J. T. 1993. *Antarctic fish biology. Evolution in a unique environment.* Academic Press, New York.

Echelle, A. A., E. W. Carson, A. F. Echelle, R. A. Van Den Bussche, T. E. Dowling, and A. Meyer. 2005. *Copeia 2005:* 320–339.

Fryer, G., and T. D. Iles. 1972. *The cichlid fishes of the great lakes of Africa: Their biology and evolution.* TFH, Neptune City, NJ.

Golani, D. 1993. The sandy shore of the Red Sea—launching pad for Lessepsian (Suez Canal) migrant fish colonizers of the eastern Mediterranean. *J. Biogeogr. 20*(3): 57–61.

Gon, O., and P. C. Heemstra (Eds.). 1990. *Fishes of the Southern Ocean.* J. L. B. Smith Institute of Ichthyology, Grahamstown, South Africa.

Grande, L. 1999. The first *Esox* (Esocidae: Teleostei) from the Eocene Green River Formation, and a brief review of esocid fishes. *J. Vert. Paleontol. 19:* 271–292.

———, and W. E. Bemis. 1991. Osteology and phylogenetic relationships of fossil and recent paddlefishes (Polyodontidae) with comments on interrelationships of the Acipenseriformes. *Soc. Vert. Paleontol. Mem. No. 1.*

———, and———. 1998. A comprehensive phylogenetic study of amiid fishes (Amiidae) based on comparative skeletal anatomy. An empirical search for interconnected patterns of natural history. *J. Vert. Paleontol. 18, Suppl. 1 (Soc. Vert. Paleontol. Mem. No. 4).*

———, and———. 1999. Historical biogeography and historical paleoecology of Amiidae and other halecomorph fishes, pp. 413–424. In *Mesozoic fishes 2: Systematics and fossil record,* G. Arratia and H.-P. Schultze (Eds.). Verlag Dr. Friedrich Pfeil, München, Germany.

Greenwood, P. H. 1983. The zoogeography of African freshwater fishes: Bioaccountancy or biogeography, pp. 179–199. In *Evolution, time, and space: The emergence of the biosphere,* R. W. Simms, J. H. Price, and P. E. S. Whalley (Eds.). Academic Press, New York.

———, D. E. Rosen, S. H. Weitzman, and G. S. Myers. 1966. Phyletic studies of teleostean fishes, with a provisional classification of living forms. *Bull. Am. Mus. Nat. Hist. 131:* 339–456.

Grove, J. S., and R. J. Lavenberg. 1997. *The fishes of the Galapagos Islands.* University of California Press, Berkeley.

Gunther, A. C. L. G. 1880. *An introduction to the study of fishes.* Adam and Charles Black, Edinburgh, UK.

Haedrich, R. L. 1983. Estuarine fishes, pp. 183–207. In *Ecosystems of the world 26, Estuaries and enclosed seas,* B. H. Ketchum (Ed.). Elsevier, New York.

———. 1997. Distribution and population ecology, Chapter 3, pp. 79–115. In *Deep sea fishes,* D. J. Randall and A. P. Farrell (Eds.). Academic Press, San Diego.

Hartel, K. E., D. B. Halliwell, and A. E. Launer. 2002. *Inland fishes of Massachusetts.* Massachusetts Audubon Society, Natural History of New England Series, Lincoln, MA.

Hastings, P. A. 2000. Biogeography of the tropical eastern Pacific: Distribution and phylogeny of chaenopsid fishes. *Zool. J. Linn. Soc. 128:* 319–335.

———, and D. R. Robertson. 2001. Systematics of tropical eastern Pacific fishes. *Rev. Biol. Trop. 49*(Suppl. 1): Preface.

Hastings, R. W. 1979. Origin and seasonality of the fish fauna on a new jetty in the northeastern Gulf of Mexico. *Bull. Fla. St. Mus. Biol. Sci. 24*(1): 1–124.

Hocutt, C. H., and E. O. Wiley (Eds.). 1986. *The zoogeography of North American freshwater fishes.* Wiley, New York.

Horn, M. H., and L. G. Allen. 1978. A distributional analysis of California coastal marine fishes. *J. Biogeogr. 5:* 23–42.

Hrbek, T., and A. Meyer. 2003. Closing of the Tethys Sea and the phylogeny of Eurasian killifishes (Cyprinodontiformes: Cyprinodontidae). *J. Evol. Biol. 16:* 17–36.

Hubbs, C. L. 1952. Antitropical distribution of fishes and other organisms. Symposium on problems of bipolarity and of pantemperate faunas. *Proc. 7th Pac. Sci. Congr. 3:* 324–329.

———. 1959. Initial discoveries of fish faunas on seamounts and offshore banks in the eastern Pacific. *Pac. Sci. 13:* 311–316.

———. 1961. The marine vertebrates of the outer coast, pp. 137–147. In Symposium: The biogeography of Baja California and adjacent seas, Part 2, Marine biotas. *Syst. Zool. 9*(3–4).

Humphries, C. J., and L. R. Parenti. 1999. *Cladistic biogeography. Interpreting patterns of plant and animal distributions* (2nd ed.). Clarendon Press, Oxford, UK.

Jordan, D. S. 1901. The fish fauna of Japan, with observations on the geographical distribution of fishes. *Science New Ser. XIV:* 545–567.

———. 1928. The distribution of freshwater fishes. *Ann. Rept. Smithson. Inst. 1927:* 335–385.

Kiener, A. 1965. Contributions a l'étude écologiques et biologiques des eaux saumatres malagaches. *Vie et Milieu C 16*(2): 113–149.

Kinziger, A. P., R. M. Wood, and D. A. Neely. 2005. Molecular systematics of the genus *Cottus* (Scorpaeniformes: Cottidae). *Copeia 2005:* 303–311.

Kozhov, M. 1963. *Lake Baikal and its life* (Monographiae Biologicae, Vol. XI). Junk, The Hague.

Krebs, C. J. 1994. *Ecology: The experimental analysis of distribution and abundance* (4th ed.). HarperCollins, New York.

Lessios, H. A., B. D. Kessing, and D. R. Robertson. 1997. Massive gene flow across the world's most potent marine biogeographic barrier. *Proc. Roy. Soc. Lond. B 265:* 583–588.

Lomolino, M. V., B. R. Riddle, and J. H. Brown. 2005. *Biogeography* (3rd ed.) Sinauer, Sunderland, MA.

Love, M. S., M. Yoklavich, and L. Thorsteinson. 2002. *The rockfishes of the northeast Pacific.* University of California Press, Berkeley.

Lovejoy, N. R., and B. B. Collette. 2001. Phylogenetic relationships of New World needlefishes (Teleostei, Belonidae) and the biogeography of transitions between marine and freshwater habitats. *Copeia 2001:* 324–338.

Lowe-McConnell, R. H. 1975. *Fish communities in tropical freshwaters.* Longmans, London.

Lundberg, J. G. 1993. Freshwater fishes, pp. 156–190. In *Biological relationships between Africa and South America,* P. Goldblatt (Ed.). Yale University Press, New Haven, CT.

———. 1998. The temporal context for the diversification of neotropical fishes, pp. 49–68. In *Phylogeny and classification of neotropical fishes,* L. R. Malabarba, R. E. Reis, R. P. Vari, Z. M. S. Lucena, and C. A. S. Lucena (Eds.). Editoria Universitaria, Pontificia Universidade Catolica do Rio Grande do Sul, Porto Alegre, Brazil.

———, M. Kottelat, G. R. Smith, M. Stiassny, and T. Gill. 2000. So many fishes, so little time: An overview of recent ichthyological discoveries in freshwaters. *Ann. Missouri Bot. Garden 87*(1): 26–62.

———, A. Machado-Allison, and R. F. Kay. 1986. Miocene characoid fishes from Colombia: Evolution or extirpation. *Science 224:* 208–209.

———, L. G. Marshall, J. Guerrero, B. Horton, M. Claudia, S. L. Malabarba, and F. Wesselingh. 1998. The stage for neotropical fish diversification: A history of tropical South American rivers, pp. 14–48. In *Phylogeny and classification of neotropical fishes,* L. R. Malabarba, R. E. Reis, R. P. Vari, Z. M. S. Lucena, and C. A. S. Lucena (Eds.). Editoria Universitaria, Pontificia Universidade Catolica do Rio Grande do Sul, Porto Alegre, Brazil.

———, and L. McDade. 1990. Systematics, pp. 65–108. In *Methods for fish biology,* C. B. Schreck and P. B. Moyle (Eds.). American Fisheries Society, Bethesda, MD.

Malabarba, L. R., R. E. Reis, R. P. Vari, Z. M. S. Lucena, and C. A. S. Lucena (Eds.). 1998. *Phylogeny and classification of neotropical fishes.* Editoria Universitaria, Pontificia Universidade Catolica do Rio Grande do Sul, Porto Alegre, Brazil.

Martens, K. 1997. Speciation in ancient lakes. *Trends Ecol. Evol. 12*(5): 177–182.

Matthews, W. J. 1998. *Patterns in freshwater fish ecology.* Chapman and Hall, New York.

Mayden, R. L. (Ed.). 1992. *Systematics, historical ecology, and North American freshwater fishes.* Stanford University Press, Stanford, CA.

———, B. M. Burr, L. M. Page, and R. R. Miller. 1992. The native freshwater fishes of North America, pp. 827–863. In *Systematics, historical ecology, and North American freshwater fishes,* R. L. Mayden (Ed.). Stanford University Press, Stanford, CA.

———, and B. R. Kuhajda. 1996. Systematics, taxonomy, and conservation status of the endangered Alabama sturgeon, *Scaphirhynchus suttkusi* Williams and Clemmer (Actinopterygii, Acipenseridae). *Copeia 1996:* 241–273.

———, and E. O. Wiley. 1992. The fundamentals of phylogenetic systematics, pp. 14–185. In *Systematics, historical ecology, and North American freshwater fishes,* R. L. Mayden (Ed.). Stanford University Press, Palo Alto, CA.

———, and R. M. Wood. 1995. Systematics, species concepts, and the ESU in biodiversity and conservation biology, pp. 58–113. In *Evolution and the aquatic ecosystem: Defining unique units in population conservation,* J. Nielson (Ed.). American Fisheries Society, Bethesda, MD.

McDowall, R. M. 1988. *Diadromy in fishes.* Timber Press, Portland, OR.

———. 1990. *New Zealand freshwater fishes, a natural history and guide* (rev. ed.). Heinemann Reed, Auckland, New Zealand.

McGinnis, R. F. 1982. Biogeography of lanternfishes (Myctophidae) south of 30°S, pp. 1–110. In *Antarctic research series,* Vol. 35, Biology of the Antarctic seas XII, D. L. Pawson (Ed.). American Geophysical Union, Washington, DC.

McHugh, J. L. 1967. Estuarine nekton, pp. 581–620. In *Estuaries,* G. Lauff (Ed.). American Association for the Advancement of Science, Washington, DC.

Mecklenburg, C. W., T. A. Mecklenburg, and L. K. Thorsteinson. 2002. *Fishes of Alaska.* American Fisheries Society, Bethesda, MD.

Miller, D. J., and R. N. Lea. 1976. Guide to the coastal marine fishes of California. *Calif. Dept. Fish Game Fish Bull. 157* (2nd printing, with addendum). Division of Agricultural Sciences, University of California, Richmond.

Miller, P. J. 1973. Gobiidae, pp. 483–515. In *Checklist of the fishes of the northeastern Atlantic and of the Mediterranean,* J. C. Hureau and Th. Monod (Eds.). CLOFNAM, UNESCO, Paris.

Miller, R. R. 1966. Geographical distribution of Central American freshwater fishes. *Copeia 1966:* 773–802.

———, W. L. Minckley, and S. M. Norris. 2005. *Freshwater fishes of Mexico.* University of Chicago Press, Chicago.

Minckley, W. L., D. A. Hendrickson, and C. E. Bond. 1986. Geography of western North American freshwater fishes: Description and relationships to intracontinental tectonism, pp. 519–614. In *The zoogeography of North American freshwater fishes,* C. H. Hocutt and E. O. Wiley (Eds.). John Wiley, New York.

Moore, J. A., M. Vecchione, B. B. Collette, R. Gibbons, and K. E. Hartel. 2004. Selected fauna of Bear Seamount (New England Seamount Chain), and the presence of "natural invader" species. *Arch. Fish. Mar. Res. 51*(1–3): 241–250.

Moritz, C. 2002. Strategies to protect biological diversity and the evolutionary processes that sustain it. *Syst. Biol. 51:* 238–254.

Moyle, P. B. 1976. *Inland fishes of California.* University of California Press, Berkeley.

———. 2002. *Inland fishes of California* (rev. ed.). University of California, Berkeley.

———, and R. M. Leidy. 1992. Loss of biodiversity in aquatic ecosystems: Evidence from the fish faunas, pp. 127–169. In *Conservation biology: The theory and practice of nature conservation,* P. L. Fiedler and S. K. Jain (Eds.). Chapman and Hall, New York.

————, and J. L. Williams. 1990. Biodiversity loss in the temperate zone: Decline of the native fish fauna of California. *Conserv. Biol. 4:* 275–284.

Myers, G. S. 1938. Freshwater fishes and West Indian zoogeography. *Smithson. Rept. 1937, Publ. 3645:* 339–364.

————. 1941. The fish fauna of the Pacific ocean, with special reference to zoogeographical regions and distribution as they affect the international aspects of the fisheries. *Proc. 6th Pac. Sci. Congr. 3:* 201–210.

————. 1949. Salt tolerance of fresh-water fish groups in relation to zoogeographical problems. *Bijdr. Dierk. 28:* 315–322.

————. 1951. Freshwater fishes and East Indian zoogeography. *Stanford Ichthyol. Bull. 7*(3). 11–21.

Myers, R. F. 1999. *Micronesian reef fishes. A field guide for divers and aquarists* (3rd ed.). Micrographics, Guam.

Nakabo, T. (Ed.). 2002. *Fishes of Japan with pictorial keys to the species* (English ed., 2 Vols.). Tokai University Press, Tokyo.

Near, T. J., and M. F. Benard. 2004. Rapid allopatric speciation in logperch darters (Percidae: *Percina*). *Evolution 58:* 2798–2808.

————, and B. P. Keck. 2005. Dispersal, vicariance, and timing of diversification in *Nothonotus* darters. *Mol. Ecol. 14:* 3485–3496.

Nelson, J. S. 1994. *Fishes of the world* (3rd ed.). Wiley, New York.

Nordlie, F. G. 1990. Rivers and springs, pp. 392–425. In *Ecosystems of Florida*, R. L. Myers and J. J. Evel (Eds.). University of Central Florida Press, Orlando.

Parenti, L. R. 1984. A taxonomic revision of the Andean killifish genus *Orestias* (Cyprinodontiformes, Cyprinodontidae). *Bull. Amer. Mus. Nat. Hist. 178:* 107–214.

————, and K. D. Louie. 1998. *Neostethus djaoorum*, new species, from Sulawesi, Indonesia, the first phallostethid fish (Teleostei: Atherinomorpha) known from East of Wallace's Line. *Raffles Bull. Zool. 46*(1): 139–150.

Parin, N. V., and T. N. Belyanina. 2002. Flying fishes of the genus *Fodiator* (Exocetidae): Systematics and distribution. *J. Ichthyol. 42:* 357–367. (Translated from *Vopr. Ikhtiol. 42:* 293–303.)

Patterson, C. 1975. The distribution of Mesozoic freshwater fishes. *Mem. Mus. Natl. Hist. Nat. (Paris) Ser. A (Zool.) 87:* 156–173.

————. 1981. The development of the North American fish fauna—a problem of historical biogeography, pp. 265–281. In *The evolving biosphere*, P. Forey (Ed.). British Museum (Natural History), London.

Paxton, J. R., and W. N. Eschmeyer. 1998. *Encyclopedia of fishes* (2nd ed.). Academic Press, San Diego.

Perry, A. L., P. J. Low, J. R. Ellis, and J. D. Reynolds. 2005. Climate change and distribution shifts in marine fishes. *Science 208:* 1912–1915.

Rahel, F. J. 2002. Homogenization of freshwater faunas. *Ann. Rev. Ecol. Syst. 33:* 291–315.

Rainboth, W. J. 1991. Cyprinids of South East Asia, pp. 156–210. In *Cyprinid fishes: Systematics, biology, and exploitation*, I. J. Whitfield and J. S. Nelson (Eds.). Chapman and Hall, New York.

————. 1996a. *Fishes of the Cambodian Mekong. FAO Species identification guide for fishery purposes.* Food and Agricultural Organization, United Nations, Rome.

————. 1996b. The taxonomy, systematics, and zoogeography of *Hypsibarbus*, a new genus of large barbs (Pisces, Cyprinidae) from the rivers of southeastern Asia. *Univ. Calif. Publ. Zool. 129:* 1–199.

Randall, J. E. 1982. Examples of antitropical and antiequatorial distribution of Indo-West Pacific fishes. *Pac. Sci. 35*(3): 197–209.

————. 1998. Zoogeography of shore fishes of the Indo–Pacific region. *Zool. Stud. (Taiwan) 37:* 227–268.

————, and K. K. P. Lim. 2000. A checklist of the fishes of the South China Sea. *Raffles Bull. Zool. 2000*(Suppl. 8): 569–667.

Regan, C. T. 1906–1908. Pisces, pp. 1–203. In *Biologia Centrali Americana*. Porter, London.

Ricklefs, R. E., and G. L. Miller. 1999. *Ecology* (4th ed.). W. H. Freeman, New York.

Roberts, T. R. 1989. The freshwater fishes of western Borneo (Kalimantan Barat, Indonesia). *Mem. Calif. Acad. Sci. No. 14.*

Roberts, Y. R. 1975. Geographical distribution of African freshwater fishes *Zool. J. Linn. Soc. 57:* 249–319.

Robertson, D. R., J. S. Groves, and J. E. McCosker. 2005. Tropical transpacific shore fishes. *Pac. Sci. 58:* 507–565.

Rocha, L. A. 2003. Patterns of distribution and processes of speciation in Brazilian reef fishes. *J. Biogeogr. 30:* 1161–1171.

Rosen, D. E. 1975. A vicariance model of Caribbean biogeography. *Syst. Zool. 24:* 431–464.

Ross, S. T. 2001. *Inland fishes of Mississippi*. University of Mississippi Press, Jackson.

Rubinoff, I., and R. Rubinoff. 1971. Geographic and reproductive isolation in Atlantic and Pacific populations of Panamanian *Bathygobius*. *Evolution 25:* 88–97.

Schaefer, S. A. 1998. Conflict and resolution: Impact of new taxa on phylogenetic studies of the neotropical cascudinhos (Siluridae: Loricariidae), pp. 375–400. In *Phylogeny and classification of neotropical fishes*, L. R. Malabarba, R. E. Reis, R. P. Vari, Z. M. S. Lucena, and C. A. S. Lucena (Eds.). Editoria Universitaria, Pontificia Universidade Catolica do Rio Grande do Sul, Porto Alegre, Brazil.

Sideleva, V. G. 2003. *The endemic fishes of Lake Baikal*. Bachhuys, Leiden, the Netherlands.

Skelton, P. H. 1996. A historical review of the taxonomy and biogeography of freshwater fishes in South Africa-the past 50 years. *Trans. Roy. Soc. Africa 51:* 91–114.

Smith, G. R., T. E. Dowling, K. W. Gobalet, T. Lugaski, D. K. Shiozawa, and R. P. Evans. 2002. Biogeography and timing of evolutionary events among Great Basin fishes, pp. 175–234. In The Great Basin: Cenozoic geology and biogeography, R. Hershler and D. Curry (Eds.), *Smithson. Contr. Earth Sci. No. 33.*

Smith-Vaniz, W. F., B. B. Collette, and B. E. Luckhurst. 1999. Fishes of Bermuda: History, zoogeography, annotated checklist, and identification keys. *Am. Soc. Ichthyol. Herpetol. Spec. Publ. No. 4*, Lawrence, KS.

Soltz, D., and R. Naiman. 1978. The natural history of native fishes in the Death Valley system. *Nat. Hist. Mus. L. A. Co. Sci. Ser. No. 30.*

Sparks, J. S., and W. L. Smith. 2004. Phylogeny and biogeography of the Malagasy and Australasian rainbowfishes (Teleostei: Melanotaenioidei): Gondwanan vicariance and evolution in freshwater. *Mol. Phylogen. Evol. 33:* 719–734.

————, and ————. 2005. Freshwater fishes, dispersal ability, and nonevidence: "Gondwana life rafts" to the rescue. *Syst. Biol. 54:* 158–165.

Springer, V. G. 1982. Pacific plate biogeography with special reference to shorefishes. *Smithson. Contr. Zool. 376:* 1–182.

Stiassny, M. L., and M. C. C. de Pinna. 1994. Basal taxa and the role of cladistic patterns in the evaluation of conservation priorities: A view from freshwater, pp. 235–249. In *Systematics Association special*, Vol. 50, P. I. Forey and C. J. Humphries (Eds.). Systematics Association, London.

————, and N. Raminosoa. 1994. The fishes of the inland waters of Madagascar. *Ann. Mus. Roy. Afr. Centr. Zool. 275:* 133–149.

Strange, R. M., and B. M. Burr. 1997. Intraspecific phylogeography of North American highland fishes: A test of the Pleistocene vicariance hypothesis. *Evolution 51:* 885–897.

Swenson, R. O. 1999. The ecology, behavior, and conservation of the tidewater goby, *Eucyclogobius newberryi. Env. Biol. Fishes 55:* 99–114.

Swift, C. C., C. R. Gilbert, S. A. Bortone, G. H. Burgess, and R. W. Yerger. 1986. Zoogeography of the freshwater fishes of the southeastern United States: Savannah River to Lake Pontchartrain, pp. 213–265. In *The zoogeography of North American freshwater fishes,* C. H. Hocutt and E. O. Wiley (Eds.). Wiley, New York.

Teugels, G. G., J. F. Guegan, and J. J. Alberet. 1994. Biological diversity of African fresh and brackish water fishes. *Ann. Mus. Roy. Afr. Centr. Zool. 235:* 3–177.

Thieme, M. L., R. Abell, M. L. J. Stiassny, and P. Skelton. 2005. *Freshwater ecoregions of Africa and Madagascar. A conservation assessment.* Island Press, Covelo, CA.

Thomson, D. A., L. T. Findley, and A. N. Kerstitch. 2000. *Reef fishes of the Sea of Cortez* (rev. ed.). University of Texas Press, Austin.

Trautman, M. B. 1981. *Fishes of Ohio* (rev. ed.). Ohio State University Press, Columbus.

Underhill, J. C. 1986. The fish fauna of the Laurentian Great Lakes, the St. Lawrence lowlands, Newfoundland and Labrador, pp. 105–136. In *The zoogeography of North American freshwater fishes,* C. H. Hocutt and E. O. Wiley (Eds.). Wiley, New York.

Vari, R. P., and A. S. Harold. 2001. Phylogenetic study of the neotropical fish genera *Creagrutus* Gunther and *Piabina* Reinhardt (Teleostei: Ostariophysi: Characiformes), with a revision of the cis-Andean species. *Smithson. Contr. Zool. 613.*

————, and L. R. Malabarba. 1998. Neotropical ichthyology: An overview, pp. 2–11. In *Phylogeny and classification of neotropical fishes,* L. R. Malabarba, R. E. Reis, R. P. Vari, Z. M. S. Lucena, and C. A. S. Lucena (Eds.). Editoria Universitaria, Pontificia Universidade Catolica do Rio Grande do Sul, Porto Alegre, Brazil.

Waples, R. S. 1995. Evolutionary significant units and the conservation of biological diversity under the Endangered Species Act, pp. 8–27. In *Evolution and the aquatic ecosystem,* J. Nielsen (Ed.). American Fisheries Society, Bethesda, MD.

Warren, M. L., Jr., and B. M. Burr. 1994. Status of freshwater fishes of the United States: Overview of an imperiled fauna. *Fisheries 19*(1): 6–18.

Waters, J. M., J. Andrea Lopez, and G. P. Wallis. 2000. Molecular phylogenetics and biogeography of galaxiid fishes (Osteichthyes: Galaxiidae): Dispersal, vicariance, and the position of *Lepidogalaxias salamandroides. Syst. Biol. 49:* 777–795.

Welcomme, R. L. 1979. *Fisheries ecology of floodplain rivers.* Longman, London.

White, B. N. 1986. The isthmanian link, antitropicality, and American biogeography: Distributional history of the Atherinopsinae (Pisces: Atherinidae). *Syst. Zool. 35:* 176–194.

————. 1994. Vicariance biogeography of the open ocean Pacific. *Prog. Oceanogr. 34:* 257–284.

Wiley, E. O. 1976: The phylogeny and biogeography of the recent and fossil gars (Actinopterygii: Lepisosteidae). *Mus. Nat. Hist. Univ. Kansas Misc. Publ. 64:* 1–111.

————. 1992. Phylogenetic relationships of the Percidae (Teleostei: Perciformes): A preliminary hypothesis. In *Systematics, historical ecology, and the North American freshwater fishes,* R. L. Mayden (Ed.). Stanford University Press, Stanford, CA.

Wilson, M. V. H., and A. M. Murray. 1996. Early Cenomanian acanthomorph teleost in the Cretaceous Fish Scale Zone, Albian / Cenomanian boundary, Alberta, Canada, pp. 369–382. In *Mesozoic fishes: Systematics and paleoecology. Proceedings of the International Meeting, Eichstätt, 1993,* G. Arratia and G. Viohl (Eds.). Verlag Dr. Friedrich Pfeil, München, Germany.

————, and R. R. G. Williams. 1992. Phylogenetic, biogeographic, and ecological significance of early fossil records of North American teleostean fishes, pp. 224–246. In *Systematics, historical ecology, and North American freshwater fishes,* R. L. Mayden (Ed.). Stanford University Press, Palo Alto, CA.

Wilson, R. R., and R. S. Kaufman. 1987. Seamount biota and biogeography, pp. 355–377. In Seamounts, islands, and atolls, B. H. Keating, P. Fryer, R. Batiza, and G. W. Boehlert (Eds.), *Geophys. Monogr. 43.* American Geophysical Union, Washington, DC.

Wydowski, R. S., and R. R. Whitney. 1979. *Inland fishes of Washington.* University of Washington Press, Seattle.

————, and ————. 2003. *Inland fishes of Washington* (2nd ed.). American Fisheries Society, Bethesda, MD.

Zahuranec, B. J. 2000. Zoogeography and systematics of the lanternfishes of the genus *Nannobrachium* (Myctophidae: Lampanyctini). *Smithson. Contr. Zool. 607.*

Zenkevitch, L. 1963. *Biology of the seas of the USSR.* George Allen and Unwin, London.

30 ECOLOGY OF FISHES I: AN INTRODUCTION TO SOME BASIC ECOLOGICAL CONCEPTS

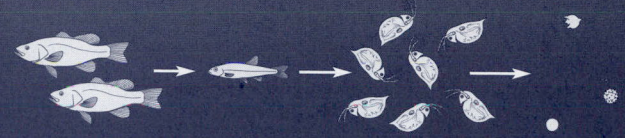

UNDERSTANDING ECOLOGY

METHODS OF ECOLOGICAL STUDY

The Abiotic Environment
The Biotic Environment

THE ABIOTIC ENVIRONMENT OF FISHES

Temperature
   Thermal Properties of Water
   Adaptations to Thermal Regimes
Density and Viscosity of Water
Hydrostatic Pressure
Dissolved Substances
   Dissolved Ions
   Dissolved Gases
Transparency of Water
   Photosynthesis and Biological Productivity

THE BIOTIC ENVIRONMENT OF FISHES

Energy: The Driving Force of Ecological Interactions
Levels of Ecological Organization
   Defining the Levels
   Ecology of Populations
   Life History Variation
   Communities
Trophic Relationships in Fish Communities: Eating and Being Eaten
   Trophic Relationships Described
   Measuring Trophic Interactions

STRUCTURE AND CONTINUITY IN ECOLOGICAL SYSTEMS

Structuring the Fish Community
   Deterministic Versus Stochastic Influences
Spatial Integrity and Continuity
Temporal Continuity

LOOKING AT THE BIG PICTURE: LANDSCAPE ECOLOGY, METAPOPULATIONS, AND MACROECOLOGY

Landscapes and Associated Metapopulations
Macroecology: The Unification of Ecology and Biogeography

Understanding the biology of fishes necessitates an understanding of the world in which they live. Fishes are defined and constrained by the physical and chemical nature of water. In this chapter, we begin our exploration of the ecological adaptations that are circumscribed by the watery realm of fishes. We will gain an appreciation for the different levels of interaction in which fishes are engaged—individual interactions with their physical environment, interactions with each other as members of the same population, and interactions with other species as members of the same community. A core element in this interaction is nutritional interrelationships—what fishes eat determines how energy and materials are passed through the community. We no longer consider aquatic ecosystems as separate and distinct from terrestrial ecosystems; what happens in one invariably influences what is going on in the other. Given the diversity of fishes in freshwater and marine communities, it should come as no surprise that they are integral to our understanding not only of the nature of aquatic ecological systems, but of the nature of the entire biosphere.

UNDERSTANDING ECOLOGY

The "Wal-Mart" phenomenon has struck the fish fauna of the continental United States. Just as individual stores and shops, reflecting local culture and characteristics—the "mom and pop" operations, so to speak—are giving way to a much more uniform merchandising of goods on a massive scale, the freshwater fishes of the United States are experiencing an unprecedented homogenization. Urban development exerts a dramatic impact on streams, transforming those with clear running water and coarsely textured beds to ones that are more turbid, with substrates composed of finer sediments (Walters et al., 2003; see Chapter 39). Regionally adapted and distinct fish faunas have become replaced with a more homogeneous assemblage that is better adapted to these degraded stream conditions. As Frank Rahel's (2000) study has shown, the widespread introduction of highly adaptable fish species that can often outcompete the native fauna has had a devastating impact on regional ichthyofaunas. More than half of the fish species found in Nevada, Utah, and Arizona are nonnative forms (Fig. 30.1).

On the other side of the globe, attempts to improve a local fishery through the introduction of Nile perch (*Lates niloticus*) into Lake Victoria have had a dramatic impact on the incredibly diverse cichlid populations in the African Great Lakes system (see Chapter 32, Going Deeper). Migratory fishes worldwide are just not coming back to home streams anymore. As mentioned in Chapter 9, eels (*Anguilla*) have virtually disappeared from the Iberian peninsula. The once-mighty runs of Pacific salmon (*Oncorhynchus*) up the Columbia River have been reduced to a few stragglers. What does this all mean? How can we understand what is going on? Perhaps the systematic and widespread replacement of highly endemic forms with more cosmopolitan species across North America is just nature's way of weeding out the losers and replacing them with hardier forms—ones that are actually more useful to us. This contention should strike us as patently absurd. But how do we answer the critics of any initiative that seeks to protect biodiversity at the expense of development? Answer them we must—and our answers had better be grounded in good science. Ecology is just the sort of science that can give us the answers we seek. The next several chapters introduce aspects of ecology that are pertinent to our understanding of the biology of fishes.

METHODS OF ECOLOGICAL STUDY

Through ecology, we seek to understand the nature of organisms and the ways they interact with each other and with the environment. Implicit in the definition of ecology is the recognition that it is a scientific discipline that seeks, through observation and experimentation, to identify patterns in the natural world. Rodriguez and Magnan (1995)

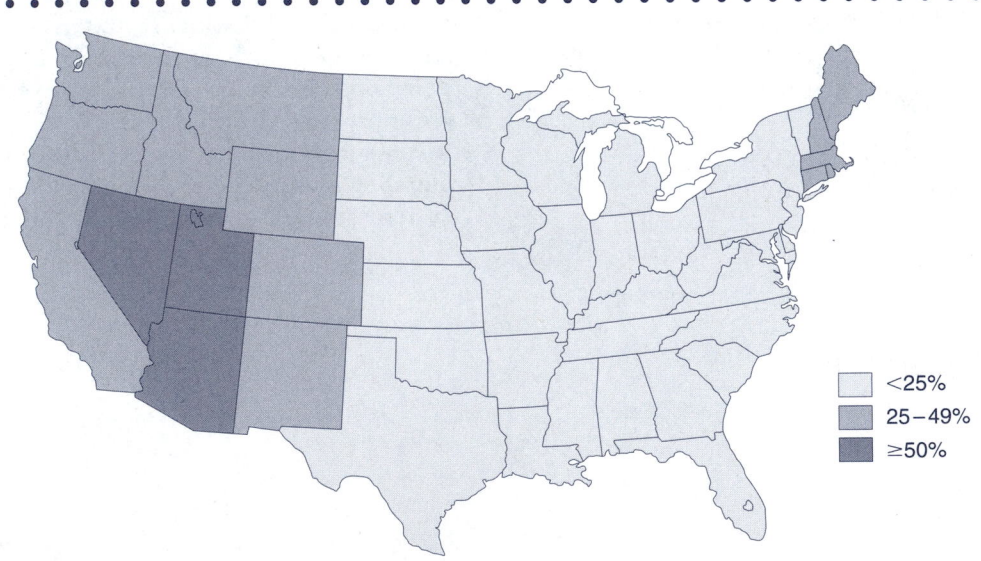

	<25%
	25–49%
	≥50%

FIGURE 30.1

Geographic distribution of introduced species across the continental United States. Degrees of shading represent the percentage of a state's current fish fauna composed of introduced species that have established reproducing populations (Source: Rahel, 2000).

have identified four methods used to study organisms in an ecological context: (1) *observational methods,* where an attempt is made to correlate a response to a number of environmental variables; (2) *"natural experiments,"* where the researcher selects comparison sites that naturally differ in the variables to be investigated; (3) *field experiments,* where the researcher actively manipulates variables of interest in the field; and (4) *laboratory experiments,* where variables can be rigorously controlled in a simplified setting. We can demonstrate these methods by applying them to a particular ecological problem—determining the optimal spawning habitat for a given species of darter, for example. After donning a wet suit, mask, and snorkel, the researcher slips into the stream to directly observe spawning activity and correlate it with the stream flow velocity measured at spawning sites (Method 1). Comparing spawning activity in a high-elevation stream, with an abundance of riffles, with that observed in a lowland stream—where pool environments predominate—demonstrates Method 2. Recognizing that sculpins may be significant predators on the eggs of darters and hence observing the effects of their removal from one of the study sites is an example of Method 3. Finally, the researcher can remove a sample of the darters and establish breeding populations in laboratory aquaria, so that variables can be much more rigidly controlled (Method 4). From these investigations, it should become apparent that the aspects of the environment that might influence the organisms of study can be divided into two categories—**abiotic** and **biotic**.

The Abiotic Environment

Ecology is a multifaceted discipline. The kinds of questions that are asked—and the kinds of experimental analyses that are applied—depend on the area of ecology under consideration. One goal of ecological study is the discernment of the relationship between organisms and their physical environment. Physical features of the environment that are known to affect the distribution and abundance of species are termed **abiotic** factors. The branch of ecology known as **physiological ecology** seeks to understand the functional adaptations of organisms to physical features of their immediate environment. (If the researcher pursuing these studies was trained in physiology, she might call this discipline environmental physiology.)

Let us consider mudskippers (Gobiidae, subfamily Oxudercinae) as a case study in physiological ecology. As mentioned in Chapter 16, mudskippers are capable of extensive terrestrial sojourns. Ishimatsu et al. (1999) have investigated the extent to which terrestrial conditions affect cardiovascular function in one mudskipper species, *Periophthalmodon schlosseri.* This species is remarkably tolerant of

atmospheric conditions, showing little change in blood O_2 or CO_2 levels, blood pressure, or heart rate. Forcible submersion of these fishes caused a rapid decline in blood O_2 saturation, however, suggesting that mudskippers are better adapted to terrestrial conditions than they are to aquatic ones (Fig. 30.2). Mudskippers will also gulp air while out on the mudflats at low tide and transport these parcels of air into J-shaped burrows, providing an atmospheric oxygen phase to supplement the available oxygen that is dissolved in the burrow water (Ishimatsu et al., 1998; Lee et al., 2005).

Another demonstration of adaptation to the physical constraints of the environment is provided by fishes that inhabit tropical coral reefs (see Chapter 33). We generally do not perceive the shallow, well-agitated waters surrounding coral reefs as being particularly prone to hypoxia (low dissolved oxygen concentration), yet shallow-water reef platforms may, in fact, become quite deficient in dissolved oxygen. This is especially true at night, when the metabolic uptake of oxygen by corals and associated animals cannot be compensated by photosynthetic activity. Measurements of hypoxia tolerance in several species of fishes that reside in these reef microhabitats show remarkable tolerance to conditions of low dissolved oxygen (Nilsson and Östlund-Nilsson, 2004; Nilsson et al., 2004).

Physiological ecology also provides insight into how the environment can affect the functional integrity of an organism—sometimes in an adverse way. Organochlorines are a category of synthetic compounds that have been widely introduced into aquatic ecosystems. Environmental physiologists have discovered that these compounds have a profound impact on aquatic organisms, because they disrupt normal endocrine function (see Chapter 26). As a consequence, growth, development, and normal metabolic processes may be affected (Colborn and Thayer, 2000). These endocrine disturbances act in a most sinister fashion, as adults may appear healthy but their offspring are severely affected by the contaminants.

The Biotic Environment

Part of the environment of a given species is, of course, other organisms. Environmental features influencing species distribution and abundance that can be attributed to the organisms themselves are classified as **biotic**. Whereas abiotic processes function in a **density-independent** manner—that is, their impact is independent of the number of organisms present—the impact magnitude of biotic factors is a function of the density of the organisms. For example, a species-specific communicable disease can be viewed as a biotic process; as the population density of the susceptible species increases, so does the rate of infection

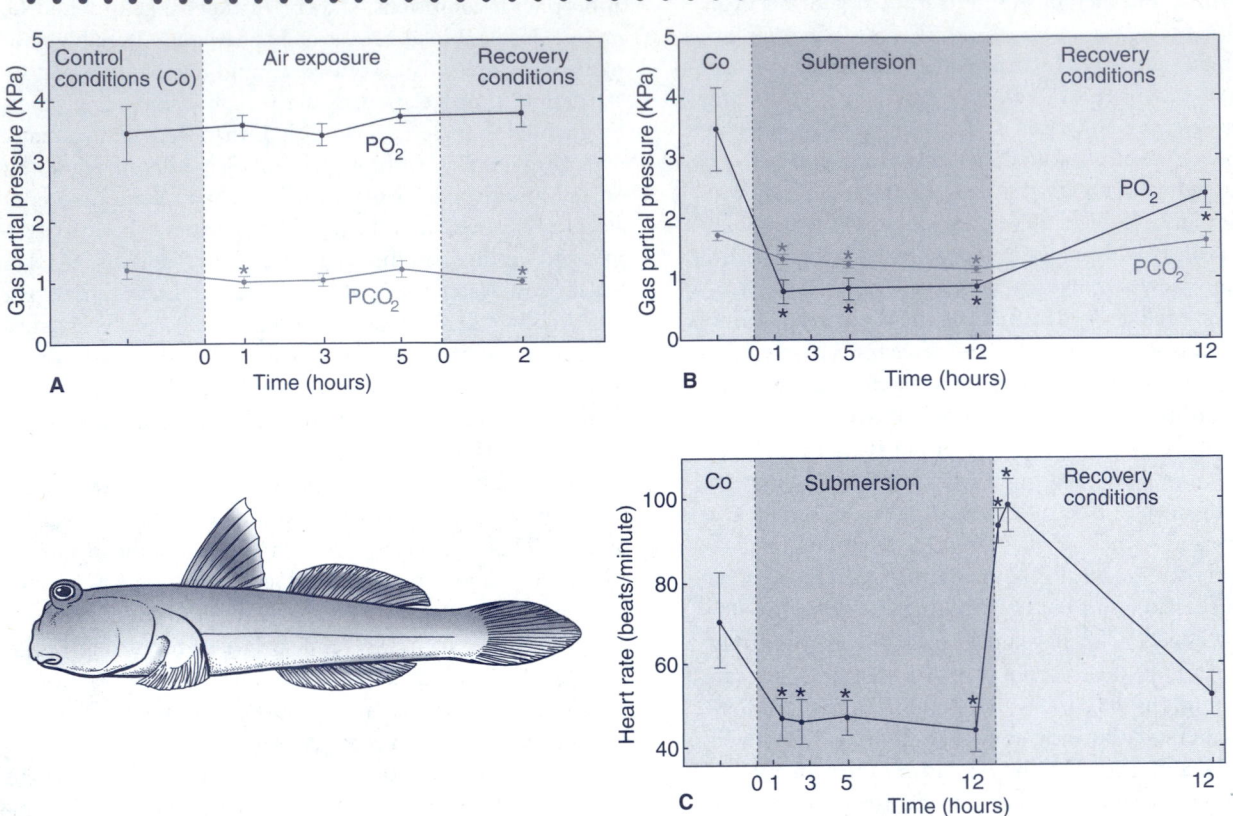

FIGURE 30.2

Physiological adaptation of the mudskipper *Periophthalmodon schlosseri* to emergent and submergent conditions: **A,** Exposure to air has little effect on the arterial partial pressure of oxygen (PO_2) or carbon dioxide (PCO_2); **B,** Forced submersion in normoxic (gas-saturated) water without access to air causes a drop in arterial PO_2 and PCO_2; **C,** Forced submersion causes a sharp drop in heart rate, followed by a rapid rebound upon emersion. *Co* indicates control conditions (Source: Ishimatsu et al., 1999).

and, hence, the mortality. This is an example of an **intra-specific** limiting factor, as the numbers of a given species affect their own future abundance and distribution. This is of serious concern to aquaculturists who rear fishes in large numbers under confined conditions (see Chapter 39). **Interspecific** limiting factors are those in which one species has an impact on the distribution and abundance of another species. Predator–prey relationships are examples of inter-specific limiting factors.

THE ABIOTIC ENVIRONMENT OF FISHES

The physical environment—which includes such things as temperature, dissolved oxygen concentration, salinity, depth, and amount of cover—is of paramount importance in governing the distribution and abundance of fishes. The adaptations of fishes to aquatic environments can best be understood by first considering some of the basic properties of the medium in which fishes have originated and diversified.

The relationships between the structure and function of fishes and the physical aspects of their aquatic environment are so fundamental that it should be intuitively obvious that the essential identity of this group of vertebrates is derived from its origins in and adaptations to water. The density of water governs the shape of the fishes that must move through it; its ionic content determines the ways in which fishes regulate the ionic composition of their own body fluids; and the amount of gases that can be dissolved in water is reflected in the structure and function of the respiratory apparatus of fishes. Yet the diversity of fishes in the aquatic realm is such that many species have escaped the confines of a watery existence. Amphibious species, such as the walking catfish, the climbing perch, and the mudskippers have become, to a greater or lesser extent, participants in terrestrial ecosystems.

Temperature

Thermal Properties of Water

Temperature is the most frequently cited example of an abiotic controlling factor. As fishes have successfully invaded most aquatic environments, it is not surprising that they cover a broad spectrum of thermal tolerances. Although the temperature of a given body of water may act as a constraint on the dispersal and distribution of fishes, water also has certain properties that facilitate the lives of fishes. The high specific heat of water (the calories required to raise the temperature of 1 g of water by 1°C) entails that water is more resistant to temperature changes than air. Aquatic organisms are therefore insulated from the sometimes extreme fluctuations in ambient temperature that prevail on land. Because water achieves its maximum density at 4°C, ice forms at the surface of bodies of water rather than at the bottom. This is especially important for freshwater fishes of higher latitudes—the ice acts as an insulating barrier between the fishes and the overlying atmosphere during the winter. Aquatic biota that inhabit lakes and ponds at high latitudes thus gain a measure of protection from the punishing extremes of temperature that terrestrial organisms must contend with during the winter months.

Surface currents flowing north or south away from polar or equatorial waters can have a dramatic effect on the dispersal and distribution of fishes. The Gulf Stream, flowing north along the Atlantic coast of North America, may transport fishes with tropical affinities, such as the tetraodontiform pufferfishes, filefishes, and triggerfishes, as far north as the Gulf of Maine (Bigelow and Schroeder, 1953).

Adaptations to Thermal Regimes

Fishes are classified as **ectotherms**, meaning that their body temperature is dependent on an external source. In most cases, this source is the sun, but in some unusual circumstances, hydrothermal vents, resulting from the tectonic activity of the earth's crust, increase the temperature of the surrounding water by several degrees (see Chapter 35). The term **poikilotherm** has also been used in reference to fishes; this means that their body temperature reflects the temperature of their surroundings. To appreciate the subtle difference between these two terms, consider the case of reptiles that engage in basking behavior to elevate and regulate their body temperature. Because they use the sun to warm their bodies, they may be considered ectotherms, but their thermoregulatory abilities exclude them from being

considered poikilotherms. Mammals are considered **endotherms,** because the heat generated from metabolic processes is used to warm the body, and **homeotherms,** because their body temperature can be regulated quite precisely. The lamnid sharks and the scombroids (tunas and their relatives) represent unusual exceptions to our thermal characterization of fishes—they use sophisticated circulatory mechanisms in conjunction with elevated levels of metabolism to generate and maintain body temperatures that are several degrees above the surrounding water (see Chapter 24).

Marine fishes at high latitudes often live in water that is below the freezing point of their tissues. The freezing point of seawater (-1.9°C) is depressed relative to that of freshwater due to the concentration of dissolved substances. Because the tissue osmoconcentrations of marine fishes fall between those of freshwater and full-strength seawater, fishes living at extremely high latitudes are faced with the threat of tissue destruction by freezing. This is counteracted by the presence in their body tissues of high-molecular-weight substances—chiefly glycoproteins—that act as a natural antifreeze by retarding the formation of ice crystals. The notothenioid fishes of the Antarctic exemplify this remarkable adaptive feature (see Chapters 24 and 35).

Just as remarkable are those fishes that are naturally exposed to high temperatures. Pupfishes (Cyprinodontidae) that live in springs and marshes of the deserts of western North America may routinely experience temperatures as high as 42°C (Soltz and Naiman, 1978; see Chapter 31). Some bony fishes—the lungfishes (see Chapter 7) and lepidogalaxiid salamanderfishes (see Chapter 11), for example—withstand the drying up of their ephemeral pool and stream habitats during the heat of the summer by resorting to aestivation.

Fishes that live in environments with a comparatively stable thermal regime and, hence, have narrow tolerances to thermal fluctuation are termed **stenothermic.** Glacial streams fed by snowmelt and the ocean depths are examples of aquatic environments where we might expect to find stenothermic fishes. Assuming an animal is stenothermic simply because it inhabits a realm of thermal stability may sometimes be presumptuous, however. The Antarctic nototheniid *Pagothenia borchgrevinki* has been demonstrated to possess sufficient plasticity in metabolic and cardiovascular functions that it is able to maintain locomotor performance when exposed to elevated temperatures (Seebacher et al., 2005).

Other aquatic environments routinely experience dramatic diurnal or seasonal fluctuations in temperature. Temperatures may fluctuate as much as 20°C in a day in the desert streams that are inhabited by the aforementioned pupfishes. Fishes that inhabit the intertidal zone may also experience broad fluctuations in temperature as the tides advance and retreat (see Chapter 34). Fishes that can tolerate such fluctuating temperatures are termed **eurythermic**.

Density and Viscosity of Water

Because water is about 800 times denser than air, successful locomotion depends on the development of effective streamlining. This is best exemplified in the fusiform shape of such pelagic (open ocean) species as tunas and billfishes (see Chapter 19, Going Deeper). Benthic (bottom-dwelling) species that live where water is moving at high velocity must also present a low profile or a smooth contour in order to maintain their position in torrents or riffles. Because the specific gravity of water (1 g/ml at 4°C) is slightly less than that of fish tissue, neutral buoyancy is achieved by the incorporation of low-density, fatty tissue or gas bladders (see Chapter 19). This relieves fishes of the necessity to expend excessive amounts of energy in maintaining their vertical position. As might be expected, the swim bladder of benthic fishes is normally reduced or completely absent.

Hydrostatic Pressure

Hydrostatic pressure increases about 1 atm for every 10 m of depth. For fishes dwelling in the mesopelagic zone, this means exposure to pressures as high as 100 atm. As might be expected, deep-dwelling fishes are well adapted to such extreme pressures. In order to avoid the problems associated with the compressibility of gases and the associated changes in buoyancy with increasing depth, the swim bladders of many deep-ocean species are infused with a large quantity of incompressible fats rather than gases (see Chapter 19). The deepest dwelling pelagic species generally have a body consisting of flabby, gelatinous tissue and skeletons that are weakly ossified. This facilitates neutral buoyancy in the absence of swim bladders. The elimination of gas bladders in the deepest dwelling marine species is not necessarily a consequence of the extremely high ambient pressures at that depth; rather, it is a consequence of their extremely reduced musculoskeletal systems—their body tissue density is so low that they are able to maintain neutral buoyancy even without the contribution of a gas bladder (Marshall, 1971).

That the loss of the swim bladder is not related to depth and the consequent pressure rise is also demonstrated by the finding that many deep benthopelagic species, such as the macrourid rattails, have well-developed swim bladders. In these cases, the length of the capillaries of the rete mirabile, which function to concentrate gases before they are secreted into the swim bladder, increases with the depth of habitat (Marshall, 1971; Pelster, 1997).

One way to appreciate the myriad ways in which deep-dwelling fishes may be adapted to extreme pressures is to consider the effects of pressure increases (hyperbaric conditions) on shallow-water forms. Experimental exposure of shallow-water fishes to elevated pressures results in increased levels of excitation, leading to tremors, convulsions, and eventually death. Elevated levels of oxygen uptake, suppression of enzymes associated with ion regulation, and an elevated heart rate have all been found to occur in shallow-water fishes on exposure to hyperbaric conditions (Sébert, 1997).

Dissolved Substances

The dissolving capabilities of water are such that chemists have termed it the "universal solvent." As it falls as rain, percolates through the ground, or runs along the surface of the earth, water collects numerous substances, including chlorides, sulfates, and carbonates of calcium, sodium, magnesium, and potassium. Silicon and phosphorus compounds as well as many organic substances are among the biologically important materials carried by water (Table 30.1). The chemistry of water is indeed complex and has a significant impact on the adaptations of fishes to aquatic environments.

Dissolved Ions

Dissolved salts, such as those encountered in the ocean or in various lakes and mineral springs, have influenced the evolution of osmoregulatory capacity in fishes (see Chapter 25). Although some species may be termed **stenohaline** due to their limited tolerance to fluctuations in the osmoconcentration of their surrounding environment, others are **euryhaline** in that they can tolerate such fluctuations. The all-time champion osmoregulator among the vertebrates is the sheepshead minnow (*Cyprinodon variegatus*). Populations of this species have been found in bodies of freshwater if the level of dissolved calcium is sufficiently elevated, and they have also become established in hypersaline lagoons, where the salinity may exceed twice that of full-strength seawater (Guillory and Johnson, 1986; Martin, 1972; Nordlie et al., 1991). In general, freshwater species regulate their salt concentration at levels much higher than that of the surrounding medium, whereas marine fishes regulate theirs at levels well below that of the surrounding medium (Table 30.1).

Dissolved Gases

Water falling as rain collects both oxygen and carbon dioxide from the atmosphere. Although the solubility of oxygen is relatively low, it dissolves in sufficient concentration that it can be extracted by the respiratory apparatus of fishes. Air contains nearly 21 percent oxygen by volume, but the amount of oxygen that can be dissolved in water is only about

TABLE 30.1   OSMOTIC AND MAJOR SOLUTE CONCENTRATIONS IN SELECTED BODIES OF WATER AND IN REPRESENTATIVE SPECIES OF TELEOSTS

Location	Average osmotic concentration (mOsm/liter)	Major Ions (mmol/l)					
		Na^+	K^+	Ca^{2+}	Mg^{2+}	Cl^-	SO_4^{2-}
Body of Water							
Average ocean	1,000	470	10	10	54	548	38
Average river	~1	~0.08	~0.01	~0.3	~0.09	~0.05	~0.08
Little Manitou Lake, Canada	2,000	780	28	14	500	660	540
Great Salt Lake, Utah	6,000	3,000	90	9	230	3,100	150
Fish Species							
Cyprinus carpio (freshwater)	274	130	2.9	2.1	1.2	125	–
Lophius piscatorius (marine)	452	180	5.1	2.8	2.5	196	2.7
Oncorhynchus tshawytscha (anadromous, freshwater phase)	288	161	0.3	2.7	1.2	114	0.5
O. tshawytscha (anadromous, marine phase)	332	179	1.0	1.0	0.9	139	0.3

Source: Evans, 1993; Holmes and Donaldson, 1969; Hutchinson, 1957; Sverdrup et al., 1947.

1/20th that of air. Whereas the proportion of oxygen in the atmosphere is fairly constant, the amount of oxygen that can be dissolved in water is inversely proportional to temperature and salinity. Freshwater can dissolve 10.23 ml/l of oxygen at 0°C, but seawater at 30 ‰ (parts per thousand, or grams/liter) holds only about 8.8 ml/l at 0°C. As might be expected, fishes that inhabit warmer waters are more tolerant of decreased levels of dissolved oxygen. A relatively diverse freshwater fish fauna can exist at a dissolved oxygen concentration as low as 3.5 ml/l, or about 55 percent saturation at 20°C. Cool-water species do better if maintained at concentrations nearer saturation (about 6.4 ml/l at 20°C).

Respiratory tissues—be they gills, lungs, or any other epithelial surface—are designed for the optimal movement of oxygen and carbon dioxide. In this context, it is important to remember that the movement of respiratory gases is exclusively a passive process, depending on the establishment and maintenance of a concentration differential across a membrane. The amount of gas that can be exchanged is thus a function of the amount of available surface area. The structure of gas exchange surfaces in fishes—especially the gills—reflects this: Relatively sluggish fishes with decreased metabolic demands for oxygen have less gill surface area than more active ones (see Chapter 24). The evolution of air breathing and the ensuing terrestriality in several groups of fishes—including perhaps those lineages that gave rise to the first tetrapods—is believed to be largely a consequence of extremely low concentrations of dissolved oxygen, as can

occur in swamps or even in some marine environments (Martin and Bridges, 1999; Packard, 1974).

Carbon dioxide has a much higher solubility in water—some 30 times greater than that of oxygen. Carbon dioxide reacts with water to form carbonic acid according to the following equilibrium reaction:

$$CO_2 + H_2O \leftrightarrow H_2CO_3 \leftrightarrow H^+ + HCO_3^- \leftrightarrow H^+ + CO_3^{2-}$$

Local biological and geological processes can affect the total concentrations of carbonate and bicarbonate. For example, compared to other types of rock, limestone is fairly soluble. Consequently, water flowing over a limestone substrate is quite "hard," as it contains higher amounts of dissolved calcium bicarbonate. As water vapor condenses and passes through the atmosphere, it picks up dissolved carbon dioxide. According to the preceding equation, the net result of this will be that pure rainwater is naturally slightly acidic, with a pH of about 5.6. The buffering capacity of bicarbonate means that seawater, on the other hand, is slightly alkaline, with a pH of about 8.2. The equilibrium reaction described in the preceding equation also exists in animal systems. Carbon dioxide produced by metabolic processes in cells and tissues is transferred to the circulatory fluids and ultimately released at gas exchange sites, such as gills and lungs (see Chapter 24). Although CO_2 is taken up and released in the form of carbon dioxide, because of its reaction with water, the bulk of carbon in the blood exists in the form of bicarbonate. In living systems, where the blood contains protein compounds such as hemoglobin to facilitate the transport of oxygen, CO_2 may also react with the NH_2 groups on these proteins to form carbamino compounds. Because of differences in the structure of the hemoglobin protein chains among the different classes of vertebrates, however, fishes and amphibians are less capable than amniotes of CO_2 transport via carbamino compounds (Randall et al., 2002).

An unfortunate consequence of industrial emissions—particularly sulfuric and nitric acids—is the increased acidification of the world's water supplies. The rainfall in heavily industrialized regions of the world, such as the northeastern United States, may have a pH as low as 4. This increased acidification results in the release of aluminum from sediments in quantities that are toxic to aquatic life. In northern Europe and in the northeastern United States and Canada, with largely granitic and insoluble strata, bodies of water are soft and thus lacking in the buffering capacity of other regions. In these places, acid precipitation exerts its greatest impact. As much as 50 percent of the species in some taxonomic groups have been eliminated from bodies of water in the Adirondacks, the Poconos and Catskills, and in southern New England. Among the most severely affected fish families are the Cyprinidae, Salmonidae, and Centrarchidae (Schindler et al., 1989).

Although nitrogen is an inert gas, its concentration in water can reach conditions of supersaturation under certain circumstances, with potentially harmful effects. Where the spillways of large dams allow air to be carried deep into plunge pools, fishes may become subject to **gas bubble disease**—a debilitating and often lethal condition caused by emboli that form in the tissues.

Transparency of Water

The transmission of light waves from different parts of the spectrum is a physical feature of water that has enormous consequences for biological processes. Although pure water is highly transparent, light absorption is rapid and differential. Only a fraction of the blue end of the spectrum penetrates to depths exceeding 1,000 m. It is these wavelengths that, when reflected back to our retinas, give oceans their characteristic blue color. Most red wavelengths are absorbed in the upper 5 m, and orange is mostly gone at 15 m. Green and yellow penetrate to about 20 m (Fig. 30.3).

An interesting consequence of this differential penetration of light of different wavelengths is the effect it has on fish coloration. Red to orange colors are frequently observed in shallow-water fishes, where they may have important adaptive functions in signaling and species recognition (see Chapters 18 and 37). The rapid extinction of these colors with distance of penetration means that they function best at close range. This may explain their widespread use among breeding fishes. Rockfishes that appear bright red when they are hauled up on the deck of a trawler appear a drab gray at the depth where they normally occur. The absence of the red-orange spectrum of color at that depth means that their bold pigmentation, which may be a consequence of their diet, actually functions as a form of cryptic coloration. Fishes living in oceanic realms that are not penetrated by light of any wavelength are uniformly black or dark brown, but they may possess highly reflective surfaces to facilitate the emission of light from photophores (see Chapter 18).

Photosynthesis and Biological Productivity

In both aquatic and terrestrial ecosystems, the energy necessary to drive biochemical reactions and, hence, to sustain life is initially made available through the process of photosynthesis. Through **photosynthesis,** the carbon dioxide that is available in the Earth's atmosphere and readily dissolved in the water is converted into energy-rich, carbon-based compounds that are then available to sustain those forms of life that cannot engage in the process of converting the energy of the sun into useful forms of chemical energy. The term **primary productivity** refers to the rate of energy accumulation through the actions of photosynthesis. The amount of organic matter thus accumulated is termed **biomass.**

Light with a wavelength between 400 and 700 nm is absorbed by the pigments of phytoplankton and other photosynthetic organisms. The **photic zone**—that portion of the aquatic realm where photosynthesis takes place—is usually restricted to the upper 50 m or so of depth, but the high intensity of light right at the water's surface is often inhibitory to photosynthesis. The red algae (**Rhodophyta**) have photosynthetic pigments that are especially well adapted for capturing and using light that penetrates deeper—the depth record for photosynthetic activity in the ocean belongs to a species of red alga that was discovered at 268 m in the Bahamas (Littler et al., 1985). In highly turbid waters—as often occur on wave-washed shores or in muddy, turbulent rivers—photosynthetic activity may be restricted to just a few upper meters.

THE BIOTIC ENVIRONMENT OF FISHES

Energy: The Driving Force of Ecological Interactions

Energy is the feature of the physical environment that makes the organization of living systems possible. In most cases, the ultimate source of energy is the sun. There are some unusual

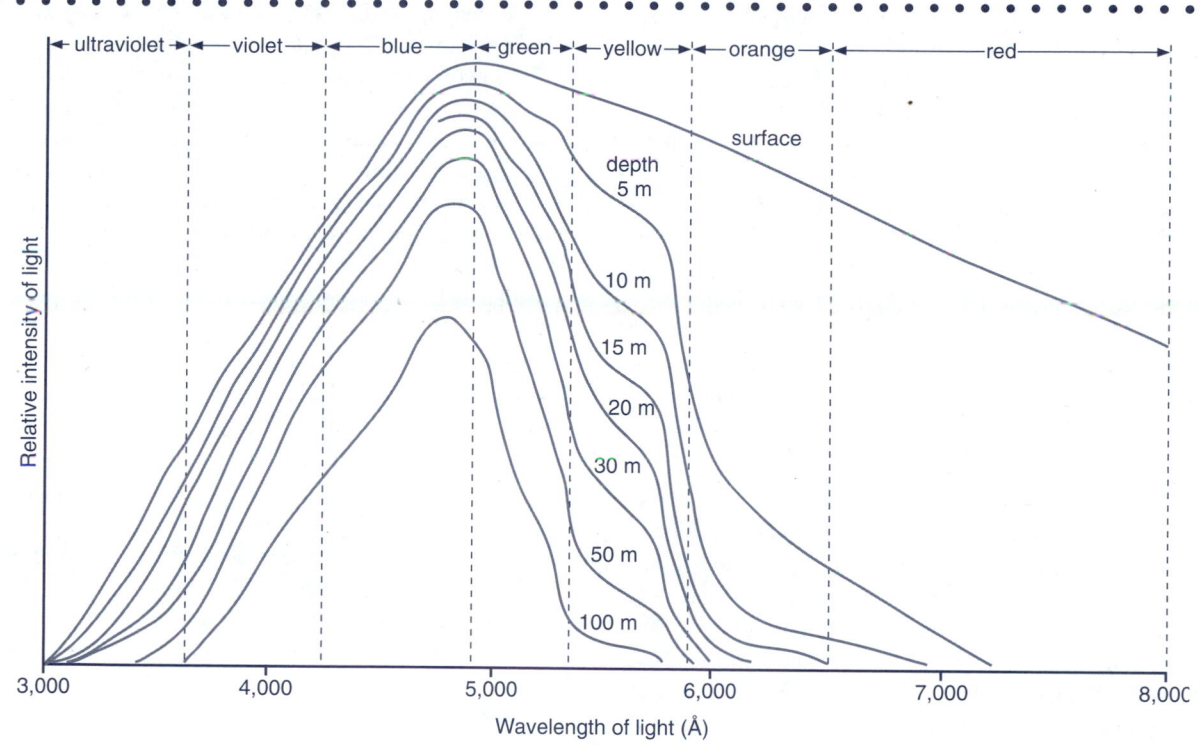

FIGURE 30.3
Selective absorption of light of different wavelengths with increasing depth. Whereas coral reef-dwelling fishes are fully visible across the color spectrum in their shallow-water habitats, an orange-colored rockfish appears gray at the depth it is normally found, owing to the attenuation of light in the red to orange wavelengths.

exceptions, of course, such as biological systems that are energized by heat arising from radioactive decay deep beneath the Earth's crust. Remarkable deep-ocean ecosystems, in which fishes play an integral part, have resulted, and we will have more to say about these in Chapter 35. **Ecological energetics** is the branch of ecology that deals with questions of energy use by living systems.

Life at all levels of organization—from the individual and its component cells through populations, communities, and ecosystems to the entire biosphere itself—is sustained by the one-way flow of energy through organisms. Whereas energy flows through life forms connected via nutritional interdependencies, the actual stuff of life—materials such as carbon, nitrogen, oxygen, and hydrogen—is constantly being recycled. Because of this, the carbon found in the proteins that make up the muscles of your arm may have, at some earlier time, existed in the swim bladder of a herring swimming in the North Atlantic Ocean.

Environments can be characterized in terms of their potential for capturing energy and making it available to organisms. As might be expected, the photosynthetic activity that

is characteristic of the photic zone of the aquatic realm imbues it with increased biological productivity. In subsequent chapters, we will explore the productivity of various aquatic environments and its impact on the distribution, abundance, and adaptations of their resident ichthyofauna. For now, a brief example of how biological productivity can influence form and function in fishes might prove instructive.

Deep-ocean predators such as the ceratioid anglerfishes (see Chapter 13) are generally small creatures, with a tissue structure that appears less developed when compared with other fishes. They adopt a "lie in wait" strategy, using bioluminescent lures and capacious mouths to attract and grab anything that comes their way (see Plates 13.11 and 35.2). A sophisticated musculoskeletal system and the associated elevated level of metabolism necessary to move about swiftly in pursuit of prey are simply not possible, given the paucity of food resources in the depths of the pelagic realm.

In the benthic realm, however, the continuous buildup of sediments causes an abundance of burrowing forms (termed **infauna**) to be available to predators; therefore, bottom-dwelling fishes are comparatively more robust. Witness, for example, the macrourid rattails; these large, active fishes, with well-developed eyes and swim bladders, are among the most common fishes that can be observed foraging over the deep-ocean bottom (see Chapter 13). Fish species that are normally associated with the poorly productive midwater realms may be attracted to the bottom because of increased feeding opportunities (Gartner et al., 1997). Deep pelagic forms with robust body structure are also found in dense concentrations in oceanic regions that are subject to localized enhancement of biological productivity (see Chapter 35, Going Deeper).

Levels of Ecological Organization

Defining the Levels

Understanding that living things are organized in a series of interacting levels under control of both abiotic and biotic processes helps to make ecological study more accessible. Ecological study is applicable at all levels of biotic organization—the individual, the population, the community, the ecosystem, and the biosphere. Interdisciplinary approaches have also provided tremendous insight into ecology. For example, through evolutionary ecology, we have come to recognize that evolutionary theory has applications not just to the individual but to all of the aforementioned levels.

Individuals of the same species that occupy roughly the same location at the same time form a **population**. Because individuals in a given population interbreed among themselves, they share a common **gene pool**. A **community** is defined as an association of interacting populations. The community and the physical environment with which it interacts form the **ecosystem**. Examples of ecosystems that we will investigate in the context of associated fish species include coral reefs, the open ocean, lakes, and rivers. All ecosystems, however, are interconnected to varying degrees. Mangrove ecosystems, for example, serve as nurseries for many species of coral reef fishes—hence, the community structure of coral reef fishes is strongly influenced by the health of neighboring mangrove forests (Mumby et al., 2004). The interconnected assemblage of ecosystems that comprises all the habitable regions of the Earth is known as the **biosphere**.

Ecology of Populations

The goal of **population ecology** is an understanding of how individuals that constitute populations interact among themselves and with the physical environment. Population ecologists study factors that affect the growth and regulation of populations or the means by which populations maintain their genetic integrity. For example, fishery biologists attempting to determine optimum stocking densities of bluegill in farm ponds are practicing population ecology.

Recently, there has been an increased recognition that the distribution of populations reflects the nonuniform distribution of habitats and that several local populations occupying a set of habitat patches may share a gene pool through the movement of individuals from one population to another. Aggregates of local populations thus form what is termed a **metapopulation**—a concept that has proven valuable in studying the dynamics of subdivided populations. Nearshore benthic marine fishes, such as gobies and blennies, are ideal candidates for metapopulation study. Although populations of these species may appear spatially discrete along a given coastline, the connectivity afforded them because they share a common marine habitat would qualify the assemblage of populations as constituting a metapopulation within a defined portion of the species range. Although the adults of these species of limited mobility may not migrate far beyond their home range, the pelagic larvae may be dispersed over great distances, so that the genetic integrity of the resultant metapopulations is maintained through larval dispersal. Molecular genetic studies (see Chapter 4) have proven especially useful in the characterization of metapopulations.

Life History Variation

Metapopulation studies may reveal more than genetic variation among populations. Fishes exhibit variation in the ways in which they adapt to local conditions and resources. Two cases of "successful" introductions of fishes into new environments during the 19th century—the carp (see Chapter 10) and the striped bass (see Chapters 16 and 39)—illustrate the adaptability of fishes to novel surroundings. Such adaptation is possible through both morphological modification (see Chapters 4, 22, and 32) and modification of the ways that fishes live—changes in food and feeding behavior, reproductive habits, and other life history traits. In some cases, changes in the environment have been documented as influencing the course of evolution by selecting for certain life history attributes that are optimal in certain environmental situations (see Bruton, 1990).

Body size is among the most important characteristics of an organism, as it can influence a number of life history attributes. **Cope's rule,** which was first applied to the evolution of mammals, states that there is a general trend toward increased body size in the evolution of a given taxon. Studies on a number of groups have suggested a broader application of this rule, yet fishes seem to be an exception. In an assessment of the size trends among North American freshwater fishes, it appears that larger species tend to occupy more basal positions in a phylogeny, whereas the more derived species consist of smaller individuals (Knouft and Page, 2003). This may be a reflection of the prevalence of freshwater habitat (specifically, smaller streams) that is available for smaller-bodied fishes—a specific example of the way in which the environment selects for certain morphological or life history attributes.

Evolutionary history also exerts an overriding influence on life history patterns. For example, although elasmobranchs and bony fishes share similar marine habitats, the former have evolved modes of reproduction that usually include internal fertilization and the production of a few large eggs, whereas the latter typically produce large numbers of small eggs that are usually fertilized externally. The modification of life history features must therefore be framed within the context of the phylogeny of the taxon in question. Even closely related species may evolve profoundly different life history patterns. The genus *Oncorhynchus* includes the Pacific salmons, which are semelparous (see Chapter 27), and the largely freshwater trouts, which are repeat spawners (iteroparous). The Atlantic salmon of the closely related genus *Salmo* is also an iteroparous species, with many adults surviving to breed again.

MacArthur and Wilson (1967) developed an early model for elucidating life history traits that remains a valuable tool for the assessment of adaptive capacity. According to their model, life histories can be conceived of as conforming to one of two strategies, termed *r*-selection and *K*-selection (Table 30.2).

The increase in numbers of a given species is described by a growth curve that is characteristic for that species and is defined by the logistic growth equation (Fig. 30.4). According to this equation, *r* refers to the net reproductive rate intrinsic to a given species and *K* refers to the number of individuals at the carrying capacity of the environment. An *r*-selected species is adapted to an environment of high unpredictability, in which the number of individuals is less likely to be limited by the available resources. Its population growth is reflected in *r*, its intrinsic rate of increase. The adaptive response to this situation is to produce the largest possible number of offspring; an abundance of resources means that parents need not invest much in the care of offspring to ensure their survival.

A *K*-selected individual, more influenced by the carrying capacity of the environment, has a life history pattern that is adapted to a stable, more predictable environment, in which competition for resources is more intense. In such cases, it is evolutionarily more prudent to produce fewer young and to invest a greater amount of energy in parental care to ensure the survival of those offspring that are produced. For fishes, this means producing fewer, larger eggs and investing greater parental care in them. Fishes such as sticklebacks and cichlids that construct nests and exhibit brooding behavior exemplify this strategy. Larger eggs give rise to larger larvae. Because mouth size, swimming capacity, and sensory abilities increase with size, larger larvae would seem to have a greater chance of survival. This implies a trade-off, because fecundity is inversely correlated with egg size in fishes (Wootton, 1992).

In terms of fecundity, cod and salmon can be viewed as polar opposites on the spectrum of life history strategies, and each demonstrates a strategy that is appropriate for its particular habitat. Although they are of approximately the same size when sexually mature, the cod produces up to a million small-diameter eggs that are broadcast into the ocean waters, whereas the salmon produces far fewer, larger eggs that are buried in a nest. For the salmon, the evolution of larger egg size and greater egg care ensures an increased survival of early life history stages in a stream environment characterized by intense competition. Egg care, however, extends only to the building of a suitable nest, the depositing of the eggs therein, and a brief period of defense against intruders. Many species of fishes are much more vigilant, tending the eggs and remaining with the larvae long after they hatch. In contrast, cod spawning in the open ocean broadcast millions of eggs and exhibit no parental care; consequently, only a

TABLE 30.2 CONTRASTING *r*-SELECTION WITH *K*-SELECTION STRATEGIES IN LIFE HISTORY TRAITS

Trait	r-strategist	K-strategist
Population growth	rapid, variable, below carrying capacity	constant, near or at carrying capacity
Competitive interactions	weak	strong
Selection favors	high fecundity, little parental care	low fecundity, much parental care
	semelparity	iteroparity
	early maturity and reproduction	late maturity and reproduction
	small body size	larger body size
	short life span	longer life span
Adaptive consequence	high productivity	high efficiency

Source: Pianka, 1970; Ricklefs and Miller, 1999.

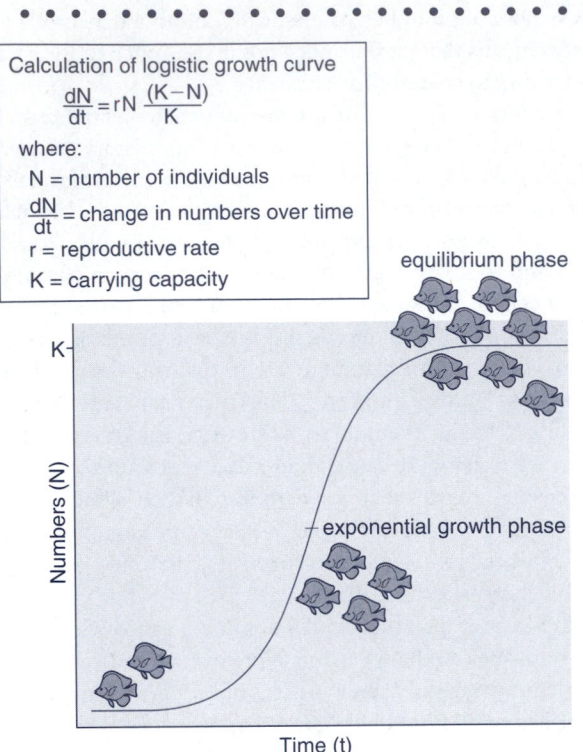

Calculation of logistic growth curve

$$\frac{dN}{dt} = rN \frac{(K-N)}{K}$$

where:

N = number of individuals

$\frac{dN}{dt}$ = change in numbers over time

r = reproductive rate

K = carrying capacity

FIGURE 30.4

Model for logistic growth. The growth curve shown has an exponential phase, where population growth is less limited by the available resources, and an equilibrium phase, where population numbers are at or near the carrying capacity.

tiny fraction of the eggs spawned produce individuals that will survive to the point that they become sexually mature. Yet this is a perfectly viable strategy, as attested by the fact that the cods have flourished for millions of years—at least until humans discovered that they were a delicious source of protein. The intense fishery pressure that preceded the collapse of populations of the Atlantic cod (*Gadus morhua;* see Chapter 39) acted as a selective force that favored the rapid evolution of individuals that matured early at smaller sizes (Olsen et al., 2004). Given the consequent impact on fecundity (larger, older females produce more eggs), it is understandable that such an artificial selection process could have contributed to the catastrophic collapse of a once-thriving fishery.

Through modification of aspects of their life histories, fishes are able to fine-tune their adaptation to local environmental conditions. Within a particular genus of fishes, variation can occur along a continuum ranging from *r*-strategists to *K*-strategists. The darters (*Etheostoma, Percina*), for example, have been shown to exhibit a range of reproductive

capacities (the measurable parameters being the number and size of eggs produced) that correlate with body size, latitude of occurrence, and spawning habits (Paine, 1990). These kinds of studies may enable us to determine why some species of a given genus are so rare in some habitats, whereas others may be exceedingly common.

As one might expect, life history strategies are not so simple that they conform exclusively to the dictates of the *r*- and *K*-selection model. A key feature that this model does not consider is the possibility of differential mortality of young versus adult stages. An alternative model, termed "bet hedging," assumes a survival rate for immature individuals that is much different from that for adults. If changes in the environment dictate a shift in the optimal number of offspring produced from year to year, selection would favor a bet-hedging strategy, in which a smaller than optimal number of offspring would be produced. This model may better reflect the adaptation of life history patterns to environmental conditions than the classical *r*- versus *K*-selection models, because it considers the consequences of spatial and temporal variation in environmental factors that influence the survival, growth, and reproduction of individuals (Stearns, 1976; Stearns and Crandall, 1984; Winemiller and Rose, 1993; Wootton, 1990).

Perhaps the most intriguing life history studies are those in which we can detect specific environmental factors that are promoting the modification of a particular life history trait. Physicochemical features of the environment, such as variations in stream flow, have been demonstrated to function as selective factors in the determination of life history traits such as fecundity and egg size (Garrett, 1982; Heins, 1979; Heins and Baker, 1987). Biotic factors such as the threat of predation also influence life histories, as documented in classic field studies by David Reznick, John Endler, and their colleagues on guppies in Trinidad streams (see Chapter 4). The adaptive coloration of males in these populations represents a delicate balance. Males rely on courtship displays and bold coloration to attract the larger, but more drably colored females. Conspicuous behavior and coloration can result in one of two outcomes: successful mating—or a quick snack for a passing pike cichlid.

Communities

The species that compose a given community are identifiable not only by their taxonomic characterization, but also by the "role" they play in the community—the habitat they occupy and the resources they exploit. This is embodied in the concept of the *ecological niche*, originally defined by Hutchinson (1957) as the activity range of each species along every dimension of the environment. Members of a community are often identified as being members of a given assemblage—a co-occurring group of taxonomically related species. Populations that are similar in their use of resources, occupying similar ecological niches, are termed *guilds*. The application of the concepts of assemblages and guilds to fish communities is explored in greater detail in Chapter 31.

Ecological communities undergo a process of development and transformation, sometimes in ways that are somewhat predictable. This is termed **ecological succession.** Depending on the community, succession can take place in a matter of a few weeks or of several thousand years. Fifteen thousand years ago, as the glaciers receded across the North American landscape, great shallow depressions were carved out and filled with meltwater. These lakes were invaded by a host of living organisms, establishing complex interrelationships based on nutritional interactions. As these assemblages acquired more structure and complexity, they may have eventually incorporated fishes that found their way to these new surroundings and flourished if the conditions were suitable.

Humans all too often play a significant role in the processes of succession. When a free-flowing river becomes dammed up, the fish fauna undergoes a dramatic transformation. The shift in physical environment elicits a change in the biota, in which a new assemblage of creatures that are better adapted to slower moving waters gradually accumulates. In North America, species such as trouts and darters, which require well-oxygenated, moving water, are replaced by species such as catfishes and sunfishes, which are better adapted to quiet waters. Beavers, of course, are well known for their damming activities in North American and Eurasian streams. Their engineering projects do not begin to approach the impact magnitude that we impose on the environment, however (see Chapter 39).

Ultimately, a point in time is reached when the community of organisms is no longer transformed through succession, and what is known as the **climax community** emerges. In a climax community, no further change in the species composition takes place, but the community is still changing as young individuals are recruited to the community and older ones perish. But the complexity that arises in ecological systems confers stability on it as well. Coral reef systems are the richest ecological assemblages on the face of the Earth. By some estimates, up to 40 percent of all described fish species are coral reef associates (Diana, 1995; Erlich, 1975). The sheer density of component species confers remarkable resilience on coral reef systems. However, that does not make them immune from the gross ecological perturbations brought about by human activity (see Chapter 39).

It should be pointed out that the concept of the climax community has its origins in terrestrial plant ecology; aquatic ecosystems may be less amenable to consideration in this context. As will be discussed later in this chapter, ecological disturbance—either natural or human-induced—

may preclude the establishment of climax communities in the classical sense.

Most of what we understand about the structure and function of communities has focused on what we may characterize as negative interactions, such as predation, competition, and disturbance effects. Hay et al. (2004) pointed out that positive interactions, such as mutualism, may also be instrumental in our understanding of the nature of the community. Hay et al. took a less restrictive view of mutualistic relationships, claiming that they need not necessarily involve obligate, coevolved interactions between partners. Among the examples provided by Hay et al. (2004) are the symbiotic relationship between coral polyps and algae that has provided the setting for the evolution of the most complex fish communities in the ocean (see Chapter 33) and the cleaner–client relationships (see Chapter 37) that are prevalent among fish communities, especially in coral reef environments.

Trophic Relationships in Fish Communities: Eating and Being Eaten

Trophic Relationships Described

Negative though it may seem, it is nevertheless an ecological fact of life that the structure and function of communities is a consequence of the sacrifice of one's own tissue for the perpetuation of another. More than any other feature, feeding and being fed on is what defines the community. Ecologists have developed a framework for understanding these nutritional interactions; it is within this context that we can define the *food chain* and *food web.*

Food Chains

Included in the range of activities of a species, as articulated in the concept of the ecological niche, is the nutritional role that a given organism might play as predator or prey. Feeding interrelationships of organisms are often depicted as **food chains,** in which the roles of component organisms as either predator or prey are indicated. A food chain is a linear depiction of nutritional relationships that shows the movement of nutrients from prey species to predators. Energetic constraints usually limit a food chain to no more than five or six links, starting at the lowest level with the primary production of photosynthetic organisms and ending with the top carnivores. The trophic relationships of fishes can be exceedingly complex. Fishes may occupy different levels of the food chain at different stages of their life histories. Planktonic larvae that feed on microscopic organisms will shift to larger prey items upon metamorphosis or maturation; this will involve modifications in oral anatomy and digestive structure and function to accommodate the dietary shift (see Chapter 23).

In the aquatic realm, fishes assume the roles of both predators and prey, and this becomes an integral feature of their ecological niche. For example, a pelagic food chain may consist of clupeoids (herrings and sardines) consuming both phyto- and zooplankton; cods consuming the clupeoids; and sharks consuming the cods. But some sharks, such as the whale shark (*Rhincodon*), basking shark (*Cetorhinus*), or "megamouth" (*Megachasma*), feed lower on the food chain by being essentially planktivorous. This enables those species to gain access to a much larger biomass of available prey, which may have contributed to their being among the largest of all fishes.

Food Webs

Perhaps the best way to envision the ecological relationships of organisms is through the construction of **food webs**. These are models that depict the trophic interrelationships of the entire community—as if a number of food chains were woven together (see Fig. 34.2). Such models make it easier to recognize the impact that changes in the abundance or distribution of one member of the food web can have on other members. The dynamics of trophic interactions are complicated by the fact that the feeding habits of fishes change as they get older and larger. Consequently, their "niche" in the food web may be only a temporary one that is defined by their life history stage.

The consequences of a loss of biodiversity on food web interactions and hence on the overall health of communities has been the focus of considerable research. Species often go extinct in a somewhat predictable manner, reflecting their differing levels of tolerance to environmental degradation. Because the surviving species tend to have greater resistance to the factors that contributed to the extinction of other community members, some researchers (Ives and Cardinale, 2004) have claimed that such "ordered extinctions" may result in a community in which resistance is conserved through the greater tolerance of the surviving species.

Regulation of Trophic Interactions

Overall control of the abundance of individual members of the food web can be perceived as being either from the "top down" or from the "bottom up." If abiotic factors—such as light, temperature, salinity, or nutrient availability—are more important, the community is said to be subject to **bottom-up regulation** (Wootton, 1990). In a pelagic community, on the other hand, it has been suggested that species composition is largely controlled by higher trophic levels—an example of **top-down regulation**. Brooks and

Dodson (1965) provided a classic demonstration of top-down regulation in their study of the impact of the planktivorous alewife (*Alosa pseudoharengus*) on the size distribution of prey species. Larger species of zooplankton were only found in those lakes that lacked populations of alewives. Blumenshine et al. (2000) also demonstrated a shift to smaller individuals in benthic communities subject to predation by fishes. In their study, however, the size distribution of prey species—in this case, benthic annelids, arthropods, and mollusks—responded to a *gradient* of predation and not just to a simple presence or absence of fish predators. Top-down regulation seems to be most important in simple food webs, where species diversity is low, few trophic levels exist, or herbivores are dominant—situations that seem to characterize most pelagic communities (Power, 1992; Strong, 1992). Yet the study by Blumenshine et al. (2000) demonstrates that top-down regulation can affect a comparatively complex, invertebrate-dominated benthic community as well. On a much larger scale, however, measurements of oceanic primary production and zooplankton density correlate with fish catch in the coastal waters of the Northeast Pacific Ocean; this reveals an ecosystem that is primarily bottom-up in the structuring of trophic linkages (Ware and Thomson, 2005).

The concept of the **trophic cascade** has been developed mainly from aquatic studies, especially those by Paine (1980, 1988) in marine ecosystems and Carpenter et al. (1985, 1987) in freshwater ecosystems. Persson (1999) defined a trophic cascade as, "the propagation of indirect mutualism between non-adjacent levels in a food chain." For example, the presence of piscivorous fishes of a certain size may affect the abundance and diversity of primary producers (algae) through the impact they may have on the intervening trophic levels that separate the primary producers from the piscivores (Fig. 30.5). In an investigation of trophic cascades in upland streams in Puerto Rico, Pringle et al. (1999) determined that the presence of fishes had a dramatic impact on the availability of fine particulate organic matter to other members of the food web. In streams where predatory fishes existed, a greater concentration of organic matter was available, owing to the decrease in processing that resulted from predation on shrimp populations. Gelwick and Matthews (1992) and Gelwick et al. (1997) have demonstrated that fishes—in this case, periphyton-grazing stoneroller minnows of the genus *Campostoma* (periphyton being substrate-attached photosynthetic organisms, such as benthic diatoms)—can also affect the size distribution of particulate organic matter in streams. Trophic cascades are best known from such studies on small-scale, more localized food webs, but they have also been observed in large-scale, open-ocean ecosystems. The overfishing that has resulted in the removal of cod as a top

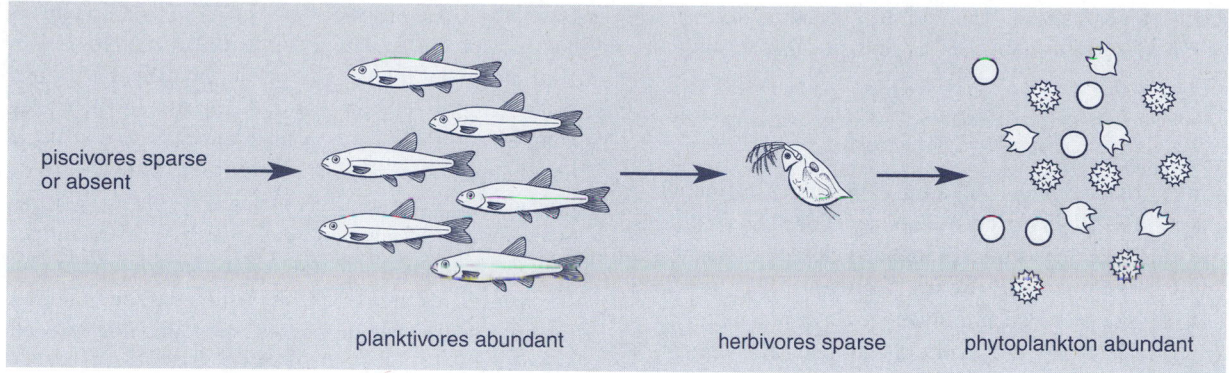

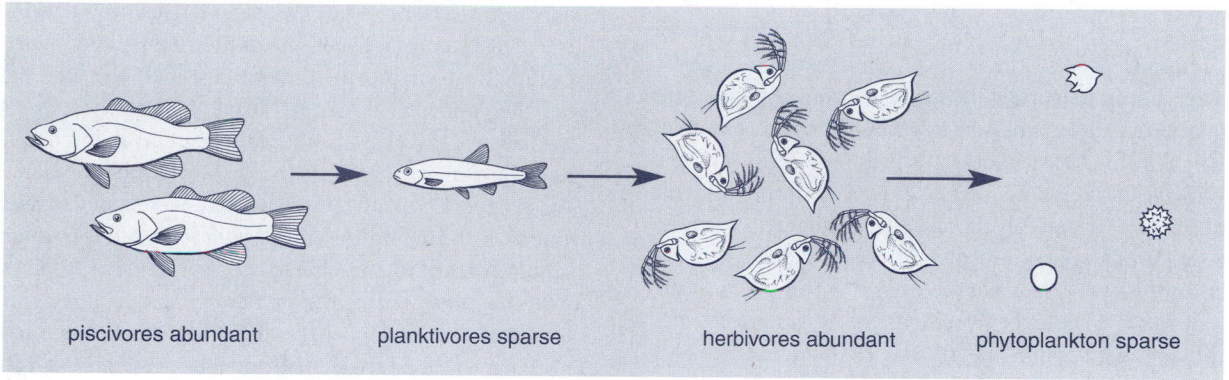

FIGURE 30.5

Example of a trophic cascade in which piscivore abundance influences the abundance of phytoplankton.
(Reproduced from Strong, 1995, with permission from McGraw-Hill Co.)

carnivore in the North Atlantic ecosystem (see Chapter 39) has dramatically altered the movement of nutrients in the trophic cascades that exist there (Frank et al., 2005).

Measuring Trophic Interactions

Interactions of Predators and Prey

The classical models for quantifying the interactions of predators and prey were developed by the population biologists A. J. Lotka and V. Volterra in the 1920s. These early mathematical models explained how populations of predators and prey oscillated with respect to each other. Lotka and Volterra also used equations developed for population growth as a basis for modeling the competition among species in a community. These models have become the basis for understanding how trophic interactions regulate population numbers in ecosystems, and they have special application to the understanding of the population dynamics of commercially exploited fish populations (see Chapter 39). The evolutionary implications of the predator–prey relationship

must also be considered. Abrams (2000) observed that prey species tend to exhibit greater adaptation to predators than predators do toward their prey.

Predator preference for prey is an aspect of behavioral adaptation that readily lends itself to quantification. Researchers have long understood that predators vary in their degree of selectivity toward different prey items, and this may be quantified in the form of the electivity index (E). The electivity index is calculated as follows:

$$E = (R_i - P_i)/(R_i + P_i),$$

with i = the prey type

R_i = the relative abundance of that particular prey in the diet

P_i = the relative abundance of prey in nature.

For example, a strongly piscivorous species, like the northern pike (*Esox lucius*), may show little preference for invertebrates (a low *E* value) and a strong preference for sunfishes (a high *E* value) as an adult, even though invertebrates may be just as abundant as sunfishes in the habitat of the pike.

Feeding Efficiency

Fishes, like other animals, will adjust their foraging tactics as a function of shifting prey availability. As a rule, feeding niches tend to broaden as food availability decreases (Diana, 1995). **Optimal foraging theory** postulates that animals will feed in a manner that minimizes their vulnerability to predation, while at the same time maximizing their nutritional intake (see Chapters 32 and 37). Optimal foraging means maximizing one's caloric intake while minimizing one's energy expenditure, and fishes have been shown to be remarkably versatile in this respect. This may mean electing to feed only at certain times of the day, or choosing larger prey items over smaller ones when both are available. J. R. Brett, in his classic work on the energetics of sockeye salmon (*Oncorhynchus nerka*), demonstrated that growth is optimized by a daily cycle of vertical migration, which brings the young sockeye to the surface of the lake to feed at dusk and dawn. When not feeding, the salmon conserve energy and thus maximize growth potential by remaining at depths below the thermocline (Brett, 1971). The ecological application of certain behaviors is seen in another example, that of longnose dace (*Rhinichthys cataractae*). Whereas most minnow species are diurnal or crepuscular feeders, dace have shifted to nocturnal foraging as a means of optimizing their prey consumption and avoiding salmonid predators and cyprinid competitors (Culp, 1989).

It is important to recognize that for many species of fishes, resource exploitation is a social activity (see Chapter 37). Resources usually are not evenly distributed across the environment, but occur in patches. The quality of a given patch of resource will vary depending on the number of consumers that have discovered it and are in the process of exploiting it. Productive patches may also attract higher order predators, which render these locations less attractive to consumers. Individuals will distribute themselves in accordance with the availability of resources in what is known as the **ideal free distribution**. (See Chapter 37 for a discussion of the behavioral ramifications of optimal foraging theory and ideal free distribution.)

Stable Isotope Analysis

Trophic relationships have recently been quantified with an unprecedented degree of precision through the application of a novel investigative technique—**stable isotope analysis**. The measurement of naturally occurring stable isotopes, particularly $\delta^{13}C$ and $\delta^{15}N$, in organisms has become a valuable tool in physiological and ecological studies. Within a given food chain, the abundance of $\delta^{15}N$ increases in predators by about 3 percent relative to their prey. The abundance of $\delta^{13}C$ shows little enrichment as the trophic level increases. Therefore, $\delta^{15}N$ is useful to define the trophic level of an organism, whereas $\delta^{13}C$ may indicate sources of production (Harrigan et al., 1989; VanderZanden et al., 1997). Although whole-animal isotope content can be measured in fishes, more reliable results can be obtained from the analysis of specific tissues, such as white muscle (Pinnegar and Polunin, 1999).

In a trophic context, anadromous salmon function to deliver nutrients of a marine origin to freshwater and terrestrial ecosystems (see Jonsson and Jonsson, 2003), and stable isotope analysis has proven to be valuable in demonstrating the connectivity of terrestrial and aquatic ecosystems. Szepanski et al. (1999) and Darimont and Reimchen (2002) have documented this in the diets of wolf populations in Alaska and British Columbia. They were able to assess the extent to which different populations fed on migrating salmon owing to the characteristic marine isotopic signature that appeared in the predator's hair or bone collagen. Collagen is a useful source in trophic studies because it is primarily derived from the protein of prey and has a characteristically long metabolic turnover rate.

Analysis of $\delta^{15}N$ content in sediments of Alaskan lakes supporting runs of sockeye salmon (*Oncorhynchus nerka*) has been used to demonstrate significant fluctuations in population abundance over the past 300 years. Studies by Finney et al. (2000) have revealed a strong correlation between run strength and temperature—warming trends are reflected in increased concentrations of sedimentary $\delta^{15}N$, pointing to a greater number of salmon returning to spawn at that time. This sedimentary record also recorded the impact of the increased salmon harvest in the past century.

Beaudoin et al. (1999) demonstrated the value of stable isotopic analysis in the corroboration of the electivity studies previously discussed. Their studies showed that isotopic signatures could determine the extent to which pike fed on invertebrates. In one of the most intuitive applications of stable isotope analysis yet published, Carpenter et al. (2003) have reconstructed the life history of a Late Cretaceous fish (†*Vorhisia vulpes*) from fossil otoliths. They determined that this species apparently was semelparous—after hatching in brackish waters, these fishes migrated to fully marine habitats, where they spent three years before returning to the estuary to spawn and die.

STRUCTURE AND CONTINUITY IN ECOLOGICAL SYSTEMS

Structuring the Fish Community

Because the majority of fishes are carnivores, we might assume that they are especially important as top-down regulators of community composition. Fish predation can alter the scope of the habitat available to prey species—the prey often including smaller individuals of those same predator species. The concept of a **keystone species**—a species whose presence or absence profoundly influences the composition of the community—is applicable to fish communities, especially in systems with comparatively well-defined boundaries, such as lakes and streams. For example, the introduction of the piscivorous Sacramento pikeminnow (*Ptychocheilus grandis*) into the Eel River in northern California caused significant shifts in microhabitat preferences of the juveniles of several co-occurring species (Brown and Moyle, 1991).

Deterministic Versus Stochastic Influences

The structuring of a fish community may be the result of a complex interaction of physical and biological factors, including predation. Large and complex communities, like the one observed in the Orinoco floodplain (see Chapter 31), may demonstrate a **deterministic** structure, where the community composition develops according to defined and predictable patterns and a strong correlation with environmental factors is evident. In the case of the fish community of the Orinoco floodplain, there is a clear connection between regional hydrological and geomorphological phenomena and the composition of the fish community.

Studies on ecological relationships in communities defined by a high degree of physical instability—such as those found in the headwaters of small streams or rocky intertidal shores—have produced conflicting schools of thought as to how these communities achieve their characteristic composition, if in fact it is ever possible to achieve such a thing. Grossman (1982; Grossman et al., 1990, 1998) and others have offered an alternative to ecological determinism, proposing that some fish communities are **stochastic,** in the sense that chance variation in the physical environment results in large variations in species composition and abundance. Stochastic community structure most likely occurs where population numbers are small (Ricklefs and Miller, 1999). Communities under the influence of stochastic events never reach an equilibrium state (in the sense of the climax community discussed earlier) because they are constantly experiencing perturbation. Consequently, competitive exclusion and resource limitation rarely figure in the

structuring of such communities. These bold assertions have not gone unchallenged, however (see Herbold, 1984).

What is apparent after decades of study is that no single variable or process can account for the diversity and complexity that is encountered in the structure of fish communities. Recent syntheses of ideas on the regulation of populations, reinforced with powerful statistical methodologies, have acknowledged that both stochastic and deterministic processes may be simultaneously influencing population structure, with effects that vary depending on the age of the organism or the developmental state of the population (Bjørnstad and Grenfell, 2001). The age-structured dynamics of cod populations (see Chapter 39) are a good case in point. The subtle interplay of a number of variables in both time and space is something we are only beginning to appreciate.

Spatial Integrity and Continuity

A potentially valuable approach to the ecological study of ecosystems, particularly in the conception of trophic interrelationships, is to view aquatic ecosystems as existing along a continuum based on the degree of permanence of habitat. Wellborn et al. (1996) have constructed a model habitat gradient, in which freshwater community structure is determined by the coupled effects of physical and biotic factors. In their model (Fig. 30.6), two major transitions are perceived—one dividing ephemeral environments that periodically dry up from those that are more permanent, and one between more or less permanent habitats that are distinguished by the presence or absence of fishes as significant predators. Community membership along this gradient is largely determined by the capacity for resident species to withstand the physical stress associated with ephemeral environments. **Vernal pools** are a good case in point (see Colburn, 2004). These are aquatic habitats created during rainy seasons. They may last anywhere from a few weeks to several months before eventually drying up. Although many invertebrate organisms have become well adapted to the rigors of such an environment, only a few species of fishes—such as the so-called "annual fishes" of the family Cyprinodontidae—are associated with such an environment (see Chapter 32).

As aquatic environments acquire a greater degree of permanence, complex community structure, characterized by a greater degree of regulation via biotic interactions—especially involving fishes—becomes the norm. The importance of recognizing that organisms and their associated habitats exist and interact along a continuum is embodied in the **river continuum concept** discussed in Chapter 31.

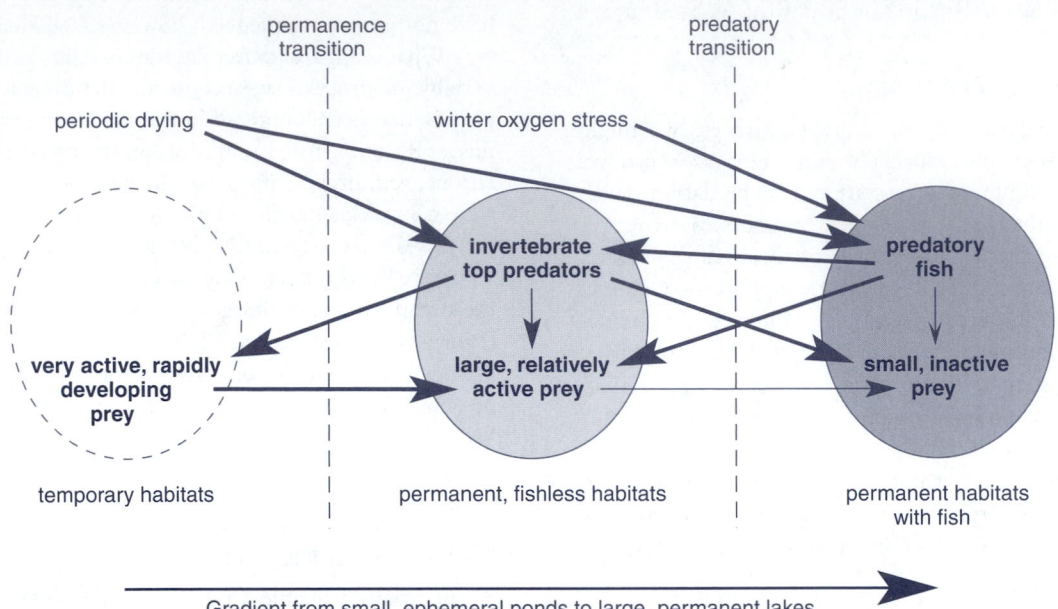

FIGURE 30.6
Model of mechanisms generating community structure along the freshwater habitat gradient. Arrows indicate the direction of negative effects. Bold arrows indicate very strong effects that act to constrain the distribution of affected species. Thinner arrows indicate weaker trophic interactions that do not prevent the coexistence of interacting species. Major transition points indicated are those between ephemeral habitats and permanent ones, and, within permanent habitats, the transition from fishless habitats to those with fishes as the dominant predators (Source: Wellborn et al., 1996).

Although marine ecosystems may also demonstrate a continuity of physical and biotic features, resulting in a gradient of habitats, they are much more insulated from the seasonality that results in the intermittent availability of water. Tidal fluctuation of the sea level, however, may impose severe physical constraints on the inhabitants of some nearshore environments (see Chapter 34).

Temporal Continuity

One of the benefits of a more integrated approach to investigating ecosystems is the emergence of the field of study known as **historical ecology**. This discipline recognizes that the structure and function of modern ecosystems are best understood in the context of their origins—their history, so to speak. Two levels of application of historical ecology have emerged, however. One seeks to understand ecosystems from the perspective of the phylogenetic and biogeographic histories of participants; the other focuses on much more recent phenomena, particularly human-induced impact on ecosystems.

The techniques of historical ecology, as articulated by Brooks (1985), have been applied in studies of the distribution of fish species and their community composition. Gorman (1992) described historical ecology as, "a research program melding phylogenetic systematics, biogeography, and ecology." Historical ecology seeks to explain present-day distribution patterns and ecological associations by combining phylogenetic studies with geological investigations. Having similar goals to evolutionary ecology, historical ecology emphasizes phylogenetic histories rather than ecological processes as a means of explaining species distribution patterns (Gorman, 1992; Mayden, 1992).

The term *historical ecology* has also been defined as historically informed environmental analysis that permits us to appreciate how populations, communities, or ecosystems have become transformed as a consequence of the spread of humanity. In this sense, historical ecology is viewed as an integration of the social and natural sciences, bridging the disciplines of anthropology and ecology (see Balée, 1998). In Chapter 39, we explore in more detail the dramatic impact that humans have had on fishes through habitat alteration and the exploitation of commercially important populations. Through an understanding of the history of the exploitation of the Pacific salmons, for example, we can better appreciate the role they currently assume in stream communities of northwestern North America.

LOOKING AT THE BIG PICTURE: LANDSCAPE ECOLOGY, METAPOPULATIONS, AND MACROECOLOGY

Landscapes and Associated Metapopulations

Ecologists have only lately come to appreciate something that artists have understood for generations—the landscape. The artist sets up her easel on a hillside, gazes at the view beyond, and appreciates and expresses it in a work that reflects the balance and integration of each of the parts rendered—the brook, the trees, the meadows. The notion that the landscape was a natural object of scientific study first emerged in Europe after World War II (the need to repair and restore the devastation obviously was a significant impetus). It was not until the early 1980s that the landscape approach took root in the United States. In a seminal workshop sponsored by the National Science Foundation and conducted in 1983, the goals of landscape ecology were articulated. In the mission statement, it was recommended "that landscape ecology consider the development and dynamics of spatial heterogeneity, spatial and temporal interactions and exchanges across heterogeneous landscapes, influences of spatial heterogeneity on biotic and abiotic processes, and the management of spatial heterogeneity" (Barrett and Peles, 1999, p. 2).

The landscape has become recognized as integral to our understanding of ecological processes that operate in the context of interacting hierarchies. The set of interacting ecosystems that might constitute a given landscape does not exist in a uniform, homogeneous manner over large areas. Nor does it exist in a uniform manner through time. Patches of ecosystem have temporal and spatial boundaries. The boundary zones of ecosystems are termed ecotones. In the landscape perspective, these boundaries may be fluid, expanding and contracting as the neighboring ecosystems change their character. Within the context of the landscape, local populations become conjoined and fragmented as the landscape changes.

Recall that we recognized that populations exist in larger aggregates, termed metapopulations, and that these metapopulations exhibit varying degrees of gene flow depending on the extent of migration into and out of their constituent population subunits. One real concern is what happens to local populations when natural landscapes become fragmented through human activity. The dynamic of the metapopulation is disrupted, ultimately leading to declines in species abundance and diversity (Gonzalez et al,. 1998). As one might expect, the concepts of landscapes and metapopulations have been most extensively applied to terrestrial ecosystems. Their application to aquatic systems has not been fully realized. Landscapes, however, are invariably identified and characterized in terms of a given watershed—the land surrounding a given body of water, into which the water running off that land flows. Here lies the connection between terrestrial and aquatic ecosystems: The landscape encompasses them both. Ultimately, we may see an increased application of these concepts to the aquatic component. The watershed, for example, plays a dominant role in the modulation of aquatic responses to terrestrial disturbance (Carigan and Steedman, 2000).

Macroecology: The Unification of Ecology and Biogeography

We have deliberately placed the chapter on the zoogeography of fishes before this one, so that you might have a bit of insight into the geographical distribution of animals before considering its ecological implications. These implications have recently been articulated in the concept of macroecology, first advanced by James Brown and Brian Maurer (1989) and elaborated upon in later books by Brown (1995) and Maurer (1999). Macroecology seeks to connect the composition of local assemblages of species to large-scale phenomena. Macroecology is, in a sense, an amplification of traditional community ecology, as it seeks to understand communities in the context of the mechanisms that determine their structure. It examines how communities respond to changes in their environments and seeks to understand the broad causal mechanisms underlying the division of food and space in continental biota. Macroecological approaches may be instrumental in understanding the dynamics of fish assemblage structure, including shifts in geographic range, body size, and the impact of the introduction of exotic species (Matthews, 1998).

One research technique that has potential application to macroecological study is geographic information systems (GIS). GIS allows the integration of massive quantities of geographic, distribution, and environmental data by generating databases constructed along certain themes. Each theme is perceived as a layer and, through GIS, these layers are superimposed to permit spatial analysis and the detection of correlations. With GIS, the attributes of a given location can be identified, locations that meet specified criteria can be determined, and predictions can be made concerning the impact of changes in the attributes of a given location. Isaak and Hubert (1997) provided an example of a fisheries application. Using GIS, a biologist with access to the bathymetric contours of a reservoir, stored as one layer of data, and telemetry data on largemouth bass location, stored as another layer, could rapidly map the horizontal and vertical distribution of the fishes, from which the habitat consequences of a reservoir drawdown could be predicted.

The complexities that are inherent in macroecological approaches may be much easier to handle with the opportunity afforded by GIS to visualize layer upon layer of environmental and geographic data for a given region. GIS and other tools of technology have given us the means of greatly expanding the breadth and depth of our ecological knowledge. The only limitation seems to be our commitment to face up to the consequences of our disturbing the natural order of things.

KEY POINTS AND CONNECTIONS

- The factors that influence the abundance and distribution of animals can be classified as either abiotic or biotic. Abiotic factors include physical features of the environment, which affect organisms in a density-independent manner. The discipline of physiological ecology is concerned with the ways in which organisms functionally adapt to the abiotic features of their environment. Biotic factors, on the other hand, vary in their impact in relation to the number of organisms present and the ways in which they interact.

- Abiotic factors include such variables as temperature, dissolved oxygen concentration, salinity, depth, and amount of cover. These physical attributes of the aquatic realm play an essential role in governing the distribution and abundance of fishes.

The process of osmoregulation, an example demonstrating the adaptation of fishes to their physical environment, is discussed in Chapter 25.

- Living things are organized in a series of interacting levels under the control of both abiotic and biotic processes; these levels typically include individuals, populations, metapopulations, communities, ecosystems, and the biosphere. Fishes demonstrate the classic divergence of life histories into *r*- and *K*-strategies. The life histories of fishes reflect the adaptation of populations to local conditions, however — a testament to the adaptive plasticity often seen in fishes. In some cases, the feature of the environment that is operating as a selective factor is discernible in its effect on the organism and its life history.

Morphological plasticity of the type known as character displacement, a demonstration of adaptive variation in feeding strategy, is discussed in Chapter 4.

- Ecological interactions among organisms are defined by the movement of energy in a one-way path through the communities that constitute the ecosystem and the recycling of the materials necessary for biological construction. The nutritional interactions of communities are defined as an interconnecting network of food chains, termed a food web.

The range of feeding adaptations that has enabled fishes to occupy such a diversity of ecological niches in food webs is discussed in Chapter 23.

- Fishes are integral to our understanding of the trophic structure of communities. Fishes participate as members of food chains; these food chains are cross-linked to form food webs. Stable isotope analysis has proven to be a valuable method for quantifying the nutritional dynamics of communities.

Our focus on nutritional dynamics here concerns the transfer of energy and materials among the members of a community; the assessment of the role of nutrition in the metabolism and growth of individuals is discussed in Chapter 23.

- Investigations of large-scale processes in ecology — so-called macroecological processes — seek to understand the metapopulation structure of species in the context of the large-scale phenomena operating at the level of the ecosystem. An appreciation for such large-scale processes is evident in our embracing the construct of landscape ecology, in which aquatic ecosystems are viewed as interacting with terrestrial ecosystems in the structure and function of watersheds.

BUILDING AN ICHTHYOLOGY LIBRARY

Four useful texts with a focus on the ecology of fishes:

Diana, J. S. 1995. *Biology and ecology of fishes.* Biological Sciences Press, Cooper, Carmel, IN.
Gerking, S. D. 1994. *Feeding ecology of fish.* Academic Press, San Diego.
Jobling, M. 1995. *Environmental biology of fishes.* Chapman and Hall, London.
Wootton, R. J. 1990. *Ecology of teleost fishes.* Chapman and Hall, London.

Three sources that will prove helpful in understanding "the big picture":

Brown, J. H. 1995. *Macroecology.* University of Chicago Press, Chicago.
Gaston, K., and T. Blackburn. 2000. *Pattern and process in macroecology.* Blackwell Science, Cambridge, UK.
Maurer, B. A. 1999. *Untangling ecological complexity: The macroscopic perspective.* University of Chicago Press, Chicago.

REFERENCES

Abrams, P. A. 2000. The evolution of predator–prey interactions: Theory and evidence. *Ann. Rev. Ecol. Syst. 31:* 79–105.
Balée, W. L. (Ed.). 1998. *Advances in historical ecology.* Columbia University Press, New York.

Barrett, G. W., and J. D. Peles. 1999. Small mammal ecology: A landscape perspective, pp. 1–8. In *Landscape ecology of small mammals*, G. W. Barrett and J. D. Peles (Eds.). Springer Verlag, New York.

Beaudoin, C. P., W. M. Tonn, E. E. Prepas, and L. I. Wassenaar. 1999. Individual specialization and trophic adaptability of northern pike (*Esox lucius*): An isotope and dietary analysis. *Oecologia 120:* 386–396.

Bigelow, H. B., and W. C. Schroeder. 1953. Fishes of the Gulf of Maine. *U.S. Fish Wildlife Serv. Fish Bull. 74.*

Bjørnstad, O. N., and B. T. Grenfell. 2001. Noisy clockwork: Time series analysis of population fluctuations in animals. *Science 293:* 638–643.

Blumenshine, S. C., D. M. Lodge, and J. R. Hodgson. 2000. Gradient of fish predation alters body size distribution of lake benthos. *Ecology 81*(2): 374–386.

Brett, J. R. 1971. Energetic response of salmon to temperature. A study of some thermal relations in the physiology and freshwater ecology of sockeye salmon (*Oncorhynchus nerka*). *Amer. Zool. 11:* 99–113.

Brooks, D. R. 1985. Historical ecology: A new approach to studying the evolution of ecological associations. *Ann. Missouri Bot. Gard. 72:* 660–680.

Brooks, J. L., and S. I. Dodson. 1965. Predation, body size, and composition of plankton. *Science 150:* 28–35.

Brown, J. H. 1995. *Macroecology.* University of Chicago Press, Chicago.

———, and B. A. Maurer. 1989. Macroecology: The division of food and space among species on continents. *Science 143:* 1145–1150.

Brown, L. R., and P. B. Moyle. 1991. Changes in habitat and microhabitat partitioning within an assemblage of stream fishes in response to predation by Sacramento squawfish (*Ptychocheilus grandis*). *Can. J. Fish. Aquat. Sci. 48:* 849–856.

Bruton, M. N. (Ed.). 1990. Alternative life history styles in fishes. Kluwer, Hingham, MA.

Carigan, R., and R. J. Steedman. 2000. Impacts of major watershed perturbations on aquatic ecosystems. *Can. J. Fish. Aquat. Sci. 57*(Suppl. 2): 1–4.

Carpenter, S. J., J. M. Erickson, and F. D. Holland, Jr. 2003. Migration of a Late Cretaceous fish. *Nature 423:* 70–74.

Carpenter, S. R., J. F. Kitchell, and J. R. Hodgson. 1985. Cascading trophic interactions and lake productivity. *Bioscience 35:* 634–639.

———, ———, ———, P. A. Cochran, J. J. Elser, M. M. Elser, D. M. Lodge, D. Kretchmer, X. He, and C. N. Ende. 1987. Regulation of lake primary productivity by food web structure. *Ecology 68:* 1863–1876.

Colborn, T., and K. Thayer. 2000. Aquatic ecosystems: Harbingers of endocrine disruption. *Ecol. Appl. 10*(4): 949–957.

Colburn, E. A. 2004. *Vernal pools: Natural history and conservation.* McDonald & Woodward, Granville, OH.

Culp, J. M. 1989. Nocturnally constrained foraging of a lotic minnow (*Rhinichthys cataractae*). *Can. J. Zool. 67:* 2008–2012.

Darimont, C. T., and T. E. Reimchen. 2002. Intra-hair stable isotope analysis implies seasonal shift to salmon in grey wolf diet. *Can. J. Zool. 80:* 1638–1642.

Diana, J. S. 1995. *Biology and ecology of fishes.* Biological Sciences Press, Cooper, Carmel, IN.

Erlich, P. R. 1975. The population biology of coral reef fishes. *Ann. Rev. Ecol. Syst. 6:* 211–247.

Evans, D. H. 1993. *The physiology of fishes.* CRC Press, Boca Raton, FL.

Finney, B. P., I. Gregory-Eaves, J. Sweetman, M. S. V. Douglas, and J. P. Smol. 2000. Impacts of climatic change and fishing on Pacific salmon abundance over the past 300 years. *Science 290:* 795–799.

Frank, K. T., B. Petrie, J. S. Choi, and W. C. Leggett. 2005. Trophic cascades in a formerly cod-dominated ecosystem. *Science 308:* 1621–1623.

Garrett, G. P. 1982. Variation in the reproductive traits of the Pecos pupfish *Cyprinodon pecoensis. Amer. Midl. Nat. 108:* 355–363.

Gartner, J. V., Jr., R. E. Crabtree, and K. J. Sulak. 1997. Feeding at depth, pp. 115–193. In *Deep-sea fishes*, D. J. Randall and A. P. Farrell (Eds.). Academic Press, San Diego.

Gelwick, F. P., and W. J. Matthews. 1992. Effects of an algivorous minnow on temperate stream ecosystem properties. *Ecology 73:* 1630–1645.

———, M. S. Stock, and W. J. Matthews. 1997. Effects of fish, water depth, and predation risk on patch dynamics in a north-temperate river system. *Oikos 80:* 382–389.

Gonzalez, A., J. H. Lawton, F. S. Gilbert, T. M. Blackburn, and I. Evans-Freke. 1998. Metapopulation dynamics, abundance, and distribution in a microecosystem. *Science 281:* 2045–2047.

Gorman, O. T. 1992. Evolutionary ecology and historical ecology: Assembly, structure, and organization of stream fish communities, pp. 659–688. In *Systematics, historical ecology, and North American freshwater fishes*, R. L. Mayden, (Ed.). Stanford University Press, Stanford, CA.

Grossman, G. D. 1982. Dynamics and organization of a rocky intertidal fish assemblage: The persistence and resilience of taxocene structure. *Am. Nat. 119:* 611–637.

———, J. R. Dowd, and M. Crawford. 1990. Assemblage stability in stream fishes: A review. *Env. Manag. 14*(5): 661–671.

———, R. E. Ratajczak, Jr., M. Crawford, and M. C. Freeman. 1998. Assemblage organization in stream fishes: Effects of environmental variation and interspecific interactions. *Ecol. Monogr. 68:* 395–420.

Guillory, V., and W. E. Johnson. 1986. Habitat, conservation status, and zoogeography of the cyprinodont fish *Cyprinodon variegatus hubbsi* (Carr). *Southw. Nat. 31:* 95–100.

Harrigan, P., J. C. Zieman, and S. A. Macko. 1989. The base of nutritional support for the gray snapper (*Lutjanus griseus*): An evaluation based on a combined stomach content and stable isotope analysis. *Bull. Mar. Sci. 44:* 65–77.

Hay, M. E., J. D. Parker, D. E. Burkepile, C. C. Caudill, A. E. Wilson, Z. P. Hallinan, and A. D. Chequer. 2004. Mutualisms and aquatic community structure: The enemy of my enemy is my friend. *Ann. Rev. Ecol. Evol. Syst. 35:* 175–197.

Heins, D. C. 1979. *A comparative life history of a closely related group of minnows (*Notropis: Cyprinidae*) inhabiting streams of the Gulf Coastal Plain.* PhD Dissertation, Tulane University, New Orleans.

———, and J. A. Baker. 1987. Analysis of factors associated with intraspecific variation in propagule size of a stream-dwelling fish, pp. 223–231. In *Community and evolutionary ecology of North American stream fishes*, W. J. Matthews and D. C. Heins (Eds.). University of Oklahoma Press, Norman.

Herbold, B. 1984. Structure of an Indiana stream fish association: Choosing an appropriate model. *Am. Nat. 124:* 561–572.

Holmes, W. N., and E. M. Donaldson. 1969. The body compartments and the distribution of electrolytes, Part 1, pp. 1–89. In *Fish physiology* (Vol. I), W. S. Hoar and D. J. Randall (Eds.). Academic Press, New York.

Hutchinson, G. 1957. *A treatise on limnology.* Wiley, New York.

Isaak, D. J., and W. A. Hubert. 1997. Integrating new technologies into fisheries science: The application of geographic information systems. *Fisheries 22*(1): 6–10.

Ishimatsu, A., Y. Hishida, T. Takita, T. Kanda, S. Oikawa, T. Takeda, and K. H. Khoo. 1998. Mudskippers store air in their burrows. *Nature 391:* 237–238.

———, N. M. Aguilar, K. Ogawa, Y. Hishida, T. Takeda, S. Oikawa, T. Kanda, and K. K. Huat. 1999. Arterial blood gas levels and cardiovascular function during varying environmental conditions in a mudskipper, *Periophthalmodon schlosseri*. *J. Exp. Biol. 202:* 1753–1762.

Ives, A. R., and B. J. Cardinale. 2004. Food-web interactions govern the resistance of communities after non-random extinctions. *Nature 429:* 174–177.

Jonsson, B., and N. Jonsson. 2003. Migratory Atlantic salmon as vectors for the transfer of energy and nutrients between freshwater and marine environments. *Freshw. Biol. 48:* 21–27.

Knouft, J. H., and L. M. Page. 2003. The evolution of body size in extant groups of North American freshwater fishes: Speciation, size distributions, and Cope's rule. *Am. Nat. 161:* 413–421.

Lee, H. J., C. A. Martinez, K. J. Hertzberg, A. L. Hamilton, and J. B. Graham. 2005. Burrow air phase maintenance and respiration by the mudskipper *Scartelaos histophorus* (Gobiidae: Oxudercinae). *J. Exp. Biol. 208:* 169–177.

Littler, M. M., D. E. Littler, S. M. Blair, and J. N. Norris. 1985. Deepest known plant life discovered on an uncharted seamount. *Science 227:* 57–59.

MacArthur, R. H., and E. O. Wilson. 1967. *The theory of island biogeography*. Princeton University Press, Princeton, NJ.

Marshall, N. B. 1971. *Explorations in the life of fishes*. Harvard University Press, Cambridge, MA.

Martin, F. D. 1972. Factors influencing the local distribution of *Cyprinodon variegatus* (Pisces: Cyprinodontidae). *Trans. Am. Fish. Soc. 101:* 89–93.

Martin, K. L. M., and C. R. Bridges. 1999. Respiration in water and air, pp. 54–78. In *Intertidal fishes: Life in two worlds*, M. H. Horn, K. L. M. Martin, and M. A. Chotkowski (Eds.). Academic Press, San Diego.

Matthews, W. J. 1998. *Patterns in freshwater fish ecology*. Chapman and Hall, New York.

Maurer, B. A. 1999. *Untangling ecological complexity: The macroscopic perspective*. University of Chicago Press, Chicago.

Mayden, R. L. 1992. Explorations of the past, and the dawn of systematics and historical ecology, pp. 3–17. In *Systematics, historical ecology, and North American freshwater fishes*, R. L. Mayden (Ed.). Stanford University Press, Stanford, CA.

Mumby, P. J., and 11 coauthors. 2004. Mangroves enhance the biomass of coral reef fish communities in the Caribbean. *Nature 427:* 533–536.

Nilsson, G. E., and S. Östlund-Nilsson. 2004. Hypoxia in paradise: Widespread hypoxia tolerance in coral reef fishes. *Proc. Roy. Soc. Lond. B 271:* S30–S33.

———, J.-P. Hobbs, P. L. Munday, and S. Östlund-Nilsson. 2004. Coward or braveheart: Extreme habitat fidelity through hypoxia tolerance in a coral-dwelling goby. *J. Exp. Biol. 207:* 33–39.

Nordlie, F. G., S. J. Walsh, D. C. Haney, and T. F. Nordlie. 1991. The influence of ambient salinity on routine metabolism in the teleost *Cyprinodon variegatus* Lacepède. *J. Fish Biol. 38:* 115–122.

Olsen, E. M., M. Heino, G. R. Lilly, M. J. Morgan, J. Brattey, B. Ernande, and U. Diekmann. 2004. Maturation trends indicative of rapid evolution preceded the collapse of northern cod. *Nature 428:* 932–935.

Packard, G. C. 1974. The evolution of air-breathing in Paleozoic gnathostome fishes. *Evolution 28:* 320–325.

Paine, M. D. 1990. Life history tactics of darters (Percidae: Etheostomatiini) and their relationship with body size, reproductive behavior, latitude, and rarity. *J. Fish Biol. 37:* 473–488.

Paine, R. T. 1980. Food webs: Linkage, interaction strength, and community infrastructure. *J. Anim. Ecol. 49:* 667–685.

———. 1988. Food webs: Road maps of interactions or grist for theoretical development? *Ecology 69:* 1648–1654.

Pelster, B. 1997. Buoyancy at depth, pp. 195–237. In *Deep-sea fishes*, D. J. Randall and A. P. Farrell (Eds.). Academic Press, San Diego.

Persson, L. 1999. Trophic cascades: Abiding heterogeneity and the trophic level concept at the end of the road. *Oikos 85:* 385–397.

Pianka, E. R. 1970. On *r* and *K* selection. *Am. Nat. 104:* 592–597.

Pinnegar, J. K., and N. V. C. Polunin. 1999. Differential fractionation of δ^{13}C and δ^{15}N among fish tissues: Implications for the study of trophic interactions. *Funct. Ecol. 13:* 225–231.

Power, M. E., 1992. Top-down and bottom-up forces in food webs: Do plants have primacy? *Ecology 73:* 733–746.

Pringle, C. M., N. Hemphill, W. H. McDowell, A. Bednarek, and J. G. March. 1999. Linking species and ecosystems: Different biotic assemblages cause interstream differences in organic matter. *Ecology 80:* 1860–1872.

Rahel, F. J. 2000. Homogenation of fish faunas across the United States. *Science 288:* 854–856.

Randall, D., W. Burggren, and K. French. 2002. *Eckert animal physiology: Mechanisms and adaptations* (5th ed.). W. H. Freeman, New York.

Ricklefs, R. E., and G. L. Miller. 1999. *Ecology* (4th ed.). W. H. Freeman, New York.

Rodriguez, M. A., and P. Magnan. 1995. Application of multivariate analyses in studies of the organization and structure of fish and invertebrate communities. *Aquat. Sci. 57*(3): 199–216.

Schindler, D. W., S. E. M. Kasian, and R. H. Hesslein. 1989. Losses of biota from American aquatic communities due to acid rain. *Env. Monit. Assess. 12:* 269–285.

Sébert, P. 1997. Pressure effects on shallow water fishes, pp. 279–323. In *Deep-sea fishes*, D. J. Randall and A. P. Farrell (Eds.). Academic Press, San Diego.

Seebacher, F., W. Davison, C. J. Lowe, and C. E. Franklin. 2005. A falsification of the thermal specialization paradigm: Compensation for elevated temperatures in Antarctic fishes. *Biol. Lett. 1:* 151–154.

Soltz, D. L., and R. J. Naiman. 1978. The natural history of native fishes in the Death Valley system. *Nat. Hist. Mus. L.A. Co. Sci. Ser. 30:* 1–76.

———, and R. E. Crandall. 1984. Plasticity for age and size at sexual maturity: A life-history response to unavoidable stress, pp. 13–33. In *Fish reproduction: Strategies and tactics*, G. W. Potts and R. J. Wootton (Eds.). Academic Press, London.

Strong, D. R. 1992. Are trophic cascades all wet? Differentiation and donor-control in speciose ecosystems. *Ecology 73:* 747–754.

———, 1995. Population and community ecology, pp. 319–321. In *McGraw-Hill Yearbook of Science and Technology*. McGraw-Hill, New York.

Sverdrup, H., M. W. Johnson, and R. H. Fleming. 1947. *The oceans: Their physics, chemistry, and general biology*. Prentice Hall, Englewood Cliffs, NJ.

Szepanski, M. M., M. Ben-David, and V. VanBallenberghe. 1999. Assessment of anadromous salmon resources in the diet of the Alexander Archipelago wolf using stable isotope analysis. *Oecologia 120:* 327–335.

VanderZanden, M. J., G. Cabana, and J. B. Rasmussen. 1997. Comparing trophic position of freshwater fish calculated using stable nitrogen isotope ratios (^{15}N) and literature dietary data. *Can. J. Fish. Aquat. Sci. 54:* 1142–1158.

Walters, D. M., D. S. Leigh, and A. B. Bearden. 2003. Urbanization, sedimentation, and the homogenization of fish assemblages in the Etowah River Basin, USA. *Hydrobiology 494:* 5–10.

Ware, D. M., and R. E. Thomson. 2005. Bottom-up ecosystem trophic dynamics determine fish production in the Northeast Pacific. *Science 308:* 1280–1284.

Wellborn, G. A., D. K. Skelly, and E. E. Werner. 1996. Mechanisms creating community structure across a freshwater habitat gradient. *Ann. Rev. Ecol. Syst. 27:* 337–363.

Winemiller, K. O., and K. A. Rose. 1993. Why do most fish produce so many tiny offspring? *Am. Nat. 142:* 585–603.

Wootton, R. J. 1990. *Ecology of teleost fishes.* Chapman and Hall, London.

———. 1992. *Fish ecology.* Chapman and Hall, New York.

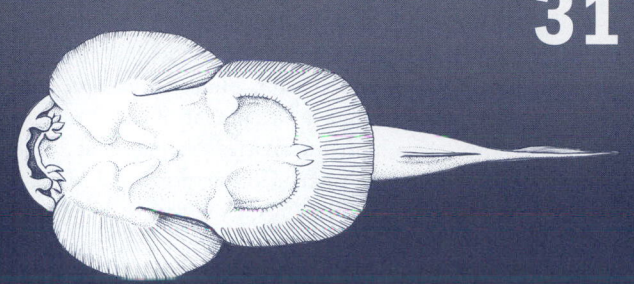

31

ECOLOGY OF FISHES II: FRESHWATER FISHES IN FLOWING WATERS

WATER—ABUNDANT AND ESSENTIAL

STREAM ENVIRONMENTS IN TEMPERATE ZONES

Upland Stream Environments
  Adaptations of Ichthyofauna of Upland Streams
  Assemblages of Temperate Upland Stream Fishes
Temperate Lowland Rivers and Streams: Habitat and Fish Assemblages
  North American Lowland Stream Fish Assemblages
  Eurasian Temperate Fish Assemblages

TROPICAL RIVERS AND STREAMS

Tropical Fish Assemblages
  Fish Fauna of Asia and Africa
  Neotropical Fish Fauna

ENERGETICS AND TROPHIC STRUCTURE OF RIVERINE FOOD WEBS

Trophic Adaptations in Stream Fishes
Riverine Trophic Structure and the Role of Fishes

THE LANDSCAPE CONNECTION

The landscape approach to ecosystem study can be traced to the pioneering investigations of F. Herbert Bormann, Gene E. Likens, Noye M. Johnson, and Robert S. Pierce. In 1963, these researchers initiated studies on a small watershed, Hubbard Brook, in the White Mountains of New Hampshire. The Hubbard Brook Ecosystem Study (HBES) remains the definitive study of nutrient flux and cycling in the forest and its associated aquatic ecosystem. In the landscape approach, the watershed is studied as an integrated whole, with terrestrial and aquatic ecosystems interacting in the movement of nutrients and materials—and this movement possesses a distinct directionality. The forces of gravity that put water in motion provide a distinct array of aquatic habitats for fishes. Moving water—be it in the form of large river systems or tiny rivulets—is one of the reasons why bony fishes have become so successful in freshwaters. Generally well oxygenated and providing ample shelter as the water courses over bottom topographies of varying composition, flowing waters provide shelter and sustenance for an array of fishes ranging from tiny darters to enormous sturgeons. Because watersheds usually represent geographically isolated entities, the opportunities for allopatric speciation are especially rich, with some families displaying remarkable adaptive radiations. This chapter represents the first of several in which we will partition the aquatic world of fishes into manageable components, starting off with a look at the fish fauna of flowing freshwaters.

WATER—ABUNDANT AND ESSENTIAL

Water is distinctive because it is capable of existing in all three states—liquid, solid, and vapor—within the Earth's thermal regime. The hydrological cycle is fundamental to ecology. Water falls to Earth in liquid or solid form, evaporates back into the atmosphere from land and sea, or is recycled by plants back into the atmosphere through the process of transpiration, in an unceasing cycle. E. C. Pielou (1998) presented a good visualization of the total quantity of water on Earth, estimated at approximately 1.4 billion km^3: If solidified into a cube, this cube would be more than 1,100 km along any edge, or about twice the length of Lake Superior. Water in its liquid state represents the medium in which fishes can exist. (Microorganisms are not so constrained, existing as they do in water vapor or in solid ice.) The total amount of liquid and frozen water is usually referred to by ecologists as the **hydrosphere**. The distribution of the total water quantity of the hydrosphere is shown in Table 31.1. The amount of freshwater amounts to only about 2.6 percent of the total, and only about 0.01 percent exists in a form that constitutes suitable freshwater fish habitat. Recall from Chapter 8 that fully 41 percent of the known fish fauna is packed into this tiny portion. Until recently, scientists believed that the hydrological cycle was essentially a closed one, and that the amount of water on the planet had remained virtually constant. There is increasing evidence to the contrary. It appears that we are being subjected to a constant barrage of house-sized snowballs, termed "small comets," that vaporize as they enter the atmosphere. These may account for a net gain of as much as 3 trillion tons of water over the past 10,000 years (Pielou, 1998).

Never before has the distribution and abundance of water resources been so critical to the continued vitality of our planet. The conflicting demands for available water continue to escalate. As much as 87 percent of the world's freshwater is used in agriculture. The consequent degradation of aquatic habitat through the introduction of pesticides and fertilizers presents a serious challenge to fish habitats worldwide. In the Northern Hemisphere, almost 80 percent of the freshwater discharge is affected by human activities (Naiman et al., 1998). The Magnuson-Stevens Fishery Conservation and Management Act, reauthorized by the U.S. Congress in 1996, mandated the establishment of protocols for the maintenance of **essential fish habitat (EFH)**. Although intended to ensure the sustainability of valuable fishery species, EFH guidelines naturally benefit all associated species. In developing countries, skyrocketing human populations have exacerbated the problem, because food production, by necessity, takes precedence over the quality of aquatic habitat. An understanding of the ecology and adaptations of ichthyofauna is essential to the resolution of conflicting demands on our precious freshwater resources.

Compared to seawater, the chemistry of inland waters is extremely variable (see Table 30.1). Traditionally, inland waters have been classified as either **lotic** (moving water) or **lentic** (still water). The ionic composition of lotic waters will largely reflect the substrate over which they flow. Likewise, for lentic waters, their ionic composition is a reflection of the geological composition of the strata in which the impoundment is found. Freshwater environments are subject to seasonal variation and, over a greater expanse of time, have experienced dramatic transformation by geological and biological processes. Changes in small streams and ponds may be readily observable over the course of a few years. Larger bodies of water change much more slowly but, from a geological perspective, are relatively young, short-lived phenomena. The North American Great Lakes, for example, can be dated only from the most recent glaciation events of North America—approximately 10,000 to 15,000 years ago.

TABLE 31.1   DISTRIBUTION OF WATER IN THE HYDROSPHERE (FROM JEFFRIES AND MILLS, 1990)

Source of water	Volume (10^3 km^3)	% of total water
Oceans	1,320,000 – 1,370,000	97.3
Freshwater		
Icesheets/glaciers	24,000 – 29,000	2.1
Atmosphere	13 – 14	0.001
Groundwater (to 4,000 m)	4,000 – 8,000	0.6
Soil moisture	60 – 80	0.006
Rivers	1.2	0.00009
Saline lakes	104	0.007
Freshwater lakes	125	0.009

Ancient bodies of freshwater, exceeding one million years of age (such as Lake Baikal, the world's deepest freshwater body, or the African Great Lakes System; see Chapter 32, Going Deeper), are most unusual.

In this and the next chapter, we will focus our attention on the range of freshwater environments in which fishes are found. Although characterizing fishes according to their habitats may be an efficient way to get a grasp on the ecology of such a diverse group, we must remember that fishes may not necessarily be always constrained by the habitats we delineate. Many species—anadromous ones, for example—transcend the conventional classification of aquatic environments employed here by passing from one aquatic ecosystem to another in the course of their life histories.

STREAM ENVIRONMENTS IN TEMPERATE ZONES

As water drains off a watershed, a natural progression of flowing water habitats develops. Small rivulets join to form brooks that may be seasonally intermittent in their flow. These brooks join to form larger streams that combine to form rivers and river systems, which ultimately empty into interior lakes or into the world's oceans. This progression is one of decreasing elevation and is generally marked by changes in velocity, as water progresses from high-gradient torrents and riffles at higher elevations to slower, meandering streams and rivers at lower elevations near sea level. A popular ecological classification scheme—one that is especially useful in correlating fish abundance and distribution with stream size—is to designate flowing waters along an ordinal gradient, with the smallest headwater streams as *first-order* streams. These join to produce a *second-order* stream. Two second-order streams merge to form a *third-order* stream—and so on. Fifth- and higher-order streams would constitute the major river systems of a continent. Although somewhat simplistic in terms of ecological context, the stream order concept is a useful means of categorizing the continuum of stream types and their associated fauna (Horton, 1945; Kuehne, 1962). The abundance of smaller, lower-order streams available as fish habitat may account for the general evolutionary trend toward smaller body size among North American freshwater fishes (see Chapter 30).

The physical and hence the biological attributes of flowing freshwaters exist along a continuum and are not the discrete entities implied by the stream order classification system. This is the fundamental precept of the **river continuum concept** (**RCC;** Vannote et al., 1980). According to the RCC, a longitudinal gradient of physical features exists along the river, and the structure of the lotic ecosystem parallels this continuum. Transport processes that deliver nutrients and other materials to the resident biota are instrumental in the regulation of what is perceived, according to the RCC, as a single continuous ecosystem—one in which organismal assemblages possess a certain degree of predictability. This concept is not without its detractors, however. Poole (2002) argued that each river system represents "a unique, patchy discontinuum from headwaters to mouth." Walters et al. (2003) proposed that river systems are best described by a *"process domains concept,"* in which community structure is primarily determined by local-scale geomorphic processes and disturbance regimes. Ward and Tockner (2001) weighed in with a proposal that holistic interpretations of stream ecology should incorporate landscape biodiversity as the integrative concept; they claimed that the classical perception of a river ecosystem is too constrained, as it does not give sufficient consideration to the connection of flowing surface waters to floodplains or contiguous groundwater aquifers.

Upland Stream Environments

Upstream environments are commonly referred to as **headwaters** or **rhithron**—the latter term being more commonly encountered in the European literature (Wootton, 1992). Headwaters are characterized by swift currents and a range of substrate types. Oxygen levels vary with the velocity of the water. A typical stretch of upland stream will consist of riffles, in which the water moves swiftly and is saturated with dissolved oxygen, and shallow pools, in which biological activity results in the depletion of oxygen. The scouring action of swift water in the riffles results in a substrate type that typically consists of rocks and cobble of varying texture, depending on local geological conditions; whereas fine-grained sediments and detritus will collect in the pools.

The flow dynamics of upland streams is extremely variable, with a range of flow velocities experienced by the resident fauna. Although the flow may appear to move downstream in a unidirectional pattern, in straight stretches of the stream, *helical currents* will tend to form: Water flows from the banks to the center, where it sinks and returns to the banks along the bottom (Fig. 31.1). These converging currents may leave a lane of flotsam at the center of the stream, where the currents converge. A rapid decrease in current velocity, caused by obstructions such as boulders in the stream bed, results in *hydraulic jumps*. These take the form of standing waves that form downstream from the obstruction. *Hydraulic drops,* sudden increases in current velocity, also occur, but they are not as conspicuous. All this agitation results in a well-oxygenated, though turbulent habitat for fishes living in swiftly flowing streams.

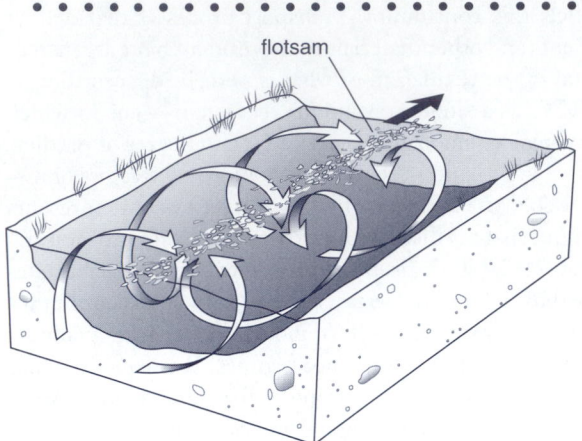

FIGURE 31.1
Helical current patterns established in a straight stretch of a stream. Note narrow accumulation of flotsam at the center of the stream. (From Pielou, 1998.)

Temperatures tend to be lower in upstream sections compared with downstream environments of the same watershed. Exposure to sunlight may be great in small streams above the timber line, but brooks in forested areas can be covered by a canopy of vegetation, which will effectively limit solar input reaching the water and thus hold temperatures down. Depending on their latitude, elevation, and exposure, upland streams may be subject to ice and snow cover for part of the year. Seasonality in precipitation patterns can lead to an intermittent flow in these streams as well. Upland streams in mediterranean climate regions experience a physically rigorous environmental regime that fluctuates in a fairly regular, seasonal manner. As the seasons progress and the hydrology changes from flooding to drought conditions, abiotic controlling factors alternate with biotic ones in influencing the dynamics of population and community structure (Gasith and Resh, 1999). Physical conditions, such as depth and bottom topography, can have a significant impact on the growth and survival of fishes (Lonzarich and Quinn, 1995), and this may be especially apparent in headwater environments that experience seasonal hydrological variation. During drought conditions, a free-flowing stream may be reduced to a series of isolated pools—consequently, the available oxygen and food resources become depleted. Nonetheless, these pools may play a critical role as refugia for the fish fauna resident in the stream (Magoulick and Kobza, 2003). The onset of flow intermittency has been shown to influence patterns of resource use in headwater fishes (Holomuzki and Stevenson, 1992; Magalhães, 1993), and can also influence gene flow in local populations.

Continued gene flow can sometimes constrain local adaptive evolution. Salamander populations that must contend with sunfish predation in intermittent streams exhibit delayed hatching and a greater degree of predator avoidance behavior in the larval stages. Storfer (1999) demonstrated that gene flow into such populations increased their susceptibility to predation.

The physical regime characteristic of a major faunal region can affect overall species diversity. The harsh physical environment and the comparatively simple bottom topography throughout the Platte River system of the Great Plains of North America have resulted in a strikingly depauperate fish fauna (Fausch and Bestgen, 1996). Changes in the water chemistry of prairie river systems that occur with the onset of flow intermittency favor species such as cyprinodontids of the genus *Cyprinodon* that are also well-known inhabitants of coastal and inland brackish waters (Ostrand and Wilde, 2004; see further and Chapter 32). The southeastern United States, with its numerous drainage systems displaying a range of habitat types, has an extraordinarily high species diversity for a temperate region, with more than 660 native freshwater species reported (Warren et al., 2000). The biological communities of stony streams tend to be quite similar worldwide; the greatest differences are seen as one moves from the tropics to the poles (Hynes, 1970). Of the three mechanisms that appear to have the greatest influence on assemblage structure and function—available resources, predation, and environmental variability—Poff and Allan (1995) and Grossman et al. (1998) have determined that environmental variability, in the form of flow regimes, had the greatest impact. Studies on the fish fauna of the Rhône River basin have demonstrated that stream hydraulics is the most influential variable in the generation of models for predicting fluvial fish community composition (Lamouroux et al., 1999).

Adaptations of Ichthyofauna of Upland Streams

Ecomorphological adaptations of fishes are influenced by at least three factors: (1) phylogenetic history; (2) hydrodynamic conditions that affect body and fin shape and size; and (3) head, jaw, and locomotor adaptations associated with feeding (Matthews, 1998). For the inhabitants of streams with fairly swift current, this means a comparatively small, fusiform body shape, well-developed fins for burst swimming or quick maneuverability in currents, and a diversity of head and jaw structures to exploit a range of feeding opportunities. Here we will see piscivorous forms with large jaws; small-jawed microcarnivores adept at capturing small insects among the rocks or drifting in the current; and bottom feeders that can suck up organic matter that settles in the interstices. Herbivores, such as stoneroller minnows

(*Campostoma*) and chiselmouths (*Acrocheilus*), may be locally abundant. These herbivores often have a cartilaginous shelf on the lower jaw that they use to scrape algae and diatoms from the substrate.

In higher latitudes of both the Old and the New World, salmonids are a conspicuous family of upland stream fishes, but they can range further downstream if temperature and oxygen conditions permit. These are active, streamlined fishes with strong locomotor capabilities. Species such as the rainbow trout (*Oncorhynchus mykiss*), the cutthroat trout (*O. clarkii*), the brown trout (*Salmo trutta*), and chars of the genus *Salvelinus* can enter streams that are essentially all rapids, with gradients up to 75 m/km. Typically, they will reside in stream pools, catching food as it drifts by. Salmonids also make occasional feeding forays into rapids, and they will invade intermittent streams on a seasonal basis. Salmonids will share this habitat with an assortment of other species. In North America, these will include dace of the genus *Rhinichthys*, creek chubs (*Semotilus*), and several species of suckers, such as the northern hog sucker (*Hypentelium nigricans*), the torrent sucker (*Moxostoma rhothoecum*), and the mountain sucker (*Catostomus platyrhynchus*).

The two most abundant and diverse groups of benthic fishes in upland streams are sculpins (Cottidae) and darters (Percidae, subfamily Etheostomatinae). These are small, bottom-dwelling fishes that typically have cylindrical to dorsoventrally compressed bodies with flattened heads, broad pectoral fins, and reduced or absent swim bladders—all adaptations enabling them to hold position on the bottoms of upland streams, habitats that may be periodically subjected to torrential conditions. In western North America, sculpins, particularly of the genus *Cottus*, are especially abundant, with species such as the Paiute sculpin (*Cottus beldingi*) and the shorthead sculpin (*C. confusus*) invading the headwaters of streams. In eastern North America, however, sculpins are not nearly as speciose. Rather, it is the darters that are the most conspicuous members of the benthic stream fish fauna. Geographical isolation, coupled with their limited mobility, has resulted in a high degree of endemism among the benthic fish fauna inhabiting upland streams—especially in the case of the darters, with more than 140 species described so far.

Although darter species, such as those of the genus *Etheostoma*, are normally associated with clear, swift streams, many species have successfully invaded backwaters and sloughs. The tiny, translucent darters of the genus *Ammocrypta* hide in the sand of broad, shallow, slowly moving streams. Studies have demonstrated considerable microhabitat partitioning among coexisting species of sculpins and darters. When two or more species are present, they will segregate according to current speed, substrate type, or depth (Daniels, 1987; Finger,

1982; Fisher and Pearson, 1987; Hlohowskyj and Wissing, 1986; Matthews, 1985; Ultsch et al., 1978).

The reproductive adaptations of fishes in high-gradient streams include burying their eggs in gravel or adhering them to suitable substrates in order to ensure that they will not be swept away by the current. Trouts, chars, and salmons seek out gravelly bottoms at the heads of riffles or in springs, where a nest is dug by the female. The eggs are temporarily adhesive, so that they remain in place until they are covered. Whitefish, grayling, and suckers spawn with vigorous activity that disturbs the gravel sufficiently to ensure that most of the eggs become lodged in interstices among the rocks. Sculpins and some darters attach adhesive clumps of eggs to the underside of stones, and these eggs may be vigorously guarded, usually by male parents.

The headwaters of upland streams are often spring-fed at their source. Headwaters with springs at their source may be more thermally stable and, hence, differ in biotic composition from neighboring drainages. Local adaptation to the prevailing conditions is evident in some spring-dwelling species that appear to have evolved a more restricted thermal tolerance regime than individuals of the same species living in more thermally labile headwaters (Walsh et al., 1997).

Some of the most remarkable freshwater fishes are those species that are adapted to conditions of torrential flow. Asiatic hill-stream loaches (Balitoridae; Fig. 31.2) have flattened ventral surfaces, with the pectoral and pelvic fins expanded to form a broad, adhesive disc that enables the fish to hold its position in extremely fast-flowing water. The family Gyrinocheilidae (the popular "algae eaters" of the tropical fish trade) have a more conventional body form, but they possess a mouth that can act as a sucker for holding on to rocks. Other torrent fishes are the South American catfish family Loricariidae (which includes species popular with aquarists), *Cheimarrichthys* (Cheimarrhichthyidae) of New Zealand, and the Rhyacichthyidae of Asia. Lampreys use their jawless mouths in much the same fashion, gripping stones as they advance on their reproductive journey upstream (the family name Petromyzontidae literally means *stone sucker*).

Assemblages of Temperate Upland Stream Fishes

The Assemblage Concept
Through the first half of the 20th century, limnologists paid little attention to the impact that fishes had on river and lake ecosystems. With the emergence of the trophic cascade concept (see Chapter 30), we now understand that fish populations have a dramatic effect on freshwater ecosystem structure (see Brooks and Dodson, 1965). The studies by William Matthews and Mary Power on the impact of stoneroller minnows on algal density in Oklahoma streams are classic

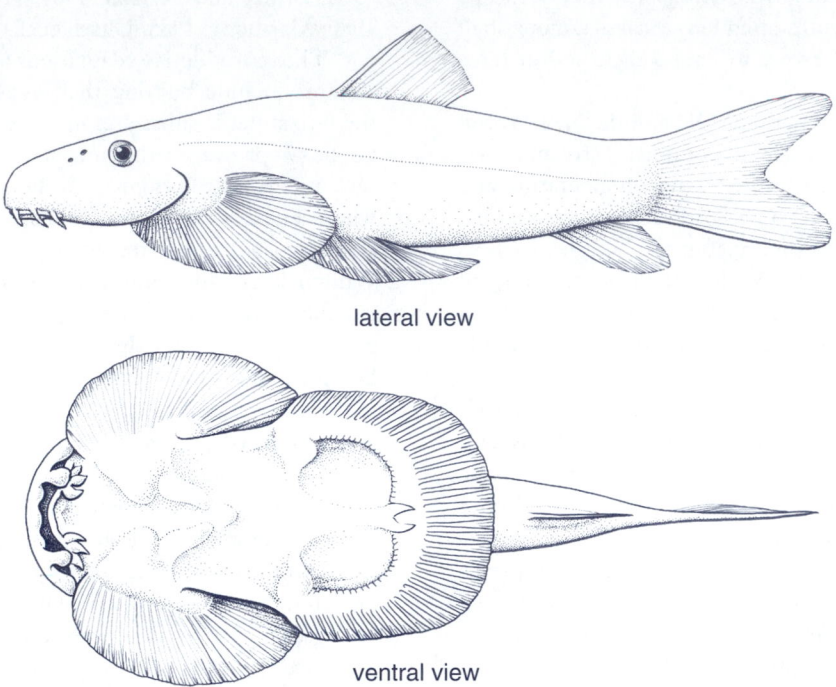

lateral view

ventral view

FIGURE 31.2
Hill-stream loach of the family Balitoridae. Note flattened profile and pectoral and pelvic fins modified to form a sucking disc.

demonstrations of the role that fishes play in the structuring of aquatic ecosystems (Gelwick and Matthews, 1992; Gelwick et al., 1997; Power and Matthews, 1983; Power et al., 1985). Most freshwater ecosystems will thus have an assemblage of fishes as a vital component. It is important to understand exactly what is meant by the term *assemblage* as it is used here. An **assemblage** constitutes those individuals found in a given locality. Perhaps the best way to define *locality* would be that portion of a lake or stream that would be visited and collected by a fish biologist.

An assemblage of fishes does not necessarily constitute a community in the ecological sense. Integral to the concept of community is some degree of trophic interaction. The term **guild** has been used to denote a subset of a given assemblage that exploits the same class of resources in a similar manner (Wootton, 1990). A given assemblage of fishes may or may not engage in significant trophic interactions. For example, a single collection may turn up nocturnal, foraging madtom catfishes and diurnal, herbivorous stoneroller minnows, but their trophic association is tenuous at best. An assemblage might consist of individuals, such as sculpins or darters, that confine themselves to a few meters of the stream, as well

as wide-ranging species such as trouts. Assemblages in the riffle habitat may constitute a subset of a larger assemblage found in the pools—especially in small stream localities (Taylor, 2000). The assemblage may be temporally variable, as seasonal visitors, such as anadromous species, arrive and depart. In one study of stream fish assemblages, seasonality was more evident in the riffle fish assemblage than in the pool assemblage (Fuselier and Edds, 1996).

Assemblage structure is a useful way to define the patterns of zonation that occur along a river gradient. Most studies of stream fishes and their ecology have been reported in terms of associations of fishes; stream zones have been named for the dominant species or associations found there. Often, these zones or associations have mainly local applications, but in Europe, where climatic and physiographic features have permitted a broadly similar fish fauna over a wide area, a zonation of streams proposed by Huet (1959) seems to have widespread application (Fig. 31.3; Wootton, 1992). Li et al. (1987) have developed a similar zonation pattern for the salmonid- and cottid-rich rivers of the Pacific Northwest, and similar ichthyofaunal zonations have been developed for other regions of the country.

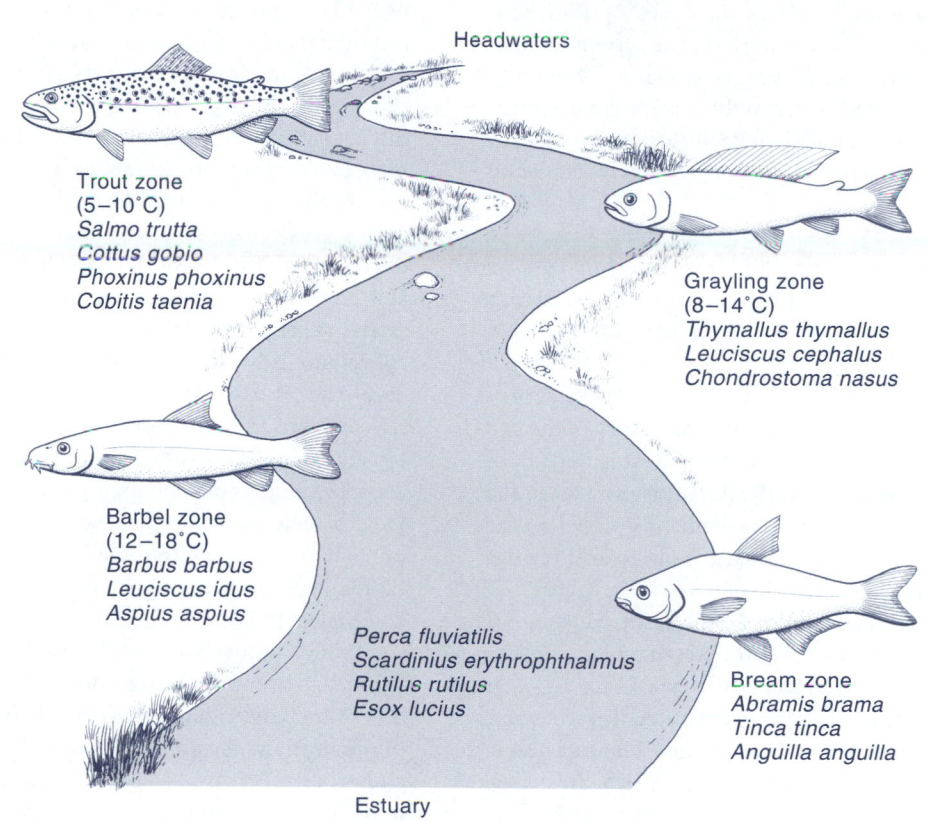

Headwaters

Trout zone
(5–10°C)
Salmo trutta
Cottus gobio
Phoxinus phoxinus
Cobitis taenia

Grayling zone
(8–14°C)
Thymallus thymallus
Leuciscus cephalus
Chondrostoma nasus

Barbel zone
(12–18°C)
Barbus barbus
Leuciscus idus
Aspius aspius

Perca fluviatilis
Scardinius erythrophthalmus
Rutilus rutilus
Esox lucius

Bream zone
Abramis brama
Tinca tinca
Anguilla anguilla

Estuary

FIGURE 31.3
Huet's zonation of fish assemblages of European rivers. Species representative of each zone are given, as is the prevailing thermal regime of each zone. (From Huet, 1959; Wootton, 1992.)

Freshwater Fishes at High Latitudes

In spite of the rigorous environment that they may impose, streams originating in either high latitudes or high elevations may contain several species of fishes. The headwaters of these streams are usually fed by glacial meltwater, with a peak in discharge in July (in the Northern Hemisphere). A diel fluctuation in flow is also characteristic, with peaks late in the afternoon. The dissolved ion content of meltwater is typically very low, and hence the specific conductance is usually less than 50 μS cm⁻¹ (Milner and Petts, 1994). Summer temperatures are typically below 10°C, and turbidity is high due to large concentrations of fine sediments (Milner and Petts, 1994). Sediments of glacial origin, termed **glacial flour** because of their characteristic color and extremely fine texture, impart a distinctive, milky white color to some streams. Although turbidity limits the amount of primary productivity, streamside boulders may be coated with a film

of diatoms. This is an important source of nutrients for invertebrates due to the paucity of allochthonous (i.e., arising from outside the stream ecosystem; discussed later) organic input (Milner and Petts, 1994). The invertebrate community is dominated by a few genera of chironomid midges. The physical environment, with an unstable stream bed and few pools, and the paucity of feeding opportunities make glacial streams a less than ideal fish habitat. Small streams may freeze solid in winter, thus restricting their availability to limited times of the year.

A total of 42 freshwater fish species have been recorded in Canadian waters above the Arctic Circle. Of these, 16 are in the family Salmonidae (Reist, 1997). This should underscore the significance of anadromy (see Chapter 36) in the lives of freshwater fishes in high latitudes. The origin of anadromy has long been debated in the ichthyological community. A high proportion of anadromous species are to be found among members of the families Petromyzontidae, Acipenseridae, Salmonidae, Osmeridae, Cottidae, and Gasterosteidae (see Chapter 4). From the range of aforementioned families, it should be apparent that anadromy is especially prevalent in more basal lineages of fishes. The ability to move readily from freshwater to the marine environment undoubtedly was a valuable adaptation during the Pleistocene epoch, when much of North America was covered with ice (Power, 1997).

Fishes and Fish Habitats in North American Deserts

Patterns of distribution and abundance of freshwater fish species in rivers and streams in the eastern part of North America differ significantly from those observed in the western part of the continent, due to the profound differences in the geological and environmental history of the two regions. The freshwater fish fauna of the West is but a fraction of that seen in the geologically older, eastern part of the continent (Smith, 1981). Given the demands placed on the limited water that is available, aquatic biota in arid regions are especially vulnerable.

Although the fish fauna of the Great Basin, Mojave, Sonoran, and Chihuahuan deserts of the western United States may not be as diverse as in regions that experience a greater amount of precipitation, they are remarkable in their scope of adaptations to extremely arid conditions. They are also in possession of a unique evolutionary history, which is characterized by a high degree of endemism. Their highly fragmented population structure, characterized by comparatively low numbers of individuals, makes desert fishes especially vulnerable to extinction (Fagan et al., 2002).

Considering their arid nature, the North American deserts are still surprisingly rich in aquatic biota. Desert waters consist usually of two types: (1) rivers and streams, often with intermittent flow; and (2) springs, which represent surface extrusions of groundwater rising up from geological faults. Marshes sometimes form in association with these sources of water, providing important habitat for aquatic organisms and a destination for many terrestrial animals. Those waters that are located along the flyways of migratory birds are especially critical, because they provide sources of water and food en route to and from nesting and wintering areas. Only a few lakes are present, exclusively in the Great Basin (Soltz and Naiman, 1981). The desert environments are all of relatively recent origin. Basically, they are relics of a much more extensive aquatic ecosystem that was present during the last great pluvial period, which ended 10,000 to 12,000 years ago (Soltz and Naiman, 1981). Consequently, the evolution of the resident fish fauna isolated in the various remnant springs and rivers is a relatively recent phenomenon. The habitats of desert fishes may experience extremes in temperature, dissolved oxygen, and salinity. Killifishes and pupfishes of the family Cyprinodontidae are well adapted to these conditions—their tolerance of environmental extremes is well known (see Hillyard, 1981).

In comparing spring-dwelling species of fishes with those living in larger bodies of water, a few generalizations can be made. Spring dwellers tend to have a shorter, stubbier body profile; their breeding season is more protracted, with a shorter time to first breeding; and they have a low fecundity, with a greater parental investment per offspring. A greater degree of territorial behavior, bolder reproductive coloration, and increased presence of alternative mating strategies also seem to characterize spring dwellers (Constantz, 1981). The cyprinodontids best represent these traits. Approximately 30 of the 50 or so known species of the genus *Cyprinodon* have evolved in the North American deserts, where they are largely limited to small streams or springs (see Echelle et al., 2005). Examples of such highly endemic species include the Devil's Hole pupfish (*Cyprinodon diabolis*) and the Owens pupfish (*C. radiosus*). In these two species, their entire natural range is restricted to one or two isolated springs (Sigler and Sigler, 1987).

Many other North American families of freshwater fishes have representative species in the western deserts. These include salmonids, such as the landlocked sockeye salmon variant known as the kokanee, whitefish (*Prosopium*), and trout; cyprinids (especially the genus *Gila*); catostomids; ictalurids; and centrarchids (Sigler and Sigler, 1987). Cichlids have been introduced in some areas and are of some concern because of their potentially adverse impact on the native biota. Studies on the fish assemblages in desert springs in Australia indicate that isolated and comparatively simple ichthyofaunal communities such as these are among the most structured and deterministic communities known (Kodric-Brown and Brown, 1993).

• •
Temperate Lowland Rivers and Streams: Habitat and Fish Assemblages

The lower reaches of streams, termed the **potamon** by Wootton (1992), possess a greater variety of habitats, and these habitats are less subject to seasonal fluctuation in physicochemical parameters. Streams here may be characterized by more pools and runs than riffles, and the current velocity will be more variable but slower than upstream locales. Straight streams are comparatively rare in nature. The course of a river or stream usually takes the form of **meanders**. Within these meanders, one bank may be scoured away, while the transported sediment builds up the other bank. The scoured bank may undercut streamside vegetation in a manner that exposes root masses, which provide shelter for fishes. As mentioned in Chapter 13, the pirate perch (*Aphre-doderus sayanus*) has evolved a most unusual reproductive morphology that has enabled it to exploit these root masses as spawning sites.

The patterns of erosion and deposition may cause a meander to actually be displaced downstream while retaining its same general morphology (Fig. 31.4). In some instances, the sediment transport may eventually cut off a meander from the main stem of a river; in these situations, distinctive streamside lakes termed **oxbows** will develop. Substrate size and texture become more variable downstream, as do the kinds of plants. The increased amount of finer sediments permits the establishment of a greater diversity of rooted plants. These provide greater microhabitat and feeding opportunities for the resident fish fauna, which tend to exhibit a greater diversity than that seen in upland streams (see Ebert and Filipek, 1988). In rivers that run through forested areas, woody debris has also been shown to contribute to habitat complexity and, hence, to fish diversity (Thévenet and Statzner, 1999). Because the distribution of nutrients and preferred habitat such as woody debris is not uniform, fish distribution is patchy in both upland and lowland streams. A continuous hydroacoustic survey of a lower stretch of the River Thames revealed patchy distributions in the form of assemblages of fishes that became specifically associated with sewage outfalls (Duncan and Kubecka, 1996).

North American Lowland Stream Fish Assemblages

The preponderance of slow-moving waters in lowland streams results in fish assemblages that resemble those seen in lakes. Members of the family Cyprinidae are abundant in both upland and lowland streams. In the Mississippi drainage, genera such as *Notropis* (the most speciose genus of North American cyprinids), *Hybopsis*, *Semotilus*, and *Pimephales* are representative. Predatory pikeminnows (*Ptychocheilus*), among the largest known cyprinids, are encountered in various western North American drainages, as are the roach (*Hesperoleucus*), chubs (*Gila*), the peamouth (*Mylocheilus*), and the redsides (*Richardsonius*). Several species of suckers, such as *Catostomus commersoni* and *C. macrocheilus*, abound in the middle to lower stretches of streams, where greater opportunities for benthic feeding exist.

North American running waters are typically classified as either warm-water streams, which occur in lower latitudes and elevations, or cool-water streams, which are seen at higher latitudes or elevations. Streams at the colder end of

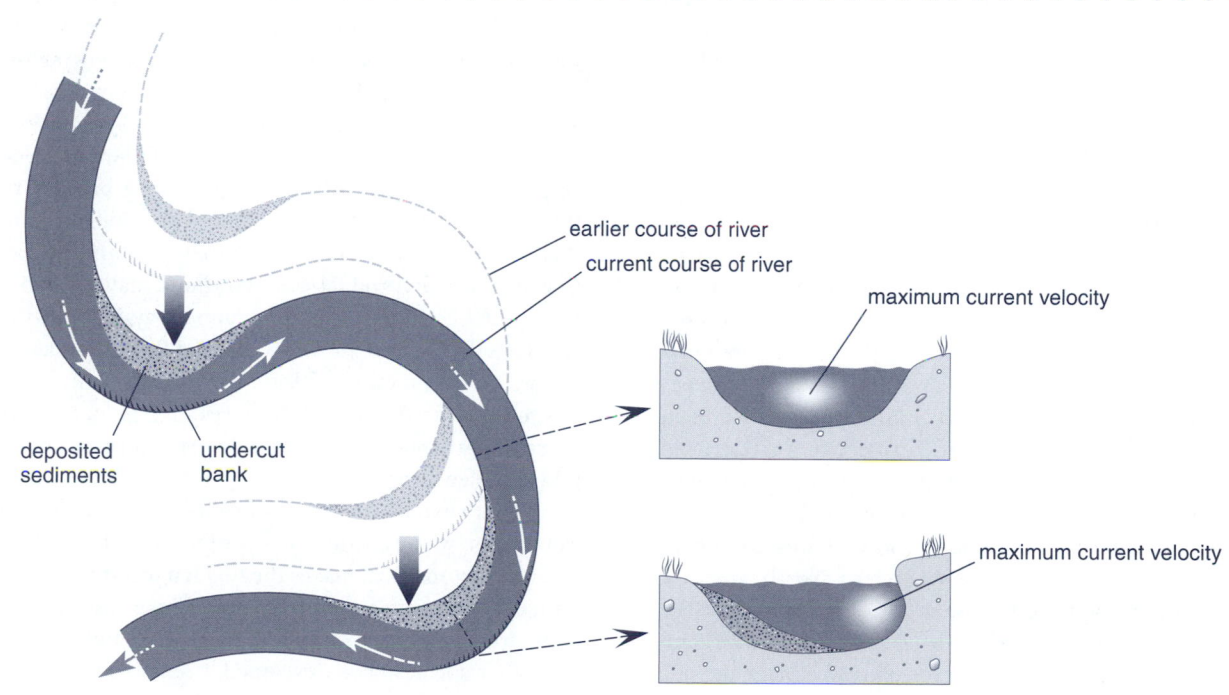

FIGURE 31.4

Downstream displacement of a river meander. Note locations of undercutting and sediment deposition. (From Pielou, 1998.)

the thermal spectrum are often identified as "trout streams" (see Cushing and Allan, 2001); their ability to support populations of salmonids makes them a subject of particular interest among the angling community. Although the top carnivores in cool-water streams are usually salmonids or larger members of the perch family, such as the walleye (*Sander*), warmwater streams will have a good representation of members of the sunfish family Centrarchidae. Whereas elongate, fusiform fishes were prevalent in swifter waters, the comparatively deep-bodied profile of sunfishes is better adapted to quieter waters. This relatively diverse family includes planktivores and microcarnivores, such as crappie (genus *Pomoxis*) and bluegill (*Lepomis macrochirus*), as well as large, piscivorous game fishes, such as the basses of the genus *Micropterus*.

As the stream order increases, current velocity decreases, depth increases, bottom materials become finer, and the water correspondingly becomes more turbid. Several species of large bottom feeders do well in such places. Suckers, including the genera *Carpiodes* and *Ictiobus,* and catfishes of the genus *Ictalurus* forage among the fine-grained sediments and rooted plants of lowland streams and rivers. Ictalurids (family Ictaluridae) are the only catfish family native to North America; they include some of the largest fishes found in freshwater. Although smaller ictalurids, such as the bullhead catfishes (*Ameiurus*) and the madtoms (*Noturus*), abound in pools and riffles in smaller streams, some species, such as the blue catfish (*Ictalurus furcatus*) and the flathead catfish (*Pylodictes olivaris*), prefer large pools or channels of large river systems. Here, they may reach sizes in excess of 50 kg. The numerous small cyprinid species in the lower reaches of streams serve as forage for predatory pikes (*Esox*) and centrarchids.

Eurasian Temperate Fish Assemblages

Moyle and Herbold (1987) have made some interesting comparisons of the life history patterns of New and Old World temperate fish communities. Communities of cold headwater streams in Europe and North America are quite similar, each comprising salmonids, sculpins, and a few species of cyprinids (catostomids abound in upland streams in North America, but their old world distribution is restricted to western Asia), whereas communities at lower elevations and latitudes tend to vary. In fact, European fish communities bear a greater resemblance to those of western North America than they do to those of eastern North America.

In Europe, cyprinids found in lotic waters include the barbel (*Barbus barbus*), the chub (*Leuciscus cephalus*), and the dace (*L. leuciscus*). Typical downstream forms include the bream (*Abramis brama*), the carp (*Cyprinus carpio;* a species that has had an adverse impact on the native biota of North America since it was introduced in the 19th century), and the tench (*Tinca tinca;* another, more recent introduction

to North America). Radiotelemetry studies on the chub (*L. cephalus*) have demonstrated the importance of cover for some species of Old World minnows that inhabit larger, slower-flowing rivers (Allouche et al., 1999). Although the loaches (Cobitidae) are most abundant and diverse in Asia, three species can be found in European freshwaters.

TROPICAL RIVERS AND STREAMS

When one considers the latitudinal distribution of freshwater discharge, two things become apparent: (1) the preponderance of landmasses in the Northern Hemisphere results in a much greater discharge there; and (2) most of the discharge takes place in the tropics (Table 31.2). The latter should not come as a surprise, as the greatest amount of rainfall occurs at the lowest latitudes. Although the "typical" temperate stream system can be conveniently divided into high-gradient, torrential upper reaches; moderate-gradient regions with a flatter profile; and slow-moving rivers and streams of the lowlands, no such simple classification system exists for rivers in tropical regions. Variations in topography have led to some rivers originating in swampy regions, with large falls and long stretches of rapids occurring at widely spaced intervals (Lowe-McConnell, 1987). Seasonal patterns of precipitation result in vast stretches of watershed being a component of the aquatic ecosystem during the rainy season but a part of the terrestrial ecosystem at other times of the year. In the Amazon system, lowland forests may be flooded for up to half of the year, permitting many species of fish to forage over a broader range; this especially facilitates the feeding habits of those species that subsist on seeds, fruits, and other terrestrial vegetation (see further). Reproductive migrations may also coincide with peak flooding, so that the young may be reared in the relative safety of the shallow reaches of flooded areas. These factors, in conjunction with the characteristically high rate of speciation in the tropics, have resulted in an exceedingly diverse ichthyofauna in tropical rivers and streams of the Old and New World.

Where extensive flooding occurs, as in the neotropical rainforests, water chemistry is profoundly altered. Decomposing plant material makes the flooded forest waters acidic, with low levels of dissolved oxygen. The forest canopy prevents wind mixing, resulting in stagnant waters. Characins and other species have developed a remarkable respiratory adaptation to these conditions. Under conditions of severe depletion of dissolved oxygen, many species develop protrusions of the lower lip that appear to facilitate aquatic surface respiration—that is, the uptake of oxygen from the thin

TABLE 31.2   LATITUDINAL DISTRIBUTION OF FRESHWATER DISCHARGE (FROM ALLAN, 1995, BASED ON MILLIMAN, 1990)

Latitude	Volume of runoff (km³)	Runoff (%)	Land Area (%)
60 – 90 N	3,551	8.9	11.6
30 – 60	8,252	20.8	31.4
0 – 30	12,597	31.8	24.5
0 – 30	11,746	29.6	19.6
30 – 60	1,567	3.9	9.4
60 – 90 S	1,987	5.0	9.4

layer of water at the surface (Saint-Paul and Bernardinho, 1988; Winemiller, 1989). This may permit herbivores to remain in the flooded forests as the water stagnates, whereas other species, including predators, are forced to return to the more richly oxygenated river channel.

If flooding occurs in tropical savanna regions, however, the conditions are quite different. In the absence of canopy cover, flooded savannas rapidly heat up. There is abundant light available for photosynthetic activity, and oxygen is replenished with wind mixing. In many tropical regions, these types of floodplains support important fisheries (Lowe-McConnell, 1987).

Tropical Fish Assemblages

Fish Fauna of Asia and Africa
The tropical fish fauna of the Old World includes a diversity of species, mainly centered in Africa and Asia. Africa has more than 2,000 species of indigenous freshwater fishes. The most striking feature of the African freshwater fish fauna is its high degree of endemism, especially in the cichlids of the African Great Lakes (see Chapter 32, Going Deeper). Fishes of the family Cyprinidae are also well represented, especially in the four large river systems—the Niger, Nile, Zaire, and Zambezi—that drain the continent. The Zaire typifies the complexity of equatorial African river systems, consisting of large rapids, swamps, main river channels, shallow stretches, and seasonally flooded areas. Large lateral lakes are also part of the Zaire system. Rivers show considerable seasonality in flow; those within the humid equatorial zone, such as the Zaire, show two high-water periods per year. Rivers to the north and south, such as the Niger and the Senegal that border on arid zones, show a single high-water interval (Lévêque, 1997). With such a diversity of habitat, the African freshwater fish fauna is correspondingly rich and varied. In addition to the aforementioned cichlids and cyprinids, rivers may be populated with bichirs (Polypteridae), elephantnose fish (Mormyridae), characins

(Characidae), and a variety of catfish families. Lungfishes (Protopteridae) occur in areas that are subject to seasonal drought; their remarkable aestivating abilities enable them to survive several months of desiccation.

Southeast Asia is another region that is extremely complex from a biogeographical perspective. For example, Borneo has more than 300 species of primary freshwater fishes in 17 families. (**Primary** freshwater fishes are those with virtually no tolerance to seawater that hence have dispersed entirely through freshwater corridors, whereas **secondary** freshwater fishes display a limited tolerance to seawater and may include species with marine affinities; see Chapter 29.) In contrast, just 140 km to the east, Sulawesi has only two primary freshwater species (Lowe-McConnell, 1987). In areas such as this, where primary freshwater fishes are conspicuous by their absence, atherinomorph fishes are particularly well established. Cyprinids dominate the tropical Asian fish fauna, as they do in many other biogeographical regions. Typical of the cyprinids is the genus *Barbus*, with 11 species recorded from Sri Lanka (Kortmulder, 1987). Many species of catfishes, including some of the largest in the world, are also found in the main channels of the river systems that drain tropical Asia. For example, *Pangasius (= Pangasianodon) gigas*, an enormous herbivore measuring more than 2 m in length, is found in the main channel of the Mekong River. An enormous male that was recently caught weighed 646 lbs (293 kg) and was almost 9 ft in length, claiming the world record for the largest freshwater fish ever captured. Another remarkable group native to the lowland swamps are the archerfishes (Toxotidae). Members of this family are able to obtain insect prey from overhanging vegetation by shooting at them with jets of water.

Neotropical Fish Fauna
The fish fauna of neotropical river systems is the richest and most diversified in the world, with approximately 3,000 species described. It is also the most poorly understood. In the past decade, an increased research focus on neotropical freshwater fishes has resulted in a corresponding increase in

the rate of new species descriptions. It has been estimated that the ichthyofauna of South and Central America may ultimately number as many as 8,000 species, or 25 percent of the total marine and freshwater fish diversity worldwide (Vari and Malabarba, 1998). A few river systems have been subjected to intensive collection efforts, so that patterns of longitudinal zonation and habitat preferences can be discerned (see Ibarra and Stewart, 1989; Rodríguez and Lewis, 1997; Fig. 31.5). One such study, a comprehensive survey along the main channels of the Amazon River, has shown that the diversity of some fish groups, such as electric fishes of the order Gymnotiformes, is positively correlated with proximity to tributaries (Fernandes et al., 2004).

Compared to the Old World tropical ichthyofauna, that of the New World is derived from fewer basic stocks. For example, primitive families, such as the osteoglossomorphs, are poorly represented, whereas characoids and

siluroids have experienced explosive adaptive radiations. Fully 85 percent of the fishes of the Amazon Basin are ostariophysans, compared to 54 percent in the Zaire River system (Lowe-McConnell, 1987). Because so many South American river systems have yet to be thoroughly explored, few generalizations concerning the ecology of the ichthyofauna can be made. The fishes that have been the subject of extensive life history investigation have usually been the larger species, which may be commercially exploitable, or those species that are remarkable for some aspect of their life history, such as the fruit- and seed-eating characoids (Araujo-Lima and Goulding, 1997; Barthem and Goulding, 1997; Goulding, 1980).

ENERGETICS AND TROPHIC STRUCTURE OF RIVERINE FOOD WEBS

Trophic Adaptations in Stream Fishes

Of foremost importance in structuring any aquatic community is its productivity—its capacity to produce food for its component species. In water, as in terrestrial habitats, the bulk of food production depends on the photosynthetic capacity of the system. In high-gradient streams, there is little opportunity for rooted plants to develop, so their primary productivity rests largely with the algae—especially diatoms—that coat the bottom. Algae form a food base on which the primary and secondary consumers of the aquatic system may subsist. As mentioned earlier, the river continuum concept has come to dominate discussions of the structure and function of riverine ecosystems. Recently, this concept has gained a lateral dimension in the elucidation of the **flood pulse concept**. The flood pulse concept emphasizes the importance of seasonally inundated floodplains to tropical and temperate river systems. Floodplains may provide nutrients and habitat for ichthyofauna on a seasonal basis. Junk et al. (1989) claimed that the main channels of rivers function as little more than corridors leading the resident fishes to feeding and habitat resources at the channel margins and beyond into the floodplains. An important consideration in the application of the flood pulse concept is the degree to which the river system has been subject to human management of its flow (Eggleton and Schramm, 1999).

Ecologists now recognize that fish assemblages are key players in the structure and function of local ecosystems. Although most fishes are secondary consumers, feeding on the creatures that feed on the diatoms, herbivory is known in some stream-dwelling species. The aforementioned hillstream loaches and gyrinocheilids are vegetation scrapers.

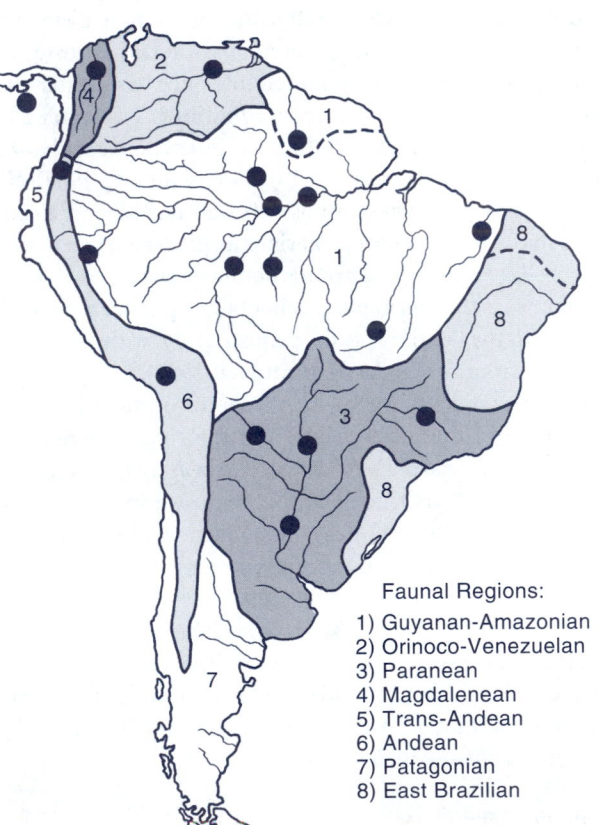

Faunal Regions:
1) Guyanan-Amazonian
2) Orinoco-Venezuelan
3) Paranean
4) Magdalenean
5) Trans-Andean
6) Andean
7) Patagonian
8) East Brazilian

FIGURE 31.5
Faunal regions of South America. Locations of extensive ichthyofaunal study are indicated with solid circles. (From Lowe-McConnell, 1987.)

The algivorous/omnivorous stoneroller minnows (*Campostoma*) and the bluntnose minnow (*Pimephales notatus*) are among the most abundant cyprinids in upland streams of eastern North America. These two genera graze on diatoms and other attached plant material (known as **aufwuchs**) and can outnumber all of the other species combined in some especially productive stream stretches. Perhaps the most remarkable examples of herbivory in fishes are those species—especially in the Amazonian basin—that subsist on seeds and fruits dropped from overhanging trees. The diet of the large characins commonly referred to as pacu that inhabit the Amazon River and its tributaries may consist almost entirely of seeds and fruits of several species of trees that form the streamside canopy. The invasion of seasonally flooded lowlands by these fishes is as clear a demonstration of the validity of the flood pulse concept as one could get.

As the gradient of the stream diminishes, there is a greater opportunity for biotic diversification, and consequently the food base becomes enhanced. Fish diversity increases in upstream sections where there is a gradient low enough to permit the establishment of alternating pools and riffles. The accumulation of soft sediments and the characteristically lower dissolved oxygen levels in the pools means that fewer feeding opportunities exist when compared to adjacent riffles.

Terrestrial organisms constitute a significant nutritional input in both upland and lowland streams. Ecologists refer to the organic resources that enter aquatic ecosystems from terrestrial ones as **allochthonous**, whereas those resources that arise from within the aquatic realm are termed **autochthonous**. Upland streams tend to receive a greater proportion of organic material from allochthonous sources, whereas autochthonous sources such as rooted aquatic plants and phytoplankton contribute a greater share downstream. Tropical Asian species, living in highly seasonal conditions influenced by monsoon rains, exhibit pronounced seasonal feeding activity in which allochthonous food resources are extensively used (Dudgeon, 2000). The abundant and diverse invertebrate community commonly referred to as **shredders** or **detritivores** subsist on leaf litter that falls into streams. These detritivores then become an important food source for fishes.

. . . .
Riverine Trophic Structure and the Role of Fishes

From the aforementioned trophic adaptations of stream-dwelling fishes, some general conclusions can be drawn. What is most apparent is the trophic versatility of stream fishes. Fish assemblage diversity increases with the availability of feeding niches. In streams of higher latitudes, invertebrate populations are the key. Where they are abundant, a diverse population of first-order carnivores—particularly

small cypriniform species—can be sustained. These fishes in turn support higher-order piscivores. Although a few algivorous species, such as the aforementioned bluntnose and stoneroller minnows, may be locally abundant, herbivores are in general poorly represented in temperate streams. Studies on the total energy budget of neotropical stream systems, however, indicate that opportunistic (omnivorous) species predominate and that approximately half of their diet consists of plant matter (Penczak et al., 1999).

The fish fauna of the Orinoco floodplain is an excellent illustration of the role that fishes play in an aquatic ecosystem. With a discharge that averages 38,000 m³s⁻¹, the Orinoco River ranks third behind the Amazon and Zaire rivers. The watershed covers more than 1.1 million km² (Fig. 31.6A) and is of inestimable value to neotropical studies because it has remained relatively unaffected by human activity (Lewis et al., 2000). Like other neotropical drainages, different tributaries of the Orinoco may vary dramatically in sediment load and water chemistry depending on their point of origin in the watershed. Tributaries that drain off the Andes carry large amounts of suspended solids due to the rapid weathering in the mountains of the Andes. Much like glacial streams of higher latitudes, these drainages have water with an opaque, milky appearance, and they are hence termed **white waters**. Streams arising in the east and south flow from the already extensively weathered Precambrian shield and hence carry little suspended and dissolved solids. These waters are stained dark from allochthonous organic matter, and they are therefore termed **black waters** (Fig. 31.6B). The transparency of the black waters permits blooms of phytoplankton and therefore supports abundant populations of planktivorous species (Lowe-McConnell, 1987).

The Orinoco floodplain—a complex network of lakes, ponds, and backwaters of the main stem of the river—is seasonally inundated. This flooding flushes massive amounts of streamside organic matter into the adjacent waters (Fig. 31.6C). Fishes tend to be relatively sparse in the main channels but congregate near banks and beaches, where productivity and terrestrial nutrient input are greater. The amazingly diverse fish assemblage that characterizes neotropical freshwaters is concentrated in comparatively few major taxa, and the resulting food web is dense with omnivores exploiting this tremendous organic input. The Orinoco assemblage is dominated by characins (60 percent of the total species), catfishes (20 percent of species), and New World knifefishes (10 percent of species). Cichlids constitute only 4 percent of the species (Rodríguez and Lewis, 1990, 1997). Riverine cichlids, like their better known brethren in the African Great Lakes, demonstrate much ecomorphological modification associated with trophic specialization

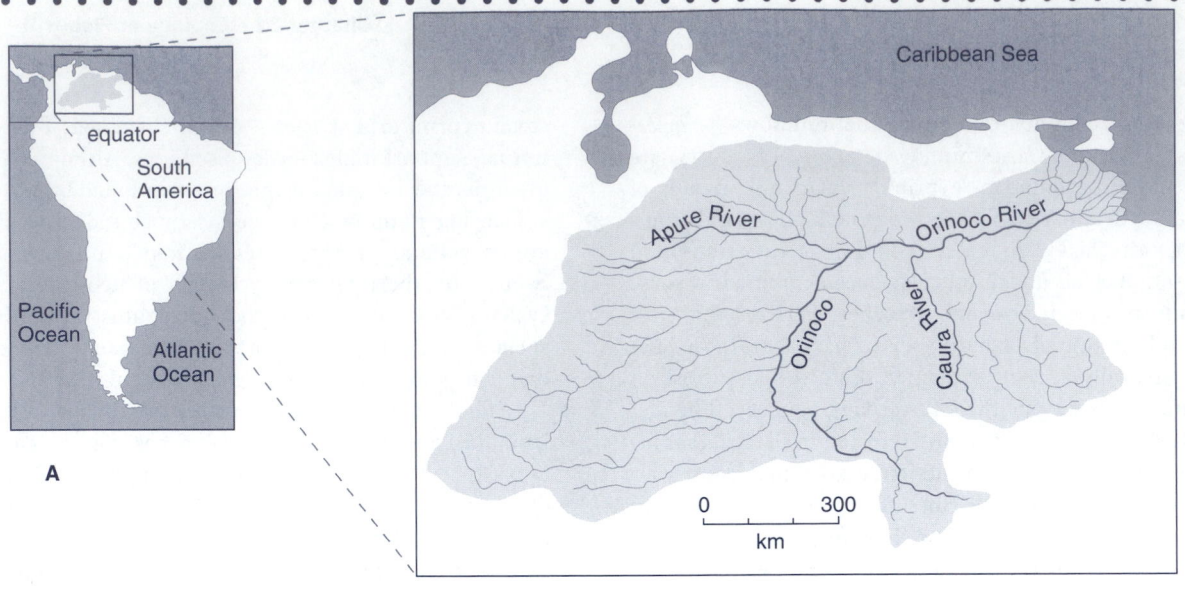

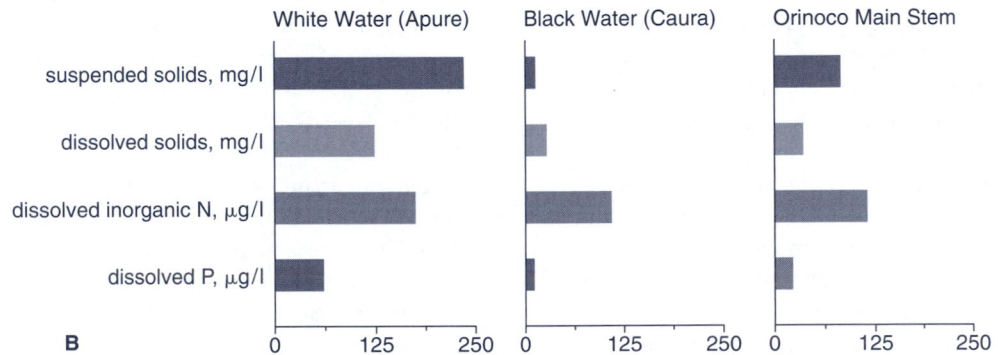

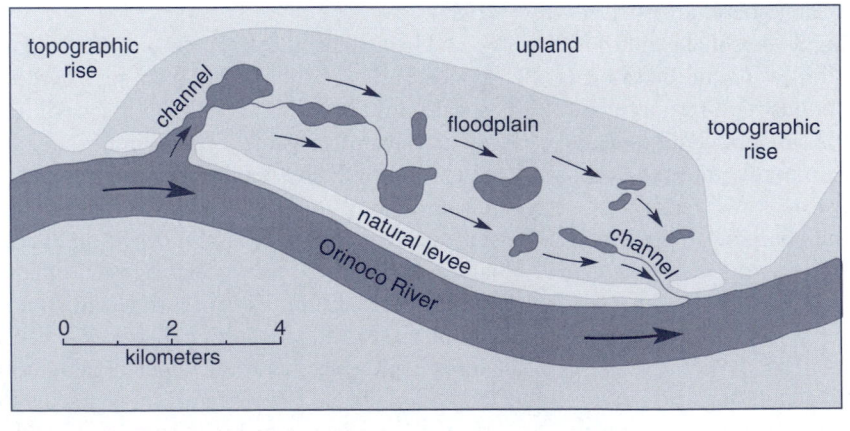

FIGURE 31.6

A, Map of the Orinoco watershed; **B,** Contrasts in dissolved and suspended sediment load between a white-water tributary (Apure River), a black-water tributary (Caura River), and the main stem Orinoco River; **C,** Hydrology of a floodplain adjacent to the Orinoco River. (From Lewis et al., 2000.)

(Winemiller et al., 1995). Yet they have not been nearly as successful as the ostariophysans in exploiting detritivore and herbivore niches.

The community structure, at first glance, seems complex beyond comprehension, but Lewis et al. (2000) claimed that the Orinoco fish community composition, like that of the rest of the floodplain ecosystem, is highly deterministic, with readily discerned causal connections with the geomorphology and hydrology of the region. The vicissitudes of biogeography may have brought characins and catfishes into the evolving environments of the neotropics, but their suite of adaptive features has enabled their successful integration into an ecosystem with a productivity heavily influenced by both in-stream and streamside vegetation.

THE LANDSCAPE CONNECTION

The Orinoco serves as an excellent example of the need to consider the landscape approach, in which the aquatic and terrestrial ecosystems must be viewed as intensively interacting components (see Chapter 30). Temperate riparian stream systems have also demonstrated the value of the landscape approach. In one such study, the fish community composition and age structure of the brown trout (*Salmo trutta*), the top carnivore of French upland streams, differed in streams with a dense canopy of streamside deciduous trees when compared with streams lacking such vegetation (Maridet et al., 1998).

As mentioned in Chapter 30, anadromous fishes serve to connect marine ecosystems with freshwater ecosystems by importing nutrients from the ocean into coastal drainages. The Japanese were among the first to recognize the interconnectedness of ecological systems and, in this sense, were pioneers in the landscape approach to ecological studies. In the early 20th century, they initiated regulations to protect *uo-tsuki-rin* (literally, "fish-attracting-forest"). Not only are migrating salmon a vital nutrient for terrestrial carnivores that forage along the stream, but their eggs and carcasses provide energy, carbon, nitrogen, and phosphorus for both aquatic and neighboring terrestrial ecosystems (Cederholm et al., 1999).

Efforts to preserve species are shifting from life history studies of individual organisms to investigations of the entire food web in which that species participates (Wootton et al., 1996). This integrated approach has great merit and is destined to shape the future of ecology and conservation biology. Future fish ecologists have unprecedented opportunities to reveal the underlying truths of the world's ecosystems. In the past few years, great advances have been made in understanding the ecology of rivers and streams, but much work remains to be done.

KEY POINTS AND CONNECTIONS

- Fully 41 percent of the known fish fauna can be found in the tiny fraction (about 0.01 percent) of the Earth's water supply that would be considered suitable habitat for freshwater fishes. Inland waters are classified as either lotic (moving water) or lentic (still water). Owing to the greater impact of the surrounding terrestrial landscape, inland waters experience a much greater degree of variation in chemical composition than marine waters.

- Flowing waters are classified along an ordinal gradient, starting with first-order streams, which represent the points of origin of lotic waters in upland locales, and culminating in fifth- and higher-order rivers, which constitute the major rivers draining the continents. According to the river continuum concept, the habitats of flowing waters are not discrete entities but exist along a continuum. Local geomorphic features, however, may result in a discontinuity of certain habitat features.

- Upland streams typically consist of well-oxygenated riffles interspersed with shallow pools, in which the dissolved oxygen content may be depleted by biological activity. Upland streams in temperate zones possess fish assemblages that are characterized by adaptations to sometimes swiftly flowing water. These adaptations include a comparatively small, fusiform body shape, large pectoral fins for burst swimming or agility, and a range of mouth designs to exploit a diversity of feeding opportunities.

The range of foraging activities and associated feeding structures of stream-inhabiting fishes are discussed in Chapter 23.

- The flowing waters in lowland temperate regions are less subject to seasonal fluctuation in the physicochemical environment than upland waters. The slower-moving waters result in fish assemblages that are more similar to those seen in lakes. Stream bottoms with a greater preponderance of fine sediments result in a greater diversity of rooted plants. Because of the abundance of microhabitats and feeding opportunities, fish assemblages tend to be more diverse in lowland temperate streams than in upland ones.

- Most of the freshwater discharge from continental landmasses comes from the enormous river systems of the tropics. Unlike temperate lotic systems, streams of the tropics defy conventional systems of classification. The fish fauna of neotropical river systems is the richest and most diversified of all freshwater ecosystems. As is the case with temperate river systems, the ostariophysans are the dominant ichthyofauna of tropical stream systems.

Several of the families of ostariophysan fishes found in the tropics are discussed in Chapter 10.

- The trophic structure of riverine ecosystems presents an excellent case for the landscape approach to ecological study. The ecosystem of a watershed is characterized by an intimate exchange of energy and materials between aquatic and terrestrial communities, with fishes often receiving the benefit in the form of allochthonous inputs of nutrients from the surrounding terrain.

Chapters 32–35 also include discussions of the trophic structure of aquatic ecosystems, including the roles of fishes.

FISH LINKS

http://www.hubbardbrook.org/ Gateway link to three websites dedicated to the Hubbard Brook Ecosystem Study.

http://www.desertfishes.org/ Home page of the Desert Fishes Council, dedicated to the preservation of desert ecosystems and their associated biota.

BUILDING AN ICHTHYOLOGY LIBRARY

Three valuable references on stream ecology:

Allan, J. D. 1995. *Stream ecology: Structure and function of running waters.* Chapman and Hall, London.
Cushing, C. E., and J. D. Allan. 2001. *Streams: Their ecology and life.* Academic Press, San Diego.
Hynes, H. B. N. 1970. *The ecology of running waters.* University of Toronto Press, Toronto.

Three essential texts on temperate freshwater fishes:

Matthews, W. J. 1998. *Patterns in freshwater fish ecology.* Chapman and Hall, New York.
Matthews, W. J., and D. C. Heins (Eds.). 1987. *Community and evolutionary ecology of North American stream fishes.* University of Oklahoma Press, Norman.
Mayden, R. L. (Ed.). 1992. *Systematics, historical ecology, and North American freshwater fishes.* Stanford University Press, Stanford, CA.

Two valuable references on tropical freshwater fishes:

Lévêque, C. 1997. *Biodiversity, dynamics, and conservation: The freshwater fish of tropical Africa.* Cambridge University Press, Cambridge, UK.
Lowe-McConnell, R. H. 1987. *Ecological studies in tropical fish communities.* Cambridge University Press, Cambridge, UK.

REFERENCES

Allan, J. D. 1995. *Stream ecology: Structure and function of running waters.* Chapman and Hall, London.
Allouche, S., A. Thévenet, and P. Gaudin. 1999. Habitat use by chub (*Leuciscus cephalus* L. 1766) in a large river, the French Upper Rhône, as determined by radiotelemetry. *Arch. Hydrobiol.* 145: 219–236.
Araujo-Lima, C., and M. Goulding. 1997. *So fruitful a fish: Ecology, conservation, and aquaculture of the Amazon's tambaqui.* Columbia University Press, New York.
Barthem, R., and M. Goulding. 1997. *The catfish connection: Ecology, migration, and conservation of Amazon predators.* Columbia University Press, New York.

Brooks, J. L., and S. I. Dodson. 1965. Predation, body size, and composition of plankton. *Science 150:* 28–35.
Cederholm, C. J., M. D. Kunze, T. Murota, and A. Sibatani. 1999. Pacific salmon carcasses: Essential contributions of nutrients and energy for aquatic and terrestrial ecosystems. *Fisheries 24*(10): 6–15.
Constantz, G. D. 1981. Life history patterns of desert fishes, pp. 237–290. In *Fishes in North American deserts,* R. J. Naiman and D. L. Soltz (Eds.). Wiley, New York.
Cushing, C. E., and J. D. Allan. 2001. *Streams: Their ecology and life.* Academic Press, San Diego.
Daniels, R. A. 1987. Comparative life histories and microhabitat use in three sympatric sculpins (Cottidae: *Cottus*) in northeastern California. *Env. Biol. Fishes 19:* 93–110.
Dudgeon, D. 2000. The ecology of tropical Asian rivers and streams in relation to biodiversity conservation. *Ann. Rev. Ecol. Syst.* 31: 239–263.
Duncan, A., and J. Kubecka. 1996. Patchiness of longitudinal fish distributions in a river as revealed by a continuous hydroacoustic survey. *ICES J. Mar. Sci. 53:* 161–165.
Ebert, D. J., and S. P. Filipek. 1988. Fish community structure and zonation related to stream habitat. *Proc. Ann. Conf. Southeast. Assoc. Fish Wildl. Agencies 42:* 234–242.
Echelle, A. A., E. W. Carson, A. F. Echelle, R. A. van den Bussche, T. E. Dowling, and A. Meyer. 2005. Historical biogeography of the New-World pupfish genus *Cyprinodon* (Teleostei: Cyprinodontidae). *Copeia 2005:* 320–339.
Eggleton, M. A., and H. L. Schramm, Jr. 1999. The "flood-pulse" concept in large temperate rivers: Modifying a paradigm. *Abstr. South. Div. Amer. Fish. Soc.* Midyear Meeting, Chattanooga, TN.
Fagan, W. F., P. J. Unmack, C. Burgess, and W. L. Minckley. 2002. Rarity, fragmentation, and extinction risk in desert fishes. *Ecology 83:* 3250–3256.
Fausch, K. D., and K. R. Bestgen. 1996. Ecology of fishes indigenous to the central and southwestern Great Plains, pp. 131–166. In *Ecology and conservation of Great Plains vertebrates,* F. L. Knopf and F. B. Samson (Eds.). Springer Verlag, New York.
Fernandes, C. C., J. Podos, and J. G. Lundberg. 2004. Amazonian ecology: Tributaries enhance the diversity of electric fishes. *Science 305:* 1960–1962.
Finger, T. R. 1982. Interactive segregation among three species of sculpins (*Cottus*). *Copeia 1982:* 680–694.
Fisher, W. L., and W. D. Pearson. 1987. Patterns of resource utilization among four species of darters in three central Kentucky streams, pp. 69–76. In *Community and evolutionary ecology of North American stream fishes,* W. J. Matthews and D. C. Heins (Eds.). University of Oklahoma Press, Norman.
Fuselier, L., and D. Edds. 1996. Seasonal variation of riffle and pool fish assemblages in a short mitigated stream stretch. *Southw. Nat.* 41: 299–306.
Gasith, A., and V. H. Resh. 1999. Streams in mediterranean climate regions: Abiotic influences and biotic responses to predictable seasonal events. *Ann. Rev. Ecol. Syst. 30:* 51–81.
Gelwick, F. P., and W. J. Matthews. 1992. Effects of an algivorous minnow on temperate stream ecosystem properties. *Ecology 73:* 1630–1645.
———, M. S. Stock, and W. J. Matthews. 1997. Effects of fish, water depth, and predation risk on patch dynamics in a north-temperate river system. *Oikos 80:* 382–389.
Goulding, M. 1980. *The fishes and the forest: Explorations in Amazonian natural history.* University of California Press, Berkeley.

Grossman, G. D., R. E. Ratajczak, Jr., M. Crawford, and M. C. Freeman. 1998. Assemblage organization in stream fishes: Effects of environmental variation and interspecific interactions. *Ecol. Monogr. 68:* 395–420.

Hillyard, S. D. 1981. Energy metabolism and osmoregulation in desert fishes, pp. 385–409. In *Fishes in North American deserts*, R. J. Naiman and D. L. Soltz (Eds.). Wiley, New York.

Hlohowskyj, I., and T. E. Wissing. 1986. Substrate selection by fantail (*Etheostoma flabellare*), greenside (*E. blennioides*) and rainbow (*E. caeruleum*) darters. *Ohio J. Sci. 86:* 14–129.

Holomuzki, J. R., and R. J. Stevenson. 1992. Role of predatory fish in community dynamics of an ephemeral stream. *Can. J. Fish. Aquat. Sci. 49:* 2322–2330.

Horton, R. E. 1945. Erosional development of streams: Quantitative physiography factors. *Bull. Geol. Soc. Amer. 56:* 275–370.

Huet, M. 1959. Profiles and biology of western European streams as related to fish management. *Trans Am. Fish. Soc. 88:* 153–163.

Hynes, H. B. N. 1970. *The ecology of running waters.* University of Toronto Press, Toronto.

Ibarra, M., and D. J. Stewart. 1989. Longitudinal zonation of sandy beach fishes in the Napo River Basin, eastern Ecuador. *Copeia 1989:* 364–381.

Jeffries, M., and D. Mills. 1990. *Freshwater ecology: Principles and applications.* Belhaven Press, London.

Kodric-Brown, A., and J. H. Brown. 1993. Highly structured fish communities in Australian desert springs. *Ecology 74:* 1847–1855.

Kortmulder, K. 1987. Ecology and behavior in tropical freshwater fish communities. *Arch. Hydrobiol. Beih. 28:* 503–513.

Kuehne, R. A. 1962. A classification of streams, illustrated by fish distribution in an eastern Kentucky creek. *Ecology 43:* 608–614.

Lamouroux, N., J.-M. Olivier, H. Persat, M. Pouilly, Y. Souchon, and B. Statzner. 1999. Predicting community characteristics from habitat conditions: Fluvial fish and hydraulics. *Freshw. Biol. 42:* 275–299.

Lévêque, C. 1997. *Biodiversity, dynamics, and conservation: The freshwater fish of tropical Africa.* Cambridge University Press, Cambridge, UK.

Lewis, W. M., Jr., S. K. Hamilton, M. A. Lasi, M. Rodríguez, and J. F. Saunders, III. 2000. Ecological determinism on the Orinoco floodplain. *BioScience 50:* 681–692.

Li, H. W., C. B. Schreck, C. E. Bond, and E. Rexstad. 1987. Factors influencing changes in fish assemblages of Pacific northwest streams, pp. 193–202. In *Community and evolutionary ecology of North American stream fishes,* W. J. Matthews and D. C. Heins (Eds.). University of Oklahoma Press, Norman.

Lonzarich, D. G., and T. P. Quinn. 1995. Experimental evidence for the effect of depth and structure on the distribution, growth, and survival of stream fishes. *Can. J. Zool. 73:* 2223–2230.

Lowe-McConnell, R. H. 1987. *Ecological studies in tropical fish communities.* Cambridge University Press, Cambridge, UK.

Magalhães, M. F. 1993. Feeding of an Iberian stream cyprinid assemblage: Seasonality of resource use in a highly variable environment. *Oecologia 96:* 253–260.

Magoulick, D. D., and R. M. Kobza. 2003. The role of refugia for fishes during drought: A review and synthesis. *Freshw. Biol. 48:* 1186–1198.

Maridet, L., J.-C. Wasson, M. Philippe, C. Amoros, and R. J. Naiman. 1998. Trophic structure of three streams with contrasting riparian vegetation and geomorphology. *Arch. Hydrobiol. 144:* 61–85.

Matthews, W. J. 1985. Critical current speeds and microhabitats of the benthic fishes *Percina roanoka* and *Etheostoma flabellare. Env. Biol. Fishes 12:* 303–308.

———. 1998. *Patterns in freshwater fish ecology.* Chapman and Hall, New York.

Milliman, J. D. 1990. Fluvial sediment in coastal seas: Flux and fate. *Nat. Resour. 26:* 12–22.

Milner, A. M., and G. E. Petts. 1994. Glacial rivers: Physical habitat and ecology. *Freshw. Biol. 32:* 295–307.

Moyle, P. B., and B. Herbold. 1987. Life-history patterns and community structure in stream fishes of western North America: Comparisons with eastern North America and Europe, pp. 25–32. In *Community and evolutionary ecology of North American stream fishes,* W. J. Matthews and D. C. Heins (Eds.). University of Oklahoma Press, Norman.

Naiman, R. J., J. J. Magnuson, and P. L. Firth. 1998. Integrating cultural, economic, and environmental requirements for fresh water. *Ecol. Appl. 8:* 569–570.

Ostrand, K. G., and G. R. Wilde. 2004. Changes in prairie stream fish assemblages restricted to isolated streambed pools. *Trans. Am. Fish. Soc. 133:* 1329–1338.

Penczak, T., A. A. Agostinho, N. S. Hahn, R. Fugi, and L. C. Gomes. 1999. Energy budgets of fish populations in two tributaries of the Paraná River, Paraná, Brazil. *J. Trop. Ecol. 15:* 159–177.

Pielou, E. C. 1998. *Fresh water.* University of Chicago Press, Chicago.

Poff, N. L., and J. D. Allan. 1995. Functional organization of stream fish assemblages in relation to hydrological variability. *Ecology 76:* 606–627.

Poole, G. C. 2002. Fluvial landscape ecology: Addressing uniqueness within the river discontinuum. *Freshw. Biol. 47:* 641–660.

Power, G. 1997. A review of fish ecology in Arctic North America. *Am. Fish. Soc. Symp. 19:* 13–39.

Power, M. E., and W. J. Matthews. 1983. Algae-grazing minnows (*Campostoma anomalum*), piscivorous bass (*Micropterus* spp.), and the distribution of attached algae in a small prairie-margin stream. *Oecologia 60:* 328–332.

———, W. J. Matthews, and A. J. Stewart. 1985. Grazing minnows, piscivorous bass, and stream algae: Dynamics of a strong interaction. *Ecology 66:* 1448–1456.

Reist, J. D. 1997. The Canadian perspective on issues in Arctic fisheries management and research. *Am. Fish. Soc. Symp. 19:* 4–12.

Rodríguez, M. A., and W. M. Lewis, Jr. 1990. Diversity and species composition of fish communities of Orinoco floodplain lakes. *Nat. Geog. Res. 6:* 319–328.

———, and ———. 1997. Structure of fish assemblages along environmental gradients in floodplain lakes of the Orinoco River. *Ecol. Monogr. 67:* 109–128.

Saint-Paul, U., and G. Bernardinho. 1988. Behavioral and ecomorphological responses of the neotropical pacu *Piaractus mesopotamicus* (Teleosteii, Serrasalmidae) to oxygen-deficient waters. *Exp. Biol. 48:* 19–26.

Smith, G. R. 1981. Late Cenozoic freshwater fishes of North America. *Ann. Rev. Ecol. Syst. 12:* 163–193.

Soltz, D. L., and R. J. Naiman. 1981. Fishes in deserts: Symposium rationale, pp. 1–9. In *Fishes in North American deserts*, R. J. Naiman and D. L. Soltz (Eds.). Wiley, New York.

Storfer, A. 1999. Gene flow and local adaptation in a sunfish–salamander system. *Behav. Ecol. Sociobiol. 46:* 273–279.

Taylor, C. M. 2000. A large-scale comparative analysis of riffle and pool fish communities in an upland stream system. *Env. Biol. Fishes 58:* 89–95.

Thévenet, A., and B. Statzner. 1999. Linking fluvial fish community to physical habitat in large woody debris: Sampling effort, accuracy and precision. *Arch. Hydrobiol. 145:* 57–77.

Ultsch, G. R., H. Boschung, and M. J. Ross. 1978. Metabolism, critical oxygen tension, and habitat selection in darters (*Etheostoma*). *Ecology 59:* 99–107.

Vannote, R. L., G. W. Minshall, K. W. Cummins, J. R. Sedell, and C. E. Cushing. 1980. The river continuum concept. *Can. J. Fish. Aquat. Sci. 37:* 130–137.

Vari, R. P., and L. R. Malabarba. 1998. Neotropical ichthyology: An overview, pp. 2–11. In *Phylogeny and classification of neotropical fishes,* L. R. Malabarba, R. E. Reis, R. P. Vari, Z. M. S. Lucena, and C. A. S. Lucena. Editoria Universitaria, Pontificia Universidade Catolica do Rio Grande do Sul, Porto Alegre, Brazil.

Walsh, S. J., D. C. Haney, and C. M. Timmerman. 1997. Variation in thermal tolerance and routine metabolism among spring- and stream-dwelling freshwater sculpins (Teleostei: Cottidae) of the southeastern United States. *Ecol. Freshw. Fish 6:* 84–94.

Walters, D. M., D. S. Leigh, M. C. Freeman, B. J. Freeman, and C. M. Pringle. 2003. Geomorphology and fish assemblages in a Piedmont river basin, U.S.A. *Freshw. Biol. 48:* 1950–1970.

Ward, J. V., and K. Tockner. 2001. Biodiversity: Towards a unifying theme for river ecology. *Freshw. Biol. 46:* 807–819.

Warren, M. L., B. M. Burr, S. J. Walsh, H. L. Bart, Jr., R. C. Cashner, D. A. Etnier, B. J. Freeman, B. R. Kuhajda, R. L. Mayden, H. W. Robison, S. T. Ross, and W. C. Starnes. 2000. Diversity, distribution, and conservation status of the native freshwater fishes of the southern United States. *Fisheries 25*(10): 7–31.

Winemiller, K. O. 1989. Development of dermal lip protuberances for aquatic surface respiration in South American characid fishes. *Copeia 1989:* 382–390.

———, L. C. Kelso-Winemiller, and A. L. Brenkert. 1995. Ecomorphological diversification and convergence in fluvial cichlid fishes. *Env. Biol. Fishes 44:* 235–261.

Wootton, J. T., M. S. Parker, and M. E. Power. 1996. Effects of disturbance on river food webs. *Science 273:* 1558–1560.

Wootton, R. J. 1990. *Ecology of teleost fishes.* Chapman and Hall, New York.

———. 1992. *Fish ecology.* Chapman and Hall, New York.

32

ECOLOGY OF FISHES III: FRESHWATER FISHES IN STILL WATERS

ENVIRONMENTAL FEATURES OF LENTIC SYSTEMS

Origins of Lakes
Lake Environments as Habitat for Fishes
Temperate Lentic Environments
Tropical Lentic Environments
    World Distribution of Tropical Lakes

FISH ASSEMBLAGES OF LENTIC WATERS

Temperate Lakes
    Origins of Temperate Fish Assemblages
Tropical Lakes

TROPHIC ADAPTATIONS

Ecomorphology and Dietary Versatility
Species Flocks
Optimal Foraging Theory

HYPOGEAN SYSTEMS: AN UNUSUAL AQUATIC HABITAT

INLAND SALINE WATERS AND ASSOCIATED FISH ASSEMBLAGES

In the previous chapter, we recognized the importance of the landscape approach to ecological study and considered flowing waters as an integral element of the watershed. The destination of these flowing waters may be the ocean, or it may be a basin that has arisen either as a consequence of geological processes or through human intervention—the impoundment of river systems. In this chapter, we will discuss freshwater fish communities that are adapted to conditions quite different from those observed in flowing waters. Because they are the recipient of sediments transported to them by flowing waters, and because of their own capacity for organic production, *lakes* are destined to be comparatively short-lived aquatic phenomena. Over the course of several hundred or thousand years, depending on the size of the impoundment, the lake will experience a process of *succession,* in which an open body of water becomes a marsh, then a swampy meadow, and eventually is reclaimed by terrestrial plant and animal communities. A host of different families of fishes are adapted to the different stages in this successional process. In this chapter, we will investigate the nature of these *lentic* (still water) ecosystems and the fishes that characterize them.

•
ENVIRONMENTAL FEATURES OF LENTIC SYSTEMS
•

Origins of Lakes

Bodies of still water can form across the landscape in several ways, most of which can be classified as one of the following: (1) glacial, (2) tectonic, (3) landslide, (4) volcanic, (5) solution, (6) fluviatile, or (7) shoreline (Cole, 1994). The University of Wisconsin, a wellspring of limnological study in the early 20th century, is located on the shores of Lake Mendota—an example of a glacial lake. Glacial events have figured significantly in the origins of many lakes, especially in Europe and North America. Best known of these *glacial* lakes are the Great Lakes, created when glacial scour left basins filled with meltwater. A glance at a map of North America reveals a chain of massive glacial lakes that runs from Great Bear Lake southeast to the Great Lakes in a line that approximates the retreat of the Laurentian ice sheet approximately 10,000 years ago. Great Slave Lake, in the Canadian Northwest Territory, is the deepest lake in North America, at more than 600 m.

Tectonic lakes are the result of instability in the earth's crust. Buckling and shifting of the crust has created depressions that fill with water. One of the most dramatic examples of these is Reelfoot Lake on the Kentucky–Tennessee border. This may have been one of the most rapidly created lakes, as it came into existence as a result of a series of massive earthquakes in 1811–1812. The quakes changed the course of the Mississippi River, which flooded a series of depressions created by the earthquake. Mountain Lake in Virginia, one of the few natural lakes in the Appalachian Mountains, was created in a valley dammed by a *landslide*. Similar phenomena have been identified in the highlands of Australia and Costa Rica. Crater Lake, at 589 m the deepest lake in the United States, is *volcanic* in origin. It was formed by the explosion and subsequent collapse of Mount Mazama about 6,800 years ago.

Solution lakes are formed in basins created by the dissolution and removal of substrate, usually carbonates that are more readily soluble in mildly acidic waters. Limestone topography that is prone to dissolution is commonly termed **karst**. Among the best known karst lakes are those of central Florida. An endemic subspecies of euryhaline cyprinodont (*Cyprinodon variegatus hubbsi*) has become adapted to the elevated concentrations of dissolved calcium characteristic of these Florida solution lakes (Guillory and Johnson, 1986). In the Yucatán, karst topography has created a series of steep-sided basins termed *cenotes*. Many are filled with the bones from human sacrifices performed in the times of the Mayas. Karst topography is also the source of extensive cave systems. The fish fauna of these caves will be given special consideration later.

Fluviatile lakes are essentially wide stretches of a river where the current slows to a near standstill. Lake Pepin, on the Mississippi River, is a good example. It was once noted for its large populations of paddlefish (*Polyodon*). *Levee* lakes and *oxbow* lakes are created when parts of the floodplain are flooded and isolated from the adjacent river when the waters recede. In Australia, they are known by the aboriginal term *billabong*. *Shoreline* lakes are created through oceanic wave action, which creates dunes and sand spits that may isolate bodies of water from the sea. These lakes may vary in the extent of marine versus freshwater influence on their biotic composition. In the Bahamas, sand dunes became cemented into series of ridges that trapped water behind them (Fig. 32.1). Euryhaline species, such as gobiids, cyprinodonts, and poeciliids, abound in these waters (Barton and Wilmhoff, 1996). Because of their marine affinities, the shoreline lakes will be treated separately.

Once a lake has been created by one of the many geological processes that can cause basins to form in the surface of the Earth, it begins to fill up with materials washed in by streams, blown in by winds, or produced in the lake itself. Usually, lakes have a life span measured in the thousands of years—a mere instant in geological time. Lakes in higher latitudes are comparatively young, as they were formed after the retreat of the last glaciers. Material deposited by retreating glaciers is termed **moraine**; this material may trap and retain water, forming shallow lakes. Small, shallow lakes, such as those formed from moraine or caused by a landslide damming a small valley, may become extinct in a few centuries. The natural progression is from lake to pond to swamp and finally to dry ground, if the processes of lake senescence are allowed to proceed unabated. Reelfoot Lake, which is popular for its sport fishery, may become one of the shortest-lived lakes. Erosion from adjacent farmlands is causing the lake to rapidly turn into marshland.

Lakes that owe their origin to major geological phenomena on the Earth's surface have a greater degree of permanence. Lakes such as Lake Baikal in Siberia or the deep rift lakes of Africa have life spans measured in the millions of years (see Going Deeper). Yet in these lakes, too, the inexorable forces of environmental and, hence, biotic transformation are still occurring, albeit at a much slower rate.

•
Lake Environments as Habitat for Fishes

The fish assemblages that inhabit inland waters are the consequence of their own requisite adaptations to lentic conditions and of the happenstance of biogeography. Species that arrived in lacustrine environments at a time in their ecological succession when suitable niches were available became established. Species that were less suited to that succession

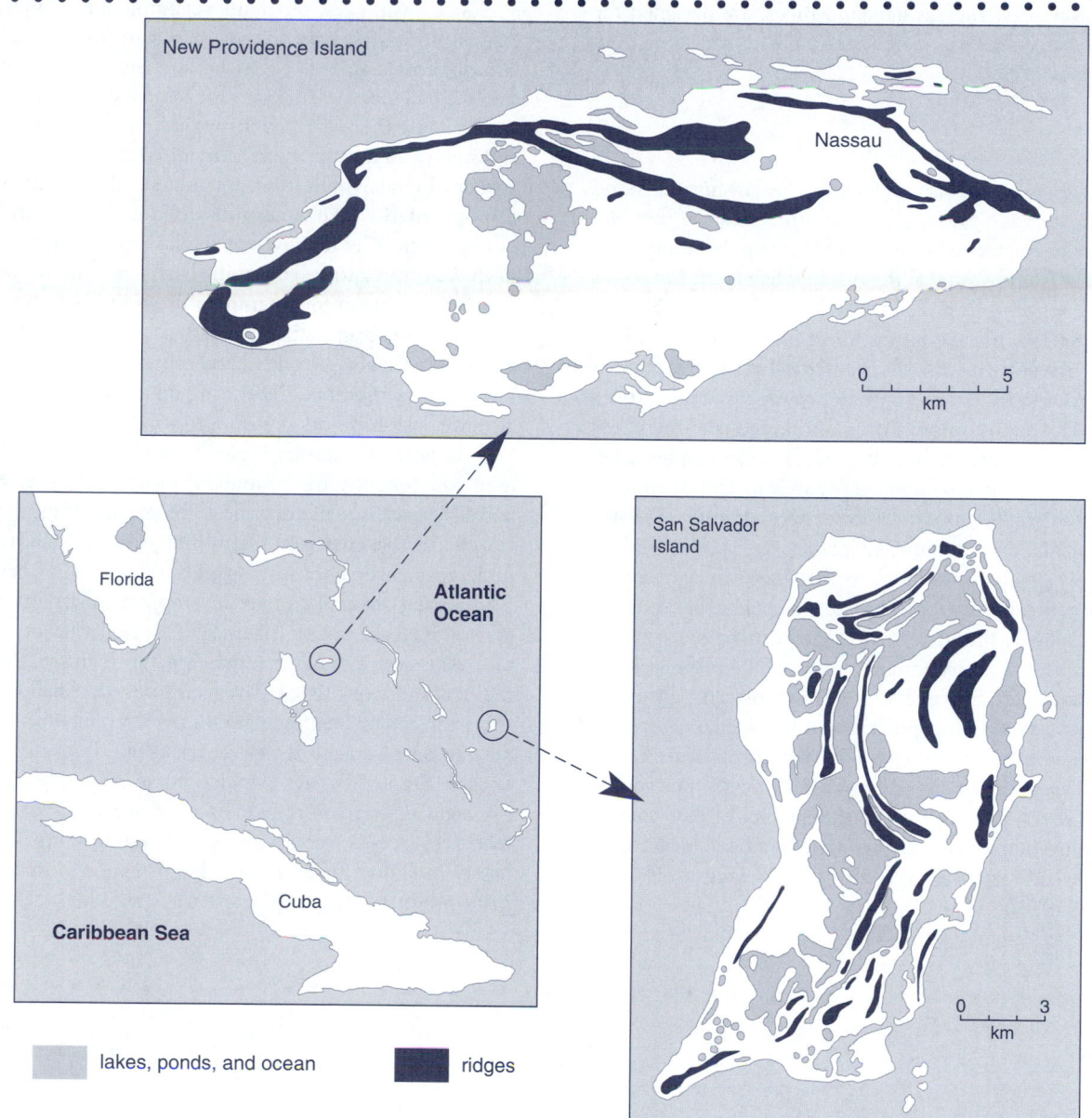

FIGURE 32.1

The maps of New Providence and San Salvador islands, showing formation of coastal lakes.
(From Sealey, 1985.)

stage were consigned to oblivion. Lakes, like oceans, provide habitats for **pelagic** species, which are adapted to existence in the water column, and **benthic** species, which are adapted to life on or near the substrate. A range of feeding opportunities is available to lake-dwelling fishes. Still waters permit the development of a plankton community. In contrast to stream-dwelling communities, planktonic organisms are abundant in still waters, and planktivory figures significantly in lentic fish assemblages. A range of body morphologies can be observed. Deep-bodied forms, such as centrarchids, are well adapted to existence in dense shoreline vegetation, where agility is more important than sustained swimming. Smaller forage fishes, such as clupeids and cyprinids, may be more fusiform to enhance their ability to escape from predators.

Larger piscivores, such as gars and pikes, are designed for short bursts of speed associated with ambush attacks on prey (see Chapter 19, Going Deeper).

Temperate Lentic Environments

Young lakes in temperate regions are low in nutrients and organic materials. These **oligotrophic** lakes, as they are called, are characterized by an abundant supply of dissolved oxygen at all depths, so that the resident fish fauna has access to the entire body of water. Thermal stratification, which, in summer, layers the lake into a warm, surface **epilimnion**; a middle **mesolimnion**, characterized by the presence of the **thermocline** (a zone of rapidly changing temperature); and a lower, cool **hypolimnion**; permits the year-round existence of cold-water fishes (Fig. 32.2). Oligotrophic lakes support a relatively low biomass per unit of area or volume. Steep, rocky banks support a sparse representation of rooted aquatic plants, and the nutrient-poor water supports a limited supply of phytoplankton, so that the primary productivity is generally low. Because the recycling of nutrients proceeds slowly, the growth of fishes can be slow in this condition, and the replacement of any fish biomass removed may take considerable time. One of the most extreme examples of this was seen in populations of brook char (*Salvelinus fontinalis*) stocked in an extremely oligotrophic lake in the Sierra Nevada mountains of eastern California. The combination of low temperatures and the paucity of food sources resulted in stunted populations that lived far longer than the normal life span for the species—as long as 24 years (Reimers, 1979).

As a basin ages, nutrients accumulate by minerals entering into solution, by organic matter washing or falling in, and by various other processes. Nutrients can be trapped in the lake by incorporation into the biomass of the lake. These materials are then recycled through the actions of scavengers and microbial decomposers. Any nutrients reaching a soluble state in the hypolimnion are not likely to be redistributed throughout the lake until there is a general recirculation of the water. The turnover of the lake occurs upon cooling in the autumn and warming in the spring, the result being that a generally uniform temperature is reached throughout the lake. This biannual circulation pattern redistributes the nutrients, dissolved oxygen, and other materials in the lake, but it can be modified depending on such factors as depth, altitude, latitude, and exposure to wind.

As bottom deposits increase and the nutrient content (and consequently the productivity) of the lake grows, the body of water is generally said to be passing from the oligotrophic to the **eutrophic** condition, and the ichthyofauna undergoes a process of succession, as discussed in Chapter 30. The greater diversity of feeding opportunities promotes a general increase in fish diversity. Accumulation of soft bottom deposits, including much organic material, enhances the development of rooted vegetation in the shallow waters and provides niches for many burrowing organisms, which may become available as forage for fishes. The enriched water can support denser populations of phytoplankton, and the zooplankton community is thus provided with greater food resources. Overall species richness in temperate lakes has been shown to be positively correlated with both primary productivity and lake surface area. The surface area is

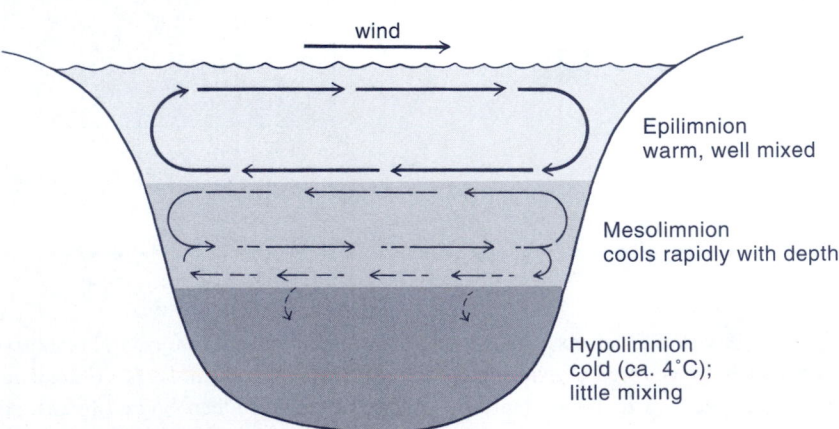

FIGURE 32.2
Zonation of a temperate oligotrophic lake, showing thermal stratification.

especially important in determining the species richness of fishes. A peak productivity value has been calculated at somewhere between 30 and 300 g C/m² yr. Increasingly eutrophic conditions will have the effect of decreasing species diversity (Dodson et al., 2000).

Rapid recycling of nutrients can result in the production of a large fish biomass. Because eutrophic lakes are capable of rapid replacement of biomass that is removed, they are suitable environments in which to develop sustainable fisheries for desirable sport and commercial fishes.

There are some trade-offs involved in the eutrophication process. The decomposition of organic material requires much oxygen, and the hypolimnetic waters of a rich lake can become anoxic. This can put some fishes in an uncomfortable squeeze during the warmer months. As surface waters warm, fishes may descend into the cooler depths, only to encounter insufficient concentrations of dissolved oxygen. Massive fish kills can result. Salmonid populations in some eutrophic lakes become confined within a narrow stratum in the mesolimnion, thus restricting their feeding opportunities. Color and transparency of the water can be affected during eutrophication due to organic compounds in solution and suspended materials, including plankton. The amount of carbon dioxide can increase significantly, as can the amount of ammonia and ammonium compounds. All these changes can be intensified and accelerated artificially by fertilization or organic pollution.

The natural succession of still-water habitats ultimately leads to the extinction of open water, with the final stages being bog ponds and swamps. These environments are characterized by shallow, acid water; emergent vegetation; and soft, organically enriched bottoms. These waters are usually well past the peak of productivity as far as fish are concerned, and in North America, they do not usually support a varied fish fauna. Some waters are termed *dystrophic* if they lack essential elements such as calcium. This results in lower populations of bacteria and, hence, inhibited rates of decomposition. Acidified bogs result from the accumulation of dead organic matter. Species that can tolerate the lower pH resulting from these conditions may be found here (Matthews, 1998). These include the aptly named swampfish (*Chologaster cornuta*), the pirate perch, *Aphredoderus sayanus* (see Chapter 13), and the pygmy sunfishes of the family Elassomatidae (see Chapter 15).

The acidification of lacustrine environments and the consequent diminution of bacterial decomposition may also come about through the introduction of atmospheric pollutants via "acid rain"—particularly in higher latitudes. Lakes that are located in granite bedrock, such as in the northeastern United States, southeastern Canada, and Scandinavia, have diminished buffering capacity and, hence, are more vulnerable to the effects of acid precipitation.

Tropical Lentic Environments

It is an artifact of history, what with limnological study originating in the Northern Hemisphere, that temperate lakes are so much better understood than tropical ones. This is ironic, given the complexity that exists in some tropical lake ecosystems. The diversity of fish species alone makes it difficult to generalize about the ecology of tropical lakes. Lakes in both temperate and tropical regions can be classified as natural or anthropogenic (i.e., of human origin). The magnitude of anthropogenic impact is evident from the fact that more than half of the world's large river systems have been modified with dams (Nilsson et al., 2005). In the temperate zone, natural bodies of water prevail, although there are vast tracts of land that have been transformed into lakes and reservoirs for irrigation, flood control, or power generation. The system of hydroelectric reservoirs created under the authority of the Tennessee Valley Authority in Tennessee and Kentucky is one example. Another is the dam building activity that took place in the American Northwest during the middle of the 20th century, which transformed the Columbia River into a series of lakes. In the tropics, large river systems are the predominant bodies of freshwater, and most of what is understood about the ecology of tropical freshwater fishes pertains to species that inhabit these river systems. However, many examples of large lakes—both natural and anthropogenic—do exist. Exploding human populations in the developing countries of the tropics are responsible for the massive hydrological transformation of many of the largest free-flowing river systems into impoundments (see Chapter 39).

Like their temperate counterparts, tropical lakes vary tremendously in size and, consequently, in habitat characteristics, ranging from small pools that seasonally dry up to Lake Victoria in East Africa, with a total surface area of more than 68,000 km². Reservoirs, such as the Aswan Lake on the Nile or Lake Kossou on the Bandama River of the Ivory Coast, generally experience thorough flushing, so that the water volume may be exchanged several times in a year, thus ensuring a constant supply of nutrients from adjacent watersheds. The waters of tropical lakes stratify like those of temperate lakes do in the summer, and a thermocline may be observed between the epilimnion and the hypolimnion—remarkable given the small temperature difference between the surface and deep waters that are necessary to establish a thermocline. Whereas temperate lakes may experience uniform mixing with seasonal turnover, deep tropical lakes, such as those of the African Rift Valley may be permanently stratified. However, shallow equatorial lakes, which are exposed to windy conditions throughout the year, such as Lake George in Uganda, experience a continuous upwelling of nutrient-rich bottom water (Lowe-McConnell, 1987).

World Distribution of Tropical Lakes

The African continent has, by far, the greatest diversity and abundance of lentic waters (Lévêque, 1997). The best known are the lakes of the East African Great Lakes system—so remarkable that they merit special consideration (see Going Deeper). Enormous areas of swamplands occur throughout Africa as well. Here, wetland vegetation, in the form of papyrus or other emergent plants, is an integral part of the aquatic ecosystem, providing large quantities of energy and materials. As elsewhere in the world, giant hydroelectric or water-storage projects have transformed many of the free-flowing rivers of Africa into lakes. The aforementioned Aswan Lake on the Nile and Kariba Lake on the Zambezi River are just two examples. In India and Southeast Asia, most lentic fish production is associated with lakes of human origin. The Grand Lac of the Tonle Sap, a lateral lake associated with the Mekong River, is an example of a natural lake that has supported a large freshwater fishery.

South America, with the world's greatest total discharge of freshwater from its Amazon system, possesses few lakes of any appreciable size. Perhaps the most noteworthy are the large lakes of Nicaragua, which have a cichlid-dominated fish fauna like that seen in African lakes. These lakes are also home to one of the most unusual fishes to be found in freshwaters, the sawfish (*Pristis perrotteti*). Pristids are more commonly associated with shallow, marine waters. The bull shark (*Carcharhinus leucas*) is another elasmobranch species that has ascended Caribbean drainages and has become well established in Lake Nicaragua. The mouths of tributaries to the major South American drainages, with their enormous surface area, and the numerous floodplain lakes of these drainages (see Fig. 31.6A), might also be considered lacustrine environments.

FISH ASSEMBLAGES OF LENTIC WATERS

Temperate Lakes

Salmonoids (trouts, whitefishes, and chars) have become the dominant open-water fish fauna of oligotrophic lakes in the Northern Hemisphere. They have also been widely introduced into similar lacustrine situations in the Southern Hemisphere. Fishes inhabiting the open waters of lakes are typically strong swimmers that can seek out and capture the crustacean or insect prey that forms most of the food base. Some of the whitefishes (*Prosopium*) and the sockeye salmon (*Oncorhynchus nerka*) are equipped with long, fine gill rakers for capturing small planktonic organisms. The young of the latter species spend their early life in lakes before migrating to the ocean. A smaller freshwater variant, called the kokanee (see later), completes its life cycle entirely in freshwater. Large, often piscivorous predators are frequently of the same species seen in larger rivers and streams. These predators include the rainbow trout (*Oncorhynchus mykiss*) and the lake trout (*Salvelinus namaycush*). The northern pike (*Esox lucius*), saugers and walleye (*Sander*), and yellow perch (*Perca flavescens*) are typical shallow-water fishes in oligotrophic lakes, but can be found in warmer water than salmonids.

The fish communities found in lakes that have progressed beyond oligotrophic, salmonid-supporting environments vary with climate and faunal region. In higher latitudes, pike, perch, sauger, and walleye can coexist with or succeed salmonids. In the warmer lentic waters of North America, many centrarchid species that are also found in river systems are encountered. These include black basses (*Micropterus*), crappies (*Pomoxis*), sunfishes (*Lepomis*), and rock basses (*Ambloplites*). Most centrarchids are compressiform, deep-bodied fishes adapted to life around cover, such as vegetation or submerged logs and brush. Some tend to be solitary, whereas other species move in schools, but they tend not to be wide-ranging in habit. Trophic versatility characterizes the centrarchids. The young of many species are planktivores. Adults of some species, such as the bluegill sunfish, are trophic generalists that continue to feed extensively on zooplankton, whereas crappies and black basses are piscivorous. The redear sunfish (*Lepomis microlophus*) and the pumpkinseed (*L. gibbosus*) have become specialists focusing primarily on molluscs. This has earned the redear sunfish the common name "shellcracker." Pumpkinseeds and bluegill sunfish commonly co-occur in lakes throughout their range in northeastern North America. Pumpkinseeds demonstrate a remarkable example of adaptive morphological differentiation in Adirondack lakes where bluegill sunfishes are absent. Here, two forms of pumpkinseeds have evolved—one that persists as a mollusc feeder and another that has become planktivorous (Robinson et al., 1993).

Typical bottom feeders in warm lakes and ponds are those species of ictalurid catfishes, buffalofishes (*Ictiobus*), or other catostomid suckers that also frequent larger rivers and streams. Some species of small, benthic families, such as the darters, also invade lake bottoms if a suitable substrate is available. Fish assemblage composition may seasonally vary as species adjust their vertical distribution in order to locate the optimum combination of temperature, dissolved oxygen, and available food (Fig. 32.3). The shads, herrings, and alewives (*Alosa* and *Dorosoma*) of the primarily marine family Clupeidae and numerous species of cyprinids form an important part of the food base for predatory species. However, the introduction of clupeids as a food source to enhance sport fisheries is somewhat controversial. Some studies have suggested that the introduced clupeids, in addition to providing forage for larger piscivores, will compete with the planktivorous young of these species, thus stunting their growth and diminishing the production capacity for desirable species in certain impoundments. This is especially true for the gizzard shad (*Dorosoma cepedianum;* Lagler, 1956; Pflieger, 1975). The widely-introduced carp (*Cyprinus carpio*) can be very numerous in North American lakes and often disrupts the natural ecological relationships of native species.

At lower latitudes, mudminnows (*Umbra*), bullhead catfishes, some sunfish species, and, especially in warmer regions, gars (*Lepisosteus*) and the bowfin (*Amia calva*) inhabit swamps and backwaters. At higher latitudes, the Alaska blackfish (*Dallia pectoralis*) is found in bog environments and shallow tundra ponds. These bodies of water may freeze to the bottom, trapping the fishes, which survive if their body temperature does not drop much below 0°C. Sticklebacks (Gasterosteidae) inhabit swampy waters in the Northern Hemisphere, but they are also known from marine and brackish waters.

Origins of Temperate Fish Assemblages

As mentioned earlier, the fish assemblage that characterizes a given lake may be the result of biogeographical phenomena (see Chapter 29). In terms of evolutionary history, lakes can be viewed much like islands: In both cases, the resident biota is relatively isolated, with limited avenues of dispersal. Fish assemblage composition may also result from the inherent adaptive features of a given species that might enable it to inhabit a given body of water. In a classic study of factors affecting assemblage composition, Bill Tonn and John Magnuson studied the comparative impact of biogeographical versus environmental influences in determining the species composition of a series of Wisconsin lakes. Tonn and Magnuson determined that the lake fish fauna could be categorized as one of two basic types: (1) an assemblage dominated by the mudminnow and different cyprinid species, or (2) one dominated by centrarchid sunfishes and pikes. They concluded that biotic conditions, such as predators and competitors, and abiotic conditions, such as seasonal thermal regimes and dissolved oxygen, took precedence over biogeography in determining the species composition of lake assemblages (Tonn and Magnuson, 1982; see also Diana, 1995). However, biogeography cannot be discounted as an influence on assemblage composition. This is especially apparent when considering the location and composition of fish fauna in refugia that existed during Pleistocene glacial events, and their routes of dispersal from these refugia to lakes that formed when the glaciers receded (see Chapters 4 and 29).

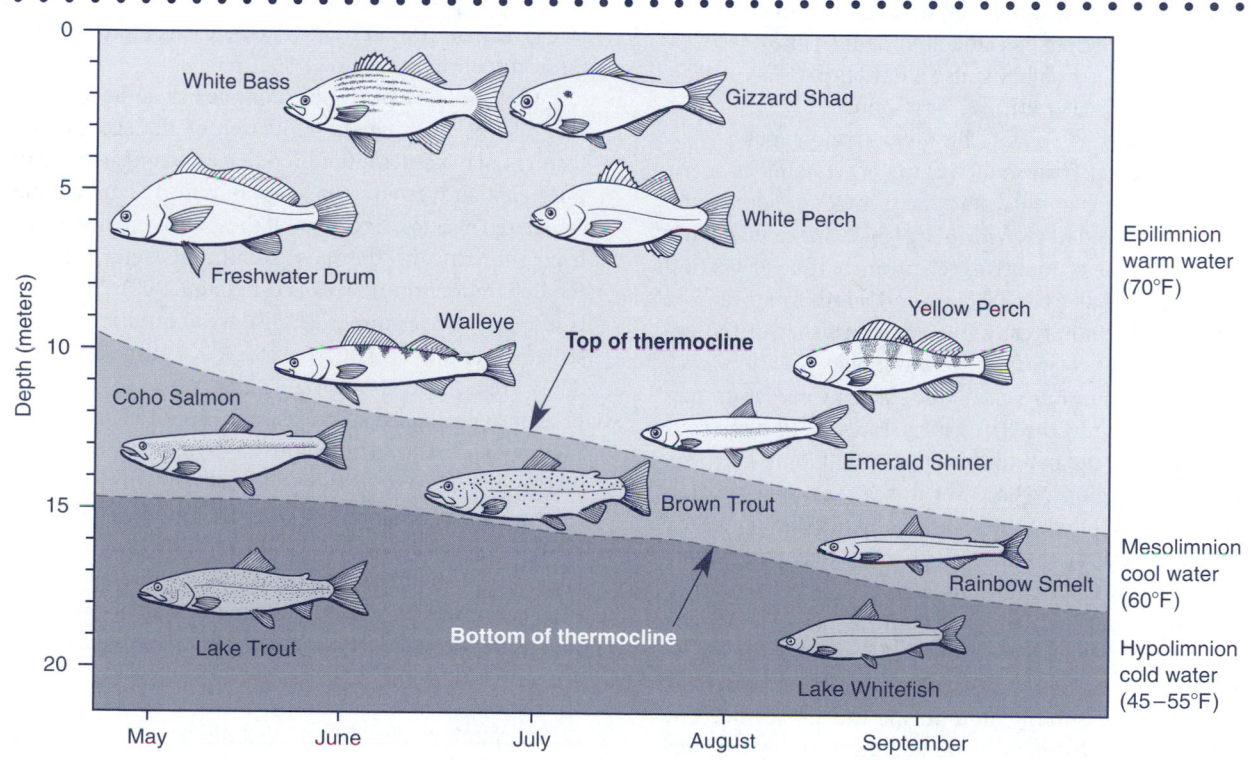

FIGURE 32.3

Distribution of fishes in Lake Erie in relation to summer temperature profile. (From Herdendorf et al., 1992.)

Tropical Lakes

As is the case with large river systems, fish assemblages in tropical lakes are dominated by ostariophysans and cichlids. Catfishes are also widespread in both the Old and New Worlds. Whereas the characoids are prevalent in Central and South America, cyprinids dominate lacustrine environments in Africa and Southeast Asia. Some African lakes (e.g., Chad and Turkana) may be home to a number of characoid species, but characoids are virtually absent in Asia. Cichlids, so abundant in South America and especially in Africa, are insignificant members of the Asian fish fauna. Some species of cichlids have also been widely introduced for pond culture. The fish assemblages of tropical lakes largely reflect the species composition seen in the riverine systems that drain into or out of the lake. In a natural lake, such as Lake Chad, there may be little difference between the riverine and lacustrine fish communities (Lowe-McConnell, 1987). Artificial impoundments, however, have experienced some successional transformation after the construction of dams for irrigation or hydroelectrical purposes. In 1968, the Kainji reservoir was created by impounding part of the Niger River. As a consequence, some fish families that were adapted to flowing water, such as the mormyrids, initially declined but later rebounded as the predator–prey dynamics changed in the ensuing years. Herrings became abundant in the open waters, and introduced cichlids also flourished (Blake, 1977; Lelek, 1973; Lewis, 1974; Lowe-McConnell, 1987).

The cichlids are by far the most conspicuous percomorphs in tropical freshwaters. Two other families merit brief mention: the Nandidae and the Centropomidae. Nandids are small predators with a leaf-like appearance that permits them to drift unobtrusively near their unsuspecting prey. They are found in Asia, Africa, and South America. The centropomids include freshwater-adapted fishes of the genus *Lates* as well as the coastal marine snooks (*Centropomus*). Fishes of the genus *Lates* are large African predators that were introduced into the Rift Valley lakes, with disastrous consequences for the native fish fauna (see Going Deeper).

In tropical regions, where wet and dry seasons alternate, numerous kinds of swamp-dwelling fishes have evolved habits and structures that permit them to occupy environments that are often lacking in dissolved oxygen, or even water. Several methods of air breathing have developed in swamp fishes; some are capable of moving overland to escape desiccation; and the African and South American lungfishes burrow into the mud during the dry season. The African lungfishes form a cocoon and enter into prolonged periods of aestivation in time of drought.

The so-called **annual fishes** of the family Aplocheilidae (the African genera *Fundulopanchax* and *Nothobranchius*) and the Rivulidae (the South American genera *Cynolebias, Pterolebias,* and others) respond to the periodic drying of their environments by laying large numbers of drought-resistant eggs, which develop over the dry season and hatch when the rains return (Berra, 1981; Breder and Rosen, 1966; Simpson, 1979). This unusual life history trait has enabled biological supply houses to market "instant fish" kits, consisting of eggs from the genus *Nothobranchius* that can be held in a suspended state until needed.

TROPHIC ADAPTATIONS

Compared to riverine ecosystems, lakes are much more limited in the degree to which faunal exchange takes place. Lakes may form in **endorheic** basins, where drainages flow only into the lake, or they may be **exorheic**—that is, their drainages ultimately flow to the sea. Unless the lake is part of a large river system and hence experiences a continuous flushing, it will be much more limited in terms of nutrient influx. Community structure and the associated trophic interdependence is much more self-contained in lakes than in streams. The input of energy and materials into streams is largely dependent on terrestrial sources (*allochthonous*); a lake's primary productivity, and hence its food web, is much more autochthonous.

One measure of trophic complexity is the length of the food chains forming the food web of the ecosystem. Two features may contribute to food chain length—overall productivity and ecosystem size, as reflected in habitat and associated species diversity. Of these two, ecosystem size has been shown to be the most significant factor determining food chain length in lakes (Post et al., 2000). Some authors (Lampert and Sommer, 1997) have questioned whether streams should even be considered a distinct ecosystem: They are a more appropriate subject for the landscape approach to ecological study, as they cannot be divorced from their associated watersheds. Lakes constitute a much more distinctive ecosystem unto themselves, and hence studies on trophic relationships in lakes are more feasible. From an evolutionary perspective, the degree of isolation afforded by lakes results in greater opportunities for endemism. Only in the headwaters of streams do we see similar opportunities for speciation as a consequence of restricted gene flow.

Ecomorphology and Dietary Versatility

Trophic polymorphism and **ecological character displacement**, in which dietary versatility results from the evolution of distinct morphs within and among species, respectively, are well known among fishes. They seem to be an adaptive feature of existence in relatively closed ecosystems, as they are most frequently seen in freshwater and anadromous forms that inhabit lakes (Smith, 1996). Trophic polymorphism is especially prevalent in recently glaciated lakes at high latitudes—a testament to the rapidity with which such adaptations evolve. For the inhabitants of these lakes, mainly trouts and salmon, chars, whitefish, and smelts, trophic polymorphisms take the form of modifications of the gill rakers and associated features to enable selective foraging on benthos, plankton, or other fishes.

Perhaps the best known example is that of the sockeye salmon (*Oncorhynchus nerka*), which exists in two forms—the anadromous form that is usually associated with river systems that have lakes, and the kokanee that becomes a permanent resident of those lakes. Kokanee mature earlier and have a smaller body size than their anadromous counterparts. The kokanee variant has independently arisen from different sockeye populations a number of times (Smith, 1996). Sticklebacks (*Gasterosteus*) inhabit many coastal waters, including lakes, streams, and estuaries. They have shown repeated evolution of divergent body shapes to facilitate feeding niche specializations (see Chapters 4 and 23).

Perhaps the most bizarre example is that of the cichlid genus *Perrisodus* of Lake Tanganyika. Its jaw morphology, which is adapted for snatching scales from unsuspecting fishes, shows distinct left- or right-handedness, enabling individuals to specialize in the side from which they approach their prey (Hori, 1993).

Trophic polymorphism naturally gives rise to partitioning of the lake habitat. The omul (*Coregonus autumnalis migratorius*) exists in Lake Baikal in three morphotypes—nearshore, open-water, and deep-water forms (Bronte et al., 1999). These examples serve to indicate that trophic polymorphism is not exclusively seen in young lakes but occurs in ancient ones as well.

Species Flocks

Ecological character displacement may be an incipient stage that eventually culminates in the evolution of *species flocks*. The cichlids of the African Great Lakes are such a dramatic case that they deserve special mention (see Going Deeper). Many other examples are known from lacustrine environments. By far the oldest lake on Earth is Lake Baikal in Siberia, with an age estimated at 25 to 30 million years (Martens, 1997). Within this lake, the sculpin subfamily Cottocomephorinae has diversified to include both benthic and open-water members. The benthic members are adapted to various bottom types and include some of the deepest-occurring freshwater fishes. They divide a food resource

consisting largely of midge and caddisfly larvae and numerous species of copepods. Pelagic members of the subfamily feed on planktonic copepods, with each fish species focusing on selected prey species. The related Comephoridae are pelagic fishes that undertake diurnal vertical migrations from depths up to 1,000 m. They are colorless and are distinguished by large mouths, extremely thin bones, and broad pectoral fins. Being viviparous and feeding on pelagic crustaceans, they have adapted completely to an open-water existence (Berg, 1949; Nelson, 1994).

Species flocks are also known among the Cyprinodontidae. As many as five species of the genus *Cyprinodon* have evolved in Laguna Chichancanab in the Yucatan Peninsula of Mexico (Humphries, 1984; Humphries and Miller, 1981; Stevenson, 1992), and another *Cyprinodon*-derived species flock has been identified in brackish inland waters of San Salvador Island, Bahamas (Bunt et al., 1999). Other well-known species flocks include an assemblage of 18 species of cyprinids in Lake Lanao in the Philippines (Kornfield and Carpenter, 1984), 13 species of cyprinids in Lake Tana in Ethiopia (Nagelkerke et al., 1994), and a complex assemblage of species flocks of killifishes in Lake Titicaca and surrounding lakes in the Andes (Parenti, 1984).

Optimal Foraging Theory

As discussed in Chapter 30, optimal foraging theory explains patterns of resource use in animals by demonstrating that they will maximize energy gain, based on the sizes and kinds of food selected, while seeking to minimize the time expended in acts of foraging—thus lessening the threat of predation (see also Chapter 37). Lacustrine fishes have provided some of the classic demonstrations of optimal foraging theory. Earl Werner and Donald Hall investigated the prey (*Daphnia*) size preferences of bluegill (*Lepomis macrochirus*) at different prey densities. When prey density was low, bluegill consumed all size classes of prey; as prey abundance increased, bluegill focused more attention on the largest individuals. In this way, the fishes were maximizing their intake while minimizing the time devoted to foraging (Werner and Hall, 1974; see also Diana, 1995). *Daphnia* are of special importance in lake ecosystems, owing to their pivotal role in trophic cascades. Planktivorous fishes, including the young of species that will eventually occupy different feeding niches when they mature, control *Daphnia* abundance in lakes. This, in turn, affects the density of phytoplankton that are grazed on by *Daphnia*. A complex, seasonal pattern of oscillation in the populations of phytoplankton and *Daphnia* is characteristic of temperate, shallow lakes (Scheffer, 1998). The presence of planktivorous fish species can have a profound effect on these population cycles.

Because lake systems are more self-contained and their resources are largely autochthonous, trade-offs between body and gonad growth may be critical. Under conditions where population size increases yet resource availability remains the same, stunting may result—an example of density-dependent regulation. As resources become more limited, less energy is available for growth. Fecundity may decline with a diminution of resources. Resource-constrained populations of yellow perch show earlier maturation, which may further influence growth, as less energy is available for growth now that a

GOING DEEPER • The Cichlid Fishes of the East African Great Lakes System

Since the time of Charles Darwin's visit to the Galápagos Islands, the phenomenon of adaptive radiation has been an integral component of the theory of evolution by natural selection. As any student of biology knows, Darwin discovered several similar but distinctly different forms of finches inhabiting the different islands. It was not until Darwin was back in England and had a chance to study his bird collections with an ornithologist colleague, John Gould, that he realized the significance of his discovery. Had Darwin been collecting fishes in the African Great Lakes, he probably would have had a nervous breakdown. Here resides an assemblage of species, all of the family Cichlidae, that is by far the greatest example of adaptive radiation known in the animal kingdom.

In eastern Africa, there are two long rift valleys, in which lake basins have formed (Fig. 32.4). Most ichthyofaunal studies have focused on three associated lakes—Victoria, Tanganyika, and Malawi. Although most of these bodies of water are found in the deep rift valleys, Lake Victoria is situated on an uplifted plateau between the two valleys. Lake Victoria is the largest in terms of surface area, and Lake Tanganyika is the deepest (Table 32.1). The fish fauna of each of these lakes is representative of the associated drainage system, but all of

these lakes are dominated to varying degrees by cichlids. Two of the northernmost rift valley lakes—Albert and Turkana—are unusual in that their fish faunas most closely represent the faunal characteristics of their drainage (the Nile River) and hence have fewer cichlids. Although Coulter (1991) claimed that Lake Tanganyika has the highest biodiversity of any lake on Earth, such assertions are difficult to prove, given the richness of the aquatic biota in this region and the limited state of our knowledge of them. Of the African Great Lakes, Tanganyika has the greatest number of fish families, with 20 of the Zaire River's 24 families adapted to the lake. Lake Malawi has only 9 of the Zambezi's 18 fish families, whereas Lake Victoria in the Nile drainage has 12 of the Nile's 17 families in residence (Lowe-McConnell, 1996). Ribbink and Eccles (1988) estimated that as much as 50 percent of the recognized cichlid taxa of Lakes Malawi and Victoria have yet to be formally described.

Complicating the picture is the fact that the East African Great Lakes are vastly different in age. Lake Tanganyika, the oldest, counts among its fish fauna 165 endemic species of cichlids. Lake Malawi may be the most speciose, with more than 500 species of cichlids, many of them as yet undescribed. Lake Victoria is the youngest and, until the past few decades, had

as many as 500 endemic species of haplochromine cichlids. Recent studies have suggested that Lake Victoria, in its present configuration, may be less than 15,000 years old, as it apparently dried up completely during the Late Pleistocene. Thus, the rate of evolution, at first glance, might seem astounding, as all these endemic cichlid species had to have evolved in the past 15,000 years (Fryer, 1997; Johnson et al., 1996; Won et al., 2005). Seehausen (2002) has suggested that this rate is consistent with that observed in other young lakes. But studies by Verheyen et al. (2003) have suggested that the history of the Lake Victoria haplochromines is much more complicated, and that a much longer evolutionary history is possible. The Lake Victoria haplochromines form part of a "superflock" that also ranges into neighboring Lakes Edward, George, and Kivu. These lakes may have served as a source for Lake Victoria cichlids, so that the desiccation of the lake may not correlate with the actual time involved in the speciation of the fish fauna— these neighboring lakes probably functioned as refugia for already highly differentiated stocks of haplochromines. The purported age of Lake Victoria and the consequent age of speciation events remains a subject of some debate (Stager et al., 2004; Verheyen et al., 2004).

TABLE 32.1 PHYSICAL AND ICHTHYOFAUNAL CHARACTERISTICS OF THREE MAJOR LAKES OF THE AFRICAN GREAT LAKES SYSTEM

Drainage	Lake	Age (My)	Surface area (km²)	Depth (m)	No. of fish families	Percentage cichlids
Nile	Victoria	0.75*	70,000	70	12	84
Zaire	Tanganyika	9–12	32,600	1,470	20	55.1–90
Zambezi	Malawi**	4.5–8.6	30,800	785	9	82.6

* Evidence suggests that the present Lake Victoria may have completely dried up approximately 14,000 years ago.

** Also sometimes referred to as Lake Nyasa.

Source: Lévêque,1997; Lowe-McConnell, 1987; Martins, 1997; Ichthyology Web Resources (www.biology.ualberta.ca/jackson.hp/iwr/iwr.html).

portion must be given over to gonadal development (Spangler et al., 1977). Perch populations in lakes at high latitudes show a periodic release from resource limitation, as these lakes are prone to seasonal winterkill, in which up to 90 percent of the population may be eliminated (Diana, 1995).

HYPOGEAN SYSTEMS: AN UNUSUAL AQUATIC HABITAT

Among the vertebrates, only salamanders rival the fishes in terms of their success in the inhabitation of subterranean (**hypogean**) environments. Many species of fishes inhabit

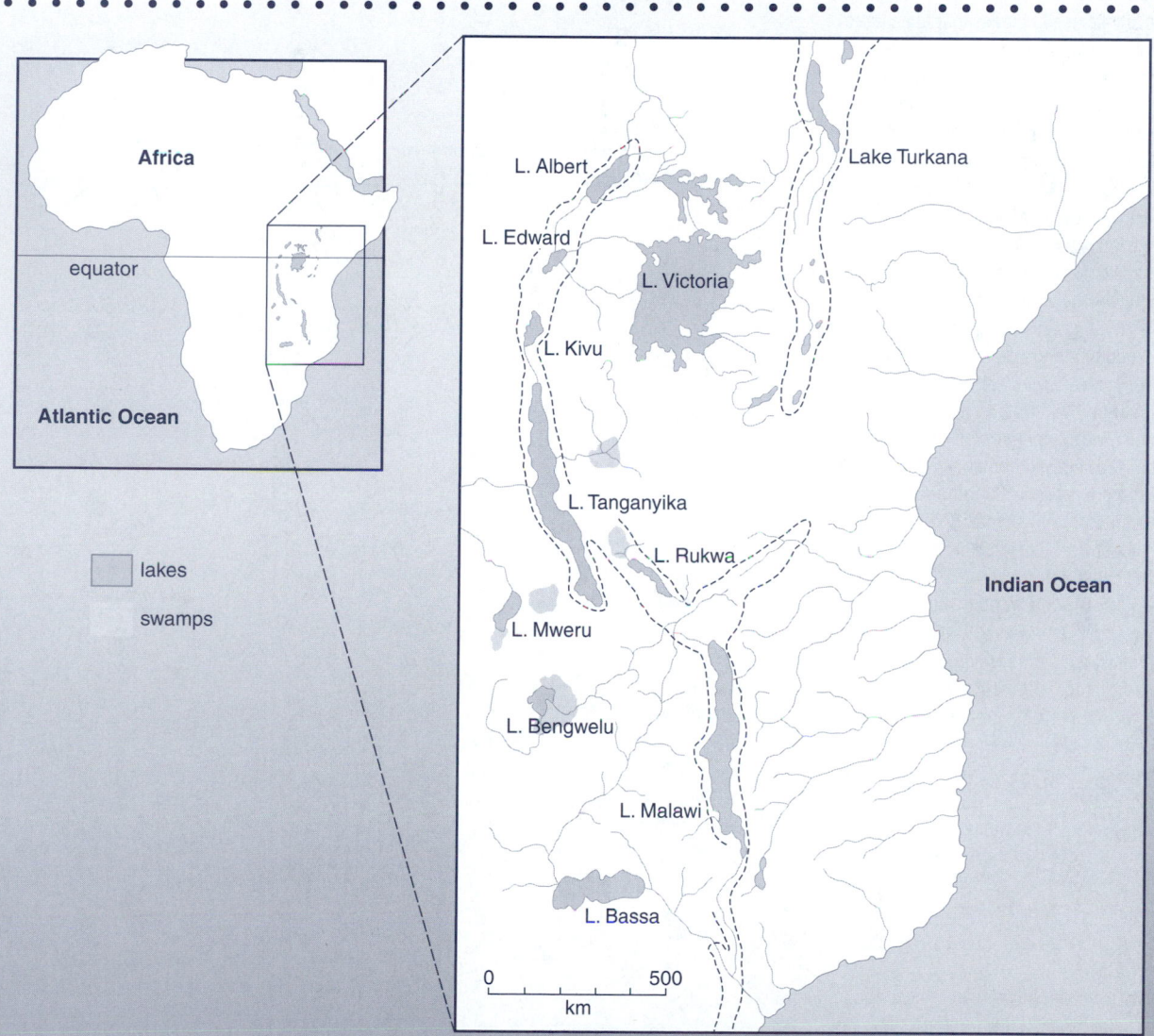

FIGURE 32.4
Map of the African Great Lakes system. Dashed lines indicate location of Rift Valley regions. (From Lévêque, 1997.)

waters beneath the ground that are partially or permanently deprived of light. As might be expected, these hypogean fishes display a range of remarkable adaptations to the unusual conditions associated with living in light-deprived areas (see Romero, 2001). Freshwater hypogean habitats usually consist of underground streams that run through karst-associated cave systems. Karst cave systems are especially prevalent in the southeastern United States, Mexico, and China. A gradation of aquatic habitats is usually present, ranging from conditions similar to noncave environments

Because the African Great Lakes differ in age, the cichlid assemblages that are characteristic of each lake can be inferred to represent different stages in the process of adaptive radiation (Sturmbauer, 1998). Cichlids are not the only family to have formed species flocks in the African Great Lakes. Catfishes of the families Clariidae, Bagridae, and Mochokidae, and spiny eels (Mastacembelidae) have also diversified in the lakes (Lowe-McConnell, 1996).

Cichlids have radiated to fill just about any ecological niche present in the lakes. Examples are phytoplankton eaters of the genus *Tilapia;* zooplankton feeders (*Haplochromis intermedius* and *Limnochromis permaxillaris*); mollusc eaters (*Macropleurodes bicolor*); and piscivores (*Rhamphochromis*). Other species feed on algae scraped from rocks (**aufwuchs**); some eat higher plants; and others filter organic matter from bottom deposits. Body form and feeding structures, including dentition and gill rakers, have become modified in a myriad of ways (Fig. 32.5). One group of haplochromine cichlids displays a remarkable specialization on different stages of development of other species of cichlids: Different species consume recently fertilized eggs, embryos, early larvae, late larvae, shoaling groups of young, and older individuals of different sizes (Greenwood, 1974; Kortmulder, 1987; Witte, 1981).

Studies on the community ecology and trophic structure of these complex communities reveal them to be consistently stable, characterized by long-term persistence and resilience of

FIGURE 32.5
Explosive adaptive radiation in body form and dentition in African Rift Valley lakes. (From Fryer and Iles, 1972, in Lowe-McConnell, 1987.)

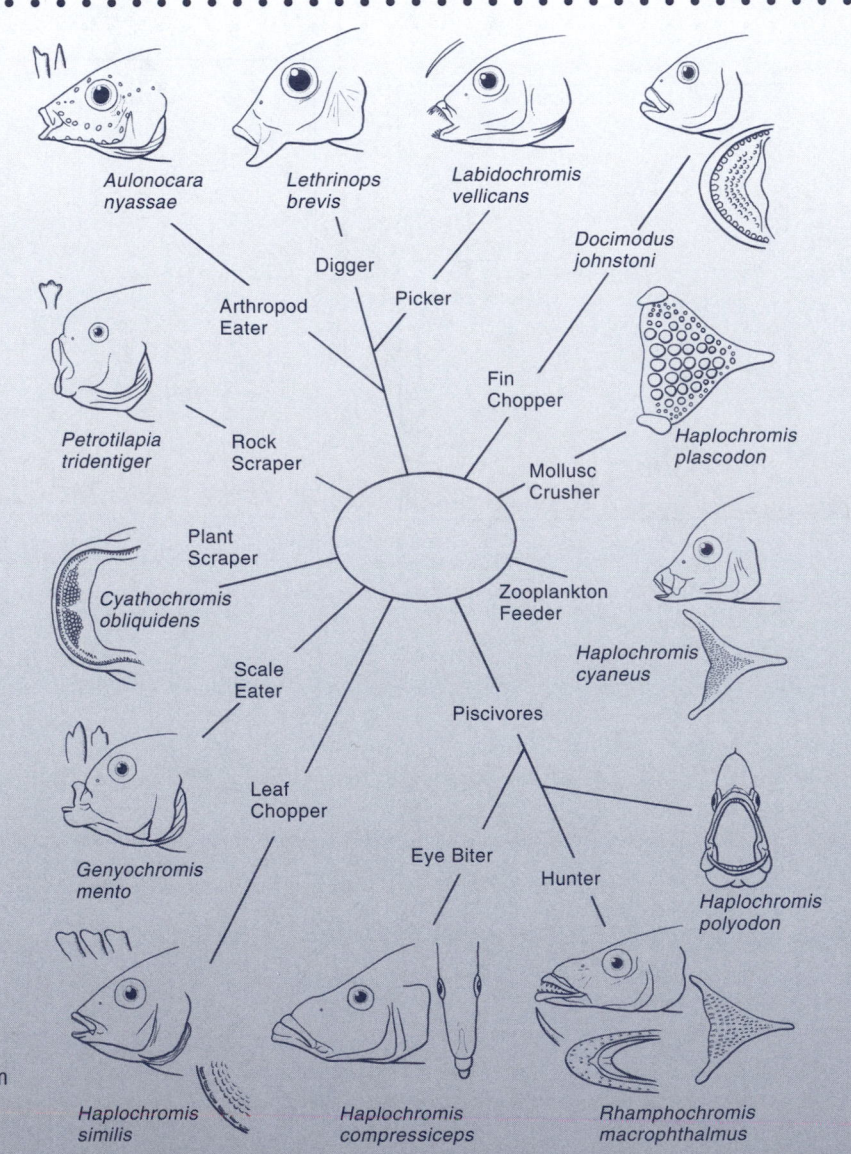

near the mouth, to a twilight area of decreasing light and minimal temperature fluctuations, to the lightless interior where temperatures show little if any seasonal fluctuation. Virtually all organic matter in the inner cave habitat must originate from outside. Streams flowing through may bring debris and plants, which either directly provide food for invertebrates or nurture fungi on which the invertebrates feed. In some caves, bat droppings provide considerable organic material. One remarkable exception is an entirely autochthonous cave ecosystem discovered in Romania, where a

structure (Hori, 1991). The rock-dwelling species, commonly referred to as *mbuna*, found in Lake Malawi are especially interesting owing to their extremely restricted distributions. These are termed **stenotopic** (as opposed to **eurytopic**) as they use finely partitioned resources in narrowly defined habitats and ranges. Their sedentary nature, coupled with the fact that their large, yolky eggs enable the young to hatch and remain feeding in the same habitat as the parents, reinforces their stenotopic character (Lowe-McConnell, 1987). An assemblage of more than 200 species of stenotopic, rock-dwelling cichlids has recently been described in Lake Victoria as well (Seehausen and Bouton, 1998).

There is one question that begs to be answered: What could account for such incredible speciation? What were the forces at work that culminated in this unprecedented event in the evolution of vertebrates? Biogeographical history, coupled with the inherent phylogenetic capacity of cichlids, must be considered. Morphological, physiological, and behavioral adaptations must be numbered among those features said to constitute phylogenetic capacity. We have already touched on biogeographic history when considering the impact of major lake level changes through the Pleistocene and possibly earlier. The comparative youth of Lake Victoria may or may not be significant in dating cichlid speciation. The stratigraphy and seismic history of the ancient Lake Tanganyika is also fairly well understood. During the Pleistocene, an extreme lowering of water levels caused the basin to be split into three isolated basins. Mitochondrial phylogenetic studies of the rock-dwelling cichlid assemblages have confirmed the impact that these geological events had on speciation events (Sturmbauer et al., 1997;

Verheyen et al., 1996). This suggests that the speciation events were *allopatric*—that is to say, they came about through physical isolation of the gene pools (see Chapter 4).

Another school of thought, though not discounting the influence of allopatry, claims that the extreme levels of diversity, coupled with their highly stenotopic consequences, could also come about through *sympatric* modes of speciation (see Chapter 4). According to this model, the cichlids achieved a level of genetic differentiation and isolation of their gene pools, ultimately culminating in the evolution of a diverse array of species, while they were still in contact. Either or both of these modes, repeated time and again and involving numerous partitioning events of the gene pool, ultimately gave rise to the staggering diversity of fishes seen among the African Great Lakes.

To be the beneficiaries of such evolutionary largesse, cichlids themselves must be bringing something to the table. Cichlids are members of one of the most successful groups of bony fishes, the suborder Labroidei. What we have learned of the labroids is that much of their adaptive capacity rests in the versatility of their jaw structure and function (see Chapter 16). Karel Liem (1973, 1975, 1980) suggested that the evolution of cichlid jaw structure was one of the uncoupling of component elements. Uncoupling these elements, thereby making them more independently functioning entities, allows for greater trophic versatility, and cichlids certainly show this. R. Craig Albertson of the Forsyth Institute in Boston has investigated the genetic foundations for the development of cichlid jaws and discovered that the dimensions of the lower jaw—and hence its feeding functions—may be controlled by a single gene. Such genetic streamlining directed at structures

that are critical in feeding processes makes the observed diversification of cichlids all that more explicable (Albertson et al., 2003; Pennisi, 2004).

Cichlids also offer good case histories for demonstrating the role of sexual selection. Mate choice, based on color preferences, plays an especially important role in the cichlids. Cichlids show pronounced sexual dimorphism, with the male in most species being significantly larger (Erlandsson and Ribbink, 1997). As is the case with most vertebrates, females are less conspicuously colored, yet they show a pronounced preference for bold coloration in potential partners. Coupled with extremely limited home ranges within narrowly defined microhabitats—as demonstrated especially among the rock-dwelling forms—rapid speciation of an allopatric or sympatric nature is feasible. There is evidence of incipient speciation in one cichlid species (*Pseudocrenilabrus philander*) in which allopatric populations can be differentiated based on subtle color variations (Twentyman-Jones et al., 1997). Incipient sympatric speciation among color morphs of *Neochromis omnicaeruleus* appears to be a consequence of sexual selection (Seehausen et al., 1999), and sexual selection also appears to play a significant role in the incipient speciation occurring among geographically distinct populations of *Pseudotropheus zebra* (Knight and Turner, 2004).

Finally, evolution of the community structure must have figured significantly in the resultant assemblages that characterize each of the African Great Lakes. The appearance of predators feeding in certain habitats or focusing on prey of certain sizes undoubtedly had an effect on the assemblage composition. The appearance of large predatory centropomids of the genus

total of 48 species of cave-adapted terrestrial and aquatic invertebrates—33 of which are endemic—form a food web that is ultimately dependent on populations of chemautotrophic bacteria (Sarbu et al., 1996).

Some fish species that normally live outside caves occasionally enter them for food or shelter. These are called **trogloxenes** and may include members of several families, including the Ictaluridae, Cyprinidae, and Cyprinodontidae. Fishes that normally spend part of their life cycle in and part of it outside of caves are termed **troglophiles;** they include the springfish (*Chologaster agassizi*) of North America and species of the genus *Chondrostoma* in Europe. Troglophiles have comparatively well-developed eyes. Those species that are confined to caves and live in constant darkness are known as **troglobites.** Most of these cave-adapted taxa are ostariophysans. One of the better known non-ostariophysan cave-dwelling families are the Amblyopsidae of North America, including the genera *Amblyopsis* (the first hypogean fish species described was *Amblyopsis spelaea*, by James DeKay in 1842) and *Typhlichthys* (see Plate 13.4).

Other families with troglobitic members are the North American ictalurids (two genera, *Satan* and *Trogloglanis*,

were recovered from deep artesian wells in Texas, up to 500 m beneath the surface; see Chapter 10); Pimelodidae (*Typhlobagrus*) of Brazil; Cyprinidae (*Aulopyge*) of Europe; and Synbranchidae (*Pluto*) of the Yucatan. *Poecilia sphenops*, a livebearer from Mexico, has been noted as having normally sighted forms outside a cave near Tabasco, with some sightless populations and intergrades inside. Several species of blind cave-dwelling fishes are known from Africa. The 13 known species are mainly in the families Cyprinidae and Clariidae, but the families Gobiidae, Eleotriidae, Synbranchidae, and Mastacembelidae also include cave-adapted representatives (Lévêque, 1997). Of the 72 hypogean species discovered so far, more than half are known from a single cave ecosystem in China (Proudlove, 1997).

Some hypogean fishes are born without sight, whereas others are secondarily blinded by the growth of circumorbital tissue over the eyes as the individuals mature. In all but a couple of species of African cavefishes, some rudiment of eye tissue remains into the adult stage (Wilkens, 2004). A blind cave characin popular with tropical fish hobbyists has been designated the genus *Anoptichthys*, yet it is, in fact, a variant of the Mexican species *Astyanax mexicanus*. Eyeless and eyed

---

Lates and of clupeids in the open waters of the evolving Lake Tanganyika ecosystem were undoubtedly contributing factors in the evolution of a resident cichlid fauna that is largely benthic or confined to rocky zones. In Lake Malawi, on the other hand, open-water planktivorous and piscivorous cichlids abound, owing to the absence of clupeids or other large predatory forms (Lowe-McConnell, 1996).

The consequences of predation are amply demonstrated in the environmental effects that the introduction of the Nile perch (*Lates niloticus*) has had on the Lake Victoria ecosystem. Increased human population surrounding the lake resulted in increased fishery pressure on resident cichlid populations, and *Lates* was introduced in the early 1950s—around the same time that nonnative tilapias were introduced to compensate for the declining fishery of native cichlids. At the time, this was undoubtedly perceived as the optimal solution—to introduce a piscivore that could avail itself of the

teeming cichlid populations and rapidly grow into a large, highly desirable food item. *Lates* had long been fished in Lake Tanganyika, where they naturally occur. About 30 years after their introduction to Lake Victoria, *Lates* underwent a population explosion. During this time, up to two thirds of the known cichlid taxa (possibly as many as 200 species) were driven to extinction. One of the introduced tilapiine species, *Oreochromis niloticus,* has apparently driven native tilapiines in Lake Victoria to extinction (Goudswaard et al., 2002).

Increased nutrient loading due to the burgeoning human population has resulted in cultural eutrophication at the same time that the ichthyofaunal assemblages were becoming transformed. Cyanobacteria, which rapidly bloom and deplete the dissolved oxygen, have replaced diatoms as the dominant algae, and the lake has become permanently stratified, with waters below 20 m being virtually anoxic. This has eliminated a vast amount of potential

habitat for the resident fauna (Kitchell et al., 1997; Lowe-McConnell, 1996). The increased turbidity has had an unanticipated evolutionary consequence: The species integrity in these highly visual and sexually selective fishes is being compromised. Increased rates of hybridization are being observed, and further speciation is being jeopardized (Galls and Metz, 1998; Seehausen et al., 1997).

The intense fishery pressure on *Lates* is beginning to achieve at least a partial compensation for the devastation wreaked in earlier years. As predation pressure has diminished, the recovery of some cichlid populations has been observed (Kitchell et al., 1997). Whether a complete recovery of the Lake Victoria ecosystem can be achieved remains to be seen. Extinction, however, is forever. All that remains of many of the cichlid species swallowed up by Nile perch are the contents of a few jars on the shelves of museums.

forms readily interbreed and produce fertile offspring (Avise and Selander, 1972; Schemmel, 1980). A cross of blind individuals from different populations will produce eyed forms, suggesting that more than one allele is responsible for the eyeless condition.

Eyeless forms apparently can reacquire eyes, as shown in an unusual population inhabiting a lake beneath a karst window far upstream from the entrance to a cave in Mexico. Karst windows, arising from the collapse of a cave roof, enable the illumination of previously dark portions of the underground stream. In a lake beneath one such karst window, both eyed and eyeless variants of *Astyanax fasciatus* were discovered, with the eyed forms preferring the illuminated side of the lake and the blind individuals the dark side (Espinasa and Borowsky, 2000). Genetic studies of *Astyanax* have revealed the role of a group of homeotic (*hox*) genes (see Chapters 2 and 28) that produce molecules, termed **hedgehog (*hh*) proteins,** that are instrumental in controlling developmental sequences. The expression of *hh* proteins along the anterior midline of embryonic individuals results in the suppression of eye growth and development (Yamamoto et al., 2004). Anyone who has kept *Astyanax* in a home aquarium soon loses any initial feelings of sympathy for these sightless fishes, because they are invariably the first to find the food when sprinkled in the tank.

In addition to the absence of well-developed eyes, true hypogean fishes also lack pigment in the skin. The lack of vision is compensated for by the development of other senses, particularly the acousticolateralis system and olfaction. The cave-dwelling catfishes use the sense of touch (and possibly taste) concentrated on their barbels. Species living in the dark depend on tactile stimulation during courtship behavior rather than on the typical visual stimulation generally seen in epigean species (Barr, 1968; Culver, 1982; Lee et al., 1980). Cephalic lateralis canals are modified in the cave dwellers and are enlarged and partially open. The lateral-line organs of amblyopsids are set out on the surface. Due to the paucity of predators in cave ecosystems, cave-adapted fishes tend not to display escape responses. Although no seasonality exists in the cave environment, some species show an annual reproductive cycle. Amblyopsids incubate their eggs in the branchial chamber. Although predation may not exist on the adults, the evolution of such behavior suggests that the eggs and possibly the larvae are vulnerable.

Although most hypogean fishes are known from freshwater systems, one marine environment is noteworthy for cave-adapted species—the *blue holes* found throughout the Bahamas and West Indies. The *cenotes* that range throughout the Yucatan Peninsula are actually a form of blue hole. These vertical shafts through the karst may be up to 100 m in depth. Whereas inland blue holes may penetrate the freshwater aquifer, coastal blue holes may have horizontal shafts that open into the adjacent ocean, thus serving as conduits for seawater exchange and marine fauna. Other blue holes, known as *ocean holes,* may be found entirely offshore on shallow banks (Whitaker, 1998). Among the several species of marine fishes known to be blue hole associates are some that preferentially inhabit the deeper, darker recesses and hence show extremely reduced eyes. Four species of brotulids of the genus *Lucifuga* (see Plate 13.6) are known from caves in Cuba and the Bahamas.

Scuba diving in blue holes and *cenotes* is difficult and dangerous, but as increasing numbers of divers who are interested in pursuing the more extreme variants of the sport enter into these poorly understood ecosystems, undoubtedly the list of known hypogean fish species will grow.

In addition to cave environments, fishes are known to inhabit waters that lack direct communication with surface flows. The term **phreatic** is used to refer to these parts of the aquifer, where no direct connection to the surface is present. Deep artesian wells or caves that are completely beneath the water table may contain fish populations. As might be expected, given the difficulty of access to these most unusual environments, fishes inhabiting these realms are poorly understood (Fernández and dePinna, 2005).

INLAND SALINE WATERS AND ASSOCIATED FISH ASSEMBLAGES

There are many inland bodies of water worldwide that, because of elevated salinities, support fish faunas that are more representative of marine environments. Discussing them here provides us with a natural segue between freshwater and marine ecosystems, as they have properties of both. Inland saline waters are termed **athalassic** if their saline condition does not arise from any preexisting association with the sea (Bayly, 1972). Coastally situated saline lakes may have arrived at that state because of a recently pre-existing or still periodically present association with the ocean. Saline bodies of water may be as small as the numerous ponds that are scattered across the Bahamian landscape (see Fig. 32.1), or as large as Lake Turkana in Kenya—the fourth largest (by volume) of the African Rift Valley lakes, with a total surface area of 7,100 km².

So-called secondary freshwater fishes—those fishes with a biogeographical history in coastal environments (see Chapter 29)—are well represented in saline lakes. Cyprinodontiform fishes, such as the families Cyprinodontidae, Poeciliidae, and Atherinidae, are especially abundant in saline lakes, as these euryhaline fishes are noted for their osmoregulatory capacity. Sticklebacks (Gasterosteidae) and even centrarchids are also known to enter temperate saline waters. Largemouth bass can tolerate exposure to moderately elevated levels of salinity and enter brackish marshes of the Gulf Coast of Texas and Louisiana (Meador and Kelso, 1990). Many species of cichlids, such as those of the genera *Tilapia* and *Oreochromis*, live in tropical brackish waters.

One of the most unusual saline lakes is the Salton Sea in the Imperial Valley of southern California. With a surface area of about 548 km², it is the largest inland body of water entirely within California. Natural phenomena conspired with human folly to create the Salton Sea when the Colorado River flooded through an inadequately engineered irrigation canal in 1905 and filled the Salton Sink—an ancient pluvial basin located 71 m below sea level. By the time that the Colorado River was rediverted back into its main channel in 1907, a vast inland sea with a salinity close to oceanic conditions had been created. Marine species, such as bairdiella (*Bairdiella icistia*), orangemouth corvina (*Cynoscion xanthulus*), and sargo (*Anisotremus davidsoni*) were introduced and formed the basis of a sport fishery. The salty substrates over which water flows before reaching the Salton Sea have resulted in an increase in salinity and introduced concentrations of toxic selenium. The salinity has reached levels (around 45 ‰) that are intolerable to the introduced marine fish species. What was once a thriving sport fishery is no more (deBuys and Myers, 1999; Moyle, 1976).

. . . .
KEY POINTS AND CONNECTIONS
. . . .

• From a geological perspective, lakes are comparatively short-lived phenomena that can form as a result of a number of geological processes. Lake ecosystems can also result from the damming of large river systems.

The impact of the formation of artificial impoundments on fish communities is considered in Chapter 39.

• Young lakes of temperate regions are oligotrophic, with comparatively low levels of productivity; these progress to eutrophic lakes, with greater levels of organic production. The Old World tropics are quite different from the New World tropics in that several large lake systems are known from Africa, whereas South America is deficient in large lakes. Here, lacustrine ecosystems may be associated with portions of large river systems, however.

The fisheries of large tropical lakes are discussed in Chapter 39.

• A range of feeding opportunities is available to the benthic and pelagic fishes of still waters. In contrast to what is observed in river ecosystems, lakes permit the development of planktonic communities that can serve as a food base for the ichthyofaunal community.

The feeding adaptations of planktivorous fishes are discussed in Chapter 23.

• Salmonoids, percoids, and centrarchids dominate the fish assemblages of northern temperate lakes, whereas characoids, cyprinids, and cichlids are the most conspicuous members of tropical lake fish communities.

Additional discussion of the adaptive radiations of cichlid fishes is included in Chapter 16.

• The biota of lakes can be described as insular, as they are relatively isolated, with limited avenues of dispersal. Limited gene flow sometimes results in highly endemic fish communities consisting of species flocks—the cichlids of the Great Lakes of Africa undoubtedly being the best-known example. Numerous instances of ecological character displacement and trophic polymorphism have also been observed; the emergence of distinct ecomorphs for a given species enables the exploitation of different food resources within a lake.

Character displacement and trophic polymorphism are also considered in Chapter 4.

• Hypogean waters represent an unusual environment, characterized by a somewhat limited assemblage of fishes that are adapted for living in areas with little or no light. As might be expected, the sensory adaptations of these fishes demonstrate exceptional compensation for the absence of vision, chiefly through enhanced tactile, acoustico-lateralis, and olfactory capabilities.

The modifications of the mechanosensory systems of cave-adapted fishes are discussed in Chapter 21.

• Inland saline waters are termed athalassic if they have no prior connection with the ocean, whereas other inland waters show pronounced marine affiliations. Euryhaline species of primary freshwater fishes, such as some centrarchid and cichlid species, or fishes characteristic of coastal marine waters, such as the cyprinodontiforms, are the dominant ichthyofauna of inland saline waters.

Osmoregulatory adaptations of euryhaline fishes are considered in Chapter 25.

. . . .
FISH LINKS
. . . .

http://www.malawicichlids.com/ The cichlid fishes of Lake Malawi, Africa. Excellent website maintained by Michael Oliver with hundreds of photographs, bibliographic references, and other Web links.

. . . .
BUILDING AN ICHTHYOLOGY LIBRARY
. . . .

A few good references on limnology that should make it into your library:

Busch, W.-D. N., and P. G. Sly. 1992. *The development of an aquatic habitat classification system for lakes.* CRC Press, Boca Raton, FL.
Hutchinson, G. E. 1957–1993. *A treatise on limnology*, Vols. 1 (Geography, physics, and chemistry), 2 (Introduction to lake biology and limnoplankton), 3 (Limnological botany), 4 (Zoobenthos).

Wiley, New York. (Hutchinson's life spanned nearly the entire 20th century, during which he became the preeminent limnologist of his time.)

Scheffer, M. 1998. *Ecology of shallow lakes*. Chapman and Hall, London.

There are probably more books published on cichlids than any other fish group. Here are a few of the best:

Barlow, G. W. 2000. *The cichlid fishes: Nature's grand experiment in evolution*. Perseus, Cambridge, MA.

Fryer, G., and T. D. Iles. 1972. *The cichlid fishes of the great lakes of Africa: Their biology and evolution*. TFH, Neptune City, NJ.

Greenwood, P. H. 1981. *The haplochromine fishes of the East African Lakes*. Cornell University Press, Ithaca, NY.

Keenleyside, M. A. 1991. *Cichlid fishes: Behavior ecology and evolution*. Chapman and Hall, London.

REFERENCES

Albertson, R. C., J. T. Streelman, and T. D. Kocher. 2003. Directional selection has shaped the oral jaws of Lake Malawi cichlid fishes. *Proc. Natl. Acad. Sci. USA 100*: 5252–5257.

Avise, J. C., and R. K. Selander. 1972. Evolutionary genetics of cave-dwelling fishes of the genus *Astyanax*. *Evolution 26*: 1–19.

Barr, T. C., Jr. 1968. Cave ecology and the evolution of troglobites, pp. 35–102. In *Evolutionary biology*, Vol. 2, T. Dobzhansky, M. K. Hecht, and W. C. Steere (Eds.). Appleton-Century-Crofts, New York.

Barton, M., and C. Wilmhoff. 1996. Inland fishes of the Bahamas—new distribution records for exotic and native species from New Providence Island. *Bahamas J. Sci. 3*(2): 7–11.

Bayly, I. A. E. 1972. Salinity tolerance and osmotic behavior of animals in athalassic saline and marine hypersaline waters. *Ann. Rev. Ecol. Syst. 3*: 233–268.

Berg, L. S. 1949. *Freshwater fishes of the U.S.S.R. and adjacent countries* (4th ed.), Vol. 3. Israel Program for Scientific Translations, Jerusalem.

Berra, T. M. 1981. *An atlas of distribution of the freshwater fish families of the world*. University of Nebraska Press, Lincoln.

Blake, B. F. 1977. The effect of the impoundment of Lake Kainji, Nigeria, on the indigenous species of mormyrid fishes. *Freshw. Biol. 7*: 37–42.

Breder, C. M., Jr., and D. E. Rosen. 1966. *Modes of reproduction in fishes*. Natural History Press, Garden City, NY.

Bronte, C. R., G. W. Fleischer, S. G. Maistrenko, and N. M. Pronin. 1999. Stock structure of Lake Baikal omul as determined by whole-body morphology. *J. Fish Biol. 54*: 787–798.

Bunt, T. M., B. J. Turner, D. Duvernell, C. Holtmeier, and M. Barton. 1999. Molecular evidence for reproductive isolation between two sympatric trophic morphs of San Salvador pupfish (*Cyprinodon*). *8th Symp. Nat. Hist. Bahamas* (abstr.).

Cole, G. A. 1994. *Textbook of limnology*. Waveland Press, Prospect Heights, IL.

Coulter, G. W. (Ed.). 1991. *Lake Tanganyika and its life*. Oxford University Press, Oxford.

Culver, D. C. 1982. *Cave life: Evolution and ecology*. Harvard University Press, Cambridge, MA.

deBuys, W., and J. Myers. 1999. *Salt dreams: Land and water in low-down California*. University of New Mexico Press, Albuquerque.

Diana, J. S. 1995. *Biology and ecology of fishes*. Biological Sciences Press, Cooper, Carmel, IN.

Dodson, S. I., S. E. Arnott, and K. L. Cottingham. 2000. The relationship in lake communities between primary productivity and species richness. *Ecology 81*: 2662–2679.

Erlandsson, A., and A. J. Ribbink. 1997. Pattern of sexual size dimorphism in African cichlid fishes. *S. Afr. J. Sci. 93*: 498–508.

Espinasa, L., and R. Borowsky. 2000. Re-acquisition of the eyed condition in a population of blind cave fish *Astyanax fasciatus* in a karst window. *80th Ann. Meet. Amer. Soc. Ichthyol. Herpetol.*, La Paz, Baja California.

Fernández, L., and M. C. C. dePinna. 2005. Phreatic catfish of the genus *Silvinichthys* from southern South America (Teleostei, Siluriformes, Trichomycteridae). *Copeia 2005*: 100–108.

Fryer, G. 1997. Biological implications of a suggested Late Pleistocene desiccation of Lake Victoria. *Hydrobiology 354*: 177–182.

Galls, F., and J. A. J. Metz. 1998. Why are there so many cichlids? *TREE 13*(1): 1–2.

Goudswaard, P. C., F. Witte, and E. F. B. Katunzi. 2002. The tilapiine fish stock of Lake Victoria before and after the Nile perch upsurge. *J. Fish Biol. 60*: 838–856.

Greenwood, P. H. 1974. The cichlid fishes of Lake Victoria, East Africa: The biology and evolution of a species flock. *Bull. Br. Mus. Nat. Hist. (Zool.) Suppl. 6*: 1–134.

Guillory, V., and W. E. Johnson. 1986. Habitat, conservation status, and zoogeography of the cyprinodont fish, *Cyprinodon variegatus hubbsi* (Carr). *Southw. Nat. 31*(1): 95–100.

Herdendorf, C. E., L. Håkanson, D. J. Jude, and P. G. Sly. 1992. A review of the physical and chemical components of the Great Lakes: A basis for classification and inventory of aquatic habitats, pp. 109–159. In *The development of an aquatic habitat classification system for lakes*, W.-D. N. Busch and P. G. Sly (Eds.). CRC Press, Boca Raton, FL.

Hori, M. 1991. Feeding relationships among cichlid fishes in Lake Tanganyika: Effects of intra- and interspecific variations of feeding behavior on their coexistence. *Ecol. Int. Bull. 19*: 89–101.

———. 1993. Frequency-dependent natural selection in the handedness of scale-eating cichlid fish. *Science 260*: 216–219.

Humphries, J. H. 1984. *Cyprinodon veracundus*, n. sp., a fifth species of pupfish from Laguna Chichancanab. *Copeia 1984*: 58–68.

———, and R. R. Miller. 1981. A remarkable species flock of pupfish, genus *Cyprinodon* from Yucatan, Mexico. *Copeia 1981*: 52–64.

Johnson, T. C., C. A. Scholz, M. R. Talbot, K. Kelts, R. D. Ricketts, G. Ngobi, K. Beuning, S. Ssemmanda, and J. W. McGill. 1996. Late Pleistocene desiccation of Lake Victoria and rapid evolution of cichlid fishes. *Science 273*: 1091–1093.

Kitchell, J. F., D. E. Schindler, R. Ogutu-Ohwayo, and P. N. Reinthal. 1997. The Nile perch in Lake Victoria: Interactions between predation and fisheries. *Ecol. Appl. 7*: 653–664.

Knight, M. E., and G. F. Turner. 2004. Laboratory mating trials indicate incipient speciation by sexual selection among populations of the cichlid fish *Pseudotropheus zebra* from Lake Malawi. *Proc. Roy. Soc. Lond. (Biol. Sci.) 271*: 675–680.

Kornfield, I., and K. E. Carpenter. 1984. Cyprinids of Lake Lanao, Philippines: Taxonomic validity, evolutionary rates, and speciation scenarios, pp. 69–84. In *Evolution of species flocks*, A. E. Echelle and I. Kornfield (Eds.). University of Maine Press, Orono.

Kortmulder, K. 1987. Ecology and behavior in tropical freshwater fish communities. *Arch. Hydrobiol. Beih. 28*: 503–513.

Lagler, K. F. 1956. *Freshwater fishery biology* (2nd ed.). Wm. C. Brown, Dubuque, IA.

Lampert, W., and U. Sommer. 1997. *Limnoecology: The ecology of lakes and streams.* Oxford University Press, Oxford.

Lee, D. S., C. R. Gilbert, C. H. Hocutt, R. E. Jenkins, D. E. McAllister, and J. R. Stauffer, Jr. 1980. *Atlas of North American freshwater fishes.* North Carolina State University, Raleigh.

Lelek, A. 1973. Sequence of changes in fish populations of the new tropical man-made lake, Kainji, Nigeria, West Africa. *Arch. Hydrobiol. 71:* 381–420.

Lévêque, C. 1997. *Biodiversity dynamics and conservation: The freshwater fish of tropical Africa.* Cambridge University Press, Cambridge, UK.

Lewis, D. S. 1974. The effects of the formation of Lake Kainji, Nigeria, upon the indigenous fish population. *Hydrobiology 45:* 281–301.

Liem, K. F. 1973. Evolutionary strategies and morphological innovations: Cichlid pharyngeal jaws. *Syst. Zool. 22:* 425–441.

———. 1975. Biological versatility, evolution, and food resource exploitation in African cichlid fishes. *Amer. Zool. 15:* 427–454.

———. 1980. Adaptive significance of intra- and interspecific differences in the feeding repertoires of cichlid fishes. *Amer. Zool. 20:* 295–314.

Lowe-McConnell, R. H. 1987. *Ecological studies in tropical fish communities.* Cambridge University Press, Cambridge, UK.

———. 1996. Fish communities in the African Great Lakes. *Env. Biol. Fishes 45:* 219–235.

Martens, K. 1997. Speciation in ancient lakes. *TREE 12*(5): 177–182.

Matthews, W. J. 1998. *Patterns in freshwater fish ecology.* Chapman and Hall, New York.

Meador, M. R., and W. E. Kelso. 1990. Physiological responses of largemouth bass, *Micropterus salmoides,* exposed to salinity. *Can. J. Fish. Aquat. Sci. 47:* 2358–2363.

Moyle, P. B. 1976. *Inland fishes of California.* University of California Press, Berkeley.

Nagelkerke, L. A. J., F. A. Sibbing, J. G. M. van den Boogaart, E. H. R. R. Lammens, and J. W. M. Osse. 1994. The barbs (*Barbus* spp.) of Lake Tana: A forgotten species flock. *Env. Biol. Fishes 39:* 1–22.

Nelson, J. S. 1994. *Fishes of the world* (3rd ed.). Wiley, New York.

Nilsson, C., C. A. Reidy, M. Dynesius, and C. Revenga. 2005. Fragmentation and flow regulation of the world's large river systems. *Science 308:* 405–408.

Parenti, L. R. 1984. Biogeography of the Andean killifish genus *Orestias* with comments on the species flock concept, pp. 85–92. In *Evolution of species flocks,* A. E. Echelle and I. Kornfield (Eds.). University of Maine Press, Orono.

Pennisi, E. 2004. The genes that change the cichlid jaws. *Science 304:* 383.

Pflieger, W. L. 1975. *The fishes of Missouri.* Missouri Department of Conservation, Jefferson City.

Post, D. M., M. L. Pace, and N. G. Hairston. 2000. Ecosystem size determines food-chain length in lakes. *Nature 405:* 1047–1049.

Proudlove, G. S. 1997. A synopsis of the hypogean fishes of the world. *Proc. 12th Int. Congr. Speleol., 1997, Switzerland, Vol. 3, Symp. 9: Biospeleology:* 351–358.

Reimers, N. 1979. A history of a stunted brook trout population in an alpine lake: A lifespan of 24 years. *Cal. Fish Game 65:* 196–215.

Ribbink, A. J., and D. Eccles. 1988. Fish communities in the East African Great Lakes, pp. 277–301. In *Biology and ecology of African freshwater fishes,* C. Lévêque, M. N. Bruton, and G. W. Ssentongo (Eds.). ORSTOM Travaux et Documents 216, Paris.

Robinson, B. W., D. S. Wilson, A. S. Margosian, and P. T. Lotito. 1993. Ecological and morphological differentiation of pumpkinseed sunfish in lakes without bluegill sunfish. *Evol. Ecol. 7:* 451–464.

Romero, A. (Ed.). 2001. *The biology of hypogean fishes.* Kluwer, Dordrecht, the Netherlands.

Sarbu, S. M., T. C. Kane, and B. K. Kinkle. 1996. A chemoautotrophically based cave ecosystem. *Science 272:* 1953–1955.

Schemmel, C. 1980. Studies on the genetics of feeding behavior in the cave fish *Astyanax mexicanus* f. *anoptichthys. Z. Tierpsychol. 53:* 9–22.

Sealey, N. E. 1985. *Bahamian landscapes: An introduction to the geography of the Bahamas.* Collins Caribbean, London.

Seehausen, O. 2002. Patterns in fish radiation are compatible with Pleistocene desiccation of Lake Victoria and 14,600 year history for its cichlid species flock. *Proc. Roy. Soc. Lond. (Biol. Sci.) 269:* 491–494.

———, and N. Bouton. 1998. The community of rock-dwelling cichlids in Lake Victoria. *Bonn. Zool. Beitr. 47:* 301–311.

———, J. J. M. van Alphen, and R. Lande. 1999. Color polymorphism and sex ratio distortion in a cichlid fish as an incipient stage in sympatric speciation by sexual selection. *Ecol. Lett. 2:* 367–378.

———, ———, and F. Witte. 1997. Cichlid fish diversity threatened by eutrophication that curbs sexual selection. *Science 277:* 1808–1811.

Simpson, B. R. 1979. The phenology of annual killifishes. *Symp. Zool. Soc. Lond. 44:* 243–261.

Scheffer, M. 1998. *Ecology of shallow lakes.* Chapman and Hall, London.

Smith, T. B. 1996. Evolutionary significance of resource polymorphisms in fishes, amphibians, and birds. *Ann. Rev. Ecol. Syst. 27:* 111–133.

Spangler, G. R., N. R. Payne, J. E. Thorpe, J. M. Byrne, H. A. Regier, and W. J. Christie. 1977. Responses of percids to exploitation. *J. Fish. Res. Bd. Can. 34:* 1983–1988.

Stager, J. C., J. J. Day, and S. Santini. 2004. Comment on "Origin of the superflock of cichlid fishes from Lake Victoria, East Africa." *Science 304:* 963b.

Stevenson, M. M. 1992. Food habits within the Laguna Chichancanab *Cyprinodon* (Pisces: Cyprinodontidae) species flock. *Southw. Nat. 37:* 337–343.

Sturmbauer, C. 1998. Explosive speciation in cichlid fishes of the African Great Lakes: A dynamic model of adaptive radiation. *J. Fish Biol. 53*(Suppl. A): 18–36.

———, E. Verheyen, L. Rüber, and A. Meyer. 1997. Phylogenetic patterns in populations of cichlid fishes from rocky habitats in Lake Tanganyika, pp. 97–111. In *Molecular systematics of fishes,* T. D. Kocher and C. A. Stepien (Eds.). Academic Press, San Diego.

Tonn, W. M., and J. J. Magnuson. 1982. Patterns in the species composition and richness of fish assemblages in northern Wisconsin lakes. *Ecology 63:* 1149–1166.

Twentyman-Jones, V., A. J. Ribbink, and D. Voorvelt. 1997. Colour clues to incipient speciation of *Pseudocrenilabrus philander* (Teleostei, Cichlidae). *S. Afr. J. Sci. 93:* 529–536.

Verheyen, E., L. Rüber, J. Snoeks, and A. Meyer. 1996. Mitochondrial phylogeography of rock-dwelling cichlid fishes reveals evolutionary influence of historical lake level fluctuations of Lake Tanganyika, Africa. *Phil. Trans. Roy. Soc. Lond. B 351:* 797–805.

———, W. Salzburger, J. Snoeks, and A. Meyer. 2003. Origin of the superflock of cichlid fishes from Lake Victoria, East Africa. *Science 300:* 325–329.

———, ———, ———, and ———. 2004. Response to comment on "Origin of the superflock of cichlid fishes from Lake Victoria, East Africa." *Science 304:* 963c.

Werner, E. E., and D. J. Hall. 1974. Optimal foraging and the size selection of prey by the bluegill sunfish (*Lepomis macrochirus*). *Ecology 55:* 1042–1052.

Whitaker, F. F. 1998. The blue holes of the Bahamas: An overview and introduction to the Andros Project. *Cave Karst Sci. 25*(2): 53–56.

Wilkens, H. 2004. Fish. In *Encyclopedia of caves,* D. C. Culver and W. B. White (Eds.). Elsevier, Academic Press, San Diego.

Witte, F. 1981. Initial results of the ecological survey of the haplochromine cichlid fishes from the Mwanza Gulf of Lake Victoria. *Neth. J. Zool. 31:* 175–202.

Won, Y.-J., A. Sivasundar, Y. Wang, and J. Hey. 2005. On the origin of Lake Malawi cichlid species: A population genetic analysis of divergence. *Proc. Natl. Acad. Sci. USA 102*(Suppl. 1): 6581–6586.

Yamamoto, Y., D. W. Stock, and W. R. Jeffrey. 2004. Hedgehog signaling controls eye degeneration in blind cavefish. *Nature 431:* 844–847.

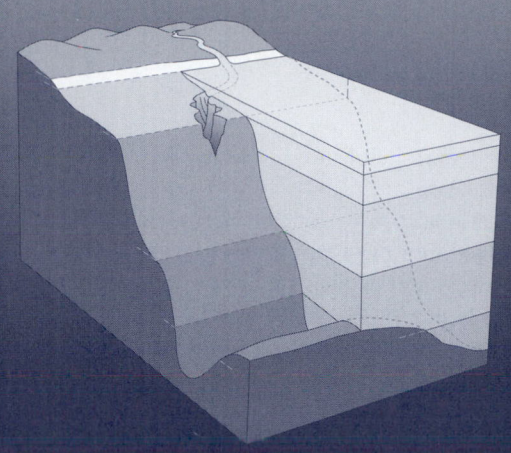

33 ECOLOGY OF FISHES IV: COASTAL MARINE ENVIRONMENTS: THE CONTINENTAL SHELF

INTRODUCTION TO THE MARINE REALM

COASTAL MARINE ENVIRONMENTS

The Continental Shelf
Pelagic Fish Fauna
Benthic Fish Fauna
  Fish Assemblages Associated with Macrophytes
  Community Ecology in Temperate Reef Areas
Coral Reefs
  Coral Reef Fishes
  Dynamics of Coral Reef Ecosystems

With more than 71 percent of the Earth's surface covered by oceans, it is not surprising that fishes have been so successful in marine environments. Although the opportunities for speciation in freshwater environments have produced a far greater number of fish species relative to the quantity of aquatic habitat, the marine realm includes an abundance of fishes with far broader geographic distributions than those of their freshwater brethren. In the next three chapters, we will investigate the range of marine environments inhabited by fishes, starting with the coastal waters of the continental shelf regions. Most marine fishes, including the vast majority of commercially valuable species, are found in shallow coastal waters. Here, a diversity of fish communities have adapted to the range of environments encountered in shelf regions. Of these coastal shallow marine environments, coral reefs merit special consideration, as they include the most diverse faunal assemblages seen in the world's oceans. More species of fishes are to be found there than anywhere else. The proximity of shelf regions to landmasses also makes them most susceptible to human activity. In exchange for the vast quantities of marine resources harvested from the continental shelf regions, humans release massive quantities of pollutants that enter coastal ecosystems in concentrations far in excess of those measured in the open ocean.

INTRODUCTION TO THE MARINE REALM

The tectonic history of the Earth has resulted in an uneven distribution of land masses, so that the Northern Hemisphere is only about 61 percent ocean, whereas about 80 percent of the Southern Hemisphere is ocean. Four ocean basins are traditionally recognized: the **Atlantic, Pacific, Indian,** and **Arctic.** The Pacific Ocean is the largest and has the deepest known regions. The Arctic Ocean is the smallest and shallowest (see Chapter 35). Contrary to freshwater environments, the world's oceans actually constitute one vast, interconnected body of water. In the Northern Hemisphere, the preponderance of land masses more readily defines the boundaries of the oceans. In the Southern Hemisphere, a continuous body of water, usually referred to as the Southern Ocean, surrounds Antarctica at about 60° S latitude. Obviously, the evolutionary history of marine fishes reflects a significantly greater degree of gene flow among populations.

Water constantly circulates through the oceans in the form of currents or local wave action. Although the salinity may vary locally due to the comparative contributions of precipitation and freshwater runoff versus evaporation, the relative ionic composition is uniform worldwide. Marine fishes can live on or near the bottom, in what we have termed the *benthic* realm, or in the open water, the *pelagic* realm. Marine ecosystems encompass these two realms, with some species linking them together. For example, *benthopelagic* fishes, such as the cods, live in the water column itself but feed on the bottom. The usual framework in which marine ecologists work is one of zonation of the two realms mentioned. Zone boundaries are crossed by numerous animals, but because the zones are established according to such factors as light penetration, temperature, and extent of the continental shelf and slope, they have considerable biological significance.

The benthic realm is divided into the *shelf* zone, which extends down to a depth of about 200 m; the *upper slope* zone, which extends to about 1,000 m; the *lower slope,* reaching to about 3,000 m; the *abyssal plain,* which extends to about 6,000 m; and the *hadal* zone, which includes the oceanic trenches as deep as 11,000 m. The pelagic realm is divided into the *epipelagic* zone, to a depth of about 200 m (this roughly corresponds to the depth of light penetration sufficient for photosynthetic activity and the edge of the continental shelf); the *mesopelagic* zone, to about 1,000 m, which is the limit of all surface light; the *bathypelagic* zone, which is aphotic and extends to 4,000 m; the *abyssopelagic* zone, from 4,000 to 6,000 m; and the *hadopelagic* zone, in the deep trenches below 6,000 m. Some ecologists find it convenient to divide the ocean into the *neritic* system (on and

above the continental shelf) and the *oceanic* system (beyond the shelf; Fig. 33.1).

An international consortium of marine researchers is currently engaged in a most ambitious project—a 10-year initiative "to assess and explain the diversity, distribution, and abundance of marine life in the oceans—past, present, and future" (http://www.coml.org/coml.htm). To date, the Census of Marine Life project has identified more than 200,000 marine organisms, including 15,482 species of fishes, of which 106 new species were described in 2004. Researchers associated with the census believe that the final tally of marine fishes will number about 20,000. This project emerged from the concern expressed by members of the scientific community that humans may be causing permanent damage to ecosystems that we still know very little about. It is hoped that such a comprehensive survey of the biota of the ocean will greatly assist in monitoring such well-known threats as overfishing and global warming (see Chapter 39).

COASTAL MARINE ENVIRONMENTS

The Continental Shelf

The actual boundaries of most continents are not visible unless one is in a submersible. Continental margins usually include a broad expanse of shallow oceanic habitat, termed the **continental shelf.** The continental shelf extends from just below the limit of low tide out to the **continental slope.** The width of the continental shelf ranges from a few meters in some locations up to 1,300 km in others. The broadest shelf areas are seen in the Arctic Ocean and in the western Pacific Ocean. Worldwide, the average width of the continental shelf is about 70 km. The proportion of ocean floor at a depth of less than 200 m, which includes most of the continental shelf, is estimated to be about 7.6 percent. In several places, the continental shelf may be penetrated by fissures that form deep canyons. The presence of one such canyon in Monterey Bay on the coast of California enables researchers to collect deep-ocean fishes within sight of land. The average slope of the continental shelf is only about 0.2 percent—an angle of inclination so slight as to appear level to the human eye. The continental slope, on the other hand, ranges from about 3.5 to 6 percent depending on the topography of the ocean bottom.

At shallow depths, there is water motion due to wave action and tidal currents, and seasonal fluctuations of light, temperature, and salinity are normally experienced. Exceptionally high productivity is encountered where **upwelling**

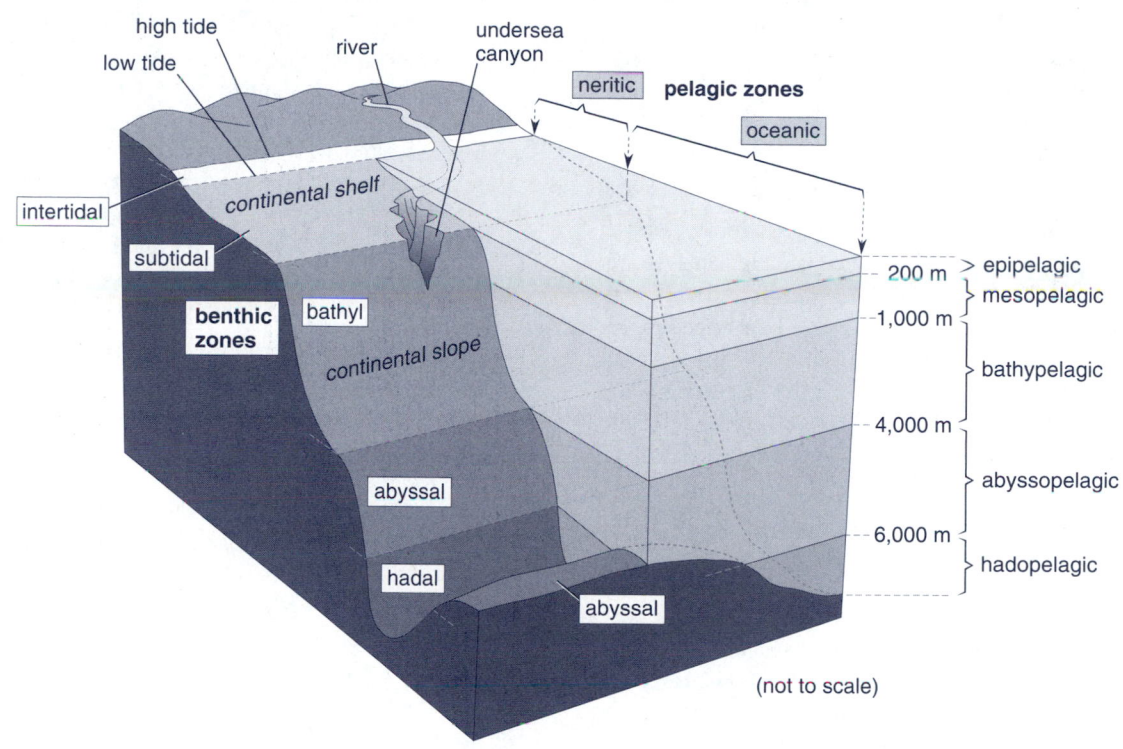

FIGURE 33.1
Profile of the ocean, showing the different life zones.

brings nutrients from the deep ocean into nearshore areas. Upwelling often determines the productivity of commercial fishing grounds in continental shelf areas. Fish faunas are at their richest in the shelf zone, especially in the upper 50 m, the "inner sublittoral" part of the shelf. Scores of families and hundreds of genera have representatives in this biome. The bottom of the continental shelf consists mostly of unconsolidated substrates—sand or finer-grained mud. Elevations that are sufficiently deep to not constitute a hazard to navigation are termed **banks**. Concentrations of commercially important fishes tend to congregate here. **Shoals** or **bars** are elevated areas of unconsolidated substrates that are shallow enough to be considered navigational hazards. Hard, consolidated substrates in shelf regions, if they are in water shallow enough to be navigational hazards, are termed **reefs**. In warmer waters, of course, these reefs may be composed of corals. From a fisheries standpoint, the continental shelf is by far the most significant oceanic realm. Here, most of the world's fisheries are concentrated, including both benthic and pelagic species (see Chapter 39).

Pelagic Fish Fauna

The fish fauna of the open waters of the continental shelves associate less with the substrate than benthic species and, hence, are generally more mobile and wide-ranging. Small planktivorous species, such as the herrings (Clupeidae) and anchovies (Engraulidae), may form enormous, commercially exploited concentrations. The waters off the coast of Peru and Ecuador, for example, are host to a fishery for the Peruvian anchovy (*Engraulis ringens*), which was until recent times the single most productive fishery in the world. Overfishing, combined with a series of unusually severe *El Niño* events, have caused a drastic decline in this fishery. Other open-water families include the flyingfishes (Exocoetidae), halfbeaks (Hemiramphidae), and the planktivorous mullets (Mugilidae). Large open-water predators include the Pacific and Atlantic salmon (Salmonidae), bluefish (Pomatomidae), temperate basses (Percichthyidae), cobia (Rachycentridae), barracudas (Sphyraenidae), and some species of tunas and mackerels (Scombridae). Many of these species, though spending their adult lives offshore, enter bays, estuaries, and—in the case of anadromous species, such as the salmon and temperate basses—freshwaters for the purpose of breeding. The larval and juvenile stages may also spend time here before moving further offshore (see Chapter 34).

Benthic Fish Fauna

The sandy or muddy bottoms of the continental shelves are populated by a diverse array of fish families that are integral components of often complex food webs. Benthopelagic or benthic families that are common to sandy or muddy bottoms of higher latitudes include the hagfishes (Myxinidae), ratfishes (Chimaeridae), skates (Rajidae), eels (Anguillidae), sablefishes (Anoplopomidae), cods (Gadidae), and flatfishes (Pleuronectidae, Bothidae, and Cynoglossidae; Allen and Smith, 1988; Rogers et al., 1999; Sasaki, 1972). Many families, such as the rockfishes (Scorpaenidae = Sebastidae) may preferentially associate with rocky outcrops. Small, cryptic families, such as the gobies (Gobiidae), sculpins (Cottidae), and poachers (Agonidae) seek shelter among the rocks and vegetation in shallower areas. Sculpins, eelpouts (Zoarcidae), and cusk eels (Ophidiidae) range into deeper rocky-bottomed areas. The bottom-dwelling ichthyofaunal community of warmer waters may include lizardfishes (Synodontidae), marine catfishes (Ariidae), and a number of perciform families, including drums or croakers (Sciaenidae), snappers (Lutjanidae), porgies (Sparidae), grunts (Pomadasyidae), and mojarras (Gerridae). Some families, such as the labrids (Labridae) and gobies (Gobiidae), may be well represented throughout temperate and tropical waters. Many species, such as stingrays (Dasyatidae), lizardfishes, and the electric stargazers (Uranoscopidae), assume a cryptic posture by partially burying themselves in the soft substratum where they occur (see Plate 33.1A–C).

In his comparison of tropical versus temperate benthic fish communities, Choat (1982) made three general observations:

1. Tropical fishes display much greater species richness, but this may not necessarily be a consequence of the evolution of specialized feeding types. For example, shell crushing may be widely practiced by tropical forms such as the balistids (see Plate 17.4), but it is also pursued by myliobatid rays and heterodontid sharks in temperate waters.

2. Although herbivorous fishes are more prevalent in tropical waters, representatives of many families, such as the Pomacentridae, Kyphosidae, Stichaeidae, and Blenniidae, consume a wide variety of plant material in temperate waters.

3. The transition from a diurnal fish community to a nocturnal one is much more defined in tropical waters. Perhaps greater predation pressure in the tropics has resulted in more rapid and orderly shelter seeking among tropical species—especially those associated with coral reefs. Benthic-feeding fishes, because of their superior mobility, orientation capacity, and ability to detect even well-hidden prey, have a dramatic impact on epi- and infaunal community structure and trophic dynamics (Choat, 1982).

Because of the commercial significance of many species, the community structure and dynamics of bottom-dwelling fishes of the continental shelf has been intensively studied for many years. Of all the members of the North Atlantic bottom fish community, Rogers et al. (1999) have determined that elasmobranchs are the most vulnerable to commercial exploitation. Several species that are harvested, including dogfishes (Squalidae), requiem sharks (Carcharhinidae), and skates (Rajidae; see Chapter 6, Going Deeper), are long-lived, slow-growing, and late-maturing species and are hence most vulnerable to overfishing. The most important species harvested in the commercial shark fishery operating in the Northwest Atlantic and the Gulf of Mexico is the sandbar shark (*Carcharhinus plumbeus*). Studies on the effects of the exploitation of this species and another commercially harvested species, the dusky shark (*Carcharhinus obscurus*), have suggested that fisheries for species with these life history characteristics can best be sustained by targeting the youngest age classes (Cortés, 1999; Simpfendorfer, 1999). In most cases, however, the current rate of exploitation of these species cannot be sustained without long-lasting impact on the fish communities.

Fish Assemblages Associated with Macrophytes

Large marine macrophytes of the phylum Phaeophyta—the kelps—are home to assemblages of fishes that seek food and shelter among the dense vegetation. Along the Pacific coast, kelp forests shelter assemblages of fish species that are quite similar to those found in association with rocky bottoms. A continuum of habitat types, ranging from the rocky bottom to which the kelp is attached to the canopy spreading out near the surface, is home to a diversity of species. Sculpins, blennies, pricklebacks, and greenlings (Hexagrammidae; see Plate 33.2A) hide among the rocks, whereas señoritas (*Oxyjulis californica*, family Labridae; see Plate 33.2B) and kelp perch (*Brachyistius frenatus*, family Embiotocidae) forage on small invertebrates associated with the canopy. Several species of rockfishes (Scorpaenidae = Sebastidae; see Plate 33.2C) are common associates of kelp forests. Topsmelt (*Atherinops*) feed on mysid shrimp and other zooplankton that congregates around the kelp (Ebeling et al., 1980). The common names of some of these ichthyofaunal associates, such as the giant kelpfish (*Heterostichus rostratus;* see Plate 33.2D), the kelp bass (*Paralabrax clathratus*), and the kelp clingfish (*Rimicola muscarum*), are indicative of their ecological affinities.

Although diurnal shifts in assemblage composition are not as conspicuous in kelp communities as in tropical coral reefs (Ebeling and Hixon, 1991), day–night variations in the activity of kelp-associated fishes have been documented (Ebeling and Bray, 1976). Large patches of kelp may break free and drift far offshore. Here they may attract pelagic species, such as yellowtail (*Seriola dorsalis*) and albacore (*Thunnus alalunga;* Mitchell and Hunter, 1970). Large aggregations of sargassum weed, as seen in the eponymous Sargasso Sea, occur in the open ocean and are also home to complex ichthyofaunal communities (see Chapter 35). Intertidal and shallow subtidal rockweeds of the genera *Ascophyllum, Pelvetia,* and *Fucus* shelter a number of small, cryptic species as well (see Chapter 34).

Community Ecology in Temperate Reef Areas

Temperate reef areas and their associated kelp forests have been the subject of intensive ecological investigation over the past 20 years. Temperate reef fishes may participate in the structuring of trophic relationships in unexpected ways. The blacksmith (*Chromis punctipinnis*, family Pomacentridae), for example, has been identified as a trophic link between the plankton and reef communities. Blacksmiths forage on zooplankton during the day and shelter in rocky reefs at night. Their nocturnal defecation serves to transfer significant amounts of organic carbon from the plankton community to the reef (Bray et al., 1981). Kelp forests have become a classic example of the significance of **keystone species**—organisms (usually predators) that exert a dominating influence on the community composition by structuring trophic relationships (see Going Deeper).

Given the quantity of algae in kelp forests and the associated shallow, rocky reefs, it is surprising that comparatively few herbivorous fishes are observed in the food web that has evolved in these communities. Seaweeds produce a number of secondary metabolites that deter predation, especially from fishes. The invertebrate grazers associated with kelp forests—amphipods, small crabs, and gastropods—have coevolved an increased resistance to these deterrent compounds and may themselves suffer less predation owing to decreased encounters with predatory fishes (Duffy and Hay, 1994). Kelp forests in temperate Australasia lack predators (specifically, sea otters; see Going Deeper) that could reduce the populations of herbivores; consequently, these kelp species have evolved much higher concentrations

of deterrent metabolites in response to higher levels of herbivory (Steinberg et al., 1995).

In the Northeast Pacific, a few species of kelp forest-associated fishes, including the opaleye (*Girella nigricans*), halfmoon (*Medialuna californiensis*), and occasionally the señorita (*Oxyjulis californica*), graze on the kelp canopy, yet fishes appear to have a negligible impact on kelp forests worldwide (Dayton, 1985). One exception is the herbivorous Australian fish species *Odax cyanomelas* (family Odacidae), which has been shown to affect the structure and dynamics of the kelp *Ecklonia radiata* (Andrew and Jones, 1990). Plant consumption is actually incidental in the señorita, as it forages on minute bryozoans that encrust on the kelp. It also consumes herbivorous isopods that may damage the individual plants and, in this way, contributes to the health of kelp beds (Bernstein and Jung, 1979).

Much study has focused on natural versus anthropogenic impacts on temperate, rocky reef ecosystems in the Northeast Pacific. The past 80 years have witnessed a dramatic increase in sea surface temperature on the West Coast of North America (McGowan et al., 1998). It is not clear if this is the result of human-induced global warming or part of a natural climate cycle. In the Southern California Bight (Fig. 33.2A), a pronounced increase in sea surface temperature was recorded in the winter of 1976–1977 (Fig. 33.2B). During this episode, the species richness of the temperate reef ichthyofauna decreased dramatically, and the fish assemblage shifted from one that was dominated by species with northern affinities to one dominated by species with southern affinities (Fig. 33.2C; Holbrook et al., 1997). The impact of global warming trends on the structure and function of tropical coral reef ecosystems has also been documented and will be discussed in Chapter 39.

• • • • • •
Coral Reefs

Although tropical coral reefs occupy only about 0.2 percent of the Earth's surface, their biodiversity is rivaled only by that of tropical rainforests. Approximately one sixth of the world's coastlines are bordered by coral reefs (Birkeland, 1997). Although coral reefs first originated 450 million years ago during the Ordovician, modern coral reef ecosystems are much more recent, dating at most to the Cenozoic (Bellwood and Wainwright, 2002). The skeleton-forming capabilities seen in the phylum **Cnidaria**—and in some species of sponges and algae—produced these conspicuous geological features in shallow seas worldwide. Corals require consistently warm temperatures (23 to 25°C) and uniform salinities; they will not form near bays and estuaries, where freshwater mixes with ocean water. They require shallow water in order to support the photosynthetic activity of the symbiotic algae that reside in their tissues.

Two phenomena are most significant in the evolution of modern coral reefs—photosymbiosis and the evolution of efficient herbivores, especially fishes. The evolution of symbiotic sheltering of photosynthetic algae by some cnidarians greatly enhanced their potential for growth and carbon fixation. Symbiotic dinoflagellates, termed **zooxanthellae,** are the best known photosynthetic associates of corals, but at least one coral species has been shown to also shelter nitrogen-fixing **cyanobacteria** that coexist with the dinoflagellates (Lesser et al., 2004).

It is important to recognize that the trophic structure of modern coral reef systems reflects to a significant degree the morphological diversification of feeding mechanisms available among the more derived perciform fishes (Wainwright and Bellwood, 2002). This is especially evident with the advent of modern groups of herbivorous fishes during the Eocene—these fishes likely had a profound impact on the evolution and diversification of corals. Herbivorous animals, including perciform fishes with their powerful and sophisticated jaw mechanisms capable of scraping, crushing, and shredding vegetation, probably promoted the proliferation of corals at the expense of competing seaweeds (Hay et al., 2004; Wood, 1998).

The heterogeneity of the reef substrate provides a wealth of niches for biota, thus making coral reefs the richest biome in the world's oceans. Early estimates of the productivity of coral reefs suggested that they were up to 20 times more productive than the surrounding ocean (Rhyther, 1959). The overwhelming complexity of the interrelationships among coral reef associates makes reliable assessments of productivity extremely difficult if not impossible to obtain (Barnes and Hughes, 1982). Coral reefs are usually located between 30° N and 30°S and have the broadest distribution and the highest diversity in the Indo–Pacific, but they are also well developed in the Caribbean and West Indies. The latitudinal range of corals tends to be broader on the eastern sides of the continents, where currents flowing away from the equator bring warm equatorial waters into higher latitudes (Fig. 33.3).

GOING DEEPER · Kelp, Fishes, and Marine Mammals—Unraveling a Once Tightly Knit Food Web

One of the most striking communities in temperate marine waters is that associated with kelp forests. Dense aggregations of brown algae (**Phaeophyta**) are common along coastlines where rocky substrates permit their attachment via specialized structures called **holdfasts.** The rapidly growing photosynthetic canopy is composed of the *stipes,* or stems, of the individual "plants" and their broad, flattened **fronds.** Some conspicuous species include the giant kelp (*Macrocystis pyrifera*) of the Pacific coasts of North and South America (individuals of which have been documented to grow as much as 50 cm per day!), *Laminaria* of the North Atlantic and Asiatic coasts of the Pacific, and *Dictyota dichotoma* from warm waters of the western Atlantic. Many species of kelps, as well as red algae (**Rhodophyta**), are commercially harvested because a number of useful products can be obtained from them. Kelps contain large amounts of starchlike compounds called **phycocolloids.** One of these, *algin,* is valued in the food industry as an emulsifier and a stabilizer; the creamy texture of commercial dairy products is largely due to algin additives. Ships designed to harvest kelp resemble giant, oceangoing lawnmowers that clip the upper layers of the rapidly regenerating canopy.

Kelps are primarily associated with cooler waters and, as such, tend to have broader distributions on the western side of continents, where currents bring polar waters toward the equator (see Chapter 35). Interestingly, the opposite is true for the distribution of coral reefs and their associated biota; these will have a broader range on the eastern sides of continents, where equatorial currents move water north or south toward higher latitudes (see Fig. 33.3).

The intricacy of the kelp forest food web has been demonstrated through the role played by keystone species. Sea otters (*Enhydra lutris*)—a fashionable item in the fur trade of the 19th century—were hunted nearly to extinction along the west coast of North America. One of the consequences of the elimination of sea otter populations was an increase in the abundance of their preferred prey items, including herbivorous sea urchins of the genus *Strongylocentrotus*. Thus, a three-tiered trophic cascade, in which sea otter abundance and macrophyte density are positively correlated, exists (Estes and Duggins, 1995; Estes and Palmisano, 1974). Urchins normally consume drift kelp that accumulates on the seafloor in inshore areas, but under conditions of rapid population increase, they begin to gnaw away on holdfast structures and stipes, with devastating consequences to the integrity of kelp communities (Tegner et al., 1995). Dense kelp forests have been reduced to "urchin deserts," where the only algae visible are encrusting coralline species. The reestablishment of sea otters and their mandated protection has helped revive kelp forests by increasing predation on sea urchins. An abrupt decline in sea otter populations has lately been observed along the Alaskan coast, however. The explanation for this, as we shall see, rests in the intimate linkage of the coastal kelp community to marine ecosystems much further offshore.

By now you should be asking yourself, "What about the fishes? Where do they fit into the equation?" Fishes, both intrinsic and extrinsic to the kelp forest community, are

Coral Reef Fishes

The total number of fish species associated—either peripherally or entirely—with coral reef ecosystems has been estimated at between 6,000 and 8,000 species, or about 35 percent of the known fish fauna (Ehrlich, 1975). The most characteristic groups can be broken down into three major assemblages: (1) *labroids*—wrasses, (Labridae), parrotfishes (Scaridae), and pomacentrid damselfishes (Pomacentridae); (2) *acanthuroids*—surgeonfishes (Acanthuridae), rabbitfishes (Siganidae), and Moorish idols (Zanclidae); and (3) *chaetodontoids*—butterflyfishes (Chaetodontidae) and angelfishes (Pomacanthidae; Sale, 1991a). The tetraodontiform fishes, including the boxfishes (Ostraciidae), triggerfishes (Balistidae), and porcupinefishes (Diodontidae), hover about the reef using their highly modified fins as their primary propulsive organs (see Chapter 19). Large predators, including several families of sharks, jacks (Carangidae), and barracudas (Sphyraenidae) periodically visit the coral reefs in search of prey. The generally restricted ranges of coral reef fishes have resulted in comparatively high degrees of endemism, making them especially vulnerable to extinction (Roberts et al., 2002). The expansion of the ornamental fish trade, which has targeted coral reef inhabitants, makes this an especially important consideration (see Chapter 39).

Conspicuous signaling coloration and deep, compressed bodies that allow rapid turns and access to narrow spaces are characteristic of fishes that forage in the water column above the coral reef. A host of other species, more cryptically colored to blend in with the bottom, make their home in the crevices and holes that abound in the reef environment. Anguilliform fishes, including the morays (Muraenidae), snake eels, and worm eels (Ophichthidae) are especially well adapted for such a cryptic existence. Among the most diverse coral reef families are blennies (Blenniidae; see Plate 16.12) and gobies (Gobiidae; see Plate 16.16). Some species avail themselves of the most unusual habitats. The shrimpfishes (Centriscidae; see Plate 15.5), so named because of their crustacean-like carapace, can insert themselves among the formidable spines of sea urchins. Anemonefishes (Pomacentridae), including 27 Indo-Pacific species of the genus *Amphiprion* (see Plate 37.8)

essential players. As discussed earlier, many species of temperate coastal fishes—especially those closely associated with rocky areas—form a fairly well-defined, kelp-associated assemblage. Of the two dominant kelp species found along the California coast, the perennial giant kelp (*Macrocystus*) harbors a greater density of kelp-associated fishes than the annual bull kelp (*Nereocystus*; Bodkin, 1986). What happens to kelp-associated fish assemblages when the kelp forest becomes decimated by urchins? As might be expected, the destruction of the kelp bed reduces the abundance of resident fish species—especially the midwater forms associated with the kelp canopy (Bodkin, 1988). The restoration of the canopy in kelp beds off the central California coast has resulted in increases in the number of canopy dwellers, such as the señorita (*Oxyjulis californica*), blue rockfish (*Sebastes mystinus*), and kelp surfperch (*Brachyistus frenatus*; Laur et al., 1988).

Kelp forests serve as vital nursery areas for many species of fishes. The temporal dynamics of kelp forests are strongly correlated with the recruitment success of several species of rockfishes (*Sebastes*; Carr, 1991). A linear relationship between *Macrocystus* density and the recruitment success of kelp bass (*Paralabrax clathratus*) has also been demonstrated (Carr, 1994). The close association of planktivorous young kelp surfperch with *Macrocystus* may affect their growth, as they are dependent on plankton availability and rate of delivery in the immediate vicinity of the kelp bed (Anderson and Sabado, 1995).

Fish assemblage structure reflects a dynamic relationship with kelp forests. As the kelp forests wax and wane owing to differences in sea surface temperature or urchin predation intensity, the fish populations fluctuate accordingly. Archaeological studies have verified the continued existence of the sea otter–fish trophic cascade for millennia. Investigations of the refuse middens of aboriginal Americans have revealed that fish bone abundance correlates positively with the density of otter bones and negatively with that of urchin tests (Ebeling and Laur, 1988; Simenstad et al., 1978).

Fish populations that are extrinsic to the kelp community also figure significantly in the maintenance of kelp forests. The recent decline in the Alaskan otter populations has been attributed to increased predation from killer whales (*Orcinus orca;* Estes et al., 1998). Killer whales and sea otters have coexisted for millennia; what caused the sudden change in the feeding preferences of killer whales? The decline of other, favored prey items, including Steller's sea lions and harbor seals, both of which have experienced dramatic declines since the late 1970s, probably prompted the killer whales' shift to sea otters. In turn, the overfishing of pollock and other benthic fish species probably precipitated the decline in the seal and sea lion populations that fed on them. (Recall from Chapter 13 that Steller's sea lion populations have been extended greater protection by closing off much of the North Pacific fishery grounds.) So there you have it: Because we absolutely could not go without fish sticks for dinner, the rich and diverse kelp forest ecosystems of the Alaskan coastline are losing a critical keystone species and may be on the brink of collapse. What were we thinking of?

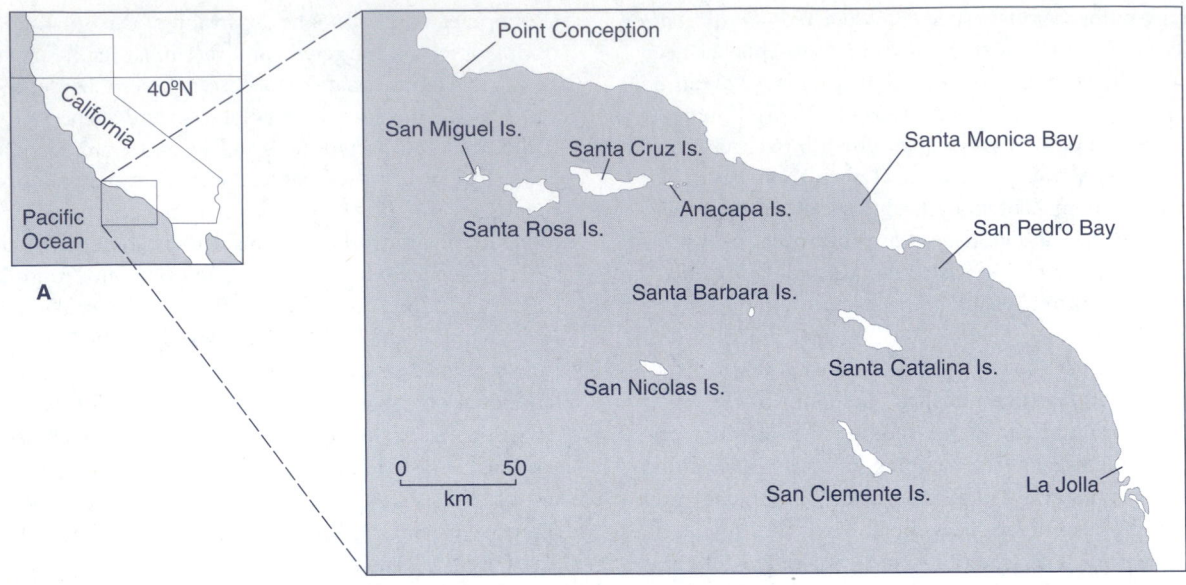

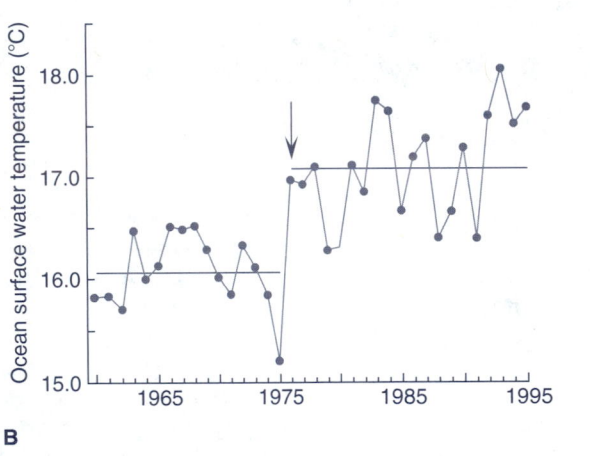

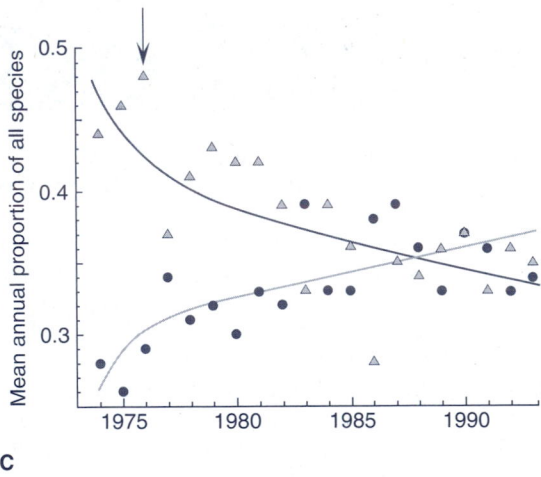

FIGURE 33.2

A, Map of the Southern California Bight; **B,** Mean annual ocean surface temperatures recorded at Scripps Pier, La Jolla, California. Arrow indicates the 1976–1977 climate shift (Source: Holbrook et al., 1997); **C,** Change in proportion of demersal fish species observed annually in Southern California. ▲ species with northern affinities; ● species with southern affinities. Arrow indicates the 1976–1977 climate shift. (Source: Holbrook et al., 1997).

and the spine-cheek anemonefish (*Premnas biaculeatus*), are well known for their habit of sheltering among the tentacles of sea anemones (see Chapter 37).

Food resources of the coral reefs are tremendous, including not only the coral polyps and other benthic invertebrates, but also algae and plankton. Modifications for food gathering range from the small mouths and long snouts of the longnose butterflyfishes (*Chelmon* and *Forcipiger* species; see Plate 33.3A) to the massive incisors forming a beaklike structure in scarid parrotfishes and balistid triggerfishes. Parrotfishes (see Plate 33.3B) use their incisors to scrape calcareous algae that can then be processed using their powerful pharyngeal dentition. At least one species (*Bolbometopon muricatus*) also

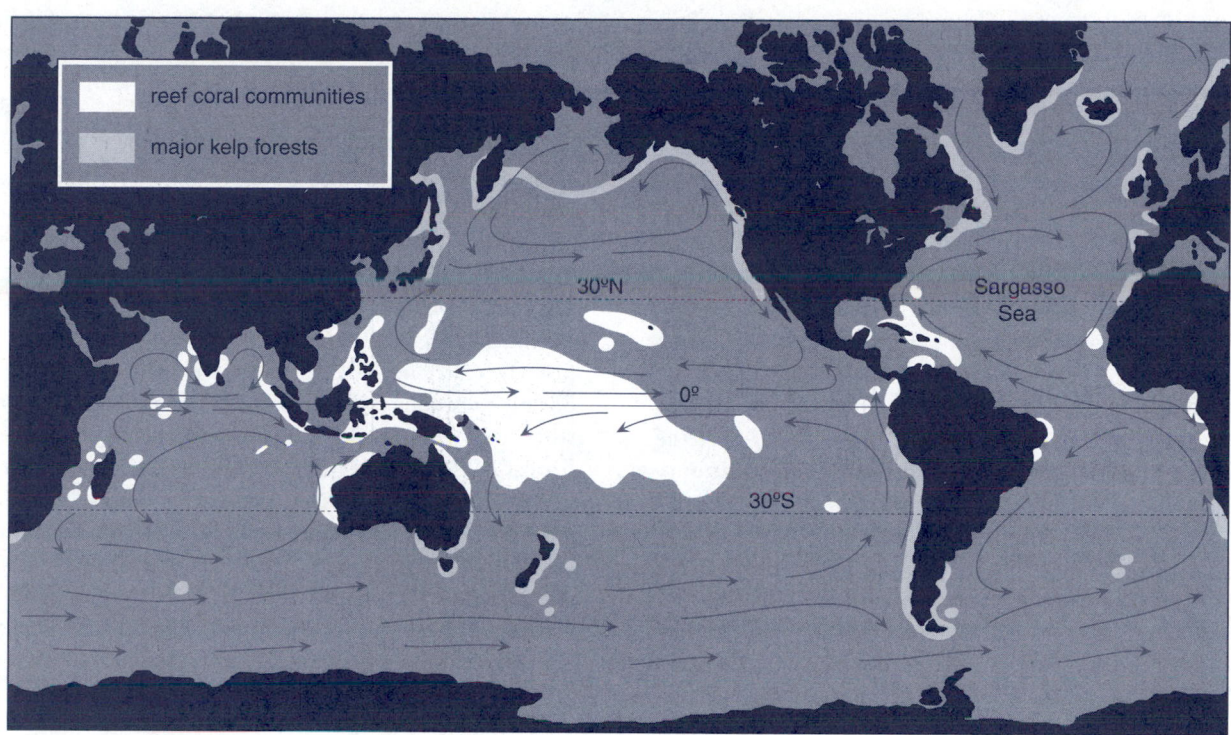

FIGURE 33.3
Current flow patterns and their impact on the distribution of coral reefs and kelp forests.

consumes live coral (Paxton and Eschmeyer, 1998). Labroid wrasses also possess well-developed pharyngeal grinding mechanisms that they use in conjunction with tusklike projecting teeth capable of snatching up small prey items (see Plate 16.8). Balistid triggerfishes use their powerful protruding incisors to grab and crush hard-shelled invertebrates (see Plate 17.4). Some triggerfishes use a clever trick to defeat the formidable spines of sea urchins: They blow jets of water at the urchin to turn it over and expose its vulnerable oral side (Paxton and Eschmeyer, 1998). Piscivores are among the most conspicuous members of the coral reef ichthyofaunal community. Predators such as the sea basses and barracudas (see Plate 33.3C) have wide mouths filled with sharp teeth suited to their carnivorous nature.

Fish families that graze on algae include the damselfishes, surgeonfishes, rabbitfishes, parrotfishes, and blennies. Herbivory is much more widespread in the tropics, and about 20 percent of the fishes inhabiting coral reefs are herbivorous (Sale, 1977). This may not reflect a disproportionate number of herbivores associated with coral reefs, however; rather, it is a function of general latitudinal trends (greater overall species diversity at lower latitudes) in reef fish assemblage structure (Meekan and Choat, 1997). To catalog the myriad adaptations encountered in the dozens of families inhabiting coral reefs is beyond the scope of this discussion, but the fishes considered here range from beautiful to bizarre, from the merely amusing to the potentially lethal. Some are adapted to life in the surf and in the channels that dissect the reef. Others seek quieter water in the lagoons that are often encircled by the reef itself.

Dynamics of Coral Reef Ecosystems

Although coral reef ecosystems are immensely complex, with a staggering diversity of associated organisms, their relative ease of accessibility, especially with the advent of scuba, has made them ideal environments to test and refine basic ecological and evolutionary concepts. In recent years, coral reefs have been the focus of studies elucidating the comparative influences of local- versus broad-scale phenomena on community structure and function (see Karlson et al., 2004). One of the few generalizations that can be made about the structure and dynamics of coral reef fish assemblages is that they are highly variable. The complex interaction of factors that result in a given fish assemblage never seems to play out in exactly the same way, so that adjacent patches of reef may have markedly different assemblages (Sale, 1991b). Ault and Johnson (1998), however, have identified certain aspects of reef fish assemblage structure that have elements of predictability.

The identification of temporal and spatial patterns of recruitment for specific groups of reef fishes, such as the labrid wrasses, also reveal predictable features (Sponaugle and Cowen, 1997). Recruitment patterns of coral reef fishes are most influenced by the fact that most species have pelagic larval stages. Molecular genetic techniques have recently been employed to define and delineate populations. Such techniques may prove helpful in resolving a troublesome paradox in the biology of coral reef fishes—how could such species diversity arise when the pelagic dispersal of larvae implies high gene flow and, consequently, little opportunity for genetic isolation? Shulman and Bermingham (1995) have identified a limited subdivision of populations in species of Caribbean reef fishes, thus supporting the claim that high gene flow is sustained by the prevailing flow patterns of surface currents. The level of divergence was far less than that observed between sister taxa that reside on opposite sides of the Panamanian isthmus. Taylor and Hellberg (2003), however, have shown that some species with extended larval life histories can retain significant local genetic integrity.

Many factors contribute to the actual assemblage structure observed on a given coral reef (see Mora et al., 2003). Doherty and Fowler (1994) have examined the role of larval recruitment in determining the structure of damselfish (Pomacentridae) populations inhabiting a series of patch reefs. Although many predictive models do not consider recruitment to be a potentially significant variable, Doherty and Fowler demonstrated that the population abundance on localized patch reefs could be explained almost entirely through the assessment of recruitment strength. A strong correlation exists between the density of new recruits to a reef site and mortality rates (Caselle, 1999; Schmitt and Holbrook, 1999). The structure and dynamics of coral reef fish assemblages is also influenced by the number and kinds of predators awaiting new recruits to the reef or by the number of potential predators that are themselves recruited (Caley, 1995a, 1995b; Hixon, 1998; Hixon and Carr, 1997). Environmental conditions, such as prevailing winds and tides at the time of spawning and larval settlement may also influence recruitment success (Robertson et al., 1999). Recruitment—specifically, its mechanism and overall impact—remains one of the more contentious issues among coral reef fish biologists (Hixon, 1998; Hixon and Webster, 2002).

The complexity of the coral reef environment, measured as the extent of available refuges for resident fishes, also influences assemblage structure (Caley and St. John, 1996).

Mechanical damage to the coral reef—as might occur with periodic hurricanes—has been demonstrated to have differential impacts on resident fish species. Algal concentrations increase on substrates where corals have been eliminated, and herbivorous species benefit, whereas species associated with live coral decline (Lewis, 1998). In the past few decades, many coral reefs have experienced dramatic increases in the abundance of the crown-of-thorns seastar (*Acanthaster planci*), a notorious coral predator. This has resulted in a greater supply of benthic algae available to herbivores; surgeonfishes (*Acanthurus nigrofuscus*) populations have experienced an enhanced growth rate on reefs where *Acanthaster* was present (Hart and Russ, 1996). Syms and Jones (2000) conducted controlled experiments on the Great Barrier Reef in which they assessed the combined effects of fish removal and habitat destruction on coral reef fish assemblages. They determined that the reef fish community structure is ultimately a function of the combined effects of deterministic factors (e.g., habitat structure and complexity) and stochastic processes (e.g., recruitment). Such conclusions probably hold true for other, less structurally and dynamically complex fish communities of the continental shelf regions.

. . . .
KEY POINTS AND CONNECTIONS
. . . .

- The continental shelf is a broad expanse of shallow oceanic habitat found at the continental margins. Ranging to a depth of 200 m, the continental shelf comprises about 7.6 percent of the ocean floor. Upwelling and coastal runoff make the continental shelf area a region of especially high productivity—as evidenced by the fact that most commercial fisheries take place in shelf waters.

- A diversity of pelagic fish families is associated with continental shelf waters. Included among these fishes are the clupeids, which constitute the most productive ocean fishery in terms of total tonnage harvested, and large, wide-ranging predatory forms, such as the anadromous salmonids and several species of scombroids.

  Migratory behavior, including that of the anadromous salmonids, is a topic of discussion in Chapter 36.

- The community structure and dynamics of bottom-dwelling fishes of the continental shelf has been the subject of intensive study; many commercially important species, including the cods, flounders, and rockfishes, are important members of shallow benthic and benthopelagic ichthyofaunal communities worldwide.

  The major pelagic and benthic fisheries of the continental shelves are discussed in Chapter 39.

- Distinctive fish communities form in close association with hard substrates in shallow waters. In temperate regions, rock outcrops provide attachment sites for dense aggregations of macrophytes; these kelp forests provide shelter and sustenance for assemblages of fishes including sculpins, blennies, greenlings, wrasses, and rock-fishes.

- The dominant hard substrate of tropical shallow waters consists of coral reefs. The most diverse fish assemblages in the world's oceans are found there, including up to 40 percent of the known fish fauna of the world. Labroids, acanthuroids, chaetodontoids, and tetraodontiform fishes are among the most conspicuous fish taxa associated with coral reefs, whereas the smaller blennioids and gobioids have been successful in exploiting the abundance of niches and crevices found among the heterogeneous substrates of tropical coral reefs.

The morphological versatility of the labroid jaw mechanism—a key feature contributing to their success in coral reef and other environments—is considered in more detail in Chapter 16. Cleaning symbiosis is an important feeding strategy that is closely associated with coral reef fishes; it is discussed in more detail in the Going Deeper box of Chapter 37.

BUILDING AN ICHTHYOLOGY LIBRARY

Caley, M. J. (Ed.). 1998. Recruitment and population dynamics of coral-reef fishes: An international workshop. *Austral. J. Ecol.* 23(4).

Sale, P. F. (Ed.). 1991. *The ecology of fishes on coral reefs.* Academic Press, San Diego.

———(Ed.). 2002. *Coral reef fishes: Dynamics and diversity in a complex ecosystem.* Academic Press, San Diego.

Of the many guides to coral reef and inshore fishes available, a few of the best are the following:

Carpenter, K. E. (Ed.). 2003. The living marine resources of the Western Central Atlantic, 3 vols. *Am. Soc. Ichthyol. Herpetol. Spec. Publ. 5.*

Humann, P. 1989. *Reef fish identification: Florida, Caribbean, Bahamas.* New World, Jacksonville, FL.

———, H. Hall, and N. McDaniel. 1996. *Coastal fish identification: California to Alaska.* New World, Jacksonville, FL.

Randall, J. E., G. R. Allen, and R. C. Steene. 1997. *Fishes of the Great Barrier Reef and Coral Sea* (2nd ed.). University of Hawaii Press, Honolulu.

Smith, C. L. 1997. *National Audubon Society field guide to tropical marine fishes.* Alfred A. Knopf, New York.

REFERENCES

Allen, M. J., and G. B. Smith. 1988. Atlas and zoogeography of common fishes in the Bering Sea and Northeastern Pacific. *NOAA Tech. Rep. NMFS 66.*

Anderson, T. W., and B. D. Sabado. 1995. Correspondence between food availability and growth of a planktivorous temperate reef fish. *J. Exp. Mar. Biol. Ecol. 189:* 65–76.

Andrew, N. L., and G. P. Jones. 1990. Patch formation by herbivorous fish in a temperate Australian kelp forest. *Oecologia 85:* 57–68.

Ault, T. R., and C. R. Johnson. 1998. Spatially and temporally predictable fish communities on coral reefs. *Ecol. Monogr. 68:* 25–50.

Barnes, R. S. K., and R. N. Hughes. 1982. *An introduction to marine ecology.* Blackwell, Oxford.

Bellwood, D. R., and P. C. Wainwright. 2002. The history and biogeography of fishes on coral reefs, pp. 5–32. In *Coral reef fishes: Dynamics and diversity in a complex ecosystem*, P. F. Sale (Ed.). Academic Press, San Diego.

Bernstein, B. B., and N. Jung. 1979. Selective pressures and coevolution in a kelp-bed canopy community in Southern California. *Ecol. Monogr. 49:* 335–355.

Birkeland, C. (Ed.). 1997. *Life and death of coral reefs.* Chapman and Hall, New York.

Bodkin, J. L. 1986. Fish assemblages in *Macrocystis* and *Nereocystis* kelp forests off central California. *Fish. Bull. 84:* 799–808.

———. 1988. Effects of kelp forest removal on associated fish assemblages in central California. *J. Exp. Mar. Biol. Ecol. 117:* 227–238.

Bray, R. N., A. C. Miller, and G. G. Geesey. 1981. The fish connection: A trophic link between planktonic and rocky reef communities? *Science 214:* 204–205.

Caley, M. J. 1995a. Community dynamics of tropical reef fishes: Local patterns between latitudes. *Mar. Ecol. Prog. Ser. 129:* 7–18.

———. 1995b. Reef-fish community structure and dynamics: An interaction between local and larger-scale processes? *Mar. Ecol. Prog. Ser. 129:* 19–29.

———, and J. St. John. 1996. Refuge availability structures assemblages of tropical reef fishes. *J. Anim. Ecol. 65:* 414–428.

Carr, M. H. 1991. Habitat selection and recruitment of an assemblage of temperate zone reef fishes. *J. Exp. Mar. Biol. Ecol. 146:* 113–137.

———. 1994. Effects of macroalgal dynamics on recruitment of a temperate reef fish. *Ecology 75:* 1320–1333.

Caselle, J. E. 1999. Early post-settlement mortality in a coral reef fish and its effect on local population size. *Ecol. Monogr. 69:* 177–194.

Choat, J. H. 1982. Fish feeding and the structure of benthic communities in temperate waters. *Ann. Rev. Ecol. Syst. 13:* 423–449.

Cortés, E. 1999. A stochastic stage-based population model of the sandbar shark in the western North Atlantic, pp. 115–136. In Life in the slow lane: Ecology and conservation of long-lived marine animals, J. A. Musick (Ed), *Am. Fish. Soc. Symp. 23.* American Fisheries Society, Bethesda, MD.

Dayton, P. K. 1985. Ecology of kelp communities. *Ann. Rev. Ecol. Syst. 16:* 215–245.

Doherty, P., and T. Fowler. 1994. An empirical test of recruitment limitation in a coral reef fish. *Science 263:* 935–939.

Duffy, J. E., and M. E. Hay. 1994. Herbivore resistance to seaweed chemical defense: The roles of mobility and predation risk. *Ecology 75:* 1304–1319.

Ebeling, A. W. and R. N. Bray. 1976. Day versus night activity of reef fishes in a kelp forest off Santa Barbara, California. *Fish. Bull. 74:* 703–717.

———, and D. R. Laur. 1988. Fish populations in kelp forests without sea otters: Effects of severe storm damage and destructive sea urchin grazing, pp. 169–201. In *The community ecology of sea otters*, G. R. Van Blaricom and J. A. Estes (Eds.). Springer Verlag, Berlin.

———, and M. A. Hixon. 1991. Tropical and temperate reef fishes: Comparison of community structures, pp. 509–563. In *The ecology of fishes on coral reefs*, P. F. Sale (Ed.). Academic Press, San Diego.

———., R. J. Larsen, and W. S. Alevizon. 1980. Habitat groups and island–mainland distribution of kelp-bed fishes off Santa

Barbara, California, pp. 403–431. In *Multidisciplinary symposium on the California islands*, D. M. Power (Ed.). Santa Barbara Museum of Natural History, Santa Barbara, CA.

Ehrlich, P. R. 1975. The population biology of coral reef fishes. *Ann. Rev. Ecol. Syst. 6:* 211–247.

Estes, J. A., and D. O. Duggins. 1995. Sea otters and kelp forests in Alaska: Generality and variation in a community ecological paradigm. *Ecol. Monogr. 65:* 75–100.

———, and J. F. Palmisano. 1974. Sea otters: Their role in structuring nearshore communities. *Science 185:* 1058–1060.

———, M. T. Tinker, T. M. Williams, and D. F. Doak. 1998. Killer whale predation on sea otters linking oceanic and nearshore systems. *Science 282:* 473–476.

Hart, A. M., and G. R. Russ. 1996. Response of herbivorous fishes to crown-of-thorns starfish *Acanthaster planci* outbreaks. III. Age, growth, mortality and maturity indices of *Acanthurus nigrofuscus*. *Mar. Ecol. Prog. Ser. 136:* 25–35.

Hay, M. E., J. D. Parker, D. E. Burkepile, C. C. Caudill, A. E. Wilson, Z. P. Hallinan, and A. D. Chequer. 2004. Mutualisms and aquatic community structure: The enemy of my enemy is my friend. *Ann. Rev. Ecol. Syst. 35:* 175–197.

Hixon, M. A. 1998. Population dynamics of coral-reef fishes: Controversial concepts and hypotheses. *Austral. J. Ecol. 23:* 192–201.

———, and M. H. Carr. 1997. Synergistic predation, density dependence, and population regulation in marine fish. *Science 277:* 946–949.

———, and M. S. Webster. 2002. Density dependence in reef fish populations, pp. 303–325. In *Coral reef fishes: Dynamics and diversity in a complex ecosystem*, P. F. Sale (Ed.). Academic Press, San Diego.

Holbrook, S. J., R. J. Schmitt, and J. S. Stephens, Jr. 1997. Changes in an assemblage of temperate reef fishes associated with a climate shift. *Ecol. Appl. 7:* 1299–1310.

Karlson, R. H., H. V. Cornell, and T. P. Hughes. 2004. Coral communities are regionally enriched along an oceanic biodiversity gradient. *Nature 429:* 867–870.

Laur, D. R., A. W. Ebeling, and D. A. Coon. 1988. Effects of sea otter foraging on subtidal reef communities off central California, pp. 151–168. In *The community ecology of sea otters*, G. R. Van Blaricom and J. A. Estes (Eds.). Springer Verlag, Berlin.

Lesser, M. P., C. H. Mazel, M. Y. Gorbunov, and P. G. Falkowski. 2004. Discovery of symbiotic nitrogen-fixing cyanobacteria in corals. *Science 305:* 997–1000.

Lewis, A. R. 1998. Effects of experimental coral disturbance on the population dynamics of fishes on large patch reefs. *J. Exp. Mar. Biol. Ecol. 230:* 91–110.

McGowan, J. A., D. R. Cayan, and L. M. Dorman. 1998. Climate–ocean variability and ecosystem response in the Northeast Pacific. *Science 281:* 210–217.

Meekan, M. G., and J. H. Choat. 1997. Latitudinal variation in abundance of herbivorous fishes: A comparison of temperate and tropical reefs. *Mar. Biol. 128:* 373–383.

Mitchell, C. T., and J. R. Hunter. 1970. Fishes associated with drifting kelp, *Macrocystis pyrifera*, off the coast of southern California and northern Baja California. *Calif. Fish. Game 56:* 288–297.

Mora, C., P. M. Chittaro, P. F. Sale, J. P. Kritzer, and S. A. Ludsin. 2003. Patterns and processes in reef fish diversity. *Nature 421:* 933–936.

Paxton, J. R., and W. N. Eschmeyer. 1998. *Encyclopedia of fishes* (2nd ed.). Academic Press, San Diego.

Rhyther, J. H. 1959. Potential productivity of the sea. *Science 130:* 602–608.

Roberts, C. M., C. J. McLean, J. E. N. Veron, J. P. Hawkins, G. R. Allen, D. E. McAllister, C. G. Mittermeier, F. W. Schueler, M. Spalding, F. Wells, C. Vynne, and T. B. Werner. 2002. Marine biodiversity hotspots and conservation priorities for tropical reefs. *Science 295:* 1280–1284.

Robertson, D. R., S. E. Swearer, K. Kaufmann, and E. B. Brothers. 1999. Settlement vs. environmental dynamics in a pelagic-spawning reef fish at Caribbean Panama. *Ecol. Monogr. 69:* 195–218.

Rogers, S. I., K. R. Clarke, and J. D. Reynolds. 1999. The taxonomic distinctness of coastal bottom-dwelling fish communities of the Northeast Atlantic. *J. Anim. Ecol. 68:* 769–782.

Sale, P. F. 1977. Maintenance of high diversity in coral reef fish communities. *Am. Nat. 111:* 337–359.

———. 1991a. Introduction, pp. 3–15. In *The ecology of fishes on coral reefs*, P. F. Sale (Ed.). Academic Press, San Diego.

———. 1991b. Reef fish communities: Open nonequilibrial systems, pp. 564–598. In *The ecology of fishes on coral reefs*, P. F. Sale (Ed.). Academic Press, San Diego.

Sasaki, T. 1972. Demersal fishes collected in the southeastern shelf waters of Alaska. *Bull. Fac. Fish. Hokkaido Univ. 22:* 281–289.

Schmitt, R. J., and S. J. Holbrook. 1999. Mortality of juvenile damselfish: Implications for assessing processes that determine abundance. *Ecology 80:* 35–50.

Shulman, M. J., and E. Bermingham. 1995. Early life histories, ocean currents, and the population genetics of Caribbean reef fishes. *Evolution 49:* 897–910.

Simenstad, C. A., J. A. Estes, and K. W. Kenyon. 1978. Aleuts, sea otters, and alternate stable-state communities. *Science 200:* 403–411.

Simpfendorfer, C. A. 1999. Demographic analysis of the dusky shark fishery in Southwestern Australia, pp. 149–160. In Life in the slow lane: Ecology and conservation of long-lived marine animals, J. A. Musick (Ed.)., *Am. Fish. Soc. Symp. 23.* American Fisheries Society, Bethesda, MD.

Sponaugle, S., and R. K. Cowen. 1997. Early life history traits and recruitment patterns of Caribbean wrasses (Labridae). *Ecol. Monogr. 67:* 177–202.

Steinberg, P. D., J. A. Estes, and F. C. Winter. 1995. Evolutionary consequences of food chain length in kelp forest communities. *Proc. Natl. Acad. Sci. USA 92:* 8145–8148.

Syms, C., and G. P. Jones. 2000. Disturbance, habitat structure, and the dynamics of a coral-reef fish community. *Ecology 81:* 2714–2729.

Taylor, M. S., and M. E. Hellberg. 2003. Genetic evidence for local retention of pelagic larvae in a Caribbean reef fish. *Science 299:* 107–109.

Tegner, M. J., P. K. Dayton, P. B. Edwards, and K. L. Riser. 1995. Sea urchin cavitation of giant kelp (*Macrocystis pyrifera* C. Agardh) holdfasts and its effects on kelp mortality across a large California forest. *J. Exp. Mar. Biol. Ecol. 191:* 83–99.

Wainwright, P. C., and D. R. Bellwood. 2002. Ecomorphology of feeding in coral reef fishes, pp. 33–55. In *Coral reef fishes: Dynamics and diversity in a complex ecosystem*, P. F. Sale (Ed.). Academic Press, San Diego.

Wood, R. 1998. The ecological evolution of reefs. *Ann. Rev. Ecol. Syst. 29:* 179–206.

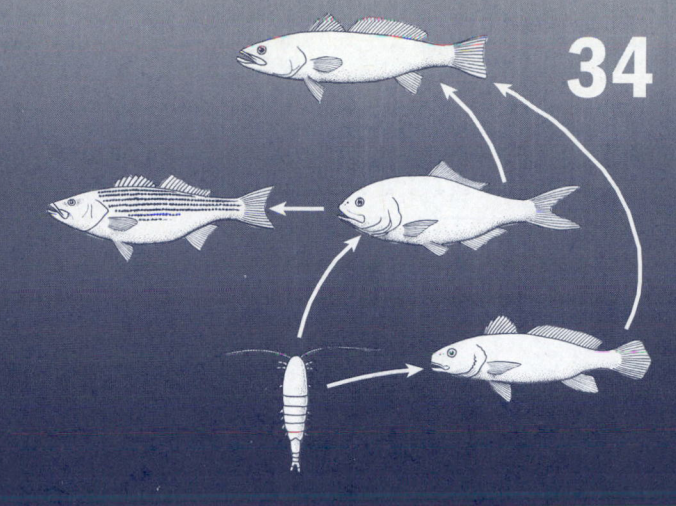

34 ECOLOGY OF FISHES V: COASTAL MARINE ENVIRONMENTS: BEACHES, ESTUARIES, AND ROCKY INTERTIDAL SHORES

THE NEARSHORE ENVIRONMENT

PROTECTED INSHORE ENVIRONMENTS: BAYS AND ESTUARIES

Temperate Estuarine Ecosystems and Associated Fish Assemblages

Tropical Estuarine Ecosystems and Associated Fish Assemblages
  Seagrass Beds as Habitat for Fishes
  Estuarine Ichthyofauna of Africa and Australia

EXPOSED INSHORE ENVIRONMENTS: SURF ZONES OF SANDY BEACHES AND ROCKY SHORES

Fish Assemblages of Surf Zones
The Fish Fauna of Rocky Shores
  Adaptations of Rocky Shore Fishes
  Intertidal Fish Assemblage Structure and Dynamics

SEASONAL VARIATION IN COASTAL FISH ASSEMBLAGES

Inshore Environments as Nurseries

Coastal waters provide an environment for fishes that is profoundly influenced by the adjacent continents. This zone of interface between land and water makes for conditions that are challenging yet remarkably accommodating. Several families of fishes have adapted to the physical and chemical challenges of this environment and have succeeded in making the productive waters of the shoreline their home. Many species of fishes have become exclusively adapted to shoreline environments, and many more species move into these realms for purposes of reproduction. Here, eggs can develop and larvae can seek refuge in shallow waters that fewer predators can invade. The rhythmic advance and retreat of the tides in coastal waters has resulted in the evolution of several fish species that have transcended, to varying degrees, the confines of aquatic existence and have become truly amphibious animals. Variations in turbulence and substrate type define the three marine realms we will investigate here—the often tumultuous waters of exposed, sandy beaches and rocky shores, and the quieter, more protected waters of bays and estuaries.

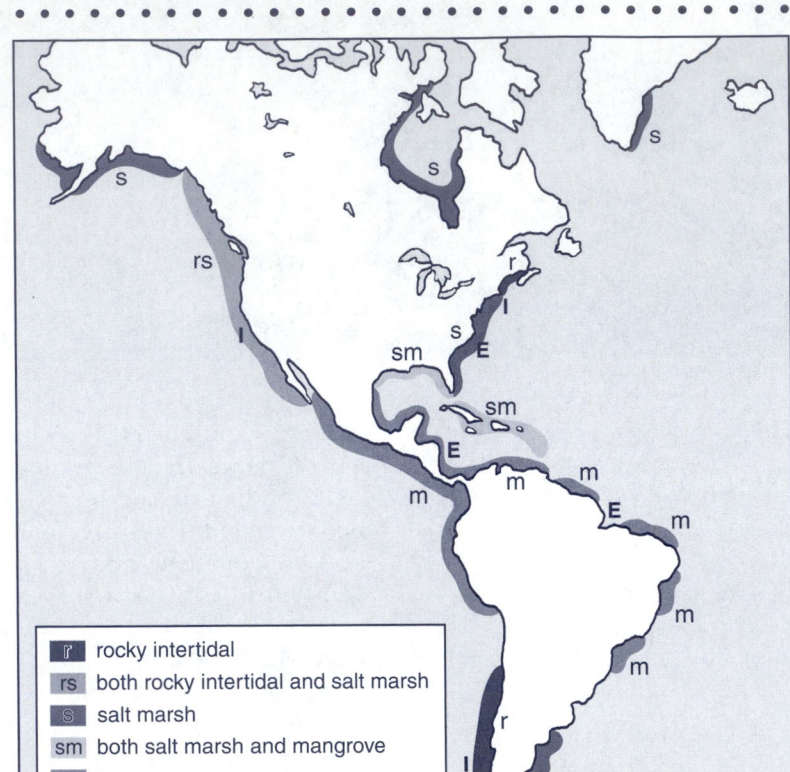

Legend:
- **r** rocky intertidal
- **rs** both rocky intertidal and salt marsh
- **s** salt marsh
- **sm** both salt marsh and mangrove
- **m** mangrove

THE NEARSHORE ENVIRONMENT

In our consideration of the oceanic realm, we have distingui-shed the neritic province from the oceanic province, with the neritic region being associated with the continental shelves (Fig. 33.1). The **nearshore** environment encompasses those neritic habitats that are immediately adjacent to the coastlines and, as such, range to only a few meters in depth. The term **littoral** is sometimes used to define this region, although this term has also been used to define shoreline environments in freshwaters. Light levels in this region are such that pho-tosynthetic activity, both by planktonic algae and by ben-thic macrophytes, is typically high. This enhanced primary productivity is also sustained by **upwelling,** which brings nutrient-enriched deeper waters into the shallows, where the nutrients can become incorporated into the biomass produced as a consequence of photosynthesis. Nutrients of terrestrial origin, delivered by freshwater drainages flowing to the coast, also are enriching. Proximity to land may also bring with it a greater seasonal impact. Freezing conditions can occur in shallow waters at high latitudes. Seasonal pre-cipitation patterns may profoundly affect ambient salinities in nearshore environments. The terms **eurythermic** (hav-ing a broad temperature tolerance) and **euryhaline** (having a broad salinity tolerance) have been used to describe the physiological capabilities of nearshore biota (see Chapter 30). In the uppermost reaches of the nearshore region, tidal fluc-tuation is most significant in influencing the distribution and abundance of biota.

Fossil evidence has suggested that the nearshore and possibly even the intertidal realm may be the ancestral home of the earliest fishes (Schultze, 1999). The fishes inhabiting the shoreline are among the most fascinating to be found in the world's oceans, as they live in a realm that is in a constant state of change. Diurnal, seasonal, and tidal fluctuations govern the lives of fishes found along the world's shorelines, and these fishes are thus by nature eurythermic and euryha-line. Those fishes living within the influence of the tides may have behavioral patterns that are cued by tidal rhythms. Yet the opportunities for existence in one of the most produc-tive zones of the ocean are such that diverse and complex communities have evolved along shorelines. By the middle

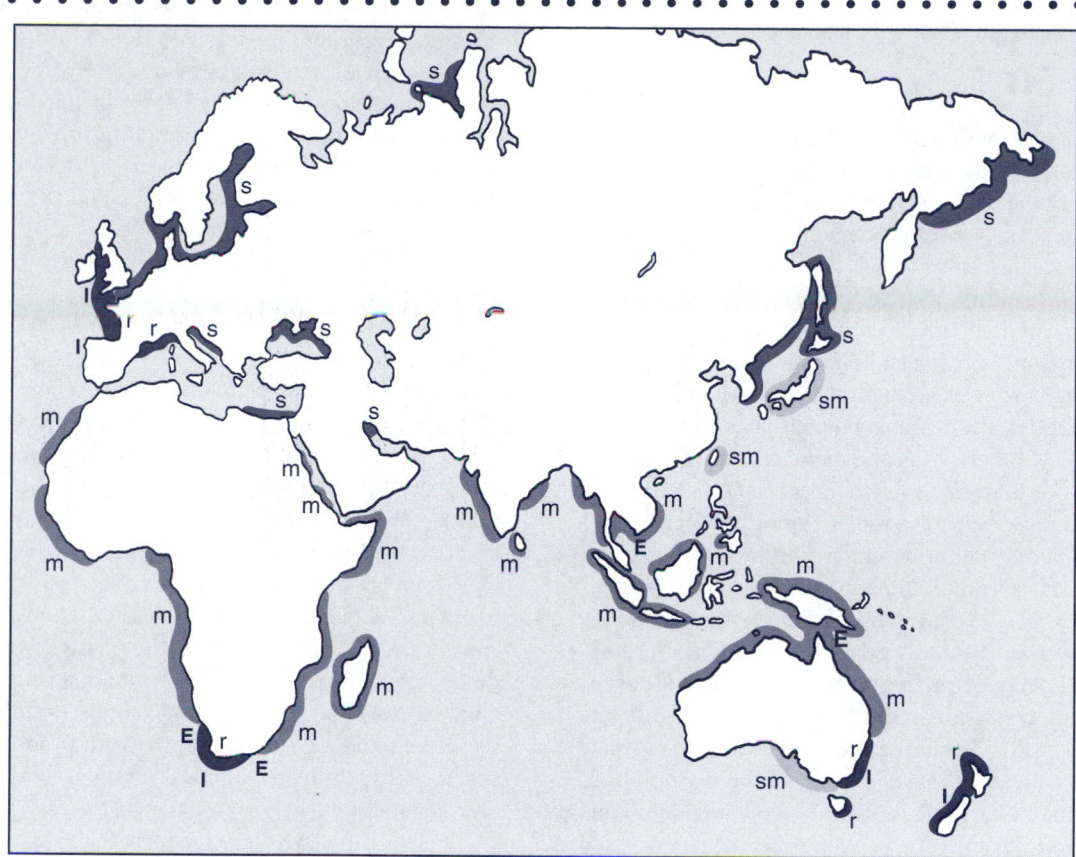

FIGURE 34.1

Global distribution of estuarine and rocky shore habitats. Areas of active estuarine research are designated with an *E;* areas of active study on rocky shore ichthyofauna are designated with an *I* (Source: Barnes and Hughes, 1982; Chapman, 1977; Horn, 1999; Prochazka et al., 1999).

of the last century, comprehensive lists of euryhaline species known to inhabit the waters of North and Middle America had been published (Gunter,1942, 1956).

In our consideration of the fishes that constitute these shoreline communities, we will distinguish ecosystems largely on the basis of their characteristic substrates. This is natural, as most species of marine animals (about 90 percent) and almost all macrophytes live in close association with the sea bottom (Sumich, 1999). Bay and estuarine communities exist in areas protected from extremes in turbulence. The shelter afforded by protected coastal embayments provides greater opportunity for the deposition of sediments of finer texture. The community structure of estuaries and their associated ichthyofaunal assemblages differ significantly, as we shall see, between tropical and temperate shores.

Exposed shorelines experience the consequences of sometimes extreme wave action. This has had decidedly dif-

ferent effects on shorelines with unconsolidated and consolidated substrates. Depending on its exposure to waves and currents, the shore can be the site of erosion or of deposition. Eroding shores are typically rocky and rugged. Wave action on sandy shores imposes extremely abrasive conditions on benthic ichthyofauna. Nevertheless, a number of fish families have become adapted to living in and above the shifting sands of wave-washed beaches. Rocky shores have evolved a distinctive community structure, based on the availability of hard substrates on which to attach. Success on rocky shores often depends on the ability to outcompete others in the colonization of exposed hard substrates. This, in general, lends itself to a much greater floral and faunal complexity. Again, distinctive ichthyofaunal assemblages characterize tropical versus temperate rocky shore environments. The global distribution of estuarine and rocky shore environments is shown in Fig. 34.1.

PROTECTED INSHORE ENVIRONMENTS: BAYS AND ESTUARIES

Bays and estuaries represent coastal environments that are partially surrounded by land, so that they are protected from the full force of the ocean. Estuaries are formed when freshwater drainages empty into embayments, creating an aquatic environment that is transitional between freshwater and fully marine. Organisms adapted to such surroundings have formed unique and complex communities, in which many species of fishes are integral components. As one might expect, estuarine organisms are both euryhaline and eurythermic. Diurnal, tidal, and seasonal fluctuations in salinity and temperature and other physical parameters, such as depth, dissolved oxygen, and light levels, define the physical environment. The complex chemistry of freshwater meeting seawater results in the deposition of much of the sediment load transported downstream. Consequently, estuarine bottoms are typically composed of sand or mud of varying texture.

Depending on the local tidal regime and the characteristic morphology of the basin, bays and estuaries experience continuous flushing, and animal life is cued to the rhythm of the tides. Terrestrial and aquatic landscapes have collaborated to create in estuaries some of the most productive ecosystems on Earth. Rivers bring tremendous amounts of organic and inorganic nutrients into the estuary. Coastal upwelling, combined with the tides, introduces nutrients from the ocean bottom offshore. The photosynthetic activity of planktonic and benthic algae and plants captures this energy, making it available to estuarine food webs. These same food webs are also heavily dependent on organic detritus flushed from the adjacent watershed.

Only a small fraction of the animal species that participate in estuarine food webs have freshwater affinities. The vast majority, fishes included, have ecological and evolutionary affinities with coastal marine species. Fishes that are permanent residents of estuaries share one physiological feature with the anadromous and catadromous fishes that pass through estuaries in the course of the completion of their life cycle—they are powerful **osmoregulators**. These species are able to maintain remarkably stable internal fluid and ionic compositions in the face of the sometimes dramatic fluctuations in salinity characteristic of the estuarine environment (see Chapter 25).

Estuarine ecosystems encompass wetlands, mudflats, seagrass beds, and channels. The abundance and diversity of estuarine ichthyofauna tends to vary among these different habitats (see Paterson and Whitfield, 2000). As the tides advance and retreat, the wetlands and mudflats experience exposure to atmospheric conditions followed by inunda-tion in a rhythmic fashion dictated by the local tidal regime. Wetland vegetation is terrestrial in origin, but its periodic immersion in seawater requires at least a partial tolerance to marine conditions. Because of this salt tolerance, plants of the wetland community are termed **halophytes**. The wetland vegetation of temperate shores is decidedly different from that encountered in the tropics, and the estuarine ecosystems also differ accordingly.

Temperate Estuarine Ecosystems and Associated Fish Assemblages

The wetlands of temperate estuarine ecosystems are dominated by two halophyte species—pickleweed (*Salicornia*) closer to the shoreline, and cordgrass (*Spartina*) growing at slightly higher elevations. Growing in the lowest parts of the marsh and therefore subject to greater inundation, pickleweed adapts to its briny environment (and earns its name) by storing large quantities of salt in its fleshy leaves. These two plants dominate the wet grassland known as the **saltmarsh**. In the fall and winter, the plants of the salt marsh die back, and the tides flush enormous quantities of their detrital remains into the water. Decomposers and detritivores recycle the carbon that, combined with products of photosynthesis from aquatic sources, makes its way into the estuarine food web.

Fishes are active participants in the temperate estuarine food web. Clupeids, such as menhaden (*Brevoortia*), harvest the plankton produced in estuarine waters, and mullets (Mugilidae) assume the role of detritivores. Microcarnivorous species, such as the spot (*Leiostomus xanthurus*, family Sciaenidae) or the flounders (Pleuronectidae, Bothidae), prey on small benthic **epifauna** (living on the mudflat) and **infauna** (hiding in burrows); higher carnivores, such as the striped bass (*Morone saxatilis*, family Moronidae) and the weakfish (*Cynoscion regalis*, family Sciaenidae) harvest the smaller forage fishes (Fig. 34.2).

Temperate estuaries are also home to a number of smaller benthic, epibenthic, and pelagic fish species that are not represented in the illustrated food web. Bays and estuaries worldwide are also invaded by a number of shark, skate, and ray species common to the open continental shelf waters. Atherinomorph fishes (see Chapter 15) are especially prevalent in estuarine waters worldwide. The upper reaches of tidal creeks that dissect the mudflats in southeastern U.S. coastal estuaries are home to sometimes dense concentrations of cyprinodonts and atherinopsids, such as the mummichog (*Fundulus heteroclitus*), sheepshead minnow (*Cyprinodon variegatus*), and tidewater silverside (*Menidia beryllina*) or Atlantic silverside (*M. menidia*). As one moves further south along the northwest Atlantic shores to the bays and estuaries

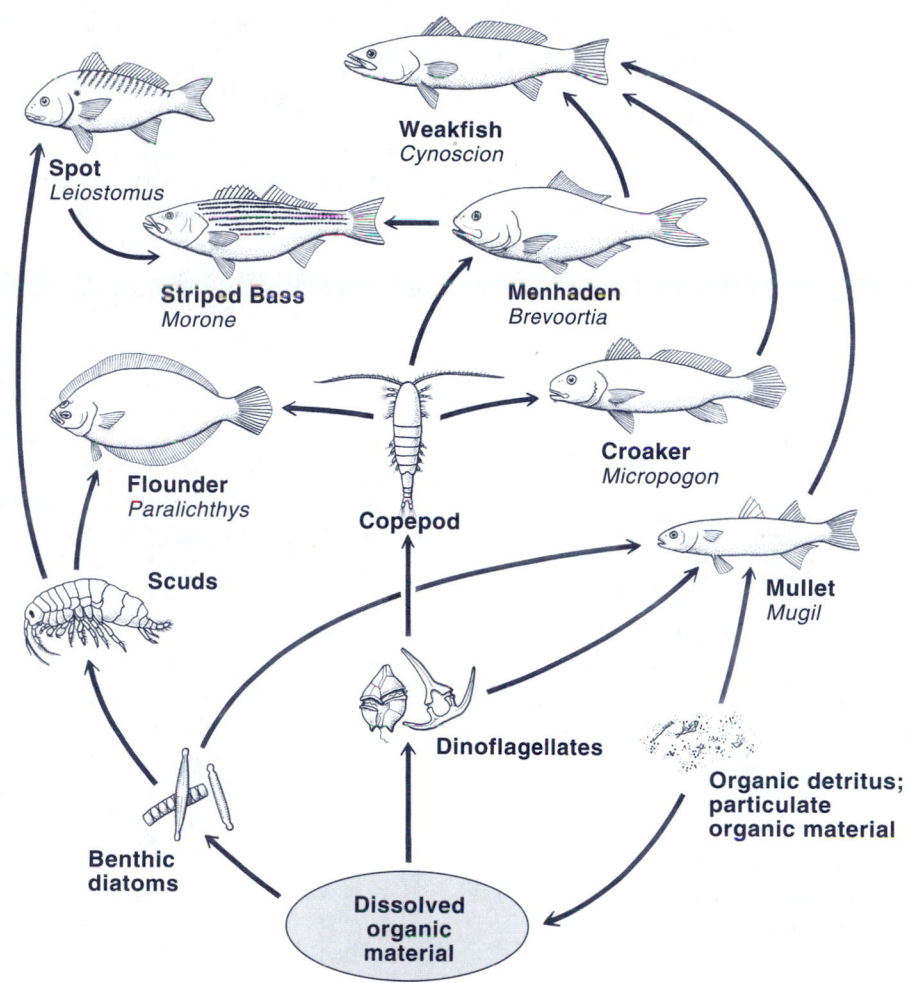

FIGURE 34.2

A simplified estuarine food web, indicating the feeding niches of some fish species common to coastal waters of the southeastern United States (Source: Smith, 1980).

of Florida and the Gulf Coast, the diversity of cyprinodontid, atherinid, and poeciliid fishes increases.

Atherinopsids, especially *Atherinops affinis,* are among the most commonly encountered fishes in Pacific coast estuaries (Horn and Allen, 1985). The tidal creeks are home to *Fundulus parvipinnis* (Fundulidae) and *Gambusia affinis* (Poeciliidae), whereas several species of viviparous surfperches, such as the white seaperch (*Phanerodon furcatus*) and the shiner perch (*Cymatogaster aggregata*) commonly forage in estuarine waters of the northeast Pacific. These waters also host a number of benthic species, including gobiids, such as the longjaw mudsucker (*Gillichthys mirabilis*), bay goby (*Lepidogobius lepidus*), and arrow goby (*Clevelandia ios*); cottids,

such as the Pacific staghorn sculpin (*Leptocottus armatus*); and batrachoidids, such as the midshipman (*Porichthys*).

Tropical Estuarine Ecosystems and Associated Fish Assemblages

The wetlands of tropical estuarine ecosystems consist of **mangals**—plant communities dominated by several species of shrubby, seawater-tolerant mangrove trees. Mangal shorelines consist of dense thickets of prop roots, which anchor the trees in the black, anoxic mud, and pneumatophores—special rootlike extensions that project up from the mud to assist in the oxygenation of the plant. These surfaces provide

718 • Part Seven Interactions of Fishes with Their Environment

a substrate for a diversity of attached flora and fauna, including polychaete worms, tree oysters (*Isognomon*), snails, crabs and shrimp, that provide ample feeding opportunities for fishes. A most unusual feeding strategy is practiced by the pufferfish (*Sphoeroides rosenblatti*). Puffers capture littorinid snails clinging to mangrove roots above the water line by partially jumping out of the water and grabbing them with their large incisors. This feeding strategy has actually been observed to affect the vertical zonation of the littorinid prey species (Duncan and Szelistowski, 1998).

The ichthyofaunal composition of tropical estuaries is dependent on a range of factors. Blaber (2000) noted the following factors as having the most influence on the structuring of tropical estuarine fish communities:

1. size, depth, and physical regime of the estuary, particularly salinity and turbidity

2. nature and depth of adjacent marine waters and, to a lesser extent, freshwaters

3. geographical location of the estuary, in terms of latitude and proximity to distinct oceanic features

These generalizations may be equally applicable to temperate estuarine ecosystems. Comparatively few families of fishes are represented in tropical estuarine communities, yet individual abundance may be quite high for those species that are able to tolerate an environment that may experience extremes in salinity or dissolved oxygen concentration. In a comprehensive study of the trophic relationships of mangrove fauna, Odum and Heald (1972) reported on the diets of 45 species of fishes collected in a south Florida estuary (Table 34.1). Several of the species reported, such as the Florida gar, American eel, and species of *Fundulus*, are commonly reported from freshwater as well.

Two families of fishes are remarkable for their adaptation to the rigors of inhabiting mangrove estuaries. The mudskippers (subfamily Oxudercinae of the family Gobiidae) are among the most conspicuous of all fishes, whereas the mangrove rivulus (*Rivulus marmoratus;* family Aplocheilidae) is known for being among the most secretive. Mudskippers (see Plate 16.17; Chapter 16) may be abundant in tropical estuaries and are conspicuous because of their terrestrial capabilities. These unusual fishes use their highly modified pectoral fins to creep about on the mudflats or prop roots of mangroves exposed at low tide. The physiological adaptations that permit such terrestrial sojourns in such an environmentally challenging environment are truly remarkable (see Chapter 30).

The distribution of the mangrove rivulus (see Plate 15.11) closely reflects the geographic distribution of the red mangrove (*Rhizophora mangle;* Davis et al., 1995). These diminutive fishes were once considered extremely rare, until their preferred habitat—the burrows of land crabs (*Cardisoma guanhumi*) above the water line—was discovered (Davis et al., 1990; Taylor et al., 1995). They have also been collected from stagnant pools and ditches. Like the mudskipper, their tolerance for extremely challenging aquatic conditions suggests remarkable physiological capabilities. *Rivulus* is also known to make limited terrestrial excursions by making short flips of its body. One of the most remarkable discoveries concerning *Rivulus* was that it is a self-fertilizing hermaphrodite (see Chapter 27).

Seagrass Beds as Habitat for Fishes

An unusual flora has evolved in inshore habitats—flowering plants that appear most closely related to the lily family and are generally referred to as **seagrasses**. Because the pollination of seagrasses is not dependent on insects, their flowers tend to be small and inconspicuous. The plants are well attached via dense mats of rhizomes to the sandy or muddy substrates on which they grow. Most genera are adapted for life at and just below the high-tide mark in protected embayments. Eelgrass (*Zostera*) is the most widely distributed, occurring in temperate bays and estuaries of the Atlantic and Pacific. Surfgrass (*Phyllospadix*) is unusual, as it inhabits open coastal environments and is well adapted to wave action, as is evident by its common name. Most seagrass species are tropical—the most common genus is turtlegrass (*Thalassia*). Seagrass beds are among the most productive patches of the estuarine environment—the plant biomass in turtlegrass beds has been measured as high as 1 kg (dry weight)/m² (Castro and Huber, 2000). Seagrasses possess a key adaptive feature that is absent in other marine macrophytes—true root systems that can absorb nutrients from the sediment. Much like the halophytes in temperate salt-marsh ecosystems, seagrasses produce large quantities of detritus, on which a shredding- and decomposition-based food web is built. A diverse epifloral community lives on the leaves of the seagrasses, whereas the sediments house a diverse community of burrowing organisms among their rhizomes. Seagrasses serve as an important food base for a number of fish species that preferentially inhabit the seagrass beds of inshore areas. Seagrass beds, developing as they do in open sand or mud flats beneath the tide line, provide significant cover—especially for larvae and juveniles.

Seagrass communities of estuaries on the northwestern Atlantic coast and in the Gulf of Mexico have been the most intensively studied (Table 34.2; Adams, 1976a, 1976b; Briggs and O'Connor, 1971; Livingston, 1982; Petrik and Levin, 2000). Although many species preferentially inhabit

TABLE 34.1   FAMILIES AND SPECIES OF FISHES INHABITING A SOUTH FLORIDA ESTUARY

Family	Species
Carcharhinidae	*Carcharhinus leucas* (bull shark)
Lepisosteidae	*Lepisosteus platyrhincus* (Florida gar)
Elopidae	*Elops saurus* (ladyfish)
Megalopidae	*Megalops atlanticus* (tarpon)
Anguillidae	*Anguilla rostrata* (American eel)
Clupeidae	*Harengula jaguana* (scaled sardine)
Engraulidae	*Anchoa mitchilli* (bay anchovy)
Ariidae	*Ariopsis felis* (sea catfish)
	Bagre marinus (gafftopsail catfish)
Synodontidae	*Synodus foetens* (inshore lizardfish)
Cyprinodontidae	*Cyprinodon variegatus* (sheepshead minnow)
	Floridichthys carpio (goldspotted killifish)
Fundulidae	*Adinia xenica* (diamond killifish)
	Fundulus confluentus (marsh killifish)
	F. grandis (gulf killifish)
	Lucania goodei (bluefin killifish)
	L. parva (rainwater killifish)
Poeciliidae	*Gambusia affinis* (mosquitofish)
	Heterandria formosa (least killifish)
	Poecilia latipinna (sailfin molly)
Atherinidae	*Menidia beryllina* (tidewater silverside)
Gobiesocidae	*Gobiesox strumosus* (skilletfish)
Batrachoididae	*Opsanus beta* (gulf toadfish)
Mugilidae	*Mugil cephalus* (striped mullet)
Centropomidae	*Centropomus undecimalis* (common snook)
Serranidae	*Epinephelus itajara* (goliath grouper)
Lutjanidae	*Lutjanus griseus* (gray snapper)
Centrarchidae	*Lepomis punctatus* (spotted sunfish)
Carangidae	*Caranx hippos* (crevalle jack)
	Oligoplites saurus (leatherjacket)
Gerreidae	*Eucinostomus argenteus* (spotfin mojarra)
	E. gula (silver jenny)
	Eugerres plumieri (striped mojarra)
Sciaenidae	*Bairdiella chrysoura* (silver perch)
	Cynoscion nebulosus (spotted seatrout)
	Sciaenops ocellatus (red drum)
Sparidae	*Archosargus probatocephalus* (sheepshead)
	Lagodon rhomboides (pinfish)
Gobiidae	*Bathygobius soporator* (frillfin goby)
	Gobiosoma robustum (code goby)
	Lophogobius cyprinoides (crested goby)
	Microgobius gulosus (clown goby)
Sphyraenidae	*Sphyraena barracuda* (great barracuda)
Achiridae	*Achirus lineatus* (lined sole)
	Trinectes maculatus (hogchoker)

Source: Odum and Heald, 1972.

seagrass beds, others seek them out simply because they are the only available shelter. Of the 39 fish species reported by Adams (1976a) in North Carolina seagrass beds, 15 species were only encountered once.

The pinfish (*Lagodon rhomboides*) is the dominant species in Atlantic seagrass beds, yet it shows strong seasonal fluctuations, so that the winter community may be dominated by other species, such as silversides (*Menidia menidia* and *Membras martinica*) and the spot (*Leiostomus xanthurus;* Adams, 1976a). Because they are so common, pinfish play an important role as a trophic link in estuarine ecosystems. They spend much of their time moving between seagrass beds and adjacent intertidal marshes and are therefore seen as providing an important link between marsh-derived secondary production and the adjacent seagrass beds (Irlandi and Crawford, 1997). (The great evolutionary biologist E. O. Wilson lost much of the vision in one eye as a child when he was struck by the dorsal spine of a hooked pinfish.)

The Atlantic silverside (*M. menidia*) and the fourspine stickleback (*Apeltes quadracus*) are the dominant species encountered in eelgrass beds of the more northerly Atlantic coastal waters of New York state (Briggs and O'Connor, 1971). Of the 106 species found by Weinstein and Heck (1979) in Caribbean *Thalassia* beds, the most common species were bucktooth parrotfish (*Sparisoma radians*), spotfin mojarras (*Eucinostomus argenteus*), and lane snappers (*Lutjanus synagris;* Table 34.2). Common inhabitants of estuarine seagrass beds of the northeast Pacific coast are also shown in Table 34.2.

Estuarine Ichthyofauna of Africa and Australia

Although we have focused our attention on estuarine and associated seagrass ichthyofaunal communities of the coastal environments of North and Central America, we would be remiss if we did not consider the ichthyofauna that inhabits protected embayments and their associated estuaries elsewhere in the world. The fish assemblages of African estuaries, for example, have been intensively studied—particularly by our South African colleagues (see Whitfield, 1999). Harrison and Whitfield (1995) have investigated the fish assemblage structure in an estuarine system that experiences seasonal interruptions in its communication with the ocean (Fig. 34.3).

TABLE 34.2 FISH SPECIES COMMONLY ASSOCIATED WITH SEAGRASS BEDS IN ATLANTIC/GULF COAST, ATLANTIC CARIBBEAN, AND NORTHEAST PACIFIC WATERS

Atlantic/Gulf Coast		Caribbean		Northeast Pacific	
Family	*Species*	*Family*	*Species*	*Family*	*Species*
Engraulidae	*Anchoa mitchilli*	Mullidae	*Pseudopeneus maculatus*	Engraulidae	*Engraulis mordax*
Syngnathidae	*Syngnathus floridae*	Gerridae	*Eucinostomus argenteus*	Syngnathidae	*Syngnathus leptorhynchus*
	S. fuscus		*E. gula*		
Atherinopsidae	*Menidia menidia*	Scorpaenidae	*Scorpaena plumieri*	Atherinopsidae	*Atherinops affinis*
	Membras martinica				
Haemulidae	*Orthopristis chrysoptera*	Haemulidae	*Haemulon aurolineatum*	Clupeidae	*Clupea harengus*
			H. bonariense		
			H. plumieri		
Sciaenidae	*Bairdiella chrysoura*	Lutjanidae	*Lutjanus analis*	Sebastidae	*Sebastes melanops*
	Leiostomus xanthurus		*L. synagris*		*S. caurinus*
			Ocyurus chysurus		
Serranidae	*Mycteroperca microlepis*	Serranidae	*Epinephelus striatus*	Osmeridae	*Hypomesus pretiosus*
Monacanthidae	*Monacanthus hispidus*	Monacanthidae	*Monacanthus ciliatus*	Cottidae	*Enophrys bison*
			M. setifer		*Leptocottus armatus*
Sparidae	*Lagodon rhomboides*	Scaridae	*Sparisoma radians*	Pholidae	*Pholis ornata*
Blenniidae	*Hypsoblennius hentzi*	Chaetodontidae	*Chaetodon capistratus*	Pleuronectidae	*Parophrys vetulus*
					Platichthys stellatus
Bothidae	*Paralichthys lethostigma*	Tetraodontidae	*Sphoeroides spengleri*	Embiotocidae	*Cymatogaster aggregata*
	P. dentatus				*Embiotoca jacksoni*
Gobiidae	*Gobionellus boleosoma*				*E. lateralis*
Batrachoididae	*Opsanus tau*				*Phanerodon furcatus*

Source: Weinstein and Heck, 1979.

Clupeids (*Gilchristella*), springers (*Elops machnata*), mullets (*Mugil, Myxus,* and *Liza*), glassperches (*Ambassis productus*), and gobies (*Glossogobius*) are the dominant estuarine fishes. Fishes of the family Cichlidae, being classified as "primary" freshwater fishes (see Chapter 29), are not normally thought of as estuarine affiliates, but cichlids (*Oreochromis*) were also reported in Harrison and Whitfield's (1995) study, and the cichlid genera *Tilapia* and *Sarotherodon* dominate the ichthyofaunal assemblages found in the tidal creeks of Nigerian mangrove estuaries.

In one of the earliest biological observations recorded from Australia, the renowned Pacific explorer James Cook named his first anchorage on the continent "Stingray Harbour," after the large numbers of smooth stingrays (*Dasyatis brevicaudata*) he observed in the shallow sandflats. The quantity of previously unknown plants collected by naturalist Joseph Banks at this location prompted Cook to reconsider his original name, changing it to what we know it as today—

"Botany Bay." Robertson and Blaber (1992) have recorded 40 species of fishes frequently encountered in Australian mangrove estuaries and their associated seagrass beds. The most common of these include *Nematalosa come* (Clupeidae), *Thryssa hamiltonii* (Engraulidae), *Ambassis gymnocephalus* (Ambassidae), and *Leiognathus equulus* (Leiognathidae).

In addition to the previously mentioned atherinomorph fishes, several other families appear to be characteristically associated with shallow estuarine waters worldwide. Stingrays of the family Dasyatidae are perhaps the most frequently encountered elasmobranchs in estuaries. Planktivorous clupeoids often form dense schools there also. Mullets (Mugilidae) are important detritivores in estuaries. Of the percomorphs, the porgies (Sparidae), drums and croakers (Sciaenidae), grunts (Haemulidae), snappers (Lutjanidae), and mojarras (Gerridae) are especially well represented in estuarine waters. Many of the aforementioned families include species that are highly esteemed food fishes. The

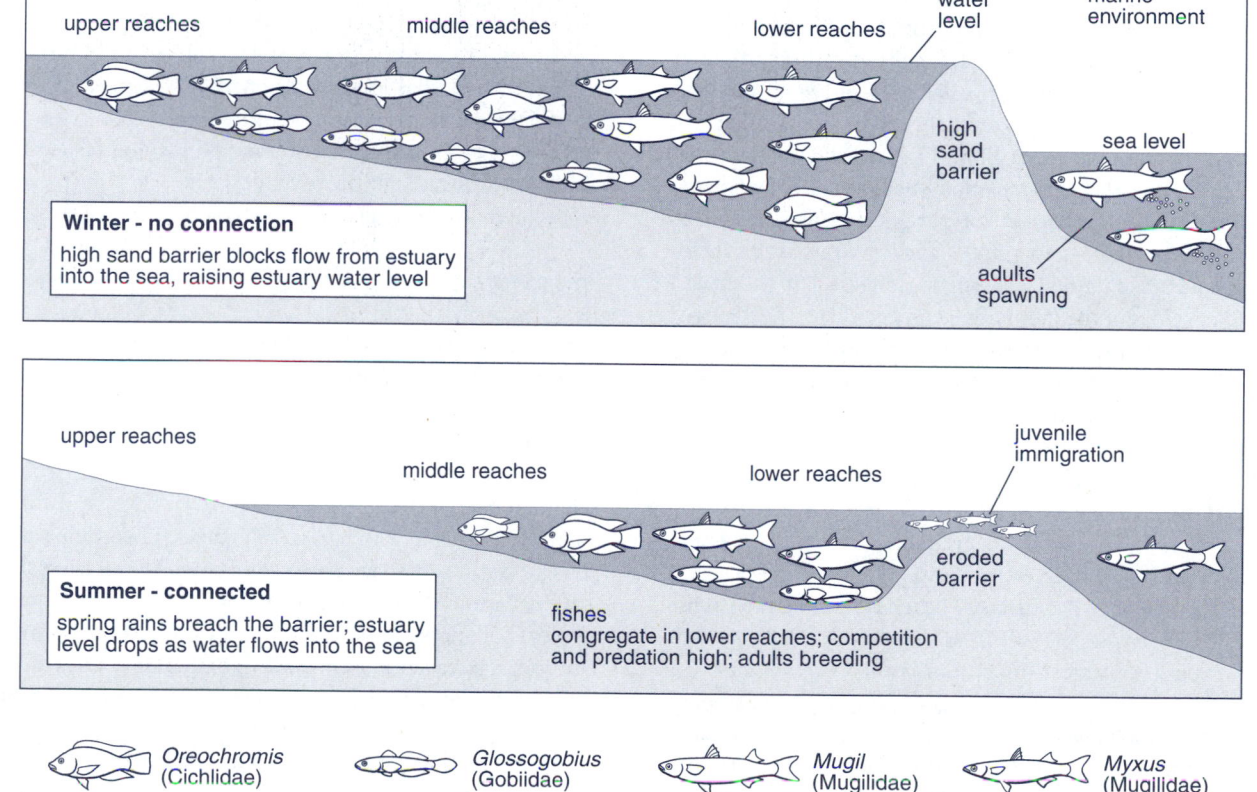

FIGURE 34.3
Seasonal dynamics of the fish assemblage in a South African estuary (Source: Harrison and Whitfield, 1995).

"usual suspects" among the bottom dwellers—the gobioids and blennioids—also count among their numbers several species found in estuaries. Several species of flatfishes also forage on the bottoms of estuaries.

EXPOSED INSHORE ENVIRONMENTS: SURF ZONES OF SANDY BEACHES AND ROCKY SHORES

Fish Assemblages of Surf Zones

Although my parents were not marine biologists, they had resided on the California coast long enough to know that it was necessary to warn us children to always shuffle our feet when entering the water at our favorite beach. This would encourage the resident population of round stingrays (*Urobatis halleri*, family Urolophidae) to move along and not surprise us with a painful sting. Stingrays are members of the somewhat depauperate fish fauna that can be found in the surf zone of northeastern Pacific sandy beaches. The surf zone presents a harsh environment of constant mechanical agitation from waves, tides, and currents, yet some fish species can be found there in abundance. Planktivorous herrings (Clupeidae) and anchovies (Engraulidae) are sometimes found schooling just beyond the surf zone. Osmerids, including the surf smelt (*Hypomesus pretiosus*) and night smelt (*Spirinchus starksi*), and atherinopsid silversides, including the celebrated grunion (see Chapter 15, Going Deeper; Plate 15.7) are well-known surf zone spawners. The aptly named surfperches (Embiotocidae)—especially the barred surfperch (*Amphistichus argenteus*) and the redtail surfperch (*A. rhodoterus*)—are favorites of surfcasters fishing on California beaches, as are the croakers and corvinas of the family Sciaenidae. On the East Coast, sciaenids are termed drums; from these common names, it should be apparent that this family is particularly adept at making sounds using their swim bladders (see Chapter 21). These fishes forage among the sands for invertebrates that may have been exposed by the surf. Flatfishes, such as the speckled sand dab (*Citharichthys stigmaeus*) and the starry flounder (*Platichthys stellatus*), frequent inshore areas where they may bury themselves in the sand. Other species that hide in the sands of Pacific coast beaches include the Pacific sandfish (*Trichodon trichodon*), which has fringed lips that allow respiratory water to pass but strain out the sand; the Pacific sand lance (*Ammodytes hexapterus*); and the smooth stargazer (*Kathetostoma averruncus*).

Atlantic coastal surf zone fish assemblages appear more diverse, with a greater representation of tropical forms, some of them seasonal visitors moving north with the Gulf Stream in the summer months. Species of anchovies, silversides, drums, porgies (Sparidae), and flatfishes are also well represented on the Atlantic coast. Inshore, uranoscopids are also represented by the southern stargazer (*Astroscopus y-graecum*), with the lancer stargazer (*Kathetostoma albigutta;* see Plate 16.11) found further offshore. A third species, the freckled stargazer (*Xenocephalus egregius*), occurs at the edge of the continental shelf (Hoese and Moore, 1998). Popular game species include the striped bass (*Morone saxatilis*), bluefish (*Pomatomus saltatrix*), and cobia (*Rachycentron canadum*). The striped bass was successfully introduced to the Pacific coast in the late 1800s, and it has become a favorite sport fish. An anadromous species, it enters rivers to spawn in the spring. Warm-water carangids, such as the permit (*Trachinotus falcatus*) and the Florida pompano (*T. carolinus*), are found along the coastal beaches of the Carolinas, Georgia, and Florida. Filefishes (Monacanthidae) and puffers (Tetraodontidae) are also occasionally seen. Of the 39 species of surf zone fishes collected by Cupka (1972), most (34) were classified as seasonal migrants or strays.

Many fish species inhabit the surf zone only as larvae or juveniles and move into deeper water as they mature. In an extensive study of larval and juvenile surf zone fishes found at the mouth of the St. Lucia estuary in South Africa, representatives of 47 families were collected. Most abundant families included the Sparidae, Haemulidae, Ambassidae, Tripterygiidae, and Chanidae (Harris and Cyrus, 1996). Studies have demonstrated that the coexistence of juveniles of closely related forms may be facilitated by differential resource allocation. For example, partitioning of the food resources has been demonstrated in the juveniles of five flatfish species encountered on the Belgian coast (Beyst et al., 1999).

The Fish Fauna of Rocky Shores

Adaptations of Rocky Shore Fishes

Rocky shores, which experience a regular cycle of emersion and immersion as the tides ebb and flow, are a harsh and stressful environment in which to live. For all of the reasons mentioned previously, rocky shores are also an environment characterized by very high productivity. The availability of hard substrate on which to attach makes for complex and intricate community structures, and their ease of accessibility has made rocky shores a favorite for illustrating various ecological principles. The classic studies of Joseph Connell on barnacles have demonstrated the comparative roles of biotic versus abiotic factors in determining community structure on rocky shores (Connell, 1961).

Among the families of fishes adapted to rocky substrates in inshore areas, many have become specifically adapted to existence in the uppermost reaches, under the influence of the tides. **Intertidal** fishes "are generally considered to be those that live their postlarval lives in the intertidal zone and possess particular morphological, physiological, and behavioral adaptations that enable them to do so" (Gibson and Yoshiyama, 1999, p. 264). The irregular substrate of rocky intertidal shores usually has depressions that retain water as the tide recedes, so that fully marine conditions can be perpetuated there when the tide recedes. These tide pools are often strewn with boulders and harbor a variety of marine algae and invertebrates. Some intertidal fishes become fully emergent when the tide recedes. As such, they have become adapted to varying degrees to terrestrial conditions. As in other inshore environments, the fish fauna found in the intertidal zone can be conveniently divided into **visitors** and **residents**. Visitors enter the intertidal zone at high tide, mainly for the purposes of exploiting the rich feeding opportunities there, but they retreat with the tides. Permanent residents of the intertidal zone possess a suite of adaptations that enable them to remain in the intertidal zone when the water retreats (Fig. 34.4; Gibson 1969, 1982).

Fishes found on exposed rocky intertidal shores must be able to withstand strong wave motion by swimming, hiding, or clinging. In temperate regions, intertidally occurring fishes must be tolerant of the fluctuations in temperature and salinity, especially if they frequent shallow pools close to the high-tide level. The complex community structure of rocky intertidal environments may be governed by an equally complex mosaic of thermal environments, in which thermal stress may be correlated with latitude (Helmuth et al., 2002).

Many intertidal fishes display anatomical features that are consistent with their benthic existence in a shallow, turbulent region. Comparatively few patterns of body shape and fin design characterize intertidal fishes, suggesting that the uniformly rigorous intertidal environment encountered worldwide has resulted in convergent evolution among different families (Horn, 1999). The fishes of rocky intertidal shores are generally small, rarely longer than 30 cm, which enables them to occupy holes, crevices, and interstices among the rocks and vegetation, although some, such as the monkeyface prickleback (*Cebidichthys violaceus;* see Plate 34.1), can grow as large as half a meter. Distinctive fin structures are also an adaptation to intertidal life. Broad, thickened pectoral fins can be used to wedge the animal into crevices. Clingfishes (Gobiesocidae) represent perhaps the most unusual morphological adaptation to the rigors of intertidal existence, especially on wave-washed shores. The paired fins and the broad, flattened body are modified into a powerful sucking disc (see Plate 34.2; Fig. 16.1D). Sharp, chisel-shaped teeth enable them to pry limpets, their favored food item, from rocks while gaining leverage from their adhesive disc (see Chapter 16).

The skin of intertidal fishes is generally tough, so that it can withstand the abrasive conditions normally encountered in their environment, yet many species, as an adaptation to periodic deprivation of aquatic sources of oxygen, engage in atmospheric gas exchange across their skin surfaces (Horn

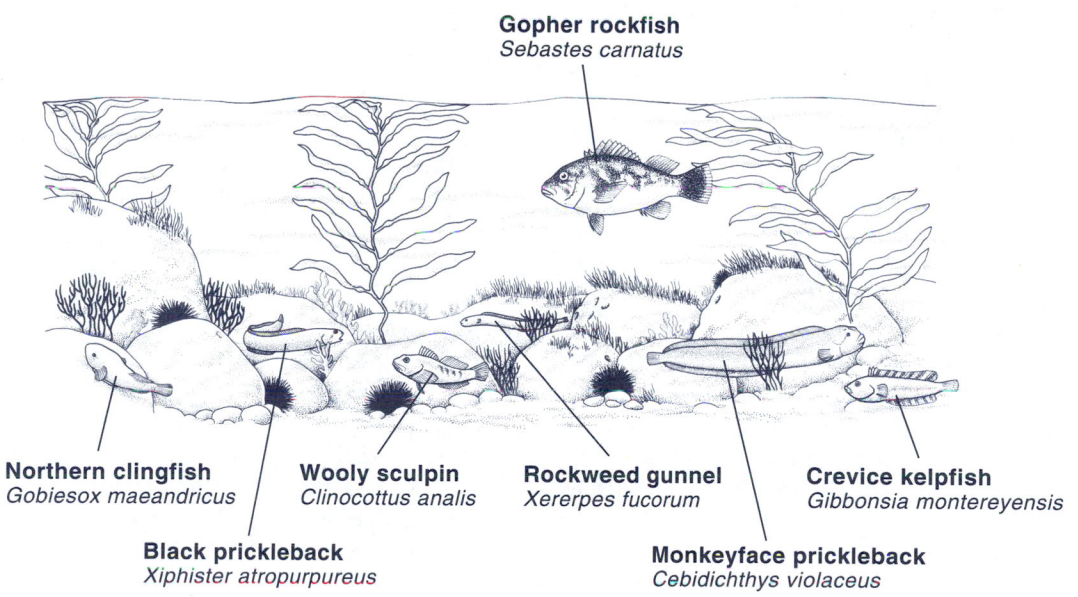

Gopher rockfish
Sebastes carnatus

Northern clingfish
Gobiesox maeandricus

Wooly sculpin
Clinocottus analis

Rockweed gunnel
Xererpes fucorum

Crevice kelpfish
Gibbonsia montereyensis

Black prickleback
Xiphister atropurpureus

Monkeyface prickleback
Cebidichthys violaceus

FIGURE 34.4
Representative intertidal fishes from the coast of California (Source: Horn and Gibson, 1988).

and Gibson, 1988; Martin, 1995, 1996; Martin and Bridges, 1999). Because intertidal fishes are generally small, their surface-to-volume ratio is comparatively high. This has consequences for both their resistance to desiccation and their respiratory capacity. A large amount of surface area, such as observed in the elongate pricklebacks (Stichaeidae) and gunnels (Pholidae), might increase evaporative water loss, yet it also provides a greater surface for cutaneous gas exchange (see Chapter 24). In order to optimize gas exchange across the skin surface, many intertidal fishes have a well-developed cutaneous vascular supply and a reduced scale coat. Atmospheric gas exchange takes place across the gills, skin, or linings of the opercular and buccal cavities (Martin and Bridges, 1999). Remarkably, marine air-breathing fishes lack the specialized organs for atmospheric gas exchange seen in their freshwater counterparts (Graham, 1976, 1997).

Intertidal fishes minimize the threat of desiccation by retreating to tide pools at low tide or by secreting themselves beneath boulders or among damp vegetation. Hypoxia, and the associated decrease in pH arising as a consequence of concentrated respiratory activity, is a common nocturnal occurrence in isolated tide pools, as the dense concentration of biota depletes dissolved oxygen, which cannot be replenished by photosynthetic activity until daylight returns. Some species will emerge from pools under such conditions. Martin (1996) has observed a range of emergent behaviors and associated survival capacity in the cottid sculpins (Fig. 34.5). Two species of labrisomid, "four-eyed" blennies that inhabit rocky shores of the Pacific coast of Central America, *Mnierpes macrocephalus* and *Dialommus fuscus,* will readily leave the water when it becomes anoxic and will also emerge when threatened by predators (Graham, 1973, 1997).

The abundance of macrophytes attached to rocks provides excellent shelter for many species, some of which are especially well adapted for inhabiting vegetation. Elongate stichaeids and pholids can withstand periodic emersion by secreting themselves in damp, dense vegetation. Cryptic coloration, including expert mimicry of the algae, is the rule in intertidal fishes (see Plate 34.3; see also Plate 16.14).

Fishes that occur in the intertidal or shallow subtidal zones of rocky shores and sandy beaches often exhibit remarkable reproductive adaptations that optimize the survival of eggs and larvae. The synchronization of spawning activities with tidal cycles, as in the grunion (see Chapter 15, Going Deeper), enables fishes to place eggs far up on the beach, where the eggs may be periodically exposed to atmospheric conditions. The increased temperatures and improved access to oxygen in these locations may enhance development, while at the same time minimizing exposure to predators. Although the adults may not necessarily show adaptations for air breathing, the eggs that they place near the

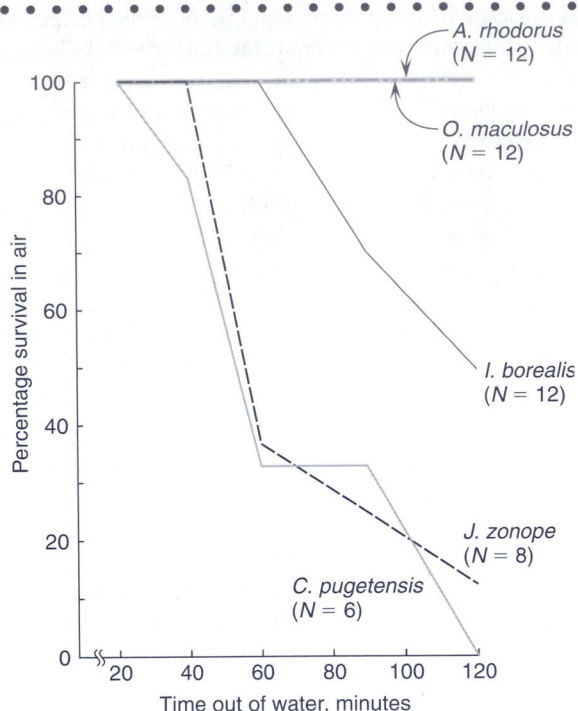

FIGURE 34.5
Tolerance to emergence, measured as percentage survival when exposed to air, in sculpins (family Cottidae). *Ascelichthys rhodorus* and *Oligocottus maculosus* live in high intertidal pools and will actively emerge from hypoxic water. *Chitonotus pugetensis* lives in deep water and does not emerge from hypoxic water (Source: Martin, 1996).

water's edge are capable of taking up oxygen from the atmosphere (Martin et al., 2004).

Intertidal Fish Assemblage Structure and Dynamics
Because the vertical range of tidal advance and retreat is generally greater at higher latitudes, studies of the community structure of intertidal fishes have generally focused on temperate families and species. Most studies of rocky intertidal fish communities have focused on the northeast Pacific coast or the coast of western Europe (Gibson and Yoshiyama, 1999). The hard substrate of intertidal shores in the tropics may be rock, as in higher latitudes, but it may also have a biological origin in the calcareous precipitation from corals. The mineral composition of reefs may, in some instances, influence the distribution of rocky shore fishes (Guidetti and Cattaneo-Vietti, 2002).

The intertidal fish community structure of temperate rocky shores tends to be a fairly well-defined entity, distinct from the subtidal ichthyofaunal community. Patchy distribution characterizes intertidal communities, and the fish assemblages are no exception. Intertidal fish communities appear to be highly resilient, stable, and deterministically structured (Gibson and Yoshiyama, 1999; Grossman, 1982, 1986; Willis and Roberts, 1996). Although the species diversity on temperate rocky intertidal shores may be quite high, comparatively few families are represented in intertidal zones worldwide. These families include the sculpins (Cottidae), gobies (Gobiidae), blennioids (families Blenniidae, Clinidae, and Tripterygiidae), zoarcoids (Stichaeidae and Pholidae), snailfishes (Liparidae), and clingfishes (Gobiesocidae). Intertidal shores in different parts of the world tend to be dominated by a particular family—the Tripterygiidae (triplefin blennies) are prominent in New Zealand, the Clinidae in South Africa, the Gobiidae in the eastern Atlantic and Mediterranean, and the Cottidae, Stichaeidae, and Pholidae along the rocky intertidal shores of the Pacific Northwest (Fig. 34.4). A few species from these families live in the supratidal zone, where the replenishment of tide pools is intermittent. In rocky areas, sculpins, gobies, blennies, and clingfishes often inhabit the supratidal "spray zone." The preferred habitat of the large Chilean clingfish (*Sicyases sanguineus*) is out of the water and well above the water line, where it will remain as long as it can be occasionally doused by wave spray (Ebeling et al., 1970). Studies on two species of clingfishes of the genus *Le-*

padogaster have shown differences in behavior that appear to reflect differences in habitat—the intertidally occurring species is less active and shows greater site fidelity than its subtidally occurring congener (Gonçalves et al., 1998).

Intertidally occurring fishes in the tropics are usually benthic taxa, such as blennies and gobies, or juveniles of species that move to offshore coral reefs as adults. As a component of larger coral reef ecosystems, the diversity of intertidal fish species may be quite high in some areas. For example, more than 60 species of intertidally occurring fishes have been recorded in Barbados (Mahon and Mahon, 1994; Prochazka et al., 1999). Tropical intertidal communities generally experience less tidal fluctuation, although wave surge from seasonal storms may significantly affect fish assemblages (Friedlander and Parrish, 1998). One of the most remarkable intertidally occurring fishes of the tropics is the frillfin goby (*Bathygobius soporator*), which is legendary for its unerring homing abilities (see Chapter 36).

The richness of intertidal habitats entails a wealth of feeding opportunities for both resident fish species and visitors that arrive with the high tide. Small invertebrates, especially copepod, amphipod, isopod, and decapod crustaceans, form the bulk of the diet (Fig. 34.6). Dietary specializations, such as the aforementioned limpet capture adaptations of clingfishes, are sometimes observed. The saddleback gun-

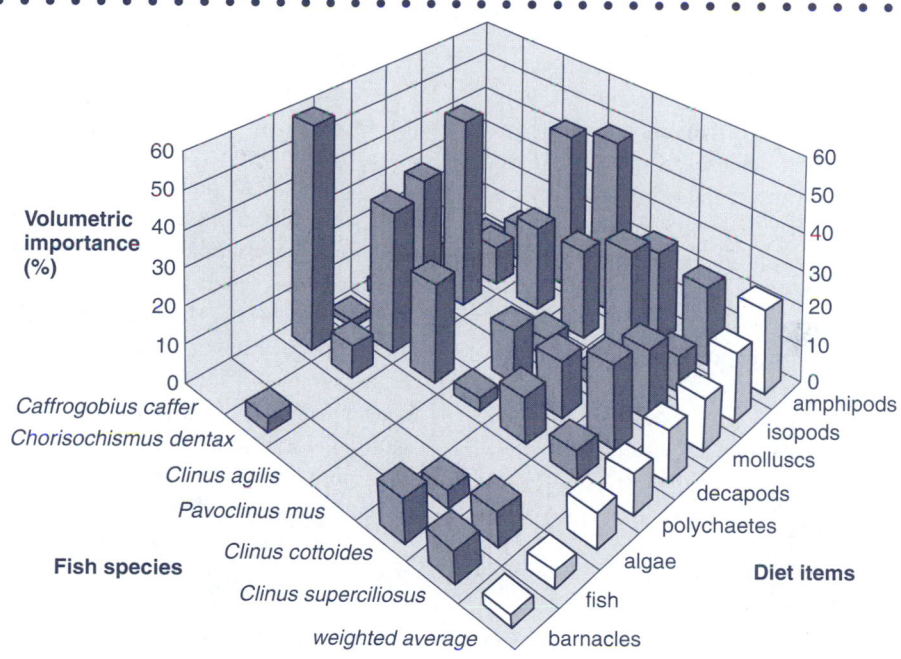

FIGURE 34.6
Diets of a rocky intertidal fish assemblage from South Africa
(Source: Data from Bennett et al., 1983; illustration from Norton and Cook, 1999).

nel (*Pholis ornata*), a resident of sheltered, rocky substrates, exemplifies another dietary specialization seen in some inshore fish species—nipping off the exposed siphon tips of bivalves (Barton, 1986). The optic modifications of the "four-eyed" blenny (*Dialommus fuscus;* see Chapter 20) permit it to forage both in and out of the water as it follows the advancing and retreating tide (Nieder, 2001).

Two of the largest intertidal species of northeast Pacific shores, the monkeyface prickleback (*Cebidichthys violaceus;* see Plate 34.1) and the rock prickleback (*Xiphister mucosus*), are carnivorous as juveniles but shift to a diet of macroalgae as they mature (Barton, 1982; Horn and Ojeda, 1999). It is puzzling that herbivorous fishes are not better represented in temperate rocky intertidal fish assemblages, especially given the abundance of benthic macrophytes. Benthic herbivorous invertebrates, such as sea urchins, may exert a profound effect on the structure of intertidal algal communities (Lubchenco, 1978; Lubchenco and Gaines, 1981). A rare instance in which a temperate herbivorous fish species has affected the population structure of associated benthic macrophytes is that of *Scartichthys viridis*, a herbivorous blenny that dominates the ichthyofaunal assemblages of the rocky coast of Chile (Ojeda and Muñoz, 1999).

SEASONAL VARIATION IN COASTAL FISH ASSEMBLAGES

Although inshore ichthyofaunal communities appear consistent in composition, deterministic, and dominated by comparatively few families characteristic of the region, a strong seasonal component to inshore ichthyofaunal assemblages is often apparent—especially in estuaries. Seasonality in assemblage composition of rocky intertidal fishes also appears to be the rule, although Prochazka (1996) observed an exception in the ichthyofaunal community of South African shores, where no seasonal variation in species composition or density was observed. Changes in the physicochemical environment, such as temperature and salinity, may make inshore habitats less suitable during certain times of the year (Hoff and Ibara, 1977). In Australian and South African estuaries, fish assemblage composition has been shown to correlate with seasonal fluctuations in turbidity (Cyrus, 1992; Cyrus and Blaber, 1992).

Inshore Environments as Nurseries

The use of inshore environments, especially bays and estuaries, for purposes of reproduction is especially significant in conferring seasonality to fish assemblage composition. Migratory activity and the exploitation of estuaries as nurseries result in strong seasonal trends in fish diversity and abundance in Atlantic coastal estuarine systems (Hoff and Ibara, 1977; McErlean et al., 1973). The abundance of available nutrients for rapidly developing eggs and larvae and the shelter afforded by macrophytes and other topographical features have prompted a number of fish species to move inshore to spawn. We have already made mention of those species adapted to spawning in the surf of sandy beaches, and of the coastal anadromous species, such as the striped bass, that move through estuaries on their way upstream to spawn in freshwater. Although young individuals may be less vulnerable to predation in the shallows of bays and estuaries, the movement of fishes toward inshore spawning habitats might serve to attract greater numbers of predators as well. The seasonal appearance of bluefish (*Pomatomus saltatrix*) and jacks (Carangidae) along the Atlantic coast may reflect improved feeding opportunities in inshore areas. The abundance of vegetation in bays and estuaries or on exposed rocky shores provides excellent sheltering opportunities for a number of species, including some that are commercially valuable. On the northwest Atlantic coast, this includes gadoids, such as the Atlantic cod (*Gadus morhua*), Atlantic tomcod (*Microgadus tomcod*), and pollock (*Pollachius virens;* Rangeley, 1994; Rangeley and Kramer, 1995a, 1995b), whereas juvenile rockfishes find shelter in rockweed along the northeast Pacific coast (Chotkowski et al., 1999). Overharvesting of rockweed (*Ascophyllum nodosum*), which is an important component of the global seaweed industry, may have an adverse impact on those species that depend on it for shelter (Rangeley, 1994).

The modification of coastal environments can sometimes affect species diversity in a positive way. With the construction of a breakwater, exposed beaches are converted into harbors. The increased protection afforded by such an artificial embayment enables the creation and maintenance of diverse and abundant fish assemblages (Allen et al., 1983). However, this can in no way offset the damage caused to fish populations by the degradation or elimination of estuaries and other coastal fish habitat. Not only are resident ichthyofauna adversely affected by the alteration and degradation of coastal habitats, but many offshore species that use inshore habitats for reproduction also suffer. The burgeoning human population will continue to place pressure on coastal areas, especially in developing countries—an ironic turn of events, as these are the areas where the ecology of the resident fish populations is least understood.

KEY POINTS AND CONNECTIONS

- Nearshore environments constitute those marine habitats that are immediately adjacent to coastlines. As such, they are influenced by both terrestrial and marine conditions. Fluctuations in temperature, salinity, and other associated physical factors characterize these regions. Fishes and other organisms that live in such conditions are termed euryhaline, as they can withstand the seasonal and diurnal fluctuations in salinity that are characteristic of this realm. Significant fluctuation in temperature is also characteristic of these nearshore regions; thus, the resident fauna is typically eurythermic as well.

The adaptations of euryhaline fishes are considered in Chapter 25; thermal tolerances of fishes are discussed in Chapter 30.

• Bays and estuaries form in coastal areas that are protected from ocean waves and currents. Freshwaters draining into these bays and estuaries create an aquatic environment intermediate between freshwater and the open ocean. The ichthyofaunal composition of estuaries is generally determined by the size and depth of the estuary, its geographic location, and the nature of adjacent marine environments and associated fishes.

• Marine environments exposed to the action of waves and currents present especially challenging conditions for existence. With respect to the resident ichthyofauna, sandy beaches are comparatively depauperate, yet several families and species are adapted for life in the surf zone. The greater opportunities for attachment to hard substrates provided by rocky shores have permitted a higher diversity of macrophytic vegetation and, hence, increased opportunities for shelter and sustenance. Here, a number of fish species can be found that have become morphologically, physiologically, and behaviorally adapted to the rhythmic advance and retreat of the tides in the intertidal zone.

Biogeographical perspectives on coastal ichthyofauna are considered in Chapter 29.

BUILDING AN ICHTHYOLOGY LIBRARY

Blaber, S. J. M. 2000. *Tropical estuarine fishes—ecology, exploitation, and conservation.* Blackwell Science, Oxford.

Graham, J. B. 1997. *Air-breathing fishes: Evolution, diversity, and adaptation.* Academic Press, San Diego.

Horn, M. H., K. L. M. Martin, and M. A. Chotkowski. 1999. *Intertidal fishes: Life in two worlds.* Academic Press, San Diego.

McLusky, D. S., and J. F. Craig (Eds.). 2002. Estuarine and lagoon fish and fisheries: The Fisheries Society of the British Isles Annual Symposium. *J. Fish Biol. 61*(Suppl. 1): 1–282.

Whitfield, A. K. 1998. *Biology and ecology of fishes in southern African estuaries.* J. L. B. Smith Institute of Ichthyology, Grahamstown, South Africa.

REFERENCES

Adams, S. M. 1976a. The ecology of eelgrass, *Zostera marina* (L.), fish communities, I. Structural analysis. *J. Exp. Mar. Biol. Ecol. 22*: 269–291.

———. 1976b. The ecology of eelgrass, *Zostera marina* (L.), fish communities, II. Functional analysis. *J. Exp. Mar. Biol. Ecol. 22*: 293–311.

Allen, L. G., M. H. Horn, F. A. Edmands, II, and C. A. Usui. 1983. Structure and seasonal dynamics of the fish assemblage in the Cabrillo Beach area of Los Angeles Harbor, California. *Bull. So. Calif. Acad. Sci. 82*(2): 47–70.

Barnes, R. S. K., and R. N. Hughes. 1982. *An introduction to marine ecology.* Blackwell Science, Oxford.

Barton, M. G. 1982. Intertidal vertical distribution and diets of five species of central California stichaeoid fishes. *Calif. Fish Game 68*: 174–182.

———. 1986. Influence of substratum on the comparative food habits of two species of estuarine stichaeoid fishes, *Anoplarchus purpurescens* (family Stichaeidae) and *Pholis ornata* (family Pholididae). *Northwest Sci. 60*(2): 125–130.

Bennett, B., C. L. Griffiths, and M. Penrith. 1983. The diets of littoral fish from the Cape Peninsula. *S. Afr. J. Zool. 18*: 343–352.

Beyst, B., A. Cattrijsse, and J. Mees. 1999. Feeding ecology of juvenile flatfishes of the surf zone of a sandy beach. *J. Fish Biol. 55*: 1171–1186.

Blaber, S. J. M. 2000. Tropical estuarine fishes: Ecology, exploitation and conservation. Blackwell Science, Oxford.

Briggs, P. T., and J. S. O'Connor. 1971. Comparison of shore-zone fishes over naturally vegetated and sand-filled bottoms in Great South Bay. *N. Y. Fish Game J. 18*(1): 15–41.

Castro, P., and M. E. Huber. 2000. *Marine biology* (3rd ed.). McGraw-Hill, Boston.

Chapman, V. J. (Ed.). 1977. *Wet coastal ecosystems.* Elsevier, Amsterdam.

Chotkowski, M. A., D. G. Buth, and K. Prochazka. 1999. Systematics of intertidal fishes, pp. 297–331. In *Intertidal fishes: Life in two worlds*, M. H. Horn, K. L. M. Martin, and M. A. Chotkowski (Eds.). Academic Press, San Diego.

Connell, J. H. 1961. The influence of interspecific competition and other factors on the distribution of the barnacle *Chthamalus stellatus.* *Ecology 42*: 710–723.

Cupka, D. 1972. A survey of the ichthyofauna of the surf zone in South Carolina. *S. C. Dept. Nat. Resources Tech. Rep. 4.*

Cyrus, D. P. 1992. Turbidity gradients in two Indo-Pacific estuaries and their influence on fish distribution. *S. Afr. J. Aquat. Sci. 18*(1–2): 51–63.

———, and S. J. M. Blaber. 1992. Turbidity and salinity in a tropical northern Australian estuary and their influence on fish distribution. *Estuar. Coast. Shelf Sci. 35*: 545–563.

Davis, W. P., D. S. Taylor, and B. J. Turner. 1990. Field observations of the ecology and habits of mangrove rivulus (*Rivulus marmoratus*) in Belize and Florida (Teleostei: Cyprinodontiformes: Rivulidae). *Ichthyol. Explor. Freshw. 1*: 123–134.

———, ———, and———. 1995. Does the autecology of the mangrove rivulus fish (*Rivulus marmoratus*) reflect a paradigm for mangrove ecosystem sensitivity? *Bull. Mar. Sci. 57*: 208–214.

Duncan, R. S., and W. A. Szelistowski. 1998. Influence of puffer predation on vertical distribution of mangrove littorinids in the Gulf of Nicoya, Costa Rica. *Oecologia 117*: 433–442.

Ebeling, A. W., P. Bernal, and A. Zuleta. 1970. Emersion of the amphibious Chilean clingfish *Sicyases sanguineus.* *Biol. Bull. 139*: 115–137.

Friedlander, A. M., and J. D. Parrish. 1998. Temporal dynamics of fish communities on an exposed shoreline in Hawaii. *Env. Biol. Fishes 53*: 1–18.

Gibson, R. N. 1969. The biology and behavior of littoral fish. *Oceanogr. Mar. Biol. Ann. Rev. 7*: 367–410.

———. 1982. Recent studies on the biology of intertidal fishes. *Oceanogr. Mar. Biol. Ann. Rev. 20*: 363–414.

————, and R. M. Yoshiyama. 1999. Intertidal fish communities, pp. 264–296. In *Intertidal fishes: Life in two worlds*, M. H. Horn, K. L. M. Martin, and M. A. Chotkowski (Eds.). Academic Press, San Diego.

Gonçalves, D. M., E. J. Gonçalves, V. C. Almada, and S. P. Almeida. 1998. Comparative behavior of two species of *Lepadogaster* (Pisces: Gobiesocidae) living at different depths. *J. Fish Biol.* 53: 447–450.

Graham, J. B. 1973. Terrestrial life of the amphibious fish *Mnierpes macrocephalus*. *Mar. Biol.* 23: 83–91.

————. 1976. Respiratory adaptations of marine air-breathing fishes, pp. 165–187. In *Respiration of amphibious vertebrates*, G. M. Hughes (Ed.). Academic Press, London.

————. 1997. *Air-breathing fishes: Evolution, diversity, and adaptation.* Academic Press, San Diego.

Grossman, G. D. 1982. Dynamics and organization of a rocky intertidal fish assemblage: The persistence and resilience of taxocene structure. *Am. Nat.* 119: 611–637.

————. 1986. Long term persistence in a rocky intertidal fish assemblage. *Env. Biol. Fishes* 15: 315–317.

Guidetti, P., and R. Cattaneo-Vietti. 2002. Can mineralogical features influence distribution patterns of fish? A case study in shallow Mediterranean rocky reefs. *J. Mar. Biol. Assoc. UK* 82: 1043–1044.

Gunter, G. 1942. A list of the fishes of the mainland of North and Middle America recorded from both freshwater and sea water. *Am. Midl. Nat.* 28: 305–326.

————. 1956. A revised list of euryhaline fishes of North and Middle America. *Am. Midl. Nat.* 56: 345–354.

Harris, S. A., and D. P. Cyrus. 1996. Larval and juvenile fishes in the surf zone adjacent to the St. Lucia Estuary mouth, KwaZulu-Natal, South Africa. *Mar. Freshw. Res.* 47: 465–482.

Harrison, T. D., and A. K. Whitfield. 1995. Fish community structure in three temporarily open/closed estuaries on the Natal coast. *Ichthyol. Bull. J. L. B. Smith Inst. Ichthyol.* No. 64.

Helmuth, B., C. D. G. Harley, P. M. Halpin, M. O'Donnell, G. E. Hofmann, and C. A. Blanchette. 2002. Climate change and latitudinal patterns of intertidal thermal stress. *Science* 298: 1015–1017.

Hoese, H. D., and R. H. Moore. 1998. *Fishes of the Gulf of Mexico: Texas, Louisiana, and adjacent waters* (2nd ed.). Texas A & M University Press, College Station.

Hoff, J. G., and R. M. Ibara. 1977. Factors affecting the seasonal abundance, composition and diversity of fishes in a southeastern New England estuary. *Estuar. Coastal Mar. Sci.* 5: 665–678.

Horn, M. H. 1999. Convergent evolution and community convergence: Research potential using intertidal fishes, pp. 356–372. In *Intertidal fishes: Life in two worlds*, M. H. Horn, K. L. M. Martin, and M. A. Chotkowski (Eds.). Academic Press, San Diego.

————, and L. G. Allen. 1985. Fish community ecology in Southern California bays and estuaries, pp. 169–190. In *Fish community ecology in estuaries and coastal lagoons: Towards an ecosystem integration*, A. Yáñez-Arancibia (Ed.). UNAM Press, Mexico.

————, and R. N. Gibson. 1988. Intertidal fishes. *Sci. Am.* 258(1): 64–70.

————, and F. P. Ojeda. 1999. Herbivory, pp. 197–222. In *Intertidal fishes: Life in two worlds*, M. H. Horn, K. L. M. Martin, and M. A. Chotkowski (Eds.). Academic Press, San Diego.

Irlandi, E. A., and M. K. Crawford. 1997. Habitat linkages: The effect of intertidal saltmarshes and adjacent subtidal habitats on abundance, movement, and growth of an estuarine fish. *Oecologia* 110: 222–230.

Livingston, R. J. 1982. Trophic organization of fishes in a coastal seagrass system. *Mar. Ecol. Prog. Ser.* 7: 1–12.

Lubchenco, J. 1978. Plant species diversity in a marine intertidal community: Importance of herbivore food preference and algal competitive abilities. *Am. Nat.* 112: 23–39.

————, and S. D. Gaines. 1981. A unified approach to marine plant-herbivore interactions. I. Populations and communities. *Ann. Rev. Ecol. Syst.* 12: 405–437.

Mahon, R., and S. D. Mahon. 1994. Structure and resilience of a tidepool fish assemblage at Barbados. *Env. Biol. Fishes* 41: 171–190.

Martin, K. L. M. 1995. Time and tide wait for no fish: Intertidal fishes out of water. *Env. Biol. Fishes* 44: 165–181.

————. 1996. An ecological gradient in air-breathing ability among marine cottid fishes. *Physiol. Zool.* 69: 1096–1113.

————, and C. R. Bridges. 1999. Respiration in water and air, pp. 54–78. In *Intertidal fishes: Life in two worlds*, M. H. Horn, K. L. M. Martin, and M. A. Chotkowski (Eds.). Academic Press, San Diego.

————, R. C. Van Winkle, J. E. Drais, and H. Lakisik. 2004. Beach-spawning fishes, terrestrial eggs, and air breathing. *Physiol. Biochem. Zool.* 77: 750–759.

McErlean, A. J., S. G. O'Connor, J. A. Mihursky, and C. I. Gibson. 1973. Abundance, diversity and seasonal patterns of estuarine fish populations. *Estuar. Coastal Mar. Sci* 1: 19–36.

Nieder, J. 2001. Amphibious behavior and feeding ecology of the four-eyed blenny (*Dialommus fuscus*, Labrisomidae) in the intertidal zone of the island of Santa Cruz (Galapagos, Ecuador). *J. Fish. Biol.* 58: 755–767.

Norton, S. F., and A. E. Cook. 1999. Predation by fishes in the intertidal, pp. 223–263. In *Intertidal fishes: Life in two worlds*, M. H. Horn, K. L. M. Martin, and M. A. Chotkowski (Eds.). Academic Press, San Diego.

Odum, W. E., and E. J. Heald. 1972. Trophic analysis of an estuarine mangrove community. *Bull. Mar. Sci.* 22: 671–738.

Ojeda, F. P., and A. A. Muñoz. 1999. Feeding selectivity of the herbivorous fish *Scartichthys viridis*: Effects on macroalgal community structure in a temperate rocky intertidal coastal zone. *Mar. Ecol. Prog. Ser.* 184: 219–229.

Paterson, A. W., and A. K. Whitfield. 2000. The ichthyofauna associated with an intertidal creek and adjacent eelgrass beds in the Kariega Estuary, South Africa. *Env. Biol. Fishes* 58: 145–156.

Petrik, R., and P. S. Levin. 2000. Estimating relative abundance of seagrass fishes: A quantitative comparison of three methods. *Env. Biol. Fishes* 58: 461–466.

Prochazka, K. 1996. Seasonal patterns in a temperate intertidal fish community on the west coast of South Africa. *Env. Biol. Fishes* 45: 133–140.

————, M. A. Chotkowski, and D. G. Buth. 1999. Biogeography of rocky intertidal fishes, pp. 332–355. In *Intertidal fishes: Life in two worlds*, M. H. Horn, K. L. M. Martin, and M. A. Chotkowski (Eds.). Academic Press, San Diego.

Rangeley, R. W. 1994. The effects of seaweed harvesting on fishes: A critique. *Env. Biol. Fishes* 39: 319–323.

————, and D. L. Kramer. 1995a. Use of rocky intertidal habitats by juvenile pollock *Pollachius virens*. *Mar. Ecol. Prog. Ser.* 126: 9–17.

————, and ————. 1995b. Tidal effects on habitat selection and aggregation by juvenile pollock (*Pollachius virens*) in the rocky intertidal zone. *Mar. Ecol. Prog. Ser.* 126: 19–29.

Robertson, A. I., and S. J. M. Blaber. 1992. Plankton, epibenthos, and fish communities. In *Tropical mangrove ecosystems* (Coastal and Estuarine

Studies 41), A. I. Robertson and D. M. Algoni (Eds.). American Geophysical Union, Washington, DC.

Schultze, H.-P. 1999. The fossil record of the intertidal zone, pp. 373–392. In *Intertidal fishes: Life in two worlds,* M. H. Horn, K. L. M. Martin, and M. A. Chotkowski (Eds.). Academic Press, San Diego.

Smith, R. L. 1980. *Ecology and field biology* (3rd ed.). Harper and Row, New York.

Sumich, J. L. 1999. *An introduction to the biology of marine life* (7th ed.). WCB/McGraw-Hill, Boston.

Taylor, D. S., W. P. Davis, and B. J. Turner. 1995. *Rivulus marmoratus:* Ecology of distributional patterns in Florida and the central Indian River Lagoon. *Bull. Mar. Sci. 57:* 202–207.

Weinstein, M. P., and K. L. Heck, Jr. 1979. Ichthyofauna of seagrass meadows along the Caribbean coast of Panamá and in the Gulf of Mexico: Composition, structure and community ecology. *Mar. Biol. 50:* 97–107.

Whitfield, A. K. 1999. Ichthyofaunal assemblages in estuaries: A South African case study. *Rev. Fish Biol. Fisheries 9:* 151–186.

Willis, T. J., and C. D. Roberts. 1996. Recolonisation and recruitment of fishes to intertidal rockpools at Wellington, New Zealand. *Env. Biol. Fishes 47:* 329–343.

35 ECOLOGY OF FISHES VI: THE PELAGIC AND BENTHIC REALM BEYOND THE CONTINENTAL SHELF; POLAR ENVIRONMENTS

LARGE MARINE ECOSYSTEMS

THE OPEN-WATER PELAGIC REALM

The Epipelagic Zone
 Adaptations of Epipelagic Fishes
 Ichthyofauna of the Epipelagic Zone
 Fishes Associated with Sargassum Beds
The Mesopelagic Zone
 Mesopelagic Fishes

THE BATHYPELAGIC REALM AND BEYOND

The Continental Slope and Abyssal Plain
Adaptations of Deep Benthic and Benthopelagic Fishes
 Feeding Adaptations
Deep Benthic Fish Assemblages
 The Deepest Dwellers
 Fish Fauna of Hydrothermal Vents

MARINE FISHES OF POLAR ENVIRONMENTS

The Nature of Polar Environments
 Origins of Polar Biota
 Adaptations of Polar Fishes
The Polar Fish Fauna
 Arctic Fishes
 Antarctic Fishes

In this chapter, we investigate the fish fauna that occupies most of the volume of the world's oceans. The waters beyond the continental shelf generally show less environmental variation than those closer to shore. Nonetheless, the fishes encountered here may be adapted to conditions that we may perceive as harsh and demanding. This is especially true for the deep ocean realm. Most of the world's ocean volume exists in a zone that is uniformly cold (about 4°C), devoid of light, and has a salinity that stays at a constant 35 ‰. Pressure dramatically increases with depth, yet several families of fishes live comfortably in this cold, dark realm where the ambient pressure could crush the hull of a submersible. As we move into higher latitudes, the thermal environment becomes even more extreme, with fishes often exposed to temperatures below the freezing point of pure water. As with the other aquatic realms we have considered, fishes can be conveniently divided into inhabitants of pelagic waters and benthic species found near or on the bottom. The most striking difference is, of course, the extreme depths of occurrence for many of the fishes that live in the marine waters beyond the continental shelf.

LARGE MARINE ECOSYSTEMS

The key feature that distinguishes most marine populations from those found in freshwater is the degree to which they are demographically open. Lakes and streams, with boundaries determined by the landforms on which they occur, are much more self-contained with respect to gene flow. As the ocean is not subject to such constraints, the recruitment of the biota appears to be largely dependent on the dispersal of planktonic larval stages from elsewhere—that is, the local production of offspring is of little significance in the determination of abundance at a given locale (Caley et al., 1996). Nowhere is this more applicable than to the open ocean beyond the continental shelf boundaries. Here, the influence of landmasses fades below the horizon, and the distributions of fish populations are far wider than those seen for coastal species.

Implicit in the concept of open marine ecosystems is the understanding that they operate on a vast scale. Fishery scientists, concerned with the rapidly diminishing stock of most commercial fish populations, have embraced the concept of **Large Marine Ecosystems (LMEs)** in the hopes that understanding ecosystems on a more global scale will lead to more effective management strategies for dwindling fishery resources (see Chapter 39). LMEs are large patches of the ocean that are characterized by distinct bathymetry, hydrography, productivity, and population structures (Sherman et al., 1990; Sherman and Duda, 1999). Currently, the LME concept has its greatest application to the biologically most productive coastal realm (neritic waters; see Chapter 33), which is the area of greatest concern to fishery biologists. As fishery researchers continue to investigate the feasibility of the commercial exploitation of the living resources of the deep ocean, this vast region ought to receive more serious attention as a distinct LME. The importance of understanding global oceanic processes is recognized in attempts to model the structure and function of ecosystems on the scale of ocean basins, so that forecasts of the abundance of economically significant species may be possible (see deYoung et al., 2004).

As with coastal ocean ecosystems, the ichthyofauna of the open ocean can be distinguished as either *pelagic* or *benthic*. The pelagic realm of the open ocean is traditionally divided into three major zones: **epipelagic, mesopelagic,** and **bathypelagic.** A **hadopelagic** zone is also recognized, consisting of the waters of the deep ocean trenches (see Fig. 33.1). Fishes that spend their entire life cycle in open waters are termed **holopelagic,** whereas **meropelagic** species may be bottom-associated at some stages in their life history, for purposes of feeding or reproduction. Although most pelagic species are associated with the more productive continental shelf regions, our concern here is with those species that have much broader distributions—some of which have a lifestyle that entails annual migrations of several thousand kilometers over the open ocean or nocturnal vertical migrations of hundreds of meters from the ocean depths to the surface.

Benthic fishes of the deep ocean range from the continental slopes down to the floors of the ocean basins. A distinction must be drawn between true **benthic** species, which remain in more or less continuous contact with the ocean floor, and **benthopelagic** species, which are bottom-affiliated but swim up in the water column. Organisms living in close association with the substrate have also been termed **demersal.**

The oceanic realm of polar latitudes has a distinctive ichthyofauna, especially in the Southern Hemisphere. It is difficult, however, to discuss polar fishes in the context of a particular ocean environment. Because the oceans of the Northern Hemisphere are largely characterized by extensive shelf regions, fishes here are technically shelf biota. This, we shall see, is much less the case in the Southern Hemisphere. The adaptations of these fishes to the rigors of life in high latitudes, and their distinctive assemblage compositions are such that they merit special consideration.

THE OPEN-WATER PELAGIC REALM

At the close of the 19th century, George Goode and Tarleton Bean produced the first definitive treatment of the ichthyofauna of the open ocean (Goode and Bean, 1896). They introduced the concept of "oceanic fishes" to describe those species that occur away from continental shelves. As such, they included pelagic, benthic, and benthopelagic species. N. V. Parin, the great Russian student of oceanic fishes considered this designation to be somewhat artificial, given the diversity of the origins and habitats of these fishes (Parin, 1984). On the whole, the diversity and abundance of pelagic fishes in the open ocean is comparatively low. This reflects the low levels of primary productivity in the open ocean compared with shelf regions that are enriched through the combined effects of upwelling and coastal runoff. Pelagic fishes, for the most part, exist in a realm lacking in topographical reference points. The major exception is when fishes associate with other biota, such as sargassum beds or other, larger marine animals. Continents may define the major oceanic regions, yet invisible boundaries, in the form of prevailing patterns of current flow (see Fig. 33.3), may also be influential in the dispersal and distribution of pelagic biota.

The Epipelagic Zone

The thin surface layer of the oceans that constitutes the epipelagic zone extends to a depth of about 200 m, although conditions typical of the epipelagic zone may disappear at shallower depths or extend deeper, depending on the location and numerous interacting physical factors. The epipelagic zone is generally the warmest layer and receives the most light, but it is subject to daily and seasonal changes in many of its physical features. Wherever light and sufficient nutrient materials coincide, photosynthesis can occur, thus producing the organic material that supports life not only in the upper layers, but indirectly in the lower layers as well. Dissolved oxygen concentration is high in the epipelagic zone—near saturation in the upper layers. Salinity ranges from about 33 to 37 ‰, depending on local conditions of precipitation or evaporation.

Close to the surface—usually in the upper 10 to 25 m in high latitudes, but sometimes reaching down to 200 m or more in the subtropics—there is a stratum of uniform temperature. This may reach or exceed 20°C in the tropics and fluctuates only a few degrees during the year. Proceeding poleward through temperate to polar latitudes, the upper layer temperatures decrease, with greater seasonal fluctuations in temperate areas than in polar regions. Below the surface layer, there is a zone of rapidly decreasing temperature, the **thermocline**, in which the temperature can drop nearly 1°C for every 10 meters of depth. The thermocline is a constant feature of warm oceans, but it develops only during the summer in high latitudes. Below the thermocline (and well below the epipelagic zone) the sea is cold, in some places reaching below 0°C.

Upwelling, usually perceived as a coastal phenomenon in which surface waters are driven offshore to be replaced by nutrient-enriched waters from the deep, also occurs in the open waters at the equator. Here, nutrient-enriched water is brought up as a result of the divergence of equatorial currents. The distribution of nutrients—and thus the primary productivity—is very patchy over the epipelagic realm, with much of the open ocean being a "wet desert," with minimal organic production.

Adaptations of Epipelagic Fishes

Buoyancy

As buoyancy is of paramount importance for a pelagic existence, it is not surprising that the gas bladder is present in almost all epipelagic fishes and is common among mesopelagic fishes as well. The inclusion of fats and oils in muscles, body cavity, gas bladder, or liver is an effective flotation device as well but, compared with gas, requires a greater mass of material. Pelagic sharks have livers that may constitute a quarter of their total weight. The specific gravity of a fish can also be brought close to that of water simply by including a large proportion of water somewhere within the confines of the skin, giving many pelagic fishes a gelatinous consistency. Many species of snailfishes (Liparidae) and the ocean sunfish (*Mola mola;* see Plate 17.6) are good examples of this. Reduction in the weight of the skeleton also confers some hydrostatic advantage. Many deep-living species that lack gas bladders have very thin, papery bones and lack scales.

A high surface-to-volume ratio is useful to reduce the rate of sinking in pelagic organisms. This is probably best illustrated in planktonic crustaceans (copepods), which possess elaborate cirri on their antennae and other appendages. Fishes as small as the larvae of pelagic species (see Figs. 17.3, 27.10, and 27.12) and as large as the manta ray probably also retard their rate of sinking through enhancement of their body surface area.

Feeding

Pelagic life requires special adaptations for food gathering. Food resources are not uniformly distributed across the open ocean but occur in patches. This necessitates some effective means of getting to the resources and processing them rapidly and efficiently once feeding commences. At the base of the food web is the phytoplankton community, on which numerous invertebrates and larval fishes (the zooplankton) feed. Clupeoids strain out both phyto- and zooplankton with fine, sievelike gill rakers, or they may pick off the zooplanktonic organisms one by one. With a long digestive tract and numerous pyloric caeca (see Chapter 23), clupeoids are very efficient at converting plankton into fish tissue.

Planktivores, such as the members of the families Clupeidae and Engraulidae, are usually more closely associated with the more productive continental shelf waters, where plankton occurs in greater densities. Some invertebrates and larval fishes (see Fig. 27.12) are predators, thus assuming the niche of top carnivores in the plankton community. The larvae of many pelagic species of the open ocean have evolved a cannibalistic lifestyle, as evidenced by a larger gape than that seen in the larvae of coastal species. This is perceived as being an adaptive response to the oligotrophic environment of pelagic species, coupled with the limited quantities of nutrients typically found in the ova of pelagic fishes (Nishimura and Hoshino, 1999).

Locomotion

Pelagic fishes are generally powerful swimmers, as this ability is necessary for the capture of prey, avoidance of predators, and migration. Members of the epipelagic suborder Scombroidei exemplify the adaptive advantages of size and power (see Chapter 19, Going Deeper). Also conspicuous among pelagic fishes are those fishes with limited means of locomotion. One definition of a planktonic organism is that it does not disperse under its own power, but rather is dependent on the ocean currents. By this definition, the ocean sunfish (*Mola mola*) is probably the largest planktonic organism, given its limited powers of locomotion. Ocean sunfishes are among those large animals mentioned earlier that serve as a reference point for other animals in the pelagic realm. They also act as habitat for a myriad of symbiotic organisms, including a number of parasitic forms, that use the sunfish's vast bulk as a source of food and shelter. Another large "planktonic" fish is the oarfish (*Regalecus;* see Plate 12.9). Large, drifting jellyfishes attract certain small fishes that can live unharmed among their tentacles. Perhaps the best known example is the man-of-war fish (*Nomeus gronovii*), which lives with the siphonophore *Physalia* (Mansueti, 1963). Smaller epipelagic fishes, such as the sauries (Scomberesocidae) and flyingfishes (Exocoetidae), are powerful swimmers, but they may also rely on currents for dispersal over the vast oceanic expanse. The genus *Schindleria* is one of the smallest ichthyological members of the plankton community, achieving an adult size of only 2 cm. This enigmatic genus of minute, paedomorphic fishes has been classified as a gobioid (Johnson and Brothers, 1993; see Chapter 16).

Reproduction

The reproductive activities of pelagic fishes—especially epipelagic species that undertake extensive migrations in the course of their lives—take place over an enormous expanse of ocean, and current patterns are instrumental in the dispersal of eggs and larvae. Lacking suitable substrates for such activities as courting and nest building, pelagic fishes tend to exhibit a minimum of reproductive behavior. The fecundities of pelagic fishes are among the highest recorded for fishes, and fertilization is accomplished simply through the broadcasting of eggs and sperm in the water. Tunas, depending on their size, release from about 1 to 10 million eggs. The ocean sunfish is reported to spawn up to 300 million eggs. This high fecundity compensates for the extremely high mortality experienced by the eggs and larvae. Eggs are released in the open ocean, mostly to drift freely with the winds and currents, but some sauries and flyingfishes attach their eggs to flotsam by means of threadlike structures on the shell. The duration of incubation is short in most drifting eggs, especially in the tropics. Usually, they hatch in one to three days. Some species—especially sauries, halfbeaks, and flyingfishes—require an incubation time of one to two weeks. Pelagic sharks, for the most part, are ovoviviparous or viviparous, giving birth to young that are well able to swim and forage. Consequently, their broods

are small. The giant whale shark (*Rhincodon*) is oviparous, depositing eggs that are surprisingly small for such a great animal (Breder and Rosen, 1966; Nikolsky, 1963).

In our earlier discussion of Large Marine Ecosystems (LMEs), it was mentioned that the pelagic dispersal of larvae in open marine populations is generally perceived as maximizing the gene flow among pelagic fish populations. Yet speciation has occurred even among widely dispersed pelagic fishes, implying some interruption of gene flow at some point in their evolutionary histories. Studies have suggested that the perception of larvae as passively drifting particles may be erroneous. They may, through controlling such things as patterns of vertical migration in response to current regimes, ensure their retention in a given area over the course of their development into adults, thus contributing to the maintenance of genetically distinct populations (Cowen et al., 2000).

Ichthyofauna of the Epipelagic Zone

The fishes of the epipelagic zone are probably the least exclusive of all the pelagic ichthyofauna. Epipelagic fishes, lacking neither the ability nor the inclination to recognize the integrity of the boundaries of the LMEs designated by fishery biologists, routinely pass into and out of neritic waters. The epipelagic ichthyofauna is a heterogeneous assemblage, with only about 150 species from four families of chondrichthyans and 15 families of teleosts that spend their entire lives within this region (Parin, 1984).

With only 70 families represented wholly or partially in the epipelagic zone, this vast region of the world's oceans encompasses less than 2 percent of all known species of fishes. These species range from the aforementioned *Schindleria* and dwarf sauries of 15 cm up to the gigantic whale shark of 18 m, but most representatives are from 30 cm to 1 m in length. The swift swimmers, such as the pelagic sharks, tunas and other mackerel-like species, and some carangids, are fusiform in body shape, with stiff, narrow, keeled caudal peduncles and crescent-shaped caudal fins (see Chapter 19, Going Deeper). This design permits maximum power output with minimum resistance imposed on the rapidly beating caudal fin. The powerful billfishes—marlins (Istiophoridae) and swordfishes (Xiphiidae)—probably derive a considerable streamlining advantage at high speeds from their elongate bills; these animals are reported to reach speeds of 130 km/hour. Many families have somewhat less of the classic fusiform shape, some retaining a lunate or forked caudal fin but having a more compressed body, and others being more elongate, even arrow shaped, such as the barracudas (Sphyraenidae).

Epipelagic fishes are remarkably uniform in coloration—a dark dorsal surface, either green or blue, giving way to silvery sides that reflect whatever the prevailing colors are in the immediate environment. This pattern, known as **obliterative countershading**, enables the fish to blend in with the background regardless of the angle from which it is viewed (see Chapter 18). Neritic fishes may sometimes add variations in the form of stripes or bars.

The fish families and species that are most frequently associated with the epipelagic realm are indicated in Table 35.1. Although the large, powerfully swimming epipelagic predators, such as the tunas and billfishes, may range widely across the world's oceans, diversity "hotspots" have been identified, particularly at mid-latitudes. Here, predator diversity

correlates with zooplankton abundance and with physical factors such as ocean temperature and dissolved oxygen (Worm et al., 2005). Some mesopelagic species, such as the myctophids *Symbolophorus, Myctophum, and Centrobranchus,* the snake mackerel *Gempylus serpens,* and the deepwater shark *Isistius brasiliensis,* make regular nocturnal forays into the epipelagic zone. The anadromous salmonids feed during most of their lives in the epipelagic zone but move back toward shore to spawn, whereas species from families that are usually associated with the continental shelf, such as mullids, hexagrammids, and bothids, may spend part of their early life history offshore in the epipelagic zone (Parin, 1984).

TABLE 35.1 FAMILIES AND REPRESENTATIVE SPECIES PRESENT IN THE EPIPELAGIC ZONE

Family	Species	Distribution
Lamnidae	White shark (*Carcharodon carcharias*)	Holopelagic, neritic
	Salmon shark (*Lamna ditropis*)	Holopelagic
Cetorhinidae	Basking shark (*Cetorhinus maximus*)	Holopelagic, neritic
Alopiidae	Thresher shark (*Alopias vulpinus*)	Holopelagic, neritic
Rhincodontidae	Whale shark (*Rhincodon typus*)	Meropelagic
Carcharinidae	Tiger shark (*Galeocerdo cuvieri*)	Holopelagic, neritic
Clupeidae	Herring (*Clupea harengus*)	Meropelagic
	Sardine (*Sardinops sagax*)	Holopelagic, neritic
Engraulidae	Anchovy (*Engraulis encrasicolus*)	Holoepipelagic, neritic
	Northern anchovy (*E. mordax*)	Holoepipelagic, neritic
Exocoetidae	Oceanic flyingfish (*Exocoetus obtusirostris*)	Holopelagic
	California flyingfish (*Cypselurus californicus*)	Holopelagic, neritic
Scomberesocidae	Pacific saury (*Cololabis saira*)	Holopelagic
	Atlantic saury (*Scomberesox saurus*)	Holopelagic
Lampridae	opah (*Lampris guttatus*)	Holopelagic
Echeneidae	remora (*Remora remora*)	Holopelagic
Carangidae	pilotfish (*Naucrates ductor*)	Holopelagic
	rough scad (*Trachurus symmetricus*)	Holopelagic, neritic
Coryphaenidae	dolphin (*Coryphaenus hippurus*)	Holopelagic, neritic
Bramidae	bigscale pomfret (*Taractes longipinnis*)	Holopelagic
Luvaridae	louvar (*Luvarus imperialis*)	Holopelagic
Gempylidae	escolar (*Lepidocybium flavobrunneum*)	Holopelagic
Scombridae	frigate mackerel (*Auxis thazard*)	Holopelagic
	albacore (*Thunnus alalunga*)	Holopelagic, neritic
Xiphiidae	swordfish (*Xiphias gladius*)	Holopelagic
Istiophoridae	white marlin (*Tetrapturus albidus*)	Holopelagic
Centrolophidae	medusafish (*Icichthys lockingtoni*)	Holopelagic
Nomeidae	silver driftfish (*Psenes maculatus*)	Holopelagic
Tetragonuridae	bigeye squaretail (*Tetragonurus atlanticus*)	Holopelagic
Stromateidae	butterfish (*Peprilus triacanthus*)	Meropelagic
Molidae	ocean sunfish (*Mola mola*)	Holopelagic
	slender mola (*Ranzania laevis*)	Holopelagic

Fishes Associated with Sargassum Beds

A special fish community of the pelagic realm consists of species that live among the dense beds of *Sargassum* weed found worldwide in the Atlantic and Pacific oceans. By far the best known *Sargassum* community is that found in the North Atlantic Ocean in a region known as the **Sargasso Sea**. The Sargasso Sea is a vast area that is encircled by the main currents of the North Atlantic Ocean (see Figs. 33.3 and 36.3). Although confined to a relatively nutrient-deficient water mass, the *Sargassum* weed community is still home to more than 50 species of fishes (from 23 families) and numerous invertebrates. Included in the fish fauna are two endemic species, the sargassumfish (*Histrio histrio*; see Plate 13.9) and the sargassum pipefish (*Syngnathus pelagicus*). Several genera of filefishes (*Stephanolepis, Alutera, Cantherhines,* and *Monacanthus*) and the young of several pelagic species, such as the carangid jacks (*Caranx, Seriola,* and *Coryphaena*), also inhabit the *Sargassum* beds (Dooley, 1972). Eels of the family Anguillidae are probably the best known *Sargassum* associates since the Sargasso Sea was discovered to be the destination of the spawning migrations of European and American eel species (see Chapter 9, Going Deeper). Leptocephali from 12 other eel families have also been collected in the Sargasso Sea (Miller and McCleave, 1994).

The *Sargassum* community of the eastern Gulf of Mexico is dominated by the filefish (*Monacanthus hispidus*; Bortone et al., 1977), whereas juvenile grunts of the genus *Haemulon* are the most common inhabitants of the seasonal *Sargassum* beds found off the southern coast of Brazil (Ornellas and Coutinho, 1998). The availability of feeding opportunities in the *Sargassum* beds attracts other larger fishes as well as several species of marine turtles and dolphins. Large predatory species, such as jacks, barracudas, tunas, and swordfishes have also been found in the vicinity of the *Sargassum* beds. *Sargassum* is currently harvested, mainly as an additive in commercial animal feeds, and there is much concern that unrestricted harvesting might adversely affect the faunal assemblages that depend on the algae for food and shelter.

The Mesopelagic Zone

The mesopelagic zone constitutes the vast, dimly lit oceanic region that extends from about 200 m to 1,000 m in depth. As such, it stretches out from the lowest part of the continental shelf zone into the slope zone of the ocean bottom (see Fig. 33.1). The mesopelagic zone is often subdivided somewhere between 600 and 700 m into shallow and deep subzones. Silvery-sided, highly reflective fishes (the sternoptychid hatchetfishes, Plate 12.3, are a good example) and partially translucent red decapod crustaceans dominate the upper mesopelagic subzone, whereas nonreflective fishes and completely red nontranslucent decapods are characteristic of the lower subzone (Angel, 1997). Although sunlight attenuates and disappears almost completely near the lower boundary of the mesopelagic zone, small populations of phytoplankton can still be found there, along with debris from epipelagic plankton. This material forms the food base for a sometimes abundant and varied zooplankton community that includes some species with the habit of moving upward into shallower depths during the night. Larger mesopelagic organisms, including many species of fishes, also migrate vertically.

Mesopelagic Fishes

As previously mentioned, the lanternfishes (Myctophidae; see Plate 12.8) are among the best known families of mesopelagic vertical migrators. In terms of numerical abundance and diversity, the myctophids and the bristlemouths (Gonostomatidae; see Fig. 12.2A) dominate the meso- and bathypelagic realms. Fishes inhabiting these depths are generally small (less than 20 cm long), dark in color (black and red are the most common colors in mesopelagic animals—red works well, as light in this part of the spectrum is rapidly extinguished with depth, so that red-colored fishes are camouflaged at depth), and endowed with photophores arranged in species-specific and sometimes gender-specific patterns on the body (Marshall, 1970, 1971, 1980).

Rapid vertical migrations of 200 m or more—one round trip each night—subject these fishes to tremendous changes of pressure. Yet many mesopelagic species with well-developed swim bladders appear to be using them to maintain buoyancy while they migrate. Not all vertical migrants move up into the surface waters. Some migrate from below the thermocline into the thermocline. These migrators form an often dense concentration of life that is readily detectable by underwater sonar—hence the name **deep scattering layer** (Farquhar, 1970).

In lieu of a gas-filled swim bladder, many mesopelagic fishes have resorted to fatty deposits or other means for buoyancy (see Chapter 19). Lipids, which are buoyant but much less compressible, often replace gases in the gas bladder. Two types of lipid-filled swim bladders have been identified in deep-dwelling fishes. In one type, the swim bladder becomes regressed and infused with fat—mainly wax esters. This type of swim bladder appears to be common among fishes that undertake extensive vertical migrations that might prove challenging if only a gas-filled bladder was available. Another type of swim bladder is one that is filled with fat yet can actively secrete oxygen, which dissolves into the fatty fraction. In this case, the lipids are mainly cholesterol, which appears to significantly reduce the diffusion of accumulated oxygen out of the swim bladder, and phospholipids (Pelster, 1997).

Mesopelagic fishes tend to have relatively large, well-developed eyes. Some species have large light-gathering structures associated with the eyes. Sunlight, however dim, still plays an important role in the lives of mesopelagic organisms. Their visual pigments are sensitive to light in the blue portion of the spectrum—the wavelength that penetrates the deepest and hence imparts the characteristic blue color to the oceans. Visual cells consist largely of rods—exclusively rods in those species that are confined to the mesopelagic zone. Luminescent photophores (see Plate 12.4) are another source of light and play a significant role in predator–prey relationships and reproductive behavior (Marshall, 1971).

Modifications of the lateralis system, apparently for greater sensitivity, appear in some mesopelagic and bathypelagic fishes. These modifications involve placement of the neuromasts on the surface of the body or actually on papillae or pedicels, as in snipe eels (Nemichthyidae) and some ceratioid anglers. Other groups, such as whalefishes (see Plate 14.1), have enlarged pores opening from large canals. The variation in the size of the sacculus and lagena—those portions of the inner ear concerned with hearing—makes generalizations about the acoustic capabilities of deep-ocean fishes difficult. Montgomery and Pankhurst (1997) suggested that hearing is of less importance in the mesopelagic realm than in the acoustically rich surface waters. The tendency to lose the gas-filled swim bladder in meso- and bathypelagic fishes eliminates an important mode of acoustic communication (see Chapter 21).

The reproductive adaptations of mesopelagic fishes are poorly understood. Most apparently release pelagic eggs, some of which float at the surface and some at density layers below the thermocline. The larvae are also pelagic and usually drift at the same level as the eggs.

Life in the mesopelagic zone is surprisingly diverse. Zooplanktonic species there usually outnumber those of the epipelagic zone, and there are about 1,000 fish species represented, including interzonal species that may enter the lower epipelagic zone during the course of vertical migrations, or range throughout the meso- and bathypelagic zones. This represents a significant increase over the number of epipelagic species. Resources are available for the establishment of many niches, and most fish species are specialized in some way. Predators in the mesopelagic and bathypelagic zones range from small stomiatoids of less than 20 cm to the daggertooths (*Anotopterus*) and lancetfishes (*Alepisaurus*) of 1 m or more. Although most species are small, they are fearsome in appearance, with large eyes and large mouths

lined with daggerlike or barbed teeth (see Plate 12.4; Plate 35.1A–D; Fig. 35.1A–C). Some species, such as the swallowers (Chiasmodontidae), which range into the bathypelagic zone, have greatly distensible stomachs and can swallow just about anything that comes their way (Fig. 35.1A). One might think that predators in what would appear to be a resource-deficient environment would be highly opportunistic, but this does not appear to be the case. Generalists that eat a broad range of prey items are actually quite rare in the mesopelagic zone (Gartner et al., 1997).

The mesopelagic species generally are representative of the soft-rayed groups that appeared early in the evolution of modern fishes (Haedrich, 1997). Representative families and species are given in Table 35.2. Though considered basal teleosts, deep pelagic fishes have evolved many highly sophisticated features enabling their continued survival in a harsh environment, in which more recently evolved spiny-rayed forms have not succeeded nearly as well. Only about 6 percent of the meso- and bathypelagic fish species of the North Atlantic Ocean are of the order Perciformes (Merrett, 1994). Some orders (e.g., Cyprinodontiformes and Tetraodontiformes) are not represented at all (Merrett and Haedrich, 1997).

THE BATHYPELAGIC REALM AND BEYOND

Below the limit of light, the environment is uniformly dark and cold. Generally, food is scarcer than in the upper zones; consequently, fewer individuals and species are found here. The fishes living here typically have flaccid bodies with reduced skeletal deposition. Gas bladders are usually absent in pelagic species below 1,000 m. The only visual cues are those associated with the luminous organs present in many deep-dwelling fishes. By far the most common fishes in the bathypelagic realm are bristlemouths (Gonostomatidae) of the genus *Cyclothone* (Marshall, 1971).

Modifications for feeding in bathypelagic fishes are truly astounding. The mouths of fishes such as the gulpers and anglerfishes are extraordinarily capacious. The gulpers (families Saccopharyngidae and Eurypharyngidae; see Plate 9.6) also possess distensible stomachs that enable them to engulf prey larger than themselves. Anglers apparently attract prey by displaying a luminous lure called the **esca**, borne on a "fishing rod" or **illicium** on the head. This is a feature characteristic of the order Lophiiformes (see Chapter 13) and, as such, is seen in many shallow-water relatives, such as the frogfishes and sargassumfish (Antennariidae) and the goosefishes (Lophiidae). Among the most bizarre of the deep-water anglers are the members of the family

FIGURE 35.1
A, Black swallower (*Chiasmodon,* family Chiasmodontidae); **B,** Bigscale (*Poromitra,* family Melamphaidae);
C, Ceratioid anglerfish (*Linophryne*).

GOING DEEPER • Seamounts—Sustaining Deep-Ocean Fisheries Along a Slippery Slope

Scattered across the deep ocean floor are numerous (more than 30,000 recorded so far), steep-sided undersea mountains, termed seamounts. Generally of volcanic origin, they constitute a unique deep-sea environment characterized by enhanced current flow and localized upwelling. Seamounts are home to a diverse array of suspension feeders, such as corals, that extract nutrients from this enriched environment (DeForges et al., 2000), suggesting the potential for a lucrative food base to be exploited by a number of foraging invertebrate and vertebrate species. Researchers studying the biogeography of these seamounts have concluded that they function much like islands, in that they demonstrate unusually high endemism for a deep-ocean habitat (DeForges et al., 2000). Large concentrations of fishes associate with seamounts, including some species that are of sufficient density to merit commercial exploitation. These seamount-associated species represent something of an anomaly for deep-sea fishes. Typical meso- and bathypelagic fishes are small and flabby and demonstrate reduced levels of metabolism when compared with shallow-water forms. This has generally been taken to be the result of reduced feeding opportunities in the deep pelagic realm. Research by Jim Childress of the University of California—Santa Barbara has suggested an alternative explanation, which has more to do with the absence of light than with decreased feeding opportunities. According to Childress, the absence of light results in a decreased need for rapid swimming capability to avoid predators, and hence a lowered metabolic rate will suffice. Precious resources can then be devoted to maintaining rapid growth (Childress et al., 1980; Koslow, 1997).

TABLE 35.2 FAMILIES AND REPRESENTATIVE SPECIES PRESENT IN THE MESOPELAGIC ZONE

Family	Species
Squalidae	broadband dogfish (*Etmopterus gracilospinus*)
	collared dogfish (*Isistius brasiliensis*)*
Synaphobranchidae	Atlantic deep sea eel (*Synaphobranchus infernalis*)
Nemichthyidae	slender snipe eel (*Nemichthys scolopaceus*)
Argentinidae	Pacific argentine (*Argentina sialis*)
Bathylagidae	California smoothtongue (*Leuroglossus stilbius*)
Opisthoproctidae	spookfish (*Macropinna microstoma*)
Alepocephalidae	slickhead (*Alepocephalus bairdi*)
Gonostomatidae	lightfish (*Gonostoma denudatum*)
	anglemouth (*Cyclothone microdon*)*
Sternoptychidae	hatchetfish (*Argyropelecus olfersi*)*
Stomiatidae	boafish (*Ichthyococcus ovatus*)
Chauliodontinae	Pacific viperfish (*Chauliodus macouni*)
Melanostomiatinae	longfin dragonfish (*Tactostoma macropus*)*
Idiacanthinae	black dragonfish (*Idiacanthus fasciola*)
Chlorophthalmidae	shortnose greeneye (*Chlorophthalmus agassizi*)
Scopelarchidae	northern pearleye (*Benthalbella dentata*)
Paralepididae	duckbilled barracudina (*Paralepis atlantica*)*
Alepisauridae	longnose lancetfish (*Alepisaurus ferox*)*
Anotopteridae	daggertooth (*Anotopterus pharao*)*
Myctophidae	lanternfish (*Myctophum punctatum*)*
	northern lampfish (*Stenobrachius leucopsaurus*)*
	numerous additional genera and species
Trachipteridae	dealfish (*Trachipterus arcticus*)
Bregmacerotidae	antenna codlet (*Bregmaceros atlanticus*)
Gempylidae	oilfish (*Ruvettus pretiosus*)

* Interzonal.

Deep-sea fishes associated with continental slopes and seamounts seem to defy our general perceptions of the adaptations of deep-ocean fishes, including Childress' provocative hypothesis. Large, powerfully swimming fishes, such as the deepwater sharks, grenadiers (Macrouridae), and codlings (Moridae) can be found in abundance. Species diversity appears to peak at about 1,500 m depth along the continental slope (Moore and Mace, 1999). Two families of deepwater fishes, the roughies (Trachichthyidae) and the oreos (Oreosomatidae) appear to exploit the enhanced productivity associated with seamounts and occur there in dense concentrations. Sedimentation and vertical migration of zooplanktonic organisms are the means of providing nutrients to the deep ocean. Seamounts represent localized beneficiaries of this nutrient influx and support a complex food web, in which roughy and oreos, consuming squid and small fishes, represent the top carnivores. They grow to large size and are active with a correspondingly high metabolic rate. Growth and reproductive efficiency is only on the order of about 5 percent (midwater species, such as lanternfishes, are much more efficient, on the order of 25 to 50 percent), because most food is consumed in activity and enhanced metabolism (Koslow, 1996, 1997). Fortunately for us, but most unfortunate for these species, their flesh is firm, rich in proteins and lipids, but low in water content. In short, they taste pretty good.

The 1980s witnessed the dramatic growth of the orange roughy (*Hoplostethus atlanticus*) and oreo fisheries in the deep waters off New Zealand and southeastern Australia. These fisheries, operating at depths between 700 and 1,400 meters, are enormously important to

Linophrynidae. These possess an enormous set of luminous barbels below the mouth as well (Fig. 35.1C). The mesopelagic viperfishes (*Chauliodus;* see Plates 12.4 and 35.1C) have an elongate first ray on the dorsal fin that appears to function as a lure, whereas the dragonfishes (Melanostomiinae and Idiacanthinae) use a luminescent chin barbel for the same purposes (see Plate 35.1D). Any of these unusual devices will assist in the capture of prey in an environment that is characteristically so low in biomass that encounters between predator and prey are relatively rare.

Males of deep-ocean species tend to be smaller than the females, and many species have pelagic eggs and larvae. Aside from these observations, little else is known about the reproductive habits of deep-ocean fishes. The ceratioid anglerfishes are well known for their obviously efficient method of ensuring that the sexes will be together at spawning time—a chancy situation, given the comparatively few individuals in the bathypelagic realm: The tiny males of many species are parasitic on the females. They grasp the female with their jaws and subsequently become permanently fused, dependent on her circulatory system for nourishment (see Plate 35.2). Most of their internal organs degenerate, save for the testes, which become the most prominent structures in the coelomic cavity of the parasitic male. Table 35.3

lists some families and representative species found in the deep pelagic zones below 1,000 m.

The Continental Slope and Abyssal Plain

Moving from the shelf onto the upper slope, the temperature decreases steadily over the 200- to 1,000-m span, and the water movement is generally slight. The last weak rays of light from the surface are eventually extinguished at about 750 to 1,000 m, but the greatly enlarged and extremely sensitive eyes of many fishes living at this depth can still make use of such low levels of ambient light intensity. Like the deeper zones, this zone depends on the shallower parts of the sea for virtually all the available food. Remains of plankton and other pelagic organisms accumulate on the bottom, where scavengers attack and recycle them.

In the lower slope and abyssal plain, physical conditions become quite uniform except for the increasing pressure with depth. These zones are cold and dark, with slow bottom currents. The sediments contain less terrigenous materials and are predominantly composed of calcareous and siliceous deposits from the pelagic realm or, in the deepest areas, of red clay. Food also originates from above. Particulate matter that reaches the bottom, if not eaten by

the economy of these two countries. Unlike most meso- to bathypelagic fish species, the orange roughy occurs in dense aggregations, making for ease of harvesting with conventional fishing gear. Sophisticated fish processing done on the high seas and overnight delivery systems mean that I can purchase orange roughy in Central Kentucky any time I want (though, considering the price of roughy compared with other species, this is not too often).

As we have depleted fisheries of the continental shelf, our attention has become more focused on species like the orange roughy that occur in the less accessible reaches of the ocean. Once considered a bait-stealing nuisance by halibut fishers, the abundant macrourids are now considered one of the most promising candidates for commercial exploitation. More than 1,500 metric tons of Pacific

grenadier (*Coryphaenoides acrolepis*) were landed along the Pacific coast of California and Oregon in 1996 (Moore, 1999). The fishery for wreckfish (*Polyprion americanus*) off the southeastern United States has grown dramatically in the last decade. Researchers are investigating the potential for developing fisheries for deep benthopelagic squaloids (*Centroscyllum* and *Centroscymnus*), chimaeras (*Hydrolagus* and *Rhinochimaera*), and bony fishes such as the smoothheads (*Alepocephalus*, a.k.a. slickheads; Moore, 1999). This is cause for some concern. Commercially desirable deep-water species, such as roughy and redfish (*Sebastes*), have exceedingly slow rates of growth (as most of their energy budget is devoted to the maintenance of an active lifestyle), are very long-lived (up to 100 years for roughy and oreos; 75 years for redfish), yet have

comparatively low fecundities. Large slow-growing fishes are especially vulnerable to overexploitation. Catches of roughy in the Australian fishery peaked in 1989–1990 with more than 34,000 tons landed (Elliott et al., 1995), and the New Zealand fishery has reduced the biomass of the roughy by an estimated 70 to 80 percent (Smith et al., 1991). The remarkable diversity of seamount fauna is just beginning to be appreciated. Exposure of these fragile, poorly understood ecosystems to heavy fishing, including damage by trawls and other fishing gear, can have devastating consequences. Perhaps the lessons learned from the demise of so many of our coastal habitats and their associated fisheries can be applied, in the hopes that the development of deep-ocean fisheries will proceed in a more prudent manner.

TABLE 35.3 REPRESENTATIVE FAMILIES AND SPECIES IN THE BATHY-, ABYSSO-, AND HADOPELAGIC ZONES

Family	Species
Nemichthyidae	snipe eel (*Nemichthys scolopaceus*)
Saccopharyngidae	whiptail gulper (*Saccopharynx ampullaceus*)
Eurypharyngidae	pelican gulper (*Eurypharynx pelecanoides*)
Bathylagidae	blacksmelt (*Bathylagus antarcticus*)
Gonostomatidae	bristlemouth (*Gonostoma bathyphilum*)
Stomiidae	loosejaw (*Malacosteus niger*)
	ribbon sawtail fish (*Idiacanthus fasciola*)
Giganturidae	gianttail (*Gigantura vorax*)
Paralepididae	slender barracudina (*Lestidium ringens*)
Evermannellidae	sabertooth (*Evermannella atrata*)
Himantolophidae	footballfish (*Himantolophus groenlandicus*)
Ceratiidae	seadevil (*Cryptopsaras couesi*)
Chiasmodontidae	black swallower (*Chiasmodon niger*)

scavengers, can be broken down by bacteria, which might then be eaten by mud-ingesting invertebrates, which in turn can be eaten by fishes.

Adaptations of Deep Benthic and Benthopelagic Fishes

Of the fishes that inhabit the deep sea, the benthic and benthopelagic taxa are quite distinct from those of the pelagic realm. Many of the benthopelagic fish families living in the oceanic depths illustrate the comparative success of an elongate body form. A common body shape consists of a large head, well-developed pectoral fins, and long dorsal and anal fins along a tapering afterbody. It is thought that the combination of an elongated snout region with an anal fin that is longer than the dorsal fin aids in positioning the body in a head-down manner to facilitate bottom feeding. Many of these fishes reach lengths of 30 to 45 cm and more; cusk eels (Ophidiidae) over 2 m long have been observed prowling the deep ocean floor.

Reproductive adaptations in some species appear to prevent the young from being displaced too far from suitable habitat. Hagfishes, skates, and chimaeras deposit large eggs, from which hatch precocious juveniles essentially like miniature adults. Some brotulids and electric rays are ovoviviparous, and a number of eelpouts are viviparous. Poachers (Agonidae) and some eelpouts deposit demersal eggs, and the latter guard their spawn. Eels, deep-sea cods, and macrourids have pelagic eggs that can drift freely with the deep currents. These species typically have very high fecundities. The eggs are small and hatch pelagic larvae that drift in the currents. These larvae are thus more vulnerable to predation and are more likely to drift away from optimum habitat. In many species, the eggs and larvae are

hydrostatically balanced, so that they float not at the surface but at some intermediate depth. In some flatfishes, eggs and larvae may drift inshore, where better nursery areas may exist. The juveniles subsequently return to deeper water. A few liparids place their eggs in the gill cavities of crabs.

The fish fauna of the benthic regions below 1,000 m contains many elements of the upper slope, but modifications for deeper life are evident. Compared with those of bathypelagic fishes, the eyes of slope-dwelling fishes are surprisingly well developed. Light gathering in the deep is facilitated through the presence of a highly reflective **tapetum lucidum** located behind the retina. In some of the deepest-dwelling species, however, the eyes may be extremely reduced—for example in *Ipnops* or *Bathypterois* (see Plate 12.6). Although bioluminescence is important in the pelagic realm at depths corresponding to the upper slope, it is relatively rare in the benthic fishes. Benthopelagic species such as the macrourids, however, may be well endowed with luminescent organs—more than half of the known benthopelagic species produce light from ventral organs (Merrett and Haedrich, 1997). Obliterative countershading (see Chapter 18) has no adaptive value at these depths. As in the mesopelagic zone, fishes tend to be uniformly dark, with red being a common color for reasons discussed earlier. Some examples are the orange roughy (*Hoplostethus atlanticus*), redfishes (*Sebastes*), and the pelagic rosy whalefish (*Barbourisia rufa*).

As is the case in shallower waters, the swim bladder tends to be absent in benthic fishes, yet benthopelagic species, requiring some buoyancy to maintain their position above the substrate, may have well-developed swim bladders. Buoyancy may also be maintained through reduction in the mass of the skeleton and scales, giving the fishes a flabby and watery consistency. The tripodfishes (*Bathypterois;* see Plate 12.6) have elongate rays on the pelvic and caudal fins, which are used to prop the animal above the soft ooze bottoms on which it lives.

As mentioned earlier, deep pelagic fishes appear to lack significant acoustic communication abilities, owing to the absence of muscles that can generate sounds in association with a gas-filled swim bladder. Although swim bladder sonic muscles (see Chapter 21) were described decades ago in benthopelagic fishes such as the grenadiers (Macrouridae) and deep-sea cods (Moridae), acoustic emissions attributable to them have only recently been recorded (Mann and Jarvis, 2004). It is interesting to note that abyssal macrourids that lack sound production mechanisms have small saccular otoliths, whereas species of the upper continental slope, in which swim bladder sonic muscles are well developed, have larger saccular otoliths (Montgomery and Pankhurst, 1997). As in deep pelagic fishes, the lateral-line systems of benthic and benthopelagic species are greatly developed, with the neuromasts often set out on stalks.

Feeding Adaptations

Faunal assessments of the deep ocean floor were pioneered by the legendary *Challenger* expedition of 1872–1876. Before these studies, it was believed that the deepest parts of the ocean were essentially biological deserts—physically harsh places with little life present. We now understand the deep ocean floor to be a complex ecosystem with a diversity of inhabitants. Energetically, the sea floor is sustained by the enormous amount of biological productivity going on in the photic zone. Organic material, ranging in size from microscopic detritus and the slightly larger particles sometimes referred to as "marine snow" to the carcasses of whales, constantly rains down from above, sustaining complex food webs as it makes its trip to the ocean floor. As is the case in ecosystems elsewhere, this nutrient influx is patchy, meaning that it is not delivered in a random fashion but aggregates in certain localities (Rex, 1981; Rice and Lambshead, 1992). The hydrothermal vent communities discussed later exemplify one kind of patchiness. Local and seasonal variation in the settling of phytodetritus (the remains of photosynthetic organisms) is another. An especially dramatic example is the arrival of huge chunks of organic matter in the form of large fish or whale carcasses. All of these processes have resulted in localized aggregations of benthic epifauna and infauna that show significant variation from one sample site to another (Rice and Lambshead, 1992).

The fact that food resources are patchy and chancy in their availability would suggest that opportunism would be the most effective feeding strategy for fishes living on or near the deep ocean floor. Studies on the feeding habits of one of the most widely distributed benthopelagic fish taxa,

the macrourids, have seemed to support this contention, leading to the general conclusion that benthic and benthopelagic fishes of the deep ocean are largely nonselective in their predatory behavior. Studies of the slope-dwelling ichthyofauna off eastern Tasmania have revealed four basic categories of feeding—pelagic piscivores, epibenthic piscivores, epibenthic invertebrate feeders, and benthopelagic omnivores (Blaber and Bulman, 1987). Gartner et al. (1997) have expanded this finding to include 10 major trophic guilds. Comparatively few species studied appear to specialize on one particular prey item. They shift their preferences based on local availability, but they show some specificity within the spectrum of available prey at a given location. Pelagic feeding appears widespread among demersal fishes, yet those feeding on the bottom tend to be larger (Merrett and Haedrich, 1997). Scavenging of large carcasses is also an important feeding strategy, especially for the ubiquitous hagfishes. Contrary to what was earlier believed, demersal fishes of the ocean depths appear to have evolved a measure of specificity in their nutritional strategies. The feeding habits of macrourids may appear broad and nonspecific simply because gadiform fishes are by nature benthopelagic opportunists. Other species appear to have evolved nutritional strategies that may be fairly specific yet flexible enough to allow them to take advantage of resources that are distributed in a nonuniform manner.

Deep Benthic Fish Assemblages

Many of the fish families from the upper slope are also encountered on the continental shelf. Flatfishes, including the halibuts (*Hippoglossus*), are common, as are hagfishes, cods, morids, scorpaenids, skates, squaloid sharks, eelpouts (Zoarcidae), seasnails (Liparidae), and cusk eels (Ophidiidae). Relatives of the cods—specifically, ophidiiform fishes (families Ophidiidae, Bythitidae, and Aphyonidae) and gadiform fishes (families Macrouridae, Merluciidae, and Moridae)—dominate the deep benthic fish fauna (Haedrich, 1997). Eelpouts, once considered a relative of the cods and cusk eels but now classified among the Perciformes, are also well represented. Many of these families exemplify the elongate body form described earlier. Other elongate fishes inhabiting the lower shelf include the chimaeras, halosaurids, and notacanthids. Some large species that inhabit the deeper slope regions or substrates associated with seamounts further offshore exist in concentrations dense enough that they are targeted by commercial fisheries (see Going Deeper). The ceratioid anglerfishes are most commonly associated with deep pelagic waters, but one genus (*Thaumatichthys*) has assumed a demersal existence, hovering just above the substrate. This ambush piscivore attracts its prey with a luminescent lure that hangs from the roof of its capacious mouth (Gartner et al., 1997). Although the diversity of deep demersal fishes is not especially high, a slightly greater number of orders are represented than in the pelagic realm—22 orders of benthic and benthopelagic fishes, compared with 18 pelagic orders (Merrett and Haedrich, 1997).

The Deepest Dwellers

Fewer than 60 species of fishes are known from the abyssal ocean floor, and fewer yet from the hadal zone. Parin (1984) noted only 3 genera of ophidiids and two genera of liparids in the hadal depths. A brotulid (*Bassogigas profundissimus*) and a liparid (*Careproctus amblystomopsis*) have been collected deeper than 7,000 m, and the deep-sea explorers Piccard and Walsh observed a "flatfish" at 10,800 m in the Marianas Trench, the deepest known point of the world's oceans. The identity of this, possibly the deepest occurring of all fishes, is not known, although Wolff (1961) suspected that it might not have been a fish at all but rather some species of holothurian (sea cucumber). If one considers the rate at which new species of deep demersal fishes have been discovered in the past few decades, we must conclude that the total number of species living near the ocean floor is still far from fully known (Haedrich, 1997; Haedrich and Merrett, 1988).

Fish Fauna of Hydrothermal Vents

In the late 1970s, oceanographers studying the deep ocean bottom about 280 km northeast of the Galapagos Islands discovered one of the most unusual ecosystems on Earth— one that is apparently energetically based on warm, chemically enriched water escaping from a series of fissures on the ocean bottom at a depth of about 2,500 m. Since that momentous discovery, these chemautotrophic ecosystems have been discovered in several other deep-ocean locations—usually in association with mid-ocean ridges at the margins of continental plates in the eastern Pacific or the north central Atlantic (VanDover, 2000). The biomass and productivity of hydrothermal vent ecosystems is enormous compared with that observed on the adjacent seafloor (Zierenberg et al., 2000). Dense aggregations of mussels, clams, and spectacular giant pogonophoran worms can be seen clustered around the vents. Many species of barnacles, crabs, and shrimp new to science have been described from these communities. A number of fish species have also been identified as associates of these deep ocean thermal springs. Cohen and Haedrich (1983) have recorded about 20 species of fishes from the Galapagos thermal vent region. The most common were three species of macrourid rattails (*Coryphaenoides*), an ophidiid (*Bassozetus*), and two or three species of zoarcid eelpouts. An unidentified member of the family Bythitidae was discovered to be actually living in

the vents. It is obvious that the trophic enrichment afforded by these thermal vents serves to attract a number of fishes in the vicinity.

MARINE FISHES OF POLAR ENVIRONMENTS

The Nature of Polar Environments

Polar environments present unusual challenges for the fishes that inhabit them. As might be expected, temperature is the overriding constraint on polar existence, yet many organisms have successfully invaded these extremely cold waters, and fishes are an integral component of the food webs that have been established in polar seas. Polar environments are those that exist roughly above 60° latitude. Some environmental features, such as cold temperatures and extreme day–night cycles, are common to both the Arctic and Antarctic. The characteristic day–night cycles, in which polar summers have essentially 24 hours of daylight, result in unusual biological production regimes, with short, intense periods of growth. Yet there are fundamental differences between the Arctic and Antarctic region, mainly pertaining to basic geological features and their consequent environmental impact.

The Arctic region consists basically of the Arctic Ocean and the adjacent water masses of the North Atlantic and the Bering Sea. The Arctic Ocean is by far the smallest ocean, with a mean depth of a little more than 1,000 m—less than a third of that seen in the Atlantic, Pacific, or Indian oceans. The depth of the Canada Basin at the center of the Arctic Ocean approximates the depth of other ocean basins, but almost half of the ocean floor is made up of broad expanses of continental shelf, thus explaining the shallower mean depth. The continental shelves of the continents bordering the Arctic Ocean extend to depths of up to 300 m—significantly deeper than most other shelf areas. The circulation of water into and out of the Arctic Ocean from the North Atlantic is partially blocked by shallow submarine ridges. Siberia and Alaska form the western boundary of the Arctic Ocean. Broad continental shelves extending from these coastlines form much of the floor of the Bering Sea, including the narrow Bering Strait that leads into the Arctic Ocean. Epipelagic fauna are constrained by the fact that most of the Arctic Ocean is permanently covered by ice. The margins of the ice cap break off and form large icebergs that drift south.

Antarctica is quite different, as it is a frozen continent surrounded by the Southern Ocean. As mentioned in Chapter 33, the Southern Ocean is unique in not being bounded by landmasses. The prevailing surface current, termed the **West Wind Drift,** flows unimpeded by continents in an

744 • Part Seven Interactions of Fishes with Their Environment

easterly direction around Antarctica. Another surface flow, called the **East Wind Drift**, flows adjacent to the continent in a westerly direction and is the dominant flow regime seen in two bodies of water, the Ross Sea and the Weddell Sea, that border Antarctica. Compared to what is observed in the Arctic, there is less shallow-water habitat in the Southern Ocean surrounding Antarctica. The continental shelf is broad, but much of it is covered with ice. As on the margins of the Arctic Ocean, the shelf areas of Antarctica are deep, averaging 500 m. The bottom has a rugged topography due to glacial scouring, and narrow trenches, some reaching depths of more than 1,000 m, course along the Antarctic shelf.

Origins of Polar Biota

Bipolar (antitropical) species distribution patterns have been found in a number of terrestrial and marine organisms (Crame, 1993; Lindberg, 1991). Many marine organisms have apparently transcended the vast oceanic barrier to become established in both Arctic and Antarctic realms. Antarctica is believed to have separated from Australia early in the Cenozoic, and the present cold-water provinces at either pole were first established between 8 and 16 million years ago. The presence of bipolar species could be explained by the convergent evolution of fauna at either pole or through transtropical gene flow enabling populations to remain in genetic contact. Darling et al. (2000) have discovered similarities in DNA sequences in the ribosomal RNA genes of Arctic and Antarctic planktonic foraminiferans (microscopic protozoans). For planktonic microbes, transtropical gene flow appears to be a distinct possibility. This seems less likely for marine fishes when we consider the duration of their pelagic early life history and the specificity of habitat requirements of adults. Indeed, the fish fauna of the Arctic appears quite dissimilar to that of the Antarctic. This may be attributed to differences in Arctic versus Antarctic environments, as discussed earlier, or it may be a consequence of long periods of evolution in isolation, in which the brief pelagic larval stage of fishes precludes any gene flow from one pole to the other. In at least one instance, however, an adult of an Antarctic fish species appears to have migrated to Arctic waters: A 70-kg Patagonian toothfish (*Dissostichus eleginoides*), a member of the Antarctic suborder Notothenioidei, was caught off the coast of Greenland (Møller et al., 2003).

Adaptations of Polar Fishes

Pioneering work by Scholander et al. (1953) suggested that polar-adapted fishes compensated for the rigors of the cold through elevated levels of metabolism. This has enabled them to remain active under conditions that would induce torpor or death in other species. More recent physiological studies

have challenged this notion. Contrary to Scholander's findings, polar fishes appear to have adapted to the decreased biological productivity and associated feeding opportunities in polar waters by decreasing their basal metabolic rates (Holeton, 1974). Compared with the Antarctic, water temperatures in the Arctic fluctuate to a greater degree. The resident fish fauna, sharing considerable phylogenetic affinity with their north temperate relatives, is more tolerant of temperature fluctuation than are Antarctic species. Deferred maturity, lower fecundity with the production of larger eggs, and increased longevity are also consistent with metabolic adaptations to polar environments—adaptations that have evolved not in response to the cold per se, but to the nature of food supply in polar environments (Clarke, 1980, 1983).

. . . .
The Polar Fish Fauna

Arctic Fishes

When compared with the Antarctic, the fish fauna of the Arctic is less distinctive, consisting largely of cold-adapted forms derived from subpolar regions of the North Atlantic and Pacific oceans. The Arctic has experienced repeated cycles of warming and cooling, with the present polar environment having existed for no longer than 0.7–2.0 million years (Eastman, 1997). More recently, during the Pleistocene, the repeated advance and retreat of the glaciers across the northern landscape and their dramatic consequences for the sea level were probably significant in influencing the dispersal of fishes into the Arctic. Consequently, less time has passed in which a distinctive ichthyofauna could have evolved in the Arctic Ocean. In the Canadian Arctic, 224 species of marine fishes (about one quarter of all marine species known from Canadian waters) have been recorded. About half of the 42 freshwater species recorded above the Arctic Circle in Canada are anadromous, entering and leaving streams that drain into the Arctic Ocean (see Chapter 31). About 416 species in 96 families have been recorded throughout the entire Arctic region. The majority of these fishes are more widely distributed boreal species—only 75 species have been recorded at the highest latitudes. Families that dominate the high Arctic include the zoarcids (21 species), salmonids (17 species), liparids (10 species), cottids (8 species), and gadids (5 species; Eastman, 1997). It is only in the Zoarcidae and the Liparidae that we see significant representation at both poles. Families recorded from the Arctic Ocean are given in Table 35.4. Physiological mechanisms to enhance resistance to freezing have evolved in members of many of these families. These mechanisms are discussed in more detail in Chapter 24.

TABLE 35.4 FISH TAXA OF THE ARCTIC REGION

Taxon	Number of Species	Habitat	Percentage of Fauna
Myxini			
Myxiniformes			
Myxinidae	1	scavenger	0.2
Cephalaspidomorpha			
Petromyzontiformes			
Petromyzontidae	3	"parasite"	0.8
Chondrichthyes			
Chimaeriformes			
Chimaeridae	2	benthopelagic	0.5
Carcharhiniformes			
Scyliorhinidae	1	benthopelagic	0.2
Triakidae	1	benthopelagic	0.2
Carcharhinidae	1	epipelagic	0.2
Lamniformes			
Cetorhinidae	1	epipelagic	0.2
Lamnidae	3	epipelagic	0.8
Hexanchiformes			
Chlamydoselachidae	1	benthopelagic	0.2
Squaliformes			
Dalatiidae	4	benthopelagic	1.1
Squalidae	1	benthopelagic	0.2
Rajiformes			
Rajidae	11	benthic	2.6
Actinopterygii			
Acipenseriformes			
Acipenseridae	4	benthopelagic	1.1
Notacanthiformes			
Notacanthidae	2	bathypelagic	0.5
Anguilliformes			
Anguillidae	2	catadromous	0.5
Synaphobranchidae	2	bathypelagic	0.5
Nemichthyidae	1	mesopelagic	0.2
Serrivomeridae	1	mesopelagic	0.2
Saccopharyngiformes			
Saccopharyngidae	1	bathypelagic	0.2
Clupeiformes			
Clupeidae*	5	epipelagic, freshwater	1.3
Cypriniformes			
Catostomidae	1	freshwater	0.2
Esociformes			
Esocidae	1	freshwater	0.2
Argentiniformes			
Argentinidae	1	mesopelagic	0.2
Microstomatidae	1	mesopelagic	0.2
Alepocephalidae	4	epibenthic	1.1
Platytroctidae	7	mesopelagic	1.7

(Continued)

TABLE 35.4 CONTINUED

Taxon	Number of Species	Habitat	Percentage of Fauna
Salmoniformes			
Osmeridae*	7	anadromous	1.7
Salmonidae	32	freshwater, anadromous	7.7
Stomiiformes			
Gonostomatidae	4	mesopelagic	1.1
Sternoptychidae	2	mesopelagic	1.4
Stomiidae	3	mesopelagic	0.8
Aulopiformes			
Notosudidae	1	benthic	0.8
Synodontidae	1	mesopelagic	0.2
Paralepididae	3	mesopelagic	0.2
Anotopteridae	1	epipelagic	0.2
Alepisauridae	1	mesopelagic	0.2
Myctophiformes			
Myctophidae	7	mesopelagic	1.7
Lampriformes			
Lampridae	1	epipelagic	0.2
Trachipteridae	1	epipelagic	0.2
Regalecidae	1	epipelagic	0.2
Ophidiiformes			
Bythitidae	1	benthic	0.2
Gadiformes			
Moridae	3	benthopelagic	1.1
Phycidae	4	benthopelagic	0.5
Merluciidae	2	benthopelagic	0.7
Gadidae*	30	benthopelagic	7.2
Macrouridae	5	benthopelagic	1.3
Lophiiformes			
Lophiidae	1	benthic	0.2
Antennariidae	1	benthic	0.2
Caulophrynidae	1	bathypelagic	0.2
Himantolophidae	1	bathypelagic	0.2
Ceratiidae	2	mesopelagic	0.5
Oneirodidae	5	mesopelagic	1.3
Linophrynidae	2	mesopelagic	0.5
Beloniformes			
Belonidae	1	epipelagic	0.2
Scomberesocidae	1	epipelagic	0.2
Stephanoberyciformes			
Melamphaidae	1	mesopelagic	0.2
Beryciformes			
Anoplogastridae	1	bathypelagic	0.2
Berycidae	1	benthopelagic	0.2
Gasterosteiformes			
Gasterosteidae	3	freshwater	0.8
Syngnathidae	2	epipelagic	0.5
Scorpaenidae	6	benthic	1.4

(Continued)

TABLE 35.4   CONTINUED

Taxon	Number of Species	Habitat	Percentage of Fauna
Triglidae	1	benthic	0.2
Hexagrammidae	2	benthic	0.5
Cottidae*	44	benthic	10.6
Hemitripteridae	2	benthic	0.5
Agonidae	12	benthic	2.9
Psychrolutidae	6	benthic	1.4
Cyclopteridae	10	benthic	2.4
Liparidae	17	benthic	4.1
Perciformes			
(Percoidei)			
Moronidae	1	anadromous	0.2
Percidae	2	freshwater	0.5
Bramidae	3	epipelagic	0.8
Caristiidae	1	epipelagic	0.2
(Zoarcoidei)			
Zoarcidae*	40	benthic	9.6
Stichaeidae	17	benthic	4.1
Cryptacanthodidae	1	benthic	0.2
Pholidae	3	benthic	0.8
Anarhichadidae	5	benthic	1.3
Zaproridae	1	benthic	0.2
(Trachinoidei)			
Chiasmodontidae	1	mesopelagic	0.2
Ammodytidae	6	benthopelagic	1.4
(Scombroidei)			
Trichiuridae	1	epipelagic	0.2
Scombridae	1	epipelagic	0.4
Xiphiidae	1	epipelagic	0.2
(Stromateoidei)			
Centrolophidae	1	epipelagic	0.2
Pleuronectiformes			
Scophthalmidae	2	benthic	0.2
Pleuronectidae*	26	benthic	6.3
Tetraodontiformes			
Molidae	1	epipelagic	0.2
Total	416		100.0

* Families in which antifreeze compounds in blood and tissues have been reported.
Bold—family found in both Arctic and Antarctic waters.
Source: Eastman, 1997; Power, 1997.

Antarctic Fishes

Although freshwater fishes were known on the Antarctic continent 180 million years ago (Eastman, 1993), only marine fishes are found there today. Lacking freshwater that is not covered by ice, freshwater or anadromous fish assemblages are entirely absent. During the Eocene (approximately 40 million years ago), the coastal fish fauna of Antarctica included many families characteristic of temperate shelf regions. This included a diverse shark fauna, including the families Hexanchidae, Squalidae, Lamnidae, and Carcharhinidae, as well as numerous teleosts, including catfishes and clupeoids (Eastman, 1993). Presently, the Southern Ocean, representing 10 percent of the total oceanic realm, contains only about 1 percent of the world's ichthyofauna (Eastman, 1995).

In its modern configuration, Antarctica is unique in its isolation. The circumpolar current, sharp temperature differences, and the absence of connection to any other continent have made Antarctica one of the most sharply defined zoogeographic regions on Earth. The ichthyofauna is highly endemic, with fully 88 percent of the species confined to the Antarctic region (Andriashev, 1987). Affinities with fauna beyond Antarctica are most evident on the Antarctic Peninsula. Because of its proximity to the southern tip of South America, the fish fauna of the Antarctic Peninsula consists of species of Patagonian as well as Antarctic affinities (Daniels and Lipps, 1982). Virtually no intertidal or shallow subtidal habitat is available for colonization by fishes, as the Antarctic shoreline is scoured by ice, and anchor ice extends to depths of more than 30 m. Likewise, the epipelagic zone of open waters is severely constrained by ice cover, and few epipelagic species are known from Antarctic waters.

Harsh as the environment is, the Antarctic presents unusual habitat opportunities. The dominant fishes of the Antarctic are members of the perciform suborder Notothenioidei (Table 35.5). This suborder has experienced an explosive diversification, with 122 species in six families occurring in the Southern Hemisphere—95 of these species are found in the Antarctic (Eastman, 1995, 1997). Notothenioids are essentially benthic fishes, although many species have secondarily evolved an epibenthic or pelagic existence (see Plate 35.3). The high rate of speciation and endemism observed in notothenioids suggests that they may be the first recognized example of a marine species flock (Eastman, 1997, 2000a, 2000b). Eastman (1993) has grouped

TABLE 35.5   FISH TAXA OF THE ANTARCTIC REGION

Taxon	Number of Species	Habitat	Percentage of Fauna
Myxini			
Myxiniformes			
Myxinidae	1	scavenger	0.4
Cephalaspidomorpha			
Petromyzontiformes			
Petromyzontidae	1	"parasite"	0.4
Chondrichthyes			
Lamniformes			
Lamnidae	1	epipelagic	0.4
Squaliformes			
Dalatiidae	2	epibenthic	0.7
Rajiformes			
Rajidae	8	benthic	2.9
Actinopterygii			
Notacanthiformes			
Halosauridae	1	benthopelagic	0.4
Notacanthidae	1	benthopelagic	0.4
Anguilliformes			
Synaphobranchidae	2	benthopelagic	0.7
Salmoniformes			
Microstomatidae	1	mesopelagic	0.4
Bathylagidae	3	mesopelagic	1.1
Alepocephalidae	4	epibenthic	1.4
Platytroctidae	1	mesopelagic	0.4
Stomiiformes			
Gonostomatidae	5	mesopelagic	1.8
Sternoptychidae	2	mesopelagic	0.7
Stomiidae	5	mesopelagic	1.8
Aulopiformes			
Scopelarchidae	2	mesopelagic	0.7

(Continued)

TABLE 35.5 CONTINUED

Taxon	Number of Species	Habitat	Percentage of Fauna
Notosudidae	1	mesopelagic	0.4
Paralepididae	4	mesopelagic	1.4
Anotopteridae	1	epipelagic	0.4
Alepisauridae	1	mesopelagic	0.4
Myctophiformes			
Myctophidae	35	mesopelagic	12.7
Lampridiformes			
Lampridae	2	epipelagic	0.7
Ophidiiformes			
Ophidiidae	1	abyssopelagic	0.4
Carapidae	1	benthic?	0.4
Gadiformes			
Muraenolepididae	4	benthic	1.4
Moridae	4	benthic	1.4
Melanonidae	1	mesopelagic	0.4
Gadidae	1	benthic	0.4
Macrouridae	11	benthopelagic	4.0
Lophiiformes			
Ceratiidae	1	mesopelagic	0.4
Oneirodidae	1	mesopelagic	0.4
Melanocetidae	1	mesopelagic	0.4
Stephanoberyciformes			
Melamphaidae	3	mesopelagic	1.1
Cetomimidae	3	mesopelagic	1.1
Zeiformes			
Oreosomatidae	1	benthopelagic	0.4
Scorpaeniformes			
Congiopodidae	1	benthic	0.4
Liparidae	31	benthic	11.3
Perciformes			
(Zoarcoidei)			
Zoarcidae*	22	benthic	8.0
(Notothenioidei)			
Bovichtidae	1	benthic	0.4
Nototheniidae*	33	benthic	12.4
Harpagiferidae	6	benthic	2.2
Artedidraconidae	24	benthic	8.7
Bathydraconidae*	16	benthic	5.4
Channichthyidae*	15	benthic	5.4
(Blennioidei)			
Tripterygiidae	1	benthic	0.4
(Scombroidei)			
Gempylidae	1	mesopelagic	0.4
Scombridae	1	epipelagic	0.4
(Stromateoidei)			
Centrolophidae	1	epipelagic	0.4

(Continued)

TABLE 35.5 CONTINUED

Taxon	Number of Species	Habitat	Percentage of Fauna
Pleuronectiformes			
Achiropsettidae	4	benthic	1.4
Totals	274		100.0

* Families in which antifreeze compounds in blood and tissues have been reported.
Bold—family found in both Arctic and Antarctic waters.
Sources: Eastman, 1993, 1997.

the benthic-derived notothenioids according to five specialized Antarctic habitats:

1. **Cryopelagic fishes.** These fishes participate in an unusual food web established just beneath the ice. Although only a small fraction of the surface irradiance reaches the underside of the ice during the austral summer, it is sufficient to sustain photosynthesis by ice-tolerant microscopic algae. A host of cryopelagic organisms, including amphipods and euphausiid shrimp, graze on these algae and are in turn preyed upon by fishes of the genus *Pagothenia*.

2. **Secondarily pelagic fishes.** These notothenioids are neutrally buoyant, permanent inhabitants of the pelagic realm and include the genera *Dissostichus*, *Pleuragramma*, and *Aethotaxis*. Lacking a swim bladder, these notothenioids accomplish neutral buoyancy by a reduction of mineral deposition in their bones and scales and by enhanced lipid deposition.

3. **Semipelagic fishes.** Fishes of this group, best exemplified by *Trematomus newnesi*, move into the water column primarily for purposes of feeding. They retain adaptations for a benthic existence.

4. **Epibenthic fishes.** Epibenthic fishes are adapted for an existence just above the seafloor. Epibenthic notothenioids differ from benthic species in being more streamlined and lacking adaptations for substrate contact.

5. **Pseudobathyal fishes.** These fishes have become adapted to the unique opportunities for deep-ocean existence in the trenches along the shelf. Again, notothenioids, especially benthic and epibenthic species typical of nearshore environments, dominate rather than the mesopelagic ichthyofauna usually encountered elsewhere in the world's oceans.

Mesopelagic fishes, especially members of the family Myctophidae, live throughout the water column in the Antarctic. Elsewhere, mesopelagic fishes typically engage in vertical migrations that bring them to the surface to feed at night. During the short, intense bouts of productivity that take place during the austral summer, lanternfishes congregate and feed near the surface at the edge of the ice pack throughout the day. In the open ocean of the Weddell Sea, six species of myctophids account for 95 percent of the biomass of mesopelagic fishes (Eastman, 1993). Here, they are an important component of the polar food chain, as marine birds and mammals feed heavily on them. A commercial fishery for *Electrona carlsbergi* has been established near the South Georgia island region (Helfman et al., 1997).

Two remarkable adaptations of Antarctic notothenioids merit a brief mention here, but they are discussed in greater detail in Chapter 24. As was described for certain species of Arctic fishes, Antarctic fishes possess a remarkable suite of compounds that confers greater resistance to freezing. Water temperatures at McMurdo Sound average −1.87°C throughout the year. With a water surface covered by sea ice and the shallow bottom with anchor ice, contact of these surfaces by fishes could seed ice formation in their tissues. Working at McMurdo Sound, Arthur DeVries (1970, 1971a, 1971b) pioneered biochemical and physiological studies on the compounds present in the blood and body tissues of Antarctic fishes that confer resistance to freezing.

Oxygen solubility exists in an inverse correlation with water temperature—the colder the water gets, the more oxygen it can hold. This, coupled with a sedentary lifestyle and the associated depressed metabolic rates seen in notothenioids, has permitted an extraordinary development in the icefishes of the family Channichthyidae. Hemoglobin and myoglobin—proteins that enhance the oxygen transport and storage capacity of the body—are not present in these fishes. Consequently, the gills of channichthyids appear very pale (see Plate 35.4).

One family of notothenioids, the plunderfishes (Harpagiferidae), displays remarkable similarities with the cottid sculpins common in North Temperate and Arctic freshwater and marine environments. Both groups are benthic, relatively sessile ambush predators that are especially sensitive to motion in prey items (mainly crustaceans). Both groups have large, spiny heads with well-developed eyes and mouths, and a small, slender body with a rounded caudal fin; and they move using robust pectoral fins. This may be taken to be an example of convergent evolution, in which the harpagiferids could be seen as Antarctic equivalents of the cottids in the Northern Hemisphere (Wyanski and Targett, 1981).

KEY POINTS AND CONNECTIONS

• The pelagic realm of the open ocean is traditionally divided into three major zones—epipelagic, mesopelagic, and bathypelagic. Although the fishes of the epipelagic zone are not especially diverse, this realm is home to the largest fishes, including species of sharks and tunas, whereas the plankton community counts among its membership the larvae of a number of species as well as a few tiny adult forms. Epipelagic fishes are usually powerful swimmers, well adapted for covering vast amounts of ocean distance. Efficient buoyancy control, involving a well-developed swim bladder, is a common characteristic of epipelagic bony fishes. Lacking significant substrates to which they can orient for purposes of reproduction, epipelagic fishes tend to display comparatively simplistic reproductive behavior—usually consisting of broadcast spawning of large quantities of eggs and sperm.

The locomotor adaptations common to epipelagic fishes and the role of the swim bladder in buoyancy control are discussed in Chapter 19.

• The mesopelagic zone is a vast, dimly lit region, generally ranging from 200 to 1,000 m in depth. Fishes here tend to have well-developed eyes to gather what little light is available; much of this light arises from bioluminescent activity among the mesopelagic fauna. Mesopelagic fishes usually show well-developed photophore patterns, indicating the importance of visual communication at these depths. The swim bladder is often present and may be useful in buoyancy control during the extensive diurnal vertical migrations that characterize many mesopelagic species. The constraints of maintaining adequate gas volume at the great pressures found at these depths have resulted in the reduction or elimination of a gas-filled bladder in several species, however.

The structure and function of bioluminescent organs are discussed in Chapter 18.

• Many species of mesopelagic fishes extend their range deeper, into the bathypelagic zone. Many other families include species that are particularly representative of this deepest of pelagic realms—a region of total darkness, save for bioluminescent activity. Fishes here are characterized by small eyes, flaccid bodies with reduced skeletons, and reduced swim bladder development. The comparative scarcity of individuals that make themselves available either as prey or potential mates has resulted in bizarre adaptations. Enormous mouths filled with elongate teeth and highly distensible stomachs enable some species to capture and ingest just about anything that comes their way. Ceratioid anglerfishes have maximized the likelihood of successful reproduction through the evolution of males that become parasitic on conspecific females.

• The deep ocean bottom consists of the continental slopes and the abyssal plains. The fish fauna seen here appears to be derived from benthic and benthopelagic fish families of shallower zones. A common body form that appears to be particularly successful in fishes of the deep demersal realm is characterized by a large head, well-developed pectoral fins, and an elongate body with long dorsal and anal fins.

Refer to Chapter 19 for a discussion of the correlation between mode of swimming and body form.

• Marine fishes of high latitudes must tolerate temperatures that may be below the freezing point of freshwater. These fishes have evolved distinct molecular compounds to combat the threat of freezing of their tissues. The polar fish fauna of the Northern and Southern Hemispheres have little in common. Whereas the fish fauna of the northern polar seas has affinities with fishes from lower latitudes and includes an anadromous component, Antarctic waters have a distinctive, more endemic fish fauna, which is dominated by members of the suborder Notothenioidei.

The distinctive features of the notothenioids are discussed in Chapters 16 and 24; in Chapter 24, adaptations that permit resistance to freezing are covered in more detail.

FISH LINKS

http://www.lifesci.ucsb.edu/~biolum/ The Bioluminescence Web Page—an excellent site maintained at the University of California Santa Barbara that provides a wealth of information and fascinating photographs.

BUILDING AN ICHTHYOLOGY LIBRARY

For fishes near the surface:

Parin, N. V. 1970. *Ichthyofauna of the epipelagic zone.* Israel Program for Scientific Translations, Jerusalem. (Hard to find, but worth having if you can locate a copy.)

For fishes in the deep ocean:

Merrett, N. R., and R. L. Haedrich. 1997. *Deep-sea demersal fish and fisheries.* Chapman and Hall, London.
Randall, D. J., and A. P. Farrell (Eds.).1997. *Deep-sea fishes.* Academic Press, San Diego.

For polar fishes:

diPrisco, G., E. Pisano, and A. Clarke (Eds.). 1998. *Fishes of Antarctica: A biological overview.* Springer Verlag Italia, Milano.
Eastman, J. T. 1993. *Antarctic fish biology: Evolution in a unique environment.* Academic Press, San Diego.
Reynolds, J. T. (Ed.). 1997. *Fish ecology in Arctic North America* (American Fisheries Society Symposium 19). American Fisheries Society, Bethesda, MD.

REFERENCES

Andriashev, A. P. 1987. A general review of the Antarctic bottom fish fauna, pp. 357–372. In *Fifth Congress of European Ichthyologists, Proceedings*, S. O. Kullander and B. Fernholm (Eds.). Swedish Museum of Natural History, Stockholm.

Angel, M. V. 1997. What is the deep sea? pp. 1–41. In *Deep-sea fishes*, D. J. Randall and A. P. Farrell (Eds.). Academic Press, San Diego.

Blaber, S. J. M., and C. M. Bulman. 1987. Diets of fishes of the upper continental slope of eastern Tasmania: Content, calorific values, dietary overlap and trophic relationships. *Mar. Biol. 95:* 345–356.

Bortone, S. A., P. A. Hastings, and S. B. Collard. 1977. The pelagic *Sargassum* ichthyofauna of the eastern Gulf of Mexico. *Northeast Gulf Sci. 1*(2): 60–67.

Breder, C. M., Jr., and D. E. Rosen. 1966. *Modes of reproduction in fishes*. Natural History Press, Garden City, NY.

Caley, M. J., M. H. Carr, M. A. Hixon, T. P. Hughes, G. P. Jones, and B. A. Menge. 1996. Recruitment and the local dynamics of open marine populations. *Ann. Rev. Ecol. Syst. 27:* 477–500.

Childress, J. J., S. M. Taylor, G. M. Caillet, and M. H. Price. 1980. Patterns of growth, energy utilization, and reproduction in some meso- and bathypelagic fishes off Southern California. *Mar. Biol. 61:* 27–40.

Clarke, A. 1980. A reappraisal of the concept of cold adaptation in polar marine invertebrates. *Biol. J. Linn. Soc. 14:* 77–92.

———. 1983. Life in cold water: The physiological ecology of polar marine ectotherms. *Oceanogr. Mar. Biol. Ann. Rev. 21:* 341–453.

Cohen, D. M., and R. L. Haedrich. 1983. The fish fauna of the Galapagos thermal vent region. *Deep Sea Res. 30:*371–379.

Cowen, R. K., K. M. M. Lwiza, S. Sponaugle, C. B. Paris, and D. B. Olson. 2000. Connectivity of marine populations: Open or closed? *Science 287:*857–859.

Crame, J. A. 1993. Bipolar molluscs and their evolutionary implications. *J. Biogeogr. 20:* 145–161.

Daniels, R. A., and J. H. Lipps. 1982. Distribution and ecology of fishes of the Antarctic Peninsula. *J. Biogeogr. 9:* 1–9.

Darling, K. F., C. M. Wade, I. A. Stewart, D. Kroon, R. Dingle, and A. J. Leigh-Brown. 2000. Molecular evidence for genetic mixing of Arctic and Antarctic subpolar populations of planktonic foraminifers. *Nature 405:* 43–47.

DeForges, B. R., J. A. Koslow, and G. C. B. Poore. 2000. Diversity and endemism of the benthic seamount fauna in the southwest Pacific. *Nature 405:* 944–947.

DeVries, A. L. 1970. Freezing resistance in Antarctic fishes, pp. 320–328. In *Antarctic ecology*, Vol. 1, M. W. Holdgate (Ed.). Academic Press, London.

———. 1971a. Glycoproteins as biological antifreeze agents in Antarctic fishes. *Science 172:* 1152–1155.

———. 1971b. Freezing resistance in fishes, pp. 157–190. In *Fish physiology*, Vol. VI, W. S. Hoar and D. J. Randall (Eds.). Academic Press, New York.

deYoung, B., M. Heath, F. Werner, F. Chai, B. Megrey, and P. Monfray. 2004. Challenges of modeling ocean basin ecosystems. *Science 304:* 1463–1466.

Dooley, J. K. 1972. Fishes associated with the pelagic *Sargassum* complex, with a discussion of the *Sargassum* community. *Contr. Mar. Sci. Univ. Texas 16:* 1–32.

Eastman, J. T. 1993. *Antarctic fish biology: Evolution in a unique environment*. Academic Press, San Diego.

———. 1995. The evolution of Antarctic fishes: Questions for consideration and avenues for research. *Cybium 19:* 371–389.

———. 1997. Comparison of the Antarctic and Arctic fish faunas. *Cybium 21:* 335–352.

———. 2000a. Fishes on the Antarctic continental shelf: Evolution of a marine species flock? *J. Fish Biol. 57*(Suppl. A): 84–102.

———. 2000b. Antarctic notothenioid fishes as subjects for research in evolutionary biology. *Antarctic Sci. 12:* 276–287.

Elliott, N. G., K. Haskard, and J. A. Koslow. 1995. Morphometric analysis of orange roughy (*Hoplostethus atlanticus*) off the continental slope of southern Australia. *J. Fish Biol. 46:* 202–220.

Farquhar, G. B. (Ed.). 1970. *Proceedings of an international symposium on biological sound scattering in the ocean*. Maury Center for Ocean Science, Department of the Navy, Washington, DC.

Gartner, J. V., Jr., R. E. Crabtree, and K. J. Sulak. 1997. Feeding at depth, pp. 115–193. In *Deep-sea fishes*, D. J. Randall and A. P. Farrell (Eds.). Academic Press, San Diego.

Goode, G. B., and T. H. Bean. 1896. Oceanic ichthyology: A treatise on the deep-sea and pelagic fishes of the world. *Spec. Bull. U.S. Nat. Mus. 2.*

Haedrich, R. L. 1997. Distribution and population ecology, pp. 79–114. In *Deep-sea fishes*, D. J. Randall and A. P. Farrell (Eds.). Academic Press, San Diego.

———, and N. R. Merrett. 1988. Summary atlas of deep-living demersal fishes in the North Atlantic Basin. *J. Nat. Hist. 22:* 1325–1362.

Helfman, G. S., B. B. Collett, and D. E. Facey. 1997. *The diversity of fishes*. Blackwell Science, Malden, MA.

Holeton, G. F. 1974. Metabolic cold adaptation of polar fish: Fact or artefact? *Physiol. Zool. 47:* 137–152.

Johnson, G. D., and E. B. Brothers. 1993. *Schindleria*: A paedomorphic goby (Teleostei: Gobioidei). *Bull. Mar. Sci. 52:* 441–471.

Koslow, A. J. 1996. Energetic and life-history patterns of deep-sea benthic, benthopelagic, and seamount-associated fish. *J. Fish Biol. 49*(Suppl. A): 54–74.

———. 1997. Seamounts and the ecology of deep-sea fisheries. *Amer. Sci. 85:* 168–176.

Lindberg, D. R. 1991. Marine biotic interchange between the northern and southern hemispheres. *Paleobiology 17:* 308–324.

Mann, D. A., and S. M. Jarvis. 2004. Potential sound production by a deep-sea fish. *J. Acoust. Soc. Am. 115:* 2331–2333.

Mansueti, R. 1963. Symbiotic behavior between small fishes and jelly-fishes with new data on that between the stromateid *Peprilus alepidotus* and the scyphomedusa *Chrysaora quinquecirrha*. *Copeia 1963:* 40–80.

Marshall, N. B. 1970. *The life of fishes*. Universe Books, New York.

———. 1971. *Explorations in the life of fishes*. Harvard University Press, Cambridge, MA.

———. 1980. *Deep sea biology: Developments and perspectives*. Garland STPM Press, New York.

Merrett, N. R. 1994. Reproduction in the North Atlantic oceanic ichthyofauna and the relationship between fecundity and species sizes. *Env. Biol. Fishes 41:* 207–245.

———, and R. L. Haedrich. 1997. *Deep-sea demersal fish and fisheries*. Chapman and Hall, London.

Miller, M. J., and J. D. McCleave. 1994. Species assemblages of leptocephali in the subtropical convergence zone of the Sargasso Sea. *J. Mar. Res. 52:* 743–772.

Møller, P. R., J. G. Nielsen, and I. Fossen. 2003. Fish migration: Patagonian toothfish found off Greenland. *Nature 421:* 599.

Montgomery, J., and N. Pankhurst. 1997. Sensory physiology, pp. 325–349. In *Deep-sea fishes*, D. J. Randall and A. P. Farrell (Eds.). Academic Press, San Diego.

Moore, J. A. 1999. Deep-sea finfish fisheries: Lessons from history. *Fisheries 24*(7): 16–21.

———, and P. M. Mace. 1999. Challenges and prospects for deep-sea finfish fisheries. *Fisheries 24*(7): 22–23.

Nikolsky, G. V. 1963. *The ecology of fishes*. Academic Press, New York.

Nishimura, K., and N. Hoshino. 1999. Evolution of cannibalism in the larval stage of pelagic fish. *Evol. Ecol. 13:* 191–209.

Ornellas, A. B., and R. Coutinho. 1998. Spatial and temporal patterns of distribution and abundance of a tropical fish assemblage in a seasonal *Sargassum* bed, Cabo Frio Island, Brazil. *J. Fish Biol. 53*(Suppl. A): 198–208.

Parin, N. V. 1970. *Ichthyofauna of the epipelagic zone*. Israel Program for Scientific Translations, Jerusalem.

———. 1984. Oceanic ichthyologeography: An attempt to review the distribution and origin of pelagic and bottom fishes outside continental shelves and neritic zones. *Arch. Fisch Wiss. 35*(1): 5–41.

Pelster, B. 1997. Buoyancy at depth, pp. 195–237. In *Deep-sea fishes*, D. J. Randall and A. P. Farrell (Eds.). Academic Press, San Diego.

Power, G. 1997. A review of fish ecology in Arctic North America. *Amer. Fish. Soc. Symp. 19:* 13–39.

Rex, M. A. 1981. Community structure in the deep-sea benthos. *Ann. Rev. Ecol. Syst. 12:* 331–353.

Rice, A. L., and P. J. D. Lambshead. 1992. Patch dynamics in the deep-sea benthos: The role of a heterogeneous supply of organic matter, pp. 469–497. In *Aquatic ecology: Scale, pattern and process*, P. S. Giller, A. G. Hildrew, and D. G. Raffaelli (Eds.). Blackwell Science, Oxford.

Scholander, P. F. , W. Flagg, W. Walters, and L. Irving. 1953. Climactic adaptation in arctic and tropical poikilotherms. *Physiol. Zool. 26:* 67–92.

Sherman, K., L. M. Alexander, and B. D. Gold (Eds.). 1990. *Large marine ecosystems: Patterns, processes, and yields*. American Association for the Advancement of Science, Washington, DC.

———, and A. M. Duda. 1999. Large marine ecosystems: An emerging paradigm for fishery sustainability. *Fisheries 24*(12): 15–26.

Smith, P. J., R. I. C. C. Francis, and M. McVeagh. 1991. Loss of genetic diversity due to fishing pressure. *Fisheries Res. 10:* 309–316.

VanDover, C. L. 2000. *The ecology of deep-sea hydrothermal vents*. Princeton University Press, Princeton, NJ.

Wolff, T. 1961. The deepest recorded fishes. *Nature 190:* 283.

Wyanski, D. M., and T. E. Targett. 1981. Feeding biology of fishes in the endemic Antarctic Harpagiferidae. *Copeia 1981:* 686–693.

Zierenberg, R. A., M. W. W. Adams, and A. J. Arp. 2000. Life in extreme environments: Hydrothermal vents. *Proc. Natl. Acad. Sci. USA 97:* 12961–12962.

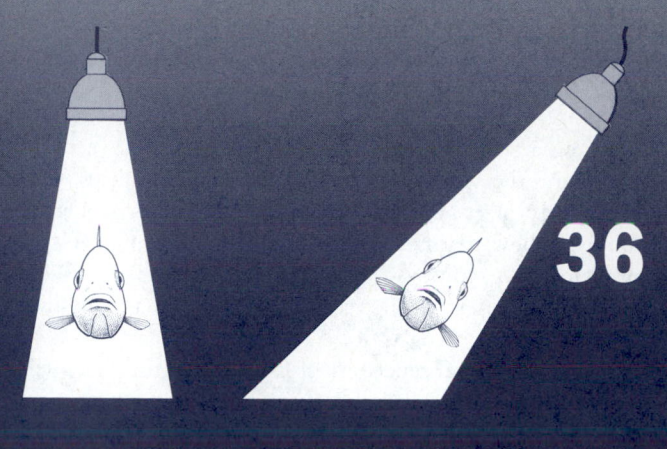

BEHAVIOR I: GETTING ALONG IN THE PHYSICAL WORLD

BEHAVIOR: THE ICHTHYOLOGICAL PERSPECTIVE

LOCOMOTOR RESPONSES TO STIMULI

Phototaxis
Geotaxis
Electrotaxis and Magnetotaxis
Thigmotaxis
Rheotaxis
 Optomotor Response
Chemotaxis

HOMING AND MIGRATION

Homing Behavior
Migration
 Diadromous Migrations
 Potamodromous Migrations
 Oceanodromous Migrations
 Vertical Migrations

As a child, your parents were always warning you to "behave," meaning to conform to an expected pattern of conduct. The assumption that behavior always has a connotation of propriety might tempt us to assume that it has less meaning for other organisms. In fact, all living things behave, and the truest meaning of the word can only be understood in an evolutionary context. *Behavior* is simply what creatures *do* in order to survive; it is the actions they pursue in order to better integrate themselves with their environment. For all creatures, behavior has consequences in terms of the achievement of anticipated outcomes. We may attempt to impose human norms of behavior on our closest animal companions ("Bad dog!"), but somehow fishes have escaped our moralistic perspective. We never accuse sharks of evil intent, even if they have just devoured our diving companion. Sharks are just doing what they do in order to ensure that they can make some contribution to their future gene pool.

In the next two chapters, we will take this perspective on behavior, trying to understand what fishes do, the actions they take in time and space, to ensure their survival. In earlier chapters, we have considered the morphological mechanisms and physiological processes that enable fishes to perceive their environment and the morphological and physiological means by which they can make appropriate responses. Here, we will investigate those actions themselves—what, for want of a better word, and free of any judgment other than that of adaptive value, we term behavior.

BEHAVIOR: THE ICHTHYOLOGICAL PERSPECTIVE

Behavior is a challenging topic, as it is both diffuse and all-encompassing. It constitutes the actions of organisms that proceed from their adaptive capacities, with the ultimate objective being the survival of the individual and its genes, the population, and the species. If a more technical definition is required, we could define **behavior** as the sum of all the motor responses of the organism to all of the external and internal stimuli it encounters. These responses would include locomotion, changes in color or appearance, or the secretion of compounds for the purposes of defense or attraction of others. For fishes, these actions are nothing if not extraordinary in their sophistication—the manifestation of an artistry that compels many of us to keep fishes in an aquarium just so we can appreciate the beauty of both their appearance *and* their action.

Early humans no doubt were practitioners of behavioral biology, for their lives depended on the ability to optimize their foraging activities like any other organism. To be able to tell when the salmon were due on spawning grounds would even become the focus of significant ceremonial activity among Native Americans of the Pacific Northwest (see Chapter 11, Going Deeper). Mass migrations of mammals, birds, fishes, and insects are conspicuous examples of behaviors that the ancients held to be important knowledge. As our abilities to discern the mysteries of fish behavior developed, we began to understand the importance of such intrinsic actions in the lives of fishes beyond our own immediate needs.

The behavior of fishes has been the subject of much classical ethological work, **ethology** being the study of behavior. In the introduction to his fine book on the topic of fish behavior, Tony Pitcher (1986) defined the essentials from which the study of behavior emerges—genes, motivation, and development. Behavioral study ultimately devolves onto the most fundamental of issues—getting food while at the same time avoiding becoming food, and getting a mate who will cooperate in getting your genes into the next generation. In many instances—particularly well studied among fishes—the two spheres are not mutually exclusive: Mating activity sometimes exacts a price in the form of increased vulnerability to predation.

Researchers recognize that behavior is a topic that can only be approached in the context of other features of the organism. Behavioral study rests on foundation disciplines such as genetics, evolution, ecology, physiology, and biochemistry. The intrinsic actions of fishes must be interpreted in the context of all these disciplines. Behavioral adaptations that promote survival in environments as fluctuating as estuaries or stream headwaters or as consistent and unvarying as the ocean depths demonstrate a connection between behavior and the structuring of ecological communities. The psychology of fishes can be explained in a physiological context—how mating behavior is a function of levels of circulating hormones—or in a biochemical context—how those hormones act at the receptor sites on the membranes of target cells.

The behavioral repertoire of fishes constitutes a bewildering array of intrinsic actions arising from a particular motivational state. Each action has a meaning that can be explained at all levels of biological investigation. Perhaps the best approach is to group the behaviors of fishes according to specific categories of action. We must not forget the essential unity of behavioral study, however—each of these topics informs our discussion of all the others.

LOCOMOTOR RESPONSES TO STIMULI

We will begin our investigation of the manifold aspects of fish behavior by considering the various kinds of stimuli that prompt responses—chiefly locomotor responses—in fishes. Fishes are attracted or repelled by a variety of external environmental cues as well as by internal signals. These stimuli result in changes in the motivational state of the animal, thus resulting in some net response, here observed as motion. The simplest and most predictable locomotor response is termed **kinesis** (plural *kineses*). This involves a simple, undirected enhancement of locomotor activity in response to some stimulus. Though most frequently associated with invertebrate phyla, instances of kinesis are recorded among vertebrates. For example, the ammocoetes larvae of lampreys, though blind, have light-sensitive cells in the skin of the tail. The larvae normally live buried in the sediment. Removal of the overlaying sediment prompts burrowing or swimming movements that will eventually return them to their proper location away from illumination. The larvae of pelagic bony fishes also exhibit kinetic responses.

Directed, nonrandom movements are termed **taxis** (plural *taxes*). Taxes are distinguished from **tropisms**—slow, growth-related orientations in response to stimuli, as witnessed in plants or sessile invertebrates (Arnold, 1974; Lyon, 1904). Taxes can be positive or negative in relation to the stimulus that initiated them. Specific taxes to be considered include **phototaxis** (response to light), **geotaxis** (gravity), **electro-** and **magnetotaxis** (electromagnetic fields), **thigmotaxis** (contact with substrate or some other object), **rheotaxis** (current flow), and **chemotaxis** (chemicals). Given the diversity of habitats occupied by fishes, it would be natural to assume that these environments differ widely in their influence on the lives of fishes. For example, thigmotaxis would be of little relevance to a solitary, pelagic predator, such as a shark, but would be essential to burrowing forms, such as garden eels (see Plate 36.1).

· · · · ·
Phototaxis

Fishes are, for the most part, highly visual creatures. It is therefore not surprising that phototaxis plays such an important role in their lives. Guthrie (1986), in his review of the role of vision in fish behavior, stated it succinctly:

> Visually mediated behavior ranges in complexity from the simple alerting or attentive state evoked by any novel but non-specific visual event, to the triggering of an elaborate fixed action pattern by means of a highly specific visual signal which consists of a precise grouping of visual properties. (p. 75)

Positive phototaxis is apparent in many species of diurnal pelagic fishes that swim toward lights at night. Some fisheries exploit this behavior by displaying lights near nets or pumps to gather fishes as they make the final phototactic response of their lives. Fish harvesters have learned that the response of the target species may vary with the intensity of illumination, or that some fishes are more attracted to blue lights than to white. Other pelagic fishes may exhibit *negative* phototaxis, as in the case of mesopelagic species—such as lanternfishes, hatchetfishes, and bristlemouths—that undertake extensive diurnal vertical migrations (see Chapter 35).

Negative phototaxis is also associated with the cryptic behavior seen in many species, such as those found in small streams. Such species as madtom catfishes (*Noturus*), sandrollers (*Percopsis*), sculpins (*Cottus*), and rock basses (*Ambloplites*), retreat from light and spend their days beneath cut-banks, rocks, vegetation, or other suitable cover. Stream pollution, in the form of discarded cans or other refuse, can often have a positive effect on species abundance by providing a greater amount of cover for such negatively phototactic species (Burr and Mayden, 1982; Kottcamp and Moyle, 1972). The larvae of one species of amphidromous goby (*Rhinogobius brunneus*) are positively phototactic at low light levels, but negatively phototactic at high light levels. This makes their downstream migration more secure, as they will rise to the surface at dawn or dusk, when light levels are low, but will stay in the nest if light levels are high (Iguchi and Mizuno, 1990, 1991). Similar phototactic responses have been observed in the ayu (*Plecoglossus altivelis*), which is also amphidromous (Koyma, 1978). On their reproductive journey, American and European eels (*Anguilla rostrata and A. anguilla*, respectively) swim at depths greater than 400 m in the daytime but ascend to 50 to 215 m at night (Tesch, 1978).

In general, light sensitivity plays a major role in governing the time of day (or night) when fishes will be active (Glass et al., 1986). Squirrelfishes (*Holocentrus*) and morays (*Muraena*) are members of the coral reef "night shift," coming out and foraging when most of the reef fish community have retired for the evening. It is interesting to note that the most conspicuous and brightly colored members of the coral reef community tend to be diurnal, and that grazing herbivores, such as surgeonfishes (Acanthuridae) and parrotfishes (Scaridae), and cleaners such as wrasses also restrict their activities to the daytime hours (Goldman and Talbot, 1976). Many species are classified as **crepuscular**—that is, with a peak activity that occurs at dusk or dawn. This is especially true of piscivorous species (Helfman, 1986). The influence of light levels on reef fish behavior is dramatically demonstrated under the conditions of a solar eclipse: Diurnal species quickly seek shelter, and nocturnal species emerge (Jennings et al., 1998).

In a few cases, fishes may seasonally shift from day-active to night-active lifestyles. This is especially true for species living at high latitudes, such as salmonids and cyprinids; decreasing temperatures prompt an increase in nocturnal activity (Erikson, 1978; Greenwood and Metcalfe, 1998; Metcalfe et al., 1999; Muller, 1978a). Fishes that demonstrate territorial behavior that is mediated by light intensity (i.e., aggressive responses to conspecifics that decrease at lower light levels) may experience a seasonal shift in territory dimensions resulting from a change in the ambient light levels (Valdimarsson and Metcalfe, 2001).

Fishes living in the mesopelagic and bathypelagic realms, where the ambient light level is very low to nonexistent, may still have well-developed visual systems, as bioluminescence plays a significant role in their lives (see Chapters 18 and 35). Mate recognition is often a function of the perception of species-specific patterns of photophore emissions. The 1989 symposium "Light and Life in the Sea" included several noteworthy contributions on the subject of bioluminescence, visual systems, and behavior of marine fishes (Herring et al., 1990).

Light functions as a *zeitgeber* (literally, "time giver") when it is responsible for the entrainment and maintenance of endogenous rhythmicity in fishes. Not all circadian rhythms are necessarily cued by light, however (Muller, 1978a, 1978b; Palmer, 1974). One of the great mysteries in biology is the mechanism by which biological rhythms are established and maintained. A major breakthrough in biorhythm research came with the discovery that a plant photopigment-like protein known as **cryptochrome** appears to be the mechanism that relays light signals to clocks controlling circadian rhythms in animals (Barinaga, 1998; Somers et al., 1998; Thresher et al., 1998; Whitmore and

Sassone-Corsi, 1999). Seasonal and diel aspects of reproductive physiology are governed at least in part by the perception of changes in ambient light in many fishes, especially those in temperate latitudes (Billard and Breton, 1978), light being detected through the pineal or parietal organ of the brain (see Chapter 26).

Geotaxis

The posture of most fishes is governed by their reaction to light stimuli acting in combination with their sensitivity to gravity. Gravity awareness is mediated through the membranous labyrinth (see Chapter 21). Fishes will tend to maintain their body position with the dorsal side up—a positive phototaxis of a sort, termed the **dorsal light reaction**. If the light source is shifted to the side, the fish will adjust its body position accordingly (Fig. 36.1). Geotaxis will prevent the fish from achieving extreme angles of inclination, but this can be overridden if the membranous labyrinth is removed. In some cases, fishes whose inner ears have been removed will orient their dorsal surface to the bottom if the light source is positioned there. The upside-down catfishes of the family Mochokidae (see Plate 10.5) contradict these general observations, because their normal mode of behavior is to maintain a body position with the ventral side up. What do you think the effect of this is on the pattern of obliterative countershading (see Chapter 18) seen in this family?

Electrotaxis and Magnetotaxis

Among the most mysterious sensory modalities that fishes possess—one that we are unable to truly appreciate, as it is generally lacking in tetrapods—is that of electroreception. Largely because of the presence of electroreceptive components in the lateral line, fishes are capable of detecting emitted electrical signals (Hopkins, 1974; see Chapter 21). As such, they display **electrotaxis**, also known as **galvanotaxis**. The knowledge that most fishes will swim toward the positive electrode of a direct current field has been used in a variety of ways, particularly in fisheries that employ direct current systems, often in conjunction with lights, to attract fishes to nets or pumps.

Although there are only a few species of fishes—such as the torpedo rays (Torpedinidae), electric catfish (Malapteruridae), and electric eels (Electrophoridae)—that generate an electrical current powerful enough to stun prey, the capacity to emit and detect weak electrical signals is much more widespread, perhaps ubiquitous, among fishes (see Chapter 21). The Australian lungfish (*Neoceratodus forsteri;* see Plate 7.2) uses electroreceptive ampullary organs to detect hidden prey (Watt et al., 1999). In some families—the elephant-fishes (Mormyridae), gymnarchid eels (Gymnarchidae), and knifefishes (Gymnotidae), for example—the emission and detection of weak electrical fields is well known as a valuable means of orientation and communication among individuals

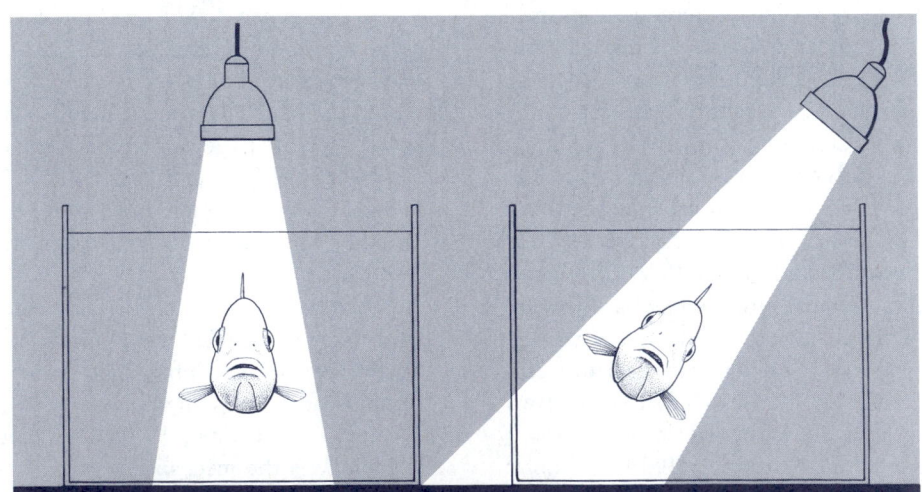

FIGURE 36.1

Representation of dorsal light reaction. When light is directly overhead, the fish assumes a normal upright position, but as the light is moved to the side, the fish maintains the same relative position in regard to the light by assuming an oblique posture.

that often live in muddy or turbid waters where visual contact is minimal (Hopkins, 1974). In the mormyrids, such impulse-generating capacity, found in modified muscle tissues, is associated with an especially large cerebellum that is instrumental in the coordination of the system. In most fishes, the brain accounts for only 0.1 to 0.3 percent of the total body weight; the mormyrid brain comprises 3 percent of the total body weight (Nilsson, 1999). Remarkably, the electroreceptive capabilities of mormyrids enable discrimination between the ohmic and capacitive properties of an object—the detection of capacitive properties being especially useful in the exploration of properties of living tissue (Heiligenberg, 1993; von der Emde, 1990). Patterns of electric organ discharge have been used to establish individual identities (Crawford, 1992), and dominance hierarchies among weakly electric fishes are established and maintained via characteristic discharge patterns (Hagedorn and Zelnick, 1989). The peculiar swimming motions of gymnotiform fishes, which incorporate a rigid body posture and rapid backward movements, apparently enhance electroreceptor scanning for potential prey items (Lannoo and Lannoo, 1993; Nanjappa et al., 2000).

The capacity of fishes to respond to *magnetic* fields is a more recent discovery—one that has provoked much debate and discussion. Training experiments have demonstrated the ability of elasmobranch fishes to orient to magnetic fields (Kalmijn, 1977, 1985). Iron-rich magnetic crystals, termed **magnetite,** have been found in the snout of yellowfin tuna (Walker et al., 1984) and in the lateral line of Atlantic salmon (Moore et al., 1990). Receptor cells incorporating magnetite have been identified in the olfactory lamellae of rainbow trout, and the configuration of these receptors in relation to the magnetite crystals has been deemed optimal for magnetoreception (Diebel et al., 2000). The possibilities are intriguing: Do fishes use magnetotaxis for orientation to the Earth's magnetic field in the course of long-distance migrations? There are many who claim that fishes—and other vertebrates, such as birds—are doing exactly that. Elasmobranchs remain among the most intensively studied fishes with respect to magnetotaxis. Although magnetic particles have been detected in the inner ear of some sharks (Hanson et al., 1990), these species do not appear to use specific magnetodetectors but rather rely on their highly tuned electroreceptors. These provide a means of magnetoreception by the detection of electrical currents generated as the fish swims through magnetic fields (Heiligenberg, 1993).

Thigmotaxis

The way that most fishes orient themselves spatially in the environment seems to suggest that they usually keep a certain amount of distance between themselves and other individuals. As we shall see in our consideration of schooling behavior in Chapter 37, this orientation is often remarkable in its precision and coordination. Most fishes, however, will eventually evince some association with a substrate of some sort. Spawning behavior will bring schooling species into contact with each other or with the substrate, if only for a short time. As we observed in Chapter 35, a few pelagic species, such as the man-of-war fish (*Nomeus*) and the sargassum fish (*Histrio histrio;* see Plate 13.9) show a special affinity for different kinds of living substrates of the pelagic realm. Benthic fishes, as might be expected, display a high affinity for particular features of a given substrate, consistent with their cryptic mode of existence. Sculpins, pricklebacks, worm eels, darters, and loaches will lurk among rocks and vegetation, whereas some species, such as skates, weeverfishes, lizardfishes, or flounders, will bury themselves in loose sediments (see Plate 33.1).

Sculpins provide a remarkable case study in substrate affinity. When placed in a rectangular glass aquarium lacking in available cover, individuals will squeeze themselves into a corner in order to achieve contact with as many surfaces as possible. When several individuals are placed in these conditions, they will pile up in an attempt to maximize contact with some other object. Some cryptic species, such as madtom catfishes, are observed clustering under a single shelter even when multiple shelters are available. This suggests that thigmotactic behavior may be preferentially directed toward conspecifics rather than the substrate. Benthic fishes are able to use their lateral-line mechanoreceptors to detect minute vibrations through the substrate that would betray the presence of prey. Mottled sculpins (*Cottus bairdi*) that were experimentally blinded were shown to be able to home in on vibrations elicited by prey. They accomplished this by placing their mandibles on the substrate and moving toward the source of vibrations in a series of short hops accompanied by biting actions directed toward the bottom (Janssen, 1990). Such behavior is obviously of benefit in nocturnal foraging species.

The intimate association between fishes and their immediate surroundings may figure significantly in courtship behavior as well. The transmission of acoustic emissions through the substrate during courtship may be more important than the sounds that travel through the water (Whang and Janssen, 1994). Thigmotactic behavior may be used in conjunction with rheotaxis (discussed later) to assist benthic fishes in maintaining optimal position in flowing waters.

Rheotaxis

Reaction to a current of water is commonly observed among stream fishes and is even recognized in some still-water forms (Arnold, 1974; Lyon, 1904). Some species will cruise as a school around a tank without a current, proceeding in seemingly random directions, but will break up and orient as individuals into a current when water is made to flow into the tank. Rheotaxis is an important component of the life history of anadromous species such as salmon. In fact, the orientation either toward or away from the current may be genetically determined. Rainbow and cutthroat trout, obtained either from streams draining into a lake or from those draining out of a lake, were shown to possess different preferred orientations to currents, which corresponded with their natural habitat: Trout from inlet streams have to migrate downstream to reach feeding areas in the lake, whereas those in outlet streams have to travel upstream to reach the lake (Raleigh, 1971). In the catadromous *Anguilla*, a positive rheotaxis is evident in the elver stage, when the newly transformed individuals first enter freshwater. This facilitates their eventual arrival at suitable upstream habitats. Near the end of the eel's life, a negative rheotaxis helps guide the animal back downstream to the ocean. Ocean currents as powerful as the Gulf Stream undoubtedly influence the migration of coastal species by providing rheotactic cues. Various species of herrings may migrate into prevailing currents to spawning grounds, from which the eggs and larvae can drift back to the rearing areas.

Knowledge of rheotactic responses in fishes is useful for various harvesting techniques. Carp can be induced to move from one holding pond to another by lowering the water level in one pond, thus causing a flow between it and the other. Rheotaxis is exploited in salmonid hatcheries to separate brood stock, which tend to exhibit a greater rheotactic response than individuals that are not in breeding condition. Reaction to water currents has also been used to guide fish away from intakes at dams, thus increasing the survival of downstream migrants. Although rheotactic response is commonly perceived as being mediated primarily by visual and tactile cues, lateral-line receptors have been demonstrated to play a significant role as well (Montgomery et al., 1997).

Optomotor Response

Optomotor response is, in a sense, a combination of phototactic and rheotactic behaviors. Fishes exhibiting optomotor response may hold a position in a current by getting a visual fix on some object on the stream bottom or side. Thigmotactic responses may also be used, as some species may brush their fins against the bottom. Optomotor response is best seen in experimental situations, where it is often used to induce fish to swim at a given velocity. A typical test device consists of a circular tank with a transparent outside wall and a circular partition inside, forming a narrow swimway in which the fishes can move adjacent to the side of the tank.

If a circular curtain marked with vertical light and dark stripes is rotated around the tank, test fishes of many species will predictably move with a given stripe or will move at a slightly different speed than the stripes. They will always move in the direction of the stripes, however.

As might be expected, optomotor response is especially well developed in stream fishes, but it is also evident in pike, perch, and other species that inhabit still waters. Smelts and several species of marine fishes, including certain herrings, cods, and jacks, show the optomotor response as well (Shaw and Tucker, 1965). Optomotor response is not as apparent in benthic species, and young fishes show a stronger response than adults (Harden-Jones, 1963). Fishery scientists are keenly aware of the impact of the optomotor response exhibited in commercially desirable species. Fishes demonstrating this behavior can outpace and eventually avoid trawl nets; this finding suggests that the most efficient trawling gear should minimize contrast with the background, thus eliminating a visual reference point for optomotor response.

• • • • • •
Chemotaxis

Given what is known about the sensitivity of the olfactory and gustatory senses, as described in Chapter 22, it is not surprising that the chemical environment figures significantly in the lives of fishes. Chemosensation is important for the location of food, for breeding and parental care, and for avoiding danger. At least in certain parts of their range, Pacific salmons show a strong escape response in the presence of the odor of bears, seals, and humans (Brett and MacKinnon, 1954). When extracts of mammalian skin or solutions of an amine that appears to be the escape stimulus are placed in fishways or natural streams containing migrating adult salmon, many of the fish will abandon their positions and retreat downstream. The odor of a pike is known to elicit violent swimming, darting to the surface, or tonic immobility (a freezing behavior in response to a perceived threat) in minnows and poeciliids.

Although certain predator avoidance behaviors, such as schooling or fright responses to visual stimuli, are known to be innate, the responses to olfactory stimuli from potential predators appear to be the result of conditioning. Predator-naïve fathead minnows (*Pimephales promelas*) exposed to pike odors showed no response, whereas wild-caught individuals with previous pike experience did (Chivers and Smith, 1994, 1995). Predator-naïve minnows were able to learn appropriate fright response behaviors from conspecifics, but they also learned fright responses from other, sympatric species, such as sticklebacks, that had experienced predators (Mathis et al., 1996). Chemical secretions released by an individual that elicit a particular behavioral or physiological response in a conspecific are termed **pheromones** (see Chapter 26). Fishes make good use of these compounds in their lives (Hara, 1975, 1992, 1993). Characteristic of the ostariophysan fishes is a fright substance termed **schreckstoff** that is released when they are injured or severely stressed (Pfeiffer, 1962, 1977). This alarm pheromone is released from special cells in the skin when it is damaged. It can be sensed in minute quantities and usually elicits flight from the area, tightening of schooling orientation, or concealment (Heczko and Seghers, 1981). Some ostariophysan species exhibit a characteristic "fin flicking" behavior, in which the detection of alarm substances from a conspecific elicits rapid movements of the dorsal, caudal, and pectoral fins with no resulting change in body position (Brown et al., 1999). The production of alarm pheromones is widespread in the Ostariophysi, with some species producing secretions that are highly species-specific in the elicitation of a fright response, whereas other species produce substances more generally recognized among members of the superorder (Verheijen and Reuter, 1969).

It has been well documented that alarm substances involuntarily released for the purpose of alerting conspecifics to the threat of predation also serve to attract predators (Mathis et al., 1995). It has been hypothesized that the release of an alarm signal may work to the ultimate benefit of prey species if it attracts other predators to the scene. The arrival of other predators may interrupt the initial predator attack, and such interference may enable the prey species to escape (Chivers et al., 1996). Fathead minnows can even detect alarm pheromones in the feces of predators that have fed on members of their own species. One such predator, the northern pike (*Esox lucius*), apparently compensates for this by restricting defecation to areas where it does not forage (Brown et al., 1995, 1996).

Alarm signaling by means of pheromones would appear to be of greatest benefit to smaller fishes, such as the ostariophysans. Darters have also been demonstrated to possess alarm substances secreted from the skin (Commens and Mathis, 1999). Just as schreckstoff acts as a stimulus to elicit avoidance behavior, pheromones emitted by fishes may serve to attract potential mates. For example, pheromones that attract mates have been identified in the urine of masu salmon (*Oncorhynchus masou;* Yambe et al., 1999). New technologies incorporating sophisticated methods of analysis will permit a more precise isolation and identification of the specific compounds that elicit behavioral responses in fishes. Such studies will also greatly expand opportunities for investigating the role of chemical communication in the structuring of ecological systems (Zimmer, 2000; Zimmer and Butman, 2000).

HOMING AND MIGRATION

Homing Behavior

Fishes of widely divergent phylogenetic affinities, occurring across a broad spectrum of habitats, have been shown to possess a home range where they spend most of their time. In some instances, this range may be as small as a few meters of stream or a tide pool; in others, it could be as large as several kilometers of coastline or lakeshore. Home range size appears to be largely a function of energy requirements and, as such, may vary seasonally as energy demands change throughout the year (Bradbury et al., 1995). The establishment and maintenance of a home range suggests that fishes are capable of recognizing their boundaries; any or all of the aforementioned taxes may figure significantly in the ability of fishes to seek out and maintain home ranges. Fishes can successfully navigate within the home range, but they can also apply these navigational powers in returning to the home range if displaced—what we term **homing behavior.** Although many cases of homing ability in fishes are truly remarkable, we cannot assume that all fishes that establish a discrete home range may be able to relocate it if they are displaced.

Homing behavior, as in the well-known case of salmon (discussed later), may involve migrations of several thousand kilometers. This is one of the most extensively researched areas of fish ethology. Homing is an integral component of the spectrum of migratory behaviors exhibited by fishes (Wootton, 1990, 1992). The ability to return to recognizable habitats figures significantly in the successful transition from one habitat type to another (Fig. 36.2). The phenomenon of homing has been demonstrated for a diverse array of fishes in a variety of habitats, including sculpins and blennioids on intertidal rocky shores (Dooley and Collura, 1988; Gibson, 1969, 1982, 1999; Griffiths, 2003; Griffiths et al., 2004; Horn and Gibson, 1988), sunfishes in stream pools (Gerber and Haynes, 1988; Gerking, 1958; Hasler and Wisby, 1958), and rockfishes on offshore banks (Carlson and Haight, 1972).

The cues by which fishes orient and navigate probably include a combination of physical and chemical entities and simple random search patterns. Searching along the shore of a small lake would eventually take a fish home, but this does not account for movement toward the home range across a large lake or from one rock reef across deep ocean water to the home reef. Displacement experiments using marine shore fishes typically demonstrate homing capacities over fairly short distances. The cunner (*Tautogolabrus adspersus*), a species with a fairly small home range, can return to its home from a distance of 4 km (Green, 1975). The yellowtail rockfish (*Sebastes flavidus*) will return to home sites, on

• •

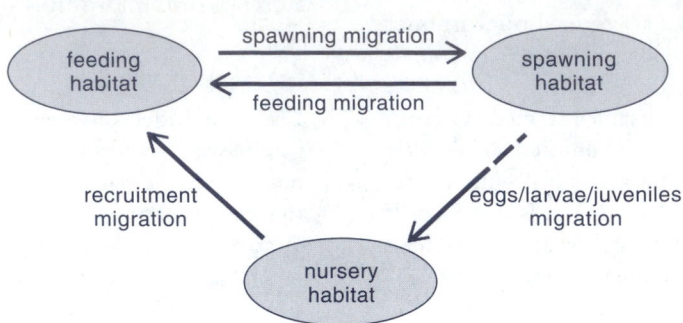

FIGURE 36.2

The role of migration in the transitions between feeding, spawning, and nursery habitats in fishes. (Adapted from Wootton, 1992.)

• •

a return journey that often involves transits across deep water, when displaced up to 22.5 km (Carlson and Haight, 1972).

Inhabitants of rocky intertidal shores are typically benthic species with very limited home ranges (Gibson, 1999; Ralston and Horn, 1986), yet they exhibit the ability to return to home areas even when displaced considerable distances. Tidepool sculpins show an ability to return to home areas when displaced up to 100 m (Green, 1971; Horn and Gibson, 1988). Shallow-water blennioids of the Canary Islands were able to locate home areas when displaced up to 200 yards (Dooley and Collura, 1988). Cunner, yellowtail rockfish, and possibly tidepool sculpins (*Oligocottus maculosus;* Green, 1971) demonstrate homing capabilities even after being held in artificial surroundings for several months. More than half of the 20 species that inhabit rock pools of temperate Australian shores exhibited homing behavior; some of these species retained the ability to return to home sites after displacement for as long as 214 days (Griffiths, 2003).

Learning and memory—specifically, its application to landmark recognition—is integral to orientation and homing in fishes (Odling-Smee and Braithwaite, 2003). A classic example are the studies by L. R. Aronson on the orientation abilities of the frillfin goby (*Bathygobius soporator*). Aronson demonstrated that this rock-pool inhabitant learned the location of pools adjacent to its home pool during high-tide foraging excursions and was able to jump to these pools at low tide when threatened (Aronson, 1951, 1971). Among the most remarkable orientation and navigation abilities are those demonstrated by the blind Mexican cavefish *Astyanax mexicanus.* The detection of near-field wave perturbations with its lateral line (see Chapter 21) gives this species a perceptual range of less than 0.05 m, but it successfully navigates home ranges of up to 30 m. This feat it apparently accomplishes by encoding a spatial map in which it is able

to learn and remember an ordered sequence of landmarks (Burt de Perera, 2004).

• • • • •

Migration

Many fishes undertake long migrations on an annual or seasonal basis for a variety of reasons. Here we adopt a broad definition of **migration** that will include any mass movement from one habitat to another with characteristic regularity in time or according to life history stage. This broad definition allows the inclusion of both active and passive mass movements, whether they are extensive seasonal or annual changes of habitat or short-term, short-distance travels. Usually, these migrations are round trips. Although migrations are best understood in an ecological and adaptive context, we normally investigate them from the perspective of the behavior of the individual migrant (Kennedy, 1985). Most migrations occur for the purposes of food gathering, reproduction, or adjustment to temperature, and are sometimes referred to simply as feeding, breeding, or wintering migrations, respectively. The linkage between breeding activities and migration is an especially intimate one (Leggett, 1985). In many fishes, migrations take place on an annual basis, but in others, individuals are involved in certain changes of habitat only at certain developmental stages. Other migrations are of a diurnal nature, such as the vertical migrations of mesopelagic fishes in the deep scattering layer (see Chapter 35).

Energetic considerations (see Chapter 23) figure significantly in discussions of migration, especially when migration is performed in the overall context of reproduction. Just as we have envisaged the propulsive mechanism of a migrating fish (see Chapter 19, Going Deeper), we must also consider its "fuel tank." Can the journey be made on only one tank of fuel, or will the distance and pace of the trip necessitate several refueling stops? The metabolic costs associated with foraging and digestion along the way will reduce resources that could otherwise be allocated to directed swimming. Fishes being ectothermic creatures, the season—and hence the prevailing water temperature—will influence their swimming speed and the distance they can cover during their migration. For example, fishes that spawn in the spring at temperate latitudes, such as shads and most cyprinids, are migrating under cool temperature conditions; early migrants have reduced swimming capacity and diminished swimming speeds (Lucas and Baras, 2001).

The most commonly used descriptive terminology for classifying migratory fishes is that of Myers (1949), as discussed by McDowall (1988):

Diadromous. These are truly migratory fishes that travel between the sea and freshwater. The three following subtypes are included:

- *Anadromous.* Diadromous fishes that spend most of their lives in the sea and migrate to freshwater to breed.
- *Catadromous.* Diadromous fishes that spend most of their lives in freshwater and migrate to the sea to breed.
- *Amphidromous.* Diadromous fishes in which migration from freshwater to the sea—or vice versa—is not for the purpose of breeding, but occurs regularly at some defined stage of the life cycle. Both marine and freshwater amphidromy exist, as first noted by McDowall (1988).

Potamodromous. These are truly migratory fishes that migrate wholly within freshwater (nondiadromous), usually from larger river systems into smaller tributaries.

Oceanodromous. These are truly migratory fishes that live and migrate wholly in the sea (nondiadromous).

The first three terms, *diadromous, anadromous,* and *catadromous,* have gained considerable usage among ichthyologists and fish biologists, whereas the latter two are not used as extensively. Several terminologies have preceded these, overlap with them, or further refine the meanings. Other terms have been introduced to classify fishes with respect to habitat salinity without regard to migration (see Table 29.1). Many of these have not been adopted widely by fish biologists because fishes vary so greatly in the degree to which they are constrained by the physical environment (Haedrich, 1983; McHugh, 1967). Much of this history is discussed by McDowall (1988), who accepted the definitions as presented in Table 29.1 and pointed out the distinction between marine and freshwater amphidromy.

Active migration requires adjustment to environmental conditions, including temperature, light, water current, salinity, alkalinity, and other sensory modalities. Major physiological changes accompany the transition from marine to freshwater and occur only at certain stages in the life cycles of diadromous fishes (see Chapter 25), so these fishes may not be euryhaline in the strictest sense of the term, as is often assumed (McDowall, 1988). It is well known that Pacific salmon orient to their natal stream largely by smell (see Going Deeper). Other fishes are known to orient to the angle of the sun in the sky, as this provides an orientation mechanism in the open sea, where olfactory and other cues are unavailable. Tunas have a well-developed pineal "eye" (see Chapter 26) on top of the skull that may function in orientation.

As discussed earlier in this chapter, the magneto- and electrotactic capacities of fishes are becoming increasingly well known. Some sharks are known to detect weak electrical fields; this could explain their migrations in deep or turbid water, where vision cannot be used. Such subtle capabilities make it possible for a diverse array of fishes to extract meaningful information aiding in migration from what initially seems to be a clueless environment.

Migrations enable organisms to take advantage of additional habitat and to avoid adverse conditions. Disadvantages of migration include not only the expenditure of energy, but exposure to predators along the migration route, losses of passively drifting eggs and larvae that end up in unsuitable regions, and the necessity to adjust to changes in the physical environment. Migratory species store extra energy in the form of fat prior to the onset of migrations. In fishes as well as birds, only a small proportion of the migrating population achieves maximum levels of lipid storage. These individuals, known as **leaders,** initiate the migration, compelling less nutritionally prepared individuals to follow (Shulman and Love, 1999). Some species, such as Pacific salmons and lampreys, mobilize all their resources and commit them irreversibly to the acts of migration and spawning, resulting in their death following spawning—the condition known as semelparity (see Chapter 27).

Ultimately, migration places individuals in the best or necessary location for given biological activities at the appropriate time. Nocturnal feeding near the sea surface by species that spend the day in the twilight of the depths places these species in contact with a high concentration of food organisms and allows them to feed when most avian and other diurnal predators are inactive. (As might be expected, several dark-colored species of sea birds also feed at night.) Predation on migratory individuals, especially by piscivorous fishes, is a highly visual phenomenon, as demonstrated in studies by Gregory and Levine (1998). Here, instances of predation were higher in clear stretches of the

Fraser River than in turbid stretches. Feeding migrations of far-ranging species allow them to reach or to follow food resources that can support far larger populations than the resources of the spawning or nursery grounds. Spawning migrations are important for species with restrictive reproduction requirements and for those whose eggs and larvae must drift to specific nursery grounds.

Diadromous Migrations

From a physiological and behavioral perspective, diadromous migrations are remarkable because fishes move from one medium to another, and the distances involved often require amazing feats of orientation, navigation, and precise recognition of home spawning areas.

Anadromy

Variation occurs in the degree of anadromy in North American salmonids and clupeids. Cutthroat trout (*Oncorhynchus clarkii*) in the Pacific Northwest often descend only into coastal estuaries before moving upstream to spawn in late fall or early spring. Steelhead trout (*O. mykiss*) range over much of the North Pacific Ocean and migrate hundreds of kilometers in the larger rivers. They often return for two or three successive years to spawn. The various species of Pacific salmon spawn from the coastal estuaries just above salt water to hundreds of kilometers upstream, depending on the species (see Going Deeper, this chapter and Chapter 11), and all die after spawning. In the Sacramento River of California, four distinct runs of Chinook salmon (*O. tsha-wytscha*) occur, one for each season (Moyle and Williams, 1990). In the eastern United States, the hickory shad (*Alosa chrysochloris*) migrates downstream to coastal areas of low salinity but has not been taken at sea. The Alabama shad (*A. alabamae*), however, has been taken far offshore in the Gulf of Mexico.

Although the sturgeons of the genus *Acipenser* are often considered anadromous, many only descend to estuarine areas and seldom venture any distance beyond. By employing

---

GOING DEEPER · The Salmonids: A Special Case of Migration and Homing Ability

Two different kinds of causative agents—**proximate** *causes* and *ultimate causes*—are invoked in studies on the seasonal movements of animals (Honore and Klopfer, 1990; Lack, 1954). Proximate causes are those that trigger and guide the migratory response. By **ultimate** causes, we mean those historical or selective factors of the environment that result in the establishment and maintenance of homing or migratory behavior. These factors will be considered in the context of what is perhaps the best known example of homing as a correlate of migratory behavior in fishes—the salmons of the genera *Oncorhynchus* and *Salmo*. What morphological and physiological mechanisms do these fishes have at their disposal to interpret the environment and make appropriate responses so that return to their home streams is possible after spending several years in the open ocean? What caused the development of such remarkable homing capabilities in salmon?

Proximate Factors

The migratory behavior of both anadromous and catadromous fishes encompasses both the freshwater and marine realms. As expected, a variety of environmental cues can be perceived by migrating species such as salmons. These cues enable them to navigate over great distances of the ocean until they find themselves in the vicinity of their natal streams. The salmons then ascend these streams, making the right choices at each tributary, until they eventually return, often to the very same stretch of gravel bed that they emerged from as alevins. It is apparent that young salmons acquire an "imprint" of their natal stream sometime during their migration downstream to the ocean. Fishes probably use their powerful olfactory capabilities here, so that the most recognizable feature of the natal environment is its smell. Researchers have demonstrated that salmon react to minute quantities of a chemical, **morpholine,** dissolved in the water. When fishes are artificially imprinted with this compound, it can be used to guide their movements to particular streams (Hasler and Scholz, 1978).

Olfactory imprinting was first advanced by A. D. Hasler and W. J. Wisby (1951) as a means of orientation. Since then, it has remained the most likely explanation for how fishes are able to distinguish different tributaries during their homeward migrations. When Atlantic salmon (*Salmo salar*) were released directly into the ocean and thus deprived of the olfactory cues that they would normally acquire during their downstream migration as smolts, they failed to return to a particular river (Mills, 1989). The mechanism that is most supported by the data on homing in streams is one in which imprinting is a sequential process, by which the smolt receives a continuous set of cues as it makes its way to the ocean. These cues then provide a continuum of reinforcing signals when the fish makes its return several years later as an adult. Other cues may prove essential in triggering upstream migration. For example, from as early as the middle of the 19th century, it has been known that Atlantic salmon do not initiate upstream movements until the stream has been freshened by a heavy fall of rain (Williamson, 1843).

In addition to olfactory cues, salmon on the high seas may employ a host of other cues, including water currents, temperature and salinity gradients, the position of the sun during the day or of the stars at night, and the pattern of polarized light in the daytime sky

a sophisticated procedure (laser ablation sampling—inductively coupled plasma—mass spectrometry, or LAS-ICP-MS) for assessing the concentration of strontium in the growth zones of skeletal elements, Veinott et al. (1999) were able to determine what proportion of the white sturgeon (*A. transmontanus*) population in the Fraser River of British Columbia actually undertakes marine migrations. They confirmed that the majority of the sturgeon population is not diadromous.

Catadromy

Catadromous migrations range from the short downstream movements of prickly sculpins in creeks tributary to the northeastern Pacific Ocean to the mid-ocean sojourns of the eels of the genus *Anguilla*. The New Zealand whitebait, or inanga (*Galaxias maculatus*), resembles the prickly sculpin in descending just to the estuarine areas, whereas the species of the tropical mullet genera *Joturus* and *Agonostomus* move long distances down large, tumultuous rivers to spawn in the coastal waters of Mexico and Central America.

The longest catadromous migrations are those of the eels of the genus *Anguilla*. Although they occur in the Indo-West Pacific and on both sides of the north Atlantic, their biology is best known from the Atlantic (see Chapter 9, Going Deeper). Geology has conspired to make the journey of the eels longer and longer, as the Atlantic seafloor has been spreading since the mid-Mesozoic. European larvae have to go much farther and delay their metamorphosis much longer than American eels (Fig. 36.3). Some fish biologists proposed that the European eel (*A. anguilla*; see Plate 9.4) was in fact not a distinct species, and that European eel populations did not spawn, owing to the distance they would have had to travel to the Sargasso Sea. Molecular studies have verified the integrity of the European eel species, however (Avise et al., 1986; Mank and Avise, 2003). Up to four or five species of anguillids occur together in the Indo–West Pacific, and their relationships are probably even more complex.

(Wootton, 1992). As discussed earlier in this chapter, the evidence for magnetoreception in fishes has become incontrovertible. Researchers have demonstrated that fishes can perceive polarized light and that specialized cone cells in the retina that are sensitive to ultraviolet light appear to be involved in this process (Hawryshyn, 1992; Land, 1991; see Chapter 20). Some researchers have suggested that salmon do not require sophisticated orientation mechanisms on the open ocean; by simply engaging in a more or less random search pattern, they will eventually locate the appropriate coastline that will take them to the outlet of their home stream.

Ultimate Factors

Although we may never be able to discover just *how* migrations and homing behavior evolved as they did in the salmon, this does not prevent us from speculating on *why* such behavior developed as it did. Fishes evolved migratory behavior—and the consequent homing capacities—for a variety of reasons (McKeown, 1984). Fishes may migrate as a means of availing themselves of highly productive feeding areas located at some distance from their natal areas. Fishes are therefore able to rapidly respond to food resources that may be seasonal in their availability. Migrations may also be timed to avoid the appearance of predators at feeding locations. Fishes—especially in temperate to polar regions—also undoubtedly migrate as a means of avoiding unfavorable climatic conditions and, consequently, optimizing reproductive success. Much of our concern about the degradation of coastal estuarine environments and the rivers that drain into them stems from the dependence of many coastal and anadromous species on estuaries and associated freshwater drainages as nursery areas. Here, egg and larval stages may be ensured a greater measure of security at critical times in the life history of the species.

We must also consider the geological and climatic history of the Earth's surface in our speculations on the origin of migrations—especially in the case of salmonid fishes. When the migratory behavior of salmonids is considered in the context of their phylogenetic history, it becomes apparent that salmonids first evolved in freshwater. Some genera—*Oncorhynchus* and *Salmo*, for example—eventually developed anadromy. The origins of anadromous behavior may have been instigated by climatic cooling during the mid-Cenozoic (Stearley, 1992). Although the widespread glaciation that covered much of the Northern Hemisphere during the Pleistocene may have had an influence on migratory routes, and certainly influenced dispersal and present-day distribution patterns, the origins of the genus *Oncorhynchus* predate the Pleistocene by several million years (Stearley, 1992). Overwintering in the ocean, with its much greater feeding opportunities, became an integral component of salmonid life history, even though these fishes continued to spawn in stream habitats. Long-distance migration, facilitated by extraordinarily acute homing capabilities, thus became the essential adaptive feature of the Salmonidae. When we consider the modern distribution of the genera *Salmo* and *Oncorhynchus*, plus the fact that *Salmo* is a repeat spawner (**iteroparous**), whereas most oceangoing species of *Oncorhynchus* die after one spawning (**semelparous**), it becomes apparent that anadromy evolved more than once within this family.

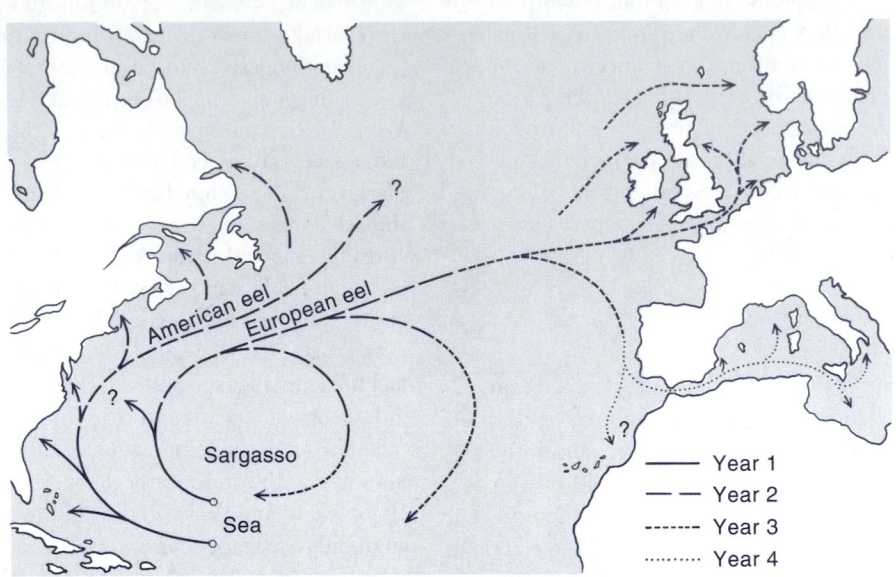

FIGURE 36.3
Movement of larval American and European eels from closely adjacent areas of hatching in the Sargasso Sea.
Note the differences in direction and larval stage duration, mostly due to the largely passive transport of the
larvae by ocean currents.

Amphidromy

Species of the families Galaxiidae, Retropinnidae, and Aplochitonidae in New Zealand, Australia, and South America have long been thought to have a diadromous lifestyle. However, McDowall (1988) believed that only *Galaxias maculatus* is truly catadromous. Most other species do not actually enter marine water enough to be considered diadromous. McDowall considered most of the other galaxiids, the retropinnids, and the aplochitonids to be amphidromous. The biology of one aplochitonid, the Tasmanian whitebait (*Lovettia seali*), is sufficiently well known that it is considered anadromous. The aplochitonids from southern South America are considered amphidromous, but they have not been well studied. McDowall (1988) also considered that these species may not be closely related to the other members of the family in Tasmania.

In Japan, the ayu (*Plecoglossus altivelis*), an annual fish, descends to lay its eggs in the lower reaches of the rivers. The larvae are carried out to sea, where they remain for a few months. They then move back into the middle and upper reaches of the rivers to grow. When mature, they migrate back downstream to spawn and die soon thereafter. This life cycle is best described as freshwater amphidromy (McDowall, 1988). Marine amphidromy is seen in the mullets (*Mugil cephalus*), which spawn at sea and whose young

spend a short period in fresh or low-salinity water before becoming fully marine.

Potamodromous Migrations

Whereas temperate river systems are characterized by several species of anadromous salmonids, tropical river systems are home to large species of characoids and catfishes that exhibit potamodromous migrations, moving hundreds of kilometers upstream to spawn. On the other hand, smaller stream inhabitants, such as minnows, suckers, trouts, and darters, may move only a few meters from feeding to spawning grounds. The **restricted movement paradigm**, which asserts that the adults of stream-dwelling fishes are largely sedentary and possess restricted home ranges, has come to dominate current research on the behavior and ecology of stream ichthyofauna (Rodríguez, 2002). Radio telemetry has proven valuable in studies of the restricted movements of stream fishes (see Bunnell et al., 1998). Many temperate fishes move downstream to deeper holes or larger streams for the winter. Bluegill sunfishes undertake seasonal migrations offshore in order to exploit available vegetation as cover from predators (Goodyear and Bennett, 1979). Often, the migrations can be in any direction, as long as the desired substrate, depth, or temperature is present. In tropical freshwater rivers, the migrations of some species are closely attuned to the seasonal changes in river flow. Opportunities for feeding and breeding are presented by flooding during the rainy season in African, South American, and Asian floodplain rivers (Goulding, 1980; Lowe-McConnell, 1975; Welcomme, 1979). Shallow, flooded expanses lateral to the river provide fruits and small animals that greatly increase the food supply for these fishes (see Chapters 31 and 32).

Oceanodromous Migrations

Migrations of this sort are usually pursued by pelagic species of the open ocean. This category encompasses species such as the clupeoids, which are more associated with the continental shelf (Fig. 36.4), and large, powerful, open-ocean migrators such as the scombroids. Two reproductive patterns, characterized by the nature of migration and larval dispersal, can be discerned in pelagic fishes (Helfman et al., 1997). Some species have a **cyclonic** pattern of migration, in which adults feed in a given area and move against the current to a spawning area, where ample food will be available

to nourish the developing larvae (Fig. 36.4). The larvae then drift back to the adult area as they grow and mature. This pattern is characteristic of species living in higher latitudes, so that spawning is a seasonal event. Herrings (*Clupea*, family Clupeidae) are a good example of this pattern of migration; they also demonstrate a pattern common to oceanodromous fishes—subdivision into a number of spawning stocks that may or may not overlap. Herrings are remarkable in the scope of migratory behaviors that they display. Not only do the adults pursue migrations of considerable distances, but they perform daily vertical migrations as well (see further). Following spawning, the larvae drift inshore, where they engage in annual migrations as juveniles, moving closer to shore in the summer and further offshore in the winter. On reaching a suitable size, they join the adult population in their open-ocean migrations (McKeown, 1984).

The **anticyclonic** pattern of migration characterizes species living at lower latitudes, such as the tropical scombroids. Here, spawning occurs repeatedly during the course

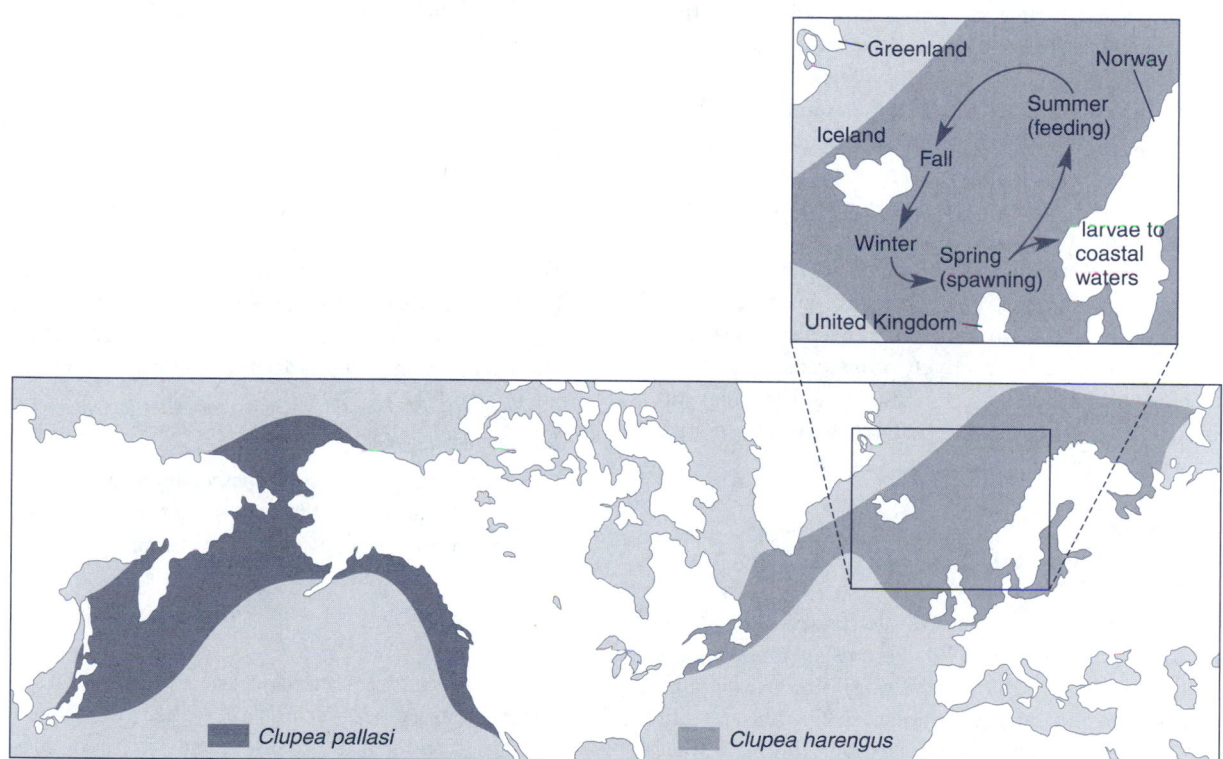

FIGURE 36.4

Geographical distribution of Atlantic (*Clupea harengus*) and Pacific (*C. pallasi*) herring. Because of similarities in anatomy and life history, some researchers do not consider Atlantic and Pacific herring to be separate species. Detail shows migration pattern of one of the three spawning stocks from the North Sea. (Adapted from McKeown, 1984.)

of the adult migration, and the eggs and larvae develop and feed in the same general areas as the adults (Helfman et al., 1997). In both cases, the recruitment of larvae into the adult population is dependent on the complex interplay of a number of environmental variables, and the year class strength fluctuates in response to these variables.

Many large marine fishes move in schools north and south on an annual basis, following seasonal temperature profiles, usually at the opposite times of the year in each hemisphere. The long north–south continental margins span a broad range of temperatures and often induce cool upwelling currents. Large pelagic fishes have adapted their life cycles to these features. The champion long-distance migrators are the tunas, whose migratory route may encompass an entire ocean basin (Nakamura, 1969). In the North Pacific, albacore (*Thunnus alalunga*) move north along the California and Oregon coast during the summer, are in mid-ocean during the winter, and move back to the coast the following June. A portion of the population may not return, but will continue west, moving northeast along the coast of Japan in May and June and returning to mid-ocean by fall. Bluefin tunas (*Thunnus thynnus*) in the Atlantic Ocean move through temperature boundaries on their migrations from Florida to Norway and between Norway and Spain. Returns of tags from Pacific populations of bluefin tunas indicate that some individuals move between Baja California and Japan and between the western coast of Australia and northern New Zealand. Tagged and recaptured marlins (*Makaira*) have been determined to have covered more than 1,850 km in a month while traveling from California to Hawaii. Blue sharks (*Prionace glauca*) tagged off New York and New Jersey have been recovered two years later off South America and off Africa—distances of 3,200 and 4,800 km, respectively. Casey et al. (1980) reported a blue shark tagged off Montauk, New York, that was recaptured 5,844 km away on the coast of Liberia nine months later—having covered an average of 21 km per day!

Vertical Migrations

Many species are known to make periodic vertical migrations between deeper and shallower waters. Although this phenomenon has been documented in freshwater fishes (Brett, 1971; Neverman and Wurtsbaugh, 1994), it is most commonly associated with marine fishes (see Chapter 35). Atlantic herrings (*Clupea harengus*) inhabit deeper water during the daytime and migrate to the surface at sunset. Around midnight, they return to the depths, only to move back to the shallows again near dawn. As the sun rises, they return back to the depths, where they spend the rest of the day (McKeown, 1984).

Nocturnal feeding migrations to the surface layers enable many mesopelagic fishes to take advantage of food organisms in the epipelagic zone, where photosynthesis sustains a higher level of productivity. Moving into shallower water at night enables these fishes to feed at a time when they are less vulnerable to predators. By descending into the cooler depths when they are not feeding, fishes reduce their metabolic rates, thus permitting a greater proportion of ingested food to be available for growth and reproduction. Many stream, lake, and marine shore fishes make similar, but much less spectacular inshore movements at night, only to return to deeper water during the day (see Chapter 37). During the reproductive season, similar nocturnal movements bring spawning adults to appropriate spawning habitats, such as gravel bars, rocks, or sandy areas. Species with pelagic eggs, such as clupeids, hiodontids, and percichthyid basses, move into shallower waters upstream, and the fertilized eggs and subsequent larvae drift passively downstream, eventually taking up residence in their juvenile habitat.

Just as we are beginning to understand the mechanisms of migration, many of our most spectacular migrating fishes are rapidly declining in numbers. Concern for their survival is stimulating drastic action to conserve the fishes that are left and providing an impetus to study them more thoroughly and thus ensure their survival (Moyle and Leidy, 1992; Moyle and Williams, 1990; Stiassny and de Pinna, 1994).

KEY POINTS AND CONNECTIONS

• The study of behavior is termed ethology. Included in ethological study are the directed responses of fishes to specific chemical and physical features of the environment. These include response to light (phototaxis), gravity (geotaxis), electromagnetic fields (electro- and magnetotaxis), physical features mediated through tactile senses (thigmotaxis), current (rheotaxis), and chemicals (chemotaxis).

Chapters 20–22 cover the sensory modalities in fishes that enable the aforementioned taxes.

• Fishes are generally perceived as possessing a home range—a limited portion of their habitat that they most frequent. Homing ability is present in many fishes as a means of facilitating migration to and from the location of their home range. Homing behavior has been demonstrated for a diverse array of fishes associated with a range of freshwater and marine habitats.

• Migration encompasses any movement from one habitat to another with characteristic regularity in time or according to life history stage. Movements may be active or passive and may include seasonal or annual changes in habitat or more limited sojourns. Diadromous movements—between freshwater and marine habitats—for the purpose of reproduction include anadromous and catadromous types of migration. Amphidromous migration includes movements between freshwater and marine habitats for reasons other than reproduction. Potamodromous migration takes place entirely within freshwater, usually from larger portions of river systems into smaller tributaries, whereas oceanodromous migration takes place entirely within the marine realm.

The evolutionary, biogeographical, and ecological perspectives on diadromy are considered in Chapters 4, 29, and 31, respectively.

BUILDING AN ICHTHYOLOGY LIBRARY

Lucas, M. C., and E. Baras. 2001. *Migration of freshwater fishes.* Blackwell Science, Oxford.

Pitcher, T. J. (Ed.). 1993. *Behavior of teleost fishes* (2nd ed.). Chapman and Hall, London.

Tricas, T., and S. H. Gruber. 2001. The behavior and sensory biology of elasmobranch fishes: An anthology in memory of Donald Richard Nelson. *Env. Biol. Fishes 60*(1–3).

REFERENCES

Arnold, L. R. 1974. Rheotropism in fishes. *Biol. Rev. 49:* 515–576.

Aronson, L. R. 1951. Orientation and jumping behavior in the gobiid fish, *Bathygobius soporator. Amer. Mus. Nov. 1486:* 1–22.

———. 1971. Further studies on orientation and jumping behavior in the gobiid fish *Bathygobius soporator. Ann. N.Y. Acad. Sci. 188:* 378–392.

Avise, J. C., G. S. Helfman, N. C. Saunders, and L. S. Hales. 1986. Mitochondrial DNA differentiation in North Atlantic eels: Population genetic consequences of an unusual life history pattern. *Proc. Natl. Acad. Sci. USA 83:* 4350–4354.

Barinaga, M. 1998. Clock photoreceptor shared by plants and animals. *Science 282:* 1628–1630.

Billard, R., and B. Breton. 1978. Rhythms of reproduction in teleost fish, pp. 31–53. In *Rhythmic activity in fishes,* J. E. Thorpe (Ed.). Academic Press, New York.

Bradbury, C., J. M. Green, and M. Bruce-Lockhart. 1995. Home ranges of female cunner, *Tautogolabrus adspersus* (Labridae), as determined by ultrasonic telemetry. *Can. J. Zool. 73:* 1268–1279.

Brett, J. R. 1971. Energetic response of salmon to temperature. A study of some thermal relations in the physiology and freshwater ecology of sockeye salmon (*Oncorhynchus nerka*). *Amer. Zool. 11:* 99–113.

———, and D. MacKinnon. 1954. Some aspects of olfactory perception in migrating adult coho and spring salmon. *J. Fish. Res. Bd. Can. 11:* 310–318.

Brown, G. E., D. P. Chivers, and R. J. F. Smith. 1995. Localized defecation by pike: A response to labeling by cyprinid alarm pheromone? *Behav. Ecol. Sociobiol. 36:* 105–110.

———, ———, and ———. 1996. Effects of diet on localized defecation by northern pike, *Esox lucius. J. Chem. Ecol. 22:* 467–475.

———, J.-G. J. Godin, and J. Pedersen. 1999. Fin-flicking behavior: A visual antipredator alarm signal in a characin fish, *Hemigrammus erythrozonus. Anim. Behav. 58:* 469–475.

Bunnell, D. B., Jr., J. J. Isely, K. H. Burrell, and D. H. VanLear. 1998. Diel movement of brown trout in a southern Appalachian river. *Trans. Am. Fish. Soc. 127:* 630–636.

Burr, B. L., and R. L. Mayden. 1982. Life history of the brindled madtom *Noturus miurus* in Mill Creek, Illinois (Pisces: Ictaluridae). *Am. Midl. Nat. 107:* 25–41.

Burt de Perera, T. 2004. Fish can encode order in their spatial map. *Proc. Roy. Soc. Lond. B 271:* 2131–2134.

Carlson, H. R., and R. E. Haight. 1972. Evidence for a home site and homing behavior of adult yellowtail rockfish, *Sebastes flavidus. J. Fish. Res. Bd. Can. 29:* 1011–1014.

Casey, J. H., H. L. Pratt, Jr., and C. Stillwell. 1980. The shark tagger. *Newsl. Coop. Shark Tagging Prog.* NOAA, NMFS, Narragansett, RI.

Chivers, D. P., and R. J. F. Smith. 1994. The role of experience and chemical alarm signaling in predator recognition by fathead minnows, *Pimephales promelas. J. Fish Biol. 44:* 273–285.

———, and ———. 1995. Free-living fathead minnows rapidly learn to recognize pike as predators. *J. Fish Biol. 46:* 949–954.

———, G. E. Brown, and R. J. F. Smith. 1996. The evolution of chemical alarm signals: Attracting predators benefits alarm signal senders. *Am. Nat. 148:* 649–659.

Commens, A. M., and A. Mathis. 1999. Alarm pheromones of rainbow darters: Responses to skin extracts of conspecifics and congeners. *J. Fish Biol. 55:* 1359–1362.

Crawford, J. D. 1992. Individual and sex specificity in the electric organ discharges of breeding mormyrid fish (*Pollimyrus isidori*). *J. Exp. Biol. 164:* 79–102.

Diebel, C. E., R. Proksch, C. R. Green, P. Nielson, and M. M. Walker. 2000. Magnetite defines a vertebrate magnetoreceptor. *Nature 406:* 299–302.

Dooley, J. K., and J. Collura. 1988. Homing behavior in tidepool fishes. *Underwater Nat. 17:* 3–6.

Erikson, L. O. 1978. Nocturnalism versus diurnalism—dualism within fish individuals, pp. 68–89. In *Rhythmic activity in fishes,* J. E. Thorpe (Ed.). Academic Press, New York.

Gerber, G. P., and J. M. Haynes. 1988. Movements and behavior of smallmouth bass, *Micropterus dolomieui,* and rock bass, *Ambloplites rupestris,* in south central Lake Ontario and two tributaries. *J. Freshw. Ecol. 4:* 425–440.

Gerking, S. D. 1958. The restricted movement of fish populations. *Biol. Rev. 34:* 221–242.

Gibson, R. N. 1969. The biology and behavior of littoral fish. *Oceanogr. Mar. Biol. Ann. Rev. 7:* 367–410.

———. 1982. Recent studies on the biology of intertidal fishes. *Oceanogr. Mar. Biol. Ann. Rev. 20:* 363–414.

———. 1999. Movement and homing in intertidal fishes, pp. 97–125. In *Intertidal fishes: Life in two worlds,* M. H. Horn, K. L. M. Martin, and M. A. Chotkowski (Eds.). Academic Press, San Diego.

Glass, C. W., C. S. Wardle, and W. R. Mojsiewicz. 1986. A light intensity threshold for schooling in the Atlantic mackerel *Scomber scombrus. J. Fish Biol. 29:* 71–82.

Goldman, B., and F. H. Talbot. 1976. Aspects of the ecology of coral reef fishes, pp. 125–154. In *Biology and geology of coral reefs,* Vol. IV, O. A. Jones and R. Endean (Eds.). Academic Press, New York.

Goodyear, C. P., and D. H. Bennett. 1979. Sun compass orientation of immature bluegill. *Trans. Am. Fish. Soc. 108:* 555–559.

Goulding, M. 1980. *The fishes and the forest: Explorations in Amazonian natural history.* University of California Press, Berkeley.

Green, J. M. 1971. High tide movements and homing behavior of the tidepool sculpin, *Oligocottus maculosus. J. Fish. Res. Bd. Can. 28:* 383–389.

———. 1975. Restricted movements and homing of the cunner, *Tautogolabrus adspersus* (Walbaum) (Pisces: Labridae). *Can. J. Zool. 53:* 1427–1431.

Greenwood, M. F. D., and N. B. Metcalfe. 1998. Minnows become nocturnal at low temperatures. *J. Fish Biol. 53:* 25–32.

Gregory, R. S., and C. D. Levine. 1998. Turbidity reduces predation on migrating juvenile Pacific salmon. *Trans. Am. Fish. Soc. 127:* 275–285.

Griffiths, S. P. 2003. Homing behavior of intertidal rockpool fishes in southeastern New South Wales, Australia. *Austral. J. Zool. 51:* 387–398.

———, R. J. West, A. R. Davis, and K. G. Russell. 2004. Fish recolonization in temperate Australian rockpools: A quantitative experimental approach. *Fish. Bull. 102:* 634–647.

Guthrie, D. M. 1986. Role of vision in fish behavior, pp. 75–113. In *The behavior of teleost fishes,* T. J. Pitcher (Ed.). Johns Hopkins University Press, Baltimore.

Haedrich, R. L. 1983. Estuarine fishes, pp. 183–207. In *Ecosystems of the world 26: Estuaries and enclosed seas,* B. H. Ketchum (Ed.). Elsevier, New York.

Hagedorn, M., and R. Zelnick. 1989. Relative dominance among males is expressed in the electric organ discharge of a weakly electric fish. *Anim. Behav. 38:* 520–525.

Hanson, M., H. Westerberg, and M. Oblad. 1990. The role of magnetic statoconia in dogfish (*Squalus acanthias*). *J. Exp. Biol. 151:* 205–218.

Hara, T. J. 1975. Olfaction in fish. *Prog. Neurobiol. 5:* 271–335.

———. 1992. *Fish chemoreception.* Chapman and Hall, London.

———. 1993. Chemoreception, pp. 191–218. In *The physiology of fishes,* D. H. Evans (Ed.). CRC Press, Boca Raton, FL.

Harden-Jones, F. R. 1963. The reaction of fish to moving backgrounds. *J. Exp. Biol. 40:* 437–446.

Hasler, A. D., and A. T. Scholz. 1978. Olfactory imprinting in coho salmon, pp. 356–369. In *Animal migration, navigation, and homing,* K. Schmidt-Koenig and W. T. Keeton (Eds.). Springer Verlag, Berlin.

———, and W. J. Wisby. 1951. Discrimination of stream odors by fishes and relation to parent stream behavior. *Am. Nat. 85:* 223–238.

———, and ———. 1958. The return of displaced largemouth bass and green sunfish to a "home" area. *Ecology 39:* 289–293.

Hawryshyn, C. W. 1992. Polarization vision in fish. *Amer. Sci. 80:* 164.

Heczko, E. J., and B. H. Seghers. 1981. Effects of alarm substance on schooling in the common shiner (*Notropis cornutus,* Cyprinidae), pp. 25–29. In *Ecology and ethology of fishes,* D. L. G. Noakes and J. A. Ward (Eds.). Junk, The Hague.

Heiligenberg, W. 1993. Electrosensation, pp. 137–160. In *The physiology of fishes,* D. H. Evans (Ed.). CRC Press, Boca Raton, FL.

Helfman, G. S. 1986. Fish behavior by day, night, and twilight, pp. 366–387. In *The behavior of teleost fishes,* T. Pitcher (Ed.). Johns Hopkins University Press, Baltimore.

———, B. B. Collette, and D. F. Facey. 1997. *The diversity of fishes.* Blackwell Science, Malden, MA.

Herring, P. J., A. K. Campbell, M. Whitfield, and L. Maddock. 1990. *Light and life in the sea.* Cambridge University Press, Cambridge, UK.

Honore, E. K., and P. H. Klopfer. 1990. *A concise survey of animal behavior.* Academic Press, San Diego.

Hopkins, C. D. 1974. Electric communication in fish. *Amer. Sci. 62:* 426–437.

Horn, M. H., and R. N. Gibson. 1988. Intertidal fishes. *Sci. Am. 256:* 64–70.

Iguchi, K., and N. Mizuno. 1990. Diel changes of larval drift among amphidromous gobies in Japan, especially *Rhinogobius brunneus. J. Fish Biol. 37:* 255–264.

———, and ———. 1991. Mechanisms of embryonic drift in the amphidromous goby, *Rhinogobius brunneus. Env. Biol. Fishes 31:* 295–300.

Janssen, J. 1990. Localization of substrate vibrations by the mottled sculpin (*Cottus bairdi*). *Copeia 1990:* 349–355.

Jennings, S., R. H. Bustamante, K. Collins, and J. Mallinson. 1998. Reef fish behavior during a total solar eclipse at Pinta Island, Galapagos. *J. Fish Biol. 53:* 683–686.

Kalmijn, A. J. 1977. The electric and magnetic sense of sharks, skates, and rays. *Oceanus 20*(3): 45–52.

———. 1985. Theory of electromagnetic orientation: A further analysis, pp. 525–563. In *Comparative physiology of sensory systems,* L. Bolis and R. D. Keynes (Eds.). Press Syndicate, University of Cambridge, Cambridge, UK.

Kennedy, J. S. 1985. Migration, behavioral and ecological, pp. 5–26. In Migration: Mechanisms and adaptive significance, M. A. Rankin (Ed.), *Contr. Mar. Sci. 27*(Suppl.). University of Texas Marine Science Institute, Port Aransas.

Kottcamp, G. M., and P. B. Moyle. 1972. Use of disposable beverage cans by fish in the San Joaquin Valley. *Trans. Am. Fish. Soc. 101:* 566.

Koyma, N. 1978. *Ecology of the ayu,* Plecoglossus altivelis. Chuokoron-sha Press, Tokyo.

Lack, D. 1954. *The natural regulation of animal numbers.* Oxford University Press, Oxford.

Land, M. F. 1991. Polarizing the world of fish. *Nature 353:* 118–119.

Lannoo, M. J, and S. J. Lannoo. 1993. Why do electric fishes swim backwards? An hypothesis based on gymnotiform foraging behavior interpreted through sensory constraints. *Env. Biol. Fishes 36:* 157–165.

Leggett, W. C. 1985. The role of migrations in the life history evolution of fish, pp. 277–295. In Migration: Mechanisms and adaptive significance, M. A. Rankin (Ed.), *Contr. Mar. Sci. 27*(Suppl.). University of Texas Marine Science Institute, Port Aransas.

Lowe-McConnell, R. H. 1975. *Fish communities in tropical freshwater.* Longmans, London.

Lucas, M. C., and E. Baras. 2001. *Migration of freshwater fishes.* Blackwell Science, Oxford.

Lyon, E. P. 1904. On rheotropism. I. Rheotropism in fishes. *Am. J. Physiol. 12:* 149–161.

Mank, J. E., and J. C. Avise. 2003. Microsatellite variation and differentiation in North Atlantic eels. *J. Heredity 94:* 310–314.

Mathis, A., D. P. Chivers, and R. J. F. Smith. 1995. Chemical alarm signals: Predator deterrents or predator attractants? *Am. Nat. 145:* 994–1005.

———, ———, and ———. 1996. Cultural transmission of predator recognition in fishes: Intraspecific and interspecific learning. *Anim. Behav. 51:* 185–201.

McDowall, R. M. 1988. *Diadromy in fishes.* Croom Helm, London.

McHugh, J. L. 1967. Estuarine nekton, pp. 581–620. In Estuaries, G. Lauff (Ed.). *Amer. Assoc. Adv. Sci. Publ. 83,* Washington, DC.

McKeown, B. 1984. *Fish migration.* Croom Helm, London.

Metcalfe, N. B., N. H. C. Fraser, and M. D. Burns. 1999. Food availability and the nocturnal vs. diurnal foraging trade-off in juvenile salmon. *J. Anim. Ecol. 68:* 371–381.

Mills, D. H. 1989. *Ecology and management of Atlantic salmon.* Chapman and Hall, London.

Montgomery, J. C., C. F. Baker, and A. G. Carton. 1997. The lateral line can mediate rheotaxis in fish. *Nature 389:* 960–963.

Moore, A., S. M. Freake, and I. M. Thomas. 1990. Magnetic particles in the lateral line of the Atlantic salmon (*Salmo salar* L.). *Phil. Trans. Roy. Soc. Lond. B 329:* 11–15.

Moyle, P. B., and R. M. Leidy. 1992. Loss of biodiversity in aquatic ecosystems: Evidence from fish fauna, pp. 127–169. In *Conservation* biology. *The theory and practice of nature conservation, preservation, and management*, P. L. Fiedler and S. K. Jain (Eds.). Chapman and Hall, New York.

———, and J. E. Williams. 1990. Biodiversity loss in the temperate zone: Decline of the native fish fauna of California. *Conserv. Biol. 4:* 275–284.

Muller, K. 1978a. The flexibility of the circadian system of fish at different latitudes, pp. 91–104. In *Rhythmic activity of fishes*, J. E. Thorpe (Ed.). Academic Press, New York.

———. 1978b. Locomotor activity of fish and environmental oscillations, pp. 1–19. In *Rhythmic activity of fishes*, J. E. Thorpe (Ed.). Academic Press, New York.

Myers, G. S. 1949. Salt tolerance of freshwater fish groups in relation to zoogeographical problems. *Bijdr. Dierk. 28:* 315–322.

Nakamura, H. 1969. *Tuna distribution and migration.* Fishing News Books, London.

Nanjappa, P., L. Brand, and M. J. Lannoo. 2000. Swimming patterns associated with foraging in phylogenetically and ecologically diverse American weakly electric teleosts (Gymnotiformes). *Env. Biol. Fishes 58:* 97–104.

Neverman, D., and W. A. Wurtsbaugh. 1994. The thermoregulatory function of diel vertical migration for a juvenile fish, *Cottus extensus. Oecologia 98:* 247–256.

Nilsson, G. 1999. The cost of a brain. *Nat. Hist. 108*(10): 66–73.

Odling-Smee, L., and V. A. Braithwaite. 2003. The role of learning in fish orientation. *Fish Fisheries 4:* 235–246.

Palmer, J. D. 1974. *Biological clocks in marine organisms.* Wiley Interscience, New York.

Pfeiffer, W. 1962. The fright reaction of fish. *Biol. Rev. Cambridge Phil. Soc. 37:* 475–511.

Pitcher, T. J. (Ed.). 1986. *The behavior of teleost fishes.* Johns Hopkins University Press, Baltimore.

———. 1977. The distribution of fright reaction and alarm substances in fishes. *Copeia 1977:* 653–665.

Raleigh, R. F. 1971. Innate control of migration of salmon and trout fry from natal gravels to rearing areas. *Ecology 52:* 291–297.

Ralston, S. L., and M. H. Horn. 1986. High tide movements of the temperate-zone herbivorous fish *Cebidichthys violaceus* (Girard) as determined by ultrasonic telemetry. *J. Exp. Mar. Biol. Ecol. 98:* 35–50.

Rodríguez, M. A. 2002. Restricted movement in stream fish: The paradigm is incomplete, not lost. *Ecology 83:* 1–13.

Shaw, E., and A. Tucker. 1965. The optomotor response of schooling carangid fishes. *Anim. Behav. 13:* 330–336.

Shulman, G. E., and R. M. Love. 1999. *Advances in marine biology, Vol. 36,* The biochemical ecology of marine fishes. Academic Press, San Diego.

Somers, D. E., P. F. Devlin, and S. A. Kay. 1998. Phytochromes and cryptochromes in the entrainment of the *Arabidopsis* circadian clock. *Science 282:* 1488–1490.

Stearley, R. F. 1992. Historical ecology of Salmoninae, with special reference to *Oncorhynchus*, pp. 622–658. In *Systematics, historical ecology, and North American freshwater fishes*, R. L. Mayden (Ed.). Stanford University Press, Stanford, CA.

Stiassny, M. L. J., and M. C. C. de Pinna. 1994. Basal taxa and the role of cladistic patterns in the evaluation of conservation priorities: A view from freshwater, pp. 235–249. In Systematics and conservation evaluation, P. I. Forey and C. J. Humphries (Eds.), *Syst. Assoc. Spec. Vol. 50,* London.

Tesch, F. W. 1978. Horizontal and vertical swimming of eels during the spawning migration at the edge of the continental shelf, pp. 378–391. In *Animal migration, navigation and homing*, K. Schmidt-Koenig and W. T. Keeton (Eds.). Springer Verlag, Berlin.

Thresher, R. J., M. H. Vitaterna, Y. Miyamoto, A. Kazantsev, D. S. Hsu, C. Petit, C. P. Selby, L. Dawut, O. Smithies, J. S. Takahashi, and A. Sancar. 1998. Role of mouse cryptochrome blue-light photoreceptor in circadian photoresponses. *Science 282:* 1490–1494.

Valdimarsson, S. K., and N. B. Metcalfe. 2001. Is the level of aggression and dispersion in territorial fish dependent on light intensity? *Anim. Behav. 61:* 1143–1149.

Veinott, G., T. Northcote, M. Rosenau, and R. D. Evans. 1999. Concentrations of strontium in the pectoral fin rays of the white sturgeon (*Acipenser transmontanus*) by laser ablation sampling—inductively coupled plasma—mass spectrometry as an indicator of marine migrations. *Can. J. Fish. Aquat. Sci. 56:* 1981–1990.

Verheijen, F. J., and J. H. Reuter. 1969. The effect of alarm substance on predation among minnows. *Anim. Behav. 17:* 551–554.

Von der Emde, G. 1990. Discrimination of objects through electrolocation in the weakly electric fish, *Gnathonemus petersii. J. Comp. Physiol. A 167:* 413–422.

Walker, M. H., J. L. Kirschvink, S. R. Chang, and A. E. Dizon. 1984. A candidate magnetic sense organ in the yellowfin tuna, *Thunnus albacares. Science 224:* 751–753.

Watt, M., C. S. Evans, and J. M. P. Joss. 1999. Use of electroreception during foraging by the Australian lungfish. *Anim. Behav. 58:* 1039–1045.

Welcomme, R. L. 1979. *Fisheries ecology of floodplain rivers.* Longman, New York.

Whang, A., and J. Janssen. 1994. Sound production through the substrate during reproduction in the mottled sculpin, *Cottus bairdi* (Cottidae). *Env. Biol. Fishes 40:* 141–148.

Whitmore, D., and P. Sassone-Corsi. 1999. Cryptic clues to clock function. *Nature 398:* 557–558.

Williamson, D. S. 1843. The statistical account of Tongland, p. 69. Cited in *Ecology and management of Atlantic salmon*, D. H. Mills, 1989. Chapman and Hall, London.

Wootton, R. J. 1990. *Ecology of teleost fishes.* Chapman and Hall, London.

———. 1992. *Fish ecology.* Chapman and Hall, New York.

Yambe, H., M. Shindo, and F. Yamazaki. 1999. A releaser pheromone that attracts males in the urine of mature female masu salmon. *J. Fish Biol. 55:* 158–171.

Zimmer, R. K. 2000. Importance of chemical communication in ecology. *Biol. Bull. 198:* 167.

———, and C. A. Butman. 2000. Chemical signaling processes in the marine environment. *Biol. Bull. 198:* 168–187.

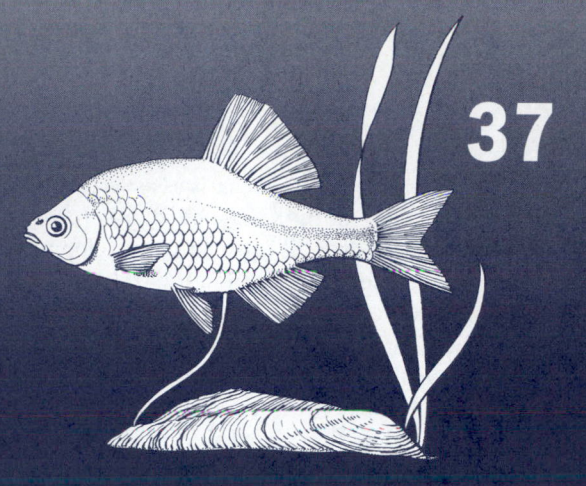

37 BEHAVIOR II: FEEDING, FOOLING AROUND, AND FINDING YOUR FRIENDS: THE SOCIAL WORLD OF FISHES

FEEDING BEHAVIOR OF INDIVIDUALS

Detection of Food
Foraging Behavior
Application of Feeding Adaptations
   Macrocarnivory
   Suction Feeding
Ecological Implications of Foraging Behavior

FISH WITH OTHER FISH: SOCIAL INTERACTIONS

Games Fishes Play
Communication: Signals and Social Behavior

REPRODUCTIVE AND PARENTAL BEHAVIOR

Courtship and Breeding Behavior
Sexual Selection and Sexual Dynamics
Parental Care

SHOALS AND SCHOOLS

Formation of Schools
Sensory Facilitation of Schooling
Adaptive Advantages of Schooling
Mixed-Species Shoals and Schools

SYMBIOSIS

Symbiotic Relationships Among Fishes
   Parasitic Fishes
Symbiotic Relationships with Invertebrates

LEARNING IN FISHES

Adaptive Behavioral Modification and Its Applications

In this chapter, we will investigate the behavioral aspects of two essential features in the lives of fishes—feeding and reproduction. We will also consider two other significant elements of their social behavior—schooling and symbiotic relationships with other animals. Although fishes often engage in group feeding, this activity is largely viewed as an individual behavior, in which fishes apply their sensory modalities to the tasks of locating and consuming food. In reproduction, the task at hand is locating a mate, accomplishing the fertilization of eggs, and for some species, providing parental care. In the cases of schooling and symbiosis, we witness the development of highly specialized forms of social behavior, manifested in remarkable demonstrations of coordination and cooperation. The development of schooling behavior is such an intrinsic feature of many species of fishes that it is difficult to think of some species, such as herrings, outside of the context of the school. The same is true for species of fishes that have evolved close associations with other fish species, or liaisons with other classes of animals. It is important that we keep in mind that the sophisticated behavioral repertoire that fishes employ in social situations is a direct consequence of the evolution of those morphological and physiological features that optimize the behavioral adaptation of fishes to their environment.

FEEDING BEHAVIOR OF INDIVIDUALS

The fundamental social unit in fishes, as in all organisms, is the individual (Keenleyside, 1979). Individual behavior is best investigated in the context of specific activities, such as feeding and reproduction. In the pursuit of these and other activities, fishes will, as a matter of course, be brought into contact with other organisms. In recent years, we have witnessed the emergence and validation of two interdisciplinary approaches to the study of behavior—*behavioral ecology* and *sociobiology*. The behavior of individual fishes has been investigated in both contexts. Through **behavioral ecology,** we address the following question: How does the behavior of the individual contribute to its adaptive capacity—that is, its ability to exist in a given environment? One of the principal proponents of **sociobiology,** Edward O. Wilson, defined it as "the systematic study of the biological basis of all social behavior" (Wilson, 1975). It is in these contexts that we attempt to evaluate the nature of those behaviors that fishes engage in when they are interacting with members of their own species (**conspecifics**) or with other species (**heterospecifics**).

Compared to those of endotherms, the energy requirements of fishes are much lower, and the onset of hunger is a more gradual process. Feeding motivation appears to arise from a combination of intrinsic contributions, including gut fullness and metabolic balance (Colgan, 1986). The significance of food as a motivational force is obvious from the amount of behavioral research in which food is used as reinforcement for learning. Given the diversity of form and function among the fishes, it is obvious that feeding behaviors are as diverse as their practitioners.

Detection of Food

The marvelous sensory arsenal of fishes generally allows them a large degree of flexibility in their feeding behavior. Certain environments and habitats favor the use of some senses over others, but usually most senses are functional to some degree and available for use. A fish might be alerted to the presence of potential food sources through the acousticolateralis system, move in the direction of the food by following a scent, and finally capture it by perceiving visual, gustatory, tactile, or electrotactic stimuli.

Visual detection of prey is of great importance in the feeding of most fishes and depends on a number of factors. An important consideration is the nature of the optical mechanisms in fishes—particularly the visual field (see Chapter 20). This involves the placement and mobility of the eye; visual acuity, which is a function of the size of the eye (larger eyes usually mean sharper vision); and brightness and color discrimination, which is associated with contrast and color perception. In the case of color perception, a correlation has been observed between the development of cone cells in the retina of young salmon and trout and the onset of feeding behavior (Noakes, 1978).

Environmental factors are significant in determining the effectiveness of visual prey detection. Feeding may be suppressed when waters become turbid as a result of rainfall or wave action. Most fishes feed during daylight hours, and their activity diminishes as light intensities approach 10^{-1} lux. Some species are crepuscular, feeding during twilight and even at intensities of 10^{-3} to 10^{-5} lux (Blaxter, 1980). J. R. Brett, in his classic work on the energetics of sockeye salmon, demonstrated that growth is optimized by a daily cycle of vertical migration that brings young fishes to the surface of the lake to feed at dusk and dawn. When not feeding, they conserve energy and thus maximize their growth potential by remaining at a depth below the thermocline (Brett, 1971). Juvenile sculpins (*Cottus extensus*), on the other hand, feed during the day in the cooler depths but move to the surface at night, where the warmer temperature improves the efficiency of digestive processes. This adaptation may be especially significant in larval and juvenile fishes, where rapid growth is essential (Neverman and Wurtsbaugh, 1994). Bat rays (*Myliobatis californica*) also engage in a form of behavioral thermoregulation, but in a horizontal fashion. This species shows a diel pattern of locomotor activity, in which its foraging behavior is coupled with movements from the warmer inner bay waters to cooler outer bay waters (Matern et al., 2000). The ecological application of certain behaviors is seen in another example, that of longnose dace (*Rhinichthys cataractae*). Whereas most species of minnows are diurnal or crepuscular feeders, dace have shifted to nocturnal foraging as a means of optimizing prey consumption and avoiding salmonid predators and cyprinid competitors (Culp, 1989).

Olfaction enables fishes to detect food beyond the limits of vision and is of primary importance to bottom-feeding species, such as catfishes, loaches, and eels. The olfactory capacities of sharks and their relation to feeding have been discussed in Chapter 22. Even such visual feeders as tunas will orient positively to the direction of food odors (Atema, 1980). Fishes often react to food odors by moving upstream toward the source of the odor and then initiating a search pattern until the source is confirmed. Externally located taste receptors, such as those on the barbels of catfishes, or vision can then aid in the final location and capture.

The lateral-line system may also assist in near orientation to food by the detection of electrical fields or acoustical disturbances emitted by the prey. In some species, feeding may be accompanied by acoustic emissions—clicks or grunts that are produced by the vibratory actions of muscles on the swim bladder, or stridulations involving skeletal elements (see Chapter 21). Streaked gurnards (*Trigloporus lastoviza*) emit repeated "growls" during competitive feeding, suggesting the establishment of a dominance hierarchy in the feeding arena (Amorim and Hawkins, 2000).

Foraging Behavior

Quests for food may involve active searching over a large area, as in pelagic predators, or they may be confined to a restricted home range. Fishes may occupy stations in the vicinity of food sources and await their chosen prey. Some species, such as anglerfishes and dragonfishes, facilitate this by the presentation of lures that induce their prey to approach (see Plate 35.1A–D). Feeding is usually not a continuous activity, however. There may be definite periods during the day (or night, in the case of nocturnal feeders) in which feeding activity is concentrated. The foraging behavior of fishes living in shallow estuaries often corresponds to the rise and fall of the tides. Staghorn sculpins (*Leptocottus armatus*) can be observed foraging immediately behind the advancing tide line, in water that is just a few centimeters in depth. Diurnal or tidal rhythmicity in feeding behavior is entrained in some species, so that individuals will display rhythms in activity and physiology even when they are held in a constant environment.

Cycles of feeding activity are often attuned to the seasons and life histories of fishes, especially in temperate waters. Fishes maximize their food intake during periods when temperatures are optimal for growth and reduce their feeding during the winter and during the hottest part of the summer. For headwater species, summertime may coincide with intermittent flow conditions, when fishes become concentrated in the remaining pool environments. Cessation of feeding at this time may simply reflect the restricted access to food.

Application of Feeding Adaptations

The behavioral "strategies" associated with individual feeding patterns naturally represent applications of a specific morphology associated with a particular feeding type. Although the range of feeding types has been explored in detail in Chapter 23, a few of the more remarkable behaviors associated with specific feeding strategies deserve mention here.

Macrocarnivory

As mentioned in Chapter 23, macrocarnivory can be expressed in a diversity of fishes that encompass a range of body forms, including fusiform pursuit predators, such as the pelagic sharks and scombroids (see Plate 16.20), and ambush predators, such as the cryptic, sessile antennariids

(see Plate 13.9) or the swift, dart-shaped gars and esocids (see Plate 12.1). What may seem at first glance to be a simple act of grabbing a large chunk of food in macrophagous forms such as sharks is in reality a complex sequence of behaviors associated with detecting, grasping, and ingesting prey (Bres, 1993; Frazzetta, 1994). White sharks (*Carcharodon carcharias*) have been observed to practice a "bite and spit" behavior, in which the prey is released after an initial bite. This apparently promotes death by blood loss, after which the shark returns to finish its meal (Klimley, 1994; McCosker, 1985). Engana and McCosker (1984) have suggested that this is the reason why North American divers have a higher rate of survival from shark attack than do their Chilean counterparts. Divers in North America are more conscientious in their practice of the "buddy system"—a companion can effect a rescue during the five- to ten-minute interval before the shark returns. The lower fat content of humans as compared to seals and sea lions also apparently makes them a less palatable prey item—humans are actually only rarely consumed by sharks (Klimley et al., 1996). White sharks also appear to be somewhat sensitive to prey freshness and will refuse to eat decomposed carcasses (Klimley, 1994). They have been known to feed on items as seemingly unpalatable as marine turtles, however (Fergusson et al., 2000).

Some pelagic predators, such as tunas and dolphins, often forage below the deep scattering layer—a concentrated band of vertically migrating marine organisms (see Chapter 35). These large predators are occasionally found with wounds of a curious origin—almost perfectly cylindrical holes of about 1 cm in depth, as if made by a cookie cutter. The culprits were determined to be a miniature, deep-dwelling but vertically migrating shark (*Isistius brasiliensis*) and a lesser known species (*I. plutodus*), which have since become known as "cookie cutter" sharks (see Plate 37.1; see Fig. 6.8B). The piranha feeds in much the same fashion, taking bite-sized chunks from larger prey. When feeding on smaller prey, such as goldfishes, a captive piranha will neatly sever the head and tail simultaneously by grasping and holding its prey crosswise in its toothy jaws. Even herbivores may employ such grasping and tearing behaviors. The monkeyface prickleback (*Cebidichthys violaceus;* see Plate 34.1), a large zoarcoid, feeds by grasping blades of seaweed and, with corkscrewlike motions, twisting them off of rocks to which they were attached.

Some fish species have large, molariform teeth, which enable them to crush otherwise indigestible food items. Another zoarcoid family, the wolf eels (Anarhichadidae), are able to feed on hard-shelled invertebrates, such as bivalves, crustaceans and sea urchins, because of this kind of dentition, whereas parrotfishes are able to bite off chunks of coral with their sharp, beaklike incisors and crush the ingested material with their powerful pharyngeal teeth. Parrotfish grazing is easily detected from the characteristic scars that they leave on the coral reef.

Suction Feeding

For many species, feeding actions are facilitated by the generation of a vacuum in the immediate vicinity of the oral cavity. By such means, food can be brought into the mouth by a sudden enlarging of the oral and buccopharyngeal cavity. A key development that greatly facilitated this mode of feeding in teleosts was the evolution of highly protrusible jaws (see Chapter 23). With these jaws, carnivorous fishes are able to "suck up" prey from some distance. Microcarnivorous or planktivorous species may also employ suction to draw small food items into their mouths, or they may feed by simply opening their cavernous mouths as they pass through dense concentrations of food. In fishes that use filter feeding, the gill rakers serve as filters to trap small prey items. Such feeding behavior is observed in species as small as anchovies and herrings or as large as the paddlefish (*Polyodon*). The largest of all fishes, the whale shark (*Rhincodon typus;* see Plate 6.3) and the basking shark (*Cetorhinus maximus*), feed in this manner. The extraordinary filter-feeding shark dubbed "megamouth" (*Megachasma pelagios*) is distinguished by a large (4.5 m), relatively flabby body and an enormous head terminating in a wide mouth lined with reflective crystals that may be associated with bioluminescent capabilities (see Chapters 6 and 18). It is surmised that this creature feeds by opening its cavernous mouth and sucking in tiny, deep-ocean crustaceans attracted by its blue-green glow (Taylor et al., 1983). The giant filter-feeding elasmobranch fishes, including the aforementioned whale shark, basking shark, megamouth, and the devil and manta rays, exemplify the range of application of suction feeding in planktivory, with some, such as the basking shark, slowly moving through and passively filtering out food, whereas the whale shark has a well-developed pharyngeal pumping apparatus at its disposal.

. . . .
Ecological Implications of Foraging Behavior

Nowhere are the applications of animal behavior to environmental adaptation more apparent than in the study of foraging. As fishes grow, their nutritional requirements change, as do the kinds of prey that they can handle. Consequently, food preferences and the behaviors associated with getting that food change. Early in the 1970s, a new approach to the study of feeding ecology, **optimal foraging theory**, became established (see Chapter 30). This theory developed from attempts to understand how animals achieve maximum net energy gain—the *optimal foraging strategy* is the one that maximizes energy intake while minimizing the metabolic costs expended in capturing and processing prey.

Models of foraging behavior typically include three components: (1) decision, (2) currency, and (3) constraints (Gerking, 1994). **Decision** involves search and location behavior that may culminate in the predator choosing to attack a given prey item. Search and location may be as slow and deliberate as that of the spot (*Leiostomus xanthurus*) searching out clams and polychaetes in the muddy substrate or the trumpetfish (*Aulostomus maculatus*) patiently stalking small fishes among Caribbean coral reefs. In what appear to be attempts to render themselves more inconspicuous, trumpetfishes often shadow other fishes while hunting (see Plate 37.2). In the case of the spot feeding on infauna, much information about prey availability must be gained from its olfactory and tactile senses. For planktivorous fishes, on the other hand, deliberate search patterns may entail only a quick and cursory search for suitable patches of prey. Piscivores and planktivores rely much more on vision while foraging. For a predator like the trumpetfish, search decisions occur in a sequential manner, as the predator encounters and evaluates each potential prey item. For planktivores, search decisions may involve having to choose among a multitude of potentially suitable prey species in the plankton community.

Currency entails a cost-benefit analysis of the energy required to obtain prey versus the potential for energy gain from the prey. Werner and Hall's (1974) study of foraging choices in planktivorous bluegill remains a classic demonstration of optimal foraging (see Chapter 32).

Constraints are those factors that influence the success of the foraging operation. They may be *intrinsic* constraints, such as the nutritional status of the fish (e.g., a starved animal may be much more motivated to feed, but it may be too weak to carry out a successful hunt), or *extrinsic* constraints, such as environmental factors that might have a bearing on the success of foraging. For example, planktivores generally are feeding in the pelagic realm, where they may be more vulnerable to predation. Optimal foraging theory dictates that they should be comparatively unselective under these conditions (Giske and Gro Vea Salvanes, 1995).

. . . .
FISH WITH OTHER FISH: SOCIAL INTERACTIONS
. . . .
Games Fishes Play

Two assumptions are fundamental to the sociobiological approach to studying the behavior of organisms: (1) that behavior is included among the heritable traits; (2) that individual fitness is measured, not in terms of the survival of the individual, but of the survival of its genes. The adaptive fitness of a certain gene pool will then be a function of the extent of successful interaction among the individuals contributing to that gene pool. This is what sociobiology seeks to explain—to what extent the social behavior of the organism contributes to its adaptive capacity. For this purpose, it is important that we take up the subject of the social behavior of fishes in more detail.

We can begin by simply continuing our discussion of feeding behavior and focusing on its social aspects. One way of alleviating competition for food among conspecifics is through the development of alternative foraging tactics—what is sometimes referred to as **resource polymorphism** (see Chapter 32). Among populations of recently emerged brook char (*Salvelinus fontinalis*), two feeding types develop. One is more sedentary and forages on crustaceans near the bottom, whereas the other is more active and feeds on insects near the surface (McLaughlin et al., 1999). This polymorphism persists through the adult stage, as adults in a given lake may be either specialists on pelagic prey or benthic prey, and the proportion of benthic specialists in a population appears to reflect the degree of competition with other benthic-feeding species (Bourke et al., 1999) Phenotypic adaptations to accommodate resource polymorphism were discussed in Chapters 4 and 32.

Cooperative foraging behaviors may also facilitate feeding. We will have much more to say about the nature of schooling behavior later, but for now, let us recognize the benefits of such behavior in foraging. Schooling by predators may make the rounding up of prey an easier task. Some bottom-feeding species may cooperate in dislodging and removing objects that solitary individuals could not successfully tackle. There is some evidence suggesting that fishes are able to identify which among them are the best at locating food. Studies on foraging behavior of bluegill sunfishes (*Lepomis macrochirus*) have revealed that individuals prefer to associate with those conspecifics with which they have had the greatest success in feeding in the past (Dugatkin and Wilson, 1992). In the past 20 years, game theory has become incorporated into a number of research endeavors, ranging from economics to biology. **Game theory** is a means of analyzing complex situations in which the strategy of a given player depends on the actions of another. Models of animal behavior are based on the assumption that animals are capable of strategic behavior as a means of optimizing their adaptation to the environment. Fishes, in being able to recognize individuals and distinguish their "value," are certainly capable of such behavioral flexibility.

The potential value of conspecifics has special meaning in the case of **predator inspection behavior**. Prey species often engage in the apparently paradoxical practice of approaching predators at a distance. The information gained from this risky procedure obviously is sufficient compensa-

tion for such behavior. It may even benefit those individuals who choose not to·engage in predator inspection but observe the outcome of one of their associates' inspections (Dugatkin and Godin, 1992; Godin, 1997; Pitcher, 1992). Although approaching and inspecting a predator in large groups obviously reduces the risk of predation (Magurran and Seghers, 1994), the apparently brazen behavior of individual inspectors also pays off in the decreased likelihood that they will be the one singled out for attack by a predator (Godin and Davis, 1995).

The decision whether to participate in predator inspection behavior or not is one example of strategic decision making by an individual. Another application, in which the behaviors of potential competitors are assessed, is manifested in the concept of the **ideal free distribution (IFD)**. Fretwell and Lucas (1970) predicted that, in an environment where resources are patchy, foragers would distribute themselves according to the distribution of the resources, in what they termed an "ideal free distribution." In achieving this state, foragers avail themselves of information such as the density and quality of the prey items or the number of potential competitors moving toward the resource (Tregenza and Thompson, 1998). They may also use alternative sources of information, such as changes in the behavioral state of competitors. Although increased levels of aggressive encounters could communicate information about patch quality and serve to attract competitors (Kennedy and Gray, 1994), this has not been shown to be the case in fishes (Koops and Abrahams, 1999). The IFD model has even been applied to the dynamics of fishing fleets competing for the same fishery resource (Gillis, 2003).

Group interactions may facilitate feeding in schooling species, to the extent that some species may be able to communicate about food resources by visual signals, such as color changes, or by other actions. In other species, feeding is such a solitary activity that conspecifics in the immediate vicinity are always regarded as intruders. Because of the relative ease with which many species of fishes can be confined in experimental situations, much more is known about the environmental factors governing territorial behavior in fishes than in other vertebrates (Maher and Lott, 2000). Herbivorous damselfishes (Pomacentridae) will cultivate gardens of algae among coral reefs and zealously guard them against intruders. In this case, we can see how the distribution of the primary production in a given environment can be a consequence of territorial behaviors.

Communication: Signals and Social Behavior

Implicit in our discussion of feeding behavior is the acknowledgement of the importance of communication. All behaviors of fishes that bring them into contact with other species will result in the elicitation of specific modes of communication. Just what is the "language" of fishes? To what extent are we able to translate it? Being mainly visual creatures, fishes depend primarily on the sense of vision for the cues, signs, and signals that trigger and maintain social behavior, but other senses can be just as significant as sight. Visual signals can range from those as simple as the perception of the characteristic form and color of a given species to complex ones that involve specific actions, such as postures, dances, approach, or flight. Some of these movements are designed to display specially colored parts of the body, including fins, bellies, mouth linings, opercular or branchiostegal membranes (see Plate 37.3). In deep-ocean fishes, the presentation of bioluminescent flashes may serve as indicators of the presence of a given species. In some species, these colors or patterns are present only during the breeding season or at other significant times of the year. A good example is the stoneroller minnow (*Campostoma*), one of the most common cyprinids in eastern North America. Breeding males of this species undergo a dramatic modification of their body morphology, including enlargement of the head, acquisition of breeding tubercles, and development of bold coloration (see Plate 10.2; Fig. 27.7A). In other species, the color may be apparent but may not be displayed until required by breeding or other behavior. Reef dwellers, such as the wrasses (Labridae) and damselfishes (Pomacentridae), are among the most conspicuous of fishes because of their brilliant coloration. They frequently show dramatic changes in these coloration patterns at different stages of their life histories.

Although significant colors and patterns are a seasonal phenomenon in some species, other species change their signals to suit the activity or "mood" of the moment—paling in fright, darkening in anger, with sexual motivation or extreme physiological stress, or changing from stripes to spots according to the stage of feeding activity. Some cichlids can display a fright pattern within a second of receiving the stimulus. A variety of patterns, including stripes, ocelli, and bars may be used. The Asian nandid *Badis badis* has about a dozen patterns that are used to signal behavior, but sometimes its activity changes faster than the patterns, so that there might be a lag between actions and the color signals associated with them (Barlow, 1963). A sand tilefish from the Philippines was observed to exhibit 24 changes in color in 15 seconds (Klauswitz et al., 1978).

Ocelli are among the most intriguing of color patterns. These "false eyes" have long been held to be directive marks, serving to direct predator attacks away from more critical areas of the body (see Plate 37.4). Reversing the perceived profile of the fish may be beneficial in other ways. Young piranhas are known to engage in fin nipping on other species.

Kirk Winemiller has demonstrated the predator-deceiving capabilities of ocelli in tests in which cichlids with ocelli near the base of the caudal fin experienced fewer attacks on their fins by young piranhas than species lacking ocelli (Winemiller, 1990; Winemiller and Kelso-Winemiller, 1993).

The presence of a variety of sound-making structures in fishes is a testament to the significance of sounds in the life of fishes (see Chapter 21). Threats, warnings, and signals to prospective mates can be produced by vibrations of the gas bladder (see Chapter 21, Going Deeper), movement of one hard part against another, or by other mechanisms. Not even Hurricane Charley, which passed through Florida in 2004, could suppress the nightly reproductive chorusing behavior of spawning fishes. Underwater acoustic data loggers placed in Charlotte Harbor recorded an even greater intensity of calling behavior during and immediately after the hurricane (Locascio and Mann, 2005). Electrical signals during certain behaviors are known from fishes with active electrical systems. Gymnotoids include bursts of electricity in their repertoire of aggressive activity, along with serpentine motions, head butts, and bites. Certain types of schooling behavior disappear in mormyrids when the nerves to the electric organs are cut (Moller, 1976).

Chemical signals may be incorporated into the behavioral repertoire of most fishes, but they have not been studied to the extent that has been accorded other means of communication. The bulk of these studies have focused on the remarkable secretions of ostariophysans, such as the *schreckstoff* mentioned in Chapters 10 and 36. Bullhead catfishes (Ictaluridae) have developed a complex hierarchy based on the characteristic odors of individuals and their ability to release pheromones that communicate social position to neighbors. Such chemical signals communicate peaceful coexistence among these normally gregarious fishes. When water from a tank in which a peaceful group coexisted was added to another tank in which agonistic behavior was present, this behavior was terminated, as the apparent perception of chemical signals from a peaceful aggregate was sufficient to instill similar behavior in the perceiving group (Todd, 1971; Todd et al., 1968). Fishes have been shown to be able to identify chemical alarm cues of other species. Pollock et al. (2003) demonstrated that fathead minnows (*Pimephales promelas*) learned to recognize chemical alarm cues of brook sticklebacks (*Culaea inconstans*).

REPRODUCTIVE AND PARENTAL BEHAVIOR

The subject of reproduction is one of great complexity, embracing as it does aspects of morphology, physiology, and behavior of the individuals to be considered. Certain reproductive behaviors, such as nest building and egg de-position, have been employed in the elucidation of phyloge-netic relationships in some groups (McLennan, 1994). With respect to behavior, a chronology of activities can be con-sidered. First, we might investigate behavioral phenomena associated with the onset of sexual maturation and ripening of the gonads, which we assume occurs simultaneously in males and females at a time propitious to the survival of the eggs and larvae. Behaviors will then be directed toward the selection of suitable breeding grounds, preparation of the site for breeding (this can be a very complex series of actions for some nest-building species), courtship and spawning activi-ties, and finally behaviors associated with the care and main-tenance of the fertilized eggs and larvae. Recognizing the tremendous diversity of fish species, we can expect consider-able variation in the extent to which each of these behavioral components is manifested. It is the diversity of reproductive modes among fishes that makes them attractive as models for reproductive studies (Amundsen, 2003). Ethologists, ecolo-gists, fishery biologists, and aquarists have all made valuable contributions to the study of reproductive behavior, with Breder and Rosen's (1966) massive treatise perhaps being the definitive work on the subject.

• • • • • • • • • • • • • • • • •

Courtship and Breeding Behavior

The onset of breeding behavior is a manifestation of physiol-ogical changes occurring within the fish. As described ear-lier for the stoneroller minnow, the onset of breeding be-havior may also be accompanied by dramatic morphological transformations. *Breeding tubercles* play an important role in reproductive behavior. They are keratinized epidermal structures used for the maintenance of body contact between the sexes, tactile stimulation of the female during spawning, and for the defense of nests or spawning territories (Wiley and Collette, 1970). At least 15 families of bony fishes develop breeding tubercles during the spawning season (see Fig. 27.7).

The gathering and movement of large numbers of fishes at a given time of the year are among the most conspicu-ous behavioral phenomena associated with reproduction. In general, reproductive migrations are synchronized with the onset of gonadal maturation, so that the fishes arrive on the spawning grounds ready to breed. Seasonal changes in the environment are detected and physiological and behavioral transformations are mediated through a series of endocrine pathways. Typically, fishes will migrate to the spawning grounds and complete the reproductive process in response to a prescribed set of environmental stimuli. For instance, the brown trout (*Salmo trutta*) migrates and spawns in the fall under the influence of decreasing light

and temperature. On the other hand, the rainbow trout (*Oncorhynchus mykiss*) over much of its range will migrate and spawn in the spring under the influence of the increas-ing photoperiod and rising water temperature.

For the Chinook salmon (*O. tshawytscha*), the situation is complicated by the apparent separation of the environ-mental cues and behaviors associated with migration from those cues and behaviors associated with the actual events of spawning. This anadromous species comprises relatively distinct races that enter freshwater at different times. The *stream-type* race apparently responds to the increasing photoperiod and warming temperatures during the spring and summer months by migrating from the ocean to its natal streams and is thus resident in freshwater months before spawning. Once in the streams, these fishes will take up residence in deep pools for the duration of the summer, during which gonadal development will take place. Spawn-ing occurs in late August and September, during a time of decreasing day length and cooling temperatures. The *ocean-type* race moves into streams in late summer and fall and spawns soon after entering freshwater. Migration into natal streams and gonadal maturation are thus apparently stim-ulated simultaneously, by the changes in photoperiod and temperature regime taking place in the fall. Age-specific sensitivity to environmental cues may also maintain the ob-served differences in stream residence time of the juveniles of the two races. Stream-type (**spring Chinook**) juveniles spend comparatively more time in freshwater than do the ocean-type juveniles (**fall Chinook;** Healey, 1991).

At the end of their migration or other movement to-ward a reproductive area, fishes select the actual arena in which their reproductive activities will be carried out. The complexity of the chosen site varies with the species and its reproductive requirements. For pelagic species, the spawn-ing site may amount to nothing more than the patch of open sea to which their migrations have brought them. Here, eggs are broadcast for external fertilization. The fertilized eggs and larvae will then drift with the currents as mem-bers of the planktonic community for a period of time. In catadromous species, such as the eels (*Anguilla*), the young will eventually drift back to the vicinity of their parents' home streams, at which time they will undergo the process of transformation into **elvers** and ascend these streams (see Chapter 9, Going Deeper).

For most species, reproduction includes the selection of some suitable patch of the bottom for the purpose of de-fining a territory or engaging in nest building. Territorial behavior in fishes is more frequently manifested in the pur-suit of reproductive activities than it is in the pursuit of food (Grant, 1997). The selection of a suitable spawning site may entail an evaluation based on a number of environmental variables, such as depth, stream flow, substrate texture, availability of nest-building materials, or availability of cover. Fishes will test the environment in several ways, exploring nooks and crannies, mouthing potential nesting materials, or settling on the bottom to feel it and make test excavations. Sometimes, a partially cleaned or excavated nest site will be abandoned if deemed unsuitable for any reason.

Mate selection or courtship activity may accompany nest preparation, or it may occur after a suitable site has been prepared. In many species, breeding is communal or polygamous, with no distinct pairing or nest preparation. In other species, males may wait at the spawning site and approach en masse any female of their species that appears. In general, courtship is a highly visual process, with colors, contrasting patterns, and configuration of body and fins providing cues. For deep-ocean species existing in virtual darkness, vision may still be an essential sensory modality, as photophore placement may give cues about species and sex. In those fishes in which pairing takes place, or in those cases in which a male entices and spawns with several females in succession, visual stimuli may be coupled with ritualized actions or postures as well as olfactory, auditory, or other cues. In some environments where fishes may be highly substrate-associated and possess limited home ranges, monogamous pairing has evolved. Barlow (1984) has identified 14 families of monogamous coral reef fishes. The vast majority of the 18 marine fish families identified by Whiteman and Côté (2004) as practicing monogamy have tropical affinities. The largest marine fish family, the Gobiidae, includes several monogamous species (Reavis, 1997a, 1997b; Reavis and Barlow, 1998).

Sex pheromones have been demonstrated to promote gonadal maturation in several species of fishes (see Chapters 26 and 36). These compounds have all been identified as either sex steroids or prostaglandins, and electrophysiological studies have demonstrated specific actions on the olfactory system (Van Weerd and Richter, 1991). Chemical cues are important in triggering nest building in certain members of the Belontiidae (bettas or Siamese fighting fishes), a family containing many bubble nest builders and mouth brooders (see Chapter 16). Although the olfactory and gustatory senses have been demonstrated to be significant in the courting ritual of the stickleback (Segaar et al., 1983), these fishes of the family Gasterosteidae are noteworthy in the simplicity of their olfactory structure. Olfaction does not appear to play a key role in their feeding behavior (Hara, 1975, 1993; Kleerekoper, 1969), nor does it figure in the location of siblings for purposes of schooling (Steck et al., 1999). Cods, croakers, toadfishes, and others employ sound in their reproductive activity, and the midshipman (*Porichthys*) adds the light of photophores to his nuptial noises (see Chapter 21,

Going Deeper). Courtship posturing and other rituals serve as stimuli promoting the pairing of mates and have become so ritualized that they are recognized as important mechanisms in ensuring the reproductive isolation of closely related species; environmental situations that compromise the effectiveness of such signals may promote the breakdown of species integrity (see Chapter 32, Going Deeper).

Courting behavior usually begins with the approach, the male sidling toward the female or otherwise displaying his form, color, or posture. Sound production figures significantly in the courtship and breeding behavior of some species. If the female is receptive and is ready to spawn or copulate, she responds by approaching or by allowing the male to approach. Nuzzling, butting, lateral contact, "dancing," or other requisite acts may ensue before mating is consummated. The aforementioned breeding tubercles are important in facilitating these activities. If a nest or other prepared site is involved, the male must cause the female to follow him to it. The well-documented "zig-zag" dance of the threespine stickleback (*Gasterosteus aculeatus;* see Plate 15.2) is a classic example. Many fishes pair off before the nesting site is prepared, and both sexes may work at building or cleaning the spawning site, as in the Cichlidae. In the salmonids, the females usually dig the nest, called a **redd,** in suitable gravel of the stream bed, while the males spend their time warding off intruders.

Territorial behavior is usually enhanced during the breeding season, with many normally nonterritorial species acquiring and defending territories. Such behavior is usually reinforced by the presence of bright colors and threatening postures that might include displays of the mouth lining, flaring of the opercula, or spreading of the fins. Although actual contact rarely occurs, intruders may be nipped, butted, or grabbed by the defender of the territory. Male–male agonistic encounters may sometimes involve the locking of jaws, with attendant pushing, twisting, or tugging in an apparent test of strength. Cichlids may even employ mouth-to-mouth contact and jaw locking as a means of mate selection. Such contests are usually of short duration, with the loser quickly fleeing the scene. In the case of the bettas, however, males will fight to the death. In some cultures, the fighting ability of individual fishes is highly esteemed, and wagers are placed on the outcome of such contests, much in the same way as they are in cockfights or dogfights. Some species of fighting fish are so belligerent that they will attack and kill the female after all the eggs have been laid in the bubble nest if she does not immediately vacate the premises.

Pacific salmon engage in aggressive encounters on the spawning grounds in defense of their territories. The **kype** of the male, with its arching curve and large, recurved teeth, can inflict potentially serious injury on an intruder. A favorite ploy is to grasp the opponent's caudal peduncle and twist it violently. Agonistic encounters may be terminated by the elicitation of so-called **displacement behavior** on the part of the loser. Such activity takes the form of some action unrelated to fighting, such as nest building, that either appears to allow some relief to the frustrated combatant or may actually be a signal of submission.

Although we usually think of sex roles as being comparatively fixed, the two-spotted goby (*Gobiusculus flavescens*) exhibits what is believed to be the first example of a sequential reversal in sex roles in a vertebrate. The comparatively brief breeding season of this goby begins with the expected situation of males vigorously courting females and engaging in intense male–male combat, but the roles then become reversed, with females actively courting males and contesting other females. It is believed that this role reversal is brought about by a decrease in the abundance of males (Forsgren et al., 2004). Another species of reef-dwelling goby, *Gobiodon erythrospilus*, exhibits an even more remarkable situation, where sex determination itself appears to be socially induced. Unpaired individuals are more likely to be immature than paired individuals, and sexual maturation is induced by the presence of an adult—if the adult is male, then the immature individual becomes female, and vice versa (Hobbs et al., 2004).

The act of spawning can be triggered by a variety of releasing mechanisms, acting alone or in concert. Tactile stimulation of the female by the male is probably the most common mechanism. Head butts, nips, lateral strokes and quivers, and passes beneath with the male's dorsal fins in contact with the ventral region of the female have all been noted in one or more species. The sight of a male or of the nest may be enough to induce the spawning sequence. An interesting form of deception is practiced among some genera of cichlid fishes—that of *egg mimicry*. Female haplochromine cichlids brood the eggs in their mouths. Males possess spots on their anal fin that are, to the human eye, virtually indistinguishable from newly laid eggs. When the female approaches the anal fin presented by the male, presumably intending to collect eggs, he will eject sperm, which will then fertilize the eggs held in the female's mouth (Goldschmidt, 1991). Not only do chemical secretions induce gonadal maturation, as previously mentioned, but they may also serve as cues to spawning behavior; secretions from either sex can have a stimulatory effect on others of the same or opposite sex.

- -
Sexual Selection and Sexual Dynamics

Where sex and procreation are concerned, energy is expended by males and females in decidedly different ways. Females must devote a considerable amount of "effort" to the generation and maintenance of eggs. For the female, the behavioral tasks at hand are locating potential mates and then choosing among them (Dugatkin and Godin, 1998). With the exception of the two-spotted goby mentioned earlier, nature has relieved her of the onerous duties of attracting and holding the attention of a potential mate. That is the job of the male. He must be bold, conspicuous, and convincing as to the quality of his genes.

Charles Darwin proposed the concept of **sexual selection**, in which the reproductive (and hence evolutionary) success of the species is largely dictated by decisions made by the female in regard to the suitability of potential mates (see Chapter 4). The expression of costly traits—boldness in appearance or aggressive and persistent courtship patterns, for example—correlates with overall individual fitness. Successful reproduction thus requires considerable preparatory effort in securing the necessary biochemical tools (Olson and Owens, 1998). Guppies (*Poecilia reticulata*) have long been a favorite of researchers for studying the nuances of the "dating game." Male guppies from populations that experience low predation pressure are more boldly colored and engage in more elaborate, time-consuming courtship rituals than individuals from populations where predation intensity is high (Andersson, 1994). Novelty appears to be an important criterion for mate selection in female guppies, as they mate more frequently with individuals presenting a color pattern they have not yet experienced (Hughes et al., 1999). In the related mosquitofishes (*Gambusia*), a trade-off governing the size of the sperm transfer organ (gonopodium) of the males is apparent. Whereas females prefer males with larger gonopodia, predation pressure favors the reverse, as smaller gonopodia contribute to greater speed and agility in escape behavior (Langerhans et al., 2005).

Carotenoids are essential in the creation of bold, highly visible orange displays, yet they are relatively rare in animal diets. It has been suggested that the presence of orange coloration in males reflects positively on their vigor, rendering them more attractive to females, and hence improving their overall fitness (see Chapter 18). An alternative proposal, termed the **sensory-bias hypothesis**, suggests that females exhibit a preference for orange coloration in males because they have a predilection for the color orange based on its prior association with preferred food items (Rodd et al., 2002). Bold colors generally accompany bold behaviors, measured as the willingness to approach and inspect predators. Female Trinidadan guppies prefer to mate with males that exhibit such conspicuousness in color and behavior, as these individuals are most inclined to seek information about the environment, including potential dangers—obviously a desirable trait (Godin and Dugatkin, 1996). The cost of sexual attractiveness in males may extend beyond their increased vulnerability to predators and the likelihood of passing these attributes to the next generation. In guppies, genes for attractive display appear to be linked with genes that contribute to decreased survivorship in offspring, even before they reach sexual maturity and acquire ornamentation (Brooks, 2000).

The females of many species of fishes show a decided preference for nests in which eggs have already been deposited. This may be interpreted as a copying behavior, in which a female chooses a mate based on his prior success at fertilizing another female. Females may choose such "experienced" mates not because of any demonstrated capabilities, but merely because placing eggs in a nest that already contains eggs may enhance their own survival (Jamieson, 1995).

Although courtship generally includes a somewhat well-prescribed series of behavioral repertoires, fishes have maximized their adaptive capacity through the evolution of alternative reproductive strategies (Alonzo and Warner, 2000; Henson and Warner, 1997). For males, this often involves "sneaky" behavior. In some species, a given population may consist of a few dominant males that court most of the females and establish nesting sites. Other, smaller males, termed **sneakers,** will grab a quick spawning while the attention of the dominant male is elsewhere (see Chapter 21, Going Deeper). In poeciliids like the mosquitofish (*Gambusia affinis*), the males are typically much smaller than the females. Studies have shown that the smaller male mosquitofish employing sneaking behavior are more successful at inseminating females than are the larger males (Pilastro et al., 1997). Selection for small size has resulted in some species exhibiting a coexistence of genetically determined small males that employ sneaking behavior with larger males that display the usual courtship behavior. The clownfish (*Amphiprion percula*), best known for its unusual symbiotic relationship with sea anemones, has evolved a remarkable reproductive hierarchy. Groups form in which only one breeding pair exists—the female assuming the position of "top dog" in the hierarchy. With her demise, her mate changes into a female, and the first male in line becomes his—now her—consort. Others move up in the queue, and order is maintained through a strict control of growth rate in subordinates, as a well-defined range of size differences reduces the chances of conflict (Buston, 2003, 2004).

· · · · · · ·
Parental Care

Parental care in fishes ranges from virtual indifference to eggs once they are spawned to intense, complex behaviors that have evolved to optimize the survival of the young.

For teleosts, 78 percent of the families exhibit no parental care, 11 percent exhibit care by the male parent only, 7 percent exhibit care by the female only, and 4 percent exhibit biparental care (Sargent and Gross, 1986, 1993). Pelagic species, lacking a substratum on which to orient, build nests, or bury eggs, typically exhibit the least amount of parental care. Parental care in salmonids extends to the building of redds in which the eggs are buried. Other species—cichlids, for example—engage in a variety of egg and larval maintaining behaviors. In addition to vigilant guarding of the eggs and larvae against predators, cichlids will fan the eggs to ensure adequate oxygenation, remove dead eggs or eggs infected with fungus, and continue to clean and maintain the nesting area. Although it is generally held that visual cues are the most important in eliciting egg care, recent studies on nocturnal egg care in cichlids have suggested that olfactory or gustatory cues can trigger and maintain egg care in the absence of visual cues (Reebs and Colgan, 1992).

Many species of fishes brood the eggs on some part of their anatomy. Male humpheads (Kurtidae) carry eggs about on a peculiar hook that projects forward above their head. Male pipefishes and seahorses carry eggs in brood pouches. Nine families of teleosts, including sea catfishes (Ariidae), jawfishes (Opisthognathidae), cardinalfishes (Apogonidae), and some cichlids, incubate their eggs in the buccal cavity (Blumer, 1982). According to the **safe harbor hypothesis,** egg size in fishes evolves in response to the nature of parental care—the evolution of mouth brooding being seen as an example of maximal parental investment (Shine, 1978). Reduced fecundity, larger eggs, and higher larval survival would all result from increased parental investment. A study of the evolution of mouth brooding in the fighting fish (*Betta*) has suggested that the correlations predicted by the safe harbor hypothesis might not always be reflected in the evolution of a reproductive strategy. In the case of *Betta*, mouth brooding has apparently evolved in more than one phyletic line—a phenomenon that has also been documented in cichlids (Rüber et al., 2004).

One problem with mouth brooding should be apparent—how does the parent breathe? Apparently, gas exchange is compromised, but, given the level of activity of brooding parents, this is not critical. Studies on the respiratory consequences of mouth brooding in cardinalfishes (*Apogon*) have indicated that when faced with hypoxic conditions, fishes will expel their eggs (Östlund-Nilsson and Nilsson, 2004).

Once hatched, the larvae may continue to swarm about the parents, which continue to guard them vigilantly. Because many species of cichlids are easily reared and maintained in captivity—a feature that contributes to their popularity with home aquarists—the parenting activities of this family are probably the best documented of all fishes. Cichlids are generally aggressive by nature, especially in defense of their nests and young. In mouth-brooding species, the young can find a safe haven in the buccal cavity of the parents, and the parents are apparently capable of distinguishing their own young by odor. Some species will even provide nourishment in the form of mucous secretions from the skin surface.

The obvious benefit of reproductive behaviors is that they optimize the number of young that survive. Although much of the behavior is tied to launching a great number of young of the right size at the right time, other behaviors help to prevent flooding an area with young in numbers that exceed the carrying capacity of the habitat. Some types of behavior ensure that the most suitable males—presumably with the genome most conducive to survival—have the best chance of spawning. Other behaviors protect the genetic integrity of the species, preventing hybridization or preventing the waste of gametes in infertile alliances. Noakes (1986) provided a good introduction to this rapidly developing area of study. Some perciform fishes have been observed actually consuming eggs from nests that they are tending. Such "filial cannibalism" might seem counter-adaptive but, in fact, is believed to promote the survival of a greater number of offspring and, hence, sustain the species (Elgar and Crespi, 1992; Lindström, 1998; Okuda, 1999). Males of a pomacentrid species, the garibaldi (*Hypsypops rubicundus*), will consume eggs in later stages of development from the nest in order to promote continued visitations and spawning by females. In this way, the use of the spawning substratum, which in this case consists of a mat of carefully tended red algae, is maximized (Sikkel, 1989, 1994). Some forms of postspawning behavior serve to lure adults away from the vicinity of hatching young, thus preventing intraspecific predation and competition. Intraspecific competition may also be avoided by distinct differences in feeding behaviors between adults and their young.

A behavior that is probably adaptive—though in a somewhat indirect manner—is the defense of the young of others. Certain Central American cichlids are known to adopt the young of other parents and even to kidnap young from nearby broods. This tends to decrease the possibility that attacks on the young would harm the offspring of the guarding pair (McKaye and McKaye, 1977). The cichlid *Cichlasoma nicaraguense*, a herbivore, has been noted guarding the young of a predator, *C. dovii*. The advantage to *C. nicaraguense* appears to be that *C. dovii* is also a predator on the major competitors of the herbivorous species (McKaye, 1977). An even more unusual relationship is the guarding of the young of cichlids by bagrid catfishes; the cichlids are kept on the periphery of the school of bagrid

young, where they presumably are more vulnerable to attack (McKaye and Oliver, 1980).

SHOALS AND SCHOOLS

It is important, at the outset, to distinguish between *shoals* and *schools*, as the two terms have been used interchangeably, causing some confusion. A group of fishes brought together so that they constitute a social group can be termed a **shoal**, in the same manner that the term *flock* defines a similar situation for birds. As such, the term makes no implications about the characteristic structure or function of the social grouping. This structure is explicit in the definition of **schools** as entities in which the participants are defined by their orienting and swimming with a particular polarity and synchrony (Pitcher, 1986; Pitcher and Parrish, 1993). The terms are unavoidably interchangeable, as fish species themselves modify their social organization (Fig. 37.1). Fish schools have been likened to an exclusive club, in which the members gain certain benefits by virtue of association with others like them (Landa, 1998). One feature that readily distinguishes fish schools from other animal aggregations is the absence of an established hierarchy. Schools sometimes demonstrate an astounding degree of coordination—remarkable in view of the fact that the school is an entirely decentralized aggregation, in which an individual can act as a leader or a follower.

Formation of Schools

In models developed to simulate the movements of fishes in schools, three basic behavioral patterns are recognized—attraction, repulsion, and parallel orientation or imitation (Huth and Wissel, 1992, 1993). Schooling is a behavior that characterizes both ancestral and more derived fish lineages (Shaw, 1978). The tendency to approach and orient to other members of the same species appears early in many fishes. For example, in the ontogeny of the schooling behavior of the European minnow (*Phoxinus phoxinus*), individuals become attracted to each other as soon as the larvae become free swimming. It is only after three weeks, however, that the schooling response occurs as a reaction to the threat of predation (Magurran, 1990). This behavior, often termed **biotaxis,** is largely dependent on visual stimuli. The size, shape, and color patterning of conspecifics attract the very young of schooling species to one another, yet there is little evidence to suggest that natural shoals of fish are necessarily composed of related individuals (Steck et al., 1999). Larval herrings will school prematurely in the presence of

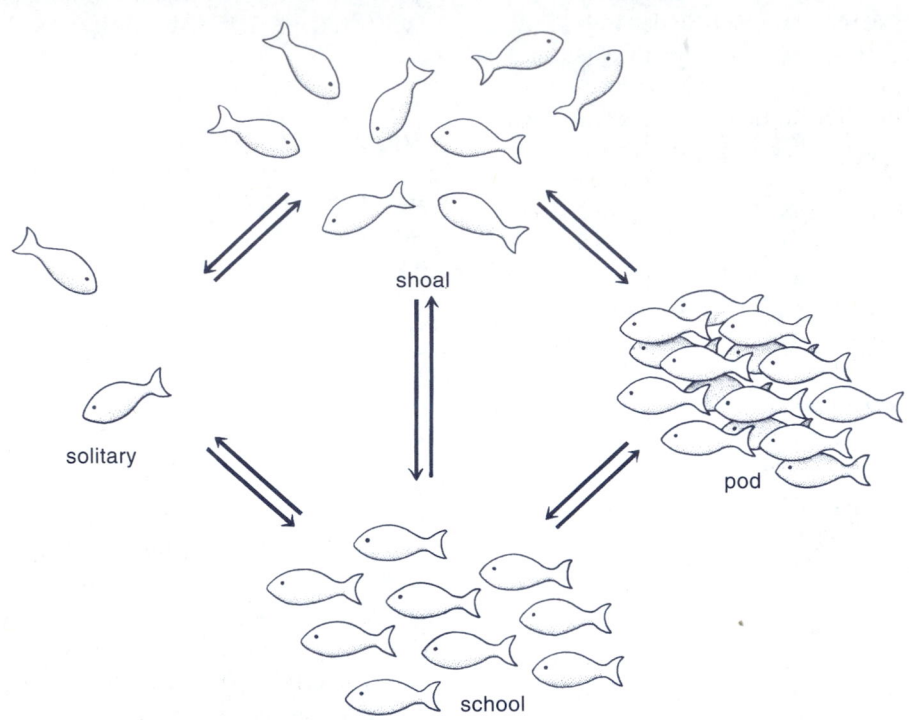

FIGURE 37.1
Relationships between various forms of fish grouping. Fishes may range from solitary existence at one extreme to the formation of pods or clumps at the other extreme. Intermediate associations include shoals and schools, with varying amounts of social cohesion and communication implied. (Adapted from Breder, 1959.)

more advanced conspecifics, even when this places them at a disadvantage in terms of competition for food. Apparently, the security afforded individuals in a school outweighs any nutritional disadvantages (Gallego et al., 1995).

Sensory Facilitation of Schooling

Although fishes are well equipped with physical senses other than vision, including keen chemical senses, vision appears to be vital in the formation and maintenance of schools. Cues to speed and directional changes come especially from the lateral visual fields. Several species have been shown to cease schooling at certain light intensities (Whitney, 1969). Blinded individuals of most species tested can school poorly if at all. To anyone watching a group of fishes schooling in a small stream pool, it becomes readily apparent which individual is afflicted with a severe case of eye fungus, as it

is unable to maintain its proper orientation relative to its schoolmates. Other physical senses are useful in schooling and may compensate for the lack of visual cues. Saithe (*Pollachius virens*), for example, still demonstrate schooling when temporarily blindfolded (Pitcher et al., 1976). Sound production may be instrumental in reinforcing the integrity of the school in the dark, and electric fishes may rely on the emission and detection of electromagnetic information as a way of keeping in contact in dark or turbid waters (see Chapter 21).

Adaptive Advantages of Schooling

It is obvious that shoaling and schooling behavior confers some adaptive advantage to the species, but the nature and extent of this advantage has been the subject of much debate (Cushing and Hardin Jones, 1968; Pitcher, 1986; Radakov,

1973). The most obvious advantage to schooling in prey species is that it affords a greater degree of protection from predation (Burgess and Shaw, 1979; Magurran, 1990; Shaw, 1978). Fishes in schools are more vigilant in their monitoring for the presence of predators (Magurran, 1990). Schools show predators multiple and shifting targets. Schools may engage in a particular avoidance pattern that reduces their vulnerability (Fig. 37.2). Prey species such as herrings form dense, writhing pods or "balls" that appear to confound a predator, so that it will only strike at individuals when they become detached from the mass. Anything that distinguishes an individual in a school singles it out for a greater likelihood of being attacked (Landeau and Terborgh, 1986). In experiments, predatory fish have been noted to consume more prey fish when they are presented singly or in small numbers rather than as a large group (Lim, 1981; Major, 1978).

Schooling can bring advantages to fishes seeking food. In tropical reef communities, herbivores such as parrotfishes (Scaridae) form schools to facilitate the invasion of the territories of competitors. Surgeonfishes (Acanthuridae)—a family that normally does not form schools—will also school as a means of invading the feeding areas of competitors (Barlow, 1974; Robertson et al., 1976; Vine, 1974).

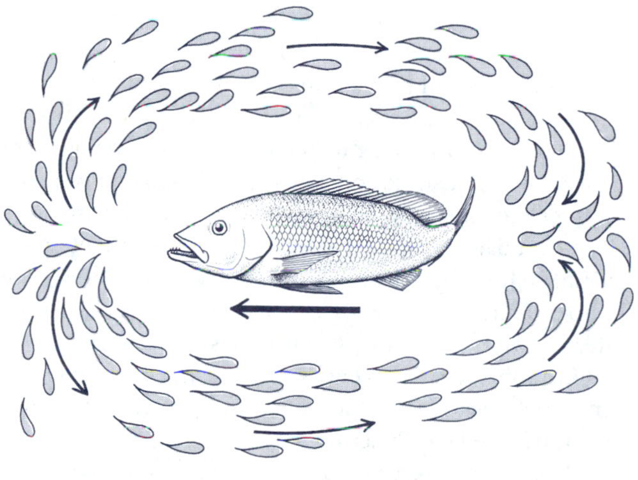

FIGURE 37.2
Illustration of the response of a schooling prey fish (*Lutjanus monostigma*) to a predator (*L. bohar*). The school retreats from in front of the predator and cascades around it while maintaining a fixed distance. The school then reforms behind the predator. (From Potts, 1970.)

The constant communication among members of a foraging school facilitates the localization of prey for a greater number of individuals. Cooperative foraging is apparent in schooling planktivores, which appear to be more efficient at feeding among dense patches of plankton in schools than they are as individuals. Schooling predators can engage in "wolf pack" behavior—cooperating in encircling prey or, as in the case of the voracious bluefish, driving prey schools into the shallows and even up onto the shore.

Laboratory experiments have demonstrated that young fishes allowed to feed in a group tend to grow faster than those eating by themselves. Possibly, fishes in shoals expend less energy on searching for prey or looking out for predators, so that more energy can be devoted to growth. The advantages of close social cohesion as a means of improving foraging efficiency do have their limits, however. In some cases—as when food resources are less abundant—solitary foraging may be most effective. Some species—the medaka (*Oryzias latipes*), for example—display aggression in competition for food and develop hierarchies when food is limited (Magnuson, 1962). Apparently, this species can rapidly evolve modifications in social behavior in response to changes in resource availability (Ruzzanti and Doyle, 1991).

Reproduction can also be facilitated by schooling and shoaling behavior. Many fishes are communal spawners, and both sexes are represented in a single school, so that at the proper season and place, they have only to release their gametes en masse in order to achieve a high rate of fertilization. Single-sex schools must, of course, commingle at spawning areas. In some species, the competitive aspect of reproductive behavior subverts the cooperation inherent in the school. During the spawning season, individual males of white cloud minnows (*Tanichthys albonubes*) defect from the school and establish territories among vegetation. If available spawning sites are limited, conflict ensues among members of the school (Magurran and Bendelow, 1990). For pelagic species, showing up as a school and shedding gametes in the vicinity of members of the opposite sex may be the extent of their reproductive behavior. For other species, as we have seen, the act of reproduction involves a much greater array of behaviors.

Researchers have demonstrated that schools and shoals do not consist of randomly distributed individuals. Rather, fishes will segregate according to size in the formation of schools and shoals (Hoare et al., 2000; Krause et al., 2000). Minnows and sardines exhibit size-segregated shoaling behavior. Individual sticklebacks, when faced with the choice of joining a shoal in which all individuals are the same size versus one in which the choosing individual would be the sole member of a different size, always prefer to shoal with individuals of their own size (Peuhkuri, 1999). The threat of predation often results in different size groupings coming together in order to form a larger body of individuals (Pitcher, 1986). Fishes do not appear to move randomly within the school but have defined fields of activity, so that some individuals may tend to be near the edge of the school, whereas others concentrate near the center (Healey and Prieston, 1973). A change of direction by a school seems to be a concerted, instantaneous action on the part of all members, so that it seems that they are of the same mind. By means of high-speed video recording and other techniques, researchers have discovered that the direction change can be initiated by individuals well within the mass of fishes. The response to the initiator's directional change comes about in a fraction of a second.

Many observations of schooling behavior have suggested that there may be some hydrodynamic advantage to fishes moving in schools (see Chapter 18). The ease with which a school seems to glide through the water seems to support such a contention, as does the observation that the role of "leading" fish in a school changes with every change in direction. Researchers studying the aerodynamics of formation flying in birds have claimed that the lead position is traded within the formation as a means of sharing the arduous duty of trail breaking, as the leading individual gains none of the aerodynamic advantage that formation flying affords to other members of the group. The mucous secretions of fishes have been suggested to enhance the movement of a school through the water, as they surely assist in drag reduction for the individual fish. Hydrodynamic advantages to schooling, intuitively appealing as they may seem, have not been adequately demonstrated experimentally. What experimental evidence does exist seems to contradict the notion that fishes move easier through the water as a group (Pitcher, 1986). In spite of this evidence to the contrary, Landa (1998) has asserted that hydrodynamic benefits accrue to members of a school.

· · · · · · · · · · · · · · · · ·
Mixed-Species Shoals and Schools

The advantages of shoaling and schooling behavior do not necessarily confine these activities to single-species aggregations. There are many instances in which fishes form mixed-species shoals. In freshwater, cyprinids and catostomids often form foraging-associated shoals among the rocks and cobbles of stream bottoms. In the Columbia River drainage, as many as five different species of minnows and suckers may compose these shoals. Such formations are

more common among juveniles, in which morphological and behavioral adaptations for different kinds of resources have yet to develop. In highly structured benthic environments, mixed-species shoaling is relatively common. Shoaling among heterospecifics has been most intensively studied in coral reefs, where some species of snappers (Lutjanidae) and grunts (Pomadasyidae) have been reported to form multispecies schools when not actively feeding (Wolf, 1983, 1985). These species tend to be crepuscular to nocturnal feeders, which school together during the day as a means of increasing their security against diurnal piscivores. Mixed-species shoaling appears to be relatively rare in open-water pelagic species, however. Commercially important fishes, such as clupeids, gadoids, and scombroids, usually form single-species schools, although mixed-species schools of hake (Merluciidae) are known to occur (Pitcher, 1986). In schools containing more than one species, it is often the rarer species that suffers greater instances of predator attack (Hobson, 1968).

Mark-and-recapture studies (see Ward et al., 2002) have demonstrated that fishes form partner associations that reinforce shoal fidelity. This apparently serves to enhance awareness of potential predators and to reduce deleterious aggressive competitive interactions. Such preferential associations may also extend to heterospecifics in the case of mixed-species shoals—cyprinids have been demonstrated to prefer associating with familiar heterospecifics from their own shoal rather than with conspecific strangers from a different social group (Ward et al., 2003).

Although schooling behavior appears to have numerous adaptive advantages—some obvious and others more subtle—one must consider one of its chief liabilities as well. Sociality brings with it the increased likelihood of transmission of disease and parasites (see Chapter 38). Because of this, it would be advantageous if organisms were able to evaluate the level of infirmity or infestation of group members. Juvenile sticklebacks (*Gasterosteus aculeatus*) have been shown to avoid schools in which conspecifics have been infected with crustacean ectoparasites (Dugatkin, 1994). Fishes probably are relying chiefly on the visual sense to make such evaluations, although stress-related secretions may also be detected by group members.

The classical perception has been that fishes school because the consequent aggregating behavior confers greater survivability on the individual. Lately, this perception has been challenged by the proponents of **complexity theory**, who view the attributes of fish schools as not necessarily reflecting the adaptive advantages accrued to the individual. According to this perception, the aggregations form as a consequence of the innate associative tendencies of animals, but the attributes displayed by the aggregation, such as pattern formation and information storage, are intrinsic to the school. In other words, evolution may explain *why* fishes school, but not *how* the schools develop (Parrish and Edelstein-Keshet, 1999, 2000).

SYMBIOSIS

Symbiotic Relationships Among Fishes

In the aquatic realm, there exist many instances of fishes sustaining close relationships among themselves or with other organisms. Such *symbiotic* relationships may be mutually beneficial (**mutualism**), beneficial to only one member of the association (**commensalism**), or beneficial to one member to the detriment of its associate (**parasitism**). There are numerous examples of social relationships involving two or more species of fishes. These relationships are usually commensalistic in nature—that is, only one of the individuals involved benefits. These benefits usually take the form of enhanced feeding opportunities. Pilotfishes (*Naucrates ductor*) and bar jacks (*Caranx ruber*) often accompany sharks and rays, with the intent of grabbing a free meal with little expenditure of effort (see Plate 37.5). Such behavior on the part of pilotfishes is so inborn that they tend to accompany any large, moving object, such as turtles or boats. Remoras (Echeneidae) will attach themselves to their host by means of powerful adhesive disks, which are in fact highly modified dorsal fins. Remoras get a free ride from their host—usually sharks or other large fishes, turtles, or whales—in addition to a free meal, which usually takes the form of scraps from the jaws of their host. Cleaning symbiosis is such a well-documented example of a mutualistic relationship involving feeding that it merits special consideration (see Going Deeper).

Some fishes exploit the foraging behavior of other species, even if their own feeding behaviors and food choices are quite different. Sea basses (Serranidae) follow herbivorous reef dwellers as they forage, snatching the inevitable small fishes and crustaceans that become dislodged as a result of algal grazing by the herbivores. Goatfishes are bottom feeders among the coral reef community, and their disturbance of the sandy sediment over which they feed attracts a number of other fish species. Less assertive species, such as sea basses, may follow more fearsome predators, such as morays, and catch small fishes as they attempt to flee (DeLoach, 1999; Karplus, 1978; Montgomery, 1975). Some predators, such

as the previously mentioned trumpetfishes, may attempt to render themselves more obscure by shadowing other predators. In Lake Malawi, the explosive radiation of cichlid species has resulted in similar feeding relationships among different species. The cichlid *Cyrtocara moori* follows several different species of bottom-feeding cichlids and will attack conspecifics and other species if they attempt to approach its feeding partner (Kocher and McKaye, 1983).

Parasitic Fishes

Although they may be parasitized by a bewildering array of invertebrate organisms, fishes themselves rarely engage in parasitic associations with other species. In the strictest definition of the term, *parasites* are prudent in that they do not kill their hosts, because this would result in their own demise. Lampreys are often seen as one of the few cases of nutritional parasitism among the vertebrates, yet their actions—consisting of rasping a wound on the host and sucking out its body fluids—actually are more like a specialized form of predation. If a lamprey attack is not fatal—as is often the case—its actions may seem more parasitic than predatory. The deep-living pugnose eel (*Simenchelys parasitica*) has been known to parasitize the tissues of large halibut. In what is perhaps the only known case of a vertebrate endoparasite, pugnose eels with guts full of blood have been found in the heart of a lamnid shark (Caira et al., 1997).

The trichomycterid candiru (*Vandellia*) of South America is parasitic on larger species, usually living within or attached to the branchial chamber of the host. Its apparent attraction to nitrogen-rich environments has resulted in some instances of penetration of the urethra of humans—a phenomenon so bizarre and painful even to hear about that many dismissed it as myth, until a South American urologist provided a graphic description of a surgical procedure performed in 1997 in which he removed a candiru from a patient's urethra. Spotte (2002) has produced the definitive work on this most bizarre of animals.

A kind of nest parasitism occurs among some North American minnows, which deposit their eggs in nests made by other species of minnows or by centrarchids. The survival of the brood appears to be enhanced by having another, perhaps larger individual guard the eggs.

Symbiotic Relationships with Invertebrates

The most common symbiotic relationships of fishes with invertebrates are those defined as parasitic. This is a special category, which is of particular concern for fish health, and will be considered in Chapter 38. One unusual invertebrate parasitic adaptation that deserves mention here is that of the freshwater mussels of the family Unionidae. The larvae of these bivalve molluscs, termed **glochidia**, are obligate parasites on the gills of fishes. Many mussel species possess mantle tissue that acts as a lure to attract fishes. In the genus *Lampsilis*, the mantle edges of the females are modified to mimic a fish. When a predatory fish approaches the lure, the mussel blows glochidia out of its excurrent siphon. These glochidia then anchor in the gill cavity of the host fish (Fig. 37.3A; see Plate 37.6); this provides a convenient means of dispersal for the otherwise sedentary bivalve. Laboratory studies have confirmed the aggressive response of fishes to the mantle tissue lures of unionids (Haag and Warren, 1999; Haag et al., 1999). Some species of mussels produce fusiform packets of glochidia, termed **superconglutinates,** that are tethered by thin mucus strands. These also deceive fishes, which perceive them as an easy meal

(Haag et al., 1999). Bitterlings (*Rhodeus*) turn the tables in this relationship, as they can lay eggs inside these bivalves using an elongate ovipositor (Fig. 37.3B).

Some remarkable symbiotic relationships have evolved between cnidarians and fishes. Members of the stromateoid genus *Nomeus* are known as man-of-war fishes because they live with the large siphonophore *Physalia*, swimming among the tentacles without injury (see Plate 37.7). The medusafish (*Icichthys lockingtoni*), another stromateoid genus, has similar habits. Larvae and small juveniles of other marine families, such as the carangids and gadids, are known to shelter under jellyfish (Mansueti, 1963).

Perhaps the best-known example of symbiotic association involving a fish and an invertebrate host is that of the pomacentrid genera *Amphiprion*, *Dascyllus*, and *Premnas*, commonly referred to as anemonefishes or clownfishes for their bold coloration. These fishes associate closely with sea

GOING DEEPER • Another Look at Cleaning Symbiosis

Several years ago, while snorkeling in Hawaii, I had the good fortune to observe a cleaning station in operation on a small patch reef. Looking much like your local gas station in the midst of a fuel crisis, it had fishes patiently queuing up and waiting for a small wrasse to attend to their dermatological needs. Fishes coming to the stations to be cleaned cooperate fully with their cleaners—remaining still, opening their mouth and gill covers when necessary, and, most important, refraining from eating the cleaner. Large barracudas and groupers allow the smaller fishes to swim around their mouths and gills with impunity. Cleaning symbiosis may be the most highly developed instance of interspecific communication known in the marine world.

The relationship between a cleaner and its client is usually interpreted as an example of mutualism—but, as we shall see, this may not always be the case. In their list of documented cases of cleaning behavior, Poulin and Grutter (1996) included two genera of decapod crustaceans, eight of tropical marine fishes, and three of birds. Most of the fish species that engage in cleaning are wrasses (Labridae) or

gobies (Gobiidae). Butterflyfishes (Chaetodontidae) and angelfishes (Pomacanthidae) have also been observed cleaning. Even though there does not appear to be any significant difference between the average parasite load of marine and freshwater fishes, the vast majority of instances of cleaning symbiosis are known from the marine environment—especially from coral reef habitats (Poulin, 1993). Some common inhabitants of kelp beds in northeast Pacific temperate waters, such as the kelp perch (*Brachyistius frenatus*) and the señorita (*Oxyjulis californica*) are known to engage in cleaning behavior (Hobson, 1971). Östlund-Nilsson et al. (2005) have recently documented two common species of shrimp of the genus *Palaemon* removing ectoparasites on plaice (*Pleuronectes platessa*)—the first report of temperate marine shrimps engaging in cleaning behavior. A few instances of cleaning symbiosis are known in freshwater environments. Juvenile bluegill sunfishes (*Lepomis macrochirus*) often clean larger adults, and adults will clean largemouth bass. This could be a most precarious relationship, as bluegill are among the preferred

food items for bass. Hippopotamuses that congregate in African pools and springs benefit from cleaning by cyprinids of the genera *Labeo*, *Barbus*, and *Garra*.

When considering the origin of cleaning symbiosis in fishes, certain families appear to possess the best suite of characteristics to make for a good cleaner. Wrasses are generally small, agile fishes with terminal mouths and dentition that is ideally suited for grasping small invertebrates. The typical wrasse diet consists of small benthic crustaceans. Despite the abundance of noncrustaceans parasitizing the surfaces of client fishes, cleaner wrasses still show a decided preference for epizootic crustaceans (Poulin and Grutter, 1996). The cleaners may not engage in this activity on a full-time basis, and they probably do not depend on it for a complete livelihood; some species clean other fishes only when they are juveniles and forsake the habit as adults. In those species for which cleaning provides a significant source of food, special color patterns and behaviors have evolved to advertise their services. In a Caribbean species of cleaner goby

anemones, seeking shelter among their stinging tentacles (see Plate 37.8). This constant and intimate association has resulted in the anemonefish being able to suppress or survive nematocyst discharges against itself. Just how this feat is accomplished is unclear. It may be that the host anemone becomes habituated to the constant presence of the fish. It is also possible that anemonefishes secrete substances that either actively inhibit nematocyst discharge or render the nematocyst toxins less effective. Some studies have suggested that the relationship is possible because anemonefishes transfer mucous secretions from their anemone host to their own bodies, resulting in an inhibition of nematocyst discharge. Researchers have also discovered that certain compounds secreted by the anemone induce anemonefishes to seek them out (Arvedlund and Nielsen, 1996; Murata et al., 1986). Although it has been suggested that this relationship is mutualistic because the anemone might

nutritionally benefit from food scraps or fecal wastes provided by the anemonefish, the anemones probably derive a greater benefit from the protection that these belligerent pomacentrids provide in the defense of their hosts from attack by butterflyfishes that prey on anemone tentacles (Fautin and Allen, 1992). Some studies (Fautin, 1991) have also suggested that photosynthetic activity by symbiotic algae is enhanced in anemones that are protected by anemonefishes.

Anther example of a mutualistic relationship between fishes and cnidarians, in this case involving stony corals, is known. Pomacentrid damselfishes dwelling among stony corals have been observed to engage in an especially remarkable form of mutualistic behavior. Many species of corals experience localized oxygen deprivation where the extensive branching patterns of the coral polyps might impede water circulation. In exchange for secure sleeping quarters

(*Elacatinus evelynae*), pronounced sex differences in cleaning behavior have been observed. This species forms monogamous pairs, in which the female engages in significantly more cleaning behavior; this is reflected in the assessment of gut contents. The female dominance of the cleaning station may reflect their greater energetic requirements for egg production (Whiteman and Côté, 2002).

In cleaning behavior, communication is essential. In a typical cleaning relationship, the cleaner will display itself at a prominent coral head, rock outcrop, sponge, or other conspicuous place where the cleaning activity will be carried out. This focal point may serve as a base from which the cleaner will range out over a larger territory that it has established and may aggressively defend against potential competitors. Such intraspecific aggression results in the partitioning of the reef into a number of non-overlapping, exclusively maintained cleaning territories. As in the case of the aforementioned goby species, cleaners may operate a station in pairs or in small groups. Two different species have sometimes been observed cleaning the

same client. A fish may "request" cleaning by a characteristic head-down posture or by gaping its mouth. The cleaner then approaches, engaging in behavior that indicates cooperation. Bluegill will darken when approaching a bass to be cleaned; this color is believed to be an indication of submissiveness (Sulak, 1975).

The behavior of a client toward a cleaner may be learned, through positive reinforcement received from the mitigation of a parasite load, or it may be innate. Although visual cues seem most important in initiating the cleaner–cleanee relationship, tactile cues are important in maintaining it (Losey, 1978). Cleaners appear to be able, through tactile stimulation, to get their clients to pose in a quiescent manner during cleaning. (Everybody loves a good massage, so why cannot fishes also enjoy a bit of pleasure from the grooming attentions of a cleaner?) Some species of cleaner fishes may be able to determine the relative satiation state of their potential clients, as it might be imprudent to enter the mouth of a client fish if it is especially hungry. These cleaners appear to direct greater amounts of tactile "dancing" behavior toward

hungry piscivorous clients than toward ones that are perceived as satiated (Grutter, 2004). It is possible that the symbiotic relationship is of sufficient benefit to the client that selection has resulted in innate posing behaviors in the presence of recognized cleaner species. In one study (Losey et al., 1995), pomacentrid damselfishes reared from eggs—and thus having no prior experience with cleaners—assumed the client posture when seeing a cleaner for the first time. Because fish size correlates with ectoparasite load, it might be expected that cleaners will exhibit preferences for larger individuals. The strength of the cleaner–client relationship, however, may be a function of factors other than the mere size of the client (Grutter and Poulin, 1998). Cleaners are known to "cheat" by consuming skin or mucus from their client. The consequences—aggression against the cleaner or avoidance of its cleaning station—appear sufficient to induce the cleaners to focus more on ectoparasites and less on the client's own tissues (Bshary and Grutter, 2005). Such "cheating" may be a consequence of ectoparasite abundance—where clients carry a heavy

among the branching corals, damselfishes of the genera *Dascyllus* and *Chromis* engage in rapid, high-frequency swimming motions with their fins while sleeping. Not only may this ensure adequate oxygen supplies for the fishes, but it also improves the oxygenation of the polyps and has been demonstrated to enhance their growth and reproductive potential (Goldschmid et al., 2004). Corals that shelter fishes show faster rates of growth and harbor greater amounts of nitrogen and zooxanthellae (Hay et al., 2004).

Juveniles of the sculpin genus *Artedius* have also been observed in a symbiotic relationship with anemones. Another Pacific coast sculpin (*Clinocottus globiceps*) appears to be a parasitic associate of sea anemones, as it feeds extensively on the tissues of its host. The reciprocal acclimatory behavior of the fish and its host, as observed in the anemonefishes, is not observed here (Yoshiyama et al., 1996a, 1996b). Scorpionfishes of the genus *Minous* exploit

another group of cnidarians by carrying a partial covering of hydroids on their skin, apparently for purposes of concealment. Many hydroid species are stinging, thus conferring deterrence in addition to camouflage.

Some gobies seek shelter in burrows excavated by shrimp or in the cavities created by the siphon tubes of clams. The blind goby (*Typhlogobius californiensis*) usually shares the burrow of the ghost shrimp *Calianassa*. A closer relationship is seen between crustaceans and the liparid genus *Careproctus*, members of which deposit eggs on or in the branchial chambers of large crabs. The pearlfishes (Carapidae; see Plate 13.5) are often associated with echinoderms and molluscs. Most species hide inside sea cucumbers, entering through the anus tailfirst in a corkscrew fashion. The inhospitable array of spines presented by sea urchins makes them a safe haven for shrimpfishes (Centriscidae; see Plate 15.5) and clingfishes (Gobiesocidae; see Plate 34.2).

ectoparasite load, the temptation to consume the client's tissues is less. This suggests that cleaning behavior exists along a continuum ranging from genuine mutualism to opportunistic parasitism, depending on the availability of resources (Cheny and Côté, 2005).

Conrad Limbaugh's (1961) pioneering studies on cleaning behavior in fishes instilled the perception that cleaning symbiosis represents a perfectly harmonious and mutually beneficial relationship. As the aforementioned studies suggest, this does not always seem to be the case. It has been difficult to demonstrate that the presence of cleaners directly benefits reef fish populations. Several studies in which cleaners were removed from the reef fish community have produced inconclusive results. Limbaugh's study claimed that cleaner removal resulted in the emigration of clients from the reef and higher levels of infestation in the remaining population. Subsequent studies (Gorlick et al., 1987; Grutter, 1997; Losey, 1972) have demonstrated no short-term negative impact of cleaner removal, however.

Rather than acting in a mutually beneficial manner, cleaner fish have been labeled "behavioral parasites" (Losey, 1987) that use skillful displays and tactile stimulation for their own ends—quick and easy access to food items on the skin of host fishes. Interesting enough, this accusation has also been leveled against one of the best-known terrestrial cleaners—oxpeckers (*Buphagus*). Studies have indicated that these birds, long perceived as benefiting their ungulate hosts by removing skin parasites, are not the most considerate of guests, because much if not most of their "cleaning" activity actually consists of keeping wounds fresh for the purpose of feeding on the blood of their hosts (Weeks, 2000). Not all client fishes are ideally cooperative, either. "Temperate seas make for intemperate behavior" might be an appropriate motto, as kelp perch do not seem to enjoy the relative immunity from predation by their clients seen in most tropical reef cleaners. Even in tropical situations, client groupers have been observed preying on cleaner wrasses (Francini-Filho et al., 2000). Yet cleaner fishes

have been demonstrated in some instances to have a measurable impact on the parasite load of their clients (Grutter, 1999), and the clients do keep coming back, lining up at the cleaning stations with the same innate attraction that children have to an ice cream stand on a summer's day. Cleaning behavior also has been shown to have practical applications to fish culture. Wrasses are used to keep parasite infestations in check in pen-reared salmon and tilapia (Bjordal, 1991; Cowell et al., 1993; Deady et al., 1995).

Perhaps the best argument in favor of cleaning symbiosis being a truly mutualistic relationship is the evolution of *cleaner mimics*—con artists that practice morphological and behavioral deception to gain access to the skin surfaces of client species (see Chapter 16, Plate 16.13A–B). Cleaning behavior—the classic demonstration of symbiosis in the ocean realm—surely must have been sustained for a long time in order to foster the evolution of deceptive intervention by cleaner mimics.

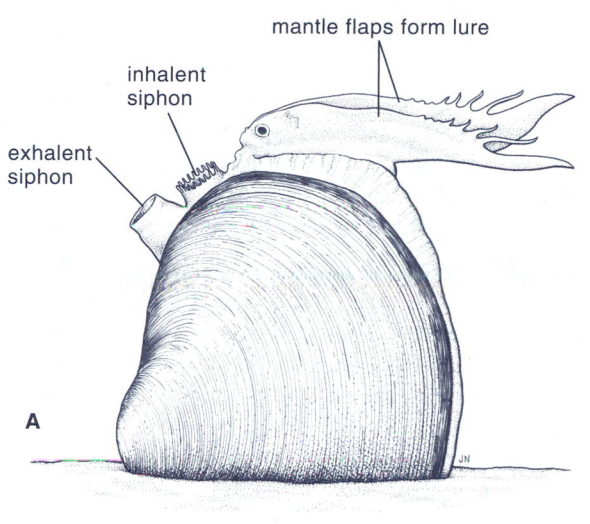

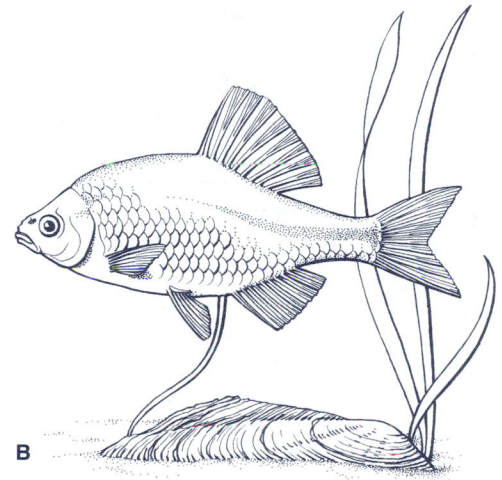

FIGURE 37.3
Reciprocal exploitation between bivalve molluscs and fishes. **A,** The mantle tissue of *Lampsilis ovata* mimics a small fish for the purpose of attracting a larger predator, to which the ejected glochidia larvae may then attach; **B,** Bitterling (*Rhodeus*) depositing eggs in the mantle cavity of a mussel (Source: Wickler, in Barth and Broshears, 1982).

LEARNING IN FISHES

Fishes engage in a variety of activities in which *learning*—defined as a change in behavior acquired through experience—plays a major role. Among the behaviors that learning plays a role in are homing and migration (see Chapter 36), foraging, social interactions, recognition processes, and avoidance of threatening situations (Kieffer and Colgan, 1992; Odling-Smee and Braithwaite, 2003). Obviously, learning is an integral component of the total adaptive response of an organism to its environment, and fishes have proven themselves adept at modifying their behavior in response to environmental circumstances. Even general observation of the lives of fishes leads to the conclusion that fishes learn quickly and have reasonably good memories. The appreciation of this aspect of fish behavior gives us insight into its application in the natural world and facilitates our manipulation of fishes in practical situations such as hatchery rearing.

Adaptive Behavioral Modification and Its Applications

Recent studies on the ecology of fishes have underscored the adaptive significance of behavioral plasticity in optimizing the use of aquatic resources by fishes. Studies on the feeding dynamics of fishes inhabiting small streams have demonstrated the ability of prey species to detect the presence of predators and to adjust their feeding regimen either spatially or temporally. Stoneroller minnows (*Campostoma*) confine their algal grazing activities to pools where predatory bass are absent (Power, 1987). The feeding behavior of the longnose dace (*Rhinichthys cataractae*), a nocturnally foraging species, can be experimentally modified using different light intensities (Beers and Culp, 1990).

Psychobiologists have investigated the behavioral adaptations of fishes not only because of their intrinsic interest, but also because they make convenient subjects for the study of certain brain functions. The **conditioned response** is well known in fishes and is widely employed in a variety of studies. An experimental animal can be conditioned through the application of reward or punishment to respond to a variety of stimuli. This response can take the form of obvious behaviors, such as flight or the initiation of feeding behavior, or more subtle ones such as changes in heart or respiratory rate. Conditioning paves the way for studying the ability of the fish to discriminate among colors, visual patterns, or small differences in sounds, odors, or temperature. Fishes can be trained to carry out tasks that require a series of actions. Once a subject is conditioned, memory can be investigated; and once memory is understood, the researcher can study the effects of chemicals, elapsed time, or other factors, such as the removal of parts of the brain on retention. The removal of the forebrain does not make experimental fish incapable of learning or remembering, although some functions are slowed or altered. Certain chemicals appear to prolong memory in goldfishes, but others are known to interfere with retention. For instance, DDT seems to reduce retention time in salmonids (Anderson and Peterson, 1969).

Conditioning has made the artificial rearing of certain species of fishes much easier. Salmonids and many other species can be taught to eat artificially formulated diets instead of natural foods that might be prohibitively expensive to furnish. Because marine species such as eels (*Anguilla japonica*) and yellowtail (*Seriola quinqueradiata*) cannot be spawned in captivity, their young must be rounded up from wild stocks. It is possible to pen-rear them in high concentrations, however, because they can be trained to consume commercially prepared diets. Not only do these fishes learn to accept unfamiliar foods, but they learn rapidly where and at what time the foods will be given. The ayu (*Plecoglossus altivelis*) normally feeds by scraping diatoms off submerged rocks, but it readily learns to take dry food from the surface. Channel catfishes (*Ictalurus punctatus*) are normally subsurface feeders, but are fed floating food pellets in fish farms.

The mutability of fish behavior enables fish culturists to breed out traits that would render a certain species unsuitable for high-density rearing situations. An example from one of the most intensively cultured species is a case in point. In the wild, trout that are unwary of objects moving on the bank or overhead have a greater chance of falling victim to predators, whereas the wary ones that retreat from such disturbances are more likely to grow to reproductive age. In the hatchery, however, little premium is placed on wariness. In fact, the wariest trout may not get its fair share of food, will grow poorly, and will be subject to greater stress. The least wary individuals are better fed and less subject to panic and injury brought on by reaction to overhead disturbances. Hatchery strains of trout that have been under domestication for 40 or 50 years have behavior patterns quite different from those of wild trout of the same species. Fishes that are only one generation removed from the wild have been seen to exhibit changes in behavior.

The behavior of fishes is indeed fascinating in its diversity. It is a challenging topic, if only because of the difficulty in extracting reliable information from a realm entirely different from our own. The challenge also lies in discerning the significance of behavioral adaptations to the environment in such a diverse and successful group of vertebrates. As such, studies of the comparative ethology of fishes make valuable contributions to understanding the basic principles of ecology and evolution.

KEY POINTS AND CONNECTIONS

- The emergence of two disciplines—behavioral ecology and sociobiology—underscores the importance of an interdisciplinary approach to the study of behavior in animals. The actions of feeding and reproduction present opportunities to investigate the behavioral adaptations of individuals to their environment—specifically, how they apply their sensory modalities to the problems of detecting and consuming food and engaging members of the opposite sex in the event of procreation.

- Fishes incorporate receptors of the acousticolateralis , olfactory, gustatory, tactile, electrotactile, and visual systems in the location and capture of food. Given the range of feeding types among fishes, the degree of variation in the application of these different sensory systems to feeding behavior should not come as a surprise. Certain feeding modes, such as macrocarnivory and suction feeding, as practiced in planktivores, are presented as examples.

The sensory systems that are important in feeding behavior are discussed in Chapters 20–22, whereas feeding adaptations themselves are covered in Chapter 23.

- The ecological significance of feeding behavior is illustrated in optimal foraging theory, which attempts to explain how animals maximize their net energy gain while minimizing their metabolic costs. Fishes make discrete choices in feeding, as measured in studies of dietary selectivity under different conditions of prey availability or predator presence.

Optimal foraging theory is also discussed in Chapters 30 and 32.

- Through sociobiology, we seek to explain the adaptive advantages of certain forms of social behavior. Those actions of fishes that bring them into contact with other organisms, especially other fishes, will elicit certain forms of communication, including visual, chemical, and electrical modes.

The nature of those cellular components of the skin that are involved in the production and control of color and their application in visual communication are discussed in Chapter 18.

- Behavior in fishes that is directed toward the process of reproduction ranges from very simple to most elaborate. Although pelagic schooling forms tend to exhibit a minimum of reproductive behavior, species that are more closely associated with the benthic realm may engage in a variety of behaviors associated with the selection and preparation of the breeding site, courtship and spawning, and care of the fertilized eggs. Sexual selection demonstrates the significance that behavior may have in the evolution of optimal fitness in species.

Reproductive adaptations of fishes are the focus of Chapter 27.

- A shoal is defined as a group of fishes brought together so that they constitute a social group. A school is defined as a social grouping of fishes that orient and swim with a particular polarity and synchrony.

Adaptive advantages of shoaling and schooling behavior include increased protection against predators and facilitation of feeding. Schooling and shoaling behavior may also accompany the process of reproduction in many species.

The hydrodynamics of swimming, which is integral to and possibly even enhanced by schooling behavior, is discussed in Chapters 18 and 19, especially in the Going Deeper boxes of those chapters.

• Fishes engage in a range of symbiotic behaviors, including mutualism, commensalism, and, to a limited degree, parasitism. Most of the social relationships involving different species of fishes are commensal in nature.

FISH LINKS

http://www.louisville.edu/~laduga01/dugatkin.htm Home page of Dr. Lee Alan Dugatkin of the University of Louisville. Dr. Dugatkin has diverse interests in the area of animal behavior. Many of his insights have come from studies of the behavior of fishes, especially poeciliids.

BUILDING AN ICHTHYOLOGY LIBRARY

Almada, V. C., R. F. Oliviera, and E. J. Gonçalves (Eds.). 1998. *Behavior and conservation of littoral fishes.* Instituto Superior de Psicologia Aplicada, Lisbon, Portugal.

Godin, J.-G. J. (Ed.). 1997. *Behavioural ecology of teleost fishes.* Oxford University Press, Oxford.

Houde, A. E. 1997. *Sex, color, and mate choice in guppies.* Princeton University Press, Princeton, NJ.

REFERENCES

Alonzo, S. H., and R. R. Warner. 2000. Female choice, conflict between the sexes, and the evolution of male alternative reproductive behaviors. *Evol. Ecol. Res. 2:* 149–170.

Amorim, M. C. P., and A. D. Hawkins. 2000. Growling for food: Acoustic emissions during competitive feeding of the streaked gurnard. *J. Fish Biol. 57:* 895–907.

Amundsen, T. 2003. Fishes as models in studies of sexual selection and parental care. *J. Fish Biol. 63*(Suppl. A): 17–52.

Anderson, J. M., and M. R. Peterson. 1969. DDT: Sublethal effects on brook trout nervous system. *Science 164:* 440–441.

Andersson, M. 1994. *Sexual selection.* Princeton University Press, Princeton, NJ.

Arvedlund, M., and L. E. Nielsen. 1996. Do the anemonefish *Amphiprion ocellaris* (Pisces: Pomacentridae) imprint themselves to their host sea anemone *Heteractis magnifica* (Anthozoa: Actinidae)? *Ethology 102:* 197–211.

Atema, J. 1980. Chemical senses, chemical signals, and feeding behavior in fishes, pp. 57–101. In Fish behavior and its use in the capture and culture of fishes, J. E. Bardach, J. J. Magnuson, R. C. May, and J. M. Reinhart (Eds.), *ICLARM Conf. Proc. No. 5,* Manila.

Barlow, G. M. 1963. Ethology of the Asian teleost *Badis badis.* II. Motivation and signal value of the colour patterns. *Anim. Behav. 11:* 97–105.

———. 1974. Extraspecific imposition of social grouping among surgeonfishes. *J. Zool. Lond. 174:* 333–340.

———. 1984. Patterns of monogamy among teleost fishes. *Arch. Fisch Wiss. 35:* 75–123.

Barth, R. H., and R. E. Broshears. 1982. *The invertebrate world.* W. B. Saunders, Philadelphia.

Beers, C. E., and J. M. Culp. 1990. Plasticity in foraging of a lotic minnow (*Rhinichthys cataractae*) in response to different light intensities. *Can. J. Zool. 68:* 101–105.

Bjordal, A. 1991. Wrasse as cleaner fish for farmed salmon. *Prog. Underw. Sci. 16:* 17–28.

Blaxter, J. H. S. 1980. Vision and feeding of fishes, pp. 32–56. In Fish behavior and its use in the capture and culture of fishes, J. E. Bardach, J. J. Magnuson, R. C. May, and J. M Reinhart (Eds.), *ICLARM Conf. Proc. No. 5,* Manila.

Blumer, L. S. 1982. A bibliography and categorization of bony fishes exhibiting parental care. *Zool. J. Linn. Soc. 76:* 1–22.

Bourke, P., P. Magnan, and M. A. Rodríguez. 1999. Phenotypic responses of lacustrine brook charr in relation to the intensity of interspecific competition. *Evol. Ecol. 13:* 19–31.

Breder, C. M., Jr. 1959. Studies on social groupings in fishes. *Bull. Am. Mus. Nat. Hist. 117:* 393–482.

———, and D. E. Rosen. 1966. *Modes of reproduction in fishes.* Natural History Press, Garden City, NY.

Bres, M. 1993. The behavior of sharks. *Rev. Fish Biol. Fisheries 3:* 133–159.

Brett, J. R. 1971. Energetic response of salmon to temperature. A study of some thermal relations in the physiology and freshwater ecology of sockeye salmon (*Oncorhynchus nerka*). *Amer. Zool. 11:* 99–113.

Brooks, R. 2000. Negative genetic correlation between male sexual attractiveness and survival. *Nature 406:* 67–70.

Bshary, R., and A. S. Grutter. 2005. Punishment and partner switching cause cooperative behavior in a cleaning mutualism. *Biol. Lett.* (*FirstCite Early Online Publ.*).

Burgess, J. W., and E. Shaw. 1979. Development and ecology of fish schooling. *Oceanus 22:* 11–17.

Buston, P. 2003. Social hierarchies: Size and growth modification in clownfish. *Nature 424:* 145–146.

———. 2004. Territory inheritance in the clown anemonefish. *Proceedings of the Royal Society, Ser. B (Suppl.) 271:* S252–S254.

Caira, J. N., G. W. Benz, J. Borucinska, and N. E. Kohler. 1997. Pugnose eels, *Simenchelys parasiticus* (Synaphobranchidae) from the heart of a shortfin mako, *Isurus oxyrinchus* (Lamnidae). *Env. Biol. Fishes 49:* 139–144.

Cheny, K. L., and I. M. Côté. 2005. Mutualism or parasitism? The variable outcome of cleaning symbiosis. *Biol. Lett. 1:* 162–165.

Colgan, P. 1986. Motivational basis of fish behavior, pp. 23–46. In *The behavior of teleost fishes,* T. J. Pitcher (Ed.). Johns Hopkins University Press, Baltimore.

Cowell, L. E., W. O. Watanabe, W. D. Head, J. J. Grover, and J. M. Shenker. 1993. Use of tropical cleaner fish to control the ectoparasite *Neobenedenia melleni* (Monogenea: Capsalidae) on seawater-cultured Florida red tilapia. *Aquaculture 113:* 189–200.

Culp, J. M. 1989. Nocturnally constrained foraging of a lotic minnow (*Rhinichthys cataractae*). *Can. J. Zool. 67:* 2008–2012.

Cushing, D. H., and F. R. Hardin Jones. 1968. Why do fish school? *Nature 218:* 918–920.

Deady, S., S. J. A. Varian, and J. M. Fives. 1995. The use of cleaner-fish to control sea lice on two Irish salmon (*Salmo salar*) farms with particular reference to wrasse behavior in salmon cages. *Aquaculture 131:* 73–90.

DeLoach, N. 1999. *Reef fish behavior: Florida, Caribbean, Bahamas.* New World, Jacksonville, FL.

Dugatkin, L. A. 1994. Juvenile three-spined sticklebacks avoid parasitized conspecifics. *Env. Biol. Fishes 39:* 215–218.

———, and J.-G. J. Godin. 1992. Predator inspection, shoaling and foraging under predation hazard in the Trinidadian guppy, *Poecilia reticulata. Env. Biol. Fishes 34:* 265–276.

———, and———. 1998. How females choose their mates. *Sci. Am. 278*(4): 56–61.

———, and D. S. Wilson. 1992. The prerequisites for strategic behavior in bluegill sunfish, *Lepomis macrochirus. Anim. Behav. 44:* 223–230.

Elgar, M., and B. Crespi (Eds.). 1992. *Cannibalism: Ecology and evolution among diverse taxa.* Oxford University Press, New York.

Engana, A. C., and J. E. McCosker. 1984. Attacks on divers by white sharks in Chile. *Calif. Fish Game 70:* 173–179.

Fautin, D. G. 1991. The anemonefish symbiosis: What is known and what is not. *Symbiosis 10:* 23–46.

———, and G. R. Allen. 1992. *Field guide to anemone fishes and their host sea anemones.* Western Australian Museum, Perth.

Fergusson, I. K., L. J. V. Compagno, and M. A. Marks. 2000. Predation by white sharks *Carcharodon carcharias* (Chondrichthyes: Lamnidae) upon chelonians, with new records from the Mediterranean Sea and a first record of the ocean sunfish *Mola mola* (Osteichthyes: Molidae) as stomach contents. *Env. Biol. Fishes 58:* 447–453.

Forsgren, E., T. Amundsen, Å. A. Borg, and J. Bjelvenmark. 2004. Unusually dynamic sex roles in a fish. *Nature 429:* 551–554.

Francini-Filho, R. B., R. L. Moura, and I. Sazima. 2000. Cleaning by the wrasse *Thalassoma noronhanum*, with two records of predation by its grouper client *Cephalopholis fulva. J. Fish Biol. 56:* 802–809.

Frazzetta, T. H. 1994. Feeding mechanisms in sharks and other elasmobranches, pp. 31–57. In *Advanced comparative environmental physiology 18: Biomechanics of feeding in vertebrates*, V. L. Bels, M. Chardon, and P. Vandewalle (Eds.). Springer Verlag, Berlin.

Fretwell, S. D., and H. L. Lucas. 1970. On territorial behavior and other factors influencing habitat distribution in birds. *Acta Biotheor. 19:* 16–36.

Gallego, A., M. R. Heath, and R. J. Fryer. 1995. Premature schooling of larval herring in the presence of more advanced conspecifics. *Anim. Behav. 50:* 333–341.

Gerking, S. D. 1994. *Feeding ecology of fishes.* Academic Press, San Diego.

Gillis, D. M. 2003. Ideal free distributions in fleet dynamics: A behavioral perspective on vessel movement in fisheries analysis. *Can. J. Zool. 81:* 177–187.

Giske, J., and A. Gro Vea Salvanes. 1995. Why pelagic planktivores should be unselective feeders. *J. Theor. Biol. 173:* 41–50.

Godin, J.-G. J. 1997. *Behavioral ecology of teleost fishes.* Oxford University Press, Oxford.

———, and S. A. Davis. 1995. Who dares, benefits: Predator approach behavior in the guppy (*Poecilia reticulata*) deters predator pursuit. *Proc. Roy. Soc. Lond. B 259:* 193–200.

———, and L. A. Dugatkin. 1996. Female mating preference for bold males in the guppy, *Poecilia reticulata. Proc. Natl. Acad. Sci. USA 93:* 10262–10267.

Goldschmid, R., R. Holzman, D. Weihs, and A. Genin. 2004. Aeration of corals by sleep-swimming fish. *Limnol. Oceanogr. 49:* 1832–1839.

Goldschmidt, T. 1991. Egg mimics in haplochromine cichlids (Pisces, Perciformes) from Lake Victoria. *Ethology 88:* 177–190.

Gorlick, D. L., P. D. Atkins, and G. S. Losey. 1987. Effect of cleaning by *Labroides dimidiatus* (Labridae) on an ectoparasite population infecting *Pomacentrus vaiuli* (Pomacentridae) at Enewetak Atoll. *Copeia 1987:* 41–45.

Grant, J. W. A. 1997. Territoriality, pp. 81–103. In *Behavioural ecology of teleost fishes*, J.-G. J. Godin (Ed.). Oxford University Press, Oxford.

Grutter, A. S. 1997. Effect of the removal of cleaner fish on the abundance and species composition of reef fish. *Oecologia 111:* 137–143.

———. 1999. Cleaner fish really do clean. *Nature 398:* 672–673.

———. 2004. Cleaner fish use tactile dancing behavior as a preconflict management strategy. *Current Biol. 14:* 1080–1083.

———, and R. Poulin. 1998. Cleaning of coral reef fishes by the wrasse *Labroides dimidiatus:* Influence of client body size and phylogeny. *Copeia 1998:* 120–127.

Haag, W. R., and M. L. Warren, Jr. 1999. Mantle displays of freshwater mussels elicit attacks from fish. *Freshw. Biol. 42:* 35–40.

———, ———, and M. Shillingsford. 1999. Host fishes and host-attracting behavior of *Lampsilis altilis* and *Villosa vibex* (Bivalvia: Unionidae). *Am. Midl. Nat. 141:* 149–157.

Hara, T. J. 1975. Olfaction in fish. *Prog. Neurobiol. 5:* 271–335.

———. 1993. Chemoreception, pp. 191–218. In *The physiology of fishes*, D. H. Evans (Ed.). CRC Press, Boca Raton, FL.

Hay, M. E., J. D. Parker, D. E. Burkepile, C. C. Caudill, A. E. Wilson, Z. P. Hallinan, and A. D. Chequer. 2004. Mutualisms and aquatic community structure: The enemy of my enemy is my friend. *Ann. Rev. Ecol. Syst. 35:* 175–197.

Healey, M. C. 1991. Life history of chinook salmon, pp. 311–393. In *Pacific salmon life histories*, C. Groot and L. Margolis (Eds.). University of British Columbia Press, Vancouver.

———, and R. Prieston. 1973. The interrelationships among individuals in a fish school. *Fish. Res. Bd. Can. Tech. Rep. No. 389.*

Henson, S. A., and R. R. Warner. 1997. Male and female alternative reproductive behaviors in fishes: A new approach using intersexual dynamics. *Ann. Rev. Ecol. Syst. 28:* 571–592.

Hoare, D. J., J. Krause, N. Peuhkuri, and J.-G. J. Godin. 2000. Body size and shoaling in fish. *J. Fish Biol. 57:* 1351–1366.

Hobbs, J.-P. A., P. L. Munday, and G. P. Jones. 2004. Social induction of maturation and sex determination in a coral reef fish. *Proc. Roy. Soc. Lond. B 271:* 2109–2114.

Hobson, E. S. 1968. Predatory behavior of some shore fishes in the Gulf of California. *U.S. Fish. Wildl. Serv. Res. Rep. 73.*

———. 1971. Cleaning symbiosis among California inshore fishes. *Fish. Bull. 69:* 491–523.

Hughes, K. A., L. Du, F. H. Rodd, and D. N. Reznick. 1999. Familiarity leads to female mate preference for novel males in the guppy, *Poecilia reticulata. Anim. Behav. 58:* 907–916.

Huth, A., and C. Wissel. 1992. The simulation of the movement of fish schools. *J. Theor. Biol. 156:* 365–385.

———, and———. 1993. Analysis of the behavior and the structure of fish schools by means of computer simulations. *Comm. Theor. Biol. 3:* 169–201.

Jamieson, I. J. 1995. Do female fish prefer to spawn in nests with eggs for reasons of mate choice copying or egg survival? *Am. Nat. 145:* 824–832.

Karplus, I. 1978. A feeding association between the grouper *Epinephelus fasciatus* and the moray eel *Gymnothorax griseus. Copeia 1978:* 164.

Keenleyside, M. H. A. 1979. *Diversity and adaptation in fish behavior.* Springer Verlag, Berlin.

Kennedy, M., and R. D. Gray. 1994. Agonistic interactions and the distribution of foraging organisms: Individual costs and social information. *Ethology 96:* 155–165.

Kieffer, J. D., and P. W. Colgan. 1992. The role of learning in fish behavior. *Rev. Fish Biol. Fisheries 2:* 125–143.

Klauswitz, W., J. E. McCosker, J. E. Randall, and H. Zetsche. 1978. *Hoplolatilus chlupatyi* n. sp., un nouveau poisson marin des Phillippines (Pisces, Perciformes, Percoidei, Branchiostegidae). *Rev. Fr. Aquariol. 2:* 41–48.

Kleerekoper, H. 1969. *Olfaction in fishes.* Indiana University Press, Bloomington.

Klimley, A. P. 1994. The predatory behavior of the white shark. *Am. Sci. 82:* 122–133.

———, P. Pyle, and S. D. Anderson. 1996. The behavior of white sharks and their pinniped prey during predatory attacks, pp. 175–191. In *Great white sharks: The biology of* Carcharodon carcharias, A. P. Klimley and D. G. Ainley (Eds.). Academic Press, San Diego.

Kocher, T. D., and K. R. McKaye. 1983. Defense of heterospecific cichlids by *Cyrtocara moorii* in Lake Malawi, Africa. *Copeia 1983:* 544–547.

Koops, M. A., and M. V. Abrahams. 1999. Assessing the ideal free distribution: Do guppies use aggression as public information about patch quality? *Ethology 105:* 737–746.

Krause, J., D. J. Hoare, D. Croft, J. Lawrence, A. Ward, G. D. Ruxton, J.-G. J. Godin, and R. James. 2000. Fish shoal composition: Mechanisms and constraints. *Proc. Roy. Soc. Lond. B 267:* 2011–2017.

Landa, J. T. 1998. Bioeconomics of schooling fishes: Selfish fish, quasi-free riders, and other fishy tales. *Env. Biol. Fishes 53:* 353–364.

Landeau, L., and J. Terborgh. 1986. Oddity and the "confusion effect" in predation. *Anim. Behav. 34:* 1372–1380.

Langerhans, R. B., C. A. Layman, and T. J. DeWitt. 2005. Male genital size reflects a tradeoff between attracting mates and avoiding predators in two live-bearing fish species. *Proc. Natl. Acad. Sci. USA 102:* 7618–7623.

Lim, T. M. 1981. Effects of schooling prey on the hunting behavior of a fish predator, p. 132. In *Ecology and ethology of fishes,* D. L. G. Noakes and J. A. Ward (Eds.). Junk, The Hague.

Limbaugh, C. 1961. Cleaning symbiosis. *Sci. Am. 205*(2): 42–49.

Lindström, K. 1998. Effects of costs and benefits of brood care on filial cannibalism in the sand goby. *Behav. Ecol. Sociobiol. 42:* 101–106.

Locascio, J. V., and D. A. Mann. 2005. Effects of Hurricane Charley on fish chorusing. *Biol. Lett. 1:* 362–365.

Losey, G. S. 1972. The ecological importance of cleaning symbiosis. *Copeia 1972:* 820–833.

———. 1978. The symbiotic behavior of fishes, pp. 1–31. In *The behavior of fish and other aquatic organisms,* D. I. Mostofsky (Ed.). Academic Press, New York.

———. 1987. Cleaning symbiosis. *Symbiosis 4:* 229–258.

———, J. L. Mahon, and B. S. Danilowicz. 1995. Innate recognition by host fish of their cleaning symbiont. *Ethology 100:* 277–283.

Magnuson, J. J. 1962. An analysis of aggressive behavior, growth, and competition for food and space in medaka (*Oryzias latipes,* Pisces, Cyprinodontidae). *Can. J. Zool. 40:* 313–363.

Magurran, A. E. 1990. The adaptive significance of schooling as an anti-predator defence in fish. *Ann. Zool. Fennici 27:* 51–66.

———, and J. A. Bendelow. 1990. Conflict and cooperation in white cloud mountain minnow schools. *J. Fish Biol. 37:* 77–83.

———, and B. H. Seghers. 1994. Predator inspection behavior covaries with schooling tendency amongst wild guppy, *Poecilia reticulata,* populations in Trinidad. *Behavior 128*(1–2): 121–134.

Maher, C. R., and D. F. Lott. 2000. A review of ecological determinants of territoriality within vertebrate species. *Am. Midl. Nat. 143:* 1–29.

Major, P. F. 1978. Predator–prey interaction in two schooling fishes, *Caranx ignobilis* and *Stolephorus purpureus. Anim. Behav. 26:* 760–777.

Mansueti, R. 1963. Symbiotic behavior between small fishes and jellyfishes with new data on that between the stromateid *Peprilis alepidotus* and the scyphomedusa *Chrysaora quinquecirrha. Copeia 1963:* 40–80.

Matern, S. A., J. J. Cech, Jr., and T. E. Hopkins. 2000. Diel movements of bat rays, *Myliobatis californica,* in Tomales Bay, California: Evidence for behavioral thermoregulation? *Env. Biol. Fishes 58:* 173–182.

McCosker, J. E. 1985. White shark attack behavior: Observations and speculations about predator and prey strategies. *Mem. So. Calif. Acad. Sci. 9:* 123–135.

McKaye, K. R. 1977. Defense of a predator's young by a herbivorous fish: An unusual strategy. *Am. Nat. 111:* 301–315.

———, and N. M. McKaye. 1977. Communal care and kidnapping of young by parental cichlids. *Evolution 31:* 674–681.

———, and M. K. Oliver. 1980. Geometry of a selfish school: Defence of cichlid young by bagrid catfish in Lake Malawi, Africa. *Anim. Behav. 28:* 1287.

McLaughlin, R. L., M. M. Ferguson, and D. L. G. Noakes. 1999. Adaptive peaks and alternative foraging tactics in brook char: Evidence of short-term divergent selection for sitting-and-waiting and actively searching. *Behav. Ecol. Sociobiol. 45:* 386–395.

McLennan, D. A. 1994. A phylogenetic approach to the evolution of fish behavior. *Rev. Fish Biol. Fisheries 4:* 430–460.

Moller, P. 1976. Electric signals and schooling behavior in a weakly electric fish, *Marcusenius cyprinoides* L. (Mormyriformes). *Science 193:* 697–699.

Montgomery, W. L. 1975. Interspecific associations of sea basses (Serranidae) in the Gulf of California. *Copeia 1975:* 785–787.

Murata, M., K. Miyagawa-Kohshima, K. Nakanishi, and Y. Naya. 1986. Characterization of compounds that induce symbiosis between sea anemone and fish. *Science 234:* 585–587.

Neverman, D., and W. A. Wurtsbaugh. 1994. The thermoregulatory function of diel vertical migration for a juvenile fish, *Cottus extensus. Oecologia 98:* 247–256.

Noakes, D. L. G. 1978. Ontogeny of behavior in fishes: A survey and suggestions, pp. 103–125. In *The development of behavior: Comparative and evolutionary aspects,* G. M. Burghardt and M. Bekoff (Eds.). Garland STPM Press, New York.

———. 1986. Genetic basis of behavior, pp. 3–22. In *The behavior of teleost fishes,* T. J. Pitcher (Ed.). Johns Hopkins University Press, Baltimore.

Odling-Smee, L., and V. A. Braithwaite. 2003. The role of learning in fish orientation. *Fish Fisheries 4:* 235–246.

Okuda, N. 1999. Female mating strategy and male brood cannibalism in a sand-dwelling cardinalfish. *Anim. Behav. 58:* 273–279.

Olson, V. A., and I. P. F. Owens. 1998. Costly sexual signals: Are carotenoids rare, risky, or required? *TREE 13:* 510–514.

Östlund-Nilsson, S., and G. E. Nilsson. 2004. Breathing with a mouth full of eggs: Respiratory consequences of mouthbrooding in cardinalfish. *Proc. Roy. Soc. Lond. (Biol. Sci.) B 271:* 1015–1022.

———, J. H. A. Becker, and G. E. Nilsson. 2005. Shrimps remove ectoparasites from fishes in temperate waters. *Biol. Lett. (FirstCite Early Online Publ.).*

Parrish, J. K., and L. Edelstein-Keshet. 1999. Complexity, pattern, and evolutionary trade-offs in animal aggregation. *Science 284:* 99–101.

———, and ———. 2000. Response: Benefits of membership. *Science 287:* 804–807.

Peuhkuri, N. 1999. Size-assorted fish shoals and the majority's choice. *Behav. Ecol. Sociobiol. 46:* 307–312.

Pilastro, A., E. Giacomello, and A. Bisazza. 1997. Sexual selection for small size in male mosquitofish (*Gambusia holbrooki*). *Proc. Roy. Soc. Lond. B 264:* 1125–1129.

Pitcher, T. J. 1986. Functions of shoaling behavior in teleosts, pp. 294–337. In *The behavior of teleost fishes,* T. J. Pitcher (Ed.). Johns Hopkins University Press, Baltimore.

———. 1992. Who dares, wins: The function and evolution of predator inspection behavior in shoaling fish. *Neth. J. Zool. 42:* 371–391.

———, and J. K. Parrish. 1993. Functions of shoaling behavior in teleosts, pp. 363–440. In *The behavior of teleost fishes* (2nd ed.), T. J. Pitcher (Ed.). Chapman and Hall, London.

———, B. L. Partridge, and C. S. Wardle. 1976. A blind fish can school. *Science 194:* 963–965.

Pollock, M. S., D. P. Chivers, R. S. Mirza, and B. D. Wisenden. 2003. Fathead minnows, *Pimephales promelas,* learn to recognize chemical alarm cues of introduced brook stickleback, *Culaea inconstans. Env. Biol. Fishes 66:* 313–319.

Potts, G. W. 1970. The schooling ethology of *Lutianus monostigma* (Pisces) in the shallow reef environment of Aldabra. *J. Zool. Lond. 161:* 223–235.

Poulin, R. 1993. A cleaner perspective on cleaning symbiosis. *Rev. Fish Biol. Fisheries 3:* 75–79.

———, and A. S. Grutter. 1996. Cleaning symbiosis: Proximate and adaptive explanations. *BioScience 46:* 512–517.

Power, M. E. 1987. Predator avoidance by grazing fishes in temperate and tropical streams: Importance of stream depth and prey size, pp. 333–351. In *Predation: Direct and indirect impacts in aquatic communities,* W. C. Kerfoot and A. Sih (Eds.). University Press of New England, Dartmouth, NH.

Radakov, D. V. 1973. *Schooling in the ecology of fish.* Halstead Press, New York.

Reavis, R. H. 1997a. The natural history of a monogamous coral-reef fish, *Valencienna strigata* (Gobiidae): 1. Abundance, growth, survival and predation. *Env. Biol. Fishes 49:* 239–246.

———. 1997b. The natural history of a monogamous coral-reef fish, *Valencienna strigata* (Gobiidae): 2. Behavior, mate fidelity, and reproductive success. *Env. Biol. Fishes 49:* 247–257.

———, and G. W. Barlow. 1998. Why is the coral-reef fish *Valencienna strigata* (Gobiidae) monogamous? *Behav. Ecol. Sociobiol. 43:* 229–237.

Reebs, S. G., and P. W. Colgan. 1992. Proximal cues for nocturnal egg care in convict cichlids, *Cichlosoma nigrofasciatum. Anim. Behav. 43:* 209–214.

Robertson, D. R., H. P. A. Sweatman, E. A. Fletcher, and M. G. Cleland. 1976. Schooling as a mechanism for circumventing the territoriality of competitors. *Ecology 57:* 1208–1220.

Rodd, F. H., K. A. Hughes, G. F. Grether, and C. T. Baril. 2002. A possible non-sexual origin of mate preference: Are male guppies mimicking fruit? *Proc. Roy. Soc. Lond. (Biol. Sci.) B 269:* 475–481.

Rüber, L., R. Britz, H. H. Tan, P. K. L. Ng, and R. Zardoya. 2004. Evolution of mouthbrooding and life-history correlates in the fighting fish genus *Betta. Evolution 58:* 799–813.

Ruzzanti, D. E., and R. W. Doyle. 1991. Rapid behavioral changes in medaka (*Oryzias latipes*) caused by selection for competitive and noncompetitive growth. *Evolution 45:* 1936–1946.

Sargent, R. C., and M. R. Gross. 1986. Williams' principle: An explanation of parental care in teleost fishes, pp. 275–293. In *The behavior of teleost fishes,* T. J. Pitcher (Ed.). Johns Hopkins University Press, Baltimore.

———, and ———. 1993. Williams' principle: An explanation of parental care in teleost fishes, pp. 333–362. In *The behavior of teleost fishes* (2nd ed.), T. J. Pitcher (Ed.). Chapman and Hall, London.

Segaar, J., J. P. C. deBruin, and M. E. van der Meche-Jacobi. 1983. Influence of chemical receptivity on reproductive behavior of the male three-spined stickleback (*Gasterosteus aculeatus* L.). *Behavior 86:* 100–166.

Shaw, E. 1978. Schooling fishes. *Am. Sci. 66:* 166–175.

Shine, R. 1978. Propagule size and parental care: The "safe harbor" hypothesis. *J. Theor. Biol. 75:* 417–424.

Sikkel, P. C. 1989. Egg presence and developmental stage influence spawning site choice by female garibaldi. *Anim. Behav. 38:* 447–456.

———. 1994. Filial cannibalism in a paternal-caring marine fish: The influence of egg developmental stage and position in the nest. *Anim. Behav. 47:* 1149–1158.

Spotte, S. 2002. *Candiru: Life and legend of the bloodsucking catfishes.* Creative Arts, Berkeley, CA.

Steck, N., C. Wedekind, and M. Milinski. 1999. No sibling odor preference in juvenile three-spined sticklebacks. *Behav. Ecol. 10:* 493–497.

Sulak, K. J. 1975. Cleaning behavior in the centrarchid fishes, *Lepomis macrochirus* and *Micropterus salmoides. Anim. Behav. 23:* 331–334.

Taylor, L. R., L. J. V. Compagno, and P. J. Struhsaker. 1983. Megamouth—a new species, genus, and family of lamnoid shark (*Megachasma pelagios,* Family Megachasmidae) from the Hawaiian Islands. *Proc. Calif. Acad. Sci. 43*(8): 87–110.

Todd, J. H. 1971. The chemical language of fishes. *Sci. Am. 244*(5): 98–108.

———, J. Atema, and J. E. Bardach. 1968. Chemical communication in social behavior of a fish, the yellow bullhead (*Ictalurus natalis*). *Science 158:* 672–673.

Tregenza, T., and D. J. Thompson. 1998. Unequal competitor ideal free distribution in fish? *Evol. Ecol. 12:* 655–666.

Van Weerd, J. H., and C. J. J. Richter. 1991. Sex pheromones and ovarian development in teleost fish. *Comp. Biochem. Physiol. 100A:* 517–527.

Vine, P. J. 1974. Effects of algal grazing and aggressive behavior of the fishes *Pomacentrus lividus* and *Acanthurus sohal* on coral reef ecology. *Mar. Biol. 24:* 131–136.

Ward, A. J. W., S. Axford, and J. Krause. 2003. Cross-species familiarity in shoaling fishes. *Proc. Roy. Soc. Lond. B 270:* 1157–1161.

———, M. S. Botham, D. J. Hoare, R. James, M. Broom, J.-G. J. Godin, and J. Krause. 2002. Association patterns and shoal fidelity in the three-spined stickleback. *Proc. Roy. Soc. Lond. B 269:* 2451–2455.

Weeks, P. 2000. Red-billed oxpeckers: Vampires or tickbirds? *Behav. Ecol. 11:* 154–160.

Werner, E. E., and D. J. Hall. 1974. Optimal foraging and the size selection of prey by the bluegill sunfish (*Lepomis macrochirus*). *Ecology 55:* 1042–1052.

Whiteman, E. A., and I. M. Côté. 2002. Sex differences in cleaning behavior and diet of a Caribbean cleaning goby. *J. Mar. Biol. Assoc. UK 82:* 655–664.

———, and ———. 2004. Monogamy in marine fishes. *Biol. Rev. 79:* 351–375.

Whitney, R. R. 1969. Schooling of fishes relative to available light. *Trans. Am. Fish. Soc. 98:* 497–504.

Wiley, M. L., and B. B. Collette. 1970. Breeding tubercles and contact organs in fishes: Their occurrence, structure, and significance. *Bull. Am. Mus. Nat. Hist. 143:* 145–216.

Wilson, E. O. 1975. *Sociobiology: The new synthesis.* Harvard University Press, Cambridge, MA.

Winemiller, K. O. 1990. Caudal eyespots as deterrents against fin predation in the neotropical cichlid *Astronotus ocellatus. Copeia 1990:* 665–673.

———, and L. C. Kelso-Winemiller. 1993. Fin-nipping piranhas: Predatory response of piranhas to alternative prey. *Nat. Geog. Res. Explor. 9:* 344–357.

Wolf, N. G. 1983. *The behavioral ecology of herbivorous fishes in mixed species groups.* PhD thesis, Cornell University, Ithaca, NY.

———. 1985. Odd fish abandon mixed-species groups when threatened. *Behav. Ecol. Sociobiol. 17:* 47–52.

Yoshiyama, R. M., A. L. Knowlton, J. R. Welter, S. Comfort, B. J. Hopka, and W. D. Wallace. 1996a. Laboratory behavior of mosshead sculpins *Clinocottus globiceps* toward their sea anemone prey. *J. Mar. Biol. Assoc. UK 76:* 793–809.

———, W. D. Wallace, J. L. Burns, A. L. Knowlton, and J. R. Welter. 1996b. Laboratory food choice by the mosshead sculpin, *Clinocottus globiceps* (Girard) (Teleostei: Cottidae), a predator of sea anemones. *J. Exp. Mar. Biol. Ecol. 204:* 23–42.

PART EIGHT

FISHES AND HUMANITY

In the two final chapters of this book, we will investigate those aspects of the biology of fishes that have an immediate impact on the human condition. I do not believe it is an exaggeration to claim that fishes are the most important vertebrates in our lives. Ensuring their survival, more than that of any other natural populations of animals, will ensure our own health and well-being. Fishes inform and educate us about the essential character and the beauty of the natural world. Because we find their glorious colors and fascinating behavior so pleasing, a fish tank has become, for many of us, a necessary part of our home or office environment. For some, the fish is a worthy adversary to stalk with rod and reel. Most important, we have become painfully aware of the fact that the ocean's capability to feed humanity is no longer inexhaustible. As we have become increasingly dependent on the world's fishery resources—both marine and freshwater—the need to conserve these resources and ensure their continued health becomes a critical issue. For countries and cultures that are heavily dependent on fishery resources, it literally has become a matter of life and death. And so we need to understand, more than ever, the fishes with which we share this planet. The biosphere may be divided into two realms—the watery one for the fishes, and the terrestrial one for us—but the destinies of both are inextricably woven together. Because fishes take care of us in so many ways, it is time to return the favor.

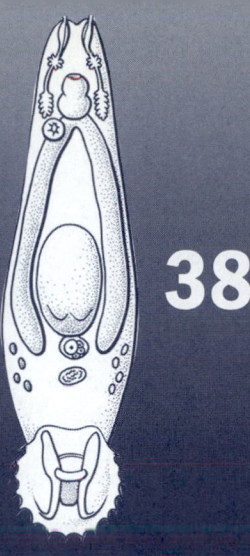

38

PARASITES
AND DISEASES
OF FISHES

PARASITES AS EXTREMELY SYMBIOTIC CREATURES

EVOLUTION OF PARASITIC FORMS AND LIFE HISTORIES

Evolutionary Considerations and Ecological Consequences
Parasite Communities

**SURVEY OF PARASITIC ORGANISMS AND THE DISEASES THEY
CAUSE**

Viruses
Eubacteria (Monera)
Protista
    Funguslike Protists
    Plantlike Protists
    Animallike Protists
Fungi
Animalia
    Myxozoans
    Metazoans

**IMPACT OF PARASITES ON EVOLUTION AND ADAPTATION OF
HOST SPECIES**

Dynamics of Host–Parasite Associations
Behavior and Susceptibility to Parasites
    Parasite- and Pathogen-Mediated Sexual Selection
Parasites as Biological "Tags" of Fish Populations

NONPARASITIC FISH DISEASES

Environmentally Induced Diseases
Dietary Diseases
Genetic Disorders

FISHES AS CARRIERS OF HUMAN PARASITES AND DISEASES

TOXIC FISHES

Ciguatera
Scombroid Poisoning
Tetrodotoxin Poisoning
Other Fish Poisons

Fishes are no different from any other organism in that their evolutionary history has been significantly influenced by that most intimate of symbiotic relationships—parasitism. Its consequences are manifested in an array of diseases that adversely affect fish populations. Not all diseases, as we shall see, arise from attack by pathogenic organisms. Many arise from nutritional deficiencies or from genetic defects in the organisms themselves. In this chapter, however, we shall focus on those pathogenic organisms that have targeted fishes. We need to know all about these organisms, not only as biological curiosities, but because they can have such a devastating impact on fish populations—both naturally occurring and cultured—that we depend on as sources of nutrition. The topic of parasites and diseases of fishes is vast in scope, so that we can hardly do justice to it in a single chapter. But we would be remiss in our attempt to inform you about the lives of fishes if we did not provide at least this brief, cursory introduction to the parasites and diseases that are integral to their existence. Issues of fish health and human health are often intertwined; although this chapter focuses on parasites and diseases that affect fishes, we will also consider some aspects of fish biology that have an immediate impact on human health—specifically, pathogens or other substances that fishes harbor that may cause harm to humans.

PARASITES AS EXTREMELY SYMBIOTIC CREATURES

Parasitic organisms represent the most intimate form of symbiosis. Parasitism is a relationship in which the symbiont is dependent on its host for nourishment and habitat—usually with a detrimental impact on the host. The basic requirements of a parasite are no different from those of free-living organisms—suitable habitat, a source of food, and at least the opportunity for the completion of its life cycle, including successful reproduction. In the strictest definition of the term, an organism can be considered a **parasite** only if it negatively affects the intrinsic growth rate of its host organism (Anderson and May, 1978). Parasitism is practiced by an astonishing diversity of organisms that encompass a variety of lifestyles and show varying degrees of host exploitation.

Authors tend to distinguish microbial parasitism—parasitism causing disease in host organisms, with the causal agents being mainly viruses and organisms from the kingdoms Monera, Protista, and Fungi—from parasitism caused by members of the kingdom Animalia. Parasites are also distinguished as **ectoparasites,** living on the epithelial surfaces of the host, and **endoparasites,** living inside the host—either embedded within its tissues or taking up residence in its body cavities. The degrees of association of parasite with host range from creatures that become intimate and permanent associates of the host tissues to tissue-ingesting or blood-sucking forms that make only occasional feeding visits on their host. Some authors do not classify blood-feeding organisms as true parasites (Price, 1980)—an especially relevant consideration for the few fish species that have been termed "parasitic" (see Chapter 37). True parasites are by nature "prudent" organisms. In an ideal world, parasites live as more or less permanent associates with their hosts, but limit their exploitation of the hosts, so that they do not kill them off. Overexploitation of the host kills it, thus eliminating the parasite's habitat and source of nutrients. We often speak of the **parasite load** of the host—a reference to the abundance and diversity of symbionts that make that host their home, or of the **parasitic intensity**—the number of parasites per parasitized host. Organisms that are the unfortunate hosts of a large diversity of parasites may be debilitated to the point of their eventual demise. In matters of host exploitation, therefore, parasites are always treading a fine line. Ultimately, they depend on extremely rapid rates of proliferation or marvelously effective modes of dispersal to ensure the completion of their life cycles and the perpetuation of their species.

The evolution of a parasitic mode of existence has a profound impact on life history and body form. The simplest taxa, such as eubacteria and protists, are among the most ruthlessly efficient. They exploit their phenomenal rates of replication to rapidly infect a host, often overwhelming its resistance mechanisms. With a comparatively simple genetic system, they are capable of rapid adaptive response to the selective factors imposed by their host environments. Parasites that have evolved from more complex taxa, such as animals, tend to demonstrate a profound reduction or simplification of body form. Many helminth parasites, for example, have lost their digestive tracts, relying on direct absorption of nutrients across their epithelial surfaces. Locomotor adaptations are of less significance when a parasite is attached or embedded in the tissues of a host organism. We have already considered a remarkable example of this in the extreme sexual dimorphism seen in deep-ocean anglerfishes, in which the males parasitize females in order to ensure proximity during the reproductive cycle (see Chapters 13 and 35; Plate 35.2). Body forms may be simple in parasites, but the life history strategies to ensure successful completion of the life cycle may be exceedingly complex, often using intermediate hosts (Fig. 38.1).

Many fish diseases—primarily those of viral or bacterial origin—are being successfully combated through vaccination. The vaccines are usually of two types: *dead vaccines,* composed of inactivated pathogens or extracts derived from them, and *live vaccines*—attenuated pathogens with diminished virulence. Vaccines have been developed to treat many of the diseases to be mentioned in this chapter (Ellis, 1988). Molecular genetic techniques have shown great potential in vaccination research—rather than injecting cultures of the pathogen or pathogen products to induce immune response, the feasibility of injecting DNA extracts encoding protective antigens for certain pathogens is being investigated (Lorenzen et al., 2000).

EVOLUTION OF PARASITIC FORMS AND LIFE HISTORIES

There is a constant evolutionary "arms race" going on between the parasite and its host. Host–parasite relationships embody the **Red Queen hypothesis** first put forth by Van Valen (1973). According to this hypothesis, the fitness of a given organism will necessarily be affected in a negative manner by the presence of others in its environment. Like the Red Queen whom Alice meets in Lewis Carroll's *Through the Looking Glass,* who must constantly be running just to stay in place, species must be constantly evolving and adapting just to avoid extinction. Parasites continually evolve more effective ways to disperse, locate, infest, and extract nutrients from their hosts. Hosts, for their part, are continually evolving novel ways to fight off infec-

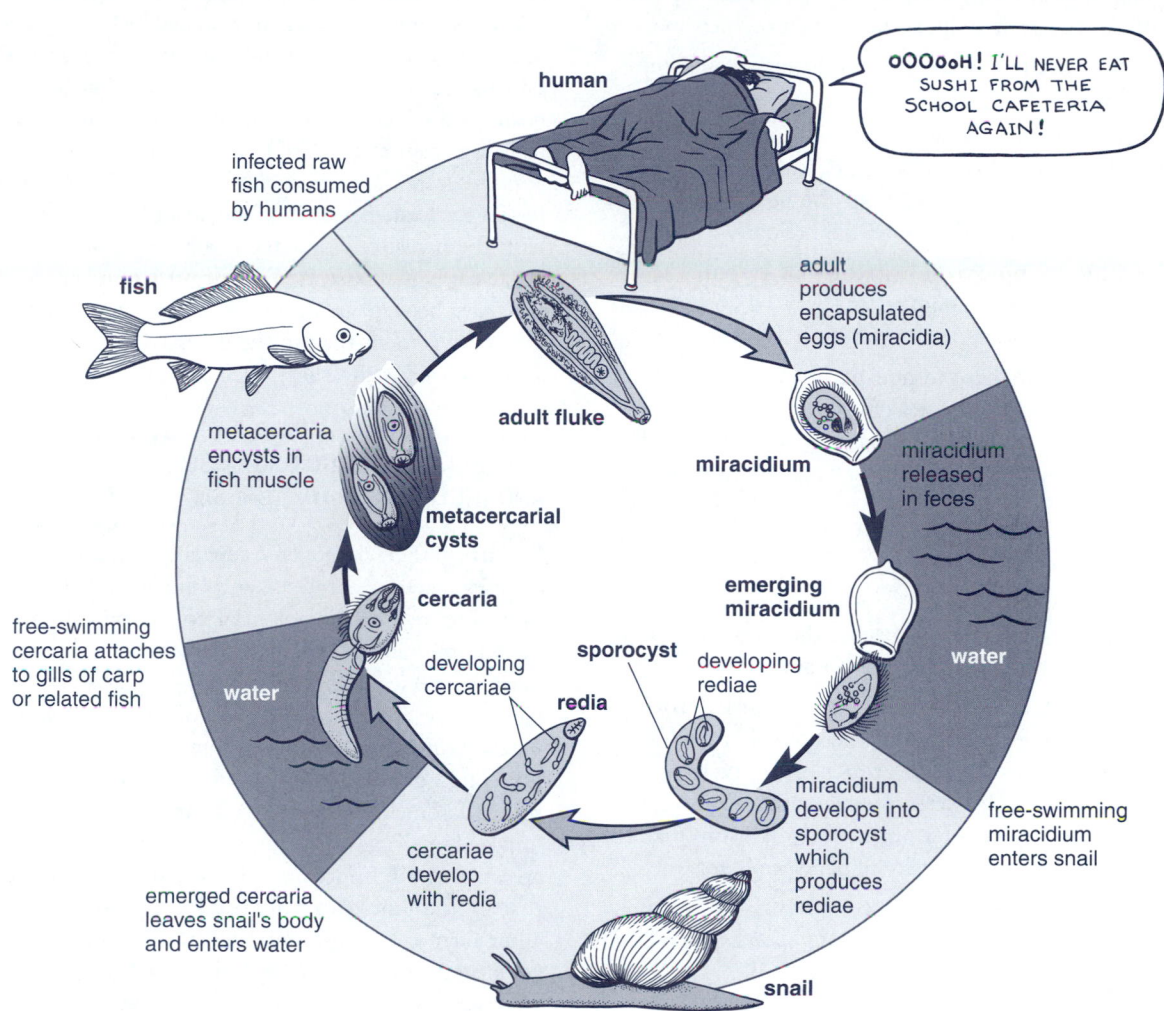

FIGURE 38.1
Life cycle of the human liver fluke (*Opisthorchis sinensis*). The snail intermediate host is usually in the genus *Bithynia,* and the fish is usually a member of the family Cyprinidae.

tions and infestations. The vertebrate immune system is a marvel of evolutionary ingenuity, as it enables the rapid identification of foreign invaders and brings to bear an array of sophisticated molecular devices to combat infection (see Chapter 24). Yet parasites are constantly evolving ways to circumvent these immunological roadblocks. The battle, once engaged, continues to rage.

Evolutionary Considerations and Ecological Consequences

Peter Price (1980) has identified three basic concepts governing the evolution of parasites, and three ecological consequences:

1. Rates of evolution and speciation can be high. The parasitic lifestyle promotes inbreeding and asexual reproduction, which fosters rapid population divergence and speciation.

2. Extensive adaptive radiation has occurred. The degree of development of this adaptive radiation is largely dependent on the nature and availability of host taxa, the length of time for evolution to occur, and the selective pressure for coevolutionary modification.

3. Allopatric speciation need not be the only speciation mode governing the evolution of parasites. Extremely

minor alterations in the genome of parasites may be sufficient to promote the establishment of new races.

The ecological consequences of these evolutionary phenomena are as follows:

1. Parasites are best adapted for the exploitation of limited, discontinuous environments. From the perspective of the parasite, its habitat (the host) is discrete and isolated from other potential habitats.

2. Parasites exemplify the extremes of resource specialization. They cannot in any way be considered trophic generalists, as many free-living forms can.

3. Parasites must adapt to nonequilibrium conditions. The life history strategies characteristic of parasites are those in which the probability of a given individual successfully completing its entire life cycle is very low.

Parasite Communities

A few broad generalizations concerning the parasitic infestation of fishes can be made courtesy of Poulin (1995):

1. Parasite community richness is most closely correlated with the body size of the host. Larger hosts offer more space and a greater variety of niches for parasites. Also, larger hosts tend to consume a greater variety of food organisms, exposing themselves to a greater diversity of parasitic fauna.

2. The richness of the parasite community tends to increase in proportion to the amount of animal food in the diet.

3. Parasite community richness tends to be greater in marine fishes than in freshwater fishes. The colonization of freshwater habitats by marine species results in a slight tendency to lose parasites.

Of particular ecological and evolutionary significance is the degree of specificity in the habitat and nutritional requirements of parasites. This may contribute to the phenomenon of **demographic release** from parasites in introduced species. Studies on a range of host species, including fishes, have shown that introduced populations are less heavily parasitized than native populations, both in terms of the diversity of parasites infesting hosts and of the percentage of the host population infected. Undoubtedly, this relative freedom from parasitism causes the populations of many invasive species to grow to the point that they become pests (Torchin et al., 2003). In some cases, the parasites borne by invasive species can have most unfortunate consequences for native species. The European cyprinid *Leucaspius delineatus*, already classified as endangered, has experienced increased

mortality and inhibited spawning caused by an intracellular eukaryote parasite that was introduced by healthy individuals of the invasive Asian cyprinid *Pseudorasbora parva*. This parasite has been shown to be capable of infecting other species, posing a distinct threat to the diversity of native European fishes (Gozlan et al., 2005). The rapid adaptive capacity of parasites, especially microbial forms, makes them formidable foes of the host, especially if the host has no prior experience with the parasite. The effectiveness of a parasitic mode of existence is demonstrated in the vast numbers of parasites that have become associated with host animals. Nowhere is this more evident than in the diverse array of parasites that are associated with fishes.

SURVEY OF PARASITIC ORGANISMS AND THE DISEASES THEY CAUSE

Because of their pathogenic capacities, microbial symbionts (viruses, bacteria, protists, and fungi) are the most significant in terms of their potential impact on fish populations, causing millions of dollars of damage in commercially important species. Their actions are usually more direct, parasitizing the cells and tissues of a single, definitive host. Their virulence is attributed to a simple genome that permits rapid population growth in host cells and tissues and responds quickly to selective changes in the environment imposed by drugs and other treatments. Metazoan parasites are structurally more complex organisms, with life cycles that typically involve **intermediate hosts**. Normally, the adult form ceases proliferation by the time it has become associated with the final or **definitive host**.

Viruses

Viruses have been implicated in more than 60 fish diseases (Bruno and Poppe, 1996), and they affect their host fishes in a number of ways. Some viruses are relatively benign, whereas others are virulent and life-threatening. Most of the viruses that are pathogenic to fishes are classified as **rhabdoviruses, herpesviruses,** or **birnaviruses**—differentiated by the structure of their enclosure (*capsid*) and genome (RNA or DNA). As is the case with most parasites and the diseases they cause, most research has focused on commercially significant species. Some of the more notorious virally induced diseases are **viral hemorrhagic septicemia (VHS)** and **infectious hematopoietic necrosis (IHN)**, both caused by rhabdoviruses, and **infectious pancreatic necrosis (IPN)**, caused by a birnavirus. VHS is perhaps the most widespread and contagious viral disease affecting salmon and trout (Bruno and Poppe, 1996). Severe hemorrhage and necrosis of organs such as the kidney and liver are associated with this disease. IHN is primarily known from Pacific salmons, but can be transmitted to Atlantic salmons under experimental conditions (Bruno and Poppe, 1996). External symptoms include darkening of the skin and cutaneous hemorrhaging, and lesions and necrosis of internal organs eventually occur. Viruses have also been implicated in the formation of **papillomas** (see Genetic Diseases section)—an example of external agents affecting the genetic control mechanisms of host cells and tissues.

Eubacteria (Monera)

Bacterial symbionts are the cause of an astounding diversity of fish diseases. Carl Sindermann (1990) has classified fish-associated bacteria into four categories: (1) primary pathogens; (2) secondary invaders that attack already debilitated hosts; (3) forms that invade dying hosts but can have a pathogenic impact on healthy individuals if experimentally injected in sufficient quantities; and (4) free-living bacteria that occur on the body surfaces of the host but are usually not pathogenic. As is the case with any infectious disease, it is essential to rapidly identify the potential pathogen. If the pathogen is suspected to be bacterial, **Gram staining** of a culture will rapidly distinguish two categories of bacteria—*Gram-positive* and *Gram-negative*—with different properties of the cell wall. Most bacterial fish pathogens are Gram-negative. Some bacterial species, such as *Vibrio anguillarum* and *Aeromonas salmonicida*, are highly virulent and cause septicemic conditions (blood poisoning), much like those observed in some viruses. **Vibriosis** is an economically important disease, usually caused by *V. anguillarum*, but other species of *Vibrio* have been identified. Salmonids farmed in seawater are especially vulnerable to vibriosis. The peculiar, straight or curved rod-shaped bacteria cause anemia, necrosis, and severe skin lesions. Vibrios demonstrate the adaptive versatility of the Eubacteria. For example, another species (*Vibrio fischeri*) is a bioluminescent symbiont that colonizes light-emitting organs in squids (Nyholm et al., 2000).

Lesions and ulcerations also characterize the disease **furunculosis** (see Plate 38.1), caused by the bacteria *Aeromonas salmonicida*. *Chlamydia* and *Rickettsia*, unusual Gram-negative microorganisms that are obligate intracellular pathogens, can affect fish populations as well. Gram-positive bacteria are also known among fish pathogens. *Streptococcus* has been implicated in mass mortalities of Gulf of Mexico fishes, including Gulf menhaden (*Brevoortia patronus*), Atlantic croaker (*Micropogon undulatus*), striped mullet (*Mugil cephalus*), and spot (*Leiostomus xanthurus;* Cook and Lofton, 1975). **Bacterial kidney disease (BKD)** is caused by the Gram-positive *Renibacterium salmoninarum.*

Protista

Although some pathologists may consider viruses and bacteria to be pathogenic but not parasitic, the numerous phyla of protists are widely considered parasitic in their mode of action. Indeed, viruses and bacteria may be more predatory in behavior, as their life cycles do not appear to manifest the prudence normally thought of as characteristic of true parasites. Yet many species of protists can, in extreme cases of infestation, cause the death of their hosts and, as such, may be considered pathological rather than parasitic. Indicative of the evolution of an extremely sophisticated parasitic mode of existence, protists have evolved life cycles with multiple stages, some parasitic and some free-living, and often use intermediate hosts. Protists have long been considered a "taxonomic wastebasket" by researchers, as they include a bewildering variety of organisms, some with attributes of fungi, others plantlike, whereas still others resemble members of the animal kingdom. Recent classifications of living organisms, based on gene sequences and mitochondrial structure, have challenged many of the earlier systematic analyses of protists based largely on morphological features such as organelles of locomotion (Prescott et al., 1999). As our understanding of the actual phylogenetic relationships of what we commonly refer to as protists improves, the following classification will undoubtedly appear more one of convenience than one representing actual evolutionary interrelationships.

Funguslike Protists

Chief among the funguslike protists are the water molds (phylum **Oomycota**). This group includes the potato blight responsible for the catastrophic famine in Ireland during the 19th century. The class **Saprolegniales** includes the genus *Saprolegnia*, which frequently infests fish eggs in nests and is one of the organisms that much parental egg care is directed against. *Saprolegnia* forms a cottony mass of filaments (**hyphae**), termed a **mycelium**, on dead or decaying matter, but it will quickly overwhelm and kill healthy eggs if infected eggs are not removed from the nest (Hoffman, 1999). The organism also infests adult fishes. The presence of mycelia has caused some biologists to classify the Oomycota as true fungi.

Plantlike Protists

These protists, sometimes referred to as "algae," are plantlike in their ability to practice photosynthesis. Many of them also have a reproductive cycle that resembles that seen in plants. Two groups of these "algae" include forms that parasitize fishes. Among the green algae (phylum **Chlorophyta**), the genera *Chlorella* and *Cladophora* have been documented to parasitize fishes (Hoffman, 1999). The dino-flagellates (phylum **Dinoflagellata**) are much more notorious for the deleterious impact they have on fishes, both as parasites and as free-living forms. Although about half of the dinoflagellates contain chlorophyll and photosynthesize (including the symbiotic **zooxanthellae** that enable photosynthesis to take place within the tissues of corals), some authors (see Pechenik, 2005) have classified dinoflagellates in the protozoan phylum **Dinozoa,** based on similarities in cell membrane structure.

Under certain conditions, the free-living dinoflagellate genera *Gymnodinium* and *Gonyaulax* experience explosive population increases. The term **red tide** comes from the reddish hue imparted to coastal waters by the enormous concentrations of these dinoflagellates. They produce a toxin that is accumulated in bivalves that filter large quantities of dinoflagellates out of the water. Human consumption of shellfish during times of red tide is indeed hazardous, as **paralytic shellfish poisoning** can result. This condition is life-threatening, as the paralysis affects the respiratory muscles. One of the most noxious consequences of red tides is the massive fish kills that are often associated with them. (In my early days as a surfer in California, I remember once being faced with a horrific dilemma—perfect waves moving through an ocean filled with bloated, stinking fish carcasses.) Not only are the dinoflagellate neurotoxins responsible for the fish kills, but the respiratory processes of the algae deplete the surface waters of oxygen at night, quickly suffocating the fishes. Another group of dinoflagellates produces a toxin that accumulates in the tissues of certain species of fishes; although manifesting no apparent ill effects in its fish hosts, this protist is responsible for the disease **ciguatera** that is contracted by humans eating infected fishes (see further).

One species of dinoflagellate that has received extraordinary attention in the past decade is *Pfiesteria piscicida.* First discovered by researchers at North Carolina State University in 1988, this species is known to have an extremely complex life history, in which it can assume as many as 24 different forms, some of them capable of producing toxins (see Plate 38.2A). *Pfiesteria* differs from most dinoflagellates that derive at least part of their energy needs from photosynthesis. It is usually found as a free-living heterotroph. Apparently, concentrations of fishes stimulate the release of toxins that incapacitate the fishes and promote the formation of lesions on their skin. *Pfiesteria* then feeds on the blood and tissues of the incapacitated fishes. In this manner, the fishes are not so much a host to *Pfiesteria* as they are its prey (see Plate 38.2B). *Pfiesteria* is widespread in brackish waters along the Atlantic and Gulf coasts. A few other

Pfiesteria-like dinoflagellates have also been discovered but have not yet been formally described (http://www.epa.gov/OWOW/estuaries/pfiesteria/fact.html).

Animallike Protists

These protists are commonly grouped in the subkingdom **Protozoa.** As might be expected, there is no consensus on the classification of such a diverse assemblage of microscopic creatures. Although recent molecular studies have challenged the conventional classification schemes used here, the four phyla presented here are recognized mainly on the basis of their mode of locomotion—the **Rhizopoda, Mastigophora, Ciliophora,** and **Sporozoa.** All of these include forms parasitic on fishes. The protozoans are probably responsible for more diseases in fish culture than any other group.

The Rhizopoda (a.k.a. Amoebozoa) include amoeboid forms that move using pseudopodia. Parasitic forms are known from the intestinal tract and gills of fishes, where they form cysts in the tissues. Adults probably feed on bacteria and other organic matter.

The Mastigophora move by means of one or more flagella. Pathogenic mastigophorans include *Ichthyobodo* (Fig. 38.2A) and *Trypanosoma* (Fig. 38.2B). The latter are blood parasites that occur in fishes but are probably best known for causing African sleeping sickness in humans. Many classification schemes place these genera in the phylum Euglenozoa (Pechenik, 2005).

Most species of Ciliophora are free-living, but many are commensal on the skin surfaces of fishes, and some are true parasites. Possibly the best known fish parasite, the scourge of tropical fish hobbyists, is *Ichthyophthirus multifiliis,* commonly known as "Ich" or "whitespot." This protist has been reported from more species of fishes than any other parasite (Hoffman, 1999). *Ichthyophthirus* are readily identified by their large, horseshoe-shaped nuclei (Fig. 38.2C). The ciliates of the family Trichodinidae are unusual parasites that are commonly found on the gills and body surfaces of fishes. They are flattened, highly symmetrical cells with a circlet of hooked teeth on the aboral surface (see Plate 38.3).

The Sporozoa (= Apicomplexa) were at one time classified as protozoans lacking any discernible locomotor apparatus, but locomotor organelles have been found in some life stages in some species, necessitating a revision of the classification of the group (Lee et al., 1985). The genus *Eimeria* (Fig. 38.2D) typifies the sporozoans that parasitize fishes. When an oocyst containing sporocysts is ingested by the host, the sporozoites are released and attack cells of the intestinal epithelium and other organs. After several

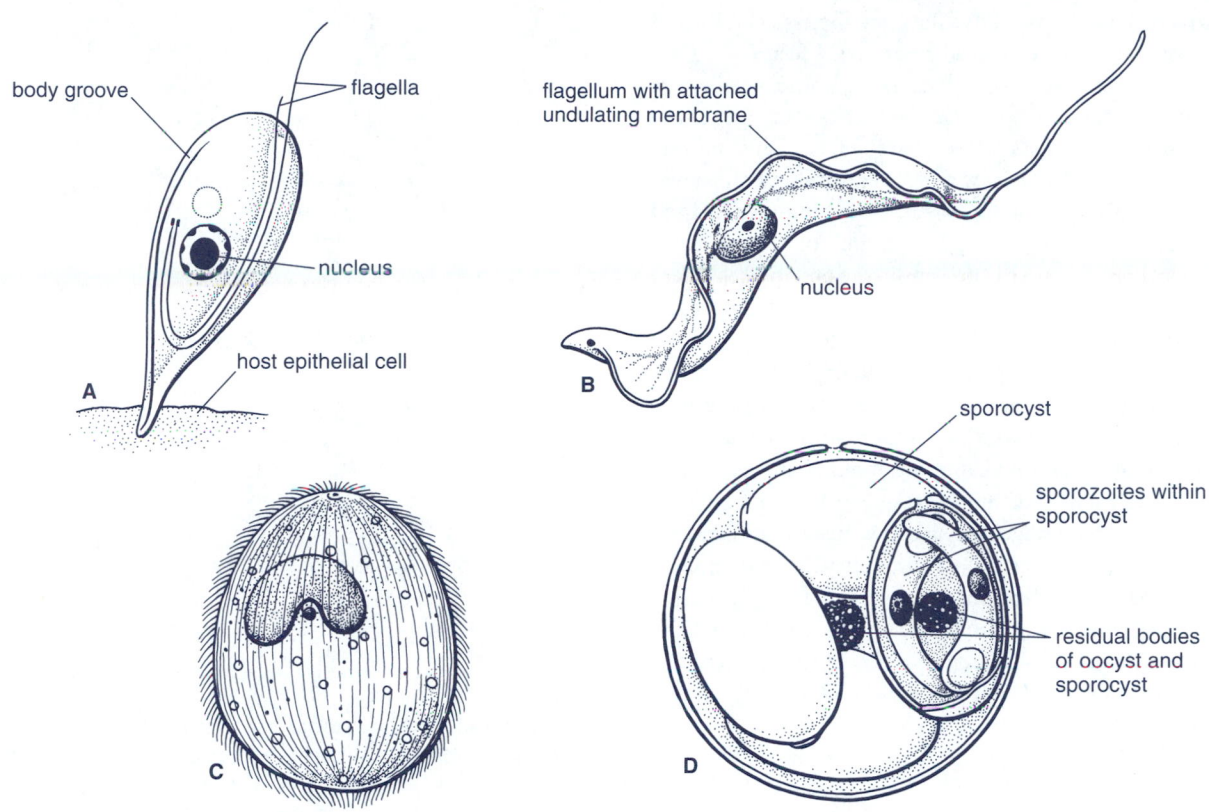

FIGURE 38.2
A, Pathogenic mastigophoran of the genus *Ichthyobodo* (from Hoffman, 1999); **B,** Pathogenic mastigophoran of the genus *Trypanosoma* (from Hoffman, 1999); **C,** Pathogenic ciliate of the genus *Ichthyophthirus* (from Hoffman, 1999); **D,** Oocyst of pathogenic sporozoan of the genus *Eimeria,* containing sporocysts. Each sporocyst contains a number of sporozoites. Residual bodies of oocyst and sporocysts also shown. (From Hoffman, 1999.)

additional life stages, gametogenesis occurs, and a zygote forms that is released as a sporulated oocyst.

Fungi

Among the "true" fungi—as distinguished from the Oomycota discussed earlier—are infectious forms classified as **Deuteromycetes** (or Fungi Imperfecti—"imperfect" because no sexual reproduction is known in the group) as well as a number of groups exhibiting sexual reproduction. One of the better known genera of "perfect" fungi, *Ichthyophonus* (phylum Zygomycota), is primarily known from marine fishes but has been known to infect salmonids cultured in freshwater when they are fed diets containing marine fish by-products (Bruno and Poppe, 1996). It has been found in

a number of internal organs and, in severe cases, can cause spinal deformities.

Animalia

Myxozoans

Parasitic forms have evolved from an astounding diversity of animals, including forms so rudimentary as to have originally been classified as protozoans, as well as several phyla of "higher" animals. Modern classifications of the kingdom Animalia have featured two lineages—the **Myxozoa** and the **Metazoa** (Prescott et al., 1999). The Myxozoa (= Myxospora) include some 1,200 species that until recently were classified as protozoans. Siddall et al. (1995) presented evidence suggesting that the myxozoans actually traced their origins from cnidarians (jellyfishes, anemones, corals, and related forms) that have evolved a parasitic lifestyle.

A myxozoan species that causes significant damage to salmonid fishes is *Myxobolus cerebralis*—the agent that causes **whirling disease** (see Plate 38.4). This minute organism, native to Eurasia, was first reported in Pennsylvania in 1956. Since then, it has spread throughout the United States wherever self-sustaining populations of rainbow trout (*Oncorhynchus mykiss*) exist. Presently, the only places where it is not reported are the Upper Midwest and the desert Southwest (Bergersen and Anderson, 1997). The parasite penetrates the head and spinal cartilage of fingerling trout. Rapid proliferation in the vicinity of the otic region causes damage to the organs of equilibrium, causing the characteristic disorientation and whirling observed in infected individuals (http://www.whirling-disease.org). The hardiness of the dormant spore stage, which can withstand desiccation and freezing, makes this a most difficult pest to eliminate. The parasite is dependent on an intermediate host, tubifex worms (*Tubifex tubifex*, phylum Annelida), which should be familiar to tropical fish hobbyists, as they are commercially freeze-dried to provide a protein-rich food supplement for aquarium fishes.

Metazoans

Parasitic species can be found in several metazoan phyla, but three phyla—**Platyhelminthes, Acanthocephala,** and **Nematoda**—have been especially successful as parasites. The phylum **Arthropoda**—the most diverse metazoan phylum, containing about 75 percent of all described animal species—owes its success to the broad range of body forms and lifestyles that have evolved. Among the arthropods, the primarily aquatic class **Crustacea** includes a diverse array of important fish parasites.

Of the platyhelminths, only about 16 percent are members of the predominantly free-living class Turbellaria. The rest—in the classes **Monogenea, Trematoda,** and **Cestoda**—are parasitic. Most monogeneans are ectoparasitic, being found on the skin or gills of fishes. They are readily identified from their prominent attachment organs—the anterior **prohaptor** and posterior **opisthaptor**—seen in the adult stages (Fig. 38.3). Intermediate hosts are lacking, so that the life cycle is one in which the fertilized egg develops into a larval **oncomiracidium**, which attaches to the host fish. As a demonstration of the extremes of host specificity seen in monogeneans, one species lives only at the base of the host fish's gills, whereas another species is only found at the tips of the same filaments (Pechenik, 2000).

The trematodes (flukes) are among the most specialized parasites in terms of their use of intermediate hosts.

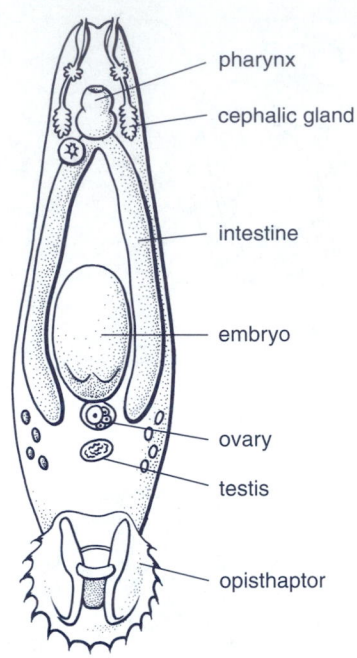

FIGURE 38.3
Monogenean of the genus *Gyrodactylus,* showing pronounced opisthaptor with hooks for anchoring to host. (From Hoffman, 1999.)

Theirs is a life cycle that—were it not for the damage they wreak on vertebrate hosts—would be inspiring in its adaptive capacity. Most trematodes are **digeneans** ("two births"), the name implying the presence of at least one intermediate host. The typical digenean life cycle includes a free-living, larval **miracidium** that hatches from a fertilized egg. These miracidia typically enter a molluscan host and become transformed into a **sporocyst,** which produces **rediae,** which in turn produce **cercariae.** The cercariae rupture through the mollusc host's tissues and assume a free-swimming existence until they encounter a fish that serves as the secondary intermediate host. They bore into the tissues and become encysted in the muscle. ("Swimmer's itch" is caused by misdirected cercariae of the blood flukes of birds attempting to bore through the skin of humans.) The encysted stage are known as **metacercariae.** Here they await consumption of the infected fish's tissues by their definitive host (Fig. 38.1). In this manner, humans who eat raw or undercooked fish run the risk of becoming infected by liver flukes (*Opisthorchis sinensis*). One of the most dreaded parasitic diseases of humans, *schistosomiasis,* is caused by flukes of the genus *Schistosoma.* Like *Opisthorchis, Schistosoma* uses snails as intermediate hosts, but differs in that it does not employ fishes in this capacity.

Most digenean trematodes that parasitize fishes as adults are found in the intestine, where they apparently have little adverse impact (Hoffman, 1999). The larval forms are most frequently encountered in fishes at the metacercarial stage. Tissue damage and hemorrhaging can occur if large numbers of cercariae penetrate and encyst in a fish host. Trematode outbreaks can result in significant losses of fishes in aquaculture facilities. The black carp (*Mylopharyngodon piceus*) is a voracious consumer of snails; for this reason, it has been touted as beneficial for the control of trematodes that use snails as intermediate hosts. Understandably, there is great concern about the possibility that fertile individuals may escape from aquaculture facilities and devastate native mollusc populations in North America (Ferber, 2001). Fishes serve as the intermediate host for a species of gill trematode (*Centrocestus formosanus*) introduced into North America. This parasite is the object of serious concern to fishery scientists, as it has caused significant losses of fish stock among tropical fish producers. It is also believed to be capable of damaging native fish populations. Whereas the definitive hosts are mainly birds, the initial intermediate host is the redrim melania (*Melanoides tuberculatus*), a hardy Asian snail widely introduced into North American waters—yet another case of adverse parasitic consequences of the introduction of exotic species (Mitchell et al., 2005).

The cestodes (tapeworms) are, in many respects, the most unusual of the platyhelminthine parasites. The adults are primarily known as gut parasites in vertebrates, but the larvae may use intermediate hosts—usually small crustaceans, such as copepods, amphipods, and isopods—in their life cycle. The adult is characterized by a head region, called a **scolex,** which anchors the worm to its host by means of a circlet of hooks. Numerous segments, termed **proglottids,** bud from the narrow neck region behind the scolex (Fig. 38.4). Each proglottid contains mainly reproductive organs (both male and female) and can produce up to 50,000 eggs. Lacking a digestive tract, tapeworms feed by the absorption of nutrients from the digestive tract of their hosts, using a highly modified epithelial surface. Larval cestodes can cause structural damage as they migrate through their hosts. Once the adult stage is anchored, its most significant impact on its host is in robbing it of necessary nutrients. Heavy tapeworm infestations can cause diminished growth and intestinal blockages in its host fish.

Acanthocephalans are easily recognized by their eversible proboscis that bears an array of hooks used to anchor the worm into its host, causing hemorrhage and tissue damage (see Plate 38.5). Eggs are shed in the host feces and ingested by small crustaceans, in which the larval stages develop. The life cycle is completed when the intermediate hosts are consumed by fishes. As with the cestodes, no digestive tract is present, and dissolved nutrients must be absorbed across the epidermis. In an unusual instance of parasites parasitizing other parasites, acanthocephalans have been found anchored in tapeworms that are parasitic in lake trout (*Salvelinus namaycush;* Miller, 1946).

Nematodes (roundworms) are an extremely successful group of helminths, and parasitism has independently

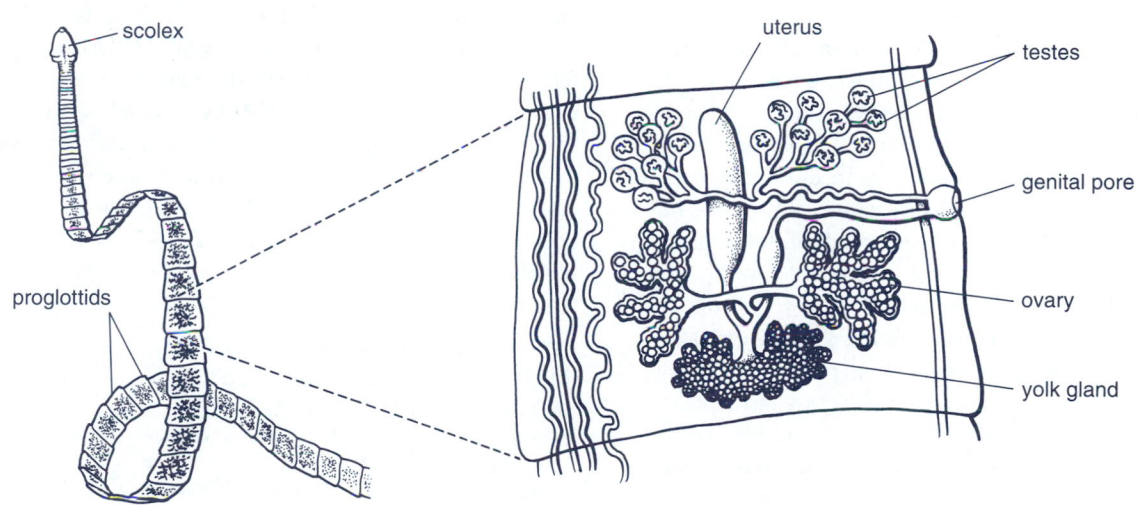

FIGURE 38.4
Cestode of the genus *Marsipometra,* with details of scolex and proglottid. (From Hoffman, 1999.)

evolved at least seven times within the group (Blaxter et al., 1998). Nematodes, especially the free-living forms, occur in such extraordinary concentrations that they may be the most abundant multicellular animals in existence. Long considered to be allied with pseudocoelomate phyla, nematodes are now considered to be more closely related to the arthropods, chiefly because of their shared possession of a cuticle that is molted (Pechenik, 2005). Although some of the better-known human parasites—such as *Trichinella*, which causes trichinosis, and *Wuchereria*, which causes elephantiasis—inflict significant damage on their hosts, those nematodes that parasitize fishes usually live a relatively benign existence in the gut of their hosts. Larval stages are found in most tissues, especially the mesenteries, liver, and musculature. Arthropods are the most common intermediate host for the species that use fishes as a definitive host. Many species that inhabit birds and mammals as the definitive host use fishes as a second intermediate host, after spending time in a crustacean first intermediate host.

Crustacean fish parasites are known from the subclasses **Branchiura** and **Copepoda**, and from the malacostracan order **Isopoda**. The branchiurans are represented by the genus *Argulus* (see Plate 38.6). About 125 species are known, all of them as ectoparasites on the skin surface, where they feed on host blood and tissue fluids by means of a proboscislike mouth inserted through the skin. They usually reach a size of about 1 cm. Copepods are among the most abundant animals on Earth, most of them being members of the planktonic or benthic communities of marine and freshwaters. About 25 percent of the known species of copepods are parasitic, including several orders that are exclusively parasitic (Pechenik, 2000). They are readily identified by unusual modifications of the head or by appendages for anchoring into or attaching onto their hosts (Fig. 38.5A, B). Parasitic copepods are generally larger than their free-living relatives. The genus *Penella*, which parasitizes pelagic fishes and whales, may reach sizes up to 25 cm. In another demonstration of the complexity of parasitic communities, *Penella* may themselves be parasitized by individuals of the barnacle genus *Chondroderma* (Barnes, 1968).

Females of both free-living and parasitic copepods are identifiable by the conspicuous paired egg sacs carried on the abdomen (Fig. 38.5). Parasitic copepods, commonly termed "fish lice," are not especially damaging in the wild, unless they occur in abundance. In ocean-ranching operations, they are extremely undesirable, as their abrasive and boring behaviors can cause skin erosion and hemorrhaging (Bruno and Poppe, 1996). Cleaner wrasses (Labridae) have been employed by aquaculturists to control outbreaks of fish lice (see Chapter 37, Going Deeper).

The order Isopoda is one of the most diverse groups of crustaceans, and most of them are scavengers, including the terrestrial pill bugs. Several isopod species are parasites on fishes. When diving on coral reefs in the Bahamas, I cannot help but feel a twinge of sympathy on seeing a soldierfish (Holocentridae) carrying about a large, "unsightly" cymothoid isopod (*Anilocra*) right between its eyes (see Plate 38.7). These isopods are very selective in the location of attachment and choice of host fish; their specific names are often derived from host species. For example, *Anilocra myripristis* attaches between the eyes of the soldierfish (*Myripristis jacobus*); *A. abudefdufi* always attaches beneath the eyes of the sergeant major (*Abudefduf saxatilis*); and *A. acanthuri* fastens itself just beneath the pectoral fins of acanthurids, such as the doctorfish (*Acanthurus chirurgus*) and the ocean surgeonfish (*Acanthurus bahianus;* DeLoach, 1999). Although many species of parasitic isopods damage their host by burrowing into and feeding on their tissues, those of the genus *Anilocra*, large and conspicuous though they may be, appear to have retained the basically omnivorous feeding habits of their free-living brethren. Most of the damage they inflict on their hosts is caused by their attachment via hooklike appendages.

Many other phyla have members that parasitize fishes. Most members of the phylum Annelida are free-living, but the subclass Hirudinea (the leeches) have a lifestyle—occasional attachment for the purposes of extracting a large blood meal—that may be interpreted as parasitic. Leeches are also known to be vectors for the transmission of the virus that causes infectious hematopoietic necrosis and of the bacterium *Aeromonas*, the causative agent for furunculosis (Hoffman, 1999). Cnidarians are not usually thought of as parasitic, yet one aberrant species (*Polypodium hydriforme*) is known to parasitize the eggs of sturgeons and paddlefishes (Hoffman, 1999). As mentioned in Chapter 37, the glochidia larvae of many freshwater clams parasitize fishes, forming cysts on the gills or fins. The larvae of several genera of mites (class Arachnida, order Acari) have also been reported as parasitic on the gills, skin, and esophagus of fishes.

IMPACT OF PARASITES ON EVOLUTION AND ADAPTATION OF HOST SPECIES

Dynamics of Host–Parasite Associations

Parasite assemblages are arranged in a complex hierarchical system that allows patterns of community richness and structure to be studied at various scales. These range from

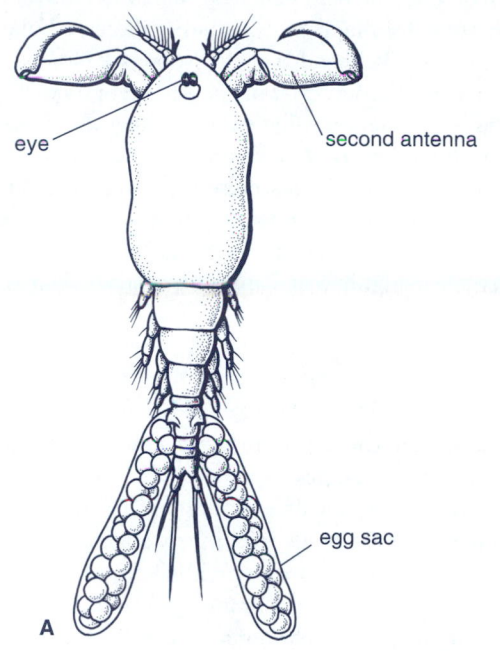

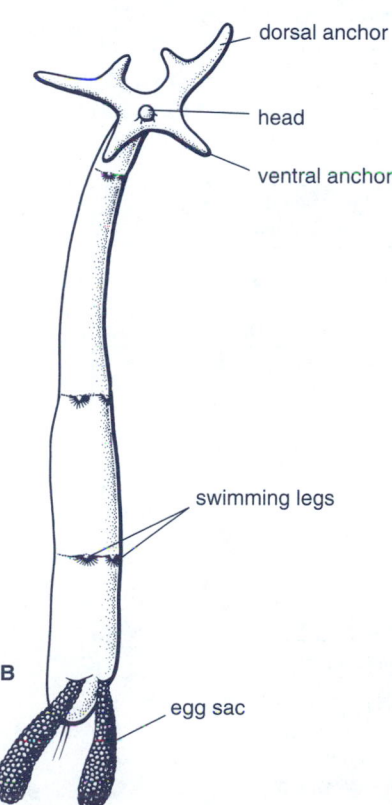

studies on the parasite community inhabiting one individual of a given species to the consideration of all parasite species exploiting a host species across its geographic range (Poulin, 1997). Assemblage dynamics become even more complex when we recognize that parasites with intermediate hosts may affect more than one community. The parasite population, together with all of the host populations with which it may interact, is often termed the **parasitic system** (Granovitch, 1999). At some time during the phase of their life cycles, when they are free-living organisms, parasites may also be integral to the community dynamics of other free-living organisms, either as prey or as competitors.

The most intuitively obvious dynamic is the one that defines the impact of the parasite on the host. Pectoral fin size in sticklebacks (*Gasterosteus aculeatus*) is associated with reproductive success, as the males oxygenate eggs in the nest with fanning motions of their fins. Smaller pectoral fin size—and hence diminished reproductive success—has been correlated with infection by intestinal parasites (Bakker and Mundwiler, 1999). The life cycle of the parasite may extend its influence beyond the immediate realm of the definitive host, however. The subtlety and complexity of host–parasite interactions is vividly demonstrated by the way in which parasites may alter features of their intermediate hosts in order to enhance transmission to a restricted range of definitive hosts.

In the neverending quest to improve their own survivability, parasites may exert profound effects on the biology of their host species. There are many examples of intermediate stages of parasites that alter the behavior of their hosts in order to render them more vulnerable to predation and thus enhance the chances for successful completion of their life cycles (Holmes and Bethel, 1972). Larval digenean trematodes (*Euhaplorchis californiensis*) parasitize the brain of the Pacific killifish (*Fundulus parvipinnis*), causing an increase in instances of conspicuous behaviors, such as contorting, jerking, and flashing (Lafferty and Morris, 1996). This seriously compromises the adaptive capacity of the host fish by attracting the attention of avian predators, which are the definitive hosts of the adult form of the trematode. An even more subtle demonstration of the diminution of host adaptive capacities for the purpose of

FIGURE 38.5

A, Parasitic copepod of the genus *Ergasilus,* showing paired egg sacs and second antennae modified into clasping hooks (from Barnes, 1968); **B,** Parasitic copepod of the genus *Lernaea,* showing paired egg sacs and pronounced hooks on head for anchoring into host (from Hoffman, 1999).

enhancing parasite survival in seen in the ability of acanthocephalan parasites to influence predation on their isopod intermediate hosts by the fish species that constitute their definitive hosts (Lyndon, 1996). The freshwater isopod *Asellus aquaticus* is the intermediate host for two species of acanthocephalans—*Acanthocephalus anguillae* and *A. lucii.* Both of these species have as their definitive hosts a number of European fish species, including the chub (*Leuciscus cephalus*), barbel (*Barbus barbus*), perch (*Perca fluviatilis*), and stickleback (*Gasterosteus aculeatus*). The cystacanth stage of *A. anguillae* infects *Asellus* at the same frequency as *A. lucii.* The previously mentioned fish species all feed on *Asellus,* yet the frequency of infection by the two species of acanthocephalan differs dramatically. Chubs and barbels are definitive hosts primarily for *A. anguillae,* whereas sticklebacks and perches host mainly *A. lucii.* The two species of acanthocephalans affect their isopod intermediate host in different ways, which results in differential susceptibility to predators. Isopods infected with cystacanths of *A. anguillae* show increased locomotor attraction to disturbance in the water and are also significantly more photophilic. Attraction

to disturbance increases their vulnerability to the nocturnal, bottom-feeding barbels, whereas increased photophilia increases the likelihood that the isopods will be swept into currents and become part of the invertebrate drift favored by chubs that hold feeding stations along the course of the stream. Infection by *A. lucii,* on the other hand, causes the isopod hosts to acquire distinctive melanization patterns on the moving respiratory opercula used to irrigate their gills, thus causing them to be more visible and making them more vulnerable to highly visual diurnal predators such as perches and sticklebacks (Fig. 38.6).

Behavior and Susceptibility to Parasites

Individuals appear to vary in their susceptibility to infestation by a given species of parasite. Within a given population, some individuals may harbor fewer parasites than expected, whereas other individuals appear to be carrying far more than their share. Individual susceptibility may be due to a number of factors, ranging from differences in genetic composition to behavioral differences that render some

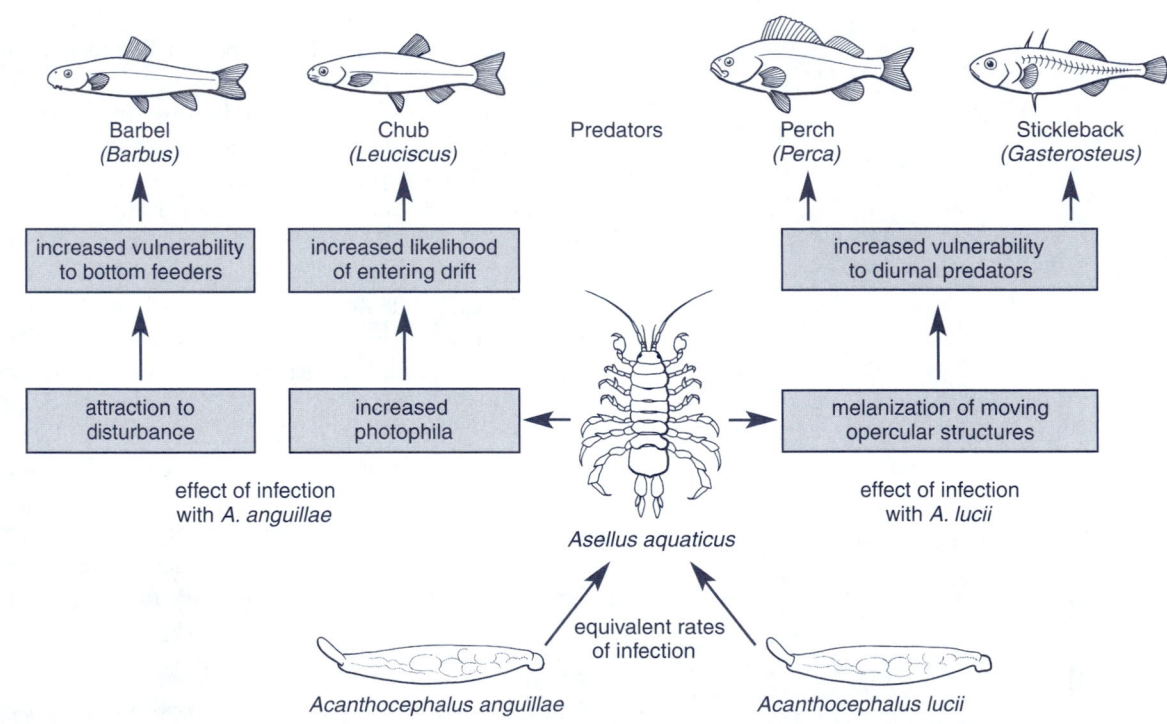

FIGURE 38.6
Morphological and behavioral modification of isopod *Asellus* by two species of *Acanthocephalus,* showing differential consequences for infestation of preferred definitive host fish species. (Based on studies by Lyndon, 1996.)

individuals more prone to infestation than others. Parasite susceptibility has had an undeniable role in the evolution of social behavior in animals (Côté and Poulin, 1995; Moore, 2002). One might expect a degree of correlation between shoal size in schooling fishes and the level of parasite infestation—the larger the school, the greater the opportunity for communicability of parasites and any associated diseases. This is true for parasites that are highly contagious but whose dispersal abilities are limited and dependent on the density of their hosts. Sometimes, the reverse is true, however. In situations of exposure to parasites with highly mobile infectious stages, schooling may confer the same benefits that it does against predators.

In some cases, parasites may exert conflicting selection pressures on shoal size because of differences in their life history and dispersal capabilities. This was demonstrated in mixed-species shoals of sticklebacks infected with crustacean ectoparasites of differing mobility (Poulin, 1999). In shoals composed of threespine sticklebacks (*Gasterosteus aculeatus*) and blackspotted sticklebacks (*G. wheatlandi*), shoal size correlated positively with infection levels of the copepod *Thersitina gasterostei*, which is transmitted via a short-lived planktonic stage and remains with its host for its entire adult life. Contrary to predictions based on laboratory studies, infection levels by the larger, more mobile crustacean *Argulus funduli* did not decrease as shoal size increased. This parasite readily moves from one host to another. Contrasting modes of transmission resulted in one parasite having a greater rate of transmission in large shoals, whereas the success of the other parasite is apparently not influenced by shoal size.

Parasite- and Pathogen-Mediated Sexual Selection

As mentioned in Chapters 18 and 37, carotenoids are dietary components essential in the creation of the bold coloration that females find so attractive. Vigorous, parasite-free males will be more successful in courting behavior, as they can maximize their carotenoid uptake and cutaneous display. Intestinal parasites can interfere with the absorption of dietary carotenoids; ectoparasites may interfere with the display of carotenoids on the skin surface; and all parasites stimulate the host immune system, thus draining the body of reserves of carotene (Olson and Owens, 1998). Parasitism may thus exert a significant impact on the process of sexual selection, as females are less likely to choose parasitized males (Hamilton and Zuk, 1982). However, one can reasonably assume that males and females suffer essentially equivalent parasite infestation. What are the consequences of parasitism on the female? Parasitized females may be less choosy than healthy females, so that the influence of sexual selection may be diminished (Poulin, 1994; Poulin and Vickery, 1996). Rolff

(1998), however, claimed that parasitized females, considered "low quality" in terms of fitness, tend to mate with "low-quality" males, so that the sexual selective force of parasites is valid and is maintained by the actions of healthy and, hence, choosy females.

Reusch et al. (2001) have presented evidence of a remarkably subtle form of sexual selection in sticklebacks. Females will choose mates that provide their offspring with the greatest diversity of histocompatibility complex genes. The **major histocompatibility complex (MHC)** is a multigene system that controls the vertebrate immune system and hence is important in conferring resistance to pathogens and parasites. Although sticklebacks are not known for especially well-developed olfactory capabilities (see Chapters 22 and 37), they apparently can determine MHC heterozygosity because of its association with a characteristic odor, and females prefer to mate with males with a greater number of MHC alleles. High allelic diversity results in the ability to recognize and combat a broader spectrum of pathogens—something that females apparently are capable of effecting through mate choice.

Parasites as Biological "Tags" of Fish Populations

The identity and genetic integrity of populations of organisms can be ascertained using well-known morphometric, meristic, and genetic methodologies. Parasites have also long been known to be of value as biological indicators in the study of fish populations. Williams et al. (1992) provided a comprehensive overview of the ways in which parasites have been used to "tag" fish populations. The specificity of the host–parasite relationship, the potential for levels of infection to vary geographically, and the ease with which many parasites can be detected make them valuable tools for fish studies. For example, Frimeth (1987a, 1987b) was able to distinguish anadromous from nonanadromous stocks of brook char (*Salvelinus fontinalis*) by the presence of metacercarial cysts of a marine digenean that produces the condition known as "black spot." Parasites have also been used to identify the recruitment migration patterns of juvenile fishes—that is, the movement from nursery grounds to adult feeding and spawning grounds (Williams et al., 1992). Parasites also provide information on the diet and feeding behavior of host fishes. Based on the differences in level of infestation by acanthocephalans versus cestodes in host flounders, Möller (1984) was able to infer where and on what the host fish was feeding. These dietary preferences were later confirmed through stomach content analysis. The specificity of the host–parasite relationship has made phylogenetic study of some host species possible. Using molecular techniques, three species of bagrid catfishes of the genus *Chrysichthys* have been identified. Each host species was also characterized by its own species of monogenean parasitizing the gills (Euzet et al., 1989).

The role of parasites as biological indicators in population studies demonstrates their manifold impact on aquatic ecosystems. Parasites may define trophic interactions in ways that dramatically influence the structure of food webs (Marcogliese and Cone, 1997). In Chapter 33 (Going Deeper box), the role of keystone species such as sea otters and urchins of the genus *Strongylocentrotus* was discussed. In this context, the parasitic amoeba (*Paramoeba invadens*) can be viewed as a "keystone parasite," as it infects the green sea urchin (*S. droebachiensis*), causing mass mortalities that permit the rapid recovery of kelp beds and their associated fauna (Jones and Scheibling, 1985). When we consider how much there is yet to be understood about the diversity, adaptations, and ecology of the world's ichthyofauna, we must recognize that our knowledge of the parasite communities they harbor is even less. We have barely begun to investigate the impact that this diverse and remarkable assemblage of creatures has on the fish fauna of the world.

NONPARASITIC FISH DISEASES

Environmentally Induced Diseases

A number of fish diseases can be attributed to causes other than parasitic organisms. Environmental conditions can contribute to health problems in fishes. One of the best known examples is that of **gas bubble disease** or "**popeye.**" This is a problem of special concern in freshwater drainages where dams have been constructed to regulate water flow. The turbulence associated with releasing water over spillways causes the water to become supersaturated with atmospheric gases—the main one, of course, being nitrogen. Fishes approach this environment and, after a short period of time, their blood and tissues are gas-supersaturated as well. When the fish eventually departs this supersaturated environment, the dissolved gases revert to their gaseous state, causing serious disruption of the tissues. The term *popeye* comes from the formation of gas bubbles behind the eye socket, causing the eye to protrude. This condition is similar to caisson disease (the "bends") that scuba divers must take precautions against.

Pathological anomalies in fishes can be attributed to excess acidity and the introduction of toxic substances into waterways—the unfortunate consequence of living in an industrialized society. Waterways that are especially sensitive to acidic precipitation, such as the lakes and streams

of New England, have experienced drastic declines in the abundance and diversity of aquatic organisms (Schindler et al., 1989). Habitat acidification results in increased mucous discharge, inflammations of the skin, and hypertrophy of gill epithelial cells (Bruno and Poppe, 1996). Fishes reared in the high densities that are often associated with aquaculture conditions may experience adverse effects of excessive concentrations of excreted ammonia. Degeneration of the liver, spleen, and gill tissues are associated with ammonia toxicity (Bruno and Poppe, 1996). The range of heavy metals, pesticides, and other industrial by-products that are released into waterways (McKim et al., 1973) continues to present a serious threat to the overall health of aquatic ecosystems and their associated fish faunas. Bioaccumulation of toxic substances (see Chapter 39) has been demonstrated to cause immunosuppression, thus leading to increased susceptibility to infectious diseases (Arkoosh et al., 1998).

Dietary Diseases

Some of the dietary diseases known in fishes (see Chapter 23) are common to all vertebrates. Chronic deficiencies in vitamin C cause a degradation of skeletal tissues, whereas anemia in fishes is caused by vitamin B_{12} deficiency. Goiters are caused by a deficiency of dietary iodine, and cataracts may arise through a deficiency of trace elements such as zinc (Bruno and Poppe, 1996). Insufficient amounts of biotin cause **blue slime disease** in trout, and a deficiency of pantothenic acid causes disorders of the gills (Royce, 1996). Farmed salmonids that are fed excessive polyunsaturated fatty acids but inadequate amounts of antioxidants, such as vitamin E, suffer from **hepatic lipoidosis,** which may manifest itself as anemia, tissue necrosis, and lesions—especially of the liver (Bruno and Poppe, 1996).

Genetic Disorders

Failure of the genetic control systems that normally operate in a healthy organism can account for a number of diseases and disorders in fishes. **Neoplasia** refers to genetic failure in the context of the creation of tumors, also known as **neoplasms.** These may be relatively benign, or they may grow uncontrollably, eventually killing the afflicted individual. Epithelial neoplasms are the most frequently encountered—possibly because of their prevalence in commercially significant species, such as eels and flatfishes (Sindermann, 1990). **Papillomas** are among the most conspicuous of the epithelial tumors (see Plate 38.8); certain viruses are often the causative agent. The discovery that elasmobranchs appear to have a greater resistance to cancer has unfortunately created a market for shark-derived products that are touted as cancer cures. Sharks, like all other vertebrates, are susceptible to cancers, although the incidence of neoplasia among elasmobranchs is lower than in any other vertebrate group (Ballantyne, 1997). Ostrander et al. (2004) have noted that the misguided perception that the ingestion of crude extracts of elasmobranch cartilage can somehow prevent cancer has had two undesirable consequences: additional stress on already decimated shark populations, and the diversion of patients from cancer therapy that can actually do some good. There is some irony in this situation, as elasmobranch cartilage, which has been promoted for its curative powers, is not even capable of repairing its own structural damage, unlike the bone of bony vertebrates, which can repair skeletal injury (Ashhurst, 2004).

Recently, there has been an increased interest in research that seeks to understand the connection between the genetic predisposition to certain forms of cancer and the role of infectious agents. Symbiotic microbes, long believed to have little role in many diseases that were considered noninfectious, are now seen as key players in such human ailments as heart disease, peptic ulcers, and cervical cancer (see Ewald, 2000). This new perspective may prove valuable in the diagnosis and treatment of some fish diseases that are normally attributed to genetic composition or environmental circumstances.

FISHES AS CARRIERS OF HUMAN PARASITES AND DISEASES

One negative aspect of the relationship between fishes and humans that is important in some geographical areas is the ability of various fish species to harbor parasites that can infect humans. The term **zoonosis** refers to any disease that humans can contract from other animals. Shapiro (http://medicine.bu.edu/dshapiro/zoo1.htm) listed 17 parasites and diseases that can be contracted from fishes. The parasites are usually of concern in areas where freshwater fishes are eaten raw or without sufficient processing, although there are some parasites that can be transmitted to humans by marine fishes.

Most of the parasites involved are worms—nematodes (roundworms), cestodes (tapeworms), or trematodes (flukes). One potentially dangerous nematode is the kidney worm (*Dioctophyma renale*), which is known mainly from Asia. Marine nematodes that are occasionally found in humans, especially in regions where raw herring is consumed, include the genus *Anisakis* and its relatives. These worms can cause illness if a person becomes heavily infested with them. One reason for refraining from the practice of swallowing live minnows (Believe it or not, college students used to engage in such silly pranks as swallowing live goldfish!) is the severe consequences of perforation of the intestinal wall by the larvae of nematodes of the genus *Eustrongyloides* (Centers for Disease Control, 1982).

Tapeworms (Cestoda) are potentially serious parasites of humans, and probably the best known fishborne cestode affecting humans is the broad tapeworm (*Diphyllobothrium latum*), which is common in some freshwater fishes in northern European countries and parts of Asia and North America. A relative (*D. pacificum*) carried by marine fishes can also infect humans (Higashi, 1985). In North America, *D. latum* is known primarily from the Great Lakes area. Humans infested with this worm generally suffer from anemia. The worm is transmitted to humans through uncooked freshwater fishes.

Among the trematodes are a few species that may be transmitted from fishes to humans. Some are intestinal parasites, but one family (Opisthorchidae) contains liver parasites that can cause serious ill effects. Infestations by the liver fluke *Opisthorchis* (= *Clonorchis*) *sinensis* are known mainly from Asia, where 20 million people are estimated to be infected (Higashi, 1985). Snails are perhaps the best known intermediate hosts for trematode parasites (see Fig. 38.1). An example is seen in the parasite dynamics involving snails, flukes, fishes, and dogs in coastal portions of the Pacific Northwest. The snail *Oxytrema silicula* harbors the early stages of the fluke *Nanophyetus salmonis*, the cercariae of which are carried mainly by salmon and trout. The adult fluke is found in various carnivores, including skunks, raccoons, and canines. The parasite itself is not especially dangerous, but it carries a rickettsial disease (caused by a form of bacteria) that is extremely dangerous to dogs. The fluke's cercariae are so common in salmonids of the region that before the true nature of the disease was discovered, it was generally believed that salmon were poisonous to dogs. Methods of prevention and treatment are now known, so the mortality rate of dogs that eat raw salmon has been considerably reduced (Baldwin et al., 1967). This fluke has now been shown also to cause illness in humans (Fritsche et al., 1989). The cercariae can be transferred by the ingestion of undercooked salmonids or, in one reported case, even by handling the fish (Harrell and Deardorf, 1990).

Bites, punctures by spines, or scratches by fins or scales are also sources of infections in humans. Various species of potentially infective *Vibrio* have been isolated from white shark teeth (Buck et al., 1984). Other pathogenic species of *Vibrio* include *V. damsa* (Fouz et al., 1992), *V. parahaemolyticus* (Ghittino, 1972), and a toxigenic strain of *V. cholerae* that was discovered in fishes and shellfishes from Mobile Bay, Alabama, in 1991 (Anonymous, 1991). Other kinds of microorganisms that are carried by fishes and are potentially infective to humans include *Aeromonas, Mycobacterium, Shigella, Salmonella, Clostridium,* and *Erysipalothrix* (Ghittino, 1972; Janssen, 1970). A painful skin disease called "fish handler's disease" is caused by *E. rhusiopathiae* (Sonnenworth et al., 1980). *M. marinum* can cause subcutaneous abscesses after gaining entry through broken skin (Wolinsky, 1980).

Home aquaria and fish culture facilities can be sources of infection. For instance, *Edwardsiella tarda* has been implicated in bouts of diarrhea in a situation where the only source of the infective agent appeared to be an aquarium (Vandepitte et al., 1983). *Salmonella* has been found in eel culture ponds in Japan (Saheki et al., 1989).

TOXIC FISHES

Aside from the harm that fishes can cause through the transmission of zoonoses, their ingestion can also result in significant distress because of intrinsic properties of their own tissues. Russell (1969) estimated the number of toxic fish species at about 1,000, but a more current assessment would undoubtedly place that number significantly higher. Included among these fishes are those that can cause illness or death when eaten (**poisonous fishes**) as well as those that have evolved mechanisms to actively deliver toxic substances (**venomous fishes;** see Chapter 18). An extensive vocabulary has been developed pertaining to the toxins of fishes and the conditions they cause in humans (Halstead, 1970, 1988; Table 38.1). A few of the more notorious examples of fish poisoning that arise through the ingestion of their tissues are discussed here.

Ciguatera

The most widely known type of poisoning caused by fishes is **ciguatera,** a type of intoxication that causes a variety of symptoms such as nausea, vomiting, abdominal pain, reversal of hot and cold sensation, and numbness of the mouth. Various other symptoms may include headache, muscular aches, dizziness, and, occasionally, blistering and loss of skin on hands and feet. A diverse array of tropical marine fishes is suspected as sources of ciguatera; some species may be toxic in some geographical areas and not in others. About 20 families, several of which contain normally safe and popular food fishes, are considered ciguatoxic (Halstead et al., 1990). Toxicity appears to increase with size in a given species. The toxin appears to accumulate through the food chain, originating in certain dinoflagellates living on macroalgae that grow around coral reefs. There still remains some uncertainty as to whether the toxins present in affected fish are the same as those produced by cultures of dinoflagellates (Anderson and Lobel, 1987).

Among the several genera of dinoflagellates suspected of being the source of ciguatera, two (*Amphidinium* and *Gambierdiscus*) have been implicated as producers of especially potent toxins (Withers, 1988; Yasumoto et al., 1987). Bomber and Aikman (1989) identified at least six toxins from 12 species of suspected ciguateric dinoflagellates. Three of these—ciguatoxin, maitotoxin, and scaritoxin—have been recovered from the flesh of fishes. Randall (1958) first suggested that ciguatoxins were passed up the food chain from unknown sources, but almost 20 years passed before this was proven to be the case, with the identification of dinoflagellates as the culprit (Anderson and Lobel, 1987). Ciguatoxin is extremely potent, having a lethal dose at which 50 percent of the test subjects die (LD_{50}) of 0.45 g kg^{-1} in mice when injected intraperitoneally. Maitotoxin, produced especially by *Gambierdiscus toxicus,* is even more potent, with an LD_{50} of 0.13 g kg^{-1} in mice. This is possibly the most potent toxin of marine origin (Ohizumi and Kobayashi, 1990). Hashimoto (1979) mentioned two genera of blue-green algae (*Schizothrix* and *Microcoleus*) as being implicated in ciguatera. Dinoflagellates with the contained toxins are thought to be more prevalent on dead coral reef areas than on undisturbed reefs (Bagnis, 1981). The toxic substances are eaten by herbivorous fishes, which then become the food of carnivores. Large carnivores concentrate the poison to the point

TABLE 38.1 TERMINOLOGY FOR TOXINS ASSOCIATED WITH FISHES

Term	Definition
Ichthyotoxin	Generally, any poison originating from fishes
Ichthyosarcotoxin	Poison found in the flesh of fishes, excluding poisons due to bacterial action
Ichthyohemotoxin	Poison found in the blood of fishes
Ichthyootoxin	Poison found only in the roe of fishes
Ichthyoacanthotoxin	Poison secreted at the site of a venom apparatus, such as spines, stings, or teeth of fishes
Ciguatera	A particular ichthyosarcotoxism caused by eating various marine fishes of tropical and subtropical areas
Scombroid poisoning	An ichthyosarcotoxism caused by eating improperly preserved scombroid fishes
Tetrodotoxin	The poison in the viscera of puffer fishes

that they become dangerous to eat—a phenomenon known as **biomagnification.**

Fishes that are most often implicated in ciguatera are morays (Muraenidae), barracuda (Sphyraenidae), snappers (Lutjanidae), groupers (Serranidae), and jacks and their close relatives (Carangidae). All these families contain excellent food fishes, which are frequently eaten in tropical areas. More than 400 species are known to cause ciguatera (Halstead and Vinci, 1988). Bagnis (1981) identified 53 species (13 families) from the Caribbean and 74 species (16 families) from the Indo–Pacific as being ciguatoxic. Randall (1980) found that more than one third of the specimens examined from Enewetok and one fifth of the specimens from Bikini showed ciguatoxicity.

Additional families of fishes, some of which are herbivores or planktivores, that have been implicated in one or more ciguatera poisonings include bonefishes (Albulidae), milkfishes (Chanidae), tarpons (Elopidae), herrings (Clupeidae), anchovies (Engraulidae), lizardfishes (Synodontidae), conger eels (Congridae), flyingfishes (Exocoetidae), squirrelfishes (Holocentridae), surgeonfishes (Acanthuridae), butterflyfishes (Chaetodontidae), mackerels and tunas (Scombridae), plus a number of other perciform fishes and some of the Tetraodontiformes (filefishes and relatives).

There have been mentions of poisonings from ciguatera dating back to the 7th century CE in China and the 17th century in Europe. The explorer Captain James Cook was apparently affected by ciguatera on his voyage of 1774 (Halstead, 1978). Reports of ciguatera poisoning usually tell of incidents involving a ship's crew or a family of islanders, but some mention 50 or 60 persons taken ill, and at least one report mentions 1,500 persons affected. Relatively few deaths (less than 10 percent) are reported in most instances, but there is a report of more than 400 deaths among Marshall and Caroline islanders during 1940 and 1941. The annual number of ciguatera cases worldwide is estimated at between 10,000 and 50,000 (Hilgerd, 1983), with about 240 cases reported in the United States from the years 1973 to 1980 (Hughes and Potter, 1991).

There is no effective remedy for ciguatera at this time. Cooking does not destroy the toxin. No immunity is imparted by a prior attack (Halstead, 1978, 1988). Recently, a breakthrough in ciguatoxin research was achieved with the first total synthesis of one form of the toxin (Hirama et al., 2001).

Scombroid Poisoning

Consumption of the flesh of tunas, their relatives, and certain other species of fishes that have not been properly processed may result in another common toxic response.

Hughes and Potter (1991) claimed that scombroid poisoning is about as common in the United States as ciguatera. Several fishes other than scombroids have been implicated in this type of poisoning, including members of the families Clupeidae, Engraulidae, Scomberesocidae, Carangidae, Coryphaenidae, Arripidae, and Pomatomidae (Auerbach, 1988). Taylor and Bush (1988) listed 45 species (in seven families) that may be implicated in scombrotoxism. Many of the species involved have a large proportion of dark muscle and maintain a temperature higher than ambient. The flesh of these fishes can undergo rapid bacterial decomposition if not quickly refrigerated. Tuna held at 20 to 25°C for several hours acquire toxicity (Halstead, 1978). Bacterial action converts free histidine, which is common in scombroid muscle, to histamine and derivatives such as saurine, the phosphate salt of histamine (Taylor and Bush, 1988). Convincing evidence that histamine is the toxic agent causing scombroid poisoning was presented by Morrow et al. (1991).

Strong allergic reactions can be caused by the ingestion of histamine-rich fish. These reactions include flushing of the skin, dizziness, nausea, diarrhea, vomiting, thirst, and palpitations (Auerbach, 1988; Halstead, 1978, 1988). Because the problem of histamine in fish flesh is not limited to scombroids, and the symptoms are allergylike, the poisoning has been referred to as "histamine (scombroid fish) poisoning" by Taylor and Bush (1988) and as "pseudo–allergic fish poisoning" by Prescott (1984).

Tetrodotoxin Poisoning

Tetrodotoxin is one of the most potent toxins that are naturally produced by animals. It is classified as a neurotoxin, as it acts by blocking the flow of sodium ions necessary for normal propagation of nerve impulses. The poison found in puffers and their relatives has been isolated and its chemical structure determined. It is identical to *tarichatoxin,* a poison found in newts of the genus *Taricha* (Halstead, 1978).

Puffers (Tetraodontidae), porcupinefishes (Diodontidae; see Plate 17.5), molas (Molidae; see Plate 17.6), and the goby *Gobius criniger* (= *Rhinogobius nebulosus;* Hashimoto, 1979) have been implicated in tetrodotoxism. These are widely distributed in tropical and warm temperate waters and have been the cause of illness and death throughout their ranges, especially in Asia. Much has been written about puffer poisoning (tetrodotoxism) caused by eating the viscera of various tetraodontiform fishes, mostly of the genus *Takifugu* (= *Fugu*), which are highly esteemed table fare in Japan (Halstead, 1978, 1988). Even though tetrodotoxin poisoning has a fatality rate higher than 50 percent, the flesh of puffers is of such high quality that the Japanese maintain a fishery for them and license chefs to prepare the fish so that only the nontoxic portions reach the table (most of the time). The ovary and liver are the most toxic parts of the fish, with the stomach and intestines being nearly as virulent. The eyes and kidneys are toxic as well. The skin, subcutaneous tissue, and testes are only moderately poisonous in some species (Halstead, 1978; Hashimoto, 1979). Other poisonous genera include *Arothon, Lagocephalus, Sphaeroides,* and *Tetraodon* (Halstead, 1992).

Other Fish Poisons

Other types of fish poisoning are known, but do not appear to be as common as the aforementioned ones. Lampreys and hagfishes have been implicated in what is called "cyclostome poisoning," which, in the case of the lampreys, might be due to an excess of bile salts in the flesh (McDowall, 1988). Several families of sharks have toxic livers or flesh. Elasmobranch poisoning might be due to three different kinds of toxins (including ciguatera). The flesh of the Greenland shark (*Somniosus macrocephalus*) often contains a toxin that can be removed by drying or thoroughly washing strips of the meat. Many tropical sharks have toxic livers, a condition apparently not related to hypervitaminosis A, which results from eating the vitamin-rich livers.

Oilfishes (Gempylidae) have been called "purgativefishes" by seafarers because of the diarrhea caused by eating their oily flesh. A few fishes are known to produce toxins from skin glands that are not associated with any kind of stinging structure. These poisons, called *ichthyocrinotoxins,* are harmful if ingested by humans. Fishes known to have poisonous skin or slime or to secrete toxins on being disturbed include hagfishes, lampreys, morays, soapfishes (Grammistidae), the Moses sole (*Pardachirus marmoratus*), puffers, and their near relatives (see Chapter 18).

KEY POINTS AND CONNECTIONS

• A diverse assemblage of viruses, bacteria, protists, fungi, and animals has evolved parasitic modes of existence; these creatures consequently have a significant impact on the lives of fishes. Parasitic forms are usually distinguished as either ectoparasites, which live on the epithelial surfaces of their hosts, or endoparasites, which parasitize internal tissues.

• The relationship between the parasite and its host is a demonstration of the "Red Queen hypothesis," which states that parasites and their hosts are engaged in a constant evolutionary arms race just to maintain their adaptive position in the host-parasite relationship. This has resulted in characteristic evolutionary patterns in parasites that have enabled their adaptation to certain ecological conditions, such as discontinuous habitats, extreme resource specialization, and nonequilibrium situations.

• Because of their simplicity and rapid rates of reproduction, the viruses, eubacteria, and protists are among the most effective parasites. A number of diseases—many of significance in fish culture operations—are attributed to these organisms.

The environmental impacts of fish culture and the genetic manipulation of cultured fishes—for example, to improve disease resistance—are discussed in Chapter 39.

• The myxozoans are a subgroup within the animal kingdom consisting of minute organisms that were previously classified as protozoans. One of these is the causative agent for whirling disease, which has had a significant impact on salmonid populations.

As whirling disease is an otic disorder, refer to Chapter 21 for a discussion of the structure and function of this system in fishes.

• Among the metazoan members of the animal kingdom, wormlike forms, such as the platyhelminths, acanthocephalans, and nematodes, have been especially successful as parasites of fishes. A number of crustaceans, especially the copepods, are important ectoparasites of fishes.

• A zoonosis is a disease that humans can contract from other animals. Fishes are known to harbor many parasites and diseases that infect humans. Fishes are also the source of human diseases, such as ciguatera, that are the result of the ingestion of tissues that have accumulated toxic substances from the environment. Many species of fishes possess tissues and organs that are inherently toxic to humans; the puffers are perhaps the most notorious example, as they produce one of the most potent of vertebrate poisons, tetrodotoxin.

Additional sources of toxicity in the skin and associated structures of fishes are explored in Chapter 18.

FISH LINKS

http://www.lsc.usgs.gov/FHL/Lsc-fhl.htm Leetown Science Center—National Fish Health Laboratory.

http://www.aphis.usda.gov/vs/aqua/ Website for the U.S. Department of Agriculture's Animal and Plant Health Inspection Service (APHIS) dedicated to health issues pertaining to the aquaculture industry (National Aquaculture Program).

http://www.biosci.ohio-state.edu/Dparasite/home.html Catalog of parasite images from Ohio State University College of Biological Sciences.

BUILDING AN ICHTHYOLOGY LIBRARY

Halstead, B. W. 1988. *Poisonous and venomous marine animals of the world* (2nd ed.). Darwin Press, Princeton, NJ.
Hoffman, G. L. 1999. *Parasites of North American freshwater fishes* (2nd ed.). Comstock, Cornell University Press, Ithaca, NY.

(First published in 1967, this remains the "bible" of North American fish parasitology, even though the systematics and taxonomy are somewhat outdated.)

Noga, E. J. 1995. *Fish disease: Diagnosis and treatment.* Mosby-Year Book, St. Louis.

Woo, P. T. K., J. F. Leatherland, and D. W. Bruno (Eds.). 1994. *Fish diseases and disorders,* Vol. 1, Protozoan and metazoan infections. CABI, New York.

Leatherland, J. F., and P. T. K. Woo (Eds.). 1999. *Fish diseases and disorders,* Vol. 2, Non-infectious disorders. CABI, New York.

Woo, P. T. K., and D. W. Bruno (Eds.). *Fish diseases and disorders,* Vol. 3, Viral, bacterial, and fungal infections. CABI, New York.

Sindermann, C. J. 1990. *Principal diseases of marine fish and shellfish,* Vol. 1, Diseases of marine fish. Academic Press, San Diego.

REFERENCES

Anderson, D. M., and P. S. Lobel. 1987. The continuing enigma of ciguatera. *Biol. Bull. 172:* 89–107.

Anderson, R. M., and R. M. May. 1978. Regulation and stability of host–parasite population interactions. I. Regulatory processes. *J. Anim. Ecol. 47:* 219–247.

Anonymous. 1991. Cholera found in Mobile Bay. *Fisheries News 2*(11): 1.

Arkoosh, M. R., E. Casillas, E. Clemons, A. N. Kagley, R. Olson, P. Reno, and J. E. Stein. 1998. Effect of pollution on fish diseases: Potential impacts on salmonid populations. *J. Aquat. Anim. Health 10:* 182–190.

Ashhurst, D. E. 2004. The cartilaginous skeleton of an elasmobranch fish does not heal. *Matrix Biol. 23:* 15–22.

Auerbach, P. S. 1988. Clinical therapy of marine envenomation and poisoning, pp. 493–565. In *Handbook of natural toxins,* Vol. 3, Marine toxins and venoms, A. T. Tu (Ed.). Marcel Dekker, New York.

Bagnis, R. 1981. L' ichtyosarcotoxisme de type ciguatera: Phénomène complexe de biologie marine et humaine. *Oceanol. Acta 4:* 375–387.

Bakker, T. C. M., and B. Mundwiler. 1999. Pectoral fin size in a fish species with paternal care: A condition-dependent sexual trait revealing infection status. *Freshw. Biol. 41:* 543–551.

Baldwin, N. L., R. E. Millemann, and S. E. Knapp. 1967. "Salmon poisoning disease." III. Effect of experimental *Nanophyetus salmincola* infection on the fish host. *J. Parasitol. 53:* 556–564.

Ballantyne, J. S. 1997. Jaws: The inside story. The metabolism of elasmobranch fishes. *Comp. Biochem. Physiol. 118B:* 703–742.

Barnes, R. D. 1968. *Invertebrate zoology* (2nd ed.). W. B. Saunders, Philadelphia.

Bergersen, E. P., and D. E. Anderson. 1997. The distribution and spread of *Myxobolus cerebralis* in the United States. *Fisheries 22*(8): 6–7.

Blaxter, M. L., P. DeLey, J. R. Garey, L. X. Liu, P. Scheldeman, A. Vierstraete, J. R. Vanfleteren, L. Y. Mackey, M. Dorris, L. M. Frisse, J. T. Vida, and W. K. Thomas. 1998. A molecular evolutionary framework for the phylum Nematoda. *Nature 392:* 71–75.

Bomber, J. W., and K. E. Aikman. 1989. The ciguatera dinoflagellates. *Biol. Oceanogr. 6:* 291–311.

Bruno, D. W., and T. T. Poppe. 1996. *A colour atlas of salmonid diseases.* Academic Press, London.

Buck, J. D., S. Spotte, and J. J. Gadbaw, Jr. 1984. Bacteriology of the teeth from a great white shark: Potential medical implications for shark bite victims. *J. Clin. Microbiol. 20:* 840–851.

Centers for Disease Control. 1982. Intestinal perforation caused by larval *Eustrongyloides*-Maryland. *Morb. Mortal. Weekly Rept. 32*(28): 383.

Cook, D. W., and S. R. Lofton. 1975. Pathogenicity studies with a *Streptococcus* sp. isolated from fishes in an Alabama-Florida fish kill. *Trans. Am. Fish. Soc. 104:* 286–288.

Côté, I. M., and R. Poulin. 1995. Parasitism and group size in social animals: A meta-analysis. *Behav. Ecol. 6:* 159–165.

DeLoach, N. 1999. *Reef fish behavior: Florida, Caribbean, Bahamas.* New World, Jacksonville, FL.

Ellis, A. E. (Ed.). 1988. *Fish vaccination.* Academic Press, London.

Euzet, L., J. F. Agnese, and A. Lambert. 1989. Value of parasites as criteria for identifying host species. Convergent demonstration by a parasitological study of branchial monogeneans and by genetic analysis of their hosts. *C. R. Hebd. Séanc. Acad. Sci. Paris III 308:* 385–388.

Ewald, P. 2000. *Plague time: How stealth infections are causing cancers, heart disease, and other deadly ailments.* Free Press, New York.

Ferber, D. 2001. Will black carp be the next zebra mussel? *Science 292:* 203.

Fouz, B., R. F Conchas, A. B. Toranzo, and C. Amaro. 1992. *Vibrio damsela* strain virulence for fish and mammals. *Fish Health Sect. Am. Fish. Soc. Newsletter 20*(1): 3–4.

Frimeth, J. P. 1987a. A survey of the parasites of nonanadromous and anadromous brook charr (*Salvelinus fontinalis*) in the Tabusintac River, New Brunswick, Canada. *Can. J. Zool. 65:* 1354–1369.

———. 1987b. Potential use of certain parasites of brook charr (*Salvelinus fontinalis*) as biological indicators in the Tabusintac River, New Brunswick, Canada. *Can. J. Zool. 65:* 1989–1995.

Fritsche, T. R., R. L. Bastburn, L. H. Wiggins, and C. A. Terhune, Jr. 1989. Praziquantel for treatment of human *Nanophyetus salmincola* (*Troglotrema salmincola*) infection. *J. Infect. Dis. 160:* 896–899.

Ghittino, P. 1972. Aquaculture and associated diseases of fish of public health importance. *J. Am. Vet. Med. Assoc. 161:* 1476–1485.

Gozlan, R. E., S. St-Hilaire, S. W. Feist, P. Martin, and M. L. Kent. 2005. Disease threat to European fish. *Nature 435:* 1046.

Granovitch, A. I. 1999. Parasitic systems and the structure of parasite populations. *Helgol. Mar. Res. 53:* 9–18.

Halstead, B. W. 1970. *Poisonous and venomous marine animals of the world,* Vol. 3. U.S. Government Printing Office, Washington, DC.

———. 1978. *Poisonous and venomous marine animals of the world* (rev. ed.). Darwin Press, Princeton, NJ.

———. 1988. *Poisonous and venomous marine animals of the world* (2nd ed.). Darwin Press, Princeton, NJ.

———. (in collaboration with P. S. Auerbach.) 1992. *Dangerous aquatic animals of the world. A color atlas.* Darwin Press, Princeton, NJ.

———, P. S. Auerbach, and D. Campbell. 1990. *A colour atlas of dangerous marine animals.* Wolfe, London.

———, and J. M. Vinci. 1988. Biology of poisonous and venomous marine animals, pp. 1–30. In *Handbook of natural toxins,* Vol. 3, A. Tu (Ed.). Marcel Dekker, New York.

Hamilton, W., and M. Zuk. 1982. Heritable true fitness and bright birds: A role for parasites? *Science 218:* 384–387.

Harrell, L. W., and T. L. Deardorf. 1990. Human nanophyetiasis: Transmission by handling naturally infected coho salmon (*Oncorhynchus kisutch*). *J. Infect. Dis. 16:* 146–148.

Hashimoto, Y. 1979. *Marine toxins and other bioactive marine metabolites.* Japanese Scientific Society Press, Tokyo.

Higashi, G. I. 1985. Foodborne parasites transmitted to man from fish and other aquatic foods. *Food Technology, March 1985:* 69–74, 111–112.

Hilgerd, T. 1983. Ciguatera food poisoning: A circum-tropical fisheries problem, pp. 1–7. In *Natural toxins and human pathogens in the marine environment,* R. Colwell (Ed.). Maryland Sea Grant, College Park.

Hirama, M., T. Oishi, H. Uehara, M. Inoue, M. Maruyama, H. Oguri, and M. Satake. 2001. Total synthesis of ciguatoxin CTX3C. *Science 294:* 1904–1907.

Hoffman, G. L. 1999. *Parasites of North American freshwater fishes* (2nd ed.). Comstock, Cornell University Press, Ithaca, NY.

Holmes, J. C., and W. M. Bethel. 1972. Modification of intermediate host behavior by parasites. *Zool. J. Linn. Soc.* 51(Suppl. 1): 123–149.

Hughes, J. M., and M. E. Potter. 1991. Scombroid-fish poisoning. *New Engl. J. Med. 324:* 766–768.

Janssen, W. A. 1970. Fish as potential vectors of human bacterial disease, pp. 284–290. In A symposium on diseases of fish and shellfishes, S. F. Snieszko (Ed.), *Am. Fish. Soc. Spec. Publ. 5.*

Jones, G. M., and R. E. Scheibling. 1985. *Paramoeba* sp. (Amoebida, Paramoebidae) as the possible causative agent of sea urchin mass mortality in Nova Scotia. *J. Parasitol. 71:* 559–565.

Lafferty, K. D., and A. K. Morris. 1996. Altered behavior of parasitized killifish increases susceptibility to predation by bird final hosts. *Ecology 77:* 1390–1397.

Lee, J. L., S. H. Hutner, and E. C. Bovee. 1985. *Illustrated guide to the Protozoa.* Society of Protozoologists, Lawrence, KS.

Lorenzen, E., K. Einer-Jensen, T. Martinussen, S. E. LaPatria, and N. Lorenzen. 2000. DNA vaccination of rainbow trout against viral hemorrhagic septicemia virus: A dose-response and time-course study. *J. Aquat. Anim. Health 12:* 167–180.

Lyndon, A. R. 1996. The role of acanthocephalan parasites in the predation of freshwater isopods by fish, pp. 26–32. In *Aquatic predators and their prey,* S. P. R. Greenstreet and M. L. Tasker (Eds.). Fishing News Books, Oxford.

Marcogliese, D. J., and D. K. Cone. 1997. Food webs: A plea for parasites. *TREE 12:* 320–325.

McDowall, R. M. 1988. *Diadromy in fishes.* Timber Press, Portland, OR.

McKim, J. M., G. M. Cristensen, J. H. Tucker, D. A. Benoit, and M. J. Lewis. 1973. Effects of pollution on freshwater fish. *J. Water Poll. Cont. Fed. 45:* 1370–1407.

Miller, R. B. 1946. Cestode "parasitized" by acanthocephalan. *Science 103:* 762.

Mitchell, A. J., R. M. Overstreet, A. E. Goodwin, and T. M. Brandt. 2005. Spread of an exotic fish-gill trematode: A far-reaching and complex problem. *Fisheries 30:* 11–16.

Möller, H. 1984. *Daten zur biologie der Elbefische.* Möller, Kiel, Germany.

Moore, J. 2002. *Parasites and the behavior of animals.* Oxford University Press, New York.

Morrow, J. D., G. R. Margolies, J. Rowland, and L. J. Roberts, II. 1991. Evidence that histamine is the causative toxin of scombroid-fish poisoning. *New Engl. J. Med. 324:* 716–720.

Nyholm, S. V., E. V. Stabb, E. G. Ruby, and M. J. McFall-Ngai. 2000. Establishment of an animal–bacterial association: Recruiting symbiotic vibrios from the environment. *Proc. Natl. Acad. Sci. USA 97:* 10231–10235.

Ohizumi, Y., and M. Kobayashi. 1990. Co-dependent excitatory effects of maitotoxin on smooth and cardiac muscle, pp. 133–143. In Marine toxins: Origin, structure and molecular pharmacology, S. Hall and C. Strichartz (Eds.), *ACS Symp. Ser. 418.*

Olson, V. A., and I. P. F. Owens. 1998. Costly sexual signals: Are carotenoids rare, risky, or required? *TREE 13:* 510–514.

Ostrander, G. K., K. C. Cheng, J. C. Wolf, and M. J. Wolfe. 2004. Shark cartilage, cancer and the growing threat of pseudoscience. *Cancer Res. 64:* 8485–8491.

Pechenik, J. A. 2000. *Biology of the invertebrates* (4th ed.). McGraw-Hill, Boston.

———. 2005. *Biology of the invertebrates* (5th ed.). McGraw-Hill, Boston.

Poulin, R. 1994. Mate choice decisions by parasitized female upland bullies, *Gobiomorphus breviceps. Proc. Roy. Soc. Lond. B 256:* 183–187.

———. 1995. Phylogeny, ecology, and the richness of parasite communities in vertebrates. *Ecol. Monogr. 65:* 283–302.

———. 1997. Species richness of parasite assemblages: Evolution and patterns. *Ann. Rev. Ecol. Syst. 28:* 341–358.

———. 1999. Parasitism and shoal size in juvenile sticklebacks: Conflicting selection pressures from different ectoparasites? *Ethology 105:* 959–968.

———, and W. L. Vickery. 1996. Parasite-mediated sexual selection: Just how choosy are parasitized females? *Behav. Ecol. Sociobiol. 38:* 43–49.

Prescott, B. D. 1984. "Scombroid poisoning" and bluefish: The Connecticut connection. *Conn. Med. 48:* 110.

Prescott, L. M., J. P. Harley, and D. A. Klein. 1999. *Microbiology* (4th ed.). WCB-McGraw-Hill, Boston.

Price, P. W. 1980. *Evolutionary biology of parasites.* Princeton University Press, Princeton, NJ.

Randall, J. E. 1958. A review of ciguatera, tropical fish poisoning, with a tentative explanation of its cause. *Bull. Mar. Sci. Gulf Caribb. 8:* 236–267.

———. 1980. A survey of ciguatera at Enewetak and Bikini, Marshall Islands, with notes on the systematics and food habits of ciguatoxic fishes. *Fish. Bull. 78:* 201–249.

Reusch, T. B. H., M. A. Häberli, P. B. Aeschlimann, and M. Milinski. 2001. Female sticklebacks count alleles in a strategy of sexual selection explaining MHC polymorphism. *Nature 414:* 300–302.

Rolff, J. 1998. Parasite-mediated sexual selection: Parasitized non-choosy females do not slow down the process. *Behav. Ecol. Sociobiol. 44:* 73–74.

Royce, W. F. 1996. *Introduction to the practice of fishery science* (rev. ed.). Academic Press, San Diego.

Russell, F. E. 1969. Poisons and venoms, pp. 401–449. In *Fish physiology,* Vol. III, W. S. Hoar and D. J. Randall (Eds.). Academic Press, New York.

Saheki, K., S. Kobayashi, and T. Kawanishi. 1989. *Salmonella* contamination of eel culture ponds (Engl. abstr.). *Jpn. Soc. Sci. Fisheries 55:* 675–679.

Schindler, D. W., S. E. M. Kasian, and R. H. Hesslein. 1989. Losses of biota from American aquatic communities due to acid rain. *Env. Monit. Assess. 12:* 269–285.

Siddall, M. E., D. S. Martin, D. Bridge, D. M. Cone, and S. S. Desser. 1995. The demise of a phylum of protists: Myxozoa and other parasitic Cnidaria. *J. Parasitol. 81:* 961–967.

Sindermann, C. J. 1990. *Principle diseases of marine fish and shellfish,* Vol. I, Diseases of marine fish. Academic Press, San Diego.

Sonnenworth, A. C., Z. A. McGee, and B. D. Davis. 1980. Other pathogenic microorganisms: L-phase variants, pp. 789–805. In *Microbiology* (3rd ed.), B. D. Davis, R. Dulbecco, S. N. Eisen, and H. S. Ginsberg (Eds.). Harper and Row, Hagerstown, MD.

Taylor, S. L., and R. K. Bush. 1988. Allergy by ingestion of seafoods, pp. 149–183. In *Marine toxins and venoms,* A. T. Tu (Ed.). Marcel Dekker, New York.

Torchin, M. E., K. D. Lafferty, A. P. Dobson, V. J. McKenzie, and A. M. Kuris. 2003. Introduced species and their missing parasites. *Nature 421:* 628–630.

Van Valen, L. 1973. A new evolutionary law. *Evol. Theory 1:* 1–30.

Vandepitte, J., P. Lemmens, and L. De Swert. 1983. Human edwardsiellosis traced to ornamental fish. *J. Clin. Microbiol. 17:* 165–167.

Williams, H. H., K. MacKenzie, and A. M. McCarthy. 1992. Parasites as biological indicators of the population biology, migrations, diet, and phylogenetics of fish. *Rev. Fish Biol. Fisheries 2:* 144–176.

Withers, N. W. 1988. Ciguatera fish toxins and poisoning, pp. 31–61. In *Marine toxins and venoms,* A. T. Tu (Ed.). Marcel Dekker, New York.

Wolinsky, E. 1980. Mycobacteria, pp. 724–742. In *Microbiology* (3rd ed.), B. D. Davis, R. Dulbecco, S. N. Eisen, and H. S. Ginsberg (Eds.). Harper and Row, Hagerstown, MD.

Yasumoto, T., N. Seino, Y. Murakami, and M. Murata. 1987. Toxins produced by benthic dinoflagellates. *Biol. Bull. 172:* 128–131.

39

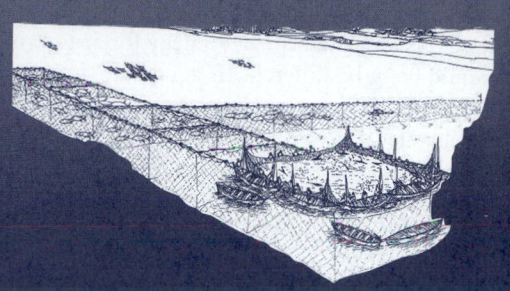

FISHES AND FISHERY RESOURCES: THEIR USE AND CONSERVATION

THE VALUE OF FISHES

FISHERY RESOURCES
A Chronology of Fishery Interests
Some Fishery Statistics

FRESHWATER AND ANADROMOUS FISHERIES
The Great Lakes: Decline of a Temperate Lake Fishery
Anadromous Fisheries
Tropical Freshwater Fisheries

MARINE FISHERIES
Pelagic Fisheries
  Clupeoids
  Scombroids
Demersal Fisheries
  The Gadoid Fishery

THE SCIENCE AND TECHNOLOGY OF FISHERIES

SCIENTIFIC MANAGEMENT OF FISHERIES

RECREATIONAL FISHERIES

THE ORNAMENTAL FISH TRADE

AQUACULTURE

CONSERVATION OF HABITAT AND BIODIVERSITY
Chemical Pollutants
  The Threat of Bioaccumulation and Biomagnification
  Impact at the Ecosystem Level
Extinctions and the Impact of Exotic Species
Hydrological Modifications
Global Environmental Perturbations

ARE THERE FISH IN OUR FUTURE?
Preserving Fish and Fisheries
Redefining Stewardship

No group of animals is more important to humans than fishes. A vital source of protein, they have become a dietary staple, shaping cultures and their characteristic cuisines. For many nations, fishery resources provide a crucial economic foundation. For such an extraordinary diverse group of animals, comparatively few species of fishes have that combination of desirability and requisite life history features that make them exploitable resources of any consequence. But these few are of inestimable value—some of them sustaining fisheries of global proportions. Fishes are also the "canary in the coal mine," providing us with forewarning of the degradation of our precious water resources. Our relationship to fishes is indeed paradoxical. Exploitation to the point of extinction of those species that have provided sustenance for humanity for millennia continues unabated, while at the same time the destruction of precious aquatic habitat spells doom for all species. These are indeed hard times for fishes, and consequently, they are hard times for us too. It is only through the prudent management of fishery resources and a greater recognition of the value of aquatic habitats that we can expect to sustain fish populations for future generations.

THE VALUE OF FISHES

"There are no places left that don't fall under humanity's shadow."

—Richard Gallager and Betsy Carpenter

This observation was made in a special issue of *Science* (25 July, 1997) that focused on human-dominated ecosystems. It is an entirely appropriate characterization of most of the aquatic ecosystems that fishes call home. The impact of humans on these systems is so vast and all-encompassing that a comprehensive analysis of it is well beyond the scope of this book. Rather, we shall attempt here in a few pages to give the reader a brief introduction to some of the major impacts that we humans have made on the fish fauna with which we share this planet. The sad state of affairs in fisheries—especially ocean fisheries—is a classic demonstration of what has come to be known as the **tragedy of the commons**: Individuals act in their own self-interest in exploiting resources and the environment for their immediate benefit but share the consequences of such selfish behavior when resources fail or the environment is irreversibly damaged (Hardin, 1968, 1998). The enormous, often devastating impact that humans have had on fishes stems from their own intrinsic commercial value, which we readily recognize, while at the same time disregarding the intrinsic value of their habitat. We shall attempt to review human impacts on fishes in general by considering the effects of human activity on fish species deemed valuable largely because of their utility as food resources. We will also consider conservation issues as they pertain to the maintenance of habitat quality and the associated species diversity for all fishes. In the latter case, the intrinsic value of the affected fish species is not gained from their palatability or their utility, but rather from the recognition that our quality of life is directly correlated with the number of species besides ourselves with which we are capable of sharing our living space.

With respect to harvested fish resources, Williams (1997) has crystallized the dilemma into four salient questions:

1. What are the ecological limits of fishery resources?

2. What costs, benefits, and social consequences must be considered in the allocation of fishery resources?

3. How can an adverse environmental impact, such as damage to spawning grounds, be economically remedied? At whose expense?

4. At what point should collateral damage caused by the exploitation of other resources, such as timber harvesting, hydroelectric power generation, or offshore oil exploration, be restrained?

Caddy (1999) provided a useful summary of the sometimes bewildering array of approaches to the management of fish resources and the limits of their effectiveness. Clearly, an effective management strategy for anadromous salmonids would be quite different from one for the demersal cod fishery. Some management strategies are best carried out at the level of local communities that are immediately affected by those resources, whereas international agreements are needed for highly migratory species that do not recognize territorial boundaries.

FISHERY RESOURCES

A Chronology of Fishery Interests

Three interest groups, each with its own management agenda, are currently competing for available fishery resources. These groups have developed in a sort of chronological fashion, which reflects the evolution of human society as a whole. Those groups that harvest fishes for *subsistence* have the longest association with any given fishery. The culture of aboriginal peoples naturally reflects the resources available to them. Resources that are deemed essential for the survival of the group may evolve a ceremonial significance, as is the case with the Native American peoples of the Pacific Northwest (see Chapter 11, Going Deeper).

When fishery products can be harvested in sufficient quantity and when their desirability extends beyond the people that harvest them, they become items of *commerce* and figure significantly in the development of economic interchange among different societies. The cod fishery discussed later is the definitive example of a fishery that has achieved commercial proportions. Because of their enormous impact, our attention will focus mostly on commercial fisheries and the species they exploit.

Finally, when the economy of a given society has developed to the point that the entire day's activities need not be centered on obtaining sustenance, *recreational* activities become increasingly important. Sport and recreational fishing has itself become a growth industry. As a member of the "baby boomer" segment of society, looking at retirement in the not too distant future, I eagerly anticipate the opportunity to make such recreational activity a full-time pursuit.

Reconciling the somewhat disparate agendas of each interest group is a challenge for today's fishery manager. The target of any fishery should be **sustainability**—the ability to continue harvesting without long-term detriment to the exploited population, yet as demands for fishery products increase, that goal gets harder to achieve—even when the most sophisticated tools of modern fishery science and technology are brought to bear.

Some Fishery Statistics

The United Nations Food and Agricultural Association (FAO) maintains comprehensive statistics on all aspects of world food production, including fisheries and aquaculture. In reviewing these statistics, one thing is apparent—the rate at which fishery technology continues to develop and be applied, especially in developing countries, far exceeds the rate at which the targeted species can replenish themselves. From 1950 to 1970, total fisheries production grew at a substantial 6 percent per year. During the 1970s and 1980s, this rate declined to 2 percent per year. By the 1990s, the rate of increase of capture fisheries had reached a plateau, with subsequent increases in fishery production accounted for largely through the development of aquaculture (FAO, 2001; Fig. 39.1). Since 1990, aquaculture production has increased at an average rate of 10 percent per year. China produces by far the most fishery products, followed by Japan and the United States (Fig. 39.2).

The FAO is presently the only institution engaged in the collection of global fishery statistics. Given that the FAO is dependent on data provided by member nations, concern has been expressed that some countries may be systematically underreporting catches, resulting in compromises in the management of valuable fish stocks (Watson and Pauly, 2001). The greatest proportion of fishery production is in the Northern Hemisphere, yet the greatest increase over the past two years has been in the Southeast Pacific (Fig. 39.3). Landings of demersal (benthic or benthopelagic; see Chapter 35) fisheries in the Northwestern Atlantic have plummeted in the past three decades, whereas the production in the Northwest Pacific seems to have reached a plateau. Almost half of the world's fishery stocks exist in a state of full exploitation, and about 22 percent are considered overexploited (Botsford et al., 1997; Fig. 39.4). The greatest potential for production increase appears to be in the Western Central Pacific Ocean and the Indian Ocean, as fewer fully exploited or overexploited stocks are to be found here.

The dominant species that constitute the marine capture fisheries are given in Figure 39.5. Fish species classified under the Magnuson-Stevens Fishery Conservation Management Act as being overfished are given in Table 39.1 (Iudicello et al., 1999).

FRESHWATER AND ANADROMOUS FISHERIES

Freshwater capture fisheries currently account for about 6.5 percent of the world fishery production. Although this may not seem significant, we must remember that the rivers, lakes, and wetlands that contribute to this fishery are but a miniscule proportion of the total hydrosphere (see Chapter 31). Freshwater aquaculture—rearing fishes in high densities,

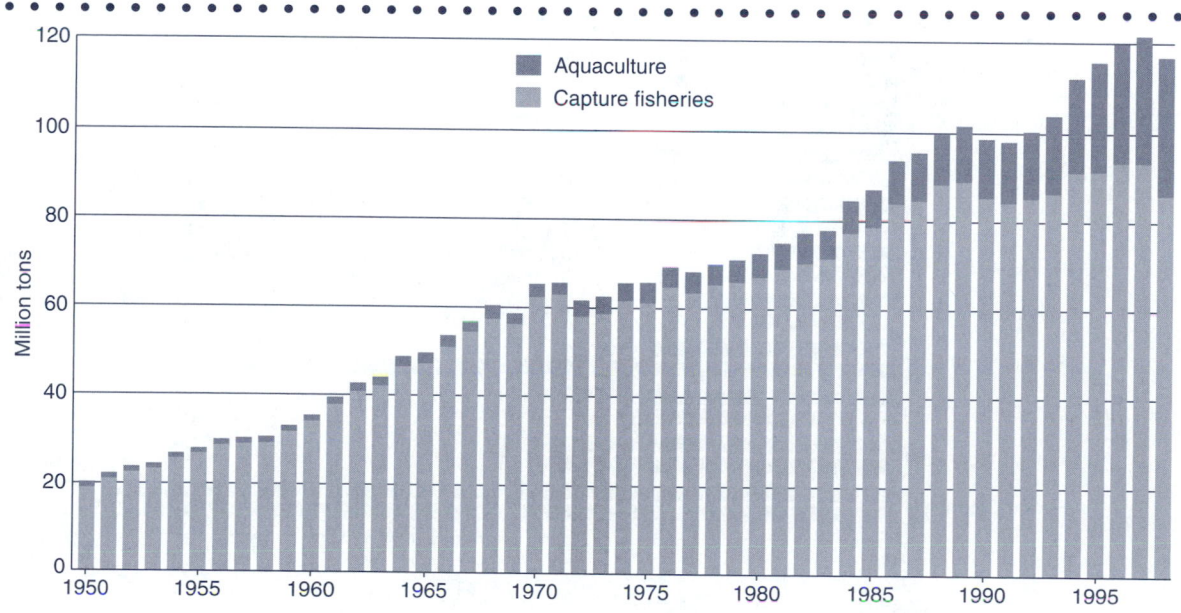

FIGURE 39.1

Growth of capture fisheries and aquaculture production. Aquaculture data prior to 1984 are estimates (Source: FAO, 2001).

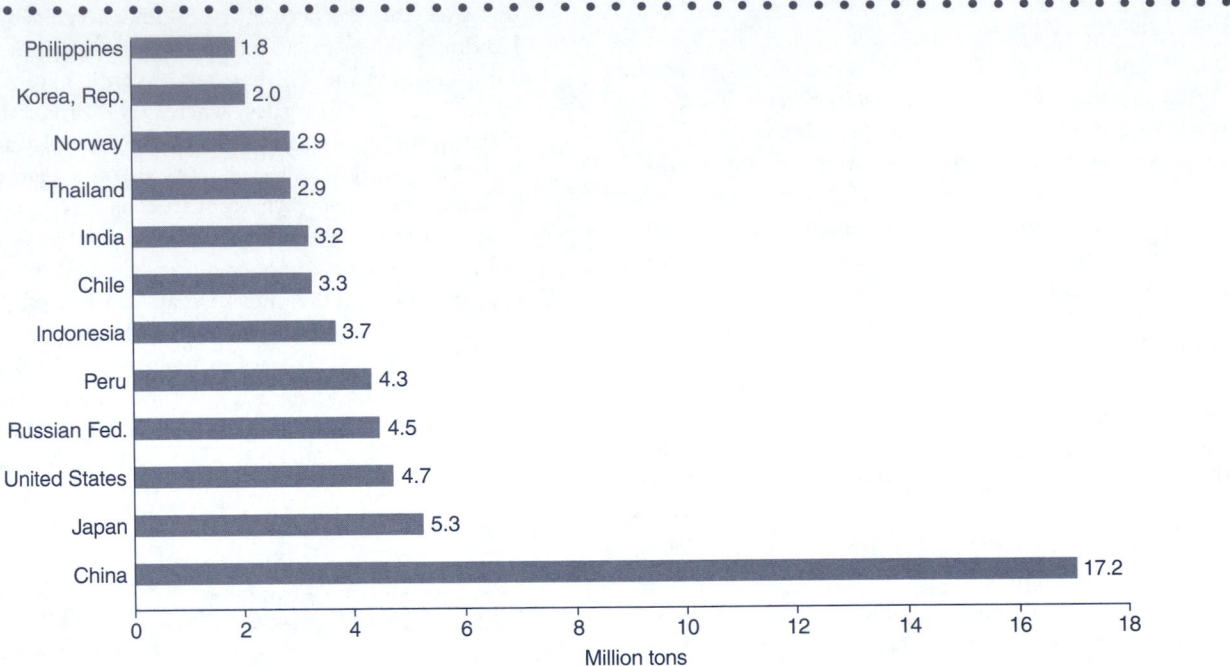

FIGURE 39.2
Freshwater and marine fisheries production in the top 12 producing countries
(Source: FAO, 2001).

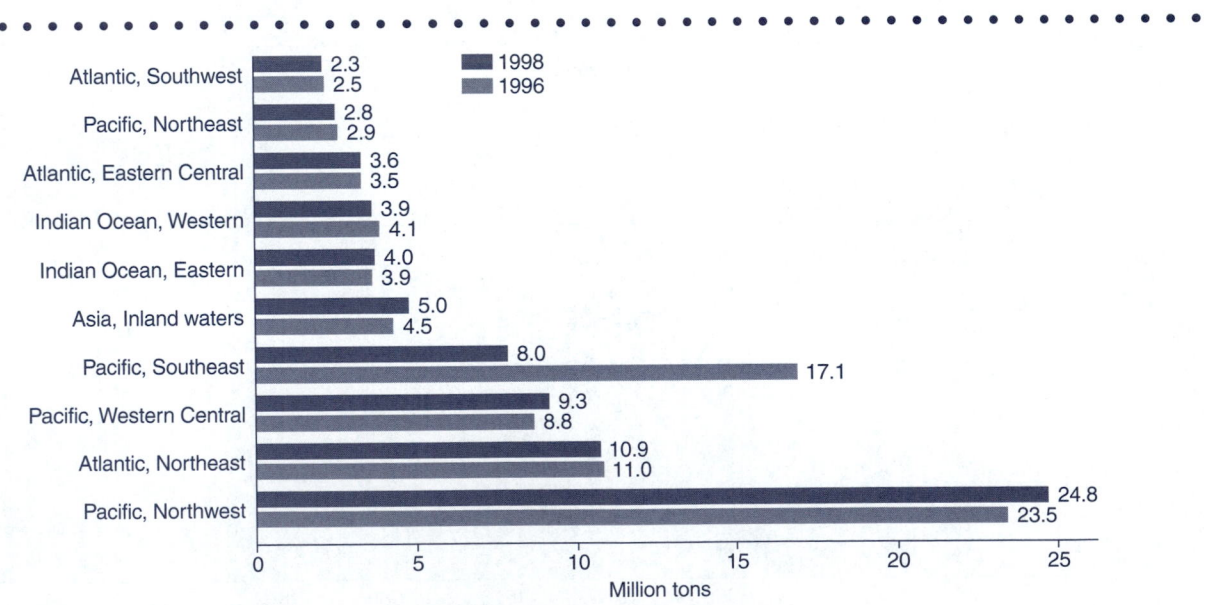

FIGURE 39.3
Capture fisheries production in 1996 compared with 1998, ranked by principal fishing areas
(Source: FAO, 2001).

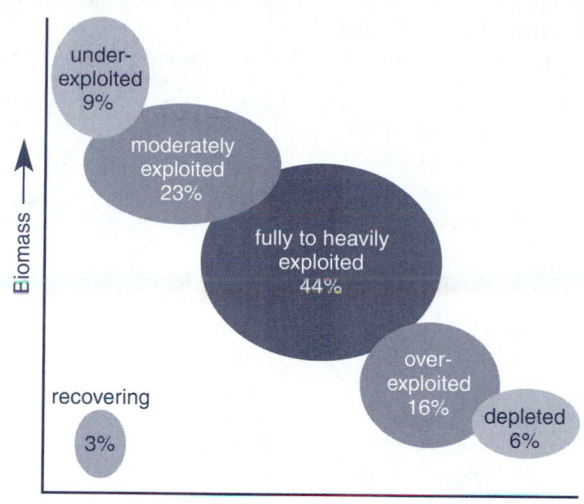

FIGURE 39.4

Distribution of exploitation in world fisheries, ranked in terms of relative biomass and fishing mortality rate
(Source, Botsford et al., 1997).

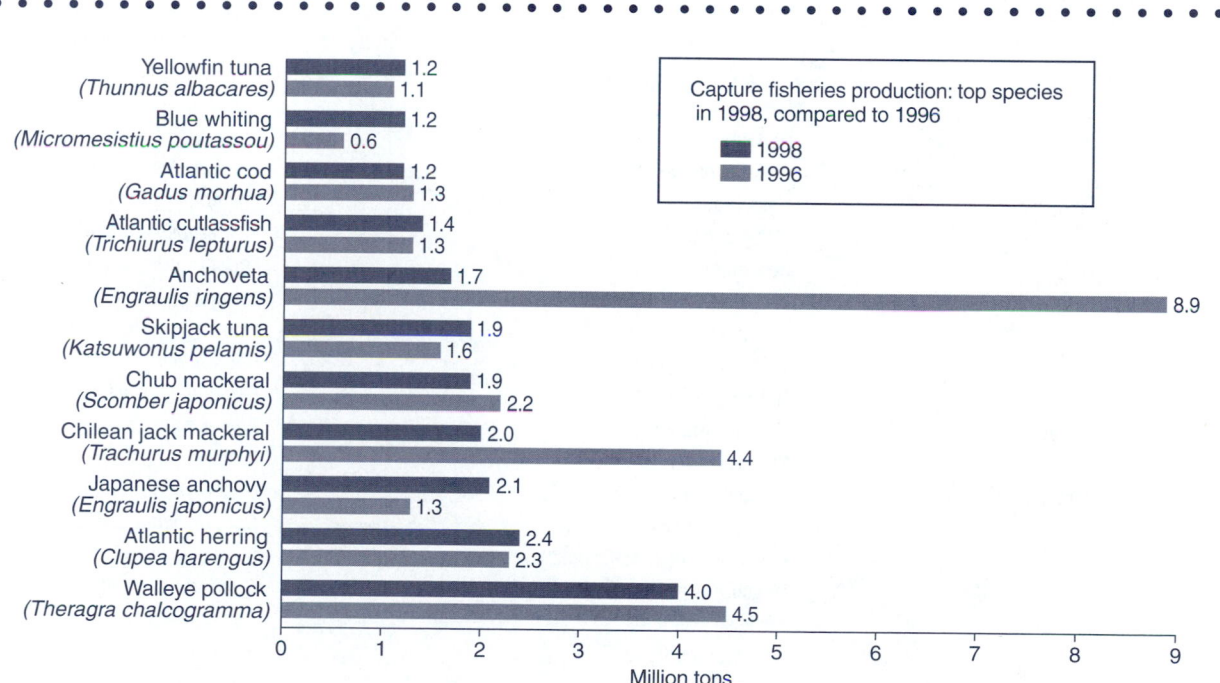

FIGURE 39.5

Dominant species in ocean capture fisheries, 1996 versus 1998
(Source: FAO, 2001).

TABLE 39.1   FISH SPECIES CURRENTLY CLASSIFIED AS OVERFISHED ACCORDING TO MANAGEMENT PLANS DEVELOPED UNDER THE MAGNUSON-STEVENS FISHERY CONSERVATION ACT

Region	Common name	Species name
New England	Atlantic salmon	*Salmo salar*
	Atlantic cod	*Gadus morhua*
	Haddock	*Melanogrammus aeglefinus*
	Silver hake	*Merluccius bilinearis*
	Red hake	*Urophycis chuss*
	American plaice	*Hippoglossoides platessoides*
	Witch flounder	*Glyptocephalus cynoglossus*
	Windowpane flounder	*Scophthalmus aquosus*
Middle Atlantic	Black sea bass	*Centropristis striata*
	Bluefish	*Pomatomus saltatrix*
	Scup	*Stenotomus chrysops*
	Summer flounder	*Paralichthys dentatus*
	Yellowtail flounder	*Limanda ferruginea**
South Atlantic	Goliath grouper**	*Epinephelus itajara*
	Snowy grouper	*Epinephelus niveatus*
	Speckled hind	*Epinephelus drummondhayi*
	Warsaw grouper	*Epinephelus nigratus*
	Black sea bass	*Centropristis striata*
	Gag	*Mycteroperca microlepis*
	Scamp	*Mycteroperca phenax*
	Golden tilefish	*Lopholatilus chamaeleonticeps*
	Red snapper	*Lutjanus campechanus*
	Vermilion snapper	*Rhomboplites aurorubens*
	White grunt	*Haemulon plumieri*
	Red porgy	*Pagrus pagrus*
	Red drum	*Sciaenops ocellatus*
	Goliath grouper**	*Epinephelus itajara*
Gulf of Mexico	Nassau grouper	*Epinephelus striatus*
	Gag grouper	*Mycteroperca microlepis**
	Red snapper	*Lutjanus campechanus*
	Red drum	*Sciaenops ocellatus*
	King mackerel	*Scomberomorus cavalla*
	Vermilion snapper	*Rhomboplites aurorubens**
	Goliath grouper**	*Epinephelus itajara*
Caribbean Sea	Nassau grouper	*Epinephelus striatus*
	Chum salmon	*Oncorhynchus keta*
Washington, Oregon	Chinook salmon†	*Oncorhynchus tshawytscha*
California	Coho salmon‡	*Oncorhynchus kisutch*
	Pacific whiting (Pacific hake)	*Merluccius productus**
	Lingcod	*Ophiodon elongatus**
	Rockfish species	*Sebastes* sp.*
	Pelagic armorhead	*Pentaceros pectoralis*
Western Pacific Pelagic	Squirrelfish snapper	*Etelis carbunculus*
	Longtail snapper	*Etelis coruscans*
	Nurse shark	*Ginglymostoma cirratum*
Atlantic Migratory Species	Whale shark	*Rhincodon typus*

(Continued)

TABLE 39.1 CONTINUED

Region	Common name	Species name
Atlantic Migratory Species (*cont.*)		
	Sand tiger shark	*Odontaspis taurus*
	Bigeye sand tiger shark	*Odontaspis noronhai*
	Basking shark	*Cetorhinus maximus*
	White shark	*Carcharodon carcharias*
	Sandbar shark	*Carcharhinus plumbeus*
	Blacktip shark	*Carcharhinus limbatus*
	Dusky shark	*Carcharhinus obscurus*
	Spinner shark	*Carcharhinus brevipinna*
	Silky shark	*Carcharhinus falciformis*
	Bull shark	*Carcharhinus leucas*
	Bignose shark	*Carcharhinus altimus*
	Narrowtooth shark	*Carcharhinus brachurus*
	Galápagos shark	*Carcharhinus galapagensis*
	Night shark	*Carcharhinus signatus*
	Caribbean reef shark	*Carcharhinus perezi*
	Finetooth shark	*Carcharhinus isodon**
	Tiger shark	*Galeocerdo cuvieri*
	Lemon shark	*Negaprion brevirostris*
	Scalloped hammerhead shark	*Sphyrna leweni*
	Smooth hammerhead shark	*Sphyrna zygaena*
	Atlantic bigeye tuna	*Thunnus obesus**
	Swordfish	*Xiphias gladius*
	Blue marlin	*Makaira nigricans*
	White marlin	*Tetrapturus albidus*

* Additions to list of overfished species; overfishing initiated between 1997 and 2002.

** Previously known as jewfish; although this name does not apparently have a history of derogatory connotation, the suggestiveness of it merited a change in common name (Nelson et al., 2001).

†Runs associated with selected river systems.

‡Runs entering Strait of Juan de Fuca.

Sources: NMFS, 1997, in Iudicello et al., 1999; NMFS, 2003.

often in artificial impoundments—accounts for almost 16 percent (Table 39.2).

Overfishing and loss of habitat continue to endanger freshwater fishes, the most vulnerable group of vertebrates harvested by humans. About 20 percent of the world's freshwater fishes are considered threatened, endangered, or have been rendered extinct in historic times (Bruton, 1995). Currently, the largest capture fisheries targeting freshwater species are centered on large lakes and river systems in the tropics, where the rapid population growth is mirrored by the rate of exploitation of native fishes.

The Great Lakes: Decline of a Temperate Lake Fishery

At one time, the Great Lakes of North America were the site of a thriving commercial fishery. Commercial harvesting of fishes on the Great Lakes began around 1820. By the late 19th century, harvests of up to 147 million tons per year were recorded. At about this time, signs of overfishing of the most desirable species began to be noticed. The Great Lakes fishery was based on the harvest of coregonids, including the lake herring (*Coregonus artedi*), bloater (*C. hoyi*), and lake whitefish (*C. clupeiformis*), yellow perch (*Perca flavescens*), and a number of top carnivores, including lake trout (*Salvelinus namaycush*), pikes (*Esox*), walleye and sauger (*Sander*, formerly known as *Stizostedion*). Pollution and the elimination of spawning habitat in tributary streams contributed to the decimation of the once thriving commercial fisheries.

Canals dug to improve access to the Great Lakes by maritime shipping facilitated the establishment of the sea lamprey (*Petromyzon marinus*), which wreaked havoc on many of the larger fish species, especially the prized lake trout. Once the lake trout was eliminated, the alewife (*Alosa*

TABLE 39.2   DISTRIBUTION OF WORLD FISHERIES PRODUCTION, 1999

Source	Production (Millions of Tons)	Relative Percentage
Inland Waters		
Capture Fisheries	8.2	6.5
Aquaculture	19.8	15.8
Marine Waters		
Capture Fisheries	84.1	67.2
Aquaculture	13.1	10.5

Source: FAO, 2001.

pseudoharengus) underwent a population explosion, reaching nuisance proportions. With the advent of successful lamprey control measures, largely involving the selective application of lampricides in tributary streams where the larval ammocoetes congregated, and the introduction of more stringent pollution control measures, the Great Lakes were ready for rehabilitation. Having a well-established forage fish base in the alewife, hatchery-reared coho salmon (*Oncorhynchus kisutch*) and Chinook salmon (*O. tshawytscha*) were successfully introduced, and they are now an important component of the revived sport fishery. What with overfishing and habitat destruction, and many species harboring toxic pollutants, the commercial fishery exists, but only as a shadow of its former self (http://www.epa.gov/glnpo/atlas/intro.html).

Anadromous Fisheries

Anadromous fisheries in the Northern Hemisphere have been among the most lucrative of fisheries, with enormous significance to all three of the aforementioned interest groups. They are also among the most endangered, due to overfishing, pollution, and loss of spawning habitat. The last factor is especially critical to anadromous fishes, as their annual return to natal streams places them squarely in the path of human "progress." In North America, three groups of fishes dominate the anadromous fishery—clupeids, salmonids, and moronids. A clue to the desirability of the American shad (*Alosa sapidissima*) is in the etymology of its taxonomic name—*sapidissima* means "most delicious." This highly esteemed food and sport fish, large by clupeid standards, with individuals commonly reaching sizes of up to 40 cm, supports a small commercial fishery but a large sport fishery on streams of the northeastern coast of North America. Its roe is also considered a somewhat passable grade of caviar (see Chapter 8, Going Deeper). Shad fishers, who pursue their quarry with artificial lures called "shad darts," like to think of themselves as a chosen few (McPhee, 2000). The striped bass (*Morone saxatilis*), an anadromous

moronid (see Chapter 16) native to the Atlantic coast of North America but successfully introduced to the Pacific coast, is another target of an important sport fishery.

In terms of cultural, economic, and recreational significance, the salmonids are without peer. When their delicate pink flesh is roasted on alder splints over an open fire, or delicately poached, all other fish pale in comparison. Several popular books have been written about the current crisis in the management of salmonids (see Cone, 1996; Lichatowich, 1999; Taylor, 1999). The term *salmon* embraces two genera. The Atlantic salmon (*Salmo salar*) is an iteroparous (repeatedly spawning) species, which runs up streams of eastern North America and Europe, and is related to the Eurasian brown trout (*S. trutta*) widely introduced into North America. Atlantic salmon have been collected from coastal marine waters and streams of British Columbia and appear capable of natural reproduction in streams where there is less competition with native salmonids (Volpe et al., 2001). The several species of the genus *Oncorhynchus* are taxonomically allied with the western trouts (see Chapter 11). These, of course, are best known for their semelparous reproductive lifestyle, spawning only once.

At the time when European Americans began their initial forays into Native American fishing grounds, salmon production in the Pacific Northwest had been estimated at 228 million to 351 million fishes annually, with more than half returning to Alaskan streams and the rest roughly divided between rivers in British Columbia and the states of California, Oregon, and Washington. Currently, up to 90 percent of the greatly diminished salmon population of the Northeastern Pacific Ocean ecosystem returns to Alaskan rivers, and not more than 1.5 percent returns to the rivers of California, Oregon, and Washington (Gresh et al., 2000). Salmon production in the Columbia River—once the source of one of the largest salmon runs on the west coast of North America—is less than 10 percent of its historic levels (Independent Scientific Group, 1999). Some runs of Pacific salmon have become so imperiled that they have merited protection under the Endangered Species Act. From an economic standpoint, this has had a catastrophic impact on the Native American, commercial, and sport fisheries of the Pacific Northwest. From an ecological perspective, the anadromous salmon runs represent a massive infusion of marine-produced nutrients returned to freshwater and terrestrial ecosystems that has simply disappeared in the regions south of Alaska. The state of Atlantic salmon runs, in streams with a significantly longer history of adverse human impact, is even more precarious. In the past 25 years, stocks of Atlantic salmon have decreased by about 90 percent. Tagging studies carried out under the auspices of the North Atlantic Salmon Conservation Organization have revealed a rate of exploitation of between 60 and 80 percent—indicating a seriously overexploited fishery (NFSC, 2000). In November of 2000, Atlantic salmon native to river systems in Maine were declared "endangered" (http://www.nero.nmfs.gov/atsalmon/).

Native American tribes, sportfishers, and many fishery managers are in agreement that one of the prime causes of the diminution of salmon run strength has been the impoundment of rivers and the consequent flooding of spawning habitat by dams. In the quest for cheap hydroelectric power, the once mighty and free-flowing Columbia River has been reduced to a series of large impoundments. Fish ladders, constructed to facilitate the passage of migratory fishes, are often not up to the task of maintaining run strength (Lauritzen, 2002). Many fishery and conservation groups have advocated the removal of dams of questionable utility (Clausen, 2000). Conservation groups have pushed for breaching four dams on the Snake River to preserve endangered salmon runs, but this controversial proposal has met with strong resistance, particularly from labor groups fearing that this would jeopardize manufacturing jobs in the Pacific Northwest. The critical power shortages experienced in the winter of 2001 in the western United States have caused many to reconsider dismantling any dams.

The timber industry has long been another source of conflict, as contemporary logging practices have also caused the devastation of critical spawning habitat for species of *Oncorhynchus* (Chamberlin et al., 1991). The conflicting demands of resource-based industries such as fishing and logging exacerbate the problems faced by the Pacific salmon, yet modern technology can provide some assistance. Landsat satellite imaging of logged areas with critical salmon spawning habitat (deemed "essential fish habitat" by the Magnuson-Stevens Act; see Chapter 31), coupled with geographic information system (GIS) analysis (see Chapter 30) permit the precise location and characterization of these critical portions of the watershed (Cohen et al., 1998; Kelly et al., 1999).

Tropical Freshwater Fisheries

Tropical freshwater fisheries in Asia and Africa have been dramatically transformed by the introduction of exotic species and by the impoundment of large river systems. The impact of the introduction of an exotic predator, the Nile perch (*Lates niloticus*), on native species of cichlids in the African Great Lakes has been discussed in Chapter 32. The traditional fishery of Lake Victoria used to focus on several native cichlid species but now concentrates on *Lates;* introduced cichlids, especially tilapias (*Oreochromis*); and a pelagic cyprinid (*Rastrineobola argentea;* Ogutu-Ohwayo, 1990). The fishery is largely conducted with gill nets, so that the best way to manage the existing fish community, including the harvesting of exotic species, is through restrictions on the mesh size of the nets (Schindler et al., 1998).

Leaders in many developing countries have staked the future of their nations on large, costly, and ostensibly prestigious dams constructed to promote agricultural and industrial development. Lake Kariba was formed in 1958 with the damming of the Zambezi River. Cyprinids of the genus *Labeo* were important commercial species that made annual spawning migrations into shallow tributaries of the Zambezi; the closure of the dam eliminated this fishery. The creation of a lentic pelagic zone in the reservoir resulted in an increase in small planktivorous clupeids and cichlids, which became forage for large predators, such as the tigerfish (*Hydrocynus vittatus;* Lévêque, 1997). The damming of the Niger River to create Lake Kainji resulted in a dramatic decrease in the abundance of mormyrids, which were one of the most frequently caught fishes in the river fishery (see Chapter 31). The gill-net fishery shifted to the harvest of the citharinid *Citharinus citharus.* Planktivorous clupeids and schilbeid catfishes rapidly proliferated and provided forage for *Lates* and *Hydrocynus*, and commercial catches on the lake rose dramatically. The boom was short, and a rapid decline followed (Lévêque, 1997). Although impoundment inevitably results in a loss of species diversity, as flowing stretches of the river are lost, the transformation of the fish community may not necessarily spell doom for a fishery, as these examples have demonstrated. Understanding the nature of the transformation of the lotic, riverine community to a lentic, reservoir one will enable the development of a management strategy that ensures sustainable harvests in the reservoir-based fishery that develops.

Perhaps the fish with the longest history of aboriginal and commercial exploitation in the tropical Americas is the pirarucú (*Arapaima gigas*). Long fished by aboriginal peoples, these enormous osteoglossids, when salted and dried by the Portuguese colonists, became a substitute for the cod that was a staple in their cuisine (Araujo–Lima and Goulding, 1997).

South American fisheries have targeted ostariophysans that inhabit the larger river systems, especially in the Amazon Basin. The tambaqui (*Colossoma macropomum*) is one of the most important fishery species (Araujo-Lima and Goulding, 1997). These large characoids, relatives of the piranhas and pacus, commonly achieve weights of up to 20 kg. Large pimelodid catfishes form an abundant, yet currently poorly regulated resource in the Amazon Basin (Barthem and Goulding, 1997). The dourada (*Brachyplatystoma flavicans*), piramutaba (*B. vaillantii*), and piraíba (*B. filamentosum*) form the basis of substantial gill-net, seine, and trotline fisheries on the lower reaches of the Amazon River. A small commercial fishery for gobies (*Gobioides*) has developed, as these fishes are the preferred bait for the trotlines used to catch catfishes. Interesting enough, many of the indigenous peoples of the Amazon hold catfishes in low esteem, and they do not form a significant part of their diet. They believe that catfishes are excessively fatty and, hence, the source of certain diseases (Dr. Adalberto LuisVal, personal communication). The fishery for these large catfish species seems largely driven by the export market (Barthem and Goulding, 1997).

Although catfishes and other large, commercially important species undertake periodic, sometimes lengthy migrations up and down the rivers (potamodromy), true diadromous fishes are not to be found in the Amazon—an especially interesting circumstance given the size of the drainage. Understanding the migration patterns of species important to local fisheries is essential for their successful harvest.

MARINE FISHERIES

The fishes that constitute the greatest proportion of the oceanic capture fishery are open-water pelagic or benthopelagic forms that are fished on or near the continental shelves, usually in regions where productivity is enhanced through processes such as upwelling (see Fig. 39.5). The continuing quest for harvestable fishery resources has pushed the open-water fisheries into deeper water, where species such as the orange roughy and rattails are becoming heavily exploited (see Chapter 35, Going Deeper). A few key fisheries merit special consideration; they are discussed hereafter.

Pelagic Fisheries

Clupeoids
In terms of the total tonnage of fishery product landed, clupeoids (sardines, herrings, and anchovies) are by far the largest fishery worldwide. In the late 1960s and early 1970s, the Peruvian anchoveta (*Engraulis ringens*, family Engraulidae)

was the single largest fishery, contributing about 20 percent of the total world production (Royce, 1996). Overexploitation, combined with adverse climatic conditions attributed to the El Niño phenomenon (see later), caused a catastrophic decline in the fishery. Being planktivorous, and thus feeding low on the food chain, clupeoids form enormous concentrations of fishes that, in addition to the fishery off the western coast of South America, are harvested in the northern Atlantic Ocean, western North Pacific, off Australia, and in the Mediterranean and Black seas (Royce, 1996). Clupeoids are usually caught near the surface with large **purse seines**. These seines are deployed so that the fish school is surrounded; the bottom line supporting the net is then drawn up, like the strings on a purse, to prevent the school from escaping before the catch is hauled aboard the fishing boat.

Much of the demand for clupeoids comes from their utility in the production of fish oils and meals for animal feeds and other industrial applications. Earlier in the last century, the Pacific sardine (*Sardinops sagax*) supported the largest fishery in North America. World War I created a huge demand for a convenient source of protein, and canned sardines readily filled the bill. By the mid-1930s, almost 800,000 short tons (a short ton equals slightly more than 907 kg) of sardines were being processed at canneries in central California—especially those in the Monterey Bay region (Wolf, 1992). The culture of the fishing and cannery industry during this time was wonderfully depicted in John Steinbeck's *Cannery Row*. Overfishing and adverse environmental conditions conspired to eliminate the fishery, so that a moratorium on sardine fishing was enacted in 1974. The once bustling Cannery Row became derelict and deserted. The moratorium permitted a limited recovery of the sardine populations, so that the fishery may eventually be restored, hopefully with a greater emphasis on scientific management (Wolf, 1992). Monterey's Cannery Row was reborn as a tourist attraction, with shops, restaurants, and—housed in one of the old cannery buildings—one of the finest aquariums in North America (http://www.mbayaq.org/).

Scombroids
Being large, swift, highly migratory open-water species, the scombroids have been targeted by one of the most sophisticated fisheries. Tunas, billfishes, and related forms are usually caught with longlines (see further) or huge purse seines. The investment in fishing boats and gear for this fishery is tremendous, but given the value of the targeted species, it is obviously worth the effort. Pound for pound, the northern and southern bluefin tunas (*Thunnus thynnus* and *T. maccoyii*) are perhaps the most valuable fishes in the ocean. Studies on the migratory behavior of *T. thynnus* have revealed that there is considerable intermixing of eastern

and western Atlantic populations, but that they may segregate when migrating to spawning grounds in the Gulf of Mexico or the Mediterranean (Block et al., 2001; Magnuson et al., 2001). For more than a thousand years, Sicilians have constructed large, elaborate trap nets called *tonnaras* (Fig. 39.6) to catch Mediterranean populations of migrating northern bluefin tuna (Maggio, 2000). Once the frozen carcass of a single bluefin tuna, weighing several hundred kilograms, hits the floor of a Japanese fish market, the bidding frenzy starts. So esteemed is the sashimi produced from these leviathans that a single fish will sell for up to a million yen (more than $10,000). Regrettably, fishery pressure has caused the southern bluefin tuna to be listed as critically endangered by the World Conservation Union (IUCN; Matsuda et al., 1998). The transformation of high-seas fisheries into large-scale industrial operations (see further) has had a disproportionate impact on larger, predatory species near the top of oceanic food chains. Studies have indicated that the current biomass of large migratory predators such as the scombroids and bottom-dwelling macrocarnivores such as the cods is only about 10 percent of the levels that existed before the advent of large-scale industrial fisheries (Myers and Worm, 2003).

Demersal Fisheries

A large proportion of the world's fishery harvest comes from the bottom fishery for a number of benthic or benthopelagic fishes, also termed **groundfishes**. The technology of

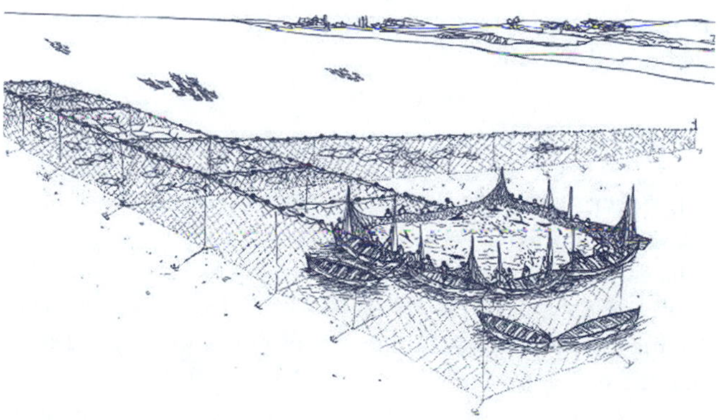

FIGURE 39.6
Illustration of a *tonnara* net used by Sicilians to trap bluefin tuna (Source: National Oceanic & Atmospheric Administration [NOAA]).

trawling—essentially, dragging a net close to the ocean bottom—took a quantum leap in the late 1800s with the invention of the **otter trawl**. These large nets are kept open through the use of stout, wooden or metal "otter doors" at either end that keep the net open as it is dragged across the ocean floor. A "tickler chain" stretched across the mouth of the net bounces along the bottom, scaring up fishes and other creatures which are then swept into the net. Fishes that live near the bottom, such as cods, flatfishes, and rockfishes, are the primary targets of the trawling industry. As indicated in Table 39.1, many of these species are considered overfished.

Along the Pacific coast of North America, a fishery for rockfishes (*Sebastes;* see Chapter 17) has existed since prehistoric times (Love et al., 2002). Technological developments after World War II—specifically, the advent of the balloon trawl, which could be used to fish over shallow, rocky reefs where rockfishes abound—provided the incentive for the development of a thriving commercial fishery. Owing to the large number of species that are targeted, this fishery is managed as a "mixed-species" fishery—one that is further complicated by the fact that the species composition of the fished stocks changes with latitude. Rockfishes are extraordinarily long-lived, with some species believed to attain ages of up to 150–200 years (Love et al., 2002). Of the 16 species of rockfishes on which the National Marine Fisheries Service has adequate fishery data, seven have been declared overfished. In recent years, overfishing has caused the imposition of severe restrictions on the harvesting of Northeastern Pacific coastal rockfishes and other demersal species (http://www.californiafish.org/AFSrockpol.html).

One problem that has potentially serious consequences for the management of mixed-species stocks such as the rockfishes and other commercially fished species from families with high species diversity is *misidentification*. As indicated in Table 39.1, the red snapper (*Lutjanus campechanus*), a highly desirable species fished in warmer waters surrounding coral reefs and other rocky outcrops, is currently in a precarious state of overexploitation. Using molecular phylogenetic techniques on samples of fishes from a number of vendors, Marko et al. (2004) determined that up to 77 percent of the fishes marketed as red snappers were in fact other species. Mislabeling of fishery products—either deliberately or inadvertently by uninformed fishery workers—might give consumers the false impression that some fish stocks, such as the red snapper, are not nearly as endangered as they actually might be.

The Gadoid Fishery

It is safe to say that no fishery product has had a greater influence on the course of human events than the cod (*Gadus morhua*). History books often overlook the pivotal role that this fish has played in the social and economic development of the countries bordering the North Atlantic Ocean. By the time of the Roman conquest, Northern Europeans had developed a commercial fishery for cod in the Baltic and North seas. Archaeologists have documented the shift in consumption of primarily freshwater fishes to marine fishes, mainly herring and cod, to have taken place in Europe around CE 1000. The overexploitation of freshwater fishes and human-induced decline in freshwater ecosystem quality may have prompted this dietary shift (Barrett et al., 2004). These large, slow-moving, benthopelagic, omnivorous predators had a palatable flesh that was just about perfect for preservation. The muscle tissue of the cod is pale, firm, and flaky—very low in fat but rich in protein. Vikings would preserve the meat simply by leaving it to dry in the frigid air—an early application of the freeze-drying principle. Curing the fish in salt greatly improved its "shelf life." Salt-cured cod became the chief source of protein in the Middle Ages. Within a few centuries after the Vikings abandoned their settlements on the North American coastline, the Basques—a distinct ethnic group that inhabits the mountainous regions along the border between northern Spain and France—had expanded the cod fishery beyond Icelandic waters into the enormously productive banks off the Maritime Provinces of Canada and New England (Kurlansky, 1997). This seemingly inexhaustible resource sustained the early settlements of New England and became an essential item of commerce in the fledgling American colonies. Many New England merchants, though expressing disdain for the practice of slavery, eagerly participated in the lucrative trade in salt-cured cod that became an essential foodstuff for the thousands of slaves laboring on plantations in the southern colonies and the Caribbean. A peculiar vestige of this trade lingers today, as codfish—salted, dried, and far from its North Sea home—remains an essential ingredient in the cuisine of the Caribbean region.

Aside from the palatability of their flesh, gadoids possess a number of attributes that contribute to their desirability in world fisheries. The extraordinary fecundity of the cod caused Alexandre Dumas to remark that if all the cod that hatched survived, it would be possible to walk dry-shod on their backs across the Atlantic in only three years. Cod make no pretense at being a sport-fish—they are easily hooked, refuse to fight, and quickly fatigue when pursued by a trawl net. In the early days of the fishery, individuals well in excess of a meter in length were commonly caught. Congregating on the banks at depths where they could easily be fished with handlines, cod stocks seemed, at first, impossible to deplete. By the close of the 19th century, however, it was apparent that many of the North Sea cod stocks were in a state of serious depletion. Icelandic and Newfoundland stocks eventually followed. The situation reached a crisis point when it appeared likely that the Newfoundland cod stocks might be exterminated altogether. With the population reduced to just 1 percent of its original size by 1992, the Canadian government took the drastic action of closing the fishery, causing the loss of 35,000 jobs (Barinaga, 1995; MacKenzie, 1995).

As might be expected, the population biology of cod is of enormous interest to fishery workers. The impact of the intense fishery pressure on cod populations must be evaluated in the context of the factors influencing the year class strength of the species. Both stochastic and deterministic factors (see Chapter 30) appear to be influential. Age structure dynamics are influenced in early life history stages by competition and cannibalism as well as by environmental fluctuations that culminate in the expatriation of eggs or starvation of larvae. Such phenomena can result in dramatic oscillations of cod populations (Bjørnstad and Grenfell, 2001). Regulation of the cod fishery depends on being able to assess the genetic integrity of different stocks. Using microsatellite genetic markers, Nielsen et al. (2001) have been able to determine the population of origin of individual cods from three different populations in the northeast Atlantic Ocean.

The gadoid fishery extends far beyond *Gadus morhua* and the North Atlantic. Three other species of *Gadus*, the Greenland cod (*G. ogac*), the Pacific cod (*G. macrocephalus*), and European whiting (*G. merlangus*) are also targeted by fisheries. Haddock (*Melanogrammus aeglefinus*), pollock (*Pollachius pollachius* and *P. virens*), and tomcods (*Microgadus tomcod* and *M. proximus*) are fished in temperate to polar waters of the Atlantic and Pacific oceans (Rounsefell, 1975). Although their flesh is softer and somewhat less palatable than that of the cods, the hakes (*Merluccius*, family Merluciidae) have become an important fishery at lower latitudes than the cod fishing grounds, where the water is warmer (Rounsefell, 1975).

THE SCIENCE AND TECHNOLOGY OF FISHERIES

The earliest fisheries pursued by humans are sometimes referred to as **artisanal fisheries**—small-scale subsistence fishing, carried out for the benefit of a family or village. Fishing gear may consist of nets, spears, handlines, or other handmade devices, and fishing craft, if used, are small. The most sophisticated technology in these fisheries may consist of small outboard motors. Developing countries are especially interested in expanding fisheries beyond the artisanal stage to that of commercial fishery. As described earlier, a fishery is classified as commercial when the harvest becomes a commodity to be bought and sold. In their earliest stages of development, commercial fisheries may be owner-operated, with a significant investment in a mechanized trawler or longline fisher. As a consequence of the technological improvements that accrued to commercial fishery, many stocks began to feel the pressure of overfishing. As technology has advanced in the past 50 years, and as the market for fishery products has grown, commercial fishing operations have made the transition to what we now term **industrial fisheries**. Intensively capitalized and bristling with the latest technology for finding, catching, and onboard processing of fish, the modern factory fishing vessel exerts a massive impact on the fishery grounds. Much like modern agricultural practices on land, the economy of scale demands the concentration of the fishery into fewer and larger operations. The *Alaska Ocean,* based in Anacortes, Washington, represents the state of the art in factory trawlers. This 376-foot-long ship is one of the largest fishing boats afloat. She spools out a trawl net that is more than half a mile in length, with a gape of up to 400 feet (120 m). More than 600 metric tons of pollock, used in the making of *surimi* for the imitation crab and lobster market, can be harvested in a day (Parfit, 1995).

Enormous trawl nets are but one of the techniques employed by industrial fisheries. **Longlines**—strings of baited hooks, sometimes up to 15 km in length—are used to catch larger pelagic species. The energetic and indefatigable Thomas Henry Huxley—prominent defender of Darwinism, and a pioneer scientific educator of the 19th century—sat on a number of scientific committees, including those managing British fisheries. When fishermen requested more stringent management of longline fisheries, claiming they were damaging the fishery, Huxley dismissed their complaints as unscientific and contrary to the proper development of the fishery. In doing so, he established the precedent of authorities tending to ignore management advice from the individuals most likely to know something about the resource—those directly involved in its harvest (Kurlansky, 1997).

Longline fisheries may also inadvertently catch other animals, such as birds that attempt to steal the bait before it sinks. Croxall et al. (2005) provided evidence that longline fisheries have adversely affected already threatened populations of oceanic albatrosses.

Drift nets, measuring up to 15 m deep and more than 50 km in length, are deployed to fish near the surface, where they catch and entangle anything that attempts to pass through them. Drift nets are a form of **gill net**—these nets snag fishes behind their opercula as they attempt to push through the net. Constructed of transparent nylon or monofilament mesh, drift nets have been used to catch squid, salmon, tuna, and billfishes, but they will indiscriminately trap anything that blunders into them, including turtles, dolphins, and whales, which quickly drown when entangled in the mesh. These nets often become detached from their moorings and lost, yet these "ghost nets" continue to fish indiscriminately for years as they drift on the high seas. Japan, Korea, and Taiwan currently have the largest drift net fisheries, deploying more than 37,000 km of nets each day (Thurman, 1994). Widely considered among the most destructive of modern fisheries, drift netting on the high seas is subject to a moratorium imposed by the United Nations in 1993, but pirate drift netters continue to wreak havoc on the open-ocean ecosystem (http://www.earthtrust.org/dnw.html).

SCIENTIFIC MANAGEMENT OF FISHERIES

Information on the distribution, fecundity, growth, and feeding of fish populations permits the fishery scientist to develop models from which she can determine what proportion of the population can be harvested. A few observations demonstrate the significance of understanding life history features for fishery managers: Fishing has a disproportionate impact on slower-growing, larger species that mature later (see Chapter 6, Going Deeper; Jennings et al., 1999; Musick, 1999). The general perception that highly fecund species, such as codfishes, may be more resistant to depletion owing to their greater production of offspring may be erroneous (Sadovy, 2001). The removal of larger, older females from harvested stocks is detrimental to the fishery, not only because of their greater fecundity, but also because older females appear to produce larvae with greater fitness (Berkeley et al., 2004). At the core of fishery science is the need to understand the dynamics of exploited populations. Because a correlation exists between certain life history parameters of exploited fishes and the resilience of their populations, modern fishery managers need to incorporate basic evolutionary and ecological principles in the management of freshwater and marine fisheries.

The quantity of a fish stock that can be harvested is a function of the *recruitment* of juveniles into the fishery. The fundamental relationship between the recruitment of juveniles and the harvestable stock—a cornerstone of fishery science—arose from early ecological studies on the concept of density dependence (Frank and Leggett, 1994). **Recruitment** is defined as the number of fishes of a single age group that enter the fishery (Royce, 1996). Favorable environmental conditions at the time of spawning may result in higher survival of larval stages and, ultimately, in a strong year class recruited into the fishery. The ultimate goal in fishery development is the sustainability of the harvest of the resource. Fish stocks that are subject to overfishing may experience an impaired ability to reproduce, or they may suffer disproportionate predation as a consequence of their extreme reduction in numbers. The consequence is a phenomenon known as **depensation**—the inability of the fish stock to successfully replenish itself. The fact that this phenomenon has been observed in only 3 of 128 overfished stocks studied suggests that there is some hope for overexploited fisheries. Most fisheries appear to experience greater survival at lower population levels (Myers et al., 1995). Severely restricting entry into a fishery, or imposing a moratorium on all fishery activity targeting an overexploited species, will likely bring about the recovery of most targeted species. Our optimism about the absence of depensation in most overfished stocks should be tempered, however, by the recognition that the recovery time for overfished stocks appears to be a lengthy process (Hutchings, 2000).

Contemporary fishery practices not only affect the targeted species, but they may also influence the distribution and abundance of organisms that are associated with the target species—the so-called **bycatch**. The bycatch includes the portion of the catch that is discarded because it is deemed worthless, but it may also include species of economic value that are caught incidental to the targeted fishery—such as a swordfish caught in a net set for herring. Hall (1996) proposed to restrict the definition of bycatch to those species that are discarded because they have no commercial value or because their retention is prohibited by law.

As discussed earlier, the nonselectivity of drift nets can have a devastating impact on the pelagic community. As fishery technology has moved beyond the use of stationary traps and small-scale, mobile fishing gear, its impact on the benthic community has correspondingly grown. Bottom trawls remove a large portion of the benthic community, of which only one or a few species are kept. The rest of the catch is dumped overboard, and the consequent mortality experienced in the bycatch surely has an adverse effect on the community structure of the ocean bottom in trawl zones. Most studies on the impact of trawls and dredges report an immediate reduction in the density of nontarget species (Auster and Langton, 1999). This may result in changes in the structure and dynamics of the benthic food web of which the target species is a member. A reduction in the topographical complexity of the ocean bottom also occurs as biogenic structures, such as sponges, corals, bryozoans, and shell aggregates, and sedimentary structures, such as sand waves and depressions, are eliminated (Auster, 1998; Auster et al., 1996).

The disturbance to the seabed caused by modern trawling has been likened to clearcutting—a forestry practice that also threatens biodiversity (Norse and Watling, 1999; Watling and Norse, 1998). The elimination of the natural bottom topography may render the habitat less suitable for target species and the organisms they may feed on. Regrettably, the intensity with which trawling has been conducted has probably eliminated pristine, intact bottom communities, so that we may never be able to properly study them (Dayton, 1998). Improvements in fishing gear technology must not only be directed toward maximizing the efficiency of harvesting the targeted species; they must at the same time minimize adverse environmental impacts on the fragile ecosystems of the ocean bottoms. The tuna purse seine fishery stands out as one that has successfully dealt with the problem of bycatch. Because dolphins often prefer to school with tuna, many of them suffocated when they were entangled in the nets. Significant improvements in gear technology and fishery practice have resulted in dramatic reductions of bycatch in this fishery (Hall, 1996; Nishida and Fluharty, 1999).

One unanticipated consequence of bycatch reduction has been its impact on marine bird populations. Everyone is familiar with the flocks of seabirds that are attracted to fishing boats because of the discards tossed overboard. This practice has become a key resource for many seabird populations—changes in the rate of discard production have been shown to affect bird populations. The decline of the demersal fishery in the North Sea, and the consequent decline in discard availability, has prompted the great skua (*Stercorarius skua*), a generalist predator and scavenger, to shift its dietary preferences to include greater numbers of other birds (Votier et al., 2004).

One ecological phenomenon of great concern to fishery ecologists is the practice of "fishing down the food web." Industrial fishing, which initially targeted macrocarnivorous benthopelagic and pelagic species, has moved down the food chain and currently focuses on microcarnivores and planktivores, both in inland and in ocean fisheries. This trend is viewed as dangerous and unsustainable (Pauly et al., 1998, 2000). Daniel Pauly, a leading conservation biologist with expertise in global fisheries management issues, has advocated abandoning the single-species model of fisheries management in favor of analyses that consider the target species in the context of its entire ecosystem (Pauly et al., 2003; Schiermeier, 2002). The goal of **ecosystem-based fishery management (EBFM)** is to maintain the health of the ecosystem; if it is healthy, then the fisheries it embraces will continue to be productive (Pikitch et al., 2004).

One method of restoring fisheries that has attracted considerable interest is to close certain areas to all fishing activities, permitting the reestablishment of the natural food web (see Shipley, 2004). The first of such *no-take zones* in United

States waters was established in 1997 in the Florida Keys National Marine Sanctuary. It is hoped that ecological reserves such as these, also referred to as **marine protected areas (MPAs)**, will contribute to neighboring depleted fisheries by furnishing them with eggs, larvae, and juveniles (Allison et al., 1998; Lauck et al., 1998; Pauly et al., 2002; Schmidt, 1997). Fishery reserves in Florida and St. Lucia that have been in existence for only a few years are already enhancing catches by adjacent sport and artisanal fishers (Roberts et al., 2001). The perception that MPAs are a cure-all for depleted fisheries, however, has not gone unchallenged. Shipp (2003) claimed that their use is not applicable to most fishery species—in other words, that traditional fisheries management approaches work best in most cases.

RECREATIONAL FISHERIES

The increased availability of leisure time has caused more individuals to pursue pastimes that significantly affect fish populations. The number of recreational anglers has steadily increased in the past several decades, and the nature of their fishery has become transformed. Although recreational landings only account for about 4 percent of the total haul of marine fishes in the United States, recreational fishery has a disproportionate impact on certain locations and species, including many that are currently overfished. When the industrial fishery harvest is excluded, the take from recreational landings increases to 10 percent; in the Gulf of Mexico, it accounts for 64 percent of the landings and includes such species of concern as the red drum (*Sciaenops ocellatus*) and red snapper (*Lutjanus campechanus*; Coleman et al., 2004).

Some species, such as the salmonids, may be actively pursued by artisanal, commercial, and sport fisheries alike—thus, developing a management plan that will satisfy all of these competing interests can be challenging indeed. The modification of natural habitats, such as streams, estuaries, and wetlands, has resulted in a greater emphasis on reservoir fisheries. The increased urbanization of society has caused fishery managers to consider ways to improve access to quality fishing in settings that are not too distant from large metropolitan areas (Schramm and Edwards, 1994). A recently published guide to recreational fishing in New Jersey—a state not often mentioned in the context of rural recreational opportunities—is perhaps indicative of the increased interest in urban fisheries (Piehler, 2000). Lakes and reservoirs often host competitive fishing events, and the number and diversity of species taken in these events continues to increase. These events, taking the form

of an intense, highly localized fishery, can have a number of adverse effects on the target population. Hooking and handling mortality, disruption of spawning activities, and the displacement of caught fish away from home areas all may exert adverse impacts on populations subjected to competitive fishing events (Hayes et al., 1995).

Scuba diving is another recreational activity that may affect fish populations. Most scuba diving is conducted in coral reef areas, and studies have demonstrated a correlation between diving intensity on coral reefs and the extent of physical destruction. A transformation of the coral community has also been observed—something that may foster a transformation of the associated coral reef fish community (Hawkins et al., 1999). One tourism development that, at the outset, might seem appalling to any environmentalist is the location of bungalows on pilings directly over the coral reefs. Such facilities appear to have had a positive effect on reef fish populations, however, as the increased availability of food and shelter associated with these structures appears to enhance fish recruitment (Planes and Doherty, 1995).

THE ORNAMENTAL FISH TRADE

For many, the value of fishes lies not in their consumption, but in their appreciation. The past few years have witnessed dramatic improvements in the technology associated with maintaining captive fishes, both at home and in public aquariums. It is now possible to successfully replicate ecosystems as diverse as flowing streams, coral reefs, rocky shores, and estuaries inside aquaria (Adey and Loveland, 1998). This technology is integral to the development and construction of public aquaria, which are now seen as instrumental in the revitalization of downtown areas intent on providing greater tourism opportunities.

The retail trade in ornamental fishes currently amounts to about 3 billion dollars per year, with Asia accounting for about half of the world's supply (FAO, 2001). Perhaps the most impressive developments have been in the technology associated with maintaining living coral reef communities in the home aquarium. The ability to successfully model ecosystems as complex as coral reefs on a small scale has promoted the rapid growth of the marine ornamental fish industry. An estimated 2 million people worldwide maintain home marine aquaria, and the global trade in marine animals has been estimated to be worth 200 to 300 million dollars annually (Wabnitz et al., 2003). Currently, more than 1,400 species of marine fishes are collected for aquaria, with damselfishes (Pomacentridae) making up almost half of the trade (Wabnitz et al., 2003).

For decades, tropical freshwater fishes have been the foundation of the home aquarium trade. Many, including several Asian species, such as the arowana (*Scleropages formosus*), bala shark (*Balantiocheilos melanopterus;* I once came across these popular cyprinids in a public aquarium display that featured them as an example of a freshwater-adapted elasmobranch!), and pygmy loach (*Botia sidthimunki*), have been subjected to serious overexploitation (Ng and Tan, 1997). Understanding the life history requirements of many freshwater species has permitted them to be bred in captivity, thus relieving pressure on the natural populations. Presently, only a few species of marine fishes can be bred and reared in captivity. Although the trade in aquarium fishes may not contribute the greatest impacts on fish resources, a balanced and comprehensive management strategy must take them into account.

AQUACULTURE

Aquaculture differs from capture fisheries in two fundamental ways. First, the culture of fishes emphasizes maximization of growth and reproduction. Second, capture fisheries are sustained from free-ranging stock, which is owned by no individual or country (although access to fisheries within territorial waters may be governmentally controlled), whereas cultured fishes are privately owned. Global aquaculture output more than doubled between 1986 and 1996. Currently, it accounts for more than one quarter of all fishes consumed by humans (Naylor et al., 1998, 2000). Although technological advances have permitted the startling growth of the aquaculture industry in the past few decades, it is important to remember that humans have practiced aquaculture for hundreds, possibly thousands of years. The common carp (*Cyprinus carpio*) is the earliest known fish to have been subjected to domestication (Balon, 2004). Carp culture in Asia may be as old as 2,500 years. The Chinese have developed a sophisticated polyculture of carps, in which several species, each feeding within its own niche, are reared together (Bardach et al., 1972). Tilapias are second only to carps as aquaculture species of nutritional significance. Experiments in the pond culture of tilapia were carried out as early as the 1920s in Africa (Bardach et al., 1972). An unfortunate consequence of the suitability of tilapias for pond culture is that they are now virtually ubiquitous in temperate and tropical fresh and brackish waters, sometimes causing significant damage to native fish populations. Salmonids have long been viewed as the most desirable freshwater sport and food fishes, and it is not surprising that attempts to culture them have been made as early as 1853 in the United States, and possibly earlier in Europe (Bardach et al., 1972). Although the commercial culture of ictalurid catfishes in North America did not get underway until the 1960s, it soon became a phenomenal growth industry, especially in the southern United States (Bardach et al., 1972). North American catfish species—especially the channel catfish (*Ictalurus punctatus*), blue catfish (*I. furcatus*), and flathead catfish (*Pylodictis olivaris*), are also valuable components of recreational and commercial fisheries in large rivers, lakes, and reservoirs (Michaletz and Dillard, 1999).

Currently, more than 220 species of finfishes and shellfishes are farmed (Naylor et al., 2000). The development of technology that has permitted large-scale aquaculture has fostered the growth of "aqua-business" as the underwater counterpart to "agri-business." Shrimp and salmon are two of the most lucrative fishery products. Although accounting for only about 5 percent of the total aquaculture production, they represent almost one fifth of the total market value of aquaculture products (Naylor et al., 1998). Aquaculture is largely held to be beneficial, as it compensates for the decreased yield of capture fisheries. Yet contemporary aquaculture practices are not without problems. For one thing, the perception that they relieve pressure on wild fisheries is generally true for detritivorous and herbivorous species, such as carps and tilapias, but most cultured species feed high on the food chain, necessitating a diet rich in fish meals and oils—a requirement that puts a severe strain on the clupeoid fisheries. The European salmon farming industry alone requires nutritional support from an area equivalent to about 90 percent of the total primary production of the North Sea (Naylor et al., 1998).

Concomitant with the growth of the aquaculture industry, there has been an increase in the introduction of exotic species (see further). This threat concerns not only the exotic species themselves—for example, the widespread introduction of Asian carps throughout the Mississippi Basin—but also the parasites that they may introduce to native fish populations (Naylor et al., 2001). Coastal ecosystems, which often serve as nursery areas for valuable marine fish species, are eliminated with the construction of ponds for large-scale aquaculture operations. The intense feeding regimen produces large quantities of eutrophicating effluents, which tax the assimilation capacity of adjacent coastal ecosystems. In January of 2000, the United States Environmental Protection Agency initiated nationally applicable standards for water discharged from aquaculture facilities. Some aquaculture operations are researching creative uses for their effluents. For example, the effluent from catfish and tilapia culture operations in Arizona is used to irrigate adjoining cotton fields (Kreeger, 2000).

The "green revolution" fostered by advances in agricultural biotechnology has permitted the development of plants and domesticated animals with tremendously enhanced production potential; a corresponding "blue revolution" is well underway in the world of aquaculture. Fishes that are deemed suitable for aquaculture have been subjected to much the same artificial selection as that seen in domesticated terrestrial animals. Compatibility with high-density living situations, enhanced food conversion efficiency, increased resistance to disease, and the ability to be bred in captivity are all seen as desirable traits in cultured species. Classical genetic improvement programs have been applied to a number of freshwater and marine species to improve their aquaculture potential. There has long been a concern that releasing species that have been subjected to artificial breeding programs into the wild as a means of supplementing natural populations might compromise the genetic integrity and adaptability of wild stocks. For example, hatchery trout are bred to be compatible with pond mates and to readily come to the surface for their daily ration of pelleted trout food. Trout in the wild are wary and territorial—traits that would not permit them to survive in the confines of the trout pond.

Hatchery rearing may, in some cases, have the potential for compromising the adaptability of fishes in the wild. Nickelson (2003) claimed that the productivity of wild coho salmon (*Oncorhynchus kisutch*) populations may be reduced by large numbers of hatchery escapees, which attract greater numbers of predators, yet Rhodes and Quinn (1999) demonstrated that the early rearing environment (hatchery versus natural conditions) does not appear to affect the growth and survivability of juveniles of genetically similar stocks of coho salmon when released into the wild. But what would be the consequences of the introduction of genes from hatchery stock subjected to artificial selection (domestication) into the gene pool of wild stocks? Concerns have been expressed that hatchery genotypes may compromise the adaptability of the wild populations (Hindar et al., 1991). Natural selection has favored the evolution of comparatively large egg size in Chinook salmon (*Oncorhynchus tshawytscha*), but rearing in hatcheries, where selection pressures are relaxed, results in the rapid evolution of reduced egg size—a feature that is disadvantageous to the species in the wild (Heath et al., 2003). Interbreeding of wild Atlantic salmon (*Salmo salar*) with hatchery-reared individuals appears to produce an overall depression of fitness (McGinnity et al., 2003). A program of continuing incorporation of wild-born individuals into the brood stock of domesticated species has been proposed to minimize the genetic differences between hatchery stock and wild fishes (Harada et al., 1998).

This may alleviate concerns about the consequences of interbreeding of hatchery escapees with wild individuals.

Lately, aquaculture breeding programs have been augmented with genetic engineering techniques, which permit a much more rapid development of desirable attributes (see Chapter 28). The DNA sequences that code for growth hormone have been identified in some species, including Atlantic salmon and gilthead seabream (*Sparus aurata*), and programs to produce transgenic strains, in which the potential for growth is greatly enhanced, are underway (Knibb, 1997). In one study, transgenic Atlantic salmon demonstrated behavior modifications consistent with enhanced growth—they showed a much greater rate of food consumption, while at the same time demonstrating significantly less sensitivity to the presence of predators (Abrahams and Sutterlin, 1999).

The "Frankenstein" phenomenon continues to haunt efforts to develop transgenic organisms, as many people are convinced that genetic manipulation will unleash hordes of mutant monsters, causing irreparable damage to the ecosystem. In a review of the risks of the genetic modification of marine fishes, Knibb (1997) concluded that, barring the sustained release of large numbers of genetically manipulated fishes, the impact of such individuals on wild gene pools would be negligible. When sexually mature, farm-reared Atlantic salmon were released into the wild, their lifetime reproductive output was only 16 percent of that of wild stocks (Fleming et al., 2000). Studies on the comparative impact of transgenic growth hormone on wild versus domesticated strains of fishes have indicated that there are limits to the extent of growth enhancement in transgenic fishes. When growth hormone transgenes were inserted into wild strains of rainbow trout, they demonstrated dramatic increases in growth rate, whereas similar treatment of domesticated strains selected for enhanced growth did not show any additional enhancement of growth. Furthermore, transgenic treatment adversely affected viability in both domestic and wild strains (Devlin et al., 2001). In spite of the demonstrated advantages of culturing transgenic strains, several researchers, not to mention the general public, remain skeptical about their environmental impact (Hoag, 2003; Stokstad, 2002).

Genetic manipulation of cultured fish stocks has been shown to be useful for purposes other than maximizing growth and enhancing disease resistance. In the future, fishes may become donors for certain cells and cell products. Scientists have cloned and modified the gene that produces insulin in tilapia, in the hope that they can develop a strain that will express only the human form of insulin. This will greatly enhance the mass production of this critical compound (MacKenzie, 1996).

CONSERVATION OF HABITAT AND BIODIVERSITY

Apart from issues of conservation related to fisheries, there remains the question of the value of intact, pristine ecosystems, where biodiversity has not yet been compromised. The total number of fish species that are in some way useful to us is miniscule compared to the wealth of species that ought to be appreciated merely as remarkable examples of adaptation to aquatic environments. A diverse assemblage of fishes is the best measure of a healthy aquatic environment. Environmental quality is compromised by four general categories of "pollutants"—chemical pollutants, introductions of exotic species, hydrological modification of the landscape, and global environmental perturbations.

Chemical Pollutants

The statistics on aquatic pollution released by the United States Environmental Protection Agency are most discouraging. Pollution advisories recommending people to limit their intake of fish from affected waterways have been issued for more than one third of the nation's lake acreage and one fourth of its river miles. The only states in which advisories have not been issued are Alaska and Wyoming (http://www.epa.gov/waterscience/fish/).

Chemical pollutants are introduced into the environment as either **point-source pollutants** or **non-point-source pollutants**. Point-source pollutants can be traced to a specific location of origin, whereas non-point-source pollutants enter waterways from several different locations. Manufacturing processes, such as oil refineries and smelting operations, are generally easy to identify as point-source polluters. Non-point-source pollution occurs when pollutants are introduced as a result of rainfall, snowmelt, or irrigation. Pollutants deposited into rivers, lakes, coastal waters, or groundwater in this manner can cause major disruptions to aquatic ecosystems. Non-point-source pollutants include such things as microbial agents that can cause disease, toxic compounds, agricultural products (fertilizers, herbicides, and pesticides), petroleum by-products, heavy metals, and a host of other products that can damage aquatic ecosystems. Toxicologists assessing the impact of specific pollutants rely on a standard research protocol, the **bioassay,** to assess the impact of pollutants. Several species of common, widely distributed fishes, including goldfishes (*Carassius*) and pupfishes (*Cyprinodon*), are popular subjects in aquatic toxicological studies (see Bengtson, 1980; Young et al., 1990).

In terms of total surface area of the landscape affected, acid precipitation is the most pervasive non-point-source pollutant (see Chapter 32). The physiological consequences of acid stress on fishes are well known (Fromm, 1980), and the acidification of streams and lakes in poorly buffered watersheds contributes to a reduction in abundance and diversity of fishes (Baldigo and Lawrence, 2000). Some non-point-source pollutants, such as phosphates and nitrates from fertilizers, upset the natural ecosystem by promoting an enhancement of primary productivity. This phenomenon, known as **eutrophication,** has been implicated in the damage to coastal fish population structures in both tropical and temperate areas. Eutrophication is suspected to have contributed to the decline of littoral fish populations in the northern Baltic Sea (Rajasilta et al., 1999) and the Red Sea (Fishelson, 1995). The huge nutrient load carried downstream by the Mississippi River enriches Gulf Coast waters, contributing to eutrophication and promoting the seasonal development of extensive hypoxic zones. These hypoxic areas can have an adverse impact on bottom fisheries by eliminating benthic food sources, forcing the emigration of targeted fishery species, or by killing vulnerable egg and larval stages. Planktivorous species may benefit from the enhanced primary productivity, however (Chesney et al., 2000).

The Threat of Bioaccumulation and Biomagnification

Pollutants may become selectively concentrated in the tissues of organisms, which is known as **bioaccumulation.** These potentially toxic compounds may become more concentrated as they move from one link in the food chain to another. Fishes that feed high on the food chain, such as tunas and swordfishes, accumulate mercury at potentially dangerous levels; this is an example of **biomagnification.** Landed fish are routinely tested, and those with excessive mercury content are not used for human consumption. The pollution of waters by various heavy metals and organic compounds because of agricultural and manufacturing practices has become a serious problem in many areas of the world. Pollutants such as DDT and mercury have accumulated in fishes to the point that advisories have been issued concerning the danger of consuming fishes from contaminated waters (Foran et al., 1989; Jakus et al., 2002). Industrial chemicals, such as the polychlorinated biphenyls (PCBs), have exceeded U.S. Food and Drug Administration tolerance levels in fishes from many waters. In 1989, 37 states in the United States had advisory programs of some kind to inform the people of the risks of consuming fishes from waters with unacceptably high levels of contaminants (Reinert et al., 1991). More stringent regulation of the release of toxic substances into coastal waters has resulted in at least a partial recovery of some Atlantic coastal commercial and sport fishery species in the past few years. The replacement of pollution-sensitive shore-zone fish species, such as menhaden (*Brevoortia*) and silversides

(Atherinopsidae), with more pollution-tolerant forms, such as the killifishes (Fundulidae), has also been documented, however (http://www.epa.gov/maia/html/es-condition.html). The worldwide impact of bioaccumulation and biomagnification of harmful chemicals is illustrated by the presence in the coelacanth (*Latimeria*) of organochlorines including PCBs and DDT (Hale et al., 1991). There is international concern about the levels of these contaminants from both ecological and public health standpoints (Borlakoglu and Dils, 1990).

Impact at the Ecosystem Level

Two examples serve to demonstrate the insidious and widespread impact that introduced pollutants may have on aquatic ecosystems—the toxic contamination of fishes in the Great Lakes, and the tragic destruction of the Aral Sea ecosystem. With the development of canals linking the Great Lakes to the Atlantic Ocean, culminating in the completion of the St. Lawrence Seaway in 1959, the cities situated on the Great Lakes could function as maritime ports. Population growth and the consequent industrial development along the shores of the Great Lakes eventually led to severe pollution, contributing to the demise of the once-thriving commercial fisheries. The introduction of more stringent water quality legislation has brought about a partial restoration of the Great Lakes ichthyofauna, yet the legacy of pollution remains. The persistence of industrial chemicals such as PCBs and dioxin, heavy metals such as mercury, and pesticides such as chlordane in the tissues of sport and commercial fish species has prompted states and provinces bordering the lakes to issue fish consumption advisories.

Citizen awareness of the environmental degradation in the Great Lakes prompted the governments of Canada and the United States to take the necessary steps to avoid an environmental catastrophe. Unfortunately, the same could not be said for the government of the former Soviet Union when dealing with the demise of the Aral Sea. Under Soviet policy, traditional agricultural practices were replaced with collectivization, and an increased emphasis was placed on exportable agricultural products. The arid lands surrounding the Aral Sea were ideal for growing cotton—if they were sufficiently irrigated. Canal construction diverted large amounts of water into the Turkmenistan desert to fuel the cotton boom. In 1960, the Aral Sea was the fourth largest inland body of water in the world. With virtually all of the inflow of the Amu Darya and Syr Darya rivers diverted for agriculture, it has shrunk dramatically.

Commercial fisheries for sturgeon, carp, bream, roach, and pikeperch once thrived in the Aral Sea. Increasing salinity and contamination of the water by pesticides had eliminated these fisheries by the early 1980s. As the sea has

shrunk, pesticides and salt have contaminated the former seabed. Winds blowing across the former seabed pick up tons of salt- and pesticide-contaminated sediment, which then rains down on the surrounding populace, causing an increase in respiratory disorders. By the beginning of the 1990s, the surface area of the Aral Sea had shrunk by nearly half, the volume was down by 75 percent, and the mineral content of the water had increased fourfold. It remains but one of the many environmental catastrophes that are the legacy of Soviet misrule (Kindler, 1998; http://www.dfd.dlr.de/app/land/aralsee/). Lest we conclude that environmental neglect is an attribute only of socialist governments, however, let us not forget the environmental disaster that has befallen the once-thriving fish populations of the Salton Sea in the interior desert of Southern California (see Chapter 32). One fundamental difference should be recognized, however: Created out of water mismanagement, the Salton Sea is not a natural body of water. Nevertheless, it has experienced many of the same problems—albeit on a much smaller scale—as those seen in the Aral Sea.

Extinctions and the Impact of Exotic Species

The rate at which species are going extinct is unprecedented. Unlike earlier widespread extinction events, which appear to have been caused by catastrophic natural phenomena, the blame for this current bout of extinction must be placed squarely at our feet. The causes of extinction are many; chief among them is the elimination of habitat. Freshwater environments have experienced the greatest extinction rates. The enormous size of the world's oceans has protected them from widespread extinction both in the present and throughout geological history (McKinney, 1998). No species of marine fish is known to have experienced global extinction in modern times (Carlton et al., 1999), yet some researchers have claimed that the extinction of as many as 1,200 species—chiefly endemic and undescribed coral reef inhabitants—may have gone completely unnoticed (Malakoff, 1997). As the case of the barndoor skate attests (see Chapter 6, Going Deeper), we may be on the verge of documenting the extinction of a large and once common marine fish. A comprehensive review of the most vulnerable stocks of marine, estuarine, and diadromous fish stocks has been provided by Musick et al. (2000).

Species of both freshwater and marine organisms vary in their vulnerability to extinction. Attempts at defining the attributes of extinction-prone fish species have revealed the most vulnerable species to be those with limited geographic distributions or highly specialized ecological adaptations (Angermeier, 1995; Moyle and Williams, 1990). Extinction in these cases is often a function of the replacement by exotic species that exhibit a greater adaptive versatility. Exotic species have virtually transformed the ichthyofaunal composition of continents (see Chapter 30). The introduction of exotic freshwater fish species is exacerbated by the hydrological transformation of freshwaters, as discussed later. In a study of colonization patterns by exotic fish species that encompassed 125 North American drainage systems, Gido and Brown (1999) determined that drainages with a large number of impoundments had the greatest number of introduced species. Drainages with a comparatively low number of native species exhibited significant invasion by exotic species, whereas species-rich drainages, such as those in the southeastern United States, contained relatively few exotic invaders (Gido and Brown, 1999). This may explain the pattern described by Rahel (2000), in which the comparatively depauperate fish fauna of the American Southwest is dominated by exotic species. The fish fauna of the southeastern United States, though not as vulnerable to competition by exotic species, still includes many species threatened with extinction as a consequence of the pollution and modification of waterways (Warren et al., 2000).

Aquaculture (mainly with the introduction of salmonids and cichlids), the tropical fish trade (ostariophysans and atherinomorphs), and deliberate introductions to enhance sport or commercial fishing (such as brown trout and striped bass) have all been implicated in the spread of exotic species. Many of these introductions have had undesirable consequences—the introduction of the common carp (*Cyprinus carpio*) into North American waters in the early 1800s, or the appearance of the walking catfish (*Clarias batrachus*) in Florida waters in the 1960s remain classic examples of the undesirability of exotic species (Lachner et al., 1970; Lever, 1996). In some instances, sport fisheries have benefited from the introduction of exotic species. One of the most widely introduced species, the brown trout (*Salmo trutta*), was established in North America by the late 1800s. Striped bass (*Morone saxatilis*) were first introduced to the Pacific coast with the release of several individuals into the Sacramento–San Joaquin estuary in 1879 and 1882. The fishes soon expanded their range north to British Columbia and south to the Mexican border, and a thriving commercial fishery had become established by the early 1900s. In recent years, striped bass and another exotic, the northern pike (*Esox lucius*), have become the subject of fish control efforts in California, as illegal introductions into some reservoirs threaten downstream salmonid populations (Courtenay and Moyle, 1992). A chronology of introductions of nonindigenous fish species into United States waters has been provided by Nico and Fuller (1999), and Semmens et al. (2004) have discussed the role of the aquarium trade in the spread of marine exotics.

Global ship traffic has greatly accelerated the rate of introduction of marine exotic species, largely through the release of ballast water (Drake and Lodge, 2004; Ruiz et al., 2000). Ballast water sampled from five container ships entering Hong Kong waters was found to contain a total of 81 species (Chu et al., 1997). Some marine exotics may become established through the exploitation of unoccupied niches, yet others may prey on or outcompete native species. Gobies appear to be among the most likely ichthyofaunal passengers to be dispersed via ballast water. The yellowfin goby (*Acanthogobius flavimanus*), a native of marine, brackish, and freshwaters of Asia, has become well established in bays, estuaries, and coastal drainages of California, where they threaten rare endemic gobies such as *Eucyclogobius newberryi* (Lafferty and Swift, 1996; Lever, 1996; Swift et al., 1989). In an ironic application of ecological justice, another Asian goby (*Tridentiger trigonocephalus*), with an introduced range similar to the yellowfin goby, is experiencing a dramatic population decline—possibly due to predation by yellowfin gobies (Fuller et al., 1999; Meng et al., 1994). The enhanced role that Great Lakes cities play as international maritime ports is reflected in the introduction of the round goby (*Neogobius melanostomus*) and the tubenose goby (*Proterorhinus marmoratus*) from their native ranges in the Caspian region of Eurasia. The tubenose goby has expanded its range to include Lake St. Clair, the upper Detroit River, and western Lake Erie. In another demonstration of Mother Nature's ironic sense of humor, the tubenose goby is considered endangered in its native range (Fuller et al., 1999). The larger and more aggressive round goby, first discovered in 1990 in the St. Clair River, has experienced explosive population growth, expanding its range throughout the five Great Lakes and becoming one of the most common benthic fishes in some locales (MacInnis and Corkum, 2000a, 2000b). Its rapid growth rate and extended spawning season has caused it to become a threat to native benthic species, such as sculpins, logperch, and darters (Lever, 1996). One potential benefit to their introduction appears to be the fact that they include another noxious exotic, the zebra mussel (*Dreissena polymorpha*), in their diet (Lever, 1996).

The control of exotic species depends largely on an understanding of their basic biology (Sakai et al., 2001). Recent success with biological control—the practice of controlling pest species with their natural enemies—has prompted many to call for the expansion of the introduction of exotics as a means to control undesirable species. Some such introductions have had unfortunate consequences, however. Black carp (*Mylopharyngodon piceus*), introduced into commercial catfish ponds for the purpose of snail control, currently poses a threat to mollusc species native to the Mississippi drainage (Fuller et al., 1999; Strong and Pemberton, 2000).

Before an exotic species is inserted into a local food web, it is imperative that the food web in question be thoroughly understood.

Hydrological Modifications

Population growth has caused an unprecedented modification of the hydrosphere. Water consumption in the United States has doubled in the last 40 years (Naiman et al., 1998; Naiman and Turner, 2000), and the massive hydroelectric projects that are underway worldwide are deemed necessary to feed our insatiable demand for energy. Freshwater ecosystems suffer from the depletion of groundwater aquifers, the construction of more and larger dams, and the diversion and channelization of waterways. All of these contribute to a diminution of species diversity. Stream ecosystems have been especially afflicted by the ever increasing rate of urbanization of the landscape (Paul and Meyer, 2001). The celebrated Devil's Hole pupfish (*Cyprinodon diabolis*) represents perhaps the best-known case of the adverse consequences of excess diversion of water from an aquifer (Anderson and Deacon, 2001). The shallow springs inhabited by this pupfish run the risk of drying up because of the diversion of groundwater. The Devil's Hole pupfish is but one of a host of fish species living in springs in the deserts of the American Southwest that are currently threatened by aquifer depletion (Cole, 1981; Williams, 1981).

Hydrological modifications are increasingly viewed as having a global impact. Reservoirs, for example, are seen as a significant source of greenhouse gases (Rosenberg et al., 2000). Dam construction has a long history in temperate countries, and the effects of modifications to freshwater ecosystems are correspondingly well understood. The United States has the most dams per land area, with more than 5,500 large (defined as higher than 15 m) dams in place (Pringle et al., 2000). Almost 1,900 stocks of Pacific salmon are currently known to exist in the waters of the United States and Canada, and the impact of dams, such as those along the Columbia River, is well known. Likewise, dam construction, with its attendant impediment to migration and modification of stream habitat, is known to affect more than half of the 67 species of European freshwater fishes that are currently considered threatened (Northcote, 1998). The damming and channelization associated with the Gabčíkovo Water Project has eliminated most wetland habitat along the middle Danube River (Balon and Holčík, 1998, 1999). A massive hydroelectric project that some claim will result in the largest artificial body of water in Europe is underway in Portugal. At the same time, introduced fish species are expanding their range across the Iberian Peninsula, whereas native species are in decline (Aparicio et al., 2000).

Contrary to what is known from the temperate zone, the potential impact of dam construction on tropical ichthyofauna is poorly understood, largely because the tropical fish fauna is so much more diverse and was, until recently, not well understood. A long history of climatic stability in the tropics has resulted in greater diversity and correspondingly greater endemism. Consequently, the impact of large-scale hydrological modifications on species diversity can be expected to be much greater than what has been experienced in the temperate fish fauna. As discussed earlier, there is a high degree of migratory behavior among tropical fishes inhabiting larger river systems, and many of these species depend on seasonal floodplain inundation for successful feeding and breeding. It is essential that these life history attributes be factored into dam construction in the tropics. Although the rate of dam construction in developing countries is escalating, the philosophy behind such hydrological modification seems to be turning in favor of conservation in developed countries. Many of the small dams that fueled the industrial economy of 19th-century New England are slated for removal, and the impact that the large dams along the Columbia River drainage have had on the region's salmonid populations has caused many to call for their removal as well.

Hydrological modifications in the form of canal construction have promoted the introduction of exotic species. One of the best documented instances of this is the so-called **Lessepsian migration,** initiated with the completion of the Suez Canal, which has profoundly affected the eastern Mediterranean ecosystem (see Chapter 29). A total of 54 fish species have been introduced from the Red Sea, including 13 families new to the Mediterranean. Swimmers might be concerned that one of these immigrants is the venomous lionfish (*Pterois miles;* Golani, 2001).

Global Environmental Perturbations

Finally, brief consideration should be given to global environmental impacts on fish abundance and diversity. In the past 25 years, the climatic condition known as El Niño has become increasingly understood to be a global phenomenon coupled with a periodic oscillation in atmospheric pressure in the Pacific. The combined event, termed **El Niño-Southern Oscillation (ENSO)** has had dramatic effects on global fisheries, such as the aforementioned Peruvian anchoveta fishery, and its effects have even been measured in the biotic composition of the deep-ocean benthos (Ruhl and Smith, 2004).

Whereas ENSO is perceived as a natural oscillation in global climate, **global warming** is widely attributed to human activity, as it is brought about through atmospheric modifications that are a consequence of increased industrialization. Global warming has enormous potential for the modification of both marine and freshwater habitats. The contemplated effects of global warming include a recession of polar environments, as tropical and temperate environments expand, and the replacement of forested and grassland habitats with deserts. As the ice caps melt, coastal habitats would be inundated. Much current study focuses on the impact of global warming on specific ecosystems. For example, the warming of the Northeast Atlantic sea surface has brought about an increase in the abundance of phytoplankton in cooler regions and a decrease of phytoplankton in warmer regions—thus, a consequential modification of the pelagic food web is possible (Richardson and Schoeman, 2004). Sea surface temperatures that are currently increasing at a rate of about 1–2°C per century have caused some temperate fish communities to become more tropical in their species composition (Parker and Dixon, 1998). Distribution and migration patterns of cod stocks have been correlated with temperature changes (Brander, 1997). Although the recruitment of Arctic cod stocks appears to increase in warmer years (Brander, 1997), the enhanced recruitment of North Sea cod stocks has been associated with below-average sea surface temperatures, and the current trend of increasing temperature appears to correlate with weak recruitment (O'Brien et al., 2000). Perry et al. (2005) have reported that both exploited and nonexploited species of fishes in the North Sea have responded to increases in temperatures with a pronounced northward shift in their distributions.

A similar phenomenon has been observed among salmonid species in the North Pacific. Using lake sedimentary records, Finney et al. (2000, 2002) have demonstrated that Alaskan sockeye salmon populations and stocks of Pacific sardine and Northern anchovy have been subject to dramatic population fluctuations long before the development of commercial fisheries. Although warming trends have been correlated with an increased abundance of Alaskan sockeye salmon (see Chapter 30), global warming could result in the extinction of other salmonid populations further to the south (Welch et al., 1998). Coral bleaching, which is symptomatic of thermal stress, has caused a significant diminution of coral species diversity worldwide (Hoegh-Guldberg, 1999). Increased temperatures are perceived as having a deadlier impact on corals than more localized chemical pollution (Normile, 2000). This could result in the global elimination of habitat for many species of coral reef inhabiting fishes. Wood and McDonald (1997) have edited a valuable overview of the potential impact of global warming, in which the contributing authors have considered the means by which fishes might adapt to temperature increases.

Ozone layer depletion is another global phenomenon that is the source of much concern among environmental scientists. The depletion of the ozone layer, permitting the increased penetration of mutagenic ultraviolet (UV) radiation, has been implicated in increased mortalities of the egg and larval stages of many aquatic species. Fish species with pelagic eggs and larvae, such as the gadids, might be especially vulnerable. Experimental studies of the effect of ultraviolet radiation on the developing eggs and larvae of cod have led some researchers to conclude that ozone depletion may be contributing to the demise of the cod fishery (Béland et al., 1999; Lesser et al., 2001). Walters and Ward (1998) have proposed that the precipitous decline of both Pacific and Atlantic species of salmonids, especially in sunny regions, may be in part attributed to UV radiation exacerbating the already stressful conditions of smolting and entry into the ocean.

ARE THERE FISH IN OUR FUTURE?

Preserving Fish and Fisheries

The virtual elimination of commercially valuable species through overfishing was the earliest example of adverse human impact on coastal ecosystems, preceding the pollution of coastal waters and human-induced climate change (Jackson et al., 2001). The consensus opinion among many marine scientists is that the only option left for the preservation of the large numbers of marine species that are threatened with extermination from overfishing is the establishment of a large network of marine preserves that would effectively eliminate fishing from much of the ocean surface. The concept of the preservation of natural areas has a long and distinguished history worldwide as far as it pertains to terrestrial ecosystems. The diverse array of freshwater ecosystems that are now perceived as integral components of the landscape has been an incidental beneficiary of these protection measures (see Chapter 30). Currently, however, only a tiny fraction of the world's oceans are set aside as marine preserves. The United States has pioneered these efforts through the establishment of preserves such as the Florida Keys National Marine Sanctuary and the Hawaiian Islands Coral Reef Ecosystem Reserve. It is hoped that these efforts will be the start of a growing international movement to set aside large tracts of the ocean for the preservation of all species, not just those deemed commercially significant. As discussed earlier, however, the concept of marine protected areas has yet to receive universal endorsement from fishery managers.

Both freshwater and marine fishes suffer in one key aspect pertaining to the measures taken to ensure their survival—living as they do in a medium foreign to our own, they are largely invisible, except to the comparatively few individuals that take the time to seek them out. Because fishes are less conspicuous, it takes more effort to understand their basic needs. Humans have historically harvested enormous numbers of many of these species without giving a thought to the consequences. That the overexploitation of fishery resources can have impacts on ecosystems far beyond their immediate environs has been graphically demonstrated in West Africa. Here, the depletion of fish stocks by heavily subsidized fishery interests from the European Union (EU) has caused Ghanans to look elsewhere for protein. This elsewhere has unfortunately become "bushmeat"—native mammals, whose serious state of decline is a regrettable consequence of the lack of availability of fish resources (Brashares et al., 2004).

An essential component of any conservation strategy is the need to preserve the genetic integrity and diversity of species to be protected. In this respect, conservation strategies for the protection of freshwater species differ from those directed toward marine ones. Freshwater fish populations—as distinct genetic entities arrayed across a landscape that is experiencing ever increasing demands for freshwater from an exponentially increasing human population—face conservation challenges quite unlike those faced by fishes in marine environments. With fewer geographical boundaries in the marine realm, and with greater opportunities for the dispersal of eggs and larvae, marine species have different requirements for genomic preservation (Ryman et al., 1994). The implications for the successful protection of marine species are obvious: Marine refuges, if they are to be of any benefit in species preservation and fishery enhancement, will have to be of appreciable size.

Redefining Stewardship

One final observation has to do with the concept of ownership versus stewardship in the management of fishes and the maintenance of their habitat. Garrett Hardin's (1968, 1998) definition of resource mismanagement as being a case of "the tragedy of the commons" has generated controversy, owing to Hardin's claim that commonly held resources are inevitably overexploited because no one assumes responsibility for their preservation. In Hardin's perspective, the only way to guarantee the continued exploitation of valuable marine resources is to impose new criteria for ownership. This view arises from one of the oldest of fishery traditions—fishes, especially those in the ocean, are owned by no one and hence are subject to harvest by anyone. One of the concessions forced on King John by his rebellious nobles, and codified in the Magna Carta, is that the king could no longer grant

exclusive fishery rights to favored individuals. In 1609, the Dutch jurist Hugo Grotius argued that ownership entailed an ability to defend one's property. In this, the doctrine of *mare liberum* ("freedom of the seas"), the seas and all they held could not be claimed by anyone (Pearse, 1994). The case of aquaculture, as discussed earlier, represents a fundamental difference in management philosophy, as the resources produced from aquaculture are entirely within the domain of the agents of production. But if a fish hatchery rears salmon to smolt size and then releases them into the wild, the salmon may move beyond territorial waters and hence no longer be beneficiaries of legislation enacted to protect them.

For the purpose of defense, nations felt the need to exercise jurisdiction over their coastal waters for some designated distance offshore. In 1702, Cornelius van Bynkershonk provided the first reasonably satisfactory compromise when he proposed that countries have jurisdiction over that stretch of ocean that could be protected by cannon shot from shore. This effectively assured nations of control over the resources on their adjacent continental shelves. In 1772, the British claimed this distance to be one league (3 nautical miles), and with the signing of a treaty in 1822, most European countries accepted this as the territorial domain of a nation.

Increased awareness of the potential value of living and mineral resources in waters beyond the domain of any one country prompted President Truman to declare that the United States had jurisdiction over the waters of its continental shelf. A series of United Nations **Law of the Sea** (**LOS**) conferences, starting in 1958, sought to resolve the thorny question of territorial claims on the high seas. The goal of these conferences was to ensure a fair and equitable allocation of ocean resources.

The LOS convention proposed a uniform 12-mile (19.2 km) coastal territorial domain and a 200 nautical mile (370 km) **exclusive economic zone (EEZ)**. In 1983, President Reagan unilaterally declared the establishment of the 200 nautical mile EEZ for the coastal waters of the United States. Other nations may fish within this zone, providing they meet the stipulations set out by the United States. In cases where the continental shelf extended beyond the 200 nautical mile limit, the EEZ was extended up to 350 statute miles (564 km or about 305 nautical miles) offshore (Thurman, 1994). The National Marine Fisheries Service supervises the **National Observer Program**, which stations trained individuals on boats that are permitted to fish within the territorial waters of the United States. It is an excellent way for young aspiring fishery scientists to get "on the job" experience in fishery management (http://www.st.nmfs.gov/st4/nop/index.html).

One criticism of Hardin's (1968, 1998) thesis arises from the assertion that the mismanagement of common resources is a feature of industrial fisheries and does not pertain to fisheries developed and maintained at a more local, artisanal level. Indeed, the historical Native American approach to salmon management is a good case in point (See Chapter 11, Going Deeper). Here, local groups, more sensitive to the significance that the resource has for their own well-being, exercise constraints that effectively prohibit overexploitation (Fairlie et al., 1995; Jensen, 2000). Most current fishery management arises from a top-down decision-making process. It is inevitable, especially in a democracy, that political expediency may overrule the scientific understanding of problems associated with resource exploitation (Caddy, 1999). The development of industrial fisheries in the context of free access to resources on the high seas—the sort of situation that leads to the "tragedy of the commons"—is a situation in which resource exploitation is driven by outside interests—by a market economy.

In his essay on fishing rights and fishing policy, Peter Pearse (1994) stated that "the long tradition of open access to fisheries is ending" (p. 89). Regulatory agencies have traditionally demanded rigorous risk management procedures in such areas of endeavor as nuclear power production and pharmaceutical manufacture, yet fishery workers, traditionally averse to any form of governmental restriction, have until now escaped such regulatory oversight (Dayton, 1998). As the global industrial fishery developed, fishery workers never felt the need to self-regulate in an environment where no one could claim ownership of the resource and, hence, no one could assume responsibility for its misuse. Caddy (1999) has advocated increased delegation of the management of local (inshore) fisheries to local communities. This seems a reasonable starting point, but, if we are to have fish in our future, they must become the objects not of ownership, but of stewardship. We must assume responsibility for the future of fish species that we consume as well as for those that we simply enjoy because they share our world.

· ·
KEY POINTS AND CONNECTIONS
· ·

- The management of fishery resources reflects the interests of different groups participating in their exploitation. These include aboriginal groups practicing subsistence fisheries, commercial interests that have developed to the level of present-day industrial fisheries, and individuals engaged in sport and recreational fishing.

- The take from capture fisheries reached a plateau in the 1990s, and the expansion of fishery production since then has been the result of aquaculture, which has increased at a rate of 10 percent per year. In terms of fishery output, China leads all other nations, followed by Japan and the United States.

- About 6.5 percent of world fishery production comes from capture fisheries conducted in freshwater. Tropical freshwater fisheries have experienced dramatic transformations in recent decades owing to the impoundment of large river systems and the introduction of exotic species.

The impact of the introduction of nonnative cichlids into the African Great Lakes is discussed in the Going Deeper box in Chapter 32.

- The management of anadromous fisheries is a special case, as it involves fishes that spend part of their life cycles in freshwater and part in the ocean. This fishery, particularly for salmonids in the Northern Hemisphere, has been an especially lucrative one, but overfishing and habitat degradation have conspired to decimate the fishery in recent years. Other anadromous species that have been become favorites of sport fishers are the shad and striped bass.

The biology of anadromous members of the families Clupeidae, Salmonidae, and Moronidae are discussed in Chapters 9, 11, and 16, respectively. The evolutionary and behavioral significance of anadromy are considered in Chapters 4 and 36.

- Clupeoids constitute the largest fishery in the world. The fishery for scombroids is another important fishery, but one that concentrates on top carnivores of the pelagic realm. These are mainly harvested using longlines or purse seines. For hundreds of years, the cod fishery has provided people living in regions bordering the North Atlantic Ocean with their main source of protein. The cod fishery is one of many important demersal or ground fisheries, in which fishes are harvested using large trawls.

The gadoids are covered in more detail in Chapter 13, whereas information on the scombroids can be found in Chapter 16. Pelagic, benthopelagic, and benthic fish assemblages of the continental shelf zone are discussed in Chapter 33.

- Technological developments have facilitated the transformation of fisheries from the artisanal to the present-day industrial operations that take place on the high seas. The scientific management of fisheries has correspondingly progressed beyond the level of single-species management strategies to ecosystem-based fishery management, in which the health of a given fishery is viewed in the context of the overall health of the entire ecosystem of which it is a part.

Additional information on ecosystem approaches to the management of marine fisheries is available in the discussion of large marine ecosystems (LMEs) in Chapter 35.

- Recreation and sport fisheries have the potential to exert a significant impact on some fish populations. The rise in popularity of recreational fishing has placed a greater emphasis on the development of reservoir fisheries. Because improvements in home aquarium technology have resulted in an increased demand for tropical marine ornamental fishes, management strategies need to be developed to prevent the excessive harvesting of coral reef fishes.

Both tropical and temperate reef ecosystems contain fish assemblages that are popular with public and private aquariums; refer to Chapter 33 for a discussion of these systems.

• Aquaculture is providing an increasing contribution to our total consumption of fishery products. More than one quarter of the fishes consumed by humans are cultured species. Although aquaculture may relieve the pressure on naturally occurring fish populations, there is a price to be paid in terms of its environmental impact. Effluents from aquaculture facilities may cause eutrophication of the immediate environment. Selective breeding and genetic manipulation has resulted in the production of fishes that are better adapted to aquaculture conditions than they are to the wild. Consequently, there is a potential for compromising the genetic integrity of naturally occurring fish populations if they come into contact with individuals from hatcheries or other aquaculture facilities.

Genetic applications in fisheries management and aquaculture are discussed in Chapter 28.

• As much as a conservation ethic should inform issues of fishery management, it should also promote the value of pristine ecosystems in which biodiversity has not been compromised. Comparatively few extant species of fishes are of value to humans as exploitable resources, yet all fishes should be valued, because a diverse assemblage of fishes is the best measure of a healthy aquatic environment. The release of toxic compounds, their cumulative effect through the phenomenon of bioaccumulation, the introduction of exotic species, the large-scale hydrological modification of waterways, and global warming all contribute to the degradation of aquatic habitats.

FISH LINKS

http://www.fisheries.org/ Home page for the American Fisheries Society.

http://www.nanfa.org/ Home page for the North American Native Fishes Association, an organization dedicated to the preservation of native fish habitat and the maintenance and enjoyment of native fishes in the home aquarium.

http://www.glfc.org/ Home page for the Great Lakes Fishery Commission.

http://www.noaa.gov/fisheries.html Links to all of the Regional Fishery Centers of the National Marine Fisheries Service.

http://www.was.org/ World Aquaculture Society. Site provides hundreds of links to other aquaculture-related sites.

http://www.invasivespecies.gov/ The Invasive Species website, maintained by the National Biological Information Infrastructure of the United States Geological Survey.

http://www.aquariumcouncil.org/ The Marine Aquarium Council. This organization, intended for those involved in the collection and care of marine animals, promotes the sustained use of coral reefs and other marine habitat by educating individuals on conservation issues relevant to the marine aquarium trade.

http://www.flmnh.ufl.edu/fish/Sharks/isaf/isaf.htm The International Shark Attack File, maintained by the Florida Museum of Natural History, University of Florida.

BUILDING AN ICHTHYOLOGY LIBRARY

Three valuable introductions to fishery science and current fishery-related problems:

Hart, P. J. B., and J. D. Reynolds (Eds.). 2002. *Handbook of fish biology and fisheries,* Vol. 1, Fish biology; Vol. 2, Fisheries. Blackwell, Malden, MA.

Mooney, H. A. (Ed.). 1998. Ecosystem management for sustainable marine fisheries. *Ecol. Appl. 8*(Suppl. 1). Ecological Society of America, Ithaca, NY.

Royce, W. F. 1996. *Introduction to the practice of fishery science.* Academic Press, San Diego.

Two important publications that focus on marine protected areas and their role in fisheries management:

Multiple Authors. 2003. The science of marine reserves. *Ecol. Appl. 13*(Suppl. 1). Ecological Society of America, Ithaca, NY.

Shipley, J. B. (Ed.). 2004. Aquatic protected areas as fisheries management tools. *Proc. Am. Fish. Soc. Symp. 42.* American Fisheries Society, Bethesda, MD.

Three excellent introductions to the problems associated with the salmon fishery of the Pacific Northwest:

Cone, J. 1996. *A common fate: Endangered salmon and the people of the Pacific Northwest.* Oregon State University Press, Corvallis.

Lichatowich, J. 1999. *Salmon without rivers: A history of the Pacific salmon crisis.* Island Press, Washington, DC.

Taylor, J. E., III. 1999. *Making salmon: An environmental history of the Northwest fisheries crisis.* University of Washington Press, Seattle.

Three valuable introductions to the practice of aquaculture:

Bardach, J. E., J. H. Ryther, and W. O. McLarney. 1991. *Aquaculture: The farming and husbandry of freshwater and marine organisms.* Wiley Interscience, New York.

Billard, R. 1999. *Carp: Biology and culture.* Springer/Praxis, Chichester, UK.

Stickney, R. R. 2000. *Encyclopedia of aquaculture.* Wiley, New York.

Useful references on exotic species:

Courtenay, W. R., and J. R. Stauffer, Jr. 1984. *Distribution, biology, and management of exotic fishes.* Johns Hopkins University Press, Baltimore.

Fuller, P. L., L. G. Nico, and J. D. Williams. 1999. Nonindigenous fishes introduced into inland waters of the United States. *Am. Fish. Soc. Spec. Publ. 27.*

Lever, C. 1996. *Naturalized fishes of the world.* Academic Press, San Diego.

REFERENCES

Abrahams, M. V., and A. Sutterlin. 1999. The foraging and antipredator behavior of growth-enhanced transgenic Atlantic salmon. *Anim. Behav. 58:* 933–942.

Adey, W. H., and K. Loveland. 1998. *Dynamic aquaria: Building living ecosystems* (2nd ed.). Academic Press, San Diego.

Allison, G. W., J. Lubchenco, and M. H. Carr. 1998. Marine reserves are necessary but not sufficient for marine conservation. *Ecol. Appl. 8*(Suppl. 1): S79–S92.

Anderson, M. E., and J. E. Deacon. 2001. Population size of Devils Hole pupfish (*Cyprinodon diabolis*) correlates with water level. *Copeia 2001:* 224–228.

Angermeier, P. L. 1995. Ecological attributes of extinction-prone species: Loss of freshwater fishes of Virginia. *Conserv. Biol. 9:* 143–158.

Aparicio, E., M. J. Vargas, J. M. Olmo, and A. deSostoa. 2000. Decline of native freshwater fishes in a mediterranean watershed on the Iberian Peninsula: A quantitative assessment. *Env. Biol. Fishes 59:* 11–19.

Araujo-Lima, C., and M. Goulding. 1997. *So fruitful a fish: Ecology, conservation, and aquaculture of the Amazon's tambaqui.* Columbia University Press, New York.

Auster, P. J. 1998. A conceptual model of the impacts of fishing gear on the integrity of fish habitats. *Conserv. Biol. 12:* 1198–1203.

———, and R. W. Langton. 1999. The effects of fishing on fish habitat. *Am. Fish. Soc. Symp 22:* 150–187.

———, R. J. Malatesta, R. W. Langton, L. Watling, P. C. Valentine, C. L. S. Donaldson, E. W. Langton, A. N. Shepard, and I. G. Babb. 1996. The impacts of mobile fishing gear on seafloor habitats in the Gulf of Maine (northwest Atlantic): Implications for conservation of fish populations. *Rev. Fish. Sci. 4*(2): 185–202.

Baldigo, B. P., and G. B. Lawrence. 2000. Composition of fish communities in relation to stream acidification and habitat in the Neversink River, New York. *Trans. Am. Fish. Soc. 129:* 60–76.

Balon, E. K. 2004. About the oldest domesticates among fishes. *J. Fish Biol. 65*(Suppl. A): 1–27.

———, and J. Holčík. 1998. An essay on the ecological calamity in the middle Danube. *Int. Rev. Hydrobiol. 83:* 51–64.

———, and ———. 1999. Gabčíkovo river barrage system: The ecological disaster and economic calamity for the inland delta of the middle Danube. *Env. Biol. Fishes 54:* 1–17.

Bardach, J. E., J. H. Ryther, and W. O. McLarney. 1972. *Aquaculture: The farming and husbandry of freshwater and marine organisms.* Wiley Interscience, New York.

Barinaga, M. 1995. New study provides some good news for fisheries. *Science 269:* 1043.

Barrett, J. H., A. M. Locker, and C. M. Roberts. 2004. The origins of intensive marine fishing in medieval Europe: The English evidence. *Proc. Roy. Soc. Lond. B 271:* 2417–2421.

Barthem, R., and M. Goulding. 1997. *The catfish connection: Ecology, migration, and conservation of Amazon predators.* Columbia University Press, New York.

Béland, F., H. I. Browman, C. A. Rodriguez, and J.-F. St. Pierre. 1999. Effect of solar ultraviolet radiation (280–440 nm) on the eggs and larvae of Atlantic cod (*Gadus morhua*). *Can. J. Fish. Aquat. Sci. 56:* 1058–1067.

Bengtson, D. A. 1980. A partial bibliography of *Cyprinodon variegatus. Gulf Res. Rep. 6:* 349–357.

Berkeley, S. A., C. Chapman, and S. M. Sogard. 2004. Maternal age as a determinant of larval growth and survival in a marine fish, *Sebastes melanops. Ecology 85:* 1258–1264.

Bjørnstad, O. N., and B. T. Grenfell. 2001. Noisy clockwork: Time series analysis of population fluctuations in animals. *Science 293:* 638–643.

Block, B. A., H. Dewar, S. B. Blackwell, T. D. Williams, E. D. Prince, C. J. Farwell, A. Boustany, S. L. H. Teo, A. Seitz, A. Walli, and D. Fudge. 2001. Migratory movements, depth preferences, and thermal biology of Atlantic bluefin tuna. *Science 293:* 1310–1314.

Borlakoglu, J., and R. Dils. 1990. Polychlorinated biphenyls (PCBs) and marine food chains. *Biologist 37*(5): 145–147.

Botsford, L. W., J. C. Castilla, and C. H. Peterson. 1997. The management of fisheries and marine ecosystems. *Science 277:* 509–515.

Brander, K. 1997. Effects of climate change on cod (*Gadus morhua*) stocks, pp. 255–278. In *Global warming: Implications for freshwater and marine fish,* C. M. Wood and D. G. McDonald (Eds.). Cambridge University Press, Cambridge, UK.

Brashares, J. S., P. Arcese, M. K. Sam, P. B. Coppolillo, A. R. E. Sinclair, and A. Balmford. 2004. Bushmeat hunting, wildlife declines, and fish supply in West Africa. *Science 306:* 1180–1183.

Bruton, M. N. 1995. Have fishes had their chips? The dilemma of threatened fishes. *Env. Biol. Fishes 43:* 1–27.

Caddy, J. F. 1999. Fisheries management in the twenty-first century: Will new paradigms apply? *Rev. Fish Biol. Fisheries 9:* 1–43.

Carlton, J. T., J. B. Geller, M. L. Reaka-Kudla, and E. A. Norse. 1999. Historical extinctions in the sea. *Ann. Rev. Ecol. Syst. 30:* 515–538.

Chamberlin, T. W., R. D. Harr, and F. H. Everest. 1991. Timber harvesting, silviculture, and watershed processes, pp. 181–206. In Influences of forest and rangeland management on salmonid fishes and their habitats, W. R. Meehan (Ed.), *Am. Fish. Soc. Spec. Publ. 19.*

Chesney, E. J., D. M. Baltz, and R. G. Thomas. 2000. Louisiana estuarine and coastal fisheries and habitats: Perspective from a fish's eye view. *Ecol. Appl. 10:* 350–366.

Chu, K. H., P. F. Tam, C. H. Fung, and Q. C. Chen. 1997. A biological survey of ballast water in container ships entering Hong Kong. *Hydrobiology 352:* 201–206.

Clausen, J. 2000. Extinction is forever: Debating dams and dollars in the Northwest salmon crisis. *Dollars and Sense 229:* 20–37.

Cohen, W. B., M. Fiorella, J. Gray, E. Helmer, and K. Anderson. 1998. An efficient method for mapping forest clearcuts in the Pacific Northwest using Landsat imagery. *Photo. Eng. Remote Sens. 64:* 293–300.

Cole, G. A. 1981. Habitats of North American desert fishes, pp. 477–492. In *Fishes in North American deserts,* R. J. Naiman and D. L. Soltz (Eds.). Wiley, New York.

Coleman, F. C., W. F. Figueira, J. S. Ueland, and L. B. Crowder. 2004. The impact of United States recreational fisheries of marine fish populations. *Science 305:* 1958–1960.

Cone, J. 1996. *A common fate: Endangered salmon and the people of the Pacific Northwest.* Oregon State University Press, Corvallis.

Courtenay, W. R., Jr., and P. B. Moyle, 1992. Crimes against biodiversity: The lasting legacy of fish introductions, pp. 365–372. *Trans. 57th N. Amer. Wildl. Nat. Res. Conf. Spec. Sess. 6. Biol. Diversity Aquat. Mgmt.*

Croxall, J. P., J. R. D. Silk, R. A. Phillips, V. Afanasyev, and D. R. Briggs. 2005. Global circumnavigations: Tracking year-round ranges of nonbreeding albatrosses. *Science 307:* 249–250.

Dayton, P. K. 1998. Reversal of the burden of proof in fisheries management. *Science 279:* 821–822.

Devlin, R. H., C. A. Biagi, T. Y. Yesaki, D. E. Smailus, and J. C. Byatt. 2001. Growth of domesticated transgenic fish. *Nature 409:* 781–782.

Drake, J. M., and D. M. Lodge. 2004. Global hot spots of biological invasions: Evaluating options for ballast-water management. *Proc. Roy. Soc. (Biol. Sci.) 271:* 575–580.

Fairlie, S., M. Hagler, and B. O'Riorden. 1995. The politics of overfishing. *Ecologist 25(2–3):* 46–73.

FAO (Food and Agriculture Organization of the United Nations). 2001. *The state of world fisheries and aquaculture 2000,* Part 1, World review of fisheries and aquaculture. FAO Information Division, Rome.

Finney, B. P., I. Gregory-Eaves, J. Sweetman, M. S. V. Douglas, and J. P. Smol. 2000. Impacts of climatic change and fishing on Pacific salmon abundance over the past 300 years. *Science 290:* 795–799.

———, ———, M. S. V. Douglas, and J. P. Smol. 2002. Fisheries productivity in the northeastern Pacific Ocean over the past 2,200 years. *Nature 416:* 729–733.

Fishelson, L. 1995. Elat (Gulf of Aqaba) littoral: Life on the red line of biodegradation. *Israel J. Zool. 41:* 43–55.

Fleming, I. A., K. Hindar, I. B. Mjølnerød, B. Jonsson, T. Balstad, and A. Lamberg. 2000. Lifetime success and interactions of farm salmon invading a native population. *Proc. Roy. Soc. Lond. B 267:* 1517–1523.

Foran, J. A., M. Cox, and D. Croxton. 1989. Sport fish consumption advisories and projected cancer risks in the Great Lakes basin. *Am. J. Publ. Health 79:* 322–325.

Frank, K. T., and W. C. Leggett. 1994. Fisheries ecology in the context of ecological and evolutionary theory. *Ann. Rev. Ecol. Syst. 25:* 401–422.

Fromm, P. O. 1980. A review of some physiological and toxicological responses of freshwater fish to acid stress. *Env. Biol. Fishes 5:* 79–93.

Fuller, P. L., L. G. Nico, and J. D. Williams. 1999. Nonindigenous fishes introduced into inland waters of the United States. *Am. Fish. Soc. Spec. Publ. 27.*

Gido, K. B., and J. H. Brown. 1999. Invasion of North American drainages by alien fish species. *Freshw. Biol. 42:* 387–399.

Golani, D. 2001. Impact of Red Sea fish migrants through the Suez Canal on the aquatic environment of the eastern Mediterranean. *Yale For. Env. Sci. Bull. 103:* 375–387.

Gresh, T., J. Lichatowich, and P. Schoonmaker. 2000. An estimation of historic and current levels of salmon production in the northeast Pacific ecosystem: Evidence of a nutrient deficit in the freshwater systems of the Pacific Northwest. *Fisheries 25(1):* 15–21.

Hale, R. C., J. Greaves, J. L. Guoderson, and R. F. Mothershead, II. 1991. Occurrence of organochlorine contaminants in tissues of the coelacanth *Latimeria chalumnae,* pp. 361–367. In The biology of *Latimeria chalumnae* and the evolution of coelacanths, J. A. Musick, M. N. Bruton, and B. K. Balon (Eds.), *Env. Biol. Fishes 32(1–4).*

Hall, M. A. 1996. On bycatches. *Rev. Fish Biol. Fisheries 6:* 319–352.

Harada, Y., M. Yokota, and M. Iizuka. 1998. Genetic risk of domestication in artificial fish stocking and its possible reduction. *Res. Popul. Ecol. 40:* 311–324.

Hardin, G. 1968. The tragedy of the commons. *Science 162:* 1243–1268.

———. 1998. Extensions of "The tragedy of the commons." *Science 280:* 682–683.

Hawkins, J. P., C. M. Roberts, T. Van't Hof, K. DeMeyer, J. Tratalos, and C. Aldam. 1999. Effects of recreational scuba diving on Caribbean coral and fish communities. *Conserv. Biol. 13:* 888–897.

Hayes, D. B., W. W. Taylor, H. L. Schramm, Jr. 1995. Predicting the biological impact of competitive fishing. *N. Amer. J. Fish. Mgmt. 15:* 457–472.

Heath, D. D., J. W. Heath, C. A. Bryden, R. M. Johnson, and C. W. Fox. 2003. Rapid evolution of egg size in captive salmon. *Science 299:* 1738–1740.

Hindar, K., N. Ryman, R. Utter. 1991. Genetic effects of cultured fish on natural populations. *Can. J. Fish. Aquat. Sci. 48:* 945–957.

Hoag, H. 2003. Transgenic salmon still out in the cold in United States. *Nature 421:* 304.

Hoegh-Guldberg, O. 1999. Climate change, coral bleaching and the future of the world's coral reefs. *Mar. Freshw. Res. 50:* 839–866.

Hutchings, J. A. 2000. Collapse and recovery of marine fishes. *Nature 406:* 882–885.

Independent Scientific Group. 1999. Scientific issues in the restoration of salmonid fishes in the Columbia River. *Fisheries 24(3):* 10–19.

Iudicello, S. W. Weber, and R. Wieland. 1999. *Fish, markets, and fishermen: The economics of overfishing.* Island Press, Washington, DC.

Jackson, J. B. C., M. X. Kirby, W. H. Berger, K. A. Bjorndal, L. W. Botsford, B. J. Bourque, R. H. Bradbury, R. Cooke, J. Erlandson, J. A. Estes, T. P. Hughes, S. Kidwell, C. B. Lange, H. S. Lenihan, J. M. Pandolfi, C. H. Peterson, R. S. Steneck, M. J. Tegner, and R. R. Warner. 2001. Historical overfishing and the recent collapse of coastal ecosystems. *Science 293:* 629–637.

Jakus, P., M. McGuinness, and A. Krupnick. 2002. The benefits and costs of fish consumption advisories for mercury. *Resourc. Fut. Disc. Pap. 02-55.* Resources for the Future, Washington, DC.

Jennings, S., S. P. R. Greenstreet, and J. D. Reynolds. 1999. Structural change in an exploited fish community: A consequence of differential fishing effects on species with contrasting life histories. *J. Anim. Ecol. 68:* 617–627.

Jensen, M. N. 2000. Common sense and common-pool resources. *Bioscience 50:* 638–644.

Kelly, N. M., D. Field, F. A. Cross, and R. Emmett. 1999. Remote sensing of forest-clearing effects on essential fish habitat of Pacific salmon. *Am. Fish. Soc. Symp. 22:* 252–267.

Kindler, J. 1998. Linking ecological and development objectives: Trade-offs and imperatives. *Ecol. Appl. 8:* 591–600.

Knibb, W. 1997. Risk from genetically engineered and modified marine fish. *Transgen. Res. 6:* 59–67.

Kreeger, K. 2000. Down on the fish farm: Developing effluent standards for aquaculture. *Bioscience 50:* 949–953.

Kurlansky, M. 1997. *Cod: A biography of the fish that changed the world.* Penguin Books, New York.

Lachner, E. A., C. R. Robins, and W. R. Courtenay, Jr. 1970. Exotic fishes and other aquatic organisms introduced into North America. *Smithson. Cont. Zool. 59:* 1–29.

Lafferty, K. D., R. O. Swenson, and C. Swift. 1996. Threatened fishes of the world: *Eucyclogobius newberryi* Girard, 1857 (Gobiidae). *Env. Biol. Fishes 46:* 254.

Lauck, T., C. W. Clark, M. Mangel, and G. R. Munro. 1998. Implementing the precautionary principle in fisheries management through marine reserves. *Ecol. Appl. 8(Suppl. 1):* S72–S78.

Lauritzen, D. V. 2002. Preferences, behaviors, and biomechanics of Pacific salmon jumping up waterfalls and fishladders. *Diss. Abstr. Int. B: Sci. Eng. 63(3):* 1109.

Lesser, M. P., J. H. Farrell, and C. W. Walker. 2001. Oxidative stress, DNA damage and p53 expression in the larvae of Atlantic cod (*Gadus morhua*) exposed to ultraviolet (290–400 nm) radiation. *J. Exp. Biol. 204:* 157–164.

Lévêque, C. 1997. Biodiversity, dynamics, and conservation: The freshwater fish of tropical Africa. Cambridge University Press, Cambridge, UK.

Lever, C. 1996. *Naturalized fishes of the world.* Academic Press, San Diego.

Lichatowich, J. 1999. *Salmon without rivers: A history of the Pacific salmon crisis.* Island Press, Washington, DC.

Love, M. S., M. Yoklavich, and L. Thorsteinson. 2002. *The rockfishes of the Northeast Pacific.* University of California Press, Berkeley.

MacInnis, A. J., and L. D. Corkum. 2000a. Fecundity and reproductive season of the round goby *Neogobius melanostomus* in the upper Detroit River. *Trans. Am. Fish. Soc. 129:* 136–144.

———, and———. 2000b. Age and growth of round goby *Neogobius melanostomus* in the upper Detroit River. *Trans. Am. Fish. Soc. 129:* 852–858.

MacKenzie, D. 1995. The cod that disappeared. *New Scientist 147:* 24–29.

———. 1996. Doctors farm fish for insulin. *New Scientist 152:* 20.

Maggio, T. 2000. *Mattanza: Love and death in the Sea of Sicily.* Perseus, Cambridge, MA.

Magnuson, J. J., C. Safina, and M. P. Sissenwine. 2001. Whose fish are they anyway? *Science 293:* 1267–1268.

Malakoff, D. 1997. Extinction on the high seas. *Science 277:* 486–488.

Marko, P. B., S. C. Lee, A. M. Rice, J. M. Gramling, T. M. Fitzhenry, J. S. McAlister, G. R. Harper, and A. L. Moran. 2004. Mislabeling of a depleted reef fish. *Nature 430:* 309–310.

Matsuda, H., Y. Takenaka, T. Yahara, and Y. Uozumi. 1998. Extinction risk assessment of declining wild populations: The case of the southern bluefin tuna. *Res. Popul. Ecol. 40:* 271–278.

McDowall, R. M. 1990. *New Zealand freshwater fishes: A natural history and guide* (2nd ed.). Heinemann-Reed, Auckland.

McGinnity, P., P. Prodöhl, A. Ferguson, R. Hynes, N. Ó. Maoiléidigh, N. Baker, D. Cotter, B. O'Hea, D. Cooke, G. Rogan, J. Taggart, and T. Cross. 2003. Fitness reduction and potential extinction of wild populations of Atlantic salmon, *Salmo salar,* as a result of interactions with escaped farm salmon. *Proc. Roy. Soc. (Biol. Sci.) 270:* 2443–2450.

McKinney, M. L. 1998. Is marine biodiversity at less risk? Evidence and implications. *Diversity Dist. 4:* 3–8.

McPhee, J. 2000. A selective advantage: How to think like a fish, from the Connecticut River to the Danube. *The New Yorker, Sept. 11:* 70–82.

Meng, L., P. B. Moyle, and B. Herbold. 1994. Changes in abundance and distribution of native and introduced fishes in Suisun Marsh. *Trans. Am. Fish. Soc. 123:* 498–507.

Michaletz, P. H., and J. G. Dillard. 1999. A survey of catfish management in the United States and Canada. *Fisheries 24*(8): 6–11.

Moyle, P. B., and J. E. Williams. 1990. Biodiversity loss in the temperate zone: Decline of the native fish fauna of California. *Conserv. Biol. 4:* 275–284.

Musick, J. A. 1999. Ecology and conservation of long-lived marine animals. *Am. Fish. Soc. Symp. 23:* 1–10.

———, M. M. Harbin, S. A. Berkeley, G. H. Burgess, A. M. Eklund, L. Findley, R. G. Gilmore, J. T. Golden, D. S. Ha, G. R. Huntsman, J. C. McGovern, S. J. Parker, S. G. Poss, E. Sala, T. W. Schmidt, G. R. Sedberry, H. Weeks, and S. G. Wright. 2000. Marine, estuarine, and diadromous fish stocks at risk of extinction in North America (exclusive of salmonids). *Fisheries 25*(11): 6–30.

Myers, R. A., N. J. Barrowman, J. A. Hutchings, A. A. Rosenberg. 1995. Population dynamics of exploited fish stocks at low population levels. *Science 269:* 1106–1108.

———., and B. Worm. 2003. Rapid worldwide depletion of predatory fish communities. *Nature 423:* 280–283.

Naiman, R. J., J. J. Magnuson, and P. L. Firth. 1998. Integrating cultural, economic, and environmental requirements for fresh water. *Ecol. Appl. 8:* 569–570.

———, and M. G. Turner. 2000. A future perspective on North America's freshwater ecosystems. *Ecol. Appl. 10:* 958–970.

Naylor, R. L., R. J. Goldburg, H. Mooney, M. Beveridge, J. Clay, C. Folke, N. Kautsky, J. Lubchenco, J. Primavera, and M. Williams. 1998. Nature's subsidies to shrimp and salmon farming. *Science 282:* 883–884.

———, R. J. Goldburg, J. H. Primavera, N. Kautsky, M. C. Beveridge, J. Clay, C. Folke, J. Lubchenco, H. Mooney, and M. Troell. 2000. Effect of aquaculture on world fish supplies. *Nature 405:* 1017–1024.

———, S. L. Williams, and D. R. Strong. 2001. Aquaculture—a gateway for exotic species. *Science 294:* 1655–1656.

Nelson, J. S., E. J. Crossman, H. Espinoza-Pérez, L. T. Findley, C. R. Gilbert, R. N. Lea, and J. D. Williams. 2001. Recommended change in the common name for a marine fish: Goliath grouper to replace jewfish (*Epinephelus itajara*). *Fisheries 26*(5): 31.

NFSC (Northeast Fisheries Science Center). 2000. Status of the fishery resources of the Northeastern United States. *NOAA Technical Memorandum NMFS-NE-115.*

Ng, P. K. L., and H. H. Tan. 1997. Freshwater fishes of Southeast Asia: Potential for the aquarium fish trade and conservation issues. *Aquar. Sci. Conserv. 1:* 79–90.

Nickelson, T. 2003. The influence of hatchery coho salmon (*Oncorhynchus kisutch*) on the productivity of wild coho salmon populations in Oregon coastal basins. *Can. J. Fish. Aquat. Sci. 60:* 1050–1056.

Nico, L. G., and P. L. Fuller. 1999. Spatial and temporal patterns of nonindigenous fish introductions in the United States. *Fisheries 24*(1): 16–27.

Nielsen, E. E., M. M. Hansen, C. Schmidt, D. Meldrup, and P. Grønkjær. 2001. Population of origin of Atlantic cod. *Nature 413:* 272.

Nishida, T., and D. Fluharty. 1999. Toward sustainable and responsible tuna fisheries. *Rev. Fish. Sci. 7:* 281–302.

NMFS (National Marine Fisheries Service). 2003. *Sustaining and rebuilding: National Marine Fisheries Service report to Congress, the status of U.S. fisheries.* http://www.nmfs.noaa.gov/sfa/reports.html#sos

Normile, D. 2000. Warmer waters more deadly to coral reefs than pollution. *Science 290:* 682–683.

Norse, E. A., and L. Watling. 1999. Impacts of mobile fishing gear: The biodiversity perspective. *Am. Fish. Soc. Symp. 22:* 31–40.

Northcote, T. G. 1998. Migratory behavior of fish and its significance to movement through riverine fish passage facilities, pp. 3–18. In *Fish migration and fish bypasses,* M. Jungwirth, S. Schmutz, and S. Weiss (Eds.). Fishing News Books, Oxford.

O'Brien, C. M., C. J. Fox, B. Planque, and J. Casey. 2000. Climate variability and North Sea cod. *Nature 404:* 142.

Ogutu-Ohwayo, R. 1990. The decline of the native fishes of Lakes Victoria and Kyoga (East Africa) and the impact of introduced species, especially the Nile perch, *Lates niloticus,* and the Nile tilapia, *Oreochromis niloticus. Env. Biol. Fishes 27:* 81–96.

Parfit, M. 1995. Diminishing returns: Exploiting the ocean's bounty. *Nat. Geog. 188*(5): 2–37.

Parker, R. O., and R. L. Dixon. 1998. Changes in a North Carolina reef fish community after 15 years of intense fishing—global warming implications. *Trans. Am. Fish. Soc. 127:* 908–920.

Paul, M. J., and J. L. Meyer. 2001. Streams in the urban landscape. *Ann. Rev. Ecol. Syst. 32:* 333–365.

Pauly, D., V. Christensen, J. Dalsgaard, R. Froese, and F. Torres, Jr. 1998. Fishing down marine food webs. *Science 279:* 860–863.

————, ————, R. Froese, and M. Palomares. 2000. Fishing down aquatic food webs. *Amer. Sci.* 88(1): 46–51.

————, ————, S. Guénette, T. J. Pitcher, U. R. Sumaila, C. J. Walters, R. Watson, and D. Zeller. 2002. Towards sustainability in world fisheries. *Nature* 418: 689–695.

————, J. Alder, E. Bennett, V. Christensen, P. Tyedmers, and R. Watson. 2003. The future for fisheries. *Science* 302: 1359–1361.

Pearse, P. H. 1994. Fishing rights and fishing policy: The development of property rights as instruments of fisheries management, pp. 76–90. In *The state of the world's fisheries: Proceedings of the World Fisheries Congress, plenary sessions,* C. W. Voightlander (Ed.). Oxford and IBH, New Delhi.

Perry, A. L., P. J. Low, J. R. Ellis, and J. D. Reynolds. 2005. Climate change and distribution shifts in marine fishes. *Science 308:* 1912–1915.

Piehler, G. R. 2000. *Exit here for fish! Enjoying and conserving New Jersey's recreational fisheries.* Rutgers University Press, New Brunswick, NJ.

Pikitch, E. K., C. Santora, E. A. Babcock, A. Bakun, R. Bonfil, D. O. Conover, P. Dayton, P. Doukakis, D. Fluharty, B. Heneman, E. D. Houde, J. Link, P. A. Livingstone, M. Mangel, M. K. McAllister, J. Pope, and K. J. Sainsbury. 2004. Ecosystem-based fishery management. *Science 305:* 346–347.

Planes, S., and P. J. Doherty. 1995. Effet des bungalows sur pilotis des hôtels en milieu corallien sur le peuplement ichtyologique. *Oceanol. Acta 18:* 123–128.

Pringle, C. M., M. C. Freeman, and B. J. Freeman. 2000. Regional effects of hydrologic alterations on riverine macrobiota in the New World: Tropical–temperate comparisons. *Bioscience 50:* 807–823.

Rahel, F. J. 2000. Homogenation of fish faunas across the United States. *Science 288:* 854–856.

Rajasilta, M., J. Mankki, K. Ranta-Aho, and I. Vuorinen. 1999. Littoral fish communities in the Archipelago Sea, SW Finland: A preliminary study of changes over 20 years. *Hydrobiology 393:* 253–260.

Reinert, R. E., B. A. Knuth, N. A. Kamrin, and Q. J. Stober. 1991. Risk assessment, risk management, and fish consumption advisories in the United States. *Fisheries 16*(6): 5–12.

Rhodes, J. S., and T. P. Quinn. 1999. Comparative performance of genetically similar hatchery and naturally reared juvenile coho salmon in streams. *N. Amer. J. Fish. Mgmt. 19:* 670–677.

Richardson, A. J., and D. S. Schoeman. 2004. Climate impact on plankton ecosystems in the Northeast Atlantic. *Science 305:* 1609–1612.

Roberts, C. M., J. A. Bohnsack, F. Gell, J. P. Hawkins, and R. Goodridge. 2001. Effects of marine reserves on adjacent fisheries. *Science 294:* 1920–1923.

Rosenberg, D. M., P. McCully, and C. M. Pringle. 2000. Global-scale environmental effects of hydrological alterations: Introduction. *Bioscience 50:* 746–751.

Rounsefell, G. A. 1975. *Ecology, utilization, and management of marine fisheries.* C. V. Mosby, St. Louis.

Royce, W. F. 1996. *Introduction to the practice of fishery science.* Academic Press, San Diego.

Ruhl, H. A., and K. L. Smith, Jr. 2004. Shifts in deep-sea community structure linked to climate and food supply. *Science 305:* 513–515.

Ruiz, G. M., P. W. Fofonoff, J. T. Carlton, M. J. Wonham, and A. H. Hines. 2000. Invasion of coastal marine communities in North America: Apparent patterns, processes, and biases. *Ann. Rev. Ecol. Syst. 31:* 481–531.

Ryman, N., F. Utter, and L. Laikre. 1994. Protection of aquatic biodiversity, pp. 92–115. In *The state of the world's fisheries: Proceedings of the World Fisheries Congress, plenary sessions,* C. W. Voightlander (Ed.). Oxford and IBH, New Delhi.

Sadovy, Y. 2001. The threat of fishing to highly fecund fishes. *J. Fish Biol. 59:* 90–108.

Sakai, A. K., F. W. Allendorf, J. S. Holt, D. M. Lodge, J. Molofsky, K. A. With, S. Baughman, R. J. Cabin, J. E. Cohen, N. C. Ellstrand, D. E. McCauley, P. O'Neil, I. M. Parker, J. N. Thompson, and S. G. Weller. 2001. The population biology of invasive species. *Ann. Rev. Ecol. Syst. 32:* 305–332.

Schiermeier, Q. 2002. How many more fish in the sea? *Nature 419:* 662–665.

Schindler, D. E., J. F. Kitchell, and R. Ogutu-Ohwayo. 1998. Ecological consequences of alternative gill net fisheries for Nile perch in Lake Victoria. *Conserv. Biol. 12:* 56–64.

Schmidt, K. F. 1997. 'No-take' zones spark fisheries debate. *Science 277:* 489–491.

Schramm, H. L., Jr., and G. B. Edwards. 1994. The perspectives on urban fisheries management: Results of a workshop. *Fisheries 19*(10): 9–15.

Semmens, B. X., E. R. Buhle, A. K. Salomon, and C. V. Pattengill-Semmens. 2004. A hotspot of non-native marine fishes: Evidence for the aquarium trade as an invasion pathway. *Mar. Ecol. Prog. Ser. 266:* 239–244.

Shipley, J. B. (Ed.). 2004. Aquatic protected areas as fisheries management tools. *Proc. Am. Fish. Soc. Symp. 42.* American Fisheries Society, Bethesda, MD.

Shipp, R. L. 2003. A perspective on marine reserves as a fishery management tool. *Fisheries 28*(12): 10–21.

Stokstad, E. 2002. Engineered fish: Friend or foe of the environment? *Science 297:* 1797, 1799.

Strong, D. R., and R. W. Pemberton. 2000. Biological control of invading species—risk and reform. *Science 288:* 1969–1970.

Swift, C. C., J. L. Nelson, C. Maslow, and T. Stein. 1989. Biology and distribution of the tidewater goby, *Eucyclogobius newberryi* (Pisces: Gobiidae) of California. *Nat. Hist. Mus. L.A. Co. Cont. Sci. 404:* 1–19.

Taylor, J. E., III. 1999. *Making salmon: An environmental history of the Northwest fisheries crisis.* University of Washington Press, Seattle.

Thurman, H. V. 1994. *Introductory oceanography* (7th ed.). Macmillan, New York.

Volpe, J. P., B. R. Anholt, and B. W. Glickman. 2001. Competition among juvenile Atlantic salmon (*Salmo salar*) and steelhead (*Oncorhynchus mykiss*): Relevance to invasion potential in British Columbia. *Can. J. Fish. Aquat. Sci. 58:* 197–207.

Votier, S. C., R. W. Furness, S. Bearhop, J. E. Crane, R. W. G. Caldow, P. Catry, K. Ensor, K. C. Hamer, A. V. Hudson, E. Kalmbach, N. I. Klomp, S. Pfeiffer, R. A. Phillips, I. Prieto, and D. R. Thompson. 2004. Changes in fisheries discard rates and seabird communities. *Nature 427:* 727–730.

Wabnitz, C., M. Taylor, E. Green, and T. Razak. 2003. *From ocean to aquarium: The global trade in marine ornamental species.* UNEP-WCMC, Cambridge, UK.

Walters, C., and B. Ward. 1998. Is solar radiation responsible for declines in marine survival rates of anadromous salmonids that rear in small streams? *Can. J. Fish. Aquat. Sci. 55:* 2533–2538.

Warren, M. L., Jr., B. M. Burr, S. J. Walsh, H. L. Bart, Jr., R. C. Cashner, D. A. Etnier, B. J. Freeman, B. R. Kuhajda, R. L. Mayden, H. W. Robison, S. T. Ross, and W. C. Starnes. 2000. Diversity, distribution, and conservation status of the native freshwater fishes of the southern United States. *Fisheries 25*(10): 7–31.

Watling, L., and E. A. Norse. 1998. Disturbance of the seabed by mobile fishing gear: A comparison to forest clearcutting. *Conserv. Biol. 12:* 1180–1197.

Watson, R., and D. Pauly. 2001. Systematic distortions in world fisheries catch trends. *Nature 414:* 534–536.

Welch, D. W., Y. Ishida, and K. Nagasawa. 1998. Thermal limits and ocean migrations of sockeye salmon (*Oncorhynchus nerka*): Long-term consequences of global warming. *Can. J. Fish. Aquat. Sci. 55:* 937–948.

Williams, C. D. 1997. Sustainable fisheries: Economics, ecology, and ethics. *Fisheries 22*(2): 6–11.

Williams, J. D. 1981. Threatened desert fishes and the Endangered Species Act, pp. 447–475. In *Fishes in North American deserts*, R. J. Naiman and D. L. Soltz (Eds.). Wiley, New York.

Wolf, P. 1992. Recovery of the Pacific sardine and the California sardine fishery. *CalCOFI Rep. 33:* 76–86.

Wood, C. M., and D. G. McDonald (Eds.). 1997. *Global warming: Implications for freshwater and marine fish.* Cambridge University Press, Cambridge.

Young, D. R., D. J. Baumgartner, S. C. Snedaker, L. Udey, M. S. Brown, and E. F. Corcoran. 1990. Effects of wastewater treatment and seawater dilution in reducing lethal toxicity of municipal wastewater to sheepshead minnow (*Cyprinodon variegatus*) and pink shrimp (*Penaeus duorarum*). *Res. J. Water Poll. Cont. Fed. 62:* 763–770.

GREEK AND LATIN WORD ROOTS AND TERMS

a- (Gr)   without, not, absence of something

acanth (Gr)   thorn, spine

acipenser (L)   sturgeon

actin (Gr)   ray or beam

ala (L)   wing

alb (L)   white

ali (L)   wing, other

-alis (L)   pertaining to

alope (Gr)   fox

ambly- (Gr)   blunt

amia (Gr)   a kind of fish

ammo (Gr)   sand

amphi (G)   double, on both sides

an- (Gr)   without, not

anabas (Gr)   gone up

anguilla (L)   eel

anoplo (Gr)   unarmed

antenna (NL)   feeler

anti (Gr)   opposed, against

aphritis (Gr)   a fish

aplo (or haplo) (Gr)   single, simple

-arch (Gr)   anus, rectum

arch-, archi- (Gr)   ancient, first, primitive; important, chief

argent- (L)   silvery

arthro (Gr)   a joint

arti (Gr)   entire, complete (in Hyperoartia, not pierced), even numbered

aspis, aspidos (Gr)   a shield

aster (Gr)   a star

atherin (Gr)   a kind of smelt

aulo (Gr)   a tube or pipe

aur, -at, -ic (L)   gold, golden

aur, -is (L)   ear

bagr (L)   a kind of fish

balist (L)   catapult

barb (L)   a beard

bathy (Gr)   deep

batrach (Gr)   a frog

belon (Gr)   a needle, dart

beryc (L)   a kind of fish

boös (G)   an ox or bull

bov (L)   cow, ox

brachi (Gr)   the arm

brad (Gr)   slow

branch (Gr)   gill

bun (Gr)   a mound or hill

calam (L)   a reed

callo (Gr)   beautiful, (L) hard

calva (L)   bald or smooth

camp (Gr)   a sea creature, a bending, (L) a field or plain

carchar (Gr)   jagged, sharp

carin (L)   keel

cato (Gr)   downwards, inferior

caud (L)   tail

caul (L <Gr)   cabbage, a stalk

cent (Gr)   puncture, point, center

cephal (Gr)   the head

ceps (L)   the head

cerat (Gr)   a horn

cerd (Gr)   a fox

cet (Gr)   whale

chano (Gr)   the open mouth, yawn

charac (Gr)   a kind of (sea) fish

chauliod (Gr)   having protruding teeth

cheimarr (Gr)   a winter torrent

chiasm (Gr)   diagonal, marked with a cross

chil (Gr)   lip

chimaera (Gr)   a mythical monster

chir (Gr)   a hand

chit -on, -in (Gr)   outer coat or covering

chlamyd (Gr)   cloak or mantle

chondr (Gr)   cartilage, also grain

cipit (L)   the head

cirr (Gr)   yellow; (L) curl

clad (Gr)   branch

clistic (Gr)   enclosed

cobit (Gr)   gudgeon-like fish

cochl (Gr)   spiral-shelled mollusk

coelo (Gr)   a hollow

coelum (L)   the heavens

coet (Gr)   bed

cotto (Gr)   head

cottus (Gr)   sculpin

cranio (Gr)   head

crosso (Gr)   fringe

cteno (Gr)   comb

cybium (Gr)   tuna

cyclo (Gr)   circle

cyprin (Gr)   carp

dasy (Gr)   with much hair

dent (L)   tooth

derm (Gr)   skin

di- (Gr)   two

distal (Eng.)   at a point away from the center

echene (Gr)   holding ships, remora

echino (Gr)   like a hedgehog; (L) prickly

elasm (Gr)   a plate

ele (Gr)   a swamp

eleuthero (Gr)   free

elop (Gr)   name of a marine fish

en- (Gr)   in

endo (Gr)   within

engraulis (Gr)   a small fish

epi- (Gr)   over or upon

erp (Gr)   creeper

esox (L)   pike

ethmos (Gr)   a strainer

eu- (Gr)   true, good

exo- (Gr)   external

flavi (L)   yellow

fontinalis (L)   of a spring

formes (L)   in the shape of

gado (Gr)   cod

galax (Gr)   milky

galeo (Gr)   shark-like

gaster (Gr)   stomach

gempyl (NL)   a mackerel-like fish

genyp (Gr)   jaw

giga (Gr)   giant

gladius (L)   sword

glanid (L, Gr)   sheatfish

glosso (Gr)   tongue

gnatho (Gr)   jaw

gobi (L)   a fish of small value

gono (Gr)   seed or offspring

grammus (Gr)   writing

gymno (<Gr)   naked

haemal (Gr)   blood red

halo (Gr)   salt

helo (Gr)   marsh

hept (Gr)   seven

heteros (Gr)   different

hex (Gr)   six

hippo (Gr)   horse

holo (Gr)   whole

hybo (Gr)   hump-backed

hyo- (<Gr)   hog

hyper (<Gr)   beyond

hypnos (Gr)   sleep

icost (Gr)   twenty

ict- (L)   fish

-icus (L)   belonging to

inio (Gr)   back of the head

intercalary (L)   inserted

iso (Gr)   equal

istio (Gr)   a sail

korso (Gr)   side of the forehead; the temple

kurt (Gr)   curved

lachrymos (L)   tearful

laemus (Gr)   throat

lamni (Gr)   a voracious fish

lamprid (Gr)   bright

lati (L)   the side

lepido (<Gr)   scale or peel

lepto (Gr)   thin or delicate

leucas (<Gr)   white

levator (Gr)   a lifter

lobi (Gr) a lobe
loph (Gr) a crest
lucio (L) light
lupus (L) wolf
lys (Gr) loose

macro (Gr) long (or large)
mala (L) jaw (or cheekbone)
mandibul (L) jaw
mantellum (L) a cloak
masta (Gr) a breast
mega (Gr) great
melano (Gr) black
mere (<Gr) part
meso (Gr) middle
meta (Gr) between
mimi (L) mimicry
mira (L) marvelous and strange
mixi (Gr) mingling
momo (Gr) ridicule
mono (Gr) one
mormyro (L) a sea fish
morph (Gr) form or shape
muraeno (L) a moray
mycto (Gr) nose
mylio (<Gr) grinding
myo (<Gr) muscle
myri (Gr) infinite
myxin (Gr) mucus
myzon (Gr) suck

nark (<Gr) torpid
nect (<Gr) swimming
nema (<Gr) thread
neo (Gr) new
nesthid (Gr) hungry
neuro (Gr) nerve
not (Gr) the back or (L) not
nym, onym (Gr) a name

ocell (L) a little eye
odons (Gr) tooth
odus (Gr) tooth
ogco (Gr) a protruberance
oidei (Gr) form of or type of
omo (Gr) the shoulder
onco (Gr) hook
ophido (Gr) serpent (actually, the root is probably ophio)
opistho (<Gr) behind

opleg (Gr) armor or tool (the root may be opl)
ops (Gr) aspect or late
orbito (L) a circle or ring
orecto (Gr) stretched out
osmer (Gr) smell or odor
osphro (Gr) to smell
ostei (Gr) a door
osteo (Gr) bone
otic (Gr) of the ear
oxy (Gr) sharp

paedia (Gr) child
paleao (Gr) ancient
panto (Gr) all
para (Gr) near or beside
parietal (Gr) walls
pegas (Gr) solid or strong
pemph(igo) (Gr) a bubble
peri (Gr) near
petro (Gr) a rock
phallo (Gr) penis
pharyngo (Gr) the pharynx
phidi (<Gr) thrifty or stingy
phili (Gr) love
pholi (Gr) a hole
pholido (Gr) a scale or spot
phor (Gr) a thief
phore (Gr) weaving
phracto (Gr) fenced in or protected
phthalmo (Gr) the eye
phyllo (Gr) a leaf
physi (Gr) a bladder
pimelod (Gr) fat and soft
pinna (L) a feather or wing
placo (Gr) flat and wide
platy (Gr) broad and flat
pleco (Gr) to weave
plect (Gr) plaited or twisted
pleuro (Gr) a rib or the side
plio (Gr) more
pnoi (Gr) breath
pnous (Gr) breathing
pogon (Gr) bearded
poly (Gr) many
pomatom (Gr) a cover
pomi (L) a fruit tree
post (L) after
pre (L) before

pria (Gr) a saw
pristio (Gr) a file
pro (L) before
proximal (L) nearest
pseph (Gr) a pebble
psett (Gr) a flatfish
pseud (Gr) false
pter (Gr) fin
pterygo (Gr) fin or wing
ptycho (Gr) a fold or layer
ptyct (Gr) folded
pungi (L) puncture
pycno (Gr) dense
pylo (Gr) grate or orifice

quadrate (Gr) squared

raj (L) a flat fish or ray
ramph (Gr) a crooked beak
retro (L) backward
rhina (Gr) a file or rasp
rhino (Gr) nose
rhyac (Gr) a brook
rhyncho (Gr) a beak
rostrat (L) a snout or beak

sacco (L) sack-like
salang (Gr) a kind of fish
salmo (L) salmon
sapid (L) savory
saur (Gr) lizard
scaen (Gr) clumsy, crooked
scapano (Gr) a spade
scombro (Gr) mackerel
scopus (Gr) a watcher
scorp (Gr) scorpion
scylio (Gr) dogfish
selachi (Gr) a cartilaginous fish
serri (L) a saw
siluri (L) a kind of river fish
siren (L) a mermaid-like creature who lured sailors to their deaths with singing
soleo (L) the bottom of the foot, sandal
spatula (L) a spoon
sphyrae (Gr) hammers
sphyrno (L) hammer-like
spondyl (Gr) vertebra
squalo (L) a kind of sea fish

squat (L) skate
stego (Gr) roof
sten (Gr) a narrow, confining space
stephano (Gr) a crown
stetho (Gr) breast or chest
stichae (Gr) rows
stomato (Gr) mouth
stroma (Gr) a mattress
stygnos (Gr) hated
sub- (L) under
supra- (L) above
sym- (Gr) together
syn- (Gr) together

taenio (Gr) ribbon
tecto (Gr) molten
teleo (Gr) perfect
tera (Gr) a monster
tetra (Gr) four
thelo (Gr) nipple
thenoid (Gr) palm of the hand
tho (Gr) quick
thriss (Gr) a kind of anchovy
tome (L) part or book
torp (L) numb
trabecular (L) marked with cross bars
trachino (L) horse mackerel
trans- (L) across
treti (Gr) pierced
triakis (Gr) three-pointed
tricho (Gr) hair
troctes (Gr) gnawer
trypao (Gr) to bore
typhlos (Gr) blind

uro (Gr) the tail

velifer (L) veiled
vulpes (L) fox

xena (Gr) a stranger
xiph (Gr) sword

zanc (Gr) sickle
zei (Gr) a kind of fish
zoarco- (Gr) life-supporting
zoön (Gr) animal
zygon (Gr) a pair, yoke

GLOSSARY

Adaptation   A structure, function, or behavior that makes the fish more fit, or a process (some form of natural selection) that makes a structure, function, or behavior more fit.

Acanthopterygian   Refers to spiny-rayed teleosts.

ACTH   Adrenocorticotropic hormone.

Aestivation   The dormant state of certain animals during periods of drought or high temperatures.

Allele   One of two or more alternative forms of a gene at a chromosome locus; one use of the term *gene*.

Allele frequency   The relative number of a particular allele at a locus in a population. A population of diploid (2N) organisms would have 2N alleles at a locus. The frequency of a particular allele is its relative proportion of that total. Often referred to as *gene frequency*.

Allopatric   Distributions that do not overlap geographically.

Allopolyploid   A polyploid derived from complete chromosomal set additions (e.g., an increase from 2N to 4N) from chromosome sets of two different species. See *autopolyploid*.

Allozymes   Alleles for a protein encoded by a locus that can be separated electrophoretically. Because both alleles are ordinarily expressed in a heterozygote, the expression is usually codominant.

Amino acid   The fundamental building block of proteins specified by nucleotide codons in DNA (or messenger RNA).

Arcualia   Bow-shaped components of vertebral arches.

Artificial propagation   Fish cultural activity, generally involving modification of natural spawning, incubation, or rearing environments.

Artificial selection   Fish cultural activity in which breeders are chosen on the basis of heritable traits (whether purposeful or inadvertent).

Autonomic nervous system   Those efferent motor fibers and their ganglia that regulate bodily functions not under voluntary (conscious) control. The system is composed of two antagonistic parts: parasympathetic and sympathetic.

Autopolyploid   A polyploid derived from complete chromosomal set additions (e.g., an increase from 2N to 4N) involving two species. See *allopolyploid*.

Backcross   A mating in which a hybrid is crossed with a parental type.

Batesian mimicry   Mimicry in which an uncommon but palatable species resembles a more common but unpalatable species.

Biomass   The total weight of all members of a species (taxon) or group of species (taxa) in a given area at an instant in time. ("Total weight of all organisms in a particular habitat or area; the term is also used to designate the total weight of a particular species or group of species.")

Bowman's capsule   The expanded proximal end of the kidney tubule surrounding the glomerulus.

Broodstock   Adult fish used for artificial propagation.

Chromosomal complement   The entire set of chromosomes in a cell or that characterize a species. Normally the complement is diploid (2N), but polyploids may have additional haploid (N) sets and be triploid (3N), tetraploid (4N), and so on.

Chromosome   A structure comprised mostly of DNA and protein and found in nuclei of cells that organizes the genetic information. Genes are arranged linearly along chromosomes. During meiosis and mitosis, chromosomes condense to facilitate apportioning of the genetic material between daughter cells.

Circadian   Pertaining to 24-hour biological cycles.

Clade   A branch of the tree of life (see *monophyletic*); a natural group.

Cladistic tree   A tree diagram showing the sequence of evolutionary branching of a study group.

Cladistics   A method of phylogenetic analysis formalized by Willi Hennig (1966), in which monophyletic (natural) groups are diagnosed by shared, derived characters called synapomorphies.

Cladogram   A cladistic tree diagram, showing branching sequence.

Codominant   When both alleles are expressed in a heterozygous (Aa) individual so the individual is distinguishable from both homozygous (AA and aa) types. Most allozyme electrophoretic banding patterns reflect codominant inheritance.

Codon   Three contiguous nucleotides in a nucleotide sequence (gene) that specify a particular amino acid in the protein encoded.

Commensalism   A symbiotic relationship in which one species benefits and the other neither suffers nor benefits.

Commissure   A linking or connecting of parts of the nervous or lateral-line systems, usually from one side to the other.

Congruence The consistency of change in character states of different characters with each other over the branching sequence of a cladogram.

Convergence The attainment of functionally similar structures by distantly related taxa.

Cristae Patches of sensory cells (neuromasts) at the juncture of the semicircular canals and the utriculus.

Cross A mating.

Cryophylic Refers to organisms that thrive at relatively low temperatures.

Dam Female parent in a cross. See *Sire*.

Defining characters These are assigned without benefit of a cladistic attempt to discover the monophyly or individuality of the group being defined.

Demersal Refers to aquatic organisms living on or in close association with the substrate (bottom).

Derived Modified relative to the primitive condition.

Diagnostic characters These are discovered in a cladistic analysis.

Diapause The state of suspended development.

Diastole The dilation phase of the heart action.

Diploid (2N) Having a chromosomal complement consisting of two homologous chromosomes of each type—that is, two haploid (N) complements.

Diverticulum Any blind sac or pouch connected to a larger cavity.

DNA (deoxyribonucleic acid) Linear, double-stranded molecule consisting of (deoxyribo) nucleotides, which encode the genetic information of an organism. The code lies in the specific order of the nucleotides. The strands are held together by complementary A-T and G-C nucleotide pairing. See *Nucleotide* and *Replication*.

Dominant An allele that is expressed in the phenotype of both homozygous (AA) and heterozygous (Aa) individuals. It can also refer to the phenotype expressed. The other allele (phenotype) is recessive.

Ecophenotypic A trait whose variation is controlled by environmental conditions (e.g., high meristic counts caused by lower developmental temperatures at high latitudes).

Electrophoresis A technique for separating molecules based on their intrinsic charge. It is applied to proteins to obtain allozyme data and to DNA fragments to estimate their size.

Emmetropic Refers to normal ocular vision (i.e., not near- or farsighted).

Endemic Native or confined (restricted) to a particular geographical region.

Endogenous Originating within. Produced from within the body, an organ, or a geographical area.

Endolymph Fluid contained within the inner ear (membranous labyrinth).

Epistasis Differential expression of a phenotype as a result of interaction between alleles at more than one locus.

Epithelium The thin layer of tissue covering internal and external body surfaces.

Exaptation Restrictive term for adaptations that began in a different adaptive context. For example, if the swimbladder originated as a respiratory organ, one might say it is an exaptation for buoyancy and hearing.

Extensive aquaculture Cultured organisms released into the natural environment for part of their life cycles. See *Intensive aquaculture*.

Fecundity An organism's capacity to produce offspring. (In fishes, often expressed as the number of eggs produced per female.)

Fenestra An aperture in a bone or a transparent portion of a membrane.

Fimbria A bordering fringe.

Fitness The productivity of a particular phenotype in a population relative to the most productive phenotype in that population in that environment. If there is a genetic basis for the phenotype, fitness-based selection can result in evolutionary change.

Follicle Any small sac or pit.

Fontanelle A membrane-covered opening in a bone.

Foramen A small opening or perforation in any body structure.

Gametes Mature, haploid, male or female reproductive cells.

Ganglion A concentration of nerve cell bodies located outside the central nervous system.

Gene The basic unit of inheritance, the nucleotide sequence that carries encoded information for a particular trait. Genes are linearly arranged on chromosomes. *Gene* is often used in a broader context to refer to loci or alleles.

Gene flow Exchange of alleles between populations (in one or both directions). See *Migration*.

Gene pool The aggregate genetic composition of a population. Often quantified in terms of allele frequencies.

Genetic diversity The genetic variation that exists within a population or unit of interest.

Genetic drift Random variation of allele frequency in a population from one generation to another that results from random errors in sampling gametes. Because such errors result from a finite sample of gametes of their parents, the genetic composition of progeny may differ from that of their parents.

Genetic integrity The degree to which the genetic composition of a population resembles its natural state, uninfluenced by anthropogenic causes.

Genetic marker An allele that may characterize a population or group of populations. The marker may be qualitative (there or not) or quantitative (present at greater or lesser frequency in the marked group). Genetic markers are the basis of genetic stock identification and have been used to determine parentage.

Genetic stock identification (GSI) Use of genetic differences occurring among populations or aggregates of populations to estimate proportionate contributions to mixtures, such as mixed stock fisheries.

Genome The entire set of genetic information (nucleotide sequences) of an individual.

Genotype The allelic composition of an individual. *Genotype* can refer to a single locus, to several loci, or to the total genomic content.

Glomerulus A knot of small blood vessels contained within Bowman's capsule at the proximal end of a kidney tubule.

Glycoproteins Organic molecules composed of carbohydrates and proteins.

Gnathostomes Jawed vertebrates.

Gonadotropic hormones Hormones, secreted by the adenohypophysis, that induce the development of gametes.

Gonopodium A modified anal fin that functions as a copulatory organ.

Gynogen An offspring possessing only maternal chromosomes. If males are the heterogametic sex, gynogens are exclusively female.

Haploid (N) Having a chromosomal complement consisting of a single chromosome of each type. Gametes of normal diploid (2N) organisms are haploid.

Hemopoesis The formation of blood cells.

Heritability The degree to which offspring resemble their parents as a result of their shared genetic background. Heritability can be in the narrow sense (h2), which takes into account only the genetic determinants of similarity that can be bred for predictability (the additive genetic variation VA), or in the broad sense (H2), which takes into account all the genetic determinants (both additive variation, VA, and dominance variation, VD).

Heritable variation That portion of an organism's variation that is controlled by genetic pathways rather than environment.

Hermaphrodite An organism with both male and female sex organs. Sequential hermaphrodites do not simultaneously possess both sets. See *Protandry* and *Protogyny*.

Heterogametic The sex that has two different sex chromosomes (e.g., XY or WZ) and can produce two different gametic types, X or W carrying and Y or Z carrying.

Heterozygous Refers to a state in which two alleles at a locus in a diploid (2N) individual differ (e.g., Aa).

Holarctic A biogeographic region that includes the arctic and north temperate zones. It includes mostly the northern parts of Eurasia and North America.

Homologous (1) Refers to characters, in different taxa, that are structurally similar due to common evolutionary origin. (2) Refers to chromosomes that carry information for the same set of traits and pair during meiosis. Although a particular gene that they carry may specify the same trait, they may be alternatives (different alleles) for that trait.

Homology Similarity due to common ancestry. Usually hypothesized by similarity of position, composition, and embryological origin. A synapomorphy may be confirmed to be a homology by demonstration that it is uniquely diagnostic of a clade, as shown by congruence with other characters in a cladistic tree analysis (e.g., the Weberian apparatus of Otophysi).

Homoplasy Similarity due to convergent, parallel, or reversed evolution (e.g., spines of catfish, goldfish, and sunfish), in contrast to homology.

Homozygous A state in which two alleles at a locus in a diploid (2N) individual are the same (e.g., AA or aa).

Hormones Chemical substances that are released from endocrine glands, are transported via the circulatory system, and regulate a wide range of physiological functions.

Hybrid Offspring of a cross between two genetically dissimilar individuals. Intraspecies hybrids result from crossing individuals of two different strains. Interspecies hybrids may result from crossing individuals of two species.

Illicium Modified first dorsal fin ray of angler fishes, used to attract prey.

Individual lineage or clade In the restricted philosophical sense, a taxon meeting strict criteria of genetic continuity within spatial and temporal boundaries, irreplaceable for purposes of conservation.

Intensive aquaculture Cultured organisms maintained in captivity during their life cycle. See *Extensive aquaculture*.

Intergrades As used in this book, individuals that are, for whatever reason, intermediate between two species (or other taxa).

Introgression The introduction of genes from one population (or species) into the gene pool of another. Introgression may compromise genetic integrity.

Intromittent organ Male copulatory organ.

Iteroparous Spawning more than once before dying. See *Semelparous*.

Lineage A genetically continuous population or group.

Linkage Genes physically connected on the same chromosome may be transmitted together in a gamete and are, therefore, linked and do not follow Mendel's law of random assortment.

Locus (plural, loci) The site on a chromosome of the nucleotide sequence that encodes information for a particular trait. A gene.

Maculae Patches of sensory cells (neuromasts) in the utriculus, sacculus, and lagena.

Meiosis Two-stage nuclear division and chromosomal allocation process that reduces the diploid (2N) chromosome complement by half. The resultant gametes that are formed are thus termed *haploid* (N).

Mendelian trait A trait resulting from the expression of alleles at a single locus. The phenotypes of these traits usually have a discrete distribution.

Meninges Protective membrane enclosing (surrounding) the brain and spinal cord.

Metamorphosis The stage in development during which an animal undergoes a radical change in form and function.

Microphagous Refers to organisms that feed on relatively small food items.

Micropyle An aperture in the vitelline membrane of an egg, through which the sperm enters.

Migration Movement of individuals or populations from one geographic location to another. In genetics, the movement of genes from one gene pool to another (gene flow).

Mimicry Imitation of another organism or object in the environment (in form, color, and/or behavior).

Mitochondrion A subcellular organelle involved in aerobic metabolism. Mitochondria possess DNA that carries information for some of their functions. Mitochondria are haploid and are believed to be maternally inherited.

Mitosis Type of nuclear division and allocation of chromosomes that accompanies cytokinesis (cell division) in which the diploid (2N) number of chromosomes is maintained. Occurs during growth and development.

Monophyletic   A natural group including all descendants of a common ancestor (i.e., a clade). Also called *holophyletic*.

MSH   Melanophore-stimulating hormone; can cause either dispersion or aggregation of pigment.

Müllerian mimicry   Mimicry in which several unrelated animal species distasteful to their predators resemble each other.

Mutation   An alteration in the nucleotide sequence of a gene. Allelic differences result from mutation. Mutation is the ultimate source of all genetic variation.

Mutualism   A symbiotic relationship in which both partners benefit.

Myelinate   Refers to nerves sheathed by a fatty membrane.

Myoid   Contractile segment of visual cell.

Native   Organisms historically indigenous to an area.

Natural selection   Conceived by Charles Darwin as the force that exerts differential pressures on organisms, thus resulting in their generational change through time. Because of this, organisms with varying genetically based phenotypes may contribute differentially to subsequent generations (see *Fitness*).

Neuromasts   Mechanoreceptors of the acousticolateralis system.

Nictitating membrane   The third "eyelid" present in many vertebrates (aids in cleaning and protecting the eye).

Nucleotide   The fundamental subunit of DNA (and RNA) structure. Nucleotides are comprised of the nitrogenous ring structures adenine [A], thymine [T] (uracil [U] in RNA), cytosine [C], and guanine [G] attached to a deoxyribose sugar (ribose in RNA), which is attached to a phosphate. Alternating (deoxy)-ribose and phosphates form the scaffolding of DNA and RNA polymers.

Nuptial   Refers to breeding.

Ocellus   An eyelike marking.

Oligotrophic   Refers to bodies of water, especially lakes, with low biological productivity.

Ontogeny   Development through the life cycle.

Otoliths   Calcareous nodules in the utriculus, sacculus, and lagena of the membranous labyrinth.

Outgroups   The relatives, outside the study group, included in a cladistic analysis as a source of information about the basal point of origin of character states.

Oviparous   Refers to egg-laying animals.

Ovoviviparity   Condition of retention and incubation of eggs within the ovary or oviduct. The young receive no (or little) nourishment from the female. Oviparous and ovoviviparous animals in which the nutrient for the offspring comes from the yolk deposited with the egg are termed *lecithotrophic*.

Paedophagous   Refers to an animal that eats larvae or juveniles.

Parabranchial cavity   Cavity bounded by the operculum and branchiostegal membrane; receives water that has passed through gills.

Paraphyletic   An unnatural assemblage defined in such a way as to exclude part of a clade—for example, Pisces (usually excludes Sarcopterygii) or Perciformes (usually excludes Scorpaeniformes, Pleuronectiformes, etc.).

Parasitism   A symbiotic relationship in which one member benefits to the disadvantage of the other. Usually, a symbiotic relationship that involves one member feeding on the other.

Parr   That stage in anadromous salmonids between yolk sac absorption and transformation to the smolt stage prior to seaward migration.

Parsimony   In cladistics, the principle dictating the choice of cladistic hypotheses requiring the fewest ad hoc assumptions about character convergence, parallelism, and reversal.

Parthenogenesis   Production of offspring by a female with no male contribution.

Pelagic   Refers to organisms that inhabit open waters of the oceans (or large lakes).

Peritoneum   Membrane lining the coelom.

Phagocytosis   The process by which certain cells engulf other cells or foreign particles.

Phenetics   The study of patterns of similarity among organisms. Also, the estimation of relationship by overall similarity rather than special (shared, derived) similarities.

Phenotype   The observed expression of a trait or characters of an organism, determined by interaction of its genes and the environment; *phenotype* can refer to a single Mendelian character or to the entire multi-locus organism. Species diagnoses must be based on parts of the phenotype that are distinctive due to inheritance rather than environmental effects.

Photophores   Light-producing organs.

Phylogenetic tree   A cladistic branching diagram with added information about times of branching and morphological change.

Phylogeny   Evolutionary relationships among species based on their descent from a common ancestor.

Placenta   An intimate association of embryonic and maternal membranes through which gases, nutrients, and wastes are exchanged.

Pleiotropy   The effect of a single locus on more than one trait. See *Polygenic trait*.

Polyandrous   Refers to species in which one female mates with more than one male.

Polyculture   The use of two or more species to participate in two or more trophic levels within a culture system. (The same effect can be obtained in some instances by the use of distinct life stages of one species.)

Polygenic trait   A trait resulting from the combined expression of alleles at numerous loci. The phenotype of these traits generally has a continuous distribution and may be referred to as quantitative or metric traits.

Polyphyletic   An unnatural group defined to include two or more groups that are not closely related (not of the same immediate line of descent)—for example, Apodes (for unrelated eel-shaped animals).

Polyploid   Possessing three or more haploid sets of chromosomes. See *Allopolyploid, Autopolyploid, Triploid,* and *Tetraploid*.

Population   A group of organisms belonging to the same species that occupy the same locality at the time of reproduction, have a reasonable chance of interbreeding, and produce progeny that will generally interbreed with progeny of other members of the population.

Primary production The rate at which plants and other autotrophs accrue biomass or energy and nutrients. Expressed as weight/area/time or calories/area/time.

Productivity The relative contribution of offspring to the next generation by a unit (e.g., an individual, phenotype, population, etc.). If differential productivity (whether it is attributable to increased fecundity, survival, or other factors) is heritable, it reflects fitness and, through natural selection, may lead to evolutionary changes.

Protandry A condition in which sequential hermaphrodites that are first males become females later in life.

Protogyny A condition in which sequential hermaphrodites that are first females become males later in life.

Recessive An allele that is expressed only when homozygous (aa). It can also refer to the trait expressed by homozygous recessive alleles.

Recombination The occurrence of combinations of genes in an organism that were in neither parental type but that can be transmitted to the organism's progeny. Recombination is a result of random assortment of unlinked loci or physical exchange between chromosomes of linked loci during meiosis.

Replication Synthesis of DNA based on the complementarity of A-T and G-C nucleotide pairs. At the end of replication, there are two complete copies of the original DNA molecule.

Ribosome The site of protein synthesis (translation) in cells. Constructed of protein and RNA.

RNA (ribonucleic acid) Linear molecule consisting of (ribo)nucleotides transcribed from the DNA sequences that encode the genetic information of an organism. The code lies in the specific order of the nucleotides. The information specifies the amino acid sequence in a protein (messenger RNAs) or other RNAs involved in cell operations.

Semelparous Spawning once and then dying, as do Pacific salmon. See *Iteroparous.*

Sexual dimorphism Morphological variation within a species correlated with the sex of the individual.

Sire Male parent in a cross. See *Dam.*

Sister chromatids At cell division following DNA replication, each chromosome has been duplicated. Identical copies of the same DNA sequence are sister chromatids. Each homolog produces a pair of sister chromatids.

Smolt The seaward migrating stage of anadromous salmonids.

Stegural A bone flanking the uroneurals in the dorsal tier of caudal fin supports. The family Salmonidae is diagnosed by a unique fan shape of this bone.

Stenophagy A narrow range of preferred foods.

Steroids A class of organic compounds composed of four interlocking carbon rings. Cholesterol and male and female sex hormones are included.

Stock A term that varies with context and ranges in meaning from a discrete, largely reproductively isolated subpopulation to an aggregation of populations managed as a unit. Alternatively, a genetic strain.

Strain A group of individuals derived from a common genetic origin; a lineage.

Symbiosis An intimate living arrangement (relationship) between two species. See *Parasitism, Mutualism,* and *Commensalism.*

Sympatric Two or more species that occupy the same geographic range.

Sympatry Temporal and spatial overlap of the ranges of two or more species.

Symplesiomorphy Shared, primitive character.

Synapomorphy Shared, derived character.

Systole The contraction phase of the heart.

Telolecithal Refers to eggs in which the yolk is concentrated at the vegetal pole. Also known as *macrolecithal.*

Tetraploid (4N) A polyploid that has four complete haploid (N) sets of chromosomes.

Trait Manifestation of a genetically determined character. A trait may be Mendelian or polygenic.

Transcription Synthesis of RNA from DNA based on the complementarity of A-(T or U) and G-C nucleotide pairs. The RNA product is complementary to the DNA strand that served as a template.

Transgenic A genetically manipulated organism that received a portion of its genetic information from another species. The genetic transfer often involves molecular genetic or recombinant DNA technology.

Translation Protein is synthesized at the ribosome from information encoded in the nucleotide sequence of the messenger RNA molecule under direction of the amino acid–specifying codons.

Triploid (3N) A polyploid that has three complete haploid (N) sets of chromosomes.

Trophic Refers to nutrition or to ecological levels or mode of feeding.

Triturate To grind.

Velum In hagfishes, a scroll-like pharyngeal membrane that acts as a respiratory pump. In adult lampreys, the fleshy, tentacled flap guarding the opening to the respiratory tube.

Viviparity The condition in which fertilized ova (eggs) are retained within the female and derive nourishment from the female via placenta or from secretions. Also known as *matrotrophy.*

Wild Fishes naturally produced, not artificially propagated or cultured. Often refers to native, self-sustaining populations.

Zymogen The inactive precursor of an enzyme.

SUBJECT INDEX

Note: Common and scientific names of fish, fish groups and fish taxa are found in the Systematic Index. *Italic* page numbers indicate illustrations; (t) indicates material in a table.

abducens nerve, 498
Abe, Tokiharu, 12
abiotic environment, 641, 642–46
   and reproduction, 521, 522, *523*, 524
abyssal plain, 740–41
abyssopelagic zone, 702
acanthocephalans, 809
accommodation, visual, 352–54
acid rain, 646, 685, 814
acoustic nerve, 498
acrodine, 415
acrosome, 520
actinotrichia, 62
active metabolic rate (AMR), 408
adaptation, 72–74
   behavior modification, 791–92
   effect of parasites, 810–12, *812*
   life history variation, 648–50
   physiological ecology, 641
adaptive radiation, 79, 690–94, *692*
adductor mandibulae, 427
adelphophagy, 529–30
adenohypophysis, 501
"adipose eyelids," 324, 348
adipose fins, 37, 195, 218
adrenaline, 480, 506
adrenocorticotropin (ACTH), 504
advertisement, 297–98, 304
aestivation, 137–38, 198, 451–53
afferent neurons, 497
Agassiz, Louis, 8
age and growth studies, 411
aggressive mimicry, 298–99
agnathan fishes. *See* jawless fishes
air breathing, 151–52, 451–53
alarm substances, 178, 397, 761
alevins, 537

alimentary canal, 64–65, 414–21, *420*
alleles, 555–57
allelochemicals, 390
allochthonous, 675
allometric growth, 13
allopatric speciation, 74–75, *76–77*
allozymes, 83, 559–60
Amazon River Basin, 187, 619–20, 674, 832
American Fisheries Society, 14, 15
American Society of Ichthyologists and Herpetologists, 12, 14
amiiform swimming, 326
ammocoetes, *22*, 94, 442, 756
amniote condition, 142–43
amphicoelous vertebra, 59
amphidromous fishes, 763, 766
amphioxus, 21, *22*, 23
amphistyly, 53, *54, 109*–10
ampulla (of archinephric duct), 519
ampulla (of inner ear), 365
ampullae of Lorenzini, 368, *369*
ampullary electroreceptors, 368–69
anadromous fishes, 763–65
   osmoregulation, 478–79
   rapid speciation in, 75
anaerobic metabolism, 408–9
anal fins, 37, 39–40
androgenesis, 536
androgens, 507
anemia, 814
angiotensin, 508
angle of attack, 324
anguilliform body, 35, *38*
anguilliform swimming, 318–20, *320*
angular acceleration, 373–74
angular bone, 58
anguloarticular bone, 58
annual fishes, 688
annuli (of scales), 48
Antarctica, 626, 632–33, 743–44, 748–50, 748(t)
anterior chamber, of eye, 346

anterohyal bone, 58
anterohyal cartilage, 52–53
anticyclonic migration, 767
antifreeze proteins, 461, 591, 643
aphakic apertures, 354
apomorphies, 81
aposematic coloration, 298
appendiculae, 529
appendicular muscles, 316, *317*
appendicular skeleton, 60–64
aquaculture, 838–39
   all-female stocks, 570, *571*
   ammonia toxicity, 814
   Atlantic salmon, 198
   behavior modification, 792
   channel catfish, 188
   diet formulation, 412, 434
   and exotic species, 842
   milkfish, 179
   mudskippers, 261
   mullets, 240
   reproductive readiness, 522
   salinity studies, 480
   selective breeding, 533–34, 575–77, 839
   sturgeon, 154
   tilapiines, 255
   triploid stocks, 570, *571*
   world production, *825*
aquaporins, 524
aquaria, public, 16
aquarium trade, 306, 837–38
aquifer depletion, 246, 843
Aral Sea, 841
Arctic waters, 626, 632–33, 669–70, 743, 745(t)
Arctogea, 609
arcualia, 59
arginine vasotocin (AVT), 505
Aristotle, 5
arm number, of chromosomes, 568–69(t)
Artedi, Peter, 6
arteries, 453
articular bone, 51

articular element, 58
artisanal fisheries, 834–35
aspect ratio, 321
assemblages, 667–68, *669*
asteriscus, 365
asterospondyly, 59
athalassic lakes, 695
atrium, 454
auditory systems, 364–66, 370–74
aufwuchs, 675
Australian region fishes, 623–24
autapomorphies, 81
autochthonous, 675
autocrine secretions, 500
autodiastyly, 110
autonomic nervous system, 498
autopaladine bone, 58
autostyly, 53, 109
axial skeleton, 48–60
Ayres, William O., 9

bacteria, as pathogens, 805, 815–16
bacterial kidney disease (BKD), 805
bacterial luminescence, 302–3
Bailey, Reeve, 10
Baird, Spencer Fullerton, 9
balistiform swimming, 326
barbels, 33–34
basal cartilage, 60
basal metabolic rate (BMR), 408
basal plate, 50
basapophyses, 59
basibranchial, 50
basihyal bone, 58
basihyal cartilage, 51
basioccipital bone, 54
basipterygia, 62
basisphenoid bone, 58
Batesian mimicry, 298
bathypelagic marine fishes, 737–40
bathypelagic zone, 702
BCF gait (body and caudal fin), 318, 319(t)
behavioral biology, 755–92
Belon, Pierre, 5

benthopelagic marine fishes, 732, 741
Berg, Lev Semёnovich, 11
"bet hedging" model, 650
biliverdin, 291
bioaccumulation, 840–41
bioassays, 840
biochromes, 292
biodiversity, 841–43
biogeography, 602–33, *603, 626, 627*
    antitropical distributions, 625, 744
    and genetics, 585–86
biological productivity, 646
bioluminescence, 300–305, 301(t), *302*
biomagnification, 817, 840–41
biomass, 646
biosphere, 648
biotaxis, 783
biotic environment, 641–42, 646–54
bipolar (antitropical) species, 625, 744
blastoderm, 536
blastopore, 21
Bleeker, Pieter, 11
blindness
    in electrogenic fishes, 376
    and extraocular photoreceptors, 359–60
    in hypogean fishes, 694–95
    and navigation, 762
Bloch, Marc Elisier, 7
blood, 461–64
blood pressure, 458, *458*
blue slime disease, 814
body regions, *33, 40*
body shapes, 35, *38, 39*
    and hydrodynamic drag, 289
body size, and Cope's rule, 648–49
Bohr effect, 336, 463, *464*
bony fishes, 28–29, 129–38, 147–278
    external features, *33,* 33–44, *38*
    jaws and branchial apparatus, 427–29
    reproductive anatomy, 519–20, *520*
    skull, 54–59
bottom-up regulation, 652
Boulenger, Georges, 9
boundary layer, 287–88
Bowman's capsule, 482
brachiopterygium, 42
brain, 494–97, *495, 496*
brain stem, 494
branchial arches, 50–51
    in elasmobranchs, 54
    and evolution of jaws, 106

external *vs.* internal, 106
    in teleosts, 59
branchial arteries, 456, *457*
branchial basket
    in hagfishes, 92
    in lampreys, 51, 92
branchial circulation, 456–58, *457*
branchial gas exchange, 446–49
branchial irrigation, 449–51, 449(t)
branchial sieve, 446
branchiocranium, 50
branchiostegal bone, 59
branchiostegal membrane, 33
branchiurans, 810
breeding tubercules, 178, 779
bulbus arteriosus, 65, 94, 454
buoyancy, regulation of, 332–37
    deep-water adaptations, 741–42
    mesopelagic adaptations, 736–37
    pelagic adaptations, 733
burglar alarm theory, 305
burst speed, 328, *329*
bycatch, 836

calcitonin, 506
calcium, dietary, 413
California coastal fishes, 628–29
calories, 408
camptotrichia, 64
canal construction, 843
canal neuromasts, 367
capillaries, 453
carangiform swimming, *320,* 321
carbohydrates, dietary, 412
carbon dioxide
    in blood, 336
    and carbonic acid, 645
    and eutrophication, 685
    gustatory response, 398
    and photosynthesis, 646
    solubility in water, 442(t)
carbonic acid, 645
cardiac muscle, 64
cardiform teeth, 415
cardinal heart, 93, 454
cardinal veins, 458
carinal muscles, 316
carotenoids, 292
cartilaginous fishes
    agnaths, 27
    gnathostomes, 27–28, 109–25
catadromous fishes, 763
    migration, 765
    osmoregulation, 478–79
catecholamines, 506
Catesby, Mark, 6
caudal fins, 37, 40–41, *43*
caudal heart, 93, 454, 459
caudal neurosecretory system, 508
caudal peduncle, 35, 36

and hydrodynamics, 289
    tendons and flexure, 322–23, *325*
caudal vein, 458
cave-dwelling fishes. *See* hypogean fishes
caviar, 152–53, 272
central nervous system (CNS), 494–98, *495*
centromere, 555
cephalic fields, 100
cephalic fins, 125
ceratobranchial, 50, 51
ceratohyal bone, 58
ceratohyal cartilage, 51, 52–53
ceratotrichia, 62, 114
cercariae, 808
cerebellum, 497
cerebrum, 494–95
cestodes, 809, *809*
character data, in DNA studies, 83
character displacement
    ecological, 79–80
    reproductive, 75
cheek, 33
chemautotrophic ecosystems, 743
chemical pollutants, 840–41
    acid rain, 646, 685, 814
    and endocrine pathways, 509
    and reproduction, 522
chemosensation
    common chemical sense, 399
    and communication, 778
    olfaction, 390–97
    solitary chemosensory cells, 399–400
    taste, 397–99
chemotaxis, 760–61
chimaeriform body, 35
chloride cells, 472, *472*
choanae, 140–41, *142*
cholecystokinin, 431
chondrocranium, 50, *50*
chondroneurocranium, 50
choroid body, 346
choroid layer, of eye, 346
choroid rete, 460–61
chromaffin tissue, 506
chromatic aberration, 352
chromatic mimesis, 298
chromatophores, 291, 292–93, *294*
chromosomes, *553,* 555, 567–72
    configurations (numbers), *557,* 568–69(t)
    and sex determination, 530
ciguatera, 806, 816–17
ciguatoxin, 816
ciliated receptor cells, 391
ciliates (Ciliophora), 806, *807*
circadian rhythms, 757–58
circulatory system, 453–61

circuli (of scales), 48
circumorbital bone, 58
circumorbital sulcus, 346
cirri, 34
clades, 15, 81
cladistics, 81
cladistic trees, *84*
cladograms, 81, *82,* 84
Clark, Eugenie, 12
claspers, 114, 519, 527, *528*
classification and categories, 14–15
cleaning symbiosis, 788–90
cleithrum, 62
climax communities, 651
cloaca, 65, 419
club cells, 284
coastal territorial domains, 845
codominance, 559
coelom, 20–21, 33
Collette, Bruce B., 11
color, 291–300
    and advertisement, 778, 781–82
    and bioluminescence, 304
color vision, 351
Columbia River, 831
commercial fisheries, 825–37
    Large Marine Ecosystem concept, 732
    sustainability, 824
    world production, 825, *825, 826, 827*
common chemical sense, 399
common names, 15
communication, among fishes, 777–78
communities, of fishes, 648, 650–51, 655–57
compressiform body, 35, *38*
concealment, 296–97, 304
condition factor, 409–10
cones, of eye, 348–50
congruence, 81
conservation
    environmental quality, 840–44
    marine preserves, 844–45
    marine protected areas, 836–37
conservation genetics, 589
conspecifics, defined, 774
continental drift, 605–8, *607,* 624–25
continental islands, 607
continental shelf, 702
continental slope, 702, 740–41
contractile iris, 355–56
conus arteriosus, 454
convergent evolution, 81
Cope, Edward Drinker, 9
copepods, 810, *811*
Cope's rule, and fishes, 648–49
coracoid cartilage, 61
coral reef environments, 705–10

cornea, 346
    pigmented, 296, 356
coronoid bone, 58
corpus cerebelli, 497
corpuscles of Stannius, 507–8
corticosterone, 505
cortisol, 505
cortisone, 505
cosmine, 48
cosmoid scales, 48, 286
countercurrent exchange, 446
counterillumination, 304
countershading, 294, 296, 735
    reverse, 187
courtship, 524, 779–80
cranial nerves, 497–98, 499(t)
Crater Lake, 682
crenate scales, 287
crepuscular activity, 757
crista, of ear, 365
critical swimming speed, 327, 329
crumenal organ, 194, 416
crus communis, 365
crustacean parasites, 810
crypsis, 304
cryptochrome, 757
C-starts, 327, 328
ctenoid scales, 48, 49, 287, 292, 293
"cucumber odor," 196
cupula, of ear, 365
cutaneous respiration, 451
cuticle (of epidermis), 46–47
Cuvier, Georges, 7, 7
cyanobacteria, 706
cycloid scales, 48, 49, 287
cyclonic migration, 767
cyclospondyly, 59
cyclostome poisoning, 818
cystovarian, defined, 520

Dahlgren cells, 508
dams, 831, 843
dark adaptation, 354–55, 355, 356
Darwin, Charles, 72
Day, Francis, 11
Dean, Bashford, 10
deep scattering layer, 736
deep-sea fishes, 633
definitive host, 804
demersal coloration, 296
demersal eggs, 524
demographic release, 804
dentition, 414–15, 416
    in early tetrapods, 139–40
    labyrinthodont, 141
    pharyngeal teeth, 428–29, 429
    polyplocont, 141
deoxyribonucleic acid. See DNA
depensation, 836
depressiform body, 35, 38
derived state, 81

dermal bones, 54
dermal denticles, 48, 114
    and drag reduction, 290
    phylogeny, 285
dermal mesethmoid bone, 58
dermal skeleton, 96
dermatocranium, 51
dermatomes, 284
dermis, 48, 284–85
desert environments, 670
detritivores, 675
detritivorous fishes, 426, 432
deuterostomes, 21
diadromous fishes, 763–66
    osmoregulation, 478–79,
    479(t)
diapause, 244, 536
diencephalon, 496
diet
    dietary requirements, 412–13
    diversity of, 423–26
    and pigmentation, 296
dietary diseases, 814
digeneans, 808
digestion, 430–34, 433(t)
digital particle image velocimetry
    (DPIV), 287, 289
dinoflagellates, 806, 816
diphycercal tail, 41
diploid, defined, 555
diplospondyly, 59
directional selection, 72, 73
diseases of fishes
    dietary, 814
    environmental, 814
    genetic, 814–15
    parasitic, 804–10
disguise, 298, 299, 304
dispersal, 604
disruptive coloration, 297, 297
disruptive selection, 72, 73
dissolved gases, 442(t), 443(t),
    644–46
dissolved salts. See salinity
distance data, in DNA studies, 83
distribution, geographic, 602–33
diversity patterns, 609
DNA (deoxyribonucleic acid), 552,
    553
    analysis methods, 562, 563,
    563–65, 564
    and archived specimens, 592
DNA-DNA hybridization
    studies, 83
dominant alleles, 556–57
dorsal aorta, 456
dorsal fins, 37–39, 41
dorsal light reaction, 758, 758
dorsal rib, 59
drift nets, 835
dual innervation, 316

ducts of Cuvier, 65, 458
dynamic equilibrium, 373–74

ear, 364–66, 365, 370–74
East Wind Drift, 744
ecological character displacement,
    79–80
    in lacustrine environments,
    689
ecological niches, 650
ecological succession, 651
ecology, 639–751
    extinctions, 841–42
    human impacts, 824–44
    optimal foraging theory, 654,
    776
ecomorphology, 430
ecosystem-based fishery manage-
    ment, 836
ecosystems, 648
ecotones, 657
ectoderm, 20, 536
ectoparasites, 802
ectopterygoid bone, 58
ectotherms, 409, 643
efferent neurons, 497
egg bearers, 530
egg mimicry, 781
eggs
    demersal vs. palagic, 524
    fecundity vs. size, 649
    hatching, 536
    internal gestation, 526–30
    macro- vs. microlecithal,
    535–36
    oviparous fishes, 524–26
    parental care of, 526, 649–50,
    782–83
    size, 520, 524
    and water salinity, 479–80
Eigenmann, Carl H., 10
elasmoid scales, 48, 287
electivity index, 653
electric organ discharge (EOD),
    376, 379–80
electric organs, 123, 377–80
electrocytes, 123
electrogenic fishes, 376–80
    and communication, 778
    electric field, 375
    jamming avoidance response
    (JAR), 375
electrolocation, 375–76
electromyogram (EMG), 318
electrophoresis, 83, 559–60, 560,
    561
electroplaques, 123, 377
electroreception, 368–70, 374–76
electrotaxis, 758–59
El Niño–Southern Oscillation
    (ENSO), 173, 843

elvers
    commercial fishery, 171
    life history, 169
    rheotaxis, 760
embryology, 534–36, 537
    brain, 494, 496
    deuterostomes vs. protostomes, 21
    gonads, 518
    kidneys, 482–83
    syncranium, 50
eminentiae granulares, 497
enamel, 284
Endangered Species Act, 246, 589
endemism, 605
    in Middle American fishes, 620
    in North American fishes, 618
endochondral bone, 54, 96
endocranium, 50
endocrine system, 500, 500–508
    disruption, 509
    and osmoregulation, 480–81
    and reproduction, 522, 532–33
endoderm, 20, 536
endolymphatic duct, 365
endoparasites, 802
endopterygoid bone, 58
endoskeleton, 96
endotherms, 643
energetics, 408, 408(t)
enterocytes, 419
entoderm. See endoderm
enzymes, digestive, 430–31, 432–33
epaxial muscles, 312
epibranchial, 50, 416
epibranchial organ, 194
epicentrals, 59
epidermal cells, 284
epidermis, 44–47, 284, 285
epididymus, 519
epihyal bone, 58
epilimnion, 684
epineprhine, 295
epineurals, 59
epiotic bone, 56
epipelagic fishes, 733–36, 735(t)
epipelagic zone, 702
epiphysis. See pineal organ
epipleurals, 59–60
equilibrium, 373–74
erythrocytes, 448, 461–62
erythrophores, 292
esca, 224
Eschmeyer, William, 10
esophagus, 418
essential fish habitat (EFH), 664
estrogens, 507
estuarine environments
    African, 720–21, 721
    Australian, 721
    temperate, 716–17, 719
    tropical, 717–18

ethanol, 14
Ethiopian region fishes, *616,*
  622–23
ethmoid bones, 58
ethmoid plate, 50, *50*
ethology, defined, 756
eubacteria, 805
eucoelomates, 20–21
euphysoclistic fishes, 336
euryhaline fishes, 602, 644
  osmoregulation, 478–79
eurytherms, 644
eutrophication, 684–85, 838, 840
Evermann, Barton Warren, 9
evolution, 72–80
  convergent evolution, 81
  and "heat shock" proteins,
    85–86
  Lamarck's theory, 85
  of parasites, 802–4
evolutionarily significant unit
  (ESU), 589
excess post-exercise oxygen consumption (EPOC), 409
exclusive economic zones (EEZ),
  845
exoccipital bone, 54
exoskeleton, 96
exotic species, 842–43
  and aquaculture, 838
  carp, 181, 672
  gobies, 261
  introduced, in United States,
    640
  lionfish, 306–7
  mosquitofish, 245
  Nile perch, 694–95
  and parasites, 804
  redrim melania, 809
  sea lamprey, 829–30
  snakeheads, 264
  walking catfish *(Clarias batra-*
    chus), 44, 187
extinctions, 841–42
extracellular luminescence, 303
extraocular photoreception, 359–
  60, 501
extraosseus tooth replacement, 415
eyes, 346–48, *347, 350, 354*
  dual-purpose, 357–59, *358*
  in flatfishes, 273
  in "four-eyed" fishes, 245
  light adaptation, 354–56, *355*
  photic *vs.* scotopic, 354
  visual cells, 348–52
eyestalks, 131

facial nerve, 498
false eyespots (ocelli), 157, 297,
  298, 778

fast (white) muscle fibers, 314–15,
  315
fecundity, 518, 521
feeding, 421–26, *422,* 774–76
  cooperative foraging, 777
  digestion, 430–34
  food selection, 653–54
  and gustation, 399
  mastication, 429
  and olfaction, 395–96
  resource polymorphism, 777
  suction feeding, 428
  trophic relationships, 651–54
feeding apparatus
  in agnathans, 426
  in bony fishes, 427–30
  in chondrichthyans, 426
field methods, 13–14
filial cannibalism, 783
filiform body, 35, *38*
fimbria, 34
finlets, 40
fin rays, 62–64, *64*
fins, 37–44, *40, 41, 45*
  angle of attack, 324
  aspect ratio, 321
  median (unpaired), 60–61
  musculature, 316, *317*
  paired, 61–62
  skeletal support, 60–64, *62*
fin-swimming, 326–27
fisheries management, 835–37
  genetic techniques, 587
  sustainability, 824
fishery resources
  anadromous, 830–31
  marine, 832–34
  over-exploitation, *827,* 828(t),
    829–30
  tragedy of the commons,
    824–25
  tropical freshwater, 831–32
  world production, 825, *825,*
    826, 827
fishery technology, 834–35
  purse seines, 832
  trawling, 833
fishes
  bony, 28–29, 129–38, 147–278
  classification, 25–29, 27(t)
  evolution and speciation, *20,*
    74–79, 85–86
  jawed cartalaginous, 27–28,
    105–25
  jawless, 27, 89–101
  nomenclature, 14–15
  origin of, 20–29, 90, *91*
  value of, 824
fish farming
  behavior modification, 792

dietary diseases, 814
  and selective breeding, 577
fish handler's disease, 816
fission (chromosomes), 567
fitness, 72, 552
flood pulse concept, 674
flukes, *803,* 808–9, 815
fluviatile lakes, 682
flying, 44
focal innervation, 316
focus (of scales), 48
focus (vision), 352
fontanelles, 51
food chains, 651–52
food webs, 652
  continental shelves, 704
  estuarine, *717*
  kelp forest, 706–7
  riverine, 674–77
foraging, 421–26
  cooperative, 777
  and schooling behavior, 785
foramen magnum, 52, 494
forebrain, 494–96
Formalin, 14, 589
Forsskål, Peter, 7
founder principle, 80
"four-eyed" fishes, 245, 357–59,
  358
fovea, 354
freezing point depression, 470,
  470(t)
frenum, 33
freshwater fishes
  biogeographic regions, *603,*
    609–24
  primary *vs.* secondary, 673
  salinity tolerance, 604(t)
frontal bone, 58
fugu, 276, 817
fulcra, 41
fundibular cells, 501
fungi, as pathogens, 807
furunculosis, 805
fusiform body, 35, *38*
fusion (chromosomes), 567

gaits, swimming, 318
galvanotaxis, 758–59
game theory, 777
gametogenesis, 521–22, *522*
ganoid scales, 48, *49,* 286
ganoin, 48, 284
Garman, Samuel, 10
gas bladders. *See* swim bladders
gas bubble disease, 646, 814
gas exchange, branchial, 446–49
gas glands, 336
gas resorption, 335
gas secretion, 336–37

gastric secretions, 430–31
gastrin, 507
gastroenteric mucosa, 507
gastrulation, 536
GenBank, 85
gene expression, 552–55, *554*
gene mapping, 565
gene pool, 74
genetic disorders, 814–15
genetic engineering, 590–91
genetics, 552–91
  and aquaculture, 839
  and biogeography, 585–86, 605
  fishes as vertebrate models,
    534–35, 589–90
  hybridization, 533, 570–72
  and phylogenetics, 23–24,
    577–83
gene trees, 585–86, *586*
genital ridge, 518
genome, 552
genotype, 552
geographic information systems
  (GIS), 657–58
geologic history, 605–8, 624–25
geotaxis, 758
gephyrocercal tail, 41
germinative layer, 46
Gilbert, Charles Henry, 9
Gill, Theodore Nicholas, 10
gill filaments, 446–49, *448, 449*
gill nets, 835
gill rakers, 415, *417,* 428–29, *429*
  "coughing," 423
  and filter feeding, 776
  shedding, in basking shark, 119
  trophic polymorphism, 79
gills, 36–37, *445*
  in agnathans, 442–43, *444*
  blood circulation in, 456–58,
    457
  branchial irrigation, 449–51
  flow velocities, 449(t)
  in gnathostomes, 443–46, *444*
  respiratory function, 446–49
  surface area, 447(t)
Girard, Charles, 9
gizzards, 431
glacial flour, 669
glacial lakes, 682
glaciation, *608,* 608–9
  and Holarctic fish distribution,
    610–11
  and Palearctic fish distribution,
    615–19
gliding locomotion, 44
global warming, 843–44
globiform body, 35, *38*
glochidia, 788, *791*
glomeruli, 482, 486, 488(t)

glossohyal bone, 58
glossopharyngeal nerve, 498
glucagon, 506
gnathostomes. *See* jawed fishes
goblet cells, 284
goiters, 814
gonadosomatic index, 521
gonadotropic hormones (GTH), 504, 522
gonadotropin releasing hormone (GnRH), 522
gonads, 65, 518–20, *520*
    hormone secretions, 507
Gondwana, 606, 624, *624*
gonopodium, 40, 527, 781
Goode, George Brown, 10
Gram-staining, 805
granular cells, 284
granulocytes, 463–64
gravistatic function, 374
gray matter, 498
Gray's paradox, 290–91
Great Lakes
    chemical pollution, 841
    fishery decline, 829–30
    origin of, 682
Great Slave Lake, 682
groundwater depletion, 843
groupings of fishes, 783–87, *784*
growth, 409–14, *410, 411*
growth hormone (GH), 504
guarders, defined, 530
guilds, 530, 668
Gunther, Albert, 9
gustation, 397–99
    and feeding, 423
    taste buds, *398*
gymnotiform swimming, 326
gymnovarian, defined, 520
gynogenesis, 533, 536

habitat, loss of, 246, 831, 843
hadopelagic fishes, 743
hadopelagic zone, 702
hair cells, 364, 365, 367
half-gills, 443
halophytes, 716
Hamilton, Francis, 11
haploid, 555
head
    in bony fishes, 33–35, *36*
    in hammerhead sharks, 120
head kidney, 65
headwaters environments, 665–70
hearing, 370–73
heart, 454–56, *455*
"heat shock" proteins, 85–86
hedgehog proteins, 695
helical currents, in streams, 665, *666*
hemal arch, 59

hemal arch pump, 459
hematocrit, 462
hematopoiesis, 462–63
hemibranch, 443
hemoglobin, 461–62
Hennig, Willi, 81
hepatic lipoidosis, 814
hepatic portal system, 454
hepatopancreas, 65, 162, 421
herbivores, 425, 431–32, 674–75
heritability, 575–77, *576*
hermaphroditism, 531–33, 573–74
heterocercal fins, 41, *43*
heterosis, 534
heterospecifics, defined, 774
heterothermy, 459–61
hindbrain, 494, 497
histamine, 817
historical ecology, 656–57
Holarctic region fishes, 154, 610–14
holoblastic cleavage, 536
holobranch, 443
holopelagic fishes, 732
"Holostei," 156
holostyly, 53, *54*, 111
holotypes, 14
homeobox genes. *See* HOX gene clusters
homeostasis, 32
homeotic genes, 23
homing behavior, 761–62
    in salmon, 396
homocercal fins, 40, *43*
homoplasy, 81
Hora, Sunder Lal, 11
hormones. *See* endocrine system; reproduction
HOX gene clusters, 23–24, 101, 106, *578*, 579–81
Hubbs, Carl, 10
human chorionic gonadotropin (hCG), 522
human diseases, fish-borne
    infections, 815–16
    parasites, *803*, 808, 815
    poisoning, 816–18
    See also venomous fishes
hybridization, 533, 570–72
hybridogenesis, 533
hybrid vigor, 534
hydraulic jumps, 665
hydrodynamics, 287–91
    angle of attack, 324
    and mechanosensory organs, 367
    quick starts *vs.* speed, 322–24
    and schooling, 786
hydrological changes, 843
hydrosphere, 664–65
hydrostatic pressure, 644

hydrothermal vents. *See* chemautotrophic ecosystems
hyoid arch, 51
hyomandibular bone, 56, 58
hyomandibular cartilage, 51
hyostylic jaw suspension, 53, *54*
hyostyly, 110–11
hypaxial muscles, 312
hyperosmotic *vs.* hypo-osmotic, 470, *471*
hypobranchial, 50
hypocercal tail, 41
hypogean fishes, 691–95
hypohyal bone, 58
hypolimnion, 684
hypophyseal portal system, 454
hypophysis. *See* pituitary gland
hypothalamo-pituitary axis, 500
hypothalamus, 496, 500–501
hypural bone, 60

ichthyocrinotoxins, 818
ichthyology
    diversification of, 12
    history of, 4–12
    methods of, 13–14
    research centers, 15–16
ideal free distribution, 654, 777
illicia, 224
immune responses, 461, 463–64
    and parasites, 813
indeterminate growth, 410
industrial fisheries, 835, 836
industrial pollutants. *See* chemical pollutants
infectious hematopoietic necrosis (IHN), 804, 805
infectious pancreatic necrosis (IPN), 804
infraorbital bone, 58
inheritance
    Mendelian, 555–65
    non-Mendelian, 565–67
inland saline waters, 695–96
inner ear, 364–66, *365*, 370–74
insulin, 506
insulin-like growth factor (IGF), 506
intercalar bone, 56
intercalary plate, 59
interhyal bone, 58
intermediate host, 804
intermuscular bones, 160
International Code of Zoological Nomenclature, 14
interopercle, 59
interorbital, 33
interorbital septum, 58
interrenal tissue, 505–6
interspecific limiting factors, 642

interspinous bone, 61
intertidal fishes, 722–23, *723, 725*
intestine, 419, 432, 432(t)
intraosseous tooth replacement, 415
intraspecific limiting factors, 642
intromittent organs, 527, *528*
    in elasmobranchs, 114
    in four-eyed fishes, 245
    modified anal fins, 40
    modified pelvic fins, 44
    in priapiumfishes, 242
ionic balance, 475(t), 485–88, *487*
iridophores, 292, 294–95
iris, of eye, 346, 355–56
iron, dietary, 413
islet organ, 506
islets of Langerhans, 506
isocercal fins, 41, *43*
isopedine, 48
isopods, 810
isosmotic solutions, 470
"Isospondyli," 162
iteroparity, 199, 523–24

jamming avoidance response (JAR), 375
jawed fishes
    bony, 28–29, 129–38, 147–278
    cartalaginous, 27–28, 105–25
    evolutionary history, *107*
    gills, 443–46
    phylogeny, 106–7, *112*
    suspensorium evolution, 109–11
    taxonomy, 107
jawless fishes, 27, 89–101, *91*
    feeding apparatus, 426
    gills, 442–43
    ionic balance, 475(t)
    skull, 51, *52*
    vertebral column, 60
jaws
    in bony fishes, *55*, 427–28
    musculature, *427*
    origin of, 106, *107*
jaw suspensions, 53–54, *54, 55*
    evolution of, 109–11, *110*
Jordan, David Starr, 9
juvenile stage, 538, *538, 539*

karst lakes, 682
karyotypes, 555, *556*
kelp forests, 704–5, 706–7, *709*
keratin, 284
keystone species, 655, 705
kidneys, 481–88, 488(t)
    hormone secretions, 508
kineses, 756
kinocilium, 365
K-selection, 649–50, 649(t)

labriform swimming, 327
labyrinthodont dentition, 139–40, *141*
Lacepède, Bernard Germain Etienne, comte de, 7
lachrymal area, 33
lachrymal bones, 58
lagena, 364
Lake Erie, *687*
lakes
  acidification, 685
  eutrophic, 684–85
  as fish habitat, 682–85
  oligotrophic, 684
  origins of, 682
  saline, 695–96
  temperate zone, 684–85
  thermal stratification, 684, *684, 685, 687*
  tropical zone, 685–86
Lake Victoria, 685, 690–94
Lamarck, Jean Baptiste de, 85
lamellae, 446–49, *448, 449*
landscape ecology, 657, 677
landslides, and lake formation, 682
lapillus, 365, 374
Large Marine Ecosystems (LME), 732
larvae, 536–43, *538, 539, 540, 541, 542*
  leptocephalus, 166, *167, 294, 539*
larviparous fishes, 529
lateral line nerve, 498
lateral line system, 36, 366–67, *367, 368, 374*
  behaviors mediated by, 374
  mesopelagic adaptations, 737
  nerves, 498
lateral recess, 367
Law of the Sea (LOS) conferences, 845
leaders (of migrations), 763
learning, 791–92
lecithotropy, 527
leeches, 810
lens, of eye, 346
  pigmented, 296, 356
lepidotrichia, 64
leptocephalus larvae, 166, *167, 294, 539*
leptocercal tail, 41
Lessepsian migration, 843
LeSueur, Charles Alexandre, 7
leucophores, 292, 295
leukocytes, 461
Leydig's gland, 519
life history studies, 648–50
light absorption, in water, 646, *647*
light adaptation, 354–56, *355*
light production, 302–3

limiting factors, defined, 642
lingual apparatus (hagfishes), 51
linkage maps, 565, *566*
linked traits, 565
Linnaeus, Carolus, 6–7, 14
lipids, dietary, 412
liver, 421
liver flukes, *803, 808*
lobe-finned fishes, 131–38
local adaptation, 586–87
locomotion, 318–31
  in actinopterygians, 148
  and hydrodynamics, 287–91
  with spines, 44, 264
  vertical movement, 334–35
locus (of genes), 555
logistic growth model, 649, *650*
longlines, 835
Losey, George, 10
Lowe-McConnell, Rosemary, 12
luciferase, 302
luminescence, 300–305
  and visual adaptations, 357
lures, 299
luteinizing hormone, 522
lymphatic system, 454, 459
lymphocytes, 464

macrocarnivores, 423–24, 775–76
macroecology, 657–58
macroevolution, 80
macrolecithal eggs, 535
macrophytes and fish assemblages, 704–5
macula, of ear, 365
magnetotaxis, 376, 759
Magnuson-Stevens Fishery Conservation and Management Act, 664
maitotoxin, 816
mandibular arch, 50–51, *54*
mandibular cartilage, 51
mangals, 717–18
Marcgrave, Georg, 6
marine environments, 702, *703*
  neritic
    continental shelf, 702–5
    estuaries and bays, *714–15,* 714–22
    reefs, 705–10
    surf zones, 722–26
  oceanic
    bathypelagic, 737–40
    deep benthic, 740–43
    epipelagic, 733–36
    pelagic, 732–37
    polar, 743–44
marker assisted selection, 577
Marshall, Norman B., 12
mass spawning, 525–26
mastication, 429

mastigophorans, 806, *807*
mate selection, 524, 779–80
matrotrophy, 527, 529
Matthiesen's ratio, 348
Mauthner neurons, 327, 497
maxillae, 33, 58
Mayr, Ernst, 74
meanders (stream), 670–71, *671*
mechanosensory systems, 367, 370, 374
Meckel's cartilage, 51
median (unpaired) fins, 37–41
  skeletal support, 60–61
medulla oblongata, 497
medullary keel, 536
Meek, Seth, 10
meiosis, 555, *558*
melanin, 292
melanocyte-stimulating hormone (MSH), 295
melanophore-concentrating hormone (MCH), 295
melanophores, 292
melanophore stimulating hormone (MSH), 504
melatonin, 295, 501
membrane bones, 54
membranous labyrinth, 364, *365*
Mendel, Gregor, 72
Mendel's laws
  independent assortment, 560–61, *561*
  segregation, 556–59, *559*
meninges, 494
mentomeckelian element, 58
mentum, 33
mercury pollution, 840
meristic characteristics, 13
Merkel cells, 509
meroblastic cleavage, 536
meroblastic eggs, 93
meropelagic fishes, 732
mesencephalon, 494, 496–97
mesocoracoid bone, 62
mesoderm, 20, 536
mesodermal somites, 536
mesolimnion, 684
mesopelagic marine fishes, 633, 736–37
mesopelagic zone, 702, 736
mesopterygium, 61
mesorchium, 518
mesovarium, 518
metabolism, 408–9
  and dietary requirements, 412–13
  and hormones, 413–14
  polar adaptations, 744
metamerism, 23
  genetic basis of, 23–24
  and mesodermal somites, 536

metamorphosis, 538–39
metapopulations, 648
metapterygium, 61
metapterygoid bone, 58
microcarnivores, 425
microevolution, 80
micropyle, 535
microsatellites, 83, 563
microvillous receptor cells, 391
midbrain, 496–97
Middle America region fishes, 619, *620*
migration, 762–68
Miller, Robert Rush, 10
milliosmol (mOsm), 470
mimicry, 298–300, *300*
minerals, dietary, 413
miracidia, 808
Mitchell, Samuel, 7
mitochondrial DNA (mtDNA), 83, 566–67, *567*
mitosis, 555, *558*
modes, of swimming, *320*
molar *vs.* molal solutions, 470
molecular clock, 83
molecular pedigree analysis, 84
molecular (DNA) studies, 81–84
monocyclic reproduction, 522
monogamy, 780
monogeneans, 808, *808*
moraine, 682
morphological color change, 295–96
morphology, 32–65
morphometric characteristics, 13
mouth brooding, 264, 300, 782–83
mouth position, 33–34, *34,* 414
MPF gait (median and paired fin), 318, 319(t)
mucous cocoons, 284
mucus
  functions of, 47
  and hydrodynamics, 290
mucus glands
  in hagfishes, 92
Müller, Johannes, 11
Mullerian mimicry, 299–300
multiple innervation, 316
muscle retes, 459–60
musculoskeletal system, 64, 312–18, *313, 314, 315, 317*
  skeletal system, 48–64
mutations, 552
myencephalon, 497
myocommata, 312
myofibers, 312
myomeres, 312
myosepta, 312
myxozoans, 807–8

nasal bone, 58
nasal capsule, 50

nasohypophyseal opening, 94
National Center for Biotechnology Information (NCBI), 85
National Observer Program, 845
national territorial limits, 845
natriuretic peptides, 509
natural selection, 72–74, *73*
Nearctic region fishes, 615–19, *618*
nearshore environments, 713–26, *714–15*
nematodes, 809–10, 815
neodarwinism, 72
Neogea, 609
neoplasia, 814
neoteny, 541
neotropical freshwater fish fauna, 673–74
Neotropical region fishes, 619–21
nephric duct, 483
nephrons, 482, *485, 487*
nephrostome, 482, *482*
nerves. *See* cranial nerves; peripheral nervous system
nest building, 526, 779–80
    salmon, 649
    sticklebacks, 238
    stoneroller minnows, 72
neural canal, 59
neural crest, 50, 90, 284, 536
neural spine, 59
neural tube, 22, 23
neurocranium, 48–50, *57*
    and jaw suspensorium, *54*
neurohypophysis, 501
neuromasts, 367, *374*
neusecretory cells, 508–9
Nicholls, John Treadwell, 10
nictitating membrane, 346–47
nidamental gland, 519
nociception, 370
node, 81
nomenclature, 14–15
nonbacterial luminescence, 303
nonguarders, defined, 530
nonswimming locomotion, 329–31
noodling, 188
noradrenaline, 506
Norman, John Richardson, 11
nostrils, 34, *37*
notochord, 22
    in agnaths, 59
    in cephalochordates, 23
    in coelacanths, 136
    in elasmobranchs, 59
    embryology, 537
    in tunicates, 22
Notogea, 609
nuchal region, 36
nucleotides, 552
nuptial tubercles, 47, 531, *532*
nutrition, 411–13

obliterative countershading, 187, 294, 296, 735
occipital arch, 50
occipital chondyles, 51–52
occiput, 36
oceanic islands, 607–8
oceanodromous fishes
    migration, *767,* 767–68
ocean surface currents, *709*
ocean surface temperature, 705, *708*
ocelli, 157, 297, 298, 778
octavolateralis system, 364
oculomotor nerve, 498
odontodes, 187
olfaction, 390–97, 395(t)
    and feeding, 774
olfactory bulbs, 391, *392,* 494
olfactory epithelium, 391
olfactory nerve, 498
olfactory rosette, 391, *393*
oligotrophic lakes, 684
omnivores, 425–26
oncomiracidia, 808
Oomycota, 805
oophagy, 529–30
opercle, 59
opercular musculature, 427, *427*
operculum, 33, 59
opisthaptors, 808, *808*
opisthocoelous vertebrae, 59, 156
opisthonephrous kidneys, 483
optical filters, 296
optic lobes, 496–97
optic nerve, 498
optimal foraging theory, 654, 776
    in lacustrine environments, 689–91
optomotor response, 760
orbital cartilage, 50
orbitosphenoid bone, 58
orbitostyly, 111
organ of Hunter, 377
organ of Leydig, 462
organ of Sachs, 377
organs of Fahrenholz, 368
Oriental region fishes, 621–22
orientation
    electroreception, 376
    homing behavior, 761–62
    magnetic fields, 759
    olfaction, 396, 764
Orinoco floodplain, 675–77, *676*
ornamental fish trade, 837–38
oscillation, 326–27
osmol (Osm), defined, 470
osmolality, 470
osmoregulation, 470–81, *474*
    in euryhaline fishes, 644
    and excretion, 485–88
    in freshwater fishes, 471–73
    in marine fishes, 473–78

osseous labyrinth, 364
ostraciiform swimming, 325–26
otarena, 365
otic capsule, 50, 54
otoconia, 365
otoliths, of ear, 364–65, *366, 374*
otophysic connection, 164
outbreeding depression, 587, *588*
outgroups, 81
ovaries, 518–20
oviducts, 519, 520, *520*
oviparous fishes, 524–26
ovoviviparous fishes, 527
oxbow lakes, 671, 682
oxygen
    aerial uptake, 451–53
    in blood, 463
    gas exchange in gills, 446–49
    solubility in water, 32, 442(t), 443(t), 644–45
ozone layer depletion, 844

paedomorphosis, 23, 541
paedophagous, 424
pain, perception of, 370
paired appendages, evolution of, 100–101
paired fins, 41–44
palaeoniscoid scales, 48
palatal organ, 416–17
palatoquadrate cartilage, 51
Palearctic region fishes, 614–15, *616*
paleostyly, 109
pancreas, 65, 162, 421
    in agnaths, 506
pancreas-peptide, 506
Pangaea, 606, *607*
papillomas, 814
parachordal cartilage, 50
paracrine secretions, 500
paralytic shellfish poisoning, 806
paraneurons, 509
parapatric speciation, 75
paraphysoclistic fishes, 336
parasite load, 802
parasites of fishes
    acanthocephalans, 809
    behavior and susceptibility, 812–13
    cestodes, 809, *809*
    crustaceans, 810, *811*
    dinoflagellates, 806
    eubacteria, 805
    fungi, 807
    glochidia, 788, *791*
    green algae, 805
    leeches, 810
    myxozoans, 807–8
    nematodes, 809–10
    platyhelminths, *808,* 808–9
    as population "tags," 813

protista, 805–7
protozoans, 806–7
trematodes, 808–9
viruses, 804–5
parasitism, 787–88
    evolution of parasites, 802–4
    host adaptations, 810–12
    and lampreys, 27, 95
    sexual, 225, 531
    and sexual selection, 811, 813
parasphenoid bone, 58
paratypes, 14
parental care, 782–83
parietal bone, 58
parietal eye, 501
parr marks, 295
parsimony, 84
pars inferior, 364
pars superior, 364
parthenogenesis, 536
partial dominance, 557–59, *559*
pearl organs, 47
pectoral fins, 41–42
    modified, 42–44, *45*
    skeletal support, 61–62, *63*
    and tetrapod origins, *140*
pelagic coloration, 296
pelagic marine fishes
    continental shelf, 703–4
    oceanic, 732–40, 735(t)
    polar, 626–28, *627,* 743–51
    Sargassum communities, 736
pelvic claspers, 114, 519, 527, *528*
pelvic fins, 44, *46*
    placement, 161
    skeletal support, 62, *63*
pericardial cavity, 33, 454
perichondral cartilage bones, 54
peripheral nervous system, 497–98
    muscular innervation, 316–18
phaeomelanophores, 292
pharyngeal jaws, 417
pharyngeal teeth, 51, 59, *417,* 428–29, *429*
    sound production, 381
pharyngobranchial, 50, *55*
pharyngocutaneous duct, 93, 442
pharyngognathous, 417
pharyngotremy, 22
pharynx, 415–17
phenetics, 81
phenotype, 552
pheromones, 390, 761, 780
photic zone, 646
photophores, 47, 302, *302*
photopic eyes, 354
photoreceptors, extraocular, 359–60, 501
photosensitive pigments, 351–52
photosynthesis, 646
phototaxis, 757–58

PhyloCode, 15
phylogenetic methods, 83–85
    DNA sequence analysis, 577–83
    gene mapping, 565
    mtDNA analysys, 566–67
phylogeny, 81, 552
phylogeography, 585
physiological ecology, 641
physoclistous, 65, 162
physostomous, 65, 162
pigmentation, 291–300
    in larvae, 541, *542*
pilaster cells, 447, *448*
pineal organ, 496, 501
        and color change, 295
        and extraocular photoreception,
            359–60
"pink" muscle, 314, *315*
pit organs, 367
pituitary gland, 501–5, *502, 503*
        and color change, 295
        and hypothalamus, 500
        and osmoregulation, 480–81
placentae, 527, 529
placoid scales, 48, *49*, 285
        and drag reduction, 291, *291*
planktivorous fishes, 776
planktivory, 173
        in lakes, 683, 686
        in pelagic zones, 733
        sharks, 119–20, 652
        and suction feeding, 776
        trophic cascades, *653*
platyhelminths, *808*, 808–9
pleomerism, 60
plesiomorphies, 81
pleural rib, 59
Pliny (the Elder), 5
ploidy manipulation, 570, *571*
pneumatic duct, 65
Poey y Aloy, Felipe, 11
poikilotherms, 409, 643
poisonous fishes, 816–18. *See also*
    venomous fishes
polar cartilage, 50
polarization of characters, 81
polarized light, 359
polar waters. *See* Antarctica; Arctic
    waters
pollution. *See* chemical pollutants
polychlorinated biphenyls (PCBs),
    840
polycyclic reproduction, 523
polygenic traits, 555, 574
polymerase chain reaction (PCR),
    83, 563–65, *564*
polyplocont dentition, 139–40, *141*
polyploidy, 568, 570, *571*
polyspermy, 535
popeye, 814
population ecology, 648

population genetics, 584–85
populations, 648
        parasites as "tags," 813
porphyropsin, 351
portal heart, 93, 454
postbranchial circulation, 458–59
posterohyal bone, 58
postlarvae, 537
postsplenial bone, 58
posttemporal bone, 62
potadromous fishes, 763
        migration, 766–67
potamon, 670
prearticular bone, 58
predator inspection behavior, 777
predator-prey interaction models,
    653–54
premaxillae, 33, 58
preopercle, 59
primitive meninx, 494
primordial germ cells, 518
proglottids, 809
prohaptors, 808
prolactin, 504
prolarvae, 537
promoters, 554
prootic bones, 54
proprioceptors, 370
prosencephalon, 494–96
protein, dietary, 412
protists, 805–7
proto-jaws, 106
protostomes, 21
protozoans, 806–7
pseudobranch, 443, 444, *446*
pseudobranchial neurosecretory
    gland, 508
pteridines, 292
pteropterygium, 61
pterosphenoid bones, 56–58
pterotic bone, 56
pterygiophores, 61
pulmonalis fold, 456
punctuated equilibrium, 80
purines, 292
purse seines, 832
pyloric caeca, 65, 419, *419*, 431

quadrate bone, 51, 58
quantitative genetics, 574–77

radii (of scales), 48
Rafinesque, Constantine, 7, *8*
rajiiform swimming, 327
ram irrigation, 450
Raney, Edward, 10
raphe, 391
Rathke's pouch, 501
Ray, John, 6
rays (of fins), 29, 37, *40*
reactor theory, 430

realized heritability, 575
recessive alleles, 557
recoil aspiration, 152
recombination (alleles), 565
recreational fisheries, 837
        and eutrophic lakes, 685
        and exotic species, 842
recruitment, 835
rectal gland, 476
rectin, 480
red blood cells, 461–63
rediae, 808
red muscle, 312, 314–15, *315*
Red Queen hypothesis, 802
red tide, 806
reef environments, 705–10
Reelfoot Lake, 682
Regan, C. Tate, 11
renal corpuscles, 482
renal portal system, 454
renin, 508
renin-angiotensin system (RAS),
    508
replication (DNA), 552
reproduction, 518–43
        deep-sea adaptations, 740, 741
        hermaphroditism and sex rever-
            sal, 531–33, 573–74
        mate selection, 524, 779–80
        and migration, *762*, 762–68,
            779
        and olfaction, 396
        parental care, 782–83
        reproductive strategies, 521–22,
            648–50
        and schooling behavior, 785
        sex determination, 572–74
        sexual selection, 78–79, 781–
            82, 813
        and sound production, 384
        territorial behavior, 780
reproductive character displace-
    ment, 75
resource polymorphism, 777. *See
    also* trophic polymorphism
respiration, 446–51
        in mudskippers, 261
        in polypteriforms, 151–52
        recoil aspiration, 152
respiratory volume, 451
respirometer, 409
restricted movement paradigm, 766
restriction fragment length poly-
    morphism (RFLP) analysis, 83,
    562, 563, *563*
rete mirabile, 336
retes, 459–61
reticulate evolution, 79
retina, 346
retinomotor mechanisms, 354–55
retractor dorsalis, 207

retroarticular bone, 58
Reynolds number, 287–88
rheotaxis, 759–60
rhithron, defined, 665
rhizopodans, 806
rhodopsin, 351
rhombencephalon, 494, 497
rhomboid scales, 286
ribonucleic acid, 554
ribosomes, 554
Richardson, John, 8
RNA (ribonucleic acid), 554
Robins, C. Richard, 11
rods, of eye, 348, 350–51
Rondelet, Guillaume, 5, *6*
Root effect, 336, 463, *464*
rostrum
        in rays, 51
        in sawfishes, 123–24, 285
        in sawsharks, 123, 285
        in sharks, *35*, 51
        in swordfishes, 263
routine metabolic rate (RMR), 408,
    409(t)
r-selection, 649–50, 649(t)

sacculus, 364
saccus vasculosus, 496
safe harbor hypothesis, 782
sagitta, 365
sagittiform body, 35, *38*
salinity
        freezing point depression, 470,
            470(t)
        inland waters, 695–96
        and marine fish distribution, 625
        of oceans, average, 470, 645(t)
        and osmotic regulation, *471*
        and reproduction, 479–80
        in selected bodies of water,
            645(t)
        and solubility of oxygen, 443(t)
        See also abiotic environment
saltatorial ontogeny, 538
saltmarshes, 716
Salton Sea, 696
salt regulation, *474*, 475(t)
        in euryhaline fishes, 478–79
        in fertilized eggs, 479–80
        in freshwater fishes, 472, 473
        in marine fishes, 473–78
Salviani, Hippolyte, 5
sapid substances, 397–99
saprophagous, 424
Sargassum communities, 628, 736
scales, 48
        of acanthodians, *131*
        in chondrichthyans, 285
        in living coelacanths, 287
        phylogeny of types, *286*
scapular cartilage, 61

scaritoxin, 816
schematochromes, 292
schooling, 783–87, *785*
    and drag reduction, 290–91, 786
    and parasites, 813
schreckstoff, 178, 397, 761
scolex, 809
scoliosis, *413*
scombroid poisoning, 817
scotopic eyes, 354
scuba diving, 837
scutes, 48, *49*
    in actinopterygians, 149
    in gasterosteiforms, 246
    in teleosts, 172
seagrass beds, 718–20, 720(t)
seamounts, 738–40
sea surface temperature, 705, *708*
secondary circulatory system, 459
selective breeding, 533–34
    and quantitative genetics, 575–77
semelparity, 199, 522–23
semicircular canals, 364
semilunar valves, 454
semiochemicals, 390
sensory-bias hypothesis, 781
sensory biology, 345–76, 389–401
sensory canals, 34, *36*
serotonin, 501
sex determination, 530, 572–73
    all-female species, 574
    hermaphroditism, 531–33, 573–74
sexual dichromatism, 531
sexual dimorphism, 44, 530–31
    in batrachoidids, 380
    in bowfins, 157
    in cichlids, 693
    in gymnotiforms, 189
    and predation, 78
sexual parasitism, 225, 531
sexual selection, 78–79, 781–82, 813
shark attacks, 27–28, 119, 120
shell gland, 519
shoals (of fish), 783, 786
shore and shelf fishes
    African coast, 629
    Antarctic region, 632–33
    Arctic region, 632
    Atlantic coast, North America, 629
    East Indian region, 629–30
    Florida peninsula, 629
    Japan and Taiwan, 631
    Pacific coast, California, 628–29
    western Pacific, 631
shoreline (coastal) lakes, 682, *683*
sigmoid growth curve, 410, *410*

simple tandem repeat polymorphisms (STRPs), 83
sinus venosus, 454
skeletal system, 48–64
skeletal tissue preparation, 14
skin, 44–48, *47*, 284–85, *285*
    and hydrodynamics, 289–90
    scales, 285–87
skull, 48–50
    in actinopterygians, 148
    in bony fishes, 54–59, *55*, *56*
    in clasmobranchs, 51–54, *53*
    in living agnaths, 51, *52*
slime, epidermal, 47
Smith, J. L. B., 11
smooth muscle, 64
Snake River, 831
snout, 33
Snyder, John O., 9
soft-rayed fishes, 29, *40*
solitary chemosensory cells, 399–400
solubility, of atmospheric gases, 442(t)
solution lakes, 682
somatolactin, 504
somatostatin, 506
sound detection, 370–73
sound location, 373
sound production, 380–84
    miscellaneous noises, 383–84
    stridulation, 380, 381, *383*
South American fish fauna, 673–74, *674*
Southern Ocean, 743–44
spawning, 525–26, 781
    finding mates, 524
    iteroparity, 523–24
    reproductive strategy and, 521–22
    semelparity, 522
speciation, 74–79
    allopatric, *76, 77*
    genetics, 583–84
species, 74–79
species flocks, 75–78, 689
specific dynamic action (SDA), 408
specific gravity, 332
    and swim bladders, 332
specimen, type, 14
specimen collection and preservation, 13–14
spectral sensitivities, 352(t), 353(t), 359
speed, of swimming, 327–29
spermatozoa, 519–20, 535
sperm sac, 519
sphenotic bone, 56
spherical aberration, 352
spinal cord and nerves, 498
spines, 34, *36*, 37

for locomotion, 44, 264
    venomous, 44, 47, 305–7, *307*
spinoid scales, 287, *288*
spinous scales, *293*
spiny-rayed fishes, 29, *40*
spiny scales, 287, *288*, *292*
spiracles, 34–35, 51
spiral valve, 65, 419
splanchnocranium, 50–51, *53*
spleen, 421
splenial bone, 58
sporocysts, 808
sporozoans, 806, *807*
sport fisheries. *See* recreational fisheries
squalamine, 122
squaline, 333
squamous cells, 46
S-starts, 327, *328*
stabilizing selection, 72, *73*
stable isotope analysis, 654
standard metabolic rate (SMR), 408
stanniocalcin, 508
Starks, Edwin C., 9
static equilibrium, 374
statistical methods, 13
Steindachner, Franz, 11
Steller, Georg, 8
stenohaline fishes, 644
stenotherms, 643
stereocilia, 365
stewardship, 845–46
stomach, *418*, 418–19
    digestive functions, 430–31
    vs. gizzard, 431
stratum argentum, 294, 356
stratum compactum, 48, 284–85
stratum spongiosum, 48
stream capture, 605–6
stream environments
    food webs, riverine, 674–77
    freshwater discharge, by latitude, 673(t)
    temperate zone, 665–72
    tropical, 672–74
streamlining, 287
stride length, of swimming, 329
stridulation, 380, 381, *383*
subcarangiform swimming, *320*, 320–21
subcutis, 48
subopercle, 59
suborbital stay, 270
subsistence fisheries, 200–203, 824, 846
substance P, 507
suction feeding, 428, 776
superfetation, 529
superficial neuromasts, 367
superscapular cartilage, 61

supraangular bone, 58
supracleithrum, 62
supraethmoid bone, 58
supramaxillae, 33
supraoccipital bone, 54
surf zones, 722–26
suspensorium, *55*
sustainability, 824
Suttkus, Royal, 10–11
swim bladders, 65, 332–37, *337*
    for aerial respiration, 453
    and habitat, 13
    otophysic connection, 164
    physostomous v. phsyoclistous, 162
    and sound production, 380–82, *383*
    and vertical migration, 736–37
swimming, 318–29, *320*
    and hydrodynamics, 287–91
    "kick-and-glide," 288–89
    schooling and drag reduction, 290–91
symbiosis, 787–90
    cleaner fish, 788–90
    and parasitism, 802
    spinous cells on *Coryphaena hippurus*, 284
sympatric speciation, 74, 75–79
symplectic bone, 58
synapomorphies, 81
synchronous spawners, 520
syncranium, 48–50
synotic tectum, 50
synplesiomorphies, 81
syntenies, 565
systematics, 80–84
    computational phylogenetics, 84–85
systemic heart, in hagfishes, 454
systolic pressure, *458*

taeniform body, 35, *38*
tapetum lucidum, reflective, 356, 741
tapeworms, 809, *809*, 815
taste, 397–99
    taste buds, *398*
taxes (movements), 756–61
taxonomy, 5, 6–7, 15, 15(t)
    equivalence of taxa, 610–11
    and systematics, 12, 80
tectonic lakes, 682
tectonic plates, 606, *630*
tectospondyly, 59
teeth, 414–15, *416*, *417*
telencephalon, 494
temperate lentic environments, 684–85
temperate stream environments, 665–72

tenaculae, 113
terminal innervation, 316
terminal nerve, 498
territorial behavior, 780
tesserae, 108
testes, 65, 518–20
tetradontiform swimming, 326
tetrads, 555
tetrapods
    origin of, 138–42, *140*, 488–89
    phylogenetic studies, 581–83,
      582
    terrestrial adaptations, 142–43
tetrodotoxin, 276, 817–18
thermal regulation, 409, 459–60
    adaptations, 643–44
thermocline
    in lakes, 684
    in oceans, 733
thigmotaxis, 759
thread cells, 92
thrombocytes, 464
thunniform swimming, *320*, 321–25
thymus gland, 463, 508
thyroid gland, 505
thyrotropin (TSH), 504
thyroxine, 505
tonic (red) muscle fibers, 314–15
top-down regulation, 652
torus longitudinalis, 497
torus semicircularis, 497
total phenotypic variability, 575,
    576
toxins, fish-borne, 816(t)
trabecula, 50
transcription (RNA synthesis),
    554, *554*
transgenic research, 590–91, 839
trawling, 833, 836
trematodes (flukes), *803*, 808–9,
    815
Trewavas, Ethylwynn, 12
trigeminal nerve, 498
tri-iodothyronine, 505
tripartite condyle, 207
triploblasty, 20, 536
triploidy, 570, *571*

trochlear nerve, 498
troglobites, 694
troglophiles, 694
trogloxenes, 694
trophic cascades, 652–53, *653*
trophic polymorphism, 79–80
    in lake habitats, 688–89
trophic relationships, *422*, 651–54
    riverine environments, 674–77
trophonemata, 527
trophotaeniae, 245, 530
tropical freshwater fisheries,
    831–32
tropical lentic environments,
    685–86
tropical stream environments,
    672–74
trunk, 33
    musculature, 312–16
truss diagrams, 13, *13*
tubal bladder, 65
tuberous electroreceptors, 369
twitch (white) muscle fibers,
    314–15, *315*
type specimens, 14
typhlosole, 65, 419

ultimobranchial gland, 506
ultraviolet (UV) radiation
    damaging effects, 296
    visual sensitivity, 359
unculi, 178
undulation, 326
unicellular glands, 284
United Nations Food and Agricul-
    tural Association (FAO), 825
urinary system, 481–88, *487*
urogenital organs, 65
urogenital sinus, 65, 519
urogenital system. *See* reproduction;
    urinary system
urohyal bone, 59, 160
uroneural bone, 60
urophysis, 508
urostyle, 60
urotensin I and II, 508
utriculus, 364

vagus nerve, 498
Valenciennes, Achille, 7
valvula cerebelli, 497
variation and adaptation, 72
vas deferens, 519
veins, 453
velar chamber, 443
venomous fishes, 305–7
    epidermal venoms, 44, 47
ventral aorta, 456
ventral fins. *See* pelvic fins
ventricle, 454
vertebrae, 59–60, *60*
    in eels, 170, 171, 172, 189
    opisthocoelus, 156
    in teleosts, 161
vertebral centra, 59
vertebral column, 59–60
vertebrates
    evolutionary history of, 90
    origin of, 20–24
    phylogenetic studies, 579–81,
      580
    phylogeny, *90*
vertical migration, 768
vertical movement, 334–35,
    736–37
vesicles of Savi, 370
vesicula seminalis, 519
vestibule, of inner ear, 364
vexilla, 223
vibriosis, 805, 815
vicariance, 74–75, 604
viral hemorrhagic septicemia
    (VHS), 804–5
viruses, as pathogens, 804–5
viscera, 64–65
vision, 345–59, 352(t), 353(t)
    and communication, 778
    and foraging, 774–75
    and schooling, 784
visual field, 348, *349*
vitamins, dietary, 412–13
vitreous chamber, 346
vitrodentine, 48
viviparous fishes, 527–30
    Baikal oilfishes, 272

brotulas, 223
eelpouts, 257
sharks, 734
surfperches, 255
topminnows, 245
volcanic lakes, 682
vomer, 58
vulvar velum, 442

Walbaum, Johann Julius, 8
Wallace's Line, 610, *616*, *617*,
    621
water
    atmospheric gas solubility, 32,
      442(t), 443(t), 644–46
    density, 32, 644
    distribution in hydrosphere,
      664, 664(t)
    freezing point depression, 470,
      470(t), 643
    light absorption, 646, *647*
    optical properties, 348, *349*
    specific gravity, 644
    specific heat, 643
    See also abiotic environment
water consumption, human,
    843
water pollution. *See* chemical
    pollutants
watersheds, 657
Weberian apparatus, 178
West Wind Drift, 743
whirling disease, 808
white blood cells, 463–64
Whitehead, Peter, 11–12
white muscle, 312, 314–15, *315*
Willoughby, Francis, 6

xanthophores, 292

Zaproridae (prowfish), 258
zoogeography. *See* biogeography
zoonoses, 815–16
zooxanthellae, 706
zygapophyses, 59
zygotes, 555

SYSTEMATIC INDEX

Note: *Italic* page numbers indicate illustrations; (t) indicates material in a table.

Abramis brama (bream), 672
Abyssobrotula galatheae, 223
Abyssocottidae, 272, 614
Acadian redfish. *See Sebastes fasciatus*
Acanthocybium (wahoos), 263, 321
    A. solandri, 328, 329(t)
Acanthodiformes, 131
Acanthodii, 130–31, *131*
Acanthogobius flavimanus (yellowfin goby), 842
Acanthomorpha, 212–13
Acanthonus armatus, 223
Acanthopterygii, 229–46
    phylogeny, *231*
Acanthostega, 139, 140, 141
Acanthothoraciformes, 108
Acanthuridae (surgeonfishes), 262, 707
Acanthuroidei, 261–62, 707
Acanthurus (surgeonfishes), 262
    A. olivaceus, 425
    habitat, 710
    parasites of, 810
    stridulation, 381
Achiridae (American soles), 274
Achoania, 131
Acipenser (sturgeons)
    A. brevirostrum (shortnose sturgeon), 154
    A. fulvescens (lake sturgeon), 154
    A. gueldenstaedti (Russian sturgeon), 152, 154
    A. medirostris (green sturgeon), 154
    A. naccarii, 568(t)
    A. oxyrhynchus (Atlantic sturgeon), 84, 154
    A. ruthenus (sterlet), 152
    A. stellatus, 152
    A. sturio (European sturgeon), 84

    A. transmontanus (white sturgeon), 154
    gills, *445*
    kidneys, 484
    and migration, 765
    spiracles, 34
Acipenseridae (sturgeons), 153–54, *155*
    caudal fins, *43*
    commercial fishery, 154
    distribution, 611, 614
    feeding, 154
    pelvic fin placement, *46*
    pyloric caeca, 419
    scales, 48
    scutes, 48, *49*
Acipenseriformes, 153–56, *155*
Acipenseroidei, 153–56
Acrocheilus (chiselmouths), 425, 667
Acropomatidae (lanternbellies), 303
Actinistia, 28
Actinopterygii, 28, 147–57
    feeding versatility, 148
    origin of, 148
    phylogeny, *150*
    pituitary gland, 504
    scales, 48
    teeth, 415
Adrianichthyidae, 243
Adrianichthyoidei, 243
Adrianichthys, 243
Aetobatus (rays), 126
African lungfish. *See Protopterus*
Agnatha, 25, 27, *91*
    chromosome and arm numbers, 568(t)
    feeding apparatus, 426
    gills, 442–43, *444*
    ionic balance, 475(t)
    islet organ, 506
    pituitary gland, 502–3
Agonidae, 35, 272
    habitat, 704, 741
    scutes, 48
Agonostomus (mullets), 623, 765

Alabama shad. *See Alosa alabamae*
Alabama sturgeon. *See Scaphirhynchus suttkusi*
Alabes, 259
Alabetidae, 258
Alaskan blackfish. *See Dallia pectoralis*
albacore. *See Thunnus alalunga*
Albula (bonefishes), *167*
    A. nemoptera, 167
    A. vulpes, 167
Albulidae, 167
Albuliformes, 167, *167*
Alepisauridae (lancetfishes), 211, 739(t)
Alepisaurioidei, 211–12
Alepisaurus (lancetfishes), *211,* 737, 739(t)
Alepocephalidae, 196, 739(t)
Alepocephaloidei, 196
alewife. *See Alosa pseudoharengus*
alligatorfish. *See Agonidae*
alligator gar. *See Atractosteus spatula*
Alloclinus holderi, 585
Allotriognathi, 212
Alopias (thresher sharks), 119, 735(t)
Alopiidae (thresher sharks), 119, 735(t)
Alosa (shads), *168*
    A. alabamae (Alabama shad), 764
    A. chrysochloris (hickory shad), 764
    A. pseudoharengus (alewife)
        chromosome and arm numbers, 568(t)
        feeding modes, 425
        lateral line system, 374
        thiaminase in, 413
        trophic interactions, 652
    A. sapidissima (American shad), 174, 830
    distribution, 613
    habitat, 686
    hearing, 372

Aluterus (filefishes), 298, 736
    A. scriptus (scrawled filefish), 276
Amarsipidae, 263
Amarsipus carlsbergi, 263
Amazon molly. *See Poecilia formosa*
Amazonsprattus scintilla, 174
Ambassidae, 722
Ambassis
    A. gymnocephalus, 721
    A. productus (glassperch), 721
Ambloplites (rock basses), 686, 757
Amblycipitidae (catfishes), 621
Amblyopsidae, 222, 615
Amblyopsiformes, 222
Amblyopsis (cavefishes), 222, 694
Ameiurus (bullheads), 188, 399
    A. natalis, 397
    A. nebulosus (brown bullhead), 35, 371
    digestion, 433(t)
    glomeruli, 488(t)
    habitat, 672, 687
    internal organs, *420*
    pectoral fins, 44
American eel. *See Anguilla rostrata*
American plaice. *See Hippoglossoides platessoides*
American shad. *See Alosa sapidissima*
American sole. *See Achiridae*
Amia calva (North American bowfin), *155,* 157
    aestivation, 453
    air breathing, 453
    amiiform swimming, 326
    distribution, 615
    gonads, 519, *520*
    habitat, 687
    inner ear, 364
    kidney, 484
    spiral valve, 419
    vertebrae, 59
Amiidae (bowfins), 157
    distribution, 615
    fins, 41, *43*
Amiiformes, *155,* 156, 157

Ammocrypta (darters), 667
Ammodytes (sandlances), *256*, 722
Ammodytidae (sandlances), 257, 320
Amphiliidae, 622
Amphipnous cuchia, 452, 453
Amphiprion (anemonefishes)
  breeding behavior, 782
  hermaphroditism, 573–74
  mucus, 47
  symbiosis, 708, 788–89
Amphistichus
  A. argenteus (barred surfperch), 722
  A. rhodoterus (redtail surfperch), 722
Amur pike. *See Esox reicherti*
Anabantidae (climbing perches), 264
  distribution, 614
  sound production, 384
Anabantoidei, 264
Anabantomorpha, 264–65
Anabas (climbing perches), *260*, 264
  A. testudineus, 447
  aestivation, 453
  air breathing, 452
  terrestrial locomotion, 330
Anablepidae (four-eyed fishes), 245, 619
Anableps, 351–52, 358, *358*
Anarhichadidae (wolffishes), 258, 330, 775
Anarhichas lupus, 258
Anarrhichthys (wolf eels), 415, *416*, 427
  A. ocellatus, 258
Anaspida, *99*, 99–100
Anchariidae, 623
anchoveta. *See* Engraulidae
anchovy. *See* Engraulidae
anemonefishes, 47. *See also* Pomacentridae
angelfish. *See* Pomacanthidae
angel shark. *See* Squatiniformes
Anglaspis, 98
angler catfish. *See Chaca chaca*
anglerfish. *See* Ceratioidei; Lophiiformes
Anguilla, 169–71
  A. anguilla (European eel), 170
  A. japonica, 508
  A. rostrata (American eel), 170, 330, *393*, 568(t)
  A. vulgaris, 328, 329(t)
  air breathing, 452
  anguilliform swimming, 319–20
  balistiform swimming, 326
  body shape, 35
  clades, 170
  commercial fishery, 171

cutaneous respiration, 451
elver stage, 169, 171, 760
eyes, 353
gas resorption, 336
larval, *539*
life history, 169–70
migrations, 765, *766*
olfaction, 394
olfactory organ, 391
olfactory threshold, 395, 395(t)
osmoregulation, 481
phototaxis, 757
pigmentation, 295–96
plasma osmolality, 479(t)
retial capillaries, 336
skin, 473
terrestrial locomotion, 330
Anguillidae, 170
  body shape, *38*
  habitat, 704, 736
Anguilliformes, *168*, 169–72
Anguilloidei, 170
Anodus, 184
Anomalopidae (flashlightfishes), 233, 300
Anomalops kaptotron, 303
Anoplogaster (fangtooths), 327
Anoplogastridae (fangtooths), 233
Anoplopoma fimbria (sablefish), 271, 626
Anoplopomatidae (sablefishes), 271, 704
Anoptichthys (Mexican blind cave-fishes), 694–95
Anostomidae, 184
Anotophysi, 178–79
Anotopteridae (daggertooths), 211, 739(t)
Anotopterus pharao (daggertooth), 211, 737, 739(t)
Antarctic dragonfish. *See* Bathydraconidae
Antarctic sculpin. *See* Harpagiferidae
antenna codlet. *See Bregmaceros atlanticus*
Antennariidae (frogfishes), 224, 737
Antennarioidei, 224–25
Antennarius nummifer, 224, 428
Antiarchiformes, 108, *109*
Apeltes quadracus (fourspine stickle-back), 238, 720
Aphanius, 625
Aphanopus carbo (black scabbard-fish), 335
Aphiosemion, 244
Aphos, 224
Aphredoderidae, 218–19, 615
Aphredoderus sayanus (pirate perch), 218–19, 671

Aphyonidae, 742
Aphyosemion, 536
Aplocheilidae, 244, 622, 688
Aplocheiloidei, 244–45
Aplochitonidae, 766
Aplochiton marinus, 198
Aplodinotus (drums)
  A. grunniens, 380, 382, 384
Apogon, 253, 782
Apogonidae (cardinalfishes), 253
Apteronotidae (knifefishes), 189, 377, 675
Apteronotus (knifefishes), *188*, 379, 380
Arandaspida, 97
arapaima. *See Arapaima*
Arapaima (arapaimas), *165*, 453
  A. gigas, 164, 831
archerfish. *See Toxotes jaculator*
Archoplites interruptus (Sacramento perch), 253, 425, 615
Arctic char. *See Salvelinus alpinus*
Arctic flounder. *See Liopsetta glacialis*
Arctic grayling. *See Thymallus arcticus*
Arctic lamprey. *See Lampetra japonica*
Argentina sialis (Atlantic argentine), 739(t)
argentine. *See* Argentinidae
Argentinidae (argentines), 195, 739(t)
Argentiniformes, 194–96
Argentinoidei, 195, *195*
Argyropelecus (hatchetfishes)
  color filters, 304
  eye pigment, 296, 357
  habitat, 739(t)
Ariidae (marine catfishes), *383*, 704
Ariommatidae, 264
Ariommus, 264
Arius graeffei, 369
armored catfish. *See* Callichthyidae
Arothron, 818
arrow goby. *See Clevelandia ios*
Artedius, 790
Arthrodiriformes, 108–9, *109*
Aspidontus taeniatus, 257, 299
Aspidorhynchiformes, 156
Astraspida, 97
Astronesthinae (snaggletooths), 210
Astronotus ocellatus (red oscar), 255
Astroscopus (stargazers), 256
  electric organs, 376, *378*, 379
  habitat, 722
Astyanax
  A. fasciatus, 695
  A. mexicanus, 399, 568(t), 694–95, 762

Ateleaspis, 102
Ateleopodidae, 210
Ateleopodiformes, *209*, 210
Ateleopus, *209*
Atherina (silversides), 240, *241*
Atherinidae (silversides), 240, 330, 623, 695
Atheriniformes, 240–42, *241*
Atherinoidea, 240–41
Atherinomorpha, 240–46, 716
Atherinops affinis, 717
Atherinopsidae, 240, 619, 717, 722
Atherinopsis californiensis, 240
Atherinopsoidea (silversides), 240
Atka mackerel. *See Pleurogrammus monopterygius*
Atlantic argentine. *See Argentina sialis*
Atlantic bigeye tuna. *See Thunnus obesus*
Atlantic bluefin tuna. *See Thunnus thynnus*
Atlantic cod. *See Gadus morhua*
Atlantic cutlassfish. *See Trichiurus lepturus*
Atlantic eel, 381, 739(t)
Atlantic flounder. *See Pleuronectes flesus*
Atlantic gunnel, 35
Atlantic hagfish. *See Eptatretus stoutii; Myxine glutinosa*
Atlantic halibut. *See Hippoglossus hippoglossus*
Atlantic herring. *See Clupea pallasi*
Atlantic midshipman. *See Porichthys plectrodon*
Atlantic mudminnow. *See Umbra krameri*
Atlantic salmon. *See Salmo salar*
Atlantic saury. *See Cololabis saira; Scomberesox saurus*
Atlantic silverside. *See Menidia menidia*
Atlantic smoothtongue. *See Leuroglossus stilbius*
Atlantic sturgeon. *See Acipenser oxyrhynchus*
Atlantic tomcod. *See Microgadus tomcod*
Atlantic viperfish. *See Chauliodus macouni*
Atractosteus (gars), 156
  A. spatula (alligator gar), 156
Aulopidae (threadsails), 210
Aulopiformes, 210–12, *211*, 626
Aulopoidei, 210
Aulopus purpurissatus, 210, 427
Aulopyge, 694
Aulorhynchidae (tubesnouts), 238

Aulorhynchus (tubesnouts), *237*
  A. flavidus, 238
Aulostomidae (trumpetfishes), 239,
  326, 776
Australian grayling. *See Prototroctes*
  maraena
Auxis (frigate mackerels), 263
  A. thazard, 735(t)
ayu. *See Plecoglossus*

Badis badis, 778
Bagridae, 621
Baikal oilfish. *See Comephoridae*
Bajacalifornia drakei, 354
Balantiocheilos melanopterus (bala
  shark), 838
bala shark. *See Balantiocheilos*
  melanopterus
Balistidae (triggerfishes), 276
  balistiform swimming, 326
  gas bladder, 373
  habitat, 707, 709
  stridulation, 381
  tetraodontiform swimming, 326
Balistoidea, 276
Balitoridae, 182
  distribution, 614–15
  habitat, 667
  sucking disc, *668*
Balitorinae (flat loaches), 182
balloonfish. *See Diodon holocanthus*
barb. *See Puntius javanicus*
barbel. *See Barbus*
Barbourisia rufa (red whalefish),
  232, *741*
Barbourisiidae, 232
Barbus (barbels), 672, 673
bar jack. *See Caranx ruber*
barracuda. *See Sphyraenidae*
barracudina. *See Paralepididae*
barramundi perch. *See Lates*
  calcarifer
barred surfperch. *See Amphistichus*
  argenteus
barreleye. *See Opisthoproctidae*
basking shark. *See Cetorhinidae*
bass. *See Moronidae*
basslet. *See Grammatidae*
Bassogigas profundissimus, 743
Bassozetus, 743
batfish. *See Ogcocephalidae*
Bathychaunax, 225
Bathyclupeidae, 254
Bathydraconidae (Antarctic dragonfishes), 258
Bathygobius soporator (frillfin goby),
  725, 762
Bathylagidae, 195, 739(t), 741(t)
Bathylagus antarcticus (blacksmelt),
  741(t)

Bathylychnops exilis, 358, 416
Bathymasteridae (ronquils), 258
Bathypterois (tripodfishes), 210,
  288, 741
Bathysauroides, 210
Batoidea, 123
Batrachoididae (toadfishes), 35, 44,
  223–24
  sound production, 382
  venomous, 306
Batrachoidiformes, 220, 223–24
Batrachoidinae (toadfishes), 223
Batrachomoeus, 223
bay goby. *See Lepidogobius lepidus*
beardfish. *See Polymixiidae*
Bedotiidae, 240–41, 623
Belantsea, 111, *113*
Belone (needlefishes), *241*
Belonidae (needlefishes), 243
  anguilliform swimming, 320
  distribution, 621
  jumping, 331
Beloniformes, *241*, 242–44
Belontiidae, 264, 400, 780
beluga sturgeon. *See Huso huso*
Benthalbella dentata (northern pearleye), 739(t)
Benthosuchus, *141*
bent-tooth. *See Champsodontidae*
Berycidae, 233
Beryciformes, 233–34, 626
Berycoidei, 233
Betta splendens (Siamese fighting
  fish), 264, 557–58, 780, 782
bichir. *See Polypteridae*
bigeye. *See Priacanthidae*
bigeye sand tiger shark. *See Odon-*
  taspis noronhai
bigeye squaretail. *See Tetragonurus*
  atlanticus
bighead carp. *See Hypophthalmich-*
  thys nobilis
bignose shark. *See Carcharhinus*
  altimus
bigscale. *See Poromitra*
bigscale pomfret. *See Taractes*
  longipinnis
billfish. *See Istiophoridae; Xiphiidae*
bitterling. *See Rhodeus*
black arowana. *See Osteoglossum*
  ferreirai
black carp. *See Mylopharyngodon*
  piceus
blackchin. *See Neoscopelidae*
black cod. *See Anoplopoma fimbria*
  (sablefish)
blackfin reef shark. *See Carcharhinus*
  melanopterus
black mackerel. *See Scombrolabrax*
  heterolepis

black marlin. *See Istiompax*
black prickleback. *See Xiphister*
  atropurpureus
black scabbardfish. *See Aphanopus*
  carbo
black sea bass. *See Centropristis*
  striata
blacksmelt. *See Bathylagus*
  antarcticus
blacksmith. *See Chromis punctipinnis*
black surfperch. *See Embiotoca*
  jacksoni
black swallower. *See Chiasmodon*
  niger
blacktip shark. *See Carcharhinus*
  limbatus
Blenniidae, 257, 707, 725
Blennioidei, 257
  fins, 44
  homing behavior, 761, 762
  kidneys, 485
blenny. *See Blennioidei*
blind goby. *See Typhlogobius*
blue catfish. *See Ictalurus furcatus*
blue-eye. *See Pseudomugilidae*
bluefin tuna, *460*
bluefish. *See Pomatomidae*
bluegill. *See Lepomis macrochirus*
blue marlin. *See Makaira*
blue shark. *See Prionace glauca*
blue whiting. *See Micromesistius*
  poutassou
bluntnose knifefish. *See Hypopomus*
  brevirostris
bluntnose minnow. *See Pimephales*
boafish. *See Ichthyococcus ovatus*
boarfish. *See Caproidae*
bobtail snipe eel. *See Cyematidae*
bocaccio. *See Sebastes paucispinis*
Bolbometapon muricatus, 708–9
Boleophthalmus, 451, 452
  B. chinensis, 261
Bombay duck. *See Harpadon*
  nehereus
bonefish. *See Albulidae;*
  Pterothrissidae
bonito. *See Sarda*
Borax Lake chub. *See Gila*
  boraxobius
Bothidae (flounders), 274, 704, 716
Bothriolepis, *109*
Bothus
  B. lunatus (peacock flounder),
    274
  B. ocellatus (eyed flounder), 275
Botia sidthimunki (pygmy loach),
  838
Bovichtidae (thornfishes), 258, 624
bowfin. *See Amiidae*
boxfish. *See Ostraciidae* (cowfishes)

Brachionichthyidae, 225
Brachymystax (lenoks), 201
Brachyplatystoma
  B. filamentosum, 832
  B. flavicans (dourada), 832
  B. vaillantii (piramutaba), 832
bradyodonts, 111
bramble shark. *See Echinorhinus*
Bramidae (pomfrets), 253, 735(t)
Bregmaceros atlanticus (antenna codlet), 739(t)
Bregmacerotidae, 739(t)
Brevoortia (menhadens), 174, 425,
  716
  B. patronus (gulf menhaden),
    539, 542, 805
bristlemouth. *See Cyclothone*
broadband dogfish. *See Etmopterus*
  gracilospinous
"brook" lampreys, 95
brook stickleback. *See Culaea*
brook trout. *See Salvelinus fontinalis*
Brosme (cusks), 222
brotula. *See Bythitidae*
Brotulina, 220
brown bullhead. *See Ameiurus*
  nebulosus
brown trout. *See Salmo trutta*
bucktooth parrotfish. *See Sparisoma*
  radians
buffalofish. *See Ictiobus*
bullhead. *See Ameiurus*
bull shark. *See Carcharhinus leucas*
bull trout. *See Salvelinus confluentus*
burbot. *See Lota lota*
buri. *See Seriola*
burrowing goby. *See Trypauchen*
butterfish. *See Stromateidae*
butterflyfish. *See Chaetodontidae;*
  Pantodon buchholzi
Bythitidae (brotulas), 223, 742, 743
Bythitoidei, 223

calcichordates, 21
California flyingfish. *See Cypselurus*
  californicus
California grunion. *See Leuresthes*
  tenuis
California halibut. *See Paralichthys*
  californicus
Callichthyidae (armored catfishes),
  48, 187, *188*
Callionymidae (dragonets), 44, 259
Callionymoidei, 259
Callionymus (dragonets), *260*
Callorhynchidae (plownose chimae-
  ras), 114
Callorhynchus, 114
Campostoma (stoneroller minnows),
  181, 778

C. anomalum (central stoneroller), 72
grazing, 425
habitat, 667
trophic interactions, 652
candiru. *See Vandellia*
Cantherhines (filefishes), 736
Canthigaster, 485
capelin. *See Mallotus*
Caproidae (boarfishes), 233
Carangidae, 253
  carangiform swimming, 321
  and ciguatera, 817
  habitat, 707, 726, 735(t), 736
  scutes, *49*
Caranx ruber (bar jack), 787
Carapidae (pearlfishes), 223, 790
Carapinae, 223
Carassius
  C. auratus (goldfish), 181
    anaerobic metabolism, 409
    digestion, 433(t)
    hearing, 371
    intestine length, 432(t)
    maneuverability, 318
    metabolic rate, 409(t)
    olfactory threshold, 395, 395(t)
    subcarangiform swimming, 321
    swimming speed, 329(t)
    thyroid gland, 505
    urine production, 473
  C. carassius (crucian carp), 400
Carcharhinidae (requiem sharks), 120, 704, 735(t)
Carcharhiniformes, 120
Carcharhinus
  C. altimus (bignose shark), 829(t)
  C. brachurus (narrowtooth shark), 829(t)
  C. brevipinna (spinner shark), 829(t)
  C. falciformis (silky shark), 829(t)
  C. galapagensis (Galapagos shark), 829(t)
  C. isodon (finetooth shark), 829(t)
  C. leucas (bull shark), 28, 120, 473, 686, 829(t)
  C. limbatus (blacktip shark), 829(t)
  C. melanopterus (blackfin reef shark), 424
  C. obscurus (dusky shark), 829(t)
  C. perezi (reef shark), 397, 829(t)
  C. plumbeus (sandbar shark), 704, 829(t)
  C. signatus (night shark), 829(t)

Carcharodon
  C. carcharias (white shark), 119, 735(t), 775, 829(t)
  C. megalodon, 119
cardinalfish. *See* Apogonidae
Careproctus (sea snails), 526, 743, 790
Carnegiella, 184
carp. *See Cyprinus*
Carpiodes, 182, 672
carpoids, 21, *21*
carpsucker. *See Carpiodes*
catalufa. *See* Priacanthidae (bigeyes)
catfishes
  Amazon Basin fishery, 832
  ampullary elecrtoreceptors, 368
  aquaculture, 838
  brain, 497
  distribution, 621
  habitat, 673, 675, 688
  hearing, 373
  madtom, 44
  migration, 766
  nares, 394
  olfactory organ, *392*
  olfactory threshold, 395(t)
  pectoral fins, 42, 44
  phylogeny, *186*
  stridulation, 381
  taste buds, 397
  USDA 103 variety, 577
  venomous, 305–6
  See also Siluriformes
catla. *See Gibelion catla*
Catlocarpio siamensis, 181
Catostomidae, 182
  chromosome and arm numbers, 569(t)
  distribution, 611, *612*, 613, 615
  gas bladder, 333
  grazing, 425
  habitat, 670, 686
  heterospecific shoaling, 786
  nostrils, *37*
  pectoral fins, 44
  pharyngeal teeth, 417
Catostomus (suckers), *183*
  C. catostomus, 182, 432(t)
    distribution, 613
  C. commersoni, 360, 671
  C. luxatus (Lost River sucker), 182
  C. macrocheilus, *419*, 671
  C. platyrhynchus (mountain sucker), 667
  gut, *419*
  nuptial tubercles, *532*
  species distribution, *77*
  Weberian ossicles, *372*
cat sharks, 120

cavefish. *See Amblyopsis; Speoplaty-rhinus; Typhilichthys*
Cebidichthys violaceus (monkeyface prickleback), 425, 432(t), 723, 726, 775
Celebes rainbowfish. *See Telmatherina ladigesi*
central mudminnow. *See Umbra limi*
central stoneroller. *See Campostoma anomalum*
Centrarchidae, 35, 253
  body shape, *38*
  chromosome and arm numbers, 569(t)
  distribution, 615
  habitat, 670, 672, 686
  sound production, 384
Centriscidae, 239, 707, 790
Centrobranchus, 735
Centrolophidae (medusafishes), 263–64, 735(t)
Centropomidae (snooks), 253, 688
Centropomus, 253, 688
Centropristis striata (black sea bass), 432(t), 828(t)
Cephalaspidiformes, *102*
Cephalaspidomorphi, 99–101
Cephalochordata, 21
Cephaloscyllium (swell sharks), 120
Ceratiidae
  habitat, 741(t)
  lure, *41*
  sexual parasitism, 531
Ceratioidei (anglerfishes), 225, 303
Ceratodontidae, 137
Ceratodontiformes, 137
Ceratodontimorpha, *26*
Cetomimidae (flabby whalefishes), 232
Cetorhinidae (basking sharks), 119, 735(t)
Cetorhinus (basking sharks), 119, 735(t), 776
  C. maximus, 117, 829(t)
Chaca chaca (angler catfish), 299, 424
Chacidae (catfishes), 621
Chaenocephalus aceratus, 463
Chaenopsidae (pikeblennies), 257
Chaetodipterus, 298
Chaetodon (butterflyfishes), 298, 415, *416*
  C. falcifer (scythe butterflyfish), *297*
  C. lunula, 296
Chaetodontidae (butterflyfishes), 707
Champsodontidae (bent-tooths), 257

Channa, 485
  C. argus (northern snakehead), 264
Channalabes, 330
channel catfish. *See Ictalurus*
Channichthyidae (icefishes), 258, 462
  polar adaptations, 750
Channidae (snakeheads), 47, 178, 264
  aestivation, 453
  distribution, 614, 622
  habitat, 722
  mucus, 47
Channoidei, 178–79, 264
Chanos chanos (milkfish), 178–79, *180*, 330
  crumenal organ, 416
  gizzard, 431
  herbivory, 425
char. *See Salvelinus*
Characidae, 184–85, *185*, 384, 673
Characiformes, 182–85, *185*
  classification, 183
  distribution, 621, 622
  habitat, 688
  migration, 766
  phylogeny, *184*
Chascanopsetta lugubris, 275
Chasmistes, 75
  C. brevirostris (shortnose sucker), *77*
  C. cujus (cui-ui), *77*
  C. liorus (June sucker), *77*
Chaudhuriidae, 237, 622
Chauliodontinae (viperfishes), 209, 739(t)
Chauliodus (viperfishes), *209*, 350, 424, 740
  C. macouni (Atlantic viperfish), 739(t)
Chaunacidae (sea toads), 225
Chaunacoidei, 225
Chaunax, 225
Cheilopogon pinnatibarbatus (spotted flyingfish), 244
Cheimarrhichthyidae (torrent fishes), 257, 667
Cheirolepiformes, 149
Cheirolepis, 149
Chela, 330–31
Chelmon (longnose butterflyfishes), 708
chestnut lamprey. *See Ichthyomyzon castaneus*
Chiasmodon niger (black swallower), *738*, 741(t)
Chiasmodontidae (swallowers), 257, 737, 741(t)

Chiloscyllium
   C. indicum, 35
   C. plagiosum, 414–15
chimaera. *See* Chimaeriformes
Chimaeridae, 113
Chimaeriformes (chimaeras), *26, 28,*
   112–14, *113*
   body shape, 35
   chromosome and arm numbers,
     568(t)
   digestive enzymes, 433
   egg cases, *525*
   endolymphatic duct, 365
   fins, 39, 44
   gills, 443
   habitat, 704, 742
   pelvic claspers, *528*
   reproductive anatomy, 519
   venomous, 47, 305
Chinook salmon. *See Oncorhynchus*
   tshawytscha
Chirocentridae, 173
Chirocentrus (wolf herrings), 139,
   168
   C. dorab (dorab), 173
   C. nudus, 173
Chirolophus nugator (mosshead warbonnet), *299*
Chirostoma, 240
chiselmouth. *See Acrocheilus*
Chlamydoselachidae, 121
Chlamydoselachoidea, 121
Chlamydoselachus anguineus (frill
   sharks), 121, *121*
Chlopsidae, 170
Chlorophthalmidae (greeneyes),
   210, 739(t)
Chlorophthalmoidei, 210
Chlorophthalmus (greeneyes), 357
   C. agassizi (shortnose greeneye),
     739(t)
Chologaster
   C. agassizi (spring cavefish),
     222, 694
   C. cornuta (swampfish), 222
Chondrichthyes, 27–28, 107,
   109–25
   egg cases, 524
   feeding apparatus, 426
   inner ear, 364
   kidneys, 483–84, 486–88
   phylogeny, *112*
   pituitary gland, 503
   skin, 285
Chondrostei, 28, 149–56
Chondrosteoidei, 153
Chondrostoma, 694
   C. nasus (nase), 432(t)
Chongichthys, 60
Choridactylus, 330

Chorochismus, 258
Chromis, 790
   C. punctipinnis (blacksmith), 705
Chrysichthys (catfishes), 814
chub. *See Gila*
chum salmon. *See Oncorhynchus keta*
Cichlasoma nicaraguense, 783
Cichlidae, 255
   distribution, 621, 622
   habitat, 675, 688
   nostril, 394
   pigmentation, 292
   sound production, 384
   speciation in African Great
     Lakes, 690–94
Ciliata (rocklings), 222, 400
Cirrhina mrigala, 433(t)
Citharichthys, 275, 569(t), 722
Citharidae, 273
Citharinidae (moonfishes), *185,* 622
Citharininae (moonfishes), 183
Citharinus
   C. citharus, 432(t), 831
   C. congicus, 432(t)
cladodonts, 114, *116*
Cladoselache, 116
Cladoselachiformes, *116,* 117
Clarias (walking catfishes)
   C. batrachus, 44, 187, 330, 452,
     842
   C. lazera, 453
Clariidae (labyrinth catfishes), 187
   distribution, 614, 621
   habitat, 694
   venomous, 305
cleanerfishes, 297
Clevelandia ios (arrow goby), 717
Climatiiformes, 131
Climatius, 130
climbing perch. *See Anabantidae*
clingfish. *See Gobiesocidae*
Clinidae, 257, 725
Clinocottus, 528
   C. analis (woody sculpin), 585
   C. globiceps, 425, 790
Clinus superciliosus, 529
clownfish. *See Amphiprion*
   (anemonefish)
Clupea (herrings), 416, 767
   C. harengus, 174
     blood, 462
     commercial fishery, *827*
     hatching at high salinity, 480
     noise production, 381
     swimming speed, 329(t)
     vertical migration, 335, 768
   C. pallasi (Atlantic herring),
     381, 568(t)
Clupeidae, 36, 174
   distribution, 621

   habitat, 722, 735(t)
   kidneys, 485
   lateral recess, 367
   musculature, 315
   pelagic adaptations, 733
   swim bladder, 332
Clupeiformes, 162, *168,* 173–74
Clupeoidei, 173–74, 767, 832
Clupeomorpha, 172–74
Cnidoglanis macrocephalus (cobbler),
   478
cobbler. *See Cnidoglanis*
   macrocephalus
cobia. *See Rachycentridae*
Cobitidae (loaches), 182
   distribution, 614, 622
   habitat, 672
Cobitus, 183
Coccostei, 108
Coccosteus, 109
cod. *See Gadidae*
codling. *See Moridae*
coelacanth. *See Latimeria*
Coelacanthidae, 134–35
Coelacanthiformes, *133,* 134–36
Coelacanthimorpha, 134–36
coffinfish. *See Chaunacidae* (sea
   toads)
coho salmon. *See Oncorhynchus*
   kisutch
Cololabis saira (Atlantic saury), 243,
   735(t)
Colorado pikeminnow. *See Ptycho-*
   cheilus lucius
Colossoma macropomum (tambaqui),
   185, 832
Comephoridae (Baikal oilfishes),
   272, 614, 689
Comephorus, 272
common carp. *See Cyprinus carpio*
common shiner. *See Luxilus cornutus*
conger. *See Conger conger*
Conger conger (conger), 171
Congridae, 171, 330
Congroidei, 171–72
conodonts, 24, 92
convict blenny. *See Pholidichthyidae*
cookie cutter shark. *See Isistius*
Copella arnoldi, 526
Coregoninae, 198
Coregonus (whitefishes), 201, 569(t),
   829
   C. autumnalis migratorius, 689
cornetfish. *See Fistulariidae*
Corynopoma riisei, 524
Coryphaena hippurus (mahi mahi),
   284, 391
Coryphaenidae, 735(t), 743
Cottidae (sculpins), 271
   chromaffin (adrenal) tissue, 506

   demersal coloration, 297
   distribution, 632
   habitat, 667, 704, 717, 725
   homing behavior, 761, 762
   kidneys, 485
   thigmotaxis, 759
   tolerance to emergence, 724, *724*
Cottocomephoridae, 271, 689
Cottus (sculpins), 271, *271*
   C. asper (prickly sculpin), 426
   C. bairdi (mottled sculpin), 759
   C. beldingi (Paiute sculpin), 667
   C. confusus (shorthead sculpin),
     667
   C. extensus, 774
   C. gobio, 433(t), 524
   C. pygmaeus, 569(t)
   distribution, 619
   fins, 37, 44
   gas bladder, 332
   metabolic rate, 409(t)
   nostrils, *37*
   phototaxis, 757
   spines, 34
cowfish. *See Ostraciidae*
cownose ray. *See Rhinoptera*
cow shark. *See Hexanchoidei*
Cranoglanididae (catfishes), 621
crappie. *See Pomoxis*
Creagrutus, 621
Creediidae (sandburrowers), 257
creek chub. *See Semotilus*
creek chubsucker. *See Erimyzon*
   oblongus
Crenichthys (springfishes), 245
Crenicichla alta, 73
crestfish. *See Lophotidae*
croaker. *See Sciaenidae*
Crossognathiformes, 172
Crossopterygii, 150
crucian carp. *See Carassius carassius*
Cryptacanthodes giganteus (giant
   wrymouth), 258
Cryptacanthodidae (wrymouths),
   258
Cryptopsaras couesi (seadevil), 741(t)
Ctenacanthiformes, 117
Ctenoluciidae, 184
Ctenopharyngodon idella (grass carp),
   181
   herbivory, 425
   intestine length, 432(t)
   pharyngeal teeth, 429
   triploid, 570
Ctenopoma, 264
Ctenosquamata, 212
cui-ui. *See Chasmistes cujus*
Culaea (brook sticklebacks), 238
cunner. *See Tautagolabrus adspersus*
Curimatidae, 184

cusk. *See Brosme*

cutlassfish, 35

cutthroat eel. *See* Synaphobranchidae

cutthroat trout. *See Oncorhynchus clarki*

Cyclopteridae (lumpfishes), 35, 44, 272
  body shape, *38*
  habitat, 739(t)

Cyclosquamata, 210–12

Cyclothone (bristlemouths), 209, *209*, 741(t)

Cyema (bobtail snipe eels), 172
  C. atrum, 168

Cyematidae, 172

Cyematoidei, 172

Cymatogaster aggregata (shiner perch), 316, 530, 717

Cynoglossidae (tonguefishes), 274, 704

Cynolebias, 244

Cynoscion (seatrouts), *333*
  C. regalis (weakfish), 364, 381, 716

Cyprinella, 616

Cyprinidae, *180*, 181–82
  alimentary canal, 418
  chromosome and arm numbers, 569(t)
  distribution, 610, 613, 614, *614*, 615, 622
  habitat, 671, 672, 673, 686, 688, 694
  heterospecific shoaling, 786
  kidneys, 485
  musculature, 315
  nares, *394*
  nuptial tubercles, 47
  olfactory rosette, *394*
  pharyngeal teeth, 417
  phototaxis, 757
  phylogeny, 181
  sound production, 384
  swim bladder, 332

Cypriniformes, *180*, 181–82, *183*, 614, 621

Cyprinodon, 246
  C. diabolis (Devil's Hole pupfish), 246, 670, 843
  C. macularius (desert pupfish), 480
  C. pecosensis (Pecos pupfish), 297
  C. radiosus (Owens pupfish), 670
  C. variegatus (sheepshead minnow), 426, 479, 480, 644, 716
    C. v. hubbsi, 682
  distribution, 619
  sexual dimorphism, 44
  species flocks, 689

Cyprinodontidae, 246
  alimentary canal, 419
  chromosome and arm numbers, 569(t)
  distribution, 621
  habitat, 670, 695
  sexual dimorphism, 44

Cyprinodontiformes, *241*, 244–46

Cyprinodontoidei, 245–46

Cyprinus (carps)
  aquaculture, 838
  brain, 495
  C. carpio (common carp), 181
    chromosome and arm numbers, 568(t)
    coloration, 557
    digestion, 433(t)
    glomeruli, 488(t)
    habitat, 672, 687
    ionic concentrations, 645(t)
    metabolic rate, 409(t)
    omnivory, 425
    plasma ionic concentrations, 475(t)
    skull, *56*
    white blood cell count, 463
  dietary requirements, 412
  digestive enzymes, 433
  as exotic, 842
  pharyngeal teeth, *417*
  taste receptors, 399

Cypselurus, 244
  C. californicus (California flyingfish), 244, 735(t)

Cyrtocara, 299, 424
  C. moori, 787

dace. *See Rhinichthys*

Dactylopteridae (flying gurnards), 44, 253–54, 331, 382

Dactyloscopidae (sand stargazers), 257

Daector, 223

daggertooth. *See* Anotopteridae

Dalatiidae (sleeper sharks), 122

Dallia pectoralis (Alaskan blackfish), 207
  chromosome and arm numbers, 569(t)
  distribution, *76*
  esophagus, 418
  habitat, 687

Dalliidae, 207, 611, *612*, 613

Danio rerio (zebrafish), 329(t), 534–35, 590

darters
  chemotaxis, 761
  habitat, 686
  See also Etheostoma; Percina

Dascyllus, 47, 359, 788–89, *790*

Dasyatidae (stingrays), 125, 704

  spines, 47
  venomous, 305

Dasyatis, 124
  D. brevicaudata (smooth stingray), 721
  D. garouaensis, 369
  habitat, 125
  spines, 28, 47

dealfish. *See Trachipterus arcticus*

deep-sea bonefish. *See* Pterothrissidae

deep-sea smelt. *See* Bathylagidae

deepwater redfish. *See Sebastes mentella*

Deltistes luxatus (Lost River sucker), 75, *77*

Denticeps clupeoides (denticle herring), 173

Denticipidae, 622

Denticipitidae, 173

Denticipitoidei, 173

denticle herring. *See Denticeps clupeoides*

desert pupfish. *See Cyprinodon macularius*

Devil's Hole pupfish. *See Cyprinodon diabolis*

Diabolepis, 142

Dialommus fuscus (blenny), 724
  air breathing, 452
  eyes, 356, 358, *358*
  habitat, 726

Diaphus, 212

Dicentrarchus labrax, 253

Diodon
  D. holocanthus (balloonfish), 276
  kidneys, 485
  olfactory organ, 392

Diodontidae (porcupinefishes), 276–77, 707, 817

Diplacanthus, 130

Diplomystidae (velvet catfishes), 186–87, 621

Diplorhina, 97

Dipnoi (lungfishes), 28, 136–38
  air breathing, 453
  ampullary electroreceptors, 368
  choanae, 392
  distribution, 28, 622
  electroreceptors, 368
  fin rays, 64
  habitat, 688
  heart, 456
  neurohypophyseal hormones, 505
  nostrils, *142*
  pituitary gland, 503–4
  relationship to tetrapods, 28, 141–42
  skull, *133*
  spiral valve, 419

  teeth, 415
  testes, 519

Dipnomorpha, *133*, 136–38

Dipterus, 133
  D. valenciennesi, 137

Diptychus dipogon, 568(t)

Diretmidae (spinyfins), 233

Diretmus (spinyfins), 357

discu. *See Symphysodon discus*

Dissostichus eleginoides (Patagonian toothfish), 744

Distichodontinae, 183, 622

Distichodus, 425

dogfish. *See* Squalidae

Dolly Varden. *See Salvelinus malma*

dolphin, 735(t)

dorab. *See Chirocentrus dorab*

Dormitator latifrons, 452

Dorosoma, 425, 686
  D. cepedianum (gizzard shad), 418, 431, 687

dory. *See* Zeidae

Doryaspis, 98

Dorypteroidei, 152

dourada. *See Brachyplatystoma flavicans*

Draconettidae (dragonets), 259

dragonet. *See* Callionymidae; Draconettidae

dragonfish. *See Echiostoma*

drum. *See* Sciaenidae

duckbill. *See* Percophidae

duckbilled barracudina. *See Paralepis atlantica*

Dunkleosteus, 109

dusky shark. *See Carcharhinus obscurus*

dwarf shark. *See Squaliolus*

eastern mudminnow. *See Umbra pygmaea*

Echeneidae (remoras), 39, 787
  habitat, 735(t)
  hitchhiking, 331
  sucking disc, *41*

Echinorhiniformes, 122

Echinorhinoidea, 122

Echinorhinus (bramble sharks), 122
  E. cookei (prickly shark), *293*

Echiostoma (dragonfishes), 357

Ecsenius gravieri, 300

eelpout. *See* Zoarcidae

eeltail catfish. *See Plotosus anguillarus*

Eigenmannia, 379

Elasmobranchii, *26*, 27–28, 114–25
  ampullae of Lorenzini, 368, 369, *369*
  blood, 462
  cartilage, and cancer, 814–15
  chromaffin tissue, 506

circulation, 459
classification, 114, 115(t), 117
cloaca, 65
dermal denticles, 114
eyes, 347, 350, 352–53
fin rays, 62
gill filaments, 447
glomeruli, 488(t)
heart, 65, 454–55
interrenal tissue, 505
ionic balance in salt water, 476(t)
kidneys, 65, 483–84, 486 88
magnetotaxis, 376, 759
nephron, 487
notocord, 59
olfactory bulb, 391
osmoregulation, 480
osmoregulation in freshwater, 472–73
osmoregulation in salt water, 474, 475–77
otoliths, 365
pectoral fins, 61
pelvic claspers, 114
phylogeny, 114
and poisoning, 818
reproductive anatomy, 518–19
scales, 286
skull, 51–54
stomach eversion, 419
teeth, 414
urogenital sinus, 65
vesicles of Savi, 370
viviparity, 527
Elassoma (pygmy sunfish), 237
Elassomatidae, 237
Elassomatoidei, 237
electric catfish. See Malapterurus
electric eel. See Electrophorus electricus
electric ray. See Torpedinidae
Electrophoridae, 189
Electrophorus electricus (electric eel), 188, 189
air breathing, 452
electric organs, 376, 377, 378
locomotion, 330
Eleotridae, 259, 623
elephantfish. See Mormyridae
Eleutheronema, 254
Elginerpeton, 141
Elopichthys bambusa (kanyu), 432(t)
Elopidae (ladyfishes), 166
Elopiformes, 166–67, 167
Elopomorpha, 166–72
Elops (ladyfishes)
E. machnata (springer), 166
E. saurus (tenpounder), 166
larvae, 167
Embiotoca jacksoni (black surfperch), 423, 585

Embiotocidae (surfperches), 255, 722
pharyngeal teeth, 417, 429
swim bladder, 332
Empetrichthys (poolfishes), 245
Enchelyopys cimbrius, 480
Engraulidae (anchovies), 173
commercial fishery, 173
distribution, 621
habitat, 722, 735(t)
hearing, 371–72
Engraulis, 168, 735(t)
E. japonicus (Japanese anchovy), 827
E. mordax (northern anchovy), 627
E. ringens (Peruvian anchoveta), 173, 703, 827, 832
Eohiodon, 164
Eopsetta jordani (petrale sole), 275
Eosalmo, 198
Ephippidae (spadefishes), 262
Epinephelus
E. drummondhayi (speckled hind), 828(t)
E. itajara (goliath grouper), 828(t)
E. nigritus (Warsaw grouper), 288, 828(t)
E. niveatus (snowy grouper), 828(t)
E. striatus (Nassau grouper), 828(t)
Epiplatys, 244
Eptatretus (hagfishes)
E. burgeri, 359
E. cirrhatus, 458(t)
E. stoutii (Pacific hagfish), 93, 96, 568(t)
Erilepis zonifer (skilfish), 271, 626
Erimyzon oblongus (creek chub-sucker), 75, 78
Erpetoichthys calabaricus (reedfish), 151–52
cutaneous respiration, 451, 452
locomotion, 330
osmotic regulation, 473
terrestrial adaptation, 151–52
Erythrinidae, 184
escolar. See Gempylidae
Esocidae (pikes), 35, 206
alimentary canal, 419
body shape, 38
distribution, 610, 611
optomotor response, 760
Esociformes, 206–7
phylogeny, 208
Esox, 206, 829
body shape, 35
chromosome and arm numbers, 569(t)

E. lucius (northern pike), 761, 842
digestion, 433(t)
distribution, 611
habitat, 686
E. masquinongy (muskellunge), 206, 317
E. reicherti (Amur pike), 206
Etelis
E. carbunculus (squirrelfish snapper), 828(t)
E. coruscans (longtail snapper), 828(t)
Etheostoma (darters), 253
E. euzonum, 77
E. kanawhae, 77
E. osburni, 77
E. tetrazonum, 77
E. variatum, 77
gas bladder, 332
habitat, 667
reproductive capacities, 650
species distribution, 75, 616
Etmopterus, 120
E. gracilospinous (broadband dogfish), 739(t)
Etroplus suratensis (pearl spot), 255
Euchondrocephali, 111–14
Eucinostomus argenteus (spotfin mojarra), 720
Eucyclogobius newberryi (tidewater goby), 625, 842
Eudontomyzon
E. danfordi, 95
E. vladikovi, 95
eulachon. See Thaleichthys pacificus
Euleptorhamphus longirostris, 331
European eel. See Anguilla anguilla
European ling. See Molva
European sturgeon. See Acipenser sturio
European wels. See Siluris glanis
European whiting. See Gadus merlangus
Eurypharyngidae, 172, 741(t)
Eurypharynx (gulpers), 168, 445
E. pellecanoides (pelican gulper), 741(t)
Euselachii, 117
Eusthenopteridae, 138
Eusthenopteron, 139, 140
Eutaeniophorinae (tapetails), 232
Euteleostei, 193–201
and Ostariophysi, 178
phylogeny, 194
Euteleostomi, 107
Euthynnus, 335
Evermannella atrata (sabertooth), 741(t)
Evermannellidae (sabertooth fishes), 211, 741(t)

Eviota zonura, 259
Exocoetidae, 45
Exocoetidae (flyingfishes), 244
distribution, 627
flying locomotion, 331
habitat, 703, 734, 735(t)
Exocoetoidei, 243–44
Exocoetus, 244
E. obtusirostris (oceanic flying-fish), 735(t)
eyed flounder. See Bothus ocellatus

false cat shark. See Pseudotriakidae
fangtooth. See Anoplogastridae
fathead sculpin. See Psychrolutidae
featherfin knifefish. See Notopterus
fifteenspine stickleback. See Spinachia
filefish. See Monacanthidae
finback cat shark. See Proscyllidae
finetooth shark. See Carcharhinus isodon
Fistulariidae (cornetfishes), 239
flabby whalefish. See Cetomimidae
flashlightfish. See Anomalopidae
flatfish. See Pleuronectiformes
flathead. See Platycephalidae
flathead catfish. See Pylodictes olivaris
flat loach. See Balitorinae
Florida pompano. See Trachinotus carolinus
flounders
body shape, 35
osmoregulation, 481
See also Bothidae; Pleuronectidae
flyingfish. See Exocoetidae
flying gurnard. See Dactylopteridae
Fodiator (flyingfishes), 244
F. acutus, 627–28
F. rostratus, 627–28
footballfish. See Himantolophus groenlandicus
Forcipiger (longnose butterflyfishes), 415, 708
four-eyed fish. See Anablepidae
fourhorn sculpin. See Myoxocephalus quadricornus
fourspine stickleback. See Apeltes quadracus
freckled stargazer. See Xenocephalus egregius
freshwater angelfish. See Pterophyllum scalare
freshwater flyingfish. See Gasteropelecidae; Pantodon buchholzi
freshwater hatchetfish. See Gasteropelecidae
frigate mackerel. See Auxis

frillfin goby. *See Bathygobius soporator*

frill shark. *See Chlamydoselachus anguineus*

frogfish. *See* Antennariidae

Fugu, 485

Fundulidae (killifishes), 245, 613–14, *620*

Fundulus, 245
  F. heteroclitus (mummichog), 507, 716
  F. parvipinnis, 717, 811

Furcacaudiformes, 99, *99*

Gadidae (cods), 39, 44, 221–22
  caudal fins, *43*
  commercial fishery, 222
  distribution, 632
  fins, 39, 44
  and global warming, 844
  habitat, 704, 742
  history of fishery, 222, 833–34
  optomotor response, 760
  and ozone depletion, 844
  pelvic fin placement, *46*
  subcarangiform swimming, 321
  tail, 41

Gadiformes, 219–22, *220*, 221(t), 626

Gadopsidae, 623

Gadus
  G. macrocephalus (Pacific cod), 222, 834
  G. merlangus (European whiting), 329(t), 834
  G. morhua (Atlantic cod), 222
    blood pressure, 458(t)
    commercial fishery, *827*, 833–34
    energy budget, 410
    eyes, 351
    glomeruli, 488(t)
    habitat, 726
    overfished, 828(t)
  G. ogac (Greenland cod), 834
  G. virens (saithe), 289

Gaidropsarus mediterraneus, 400

Galapagos shark. *See Carcharhinus galapagensis*

Galaxias maculatus (inanga), 198, 479, 765, 766

Galaxiidae, 197–98
  and diadromy, 766
  distribution, 623–24

Galaxiinae, 198

Galaxioidea, *195*, 197–98

Galeaspida, 100

Galeichthys, 382

Galeocardo cuvieri (tiger shark), 120, 735(t), 829(t)

Galeoidea, 118

Galeomorphii, 117, *118*

Galeorhinus (topes), 120

Galeus zyopterus (soupfin shark), 120

Gambierdiscus toxicus, 816

Gambusia, 781
  G. affinis, 245, 717
  G. holbrooki, 245

gar. *See* Lepisosteidae

garibaldi. *See Hypsypops rubicundus*

Gasteropelecidae, 42–44, 184, 331

Gasteropelecus, *45*, 184

Gasterosteidae (sticklebacks), 238
  chromosome and arm numbers, 569(t)
  habitat, 687, 695–96
  nostrils, 394
  scutes, 48, *49*

Gasterosteiformes, *237*, 238–39

Gasterosteoidei, 238

Gasterosteus (sticklebacks), 238
  breeding behavior, 780
  G. aculeatus (threespine stickleback)
    distribution, 611
    genetic divergence, 585
    osmotic regulation, 473
    schooling behavior, 786
    trophic polymorphism, 430, 689
  parasites, 811, 813
  sexual selection, 79
  stridulation, 381

Gazza, 303

Gempylidae (escolars), 262, 735(t), 739(t)

Gempylus serpens (snake mackerel), 735

Gemuendina, 109

Genyonemus lineatus (white croaker), *542*

Genypterus, 223

Geotria australis (pouched lamprey), 94, 291

Geotriidae (lampreys), 94, 623

Gephyrochromis, 425

Gerridae (mojarras), 415, 704, 721

ghost pipefish. *See Solenostomus*

giant gourami. *See Osphronemus goramy*

giant kelpfish. *See Heterostichus rostratus*

gianttail. *See Gigantura vorax*

giant wrymouth. *See Cryptacanthodes giganteus*

gibberfish. *See Gibberichthys*

Gibberichthyidae, 230–32

Gibberichthys (gibberfishes), 230–32, *232*

Gibelion catla (catla), *180*, 181, 433(t)

Gigantactis (whipnose anglerfishes), 225

Gigantura vorax (gianttail), 741(t)

Giganturidae (telescopefishes), 210, 741(t)

Giganturoidei, 210

Gila (chubs), 670, 671
  G. bicolor (tui chub), *36*, 426, *429*
  G. boraxobius (Borax Lake chub), 521, 522

Gilchristella, 721

Gillichthys mirabilis (longjaw mudsucker), 261, 414, 452, 717

Gilpichthys, 92

Ginglymostoma cirratum (nurse shark), 118, 426, 828(t)

Girella nigricans (opaleye), 705

gizzard shad. *See Dorosoma cepedianum*

"glass eels," 170

glassfish. *See* Salangidae

glassperch. *See Ambassis productus*

Glossogobius (gobies), 721

Glyptocephalus
  G. cynoglossus (witch flounder), 828(t)
  G. zachirus (rex sole), 275

Glyptothorax (catfishes), 42, *45*

Gnathonemus (elephantfishes), 166

Gnathostomata, 25, 27–29, 107, *112*

goatfish. *See Mulloides auriflamma*

Gobiesocidae (clingfishes), 44, 258–59, 723, 725
  epidermis, 47
  modified pelvic fins, *46*
  pelvic fins, 44
  symbiosis, 790

Gobiesocoidei, 258–59

Gobiesox (clingfishes), *256*
  fins, 44
  G. maeandricus (northern clingfish), 258–59

Gobiidae, 44, 259–61
  brackish-adapted, 625
  cleaning symbiosis, 788
  distribution, 623, 625
  habitat, 704, 707, 717, 725
  modified pelvic fins, *46*
  monogamy, 780
  pectoral fins, 44
  sexual dimorphism, 44

Gobiodon erythrospilus, 531, 781

Gobiogobio (gudgeons), 384

Gobioidei, 259, 485

Gobioides (gobies), 832

Gobiomorus dormitor, 259

Gobionellus shufeldti, 573

Gobiosoma bosci (naked goby), 384

Gobius
  G. cobitis, 433(t)
  G. criniger, 817
  G. exanthonemus, 462

Gobiusculus flavescens (two-spotted goby), 781

goblin shark. *See* Mitsukurinidae

goby. *See* Gobiidae

golden redfish. *See Sebastes norvegicus*

goldeye. *See Hiodon alosoides*

goldfish. *See Carassius auratus*

goliath grouper. *See Epinephelus itajara*

Gonorynchidae, 179

Gonorynchiformes, 178–79, *180*

Gonorynchoidei, 179

Gonorynchus (sandfishes), *34*, 179, *180*

Gonostoma denudatum (lightfish), 739(t)

Gonostomatidae, 209
  distribution, 627
  habitat, 736, 739(t), 741(t)

Gonostomatoidei, 209

Goodeidae, 245–46, 619, *620*

Goodeinae (splitfins), 245

goosefish. *See* Lophiidae

gourami. *See Trichogaster*

Grammatidae (basslets), 253

Grammistidae (soapfishes), 818

grass carp. *See Ctenopharyngodon idella*

graveldiver. *See* Scytalinidae

grayling. *See Thymallus*

great barracuda. *See Sphyraena barracuda*

greeneye. *See* Chlorophthalmidae

Greenland cod. *See Gadus ogac*

Greenland shark. *See Somniosus macrocephalus*

greenling. *See* Hexagrammidae

green sturgeon. *See Acipenser medirostris*

green swordtail. *See Xiphophorus helleri*

grenadier. *See* Macrouridae

grideye. *See Ipnops*

grouper. *See* Serranidae

grunion. *See Leuresthes tenuis*

grunt. *See* Haemulidae; Pomadasyidae

gudgeon. *See Gobiogobio*

guitarfishes, 124

gulf menhaden. *See Brevoortia patronus*

gulper. *See Eurypharynx*

gulper eel. *See* Eurypharyngidae; Monognathidae

gunnel. *See* Pholidae
guppy. *See* Poecilia
Gymnachirus melas (naked sole), 298
Gymnarchidae, 166
  distribution, 622
  electric organs, *378*
  electrotaxis, 758
  habitat, 379
Gymnarchus
  amiiform swimming, 326
  electric organs, 379, 380
  G. niloticus, 166, 379
Gymnocephalus cernua, 428
Gymnorhamphichthys, 379
Gymnothorax, 391, 485
Gymnotidae, 189
  electrotaxis, 758
  fins, 39
Gymnotiformes, *188,* 189
  distribution, 621
  tuberous organs, 375
Gymnotoidei, 189
Gymnotus, 379
Gyrinocheilidae, 42, 44, 182, 667
Gyrinocheilus, 451

haddock. *See Melanogrammus*
  aeglefinus
Haemulidae (grunts), 721, 722, 736
Haemulon plumieri (white grunt),
  828(t)
hagfishes, *26, 27,* 92–94.
  alimentary canal, 418–19
  anguilliform swimming, 318
  blood pressure, 458
  branchial irrigation, 450
  branchial ventilation, 450
  circulation, 458
  commercial uses, 48
  and conodonts, 92
  cyclostome poisoning, 818
  digestion and excretion, 432
  eggs, 93, *93*
  epidermis, 47
  feeding, 92–93, 424
  fin rays, 62
  gills, 36–37, 442–43, *444*
  habitat, 93, 742
  heart(s), 93, 454
  ichthyocrinotoxins, 818
  kidney, 483, 486
  lingual apparatus, 51, 426
  liver, 421
  locomotion, 330
  membranous labyrinth, 366
  mucus glands, 47
  nephron, *487*
  neural crest, 90
  neurohypophyseal hormones,
    505

  notocord, 59
  olfaction, 396
  olfactory threshold, 395(t)
  osmoregulation, 475, 475(t)
  pancreas, 421
  phylogeny, 90–92, *91*
  pituitary gland, 502–3
  reproductive anatomy, 518
  skull, 51, *52*
  thread cells, 92
  thyroid gland, 505
  vascular system, 93
  velum, 51
  See also Eptatretus;
    Myxiniformes
Haikouella, 25
Haikouichthys, 25
hairtail. *See Trichiurus*
hairyfish. *See Mirapinna esau*
hake. *See* Merlucciidae
halfbeak. *See* Hemiramphidae
halfmoon. *See Medialuna*
  californiensis
halibut. *See Hippoglossus*
Halosauridae, 35, 167–69, 742
Hamiltonichthys, 116
hamlet. *See Hypoplectrus*
hammerhead shark. *See* Sphyrnidae
Haplochromis, 255
Haplopagrus guentheri, 391
Hardistiella, 94
Harpadon nehereus (Bombay duck),
  212
Harpagiferidae (plunderfishes), 258,
  750–51
hatchetfish. *See Argyropelecus;*
  Sternoptychidae
Helicophagus, 187
Helostomatidae, 264
Hemiemblemaria simulus, 257
Hemigaleidae (weasel sharks), 120
Hemiodontidae, 184
Hemiodontinae, 184
Hemipteronotus pavo (labrid), 298
Hemiramphidae (halfbeaks), 243–
  44, 331, 703
Hemitripteridae, 272
Hemitripterus americanus (sea raven),
  272
Hepsetidae, 184, 622
Hepsetus odoe (Kafue pike), 184
Heptranchidae (seven-gill sharks),
  121–22
Herichthys cyanoguttatum (Rio
  Grande perch), 255
herrings.
  carangiform swimming, 321
  genetic divergence, 585
  habitat, 688
  hearing, 371–72

  lateral line, 36
  migration, 767, *767*
  optomotor response, 760
  See also Clupea; Sardinella
Hesperoleucus (roaches), 671
Heterodontidae, 117–18
Heterodontiformes, 117–18
Heterodontoidea, 117–18
Heterodontus (horn sharks), 117–18,
  118, 590
Heteropneustes fossilis, 360
Heteropneustidae, 187, 305
Heterostichus rostratus (giant kelp-
  fish), 705
Heterostraci, 97–98
Heterotidinae, 164
Heterotis niloticus, 164
Hexagrammidae (greenlings), 36,
  271, 296
Hexagrammos (greenlings), 36, 271,
  271
  H. octogrammus, 569(t)
  H. stelleri, 356
Hexanchidae, 121
Hexanchiformes, 121–22
Hexanchoidei (cow sharks), 121–22
Hexanchus (six-gill sharks), *121,*
  122
  H. griseus, 122
Hexatrygonidae, 125
hickory shad. *See Alosa chrysochloris*
hill stream catfish. *See* Sisoridae
Himantolophidae, 741(t)
Himantolophus groenlandicus (foot-
  ballfish), 741(t)
Hiodon
  H. alosoides (goldeye), 164,
    568(t)
  H. tergisius (mooneye), 164
Hiodontidae, 164, 615
Hiodontiformes, 164
Hippocampus (seahorses), 238
Hippoglossoides platessoides (Ameri-
  can plaice), 828(t)
Hippoglossus (halibuts), 275, 742
  H. hippoglossus (Atlantic hali-
    but), 275, 421
  H. stenolepis (Pacific halibut),
    275
Histrio histrio (sargassumfish), 224,
  299
  distribution, 628
  drifting, 331
  habitat, 736
  kidneys, 485
  locomotion, 330
  thigmotaxis, 759
Holacanthus passer, 431
Holarctic capelin. *See Mallotus*
  villosus

Holcentroidei, 233–34
Holocentridae, 233–34
Holocentrus (squirrelfishes), 757
Holocephali, *26,* 28, 47, 53, 111–12,
  633
Holocephalimorpha, 112
Holoptychius, 133
Homalopterus, 182
Hoplosternum, 453
Hoplostethus atlanticus (orange
  roughy), 233, 741
horn shark. *See Heterodontus*
houndfish. *See Tylosurus crocodilus*
huchen. *See Hucho*
Hucho (huchens), 201
Huso huso (beluga sturgeon), 152,
  154
Hybodontiformes, 114, *116,* 117
Hybognathus (silvery minnows), 583
Hybopsis, 671
Hydrocynus (tigerfishes), 185, *185,*
  831
Hydrolagus, 113
  chromosome and arm numbers,
    568(t)
  egg cases, *525*
  H. colliei, 519
  osmoregulation, 477(t)
  pelvic claspers, *528*
Hynerpeton, 142
Hypentelium nigricans (northern hog
  sucker), 667
Hyperotreti, 90
Hyphessobrycon eques (jewel tetra),
  557
Hypnosqualea, 123
Hypomesus pretiosus (surf smelt), 722
Hypophthalmichthys
  H. molotrix (silver carp), 181,
    432(t)
  H. nobilis (bighead carp), 181
  planktivory, 425
Hypoplectrus (hamlets), 574
Hypopomidae, 189
Hypopomus brevirostris (bluntnose
  knifefish), 452
Hypoptychidae (sand eels), 238
Hypsypops rubicundus (garibaldi),
  783
Hysterocarpus traski, 255

icefish. *See* Channichthyidae;
  Salangidae
Ichthyoboridae, 424, 622
Ichthyococcus ovatus (boafish), 739(t)
Ichthyodectiformes, 161
Ichthyomyzon castaneus (chestnut
  lamprey), 95
Ichthyostega, 141

Icichthys lockingtoni (medusafish),
    263–64, 735(t)
    distribution, 627
    symbiosis, 788
Icosteidae, 258
Icosteoidei, 258
Icosteus (ragfishes), *256*
    I. aenigmaticus (ragfish), 258
Ictaluridae (catfishes), 44, 187–88
    alimentary canal, 419
    body shape, *39*
    chromosome and arm numbers,
        569(t)
    distribution, 615
    eggs, 521
    habitat, 670, 672, 686
Ictalurus (channel catfishes)
    behavior modification, 792
    blood pressure, 458(t)
    diet, 425
    dietary requirements, 412
    gas bladder, *333*
    habitat, 672
    I. furcatus (blue catfish), 187,
        672, 838
    I. punctatus, 187, 557, 838
Ictiobus (buffalofishes), 182, *183*,
    672, 686
Idiacanthinae (stalkeyes), 210,
    739(t), 740
Idiacanthus fasciola, 739(t), 741(t)
ilisha. *See Ilisha;* Pristigasteridae
Ilisha (ilishas)
    I. africanus, 173
    I. elongatus, 173
inanga. *See Galaxias maculatus*
Indian mullet. *See Rhinomugil corsula*
Indostomidae, 239
Inimicus, 330
Iniopterygi, 111, *113*
Iniopterygiformes, 111
Ipnopidae, 210
Ipnops (grideyes), 210, 741
Ischnacanthiformes, 131
Isistius (cookie cutter sharks), *121*,
    122–23
    feeding behavior, 775
    I. brasiliensis, 305, 739(t)
    I. plutodus, 775
Isonidae, 240
Istiompax (black marlins), 263
Istiophoridae, 263
    distribution, 627
    habitat, 735(t)
    locomotion, 734
Istiophorus (sailfishes), 263, 321, 328
Isurus (mako sharks), 119

jack mackerel. *See Trachurus*
Jamoytius, 99

Janassa, 111
Japanese anchovy. *See Engraulis*
    japonicus
Japanese dace. *See Tribolodon*
Jenynsia, 245
jewel tetra. *See Hyphessobrycon eques*
John Dory. *See Zeus faber*
Joturus (mullets), 765
June sucker. *See Chasmistes liorus*

Kafue pike. *See Hepsetus odoe*
Kamoharaia megastoma, 275
kanyu. *See Elopichthys bambusa*
kasidorons, 230–32, *232*
Kathetostoma
    K. albigutta (lancer stargazer),
        722
    K. averruncus (smooth star-
        gazer), 722
Katsuwonus pelamis (skipjack tuna),
    263
    commercial fishery, *827*
    digestion, 433(t)
    gill apparatus, 449(t)
    musculature, *314*, 316
Kaupichthys nuchalis, 170
kelp bass. *See Paralabrax clathratus*
kelp clingfish. *See Rimicola mus-*
    carum
Kenichthys, 140, *142*
Kern brook lamprey. *See Lampetra*
    hubbsi
killifish. *See* Fundulidae; Profun-
    dulidae
king-of-the-salmon. *See Trachipterus*
    altivelus
Kneria auriculata (shell-ear), 179,
    180
Kneriidae, 179, 622
knifefish. *See* Apteronotidae
kokanee. *See under Oncorhynchus*
    nerka
Kraemeriidae (sand gobies), 261
Kryptopterus, 356
Kurtidae, 261, 782
Kurtoidei, 261
Kurtus (nurseryfishes), *260*, 261, 526
Kyphosus elegans (rudderfish), 298,
    434

Labeo horie, 432(t)
Labidesthes sicculus, 240
labrid. *See Hemipteronotus pavo*
Labridae, 254, 704, 707
    cleaning symbiosis, 788
    eyes, 355
    labriform swimming, 327
Labrisomidae, 257
Labroidei, 254, 707
Labroides dimidiatus, 299, 573

labyrinth catfish. *See* Clariidae
Lacantunia enigmatica, 187
ladyfish. *See* Elopidae
Lagocephalus, 818
Lagodon rhomboides (pinfish), 399,
    720
lake sturgeon. *See Acipenser*
    fulvescens
lake trout. *See Salvelinus namaycush*
Lamna ditropis (salmon shark),
    735(t)
Lamna nasus (mackerel shark), 119,
    627
Lamnidae (mackerel sharks), 119
    habitat, 735(t)
    thunniform swimming, 321
Lamniformes (mackerel sharks),
    119–20
Lampanyctus, 212
Lampetra (lampreys)
    L. ayresi (river lamprey), 95, 424
    L. fluviatilis, 95
        kidney, 483, 508
        plasma ionic concentrations,
            475(t)
    L. hubbsi (Kern brook lam-
        prey), 95
    L. japonica (Arctic lamprey), 95
    L. lethophaga (Pit-Klamath
        brook lamprey), 95
    L. minima, *35*, 95, 424
    L. planeri, 95
    L. richardsoni (western brook
        lamprey), 95
    L. tridentata (Pacific lamprey),
        95, *365*
    phylogenetic studies, 583
lampreys, *26*, *27*, 94–96
    alimentary canal, 418, 419
    anguilliform swimming, 318
    blood, 462
    brain, *495*
    branchial arteries, *457*
    branchial basket, 51
    branchial irrigation, 450
    branchial ventilation, 450
    "brook" form, 95
    circulation, 458
    cyclostome poisoning, 818
    electroreceptors, 368–69
    eyes, 347, 352
    feeding, 95, 426
    fin rays, 62
    fins, 39
    gills, 36, 442–43, *444*
    habitat, 94
    heart, 454
    hitchhiking, 331
    ichthyocrinotoxins, 818
    kidneys, 483, 486

larvae, 94, 95
    larval, *22*
    life history, 94
    membranous labyrinth, 366
    metamorphosis, 538–39
    musculature, 312, *313*, *314*
    nasohypophyseal opening, 94
    nephron, *487*
    neurohypophyseal hormones,
        505
    notocord, 59
    olfaction, 394, 396
    oral disc, 34, *35*
    osmoregulation in freshwater,
        472
    osmoregulation in salt water,
        475, 475(t)
    "paired" species, 95
    pancreatic tissue, 421
    "parasitic" form, 95
    pheromones, 524
    phylogeny, 90–92, *91*
    pituitary gland, 502–3
    reproductive anatomy, 518
    skull, 51, *52*
    spinal cord, 498
    suck-and-hitch locomotion, 330
    thyroid gland, 505
    See also Lampetra;
        Petromyzontiformes
Lampridae, 213, 627, 735(t)
Lampridiformes, 212–13
Lampris (opahs)
    L. guttatus, 213, 627, 735(t)
    L. immaculata, 213
Lamproidei, 213
lancer stargazer. *See Kathetostoma*
    albigutta
lancetfish. *See* Alepisauridae
lane snapper. *See Lutjanus synagris*
lanternbelly. *See* Acropomatidae
lanternfish. *See* Myctophidae
largemouth bass. *See Micropterus*
    salmoides
Lates, 688
    L. calcarifer (barramundi perch),
        253
    L. niloticus (Nile perch), 253,
        694, 831
Latidae, 253
Latimeria (coelacanths), 134–36
    buoyancy, 333
    distribution, 28
    eggs, 529
    extraocular photoreception, 359
    feeding, 136
    fins, 42, 60–61, 135–36
    kidney, 488
    L. chalumnae, *133*, 583
    L. menadoensis, 583

living, discoveries of, 134–35
nostrils, *142*
olfaction, 394
olfactory bulb, 391
osmoregulation, 477, 477(t)
ovoviviparity, 136
pituitary gland, 503
pollutants, 841
relationship to tetrapods, 136
scales, 135
spiral valve, 419
swim bladder, 136
viviparity, 529
leaffish. *See Monocirrhus polyacanthus*
leafy seadragon. *See Phycodurus eques*
Lebiasinidae, 184
Leedsichthys, 148, *149*
Leiognathidae (slipmouths), 300, 303, 304, 415, *416*
Leiognathus equulus, 721
Leiostomus xanthurus (spot), 716, 720, 776, 805
lemon shark. *See Negaprion brevirostris*
lenok. *See Brachymystax*
Lepadogaster, 725
Lepidogalaxias (salamanderfishes)
L. salamandroides, 197–98, 453
Lepidogalaxiidae, 197
Lepidogobius lepidus (bay goby), 717
Lepidomeda, 619
Lepidopsetta bilineata (rock sole), 275
Lepidosiren (South American lungfishes), 42, 456
kidneys, 484
L. paradoxa, 138, 446
Lepidosirenidae, 138
Lepidosireniformes, 137–38
Lepisosteidae (gars), 35, 156–57
distribution, 615
fins, 41
scales, 48
vertebrae, 59
Lepisosteiformes, *155*
Lepisosteus (gars), *155*, 156
body shape, 35
Dahlgren cells, 508
habitat, 687
kidneys, 484
L. platostomus, 568(t)
pancreas, 421
spiral valve, 419
Lepomis (sunfishes), 35
body shape, 35
habitat, 686
homing behavior, 761
L. gibbosus (pumpkinseed), 686

L. macrochirus (bluegill), 253
conspecific foraging, 777
habitat, 672
male dimorphism, 507
mating tactics, 532
migration, 766
prey size preferences, 689
L. microlophus (redear sunfish), 686
olfactory orientation, 396
Leptacanthichthys, 531
Leptochariidae, 120
Leptocottus armatus (staghorn sculpin), *393*, 717, 775
Leptolepidiformes, 161
Leptoscopidae, 257
Lestidium ringens (slender barracudina), 741(t)
Leucaspius delineatus, 804
Leuciscinae, 181
Leuciscus
L. cephalus (chub), 672
L. leuciscus (dace), 672
Leuresthes tenuis (California grunion), 240, 722
life history, 242–43
Leuroglossus
L. schmidti, *540*
L. stilbius (Atlantic smoothtongue), 739(t)
lightfish, 739(t)
lighthousefish. *See* Photichthyidae
Limanda
L. aspera (yellowfin sole), 275
L. ferruginea (yellowtail flounder), 828(t)
Limnichthytes fasciatus, 357
ling cod. *See Ophiodon elongatus*
Linophryne (anglerfishes), *738*
Linophrynidae, 740
Linophymidae, 531
lionfish. *See Pterois*
Liopsetta glacialis (Arctic flounder), 632
Liparidae (snailfishes)
buoyancy, 733
distribution, 632
habitat, 725, 742, 743
modified pelvic fins, *46*
nostrils, 394
pyloric caeca, 419
Liparis (snailfishes), 297
Lipophrys pholis, 433(t)
Liza (mullets), 721
lizardfish. *See* Synodontidae
loach. *See* Balitoridae; Cobitidae
loach goby. *See* Rhyacichthyidae
Lobotes (tripletails), 298, *299*
longfin dragonfish. *See Tactostoma macropus*

longjaw mudsucker. *See Gillichthys mirabilis*
longnose butterflyfish. *See Chelmon; Forcipiger*
longnose chimaera. *See* Rhinochimaeridae
longnose dace. *See Rhinichthys cataractae*
longtail snapper. *See Etelis coruscans*
loosejaw. *See* Malacosteinae
Lophichthyidae, 225
Lophichthys boschmai, 225
Lophiidae (goosefishes), 35, 224, 737
Lophiiformes (anglerfishes), 39, 224–35
Lophioidei, 224
Lophioides, 224
Lophiomus, 224
Lophius (goosefishes), 35, 224
L. americanus, 224
L. budegassa, 224
L. piscatorius, 224
ionic concentrations, 645(t)
kidneys, 485
plasma ionic concentrations, 475(t)
Lopholatilus chamaeleonticeps, 254, 828(t)
Lophotidae (crestfishes), 60, 213
Loricariidae, 48, 187, 425, 667
Lost River sucker. *See Catostomus luxatus; Deltistes luxatus*
Lota lota (burbot), 222, 391, 611
Lotidae, 222
louvar. *See Luvarus imperialis*
Lovettia, 197, 766
Lucania parva (rainwater killifish), 245
Lucifuga, 223, 695
Luciocephalidae, 264, 622
Luciocephalus, 373
L. pulcher (pikehead), 264, 452
lumpfish. *See* Cyclopteridae
lungfish. *See* Dipnoi
Lutjanidae (snappers), 35, 253
body shape, 35
and ciguatera, 817
habitat, 721
heterospecific schooling, 786
Lutjanus
L. campechanus (red snapper), 828(t), 833, 837
L. griseus, 488(t)
L. synagris (lane snapper), 720
Luvaridae, 262, 735(t)
Luvarus imperialis (louvar), 262, 735(t)
Luxilus (shiners), 75
genetic studies, 583

hybridization, 572
L. chrysocephalus (striped shiner), 75
L. cornutus (common shiner), 75

mackerel. *See Scomber*
mackerel shark. *See* Lamniformes
Macropinna (spookfishes), 40, *195*, 739(t)
Macropodus (paradise fishes), 264
Macrorhamphosidae (snipefishes), *237*, 239
Macrouridae (grenadiers), 35, 221, 303, 337, 742
Macrourus (grenadiers), 35, *220*
madai. *See Pagrus major*
madtom. *See Noturus*
mahi mahi. *See Coryphaena hippurus*
mahseer. *See Tor tor*
Makaira (blue marlins), 263, 768, 829(t)
mako shark. *See Isurus*
Malacanthidae (tilefishes), 254
Malacosteinae (loosejaws), 210
Malacosteus niger (loosejaw), 357, 741(t)
Malapteruridae, 187, 622
Malapterurus (electric catfishes), 187
electric organs, 376, 377, *378*, 379
Mallotus (capelins)
M. villosus (Holarctic capelin), 196
mangrove rivulus. *See Rivulus marmoratus*
man-of-war fish. *See* Nomeidae
manta. *See Manta*
Manta (mantas), 123, 125
body shape, 35
cephalic fins, 125
feeding, 125
M. birostris, *124*, 125
masou salmon. *See Oncorhynchus masou*
Mastacembelidae (swamp eels), 236–37, *237*
distribution, 614, 622
nostrils, *37*
Mastacembeloidei (spiny eels), 236–37
Mayomyzon, 94
medaka. *See Oryzias*
Medialuna californiensis (halfmoon), 585, 705
medusafish. *See* Centrolophidae
Megachasma pelagios (megamouth shark), 117, *118*, 119–20, 776
Megachasmidae (megamouth sharks), 119
Megalomycteridae, 232

Megalopidae, 166–67
Megalops (tarpons), *167*
    M. atlanticus, 166–67
    M. cyprinoides (oxeye), 166
megamouth shark. *See Megachasma pelagios*
Meiacanthus (blennies)
    M. nigrolineatus (sabertooth blenny)
        mimicry, *300*
        pigmentation, 298
        trauma from, 307
Melamphaidae, 230
Melanogrammus aeglefinus (haddock), 222, 828(t), 834
Melanostigma pammelas (eelpout), 327, 328
Melanostomiatinae, 739(t)
Melanostomiinae, 210, 740
Melanotaeniidae (rainbowfishes), 241, 623
Membras martinica, 720
menhaden. *See Brevoortia*
Menidia (silversides)
    M. beryllina, 716
    M. clarkhubbsi (Texas silverside), 574, 583
    M. menidia (Atlantic silverside), 573, 716, 720
Merlucciidae (hakes), 41, 221, 742, 786
Merluccius, 424
    M. bilinearis (silver hake), 828(t)
    M. productus (Pacific hake), 221, 828(t), 834
Metynnis, 185
Mexican blind cavefish. *See Anoptichthys*
Microcyprini, 244
Microdesmidae (wormfishes), 261
Microgadus tomcod (Atlantic tomcod), 726, 834
Micromesistius poutassou (blue whiting), 222, *827*
Micropogon, 382, 805
Micropterus (bass)
    habitat, 672, 686
    hybridization, 572
    M. salmoides (largemouth bass), 253
        digestion, 433(t)
        gill apparatus, 449(t)
        habitat, 696
        pharyngeal teeth, 429
        seasonal feeding, 411
        teeth, 415
Microstomatidae, 195
midshipmen. *See Porichthys*
milkfish. *See Chanos chanos*
minnow. *See Phoxinus*

Minous (scorpionfishes), 790
Mirapinna esau (hairyfish), 232
Mirapinnidae, 232
Misgurnus (weatherfishes), 182
    kidneys, 485
    M. anguillicaudatus, 556, 569
    M. fossilis (squeaker), 384
Mitsukurina owstoni (goblin shark), 119
Mitsukurinidae (goblin sharks), 119
Mnierpes macrocephalus, 724
Mobula, 35, 125
Mobulidae, 35, 330
Mochokidae, 187
    distribution, 622
    electric discharge, 379
    geotaxis, 758
mojarra. *See* Gerridae
mola. *See* Molidae
Mola mola (ocean sunfish), 277, *278,* 735(t)
    buoyancy, 333, 733
    distribution, 627
    larval transformation, *277*
    locomotion, 734
Molidae (molas), 277, 817
    habitat, 735(t)
    tail, 41
Molva (European lings), 222
Monacanthidae (filefishes), 276, 381, 722
    balistiform swimming, 326
    stridulation, 381
Monacanthus (filefishes), 736
monkeyface prickleback. *See Cebidichthys violaceus*
monkfish. *See Lophius* (goosefishes)
Monocentridae (pineconefishes), 233
Monocentrus japonicus (pineconefish), 303
Monocirrhus polyacanthus (leaffish), 298
Monognathidae, 172
Monopterus albus (rice eel), 236
    aerial respiration, 453
    esophagus, 418
    gill openings, 37
    pituitary gland, 504
    suction feeding, 428
Monorhina, 99
mooneye. *See Hiodon tergisius*
moonfish. *See* Citharinidae
Moorish idol. *See* Zanclidae
moray. *See* Muraenidae
Mordacia, 94
    M. mordax, 95
    M. praecox, 95
Mordaciidae (lampreys), 94
    distribution, 623

Moridae (codlings), 221, 742
    hearing, 371
Moringuidae (spaghetti eels), 330
Mormyridae, 165–66
    brain, 494, 497
    distribution, 622
    electric organs, 369, *378,* 380
    electrotaxis, 758, 759
    habitat, 379, 673
    hearing, 371
Mormyriformes, 375
Mormyrus (elephantfishes), *165*
Morone
    distribution, 613
    M. saxatilis (striped bass), 57, *63,* 253, 842
        habitat, 716, 722
        hermaphroditism, 531
        metabolic rate, 409(t)
        sport fishery, 830
Moronidae (basses), 253
    caudal fins, *43*
    pelvic fin placement, *46*
    subcarangiform swimming, 321
Moses sole. *See Pardachirus marmoratus*
mosshead warbonnet. *See Chirolophus nugator*
mottled sculpin. *See Cottus bairdi*
mountain sucker. *See Catostomus platyrhynchus*
mountain whitefish. *See Prosopium williamsoni*
Moxostoma rhothoecum (torrent sucker), 667
Mozambique tilapia. *See Oreochromis mossambicus*
mudfish. *See Neochanna*
mudminnow. *See* Umbridae
mudskipper. *See* Oxudercinae
Mugil (mullets)
    habitat, 721
    jumping, 330
    M. cephalus (striped mullet), 239–40
        chromosome and arm numbers, 569(t)
        eyes, 353
        marine amphidromy, 766
        pathogens, 805
    M. curema, 569(t)
    stomach, 418
Mugilidae (mullets), 239–40, *418,* 703, 716
Mugiliformes, 239–40
Mugilomorpha, 239–40
mullet. *See* Mugilidae
Mulloides auriflamma (goatfish), 432(t), 787

mummichog. *See Fundulus heteroclitus*
Muraenesocidae, 171
Muraenidae (morays), 170–71
    and ciguatera, 817
    habitat, 707
    ichthyocrinotoxins, 818
    locomotion, 329
    phototaxis, 757
    trauma from, 307
Muraenoidei, 170–71
muskellunge. *See Esox masquinongy*
Mustelus canis, 488(t)
Mycteroperca
    M. microlepis, 828(t)
    M. phenax, 828(t)
Myctophidae (lanternfishes), 211, 212
    chromosome and arm numbers, 569(t)
    distribution, 626, 627
    gas bladder, 337
    habitat, 736, 739(t)
    kidneys, 485
    vertical migrations, 335
Myctophiformes, *211,* 212
Myctophum, 212, *541,* 735
Myleus, 425
Myliobatidae, 35, 330
Myliobatiformes (rays), *124,* 125
Myliobatis (rays), 35
    feeding, 125, 774
    M. californica, 774
    rostrum, 51
Mylocheilus (peamouths), *532,* 671
Mylopharyngodon piceus (black carp), 809, 842
Myoxocephalus quadricornus (fourhorn sculpin), 611, 632
Myripristis (soldierfishes), 372
    M. jacobus, 810
Myrophis (worm eels), *37*
Myxine, 458(t)
    M. glutinosa (Atlantic hagfish), 93
        chromosome and arm numbers, 568(t)
        plasma ionic concentrations, 475(t)
Myxini, 27
Myxinidae, 27, 93, 704
Myxiniformes (hagfishes), *26,* 27, 92–94
Myxinikela, 92
Myxocyprinus, 182, 613
Myxus (mullets), 721

naked goby. *See Gobiosoma bosci*
naked sole. *See Gymnachirus melas*
Nandidae, 622, 688

Nannopercidae (pygmy perches), 623

Narcine brasiliensis, 568(t)

narrowtooth shark. *See Carcharhinus brachurus*

nase. *See Chondrostoma nasus*

Naso (unicornfishes), 262

Nassau grouper. *See Epinephelus striatus*

Naucrates ductor (pilotfish), 735(t), 787

Nautichthys, 326

needlefish. *See* Belonidae

Negaprion brevirostris (lemon shark), 395(t), 434, 829(t)

Nemacheilus (loaches), 182
  N. barbatula, 371

Nematalosa come, 721

Nematognathi, 186

Nemichthyidae (snipe eels), 35, 171–72
  body shape, 38
  habitat, 737, 739(t), 741(t)
  vertebrae, 60

Nemichthys (snipe eels)
  N. scolopaceus, 35, 739(t), 741(t)

Neobythitinae, 223

Neoceratiidae, 531

Neoceratodus (lungfishes), *133*
  fins, 42, *42*
  gills, 446
  kidneys, 484
  N. forsteri (Queensland lungfish), 137
    distribution, 623
    electrotaxis, 758
    heart, 456

Neochanna (mudfishes), 451, 453

Neocyema (bobtail snipe eels), 172

Neogobius melanostomus (round goby), 842

neon tetra. *See Paracheirodon innesi*

Neopterygii, *26, 28,* 156–57

Neoscopelidae (blackchins), 212

Neoselachii, 117

Neostethus (priapiumfishes), *241*

Neoteleostei, 207–12
  phylogeny, *208*

Nerophis (pipefishes), 419

Nettenchelys, 394

neurohypophyseal hormones, 505

night shark. *See Carcharhinus signatus*

night smelt. *See Spirinchus starksi*

Nile perch. *See Lates niloticus*

Nimbochromis livingstonii, 255

ninespine stickleback. *See Pungitius*

Nomeidae (man-of-war fishes), 264, 735(t)

Nomeus (man-of-war fishes), 264

esophagus, 418
symbiosis, 734, 788
thigmotaxis, 759

noodlefish. *See* Salangidae

North American bowfin. *See Amia calva*

North American sauger. *See Stizostedion canadense*

northern anchovy. *See Engraulis mordax*

northern clingfish. *See Gobiesox maeandricus*

northern hog sucker. *See Hypentelium nigricans*

northern lampfish. *See Stenobrachius leucopsarus*

northern pearleye. *See Benthalbella dentata*

northern pike. *See Esox lucius*

northern pikeminnow. *See Ptychocheilus oregonensis*

northern sea robin. *See Prionotus carolinus*

northern snakehead. *See Channa argus*

Notacanthidae, 169, 742

Notacanthiformes, 167–69, *168*

Notacanthus, 35, *168*

Nothobranchius, 244
  diapause, 536
  habitat, 688
  N. rachowi, 569(t)

Nothonotus (darters), 615

Notidanoidea, 121–22

Notocheiridae (surf silversides), 240

Notopteridae, 165
  ampullary electroreceptors, 368
  distribution, 622
  gymnotiform swimming, 326
  inner ear, 364
  swim bladder, 332

Notopteroidei, *165,* 165–66

Notopterus (featherfin knifefishes), *165*

Notorhynchus (seven-gill sharks), 121

Notoscelopus
  N. japonicus, 288
  N. kroyeri, 335

Notosudidae (paperbones), 210

Nototheniidae, 258

Notothenioidei, 258, 632, 748–51

Notropis (shiners), *180,* 181, 671

Noturus (madtoms), 188
  distribution, 615
  habitat, 672
  phototaxis, 757
  thigmotaxis, 759
  venomous, 44, 47, 306

Novumbra, 207

N. hubbsi (Olympic mudminnow), 76

nurseryfish. *See Kurtus*

nurse shark. *See Ginglymostoma cirratum*

oarfish. *See* Regalecidae

oceanic flyingfish. *See Exocoetus obtusirostris*

ocean sunfish. *See Mola mola*

Odacidae, 255

Odax cyanomelas, 705

Odontaspididae (sand tiger sharks), 119

Odontaspis
  O. noronhai (bigeye sand tiger shark), 829(t)
  O. taurus (sand tiger shark), 829(t)

Odontobutidae, 259

Ogcocephalidae (batfishes), *45,* 225, 330

Ogcocephalioidei, 225

Ogcocephalus (batfishes), 44

Ogilbia, 223

oilfish. *See Ruvettus*

Oligocottus maculosus (tidepool sculpin), 762

Oligoplites saurus, 542

olive flounder. *See Paralichthys olivaceus*

Olympic mudminnow. *See Novumbra hubbsi*

Omosudidae, 211

Ompok, 356

Oncorhynchus, 35, 198
  brain, *495*
  chromosome and arm numbers, 569(t)
  eyes, 351, 355
  genetic engineering, 591
  genetic studies, 583
  habitat destruction, 831
  homing and olfaction, 396
  hybridization, 572
  migration, 764–65
  O. clarki (cutthroat trout)
    coastal ESU, 589
    habitat, 667
    inner ear, *365*
    life history, 199
    migration, 764
    pigmentation, 296–97
  O. gorbuscha (pink salmon), 200
    diploid, 570
    genetic divergence, 585–86
    gene tree, *586*
  O. keta (chum salmon), 200
    hatching in sea water, 480
    osmoregulation, 478
    overfished, 828(t)

O. kisutch (coho salmon), 198, 199–200
  in Great Lakes, 830
  osmoregulation, 481
  overfished, 828(t)
  swimming speed, 329(t)
  weight heritability, 577

O. masou (masou salmon), 518, 761

O. mykiss (rainbow trout), 198
  albinism, 557
  appendicular skeleton, *62*
  blood pressure, 458(t)
  digestion, 433(t)
  gill filaments, *448*
  habitat, 667, 686
  life history, 199, *779*
  migration, 764
  olfactory rosette, *393*
  osmoregulation, 481
  otolith, *366*
  pituitary gland, *502*
  salt tolerance, 478
  scoliosis, *413*
  skull, *55*
  swimming speed, 328, 329(t)
  taste buds, *398*
  whirling disease, 808

O. nerka (sockeye salmon), 80, 200
  commercial fishery, 200
  crepuscular feeding, 774
  fry migration, 587, 654
  habitat, 686
  kokanee variant, 200, 670, 689
  life history, 199
  population studies, 654
  Redfish Lake ESU, 589
  river *vs.* lake spawning, 80

O. tshawytscha (Chinook salmon), 198, 199, 200
  aquaculture, 839
  in Great Lakes, 830
  habitat, 735(t)
  ionic concentrations, 645(t)
  migration, 764, 779
  musculature, *314*
  overfished, 828(t)
  plasma osmolality, 479(t)
  Snake River ESU, 589
  stream-type *vs.* ocean-type, 779
  triploid, 570
  UV sensitivity, 296

semelparity, 522–23

swimming speed, 328

taste buds, 397

territorial behavior, 780–81

Onychodontiformes, 132, *133*

Opaeophacus (eelpouts)
   O. acrogeneius, 357
opah. *See Lampris*
opaleye. *See Girella nigricans*
Ophichthidae, *37*, 171, 329, 707
Ophidiidae, 223, 330
   habitat, 704, 741, 742, 743
Ophidiiformes, *220*, 222–23
Ophidioidei, 223
Ophiodon
   O. elongatus (ling cod), 271,
      458(t), 828(t)
   O. marginatum, 381
Opisthoproctidae, 40, 195, 303,
   739(t)
Opsanus (toadfishes), 223, 382, *383*
   O. tau, 382, 479
orange roughy. *See Hoplostethus
   atlanticus*
Orectolobidae, 118
Orectolobiformes, 118–19
Orectolobus (wobbegons), 118
Oreochromis, 696, 721, 831
   O. hornorum, 591
   O. mossambicus (Mozambique
      tilapia), 255
      adaptation to salt water, 479,
         481
      gastric enzymes, 431
      ion balance, *472*
      noise production, 381
   O. nilotica, 255
Orestias, 621
Orthopristis, 399
Oryzias (medakas), 243
   albinism, *559*
   kidneys, 485
   social behavior, 785
   spinal abnormalities, 597
Osmeridae (smelts), 196–97
   distribution, 611, 632
   optomotor response, 760
Osmeroidea, 196–97
Osmeroidei, *195*, 196
osmoregulation, 477
Osphronemidae, 39, 47, 264
Osphronemus goramy (giant
   gourami), 264
Ostariophysi, 178–89
   chemotaxis, 761
   distribution, 621
   habitat, 674, 677, 688, 694
   hearing, 371, 373
   inner ear, 364
   phylogeny, *179*
"Osteichthyes," 28, 130, 138
Osteoglossidae, 164, 622, *624*
Osteoglossiformes, 164
Osteoglossoidei, 164–65, *165*
Osteoglossomorpha, 162–66

Osteoglossum
   O. bicirrhosum (silver arowana),
      164
   O. ferreirai (black arowana), 164
Osteolepiformes, 138, *139*
Osteolepimorpha, 138
Osteostraci, 100, *102*
Ostraciidae (cowfishes), 276
   body shape, 35, *39*
   habitat, 707
   ostraciiform swimming, 325–26
Ostracion, 485
ostracoderms, 96–101
Ostracoidea, 276
Otocephala, 173
Otophysi, 179–89
Owens pupfish. *See Cyprinodon
   radiosus*
oxeye. *See Megalops cyprinoides*
Oxudercinae (mudskippers), 261,
   330, 718
Oxyeleotris marmoratus, 259
Oxyjulis californica (señorita), 705

Pachycormiformes, 156
Pachypanchax, 244
Pachystomias, 357
Pacific bluefin tuna. *See Thunnus
   orientalis*
Pacific cod. *See Gadus macrocephalus*
Pacific hagfish. *See Eptatretus
   stoutii*
Pacific hake. *See Merluccius
   productus*
Pacific halibut. *See Hippoglossus
   stenolepis*
Pacific lamprey. *See Lampetra
   tridentata*
Pacific Ocean perch. *See Sebastes
   alutus*
Pacific salmon. *See Oncorhynchus*
Pacific sandfish. *See Trichodon
   trichodon*
Pacific sardine. *See Sardinops sagax*
Pacific whiting. *See Merluccius pro-
   ductus* (Pacific hake)
pacu. *See Piaractus mesopotamicus*
paddlefish. *See Polyodontidae*
Pagothenia borchgrevinki, 643
Pagrus
   P. major (madai), 412, 433
   P. pagrus (red porgy), 828(t)
Paiute sculpin. *See Cottus beldingi*
Palaeacanthaspidoidei, 108
Palaeonisciformes, 48, *151*, 152
Palaeoniscoidei, 152
Palaeoselachii, 117
pallid sturgeon. *See Scaphirhynchus
   albus*
Palonia castelnaui (herring), 174

Pampus, 418
Panderichthyidae, 138
Panderichthys, *139*
Pangasiidae, 621
Pangasius gigas, 187, 673
Pantodon buchholzi (butterflyfish),
   42, 164–65
   jumping, 331
   pectoral fins, 42–44, *45*
Pantodontidae, 164–65, 622
Pantosteus, 619
paperbones. *See Notosudidae*
Parablennius sanguinolentus, 433(t)
Paracanthopterygii, 217–25
Paracheirodon innesi (neon tetra),
   294–95
paradise fish. *See Macropodus*
Parahucho, 201
Paralabrax clathratus (kelp bass),
   705
Paralepididae (barracudinas), 211,
   739(t), 741(t)
Paralepis atlantica (duckbilled bar-
   racudina), 739(t)
Paralichthyidae, 275
Paralichthys
   P. californicus (California hali-
      but), 275
   P. dentatus (summer flounder),
      275, 828(t)
   P. olivaceus (olive flounder), 275
Parapriacanthus, 302
Paraselachii, 111
Parasilurus asotus, *392*
Pardachirus marmoratus (Moses
   sole), 274, 305, 818
Paretroplus, 623
Parexocoetus mento, 244
Parodon affinis, 573
parrotfish. *See Scaridae*
Patagonian toothfish. *See Dissosti-
   chus eleginoides*
peacock flounder. *See Bothus lunatus*
peamouth. *See Mylocheilus*
pearleye. *See Scopelarchidae*
pearlfish. *See Carapidae*
pearl spot. *See Etroplus suratensis*
Pecos pupfish. *See Cyprinodon
   pecosensis*
Pegasidae (seamoths), 35, *39*, 238
pelican gulper. *See Eurypharynx pellecanoides*
pencil catfish. *See Trichomycteridae*
pencilsmelt. *See Microstomatidae*
Pentaceros pectoralis, 828(t)
Peprilus, 735(t)
Peprilus (butterfishes), *260*
Perca (perches)
   muscle fiber types, 314, 316
   optomotor response, 760

   P. flavescens (yellow perch), 34,
      253
      eggs, 526
      glomeruli, 488(t)
      habitat, 686
      ovary, 520
      pyloric caeca, 419
   P. fluviatilis, 253
Percesoces, 239
perch. *See Perca*
Percichthyidae, 253
   distribution, 623
   habitat, 703
Percichthys trucha, 79
Percidae, 253
   chromosome and arm numbers,
      569(t)
   distribution, 610, 611, 613, 615
Perciformes, 251–65, *256*, *260*, 270
Percilia gillissi, 526
Percina (darters), 650
Percoidei, 252–54, 485
Percomorpha, 230–34
   classifications, 236
   phylogeny, 230
Percophidae (duckbills), 257
Percopsidae (sandrollers), 46, 218,
   615
Percopsiformes, 218–19, *220*
Percopsis (troutperches)
   P. omiscomaycus, 218, *220*
   P. transmontana (sandroller), 218
   phototaxis, 757
Periophthalmodon schlosseri (mud-
   skipper), 641, *642*
Periophthalmus, 261
   cutaneous respiration, 451, 452
   eyes, 358, *358*
   gills, 447
permit. *See Trachinotus falcatus*
Perrisodus, 689
Peruvian anchoveta. *See Engraulis
   ringens*
Petalichthyformes, 108
Petalodontiformes, 111, *113*
petrale sole. *See Eopsetta jordani*
Petromyzon
   brain, *495*
   P. marinus (sea lamprey), 351
      chromosome and arm
         numbers, 568(t)
      invasion of Great Lakes, 95,
         424, 829
      plasma ionic concentrations,
         475(t)
      plasma osmolality, 479(t)
Petromyzontidae, 94, 611, 614
Petromyzontiformes (lampreys), *26*,
   27, 94–96
Petrotilapia tridentiger, 426

Phallostethidae (priapiumfishes), 242

Phanerodon furcatus (white seaperch), 433(t), 717

Phanerorhynchiformes, 153

Phareodus, 164

Pholidae (gunnels), 35, 258, 724, 725
  body shape, *38*
  locomotion, 330
  pigmentation, 297

Pholidichthyidae (convict blennies), 257

Pholidichthyoidei, 257

Pholidophoriformes, 160

Pholis
  P. gunnelus, 329(t)
  P. ornata (saddleback gunnel), 726

Photichthyidae (lighthousefishes), 209, 210

Photichthyoidei, 209–10

Photoblepharon palpebratum, 303

Phoxinus (minnows)
  chemotaxis, 760
  hearing, 371
  olfactory threshold, 395(t)
  P. erythrogaster (redbellied dace)
    distribution, 616
    schooling behavior, 783
    taste threshold, 399

Phractolaemidae, 622

Phreatichthys andruzzii (Somalian cavefish), 359, 360

Phycodurus eques (leafy seadragon), 239

Phyllolepiformes, 108

Phyllopteryx foliatus (weedy seadragon), 239

Piaractus mesopotamicus (pacu), 675

pickerel. *See* Esox

Pikaia, 24

pike. *See* Esocidae

pikeblenny. *See* Chaenopsidae

pike conger. *See* Muraenesocidae

pikehead. *See* Luciocephalus pulcher

pikeminnow. *See* Ptychocheilus

pilotfish. *See* Naucrates ductor

Pimelodidae (catfishes), 187

Pimephales (bluntnose minnows)
  chemotaxis, 760
  habitat, 671
  P. promelas, 433(t)

pineconefish. *See* Monocentridae

pinfish. *See* Lagodon rhomboides

Pinguipedidae (sandperches), 257

pink salmon. *See* Oncorhynchus gorbuscha

pipefishes
  amiiform swimming, 326

kidneys, 485
  See also Nerophis; Syngnathus

Pipiscus, 94

piramutaba. *See* Brachyplatystoma vaillantii

piranha. *See* Serrasalminae

pirate perch. *See* Aphredoderus sayanus

Pit-Klamath brook lamprey. *See* Lampetra lethophaga

Pituriaspida, 100–101

Placodermi, 107–9

Plagiotremus townsendi, 300, *300*

plaice. *See* Pleuronectes platessa

plainfin midshipman. *See* Porichthys notatus

Platax, 298

Platichthys
  P. flesus, 328, 478
  P. stellatus (starry flounder), *274*, 275
    eye migration, 273
    habitat, 722

Platybelone (needlefishes), 243

Platycephalidae (flatheads), 270

platyfish. *See* Xiphophorus maculatus

Platyrhinidae, 125

Platysomoidei, 152

Platytroctidae, 196, 303

Plecoglossus (ayus), *195*
  grazing, 425
  kidneys, 485
  P. altivelis (ayu), 196–97
    behavior modification, 792
    digestive enzymes, 433
    distribution, 614
    fishery, 196–97
    migration, 766
    phototaxis, 757

Pleurogrammus monopterygius (Atka mackerel), 271, 626

Pleuronectes
  P. flesus (Atlantic flounder), *393*
    plasma ionic concentrations, 475(t)
  P. platessa (plaice), 275

Pleuronectidae (flounders), 35, 39, 275
  habitat, 704, 716
  kidneys, 485

Pleuronectiformes (flatfishes), 273–75, *274*
  gas bladder, 332
  phylogeny, 272
  pyloric caeca, 419

Pleuronectinae, 275

Pleuronectoidei, 273–75

Pliotrema (saw sharks), 51, *121*, 123

Plotosidae, 305

Plotosus, 485

P. anguillarus (eeltail catfish), 298

plownose chimaera. *See* Callorhynchidae

plunderfish. *See* Harpagiferidae

Pluto, 694

Poecilia (guppies), 72–74, *241*, 760, 781
  P. formosa (Amazon molly), 533
  P. reticulata, 72–74, 245, 297, 350, 478
  P. sphenops, 694

Poeciliidae, 245
  chromosome and arm numbers, 569(t)
  distribution, 619
  gonopodium, 40
  habitat, 695
  viviparity, 529

Poeciliopsis, 245, 529

Pollachius (pollocks)
  commercial fishery, 222, 834
  P. pollachius, 834
  P. virens, 222
    chromosome and arm numbers, 569(t)
    habitat, 726
    schooling behavior, 784

pollock. *See* Pollachius

Polydactylus, 254

Polymixia (beardfishes), *211*, 213

Polymixiidae (beardfishes), 213

Polymixiiformes, *211*, 213

Polymixiomorpha, 213

Polynemidae (threadfins), 44, *45*, 254

Polynemus, 254

Polyodon spathula (paddlefish), 154–56, *155*
  chromosome and arm numbers, 568(t)
  spiracles, 34
  suction feeding, 776

Polyodontidae (paddlefishes), 154–56
  distribution, 611, *612*
  scales, 48

Polyplocodus, *141*

Polypteridae (bichirs), 28, 150–52, 368
  distribution, 622
  dorsal fins, 39
  habitat, 673
  scales, 48
  spiral valve, 419

Polypteriformes, 149–52

Polypterus (bichirs), 151, *151*
  Dahlgren cells, 508
  fins, 42, *42*
  kidney, 484

P. senegalus, 152
  pyloric caecum, 419
  spiracles, 34
  testes, 519

Pomacanthidae (angelfishes), 707

Pomacentridae (anemonefishes), 255, 394, 427, 707, 777

Pomacentrus (anemonefishes)
  P. partitus, 384

Pomadasyidae (grunts), 704, 786

Pomalobus pseudoharengus, 329(t)

Pomatomidae (bluefishes), 703

Pomatomus saltatrix (bluefish)
  habitat, 722, 726
  overfished, 828(t)
  swimming speed, 328

pomfret. *See* Bramidae

Pomoxis (crappies), 253, 672, 686

poolfish. *See* Empetrichthys

popeye catalufa. *See* Pristigenys serrula

porbeagle shark, *460*

porcupinefish. *See* Diodontidae

Porichthyinae, 224

Porichthys (midshipmen), *220*, 224
  habitat, 717
  P. notatus (plainfin midshipman), 302, 305
  P. plectrodon (Atlantic midshipman), 224, 304
  vocal adaptations, 380–81

Porolepiformes, *133*, 136

Porolepimorpha, 136

Poromitra (bigscales), *738*

Potamotrygon, 473

Potamotrygonidae (stingrays), 125, 305, 621

pouched lamprey. *See* Geotria australis

Preitella, 189

Premnas, 47, 708, 788–89

Priacanthidae (bigeyes), 253, 289

priapiumfish. *See* Phallostethidae

prickleback. *See* Stichaeidae

pricklefish. *See* Stephanoberycidae

prickly sculpin. *See* Cottus asper

prickly shark. *See* Echinorhinus cookei

Prionace glauca (blue shark), 120, 627, 768

Prionotus carolinus (northern sea robin), 432(t)

Prionurus, 262

Pristidae (sawfishes), 123–24

Pristiformes, 123–24

Pristigasteridae, 173

Pristigenys serrula (popeye catalufa), *292*

Pristiophoridae (sawsharks), 123

Pristiophoriformes, 123

Pristiophorus (sawsharks), 123

Pristis (sawfishes), 51

    P. pectinata (smalltooth sawfish), 473

    P. perrotteti, 686

Probolobus heterostomus, 299

Prochilodontinae, 184

Prochilodus scrofa, 426

Profundulidae (killifishes), 245, 619, *620*

Profundulus, 245

Proscyllidae (finback cat sharks), 120

Prosopium (whitefishes), 75, 201

    chromosome and arm numbers, 569(t)

    habitat, 670, 686

    P. cylindraceum (round white-fish), 75

    P. williamsoni (mountain white-fish), 75

    P. williamsoni (Rocky Mountain whitefish), 201

Proterorhinus marmoratus (tubenose goby), 842

Protacanthopterygii, 194, 206

Protopteridae, 138

    aestivation, 47

    distribution, 622

    gill apparatus, 445–46

    habitat, 673

Protopterus (African lungfishes), 42, 396, 421, 456, 473

    kidneys, 484

    P. aethiopicus, 138, 531

    P. amphibius, 138

    P. annectans, 138, 539

    P. dolloi, 138

Protospinaciformes, 123

Prototroctes

    P. maraena (Australian grayling), *195*, 197

    P. oxyrhynchus, 197

Psarolepis, 131

Psenes maculatus (silver driftfish), 735(t)

Psephurus gladius (paddlefish), 154

Psetta maxima (turbot), 275

Psettodes, 273

Psettodidae, 273

Psettodoidei, 273

Pseudocarchariidae, 119

Pseudomugilidae (blue-eyes), 241, 623

Pseudopleuronectes americanus, 461, 488(t)

Pseudorasbora, 399

    P. parva, 804

Pseudoscaphirhynchus (shovelnose sturgeon), 154, *155*

P. hermanni, 154

    phylogeny, 153–54

Pseudotriakidae (false cat sharks), 120

Pseudotropheus, 425

    P. zebra, 426

Psychrolutes phrictus, 272

Psychrolutidae (fathead sculpins), 272

Pteraspidomorphi, 97–99, *98*

Pteraspis, 97–98, *98*

Pterichthyes, 108

Pteroinae, 270

Pterois (lionfishes), 270, 298, 306–7

Pterolebias, 244

Pterophyllum scalare (freshwater angelfish), 255

Pterothrissidae, 167

Pterothrissus, 167

Ptilichthyidae (quillfishes), 258

Ptychocheilus (pikeminnows), 182

    habitat, 671

    P. grandis (Sacramento pikemin-now), *76*, 655

    P. lucius (Colorado pikemin-now), *76*, 182

    P. oregonensis (northern pike-minnow), *76*, 182, 432(t)

    P. umpquae, *76*

    pharyngeal teeth, *417*, 429

    species distribution, 75, 619

Ptyctodontiformes, 108

puffer. *See* Tetraodontidae

pumpkinseed. *See Lepomis gibbosus*

Pungitius (ninespine sticklebacks), 238

Puntius javanicus (barb), 425

Pycnodontiformes, 156

Pycnosteus, 98

pygmy loach. *See Botia sidthimunki*

pygmy perch. *See* Nannopercidae

pygmy sunfish. *See* Elassoma

Pygocentrus (piranhas)

    P. nattereri (red-bellied piranha), 185

Pylodictes olivaris (flathead catfish), 187–88, 672, 838

Pyrrhulina filamentosa, 432(t)

Queensland lungfish. *See Neocera-todus forsteri*

quillback. *See Carpiodes*

quillfish. *See* Ptilichthyidae

rabbitfish. *See* Chimaeriformes (chimaeras)

Rachycentridae (cobias), 703, 722

Raconda, 173

ragfish. *See Icosteus*

rainbowfish. *See* Melanotaeniidae

rainbow trout. *See Oncorhynchus mykiss*

rainwater killifish. *See Lucania parva*

Raja, 379, 458(t), 488(t), 524, 568(t)

Rajidae (skates), *38*, 124–25

    commercial fishery, 125, 704

    distribution, 632

    electric organs, 376, *378*

    electricreceptors, 369

    habitat, 704

Rajiformes, 124–25

Rajomorphii, 123–25, *124*

    classification, 122(t)

Ranzania laevis (slender mola), 735(t)

Rastrelliger, 263

Rastrineobola argentea, 831

ratfish. *See* Chimaeriformes (chimaeras)

rays, 26, 28

    classification, 115(t), 122(t)

    electroreceptors, 369

    endolymphatic duct, 365

    fins, 39, 44

    gills, 37, 443

    kidneys, 483–84

    musculature, 316

    negative bouyancy, 332

    neurohypophyseal hormones, 505

    olfaction, 393–94

    rajiform locomotion, 327

    rostrum, 51

    skull, 51

    taxonomy, 123–25

    See also Myliobatiformes

razorback sucker. *See Xyrauchen texanus*

redbellied dace. *See Phoxinus eryth-rogaster*

red-bellied piranha. *See Pygocentrus nattereri*

red drum. *See Sciaenops ocellatus*

redear sunfish. *See Lepomis microlophus*

Redfieldioidei, 152

redfish. *See* Sebastinae

redmouth whalefish. *See Rondeletia*

red oscar. *See Astronotus ocellatus*

red porgy. *See Pagrus pagrus*

redside. *See Richardsonius*

red snapper. *See Lutjanus campechanus*

redtail surfperch. *See Amphistichus rhodoterus*

red whalefish. *See Barbourisia rufa*

reedfish. *See Erpetoichthys calabaricus*

reef shark. *See Carcharhinus perezi*

Regalecidae (oarfishes), 213

Regalecus glesne (oarfish), 213, 734

remora. *See* Echeneidae

Remora remora (remora), 39, 735(t)

requiem shark. *See* Carcharhinidae

Retropinna, 197

Retropinnidae, 197, 760, 766

rex sole. *See Glyptocephalus zachirus*

Rhamphichthyidae, 39, 189

Rhenaniformes, 108, *109*

Rhincodontidae (whale sharks), 118–19, 735(t)

Rhincodon typus (whale shark), 28, 117, 118–19

    habitat, 735(t)

    overfished, 828(t)

    suction feeding, 776

Rhinecanthus aculeatus, 329(t)

Rhinichthys (daces)

    distribution, 619

    habitat, 667

    mouth position, *34*

    R. cataractae (longnose dace), 334, 654, 774

Rhinidae, 124

Rhiniformes, 124

Rhinobatidae, 124

Rhinobatiformes, 124

Rhinochimaeridae (longnose chi-maeras), 113–14

Rhinogobius brunneus, 757

Rhinomugil corsula (Indian mullet), *358*, 358–59

Rhinoptera (cownose rays), 125

Rhipidistia, 138

Rhizodontimorpha, 138

Rhodeus (bitterlings), 788, *791*

Rhomboplites aurorubens (vermilion snapper), 828(t)

Rhyacichthyidae (loach gobies), 259, 667

Rhyncolepis, 99

ribbonfish. *See* Trachypteridae

rice eel. *See Monopterus albus*

Richardsonius (redsides), 671

Rimicola muscarum (kelp clingfish), 705

Rio Grande perch. *See Herichthys cyanoguttatum*

river lamprey. *See Lampetra ayresi*

Rivulidae, 244–45, 688

Rivulus

    R. hartii (killifish), 73

    R. marmoratus (mangrove rivu-lus), 244–45, 531, 718

roach. *See Hesperoleucus*

rock bass. *See Ambloplites*

rock cod. *See* Sebastinae

rockfish. *See* Sebastidae

rockling. *See* Lotidae

rock prickleback. *See Xiphister mucosus*
rock sole. *See Lepidopsetta bilineata*
Rocky Mountain whitefish. *See Prosopium williamsoni*
Rondeletia (redmouth whalefishes), 232
Rondeletiidae, 232
ronquil. *See Bathymasteridae*
rougheye rockfish. *See Sebastes aleutianus*
rough pomfret. *See Taractes asper*
rough scad. *See Trachurus symmetricus*
round goby. *See Neogobius melanostomus*
round stingray. *See Urobatis halleri*
round whitefish. *See Prosopium cylindraceum*
rudderfish. *See Kyphosus elegans*
Russian sturgeon. *See Acipenser gueldenstaedti*
Ruvettus (oilfishes), 290, 739(t)

sabertooth. *See Evermannella atrata*
sabertooth blenny. *See Meiacanthus nigrolineatus*
sabertooth fish. *See Evermannellidae*
sablefish. *See Anoplopomatidae*
Sacabambaspis, 25, 96, *96*, 97
Saccobranchus fossilis, 453
Saccopharyngidae, 172, 741(t)
Saccopharyngiformes, 172
Saccopharyngoidei, 172
Saccopharynx ampullaceus (whiptail gulper), 741(t)
Sacramento perch. *See Archoplites interruptus*
Sacramento pikeminnow. *See Ptychocheilus grandis*
saddleback gunnel. *See Pholis ornata*
sailbearer. *See Veliferidae*
sailfish. *See Istiophorus*
saithe. *See Gadus virens*
salamanderfish. *See Lepidogalaxias*
Salangidae, 197
   commercial fishery, 197
   distribution, 614
   paedomorphosis, 541
Salarias, 257
Salmo, 198
   life history, 199
   migration, 764–65
   S. salar (Atlantic salmon), 198, 735(t)
      aquaculture, 839
      body shape, 35
      genetic engineering, 591
      magnetotaxis, 759

osmoregulation, 481
overfished, 828(t)
plasma osmolality, 479(t)
UV sensitivity, 296
S. trutta (brown trout), 198, 667, 779, 842
   metabolic rate, 409(t)
   spotting, 559, *559*
swimming speed, 328
salmon
   aquaculture, 480, 838
   chemotaxis, 760
   dietary requirements, 412, 413
   gill raker morphology, 416
   and global warming, 844
   homing behavior, 396
   migration, 764
   musculature, *313*
   Native American fishery, 200–203
   North Atlantic fishery, 831
   and ozone depletion, 844
   Pacific Northwest fishery, 200–203, 830–31
   pigmentation, 295
   rheotaxis, 760
   sport fishing, 837
   vertebrae, 60
   See also Oncorhynchus; Salmo; Salmonidae
Salmonidae, 198–201
   alimentary canal, *418*
   distribution, 611, 626
   fungal infections, 807
   habitat, 667, 669, 670, 703
   kidneys, 485
   olfactory threshold, 395
   phototaxis, 757
   phylogeny, *199*
   pyloric caeca, 419
   rheotaxis, 760
   vibriosis, 805
   whirling disease, 808
Salmoniformes, 196–201
Salmoninae, 198–201
Salmonoidei, 198–201, 686
salmon shark. *See Lamna ditropis*
Salmopercae, 218
Salvelinus (chars), 201
   chromosome and arm numbers, 569(t)
   distribution, 626
   habitat, 667
   S. alpinus (Arctic char), 395, 396, 632
      metabolic rate, 409(t)
   S. confluentus (bull trout), 201
   S. fontinalis (brook trout), 201, 777, 813
   S. malma (Dolly Varden), 201

S. namaycush (lake trout), 201, 829
   habitat, 686
Samaridae, 274
sandbar shark. *See Carcharhinus plumbeus*
sandburrower. *See Creediidae*
sanddab. *See Citharichthys*
sand diver. *See Trichonotidae*
sand eel. *See Hypoptychidae*
Sander, 253, 686, 829
   S. vitreus, 409(t)
sandfish. *See Trichodontidae*
sand goby. *See Kraemeriidae*
sandlance. *See Ammodytidae*
sandperch. *See Pinguipedidae*
sandroller. *See Percopsidae*
sand stargazer. *See Dactyloscopidae*
sand tiger shark. *See Odontaspididae*
Sanopus splendidus, 223
Sarcopterygii, 28, 131–38, *133*
   classification, 131–32
   fins, 60–61
   phylogeny, *132*
Sarda (bonitos), 263
sardine. *See Sardinops*
Sardinella (herrings), 425
Sardinops (sardines), 735(t)
   carangiform swimming, 321
   S. sagax (Pacific sardine), 174
      commercial fishery, 832
      distribution, 627
sargassumfish. *See Histrio histrio*
sargassum pipefish. *See Syngnathus pelagicus*
Sarotherodon, 721
Satan, 189, 694
Saurichthyiformes, 153
Sauripterus, 141
saury. *See Scomberesocidae*
sawfish. *See Pristidae*
saw shark. *See Pristiophoridae*
Scaphirhynchinae (sturgeons), 611
Scaphirhynchus (sturgeons), 154
   S. albus (pallid sturgeon), 154
   S. platorhynchus, 154, 568(t)
   S. suttkusi (Alabama sturgeon), 154
Scaridae (parrotfishes), 254–55, 425, 707
   alimentary canal, *418*
   beak, 415, *416*
   mucous envelope, 47
Scartichthys viridis, 726
Scarus schlegeli, 329(t)
scat. *See Scatophagidae*
Scatophagidae (scats), 262
Schilbeidae, 621
Schindleria

S. brevipinguis, 261
   locomotion, 734
   neoteny, 541
S. praematura, 261
Schindleriidae, 261
Sciaenidae, 253, 254
   distribution, 621
   habitat, 704, 721
   sound production, 382
Sciaenops ocellatus (red drum), 828(t), 837
Scleropages
   S. formosus, 164, 838
   S. jardini, 164, 623
   S. leichardti, 164, 623
Scoloplacidae, 187
Scomber (mackerels), 59, 335, 462
   S. japonicus, 263, *827*
   S. scombrus, 263, 328, 329(t), 627
Scomberesocidae (sauries), 243
   alimentary canal, 418
   anguilliform swimming, 320
   distribution, 627
   habitat, 734, 735(t)
Scomberesox (sauries), *241*
   S. saurus (Atlantic saury), 735(t)
Scomberomorus (Spanish mackerels), 263, 334
   S. cavalla, *325*, 828(t)
Scombresocidae (sauries), 40
Scombridae, 263
   body shape, *38*
   commercial fishery, 263
   finlets, 40
   fins, 39
   habitat, 703, 735(t)
   jumping, 330
   kidneys, 485
   musculature, 322–24
   pyloric caeca, 419
Scombroidae, 767
Scombroidei, 262
   commercial fishery, 832–33
   drag reduction, 289
   pelagic adaptations, 733–34
Scombrolabracidae, 262, 373
Scombrolabracoidei, 262
Scombrolabrax heterolepis (black mackerel), 262
Scopelarchidae (pearleyes), 210, 739(t)
Scopelarchus (pearleyes), 357
Scopelomorpha, 212
Scophthalmidae, 275
Scophthalmus aquosus (windowpane flounder), 828(t)
Scorpaenidae, 270
   habitat, 704, 742
   larviparous, 529
   venomous, 306

Scorpaeniformes, *271*

Scorpaenoidei, 270–72

scorpionfish. *See* Sebastinae

scrawled filefish. *See Aluterus scriptus*

sculpin. *See* Cottidae

scup. *See Stenotomus chrysops*

Scyliorhinidae, 120, *525*

Scyliorhinus canicula (spotted dogfish), 314, 316, 373

Scytalinidae (graveldivers), 258, 330

scythe butterflyfish. *See Chaetodon falcifer*

seadevil. *See Cryptopsaras couesi*

seahorses. *See also Hippocampus;* Syngnathidae
  amiiform swimming, 326
  body shape, 35
  kidneys, 485

sea lamprey. *See Petromyzon marinus*

seamoth. *See* Pegasidae

sea raven. *See Hemitripterus americanus*

searobin. *See* Triglidae

sea snail. *See Careproctus*

sea toad. *See* Chaunacidae

seatrout. *See Cynoscion*

Sebastes (rockfishes), *563*
  coloration, 741
  commercial fishery, 833
  homing behavior, 761
  overfished, 828(t)
  S. aleutianus (rougheye rockfish), 411
  S. alutus (Pacific Ocean perch), 270
  S. borealis (shortraker rockfish), 411
  S. diploproa, 349–50
  S. fasciatus (Acadian redfish), 270
  S. flavidus (yellowtail rockfish), 761–62
  S. mentella (deepwater redfish), 270
  S. norvegicus (golden redfish), 336–37
  S. paucispinis (bocaccio), 429
  S. serriceps (treefish), *297*
  S. wilsoni, 429

Sebastidae (rockfishes), *36*

Sebastinae, 270

Selene vomer, *539, 541*

Semionotiformes, 156–57

Semotilus (creek chub), 667, 671

señorita. *See Oxyjulis californica*

Seriola (buris), 705
  S. quinqueradiata, 412

Serranidae (groupers), 253, 787, 817

Serrasalminae (piranhas), 184–85, 775

Serrasalmus (piranhas), 415
  S. elongatus, 424

seven-gill shark. *See* Heptranchidae

shad. *See Alosa*

sharks, *26*, 27–28
  age and growth, 411
  angel shark, 35
  anguilliform swimming, 318
  appendicular skeleton, *62*
  blood, 463
  brain, *495*
  branchial arteries, *457*
  and cancers, 815
  classification, 114, 115(t), 117, 120
  commercial uses, 119, 120
  dermal denticles, 48
  drag reduction, 290–91
  egg cases, *525*
  electrosensitivity, 376
  endolymphatic duct, 365
  eyes, 347–48
  fins, 44
  gills, 37, 443, *444, 445*
  habitat, 707
  heart, *455*
  heat exchange, 460
  jaw suspension, 53
  kidneys, 483–84
  liver, 421
  mouth, *35*
  negative buoyancy, 332
  neurohypophyseal hormones, 505
  olfaction, 393–94, 395–96
  placentae, 527, *527*
  placoid scales, 48
  planktivorous, 117, 118, 119
  and poisoning, 818
  reproductive anatomy, 518–19
  skull, 51, *53*
  sound location, 373
  spiracular organs, 370
  spiral valve, 419
  taste preferences, 397–98
  teeth, 415
  thunniform swimming, 321
  urogenital organs, *484*
  viscera, *420*
  viviparous, 527, 529

sheatfish. *See* Siluridae

sheepshead minnow. *See Cyprinodon variegatus*

shell-ear. *See Kneria auriculata*

shiner. *See Luxilus; Notropis*

shiner perch. *See Cymatogaster aggregata*

shorthead sculpin. *See Cottus confusus*

shortnose greeneye. *See Chlorophthalmus agassizi*

shortnose sturgeon. *See Acipenser brevirostrum*

shortnose sucker. *See Chasmistes brevirostris*

shortraker rockfish. *See Sebastes borealis*

shovelnose sturgeon. *See Pseudoscaphirhynchus*

Siamese fighting fish. *See Betta splendens*

Sicyaces sanguineus, 259, 725

Sicydiinae, 261

Siganidae (spinefoots), 261, 707

Siganus, 434

silky shark. *See Carcharhinus falciformis*

Siluridae (sheatfishes), 187, 614, 621

Siluriformes, 186–89, *188*
  distribution, 621
  phylogeny, *186*

Siluris (wels), *188*
  S. glanis (European wels), 187

silver arowana. *See Osteoglossum bicirrhosum*

silver carp. *See Hypophthalmichthys molotrix*

silver driftfish. *See Psenes maculatus*

silver hake. *See Merluccius bilinearis*

silversides. *See* Atherinidae

silvery minnow. *See Hybognathus*

Simenchelys parasitica (snubnose eel), 171, 787

Siphamia, 303

Sisoridae (hill stream catfishes)
  distribution, 614, 621
  pectoral fins, 42

six-gill shark. *See Hexanchus*

skates, *26*, 28, 124–25. *See also* Rajidae
  classification, 122(t)
  fins, 39
  habitat, 742
  pelvic claspers, 527
  punting locomotion, 330
  rajiform locomotion, 327

skilfish. *See Erilepis zonifer*

skipjack tuna. *See Katsuwonus pelamis*

Sladenia, 224

sleeper shark. *See* Dalatiidae

slender barracudina. *See Lestidium ringens*

slender mola. *See Ranzania laevis*

slickhead. *See* Alepocephalidae

slipmouth. *See* Leiognathidae

smalleye squaretail. *See Tetragonurus cuvieri*

smalltooth sawfish. *See Pristis pectinata*

Smegmamorpha, 236

smelt. *See* Osmeridae; Retropinnidae

smooth dogfish. *See* Triakidae

smooth stargazer. *See Kathetostoma averruncus*

smooth stingray. *See Dasyatis brevicaudata*

snaggletooth. *see* Astronesthinae

snailfish. *See* Liparidae

snake eel. *See* Ophichthidae

snakehead. *See* Channidae

snake mackerel. *See Gempylus serpens*

snapper. *See* Lutjanidae

snipe eel. *See* Nemichthyidae

snipefish. *See* Macrorhamphosidae

snook. *See* Centropomidae

snowy grouper. *See Epinephelus niveatus*

snubnose eel. *See Simenchelys parasitica*

soapfish. *See* Grammistidae

sockeye salmon. *See Oncorhynchus nerka*

soldierfish. *See Myripristis*

sole. *See* Soleidae

Solea (soles), *274*
  S. solea, 274

Soleidae (soles), 274, 621

Solenostomidae, 239

Solenostomus (ghost pipefishes), 239

Somalian cavefish. *See Phreatichthys andruzzii*

Somniosus macrocephalus (Greenland shark), 120, *495, 818*

soupfin shark. *See Galeus zyopterus*

South American lungfish. *See Lepidosiren*

spadefish. *See* Ephippidae

spaghetti eel. *See* Moringuidae

Spanish mackerel. *See Scomberomorus*

Sparidae, 35, 704, 721, 722

Sparisoma
  S. radians (bucktooth parrotfish), 720
  S. viride, 507

speckled hind. *See Epinephelus drummondhayi*

Speoplatyrhinus (cavefishes), 222

Sphaerichthyes osphromenoides, 569(t)

Sphoeroides, 818
  S. rosenblatti (pufferfish), 718

Sphyraena (barracudas)

and ciguatera, 817
kidneys, 485
S. barracuda (great barracuda), 262
teeth, 415
Sphyraenidae (barracudas), 262, 703, 707
Sphyrna (hammerhead sharks), 120
  S. leweni, 829(t)
  S. mokarran, 120
  S. zygaena, 120, 829(t)
Sphyrnidae (hammerhead sharks), 120
spikefish. *See* Triacanthoidei
Spinachia (fifteenspine sticklebacks), 238
spinefoot. *See* Siganidae
spinner shark. *See* Carcharhinus brevipinna
spiny eel. *See* Mastacembeloidei
spinyfin. *See* Diretmidae
Spirinchus starksi (night smelt), 722
splitfin. *See* Goodeinae
spookfish. *See* Macropinna
spot. *See* Leiostomus xanthurus
spotfin mojarra. *See* Eucinostomus argenteus
spotted dogfish. *See* Scyliorhinus canicula
spotted flyingfish. *See* Cheilopogon pinnatibarbatus
sprat. *See* Sprattus sprattus
Sprattus sprattus (sprat), 174
spring cavefish. *See* Chologaster agassizi
springer. *See* Elops machnata
springfish. *See* Crenichthys
Squalea, 120, *121*
Squalidae (dogfishes), 704, 739(t), 742
Squaliformes, 122–23
Squaliolus (dwarf sharks), 28
  S. laticaudus, 122
Squaloidea, 122–23
Squalus (dogfishes)
  blood pressure, 458(t)
  commercial fishery, 122
  S. acanthias, 122
    chromosome and arm numbers, 568(t)
    gill filaments, *449*
    ionic balance in salt water, 476(t)
  squalamine antibiotic, 122
  venomous, 305
squaretail. *See* Tetragonuridae
Squatina (angel sharks), 35, 123
Squatinidae, 35
Squatiniformes (angel sharks), 123

squeaker. *See* Misgurnus fossilis
squirrelfish. *See* Holocentrus
squirrelfish snapper. *See* Etelis carbunculus
staghorn sculpin. *See* Leptocottus armatus
stalkeye. *See* Idiacanthinae
stargazer. *See* Uranoscopidae
starry flounder. *See* Platichthys stellatus
Steatogenys, 379
steelhead trout. *See* Oncorhynchus mykiss
Stegostoma (zebra sharks), 118
Steindachneriidae, 303
Stemoptychidae, 739(t)
Stenobrachius leucopsaurus (northern lampfish), 739(t)
Stenodus, 201, 626
Stenopterygii, 207–10
Stenotomus chrysops (scup), 828(t)
Stephanoberycidae (pricklefishes), 232
Stephanoberyciformes, 230–32
Stephanolepis (filefishes), 736
sterlet. *See* Acipenser ruthenus
Sternoptychidae, 189, 209, 627
Sternopygoidei, 189
Sternopygus, 379, 380
Stichaeidae (pricklebacks), 35, 258
  cutaneous respiration, 724
  habitat, 725
  locomotion, 330
  pigmentation, 297
stickleback. *See* Gasterosteidae
stingray. *See* Dasyatidae; Potamotrygonidae
Stizostedion
  S. canadense (North American sauger), 253
  S. lucioperca, 253
  S. vitreum (walleye), 253 ·
Stokellia, 197
Stomiatidae, 739(t)
Stomiidae, 209–10, 335, 741(t)
Stomiiformes, 207–10, *209*
  distribution, 626
  luminescence, 47, 300
  teeth, 415
Stomiinae, 209
stonefish. *See* Synanceiidae
stoneroller minnow. *See* Campostoma
streaked gurnard. *See* Trigloporus lastoviza
striped bass. *See* Morone saxatilis
striped mullet. *See* Mugil cephalus
striped shiner. *See* Luxilus chrysocephalus

Stromateidae (butterfishes), 264
  esophagus, 418, 430
  habitat, 735(t)
Stromateidoidei, 416
Stromateoidei, 263–64
Strongylura kreffti, 243, 520
Strunius, 132, *133*
sturgeons
  commercial uses, 152–54
  mouth position, *34*
  spiral valve, 419
  See also Acipenseridae; Scaphirhynchinae
Stylephoridae (tube-eyes), 213
Stylephoroidei, 213
Stylephorus chordatus (tube-eye), 213
sucker. *See* Catostomidae
summer flounder. *See* Paralichthys dentatus
Sundasalanx, 197
sunfish. *See* Lepomis
surfperch. *See* Embiotocidae
surf silversides. *See* Notocheiridae
surf smelt. *See* Hypomesus pretiosus
surgeonfish. *See* Acanthuridae
swallower. *See* Chiasmodontidae; Saccopharyngidae
swamp eel. *See* Mastacembelidae; Synbranchidae
swampfish. *See* Chologaster cornuta
swell shark. *See* Cephaloscyllium
swordfish. *See* Xiphiidae
Symbolophorus, 735
  S. californiense, *540*
Symmoriiformes, 117
Symphurus, 274
Symphysodon discus (discus), 255
Synanceia horrida (stonefish), 307, *307*
Synanceiidae (stonefishes), 270–71
  pectoral fins, 44
  spines, *47*
  venom, 307
Synanceinae (stonefishes), 297
Synaphobranchidae, 171, 739(t)
Synaphobranchus, 337
Synbranchidae (swamp eels), 37, 236, *237*, 330, 622
Synbranchiformes, 236–37, *237*
Synbranchoidei, 236
Synbranchus, 236, 452, 453
Synentognathi, 242
Syngnathidae, 35, 238–39
  alimentary canal, 418
  body shape, *39*
  osmoregulation and brooding, 479
  pigmentation, 297

Syngnathoidei, 238–39
Syngnathus (pipefishes), 39, 238
  S. pelagicus (sargassum pipefish), 736
Synodontidae (lizardfishes), 211, 212, 704
Synodus luciocaps, 569(t)

Tachysurus, 526
Tactostoma macropus (longfin dragonfish), 739(t)
Takifugu rubripes (pufferfish), 275, 590, 817
tambaqui. *See* Colossoma macropomum
Tanichthys albonubes (white cloud minnows), 785
tapetail. *See* Eutaeniophorinae
Taractes
  T. asper (rough pomfret), 289, *292*
  T. longipinnis (bigscale pomfret), 735(t)
Tarletonbeania, 212, 335
tarpon. *See* Megalopidae
Tarrasiiformes, 153
Tautogolabrus adspersus (cunner), 431, 761, 762
Teleostei, *26*, 28–29, 159–74
  appendicular skeleton, 61–62, *63*
  blood, 462, 463
  branchial arches, 59
  branchial arteries, *457*
  branchial pumps, 450
  caudal heart, 459
  cerebellum, 497
  eggs, *525*
  embryonic stages, *537*
  endocrine system, *500*, 500–501
  eyes, 347, 353–54, *355*
  fin rays, 64
  gills, 443, 445, *445*, 446, *448*
  gill surface area, 447(t)
  glomeruli, 488(t)
  gonads, *520*
  gustatory responses, 398
  heart, 455–56
  inner ear, 364
  ionic balance, 475(t)
  jaws, 427–28
  kidneys, 484–85, *485*, 486
  lateral line system, 367
  nephron, *487*
  neurohypophyseal hormones, 505
  olfaction, 394
  olfactory bulb, 391
  osmoregulation, 477, 480

Teleostei (*continued*)
  osmoregulation in freshwater, 473, *474*
  osmoregulation in salt water, 473–75, *474*, 477–78
  pancreatic islets, 506
  phylogeny, *163*
  pituitary gland, 504
  reproductive anatomy, 519–20
  scales, 48, *286*, *287*
  skull, 54–59
  teeth, 415, *416*
  tubal bladder, 65
  vertebrae, 59, *60*
  vertebral column, 60
Teleostomi, 28–29, 129–38
telescopefish. *See* Giganturidae
Telmatherina ladigesi (Celebes rainbowfish), 241–42
Telmatherinidae, 241–42
tench. *See Tinca tinca*
tenpounder. *See Elops saurus*
tetra. *See* Characidae
Tetrabrachiidae, 224–25
Tetrabrachium ocellatum, 224–25
Tetragonuridae (squaretails), 264, 735(t)
Tetragonurus (squaretails), 289–90
  T. atlanticus (bigeye squaretail), 735(t)
  T. cuvieri (smalleye squaretail), 264, *293*
Tetraodon, 392, 818
Tetraodontidae (puffers), 276, 326, 722, 817
Tetraodontiformes, 275–77
  adductor mandibulae, 427
  habitat, 707
  phylogeny, 272
Tetraodontoidea, 276–77
Tetraodontoidei, 276–77
Tetrapturus albidus (white marlin), 735(t), 829(t)
Texas silverside. *See Menidia clarkhubbsi*
Thalassoma duperreyi, 573
Thalassophryne, 223
Thalassophryninae (toadfishes), 224
Thaleichthys pacificus (eulachon), 196, 333
Thallassenchelys coheni, *294*
Thaumatichthys, 742–43
Thelodonti, 98–99, *99*, 106–7
Thelodus parvidens, 98
Theragra, 424
  T. chalcogramma (walleye pollock), 222
    commercial fishery, *827*
    distribution, 626
    larvae, 542

Thoracocharax, 184
thornfish. *See* Bovichtidae
threadfin. *See* Polynemidae
threadsail. *See* Aulopidae
threefin. *See* Tripterygiidae
threespine stickleback. *See Gasterosteus aculeatus*
threetooth puffer. *See Triodon macropterus*
thresher shark. *See* Alopiidae
Thryssa hamiltonii, 721
Thunnus (tunas), 735(t)
  blood, 462
  commercial fishery, 832–33
  and scombroid poisoning, 817
  T. alalunga (albacore), 263, *325*, 335
    distribution, 627
    habitat, 705
    migration, 768
  T. albacares (yellowfin tuna), 263
    commercial fishery, *827*
    magnetotaxis, 376, 759
    swimming speed, 329(t)
  T. obesus (Atlantic bigeye tuna), 829(t)
  T. orientalis (Pacific bluefin tuna)
    distribution, 627
  T. thynnus (Atlantic bluefin tuna), 263
    distribution, 627
    migration, 768
    swimming speed, 328, 329(t)
  thunniform swimming, 321
  vertebral column, 59
Thymallinae, 198
Thymallus (graylings), 201
  olfactory threshold, 395(t)
  T. arcticus (Arctic grayling), 201, 395
tidepool sculpin. *See Oligocottus maculosus*
tidewater goby. *See Eucyclogobius newberryi*
tigerfish. *See Hydrocynus*
tiger shark. *See Galeocardo cuvieri*
Tilapia
  aquaculture, 838
  dietary requirements, 412
  habitat, 696, 721
  hearing, 373
  T. galilaea, 428
  T. nilotica, 433(t)
  T. zilli, 255, 399
tilefish. *See* Malacanthidae
Tinca tinca (tench), 672
toadfishes
  body shape, 35

fins, 44
  See also Batrachoididae; *Opsanus*
tonguefish. *See* Cynoglossidae
tope. *See Galeorhinus*
topminnow. *See* Goodeinae (splitfins)
Torpedinidae (electric rays), 123, 376
Torpediniformes, 123, *124*
Torpedo (electric rays), 123, *124*
  electric organs, 376, *377*, *378*, 379
torrent fishes, 257, 667
torrent sucker. *See Moxostoma rhothoecum*
Tor tor (mahseer), 181
Toxotes jaculator (archerfish), 359
Toxotidae, 622, 673
Trachichthyidae, 233, 303
Trachichthyoidei, 233
Trachinidae (weevers), 47, 256, 307, *307*
Trachinoidei, 255–57
Trachinotus
  T. carolinus (Florida pompano), 722
  T. falcatus (permit), 298, 722
Trachinus (weevers), 47, *256*, 298
Trachipteroidei, 213
Trachipterus
  eyes, 346
  fins, 39
  T. altivelus (king-of-the-salmon), 213
    fins, 39
  T. arcticus (dealfish), 213, 739(t)
Trachurus (jack mackerels)
  fins, 40
  T. mediterraneus, 329(t)
  T. murphyi
    commercial fishery, *827*
  T. symmetricus (rough scad), 328, 735(t)
Trachypteridae (ribbonfishes), 213, 290, 739(t)
treefish. *See Sebastes serriceps*
Trematomus, 462
Triacanthodidae, 276
Triacanthoidei, 276
Triakidae (smooth dogfishes), 120
Tribolodon (Japanese daces), 349
Trichiuridae, 35, 262, 326
Trichiurus (hairtails), 35, 332
  T. lepturus (Atlantic cutlassfish), *827*
Trichodontidae (sandfishes), 257
Trichodon trichodon (Pacific sandfish), 722
Trichogaster (gouramis), 44, 47
Trichomycteridae, 187

Trichonotidae (sand divers), 257
Tridentiger trigonocephalus, 842
triggerfish. *See* Balistidae
Triglidae (searobins), 271
  chemosensory cells, 400
  locomotion, 330
  pectoral fins, 44, *45*
  sound production, 382
Trigloporus lastoviza (streaked gurnard), 775
Trimmatom nanus, 259
Triodon macropterus (threetooth puffer), 276
Triodontidae, 276
tripletail. *See Lobotes*
tripodfish. *See Bathypterois*
Tripterygiidae (threefins), 257, 722, 725
Trogloglanis, 189, 694
trout
  aquaculture, 839
  diet, 412
  dietary diseases, 814
  mechanosensory functions, 374
  mouth position, *34*
  olfactory organ, *392*, *393*
  pigmentation, 295
  rheotaxis, 760
  subcarangiform swimming, 321
  See also Oncorhynchus; Salmo
troutperch. *See Percopsis*
trumpetfish. *See* Aulostomidae
trunkfish. *See* Ostraciidae
Trypauchen (burrowing gobies), 261
Tselfatoidei, 161
tube-eye. *See* Stylephoridae
tubenose goby. *See Proterorhinus marmoratus*
tubeshoulder. *See* Platytroctidae
tubesnout. *See* Aulorhynchidae
tui chub. *See Gila bicolor*
Tulerpeton, 141
tunas
  body shape, *38*
  drag reduction, 289
  growth rate, 325
  heat exchange, 460
  musculature, 313
  ram irrigation, 451
  swimming speed, 328
  thunniform swimming, 321, 324
  See also Thunnus
tunicates, 21–23, *22*
turbot. *See Psetta maxima*
two-spotted goby. *See Gobiusculus flavescens*
Tylosurus crocodilus (houndfish), 243

Typhilichthys (cavefishes), 222, 694
Typhlobagrus, 694
Typhlogobius (blind gobies), 259
 T. californiensis, 261, 790

Umbra
 habitat, 687
 species distribution, 74–75, *76*
 U. krameri (Atlantic
  mudminnow)
  distribution, *76*, 611
 U. limi (central mudminnow),
  207, 453
  chromosome and arm
   numbers, 569(t)
  distribution, *76*, 611
 U. pygmaea (eastern mudminnow), *76*, 611
Umbridae (mudminnows), 206–7,
 611
unicornfish. *See Naso*
Uranoscopidae (stargazers), 256,
 297, 704
Urobatis halleri (round stingray),
 722
Urochordata, 21–23
Urolophus halleri (stingray), *306*
Urophycis, 400, 828(t)
Urotrygonidae, 125

Vandellia (candirus), 187, 787
Veliferidae (sailbearers), 213
Veliferoidei, 213
velvet catfish. *See Diplomystidae*
vermilion snapper. *See Rhomboplites
 aurorubens*
Vinciguerria, 210
viperfish. *See Chauliodontinae*

wahoo. *See Acanthocybium*
walking catfish. *See Clarias*
Wallagonia attu, 187
walleye. *See Stizostedion vitreum*
walleye pollock. *See Theragra
 chalcogramma*
Warsaw grouper. *See Epinephelus
 nigritus*
weakfish. *See Cynoscion regalis*
weasel shark. *See Hemigaleidae*
weatherfish. *See Misgurnus*
weedy seadragon. *See Phyllopteryx
 foliatus*
weever. *See Trachinidae*
wels. *See Siluris*
western brook lamprey. *See Lampetra richardsoni*
whale shark. *See Rhincodontidae*
whipnose anglerfish. *See Gigantactis*
whiptail gulper. *See Saccopharynx
 ampullaceus*
white cloud minnow. *See Tanichthys
 albonubes*
white croaker. *See Genyonemus
 lineatus*
whitefish. *See Coregoninae*
white grunt. *See Haemulon plumieri*
white marlin. *See Tetrapturus albidus*
white seaperch. *See Phanerodon
 furcatus*
white shark. *See Carcharodon
 carcharias*
white sturgeon. *See Acipenser
 transmontanus*
windowpane flounder. *See Scophthalmus aquosus*
witch flounder. *See Glyptocephalus
 cynoglossus*

wobbegon, 118
wolf-eel. *See Anarrhichthys*
wolffish. *See Anarhichadidae*
wolf herring. *See Chirocentrus*
woody sculpin. *See Clinocottus a
 nalis*
worm eel. *See Moringuidae;
 Ophichthidae*
wormfish. *See Microdesmidae*
wrymouth. *See Cryptacanthodidae*

Xenacanthiformes, 117
Xenisthmidae, 261
Xenocephalus egregius (freckled stargazer), 722
Xenocharax spirulus, 432(t)
Xenomystis, 165
Xenopoecilus, 243
Xiphactinus, 161
Xiphias gladius (swordfish), 263
 distribution, 627
 gas bladder, 335
 habitat, 735(t)
 larva, *540*
 overfished, 829(t)
Xiphiidae (swordfishes), 263
 habitat, 735(t)
 locomotion, 734
 scales, 287
Xiphister
 X. atropurpureus (black prickleback), 432(t)
 X. mucosus (rock prickleback),
  425, 432(t), 726
Xiphophorus
 locus linkage map, *566*
 thyroid gland, 505
 X. helleri (green swordtails), 597

X. maculatus (platyfish), 573,
 597
Xyrauchen, 75
 X. texanus (razorback s
  ucker), *77*

yellowfin goby. *See Acanthogobius
 flavimanus*
yellowfin sole. *See Limanda aspera*
yellowfin tuna. *See Thunnus
 albacares*
yellow perch. *See Perca flavescens*
yellowtail flounder. *See Limanda
 ferruginea*
yellowtail rockfish. *See Sebastes
 flavidus*
Yongeichthys nebulosus, 288

Zanchus cornutus (Moorish idol),
 262, 708
Zanclidae (Moorish idols), 262,
 707
Zanobatidae, 125
Zantodon buchholzi, 370
zebrafish. *See Danio rerio*
zebra shark. *See Stegostoma*
Zeidae (dories), 233, 415
Zeiformes, 232–33
Zeus faber (John Dory), 233, 382
Zoarces viviparus, 328, 329(t), 530
Zoarcidae (eelpouts), 257–58
 anguilliform swimming, 320
 distribution, 632
 habitat, 704, 742, 743
 locomotion, 330
 nostril, 394
Zoarcoidei, 257–58

Selected Coastal Marine Fish Families and Their Continental Affiliations*

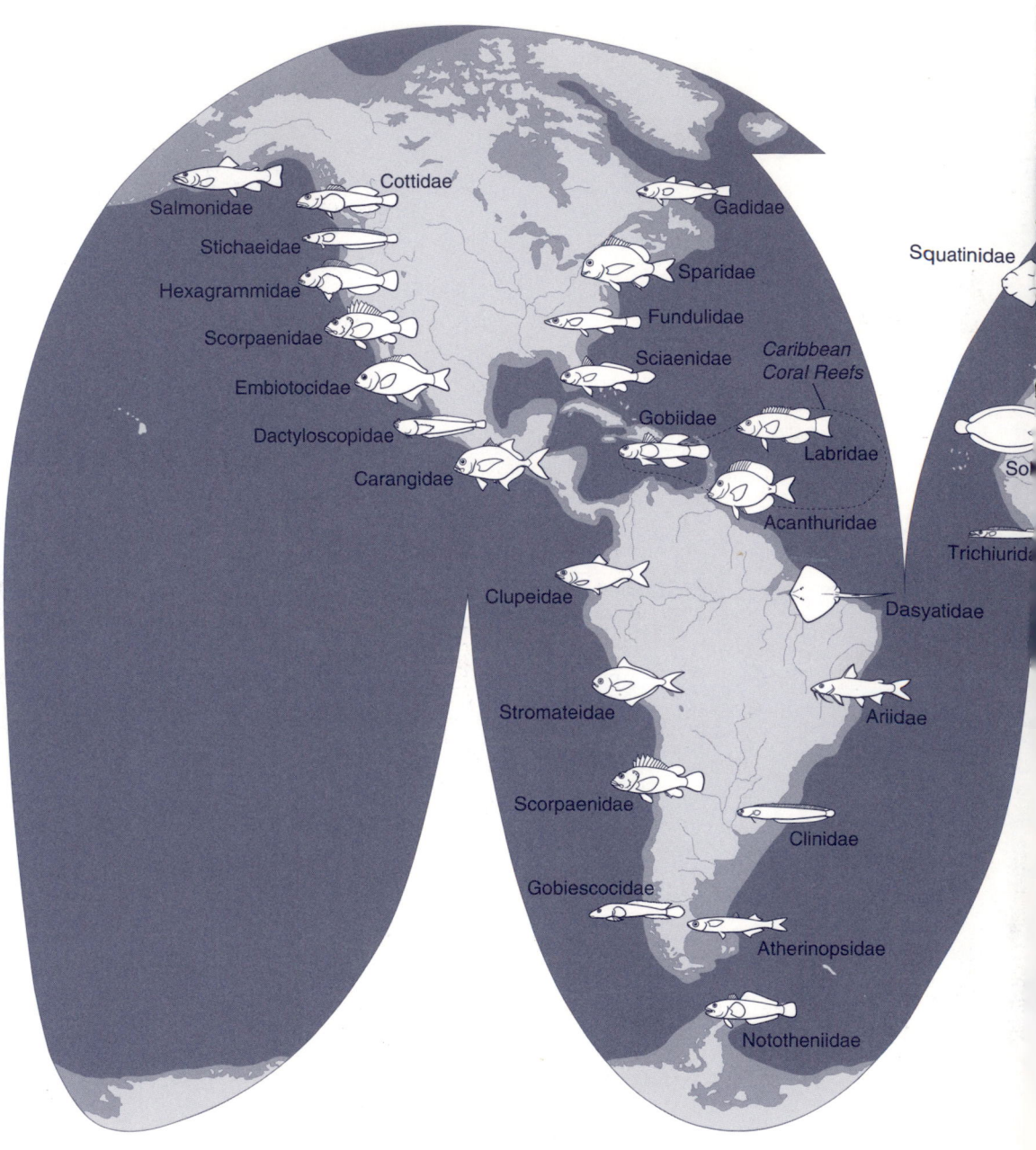